MARSH
D0082962

GENES & GENOMES

A CHANGING PERSPECTIVE

GENES & GENOMES

A CHANGING PERSPECTIVE

MAXINE SINGER

President, Carnegie Institution of Washington
Scientist Emeritus, National Institutes of Health

PAUL BERG

Willson Professor of Biochemistry
Director, Beckman Center for Molecular and Genetic Medicine,
Stanford University School of Medicine

415 - 383-1430

UNIVERSITY SCIENCE BOOKS
MILL VALLEY, CALIFORNIA

University Science Books
20 Edgehill Road
Mill Valley, CA 94941
Fax: (415) 383-3167

Production manager: Mary Miller
Manuscript editor: Carol Dempster
Copy editor: Sylvia Stein Wright
Text designer: Gary Head
Jacket and cover designer: Robert Ishi
Art director: Charlene Kornberg
Indexer: Maria Coughlin
Compositor: Syntax International
Printer: R. R. Donnelley & Sons

Library of Congress Catalog Number: 90-070-698

ISBN 0-935702-17-2

Printed in the United States of America

10 9 8 7 6 5 4 3 2 1

Abbreviated Contents

Contents

Chapter 1
The Genetic Molecules 35

Chapter 2
Replication, Maintenance, and Modification of the Genome 73

Chapter 8
The Structure and Regulated Expression of Eukaryotic Genes 455

Chapter 9

The Molecular Anatomy of Eukaryotic Genomes 621

Chapter 10

Genomic Rearrangements 713

PART IV
Understanding and Manipulating Biological Systems 859

Preface

This book is about the molecular structures and mechanisms that underlie the utilization of genetic information by complex organisms. We hope, in the presentation, to capture the sense of discovery, understanding, and anticipation that has followed from the recombinant DNA breakthrough.

To introduce these developments, Perspective I and Chapters 1 through 3 emphasize the classical genetic and biochemical studies from the beginning of this century to circa 1972. These studies, carried out largely with microorganisms, defined our knowledge of the structure and function of genes and genomes. The emergence of the recombinant DNA methods in the early 1970s and the associated advance in rapid DNA sequencing technology soon thereafter provided the means for a molecular approach to the analysis of more complex genomes. Thus, the core of this book deals with research covering less than two decades and reveals the extraordinary depth and breadth of opportunity that these methods provide to biology.

The book has been a long time in the works. The idea for it originated with a series of public lectures one of us (P.B.) gave in 1979 at the University of Pittsburgh. The efforts to prepare the lectures for ublication were frustrating. In speaking to an audience of nonbiologists, many interesting and exciting details of the science were omitted. A book that lacked such detail would surely be unsatisfactory. As the concept of the book expanded, the need to share the effort became apparent. Thus, the two of us set to work. With only two authors, both working over all the material, we hoped to obtain a unified, coherent approach and style. But we also recognized that it would be foolhardy to attempt the comprehensive coverage attained by larger groups of authors in several excellent recent texts. We also faced the continuing task of updating previously written sections in order to have the final manuscript reasonably current. To avoid the risk of being overwhelmed by the explosive pace at which new information emerged, we limited our goals. Thus, we have focused on selected specific areas that have already been studied in some depth and that illustrate the progress that is being made.

Our overall plan was to emphasize by depth and experimental approach the science that has come to be called molecular genetics, as it is applied to eukaryotes. To accomplish this, and yet have a text that can be used by readers with only limited prior knowledge of biochemistry, cell biology, and genetics, we have provided background materials in two ways. First, Chapters 1, 2, and 3, respectively, summarize the essential information on the structures of DNA, RNA, and proteins; the various cellular transactions involving DNA (replication, repair, and recombination); and the fundamental mechanisms of transcription, translation, and control of gene expression. Readers who are well versed in this material may want to skip the first three chapters. Second, the essays denoted Perspectives I, II, and III provide a historical introduction and overview of the concepts described in the chapters comprising Parts I, II, and III of the book, respectively. These essays do not say much about how these concepts were discovered and verified. Rather, they attempt to convey how the intellectual framework of modern biology was constructed from diverse studies involving chemistry, genetics, microbiology, cell biology, and evolutionary biology. Thus, Perspective I, which introduces Chapters 1, 2, and 3, traces the historical path to our present views on inheritance. It introduces, along the way, the concepts of a gene, the transmission and segregation of genes, the rationale for the early mapping of genetic determinants to specific chromosomal locations, the identification of genes as segments of nucleic acid, and the emergence of the informational relationships between DNA, RNA, and proteins.

The four chapters in Part II describe tools and methods for constructing, cloning, selecting, and characterizing recombinant DNA molecules as well as other experimental methods in molecular genetics. The emphasis is on the logic of the methods; the chapters do not constitute a laboratory manual. Readers interested in detailed procedures are directed, by reference, to the several excellent manuals published during the last decade. The Perspective II essay that precedes these chapters stresses the historical development of the tools and methods from a large and disparate body of fundamental research in enzymology, bacterial and bacteriophage genetics, and nucleic acid chemistry.

The principal aim of the book is realized in Part III, which describes the major concepts that define current understanding of the genetic information systems of eukaryotes. The Perspective III essay includes an introduction to the distinctive properties of eukaryotic genomes, including introns and the complex signals that regulate transcription, as well as the abundance of sequence repetition in the genome, and the related question of the role of reverse transcription in the origin of eukaryotic DNA segments. The essay ends with a description of the unifying concepts provided by thinking about biological evolution as a process that fundamentally involves the reshaping of nucleic acid and, consequently, protein structure.

The first of the three chapters in Part III (Chapter 8) examines the structure of eukaryotic genes and our present understanding of their expression, including (1) the complex signals for regulating transcription and (2) the origins, locations, and structures of introns and the mechanisms by which introns are removed from primary transcripts by splicing. Fundamental to these descriptions is the use of reverse genetics—the design and construction of specifically mutated DNA segments—for the analysis of structure-function relationships in eukaryotic genes. Chapter 9 focuses on the way genetic information is organized in complex genomes. The emphasis is on the distribution of genes and other DNA sequences, including their relation to chromosome morphology. The concept of the genome as a record of evolutionary history threads throughout. The chapter concludes with a description of the genomes of the intracellular organelles—mitochondria and chloroplasts. The diverse mechanisms involved in random and nonrandom rearrangements of genomic DNA are described in Chapter 10. These include amplifications, deletions, and transpositions, both those that are unprogrammed and lead to mutagenesis and those such as yeast mating type switches and immunoglobulin gene construction, which are programmed and regulate gene expression in precise and essential ways.

This volume closes with the Perspective IV essay, which illustrates briefly how the general concepts presented earlier apply to specific and complex biological systems. Perspective IV provides an introductory overview to Part IV of the book, which will appear in a separate volume. Part IV emphasizes that genes operate in complex, multicomponent, interacting systems. In each system, the general operating principles are applied in different ways, leading to the great diversity of living things. In the past, much of this diversity was described in phenomenologic terms. Now the phenomenology can be reinterpreted with molecular descriptions of the regulation, in time and place, of gene expression. Part IV also describes how, as a consequence of the recombinant DNA methods, biology has changed from a descriptive to a manipulative science. The genotypes, and thus the phenotypes of individual proteins and of whole cells and organisms, can be altered, providing future opportunities to investigate fundamental biological processes as well as to address critical problems facing our species and the planet we inhabit. Many of the opportunities will be realized only after we acquire greater understanding of the structure of genomes. This includes locating the positions of genes and extensive sequence information. Part IV introduces the concepts fundamental to current efforts to map and sequence the genomes of several species, including humans.

In 1980, when we first began thinking about this book, we, and many other biologists, were just emerging from a period when public policy issues vied with science for daily attention. The exciting research stemming from the recombinant DNA methodology was only one aspect of what has been called a "revolution in biology." Besides initiating a new era in our ability to understand living things, the revolution initiated unprecedented social concern over the impact of biological research.

The earliest public concerns reflected questions raised by the community of biological scientists and centered on the safety of recombinant DNA experiments. Laboratory safety has always concerned people working with pathogenic microorganisms—viruses and bacteria. These concerns were renewed about 1971 as more work was carried out worldwide on viruses that cause tumors in experimental animals. The distinctive life cycle of these viruses compared to the more familiar viruses (for example, those that cause measles and polio) and the possibility that some of them might cause human cancer prompted a serious look at the possible hazards to scientists and their students. It was in this context that the recombinant DNA methods emerged. The general atmosphere of concern was also influenced by the then pervasive sense of social responsibility in the United States, a sense that grew out of the prior decade's turmoils over civil rights and the war in Vietnam.

The earliest prototypes for recombinant DNA experiments involved the joining of DNA from a virus causing tumors in small laboratory rodents to parts of the DNA of a well-studied bacterial virus. Plans to introduce the new DNA molecule into bacterial cells aroused serious reservations, but these were stilled by the decision to defer such attempts. The next major step involved constructing recombinant DNA with a gene that makes bacterial cells resistant to antibiotics. The ability to transform living cells genetically by introducing such novel DNA molecules into bacteria raised questions about whether such bacteria might cause tumors or acquire natural resistance to medically important antibiotics. The issues were raised at scientific meetings and in private conversations. At one such meeting in June, 1973, the participating scientists called attention to this promising line of research and asked the National Academy of Sciences to undertake a focused study of the possible risks of recombinant DNA experimentation. To publicize this request, their letter was published in *Science*.

The Academy responded (as academies typically do) by forming a committee, this one of scientists actively engaged in recombinant DNA and related work. This committee met in April, 1974, and proposed two major steps that became, to their surprise, front page news. First, they recommended a worldwide moratorium on those recombinant DNA experiments that involved tumor virus DNAs or introduced genes for potent toxins or antibiotic resistance into bacteria that normally did not have such genes. Second, they called for an international, broad-gauged discussion of the issues, to be held at a meeting the following winter.

Although there were rumblings of discontent and even accusations that U.S. scientists were trying to slow up everyone else's work so they could win a race to the major discoveries promised by the new methods, the moratorium was, as far as anyone knows, universally honored. Molecular biologists, virologists, microbiologists, and biochemists from the United States and abroad, as well as science administrators, journalists, and even lawyers, readily accepted invitations to the proposed conference.

That gathering took place in February, 1975, at Asilomar, a state of California meeting center on the edge of the Pacific, a few miles south of Stanford University's Hopkins Marine Station in Pacific Grove, where these words are being written. The Organizing Committee had arranged for prior working group sessions and working paper preparation by experts in several different areas for presentation. There was a good deal of discussion, some heated, about the reality of the dangers being discussed. There was broad agreement but not unanimity that if risks of constructing hazardous organisms existed, their likelihood was very low. But consensus on a future course of action seemed remote until the lawyers took their turn. They underscored the personal legal responsibility of scientists, even for highly unlikely events. They also reminded us that the public can be very restrictive about remote risks if fear of consequences, however unlikely, is great. The message was clear. Continuing a responsible stance was the only sensible course. On the last day, the Organizing Committee's proposed report was debated vigorously but finally accepted. The Asilomar Conference's recommendations were widely reported in the press and later published in several scientific journals. Similar conclusions had been reached by a British government committee under Lord Ashby a few weeks earlier.

The recommendations made at Asilomar provided the framework and starting point for official U.S. action, which began the day after the conference closed. A committee organized by the National Institutes of Health (NIH) began work on guidelines to govern all recombinant DNA experiments carried out in institutions with their funding. The original guidelines, published in June, 1986, were intentionally strict, with the expectation that, as experience and knowledge accumulated, they could be revised. To this day, no untoward events to laboratory personnel or the public are known to have originated from the tens of thousands of recombinant DNA experiments that have been conducted. The containment requirements for most routine recombinant DNA experimentation have been eliminated or relaxed. The only experiments that still require strict containment within the NIH guidelines (or those in most other countries) involve recombinants that include extensive DNA regions from highly pathogenic organisms. Interestingly, recombinant DNA techniques have made the study of some important but very dangerous infectious agents of humans and animals feasible and safe.

Soon after Asilomar, and with varying intensity until the present, public interest and concern have continued over recombinant DNA experiments and their extension to the genetic engineering of whole organisms—bacteria, plants, and animals. At first, the focus was the same as that of the majority of the scientific community—

the potential of the experiments to create disease-producing agents. Local and state governments passed laws and ordinances mandating compliance with the NIH guidelines or somewhat more restrictive rules. Bills proposing to make the NIH guidelines law and mandating punishments for noncompliance were introduced and debated in Congress, but none ever passed. These numerous independent debates, in various locales and at many levels, ultimately validated the NIH approach to the problem—both the scientific risk analysis and the administrative organization.

Later, and to this day, other kinds of issues dominate the public debate. A few scientists raised questions about the possible evolutionary consequences of passing DNA across species boundaries, for example, inserting human DNA into bacteria. These questions attracted attention from many nonscientists, but the issue is now viewed as inconsequential. There are myriad opportunities in nature for such exchanges of DNA, and they have probably occurred many times. Recombinant DNA experiments do not significantly increase the opportunities. Moreover, most organisms harboring recombinant DNA molecules are unlikely to thrive except under very special laboratory conditions. The latter argument is also pertinent to the recent discussions regarding government regulation of the deliberate release of genetically engineered organisms into the environment for various agricultural purposes. Although careful consideration of the potential problems associated with such releases is essential, excessive requirements, including protective garb like astronaut suits, are, from a scientific standpoint, unwarranted.

The possible application of recombinant DNA techniques to the development of biological warfare agents is raised continually; the United States is party to a 1972 international convention prohibiting such work. The Defense Department continues to support work aimed at defense against biological weapons that others might develop. This too has been debated because the differences between research intended for offensive and defensive purposes are not easily distinguished. This debate, properly, attracts scientists and nonscientists alike because it is fundamentally a political and social question, not a scientific one, as are discussions by the public of other questions arising out of the new biology. What are the advantages and disadvantages of gene therapy for humans should it ever become feasible? How should society, employers, and individuals deal with the growing amount of information about people's genes that will be acquired as a result of the new technologies? Are there valid ethical concerns associated with introducing human genes into animals? Should genetically engineered animals and plants be patentable?

The early public fears about the biological revolution engendered many negative attitudes about the research. Biologists then feared the worst: highly restrictive laws or regulations that would seriously hinder further experimentation and its promise of new knowledge and beneficial applications to medicine, agriculture, and industry. Many scientists regretted the initial open discussion of the issues in the face of successful demagoguery by critics and the tendency of newspapers to build hype rather than carefully explore difficult issues. Yet it has turned out well. The science described in this book attests to that, as do the growing number of important products being produced by the young and energetic biotechnology industry. Perhaps the moratorium and initially restrictive guidelines held things up, but only briefly. The early caution in the face of ignorance was prudent, even though hindsight suggests that the risk scenarios were far less likely than we had supposed.

MAXINE SINGER AND PAUL BERG

Acknowledgments

Contrary to the widely held image, science is a very social activity, and most scientific work depends on a good deal of collaboration and consultation. This book is no exception. We are grateful for the cooperation and constructive, patient help of our colleagues who read and criticized sections or chapters of the text: David Finnegan, Claude Klee, Arthur Kornberg, I. R. Lehman, Howard Nash, Bruce Paterson, Carl Schmid, and Robert Tjian. We also thank those who provided unpublished information and manuscripts.

Five colleagues read almost the entire book in an early draft. The final product is much improved for their careful and detailed criticisms. We are thankful for the efforts of Barbara H. Bowman, David Dressler, Paul Schimmel, Jean O. Thomas, and William B. Wood. We were very fortunate to have the wise help of Carol Dempster, one of those rare individuals who can read in many scientific disciplines with enthusiasm, perception, and insight into the clarity of presentation. Bruce Armbruster, the publisher, president of University Science Books, was responsible for obtaining Carol Dempster's assistance, as he was for the other fine members of the group that helped assemble this book: Sylvia Stein Wright, copy editor; Gary Head, text designer; and Mary Miller, production manager. But more than this, Bruce's patience and unflagging enthusiasm during the long years when other matters slowed our progress, sometimes almost to a standstill, were extraordinary; and we are most grateful.

Molecular geneticists think about genes and genomes visually. We "see" DNA and schemes describing how DNA functions in our minds' eyes. Thus, the diagrams in scientific papers and textbooks like this one are critical to our understanding and memory. The wonderful pictures and diagrams in this book were prepared by Charlene Kornberg and her colleagues at Stanford University Medical School. They are an integral part of the narrative, and in many instances, they clarify and amplify the text. Our own naivete about the difficult process of preparing the drawings and our insistence on an abundance of pictures made the job burdensome and long. We thank all those who worked so diligently under Charlene Kornberg's direction: Meryle Colten, Butch Colyear, Mike Maystead, Eunice Ockerman, Lois Schoen, Kelly Solis-Navarro, Karen Sullivan, and Linda Toda. But it is we, not they, who are responsible for any errors that may remain.

Among the splendid advantages available to modern biologists are the libraries wisely established by earlier generations. During our work, we have spent time reading and writing at three such places, all conveniently remote from telephones and daily responsibilities. The librarians and other staff members of the Marine Biological Laboratory at Woods Hole, Massachusetts, of The Jackson Laboratory at Bar Harbor, Maine, and of the Hopkins Marine Station of Stanford University at Pacific Grove, California, were all hospitable and helpful.

Harry Woolf, then director of the Institute for Advanced Studies at Princeton, New Jersey, graciously provided opportunities to work at that quiet place. His kind interest, hospitality, and friendship, and the fine food offered by the Institute's eclectic chef, Franz Moehn, made those days memorable. Curiously, our ability to work uninterrupted at Princeton was assured by the almost universal disinterest of the institute's physicists and mathematicians in talking to biologists.

Quiet, uninterrupted days for work were also assured during Paul Berg's tenure as a fellow of Clare Hall, Cambridge. The lively intellectual community there, presided over by Michael Stoker, was congenial and stimulating.

Over the years, the science we hoped to capture here progressed at an extraordinary and unprecedented pace. Continual revisions put heavy loads on the good humored and expert people who typed and retyped innumerable drafts. Without Eleanor Olson and Dot Potter at Stanford and May Liu and Gail Gray in Bethesda, nothing would have been accomplished.

Millie Berg and Dan Singer have been understandably impatient at times, but their support was always clear and always welcomed. So, too, the younger generations of the Berg and Singer families, though none is a biochemist or molecular biologist, have bolstered us with their curiosity, enthusiasm, pride, and love.

GENES & GENOMES

A CHANGING PERSPECTIVE

PERSPECTIVE

Molecular Basis of Heredity: An Overview

PART I

The disciplined study of heredity began with genetically complex plants and animals. Emerging from these early investigations was the concept of an indivisible gene as the fundamental unit of inheritance and a recognition that the transmission of genes from one generation to the next is strongly influenced by random events. But insights into the chemical nature of genes or how they act were nonexistent. Explorations into the nature of the genetic molecules and the detailed mechanisms governing heredity began only when bacteria and viruses—whose existence was unsuspected by early geneticists—were adopted as experimental models. Indeed, it was in these organisms that **deoxyribonucleic acid (DNA)**, **ribonucleic acid (RNA)**, and **protein** were first recognized as the universal determinants of genetic behavior. The speed and decisiveness of the advances that followed were possible because these microorganisms' biological characteristics facilitate the manipulations needed to analyze their genetic structures. Because comparable analytic capabilities for examining more complex genetic systems did not exist; these insights could not be extended to animals and plants. The emergence of recombinant DNA techniques erased a formidable array of technical and conceptual barriers to deciphering and understanding complex genetic systems. Not surprisingly, our vision of the structure and function of genes, chromosomes and genomes is now profoundly modified, and this new understanding is, in turn, radically altering broad areas of biology.

Some perspective on these achievements can be obtained by examining the origins and successive revisions of our fundamental concepts about inheritance. Historically, the extraordinary diversity of living things was a major obstacle to the construction of unifying principles about heredity. Theophrastus, a student of Aristotle, first recognized the analogy between animal and plant reproduction and coined the words *male* and *female* to describe the participants in sexual reproduction. Before Theophrastus, fifth-century Greek philosophers challenged strong cultural influences by concluding that, because offspring resemble both parents, both sexes must contribute to the formation of a new individual. They also believed that these contributions were information of some kind, collected into the male or female "semen" from parts of mature individuals. Democritus, whose view did not prevail, suggested that the information was carried in particles whose shape, size, and arrangement influenced the offspring's properties.

Early in the nineteenth century, following the introduction of improved microscopes, the concept of the cell provided a major unifying simplification for biology. All organisms could be viewed as single, free-living cells or collections of cells. Continuing refinements in microscope optics and innovative methods for preparing and staining materials provided increasingly detailed descriptions of the insides of cells (Figure I.1). Research showed that cells are bounded by a membrane with a defined structure, that cell shapes vary, and that cells contain definite structures, the most striking being the **nucleus**, which itself contains small brightly stainable structures called **chromosomes**. Later it was discovered that certain microorganisms (bacteria) do not contain discrete nuclei. New cells were found to arise only by the division of preexisting cells.

Today, living things are divided into two groups. One, the **eukaryotes**, has nuclei within which the chromosomes, the repositories of genetic information, are sequestered; the second group, the **prokaryotes**, comprises the single-celled bacteria that lack nuclei and contain their chromosome(s) in the cytoplasm. With only a few specific exceptions, every cell in a multicellular organism carries the same complete set of chromosomes. Eukaryotic organisms are more complex than prokaryotes and generally contain more genetic information. Furthermore, eukaryotes are capable of true sexual reproduction, and, for many, this mode is obligatory for the production of offspring. One corollary of sexual reproduction is that each nucleus contains two copies of each chromosome; such eukaryotic cells are called **diploid**. Prokaryotes, with only a single chromosome, are called **haploid**. Under some circumstances, prokaryotes engage in pseudosexual activities, the consequences of which render the participants partially diploid, a property that is useful to geneticists.

Soon after the cell theory was defined, three approaches to the study of living things were initiated: the study of chromosomes, the statistical analysis of the inheritance of single traits, and the chemical isolation and characterization of the chromosomal constituents. These approaches proceeded in parallel and developed into important scientific disciplines before merging in the middle of the present century.

Chromosomes

The second half of the nineteenth century witnessed the construction of an increasingly detailed picture of chromosome morphology and behavior. All cells of a given species, with one important exception, have the same characteristic number of chromosomes. For example, flies of the species *Drosophila melanogaster* have 8 chromosomes; humans and evening bats have 46, corn 20, and rhinoceri 84. The chromosomes can be grouped into homologous pairs on the basis of similar shapes: 4 pairs in *D. melanogaster*, 23 pairs in humans, and so forth. Microscopic examination of fixed and stained cells gave only static pictures, but these could be ordered in a time sequence, beginning with the formation of a cell by division and ending with its division into two cells. It became apparent that a duplicate of each chromosome is made in the course of cell division, resulting in a doubling of the number of chromosomes (Figure I.2, pages 6–7). Upon division, the duplicate sets of chromosomes separate so that each of the two daughter cells has the same number and kind as the parent cell. The entire process is called **mitosis**.

The Cell Cycle

The events that occur from one cell division to the next are termed the **cell cycle** (Figure I.3, page 8). The mitotic phase (**M-phase**) of the cycle encompasses the period when both chromosome and cell division occur. Following cell separation (**cytokinesis**), each daughter cell enters a period of accelerating biosynthetic activity, a phase called $\mathbf{G_1}$ (for "gap"). G_1 ends with the onset of chromosome doubling—in molecular terms, the duplication of chromosomal DNA; this period of genome replication is called the synthesis phase (**S-phase**). With completion of the S-phase, cells initiate the events characteristic of mitotic prophase, the part of the cycle named $\mathbf{G_2}$. Finally, mitosis and cytokinesis (the M-phase) recur, and the cycle repeats. Generally, the G_1 S, and G_2 periods, which together constitute the **interphase**, consume about 90 per cent of a typical cell cycle, and the M-phase takes up less than 10 percent. The total time for a complete cell cycle varies considerably (from minutes to days) in different cell types and under different growth conditions. Duration of the G_1 phase is the principal determinant of cycle length. For example, resting cells that are deprived of an essential nutrient and cannot grow exponentially are arrested in G_1 and are said to be in a $\mathbf{G_0}$**-phase**. When the nutrient is supplied, they leave G_0 and resume processes characteristic of the end of G_1, processes that lead to S-phase.

Meiosis and the Formation of Gametes

The formation of eggs and sperm uniquely involves a reduction in the normal number of chromosomes by precisely half, a process called **meiosis** (Figure I.4, page 9). These germ cells or **gametes** con-

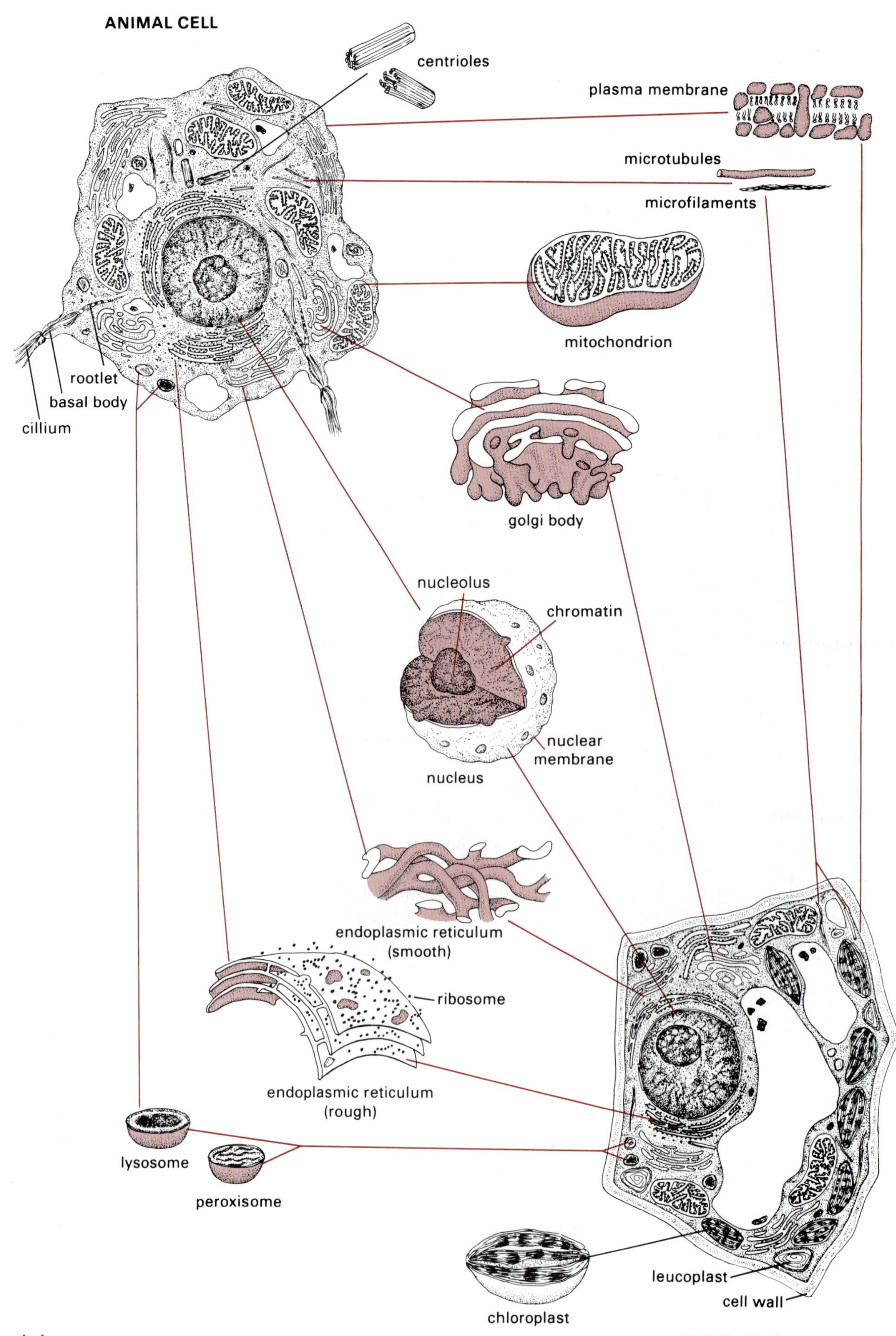
ANIMAL CELL
centrioles
plasma membrane
microtubules
microfilaments
mitochondrion
rootlet
basal body
cillium
golgi body
nucleolus
chromatin
nuclear membrane
nucleus
endoplasmic reticulum (smooth)
ribosome
endoplasmic reticulum (rough)
lysosome
peroxisome
leucoplast
cell wall
chloroplast
PLANT CELL

(a)

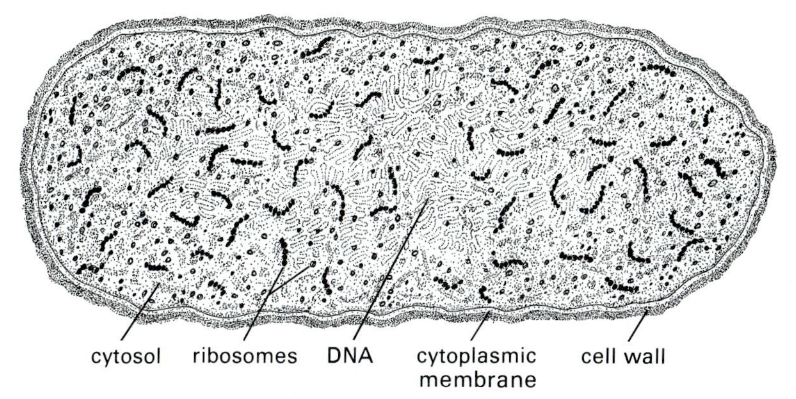

(b)

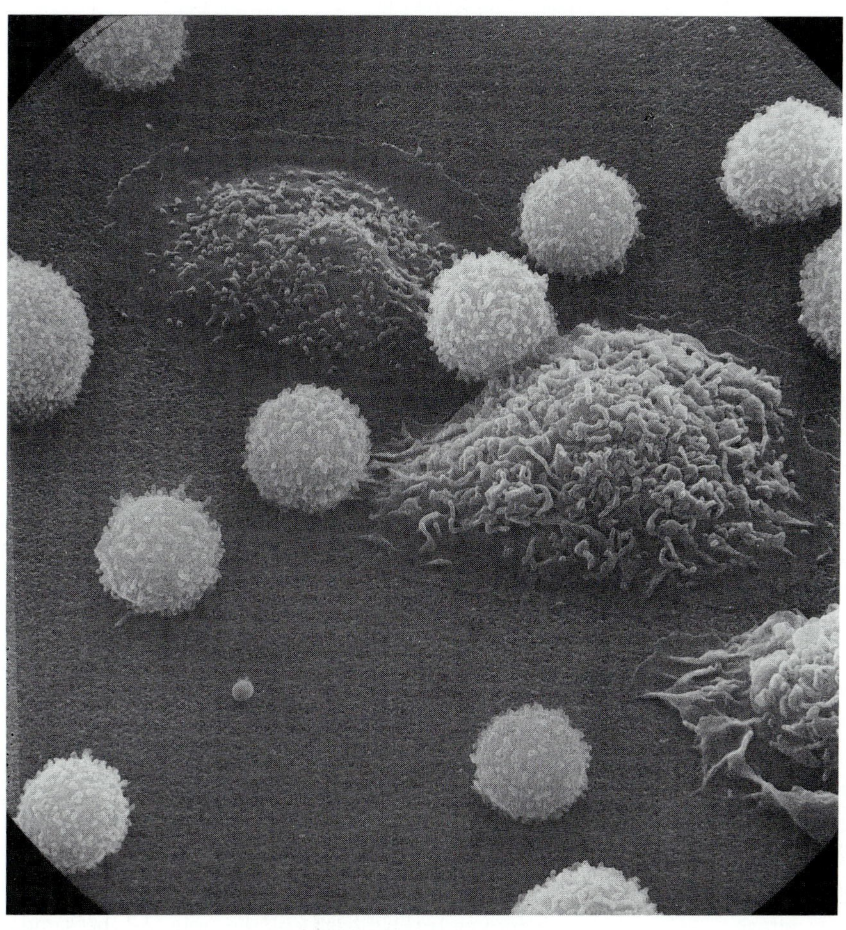

(c)

Figure I.1

(a) PAGE 4: Schematic diagram of typical animal and plant cells.
(b) Schematic diagram of a bacterial cell. (c) Macrophage (larger cells) and lymphocytes (smaller cells) photographed through a scanning electron microscope. The magnification is 6000 times. Courtesy of Emma Shelton and Jan Ornstein.

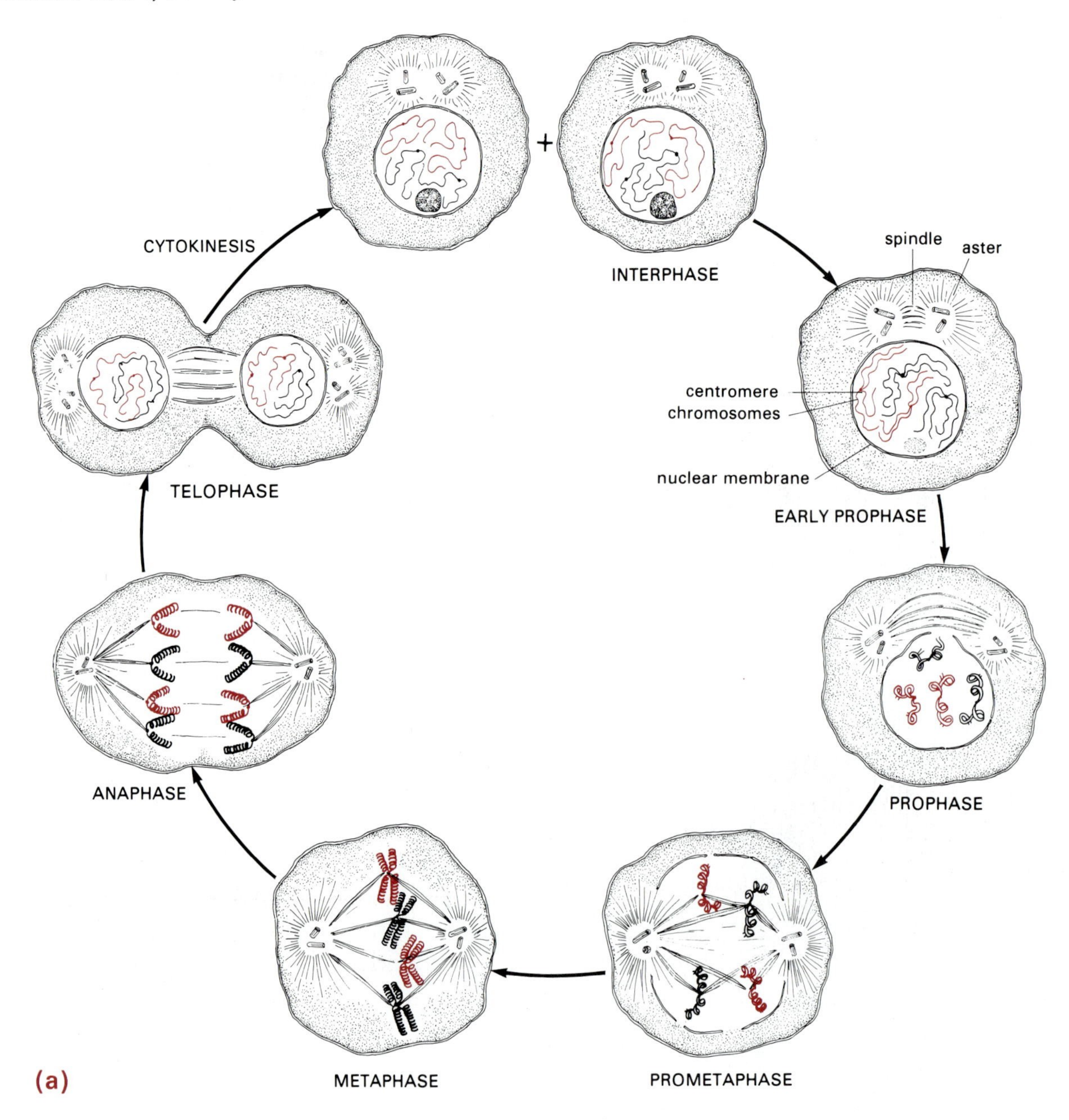

Figure I.2
Mitosis: the stages leading to division of a diploid cell. (a) The diagram shows two pairs of homologous chromosomes (each pair a different color). Each member of a pair goes through mitosis as an independent entity. During interphase, the chromosomes are fine, diffuse strands that are normally difficult to see; chromosomes are duplicated during this period. The replicated chromosomes then condense into distinct structures that are visible at prophase. At prometaphase, the chromosomes condense further, the nuclear envelope disintegrates, and the chromosomes become associated with the spindle. The duplicated structures remain attached to one another at a region called the centromere (dots). Except at the centromere, the two copies of each chromosome, the sister chromatids, separate. By metaphase, the sister chromatid pairs are aligned near the center of the spindle with their centromeres still associated. During anaphase, the sister chromatid pairs separate, and one member of each pair moves toward each spindle pole. At the same time, both the spindle fibers and the cell begin to elongate. When the chromatids reach opposite poles at telophase, a new nuclear envelope is formed around each set, and the chromosomes begin to decondense. Finally, the plasma membrane separates the two nuclei and surrounding cytoplasm into two cells: cytokinesis. The chromosomes take on the extended diffuse shape typical of interphase nuclei, and the process can begin again.

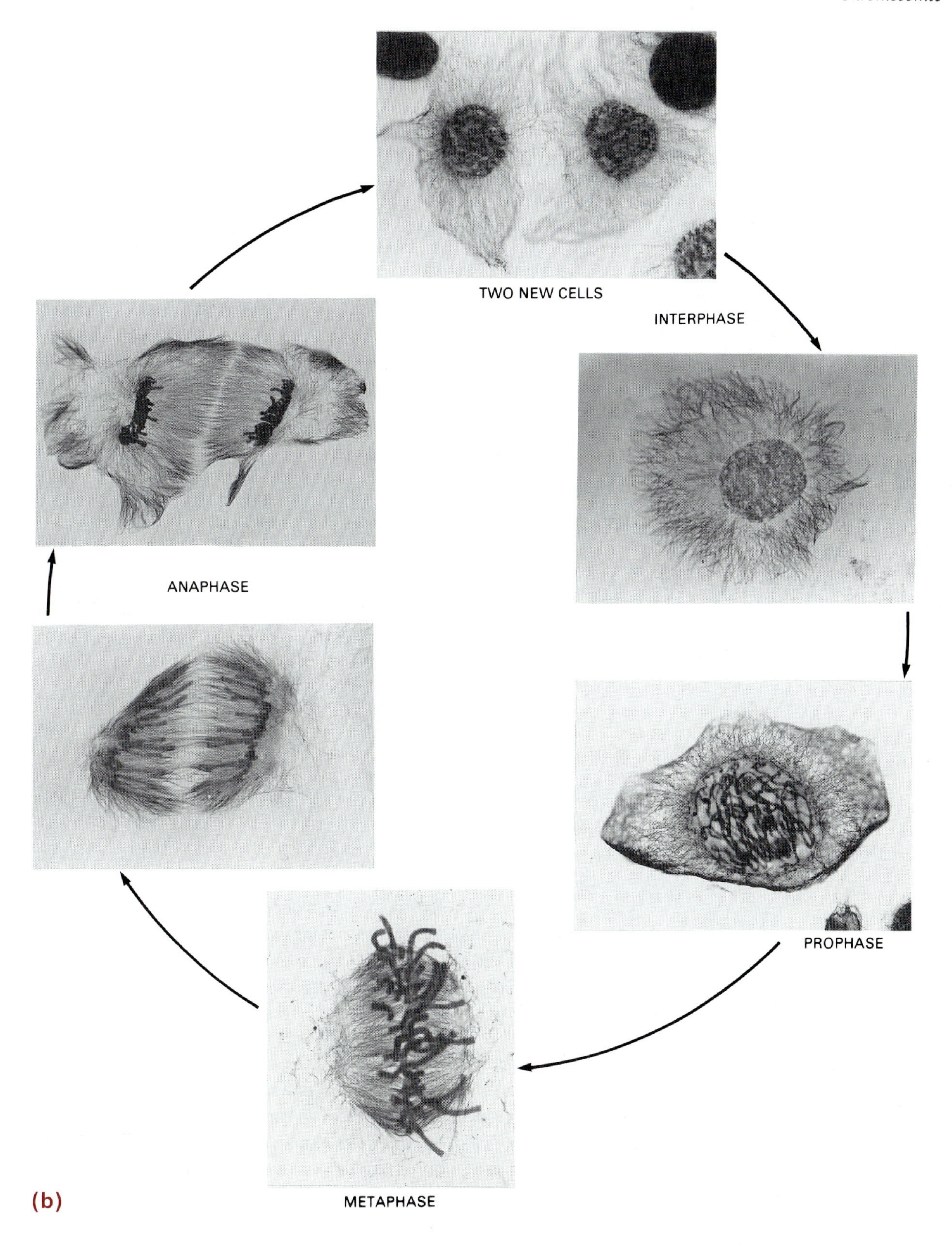

Figure I.2
(b) ABOVE: Photomicrographs of mitosis in the lily, *Haemanthus katherinae*. Cells were stained by an immunogold/silver method. Magnification is 600 times. Courtesy of A. S. Bajer.

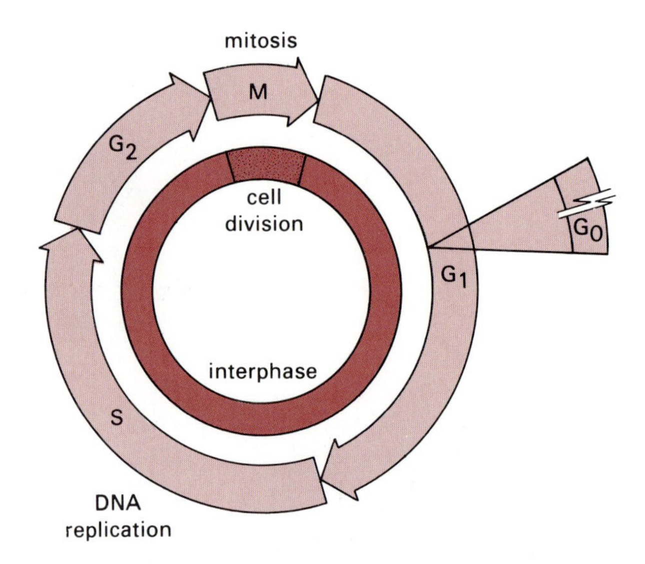

Figure I.3
The cell cycle: mitosis and cytokinesis (cell division) comprise the M-phase of the cycle, which culminates in the formation of two daughter cells. Each daughter cell enters the G_1 portion of the interphase and may initiate a new cell cycle. G_1 is followed by S-phase, during which DNA and chromosomes are duplicated, and then by G_2. The beginning of mitosis marks the end of interphase. Quiescent cells are blocked in the G_1 phase and are said to be in G_0. Typically, eukaryotic cells that do not stop in G_0 take about 24 hours to complete a cycle.

Figure I.4 (PAGE 9)
Meiosis: the stages leading to the division of a diploid cell into four haploid daughter cells. The process differs from mitosis by including two cell divisions and only one round of chromosome replication. The schematic diagram shows two pairs of homologous chromosomes (each pair a different color). During interphase, the chromosomes are fine, diffuse strands. After replication, the sister chromatids remain closely associated and begin to condense, marking the start of prophase. Then homologous pairs of sister chromatids become closely associated to form a tetrad; the process is called synapsis. The beginning of meiotic metaphase I is marked by further condensation of the chromosomes and disintegration of the nuclear membrane. At anaphase I, each member of a homologous pair of sister chromatids begins to move to a different end of the elongating cell. By the end of telophase I and cell division I, there are two daughter cells, each of which has one of the homologous pairs of sister chromatids. The second round of cell division occurs without additional chromosome duplication and begins with prophase II, followed by metaphase II. At anaphase II, the two sister chromatids that have remained in association until now begin to move to opposite ends of the elongating cell. After telophase II and cell division II, there are four haploid cells, the precursors to germ cells. Only one of the original pair of homologous chromosomes occurs in each daughter cell.

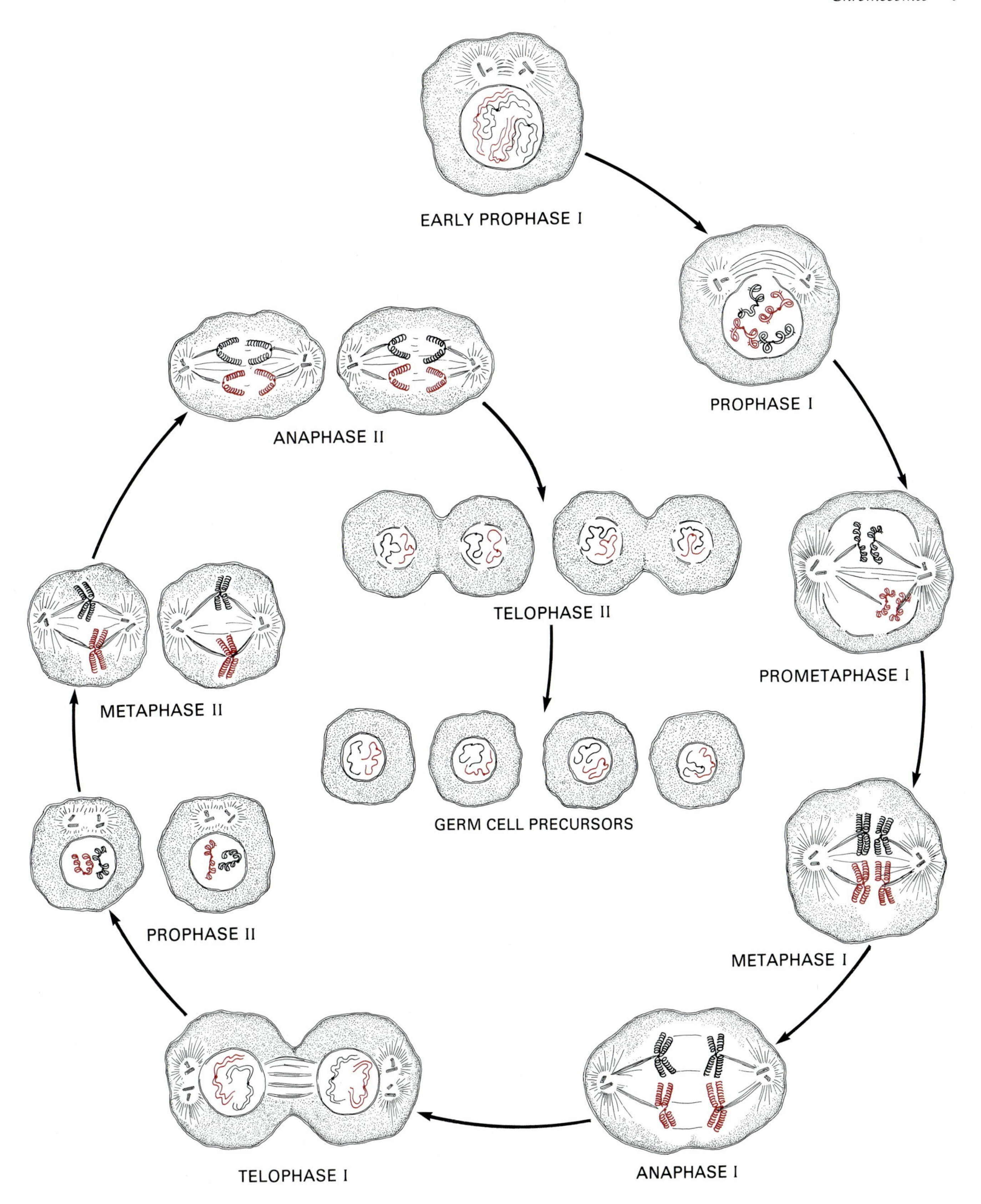
EARLY PROPHASE I
PROPHASE I
PROMETAPHASE I
METAPHASE I
ANAPHASE I
TELOPHASE I
PROPHASE II
METAPHASE II
ANAPHASE II
TELOPHASE II
GERM CELL PRECURSORS

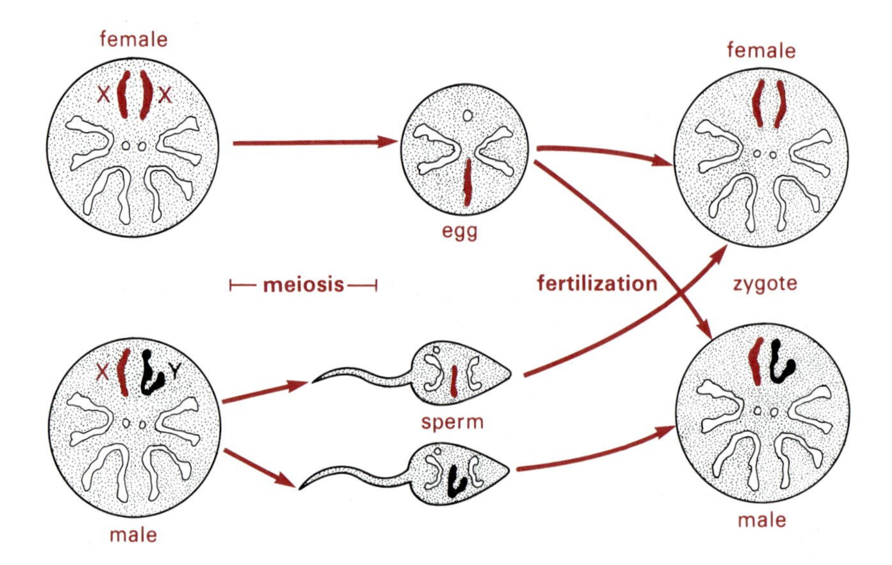

Figure I.5
The formation of haploid gametes during meiosis and the fusion of two gametes to yield a diploid cell at fertilization. Note that in this example (the fly, *D. melanogaster*), as in many other organisms, including mammals, two chromosomes, termed X and Y, are not homologous in the male. During meiosis, two types of sperm are formed, one carrying an X and one a Y chromosome. Meiosis in the female, which has a pair of X chromosomes, gives rise to one type of gamete. The sex of the offspring is determined by whether the fertilizing sperm carries an X or Y chromosome. In some other organisms (e.g., birds), the female carries the nonhomologous, sex-determining chromosome.

tain only one member of each pair of homologous chromosomes and thus only one-half the number of chromosomes found in the other or **somatic** cells of the corresponding organism; therefore, gametes are haploid. Chromosome sorting during meiosis is random, so either member of each homologous pair can end up in the newly formed germ cells.

Upon fertilization, the haploid sets of chromosomes from the sperm and egg mingle in the egg (Figure I.5). The diploid state of the fertilized egg (**zygote**) is then maintained by mitotic division in the many somatic cells that ultimately give rise to the mature individual during development. Fertilization thus reestablishes the full set of homologous chromosome pairs, one member of each pair being derived from the egg and one from the sperm of the respective parents. The two members of a homologous chromosome pair have a very important functional relationship—they contain duplicate sets of genetic information. Thus, complete organisms can sometimes develop from unfertilized, haploid eggs or from fertilized eggs that contain only a partial complement of maternal chromosomes. As already pointed out, either member of a homologous chromosome pair can occur in a functional gamete. Chromosome reduction during meiosis sorts one member of each pair into each mature egg or sperm.

Chromosome Structure

The easiest chromosomes to see in a microscope, and the ones normally depicted in photographs and drawings, are metaphase chromosomes (Figure I.6). At this stage, the chromosomes are highly condensed and form discrete structures. In many organisms, individual chromosomes and their homologs are readily recognized by size and shape. Each metaphase chromosome is actually composed of two identical portions, called **sister chromatids**, because the duplication of chromosomal DNA occurs well before metaphase in the S-phase of the cell cycle.

Chromosomes possess a constriction called a **centromere**. Its position is fixed for each chromosome. Very specific chromosomal functions are associated with centromeres. They are the last site of association of sister chromatids before they separate during mitotic division and meiotic division II (Figures I.2 and I.4). The remainder of the sister chromatids, the arms, are clearly visible as separate entities long before the centromeres separate at anaphase.

Another difference between centromeric regions and the arms of chromosomes becomes apparent after treatment with certain types of dyes. Centromeres can appear dense and compact compared to the appearance of the arms after staining (Figure I.7, page 14). Such densely staining chromosomal regions are called **heterochromatic**. Centromeric heterochromatin is often visible after staining even in otherwise obscure interphase chromosomes. The regions of chromosomes that are not heterochromatic are generally referred to as **euchromatic**. (The prefix "eu," which appears in the term *euchromatic* as well as in the word *eukaryote*, has a Greek origin and means good or well. Thus, euchromatic regions stain well or normally, and heterochromatic regions stain other than normally. Similarly, eukaryotes have a "good" nucleus, and prokaryotes do not.)

Additional morphological features that are used to describe chromosomes include the ends of the chromatids, the **telomeres**. These too are often heterochromatic. Frequently, but not always, small constrictions called **nucleolar organizer regions** are visible on mitotic chromosomes (Figure I.8, page 14). These regions look like knobs on meiotic chromosomes. Within a species, nucleolar organizer regions may occur on one or more specific chromosomes (and their homologs); when they do, they are always at the same location. By the G_1 phase of the cell cycle, the several nucleolar organizers have begun to expand; and, if there is more than one, the expanded regions may coalesce into one or more large, roughly spherical structures, the **nucleoli** (Figure I.1). In interphase nuclei, nucleoli are often the only clearly visible structures, but as prophase begins, they begin to disappear.

Special dyes and staining techniques developed over the last few decades have revealed additional details of prometaphase and metaphase chromosome structure, even in the rather small chromosomes of mammals. Thus, after staining, each chromosome displays a unique pattern of light and dark bands; homologous chromosomes have identical patterns (Figure I.7). The pattern is sufficiently reproducible to identify each chromosome in the set. Figure I.9 (page 15)

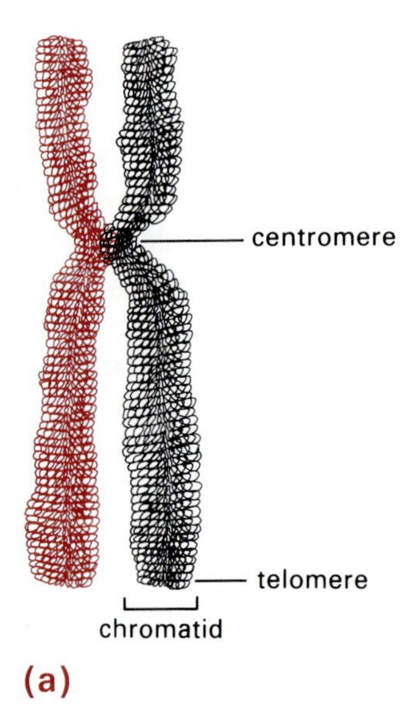
centromere
telomere
chromatid
(a)

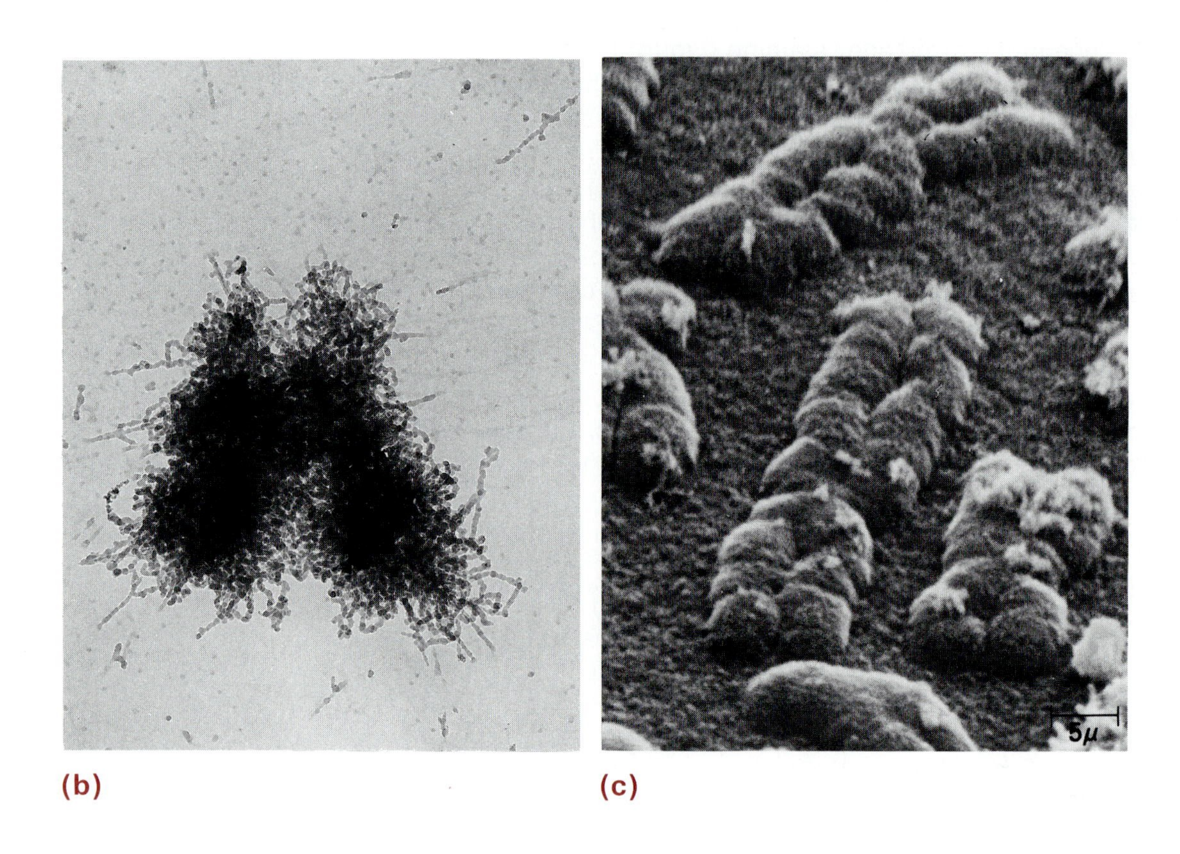
5μ
(b)
(c)

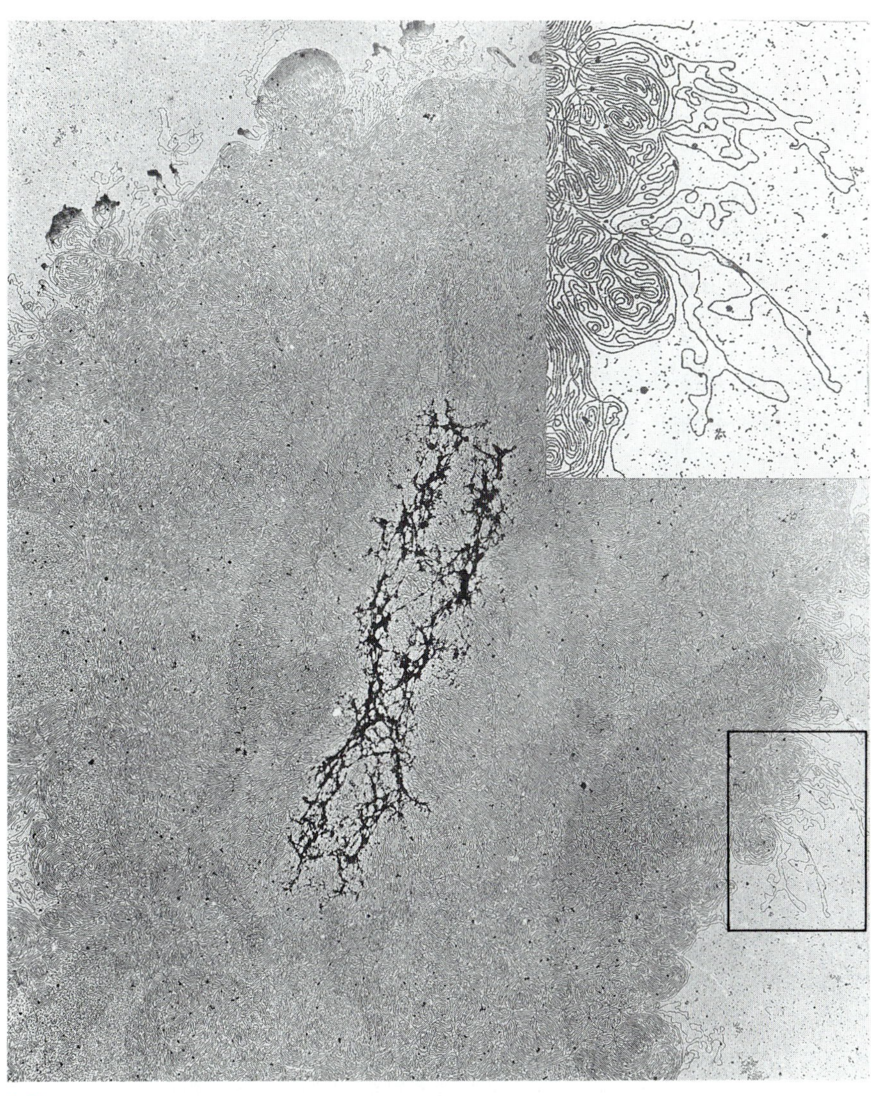

(d)

Figure I.6
Properties of eukaryotic metaphase chromosomes. (a) PAGE 12: Schematic drawing showing the two copies or chromatids of a duplicated chromosome. The two chromatids are held together at the centromere, which is, in this example, near the chromosomal center (a metacentric chromosome). (b) PAGE 12: An electron micrograph of a submetacentric chromosome; magnification is 30,000 times. Courtesy of G. F. Bahr. (c) PAGE 12: Photograph taken with a scanning electron microscope of several human chromosomes, showing their coiled structure. Courtesy of J. B. Rattner. See J. B. Rattner and C. C. Lin, *Cell* 42 (1985), p. 291. (d) This chromosome has been depleted of histone proteins, revealing a spaghettilike mass of DNA. The insert shows the DNA strands at higher magnification. Micrograph is by U. K. Laemmli, as published in D. W. Fawcett, *The Cell*, 2nd ed. (Philadelphia: W. B. Saunders, 1981), p. 237.

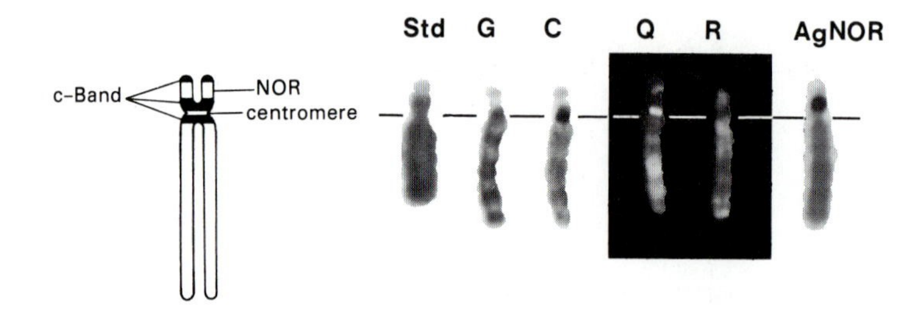

Figure I.7
Photographs of human metaphase chromosome number 13 showing overall morphology (Std), banding patterns in euchromatic arms as observed after three different staining procedures (G-, Q-, and R-banding), heterochromatic region at the centromere as observed after a special staining technique (C-banding), and the nucleolar organizer region highlighted by a specific staining method (AgNOR). The position of the centromere on this acrocentric chromosome is marked by the horizontal line. Courtesy of T. A. Donlon.

displays the full complement of prometaphase chromosomes in a human cell; this representation, called the human **karyotype**, reveals their relative size and shape as well as the positions of the centromeres and the characteristic banding patterns.

During interphase, the chromosomes are stretched out and generally not visible. There are, however, important exceptions that have been studied intensively for many years. Secretory cells in certain insect larvae (such as those of *D. melanogaster*) grow to very large sizes and go through many S-phases in the absence of mitosis and cell division. Multiple chromatids, sometimes numbering as many as a thousand, are formed and remain linked and aligned

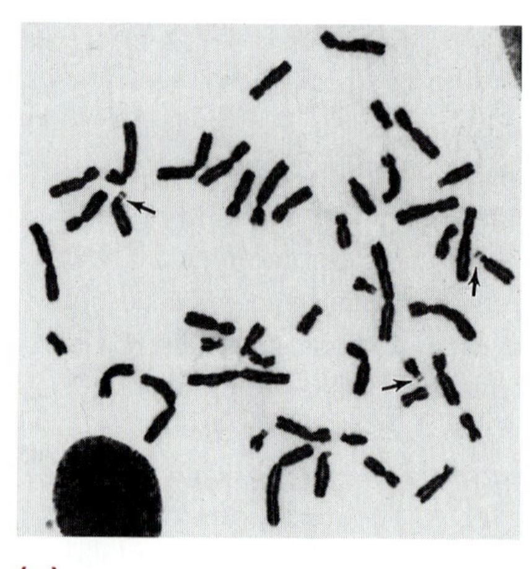
(a)

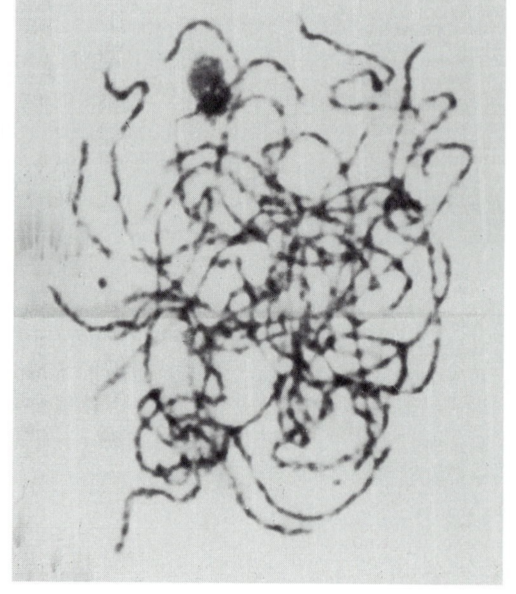
(b)

Figure I.8
Photographs showing nucleolar organizer regions (NOR) on chromosomes. (a) Human metaphase chromosomes stained to leave NORs unstained (arrows). Not all the NORs are apparent. Courtesy of T. C. Hsu. (b) Meiotic human chromosomes showing a knoblike nucleolar organizer region after staining with silver. From C. Mirre, M. Hartung, and S. Stahl, *Proc. Natl. Acad. Sci. USA* 77 (1980), p. 6019.

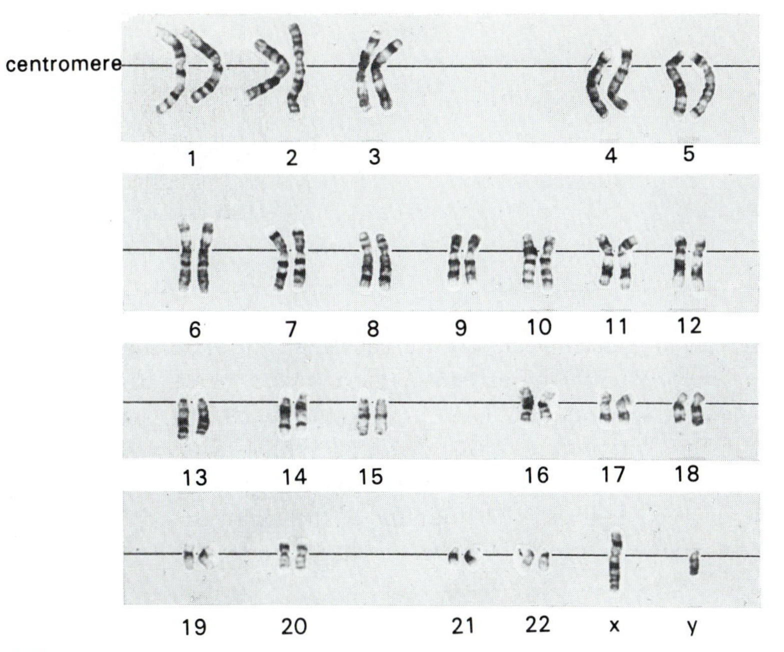

(a)

Figure I.9
The human karyotype. A horizontal line is drawn through the centromeres, and the chromosomes are numbered according to decreasing length.
(a) Photograph of a full set of male prometaphase chromosomes stained to reveal distinctive banding patterns. The two members of each homologous pair are shown. Courtesy of Uta Francke.
(b) Schematic representation (ideogram) summarizing the information in (a). Heavy hatching marks the centromeres; light hatching marks variable intensity bands. From D. G. Harnden and H. P. Klinger, eds., *An International System for Human Cytogenetic Nomenclature* (Basel: S. Karger, AG., 1985), p. 50.

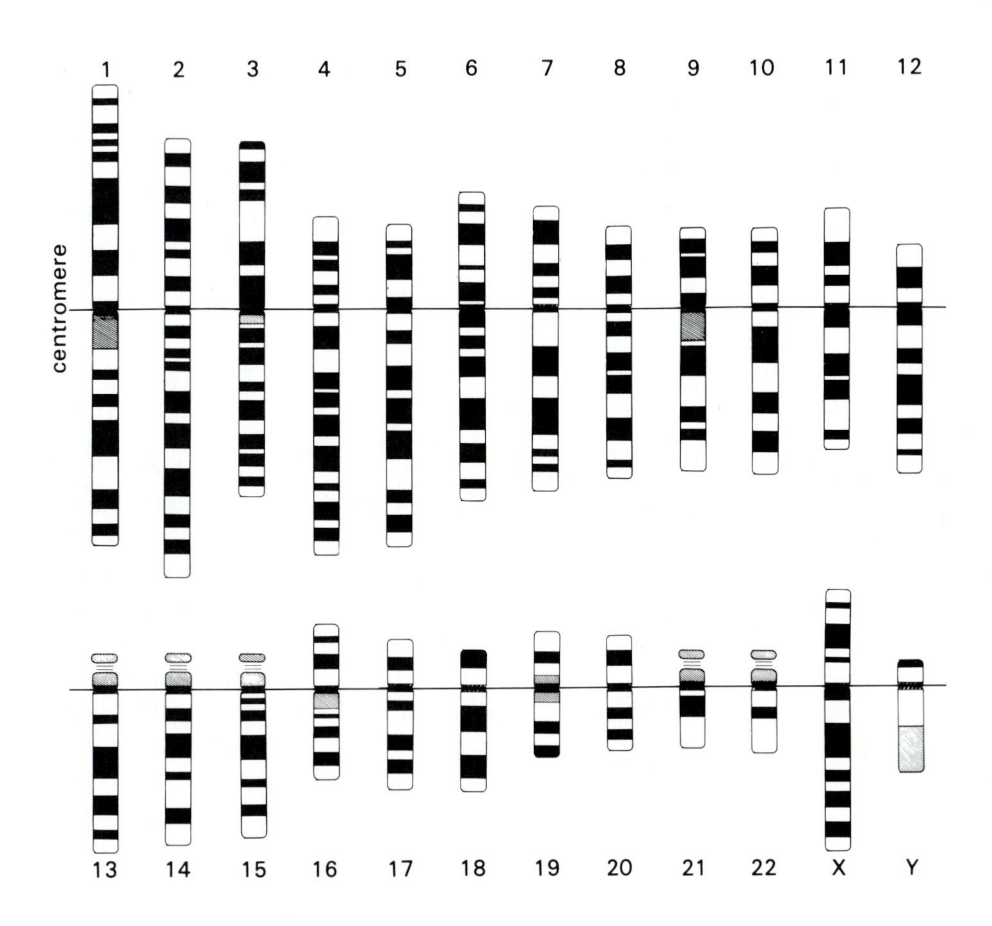

(b)

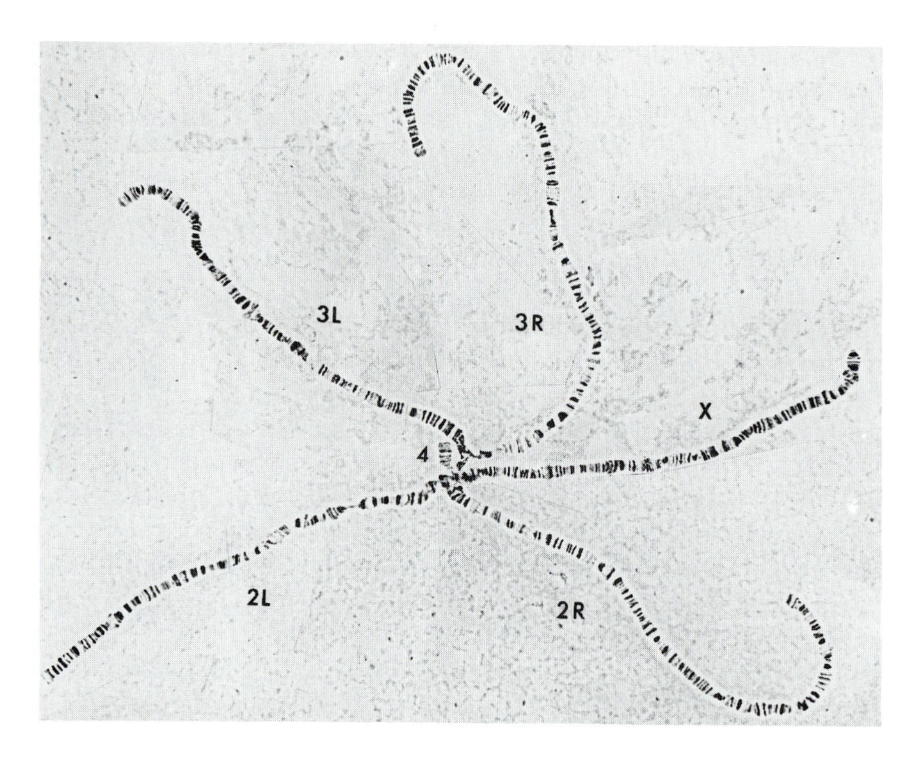

Figure I.10
Composite photograph showing the complete set of giant polytene chromosomes found in *D. melanogaster* salivary gland cells. All four chromosomes are joined near their centromeres (chromocenter). Each of these interphase chromosomes is a tight bundle of many chromatids that are replicas of both members of a homologous pair aligned with one another. The bands are revealed by staining, and the patterns and relative thickness of the bands are characteristic of each chromosome. Note that each band extends across the entire bundle. From G. Lefevre, Jr., as published in Chapter 2, Figure 3, in M. Ashburner and E. Novitski, eds., *The Genetics and Biology of Drosophila,* volume 1a (London: Academic Press, 1976).

side by side, resulting in fat and untangled bundles of chromatids that are called **polytene** chromosomes (Figure I.10). Like all interphase chromosomes, polytene chromosomes are also stretched out rather than condensed like metaphase chromosomes. Staining of the polytene chromosomes with special dyes gives a distinctive pattern of dark and light bands. Unlike the bands visible in the highly condensed metaphase chromosomes, these bands are very numerous. For example, about 5000 dark bands are visible on the 4 *D. melanogaster* polytene chromosomes; at most, 2000 are visible on the full set of 23 human metaphase chromosomes.

The distinctive morphological features of individual prometaphase and polytene chromosomes are highly reproducible from one individual of a species to another. Unusual shapes or banding patterns signal aberrant chromosomal material, as do atypical numbers of chromosomes (Figure I.11). The presence of such chromosomes is frequently associated with disease. For example, portions of one chromosome are sometimes translocated to a completely unrelated chromosome, a rearrangement that is readily detected by an unusual size or banding pattern. Translocations are sometimes reciprocal, with two unrelated chromosomes exchanging segments. Other examples of aberrant chromosomes include deletions of parts of the normal chromosomes, duplications of certain regions, and even inversions of segments. Sometimes there are missing or extra chromosomes. For example, the human disease known as **Down's syndrome** is associated with three copies of chromosome 21 (termed chromosome 21 trisomy) rather than the usual diploid state.

Fruitful analysis of chromosome structure depended on the choice of suitable experimental organisms. Thus, the bulky polytene chromosomes of *D. melanogaster* became a favorite experimental system early in the development of the field termed **cytogenetics**; systematic analysis of human and other small mammalian chromosomes awaited technical improvements made after 1950. The small chromosomes of prokaryotes are not visible in the light microscope, and, similarly, the small and diffuse chromosomes of such simple eukaryotes as yeast and the trypanosomes have eluded light microscope analysis.

The Inheritance of Single Traits

The concept of a gene began with Gregor Mendel in the 1860s, although the word itself was not coined until others had repeated and extended his findings during the beginning of the twentieth century. The word **gene**, which was introduced by W. Johannsen in 1910, referred to a hypothetical unit of information that governs the inheritance of an individual characteristic in an organism. The existence of genes was inferred from the statistical distribution of simple heritable traits among the offspring of known parents over several generations. These early studies dealt with genes as abstract and statistical entities because there was no information regarding the chemical basis of the traits being studied. For example, the texture or color of seeds or flowers could be followed as a discernible, heritable trait without regard for the chemical or metabolic rationale of that property. Nevertheless, the logical intellectual framework established by Mendel and those who followed is consistent with our current views of the chemical structure of genes and how that structural information determines an organism's attributes.

Independent Segregation and Independent Assortment

The Mendelian view of inheritance in eukaryotes was defined by two major findings. The first is the occurrence of independent segregation. Organisms contain a pair of genes for any single inherited trait; each member of the pair is either maternal or paternal in origin. In each generation, members of each pair of genes segregate during formation of new eggs or sperm, and, upon fertilization, a new gene pair is established. Each member of a pair of genes is now called

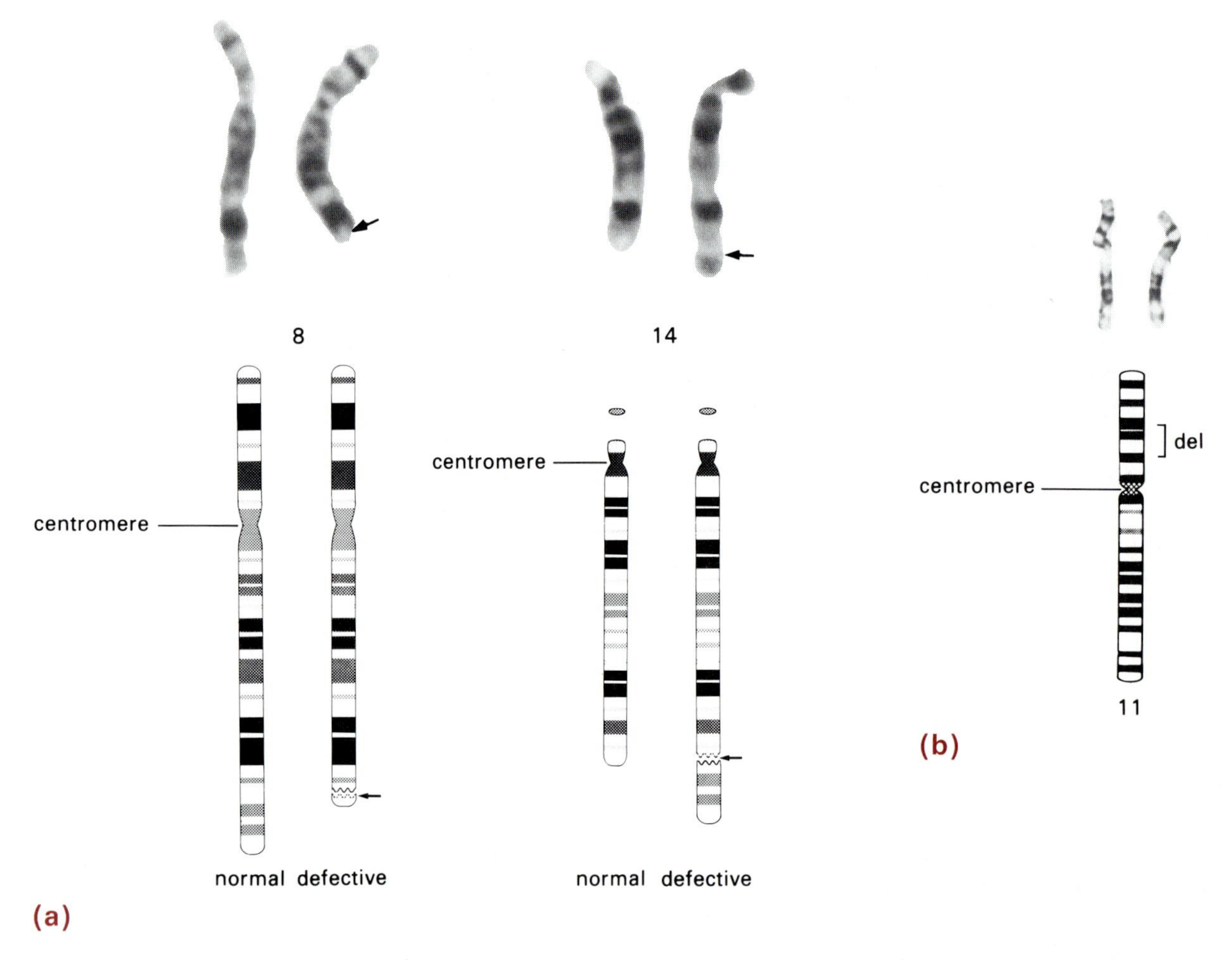

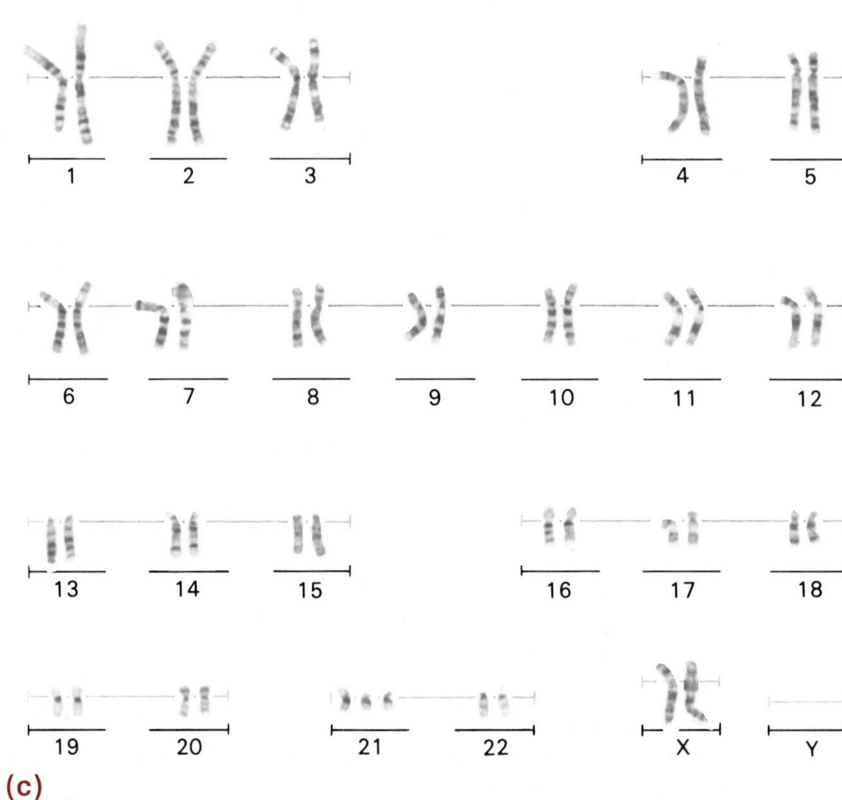

Figure I.11
Aberrant chromosomes associated with human disease. In (a) and (b), photos of banded normal (left) and aberrant (right) chromosomes are shown with corresponding ideograms below. These ideograms display more detailed band patterns than those in Figure I.9. (a) Translocation (arrow) from chromosome 8 to 14 in a human Burkitt's lymphoma. (b) Deletion in part of human chromosome 11 in cells from a Wilms's tumor. (a) and (b) are courtesy of U. Francke. (c) Three copies of chromosome 21 in the karyotype of a female Down's syndrome patient. Courtesy of T. A. Donlon.

an **allele**, and the particular form of the trait specified by each allele may be the same (the organism carrying identical alleles is said to be **homozygous**) or different (**heterozygous**). Thus, an allelic pair determining trait a may be a^1a^1 or a^1a^2 or a^2a^2 in different individuals. Sperm or eggs will then be either a^1 or a^2. Although an individual organism can contain at most only two different alleles of a particular gene, many different alleles can exist in the entire population of a species. For example, there may be multiple forms of gene a: a^1, a^2, a^3, a^4, and so forth. Individual organisms may then contain various pairs, such as a^1a^2, a^2a^2, a^3a^2, a^1a^4, a^4a^5, a^4a^4.

The second major Mendelian finding concerns the independent assortment of different allelic pairs of genes, each of which specifies a different trait. For example, in an organism that contains the allelic pairs a^1a^2 for trait a and b^1b^2 for trait b, the eggs or sperm may contain a^1b^1, a^1b^2, a^2b^1, or a^2b^2. During the formation of gametes, the segregation of allelic pair a is independent of the segregation of allelic pair b and so on for other allelic pairs.

The Relation Between Genes and Chromosomes

The correlation between the physical behavior of chromosomes and the insights of Mendelian genetics was discerned early in the twentieth century. Each member of an allelic pair of genes could be associated with one of a pair of chromosomes, and independent assortment of alleles could be explained if different allelic pairs were on different chromosomes. Thomas Hunt Morgan and his colleagues proved that, in *D. melanogaster*, genes are associated with the fly's chromosomes. They had adopted this organism because its short generation time and the large numbers of offspring produced from each mating make genetic analysis convenient and precise; *D. melanogaster* chromosomes are also readily observable in the light microscope. The experiments revealed that the inheritance of an allele producing white rather than the usual red eyes was always associated with inheritance of an X-chromosome, never with a Y-chromosome. Other alleles, governing other traits, were also associated with the inheritance of X-chromosomes, and still others were frequently inherited together in linked groups unrelated to the X-chromosome. Subsequently, it was evident that the number of groups of linked alleles equaled the number of chromosome pairs. These studies established that alleles associated with different chromosomes assort independently among the progeny, but those associated with the same chromosome remain together in the offspring.

Recombination

Almost as soon as the existence of linkage groups was established, surprising exceptions to linkage were noticed. For example, a grouping of alleles such as a^1b^1 or a^2b^2 might generally be linked in the offspring; but occasionally the pairings were reassorted, and new linkages such as a^1b^2 and a^2b^1 appeared and were inherited in subsequent generations. Because cytogenetic analysis revealed that

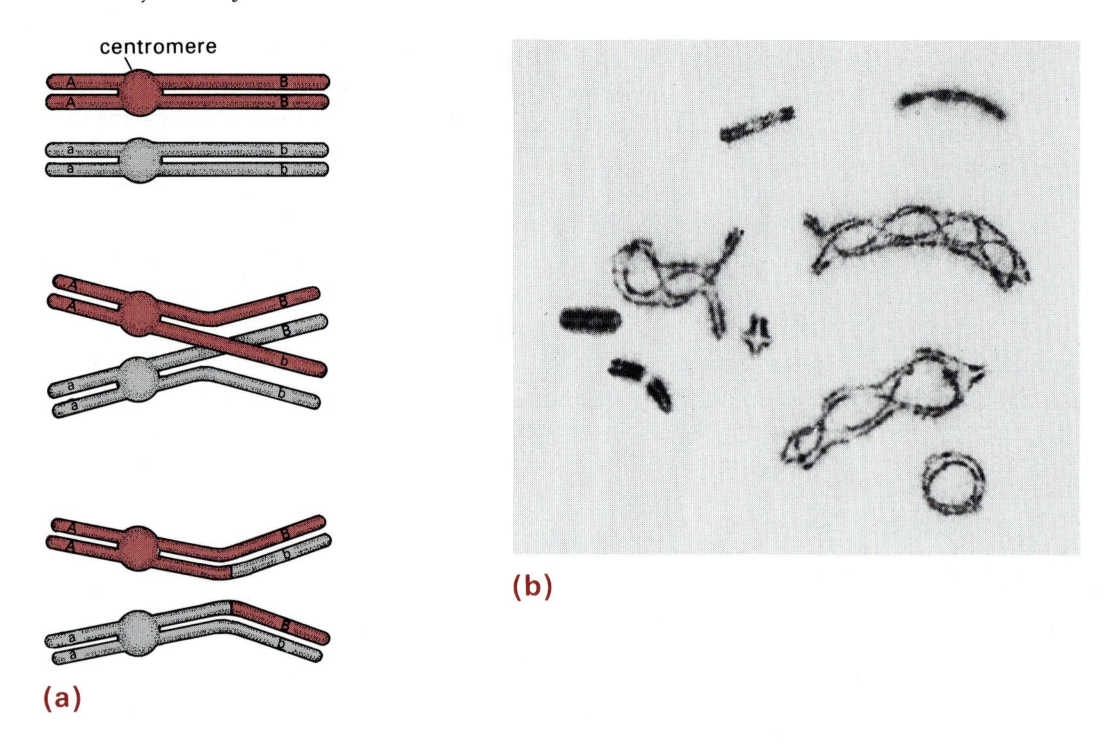

Figure I.12
Diagram and photograph of meiotic chromosomal crossing-over. (a) On the diagram, A and a and B and b are pairs of alleles. (b) The photograph shows grasshopper (*Chorthippus parallelus*) chromosomes in the diplotene stage of meiosis (prophase I). Homologous pairs of already duplicated chromosomes are visible, as illustrated in the drawing. Several of the pairs show multiple crossover sites (chiasmata). Some pairs have already condensed in preparation for metaphase I. Courtesy of B. John.

homologous chromosomes can wrap around one another during meiosis, Morgan surmised that they might exchange portions of themselves, thereby creating new combinations of linked alleles (Figure I.12). This process was termed **crossing-over** or **recombination**. Without knowing anything of the underlying chemistry, geneticists used the phenomenon of recombination as the basic tool of genetic research. Measurements of recombination frequencies between linked allelic pairs in *D. melanogaster* led to three important conclusions: Genes are arranged in a linear order with the members of an allelic pair usually occupying the same relative position on homologous chromosomes. Recombination occurs only within the same linkage group (i.e., between homologous chromosomes). The frequency with which two different linked alleles cross over together (as a bundle, so to speak) is a function of how close they are on the chromosome (the closer together, the higher the frequency). The relative chromosome positions of different genes in *D. melanogaster*, and eventually in other organisms, were mapped using these ideas. By 1922, Morgan and his colleagues could order several hundred genes on the four *D. melanogaster* chromosomes (Figure I.13).

The Relation Between Genes and Proteins

The first clue as to how the information in genes is expressed to give specific cellular and organismal characteristics preceded even the invention of the word *gene*. In the first decade of the twentieth century, an English physician named Archibald Garrod noted that the inheritance of certain human metabolic idiosyncrasies and disorders followed Mendelian rules. He also surmised that such inherited disorders stemmed from the deficiency or absence of a particular en-

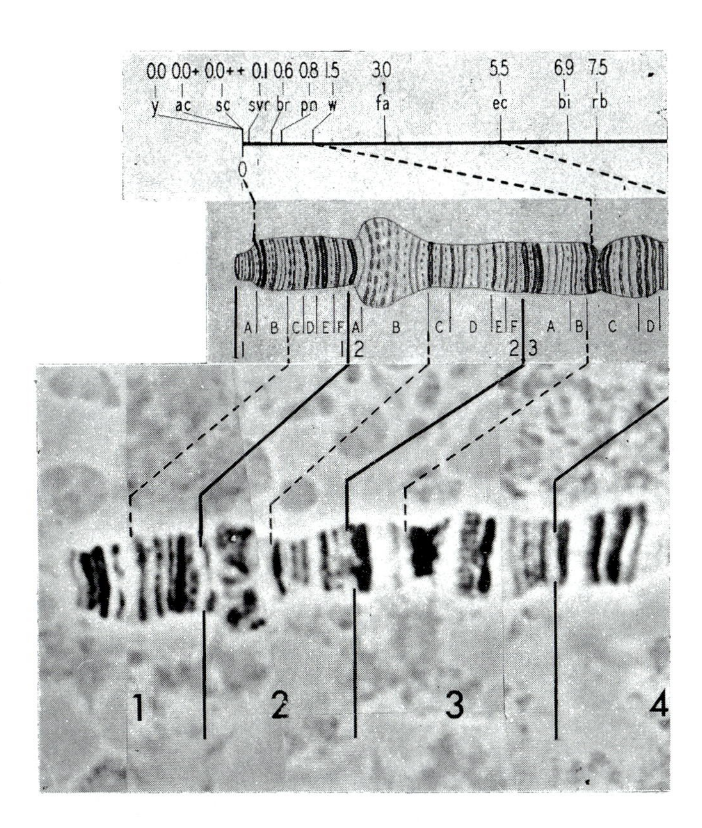

Figure I.13
A map of genes arranged along part of the X chromosome (polytene) of *D. melanogaster*. Note that the position of the genes can be described relative to the presence of a specific band. Both a diagram and a photograph are shown. Genes are indicated at the top. For example, mutation in gene *w* (region 3C) leads to white eyes. From C. B. Bridges, *Journal of Heredity 26* (1935), p. 60.

zyme required for normal metabolism. Garrod then suggested that the determinants of heredity controlled the production of enzymes. Thus, the ability to produce a particular enzyme, or even the quality of the enzyme, was attributed to genes. Even in the absence of experimental proof, the proposal was attractive because it related the then contemporary genetic studies on flies and plants to human biology.

To pursue Garrod's idea further, a new experimental approach was needed. That was provided by the adoption of microorganisms as experimental subjects in the late 1930s. Initially, simple fungi from the genera *Aspergillus* and *Neurospora* were the focus of attention. These organisms grow readily in simple, defined culture conditions and reproduce quickly enough to yield many generations of descendants in a short time. Sufficient genetic and biochemical data were obtained and correlated by the mid 1940s to establish that the presence or absence of an enzyme was heritable and governed by the expression of a single gene. George Beadle and Edward Tatum generalized the relation between enzyme and gene as the "**one enzyme–one gene**" dictum. Because enzymes are proteins and many proteins contain more than one kind of polypeptide chain, the notion evolved to "one polypeptide–one gene." Research has since shown that some genes specify proteins that are not enzymes (e.g., hormones and structural proteins), and some control the production of RNA molecules needed to assemble proteins.

The two components in this informational relationship are often described as the **genotype** and **phenotype**, referring respectively to a gene and the protein (or RNA) it encodes. More generally, genotype sometimes refers to the totality of genetic information in a specific

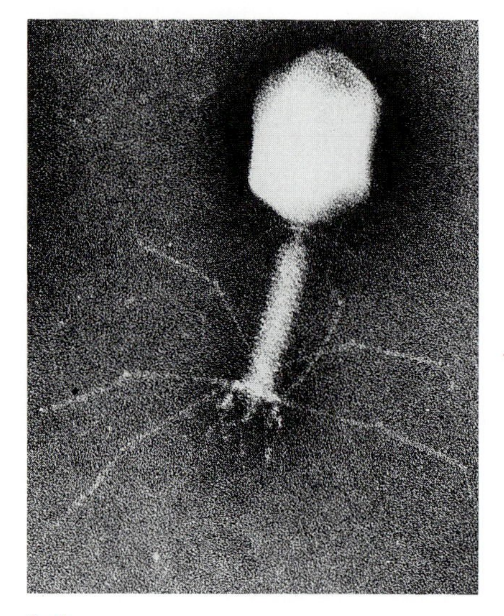

(a)

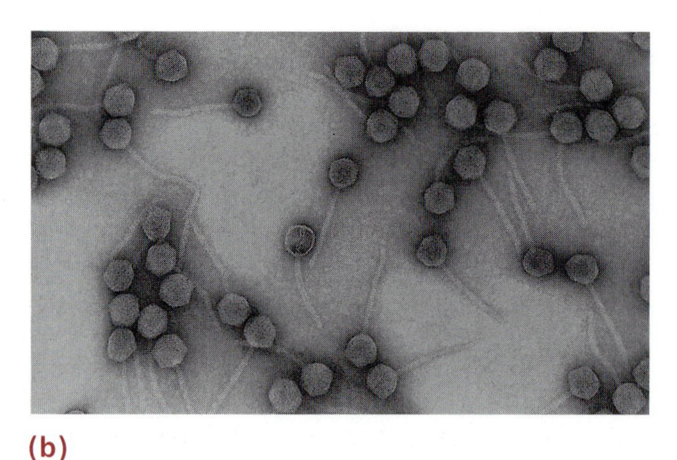

(b)

Figure I.14
(a) Electron micrograph of bacteriophage T4, magnification, × 200,000. Courtesy of R. C. Williams.
(b) Electron micrograph of bacteriophage λ, magnification, × 77,000. Courtesy of A. D. Kaiser.

cell or organism. Similarly, phenotype is used more globally to describe the observable characteristics of a cell or organism, be they particular proteins or functions or morphological or behavioral attributes. Phenotypes are frequently the result of an interplay between the genetically encoded information and the environment in which it is expressed. The term **genome** is used to describe the totality of the chromosomes (in molecular terms, DNA) unique to a particular organism (or any cell within the organism), as distinct from genotype, which is the information contained within those chromosomes (or DNA).

By 1950, an even more attractive system for examining the relationship between genes and cellular functions had been developed. The common intestinal bacterium, *Escherichia coli* (*E. coli*), has simple nutritional requirements and divides every 20 to 60 minutes (depending upon culture conditions) to yield large numbers of cells (10^9 per ml). Moreover, a large number of readily measurable physiological characteristics was found to be genetically controlled. Furthermore, readily obtainable and characterizable mutants of *E. coli* enabled the identification of genes for specific cellular functions. The way was opened to a more formal genetic analysis and the construction of a genetic map of *E. coli*'s single chromosome. An added bonus was that *E. coli* is a host for several viruses (bacteriophages) (Figure I.14) that themselves display genetic variabilities in their infectious behavior.

Bacteriophages, or phage for short, provided even simpler genetic systems for analysis. Two or more phage inside an infected cell can undergo exchanges between parts of their homologous genomes to yield progeny phage with new genetic properties (Figure I.15). Phage genomes can even insert themselves, reversibly, into bacterial chromosomes. When a phage genome is excised, it occasionally carries part of the bacterial genome with it; thus, phage can become carriers of bacterial genes. Analysis of these genetic exchanges indicated that even such primitive organisms contain an organized genome and that individual genes could be ordered on a genetic map.

Genes and DNA

Today's genetic chemistry has its origin in Friedrich Miescher's discovery of DNA in 1869. He recognized that material extracted from pus cells and cell nuclei was chemically different from proteins in both its content of organic phosphorous and its resistance to destruction by proteolytic enzymes. During the next 85 years, new methods were developed to isolate DNA, to establish the nature of its chemical substituents, and to determine the linkages between them. These investigations culminated in the identification of the basic structural unit of DNA: a phosphorylated sugar, deoxyribose phosphate, joined to a nitrogenous base, either one of the purines—adenine or guanine—or one of the pyrimidines—cytosine or thymine (Figure I.16, page 26). Meanwhile, biophysical studies established that DNA molecules are very long chains, the backbone being composed of deoxyribose phosphate units joined to one another through **phosphodiester** bridges; attached to each deoxyribose in the chain is either a purine or a pyrimidine base (Figure I.17, page 26). In 1953, the existing chemical and physical information about the composition and structure of DNA was assimilated by James Watson and Francis Crick into their now classical double helical structure. The significance and impact of their discovery was magnified by the growing acceptance of the discoveries by Oswald Avery and his colleagues and by Alfred Hershey and Margaret Chase that DNA alone was the carrier of genetic information. The central role in heredity attributed to chromosomes could now be assigned to the DNA they contained.

DNA was not the only nucleic acid (or polynucleotide) found in cells. Closely related molecules, ribonucleic acids (RNA), differ chiefly by having the sugar ribose in place of the deoxyribose in DNA and by occurring most frequently as single strands.

The solution of the structure of DNA and the recognition of its central role in heredity represent the crowning achievements in the transition of genetics from a statistical and phenomenological science to one with an overriding chemical and molecular perspective. The immediate, exuberant response to the double helix reflected a sense of its correctness. Not only did the model fit the chemical and physical data, but the structure was perfectly suited to the functions required of the genetic material. It could encode a large amount of information in the linear sequence of the four purines and pyrimidines, and, in principle, could direct its own replication. Solving the double helical structure of DNA illuminated every aspect of biology and provided a foundation upon which the protean observations of preceding centuries could settle. It defined a fundamental unity for interpreting the enormous diversity of life. All at once, heredity could be equated with molecular structure.

Genetics—a Molecular Science

Beginning in the early 1950s, the longstanding questions concerning the mechanisms of transmission, segregation, mutation, reassort-

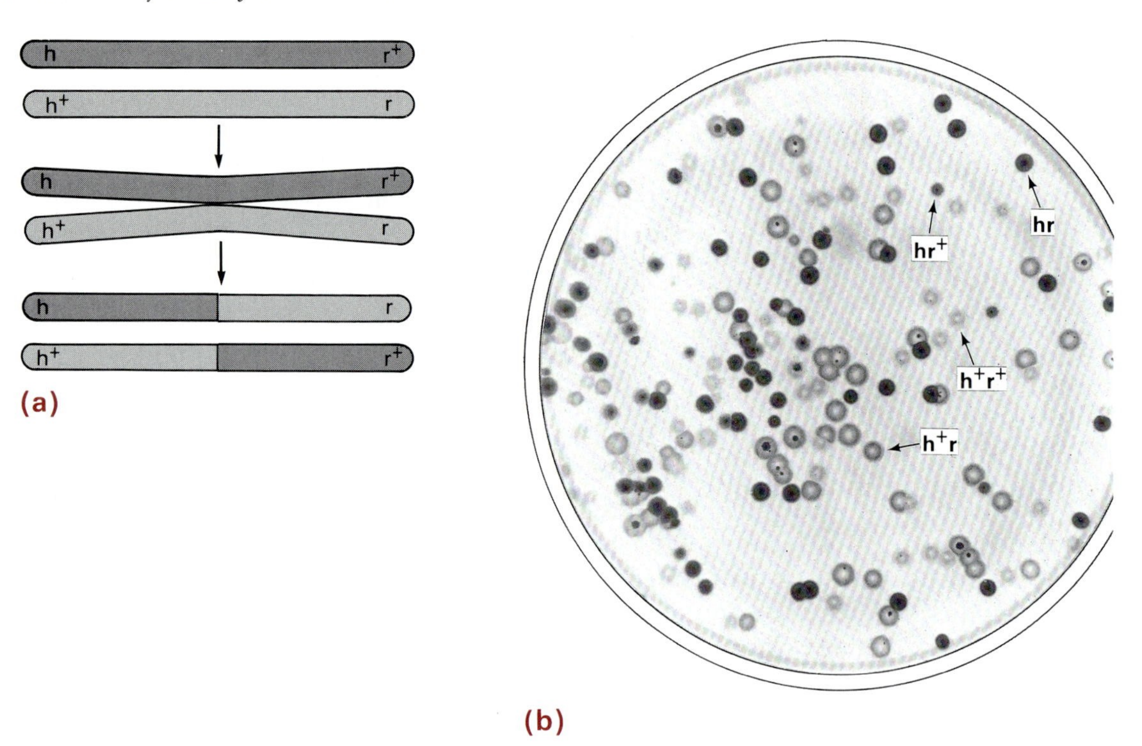

Figure I.15
Alternate fates of phage chromosomes after infection. (a) Two phage chromosomes carrying different pairs of alleles for the genes h and r (hr^+ or h^+r) infect a single bacterial cell. Four different kinds of chromosomes occur in the progeny phage after recombination: the two parentals, hr^+ and h^+r (no recombination) and two recombinants, hr and h^+r^+. (b) Each type of chromosome gives phage plaques with a different morphological phenotype. The plaques are visible as relatively clear areas on a lawn of bacteria growing on agar. Each plaque arises when a single, initially infected cell releases phage, which then infect the surrounding cells, leaving a clear area of dead cells on the lawn. From G. S. Stent, *Molecular Biology of Bacterial Viruses* (San Francisco: W. H. Freeman, 1963), p. 335. (c) PAGE 25: Recombination between phage and bacterial chromosomes. A single break in each chromosome allows them to join broken ends. The result is the insertion of the phage chromosome into the bacterial chromosome. The process can be reversed, causing excision of the phage chromosome. Occasionally, excision is imprecise. The excised phage DNA then carries with it one or more bacterial genes, leaving phage genes behind in the bacterial chromosome.

ment, and expression of genetic characteristics were reformulated in molecular and chemical terms. How do DNA molecules replicate and recombine? Do mutations correspond to alterations in DNA structure? How are they maintained in succeeding generations? How does information encoded in DNA determine the production of phenotypic products, the proteins? How is the readout of DNA information regulated in cell growth and during development and other physiological states? And how are these processes altered in disease? These and other questions have formed the agenda for research in molecular

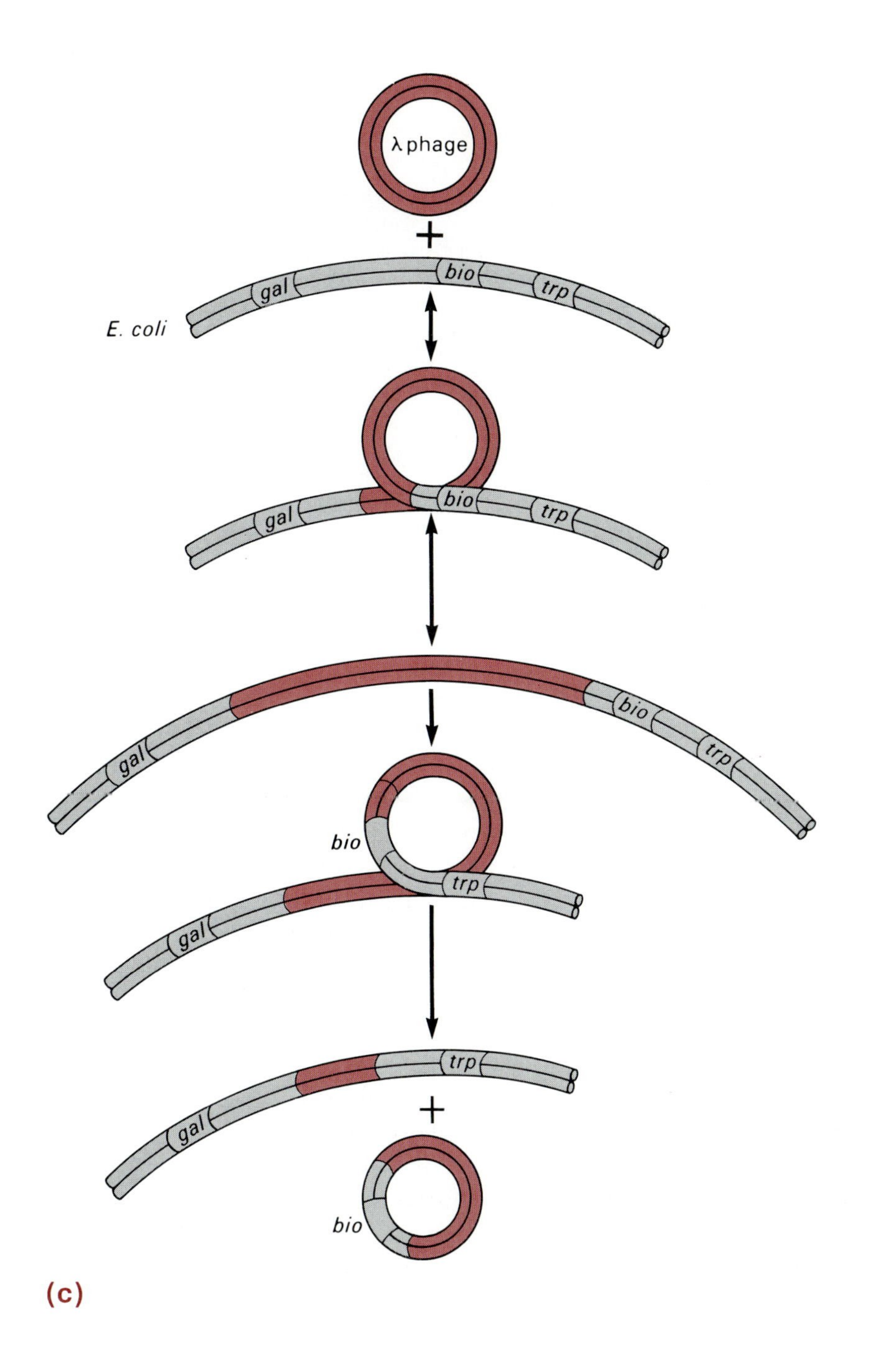

genetics during the last 35 years. The explosive progress during the first 20 of those years, primarily in studies with prokaryote systems, identified the molecular structures involved in the storage, maintenance, transmission, and utilization of genetic information.

The Intracellular Flow of Genetic Information

The informational relationships between the genetic molecules, DNA, RNA, and protein, are now firmly established (Figure I.18, page 27).

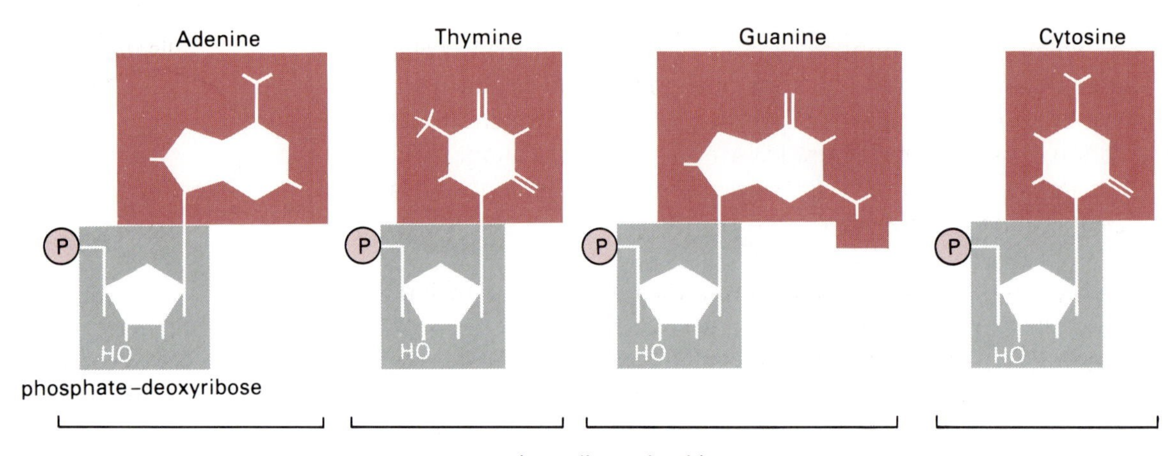

Figure I.16
The basic structural units in DNA—the deoxyribonucleotides.

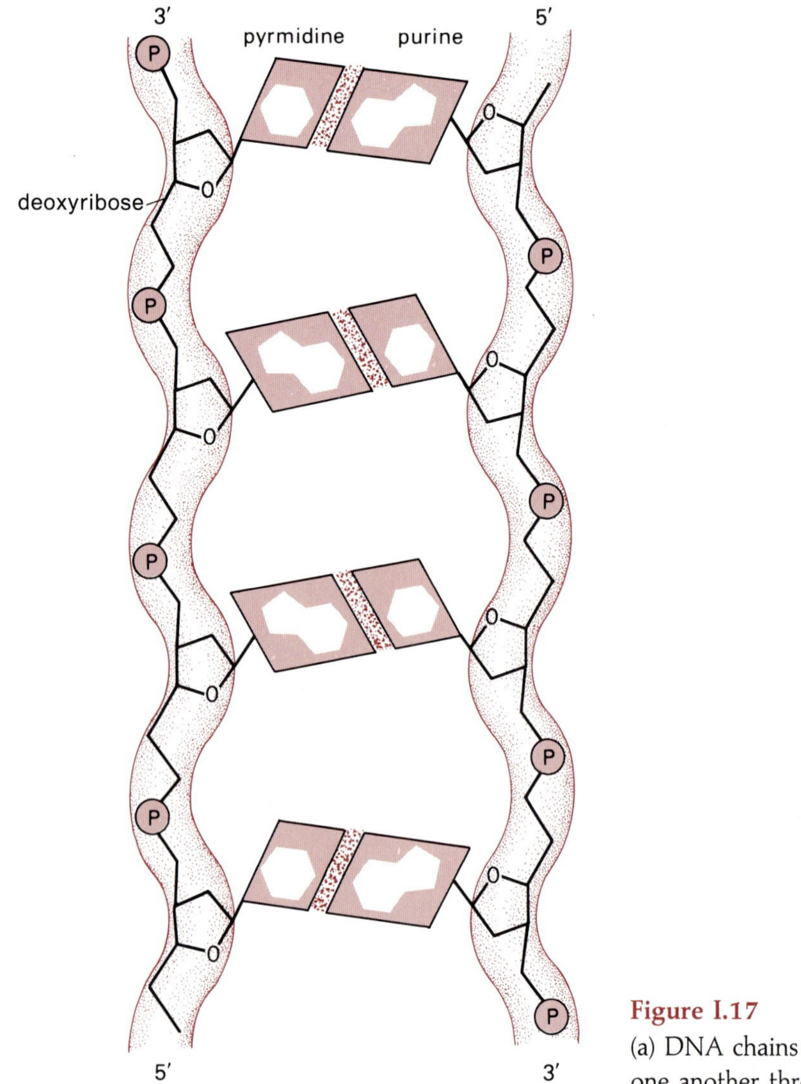

C-G
T-A
A
C-G
A-T
C-G
C-G
G
T-A
G-C
C-G
T-A
A
C-G
A-T
C-G
C-G
G
T-A
A-T
T-A

(b)

Figure I.17
(a) DNA chains are formed from deoxyribonucleotides joined to one another through sugar-phosphate linkages. (b) Two such chains wind around a common axis to form double helical DNA.

Replication, which creates identical copies of the parent DNA molecule, maintains genetic continuity from one generation to the next. **Transcription** of DNA information into RNA permits the **translation** of that information into proteins. Thus, DNA performs two critical functions. One is its role in directing its own replication. The second is its role in defining the phenotype via the formation of RNA molecules involved in the translation of DNA information into protein. Uniquely in eukaryotes, at least as far as is known, information can also flow from RNA into DNA by a process called **reverse transcription**.

A universal feature of information transfer from DNA to DNA or to RNA and from RNA to DNA is the involvement of a nucleic acid as a **template**. Nucleic acids direct the assembly of identical or related molecules and participate directly in protein synthesis. So far as is known, information does not pass from proteins to nucleic acids (i.e., reverse translation is unknown). However, proteins have a critical role in catalyzing and facilitating information transfer between nucleic acids as well as in assembling proteins themselves.

What follows is a brief synopsis of the key features of the genetic machinery and its operation: the structural features of the principal molecular components—DNA, RNA, and protein—and how they function to maintain the integrity of the genome and to translate an organism's genotype into its phenotype. Chapters, 1, 2, and 3, which make up Part I, provide details of these matters.

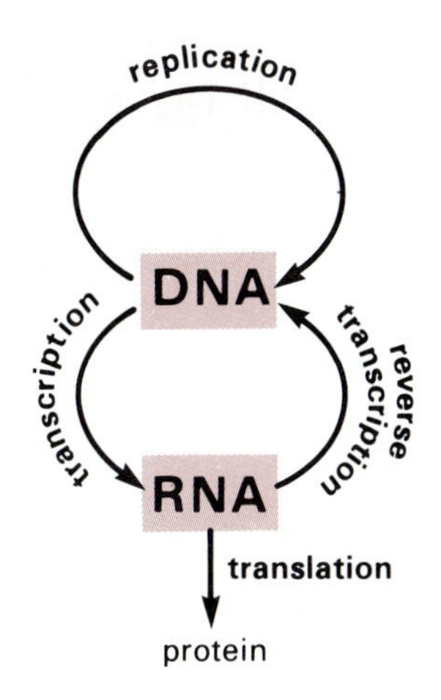

Figure I.18
The informational relationships between DNA, RNA, and protein and the processes that mediate the information flow. Note that reverse transcription has been observed only in eukaryotic systems.

The Structure and Maintenance of DNA Genomes

All cellular DNA consists of two polynucleotide chains wound around a common axis: the DNA double helix (Figure I.19). Tracing the outside of the helix are the backbones of each chain formed by very long stretches of repeating deoxyribose-phosphate residues. The two chains are held together by hydrogen bonds formed between purine bases on one strand and pyrimidine bases on the other: adenine always pairs with thymine, and guanine always pairs with cytosine. A consequence of these essentially invariant base pairs is that the sequence of bases on one strand uniquely defines the sequence of bases on the other strand, and the two DNA chains in a double helix are thus said to be **complementary**.

DNA molecules play two different roles in their own replication. First, the sequence of purine and pyrimidine bases on each chain provides a template from which a new strand is copied. Second, genes encoded in DNA direct the synthesis of enzymes and other proteins that are required for new DNA synthesis. Replication involves unraveling the two strands at a particular site in the DNA double helix. Each strand then acts as a template to direct the formation of a new complementary strand (Figure I.19). Thus, each of the two daughter helices inherits one of its chains from the parental DNA helix, the other chain being derived by *de novo* synthesis. Although simple in its logic, DNA replication is actually a complex process requiring many proteins, central among which are enzymes called **DNA polymerases**. Their role in replication is to assemble chains of polynucleotides from individual mononucleotides (deoxy-

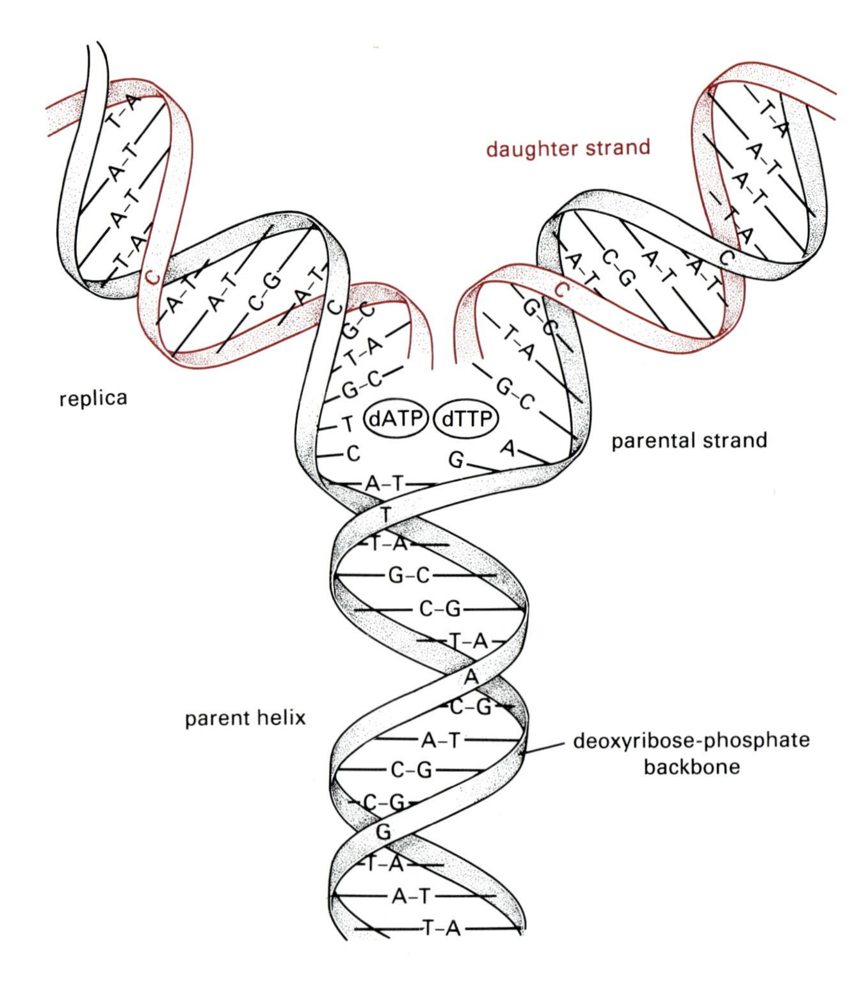

Figure I.19
Diagram of a replicating DNA molecule. The bottom portion is still double helical and base paired. The top portion is unwound, and new DNA strands are associated with each parental strand by base pairing.

nucleoside triphosphates). All DNA polymerases elongate chains one deoxynucleotide at a time. The choice of the deoxynucleotide to be added is determined by the ability of its base to form a complementary pair with the next available base in the template strand. The high fidelity of the replication process assures a nearly error-free transmission of genetic information from one generation to the next.

One of the revelations that emerged from studies of simple genomes is that they encode the machinery for their perpetuation and maintenance (i.e., replication and repair of the DNA genome). Moreover, the genetic program also promotes rearrangements of DNA, oftentimes producing disadvantageous arrangements but occasionally creating new combinations of genes that allow for evolutionary experimentation. All genomes encode the information for the synthesis of RNA, enzymes, and other proteins that participate in these processes. One of these processes, genetic recombination, produces exchanges between segments of homologous chromosomes. Earlier we noted that such genetic exchanges frequently correlate with the occurrence of meiotic chromosome pairing and visually distinctive crossovers between the involved chromosomes. In molecular terms, the recombinations occurring at crossover sites result from the breaking and rejoining of strands within corresponding (homologous) regions of the DNA helices in the recombining chromosomes (Figure

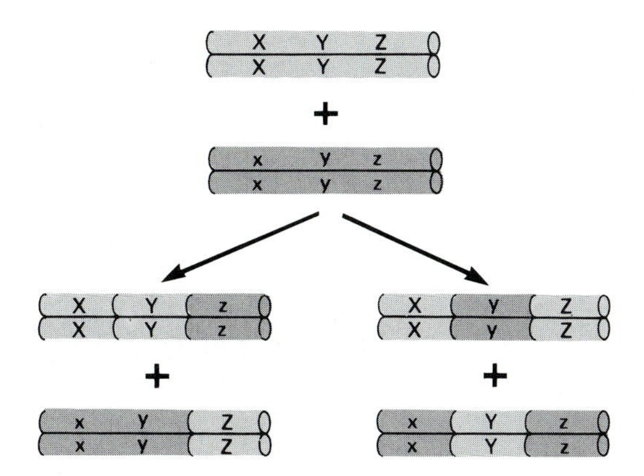

Figure I.20
Recombination between two homologous parental helices carrying different alleles (capital and lowercase letters) for the genes *X*, *Y*, and *Z*. Various possible products of recombination are shown.

I.20). Certain types of recombinations, also programmed by chromosomal information, occur between special DNA segments in otherwise nonhomologous chromosomes; as a result, novel associations and rearrangements of genetic structures are produced. The various processes of recombination uncovered in prokaryotes require a panoply of enzymes to promote the matching of homologous or specific sequences and to catalyze the chain scissions and rejoinings.

There are also special mechanisms for the repair of damage to the DNA structure. Exposure of cells to ultraviolet light, X rays, or a variety of chemical agents produces a wide assortment of lesions in the bases or backbone of the DNA helix. DNA encodes repair enzymes and proteins that maintain the integrity of an organism's genome.

Expression and Regulation of a Genome's Phenotype

Proteins are the principal determinants of an organism's phenotype. They make up both the enzymatic machinery for the metabolic, energetic, and biosynthetic activities of all cells and the regulatory elements that coordinate these activities in response to endogenous and exogenous stimuli. Proteins also contribute many of the structural elements that characterize a cell's morphology and movement. In short, organisms are what they are because of the array of proteins they manufacture.

The one polypeptide–one gene theory of gene action provided an attractive conceptual framework for relating an organism's genotype to its phenotype. But not until the structures of proteins and DNA were solved, which occurred in the early 1950s, did the theory acquire a molecular reality. Newly developed methods for analyzing protein structures established that each protein possesses a unique linear amino acid sequence along its polypeptide chain (Figure I.21a). This unique amino acid sequence, called the **primary structure**, directs the folding of the linear polypeptide chain into a characteristic and biologically functional three-dimensional shape (Figure I.21b and c). Therefore, research indicated that genes determine protein structure by specifying amino acid sequence. The finding that mutations change the amino acid sequence of the affected proteins substantiated this. Moreover, the sites of mutations in genes and the positions at which amino acids in the corresponding proteins were changed were colin-

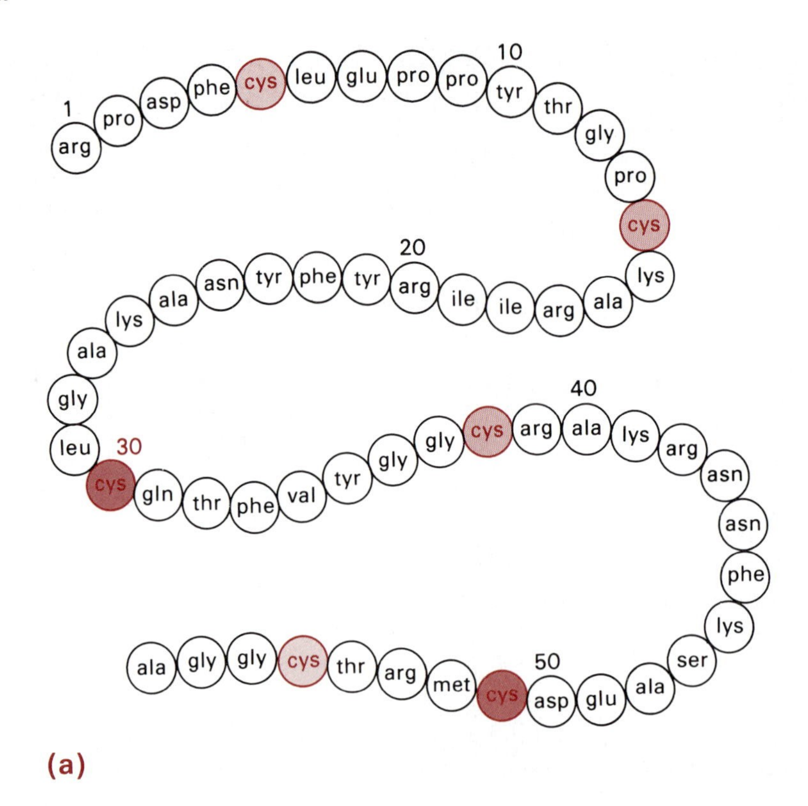

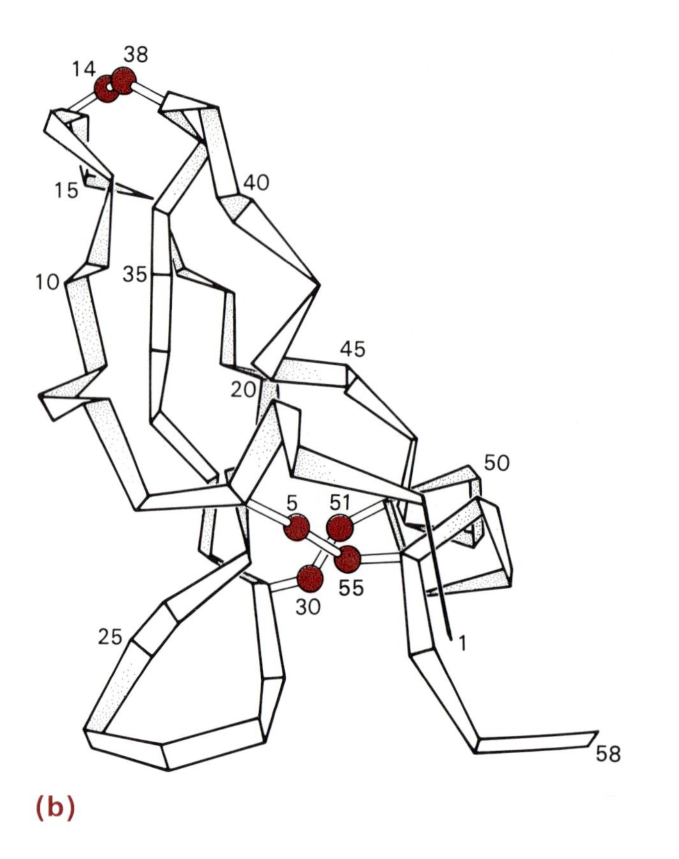

Figure I.21
(a) The primary structure of the bovine protein, pancreatic trypsin inhibitor. Names of individual amino acids are abbreviated. (b) The polypeptide is folded into its characteristic three-dimensional structure. The drawing is a "ribbon" diagram. It shows the localized folding or secondary structure (e.g., the helical region from residue 47 to 56) and the long range folding or tertiary structure involving interactions between amino acids that are not near-neighbors in the primary structure. Disulfide bonds between pairs of the amino acid cysteine (colored balls) stabilize the folded structure. (c) PAGE 31, TOP: Photograph of a model of the protein. The polypeptide backbone is light. The ribbon diagram in (b) is equivalent to the view from the base of the model in (c). Courtesy of T. E. Creighton, *J. Mol. Biol.* 95 (1975), p. 167.

(c)

ear (i.e., in the same order). These findings indicated that the linear arrangement of nucleotides in DNA and of amino acids in proteins were related. Thus, a form for a genetic code could be forecast.

The notion of a genetic code implies that there is a mechanism for translating the nucleotide sequences in DNA into the amino acid sequences of proteins. Research from the mid 1950s through the early 1960s unraveled the molecular features of the genetic code and how it is deciphered in assemblying polypeptide chains. Breaking the genetic code represents one of the monumental achievements in contemporary molecular genetics. Unexpectedly, the code proved to be quite straightforward and virtually the same in all forms of life. Moreover, the general rules for translating genetically coded messages are also universal.

The genetic dictionary contains 64 **codons**, each comprising three adjacent nucleotides in a DNA chain (Figure I.22). Sixty-one of the

Ala	Arg	Asp	Asn	Cys	Glu	Gln	Gly	His	Ileu	Leu	Lys	Met	Phe	Pro	Ser	Thr	Trp	Tyr	Val	stop
	AGA									TTA					AGT					
	AGG									TTG					AGC					
GCA	CGA						GGA			CTA				CCA	TCA	ACA			GTA	
GCG	CGG						GGG		ATA	CTG				CCG	TCG	ACG			GTG	TAA
GCT	CGT	GAT	AAT	TGT	GAA	CAA	GGT	CAT	ATT	CTT	AAA		TTT	CCT	TCT	ACT		TAT	GTT	TAG
GCC	CGC	GAC	AAC	TGC	GAG	CAG	GGC	CAC	ATC	CTC	AAG	ATG	TTC	CCC	TCC	ACC	TGG	TAC	GTC	TGA

Figure I.22
The genetic code.

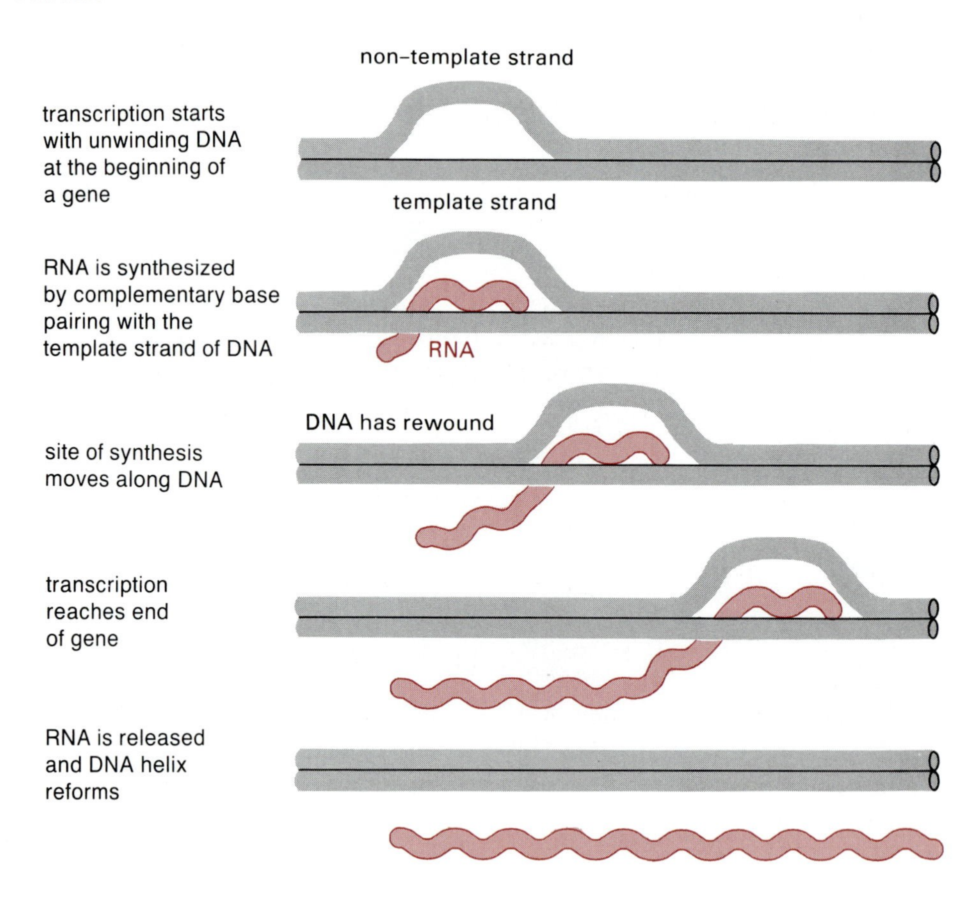

Figure I.23
The basic features of transcription. The RNA has the same nucleotide sequence as the nontemplate strand of DNA.

64 triplets specify amino acids, and each triplet encodes only one amino acid. One of these triplets has a dual function: it encodes the amino acid methionine, and it marks the beginning of protein coding stretches on DNA. Any of the remaining three triplets can signal the end of a protein coding sequence. The code is said to be "degenerate" because more than one codon can specify the same amino acid; but the code is not ambiguous because a particular codon never specifies more than one amino acid. If one knows the codon dictionary, the translation on paper of a gene sequence into its corresponding protein product is a trivial exercise.

The expression of a gene as protein requires that its DNA be transcribed into RNA (Figure I.23). This process is catalyzed by **RNA polymerases**, enzymes that assemble RNA chains by copying the nucleotide sequence in one strand of DNA by complementary base pairing. Genes that encode proteins yield messenger RNAs (**mRNAs**), so called because they carry the genetic message in DNA and function directly in protein assembly. Some genes do not encode proteins. Rather than mRNA, their transcription produces RNA molecules that are needed for the maturation of RNAs and the translation of mRNA sequences into proteins.

Studies of the interactions of RNA polymerases and other accessory transcription proteins with DNA have broadened our understanding of the specificity and strength of macromolecular interactions. Indeed, precise molecular contacts between proteins and specific groupings of nucleotides in DNA have been established,

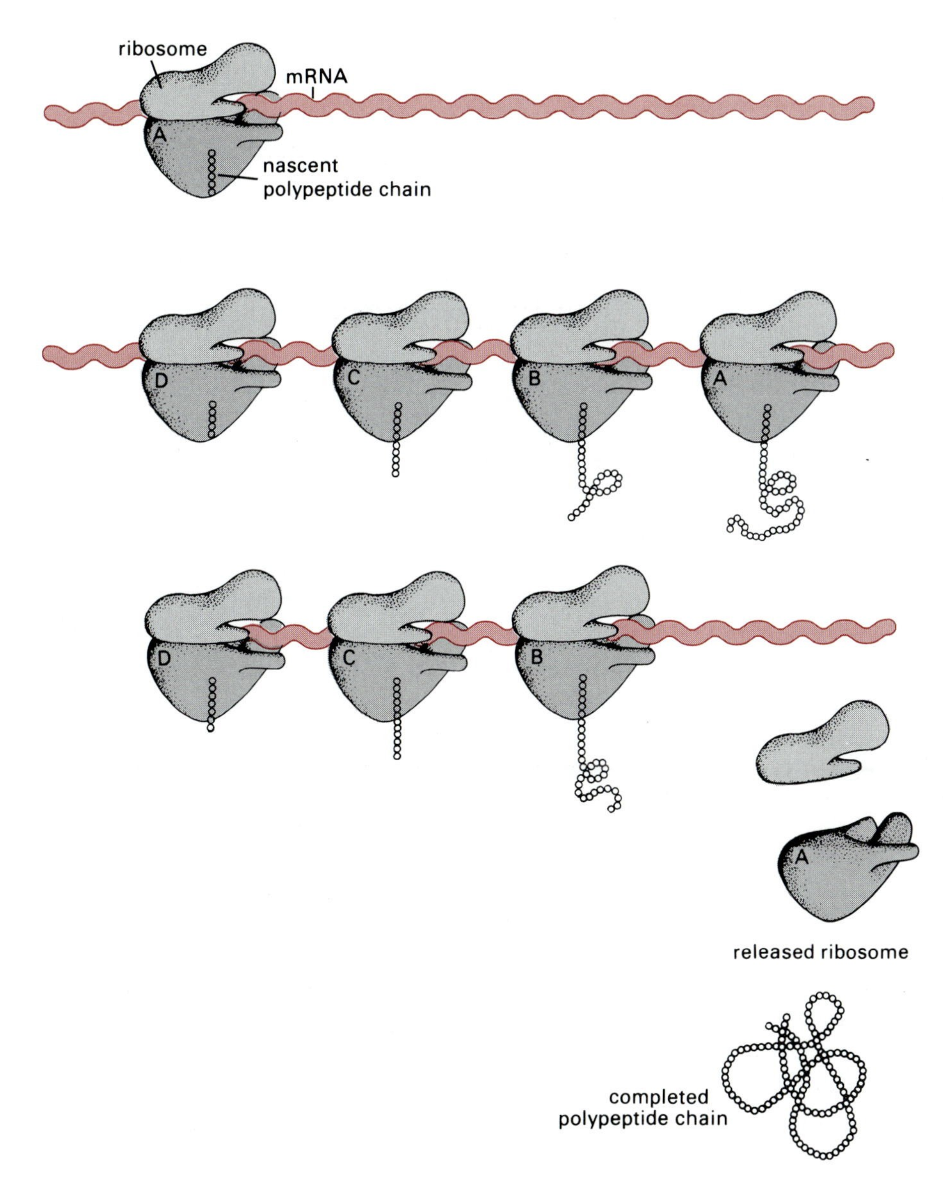

Figure I.24
The basic features of the translation of messenger RNA into protein. The first ribosome has already engaged the messenger RNA chain and carries a short polypeptide representing the first part of the encoded message. Polypeptide and messenger RNA chains are not drawn to scale. A few seconds later (second line), the ribosome has moved along the messenger RNA, which directs, triplet codon by triplet codon, the addition of more amino acids to the nascent polypeptide chain. Meanwhile, a second, third, and fourth ribosome have successively engaged the messenger RNA chain and initiated the assembly of additional polypeptide chains, forming a polyribosomal cluster on the messenger RNA. Somewhat later (third line), the first ribosome has completed assembly of its polypeptide chain, has released that chain, and has itself been released from the messenger RNA chain and from the polyribosomal cluster. Other ribosomes are nearing completion of their polypeptide chains. The nascent polypeptide chains begin to fold into their secondary and tertiary structures even before they are complete.

thereby providing a new perspective on how genes are expressed and regulated. We shall comment shortly on how such interactions are modulated to regulate gene function.

It is beyond the scope of this essay to describe the elaborate process by which the sequence of nucleotides in a messenger RNA is translated into a protein chain. The process is complex and involves a very large number of repetitive steps (Figure I.24). Ribonucleoprotein particles (**ribosomes**), containing more than 50 different proteins and 3 kinds of RNA molecules, catalyze the translation of mRNAs into proteins. The assembly of a protein chain begins when ribosomes attach to a messenger RNA. The protein chain elongates one amino acid at a time as the ribosome moves along the messenger RNA one codon at a time. The key element in translation, the conversion of the genetic information encoded in the triplet codons in messenger RNA into specific amino acids, depends on complementary base pairing. Each amino acid is attached to a special cognate **transfer RNA** (tRNA) that contains a triplet (an **anticodon**) that is complementary to the coding triplet on the messenger RNA. Base pairing between the codon and the anticodon on the tRNA puts the right amino acid in place and facilitates the joining of the amino acids to the growing ends of the protein chains. One pass of the ribosome along the length of the messenger RNA's protein coding sequence produces one molecule of the protein.

Learning how genes are expressed is only one aspect of the study of the mechanism of gene function. Another concerns the regulatory processes that control the rate and amount of gene expression under a wide variety of circumstances. Not surprisingly, progress in understanding the mechanics of transcription and translation illuminated the subject of regulation. Thus, it was discovered that bacteria differentially regulate the expression of their genes. Indeed, under certain conditions, many genes are not expressed at all, and the rates of expression of others differ by orders of magnitude. However, a change in conditions can trigger the activation of previously silent genes and the repression of previously active ones. This capability provides cells with a broad ranging flexibility to adjust their phenotypes.

Gene expression is most frequently regulated at the level of RNA production. Generally, transcription initiation is the regulated event, the control being mediated either by **repressor** proteins, which prevent transcription, or by **activator** proteins, which are needed to start transcription. In the former case, expression occurs when repression is reversed by modification of the repressor protein. In the latter, the gene is transcribed only if the activator protein is in the proper functional state. Repressor and activator proteins are not the sole means for regulating gene transcription. In some instances, the protein product of a gene's expression is itself the regulator of that gene's transcription. There are also examples where structural changes in the DNA affect the efficiency of a gene's transcription. RNA production can also be regulated by controlling the rate of elongation or by determining whether transcription continues through the entire gene or is terminated at specific stop-go signals within it. Gene expression can also be regulated during the translation of messenger RNA into protein. Here, too, specific regulation is most often achieved in the initiation step of the decoding process.

In this essay, we have sketched the birth of genetics and its evolution to a molecular science. Much of the information that provides the basis for this molecular understanding is described in the three chapters of Part I. These insights, which were obtained largely from studies with microorganisms (bacteria and their viruses), provide the background for the analyses and discussions of the more complex genomes in eukaryotic organisms.

CHAPTER 1

The Genetic Molecules

The genetic information of all cells is encoded in the variety and arrangement of nucleotides in their deoxyribonucleic acid (DNA). That information is expressed first via the formation of a related nucleic acid, ribonucleic acid (RNA), which in turn participates in the production of specific proteins. Each organism's phenotypic characteristics are ultimately determined by the number and variety of proteins expressed from its DNA. The informational relationships between the genetic molecules, DNA, RNA, and protein, are summarized in Figure 1.1.

Genetic continuity from one cell generation to the next requires **DNA replication**, the process by which the parental DNA molecules are duplicated and passed on to each of the offspring. This replication process must occur with high fidelity; moreover, damage and inadvertent errors introduced into DNA between and during replication cycles must be rectified and not be passed on to the progeny genomes. Genetic information must also be expressed to create the phenotype. In all cellular organisms, gene expression requires that DNA be copied into RNA (**transcription**), followed by **translation** of the RNA into proteins. As we shall see in Chapter 3, DNA is transcribed into several kinds of RNA, one of which, **messenger RNA** (mRNA), encodes protein structures; others are involved in the various processes that are essential for protein assembly. Besides encoding the cell's enzymatic machinery, DNA is also the substrate for repair and, under certain circumstances, its rearrangement. These DNA transac-

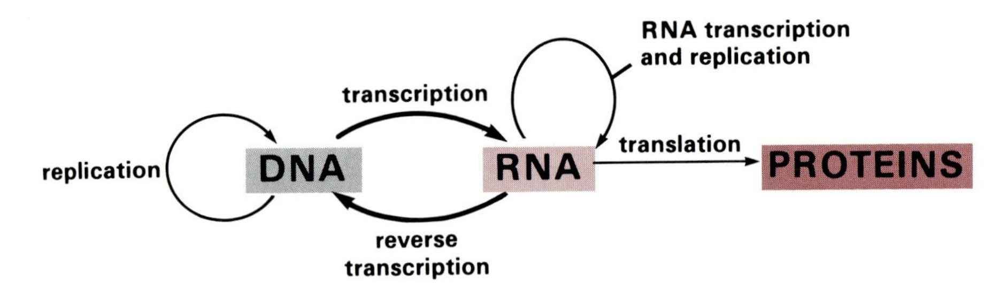

Figure 1.1
Established informational relationships between DNA, RNA, and protein.

tions—replication, repair, and rearrangement—are the key processes by which organisms maintain and diversify their genotype.

The genetic information of many viruses is also encoded in DNA; and the way their DNA is replicated, repaired, rearranged, and expressed is analogous to that used in cells. Some viruses have RNA rather than DNA genomes. They translate their genomic RNA directly into proteins, or their RNAs provide the genetic information for synthesizing related RNA molecules that are then translated into proteins. Those viral genomes that remain as RNA throughout their life cycle must replicate the parental RNA genomes in order to produce progeny virus. There is also a class of viruses called **retroviruses** whose RNA genomes begin their reproductive cycles by a **reverse transcription** of their genomic information from RNA into DNA. Such DNA copies, or proviruses, are replicated and expressed only after integration into the cellular chromosomal DNA. In that form, the viral genomes are replicated in concert with the host DNA, and the same transcription machinery used for cellular genes synthesizes the progeny RNA genomes as well as the mRNA needed to make the viral proteins.

A key feature of information transfer among nucleic acids, whether it be replication, transcription, or reverse transcription, is the involvement of the nucleic acid as a **template** to direct the assembly of identical or related structures. So far as is known, information stored in proteins is not used to assemble corresponding nucleic acids; thus, reverse translation is unknown. Nevertheless, proteins are critical participants in the processes that transfer information between nucleic acids and subsequently to proteins as well.

This chapter considers the chemical and physical features of DNA, RNA, and proteins because the secrets of their genotypic and phenotypic functions lie in their molecular structures and properties. The details of template-directed replication, repair, and recombination of nucleic acids appear in Chapter 2; Chapter 3 examines the mechanisms of transcription and translation in the expression of genetic information.

1.1 Structure and Behavior of DNA

a. Constituents and Chemical Linkages in DNA

Both chemical and physical studies established that DNA is a polymer composed of four different but related monomer units. Each monomer—a

nucleotide—contains a distinctive heterocyclic nitrogenous base, either adenine (A), guanine (G), cytosine (C), or thymine (T), joined to a deoxyribose-phosphate (Figure 1.2). Long polynucleotide chains result from joining the deoxyribose units of neighboring nucleotides through phosphodiester linkages (Figure 1.3). Each phosphate links the hydroxyl group (OH) on the 3′ carbon atom of a deoxyribose of one nucleotide to the OH on the 5′ carbon atom in the deoxyribose group of the adjacent nucleotide. Thus, the deoxyribose-phosphate backbone of the polynucleotide chain has a specific directionality of phosphodiester linkages.

The frequency of any two nearest neighbor bases in bacterial, bacteriophage, and yeast DNAs reflects the abundance of those bases in the DNA (Table 1.1). For example, the frequency of the nearest neighbors, 5′-CG-3′ and 5′-GC-3′ in prokaryote DNAs is about the same and nearly random;

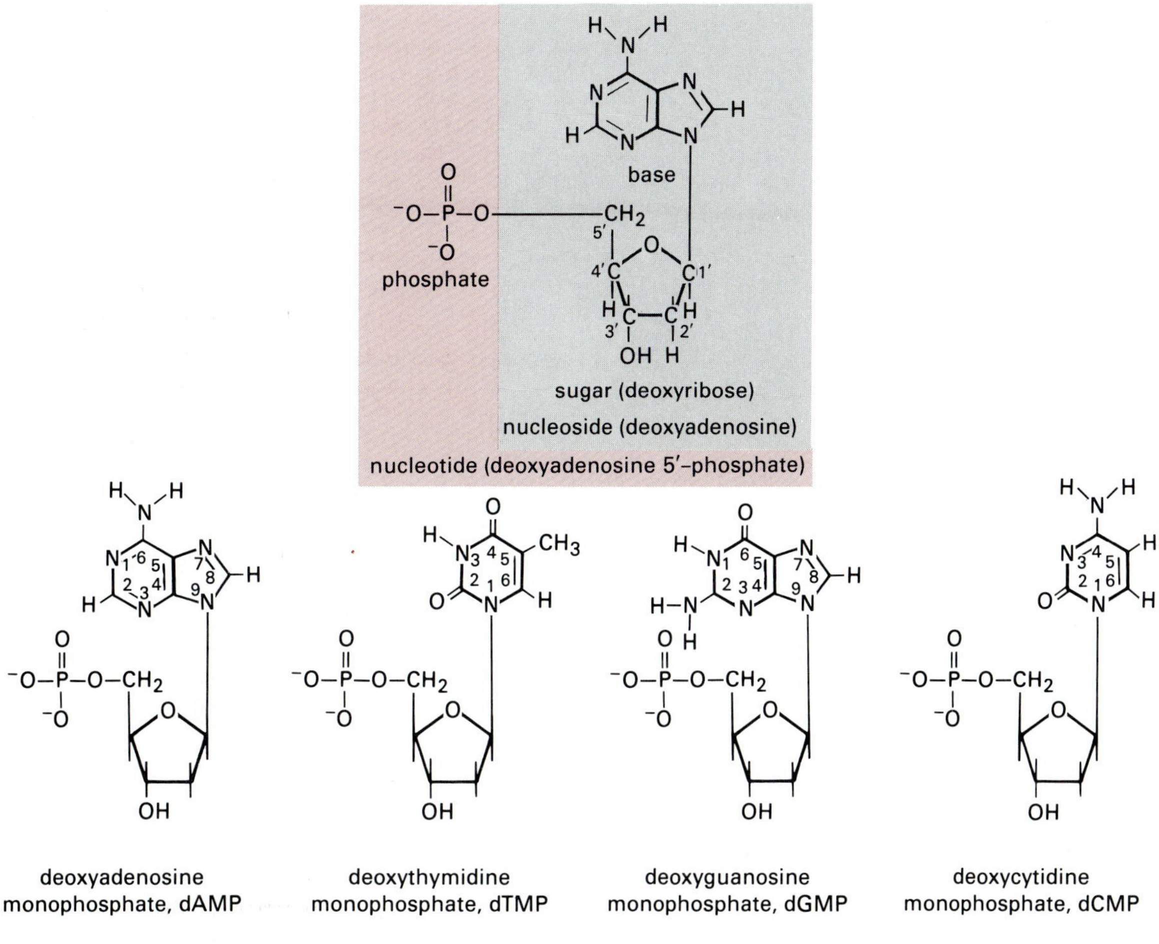

Figure 1.2
Deoxynucleotides in DNA. Shown are the conventional numbering for various positions in the purine (adenine and guanine) and pyrimidine (thymine and cytosine) heterocyclic rings and for the carbon atoms in deoxyribose.

Figure 1.3
The linkages between adjacent deoxynucleotides in a polydeoxynucleotide chain. The lower right-hand portion of the diagram shows several shorthand ways to designate a chain's deoxynucleotide sequence. Nucleotide sequences are usually displayed from left to right, from the 5′-to-3′ end.

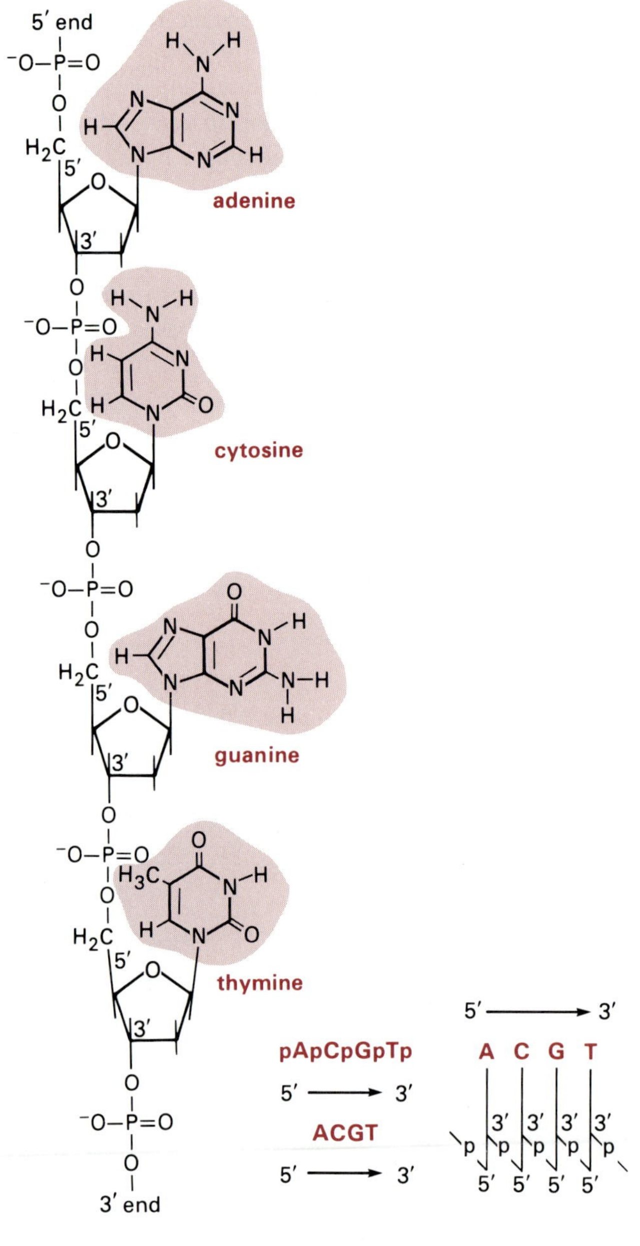

the same is true for the dinucleotide nearest neighbors 5′-GA-3′ and 5′-AG-3′. However, in animal, animal virus, and plant DNAs, the frequency of 5′-CG-3′ is between one-half to one-fifth that of 5′-GC-3′. Thus, the 5′-CG-3′ sequence occurs relatively infrequently in the DNA of higher eukaryotes. This feature is related to the use of this dinucleotide as a signal for methylation and its role in the regulation of gene expression.

After DNA is synthesized, some of the purine and pyrimidine bases may be chemically modified. For example, 5-methylcytosine, 5-hydroxymethylcytosine, 5-hydroxymethyluracil, and N^6-methyladenine have been identified as constituents of some DNAs (Figure 1.4). In certain bacteriophage DNA, a mono- or disaccharide is linked to the hydroxymethyl group of hydroxymethylcytosine by a glycosidic bond.

Table 1.1 Some Nearest Neighbor Frequencies in Various DNAs

Source of Template	Ratio 5′-CG-3′/5′-GC-3′	5′-AG-3′/5′-GA-3′
Bacterial viruses		
λ	1.02	0.90
T2	0.82	0.96
Bacteria		
Escherichia coli	0.95	1.0
Micrococcus lysodeikticus	1.15	0.75
Bacillus subtilis	0.82	0.87
Animal Viruses		
SV40	0.14	1.35
Polyoma	0.35	1.28
Shope papilloma	0.44	1.16
Unicellular eukaryotes		
Saccharomyces cerevisiae	0.87	0.96
Chlamydomonas	0.68	1.36
Multicellular eukaryotes		
Chicken	0.21	1.28
Human	0.23	1.15
Mouse	0.23	1.15
Cow	0.36	1.10
Wheat	0.78	0.97

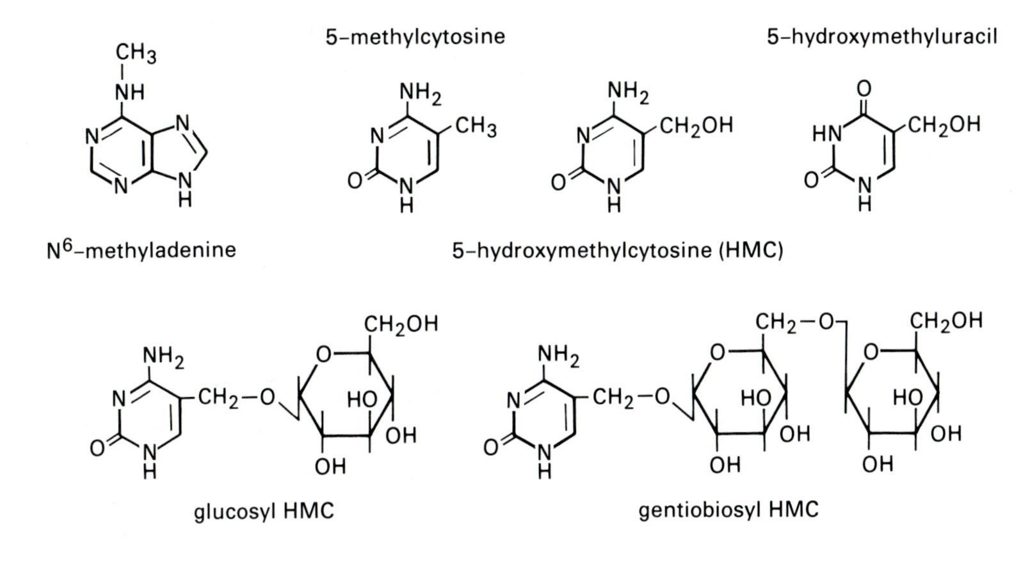

Figure 1.4
Structures of modified purines and pyrimidines found in some DNAs.

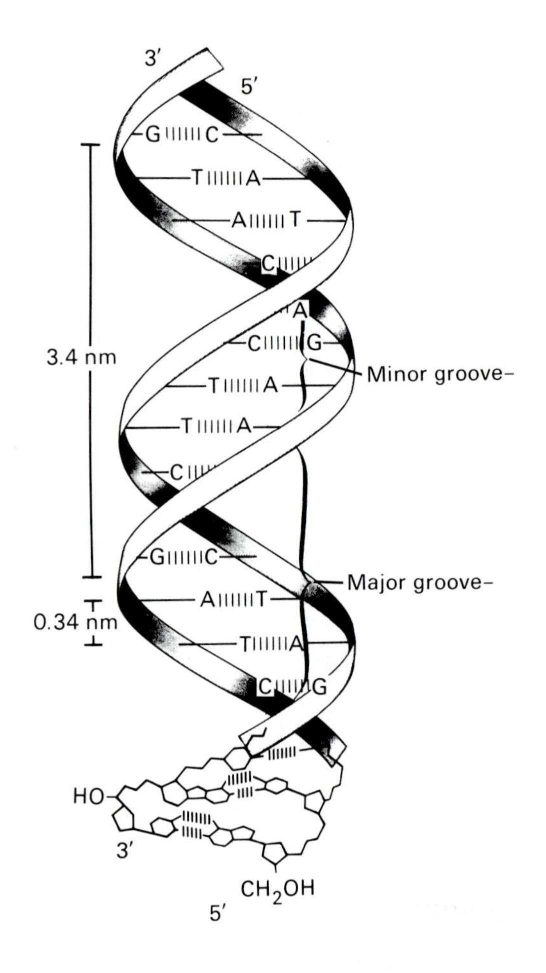

Figure 1.5
A diagrammatic representation of the B form of the DNA double helix showing the major and minor grooves, the spacing between adjacent base pairs, and the distance between turns of the helix. From A. Kornberg, *DNA Replication* (San Francisco: W.H. Freeman, 1980).

Most simple eukaryotes and invertebrates contain relatively small amounts of 5-methylcytosine and N^6-methyladenine. However, methylation of bases is more extensive in vertebrates and 5-methylcytosine is the most common. Indeed, more than 95 percent of the methyl groups in vertebrate DNAs occur at the C residue of rare CG dinucleotides, and more than 50 percent of all CG dinucleotides may be methylated. There is strong evidence (Section 8.7) that the extent of methylation of certain CG-containing sequences is a determining factor in regulating the expression of particular genes. In plants, the 5-methylcytosine residues are contained within CG dinucleotides and CNG trinucleotides, where N can be C, A, or T.

b. The Double Helical Structure of DNA

Physicochemical studies, electron microscopy, and X-ray diffraction analyses established that most DNA molecules are long, flexible, threadlike structures. These methods also verified that the DNA fiber has a nearly constant diameter with regularly spaced and repeated structures, irrespective of the base composition. Thus, unlike proteins, whose two- and three-dimensional structures depend critically upon the composition and order of amino acids (Section 1.3), DNA has, under usual conditions, a quite regular and nearly identical structure regardless of the composition or order

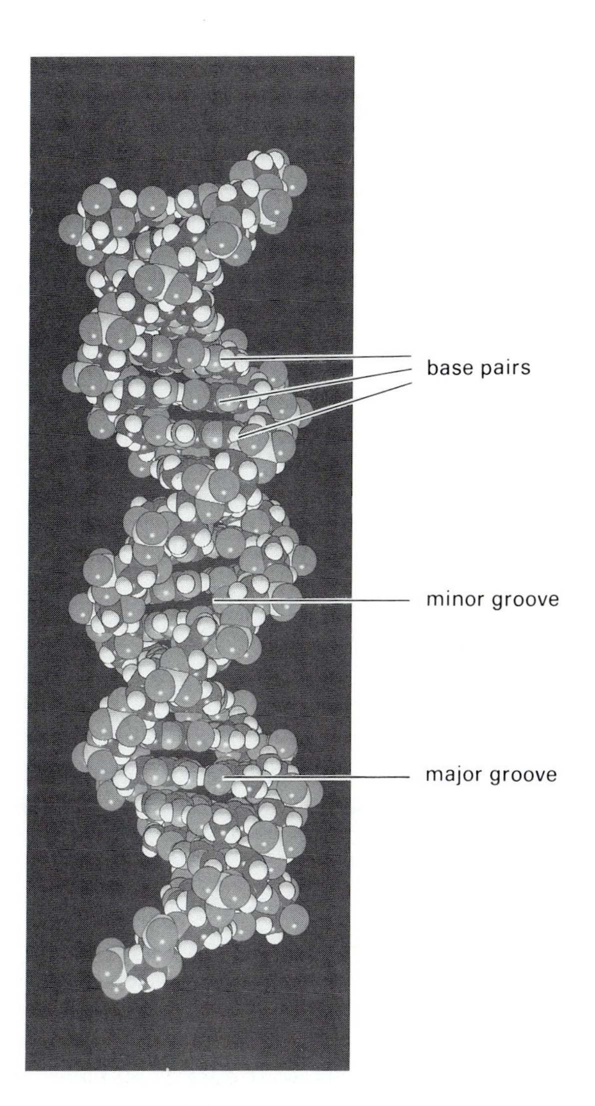

Figure 1.6
A space filling model of the B form of the DNA double helix. (Courtesy of Dr. Melson Max, Lawrence Livermore National Labs.)

of the four bases. These somewhat paradoxical chemical and physical features are explained by the structure of the DNA double helix.

The predominant form of DNA is a double helical structure formed from two separate polynucleotide chains wound around each other. The two deoxyribose-phosphate backbones, which trace the periphery of the molecule, have antiparallel (i.e., opposite) 5′-to-3′ orientations (Figures 1.5 and 1.6). In the usual form of DNA, the purine and pyrimidine rings in each chain are stacked 0.34 nm apart in the interior, with the plane of their rings nearly perpendicular to the helical axis. The helix makes one complete turn every 3.4 nm, that is, every ten base pairs. The bihelical structure has two external grooves, one wide (major) and the other narrow (minor).

The nitrogenous bases in the four nucleotides in DNA do not occur in 1:1 proportions as they are illustrated in Figure 1.3. On the contrary, the molar amounts of the two purines, A and G, and the two pyrimidines, T and C, are different and distinctive among DNAs from different organisms (Table 1.2). But there are certain invariant quantitative relationships among the purines and pyrimidines irrespective of the source of the DNA. The sum of the purine nucleotides, A + G, is equal to the

Table 1.2 Base Composition of Various DNAs

Source	A*	G	C	T	$\frac{A+T}{G+C}$	$\frac{A+G}{T+C}$	Mole % G + C
Bacteriophage λ	26.0	23.8	24.3	25.8	1.08	0.99	48
Bacteriophage T2	32.5	18.2	16.7**	32.6	1.86**	1.03**	35**
Escherichia coli	23.8	26.0	26.4	23.8	0.91	0.99	52
Bacillus subtilis	29.0	20.7	21.3	29.0	1.38	0.99	42
Shope papilloma virus	26.6	24.5	24.2	24.7	1.05	1.04	49
Saccharomyces cerevisiae	31.3	18.7	17.1	32.9	1.79	1.00	36
Chlamydomonas	19.6	30.2	30.0†	19.7	0.65†	0.99†	60†
Chicken	27.9	21.2	21.5†	29.4	1.34†	0.96†	43†
Mouse	28.9	21.1	20.3†	30.0	1.44†	1.00†	41†
Cow	27.3	22.5	22.5†	27.7	1.22†	0.99†	44†
Wheat	27.2	22.6	22.8†	27.4	1.20†	0.99†	45†

* The usual abbreviations for the purines and pyrimidines are used.
** As 5-hydroxymethyl cytosine.
† Includes 5-methylcytosine.

sum of the pyrimidine nucleotides, T + C; the amounts of A and T are always equal, as are the amounts of G and C. Indeed, these facts suggested that the purine and pyrimidine nucleotides are paired in DNA and led to the idea that hydrogen bonds between purines on one strand and pyrimidines on the other hold the two chains of the double helix together (Figures 1.5 and 1.6).

Two kinds of base pairs, often referred to as **complementary base pairs**, predominate in most DNAs: A with T and G with C (Figure 1.7). A·T base pairs have two hydrogen bonds, one between the amino and keto groups of the purine and pyrimidine, respectively, and the other between two ring nitrogen atoms. G·C base pairs have three hydrogen bonds, two between amino and keto groups on each base and a third between ring nitrogens. Base pairing between two purines, two pyrimidines, or the noncomplementary bases A and C or G and T is unfavored because appropriate hydrogen bonds cannot be formed, their formation is sterically hindered, or the helix geometry is disrupted. The modified purines and pyrimidines that occur occasionally in DNA (Figure 1.4) form the same hydrogen bonds as do their unmodified counterparts; consequently, they do not disrupt the base-paired helix structure. Following from these base pairing rules, the base sequence in one strand defines the base sequence in the other. Indeed, complementarity in the base sequence of the two polynucleotide chains is a profound feature of DNA and endows it with its genetic properties.

Additional stability for the bihelical arrangement of the two chains is provided by stacking interactions between aromatic rings of adjacent base pairs. Given the nearly identical dimensions of the complementary base pairs, the angle and direction of the deoxyribose-base linkage, the 0.34 nm vertical spacing, and the 36° average radial rotation between adjacent base pairs, one can account for the nearly constant diameter of the helix and the helical repeat of approximately ten base pairs per turn (Figure 1.5). Very precise information about the placement, orientation, and dimensions of the various constituents in three-dimensional space are available from X-ray diffraction analysis of model DNAs.

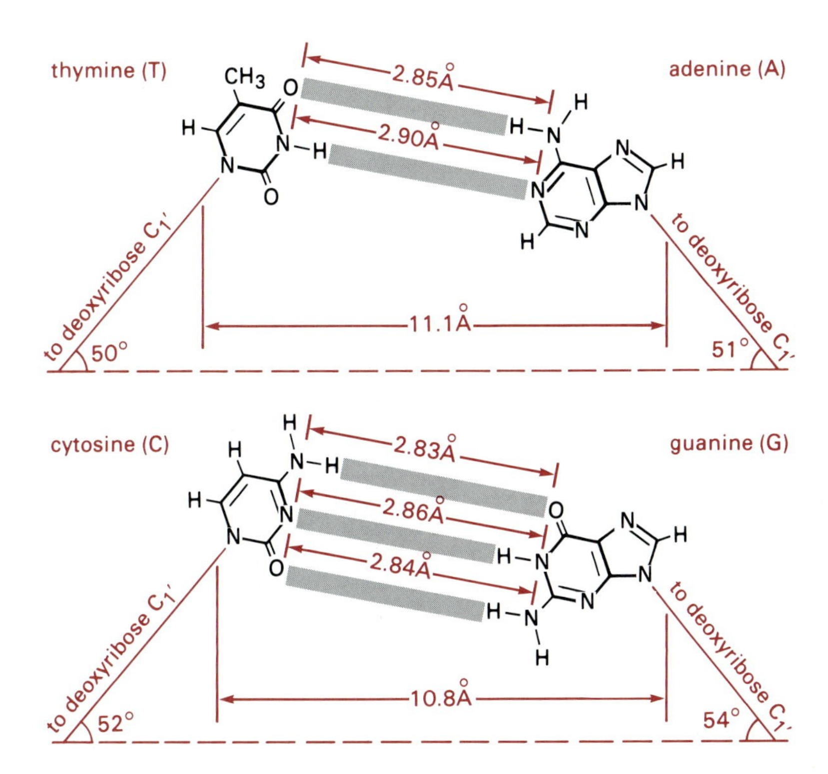

Figure 1.7
Hydrogen bonding between the complementary bases in DNA. Note that the geometry and distance between the two kinds of base pairs are almost the same. After A. Kornberg, *DNA Replication* (San Francisco: W.H. Freeman, 1980).

c. *Alternative Forms of the DNA Double Helix*

The previous description applies to the predominant or B form of the DNA double helix. Two additional, isomeric species of the DNA double helix are known. These can exist because the bond angles between the bases and deoxyribose are changeable and because the deoxyribose ring and deoxyribose-phosphate backbone are flexible enough to permit alternate configurations. The rare A form, which exists only in dehydrated states, differs from the B form in that the plane of the base pairs is tilted 20° from the perpendicular to the helix axis (Figure 1.8). This difference lowers the base pair to base pair distance in A–DNA to about 0.29 nm and increases the number of base pairs per turn to 11 to 12. No biologic function has yet been attributed to the A form.

A characteristic feature of B–DNA is that the sugar-phosphate backbones of the two chains form right-handed helices (Figures 1.5 and 1.6). However, under certain conditions, segments of DNA in which purine- and pyrimidine-containing nucleotides alternate can assume a left-handed helical conformation with a lengthened spacing between base pairs (0.77 nm for two base pair steps) and 12 base pairs per helical turn. The somewhat zigzag course of the deoxyribose-phosphate backbone in this structure accounts for its designation as Z–DNA. The question of whether Z–DNA occurs naturally and arises in discrete regions of the B helix under the influence of specific proteins that can facilitate a B helix to Z helix transition is being actively investigated (Chapter 8, Section 8.7).

d. *The Size of DNA Molecules*

Generally, DNA size is given as the number of base pairs (bp), with 1000 base pairs (kilobase pairs [kbp]) being a common unit. One kbp of

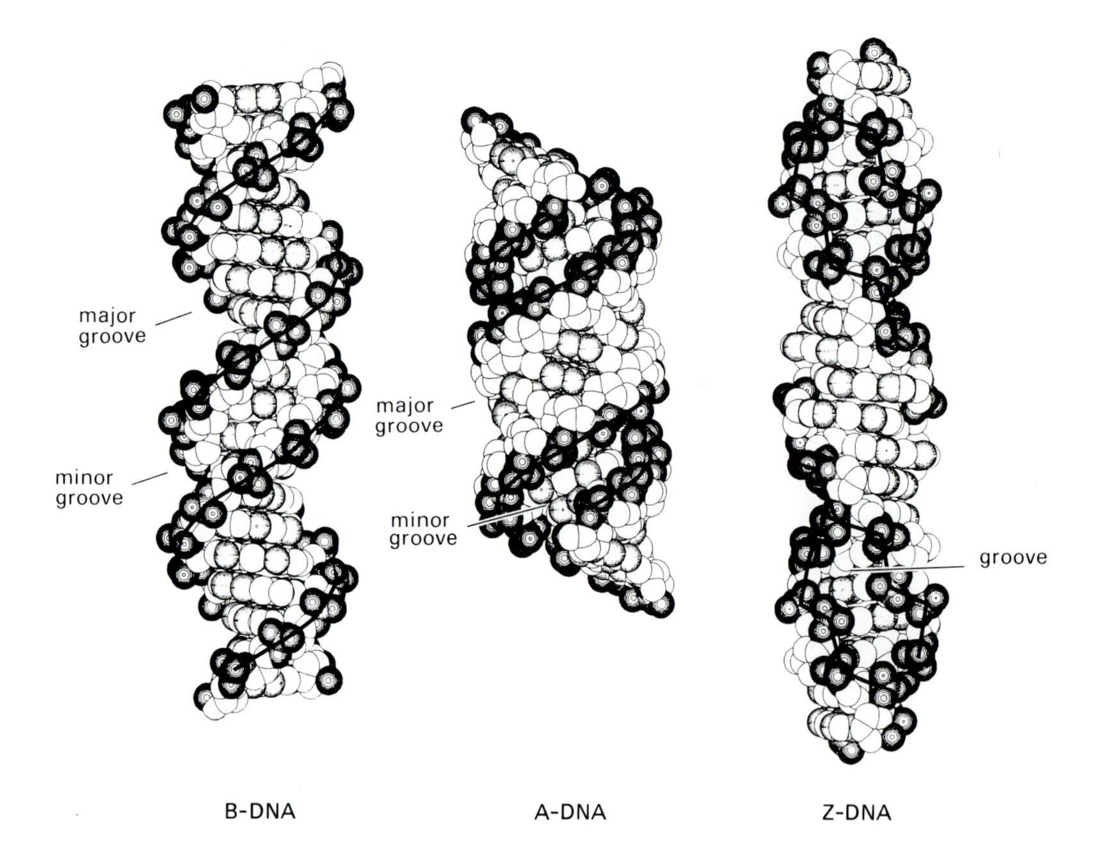

Figure 1.8
Space filling (van der Waals) representations of B, A, and Z–DNAs (each containing 20 base pairs). The phosphorous and phosphate oxygen atoms are shown as solid black circles, and the nitrogen atoms are lightly stippled. The solid line from phosphate to phosphate traces the polynucleotide chain; note the zigzag course of the backbone in Z–DNA. A–DNA is much shorter and thicker, and Z–DNA is slightly larger and thinner than B–DNA. A–DNA also has a much deeper major groove, but the minor groove is flattened out to make an almost ribbonlike surface on the outside. Z–DNA has only a single groove, somewhat deeper than the major groove in B–DNA, but not as deep as the major groove in A–DNA. Courtesy of A. Rich.

B–DNA has an average molecular weight of 6.6×10^5 and a length of 340 nm. When care is taken not to break DNA during its isolation, and gentle methods are used to assess its length, there is a striking agreement between the length of DNA and the mass contained in single small chromosomes (Table 1.3). Thus, the DNA molecules from the *E. coli* bacteriophages λ and T4 and from the adeno- and herpesviruses have lengths corresponding to the number of base pairs in the single chromosome that constitutes each virus genome. Indeed, the entire *E. coli* genome (about 4×10^6 base pairs) is a single DNA molecule, 1.4 mm long. There is also reason to believe that each chromosome from yeast, *Drosophila,* and even humans contains a single DNA molecule with lengths ranging from tens of thousands to many millions of base pairs.

e. Variations in DNA Shape and Strandedness

The view stated earlier that B–DNA is a perfectly regular double helix in which the periodic geometry of the interwound sugar phosphate backbones is identical, irrespective of the base pair sequence, is not entirely correct. Detailed X-ray structural analysis, model building, and thermodynamic considerations have established that the planes of adjacent base pairs in the helix are not strictly parallel. Each complementary base pair thus serves as a local wedge inclining the helix axis in one direction or another. The greatest wedge occurs when 2 neighboring As on one strand pair with 2Ts on the other. As a consequence, runs of As paired with runs of Ts produce a localized curve in the helix. If such runs occur periodically, at approximately ten base pair intervals (i.e., one turn of the helix), the DNA molecule assumes a distinctly curved structure. Curved

Table 1.3 Molecular Weights, Lengths and Structures of DNA from Various Sources

Source	Molecular Weight	Length	Number of Base Pairs	Structure
Bacteriophage ϕX174	1.6×10^6	1.6 μm	5×10^3 *	circular single strand
SV40	3.5×10^6	1.1 μm	5.2×10^3	circular double strand
Bacteriophage T2	1.2×10^8	50 μm	2×10^5	linear double strand
Hemophilus influenzae chromosome	7.9×10^8	300 μm	1.2×10^6	not known
Escherichia coli chromosome	2.6×10^9	1 mm	4×10^6	circular double strand
Saccharomyces cerevisiae				
chromosome 1	1.4×10^8	50 μm	2.1×10^5	linear double strand
chromosome 12	1.5×10^9	500 μm	2.2×10^6	linear double strand
Drosophila melanogaster				
chromosome 2	4×10^{10}	15 mm	6.0×10^7	linear double strand
chromosome 3	4.2×10^{10}	16 mm	6.3×10^7	linear double strand
chromosome 4	4×10^9	1.5 mm	6×10^6	linear double strand

* In this instance, bases instead of base pairs.

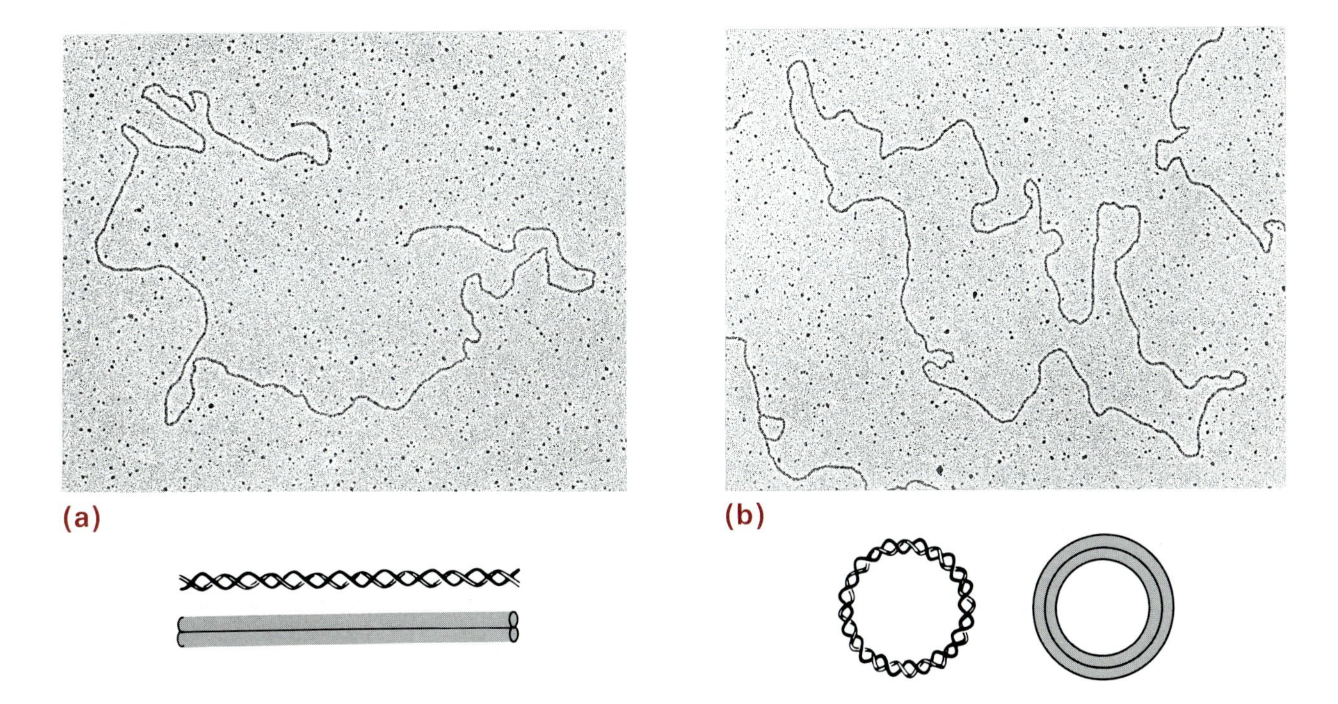

Figure 1.9
Diagrams and electron micrographs of linear and circular (monomer length), double strand λ phage DNA. Electron micrographs courtesy of L. Chow. The diagrams show the DNA in its double helical form and in the way double strand DNA is shown throughout this book's figures.

or bent DNA has been detected, for example, in linear fragments from the kinetoplast DNA of the trypanosome *Leishmania tarentolae* by their anomalously reduced mobility during electrophoresis in polyacrylamide gels. The bends in this DNA are localized around a sequence with an unusual frequency of periodically repeated $(A \cdot T)_{5-6}$ tracts separated by 4 to 6 G·C rich base pairs. The full significance of DNA curving or bending is not yet certain. A sequence-dependent predisposition to bending may influence how DNA is wrapped around histone octamers in chromatin (Section 1.1g). Bending may also be important in the way specific DNA binding proteins regulate gene expression.

DNA occurs as both linear and circular molecules (Figure 1.9). Bacterial plasmids, some bacterial chromosomes, many mitochondrial and chloroplast DNAs, and mammalian virus genomes consist of a single, covalently joined, circular duplex DNA molecule. The λ bacteriophage chromosome in different stages of its life cycle exists as either a linear molecule or a closed circular structure or as a circle with breaks in the phosphodiester backbone. There appears to be no upper limit to the size of circular double helical DNA.

The DNA inside cells is normally associated with proteins (see Section 1.1g). Bound protein unwinds the helix slightly; consequently, the number of helical turns per unit length is less than it is in free B–DNA. When the protein is removed from DNA, the usual number of right-handed or positive helical turns is restored. With linear DNA, this occurs readily because the two strands are free to rotate around each other. However, in closed circular DNA, the total number of helical turns is topologically fixed, that is, the number of turns of one strand around the other cannot be changed without compensatory twists elsewhere. Thus, when natural circular duplexes are freed of the proteins with which they are often associated *in vivo,* two things happen: (1) The number of right-handed (positive) helical turns is increased to the value characteristic of B–DNA. (2) The duplex itself is twisted an equal number of times in the opposite (or negative) sense to compensate for the increase in the number of helical turns. Such molecules are said to be negatively **supercoiled** (Figure 1.10). A single nick in a supercoiled circular DNA eliminates the supercoiling and relaxes the circular structure because the molecule is no longer topologically constrained. Any chemical or physical changes that reduce the number of helical turns per molecule will lower or eliminate negative supercoils in closed circular DNA.

Not all naturally occurring DNA is double stranded. The genomes of some small bacterial, plant, and animal viruses consist of covalently closed circles containing only single strands (Figure 1.11). All known single strand circular DNAs are relatively small: for example, the DNA of the *E. coli* bacteriophages ϕX174 and M13 contain about 5300 and 6000 nucleotides, respectively, and are between 1.5–2 μm in length; those of animal parvoviruses and certain plant viruses are about two-thirds to one-half that size, respectively. However, replication of each of these viral DNAs involves conversion of the single strand circle to the corresponding double strand circle, from which single strand progeny circles are then produced (Chapter 2). Moreover, expression of the genetic information in such genomes always occurs via the double strand form because that form is the substrate for transcription of the DNA sequence into RNA.

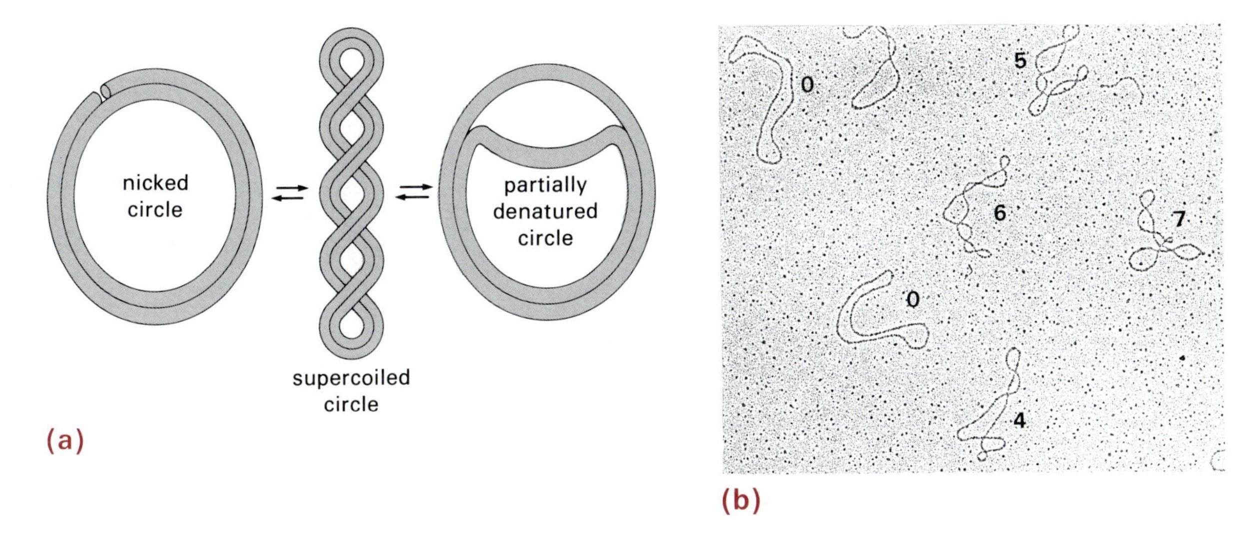

Figure 1.10
(a) Diagrammatic representation of supercoiled circular DNA and the relaxed circular forms that are obtained from it by nicking either of the two strands or by separating the two strands over a short stretch. (b) The electron micrographs show M13 phage double strand circular DNA with different extents of supercoiling. The numbers next to each molecule indicate the estimated number of superhelical turns. Courtesy of L. Chow.

f. Denaturation and Renaturation of DNA

The hydrogen bonds and stacking interactions that maintain the double helical structure are weak forces, and relatively small amounts of energy separate the two strands—a process called **denaturation** or melting (Figure 1.12a). A solution of double helical DNA is readily melted to single polynucleotide chains by heating close to 100°C. Denaturation also occurs if the hydrogen bonds between bases on the two strands are disrupted by raising the pH of the solution. A variety of factors (such as monovalent and divalent cations, polyamines, and proteins) influences the denaturation reaction by, for example, neutralizing or masking the repulsive forces of negatively charged phosphate groups in the backbone. With homogeneous double strand DNA, the temperature or pH range over which the two strands separate is sharp, the T_m or pH_m being the midpoint of the transition (Figure 1.12b). Because less energy is required to disrupt the two hydrogen bonds of an A·T base pair than is needed to break three hydrogen bonds in a G·C base pair, the ease with which double helices denature, whether by heat or increased pH, depends on their base compositions. The higher the fraction of G·C base pairs, the greater the T_m or pH_m (Figure 1.13).

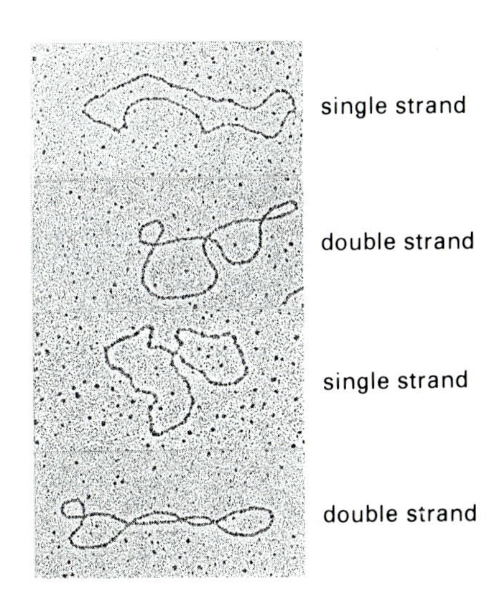

Figure 1.11
Electron micrographs of single and double strand circular forms of M13 phage DNA. The double strand form appears smoother and more extended and is more easily visualized than single strand DNA. Courtesy of L. Chow.

Denaturation is a reversible process, and reformation of double strand DNA can occur even when the two strands have been completely separated. That process, called **renaturation**, **reassociation**, or **annealing**, occurs if the temperature or pH are lowered (Figure 1.12). If the change in temperature or pH is gradual, the two strands align properly and reform all the original base pairs. If the temperature or pH is lowered abruptly, the proper alignment of complementary base pairs is prevented or impeded

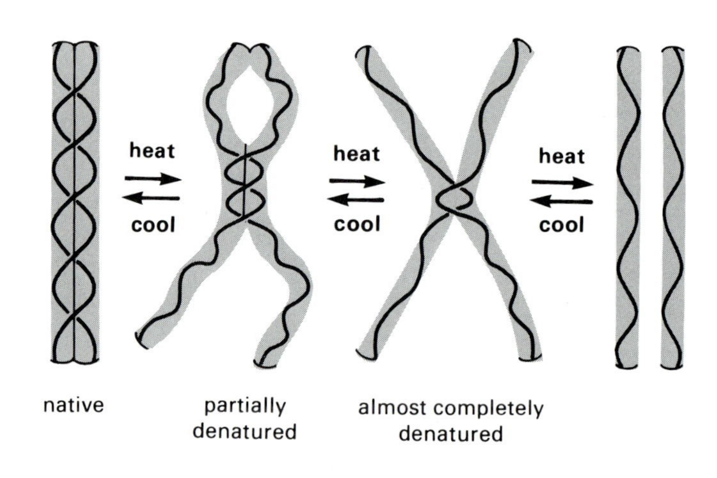

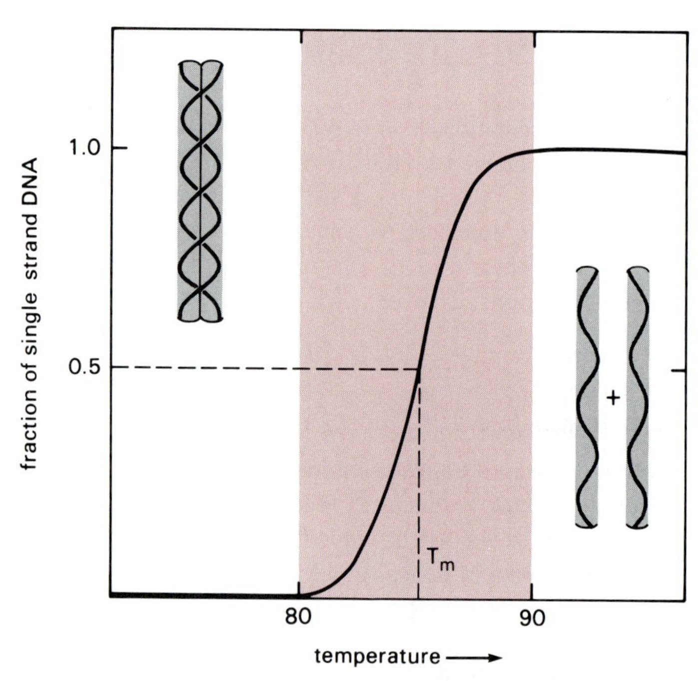

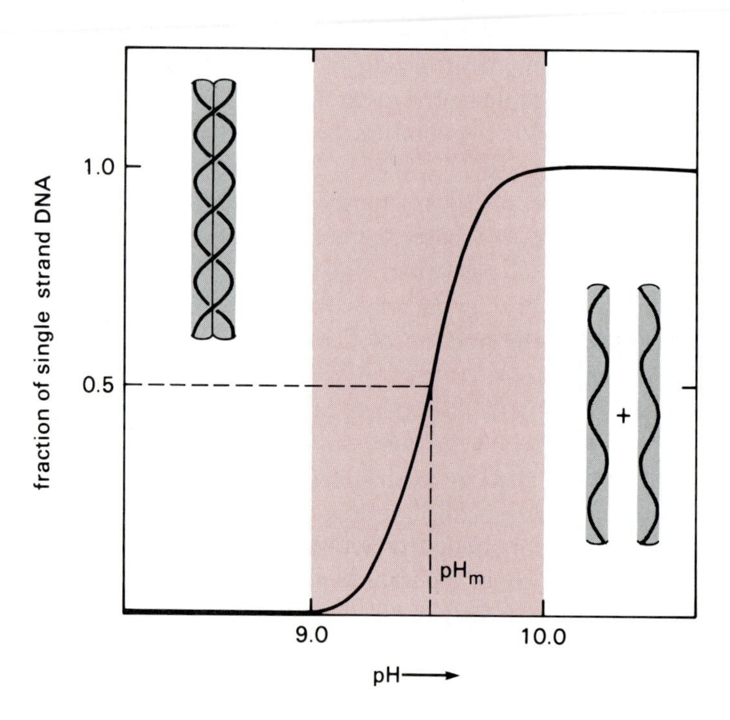

Figure 1.12
(a) The denaturation (dissociation) of double strand DNA into two single strands by increasing the temperature and the renaturation (reassociation) of the two complementary strands by cooling. (b) The denaturation-renaturation curves of a typical double strand DNA as a function of temperature and pH. The T_m and pH_m designate the temperature and pH, respectively, at which the DNA is half-denatured or renatured.

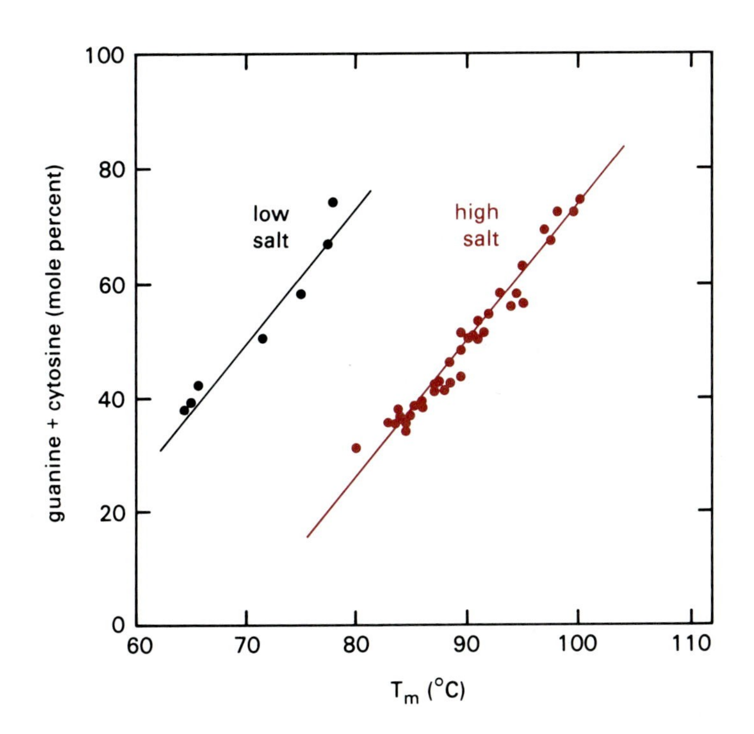

Figure 1.13
The relationship between the T_m and the mole percentage of guanine plus cytosine in DNA at low and high salt concentrations. The points refer to individual DNAs from bacteria, bacteriophages, yeast, plants, and animals. After J. Marmur and P. Doty, *J. Mol. Biol.* 5 (1962), p. 109.

because of the interaction between partially complementary regions within a strand or between different stretches on the two strands (Figure 1.14).

Dissociation (denaturation) and reassociation (renaturation) of DNA strands in solution are, in fact, artificial reenactments of processes critical to a wide variety of normal biologic functions. Moreover, and most important for the purposes of the forthcoming presentations, the ability of two complementary single strands of nucleic acid to reconstitute a double strand structure is essential for joining DNA molecules together *in vitro* and for isolating, comparing, and identifying specific species of nucleic acids. Indeed, the unique capability of nucleic acid chains to form double strand helices by the association of complementary single strands is critical to every area of genetic chemistry.

g. Packaging of DNA in Chromosomes

DNA probably never occurs free, in extended form, in cells or viruses. Rather, it exists associated with low molecular weight cations—divalent metals or di- and polyamines or with proteins, but probably with combinations of these. The interactions are electrostatic, the negatively charged internucleotide phosphate anion being neutralized by positively charged metals and polyamines or by the basic amino acid residues in proteins. A consequence of many diverse types of these interactions is the condensation of the DNA, sometimes several thousandfold. The 1.4 mm long circular DNA from *E. coli* fits into a rodlike cell with dimensions of about 1 μm diameter and 2 μm in length; in the interphase eukaryotic cell, about 2 m of nuclear DNA is confined to a nucleus of less than 10 μm in diameter. The nuclear DNA in mitotic cells is condensed even further and can be seen by light microscopy as highly compacted chromosomes (Figure I.7).

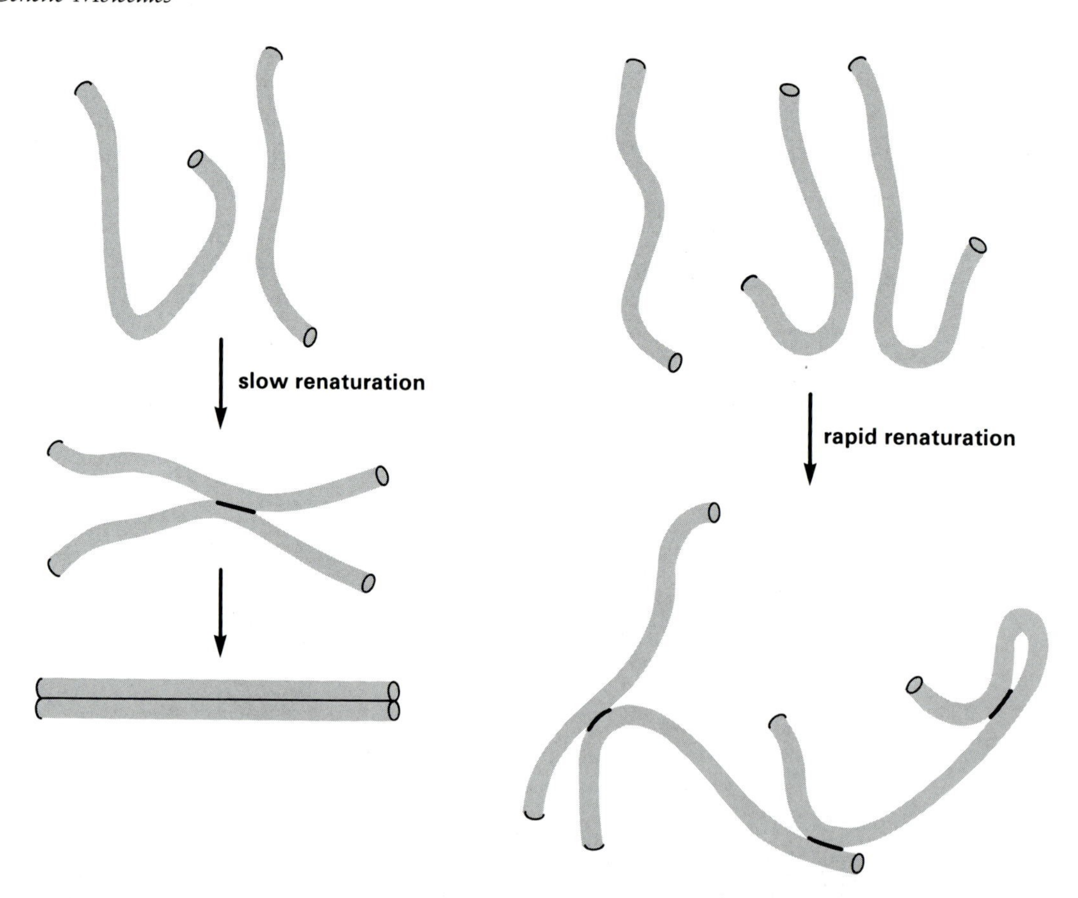

Figure 1.14
The outcome of slow and rapid renaturation of complementary DNA strands.

Eukaryotic chromosomes The chromosomes of eukaryotic cells are composed largely of chromatin, a complex of double strand DNA and five histone proteins called **H1**, **H2A**, **H2B**, **H3**, and **H4** (Table 1.4). The histones may be modified by acetylation, methylation, phosphorylation, poly ADP ribosylation, and, in histone H2A and H2B, by the covalent addition of a protein called ubiquitin. However, the full significance of the effect of these substituents on the structure or function of the histone is obscure. Mammalian histone H1 contains about 215 amino acids; the other histones range in size from 100 to 135 amino acids. All have unusually large amounts of the positively charged amino acid lysine; H3 and H4 are distinguished from the others by also having high levels of positively charged arginine. The chromatin of small eukaryotes such as yeast and molds contains H2A, H2B, H3, and H4 and in the same proportions as mammalian chromatin.

Depending upon the conditions of extraction and spreading, chromatin may appear in the electron microscope as a long 10 nm diameter fiber or, more frequently, as a more extended fiber with "beads" of about 10 nm diameter arrayed along its length at regular intervals (Figure 1.15). Approximately 145 base pairs of chromosomal DNA are wound around each **nucleosome** core (the beads seen in Figure 1.15b), which consists of an octamer of histones containing two molecules each of H2A, H2B, H3, and H4 (Figure 1.16). Thus, the one-and-three-quarter turns of double strand DNA that is wound around the perimeter of the nucleosome forms a "super helix."

The fifth histone, H1, is not included in the nucleosome core nor is it involved in the coiling of the DNA helix about the histone octamer.

Table 1.4 Typical Characteristics of Mammalian Histones

Type	Number of Amino Acids	Molecular Weight (kD)	Number of Basic Amino Acids	Ratio Lys/Arg	Number of Acidic Amino Acids
H1 (rabbit)	213	23.0	65	21	12
H2A (cow)	129	14.0	26	1.2	20
H2B (cow)	125	13.8	28	2.5	16
H3 (cow)	135	15.3	32	0.7	18
H4 (cow)	102	11.3	26	0.8	10

Rather, the H1 histone is bound to the DNA where the double helix enters and leaves the nucleosome core (Figure 1.17). In this structure, 168 base pairs of helical DNA are associated with one histone octamer and a molecule of histone H1. As noted previously, chromatin fibers often appear in the electron microscope in two alternative forms: a fiber with well-separated nucleosomes or a 10 nm fiber in which the nucleosomes are loosely packed along the length of the fiber (Figure 1.15). The 10 nm

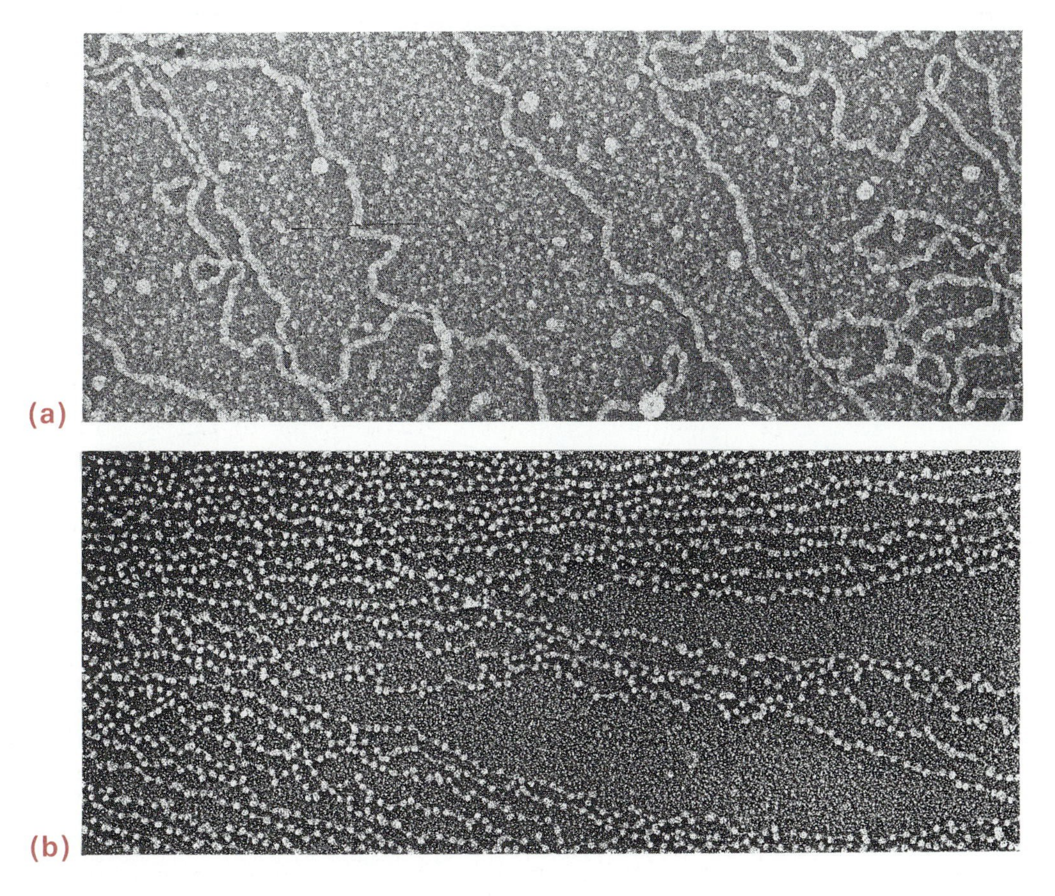

Figure 1.15
Electron micrographs of chromatin. (a) The 10 nm chromatin fiber prepared from CV1 monkey kidney cells. Courtesy of J. Griffith. (b) The beads-on-a-string form of chicken erythrocyte chromatin. Courtesy of H. Zentgraf.

fiber can undergo another condensation, which creates a higher order packing. In this structure, the nucleosomes are believed to be arranged in a solenoid to produce a fiber of 30 nm diameter (Figure 1.18).

The DNA-histone interactions convert a double helix of 168 base pairs of DNA with overall diameter of 2 nm and a length of 57 nm into a 10 nm diameter coil about 5 nm long (Figure 1.18). The additional helical coiling of the 10 nm diameter fiber to produce the 30 nm diameter fiber increases the condensation another 6-fold. Thus, the packaging of duplex DNA with the five histones condenses DNA about 50-fold. However, this degree of folding still does not account for the nearly 5000-fold packing that is needed to achieve the very dense organization of DNA in metaphase chromosomes.

Eukaryotic chromatin also contains other proteins generally referred to as nonhistone proteins. Some, like the enzymes that are necessary for the replication and expression functions of DNA, may be associated with the chromatin only transiently. Those involved in regulation may be associated in only specific tissues or at particular stages of differentiation. These aspects and the role of alternate states of chromatin organization in DNA replication and expression are discussed in more detail in later chapters.

Prokaryotic chromosomes Only two or three proteins have been implicated so far in the condensation or packaging of prokaryotic DNA genomes. But less is known about these proteins, the nature of their associations with DNA, and the structure of the condensed protein-nucleic acid complexes. In *E. coli*, there appears to be one or a class of DNA binding proteins, the HU proteins; these resemble the eukaryotic histone H2A in size, lysine and arginine content, and antigenicity. Another protein, protein II, present in *E. coli* and cyanobacteria as well, also resembles a eukaryotic histone in its high lysine content and DNA binding properties. Proteins HU and II occur in high enough levels to complex nearly half the *E. coli* DNA and, perhaps in conjunction with polyamines and yet

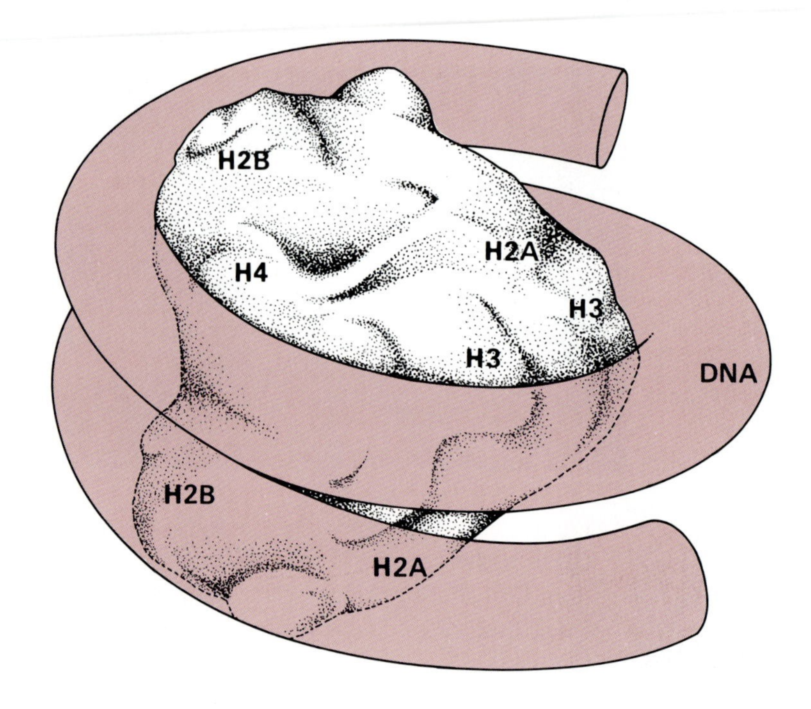

Figure 1.16
Model for the nucleosome core particle deduced from low and high resolution crystallographic analysis. The DNA (145 bp), shown as a tube, is wound one and three-quarter turns around the histone octamer. After R. D. Kornberg and A. Klug, *Sci. Amer.* 244 (2) (1981), p. 52.

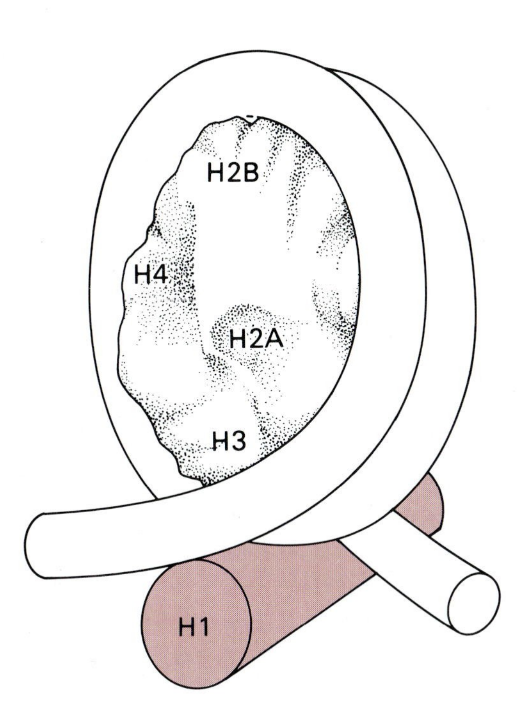

Figure 1.17
Histone H1 cross links the DNA as it enters and leaves its association with nucleosomes. After A. Klug, *Les Prix Nobel* (Stockholm, Sweden: Nobel Foundation, 1982), p. 93.

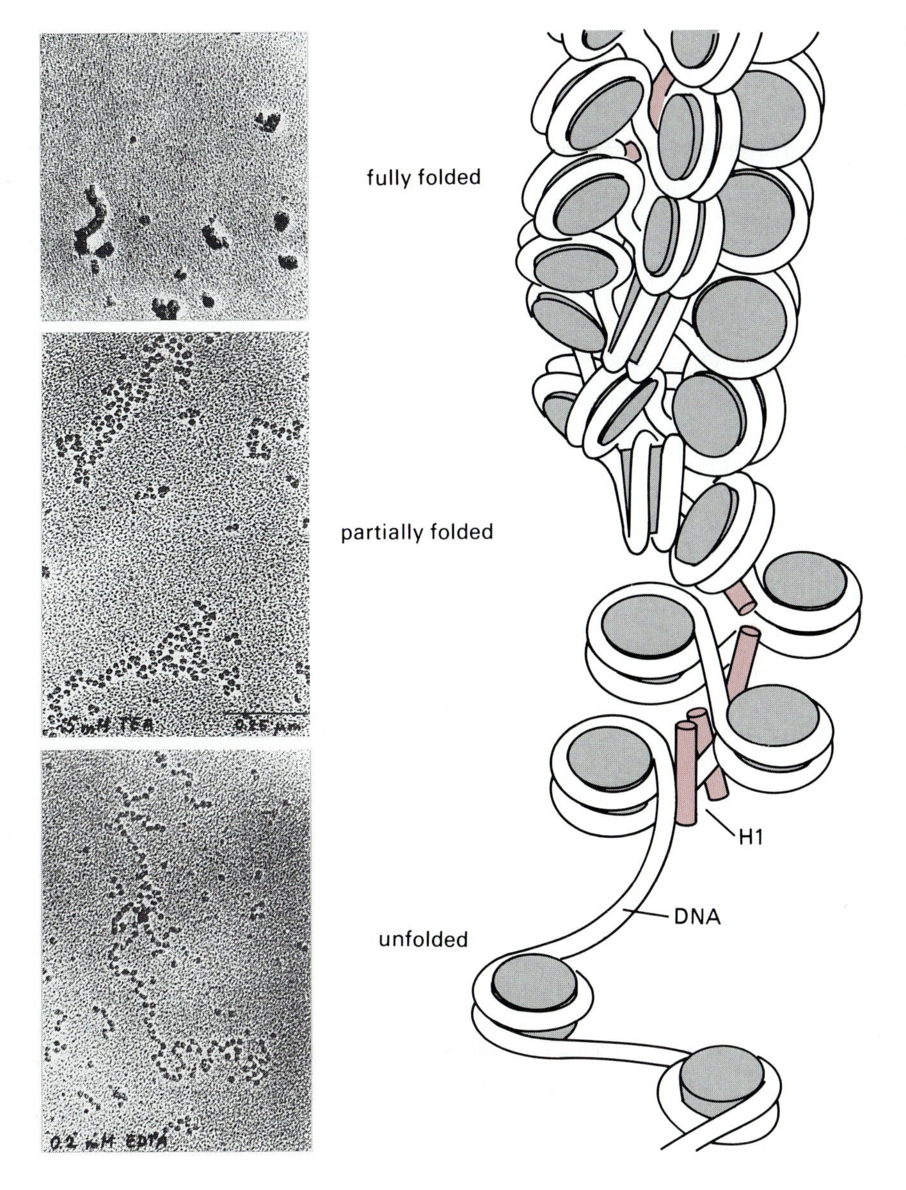

Figure 1.18
A model of the various folded chromatin structures. The extended, beads-on-a-string chromatin structure is shown in the drawing and electron micrograph at the bottom of the figure. The partially folded form, the 10 nm diameter fiber, is shown in the middle, and the more highly condensed helical array of the 10 nm fiber to form a 30 nm diameter solenoid structure is illustrated at the top. Note how the H1 histones, associated with each nucleosome, interact to facilitate the folding of the 10 nm fiber into the more condensed 30 nm fiber. The electron micrographs are courtesy of A. Klug. The drawing is after R. D. Kornberg and A. Klug, *Sci. Amer.* 244 (2) (1981), p. 52.

to be discovered proteins, may perform the same functions that the five eukaryotic histones have in condensing and packing eukaryotic nuclear DNA.

1.2 Structure and Behavior of RNA

a. *Types of RNA and Their Occurrence*

RNA constitutes the bulk of the nucleic acid in all cells, being five to ten times more abundant than DNA. Its principal and best understood role is in translating genetic information into proteins. However, RNAs participate in certain specialized endonuclease functions that probably regulate various steps in gene expression. Certain viruses—retroviruses and a variety of single and double stranded animal, plant, and insect viruses—have genomes composed of RNA.

Several types of RNA occur in all cells: **ribosomal RNA** (rRNA), **transfer RNA** (tRNA), and messenger RNA (mRNA). Most, if not all cells also contain a variety of other small cytoplasmic RNAs (scRNAs), and eukaryotic cells contain in addition a variety of small nuclear RNAs (snRNAs) (Table 1.5). About 80 percent of the cellular RNA is composed of the three or four species of rRNAs, and the nearly hundred kinds of tRNAs constitute about 15 percent. The several thousand different messenger RNAs account for less than 5 percent of the cellular RNA. The as yet uncounted number of small nuclear and cytoplasmic RNAs amount to less than 2 percent of the total.

b. *Constituents and Chemical Linkages in RNA*

RNAs are polynucleotides that range in size from as few as 70 or so nucleotides in some tRNAs to over 10,000 in some mRNAs. The four commonly occurring nucleotides in RNA are shown in Figure 1.19. Two purines (adenine and guanine) and one pyrimidine (cytosine) are common

Table 1.5 Some Important RNAs

Types of RNA	Approximate Number of Different Kinds in Cells	Approximate Length in Nucleotides	Distribution*
Transfer RNA (tRNA)	80–100	75–90	P, E
5S Ribosomal RNA (rRNA)	1–2	120	P, E
5.8S Ribosomal RNA (rRNA)	1	155	E
16S Ribosomal RNA (rRNA)	1	1600	P
23S Ribosomal RNA (rRNA)	1	3200	P
18S Ribosomal RNA (rRNA)	1	1900	E
28S Ribosomal RNA (rRNA)	1	5000	E
Messenger RNA (mRNA)	thousands	vary	P, E
Heterogeneous nuclear RNA (hnRNA)	thousands	vary	E
Small cytoplasmic RNA (scRNA)	tens	90–330	P, E
Small nuclear RNA (snRNA)	tens	58–220	E

* P = prokaryotic, E = eukaryotic.

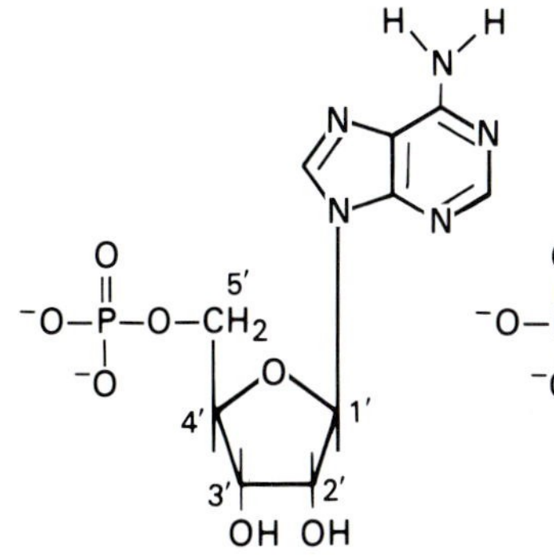

adenosine 5′-phosphate (AMP)

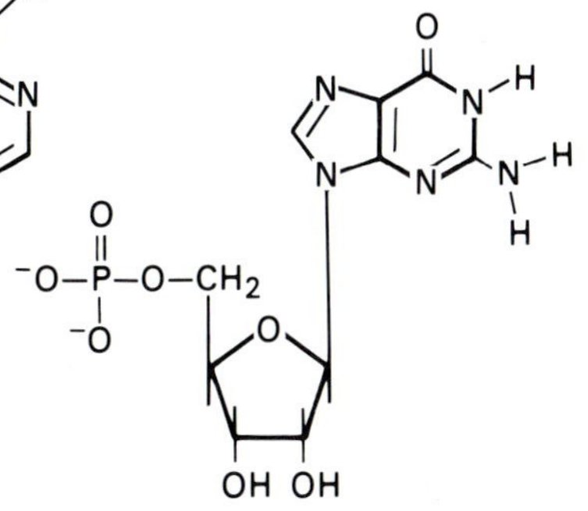

guanosine 5′-phosphate (GMP)

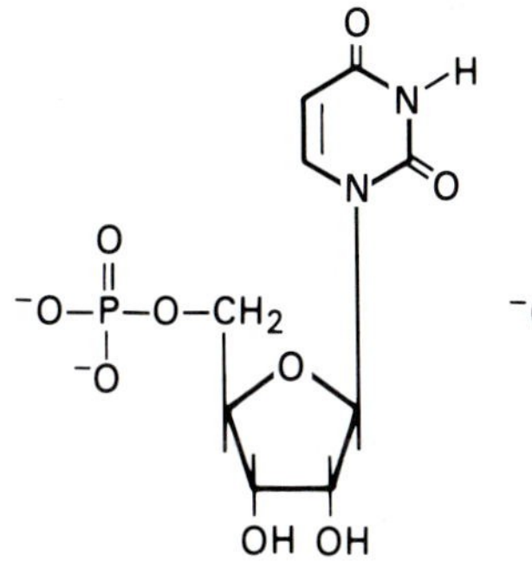

uridine 5′-phosphate (UMP)

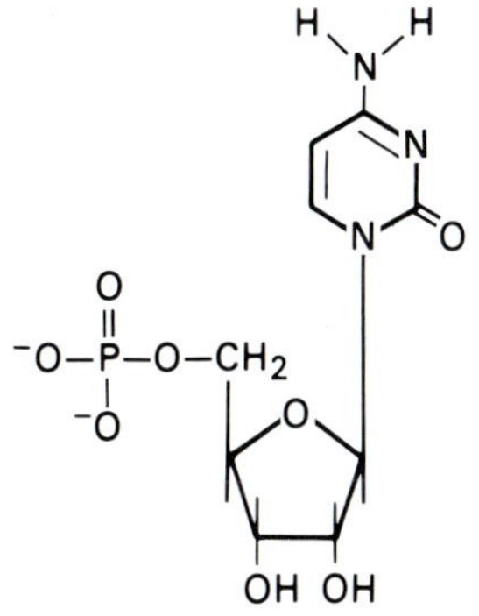

cytidine 5′-phosphate (CMP)

Figure 1.19
Structures of the four ribonucleotides that commonly occur in RNA.

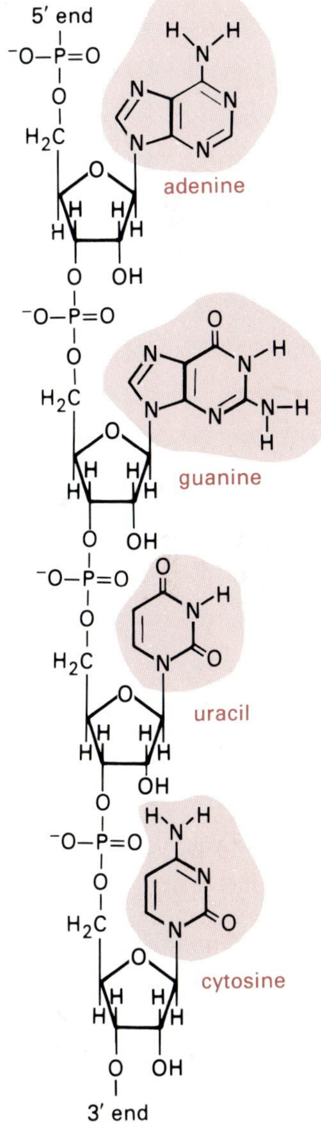

to RNA and DNA nucleotides. But thymine, the 5-methyldiketopyrimidine in DNA, is substituted in RNA by uracil, which lacks the 5-methyl substituent. The nucleotides in RNA are joined into chains by the same 5′-to-3′ phosphodiester linkages that occur in DNA (Figure 1.20). The presence of the 2′-OH adjacent to the internucleotide phosphodiester linkage renders the P-O bond sensitive to alkali and to enzymes that cleave RNA.

Some RNAs contain methyl, thiol, hydrogen, and isopentenyl substituents on certain purines and pyrimidines, 2′-O-methyl modifications on specific ribose residues, and even an altered linkage between special uracils and their ribose moieties (e.g., pseudouridylic acid) (Figure 1.21). Nucleotides containing such modifications occur only rarely in rRNAs and mRNAs but very frequently in tRNA. Generally, modifications of the bases and ribose residues occur after RNA synthesis and not at the level of the biosynthetic precursors. In only a few instances is their functional significance known.

5′ end

adenine

guanine

uracil

cytosine

3′ end

Figure 1.20
A short length of polyribonucleotide in RNA.

c. *RNA Structure*

Most cellular RNA is single stranded, although some animal virus genomes (e.g., reoviruses) consist of double stranded RNA molecules resembling the A form of DNA. The single strands almost invariably form

Figure 1.21
The structure of modified purine and pyrimidine nucleosides that occur in small quantities in some RNAs.

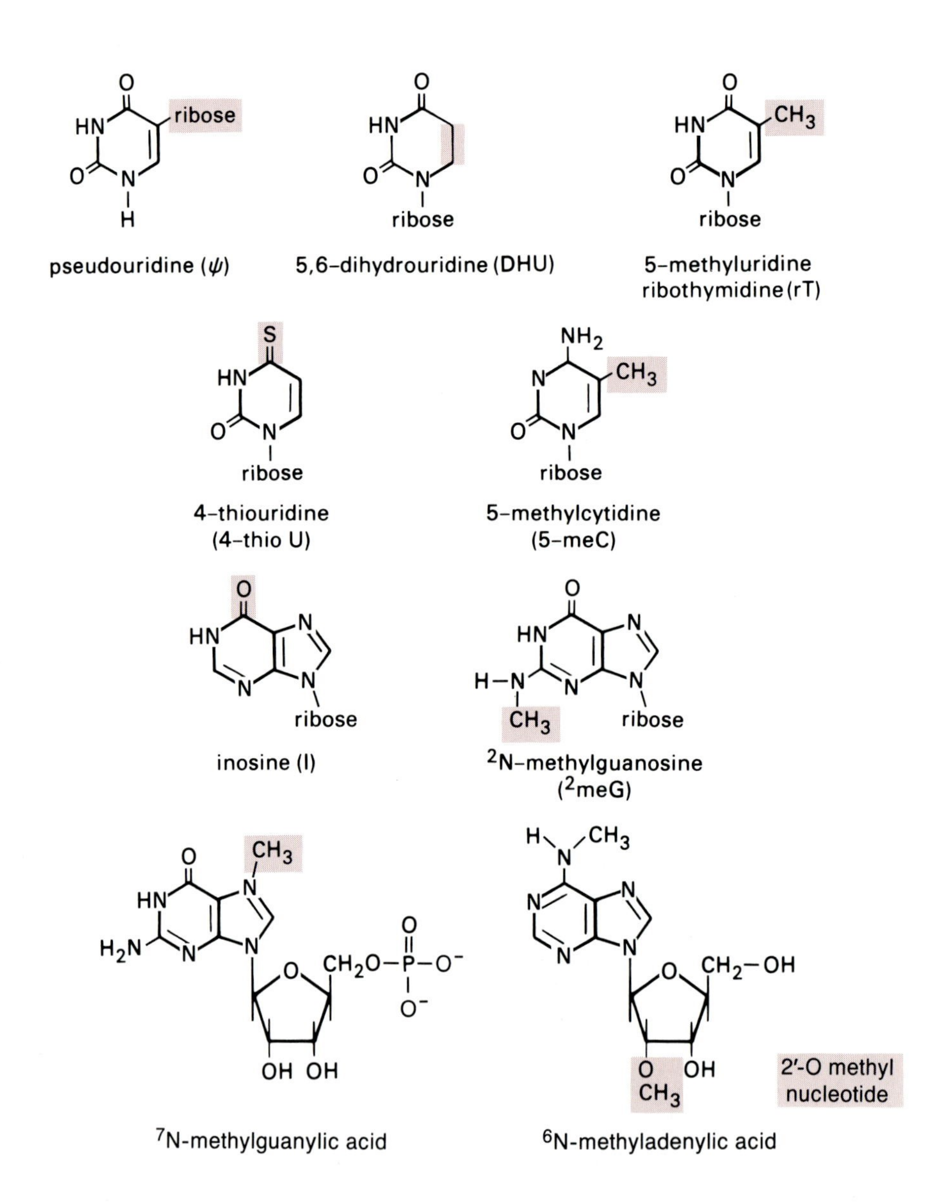

intramolecular, short, double helical stretches. These arise because most RNA chains have short regions of complementary sequences that allow the chain to loop back on itself to form localized helical regions (Figure 1.22). In the double strand regions, A pairs with U and G pairs with C; G can also form base pairs with U, but G·U base pairs are less stable than the standard G·C pair because they make two instead of three hydrogen bonds. Double helical segments formed in this way are usually short and interrupted because the base sequences on the two interacting regions are rarely perfectly or continuously complementary.

Most RNA chains can fold in more than one way, but the biological significance of these isomers is known in only a few cases. For example, proper folding is critical for gene expression of certain viral RNAs because key regulatory signals may be available or obscured in one or another folded structures (Section 3.11f). This dependence of function on proper RNA folding is best understood in tRNA. In spite of their distinctive nucleotide sequences, the three-dimensional structures of tRNAs are very similar, and the maintenance of their structures is critical for their

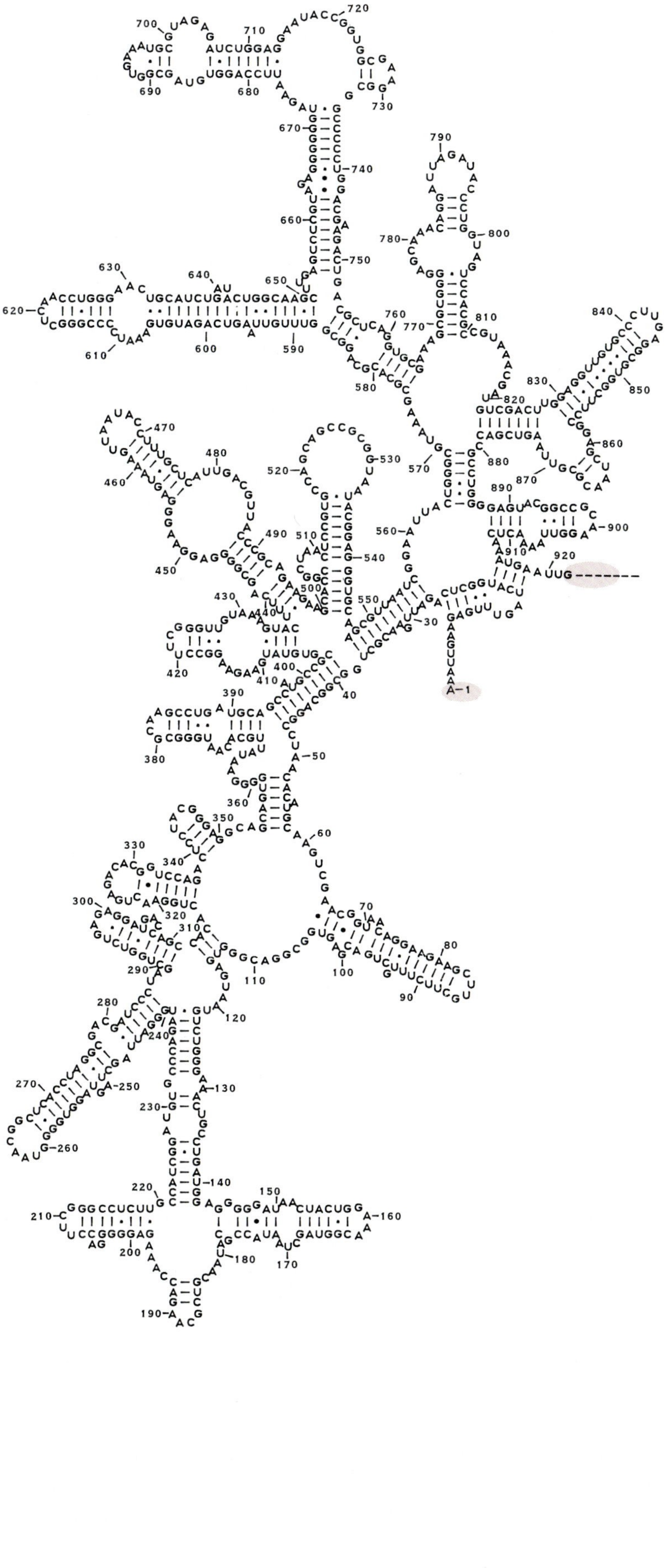

Figure 1.22
The 5′ proximal portion of the *E. coli* 16S rRNA chain showing how intramolecular base pairing is believed to fold the chain. Note that G·U base pairs are acceptable for maintaining an RNA duplex and that bases may loop out from a base paired region where their complementary base is absent from the opposite strand. The shaded regions designate both the first nucleotide in the chain and the point at which the chain continues on to the 3′ end. Courtesy of H. Noller.

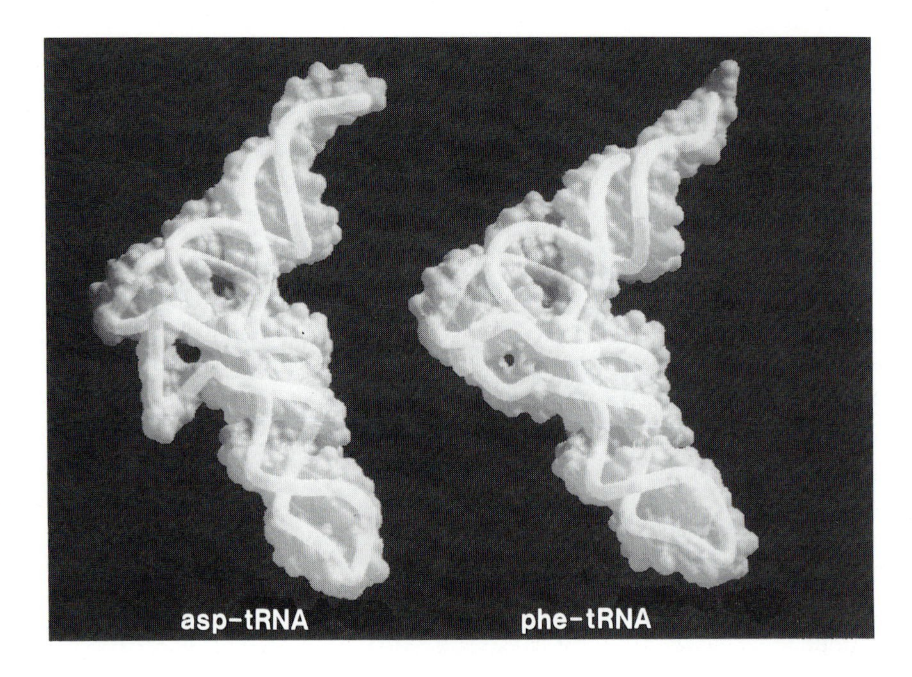

Figure 1.23
Computer graphic representation of the structure of $tRNA^{Asp}$ and $tRNA^{Phe}$ The backbone of each RNA chain is shown as a solid tube superimposed over the space filling image. The chains are held in this folded form by base pairing between bases that are distant from each other in the tRNA sequence. The 5′ and 3′ ends of the RNA chains are at the top right of each molecule. The models and photographs were provided by Arthur J. Olson, Ph.D. Copyright 1987, Research Institute of the Scripps Clinic.

biological function (Figure 1.23). The proper termination of RNA synthesis and the processing of the transcript also frequently depend on the structure of the folded RNA.

d. RNA Denaturation and Renaturation

Just as in DNA, double helical regions in RNA are disrupted by elevated temperatures and high pH; but unlike in DNA, the phosphodiester bonds in RNA are cleaved at high pH. Because the lengths of helical regions in single strand RNA are short and frequently imperfect, these regions are readily disrupted. However, fully complementary double strand RNA melts sharply over a narrow temperature range, as does double strand DNA. Also, as with DNA, denaturation of double strand RNA yields two complementary single strands that can reassociate when the temperature is lowered slowly. With single strand RNA, after denaturation, it is more difficult to reform the same base paired regions, and several alternative structures may be formed.

e. RNA–DNA Hybrid Helices

RNA and DNA polynucleotide chains that have complementary base sequences form antiparallel RNA-DNA double helices. Their structures resemble that of the A–DNA double helical structure (Figure 1.8). The base-pairing relationships in such structures conform to the Watson-Crick rules for DNA: dA pairs with rU, rA with dT, dG with rC, and dC with rG. In such RNA-DNA duplexes, the rG·dC base pair is more stable than the dG·rC base pair, and both are more stable than the dG·dC base pair of DNA; both the dT·rA and rU·dA base pairs are less stable than dA·dT base pairs in DNA. Such hybrid helices or **heteroduplexes** are denaturable and renaturable in the same way as DNA. In aqueous solution, with moderate concentration of salt, double strands formed from

RNA and DNA are more easily denatured than the corresponding helices of exclusively RNA or DNA. However, in solutions containing certain concentrations of formamide and salt, the RNA-DNA duplex is more stable than a DNA duplex. These conditions can be used to produce and specifically maintain the hybrid duplex.

Polynucleotide chains, whether RNA or DNA, are complementary if they can form a continuous double helix with Watson-Crick base pairs. Two nucleic acids are said to be **homologous** if their nucleotide sequences are identical or very closely related. By measuring the extent to which their respective single strands form RNA or DNA double helices or RNA-DNA hybrid helices, the extent of homology between two nucleic acids can be determined. The formation of DNA, RNA, and DNA-RNA hybrid helices can be detected by a variety of procedures. Single strand DNA is most often used as a probe to detect and quantitate homologous RNA molecules. Thus, the extent of conversion of single strand RNA to a double strand form in the presence of a second nucleic acid is a measure of their sequence homology. With appropriate **hybridization** partners, the amount of a specific nucleic acid sequence and the extent of nucleotide sequence homology between two DNAs, DNA and RNA, or two RNAs can be estimated.

1.3 Structure of Proteins

a. Constituents and Chemical Linkages in Proteins

Protein molecules contain one or more **polypeptides**, each consisting of a long unbranched polymer of amino acids. All polypeptides, whether from viruses or humans, are constructed from a repertoire of only 20 different amino acids. Amino acids have common and unique chemical groupings: a carbon atom (the α-carbon atom) that bears a carboxyl group, an amino group, and a distinctive substituent characteristic of each amino acid (Figure 1.24). These "side chains," often referred to as the R groups, differ in size, shape, charge, and chemical reactivity.

The backbone of every protein chain is composed of amide (or peptide) bonds that are formed by joining the amino group of one amino acid to the carboxyl group of its neighbor. Figure 1.25 shows the simplest unit, a dipeptide, and Figure 1.26 shows a more realistic representation of the bond angles and various substituents in a heptapeptide. Thus, polypeptides are long chains with regularly repeating peptide bonds and an assortment of distinctive side chains arrayed along the backbone. Moreover, polypeptide chains have a direction. One end contains the free amino group (under neutral conditions, this group is protonated to the form NH_3^+), and that amino acid is designated the N terminal residue and that end of the chain, the amino terminus. The carboxyl group at the other end of polypeptide chains generally exists as the COO^- anion, the amino acid at that location is referred to as the C terminal residue, and that end is called the carboxyl terminus. In some proteins, the cysteine residues form intrachain disulfide linkages to join two parts of a chain together and thereby constrain intervening portions of the structure (Figure 1.27). Such disulfide bridges can also join two separate polypeptide chains, which may or may not be iden-

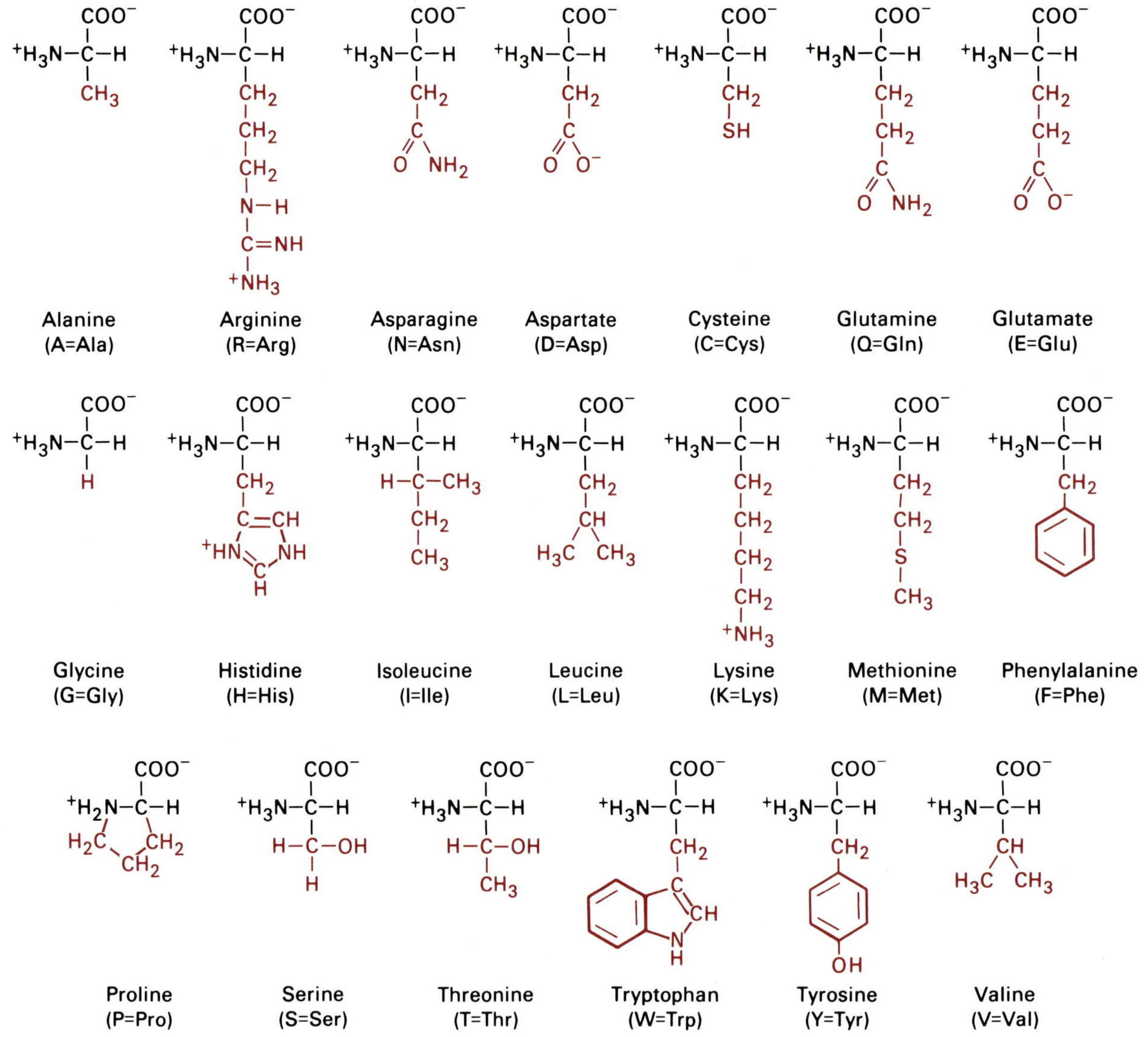

Alanine (A=Ala) | Arginine (R=Arg) | Asparagine (N=Asn) | Aspartate (D=Asp) | Cysteine (C=Cys) | Glutamine (Q=Gln) | Glutamate (E=Glu)

Glycine (G=Gly) | Histidine (H=His) | Isoleucine (I=Ile) | Leucine (L=Leu) | Lysine (K=Lys) | Methionine (M=Met) | Phenylalanine (F=Phe)

Proline (P=Pro) | Serine (S=Ser) | Threonine (T=Thr) | Tryptophan (W=Trp) | Tyrosine (Y=Tyr) | Valine (V=Val)

Figure 1.24
Structures of the amino acids commonly found in proteins. The variable portion of each amino acid is shown in color. The complete name and the one- and three-letter designations of each amino acid are shown below each structure. Many representations of peptides and polypeptides show the side chains as *R* groups.

tical. Certain oligomeric proteins have their polypeptide subunits joined in this manner.

Some proteins contain small quantities of modified forms of the natural amino acids (Figure 1.28). In virtually all cases, the alteration in the amino acid structure occurs after it has been incorporated into peptide linkage: hydroxyproline and hydroxylysine arise by hydroxylation of protein-bound proline and lysine, respectively; γ-carboxyglutamate is formed by carboxylation of glutamate and phospho-amino acids by the phosphorylation of the hydroxyl groups of serine and threonine or the phenolic group of tyrosine.

Members of a major group of proteins, called glycoproteins, are important constituents of eukaryotic cells and many viruses. They contain complex carbohydrates covalently linked to peptide-bound asparagine, hydroxylysine, serine, and threonine residues. As in the case of the amino acid modifications previously cited, the various sugars are added in discrete steps following the assembly of the polypeptide chain (Section 3.10).

a dipeptide

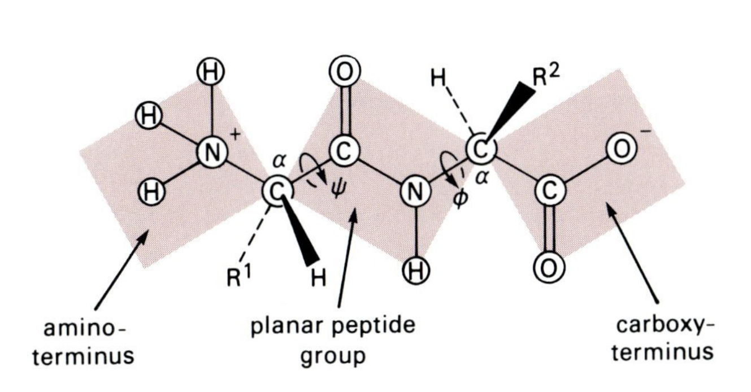

Figure 1.25
Structure of a dipeptide. The top representation shows the chemical linkages of the peptide bond, the ionized terminal amino and carboxyl groups, and the α carbon atoms to which the side chains are attached. The lower diagram shows the angular features of the peptide bond. These include the angle formed by the bond between the α carbon and the amide group (ϕ), between the α carbon and the carboxyl group (ψ), and the projections of the alternating *R* groups and hydrogen atoms from the α carbon atoms.

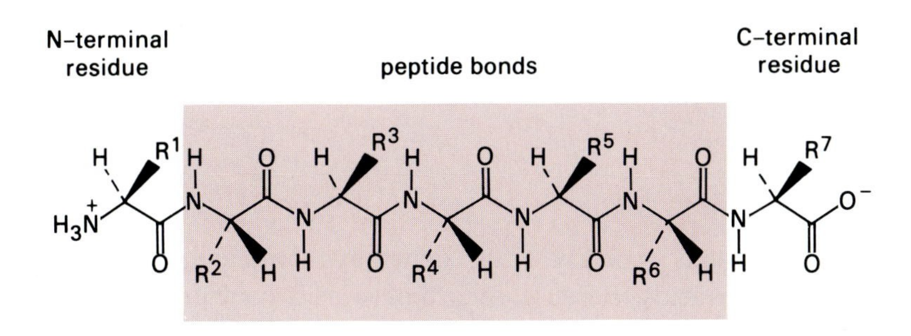

Figure 1.26
Structure of a heptapeptide. The symbols and conventions are as described in Figure 1.25.

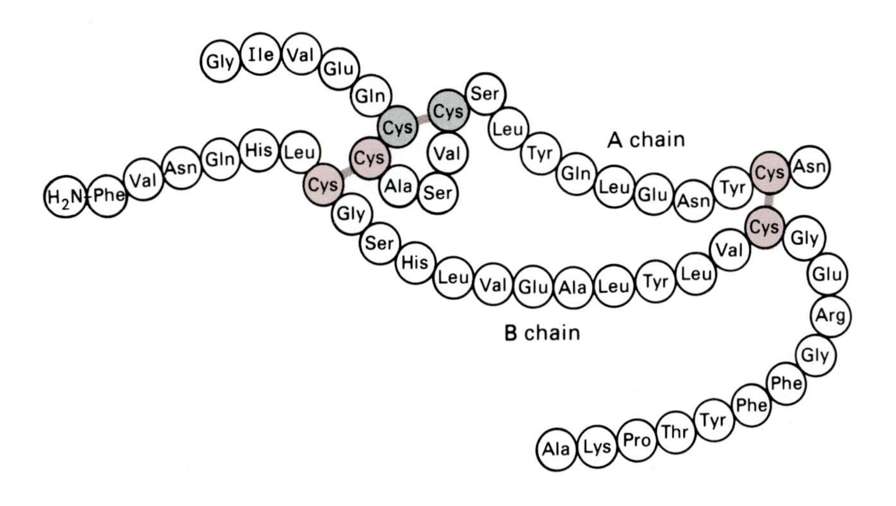

Figure 1.27
Interchain and intrachain disulfide linkages in bovine insulin. The A chain contains two neighboring cysteine (Cys) residues that form one intrachain disulfide bond; two disulfide bridges link cysteines in the A and B chains. The amino terminal phenylalanine (Phe) of the B chain and the carboxyl terminal asparagine (Asn) of the A chain are the amino and carboxyl termini, respectively, of the insulin precursor polypeptide.

Figure 1.28
Some modified amino acids in proteins.

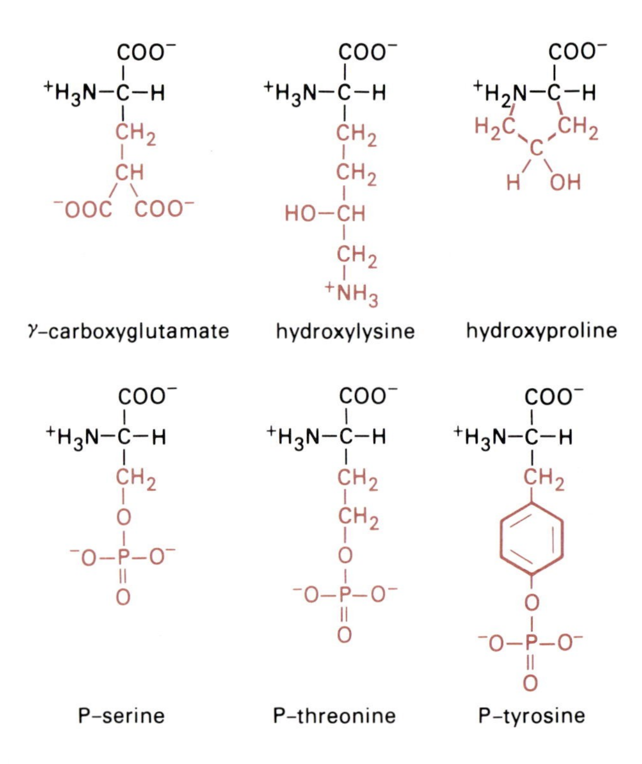

Such glycosylation reactions play an important role in transporting proteins from the sites of synthesis to specific intracellular organelles and to the cell's exterior; glycoproteins also provide a distinctive signature to a cell's surface.

b. Protein Size and Shape

Polypeptides come in many sizes, generally containing 50 to several thousand amino acids. Because each amino acid contributes about 110 units (Daltons) of molecular weight, proteins range in size from 10^4 to 10^6 Daltons. Some proteins consist of only a single polypeptide chain, others contain more than one chain of the same type, and still others are composed of different types of chains (Table 1.6). Myoglobin is an example of a protein with a single polypeptide chain; bacterial glutamine synthetase has 12 identical polypeptide chains. Hemoglobin is a tetramer, having two different kinds of chains, and the functional form of *E. coli* RNA polymerase contains at least four different kinds of polypeptide chains.

Proteins have well-defined conformations or three-dimensional shapes in solution. Almost invariably, the biological activity of proteins, whether it be catalysis, structure, transport, support, movement, or regulation, depends on maintenance of the native or active conformation. Most protein conformations fall into two general categories. The **globular** proteins are compact and roughly spherical structures, resulting from tight and irregular folding of the polypeptide chains (Figure 1.29). **Fibrous** proteins have their polypeptide chains in parallel arrangements, yielding long fibers or sheets (Figure 1.30). Most enzymes and soluble proteins have globular structures; proteins such as collagen and keratin, that occur in structural or connective tissues, assume the fibrous conformation. Other proteins, such as fibrinogen and muscle myosin, include both types of conformations.

Table 1.6 Size and Subunit Compositions of Some Globular Proteins

Protein	Source	Molecular Weight of Protein (kD)	Number of Subunits	Molecular Weight of Subunits (kD)
DNA ligase	*E. coli*	75	1	—
DNA polymerase I	*E. coli*	109	1	—
Alkaline phosphatase	*E. coli*	86	2	43
Lac repressor	*E. coli*	160	4	40
β-galactosidase	*E. coli*	544	4	135
Glutamine synthetase	*E. coli*	592	12	49
Hemoglobin	mammalian	64	2α	16
			2β	16
Tryptophan synthetase	*E. coli*	148	2α	29
			2β	45
Aspartic transcarbamylase	*E. coli*	310	6 (C)[1]	33
			6 (R)[2]	17
RNA polymerase core	*E. coli*	400	2 (α)	40
			1 (β)	155
			1 (β')	165
Pyruvate dehydrogenase complex	*E. coli*	4500	24[3]	91
			24[4]	70
			12[5]	56

[1] The catalytic subunits
[2] The regulatory subunits
[3] Pyruvate decarboxylase subunits
[4] Dihydrolipoyl transacetylase subunits
[5] Dihydrolipoyl dehydrogenase subunits

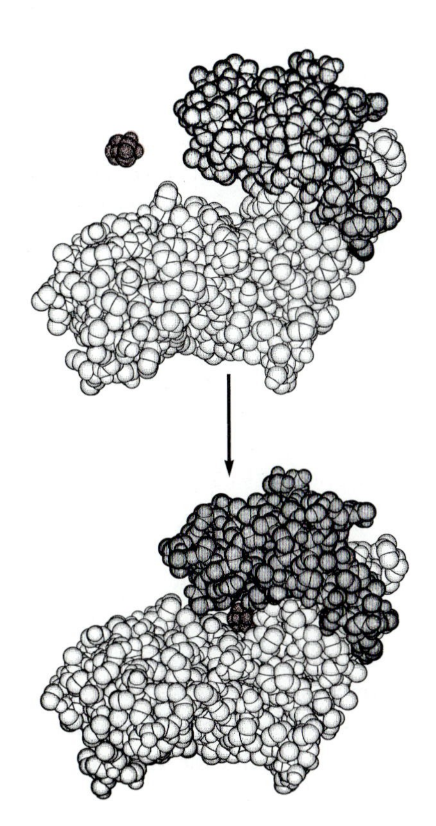

Figure 1.29
Structure of the globular protein, hexokinase. The model at the top shows the hexokinase molecule and one of its substrates, D-glucose, in the uncomplexed forms. Upon binding glucose, the darkly and lightly shaded domains of the protein move to close the binding pocket. From W. S. Bennett, Jr., and T. A. Steitz, *J. Mol. Biol.* 140 (1980), p. 211.

(a)

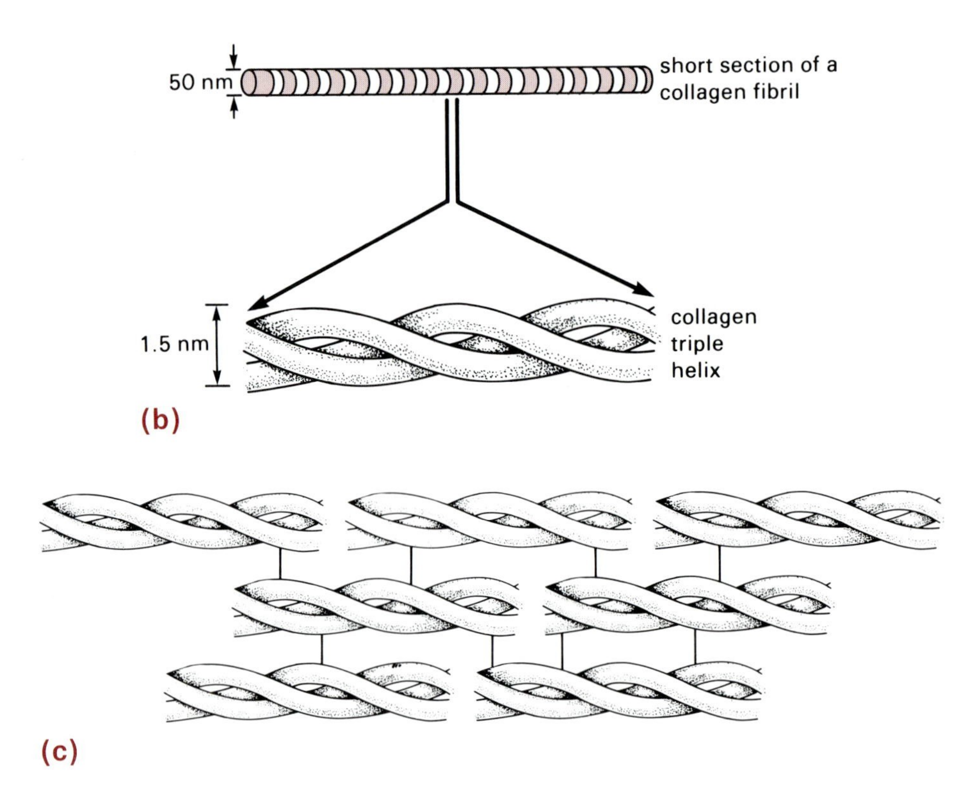

Figure 1.30
Collagen structure. (a) Electron micrograph of intact collagen fibrils from skin. Courtesy of J. Gross. (b) Collagen fibrils contain tropocollagen fibers composed of three polypeptide chains organized in a triple helix. (c) The fibrils are composed by interactions between multiple, overlapping tropocollagen triple helices.

c. *Determinants of Protein Conformation*

Protein structure can be considered at several different levels of organization, which are designated primary, secondary, tertiary, and quaternary. The first three refer to structural features of polypeptide chains. The fourth, quaternary structure, refers to the structure of oligomeric proteins—those with two or more polypeptide chains.

Each protein possesses a unique amino acid order along its polypeptide backbone (see Figure 1.27 for the amino acid sequence of the two chains of bovine insulin); this defined order of amino acids is called the **primary structure**. The uniqueness of a primary structure was first established for bovine insulin but has been confirmed by analysis of literally hundreds of other proteins of various sizes. A protein's primary structure is determined by the structure of the corresponding gene. Therefore, a change in the nucleotide sequence of a gene coding for a protein can alter the primary structure of the protein.

Polypeptide chains can fold into regularly repeating structures called the **secondary structures**. The most prevalent periodic conformations in proteins are the **right-handed α helix** and the **β pleated sheet**. In the α helix, the backbone assumes the configuration of a helical coil with a periodicity of 0.54 nm and about 3.6 residues per turn (Figure 1.31). Such a helical arrangement is stabilized by intrachain hydrogen bonds between the hydrogen atom on the nitrogen of one peptide bond and the carboxyl oxygen of the fourth amino acid away in the helix coil. In the α helix, the amino acid side chains extend outward from the helical backbone (Figure 1.32).

The extent to which a given polypeptide chain can adopt the α helical conformation depends upon the composition and order of amino acids in the peptide chain. Some amino acids or sequences destabilize the α helix,

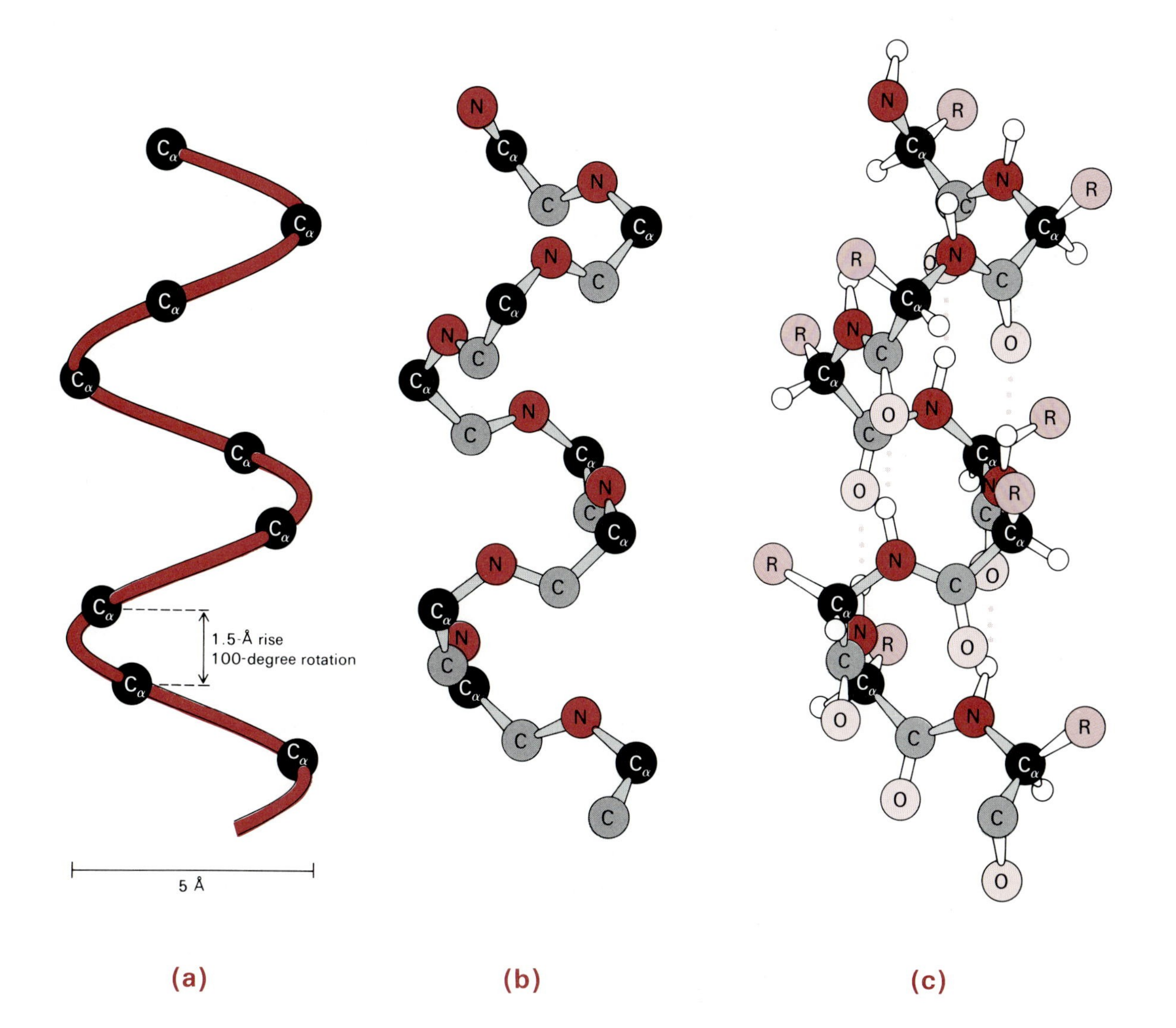

Figure 1.31
Models of a right-handed α helix. (a) Only the α carbon atoms are shown on a helical thread. (b) Only the backbone nitrogen (N), α-carbon (Cα), and carbonyl carbon (C) atoms are shown. (c) The entire helix. Hydrogen bonds (shown in part c) between NH and CO groups stabilize the helix. From L. Stryer, *Biochemistry* (San Francisco: W.H. Freeman, 1981).

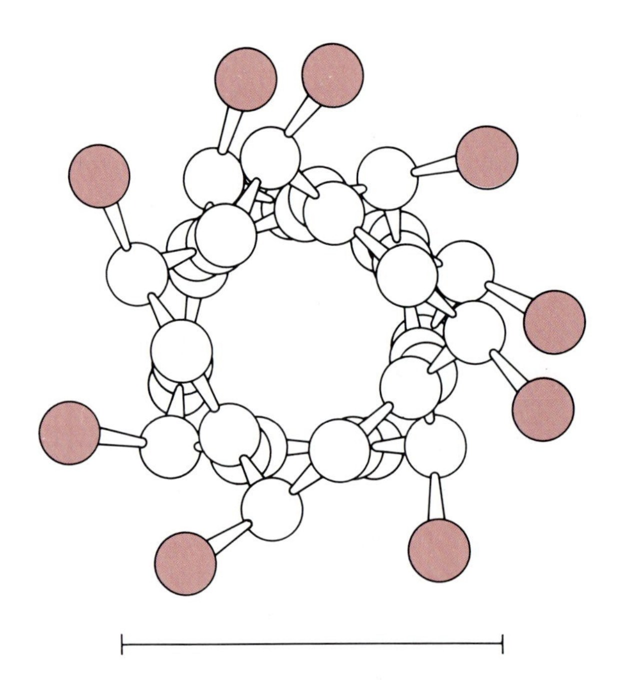

Figure 1.32
Cross section through an α helix. The *R* groups of each amino acid radiate to the outside of the helix. The space through which the axis of the helix passes has been enlarged for illustrative purposes. From L. Stryer, *Biochemistry* (San Francisco: W.H. Freeman, 1981).

and two, proline and hydroxyproline, break or interrupt α helices because of restricted rotation about the peptide bond and the lack of the H atom for hydrogen bond formation. Usually, α helical regions are relatively short, involving on average 10 to 20 amino acids. Occasionally, hydrogen bonds between amino acid side chains in two α helical segments will stabilize a side-by-side pairing of helix coils from the same or different polypeptide chains. There are also instances, for example in fibrinogen, in which multiple α helical regions of a polypeptide chain form interlocked bundles. Some proteins, such as the enzyme chymotrypsin, are devoid of α helical segments; in others, for example, the α and β subunits of hemoglobin, the α helix recurs frequently along each of the chains.

An alternate secondary structure in proteins is the β pleated sheet conformation, so called because the structure resembles a sheet. Instead of the tightly coiled chain and rodlike structure of the α helix, the polypeptide chain in a β pleated sheet is extended. Moreover, hydrogen bonds occur between amino acids distant from each other along the polypeptide chain or between amino acids of different chains (Figure 1.33). Rows of two to five such hydrogen-bonded chains, organized in parallel or antiparallel arrangements, form pleated sheet structures.

Certain proteins (e.g., collagen) have an unusually high percentage of glycine and proline, two amino acids that destabilize or break α helices, and of two unusual amino acids, hydroxyproline and hydroxylysine. Such proteins have a third type of secondary structure formed from three long left-handed helical chains wound around one another to form a superhelical cable (Figure 1.34). This structure is also stabilized by hydrogen bonds formed between residues on adjacent chains in the cable.

Virtually all enzymes and regulatory proteins have globular or compactly folded three-dimensional structures; this constitutes the **tertiary structure**. Almost invariably, the polar amino acid side chains are on the surface interacting with the solvent, and the nonpolar residues are buried in the interior, shielded from the aqueous phase. The highly convoluted polypeptide chain contains variable amounts of α helical and β pleated

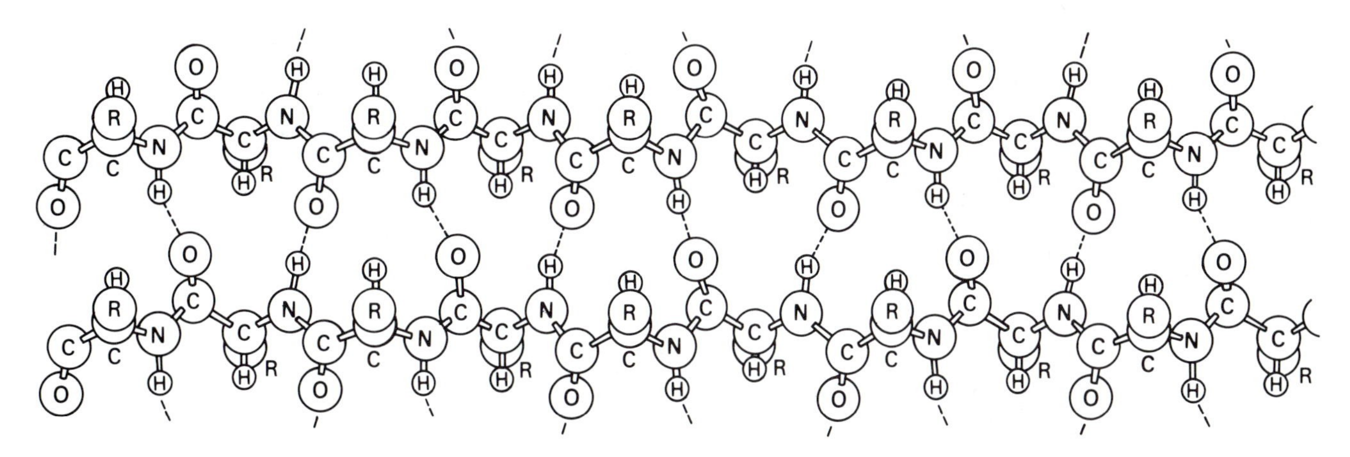

Figure 1.33
Structure of a β pleated sheet. β pleated sheet structures are formed by the interaction of more than one polypeptide chain. Hydrogen bonds between the hydrogen of an amide group (NH) of one chain and carbonyl oxygens (CO) of parallel chains stabilize the structure. Note that intrachain hydrogen bonds characteristic of the α helix are replaced by interchain hydrogen bonds of the β pleated sheet. The structure shown here has the parallel arrangement of chains, but antiparallel arrangements also exist in some β pleated sheet structures.

sheet structures interspersed with unstructured regions that generally occur at or near bends of the backbone (Figures 1.29 and 1.35). The biologically active, tertiary structure of the protein is stabilized by a variety of interactions between amino acids in the chain: (1) those that contribute to the formation of α helical and β pleated sheet structures, (2) hydrogen bonds between some amino acid side chains, (3) ionic bonds between oppositely charged side chains, (4) covalent disulfide linkages between distantly located cysteines, and (5) hydrophobic side chains that interact with one another in the interior rather than with the aqueous phase on the exterior of the protein. The relative contributions of these stabilization modes vary among different proteins.

After a native globular structure is unfolded by heating or by raising or lowering the pH, the **denatured** chain can reform its initial three-dimensional or tertiary structure. This involves reestablishing the chemical and physical interactions that stabilized the original folded conformation, a process called **renaturation** or **refolding**. The mechanism by which the polypeptide chain is properly folded and the intermediate stages that are encountered in the process are active subjects of investigation.

Many globular proteins are **oligomeric** (composed of two or more identical or nonidentical polypeptide chains). Hemoglobin (Figure 1.36) illustrates one such protein, and an immunoglobulin (Figure 1.37, page 70) represents another. The **quaternary structure** of such proteins is defined by the way the individual folded polypeptide chains are organized and interact in the oligomeric molecule. In hemoglobin, for example, there are contacts between amino acids on the α and β polypeptides to form the $\alpha\beta$ dimer and additional interactions between $\alpha\beta$ dimers to form the $\alpha_2\beta_2$ tetramer (Figure 1.36). Detailed structural studies of hemoglobin have also clarified how these contacts change as the protein binds and releases oxygen.

The secondary, tertiary, and quaternary conformations of proteins are intimately interdependent and ultimately determined by the primary structure of one or more polypeptide chains. The ramifications of this interdependency are profound: The information to fold a protein correctly into its biologically active state is inherent in the amino acid sequence of the polypeptides. Consistent with this principle is the finding that chemical modifications and mutational alterations of amino acid sequences in polypeptides strikingly alter their renaturation rates and their ability to

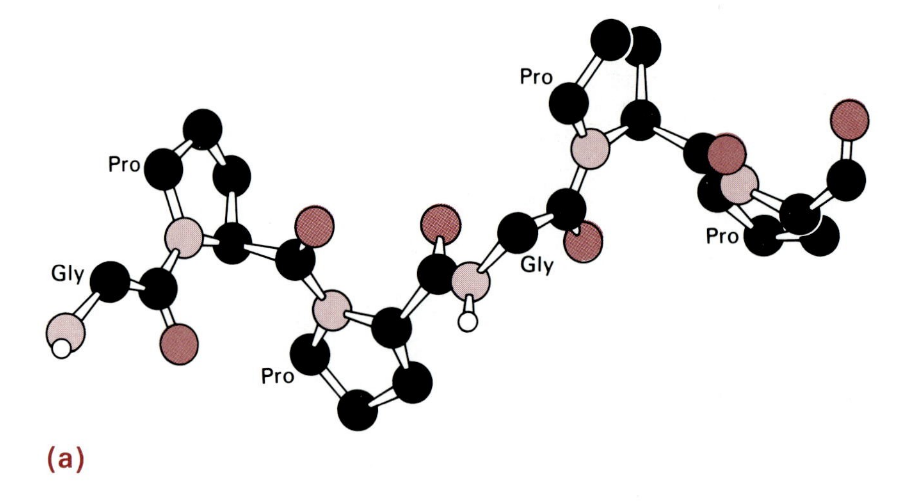

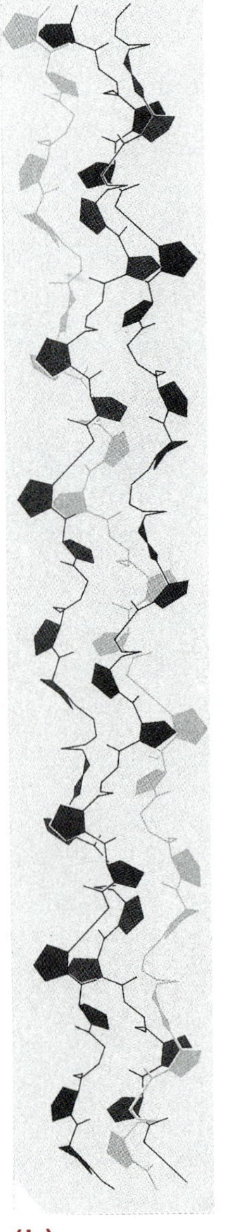

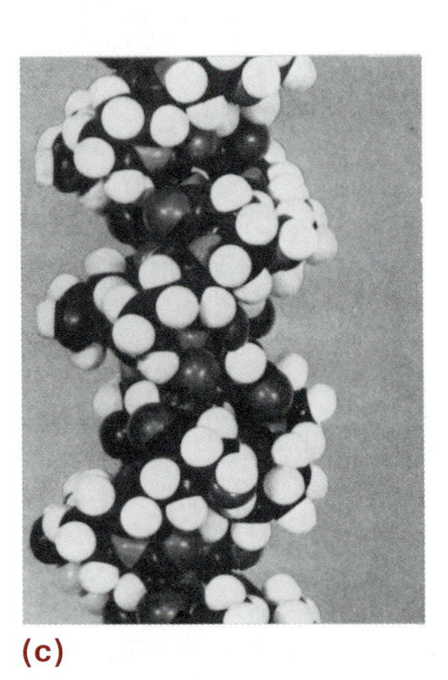

Figure 1.34
Molecular details of the collagen structure. (a) The basic repeating structure composed of glycine, proline, and proline. (b) Skeletal model of the repeating glycine and prolines in the triple strand collagen helix. From L. Stryer, *Biochemistry* (San Francisco: W.H. Freeman, 1981). (c) A space filling model of collagen II with a repeating sequence of glycine, proline, and hydroxyproline. The helix repeat distance is 2.86 nm. From A. Rich and F. H. C. Crick, *J. Mol. Biol.* 3 (1961), p. 483.

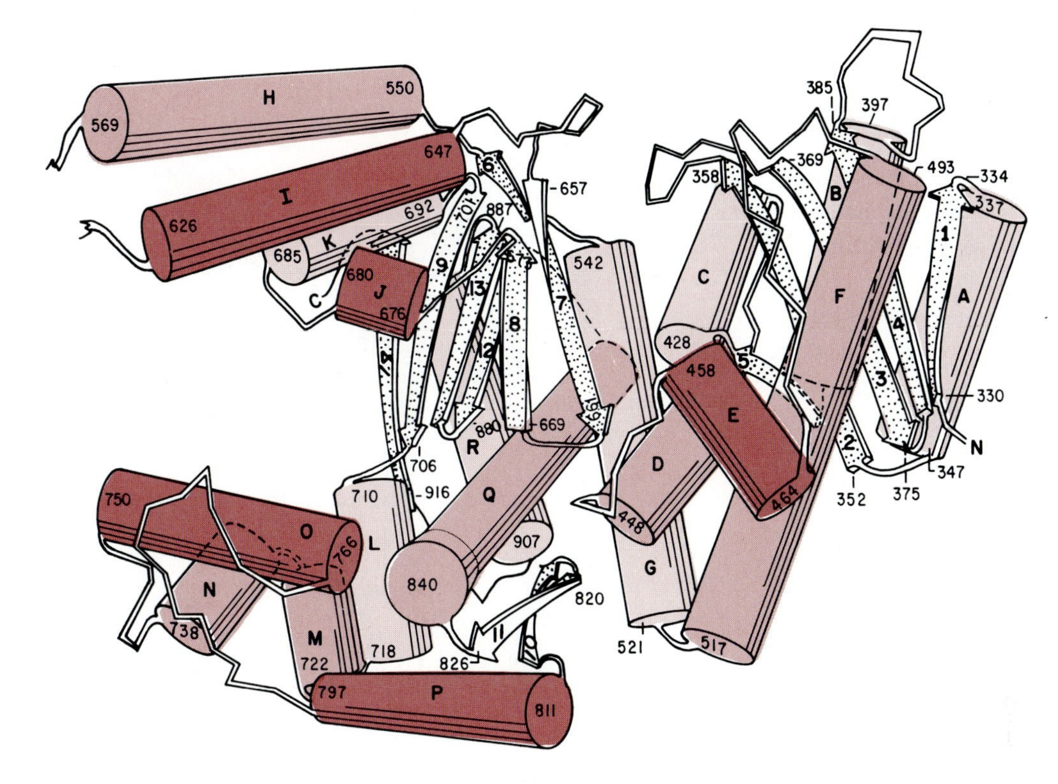

Figure 1.35
Schematic representation of part of the structure of *E. coli* DNA Polymerase I. The cylinders represent α helices, and the arrows represent β pleated sheets. The more intensely colored the region, the closer it is to the viewer. After D. L. Ollis et al., *Nature* 313 (1985), p. 762.

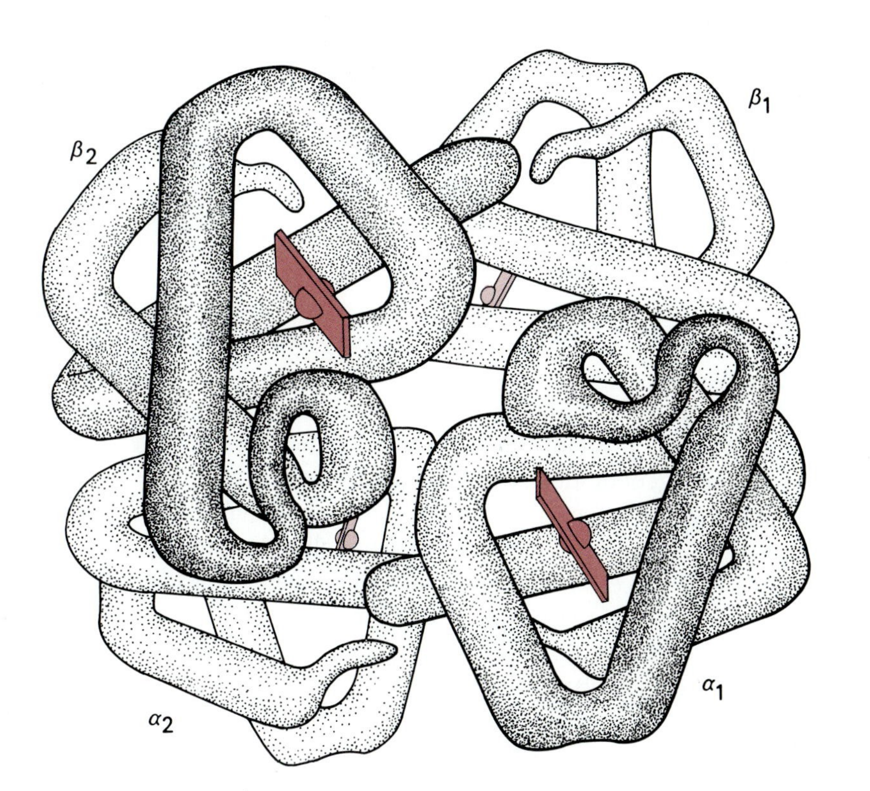

Figure 1.36
Model of hemoglobin at low resolution. Hemoglobin is a tetrameric protein composed of two α polypeptide chains (designated $\alpha 1$ and $\alpha 2$) and two β polypeptide chains (designated $\beta 1$ and $\beta 2$). Each chain has a ferro heme group shown in color; the planar structure is the heme moiety and the sphere represents the iron atom. The more intense the stippling, the closer the structure is to the viewer. Copyright Irving Geis.

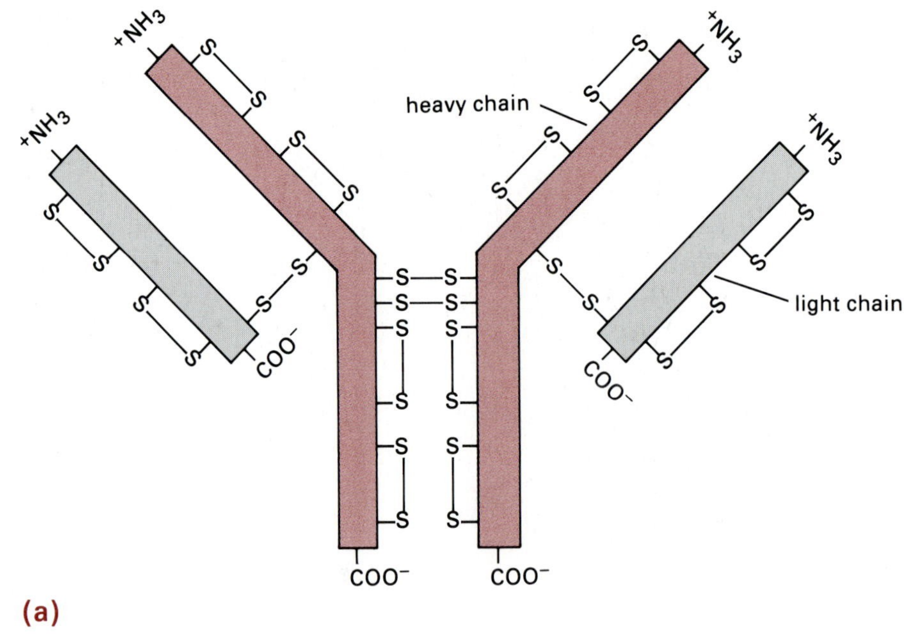

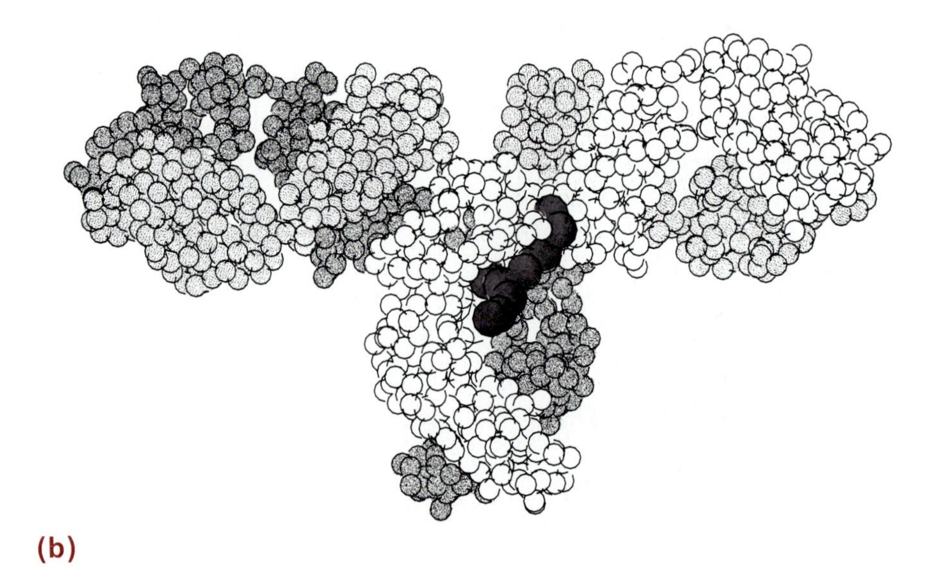

Figure 1.37
The structure of antibody molecules. (a) Immunoglobulin molecules are composed of four polypeptide chains, two identical light chains, and two identical heavy chains. Disulfide bonds between the two heavy chains and between a heavy and light chain hold the four chains together. Intrachain disulfide linkages also occur within the heavy and light chains. (b) Model of an immunoglobulin molecule shows how the light (lightly stippled) and heavy (heavily stippled and unstippled) chains are intertwined. From E. W. Silverton, M. A. Navia, and D. R. Davies, *Proc. Natl. Acad. Sci. USA* 74 (1977), p. 5140.

form the secondary, tertiary, and quaternary structures with full biological activity.

Sickle cell anemia is one of many convincing examples of the dependence of normal protein structure and function on correct amino acid sequence. The genetic defect in this disease results from the replacement of glutamic acid, the sixth amino acid from the amino terminus of the β chain, by valine in sickle hemoglobin. That alteration in the β globin primary structure places an aberrant hydrophobic amino acid on the surface of the protein, causing deoxygenated hemoglobin to aggregate and form higher order oligomeric structures (Figure 1.38). As a consequence, the shape and deformability of the red blood cells are altered, and blood flow through the capillaries and small venules is impeded or interrupted. The fundamental implication of this classic example is that a single mutation,

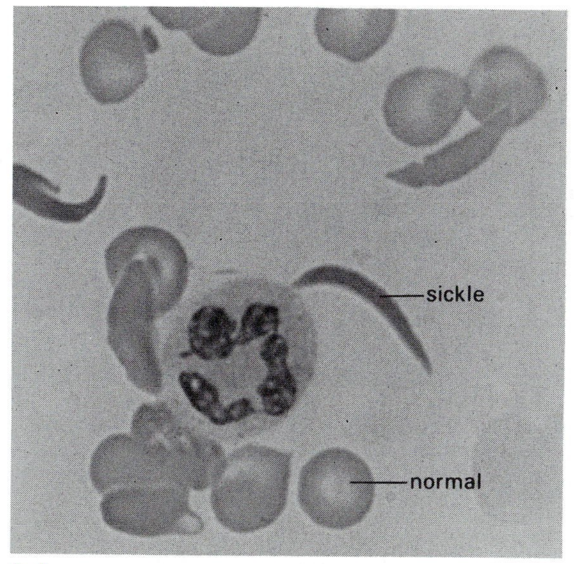

(a)

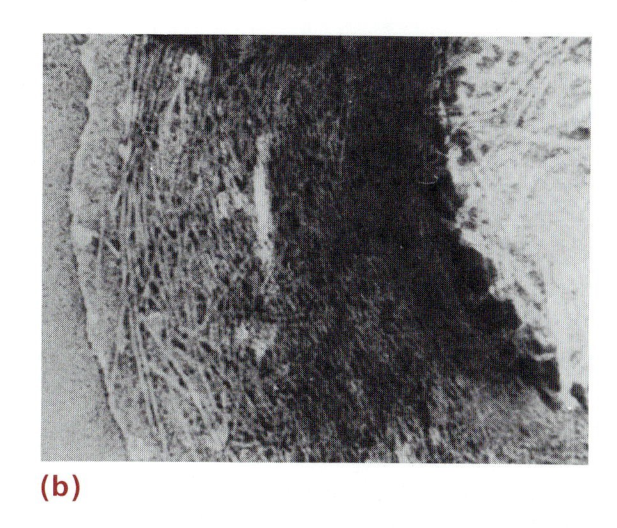

(b)

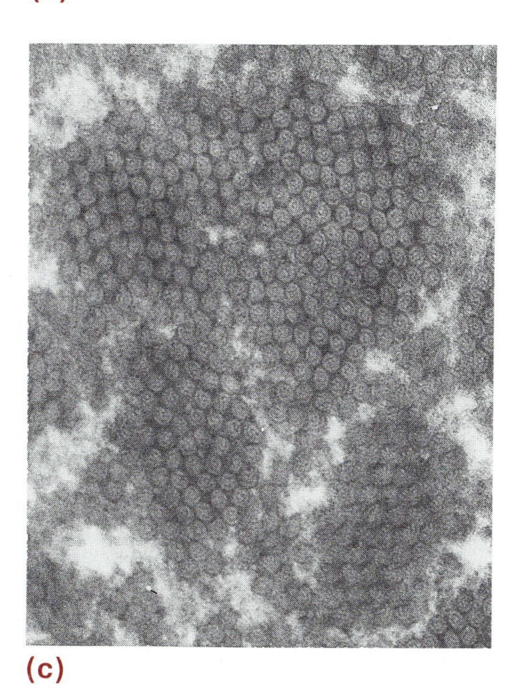

(c)

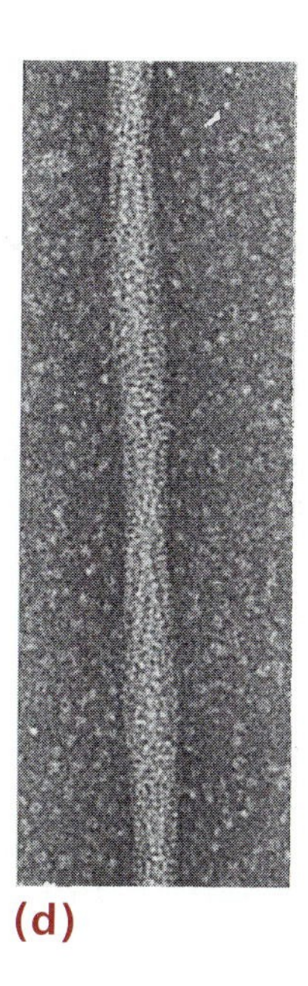

(d)

Figure 1.38
Cellular and molecular defects associated with sickle cell disease. (a) Micrograph showing normal and sickled erythrocytes. Courtesy of S. Schreier. (b) Micrograph showing sickle hemoglobin fibers in a disrupted sickled erythrocyte. (c) Cross section through hemoglobin fibers formed by deoxygenation of sickle hemoglobin. (b) and (c) courtesy of S. J. Edelstein. (d) Electron micrograph of a negatively stained fiber of sickle hemoglobin. From G. Dykes, R. H. Crepeau, and S. J. Edelstein, *Nature* 272 (1978), p. 509.

a change of one nucleotide in the DNA sequence, resulting in the replacement of one amino acid for another at a specific location in a polypeptide chain, can have a profound effect on the conformation of a protein and cause drastic alterations in its physiological function. In the most general sense, it is this mechanism that links the phenotypes of all organisms to their genotypes.

CHAPTER 2

Replication, Maintenance, and Modification of the Genome

A cellular organism's genetic program is specified in its DNA's nucleotide sequence. Consequently, faithful replication of that sequence at each generation is essential for preserving the organism's unique characteristics. *E. coli*, for example, must duplicate its entire genome of 4×10^6 nucleotide pairs virtually without error at each generation; and the nearly 4×10^9 nucleotide pairs in a human's 23 pairs of chromosomes need to be copied accurately between each cell division.

One of the wonders of DNA is that it encodes the complete machinery and instructions needed for its own duplication: some genes code for enzymes that synthesize the nucleotide precursors of DNA, and others specify proteins that assemble the activated nucleotides into polynucleotide chains. There are also genes for coordinating the replication process with other cellular events; and still others that encode the proteins that package DNA into chromatin. Another extraordinary property of DNA is that it functions as a template and directs the order in which nucleotides are assembled into new DNA chains. Provided with precisely the same synthetic machinery, different DNAs direct only the formation of replicas of themselves.

The genetic program also specifies an enzymatic machinery that rectifies the errors that occur occasionally during DNA replication, as well as enzymes that repair damage to the bases or helical structure caused by exposure to X rays, ultraviolet light, chemicals, and even certain diseases. The genetic program also provides opportunities for genome variation and evolutionary change. Certain genes encode proteins that promote strand exchanges between DNA molecules and thereby create new combinations of genetic information for the progeny. Other proteins cause genome rearrangements by catalyzing translocations of small segments or even large regions within and among DNA molecules. Such recombinations and translocations provide some of the substrates for evolution's experiments, but some rearrangements cause disease. By contrast, the proper functioning of several genetic programs actually depends on specific DNA rearrangements (Chapter 10).

In this chapter, we examine the mechanisms of DNA replication, repair, recombination, and rearrangement—in essence, the genetic transactions of DNA. We also include the replication of viral RNA genomes because features of that process mimic DNA replication.

2.1 DNA Replication

a. DNA's Template Function During Replication

In announcing the structure of DNA in 1953, James Watson and Francis Crick stated: "It has not escaped our notice that the specific (base) pairing we have postulated suggests a possible copying mechanism for the genetic material." A month later they expanded the idea: "If the actual order of the bases on one of the pair of chains were given, one could write down the exact order of the bases on the other, because of the specific base pairing. Thus, one chain is, as it were, the complement of the other, and it is this feature which suggests how DNA might duplicate itself."

Watson and Crick suggested that DNA duplication requires breaking the hydrogen bonds holding the helical duplex together and unwinding the two chains. They supposed that each chain could act as a **template** to direct the formation of a complementary chain, producing two pairs of chains where only one had existed before (Figure 2.1). In this way, the sequence of nucleotide pairs in a double helix could be duplicated exactly. Watson and Crick suggested that DNA replication could proceed by a spontaneous, nonenzymatic mechanism, but this proved to be incorrect. Nevertheless, their proposal that DNA replication involves assembling and joining an array of nucleotides complementary to those on each of the two chains of the helix solved the conceptual problem of how genes are faithfully replicated.

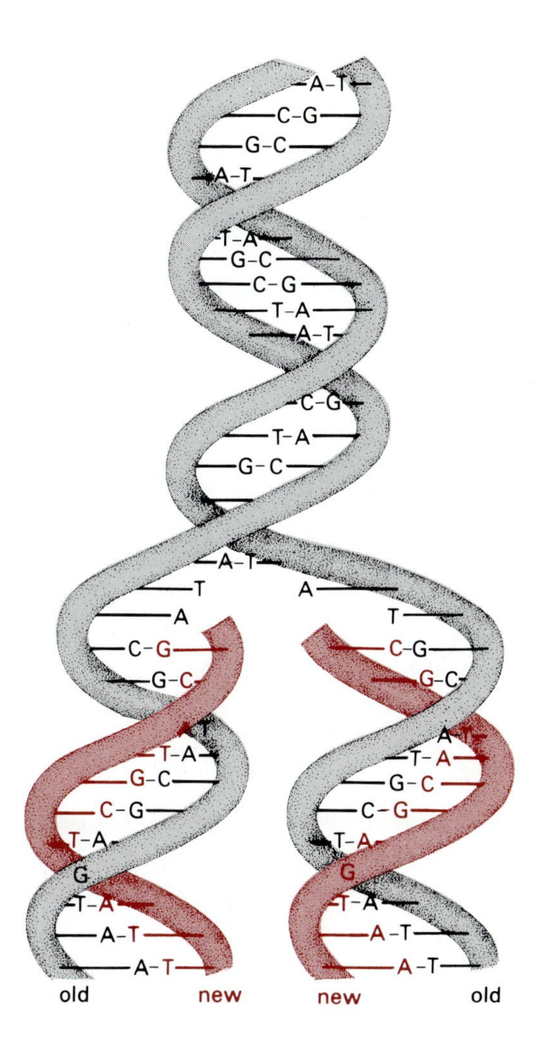

Figure 2.1
DNA duplication requires breaking the hydrogen bonds that hold the double helix together and unwinding the chains. Each chain acts as a template for the formation of a complementary chain. After J. D. Watson and F. H. C. Crick, *Cold Spring Harbor Symposium* 18 (1953), p. 123.

Since its proposal, the basic nature of the template mechanism has been verified by a large body of evidence obtained from many different types of organisms, *in vivo* and *in vitro*. As the model predicted, replication of all double strand DNA is **semiconservative** (Figure 2.2). Alternatives such as conservative or dispersive modes of double strand DNA replication are not known to occur. Thus, after one round of replication, one of the two polynucleotide chains of a parental duplex DNA is always conserved intact, base paired to a newly synthesized chain, in each of the two daughter molecules. Where genomes consist of only one DNA strand (e.g., in certain viruses), that strand provides the template to produce a duplex containing the complementary chain. Then the duplex directs the formation of either progeny duplexes or single strand copies of one of the template chains (Figure 2.3).

b. Replication Initiates at Discrete Locations

Origins of replication Replication does not begin just anywhere in a DNA molecule. Initiation occurs at specific locations called **replication origins**, and copying proceeds via **replication forks** in one or both directions until the DNA is completely duplicated (Figure 2.4). With circular duplex

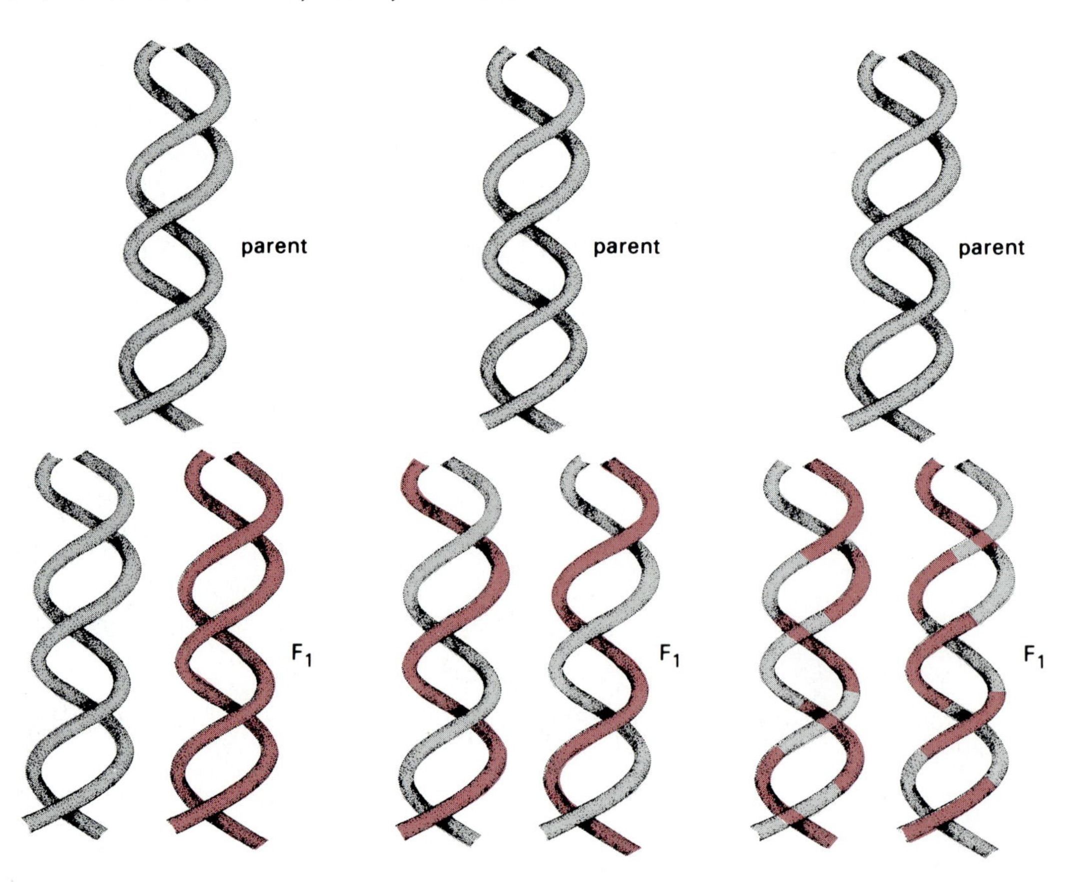

Figure 2.2
Possible models for replication of duplex DNA. The middle diagram shows the actual mechanism, semiconservative replication. Alternative possibilities, conservative (left) and dispersive (right) replication, are not known to occur.

Figure 2.3
Replication of single strand DNA genomes of viruses. *V* indicates the viral genome and *C*, its complement.

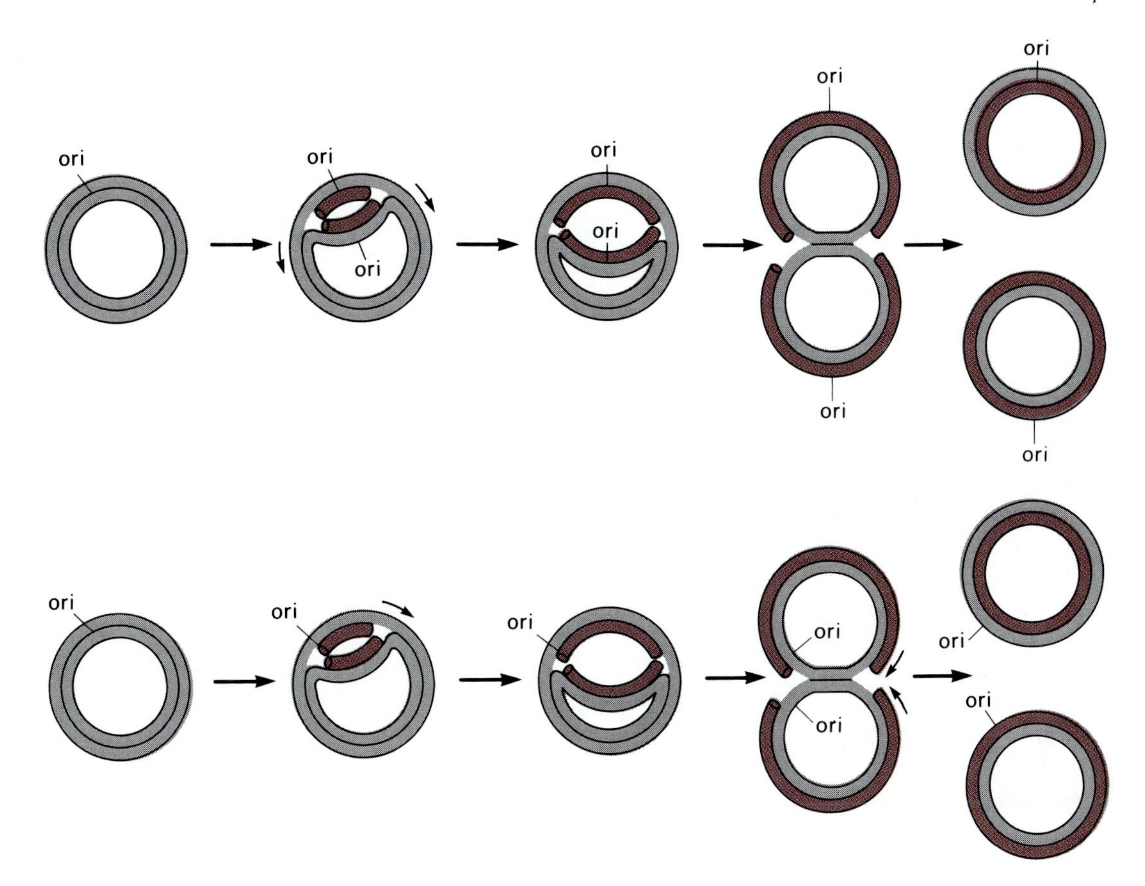

Figure 2.4
Replication initiates at specific DNA sequences called replication origins (ori). The growing chains form replication forks. New duplex DNA synthesis occurs at replication forks that move either bidirectionally (top) or unidirectionally (bottom), depending on the nature of the ori.

DNAs, the newly synthesized chains are covalently joined where the growing forks meet or where a single fork returns to the origin. The progeny molecules usually separate before another round of replication begins. For example, genomes of such widely differing size as SV40 (5.2 kbp), λ bacteriophage (48.5 kbp), and *E. coli* (4×10^3 kbp) are each copied after a single initiation event that occurs at a unique location.

Novel variations on this theme occur among pro- and eukaryotic organisms. Animal mitochondria (15 kbp) and bacterial colE1 plasmid (6 kbp) DNAs have different origins of replication for copying each strand of the parental helix (Figure 2.5). The synthesis of the complementary strand of several small single strand bacteriophage genomes begins at one specific sequence, but replication of the resulting duplex may be initiated at a different sequence (Figure 2.6). Replication of linear duplex DNAs is also initiated at unique sites. For example, T7 bacteriophage DNA (40 kbp) is replicated toward each end from a primary replication origin (Figure 2.7) but each of the two strands of human adenovirus DNAs (30–38 kbp) is always replicated sequentially starting from its 3′ end (Figure 2.8).

Multiple replication origins, spaced along the chromosome about 20 kbp apart, are common in eukaryotic cell genomes (Figure 2.9 on page 80). Following initiation, replication proceeds bidirectionally from each bubble until the forks from adjacent origins merge. Therefore, the very long DNA chains in each daughter chromosome arise by joining shorter stretches of independently initiated, newly synthesized strands.

Rate of genome replication In most instances, the genome replication rate is controlled by regulating the frequency of the initiation event. For ex-

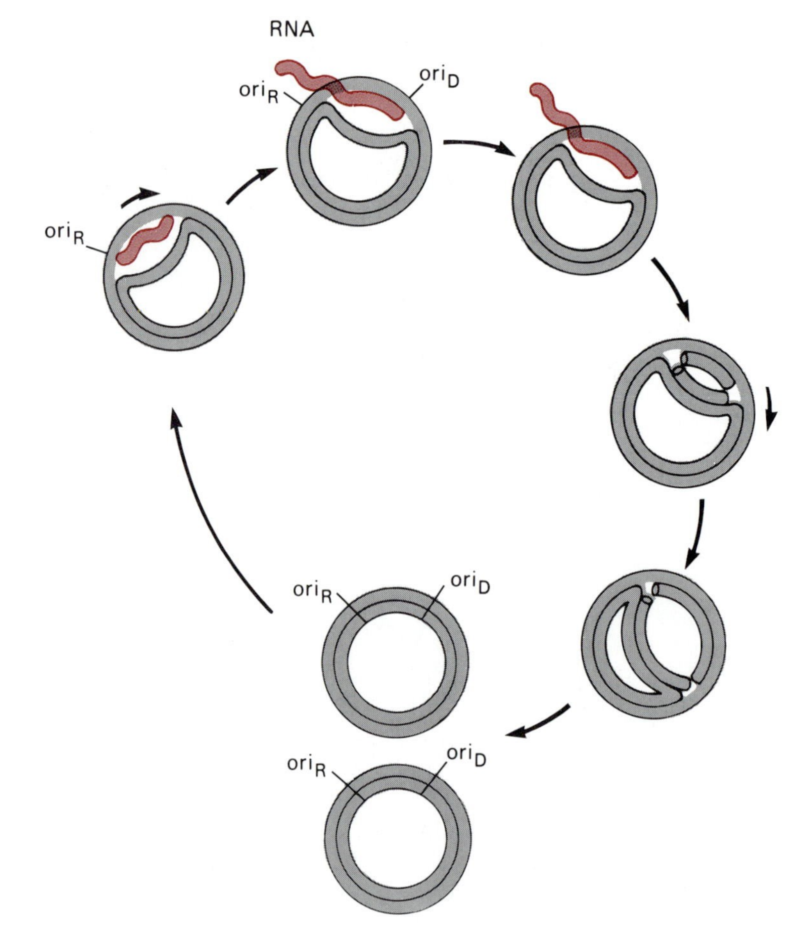

Figure 2.5
Some circular genomes utilize different replication origins for each strand. Animal mitochondrial DNA replication is shown here as an example. Synthesis of one strand begins at the region ori_R. When the new strand reaches ori_D, synthesis of the second strand starts. Synthesis is initiated by formation of an RNA primer, as described in detail in Section 2.1g.

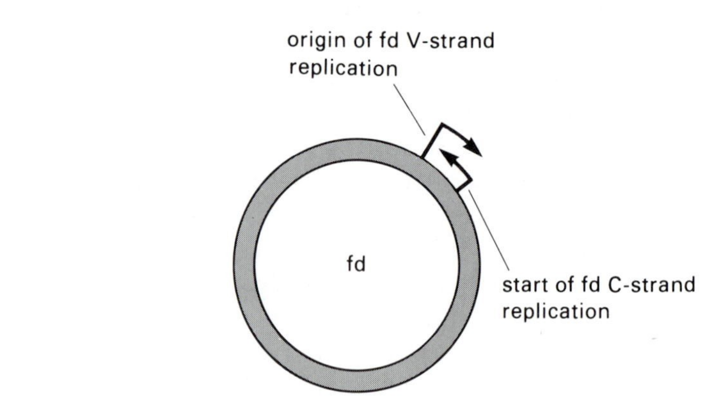

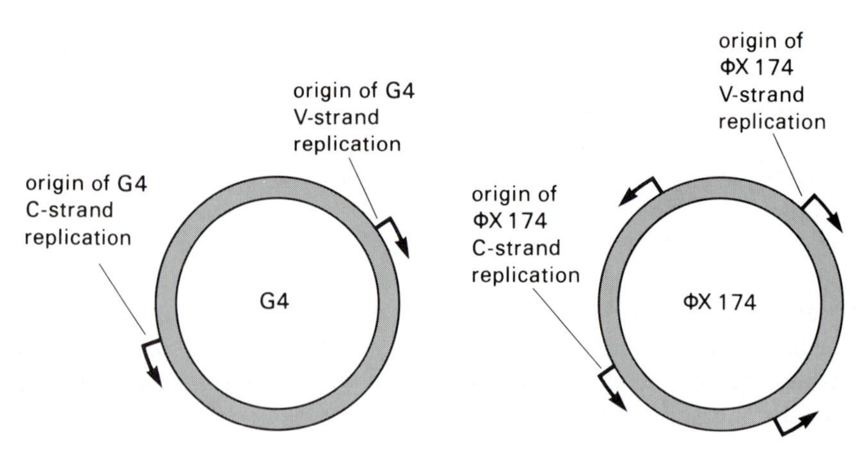

Figure 2.6
Origins of replication in single strand bacteriophage genomes. Each of the three phage DNAs, fd, G4, and ϕX174, has a single ori for initiating genomic DNA (V-strand) synthesis from the duplex DNA that is a replication intermediate. Complementary strand (C-strand) synthesis initiates at one (fd and G4) or multiple (ϕX174) origins.

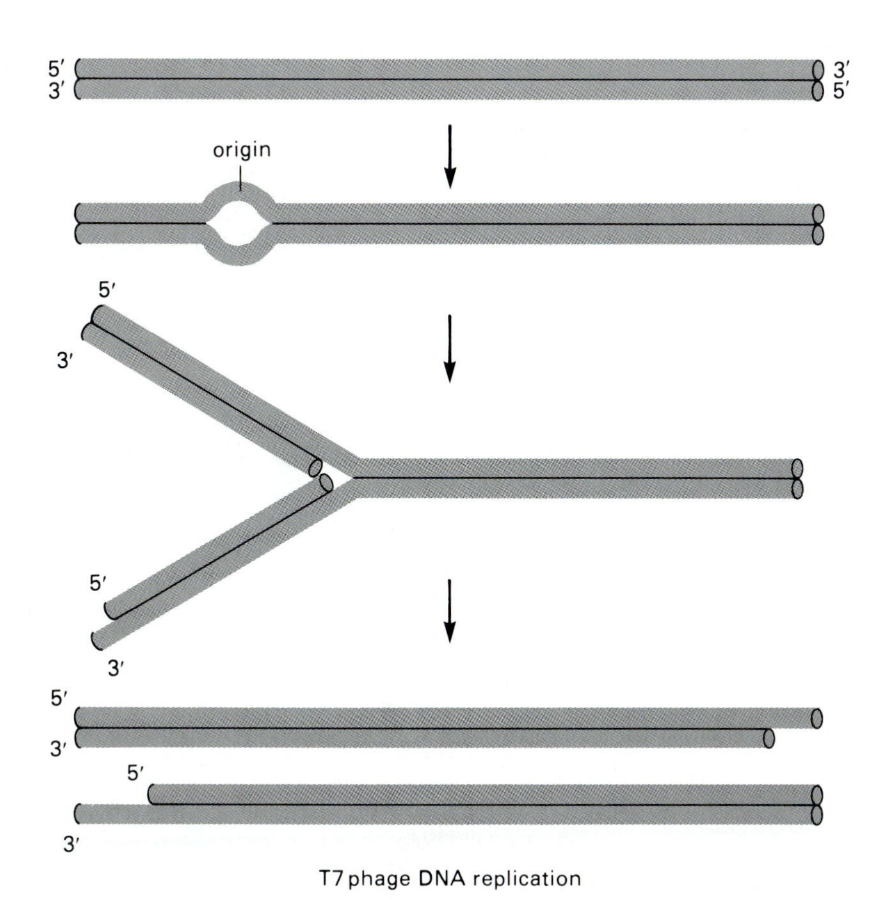

Figure 2.7
Replication of the T7 bacteriophage linear duplex genome. DNA synthesis proceeds in both directions from a primary origin. As a consequence of the details of DNA replication mechanisms, the 5′ ends of the newly synthesized strands are incomplete. Additional reactions are required to form fully duplex phage genomes (see Section 2.1h).

ample, in *E. coli*, the copying rate at each replication fork is constant at about 1500 nucleotide pairs per second; consequently, the entire genome of 4×10^6 pairs is replicated in about 40 minutes. When the chromosome is replicated more rapidly, the frequency of initiations at the same origin increases, but the rate of copying remains nearly the same. Indeed, when *E. coli* divides every 20 minutes, DNA replication is initiated on chromosomes that are still replicating from a prior initiation event.

Although the rate of replication fork movement is considerably slower in eukaryotic cells (10–100 nucleotide pairs per second), simultaneous initiations at multiple replication origins can complete chromosomal replication within reasonable times. Thus, the rate of chromosomal replication is controlled by the number and spacing of origins. Early *Drosophila* embryos, for example, replicate their chromosome complement every three minutes by nearly simultaneous initiations at replication origins spaced 7000–8000 nucleotide pairs apart. The much slower rate of chromosome duplication in *Drosophila* cells in culture stems from a markedly reduced frequency of initiations at only a subset of replication origins spaced about 40,000 nucleotide pairs apart. Therefore, with a fixed rate of DNA chain growth, multiple initiations permit faster replication rates and shorten the time needed to duplicate long chromosomal stretches.

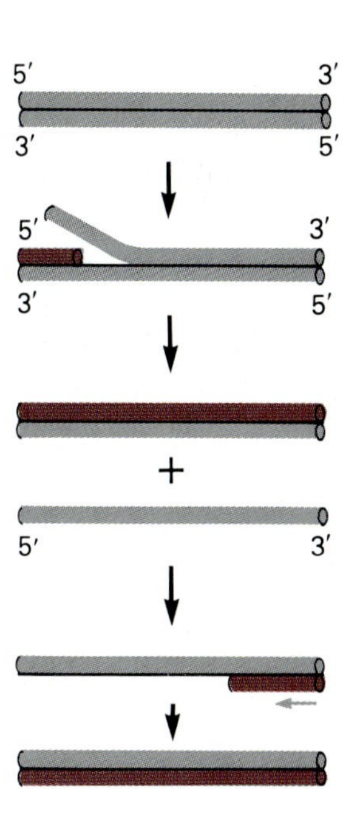

Figure 2.8
Replication of the human adenovirus linear duplex genome. The two strands are synthesized sequentially. In each case, copying of the template strand begins at its 3′ end.

The structure of replication origins DNA segments specifying an origin of replication have been isolated from *E. coli* and several coliphages and plasmids as well as from yeast and a number of eukaryotic viruses. In some instances, the nucleotide sequence of a replication origin may cause

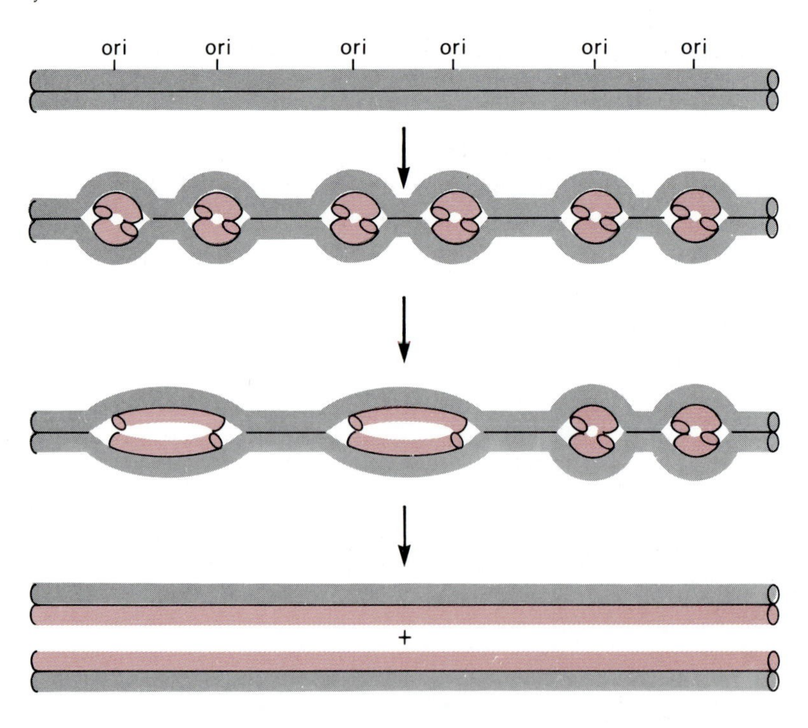

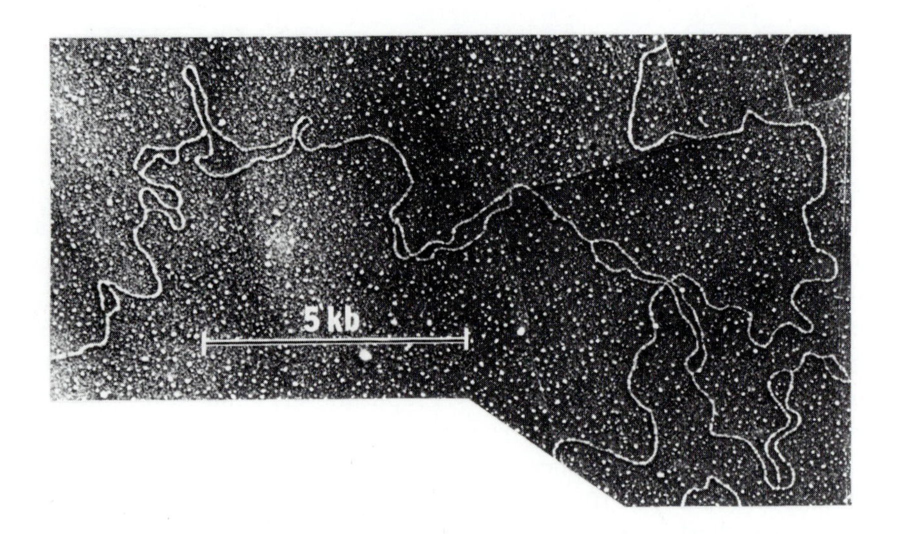

Figure 2.9
Replication of eukaryotic chromosomal DNA. Replication proceeds bidirectionally from multiple origins forming adjacent bubbles. The diagram shows that the bubbles grow larger as the replication forks advance and eventually merge with neighboring bubbles. The electron micrograph shows bubbles of various sizes in replicating *Drosophila* DNA. The micrograph is courtesy of D. S. Hogness.

the duplex to assume an unusual structure that is recognized by proteins involved in the initiation event. Although the nature of the interactions between replication origins and proteins and the mechanism of initiation per se are only beginning to be understood (Section 2.1g), the mechanism is probably different in each case.

c. DNA Replication Is Semiconservative

Once initiated at an origin, replication forks move in one or both directions along the DNA, identifying both the sites of DNA synthesis and the extent of the copying process. Before examining the mechanisms of fork initiation and movement suggested by studies with model prokaryotic replication systems (Section 2.1g), we need to consider the basic reaction occurring at the replication fork during the copying of double strand DNA.

DNA strands are always synthesized by the addition of the 5′-deoxynucleotidyl units of deoxynucleoside triphosphates to the 3′-hydroxyl end of an already existing chain (the **primer**) (Figure 2.10). Consequently, DNA chains grow in the 5′-to-3′ direction along a template strand that is oriented in the antiparallel 3′-to-5′ direction (Figure 2.11). Because DNA strands are never extended in the 3′-to-5′ direction, the two newly synthesized strands must be elongated in opposite directions at each replication fork (Figure 2.12). Synthesis of one strand (the **leading strand**) is continuous; formation of the other (the **lagging strand**) is discontinuous. This replication mode is termed **semidiscontinuous**. The leading strand grows 5′-to-3′ in the same direction as the replication fork and requires only a single initiation event. Elongation of the lagging strand also occurs in the 5′-to-3′ direction, but away from the replication fork. Synthesis of the lagging strand necessitates multiple initiations, resulting in the tran-

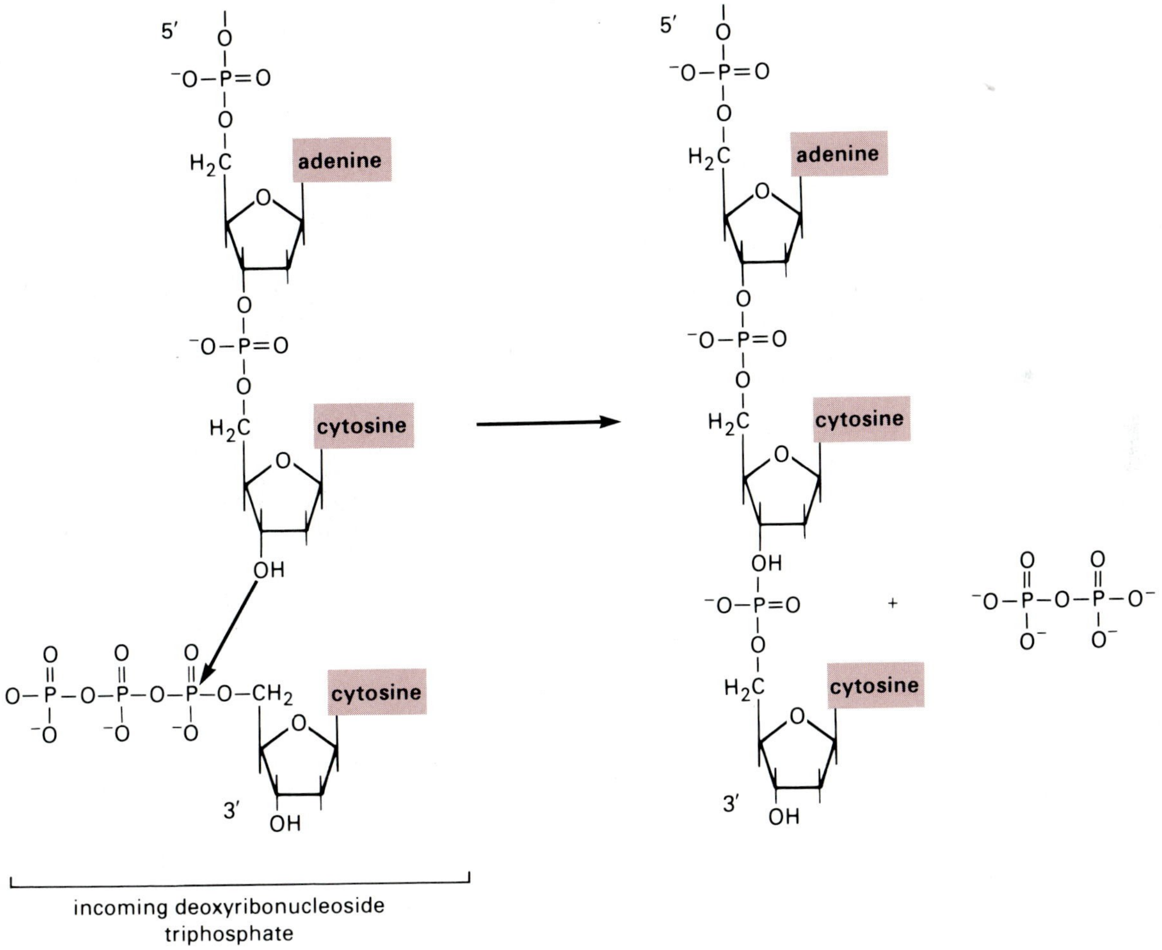

adenine
cytosine
cytosine
incoming deoxyribonucleoside triphosphate
adenine
cytosine
cytosine

Figure 2.10
DNA strands grow by adding deoxynucleotide units to their terminal 3′-hydroxyl groups. 5′-deoxynucleoside triphosphates are the donors of new deoxynucleotide units. One molecule of inorganic pyrophosphate is released for each unit added.

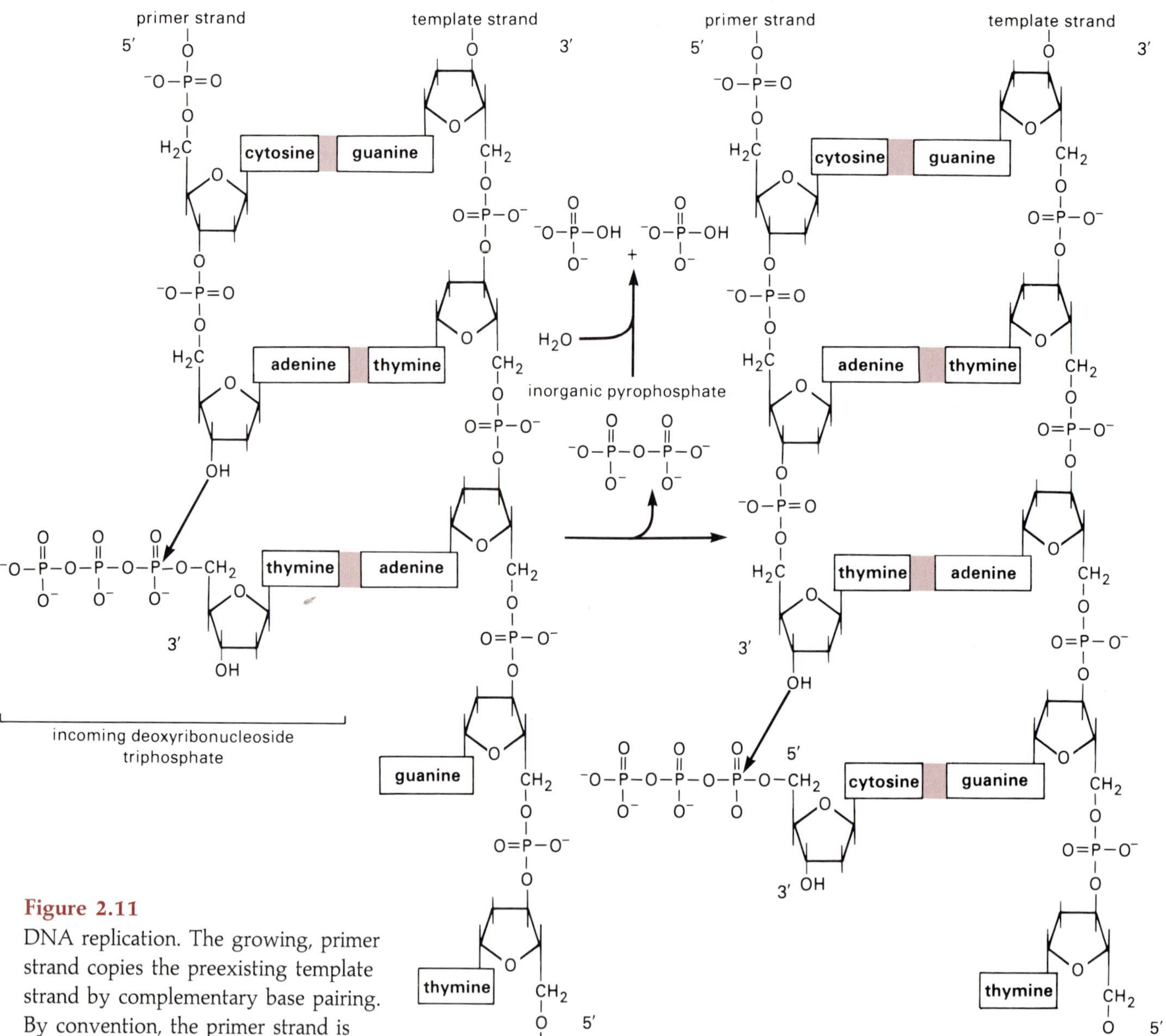

Figure 2.11
DNA replication. The growing, primer strand copies the preexisting template strand by complementary base pairing. By convention, the primer strand is said to grow in the 5′-to-3′ direction, always adding new deoxynucleotide units to the terminal 3′-hydroxyl. The complementary template strand goes in the opposite, antiparallel, or 3′-to-5′ direction. Hydrolysis of the inorganic pyrophosphate product helps drive the reaction.

sient appearance of short DNA chains. These fragments, named **Okazaki fragments** after their discoverer, are 1000–2000 nucleotides long in prokaryotes and 100–200 nucleotides long in replicating eukaryotic DNA. As the replication fork moves, the ends of contiguous Okazaki fragments are joined to form a continuous lagging strand.

The mechanism of initiation at replication origins and of the formation of Okazaki fragments on the lagging strand are similar in principle, though different in detail (Section 2.1g). Both involve the formation of short **RNA primers** complementary to the template DNA and their extension as DNA chains (Figure 2.13). Subsequently, DNA replaces the short stretches of RNA, and the individual segments are joined to produce a continuous lagging strand.

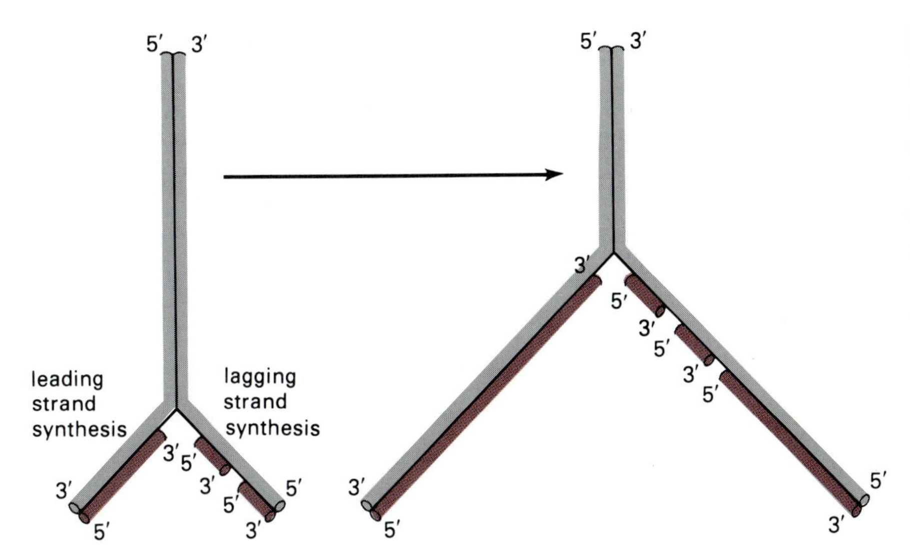

Figure 2.12
Both new DNA strands must be synthesized in the 5′-to-3′ direction at a replication fork. Consequently, the two strands elongate in opposite linear directions. One, the leading strand, is synthesized as a single chain. The other, the lagging strand, is synthesized in short segments that are subsequently joined to form a continuous chain.

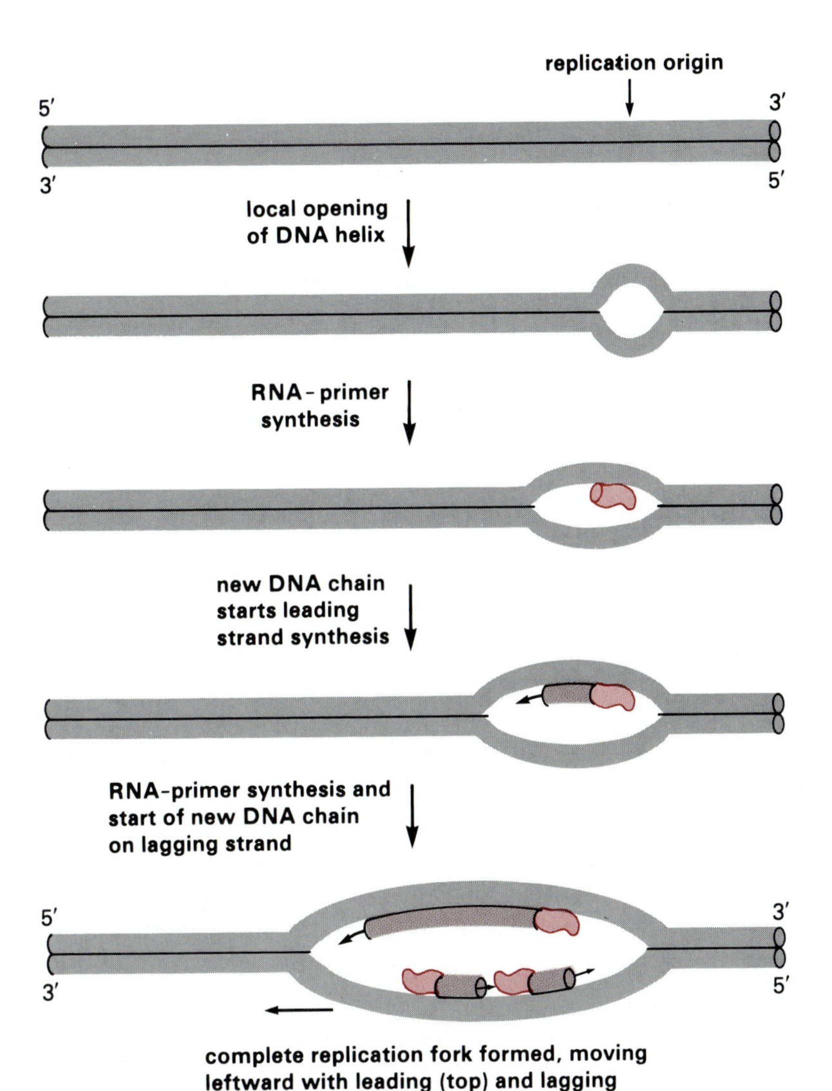

Figure 2.13
Initiation of DNA replication.

d. Complementary Base Copying, Deoxynucleotidyl Transfer, and DNA Ligation in DNA Replication

The nucleotide sequence of a template strand establishes the order of nucleotide assembly in newly formed DNA chains (Figure 2.11). Although complementarity of hydrogen bonding ensures a high fidelity in the copying process, it is not infallible. Stochastic fluctuations in the structures of the incoming or template base allow rare mispairing and nucleotide misincorporation (Figure 2.14). When such misincorporation occurs, the unmatched nucleotide usually does not serve as a primer for further chain growth (Figure 2.15). As a safeguard, the enzymes that catalyze nucleotidyl addition have an efficient editing capability that removes the incorporated but mispaired nucleotide almost immediately after its addition to the end of the growing chain.

The overall reaction for the addition of each 5′-deoxynucleotidyl group to the 3′-hydroxyl group of the primer chain's terminal nucleotide is

$$[\mathrm{dNMP}]_n + \mathrm{dNTP} \rightleftharpoons [\mathrm{dNMP}]_{n+1} + \mathrm{PP_i}$$

where dNMP designates any of the four common deoxynucleotides. In each step, the primer chain is elongated by one residue with a concomitant elimination of inorganic pyrophosphate. Although the addition reaction is energetically reversible, the reaction proceeds toward synthesis because inorganic pyrophosphate is rapidly destroyed in cells. The enzymes that catalyze primer-dependent, DNA template–directed deoxynucleotidyl addition reactions are called **DNA polymerases**. Many DNA polymerases have been isolated and characterized, and a detailed discussion of their properties and the reactions they catalyze is presented in Section 2.1e.

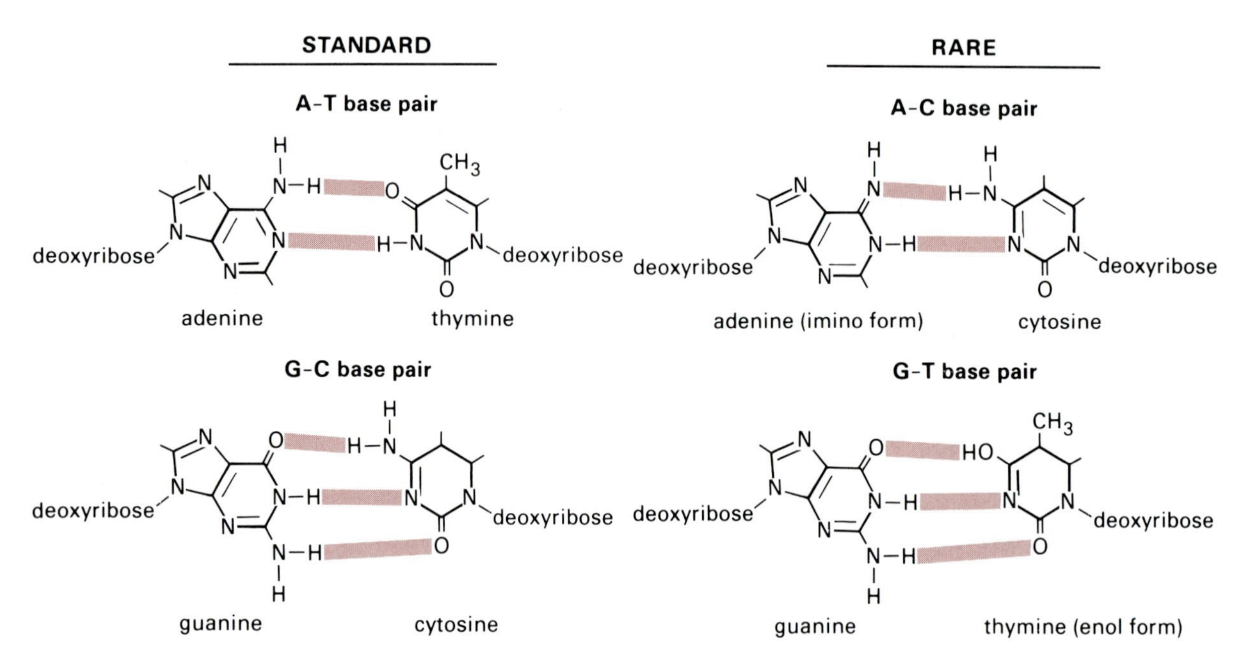

Figure 2.14
Comparison of standard and unusual, rarely formed, base pairs.

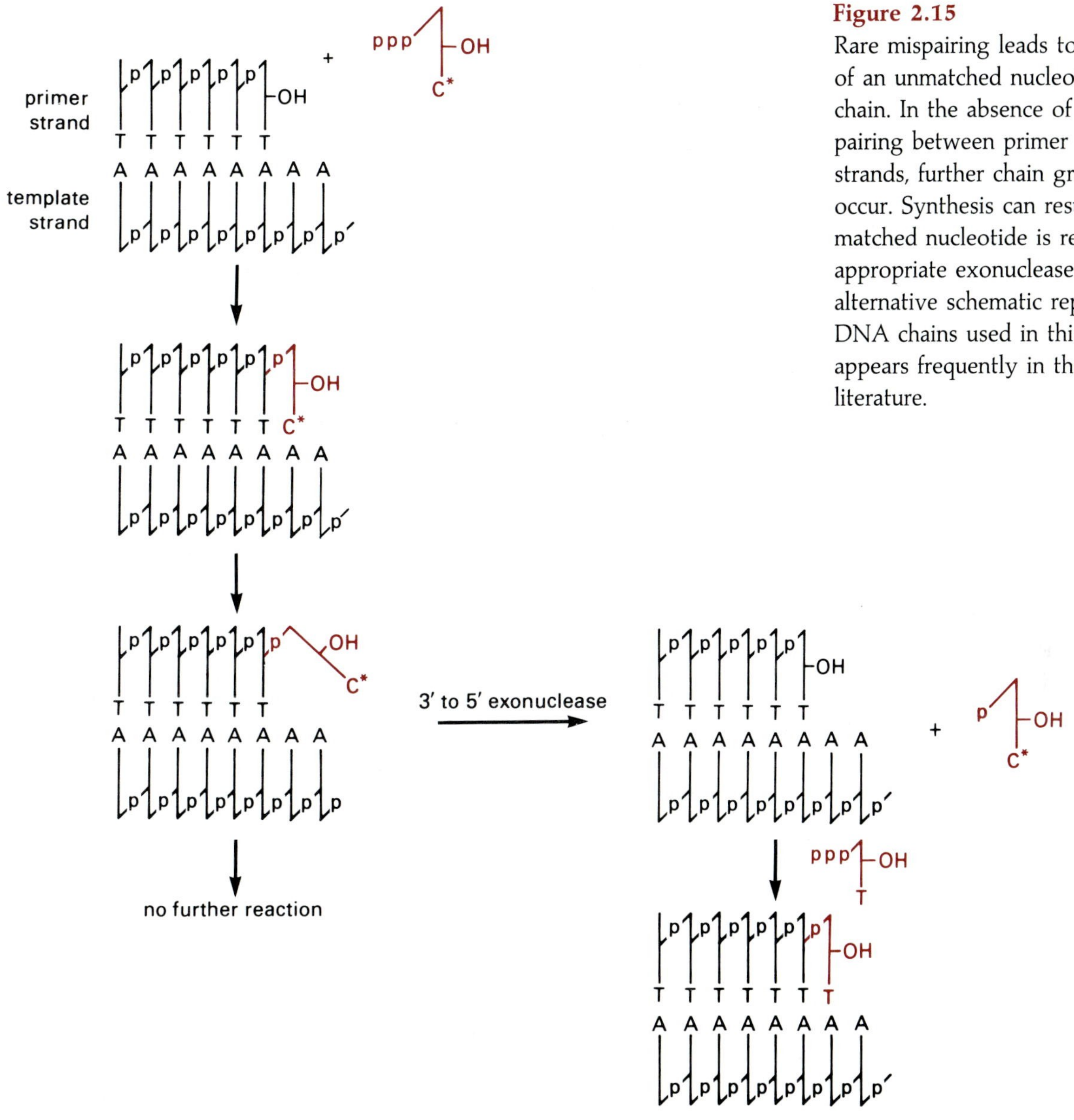

Figure 2.15
Rare mispairing leads to the addition of an unmatched nucleotide in a DNA chain. In the absence of proper base pairing between primer and template strands, further chain growth fails to occur. Synthesis can resume if the mismatched nucleotide is removed by an appropriate exonuclease. Note that the alternative schematic representation of DNA chains used in this diagram appears frequently in the earlier literature.

A necessary consequence of the discontinuous mode of DNA chain synthesis is that the separate Okazaki fragments must eventually be ligated together to form the continuous lagging strand (Figure 2.12). This reaction is accomplished by **DNA ligases**, enzymes that catalyze the covalent joining of duplex DNA strands. Besides having a critical role in DNA replication, DNA ligases are also necessary for the repair of DNA damage (Section 2.3) and in DNA recombination (Section 2.4).

e. *Key Enzymes in DNA Synthesis*

Our understanding of many of the molecular details of DNA replication derives from studies of the behavior and activities of the enzymes that constitute the replication machinery. Most of our information relates to the proteins and mechanisms involved in bacterial DNA replication, principally *E. coli* and the bacteriophages that multiply in it. Additional insights have been obtained with enzymes from yeast, *Drosophila*, and mammalian viruses and cells. The ensuing discussion examines the DNA polymerases and DNA ligases because the two act in concert during the assembly of long DNA chains.

DNA polymerases All prokaryotic and eukaryotic cells contain DNA polymerases. Moreover, many bacterial and animal viruses induce the formation of virus-specific DNA polymerases or induce proteins that facilitate or enable host DNA polymerases to replicate the viral DNA more effectively. Several prokaryotic and eukaryotic DNA polymerases have been extensively purified and characterized with respect to their physical and enzymatic attributes. Although some of their properties differ in detail, the general features of their catalytic activities are strikingly similar.

The best understood enzyme, *E. coli* DNA polymerase I (Pol I), is a single polypeptide with multifunctional activities. As a DNA polymerase, it catalyzes the transfer of the 5′-deoxynucleotidyl unit of a deoxynucleoside 5′-triphosphate to the 3′-hydroxyl group of either a DNA or RNA chain that is base paired to a DNA strand (the equation in Section 2.1d and Figure 2.11). Thus, for polymerization, the enzyme requires a primer as the deoxynucleotidyl acceptor and a template that determines the nucleotide to be added. Besides nucleotide polymerization, Pol I catalyzes two other reactions that are significant for its biological role (Figure 2.16). One reaction causes the hydrolysis of phosphodiester bonds on single strands of DNA or at an unpaired end of duplex DNA, removing one nucleotide at a time, starting at the 3′ end of the chain (**3′-to-5′ exonuclease**). The second reaction also results in the removal of nucleotides, but hydrolysis begins at the 5′ end of a duplex DNA and progresses toward the 3′ end (**5′-to-3′ exonuclease**). These distinctive enzyme activities reside in different parts of the Pol I polypeptide. If Pol I is digested with trypsin *in vitro*, the polypeptide chain is cleaved into a large and small fragment. The large carboxy terminal fragment ("Klenow fragment") retains the DNA polymerase and 3′-to-5′ exonuclease activities; the small amino terminal fragment contains the 5′-to-3′ exonuclease.

Pol I and its associated exonuclease activities are essential for the replication and repair of *E. coli* chromosomal DNA. The 3′-to-5′ exonuclease activity enables Pol I to edit each nucleotide addition and to remove an added mismatched nucleotide from the growing end of a chain (Figure 2.15). If the 3′-to-5′ exonuclease activity of the Pol I is impaired, as is the case in some mutations affecting the Pol I gene, frequent mutations due to base changes introduced into DNA during its replication occur throughout the genome.

Pol I's ability to extend the 3′ ends of chains that are base paired to a template strand enables it to fill in the gaps between segments made on the lagging strand. Pol I extends Okazaki fragments at their 3′ ends and removes the ribonucleotides that initiated the 5′ end of the adjacent fragment, an essential prelude to the formation of a continuous lagging strand (Figure 2.13). Because Pol I can extend the 3′ end of a strand at a nick in double strand DNA and remove nucleotides from the 5′ end of the same nick (a process called **nick translation**), it has a crucial role in repairing damaged DNA (Figure 2.16 and Section 2.3). The nick translation reaction has also been extremely useful in the *in vitro* preparation of radioactively labeled DNA (Section 4.6b).

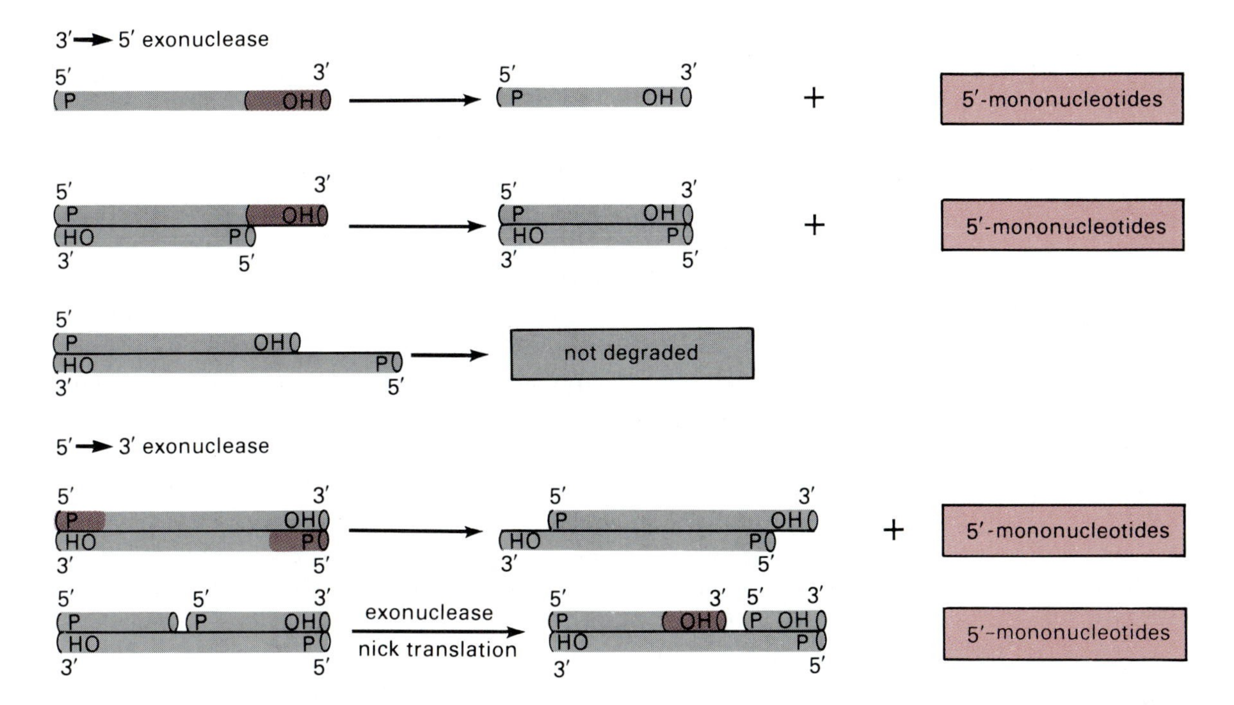

E. *coli* contains lesser amounts of two other DNA polymerases. Pol II is considerably less active than Pol I for nucleotide addition and lacks the 5′-to-3′ exonuclease activity. Consequently, Pol II can fill in gaps between DNA fragments that are base paired to a template strand, but it is unable to remove RNA from Okazaki fragments or to promote nick translation of DNA. The function of Pol II in *E. coli* chromosomal DNA replication and maintenance is obscure.

DNA polymerase III holoenzyme (Pol III holoenzyme) is the principal enzyme responsible for chromosomal DNA replication in *E. coli*. Although there are only 10–20 copies per cell of the Pol III holoenzyme, it is a critical component of the multienzyme complex that initiates the formation of replication forks at origins, continuously elongates the leading strand at the fork, and elongates RNA primers to form Okazaki fragments. But because Pol III holoenzyme lacks the 5′-to-3′ exonuclease, replication of the lagging strand requires the action of Pol I to extend the Pol III product and remove the RNA primers at the 5′ end of the Okazaki fragments.

The polypeptide substituents of the Pol III core enzyme are known, and certain of the enzyme's activities can be assigned to specific subunits in the complex. Thus, a subunit called α provides the polymerizing activity, and the ε subunit contributes the 3′-to-5′ exonuclease activity. But the complex of α and ε subunits has substantially greater polymerase and exonuclease activity than the individual subunits alone. The function of another subunit, θ, is still unknown.

Besides two equivalents of the Pol III core, the Pol III holoenzyme contains seven additional subunits: τ, γ, β, δ, δ', χ, and ψ. These latter polypeptides also occur as multiple copies, yielding a complex molecular mass of about 10^3 kD. The β subunit serves to minimize the possibility that the enzyme will dissociate from the template before completing the copying process, but the precise roles of the other subunits are not known. It may be that there are two forms of Pol III holoenzyme, each having

Figure 2.16
DNA polymerase catalyzes 3′-to-5′ and 5′-to-3′ exonuclease reactions. The 3′-to-5′ exonuclease degrades single strand DNA starting at the end with a 3′-hydroxyl. The 5′-to-3′ exonuclease degrades duplex DNA starting at a 5′ end. Together, the 5′-to-3′ exonuclease and polymerization activities catalyze DNA degradation starting at a 5′ end at a nick in a duplex and chain elongation starting at the 3′ end at the nick. Consequently, the nick moves down the chain in the 5′-to-3′ direction: nick translation. Deoxynucleoside triphosphates are required for nick translation.

a different assortment of auxiliary subunits that impart distinctive synthetic capabilities. One may synthesize the continuous leading strand and the other, the discontinuous lagging strand.

The Pol III holoenzyme catalyzes the same synthetic reactions as Pol I but about 60 times faster. Moreover, the Pol III holoenzyme has an enhanced template affinity and copying efficiency. The Pol III holoenzyme may also associate with other proteins in more complex assemblies, making the copying process more efficient by coordinating several essential enzymatic steps that occur during replication. Such higher order complexes may include proteins that unwind the helix at replication origins and forks (helicases), initiate the formation of RNA primer fragments (primase), promote their processive extension as DNA chains, terminate the replication process, and segregate the daughter DNA helices.

DNA polymerases produced by other bacteria and many bacteriophages have different physical structures and properties. Nevertheless, the reactions they catalyze are virtually identical to those from *E. coli*. All the DNA polymerases contain the corrective 3′-to-5′ exonuclease, but some lack the 5′-to-3′ exonuclease. For example, the bacteriophage T4 DNA polymerase can perform the error-correcting 3′-to-5′ exonuclease reaction, but it does not catalyze the 5′-to-3′ exonucleolytic reaction and therefore fails to promote nick translation. During the replication of T4 DNA, another phage-encoded protein catalyzes the 5′-to-3′ exonuclease reaction, thereby removing the RNA primers in preparation for the ligation of the Okazaki fragments into a continuous lagging strand. This enzyme acts coordinately with the phage DNA polymerase to permit discontinuous synthesis of the lagging strands and the repair of damage to T4 DNA.

Although a variety of DNA polymerases have been identified in eukaryotic cells, analysis of their physical and functional properties is not as advanced as that of the prokaryotic enzymes. Four DNA polymerases have been resolved from mammalian cells: α, β, and δ occur in cell nuclei, and γ resides in mitochondria. DNA polymerase α is involved in chromosomal DNA replication. Its polymerizing activity is associated with a large polypeptide, but it exists and probably functions as a multisubunit protein analogous to the *E. coli* Pol III holoenzyme. The β polymerase is a single polypeptide that provides the gap-filling function for repairing DNA damage. The mitochondrial polymerase (γ), which contains four identical polypeptides, is responsible for replicating the mitochondrial genome. The δ polymerase resembles the α polymerase in its molecular and synthetic properties and is also involved in chromosomal DNA replication. Because the mammalian α, β, and γ DNA polymerases lack the 3′-to-5′ and 5′-to-3′ exonuclease activities that are intrinsic to the *E. coli* enzymes, it is unclear how these organisms remove the inadvertent nucleotide misincorporations and the RNA primers from the ends of Okazaki fragments during replication of their DNA.

Some animal viruses (e.g., Herpes, vaccinia, and hepatitis) induce the synthesis of specialized polymerases to replicate their DNA genomes, but others produce proteins that augment the cellular DNA replication systems or perform a function unique to viral DNA replication. For example, papovaviruses produce proteins called T antigens that are essential for initiating replication at their origins. Human adenoviruses encode proteins that can prime the initiation of the synthesis of each

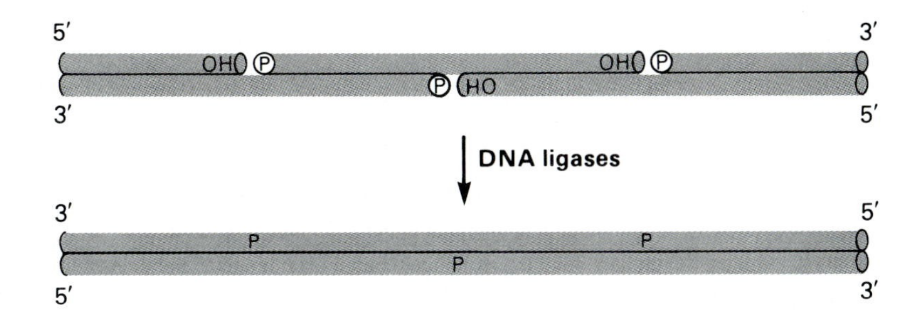

Figure 2.17
All DNA ligases join a 5′-phosphoryl and 3′-hydroxyl on adjacent deoxynucleotides surrounding a single strand nick in a DNA duplex. A new phosphodiester bond is formed, thereby sealing the nick.

strand of the linear viral DNA (Section 2.1g). They also produce specialized DNA-binding proteins that facilitate their special mode of replication.

DNA ligases DNA ligases are the principal enzymes for joining DNA chains during replication, repair, and general recombination. All known ligases can form phosphodiester bridges between the 5′-phosphoryl and 3′-hydroxyl groups of adjacent deoxynucleotides at a nick in DNA (Figure 2.17). The DNA ligase induced in *E. coli* after T4 phage infection (T4 ligase) is unique in being able to join double strand DNA fragments at their flush ends (Figure 2.18), a reaction whose physiological significance is unknown but whose practical value in recombinant DNA manipulations is inestimable (Sections 4.5 and 6.2).

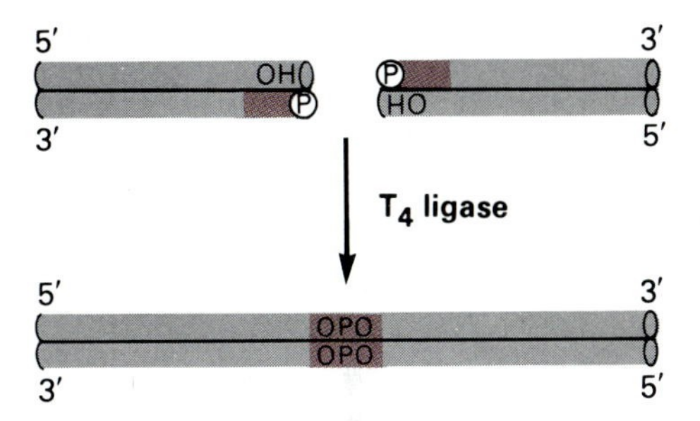

Figure 2.18
The DNA ligase encoded on the T4 phage genome joins two duplex DNA fragments end to end. A new phosphodiester bond is formed on each strand.

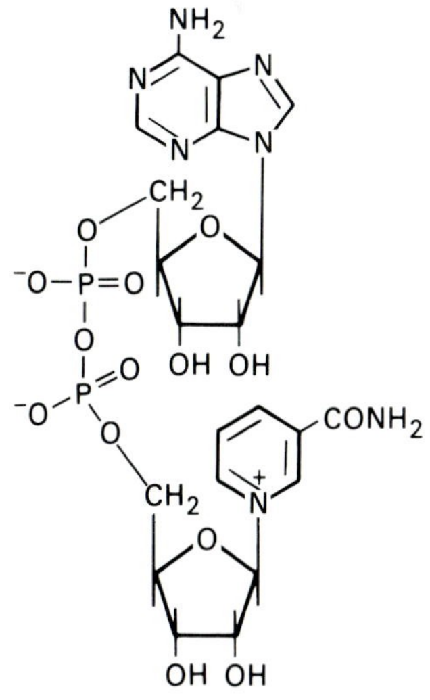

Several bacterial (*E. coli* and *B. subtilis*), bacteriophage (T4 and T7), and mammalian DNA ligases have been isolated and their structures and catalytic mechanism defined. The *E. coli*, T4, and T7 DNA ligases are known to be single polypeptide chains, but the composition of the two forms of mammalian DNA ligase is not known. In forming a phosphodiester bond between the appropriate ends of nucleic acid chains, ligases utilize the energy of either ATP or a derivative of ATP, nicotinamide adenine dinucleotide (NAD) (Figure 2.19). The reaction (Figure 2.20) occurs in discrete steps: (1) The adenylyl portion of either NAD (with the *E. coli* and *B. subtilis* ligases) or ATP (with the T4, T7 and mammalian ligases) is transferred to an ε-amino group of a lysine residue of the ligase, with a concomitant release of nicotinamide mononucleotide or inorganic pyrophosphate, respectively. (2) The adenylyl group is transferred from the protein to the 5′-phosphoryl group of the terminal residue of the DNA chain to form a pyrophosphoryl derivative, adenylyl-DNA. (3) The adenylyl residue from the 5′-phosphoryl group is displaced by the 3′-hydroxyl group of the adjacent DNA terminus. The net result of these reactions is to create a phosphodiester linkage in a DNA strand, utilizing the energy of the pyrophosphoryl linkage of NAD or ATP. With all but the T4 ligase, the nucleotide with the acceptor 3′-hydroxyl group must be base paired adjacent to the activated 5′-phosphoryl group for phosphodiester bond formation to occur. Both the *E. coli* and T4 DNA ligases can join the ends of two different duplex fragments or the ends of a strand in a linear or circular nicked DNA. Thus, either linear or circular duplex molecules can be produced by DNA ligase action.

NAD^+

Figure 2.19
Nicotinamide adenine dinucleotide (NAD).

f. Replication Requires Unwinding the Helix

Double strand DNA must be progressively unwound to permit complementary copying of each strand. But this unwinding occurs only at the

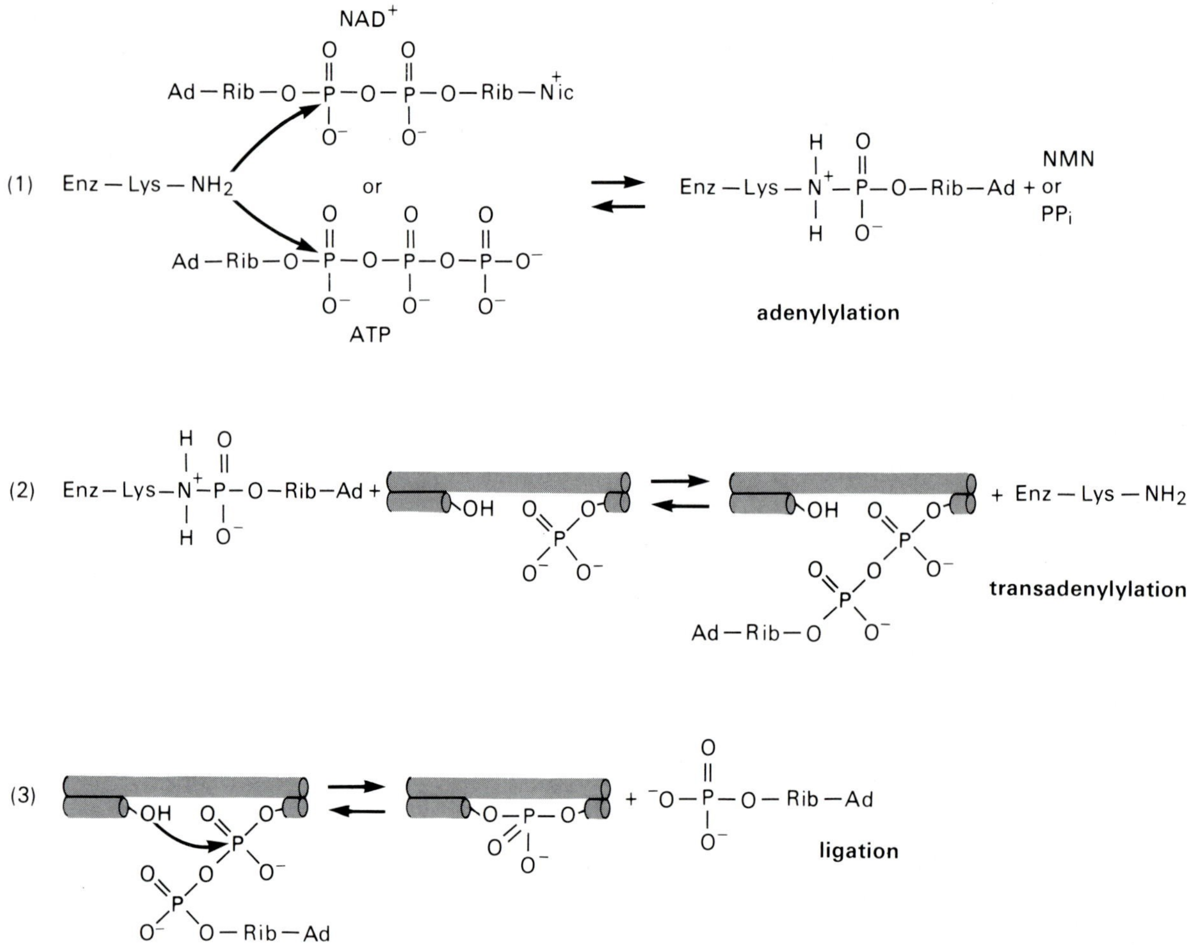

NAD$^+$

(1) Enz—Lys—NH_2 or ATP

Ad—Rib—O—P(=O)(O$^-$)—O—P(=O)(O$^-$)—O—Rib—$\overset{+}{N}$ic

Ad—Rib—O—P(=O)(O$^-$)—O—P(=O)(O$^-$)—O—P(=O)(O$^-$)—O$^-$

⇌ Enz—Lys—$\overset{+}{N}H_2$—P(=O)(O$^-$)—O—Rib—Ad + NMN or PP_i

adenylylation

(2) Enz—Lys—$\overset{+}{N}H_2$—P(=O)(O$^-$)—O—Rib—Ad + OH O P O$^-$ O$^-$ ⇌ OH O P O$^-$ O P O$^-$ Ad—Rib—O + Enz—Lys—NH_2

transadenylylation

(3) OH O P O$^-$ O P O$^-$ O—Rib—Ad ⇌ O—P—O O O$^-$ + $^-$O—P(=O)(O$^-$)—O—Rib—Ad

ligation

Figure 2.20
Mechanism of DNA ligase (Enz) catalyzed reaction. Abbreviations: Lys, lysine on the ligase; Rib, ribose; Ad, adenine; Nic, nicotinamide; NMN, nicotinamide mononucleotide; PP_i, pyrophosphate.

localized region of the replication fork. The unwinding of double helical DNA is not a spontaneous process; it is facilitated by two kinds of proteins (Figure 2.21). One group, called **DNA helicases**, uses the energy released by ATP hydrolysis to ADP to promote the separation of the two strands of double helical DNA. Helicases often function as part of a complex that acts to advance the replication fork and replicate the unwound strands (Section 2.1g). One helicase protein is generally sufficient, but more than one type of helicase may act in concert to maximize the rate of unwinding the fork. The second type of helix-destabilizing protein is **single strand binding proteins**, which facilitate the unwinding of the duplex by stabilizing the unwound single strands. Thus, helicases cause a localized unraveling of the double helix, and single strand binding protein binds cooperatively to the single strands, leaving the bases available for copying by base pairing.

In eukaryotic organisms, the need to replicate the chromatin form of DNA (Section 1.2g) introduces further complexities. To unwind the DNA strands in chromatin, the highly condensed DNA histone complex must be disassembled. Then, following the wave of replication, the two daughter helices must be repackaged into the characteristic chromatin structure. Very little is known about how the eukaryotic chromosome is unfolded in preparation for replication. Once unfolded, preexisting and new nucleosomes (assembled from newly synthesized histones) become asso-

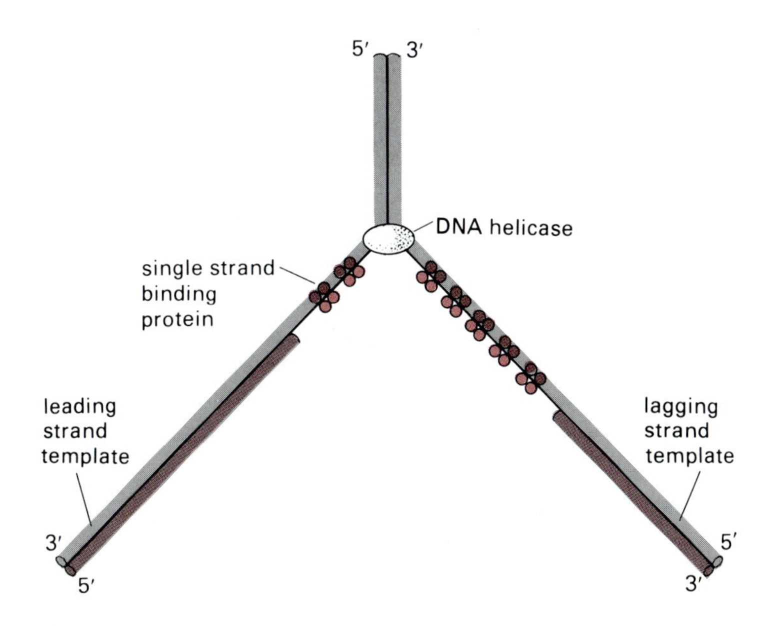

Figure 2.21
Unwinding DNA duplexes is facilitated by two proteins: helicase and single strand binding protein.

ciated with the newly synthesized duplex DNA. To preclude the occurrence of extensive lengths of free DNA during the replication process, chromatin assembly occurs promptly after the duplex DNA is formed. At present, it is not known if the parental histone octamers become associated with only one or both of the daughter helices.

Topological problems of DNA unwinding and replication The unwinding of the two strands of the double helix at replication forks creates mechanical and topological problems. With linear duplex DNA, the unwinding of the two strands can be accommodated in principle by rotating the parental helix about its axis (Figure 2.22) Nevertheless, the mechanical problem of rotating very long DNA chains about their long axes in the intracellular milieu is formidable. During replication of a closed circular DNA, unwinding the two strands at a fork poses additional problems. As the two strands unravel, the normally negatively supercoiled segment ahead of the fork becomes progressively less negatively supercoiled and eventually would become positively supercoiled. As a consequence, the fork's progress would be impeded and eventually blocked. The impediment to further replication is removed by introduction of a nick, creating a "swivel," which allows the unreplicated duplex ahead of the fork to rotate relative to the fork (Figure 2.23). Such a nick is provided by the action of a class of enzymes referred to collectively as **DNA topoisomerases**.

DNA topoisomerases These enzymes alter the extent and type of DNA supercoiling. Besides providing the swivel that allows the continued propagation of a replication fork, topoisomerases promote the separation or creation of intertwined or catenated circular DNAs and the removal of knots and tangles from long linear DNA. Topoisomerases are also essential participants in certain types of recombination (Section 2.4).

Two general classes of topoisomerases have been identified in a wide variety of organisms. One, referred to as type I topoisomerases, or nicking-closing enzymes because of their mechanism, removes one supercoil from DNA per reaction cycle. Such topoisomerases act by

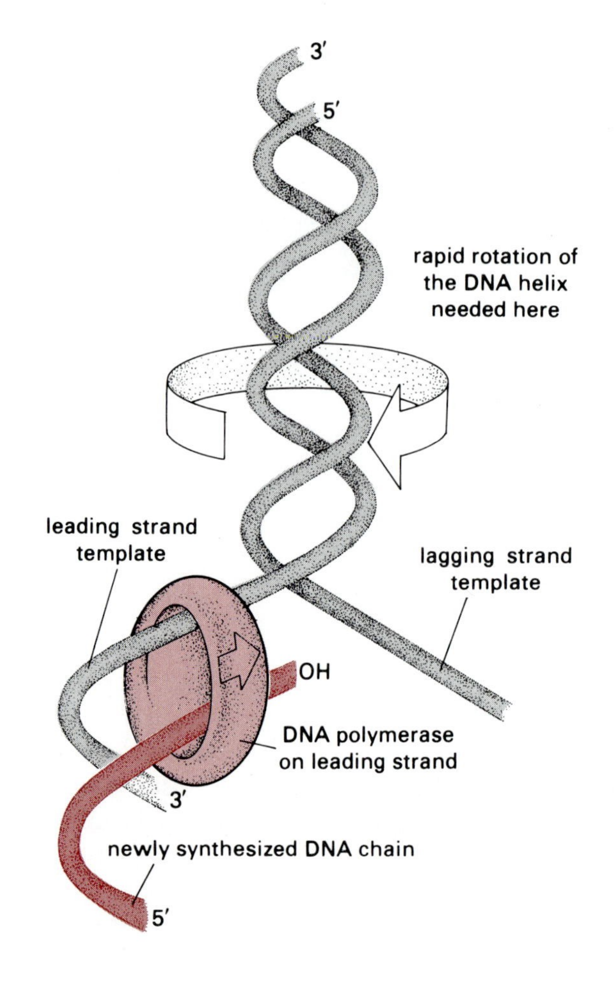

Figure 2.22
Rotation of the DNA double helix can relieve the mechanical stress caused by unwinding the two strands at a replication fork. After B. Alberts et al., *Molecular Biology of the Cell* (New York: Garland, 1983), p. 228.

nicking one of the two strands, allowing the flanking duplex regions to rotate once about the intact strand and then rejoining the ends of the broken strand (Figure 2.24). ATP is not required in type I topoisomerase reactions because the energy of the phosphodiester bond is conserved by having a tyrosine residue in the enzyme act transiently as an acceptor and a donor of the phosphoryl end of the broken chain. The single strand passes through the break spontaneously. Two interesting but probably unrelated differences between prokaryotic and eukaryotic type I topoisomerases have been noted: (1) Prokaryotic type I topoisomerases become linked to the 5′-phosphoryl end of the broken strand, and the enzymes from eukaryotic organisms are joined at the 3′-phosphoryl end. (2) Prokaryotic type I topoisomerases eliminate only negative supercoiling, but the eukaryotic counterpart removes both negative and positive supercoiling.

Type II topoisomerases remove both negative and positive supercoils from DNA. By contrast to the type I enzymes, the type II topoisomerases catalyze transient breaks in both complementary strands of a supercoiled duplex, pass a double strand segment from the same or another DNA molecule through the break, and then rejoin the ends (Figure 2.25). Type II topoisomerases also use tyrosyl residues, one on each of two subunits, to bind the 5′ end of each broken strand while the other duplex passes through the break. As a consequence of the double strand cleavage and passage of a duplex through the break, two negative or positive super-

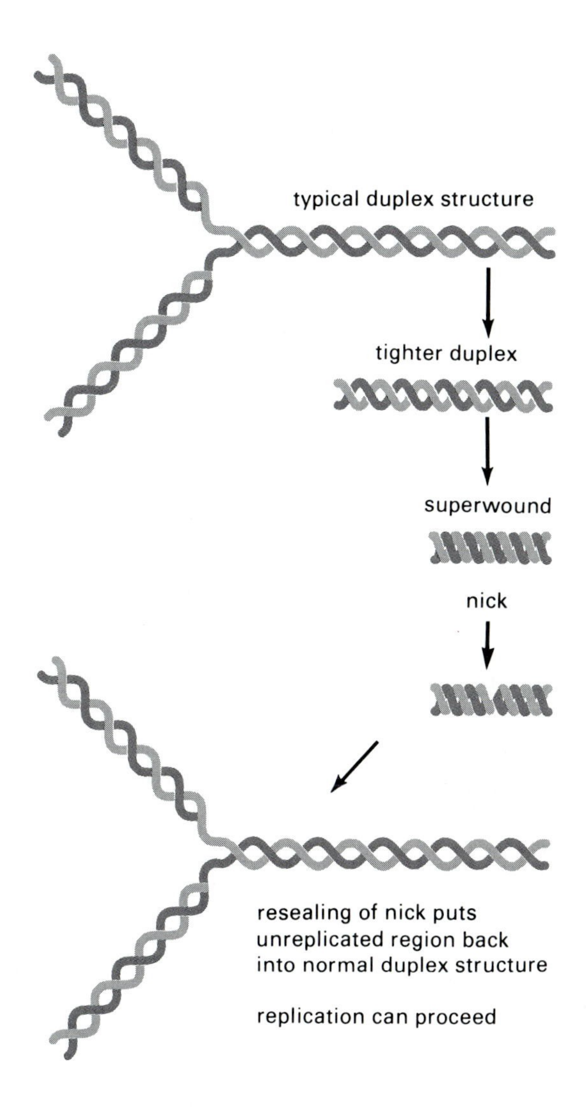

Figure 2.23
Relieving the mechanical and topological stress from closed circular duplex DNA ahead of a replication fork. Introduction of a nick allows the unreplicated, superwound DNA to rotate relative to the fork. The stress imposed by replication on long intracellular DNA chains is removed in a similar way.

coils are removed per cycle. In some circumstances, the duplex that passes through the break is another circular DNA molecule, and this leads to **decatenation** (separation of interlinked circular DNAs) or **catenation** (intertwining of circular DNAs). The same mechanism can account for unknotting and knotting or untangling and condensing large DNA duplexes.

Although type I and type II topoisomerases remove superhelical twists that accumulate during replication of circular DNA, one particular type II topoisomerase, called **gyrase**, so far found only in bacteria, introduces negative superhelical turns into relaxed circular DNAs. This requires double strand scissions and rejoining in a special way (Figure 2.26a). Thus, the action of gyrase removes positive supercoiling and introduces negative supercoiling into relaxed DNA. In bacteria at least, the balanced action of type I topoisomerases and gyrase probably regulates the extent of supercoiling in DNA and thereby influences the rate of fork movement during replication.

The mechanism by which gyrase catalyzes the introduction of negative superhelical turns into circular or otherwise topologically constrained DNA is only partially understood. The DNA gyrase of *E. coli* is a tetramer of two kinds of subunits ($\alpha 2\beta 2$), the α subunits containing the sites at which

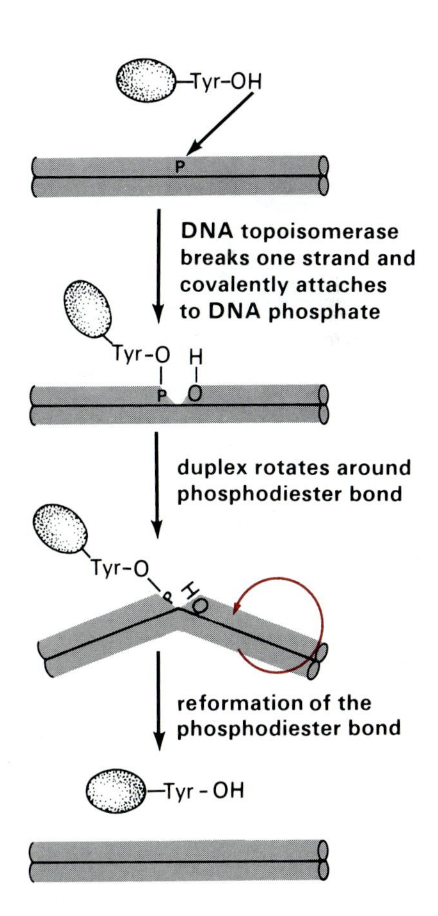

Figure 2.24
The reaction catalyzed by type I topoisomerases.

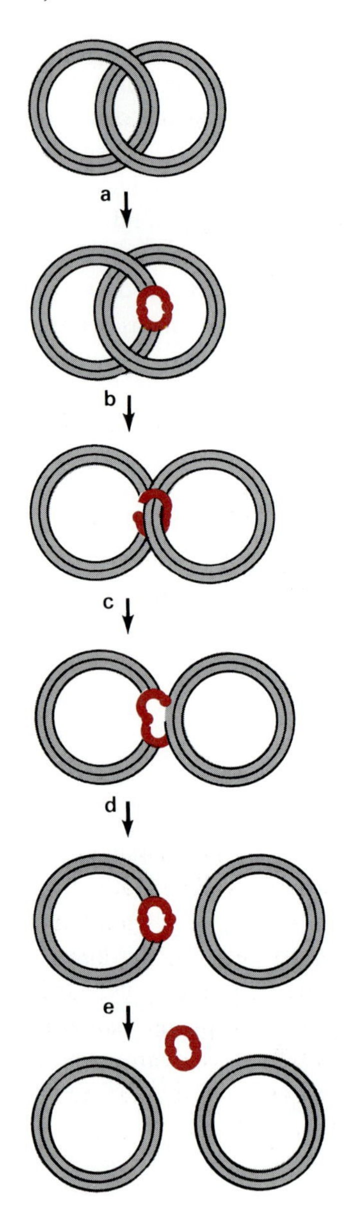

Figure 2.25
Type II topoisomerases catalyze the passage of one duplex DNA through another. After B. Alberts et al., *Molecular Biology of the Cell* (New York: Garland, 1983), p. 229.

the DNA termini are covalently bound to the enzyme. Gyrase introduces negative superhelicity by first creating positive superhelical turns at local regions of DNA associated with the enzyme (Figure 2.26b). Then the orientation of the two helical segments in that region are reversed by passing one chain through the gap created by a double strand break in the other. The end result of these actions is the creation of two negative superhelical turns per catalytic cycle. A dependence on the cleavage of ATP relates to the energy required to convert a relaxed circular DNA into the higher energy state associated with a superhelical conformation.

g. Initiation of New DNA Chains and Their Extension at Replication Forks

Initiation of new DNA chains Unlike RNA chains, which are initiated *de novo* by RNA replicases (Section 2.5) and RNA polymerases (Chapter 3), the formation of DNA chains requires a primer. Thus, DNA chains are normally begun by the synthesis of short RNA primers, which are then extended as DNA. For example, the conversion of the circular single strand M13 phage DNA to its double strand replicative form utilizes RNA polymerase holoenzyme to synthesize a short RNA segment at a replication origin on the DNA (Figure 2.27). With another single strand circular phage DNA (G4), the primer RNA is synthesized at an origin by a specialized RNA polymerase, termed a **primase**. With still a different single strand circular phage (ϕX174), a multienzyme complex containing seven discrete proteins, composed of 15 to 20 polypeptide chains (a **primosome**), activates the template strand so that a primase can synthesize the RNA primer. A primosome-primase mode of initiation is also used to initiate leading strand synthesis at *E. coli*'s replication origin as well as the formation of the Okazaki fragments in the synthesis of the lagging strand.

It seems likely that, with some variations, DNA strand synthesis is initiated by similar mechanisms in prokaryotic and eukaryotic chromosomes, plasmids, and DNA viruses. Among the likely variations are the different kinds of polymerases that synthesize the RNA primers, the accessory proteins that facilitate the initiation of the RNA transcripts, and either the nucleotide sequences *per se* or special structural features that specify where transcription initiation occurs.

The length of the RNA primers varies with the particular initiation event. In bacteriophage ϕX174 and *E. coli*, the primase catalyzes the synthesis of RNA primers about two to five nucleotides long, but in eukaryotic DNA replication, the primer RNAs are about twice that length. With the bacteriophages M13 and G4, where RNA polymerase or G4 primase, respectively, promote RNA primer synthesis, the RNA stretches are 20 to 30 nucleotides long and are complementary to hairpin loop regions in these phage DNA structures (Figure 2.27). The precise nature of the signals that specify the sites of primer RNA synthesis for initiation of replication or for lagging strand synthesis are not known.

Initiation of DNA replication in the colE1 type plasmids is also mediated by an RNA transcription event. (Chapter 5 describes the use of such plasmids as cloning vectors.) But in this instance, the primer RNA

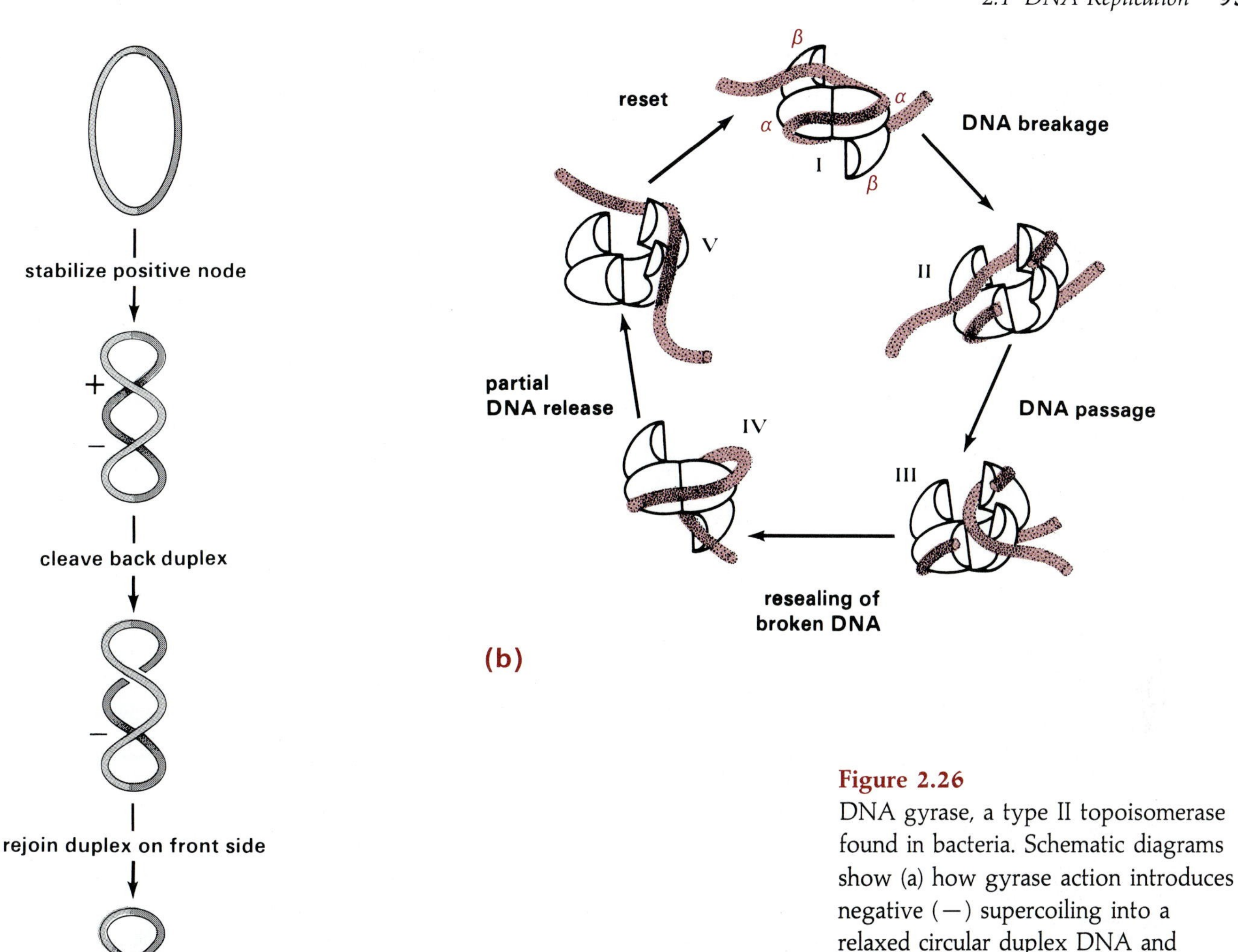

Figure 2.26
DNA gyrase, a type II topoisomerase found in bacteria. Schematic diagrams show (a) how gyrase action introduces negative (−) supercoiling into a relaxed circular duplex DNA and (b) details of the gyrase catalyzed reaction. After A. Morrison and N. R. Cozzarelli, *Proc. Natl. Acad. Sci.* U.S.A. 78 (1981) 1416–1420.

is initiated by RNA polymerase at a sequence that is 555 base pairs upstream of the origin of plasmid DNA replication (Figure 2.28). Transcription proceeds through the origin, producing an RNA chain about 500–600 nucleotides long. RNAse H then destroys most of the RNA strand. The remaining 3′ end of the RNA within the origin region serves as the primer for DNA chain synthesis by Pol I. This DNA strand becomes the leading strand for replicating the entire plasmid DNA.

In the case of human adenoviruses, the synthesis of new DNA strands is initiated by a protein rather than an RNA (Figure 2.29 on page 98). Adenovirus type 2 DNA has a protein of 55 kD covalently joined to the 5′-phosphate terminus of each strand. The first step in initiating new strands is a reaction between a virus-coded primer protein (80 kD) and the deoxynucleoside triphosphate corresponding to the first deoxynucleotidyl residue in the DNA (dCTP), to produce a protein-dCMP. The 3′-hydroxyl group of the protein-linked deoxycytidylate residue serves as the primer to elongate the DNA chain using the complementary DNA strand as a template. Eventually, the 80 kD viral initiation protein is

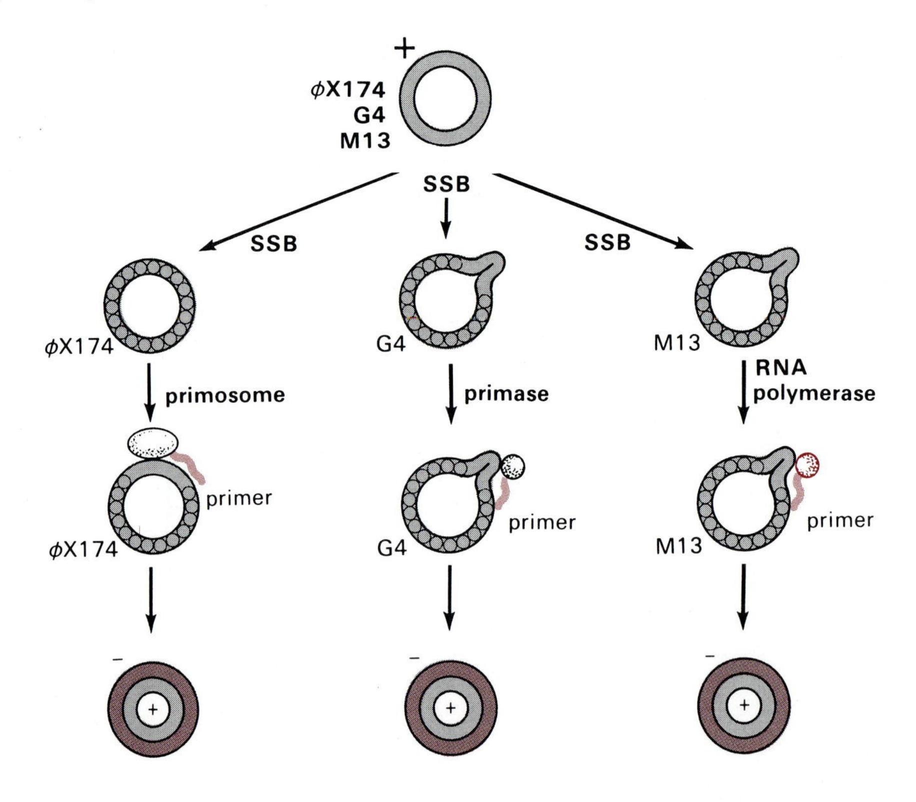

Figure 2.27
Initiation of single strand phage DNA replication. Each phage depends on a different enzyme to catalyze RNA primer synthesis at the replication origin (*C* strand origin, Figure 2.6).

cleaved to leave the 55 kD terminal protein still joined to the ends of the mature viral DNA strands. Other adenoviruses and some bacteriophages use an analogous mechanism to initiate DNA chain synthesis.

Simultaneous replication of both strands at a replication fork Studies of the mechanism of phage DNA replication with purified enzymes from *E. coli* have provided a sophisticated model for semiconservative, semidiscontinuous replication at a fork (Figure 2.30). Unwinding the helix to create the fork requires that the DNA be negatively supercoiled (Section 2.1f). The unwinding is achieved by an ATP-driven helicase and facilitated by the single strand binding protein that associates with the unraveled single strands. Once initiated at the origin, Pol III holoenzyme extends the leading strand over the entire length of the chromosome by adding deoxynucleotidyl units to the successive 3′-hydroxyl termini.

Lagging strand synthesis is initiated by a primosome-primase complex (Figure 2.31). The primosome, which contains a helicase activity, unwinds the helix and probably creates appropriate structures that allow the associated primase to produce the primer. The primosome-primase complex moves in the direction of the fork, pausing as it unwinds the duplex to synthesize the RNA primers. The short RNA segments are then extended by the Pol III holoenzyme with deoxynucleoside triphosphates to produce the 1000–2000 nucleotide long Okazaki fragments (Figure 2.30). Removing the RNA segments from the 5′ end of each Okazaki fragment and filling in those gaps between fragments relies upon the chain extension and nick translation capabilities of Pol I. When the growing 3′-hydroxyl end of each Okazaki fragment meets the 5′-deoxynucleotidyl end of the adjacent fragment, the action of DNA ligase then produces a continuous lagging strand. A more elaborate but still conjectural model

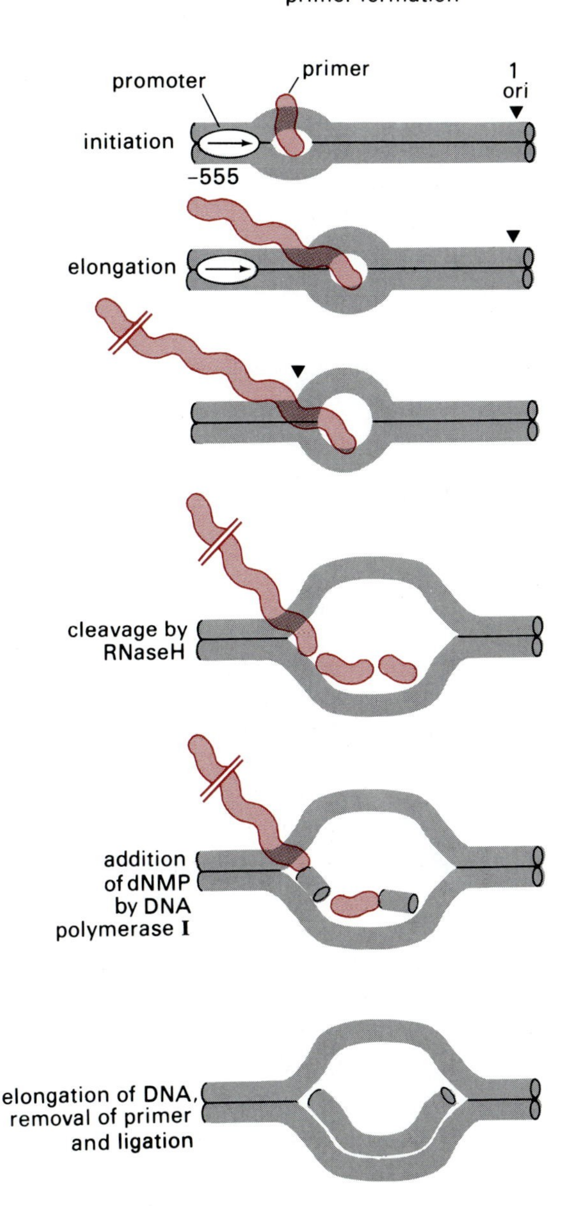

Figure 2.28
Initiation of colEl type plasmid DNA replication.

(Figure 2.32) for how the replication fork is advanced involves a higher order multienzyme complex—a **replisome** that contains a functional primosome-primase complex, the helicase, Pol III holoenzyme, and possibly gyrase. This complex can promote concurrent extension of the leading strand while priming RNA initiation and DNA extension for lagging strand synthesis. Two such replisomes operating in concert at the two oppositely moving forks on circular chromosomes is still a further embellishment of the model.

Rolling circle replication Some double strand circular chromosomes are replicated via an alternate route, termed **rolling circle replication** (Figure 2.33 on page 100). In this mode, a double strand circular DNA is nicked by a specific enzyme at a unique location in one strand (the rolling circle origin), and the 3′-hydroxyl end so created is extended via nucleotidyl additions by Pol III holoenzyme using the intact circle as the template.

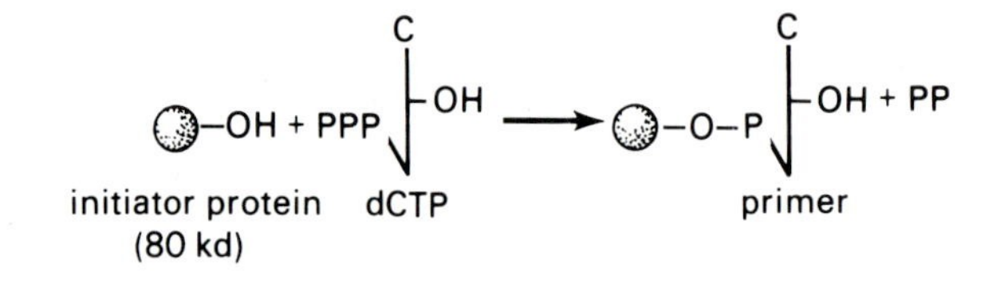

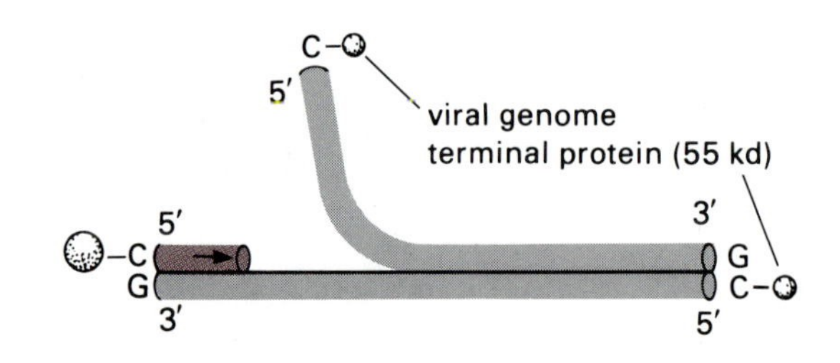

Figure 2.29
Initiation of adenovirus DNA replication. The key step is the reaction between the kD initiator protein and dCTP and the association of the protein-deoxycytidylate at the 3′ end of the template strand.

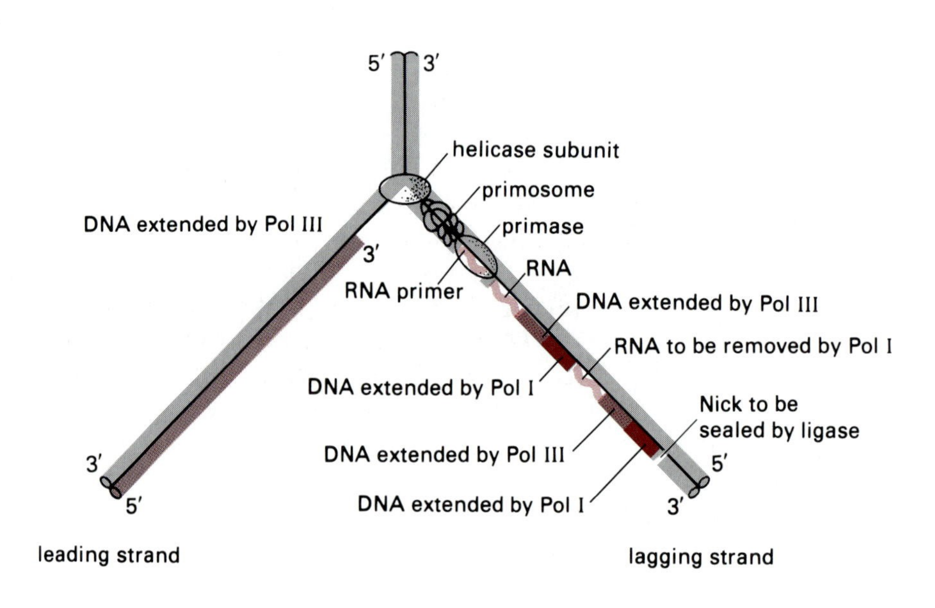

Figure 2.30
Model of semiconservative, semidiscontinuous replication at a fork in *E. coli*.

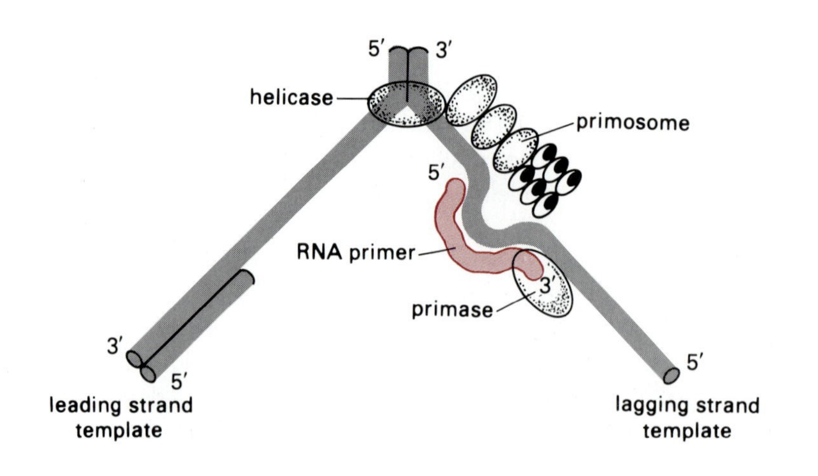

Figure 2.31
Initiation of lagging strand synthesis. Adapted from A. Kornberg, *DNA Replication, Supplement* (San Francisco: W.H. Freeman, 1982), p. S113.

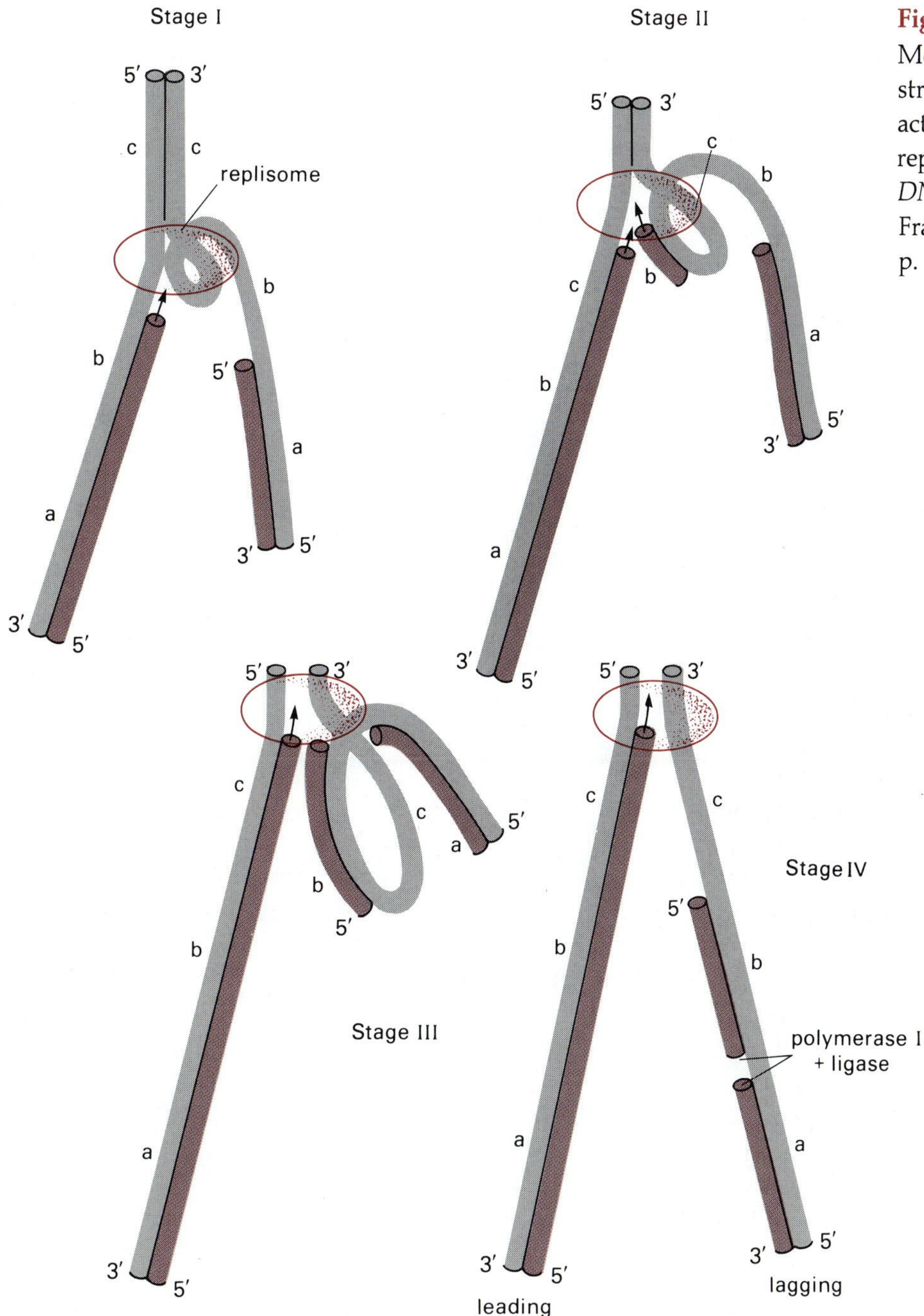

Figure 2.32
Model for advancement of both strands at the replication fork by the action of a multienzyme complex, a replisome. Adapted from A. Kornberg, *DNA Replication, Supplement* (San Francisco: W. H. Freeman, 1982), p. S125.

Thus, only the leading strand is synthesized at the fork. As the leading strand is extended, it displaces the 5′ end of the nicked circle as a single strand. As a result, the length of the leading strand may exceed that of the template, occasionally being two to five times as large.

For phages M13 or ϕX174, whose mature genomes are single strand circular DNA, this replication mode occurs late in the infection, after the infecting DNA has been converted to double strand circular DNA and amplified in that form. The continuously extruded single strand of DNA produced by rolling circle replication is cleaved at each origin and circularized to its mature form during the formation of virus particles.

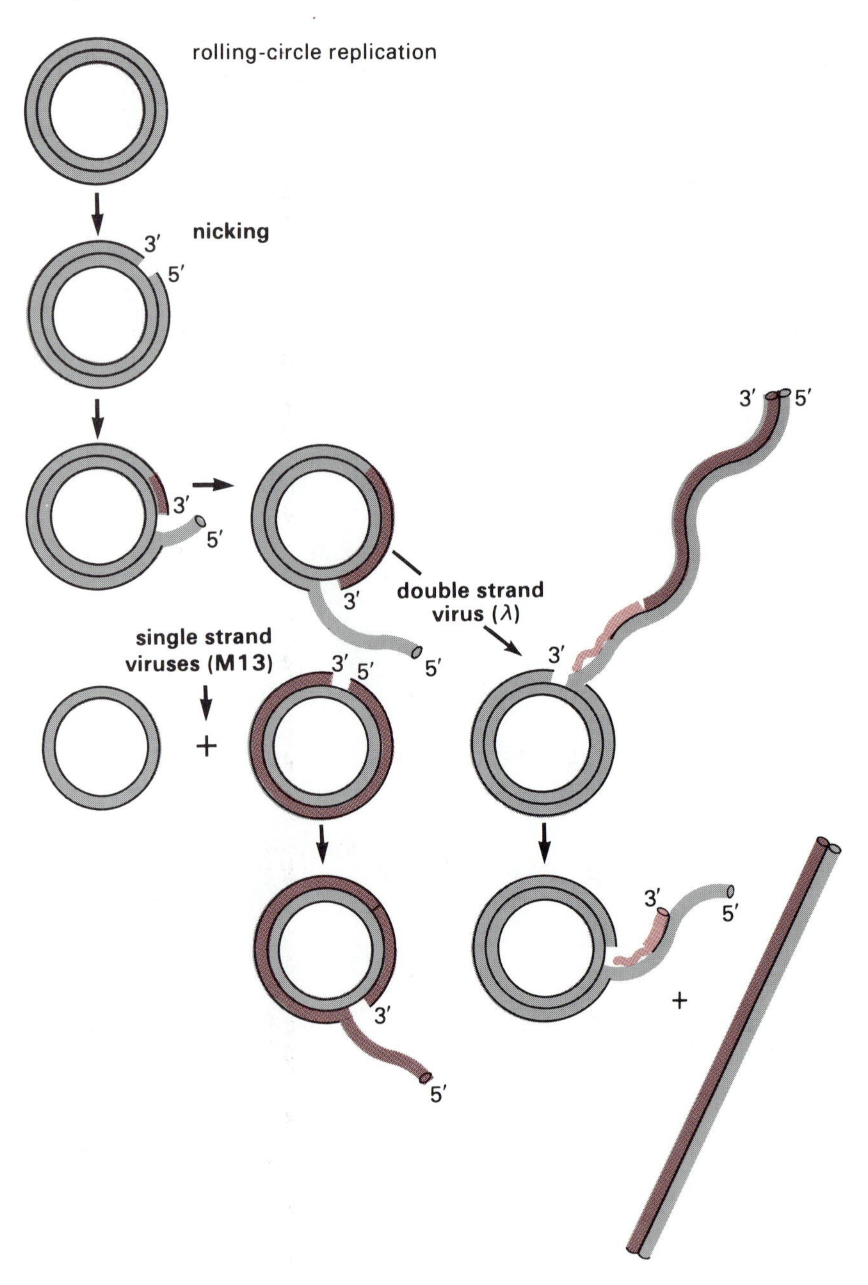

Figure 2.33
Rolling circle replication. See text.

Phage λ utilizes the rolling circle replication mode to produce double strand linear viral DNA (Figure 2.33). In this instance, the substrate template is double strand circular DNA that had been replicated after conversion of the linear viral DNA to the circular replicative form early in the infection. For the production of progeny linear DNA, the double strand circles are nicked and replicated asymmetrically as in the case of M13 and φX174 DNA replication. However, the extruded single strands are converted to double strands. First, short RNA chains are formed by primase at multiple sites along the single strand. Then these RNA chains are extended by Pol III holoenzyme, the gaps are filled, the RNA is removed by Pol I, and the short DNA fragments are eventually joined by DNA ligase. During the packaging of the DNA into bacteriophage

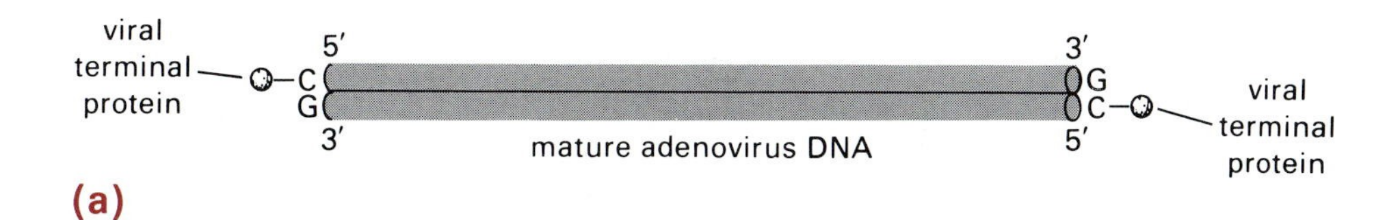

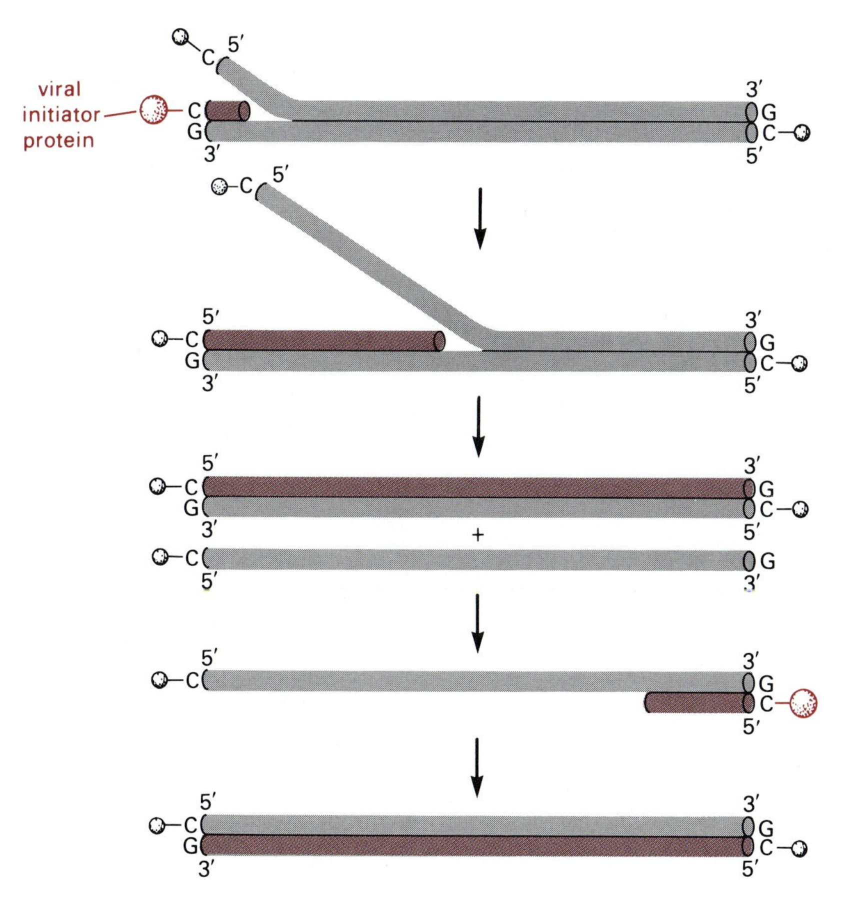

Figure 2.34
Alternate strand replication of adenovirus DNA. (a) Double strand adenovirus DNA with 55 kD protein attached to C residue at each 5′ end. (b) New DNA strands are initiated alternately at each end with an 80 kD protein-deoxycytidylate primer which is then extended to the end of the template strand. The 80 kD proteins are trimmed to 55 kD before the strands are completed.

particles, staggered cleavages are made at special sequences called **cos sites**, spaced one viral genome length apart, thereby converting the long duplex containing multiple lengths of λ DNA to the mature DNA size found in the virions.

Alternate strand replication Adenovirus DNA replication dispenses with lagging strand synthesis and therefore the necessity for multiple initiations and the formation of Okazaki fragments (Figure 2.34). Instead, each strand of the linear duplex genome is replicated alternately. First, one strand is initiated via the protein primer and is extended continuously to complete the replication process. The displaced genome length single strand then serves as the template for the synthesis of another duplex DNA by the same steps. There appears to be no specificity as to which end of the parental duplex starts the process. Adenovirus DNA replication is novel in two aspects: (1) the use of a protein to prime DNA chain synthesis starting at the end of a linear duplex and (2) the lack of discontinuous replication via RNA transcription.

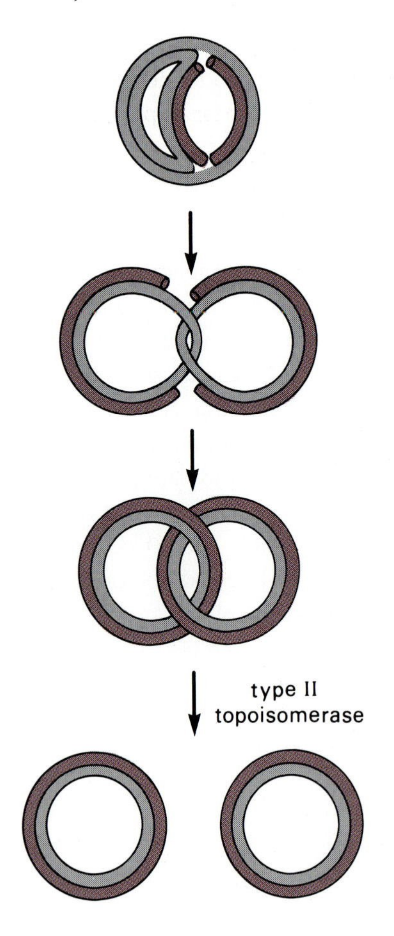

Figure 2.35
Termination of closed circular duplex DNA replication and resolution (separation) of the two progeny circles.

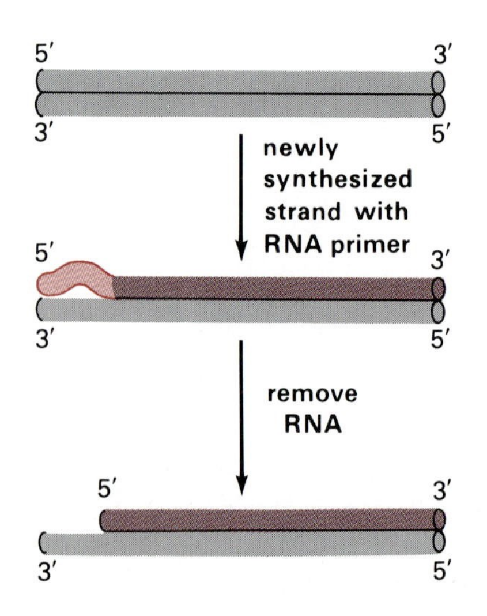

Figure 2.36
Removal of the RNA primer from the 5′ end of a newly replicated DNA strand leaves a gap at one end of the duplex.

h. Termination of DNA Replication and Resolution of the Daughter Helices

Termination and resolution in circular genomes Circularity of many DNA genomes simplifies the problem of completing the replication of the entire nucleotide sequence. Continued growth of the leading and lagging strands around a circular template inevitably juxtaposes the 3′-hydroxyl and 5′-phosphoryl ends of the same strand either at the origin or, where replication is bidirectional, about halfway around the ring (Figure 2.35). The rings are closed at these sites by DNA ligase, but this generally leads to catenated or intertwined dimer rings, which must be resolved for each genome to replicate further. Intertwined rings are resolved by the action of a type II topoisomerase (Figure 2.25).

Termination and completion in linear DNAs With the exception of adenovirus DNA replication (Figure 2.34) where a protein primer initiates new DNA strand synthesis and the entire template strand can be copied, the requirement for an RNA primer poses a special problem for completing the replication of linear duplex DNA (Figure 2.36). Note that after a new strand is initiated and the RNA primer is subsequently removed, the newly made DNA strand is left with a gap at the 5′ end. Because there is no way to extend the 5′ end of DNA chains, another means is needed to complete the replication.

Two interesting ways to accomplish this were proposed in the past. One involves DNA chains with direct repeats at their ends (Figure 2.37). After replication, the two complementary ends on two incomplete duplexes can base pair to produce nicked or gapped linear concatemers. The nucleotide sequence between the gaps or nicks can either be filled in by extending the chains from their 3′-hydroxyl termini to the 5′ ends at the nicks and be joined by DNA ligase or be joined by ligase directly at abutting termini to produce concatemeric dimers. The mature termini are produced if the concatemer is cleaved by a specific endonuclease that introduces staggered breaks with protruding 5′ termini. DNA polymerase can now extend the 3′ ends to complete the chains. Another scheme depends on the occurrence of a short inverted repeat sequence at the terminus of each DNA strand, which allows the strands to form short snap-back loops at their ends (Figure 2.38). This permits the 3′ end of each strand to serve as a primer for copying the unreplicated stretch. Thereafter, a specific break at the beginning of the inverted repeat creates a structure that can be extended at the new 3′ end to reproduce the original double strand terminal sequence.

Recent evidence suggests that telomeres, the ends of eukaryotic chromosomes, are replicated and maintained by a special mechanism, different from either of the two earlier models. The ends of yeast, invertebrate, plant, and vertebrate chromosomes have analogous and unusual structures (Section 9.6b). These contain hairpin loops that bring the 3′ and 5′ termini at the end of the DNA duplex together and many tandem repeats of a short sequence. On one strand, near the loop, there are multiple single strand nicks within the clusters of the repeats. How this sequence organization permits and participates in replicating the end of the duplex is becoming clearer. In view of the similarities in the structural features of

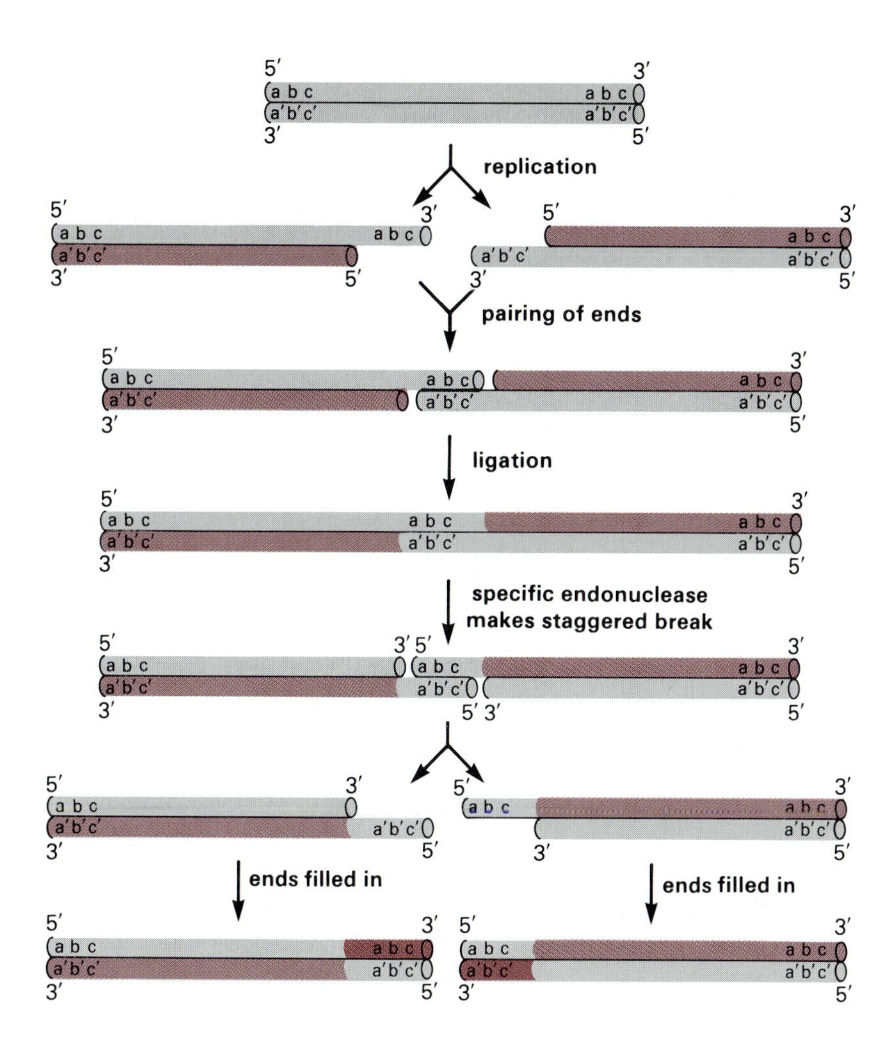

Figure 2.37
A speculative model describing how replication of linear duplex DNA might be completed. Based on J. D. Watson, *Nature*, New Biology 239 (1972), p. 197.

chromosomes ends, it is likely that there is a common mechanism for replicating the ends of all eukaryotic chromosomes (Section 9.6b).

2.2 Replication of RNA into DNA

a. *Replication of Retroviral Genomes*

Retroviral genomes are contained in a single molecule of single strand RNA. Following entry into its host cell, the genetic information encoded

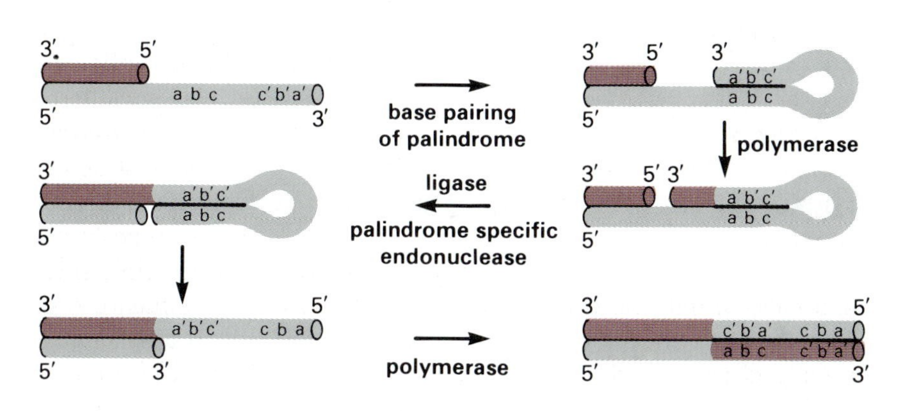

Figure 2.38
Another speculative model for completing a newly replicated linear double helical DNA. After T. Cavalier-Smith, *Nature* 250 (1974), p. 467.

in the virus' RNA genome is **reverse transcribed** into an RNA-DNA duplex and then into double strand DNA. These steps precede the expression of the virus' genes as proteins and the production of progeny RNA genomes.

The enzyme that catalyzes the complementary copying of RNA into DNA is called **reverse transcriptase**. The enzyme is an integral component of retrovirus particles (virions) and is activated after entry into cells when the viral glycoprotein-lipid envelope is removed. Very likely, auxiliary proteins present in the virion core facilitate reverse transcription because the enzymatic reaction carried out by virion cores is considerably more efficient and extensive than that obtained with only the purified enzyme. There are increasing indications that reverse transcription occurs in a variety of eukaryotic cells and that the enzyme plays an active role in the mechanism of genome rearrangements (Sections 10.3 and 10.4).

Retroviral reverse transcriptases are actually DNA polymerases and can function *in vitro* with DNA as a template. However, they are distinctive by the efficiency with which they use RNA as a template. As with all other DNA polymerases, reverse transcriptase does not initiate new DNA chains. But once synthesis is primed with an RNA or DNA having a 3′-hydroxyl end, the enzyme is efficient for template-directed DNA assembly.

Retroviruses are diploid because each virion contains two identical RNA chains of 8000 to 10,000 nucleotides. The two RNA chains are associated near their 5′ ends, but the nature of this noncovalent interaction is unknown. Each RNA chain has at its 5′ and 3′ ends the modifications characteristic of eukaryotic mRNAs (5′ caps and 3′ polyadenylated tails, Section 3.8a) (Figure 2.39). In considering the mechanism of the reverse transcription reactions, five features of the viral RNA structure must be emphasized: (1) the existence of direct repeats of nucleotides at the 5′ and 3′ ends of the RNA (R), (2) a sequence of 80–120 nucleotides adjacent to the 5′ terminal repeat (U5), (3) a sequence of 170–1200 nucleotides adjacent to the 3′ terminal repeat (U3), (4) a sequence of 15–20 nucleotides (P) at which a cell-derived tRNA is base paired to the retroviral RNA to provide a primer for synthesis of the first DNA strand, (5) a segment (Pu) just before U3 that provides the site for priming the second strand of DNA and appears to be conserved in all retroviruses of a particular type (i.e., this sequence is virtually identical in all avian viruses and different from that sequence in all murine viruses).

There are three products of the reverse transcription reaction: form A, a linear DNA duplex with the U5RU3 sequence (**long terminal repeat [LTR]**) repeated at both ends of the duplex and two circular DNA duplex derivatives of A, form B with both LTRs, and form C with only one

Figure 2.39
The RNA genome of a retrovirus. Each virion contains two identical copies of the RNA. Typically, retroviruses have three genes, *gag*, *pol*, and *env*; the *pol* gene encodes the reverse transcriptase. The RNA has a 5′ cap and a 3′ polyadenylated tail (described in Section 3.8a). The other features of this genome, *R*, U5, etc., are described in the text.

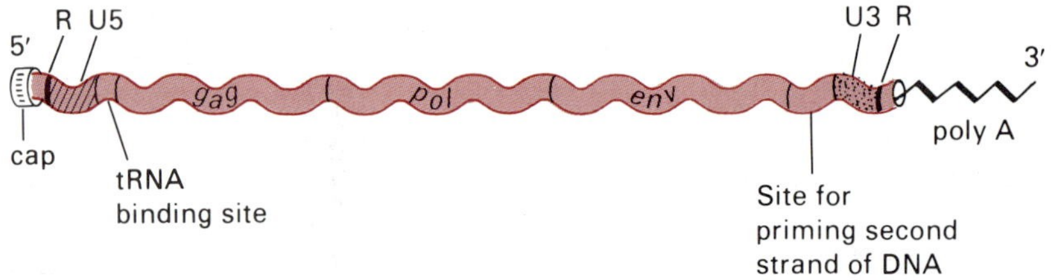

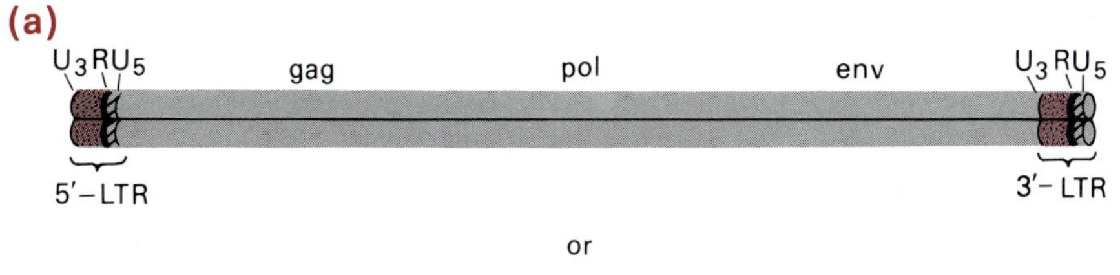

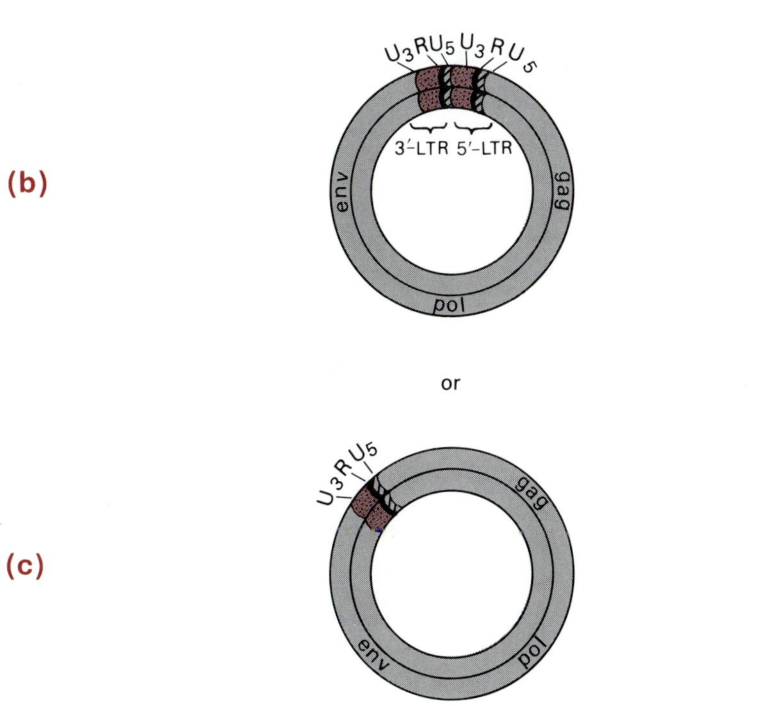

Figure 2.40
Duplex DNA forms produced by reverse transcription of retroviral RNA.

(Figure 2.40). The problem is to explain how structures (a), (b), and (c) are derived by reverse transcription of the viral RNA. The process, summarized in Figure 2.41, begins by extending the tRNA primer through U5 and R to the 5′ end (step 1). Then an RNase H specific for the RNA in an RNA-DNA hybrid duplex digests the RNA segment of the hybrid duplex (step 2). Because R is repeated at the 3′ end of the RNA, the newly made short DNA chain "jumps" to the other end of the RNA and base pairs with its complement at the 3′ end (step 3). Following that, the DNA chain is extended using the remainder of the RNA as a template (step 4). By the time the first DNA strand is completed, most of the viral RNA is degraded by RNase H. Then, synthesis of the second DNA strand is initiated at the putative primer-binding site (Pu) near U3, using the first DNA strand as the template (step 5). The primer for the second strand DNA synthesis may be RNA, but it is not known if the second strand is extended continuously or discontinuously. After replicating the tRNA binding sequence at the 5′ end of the first DNA strand (step 6), the tRNA is believed to be removed (step 7). Next, the newly synthesized second DNA strand becomes paired through the first strand's tRNA binding sequence (step 8). By extending the 3′ ends of each strand, a fully duplex DNA is produced (step 9). Note that the DNA duplex contains a direct repeat of the sequences U3RU5 at each end; these constitute the LTRs.

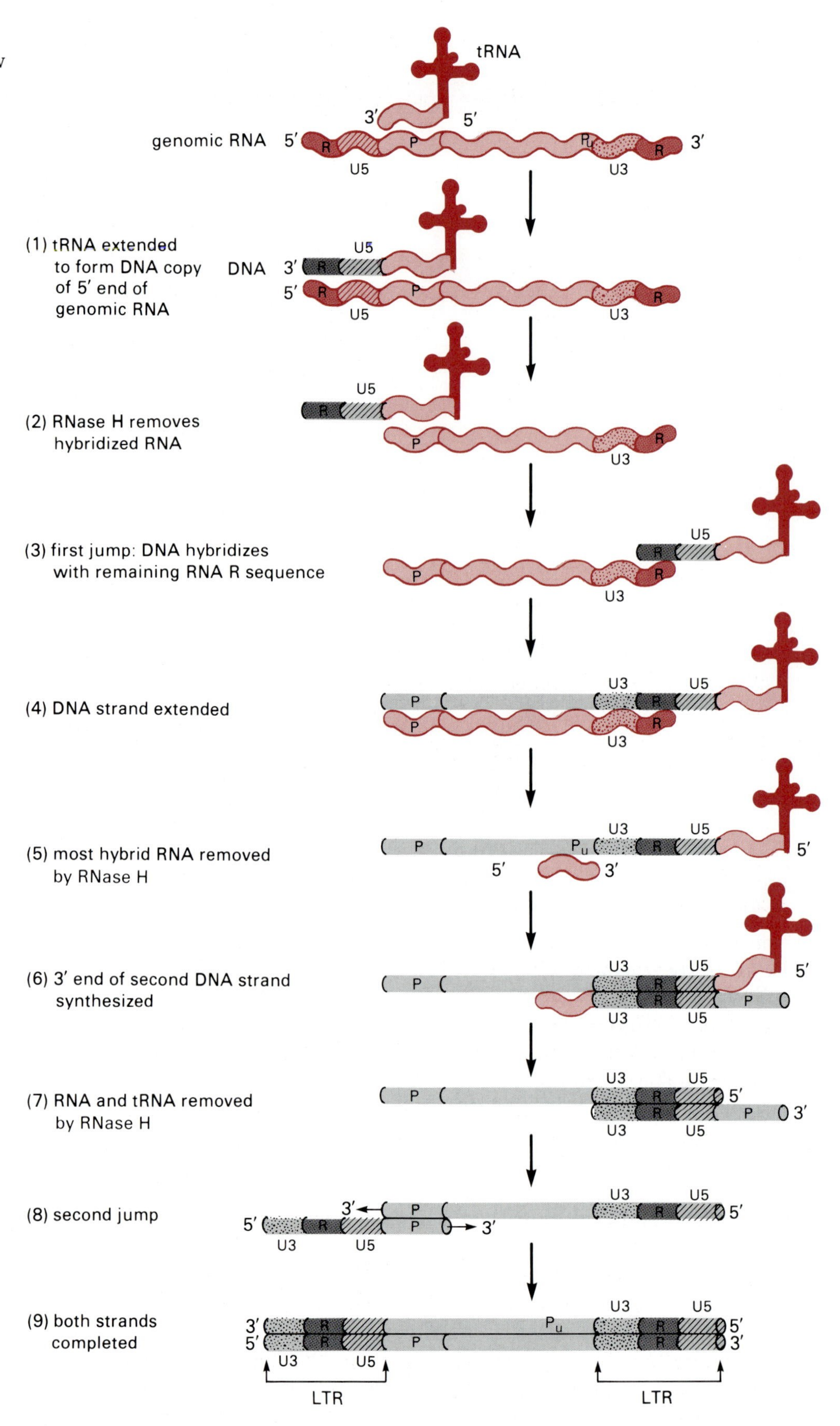

Figure 2.41
Schematic description of how reverse transcription of retroviral RNA leads to formation of a linear duplex DNA bounded by LTRs. After J. Darnell, H. Lodish, and D. Baltimore, *Molecular Cell Biology* (New York: Scientific American Books, 1986), p. 1052.

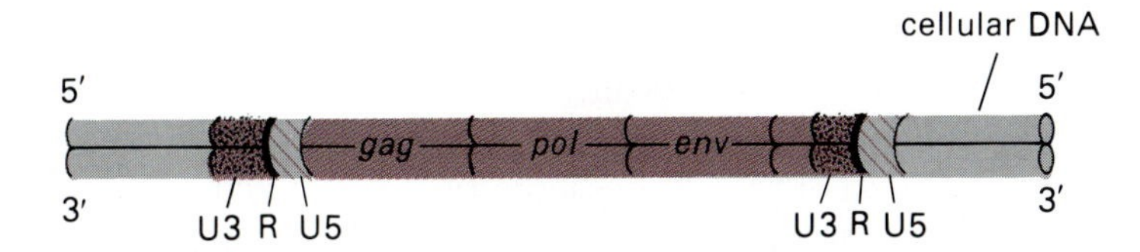

Figure 2.42
Structure of retroviral DNA integrated in a cellular genome: a provirus.

The circular DNAs shown in Figure 2.40 are presumed to arise either by ligation of the two ends of the linear DNA (b) or by homologous recombination (Section 2.4) between the LTR segments (c). It is remarkable that this complex series of reactions occurs without the apparent participation of cellular DNA replication enzymes (e.g., helicase, primase, DNA-binding protein, ligase, topoisomerase, etc.).

Replication of the double strand form of retrovirus DNA does not occur until it is integrated into the cellular DNA. The reverse transcription product shown in Figure 2.40a, the linear duplex DNA, is the substrate for the integration event. The mechanism of the recombinational insertion is not well understood. However, the product is identical to the linear DNA shown in Figure 2.40a except that the sequence is joined to cell DNA at both ends (Figure 2.42). The integration mechanism causes the loss of a few nucleotides from the ends of the LTRs and a duplication of three to ten nucleotides of cell DNA at the two ends of the integrated viral DNA. After integration, the retroviral DNA sequences are replicated as an integral part of the cellular DNA. The RNA in progeny virions is derived by transcription of the integrated viral DNA copies. RNA strands are initiated at the leftmost U3R junction and terminate at the rightmost RU5 junction (Figure 2.42).

b. Some DNA Viruses Use Reverse Transcription for Replication

Human hepatitis B virus contains a circular, interrupted double strand DNA as its chromosome. Once this virus is inside the cell, gaps in its incomplete DNA strands are filled in by a virion-associated DNA polymerase. A nearly full-length DNA strand then acts as a template to produce a genome-length RNA that serves both as a source of mRNA to express the viral proteins and as a template for reverse transcription to regenerate progeny viral DNA. A similar mechanism operates in the replication of cauliflower mosaic virus, a plant virus.

2.3 DNA Repair

DNA is the only cellular macromolecule capable of being repaired after sustaining structural damage. Moreover, it encodes the machinery for effecting a wide variety and large number of repair processes. Just as complementary base pairing is the key feature in DNA replication, it is also central to directing the restoration of the original structure after it suffers a damaging break in the backbone, undergoes a modification of one of its bases, or acquires a mismatch during recombination (Section 2.4). Events that alter both strands of a given region or cause double strand scissions are often lethal because the damage is only rarely reparable.

Thermal disruption of the purine-deoxyribose-N-glycosyl linkage is the most frequent type of damage inflicted on DNA (Figure 2.43). Between 5000 and 10,000 depurinations occur per genome per day in a human cell. If left untended, they would impair replication and gene expression. Moreover, cytosines and adenines are spontaneously deaminated to produce the corresponding uracil- and hypoxanthine-containing nucleotides, respectively, about 100 times per genome per day. This creates potential mutations unless rectified before the next wave of replication.

Environmental chemicals cause a wide variety of alterations of DNA structure. These substances include alkylating agents such as nitrogen mustards, alkylsulfonates, and nitrosoureas that modify guanine residues preferentially, but other bases as well; molecules that intercalate between the stacked base pairs and cause insertions and deletions of nucleotides during replication; and bifunctionally reactive agents that sometimes form covalent bridges between the two DNA chains, thereby blocking their separation during replication. Physical forces can be equally disruptive. Ultraviolet light absorbed by a thymine or cytosine base can cause it to form a cyclobutyl derivative with an adjacent pyrimidine (Section 2.3a); ionizing radiation, as in cosmic rays, can create reactive free radicals, which alter the substituents of DNA in myriad ways; and X rays received in the course of medical diagnosis and treatment can induce single and double strand breaks as well as other free radical–mediated modifications.

How does DNA survive these disruptive forces? What mechanisms restore the normal nucleotide structure and sequence before their effects are permanently fixed and expressed as mutations? Two general categories of repair processes are known: (1) reactions that reverse the modification or mismatch directly and do not require replication to restore the original structure and (2) reactions that require removal of the nucleotides surrounding the mismatched or altered base pair and resynthesis of that region by replication.

a. Repair by Reversing the Modification

Alkylating agents such as N-methyl-N-nitrosourea or N^1, N-dimethyl nitrosoguanidine act on DNA to produce O^6-methyl or other O^6-alkyl substituted guanine residues. These modified guanine residues can be dealkylated by enzymes in bacterial and mammalian cells. An O^6-methyl guanine-DNA alkyl transferase catalyzes the transfer of alkyl groups to the sulfhydryl group of a cysteine residue in the enzyme, the acceptor protein being inactivated by the transalkylation event (Figure 2.44). In *E. coli*, the level of alkyl transferase is increased in response to a challenge of O^6-alkyl guanine, but no such inducibility has been noted in mammalian cells.

Ultraviolet light irradiation of DNA causes the formation of cyclobutyl dimers between adjacent pyrimidines (Figure 2.45). Such structures block DNA replication and must therefore be removed for a cell to survive. One way to eliminate pyrimidine dimers is to convert them back to monomers enzymatically with the assistance of light in the 300–600 nm

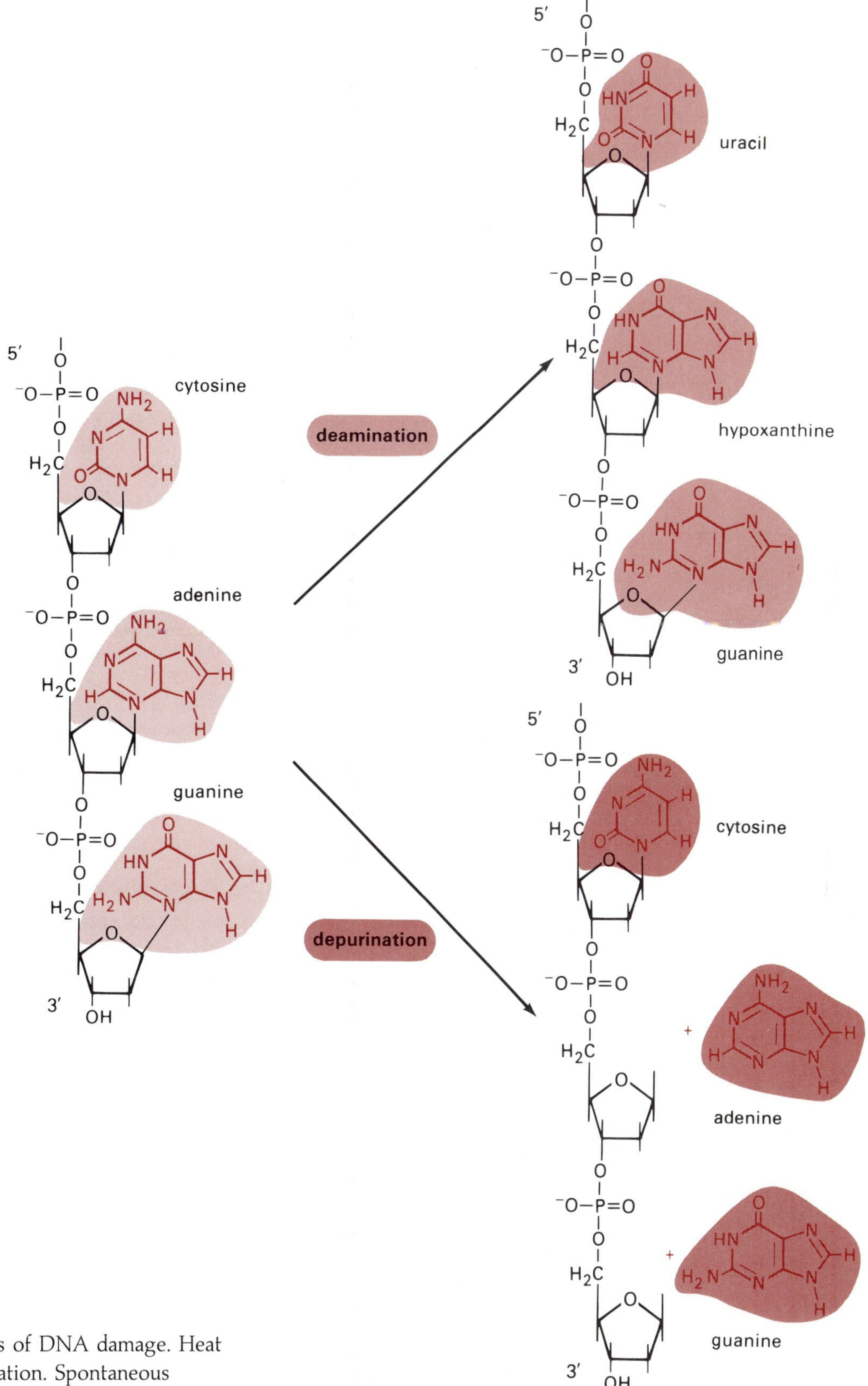

Figure 2.43
Common types of DNA damage. Heat causes depurination. Spontaneous deamination of adenine and cytosine leads to formation of hypoxanthine and uracil residues, respectively.

Figure 2.44
The dealkylation of O^6-methyl guanine residues is catalyzed by a specific DNA alkyl transferase.

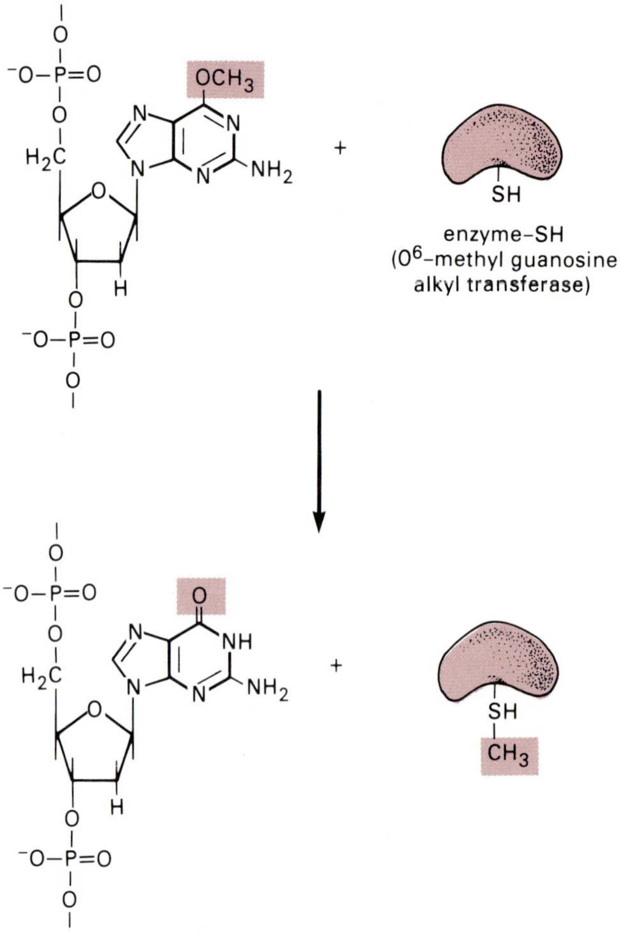

(visible) range (Figure 2.46). Such photoreactivating enzymes (photolyases) are present in bacteria and small eukaryotic organisms but have not been found in mammalian cells. The enzyme forms a stable protein-pyrimidine dimer complex and uses the absorbed light energy to effect the conversion of the dimer to its monomers without disrupting the double strands.

Figure 2.45
The formation of cyclobutyl dimers between adjacent thymines on a DNA strand.

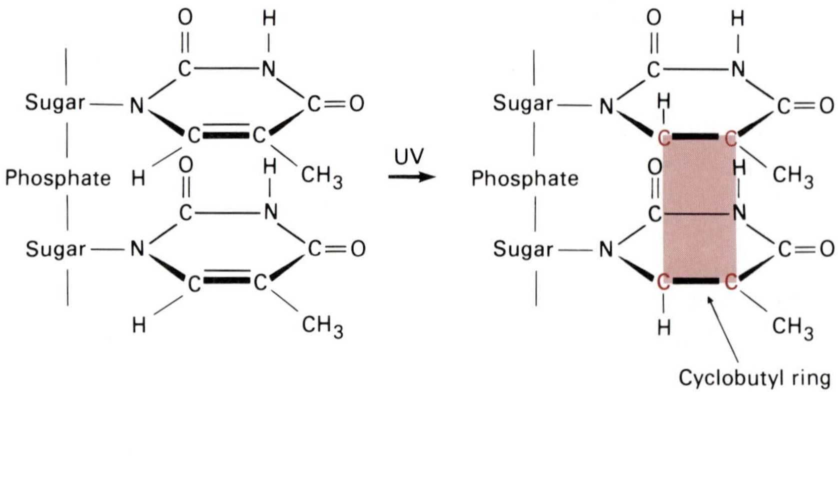

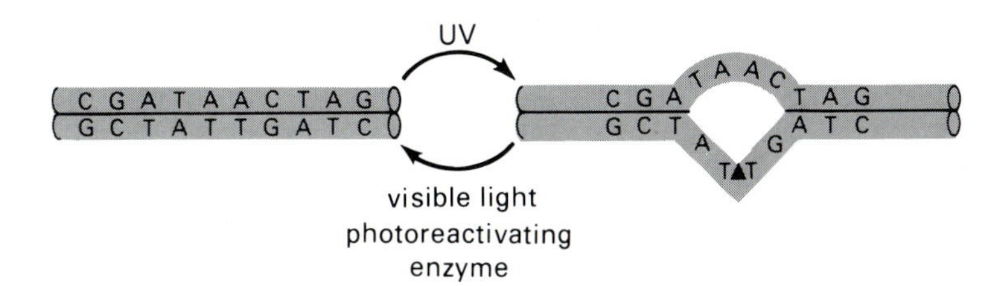

Figure 2.46
Formation of pyrimidine dimers by ultraviolet light is reversed by enzymes that depend on visible light.

b. Repair by Replacing Modified Residue(s)

Replacing a modified nucleotide generally involves four steps. First, the modification is recognized and the polynucleotide chain adjacent to the modified sequence is cut, or the glycosidic bond between the modified base and its deoxyribose is cleaved. Second, exonuclease excises the modified nucleotide and/or surrounding nucleotides leaving a small gap. Third, the excised region is resynthesized, from the 3′-hydroxyl end using the opposite strand as a template. Fourth, the ends at the nick produced by repair are ligated to restore the covalent integrity of the repaired strand (Figure 2.47).

Sites at which depurination or depyrimidination has occurred are cleaved by enzymes referred to as AP (apurinic and apyrimidinic) endonucleases (Figure 2.48a). Prokaryotic and eukaryotic cells contain a large variety of AP endonucleases, often several different types in the same cell. Some AP endonucleases cleave at the 3′ side of the AP site while others incise the diester bond on the 5′ side; in all cases, 3′-hydroxyl and 5′-phosphoryl ends are produced. The cleavage of the phosphodiester linkage on one or the other side (sometimes on both sides) permits exonucleases to remove the adjacent residues on either side of the cleavage site and allows for resynthesis of the excised sequence.

During the repair of N-alkylated purines and other modified bases, specific N-glycosylases—enzymes that cleave the glycosidic linkage between a specific modified base and the deoxyribose—act first (Figure 2.48b). The correction of changes caused by the spontaneous deamination of cytosine to uracil or of adenine to hypoxanthine, each of which is mutagenic unless rectified before replication, also involves specific N-glycosylases. These enzymes, uracil-N-glycosylase and hypoxanthine-N-glycosylase, remove uracil or hypoxanthine, respectively, from DNA, leaving apyrimidinic or apurinic sites in their place. Again, excision-resynthesis restores the original sequence on the affected strand.

The uracil-N-glycosylase also has an especially important antimutagenic role in rectifying errors produced when dUTP is used instead of dTTP during replication. Because the dUMP base pairs with the template base virtually as well as dTMP, it is not edited away by Pol III (or Pol I). Therefore, the dUMP may, unless specifically removed, cause dGMP to be incorporated into the new strand in a subsequent round of replication. Thus, uracil-N-glycosylase protects cells against the potential mutagenic consequences of the unavoidable production of dUTP.

Cyclobutyl thymine dimers can be repaired without light in either of two ways. In *E. coli*, three polypeptides, coded by the genes *uvrA*, *uvrB*, and *uvrC*, form an enzyme complex, the *uvrABC* excision nuclease. This enzyme cleaves DNA containing a pyrimidine dimer eight phospho-

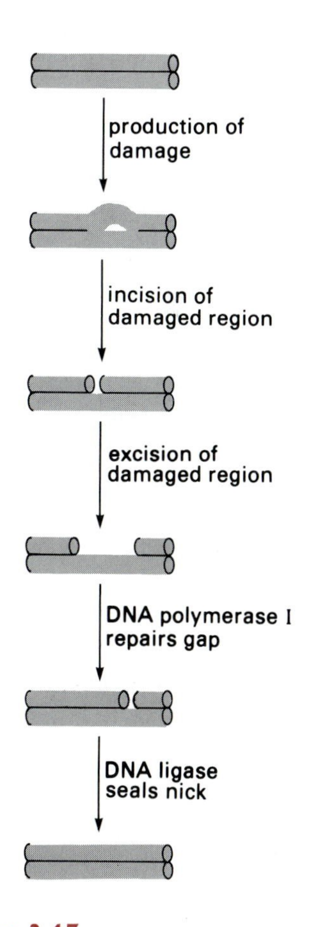

Figure 2.47
Repair of damaged DNA by replacement of modified nucleotide residues.

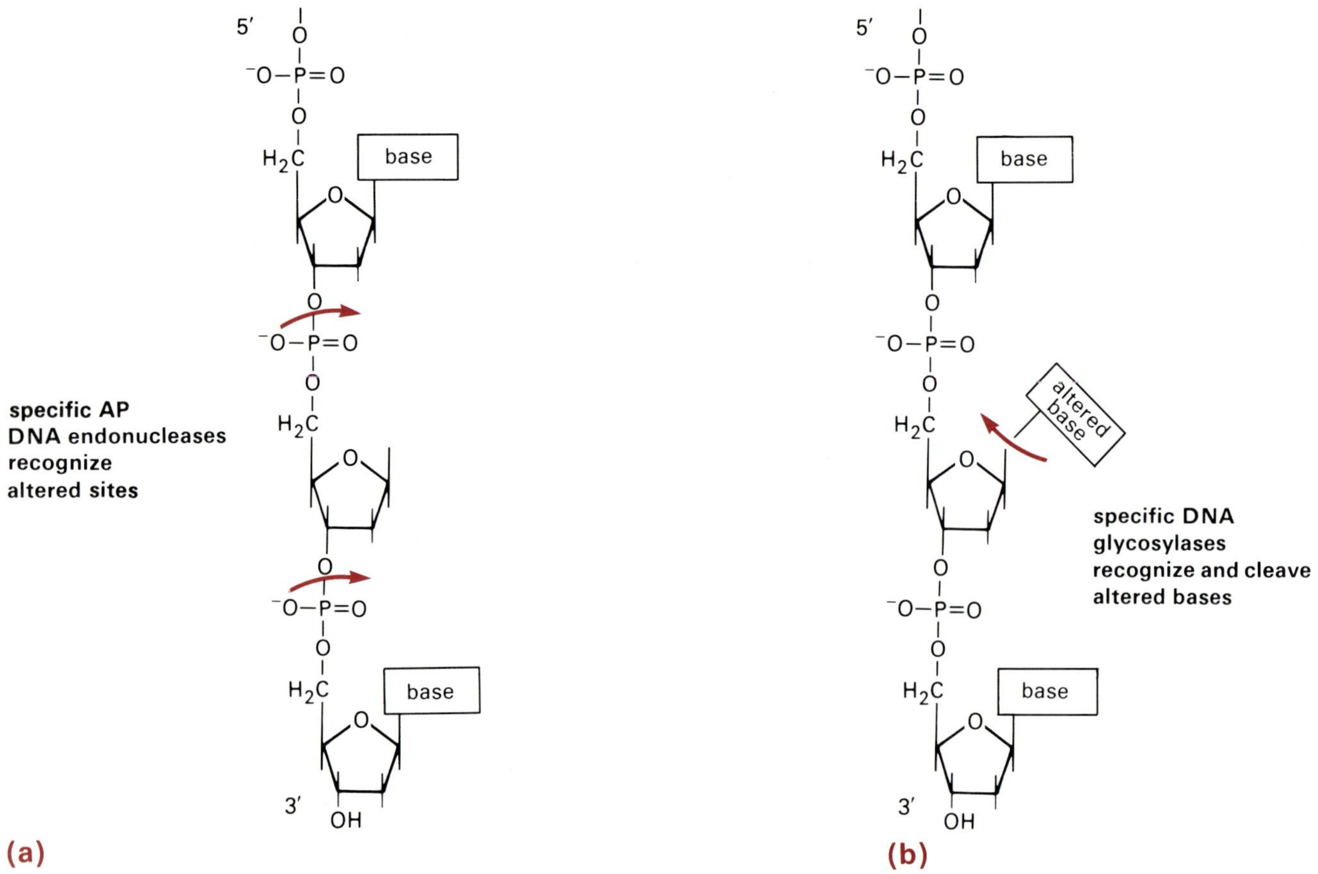

specific AP DNA endonucleases recognize altered sites

(a)

altered base

specific DNA glycosylases recognize and cleave altered bases

(b)

Figure 2.48
Cleavages in DNA at altered sites. (a) *AP* endonucleases recognize depurinated and depyrimidinated residues and hydrolyze a phosphodiester bond at either the 3′ or 5′ side of the modified residue. (b) *N*-glycosylases initiate repair of modified bases by hydrolyzing the glycosidic linkage between the base and deoxyribose.

diester bonds on the 5′ side and four or five phosphodiester bonds on the 3′ side of the thymine dimer (Figure 2.49). Removal of the damaged segment appears to be facilitated by a helicase encoded in still another uvr gene, *uvrD*. As a result of the incisions, the thymine dimer and about 12 flanking nucleotides are removed. The missing nucleotides are replaced by the gap-filling action of Pol I, whereupon DNA ligase completes the excision-repair phase by ligating the two adjacent nucleotides.

Another pathway used for repairing pyrimidine dimers in some organisms involves a pyrimidine dimer N-glycosylase, which creates an apyrimidinic site. Subsequently, the adjoining phosphodiester bond is cleaved, creating a thymine dimer at the 5′ terminus and a 3′-hydroxyl group on a terminal deoxyribose (Figure 2.50). The 3′-to-5′ exonuclease action of DNA polymerase can remove the newly created 3′ AP terminus. When this occurs, the nick translation-ligation pathway removes the nucleotide with the thymine dimer and completes the repair process.

c. *The Importance of DNA Repair*

Cells have evolved a complex machinery to minimize damage to DNA from a wide variety of chemical, physical, and replication—or recombination—associated events. It is not difficult to see why. Many of the alterations would block transmission of the genetic information to the next generation. Others, if left unattended, would be perpetuated in progeny genomes and lead to unacceptable changes in the proteins and enzymes needed to sustain the life of the cell. A loss of specific parts of

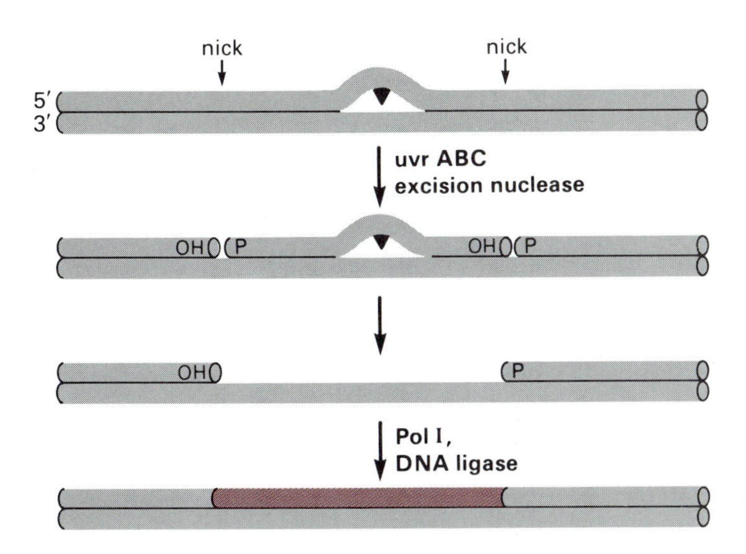

Figure 2.49
One pathway for light-independent repair of cyclobutyl thymine dimers. The *E. coli uvrABC* enzyme complex initiates repair by hydrolyzing the chain eight phosphodiester bonds to the 5′ side and four or five phosphodiester bonds to the 3′ side of the dimer.

the repair machinery makes cells especially vulnerable to certain chemical and physical agents. For example, *E. coli* that have a defective incision system for repairing thymine dimers are very sensitive to ultraviolet light. Those that cannot perform one or another N-glycosylase reaction are particularly prone to mutagenesis or killing by alkylating agents or ionizing radiation. *E. coli* that are deficient in Pol I have a markedly impaired ability to survive low doses of UV.

The small eukaryote *Saccharomyces cerevisiae* has at least five genes that are required for the incision of UV-irradiated DNA. A defect in any one of these *RAD* genes renders the cells totally defective in DNA incision and pyrimidine dimer excision. Similarly, there are yeast mutants that cannot remove interstrand cross-links even though they normally eliminate UV-induced damage. This suggests that in yeast, as in humans, there are specific and complex mechanisms for repairing UV-damaged DNA, for eliminating interstrand cross-links, and probably for a variety of other chemical modifications of DNA structures.

Humans who have the genetic defect *Xeroderma pigmentosum* are extremely sensitive to ultraviolet light and contract a variety of skin cancers from even very low exposures to sunlight. They have mutations that resemble yeast *RAD* mutants in that neither can excise pyrimidine dimers from UV-irradiated DNA. The *Xeroderma pigmentosum* disease phenotype can result from mutations in any one of at least nine different genes, indicating that the machinery needed to repair DNA containing thymine dimers in humans is quite complex. In most forms of the disease, the failure to excise the thymine dimers is the only defect; UV damage induced in cultured cells from *Xeroderma pigmentosum* patients can be repaired if an enzyme possessing thymine dimer glycosylase and AP endonuclease activities is introduced artificially into the irradiated cells.

2.4 DNA Recombination

Genetic recombination includes several related processes that create new associations of genetic information for the cells or organisms in which they occur. Recombination between closely paired homologous chromo-

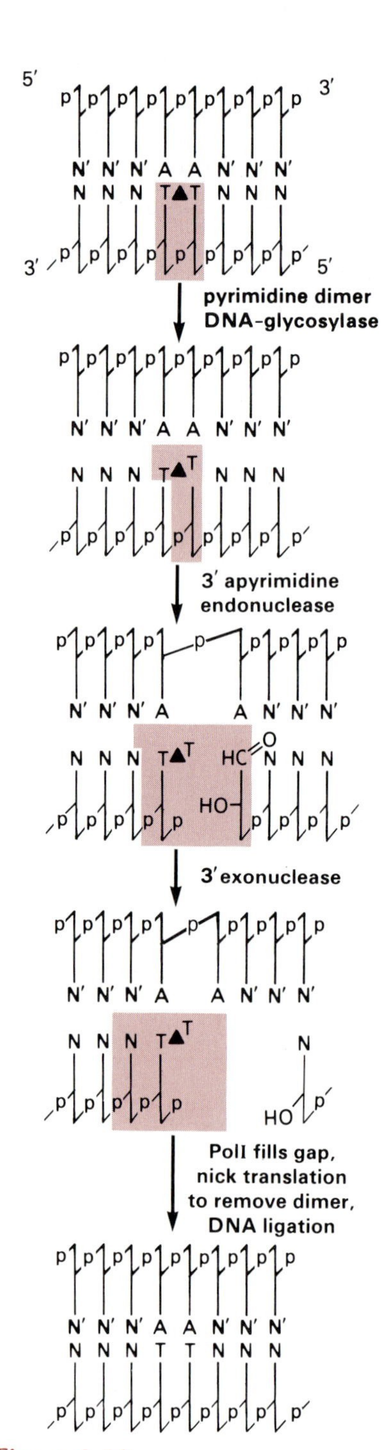

Figure 2.50
A second pathway for light-independent repair of cyclobutyl thymine dimers. A specific pyrimidine dimer *N*-glycosylase initiates repair by cleaving the glycosidic bond between one of the thymines and deoxyribose. Some glycosylases can also catalyze the subsequent endonuclease cleavage.

somes at meiosis causes extensive reshuffling of paternal and maternal genes, allowing evolutionary trials of new combinations among the offspring. Generally, recombinations occurring in somatic cells, either during or following DNA replication, are manifested as sister chromatid exchanges (Figure 2.51) with outcomes that, by themselves, do not alter the cellular genotype or phenotype. However, other recombinational events create novel rearrangements in the organism's genome. These include the loss, acquisition, or amplification of genetic information and the establishment of new linkage relationships between existing genetic elements.

In molecular terms, genetic recombination creates covalent linkages between nucleic acid sequences from two different regions of the same DNA molecule or between nucleotide sequences from two different DNA molecules. Just as all cells and many viruses encode the enzymatic machinery to synthesize and repair damage to their DNA, their genomes are also genetically programmed to produce enzymes that promote recombination. Indeed, some enzymes involved in DNA replication and repair are essential for recombination. In this section, we examine the mechanisms and enzymes for several different types of recombinational events, focusing principally on recombinations in bacteria and phages because we understand these systems best. Recombination in eukaryotic cells, though long recognized in its genetic and morphologic context, is still unexplained in molecular terms.

a. *Types of Recombination*

Genetic recombination events can be subdivided into three categories, each with distinctive characteristics: **general** or **homologous recombination**, **site-specific recombination**, and random or **nonhomologous recombination**.

General recombination Occurring most often within extended regions that share nearly identical or homologous nucleotide sequences, general recombination is frequently referred to as homologous recombination or **crossing-over**. In general recombination, two homologous DNA helices appear to have been broken and each of the ends of one segment joined to the corresponding ends of the other so that the two products contain different parts of the two contributing DNA helices (Figure 2.52). Actually, the sites at which breakage and reunion occur on each of the two strands of a helix are most often separated by a variable distance.

Although general recombination occurs between homologous and allelic regions on different DNA molecules, it can also occur between homologous but nonallelic segments on the recombining DNAs. When this occurs, one of the recombination products loses DNA, and the other gains it. Hence, such an event is referred to as **unequal crossing-over** (Figure 2.53). Recombination can also occur between nonallelic regions in the same chromosome (e.g., between repeated DNA sequences), with the consequent loss of the region between the recombination sites (Figure 2.54). Unlike the cases already mentioned, certain recombinations are nonreciprocal; as a consequence, one product of the recombination remains unchanged while the other is altered (Figure 2.55). Such events are often referred to as **gene conversions**.

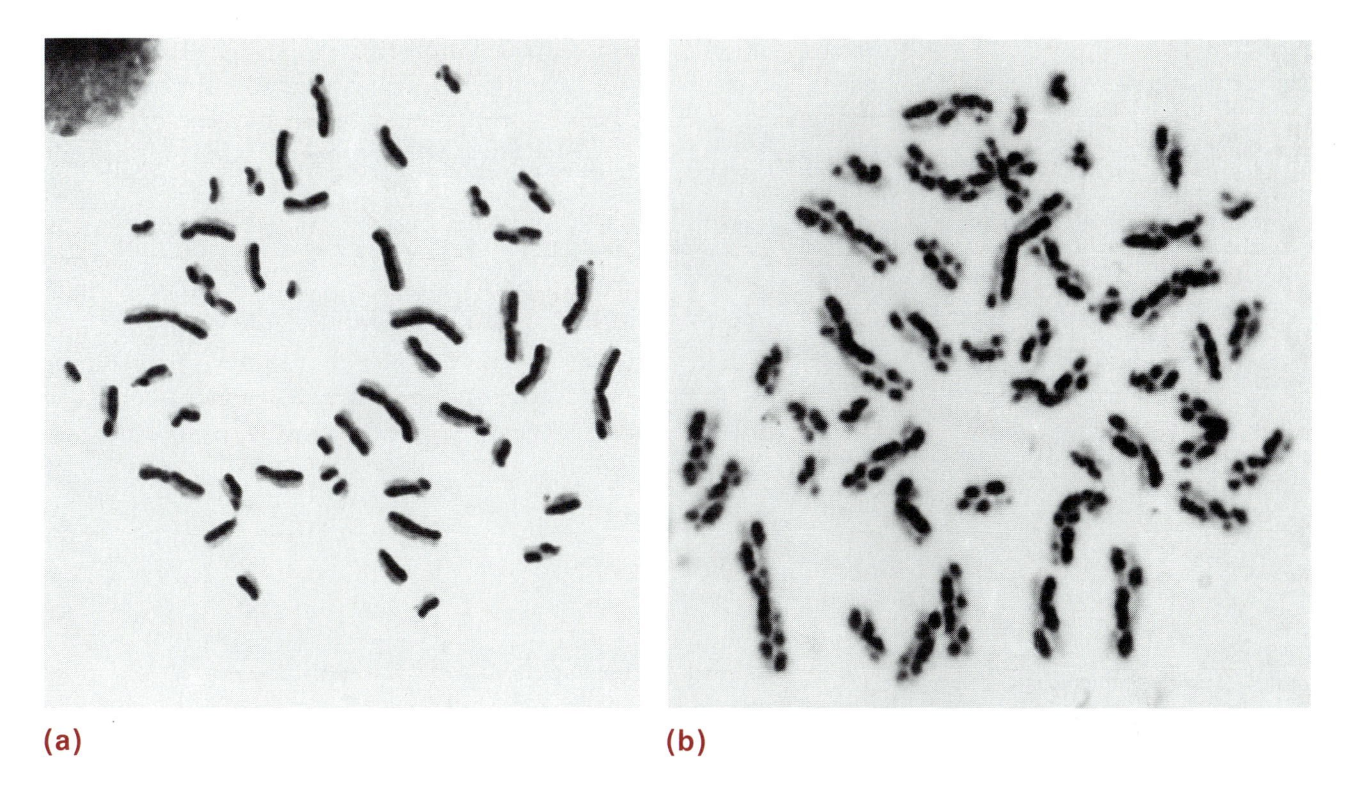

Figure 2.51
Sister chromatid exchange. Interchanges of DNA between the two products of chromosome duplication (sister chromatids) are detectable in metaphase chromosomes by special techniques. Briefly, cells are permitted to go through one cell cycle in the presence of bromodeoxyuridine and a second cycle in the absence of the thymidine analogue. The dye, Hoechst 33258, fluoresces less when bound to DNA containing bromodeoxyuridine than when bound to normal, thymidine-containing DNA. Thus, the sister chromatid products of the second cell cycle take on different intensities. The products of exchanges between portions of sister chromatids are readily visible as discontinuities in the fluorescence intensities. (a) Chromosomes from normal human peripheral lymphocytes. (b) Human peripheral lymphocytes from a patient with Bloom's syndrome. S. A. Latt and R. Schreck, *Amer. J. Hum. Genet.* 32 (1980), p. 294. Micrographs are courtesy of S. A. Latt (deceased).

Site-specific recombination When the sites of breakage and rejoining between two DNA molecules or between two segments of the same DNA molecule occur within rather short specific regions of sequence homology, usually less than 25 nucleotides, the recombination is referred to as site specific. The short specific nucleotide sequences may be present in only one of the partners and random sequences on the other (Figure 2.56), or the specific nucleotide sequences may be required in each of the partners (Figure 2.57). Transpositions by some prokaryotic and eukaryotic transposable elements exemplify the former (Sections 10.2 and 10.3), and the integration-excision processes between phage λ and *E. coli* DNA illustrate the latter (Section 2.4d). The programmed chromosomal DNA rearrangements involved in yeast-mating type alterations and the ontogeny of the immune response (Section 10.6) all involve site-specific recom-

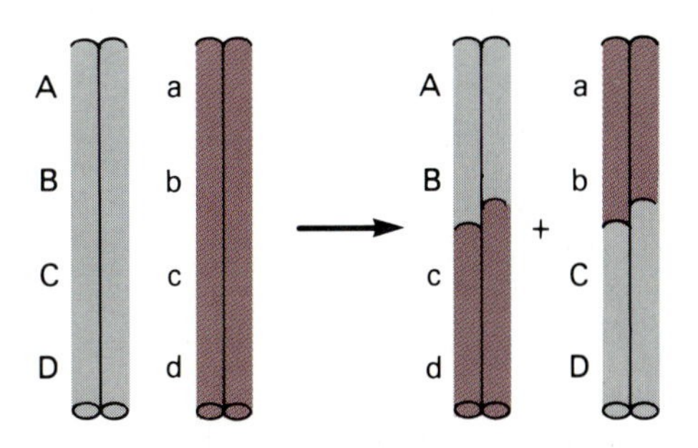

Figure 2.52
Homologous recombination or crossing-over.

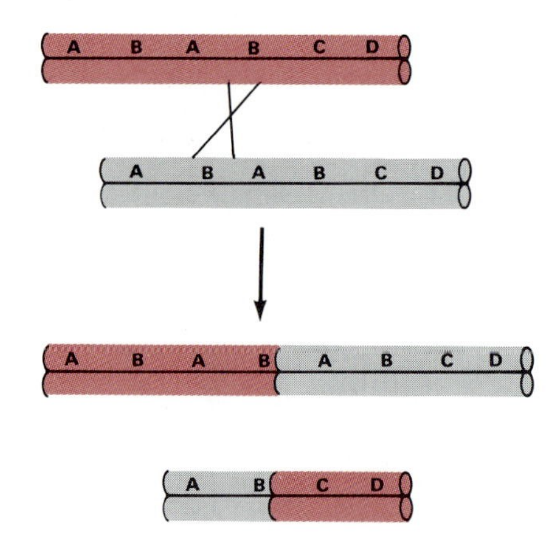

Figure 2.53
Unequal crossing-over.

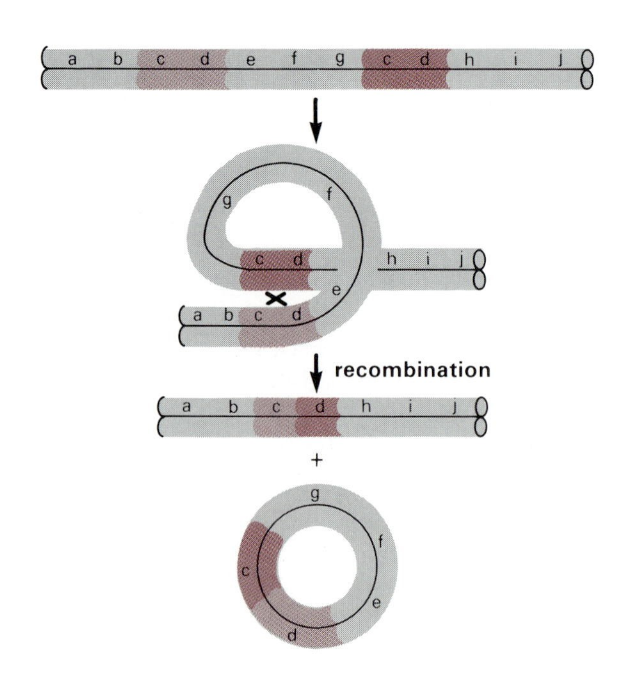

Figure 2.54
Intramolecular recombination.

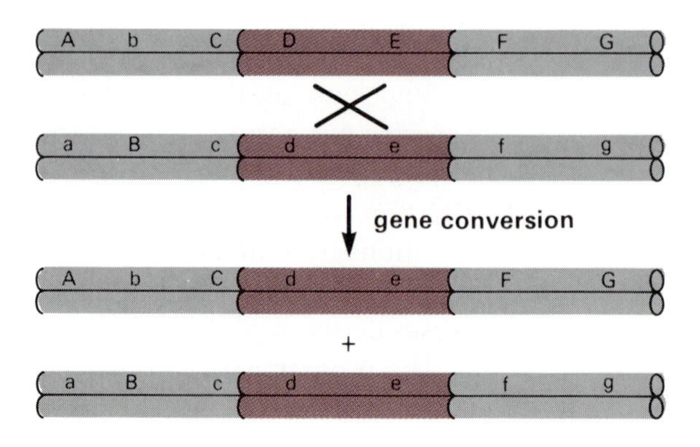

Figure 2.55
Nonreciprocal recombination or gene conversion.

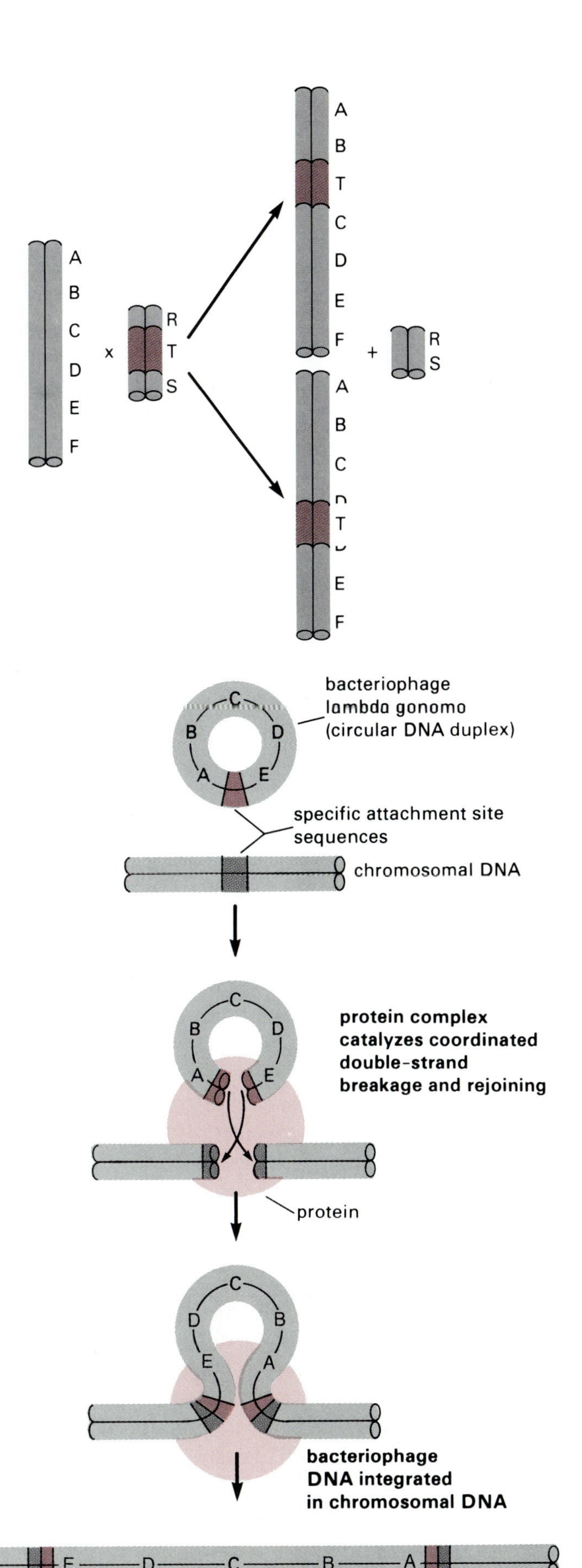

Figure 2.56

Site-specific recombination involving specificity on one DNA segment but not the other. In this example, the ends of region T are specific recombination sites, but T can recombine with random DNA sequences, for example, between B and C (top) or within region D (bottom).

Figure 2.57

Site-specific recombination requiring specific sequences on both interacting DNAs. This example shows the integration of λ phage DNA into *E. coli* chromosomal DNA. Specific recombination sites occur on both genomes. After B. Alberts et al., *Molecular Biology of the Cell* (New York: Garland, 1983), p. 248.

binations. Whereas a single set of enzymes probably suffices to mediate general recombination between any pair of homologous sequences, a special set of proteins is probably needed for each particular site-specific recombination.

Nonhomologous recombination Recombinations between nonhomologous nucleotide sequences occur only rarely in prokaryotes and yeast, but they are considerably more frequent in mammalian cells. Breakage and reunion between nonhomologous regions of DNA account for the random integrations of viral and plasmid DNAs into animal cell DNA and the rapid evolution of deletions and duplications in replicating papovavirus genomes. Moreover, the ends of broken DNA are efficiently joined even if their termini have nonhomologous sequences. In some instances, such recombinations may involve sequences with several base pairs of homology or short regions with only partial homology. But in most cases, there is no sequence homology between the recombined segments.

b. General Recombination Between Homologous DNAs

General recombination involves the coordinated breakage and reunion of strands on two helices resulting in the formation of extensive heteroduplex regions For two double strand helices to be recombined as illustrated in Figure 2.52, each of the four strands must be broken and rejoined to new partners. The stages at which the breakages and reunions are thought to occur are summarized in Figure 2.58. First, the corresponding strands of two aligned homologous DNA duplexes are nicked, and the free ends from each helix pair with the complementary regions of the other helix (a–d). The crossover is stabilized by ligating the ends on the invading strands to the free ends in the recipient helices (e). The crossover point of such a strand exchange is mobile and moves freely in either direction by a process called **branch migration** (f). Branch migration involves the simultaneous separation of strands in the original helices coupled to the reassociation of the freed strands to form new duplexes. The crossover structures shown in e and f, as well as the structure represented in g, are all referred to as **Holliday structures** after their first proponent.

The Holliday structure can be resolved into the recombinant double helices by two alternative ways to break and rejoin strands. One alternative follows from a break and reunion in the strands that crossed over. The two reciprocal products are shown in k and l and can be understood by examining the outcome of breaking and joining the crossed strands at the node in structures f and g or across the four-strand node of the isomerized Holliday structure shown in i. The length of the exchanged segment depends on the extent of the branch migration that preceded the recombination event. An exchange between both strands of the double helices, as represented by the alternate products in m and n, occurs if the nonexchanged strands in the Holliday structure, h, are broken and rejoined as indicated by j.

Because the driving force in this type of recombination is the homologous pairing of strands derived from two different DNA helices, recombination events most likely occur only at regions where the initial

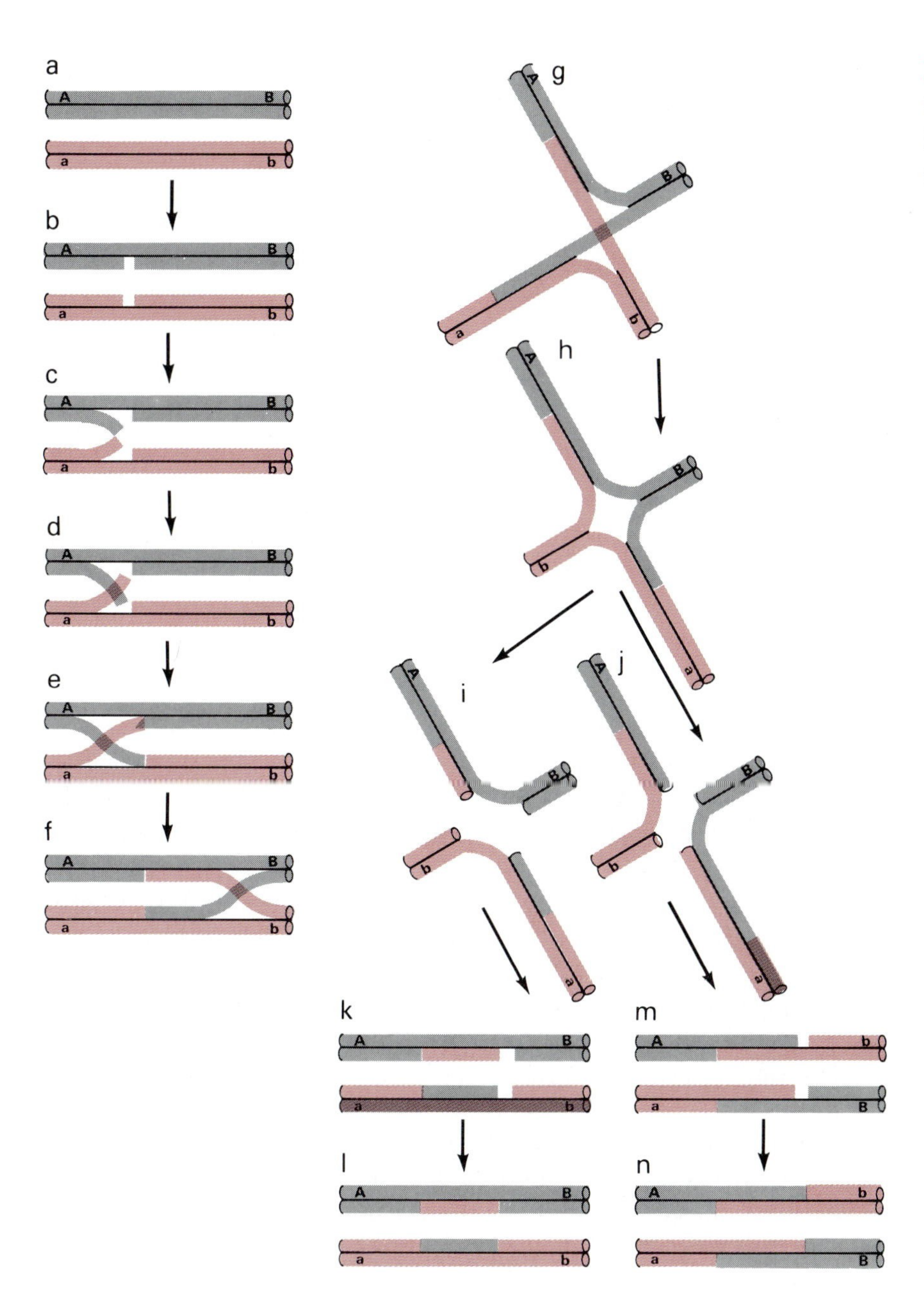

Figure 2.58
Stages in recombination between homologous duplex DNAs. See text for explanation.

base pairing is possible and where the nucleotide sequence homology is sufficient to permit branch migration to proceed within the cross-stranded structure. Accepting that constraint, we can understand why general or homologous recombination can occur between two repeated regions in the same DNA molecule or between both allelic and nonallelic occurrences of the same sequence in two different chromosomes (Figures 2.53 and 2.54).

During branch migration, the pairing of strands from the two helices forms **heteroduplexes**. Such heteroduplexes may have one or more mismatched bases within the segment between the site at which the Holliday structure was formed and the site at which crossing-over occurred (Figure 2.58). The mismatch can be removed by repair in much the same way as a modified base is removed from damaged DNA (Section 2.3b). However, because either base of the mismatched pair can be rectified, the

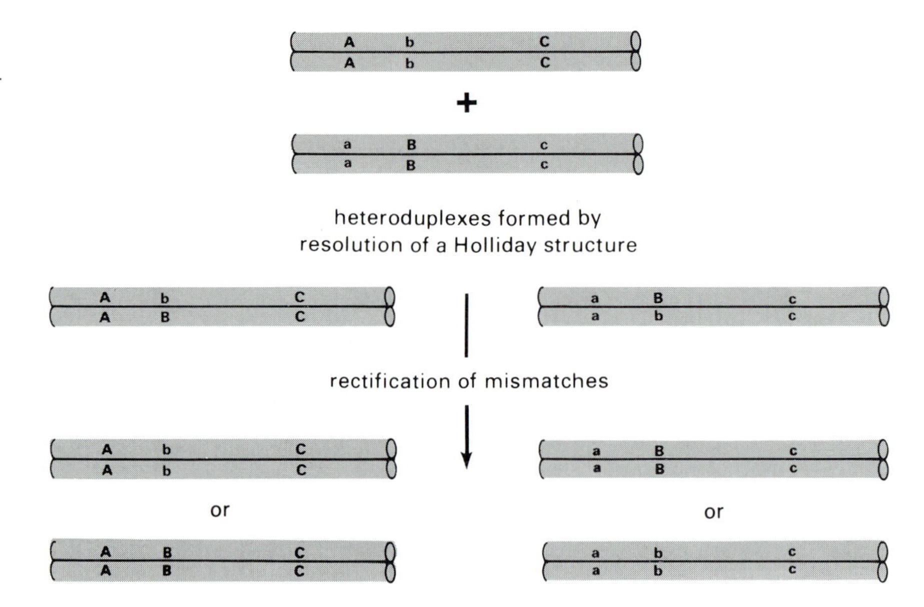

Figure 2.59
Products of mismatch repair of heteroduplexes formed during homologous recombination.

products may have the same base pair in both recombinant helices; therefore, the recombination will appear to be nonreciprocal for that site (Figure 2.59). Thus, each of the recombinant helices could resemble either of the initial duplexes at a position where they originally differed.

Double strand break model of general recombination A variant explanation for the recombination mechanism assumes that the exchange begins with a double strand break in one of the partners (Figure 2.60). The model further supposes that the action of exonucleases converts the break to a gap. Pairing of the 3′ single strand terminus at one end of the gap with its complementary strand in the intact helix creates a displacement loop. The length of the displacement loop increases as DNA polymerase extends the 3′ end of the invading strand. Eventually, the single strand terminus at the other end of the gap pairs with its complementary sequence in the displacement loop. This pairing creates a new primer-template structure, which allows DNA polymerase to synthesize a new strand across the gap. Ligation of the two growing ends to their original strands creates a double Holliday structure (i.e., a structure in which the two helices have become interconnected by two crossovers, one at each end of the gap). Branch migration at either or both crossovers moves the two interconnections in either direction, thereby creating possible mismatches between the two strands in the regions flanking the gap. Resolution of such structures can occur in two possible ways—crossover and noncrossover (Figure 2.58)—and thereby produce four possible products (Figure 2.60).

Several features of this mechanism should be noted. The creation of possible mismatches (heteroduplexes) at the regions flanking the gap provides a way to obtain reciprocal as well as nonreciprocal recombinations between genetic markers. If the double strand break occurs near or within a region where the two helices differ (base changes, deletions, insertions, inversions, etc.), the recombinants will carry the sequence information of the unbroken partner (i.e., the nucleotide sequence in one partner will

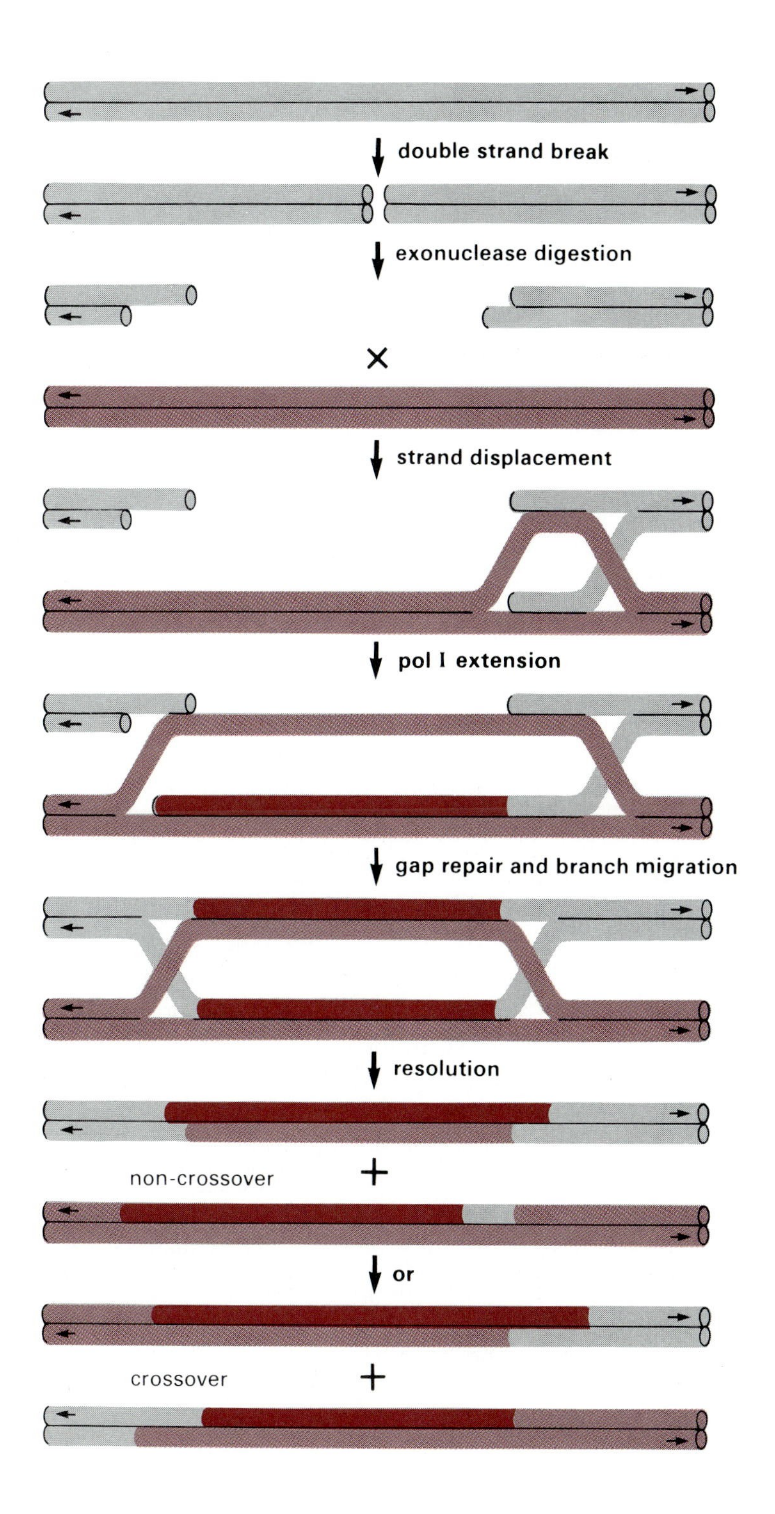

Figure 2.60
The double strand break model for general recombination.

have been transferred to that of the other partner). This mechanism explains many examples of gene conversion, particularly those in which an extended sequence in one duplex is replaced by the corresponding but different sequence in another duplex.

Nonreciprocal, general recombination also serves in repair of certain kinds of DNA damage. For example, if thymine dimers have not been excised from ultraviolet-irradiated DNA before the replication fork reaches that site, the strand complementary to that region cannot be completed (Figure 2.61). Because thymine dimers opposite a gap cannot be excised

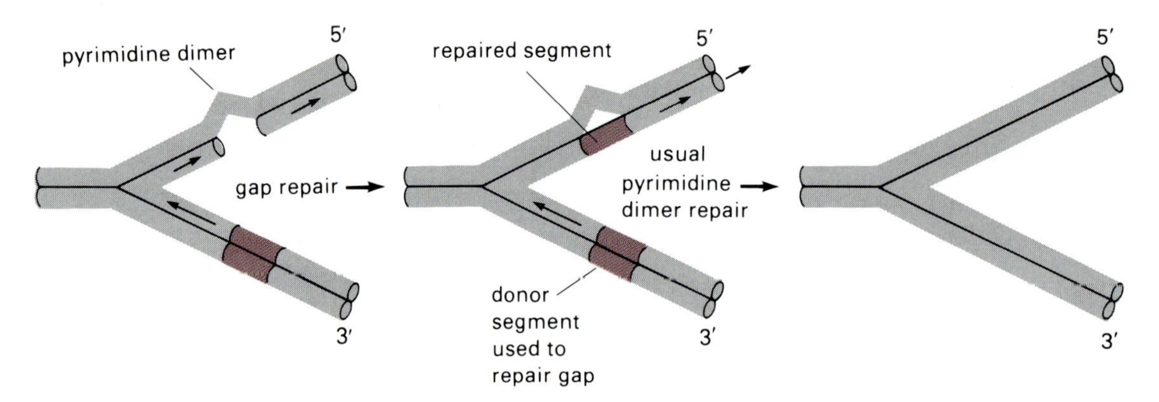

Figure 2.61
Nonreciprocal homologous recombination can foster the repair of cyclobutyl pyrimidine dimers.

and repaired, one means for salvaging that chromatid is to use the homologous information in the sister chromatid to fill in the gap. This can be accomplished by the pathway proposed for gap repair (Figure 2.60). Assimilation of an end from the gapped region can produce a displacement loop that pairs with the single strand segment of the damaged strand. This allows the usual processes of pyrimidine dimer excision and repair to proceed by any one of the pathways already described.

c. *Necessary Enzymes in General Recombination*

Two special enzymes and several others involved in DNA replication and repair are needed for general recombination. The enzymology of general recombination is known only for several prokaryotic organisms, principally *E. coli* and its phages. One of the special enzymes required for proficient homologous recombination is called the recA protein. When single strands arise (e.g., via nicks or during replication), the recA protein catalyzes the exchange of strands between such single strands and a homologous duplex at the expense of ATP hydrolysis to ADP and inorganic phosphate (Figure 2.62). The recA-mediated invasion of a duplex by single strand DNA is the first step in the recombination in both the Holliday scheme (Figure 2.58) and the double strand gap proposal (Figure 2.60). A second enzyme, containing three distinct subunits (B, C, and D) and therefore termed the recBCD nuclease, is an endo- and exonuclease as well as a helicase. Its action is not well understood, but the rec*BCD* nuclease may introduce the nicks in duplex DNA and, together with its intrinsic helicase activity and recA, initiate the strand assimilation that begins the recombination event. An enzyme that cleaves the nodes of Holliday structures has also been identified, and its action creates cohesive ends for rejoining by ligase.

Other enzymes needed in general recombination are helicases and proteins that bind to single strand DNA, both of which are needed to facilitate branch migration. Pol I is known to promote strand displacement during branch migration, and DNA ligase is essential to reestablish the continuity of the broken strands. A type I topoisomerase and probably gyrase may also be needed to manage the topological complexities of the unwinding and resolution of cross-stranded structures.

d. *Site-Specific Recombination*

Site-specific recombination occurs between specific segments of duplex DNA that do not have extensive homology. The best understood instance of this is the integration of phage λ circular DNA into the *E. coli*

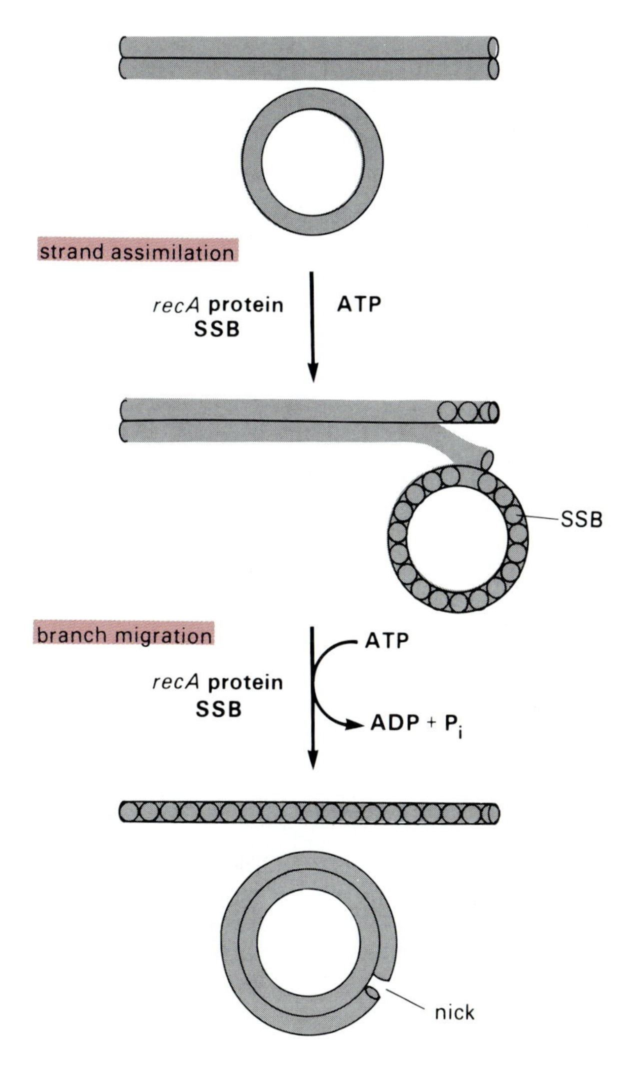

Figure 2.62
The role of the recA protein in general recombination in *E. coli.* The reactions are facilitated by the binding of single strand binding protein (SSB) to the single strands. After M. Cox and I. R. Lehman, *Proc. Natl. Acad. Sci. U.S.A.* 78 (1981), p. 3433.

chromosome and its excision from that site. Although these recombination events also involve the breaking and rejoining of two helical DNA segments, the reaction mechanism is quite different from general recombination. Recombination occurs within a specific nucleotide sequence in λ DNA (the *att*P site) and within a unique sequence in *E. coli* DNA (the *att*B site) (Figure 2.63). The nucleotide sequence of the *att*P and *att*B sites

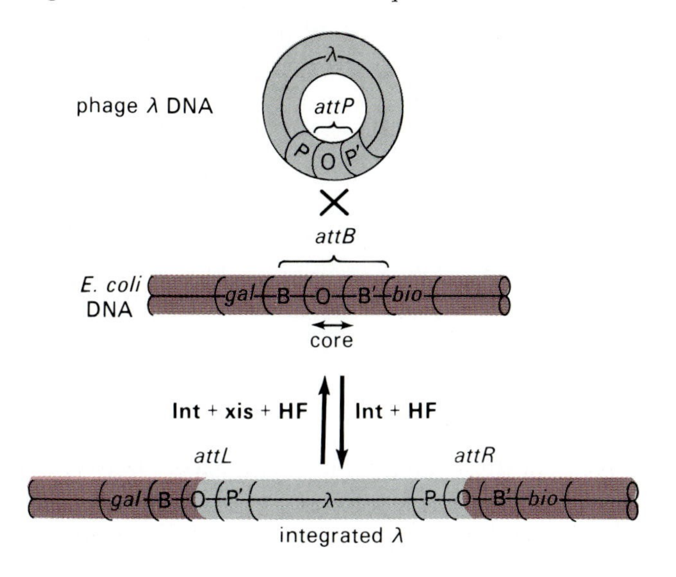

Figure 2.63
The integration and excision of circular λ phage DNA into and out of the *E. coli* chromosome. The *att*B site is between the *E. coli gal* and *bio* genes. The protein encoded by the phage *int* gene and a bacterial protein, HF, are required for integration. These two, and an additional phage encoded protein, xis, are required for excision.

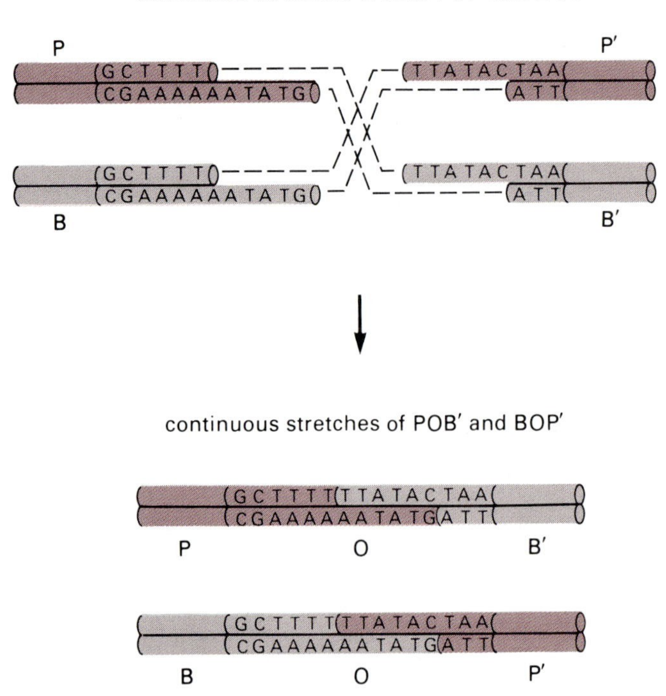

Figure 2.64
Common nucleotide sequences at *att*P and *att*B participate in site-specific recombination.

(Figure 2.64) are quite different, although they share a common core (O) of 15 nucleotide pairs. The *att*P (POP′) extends about 150 nucleotides to the left (P) and 75 nucleotides to the right (P′) of the common core; *att*B (BOB′) is only about 25 nucleotides long, including the common core. The recombination events that summarize the integration of λ DNA into the *E. coli* chromosome and its excision from the integrated state are summarized in the equation in Figure 2.63.

Because the nucleotide sequences surrounding *att*L and *att*R are different from those that flank the *att*P and *att*B sites, the mechanism of the recombinational excision of λ DNA from *E. coli* DNA must differ from the process resulting in the recombinational integration (Figure 2.63). Indeed, another phage-encoded protein, xis, and a cellular protein, HF, are needed in addition to Int for recombination between *att*L and *att*R to result in the excision of λ DNA. The recombinational excision reaction probably shares certain features with the integration reaction, but the role of the three proteins, particularly the contribution of xis to this process, is still being investigated.

2.5 RNA Replication

So far as is known, RNA replication occurs uniquely in prokaryotic and eukaryotic viruses that use RNA as their genome; all cellular RNAs are produced by DNA transcription (Chapter 3). Here we focus on how RNA is replicated and do not consider how viral RNA replication relates to the life cycles of the respective viruses.

With the exception of retrovirus RNA replication (Section 2.2), the underlying theme of RNA replication recapitulates DNA synthesis. RNA chains are assembled one nucleotide at a time, growing from the 5′ end

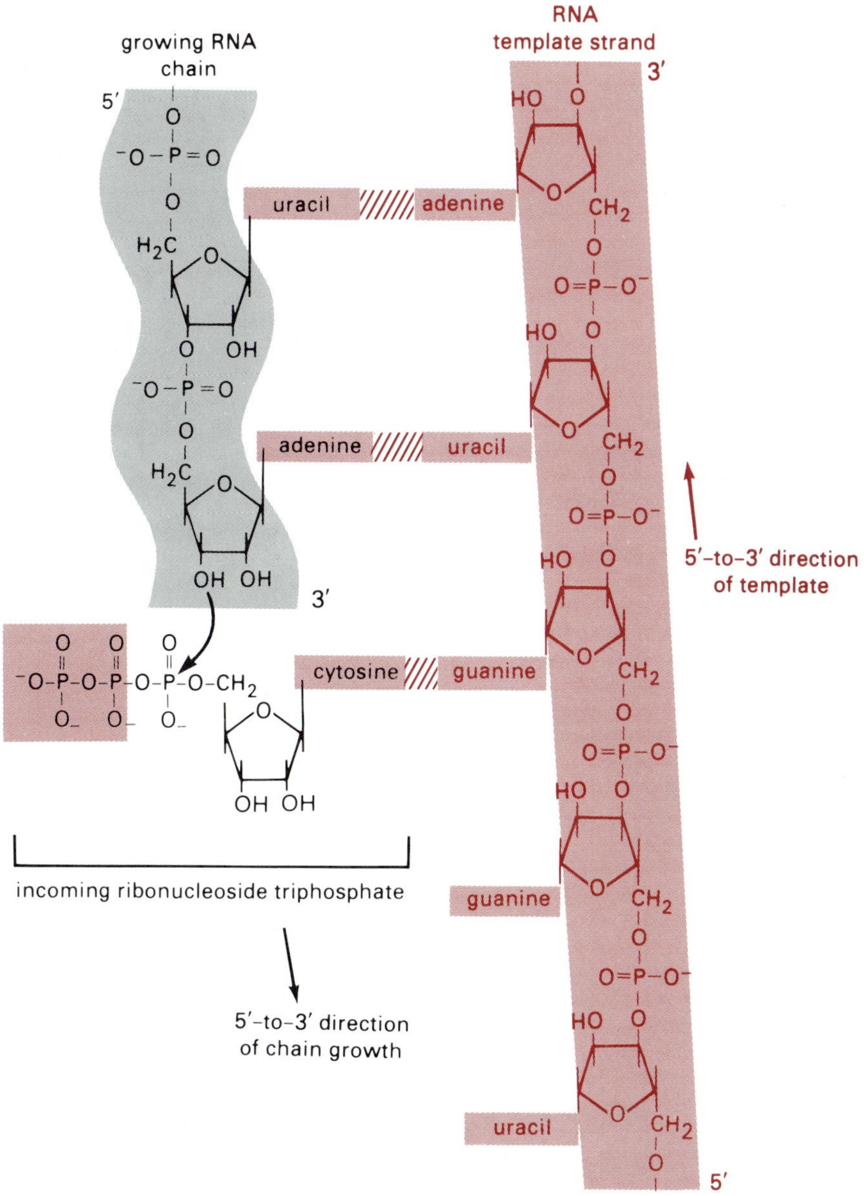

Figure 2.65
RNA-dependent RNA synthesis.

toward the 3′ end by adding the nucleotidyl moieties of ribonucleoside triphosphates to the 3′-hydroxyl end of the growing chain. As with DNA replication, the order of ribonucleotide addition is determined by complementary copying of a template—in this case, always an RNA chain (Figure 2.65). The enzymes are called RNA-dependent replicases.

The bacterial RNA viruses, R17 and MS2, and the animal viruses, polio and sindbis, are all designated (+) strand viruses because they contain an RNA genome whose nucleotide sequence is the same as that of their mRNA. Thus, the infecting genomes can serve directly as mRNA to guide the formation of some or all of the viral proteins. A specific replicase, encoded in the viral genome and produced shortly after infection, associates with one or more host cell proteins and initiates copying at the 3′ end of the (+) strand RNA, producing a complete (−) strand

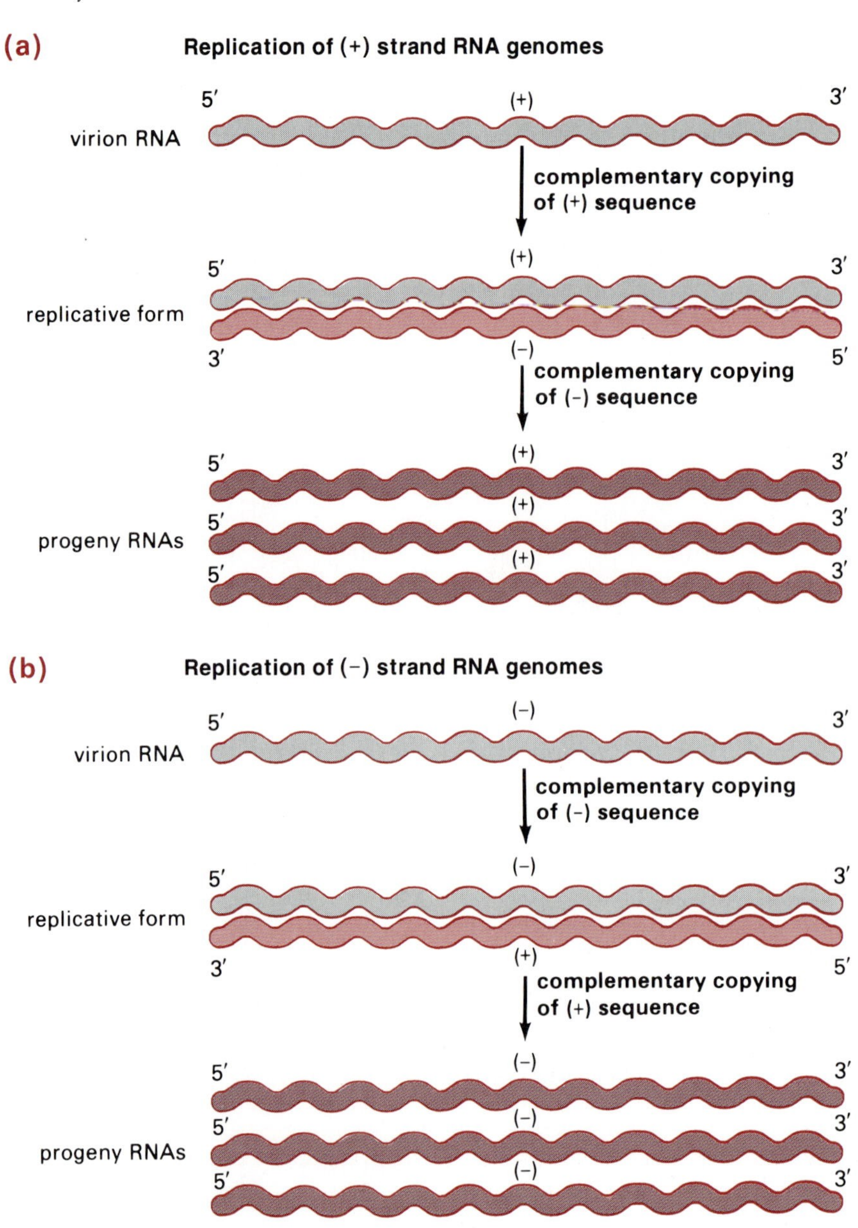

Figure 2.66
Replication of (a) (+) and (b) (−) strand viral RNA genomes by RNA-dependent RNA synthesis.

associated with the (+) strand template (Figure 2.66a). Then the same replicase, perhaps with additional proteins coded by the virus RNA, synthesizes multiple copies of (+) strand RNA using the new (−) strand as the template. Several of the new (+) RNA strands are made concurrently from a single (−) strand RNA template (Figure 2.67). With many of the (+) strand viruses, the copying of either the (+) or the (−) strand is initiated with the ribonucleoside triphosphate that pairs with the first nucleotide at the 3′ end of the template strand.

The polio virus (+) strand RNA also replicates according to the scheme outlined in Figure 2.66a, but its genomic RNA has an unusual feature—a protein linked to the 5′ end of the RNA through a phosphodiester linkage to a tyrosine residue. The (−) strand formed in the first round of the replication does not contain a protein at its 5′ end. However, all the (+) strands made using the (−) strand as the template contain the protein at their 5′ ends even when the growing (+) strands are very short. This

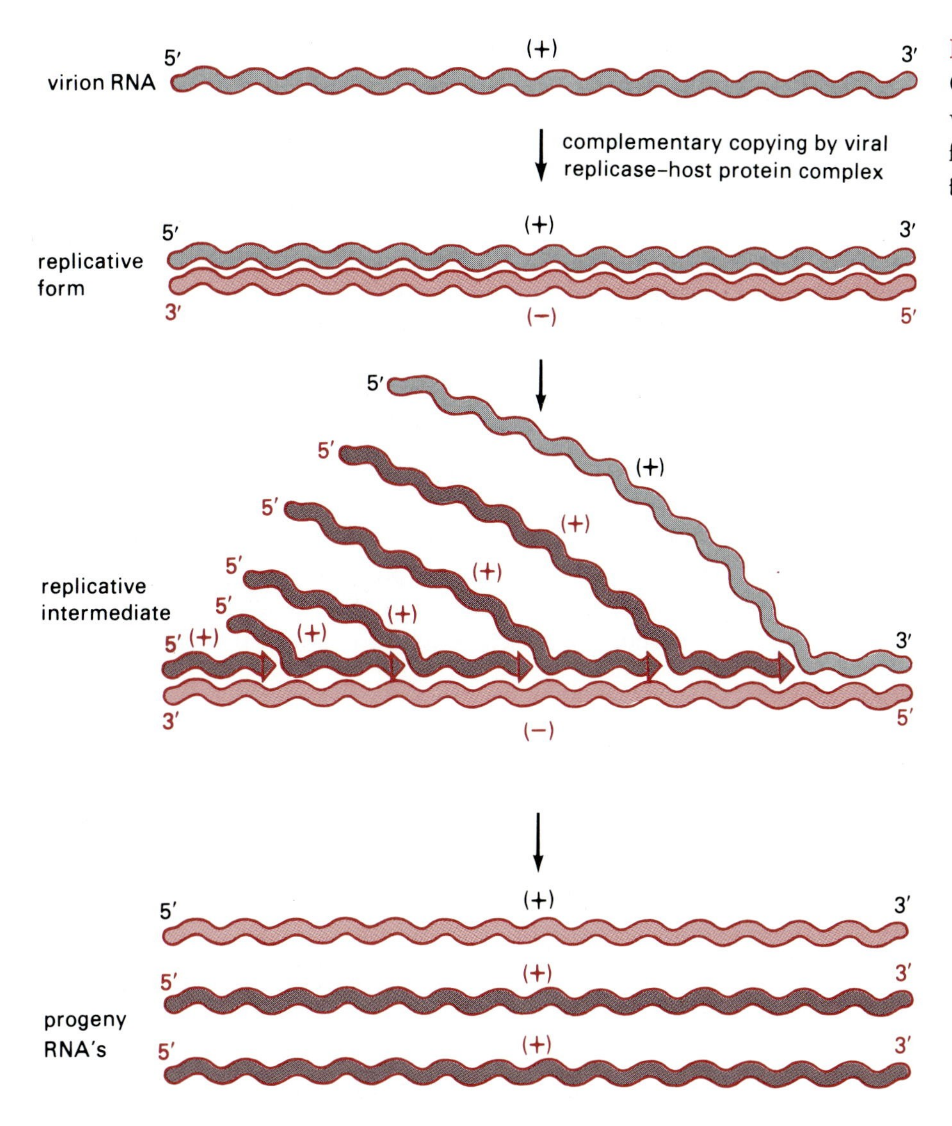

Figure 2.67
Concurrent synthesis of viral (+) RNA strands from a single (−) RNA template.

implies that (+) strand synthesis by the polio virus replicase utilizes the tyrosine hydroxyl group as a primer to initiate new chains but that the enzyme can initiate (−) strands *de novo.*

The genomes of some viruses (e.g., vesicular stomatitis) consist of a single RNA strand whose nucleotide sequence is complementary rather than identical to their mRNA. These are referred to as (−) strand genomes. There are also viruses (influenza viruses) whose genomes consist of multiple (−) strand RNA molecules, each segment corresponding to one or two viral genes. When the viral genome consists of one or more (−) strands of RNA, the virion itself contains a complex of the genomic RNA and a specific replicase. After entering the cell, the complex is activated and (+) strands are made using the (−) strands as templates (Figure 2.66b). The newly made (+) strands serve both as mRNA to produce new or additional replication proteins and as templates for the synthesis of virus progeny (−) strands. The new (−) strands are synthesized by a process akin to that shown in Figure 2.67 for (+) strand viruses.

Some viruses (e.g., reoviruses) have double strand RNA as their genome and therefore contain both (+) and (−) strands. Their replication is also initiated by enzyme(s) that are part of the virion. The virion enzymes copy the (−) strand of the RNA duplexes by a conservative process (i.e., the original double helical RNA is conserved, and the new (+) strand RNAs are released). Replicase proteins translated from these (+) strand mRNAs then produce the (−) strand complements from the accumulating (+) strands to regenerate double strand RNA progeny.

Our information about the structure and function of RNA replicases is limited. The best understood RNA replicase is produced by the RNA coliphage Qβ. It contains four nonidentical polypeptides, only one of which is encoded in the virus genome. The phage-encoded polypeptide is unable to replicate the viral (+) strands, but it can catalyze a limited reaction of ribonucleotide addition. The three proteins associated with the Qβ RNA replicase are involved in the host's machinery for protein synthesis, but their precise contribution to RNA replication is obscure. Animal virus-encoded RNA replicases also depend on proteins produced by the host cell to perform the initiation and replication reactions. A notable feature of the RNA replicases is their unusual capacity to recognize specifically the nucleotide sequences at the 3′ ends of both the (+) and (−) strands, thereby assuring correct initiation of new strands.

CHAPTER 3

The Logic and Machinery of Gene Expression

The one gene–one protein theory of gene action provided the conceptual framework for relating an organism's genotype and phenotype. However, the theory did not consider the molecular nature of genetic information or how that information would be translated from genes to proteins. Neither did it provide any insight into the mechanisms for regulating the expression of genetic information. But progress on these questions followed quickly after several key advances: (1) the identification of genes as DNA, (2) the solution of the molecular structure of DNA, (3) the recognition that the structure and function of proteins stem from their individual unique amino acid sequences, (4) the discovery that information transfer from DNA to protein is mediated by RNA, (5) the development of relatively simple bacterial genetic systems for correlating mutational alterations of genes with structural changes in the corresponding protein, and (6) the development of *in vitro* systems for studying the formation of RNA and the assembly of proteins.

DNA and its informationally related molecules, RNA and proteins, may be viewed as one-dimensional arrangements of subunits. DNA stores its information as deoxyribonucleotide pairs arrayed sequentially along the double helical chains; the same information is conveyed in RNA by the sequence of ribonucleotides along the chain. The uniqueness of a protein resides in the linear arrangement of amino acids in the polypeptide. Understanding of the informational relationship between DNA and protein emerged from genetic and biochemical analyses that linked mutations in a single gene to specific alterations in the amino acid sequence of the corresponding protein. These studies also established the existence of colinearity between the sequence of nucleotides in DNA and the sequence of amino acids in proteins. The implication of this correlation was the existence of a code—the **genetic code**—that relates the sequences of nucleotides and amino acids in the two polymers. But what is the nature of the code? Specifically, how do DNA sequences, utilizing only 4 nucleotides, specify protein sequences composed of as many as 20 amino acids?

Furthermore, what chemical processes govern the translation of the genetic code, and how are those processes regulated to create the characteristic phenotype of cells and organisms?

The nature of the genetic code is now known, and the dictionary relating nucleotide and amino acid sequences is well established. Furthermore, the basic features of the various steps in gene expression and regulation are also known, although many of the molecular details remain to be elucidated. This chapter describes the nature of the genetic dictionary and the basic mechanisms cells employ to express their genetic information. It emphasizes prokaryotic organisms but includes a brief description of comparable features in eukaryotic systems. Hypotheses and investigations at the edge of our understanding of these processes in eukaryotes are covered in Chapter 8.

3.1 A Synopsis of Gene Expression

Gene expression refers to the processes by which information encoded in DNA structure is read out into RNA and protein products. To provide an overview for the more extensive coverage and illustrations of these processes in succeeding sections, our discussion begins with a brief synopsis of the mechanism of gene expression and its regulation.

a. Transcription of DNA into RNA

The expression of all cellular genes begins with the transcription of their nucleotide sequence into RNA. In that process, a region of one of the two DNA strands is used as a template to direct the synthesis of an RNA chain by complementary base pairing. Genes that encode structural information for proteins yield mRNAs; others produce RNAs that are themselves parts of the machinery needed to translate mRNA into proteins. In prokaryotic organisms such as *E. coli*, DNA is transcribed by a single enzyme—a **DNA-dependent RNA polymerase**—to yield all the classes of RNA. In contrast, eukaryotes possess three distinctive DNA-dependent RNA polymerases, each responsible for the transcription of genes that encode different classes of cellular RNAs (Chapter 8). Although the mechanism of RNA synthesis and template copying is identical for all RNA polymerases, each enzyme recognizes unique features in the DNA template to identify specific sites for initiation, termination, and regulation of transcription.

b. Relation of Nucleotide Triplets to Amino Acids

The genetic code defines how the amino acid sequence of a polypeptide is derived from the nucleotide sequence of its corresponding mRNA. The format of the genetic code and the particular coding units that specify each amino acid (**codons**) are virtually the same in all forms of life. Moreover, the general rules and mechanisms for translation of genetically coded messages are also universal.

The genetic dictionary contains 64 codons, each comprising 3 adjacent nucleotides (a triplet) in an mRNA chain. Among the 64 codons, 61 are used to specify the 20 amino acids found in proteins, 1 specifies the beginning of most protein coding sequences, and 3 provide signals for the end of a coding sequence.

A significant feature of the genetic code is that no codon specifies more than one amino acid—that is, the code is not ambiguous. Consequently, knowing the dictionary and the rules for its use, we can translate the entire nucleotide sequence of an mRNA into an unambiguous amino acid sequence. The genetic code is also "degenerate," which means that several codons can specify the same amino acid. The existence of degeneracy in the genetic code makes it impossible to define a unique nucleotide sequence from a protein's amino acid sequence.

c. Recognition of Codes by tRNAs

Amino acids do not interact directly with their corresponding codons. Instead, each amino acid is first attached to an adapter—a **cognate tRNA**—and the resulting aminoacyl-tRNA recognizes its respective codon by base pairing. Thus, the actual decoding process is accomplished by complementary base pairing between codon triplets in mRNA and **anticodon** triplets in aminoacyl-tRNAs.

Enzymes called **aminoacyl-tRNA synthetases** catalyze the attachment of amino acids via their carboxyl groups to their cognate tRNAs. As a result of linking the amino acids to tRNA, the carboxyl group of the amino acid is chemically activated; this makes the formation of peptide bonds an energetically favorable reaction. The energy for amino acid activation via its attachment to tRNA is provided by the cleavage of a pyrophosphoryl group of ATP.

A different enzyme is needed to join each amino acid to its cognate tRNA. For example, tyrosyl-tRNA synthetase joins L-tyrosine only to those tRNA chains that can base pair with tyrosine codons. Similarly, leucyl-tRNA synthetase attaches leucine to a family of tRNAs that recognize the leucine codons. Thus, the specificity of the decoding process is set at two stages: accurate joining of each amino acid to its cognate tRNA and complementary base pairing of the aminoacyl-tRNAs to their corresponding codons in mRNA.

d. Proper Initiation of Translation

There are three possible "reading frames" by which successive nucleotide triplets in an mRNA can be decoded into amino acids. Proper initiation of translation is essential for correct readout of the genetic code. The choice of reading frame depends on which set of three contiguous nucleotides is chosen as the first codon. Examples of each of the three possible reading frames for the nucleotide sequence GUACGUAAGUAAGUAUGGACGUA are shown below.

```
Reading frame 1        GUA CGU AAG UAA GUA UGG ACG . . .
Reading frame 2  . . G UAC GUA AGU AAG UAU GGA CGU . . .
Reading frame 3  . GU  ACG UAA GUA AGU AUG GAC GUA . . .
```

Generally, only one of the frames will correspond to the amino acid sequence of the encoded polypeptide chain. Therefore, a way is needed to initiate translation in the correct reading frame. In all organisms studied so far—bacteria, viruses, and eukaryotes—the correct reading frame is set by a mechanism that recognizes a specifically located codon that defines the amino terminal amino acid of the nascent protein. Almost invariably, the initiator codon is AUG, the triplet that specifies methionine. Therefore, nascent polypeptides always have a methionine at the amino terminus, but subsequent removal of the amino terminal end of the protein usually leaves an initially internal amino acid as the amino terminus in the final protein product. Reading frame 3 in the example contains an AUG codon that could initiate translation.

e. Codon Translation and Amino Acid Assembly

The matching of aminoacyl-tRNAs to successive codons in the mRNA and the accompanying stepwise assembly of polypeptide chains involve a complex series of coordinated and interdependent reactions. One of the principal participants in this highly orchestrated process is the **ribosome**—a multienzyme particulate complex of several RNAs and many proteins. In addition, a battery of enzymes and factors catalyzes the myriad chemical events needed to effect protein synthesis.

Ribosomes carrying the special initiator methionyl-tRNA locate and bind to the AUG initiator codon of an mRNA. Then the aminoacyl-tRNA corresponding to the second codon binds to the ribosome, and a ribosome-associated enzyme activity links the methionyl group to the second amino acid still bound to its tRNA. In the process, a dipeptidyl-tRNA is formed. As the ribosome moves along the mRNA chain encountering each successive codon, the polypeptide chain is lengthened by one amino acid at a step. Elongation ceases when the ribosome reaches any one of the three termination codons. Concomitantly, the completed polypeptide chain is released from the last tRNA, and the ribosome dissociates from the mRNA.

f. Regulation of Gene Expression at Various Stages of RNA and Protein Formation

Both prokaryotic and eukaryotic cells possess the ability to regulate gene expression differentially. Thus, under a particular set of conditions, many genes may not be expressed at all, while the rates of expression of others may differ by orders of magnitude. A change in conditions can trigger the activation of previously silent genes and the repression of active genes. This capability provides cells with a broad ranging flexibility to adjust their phenotypes in response to different environmental and physiological stimuli. Differential expression of a constant genome also accounts for the ability of multicellular eukaryotes to develop myriad cell types with various specialized functions from one or a few progenitor cells.

Gene expression is most frequently regulated at the level of RNA production. Generally, transcription initiation is the regulated event, the control being mediated either by **repressor** proteins, which prevent transciption, or by **activator** proteins, which are needed to start it. In the former case, transcription begins only after the repressor protein is inactivated. In the latter case, the gene is transcribed only if the activator protein is in the proper functional state. Repressor and activator proteins are not the sole means for regulating gene transcription. In some instances, the protein product of a gene's expression is itself the regulator of its own gene's transcription. The conformational state of the DNA or RNA may also affect the efficiency of a gene's transcription. In addition, RNA production can be regulated by controlling the rate of its elongation or by determining whether transcription reads through the entire gene or is terminated at specific "stop-go" signals within the transcription unit. When modification and/or processing of the primary transcripts are needed to produce mature functional RNA, such steps also provide sites for regulation.

Gene expression can also be regulated during the translation of mRNA into protein. Here, too, specific regulation is most often achieved in the initiation step of the decoding process. However, control can also occur at several steps in polypeptide chain assembly. Furthermore, the synthesis of proteins that require posttranslational modifications or transport to specific intracellular locations may be regulated at each of these stages.

Later, as we examine these processes in more detail, we shall see that the mechanisms for regulating gene expression are diverse, numerous, and often complex. Although common themes unify many of them, the precise mechanisms are frequently unique to a gene, the physiological state of the organism, and the environment. The analysis of regulatory mechanisms in bacterial systems has provided significant insights and identified a broad range of possibilities for regulating and coordinating gene expression. But the nature of the controls that govern gene expression in eukaryotic cells is only beginning to be unraveled, and the mechanisms that guide the differentiation of multicellular organisms remain obscure.

3.2 Transcription: The Transfer of DNA Sequence Information to RNA

In this and the next section, we describe several aspects of the transcription of DNA sequence information into RNA, a process responsible for the synthesis of all species of cellular RNA in both prokaryotic and eukaryotic organisms. Much of the early work defining the properties of transcription—the nature of the reaction and its substrates, the characterization of the enzymatic machinery, the definition of the nucleotide sequence signals that specify which DNA regions will be transcribed, and some of the processing pathways that convert **primary transcripts** into mature RNAs—was carried out with prokaryotic systems. Here, the interplay of genetics and biochemistry has been the key to analyzing the transcription enzymes and mechanisms. Concurrent efforts to study the transcription and regulatory machinery of eukaryotes progressed more slowly and were less successful, mainly because the components of the transcription machinery—DNA in the form of chromatin and RNA polymerases—were ill defined. Moreover, the nature of the transcription units was unknown, and genetic approaches were not available for defining their structure.

Since the advent of molecular cloning, the situation has changed remarkably. Now, many genes constituting different types of transcription units have been isolated, sequenced, and even modified in directed ways to explore their function. Moreover, a new view of the transcriptional apparatus of a wide range of organisms—from yeast to humans—has been provided through several approaches. These include methods for introducing DNA molecules into cultured mammalian cells (Chapter 5) and even into whole animals, as well as the more traditional studies *in vitro* with purified transcription systems. Simultaneously, important progress has been made in understanding chromatin structure and in defining the nature of eukaryotic RNA polymerases. In this chapter, we will make only brief references to eukaryotic transcription systems and their signals, primarily to contrast them with the prokaryotic systems. Gene structure, transcription, and the closely related posttranscriptional events that account for RNA formation in eukaryotic species are covered in Chapter 8.

a. Copying DNA Sequences into RNA

Double strand DNA is the physiologic template for all cellular RNA synthesis. Even where single strand DNA is the natural form of a genome, as in certain viruses, conversion to double strand DNA precedes transcription. Although either of the two DNA strands in a genome may be transcribed, only one of the strands of a gene usually serves as template for the transcription (Figure 3.1). However, in some instances (e.g., ϕX174

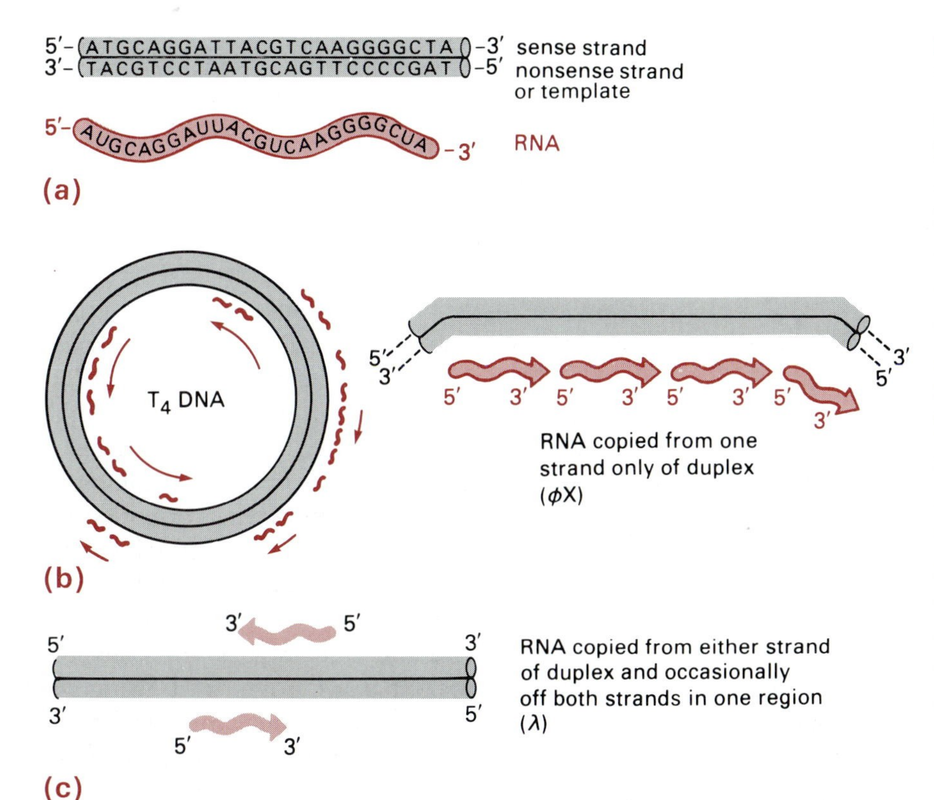

Figure 3.1
The relationship between RNA transcripts and their DNA templates. (a) RNA transcripts are complementary and antiparallel to the template strand and have the same nucleotide sequence (except for U in place of T) as the sense strand. (b) Individual genes are generally encoded on one strand of a duplex genome, although either strand may be the template for different genes (as in bacteriophage T_4). Thus, transcription, which is always 5′-to-3′, goes in either direction. Occasionally, all the transcripts are made off one strand of the genome (as in bacteriophage ϕX174). (c) In rare instances, where there are overlapping genes, both strands in the same region may be templates for RNAs transcribed in opposite directions.

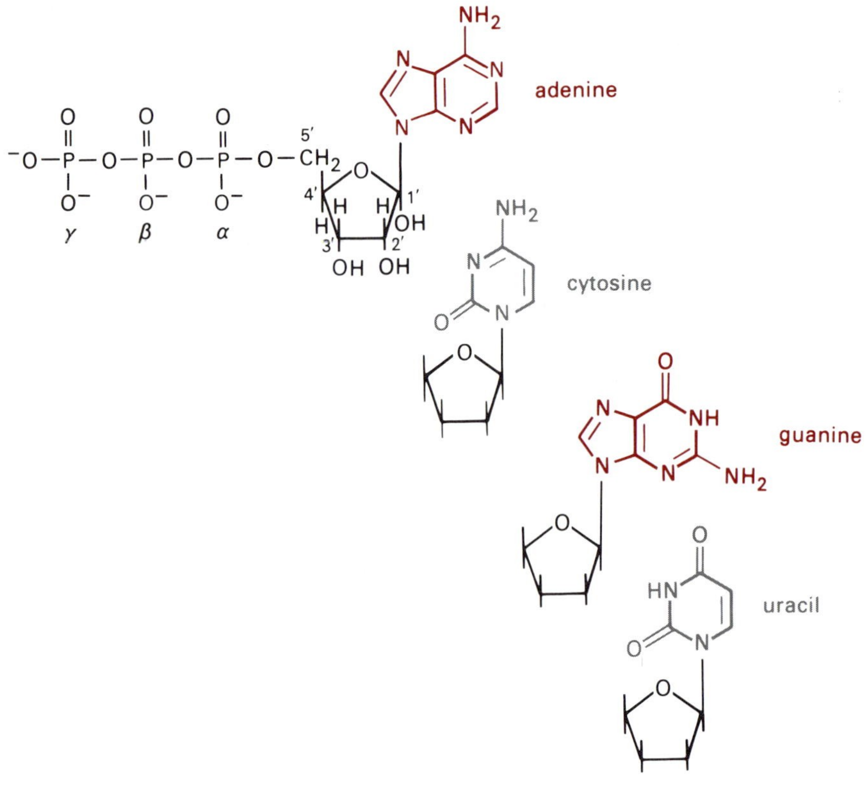

Figure 3.2
The four common ribonucleoside 5′-triphosphates: ATP, CTP, GTP, and UTP. The ribose triphosphate portion and the linkages are identical in all four.

bacteriophage), all the mRNA is transcribed from the same strand. In rare circumstances, transcription occurs from both strands in the same region, producing RNA strands that are complementary to each other; such transcriptional behavior may have special regulatory significance.

The nucleotidyl precursors for RNA synthesis are the four ribonucleoside 5′-triphosphates, ATP, GTP, UTP, and CTP (Figure 3.2). Even though many RNAs contain modified nucleotides, the alterations of the bases and ribose residues are introduced after the nucleotides have been polymerized—that is, posttranscriptionally. Nevertheless, RNA polymerases can utilize ribonucleoside 5′-triphosphates other than the four standard ones, provided that they have base pairing capability comparable to adenine, guanine, cytosine, and uracil.

RNA polymerases catalyze a reaction in which the 3′-hydroxyl group of the nucleotide at the growing end of the chain is joined to the α-phosphate of the incoming ribonucleoside 5′-triphosphate (Figure 3.3). Repetitions of this reaction result in stepwise elongation of the RNA chain. The formation of each new phosphodiester bond is accompanied by the release of inorganic pyrophosphate; rapid hydrolysis of the pyrophosphate to inorganic phosphate *in vivo* makes the formation of the phosphodiester bond energetically favorable.

Transcription mimics replication in its requirement for DNA as the template (Figure 3.3). Complementary base pairing is the guiding principle in determining the order of ribonucleotide assembly. During transcription, the DNA helix must be unwound by RNA polymerase to permit complementary matching of each successive nucleoside triphosphate to the template bases being transcribed (Figure 3.4). The growing RNA chain remains bound to the enzyme and base paired to the template strand for

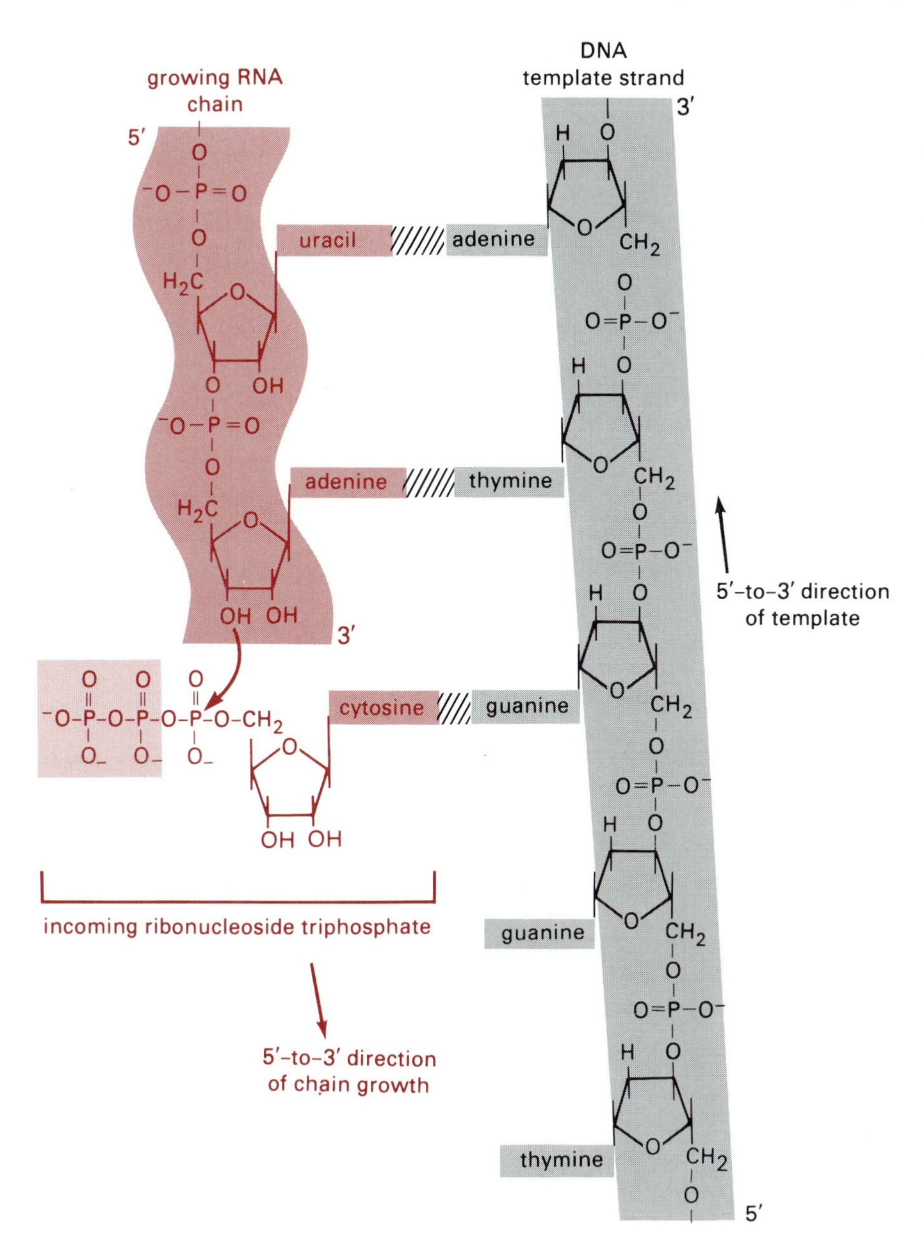

Figure 3.3
RNA chains grow one ribonucleotide residue at a time in the 5′-to-3′ direction. Complementary base pairing with the DNA template strand determines the order of ribonucleotide assembly. With each addition to the chain, one molecule of inorganic pyrophosphate is released. The reaction is catalyzed by RNA polymerases.

about 20 to 30 nucleotides at the growing end; the remainder of the nascent chain appears not to be associated with the enzyme or the DNA. As the enzyme continues transcription, the transiently separated DNA strands reassociate to reform the original duplex structure. Thus, transcription is a conservative process in that the DNA double helix is preserved, and the RNA chain is dissociated. By contrast, DNA replication is semiconservative because the two strands of the original helix are separated, each ending up in one of the two daughter helices (Figure 2.2). Another significant difference between DNA replication and transcription is that replication cannot occur without a primer chain (Section 2.1), but RNA polymerase can initiate RNA chains *de novo*, with the ribonucleoside triphosphate corresponding to the first nucleotide in the RNA chain.

RNA chains are extended in the 5′-to-3′ direction along a template strand oriented in the 3′-to-5′ or antiparallel sense (Figure 3.3). Although elongation is processive—that is, the enzyme does not dissociate from

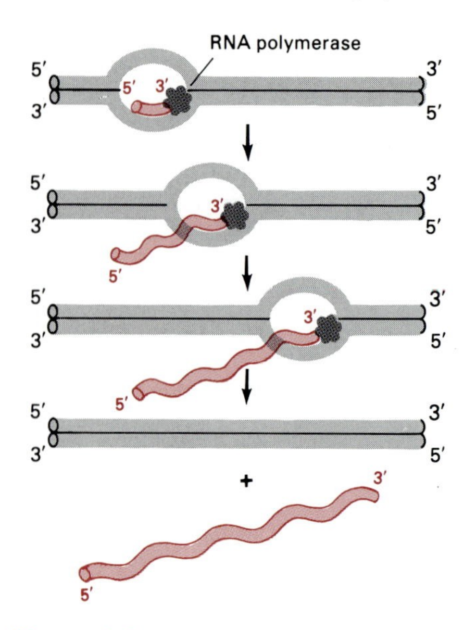

Figure 3.4
RNA polymerase unwinds the DNA duplex in the region of active transcription. As transcription proceeds, the enzyme and the unwound "bubble" move along the DNA while the RNA chain elongates.

the template during a round of transcription—the rate of elongation is not constant along the template. Indeed, the enzyme pauses at certain locations—possibly those in which the single strand DNA or the RNA itself forms intrastrand duplexes that impede polymerase progress. In fact, such pauses may, under certain circumstances, cause premature termination of transcription. We shall see shortly that specialized structures of looped back RNA are critical for normal termination and release of the completed RNA and polymerase from the template.

It is still unclear how RNA polymerase moves unidirectionally along the DNA template. Nor do we understand how the DNA duplex is unwound and rewound during the transcription process or what favors reformation of the DNA duplex rather than retention of the transient DNA-RNA duplex. However, because RNA polymerase alone performs all these functions *in vitro*, even with covalently closed circular DNA templates, these must be intrinsic attributes of the enzyme. By contrast, recall that DNA polymerases cannot initiate new DNA chains *de novo* and that helicases and topoisomerases are needed to unwind and rewind the duplexes during double strand DNA replication.

b. DNA-Dependent RNA Polymerases

In prokaryotes, the synthesis of all RNA species—mRNAs, rRNAs, and tRNAs, as well as more specialized RNAs that function in RNA processing (Section 3.3c)—is catalyzed by a single DNA-dependent RNA polymerase. (RNA primers for DNA synthesis are made by special RNA polymerases called primases; Section 2.1g.) Bacterial RNA polymerases are complex proteins containing several different kinds of subunits. The best studied enzyme, the *E. coli* **RNA polymerase holoenzyme**, contains five different polypeptide subunits (Table 3.1): two alpha chains, one beta and one beta-prime chain, a sigma and an omega chain ($\alpha_2\beta\beta'\sigma\omega$). An alternate and coexisting form, referred to as the **RNA polymerase core**, lacks the sigma subunit.

Studies with drugs that inhibit RNA polymerase and with enzymes whose individual subunits have been altered by mutation have shed

Table 3.1 Subunits of *E. coli* RNA Polymerase

Subunit	No. of Copies per Holoenzyme	Molecular Weight (Daltons)	Function
Alpha (α)	2	36,500	Not known
Beta (β)	1	150,000	Ribonucleoside triphosphate binding and RNA polymerization
Beta prime (β')	1	155,000	DNA binding
Sigma (σ)	1	70–80,000	Promoter recognition and initiation
Omega (ω)	1	11,000	Not known

some light on the role of the core subunits. The beta subunit is most likely involved in binding ribonucleoside triphosphates for both the initiation and elongation reactions. A complex of the alpha and beta-prime subunits ($\alpha_2\beta'$) is involved in nonspecific strong binding to DNA and in the specific interaction that occurs between the holoenzyme and the **promoters**—the sites that determine where transcription is initiated (Section 3.2c). However, of all the subunits in the holoenzyme, the role of the sigma subunit is best understood.

The core enzyme, which lacks sigma, catalyzes most of the reactions needed to transcribe DNA into RNA—that is, complementary copying of the template strand, phosphodiester bond formation, and RNA chain termination. But the core enzyme does not initiate RNA chains at the proper locations because it cannot recognize promoter sites (Section 3.2c). Accurate binding and initiation at promoters occur only after the core enzyme acquires the sigma subunit to form the holoenzyme (Figure 3.5). One explanation for this behavior is that the core enzyme binds very tightly but nonspecifically to DNA and therefore rarely locates the promoter. By contrast, the holoenzyme binds to nonspecific regions of DNA with a relatively low affinity and finds the promoter by a process involving successive binding and dissociation; when the holoenzyme encounters the promoter, it remains strongly bound at that sequence. Following the formation of the first few phosphodiester bonds (perhaps 5-to-10), the sigma subunit dissociates from the initiation complex, and further transcription is mediated by the core complex. Transcription proceeds processively until the enzyme reaches the transcription termination signal. Thus, sigma confers efficient promoter binding properties on the RNA polymerase holoenzyme, and its dissociation transforms the polymerase into an effective catalyst for the chain elongation phase of transcription. Meanwhile, sigma can promote initiation by specific binding to another molecule of RNA polymerase.

There is evidence that the promoter recognition properties of RNA polymerase can be altered by association with different sigma subunits

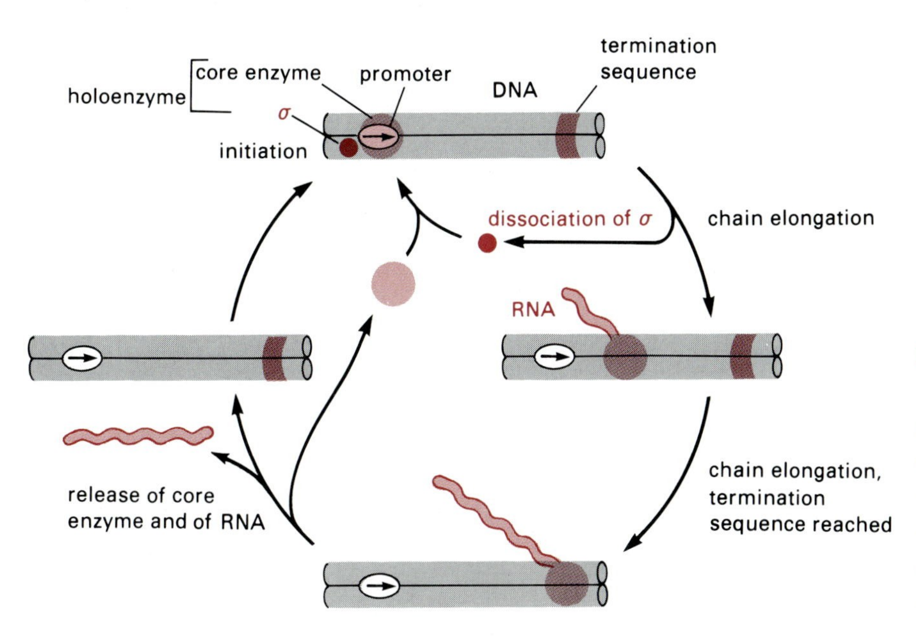

Figure 3.5
The sigma subunit cycle. RNA polymerase core enzyme interacts with a sigma (σ) subunit to form the holoenzyme that binds tightly to a promoter. After transcription is under way, sigma dissociates, and core enzyme catalyzes chain elongation.

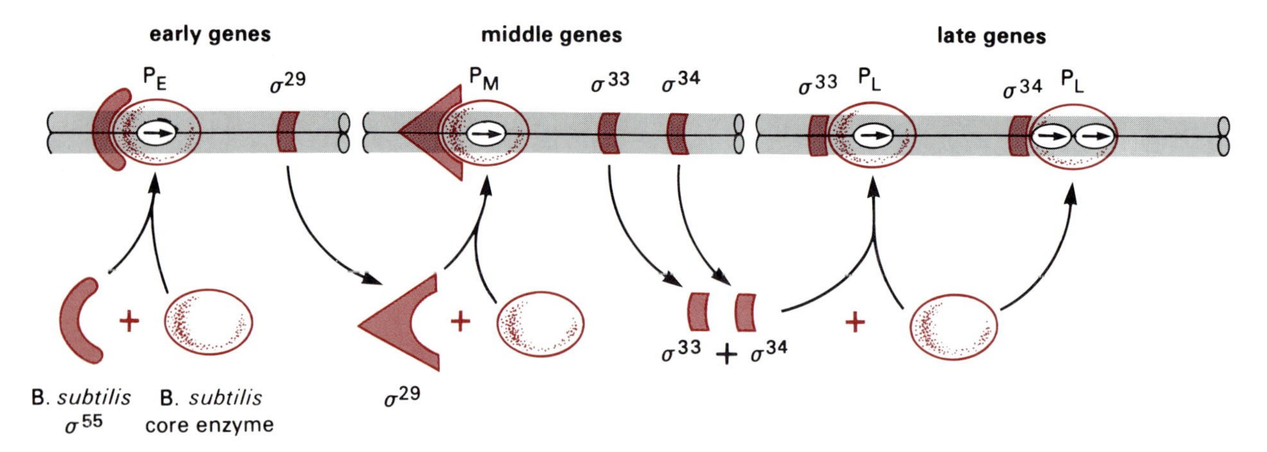

Figure 3.6
Simplified schematic diagram indicating the use of alternate sigma subunits during sporulation in *B. subtilis.* Several sigmas, including σ^{55}, occur in vegetative cells. Under conditions that trigger sporulation (e.g., starvation), several new sigmas such as σ^{29} are expressed and associate with the core RNA polymerase. The new holoenzyme binds to specific sporulation gene promoters, here labeled "middle" genes. Other specific sigmas are then expressed (e.g., σ^{33} and σ^{34}), and these, in association with the core polymerase, initiate transcription of genes whose products are involved in the late sporulation stages.

(Figure 3.6). Thus, after infection of *Bacillis subtilis* with certain bacteriophages, or in the early stages of sporulation, different sigma subunits may be expressed and thereby alter the transcription pattern of cellular and viral genes. Whether RNA polymerases in other prokaryotic organisms also utilize alternate sigma subunits to regulate their promoter specificity is still not certain.

Not all prokaryotic RNA polymerases are multisubunit enzymes. The RNA polymerases encoded by the *E. coli* bacteriophages T3 and T7 are moderately sized single polypetide chains ($\sim$100 kDaltons). These enzymes are highly specific in their ability to recognize and bind to specific promoter sites that are used for the transcription of particular sets of viral genes. These "simpler" enzymes contain all the activities of the multisubunit RNA polymerases. They catalyze DNA-directed RNA chain synthesis and terminate RNA chains properly.

c. Transcription Initiates at Characteristic Nucleotide Sequences

Transcription is initiated by the formation of a stable complex between the RNA polymerase holoenzyme and a characteristic control sequence, called the promoter, that occurs at the beginning of all transcription units (Figure 3.7). Nucleotide sequence and mutational analyses of more than 50 different prokaryotic promoter sites have revealed only two regions that are consistently similar, presumably because they are crucial for promoter recognition and function. One of the important sequences contains 6 or 7 base pairs and occurs about 10 base pairs upstream (i.e., 5′)

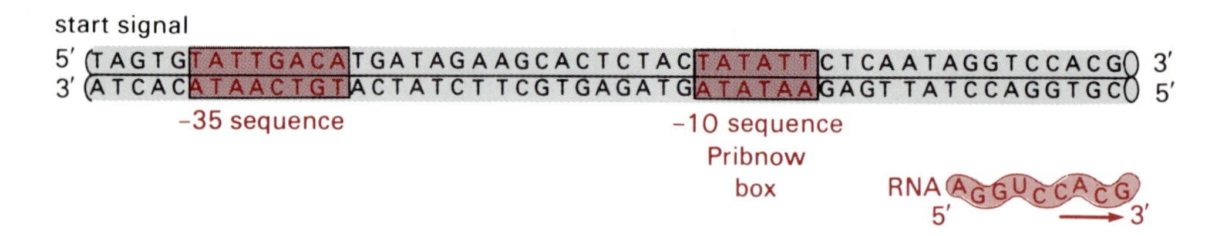

Figure 3.7
A typical *E. coli* promoter showing the -10 sequence (or Pribnow box), the -35 sequence, and the transcription initiation site.

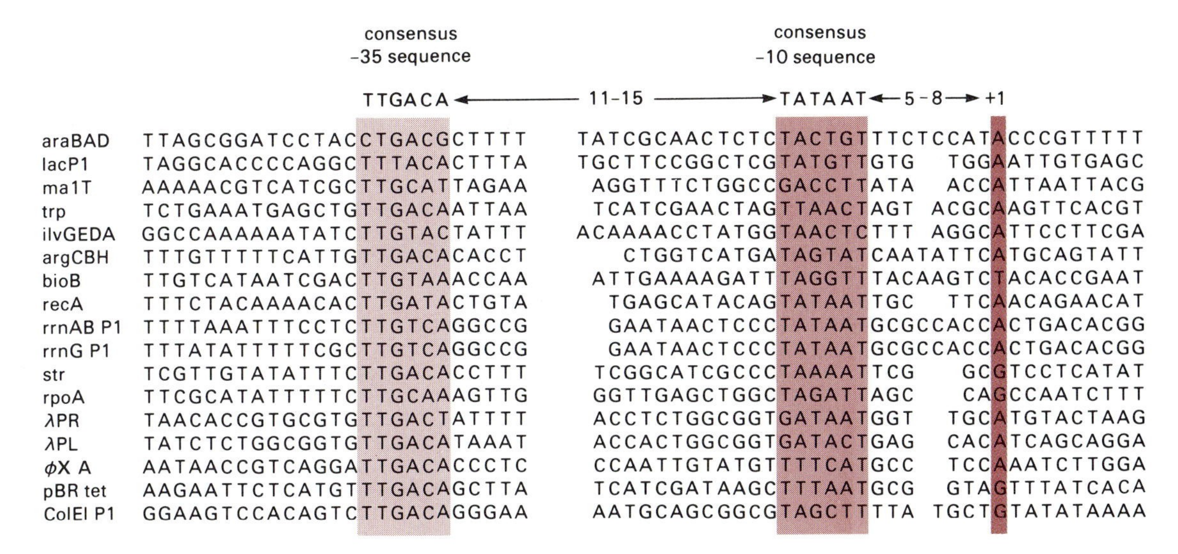

Figure 3.8
The nucleotide sequences on the sense strand of DNA upstream of the transcription initiation site of various *E. coli* and bacteriophage genes. Data taken from D. K. Hawley and W. R. McClure, *Nucleic Acid Research* 11 (1983), p. 2237.

of the nucleotide at which transcription starts (+1); this signal is commonly referred to as the −10 sequence or the **Pribnow box**, after its discoverer. Comparing the sequence of the −10 regions of nearly 50 prokaryotic promoters indicates that each varies somewhat from a consensus sequence of TATAAT (Figure 3.8). The underlined T is virtually invariant in all promoters, but one or a few variations occur in the surrounding nucleotides of any particular promoter.

A second region, typically 9 nucleotides long and clustered about 35 nucleotides upstream of the RNA initiation site (the −35 sequence), also occurs in most prokaryotic promoters. The nucleotide sequence between the −35 and −10 sequences is not critical, although the distance between the two sequences is important. The −35 sequence has been implicated in the binding of the RNA polymerase, which precedes the positioning of the enzyme at the Pribnow box. By inference, the RNA polymerase opens a short stretch of helix, probably beginning at the Pribnow box, to permit RNA chain initiation to occur (Figure 3.9).

It is not clear whether binding of the RNA polymerase at the promoter is sufficient to induce opening of the DNA duplex in the neighborhood of the RNA initiation site or if RNA polymerase itself unwinds the helix at the start site. Whatever the mechanism, formation of the open promoter complex permits RNA polymerase to mediate the base pairing of the first and second ribonucleoside triphosphates to the template strand and to catalyze the synthesis of the first phosphodiester bond.

Differences in the transcription efficiency of particular genes are in part a consequence of the structure of their promoters. Both the strength of the RNA polymerase interaction with the promoter sequence and the efficiency of forming the open promoter complex are conditioned by the particular nucleotide sequences in the −35 and −10 regions, respectively. Mutational changes in the −35 and −10 regions of many promoters produce striking alterations in their ability to promote transcription initiation (Figure 3.10). In certain instances, transcription initiation at intrinsically weak promoters is facilitated by auxilliary proteins that bind adjacent to the −35 region (Section 3.11c). The mechanism of such facilita-

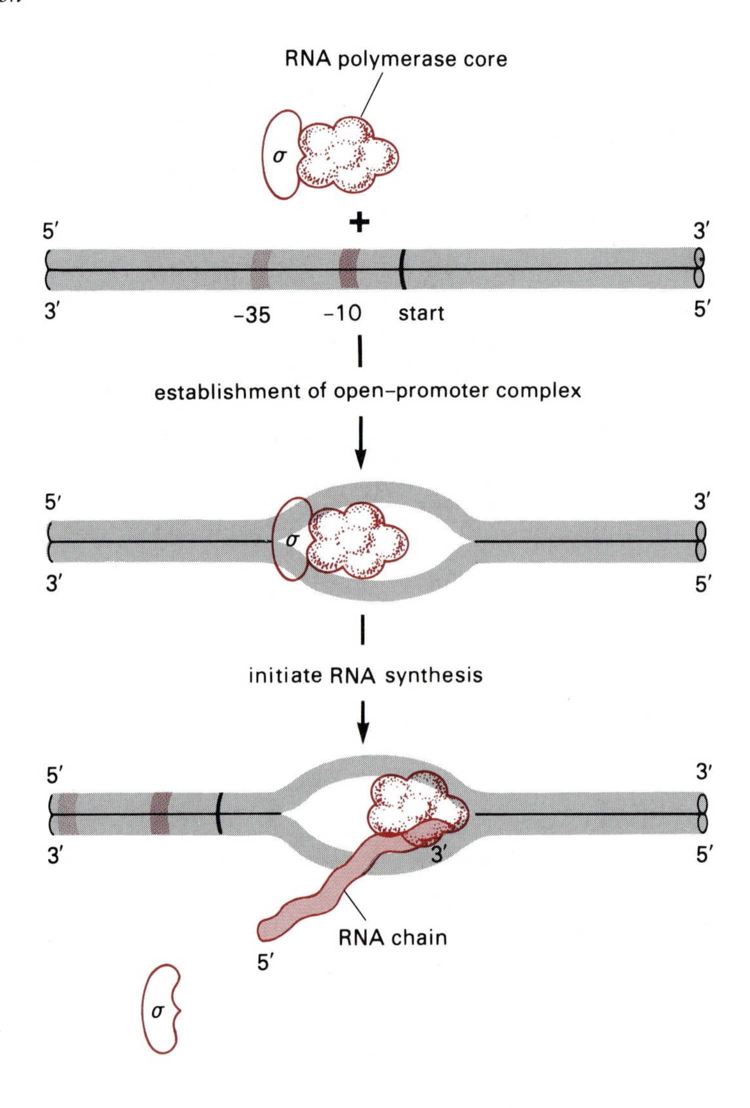

Figure 3.9
Binding of RNA polymerase holoenzyme (shown schematically) to the promoter region. The protein is thought to cover a region from slightly upstream of the −35 sequence to just within the transcribed region, a total of about 50 bases. Binding leads to unwinding (or "opening") of the helix and transcription initiation.

tion is not known in all cases, but in certain instances, it appears that the proteins bound adjacent to the −35 region increase the probability that RNA polymerase will encounter, or be bound to, the −35 sequence. Binding of the activator proteins may also alter the DNA structure and facilitate transcription initiation per se. Alterations in DNA's topological structure—particularly an increase in the extent of negative supercoiling—can also increase or decrease the efficiency of some promoters.

d. Termination and Release of RNA Chains

The DNA sequences that serve as signals to stop transcription are called **transcription terminators**. Two kinds of termination signals have been discovered, **rho-dependent** and **rho-independent** terminators. Rho-dependent and rho-independent terminators have some common characteristics (Figure 3.11). Both contain inverted repeat sequences that cause the 3′ ends of RNA transcripts to fold into **stem-loop** structures of variable lengths. The stems of rho-independent terminators are generally rich in adjacent G·C base pairs, particularly near the base of the stem, and are followed by a run of four to six uridylate and one or two adenylate residues. By contrast, rho-dependent terminators have few, if any, stretches

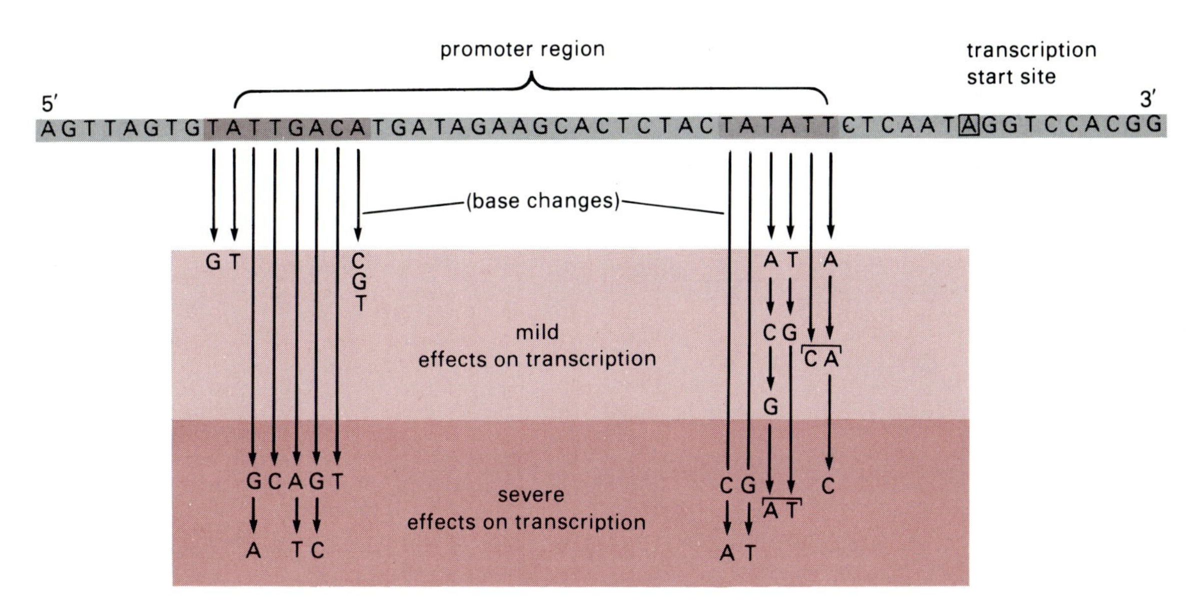

Figure 3.10
Effects on transcription of mutational changes in the −10 and −35 regions. The experiments utilized a promoter in the bacteriophage P22 genome. All mutations shown were "down mutations," that is, they decreased transcription. In several of the mutants, two bases were changed, as indicated by brackets. After P. Youderian, S. Bouvier, and M. M. Suskind, *Cell* 30 (1982), p. 843.

of neighboring G·C base pairs in their stems, and they may or may not have a run of uridylate residues at the 3′ ends.

The precise mechanism for rho-independent and rho-dependent transcription terminations are still matters of conjecture. The most prevalent speculation to account for rho-independent termination is that the polymerase pauses after the inverted repeat sequence is transcribed because of interference by the stem-loop structure. As a result of or following the pause, the RNA chain dissociates from the template strand in the vicinity of the stretch of U's, which are relatively weakly paired with the stretch of dA's in the template. This probably explains why the most frequent termini of RNA chains are uridylate or adenylate nucleotides.

Rho is an oligomeric protein that binds strongly to RNA and in that state hydrolyzes ATP to ADP and inorganic phosphate. One model of

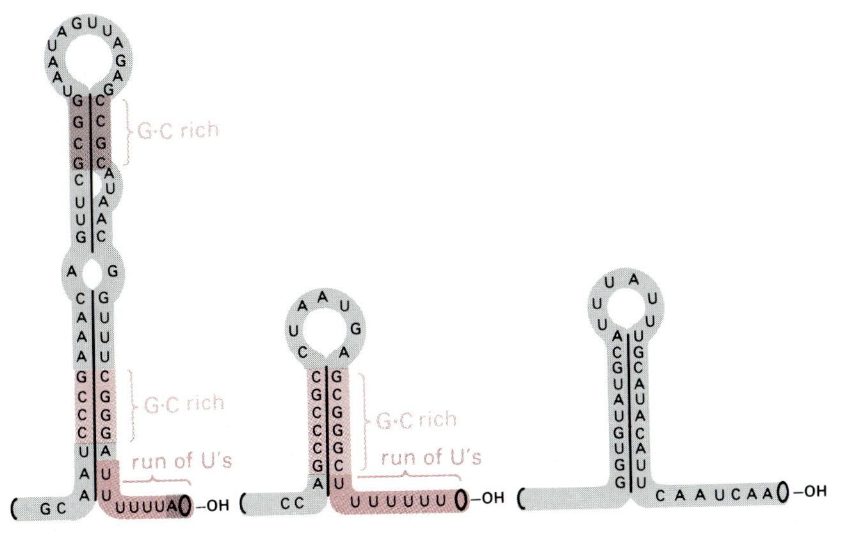

Figure 3.11
Examples of the stem-loop structures in rho-independent and rho-dependent terminators. Adapted from B. Lewin, *Genes III* (New York: Wiley, 1987), p. 250.

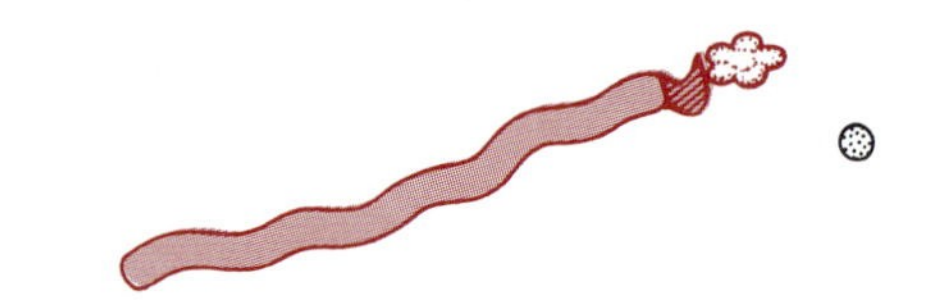

Figure 3.12
Schematic diagram indicating how rho may influence the pausing of RNA polymerase at a terminator, causing transcription to cease with release of the enzyme and nascent RNA. Adapted from B. Lewin, *Genes III* (New York: Wiley, 1987), p. 260.

rho action is that it binds to and migrates along the nascent RNA chain in the 5′-to-3′ direction toward the site of RNA synthesis; the energy for rho migration is derived from the hydrolysis of ATP (Figure 3.12). When rho encounters an incipient stem-loop structure in the RNA, it causes the polymerase to pause where it otherwise would continue transcription. Rho-induced pausing may be sufficient to cause the RNA terminus to dissociate from the template, or, alternatively, rho may promote the dissociation and release.

Transcription termination is rarely absolutely dependent or independent of rho. Some rho-independent terminators are made more efficient by rho's presence. And, under some conditions, some so-called rho-dependent terminators cause termination in the absence of rho, albeit inefficiently.

Just as the initiation of RNA synthesis can be regulated, so is termination a regulatable reaction. Proteins exist that function as **antiterminators** by preventing termination at rho-independent terminators; still other proteins are able to inhibit the action of rho and thereby permit RNA chain elongation to continue beyond these termination signals. Also, under certain circumstances, alternate stem-loop structures can be formed in the RNA at specialized sequences within a particular region of DNA. As a result, one such structure may cause abortive termination while the alternative structure allows transcription to proceed. Regulated termination as a means of controlling mRNA formation is considered in Section 3.11d.

3.3 RNA Processing in Prokaryotes

The primary transcripts produced by transcription of prokaryotic protein coding genes serve as mRNA without further modification or processing. Indeed, the translation of mRNAs frequently begins even before the 3′ end of the transcript is completed (Section 3.7a). But the situation is quite different for rRNAs and tRNAs. In these cases, clusters of rRNA or tRNA genes, or interspersions of the two types of genes, are often transcribed as a single RNA chain. Thus, although transcription of these genes begins at promoters and ends at terminators, the primary RNA transcripts must be cleaved at specific locations, and specific nucleotides must be modified to produce the mature functional forms. These molecular changes are referred to collectively as **posttranscriptional modifications** or simply **RNA processing**. The pathways and enzymes for processing rRNAs and tRNAs are best understood in *E. coli*, and we shall use that system to illustrate features of posttranscriptional RNA processing. The analogous modifications of eukaryotic RNAs are considered in Chapter 8 because, in addition to processing of rRNA and tRNA, there is an elaborate system for the maturation of transcripts destined to become mRNA.

a. *The Grouping of rRNA and tRNA Genes*

Seven discrete transcription units that encode rRNAs have been identified and mapped on the *E. coli* genome; these are referred to as *rrn*A–H in Figure 3.13. Each transcription unit yields an RNA molecule that contains about 5000 nucleotides and includes one copy of the coding sequences for the 5S, 16S, and 23S rRNAs. The direction of transcription of this region is 16S → 23S → 5S. In addition to the three rRNA sequences, the transcripts contain variable lengths of spacer nucleotides and one or more tRNA genes (Figure 3.14). The spacers may occur before, between, and following the rRNA sequences, and the tRNA sequences generally lie within the interspersed or 3′ terminal spacer segments. Processing of these transcripts is clearly necessary to form the functional mature RNAs. Modification of specific bases within the spacer, rRNA, and tRNA sequences also occurs preceding and during the processing steps.

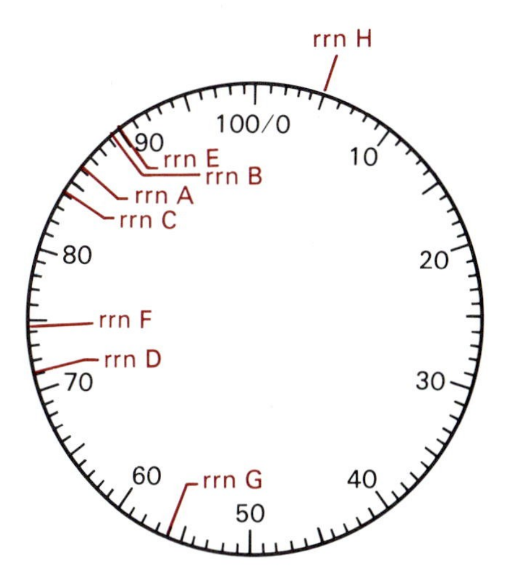

Figure 3.13
The map locations of the seven rRNA transcription units on the *E. coli* chromosome.

b. *Trimming the rRNA-tRNA Cotranscripts*

RNase III is the endonuclease responsible for the initial cleavage of the primary transcripts into fragments that contain either a tRNA or a 16S, 23S, or 5S rRNA sequence. The targets for RNase III cleavage are short RNA duplexes formed by intramolecular base pairing of sequences that straddle each of the rRNA segments (Figure 3.15). For example, stretches

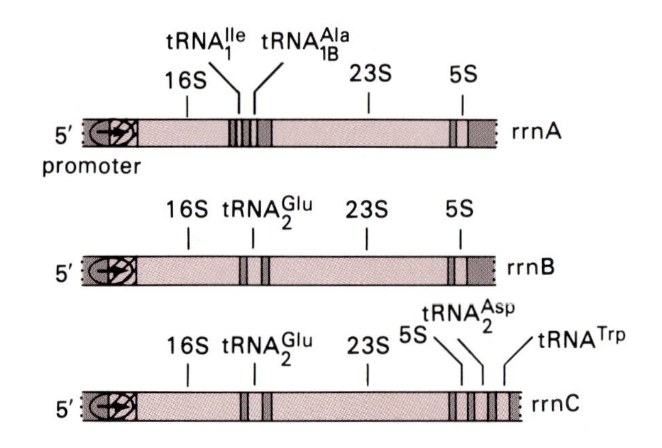

Figure 3.14
Diagram of three characteristic *E. coli* rRNA transcription units showing coding regions for 5S, 16S, and 23S rRNAs, tRNAs, and spacer segments (most intensely colored).

of complementary nucleotides in the spacer regions that flank the 16S rRNA sequence form a stem-loop structure in which the 16S rRNA sequence resides in the loop; similar stem-loop configurations occur with the 23S and 5S rRNA segments. RNase III introduces staggered cleavages within the double strand RNA stems, yielding RNA chains that contain one of the rRNA sequences flanked by short segments of spacer with 5′-phosphate and 3′-hydroxyl termini. The extra nucleotides from the spacer sequences are ultimately trimmed away, possibly by the same RNA exonuclease that performs the final step of trimming in tRNA processing (Section 3.3c). In principle, only the nucleotide sequences needed to form the stem-loop structure for each rRNA must be transcribed for the enzymatic cleavage to occur. However, processing appears to follow comple-

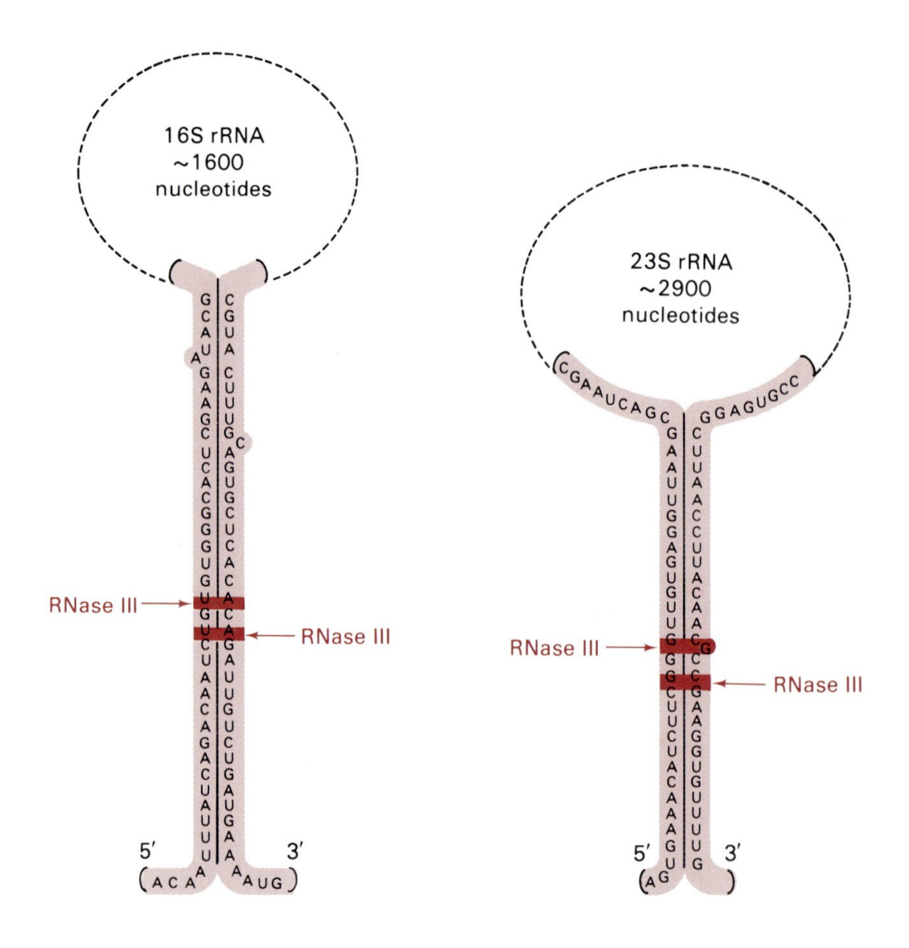

Figure 3.15
RNase III cleavage sites in rRNA precursors. Cleavage occurs in the stems formed by intramolecular base pairing of sequences in the spacer regions flanking the 16S and 23S rRNA sequences. Adapted from B. Lewin, *Genes III* (New York: Wiley, 1987), p. 453.

tion of the primary transcript because ribosomal or other proteins are probably required to enable the entire RNA transcript to fold properly for endonuclease III action. The tRNA segments that are excised from the multigene transcripts are processed the same way as are the tRNAs from single-gene transcription units (Section 3.3c).

c. Producing Mature tRNAs from Larger Transcripts

Although some tRNA genes are located within rRNA transcription units and are expressed concomitantly with rRNAs, the majority of tRNA genes occur singly or in clusters. The clusters may contain multiple copies of the same tRNA gene, for example, three copies of an identical $tRNA^{Tyr}$, while other arrays contain different and unrelated tRNA genes (Figure 3.16). In either case, each cluster is transcribed into a single large RNA molecule that is processed by successive cleavages to the mature tRNA lengths (Figure 3.17). Specific base modifications and the addition of one, two, or all three nucleotides of the 3′-CCA end may be needed to produce a fully functional tRNA.

A single endonuclease, called RNase P, is responsible for creating the 5′-phosphate terminus of all tRNAs, whether the primary transcript contains one or more tRNA sequences and whether it is embedded within an rRNA spacer region. Apparently, RNase P recognizes the folded structure of the tRNA in the precursor polynucleotide and cleaves away the leader or spacer sequences that are 5′ to the start of the mature tRNA. The 3′ end of the tRNA is created by a combination of activities. An as yet unidentified endonuclease cleaves the precursor at a hairpin loop near the mature 3′ end of the tRNA, and then an exonuclease, RNase D, gen-

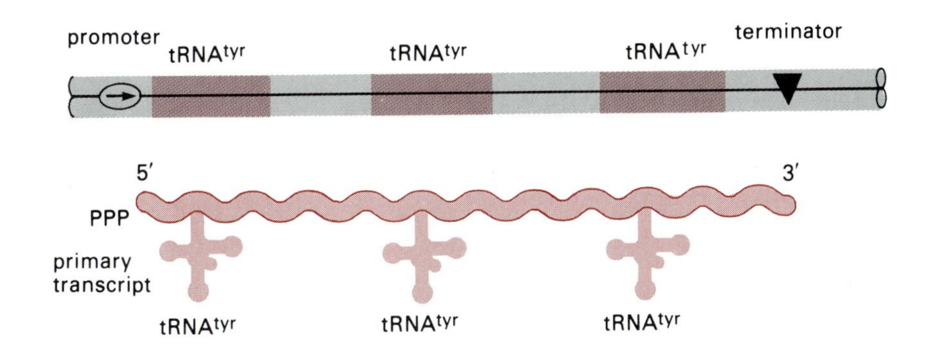

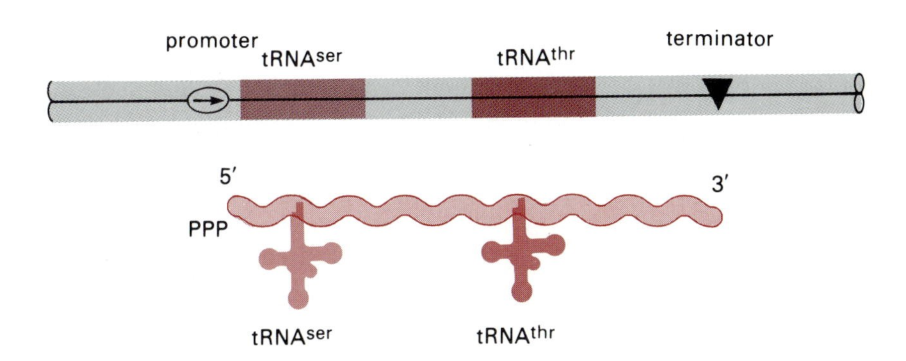

Figure 3.16
Typical clusters of *E. coli* tRNA genes and their primary transcripts.

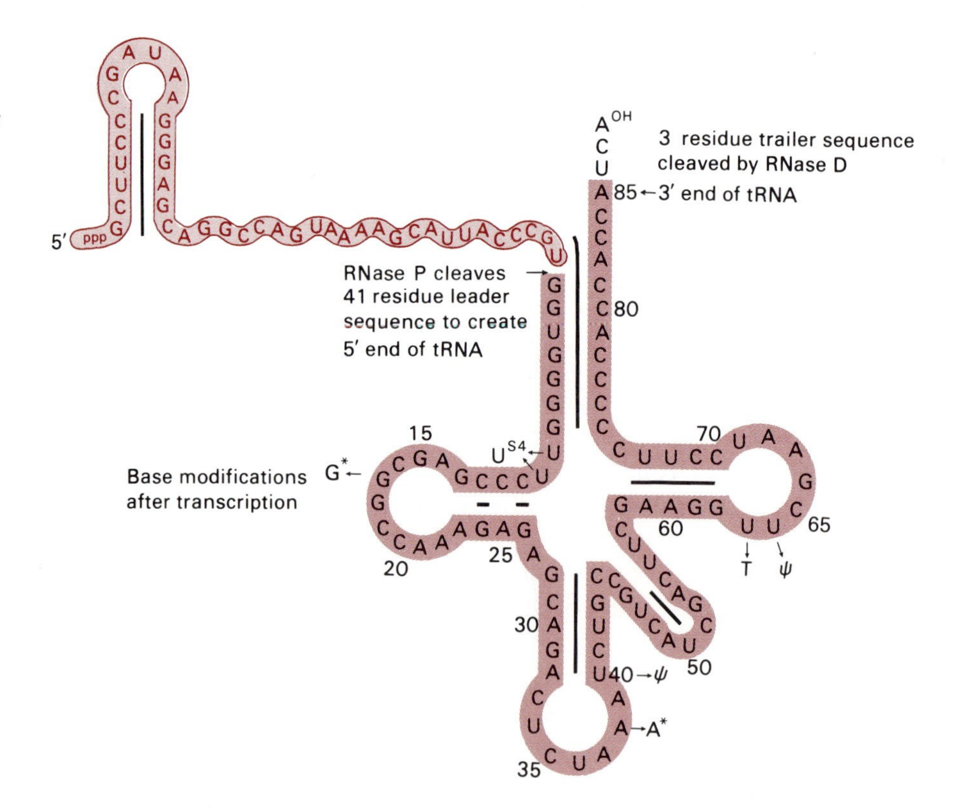

Figure 3.17
Various steps in the maturation of a tRNA from a precursor. The example shown is *E. coli* $tRNA^{Tyr}$.

erates the correct 3′ terminus. In some cases, the exonuclease digestion stops at the proper 3′-CCA terminus on the mature tRNAs (Figure 3.17), but in others, the exonucleolytic action generates an end that serves as the primer to which tRNA nucleotidyl transferase adds one or more of the invariant terminal nucleotides (C and/or A).

RNase P has an unusual specificity in that the correctly folded tRNA structure specifies the cleavage site. Nucleotide sequence changes that do not alter the folded structure do not affect processing at the 5′ end, but nucleotide changes that disrupt the folded conformation impair or prevent RNase P action. Another intriguing feature of RNase P is that the enzyme is composed of both protein and RNA. The RNA has a specific sequence of 377 nucleotides and is itself transcribed from a slightly larger gene by RNA polymerase and processed to its mature length. A surprising property of this RNA is that it alone can catalyze the same endonuclease cleavage of tRNA precursors as does the native ribonucleoprotein; by contrast, the protein alone is inactive as an endonuclease. Thus, the endonuclease activity may be an intrinsic property of the RNA, and the protein may serve only to keep the RNA folded in its maximally active configuration.

Mature tRNAs are unusual not only for their very characteristic folded conformation, but also for their content of modified bases and nucleotides (Figure 3.18). Several of the modified bases in tRNA appear to be essential for one or more of tRNA's physiological functions. Although only a few of the battery of enzymes that catalyze the myriad modifications have been characterized, most of the modifications clearly occur at both the RNA precursor stage and in the fully processed tRNA. These modifying enzymes are particularly interesting because of their unusual sequence specificity; for example, only certain of the uracil residues are converted

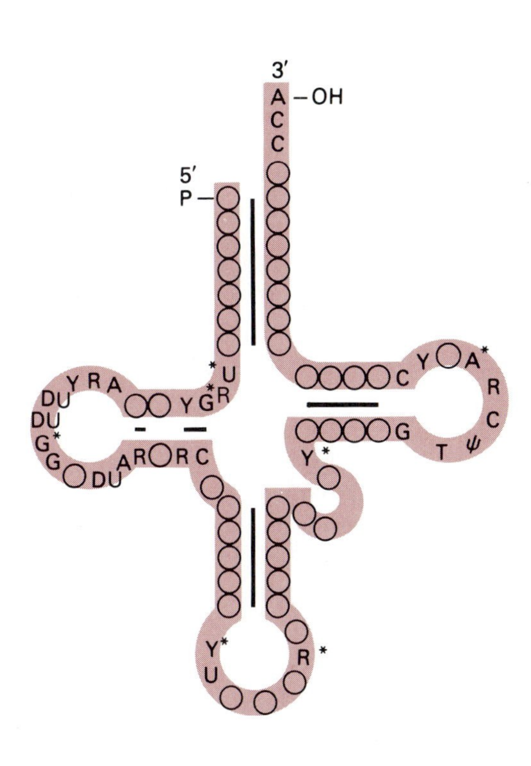

Y = pyrimidine
R = often purine
DU = frequently dihydrouridine
* = modified base
T = thymidine
ψ = pseudouridine

Figure 3.18
Schematic diagram of the cloverleaf form of a tRNA. The several kinds of modified bases occur at characteristic positions in all tRNAs.

to thiouracil, methylated to thymine, or reduced to dihydrouracil. Even more puzzling is how the usual linkage of uracil and ribose in uridylate is altered to produce pseudouridylate (Figure 1.21).

3.4 The Genetic Code

Cellular genes that code for proteins specify the amino acid sequence by the order of deoxynucleotides in the DNA but more directly by the sequence of ribonucleotides in their mRNA transcripts. The informational relationship between nucleotide and amino acid sequences is defined as the **genetic code**. The genetic dictionary, which also includes punctuation signals that mark the beginnings and ends of protein coding regions, was deduced from a variety of genetic and biochemical experiments. Except for minor variations in the usage of a few nucleotide sequences for particular amino acids in mitochondria (Section 3.4e) and certain ciliates, the genetic dictionary is universal—that is, a specific grouping of nucleotides dictates the same amino acid sequence in all living organisms.

The existence of such a coding system necessitates a mechanism for translating nucleotide sequences into unique polypeptide chains. As expected, the machinery and reactions used to perform the translation are complex. Nevertheless, in spite of the differences between prokaryotes and eukaryotes with respect to both the structure of their mRNAs (Section 3.8a) and the physical separation between their genes and the translation machinery, both types of organisms use very much the same mechanisms for decoding their genetic messages.

This section describes the characteristics of the code. We discuss the nature of the machinery and the protein assembly process in Sections 3.5–3.8.

a. *Amino Acid Sequences in Proteins Correspond to Nucleotide Sequences in Genes*

The existence of a colinear relationship between nucleotide and amino acid sequences was one of the earliest speculations on the nature of the genetic code. Initially, this was surmised from demonstrations that many mutations cause single amino acid replacements in proteins of bacteria, plants, and animals. But not until detailed genetic and biochemical analyses were done with a well-characterized gene-protein system was the colinearity hypothesis established. For example, the relative positions of amino acid changes in the α subunit of *E. coli* tryptophan synthetase were shown to match the relative map positions of the corresponding mutations in that organism's *trpA* gene. But the nucleotide sequence that encoded each amino acid could not be specified directly because of the inability to isolate and characterize the altered DNA.

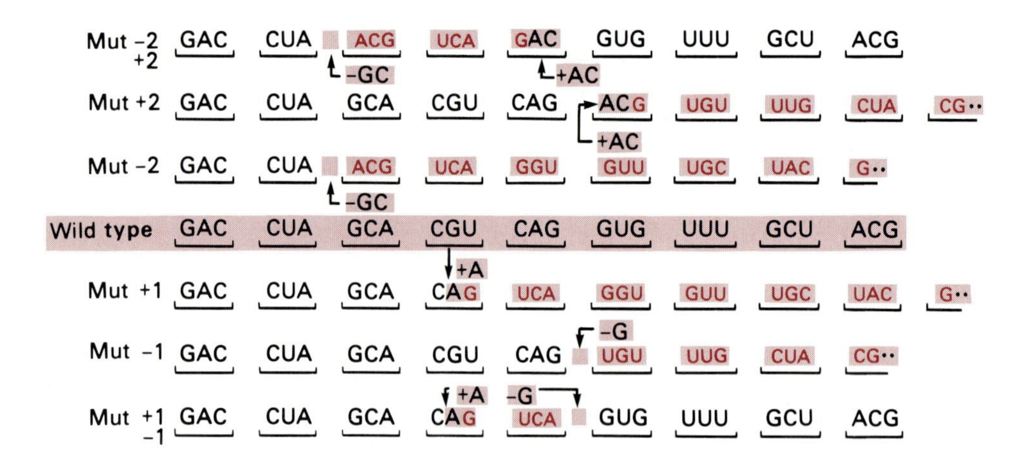

Figure 3.19
Genetic data on frameshift mutations indicated that codons are composed of three adjacent nucleotides. Frameshift mutagens (e.g., proflávin) can cause the deletion or insertion of one or two bases. The resulting mutants make inactive or shortened polypeptides because the change shifts the frame to one that encodes the wrong amino acids or brings a stop codon into the frame, respectively. Recombination between two closely spaced deletion and insertion mutants (e.g., $-1 \times +1$ or $-2 \times +2$) can, however, regenerate a functional gene if the few residual, out-of-frame codons are in a nonessential part of the polypeptide. Recombination between two closely spaced insertions (e.g., $+1 \times +1$) or deletions (e.g., -2×-2) fails to correct the mutation. In contrast, recombinations that yield genes with three closely spaced insertions (e.g., $+1 \times +2$) or three closely spaced deletions (e.g., -1×-2) do occasionally yield functional genes because the normal reading frame is resumed downstream of the altered bases.

Theoretical considerations suggested that a grouping of 3 nucleotides was the most likely size of the genetic coding unit or **codon**. This estimate was based on the following arguments. First, 4 nucleotides taken 1 at a time could specify only 4 different amino acids. Two nucleotides taken at a time would permit only 4^2 or 16 amino acids to be specified, still fewer than the 20 known to be used in proteins. But any 3 adjacent nucleotides create 64 possible codons (4^3), a number more than enough to specify 20 different amino acids. Genetic experiments with mutants that had deletions or insertions of 1, 2, or 3 nucleotides in a protein coding gene indicated that the most likely size for a codon was 3 adjacent nucleotides (Figure 3.19). Moreover, it was inferred from these studies that a nucleotide sequence is translated by successive triplets from a fixed point. These deductions, plus the information that polypeptide chains are assembled from the amino to the carboxy terminal amino acid, provided the opening wedge in solving the genetic code.

b. Matching Amino Acids to Their Codons

A perplexing aspect of the coding problem was the apparent lack of structural complementarity between nucleic acids and the different side chains of the amino acids. The conceptual impasse—how amino acids are matched up with their corresponding codons—was solved by the **adapter hypothesis**. According to this model, amino acids are first attached to RNA molecules, and these amino acid–RNA's are then positioned on mRNA by complementary base pairing between several bases in the adapter RNA and the appropriate codon (Figure 3.20). The adapter hypothesis received strong support from the discovery of tRNA and of enzymes that link amino acids to tRNAs and from the demonstration that amino acids linked to tRNA are the direct precursors for polypeptide assembly.

If tRNAs are the adapters, then amino acids should be linked to specific tRNAs, and each tRNA should be able to pair with only its corresponding codon. The first of these requirements was met by the discovery of specific enzymes—aminoacyl-tRNA synthetases—each of which links a single amino acid to one or a few cognate tRNAs (Figure 3.21). These enzymes and the reactions they catalyze are discussed in greater detail in Section 3.5a. For the present, suffice it to say that amino acids are linked to tRNAs as the first step in the decoding process. What of the second prediction—that the tRNA alone determines the position

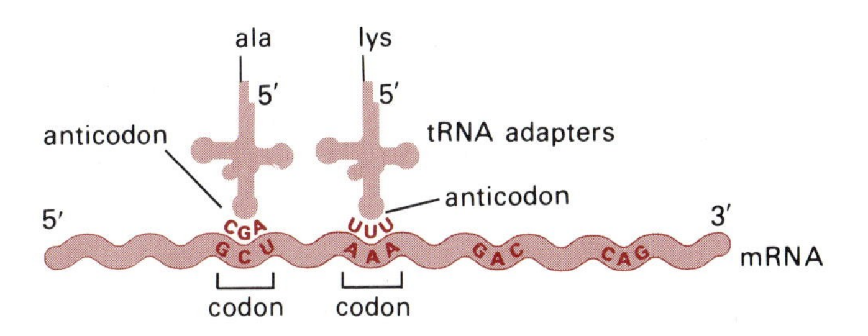

Figure 3.20
The adaptor hypothesis. Amino acids are first attached to specific adaptor RNAs (tRNAs) that are then aligned on mRNA through complementary base pairing between the codons in mRNA and anticodons in the adaptor.

Figure 3.21
Each aminoacyl-tRNA synthetase joins a specific amino acid to one or a few cognate tRNAs.

Overall Reactions

$$\text{ATP} + \text{L-Tyrosine} + \text{tRNA}^{\text{Tyr}} \xrightleftharpoons{\text{Tyrosyl-tRNA synthetase}} \text{L-Tyrosyl-tRNA}^{\text{Tyr}} + \text{AMP} + \text{PP}_i$$

$$\text{ATP} + \text{L-Leucine} + \text{tRNA}^{\text{Leu}} \xrightleftharpoons{\text{Leucyl-tRNA synthetase}} \text{L-Leucyl-tRNA}^{\text{Leu}} + \text{AMP} + \text{PP}_i$$

of its amino acid in the polypeptide chain? A simple experiment established the correctness of this aspect of the adapter theory. One amino acid was changed into another while it was still linked to its cognate tRNA. Then, after incorporation into protein *in vitro*, the position of the changed amino acid in the particular protein was checked (Figure 3.22). It was found that, after cysteinyl-tRNA^{Cys} was converted chemically to alanyl-tRNA^{Cys}, the alanyl residues linked to the tRNA^{Cys} were incorporated only into positions in the protein that normally contained cysteine and not into the sites occupied by alanine. Thus, it was clear that the tRNA to which an amino acid is linked, and not the amino acid itself, determines how codons are translated.

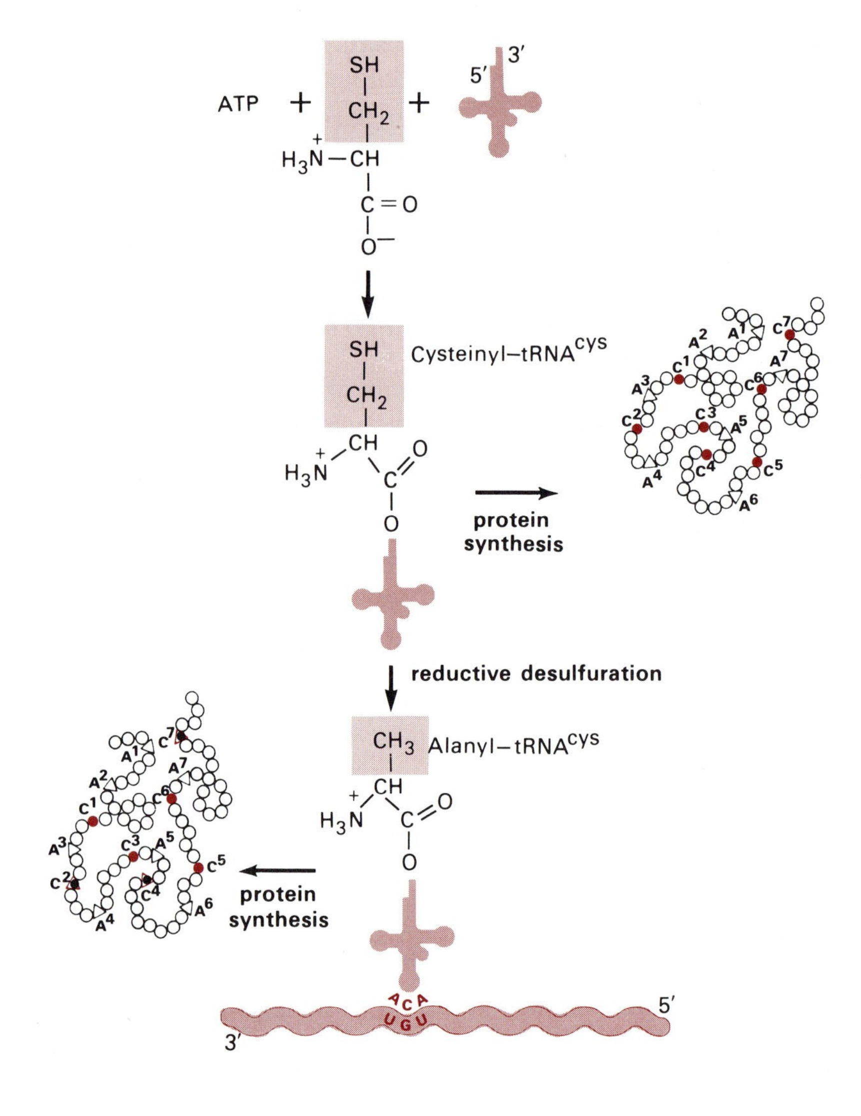

Figure 3.22
The reductive conversion of cysteinyl-tRNA^{Cys} to alanyl-tRNA^{Cys} results in the synthesis of polypeptides containing alanine at positions encoded by a cysteine codon, UGU. The positions where alanine has been incorporated are shown by black dots within colored triangles.

c. *Solving the Genetic Code*

Two advances paved the way for deciphering the code. First, it was established that mRNA was the informational intermediary between genes and proteins. Second, added mRNAs could be translated in bacterial extracts into their corresponding proteins. This breakthrough occurred when synthetic RNA polymers such as polyuridylate (poly U), polyadenylate (poly A), and polycytidylate (poly C) were translated in *E. coli* extracts into polyphenylalanine, polylysine, and polyproline, respectively. This implicated triplets of U, A, and C as codons for phenylalanine, lysine, and proline, respectively. Subsequent experiments using mixed polymers with varying ratios of two or three nucleotides provided information about the composition of codons (Table 3.2). However, although such data estimated the nucleotide composition of a codon, it could not establish the precise order of nucleotides.

All the codons were finally identified by two definitive experiments. One compared the amino acid sequences of polypeptides made *in vitro* using synthetic polynucleotide mRNAs containing repeated defined sequences of two or three nucleotides (Figure 3.23 and Table 3.3). The second identified which aminoacyl-tRNAs were bound to ribosomes in the presence of each of the possible trinucleotides (Figure 3.24). Both approaches yielded a consistent dictionary in which 61 trinucleotides code for the 20 amino acids and 3 designate the end of the coding sequence (Figure 3.25).

The code contains a potential for ambiguity that stems from the way it is translated rather than from the codon–amino acid relationships themselves. That ambiguity is caused by the existence of alternate sets of triplets or reading frames that exist inherently in any polynucleotide sequence (Figure 3.26 on page 156). Most prokaryotic genes are translated in only one continuous reading frame; indeed, the alternate reading frames generally contain a termination codon once in every 20 codons, on average.

In the decoding experiments described previously, the synthetic polymers were translated using conditions that obviate the requirements for accurate initiation. However, *in vivo*, and under the proper conditions *in vitro*, initiation occurs only in the correct reading frame. Unambiguous readout of the protein coding sequence is achieved because translation of the mRNA sequence can begin only at a specific AUG triplet in the

Table 3.2 Coding Properties of a Random Copolymer Composed of U (0.76) and G (0.24)

Triplet	Calculated Triplet Frequency	Relative Frequency	Amino Acid	Relative Incorporation into Polypeptide	Probable Codon Composition
UUU	0.44	100	Phenylalanine	100	3U
UUG	0.14	32	Valine	37	2U1G
UGU	0.14	32	Leucine	36	2U1G
GUU	0.14	32	Cysteine	35	2U1G
UGG	0.044	10	Tryptophan	14	1U2G
GUG	0.044	10			
GGU	0.044	10	Glycine	12	1U2G
GGG	0.014	3			

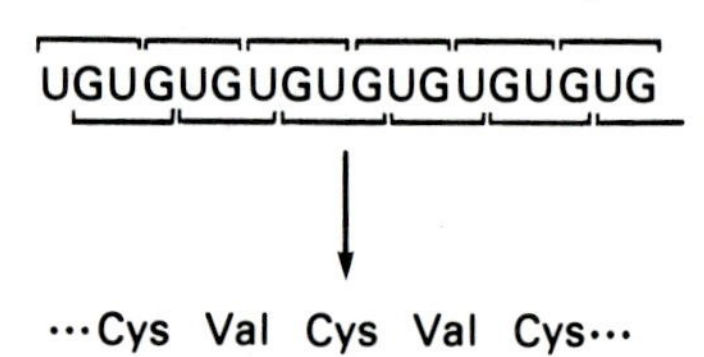

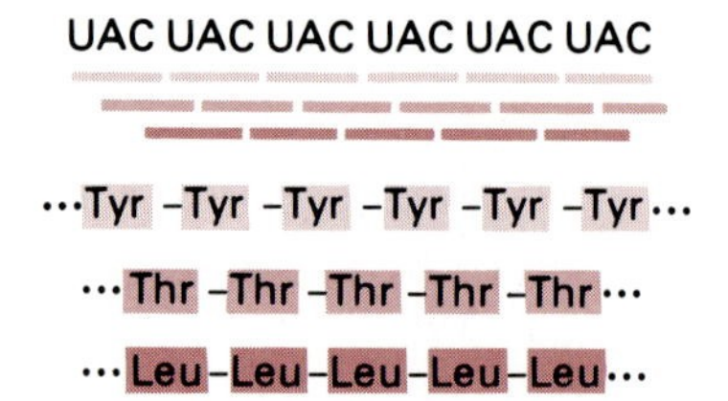

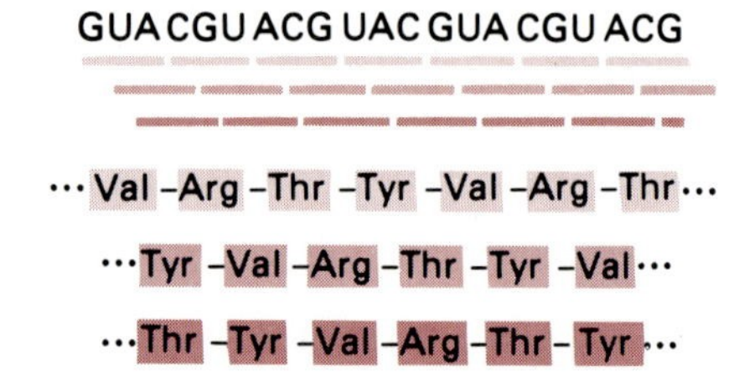

Figure 3.23
Synthetic mRNAs can specify defined polypeptides. Polyribonucleotides with defined nucleotide sequences can direct the synthesis of specific polypeptides in cell-free extracts that catalyze protein synthesis. Note that the polypeptide product that is made depends on which of the three possible sets of repeating triplets is read in poly(UAC)$_n$: UAC, ACU, or CUA.

Table 3.3 Incorporation of Amino Acids into Proteins Stimulated by Synthetic Polynucleotides Having Known Repeating Sequences

Repeating Sequence	Codons in Sequence	Amino Acids Incorporated
UC	UCU-CUC	Ser-Leu
AG	AGA-GAG	Arg-Glu
UG	UGU-GUG	Cys-Val
AC	ACA-CAC	Thr-His
UUC	UUC; UCU; CUU	Phe; Ser; Leu
AAG	AAG; AGA; GAA	Lys; Arg; Glu
UUG	UUG; UGU; GUU	Leu; Cys; Val
CAA	CAA; AAC; ACA	Gln; Thr; Asp
GUA	GUA; UAG; AGU	Val; Ser
UAC	UAC; ACU; CUA	Tyr; Thr; Leu
AUC	AUC; UCA; CAU	Ile; Ser; His
GAU	GAU; AUG; UGA	Asp; Met
UAUC	UAU-CUA-UCU-AUC	Tyr-Leu-Ser-Ile
GAUA	GAU-AGA-UAG-AUA	None
UUAC	UUA-CUU-ACU-UAC	Leu-Thr-Tyr
GUAA	GUA-AGU-AAG-UAA	None

From H. G. Khorana et al., *Cold Spring Harbor Symp. Quant. Biol., 31,* 39 (1966). Dashes separate codons in the same reading frame, while semicolons indicate codons in alternate reading frames.

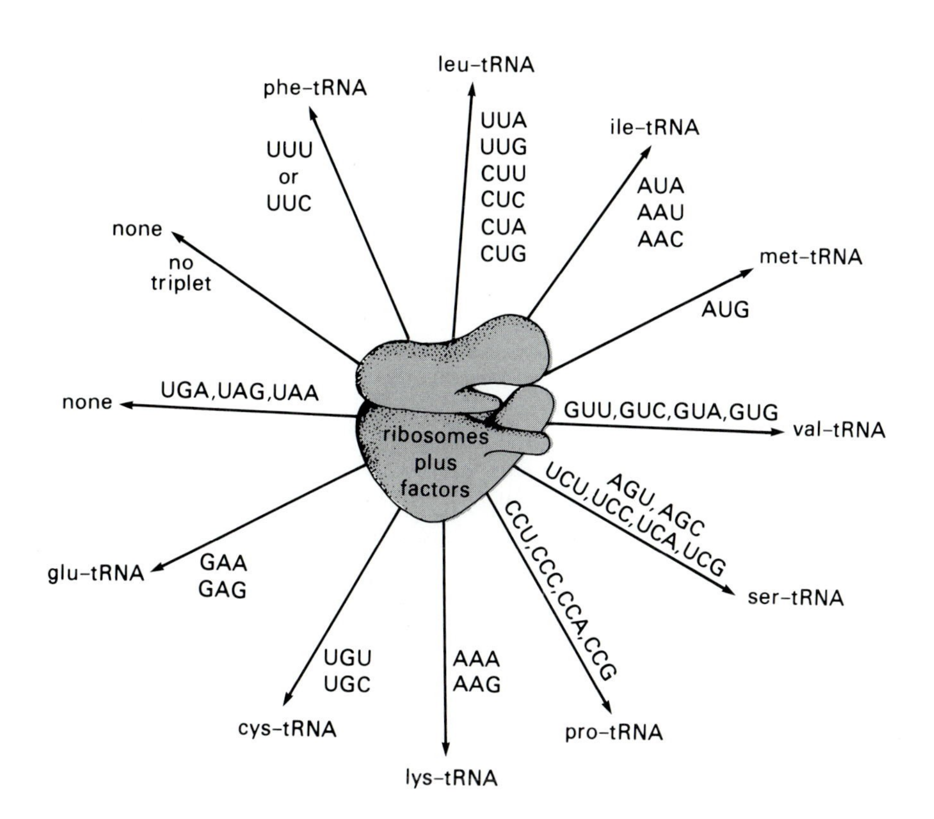

Figure 3.24
Identification of codons by measuring the binding of specific aminoacyl-tRNAs to ribosomes in the presence of synthetic trinucleotides. Ribosomes and required soluble factors were incubated with individual trinucleotides and mixtures of aminoacyl-tRNAs in which only one aminoacyl moiety was radiolabeled. Only a single kind of radiolabeled aminoacyl-tRNA is bound to each ribosome-trinucleotide complex (e.g., radiolabeled prolyl-tRNA to the ribosome-UCA complex).

The Genetic Code

First Position (5′)	Second Position: U	C	A	G	Third Position (3′)
U	PHE	SER	TYR	CYS	U
	PHE	SER	TYR	CYS	C
	LEU	SER	Stop	Stop	A
	LEU	SER	Stop	TRP	G
C	LEU	PRO	HIS	ARG	U
	LEU	PRO	HIS	ARG	C
	LEU	PRO	GLN	ARG	A
	LEU	PRO	GLN	ARG	G
A	ILE	THR	ASN	SER	U
	ILE	THR	ASN	SER	C
	ILE	THR	LYS	ARG	A
	MET	THR	LYS	ARG	G
G	VAL	ALA	ASP	GLY	U
	VAL	ALA	ASP	GLY	C
	VAL	ALA	GLU	GLY	A
	VAL	ALA	GLU	GLY	G

Figure 3.25
The coding dictionary. The upper representation of the dictionary allows one to determine the amino acid encoded by a particular triplet. Thus, the triplet CAG is translated by locating C in the left column, A across the top, and G in the right column. The dictionary at the bottom allows one to examine easily the number and types of codons that specify each amino acid.

ALA	ARG	ASP	ASN	CYS	GLU	GLN	GLY	HIS	ILE	LEU	LYS	MET	PHE	PRO	SER	THR	TRP	TYR	VAL	Stop
	AGA									UUA					AGC					
	AGG									UUG					AGU					
GCA	CGA						GGA			CUA				CCA	UCA	ACA			GUA	
GCG	CGG						GGG		AUA	CUG				CCG	UCG	ACG			GUG	UAA
GCC	CGC	GAC	AAC	UGC	GAA	CAA	GGC	CAC	AUC	CUC	AAA		UUC	CCC	UCC	ACC		UAC	GUC	UAG
GCU	CGU	GAU	AAU	UGU	GAG	CAG	GGU	CAU	AUU	CUU	AAG	AUG	UUU	CCU	UCU	ACU	UGG	UAU	GUU	UGA

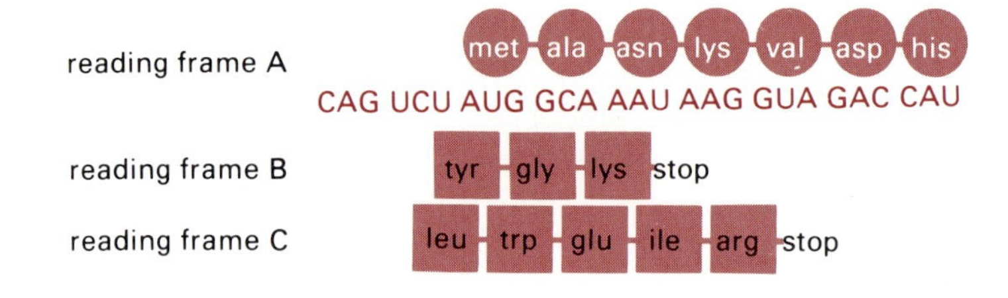

Figure 3.26
A nucleotide sequence can be translated in any of three reading frames. The nucleotide sequence is translated from 5′-to-3′, left to right. In this example, reading frames *B* and *C* are interrupted by stop codons. Only reading frame *A*, which begins with the specific start codon, AUG, is "open" throughout.

mRNA. Translation proceeds by decoding each successive triplet in turn in the 5′-to-3′ direction on the mRNA. So, for example, the coding sequence in Figure 3.27 is translated from the AUG near the 5′ end of the mRNA toward the 3′ end, where it ends at one of the three terminator codons. The recent development of rapid nucleic acid and protein sequencing techniques has made it possible to verify the code by direct comparisons of the sequences in DNA, RNA, and the proteins they encode. These comparisons also confirm that a coding sequence is read from the 5′-to-3′ end of the mRNA.

d. *Redundancy in the Genetic Code*

A striking feature of the code is that all but two amino acids are encoded by more than one codon. Amino acids that are represented in proteins relatively rarely (e.g., methionine and tryptophan) are encoded by one triplet each. Serine and leucine, which occur more abundantly in proteins, have the largest number of codons. And the moderately abundant amino acids such as cysteine, alanine, glycine, valine, and the dicarboxylic acids and their amides, are each represented by two to four codons. Because of this redundancy, different nucleotide sequences can specify the same amino acid sequence (Figure 3.28). Thus, although we can unequivocally define a protein sequence from a nucleotide sequence, we cannot do the reverse.

Any one of three codons, UAA, UAG, and UGA, signals the point on the mRNA at which protein assembly stops. The codon AUG has two functions: it specifies the amino acid methionine and, in certain sequence contexts, also marks the beginning of the protein coding sequence.

Codon redundancy has an especially interesting feature. Most of the variation among codons for the same amino acid occurs in the third position (the 3′ end of the triplet). For example, glycine, valine, proline, alanine, and threonine are each coded by four codons, and the four codons for any one of these amino acids differ from one another only in the third

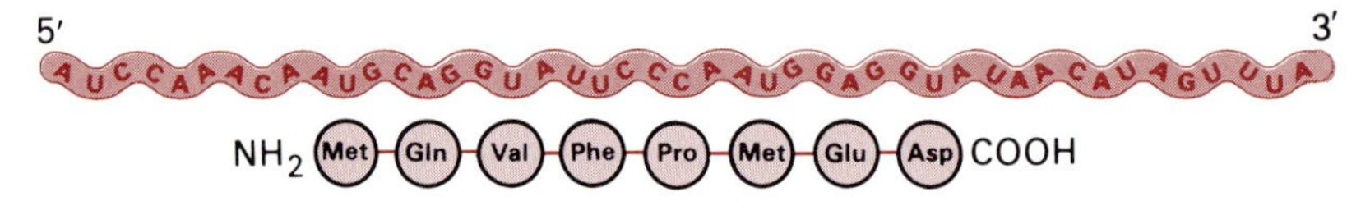

Figure 3.27
A coding sequence starts with an AUG and ends with a terminator (stop) codon. Flanking the coding sequence is a 5′ untranslated region (5′ leader) and a 3′ untranslated region (3′ trailer).

nucleotide (Figure 3.25). Where an amino acid is encoded by two different triplets, they vary in either the type of purine or the type of pyrimidine nucleotide in the third position. Only the codon groups for leucine, serine, and arginine differ from each other in the first, second, or both positions. Thus, mutations causing a nucleotide change in the third position of codons frequently do not alter the amino acid sequence. In addition, the code is such that even changes in the first or second nucleotides of some codons cause substitutions by structurally related amino acids, thereby tending to produce minimal alterations in the protein structure. For example, codons for the hydrophobic amino acids phenylalanine, leucine, isoleucine, and valine differ by only one nucleotide. A similar relationship exists in the codons for serine and threonine or for alanine and glycine.

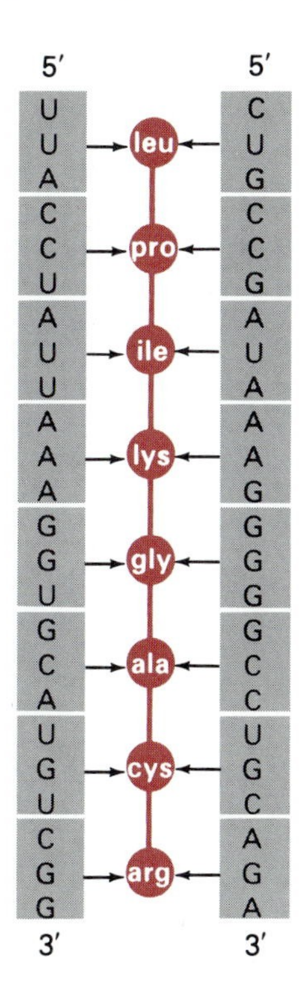

Figure 3.28
Two different nucleotide sequences can specify the same polypeptide sequence.

e. *Universality of the Genetic Code*

So far, all prokaryotic organisms—cells and viruses—as well as most nuclear genes for eukaryotes appear to use the same codon dictionary, irrespective of whether their genomes are composed of DNA or RNA. In that sense, the code is often referred to as being universal.

Although the code is virtually universal, the frequency with which synonymous codons are used varies among different organisms and even in different mRNAs. Where there is a bias in codon usage in most of an organism's protein coding sequences, that bias is reflected in the relative amounts of the different tRNAs that decode synonymous codons. Thus, if the codon for AGA appears infrequently in an organism's mRNA, the $tRNA^{Arg}$ that decodes AGA is likely to be rare. In other instances, the selection of codons for a particular coding sequence may have implications other than those related to their translatability. For example, certain codon arrangements may favor specific secondary or folded mRNA structures; in such instances, the use of particular codons may affect the availability of the mRNA for translation and thereby have regulatory significance. In some cases, where naturally occurring codons have been replaced experimentally by their synonyms, the metabolic stability of the mRNA is altered, although the translatability of the message is not appreciably changed.

Rare exceptions to the standard codon dictionary are found in certain ciliate and mitochondrial genes. For example, in mammalian mitochondria, the UGA codon is translated as tryptophan because the mitochondrial DNA encodes a $tRNA^{Trp}$ whose anticodon (UCA) matches UGA almost as well as the normal tryptophan codon, UGG. Mammalian mitochondria translate the AGA and AGG codons as termination signals. AUU, AUC, AUA and AUG serve as initiation codons, and AUA encodes methionine, instead of isoleucine. In yeast mitochondria, the CUU, CUC, CUA and CUG triplets encode threonine rather than leucine.

3.5 The Translation Machinery

The central participants in the readout of the protein coding sequences in mRNA are the aminoacyl-tRNA synthetases and tRNA, the ribosomes and various ribosome-bound and other proteins. Together, these elements act in the initiation, elongation, and termination stages of polypeptide assembly. This section examines the properties of each of these components. In the next section, we discuss how they participate in the different steps of the translation process itself.

a. *Attachment of Amino Acids to Cognate tRNAs*

tRNA's role in codon recognition depends on two unique activities: First, specific **aminoacyl-tRNA synthetases** link amino acids to the 3′ termini of cognate tRNAs. Second, the aminoacyl-tRNAs bind specifically to their corresponding codons in ribosome-mRNA complexes. The key feature of both reactions is specificity, for a breakdown in the accuracy of aminoacyl-tRNA formation or in the interaction between the aminoacyl-tRNA and its corresponding codon would create a shambles of gene expression. It is therefore important to understand the basis of this specificity, how it is assured, and the consequences of alterations in the fidelity of those two steps.

Key features of tRNA structure Each cell has a population of tRNA molecules. Typically, each tRNA contains 75 to 85 nucleotides and a unique nucleotide sequence that determines which amino acid it can accept and the codon with which it can base pair. Computer-generated intramolecular complementary base pairing patterns of more than 150 discrete tRNA species from both pro- and eukaryotes have been derived based on their nucleotide sequences. The patterns indicate that virtually all tRNAs, irrespective of their nucleotide sequence, possess a characteristic secondary structure, often referred to as the **cloverleaf** structure because of the three stem-loop domains (Figure 3.29a). The reality of the predicted structures is supported by the distinctive chemical reactivities of bases expected to be in base paired or looped out regions.

The folded forms of most tRNAs contain four discrete regions, each of which has certain invariant features, regardless of their amino acid specificity. Four nucleotides at the 3′ terminus are single strand, and the 3′ end is always 5′-CCA-3′. The amino acid acceptor stem contains both the 5′ and 3′ ends of the RNA chain, held together by complementary base pairing of seven nucleotides at the 5′ end with seven nucleotides near the 3′ end. The TΨC region, so called because it contains two unusual forms of uridine (ribothymidine [T] and pseudouridine [Ψ]), consists of a base paired stem and a seven-nucleotide-long loop structure. The stem always contains five base paired nucleotides, including a G·C pair. The TΨC trinucleotide always occurs in the same position in the loop. The **anticodon** region also consists of a stem and loop structure. Seven base paired nucleotides always make up the stem. The triplet that is complementary to the cognate codon, the anticodon, occurs within a loop of seven nucleotides. An invariant uracil residue and a modified uracil flank the 5′ end of the anticodon, and a variably modified purine, usually adenine, occurs immediately 3′ to the anticodon. The fourth region is another stem-loop structure that consists of a stem containing three or four base paired nucleotides and a variably sized loop that frequently contains a reduced form of uracil, dihydrouracil (DU).

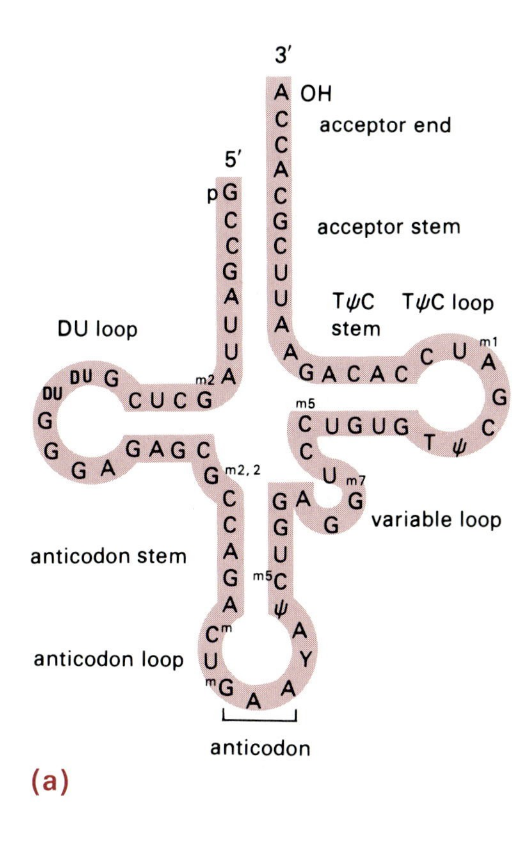

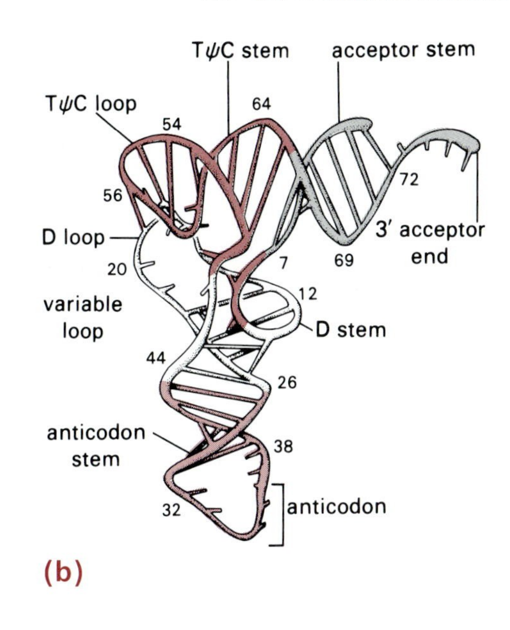

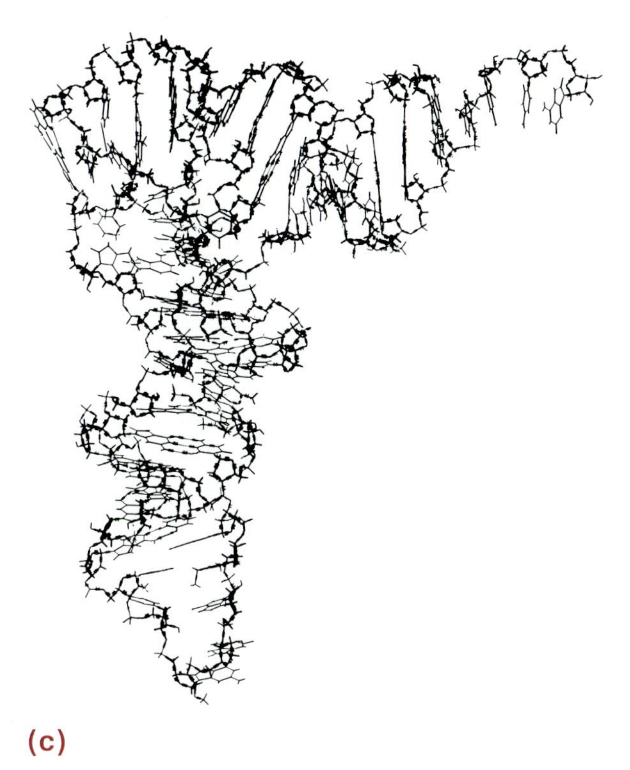

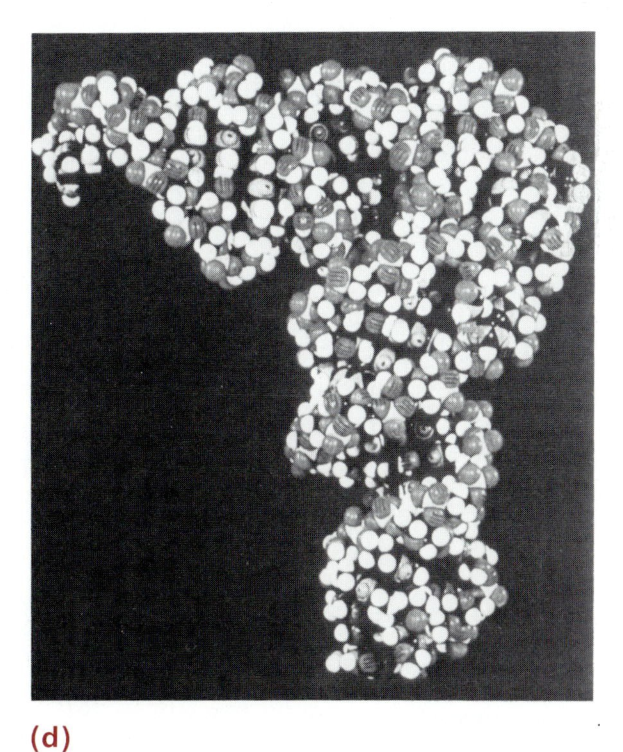

Figure 3.29
A typical eukaryotic tRNA: yeast $tRNA^{Phe}$. (a) The open cloverleaf form. (b) The folded cloverleaf with the ribose phosphate backbone shown as a continuous ribbon and the hydrogen bonds between paired bases as bars. (c) and (d) Wire and space filling molecular models of the folded cloverleaf, respectively. The tRNA structure in (d) is flipped 180° compared to the model in (b). In all pictures, the anticodon is at the bottom. (b), (c), and (d) are courtesy of S. M. Kim. See S. M. Kim et al., *Science* 185 (1974), p. 435.

The most variable features of tRNAs are the nucleotide sequences that make up the stems, the different number of nucleotides between the anticodon and TΨC regions (the variable arm), and the size of the loop and locations of the dihydrouracil residues in the DU region.

X-ray crystallographic analysis of several tRNAs has revealed that tRNAs also have a characteristic tertiary structure (Figure 3.29b–d). The proposed structure is more compact than implied by the cloverleaf imagery. It is formed by intramolecular interactions that bring the DU and TΨC regions together. This causes the overall tRNA molecule to appear as if it consists of two perpendicular components—one being the amino acid acceptor region and the other, the anticodon region. Because of its appearance, the structure is referred to as the L-configuration. The L-shaped structure is more relevant when considering how tRNA functions as an adapter during the interaction of the codon and anticodon on the ribosome.

There are usually several different tRNAs that can accept the same amino acid (iso-accepting tRNAs) (Figure 3.30), but they may have different anticodons that allow them to base pair with synonymous codons. This explains, in part, the degeneracy of the code—that is, how different codons can specify the same amino acid.

Esterification of tRNAs In order to function properly as an adapter in mRNA translation, a tRNA must acquire the amino acid consistent with its anticodon. That takes place in an ATP-dependent reaction catalyzed

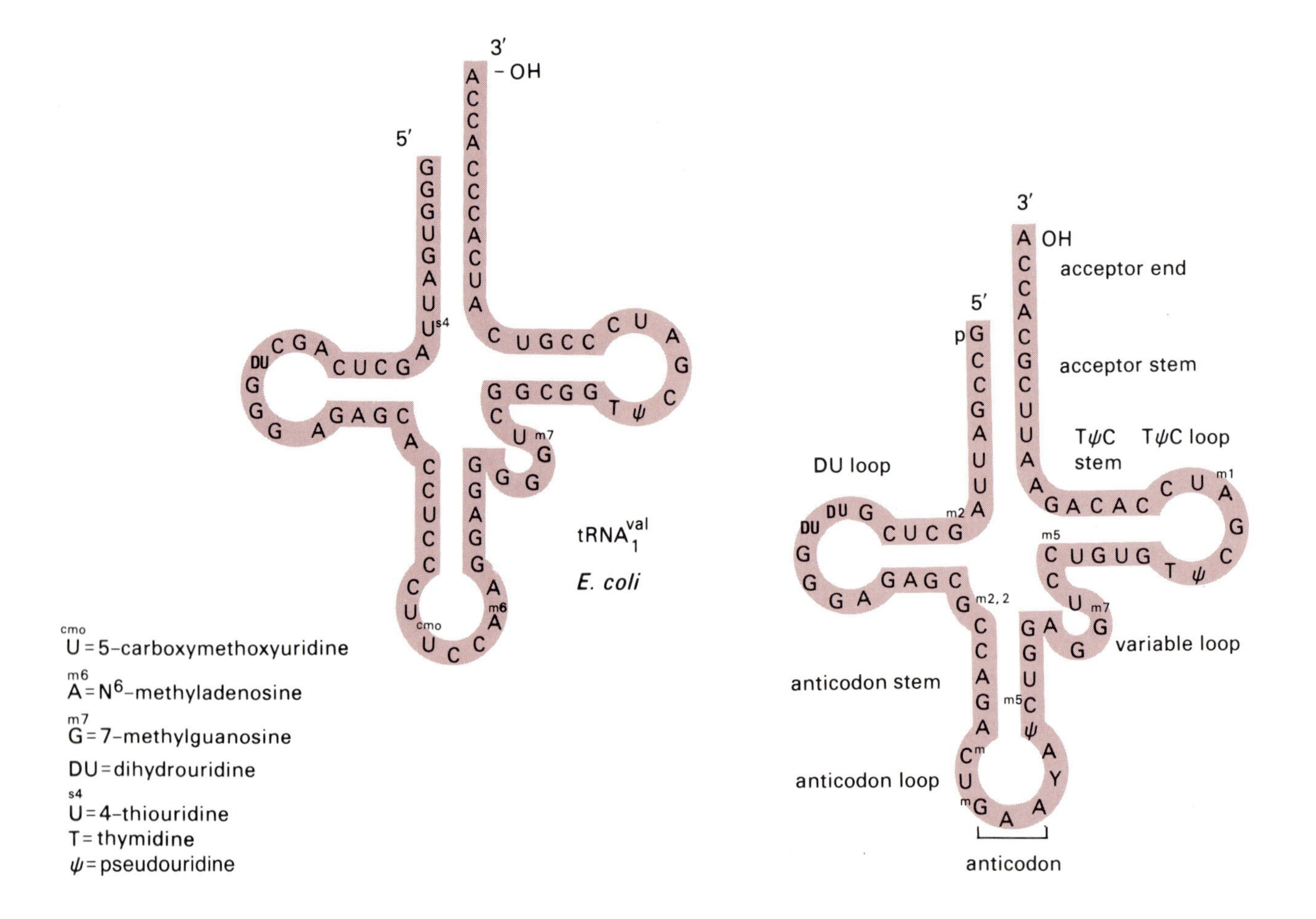

Figure 3.30
Iso-accepting tRNAs for valine in *E. coli*. The two tRNAs differ in their primary nucleotide sequences, their anticodons, and the kinds and locations of their modified bases.

by specific enzymes, aminoacyl-tRNA synthetases (Figure 3.31). In the overall reaction, ATP is cleaved into 5′-adenylic acid (AMP) and inorganic pyrophosphate (PP_i), and the energy available from that breakdown is conserved by linking the amino acid's carboxyl group to one of the hydroxyl groups of the terminal ribose at the 3′ end of the tRNA. The formation of aminoacyl-tRNAs actually occurs in two discernable steps. In the first step, the amino acid's carboxyl group is joined to the α-phosphate of ATP, eliminating inorganic pyrophosphate and forming an aminoacyl-adenylate. The aminoacyl-adenylate is extremely reactive and is stabilized by tight association with the enzyme. Transfer of the aminoacyl group from the enzyme-bound aminoacyl-adenylate to either the 2′- or 3′-hydroxyl group of the terminal ribose group of a tRNA constitutes the second step. The acyl group transfer potential of aminoacyl-tRNA is high

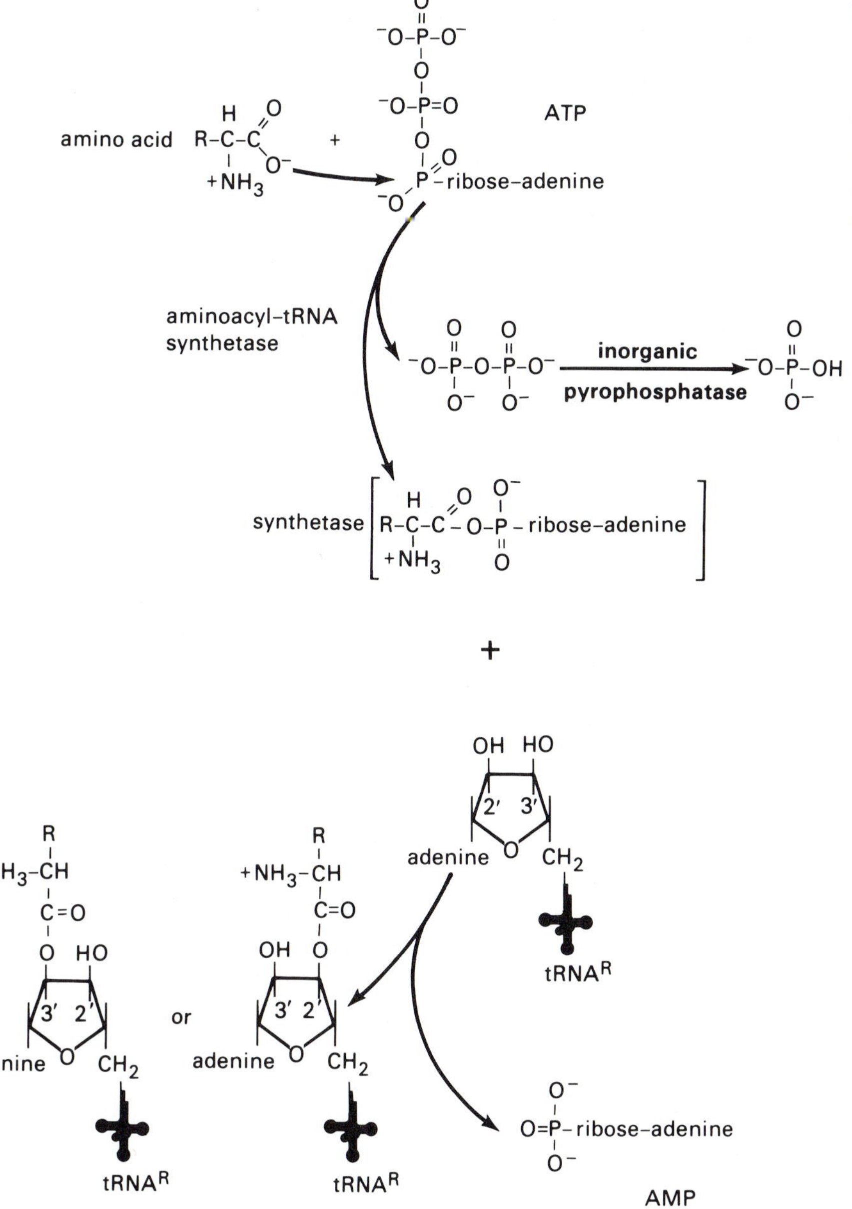

Figure 3.31
The formation of an aminoacyl-tRNA. Aminoacyl-tRNA synthetases catalyze the formation of an enzyme-bound aminoacyl-adenylate intermediate that can transfer the aminoacyl moiety to the tRNA.

enough for peptide bond formation to occur without a further input of energy.

The most significant feature of the reaction leading to the aminoacylation of tRNA is the specificity of the enzymes. Thus, there is a discrete aminoacyl-tRNA synthetase for each of the 20 amino acids that occur in proteins. Each enzyme must be able to distinguish 1 amino acid from the 19 others and transfer it to 1 or a few iso-accepting tRNAs in the presence of the nearly 75 other kinds of tRNAs. Bear in mind that certain amino acids are very closely related in structure: leucine, valine, and isoleucine; valine and threonine; aspartic and glutamic acids. Moreover, the aminoacyl-tRNA synthetases must distinguish their cognate tRNAs from all others in spite of the striking similarities in the two- and three-dimensional structures of all tRNAs. Therefore, these enzymes must be exquisitely specific in their ability to discriminate among such related structures if errors in protein synthesis are to be avoided.

Comments on the structure of aminoacyl-tRNA synthetases and their recognition of amino acids and cognate tRNAs Many aminoacyl-tRNA synthetases have been purified. Some consist of a single polypeptide chain; others contain two or four identical chains, each ranging in size from about 35 to 115 kDaltons. Some dimeric and tetrameric enzymes contain two different types of subunits. There is no apparent correlation between the size of the enzyme or its subunit complexity and its specificity.

Studies of the interactions between aminoacyl-tRNA synthetases and their cognate tRNAs do not provide a simple explanation for their specificity. Most of the available studies indicate that an enzyme's specificity results from its strong contacts with the amino acid acceptor end of the tRNA, the DU region, and the variable sized loop between the TΨC and anticodon regions. Some enzymes appear to be oblivious to the anticodon triplet and catalyze the aminoacylation reaction even if the anticodon is changed. But several enzymes show markedly reduced activity with such altered tRNAs and occasionally attach the wrong amino acid when the anticodon is changed. In some instances, therefore, interactions with the anticodon loop occur as well. Whatever the determinants, the fit must orient the amino acid acceptor end of the tRNA so that the enzyme's catalytic site can transfer the bound aminoacyl-adenylate to the terminal nucleotide of the tRNA.

Part of an enzyme's ability to attach the correct amino acid to its cognate tRNAs relies on specific binding of the amino acid. However, where absolute discrimination between related amino acids is impossible, the synthetases can correct mistakes in the charging reaction. For example, the isoleucyl-tRNA synthetase cannot exclude valine entirely from the amino acid binding site because of the similarity in size and structure between valine and isoleucine. This breakdown in specificity manifests itself in the first reaction; isoleucyl-tRNA synthetase forms an enzyme bound valyl-adenylate, albeit less efficiently than an isoleucyl-adenylate. However, the activated valine does not become linked to $tRNA^{Val}$ or to $tRNA^{Ile}$. Instead, the enzyme bound valyl-AMP is rapidly hydrolyzed in the presence of $tRNA^{Ile}$, and the formation of valyl-$tRNA^{Ile}$ is prevented. A similar mechanism permits valyl-tRNA synthetase to discriminate between valine and threonine and methionyl-tRNA synthetase to distinguish threonine from methionine. Evidently, aminoacyl-tRNA synthetases use editing mechanisms to prevent inescapable mistakes in tRNA aminoacylation. By contrast, if the wrong amino acid attaches to a tRNA, there is no mechanism for removing it. In such cases, the amino acid is incorporated into an incorrect position in the protein. The frequency of such errors is very small. In rabbit hemoglobin, for example, valine appears in positions normally occupied by isoleucine only once in 25,000 to 50,000 isoleucine residues. Thus, the first step in the pathway of gene readout provides a very accurate set of aminoacyl-tRNAs.

b. Ribosomes Match Aminoacyl-tRNAs to Codons and Assemble Protein Chains

Because of their key role in protein synthesis, all cells contain ribosomes, the number varying from 20,000 to 50,000 per cell, according to the cell's protein synthesizing activity. Ribosomes are undifferentiated with respect to the proteins they can synthesize or the cellular targets to which they direct their products. Which protein a ribosome will make in any one synthetic cycle is dictated by the mRNA with which it becomes associated. The intra- or extracellular locations to which the newly synthesized protein is targeted are determined by interactions of specialized membranes and organelles with structural features on that protein.

Prokaryotic and eukaryotic ribosomes have very similar overall structures and functions. Nevertheless, they differ in detail because of variations in the structure and organization of pro- and eukaryotic mRNAs and in the way transcription and translation are temporally and spatially coupled in prokaryotes. *E. coli* ribosomes are prototypic of prokaryotic ribosomes, and because their structure and functions are best understood, the ensuing discussion uses this model. Several structural features of eukaryotic ribosomes are also mentioned for comparison.

The composition of ribosome particles Prokaryotic ribosomes (referred to as 70S particles because of their sedimentation behavior) are composed of small (30S) and large (50S) subunits (Figure 3.32). The 30S ribosomal subunit contains a single rRNA molecule (16S) of 1542 nucleotides and 21 different proteins (S1–S21); the 50S subunit contains two rRNAs, the large RNA (23S) containing 2904 nucleotides and the smaller RNA species (5S) of 120 nucleotides, associated with 34 different proteins (L1–L34). The nucleotide and amino acid sequence of each rRNA and protein is known. Electron micrographs of 70S ribosomes (Figure 3.33) and their three-dimensional reconstructions (Figure 3.34) indicate that the small and large subunits are in contact over several regions, but the most notable feature is the existence of a groove between them, possibly to accommodate the mRNA during its translation.

Both small and large ribosomal subunits can be dissociated into their constituent RNAs and proteins. Moreover, even after each of the RNAs and proteins are separated from one another, fully native and functional ribosomal subunits can be reconstituted by mixing them under appropriate conditions. This means that all the information for assembling this multimeric complex is embodied in the structures of its constituent parts. Such

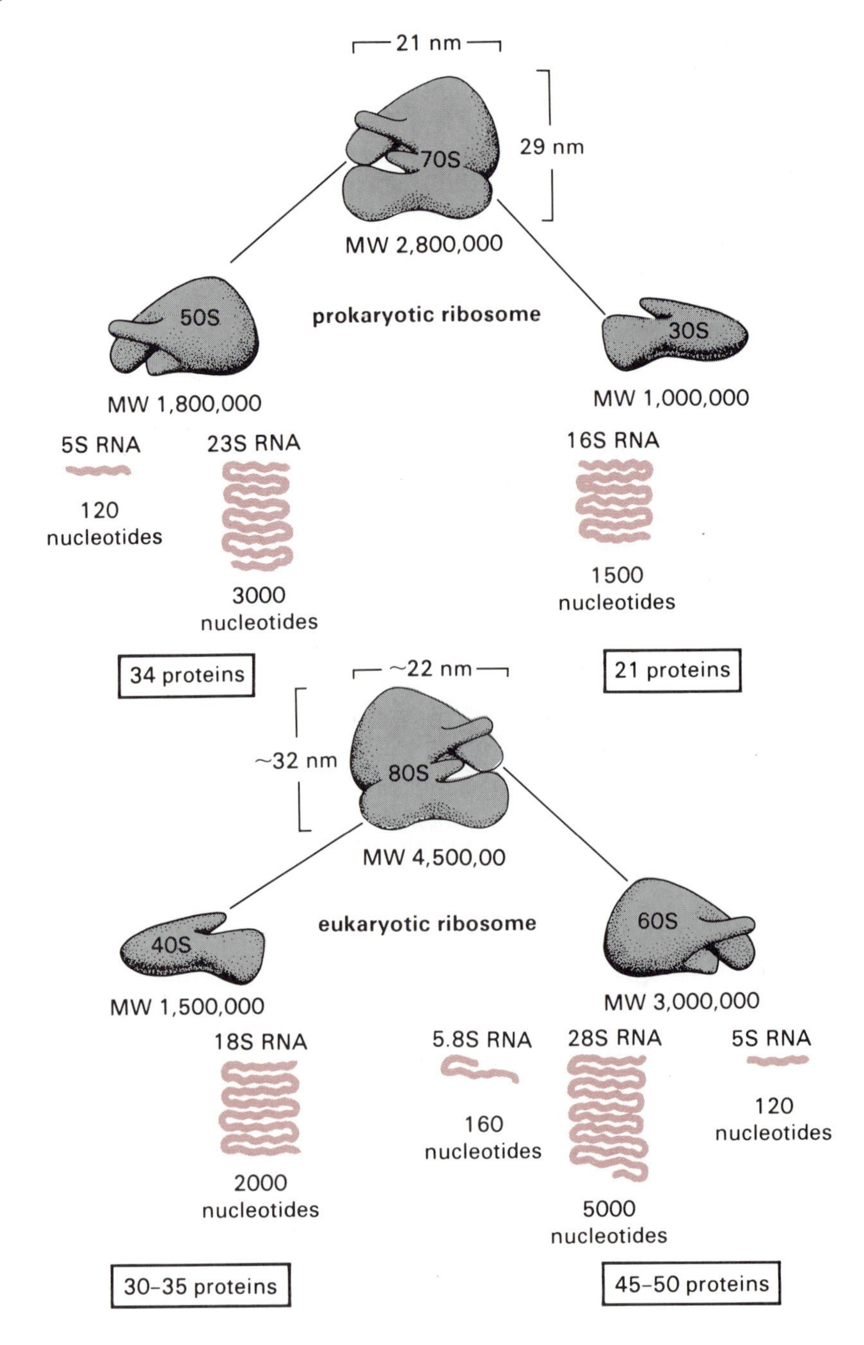

Figure 3.32
The composition of typical prokaryotic and eukaryotic ribosomes.

reconstitution experiments provide important insights into the contacts between the different components of each ribosome and the probable order in which the proteins and RNA assemble in the biological pathway. Moreover, such experiments allow one to test the compatibility of equivalent RNAs or protein subunits from different species (i.e., among distantly related prokaryotes and between prokaryotes and eukaryotes). Further, this approach can evaluate the ability of individual mutationally altered RNAs and proteins to participate in reconstitution and the various activities catalyzed by reconstituted ribosomes.

The eukaryotic cytosolic ribosome also comprises a small (40S) and a large (60S) subunit (Figure 3.32). The small subunit contains a single

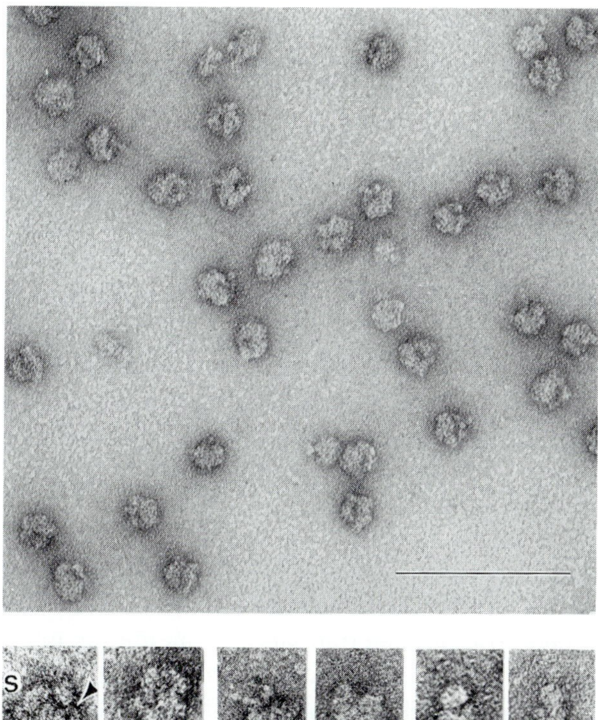

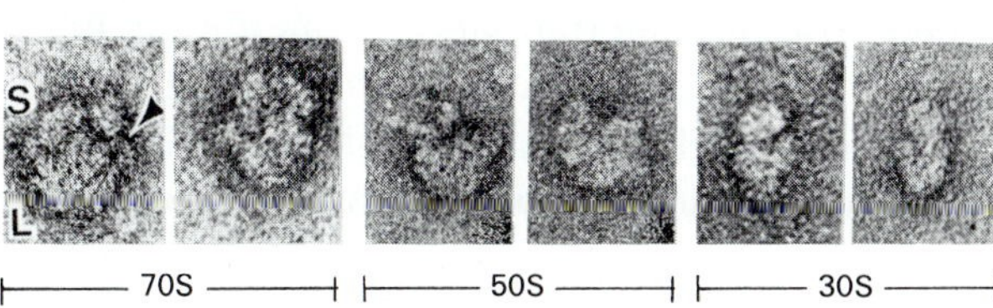

Figure 3.33
Electron micrographs of *E. coli* ribosomes. S and L shown in the bottom left panel refer to the 30S and 50S subunits, respectively. The arrow points to the groove between the two subunits. The top photograph shows 70S monosomes at a magnification of 350,000x. The gallery at the bottom shows enlarged (560,000x) images of 70S ribosomes and the large (50S) and small (30S) subunits. M. Boublík, *Cytobiologie* 14 (1977), p. 293. Courtesy of M. Boublík.

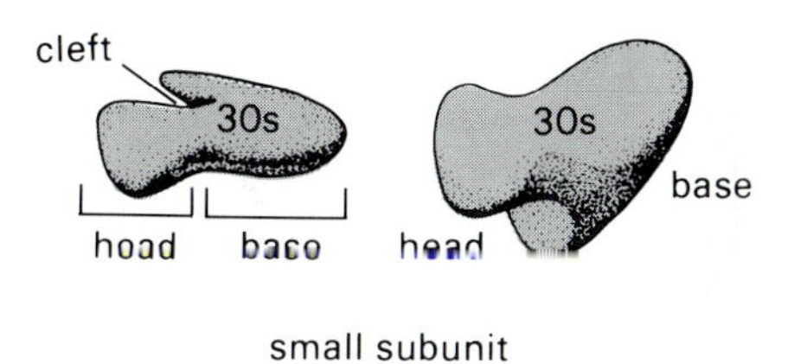

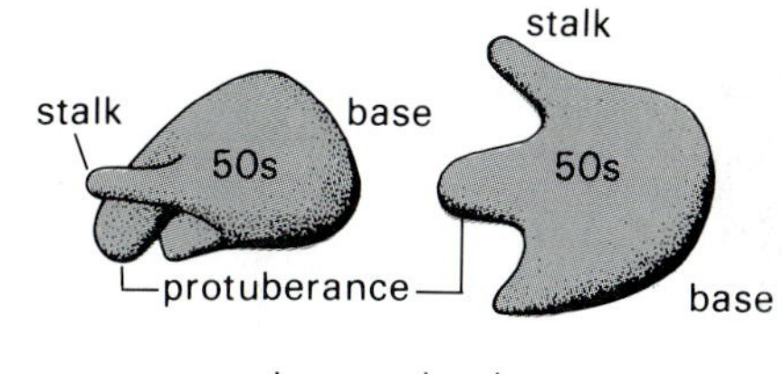

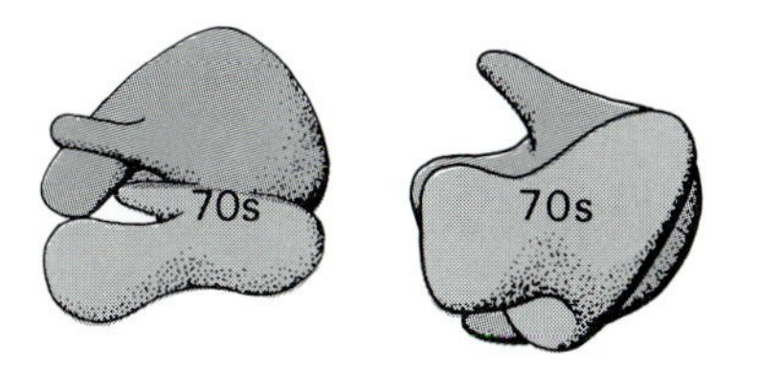

Figure 3.34
Representations of *E. coli* 30S, 50S, and 70S ribosomal particles reconstructed from electron diffraction data. Two views of each particle are shown: they are rotated 90° around the plane of the page relative to each other. After J. A. Lake, *Scientific American* 245 (1981), p. 84.

RNA (18S) of about 1900 nucleotides and 30 to 35 proteins; the large subunit contains 3 RNA chains of 120 (5S), 160 (5.8S), and about 4800 (28S) nucleotides and 45 to 50 proteins. The ribosomes in mitochondria and chloroplasts are different from those of the respective cytosolic ribosomes. Generally, they are smaller with fewer proteins and distinctive rRNAs. Physical and chemical reconstitution studies with the more complex eukaryotic ribosomes have lagged behind those from *E. coli*.

It is useful to identify two functionally important sites that are created by the association of the two subunits to form the 70S ribosome (Figure 3.35). These are the sites where two tRNAs bind, one attached to the growing protein chain (the "P site") and the other bearing the next amino acid to be added (the "A site") (Section 3.6b).

A special tRNA and several accessory proteins are involved in translation Both prokaryotic and eukaryotic organisms have two kinds of tRNA that bind methionine. In prokaryotes, one is referred to as $tRNA_F^{Met}$ and the other as $tRNA_M^{Met}$; in eukaryotes, the two corresponding species are called $tRNA_I^{Met}$ and $tRNA_M^{Met}$. Each of the $tRNA^{Met}$ species in both types of organisms is aminoacylated with methionine by the corresponding methionyl-tRNA synthetases. The $tRNA_F^{Met}$ of prokaryotes and the $tRNA_I^{Met}$ of eukaryotes have unusual properties that permit them to function as adapters in the initiation of polypeptide chain synthesis at appropriate initiator AUG codons. The $tRNA_M^{Met}$ species in both pro- and eukaryotes decode AUG codons within the protein coding sequence.

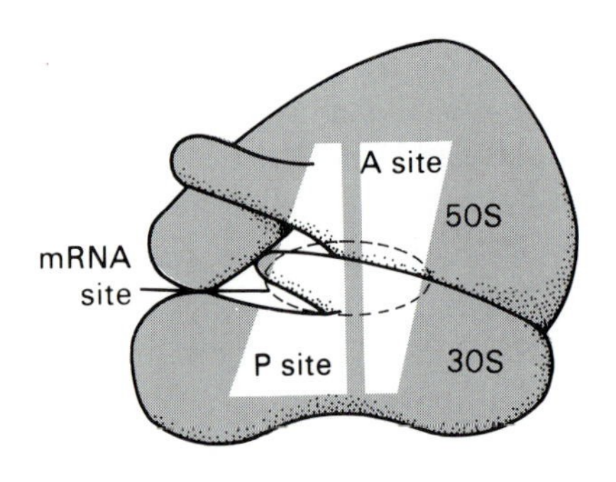

Figure 3.35
Diagram indicating how the P and A sites may be positioned in an *E. coli* 70S ribosome. Elements of both the 30S and 50S subunits probably contribute to both sites. The region believed to be the mRNA binding site on the 30S subunit is also shown.

In prokaryotes, the amino group of methionyl-$tRNA_F^{Met}$ (met-$tRNA_F^{Met}$), but not of methionyl-$tRNA_M^{Met}$ (met-$tRNA_M^{Met}$), is formylated to Fmet-$tRNA_F^{Met}$ by a specific enzyme (methionyl-tRNA-transformylase) using N^{10}-formyltetrahydrofolate as the formyl donor (Figure 3.36). Apparently, the transformylase distinguishes between met-$tRNA_F^{Met}$ and met-$tRNA_M^{Met}$ Fmet-$tRNA_F^{Met}$ is used exclusively to initiate protein chains, and met-$tRNA_M^{Met}$ is used exclusively to decode internal methionine codons. Although $tRNA_I^{Met}$ in eukaryotes serves uniquely for initiation, the methionyl group is not formylated. Evidently, the structural feature needed to endow $tRNA_I^{Met}$ with special initiator function is contained wholly within its nucleotide sequence and/or three-dimensional structure.

Several proteins are only transiently associated with ribosomes during the translation process. These perform essential functions in the initiation, elongation, and termination phases of protein chain assembly. To set the stage for the subsequent discussion of these processes (Section 3.6), we introduce these proteins here and examine briefly their properties and role in the translation machinery.

Three proteins, called initiation factors and abbreviated as IF-1, IF-2, and IF-3, are needed to initiate translation of an mRNA into protein. IF-1 and IF-3 associate with the 30S subunit to enable it to interact with a complex of IF-2 and the unique initiator aminoacyl-tRNA, Fmet-$tRNA_F^{Met}$. The complex of the 30S subunit with all three initiation factors and Fmet-$tRNA_F^{Met}$ is the entity that recognizes the appropriate initiator AUG on the mRNA. The three initiation proteins dissociate when the initiation complex is completed by acquiring the 50S subunit. Two proteins, called elongation factors EF-Tu and EF-Ts, are needed to bring successive aminoacyl-tRNAs to the translating ribosome. EF-Tu, with GTP bound to it, binds all aminoacyl-tRNAs except Fmet-$tRNA_F^{Met}$ and delivers them to the A site of the 70S ribosome-mRNA complex. In the process, EF-Tu-GDP is formed, and EF-Ts is needed to regenerate EF-Tu-GTP. Another protein,

Figure 3.36
Formylation of methionyl-$tRNA_F^{Met}$ in *E. coli.*

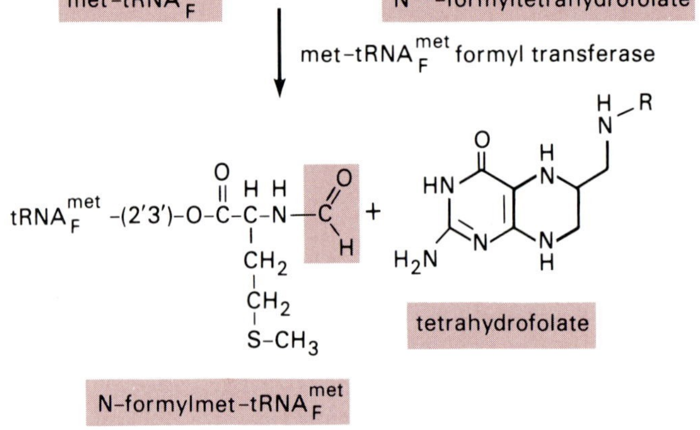

met-$tRNA_F^{met}$ + N^{10}-formyltetrahydrofolate

met-$tRNA_F^{met}$ formyl transferase

N-formylmet-$tRNA_F^{met}$ + tetrahydrofolate

EF-G, functions during elongation by promoting a translocation of the ribosome along the mRNA after each codon is translated. Termination of protein assembly at any of the three stop codons also requires special proteins that are bound only transiently to the translating complex. Three proteins are known: RF-1 causes release of the polypeptide chain in response to UAA and UAG. RF-2 acts similarly with UAA and UGA. RF-3 may facilitate the action of the other two factors.

The number and variety of accessory proteins needed in the various stages of protein assembly in eukaryotic cells are less well known, and the proteins' functions are only poorly understood. There may be more than five proteins needed for initiation, some of which may have functions analogous to those of the prokaryotic IF-2 and IF-3 proteins. The elongation machinery is also more complex, with many more proteins involved both in the binding of aminoacyl-tRNAs to ribosomes and in ribosome translocation. Termination of translation in eukaryotic protein synthesis is also less well understood; moreover, in this instance, only a single protein (eRF) seems to be needed. In this instance, GTP hydrolysis is an important concomitant of either the termination reaction or the release of the protein and mRNA chains.

3.6 mRNA Translation in Prokaryotes

With a knowledge of the participants, we can now examine the chemical events in the assembly of polypeptides—that is, the process of translation itself. Although the process undoubtedly occurs continuously from start to finish, it is customary to delineate three stages: initiation, elongation, and termination. Two general features of the mRNA directed assembly of polypeptide chains serve as a backdrop for considering each of the stages. First, polypeptide chains are assembled unidirectionally, beginning at their amino termini and proceeding toward their carboxyl ends (Figure 3.37). This implies a reaction in which the carboxyl group of the nascent polypeptide chain is joined to the amino group of the next amino acid

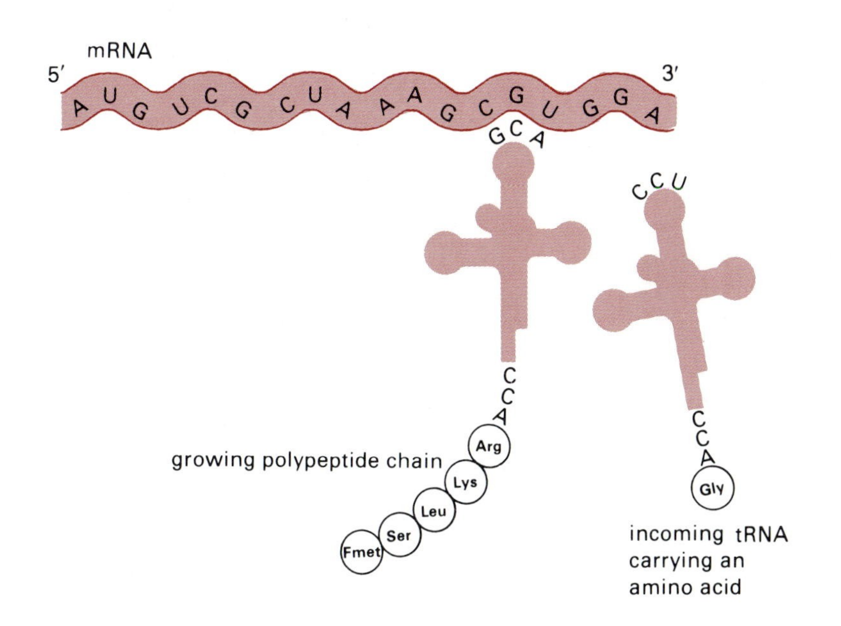

Figure 3.37
The assembly of a polypeptide chain. The mRNA codons are read one by one in the 5′-to-3′ direction, starting with the initiator codon, AUG, which binds to N-formylmethionyl-$tRNA_F^{Met}$. The polypeptide grows from the amino to the carboxyl end.

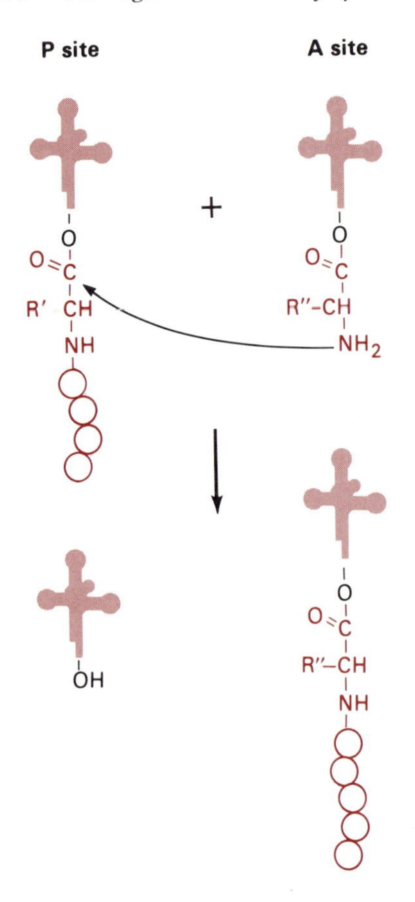

Figure 3.38
Details of the peptidyl transferase reaction. The amino N on the incoming amino acid makes a nucleophilic attack on the activated carboxyl group of the growing chain to form the new peptide bond.

to be added. This could be achieved if the carboxyl end of the growing polypeptide chain is energetically activated to drive the formation of each new peptide bond. We noted earlier that this energetic advantage is provided by the attachment of the carboxyl group of the growing polypeptide chain and of each amino acid to tRNA (Figure 3.38). A second invariant feature of translation is that mRNA decoding begins at a uniquely located AUG codon, which identifies the 5′ end of the coding sequence and specifies the amino terminal amino acid of the nascent polypeptide (Figure 3.37). During intiation, the first and second aminoacyl-tRNAs are matched to the first two codons of the message. Translation then proceeds, codon by codon in the 5′-to-3′ direction on the mRNA, until it reaches a stop signal just beyond the codon specifying the carboxyl terminal amino acid.

a. Requirements for Initiation

Although the 70S ribosome translates the mRNA sequence, it cannot initiate the process. The two initiation proteins, IF-1 and IF-3, cause the dissociation of the 70S ribosome by binding to the 30S subunit. The 30S subunit complexed with IF-1 and IF-3 binds IF-2, GTP, and Fmet-$tRNA_F^{Met}$. This entire complex binds to the region surrounding the AUG codon at the 5′ end of the mRNA's coding sequence (Figure 3.39). Apparently, IF-2 can distinguish Fmet-$tRNA_F^{Met}$ from met-$tRNA_M^{Met}$; part of the specificity derives from the N-formyl group, which is not present on met-$tRNA_M^{Met}$. The formation of a fully functional initiation complex is completed by association of a 50S subunit with the preinitiation complex.

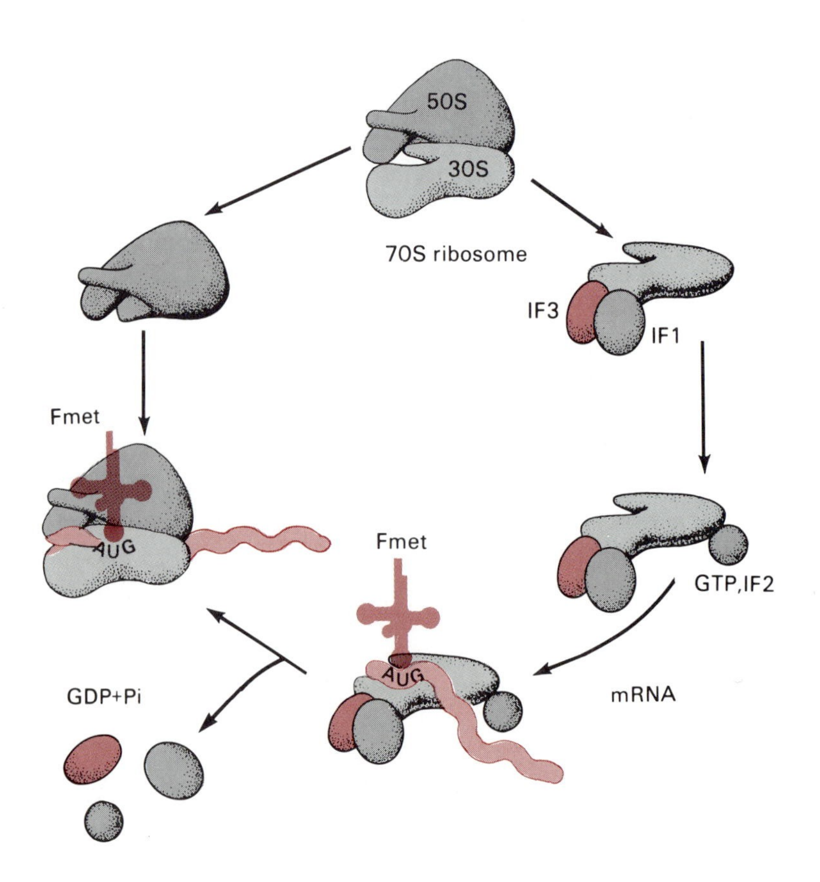

Figure 3.39
Diagram showing the steps that precede translation initiation. Dissociation of the 70S ribosome is induced by the binding of IF-1 and IF-3 to the 30S subunit. This is followed by formation of the preinitiation complex of the 30S subunit with N-formylmethionyl-$tRNA_F^{Met}$ and mRNA. Subsequent association of the complex with the 50S subunit is accompanied by release of IF-1, IF-2, and IF-3 and by hydrolysis of one molecule of GTP to GDP.

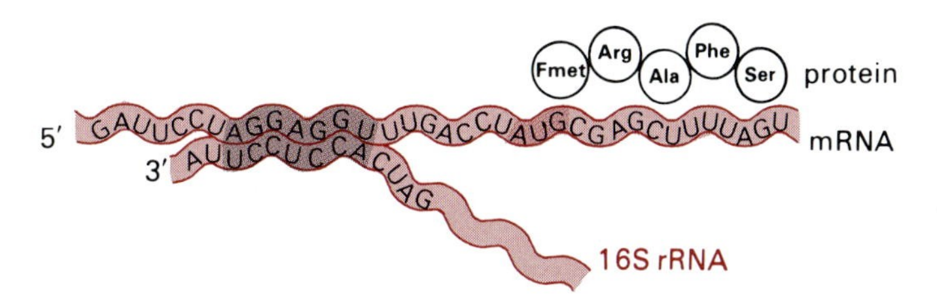

5′AAUCUUGGAGGCUUUUUUAUGGUUCGUUCU3′ ϕX174 gene-A protein
5′UAACUAAGGAUGAAAUGCAUGUCUAAGACA3′ Qβ replicase
5′UCCUAGGAGGUUUGACCUAUGCGAGCUUUU3′ R17 gene-A protein
5′AUGUACUAAGGAGGUUGUAUGGAACAACGC3′ λ gene-cro protein

Figure 3.40
The purine-rich Shine-Dalgarno sequence just upstream of the initiator AUG interacts with a sequence near the 3′ end of *E. coli* 16S rRNA by complementary base pairing. Typical sequences that serve as Shine-Dalgarno sequences in several bacteriophage mRNAs are shown below.

The formation of the functional 70S subunit triggers release of the three initiation proteins.

How is the first codon of the protein sequence recognized? Binding of the 30S subunit to the mRNA is strongly influenced by the nucleotide sequence about ten nucleotides 5′ to the initiator codon. The interaction is promoted by complementary base pairing between this five- to eight-base purine-rich sequence, termed the **Shine-Dalgarno sequence**, with a pyrimidine-rich stretch near the 3′ end of the 16S rRNA (Figure 3.40). The extent of complementarity between the Shine-Dalgarno and 16S rRNA sequences, as well as the distance of the stretch of purines from the AUG codon, strongly influence the efficiency of initiation. Indeed, this requirement, along with the constraints mentioned later, help explain differences in the efficiency of translation initiation with different mRNAs.

The mRNA secondary structure surrounding the initiator AUG is another variable in the initiation step. Initiation is blocked or inefficient if the AUG initiator codon is sequestered in intramolecular base paired regions (Figure 3.41). Indeed, the accessibility of an initiator AUG to the 30S ribosome can be regulated in this way. Thus, an initiator AUG may be inaccessible when it is base paired in a folded form of the mature mRNA but be available for initiation during transcription of the mRNA or during translation of other coding sequences on the same mRNA.

b. Polypeptide Chain Elongation

The association of the two ribosome subunits upon initiation of translation creates the two functional sites needed for protein assembly: the P site and the A site. Fmet-tRNA$_F^{Met}$ occupies the P site, and formation of the first peptide bond requires that the aminoacyl-tRNA corresponding to the next codon be in the A site. To fill the A site, aminoacyl-tRNAs must first bind EF-Tu and GTP. The resulting ternary complex (aminoacyl-tRNA-[EF-Tu-GTP]) deposits the aminoacyl-tRNA into the A site (Figure 3.42 on page 172). GTP is hydrolyzed during this process, and the complex (EF-Tu-GDP) is released from the ribosome. With both the P and A sites occupied, the peptidyl transferase activity of the 50S subunit catalyzes the transfer of the Fmet group from its tRNA to the amino group of the aminoacyl-tRNA in the A site. A dipeptidyl-tRNA now occupies the A site, and a free tRNA occupies the P site.

To read the next codon and extend the polypeptide chain with an additional amino acid, the whole series of reactions must be repeated. Before

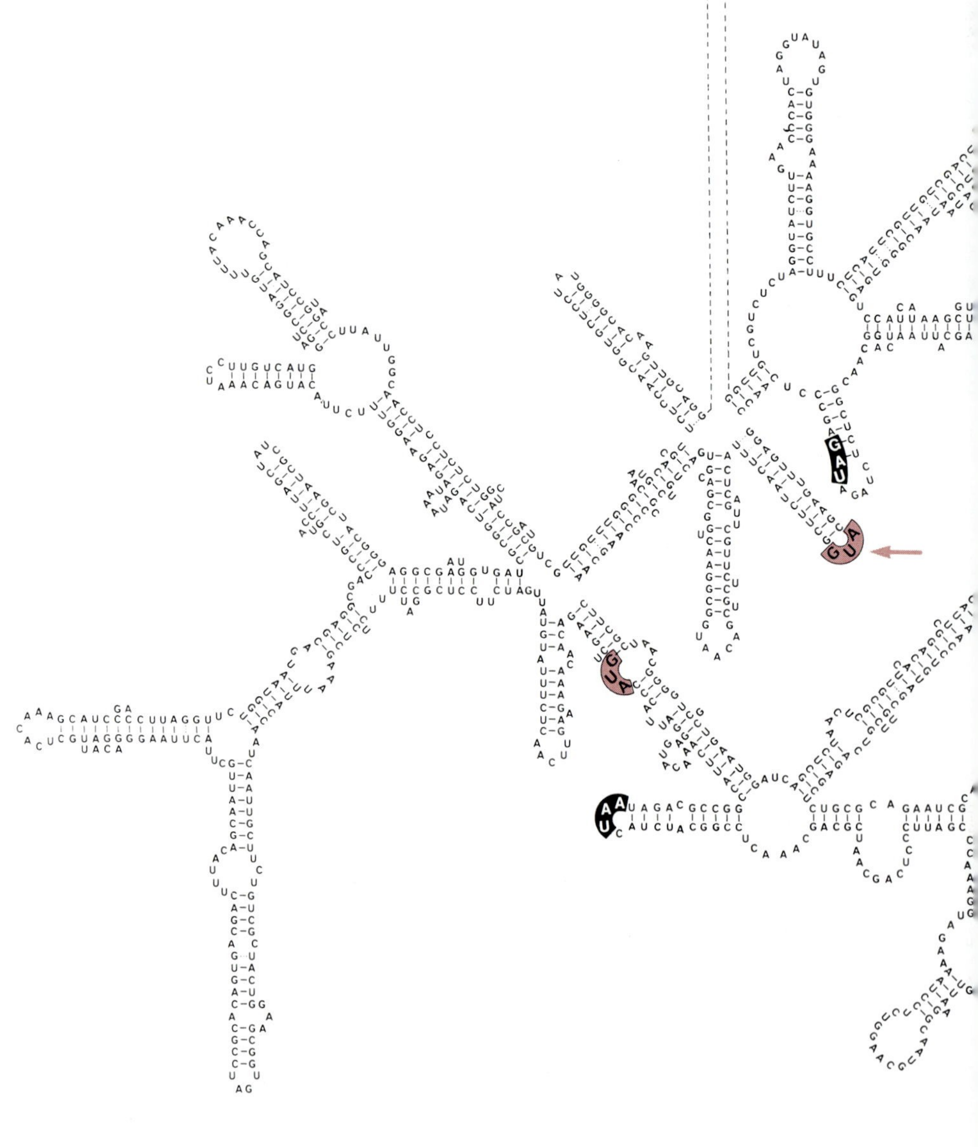

Figure 3.41
Primary and proposed secondary structure of the 5′ portion of the bacteriophage MS2 genome. Intramolecular base pairing sequesters several initiator codons (including the unusual GUG closest to the 5′ end). One AUG (arrow), which is exposed at a hairpin turn, is accessible and binds to ribosomes immediately after infection. The open reading frame following the AUG encodes the MS2 coat protein. Note that the Shine-Dalgarno sequence upstream of this AUG is also relatively accessible in this model; two G residues form relatively weak hydrogen bonds with Us. Courtesy of W. Fiers.

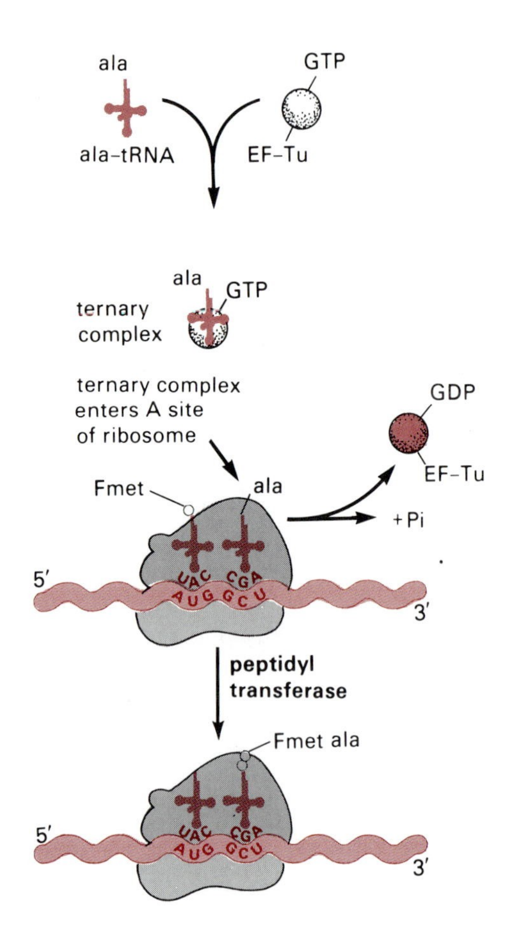

Figure 3.42
Formation of the first peptide bond. The preinitiation complex of 70S ribosome with mRNA and with N-formylmethionyl-tRNA$^{Met}_{F}$ in the P site interacts with a ternary complex of alanyl-tRNA^{Ala}, EF-Tu, and GTP. The alanyl-tRNA^{Ala} enters the A site, interacting with the second codon, GCU, through the anticodon, 5′-AGC-3′. The GTP is hydrolyzed, and a complex of EF-Tu with GDP is released. After the peptidyl transferase reaction, a dipeptidyl-tRNA occupies the A site.

that can happen, however, the tRNA in the P site must be ejected, the newly made dipeptidyl-tRNA must be shifted to the P site, and a new codon must be brought in to occupy the vacant A site (Figure 3.43). EF-G mediates this step by a GTP-dependent translocation of the ribosome. The breakdown of GTP to GDP provides the energy to move the ribosome to the next triplet in the coding sequence and to eject the free tRNA from the P site. The new codon in the A site is now available to pair with its cognate aminoacyl-tRNA.

A functional EF-Tu-GTP must be regenerated from the EF-Tu-GDP complex that is released each time an aminoacyl-tRNA is deposited into the A site (Figure 3.44). The EF-Tu-GDP reacts with the protein EF-Ts to dissociate GDP and form the complex, EF-Tu·EF-Ts. Thereafter, EF-Tu·EF-Ts reacts with GTP to regenerate EF-Tu-GTP and release EF-Ts, both of which are then available for another cycle.

Several features of the elongation process should be stressed. (1) The formation of each peptide bond consumes four equivalents of phosphate bond energy—two equivalents of ATP during the aminoacylation of tRNA and two of GTP during each cycle of chain elongation. (2) During the initiation of translation, Fmet-tRNA$^{Met}_{F}$ is recognized to the exclusion of all other aminoacyl-tRNAs by IF-2, and EF-Tu discriminates between met-tRNA$^{Met}_{M}$ and Fmet-tRNA$^{Met}_{F}$ for insertion into the A site. (3) The elongation factors EF-Tu and EF-G shuttle on and off the ribosome, depending upon whether they are bound to GTP (on) or GDP (off). (4) The growing peptide chain is always attached via its carboxyl end to a tRNA, the tRNA being the cognate of the carboxy terminal amino acid in the

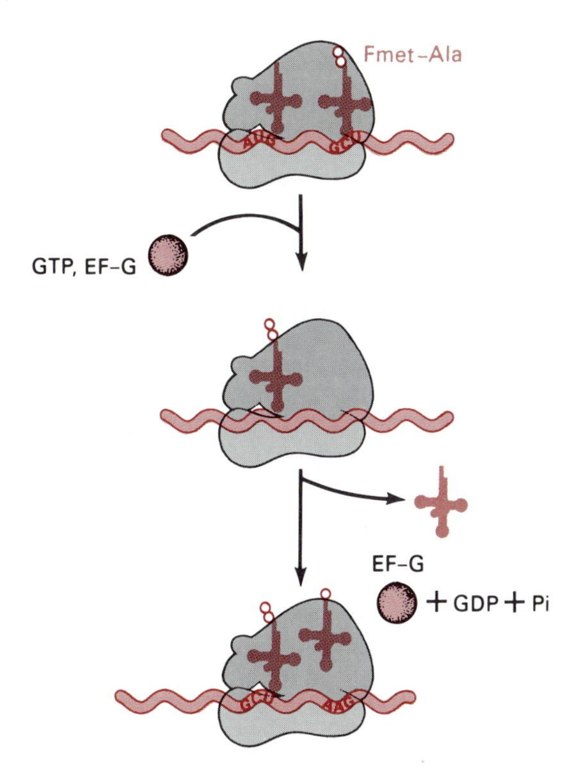

Figure 3.43
Polypeptide chain elongation. Factor EF-G mediates the GTP-dependent translocation of the 70S ribosome relative to mRNA. In the process, the GTP is hydrolyzed, the tRNA in the P site is ejected, and the peptidyl-tRNA is shifted from the A site to the vacant P site. The A site and the next codon can now receive the cognate aminoacyl-tRNA.

growing chain. (5) Peptidyl transferase catalyzes the formation of peptide bonds between the carboxy terminus of the growing chain and the amino group of an aminoacyl-tRNA.

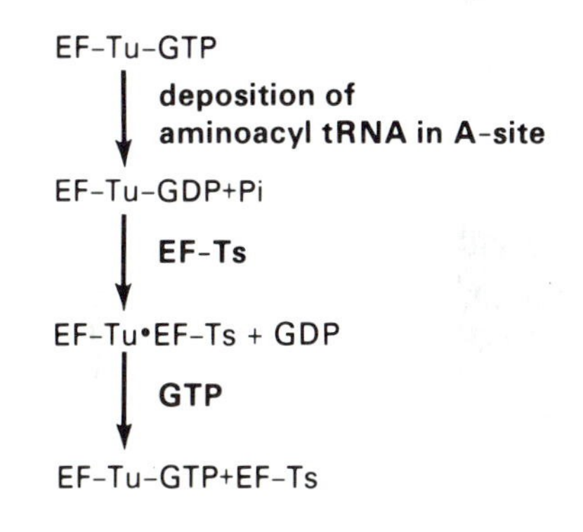

Figure 3.44
Regeneration of EF-Tu-GTP after release of EF-Tu-GDP from the complex with aminoacyl-tRNA.

c. *Termination of Polypeptide Chain Elongation*

The succession of codon translations ultimately brings one of the three termination codons, UAG, UAA, or UGA, into the A site (Figure 3.45). Lacking tRNAs to translate these codons, the polypeptidyl-tRNA remains in the P site. Specific release factors intervene and catalyze cleavage of the polypeptide chain from the tRNA, release of both from the ribosome, and dissociation of the 70S ribosome from the mRNA. Release factor RF-1 recognizes either UAA or UAG in the A site; RF-2 functions when either UAA or UGA appears in the A site; RF-3 may act to facilitate the action of the other two proteins. UAA may be more efficient in terminating the translation process because either RF-1 or RF-2 can trigger the termination with that codon. However, the efficiency of termination by any of the stop codons is influenced by the flanking sequences in the mRNA. Although the outlines and some features of the termination reaction are known, the mechanism of the termination reaction and the precise catalytic role of the release factors are unclear.

3.7 Some Notable Features of the Translation Process

The previous section detailed the events of a single translation cycle of an mRNA by a ribosome. Here the emphasis is on several features of the overall process and how its various stages may be subverted or blocked.

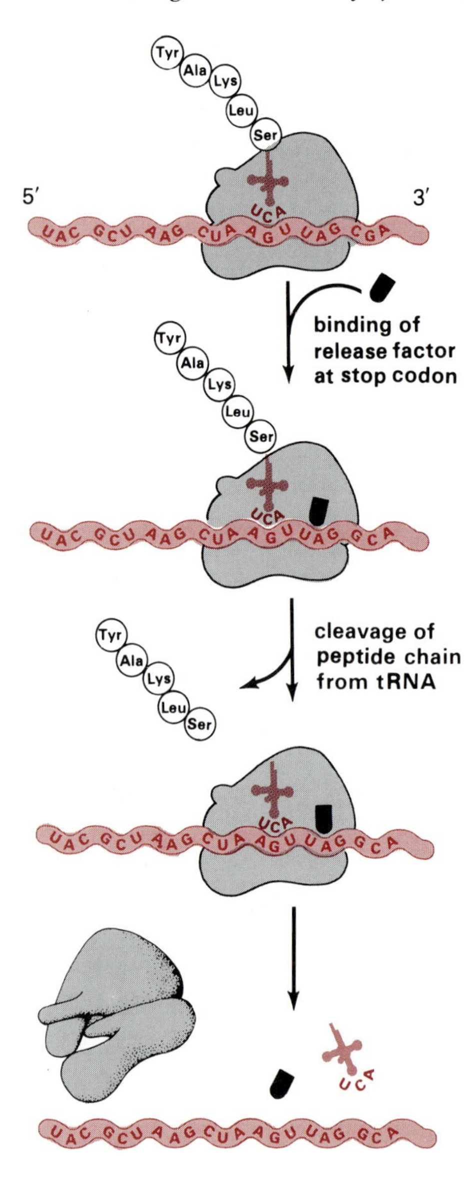

Figure 3.45
Termination of translation. When a stop codon moves into the A site, release factors initiate a series of reactions leading to release of the new polypeptide chain, the tRNA, and the mRNA from the 70S ribosome.

a. *Simultaneous Translation of mRNAs by More Than One Ribosome*

Following initiation, the 70S ribosome moves away from the initiation region as it translates each successive codon. By the time the ribosome has moved 100 to 200 nucleotides along the mRNA, the initiation region is available for a second initiation event (Figure 3.46). Furthermore, after the second ribosome translates the initiation sequence, a third initiation can occur, and so on. Thus, an mRNA may have several ribosomes translating the same protein coding sequence simultaneously. Such multiribosome translation complexes are called **polyribosomes** or **polysomes**. Each ribosome in a polysome eventually translates the entire coding sequence and yields a full-length polypeptide. Each individual ribosome in a polysome carries a length of nascent polypeptide that increases in proportion to how far the ribosome has moved from the 5′ end of the coding sequence.

b. *Translation of Bacterial mRNAs Can Occur During Transcription*

Transcription of a gene or cluster of genes produces the 5′ end of the mRNA sequence first and then progresses toward the 3′ end (Section 3.2). Consequently, the formation of the translation initiation complex can occur almost as soon as the sequence surrounding the initiation codon is transcribed (Figure 3.47). Indeed, synthesis of the polypeptide chain generally begins even before the 3′ proximal region of the mRNA is completed.

In those instances where the bacterial transcription unit contains more than one protein coding sequence, for example, the *trp* or *lac* operons (Section 3.11), ribosomes can initiate and complete translation of the first coding sequence while the subsequent coding sequences are still being transcribed. Messenger RNAs which contain multiple different protein coding sequences are often translated sequentially. Thus, initiation, elongation, and termination of the first coding sequence is followed by the same progression on the second, third, or subsequent coding segments. After ribosomes reach the termination signal of the first or subsequent coding sequences, they usually dissociate from the mRNA, and a new initiation complex is assembled at the next initiator sequence. However, under certain circumstances, ribosomes may not dissociate from the mRNA, instead migrating along the mRNA to form a new initiation complex at the next initiation sequence.

c. *Ribosomes Recycle After Translating a Coding Sequence*

We have already noted that, when a translating 70S ribosome reaches a termination codon, the completed polypeptide chain is split from the tRNA, both the polypeptide and the tRNA chains are released from the ribosome, and the 70S ribosome dissociates from the mRNA. In their 70S form, ribosomes cannot initiate new rounds of polypeptide synthesis, so they must first be dissociated into their component 50S and 30S subunits. This dissociation is driven by the initiation factor IF-3 in conjunction with

IF-1 (Figure 3.48). The "ribosome cycle" is completed by the association of the 50S subunit with the mRNA-bound 30S subunit containing IF-2, Fmet-tRNA$_F^{Met}$, and GTP to form the functional 70S translation apparatus (Figure 3.39). By modulating the amount of 30S and 50S subunits relative to their precursor 70S ribosomes, IF-3 serves as a general control on the rate of protein synthesis.

Figure 3.46
Formation of a polyribosome. After one ribosome begins to elongate the polypeptide chain, a second one can form an initiation complex at the same start site and begin to decode the mRNA. Thereafter, additional ribosomes initiate polypeptide synthesis so that multiple polypeptides are synthesized simultaneously on the same mRNA.

d. Codon-Anticodon Interactions

Most tRNAs pair with more than one codon Because codons are translated by means of the anticodon sequence of tRNAs, one might have expected that a distinctive tRNA would exist for each of the 61 codons that specify amino acids. But there are not, for example, different tRNAs for each of the four valine or glycine codons; nor are there different tRNAs to translate the two tyrosine or two lysine codons. Indeed, both *in vitro* and *in vivo* experiments have proven that some tRNAs can translate more than one codon. Thus, a single tRNA^{Tyr} species translates the codons UAU and UAC (Figure 3.49 on page 178). Because the single tRNA^{Tyr} contains the anticodon sequence 5′-GUA-3′, it can form complementary base pairs with the first two bases of either codon (remember that base pairing occurs between antiparallel sequence arrangements). Apparently, G is able to pair acceptably with both U and C in the third position of the codon; similarly, U at the 5′ end of an anticodon allows the tRNA to pair with an A or G in the 3′ end of codons. In fact, the translation of all pairs of codons that have U or C in the third (3′) position can be mediated with a single tRNA that uses G or a modified base in the first (5′) base of the anticodon. The matching of codons and anticodons at the P and A sites of the translating ribosome probably involves stabilizing interactions other than the usual complementary base pairing.

An examination of the genetic code shows, however, that there must also be specific interactions that can distinguish between codons having A or G in the third position. For example, the tRNA that decodes the AUG codon as methionine must distinguish this triplet from AUA, which specifies isoleucine. Furthermore, tRNA^{Trp} translates UGG but not the termination codon UGA. The specificity of both of these decoding operations is achieved by the use of C in the anticodon to pair with G in the codon's third position (Figure 3.49).

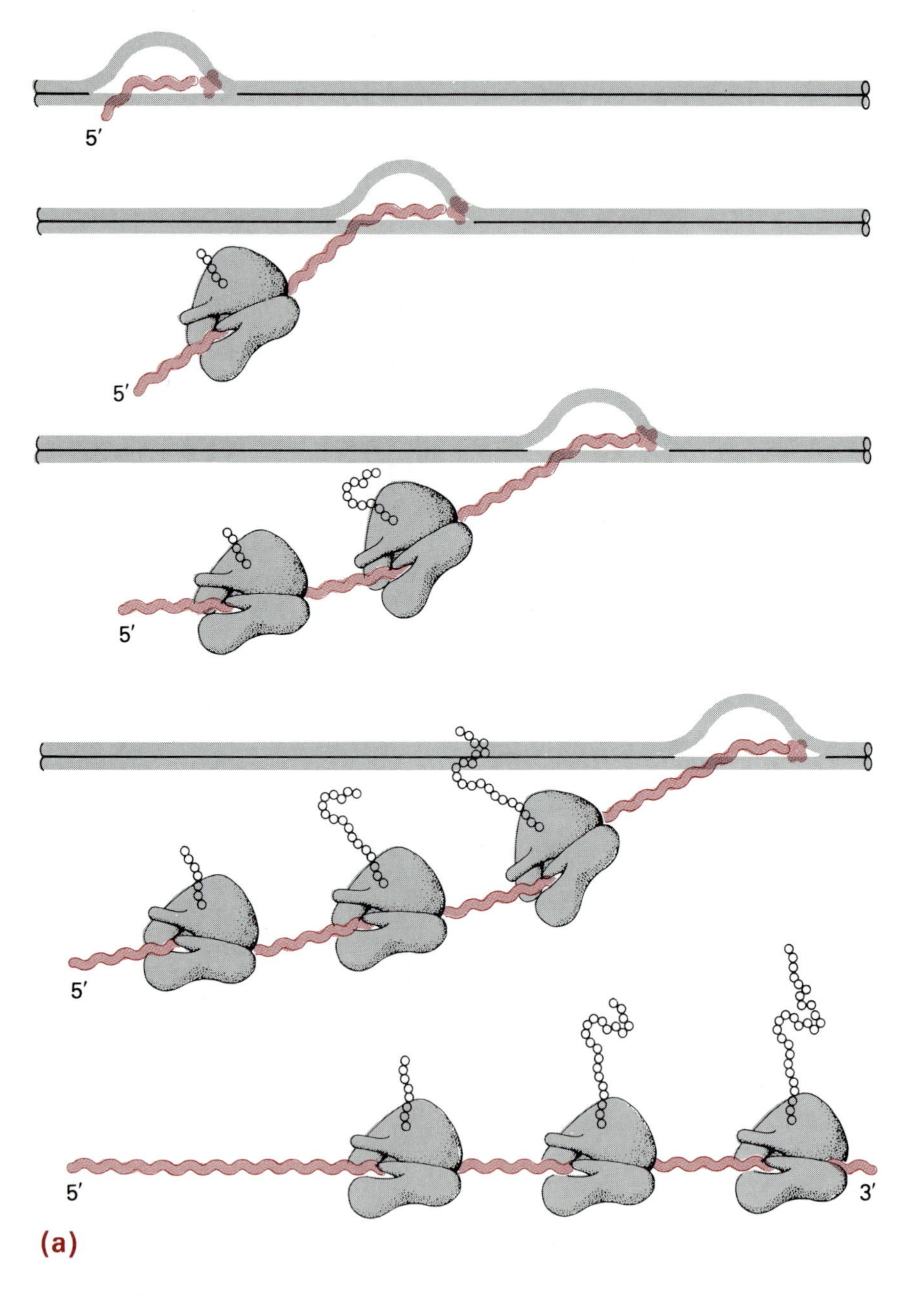

Figure 3.47
Transcription and translation in prokaryotes occur concurrently. (a) Diagrammatic representation of simultaneous transcription of DNA and translation of the mRNA by polyribosomes. (b) FACING PAGE, TOP: An electron micrograph showing multiple ribosomes associated with RNA molecules being transcribed from *E. coli* DNA. The larger RNA chains, which are farthest from the transcription start site, have a larger number of ribosomes (magnification 142,000x). From O. L. Miller, Jr., B. A. Hamkalo, and C. A. Thomas, Jr., *Science* 169 (1970), p. 392.

Modifications in anticodon bases can further restrict codon-anticodon interactions. For example, hypoxanthine (Hx) in place of adenine at the anticodon position that pairs with the third base of the codon permits pairing with codons having U, C, or A in the third position. A variety of modifications of bases in the anticodon, or more frequently adjacent to the anticodon, alters the specificity of the aminoacyl-tRNA-codon interaction. Generally, these prevent misreading of the third position of codons and promote the fidelity of the decoding process.

The base pairing rules that govern multicodon translations by single tRNAs and prevent the misreading of two noncognate codons by a single tRNA are referred to as the "wobble" rules (Table 3.4). It should be emphasized, however, that the word "wobble," which is used to describe the relaxed pairing that occurs at the third base of the codon, glosses over our ignorance about the precise chemical and structural features that guide codon-anticodon interactions in the P and A sites of the ribosome.

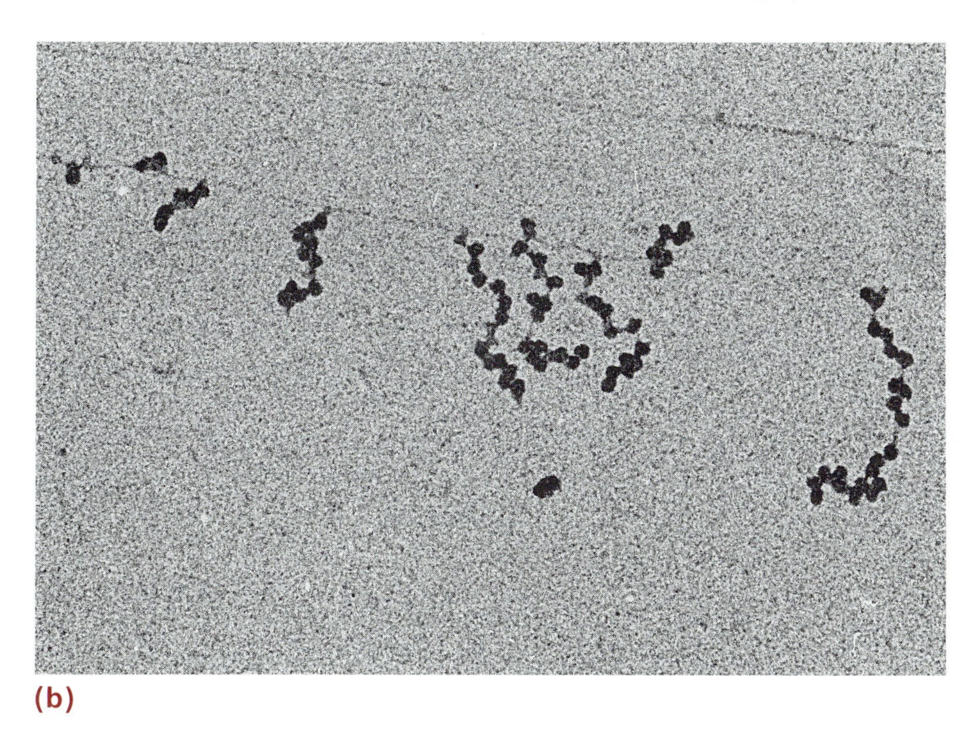

(b)

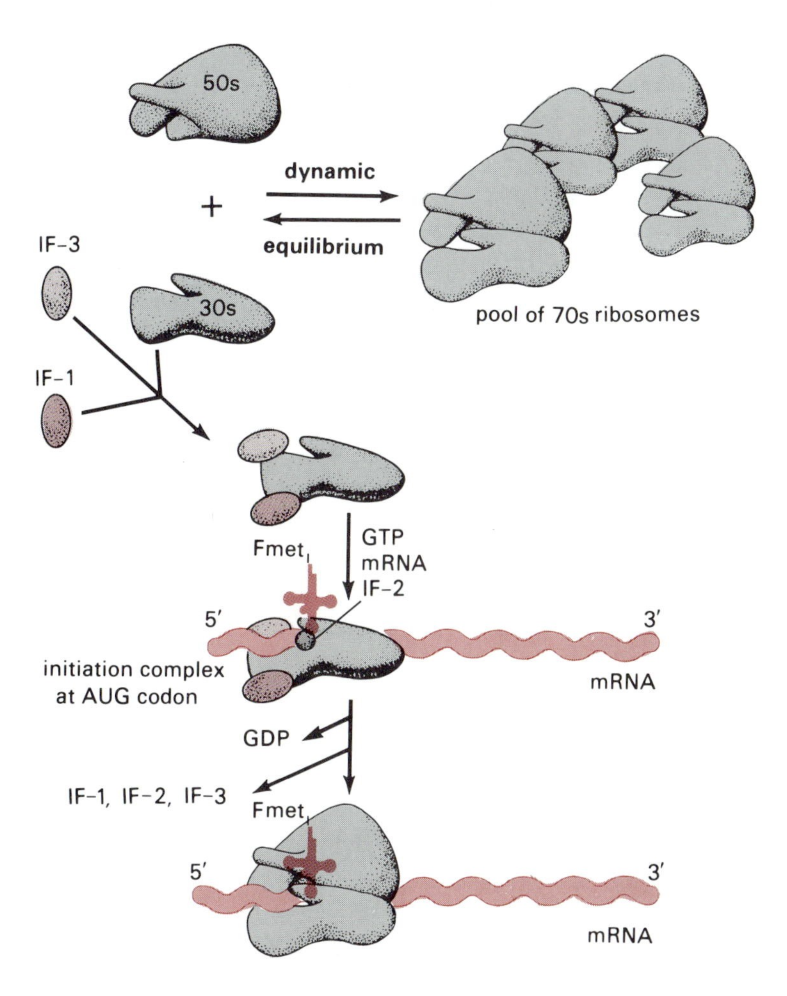

Figure 3.48
The recycling of 70S ribosomes during protein synthesis.

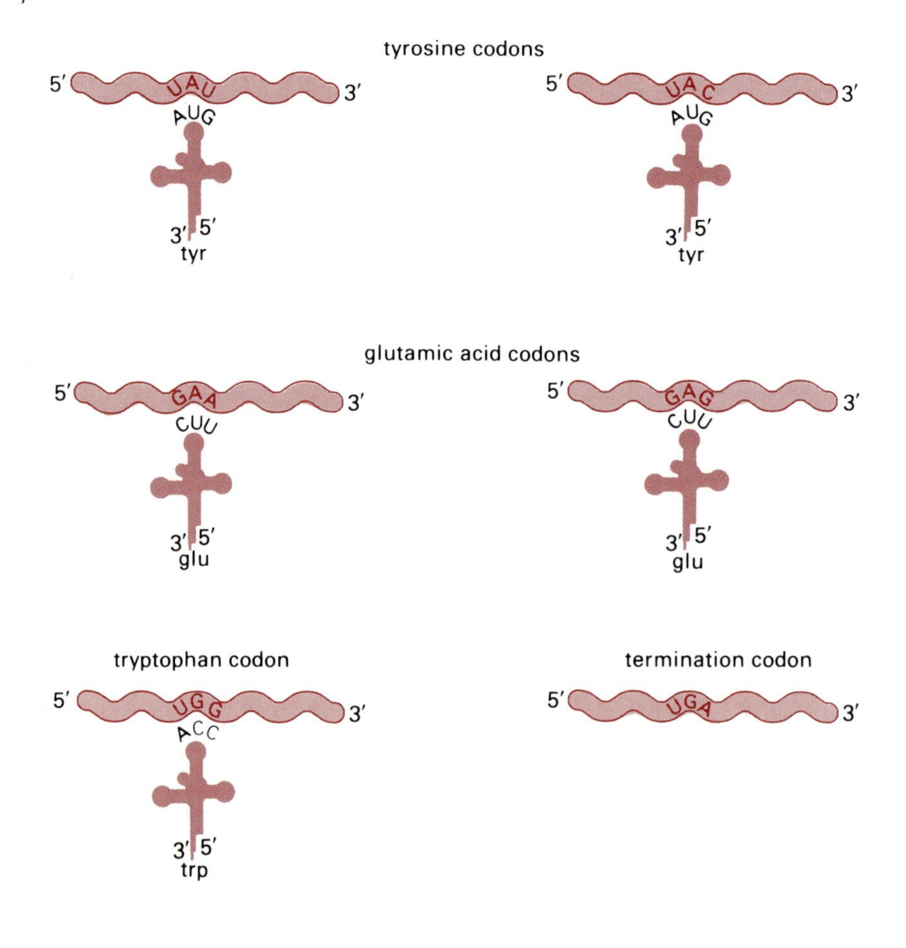

Figure 3.49
Most tRNAs pair with more than one codon. The tyrosine codons UAU and UAC are translated by a single $tRNA^{Tyr}$. Similarly, the glutamic acid codons GAA and GAG are translated by one $tRNA^{Glu}$. However, $tRNA^{Trp}$ recognizes the tryptophan codon, UGG, but not the termination codon, UGA.

Mutations in codons and anticodons Mutations that affect different components of the translation process can alter the outcome of the readout of the coding sequence. Most seriously, mutations in a protein coding gene that convert an amino acid to a termination codon prevent completion of the protein because translation is prematurely terminated at the site of the mutation (Figure 3.50). Examples include conversion of the lysine codon AAA to UAA and the glutamine codon CAG to UAG. Similarly, any mutation that converts an amino acid codon to UGA also causes premature termination of polypeptide chain synthesis. However, if a second

Table 3.4 Types of Base Pairing Proposed by the Wobble Hypothesis

Base in Anticodon That Pairs with Third Position in Codon	Base in Third Position of Codon
G	U or C
C	G
A	U
U	A or G
Hx	A or U or C

U, C, A, G refer to the usual four bases in RNA. Hx refers to the deaminated form of adenine, hypoxanthine.

wild type mRNA	AUC CUA AAA UCU GUA AUA CAG AUG GGC
	↓ mutations causing termination
mutant $mRNA_1$	AUC CUA UAA UCU GUA AUA CAG AUG GGC
	or
mutant $mRNA_2$	AUC CUA AAA UCU GUA AUA UAG AUG GGC
	↓ polypeptides produced
wild type polypeptide	ile–leu–lys–ser–val–ile–gln–met–gly–
mutant $polypeptide_1$	ile–leu–COO^-
mutant $polypeptide_2$	ile–leu–lys–ser–val–ile–COO^-

Figure 3.50
Mutations of internal codons to terminator (stop) codons cause premature termination of translation.

mutation changes the appropriate base in the anticodon of a tRNA, termination is **suppressed**, and a complete protein, albeit an altered one, is produced (Figure 3.51). For example, if $tRNA^{Tyr}$, $tRNA^{Leu}$ or $tRNA^{Ser}$ is changed in this way, UAG can be translated as an amino acid. The mutant UAA or UGA codons can also be mistranslated by a different mechanism. Mutations in tRNA genes, other than in the bases specifying the anticodon, may modify the specificity or stability of codon-anticodon interactions. As a consequence of these mechanisms, premature termination of polypeptide assembly can be prevented by translating the terminator codon as an amino acid. Such read-through, or suppression of termination, is generally inefficient; therefore, both the prematurely terminated polypeptide and the completed chain are produced. Because such translational suppression is relatively inefficient, there is no deleterious effect from occasional read-through of the terminator codon at the natural ends of mRNAs.

Missense mutations, that is, those that cause amino acid substitutions and occasionally loss of protein function, can also be reversed by suppressor mutations that cause the mutant codon to be misread (Figure 3.52). This can occur if a tRNA, which can carry the normal amino acid or any other amino acid that is acceptable at that position, acquires an anticodon that pairs with the mutant codon. Mutations that change the reading frame in a coding sequence can also be suppressed if mutationally altered tRNAs or ribosomes occasionally translate two or four bases at a time.

Thus, translation errors can reverse the effects of coding sequence changes. Mutational alteration of the anticodons of tRNAs is the most prevalent mechanism of suppression, but changes in other parts of the tRNA can cause errors in the esterification of amino acids by the aminoacyl-tRNA synthetase or mispairing at the ribosome. Mistranslation may also occur if mutations alter the protein or RNA components of the

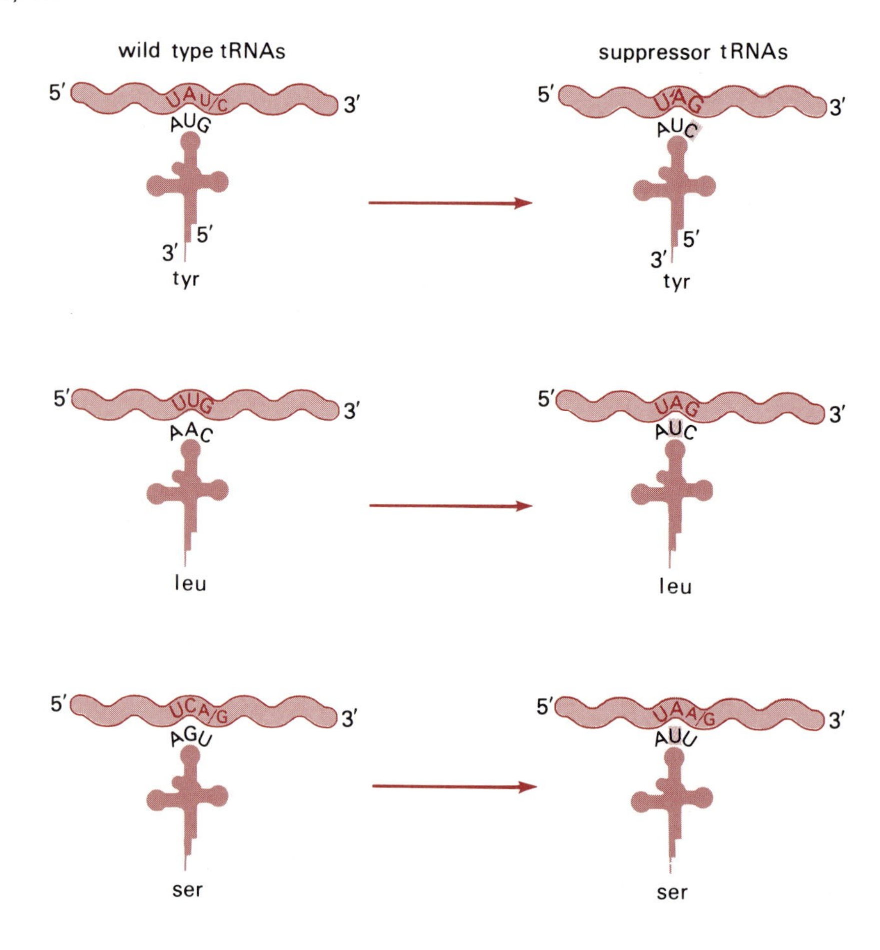

Figure 3.51
Translational suppression of terminator codons. Mutations causing a change in the anticodons of $tRNA^{Tyr}$, $tRNA^{Leu}$, or $tRNA^{Ser}$ permit translation of termination codons as amino acids.

ribosome that are involved in the codon-anticodon interactions. Also, certain chemicals such as aminoglycosides (e.g., streptomycin) bind to ribosomal proteins in the 30S subunit and can alter the fidelity of translation. In these cases, there is a more widespread breakdown in the accuracy of the translation process.

3.8 mRNA Translation in Eukaryotes

Translation of eukaryotic mRNA is basically similar to that of prokaryotic mRNA. With the exceptions already noted, the genetic code is identical, and the codons are translated successively by aminoacyl-tRNAs in conjunction with ribosomes. There are, however, three notable differences imposed by certain characteristics of eukaryotic cells. First, the transcription and translation machinery in eukaryotes are physically separated, transcription occurring in the nucleus and translation in the cytoplasm. Second, the 5′ and 3′ ends of eukaryotic mRNAs have special structures. Third, with the exception of the mRNAs transcribed from the DNA genomes of viruses, eukaryotic mRNAs usually contain only a single protein coding sequence.

At present, we know considerably less about the structures and properties of the participants in eukaryotic translation than of their counterparts in prokaryotes. Although the same three stages—initiation, elongation, and termination—are discernable in eukaryotes, each is more complex in the number of extraribosomal protein factors that are required. In spite of the differences, protein coding sequences from prokaryotes are readily translated by the eukaryotic translation system, provided that their mRNAs possess the appropriate modifications at the 5′ and 3′ termini (Section 3.8a). Conversely, eukaryotic protein coding sequences are translated efficiently in prokaryotes, provided they contain a Shine-Delgarno sequence 5′ to the initiator AUG. This means that the translation machinery of both types of organisms can contend with the nucleotide sequence arrangements in mRNAs from whatever source.

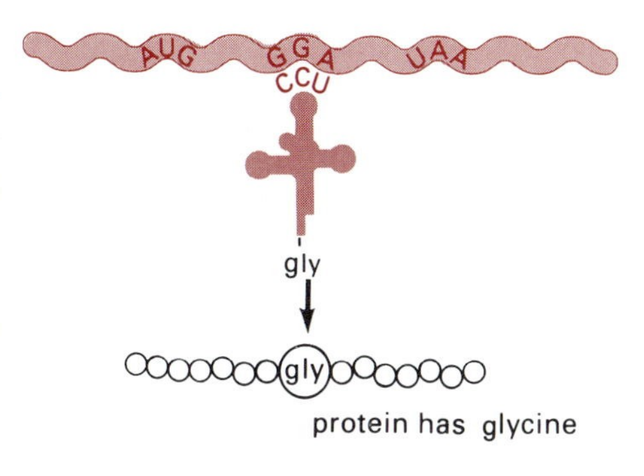

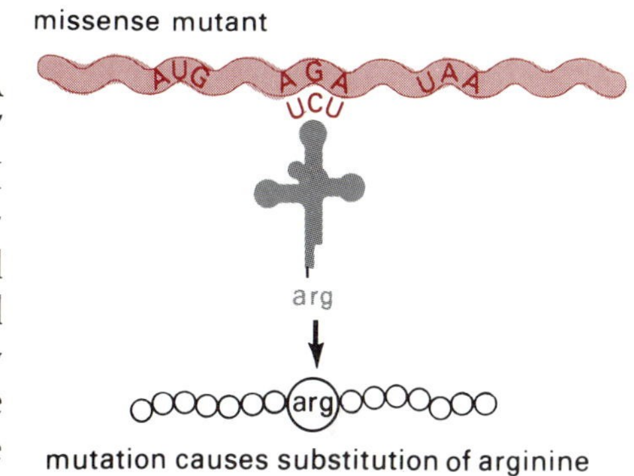

Figure 3.52
Translational suppression of missense mutations. In this example, a $tRNA^{Gly}$ acquires an altered anticodon that permits insertion of glycine at an arginine codon, AGA.

a. Special Modifications in Eukaryotic mRNAs

Eukaryotic mRNAs transcribed from nuclear or viral DNA genes by RNA polymerase II always have a modified 5′ terminus, referred to as a "cap" (Figure 3.53). But RNAs transcribed by eukaryotic RNA polymerases I (i.e., rRNAs) and III (5S and tRNAs) are not capped and retain their original 5′-triphosphate termini. Most of the mRNA produced by animal RNA viruses is also capped, even though it is produced by virus encoded RNA transcriptases. Most uncapped mRNAs are poorly translated by eukaryotic protein synthesizing systems because of inefficient ribosome binding to the mRNA. Capping occurs at the 5′ nucleoside triphosphate and shortly after RNA transcripts are initiated, well before the transcript is completed. The details of the capping process are presented in Section 8.3c.

Eukaryotic mRNAs also contain a polyadenylate sequence at their 3′ ends. This 3′ "tail," which is 50 to 200 adenylate residues long, is not encoded in the sequence of protein coding genes, but is added posttranscriptionally after cleavage of the transcript at a specific sequence beyond the translation termination signal (Section 8.3c).

b. Initiation of Translation by Small Ribosomal Subunits at the 5′ Capped Ends of mRNAs

Just as the dissociation of 70S ribosomes is an obligatory step for initiating the translation of prokaryotic mRNA, the 80S ribosomes must be dissociated before the translation of eukaryotic mRNAs can begin. The small subunit (40S), in association with an array of accessory proteins (eIFs), one or more of which are needed to dissociate the ribosome into its component subunits, binds the special initiator met-$tRNA_I^{Met}$. Here, too, GTP and a special protein—eIF-2—are needed to bind the initiator aminoacyl-tRNA. In eukaryotes, however, met-$tRNA_I^{Met}$ is not N-formylated; but as in prokaryotes, the structure of $tRNA_I^{Met}$ is different from $tRNA_M^{Met}$. The 40S complex containing met-$tRNA_I^{Met}$, GTP, and a panoply of other eIFs binds to the mRNA at or near the capped 5′ end; at least one of the factors recognizes and binds to the cap structure. The small subunit moves from the capped terminus to the first AUG downstream from the 5′ end by a mechanism that is still unknown. At present, there is no requirement for a nucleotide sequence comparable to the Shine-Delgarno sequence near the AUG. However, the efficiency with which an AUG serves as an initiator codon is influenced by certain nucleotides on both sides of the AUG. This preinitiation complex com-

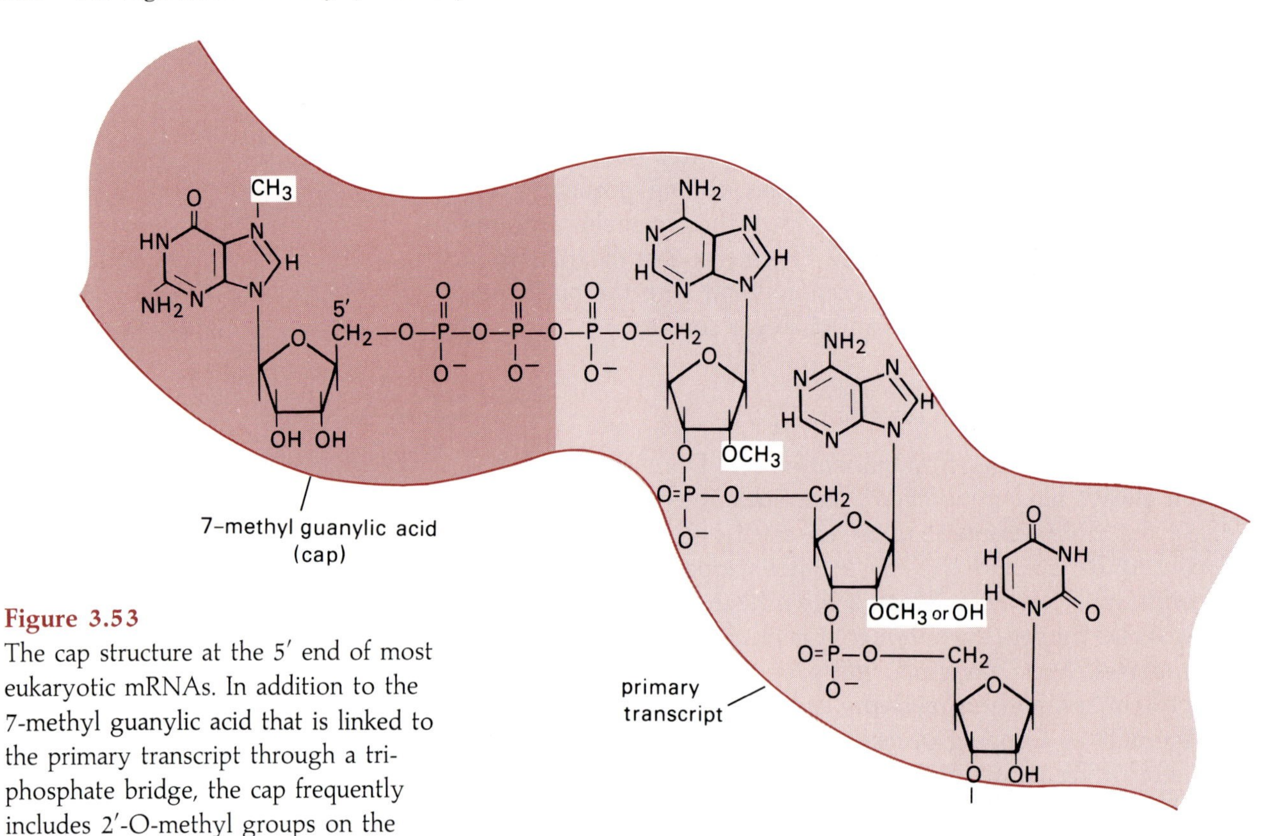

Figure 3.53
The cap structure at the 5′ end of most eukaryotic mRNAs. In addition to the 7-methyl guanylic acid that is linked to the primary transcript through a triphosphate bridge, the cap frequently includes 2′-O-methyl groups on the first or first two ribose residues in the transcript.

bines with the 60S subunit in an energy- and factor-dependent reaction to create the functional initiation complex. The initiation of protein synthesis can be regulated by phosphorylation and dephosphorylation of an eIF-2.

c. *Polypeptide Chain Elongation and Termination*

The stepwise translation of successive codons with aminoacyl-tRNAs is not substantially different in eukaryotes and prokaryotes. GTP and the elongation factor, eEF-1, which corresponds to the prokaryotic complex of EF-Tu and EF-Ts, cycle to bring aminoacyl-tRNAs to the ribosomes. GTP and eEF-2, which is functionally analogous to prokaryotic EF-G, promote the translocation operation. Termination of translation in eukaryotes also occurs at any one of the three stop codons, causing release of free polypeptide chains, tRNA, and, very likely, the 80S ribosome from the mRNA. One factor, eRF, and GTP appear to mediate the entire series of termination events.

The biogenesis of eukaryotic mRNA is discussed in detail in Section 8.3, but here we need to stress that translation does not occur until the mature mRNA reaches the cytoplasm, whereupon initiation can occur as mentioned previously. Polysomes are formed by successive initiations, and simultaneous, multiple translations of the protein coding sequence occur as in prokaryotes. A notable difference, however, is that cellular mRNAs generally contain only a single protein coding sequence. When mRNAs contain consecutive coding regions, as occurs with many animal DNA viruses, the downstream coding sequences are inefficiently or not translated.

3.9 Inhibitors of Transcription and Translation

A large number of natural or synthetic compounds have been screened for their potential antimicrobial or anticancer activity. Such searches have identified many compounds with profound inhibitory activity on either transcription or translation in pro- and eukaryotes. Most are nonspecific and therefore too toxic for therapeutic use. But some have proven to be extremely useful experimental tools for dissecting the machinery for RNA and protein synthesis. Here we mention just a few of the compounds that have been helpful.

a. *Inhibition of RNA Polymerase*

Actinomycin D, a polypeptide antibiotic produced by many species of *Streptomyces*, intercalates between adjacent G·C base pairs in double strand DNA (Figure 3.54). Although all RNA polymerases are inhibited by actinomycin D, there are substantial differences in the extent of inhibition of various enzymes. At concentrations at which actinomycin blocks transcription, DNA replication is unimpaired. Rifamycin, an antibiotic from *Streptomyces mediterranei*, or its synthetic analogue, rifampicin, block transcription in prokaryotic organisms, but neither compound affects eukaryotic transcription. These compounds block the initiation of RNA chains by binding to the β-subunit of the RNA polymerase holoenzyme. Once RNA chains are initiated, elongation is unaffected by rifampicin. Another useful antibiotic for blocking bacterial RNA polymerases is streptolydigins, from *Streptomyces lydicus*, which blocks elongation but not initiation of RNA chains.

A widely used inhibitor of transcription in eukaryotes is the bicyclic octapeptide, α-amanitin, from the fungus *Amanita phalloides*. Of the three eukaryotic RNA polymerases (Section 3.1a), RNA polymerase II is the most sensitive to α-amanitin. RNA polymerase I is not inhibited at all, and RNA polymerase III is inhibited only at high concentrations. Therefore, this compound is very useful for distinguishing among the enzymes responsible for transcribing a particular RNA species.

b. *Inhibition of Translation*

Only a few of the more widely used and understood antibiotic inhibitors of translation in both prokaryotic and eukaryotic organisms are commented upon here (Tables 3.5 and 3.6).

The highly basic trisaccharide, streptomycin, like many aminoglycosides, binds strongly to the S12 protein of the 30S subunit of bacterial ribosomes. One of the consequences of this binding is to inhibit the formation of a proper initiation complex, thereby blocking the start of polypeptide chains. Binding of streptomycin to the 30S subunit also modifies the structure of the A site, causing occasional mismatching between aminoacyl-tRNAs and codons. Bacterial mutants that are resistant to streptomycin have an altered 30S ribosomal subunit, which either fails to bind streptomycin or binds it in such a way as to thwart its inhibitory action. Because streptomycin does not bind to eukaryotic ribosomes, their function is not affected by this compound.

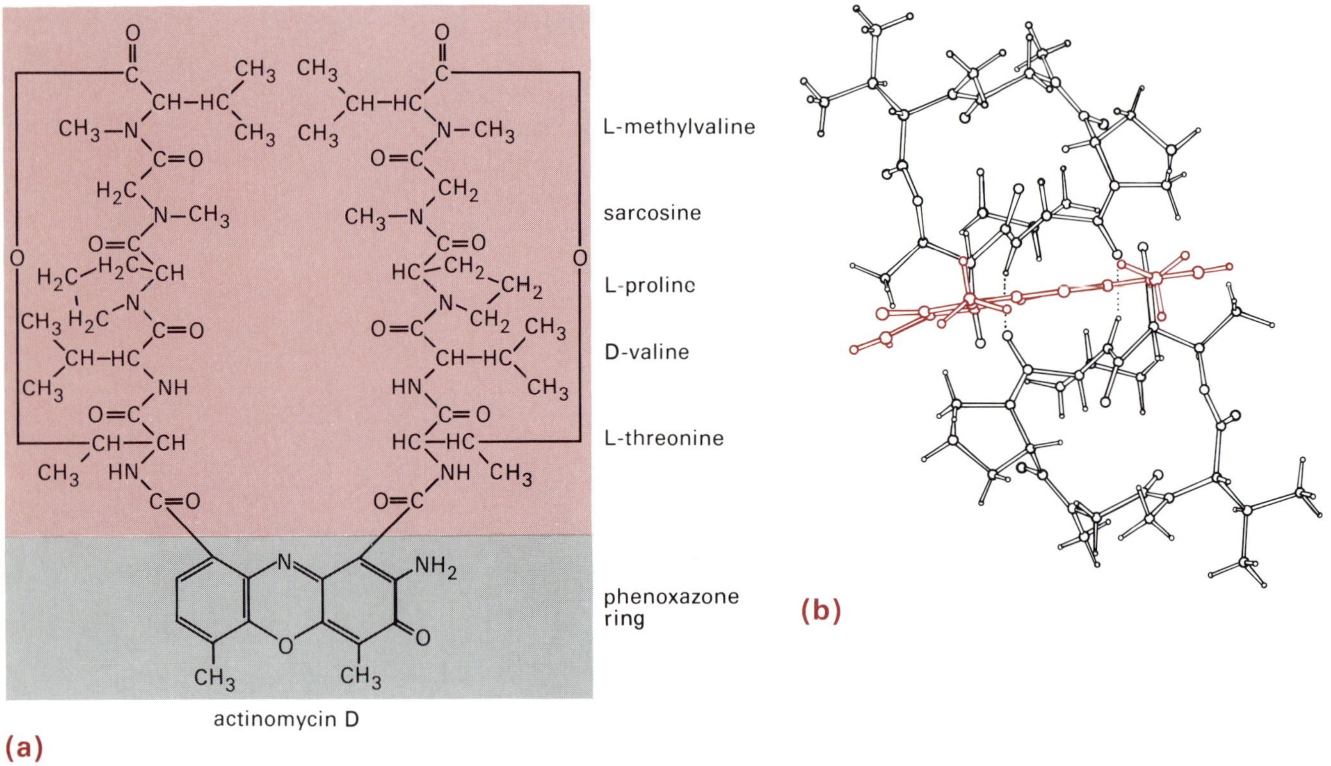

L-methylvaline

sarcosine

L-proline

D-valine

L-threonine

phenoxazone ring

(b)

actinomycin D

(a)

Figure 3.54
The structure of actinomycin D.
(a) The composition and structure are shown in a standard way. (b) A three-dimensional representation in which the plane of the phenoxazone ring is perpendicular to the plane of the paper.
(c) The structure shown in (b) is intercalated between two adjacent $G \cdot C$ base pairs in a duplex DNA. Adapted from S. C. Jain and H. M. Sobell, *J. Mol. Biol.* 68 (1972), p. 1.

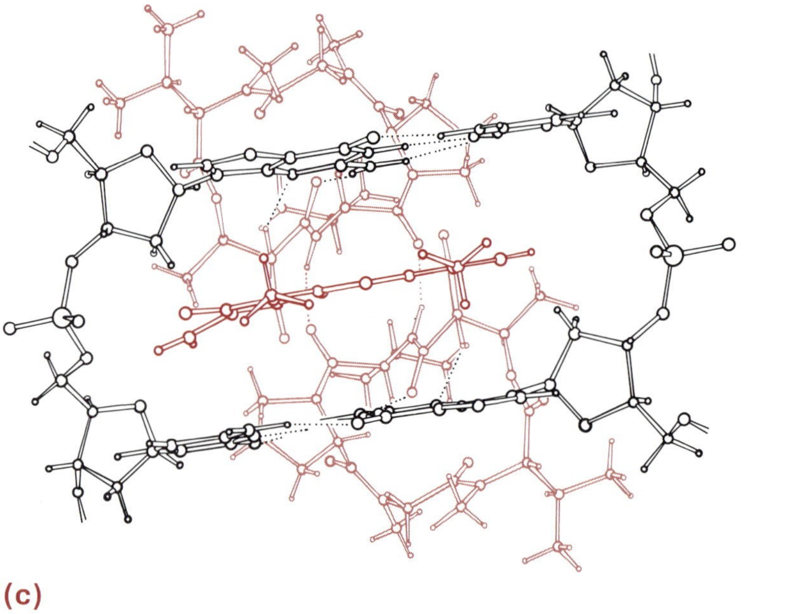

(c)

Puromycin, from *Streptomyces alboniger*, blocks polypeptide assembly in both pro- and eukaryotic ribosomes by mimicking the behavior of an aminoacyl-tRNA (Figure 3.55). When present, puromycin enters the vacant A site, where its amino group acts as an acceptor of the nascent polypeptide chain, much in the same way as the amino group of a bound aminoacyl-tRNA. Lacking any interaction with the codon, the polypeptidyl puromycin product dissociates from the ribosome, and the ribosome is separated from the mRNA. In effect, puromycin induces an abortive termination of polypeptide assembly.

Table 3.5 Inhibitors of Protein Synthesis in Prokaryotes

Antibiotic	Action
Streptomycin	Binds to the 30S ribosomal subunit and inhibits binding of Fmet-tRNA$_F^{Met}$ to the P site; also causes mistranslation by the ribosome
Neomycin, kanamycin	Same as streptomycin
Chloramphenicol	Inhibits peptidyl transferase of 70S ribosome
Tetracycline	Inhibits binding of aminoacylated tRNA to 30S particle
Erythromycin	Binds to free 50S particle and prevents formation of the 70S ribosome; has no effect on an active 70S ribosome
Puromycin	Causes premature chain termination by acting as an analogue of aminoacyl-tRNAs
Fusidic acid	Inhibits binding of aminoacyl-tRNAs to the A site by blocking the release of EF-G after translocation
Kasugamycin	Inhibits binding of Fmet-tRNA$_F^{Met}$ to the P site
Lincomycin	Inhibits peptidyl transferase
Kirromycin	Binds to EF-Tu, stimulates formation of (EF-Tu)-GTP and binding of ternary complex to ribosome but inhibits release of EF-Tu
Thiostrepton	Prevents translocation by inhibiting EF-G

Table 3.6 Inhibitors of Protein Synthesis in Eukaryotes

Inhibitor	Action
Abrin, ricin	Inhibits binding of aminoacyl-tRNA
Diphtheria toxin	Catalyzes a reaction between NAD and eIF-2 to yield an inactive factor; inhibits translocation
Chloramphenicol	Inhibits peptidyl transferase of mitochondrial ribosomes; inactive against cytoplasmic ribosomes
Puromycin	Causes premature chain termination by acting as an analogue of aminoacyl-tRNAs
Fusidic acid	Inhibits translocation by inhibiting eEF-2
Anisomycin	Inhibits peptidyl transferase
Cycloheximide	Inhibits peptidyl transferase
Pactamycin	Inhibits positioning of met-tRNA$_I^{Met}$ on the 40S ribosome
Showdomycin	Inhibits formation of the eIF-2-met-tRNA$_I^{Met}$-GTP complex
Sparsomycin	Inhibits translocation
Interferon	Blocks initiation by inducing phosphorylation of eIF-2

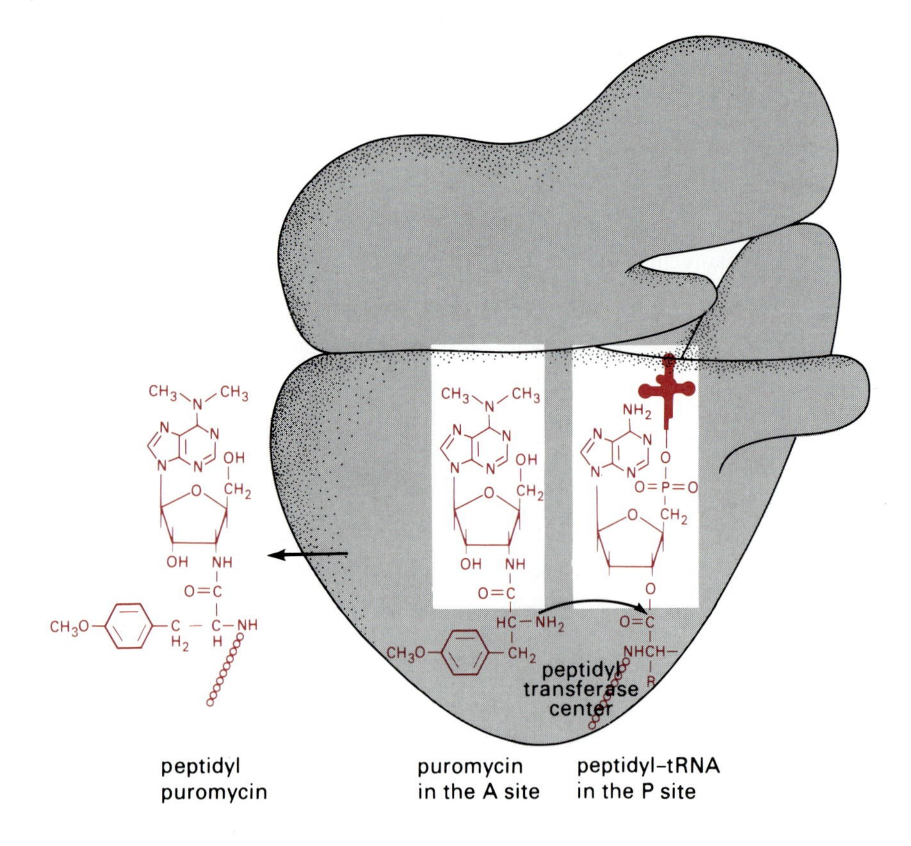

Figure 3.55
Puromycin mimics an aminoacyl-tRNA and interrupts polypeptide chain elongation.

Several antibiotics (e.g., erythromycin [from *Streptomyces erythreus*] and fusidic acid [from *Fusidium coccineum*]) block translocation of the ribosome after completion of the elongation or peptidyl transfer reaction. As a result, the ribosome cannot move to the next codon, and the peptidyl-tRNA remains bound in the A site and blocks entry of the next aminoacyl-tRNA.

Both chloramphenicol and cycloheximide inhibit the same step in protein chain assembly. Chloramphenicol acts on bacterial, mitochondrial, and chloroplast ribosomes but cycloheximide inhibits only eukaryotic cytosolic ribosomes. Both compounds act on the peptidyl transferase activity of their respective large ribosomal subunits. As a consequence, elongation of the peptide chain is blocked, and both the P and A sites are filled.

3.10 The Fate of Newly Synthesized Proteins

The assembly of amino acids into polypeptide chains is just the first step in the formation of many proteins. Polypeptide chains must be folded into their proper two- and three-dimensional structures, and, in most cases, individual polypeptides must be assembled into functional oligomeric proteins. Whether these processes are always wholly self-guided or sometimes facilitated by enzymes remains unsettled. Primary translation products often undergo other important alterations before the proteins can perform their assigned functions. A few examples will suffice to illustrate these points.

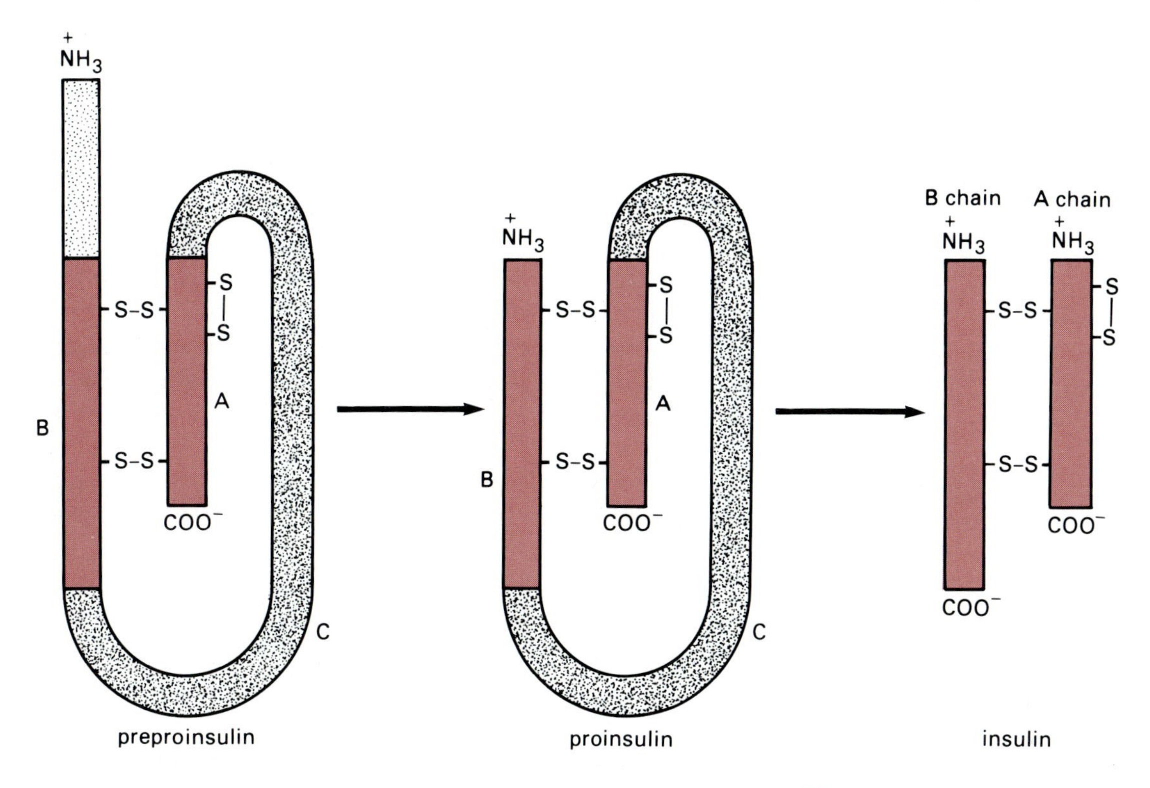

Figure 3.56
Stepwise conversion of preproinsulin to insulin. Translation of insulin mRNA yields a large precursor, preproinsulin. Functional hormone, a heterodimer, is formed by removal of the amino terminal signal segment followed by two cleavages within the proinsulin molecule that eliminate the central portion (C) of the primary translation product. The *A* and *B* chains of active insulin remain, held together by disulfide bonds that formed in the initial, long polypeptide.

a. Posttranslational Alterations in Polypeptide Chains

Functional hemoglobin is produced only after the two types of subunit chains (α and β) have assembled into an $\alpha_2\beta_2$ structure and the heme group becomes associated with the amino acid side chains of both subunits. Biotin must be covalently linked to certain amino acid side chains before the protein chains of pyruvate carboxylase and acetyl-CoA carboxylase become active as enzymes. Certain proteins involved in blood clotting require the carboxylation of specific glutamic acid side chains to produce the dicarboxylic centers needed for binding Ca^{2+}. To form collagen, specific proline and lysine residues in the polypeptide chain must be hydroxylated. A prevalent and highly significant mode of metabolic regulation is mediated by the phosphorylation and dephosphorylation of specifically located serine, threonine, and tyrosine residues by specific protein kinases and protein phosphatases, respectively. Many proteolytic proteins involved in digestion and blood clotting are synthesized as large precursors, which are then activated by trimming away a portion of the polypeptide chain. Insulin is synthesized as a preproinsulin polypeptide that is converted to mature insulin by cleavages that remove, first, an amino terminal segment and then an internal segment (Figure 3.56). Furthermore, many viral proteins, hormones, and neuropeptides are derived from a primary translation product—a polyprotein—that is cleaved at multiple sites to yield several mature sized proteins and peptides.

b. Targeting of Eukaryotic Proteins into and Through Cell Membranes

In eukaryotic cells, there are, in addition to the bounding plasma membrane, innumerable intracellular membranes that delimit the various intracellular organelles: mitochondria, chloroplasts (in plants), **endoplasmic reticulum**, **Golgi structures**, **peroxisomes**, **lysosomes**, and **secretory vesicles** (Figure I.1a). How do proteins destined for these many different membranes or the compartments they define arrive at their ultimate locations? Are the intra- or extracellular destinations encoded in a protein's structure? If so, what is the chemical nature of this information, and how is that information acted upon?

The elaborate transport machinery of eukaryotic cells Ribosomes are found in two forms in eukaryotic cells: free and bound to membranes. Proteins that are destined for certain organelles and the **cytosol**—the soluble, aqueous portion of the cytoplasm—are synthesized on free ribosomes. Proteins that end up in lysosomes, the Golgi structures, and the plasma membrane are made on ribosomes that are associated with the endoplasmic reticulum (**ER**)—a continuous intracellular membrane that encloses a space called the **ER lumen**. The ER membranes with which ribosomes are associated are referred to as the "**rough ER**"; those membranes that lack bound ribosomes are called the "**smooth ER**" (Figure 3.57). The proteins synthesized on ER bound ribosomes pass through the ER membrane into the ER lumen, with the amino terminus leading the way (Figure 3.58). In the process, a stretch of largely hydrophobic amino acids at the amino terminus of the proteins—the **signal sequence**—is cleaved by a specific endopeptidase localized in the ER lumen. This directional transfer from ribosome into the ER lumen occurs while the chain is being assembled and is therefore referred to as **cotranslational transport**.

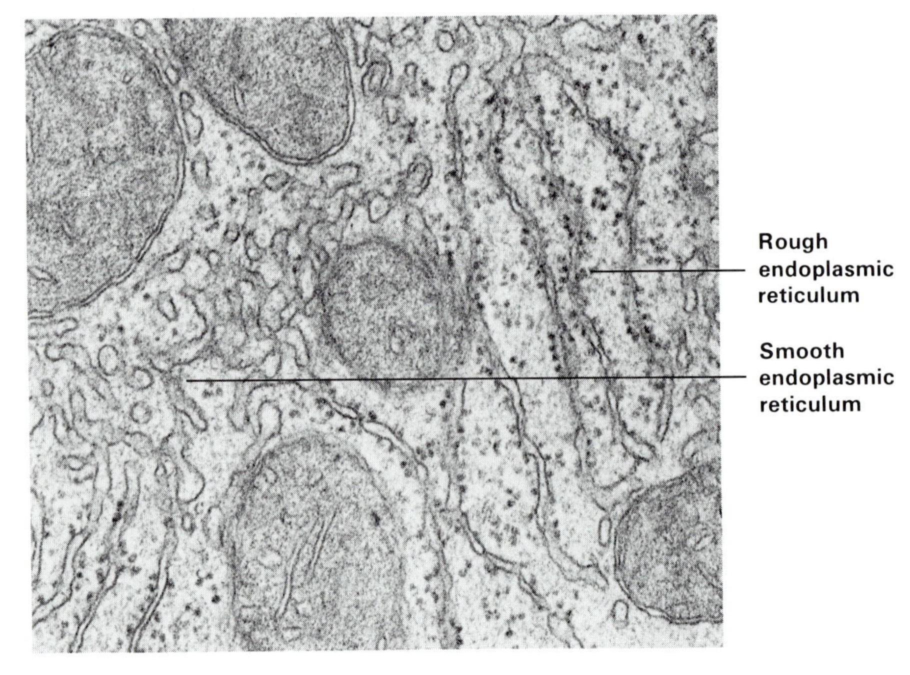

Figure 3.57
An electron micrograph of a thin section through a rat liver cell showing both smooth and rough endoplasmic reticulum (ER). The membranes of the smooth ER, shown at the left and top of the photo, are rarely associated with ribosomal particles. By contrast, the narrow, elongated, membranous structures of the rough ER, shown in the right-hand portion of the picture, are studded with ribosomes. The large, more electron dense structures are mitochondria. Courtesy of George E. Palade.

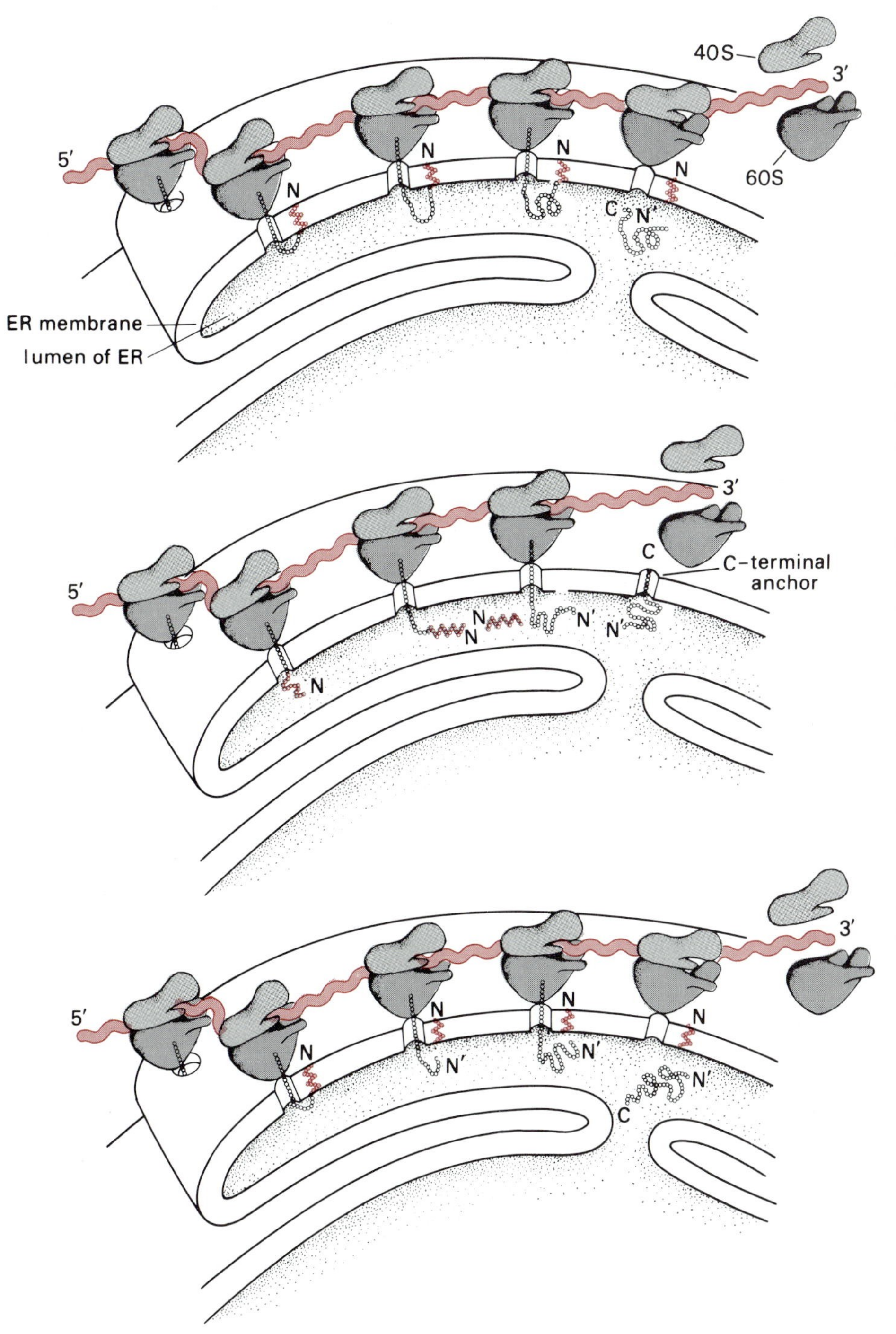

Figure 3.58
Cotranslational transport of polypeptide chains. Schematic diagram showing how polypeptides synthesized on the rough ER are directed through the membrane into the lumen. Soon after translation begins, the hydrophobic amino terminus of the nascent protein, the signal sequence, passes through the membrane. As the polypeptide is elongated, it is extruded into the lumen. Various nascent polypeptides undergo different reactions within the ER, depending on their eventual fate, although all lose their signal sequences. N′ is the new amino terminus formed by cleavage of the signal sequence. In the example at the top, the signal sequence, anchored within the membrane, is cleaved off after translation is complete; the polypeptide, with a new amino terminus, is free in the lumen. In the middle example, the signal sequence is removed before translation is complete; the carboxyl terminus of the new polypeptide is anchored in the ER membrane. In the bottom example, the signal sequence is clipped off shortly after the polypeptide enters the lumen.

Targeting nascent polypeptide chains into the ER lumen Interactions between a polypeptide signal sequence, a **signal sequence recognition particle (SRP)**, and a **signal sequence recognition particle receptor (SRP-R)** direct the transport of membrane and secreted proteins into the ER lumen. What is it that steers a ribosome to the ER if the protein it is synthesizing is destined to be secreted or to be transported to the lysosomes or plasma membrane? First and foremost, it is not an intrinsic property of the ribosome. Cytosolic proteins are synthesized on ribosomes that are

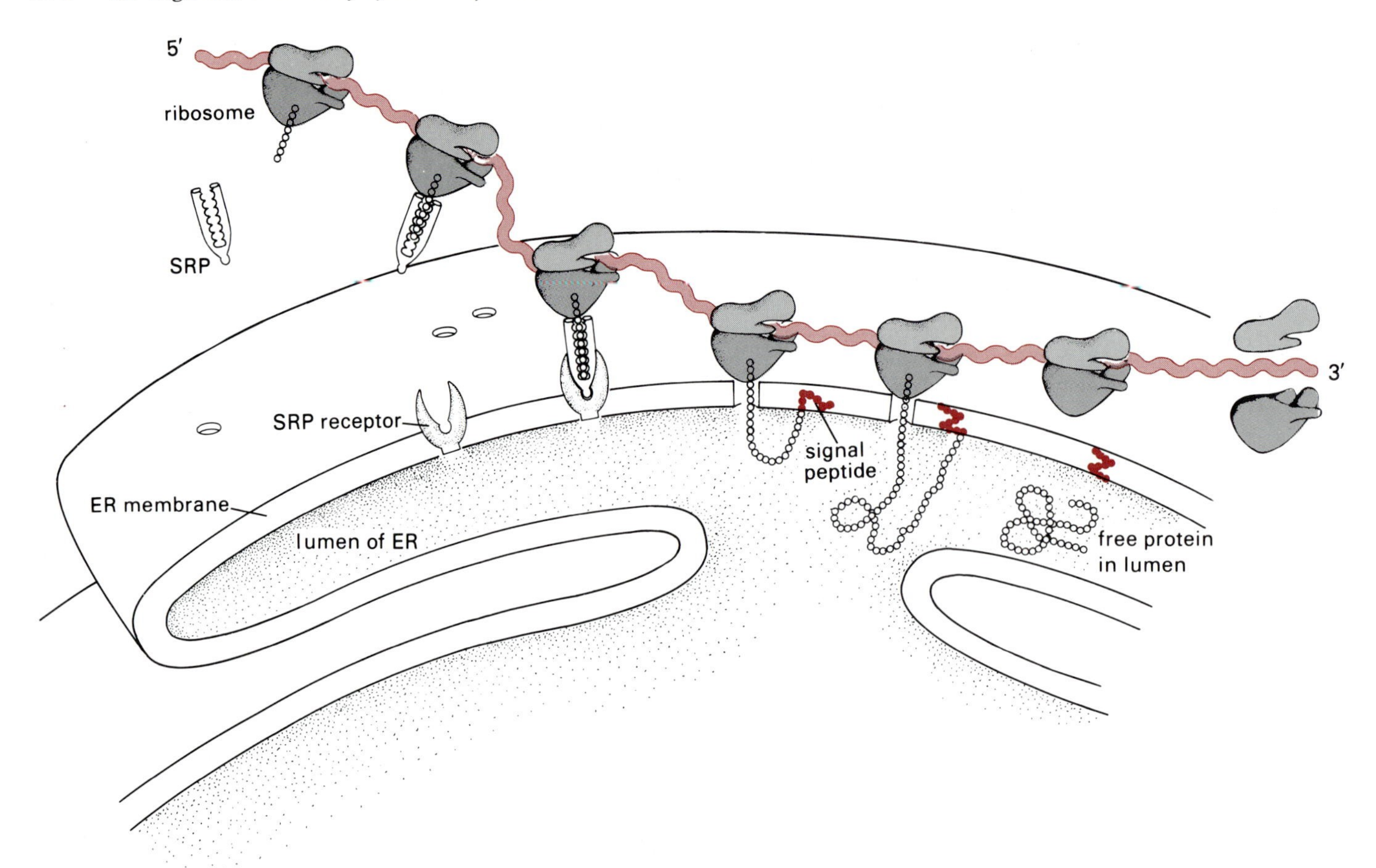

Figure 3.59
A signal recognition particle mediates entry of nascent polypeptides into the ER. Schematic diagram showing how a signal recognition particle (SRP) interacts with the amino terminal signal sequence on a growing polypeptide and then with an SRP-receptor in the ER membrane to target the polypeptide into the ER lumen.

identical to those that synthesize proteins residing in the ER. Indeed, a ribosome that synthesizes a cytosolic protein in one cycle may synthesize a membrane protein in its next cycle. The insertion of a protein into the ER results entirely from the interaction of the SRP with the signal sequence in the nascent protein chain (Figure 3.59). The SRP-signal sequence complex subsequently interacts with the SRP-R located in the ER membrane. Thus, the SRP serves as an adapter between cytoplasmic protein synthesis and the protein targeting machinery responsible for directing proteins into the ER lumen.

The SRP binds to the signal sequence just as it emerges from the ribosome, that is, after the assembly of about 70 amino acids. Polypeptide chain elongation slows until the SRP-signal sequence complex binds to the SRP-R in the ER membrane. Shortly thereafter, the SRP is released, the rate of polypeptide chain synthesis increases, and the nascent polypeptide chain traverses the membrane into the lumen. Thus, the association of the translating ribosome with the SRP-R causes a unidirectional transport of the growing polypeptide chain into the ER lumen. A signal sequence peptidase located in the ER lumen cleaves the signal sequence from the nascent protein at a specific location.

The SRP consists of a complex of six proteins ranging in molecular weight from about 10,000 to 75,000 Daltons and a single RNA molecule of about 300 nucleotides, the 7SL RNA. Neither the RNA nor the proteins by themselves can serve as an SRP. However, a fully functional SRP is reconstituted by mixing the RNA and proteins together, *in vitro*. In such experiments, the 7SL RNAs from mammalian, amphibian, and insect cells

Protein	Charged Region	Hydrophobic Region	
Lipoprotein	Met Lys Ala Thr Lys	Leu Val Leu Gly Ala Val Ile Leu Gly Ser Thr Leu Leu Ala Gly	Cys Ser Ser
β-lactamase of pBR322	Met Ser Ile Gln His Phe Arg	Val Ala Leu Ile Pro Phe Phe Ala Ala Phe Cys Leu Pro Val Phe Ala	His Pro Glu
Phage fd, major coat protein	Met Lys Lys Ser Leu Val Leu Lys	Ala Ser Val Ala Val Ala Thr Leu Val Pro Met Leu Ser Phe Ala	Ala Glu Gly
Alkaline phosphatase	Met Lys	Gln Ser Thr Ile Ala Leu Ala Leu Leu Pro Leu Leu Phe Thr Pro Val Thr Lys Ala	Arg Thr Pro
Maltose binding protein	Met Lys Ile Lys Thr Gly Ala Arg	Ile Leu Ala Leu Ser Ala Leu Thr Thr Met Met Phe Ser Ala Ser Ala Leu Ala	Lys Ile Glu
Leucine binding protein	Met Lys Ala Asn Ala Lys	Thr Ile Ile Ala Gly Met Ile Ala Leu Ala Ile Ser His Thr Ala Met Ala	Asp Asp Ile
Proalbumin	Met Lys Trp Val Thr	Phe Leu Leu Leu Leu Phe Ile Ser Gly Ser Ala Phe Ser	Arg
IgG light chain	Met Asp Met Arg Ala Pro Ala Gln	Ile Phe Gly Phe Leu Leu Leu Leu Phe Pro Gly Thr Arg Cys	
Lysozyme	Met Arg Ser	Leu Leu Ile Leu Val Leu Cys Phe Leu Pro Leu Ala Ala Leu Gly	Lys
Prolactin	Met Asn Ser Gln Val Ser Ala Arg Lys Ala Gly Thr	Leu Leu Leu Leu Met Met Ser Asn Leu	Leu
Vesicular stomatitis virus glycoprotein	Met Lys Cys	Leu Leu Tyr Leu Ala Phe Leu Phe Ile His Val Asn Cys	Lys

Figure 3.60
Characteristic signal sequences at the amino terminus of some prokaryotic and eukaryotic proteins. A short segment with one or more charged amino acids occurs at the beginning of the chain, followed by a typical hydrophobic region. After S. Michaelis and J. Beckwith, *Ann. Rev. Microbiol.* 36 (1982), p. 435, and D. P. Leader, *Trends Biochem. Sci.* 4 (1979), p. 205.

can reconstitute an SRP with mammalian proteins. This finding, plus the considerable sequence similarity between the 7SL RNAs from these species, indicates that the structure of the 7SL RNA is conserved in evolution. Whether the 7SL RNA in SRP has a direct role in recognizing the signal sequence or serves as a "scaffold" to organize the various protein subunits into the functional SRP, or both, is not known.

Signal sequences are generally located at the amino terminus of proteins destined to be exported to one of the cellular membranes or intracellular organelles. In some instances, however, the amino acid sequence that is recognized by the SRP is neither amino terminal nor is it cleaved from the translocated protein, and some secreted proteins do not possess a recognizable signal sequence.

Signal sequences are usually between 15 and 35 amino acids long and contain a preponderance of hydrophobic amino acids in the first three-quarters of the sequence (Figure 3.60). However, signal sequences do not have a unique, strictly conserved stretch of amino acids. Rather, the evidence indicates that the secondary structure of the signal sequence is important for its recognition by the SRP and behavior in the membrane. It is interesting that the signal sequences of eukaryotic proteins are recognized by the bacterial translocation machinery, and signal sequences from proteins secreted by bacteria can promote translocation through the membrane components of eukaryotic cells.

The signal sequence also specifies the cleavage site for signal peptidases. Most frequently, cleavage is favored on the carboxyl side of glycine, serine, or alanine residues. As a consequence, the amino terminal amino acids of many mature secreted and membrane proteins are adjacent to one of these amino acids in the signal sequence.

Transport of proteins from the ER to the Golgi and beyond Proteins bound for lysosomes, plasma membranes, or secretion pass through the Golgi structure—a set of closely packed, interconnecting, membrane-enclosed cisternae that lie near the nucleus (Figure 3.61). Transport to the Golgi occurs via coated vesicles that bud off from the ER membranes and fuse with the Golgi cisternae (Figure 3.62). Proteins pass through the Golgi in a specific direction: from the entry face (or cis-Golgi stacks), to the exit face (or trans-Golgi stacks). Transport of proteins through Golgi stacks is also mediated by coated transport vesicles; these originate from the donor cisternae membranes.

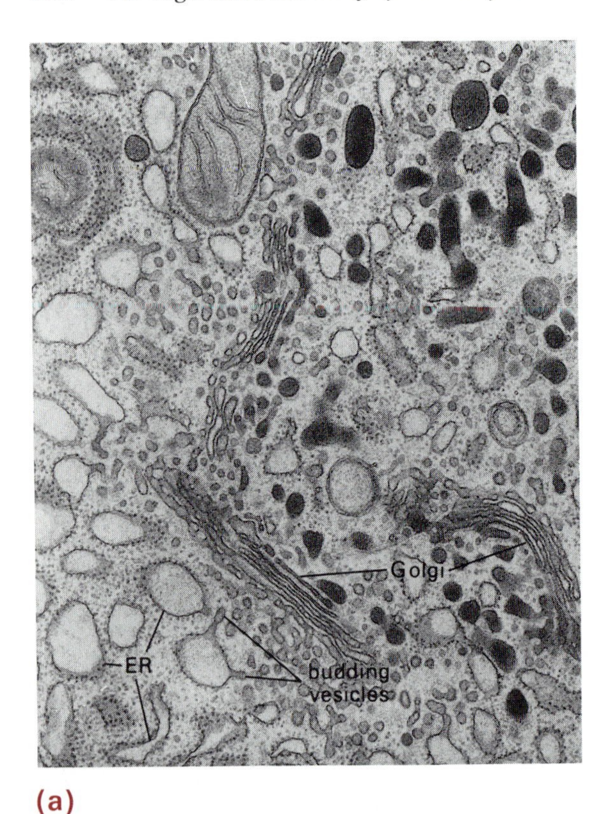

(a)

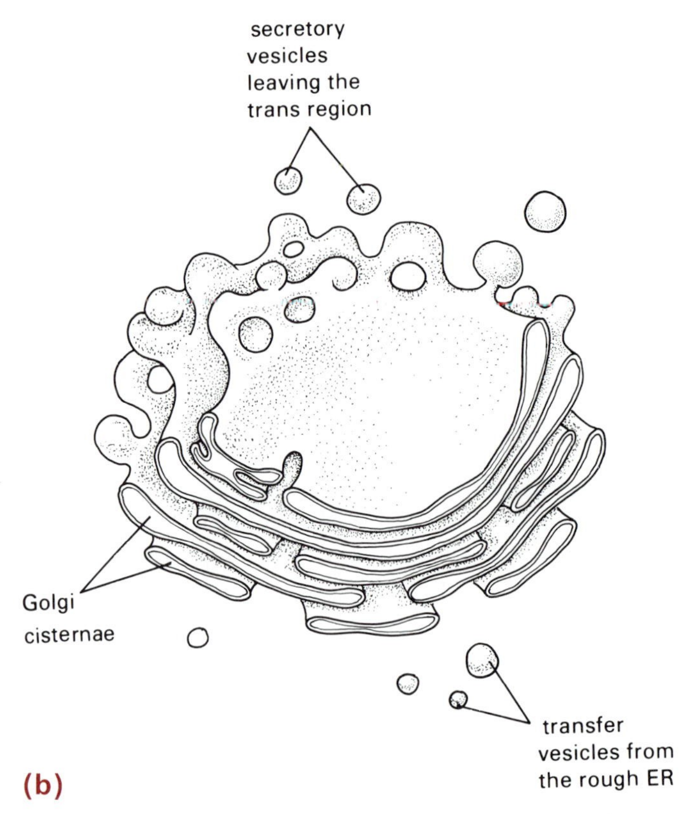

(b)

Figure 3.61
The Golgi structure. (a) Electron micrograph shows Golgi stacks adjacent to ER in an exocrine secretory cell from rat pancreas. The Golgi stack proximal to the ER is designated the *cis* Golgi, and the stack distal to the ER is the *trans* Golgi. Note vesicles budding off from some ER elements. From E. M. Merisko, M. Fletcher, and G. Palade, *Pancreas* 1 (1986), p. 95. Courtesy of M. G. Farquhar. (b) Drawing illustrating the cisternae and vesicles associated with the Golgi structure. Adapted from J. Darnell, H. Lodish, and D. Baltimore, *Molecular Cell Biology* (New York: Scientific American Books, 1986), and after a model by J. Kephart.

Proteins that are targeted for secretion are first incorporated into secretory vesicles that eventually fuse with the plasma membrane to empty their contents to the outside. In those instances where protein secretion is induced, as in the case of insulin and some neuropeptides, the cytosolic secretory vesicles fuse with the membrane and empty their contents to the outside only after induction.

At present, there is no explanation for how coated vesicles distinguish proteins destined for one or another cellular compartment. Nor is it known if a vesicle carrying a particular protein is distinctive and, if it is, how that distinctiveness is expressed.

Glycosylation Proteins are extensively modified as they pass through the ER and the Golgi stacks (Figure 3.63). An enzyme in the ER attaches a complex oligosaccharide to specific asparagine residues of transported proteins. The protein-bound oligosaccharide is subsequently remodeled by a series of specific glycosyl hydrolase and glycosylation reactions that occur in the ER and throughout the Golgi structure.

c. *Transport of Proteins to Eukaryotic Cellular Organelles*

Not all proteins that are localized to intracellular organelles are routed via the pathway involving the ER and the Golgi structure. Most of the proteins of mitochondria, nuclei, and plant chloroplasts are synthesized on free cytoplasmic ribosomes either as mature or precursor proteins and are then taken up by the appropriate organelle. Other mitochondrial and chloroplast proteins are encoded in organellar DNA and translated by an organellar protein synthesizing apparatus analogous to the cytoplasmic system described earlier.

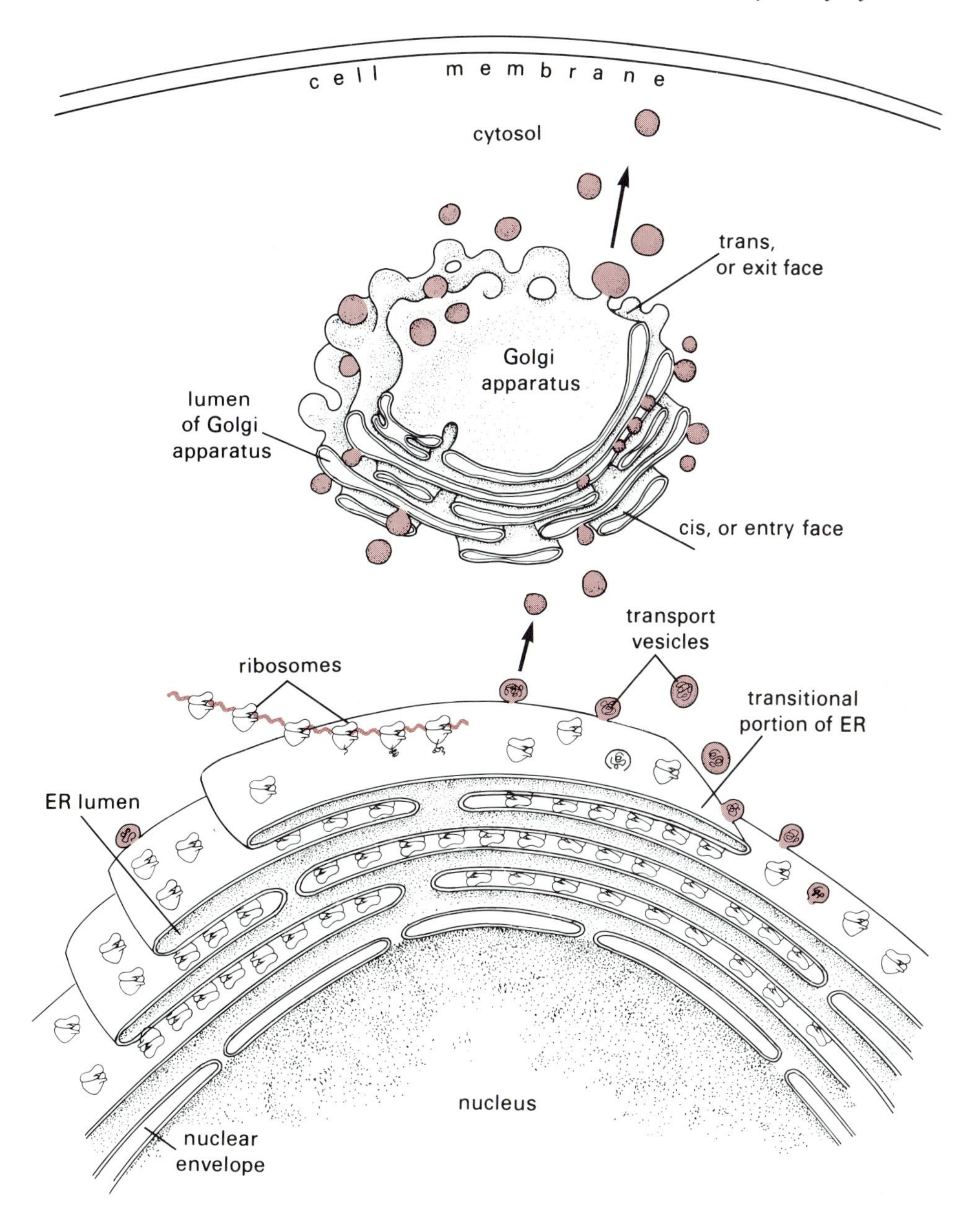

Figure 3.62
Coated transport vesicles containing polypeptides destined for lysosomes, plasma membrane, or secretion bud off the ER and fuse with the *cis* or entry face of the Golgi. The polypeptides traverse the Golgi via vesicles and emerge into the cytosol from the exit or *trans* face in transport vesicles derived from the Golgi cisternae membrane. Post-translational modifications such as proteolytic cleavage and glycosylation occur during passage through the Golgi.

The mitochondrion, for example, contains several compartments: the inner membrane-matrix compartment, the outer membrane, and an intermembrane space (Figure 3.64). Nuclear encoded proteins that occupy the inner membrane-matrix compartment are synthesized in precursor form containing an extra 25 to 60 amino acids at the amino terminus. After binding to a receptor in the outer membrane, the precursor form of the protein is translocated into the inner membrane-matrix compartment. The driving force for the transport is the electrochemical potential across the inner membrane, which is created by energy-yielding reactions occurring in the mitochondrion. Concomitant with, or immediately following, the translocation, the extra amino acids at the amino terminus are removed by a specific protease located in the matrix. Some proteins remain embedded in the inner membrane, and others pass through the membrane into the matrix.

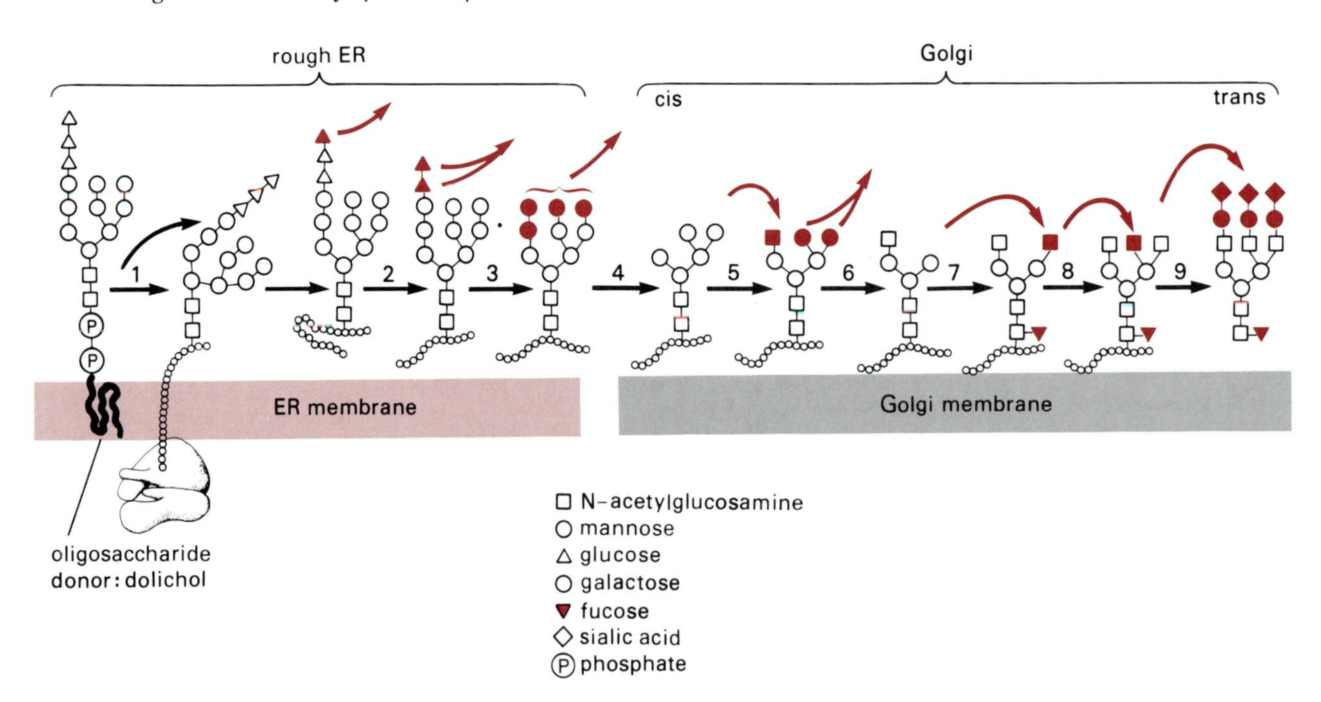

Figure 3.63
Glycosylation of proteins in the rough ER and Golgi stacks. In the ER, a branched oligosaccharide linked to the lipid dolichol through a pyrophosphate bridge is transferred to protein asparagine residues within the sequences asn-x-ser or asn-x-thr (where x is any amino acid). Thereafter, the N-linked oligosaccharide is modified by a stepwise series of specific glycosyl hydrolase and glycosylation reactions that occur either in the ER or later, as the protein passes through the Golgi. Different proteins acquire different final N-linked oligosaccharides, depending on their primary and secondary structures and the presence or absence of specific oligosaccharide processing enzymes in the cell carrying out the synthesis. After J. Darnell, H. Lodish, and D. Baltimore, *Molecular Cell Biology* (New York: Scientific American Books, 1986).

Proteins made in the cytosol and targeted for the mitochondrial intermembrane space are translocated in several different ways (Figure 3.64). Some precursor proteins possess amino terminal extensions and enter the mitochondrion as described previously. But in others, cleavage of the terminal sequence leaves the protein transiently embedded in the inner membrane, where a second processing enzyme cleaves the intermediate to release the mature protein into the intermembrane space. Proteins localized in the outer membrane enter by an energy-independent process and without conversion of a precursor to a mature protein. In some instances, conformational changes in the protein structure, induced by binding a prosthetic group, may trigger entry of the cytosolic protein into the outer membrane. Proteins are stabilized in membranes by anchoring sequences that span the lipid bilayer; these anchoring sequences may be at either the amino or carboxy terminus or within the middle of the polypeptide chain.

d. *Transport of Proteins in Prokaryotes*

Both prokaryotic and eukaryotic cells contain bounding lipoprotein membranes. Bacterial inner membranes define the interface between the cytoplasmic contents and the intermembrane or **periplasmic space**. The outer membrane separates the periplasmic and external environments. Newly synthesized proteins in bacteria must be partitioned between the cytosol, the inner or plasma membrane, the intermembrane or periplasmic space, the outer membrane, and the outside. As is the case in eukaryotes, bacterial proteins destined for the plasma membrane or more external sites are distinguished from those retained in the cytosol by having an amino terminal signal sequence (Figure 3.60). Moreover, the signal sequence is removed to create the mature forms of inner membrane or intermembrane proteins. Unlike eukaryotic membrane proteins, however, the bacterial counterparts are not glycosylated.

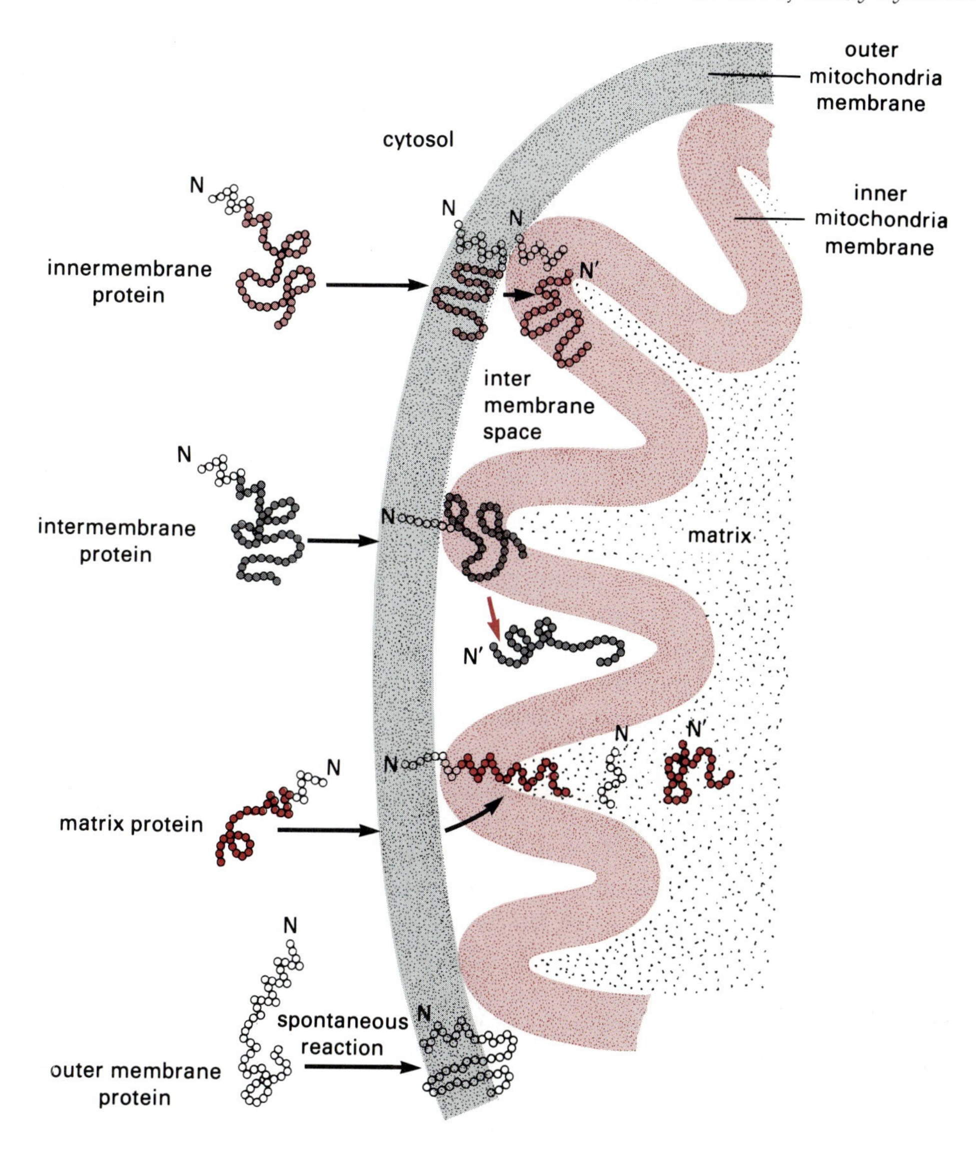

Figure 3.64
Uptake of completed proteins by a mitochondrion. Depending on the mitochondrial target site (shown at the left) of nuclear-encoded mitochondrial proteins, they take different routes and undergo different processing reactions during translocation from the cytosol.

Bacterial membrane proteins are also transported through the inner membrane cotranslationally. It is not certain if the signal sequence is directed to the inner membrane by a nucleoprotein particle analogous to the eukaryotic SRP or if that results from a direct interaction between the hydrophobic amino terminus and the membrane. In either case, the nascent polypeptide chain feeds through the inner membrane as it is assembled. A membrane associated signal protease cleaves the signal sequence away, leaving the mature protein anchored in the membrane or entirely free in the periplasm. The vectorial transport of some prokaryotic proteins across membranes can also be achieved posttranslationally; here, too, the driving force for the transfer is an energized membrane. Given the similarity in the transport processes in pro- and eukaryotic cells, it is not surprising that proteins that are exported from bacteria are also exported from eukaryotic cells and vice versa. Nor is it startling that, if an appropriate signal sequence is artificially fused to normally cytosolic proteins, such proteins are targeted to and through membranes, depending upon their secondary structure.

3.11 Regulation of Gene Expression

A fundamental characteristic of both prokaryotic and eukaryotic organisms is the capacity to regulate the expression of their genes differentially. By controlling which genes are expressed and which are silent, or by adjusting the rate at which different genes are expressed, cells adapt their phenotypes to particular extra- and intracellular environments. Genes are often expressed in a temporal sequence, the activation of one gene triggering the expression of one or more others, ultimately leading to a cascade of functions. Some genes or related groups of genes are expressed coordinately—that is, they respond simultaneously, and usually to about the same extent, to a regulatory signal. Perhaps the most elaborate regulatory systems are those that enable a multipotential **stem cell** to differentiate into the distinct cell types of a complex organism during its intricate developmental processes.

The phenotypic characteristics of different types of cells, or of similar cells in different environments, depend on the assortment and amounts of the structural, catalytic, and regulatory proteins they make. Any one or more of the several steps in the readout of genetic information into protein may serve as the target for regulation. In bacteria such as *E. coli,* the production of proteins is controlled principally by the level of mRNA available for translation. But regulating different steps in translation, or the rate of protein breakdown, provides additional means for maintaining proper levels of cellular proteins. Eukaryotic cells possess more elaborate means for controlling the amounts and array of proteins. In addition to controls at the level of transcription initiation in the nucleus, the level of cytoplasmic mRNA is regulated by processing of the primary transcripts and transport of the mature RNAs to the cytoplasm. As in prokaryotes, eukaryotic cells can also adjust the levels and types of their proteins by regulating both translation and the rates of protein transport and breakdown.

The purpose of this section is to illustrate, with a few examples, several mechanisms used by prokaryotic organisms such as *E. coli* and its bacteriophages to regulate gene expression. The more complex regulatory mechanisms used in eukaryotic gene expression are described in Chapter 8.

a. Regulation of RNA Levels During Biosynthesis

RNA formation in prokaryotes is most commonly regulated at the initiation of transcription. This is accomplished in several ways. One is modifications in the structure of RNA polymerase. For example, the beta (or beta-prime) subunit is altered following infection of *E. coli* with some bacteriophages, or a new sigma subunit is made during sporulation of certain *Bacillus* strains; both change the promoter-binding properties of the RNA polymerase and the rate of transcription of selected genes. A second way is changes in the topology of DNA affecting the ability of RNA polymerases to interact with certain promoters or to initiate RNA chain synthesis. Third, interaction between RNA polymerase and particular promoters is inhibited or enhanced by proteins that bind at or near the polymerase binding site. The binding of these regulatory proteins—**repressors** and **activators**—

is often influenced by certain metabolites, which serve as **corepressors** and **coactivators**.

The rate of elongation of RNA chains is also governed by the secondary structure of the mRNA. Signals that determine whether transcription terminates prematurely or proceeds through to the end of the transcription unit provide a prominent motif for regulating mRNA levels. In these cases, special nucleotide sequences and proteins modulate the go or no go decision for transcription.

The secondary structure of transcripts can also influence their stability, the structure of the 3′ end being particularly influential in determining the lifetime of an RNA in the cell. Some mRNAs are quite stable and participate in many rounds of translation. However, others are destroyed rapidly, even while being translated; transcriptional control is more significant with the most unstable mRNAs. As mentioned earlier, mature rRNA and tRNA molecules are processed and modified products of primary gene transcripts (Section 3.3). These posttranscriptional events are also regulated to control the production levels of functional RNAs.

b. Coordinate Regulation of Prokaryotic Gene Expression

Coordinate regulation of groups of related genes In *E. coli*, the genes that encode proteins involved in the same metabolic pathway, or genes that specify closely related functions, are frequently regulated coordinately. This means that the expression of these genes is turned on and off, or fine-tuned, as a group, usually by the same regulatory event. Genes that are coordinately regulated are often linked together in the genome and are transcribed from a promoter at the 5′ end of the gene cluster into a single RNA molecule, referred to as a **polycistronic** (or polygenic) transcript. Such a coordinately expressed set of genes is called an **operon**. The three genes encoding the enzymes responsible for metabolizing galactose in *E. coli* are organized into the *gal* operon with the promoter (P) and associated regulatory segment, the operator (O), at the 5′ end of the transcribed sequence, *gal*E-*gal*T-*gal*K (Figure 3.65). The synthesis of the enzymes needed for arabinose utilization is regulated by two adjacent transcription units

Figure 3.65
Schematic diagram of the galactose (*gal*) operon of *E. coli* showing operator (O), promoter (P), and the three genes of the operon, *gal*E, *gal*T, and *gal*K. The reactions catalyzed by three gene products, galactokinase (K), uridyl transferase (T), and UDP-galactose epimerase (E), are shown below. All three enzymes are translated from the single polycistronic *gal* mRNA. A third gene (*gal*R), which encodes the *gal* operon repressor, is not linked to the operon.

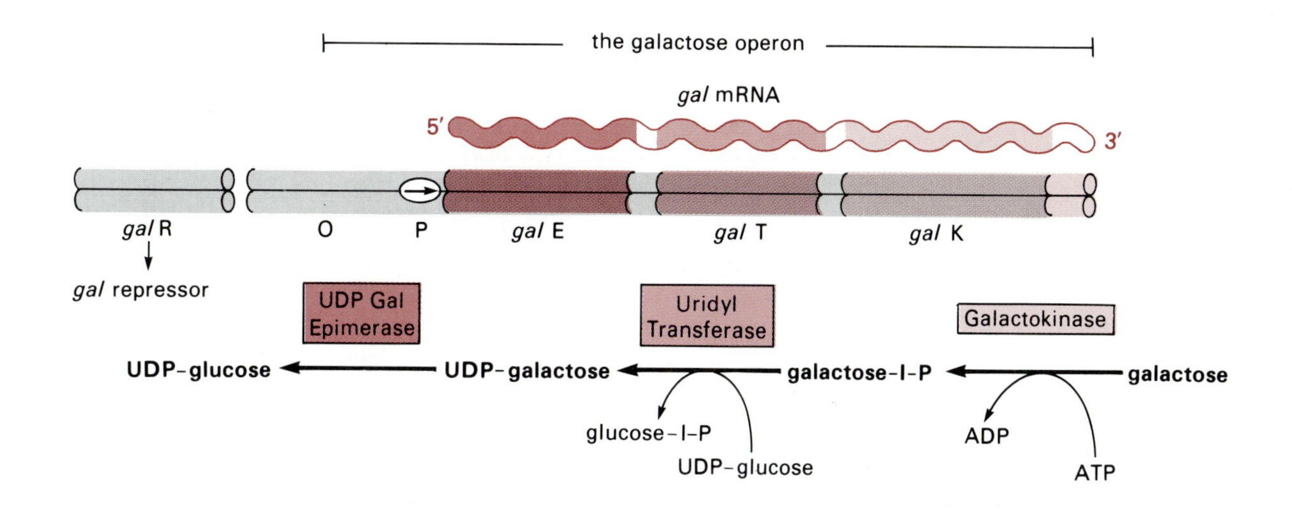

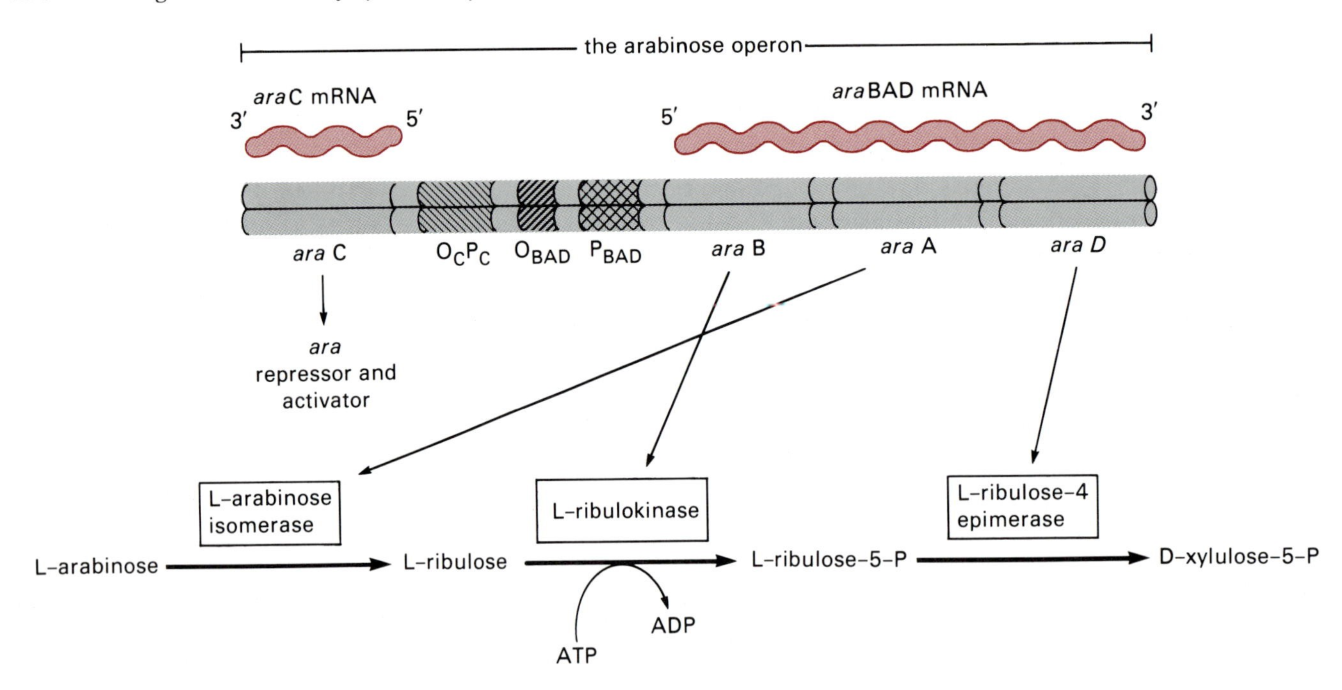

Figure 3.66
Schematic diagram showing the arabinose (*ara*) operon of *E. coli* and the adjacent *ara* C gene. The two transcription units go in opposite directions, and each has its own promoter and operator: O_C and P_C for *ara* C and O_{BAD} and P_{BAD} for the *ara* operon.

(Figure 3.66). One is the *ara* operon, which contains three genes, *ara*B, *ara*A, and *ara*D, and a 5′ control region. The second consists of the *ara*C gene, which encodes a regulatory protein that is required for the transcription of the *ara* operon.

The genes encoding a set of related functions are not always clustered in one operon. Thus, the genes that encode 30S and 50S ribosomal proteins are organized into multiple operons, some of which contain genes encoding other proteins involved in transcription and/or translation (Figure 3.67). Similarly, the *E. coli* genes required for the synthesis of arginine occupy many scattered genetic loci, only one cluster of genes (*arg*ECBH) being organized in a typical operon. Usually, separate operons specifying related functions contain common or very closely related regulatory sequences so that they can respond to a given regulatory signal in a similar manner.

Positive and negative regulation Negative control of transcription initiation—or repression—can be achieved by repressor proteins that bind to operators. Because operator sequences frequently overlap with promoter sequences, the binding of repressors to their operators limits access of the RNA polymerase to the promoter, thereby inhibiting transcription initiation. Positive control can also be achieved by the binding of specific proteins to nucleotide sequences in the general vicinity of the promoter. The present view is that the bound activator protein facilitates the association of RNA polymerase with the promoter and consequently increases the rate of transcription initiation.

The genes encoding the regulatory proteins that bind to the operator or activator sequences may or may not be located in the vicinity of the genes they control. For example, the gene encoding the repressor of the galactose operon (*gal*R) is not closely linked to the transcription unit containing the *gal*E, *gal*T, and *gal*K genes (Figure 3.65). By contrast, transcription of the arabinose operon is regulated either positively or negatively,

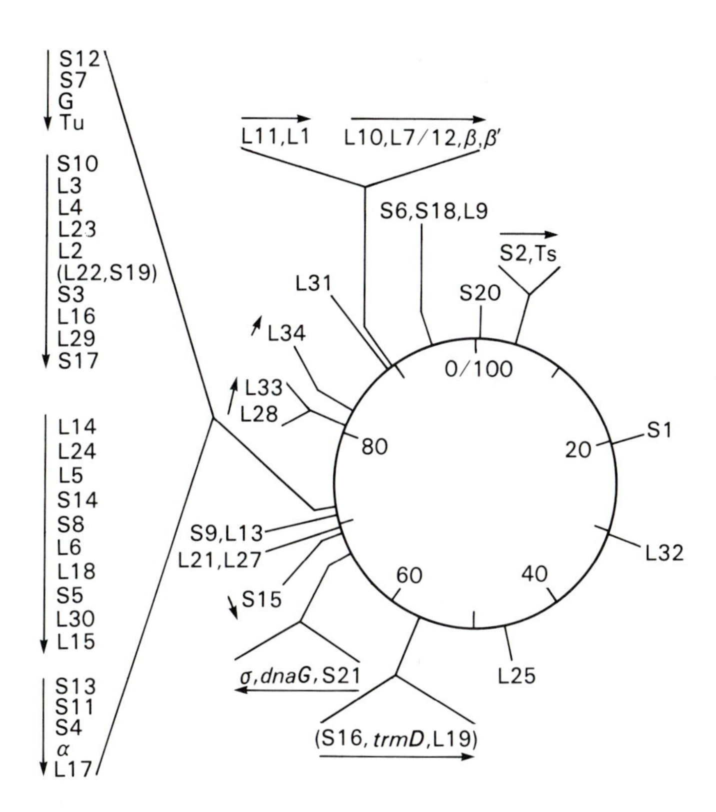

Figure 3.67
The location of genes encoding ribosomal proteins on the *E. coli* genome. Proteins on the 30S and 50S ribosomal subunits are designated S and L, respectively. The map also shows the positions of genes encoding RNA polymerase subunits α, β, β', and δ; DNA primase, *dna* G; and tRNA methyl transferase, *trm* D. The translation factors, EF-G, EF-Tu, and EF-Ts, are designated G, Tu, and Ts, respectively. When known, the direction of transcription is shown by arrows.

depending upon whether or not the protein encoded by *ara*C, a closely linked gene, is complexed with arabinose (Figure 3.66).

c. Regulation of Lactose Operon Expression

Negative control *E. coli* can use lactose as a sole source of carbon and energy because it can produce large quantities of β-galactosidase, an enzyme that cleaves lactose into glucose and galactose. When grown on other carbon sources, *E. coli* produces barely detectable levels of β-galactosidase. The gene that encodes β-galactosidase (*lac*Z) is called an "inducible" gene because the enzyme is synthesized only when the cell is provided with sugars that have a β-galactosyl substituent. Besides β-galactosidase, two additional proteins are induced by β-galactosides: a β-galactoside permease (encoded by the *lac*Y gene), which is required for the uptake of β-galactosides into the cell, and β-galactoside transacetylase (*lac*A), an enzyme whose physiologic function is obscure (Figure 3.68). The three genes, *lac*Z, *lac*Y, and *lac*A, contain the protein coding information of the *lac* operon; they are transcribed into a single polycistronic RNA, which is translated into nearly equal amounts of the corresponding proteins. In this sense, the three genes of the *lac* operon are said to be coordinately expressed.

Several types of regulatory elements are associated with the structural genes of the *lac* operon and account for their inducibility and coordinate regulation (Figure 3.68). The promoter is the nucleotide sequence at which RNA polymerase binds and begins to transcribe the three structural

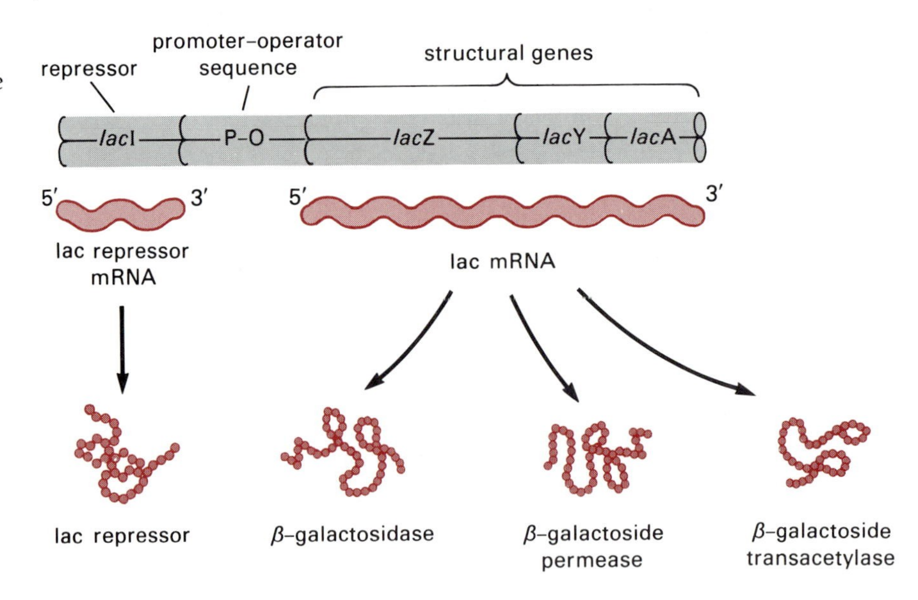

Figure 3.68
Schematic diagram of the *E. coli* lactose (*lac*) operon showing the closely linked *lac* repressor gene (*lac* I).

genes. The operator is the site at which the *lac* repressor binds to inhibit transcription of the *lac* operon. The gene *lac*I, which is not part of the *lac* operon, encodes the repressor, a polypeptide chain of 37,000 Daltons. The form in which it binds tightly to the operator is a tetramer of four identical chains.

Because the promoter and operator sequences overlap, binding of the repressor to the operator prevents RNA polymerase from binding to the promoter and blocks transcription of the structural genes. Transcription of the operon can be induced by preventing or reversing the binding of the repressor to the operator (Figure 3.69). This occurs when any one of a

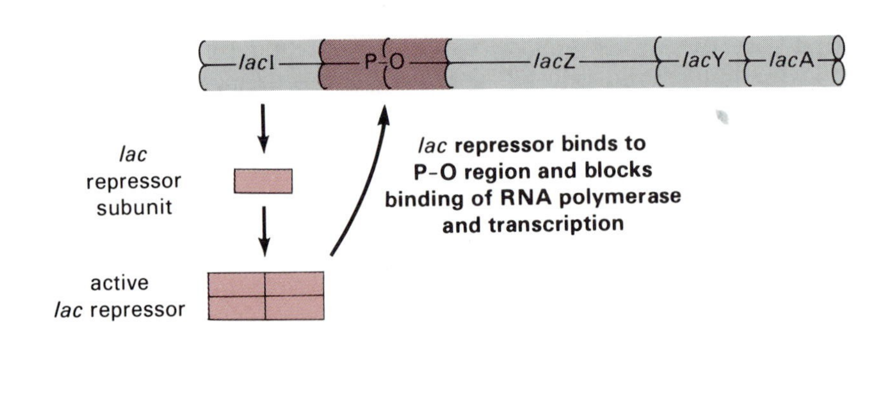

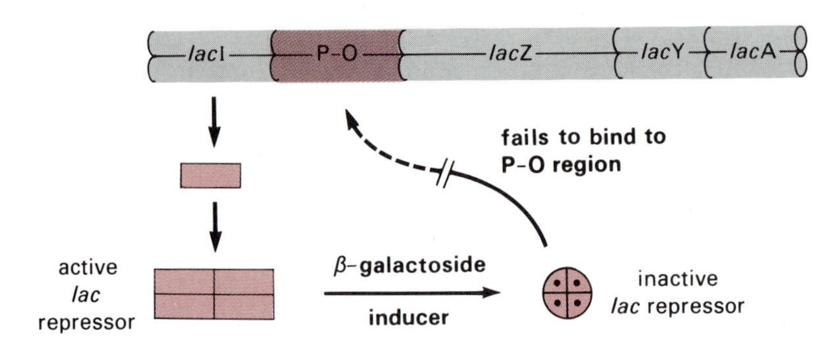

Figure 3.69
Repression of the *lac* operon by the homotetramer, *lac* repressor (top), and induction of the *lac* operon upon binding of a *β*-galactoside inducer to the repressor (bottom).

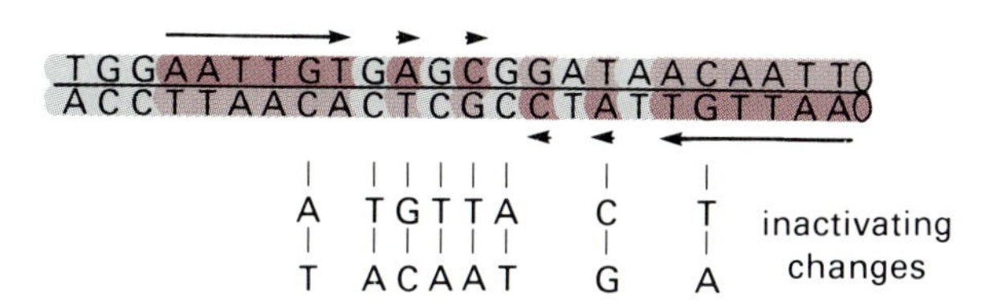

Figure 3.70
Nucleotide sequence at the repressor-binding domain of the *lac* operator. Base pair changes that lower the affinity of binding of repressor to operator are shown below. Arrows denote an inverted DNA repeat within the domain. Such inverted repeats characterize *E. coli* operator sequences. After J. D. Watson et al., *Molecular Biology of the Gene*, 4th ed. (Menlo Park, Calif.: Benjamin/Cummings, 1987).

variety of β-galactosides binds to one of the repressor subunits, altering the repressor protein's conformation and lowering its affinity for the operator. Removing the repressor from the promoter region allows the polymerase to bind and initiate operon transcription.

The identity of the nucleotide sequences that define the repressor-binding domain of the *lac* operator was determined by measuring the efficiency of repression *in vivo* with mutants having an altered repressor or operator (Figure 3.70). *In vitro* studies of the binding of purified wild-type and mutant repressor proteins with wild-type and mutant operators were also informative. Mutations that lower the repressor's affinity for the operator lead to **constitutive** synthesis of the *lac* operon enzymes—that is, expression of the *lac* enzymes in the absence of an inducer. Mutations that cause the repressor to accumulate in cells or that increase the repressor's affinity for the operator make the *lac* operon noninducible.

Positive control Besides the necessity for relieving repression, expression of the *lac* operon, as well as of other inducible operons controlling the synthesis of sugar metabolizing enzymes, requires a positive signal. The positive signal is provided by a complex of **cyclic AMP** (cAMP) and a **catabolite activator protein** (CAP) that binds to a specific sequence just upstream of the *lac* promoter (Figure 3.71). cAMP, which affects many cellular processes, is formed from ATP (Figure 3.72) in response to a variety of extra- and intracellular stimuli. When CAP, a dimer of identical 22 kDalton polypeptides, binds cAMP, it interacts with the "CAI binding site" just upstream of the promoter's −35 sequence with high affinity. This increases transcription of the *lac* operon nearly 50-fold. CAP alone fails to associate with the CAP binding site or to stimulate transcription. One explanation for how the CAP-cAMP complex stimulates transcription is that its binding adjacent to the RNA polymerase binding site increases

Figure 3.71
Nucleotide sequences that control *lac* operon expression. The repressor binding site, the −10 and −35 promoter elements, and the region that binds the catabolite activator protein CAP-cAMP complex are shown.

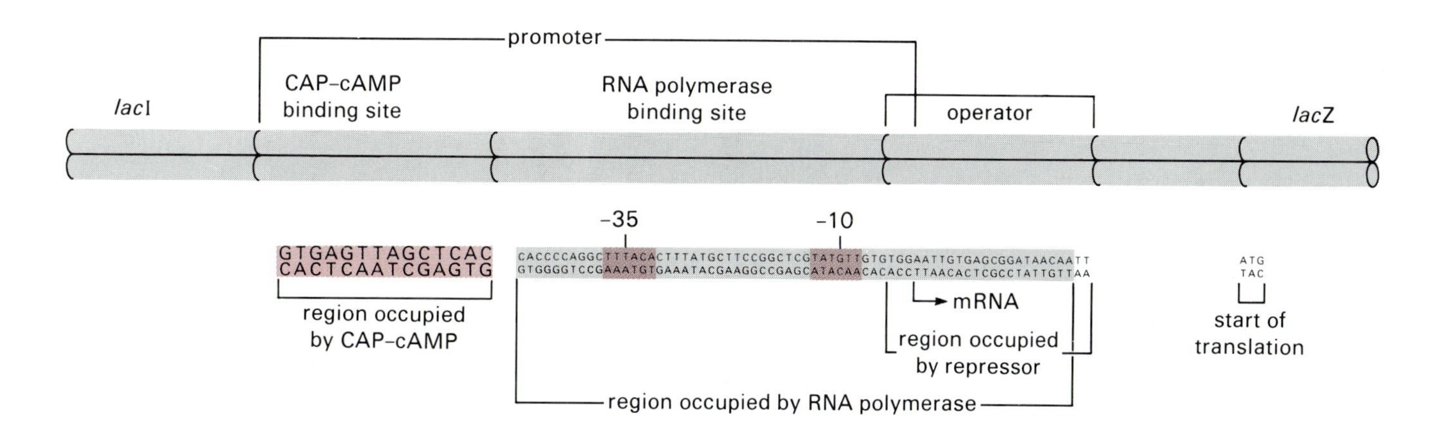

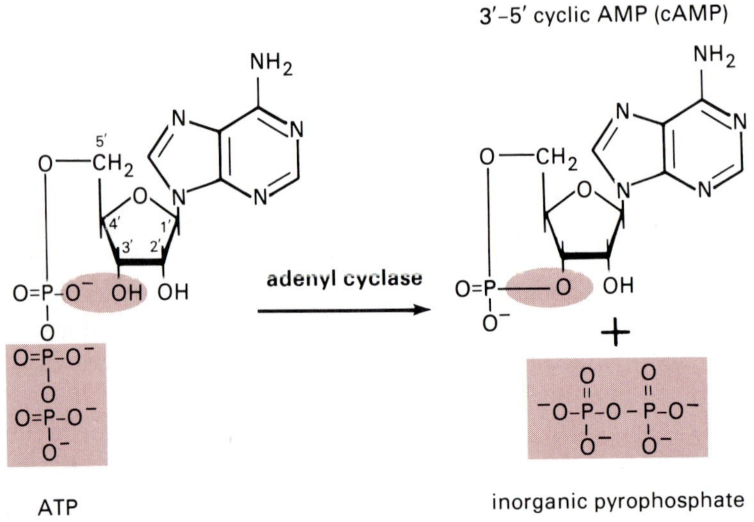

Figure 3.72
The synthesis of cyclic AMP (cAMP) from ATP is catalyzed by adenyl cyclase.

ATP

adenyl cyclase

3′-5′ cyclic AMP (cAMP)

inorganic pyrophosphate

the affinity of the enzyme-promoter interaction (Figure 3.73). Another possibility is that CAP-cAMP binding to the CAP site prevents RNA polymerase from binding to a nearby weak promoter, thereby increasing the likelihood that the polymerase will associate with the "proper" promoter site. The possibility that binding the CAP-cAMP complex alters DNA topology so that polymerase binding to the promoter or transcription is favored is now considered unlikely.

The CAP-cAMP complex also provides a positive signal for regulating the expression of other operons, particularly those that encode the enzymes for carbohydrate degradation. For example, the expression of the *ara* and *gal* operons requires both that repression be reversed by their inducers, arabinose and galactose, respectively, and that the CAP-cAMP complex be bound near their promoters. Thus, in bacteria growing with glucose, the intracellular level of cAMP and therefore also of CAP-cAMP complex is very low. Consequently, even if arabinose or galactose are in the medium, the cells do not produce the enzymes needed to metabolize these or other sugars. However, if glucose becomes limiting, the levels of cAMP and CAP-cAMP rise, and those operons whose inducers are present are expressed. This combination of positive and negative control systems serves the organism well because it prevents the wasteful production of enzymes when they are not required. The CAP-cAMP signaling system also controls operons whose enzymes degrade amino acids, purines, and pyrimidines. In these instances, the accumulation of CAP-cAMP—the "starvation" signal—reduces the expression of operons that encode enzymes that degrade amino acids, purines, and pyrimidines. Here, too, the system's logic is to prevent the production of unnecessary or wasteful enzymes in times of starvation.

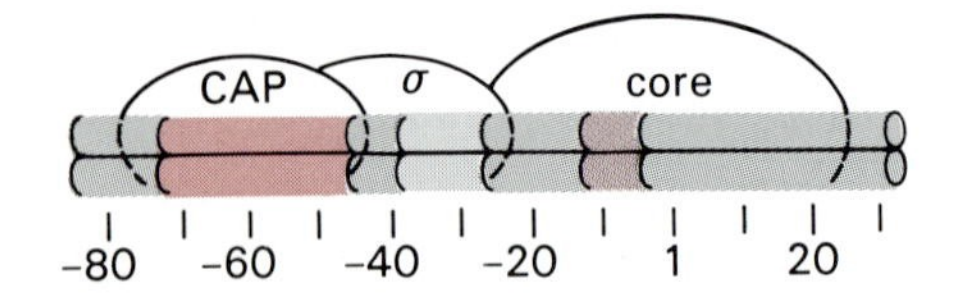

Figure 3.73
Model showing how the binding of the CAP-cAMP to its binding site might facilitate the adjacent binding of RNA polymerase to the *lac* promoter. The model implies an interaction between CAP and the holopolymerase.

d. *Regulation of Tryptophan Operon Expression*

Tryptophan is synthesized in *E. coli* from an aromatic precursor, chorismic acid, via a series of five reactions catalyzed by enzyme complexes of five proteins (Figure 3.74). The five proteins encoded in the *trp* operon are produced coordinately, in nearly equal amounts, by translation of a 7000-

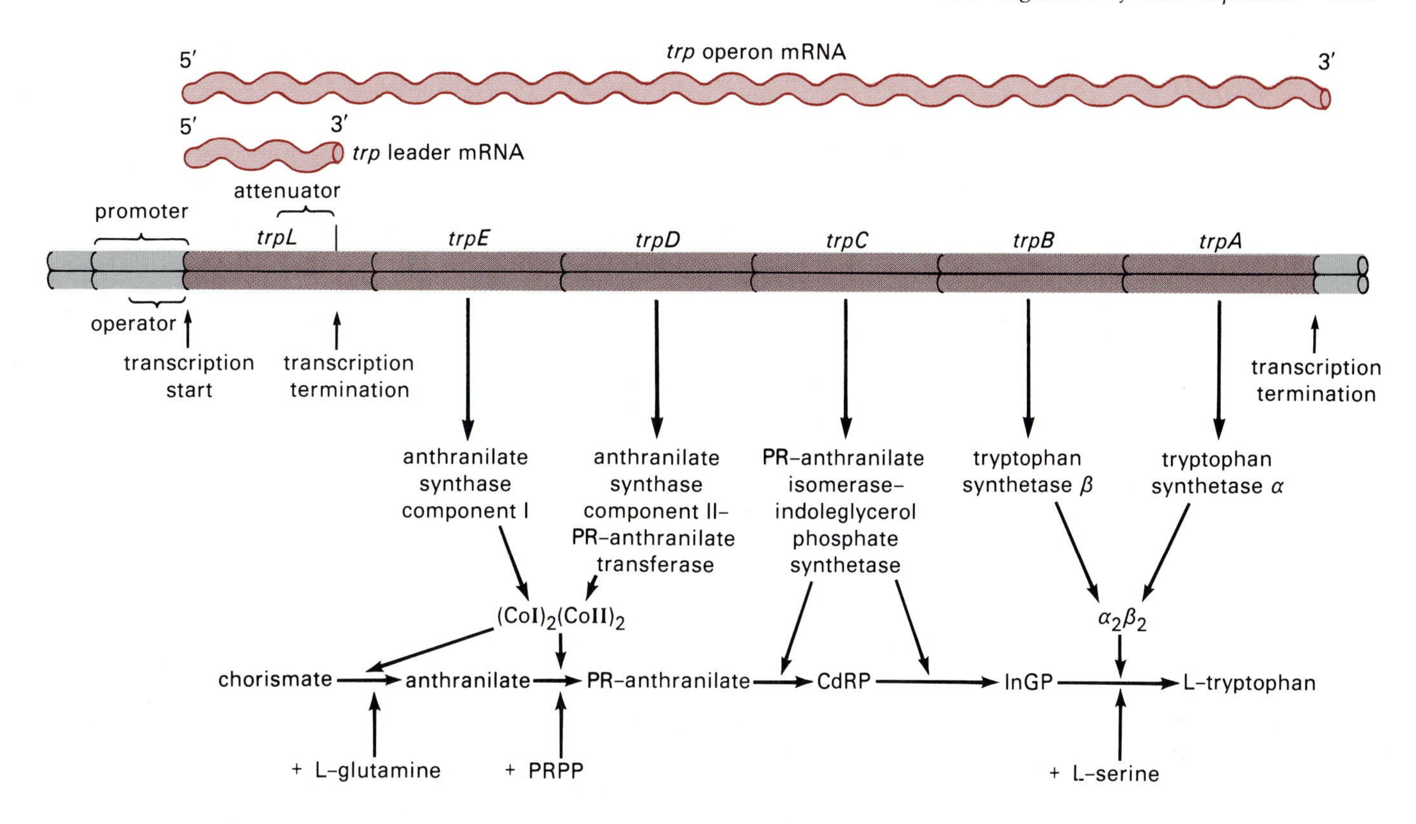

Figure 3.74
Schematic diagram of the tryptophan (*trp*) operon of *E. coli* showing the gene products and the reactions they catalyze in the conversion of chorismic acid to tryptophan. Five gene products yield four enzymes. The *trp* leader segment, *trp* L, and its transcript, the *trp* leader RNA, are shown, as are the attenuator site and the polycistronic *trp* operon mRNA. The abbreviations are PR, phosphoribosyl; PRPP, phosphoribosyl pyrophosphate; CdRP, carboxyphenyl-amino-deoxyribulose phosphate; and InGP, indole glycerol phosphate. Courtesy of C. Yanofsky.

nucleotide-long polycistronic mRNA that is transcribed from the operon's promoter. Provided with sufficient tryptophan for growth, *E. coli* produces almost undetectable levels of the tryptophan biosynthetic enzymes. However, if the cells lack tryptophan, they synthesize high levels of the five enzymes. Depending upon the intracellular level of tryptophan, *E. coli* can vary the production of this group of enzymes over a nearly 700-fold range.

This range of expression is a consequence of two nearly independent regulatory motifs, each of which adjusts the level of *trp* mRNA in response to the concentration of intracellular tryptophan. One mechanism relies on repression to modulate the rate at which operon transcription is initiated at the promoter. The second, referred to as **attenuation**, regulates transcription through a transcription termination signal located between the promoter and the beginning of the first structural gene. The effects of the two regulatory mechanisms are complementary, multiplicative, and permit the extended range in the operon's expression. The remainder of this section considers the two types of regulation and how they account for the fine-tuning of tryptophan biosynthesis.

Repression of the tryptophan operon Transcription of the *trp* operon is blocked by the interaction of a repressor with the *trp* operator sequence. The *trp*R gene, which encodes the 58,000 Dalton repressor protein, is located far from the *trp* operon. For the *trp*R protein to bind to the operator and function as a repressor, it must be associated with tryptophan. Because the expression of *trp*R is low and is not influenced by tryptophan, the concentration of active repressor reflects the concentration of intracellular tryptophan.

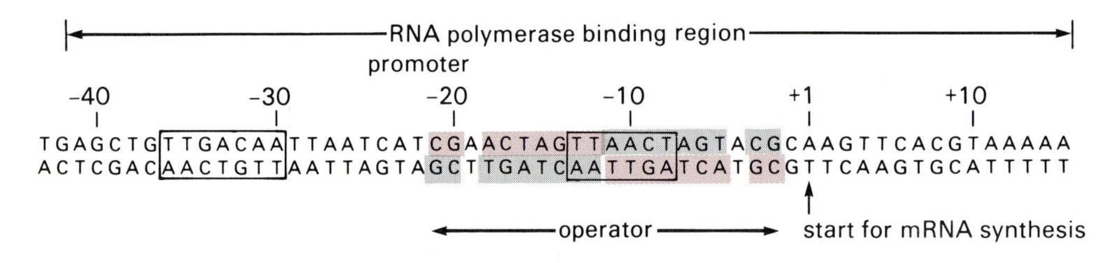

Figure 3.75
The nucleotide sequence of the overlapping promoter and operator of the *trp* operon of *E. coli*. Binding of the *trp* repressor-tryptophan complex to the operator prevents interaction of RNA polymerase with the promoter.

The nucleotide sequence of the *trp* operator and promoter overlap (Figure 3.75), and therefore the binding of the repressor-tryptophan complex to the operator prevents the proper association of RNA polymerase with the promoter. Under these circumstances, operon transcription does not occur, and the biosynthetic enzymes are not made. When tryptophan is absent, repressor binding to the operator does not occur, leaving RNA polymerase free to initiate transcription at the promoter and produce mRNA.

Attenuation of tryptophan operon expression Regulation of the *trp* operon by attenuation depends on a sequence about 100 to 140 base pairs downstream from the site where transcription begins (Figure 3.74). The so-called leader segment, *trp*L, contains the **attenuator** sequence. The attenuator's nucleotide sequence causes RNA polymerase to terminate transcription just before the beginning of the *trp*E gene and release a 141-nucleotide-long RNA, the *trp* leader mRNA. Such termination occurs when the cell contains moderate to high levels of tryptophan. However, when the level of intracellular tryptophan is low, transcription termination at this sequence is attenuated, and transcription proceeds through the terminator to produce the full-length *trp* operon mRNA.

How does the tryptophan level regulate transcription termination at the attenuator? The answer emerges from an examination of the nucleotide sequence of the attenuator region and the knowledge that translation, which occurs simultaneously with transcription, plays a part in attenuation. The first 145 nucleotides in the *trp* operon transcript contain three novel nucleotide sequence arrangements (Figure 3.76). One of them encodes a short polypeptide of 14 amino acids, 2 of which are tandem tryptophan residues. The other two contain a novel arrangement of in-

Figure 3.76
The *trp* leader mRNA encodes a short peptide and contains the transcription attenuator sequence. The leader polypeptide has two neighboring tryptophan residues.

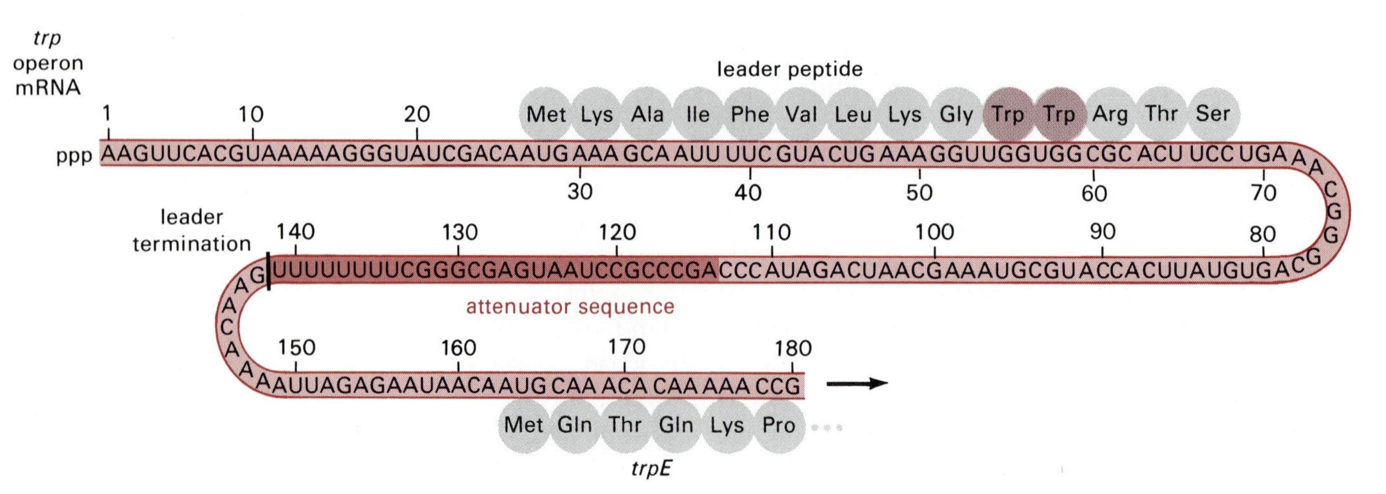

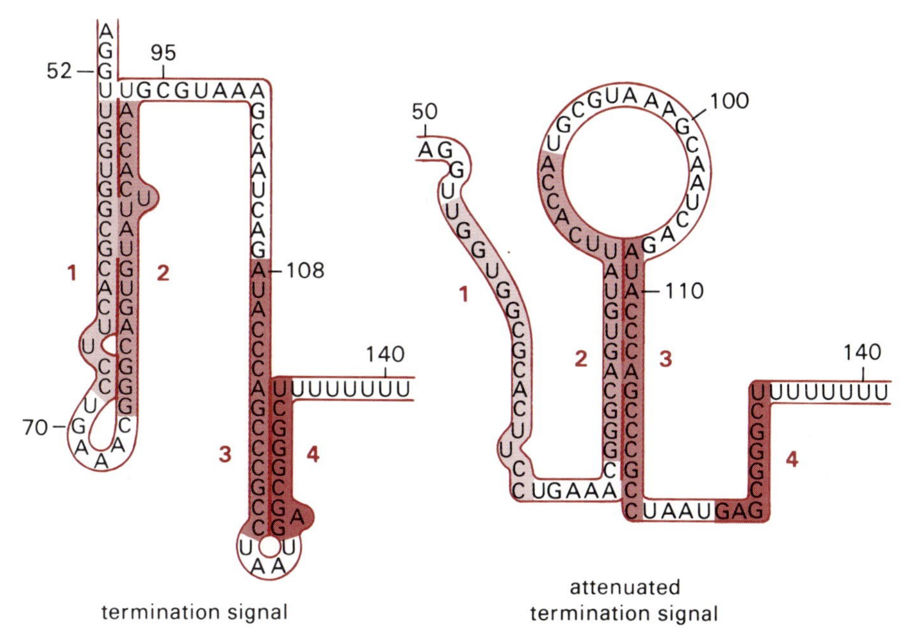

Figure 3.77
Alternative stem-loop structures for the *trp* leader mRNA. The numbering of nucleotide residues is as in Figure 3.76.

verted repeats that permits intramolecular base pairing and folding of the RNA transcript into either one of two alternative base paired stem-loop structures (Figure 3.77). One of the alternative stem-loop structures, in which segments 1 and 2 and segments 3 and 4 are base paired, respectively, promotes a rho-independent termination of transcription at the run of U's following segment 4 (Section 3.2d). The other stem-loop structure—in which segments 2 and 3 are base paired and segments 1 and 4 are single stranded—permits transcription to continue to the end of the operon.

Which of the two alternative stem-loop structures is formed during transcription of the attenuator region depends on whether or not the entire coding sequence for the 14-amino-acid-long polypeptide is translated by ribosomes. Once the leader peptide is initiated at the AUG codon (Figure 3.76), ribosomes translate the following 13 codons, provided there is an adequate supply of all the required aminoacyl-tRNAs. If tryptophanyl-tRNA is limiting, ribosomes will stall when they reach the tandem tryptophan codons. When ribosomes stall at either of the tryptophan codons, which are located in attenuator segment 1, the RNA bases in segments 2 and 3 are free to base pair and form a stem-loop structure while segment 4 remains single stranded (Figure 3.78). This structure allows transcription to continue through this region to produce the 7000-nucleotide-long, full-length *trp* mRNA. If the supply of tryptophanyl-tRNA is adequate for translation of the tandem tryptophan codons, the ribosomes progress to the termination codon for the leader peptide, and, because segment 2 is associated with the ribosome, segments 3 and 4 form a stem-loop, the structure favored for transcription termination.

Thus, the attenuator permits RNA polymerase to sense the concentration of tryptophanyl-tRNA through the ribosome's location. The translatability of the short coding sequence in the leader segment determines the secondary structure assumed by the transcript and consequently the choice between termination and read-through. Adequate tryptophanyl-tRNA

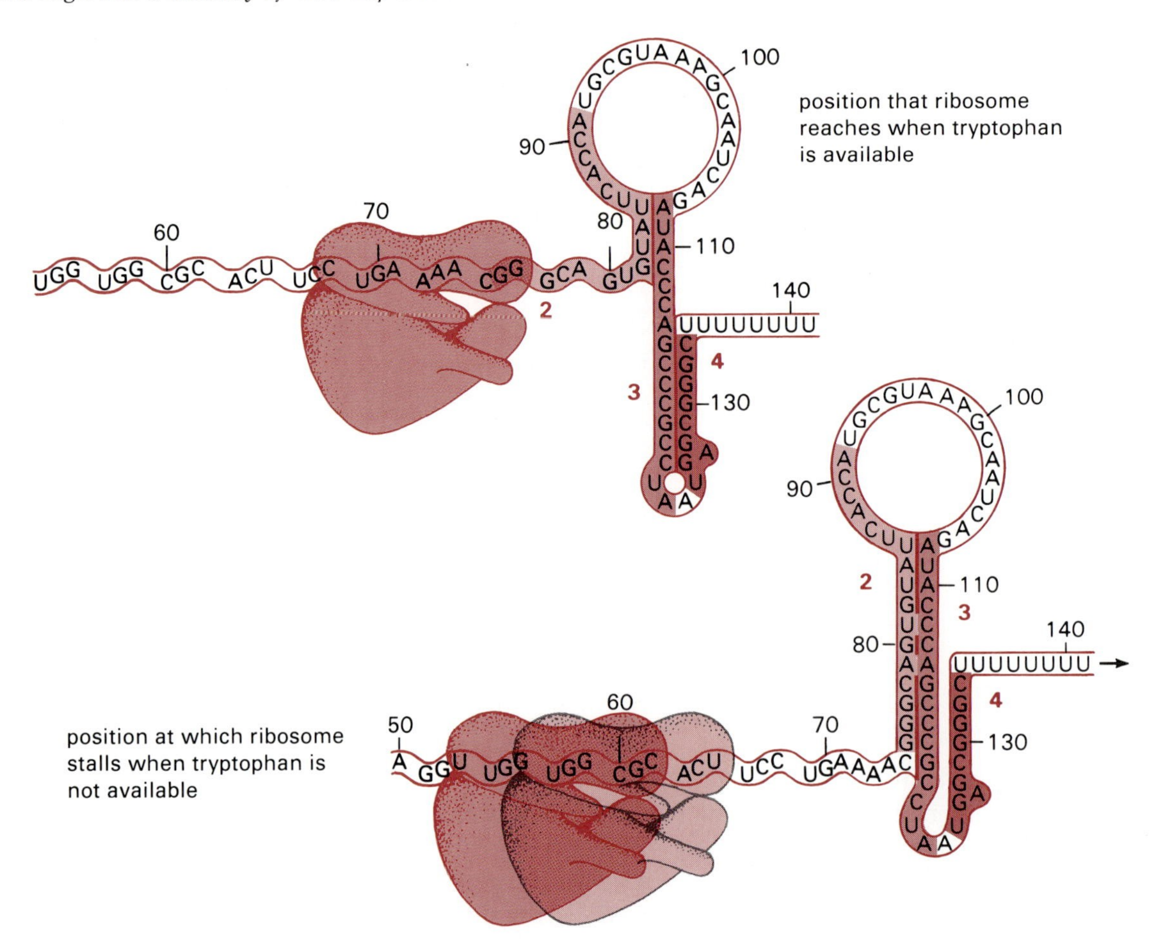

Figure 3.78
The influence of translation on the stem-loop structure of the *trp* leader mRNA. The position of the ribosome determines which of the alternative stem-loop structures exist and thus whether the transcriptional terminator structure can form.

triggers transcription termination, and inadequate tryptophanyl-tRNA allows transcription to continue through the entire operon.

The multiplicative effects of attenuation and repression Together, attenuation and repression provide for a fine-tuning of the expression of the *trp* operon over a wide range. When tryptophan is abundant, repression blocks initiation of mRNA synthesis. As the quantity of tryptophan decreases, the level of the functional repressor-tryptophan complex also decreases, and transcription begins. However, at tryptophan levels that allow the initiation of transcription, the quantity of tryptophanyl-tRNA is still high enough to cause transcription to be terminated within the leader region most of the time; under these conditions, therefore, the operon is expressed only at low levels. As the amount of tryptophan decreases further, repression is completely lifted; if the tryptophanyl-tRNA levels fall below that needed for protein synthesis, termination is attenuated, and the *trp* mRNA and enzymes are synthesized at increased rates. Thus, operon expression is maximal when repression is absent and termination is attenuated maximally; expression is at its lowest when the operon is almost completely repressed and termination is not attenuated.

Attenuation as a general mechanism Attenuation is used in regulating the expression of many genes and operons in *E. coli* as well as in other organisms. The operon that specifies the nine enzymes for histidine biosyn-

operon	leader peptide sequence	regulatory amino acids
his	Met-Thr-Arg-Val-Gln-Phe-Lys-His-His-His-His-His-His-His-Pro-Asp	His
pheA	Met-Lys-His-Ile-Pro-Phe-Phe-Phe-Ala-Phe-Phe-Phe-Thr-Phe-Pro	Phe
leu	Met-Ser-His-Ile-Val-Arg-Phe-Thr-Gly-Leu-Leu-Leu-Leu-Asn-Ala-Phe-Ile-Val-Arg-Gly-Arg-Pro-Val-Gly-Gly-Ile-Gln-His	Leu
thr	Met-Lys-Arg-Ile-Ser-Thr-Thr-Ile-Thr-Thr-Thr-Ile-Thr-Ile-Thr-Thr-Gly-Asn-Gly-Ala-Gly	Thr, Ile
ilv	Met-Thr-Ala-Leu-Leu-Arg-Val-Ile-Ser-Leu-Val-Val-Ile-Ser-Val-Val-Val-Ile-Ile-Ile-Pro-Pro-Cys-Gly-Ala-Ala-Leu-Gly-Arg-Gly-Lys-Ala	Leu, Val, Ile

Figure 3.79
Examples of amino acid biosynthetic operons that include leader mRNAs containing multiple codons for the amino acids that regulate the operon.

thesis is controlled by attenuation only; its attenuator region uses a polypeptide coding sequence containing six tandem histidine codons (Figure 3.79). Operons that are regulated by the levels of several related amino acids—threonine and isoleucine or leucine, valine and isoleucine—have the codons for these amino acids heavily represented in their leader peptide coding sequences. Antitermination, that is, attenuated termination, does not require coupling of translation and transcription. Antitermination can also be mediated by certain proteins, which allows particular gene clusters to be expressed only after the proteins that mediate the attenuation are made (Section 3.11e).

e. *Temporal Control of Gene Expression in the Life Cycle of Bacteriophage λ*

Bacteriophage λ has two alternative life-styles. In the lytic pathway, all the viral genes are expressed in a temporal sequence, leading ultimately to the production of about 100 progeny phage and lysis of the infected bacterium. The infecting virus may also enter a lysogenic pathway during which its DNA is covalently inserted into the host cell DNA at a specific chromosomal site (Section 2.4d). The integrated form of the viral genome is called a **prophage**. In lysogenic cells, the prophage DNA is replicated by the cell's machinery for many generations just as if it were a normal part of the cell's genome. However, all but one of the prophage's genes are switched off. Lysogenic cells initiate the lytic pathway either spontaneously at low frequency or by exposure to various chemical or physical agents. Irrespective of whether induction occurs spontaneously or by other means, the formerly quiescent viral genes are turned on, triggering the reactions that lead to the production of infectious phage particles.

How is the expression of viral genes regulated temporally to ensure that viral DNA replication, the production of viral proteins, and the assembly of bacteriophage particles occur in an orderly fashion during lytic infection? Furthermore, what regulatory mechanisms determine whether the initial infection follows the lytic or lysogenic pathway? These questions are worth considering because their answers illustrate the elegant interplay of different mechanisms prokaryotic organisms use to regulate their phenotypes.

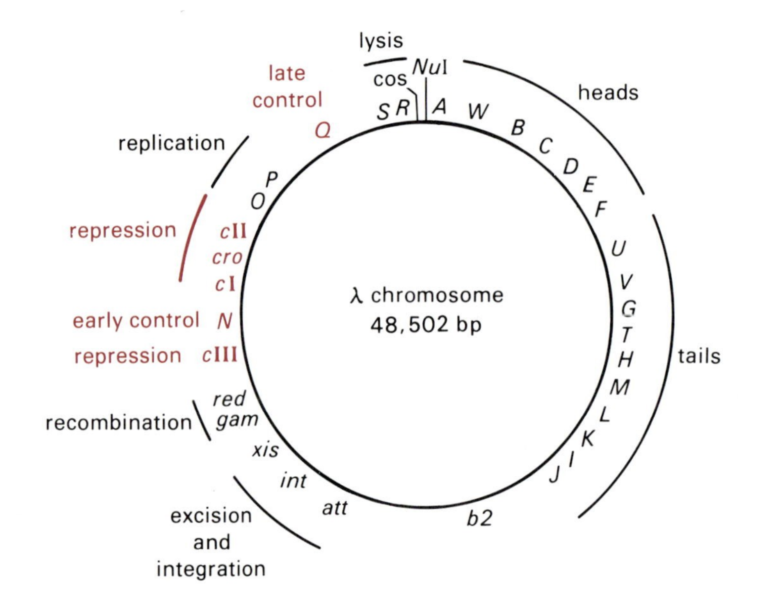

Nul
cos A W B C D E F U V G T H M L K I J b2 att int xis gam red cIII N cI cro cII O P Q S R cos

Figure 3.80
Linear and circular forms of the bacteriophage λ genome showing the positions of its genes. The circular map shown at the top is formed by annealing and ligating the single strand cohesive end of the linear form of λ DNA. The ends and the join are labeled *cos*.

Figure 3.81
Genomic positions and temporal order of expression of bacteriophage mRNAs that are formed in the lytic cycle.

The lytic pathway First, we need to consider the pattern of gene expression in the lytic infection. Figure 3.80 shows the location of the major structural and regulatory genes arrayed on the circular and linear maps of the phage's genome. The genes that encode the structural proteins (heads and tails) are clustered in one region of the DNA; the genes encoding the enzymatic (replication and recombination) and regulatory functions (repressors and antiterminators) are interspersed with one another in another region of the DNA.

Following infection, the phage DNA is circularized by ligation of the cohesive ends (Section 2.4d). Then *E. coli* RNA polymerase transcribes a class of λ mRNAs that encode the proteins required for the earliest events in the life cycle—the immediate early functions (Figure 3.81). One of the immediate early mRNAs is transcribed in the leftward direction from the P_L promoter and terminates at a sequence t_{L1}; this mRNA (*L1*) produces a regulatory protein, N, which acts as an antiterminator. The other immediate early mRNA is transcribed in the rightward direction

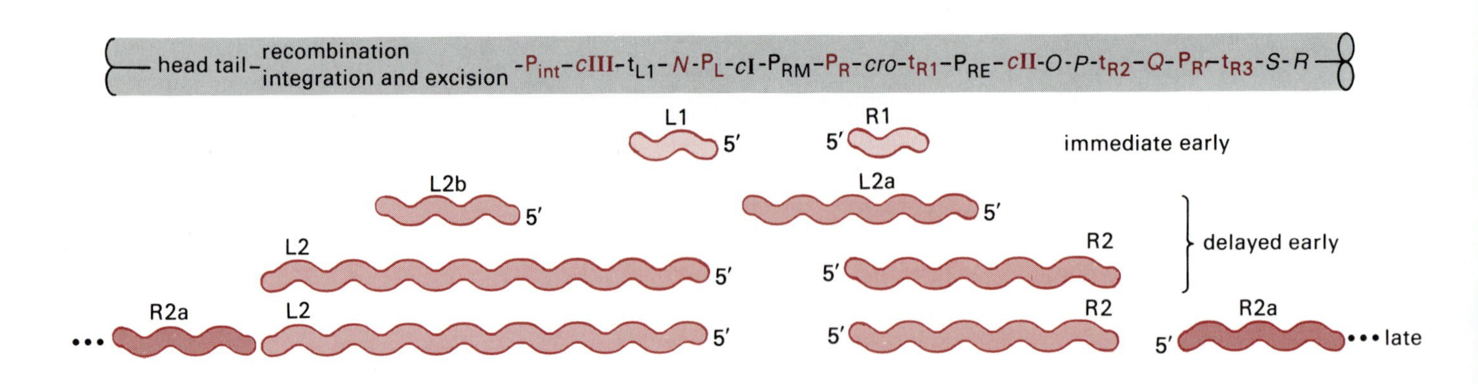

from the P_R promoter to the t_{R1} terminator, producing an mRNA that encodes only the Cro protein. We shall return shortly to consider how the Cro protein influences the lytic versus lysogenic outcome.

As the N protein accumulates, it attenuates termination at t_{L1} and t_{R1}, allowing RNA polymerase to transcribe through those sites to produce a second class of mRNAs—the delayed early transcripts. The larger transcript from P_L (*L2* in Figure 3.81) encodes the enzymatic machinery for recombination, particularly the enzymes that catalyze the integration of λ DNA into the host DNA. The delayed early transcript from P_R—the transcript *R2*—encodes the enzymes responsible for λ DNA replication (proteins O and P) and another regulatory protein, Q. Protein Q attenuates transcription termination at a termination site (t_{R3}) just downstream of the promoter, $P_{R'}$. Transcription from $P_{R'}$ to the right transcribes genes (*S* and *R*) responsible for triggering lysis of the cells. Moreover, because λ DNA is circularized very soon after infection (Figure 3.80), the *S* and *R* genes are contiguous with the genes that code for the head and tail proteins. Consequently, transcription initiating at $P_{R'}$ and reading through t_{R3} produces an mRNA that encodes the lysis proteins and all the viral structural proteins. So far, we have accounted for the production of the machinery needed for the lytic infection: the enzymes needed for replication of the viral DNA and the viral proteins needed to form mature phage particles.

The lysogenic pathway How can we explain the detour to a lysogenic pathway from a series of events geared to produce the lytic outcome? To do so, we need to upgrade our picture of the structure of P_L and P_R and to introduce several additional genes, gene products, and promoters. P_L and P_R are each actually part of complex regulatory regions in which the promoters are interspersed with operator sequences, O_L and O_R respectively (Figure 3.82). Both O_L and O_R are binding sites for two regulatory proteins, the cI repressor and the antirepressor Cro. As already mentioned, Cro protein is the product of the translation of the immediate early mRNA transcribed in the rightward direction from P_R. The repressor is encoded by the *cI* gene, which is located between P_R and P_L/O_L (Figure 3.82). The *cI* mRNA is transcribed in the leftward direction from a promoter called P_{RE}, which is to the right of the *Cro* gene. P_{RE} is itself activated by two positive regulatory proteins, cII and cIII, which are synthesized following the action of N protein in extending the mRNAs initiated at P_R and P_L, respectively.

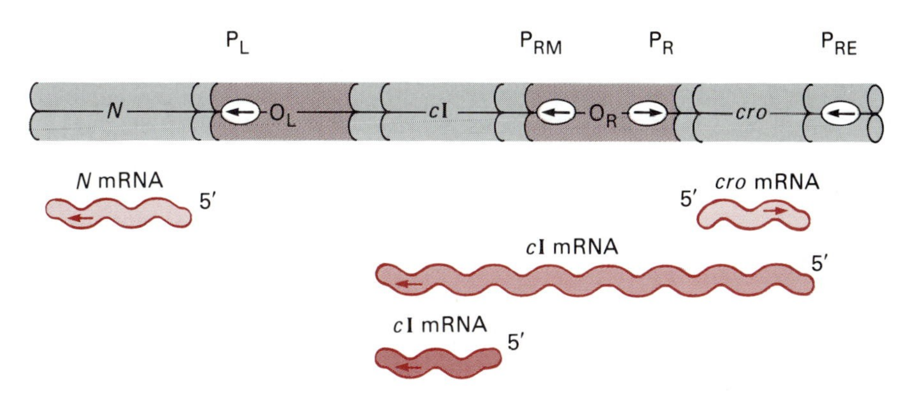

Figure 3.82
The arrangement of promoter-operator sequences that control the lytic or lysogenic outcome of a bacteriophage λ infection. Four different promoters are shown, as are several mRNAs that encode the regulatory proteins Cro, cI repressor, and N.

As the cI repressor accumulates, it binds to the left and right operator sites, shutting down the expression of all the genes that are transcribed from P_R and P_L. These circumstances favor the lysogenic pathway because production of the replication enzymes and the viral structural proteins is blocked, and the small amount of integration enzyme, Int, made from the delayed early mRNA before repression is complete catalyzes the recombination between the phage and cell DNA. Once integrated, lysogeny is maintained by the low level production of the cI repressor from an mRNA that is transcribed from the promoter, P_{RM}, which occurs immediately to the right of the *cI* structural gene.

But the lytic response does occur, and this is a consequence of both the antirepressor and activator properties of Cro. Cro also binds to sites in O_R and O_L and in doing so prevents cI binding to O_L and O_R, thereby antagonizing the repressor action of cI. As Cro accumulates, the synthesis of the cI repressor is progressively blocked while the accumulation of the replication, lysis, and phage structural proteins accelerates.

Binding of Cro and cI to O_R *and* O_L To understand the subtle antagonistic actions of Cro and cI proteins, we need to examine the nature and arrangement of the binding sites within the two operators O_R and O_L. O_L and O_R each consist of a tandem cluster of three suboperators O_{L1}, O_{L2}, and O_{L3} or O_{R1}, O_{R2}, and O_{R3}, respectively. O_L overlaps P_L, O_R overlaps P_{R1}, and O_{R3} overlaps P_{RM} (Figure 3.83). Both the Cro and cI proteins bind to all suboperator segments, but their relative affinities for each suboperator differs. Thus, Cro protein binds most strongly to O_{R3} and progressively less so to O_{R2} and O_{R1}; the affinity of binding of Cro protein to O_L is greatest for O_{L3} and decreasingly strong for O_{L2} and O_{L1}. As Cro protein accumulates early in the infection, it binds to O_{R3}, and transcription of the *cI* gene from P_{RM} is prevented. Because O_{R1} and O_{L1} are not occupied until the concentration of Cro protein rises to much higher levels, transcription from P_R and P_L is not inhibited. On the contrary, binding of Cro protein to O_{R3} and O_{L3} stimulates transcription from P_R and P_L, respectively. The cI protein also has different binding affinities for the suboperators. Suboperators O_{R1} and O_{L1} bind the cI protein most strongly; therefore, transcription from P_R and P_L is blocked at low concentrations of the repressor. Here, too, binding of the cI repressor at O_{R1} stimulates transcription of the *cI* gene from P_{RM}.

Thus, the choice between the lytic and lysogenic pathways reflects the relative rates of accumulation of four regulatory proteins: the cI repressor protein, Cro antirepressor protein, and the positive regulatory proteins cII and cIII. The lytic pathway prevails if the antirepressor function of Cro protein dominates. If repression is established first, the functions needed for the lytic pathway are blocked, and lysogeny results. Because the balance of these proteins is delicate, variations in the host, culture medium, or other conditions influence the probability of the lytic versus lysogenic response.

f. Translational Regulation of the Expression of Some Gene Products

As pointed out earlier, the expression of prokaryote genes that specify RNA molecules, rRNA and tRNA, for example, is regulated at two

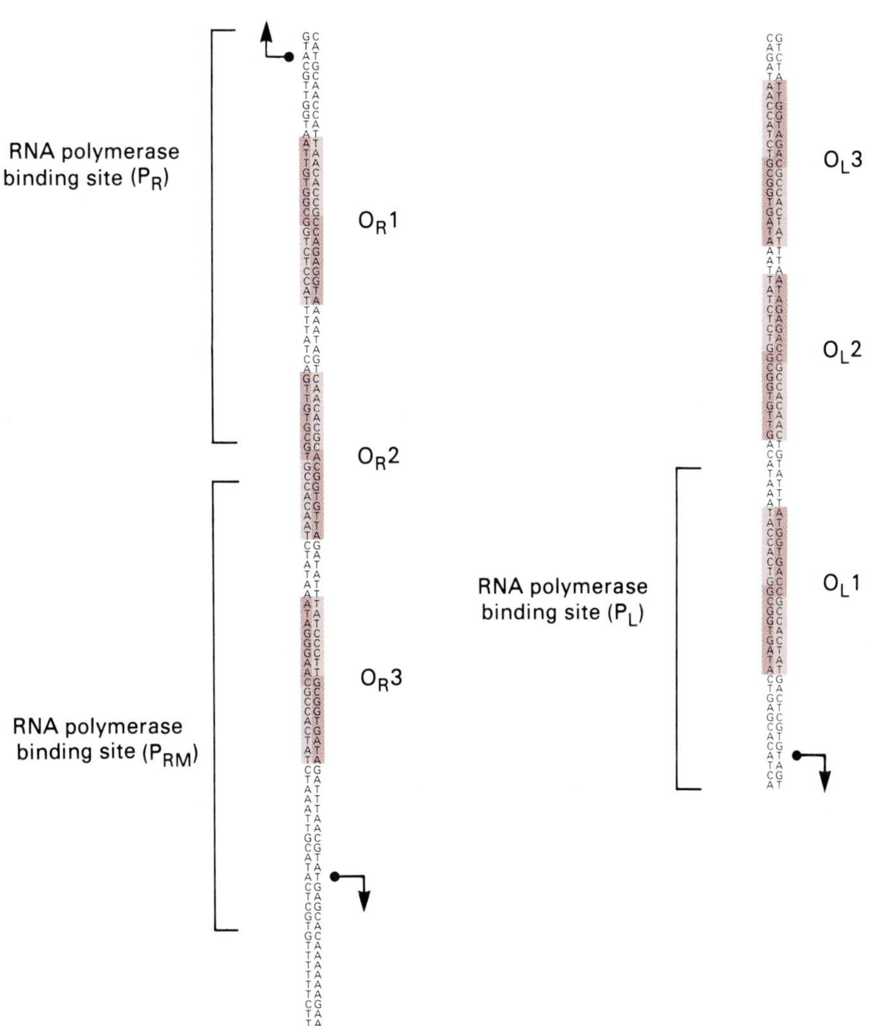

Figure 3.83
Organization of the binding sites for Cro, cI repressor, and RNA polymerase in the operator regions O_L and O_R.

stages—at transcription, and during posttranscriptional modification and processing of precursor RNAs to their mature forms. Expression of some protein coding genes is also regulated at two stages. One is the control of mRNA levels, as discussed in Sections 3.11c–e. But regulation also occurs by repression of translation. Two examples follow.

First, each of the related *E. coli* bacteriophages, MS2, R17, and Qβ, contains a single RNA molecule that functions both as its genome and as the mRNA that produces the proteins needed for virus multiplication. The bacteriophage MS2 genome, which is representative of this class of virus, has four genes (Figure 3.84). One of the genes (*rep*) encodes a subunit of the RNA replicase complex, which amplifies the infecting RNA to provide high levels of mRNA for translation of its genes, as well as providing the genomes in the progeny virus. A second gene specifies the coat protein (CP), 180 copies of which form the virus' icosohedral capsid. The third gene product, the A protein or maturase, is part of the mature virus particle and is essential for the adsorption and entry of the virus into cells. A fourth protein (Lys) promotes the lysis of infected cells and the release of the progeny virus. The sequence encoding the Lys protein begins within the 3′ end of the *CP* gene and extends into the 5′ end of the *rep* gene. However, the reading frame encoding the Lys protein is

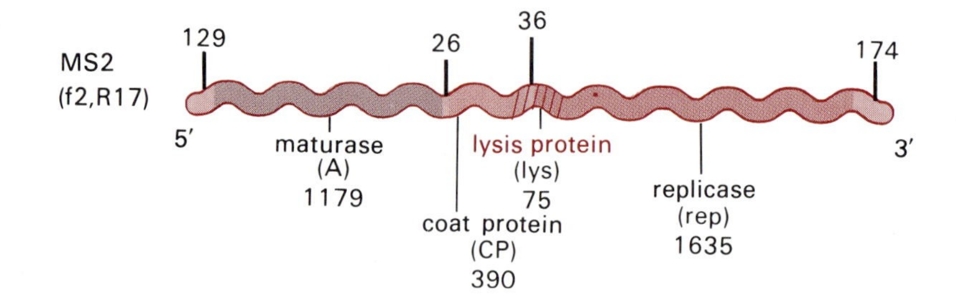

Figure 3.84
Arrangement of the four genes on the bacteriophage MS2 RNA genome (3569 nucleotides). The lengths of the 5′ leader, the genes, and the 3′ trailer are given as number of nucleotides, as are the lengths of the spacers between genes *A* and *CP* and genes *CP* and *rep*. The spacer between *CP* and *rep* is part of the *lys* gene coding region. See Figure 3.41 for the sequence of a portion of MS2 RNA.

shifted by one nucleotide from that of the CP and Rep coding sequences.

In a normal infection, large quantities of CP must be made, but much smaller amounts of Rep and Lys and very few molecules of A suffice. Differential expression of the viral genes is achieved by having only the initiator AUG codon of the *CP* gene available for ribosome attachment. The corresponding sequences of the *A*, *lys*, and *rep* genes are buried within base paired regions of the RNAs secondary structure (Figure 3.41). Ribosomes translate the *A* gene only rarely, in fact, only once, just as the 5′ end of the RNA is being synthesized. But the *CP* gene is translated many times by ribosomes that initiate at the *CP* gene's exposed initiator coding sequence AUG. In translating the CP coding sequence, the ribosomes unfold the RNA and expose the initiator AUG of the *rep* gene, allowing that coding sequence to be translated. To translate the Lys protein's coding sequence, it appears that a ribosome occasionally translocates "unnaturally" during translation of the *CP* gene sequence, and translation terminates shortly thereafter. This permits translation to begin again at the "out of phase" initiation sequence of the Lys coding sequence.

As the concentration of CP rises, it binds to a sequence that overlaps the translation initiation sequence in the *rep* gene; this prevents synthesis of the replicase but allows the formation of coat protein to continue. Thus, the translation of the *rep* gene is regulated initially by structural constraints that limit the ability of ribosomes to bind to its initiator sequence, by occasional slippage of ribosomes during translation and later by translational repression caused by binding of the coat protein to its initiation sequence.

In our second example, the synthesis of proteins that make up much of the translation machinery is regulated at the level of translation. The genes that encode the proteins of the large (L) and small (S) ribosomal subunits and some of the proteins that mediate the translation process (e.g., EF-Tu and EF-G) are intermingled among several operons (Figure 3.67). This permits the synthesis of many gene products that function in concert to be regulated coordinately. The expression of these genes is mediated coordinately during both transcription and translation. As will be evident in what follows, ribosomal protein synthesis is in part also regulated by the level of the three rRNAs and by the kinetics of ribosome assembly.

The motif for controlling the translation of the several ribosomal protein operons is the same (Table 3.7). One of the ribosomal proteins encoded in a polycistronic mRNA binds to a specific sequence at either the 5′ end of the mRNA or at the beginning of one of the coding sequences

Table 3.7 The Arrangement of Sequences Encoding Ribosomal Proteins, Translation Factors and RNA Polymerase Subunits Are Interspersed in the mRNAs Transcribed from Several Operons

Operon	Encoded Proteins (in order from 5′ end of mRNA)	Regulator	Initiator Sequence to Which Regulator Binds
str	S12 S7 EF-G EF-Tu	S7	S7
spc	L14 L24 L5 S14 S8 L6 L18 S5 L30 L15 L30	S8	L5
S10	S10 L3 L2 L4 L23 S19 L22 S3 S17 L16 L29	L4	S10
α	S13 S11 S4 α L17	S4	S13
L11	L11 L1	L1	L11
rif	L10 L7/12 β β'	L10	L10

Each mRNA is written so that its 5′ end is at the left. The regulator protein is indicated at the right, and the proteins subject to regulation are underlined. Coding sequences whose inclusion in the regulation is uncertain are indicated by broken lines.

within the mRNA. In either case, binding blocks ribosome access to the nearby translation initiator sequence. Depending upon whether the site for initiating translation of the mRNA occurs only at the 5′ proximal coding sequence or at a few internal regions, translation of all or part of the mRNA sequence is affected. This kind of feedback control, in which a gene product regulates expression of its own gene, is referred to as **autogenous regulation**. It occurs at the level of transcription (for example, the cI repressor protein of λ regulates its own transcription at P_{RM}) and in translation, as in the present case.

In the course of assembling ribosomes, certain of the ribosomal proteins bind to rRNA. This suggests a way that free rRNA could regulate the synthesis of ribosomal proteins. How would this be accomplished? When rRNA is available, newly synthesized ribosomal proteins associate with it to initiate ribosome assembly. However, when rRNA is limiting, the newly made ribosomal proteins accumulate and bind instead to their own mRNAs and block their own synthesis as well as that of related ribosomal proteins. This prevents the accumulation of free ribosomal proteins. Thus, several key ribosomal proteins are repressors that block translation of the mRNAs that encode them. Simultaneously, they block translation of the other proteins encoded by the same mRNA. The capability of a ribosomal protein to recognize both the rRNA and its own mRNA derives from the fact that the two RNAs share a related nucleotide sequence (Figure 3.85). Thus, the sequences at which the ribosomal S8 and S7 proteins bind to the 16S RNA and to their own mRNAs form comparable secondary structures and share an identical sequence in their loops.

In addition to the translational control of ribosomal protein synthesis, there is also a feedback regulation on the transcription of ribosomal protein operons. We shall not consider the transcriptional regulation of these operons, except to note that repression and attenuation of transcription by ribosomal subunits or even whole ribosomes may be involved. Most likely, however, the major control point for setting the pace of ribosome synthesis is the transcription of rRNAs. Thus, the level of rRNA controls the translation of ribosomal proteins from their mRNAs and, hence, the assembly of ribosomes.

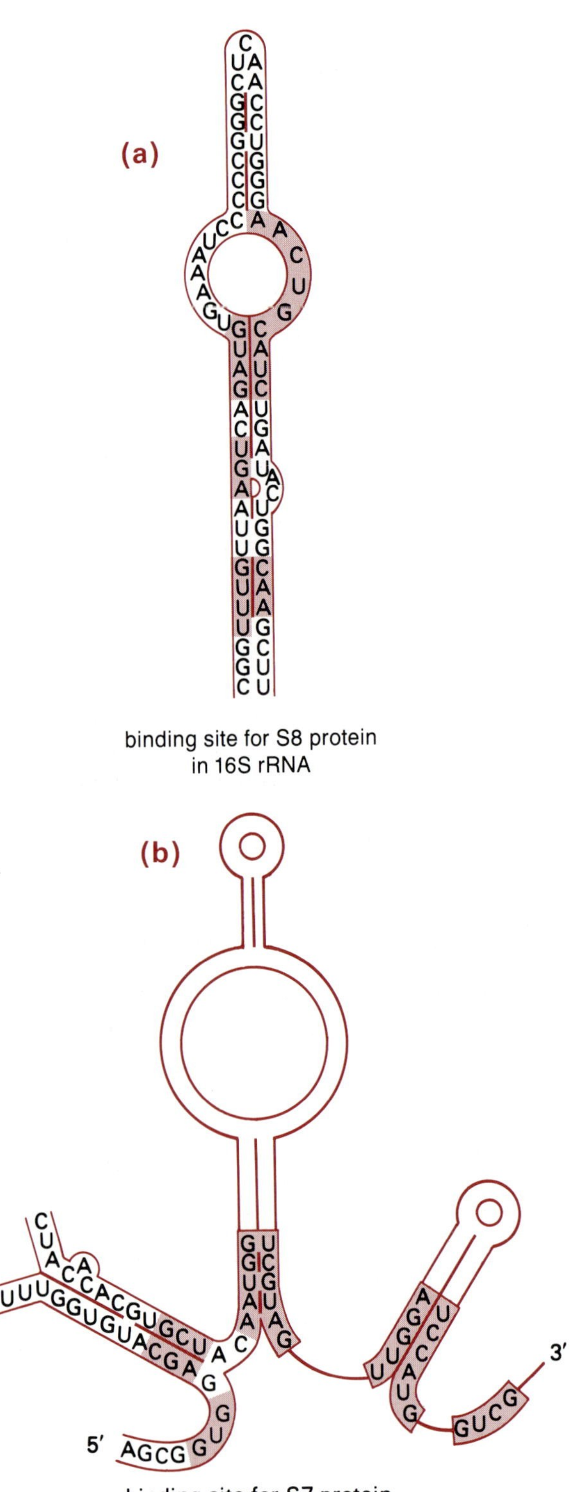

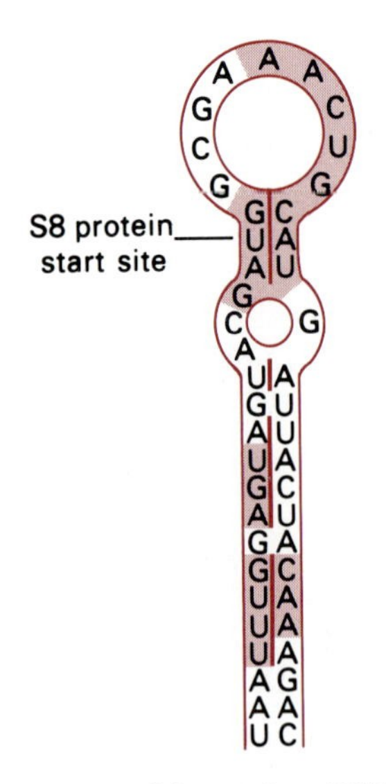

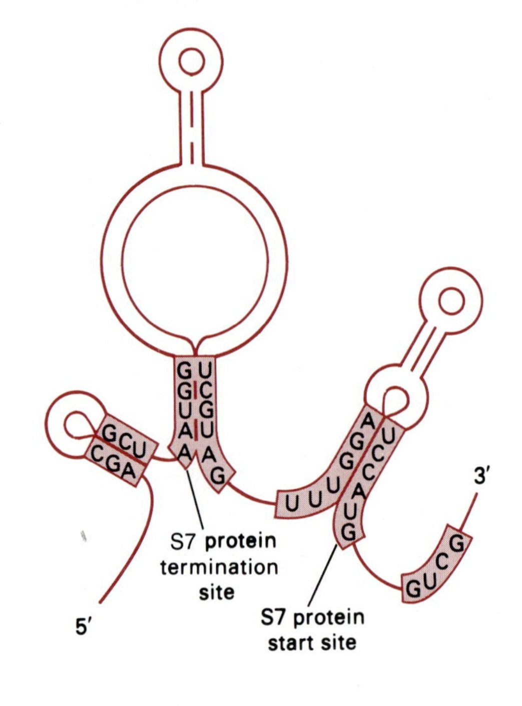

Figure 3.85

Support for a proposal whereby binding of a ribosomal protein to the mRNA inhibits translation of ribosomal protein mRNA. (a) Similarities between the primary and secondary structure of S8 mRNA (*spc* operon polycistronic mRNA) and the S8 binding site on 16S rRNA. (b) Similarities in the primary and secondary structures between the S7 protein binding site on 16S rRNA and the translation initiation site of the S7 mRNA (*str* operon polycistronic mRNA). Similar nucleotide sequences are shown in color. After M. Nomura et al., *Proc. Natl. Acad. Sci. U.S.A.* 77 (1980) p. 7084, and J. D. Watson et al., *Molecular Biology of the Gene*, 4th ed. (Menlo Park, Calif.: Benjamin/Cummings, 1987). See Figure 3.67 and Table 3.7 for more information about these operons.

References for Part I

PERSPECTIVE

The following classical articles and books in genetics and biochemistry contributed major conceptual advances to the development of molecular genetics. The list is chronological.

G. Mendel. 1866. Versuche über Pflanzen-Hybriden. *Verhandlungen der Naturforschenden Verein*, Brünn, 4 3–47 (reprinted in *Journal of Heredity* 42 3–47 [1951]). This and the other important paper by Mendel, Über einige aus künstlicher Befruchtung gewonnene Hieracium-Bastarde, are available in English translation in C. Stern and E. R. Sherwood (eds.). 1966. The Origins of Genetics. A Mendel Source Book. W. H. Freeman, San Francisco.

F. Miescher. 1871. Über die chemische Zusammensetzung der Eiterzellen. *Hoppe-Seyler*'s *Medizinish-Chemischen Untersuchungen* 4 441–460.

A. E. Garrod. 1902. Inborn Errors of Metabolism. *Lancet* 2 1616–1620.

W. S. Sutton. 1903. The Chromosomes in Heredity. *Biol. Bull.* 4 231–251.

A. E. Garrod. 1909. Inborn Errors of Metabolism. Frowde, Hodder, and Stoughton, London.

W. Johannsen. 1909. Elemente der Exakten Erblichkeitslehre. Fischer. Jena.

T. H. Morgan. 1910. Sex-Linked Inheritance in *Drosophila. Science* 32 120–122.

T. H. Morgan, A. H. Sturtevant, H. J. Muller, and C. B. Bridges. 1915. The Mechanism of Mendelian Heredity. Holt, Rinehart & Winston, New York.

B. McClintock and H. B. Creighton. 1931. A Correlation of Cytological and Genetical Crossing Over in *Zea mays. Proc. Natl. Acad. Sci.* U.S.A. 17 492–497.

B. McClintock. 1934. The Relation of a Particular Chromosomal Element to the Development of the Nucleoli in *Zea mays. A. Zellforsch. u Mikr. Anat.* 21 294–328.

G. W. Beadle and E. L. Tatum. 1941. Genetic Control of Biochemical Reactions in *Neurospora. Proc. Natl. Acad. Sci.* U.S.A. 27 499–506.

O. T. Avery, C. M. Macleod, and M. McCarty. 1944. Studies on the Chemical Nature of the Substance Inducing Transformation of Pneumococcal Types. I. Induction of Transformation by a Desoxyribonucleic Acid Fraction Isolated from *Pneumococcus* Type III. *J. Exp. Med.* 79 137–158.

E. Chargaff, R. Lipschitz, C. Green, and M. E. Hodes. 1951. The Composition of the Deoxyribonucleic Acid of Salmon Sperm. *J. Biol. Chem.* 192 223–230.

D. M. Brown and A. R. Todd. 1952. Nucleotides. Part X. Some Observations on the Structure and Chemical Behavior of the Nucleic Acids. *J. Chem. Soc.* 52–58.

A. D. Hershey and M. Chase. 1952. Independent Functions of Viral Protein and Nucleic Acid in Growth of Bacteriophage. *J. Gen. Physiol.* 36 39–56.

J. D. Watson and F. H. C. Crick. 1953. Molecular Structure of Nucleic Acids: A Structure for Deoxynucleic Acids. *Nature* 171 737–738.

J. D. Watson and F. H. C. Crick. 1953. General Implications of the Structure of Deoxyribonucleic Acid. *Nature* 171 964–967.

V. M. Ingram. 1957. Gene Mutations in Human Hemoglobin: The Chemical Difference Between Normal and Sickle Cell Hemoglobin. *Nature* 180 326–328.

M. B. Hoagland, M. L. Stephenson, J. F. Scott, L. I. Hecht, and P. C. Zamecnik. 1958. A Soluble Ribonucleic Acid Intermediate in Protein Synthesis. *J. Biol. Chem.* 231 241–257.

M. Meselson and F. W. Stahl. 1958. The Replication of DNA in *Escherichia coli. Proc. Natl. Acad. Sci.* U.S.A. 44 671–682.

A. Kornberg. 1960. Biological Synthesis of Deoxyribonucleic Acid. *Science* 131 1503–1508.

S. Brenner, F. Jacob, and M. Meselson. 1961. An Unstable Intermediate Carrying Information from Genes to Ribosomes for Protein Synthesis. *Nature* 190 576–581.

F. Jacob and J. Monod. 1961. Genetic Regulatory Mechanisms in the Synthesis of Proteins. *J. Mol. Biol.* 3 318–356.

M. W. Nirenberg and H. J. Matthaei. 1961. The Dependence of Cell-Free Protein Synthesis in *E. coli* upon Naturally Occurring or Synthetic Polyribonucleotides. *Proc. Natl. Acad. Sci.* U.S.A. 47 1588–1602.

C. Yanofsky, B. C. Carlton, J. R. Guest, D. R. Helinski, and U. Henning. 1964. On the Colinearity of Gene Structure and Protein Structure. *Proc Natl. Acad. Sci.* U.S.A. 51 266–272.

The Genetic Code. 1966. *Cold Spring Harbor Symp. Quant. Biol.* 31 whole issue.

The historical development of genetics, nucleic acid biochemistry, and enzymology is described in the following works.

J. S. Fruton. 1972. Molecules and Life. Wiley-Interscience, New York.

H. Stubbe. 1972. History of Genetics: From Prehistoric Times to the Discovery of Mendel's Laws. Translated by T. R. W. Waters. MIT Press, Cambridge, Massachusetts.

R. Olby. 1974. The Path to the Double Helix. University of Washington Press, Seattle.

F. H. Portugal and J. S. Cohen. 1977. A Century of DNA. MIT Press, Cambridge, Massachusetts.

H. F. Judson. 1979. The Eighth Day of Creation. Simon and Schuster, New York.

General Reading for Chapters in Part I

C. R. Cantor and P. R. Schimmel. 1980. Biophysical Chemistry. W. H. Freeman, San Francisco.

B. Alberts, D. Bray, J. Lewis, M. Raff, K. Roberts, and J. D. Watson. 1983. Molecular Biology of the Cell. Garland, New York.

G. Zubay. 1983. Biochemistry. Addison-Wesley, Reading, Massachusetts.

J. Darnell, H. Lodish, and D. Baltimore. 1986. Molecular Cell Biology. Scientific American Books, New York.

D. Freifelder. 1987. Molecular Biology, 2nd ed. Jones and Bartlett, Boston.

B. Lewin. 1987. Genes, 3rd ed. Wiley, New York.

J. D. Watson, N. H. Hopkins, J. W. Roberts, J. A. Steitz, and A. M. Weiner. 1987. Molecular Biology of the Gene, 4th ed. Benjamin/Cummings, Menlo Park, California.

J. D. Rawn. 1988. Biochemistry. Carolina Biological Supply Co., New York.

L. Stryer. 1988. Biochemistry, 3rd ed. W. H. Freeman, San Francisco.

References for Indicated Chapter Sections

1.1

F. H. C. Crick and J. D. Watson. 1954. The Complementary Structure of Deoxyribonucleic Acid. *Proc. Roy. Soc.* 223(4) 80–96.

F. H. C. Crick. 1976. Linking Numbers and Nucleosomes. *Proc. Natl. Acad. Sci.* U.S.A. 73 2639–2643.

W. R. Bauer, F. H. C. Crick, and J. H. White. 1980. Supercoiled DNA. *Sci. American* 243 118–133.

A. Kornberg. 1980 and 1982 supplement. DNA Replication. W. H. Freeman, San Francisco.

R. D. Kornberg and A. Klug. 1981. The Nucleosome. *Sci. American* 244 52–79.

R. E. Dickerson, H. R. Drew, B. N. Conner, R. M. Wing, A. V. Fratini, and M. L. Kopka. 1982. The Anatomy of A-, B-, and Z-DNA. *Science* 216 475–485.

A. Klug. 1982. From Macromolecules to Biological Assemblies. Nobel lecture. Nobel Foundation, Stockholm, Sweden.

Structure of DNA. 1982. *Cold Spring Harbor Symp. Quant. Biol.* 47 whole issue.

J. C. Wang. 1982. DNA Topoisomerases. *Sci. American* 247 94–109.

A. Rich, A. Nordheim, and H.-J. Wang. 1984. The Chemistry and Biology of Left-Handed Z-DNA. *Annu. Rev. Biochem.* 53 791–846.

W. Sanger. 1984. Principles of Nucleic Acid Structure. Springer-Verlag, New York.

G. Felsenfeld. 1985. DNA. *Sci. American* 253(4) 58–67.

D. E. Pettijohn. 1988. Histone-like Proteins and Bacterial Chromosome Structure. *J. Biol. Chem.* 263 12793–12796.

K. E. van Holde. 1988. Chromatin. Springer-Verlag, New York.

R. D. Wells. 1988. Unusual DNA Structures. *J. Biol. Chem.* 263 1095–1098 (a minireview).

1.2

R. W. Holley. 1968. The Nucleotide Sequence of a Nucleic Acid. *Sci. American* 214 30–39.

A. Rich and U. L. Raj Bhandary. 1976. Transfer RNA: Molecular Structure, Sequence and Properties. *Annu. Rev. Biochem.* 45 805–860.

H. F. Noller. 1984. Structure of Ribosomal RNA. *Annu. Rev. Biochem.* 55 119–162.

J. E. Darnell, Jr. 1985. RNA. *Sci. American* 253(4) 68–78.

1.3

R. E. Dickerson and I. Geis. 1969. The Structure and Action of Proteins. Benjamin/Cummings, Menlo Park, California.

T. E. Creighton. 1983. Proteins: Structure and Molecular Properties. W. H. Freeman, San Francisco.

R. E. Dickerson and I. Geis. 1983. Hemoglobin: Structure, Function, Evolution and Pathology. Benjamin/Cummings, Menlo Park, California.

C. Chothia. 1984. Principles That Determine the Structure of Proteins. *Annu. Rev. Biochem.* 53 537–572.

R. F. Doolittle. 1985. Proteins. *Sci. American* 253(4) 88–99.

R. J. Fletterick, T. Schroer, and R. J. Matela. 1985. Molecular Structure: Macromolecules in Three Dimensions. Blackwell Scientific Publications, Oxford, England.

M. E. Goldberg. 1985. The Second Translation of the Genetic Message: Protein Folding and Assembly. *Trends Biochem. Sci.* 10 388–391.

2.1

J. Cairns. 1966. The Bacterial Chromosome. *Sci. American* 214 36–44.

W. Gilbert and D. Dressler. 1968. DNA Replication: The Rolling Circle Model. *Cold Spring Harbor Symp. Quant. Biol.* 33 473–484.

J. S. Sussenbach, P. C. Van der Vliet, D. J. Ellens, and H. S. Jansz. 1972. Linear Intermediates in the Replication of Adenovirus DNA. *Nature New Biol.* 239 47–49.

J. D. Watson. 1972. Origin of Concatameric T7 DNA. *Nature New Biol.* 235 197–201.

J. D. Wolfson, D. Dressler, and M. Magazin. 1972. Bacteriophage T7 DNA Replication: A Linear Replicating Intermediate. *Proc. Natl. Acad. Sci.* U.S.A. 69 499–504.

I. R. Lehman. 1974. DNA Ligase: Structure, Mechanism, and Function. *Science* 186 790–797.

D. Bastia, M. Sueoka, and E. Cox. 1975. Studies in the Late Replication of Phage λ: Rolling Circle Replication. *J. Mol. Biol.* 98 305–320.

D. M. Rekosh, W. C. Russel, A. J. D. Bellet, and A. J. Robinson. 1977. Identification of a Protein Linked to the Ends of Adenovirus DNA. *Cell* 11 283–295.

DNA: Replication and Recombination. 1978. *Cold Spring Harbor Symp. Quant. Biol.* 43 whole issue.

S. H. Wickner. 1978. DNA Replication Proteins of *Escherichia coli*. *Annu. Rev. Biochem.* 47 1163–1191.

T. J. Kelly, Jr., and R. L. Lechner. 1979. The Structure of Replicating Adenovirus DNA Molecules: Characterization of DNA-Protein Complexes from Infected Cells. *Cold Spring Harbor Symp. Quant. Biol.* 43 721–728.

N. R. Cozzarelli. 1980. DNA Gyrase and the Supercoiling of DNA. *Science* 207 953–960.

M. Gellert. 1981. DNA Topoisomerases. *Annu. Rev. Biochem.* 50 879–910.

Y. Hirota, M. Yamada, A. Nishimura, A. Oka, K. Sugimoto, K. Asada, and M. Takanami. 1981. The DNA Replication Origin (*ori*) of *Escherichia coli*: Structure and Function of the *ori*-Containing DNA Fragment. *Prog. Nucleic Acid Res.* 26 33–48.

O. Sundin and A. Varshavsky. 1981. Arrest of Segregation Leads to Accumulation of Highly Intertwined Catenated Dimers: Dissection of the Final Stages of SV40 DNA Replication. *Cell* 25 659–669.

B. M. Alberts, B. P. Bedinger, T. Formosa, C. V. Jongeneel, and K. N. Kreuzer. 1982. Studies on DNA Replication in the Bacteriophage T4 *in vitro* Systems. *Cold Spring Harbor Symp. Quant. Biol.* 47 655–668.

B. M. Baroudy, S. Venkatesan, and B. Moss. 1982. Structure and Replication of Vaccinia Virus Telomeres. *Cold Spring Harbor Symp. Quant. Biol.* 47 723–729.

B. W. Stillman and F. Tamenoi. 1982. Adenoviral DNA Replication: DNA Sequences Required for Initiation *in Vitro*. *Cold Spring Harbor Symp. Quant. Biol.* 47 741–750.

A. Kornberg. 1983. Mechanism of Replication of the *E. coli* Chromosome. *Eur. J. Biochem.* 137 377–382.

N. G. Nossal. 1983. Prokaryotic DNA Replication Systems. *Annu. Rev. Biochem.* 52 581–615.

B. M. Alberts. 1984. The DNA Enzymology of Protein Machines. *Cold Spring Harbor Symp. Quant. Biol.* 49 1–12.

S. Dinardo, K. Voelkel, and R. Sternglanz. 1984. DNA Topoisomerase II Is Required for Segregation of Daughter Molecules at the Termination of DNA Replication. *Proc. Natl. Acad. Sci.* U.S.A. 81 2616–2620.

M. Kaguni and A. Kornberg. 1984. Replication Initiated at the Origin (*ori*C) of the *E. coli* Chromosome Reconstituted with Purified Enzymes. *Cell* 38 183–190.

J. Tomizawa. 1984. Control of ColE1 Plasmid Replication: The Process of Binding of RNA I to the Primer Transcript. *Cell* 38 861–870.

T. Kirchhausen, J. C. Wang, and S. C. Harrison. 1985. DNA Gyrase and Its Complexes with DNA: Direct Observation by Electron Microscopy. *Cell* 41 933–943.

C. S. McHenry. 1985. DNA Polymerase III Holoenzyme of *Escherichia coli*: Components and Function of a True Replicative Complex. *Mol. Cell Biochem.* 66 71–85.

B. W. Stillman and Y. Gluzman. 1985. Replication and Supercoiling of SV40 DNA in Cell-Free Extracts from Human Cells. *Mol. Cell. Biol.* 5. 2051–2060.

J. L. Campbell. 1986. Eukaryotic DNA Replication. *Annu. Rev. Biochem.* 55 733–771.

H. Echols. 1986. Multiple DNA-Protein Interactions Governing High-Precision DNA Transactions. *Science* 233 1050–1056.

A. Maxwell and M. Gellert. 1986. Mechanistic Aspects of DNA Topoisomerases. *Adv. Prot. Chem.* 38 69–107.

S. A. Wasserman and N. R. Cozzarelli. 1986. Biochemical Topology: Applications to DNA Recombination and Replication. *Science* 232 951–960.

C. M. Joyce and T. Steitz. 1987. DNA Polymerase I: From Crystal Structure to Function via Genetics. *Trends Biochem. Sci.* 12 288–292.

T. Kelly and B. Stillman (eds.). 1988. Eukaryotic DNA Replication. Cold Spring Harbor Laboratory, New York.

A. Kornberg. 1988. DNA Replication. *J. Biol. Chem.* 263 1–4.

2.2

D. Baltimore. 1970. Viral RNA-Dependent DNA Polymerase. *Nature* 226 1209–1211.

H. M. Temin and S. Mizutani. 1970. Viral RNA-Dependent DNA Polymerase. *Nature* 226 1211–1213.

E. Gilboa, S. W. Mitra, S. Goff, and D. Baltimore. 1979. A Detailed Model of Reverse Transcription and Tests of Crucial Aspects. *Cell* 18 93–100.

W. S. Mason, J. M. Taylor, and R. Hull. 1987. Retroid Virus Genome Replication. *Adv. Virus Res.* 32 35–96.

H. Varmus. 1987. Reverse Transcription. *Sci. American* 257 56–64.

2.3

P. Howard-Flanders. 1981. Inducible Repair of DNA. *Sci. American* 245 72–103.

T. Lindahl. 1982. DNA Repair Enzymes. *Annu. Rev. Biochem.* 51 61–87.

L. A. Loeb and T. A. Kunkel. 1982. Fidelity of DNA Synthesis. *Annu. Rev. Biochem.* 51 429–457.

A. Sancar and W. D. Rupp. 1983. A Novel Repair Enzyme: *UVR*ABC Excision Nuclease of *Escherichia coli* Cuts a DNA Strand on Both Sides of the Damaged Region. *Cell* 33 249–260.

A. T. Yeung, W. B. Mattes, E. Y. Oh, and L. Grossman. 1983. Enzymatic Properties of Purified *Escherichia coli uvr*ABC Proteins. *Proc. Natl. Acad. Sci.* U.S.A. 80 6157–6161.

E. C. Friedberg. 1984. DNA Repair. W. H. Freeman, San Francisco.

A. Sancar and G. B. Sancar. 1988. DNA Repair Enzymes. *Annu. Rev. Biochem.* 57 29–67.

2.4

N. Sigal and B. Alberts. 1972. Genetic Recombination: The Nature of a Crossed Stranded Exchange Between Two Homologous DNA Molecules. *J. Mol. Biol.* 71 789–793.

S. A. Latt. 1979. Sister Chromatid Exchanges. *Genetics* 92 583–595.

F. W. Stahl. 1979. Genetic Recombination: Thinking About It in Phage and Fungi. W. H. Freeman, San Francisco.

C. DasGupta, A. Wu, R. Kahn, R. Cunningham, and C. Radding. 1981. Concerted Strand Exchange and Formation of Holliday Structures by *E. coli* RecA Protein. *Cell* 25 507–516.

H. Nash. 1981. Integration and Excision of Bacteriophage λ: The Mechanism of Conservative Site Specific Recombination. *Annu. Rev. Genet.* 15 143–167.

C. Radding. 1982. Homologous Pairing and Strand Exchange in Genetic Recombination. *Annu. Rev. Genet.* 16 405–437.

J. Szostak, T. Orr-Weaver, R. Rothstein, and F. Stahl. 1983. The Double-Strand Break Repair Model for Recombination. *Cell* 33 25–35.

R. Weisberg and A. Landy. 1983. Site-Specific Recombination in Phage λ. In R. Hendrix, J. Roberts, F. Stahl, and R. Weisberg (eds.), Lambda II, pp. 211–250. Cold Spring Harbor Laboratory, Cold Spring Harbor, New York.

M. M. Cox and I. R. Lehman. 1987. Enzymes of General Recombination. *Annu. Rev. Biochem.* 56 229–262.

P. Modrich. 1987. DNA Mismatch Correction. *Annu. Rev. Biochem.* 56 435–466.

2.5

S. Spiegelman, I. Haruna, I. B. Holland, B. Beaudreau, and D. Mills. 1965. The Synthesis of a Self-Propagative and Infectious Nucleic Acid with a Purified Enzyme. *Proc. Natl. Acad. Sci.* U.S.A. 54 919–927.

T. Blumenthal and G. C. Carmichael. 1979. RNA Replication: Function and Structure of QB-Replicase. *Annu. Rev. Biochem.* 48 525–548.

3.1

F. H. C. Crick. 1958. On Protein Synthesis, Biological Replication of Macromolecules. *Symp. Exp. Biol.* 12 138–163.

J. D. Watson. 1963. The Involvement of RNA in the Synthesis of Proteins. *Science* 140 17–26.

H. Weissbach and S. Pestka (eds.). 1977. Molecular Mechanisms of Protein Biosynthesis. Academic Press, New York.

3.2

J. W. Roberts. 1969. Termination Factor for RNA Synthesis. *Nature* 224 1168–1174.

A. A. Travers and R. R. Burgess. 1969. Cyclic Reuse of the RNA Polymerase Sigma Factor. *Nature* 222 537–540

C. Lowery-Goldhammer and J. P. Richardson. 1974. An RNA-Dependent Nucleoside Triphosphate Phosphohydrolase (ATPase) Associated with Rho Termination Factor. *Proc. Natl. Acad. Sci.* U.S.A. 71 2003–2007.

M. Rosenberg and D. Court. 1979. Regulatory Sequences Involved in the Promotion and Termination of RNA Transcription. *Annu. Rev. Genet.* 13 319–353.

R. Losick and J. Pero. 1981. Cascades of Sigma Factors. *Cell* 25 582–584.

M. Chamberlin. 1982. Bacterial DNA-Dependent RNA Polymerases. In P. D. Boyer (ed.), The Enzymes, part B, vol. 15, pp. 61–108. Academic Press, New York.

R. Rodriguez and M. Chamberlin (eds.). 1982. Promoters: Structure and Function. Praeger, New York.

D. K. Hawley and W. R. McClure. 1983. Compilation and Analysis of *Escherichia coli* Promoter DNA Sequences. *Nucleic Acid Res.* 11 2237–2255.

P. H. von Hippel, D. G. Bear, W. D. Morgan, and J. A. McSwiggen. 1984. Protein-Nucleic Acid Interactions in Transcription: A Molecular Analysis. *Annu. Rev. Biochem.* 53 389–446.

W. R. McClure. 1985. Mechanism and Control of Transcription Initiation in Prokaryotes. *Annu. Rev. Biochem.* 54 171–204.

T. Platt. 1986. Transcription Termination and the Regulation of Gene Expression. *Annu. Rev. Biochem.* 55 339–372.

C. A. Brennan, A. J. Dombrowski, and T. Platt. 1987. Transcription Termination Factor Rho Is an RNA-DNA Helicase. *Cell* 48 945–952.

3.3

E. Lund, J.E. Dahlberg, L. Lindahl, R. Jaskunas, P. P. Dennis, and M. Nomura. 1976. Transfer RNA Genes Between 16S rRNA Gene in RNA Transcription Units. *Cell* 7 165–177.

J. D. Smith. 1976. Transcription and Processing of Transfer RNA Precursors. *Prog. Nucleic Acid. Res.* 16 25–73.

N. Nakajima, H. Ozeki, and Y. Shimura. 1977. Organization and Structure of an *E. coli* tRNA Operon Containing Seven tRNA Genes. *Cell* 23 239–249.

P. Gegenheimer and D. Apirion. 1981. Processing of Prokaryotic Ribonucleic Acid. *Microbiol. Rev.* 45 502–541.

S. Altman, C. Guerrier-Takeda, H. Frankfort, and H. Robertson. 1982. RNA Processing Nucleases. In S. Linn and R. Roberts (eds.), Nucleases, pp. 243–274. Cold Spring Harbor Laboratory, Cold Spring Harbor, New York.

H. D. Robertson. 1982. *Escherichia coli* Ribonuclease III Cleavage Sites. *Cell* 30 669–672.

M. Deutscher. 1984. Processing of tRNA in Prokaryotes and Eukaryotes. *Crit. Rev. Biochem.* 17 45–71.

3.4

F. H. C. Crick. 1958. On Protein Synthesis, Biological Replication of Macromolecules. *Symp. Exp. Biol.* 12. 158–163.

F. H. C. Crick, L. Barnett, S. Brenner, and R. J. Watts-Tobin. 1961. General Nature of the Genetic Code for Proteins. *Nature* 192 1227–1232.

M. W. Nirenberg and J. H. Matthaei. 1961. The Dependence of Cell-Free Protein Synthesis in *E. coli* upon Naturally Occurring or Synthetic Polyribonucleotides. *Proc. Natl. Acad. Sci.* U.S.A. 47 1588–1682.

J. F. Speyer, P. Lengyel, C. Basilico, A. J. Wahba, R. S. Gardner, and S. Ochoa. 1963. Synthetic Polynucleotides and the Amino Acid Code. *Cold Spring Harbor Symp. Quant. Biol* 28 559–568.

M. Nirenberg and P. Leder. 1964. The Effect of Trinucleotides upon the Binding of sRNA to Ribosomes. *Science* 145 1399–1407.

A. S. Sarabhai, A. O. Stretton, S. Brenner, and A. Bolle. 1964. Colinearity of the Gene with the Polypeptide Chain. *Nature* 201 13–17.

C. Yanofsky, B. C. Carlton, J. R. Guest, D. R. Helinski, and U. Henning. 1964. On the Colinearity of Gene Structure and Protein Structure. *Proc. Natl. Acad. Sci.* U.S.A. 51 266–272.

F. H. C. Crick. 1966. Codon-Anticodon Pairing: The Wobble Hypothesis. *J. Mol. Biol.* 19 548–555.

The Genetic Code. 1966. *Cold Spring Harbor Symp. Quant. Biol.* 31 whole issue.

M. Ycas. 1969. The Biological Code. Wiley-Interscience, New York.

3.5

P. Berg and E. J. Ofengand. 1958. An Enzymatic Mechanism for Linking Amino Acids to RNA. *Proc. Natl. Acad. Sci.* U.S.A. 44 78–86.

M. B. Hoagland, M. L. Stephenson, J. F. Scott, L. I. Hecht, and P. C. Zamecnik. 1958. A Soluble Ribonucleic Acid Intermediate in Protein Synthesis. *J. Biol. Chem.* 231 241–257.

F. Chapeville, F. Lipmann, G. V. Elivenstein, B. Weisblum, W. J. Ray, Jr., and S. Benzer. 1962. On the Role of Soluble Ribonucleic Acid in Coding for Amino Acids. *Proc. Natl. Acad. Sci.* U.S.A. 48 1086–1092.

R. W. Holley. 1966. The Nucleotide Sequence of a Nucleic Acid. *Sci. American* 214 30–39.

M. Nomura. 1973. Assembly of Bacterial Ribosomes. *Science* 179 864–873.

A. Rich and U. L. Raj Bhandary. 1976. Transfer RNA: Molecular Structure, Sequence and Properties. *Annu. Rev. Biochem.* 45 805–860.

A. Rich and S. H. Kim. 1978. The Three-Dimensional Structure of Transfer RNA. *Sci. American* 238 56–62.

P. R. Schimmel and D. Söll. 1979. Amino Acyl tRNA Synthetases: General Features and Recognition of tRNAs. *Annu. Rev. Biochem.* 48 601–648.

I. Wool. 1979. The Structure and Function of Eukaryotic Ribosomes. *Annu. Rev. Biochem.* 48 719–754.

J. A. Lake. 1981. The Ribosome. *Sci. American* 245 84–97.

K. H. Nierhaus. 1982. Structure, Assembly and Function of Ribosomes. *Current Topics in Microbiol. and Immun.* 97 82–155.

P. Schimmel, S. Putney, and R. Starzyk. 1982. RNA and DNA Sequence Recognition and Structure-Function of Amino Acyl tRNA Synthetases. *Trends Biochem. Sci.* 7 209–212.

H. G. Wittmann. 1982. Components of Bacterial Ribosomes. *Annu. Rev. Biochem.* 51 155–183.

H. G. Wittmann. 1983. Architecture of Prokaryotic Ribosomes. *Annu. Rev. Biochem.* 52 35–66.

M. Yarus and R. Thompson. 1983. Precision of Protein Biosynthesis. In J. Beckwith, J. Davies, and J. A. Gallant (eds.), Gene Function in Prokaryotes, pp. 23–63. Cold Spring Harbor Laboratory, Cold Spring Harbor, New York.

H. F. Noller. 1984. Structure of Ribosomal RNA. *Annu. Rev. Biochem.* 53 119–162.

D. M. Blow and P. Brick. 1985. Amino Acyl tRNA Synthetases. In F. A. Jurnak and A. McPherson (eds.), The Structure of Biological Macromolecules and Assemblies, vol. 2, Nucleic Acid Binding Proteins, pp. 442–469. Wiley, New York.

J. A. Lake. 1985. Evolving Ribosome Structure: Domains in Archaebacteria, Eubacteria, Eocytes and Eukaryotes. *Annu. Rev. Biochem.* 54 507–530.

G. R. Björk, J. U. Ericson, C. E. D. Gustafsson, T. G. Hagerrall, Y. H. Jönsson, and P. M. Wilkström. 1987. Transfer RNA Modification. *Annu. Rev. Biochem.* 56 263–288.

P. Schimmel. 1987. Amino Acyl tRNA Synthetases: General Scheme of Structure-Function Relationship in the Polypeptides and Recognition of Transfer RNAs. *Annu. Rev. Biochem.* 56 125–158.

3.6

H. M. Dintzis. 1961. Assembly of the Peptide Chain of Hemoglobin. *Proc. Natl. Acad. Sci.* U.S.A. 47 247–261.

J. Shine and L. Dalgarno. 1974. The 3'-Terminal Sequence of *E. coli* 16S rRNA: Complementarity to Nonsense Triplets and Ribosome Binding Sites. *Proc. Natl. Acad. Sci.* U.S.A. 71 1342–1346.

Y. Kaziro. 1978. The Role of Guanosine 5'-Triphosphate in Polypeptide Chain Elongation. *Biochem. Biophys. Acta* 505 95–127.

M. Grunberg-Manago. 1980. Initiation of Protein Synthesis as Seen in 1979. In G. Chambliss, G. R. Craven, J. Davies, K. Davis, L. Kahan, and M. Nomura (eds.), Ribosomes: Structure, Function and Genetics, pp. 445–478. University Park Press, Baltimore.

J. Ofengand. 1980. The Topography of tRNA Binding Sites on the Ribosome. In G. Chambliss, G.R. Craven, J. Davies, K. Davis, L. Kahan, and M. Nomura (eds.), Ribosomes: Structure, Function and Genetics, pp. 497–530. University Park Press, Baltimore.

H. Weissbach. 1980. Soluble Factors in Protein Synthesis. In G. Chambliss, G. R. Craven, J. Davies, K. Davis, L. Kahan, and M. Nomura (eds.), Ribosomes: Structure, Function and Genetics, pp. 445–478. University Park Press, Baltimore.

L. Gold, D. Pribnow, T. Schneider, S. Shinedling, B. S. Singer, and G. Starmo. 1981. Translational Initiation in Prokaryotes. *Annu. Rev. Microbiol.* 35 365–403.

R. A. Garrett and P. Wooley. 1982. Identifying the Peptidyl Transferase Centre. *Trends Biochem. Sci.* 7 385–386.

A. Johnson, H. Adkins, E. Matthews, and C. Cantor. 1982. Distance Moved by Transfer RNA During Translocation from the A Site to the P Site on the Ribosome. *J. Mol. Biol.* 156 113–140.

M. Kozak. 1983. Comparison of Initiation of Protein Synthesis in Prokaryotes, Eukaryotes and Organelles. *Microbiol. Rev.* 47 1–45.

C. T. Caskey, W. S. Forrester, and W. Tate. 1984 Peptide Chain Termination. In B. Clark and H. Petersen (eds.), Alfred Benzon Symposium, vol. 19, pp. 457–466. Munksgaard, Copenhagen.

K. Moldave. 1985. Eukaryotic Protein Synthesis. *Annu. Rev. Biochem.* 54 1109–1149.

3.7

K. Moldave. 1965. Nucleic Acids and Protein Biosynthesis. *Annu. Rev. Biochem.* 34 419–448.

F. H. C. Crick. 1966. Codon-Anticodon Pairing: The Wobble Hypothesis. *J. Mol. Biol.* 19 548–555.

Protein Synthesis. 1969. *Cold Spring Harbor Symp. Quant. Biol.* 34 whole issue.

B. A. Hamkalo and O. L. Miller, Jr. 1973. Electron Microscopy of Genetic Activity. *Annu. Rev. Biochem.* 42 379–396.

R. Haselkorn and L. B. Rothman-Denes. 1973. Protein Synthesis. *Annu. Rev. Biochem.* 42 397–438.

M. Nomura, A. Tissieres, and P. Lengyel (eds.). 1974. Ribosomes. Cold Spring Harbor Laboratory, Cold Spring Harbor, New York.

E. Beronak. 1978. Mechanisms in Polypeptide Chain Elongation on Ribosomes. *Prog. Nucl. Acid Res. Mol. Biol.* 21 63–100.

M. Grunberg-Manago, R. H. Buckingham, B. S. Cooperman, and J. W. B. Hershey. 1978. Structure and Function of the Translation Machinery. *Symp. Soc. Gen. Microbiol.* 28 27–110.

D. A. Steege and D. G. Söll. 1979. Suppression. In R. F. Goldberger (ed.), Biological Regulation and Development I, pp. 433–486. Plenum, New York.

H. Ozeki, H. Inokuchi, F. Yamao, M. Kodaira, H. Sakano, T. Ikemura, and Y. Shimura. 1980. Genetics of Nonsense Suppressor tRNAs in *E. coli*. In D. Söll, J. M. Abelson, and P. R. Schimmel (eds.), Transfer RNA, Biological Aspects, pp. 341–349. Cold Spring Harbor Laboratory, Cold Spring Harbor, New York.

J. R. Roth. 1981. Frameshift Suppression. *Cell* 24 601–602.

M. Yarus and R. Thompson. 1983. Precision of Protein Biosynthesis. In J. Beckwith, J. Davies, and J. A. Gallant (eds.), Gene Function in Prokaryotes, pp. 23–63. Cold Spring Harbor Laboratory, Cold Spring Harbor, New York.

3.8

K. Moldave. 1985. Eukaryotic Protein Synthesis. *Annu. Rev. Biochem.* 54 1109–1149.

3.9

J. H. Goldberg and P. A. Friedman. 1971. Antibiotics and Nucleic Acids. *Annu. Rev. Biochem.* 40 775–810.

H. M. Sobell. 1974. How Actinomycin Binds to RNA. *Sci. American* 231 82–91.

S. Pestka. 1977. Inhibitors of Protein Biosynthesis. In H. Weissbach and S. Pestka (eds.), Molecular Mechanisms of Protein Biosynthesis, pp. 467–553. Academic Press, New York.

R. J. Subadolnik. 1979. Naturally Occurring Nucleoside and Nucleotide Antibiotics. *Prog. Nucl. Acid Res. Mol. Biol.* 22 193–291.

E. Cundliffe. 1980. Antibiotics and Prokaryote Ribosomes: Action, Interaction and Resistance. In G. Chambliss, G. R. Craven, J. Davies, K. Davis, L. Kahan, and M. Nomura (eds.), Ribosomes: Structure, Function and Genetics, pp. 377–412. University Park Press, Baltimore.

3.10

S. Michaelis and J. Beckwith. 1982. Mechanism of Incorporation of Cell Envelope Proteins in *Escherichia coli. Annu. Rev. Microbiol.* 36 435–465.

G. Schatz and R. A. Butow. 1983. How Are Proteins Imported into Mitochondria? *Cell* 32 316–318 (a minireview).

T. J. Silhavy, S. A. Benson, and S. D. Emr. 1983. Mechanisms of Protein Localization. *Microbiol. Rev.* 47 313–344.

L. L. Randall and S. J. S. Hardy. 1984. Export of Protein in Bacteria: Dogma and Data. In B. Satir (ed.), Modern Cell Biology, vol. 3, pp. 1–20. Liss, New York.

P. Walter, R. Gilmore, and G. Blobel. 1984. Protein Translocation Across the Endoplasmic Reticulum. *Cell* 38 5–8.

J. E. Rothman. 1985. The Compartmental Organization of the Golgi Apparatus. *Sci. American* 253(3) 74–89.

R. Scheckman. 1985. Protein Localization and Membrane Traffic. *Annu. Rev. Cell Biol.* 1 115–143.

M. Schleyer and W. Neupert. 1985. Transport of Proteins into Mitochondria: Translocational Intermediates Spanning Contact Sites Between Outer and Inner Membranes. *Cell* 43 339–350.

S. R. Pfeffer and J. E. Rothman. 1987. Biosynthetic Protein Transport by the Endoplasmic Reticulum and Golgi. *Annu. Rev. Biochem.* 56 829–852.

J. E. Rothman. 1987. Protein Sorting by Selective Retention in the Endoplasmic Reticulum and Golgi Stack. *Cell* 50 521–522.

H. F. Lodish. 1988. Transport of Secretory and Membrane Glycoproteins from the Rough Endoplasmic Reticulum to the Golgi. *J. Biol. Chem.* 263 2107–2110.

D. Roise and G. Schatz. 1988. Mitochondrial Presequences. *J. Biol. Chem.* 263 4509–4511 (a minireview).

3.11

F. Jacob and J. Monod. 1961. Genetic Regulatory Mechanisms in the Synthesis of Proteins. *J. Mol. Biol.* 3 318–356.

S. Adhya and M. Gottesman. 1978. Control of Transcription Termination. *Annu. Rev. Biochem.* 47 967–996.

W. Fiers. 1979. Structure and Function of RNA Bacteriophages. *Comp. Virology* 13 69–204.

R. F. Goldberger (ed.). 1979. Biological Regulation and Development. I. Gene Expression. Plenum, New York.

J. Miller and W. Reznikoff (eds.). 1980. The Operon. Cold Spring Harbor Laboratory, Cold Spring Harbor, New York.

C. Yanofsky and R. Kolter. 1982. Attenuation in Amino Acid Biosynthesis Operons. *Annu. Rev. Genet.* 16 113–134.

C. Bauer, J. Carey, L. Kasper, S. Lynn, D. Woechter, and J. Gardner. 1983. Attenuation in Bacterial Operons. In J. Beckwith, J. Davies, and J. Gallant (eds.), Gene Function in Prokaryotes, pp. 65–89. Cold Spring Harbor Laboratory, Cold Spring Harbor, New York.

K. M. Campbell, C. D. Starmo, and L. Gold. 1983. Protein-Mediated Translational Repression. In J. Beckwith, J. Davies, and J. Gallant (eds.), Gene Function in Prokaryotes, pp. 185–187. Cold Spring Harbor Laboratory, Cold Spring Harbor, New York.

R. W. Hendrix, J. W. Roberts, F. W. Stahl, and R. A. Weisberg (eds). 1983. Lambda II. Cold Spring Harbor Laboratory, Cold Spring Harbor, New York.

O. C. Uhlenbeck, J. Carey, P. J. Romaniok, P. T. Lowary, and D. Beckett. 1983. Interaction of R17 Coat Protein with Its RNA Binding Site for Translational Repression. *J. Biomol. Structure and Dynamics* 1 539–552.

B. De Crombrugghe, S. Busby, and H. Buc. 1984. Cyclic AMP Receptor Protein: Role in Transcription Activation. *Science* 224 831–838.

S. Gottesman. 1984. Bacterial Regulation: Global Regulatory Networks. *Annu. Rev. Genet.* 18 415–442.

M. Nomura, R. Gourse, and G. Baughman. 1984. Regulation of the Synthesis of Ribosomes and Ribosomal Components. *Annu. Rev. Biochem.* 53 75–117.

C. Pabo and R. Sauer. 1984. Protein-DNA Recognition. *Annu. Rev. Biochem.* 58 293–321.

D. Raibaud and M. Schwartz. 1984. Positive Control of Transcription Initiation in Bacteria. *Annu. Rev. Genet.* 18 173–206.

M. B. Matthews (ed.). 1986. Translational Control. Current Communications in Molecular Biology. Cold Spring Harbor Laboratory, Cold Spring Harbor, New York.

M. Ptashne. 1986. A Genetic Switch, Gene Control and Phage λ. Cell Press and Blackwell Scientific Publications, Palo Alto, California.

C. Yanofsky. 1988. Transcription Attenuation. *J. Biol. Chem.* 263 609–612.

The Recombinant DNA Breakthrough

PART II

By about 1970, certain fundamental attributes of genetic systems were well established. Although many important details were lacking, the basic outlines of DNA replication, recombination, and repair were understood, and each of these processes had already been reproduced *in vitro*. The central dogma, which posited a flow of information from DNA to RNA to protein, provided the basis for molecular explanations of an organism's genotype and phenotype. Messenger RNA was identified as a key intermediary in the transfer of information from DNA to protein. The genetic code had been deciphered, and information about the cellular machinery and general mechanisms for decoding messenger RNA into protein was emerging from studies with reconstituted cellular components *in vitro*. As anticipated, transcription of DNA into RNA and translation of RNA into protein were shown to be regulated, and the evidence indicated that there are both positive and negative controls for gene function. Once the genetic code was solved, the long sought after key for relating the chemical structure of a gene with that of the protein it encodes was in hand, allowing mutations to be interpreted in terms of alterations in DNA structure. An often unacknowledged reward from the advances made during this extraordinary period of discovery was the identification of many enzymes that act on nucleic acid substrates. Their isolation and characterization greatly facilitated the analysis of nucleic acid structure and function; moreover, their availability made possible the next generation of advances; the recombinant DNA techniques described in the remainder of this book.

Although it was widely accepted that all genetic systems share the same fundamental attributes, the depth of understanding was considerably greater for prokaryotic than for eukaryotic organisms. Certainly, genetic analysis was much simpler with the smaller and less complex bacterial genomes. Mutational alterations of specific genes were readily produced and identified. The promiscuous exchange of genetic information between various bacteria and between certain bacteria and their viruses (bacteriophages) facilitated the mapping of these genes, thus leading to descriptions of the overall arrangements of bacterial and phage genomes. Even more important was the remarkable interplay between genetics and biochemistry. This two-pronged approach helped unravel the complexities of DNA replication, even leading to the *in vitro* replication of fully infectious viral genomes. The combination of genetic and biochemical techniques also led to the isolation of genes, thereby paving the way for an analysis of gene transcription and translation *in vitro* and the identification of the molecular entities mediating these processes. The same combination of strategies revealed the logic governing the regulation of gene expression: gene expression is mainly controlled by molecular interactions between specific proteins and corresponding regulatory sequences in DNA or messenger RNA.

By comparison, progress toward understanding the molecular features of eukaryotic genome structure, organization, and function languished. Quite sophisticated genetic maps of mutationally altered loci existed for those few eukaryotes whose genetic systems were experimentally manipulable (e.g., several yeasts, *Neurospora crassa*, and *D. melanogaster*). However, genetic maps for mammalian organisms such as the mouse and man were primitive by comparison. Ignorance at the molecular level was even more profound. The structure of eukaryotic genes and their organization in chromosomal DNA were virtually complete mysteries. The multiple repetition, in most eukaryotes, of certain DNA segments posed an additional enigma. Without a more thorough analysis of the molecular anatomy of these genomes, further progress was impossible.

Biochemical experiments on eukaryotic gene expression and regulation were also stymied by the inaccessibility of structural information about cellular genes. It was clear that the nuclear DNA of eukaryotes is transcribed into RNA and that messenger RNAs are translated into proteins by a cytoplasmic ribosome-tRNA mediated process much like that in prokaryotes (Chapter 3). But the nature of the transcription process and the subsequent fate of the transcripts were particularly puzzling. In many eukaryotes, only a small fraction of the nuclear RNA (<10 percent) ends up in the cytoplasm as messenger RNA, ribosomal RNA, and transfer RNA. Some is sequestered in stable form as short RNA chains within ribonucleoprotein particles, but most of it is rapidly degraded without leaving the nucleus. The nature of the rapidly turning over RNA and its origins and function were vexing issues. The question of messenger RNA biogenesis was also confounded by the fact that eukaryotic messenger RNAs differ from their prokaryotic counterparts in having modifications at both ends—a so-called "cap" and a "poly A tail" at the beginning and end of the messenger RNA, respectively. This implied that eukaryotic messenger RNAs undergo posttranscriptional modifications during their biogenesis. Why, where, and how do the modifications occur? What is their significance? What is the pathway for converting primary transcripts of DNA into mature messenger RNAs? Further, what controls the transcription and turnover of RNAs from different genes in different cell types of the same organism? How do the mechanisms of gene expression and regulation differ between eukaryotes and prokaryotes? In the absence of hard-core molecular genetic information and the methodology to obtain it, these vital questions about eukaryotes remained unanswered.

As more and more was learned about how genetic information is organized and expressed in prokaryotes, the ignorance about eukaryotes became increasingly frustrating. What was needed was a general methodology that would facilitate the molecular analysis of eukaryotic cellular genomes. Ideally, such a breakthrough would permit the isolation of discrete genes and the determination of their molecular structures and their genomic organization. The availability of such isolated genetic elements could then permit biochemical experiments to characterize the transcriptional and translational mechanisms that govern their expression. That objective finally became a reality during the first half of the 1970s, when the recombinant DNA technologies were developed. Before considering these developments, we shall examine the origins of the concepts and methodologies that paved the way for the key experiments. These origins derive primarily from experiments in bacterial genetics and specifically from the discovery that DNA molecules can be introduced into bacterial cells. The newly introduced DNA, which may be the DNA genome of a bacteriophage or DNA from other bacterial cells, then contributes to the genotype and often the phenotype of the recipient cell. Genes on the added DNA can be expressed, and the DNA can undergo recombination with the chromosomal DNA.

Introducing New Genetic Information into Bacteria

Bacteria can acquire new genetic information in several ways, and, depending on the species, the different methods work more or less efficiently. These methods include (1) **transformation**, in which cells take up DNA molecules that are added to the surrounding medium; (2) **conjugation**, in which DNA is transferred directly from one cell to another; and (3) bacteriophage-mediated **transduction**, in which new genetic information is inserted into the cell from a bacteriophage particle. Whatever the mode of entry into the recipient cell, the acquired DNA recombines with homologous regions or special sites in the recipient organism's genome or is maintained as an autonomous minichromosome, thereby altering the genotype.

Bacterial Transformation

Transformation, which refers to the alteration of a cell's genotype following the uptake of DNA molecules from the culture medium, was

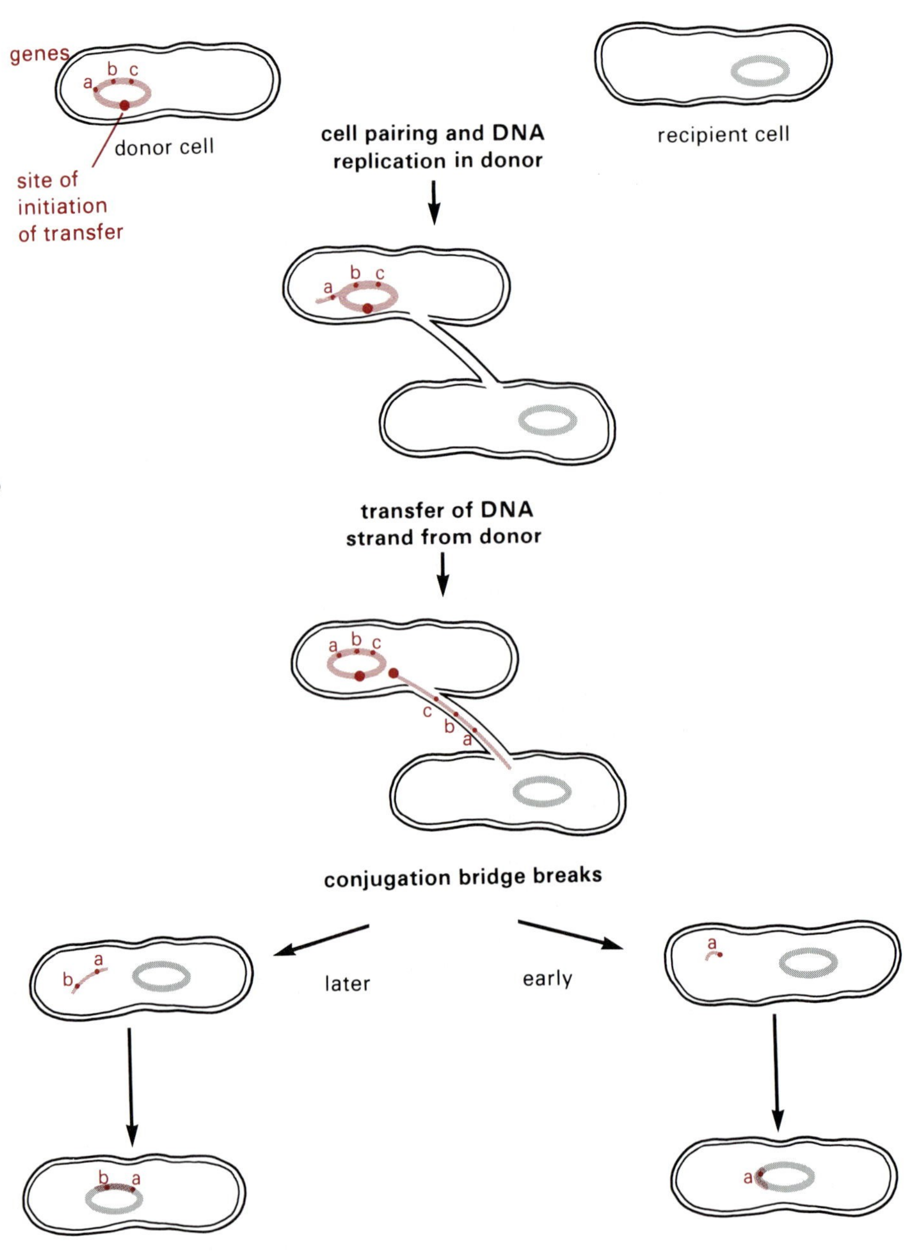

Figure II.1
Transfer of genetic information between bacteria by conjugation. Two *E. coli* cells, one capable of donating its chromosome to the other, make contact through a conjugation bridge composed of protein. DNA replication starts at a special position (dot) in the donor cell's chromosome. One of the DNA strands is transferred to the recipient cell. The transfer stops when the conjugation bridge is disrupted (e.g., by random motion of the cells or shaking the flask containing the cells). The longer the two cells are in contact before the bridge breaks, the greater the number of genes transferred. The transferred DNA can then replace the corresponding region in the recipient cell's genome, by recombination. Different strains of *E. coli* initiate transfer at different chromosomal loci.

the first way discovered to introduce new genes into bacteria and provided the first proof that DNA alone is the carrier of genetic information. In practice, exposure of an organism with a particular genetic disability (e.g., an inability to synthesize tryptophan, metabolize galactose, produce a surface polysaccharide, etc.) to DNA from cells that can perform the function often results in acquisition of the capability characteristic of the donor DNA. Such transformations are usually heritable and stable because of recombination between the functional gene(s) in the the donor DNA and the defective gene(s) in the recipient. However, DNA-mediated transformation is only marginally useful for studying prokaryotic molecular genetics because the transforming gene cannot be easily recovered; therefore, its structure cannot be determined. Nevertheless, the principle of transformation has become of major significance in other ways. For example, production of transformed cells using naked DNA molecules is an essential step in many recombinant DNA experiments. And the term transformation has been adopted in the field of eukaryotic molecular biology to refer to the permanent alteration of a cell's genotype and phenotype.

Conjugation

Conjugation involves the direct transfer of DNA from one cell to another upon contact. It is initiated, for example, between certain strains of *E. coli*, one of which acts as the DNA donor and the other as the recipient (Figure II.1). The donor cell's chromosomal DNA is transferred through a bridge formed between the two cells. Complete transfer takes about 90 minutes, but if the bridge is broken, conjugation is terminated and chromosome transfer is incomplete.

Depending on the *E. coli* strain that is the donor, transfer begins at different positions on the chromosome. Thus, different sets of *E. coli* genes are transferred early or late during conjugation with different donors. However, the genes are always transferred in one of only two orders, and these are the inverse of each other (i.e., transfer occurs in one of two directions around the *E. Coli* chromosome) (Figure II.2). These experiments first suggested that all the *E. coli* genes are on a single and circular DNA molecule, as shown in Figure II.3. And the order of genes on the *E. coli* chromosome, the genetic map, was constructed by comparing the time required for different genes to be transferred during conjugation.

Transduction

Two types of bacteriophage-mediated transduction (generalized and specialized) have been characterized. In the first, **generalized transduction**, relatively large segments of one bacterial cell's genome are transferred to another cell by bacteriophage particles that have acquired cellular DNA segments. Transducing bacteriophage particles are produced in the course of certain infections because the cell's DNA is extensively degraded, and fragments about the size of the phage genome are accidentally packaged in mature bacteriophage structures (Figure II.4). Subsequent infections by bacteriophage populations containing the transducing phages introduce the cellular DNA into the "infected" cells. Altered genotypes are produced when the newly introduced cellular DNA fragments recombine with the recipient cell's DNA. Each transducing phage particle

Donor Strain	Order of gene transfer
1	•-*thr-leu-azi-ton-pro-lac-pur-gal-trp-his-gly-str-mal-xyl-mtl-ile-met-thi*
2	•-*met-thi-thr-leu-azi-ton-pro-lac-pur-gal-trp-his-gly-str-mal-xyl-mtl-ile*
3	•-*str-mal-xyl-mtl-ile-met-thi-thr-leu-azi-ton-pro-lac-pur-gal-trp-his-gly*
4	•-*leu-thr-thi-met-ile-mtl-xyl-mal-str-gly-his-trp-gal-pur-lac-pro-ton-azi*
5	•-*his-trp-gal-pur-lac-pro-ton-azi-leu-thr-thi-met-ile-mtl-xyl-mal-str-gly*
6	•-*mtl-xyl-mal-str-gly-his-trp-gal-pur-lac-pro-ton-azi-leu-thr-thi-met-ile*

Figure II.2
The transfer of DNA during conjugation starts with different genes, depending on the donor strain. Thus, transfer with strain 1 starts at the *thr* gene, and transfer with strain 2 starts at *met*. However, groups of genes are always transferred in one of only two different orders, and these are the inverse of one another (e.g., in strains 1, 2, and 3, the order is *thr-leu-azi*, etc.; in strain 4, transfer begins between *azi* and *leu*; in strains 5 and 6, it is *azi-leu-thr*, etc.).

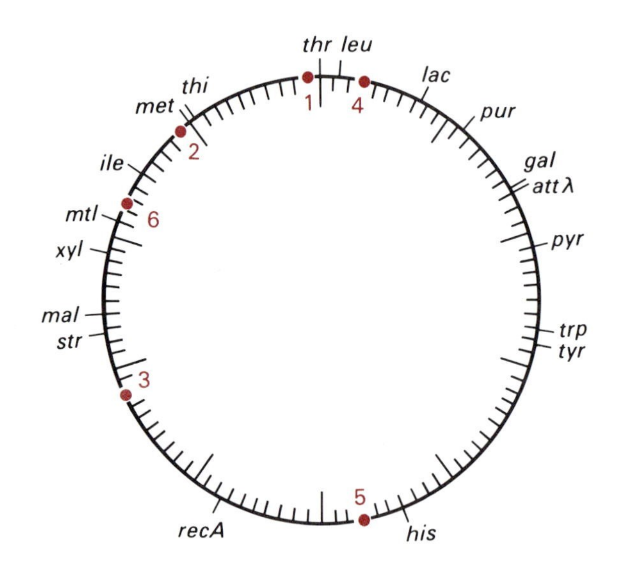

Figure II.3
Circular gene map of *E. coli* showing the locations of several representative genes. The order of the genes around the chromosome is the same as shown in Figure II.2. The dots indicate the points at which transfer was initiated in the examples shown in Figure II.2.

generally contains only a single random fragment of the original donor chromosome, and there is a nearly equal probability of it being any particular portion of that genome. However, because the transduced DNA segments are quite large (as much as 100 kbp or 2.5 percent of the *E. coli* chromosome with a certain transducing bacteriophage), the recipient cell generally acquires a group of genes in a single event. Thus, genes that are close to each other on the donor chromosome are cotransduced at a high frequency, whereas distant genes are transduced independent of one another (Figure II.5). Measurements of the frequency with which genes are cotransduced help refine genetic maps by providing an estimate of the relative distance between closely spaced genes.

The second type of transduction, **specialized transduction**, is an attribute of bacteriophages whose infectious cycle is often interrupted by the integration of the viral genome into a special location

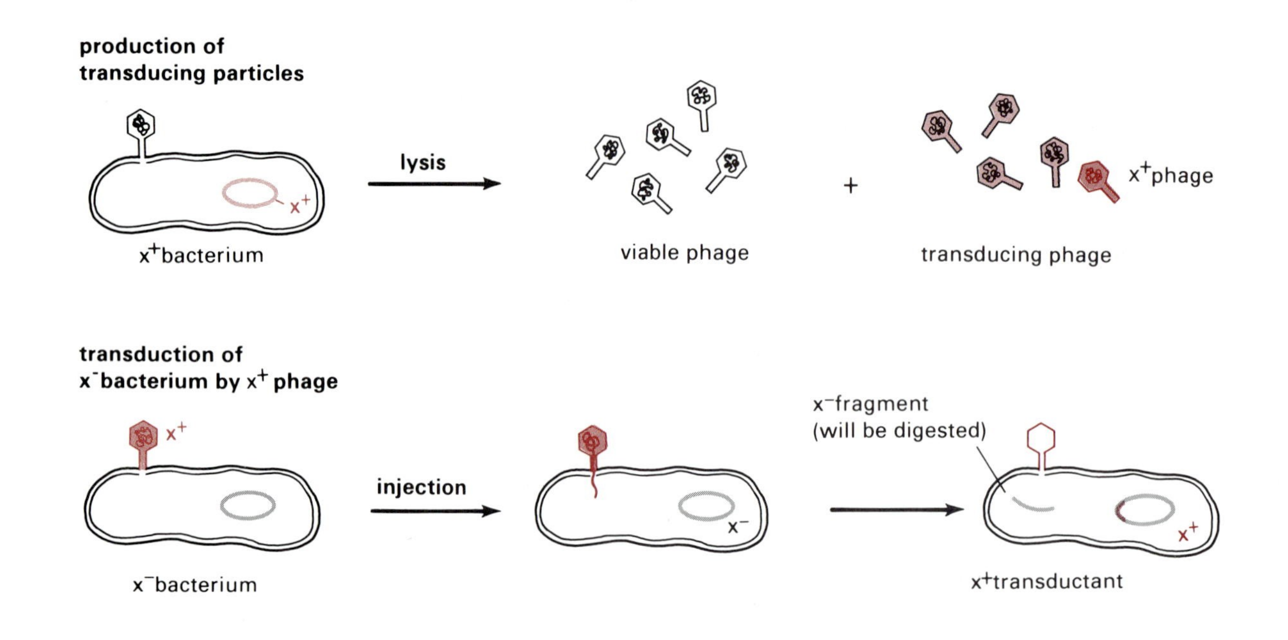

Figure II.4
The production of generalized transducing phage and the way they introduce new DNA into the cells they infect. In this example, one of the transducing phages carries the wild type bacterial gene x^+. In a subsequent infection, the x^+ phage introduces x^+ into a bacterial cell carrying a mutation, x^-, in gene x and by recombination converts the recipient to x^+.

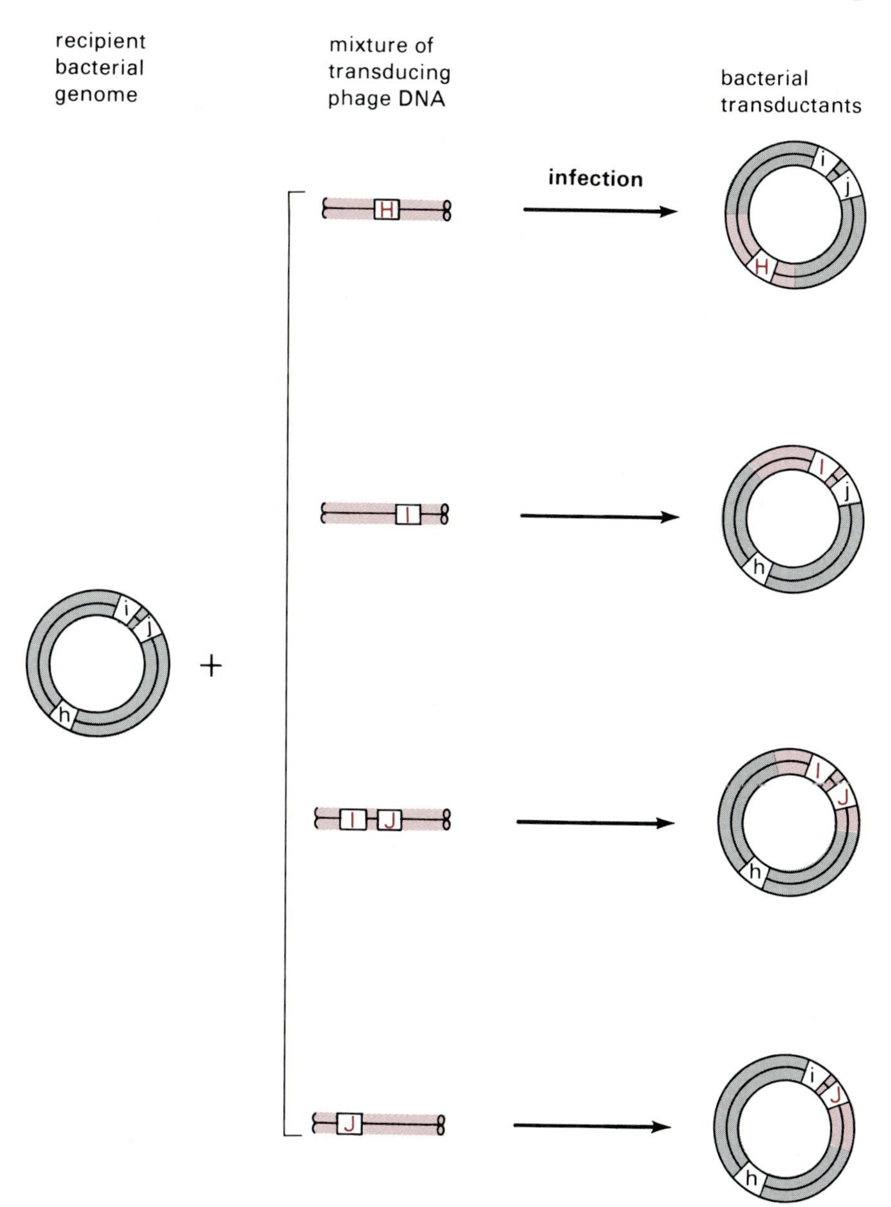

Figure II.5
The effect of distance on cotransduction. Genes *i* and *j* can be transduced independently or together because they are close to one another on the bacterial chromosome. Gene *h* is too far from either *i* or *j* to be cotransduced.

in the infected cell's chromosome (Chapter 2). Bacteria containing such integrated phage genomes are called **lysogenic**, and they contain viral genomes as heritable elements within their own chromosomes (Figure II.6). In a lysogenic cell, the viral and cellular genomes replicate as one and maintain a mutually compatible existence. Phage genome integration into cellular DNA keeps the phage from killing cells and producing infectious particles. For this reason, bacteriophage that can lysogenize are called **temperate**, in contrast to **virulent** phage. Under certain conditions—induction—the lysogenic state is disrupted, and the viral genome is excised from the cell's chromosome. The viral genome replicates, produces a burst of infectious virus, and kills the cell. Generally, viral genome excision is precise, and the resulting phage contain the original virus genome structure. Occasionally, however, phage excision is aberrant, and cellular genes adjacent to the integrated viral genome are incorporated into the progeny phage genomes in place of some viral genes

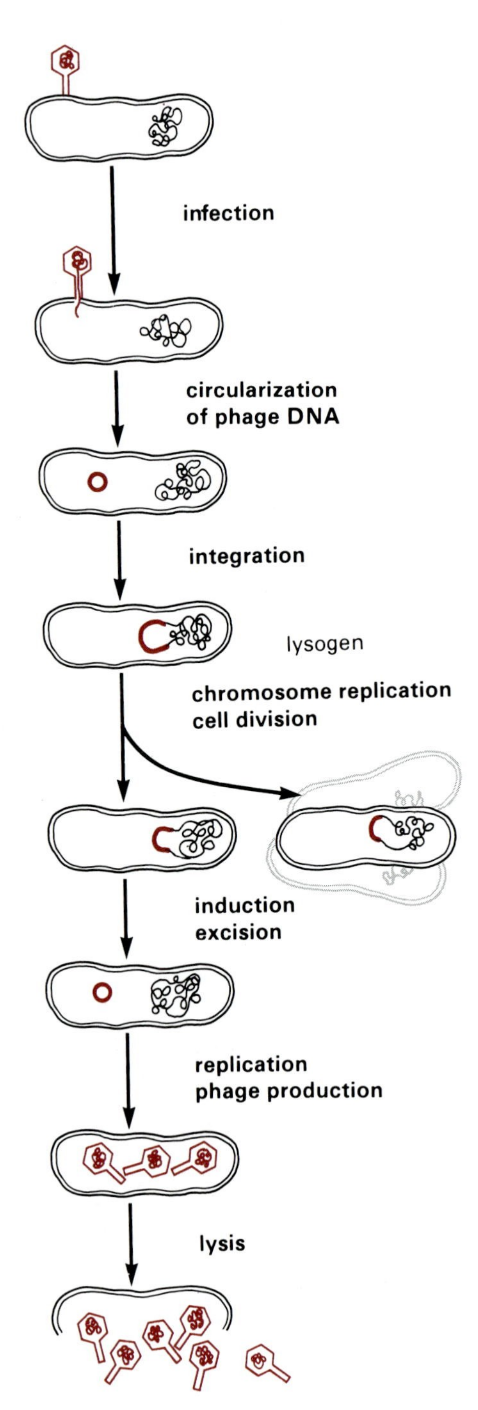

Figure II.6
General events in the establishment of lysogeny and the induction of lysogens to produce phage. The infecting phage DNA circularizes before being integrated into the bacterial chromosome and is excised as a circle upon induction.

(Figure II.7). During the next cycle of infection, the cellular genes are introduced along with phage genes into recipient cells. After the transducing phage DNA is introduced into the recipient cell genome, the cell acquires, in addition to phage genes, genetic information that originated from the phage's previous host. Thus, in specialized transduction, the phage serves as a vector for transferring genes from one cell to another; only cellular genes that are close to the viral genome integration site are transducible by this mechanism.

Because different temperate phages have distinctive chromosomal integration sites, their aberrant excisions yield phages that transduce different chromosomal genes. Thus, λ phages transduce genes responsible for galactose metabolism (λ *gal*) or genes needed to synthesize biotin (λ *bio*), and ϕ80 phages incorporate varying amounts of the gene cluster that encodes the enzymes for tryptophan biosynthesis (Figure II.8). Certain genetic tricks have been developed that enable these and other temperate phages to acquire different *E. coli* genes as well as genes from related organisms. The same strategy, with minor modifications, also yields transducing phages that contain mutant bacterial genes. Because such transducing phages can be readily identified, grown, and purified, it is possible to obtain substantial amounts of wild type or mutant alleles of *E. coli* genes in highly enriched form.

The enrichment of a bacterial gene that is afforded by its incorporation into a transducing phage genome is dramatic. Consider, for example, the *E. coli* gene encoding β-galactosidase (*lac* Z). The protein is composed of identical polypeptide chains 1173 amino acids long; therefore, the gene encoding that polypeptide is about 3600 bp long. Although the β-galactosidase gene represents about one-thousandth of the *E. coli* genome (3.6×10^3 out of 4×10^6 bp), it is about one-fifteenth of the genome of the transducing virus, λ *lac* (3.6×10^3 out of 5×10^4 bp). Thus, λ *lac* DNA is about 100 times richer a source for isolating the β-galactosidase gene than is *E. coli* DNA. This simplification in isolating the β-galactosidase gene aided the identification of its control regions and the analysis of its nucleotide sequence. In a similar way, the ϕ80*trp* transducing phages facilitated the isolation and characterization of the genes and regulatory signals that constitute the tryptophan operon (Section 3.11d). Moreover, because of the ability to transduce cells, it was possible to test the effect of various mutational alterations on the expression and regulation of genes *in vivo*.

Consider the special attributes of a viral genome that enable it to function in specialized transduction (Figure II.9). First, the genome must replicate following infection (i.e., the viral DNA must retain an origin of replication and the genes necessary to support replication). Second, it must acquire a covalently linked segment of nonviral DNA, the segment being transduced. This DNA is usually cellular in origin, but in principle it could be from any source. It may be inserted at any position in the viral genome so long as it does not impair replication of the viral DNA in the infected host cell or the DNA's ability to be packaged into a mature virion particle. Having become an integral part of the viral genome, the transduced

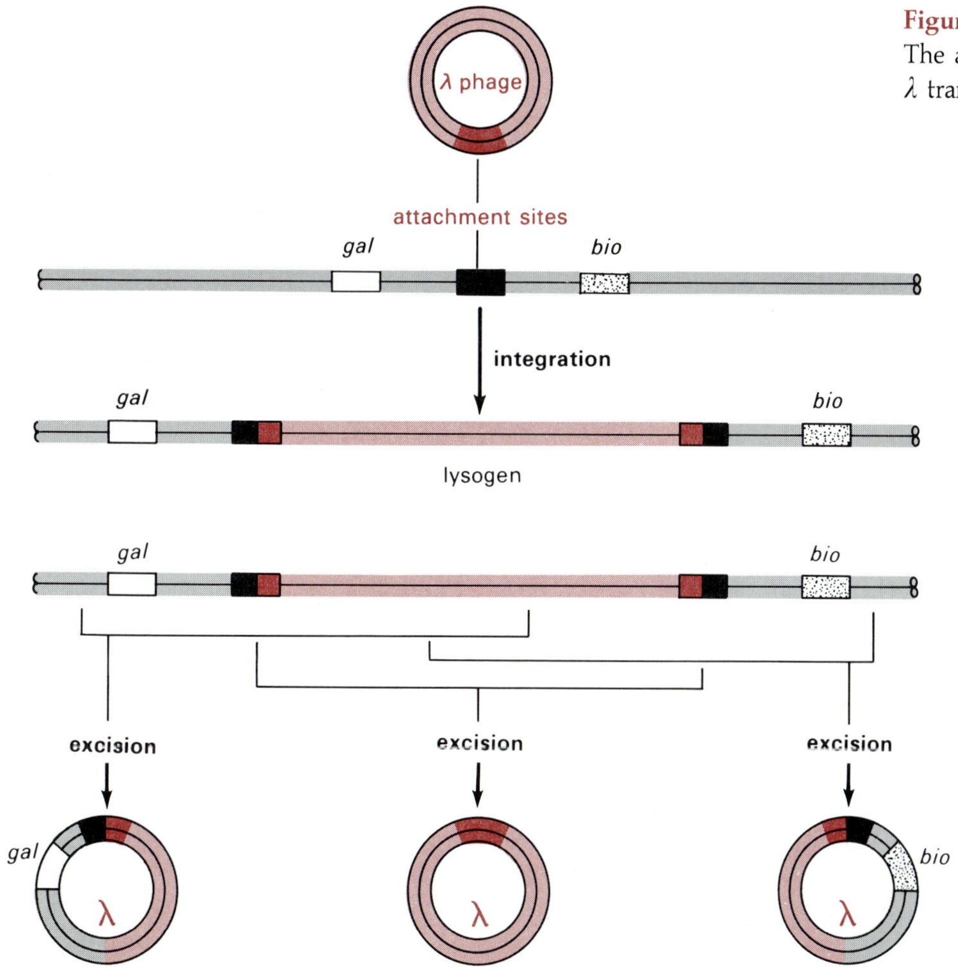

Figure II.7
The acquisition of cellular genes by λ transducing phage.

DNA segment is replicated along with the viral DNA. Third, the genes encoding virion structural proteins must be either functional or be supplied by a coinfecting virus or the host cell. Fourth, because transducing viruses are frequently accompanied by coinfecting wild type viruses, there must be a way to separate the different types of viral genomes and to identify the particular one of interest. This is generally achieved by **cloning**.

The Principle of Cloning

In order to understand how the concepts of transduction were applied to the recombinant DNA method, we need to examine the meaning of cloning. A clone of virus or of cells is a population of individuals all of which are derived from the reproduction of a single virion or a single cell, respectively. All the members of a clone, whether viruses or cells, are essentially identical to the virus or cell that initiated the production of the clone and to one another. Cloning viruses requires that the viral progeny from a single cell infected with a single viral particle multiply through many cycles of infection without becoming mixed with the progeny from other infected cells

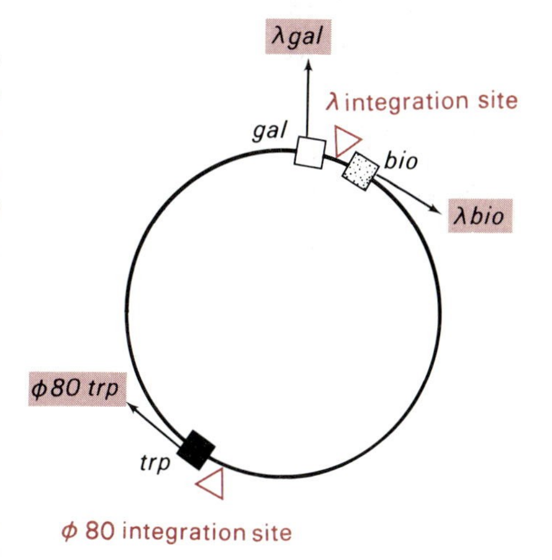

Figure II.8
Different specialized transducing phage transfer only those genes that are near their integration (attachment) sites on the bacterial genome. The integration sites of λ phage (λ*att*) and φ80 phage (φ80*att*) are shown.

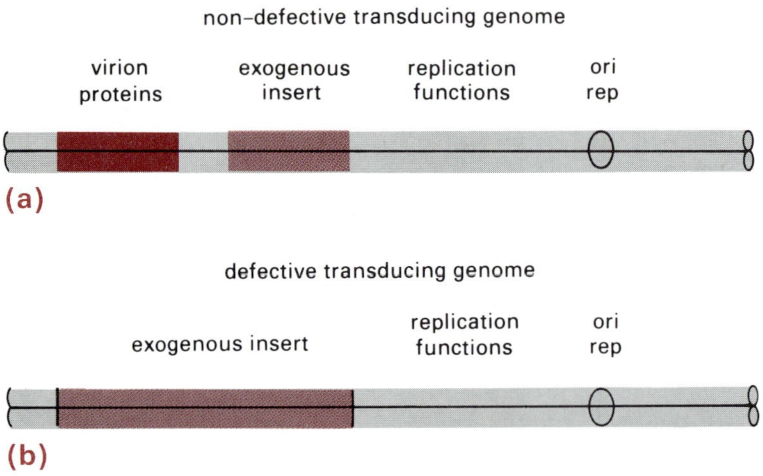

non-defective transducing genome

virion proteins
exogenous insert
replication functions
ori rep

(a)

defective transducing genome

exogenous insert
replication functions
ori rep

(b)

Figure II.9
Essential features of a transducing virus genome. (a) A nondefective genome that encodes all the functions required for virus replication and encapsidation. (b) A defective genome that does not include genes encoding virion proteins; these proteins must be provided by genes within the host cell or on a coinfecting viral genome.

(Figure II.10). Such viral clones appear as isolated clear areas (or plaques) on an otherwise normal monolayer (or lawn) of uninfected cells. Cloning of cells is achieved if the cells in a population are permitted to multiply in isolation from one another (Figure II.11). Clones of bacterial or mammalian cells are readily produced when single cells are spread sparsely on a plate so that they form isolated multicell colonies after growth on the surface of, within, or under a suitable medium.

Cloning is a means for obtaining a pure preparation of a single genotype, be it of a virus or a cell. It is also a way to obtain a pure preparation of a single genome because each individual in a clone has the same DNA. This same concept, molecular cloning, is used to obtain pure preparations of particular recombinant DNA molecules.

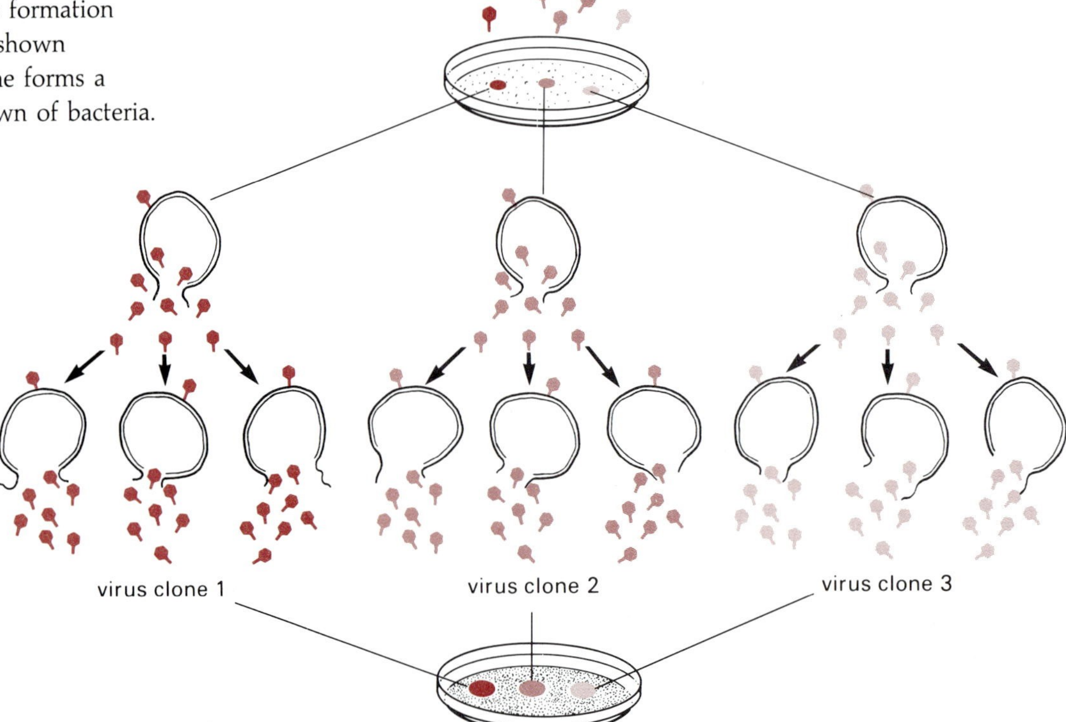

Figure II.10
A clone of viral particles is a population of viruses derived from the replication of a single particle. The formation of three viral clones is shown schematically. Each clone forms a separate plaque on a lawn of bacteria.

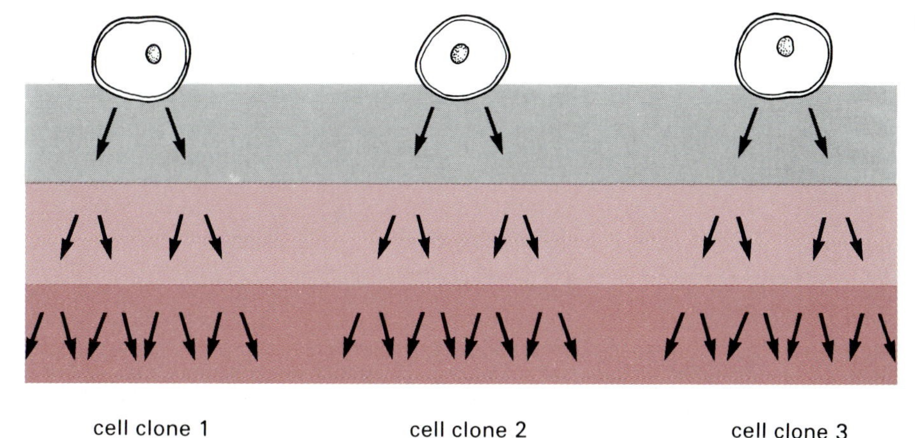

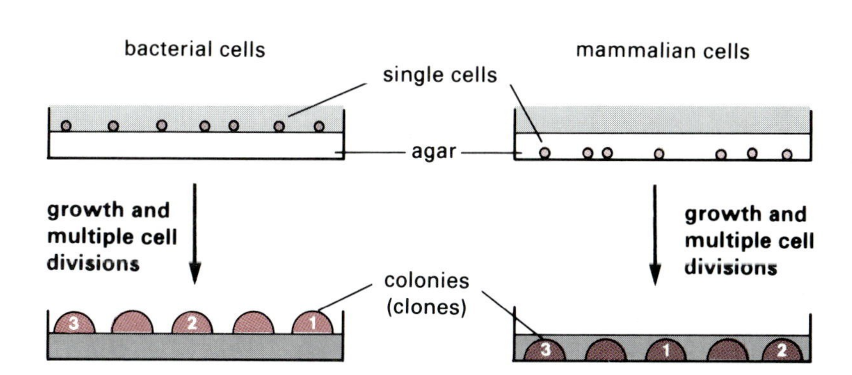

Figure II.11
A clone of cells is a population of cells derived by multiple cell divisions from a single cell. Each clone is seen as a discrete colony.

The Recombinant DNA Concept

The recombinant DNA methodology is based upon the same basic principle that underlies transduction. DNA molecules capable of replication in an appropriate cell, be they viral genomes or plasmids (see below), serve as carriers or **vectors** of "foreign" DNA segments, hereafter referred to as **DNA inserts**. However, instead of relying on cellular processes to produce the recombinant transducing genomes, appropriately modified DNA inserts and vector DNAs are joined or recombined *in vitro* through the action of an enzyme, DNA ligase (Figure II.12). Such recombinant DNAs are then introduced into appropriate cells, where they multiply through DNA replication. The cells are thereby transformed by the recombinant DNA.

This methodology's enormous potential stems not just from the construction and replication of recombinant DNAs, but from the ability to clone individual recombinant DNA molecules. Consider, for example, the consequences of joining a mixture of random segments of genomic DNA from any organism with a DNA vector (Figure II.13). The assortment of all possible recombinants is extremely varied; each contains a different segment of the original DNA. However, when the recombinant DNAs are cloned as individual virus plaques, each plaque contains a unique recombinant molecule composed of the vector DNA and a single segment of the original genome. Overall, the recombinant DNA technique makes it possible

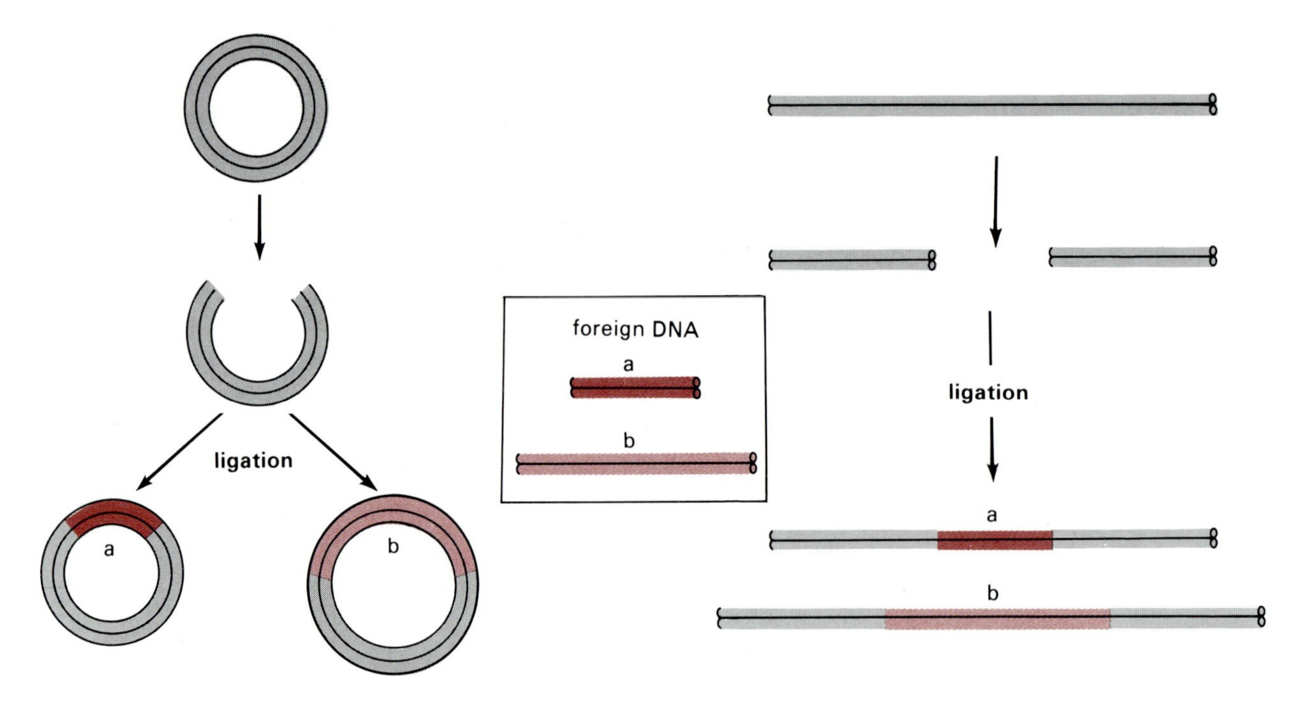

Figure II.12
Insertion of a mixture of foreign DNA fragments (a and b) into circular or linear virus DNA molecules *in vitro*.

to recover substantial yields of single DNA segments from extremely complex mixtures of segments generated from any cellular or viral genome.

An approximately 100-fold enrichment is achieved by the transfer of an *E. coli* gene from the bacterial genome to a transducing phage genome. This is a small purification compared to what is achieved by molecular cloning of DNA segments from complex organisms. For example, a mammalian gene containing 5 kbp is only about one-millionth of the mammalian genome (5 kbp out of about 3×10^6 kbp), but it is about one-tenth of a λ phage recombinant. Thus, molecular cloning provides a way to dissect even the largest and most complex genomes and to obtain discrete segments containing one or a few genes in pure form. This relatively straightforward application of the principles and methods developed to analyze the molecular genetics of prokaryotes removed the barrier that had prevented similar analysis of eukaryotic genomes.

Important Discoveries

As is often the case in scientific advances, many seemingly unconnected discoveries paved the way for the recombinant DNA methodology. One of these was transduction, as already described. Others include the discovery and characterization of bacterial plasmids and restriction endonucleases.

Bacterial Plasmids

One of the striking consequences of the use of antibiotics for treating infectious disease was the emergence of resistant strains of

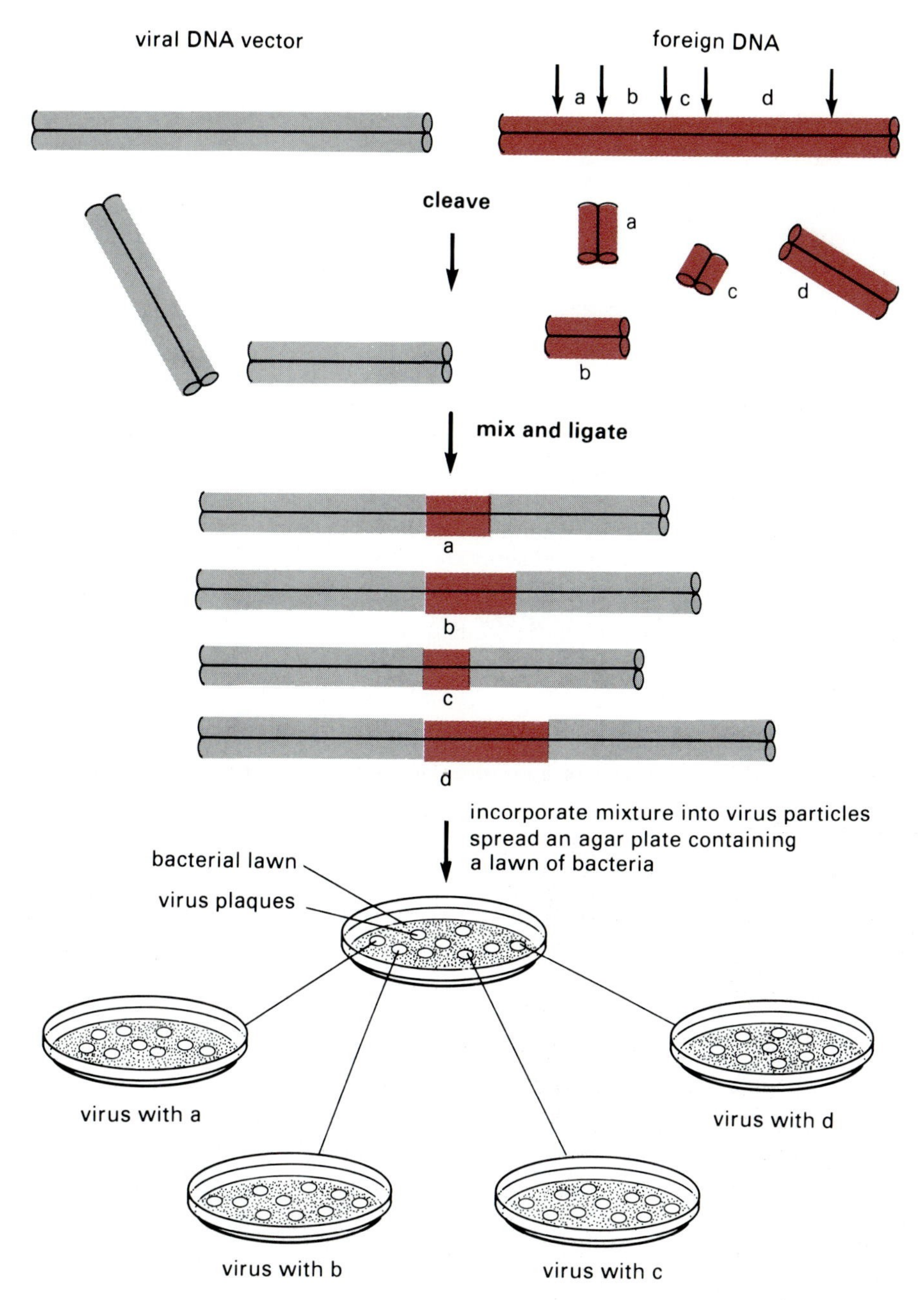

Figure II.13
Molecular cloning of foreign DNA in a viral (bacteriophage) vector.

pathogenic bacteria. The seriousness of this problem stimulated research to explain the phenomenon. Soon it became apparent that drug resistance is a relatively stable genetic trait that can be transmitted to sensitive bacterial cells in a manner akin to an infectious process. Subsequently, it was established that cell-to-cell contact is responsible for the spread of antibiotic resistance (Figure II.14). This phenomenon was explained by the discovery that antibiotic resistant cells contain genetic elements, plasmids, separate from their chromosomal DNA that can replicate independently of the chromosome and be transferred during cell-to-cell contacts. Furthermore, such extrachromosomal plasmids carry genes that confer resistance to one or more antibiotics on the cells that harbor them. These plasmids were called resistance or **R-factors**.

Figure II.14
Spread of a drug-resistance plasmid by cell-to-cell transfer.

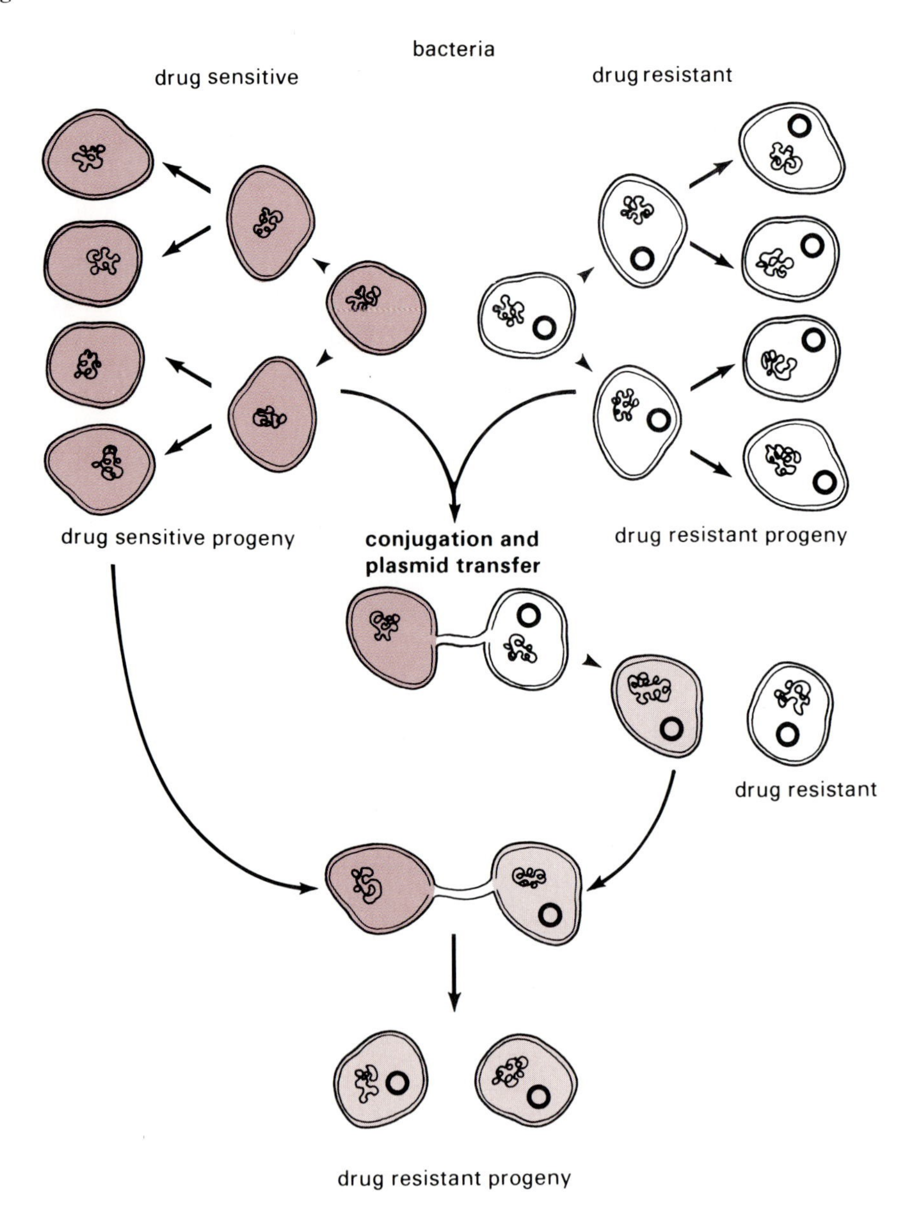

This understanding was greatly influenced by the knowledge of how genetic information is transferred from one cell to another during bacterial conjugation. Figure II.1 shows that genetic transfer is a consequence of a donor cell's ability to replicate and transfer its genomic DNA across a conjugation bridge to the recipient. *E. coli* strains that function as donors contain, in their genomes, DNA derived from a plasmid called a fertility or **F-factor**. Such chromosome donors are formed after acquisition of an F-factor and recombination between the plasmid and the cellular chromosome. The presence of the integrated F-factor sequences promotes conjugation and thereby fosters a high frequency of chromosome transfer. Transfer is initiated at the site of F-factor integration, and because the plasmid can integrate at many different sites, different strains initiate transfer at dif-

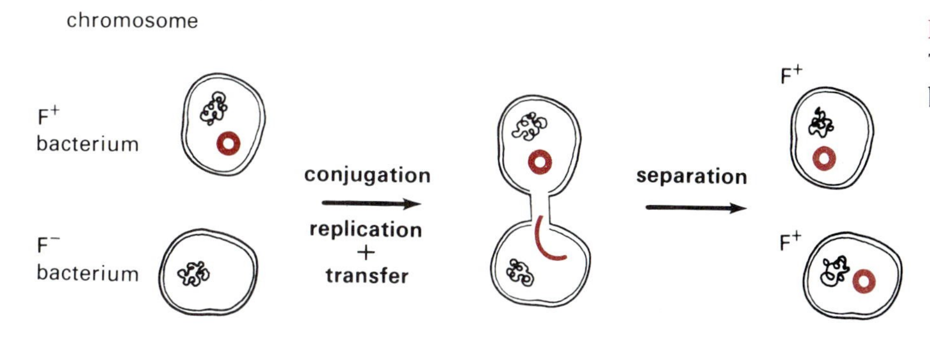

Figure II.15
Transfer of F-factor plasmids between bacteria by conjugation.

ferent sites on the *E. coli* chromosome (Figures II.1 and II.2). In some instances, the F-factor segment remains as an independent extrachromosomal element, the F-plasmid, rather than integrating (Figure II.15). Such cells, referred to as F^+, have the ability to transfer the F-plasmid itself to recipient cells, just as R-factors are transferred from cell to cell.

R-factors and F-factors are covalently closed circular duplex DNA molecules. Both contain genes that enable them to replicate as autonomous plasmids and to be transferred upon contact with appropriate recipients. R-factors also contain the genes that confer antibiotic resistance. Some of these genes alter a cell's response to an antibiotic; others induce the formation of proteins that degrade or modify particular antibiotics. Thus, an R-plasmid encodes β-lactamase, which causes resistance by degrading ampicillin. Chloramphenicol resistance is caused by acetylation of that antibiotic by a plasmid-encoded chloramphenicol transacetylase.

Because plasmid DNAs can be readily isolated from cells in substantial quantities and pure form, their potential as vectors for introducing new DNA segments into cells was explored. The joining of plasmid DNA to appropriately modified DNA inserts can be carried out *in vitro*, much like the joining of inserts to bacteriophage vectors (Figure II.12). Here, the experience with DNA-mediated cell transformation proved significant. Methods were developed to promote the uptake of purified plasmid DNA into suitable bacterial host cells. The presence of the antibiotic resistance genes provides a convenient way to select cells that acquire an R-plasmid as a stable replicating entity. Cells containing R-plasmids live and divide in the presence of the antibiotic; those lacking the plasmid do not survive. In the continuing presence of the antibiotic, the recombinant plasmid DNA is maintained as an autonomously replicating genome during subsequent host cell divisions. Cloning yields colonies of cells containing a unique recombinant DNA molecule because each host cell takes up only one plasmid. Because many plasmid vectors are only a few kilobase pairs long, enrichment of eukaryotic DNA segments is even greater than with phage vectors. These plasmids were actually the first vectors used for molecular cloning in bacteria.

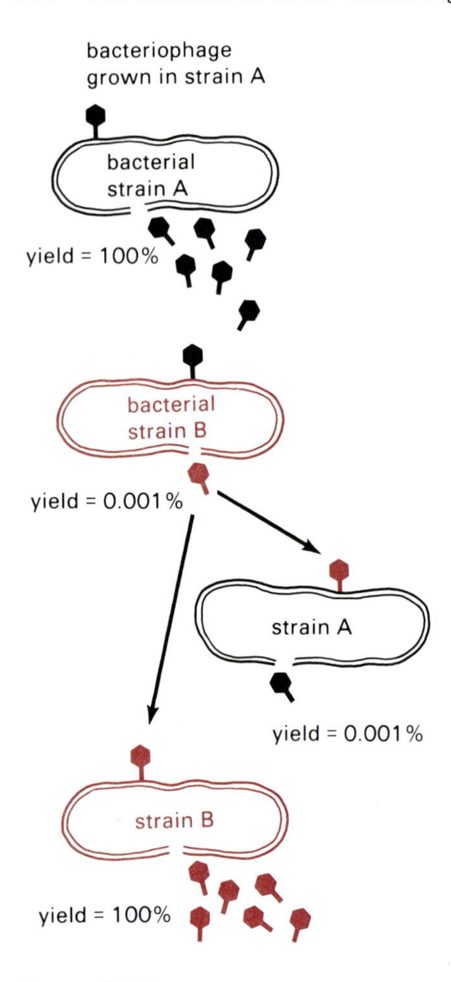

Figure II.16
Pattern of host controlled restriction of bacteriophage.

Restriction Endonucleases

One of the very important outcomes of genetic studies on bacteria and their bacteriophages was the discovery of restriction endonucleases—enzymes that recognize specific short DNA sequences and cleave both strands of the duplex at or some distance from that site. The key observation, made about 30 years ago, was that bacteriophage grown in one strain of cells often grow very poorly when they infect another strain of the same species. Moreover, although the few bacteriophage recovered from the second inefficient infection propagate very well in the host cells from which they were just obtained, they now grow poorly in the cells in which they had grown originally (Figure II.16). This phenomenon, which is unrelated to the phage genotype, was referred to as **host-controlled variation** or **host-induced restriction** because the effect is clearly correlated with a property of the host. These observations were explained by the suggestion that some phage component required for replication is modified in a strain-specific manner by the host cells, thus allowing the phage to complete replication upon reinfection of the same host strain but restricting its growth in unrelated strains that do not contain the same modification system.

The modified phage component proved to be DNA. Restriction of phage growth in an unfamiliar strain is caused by the degradation of the infecting phage DNA (Figure II.17). DNA breakdown is initiated by several highly specific cleavages, followed by massive nonspeci-

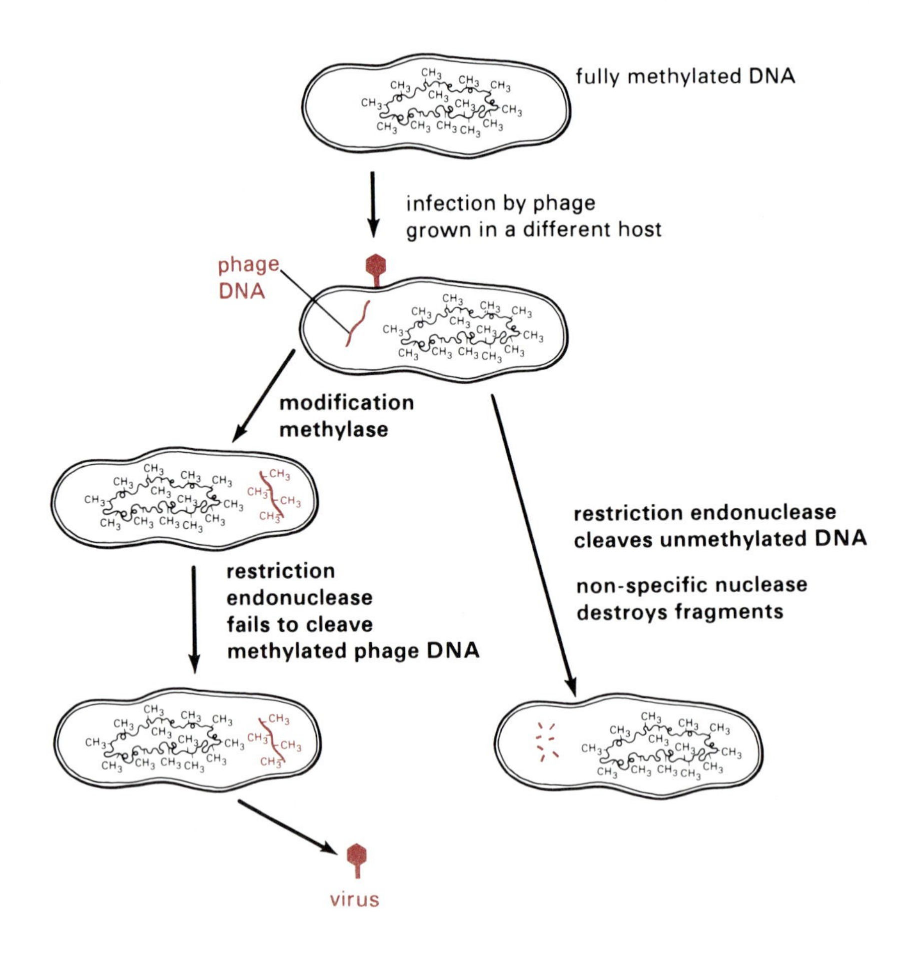

Figure II.17
Role of modification methylase and restriction endonuclease in host controlled restriction.

fic degradation. Strain-specific modification protects some of the infecting phage DNA as well as the cell's genome from breakdown by methylating the DNA. Genetic and biochemical analyses established that modification involves DNA methylation at short specific sequences; restriction is mediated by endonucleases that recognize the same short specific sequences lacking the methyl groups. Invariably, modification and restriction systems are matched, that is, methylation and cleavage are directed at the same DNA sequence. For each system, a different short DNA sequence is the target for the strain-specific restriction-modification system. Appropriately methylated DNA is not cleaved at that sequence by the cognate restriction endonuclease. Correspondingly, restriction endonucleases cleave only DNAs that are not modified at their restriction sites. Modification and restriction systems are controlled by related sets of genes, and such paired systems are encoded in many different species, in bacteriophages and in plasmids.

Two features of restriction endonucleases have proven to be extraordinarily influential in analyzing the genetic and physical organization of complex genomes. First, the enormous range of specificities displayed by different restriction endonucleases has provided many different ways to cleave DNA from virtually any source into discrete populations of fragments. These fragments can be separated according to size by electrophoresis through gels (Figure II.18). The distinctive fragment pattern created by the action of each restriction endonuclease on its special targets in a molecule or

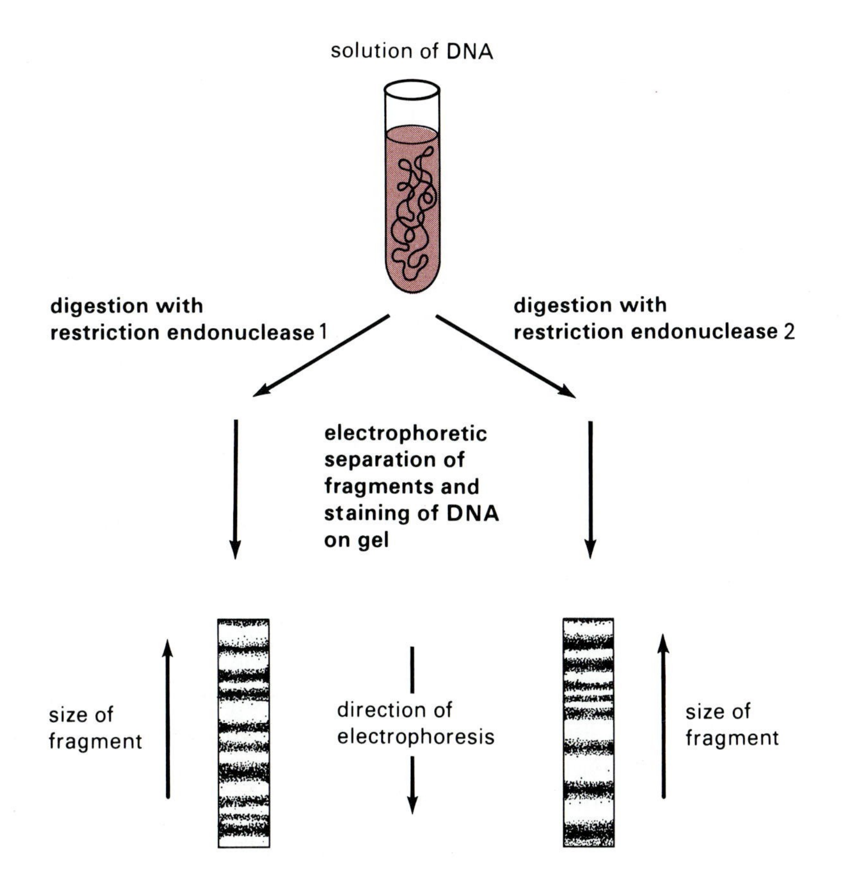

Figure II.18
Restriction endonucleases cleave a DNA molecule into a characteristic population of fragments that can be separated according to size by gel electrophoresis.

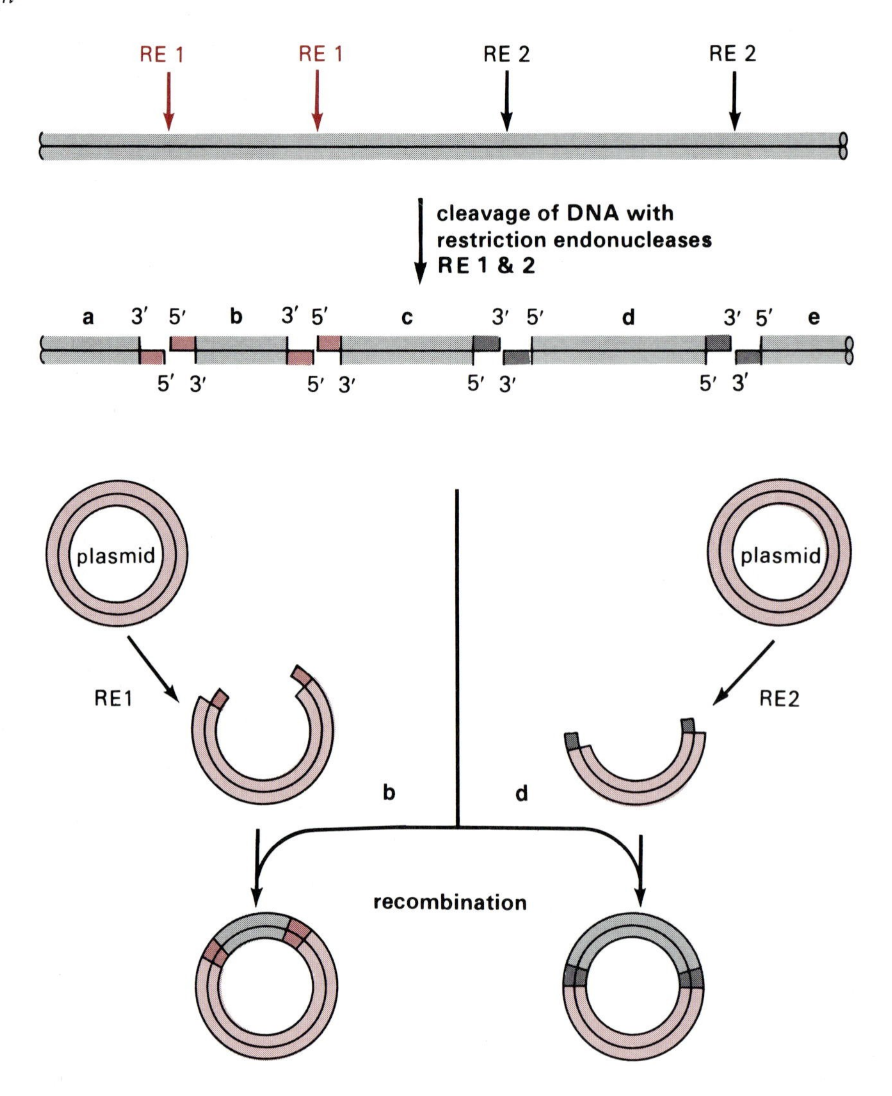

Figure II.19
Many restriction endonucleases make staggered cleavages in DNA and yield complementary single strand termini. Any two DNA segments carrying such complementary ends can be recombined *in vitro.* If one of the segments is competent to replicate within a suitable host cell, then the whole recombinant can be cloned and amplified. In this example, plasmid DNAs are joined with two of the fragments generated by the restriction endonucleases at the top of the diagram.

region of DNA provides a unique fingerprint that characterizes that DNA sequence. Second, many restriction endonucleases make staggered breaks on the two DNA strands within the restriction site, thereby creating fragments with complementary single strand ends (Figure II.19). This fact makes it possible to recombine DNAs *in vitro*. Any two DNA segments possessing matching ends can be joined. This provides a way to insert DNA fragments into phage, plasmids, or other potential vector DNAs and to recover specific vector-insert combinations in pure form by molecular cloning and amplification in suitable hosts (Figure II.12).

Upcoming chapters discuss how the unique properties of restriction endonucleases, together with molecular cloning of recombinant DNAs, led to the isolation of specific genes from even the most complex genomes. Detailed information about the structure of genes and their regulatory elements, as well as detailed maps of their arrangements in the genome, followed very quickly. The bottleneck to exploring the molecular genetics of eukaryotic genomes was broken.

The development of recombinant DNA techniques and molecular cloning would, by itself, have opened vast new areas of biological investigation. But the full realization of these advances depended upon other, simultaneous developments: fractionation methods that allow the separation of DNA fragments differing only slightly in their length (Figure II.18); relatively simple and rapid procedures for determining the sequence of thousands of nucleotides in DNA; production of specifically altered genes by mutagenesis of their cloned versions *in vitro*; and genetic transformation of cells, tissues, and whole organisms by introducing cloned genes into appropriate systems.

The four chapters in Part II describe the tools and methods for constructing, cloning, selecting, and characterizing recombinant DNA molecules. The coverage is not intended to provide a laboratory manual. Instead, the focus is on concepts. The presentation frequently emphasizes the earlier experimentation from which the method derives. The scientists who made the earlier advances could not have foreseen how their discoveries would facilitate the recombinant DNA developments. Yet the successful analysis of prokaryotic molecular genetics and the study of enzymes that synthesize and degrade nucleic acids opened the way to the molecular understanding of eukaryotic genomes.

CHAPTER 4

The Tools: Enzymes

The ability to manipulate DNA *in vitro* depends entirely on the availability of purified enzymes that cleave, modify, and join the molecules in specific ways. At present, no purely chemical methods can achieve either the selectivity or the range of enzymatic reactions for restructuring DNA. Yet only a rather small number of enzymes provides the essential tools for preparing recombinant DNA molecules. Most of them were discovered under circumstances unrelated to their present use in manipulating DNA. Indeed, each enzyme has a vital role in the genetic chemistry of the organism from which it is derived. The utility of the enzymes as tools depends in part on their availability and stability, but most important on their purity, particularly the absence of interfering enzymatic activities.

4.1 Nucleases

a. General Properties

Nucleases provide specific tools for manipulating DNA and RNA. Each enzyme can be described according to its specificity and the reactions it catalyzes. Thus, some enzymes, like the restriction endonucleases, cleave only DNA; others, like pancreatic RNase, hydrolyze only RNA (Table 4.1). Still other enzymes find the presence or absence of the 2′-OH on the ribose moiety irrelevant and use both DNA and RNA as substrates. There are nucleases that prefer either double strand or single strand polynucleotides as substrates and others that show no preference regarding strandedness. Nucleases are also characterized as being either **exonucleases** or **endonucleases**. Exonucleases require polynucleotide substrates that have ends, and they initiate cleavage at or close to the chain terminus. Exonucleases may prefer to initiate cleavage at either the 5′ or the 3′ terminus or may have no preference for a particular end. Endonucleases do not require ends on their substrates and can thus hydrolyze circular molecules; they always cleave internal phosphodiester bonds and produce polynucleotide products of variable size. Exonucleases can also yield short polynucleotide products, but many exonucleases produce nucleoside monophosphate products as they hydrolyze the chains one residue at a time. Finally, nucleases are distinguished depending on which side of the internucleotide phosphodiester bridge is hydrolyzed (Figure 4.1). Some enzymes cleave

Table 4.1 Typical Nucleases

Name	Source	Specificity	Strand Preference	Endo or Exo	Product
*Bal 31	*Altermonas espejiana*	DNA	s.s.	endo	5′-p(Np)$_n$N
		DNA	d.s.	exo {5′ to 3′, 3′ to 5′}	5′-NMP
ExoI	*E. coli*	DNA	s.s.	exo (3′ to 5′)	5′-NMP
ExoIII	*E. coli*	DNA	d.s.	exo (3′ to 5′)	5′-NMP
*ExoVII	*E. coli*	DNA	s.s.	exo {5′ to 3′, 3′ to 5′}	5′-p(Np)$_n$N
λ-exonuclease	*E. coli* (λ-phage)	DNA	d.s.	exo (5′ to 3′)	5′-NMP
*Mung bean nuclease	mung bean sprouts	DNA, RNA	s.s.	endo	5′-p(Np)$_n$N
*Neurospora nuclease	*Neurospora crassa*	DNA, RNA	s.s.	endo	5′-p(Np)$_n$N
Pancreatic DNase I	bovine	DNA	s.s., d.s.	endo	5′-p(Np)$_n$N
Pancreatic RNase	bovine	RNA	s.s.	endo	(Np)$_n{}^{C}_{U}$-p-3′
*RNase H	cellular	RNA·DNA	d.s.	endo	5′-p(Np)$_n$N
*RNase H	retrovirus	RNA·DNA	d.s.	exo (3′ to 5′)	5′-p(Np)$_n$N
*Sl nuclease	*Aspergillus oryzae*	DNA, RNA	s.s.	endo	5′-p(Np)$_n$N
Snake venom phosphodiesterase	*Crotalus adamanteus*	DNA, RNA	s.s.	exo (3′ to 5′)	5′-NMP
Spleen phosphodiesterase	bovine	DNA, RNA	s.s.	exo (5′ to 3′)	3′-NMP
Staphylococcal nuclease	*Staphylococcus aureus*	RNA, DNA	s.s.	endo	(Np)$_n$N-p-3′

Note: Enzymes marked with an asterisk (*) are described in the text. N denotes any nucleotide. (Np)$_n$ denotes a polynucleotide chain that is n residues long.

between the phosphate and the 3′-hydroxyl to yield 5′-phosphomonoester products; others cleave between the phosphate and the 5′-hydroxyl, resulting in products with 3′-phosphomonoester termini.

Most of the enzymes listed in Table 4.1 are mentioned in the course of this book. The nucleases marked with an asterisk are routine reagents in recombinant DNA experiments and are therefore described in more detail.

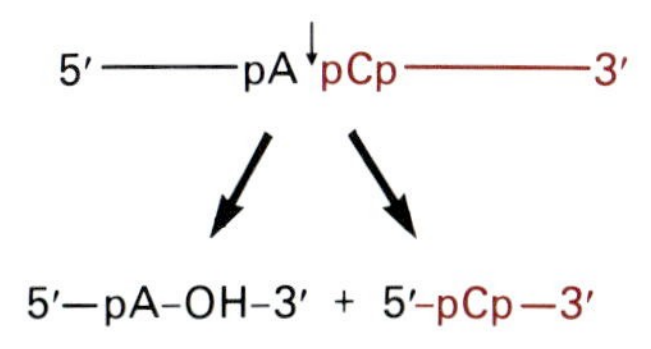

hydrolysis between the phosphate and the 3′-hydroxyl yields 5′-phosphomonoester end groups

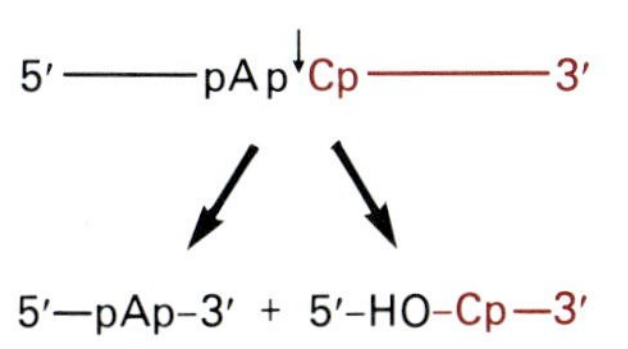

hydrolysis between the phosphate and the 5′-hydroxyl yields 3′-phosphomonoester end groups

Figure 4.1
Nucleases may cleave polynucleotides on either side of a phosphodiester bridge. Each enzyme is specific for one or the other side.

b. Single Strand Specific Nucleases

Endonucleases Some endonucleases hydrolyze single strands of DNA up to a thousand times more rapidly than duplex molecules. This specificity is experimentally useful in many different ways—for constructing DNA recombinants, for analyzing heteroduplexes, even for analyzing gene expression. Some of the reactions upon which these applications depend are outlined in Figure 4.2.

Several different single strand specific endonucleases have been sufficiently purified and characterized to be used as reagents. S1 nuclease is obtained from dried preparations of the mold *Aspergillus oryzae*; *Neurospora* and mung beans are the sources of two other widely used enzymes. Each of these three enzymes requires specific reaction conditions for optimal activity and for maximal discrimination between single and double strand molecules; all three hydrolyze both DNA and RNA. The cleavage of phosphodiester bonds by these enzymes produces 5′-monophosphate and 3′-hydroxy ends.

Exonucleases An *E. coli* exonuclease, exoVII, is specific for single strands of DNA. Its unusual exonucleolytic activity initiates digestion at both

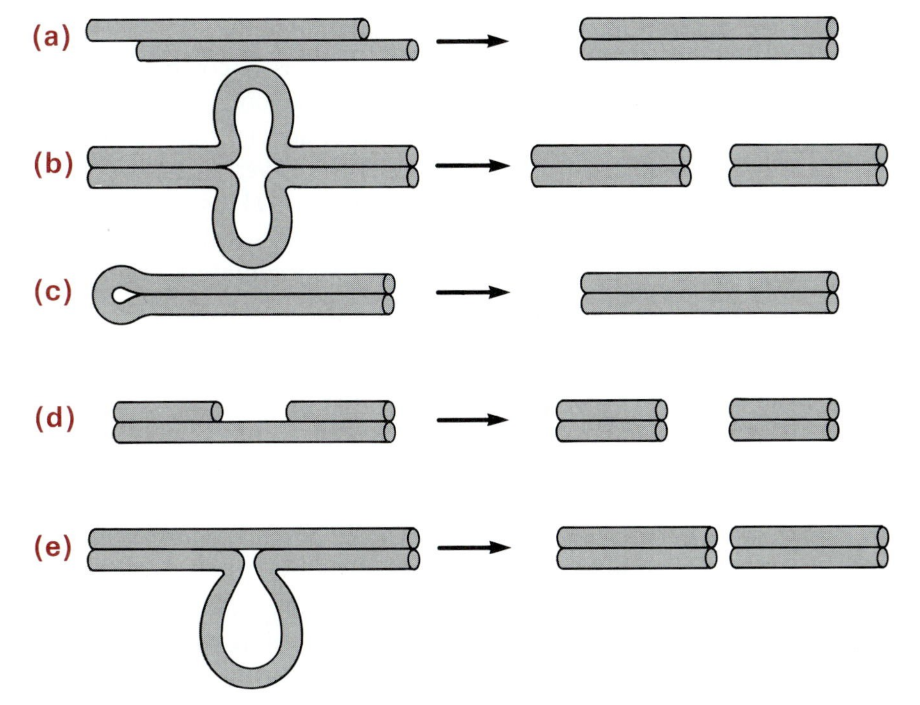

Figure 4.2
Some reactions catalyzed by single-strand-specific endonucleases. (a) Single strand tails on duplex DNA are converted to flush ends. (b) Duplex DNA with a small region of mismatched base pairs is cleaved into two fragments at the point of mismatch. (c) Duplex DNA with a hairpin turn at one end (i.e., a single duplex chain) is cleaved into a two-chain duplex. (d) A duplex containing a gap is converted into two double helical segments by destruction of the gap. (e) Duplex DNA containing a mismatched single strand loop is converted to two duplex fragments by destruction of the loop and cleavage at the resulting single strand region on the opposite strand.

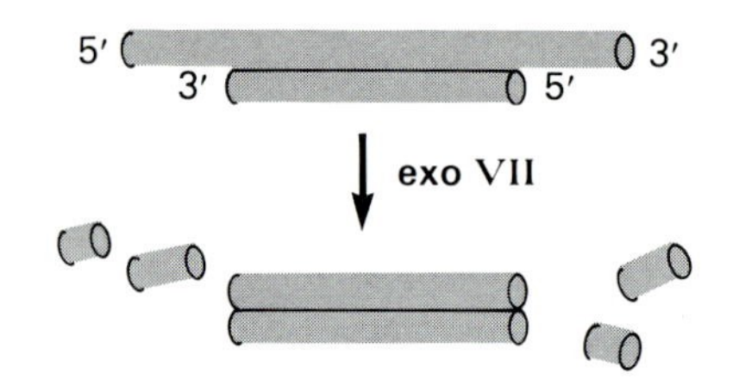

Figure 4.3
A reaction catalyzed by exoVII. Exonucleolytic digestion at both 5′ and 3′ ends yields oligonucleotides as long as 25 residues.

the 5′ and the 3′ termini of a chain; most known exonucleases work at one or the other end, but not both (but see the description of *Bal* 31 in Section 4.1c). Figures 4.2a and 4.3 illustrate the mode of action of exoVII on two different kinds of partly duplex DNAs. Although exoVII initiates its action at the ends of chains, it does not hydrolyze one nucleotide at a time. Rather, the products of exoVII digestion of single strands are oligonucleotides up to about 25 nucleotides long, and they contain 5′-phosphomonoester end groups.

c. Bal *31 Nuclease*

A Pseudomonad, *Alteromonas espejiana Bal* 31, secretes a single deoxyribonuclease called *Bal* 31. *Bal* 31 acts as an endonuclease on single strand DNA, including single strand regions on otherwise duplex DNA, in much the same way as other single strand specific endonucleases. However, with intact duplex DNA, this enzyme mimics exonuclease activity, presumably because it recognizes the partially single strand character (or duplex instability) at the termini. *Bal* 31 cleaves both chains at both ends of the duplex; that is, it degrades in the 3′-to-5′ and the 5′-to-3′ directions simultaneously. As a result, the size of the duplex molecule is progressively shortened (Figure 4.4). If the rate of shortening is determined empirically, then *Bal* 31 can be used to shorten DNA fragments a desired amount. Although the shortening of each DNA molecule in a population is not synchronous, products clustering around a chosen length are readily obtained. After exhaustive digestion with *Bal* 31 nuclease, intermediate oligonucleotide products are completely degraded to 5′-mononucleotides.

d. *RNase H*

There is a group of enzymes that are all called RNase H because they specifically degrade the RNA strand of an RNA·DNA hybrid duplex. *E. coli* RNase H is an endonuclease whose products are oligoribonucleotides with 5′-phosphomonoester termini; this enzyme is a commonly used reagent (Section 4.7, for example). Eukaryotic cells contain a similar endonucleolytic RNase H. Exonucleolytic RNase H activities are associated with *E. coli* exonuclease III (Table 4.1) and the reverse transcriptases encoded by retroviruses (Section 4.7). The RNase H associated with exonuclease III degrades the RNA to 5′-nucleoside monophosphates in the 3′-to-5′ direction. The products of reverse transcriptase RNase H are 5′-phosphorylated oligoribonucleotides generally between two and ten nucleotides long.

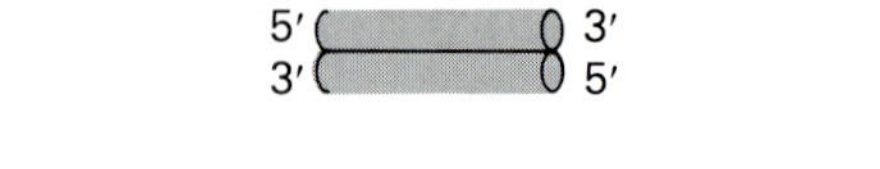

5′ 3′ 3′ 5′
Bal 31
5′ 3′ 3′ 5′
Bal 31
5′ 3′ 3′ 5′

Figure 4.4
The action of Bal 31 on double strand linear DNA molecules.

4.2 The Restriction Endonucleases

Evolution endowed different bacterial species with unique endonucleases that allow them to distinguish their own from foreign DNAs. In so doing, nature provided scientists with a large array of exquisitely specific reagents for dissecting DNA. Two significant features of **restriction endonucleases** are important for studying DNA. The first is the remarkable specificity with which the enzymes recognize short nucleotide sequences in DNA molecules. The second is the existence of many different restriction endonucleases, each recognizing a specific sequence.

a. *Three Types of Restriction Endonucleases*

Types I and III Both Types I and III enzymes are complex proteins that include the restriction endonuclease and the cognate methylase activities. They are interesting enzymes, but are not useful reagents in constructing recombinant DNA molecules. Type I enzymes bind to DNA at specific recognition sequences and then make double strand cleavages at variable distances from the recognition site. At least 400 bp and as many as 7 kbp separate the recognition site from the nonspecific cleavage site. Enzymatic hydrolysis of DNA requires the presence of Mg^{2+}, ATP, and S-adenosylmethionine; the latter activates the enzyme. ATP hydrolysis accompanies the cleavage, whereupon the enzyme becomes inactive as an endonuclease but potent as an ATPase. Thus, the type I enzymes are DNA-dependent ATPases. They are also site-specific methylases producing 6-methyladenine residues within the recognition site. For example, the recognition site for the *E. coli* K12 enzyme is

$$5'\text{-A}\dot{\text{A}}\text{CNNNNNNGTGC-}$$
$$3'\text{-TTG}\underline{\text{NNNNNN}}\text{C}\underset{\bullet}{\text{A}}\text{CG-}$$

where N can be any base and $\underline{\text{N}}$, its complement. Although endonuclease action occurs a long distance from this site, methylation to 6-methyladenine occurs within the site at the second A on the top strand and at the lone A on the bottom (dotted). Only a completely unmethylated site provides a substrate for endonuclease activity. The bipartite recognition site of the *E. coli* K12 enzyme, two short, specific oligonucleotide segments separated by six to eight unspecified base pairs, is typical of the recognition sites of type I enzymes found in different bacterial strains. Although the specific oligonucleotide sequences differ from one enzyme to another, the positions of the methylated A residues are the same.

Related type I restriction-modification systems are encoded in allelic positions in the genomes of different *Enterobacteriaceae* (e.g., *E. coli* and *Salmonella typhimurium*). Three linked genes are associated with each system: *hsdR*, *hsdM*, and *hsdS*, in the order of transcription. The polypeptide products of *hsdM* and *hsdS* are translated from a single dicistronic mRNA and together make up the methylase; the two gene products encode the methylase activity and the site-specific recognition functions, respectively. The product of the *hsdR* gene is responsible for the endonuclease activity. All three polypeptides occur together in varying proportions in active preparations of the type I enzymes.

Both nuclease and methylase activity occur together in type III enzymes as they do in type I endonucleases. However, although the type III enzymes are stimulated by S-adenosylmethionine and require ATP for cleavage of DNA, they do not catalyze ATP hydrolysis. The type III endonucleases make double strand breaks in DNA about 25 bp from their recognition sites. In the presence of both ATP and S-adenosylmethionine, the same enzymes catalyze site-specific methylations. Type III enzymes are heterodimeric proteins; the two distinct subunits are encoded by two linked genes on extrachromosomal (phage and plasmid) genomes in certain *E. coli* strains. Each gene is a separate transcription unit. One gene product is associated with specific sequence recognition and methylase activity; the other provides endonuclease function.

Because type I enzymes do not break DNA molecules into reproducible sets of fragments and both type I and III endonuclease activities are complicated by competing methylation reactions, neither is used as a reagent in recombinant DNA research.

Type II The type II restriction endonucleases are critical tools for constructing recombinant DNAs and for analyzing DNA structure. These enzymes recognize and bind to specific short nucleotide sequences in DNA, but, in contrast to type I and III endonucleases, they catalyze double strand cleavages at specific phosphodiester bonds either within or a precise short distance from the recognition sequence. Regardless of its overall length, a particular DNA molecule is always broken into the same set of fragments by a single type II enzyme. Fragment length is defined by the distance between the specific sequences recognized by that enzyme. The type II enzymes hydrolyze phosphodiester bonds between the 3′-hydroxyl group and the phosphate, producing a 5′-phosphomonoester on one side of the break and a 3′-hydroxyl group on the other. A divalent cation, usually Mg^{2+}, is essential, but neither ATP nor S-adenosylmethionine is required.

b. A Typical Type II Restriction Endonuclease

Eco*RI Endonuclease* The widely used *Eco*RI restriction endonuclease recognizes and cleaves DNA wherever the base sequence 5′-GAATTC-3′ occurs (Figure 4.5). As is characteristic of one group of type II restriction endonucleases, this sequence is a **palindrome**; the same sequence occurs precisely opposite on the complementary DNA strand, also in the 5′-to-3′ direction. Endonuclease-catalyzed hydrolysis occurs at the phosphodiester bond between the G and the A on each strand. Because *Eco*RI

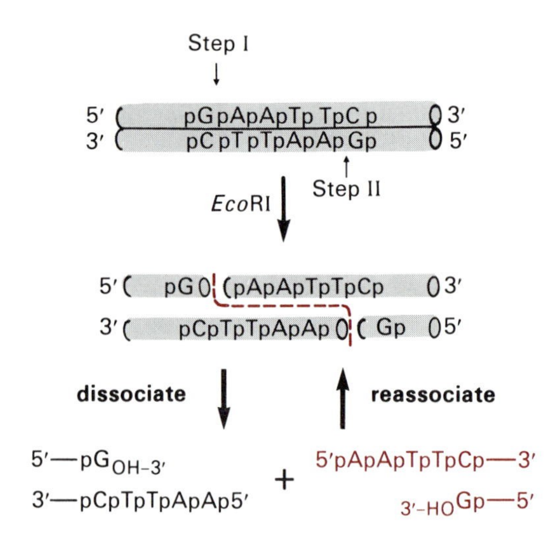

Figure 4.5
The recognition and cleavage site of restriction endonuclease *Eco*RI. Cleavage takes place in two steps; first one strand and then the other is hydrolyzed. Under the usual reaction conditions, the products are two separate DNA duplexes with complementary single strand tails. When conditions are altered to stabilize hydrogen bonding between the four nucleotide long single strands (e.g., lower temperature), the fragments can reassociate.

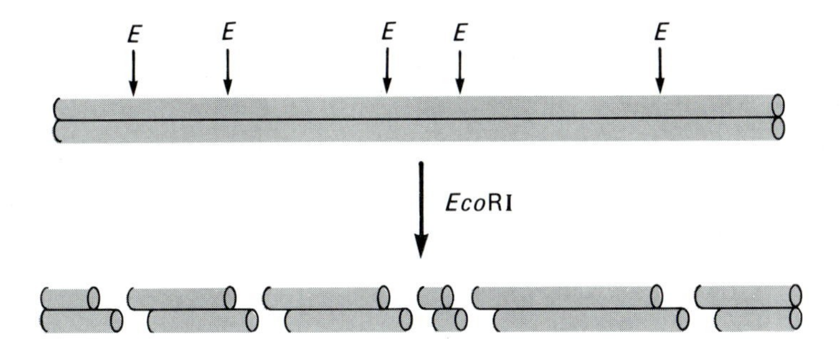

Figure 4.6
Schematic diagram of the cleavage by *Eco*RI of a long DNA duplex containing several *Eco*RI recognition sites (E).

endonuclease cleaves both strands at each 5′-GAATTC-3′ sequence it encounters, it divides a DNA molecule into a set of fragments that defines a characteristic fingerprint (Figure 4.6).

Cohesive ends Because *Eco*RI endonuclease produces staggered breaks in the two strands, the resulting DNA fragments end in short, complementary single strand tails containing the four bases 5′-AATT-3′ (Figure 4.5). Depending on the conditions of salt and temperature, these complementary tails either remain hydrogen bonded or are denatured to produce separate DNA fragments with short single strand extensions at their 5′ ends. Single strand complementary tails are readily reannealed with a change in conditions (Figure 4.5). The single strand termini produced by *Eco*RI endonuclease are referred to as **sticky** or **cohesive** ends because they can form base pairs with one another. An important consequence of this type of cleavage is that fragments, obtained by *Eco*RI endonuclease digestion of any two different DNAs (e.g., *E. coli* DNA and yeast DNA), can be annealed together at their identical cohesive ends; the sequence differences within the double helical segments do not influence the joining.

The enzyme Unlike the type I and III restriction endonucleases, type II enzymes like *Eco*RI do not catalyze methylation; the cognate methylase activity is in a separate protein. Enzymatically active *Eco*RI endonuclease is a dimer of two identical polypeptide chains of molecular weight 31,000. Because of the homodimer structure, one might expect that the enzyme would interact symmetrically with the identical sequences on the two strands in the recognition site and cleave both simultaneously. However, hydrolysis actually occurs in a stepwise fashion. The protein interacts with a total of about ten base pairs: the six within the recognition site itself as well as a few flanking base pairs. These flanking segments seem to influence the selection of the first strand to be cleaved. Similarly, flanking base pairs affect the overall cleavage rate at particular *Eco*RI recognition sites so that different sites in a single DNA are hydrolyzed at somewhat different rates. X-ray analysis of crystals of the complex between the enzyme and the oligonucleotide 5′-TCGCGAATTCGCG-3′ has given a detailed description of the way the protein and DNA interact. Hydrogen bonds are formed between two arginines and one glutamate (on each subunit) and base pairs in the recognition site. As a consequence, the B-DNA structure is partly unwound. Although less is known about the enzymatic properties of other type II restriction endonucleases, they appear to be similar to *Eco*RI in structure and function.

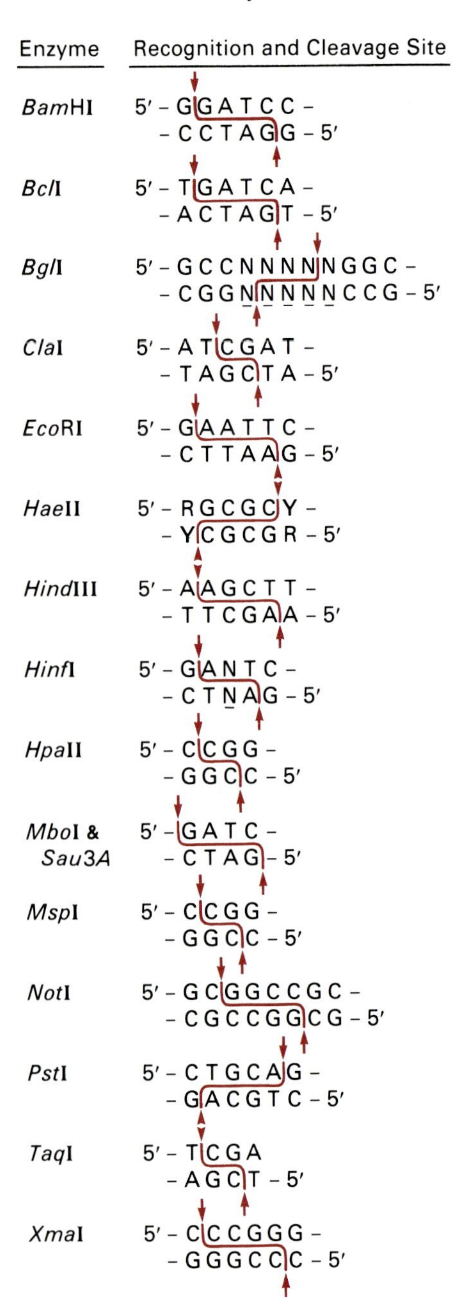

Figure 4.7
A group of type II restriction endonucleases that cleave palindromic sequences to yield cohesive ends. The arrows denote the cleavage points. The "p" designating the phosphodiester bridge is dropped from this and subsequent diagrams; it is understood to be present. Also understood is that cleavage by restriction endonucleases always yields a 5′ terminal phosphomonoester and a 3′ terminal hydroxyl (Figure 4.1). N represents any base and N̲ its complement. R represents either purine and Y either pyrimidine.

c. *Several Groups of Type II Restriction Endonucleases*

A large catalogue of reagents Figures 4.7, 4.8, and 4.9 together present a partial list of the approximately 400 type II restriction endonucleases that have been described. Among the 400, about 90 different recognition sites occur. Each of the known enzymes is derived from a particular prokaryotic organism, and its name reflects its origin. Thus, *Eco*RI is from *Escherichia coli* strain RY13; *Hae*II and *Hae*III, from *Haemophilus aegyptius*; *Bam*HI, from *Bacillus amyloliquefaciens* strain H; *Mbo*I and *Mbo*II, from *Moraxella bovis*; and so forth. Two different enzymes that have the same recognition site are called **isoschizomers**. However, isoschizomers do not necessarily cleave at the same position; compare, for example, *Xma*I and *Sma*I in Figures 4.7 and 4.8, respectively.

To facilitate the presentation of restriction sites, we use a set of symbols to represent the cleavage sites of the commonly used enzymes (Table 4.2).

Palindromic recognition and cleavage sites *Eco*RI endonuclease is one of many type II restriction endonucleases (Figure 4.7) that recognize different palindromic nucleotide sequences and cleave to produce complementary single strand tails. Within this group, several different kinds of enzymes can be distinguished. For example, *Eco*RI, *Hin*dIII, and some other endonucleases recognize six base pair sequences; *Hin*fI endonuclease recognizes five base pairs in which the central residue (marked N in Figure 4.7) may be any of the four nucleotides. *Bgl*I endonuclease recognizes six specific nucleotide residues, but these must be separated in the middle by any five other base pairs. The endonuclease *Not*I has an eight-base pair long recognition site; enzymes such as *Taq*I and *Mbo*I have tetranucleotide recognition sites. Another notable distinction concerns the nature of the single strand ends produced by different restriction endonucleases (Figure 4.7). The *Eco*RI, *Hin*dIII, *Mbo*I, and *Cla*I endonucleases produce single strand tails containing the 5′ terminal residues of each strand; but other enzymes, such as *Pst*I and *Hae*II endonucleases, yield 3′ terminal single strand tails.

Endonucleases with different recognition and cleavage sites frequently produce identical cohesive ends. For example, *Mbo*I, *Bcl*I, and *Bam*HI endonucleases yield 5′ terminal 5′-GATC-3′. Consequently, DNA segments produced by any of these enzymes will, upon mixing and annealing, be joined to one another. However, segments produced from different DNAs by cleavage with *Bgl*I are unlikely to anneal to one another because the single strand tails can contain any trinucleotide sequence.

Other type II restriction endonucleases (Figure 4.8) also recognize specific palindromic nucleotide sequences but cleave the phosphodiester bridge at the midpoint of the recognition sequence, thereby producing DNA fragments whose ends are fully base paired. Such fragments are termed **flush-ended** or **blunt-ended**.

Nonpalindromic recognition sites with cleavage at a distance Type II enzymes of still another kind (Figure 4.9) recognize a specific nucleotide sequence but hydrolyze a phosphodiester bridge outside the recognition sequence. The recognition sequence is not a palindrome. In these instances, the restriction endonucleases appear to count a precise number of base pairs

Table 4.2 Symbols for Cleavage Sites of Commonly Used Enzymes

Restriction Endonuclease	Symbol	Restriction Endonuclease	Symbol
*Alu*I	Al	*Kpn*I	K
*Ava*I	A	*Mbo*I	M1
*Bam*HI	B	*Mbc*II	M2
*Bcl*I	Bc	*Msp*I	M
*Bgl*I	Bg	*Pst*I	P
*Bgl*II	B2	*Pvu*I	Pu
*Cla*I	C	*Pvu*II	Pv
*Dde*I	D	*Sac*I	Sc
*Eco*RI	E	*Sal*I	S
*Eco*RII	E2	*Sau*3A	S3
*Eco*RV	E5	*Sma*I	Sm
*Hae*II	He	*Sph*I	Sp
*Hae*III	Ha	*Sst*I	Ss
*Hinc*II	Hc	*Taq*I	T
*Hind*III	H	*Xba*I	X
*Hinf*I	Hf	*Xho*I	Xh
*Hpa*I	HI	*Xma*III	Xm
*Hph*I	Hp		

from the recognition sequence and then cleave the chains. The counting mechanism results in hydrolysis of each of the two chains at phosphodiester bridges that are one base pair apart. The end result is the formation of DNA fragments with single strand overhangs that are only one nucleotide residue long.

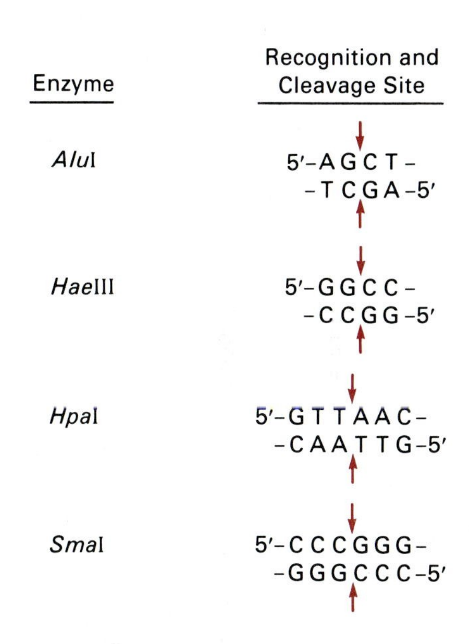

Figure 4.8
A group of type II restriction endonucleases that cleave palindromic sequences to yield flush ends.

d. *Mapping DNA Segments with Type II Restriction Endonucleases*

Restriction endonucleases provide a means to break up complex genomes or long DNA segments into reproducible sets of smaller units. The smaller units can be separated from one another on the basis of their size differences by several convenient methods. The most popular is electrophoresis of the DNA fragments through a semisolid gel made of agarose or polyacrylamide (Figure 4.10). Generally, the mobility of a duplex fragment in the electric field is inversely proportional to the logarithm of its size. Thus, given some marker fragments of known size, one can readily determine the size of a new fragment with a ruler and a simple graph. Less than a microgram of DNA is usually sufficient for such analyses because the DNA fragments are readily detected when stained with an appropriate dye such as ethidium bromide.

The set of fragments produced with a particular restriction endonuclease provides a characteristic fingerprint of that DNA. By analyzing the fragments produced by several restriction endonucleases alone and in combinations, one can frequently deduce the order of the segments within the original DNA molecule. The result is a physical map of the DNA, which, if sufficiently detailed, uniquely defines a genomic segment (Figure 4.11). Complex genomes or large DNA molecules generate complex

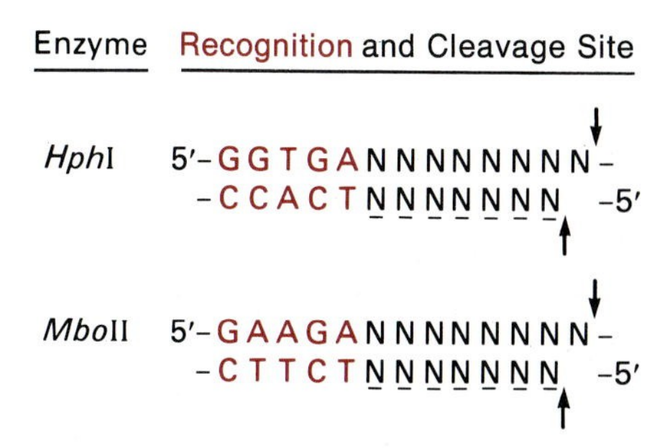

Figure 4.9
A group of type II restriction endonucleases that cleave DNA chains a precise distance from the recognition sequence.

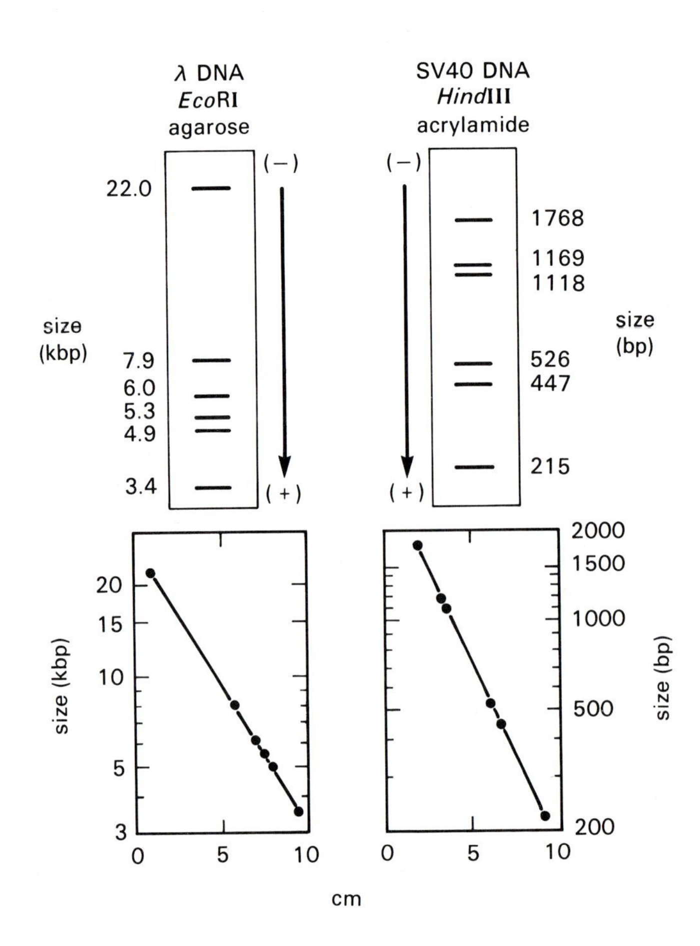

Figure 4.10
Electrophoresis of DNA fragments produced by restriction endonuclease digestion. On the left is a diagram of an agarose gel on which the fragments produced by digestion of λ bacteriophage DNA with *Eco*RI are indicated. On the right is a diagram of a polyacrylamide gel showing the separation of the fragments produced by digestion of simian virus 40 DNA with *Hind*III. Agarose gels separate fragments that are relatively large (effective range 23 to 0.3 kbp) while polyacrylamide resolves smaller fragments (effective range 6 kbp to 2 bp). Varying such components as the percentage of agarose or acrylamide in the gel allows the separation of fragments in different size ranges to be optimized. The DNA fragments are visualized after staining with ethidium bromide; as little as 50 ng can be seen as a band upon illumination with ultraviolet light. At the bottom of each gel is a plot of the log of the fragment sizes (in kbp or bp) against their relative mobility.

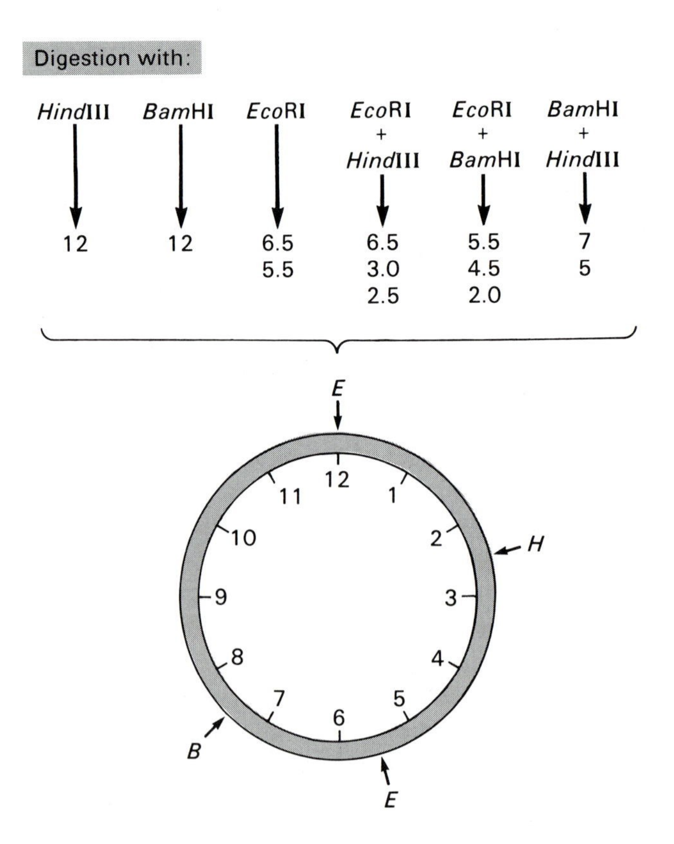

Figure 4.11
Constructing a restriction endonuclease map of a 12 kbp circular DNA molecule. The DNA was digested with the indicated endonucleases, and the size (in kbp) of the products was determined by agarose gel electrophoresis. From these sizes, the positions of the recognition sites for the three enzymes were deduced.

restriction endonuclease digests, and their analysis can be expedited by specially designed computer programs.

e. *Protection by Methylation*

Cognate methylases Genomes of the bacteria that make type II restriction endonucleases are protected from self-destruction by corresponding modification systems. Specific modification methylases recognize the same nucleotide sequences as their cognate endonucleases and transfer a methyl group from S-adenosyl methionine to specific adenine (forming N^6-methyladenine) or cytosine (forming 5-methylcytosine) residues within the site (Figure 1.4). As a result, cleavage by the endonuclease is barred. Figure II.17 illustrates the general set of reactions, and Figure 4.12 shows the characteristic methylation sites of several type II systems. A few of the restriction methylases have been purified sufficiently to be useful as tools in constructing recombinant DNAs (Section 6.2b). Methylation, like endonuclease cleavage, occurs symmetrically on both strands. However, the methylases that have been studied, including the *Eco*RI enzyme, are monomeric proteins, and methylation of the two strands occurs in two separate steps. Thus, although they recognize the same short DNA sequence and presumably evolved hand in hand, type II endonucleases and methylases are distinctly different proteins.

Identifying methylated sites in DNA Apart from restriction-modification systems, methylation is a common feature of DNA. As a result, some correct restriction sites within various DNAs resist cleavage by particular endonucleases. For example, 5′-meCG-3′ dinucleotides are common in vertebrate DNA (Section 1.1a). *Hpa*II restriction sites that include the methylated dinucleotide are not cleaved by the *Hpa*II endonuclease (Figure 4.13). This complication can be exploited to determine the state of methylation of those 5′-CG-3′ dinucleotides in complex genomes that are contained within *Hpa*II recognition sites. Thus, although *Hpa*II endonuclease does not cleave 5′-CmeCGG-3′, its isoschizomer, the *Msp*I endonuclease, does (Figure 4.13). Comparing the susceptibility of DNA fragments to cleavage by the two enzymes helps to locate a methylated cytosine on a map. Similarly, the common plant sequence 5′-CmeC$^{A}_{T}$GG-3′ can be distinguished from the unmethylated form 5′-CC$^{A}_{T}$GG-3′ by comparing the products obtained by digestion with the isoschizomeric endonucleases *Bst*N1 and *Eco*RII (Figure 4.14).

Independent methylases in E. coli When eukaryotic DNA fragments are inserted into bacterial vectors and replicated in *E. coli* or other prokaryotes, they lose their characteristic methylation patterns because they become subject to the methylation reactions of the new host cell. *E. coli* cells, for example, do not methylate 5′-CG-3′ dinucleotides, so sites previously protected from *Hpa*II endonuclease become susceptible (Figure 4.13). However, two common *E. coli* enzymes, the *dam* DNA methyltransferase that methylates A residues in 5′-GATC-3′ sequences and the *dcm* DNA methyltransferase that methylates the second C residue in

Restriction Endonuclease	Cognate Methylase Specificity
*Bam*HI	CH_3 (on C) 5′-GGATCC CCTAGG-5′ CH_3 (on C)
*Eco*RI	CH_3 (on A) 5′-GAATTC CTTAAG-5′ CH_3 (on A)
*Eco*RII	CH_3 (on C) 5′-CC$^{A}_{T}$GG GG$^{T}_{A}$CC-5′ CH_3 (on C)
*Hae*III	CH_3 (on C) 5′-GGCC CCGG-5′ CH_3 (on C)
*Hin*dIII	CH_3 (on A) 5′-AAGCTT TTCGAA-5′ CH_3 (on A)
*Hpa*I	CH_3 (on A) 5′-GTTAAC CAATTG-5′ CH_3 (on A)
*Hpa*II	CH_3 (on C) 5′-CCGG GGCC-5′ CH_3 (on C)

Figure 4.12
Cognate methylation sites of some restriction-modification systems.

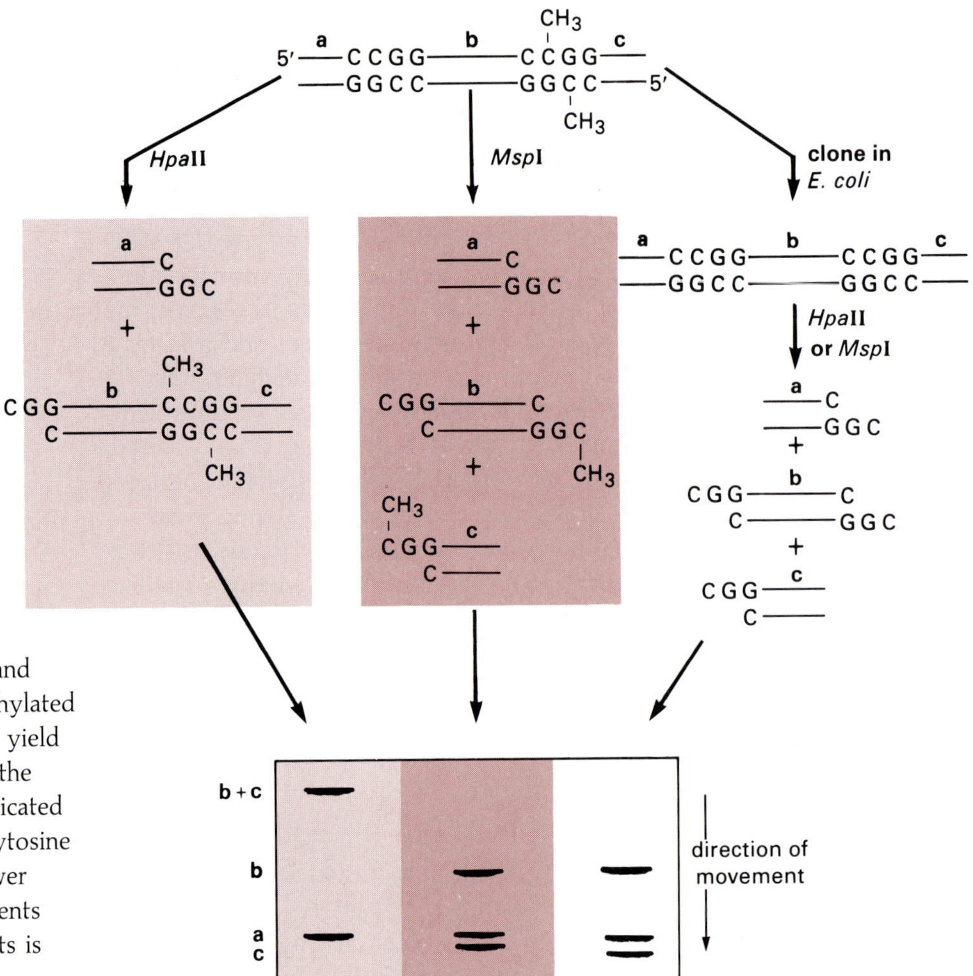

Figure 4.13
The differential action of *Msp*I and *Hpa*II on a region of partly methylated animal DNA. The two enzymes yield the same set of fragments after the region has been cloned and replicated in *E. coli* because the 5-methylcytosine is replaced by cytosine. The lower diagram shows the DNA fragments obtained after each of the digests is analyzed by gel electrophoresis.

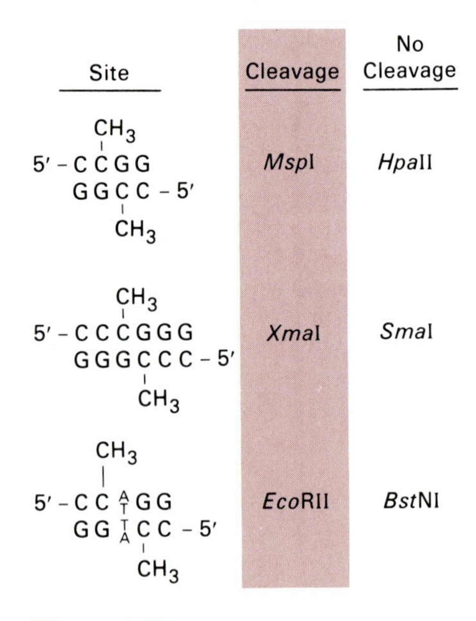

Figure 4.14
Pairs of restriction endonucleases that locate 5-methylcytosine residues.

5′-CC$^{A}_{T}$GG-3′ sequences, introduce newly resistant sites; neither of these methylases is part of restriction-modification system. Thus, eukaryotic DNA replicated in *E. coli* becomes resistant to *Mbo*I and *Bcl*I endonucleases because of the 6-methyladenine within the recognition sites. Note that the same 6-methyladenine within 5′-GATC-3′ does not interfere with the action of *Bam*HI, *Bgl*I, or *Sau*3A endonucleases; *Sau*3A cleavage is, however, inhibited if the C is methylated.

4.3 Phosphomonoesterases

Frequently in the course of recombinant DNA experiments or in the analysis of DNA structure, phosphomonoester groups must be removed from the terminus of a DNA chain (Figure 4.15). A variety of nonspecific phosphomonoesterases (phosphatases) have been known for many years. Those from *E. coli* and from calf intestine are highly purified. The phosphatases hydrolyze both 5′ and 3′ terminal phosphomonoesters on DNA and RNA.

Phosphomonoester groups at nicks in duplex DNA or those that are "hidden" by an overhanging single strand, such as those produced by

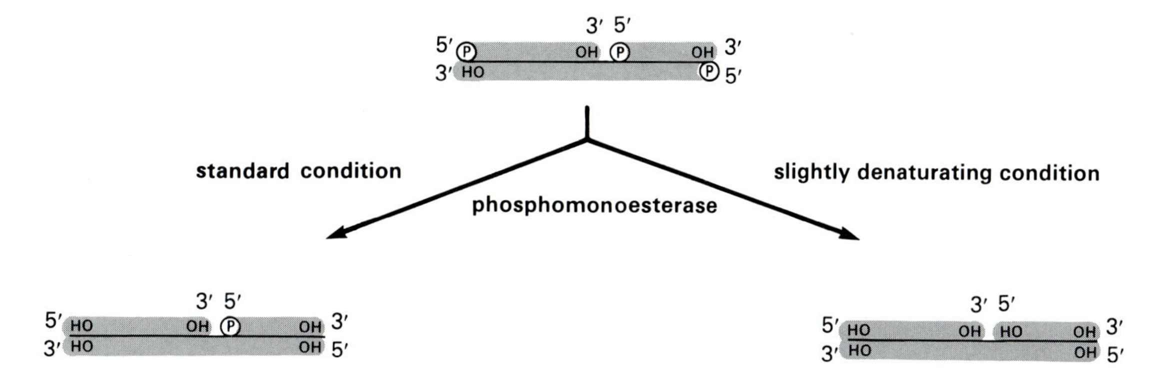

cleavage of DNA with *Pst*I endonuclease (Figure 4.7), are removed by phosphatase only under slightly denaturing conditions (e.g., elevated temperature).

Figure 4.15
The hydrolysis of terminal phosphomonoester groups by phosphatase.

4.4 Polynucleotide Kinase

Kinases constitute a large class of enzymes that catalyze the phosphorylation of many different biochemicals ranging from small molecules to very large macromolecules including polypeptides and polynucleotides. The polynucleotide kinase (Figure 4.16) that is widely used as a reagent in recombinant DNA experiments is purified from *E. coli* infected with T4-bacteriophage; the enzyme is encoded by the bacteriophage genome. ATP is the phosphate donor, and ADP is one product of the reaction. The reaction specifically phosphorylates 5′ terminal hydroxyl groups on DNA and RNA chains; 3′ terminal hydroxyls are not phosphorylated. An important use of polynucleotide kinase is to label the 5′ terminus of a polynucleotide chain with radioactive ^{32}P using ATP labeled with ^{32}P in the γ phosphate. The consecutive action of phosphatase (Section 4.3) and polynucleotide kinase replaces an unlabeled 5′ terminal phosphomonoester with a radioactive one, with no other alterations in the chain.

4.5 DNA Ligase

Recombinant DNA experiments require the joining *in vitro* of DNA segments from disparate sources. The cohesive ends that certain restriction endonucleases produce readily allow segments to be joined by annealing, but the hydrogen bonds that hold the segments together are relatively weak and not always stable under the conditions required for experimentation. DNA ligase joins the molecules covalently by catalyzing the formation of phosphodiester bonds between adjacent nucleotides (Section 2.1e); the mechanism of ligation is shown in Figure 2.20.

To have a successful ligation, the strands to be joined must be appropriately lined up with one another, as they are after annealing of cohesive ends; both strands are ligated (Figure 4.17a). A similar alignment is attained if the phosphodiester bridge to be mended is simply a nick in a duplex (Figure 4.17b), because the two strands are held in place by hydrogen bonding to the intact complementary strand. Of the two well-

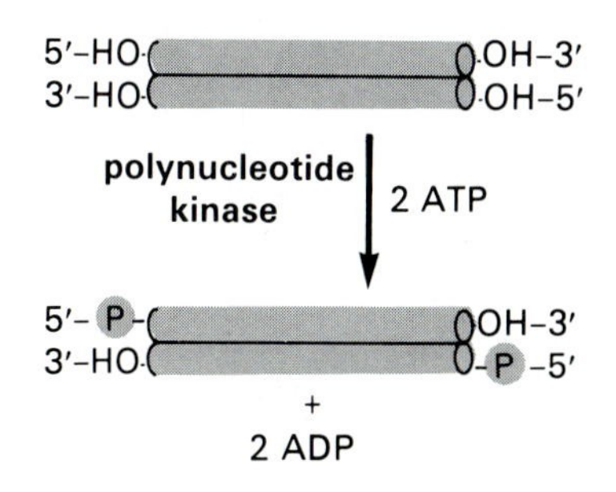

Figure 4.16
The addition of 5′ terminal phosphomonoester groups by the action of polynucleotide kinase. The γ phosphate of the ATP can be labeled with ^{32}P to yield a radioactive DNA molecule.

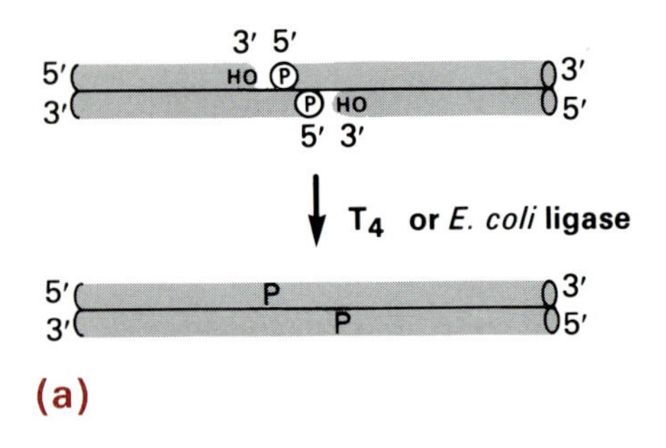

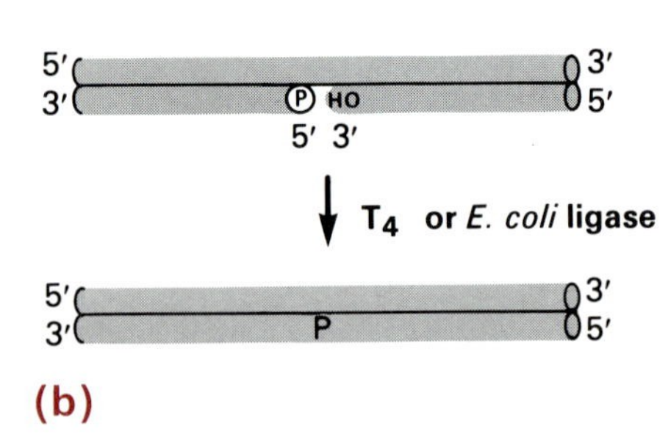

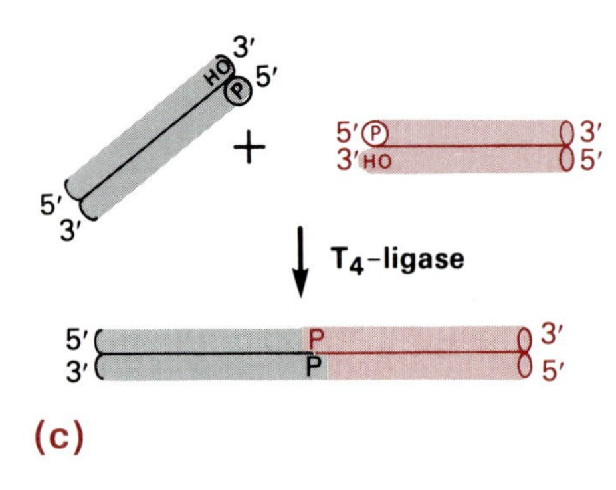

Figure 4.17
(a) Ligase covalently joins two duplexes held together by cohesive ends.
(b) Ligase catalyzes the formation of phosphodiester bonds at nicks in duplexes. (c) T4 DNA ligase joins flush-ended duplexes.

known DNA ligases, the T4–ligase, but not the *E. coli* enzyme, can even join two flush-ended duplexes (albeit inefficiently), forming both the required phosphodiester bonds (Figure 4.17c). It remains to be discovered if any special mechanism operates to bring the two flush-ended fragments together in proper alignment.

4.6 DNA Polymerase I

a. *A Versatile Enzyme*

The mechanism by which DNA polymerases catalyze the semiconservative synthesis of new DNA chains was described in Section 2.1e. DNA chains are elongated by the sequential addition of nucleotides to a primer with a free 3′-hydroxyl; at each step, the choice of nucleotide residue is directed by a DNA template. *E. coli* DNA polymerase I also catalyzes several other important reactions, two of which are of special significance in recombinant DNA experiments. One is a 3′-to-5′ exonuclease and the other, a 5′-to-3′ exonuclease. Both nucleolytic activities yield products with a 5′-phosphomonoester terminus. The two exonuclease activities are catalyzed by different regions of the DNA polymerase I protein molecule, and these can be separated physically by treating the enzyme with proteolytic enzymes. When this is done, the 3′-to-5′ exonuclease and the polymerizing activity remain associated with a large, carboxy-terminal polypeptide while the 5′-to-3′ exonuclease activity is in a second, smaller fragment.

b. *Nick Translation*

Various uses are made of the different reactions catalyzed by DNA polymerase I. One important use depends upon DNA polymerase I's ability to catalyze simultaneously both polymerization and 5′-to-3′ exonucleolytic digestion (Figure 4.18). The exonuclease degrades the chain in the 5′-to-3′ direction starting at the 5′ terminus of a single strand nick

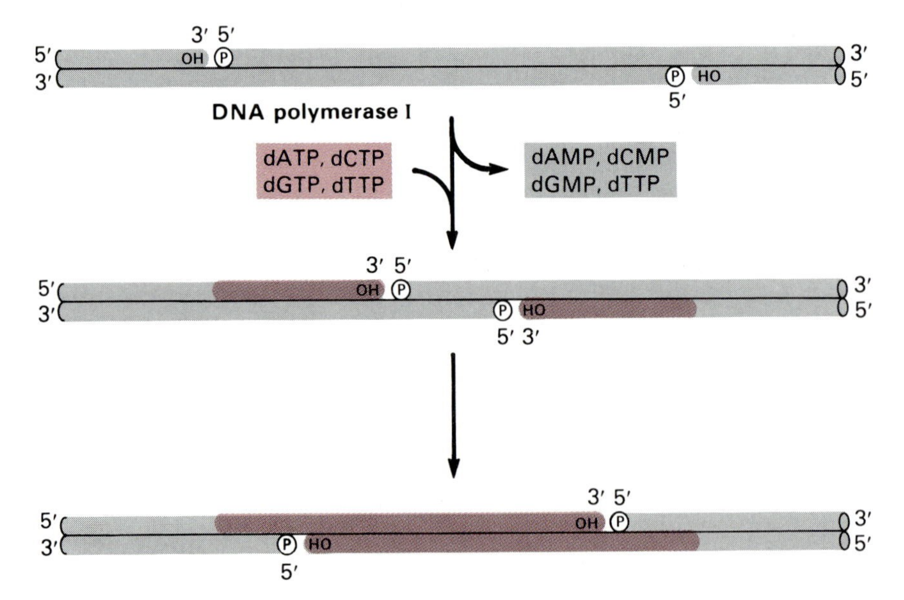

Figure 4.18
Nick translation. Starting with a nicked duplex, the combined action of the 5′-to-3′ exonuclease and polymerase activity of *E. coli* DNA polymerase I in the presence of α-^{32}P-labeled deoxynucleoside triphosphates (color) results in the movement of the nick down the DNA strand and the synthesis of ^{32}P-labeled DNA in the wake of the nick.

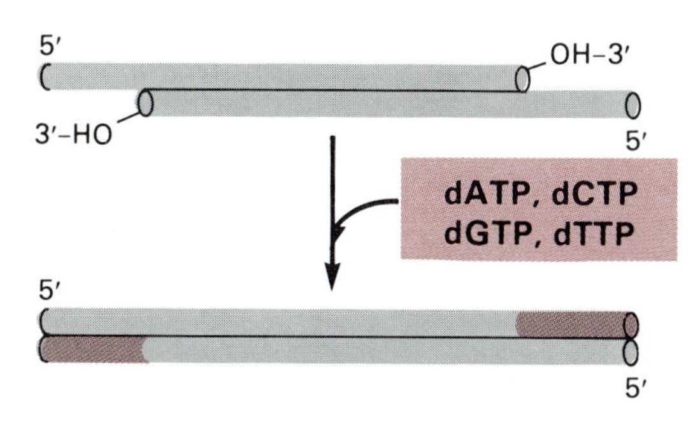

Figure 4.19
Making flush ends by filling in with the large fragment of DNA polymerase I.

in a duplex while the polymerase rebuilds the chain by adding new mononucleotide residues to the free 3′-hydroxyl that bordered the original nick. Net synthesis of DNA does not occur. The position of the nick thus progresses down the chain, which accounts for the common name for this process—**nick translation**. Nick translation with α-^{32}P-labeled deoxynucleoside triphosphates as substrates is widely used to prepare highly radioactive DNA fragments.

c. *Filling In*

DNA polymerase can convert DNA fragments with cohesive ends to fragments with flush ends if the overhang is a 5′ terminus (Figure 4.19). The 3′-hydroxyl terminus serves as a primer, the 5′ overhang as a template, and the shortened 3′ end is filled in by the addition of deoxynucleotide residues. The large carboxy-terminal fragment of DNA polymerase I produced by proteolysis is preferred for filling in in order to avoid having the 5′-to-3′ exonuclease destroy the template (Section 2.1e). Cohesive ends can also be converted to flush ends with a single-strand-specific nuclease, but with this method, several nucleotide residues are lost (Figure 4.2a).

4.7 RNA-Dependent DNA Polymerases (Reverse Transcriptases)

These enzymes are purified from RNA tumor viruses, and they permit the *in vitro* synthesis of DNA with an RNA template (Section 2.2). The reaction catalyzed by reverse transcriptases is analogous to the standard DNA polymerase reactions, and, as with other DNA polymerases, a primer is needed. The RNA template is generally a single strand to begin with, and only a single, complementary DNA strand is synthesized, yielding an RNA·DNA hybrid. The schematic diagram in Figure 4.20 demonstrates that a short polydeoxyribothymidylic acid (polydT) is often a convenient primer because it anneals to the polyriboadenylic acid tail commonly found on eukaryotic mRNAs (Section 3.8a), thus allowing the synthesis of a DNA copy (a **cDNA**) of the RNA.

The cDNA copy of the mRNA can be converted to a double strand DNA after RNase H or alkaline hydrolysis removes the original RNA template. Although the reasons are not well understood, the single strand cDNA product serves as its own primer (as well as the template) for synthesis of the complementary second DNA strand. The latter reaction may be catalyzed either by DNA polymerase I or by reverse transcriptase

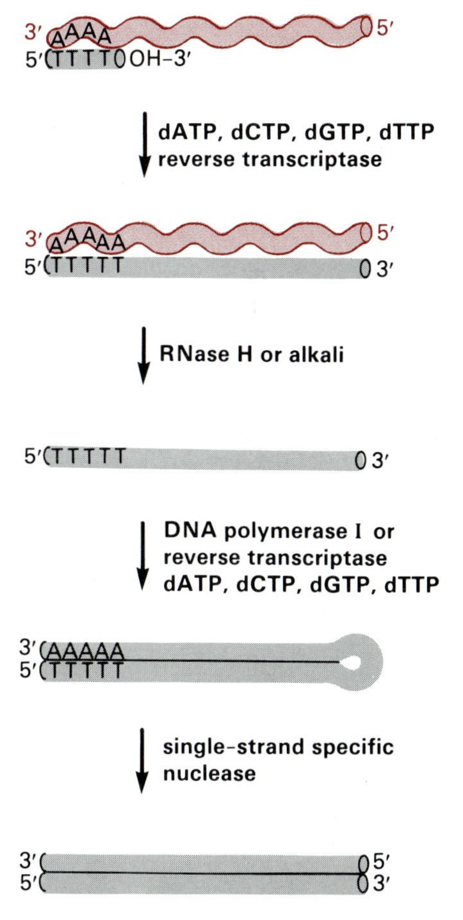

Figure 4.20
Schematic representation of the reaction catalyzed by reverse transcriptase and the synthesis of duplex cDNA.

itself. Apparently, a short hairpin loop forms near the 3′-hydroxyl end of the first strand, thereby providing the primer. The final duplex still has the hairpin turn at one end, and this is cleaved by a single strand specific nuclease (see Figure 4.2c) to yield two fully complementary single DNA strands (a duplex cDNA).

4.8 Terminal Deoxynucleotidyl Transferase

a. *Polymerization Without a Template*

In one sense, **terminal deoxynucleotidyl transferase** (often called **terminal transferase**) is a polymerase in that it catalyzes the synthesis of polydeoxyribonucleotides from deoxyribonucleoside triphosphates with the release of inorganic pyrophosphate. Like the DNA polymerases, it is unable to initiate a new polymer chain and therefore requires a primer with a free terminal 3′-hydroxyl group. But unlike the true DNA polymerases, it does not require a template and does not copy anything at all. The product of the polymerization reflects the deoxynucleoside triphosphates used as substrate (Figure 4.21). If dATP is added to the reaction, the product is a polydeoxyadenylic acid (poly dA) added to the 3′ end of the primer; if dGTP is used, poly dG is added to the primer. Because terminal deoxynucleotidyl transferase does not copy a template, the product is a single strand. If the primer molecule is itself double stranded, then similar 3′ terminal tails are produced on either end of the duplex. Actually, terminal deoxynucleotidyl transferase prefers a single strand to a double strand primer. Given a fully duplex DNA fragment to which single strand tails are to be added, a variety of methods can be used to produce single strand ends suitable for priming (Figure 4.22).

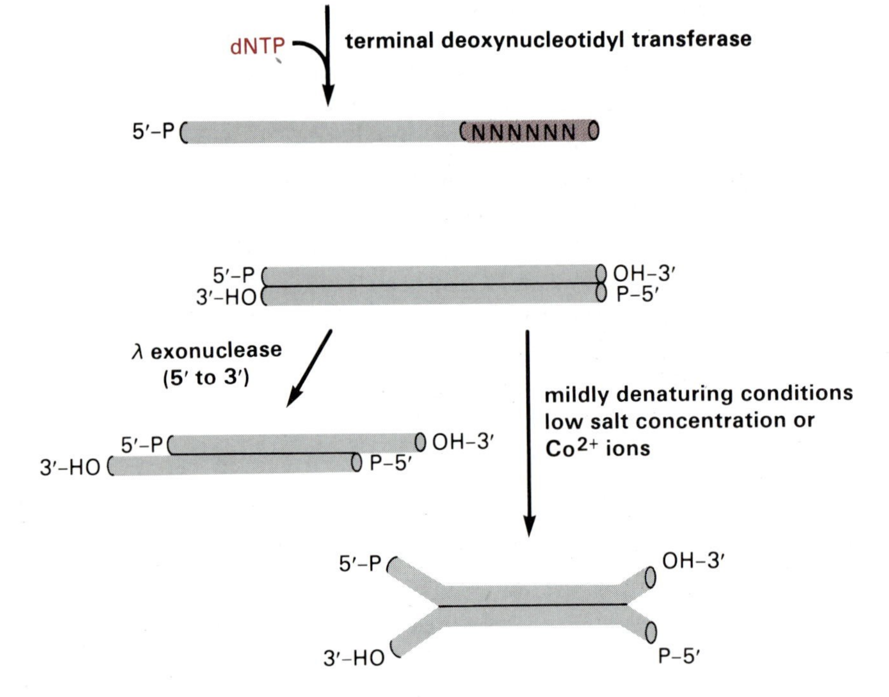

Figure 4.21
The reaction catalyzed by terminal deoxynucleotidyl transferase. N stands for any of the four deoxynucleotides A, T, G, or C.

Figure 4.22
Ways to make a suitable single strand primer for terminal deoxynucleotidyl transferase on a fully duplex DNA molecule.

b. Synthesizing Cohesive Ends

Cohesive ends can be added to DNA molecules that do not otherwise have them with terminal deoxynucleotidyl transferase (Figure 4.23). These include random fragments obtained by mechanical shearing of large DNA (e.g., by rapid stirring or forcing a DNA solution through a small orifice), fragments produced by restriction endonucleases that yield flush ends, or fragments produced by nonspecific DNases. For example, if poly dT is added to one set of DNA fragments and poly dA to another, and the two samples are mixed, the fragments anneal to one another. Any gaps that occur because of inequality in the lengths of the poly dT and poly dA tails may be filled in with DNA polymerase I. The molecules can then be joined with DNA ligase. This is the approach that was used to con-

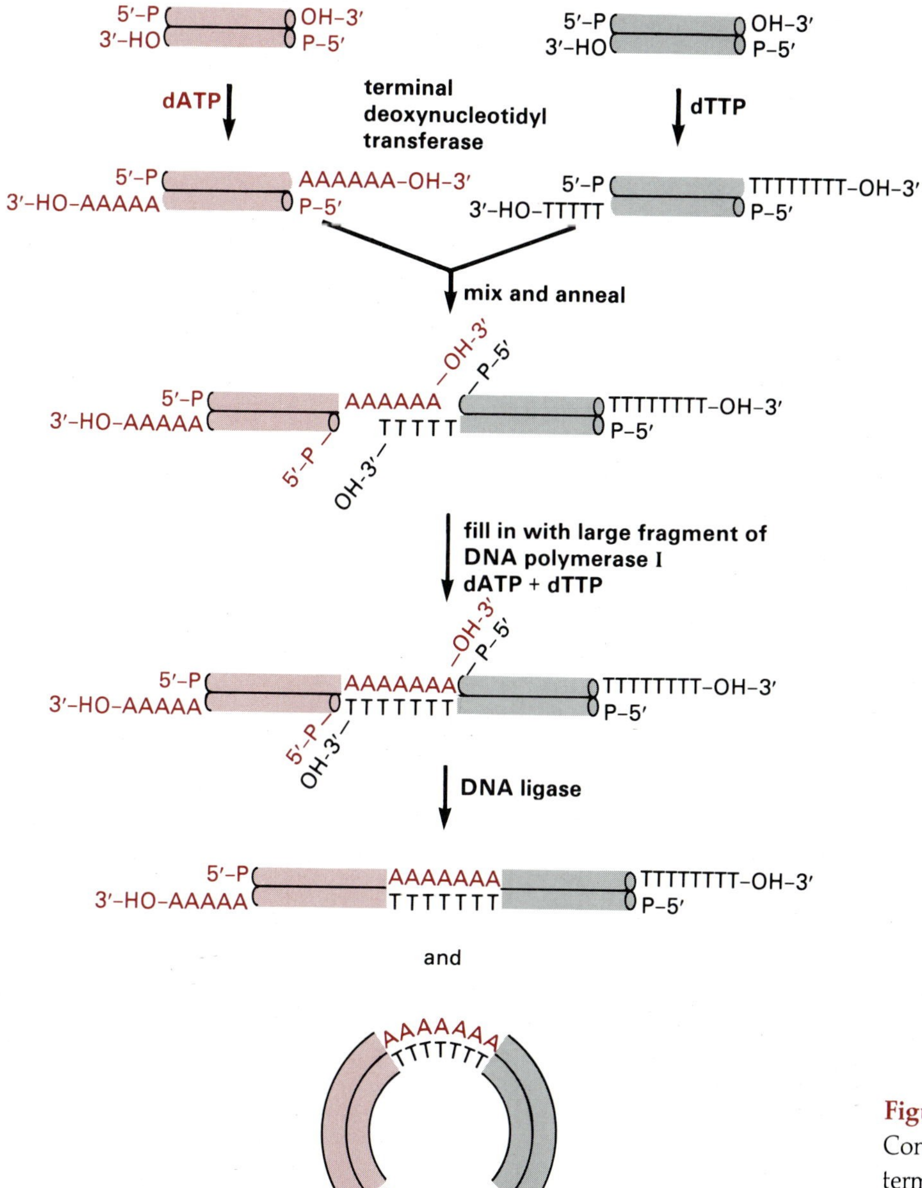

Figure 4.23
Constructing cohesive ends with terminal deoxynucleotidyl transferase. The final linear joined molecule can readily be circularized.

struct the first recombinant DNA molecules. Although the discovery that restriction endonucleases produce cohesive ends greatly simplified the *in vitro* recombination of DNA molecules, constructing cohesive ends with terminal deoxynucleotidyl transferase is sometimes still the method of choice.

4.9 Poly A Polymerase

Like terminal deoxynucleotidyl transferase, poly A polymerase adds nucleotide residues to the 3′ terminus of a chain without involving a template. However, poly A polymerase is specific for RNA. The priming strand is a polyribonucleotide, and ATP is the substrate. GTP, CTP, UTP, or dATP cannot substitute for ATP. Poly A polymerase has a highly specialized but important role in recombinant DNA experiments: it is used to add poly A tails to RNA molecules in preparation for synthesizing cDNA (Section 4.7).

CHAPTER 5

The Tools: Host-Vector Systems

Molecular cloning requires that the DNA fragment to be cloned (the insert) be joined to another DNA molecule (the vector) that can replicate in an appropriate host cell (Figure II.12). The joining is accomplished *in vitro*, and the resulting recombinant DNA molecules are then introduced into the cells. The vector molecule must contain an origin of DNA replication, and the process depends on replication functions (enzymes and other proteins) provided by the host cell or encoded on the vector itself. Any extrachromosomal small genome (like those of plasmids, phage, and viruses) is, in principle, a potential vector. Each of these genomes is found in nature within a particular species and, for the most part, replicates only within its natural host or cells from closely related species. The replication mechanism has usually evolved to optimize the extrachromosomal genome's success within the natural host and relies on the host for essential metabolites, enzymes, and other proteins as well as for the machinery for protein synthesis. Therefore, the fundamental tool of molecular cloning is always a two-component system—a compatible host-vector combination.

Recombination *in vitro* generally yields a population of DNA molecules, only some which have the desired structure. The cloning of unique constructs requires the following:

1. The conditions under which the population of recombinant DNAs is mixed with a population of recipient cells must favor the introduction of a single recombinant molecule into a recipient cell. This results in the separation of each recombinant from all the others.
2. Each recipient cell needs to be separated from all the others in the population to permit isolation of a clone of cells or viruses containing a unique recombinant.
3. Cells or viruses that receive recombinant DNAs must be distinguishable from those that do not so they can be selected for or identified by screening.
4. Cells that receive the desired recombinant must be distinguishable by screening or selection from those that contain other recombinant DNA molecules.

This chapter describes a variety of typical host-vector systems and deals with the first three requirements because the methods used for screening and selection largely depend on the particular properties of the host-vector combination being used. The techniques that allow a specific desired recombinant clone to be isolated—requirement 4—are described in Chapter 6. In general, we have selected examples that illustrate basic principles now being applied to the design of increasingly sophisticated systems for use in complex experimental situations.

The most widely used host-vector combinations involve *E. coli* strain K12 as host and *E. coli* plasmids and phage as vectors. *E. coli* K12 was an attractive choice because it had been the standard tool of microbial genetics long before the advent of recombinant DNA technology. Its genetic and physiological behavior was known in considerable detail, and many well-characterized mutants were available. Moreover, *E. coli* K12 supports the replication of a large number of bacteriophage and plasmids that are potentially useful vectors. The intimate knowledge of this bacterium and its compatible vectors assured its preeminence in recombinant DNA experiments. Other host-vector systems have also been developed, primarily as tools for specific investigations. For example, *Bacillus subtilis* systems are an important alternative where the aim is high level production and secretion of a protein encoded in a cloned gene. The extension of genetic manipulation to eukaryotic cells and in particular to studies of gene expression in yeast, plant, and animal cells also required suitable eukaryotic host-vector systems.

Shortly after recombinant DNA work began, some scientists expressed concern about the possibility of microorganisms—cells or viruses—containing "foreign" DNA inserts having unexpected and perhaps hazardous properties. Those concerns prompted a search for suitably enfeebled host-vector systems to limit the possibility of infecting laboratory workers or other living things. Certain naturally occurring or laboratory substrains of *E. coli* K12 themselves proved unlikely to survive or spread outside of very special experimental conditions not usually found in nature. Modifications of *E. coli* K12 and other strains by both classical genetic and recombinant DNA techniques enlarged the repertoire of enfeebled hosts. In some cases, the containment considerations and experimental requirements overlapped. But in others, the debilitated strains added considerably to the technical challenge of experiments. As evidence accumulated to indicate that many of the hypothetical hazards are highly improbable or nonexistent (at least for the vast majority of experiments being conducted), the need for enfeebled host-vector systems became less important.

5.1 *E. Coli* Systems—the Host Cells

a. *A Versatile Host*

Although *E. coli* K12 is uniquely different from other *E. coli* strains, it is in fact not a single entity. Rather, it is a family of related bacterial clones, all derived from a single original isolate. The present-day members of the *E. coli* K12 family differ from one another by mutations in one or many genes. Although some of the *E. coli* K12 strains were obtained by purposeful selection for specific phenotypes, most strains contain undetected allelic differences. Many of these are irrelevant to the use of a particular substrain as a host in recombinant DNA experiments, but others are important. The host cell's significant properties, which depend on the vector being used, fall into several classes: those relevant to faithful replication of the vector, to successful introduction of recombinant vectors, and to convenient selection of desired recombinants.

b. *A Hospitable Host*

The host must supply the requisite and compatible functions for vector DNA replication and not contain elements that inhibit vector replication or compromise the selection techniques. Cells that contain active restriction systems are avoided because they compromise the survival of a recombinant DNA containing susceptible cleavage sites. Cells that have the normal *E. coli dam* and *dcm* methyltransferases are avoided because they would yield replicated recombinant DNA that is resistant to useful restriction endonucleases (Section 4.2e). Host cells that are deficient in normal recombination functions help suppress undesirable alterations of the insert.

Other important host cell properties depend on whether they will be used with plasmid or phage vectors. For example, if the vector is λ bacteriophage DNA, then the host cells cannot be lysogenic for lambda. Such cells are immune to new infections (superinfections) due to the presence of the λ repressor, the protein cI, which shuts down expression of all genes required for a lytic infection (Section 3.11e). Plasmid vectors generally carry marker genes that signal their presence by imparting selectable phenotypes to the host cells; therefore, the host cells should not themselves have the relevant phenotype. For example, mutant *E. coli* cells that require a particular amino acid are useful hosts in combination with vectors carrying a functional gene for the missing biosynthetic enzyme. In a culture containing some cells with the vector and some cells without, only those *E. coli* carrying the vector will produce colonies in a medium lacking the amino acid. Similarly, cells sensitive to a particular antibiotic or toxin are useful with vectors carrying genes for resistance to the same agents. As a general matter, host cells that harbor extraneous plasmids are not used with plasmid vectors because purification of the recombinant DNA, which depends on the physical properties of plasmid DNA, would then be difficult.

c. *An Accessible Host*

The problem of introducing naked DNA molecules into viable *E. coli* cells—the process called **transfection**—was solved, empirically, early in recombinant DNA work; in the presence of $CaCl_2$, otherwise resistant cells become permeable to DNA. Efficient transfection with either plasmid or phage DNA takes only a few minutes. Although the method works well, the process remains poorly understood. However, more efficient transfer of phage genomes is achieved if they are first packaged into phage particles *in vitro*, and standard infection procedures are used (Section 5.3). The essential requirement is that the cells permit bacteriophage adsorption and replication; not all *E. coli* strains are permissive for all *E. coli* bacteriophage.

d. *Some Examples*

The special attributes of different *E. coli* K12 variants are best illustrated by considering some commonly used substrains (Table 5.1). *E. coli* K12 C600 is a standard substrain, although various clones of C600 have differing genotypes. Substrain RR1 is unable to take up galactoside sugars from the medium (the designation *lac*Y indicates the absence of a functional β-galactoside permease). A plasmid vector that supplies the β-galactoside permease gene confers on recipient cells the ability to take up β-galactosides, which are then hydrolyzed by β-galactosidase. This is a handy marker because successfully transfected cells, which produce blue colonies on agar containing 5-bromo-4-chloro-3-indolyl-β-D-galactoside (called Xgal), can be detected. Xgal is hydrolyzed by β-galactosidase to yield a blue

Table 5.1 Some Commonly Used *E. coli* K12 Host Cells

Substrain	Vector	Useful Genetic Properties*
C600	plasmids, λ phage	
RR1	plasmids	*lac*Y, *hsd*R, *end*A
HB101	plasmids	*lac*Y, *hsd*R, *end*A, *rec*A
χ1776	plasmids	*dap*
71-18	M13 phage	Δ[*lac-pro*]; F′*lac* (Δ M15)
DH1**	plasmids	*hsd*R, *end*A, *rec*A

* The *E. coli* genes listed here are described in the text.
** This strain gives especially high efficiency of transfection by plasmid vectors.

chromophore. Strain RR1 carries the mutation *hsd*R, which inactivates the endogenous *E. coli* K12 restriction endonuclease, thereby minimizing degradation of the incoming recombinant plasmid DNA. Further protection from degradation is achieved by the absence of the major *E. coli* K12 deoxyriboendonuclease (*end*A). Substrain HB101, a close relative of RR1, has the added advantage of lacking the general *E. coli* K12 recombination function (*rec*A).

Substrain χ1776 was developed especially for recombinant DNA work in response to the initial concerns over possible hazards. Standard genetic techniques were used to incorporate mutations (*dap*) that severely restrict the cell's ability to synthesize a cell wall except in the presence of an exogenous supply of an essential cell wall constituent, the rare amino acid diaminopimelic acid. This and other mutations affecting proteins essential for cell wall construction mean that χ1776 is very sensitive to some antibiotics and to lysis by bile salts in animal intestines and by ionic detergents. This same sensitivity necessitates special procedures for successful transfection.

Convenient vectors have been constructed from the phage M13 (Section 5.3), a member of the group called Ff (for F-specific filamentous) phage. These phage infect only *E. coli* cells carrying the sex factor plasmid called F (Perspective II). Therefore, host cells for M13 must be F^+, as is substrain 71-18. The *lac* operon of strain 71-18 is deleted (Δ [*lac-pro*]) from the *E. coli* genome but is present on the F-plasmid (F′*lac*). To facilitate the detection of successful infection, the F′*lac* plasmid contains a significant deletion (Δ M15) corresponding to the amino terminal end of β-galactosidase. Blue plaques, signifying an active β-galactosidase, are found on agar containing Xgal if the M13 vector contains the region (*lacZ*α) of the *lac* operon that is missing from F′*lac*. The two genomes (F′*lac* [Δ M15] and M13 *lacZ*α) each produce a different part of the β-galactosidase polypeptide, and the two interact with each other within the cell to yield active β-galactosidase.

5.2 *E. Coli* Systems—Plasmid Vectors

a. *The Modular Structures of Plasmids*

There are many different ways to classify plasmids, but it is perhaps most useful here to consider them as being constructed from modular DNA

Table 5.2 Modular Organization of Plasmids

Modules	Plasmids (approximate size in kbp) F (93)	R100 (100)	ColV (140)	ColE1 (6)	Ap201 (14)
Replication stringent (1–2 copies)	+	+	+	−	−
Replication relaxed (10–30 copies)	−	−	−	+	+
Conjugation	+	+	+	−	−
Tetracycline resistance	−	+	−	−	−
Colicin	−	−	+	+	−
Ampicillin resistance	−	−	−	−	+
Chloramphenicol resistance	−	+	−	−	−

segments. Table 5.2 summarizes the modular features of several different naturally occurring plasmids. Each modular segment can contain one or more genes or *cis-acting* elements such as a replication origin. Functionally related modules in different and independently isolated plasmids are frequently, but not always, related structurally, as if they had a common ancestral DNA. Every plasmid must have at least one of several different kinds of replication modules that permit autonomous replication. In one class, typified by F-plasmids, replication and segregation are under the same control as replication of the bacterial genome, and only one or two copies of the plasmid occur in each cell. This is referred to as **stringent replication**. A second type of replication module is free from such control and allows for many copies per cell—**relaxed replication**.

Other DNA modules, unrelated to replication, may or may not be present in any particular plasmid. Sex factors, that is, **conjugative** (or **self-transmissible**) plasmids like F-factor, have modules containing the genes and regulatory regions required for transfer of the plasmid from one cell to another (Perspective II). Sex factors that carry a DNA module derived from the bacterial chromosome are designated "prime," as in F′.

Another type of module contains genes whose protein products inactivate antibiotics. Plasmids carrying such modules are frequently termed **R-plasmids** (for resistance) (Table 5.3). A single plasmid often provides resistance to several antibiotics, and some or all of the genes may be grouped in one module. An R-plasmid may or may not function as a sex factor, depending on whether it includes a conjugation module. **Nonconjugative** plasmids (those lacking a conjugation module) are not self-transmissible, but some of them are transferable from one cell to another if they cohabit a donor cell along with a sex factor. The sex factor mediates transfer of the nonconjugative plasmid, which is then said to be **mobilizable**.

Another class of modules found on both conjugative and nonconjugative plasmids is referred to as *Col*. *Col* modules encode a gene for one of several proteins called colicins (antibacterial agents produced by bacteria). The various colicins are distinctive in their structure and mode of action. Plasmids that encode a particular colicin often also contain genes that

Table 5.3 Antibiotic Resistance Genes

Antibiotic (gene)	Encoded Proteins	Mode of Action Leading to Resistance
Ampicillin (*amp*)	β-lactamase (penicillinase)	Hydrolysis of C—N bond in β lactam ring
Chloramphenicol (*cam*)	Chloramphenicol transacetylase	Acetylation of CAM with acetyl-CoA to form O-acetyl-CAM
Tetracycline (*tet*)	Not well characterized	Involves decreased ability of cells to concentrate tetracycline
Streptomycin (*str*)	Streptomycin phosphotransferase (1) Streptomycin adenylate synthetase (2)	Phosphorylation (1) or adenylylation (2) of OH on streptomycin by ATP
Kanamycin (*kan*) Neomycin (*neo*)	Kanamycin acetyltransferase (1) Aminoglycoside phosphotransferase (2)	N-acetylation (1) or O-phosphorylation (2) of these related antibiotics

provide for immunity to that colicin, thereby assuring that the producer cell is not hoist on its own petard.

Some plasmids contain modules that encode restriction-modification systems; the *Eco*RI system is one example.

A typical plasmid The modular construction of a typical R-plasmid is shown schematically in Figure 5.1. Of approximately 10^5 base pairs, about 50 percent are homologous to portions of an F-plasmid in structure and function. This region includes genes required for conjugation (*tra*) and the stringent replication mode. An extra module that includes a set of genes specifying resistance to the antibiotic tetracycline is contained within the F-like region. A second large portion of the R-plasmid is unrelated to F and includes modules encoding resistance to streptomycin, sulfonamides, chloramphenicol, and kanamycin.

Modules as movable elements Plasmids isolated independently, even in different parts of the world, often contain closely related modules. For example, many plasmids carry the same antibiotic resistance genes. From these findings, the notion emerged that some genetic modules are exchanged between genomes as intact DNA segments. Now it is clear that a great deal of rearrangement and exchange occurs among bacterial plasmids. For example, when R-plasmids are grown in certain bacterial species other than *E. coli,* notably *Salmonella typhimurium* or *Proteus mirabilis,* they may dissociate into two separate plasmids, each containing one or the other of the two distinctive portions (Figure 5.1). The F-like sequences form an independently replicating conjugative plasmid called RTF while the remaining region forms another independently replicating but nonconjugative plasmid called simply r. Because both RTF and r are independent replicons, each must include a DNA module specifying replication functions. These rearrangements are not simply laboratory curiosities; RTF- and r-plasmids are found independently in nature. The

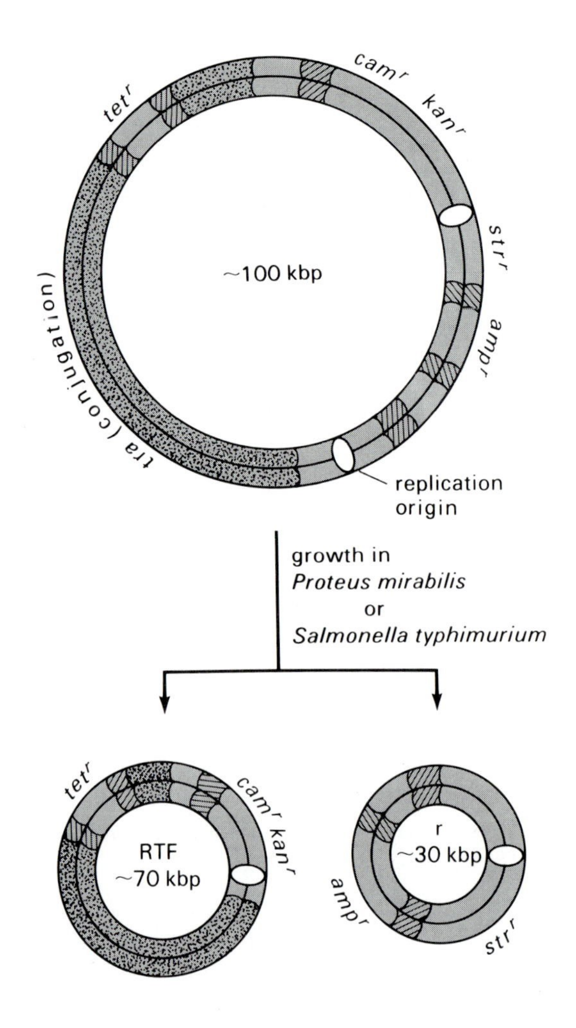

Figure 5.1
Schematic diagram of the genome of a typical conjugative R-plasmid and its dissociation into two independent replicons: the RTF- and r-plasmids. The drawing is not to scale. Antibiotic resistance genes are defined in Table 5.3. The ovals are origins of replication, the stippled regions are required for conjugation, and the slashed regions are movable elements.

rapid dispersal of antibiotic resistance genes in nature is explained by the exchange of modules among plasmids.

These and other observations made during the study of bacterial genetics converged in the identification of discrete movable elements called **insertion sequences (IS)** and **transposons** (Chapter 10), which can shuttle not only among plasmids but between plasmids and cellular genomes and within a bacterial genome itself. Typically, many of the modules in plasmids are or are flanked by movable elements (Figure 5.1).

Plasmids as vectors Many different plasmids are used as vectors for molecular cloning in *E. coli*. Plasmids originally found in nature have since been modified, shortened, reconstructed, and recombined in both cells and test tubes to enhance their utility either for general purposes or to suit particular experimental designs. The best ones are designed to include genetic markers that permit the easy selection of recombinants.

b. Designing Vectors for Selection

The efficiency of the transfection of *E. coli* K12 with recombinant plasmid DNA is low. Some percentage of the host cells escape transfection altogether; others receive a vector with no insert either because the vector

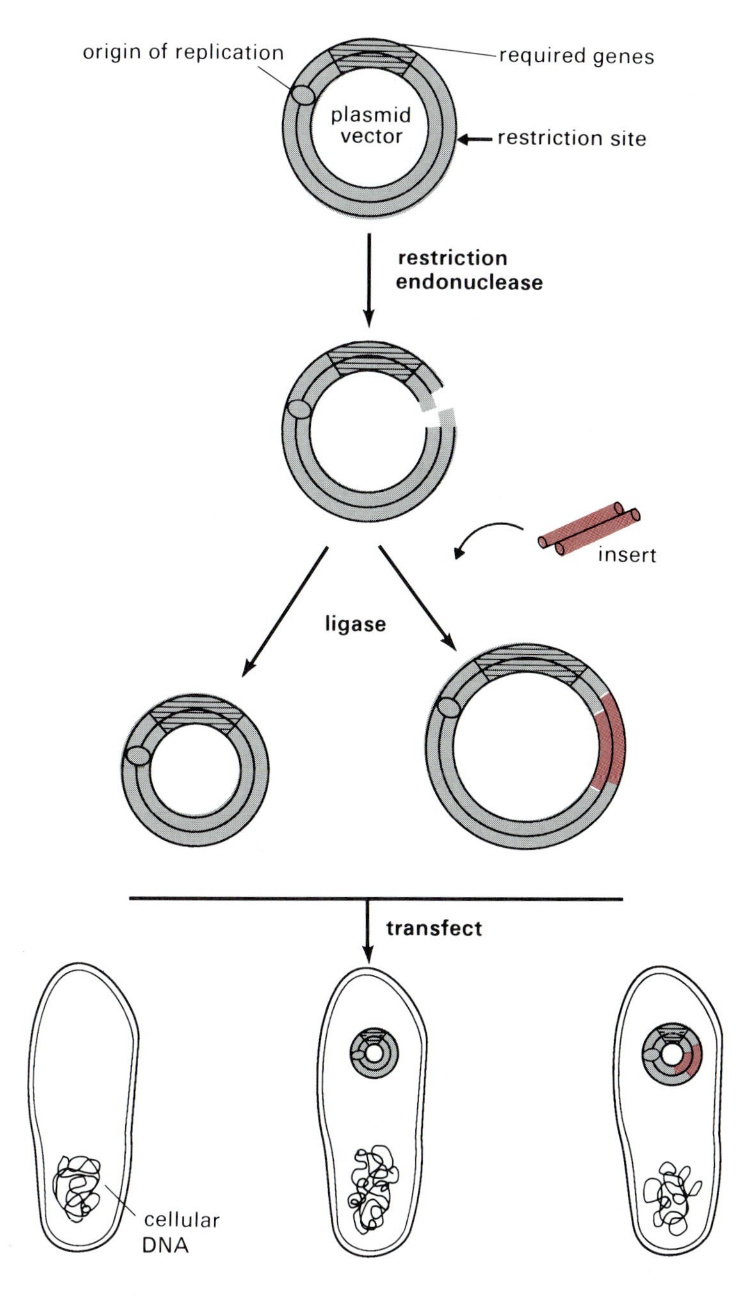

Figure 5.2
After transfection with a recombinant plasmid preparation, some cells fail to acquire a plasmid, some acquire the original vector with no insert, and some acquire the recombinant molecule.

escaped cleavage by the restriction endonuclease or because it religated without an insert (Figure 5.2). A way is needed to detect or selectively grow those cells that acquired recombinant vectors. For this purpose, genes imparting selectable phenotypic properties are incorporated into plasmid vectors. Once cells carrying recombinant vectors are obtained, those containing the insert of interest must be identified and cloned. The methods used for this final step in molecular cloning are described in Chapter 6.

A single selectable marker The earliest recombinant DNA experiments used a plasmid vector called pSC101. It contains all the information required for its replication in *E. coli* as well as genes specifying resistance to tetracycline and single recognition sites for the restriction endonucleases

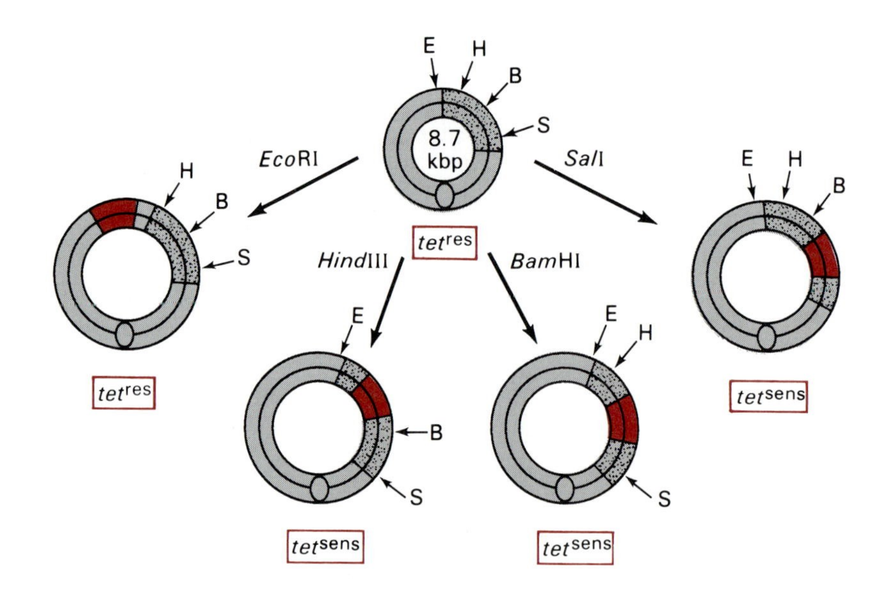

Figure 5.3
The structure of pSC101 and the phenotypic properties of *E. coli* K12 transfected with pSC101 or various recombinant derivatives of pSC101. In each derivative, the new DNA fragment was inserted at a different restriction endonuclease site. *E. coli* K12 cells that contain no plasmid are sensitive to tetracycline.

*Eco*RI, *Hin*dIII, *Bam*H1, and *Sal*I (Figure 5.3). The latter three sites are within the genetic region responsible for tetracycline resistance. When a DNA segment is inserted into the *Eco*RI site, the recombinants yield tetracycline resistant transfectants. In a growth medium containing tetracycline, the untransfected cells do not form colonies, but the transfectants do. However, so do cells that contain unaltered pSC101. A DNA fragment inserted into the *Bam*H1, *Sal*I, or *Hin*dIII site interrupts and inactivates the genes encoding tetracycline resistance, thereby resulting in the formation of tetracycline sensitive transfectants. After transfection, cells containing the recombinant are distinguishable from those transfected by pSC101 but not from the original tetracycline sensitive host cells themselves. Thus, because pSC101 contains only a single selectable marker (tetracycline resistance), it does not permit direct selection of recombinant-containing cells. Furthermore, it has other shortcomings; it contains unnecessary DNA, and, most important, its replication is regulated stringently so that only a few copies of the plasmid are produced in a host cell, with correspondingly low yields of the recombinant DNA.

Two selectable markers Having two selectable markers on a plasmid vector allows discrimination among the three kinds of cells (Figure 5.2): untransfected cells, those that acquired unaltered plasmid, and those that acquired recombinant plasmid. This principle is illustrated using resistance to the antibiotics tetracycline and chloramphenicol as examples of two selectable markers (Figure 5.4). The discrimination depends on the interruption of only one of the marker genes by the insert. In the example, the gene for tetracycline resistance is interrupted. Only cells containing a recombinant plasmid grow in the presence of chloramphenicol and fail to grow in the presence of tetracycline.

In addition to permitting the isolation of clones of bacterial cells containing recombinant plasmids, selectable markers serve another important purpose. Unless a plasmid supplies its host with an essential function, the cells usually multiply more efficiently without the burden of the extra

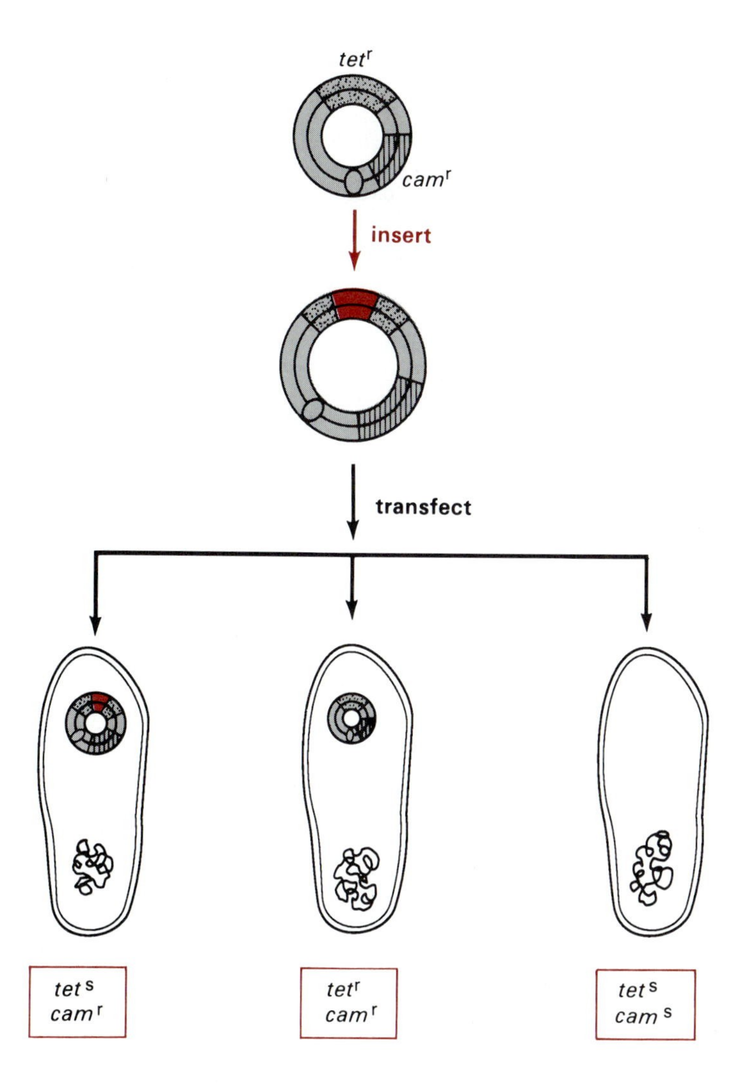

Figure 5.4
The advantages of two selectable phenotypic markers. When the insert disrupts one of the markers, all three types of transfected cells can be distinguished by their phenotypes.

DNA. Therefore, a cell that loses the plasmid rapidly outgrows the others, and the recombinant will be lost. An environment that makes the cells dependent on the plasmid avoids this problem. For example, cells that contain a recombinant vector encoding resistance to tetracycline grow in a medium that includes tetracycline, but cells that lose the vector will themselves be lost. This type of selective pressure on the host cells is almost always necessary to assure maintenance of a recombinant plasmid vector.

Properties of an ideal plasmid vector An ideal plasmid vector for molecular cloning has a minimum amount of DNA, replicates in the relaxed rather than the stringent mode to ensure a good yield of DNA, contains at least two selectable markers, and has only a single recognition site for at least one restriction endonuclease. The last specification permits cleavage of the circle at a unique site for ligation to the insert segment. For maximum convenience in selection, the unique restriction site should be within one of the two selectable marker genes. It cannot interrupt sequences that are essential for plasmid maintenance. Vectors that approach such an ideal have been constructed from naturally occurring DNA modules using both classical genetic and recombinant DNA techniques.

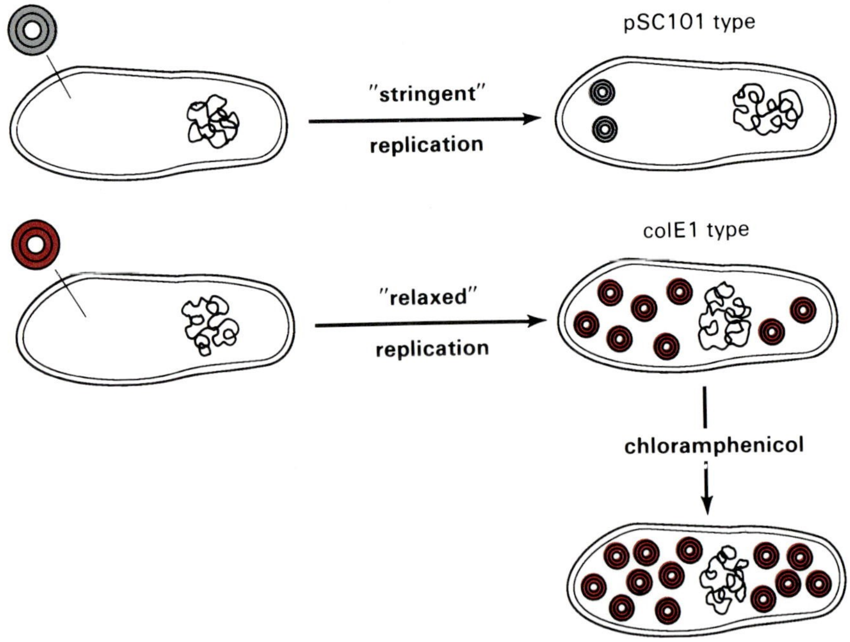

Figure 5.5
Stringent and relaxed control of plasmid replication.

Modules specifying relaxed replication occur on a variety of natural plasmids. One, the plasmid called colE1 (Table 5.2), has an added advantage in that plasmid replication continues even when protein synthesis and *E. coli* chromosomal DNA synthesis are inhibited by amino acid starvation or by the antibiotic chloramphenicol (colE1 replication is described in Chapter 2). Under the inhibitory conditions, the cells do not divide, but several thousand copies of the plasmid genome accumulate in each cell (Figure 5.5). Yields of a milligram of plasmid DNA are readily recovered from a liter of culture fluid. Although colE1 itself is not often used as a vector, the colE1 replication module has been incorporated into a large number of other plasmids, all of which retain the advantages of relaxed replication and massive accumulation of plasmid upon inhibition of protein synthesis. Some of these plasmids are listed in Table 5.4.

c. *A Plasmid Vector—pBR322*

Of the available *E. coli* vectors, pBR322 is among the most popular and most widely used (Table 5.4). It was constructed using both classical (*in vivo*) genetic techniques and recombinant DNA methodology and has many of the attributes of an ideal plasmid-cloning vector (Figure 5.6). The entire 4362 base pair long nucleotide sequence is known. pBR322 replicates by the colE1 mode, and thus high yields of the plasmid can be obtained. Also, it has two selectable markers: resistance to ampicillin and tetracycline. Recognition sites for 12 restriction endonucleases occur only once each on the molecule, making it easy to produce full length linear molecules with a variety of cohesive ends. Transfection of *E. coli* K12 cells with pBR322 can yield more than 10^8 successfully **transformed** cells per μgram of plasmid DNA, depending on the conditions and the strain of host cells used. After DNA segments are inserted into pBR322 *in vitro*, the efficiency of transfection drops one or two orders of magnitude, depending again on the strain of host cells used and the transfection conditions.

Table 5.4 Some Plasmid Vectors Used in *E. coli*

Plasmid Vector	Replication Mode	Size (bp)	Selective Markers and Some Unique Restriction Endonuclease Sites*
pSC101	stringent	8700	Tetracycline^{res}(*Hin*dIII, *Bam*H1, *Sal*I); *Eco*RI
pUC7	relaxed	2700	Ampicillin^{res} *lacZα*** (*Eco*RI, *Bam*H1, *Hin*cII, *Pst*I)
pMB9	relaxed	5500	Tetracycline^{res}(*Hin*dIII, *Bam*H1, *Sal*I) Colicin immunity†; *Eco*RI; *Hpa*I; *Sma*I
pBR322	relaxed	4362	Tetracycline^{res}(*Hin*dIII, *Bam*H1, *Sal*I); Ampicillin^{res}(*Pst*I, *Pvu*I); *Eco*RI; *Ava*I; *Pvu*II; *Cla*I
pDF41	stringent	12,800	*Trp*E gene; *Bam*H1, *Eco*RI, *Hin*dIII, *Sal*I
pRK2501	stringent	11,100	Tetracycline^{res}(*Sal*I) Kanamycin^{res}(*Hin*dIII, *Xho*I)

* Restriction endonuclease sites within genes for selectable markers are given in parentheses after the marker.

** *lacZα* contains the promoter region and amino-terminal portion of the *E. coli* gene (*lacZ*) for *β*-galactosidase (Sections 5.1d and 5.2d).

† Colicin is a toxin active against *E. coli* and encoded in some plasmids.

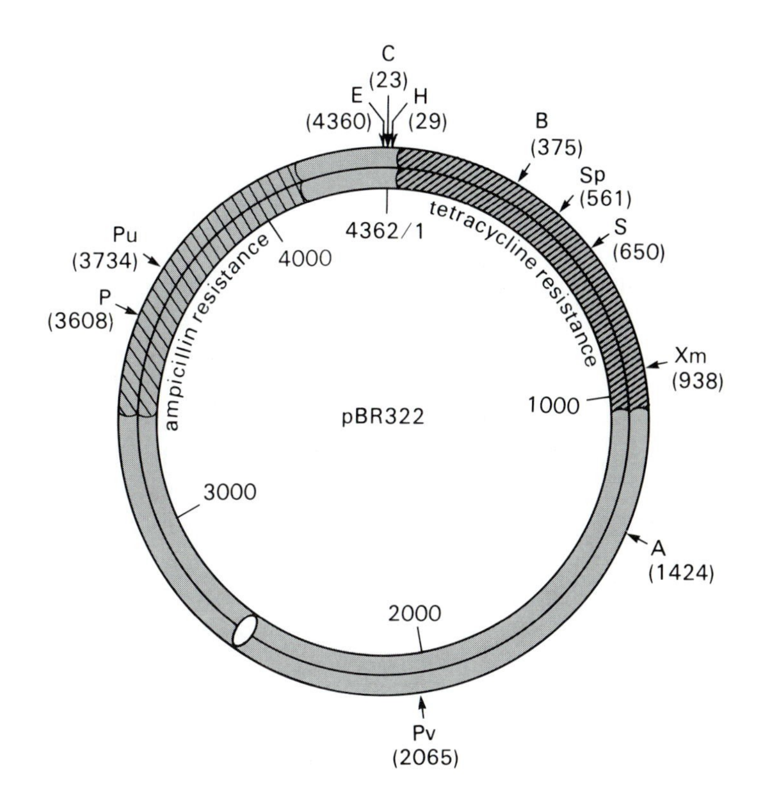

Figure 5.6
pBR322. The numbering of the base pairs is indicated, as are restriction endonuclease sites that occur only once on the molecule.

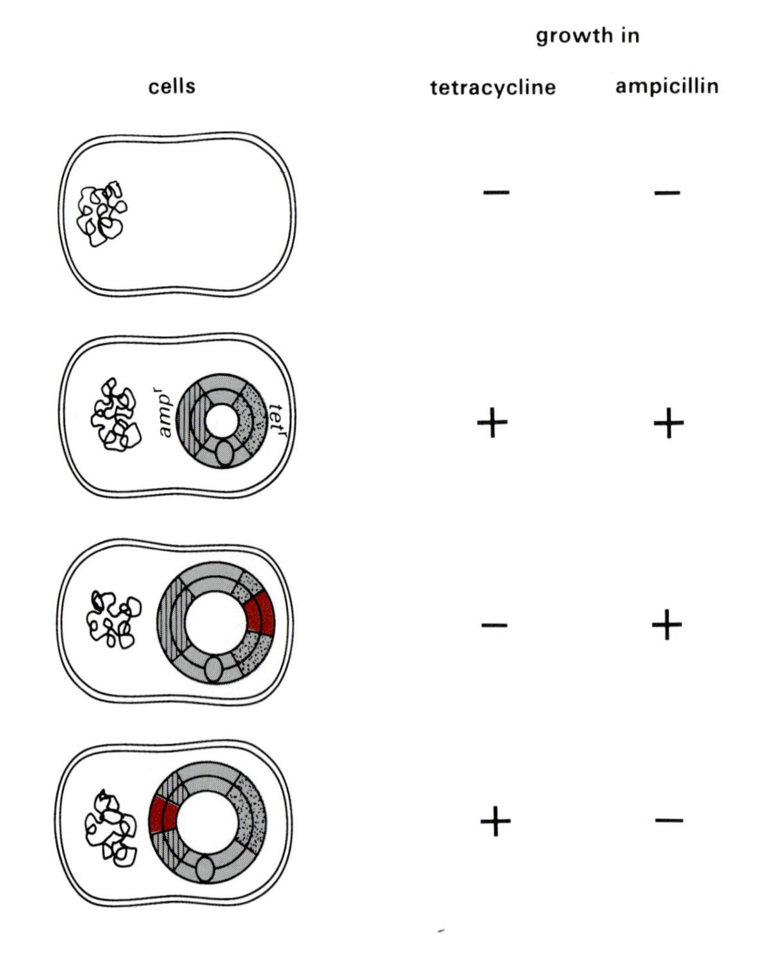

Figure 5.7
Phenotypes of cells containing pBR322 with and without inserts.

E. coli K12 cells containing pBR322 grow in media containing either ampicillin or tetracycline or both; cells lacking the plasmid do not grow in the presence of either antibiotic (Figure 5.7). When a segment of DNA is inserted into the region of pBR322 encoding resistance to tetracycline (e.g., at the *Bam*H1 or *Sal*I restriction site), transformants still grow in ampicillin, but not in tetracycline. The reverse is true when the insert is placed within the ampicillin gene (e.g., at the *Pvu*I or *Pst*I restriction site). Either type of insertion permits easy selection of only those bacterial colonies that contain the recombinant plasmid. In practice, selection is carried out by plating transfected *E. coli* K12 on an agar plate containing the antibiotic to which cells containing the plasmid are expected to be resistant. All the colonies obtained have a plasmid. Cells from these drug resistant colonies are then tested for their ability to grow in media containing the alternate antibiotic. This screening distinguishes between colonies that contain pBR322 plasmid itself (growth) and those that contain a recombinant plasmid (no growth). The whole procedure can be carried out in a day or two using the **replica-plating** technique (Figure 5.8). Even a single successfully transformed cell can be detected as a colony of cells after plating 10^8 cells. The importance of two selectable markers is evident when one compares how much easier it is to obtain clones containing recombinant plasmids using pBR322 as vector than with pSC101 (compare Figures 5.3 and 5.7).

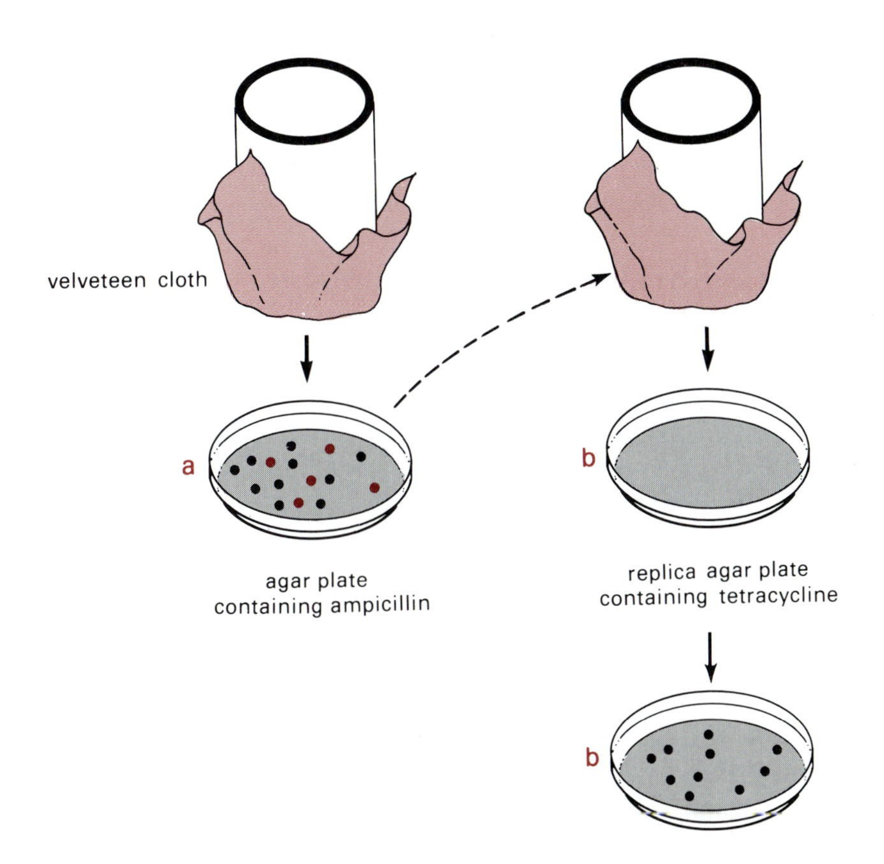

Figure 5.8
Use of replica plating to distinguish several phenotypic properties of a large number of bacterial colonies. In this example, *E. coli* K12 cells were transfected with pBR322 that had had a DNA insert fragment placed in the *Bam*H1 site, *in vitro*. All the cells that contained a plasmid formed colonies on nutrient agar medium prepared with ampicillin (plate a). A cylindrical block covered at one end with sterile velveteen was pressed onto plate a and then onto a fresh nutrient agar plate containing tetracycline but no ampicillin (plate b). In this way, a replica of the distribution of colonies on plate a is formed on plate b. However, some of the colonies (colored on plate a) fail to grow on plate b because the insert interrupted and inactivated the genes encoding tetracycline resistance.

d. Different Vectors for Different Purposes

In addition to pSC101 and pBR322, dozens of plasmid vectors have been designed and constructed for various purposes. Table 5.4 lists several of these. Although pUC7 is much smaller than pBR322, it preserves its essential parts, including the gene for ampicillin resistance. The second selectable marker is a portion of the *lac*Z gene (*lac*Zα), which, in conjunction with host cells containing the rest of *lac*Z, gives blue colonies on Xgal plates (Section 5.1d). All the unique restriction sites used to insert a DNA fragment interrupt the *lac*Zα fragment on pUC7, so cells with a recombinant vector give white colonies on Xgal. Replicating by the relaxed mode, pMB9 contains a pair of selectable markers, one of which encodes resistance to colicin E1. Cells containing pMB9 grow on a medium that contains purified colicin E1; cells without plasmid are killed. The replication and tetracycline resistance modules of pMB9 are the same as those in pBR322. A completely different type of vector, pDF41, contains a replication module derived from the conjugative F-plasmid and an *E. coli* chromosomal gene (*trp*E) for an enzyme involved in tryptophan biosynthesis. The latter allows selection of transformants when *trp*E$^-$ host cells are used and grown on a medium lacking trytophan. The low copy number of pDF41 is useful in experiments designed to study the regulation of cloned bacterial genes in *E. coli* when a high copy number might mask regulatory effects. The genome of the vector pRK2501 is derived from a plasmid called RK2, a representative of the P class of plasmids. The distinctive feature of P plasmids is their ability to be transferred by conjugation into gram-negative bacteria of many genera (e.g., *Serratia marcescens, Pseudomonas aeruginosa*). Therefore, recombinant plasmids constructed with pRK2501 can be used in studies with many bacterial species.

5.3 *E. Coli* Systems—Bacteriophage Vectors

a. *Some Differences Between Plasmid and Phage Vectors*

With phage vectors, the viable product of a recombinant DNA construction is a population of phage. Unlike the situation with plasmid vectors, in which cells are cloned and selected, the recombinant phage genome itself is cloned in the form of a plaque on a lawn of susceptible host cells (Figure II.13). Thus, foreign DNA fragments are inserted into a population of phage DNA vectors *in vitro*, and the recombinant phage genomes may then be introduced into permissive cells as naked DNA (transfection) or as reconstituted particles (infection). Either way, a cell infected with a single molecule of recombinant viral vector will produce viral progeny and finally a plaque because of the subsequent infection and lysis of neighboring cells. Formation of a viral plaque itself indicates successful infection (or transfection), but identification of recombinant phage containing an insert of interest may require additional selection conditions or screening of the plaques for particular DNA sequences (Chapter 6).

Whether one uses a plasmid or a bacteriophage vector depends on the aim of a given experiment, the size of the fragment to be inserted, and the relative abundance of the insert of interest in the available mixture of DNA fragments. In general, bacteriophage vectors are more efficient than plasmids for cloning large inserts, and screening large numbers of bacteriophage plaques is easier than screening bacterial colonies for a specific desired insert (Chapter 6). Two bacteriophage, λ and M13, are frequently employed as vectors in *E. coli*.

b. *λ Phage*

The λ genome The most frequently used bacteriophage vectors are derived from the temperate phage λ (Chapters 2 and 3). A schematic map of the genes and restriction endonuclease sites of wild type λ is shown in Figure 5.9. The entire 48,502 base pair sequence is known. The molecule contains an origin of replication and genes for viral structural proteins (head and tail) and for enzymes used in DNA replication, lysis of infected cells, and establishing lysogeny (Section 3.11e). Approximately one-third of the genome, the continuous internal segment between about 20 and 38 kbp on the map, is dispensable for successful lytic infection; it contains genes required for lysogeny. Another small region at about 43 kbp is also dispensable. However, for λ DNA to be packaged into bacteriophage particles, it must be larger than 38 and smaller than 52 kbp in length. Therefore, the essential characteristic of vectors constructed from λ phage DNA is that all or part of the central dispensable region is replaced with an insert segment of appropriate length. Generally, many, if not all, of

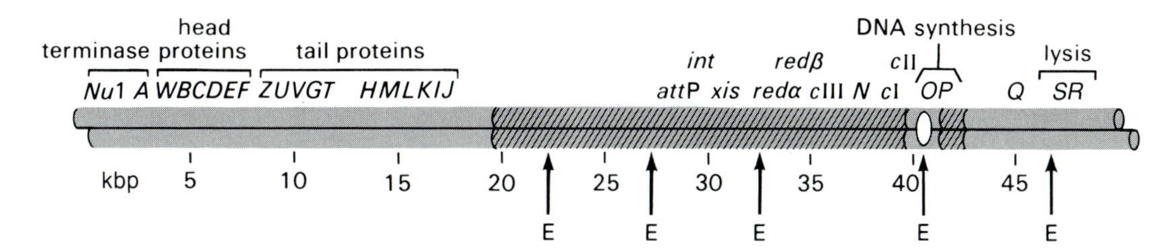

the genes required for lysogeny are removed so that infection always causes lysis of the host. Derivatives of wild type λ that lack some of the usual restriction endonuclease sites are used to simplify the construction of recombinants. The λ vectors that allow an insert segment to replace a dispensable portion of the genome without perturbing the essential regions were produced from the wild type λ genome by mutation and recombination *in vivo* as well as by recombinant DNA techniques.

Figure 5.9
The linear genome of bacteriophage λ. The dispensable regions are slashed, and the oval is the origin of replication. Individual genes are shown as single capital letters or lowercase names. The *Eco*RI (E) sites are shown.

Multiple forms of the genome During its life cycle, the λ genome exists in four different physical forms. The DNA is linear within the phage particle (Figures 5.9 and 5.10a). The ends of the linear DNA have 12 base long single strand termini that are mutually cohesive through base pairing. When the λ genome enters *E. coli* cells, the cohesive ends anneal to form a circular molecule, which is then sealed by DNA ligase (Figure 5.10b).

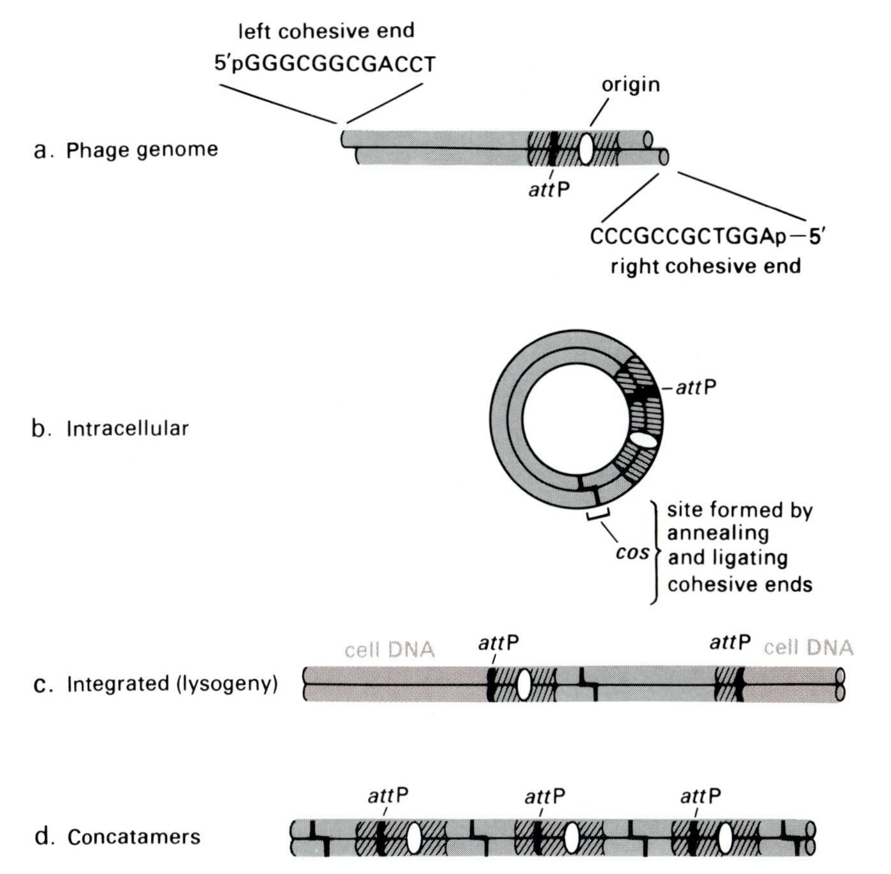

Figure 5.10
The four forms of the λ genome. The symbols are as in Figure 5.9.

The sealed cohesive ends are called cos sites. The circular form is obligatory for replication of the viral DNA and is also the precursor for lysogenic integration into the host genome (Figure I.15c and Chapter 2.4d). As a consequence of the integration, the λ genome again becomes linear (Figure 5.10c). However, because the position of the integration site on the phage DNA (*attP*) is distant from the cos site, the integrated linear prophage DNA is permuted relative to the form in the phage particle. Finally, long concatemers composed of many copies of the λ genome are produced during rolling circle replication of λ DNA (Figure 2.32) and are cleaved at the cos sites to yield mature phage DNA (Figure 5.10d).

c. λ Phage Vectors

Typical vectors A simple λ vector is shown in Figure 5.11. It is shorter than the wild type λ genome by more than 10 kbp. Both the small dispensable region around 43 kbp and a part of the central dispensable region are deleted. All the genes required for a lytic infection are intact, and the molecule is large enough to be packaged into bacteriophage particles. Only two of the five original *Eco*RI sites remain; these flank 7 kbp

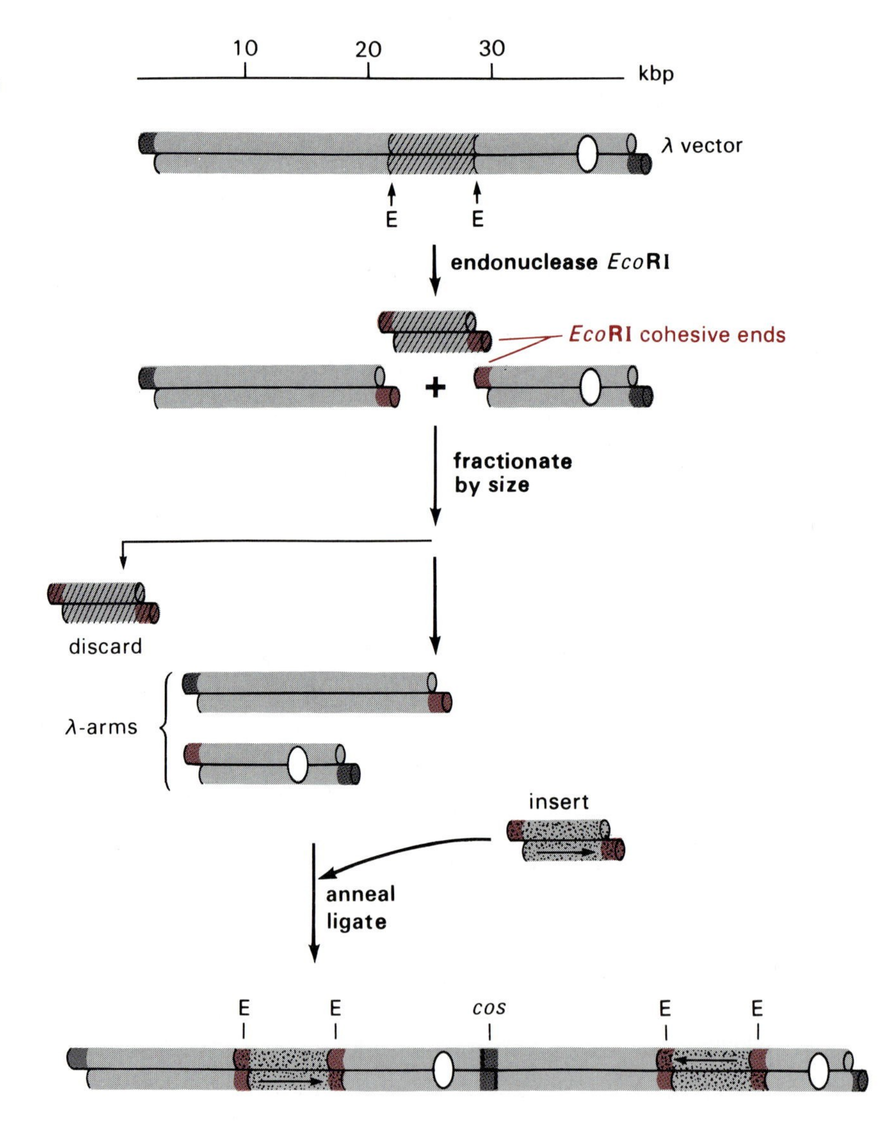

Figure 5.11
A typical λ vector. After cleavage with *Eco*RI, the dispensable center region is separated from the two arms by taking advantage of its smaller size. The arms are annealed and ligated with an insert fragment having *Eco*RI cohesive ends. The concatemer that is formed links λ arms and the insert through the *Eco*RI ends and the cos sites.

of dispensable DNA. Cleavage with *Eco*RI yields three DNA fragments, one corresponding to the left arm, another to the right arm, and the third to the dispensable center region. These three can be separated because of their different sizes. To form a recombinant, the two purified arm fragments are mixed with insert fragments, annealed, and then ligated together. The insert DNA must have *Eco*RI cohesive ends and be between about 5 and 17 kbp in length to produce a packageable molecule. Note that during annealing and ligation, two different kinds of joining occur. The insert DNA is joined to the two arms in either of two orientations through cohesive *Eco*RI sites, and the arm fragments are linked to each other through their cohesive λ ends. The product is a **concatemer**, the proper precursor for packaging viral DNA into mature phage particles.

A large array of λ vectors has been constructed for different purposes. They all share two basic characteristics: (1) The vectors themselves can be propagated as phage in a lytic infection so that vector DNA can be prepared. (2) Carefully positioned restriction endonuclease sites permit cleavage or removal of the central dispensable region and also provide sites for ligation of an insert. Many λ phage vectors use restriction endonuclease sites other than *Eco*RI to provide the essential left and right arm fragments, thus allowing for ligation to a variety of insert fragments. Also, various λ vectors accommodate different sizes of insert segments. The maximum insert size is somewhat over 24 kbp in those vectors from which the entire dispensable region is deleted.

Selecting recombinant phage As with plasmid vectors, the desired recombinant phage must be selected from among a variety of extraneous products. Plaque formation upon infection of susceptible cells provides the first selection because some possible products of ligation are not infectious. For example, with the vector shown in Figure 5.11, the side products of the joining reaction include concatemers containing only the two λ arms. However, the distance between the cos sites in these molecules is less than 38 kbp, and they are thus not packageable into phage particles. Additional selection methods are required if the λ vectors have longer arm fragments or if the dispensable λ segment is not completely removed prior to annealing and ligation because either of these situations allows phage without inserts to form plaques. To distinguish recombinant from nonrecombinant phage, some λ vectors include a DNA segment that contains the promotor, operator, and β-galactosidase gene (*lac*Z) from the *E. coli lac* operon (Section 3.11c). Most of the *lac*Z gene is discarded when the dispensable central region is removed by restriction endonuclease cleavage. Consequently, reconstituted vector DNA, which can form β-galactosidase, produces blue plaques on agar plates containing Xgal, whereas recombinant vector DNA, which does not produce β-galactosidase, generates colorless plaques on the same culture medium. Phage containing an insert are thus distinguished by their color on indicator plates designed to demonstrate the presence or absence of β-galactosidase (Section 5.1d).

d. Packaging λ Vectors into Phage Particles

In vitro packaging A great advantage of using λ DNA vectors for molecular cloning is the high efficiency with which the recombinant genomes can be transferred into *E. coli* K12 cells. In theory, wild type λ DNA can

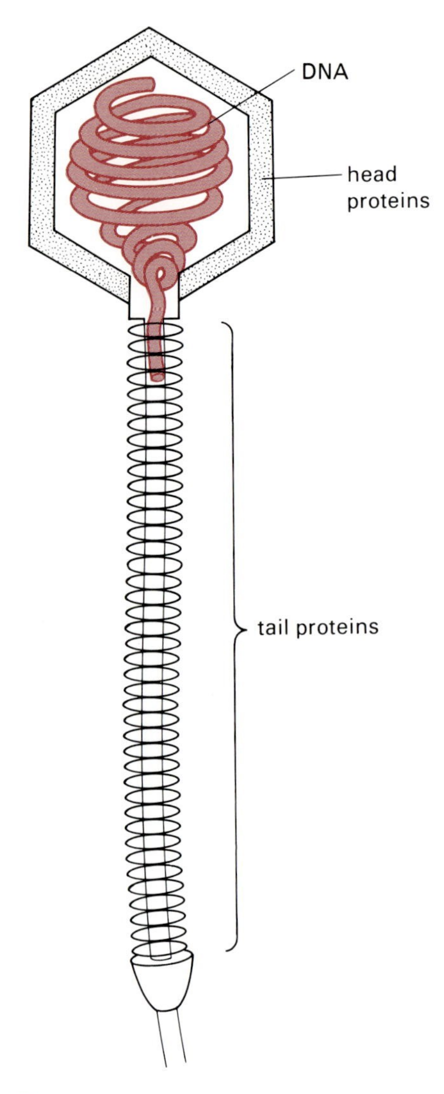

Figure 5.12
The structure of λ bacteriophage.

give about 2×10^{10} plaques per μg of DNA because the molecular weight of the genome is about 3×10^7, and each molecule yields a plaque. Cells transfected with wild type λ DNA by the $CaCl_2$ technique (Section 5.1c) produce only between 10^4 and 10^6 plaques per μg of DNA. The efficiency is several orders of magnitude less with ligated recombinant DNAs. These low yields would necessitate the use of relatively large amounts of insert and vector DNA in the ligation mixtures, and a large number of cells would have to be transfected in order to find a DNA segment of interest. However, recombinant λ vectors can be packaged into infectious bacteriophage particles *in vitro*, and infection with these gives as many as 10^8 plaques per μg of wild type λ DNA or 10^6 per μg of a recombinant λ DNA.

Constructing phage heads and tails The DNA of a λ phage particle is wound up within a hexagonal **head** made up of several proteins (Figure 5.12). Attached to the head is a **tail** composed of another set of proteins. Both head and tail proteins are encoded by the λ genome (Figure 5.9) and are produced in cells infected with λ. During a lytic infection, head proteins combine to form empty proheads (Figure 5.13). These then interact with a complex formed of concatemeric λ DNA and another λ protein, **terminase**; terminase specifically binds close to the cos sites. The DNA then enters the head, left end first, and a full length linear λ genome is generated by terminase-catalyzed endonucleolytic cleavage at the cos sites. This reaction produces the cohesive ends typical of mature linear λ DNA. Finally, preformed tails are attached, completing the phage particle.

Wild type or recombinant λ DNA is packaged *in vitro* by mixing concatemeric DNA, under appropriate conditions, with preformed empty proheads, tails, and terminase. Because, during a wild type λ infection, all these proteins are promptly consumed by the formation of new λ particles, special conditions are needed to obtain abundant quantities of heads, tails, and terminase free of DNA. This is achieved by growing separate cultures of *E. coli* infected with two different λ phage, each of which contains a mutation in a different gene (Figure 5.14). One λ mutant is unable to synthesize protein A (one of the polypeptides in terminase); the other is unable to synthesize head protein E. Both cultures contain abundant amounts of the other head proteins as well as tail proteins, but neither can produce complete phage particles. When cell-free and DNA-free extracts of the two kinds of infected cells are mixed, intact empty heads form. If wild type or recombinant λ DNA is added, it is packaged, and tails are attached, just as occurs within cells infected with wild type λ bacteriophage (Figure 5.13).

e. Phage M13

M13 is one of the filamentous bacteriophage of *E. coli*. Its genome is a single strand circular DNA molecule about 6400 nucleotides long and is packaged in a tubelike capsid constructed of at least three different phage-encoded proteins. M13 only infects *E. coli* if the cells carry a hairlike structure called an "**F-pilus**" extending from their membranes; the F-pilus is required for attachment and adsorption of the phage to the cell (Figure 5.15). Because the genes for construction of an F-pilus are encoded by

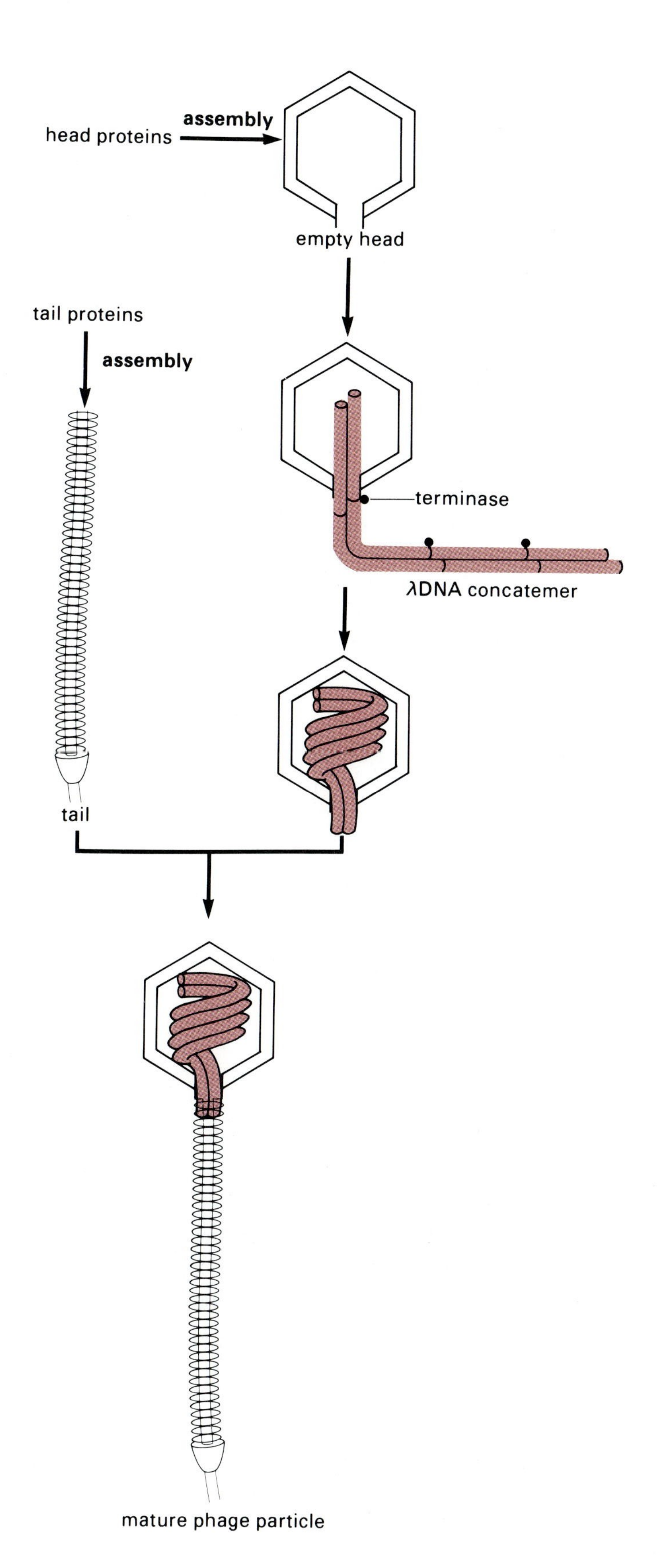

Figure 5.13
The assembly of λ DNA into phage particles.

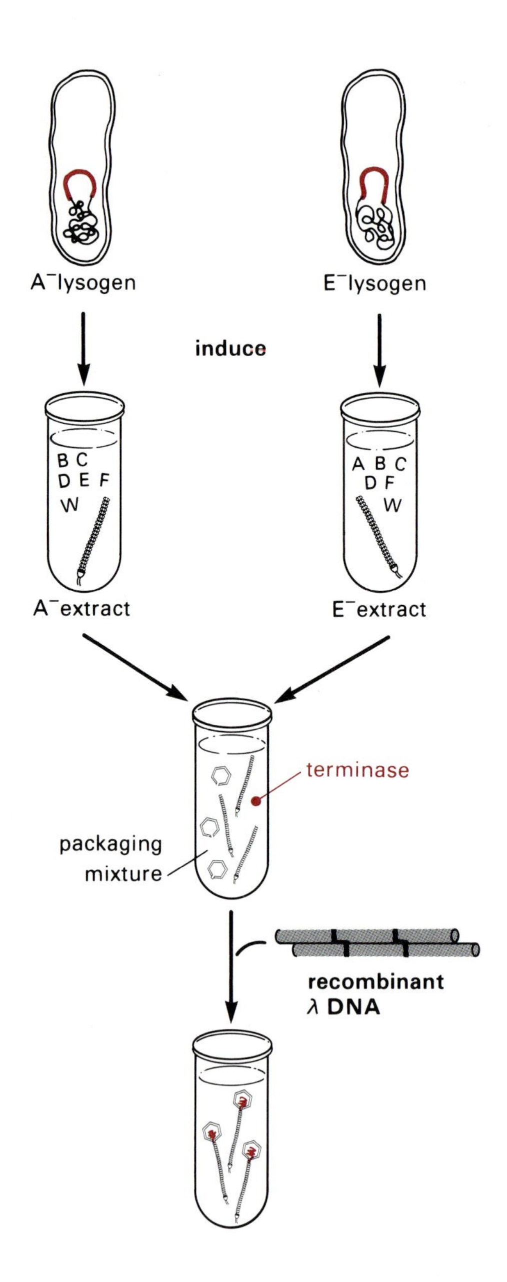

Figure 5.14
Preparing packaging extracts for the *in vitro* packaging of λ DNA. Two strains of *E. coli*, each lysogenic for a different mutant strain of λ phage, are maintained. The two mutations are in genes encoding head proteins A and E, both of which are required to package λ DNA. Separate extracts are made after inducing cultures of the two bacteria to form the bacteriophage proteins, and these are combined to form a DNA packaging mixture. Concatemeric λ DNA is added and interacts with terminase before being cleaved at cos sites and inserted into phage heads.

the sex factor F, only F^+ cells are susceptible to M13. The first step in replicating M13 in *E. coli* is converting the single strand DNA to a double strand circular form of the genome (Figure 2.3, Section 2.1g). This is accomplished by synthesis of a strand complementary to that of the infecting DNA. The duplex DNA directs the synthesis of about 300 copies of itself, which then serve as templates to produce single strand phage DNA for packaging. Concomitantly, the phage genes are transcribed and translated, and the major capsid protein is deposited in the inner cell membrane. Mature single strand phage DNA molecules pass through the membrane, picking up the capsid protein as they go, and large numbers of mature phage particles are extruded. Although the infected cells are not killed, their growth is retarded sufficiently to yield turbid plaques in a lawn of susceptible bacteria. The double strand replicative intermediate is readily isolated from infected cells for use as a vector, and the single strand genome is abundantly available in the form of phage in the growth

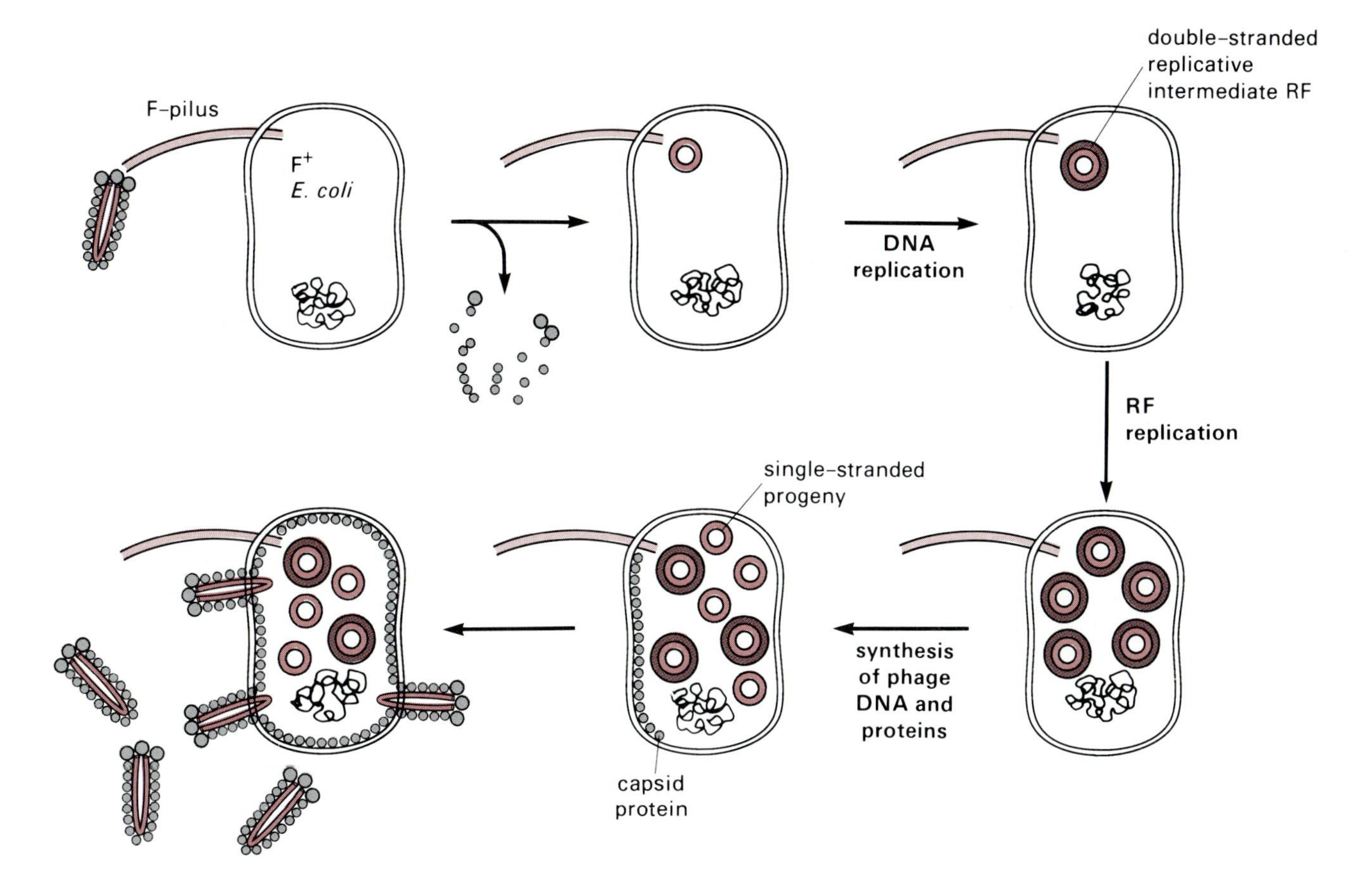

medium. Each cell in each generation produces several hundred phage. The availability of single strand recombinants is especially convenient for nucleotide sequence analysis (Chapter 7).

Figure 5.15
The life cycle of bacteriophage M13. Note that the infected cell and its progeny continually extrude new phage particles.

f. M13 Vectors

As with λ, the wild type M13 genome (Figure 5.16) has been modified to construct convenient vectors. The actual molecules used for the construction of recombinants are always the double strand replicative intermediate, as previously described. Single strand DNAs are not used as vectors because they are not generally cleaved by type II restriction endonucleases. Most of the approximately 6400 nucleotide residues in the DNA encode genes required for coat proteins, viral replication, and viral assembly. The origin of replication is a region of about 150 nucleotides contained within a noncoding segment of 507 nucleotides. An insert can be placed within this segment so that it does not interfere with M13 replication.

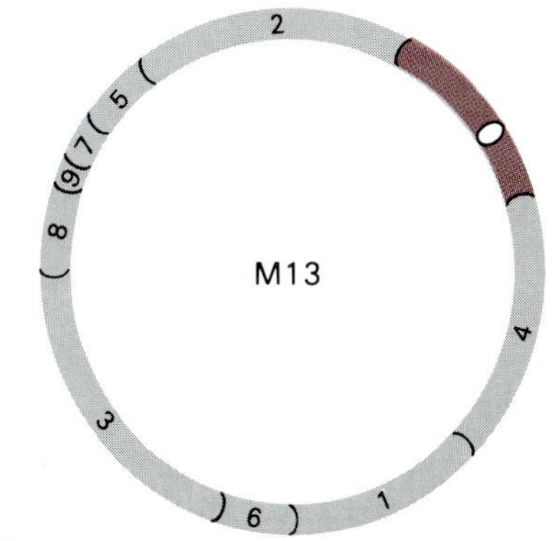

Figure 5.16
A map of the M13 genome. The numbers refer to the genes (genes 2 and 5 encode replication functions, and the remainder are involved in capsid formation and assembly), the origin is marked by an oval, and the noncoding region is colored.

A typical vector A portion of an M13 vector is shown in Figure 5.17. The wild type phage genome was adapted for convenient selection of recombinants by inserting part of the *E. coli lac* operon, including the operator, promotor, and the 5′ coding region for β-galactosidase, into a nonessential site in the 507 bp noncoding region. The short portion of the β-galactosidase gene has itself been interrupted by a 42 bp segment that contains several unique restriction endonuclease sites, any one of which

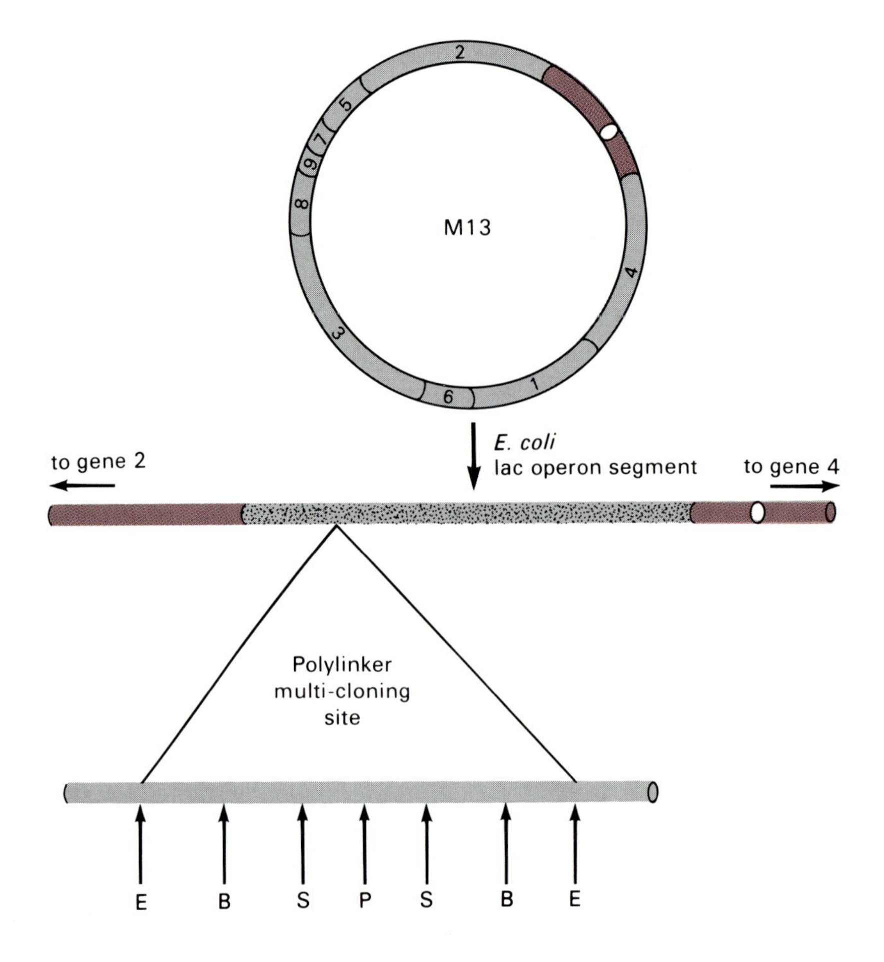

Figure 5.17
A typical M13 vector. A part of the *E. coli lac* operon was inserted into the noncoding region. Thereafter, a 42 bp segment containing multiple restriction endonuclease sites (a polylinker or multicloning site) was inserted into the *lacZ* gene.

can be used to insert DNA segments with compatible cohesive ends. Such a segment is called a **polylinker**. Neither the coding frame nor the activity of the NH_2-terminal portion of the β-galactosidase is disrupted by the additional 42 base pair segment. Lacking an insert, the vector genome directs the synthesis of the NH_2-terminal portion of β-galactosidase. With the carboxy-terminal portion of β-galactosidase produced by special *E. coli* host strains (71-18 in Table 5.1), it forms active β-galactosidase (Section 5.2d). Consequently, plaques formed by the vector itself are blue if grown on agar containing Xgal. However, when a DNA fragment is inserted into any one of the restriction endonuclease sites in the polylinker, formation of the NH_2-terminal portion of β-galactosidase is prevented, and white plaques are produced. Recombinant virus are thus easily detected among a large number of plaques produced by the vector lacking a DNA insert. Different devices for convenient detection of recombinants and different unique restriction endonuclease sites have been incorporated into other derivatives of M13. Some of these make selection dependent on antibiotic resistance; others provide genes that circumvent a nutritional requirement of an appropriate host.

Transfection As with plasmid vectors, the $CaCl_2$ technique (Section 5.1c) allows the initial transfection of M13 recombinant molecules into susceptible F^+ cells. When the desired recombinant plaque is identified and purified, it yields growing and dividing cells that continually extrude recombinant phage. Although the constructed recombinant DNA is always a duplex, the phage DNA is always a single strand.

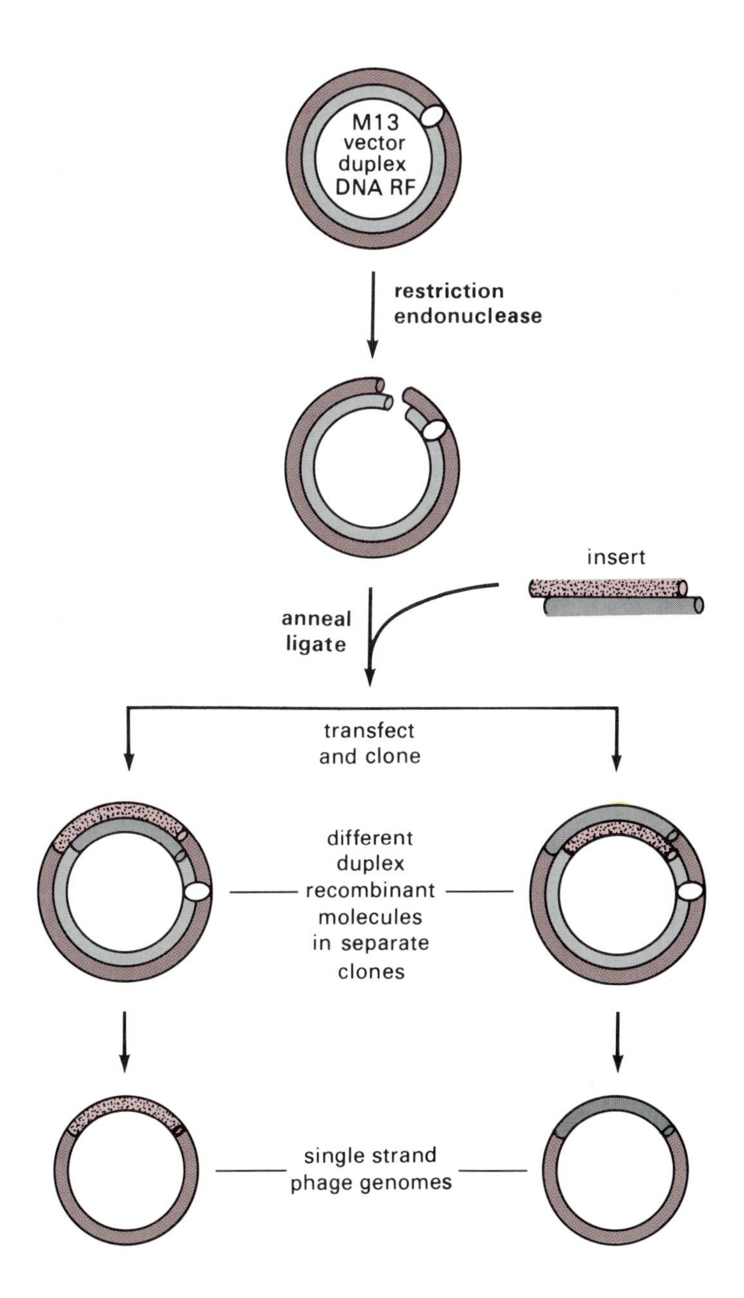

Figure 5.18
M13 vectors provide the separate strands of the insert DNA in different clones. The strand of the duplex vector that is identical to the phage genome is the colored one. (The replication process is described in Section 2.1g.)

Single strand recombinant DNAs The M13 host-vector systems have several desirable properties in common with both plasmids (e.g., cells remain viable) and λ phage (the recombinant DNA is obtained within stable bacteriophage particles). However, they also have several unique features. First, very large inserts can be cloned because packaging is independent of the size of the phage genome. Second, the double strand recombinant DNA molecules constructed with the replicative intermediate are converted to abundant single strand DNA molecules in the phage particles. Furthermore, the system provides each of the two single DNA strands of the insert fragment in different recombinant clones. M13 vector molecules accept inserts in either of two possible directions (Figure 5.18), as do plasmid and λ phage vectors. Upon transfection, each kind of recombinant will enter different cells and form separate plaques. However, because only one strand of the M13 duplex replicating form is the tem-

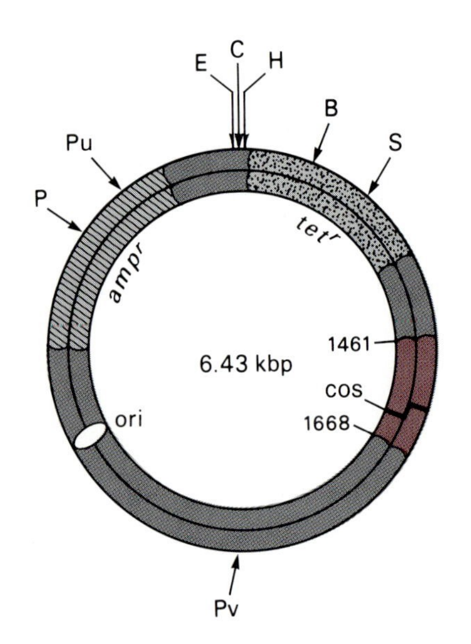

Figure 5.19
A typical cosmid. The pBR322 regions are grey. Residues 1461 through 1688 of pBR322 are missing (see Figure 5.6), and a segment of the λ genome (colored) is inserted in their place. The cos site is shown.

plate for synthesizing the single strand phage DNA, only one of the two strands of the replicative intermediate is present in all the phage in a particular plaque (Figure 2.3). Thus, cloning any insert always yields two different kinds of phage, each of which contains one of the strands of the insert. This property is very convenient for rigorous nucleotide sequence analysis and for the synthesis of specific, radiolabeled DNA probes (Chapter 7).

5.4 *E. Coli* Systems—Plasmid-Phage Combination Vectors

Because plasmid and phage vectors have distinctive advantages for molecular cloning, DNA molecules that combine selected features of the two have been developed.

a. Cosmids

Cosmids are one type of hybrid vector that replicates like a plasmid but can be packaged *in vitro* into λ phage coats. A typical cosmid (Figure 5.19) has replication functions, unique restriction endonuclease sites, and selective markers contributed by plasmid DNA, combined with a λ DNA segment that includes the joined cohesive ends (cos sites). Cosmids are constructed by recombinant DNA techniques. Appropriate restriction endonuclease digestion of λ DNA concatemers (Figure 5.10) produces the cos segments, which are inserted into a standard plasmid vector. As little as 250 bp of λ DNA is sufficient to provide the cos junction, including the sequences required for binding to and cleavage by terminase. An important feature of most cosmid cloning vectors is that they accommodate DNA inserts as large as 45 kbp.

Circular cosmid DNA is opened at a unique restriction endonuclease site, mixed with an insert DNA with matching cohesive ends, and annealed. Among the products are long concatemers (Figure 5.20). When the concatemers are mixed with a λ packaging mixture (Section 5.3d), the cos sites are cleaved, and the DNA is packaged in the usual way. This procedure selects for long inserts because the distance from one cos site to another must be between about 38 and 52 kbp to be packaged in λ heads. As with λ vectors, the mixture of concatemers is complex and can contain vector fragments with no or multiple consecutive inserts.

Recipient host cells acquire the packaged cosmids by infection with the "fake" phage particles—a much more efficient process than transfection with plasmid DNA. Between 1×10^4 and 5×10^5 colonies of transformed cells are obtained per μg of insert DNA used. Once inside the host cell, the recombinant DNA is amplified and maintained as a plasmid (Figure 5.20).

b. Phasmids

Cosmids are basically plasmid vectors that borrow the λ phage cos site to enable efficient packaging *in vitro*. **Phasmid** vectors, however, are truly

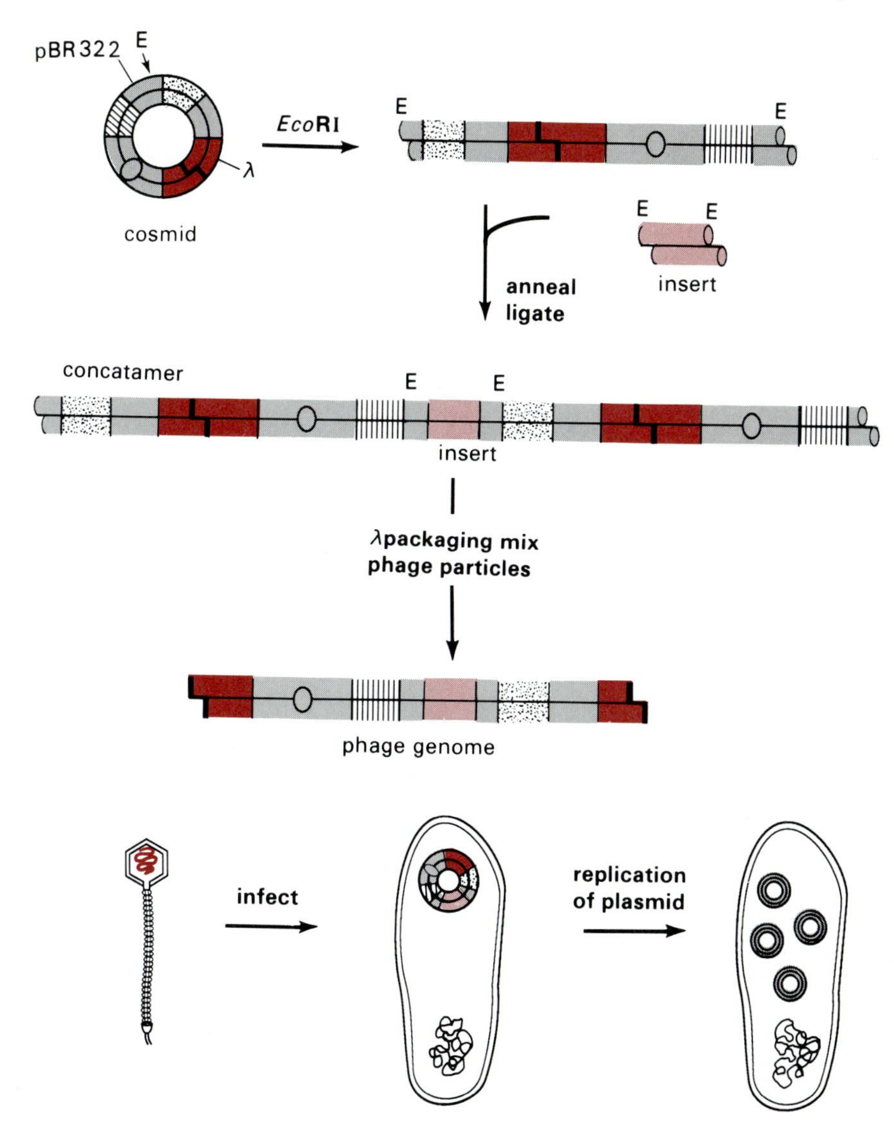

Figure 5.20
Molecular cloning with a cosmid vector. (See Figures 5.13 and 5.14 for descriptions of *in vitro* packaging.)

joint ventures between phage and plasmid. They are linear duplex DNAs whose ends are λ segments that contain all the genes required for a lytic infection and whose middle is linearized plasmid. Both the λ and the plasmid replication functions are intact. In practice, the vector usually contains several tandem repetitions of the plasmid segment in order to supply a genome long enough to be packaged into phage particles. One or more of these is replaced by a DNA insert during construction of a recombinant (Figure 5.21). As with λ phage recombinants and cosmids, phasmid recombinants are packaged *in vitro* before infection. Once inside an *E. coli* cell, the phasmid can replicate like a phage and form plaques in the normal way. However, if the vector contains the gene that encodes the λ repressor, then the phasmid replicates as a plasmid rather than as a phage. Moreover, if the repressor gene encodes a mutant cI protein that is inactive at elevated temperature (a **temperature-sensitive repressor**), the phasmid replicates as a plasmid at low temperature and as a phage when the temperature is raised a few degrees. This versatility can be quite useful in certain kinds of experiments.

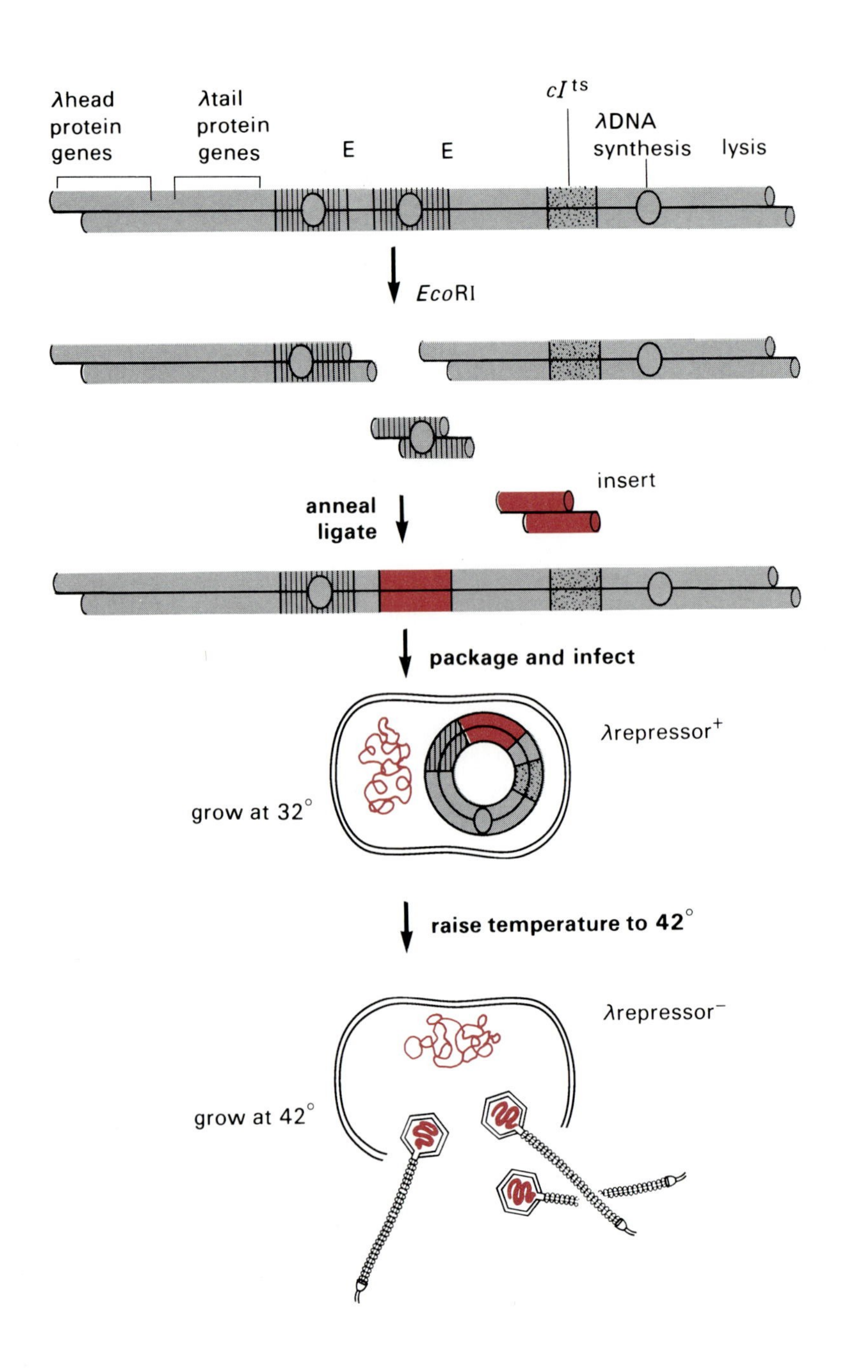

Figure 5.21
A typical phasmid. The open regions are λ DNA; the lined regions are pBR322. Origins of replication are ovals. The gene for a temperature sensitive λ repressor (cI^{ts}) is shown.

5.5 Other Prokaryotic Host-Vector Systems

Other host-vector systems are unlikely ever to compete with the *E. coli* systems for the general purpose of isolating and purifying DNA segments. Nevertheless, efficient means for cloning in other species are important for the study of genetics and gene expression in organisms that are of inherent scientific, medical, or industrial interest. Therefore, useful systems have been devised in spite of a paucity of well-characterized plasmids and bacteriophage that can serve as vectors and in spite of the difficulty experienced in introducing DNA molecules into some kinds of cells. Sometimes the problem posed by the lack of good indigenous plasmids or phages is circumvented by counterparts obtained from other species. Although many plasmids and bacteriophages are quite finicky and replicate in only one or a few related species of host cells, others are remarkably fickle and reproduce in a wide range of host species.

a. *Gram-Negative Organisms*

The R-plasmid depicted in Figure 5.1 is an example of a plasmid that can propagate in many gram-negative bacteria. Using derivatives of such plasmids, recombinant DNA molecules can be constructed, cloned, and amplified in *E. coli* and then transferred into other bacteria for specific purposes.

Another cloning system has been developed using *Hemophilus influenzae* and a naturally occurring, nonconjugative *Hemophilus* plasmid. The plasmid is about 6 kbp long, carries a gene for ampicillin resistance, and forms multiple copies in the host cells. Because *H. influenzae* is naturally competent for transformation, no special transfection procedures are required. Useful vectors are also available for *Klebsiella pneumoniae* (a nitrogen-fixing bacterium). These are derived from the small bacteriophage P4, which grows in both *E. coli* and *K. pneumoniae* cells and can, depending on conditions, replicate either as a phage or a plasmid.

b. *Gram-Positive Organisms*

Bacillus subtilis Vectors permitting genetic manipulation of *B. subtilis* are particularly important because this organism is widely used in commercial fermentation procedures, including the production of antibiotics and enzymes, and secretes proteins into the growth medium, thereby facilitating the isolation of gene products encoded on recombinant vectors. Bacteriophages and plasmids carrying useful selectable markers, such as genes for antibiotic resistance, have been isolated and developed into vectors. One useful plasmid originates from another gram-positive organism, *Staphylococcus aureus*, but replicates and is maintained in both *B. subtilis* and other *Bacillus* species (Figure 5.22). This vector is about 5 kbp long and carries two selectable markers. One confers kanamycin resistance, and the second is a *B. subtilis* gene required for tryptophan biosynthesis. The vector DNA replicates in *B. subtilis* to about 50 copies per cell. Single cleavage sites for several restriction endonucleases are present and amenable for recombination with DNA inserts. Insertion of DNA into the single *Bgl*II site inactivates the kanamycin resistance gene. Insertion into the unique *Hind*III site inactivates the gene required for tryptophan biosynthesis, thereby providing a second phenotypic marker in host cells that themselves lack the active gene. As in the *E. coli* systems, plasmid vectors with two markers permit easy identification and isolation of recombinants.

Bacillus species, unlike *E. coli*, are susceptible to transformation even in the absence of special methods. However, transfection efficiency is markedly improved when **protoplasts** are used in place of normal cells. Protoplasts lack the cell wall and are prepared by treating cells with lysozyme. After exposure to the transforming DNA, the cell walls are regenerated in a special nutrient medium. Efficiencies ranging from 100 to as high as 10^6 transformants per μg of vector DNA are obtained, depending on the particular host-vector system and the transfection method.

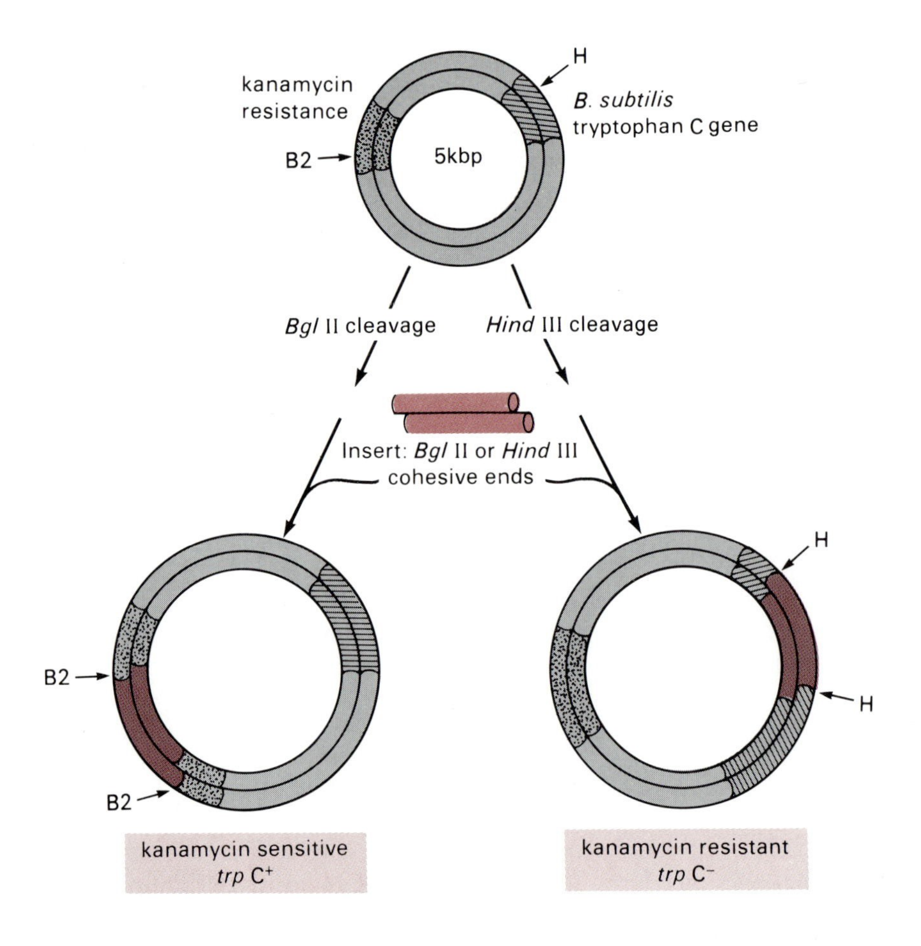

Figure 5.22
A host vector system for *Bacillus* species.

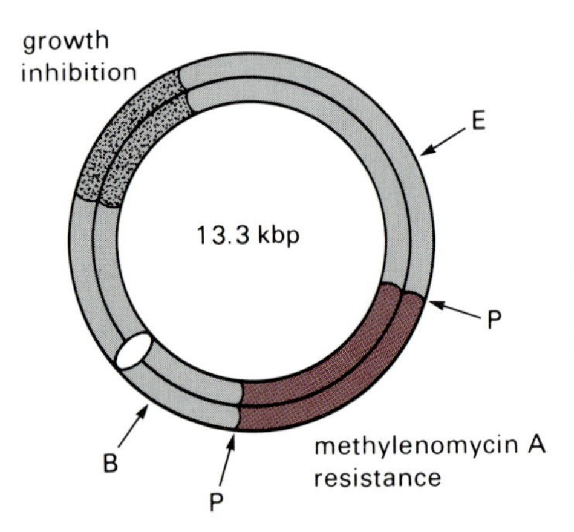

Figure 5.23
A *Streptomyces* vector. The bulk of the DNA is derived from a plasmid indigenous to *S. lividans,* including a region whose phenotype is to inhibit the growth of cells that lack the plasmid. An additional region is derived from *S. coelicolor* and encodes resistance to methylenomycin A.

Streptomyces Members of the gram-positive genus *Streptomyces* encode most of the known and useful antibiotics. Convenient host-vector systems for *Streptomyces* are therefore important for studying the biosynthesis of antibiotics, for designing new antibiotics, and for increasing production efficiency. Plasmids currently provide the best developed vector systems, and plasmids that attain either low or high intracellular copy numbers are available. Special properties of the genus have been adapted for transfection and for selection of successful recombinants. Some of the *Streptomyces* plasmid vectors are self-transmissible and can be moved from one species to another by conjugation.

Typical of the available vectors is the plasmid shown in Figure 5.23. About 11 kbp of its DNA were derived from a plasmid indigenous to *S. lividans.* The remaining DNA contains a gene that encodes resistance to the antibiotic methylenomycin A and is derived from the chromosome of an *S. coelicolor* strain. A second selectable marker is provided by the ability of cells containing the plasmid to inhibit the growth of neighboring *Streptomyces* cells that lack the plasmid. When transfected cells are distributed on a lawn of cells lacking plasmid, transformants are picked from the center of regions in which lawn growth is inhibited. This and similar vectors can be transfected into protoplasts of several species of *Streptomyces;* normal cells regenerate when the protoplasts are placed in standard medium. As many as 10^5 transformants can be obtained per μg of plasmid.

c. Shuttle Vectors

Plasmids and bacteriophages evolved synergistically with their natural hosts. As a result, their replication is usually limited to one or a small group of species. There are, however, distinct advantages to vectors that can replicate in diverse hosts. For example, cloning and isolating DNA segments for structural analysis is most conveniently achieved with *E. coli* host-vector systems. However, functional analysis of a cloned insert must usually be carried out in the species of origin. The desired DNA can always be cloned in *E. coli*, removed, and reinserted into a vector that is compatible with the host cells of interest. But it is much simpler to carry out the initial cloning with a vector that can, without extra manipulation, shuttle back and forth between *E. coli* and the alternate cells and replicate in both. **Shuttle vectors** designed to replicate in the cells of two species contain two origins of replication, one appropriate for each species, as well as any genes that are required for replication and not supplied by the host cells. These vectors are themselves constructed by recombinant DNA techniques, and many different types have been made. Some of them shuttle between two prokaryotic species, others between a prokaryote (usually *E. coli*) and eukaryotic cells (including yeast, plants, and animals). Indeed, most of the eukaryotic vectors described in the following sections are shuttles.

Shuttle vector design poses certain special problems because one portion of a shuttle vector is always a "stranger" and the other a "native" in the alternative host cells. The "stranger" must not interfere with replication functions encoded in the "native" portion of the vector. Unusual sequence arrangements that render the shuttle unstable in one of the pair of host cells need to be avoided, as do sequences that interfere with replication or transcription in the alternate hosts. Similarly, the presence of an active restriction-modification system in one or another (or both) of alternate host cells can compromise the effectiveness of a shuttle vector. This latter problem can be circumvented by using mutant cells that lack active restriction endonuclease.

An example of a shuttle vector that replicates in both *S. lividans* and *E. coli* is shown in Figure 5.24. DNA segments derived from a *Streptomyces* plasmid (Figure 5.23) contribute the signals needed for replication in *Streptomyces* and for methylenomycin A resistance while an *E. coli* plasmid supplies the replication functions and an antibiotic resistance marker for maintenance in *E. coli*. This plasmid allows cloning of *Streptomyces* DNA segments in *E. coli* and then propagation and functional tests in *Streptomyces.*

5.6 Eukaryotic Host-Vector Systems: Yeast

a. Versatility and Convenience

Laboratory strains of the common baker's yeast, *Saccharomyces cerevisiae*, are convenient hosts for recombinant DNA experiments. Although yeast can reproduce sexually (Chapter 10), the cells usually multiply asexually by budding. They grow as single cells in suspension and produce colonies in solid medium, much as do *E. coli*. In usual growth media, the cell dou-

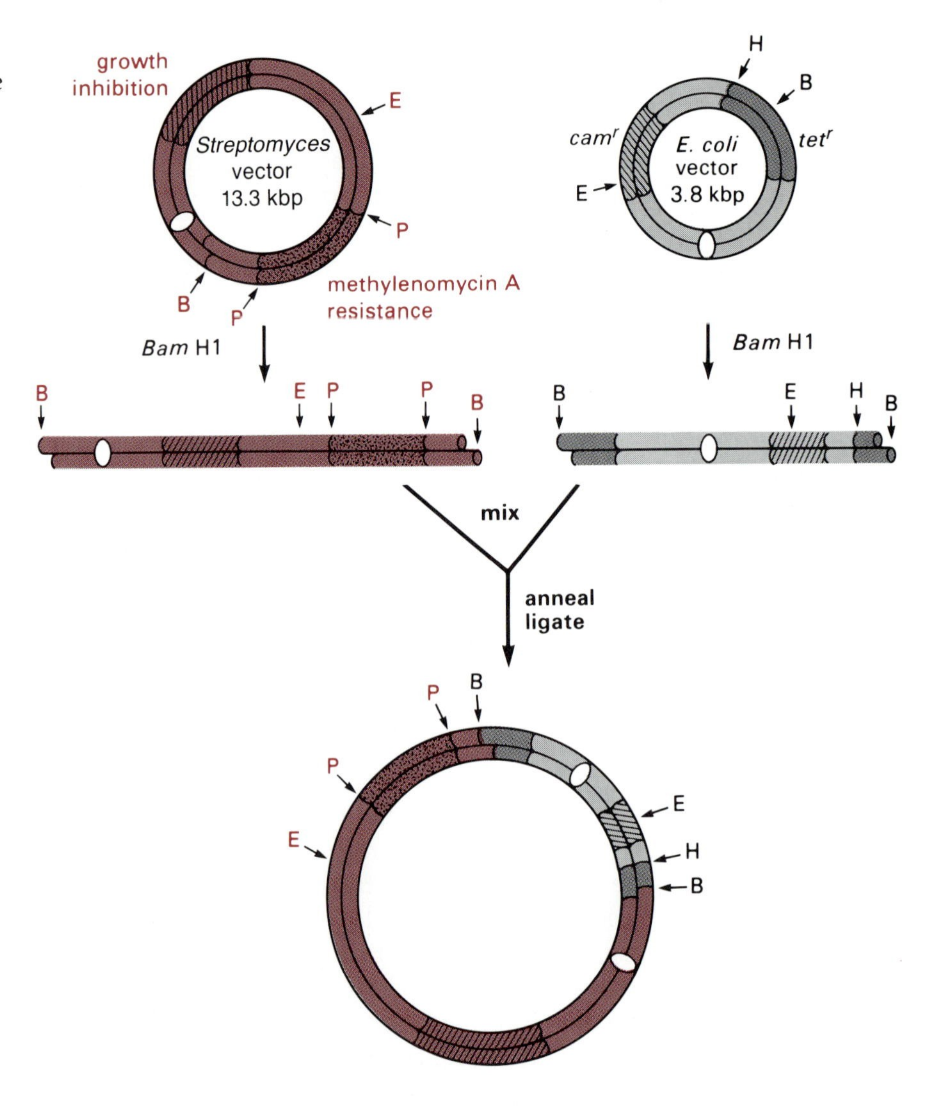

Figure 5.24
Construction of a shuttle vector by the joining of a *Streptomyces* (Figure 5.23) and an *E. coli* plasmid. The tetracycline resistance gene of the *E. coli* plasmid is inactivated in the shuttle.

bling-time is from 1.5 to 2.5 hours. Thus, large numbers of these nonpathogenic organisms can be obtained inexpensively and in a relatively short time.

The haploid genome of this unicellular eukaryote contains 1.4×10^7 base pairs, about three times as many as *E. coli*, divided among 17 (haploid) chromosomes. A large collection of metabolic, biosynthetic, and cell cycle defective mutants are known and genetically mapped. Viruses that infect yeast are not known, but the only known yeast plasmid provides a basis for some of the vectors now used to introduce recombinant DNA molecules into yeast cells. Other vectors contain yeast genomic segments that permit independent replication and, when cloned centromeric DNA is added, even behave like chromosomes in meiosis and mitosis. Moreover, cloned DNA fragments can recombine with homologous segments on the yeast genome, resulting in permanent and **site-specific transformation** of the cellular genome, independent of vector maintenance. These techniques, together with those of classical yeast genetics, permit this eukaryote to be studied with the rigor and depth previously confined to prokaryotes like *E. coli*.

The tough polysaccharide cell wall that surrounds yeast cell membranes bars the entry of DNA molecules. Before free DNA can enter yeast cells, the wall must first be removed without killing the cell. Enzymes that degrade the wall polysaccharides are used to denude cells and produce **spheroplasts** that can take up DNA following treatment with $CaCl_2$. The spheroplasts regenerate the walls when placed in special nutrient. Also, lithium acetate can permeabilize cells. As many as 10^4 transformed *S. cerevisiae* cells per μgram of DNA have been obtained with some vectors.

b. Vectors That Replicate in Yeast

Plasmid vectors Certain *S. cerevisiae* strains contain a circular duplex plasmid that is 6318 base pairs long. This so-called **2 μm** plasmid DNA contains an origin of DNA replication, another *cis*-acting region (*REP*3), and two genes (*REP*1 and *REP*2) that together permit stable maintenance of a high copy number (about 50 per cell) in yeast cells. About 50 percent of the 2 μm plasmid genome is essential for replication and maintenance, the remainder of the DNA being dispensable. The 2 μm DNA has no marker genes that permit selection of yeast cells that contain the plasmid. Therefore, neither the plasmid nor its essential half alone is a suitable vector for use with yeast cells. However, the essential portion of the 2 μm plasmid has been used to construct a shuttle vector by linking it to pBR322 that also contains a yeast gene (*HIS*3), encoding one of the enzymes required for histidine biosynthesis (Figure 5.25). When used in conjunction with a *his*3$^-$ yeast strain as host, the vector permits selection of transformed cells that grow in the absence of histidine. Actually, the same yeast *HIS*3 gene is functional in *E. coli* because its DNA contains a region of fortuitous homology to an *E. coli* promoter. Consequently, the yeast *HIS*3 gene provides a selectable marker in bacterial cells that lack the analogous gene. The vector contains several unique restriction sites that allow DNA segments to be inserted.

ARS vectors The second type of yeast vector incorporates a yeast genomic replication origin into a circular duplex DNA. These vectors are usually shuttles. A serendipitous discovery identified the presence, in the yeast genome, of chromosomal origins of DNA replication. Consider the vector shown in Figure 5.25. If the 2 μm plasmid segment is removed, the resulting plasmid is equivalent to the insertion of the *HIS*3 yeast segment into the tetracycline resistance region of pBR322. It fails to replicate autonomously in yeast, although it retains its ability to replicate in *E. coli*. However, when certain yeast chromosomal genes other than *HIS*3 were cloned in pBR322, the resulting recombinants replicated autonomously in both *E. coli* and yeast! Among the genes that conferred replication capability were *TRP*1 and *ARG*4, which encode enzymes required for biosynthesis of tryptophan and arginine, respectively. Fortuitously, a yeast chromosomal origin of replication lies close enough to each of these genes to be included in the cloned yeast segment (Figure 5.26). The replication origins are defined by 100 base pairs or less and are called **autonomously replicating sequences** (or *ARS*). Different *ARS* sequences all contain a short common oligonucleotide stretch $5'\text{-}{}^{A}_{T}TTTAT^{A}_{G}TTT^{A}_{T}\text{-}3'$

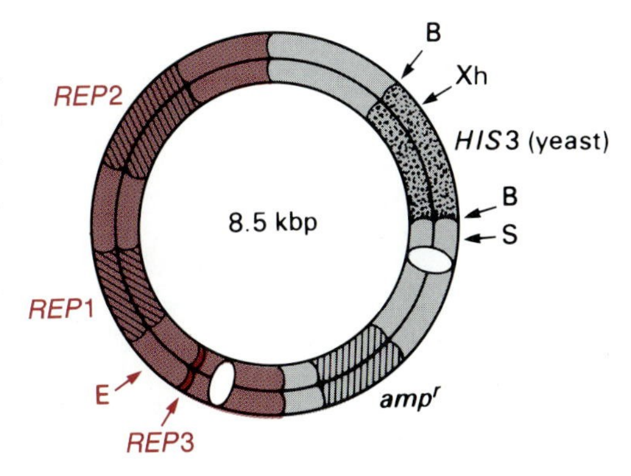

Figure 5.25
Schematic diagram of a yeast-*E. coli* shuttle vector based on the yeast 2 μm plasmid (color). The *E. coli* portion is pBR322. Origins of replication are indicated by ovals, and the *HIS*3 region derived from yeast chromosomal DNA is stippled.

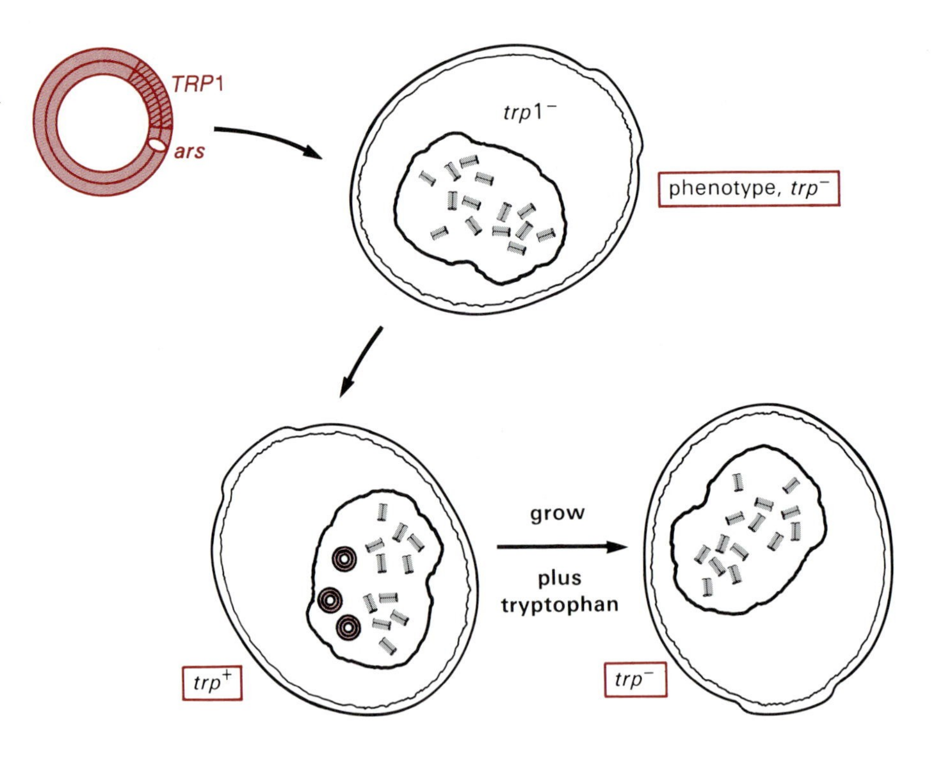

Figure 5.26
A yeast plasmid vector that includes a chromosomal origin of DNA replication. When *trp*$^+$ transformed cells are grown in the presence of tryptophan, the plasmid is lost.

that is essential for function. Other functionally important sequences surround the 11 bp stretch, but these vary from one *ARS* element to another. Any circular duplex that contains an *ARS* serves as a yeast vector and can be used directly for cloning in yeast. The molecules are stably maintained within *S. cerevisiae* cells so long as they are the sole source of a required gene (for example, of *TRP1* in a *trp1*$^-$ cell grown in the absence of tryptophan), but they are rapidly lost in the absence of such selective pressure.

Minichromosome vectors The third type of yeast vector behaves like a functional chromosome, albeit a tiny one. These minichromosome vectors replicate only once in each cell generation, and the two copies are segregated, like true chromosomes, into daughter cells during meiosis and mitosis. Three small segments of yeast chromosomal DNA are required for a functional minichromosome (Figure 5.27). One, an *ARS*, supplies an origin of DNA replication. A second segment provides for selection; in the example in Figure 5.27, the plasmid contains the *LEU2* gene for use with *leu2*$^-$ hosts. The third segment, *CEN*, controls the intracellular copy number and confers the property of segregation through many generations. The DNA segments that can impart this remarkable behavior represent functional yeast centromeres and come from the centromeric regions of yeast chromosomes (thus, *CEN*). They need be no larger than 500 base pairs in length. The relation between the structure and function of *CEN* segments is described in Chapter 9, as are the molecular features of yeast telomeres. Chapter 9 also describes how cloning these two critical chromosomal domains has permitted the reconstruction of functional yeast chromosomes using linear DNA rather than the circular DNA required for the yeast vectors described thus far.

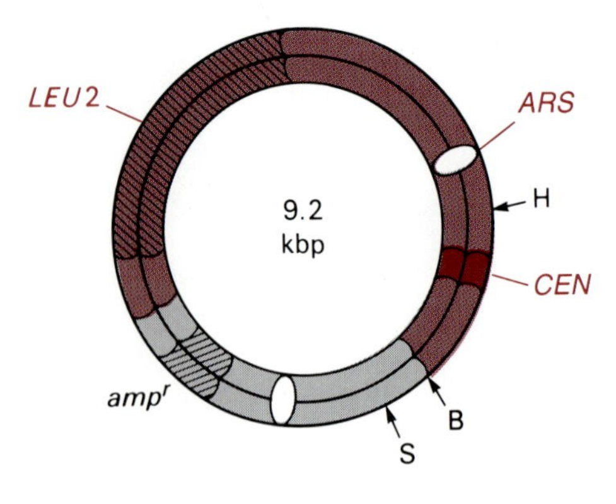

Figure 5.27
A yeast minichromosome shuttle vector. The yeast sequences are colored, and pBR322 sequences are grey.

c. *Permanent Transformation by Recombination with the Yeast Genome*

This chapter has stressed systems designed to permit the independent, extrachromosomal replication of a recombinant vector in appropriate cells. However, the ability to effect the **permanent transformation** of cellular genomes by inserting recombinant DNA molecules into chromosomal DNA is also of great significance. Here, the newly acquired genotype and phenotype are stable, independent of selective pressure or maintenance of the extrachromosomal recombinant in future generations. The inserted DNA replicates whenever the cellular genome is duplicated. Such permanent transformations are relatively rare, depending on the nature of both the donor DNA and the recipient cells.

Permanent transformation by recombinant plasmids Yeast cells can be permanently transformed, albeit at a low frequency, by recombinant plasmids that are unable to replicate in yeast for lack of a suitable origin of replication. A microgram of a pBR322 derivative carrying the yeast *LEU2* gene transforms about 1 out of 10^7 *leu2*$^-$ yeast cells to *LEU2*$^+$. When the recombinant vector also contains an *ARS* element and can replicate autonomously, permanent transformation is more efficient. The high intracellular copy number of the transforming DNA increases the likelihood of insertion into the yeast genome.

Insertion involves recombination between the donor recombinant molecule and the recipient yeast genome. The recombination may be nonspecific, with the donor DNA being inserted at random chromosomal loci, but this is the rarest situation. Most frequently, insertion is at a specific chromosomal locus homologous to yeast DNA sequences in the recombinant donor. For example, a plasmid carrying the wild type yeast *TRP1* gene and replicating in *trp1*$^-$ yeast cells recombines most often with the homologous mutant *trp1* locus in the yeast genome; the cell is permanently transformed to tryptophan independence. Two different types of homologous (and thus site-specific) recombination are observed. Crossing-over between the circular donor and the recipient results in the insertion of the entire vector into the corresponding chromosomal locus (Figure 5.28). In this case, excision by homologous recombination between the two copies of the gene can occur in subsequent generations to regenerate the normal DNA structure and either wild type or mutant cells. The other type of recombination involves gene conversion, a nonreciprocal recombination in which the wild type gene on the vector repairs the mutant segment on the genome (Section 2.4b).

Site-specific transformation by linear donors Circular plasmids that can not replicate in yeast are very inefficient donors for site-specific transformation. However, the transforming efficiency of such molecules can be significantly enhanced if they are linearized prior to transfection and if the linear duplex is homologous, at both ends, to sequences in the genomic target (Figure 5.29). The free ends on such linear molecules are especially recombinogenic. The linear transforming DNA is usually derived by appropriate restriction endonuclease cleavage of a recombinant plasmid obtained after replication in *E. coli*. Essentially, any DNA sequence at all

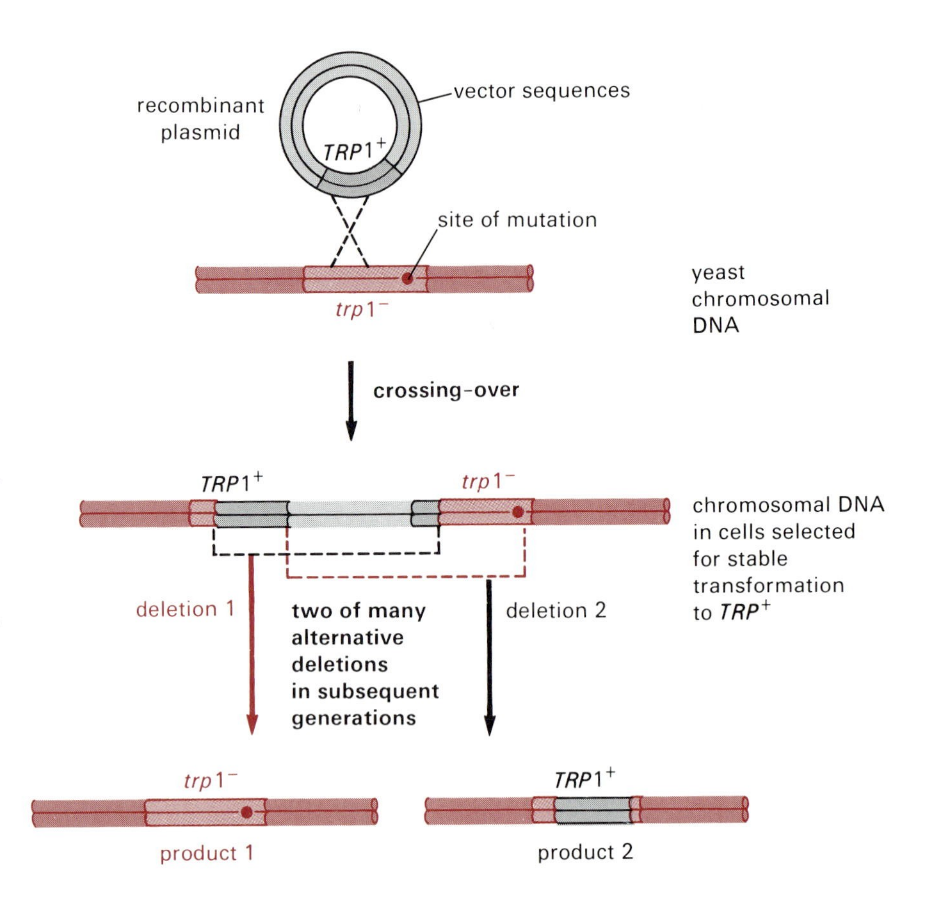

Figure 5.28
Reciprocal homologous recombination between a replicating plasmid vector and the yeast chromosome permanently transforms the yeast cell. In this example, the vector carries a wild type *TRP1* gene (as well as required replication functions). Homologous recombination at the mutant *trp1* locus on the yeast chromosome yields two copies of the *TRP1* gene, one functional and one not; vector sequences are also inserted. The resulting cells are stable for the $TRP1^+$ phenotype, even in the absence of selective pressure (i.e., presence of tryptophan). In subsequent generations, rare homologous recombinations between the two copies of the gene can delete the intervening sequences as well as one gene copy. The resulting cells may be either $TRP1^+$ or $trp1^-$.

can be between the two homologous ends. Thus, specific genes can be introduced into known genomic loci, or genes can be altered or deleted, at will. Rather than depending on the chance formation and selection of specific mutations, the study of yeast genetics can now proceed on a rational basis, independent of the chance isolation of interesting mutants.

5.7 Eukaryotic Host-Vector Systems: Animals

a. The Transformation of Animal Cells

Transient and permanent transformation The cloning of DNA segments from animal cells is most often carried out in *E. coli* cells, and recombinant DNAs isolated from the bacteria are used for structural analysis. However, functional studies require the transformation of animal cells. As with yeast, transformation takes several forms. Animal cells can be transformed by infection or transfection with a vector that replicates independently as an extrachromosomal element. Some of these vectors are viral genomes that yield virions with recombinant genomes. In other cases, they are more analogous to plasmids. Frequently, they are shuttles that use *E. coli* as an alternate host. Unlike the situation in yeast and bacteria, replicating extrachromosomal plasmidlike vectors are frequently unstable in animal cells during the cell generations subsequent to the initial transfection,

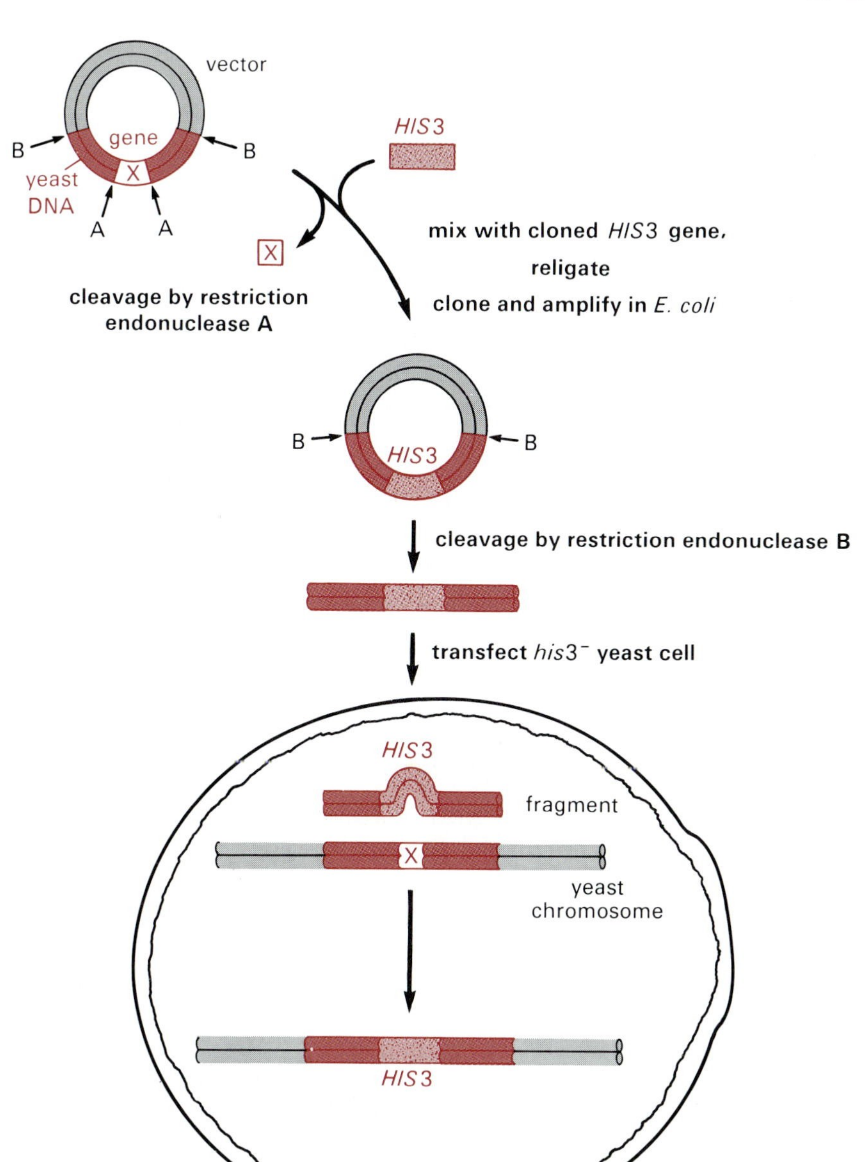

Figure 5.29
Site-specific transformation of a yeast genomic locus by a linear fragment. A segment of yeast genomic DNA from the region to be transformed is cloned in an *E. coli* plasmid vector. A portion of that segment is then altered by insertion, deletion, or mutagenesis, and the new construction is cloned and amplified in *E. coli*. In the example, gene X is deleted, and a selectable marker, the *HIS3* gene, is inserted. The plasmid is cleaved at restriction endonuclease sites marked B to release the yeast DNA fragment in a linear form with regions homologous to the genomic target site at both ends. Yeast cells that are *his3*$^-$ are transfected, and *HIS3*$^+$ colonies that lack gene X are selected. Both haploid and diploid yeast can be transformed. Only one of the two homologous chromosomes is disrupted in diploids, but the two chromosomes segregate in haploid spores and are thus separable.

even when selective pressure is imposed. Populations of such transformed cells often can be maintained only transiently (i.e., for a few days), but this time period is sufficient to study the expression of genes in the vector DNA. **Transient expression** can in fact be detected even if the vector is unable to replicate in animal cells. However, a replicating vector is advantageous because of its increased intracellular copy number, which in turn increases the level of gene product and facilitates the analysis of gene expression.

Some animal vectors do establish themselves as stable, replicating, extrachromosomal DNAs in animal cells. These include those derived from bovine papillomavirus (Section 5.7c) and others that utilize the origin of replication of the herpes virus, Epstein-Barr. In most instances, permanent transformation is associated with the integration of the transforming DNA into the cellular genome by recombination. Both replicating vectors and nonreplicating DNA can cause **insertional transformation**; integration usually involves nonhomologous recombination into random genomic positions. This differs from the situation in yeast, where transformation is most often site specific through homologous recombination between the donor DNA and the target site on the yeast genome.

Host cells Animal cells growing in what are called **cell cultures** (or, inaccurately, **tissue culture**) are the most frequent targets for transformation by recombinant DNA molecules and analysis of gene expression. *Xenopus* oocytes also provide an environment where many cloned genes, even those of mammalian origin, are efficiently expressed (Chapter 7). In recent years, techniques for inserting cloned DNA sequences into early animal embryos, including mammalian zygotes, have been developed. The DNA can insert into the genome of these embryonic cells, resulting in their permanent transformation. Viable offspring are obtained, and when the DNA is conserved in the genome of germ line cells, the newly acquired genotype and phenotype are inherited by Mendelian rules in subsequent generations. With placental mammals, development of the new offspring requires that the transformed early embryos be implanted into foster mothers. Systems for such germ line modification of animals (**transgenic** animals) are described in Perspective IV. Here we are concerned with animal cells growing in culture.

Many kinds of animal cells grow and divide in appropriate nutrient media. Some are most successful fixed to a solid surface, such as the bottom of a plastic dish; they continue to divide until the surface of the plate is covered by a one-cell-thick layer of cells (a **monolayer**). Other cell types are not inhibited by contact with neighboring cells and pile up in multiple layers on the dish; cells derived from tumors often have this property. Some animal cells do not require attachment to a surface and grow in suspension much like bacteria or yeast.

Primary and **secondary** cell cultures, those recently derived from fresh tissues, generally have a finite life span; they may divide a few times and then stop. However, **continuous cells lines**, in which the cells are essentially immortal, can be derived from such cultures, and **established cell lines** are widely used in recombinant DNA experiments. They have been obtained from a variety of organisms and organs such as rodent embryos, monkey kidneys, *Drosophila*, and rodent and human tumors.

Under proper conditions, single animal cells grow and divide to yield a discrete colony or clone, much as do bacteria and yeast. Although a cloned cell line derived from a single cell is the ideal, it is not always easy to achieve. Therefore, the population uniformity in such cultures may be problematic. Also, cell lines with a common lineage propagated in different laboratories often display different properties. The most likely explanation is that even slightly altered growth conditions can lead to the overgrowth of a cell that has had a favorable mutation. Some cell lines containing specific mutations are available and provide opportunities to use selection as a means for isolating colonies of transformed cells that have acquired a new genetic marker.

Introducing DNA into animal cells The genomes of animal viruses are frequently used as vectors. Such molecules, along with any foreign DNA insert, can often be packaged into a virion particle, which is introduced into appropriate host cells by infection. Packaging is always achieved *in vivo*; unlike the λ phage system, packaging of animal viruses *in vitro* is not now possible. Although infection with virion particles is by far the most efficient way to introduce recombinant genomes into animal cells (essentially every cell becomes infected), the structure of some recombinants prohibits packaging, and the naked recombinant DNA must then be transfected into the cells. Similarly, transfection is required with DNA fragments that are unable to replicate, such as those often used to obtain permanent insertional transformation. Just as with prokaryotes and yeast, special techniques are required to introduce naked DNA molecules into animal cells and their nuclei. Direct injection into cells and nuclei is straightforward, if technically demanding, but has the disadvantage of allowing only a relatively small number of cells to be treated. Large populations of cells can be transfected by treating the cells with particles formed by the coprecipitation of the transfecting DNA and calcium phosphate. The particles are taken up by the cells, in a poorly understood manner, and intact DNA reaches the nucleus. Generally, 10 percent or less of the cells are successfully transfected, and with certain cell types this method does not work at all. DNA can be introduced into many additional cell types by subjecting a suspension of the cells mixed with the DNA to a high-voltage electric discharge (2000–4000 volts). This process, **electroporation**, apparently works by creating reparable holes in the cell membrane through which the DNA can pass.

Phenotypic selection of transformed cells Usually less than 10^{-2} percent of a population of transfected animal cells are permanently transformed after foreign DNA sequences are introduced. Finding and isolating the rare transformants from among millions of unaltered cells requires highly discriminating selection procedures. Yet unlike the situation in *E. coli*, only a few selection systems are now available. Two general approaches are widely used: one utilizes mutant and the other normal recipient cells.

As with bacteria, if recipient cells have a mutation in a gene carried in wild type form on a cloned transforming DNA, then the desired transformants can be selected after growth under conditions that discriminate between the mutant and wild type phenotype. For example, TK^- rodent cells have a mutation affecting thymidine kinase (the enzyme that converts thymidine to thymidine-5′-monophosphate). They do not grow in the presence of thymidine if the *de novo* pathway for dTMP synthesis is blocked by the drug aminopterin, which inhibits regeneration of tetrahydrofolate (Figure 5.30). However, cells transformed by DNA containing a functional thymidine kinase gene grow into colonies in the presence of thymidine and aminopterin (Figure 5.31). Fewer than 1 in 10^6 untransformed cells produce colonies under these conditions, so this is an extremely sensitive way to detect and isolate the rare TK^+ transformants. A thymidine kinase gene isolated from herpes simplex virus DNA by molecular cloning is widely used in this way in conjunction with TK^- rodent cells. The same approach works with other types of mutations. Unfortunately, very few suitable mutant cell lines are available, so the technique is not widely applicable. This limitation is often overcome by transfecting TK^- cells with a vector that contains both the thymidine kinase gene and an unselectable gene of interest. Transformed cells selected for the TK^+ phenotype are almost always also transformed with

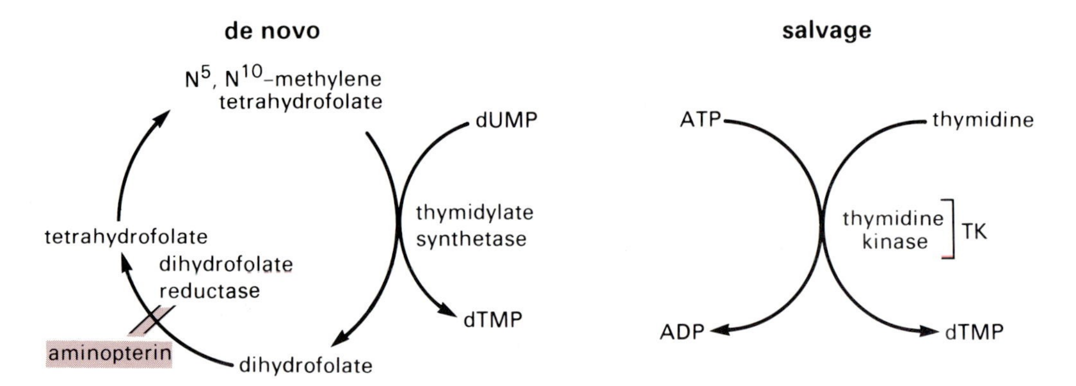

Figure 5.30
The *de novo* and salvage pathways for dTMP synthesis. *De novo* synthesis is blocked by aminopterin, making the cell dependent on thymidine and thymidine kinase.

respect to the accompanying portions of the vector molecule, including the gene of interest. In fact, such **cotransformation** does not depend on the covalent linkage of the thymidine kinase gene and the unselectable second gene. Permanent, cotransformants are formed when cells are cotransfected with a mixture of linear DNA fragments containing the two genes on separate molecules. Greater than 50 percent of the cells selected as TK^+ also contain the second gene. The two genes are usually found integrated side by side because they tend to be joined after entering the cellular environment and prior to the integration event.

Cotransformation expands the range of genes that can be effectively introduced into animal cells. However, the number of different cell types that can be studied is limited by the availability of appropriate mutant cell lines. A more general approach relies on making normal cells dependent on a transfected gene. For example, the cells of many species are unable to synthesize guanylic acid (GMP) in a medium containing mycophenolic acid, a compound that specifically inhibits inosinate (IMP) dehydrogenase (Figure 5.32). The block is circumvented by supplying xanthine and a functional *E. coli* gene (*gpt*) that encodes xanthine-guanine phosphoribosyltransferase. Animal cells do not normally use xanthine as a precursor to guanine nucleotides because xanthine is a very poor substrate for the analogous animal enzyme, hypoxanthine-guanine phosphoribosyltransferase. Therefore, cells transformed by the *E. coli gpt* gene and synthesizing functional xanthine-guanine phosphoribosyltransferase can be selectively grown in a medium containing xanthine and mycophenolic acid (in the absence of guanine). Similarly, an *E. coli* gene (*neo*) encoding

Figure 5.31
Phenotypes of TK^+ and TK^- (thymidine kinase) cells in various growth mediums. **HAT** medium contains **h**ypoxanthine, **a**minopterin, and **t**hymidine. Hypoxanthine is included in the HAT medium to bypass the block in the *de novo* path of purine synthesis, which also requires tetrahydrofolate.

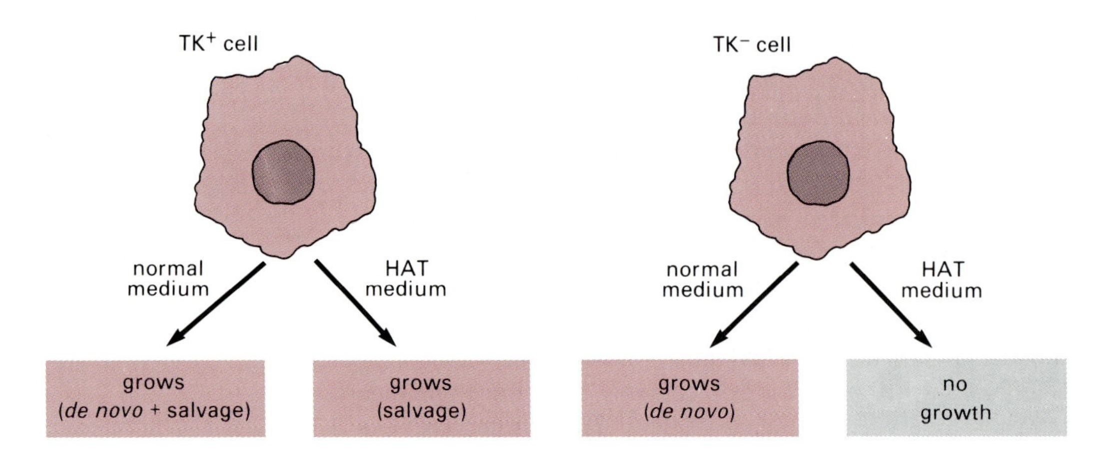

a. normal animal cells

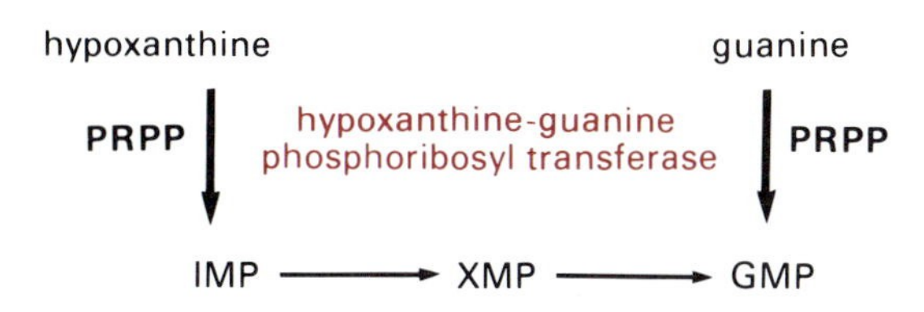

b. cells inhibited by mycophenolic acid (absence of guanine)

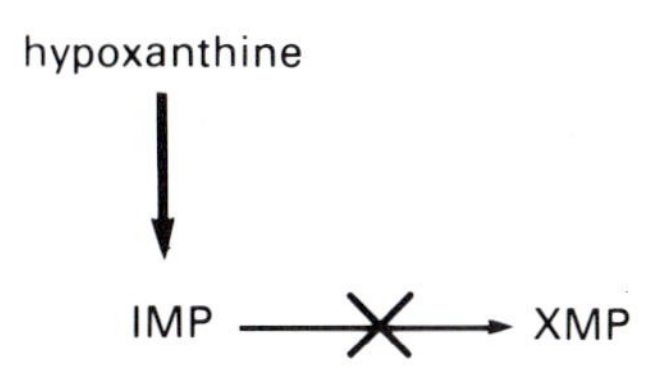

c. overcoming mycophenolic acid inhibition (absence of guanine)

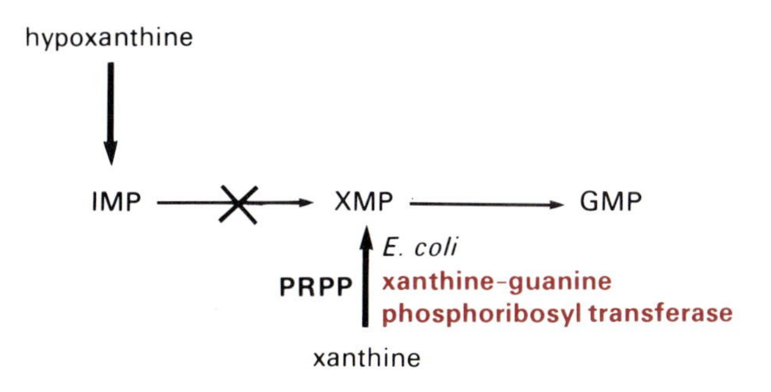

Figure 5.32
Synthesis of GMP under various conditions. PRPP is phosphoribosyl pyrophosphate.

an aminoglycoside phosphotransferase can be used as a dominant marker in conjunction with an aminoglycoside called G-418. This analogue of the antibiotics neomycin and kanamycin blocks protein synthesis in eukaryotic cells and is inactivated by the phosphotransferase (Table 5.3). The *E. coli* genes are supplied on vectors, usually shuttles, that have been cloned and replicated in *E. coli*. During construction of the recombinant, the normal *E. coli* promoters are replaced by transcriptional regulatory signals that are active in animal cells. The *E. coli gpt* and *neo* genes can be used to cotransform normal cells with a gene for which no selectable marker is available. A bacterial gene that confers resistance to hygromycin (*hph*), and a mammalian gene encoding a methotrexate-resistant form of a dihydrofolate reductase are additional dominant selective markers.

Vectors The genome of the primate papovavirus, simian virus 40 (SV40), has been used extensively as the basis for vector construction. Just as the extensive knowledge and ease of handling of *E. coli* systems made them so attractive for establishing bacterial cloning systems, so too was SV40 the logical candidate for use in mammalian cells. In the years since efforts to develop SV40 vectors began, considerable information has accumulated about the structural and genetic aspects of other eukaryotic viral genomes. Several of these viruses have now provided interesting host-vector sys-

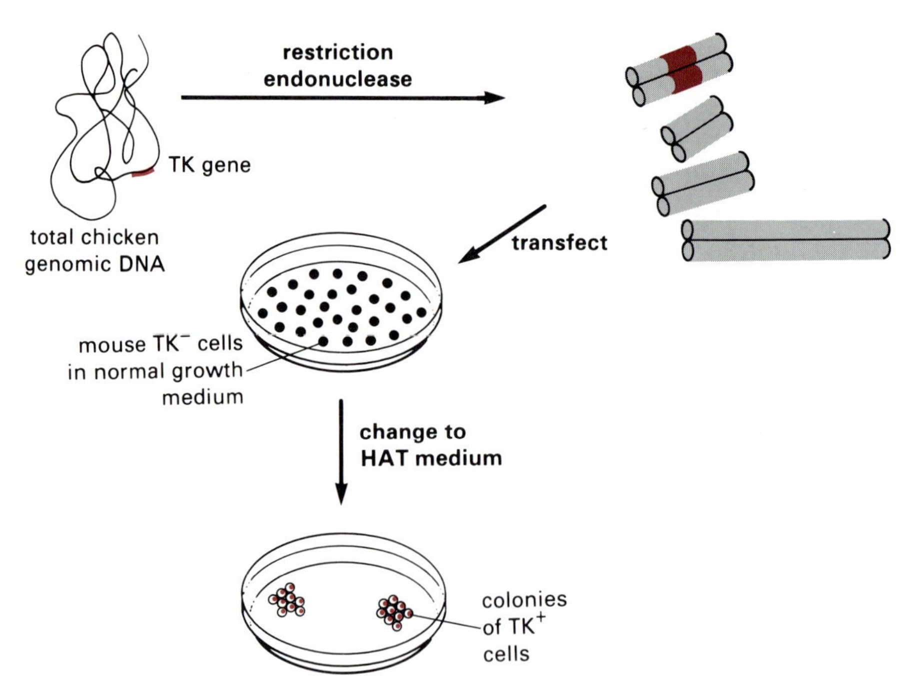

Figure 5.33
Transformation of TK^- mouse cells to TK^+ by chicken DNA. A restriction endonuclease digest of total chicken DNA is used to transfect mouse TK^- cells growing in culture. After the transfection period, the cells are placed in HAT medium. Cells that received no chicken DNA or chicken DNA fragments that do not contain the chicken thymidine kinase gene fail to grow. The rare transformed cells that integrated a chicken DNA fragment containing the gene into the mouse genome grow and form colonies. Note that if the restriction endonuclease cleaves the chicken thymidine kinase gene, no transformed cells appear.

tems, including, for mammals, papillomavirus, adenoviruses, the herpes virus called Epstein-Barr, vaccinia virus, and retroviruses and, for insects, baculoviruses.

As already pointed out, two different kinds of animal cell vectors are used. Sometimes they are analogous to bacterial plasmid or phage vector systems. After transfection or infection, they replicate and are maintained as independent extrachromosomal genomes or are packaged into infectious viral particles. In other instances, the vectors are nonreplicating and are used to introduce specific DNA segments into cells to analyze the transient expression of a cloned gene or to produce permanent transformation by integrating the vector into the host genome.

Transformation without vectors This chapter deals with vectors, that is, DNA molecules carrying specific cloned DNA inserts. However, because cells can be transformed with nonreplicating DNA segments, any DNA fragment or mixture of uncloned fragments can be used, so long as the method for selecting the desired transformants is sensitive enough. For example, when a restriction endonuclease digest of total chicken DNA is transfected into TK^- mouse cells, colonies of cells expressing the chicken thymidine kinase gene can be isolated (Figure 5.33). The chicken gene is inserted into the mouse cell DNA. It can be recovered from the mouse genome by molecular cloning using special selection techniques. Application of this approach has proven extremely useful for the cloning of many important genes, including the oncogenes that occur in tumors.

b. SV40 Vectors

The virus and its genome SV40 is a spherical virus, and its genome is a single closed circular duplex DNA molecule (Figure 5.34). The viral

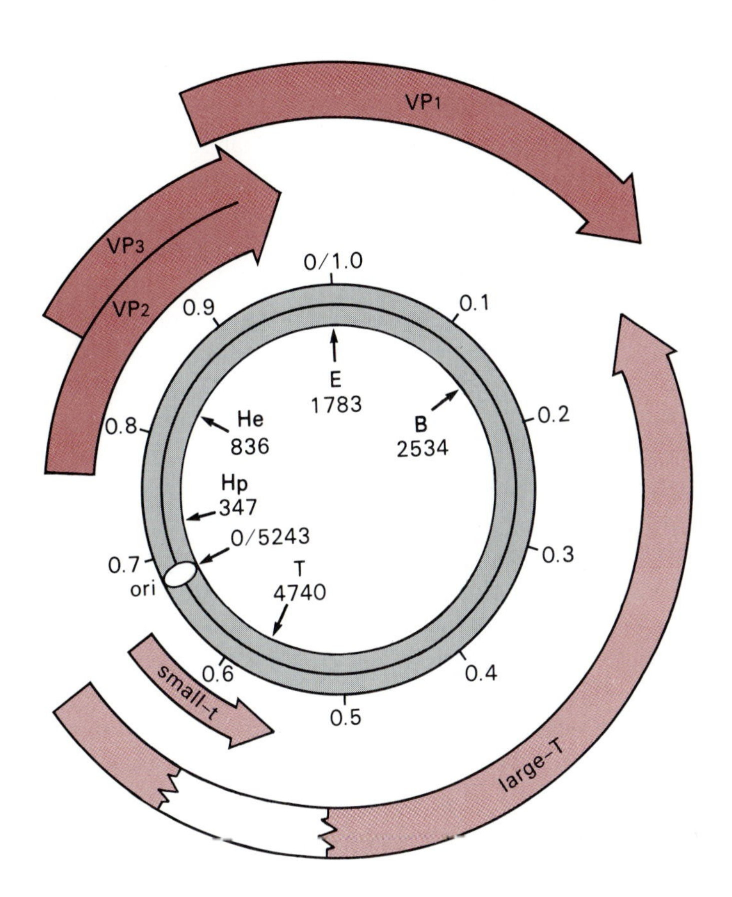

Figure 5.34
The genome of simian virus 40. Genetic map units are shown on the outside of the circle and base pair numbers on the inside. Unique restriction endonuclease sites are indicated. The regions encoding five SV40 proteins, some of which overlap, are shown around the outside. Note that large-T antigen is encoded by two separate regions (see Chapter 8 for a description of intervening sequences).

particle contains three viral coded proteins. VP1 is the major virion protein and comprises the bulk of the capsid; VP2 and VP3 are present in smaller amounts, but their function is unknown. The DNA in the viral particle is in the form of chromatin and is associated with the four nucleosomal cellular histones (Chapter 2). Although SV40 occurs naturally in rhesus monkeys (*Macaca mulatta*), the preferred laboratory host cells are usually kidney cells from the African green monkey (*Cercopithecus aethiops*). The entire sequence of the 5243 base pairs in the SV40 genome is known, as are the precise segments encoding five viral proteins: small-T, large-T, VP1, VP2, and VP3 (Figure 5.34). The SV40 DNA map is defined in terms of both genetic map units and base pair numbers; nucleotide 1 is within the origin of DNA replication (ori).

The infecting SV40 virion travels to the nucleus, where the DNA is uncoated and its genetic information is expressed in a carefully timed sequence of events. New viral particles begin to appear within about 30 hours after infection of a monolayer of monkey kidney cells, and after 4 days, all the cells are destroyed. Initially, the genes for the two "early" proteins, called small-T and large-T are transcribed in a counter-clockwise direction starting close to the origin of replication. The polyadenylation site is near map position 2550. The two proteins are then synthesized. Neither small-T nor large-T appears in mature viral particles, but large-T is essential for viral DNA replication, which begins soon after the early proteins appear. Replication starts at the origin and is bidirectional; it terminates when the two replicating forks meet about halfway around the genome, and the new duplexes are joined into covalently closed circular progeny DNA. More than 10^5 copies of the viral genome are produced in each infected cell. At about the same time as DNA replication begins, synthesis of the early proteins decreases and abundant transcription of the "late" genes occurs in a clockwise direction starting near the origin. Finally, the late proteins VP1, VP2, and VP3 appear. The accumulation of progeny DNA molecules and virion proteins culminates in the assembly and release of new virion particles accompanied by the death of the cell. The entire infectious process can also be initiated by transfection with naked SV40 DNA.

Several types of SV40 vectors The general principles used to design SV40 vectors are similar to those used with λ phage. Portions of the viral genome are removed (by appropriate restriction endonuclease cleavage) and replaced by other DNA segments. Which DNA segments are eliminated depends on the type of vector being constructed. It is convenient to distinguish three categories of SV40 vectors, each of which has distinct advantages and disadvantages for various kinds of experiments. First, there are **SV40 transducing vectors** that replicate in appropriate host cells (monkey kidney cells) and are packaged into virion particles; these are analogous to specialized transducing phage and λ phage vectors. Second, there are SV40 vectors that can replicate but cannot be packaged; these are called **SV40 plasmid vectors**. Finally, there are passive **transforming vectors** that can neither replicate nor be packaged as virions; small segments of the SV40 genome that foster gene expression are incorporated into these molecules.

SV40 transducing vectors If SV40 vectors are to produce viral particles, they must meet three important requirements. First, the vector must include the 300 base pairs from around the origin of replication to assure viral DNA synthesis and to provide the transcriptional regulatory signals for the synthesis of mRNAs. Second, the total amount of DNA in the recombinant must not exceed about 5300 base pairs or be less than about 3900 base pairs, the size range required for packaging into viral particles. Third, the proteins large-T, VP1, VP2, and VP3 must be provided, although they need not be encoded on the recombinant genome itself. It is, in fact, almost impossible to add any extra DNA to an intact SV40 genome without exceeding the maximum size for packaging. Therefore, the general scheme is to delete a coding region from the vector, replace it with the DNA fragment of interest, and provide the missing viral gene product from a gene on a helper viral genome or on the host genome (i.e., supply it in *trans*). For example, a wild type helper virus can provide sufficient T or VP1, VP2, and VP3 for itself and the recombinant. The experiment is initiated by transfecting a cell monolayer with a mixture of wild type SV40 DNA and the defective viral DNA carrying the insert (Figure 5.35). Cells that take up either wild type or both DNAs will undergo lytic infection, and progeny DNA will be packaged in virions.

Regions of infected cells on the monolayer can be identified as plaques. Note that there is no simple way to distinguish plaques containing mixed progeny from those containing only wild type SV40. Additional analyses of the virions recovered from a single plaque are required to identify stocks containing the recombinant. This disadvantage is, however, readily overcome if the vector retains at least one functional viral gene and if the helper is not a wild type SV40 but is defective in the gene provided

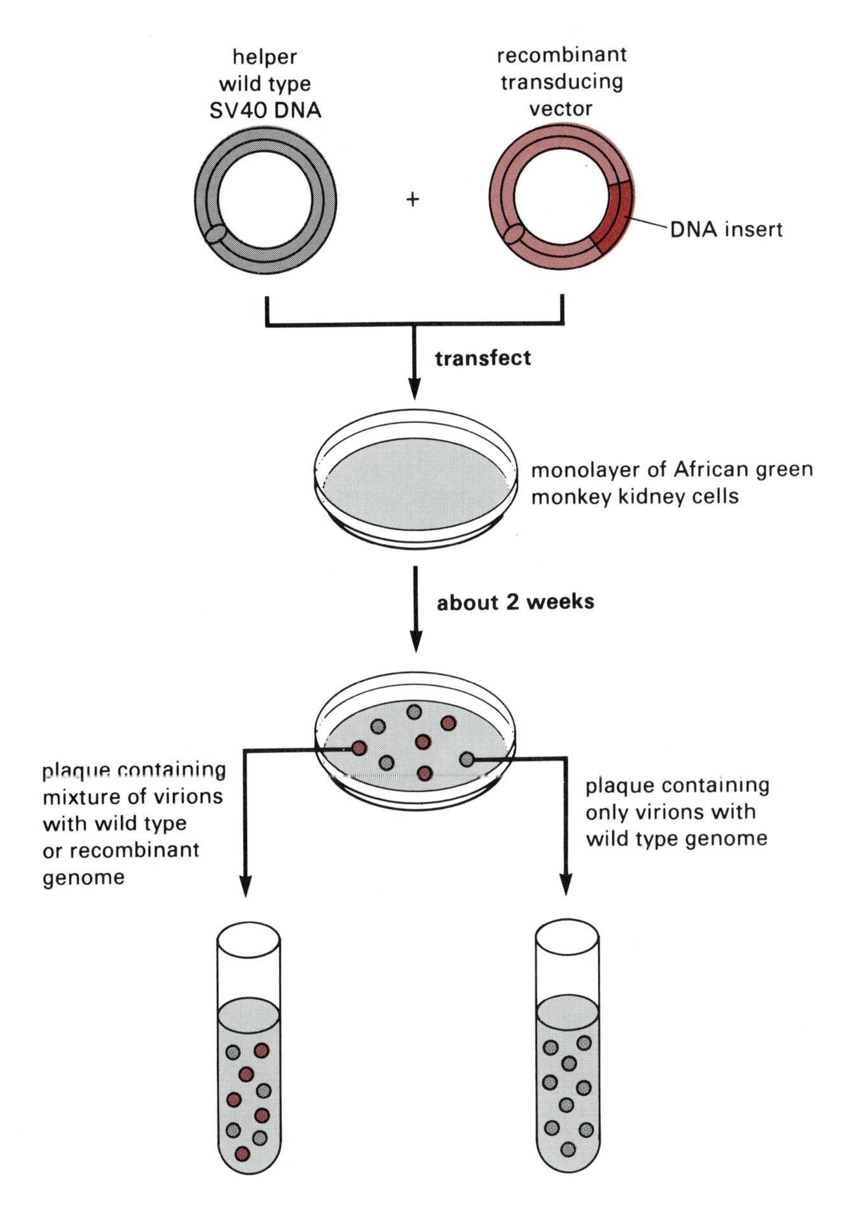

Figure 5.35
Complementation of an SV40 transducing vector with wild type SV40. A monolayer of African green monkey kidney cells is transfected with a mixture of wild type SV40 DNA and an SV40 transducing vector in which the early region (encoding large-T) of the viral DNA is replaced by a foreign DNA insert. Note that the viral origin of replication must be intact. Alternatively, the foreign DNA insert could replace the late region encoding VP1, VP2, and VP3. Some of the cells receive both types of DNA, and the viral functions missing in the recombinant are supplied by the wild type genome. Both DNAs replicate, and their progeny are encapsidated into new virions. These virions then co-infect neighboring cells. A clear area devoid of living cells—a plaque—appears on the monolayer around the initial infected cell. Material recovered from some plaques is a mixed stock of wild type and recombinant virus. Other plaques produce only wild type SV40.

by the vector. In this case, only cells transfected by both DNAs will produce plaques. For example, if the inserted DNA replaces portions of the late region that encode VP1, VP2, and VP3, an appropriate helper is one that has a mutation in the gene for large-T. The recombinant supplies large-T protein while the helper provides VP1, VP2, and VP3 (Figure 5.36). Conversely, if the inserted DNA replaces the early region of SV40, a helper that provides functional large-T but not the capsid proteins is chosen. Unlike the yields from transfections that utilize wild type helper, each plaque produced after transfection with vector and defective viral DNA yields a mixture of progeny.

In an alternative approach, the host genome supplies the missing function (Figure 5.37). COS cells are a line of African green monkey kidney cells that have a functional SV40 large-T gene integrated in the cellular genome. The cells produce sufficient large-T protein to support the replication of molecules that contain the SV40 origin of replication, about 80 base pairs. If such a transducing vector also encodes VP1, VP2, and VP3,

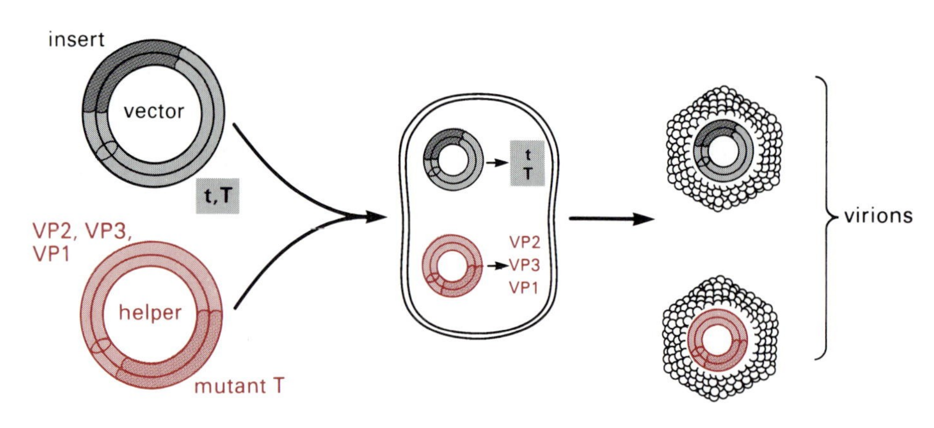

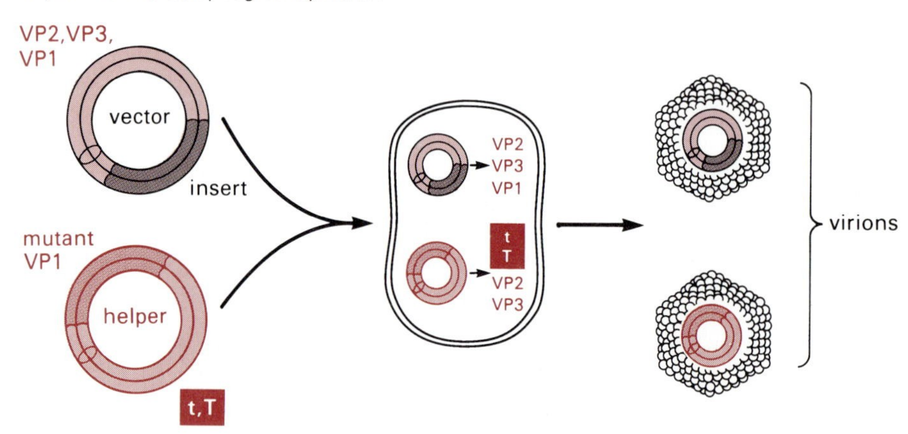

Figure 5.36
Complementation of SV40 transducing vectors with defective helpers. In the top diagram, the foreign DNA insert replaces the SV40 late gene region. Coding sequences for VP1, VP2, and VP3 are supplied by a helper genome with a mutated large-T antigen gene. In the bottom diagram, the foreign DNA insert replaces the T antigen coding region, and the helper, which has a mutation in the VP1 gene, provides a functional large-T antigen gene.

it will yield virion particles in COS cells even in the absence of a helper viral genome. Because no helper viral genome is required, the virions contain only a single type of genome, that of the transducing vector.

SV40 plasmid vectors The second category of SV40 vectors replicates in monkey cells but is not packaged into viral particles (Figure 5.38). Retention of the origin of replication is essential, as is a source of large-T; but VP1, VP2, and VP3 are not required. Either the vector itself or COS cells can supply large-T. SV40 plasmid vectors can be *E. coli*-animal shuttles because they are no longer limited to a size compatible with packaging. Recombinants are prepared in *E. coli*, and the DNA is then transfected into monkey cells. The cells can be used to study the transient expression of genes inserted in the vector. The high copy number achieved by the replicating vector increases the intracellular concentration of any gene contained on the vector, thereby facilitating analysis of the gene products. It would be advantageous for many purposes if these SV40 plasmid vectors behaved in animal cells as plasmids do in bacterial cells. Such cells would pass the recombinant genome on to daughter cells at each cell division and provide a permanent system for investigating gene expression. However, the plasmid vectors are unstable in most animal cells and cannot be maintained indefinitely. A relatively stable situation

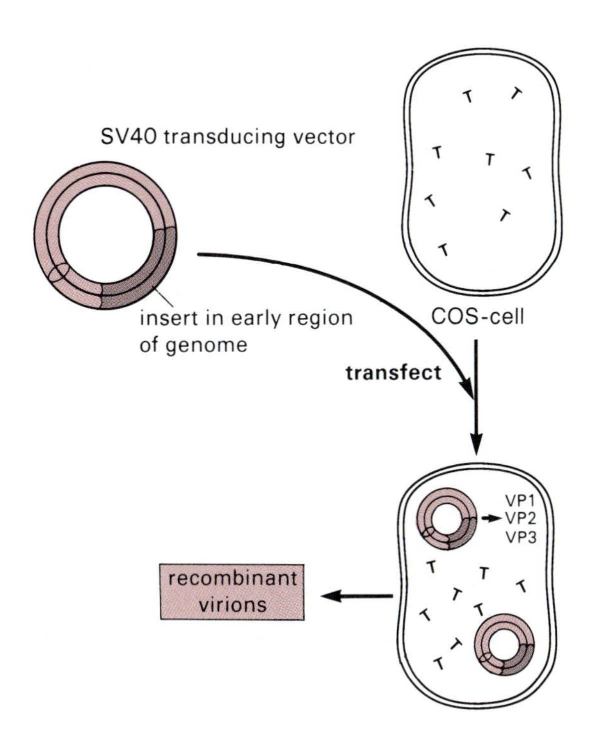

Figure 5.37
COS cells complement an SV40 transducing vector that lacks the large-T gene. A foreign DNA insert replaces the large-T coding sequences in this transducing vector. The genes for VP1, VP2, and VP3 are present and functional. The COS cell genome contains an integrated, functional large-T gene, and the cells produce large-T protein (T). After transfection, all the proteins required for replication and packaging of the vector DNA are present. Virions carrying the recombinant transducing vector are the only type of progeny produced.

can be achieved if COS cells are used to supply large-T for replication and if the vector bears a selectable marker, such as the *E. coli neo* gene, and cells are maintained and grown in G-418.

Stable transformation can be achieved if SV40 plasmid vectors integrate into the host genome. This occurs at a frequency of between 10^{-5} and 10^{-3}, depending on the system. Colonies of transformed monkey cells can be isolated if a selective marker is incorporated into the vector.

Passive transforming vectors The third type of SV40 vector does not replicate independently. Instead, it provides a vehicle for inserting cloned DNA segments into the cellular DNA. The permanently transformed cells that result replicate the new DNA as an integral part of their own genomes. These vectors are generally shuttles that are first cloned in and isolated from *E. coli* and then transfected into various mammalian cells (e.g., pSV2 in Figure 5.38). They are not restricted to use in monkey kidney cells because an environment conducive to replication is irrelevant. Selection for a marker included on the vector yields colonies of transformants. Selective markers such as the herpes virus thymidine kinase or the *E. coli gpt* and *neo* genes have been used. The SV40 DNA segments included in these vectors are generally the transcriptional regulatory signals and the polyadenylation sites that flank the genes in the SV40 genome.

c. *Bovine Papillomavirus Vectors*

The virus and its genome Warts, the benign skin tumors common in many mammals, are induced by papillomaviruses. This large group of viruses is in the papovavirus class, as is SV40, and different papillomaviruses are generally highly species specific. Virions are composed mainly of a capsid protein surrounding a closed, circular duplex DNA genome about 8 kbp

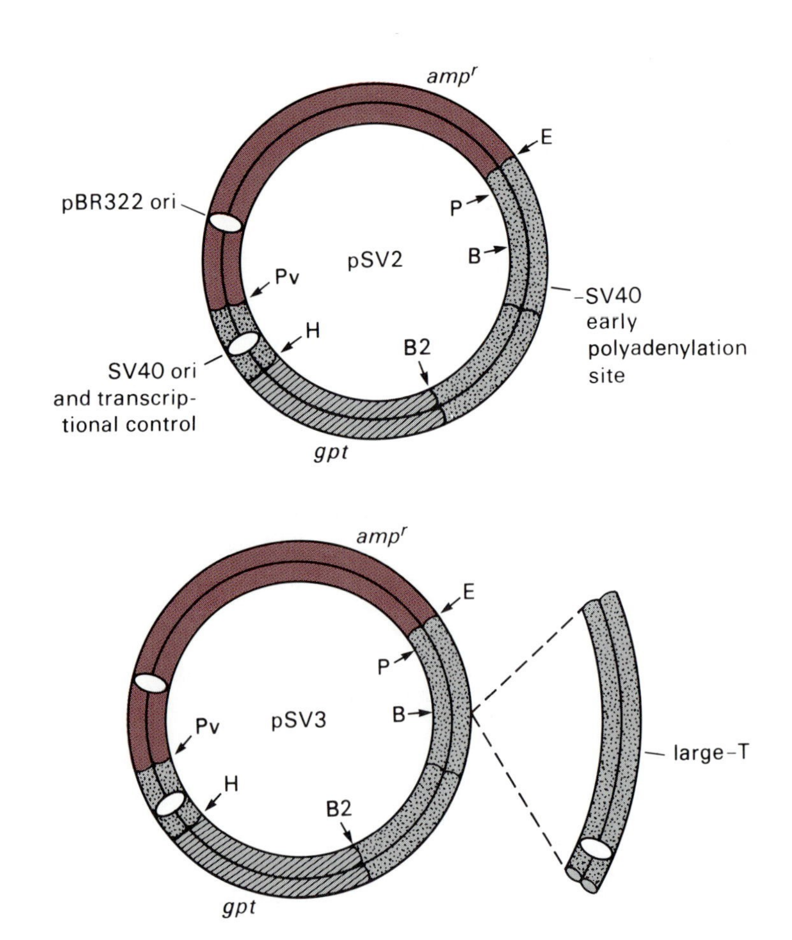

Figure 5.38
SV40 plasmid vectors. These shuttle vectors contain about 2.3 kbp of pBR322 sequences, including the origin of replication and the ampicillinase gene, for cloning in *E. coli.* The SV40 segment from about SV40 map position 5171 to 270 (counterclockwise), which includes the origin of replication and the transcriptional regulatory sequences for early transcription, is placed just to the 5′ side of the coding region for the *E. coli gpt* gene. At the 3′ end of the *gpt* gene are additional SV40 sequences, including some from around map position 2500 that have signals for polyadenylation of mRNA in animal cells and others that are required for the maturation of mRNA. The vectors were prepared by cloning in *E. coli.* The vector at the top, pSV2, is a plasmid vector in COS cells, but it cannot replicate in normal monkey cells because of the absence of large-T. In normal monkey cells or in other mammalian cells, pSV2 is used to study transient expression of the *gpt* gene or some other gene inserted at the *Bam*H1 or *Eco*RI site, for example. The vector at the bottom (pSV3) does not need COS cells to replicate; it acts as a plasmid vector in normal monkey cells because it encodes its own large-T. The SV40 large-T gene, along with its transcriptional control sequence, is inserted into the *Bam*H1 site of pSV2. Both pSV2 and pSV3 can insert into the host cell genome and effect permanent transformation. The 1 in 10^4 to 10^5 cells that is transformed grows into a colony in the presence of mycophenolic acid.

in length. The DNA, like SV40 DNA, is wound around nucleosomal cores composed of four cellular histones. Progress in understanding the biology and genetics of the papillomaviruses was frustratingly slow because no tissue culture system is known to support replication of the virus, requiring it always be isolated directly from warts. But with the advent of recombinant DNA techniques, the small amounts of papillomavirus DNA available from warts were purified, large quantities were prepared by molecular cloning in *E. coli*, and the genome structures were studied. The viral DNA has proven to be a useful and unique vector for animal cells, although the natural history of papillomaviruses remains poorly understood.

The entire nucleotide sequence of both a bovine and a human papillomavirus genome is known. The bovine virus genome (Figure 5.39) is the basis for a useful animal vector system. Unlike SV40, bovine papillomavirus (BPV) replicates as a stable plasmid in rodent and many bovine cells. No virions are produced because capsid protein expression is, as far as is known, limited to terminally differentiated bovine epidermal cells. Moreover, infected cells are not killed, and the BPV genome is passed on to daughter cells. These genomes transform the cells, causing heritable phenotypic alterations, including conversion of normal cells to tumor cells. Thus, transformants are detected as a piled up colony of cells, a **focus**, on the transfected monolayer. Clones of transformed cells can be established in culture much the same way as a colony is derived from

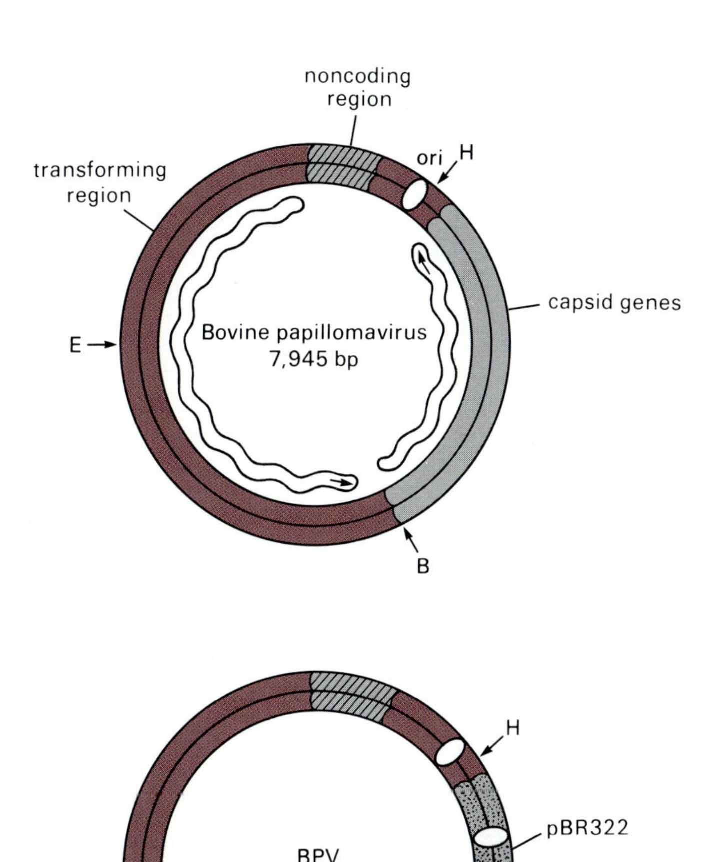

Figure 5.39
Schematic maps of the bovine papillomavirus (BPV) genome and a BPV shuttle vector. The color region is sufficient for transformation and includes a noncoding regulatory segment. The most likely position of the origin of replication is shown as an oval; a second origin may function under certain conditions. The direction and region of transcription for the transformation function(s) and for capsid protein synthesis are shown in the inner circle. Only the viral transforming sequences are included in the vector, which also carries pBR322 segments required for replication and selection in *E. coli.*

a single *E. coli* cell. Like many transformed eukaryotic cells, these grow more rapidly than their normal counterparts. Only those genes within the 69 percent of the genome marked "transforming" in Figure 5.39 are expressed or required for maintenance of the transformed state. Permanent transformation by BPV DNA does not involve integrating viral DNA into the cellular genome; the plasmidlike state is sufficient. Each transformed cell contains from 10 to 100 copies of the BPV genome.

Designing vectors Papillomavirus vectors can be constructed by ligating "foreign" inserts to the approximately 5500 bp of the "transforming" segment of the bovine viral genome (Figure 5.39). Generally, shuttle vectors are prepared by cloning and replication in *E. coli* using a plasmid vector such as pBR322. Eukaryotic DNA fragments are then inserted, and cloning in *E. coli* is repeated. Prior to transfection, the vector molecules may be cleaved by restriction endonuclease digestion to separate the pBR322 segments from the BPV (and insert) DNA because the bacterial plasmid sequences can inhibit the replication of covalently linked papilloma virus DNA. The linear DNA molecules are transfected into animal cells, where the segment containing the insert and the BPV transforming region circularizes and replicates as an independent plasmid. Many variations on the basic BPV vector have been used. For example, if the *E. coli neo* gene that is functional in animal cells is included, then transformed cells can be selected using the aminoglycoside G-418. This obviates the need to recognize a focus of piled-up cells, thereby widening the range of susceptible cells to those that do not show this morphological property.

d. Retrovirus Vectors

The retrovirus life cycle (Section 2.2a) The replication of the single strand RNA genome of retroviruses involves the formation of a double strand DNA copy of the RNA by reverse transcription, insertion of the viral DNA into the host genome, transcription of mRNAs and of new viral genomes from the integrated DNA, synthesis of viral structural proteins and reverse transcriptase, assembly of new virion particles, and extrusion of the particles into the surrounding medium (Figure 5.40). Virion particles contain, in addition to structural proteins, the reverse transcriptase that is necessary to initiate the life cycle. The integrated viral genome is called a **provirus**; its insertion causes the permanent transformation of the cells. Transformed cells are often detectable as dense foci of rapidly growing tumorlike cells, and many retroviruses are oncogenic.

Certain nucleotide sequences on the retroviral genome must be present for a productive life cycle to occur; that is, they are required in *cis*. However, the products of the viral genes, the group-specific antigens (**gag**) and envelope proteins (**env**) and the reverse transcriptase (**pol**), can be supplied in *trans*, by a helper virus or by a provirus. The structural elements that are required in *cis* are (1) those involved in reverse transcription and designated R, U5, U3, P, and Pu; (2) a short sequence, ψ, that is necessary for packaging genomes into virions; and (3) two sequences (S) required for splicing out parts of the RNA to form a functional mRNA for envelope protein synthesis (Figure 5.40). The long direct terminal repeated sequences (LTRs) at either end of the duplex DNA form of the viral genome and the provirus contain transcriptional regulatory units that foster transcription of the viral genes.

Retroviral vector design Three basic principles underlie the design of retroviral vectors. First, vectors must retain the viral genomic sequences required for replication, gene expression, and packaging. The viral proteins can be supplied in *trans*. Second, foreign DNA segments are inserted into or replace parts of the coding regions of the genome. Third, retroviral vectors are generally used as transducing viruses, the recombinant genome being packaged in a virion. In fact, retroviruses are themselves natural transducing viruses; they recombine in nature with cellular genomes and transduce cellular genes into newly infected cells as part of a provirus.

The DNA forms of many retroviral genomes have been cloned in *E. coli* host-vector systems and are thus available for further manipulation. These genomes are restructured into recombinant vectors *in vitro*, and the newly derived forms are cloned and replicated in *E. coli*. A typical vector

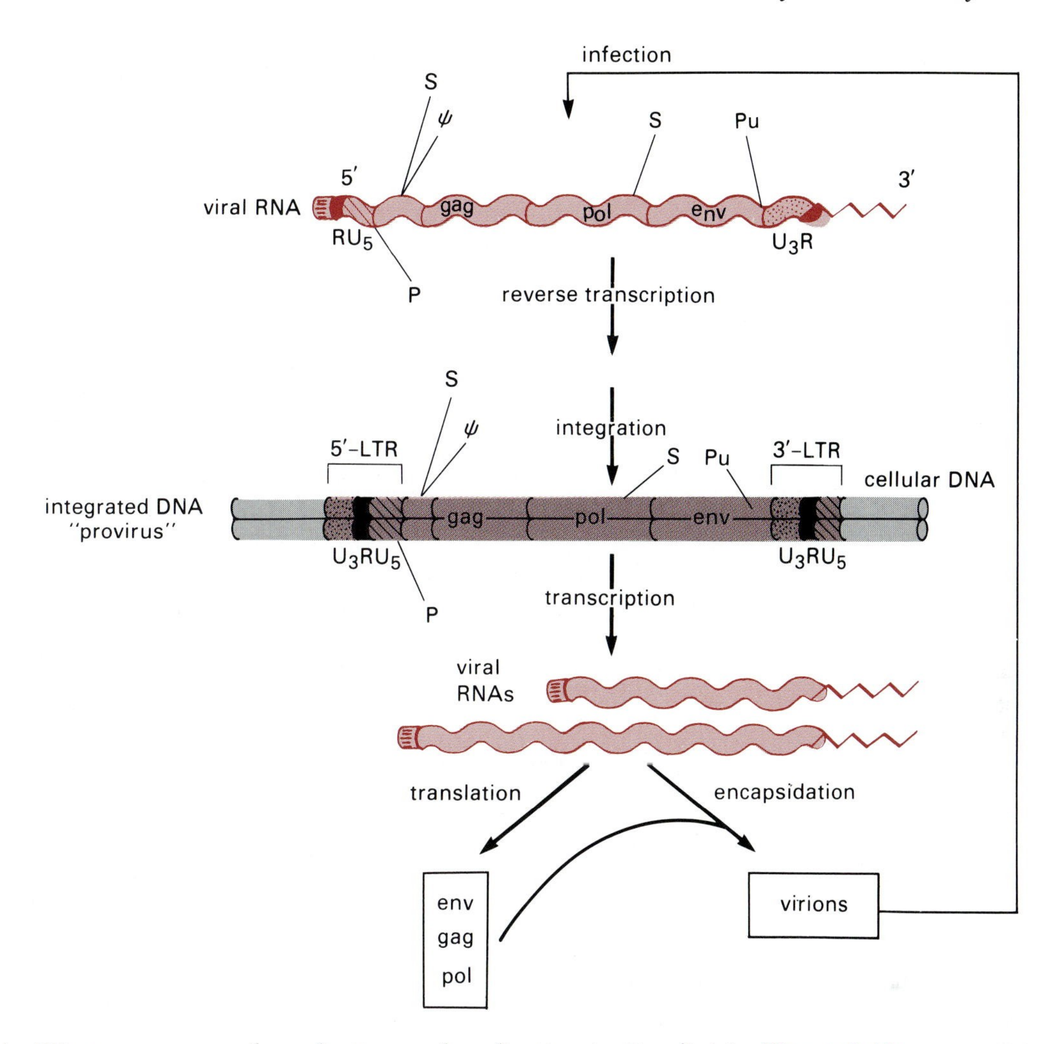

Figure 5.40
The life cycle of a retrovirus. (See Chapter 2 for details.) After infection, the RNA genome is converted to a duplex DNA by reverse transcriptase present in the virion. A provirus is formed when the duplex DNA is integrated into the cellular genome. Proviral DNA is bounded by long direct terminal repeats (5′ LTR and 3′ LTR). Transcriptional regulatory signals in the 5′ LTR promote transcription of the proviral DNA, and both mRNAs and new viral genomes are produced. The mRNAs are translated into virion proteins, gag, pol, and env, which together with the viral genomes, are assembled into virion particles that bud off from the plasma membrane. Nonviral genes inserted into the central coding region are also expressed under the control of the 5′ LTR.

contains (1) pBR322 sequences for selection and replication in *E. coli*, (2) the LTRs (long terminal direct repeats) of the viral genome, including the R, U5, and U3 segments; (3) the sequences P, Pu, ψ, and S; (4) a selectable marker such as the *E. coli neo* gene; and (5) at least one unique restriction endonuclease site that can be used to insert additional DNA segments and does not disrupt any essential viral sequences (Figure 5.41).

Vectors prepared in *E. coli* are transfected directly into appropriate animal cells. Sequences in the 5′ LTR serve as a transcriptional control element for RNA polymerase II, and a transcriptional termination signal occurs in the 3′ LTR. Consequently, an RNA is synthesized that is essentially a viral genome (Figure 5.41) except that one or more nonviral segments are included between the 5′ and 3′ end. Note that the *E. coli* vector sequences are not transcribed and do not appear in the recombinant RNA retroviral genome.

The recombinant RNA retroviral genomes can be recovered from the medium as virion particles, providing the proper viral proteins are supplied. This happens if the host cells contain a completely functional provirus that supplies the viral proteins in *trans* (Figure 5.42). In this case, the progeny will be a mixed population of recombinant and wild type helper virions. Such a stock can be used repeatedly to infect new cells, the necessary helper virus being supplied along with the recombinant. However, pure stocks of recombinant virions can be obtained if the

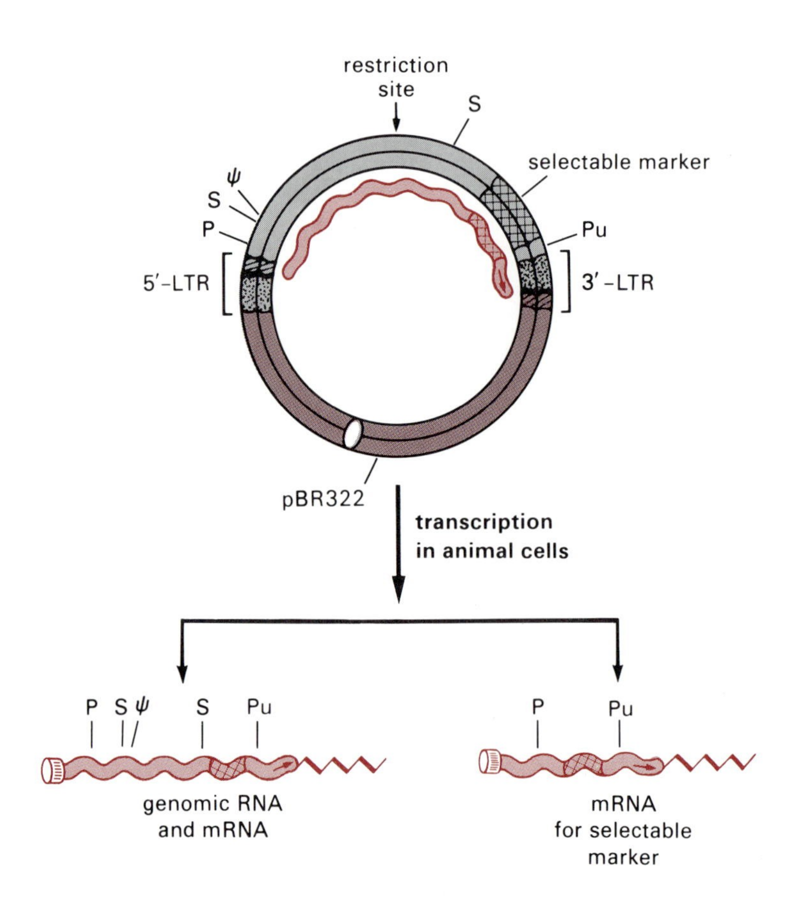

Figure 5.41
A retroviral shuttle vector. This plasmid contains pBR322 sequences that permit selection, replication, and cloning in *E. coli.* A modified retroviral genome is joined to the bacterial plasmid. Note that both LTRs are present, as are the other sequences (P, ψ, S, and Pu) required in *cis* for replication, mRNA formation, and packaging of a retrovirus. The DNA between the LTRs contains some retroviral sequences (open bar) as well as an insert encoding a selectable marker (e.g., the *E. coli neo* or *gpt* genes or the herpes virus thymidine kinase gene). Other DNA segments can be inserted at the unique restriction endonuclease site or can replace parts of the remaining retrovirus segments. After transfection into an animal cell, transcription initiates in the 5′ LTR and the RNA ends at transcription termination signals in the 3′ LTR. Some of the transcripts will serve as viral genomes and be encapsidated by viral proteins encoded either on the vector or in *trans,* either on a helper virus or in the genome of the host cells. Others will be used as mRNAs. In this example, the selectable marker gene is in the position usually occupied by the *env* gene. Transcripts destined to serve as mRNA for genes in this region lose all the sequences between the two sites S by splicing out the intervening RNA and joining the two ends at the S sites.

provirus in the initial host cells is defective for packaging. Such host cells are obtained by infecting cells with a deletion mutant missing the ψ sequences (Figure 5.43). Although the viral RNA produced from the ψ^- provirus cannot be packaged, the recombinant genomes are capable of being packaged, reinfecting new cells, and forming provirus because their ψ sequence is intact. The virion particles contain the proteins necessary for reverse transcription and insertion into a cellular genome in the next infectious cycle.

The use of retroviral vectors Transducing virus vectors always have the advantage that infection of cells is much more efficient than transfection. Once a viral stock of a newly constructed retroviral vector is prepared by transfection into cells supplying helper function, all subsequent transfers can be by infection. Moreover, the stock will not require helper if the aim of the experiment is cell transformation. The recombinant genome forms a provirus, and the cells are permanently transformed. Transformed colonies can be selected in several ways (Figure 5.44). If the vector includes a functional marker such as the *E. coli neo* or *gpt* genes, then transformed cells can be selected in the presence of G-418 or mycophenolic acid, respectively. Note that the transcriptional regulatory signals in the 5′ LTR promote the synthesis of functional mRNA for the *E. coli* genes in the animal cells. Also, if the *E. coli* (or any other) gene is inserted in place of the *env* gene (as it is in Figure 5.41), then the S sites must be intact if a functional mRNA is to be produced. Alternatively, transformed foci can be detected by their characteristic morphology if the vector retains an oncogene that was present in the original virus. An otherwise unselect-

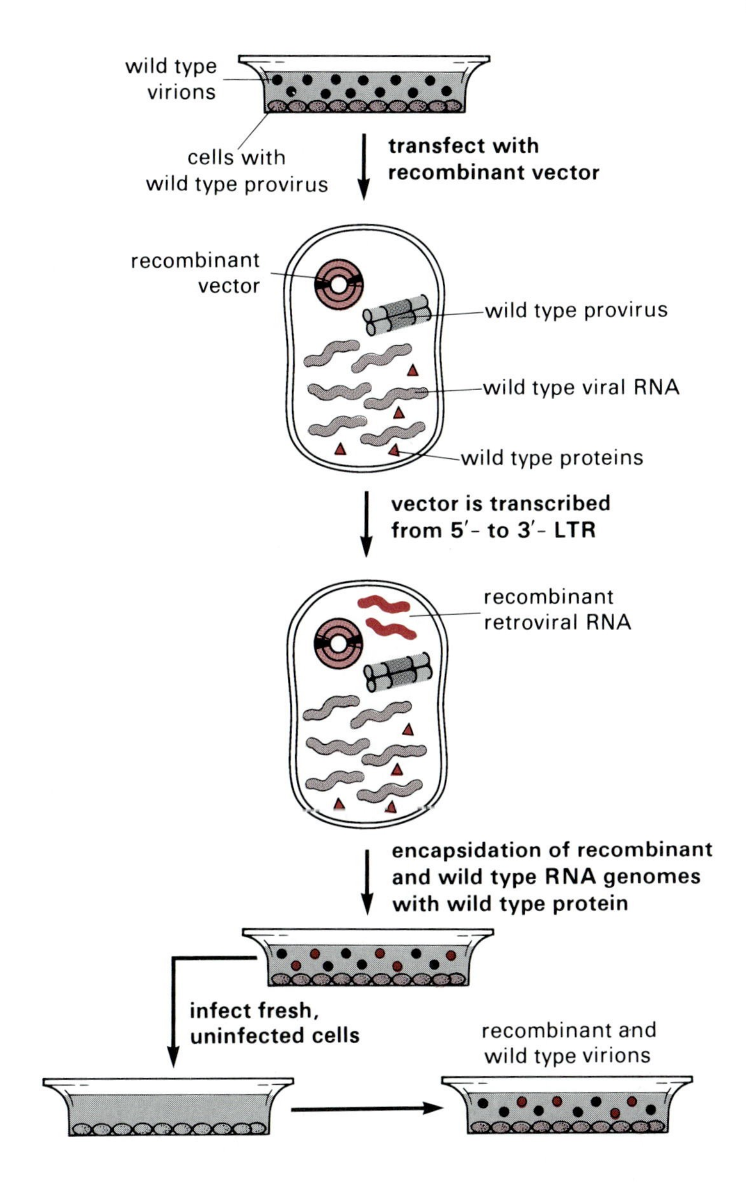

Figure 5.42
Conversion of a duplex recombinant DNA retrovirus vector into an encapsidated RNA genome using host cells containing a wild type provirus. The cells in the culture dish at the top contain integrated proviruses that produce wild type genomic RNA and viral proteins and extrude wild type virions. Within a short time after transfection with the double strand retroviral recombinant vector (Figure 5.41), recombinant RNA is produced by transcription from the 5′ to the 3′ LTR. Both recombinant and wild type RNA genomes are encapsidated in wild type proteins, and the mixture of virions collects in the medium. This fluid is then used as a stock of recombinant plus helper virus for infection of new cells; after reverse transcription, both wild type and recombinant proviruses integrate into some of the newly infected cells, which then become producers of mixed virions.

able gene inserted in the vector along with the selectable marker will also be present in the transformed cells. These properties of transducing recombinant retroviruses, in particular the efficiency with which cells are infected and transformed, make this system especially useful for the experimental transformation of cells in very early embryos and whole animals.

The construction of a retroviral recombinant vector begins with a shuttle that replicates in *E. coli* and is then transfected into animal cells. However, the *E. coli* plasmid sequences are lost in the initial transfection because transcription in animal cells proceeds only from the 5′ LTR to the 3′ LTR (Figure 5.41). The packaged RNA genomes in the virions contain no plasmid vector segments. Often this is of no consequence in subsequent experiments, but in some situations, being able to shuttle the recombinant sequences back through *E. coli* is important. This is accomplished by inserting an *E. coli* plasmid segment such as pBR322 between the LTRs in the initial construction (Figure 5.45, page 316). All forms of the viral genome, the RNA, the duplex DNA, and the provirus will carry the pBR322 segment. Thus, the retroviral vector can be rescued from animal cells by cloning directly into *E. coli*. Duplex viral DNA or genomic DNA

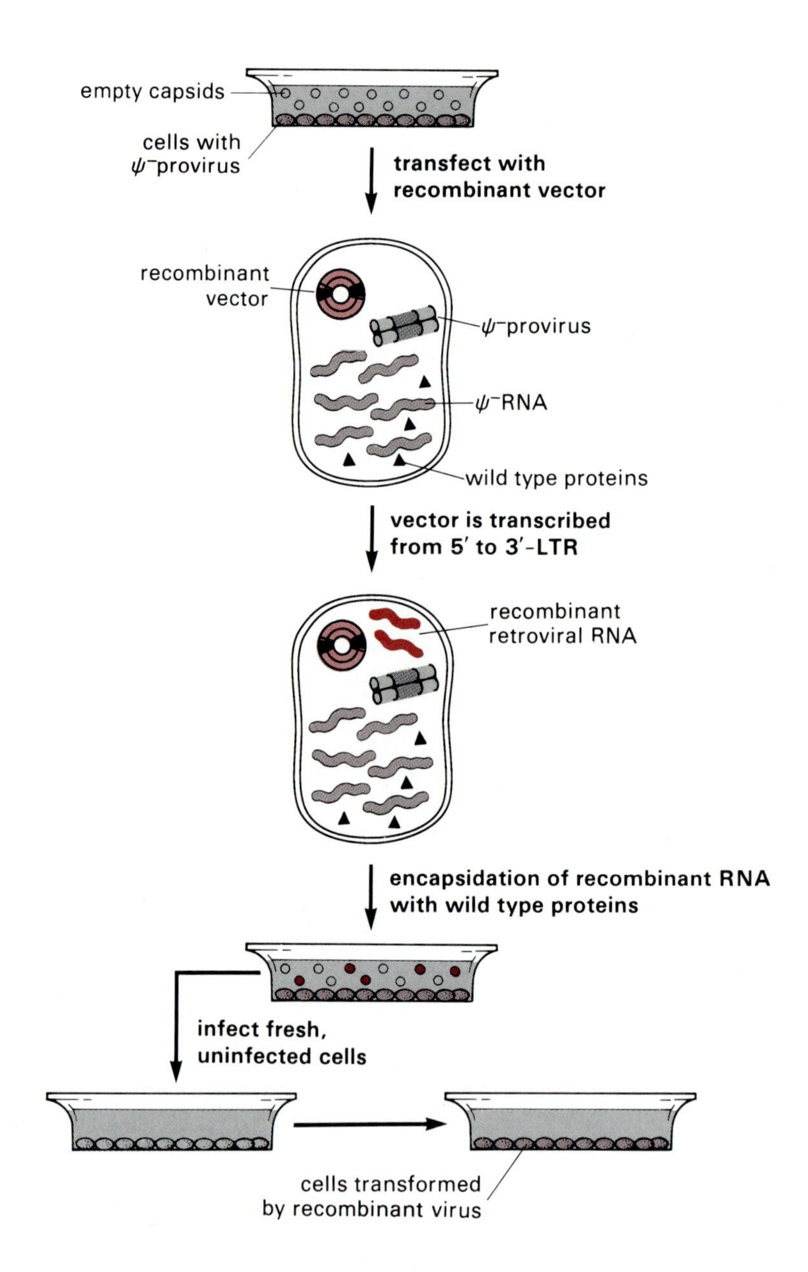

Figure 5.43
Conversion of a duplex recombinant DNA retrovirus vector into an encapsidated RNA genome using host cells carrying a ψ^- provirus. The cells in the culture dish contain an integrated provirus that has a deletion of the ψ sequence. Viral RNA and proteins are produced, but the RNA cannot be encapsidated; only empty, noninfectious capsids accumulate in the medium surrounding the cells. Within a short time after transfection with the double strand retroviral recombinant vector, recombinant RNA is produced by transcription from the 5′ to the 3′ LTR. The recombinant RNA is encapsidated in the wild type proteins, and recombinant virions collect in the medium. This stock contains only empty capsids and recombinant virions. When the stock is used to infect fresh cells, replication of the recombinant genome is initiated by the reverse transcriptase in the virions. The DNA integrates as a provirus, and the resulting transformed cells express genes contained in the recombinant provirus.

segments carrying the provirus are circularized, ligated, and transfected into *E. coli*, where they replicate. The new recombinant molecules frequently contain segments of genomic DNA that had flanked the provirus in the host genome. Because provirus insertion into or near functional genes is often mutagenic or oncogenic or both, analysis of such flanking sequences can be used to isolate the gene associated with the observed phenotype.

5.8 Eukaryotic Host-Vector Systems: Plants

a. *General Considerations*

No plasmids indigenous to plant cells are known. The independently replicating genomes of various plant viruses can, in principle, form the basis for transforming vector systems, and systems based on the genome of cauliflower mosaic virus are being developed. The best developed plant

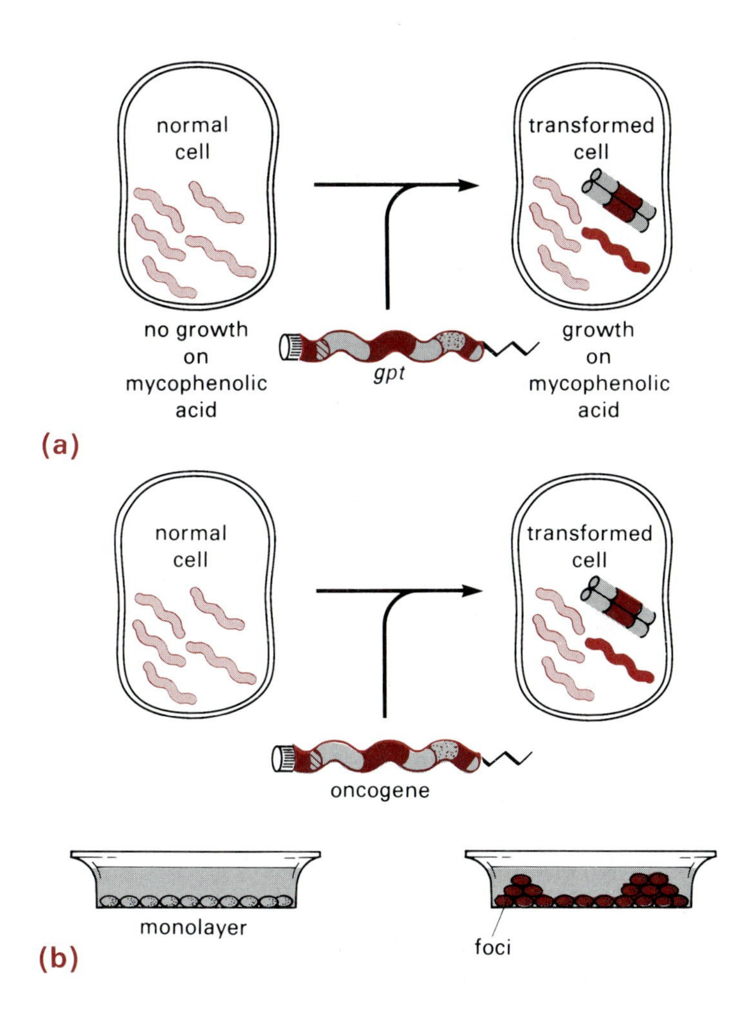

Figure 5.44
The selection of cells transformed by retrovirus vectors. (a) The recombinant genome contains the *E. coli gpt* gene. The reverse transcriptase in the recombinant virion catalyzes the formation of duplex DNA after the virion enters the cell, and the DNA is integrated into the cellular genome to form a provirus. Transformed cells are selectable because they express the *gpt* gene, which overcomes the inhibition by mycophenolic acid. If an unselectable gene is included in the recombinant, then it too is expressed.
(b) The retroviral genome contains an oncogene. Transformed cells can be selected because they form multilayer foci of cells, but untransformed cells grow as a monolayer. Again, the transformed cells also express an unselectable gene that is included in the transforming genome.

vector systems all derive from a family of unusual bacterial plasmids called **pTi**. These plasmids make up a naturally occurring transformation system that transfers plasmid DNA segments into the genomes of a wide variety of dicotyledonous plants.

In the absence of a specific vector, direct transformation of at least some plant cells is attained by transfection with foreign DNA fragments added to a culture medium. In analogy to animal cells, the plant cells take up the DNA and integrate it into the cellular genome, yielding permanently transformed cells. The efficiency of direct transformation is, however, quite low. Because only about 1 in 10^6 treated cells is transformed, a highly efficient selection procedure is required. DNA can be introduced directly into plant cells more efficiently by electroporation. Up to 1 percent of the cells may become transformed, and the procedure works with both monocotyledonous and dicotyledonous species. Various plant cells, including tobacco, petunia, tomato, and sunflower, have been transformed by recombinant DNA molecules. Protoplasts prepared by digestion of the tough cell walls with cellulase, for example, are often used because they are permeable to DNA. The cell wall regenerates when the transfected protoplasts are placed in a proper medium.

b. pTi-A Tumor-Inducing Plasmid

The pTi family of conjugative plasmids is found in the gram-negative bacteria *Agrobacterium tumefaciens*. The closed circular duplex genomes of

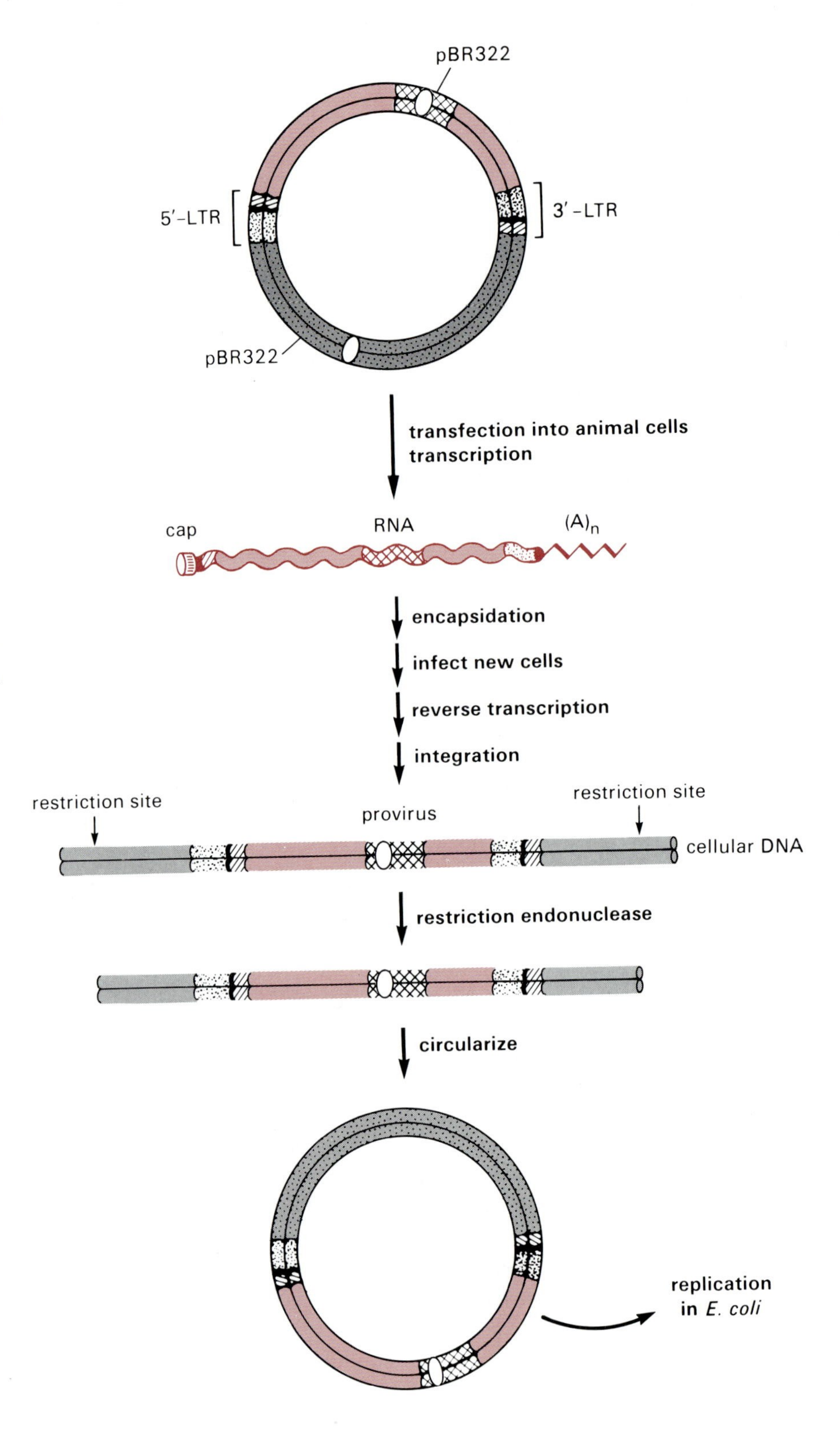

Figure 5.45
A retrovirus vector that can be recovered from animal cells and cloned in *E. coli*. The initial vector is essentially like that in Figure 5.41, except that additional pBR322 sequences, capable of supporting replication in *E. Coli*, are inserted between the 5′ and 3′ LTR, along with other inserts. After transfection, transcription from the 5′ LTR to the 3′ LTR yields a viral genome (RNA) that includes the pBR322 segment, and this is encapsidated by viral proteins supplied in *trans*. Infection of a new cell yields duplex DNA and provirus still carrying the pBR322 segment. If DNA isolated from transformed cells is cleaved with restriction endonucleases, fragments containing the provirus are produced (along with many others). Upon circularization and transfection into *E. coli*, the circle containing the provirus is uniquely capable of replication.

pTi plasmids range from 150 to 250 kilobase pairs and contain a variety of unique genes (Figure 5.46 and Table 5.5). *A. tumefaciens* cells bind to wounded plant tissue, whereupon pTi sequences enter the plant cells and recombine with cellular DNA. There is no one specific site for integration on the plant genome. The cells are transformed and produce a tumor

Table 5.5 Some of the Genes on Ti-Plasmids

Octopine Plasmids	Nopaline Plasmids
6 virulence genes (*vir*)	6 virulence genes (*vir*)
octopine synthesis (*ocs*)	nopaline synthesis (*nos*)
octopine catabolism (*occ*)	nopaline catabolism (*noc*)
exclusion of phage AP1 (*ape*)	required for conjugation (*tra*)
auxin biosynthesis (*iaa*)	auxin biosynthesis (*iaa*)

Figure 5.46
Schematic map of a typical nopaline pTi. The colored region shows the T-DNA segment that is found inserted in the DNA of crown gall tumor cells. Some genes are defined in Table 5.5 (*agr* encodes sensitivity to Agrocin 84). The arrowheads show the position of the 25 bp sequences at the right and left borders of the T-DNA segment; the right-hand 25 bp are required for integration of T-DNA into a plant genome. The position of the replication origin (for replication in *A. tumefaciens*) is indicated (*ori*).

called a **crown gall** (Figure 5.47). *A. tumefaciens* strains that lack pTi do not produce crown galls. Cells isolated from crown galls, unlike normal plant cells, grow readily in laboratory nutrient media, even in the absence of plant growth hormones. If they are grafted onto a healthy plant, a new gall forms. Sometimes the expression of the tumor phenotype is suppressed, and normal plant tissues can be regenerated from gall cells. In the laboratory, transformation is achieved by inoculating wounded plant tissues such as stems, roots, or bits of leaves with *A. tumefaciens* or by cocultivating the bacteria with plant protoplasts. Ten percent or more of tobacco protoplasts are transformed by cocultivation, for example.

pTi genomes Generally only a small and varying portion (between 13 and 25 kbp) of the pTi genome is inserted into the DNA of crown gall cells, but this always includes one particular segment called **T-DNA** (Figure 5.46). RNA is transcribed from the integrated T-DNA in the crown gall cells, and T-DNA encoded proteins are synthesized. Among these are proteins that establish the tumor phenotype of the crown gall cells. In addition, enzymes that catalyze the synthesis of a group of arginine derivatives known as **opines** are produced. Opines are not produced by untransformed cells. Ti-plasmids are classified according to the particular opine associated with them. For example, octopine Ti-plasmids induce

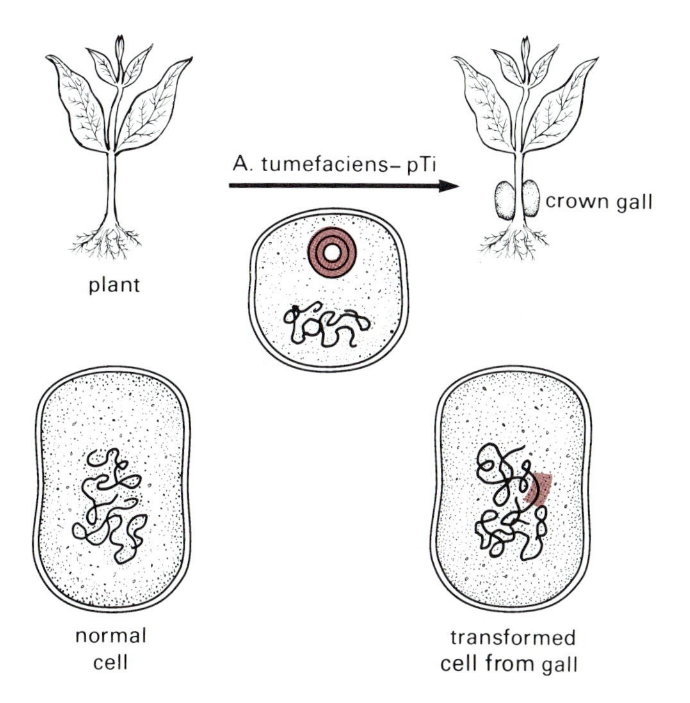

Figure 5.47
pTi-induced transformation and crown gall tumor formation.

octopine synthesis within transformed plant cells, and nopaline Ti-plasmids induce nopaline synthesis (Table 5.5). However, each Ti-plasmid also has genes that are expressed in *A. tumefaciens;* these are encoded outside the T-DNA region. Some of these, the virulence (*vir*) genes, are required for the transfer of T-DNA to the plant and are thus essential for tumor formation. Others are associated with the ability to catabolize the same opine whose synthesis is encoded by the T-DNA. Expression of the opine-metabolizing genes is induced in the bacteria by opines themselves. Ti-plasmids thus provide the bacteria with two important resources: a source of a metabolite (by transforming a plant cell) and the means to use that metabolite as a source of carbon, nitrogen, and energy. The secretion of opines by the transformed plant cells is highly advantageous to the *A. tumefaciens* in the surrounding soil because other common soil bacteria are unable to metabolize these unusual compounds.

The pTi-*A. tumefaciens* system is remarkable because it is an example in nature of the insertion of prokaryotic plasmid DNA into a eukaryotic genome. Actually, the plasmid itself seems to be a natural chimera; it carries two sets of genes, one active in bacteria and the other, in plants. The genes in the T-DNA segment are associated with transcriptional control signals that operate in plants while those in the remainder of the plasmid are under the control of bacterial promoters. Note that genes in both groups are required for tumor formation. The *vir* genes, which are expressed in the bacteria, are essential for the transfer of T-DNA to the plant genome. Other genes within the T-DNA are expressed in the plant and are responsible for the tumor phenotype of the crown gall cells. These oncogenes encode proteins that catalyze the production of the plant growth hormones auxin and cytokinin.

Insertional transformation by pTi The mechanism of insertional transformation by pTi differs from that of the other eukaryotic systems described in this chapter but has some similarities to bacterial conjugation. The *A. tumefaciens* chromosome encodes functions required for attachment of the bacteria to plant cells. pTi encodes *cis* and *trans* functions required for integration. A 25 bp segment at the right border of the T-DNA segment must be present for integration to occur; only sequences to the left of this region are transferred. A similar sequence occurs at the left border of the T-DNA. It does not appear to be required for integration, but it helps define the end of the integrated T-DNA. *Trans* functions are provided by the gene products of the *vir* genes (Figure 5.46). Among these is a site-specific endonuclease that nicks the lower T-DNA strand within both border sequences. The 3′ end of the pTi DNA next to the right-hand nick serves as a primer for DNA synthesis, displacing the lower T-DNA strand. The free T-DNA strand transfers to the plant cell, 5′ to 3′, starting with the right border. The mechanism of insertion into random sites on the plant DNA is not understood.

c. *Designing Recombinant DNA Vectors with pTi*

Exploitation of pTi as a recombinant DNA vector depends on being able to insert a DNA segment into the T-DNA region and to select and establish

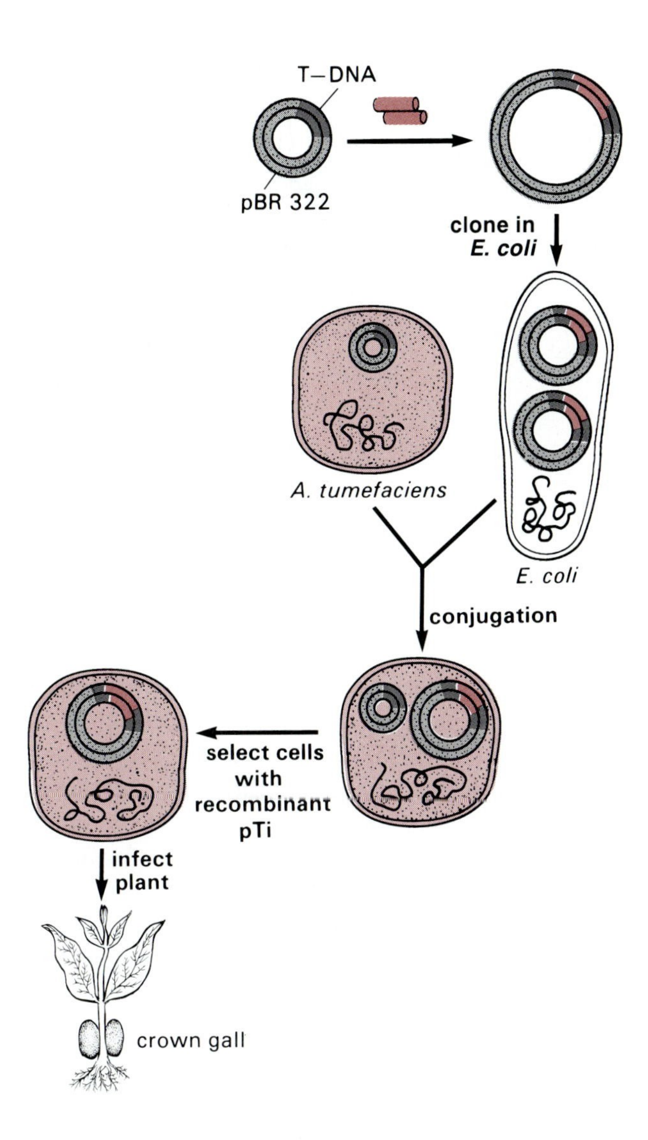

Figure 5.48
Construction and use of a pT intermediate vector called iv.

the recombinant in *A. tumefaciens* cells so that it can be transferred to a plant cell. Although the absence of unique restriction endonuclease sites in the large pTi molecule precludes direct insertion of a foreign DNA segment, there is a general route for inserting genes into the T-DNA region. A segment of the T-DNA region is itself first cloned in *E. coli* (using pBR322, for example). Desired DNA segments are inserted at convenient restriction endonuclease sites in the T-DNA segment in the isolated plasmid, and these recombinants are recloned in *E. coli* (Figure 5.48). When such plasmids are introduced by conjugation into *A. tumefaciens* that already carry a wild type pTi, homologous recombination within the cell transfers the T-DNA containing the new insert segment into the T-DNA region of the full pTi. *A. tumefaciens* cells that carry the recombinant pTi can be selected if an appropriate marker gene such as the *E. coli neo* gene is included in the foreign DNA insert. Replication of the pTi recombinant in *A. tumefaciens* followed by infection of plants yields crown gall tumors carrying the recombinant T-DNA segment.

A relatively simple pTi vector system has been designed by taking advantage of the distinction between *cis* and *trans* functions on the plasmid. A shuttle plasmid that can replicate both in *E. coli* and in a wide range of other bacteria (including *A. tumefaciens*), and carries both 25 bp border

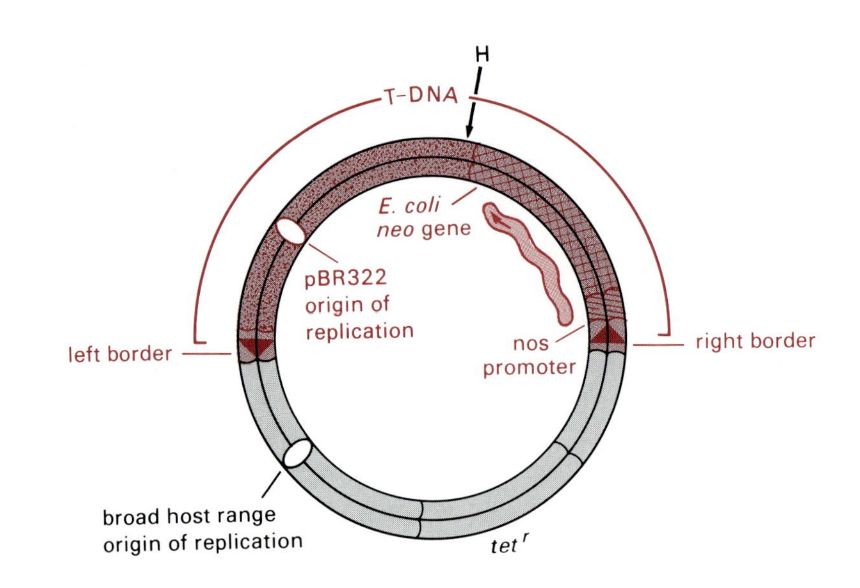

Figure 5.49
T-DNA–*E. coli* shuttle vector. The grey region is derived from an *E. coli* R plasmid (Section 5.2) that can replicate in a wide range of gram-negative bacteria, including *A. tumefaciens.* The region includes an origin of replication, all functions required for replication and for conjugation by mobilization, and a gene encoding tetracycline resistance (*tet*^{r}) for selection in *E. coli.* The arrowheads are the T-DNA border sequences that function in *cis* for integration into plant cell DNA. Between the left and right border sequences is the T-DNA region that will be transferred. This includes a pBR322 replication region and an *E. coli neo* gene that can be expressed in plant cells because it is linked to a transcriptional control region that normally promotes transcription of the *nos* gene of pTi (the arrow shows the direction of transcription). The single *Hin*dIII site provides a position for insertion of additional DNA sequences into the T-DNA region.

segments of T-DNA is the basic vector (Figure 5.49). Preparations of vector derivatives containing various DNA insertions to the left of the right-hand 25 bp border segment are readily obtained by cloning in *E. coli.* When such molecules are transferred by conjugation into *A. tumefaciens* carrying a wild type pTi, they replicate, and the gene products required for insertion into plant DNA are supplied to the recombinant plasmid in *trans.* Plant cells infected with such *A. tumefaciens* cells integrate either the wild type T-DNA or the recombinant T-DNA or both. Functional marker genes included in the recombinant T-DNA are expressed from its integrated position in the plant DNA so long as they are associated with transcriptional regulatory signals that promote RNA synthesis in eukaryotic cells. This can be achieved, for example, by placing the marker gene to the 3′-side of the regulatory signals of the *nos* gene in the recombinant T-DNA. If, instead of the wild type helper, a helper pTi plasmid is used from which the 25 bp long *cis* function or the entire T-DNA segment has been deleted, then only the recombinant T-DNA can be transferred.

Many plant cells are **totipotent** (can multiply, differentiate, and form whole and fertile plants). The pTi vector systems could, ideally, permit the formation of genotypically altered plants from transformed cells, and this has been achieved in several instances. Moreover, the experiments performed thus far have revealed fundamental information about gene expression and differentiation in several plant species.

CHAPTER 6

The Means: Constructing, Cloning, and Selecting Recombinant DNA

This chapter describes the steps essential for obtaining cloned DNA molecules. First, the DNA fragment (the "insert") is prepared in a form suitable for ligation to a vector. Next, the ligation itself is performed. These steps take place *in vitro*. Thereafter, the recombinant DNA molecules are introduced into individual cells in which they are amplified by replication. This is followed by cloning, selection, and further amplification.

6.1 Inserts

a. General Considerations

Recombinants are usually constructed using a large and complex mixture of potential inserts, thereby generating an extensive collection of clones. The desired recombinant is finally obtained by specific selection and screening procedures. Both procedures are much easier if the initial mixture is partly enriched for the segment of interest because fewer recombinant clones will need to be tested to find the desired one. Although a pure DNA fragment can be used to construct a recombinant DNA molecule, molecular cloning itself is usually the simplest and most efficient way to purify a fragment. It is also the best way to prepare most genomic DNA fragments in significant quantities.

There are three sources of inserts for cloning: (1) genomic DNA fragmented either by restriction endonucleases or by physical methods such as mechanical shearing or sonication, (2) synthetic DNA segments prepared by chemical or enzymatic methods or a combination of both, and (3) DNA segments (cDNAs) prepared by enzymatic copying of RNA templates *in vitro*. Restriction endonuclease digestion often yields fragments that can be ligated directly into either matching cohesive termini or flush-ended termini on the vector. Otherwise, appropriate terminal extensions are attached to the fragments before ligation.

b. Inserts from Genomic DNA

Restriction endonuclease digestion Using a restriction endonuclease to digest total genomic DNA from any organism or virus is a direct way to obtain inserts. This procedure has the advantage of reproducibility; the same set of fragments is produced each time a specific enzyme cleaves a particular DNA. If the endonuclease yields cohesive ends that match those on the vector, cloning is carried out directly or after enrichment of the population of fragments for the desired insert.

Enriching for selected fragments in the digest Two procedures, **electrophoresis** on semisolid supports (Section 4.2d, Figure 4.10) and **high performance reverse phase liquid chromatography**, afford extensive resolution of large and complex mixtures of DNA fragments and allow the separated fragments to be recovered for cloning. However, neither method yields pure DNA fragments from complex mixtures. Preparations are usually contaminated with many other fragments of similar size or elution properties. Nevertheless, enrichment for the desired fragment can be substantial, significantly decreasing the ultimate problem of finding the desired DNA segment in a large collection of clones.

Electrophoresis and chromatography separate DNA fragments primarily on the basis of size and base composition. When small genomes are used, the relatively few fragments are well resolved (Figure 4.10) and are easily isolated by collecting appropriate fractions of the eluate from a chromatography column or by eluting the fragments from appropriate pieces of the electrophoresis gel. When the digested genome is large, however, few if any well-separated DNA fragments are obtained. Rather, a continuous smear of DNA fragments of all possible sizes appears (Figure 6.1a). It is easy to calculate why. A typical mammalian haploid genome contains about 3×10^9 base pairs. A rough estimate of the number of fragments produced by exhaustive endonuclease digestion is obtained by dividing the genome size by the expected average distance (in base pairs) between two sites for a given restriction endonuclease (assuming a random array

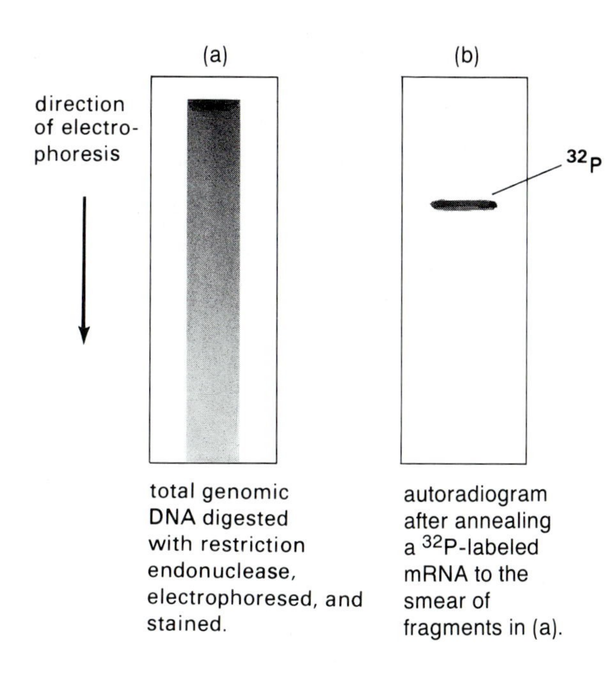

Figure 6.1
(a) Electrophoretic separation of the fragments produced by digestion of a eukaryotic genomic DNA with a restriction endonuclease. The fragments are visualized by staining with ethidium bromide. Flat gels of about 0.5 mm to several millimeters in thickness are prepared on a solid (e.g., glass) support. Gels are generally made of polyacrylamide or agarose with an appropriate buffered salt solution.
(b) The DNA on the gel in (a) was transferred to nitrocellulose by the DNA blotting technique described in Figure 6.2 and hybridized with a purified ^{32}P-labeled mRNA. The diagram shows an autoradiogram made by contact exposure of X-ray film to the nitrocellulose sheet.

of the four bases). An enzyme with a six base pair recognition site (e.g., *Eco*RI or *Hin*dIII) will find a site every 4^6 (4096) base pairs on the average. If the recognition site is four base pairs long, the number is every 4^4 (256) base pairs. Digestion of mammalian DNA with enzymes that cleave at six base pair recognition sites should yield about 7×10^5 unique fragments, digestion with a four base pair enzyme, about 12×10^6. The resulting sets of fragments yield a continuous spread of DNA molecules of all possible sizes; individual fragments are invisible because each one represents a minute amount of the total.

Finding a specific fragment The electrophoresis gel in Figure 6.1a illustrates the problem. If about 50 μgrams of genomic DNA are digested and electrophoresed, then the mass of a DNA segment 1000 base pairs long and occurring once in the genome is only 17×10^{-6} μgrams or 17 picograms. How can a specific fragment be located for elution? If a homologous DNA or complementary RNA preparation is available for use as a hybridization probe, then the fragment can be found by annealing the probe to the fragments after they are denatured. For example, if the messenger RNA corresponding to the gene to be cloned has been purified, it can be used as a probe. Sometimes a homologous gene cloned from another organism is similar enough in structure to be used as a probe. To detect the hybridization, the probe must be labeled, usually with a radioactive isotope.

Figure 6.1b shows the idealized result of annealing a ^{32}P-labeled messenger RNA probe to the (denatured) fragments in the gel of Figure 6.1a. The genomic fragment that contains sequences complementary to the messenger RNA has formed a duplex with the probe, and its electrophoretic position is revealed by the location of the ^{32}P-RNA. Once the approximate size of the desired fragment is ascertained, cloning can be carried out with material eluted from the comparable region of a similar, unhybridized gel.

Fractions collected by elution of a digest from a chromatography column can also be tested for hybridization with a radioactive probe.

Blotting Almost all experiments like that described in the previous paragraphs as well as many others involve **DNA blotting** rather than directly treating the electrophoretic gel with the radioactive probe. Blotting deserves special attention because it is widely used and particularly important, albeit quite simple in concept. Analogous procedures for detecting specific RNA and protein molecules following their electrophoretic separation exist and are described in Sections 7.7 and 6.4, respectively. These methods are often referred to by nicknames. DNA blotting is termed **Southern blotting** after its inventor, and subsequently developed techniques were dubbed **Northern blotting** (for RNA blotting) and **Western blotting** (for proteins).

Hybridization of a labeled probe to DNA fragments embedded in a gel is very insensitive. Therefore, before carrying out the annealing, fragments are transferred by blotting from the gel to a more suitable solid support, usually a paperlike sheet of nitrocellulose or nylon. First, the gel is treated with alkali to denature the DNA. The single strands, which will be efficiently trapped by the solid support, are required if the complementary sequences are to anneal with the probe. After DNA denaturation, the support is placed in contact with the gel, and a buffer solution is passed through the gel and into the matrix (Figure 6.2). The DNA is carried along during the blotting process, and the geographic location of the DNA fragments is undisturbed. Thereafter, the solid support is incubated in a solution containing the ^{32}P probe under conditions of salt concentration and temperature that stabilize hydrogen bonding between the probe and the complementary DNA fragment. The sensitivity of the method is such as to permit detection of picograms of a particular fragment, depending on the specific radioactivity of the probe. For example, ^{32}P-labeled probes prepared by nick translation (Section 4.6b) have specific activities of more than 100 cpm per picogram, sufficient for visualization by autoradiography.

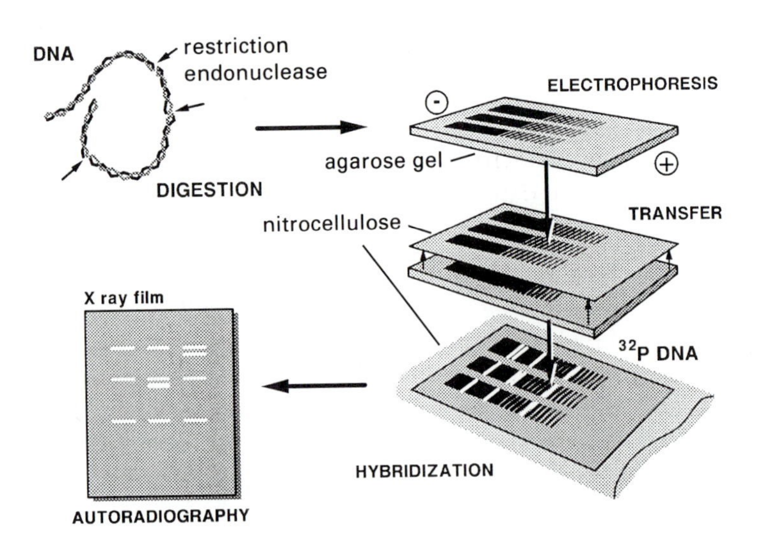

Figure 6.2
DNA blotting. A gel such as that in Figure 6.1a is placed in contact with a sheet of nitrocellulose (the same size as the gel). An appropriate solution is drawn through the gel and nitrocellulose, eluting the DNA and trapping it on the nitrocellulose sheet. Very similar procedures can be used to blot RNA and protein from suitable electrophoretic gels onto solid materials.

Random fragmentation of genomic DNA Because long DNA molecules are quite fragile, small inputs of energy readily fracture the duplex structures. Therefore, DNA molecules can be broken into random fragment sets by mechanical shearing, sonic vibration, rapid mixing, or passing a solution through a small orifice. Restriction endonucleases themselves can be used to produce random overlapping sets of fragments if only a minority of the available sites are actually cleaved during the digestion. The average size of fragments in a set can be varied by the amount of shearing energy or by adjusting the concentration of endonuclease. DNA is generally isolated from a large population of cells; therefore, the initial DNA preparations always represent a population of many identical genomes. Any particular DNA segment will occur in many sizes in the final set of fragments, making purification prior to cloning impossible. However, the random sets are more likely to include at least one copy of any desired segment all in one piece than are the sets generated by complete restriction endonuclease digestion.

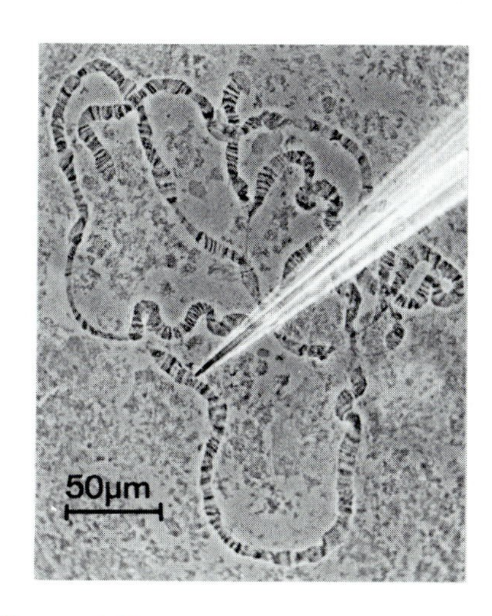

Figure 6.3
Dissection of a *Drosophila* polytene chromosome. The glass needle points to the tip of the X-chromosome. Courtesy of V. Pirotta.

Dissecting chromosomes If the position of a gene within a chromosomal region is known precisely, and if the chromosome is large enough to be manipulated, then, in principle, the region containing the gene is recoverable in a slice of the chromosome. The polytene chromosomes of *Drosophila* salivary glands meet these requirements. Each one contains more than 1000 copies of the DNA, and the DNA molecules are lined up parallel with one another so that a small region may contain 1000 copies of a particular gene. Moreover, the position of many *Drosophila* genes is known with some precision as a result of both genetic and cytogenetic analysis. Working under a phase contrast microscope, the region containing a particular gene is located and cut out with a fine needle (Figure 6.3). In this way, segments representing about 200 kbp of the genome can be obtained, cleaved with restriction endonucleases, and inserted into vectors. The enrichment over total *Drosophila* DNA is considerable because 200 kbp represents only about 0.1 percent of the 1.8×10^8 bp in the genome.

c. Synthetic Inserts

Great advances in chemical techniques make it possible to synthesize DNA molecules up to about 50 residues long directly, rapidly, and in good yield from simple nucleosides. Molecules that are otherwise difficult or impossible to obtain can then be cloned and prepared in significant amounts. For example, screening for a particular cloned gene may be impossible if no appropriate probe is available for annealing. But a completely synthetic gene can be prepared when a coding sequence can be deduced from a polypeptide amino acid sequence. This is how DNA segments that encode the polypeptide hormones somatostatin and insulin were initially cloned in *E. coli*. Synthesis was accomplished by preparing blocks of oligonucleotides and then joining these with DNA ligase. Somatostatin gene synthesis is illustrated in Figure 6.4. Eight short single strand DNA fragments were synthesized. They were designed to anneal to form a double strand structure by means of overlapping complementary terminal sequences, thereby providing appropriate substrates for DNA

Figure 6.4
Synthesis of a somatostatin gene by ligation of short chemically synthesized single strand chains. The eight individually synthesized oligodeoxyribonucleotides are indicated by brackets. The color marks the phosphomonoester groups (p′s) added with polynucleotide kinase and ATP.

ligase and for the subsequent cloning steps. Because the synthetic products all had 5′ terminal hydroxyl groups, the termini were phosphorylated with polynucleotide kinase and ATP prior to the ligation. These phosphate groups are colored in Figure 6.4. More recently, a 514 bp long interferon gene was assembled from 66 short synthetic fragments. Note that because of the degeneracy of the genetic code, predicting the actual sequence of a cellular gene from the sequence of its corresponding polypeptide is impossible. However, a functional coding sequence can nevertheless be deduced and synthesized.

Synthesis of an entire long gene remains a formidable task. But even the synthesis of a short portion of the coding sequence for a long polypeptide is very useful because sequences 15 to 20 bases long provide specific hybridization probes when other probes are unavailable. The synthetic probes can be used in conjunction with DNA or RNA blots as well as for screening populations of recombinants for the desired clone (Section 6.4). In addition, short synthetic segments are widely used as primers in determining the nucleotide sequence of long DNA segments (Section 7.2) and are also useful as primers in the enzymatic synthesis of DNA copies of RNA molecules by reverse transcriptase. Another important application of the synthetic techniques is in preparing short DNA segments that contain the recognition sites of known restriction endonucleases. These "linker" segments are ligated to flush-ended DNA fragments in preparation for insertion into a vector (Section 6.2).

Chemical synthesis of polydeoxynucleotides Two methods are widely used for the chemical synthesis of polydeoxynucleotides. Both utilize deoxyribonucleosides as starting materials and involve the stepwise coupling of mononucleotides and oligonucleotides. Moreover, both methods have been fully automated in commercially available "DNA synthesizers."

The first consideration, regardless of method, is protecting those reactive groups on the deoxynucleosides that do not participate in the coupling reactions. Thus, the amino groups on deoxyadenosine and deoxycytidine are usually benzoylated, and that on guanosine is protected by an isobutyryl group (Figure 6.5). Thymidine, which has no amino group, is used unmodified. The 5′-hydroxyl groups are usually protected by formation of an ether with a 4, 4′, di-methoxytriphenylmethyl group that is commonly called dimethoxytrityl and abbreviated $(MeO)_2Tr$ (Figure 6.6). The protecting groups are all designed for easy removal at the end of the synthesis. The free amino groups are regenerated by mild alkaline hydrolysis of the aminoacyl moieties, and the dimethoxytrityl groups are removed by gentle acid hydrolysis.

The phosphate triester (or phosphotriester) method The 5′-dimethoxytrityl protected nucleoside is suitable, without further manipulation, for constructing a phosphodiester at the 3′-hydroxyl by coupling with, for ex-

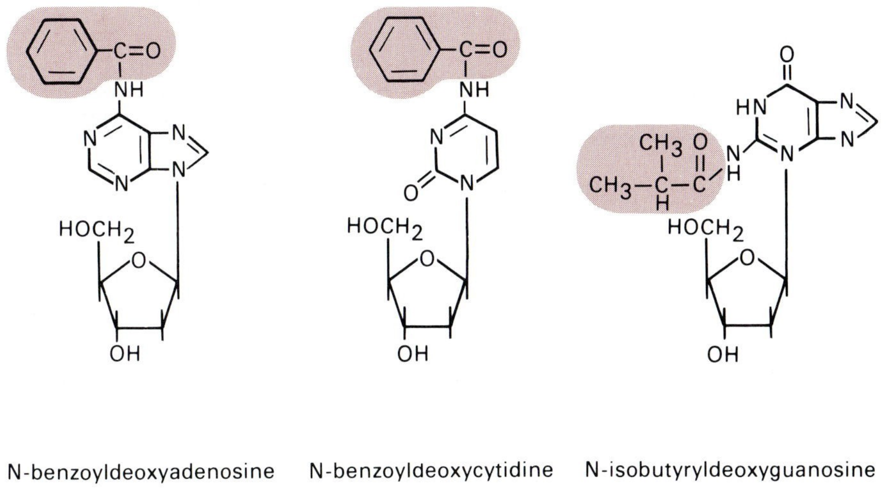

N-benzoyldeoxyadenosine N-benzoyldeoxycytidine N-isobutyryldeoxyguanosine

Figure 6.5
N-acyl protected deoxyribonucleosides are the starting materials for chemical synthesis of polydeoxyribonucleotides.

ample, p-chlorophenylphosphorodichloridate (Figure 6.7). The diester can serve directly as the precursor of the 5′ terminus of the new chain. A second such diester, destined to be the 3′ terminus of the new dinucleotide, is converted to a triester with, for example, β-cyanoethanol, and the 5′-dimethoxytrityl group is then removed by mild acid hydrolysis to give a reactive 5′-hydroxyl group (Figure 6.7). The diester and the triester are mixed in the presence of reagents that promote their coupling (Figure 6.8). Typically, these reagents are arylsulfonyl compounds such as triisopropylbenzenesulfonyl chloride. The precise mechanism of the coupling reaction is not understood. The product (Figure 6.8a) is a fully protected dinucleotide. Removal of all the protective groups leaves the simple dinucleoside monophosphate (Figure 6.8b). More important, the fully protected dinucleotide is a good starting material for the construction of still larger molecules; even the existing internucleotide phosphate is protected as a triester. Treatment with an acid such as benzenesulfonic acid or $ZnBr_2$ removes the dimethoxytrityl group and provides a dinucleotide for the 3′ end of a longer chain (Figure 6.8c). Alternatively, treatment with triethylamine preferentially hydrolyzes the cyanoethyl ester, thereby providing a reactive 3′ end on a dinucleotide that can be used as the 5′ end of a longer chain (Figure 6.8d). Coupling of the two dinucleotides yields a tetranucleotide (Figure 6.8e). In analogous ways, appropriately protected

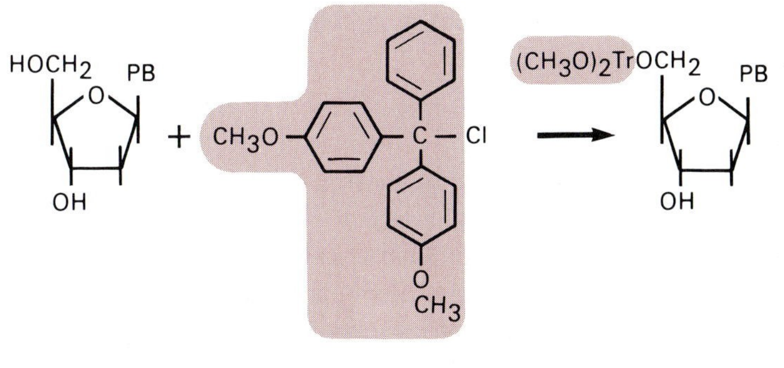

Figure 6.6
Protection of the 5′ OH of a protected nucleoside (Figure 6.5) by reaction with di-*p*-methoxytrityl chloride. In this and the following figures, PB denotes a protected purine or pyrimidine base.

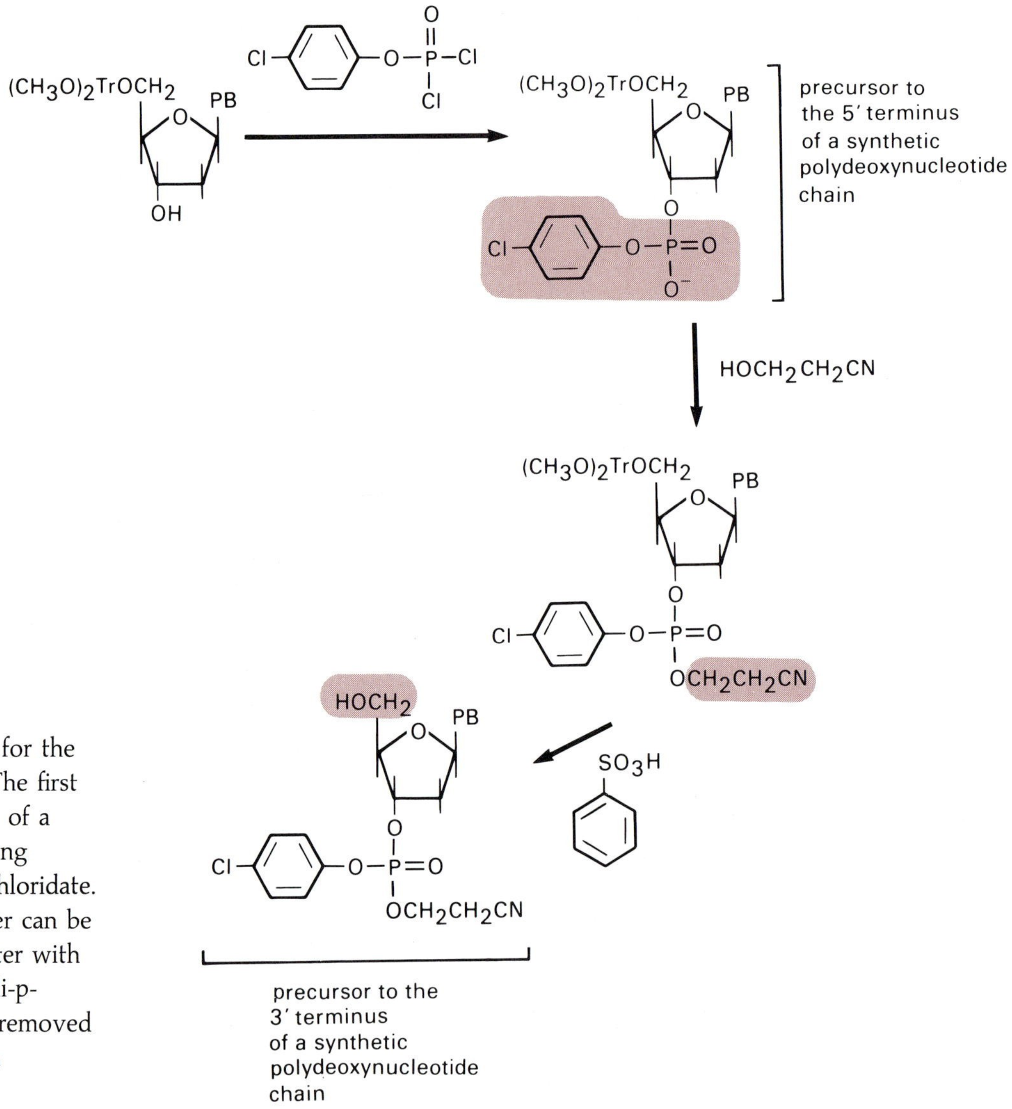

$(CH_3O)_2TrOCH_2$ PB OH
$Cl-C_6H_4-O-P(=O)(Cl)-Cl$
$(CH_3O)_2TrOCH_2$ PB
$Cl-C_6H_4-O-P(=O)-O^-$
precursor to the 5′ terminus of a synthetic polydeoxynucleotide chain
$HOCH_2CH_2CN$
$(CH_3O)_2TrOCH_2$ PB
$Cl-C_6H_4-O-P(=O)-OCH_2CH_2CN$
SO_3H
$HOCH_2$ PB
$Cl-C_6H_4-O-P(=O)-OCH_2CH_2CN$
precursor to the 3′ terminus of a synthetic polydeoxynucleotide chain

Figure 6.7
Preparation of intermediates for the phosphate triester method. The first reaction shows the synthesis of a protected phosphodiester using p-chlorophenylphosphorodichloridate. The protected phosphodiester can be converted to a phosphotriester with β-cyanoethanol, and the 5′-di-p-methoxytrityl group can be removed by mild acid hydrolysis with benzenesulfonic acid.

mononucleotides and oligonucleotides can be put together in long blocks. Depending on the choice of starting materials, the DNA chains are constructed in either the 3′-to-5′ or 5′-to-3′ direction.

The phosphate triester method is expedited and simplified if one terminal nucleotide is fixed, through a hydroxyl, to a solid support. In principle, either hydroxyl could be fixed, but in practice, things work better if the 3′-hydroxyl is anchored. This is generally accomplished by forming an ester between the 3′-OH and a carboxyl group on a support such as controlled pore glass beads (Figure 6.9). Not only is the first nucleotide

---

Figure 6.8 (PAGE 329)
Chemical synthesis of a di-deoxyribonucleotide by the phosphate triester method. (a) is a fully protected dinucleotide that can be used, after appropriate deprotection, for coupling to additional units. The structure of the coupling agent, 1-(2,4,6)-triisopropylbenzenesulfonyl chloride (TPS), is shown. Note the simplified schematic depiction of deoxyribonucleotides.

PB^1 PB^2

3′ 5′ $(CH_3O)_2TrO$ — O — P(=O)(O⁻) — O — C₆H₄Cl

3′ 5′ HO — O — P(=O) — OCH_2CH_2CN — O — C₆H₄Cl

CH_3 CH_3 CH_3 CH_3 CH_3 CH_3 O=S=O Cl

1-(2,4,6)-triisopropylbenzenesulfonyl chloride 'TPS'

(a)

$(CH_3CH_2)_3N$

$C_6H_5SO_3H$

NH_3

$C_6H_5SO_3H$ or $ZnBr_2$

$(CH_3CH_2)_3N$

B^1 B^2 HO OH O^-

(b)

PB^1 PB^2 HO OCH_2CH_2CN Cl Cl

(c)

PB^1 PB^2 $(CH_3O)_2TrO$ OH Cl Cl

(d)

TPS

PB^1 PB^2 PB^1 PB^2 $(CH_3O)_2TrO$ OCH_2CH_2CN Cl Cl Cl Cl

(e)

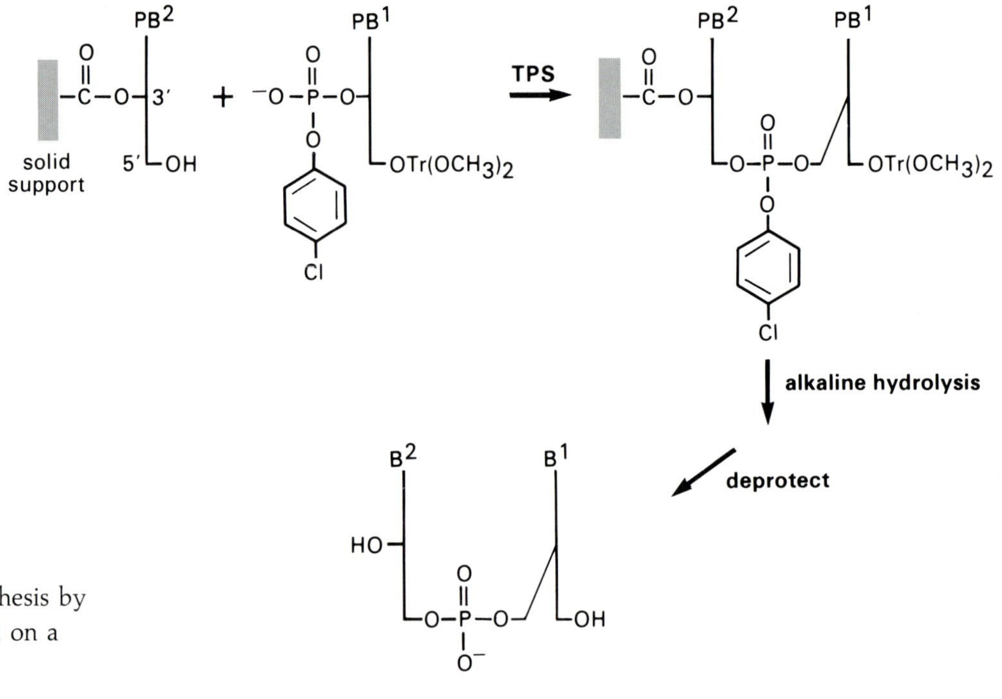

Figure 6.9
Polydeoxyribonucleotide synthesis by the phosphate triester method on a solid support.

fixed, but the 3′-OH is protected. Nucleotides or oligonucleotides protected at their 5′-OH groups by dimethoxytrityl groups are added one by one. In between each addition, the terminal dimethoxytrityl is removed (Figure 6.7), leaving the growing chain with a free 5′-OH group for the next addition. Finally, the desired polydeoxynucleotide is removed from the support by alkaline hydrolysis (Figure 6.9). This procedure eliminates the tedious purifications that must otherwise be carried out after each addition to the growing chain. And the procedure on the fixed support has been adapted for automated stepwise synthesis. The time required for synthesis of a polymer 10 to 20 nucleotides long is a few days.

The phosphite triester (or phosphoramidite) method Deoxynucleosides with protected amino groups and a 5′-dimethoxytrityl group again provide the basic materials (Figures 6.5 and 6.6). In the scheme outlined in Figure 6.10, the 3′-hydroxyl of the terminal nucleoside is protected by fixation to a solid support through an ester linkage. The precursor of the next residue is a protected deoxynucleoside 3′-phosphoramidite that is activated for coupling by the addition of tetrazole. The success and speed of the procedure depends on the phosphoramidites, which are stable and efficient coupling agents and readily synthesized (Figure 6.11). The immediate product of the coupling reaction is a phosphite; it is oxidized to a phosphate (triester) with iodine. As in the previous method, the triester protects the new internucleotide bond from destruction during subsequent, stepwise additional couplings. When the desired chain is completed, the various protecting groups are removed, and the polydeoxynucleotide is freed from the solid support by alkaline hydrolysis. The phosphite method on a silica-based solid support (or controlled pore glass beads) is used for automated sequential synthesis of oligodeoxyribonucleotides. Adding each nucleotide residue takes less than 15 minutes, and chains as long as 50 residues can be prepared in good yields.

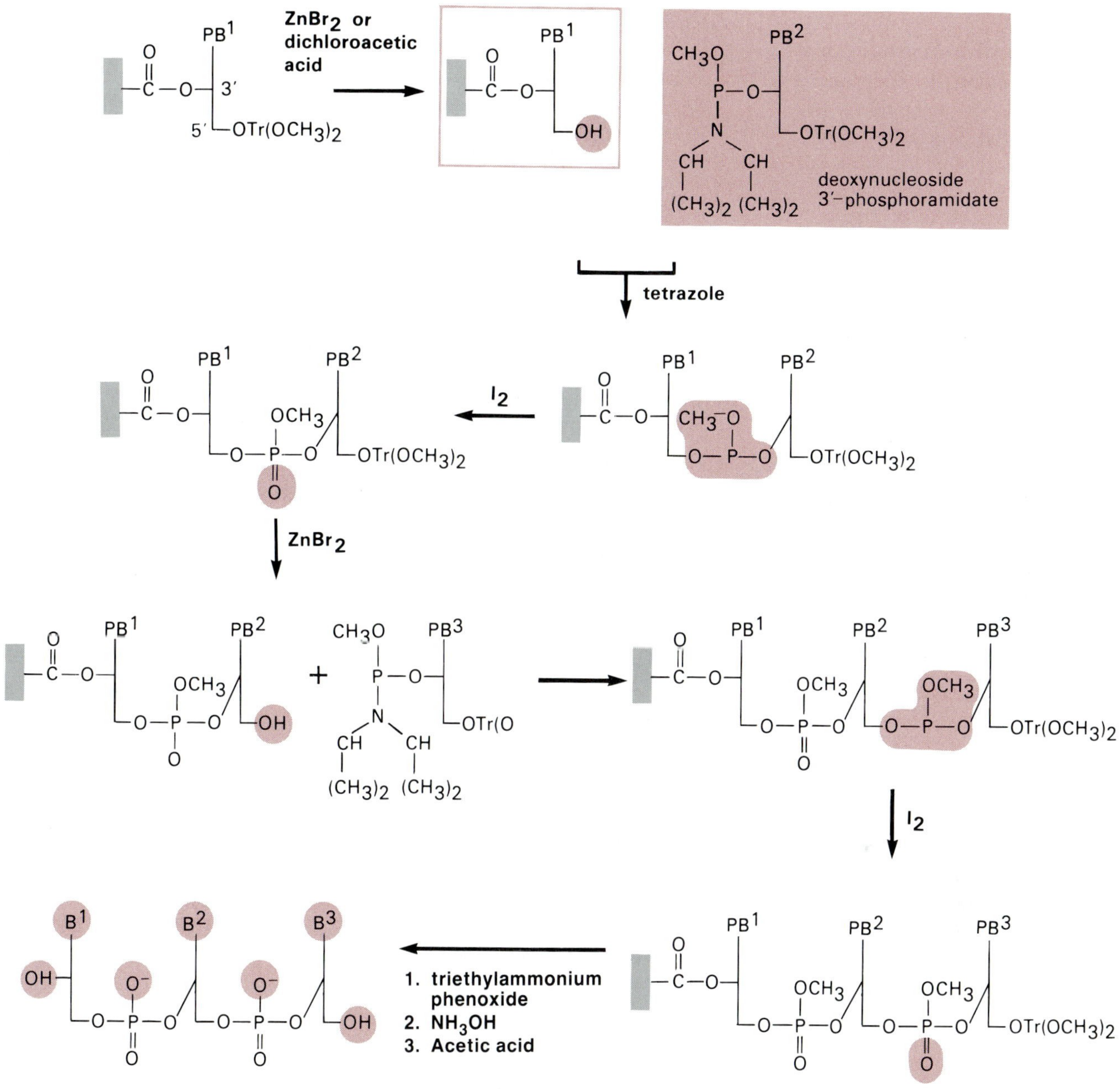

Figure 6.10
Polydeoxyribonucleotide synthesis by the phosphite triester (or phosphoramidite) method on a solid support.

Enzymatic methods Enzymatic synthesis of short polydeoxyribonucleotides of defined sequence is conceptually simpler than the strictly chemical methods, and the manipulations are easier to perform (unless an automated DNA synthesizer is available). No protective groups are required for these enzymatic reactions. However, the reactions are limited and hard to control. The methods utilize the bacterial enzyme polynucleotide phosphorylase, which is generally considered specific for ribonucleotides but will polymerize deoxyribonucleotides (though at a slow rate) when Mn^{2+} replaces

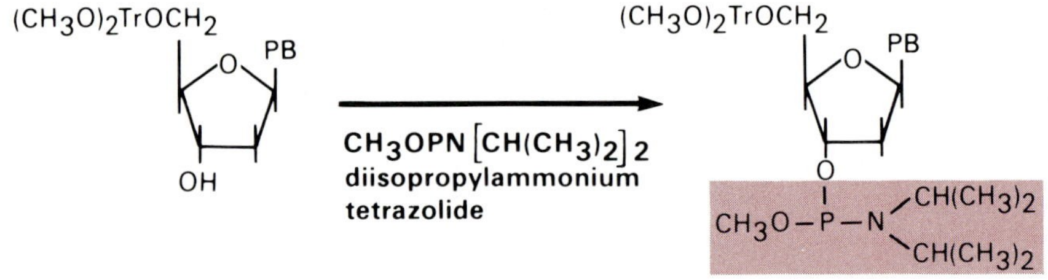

Figure 6.11
Synthesis of a deoxyribonucleoside 3′-phosphoramidite, the coupling intermediate in the phosphite triester (phosphoramidite) method.

$(CH_3O)_2TrOCH_2$ … PB … OH

$CH_3OPN[CH(CH_3)_2]_2$
diisopropylammonium tetrazolide

$(CH_3O)_2TrOCH_2$ … PB … $CH_3O-P-N(CH(CH_3)_2)_2$

Mg^{2+} in the reaction (Figure 6.12a). The substrates are always deoxyribonucleoside diphosphates, and a template strand is neither required nor used. One or two deoxyribonucleotide residues can be added at a time to primers at least three residues long. Figure 6.12b outlines the synthesis of a dodecanucleotide that represents a segment of the isocytochrome c gene of yeast.

Combined chemical and enzymatic synthesis using either DNA polymerase I or reverse transcriptase is also useful. The example in Figure 6.13 shows two partially complementary oligodeoxynucleotides synthesized by chemical procedures. The fully duplex molecule is readily obtained by using the single strand regions as templates for DNA polymerase I or reverse transcriptase (Figure 4.19).

d. Copying RNA into DNA

Methods for analyzing RNA are tedious compared to those developed for investigating DNA. The easiest way to study the primary structure of an RNA molecule is to copy it into DNA, clone the copy, and determine the DNA sequence. DNA molecules that are synthesized enzymatically by copying an RNA template are called **cDNAs** (for complementary or copy DNA). Purified RNAs or mixtures of RNA molecules such as the total mRNA from a tissue or a cell population can serve as templates.

Highly differentiated tissues and cells often produce abundant quantities of a particular mRNA and are good sources of RNA for purification. The early and successful application of recombinant DNA techniques to the study of the ovalbumin gene family (Chapter 7), for example, employed mRNAs isolated from specialized cells in which the corresponding proteins are abundantly synthesized. Once the cDNAs were cloned, they were used as probes for isolating the corresponding genes.

Copying eukaryotic mRNA The general scheme for synthesizing double strand cDNA from mRNA is outlined in Section 4.7 (Figure 4.20). First, the mRNA sequence is copied into a complementary single strand of DNA using reverse transcriptase. The RNA is the template, and the primer is a short polydeoxyribothymidylic acid (oligo dT) annealed to the polyriboadenylic acid end of the RNA. In the second step, the RNA template is degraded either by treatment with ribonuclease or by alkaline hydrolysis. Then the single strand cDNA is used as a template to synthesize the complementary DNA strand with DNA polymerase I or reverse transcriptase. No primer is needed at this step because a variable length hairpin loop is formed transiently at the 3′ end of the first strand and serves as

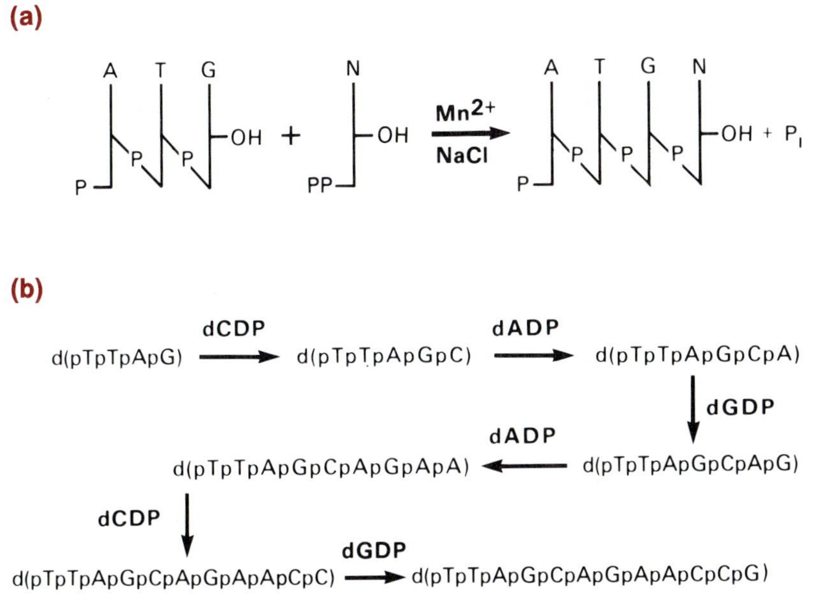

(a)

A T G + N → (Mn^{2+}, NaCl) A T G N

P–(A)–P–(T)–P–(G)–OH + PP–(N)–OH → P–(A)–P–(T)–P–(G)–P–(N)–OH + P_i

(b)

d(pTpTpApG) —dCDP→ d(pTpTpApGpC) —dADP→ d(pTpTpApGpCpA)

—dGDP→ d(pTpTpApGpCpApG) —dADP→ d(pTpTpApGpCpApGpApA)

—dCDP→ d(pTpTpApGpCpApGpApApCpC) —dGDP→ d(pTpTpApGpCpApGpApApCpCpG)

Figure 6.12
(a) Polynucleotide phosphorylase catalyzed addition of a single deoxyribonucleotide residue to an oligodeoxyribonucleotide. (b) Enzymatic synthesis of a segment of the yeast isocytochrome c gene. One or two new deoxyribonucleotide residues are added at each step. The desired product must be purified after each addition reaction.

a primer for second strand DNA synthesis. In the last step, the hairpin structure is cleaved by a single-strand-specific nuclease, producing a duplex cDNA.

Copying other RNAs Preparing cDNAs from the genomes of some RNA viruses or from other RNA molecules that do not have polyriboadenylate tails requires some special tricks. One is to add poly A tails to the 3′-OH end of the RNA with poly A polymerase (Section 4.9) and ATP and proceed as before. Alternatively, an oligodeoxyribonucleotide complementary to a region of the RNA molecule is annealed to the RNA to serve as primer.

Complications The scheme in Figure 4.20 shows an idealized outcome of cDNA synthesis. But complications arise at each step, so even starting with a pure mRNA preparation, the final product is generally a mixture of cDNAs most of which are somewhat shorter than the complete RNA. The following is a list of some of the problems:

1. The mRNA itself may be partly degraded because of ribonuclease action during isolation.

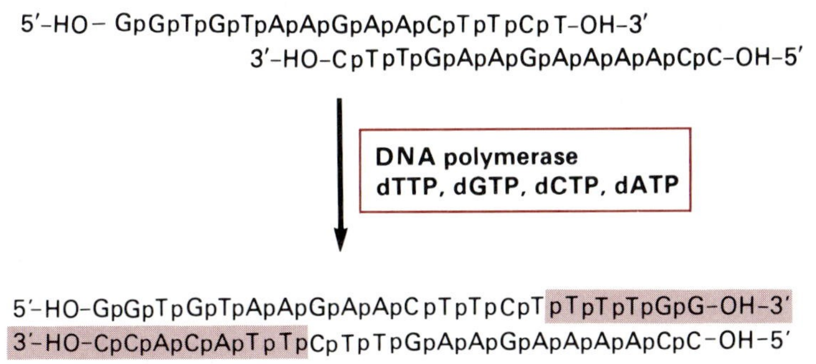

5′-HO– GpGpTpGpTpApApGpApApCpTpTpCpT-OH-3′
3′-HO-CpTpTpGpApApGpApApApApCpC-OH-5′

DNA polymerase
dTTP, dGTP, dCTP, dATP

5′-HO-GpGpTpGpTpApApGpApApCpTpTpCpTpTpTpTpGpG-OH-3′
3′-HO-CpCpApCpApTpTpCpTpTpGpApApGpApApApApCpC-OH-5′

Figure 6.13
Polydeoxynucleotide synthesis by a combination of chemical and enzymatic methods. The two partly complementary strands shown at the top are prepared by chemical synthesis and annealed together. The partial duplex is then used as a template and primer with DNA polymerase I of *E. coli* or reverse transcriptase to yield the fully duplex molecule.

2. The mRNA may be incompletely copied so that the sequence at its 5′ end is missing in the cDNA. The longer the RNA, the more likely this is to be a problem.
3. Synthesis of the second DNA strand may stop before the end of the template so that the 3′ end of the RNA will be missing. The remaining first strand will be removed in the single strand nuclease step.
4. The nuclease may, in addition to cleaving the hairpin, nibble away the ends of the duplex.

Because of these problems and because even highly purified RNA preparations are never absolutely pure, double strand cDNAs are always mixtures of different kinds of molecules. Physical and chemical methods are incapable of resolving the mixtures, but molecular cloning of the cDNAs permits isolation of single pure DNA segments completely free of others.

The serious problems encountered in obtaining full length cDNA copies of RNAs by the standard method frustrate many experiments. Much effort has therefore gone into inventing ways to circumvent the difficulties. Eliminating the nuclease step avoids some of the problem. This can be accomplished by adding (with terminal deoxynucleotidyl transferase) a polydeoxycytidylate tail to the 3′ end of the first DNA strand and then using an oligodeoxyguanylate primer to synthesize the second strand (alternatively, A and T tailing could be used).

Figure 6.14 illustrates a different method that ameliorates the chief drawback of the standard method, namely, the synthesis of incomplete copies of the RNA. First, the oligo dT primer is made an integral part of the *E. coli* plasmid vector so that the cDNA can be synthesized directly on the vector, thus eliminating separate isolation and ligation steps. Second, no single strand nuclease is used, so sequences corresponding to the 5′ end of the mRNA are not eliminated. Third, short DNA products of the first reverse transcriptase reaction (missing the sequences from the 5′ end of the RNA) tend to be eliminated in the oligo dC tailing step because the terminal transferase adds nucleotide residues poorly to a 3′-OH end that is completely protected by hydrogen-bonded base pairing (Section 4.8). Thus, the process yields a recombinant population that is enriched for complete cDNA copies of the initial RNA and ready for transfection into host cells.

A simple modification of this procedure produces plasmids containing cDNAs in a form that is expressible in eukaryotic cells. The modification involves incorporating a eukaryotic promoter and RNA processing signals into the linker segment shown in Figure 6.14.

6.2 Ligating Vector to Insert

After a vector and an insert have been prepared, constructing recombinant DNA molecules is quite straightforward. Only two double strand connections are usually required, regardless of whether the vector is a linear plasmid or viral genome or the two arms of λ DNA made into a single linear molecule by annealing through the cos sites. If the inserts have flush ends, cohesive ends are often attached before ligation to increase the efficiency of ligation.

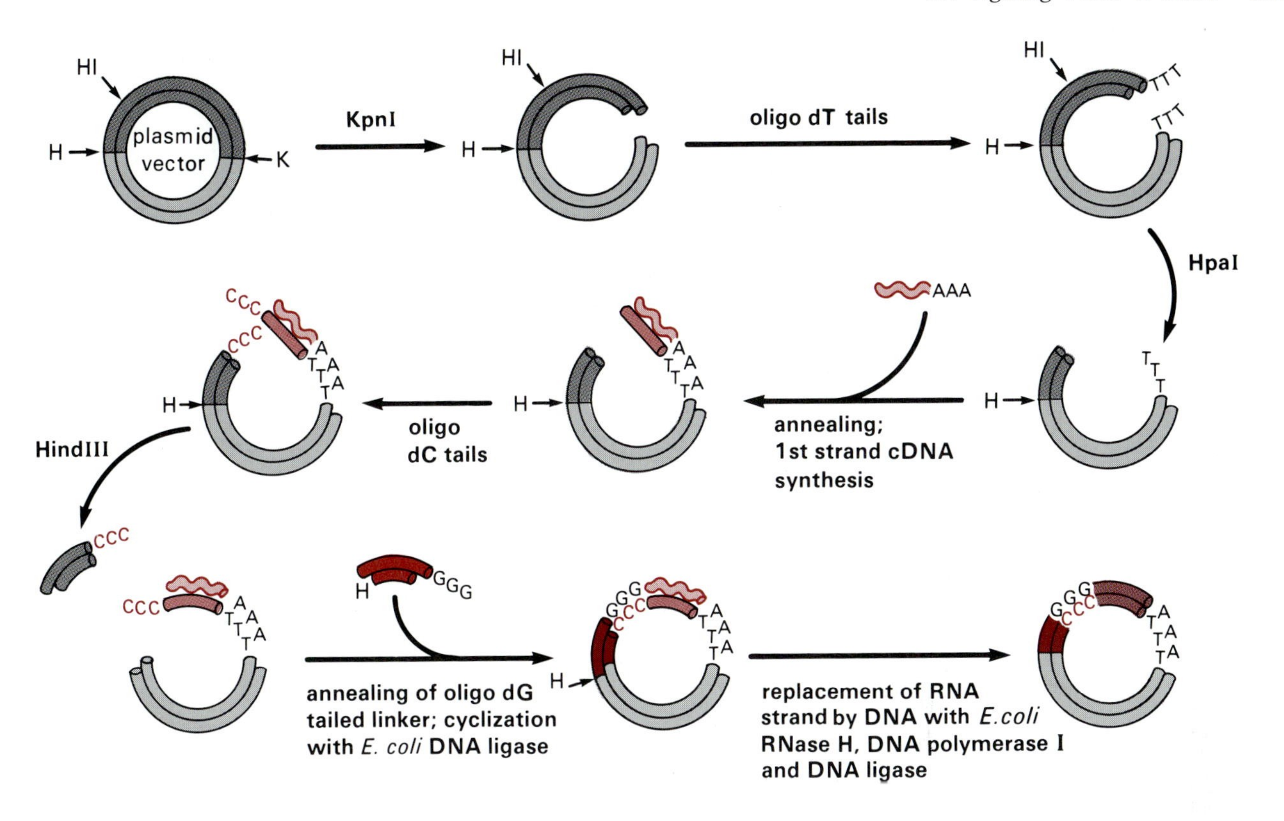

Figure 6.14

A procedure for cloning full length cDNAs. A circular, specially designed *E. coli* plasmid vector (derived from pBR322) is linearized by cleavage with a restriction endonuclease (*Kpn*I) at a single site in a nonessential region. Multiple dT residues are added to the 3′ ends with terminal deoxynucleotidyl transferase. One of the oligo dT regions is then removed with an appropriate restriction endonuclease (*Hpa*I) that leaves a blunt end. In this way, the oligo dT primer for the first reverse transcriptase reaction is an integral part of the plasmid vector. The vector also has a specially placed unique restriction endonuclease site (in the example, it is *Hin*dIII) close to the end opposite the oligo dT and within a dispensable region. After annealing of the RNA and synthesis of the first cDNA strand, tails of C residues are added to the available 3′ ends with terminal deoxynucleotidyl transferase. Then the unwanted C tail on the vector is clipped off with *Hin*dIII endonuclease and replaced, by ligation, with a separately prepared segment with a G tail on one end and a *Hin*dIII cohesive site on the other. This fragment can bridge the two ends of the construction to form a circle. Finally, the RNA strand is removed with RNAse H and replaced by DNA (the G tail provides the primer for DNA polymerase), and the circle is sealed with DNA ligase. Throughout these manipulations, it is necessary to use conditions that maintain the heteroduplex structure because only complementary base pairing holds the RNA to the vector until the last step. Note too that the RNA·DNA hybrid portion of the molecule is resistant to cleavage by *Hin*dIII; otherwise any *Hin*dIII sites that happen to be in the insert would be cleaved.

a. *Joining Ends*

Four situations are possible. The vector and insert have completely compatible cohesive ends (Figure 6.15a), two different pairs of compatible cohesive ends (Figure 6.15b), flush ends (Figure 6.15c), or a pair of compatible cohesive ends and a pair of flush ends (Figure 6.15d). A vector-insert pair containing one cohesive end and one flush end each (Figure

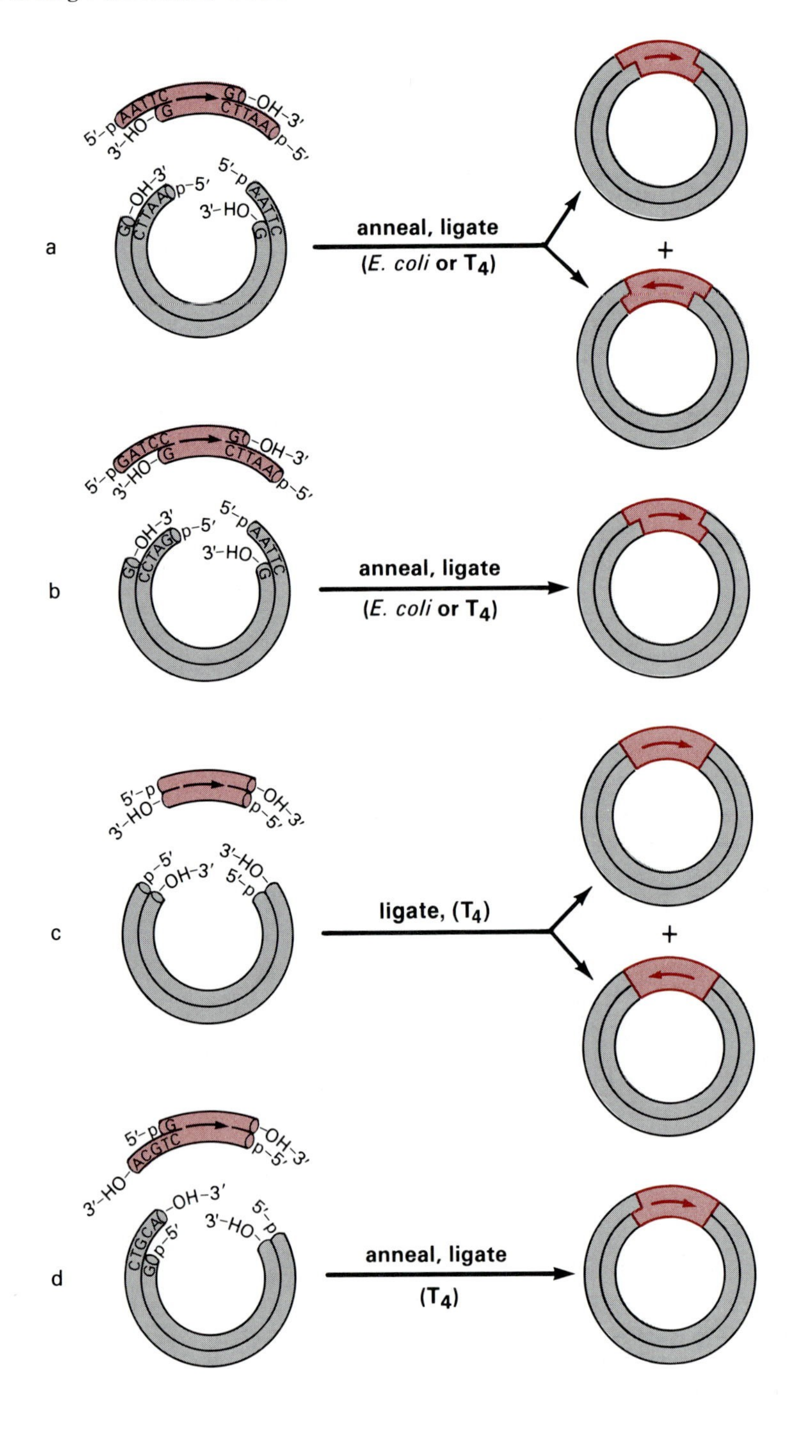

Figure 6.15
Strategies for ligation of insert and vector. Recall that the DNA ligase of *E. coli* is unable to catalyze the joining of flush-ended fragments (Section 4.5).

6.15d) joins up in only one way. The same is true if both vector and insert each have two different but compatible cohesive ends (Figue 6.15b). But when each molecule has two identical cohesive ends (Figure 6.15a) or two flush ends (Figure 6.15c), the insert can be ligated in either of two (opposite) orientations relative to the vector producing two different products. When cohesive termini are to be joined, the vector and insert are treated with DNA ligase under conditions that permit hydrogen bonds to form between the complementary single strands. For joining flush ends,

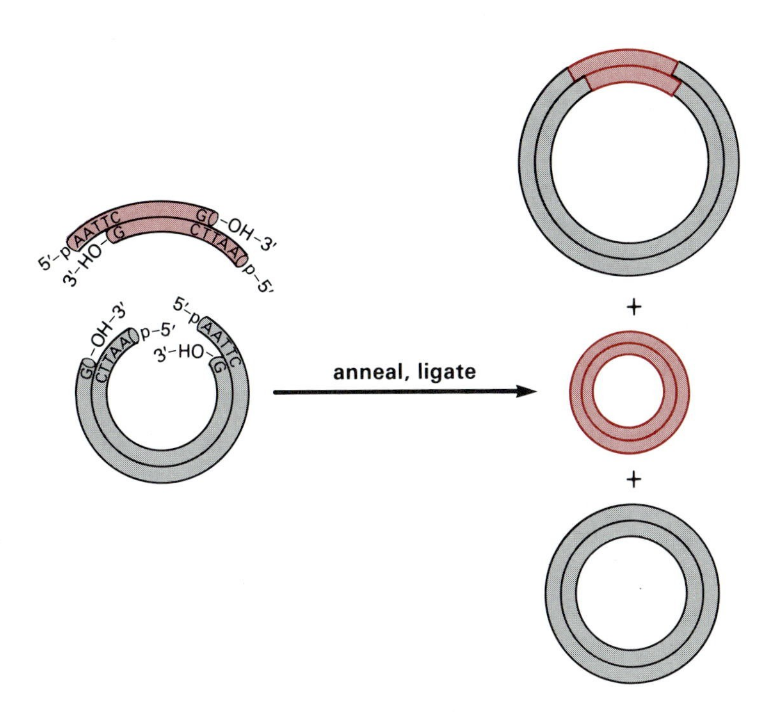

Figure 6.16
Side products of ligation: intramolecular joining.

high concentrations of both T4 DNA ligase and fragments are required because the affinity of the ligase for flush ends is low.

Intramolecular ligation of both vector and insert linear fragments competes seriously with intermolecular joining and reduces the yield of the recombinant molecule (Figure 6.16). These pesky side reactions are minimized by dephosphorylating either the vector or insert with phosphatase prior to ligation (Figure 6.17). In the absence of 5′-phosphomonoester groups, no intramolecular ligation can occur. The vector is usually dephosphorylated to minimize its amplification after transfection; a circularized insert is incapable of replication and therefore will not be a major contaminant. As shown in Figure 6.17, the recombinant molecule formed

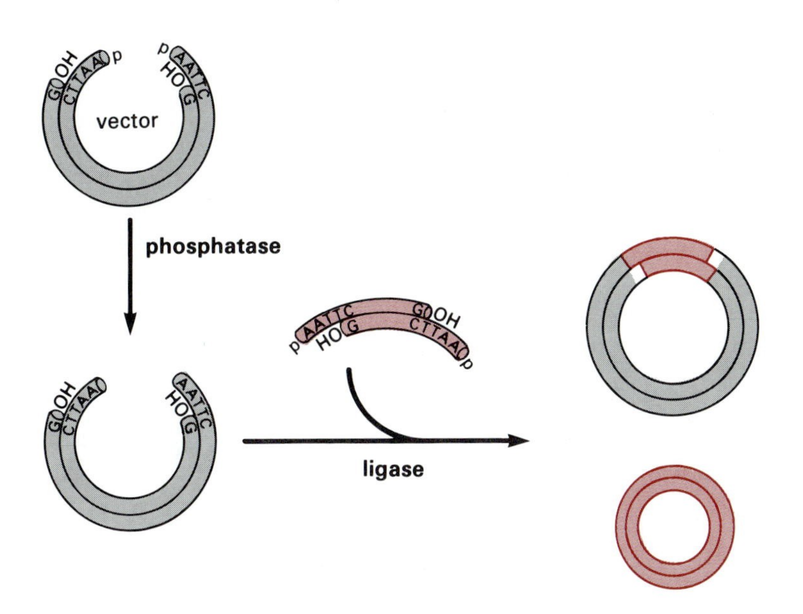

Figure 6.17
Avoiding intramolecular ligation of the vector by dephosphorylation.

with a dephosphorylated vector contains one nick in each DNA strand; nicked recombinant plasmid molecules are readily repaired after they are taken up into appropriate host cells.

b. Attaching Cohesive Ends

Inserting DNA segments into vectors is always more efficient if the segments and vector have matching cohesive ends. There are two good ways to create cohesive ends on flush-ended DNA segments such as cDNAs: tailing and attaching restriction endonuclease sites.

Tailing If terminal deoxynucleotidyl transferase is used to construct a run of nucleotides at each of the two 3′-hydroxyl termini of a double strand flush-ended insert, and runs of the complementary nucleotide are similarly added to the ends of a linear vector DNA, the tails can anneal to form a circle (Figure 6.18). If all tails are the same size, the circle is readily sealed with DNA ligase. However, that rarely occurs, so the gaps in the annealed molecules are filled in with DNA polymerase I prior to ligation. Recall too that the terminal deoxynucleotidyl transferase requires a single strand 3′-hydroxyl terminus as a primer and that special conditions are required for efficient addition to a flush-ended DNA duplex or to a molecule in which the 3′-hydroxyl terminus is recessed (Section 4.8).

Attaching restriction endonuclease sites Short flush-ended DNA fragments containing sequences that include specific restriction endonuclease sites are available through chemical synthesis. The structure of two such "linkers" and the way they are used to produce cohesive ends on flush-ended inserts are shown in Figure 6.19. Joining of linker to insert is cata-

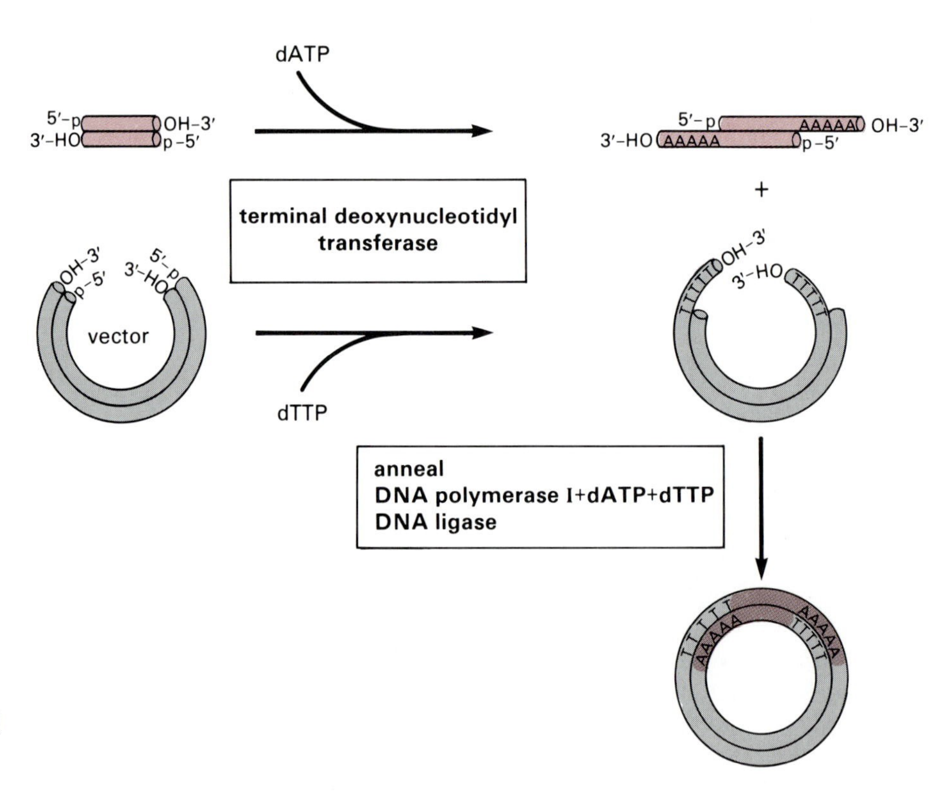

Figure 6.18
Adding cohesive ends with terminal deoxynucleotidyl transferase.

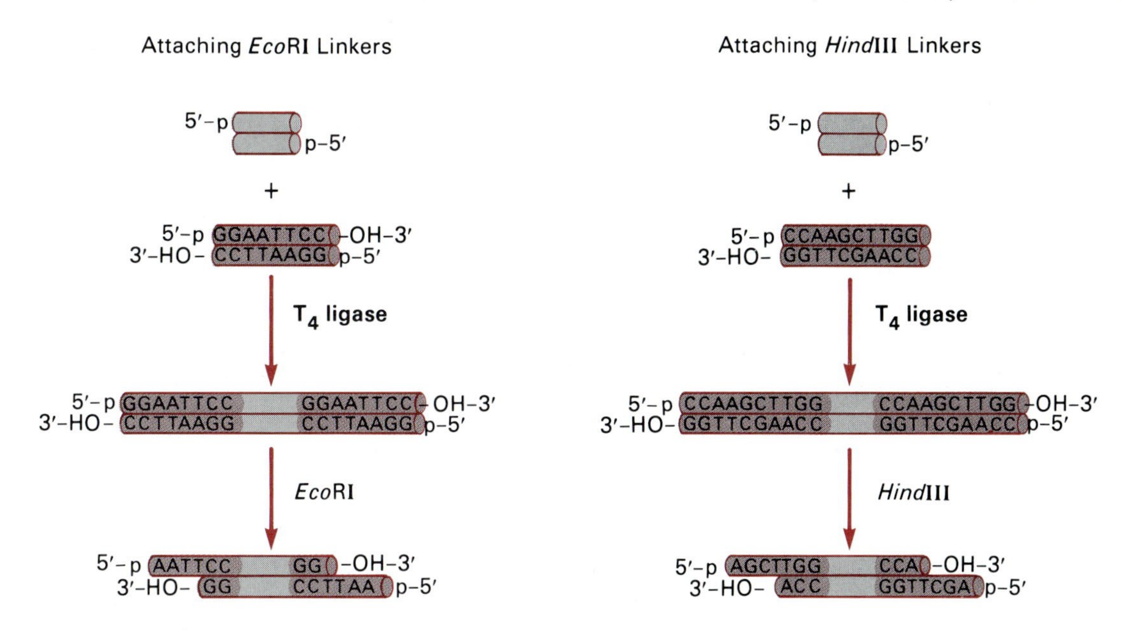

Figure 6.19
Constructing cohesive termini with "linkers."

lyzed by the T4 DNA ligase. Thereafter, the DNA is digested with the corresponding restriction endonuclease to generate the cohesive ends. This last step creates a complication. If the fragment to which the linkers are being attached contains internal recognition sites for the enzyme being used, then the generation of cohesive termini will destroy the fragment. There are two ways to avoid this problem. The first is to choose another type of linker. The second is to protect the susceptible internal sites by prior methylation with the methylase specific for the recognition site (Figure 6.20, and Section 4.2e).

Linkers are also a convenient way to arrange for compatible cohesive ends on an unmatched pair of insert and vector. Existing cohesive ends are first made flush with a single-strand-specific nuclease or by filling in the recessed ends with DNA polymerase I or reverse transcriptase; then linkers are attached.

6.3 Infection, Transfection, and Cloning

a. Moving Recombinant Molecules from Test Tube to Cell

Once constructed, recombinant DNA molecules are introduced into cells or viral particles for cloning and amplification. Several ways in which this is accomplished are given in Chapter 5; different techniques are appropriate to different host-vector systems. Briefly, recombinant DNAs built from bacterial or yeast plasmids or from eukaryotic viruses are transfected into hospitable host cells after making the cell membranes (and walls) permeable. Recombinants constructed with λ DNA vectors or cosmids are packaged into bacteriophage particles, which are then used to infect permissive *E. coli* K12 cells.

Transfection is generally not very efficient. Only a portion of the treated cells actually incorporates a recombinant genome. Growing the

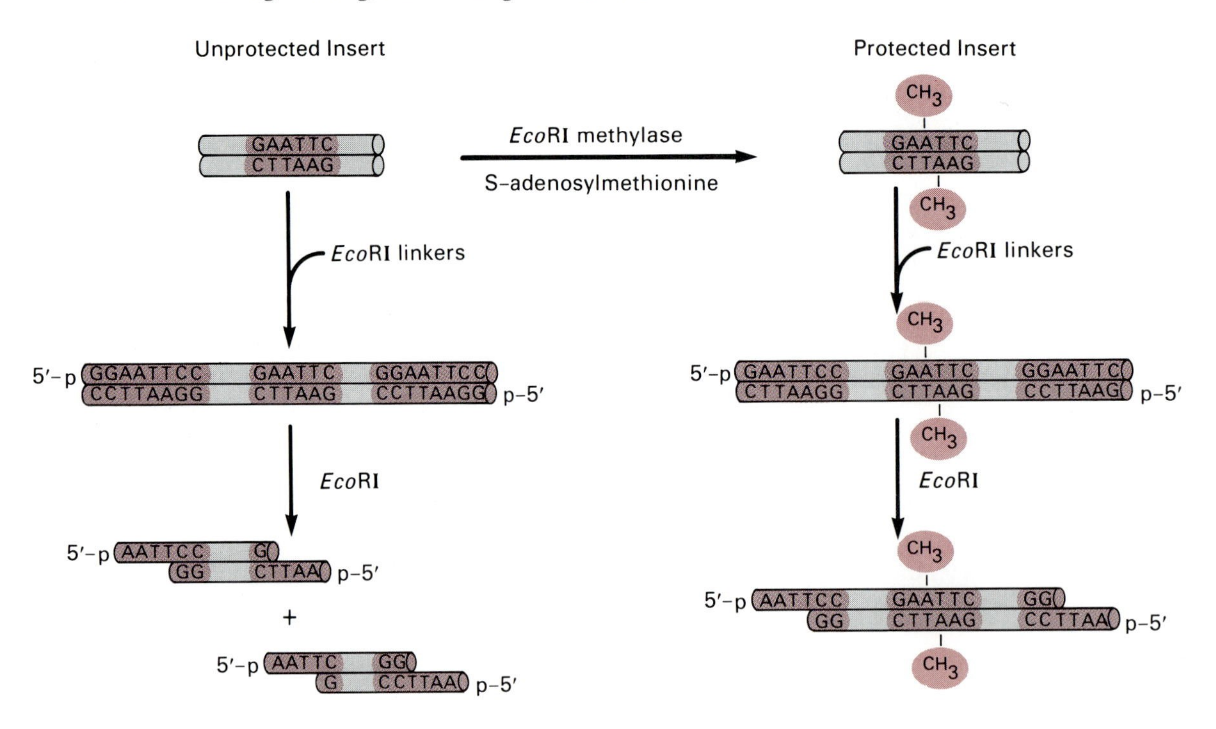

Figure 6.20
Restriction endonuclease sites within an insert fragment are protected by methylation prior to the addition of linkers.

cells under conditions that force a dependency on genes encoded within the vector is a powerful way to select and maintain recombinant vectors. This type of selective pressure is also often crucial to identifying and isolating the desired recombinant.

One DNA molecule per cell Two principles are central to the concept and feasibility of molecular cloning. First, after construction *in vitro,* single recombinant DNAs must be parceled into separable entities. Therefore, no cell should receive more than one plasmid molecule or viral particle. Second, the separated entities must be able to replicate.

The products of transfection and infection are populations of cells or populations of viruses or bacteriophage. Among the predictable members of the populations are those that contain the desired recombinants, recombinants containing inserts that are of no interest, vectors carrying no insert at all, and, in the case of plasmid and some eukaryotic viral vectors, unaltered cells. In addition, a variety of unpredictable and aberrant recombinants can occur, including vectors that have incorporated two or more inserts and recombinants in which the inserts have been altered by recombination within the host cell. For these reasons, transfection and infection mark not the end but only one step in any recombinant DNA experiment. The next jobs are cloning and identifying the recombinant of interest among the members of the population.

b. Cloning

The first requirement for cloning is to separate all the transfected or infected cells from one another. Each clone then takes the form of a single bacterial or animal cell colony, or phage or viral plaque, and each carries a cloned recombinant DNA molecule.

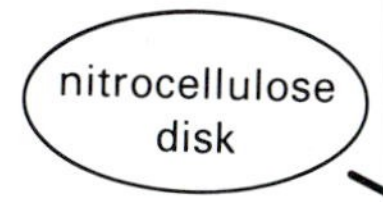

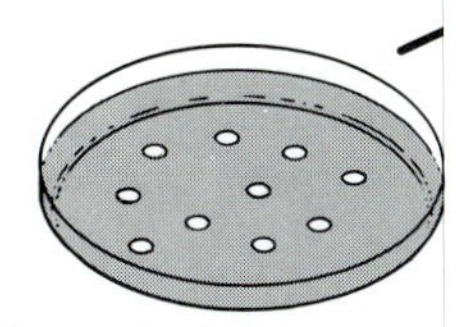

bacteriophage plaque
on a lawn of
susceptible cells

Plasmid vectors Transfected bacterial or yeast cells are thinly spread on nutrient agar plates so that single cells are well separated and each can grow into a separate single colony. The vectors usually contain at least one selectable marker so that untransfected cells fail to yield colonies and are lost.

Bacteriophage vectors Infected (or transfected) cells are spread on a lawn of susceptible cells, and bacteriophage plaques form through infection of neighboring cells. Some vector systems have built-in markers that allow easy discrimination of recombinants from reconstructed intact vectors (Section 5.3).

Animal virus vectors SV40 transducing vectors designed to produce viral particles (Section 5.7b) are cloned in much the same way as bacteriophage vectors. However, because such SV40 vectors often require a helper virus to supply missing functions, the viral plaques (areas of lysis on a monolayer of cells) may contain both recombinant and helper genomes. SV40 plasmid and passive transforming vectors, which do not produce viral particles, are first cloned as shuttle vectors in *E. coli*. These usually encode a selectable phenotype for animal cells, so clones of transformed cells can be obtained. Similarly, recombinant bovine papillomavirus vectors and retroviral vectors are cloned in *E. coli* as shuttles and then transfected into animal cells.

6.4 Screening Cloned Populations of Recombinants

a. Finding the Right Clone

Once the mixed population of recombinant DNA molecules is sorted into individual clones, a formidable task remains: finding the desired clone or clones. The search must depend on a detectable and unique property of the insert, and there are only two such properties. One is the structure of the insert itself (i.e., its nucleotide sequence). The other is its function (i.e., the product of expression of the cloned gene). These two unique properties are used in various ways to screen large numbers of clones and identify the one that is wanted. It is easy to lose sight of how extraordinarily sensitive the screening methods are. The search may be for a single clone among hundreds of thousands or a few million. Yet a single, particular recombinant clone is distinguishable.

b. Annealing with a Complementary Polynucleotide

Complementary single strand polynucleotides, whether RNA or DNA, anneal together (renature) efficiently and with high specificity regardless of the presence of a great excess of unrelated chains. This fact provides a powerful tool for finding a specific insert, and it is exploited in several ways.

Annealing with a radioactive probe Given a large collection of cloned recombinants, only those with complementary sequences become radioac-

Figure 6.21
Screening bacterial (c
for hybridization to a
labeled probe. Coloni
selection for markers
shown. Two of them,
the desired insert seg

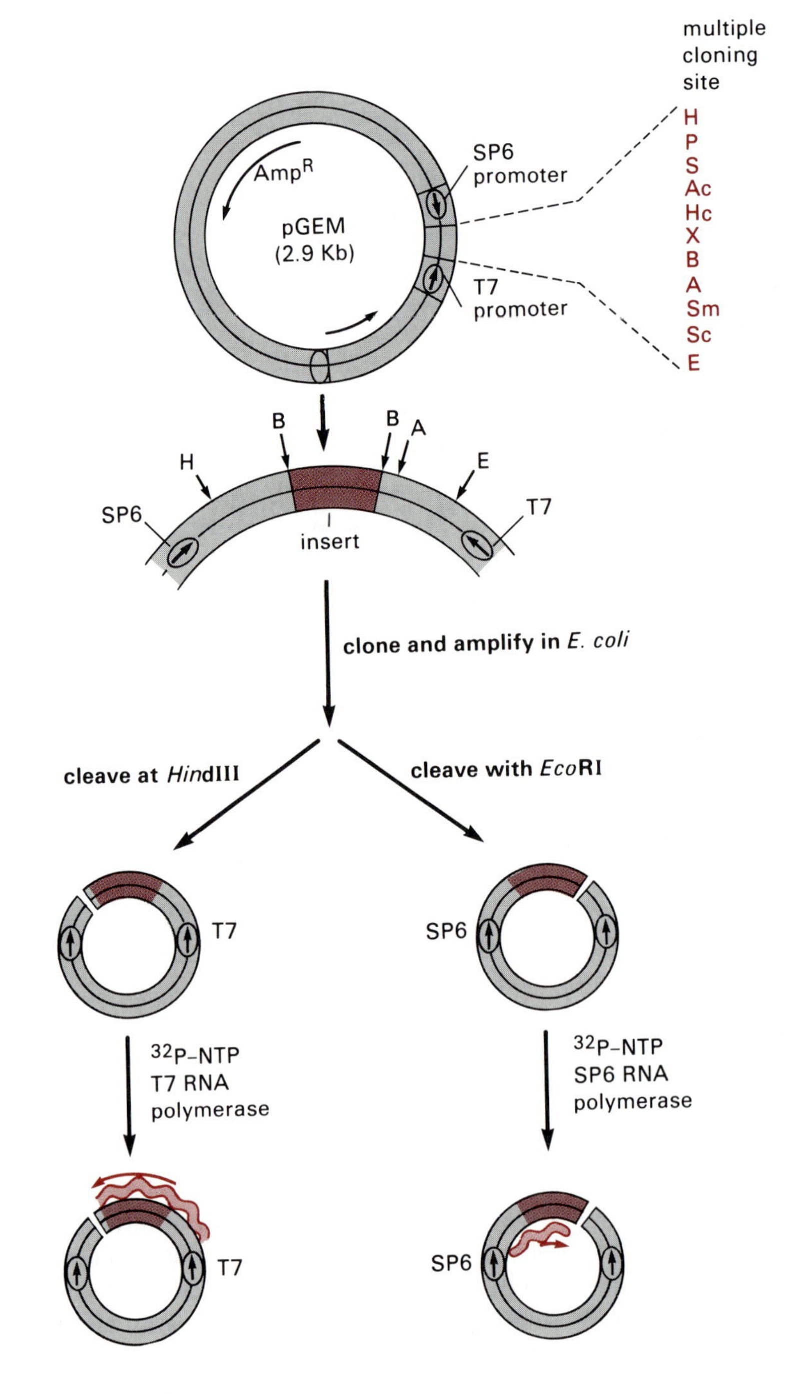

Figure 6.23
A recombinant DNA plasmid that can be used to synthesize radiolabeled single strand RNAs. The basic plasmid, pGEM, contains sequences from pBR322 and two prokaryotic promoters. The SP6 promoter is derived from the *Salmonella typhimurium* bacteriophage SP6 and functions only with the specific SP6 RNA polymerase. The T7 promoter is derived from *E. coli* bacteriophage T7 and functions only with the specific T7 RNA polymerase. Between the two promoters, there is a polylinker containing multiple restriction endonuclease sites. An insert can be placed in any one of these sites; insertion into the *Bam*H1 site is shown. Purified recombinant plasmid DNA preparations are cleaved with restriction endonuclease to form linear molecules prior to carrying out *in vitro* transcription. Depending on the cleavage site selected, which in turn depends on whether the SP6 or T7 RNA polymerase is to be used, an RNA copy of one or the other strand of the insert is obtained.

convenient because they need not be denatured prior to use and because the cloned DNAs being tested need not compete with any other hybridizing DNA strand during the annealing reaction. Messenger RNA probes are themselves single strands, and single strand cDNA probes are readily prepared by limiting copying to only one strand (Figure 4.20). Single strand RNA probes are also readily prepared from a recombinant plasmid containing the probe sequence inserted into a special vector such as pGEM (Figure 6.23). This vector includes two different, highly specific bacterial promoters that flank the insert in recombinants. By using purified, suitably linearized plasmid as a template *in vitro* with one or the other of the

peptide	glu · phe · trp · met · gly
16 possible mRNAs	5′ GAA UUU UGG AUG GGN* G C
16 possible DNA probes	3′ CTT AAA ACC TAC CCN* C G

* N is any of the 4 nucleotides

Figure 6.24
The sequence of a pentapeptide predicts several possible sequences for the corresponding region of the mRNA. Chemical synthesis of a mixture of oligodeoxyribonucleotides complementary to the possible mRNAs provides a probe for the detection of mRNAs and clones that contain segments encoding the peptide.

corresponding prokaryotic RNA polymerases, RNAs representing a copy of one or the other of the strands of the insert are synthesized.

If the amino acid sequence of the polypeptide corresponding to the gene is known, a nucleotide sequence for a portion of the gene can be deduced from the genetic code and then synthesized by chemical procedures. Even a DNA chain as short as 15 nucleotides can give a relatively specific signal when used as a probe. Because of the degeneracy of the genetic code, predicting a unique sequence for the coding region is usually impossible. The problem can be circumvented by synthesizing a mixture of oligonucleotides differing in one or more nucleotides (Figure 6.24). This is not as complicated as it sounds. Instead of using a single protected mononucleotide at a particular step in chemical synthesis, a mixture of protected mononucleotides is used.

Testing for specific polypeptide synthesis in vitro Sometimes there is simply no available probe of suitable purity for direct screening of clones. In such a case, indirect methods can still be used, although they are not generally convenient for screening massive cloned populations. Most of the commonly used indirect procedures depend on two principles. First, an mRNA can be translated, *in vitro*, into an identifiable polypeptide. Second, translation is inhibited if the single strand messenger RNA is annealed to a complementary, cloned DNA; the RNA in the RNA·DNA duplex no longer functions as a messenger.

When a mixture of mRNAs is incubated *in vitro* in the presence of the required components, polypeptides corresponding to each mRNA may be synthesized. The mixture of polypeptide products is separable on the basis of size by gel electrophoresis. Efficient *in vitro* translation requires single strand mRNAs. If, prior to translation, the mRNA is annealed with complementary DNA to give a DNA·RNA hybrid duplex, the messenger is no longer translatable. In a procedure called **HART**—for hybrid-arrested translation (Figure 6.25)—cloned recombinant DNA molecules are purified, denatured, and annealed (in solution) with an mRNA preparation prior to the *in vitro* translation reaction. DNA molecules homologous to the desired gene hybridize with the mRNA, and the corresponding polypeptide disappears from among the translation products.

A second approach to identifying a recombinant by *in vitro* translation is called **hybrid selection** (Figure 6.26). Recombinant DNA is purified from a number of recombinants, each of which is suspected to include the correct one. These DNA preparations are denatured, and each is fixed to a solid support, such as a nitrocellulose disk. Samples of a mixture of

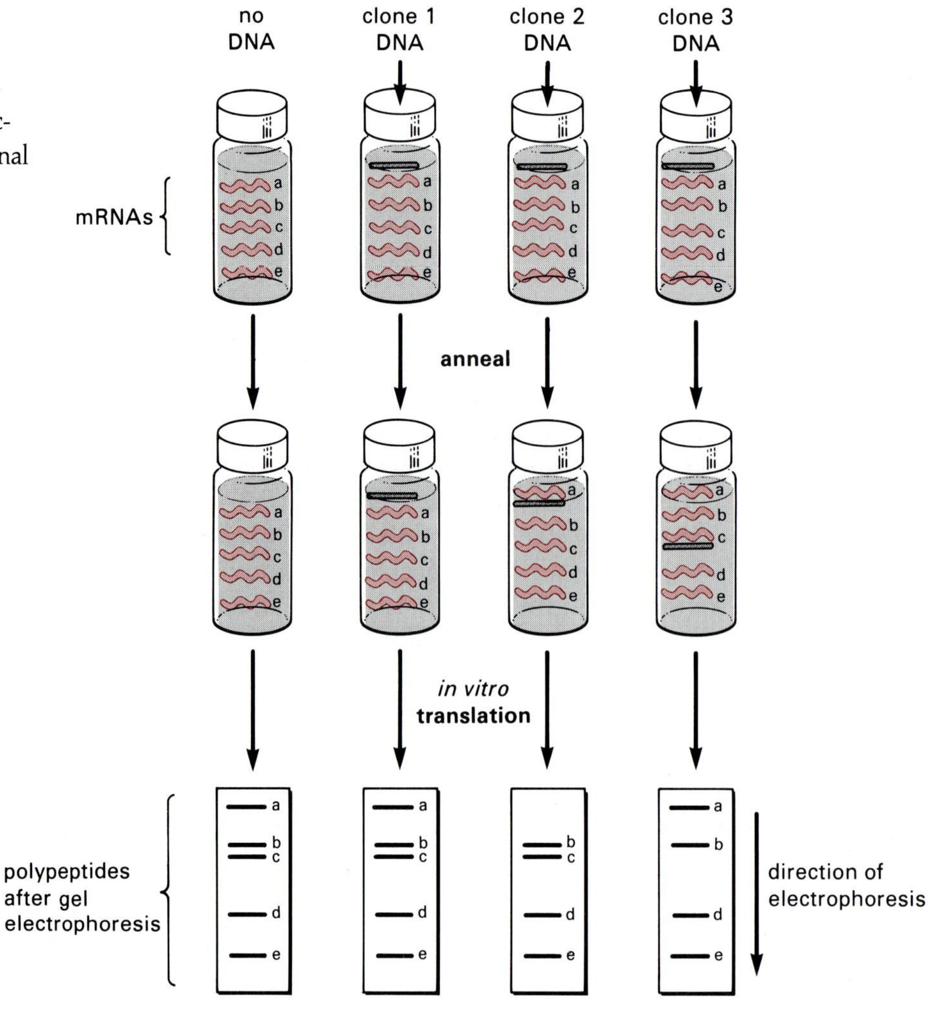

Figure 6.25
Hybrid-arrested translation (HART). Clones 2 and 3 contain DNA corresponding to mRNAs a and c, respectively. (Figure 6.27 provides additional details about *in vitro* translation.)

mRNAs (in solution) are incubated with each disk, but only an RNA that is complementary to the cloned DNA hybridizes and is bound to the disk. The other RNAs are removed by washing. The bound RNA is dissociated from the individual DNAs, recovered, and tested for its ability to yield a specific polypeptide in an *in vitro* translation system, thereby identifying the proper recombinant.

A variety of other techniques is used to facilitate or confirm the identity of the polypeptides produced by *in vitro* translation of mRNA mixtures. When the mRNA preparation is enriched for a specific messenger, the polypeptide encoded by that messenger can sometimes be identified among the translation products by size and by abundance (Figure 6.27a). If the mixture of messengers and polypeptides is complex, the polypeptide of interest may still be identifiable by its interaction with a specific antibody. One way to do this is by the protein-blotting technique (Western blotting) analogous to DNA blotting (Figure 6.27b). In an alternative approach, the polypeptide of interest is precipitated by a specific antibody from the mixture of products of *in vitro* translation (Figure 6.27c). The precipitate is dissolved and then subjected to electrophoresis. In principle, the only two proteins on the gel should be the desired polypeptide and the immunoglobulin.

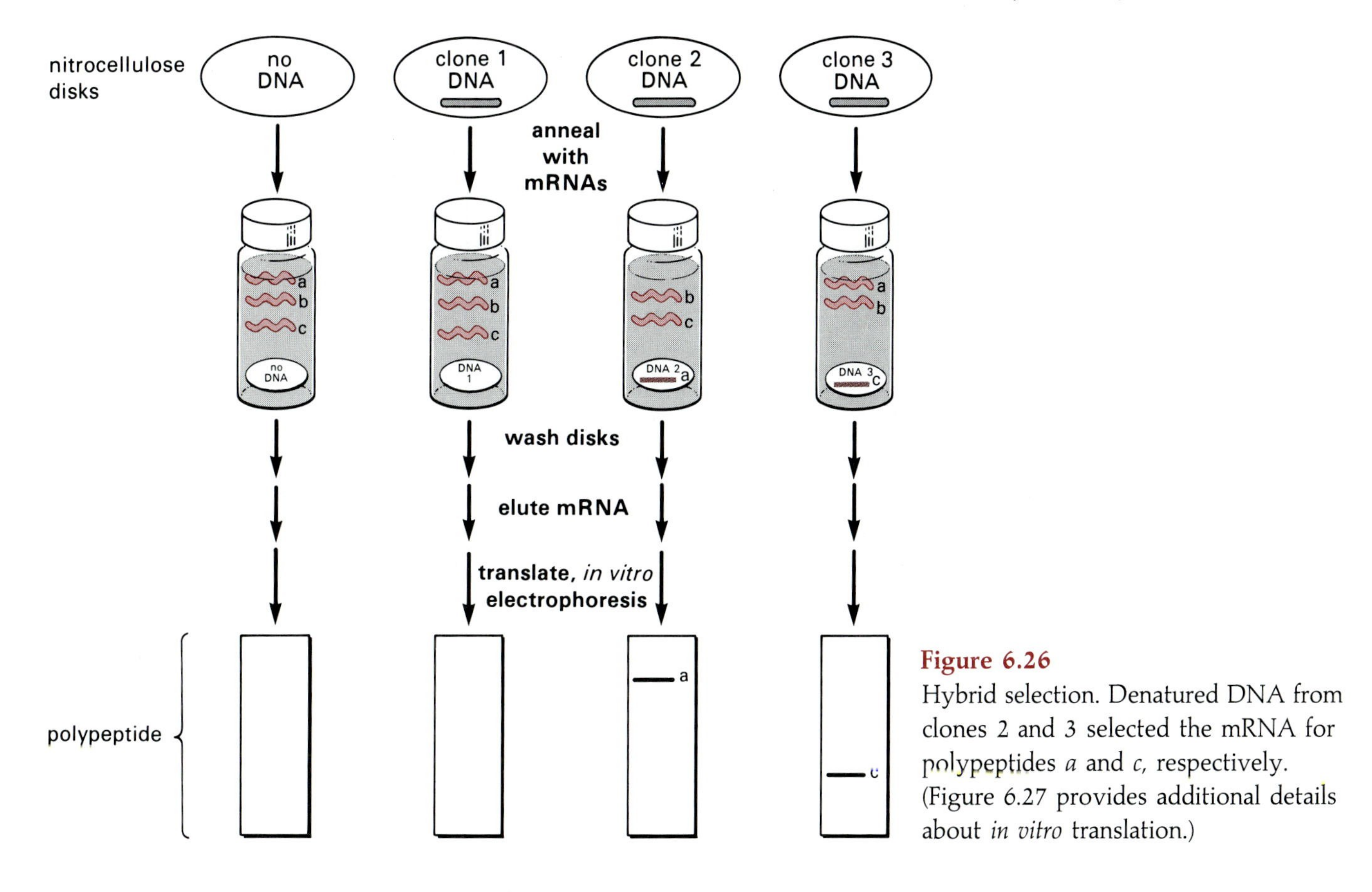

Figure 6.26
Hybrid selection. Denatured DNA from clones 2 and 3 selected the mRNA for polypeptides *a* and *c*, respectively. (Figure 6.27 provides additional details about *in vitro* translation.)

c. *Testing for Gene Expression in Cells*

The expression of a gene contained within a recombinant DNA segment will, under some circumstances, impart to a host cell a distinct and detectable phenotype. Selection techniques based on that phenotype enable identification and isolation of the clone, just as the standard selectable markers included in many vectors permit detection of transformed cells. Examples of such standard selectable marker systems described in Chapter 5 include the use of *E. coli* that lack β-lactamase (and are thus sensitive to ampicillin) to select for transformation by vectors carrying the *amp* gene and of mammalian cells that are TK^- to select for transformation by vectors carrying a thymidine kinase gene. This principle—**phenotypic selection**—can be used to select specific clones directly from populations of bacterial, yeast, or animal cells transformed with mixtures of recombinant vectors containing various bacterial, yeast, or animal genes, respectively. The feasibility of phenotypic selection depends on having host cells of a different phenotype, most often, for example, cells that lack or are deficient in the product of the gene being cloned. Other approaches to phenotypic selection used in eukaryotic cells are described later in this section and in Section 5.7. When no appropriately matched host cell is available, immunological screening is an effective alternative to phenotypic selection. Here, clones are screened with an antibody to the specific gene product.

Phenotypic selection In those situations where it is practicable, phenotypic selection from a population of transformed cells is a straightforward

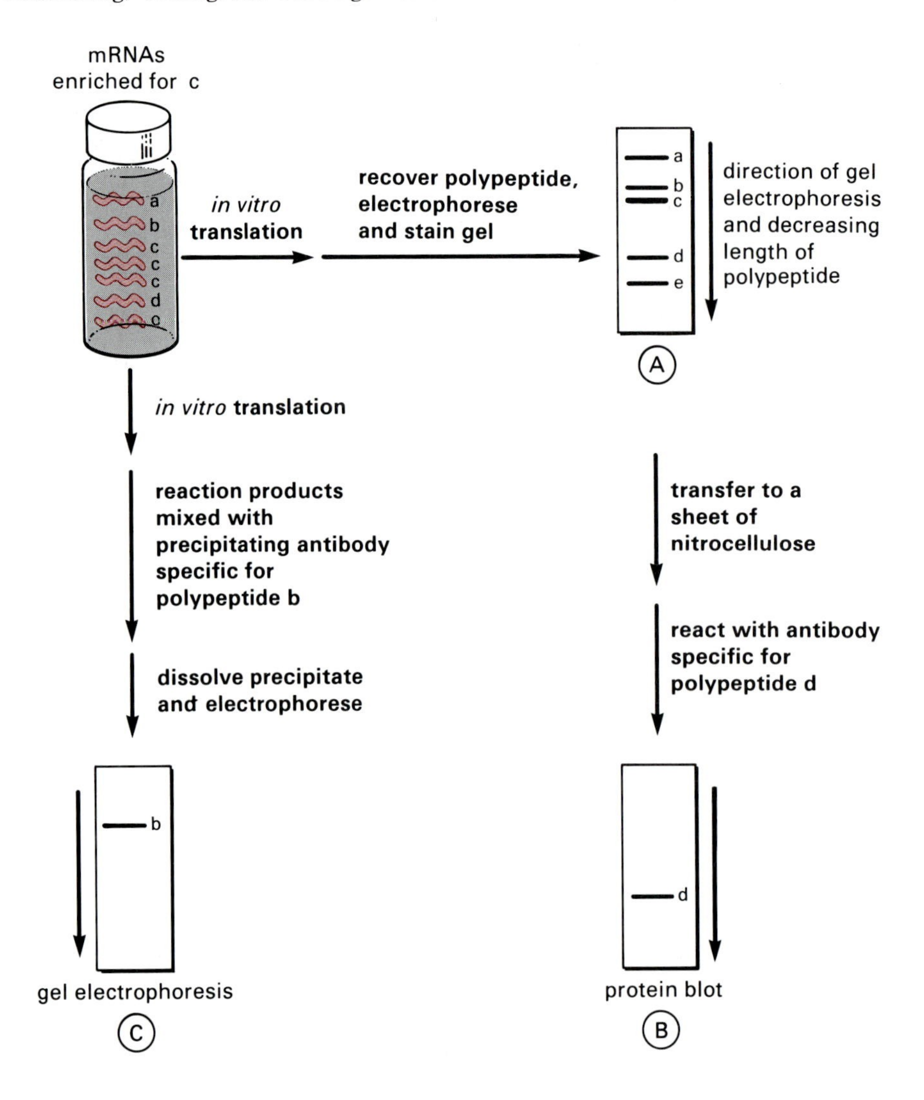

Figure 6.27

In vitro translation of a mixture of RNAs. In addition to the RNAs, the reaction mixture contains amino acids (sometimes including at least one radioactively labeled amino acid), ATP, and cell extracts containing the other components required for protein synthesis (e.g., enzymes, ribosomes, and tRNA). (a) Each mRNA yields its corresponding polypeptide. These are separated according to size by electrophoresis and visualized by staining with an appropriate dye. Alternatively, if radioactive amino acids are used, an autoradiogram can be prepared by exposing an X-ray film to the gel. In this example, mRNA c is especially abundant, and polypeptide c is correspondingly enriched in the translation products. (b) The electrophoresis gel in (a) is analyzed by protein blotting. The *in vitro* translation is carried out with unlabeled amino acids. The separated polypeptides are transferred from the gel onto an immobilizing matrix, where they are trapped by adsorption. Most often the transfer involves electroelution of the polypeptides from the gel, rather than simple blotting. The setup is much like that described in Figure 6.2, except that the sandwich of gel and matrix is immersed in a buffered salt solution with electrodes on either side. The polarity is fixed so the proteins will be eluted onto the matrix. The blot (the immobilizing matrix containing the polypeptides) is then incubated in a solution of an appropriate antibody, in this case, antibody against polypeptide d. The antibody itself can be tagged by radioactive isotopes, by a fluorescent adduct, or by conjugation to an enzyme that is easily detected. In this way, the presence and position of the polypeptide on the original gel is marked. Alternatively, the position of a bound antibody can be determined by incubating the blot with a radioactively labeled preparation of protein A of *Staphylococcus aureus*; this protein binds specifically to the immunoglobulin IgG. (c) *In vitro* translation is carried out with radiolabeled amino acids, and the polypeptide products are mixed with antibody to polypeptide b. The precipitate is dissolved and analyzed by gel electrophoresis, and an autoradiogram is made. The only radioactive product visible is polypeptide b.

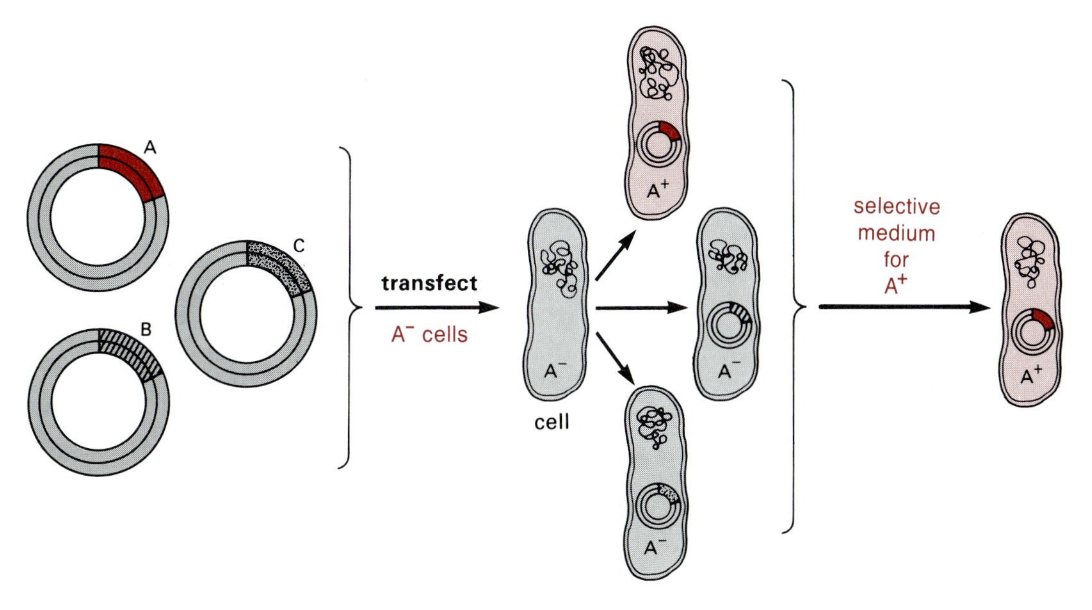

Figure 6.28
Phenotypic screening.

route to the desired clone (Figure 6.28). In practice, this method is by and large limited to cloning prokaryotic genes in prokaryotes, some yeast and other fungal genes in prokaryotes, and eukaryotic genes in eukaryotic cells. The general approach was introduced in Section 5.6b when the construction of a yeast plasmid vector was described. To clone gene A (or cDNA-A), a mixture of DNA fragments (e.g., from an entire cellular genome) is inserted into vector molecules and transfected into cells that contain mutations in gene A (A^-). The cells are grown under selective conditions that require the gene A product; only cells provided with gene A product by expression from the vector form colonies. Preferably, the A^- host cells contain deletions including all or part of gene A. Otherwise there is a problem in distinguishing revertants from transformants. Also, deletion mutants limit the possibility that the A^+ phenotype will be caused by genes other than A that suppress the A^- phenotype.

Eukaryotic gene expression in bacterial host-vector systems Neither phenotypic selection nor immunological screening is generally applicable to the isolation of *E. coli* cells containing a particular recombinant eukaryotic gene. Exceptions occur in the case of genes from yeast and other fungi, some of which are expressed in *E. coli* cells. However, most eukaryotic genes are not properly expressed in *E. coli* because they lack functional promoters and contain noncoding interruptions in their coding sequences (Chapter 8). The situation is different when the recombinants contain cDNA copies of eukaryotic mRNAs because the noncoding intervening sequences are spliced out during maturation of mRNA in eukaryotic cells. *E. coli* cells are incapable of carrying out this splicing. The cDNA segments are thus translatable into the correct polypeptides in *E. coli* (Figure 6.29). To ensure that the *E. coli* cells will transcribe and translate the cloned eukaryotic cDNA, an *E. coli* promoter and translation signals are usually included at appropriate positions on the vector. Similarly, when eukaryotic

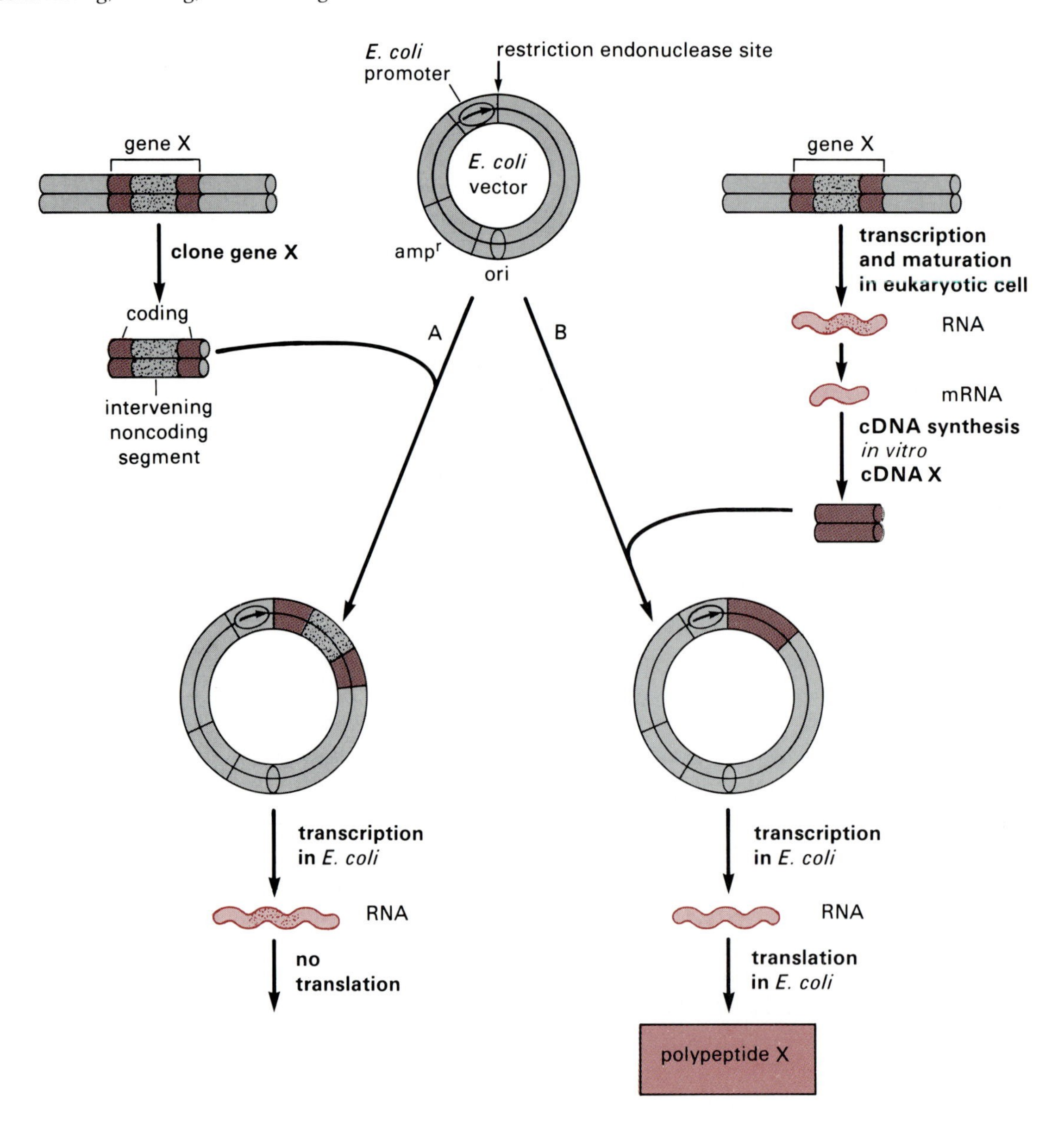

Figure 6.29

The expression of eukaryotic genes in *E. coli* host-vector systems. The eukaryotic DNA segment is inserted downstream of an *E. coli* promoter to ensure transcription in the bacteria. Eukaryotic genes (such as gene X, depicted here) are generally interrupted by noncoding segments (stippled); therefore, the transcript (RNA) of gene X that is produced in *E. coli* cannot be translated into the correct polypeptide (A). When functional mRNA is synthesized in a eukaryotic cell, the noncoding intervening sequence is spliced out (Chapter 8). Thus, a cDNA made with the mRNA corresponding to gene X as template lacks the noncoding sequence. When this cDNA is inserted into the bacterial vector and the construction is introduced into *E. coli* cells, an RNA that can be properly translated into the correct X-gene product is produced (B).

cDNA recombinants are to be expressed in eukaryotic cells, the vector must supply appropriate transcriptional regulatory signals. Because the normal genomic regulatory signals generally lie outside the transcribed portion of the gene in the 5′ or 3′ flanking region, they do not appear in the mRNA or cDNA.

Other methods for phenotypic screening in eukaryotic cells Phenotypic screening is rarely applicable in diploid eukaryotic cells because it requires a cell line in which both alleles of the desired gene are defunct. Cotransformation using a selectable marker and correspondingly mutated cells (e.g., the thymidine kinase gene and TK^- rodent cells) or a dominant marker and drug-inhibited cells (e.g., the *E. coli neo* gene encoding aminoglycoside phosphotransferase and cells treated with the aminoglycoside G-418) allows the identification of transformed cells (Section 5.7). However, it does not discriminate between the desired cotransformant and the accompanying cotransformed colonies that contain different recombinant DNAs. Thus, successful methods for phenotypic screening generally rely on specific properties expected of the desired clone. One example is the use of the morphological changes that occur when normal cells are converted to tumor cells, as described in the sections on bovine papillomavirus and retrovirus vectors in Section 5.7. Cells transformed to a tumorigenic phenotype pile up on top of one another to form a distinctive colony, or focus, instead of growing as a monolayer. This method was used, for example, to select recombinants containing cellular oncogenes from genomic animal DNA.

Another specialized screening method is used to clone animal cells expressing a recombinant gene (or cDNA) encoding a secreted polypeptide. Host cells that do not normally produce the polypeptide are used, and the medium surrounding the transformed cells is assayed for the desired product. If the product is, for example, a hormone, its presence in the medium can often be measured by a convenient and highly sensitive biological assay. Genes encoding growth factors that stimulate the proliferation of specific target cells have also been cloned in this way. If the polypeptide is not associated with an easily measurable activity, then it may be detectable in the medium with a specific antibody. This method does not permit the simultaneous screening of massive populations of cloned cells containing different recombinants because too many (perhaps millions) of individual assays would need to be performed. Instead, the mixture of transformed cells is divided into a convenient number of individual groups, and the groups are assayed. A positive group is then subdivided for assay, and again any positive subgroup is subdivided and assayed and so forth until a single positive clone is obtained.

Immunological screening Immunological techniques provide a more general method for screening a recombinant population for gene expression. Functional protein is not required; what is necessary is a specific antibody to the protein in question and host cells that do not synthesize the antigen. The antibody reacts with the protein in a single clone. The reaction is visualized by accumulation of radioactivity from either labeled antibody or secondary interaction of the immunoglobulin with a specific (labeled) reagent that recognizes immunoglobulins (e.g., the protein A of *Staphylococcus aureus*).

Several different versions of antibody screening techniques have been devised for use with *E. coli* host-vector systems. They are similar in principle, although different techniques are used depending, for example, on whether phage or bacterial clones are to be screened. In one version of this method, the antibody is spread uniformly over a solid support such as a plastic or paper disk and placed in contact with an agar layer containing lysed colonies or phage plaques. If the specific antigen was synthesized by any of the clones, it will bind to the antibody on the disk. After the disk is removed from the agar, the position of the antigen can be marked by treating the disk with additional antibody that has been radiolabeled (usually with ^{125}I), washing and preparing an autoradiogram on X-ray film. This procedure depends on the ability of the two antibodies (one coating the disk and the other ^{125}I-labeled) to interact with different antigenic determinants on the same polypeptide. Just as with screening by annealing to a DNA or RNA probe, a positive clone gives a localized dark spot on the film.

Highly specialized techniques permit immunological screening and selection of transformed animal cells if the gene in question encodes a polypeptide that becomes localized to the cell surface. Surface proteins are available for interaction with specific antibodies. If the antibodies are coupled to a fluorescent molecule, then cells synthesizing a particular protein can be distinguished from the other transformants by their fluorescence. A specially designed apparatus, a **fluorescence-activated cell sorter** (or FACS), sorts individual viable fluorescent cells away from the vast number of nonfluorescent cells. The viable cells can then be grown into cloned populations (Figure 6.30). Some surface proteins are receptors that bind specific ligands (e.g., insulin receptors for insulin). If the gene in question encodes such a receptor, then a fluorescent ligand will allow a FACS to separate transformed cells.

6.5 Libraries

Talk of "shotgun" experiments attracted attention even when the recombinant DNA technique was still quite new. The idea of amplifying very large mixtures of DNA fragments and then trying to find the single segment of interest was dramatic, even fanciful. Now it is almost commonplace. The ultimate shotgun approach is embodied in the construction of a collection of recombinants that represents an organism's entire genome (a **recombinant library**). Basically, a total unfractionated set of DNA fragments is converted into a corresponding set of stable recombinants that can be stored and used repeatedly to clone different inserts. *E. coli* host-vector systems, either plasmids or bacteriophage, are used because they are most efficient for dealing with large numbers of recombinants and because they are readily preserved.

Two different kinds of libraries are important. One, constructed from total genomic DNA, in principle contains all the organism's genes and other DNA sequences. However, this ideal is usually not attained because some DNA sequences escape cloning. A variant of a total genomic library is one representing all the DNA in only one particular chromosome.

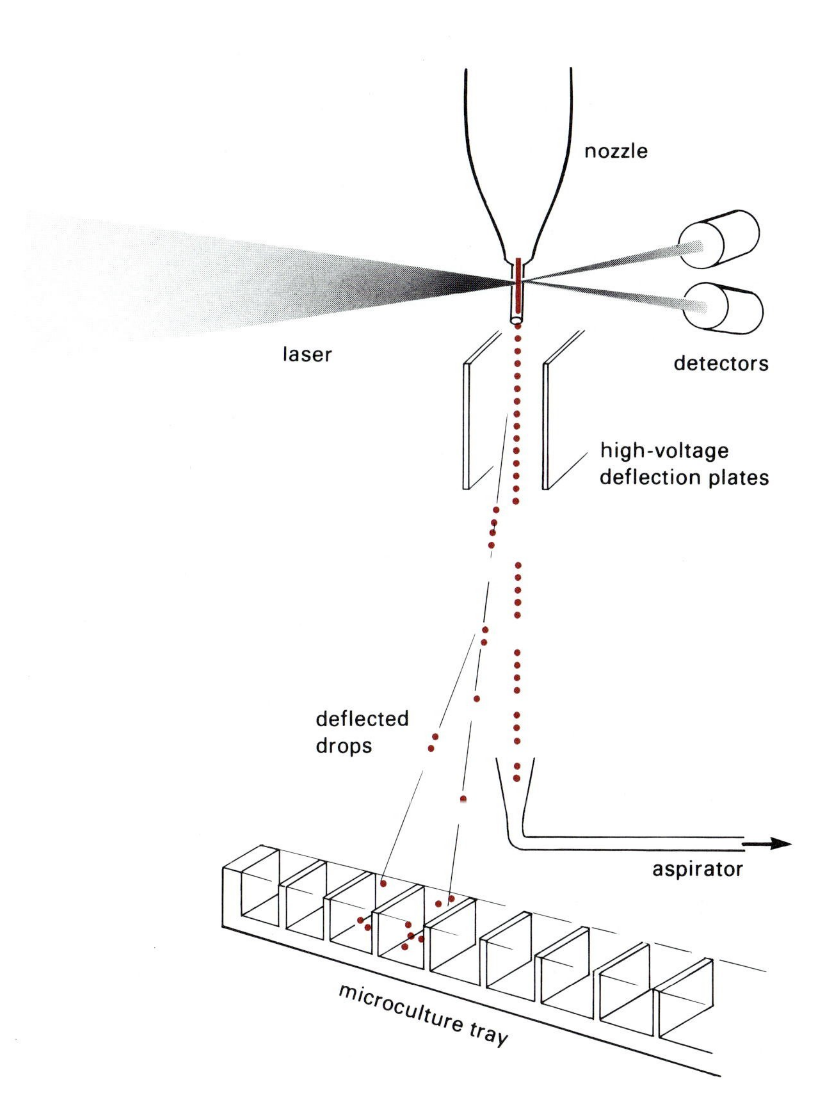

Figure 6.30
Schematic diagram of a fluorescence-activated cell sorter. A population of cells is treated with a specific antibody to a cell surface protein. The antibody is coupled to a fluorescent tag so that cells carrying the surface antigen become fluorescent. The cells are then passed, one by one in a stream, between a laser and a fluorescence detector. Cells that fluoresce above a fixed threshold value are deflected and dropped directly into wells in a cell culture dish, after which they multiply and can be cloned. Cells that do not fluoresce are aspirated away. About 10^3 cells can be screened per second. Adapted from D. R. Parks et al., *Proc. Natl. Acad. Sci. U.S.A.* 76 (1979), p. 1962.

Making such a library requires preparations of DNA from individual chromosomes. For example, human chromosomes have been fractionated by techniques similar to those used in fluorescence-activated cell sorting. The method depends on the differential uptake, by chromosomes, of fluorescent dyes, depending on chromosome size and relative content of A·T compared to G·C base pairs. A solution of stained chromosomes is then forced into a high speed stream across which a beam of laser light is focused. The chromosomes flow down the stream in single file and are caused to fluoresce in characteristic fashion by the beam. A sorting device detects the fluorescence intensity and deflects those chromosomes whose fluorescence equals a preset level into a special collector. Changing the preset level allows different chromosomes to be collected in separate runs through the device.

The DNA from individual yeast and some protozoan chromosomes can also be isolated, notwithstanding the fact that the chromosomes are not visually identifiable by cytogenetic techniques (i.e., by staining and microscopy). The DNA is gently isolated from cells to avoid breakage and subjected to electrophoresis through agarose gels under special conditions that permit separation of molecules as large as 2000 kbp (Figure 6.31).

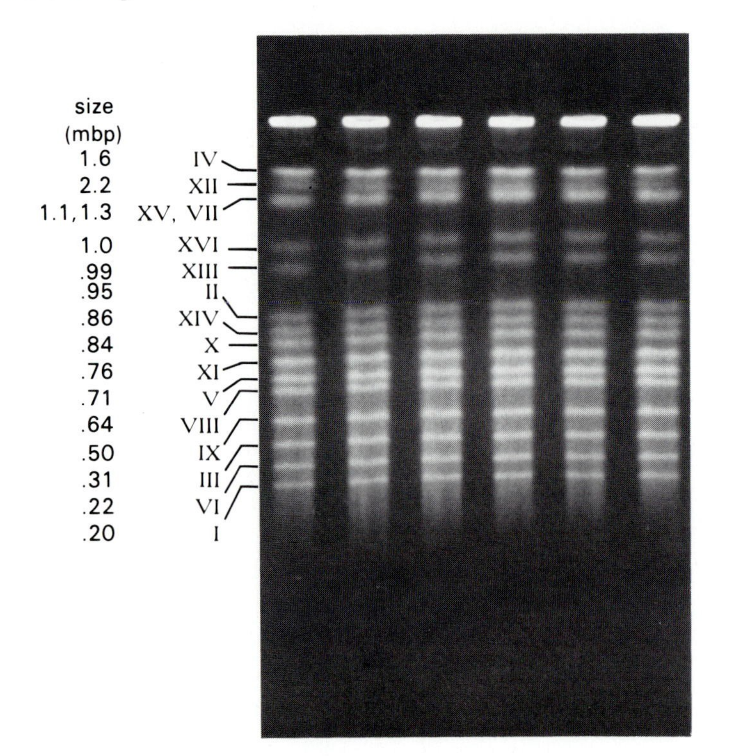

Figure 6.31
Pulsed-field gel electrophoresis. Photograph of an ethidium bromide–stained gel in which chromosomal DNA from yeast (*S. cerevisiae*) was electrophoresed for 24 hours under conditions that provide a homogeneous field (CHEF hexagonal array apparatus). The field was reoriented by an angle of 120° every 80 seconds. The yeast chromosomes (roman numerals) were identified by hybridization to previously mapped probes. Their estimated sizes are given in units of 10^6 bp (megabase pairs, mbp). After G. Chu, D. Vollrath, and R. W. Davis, *Science* 234 (1986), p. 1582.

Conventional electrophoretic techniques do not separate duplex DNAs that are much longer than 20 kbp; above this size, the molecules do not migrate in a size-dependent manner. However, if instead of the constant one-directional electrical field used in standard gels, the DNA is exposed to fields in alternating orientations, then even very large molecules can be separated according to size. This behavior most likely depends, at least in part, on the way DNA molecules move through the pores of the agarose gel. They are probably elongated in the direction of the field and must then continually reorient if they are to move in the changing fields. The time required for reorientation is likely to depend on the length of the chain and the angle between the two fields; thus, the net mobility will depend on both these parameters. The simplest **pulsed-field electrophoresis** devices pulse (automatically) between two nonhomogeneous, essentially perpendicular fields. More complex arrangements allow for homogeneous fields across the gel and optimize the angle between the two fields, resulting in improved resolution. Pulse times are generally of the order of a minute.

The second kind of library contains sequences representing all the mRNAs found in a particular cell type. In this case, the total mRNA population is converted to cDNAs, which are then cloned. **Genomic libraries** are a source of genes and DNA sequences; **cDNA libraries** represent the expression of those genes in the form of mRNA.

a. Genomic Libraries

Genomic libraries are usually constructed using λ bacteriophage or cosmid vectors and *in vitro* packaging (Figure 6.32). These vectors accept large inserts, thereby minimizing the total number of recombinants required to constitute the library. For example, if a λ library with an average insert

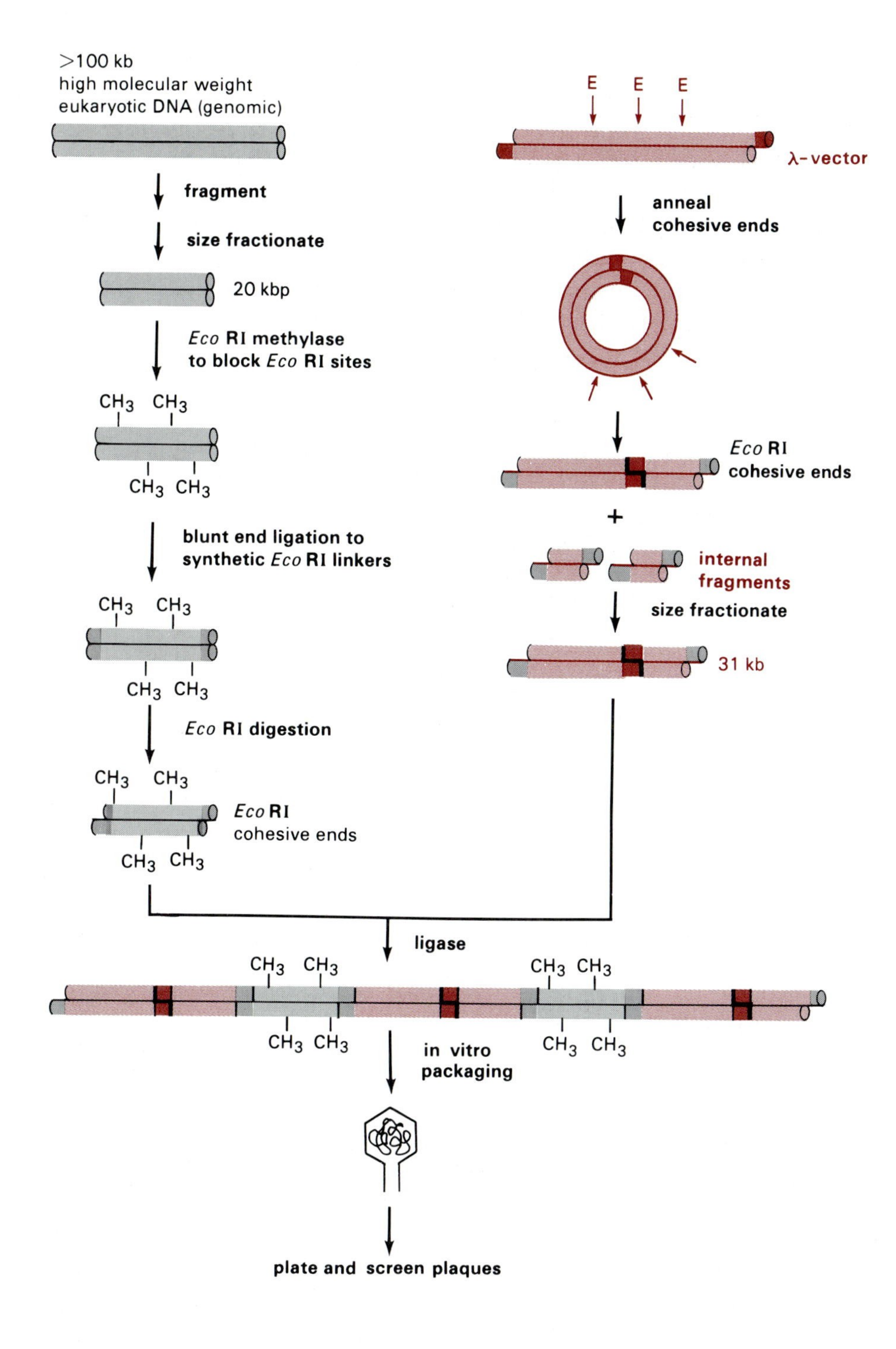

Figure 6.32
Constructing a genomic library. The high molecular weight genomic DNA is fragmented either by limited digestion with a restriction endonuclease or by physical methods. All the other steps have been described in Chapters 5 and 6. Note that the preparation of the arms of the λ vector is different from the procedure outlined in Figure 5.11. Here, the vector DNA is first circularized by annealing the cos sites together. Then the dispensable central portion is separated by restriction endonuclease cleavage and fractionated away from the annealed arms. The arms are then annealed to the insert fragments through the cohesive *Eco*RI ends. Adapted from T. Maniatis et al., *Cell* 15 (1978), p. 687.

size of 17 kbp pairs is constructed from a mammalian genome containing 3×10^9 bp (per haploid), then the whole genome is represented in 3×10^9 divided by 1.7×10^4 or 1.8×10^5 individual recombinants. Actually, about 10^6 individual recombinants are required for a library in which the chance of finding a particular unique genomic segment is greater than 99 percent because the individual fragments are ligated at random. Thus, some may be inserted in more than one vector molecule, but others may resist (or even escape) ligation or packaging. Complete cosmid libraries require fewer members because the inserts may be up to 45 kbp.

There are several ways to prepare fragments from total genomic DNA for library construction. The most convenient method involves partially digesting the DNA with a restriction endonuclease that has a recognition site of six bases (or longer) and yields cohesive ends appropriate for the chosen vector. In this way, only a limited number of the possible sites are cleaved. Because those sites that are cleaved are selected at random, and because the amount of genomic DNA used includes many replicas of any particular genomic segment, nearly every stretch of DNA should be represented in DNA segments of appropriate size for cloning. This approach does, however, produce a biased library. Regions of the genome in which the restriction endonuclease recognition sites are spaced too far apart are not incorporated into viable recombinant phage. Similarly, regions in which the sites are clustered will tend to be in very short pieces that may not be represented in the library.

In principle, a more representative library is obtained if the limited cleavage of genomic DNA is more random than can be realized by using an enzyme with a six–base pair recognition site. This can be accomplished by mechanical shearing of DNA or by very limited treatment with four–base pair recognition site endonucleases such as *Alu*I or *Hae*III. In either case, appropriate cohesive ends (linkers) must then be attached to the population of segments.

b. cDNA Libraries

Plasmid and λ phage vectors are generally used to prepare cDNA libraries. Except for the fact that a mixture of total rather than purified mRNA is used, the principles involved are as described in Section 6.1d. When prepared in special vectors or screened in special ways, cDNA libraries can be used to clone mRNA sequences even if the protein amino acid sequence and nucleotide coding sequence are entirely unknown and even if no homologous probe is available. Examples of these special tools are illustrated in the following paragraphs.

A cDNA expression library Purified cDNAs or mixed cDNA populations can be ligated into vectors that are specially designed to permit transcription and translation of the cDNA coding region. The desired clone is identified by immunological screening with an antibody specific for the encoded polypeptide. This quite useful technique permits cloning even when nothing is known about the structure of the gene or the protein. One widely used expression vector is a derivative of λ phage called λgt11 (Figure 6.33). When a population of cDNAs containing *Eco*RI linkers is ligated into the single endonuclease *Eco*RI site on λgt11, it is inserted into a region encoding the bacterial gene for β-galactosidase (*lacZ*). About one out of six insertions of any particular cDNA will be in the proper transcriptional direction and in phase with the β-galactosidase reading frame. (Copying of mRNA into a duplex cDNA is rarely perfect, and only one out of three cDNAs will likely start at an appropriate position.) When expression of the β-galactosidase is induced by a β-galactoside, transcription begins at the *lac* promoter and continues through the eukaryotic

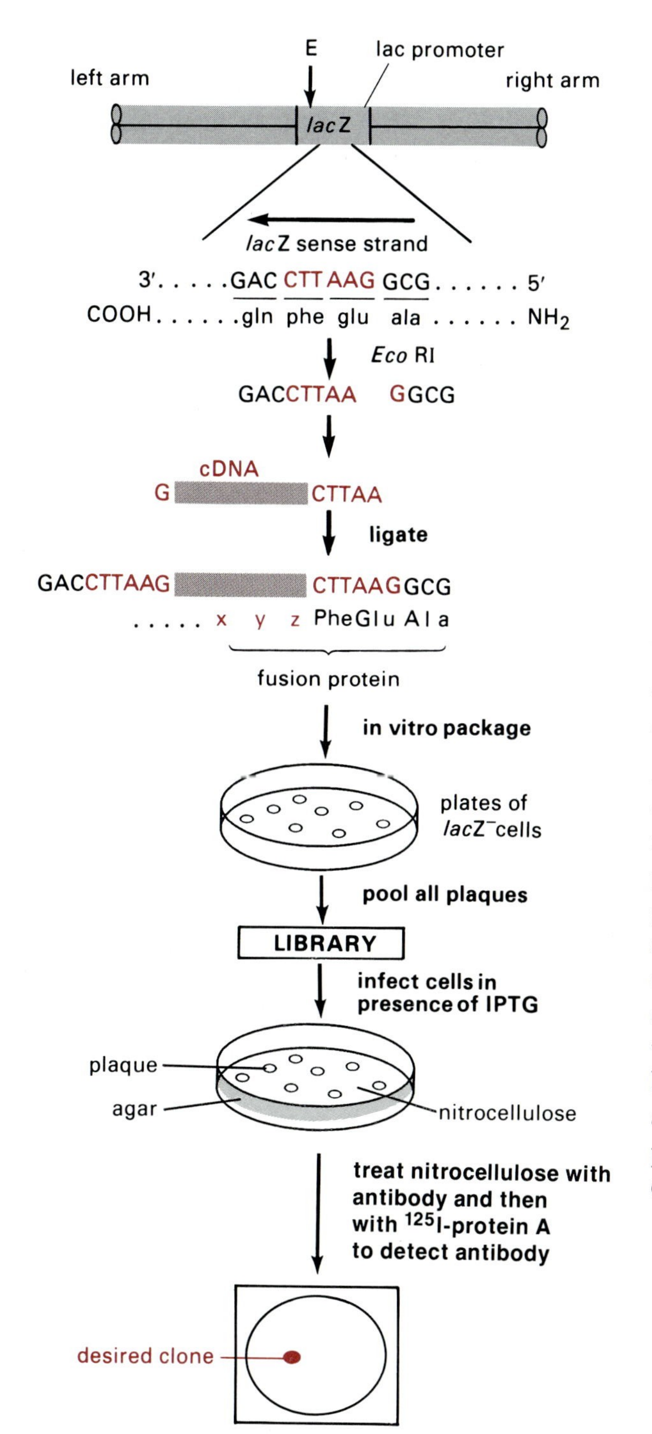

Figure 6.33
Cloning cDNAs in the expression vector, λgt11 (a modified λ phage genome that is about 44 kbp long and contains a copy of the *E. coli* gene for β-galactosidase, *lacZ*). There is a single endonuclease *Eco*RI site near the 3′ end of the *lacZ* gene. A population of cDNAs with *Eco*RI linkers is ligated into the *Eco*RI site, and the molecules are packaged into λ particles *in vitro*. Insertion disrupts the *lacZ* coding region, and no active β-galactosidase is produced. However, if the cDNA is in the right orientation and forms a continuous open reading frame with the *lacZ* coding region, a fusion protein with β-galactosidase sequences at the NH_2 end and a eukaryotic polypeptide at the COOH end can be formed. Recombinant phage plaques are readily distinguishable from λgt11 itself because they do not make active β-galactosidase. To utilize this marker, plaques are formed on host cells that have no functional *lacZ* gene. The host cells also have multiple copies of a plasmid containing the *lac* repressor gene (*lacI*) so that transcription of the λgt11 *lacZ* gene is inhibited. This ensures that no phage will be lost because the fusion protein is detrimental to the cell or phage. The phage in all the plaques are collected and constitute the library. To isolate and clone a desired recombinant, a portion of the library is used to infect fresh cells, and plaques are formed on a nitrocellulose disk laid on top of the agar. Here, plaque formation is in the presence of isopropyl thio-β-*D*-galactopyranoside (IPTG), an inducer of *lacZ* transcription. As a result, the β-galactosidase-eukaryotic fusion protein is produced. The nitrocellulose disk is then treated with a solution containing antibody to the desired polypeptide. After appropriate washing, plaques that interacted with the antibody are identified by the ability of IgG antibodies to bind ^{125}I-labeled protein A, a protein obtained from *Staphylococcus aureus*. The desired plaques are visualized by autoradiography and are then recovered from the corresponding positions in the agar layer.

segment. Translation of this RNA yields a fusion protein having β-galactosidase polypeptide in the amino portion and the eukaryotic polypeptide in the carboxy portion. A cDNA library constructed in λgt11 yields a population of plaques some of which contain fusion proteins. The plaque containing the protein (and therefore the cDNA) of interest is identified by its ability to complex the appropriate antibody. A single positive plaque among 10^4 can be identified on a single plate, and as many as 10^6 plaques are readily screened.

Selective cDNA libraries Development and differentiation of complex multicellular organisms from a single fertilized egg cell involve a complex and highly regulated program of differential gene expression. Some genes are expressed only in one or a limited number of cell types. Some are expressed for only a limited time period. As a result, the cytoplasmic mRNAs in different tissues and cell types contain distinctive populations of molecules. This differential gene expression can be used to clone cDNAs corresponding to the regulated genes, even if nothing is known about their gene products.

For example, to understand the process of early *Xenopus* development, it is important to know what genes are expressed during the initial cell divisions. An approach to this question requires identifying the mRNAs present in gastrulae but not in eggs. It is possible to do this by preparing a genomic library and screening it separately with RNA preparations from both eggs and gastrulae. Phage plaques that anneal only to gastrula RNA can then be identified and cloned. In practice, however, this procedure has the disadvantage that many of the 10^5 or more different mRNAs in the two samples are the same, and a huge number of plaques must be screened to find the few that are specific. A more efficient alternative is to prepare a cDNA library with gastrula RNA that has been enriched for mRNAs specific to that stage (Figure 6.34). Such an enrichment can be achieved as follows. A single strand cDNA is made from gastrula mRNA with reverse transcriptase. Then all the cDNA molecules that are homologous to mRNAs present in both eggs and gastrulae are removed by formation of cDNA·mRNA hydrids between the gastrula cDNA and egg mRNA. These hybrids are adsorbed to hydroxyapatite, leaving the remaining single strand cDNA in solution (Section 7.5b). The single strand cDNA, which is gastrula specific, is then converted to a duplex cDNA in the usual way and used to construct a cDNA library. The library is expected to be and often is highly enriched for cDNAs that represent gastrula-specific mRNAs. Relatively few of the recombinants anneal to *Xenopus* egg mRNAs. This same approach can be used with mRNAs from any pair of related tissues or cells. A variation on this theme involves using as a probe cDNAs enriched by the previously described method. Here a complete cDNA library is screened for those clones that hybridize with the enriched mRNA (or cDNA) probe (Figure 6.34).

6.6 Examples of Strategies Used for Cloning Genes and cDNAs

Many different approaches to constructing, cloning, and selecting specific recombinant genes have been described in this chapter. Using these procedures, numerous different strategies can be devised to obtain particular cloned genes or cDNAs. The nature of the gene and its gene product, as well as the species of origin, are pertinent to the choice of an efficient and effective strategy. The best way to get a sense of the versatility and selectivity of the recombinant DNA methods is to study the way a variety of genes and cDNAs has been cloned. Therefore, a series of cloning experiments that illustrate different strategies is given in Figure 6.35 through 6.40. Other examples are described in subsequent chapters.

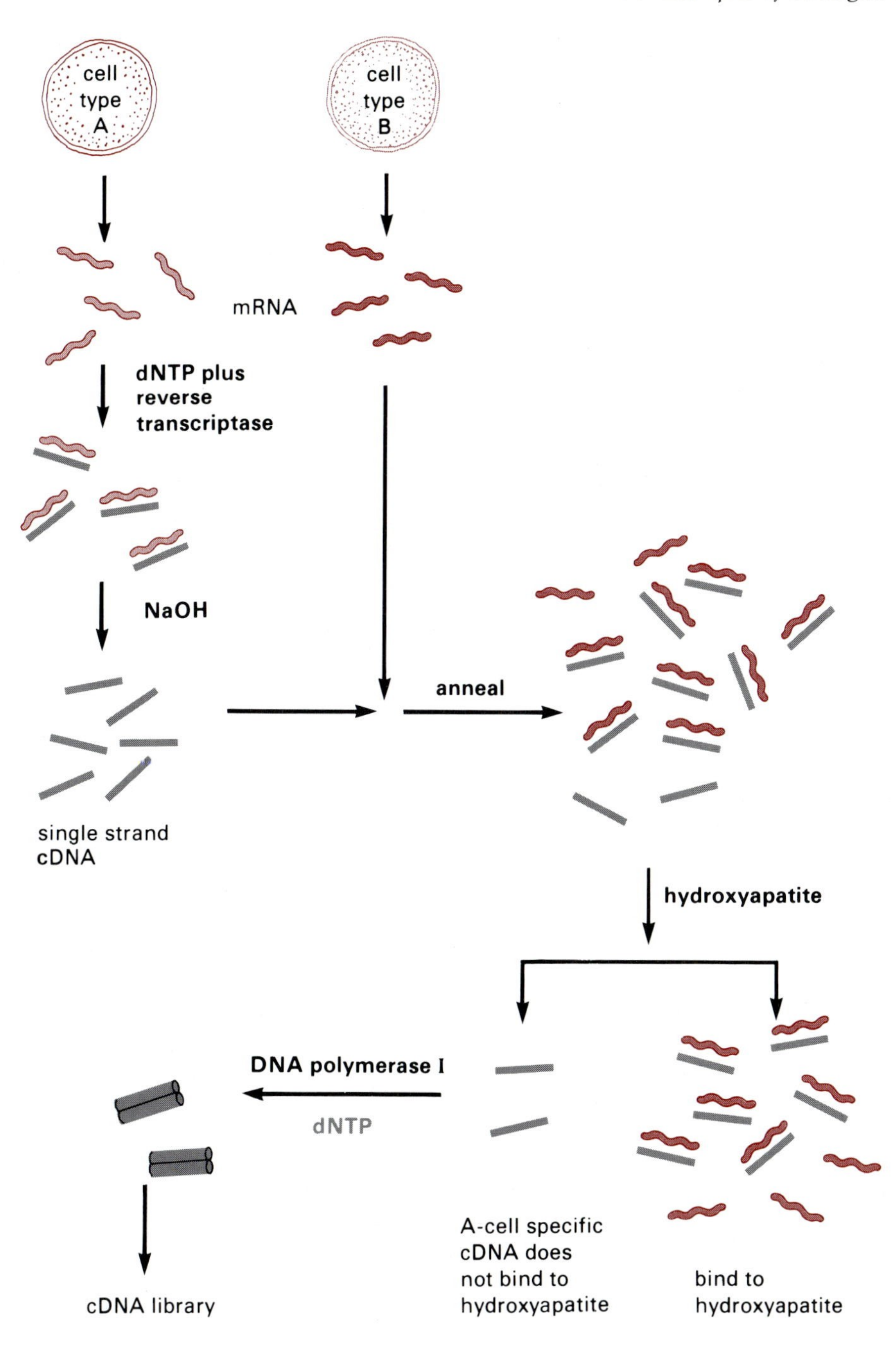

Figure 6.34
Preparation of a cDNA library representing mRNA specific to a particular cell type or tissue. Cytoplasmic mRNA is prepared from two related cell types or tissues, A and B. A specific cDNA library from A is to be constructed. The A-mRNA is copied into a single strand cDNA. Then the A-cDNA is annealed with an excess of B-mRNA. Duplex hybrids of complementary A-cDNA and B-mRNA are formed. The A-cDNA that is specific to A cells remains single stranded and is separated from the duplex hybrids and remaining RNA. This is accomplished using the adsorbent hydroxyapatite, which, under suitable conditions, binds the hybrids and free RNA but not single strand DNA (Section 7.5b). The single strand A-cDNA is then used as a template to synthesize the second cDNA strand, and a library is constructed with the duplex cDNA. The library represents almost exclusively those mRNAs specific for cell type A. Alternatively, the enriched cDNAs are used as a probe for hybridization with a complete cDNA library from cell type A mRNA to detect cDNA clones that correspond to cell type A mRNAs.

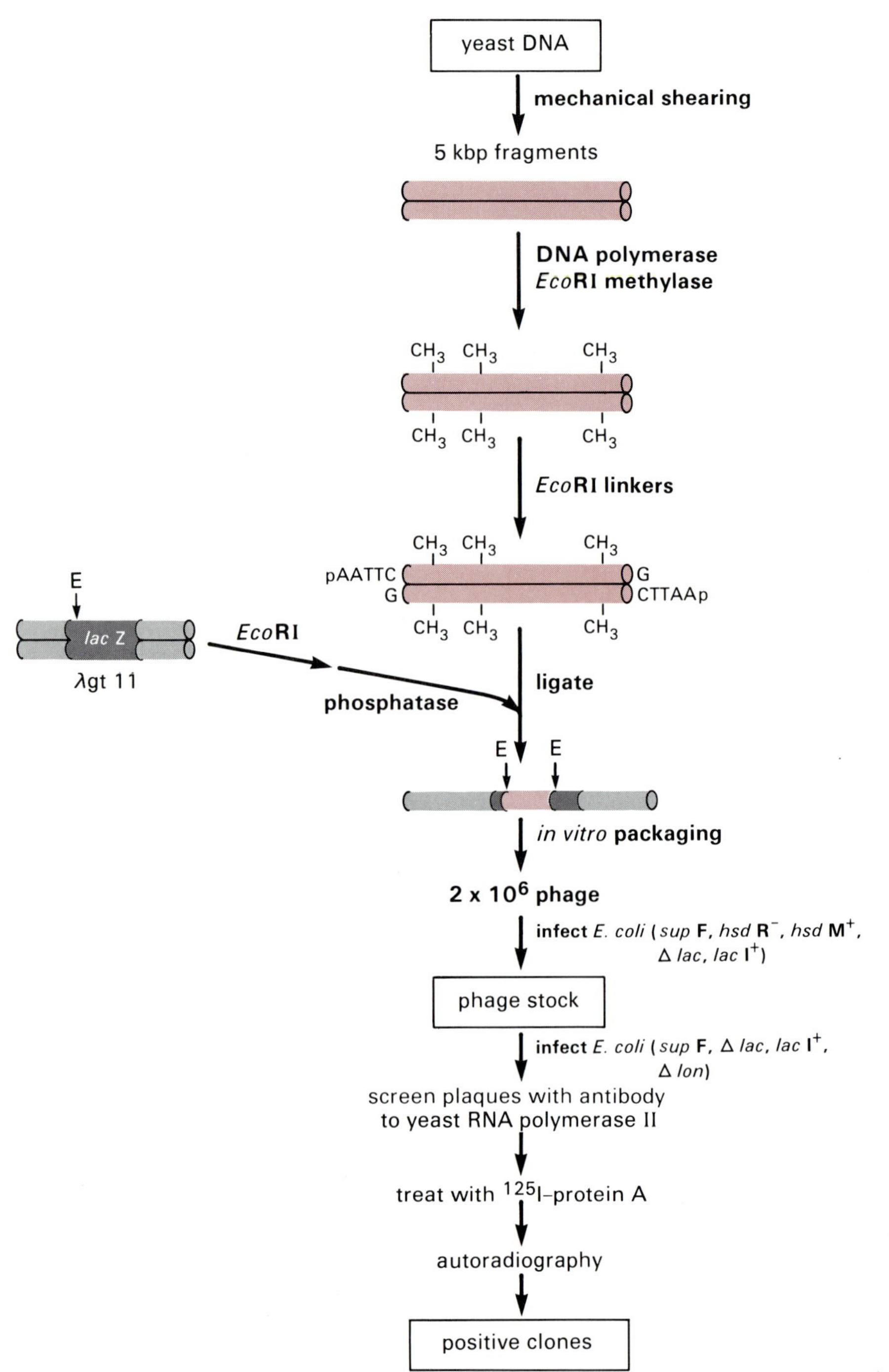

Figure 6.35
Cloning genes encoding subunits of yeast RNA polymerase II. *S. cerevisiae* DNA was sheared mechanically to give fragments averaging 5 kbp. After filling in any single strand regions with DNA polymerase, the *Eco*RI sites were methylated, *Eco*RI linkers were attached, and the fragments were ligated into λgt11 that had been cleaved with endonuclease *Eco*RI and treated with phosphomonoesterase. The recombinant DNAs were packaged into λ particles, *in vitro*, yielding 2×10^6 phage, which were used to infect *E. coli* with the following genotype: *sup*F (to suppress the amber mutation in the λ S gene), *hsd*R$^-$, *hsd*M$^+$ (to prevent restriction of the infecting DNA), Δ*lac* (to allow discrimination between plaques formed by λgt11 and those containing yeast inserts), and a pBR322 plasmid containing the *lac*I gene (which encodes the *lac* repressor and thus inhibits expression of yeast proteins that might inhibit cell or phage growth). In this step, the library phage multiply, yielding a stock. An aliquot of the stock containing 10^6 phage was used to screen for phage plaques producing proteins that bind to antibody to yeast RNA polymerase II, as described in Figure 6.33. For the screening, an *E. coli* strain containing a deletion of an important protease gene, (Δ*lon*) was used to reduce degradation of yeast proteins. Rescreening of the 60 positive plaques yielded multiple clones containing coding sequences for several different subunits of RNA polymerase II. After R. A. Young and R. W. Davis, *Science* 222 (1983), p. 778.

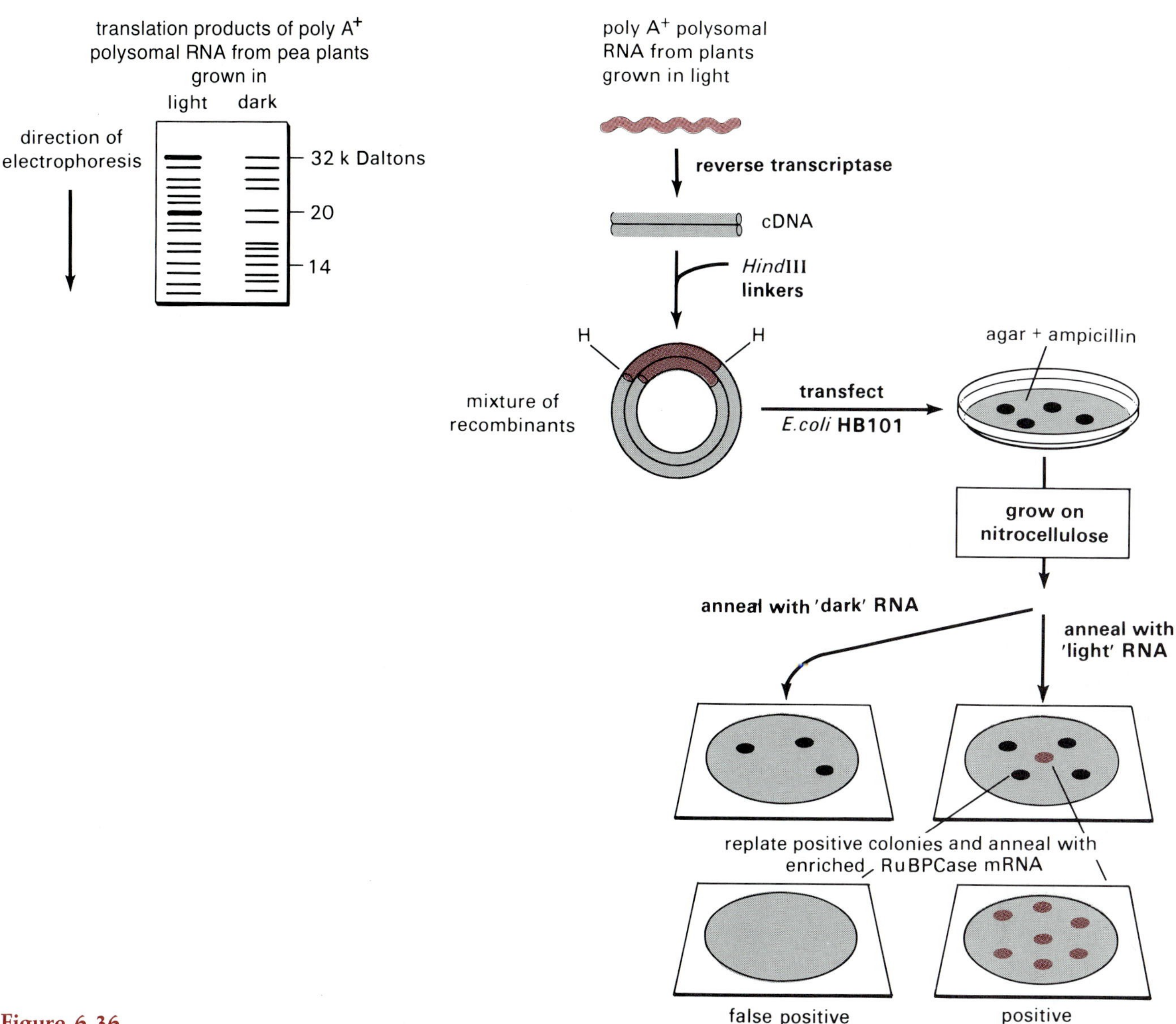

Figure 6.36

Cloning a cDNA encoding the small subunit of ribulose-1,5-bisphosphate carboxylase (RuBPCase) from pea (*Pisum satirum*) leaves. RuBPCase, probably the most abundant protein on earth, is composed of eight identical large and eight identical small subunits. The gene for the large subunit (55 kDaltons) is in the chloroplast genome (Section 9.7c), and the one for the small polypeptide (a 14 kDalton chain produced by posttranslational cleavage of a 20 kDalton precursor) is in the nuclear, chromosomal DNA. Expression of the genes depends on light, as illustrated in the left top panel. Pea plants were kept in the dark for 9 days, and then some were illuminated for 48 hours. Polysomal, polyA^+ RNA isolated from both types was translated *in vitro*; only RNA from illuminated plants synthesized the 20 kDalton precursor of the RuBPCase small subunit. The RNA from illuminated pea leaves was used to prepare double strand cDNA, to which *Hin*dIII linkers were attached. The DNA was ligated to the *Hin*dIII site of pBR322, and the recombinant plasmids were used to transform *E. coli*, strain HB101. Transformed bacterial colonies resistant to ampicillin were grown on nitrocellulose sheets laid upon nutrient agar plates. After fixing the DNA to the sheets, clones containing RuBPCase cDNA were identified by their ability to anneal with radiolabeled polyA^+, polysomal RNA from illuminated leaves, but not with RNA from dark-grown leaves. False positives were eliminated by annealing the initially positive clones to radiolabeled RNA enriched for RuBPCase mRNA. Enrichment was achieved by size fractionation on polyacrylamide gel electrophoresis. The mRNA was identified as a 900 base long chain found in illuminated but not dark-grown leaves and capable of translation, *in vitro*, into the RuBPCase small subunit. Finally, the clones' identities were confirmed by hybrid selection (Section 6.4b). After J. R. Bedbrook, S. M. Smith, and R. J. Ellis, *Nature* 287 (1980), p. 692.

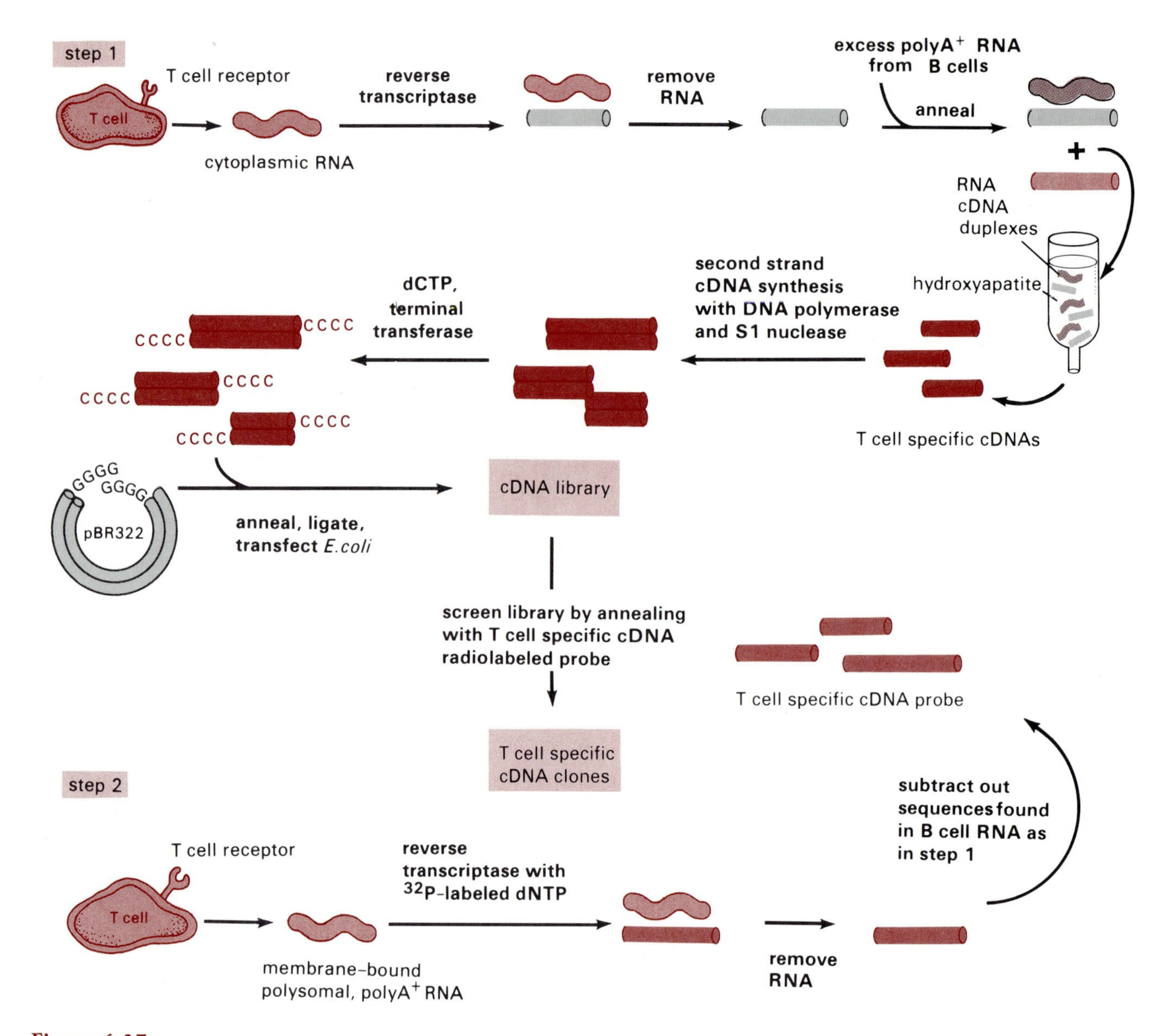

Figure 6.37

Cloning a murine cDNA encoding the β chain of the T cell receptor. T lymphocytes play a critical role in cellular and humoral immunity. Their functions are mediated by a characteristic, heterodimeric surface receptor (Section 10.6c). The first step in cloning the gene for the β chain was construction, in pBR322, of a cDNA library enriched for mRNAs specifically found in T cells. PolyA$^+$, cytoplasmic, T cell RNA was used as a template for first strand cDNA synthesis. The cDNA was annealed with polyA$^+$, cytoplasmic RNA from another type of lymphocyte, B cells, and the RNA·cDNA duplexes removed by binding to hydroxyapatite (Section 7.5b). Over 95 percent of the cDNA was removed, representing the mRNAs common to both B and T cells. Synthesis of the second cDNA strand was then carried out on the remaining, T cell–specific cDNA (5 percent). After S1 nuclease treatment, C tails were added, the tailed molecules were ligated to pBR322 that had been cleaved at the endonuclease *Pst*I site and tailed with G residues, and the recombinants were used to generate transformed *E. coli* colonies. The second step was preparation of a radiolabeled probe for screening the cDNA library. The probe was a first strand cDNA made on a template of cytoplasmic, membrane bound polysomal, polyA$^+$ RNA from T cells. (Recall that because the T cell receptor is found on the cell's plasma membrane, it is translated on ribosomes associated with the endoplasmic reticulum membrane.) To enrich this cDNA for T cell–specific transcripts, it was annealed with polyA$^+$ B cell RNA, and the RNA·cDNA duplexes removed by hydroxyapatite. The remaining radiolabeled, T cell–specific cDNA was used to screen the library. Of the 5000 library clones screened, 35 were positive. Five of these annealed to B cell RNA on RNA blots. Of the remaining 30, one proved to include sequences encoding the T cell receptor β chain. Prior to the successful cDNA cloning, the T cell receptor itself had not been isolated, and nothing useful was known of its structure. After S. M. Hedrick et al., *Nature* 308 (1984), p. 149.

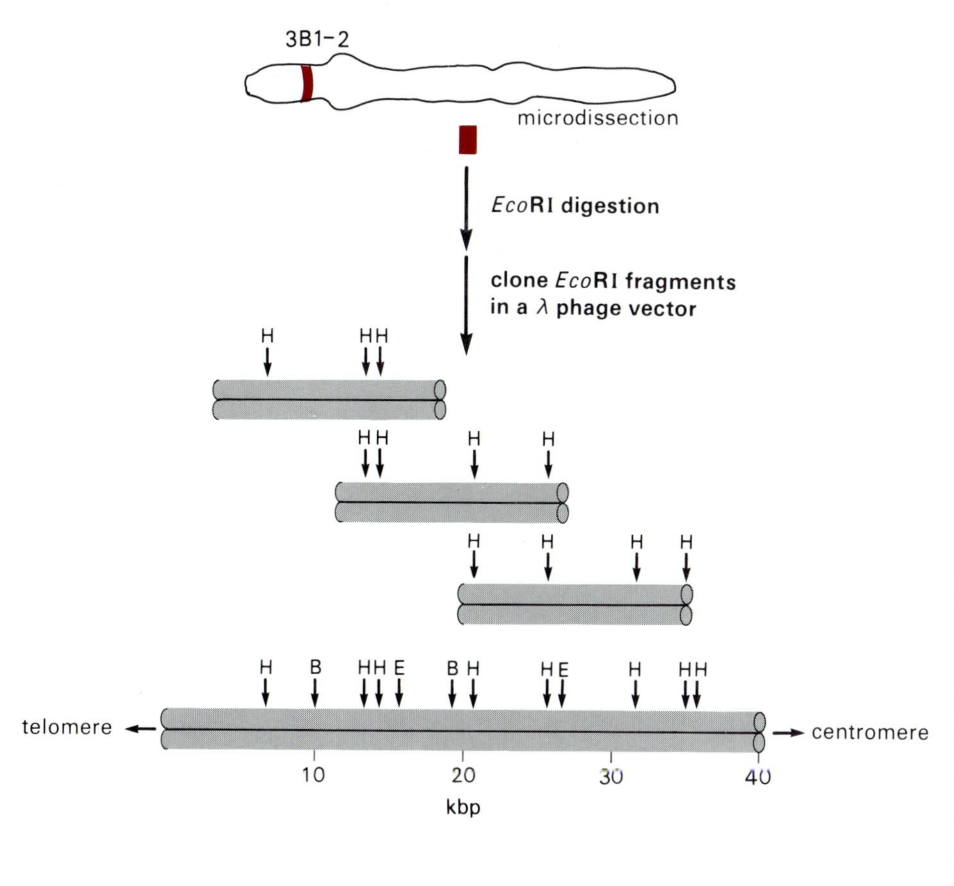

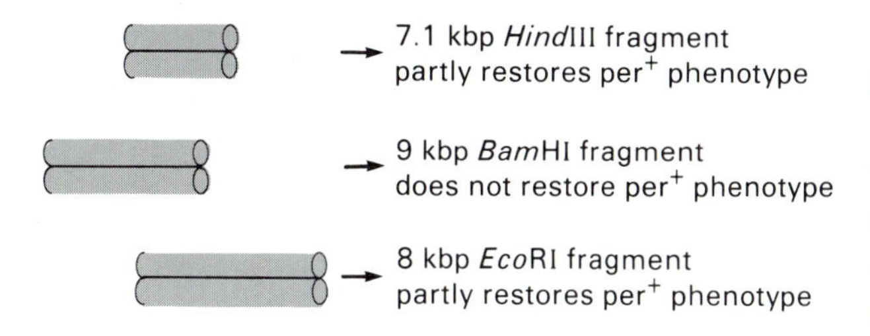

Figure 6.38
Cloning a gene that regulates rhythmic behavior in *Drosophila.* Mutations at the *per* locus on region 3B1-2 of the *D. melanogaster* X-chromosome cause abnormal circadian (24 hour) rhythms as well as changes in the repeat time of the male *Drosophila*'s courtship song. Such mutations are frequently associated with chromosomal rearrangements. DNA segments were microdissected from this portion of a normal X-chromosome, partially digested with restriction endonuclease, and cloned in a λ phage vector. Clones with overlapping sequences were identified by cross-hybridization and common restriction endonuclease site maps (not all sites are shown here). Thus, the map of a long DNA segment that includes the *per* locus was deduced. Subcloned regions of the segment were used as probes to construct maps of X-chromosomes that are rearranged near the *per* locus and known to be per^+, per^-, or per^{ab} (abnormal rhythmic behavior). In this way, loss of normal phenotype was correlated with an approximately 15 kbp DNA segment near the center of the map. Also, certain of these subclones partly restore circadian rhythm and courtship song timing to flies with *per* deletions when they are introduced by P element transduction (described in Section 10.2b). Thus, without knowing the precise location of the gene, and in the absence of any knowledge of the gene product, a functional gene that governs complex behavioral phenotypes was isolated. After T. A. Bargiello and M. W. Young, *Proc. Natl. Acad. Sci. U.S.A.* 81 (1984), p. 2142; W. Zehring et al., *Cell* 39 (1984), p. 369; and Y. Citri et al., *Nature* 326 (1987), p. 42.

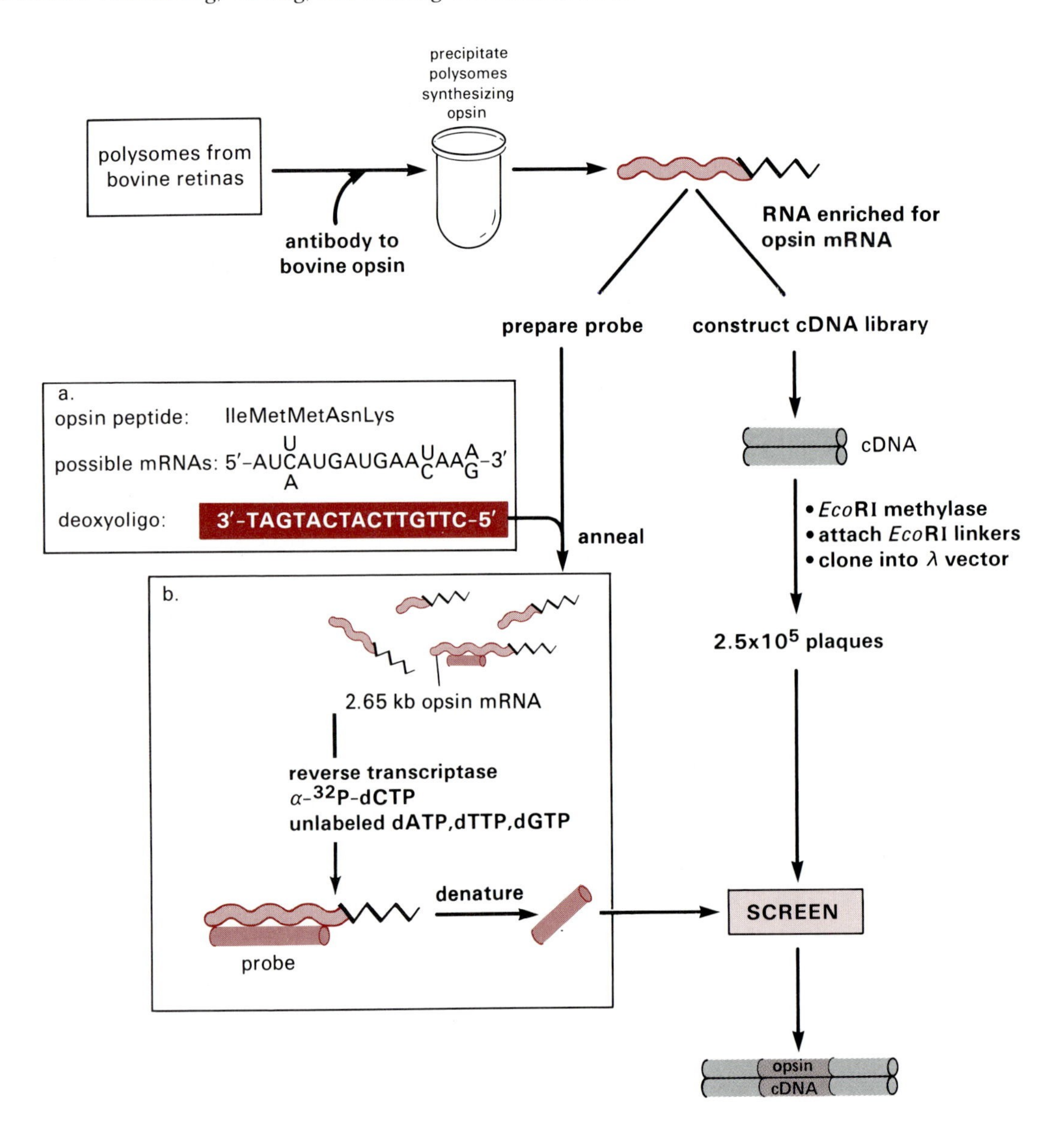

Figure 6.39

Cloning genes and cDNAs encoding rhodopsin from cows, humans, and *Drosophila*. Opsins are the apoproteins of retinal photoreceptors (visual pigments), with rhodopsin being the major photoreceptor in both vertebrate and invertebrate eyes. In vertebrates, rhodopsin is associated with rod cells; the amino acid sequence of rhodopsin is known. The bovine cDNA and gene were the first isolated. PolyA^+ RNA enriched for opsin mRNA was obtained from polysomes of bovine retina. A cDNA library was constructed in a λ phage vector by standard procedures. It was screened with a specific probe prepared by the following two steps (shown in boxes): (a) chemical synthesis of a 15 residue long deoxyoligonucleotide complementary to the mRNA predicted from five amino acids near the carboxy end of opsin and (b) use of the 15-mer as a specific primer for reverse transcription of the opsin mRNA present in the bovine retina, polysomal, polyA^+ RNA. The probe was radiolabeled using α-^{32}P-dCTP during reverse transcription. The 15-mer might have permitted positive identification of opsin clones, but the longer probe increased the specificity. Of the 2.5×10^5 clones in the library, 0.2 percent were specifically labeled by the probe. Eight of these were analyzed, and their nucleotide sequences indicated that

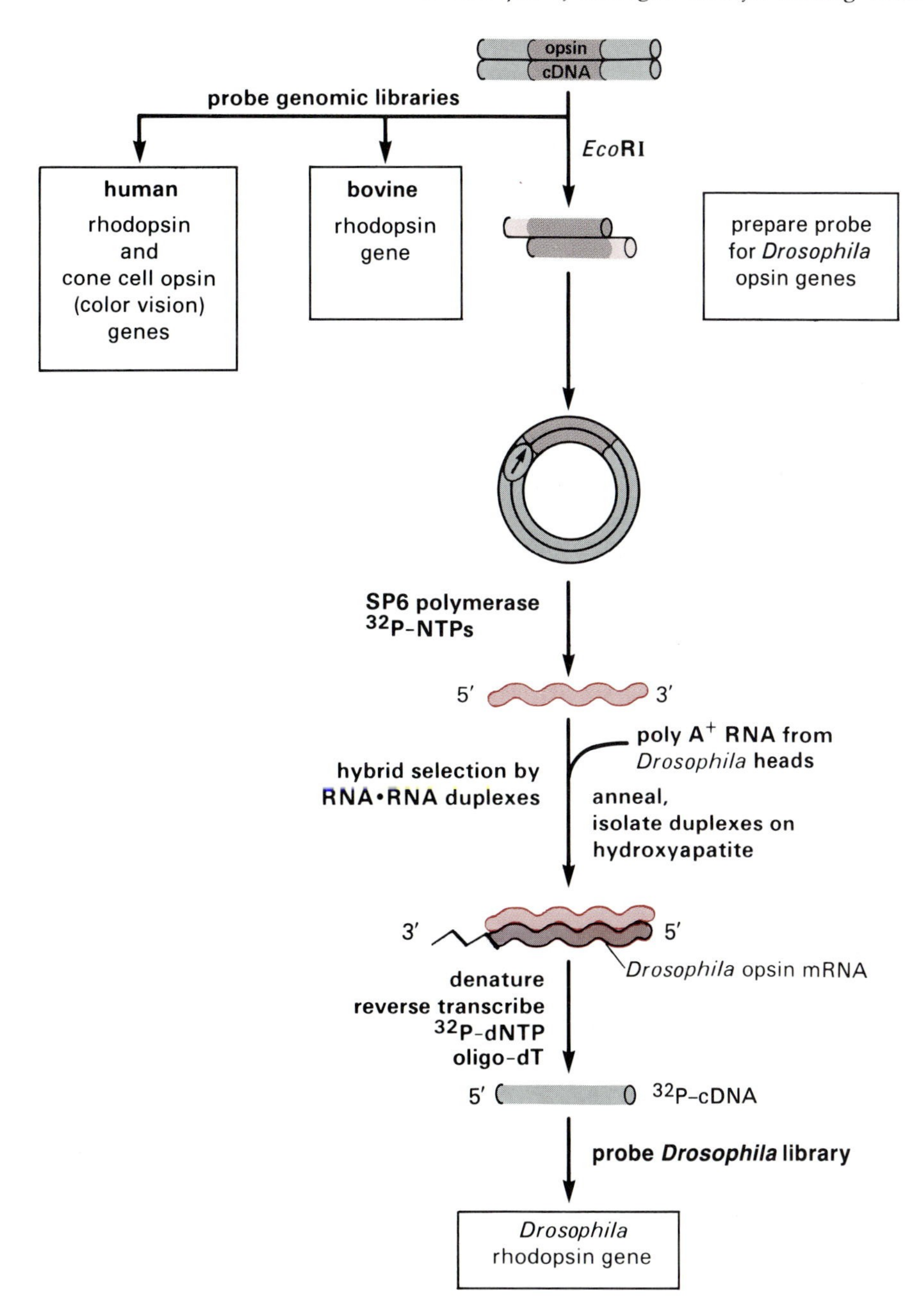

they encoded portions of opsin. The cDNA clones were used as probes to screen *Drosophila*, bovine, and human genomic libraries. No positive recombinants were detected in the *Drosophila* library, but the two mammalian genes were isolated. The distantly related *Drosophila* sequence was, however, obtained by special techniques. The bovine cDNA was excised from the λ vector DNA by restriction endonuclease cleavage and recloned into a vector like pGEM. Radiolabeled RNA corresponding to the non-sense strand was synthesized using SP6 RNA polymerase and annealed to polyA$^+$ RNA from *Drosophila* heads. The RNA·RNA duplexes containing *Drosophila* opsin mRNA were isolated on hydroxyapatite (Section 7.5b). After denaturing the duplexes, the *Drosophila* RNA was used as a template for reverse transcription with an oligodeoxy-T primer and radiolabeled deoxyribonucleotides. This radiolabeled cDNA was then used to select clones from a *Drosophila* genomic library. The success of this procedure depended on using the RNA·RNA duplexes. Mismatched RNA·RNA duplexes are much more stable than the corresponding mismatched DNA·DNA duplexes. After J. Nathans and D. S. Hogness, *Cell* 34 (1983), p. 807; and C. S. Zuker, A. F. Cowman, and G. M. Rubin, *Cell* 40 (1985), p. 851.

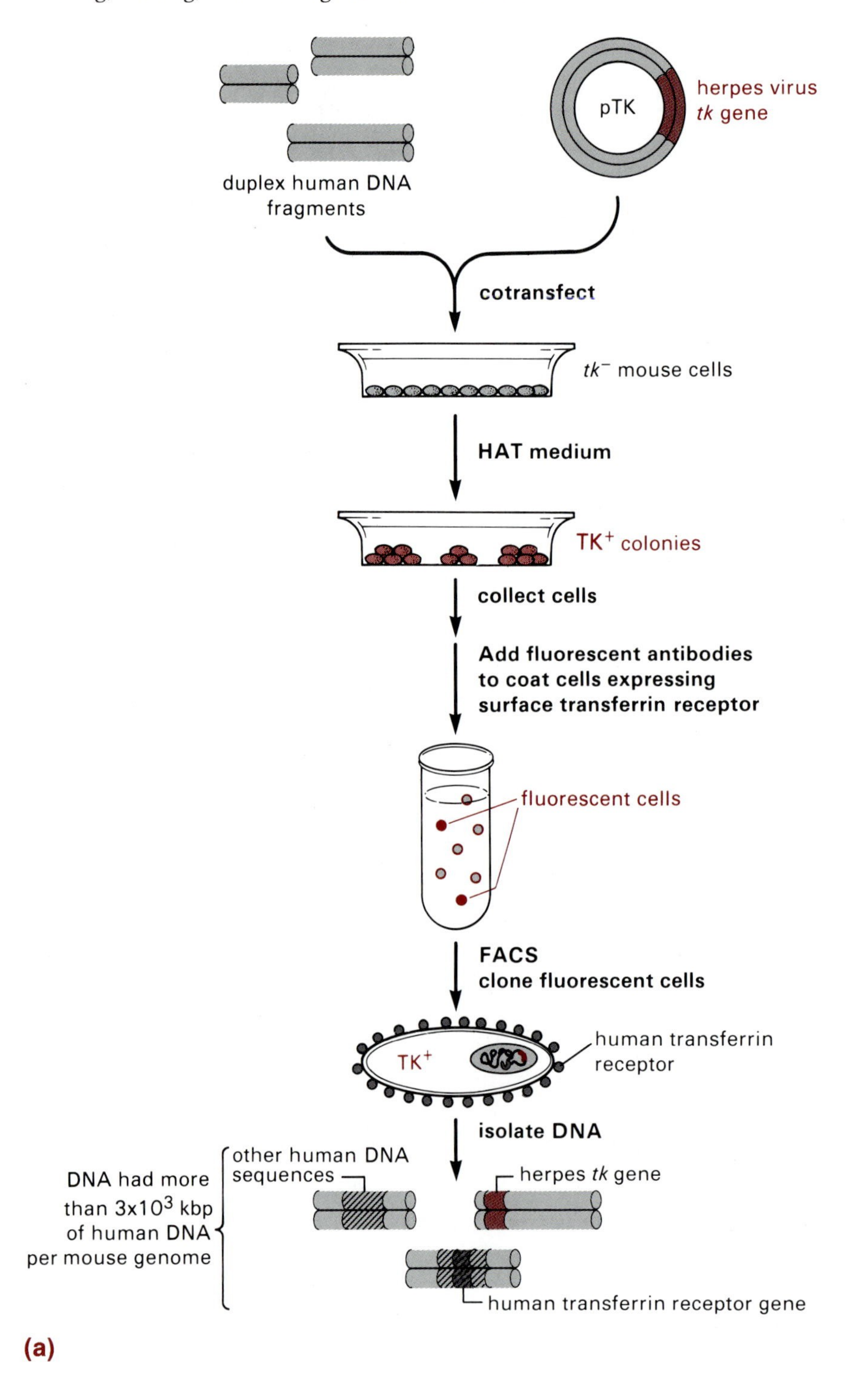

Figure 6.40
Cloning the human gene encoding the specific cell surface receptor for transferrin. Transferrin is an Fe^{3+}-binding serum protein that transports iron into cell interiors. Uptake of Fe^{3+}-transferrin complexes is mediated by a specific receptor that is a homodimeric, cell membrane glycoprotein. (a) To clone the receptor gene, high molecular weight fragments of total human DNA and a recombinant plasmid containing the herpes simplex virus thymidine kinase gene (*tk*) were cotransfected into tk^- mouse cells. After selection in HAT medium, the TK^+ cells were mixed with antibody to human transferrin receptor. Cells that bound to antibody (and thus carried the receptor on their surfaces) were reacted with a fluorescent tag (fluorescein coupled to an antibody to the antibody), recovered by fluorescence activated cell sorting, and cloned. DNA isolated from the cloned cells contained much more human DNA (more than 3×10^3 kbp per mouse genome) than required to encode a single gene. (The

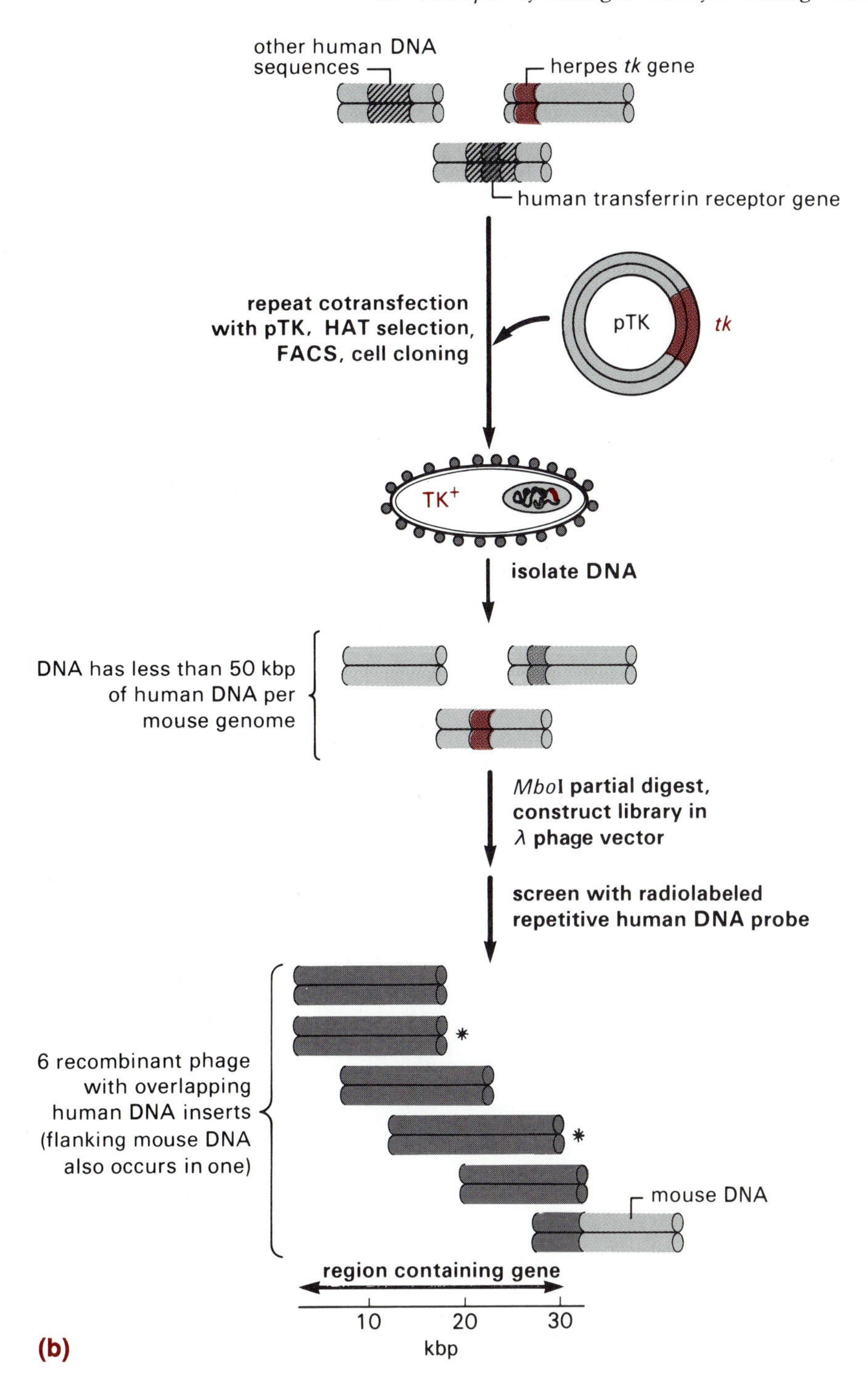

amount of human DNA was estimated in the presence of the large excess of mouse DNA by annealing with a probe containing DNA sequences that are repeated in almost 10^6 random locations interspersed throughout human DNA and do not occur in mouse DNA [Section 9.5c].) (b) To reduce the amount of unwanted human DNA, the DNA from these cells was used in a second round of cotransfection, HAT selection, FACS sorting, and cloning. The DNA of the resulting cells (secondary transformants) had less than 50 kbp of human DNA per mouse genome and were used to construct a genomic library. Six phage clones that annealed with the interspersed, highly repeated human DNA probe were selected. The six contained overlapping DNA segments. Two of the six inserts (starred) were recombined in a phage vector to reconstitute a unique, 31 kbp DNA segment that directed the synthesis of human transferrin surface receptor upon transfection into mouse cells. After L. C. Kuhn, A. McClelland, and F. H. Ruddle, *Cell* 37 (1984), p. 95.

CHAPTER 7

The Products: Characterizing and Manipulating Recombinants

After a cloned recombinant DNA is obtained by selection, the nature of the insert must be established. Full characterization requires answers to a series of questions. First and foremost, does the recombinant really contain the desired DNA segment? Also, is it complete, or does it represent only a portion of the required sequence? Does it faithfully represent the genomic DNA from which it is derived? Is it functional? This chapter describes the various ways in which cloned DNA fragments are characterized. The first step is generally to purify the recombinant DNA free from host cell DNA, RNA, and proteins. This is quite straightforward and involves extractions, physical separations, and enzymatic digestion of RNA and protein. For example, plasmid DNA can be separated from genomic DNA on the basis of its small size and circularity (a large circular genome like that of *E. coli* is usually broken into linear fragments when cells are extracted). Phage DNA can be isolated free of cellular DNA by purifying phage particles and then extracting their DNA.

7.1 The Gross Anatomy of a Cloned Insert

An insert's overall structure is initially characterized by its size and by the location within it of restriction endonuclease sites. Because cloned genomic DNA fragments are often either longer or shorter than the region of interest, the location and length of the desired segment within the insert also needs to be determined.

a. Insert Size

Because vector size is generally known, the insert's size can be deduced from the length of the recombinant molecule itself, assuming that no portion of the vector was deleted during cloning. Gel electrophoresis yields the size of many recombinants directly. Circular recombinants need to be linearized for accurate size determination because their electrophoretic mobility varies with the superhelical density of circles. Also, standard electrophoresis systems are impractical for molecules over about 15 kbp in length because larger molecules move too slowly to be well resolved. Therefore, the size of many recombinants cannot readily be determined directly by electrophoresis unless an apparatus for pulsed-field electrophoresis is available (Section 6.5a). Alternatively, the total size of the recombinant can be estimated by measuring its length on electron micrographs. Here, too, size standards are essential, and the number of base pairs in a given length varies with molecular configuration and technical factors.

Another way to determine an insert's size is to reverse the reactions used in constructing the recombinant and cut out the insert with appropriate restriction endonuclease(s) (Figure 7.1a). The size of the linear insert is then determined by gel electrophoresis. Because the map of the restriction endonuclease sites on the vector is generally known, the insert fragment is readily identifiable. If the ligation reactions eliminated the restriction endonuclease sites bordering the insert (e.g., in blunt end ligation), an alternative procedure can be used. Restriction endonuclease cleavages within the vector sequences that flank the insert will excise the insert joined to vector segments of known size (Figure 7.1b). Note that, in either

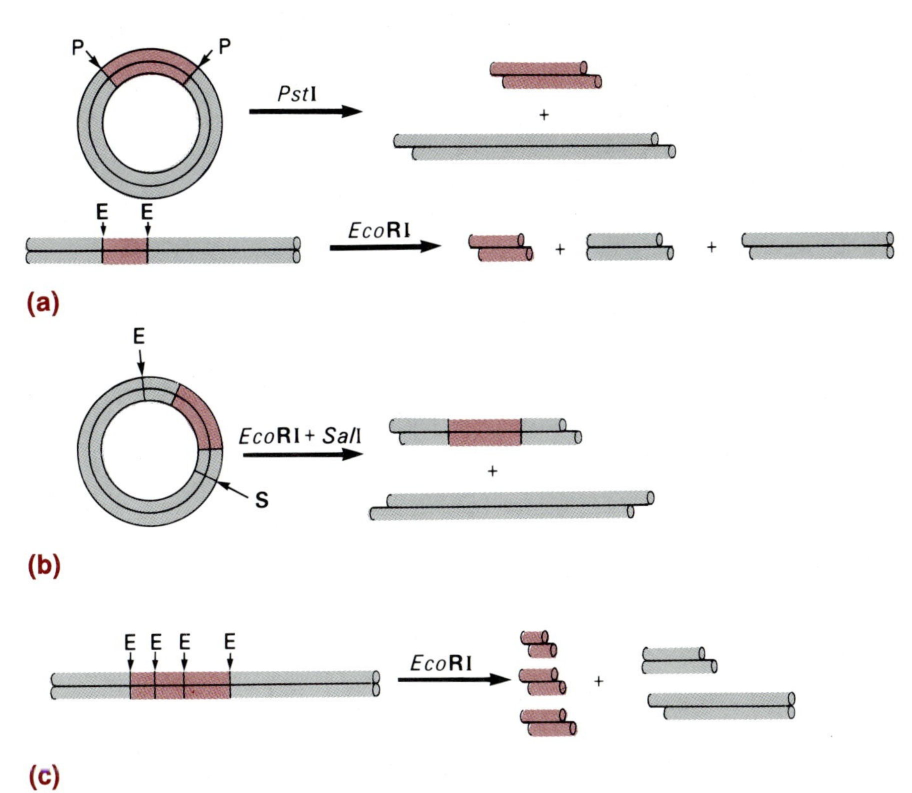

Figure 7.1
Determining insert size in a recombinant DNA molecule by (a) reversing the reaction used to ligate the vector and the insert in the cloning procedure, (b) utilizing flanking restriction endonuclease sites, and (c) summing the sizes of restriction endonuclease fragments generated by the insert.

case, the results are more complex if the insert itself contains cleavage sites for the endonuclease(s) used (Figure 7.1c). Here the sum of the sizes of all fragments produced by the insert gives the total length. And if the insert is quite large, internal cleavage and summation may in fact be required to obtain an accurate length.

b. Mapping Restriction Endonuclease Sites

The principles by which maps of the restriction endonuclease sites on a DNA segment are determined and aligned were described in Section 4.2d. If the vector is small and not complex, maps can sometimes be determined directly on the recombinant molecule itself. However, it is often necessary first to excise the insert and purify it free of otherwise confusing vector segments.

c. Subcloning

Long cloned inserts such as those obtained with λ phage or cosmid vectors are often unmanageable. Once a partial restriction endonuclease map is constructed, the insert can be divided into smaller fragments by **subcloning**. Subcloning is not different in principle from other recombinant DNA constructions; the starting genome is simply a recombinant clone. Using restriction endonucleases that have four base pair recognition sites for subcloning is often advantageous because they usually cleave DNA segments frequently.

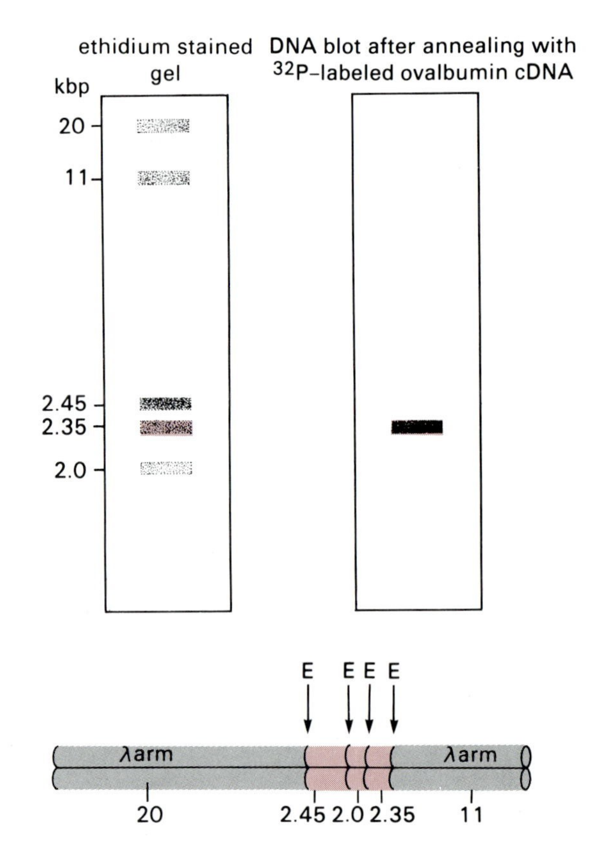

Figure 7.2
Annealing of an ovalbumin mRNA probe to an *Eco*RI digest of a λ phage vector containing chicken ovalbumin gene sequences. The digest was separated by electrophoresis and blotted. The restriction endonuclease map of the insert is shown at the bottom; the numbers are the fragment lengths in kbp.

d. Locating the Segment of Interest in the Insert

The same principles used to find specific segments in a large genome prior to cloning (Section 6.1b) can also be used to locate segments in smaller recombinant genomes. Once a restriction map of the insert is determined, precise localization is simple. The only requirement is a suitably labeled probe such as the DNA or RNA probe used to select the clone to begin with.

Figure 7.2 illustrates how this works with an example taken from experimental work on the chicken ovalbumin gene. Ovalbumin contains a single polypeptide chain (43,500 Daltons) that is synthesized in chicken oviducts after stimulation by the steroid hormone estrogen (or progesterone). The protein is deposited in egg whites. Hormone stimulation occurs naturally in laying chickens, and immature chicks are induced to form oviducts and synthesize ovalbumin by treatment with estrogen. A primary action of the hormone is to induce transcription of the ovalbumin gene; the concentration of ovalbumin mRNA goes from essentially zero to about 50,000 molecules per synthesizing cell after two to three weeks of treatment and falls to below 10 molecules per cell within two weeks after cessation of estrogen treatment. Oviducts of laying hens are a good starting material for purification of ovalbumin mRNA because it represents about 50 percent of the total mRNA. The purified mRNA is then used to prepare a radioactive probe for identifying a cloned ovalbumin gene. A radioactive single strand cDNA made directly from the mRNA with reverse transcriptase or a cloned duplex cDNA can be used. Using

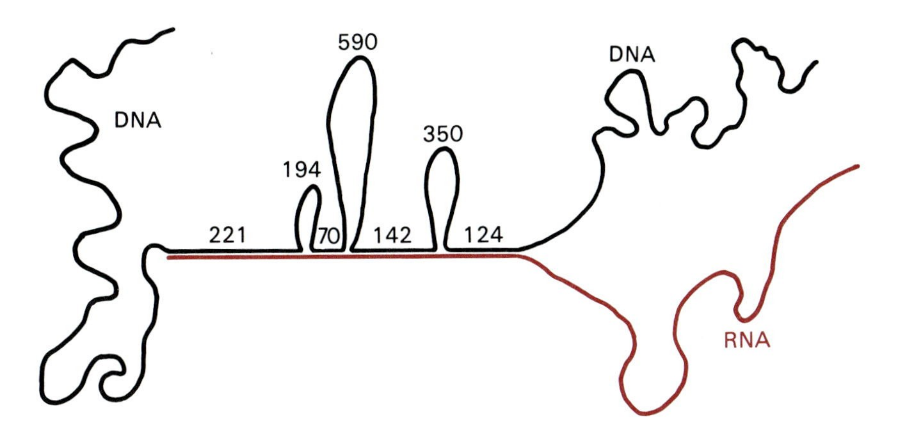

Figure 7.3
Schematic drawing of an electron micrograph showing a heteroduplex molecule formed between one strand of the recombinant described in Figure 7.2 (black) and ovalbumin mRNA (color). The lengths of the various regions are given in base pairs (or bases for the single strand regions). They are determined by comparison with the lengths of polynucleotides of known size present in the same sample.

such probes, cloned ovalbumin gene sequences were selected from a population of λ phage vectors containing chicken genomic DNA inserts that had been generated by partial *Eco*RI digestion. The clone shown in Figure 7.2 had a 6.8 kbp long insert that gave three *Eco*RI fragments; only the 2.35 kbp long fragment annealed with the ovalbumin cDNA probe.

Electron microscopy is another tool for locating a sequence within a cloned segment. A probe and recombinant DNA are mixed, denatured, and annealed prior to preparing the sample for microscopy. Complementary regions form duplexes that appear as relatively thick strands in the electron microscope while noncomplementary regions remain single stranded and appear thinner. A heteroduplex (one RNA chain, one DNA chain) formed between the ovalbumin recombinant described in Figure 7.2 and ovalbumin mRNA is shown schematically in Figure 7.3. Such heteroduplexes form if the annealing conditions are set to favor RNA·DNA hybridization over DNA·DNA interactions. Within the 2.35 kbp DNA segment that hybridized with mRNA (Figure 7.2), only about 550 bp anneal to the messenger. Thus, this heteroduplex analysis reveals much more than the location of coding sequences in the clone. It shows that, although the mRNA is about 1800 bp long, less than a third of the ovalbumin coding sequences are present in the DNA segment that was cloned. Furthermore, the 550 bp long complementary region on the cloned DNA is not in a single contiguous stretch; it is broken up into four separate regions. In between each DNA segment that hybridized to mRNA is a single strand loop of DNA. The genomic DNA coding sequences are interrupted by noncoding regions. As described in Section 8.5, such **intervening sequences** or **introns** are typical of eukaryotic genes.

7.2 The Fine Anatomy of a DNA Segment—Primary Nucleotide Sequence

Ultimately, understanding the structure, function, and evolutionary history of a cloned DNA segment depends on knowing its primary structure—the nucleotide sequence. Fast, accurate methods for determining DNA sequences were invented shortly after recombinant DNA techniques were developed. In principal, there is now no limit to the length of DNA that can be sequenced. The sequences of DNA segments hundreds, even thousands of nucleotides long are determined routinely. Yet until about 1975, obtaining information about DNA sequences was very difficult. RNA copies of a DNA strand had to be prepared with RNA polymerase and then the sequence of the cRNA determined. The fidelity and completeness of the copying was often questionable, and the RNA sequencing methods are themselves tedious. The new methods use DNA itself, and sequencing RNA molecules is now usually done by first making cDNA copies.

a. General Principles

DNA sequencing methods fall into two main classes. One depends exclusively on chemical reactions carried out directly on pure DNA segments. In the second, sequence information is generated from enzymatically produced DNA copies of pure segments. Certain general features are common to both approaches. First and foremost, the DNA fragment is usually pure, a condition that is readily met with cloned DNA. Second, both methods provide the sequence of only one DNA strand at a time. The most accurate data are obtained when each strand of a duplex molecule is separately analyzed, thereby confirming all nucleotide assignments. Data are collected for single strands by having only one strand at a time labeled with a radioactive isotope. Only the radioactive strand is detected, and only that strand is usually described in the following paragraphs. Third, both methods generate sets of radiolabeled single strands containing all possible chain lengths from one to n, where n is the total length sequenced.

In both the chemical and enzymatic approaches, the complete fragment set (sizes one to n) is actually contained in four separate collections of fragments. Ideally, each collection of fragments contains all possible molecules starting from one end of the chain and extending to every position at which a particular nucleotide occurs. One collection contains all possible chains that start at one end of the chain and terminate at the different deoxyadenosine residues (Figure 7.4). A second collection contains all possible chains that start at that same end and terminate at all the different deoxycytidine residues. The third and fourth collections contain stops at either the thymidine or deoxyguanosine residues. In practice, the enzymatic method works in this straightforward way, but the chemical procedure is somewhat more complex.

The chemical and enzymatic methods differ in the ways the four collections are generated. In the chemical approach, four samples of the prelabeled radioactive chain are cleaved in four different ways to form the collections of fragments. In the enzymatic method, the four radiolabeled fragment collections are synthesized by copying the unlabeled DNA. The final step in both procedures, separating each collection of strands according to chain length, is accomplished by electrophoresis through polyacrylamide gels under denaturing conditions; single polynucleotide chains that differ from one another by one nucleotide residue are well resolved. The radioactive isotope allows the chains to be visualized by autoradiography on X-ray films, so only very small quantities—picomoles of fragment or less—need be analyzed. All four collections making up the set derived from a single DNA fragment are electrophoresed

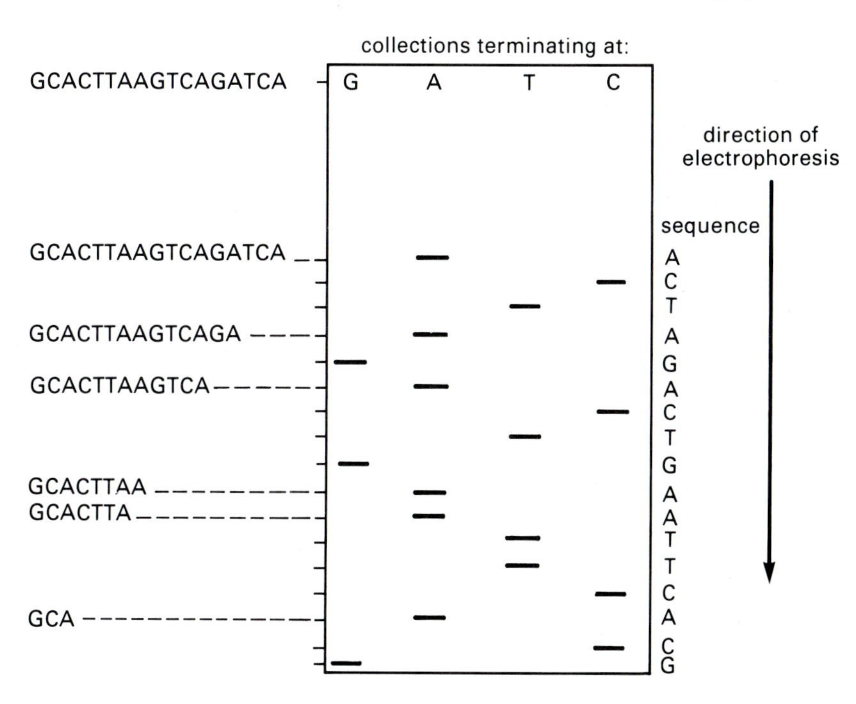

Figure 7.4
A schematic representation of separate but simultaneous electrophoresis of four collections of oligodeoxynucleotides generated from the chain at the top left. Each collection represents all possible chains extending from the 5′ end to each occurrence of a particular deoxynucleoside– deoxyguanosine, deoxyadenosine, deoxycytidine, or thymidine. The collection ending in deoxyadenosine is listed as an example at the left. These diagrams represent the type of collection generated by the enzymatic sequencing method.

simultaneously in parallel lanes on a single gel. By reading up the gel, scanning each lane, one can read the entire sequence. Special gel electrophoresis techniques allow separation of chains from 1 to 300 or more nucleotide residues in length. Molecules that are more than several hundred nucleotides long must be divided into smaller units (often by restriction endonuclease digestion or subcloning) and the sequence of each unit determined separately.

b. *Chemical Sequencing*

Generating nested sets Chemical sequencing depends on reactions that specifically modify the different purine and pyrimidine bases. The resulting modified bases are then evicted from the polydeoxynucleotide chain under conditions that preserve the glycosidic bonds between unmodified bases and deoxyribose. Finally, the relatively unstable phosphodiester bonds surrounding the eviction sites are cleaved by hydrolysis, thereby breaking the chain. A diagram of the overall scheme is shown in Figure 7.5 (using guanine residues as an example). Typical base-specific reactions are illustrated in Figure 7.6. In these examples, two reactions are each specific for only one base—either guanine or cytosine—while two others are specific either for the pair of purines or the pair of pyrimidines. The four reactions are carried out on four separate samples of the DNA segment generating the four collections of fragments required for unambiguous sequence determination. The extent of each reaction is carefully limited so that, on the average, only one of the susceptible bases on each DNA molecule in the sample reacts. Within the population of identical DNA molecules in the initial sample, the randomly placed reactions generate the complete nested collection of chains extending from the ends to every possible reactive base. The example in Figure 7.7 (PAGE 378) outlines the situation for cleavage at deoxycytidine residues only.

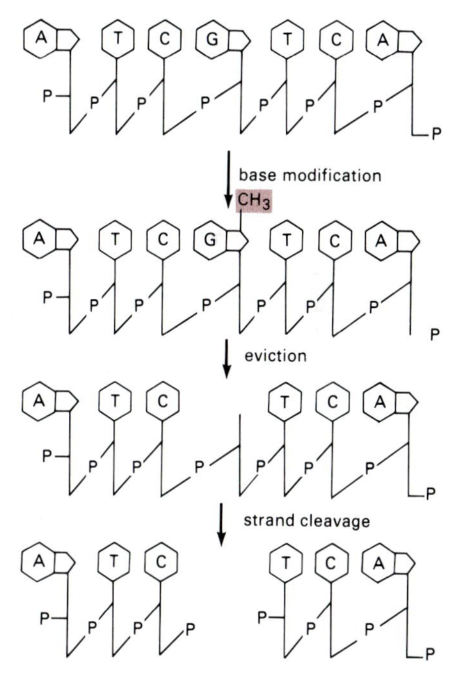

Figure 7.5
The three steps used in chemical sequencing methods: base modification, eviction of modified base, and cleavage of the chain at the "empty" deoxyribose. Reaction at a guanine is used as an example.

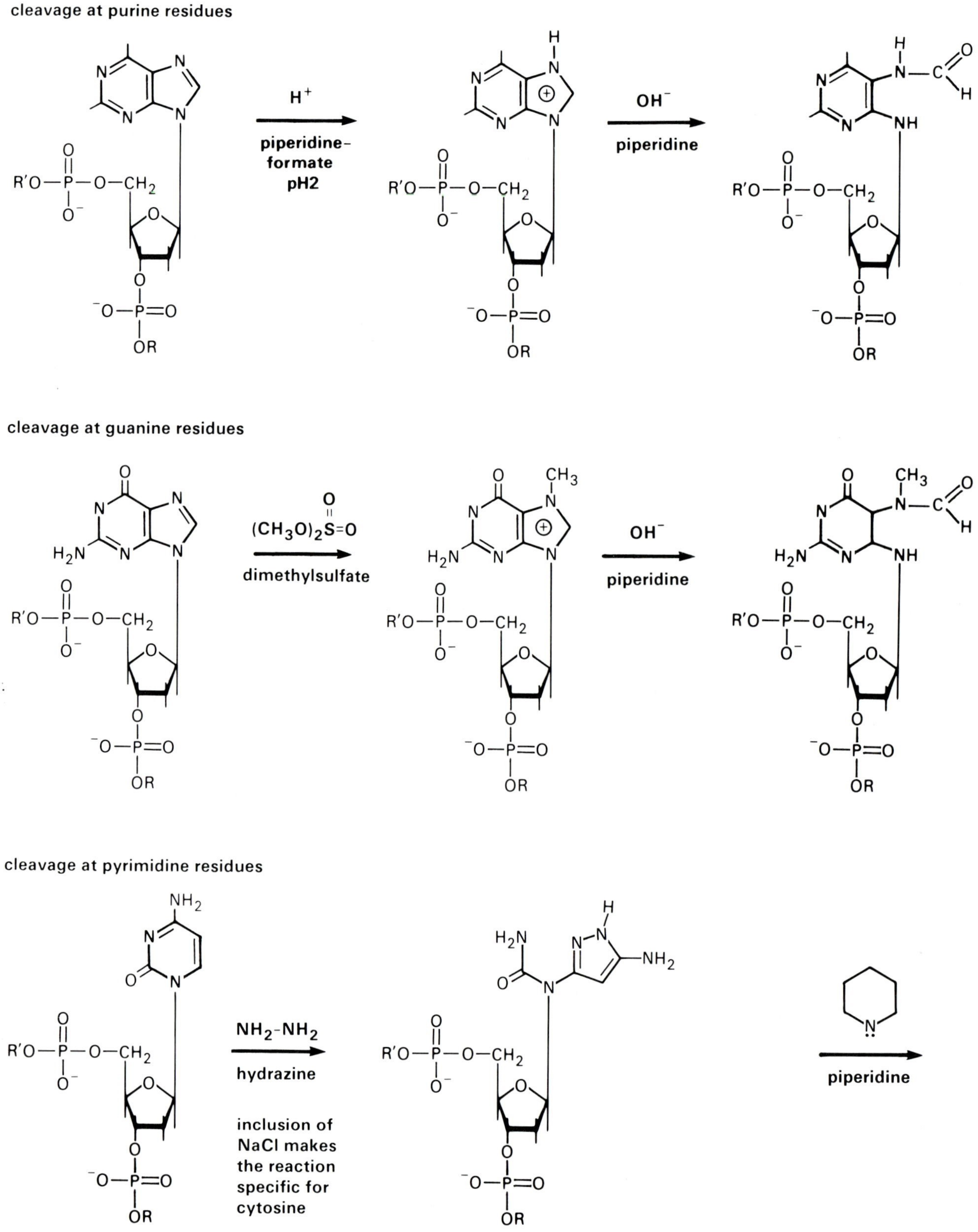

cleavage at purine residues

H^+

piperidine-formate pH2

OH^-

piperidine

cleavage at guanine residues

$(CH_3O)_2S=O$

dimethylsulfate

OH^-

piperidine

cleavage at pyrimidine residues

NH_2-NH_2

hydrazine

inclusion of NaCl makes the reaction specific for cytosine

piperidine

Figure 7.6
Reactions used for specific chemical modification of bases.

piperidine $\xrightarrow{}$ (5-formamido pyrimidine, NH_2) + $R'O-PO(O^-)-O-CH_2$ (sugar, HO, $HC{=}N^+$ piperidine; $^-O-P(=O)-OR$) $\xrightarrow[\beta\text{-elimination}]{OH^-}$ $R'-O-PO_2-O^-$ + $R-O-PO_2-O^-$ + $CH_2{=}C(OH)-CH{=}CH-CH{=}N^+$ (piperidine)

piperidine $\xrightarrow{}$ (H_2N, NH_2, O, CH_3–N–CHO pyrimidine) + $R'O-PO(O^-)-O-CH_2$ (sugar, HO, $HC{=}N^+$ piperidine; $^-O-P(=O)-OR$) $\xrightarrow[\beta\text{-elimination}]{OH^-}$ $R'-O-PO_2-O^-$ + $R-O-PO_2-O^-$ + $CH_2{=}C(OH)-CH{=}CH-CH{=}N^+$ (piperidine)

mixed fragments + $R'O-PO(O^-)-O-CH_2$ (sugar, HO, $HC{=}N^+$ piperidine; $^-O-P(=O)-OR$) $\xrightarrow[\beta\text{-elimination}]{\text{piperidine}}$ $R'-O-PO_2-O^-$ + $R-O-PO_2-O^-$ + $CH_2{=}C(OH)-CH{=}CH-CH{=}N^+$ (piperidine)

5′-pApGpCpTpTpTpCpTpGpApGpApApApCpTpGpCpTpCpTpG-3′

5′- set	3′- set
5′-pApGp	pTpTpTpCpTpGpAGpApApApCpTpGpCpTpCpTpG-3′
5′-pApGpCpTpTpTp	pTpGpApGpAGpApApApCpTpGpCpTpCpTpG-3′
5′-pApGpCpTpTpTpCpTpGpApGpApApAp	pTpGpCpTpCpTpG-3′
5′-pApGpCpTpTpTpCpTpGpApGpApApApCpTpGp	pTpCpTpG-3′
5′-pApGpCpTpTpTpCpTpGpApGpApApApCpTpGpCpTp	pTpG-3′

Figure 7.7
Generating nested collections of chains in chemical sequencing. In this example, deoxycytidine residues were modified and eliminated. Either the 5′ or the 3′ set can be used for sequence analysis, depending on whether the 5′ or 3′ end of the starting DNA is radiolabeled.

Labeling chain termini with ^{32}P As is evident in Figures 7.5, 7.6, and 7.7, each of the four reaction mixtures actually contains two collections of chains derived from the strand being analyzed. One includes all partial molecules starting at the 5′ end of the original chain, and the other includes all partial molecules starting at the 3′ end. Either collection can give the required sequence information, and the set to be used is determined by whether the starting DNA fragment has been labeled with ^{32}P at the 3′ or 5′ end prior to carrying out the base-specific chemical reactions. Unlabeled material is invisible in the autoradiogram. If the starting material is a duplex DNA, only one of the two strands can be labeled. If both contain ^{32}P, then the final cleavage products will each contain two different collections of labeled chains, and the data will be uninterpretable.

Labeling at 5′ ends is accomplished with polynucleotide kinase and γ-^{32}P[ATP] (Figure 7.8); if 5′ terminal phosphomonoester groups are present, they must first be removed with phosphomonoesterase. Single strand preparations are sequenced directly after labeling. If the sample is a duplex, it can be, after kinase treatment, separated into two complementary single strands by denaturation and electrophoresis (Section 7.2c). Alternatively, the labeled double strand fragment can be cleaved in two at an appropriately placed internal restriction endonuclease site and the resulting fragments separated by electrophoresis. In this case, the material used for the sequencing reactions is actually double stranded, but one

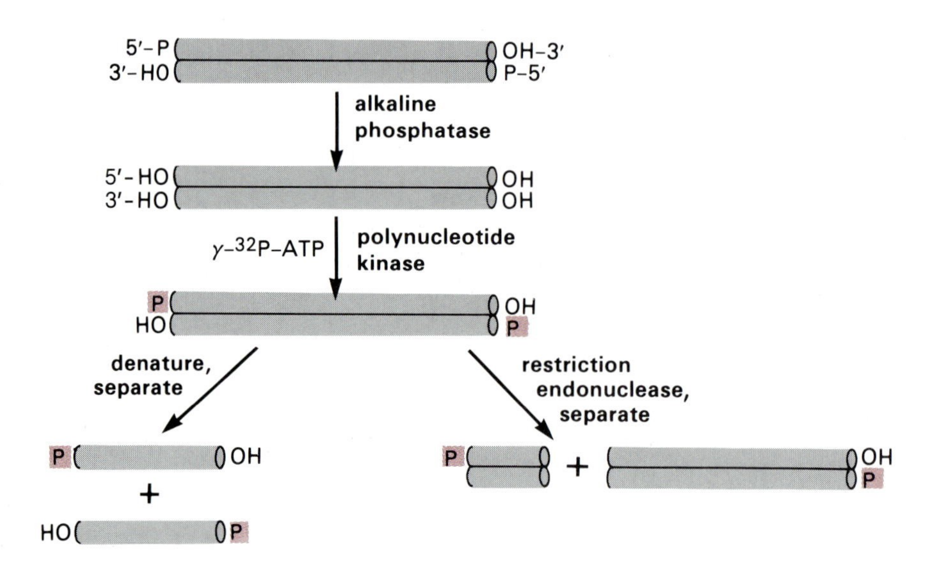

Figure 7.8
5′ terminal labeling of double strand DNA in preparation for chemical sequencing. Two methods for obtaining a fragment with the radiolabel on only a single strand are illustrated.

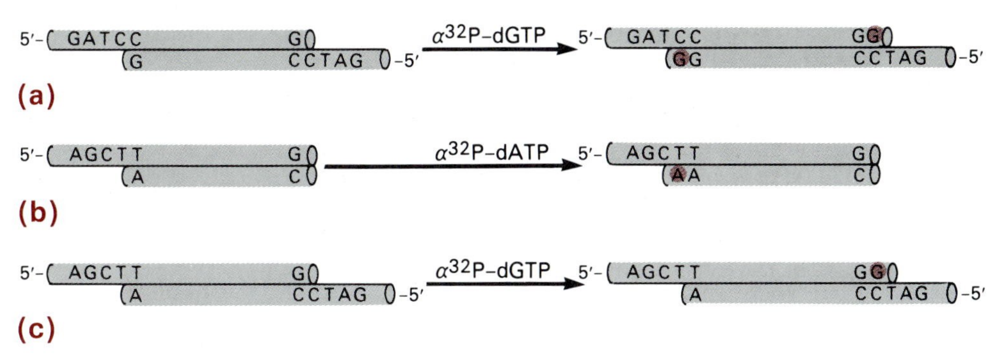

Figure 7.9
Strategies for labeling the 3′ hydroxyl termini of DNA fragments by filling in with α-^{32}P-labeled deoxynucleoside triphosphates and DNA polymerase I or reverse transcriptase.

strand is unlabeled and therefore invisible. When DNA is labeled at the 5′ termini, the autoradiograms show the fragments closest to the 5′ end as the shortest products.

Labeling the 3′ termini of duplex chains is practical when there are single strand overhangs at the 5′ termini. Either DNA polymerase I or reverse transcriptase catalyzes the extension of the recessed 3′-hydroxyl end, using the 5′ terminal overhang as a template; ^{32}P is incorporated into the chain by using an appropriate α-^{32}P-labeled deoxynucleoside triphosphate substrate (Figure 7.9a). Because the reaction requires a duplex DNA, it must be followed by either strand separation or restriction endonuclease cleavage to obtain a single ^{32}P-labeled terminus. However, if the ends of the duplex fragment are not identical, generating a single ^{32}P-labeled terminus directly is frequently possible. For example, no extension occurs at a flush end (Figure 7.9b), and careful selection of the α-^{32}P-labeled deoxynucleoside triphosphate may label only one of a pair of nonidentical cohesive ends (Figure 7.9c). If labeling is at the 3′ terminus, the shortest chains visible on the autoradiogram will be those closest to the 3′ end.

Figure 7.10 illustrates data obtained from chemical sequencing.

c. *Sequencing by Enzymatic Copying*

Generating nested sets To sequence DNA by enzymatic copying requires a single strand of DNA as well as a short polydeoxynucleotide primer complementary to a small portion of the single strand. The single strand serves as the template, and new strands are extended from the 3′-hydroxyl group on the primer using deoxyribonucleoside triphosphate substrates. Four separate copying reactions are carried out, each designed to yield a collection of strands that stop at A, C, G, or T residues. The stops themselves are caused by incorporation of a dideoxynucleotide residue that, because it lacks a 3′-hydroxyl group, prohibits further extension of the chain. In order to ensure that each collection of chains includes molecules representing all possible stop positions, the reactions are set up so that there is only a small probability of a stop at any particular residue.

DNA polymerase I (as well as reverse transcriptase) utilizes the 2′, 3′-dideoxynucleoside triphosphate analogues of its normal deoxynucleoside triphosphate substrates. As in the typical polymerization reaction, the appropriate dideoxynucleotide is added to the 3′-hydroxyl terminus of the primed chain (Figure 7.11). However, the product of such an addition

Figure 7.10
A sequencing gel generated by the chemical method. The strand was labeled at its 5′ terminus. Reading up the gel gives the sequence in the 5′ to 3′ direction. The column headings indicate the base-specific reaction used for each of the four samples. Courtesy of R. E. Thayer.

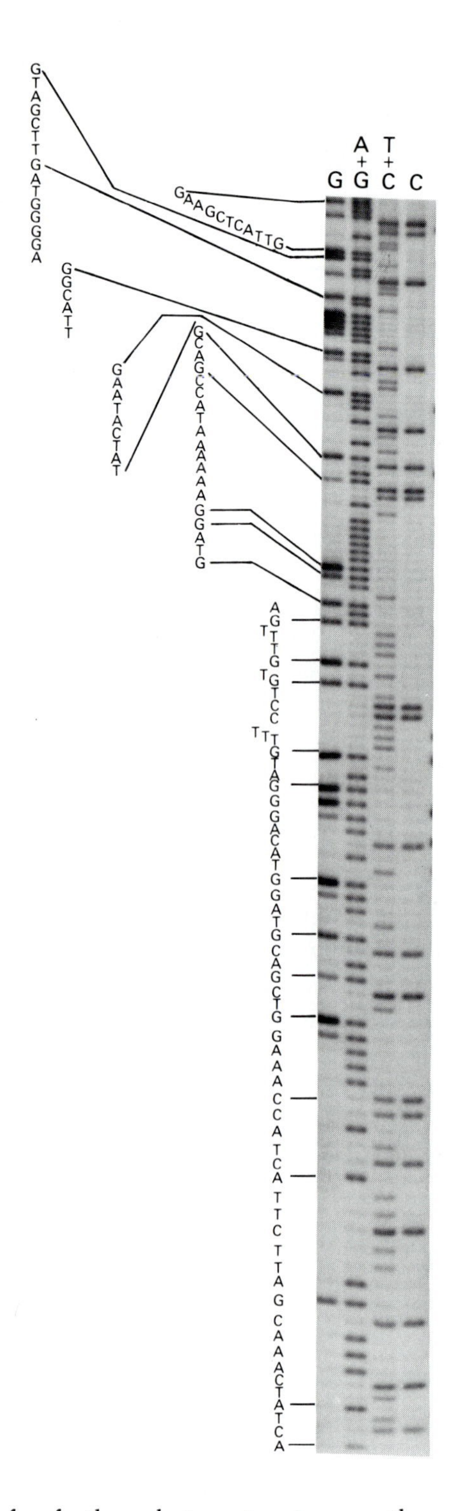

is not a primer for further chain extension, so the reaction terminates. Four separate reactions are run in which the template-primer complex is incubated with all four normal deoxyribonucleoside triphosphates plus a single dideoxynucleoside triphosphate; each reaction has a different dideoxynucleotide. For example, one reaction contains ddATP, dATP, dGTP, dCTP, and dTTP; another contains dATP, ddGTP, dGTP, dCTP, and dTTP, and so forth. Chain extension by template copying proceeds normally. Whenever a dideoxynucleotide rather than the deoxynucleotide is inserted, chain extension terminates. Because termination occurs at random, the final product is a collection of chains representing all possible lengths between the 5′ end of the primer and the positions corresponding to the added dideoxynucleotide (Figure 7.12). The newly synthesized

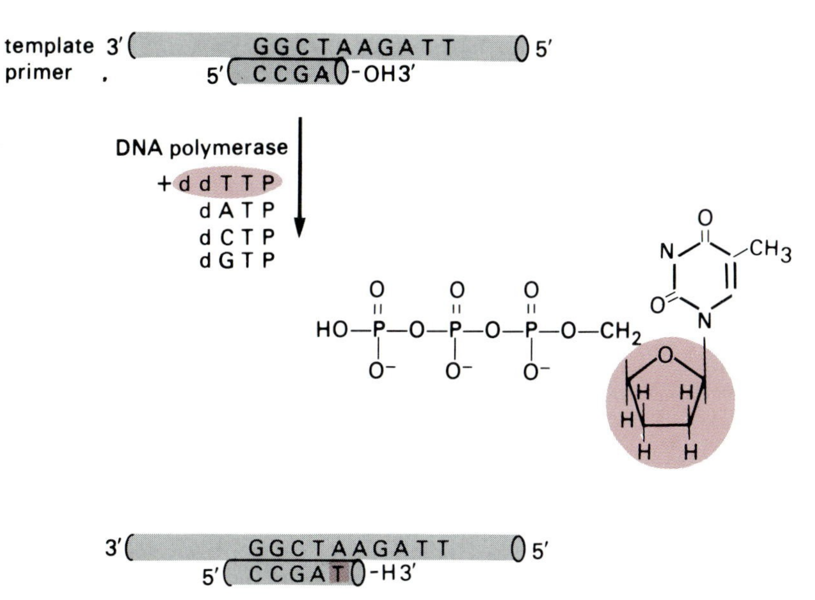

Figure 7.11
Addition of a dideoxynucleotide to a growing chain terminates synthesis; no 3′-hydroxyl group is available for further extension.

chains are radiolabeled by using at least one radioactive deoxynucleoside triphosphate in the copying reaction. Often this is an α-^{32}P-deoxynucleoside triphosphate, but a ^{35}S-(α-thio)-deoxynucleoside triphosphate can also be used (Figure 7.13). Alternatively, a radiolabeled primer can be used. The four reaction products are electrophoresed separately and simultaneously as for chemical sequencing. The sequence is read off the resulting autoradiogram, which looks like those obtained with chemical methods (Figure 7.10).

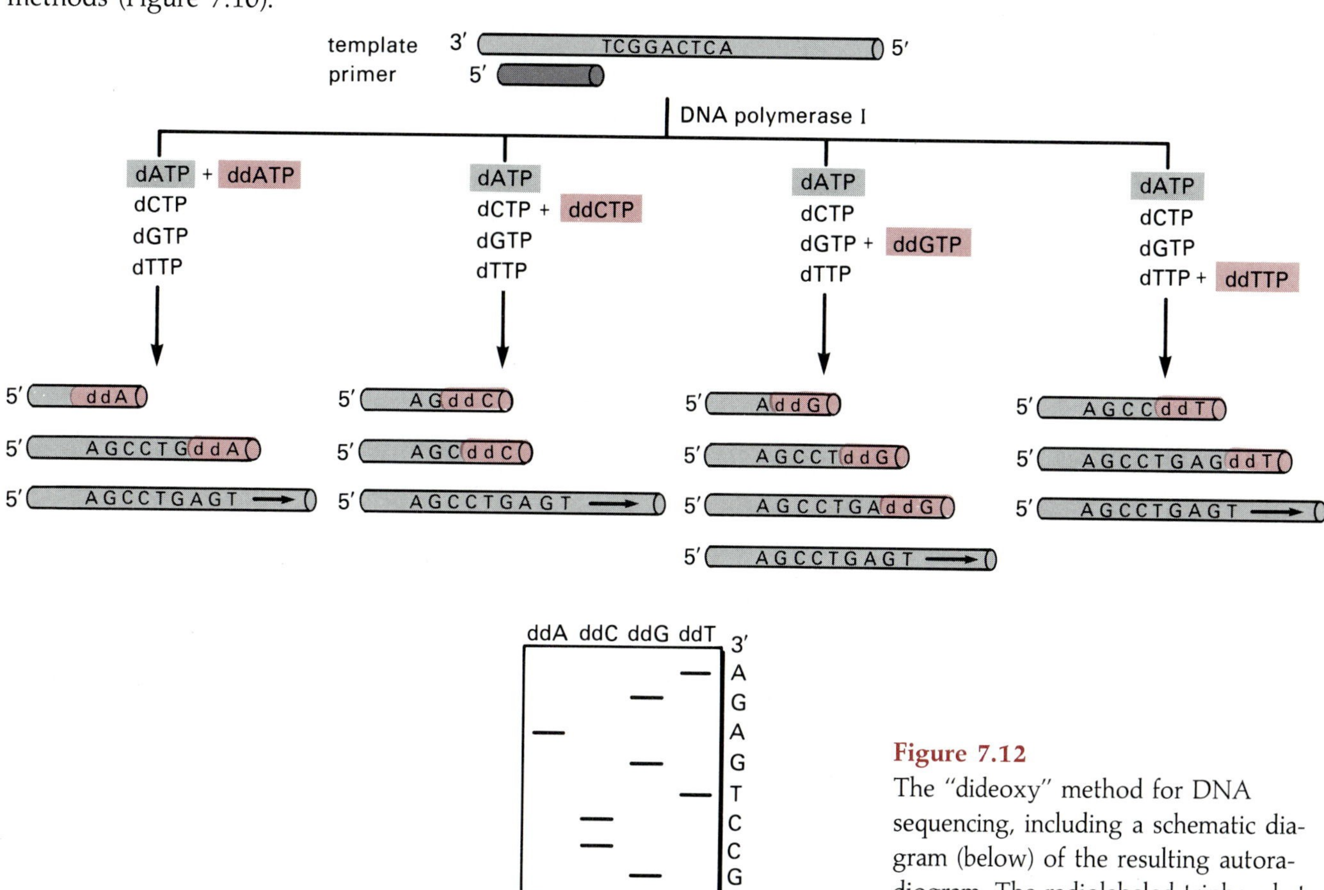

Figure 7.12
The "dideoxy" method for DNA sequencing, including a schematic diagram (below) of the resulting autoradiogram. The radiolabeled triphosphate is α-^{32}P-deoxyATP.

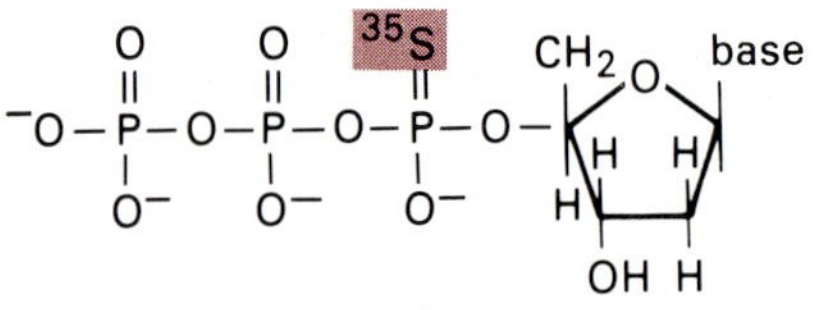

^{35}S- (α-thio)-deoxynucleoside triphosphate

Figure 7.13
One or more of the deoxyribonucleoside triphosphates in a "dideoxy" sequencing reaction is radiolabeled to permit detection of the product strands. Either an α-^{32}P-triphosphate or a ^{35}S-(α-thio)-triphosphate (shown here) can be used. Note that the presence of a sulfur atom in place of the oxygen on the α-P does not significantly alter the molecule's ability to serve as a substrate for DNA polymerase I.

Obtaining single strand DNA for "dideoxy" sequencing With the special exception of single strand bacteriophage, DNA molecules are usually double stranded. As already mentioned, physical separation of duplex strands is not always possible (Section 7.2a). Strand separation involves denaturing the duplex with heat or alkali and separating the two strands by gel electrophoresis or chromatography. Presumably, the different shapes dictated by intramolecular secondary structure make the two strands separable. When strand separation is not successful, other procedures are used.

The most effective way to prepare single strands for sequencing is to clone the DNA fragment in an M13 vector (Section 5.3f). Regardless of the cloned insert, it will always be flanked by the same vector sequence (Figure 7.14); therefore, standard primers can be used. These have been synthesized chemically, so a convenient supply can be assured. Also, because it is simple to obtain separate clones containing each of the two complementary DNA chains (Section 5.3f), both strands are readily sequenced.

The combination of cloning in M13 and dideoxy sequencing furnishes an efficient way to determine the sequence of DNA molecules that are thousands of base pairs long. The large duplex molecule is broken into segments by shearing, and the collection of fragments is cloned. Each plaque contains different fragments or strands. The sequence of the inserts in randomly selected clones is determined. Because the inserts are generated by random shearing, overlapping fragments are included. The sequence of each fragment is compared with the others using computer programs (Section 7.3), the overlaps are established, and the separate fragments are ordered. The sequences of the human mitochondrial (16,569 base pairs) and λ phage (48,502 base pairs) genomes were determined in this way.

Other applications of sequencing methods Sequencing by the dideoxyribonucleotide procedure can be adapted to the direct determination of an RNA or DNA sequence, even if the chain is present in a mixture of unrelated molecules. The essential element is a primer known to be com-

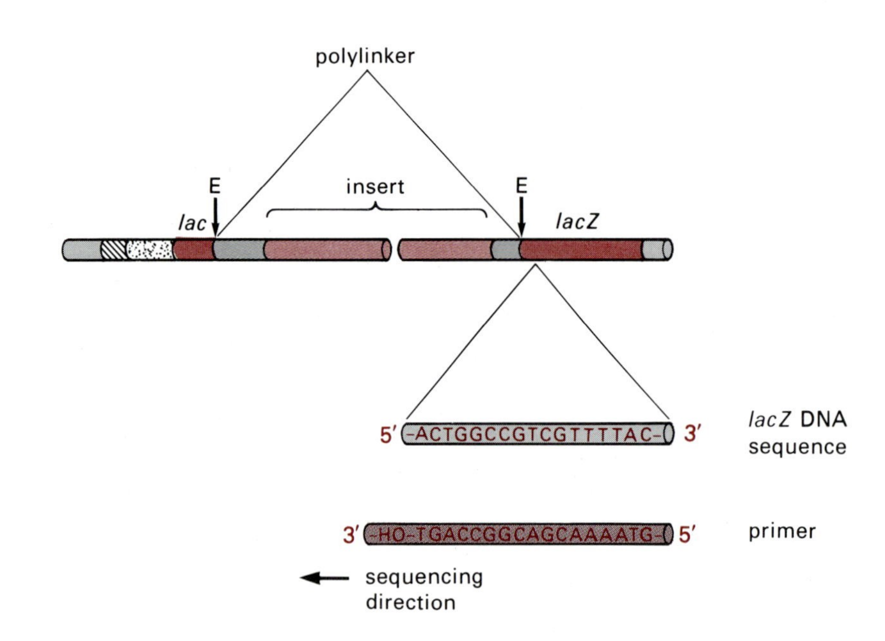

Figure 7.14
Sequencing by the "dideoxy" method of single strand DNA inserts cloned in an M13 vector. (See Section 5.3f for a description of the vector.) The same single strand primer can be used, regardless of the insert.

plementary to a small region of interest and unlikely to anneal with other molecules in the preparation. Such specificity is usually assured by a primer with 20 bases of a unique sequence. The cDNA is synthesized with reverse transcriptase. Although the method is useful only if a substantial amount is already known about the RNA or DNA, the number of applicable experimental situations is growing.

The chemical sequencing method can be applied to specific DNA sequences in a large mixture of DNAs—including even a single gene in total mammalian genomic DNA. First, the entire DNA sample is cleaved with a restriction endonuclease that will leave the segment of interest in a fragment several hundred base pairs long. Then the base-specific chemical reactions are carried out, thereby generating four collections of fragments. These collections will include all possible lengths of the region of interest starting from the restriction site used for cleavage. They will also include millions of other fragments from the rest of the genome. The fragments are electrophoresed on a sequencing gel. Thereafter, they are blotted onto a nylon membrane and annealed with a ^{32}P-labeled probe that is complementary to the target sequence in the region close to one of the restriction endonuclease cleavage sites. The probe labels all the fragments produced from one end of the target region, thereby yielding typical sequencing ladders on an autoradiogram. This method can, for example, yield the nucleotide sequence of mutant forms of a previously cloned gene. It can also be used to determine the pattern of methylation in a specific, already cloned genomic region because hydrazine reacts poorly with 5-methylcytosine residues compared to cytosine and thymine residues. The presence of 5-methylcytosine is marked on the gel by the absence of a cytosine residue that is present in the corresponding DNA fragment isolated after cloning and replication in *E. coli* (Figure 4.13)

7.3 Computer Analysis of DNA Sequences

The great ease with which DNA segments can be purified by cloning and analyzed has yielded an enormous amount of nucleotide sequence data. Although the biologist's yearnings for chemical precision are nourished by this information, it is useless until the significant features of the sequence can be deduced. Yet to the eye, long runs of bases are baffling, and patterns evade recognition. Fortunately, highly developed computer methods are readily applicable to these problems, and programs in several standard computer languages are available for routine use. Although many different programs exist, their capabilities are to a large extent overlapping. They are available on large computers as well as for mini- and microcomputers. Interactive and conversational modes increase convenience. Broadly speaking, the programs are designed to analyze both the structural and biological features of sequences.

a. Storing Primary Sequence Data

Standard editing procedures facilitate the entry and storage of sequence data for later retrieval by the investigator or by an analytical program. Most of these also provide for correcting the stored sequence should that become necessary. Sequence entries are often made by typing the sequence into a computer terminal after reading and recording the data from sequencing gels. However, there are also programs for the direct entry of the primary data into a computer. In principle, such programs minimize errors made by misreading gels or inaccurately recording the data. There are two approaches to this automation. In one, a detector automatically scans the autoradiograms, and the data enter directly into the computer. In the second, the data are automatically recorded from a sensor manipulated by hand. The second approach leaves the investigator as the arbiter of any difficult or ambiguous assignments.

b. Structural Analysis

Primary structure Given an entered single strand, most programs will generate the complementary strand, calculate the base composition, delineate regions rich in purines, pyrimidines, or other combinations of bases, and count the frequency of the different dinucleotides (Figure 7.15). Specific subsequences can be located within the segment, a feature that is frequently used to find restriction endonuclease sites (Figure 7.16). The recognition sites of the known enzymes are stored in the program, and a single command yields the location of cleavage sites for each enzyme, the number of fragments expected from digestion with each enzyme, the size of each fragment in base pairs and percentage of the total segment, and the residues corresponding to the ends of each fragment. Other programs predict the products of simultaneous digestion with two or more endonucleases. In these operations, the sequence may be treated as either a linear or circular molecule. This information can be used to confirm sequencing data by comparing the predictions with observed products of endonuclease cleavage.

The available programs also search for distinctive sequence features, including direct repeats (Figure 7.17) and inverted repeats (dyad symmetries) (Figure 7.18). Not only are perfectly matched segments reported, but those with varying degrees of mismatching can be called for by setting flexible program parameters. In the example in Figure 7.17, all repeats at least six base pairs long and at least 75 percent perfectly matched are listed, and a loop-out of two unmatched base pairs is allowed.

Homologies and partial homologies between different sequences are also discernable. Examples might include the structure of the same gene in two different organisms or of related genes in a single organism. Such comparisons are widely used in the study of evolution at the molecular level. Algorithms exist for comparing two sequences or for comparing one sequence with a group of other sequences simultaneously. The same probability analyses and limitations previously described can be imposed. Comparison of sequence information obtained in distant laboratories is facilitated by the existence of central data storage banks from which sequences can be retrieved and transferred to local systems by subscribers. As of 1989, more than 20 million base pairs of sequence were stored in the banks, representing data from many genes and organisms.

Besides its more general utility, the search for similarities between different sequences is an integral component of determining very long

```
THE NUCLEOTIDE SEQUENCE IS:

      5109        5119        5129        5139        5149        5159
TCCTTTCAAG  ACCTAGAAGG  TCCATTAGCT  GCAAAGATTC  CTCTCTGTTT  AAAACTTTAT

      5169        5179        5189        5199        5209        5219
CCATCTTTGC  AAAGCTTTTT  GCAAAAGCCT  AGGCCTCCAA  AAAAGCCTCC  TCACTACTTC

      5229        5239     1    6          16          26          36
TGGAATAGCT  CAGAGGCCGA  GGCGGCCTCG  GCCTCTGCAT  AAATAAAAAA  AATTAGTCAG

        46          56          66          76          86          96
CCATGGGGCG  GAGAATGGGC  GGAACTGGGC  GGAGTTAGGG  GCGGGATGGG  CGGAGTTAGG

       106         116         126         136         146         156
GGCGGGACTA  TGGTTGCTGA  CTAATTGAGA  TGCATGCTTT  GCATACTTCT  GCCTGCTGGG

       166         176         186         196         206         216
GAGCCTGGGG  ACTTTCCACA  CCTGGTTGCT  GACTAATTGA  GATGCATGCT  TTGCATACTT

       226         236         246         256
CTGCCTGCTG  GGGAGCCTGG  GGACTTTCCA  CACCCTAACT  GACA


THE COMPLEMENTARY STRAND:

       251         241         231         221         211         201
TGTCAGTTAG  GGTGTGGAAA  GTCCCCAGGC  TCCCCAGCAG  GCAGAAGTAT  GCAAAGCATG

       191         181         171         161         151         141
CATCTCAATT  AGTCAGCAAC  CAGGTGTGGA  AAGTCCCCAG  GCTCCCCAGC  AGGCAGAAGT

       131         121         111         101          91          81
ATGCAAAGCA  TGCATCTCAA  TTAGTCAGCA  ACCATAGTCC  CGCCCCTAAC  TCCGCCCATC

        71          61          51          41          31          21
CCGCCCCTAA  CTCCGCCCAG  TTCCGCCCAT  TCTCCGCCCC  ATGGCTGACT  AATTTTTTTT

        11           1        5234        5224        5214        5204
ATTTATGCAG  AGGCCGAGGC  CGCCTCGGCC  TCTGAGCTAT  TCCAGAAGTA  GTGAGGAGGC

      5194        5184        5174        5164        5154        5144
TTTTTTGGAG  GCCTAGGCTT  TTGCAAAAAG  CTTTGCAAAG  ATGGATAAAG  TTTTAAACAG

      5134        5124        5114        5104
AGAGGAATCT  TTGCAGCTAA  TGGACCTTCT  AGGTCTTGAA  AGGA
```

THE SEQUENCE CONTAINS 404 NUCLEOTIDES:

95	(23.5)	adenine
96	(23.8)	cytosine
110	(27.2)	guanine
103	(25.5)	thymine
0		other

THE DISTRIBUTION OF DINUCLEOTIDES IS:

	number	percent	expected percent
AA	32	7.9	5.5
AC	18	4.5	5.6
AG	24	6.0	6.4
AT	20	5.0	6.0
CA	21	5.2	5.6
CC	23	5.7	5.6
CG	9	2.2	6.5
CT	43	10.7	6.1
GA	24	6.0	6.4
GC	35	8.7	6.5
GG	44	10.9	7.4
GT	7	1.7	6.9
TA	18	4.5	6.0
TC	20	5.0	6.1
TG	33	8.2	6.9
TT	32	7.9	6.5

Figure 7.15

The sequence of a portion of the SV40 genome surrounding the origin of replication. The numbers are as in the map in Figure 5.34. Residue 1 is within the origin and is attached to residue 5243 in the circular genome. All sequences are shown 5′ to 3′, left to right, top to bottom. This is a typical computer output including the sequence recorded in the computer, the complementary strand, the base composition, and the dinucleotide frequency compared to the frequency expected for a chain with a random arrangement of the same nucleotides.

Figure 7.16
A partial computer output showing some of the restriction endonuclease sites in the SV40 DNA segment in Figure 7.15. The number (#) of sites for each enzyme is given, as is the position of the cleavage (bp). Also given are the length of each fragment produced by digestion with the endonuclease, stated both as the number of base pairs and the percentage of the total segment (in parentheses), as well as the residue numbers at the ends of the fragments.

Enzyme	Sites #	Sites bp	Fragments length bp	Fragments length %	Fragments ends	
ALU 1, OXA 1, (AGCT)	3					
		5126	277	(68.6)	5226	260
		5172	54	(13.4)	5172	5226
		5226	46	(11.4)	5126	5172
			27	(6.7)	5100	5126
AVA 3 (ATGCAT)	2					
		126	270	(66.8)	5100	126
		198	72	(17.8)	126	198
			62	(15.3)	198	260
BGL 1 (GCCNNNNNGGC)	1					
		5235	268	(66.3)	5235	260
			136	(33.7)	5100	5235
DDE 1 (CTNAG)	1					
		5228	275	(68.1)	5228	260
			129	(31.9)	5100	5228
HAE 3 (GGCC)	4					
		5191	254	(62.9)	6	260
		5234	92	(22.8)	5100	5191
		5243	43	(10.6)	5191	5234
		6	9	(2.2)	5234	5243
			6	(1.5)	5243	6
HIND 3 (AAGCTT)	1					
		5171	332	(82.2)	5171	260
			72	(17.8)	5100	5171
HINF 1 (GANTC)	1					
		5135	368	(91.1)	5135	260
			36	(8.9)	5100	5135

sequences by the dideoxy method coupled to random cloning in M13 (Section 7.2c). Using the computer to meld overlapping sequences not only makes the job less tedious, but allows sensible decisions regarding the need for additional data to be made rapidly.

Secondary structure Other programs are designed to predict stable intramolecular secondary structure in a single strand RNA or DNA. These have, for example, constructed models for folding tRNA and rRNA, and many of the predictions have been confirmed experimentally by analysis with single strand-specific nucleases. The calculations are based on assumptions including likely base pairs and thermodynamic information regarding base pair and helix stability.

c. Biological Significance

Computer programs translate nucleotide sequences into the corresponding amino acids according to the genetic code. All three possible reading frames are indicated, along with translational stop codons. If the amino acid sequence of the encoded polypeptide is known, the proper reading frame is readily identified. If not, the proper reading frame may be suggested by its length. Frames that are frequently interrupted by stop codons are not likely candidates. Figure 7.19 shows a portion of the SV40

```
   49     GAATGGGCGGA  59
   81     GA TGGGCGGA  90
0.909 a
3E-05 b
1E+00 c

   52     TGGGCGGA  59
   62     TGGGCGGA  69
1.000
4E-05
2E+00

   53     GGGCGGAACT   62
   96     GGGCGGGACT  105
0.900
9E-05
4E+00

   57     GGAACTGGGCGGAGTTAGGGGCG   79
   80     GGA  TGGGCGGAGTTAGGGGCG  100
0.913
0E+00
0E+00

   62     TGGGCGGAGTTAGGGGCGGGA   82
   83     TGGGCGGAGTTAGGGGCGGGA  103
1.000
0E+00
0E+00

 .107     TGGTTGCTGACTAATTGAGATGCATGCTTTGCATACTTCTGCCTGCTGGGGAGCCTGGGGACTTTCCACACC  178
  179     TGGTTGCTGACTAATTGAGATGCATGCTTTGCATACTTCTGCCTGCTGGGGAGCCTGGGGACTTTCCACACC  250
1.000
0E+00
0E+00

  127     TGCATGCTT TGCATACT  143
  208     TGCATACTTCTGCCTGCT  225
0.778
1E-05
5E-01

  136     TGCATACTTCTGCCTGCT  153
  199     TGCATGCTT TGCATACT  215
0.778
1E-05
5E-01

  THE NUMBER OF MATCHES IS 22

MAXPROB= 9.99E-05    EXPECT=   4         MINRATIO=7.50E-01;
                     LOOPLENGTH=  2   DISTANCE=        404;
```

Figure 7.17
A partial computer output showing some of the repeated sequences within the SV40 segment in Figure 7.15. In this search, the program was preset to eliminate repeats if they were shorter than 7 bp, less than 75 percent homologous (MINRATIO), or the probability (MAXPROB) of finding as good a match in a random sequence was greater than 1×10^{-4} (written as 9.99E-05). The actual percent homology is shown by *a* and the actual probability of the match by *b*. A maximum loop length of 2 was allowed. When the value of *b* is less than 1×10^{-5} (1E-05), *b* appears as 0 (0E + 00), and *c* is automatically also 0. The number of matches of equal percent homology expected in a random sequence of the same length and base composition is *c*. A total of 22 matches was detected (8 are shown), whereas only 4 were expected from a random arrangement of the nucleotides.

genome within the region encoding the NH_2-terminus of the small-T and large-T proteins (Figure 5.34) and the computer-generated polypeptides for all three reading frames; frame #3 corresponds to the actual coding frame. It is also possible to count the frequency with which each codon appears and thus learn whether particular codons are favored for particular amino acids. The reverse of these procedures allows deduction of possible nucleotide sequences from the known sequence of a polypeptide, although the solution is never unique because of the degeneracy of the genetic code. Central storage banks generally include the amino acid sequences of analyzed polypeptides; thus, newly determined DNA sequences can be compared with known proteins.

Because the programs allow searching for specific nucleotide sequences, they are useful for detecting biologically important segments such as promotor sequences and ribosome-binding sites. The computer programs can also assist in the discovery of significant but unsuspected sequences. For example, similar sequences in or near otherwise unrelated genes invite speculation regarding possible regulatory significance. Also, sequence homologies between coding regions in different genes raise questions regarding common evolutionary origins. Insights like this figure prominently in Part III of this book. However, it is important to recognize that statistical significance is neither necessary nor sufficient for biological

```
THE DYAD SYMMETRIES ARE:

 5177    T T T G C A A A  5184
 5172    A A A C G T T T  5165
1.000
4E-05
2E-01

    1    G C C T C G G C C T C T G     13
 5242    C G G A G C C G G A G A C   5230
1.000
0E+00
0E+00

THE NUMBER OF MATCHES IS  2
MAXPROB= 9.99E-05   EXPECT=     0
                    LOOPLENGTH=   2

MINRATIO= 7.50E-01;
LOOPDIST=     20;
```

Figure 7.18
The computer output showing inverted repeats (dyad symmetries) in the SV40 DNA segment of Figure 7.15. The symbols are as in Figure 7.17. A maximum of 20 bases was allowed between the two elements in the symmetry (LOOPDIST).

```
THE TRANSLATED SEQUENCE PHASED FROM POSITION 1

                                              5214                                            5184
5'CGC CTC GGC CTC TGA GCT ATT CCA GAA GTA GTG AGG AGG CTT TTT TGG AGG CCT AGG CTT
  ARG LEU GLY LEU END ALA ILE PRO GLU VAL VAL ARG ARG LEU PHE TRP ARG PRO ARG LEU

                                              5154                                            5124
  TTG CAA AAA GCT TTG CAA AGA TGG ATA AAG TTT TAA ACA GAG AGG AAT CTT TGC AGC TAA
  LEU GLN LYS ALA LEU GLN ARG TRP ILE LYS PHE END THR GLU ARG ASN LEU CYS SER END

                                              5094                                            5064
  TGG ACC TTC TAG GTC TTG AAA GGA GTG CCT GGG GGA ATA TTC CTC TGA TGA GAA AGG CAT
  TRP THR PHE END VAL LEU LYS GLY VAL PRO GLY GLY ILE PHE LEU END END GLU ARG HIS


THE TRANSLATED SEQUENCE PHASED FROM POSITION 2

                                              5213                                            5183
5'GCC TCG GCC TCT GAG CTA TTC CAG AAG TAG TGA GGA GGC TTT TTT GGA GGC CTA GGC TTT
  ALA SER ALA SER GLU LEU PHE GLN LYS END END GLY GLY PHE PHE GLY GLY LEU GLY PHE

                                              5153                                            5123
  TGC AAA AAG CTT TGC AAA GAT GGA TAA AGT TTT AAA CAG AGA GGA ATC TTT GCA GCT AAT
  CYS LYS LYS LEU CYS LYS ASP GLY END SER PHE LYS GLN ARG GLY ILE PHE ALA ALA ASN

                                              5093                                            5063
  GGA CCT TCT AGG TCT TGA AAG GAG TGC CTG GGG GAA TAT TCC TCT GAT GAG AAA GGC ATA
  GLY PRO SER ARG SER END LYS GLU CYS LEU GLY GLU TYR SER SER ASP GLU LYS GLY ILE


THE TRANSLATED SEQUENCE PHASED FROM POSITION 3

                                              5212                                            5182
5'CCT CGG CCT CTG AGC TAT TCC AGA AGT AGT GAG GAG GCT TTT TTG GAG GCC TAG GCT TTT
  PRO ARG PRO LEU SER TYR SER ARG SER SER GLU GLU ALA PHE LEU GLU ALA END ALA PHE

                                              5152                                            5122
  GCA AAA AGC TTT GCA AAG ATG GAT AAA GTT TTA AAC AGA GAG GAA TCT TTG CAG CTA ATG
  ALA LYS SER PHE ALA LYS MET ASP LYS VAL LEU ASN ARG GLU GLU SER LEU GLN LEU MET

                                              5092                                            5062
  GAC CTT CTA GGT CTT GAA AGG AGT GCC TGG GGG AAT ATT CCT CTG ATG AGA AAG GCA TAT
  ASP LEU LEU GLY LEU GLU ARG SER ALA TRP GLY ASN ILE PRO LEU MET ARG LYS ALA TYR
```

Figure 7.19
Computer output showing the translation of a portion of the SV40 genome (residues 5243 to 5062, Figure 5.34) into polypeptides corresponding to all three possible reading frames. The DNA strand shown has the same sequence as the mRNA for early proteins (the sense strand). The actual reading frame is number 3, and the initiation codon, ATG (AUG), is boxed. Translational stop signals are marked "end."

significance. Independent biological confirmation of the meaning of sequences is essential.

7.4 Locating Cloned Segments in Genomes

Ideally, localizing a cloned DNA segment means defining its flanking sequences on a molecular level and placing it at a specific chromosomal site. Such mapping aims ultimately at a complete description of chromosome structure.

a. Molecular Location

Characterizing a cloned segment The first goal is to demonstrate that a cloned segment is a true replica of a genomic segment. Does a homologous segment with identically placed restriction endonuclease sites exist in the genome? This is important because cloning "accidents" caused by

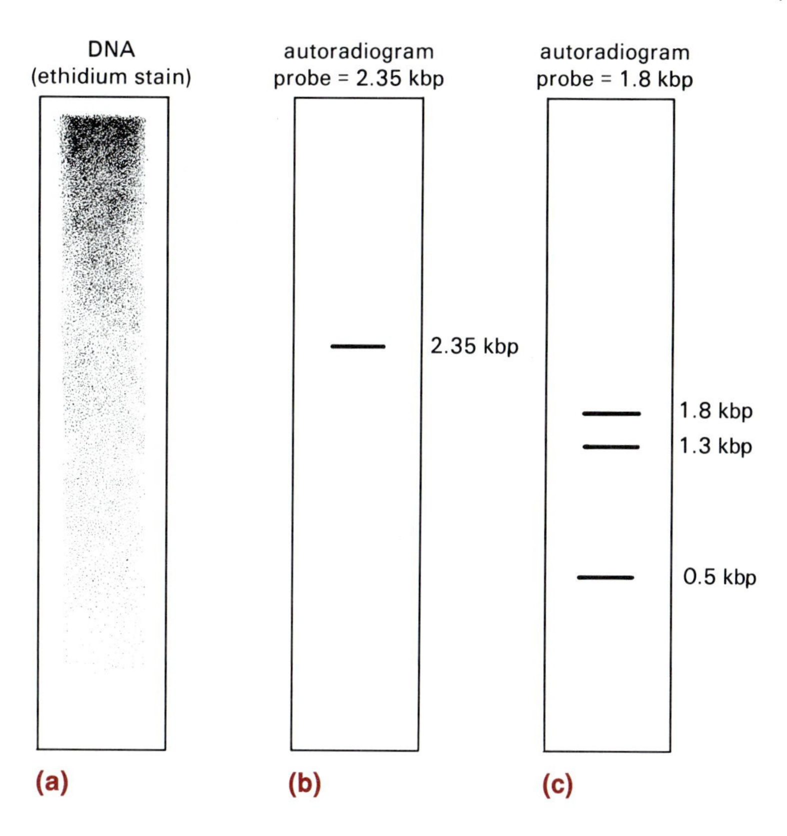

Figure 7.20
Annealing the ^{32}P-labeled 2.35 kbp (b) and 1.8 kbp (c) long cloned segments of the ovalbumin gene to the electrophoretically separated *Eco*RI fragments of total chicken DNA (a).

recombination, deletion, or mutation occur occasionally during replication in host-vector systems. The basic approach is to use the cloned DNA as a probe against restriction endonuclease digests of the total genomic DNA that was its source. This is exactly the same kind of experiment described in Section 6.1b except that the probe is now the cloned insert labeled with ^{32}P. In the experiment shown in Figure 7.20b, the cloned probe is the subcloned 2.35 kbp portion of the ovalbumin gene shown in Figure 7.2; the DNA on the gel is an *Eco*RI digest of chicken DNA (Figure 7.20a). A single 2.35 kbp band anneals to the probe. This experiment demonstrates (1) that the cloned insert is indeed a chicken DNA fragment, (2) that within the genome, the annealing segment is between two *Eco*RI sites that are 2.35 kbp apart, and (3) that no fragments of any other size anneal to the clone. Thus, the 2.35 kbp cloned fragment appears to be a true representative of a genomic sequence, and the two alleles in the diploid are identical insofar as the placement of these *Eco*RI sites is concerned.

Another part of the ovalbumin gene is found in a 1.8 kbp long *Eco*RI restriction fragment. When this was cloned and used as a probe against an *Eco*RI digest of chicken DNA, three fragments annealed (Figure 7.20c): they are 1.8, 1.3, and 0.5 kbp long, respectively. This shows that the two allelic ovalbumin genes are not identical; one has an extra *Eco*RI site that divides the 1.8 kbp fragment into two pieces. Note that this rather simple procedure provides a general method for genetic analysis. The inheritance of two or more alleles is readily followed by restriction endonuclease digestion, electrophoresis, DNA blotting, and annealing with an appropriate probe, using DNA isolated from different individuals. The only requirements are polymorphic restriction endonuclease sites in the alleles

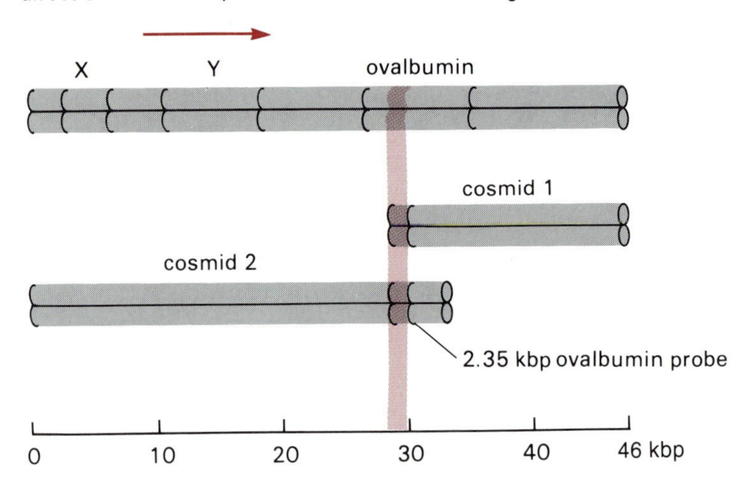

Figure 7.21
Two overlapping cosmids give a fine structure map of the chicken ovalbumin gene region. In addition to the ovalbumin gene, two related genes of unknown function, X and Y, occur within a 46 kbp segment. After A. Royal et al., *Nature* 279 (1979), p. 125.

of any gene or DNA segment. Because only a small amount of tissue or a few milliliters of blood yield sufficient DNA for these analyses, the method is applicable to most species, including humans.

Extending the molecular map Relatively short segments such as the subcloned 2.35 kbp fragment previously described enable construction of fine structure molecular maps that can be extended down to the level of nucleotide sequence. Longer cloned segments like those obtained in λ phage vectors or in cosmids frequently contain several genes as well as other DNA segments and can be used to extend the map. Thus, the 2.35 kbp subcloned ovalbumin gene fragment was used as a probe to search for the homologous sequence in cosmid libraries constructed from chicken genomic DNA. Two cosmids whose sequences overlap only in the region of the probe were selected; together they represent 46 kbp of the genome (Figure 7.21), of which 7.7 kbp are occupied by the ovalbumin gene itself (including noncoding intervening sequences). The identity of the region containing the ovalbumin gene was confirmed by heteroduplex formation with ovalbumin cDNA. Unexpectedly, two other portions of the 46 kbp segment also hybridized with the ovalbumin cDNA, although to a lesser extent. When RNA from chicken oviducts was annealed to the cosmids, two mRNAs other than the ovalbumin mRNA also hybridized to these two portions. This suggested that two additional genes are present on the 46 kbp segment and that their sequences resemble that of the ovalbumin gene. Restriction endonuclease analysis, heteroduplex analysis, and nucleotide sequence determination confirmed the suggestion. The two genes, called X and Y because their function and gene products are unknown, are, like the ovalbumin gene, transcribed in response to steroid hormones.

The methods used to map the ovalbumin, X, and Y genes on a contiguous stretch of the chicken genome constitute a general procedure for fine structure mapping called **chromosome walking**. A unique DNA segment close to one end of a cloned fragment is purified (by subcloning) and used to probe a genomic DNA library. Some of the clones that hybridize will completely overlap the original cloned segment, but others will include new sequences, thereby extending the contiguous map. The walk is lengthened by repeating the procedure. In practice, the procedure is difficult in those genomes that contain large amounts of interspersed repeated sequences because unique probes are then difficult to purify.

b. Chromosomal Location

Classical genetic mapping depends on chance mutations and the analysis of recombination frequencies (see Perspective I). With *Drosophila,* the maps were extended and confirmed by correlating the genetic analysis with chromosomal aberrations such as deletions, inversions, and translocations that are visible as altered polytene banding patterns. The methods used for *Drosophila* are, however, less rewarding with most plants and animals; genetic analysis of small populations with long generation times is difficult, polytenization is rare, and the chromosomes are often numerous, small, and difficult to identify. In mammals, for example, the correlation of phenotypic abnormalities with deletions, translocations, and inversions allow only a limited number of genes to be located on specific chromosomes or chromosomal regions. With the advent of cloned DNA segments, generally applicable mapping procedures were developed that are independent of phenotypically observable mutations. Increasing numbers of genes, including human genes, are being localized to specific chromosomal regions.

Two commonly used methods are described here. One locates a cloned DNA segment by its ability to hybridize to a specific site on essentially intact chromosomes. The second identifies the specific chromosome that carries the sequence by annealing a suitable labeled probe to hybrid cells that contain different assortments of only a few chromosomes from one species. A third method, which is described in Perspective IV, utilizes cloned DNA fragments to identify restriction endonuclease site polymorphisms (such as the one illustrated in Figure 7.20). These **RFLPs** (**restriction fragment length polymorphisms**) are then used as genetic markers to construct linkage maps of eukaryotic chromosomes in much the same way as classical genetic analysis utilizes phenotypic markers.

Determining chromosomal location by annealing in situ RNA or denatured DNA probes anneal to corresponding sequences in denatured DNA contained within morphologically recognizable chromosomes. The position of a radioactive probe is determined by coating the microscope slide with a radiosensitive photographic emulsion; dark silver grains form over the regions that annealed to the probe. A photomicrograph of the entire preparation shows the probe's location. Reasonably accurate locations are obtained with the large polytene chromosomes of *Drosophila,* especially because the position of the grains can be correlated with the banding pattern (Figure 7.22a). However, locating single copy genes in the small and numerous chromosomes of plants and vertebrates by *in situ* hybridization is difficult for two reasons.

First, identifying individual chromosomes is not always easy. It depends on chromosome size, position of the centromere, and characteristic patterns of dark and light bands seen after staining fixed metaphase chromosomes with certain dyes (Figure 7.23). Moreover, the bands them-

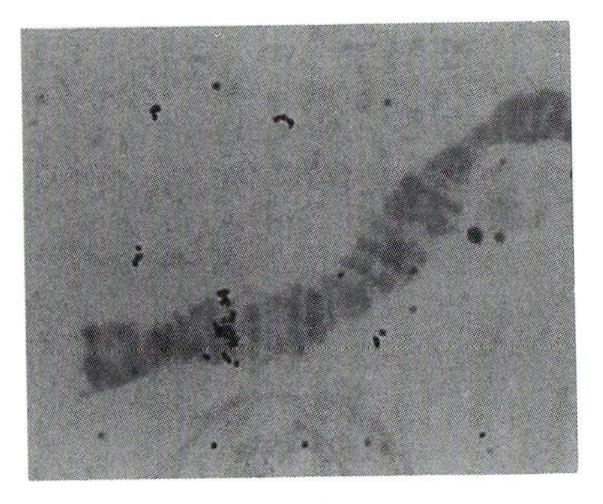

(a)

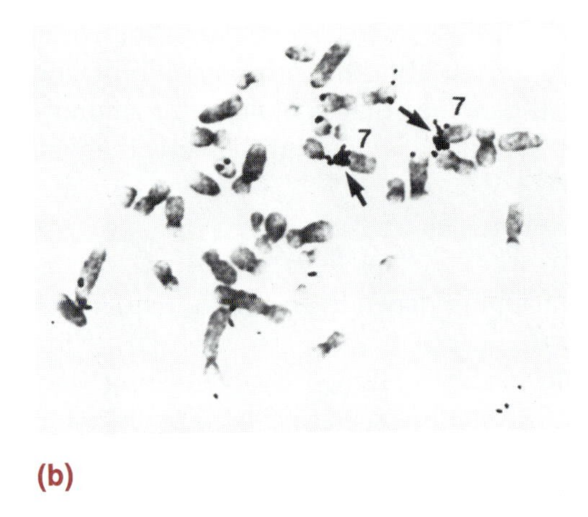

(b)

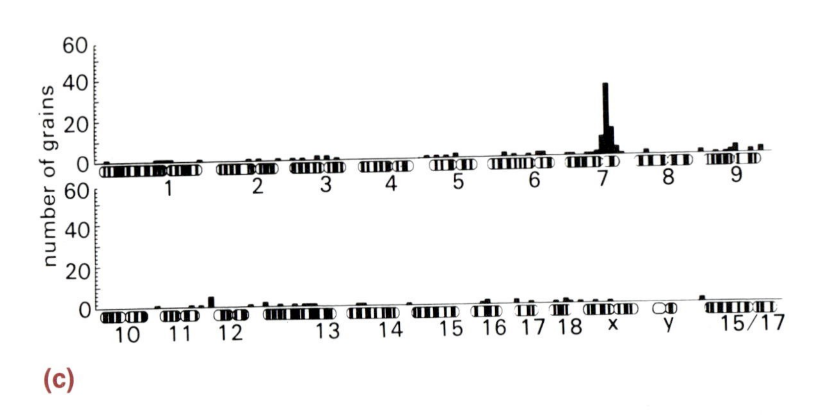

(c)

Figure 7.22
(a) Localizing a cloned myosin light chain 2 gene on *Drosophila* polytene chromosome 3 by *in situ* hybridization. The silver grains are clustered at the locus containing sequences homologous to the probe. Courtesy of J. Toffenetti, D. Mischke, and M. L. Pardue, *J. Cell. Biol.* 104 (1987), p. 19. (b) *In situ* hybridization of a single copy sequence to pig chromosome 7. The probe was a cloned portion of a gene encoding a porcine major histocompatibility locus antigen. The arrow shows the grains on chromosome 7. Inspection of a single chromosome spread does not permit reliable placement of the sequence because scattered grains appear at other positions on the spread. Counting the grains associated with each band in 100 or more spreads makes the specific hybridization site apparent. Thus, in the diagram (c) the sum of the number of grains appearing at each chromosomal site in more than 100 spreads is plotted. Chromosomes other than 7 showed only scattered grains. From M. Rabin et al., *Cytogenet. Cell. Genet.* 39 (1985), p. 206.

selves define only very large stretches of DNA; for example, the 3×10^9 base pairs in the haploid human genome are included in fewer than 1000 discernable bands (Figure I.9 shows a full human karyotype). In plants, even fewer bands are obtained. Thus, the position of silver grains affords, at best, only a regional location. Furthermore, the large size and scatter of the grains relative to the size of the chromosomes limits resolution to about 10^7 bp.

The second problem relates to the sensitivity of the probes. A single copy of a duplex DNA fragment 1 kbp long weighs approximately 10^{-12} μgrams; radioactive DNA probes prepared by nick translation (Section 4.6b) rarely have specific radioactivities much greater than 10^8 disintegrations per minute (dpm) per μgram. Annealing of such a probe to a single comparable segment in a chromosome yields only 10^{-4} dpm or one disintegration in about a week. Several special procedures must be combined to obtain the required sensitivity. For example, probes with specific radioactivities of about 10^9 dpm per μgram can be made by nick translation using [^{125}I]-5-iodo-dCTP as one substrate. Annealing is carried out in the presence of dextran sulfate, which increases the apparent rate of annealing up to 100-fold, thereby enhancing the signal. The number of grains can also be increased if the single strand vector DNA sequences that are linked to the annealed eukaryotic sequence in the cloned probe themselves anneal to excess probe molecules. Exposure times of several weeks are used to yield a significant number of silver grains (Figure 7.22b and c). In this way, single copy genes are localized to specific but broad regions on particular chromosome arms in mammalian cells.

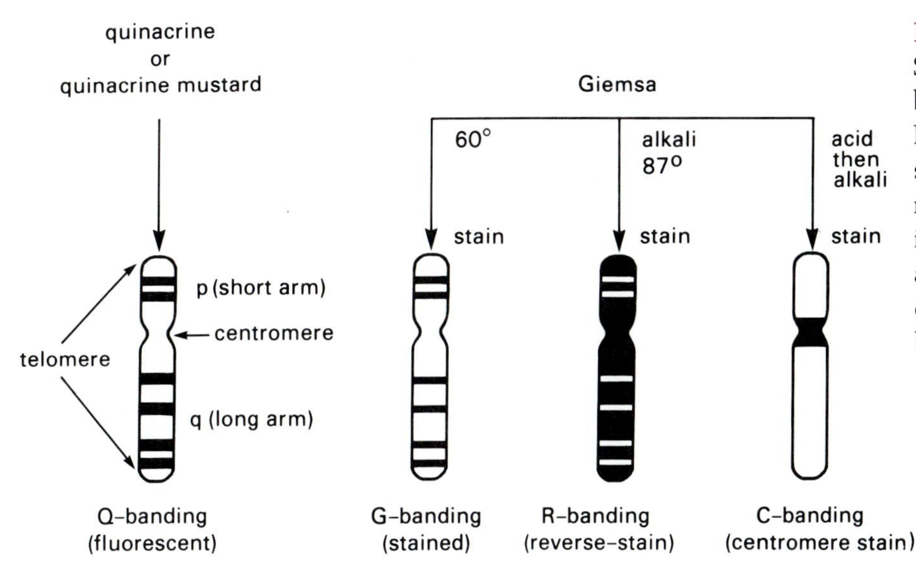

Figure 7.23
Staining methods reveal characteristic bands on mammalian chromosomes. Note that Q and G banding give the same patterns, but R banding gives the reverse pattern. The human karyotype in Figure I.9 is G banded. The acid/alkaline Giemsa procedure stains the centromeric heterochromatin– C banding. (See also Figure I.7.)

Chromosomal mapping using somatic cell hybrids Hybrid cell lines that contain chromosomes from two different species are useful for placing a cloned segment on a specific chromosome. Generally, these **somatic cell hybrids** contain a full complement of the chromosomes of one species (the recipient) but only one or a limited number of chromosomes from the second species (the donor). A gene can then be located on a particular donor chromosome by analyzing a group of different hybrid cell lines and correlating the presence of the gene with the presence of a specific chromosome.

Somatic cell hybrids are formed by (1) fusing the cells of two different species, (2) fusing the cells of one species with microcells of another that contains one or a few donor chromosomes, and (3) transfecting recipient cells with purified donor chromosomes (Figure 7.24). Appropriate selective conditions are then applied to ensure that unfused donor or recipient cells do not grow. For example, the donor cell may be sensitive to a particular drug, but the recipient cells might be mutants requiring special conditions for growth (e.g., TK^- cells do not grow in HAT medium, Figure 5.31). In the presence of the drug and HAT medium, only hybrid cells containing a functional thymidine kinase gene on a donor chromosome grow.

If hybrid cells are grown under nonselective conditions after the initial selection, donor chromosomes tend to be lost more or less at random. At any point, clones of hybrids can be prepared, and the resident donor chromosomes are identified by standard banding techniques, by the presence of previously mapped DNA sequences, by assay of already mapped enzymatic markers, or (preferably) all three. In this way, banks of various hybrid cell lines, each containing a different set of donor chromosomes, can be obtained. Such a bank is called a **mapping panel**. Alternatively, selective conditions may be maintained. By imposing selective pressure for a given gene over many cell generations (e.g., by using known mutant recipient cells and an appropriate medium), one can obtain cell lines with only one stable donor chromosome. Another way to obtain

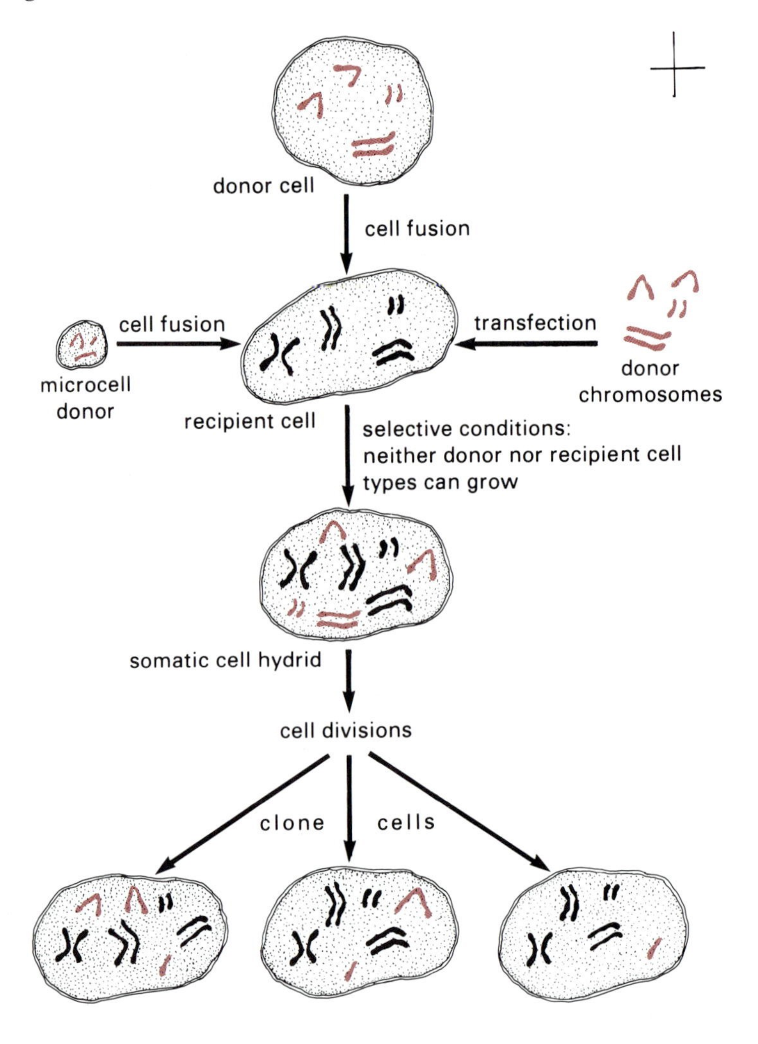

Figure 7.24
Constructing somatic cell hybrids.

hybrid cells with a single donor chromosome is to carry out the transfection with preparations enriched for a specific donor chromosome by physical separation methods.

A question that arises when constructing hybrids concerns which species will act as donor and which as recipient. The answer is generally unpredictable except on the basis of experience. Human chromosomes are usually lost from human-rodent cell hybrids.

Once the foreign chromosomes within somatic cell hybrids are identified, there are three approaches to correlating the presence of a test gene (or DNA segment) with a given chromosome. The first is to correlate the presence of the chromosome with expression of a particular gene product. This can be accomplished by measuring enzyme activity or interaction with a specific antibody, provided that the assay distinguishes between donor and recipient cell gene products. For example, the two proteins might have distinctive electrophoretic properties, as do the galactokinases of rodents and primates. The second is to anneal a cloned probe to the hybrid set of chromosomes *in situ;* this has the limitations already described. The third and most general approach is to detect the gene (or DNA segment) of interest by annealing an appropriate DNA or RNA probe to a DNA blot prepared from the hybrid cell DNA. Annealing to restriction endonuclease digests often demonstrates the presence of the donor gene in fragments of a size typical of the donor genome, even if the donor and recipient genes cross-hybridize. Because hybrids with a single donor chromosome are difficult to obtain, a mapping panel of hybrid cell lines that contain different donor chromosomes is usually examined. If the set is large enough, the presence of the probe sequence can usually be correlated with the presence of a particular chromosome.

Somatic cell hybrids that contain only a portion of a particular donor chromosome are also available. For example, some donor cell lines have a deletion in a chromosome or have a chromosome that contains a small translocation from another chromosome. Such lines are not uncommon, and over the years, many hybrid cell lines carrying deleted or composite donor chromosomes have been isolated and characterized. Locating the human immunoglobulin heavy chain genes at position 32 on the long arm of chromosome 14 (14q32), for example, depended on such hybrids. A pair of mouse-human hybrid cell lines, each containing one of a pair of reciprocal translocations between chromosome 14 and the X-chromosome were analyzed. The cloned immunoglobulin gene probe hybridized only to DNA from that hybrid containing the piece of chromosome 14 corresponding to q32.

Fine structure maps of regions of individual chromosomes are obtained by analyzing clones from libraries representing the DNA of a single chromosome. The most direct way to obtain such a library is to use purified individual chromosomes (Section 6.5a). Another method is to use DNA from a somatic cell hybrid carrying a single donor chromosome. In such a library, recombinants containing donor DNA sequences can be selected by their ability to anneal with species-specific probes.

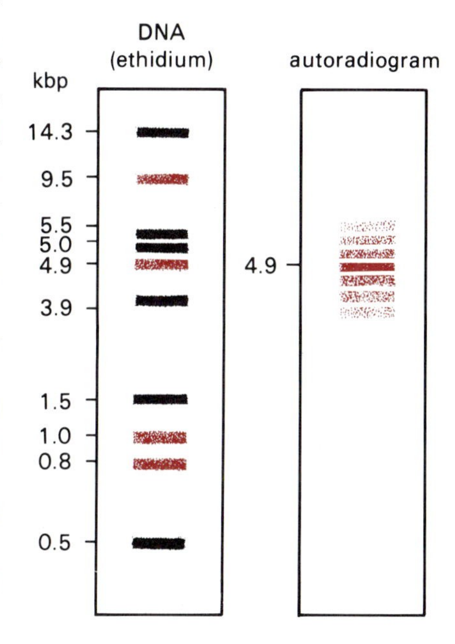

Figure 7.25
Instability during replication of a recombinant in a λ phage vector. A segment containing a tandemly repeated DNA sequence was cloned in a λ phage vector, and DNA was isolated after phage growth from a purified single plaque. The DNA was digested with restriction endonuclease, it was electrophoresed, and the gel was stained with ethidium bromide (left). A DNA blot was then prepared and annealed with a cloned probe containing the same repeated sequence. A 4.9 kbp insert fragment contained most of the repeated segment. However, additional DNA fragments both larger and smaller than 4.9 kilobase pairs annealed with the probe. These fragments occurred in submolar amounts and were not detected when the DNA in the gel was stained with ethidium. Insert fragments are colored; λ fragments are black.

c. *The Fidelity of Cloning*

Although most cloned inserts are faithful replicas of genomic sequences, sequence alterations sometimes occur during cloning. Whenever there is a discrepancy between the size of restriction endonuclease fragments in a cloned insert and those of corresponding genomic DNA segments, alteration by recombination, deletion, or insertion during cloning is likely to have occurred. Another frequent warning signal is instability in the size or restriction endonuclease map of an insert upon repeated isolation. When the cloned fragment contains tandem repetitions of a DNA sequence, the restriction endonuclease digests of the cloned recombinant DNA frequently show a complex pattern (Figure 7.25). Some fragments may, for example, be present in less than a mole equivalent (i.e., submolar). Such a pattern again indicates that rearrangements occurred during replication of the recombinant molecule, resulting in a mixed population of recombinants. The cause of this phenomenon is homologous recombination, which yields deletions and amplifications of the cloned tandem repeating units. Instability of cloned repeated sequences in *E. coli* host-vector systems is decreased but not eliminated in cells that are mutants in recombination functions (e.g., *rec*A$^-$, *rec*BC$^-$).

7.5 Determining the Number of Copies of a DNA Sequence in a Genome

The ovalbumin gene occurs only once in a haploid chicken genome. But many genes and other DNA segments occur multiple times (Chapter 9). Eukaryotic genomes are in fact typified by the reiteration of DNA sequences. Sometimes the several copies are precisely the same. Then annealing a DNA blot of restriction endonuclease–digested genomic DNA with a labeled homologous probe can give a single band that is evident even if 1 μgram or less of digested DNA is used. Usually, 10 μg or more of mammalian cell DNA are needed to see a segment present only once in the genome. If a band is evident with 1 μg or less, then that segment must be present many times. If the different copies of the sequence diverge in one or more base pairs, they may have varying restriction endonuclease sites. Then the autoradiogram will show multiple bands or even a large dark smear, depending on the copy number of the repeated sequence.

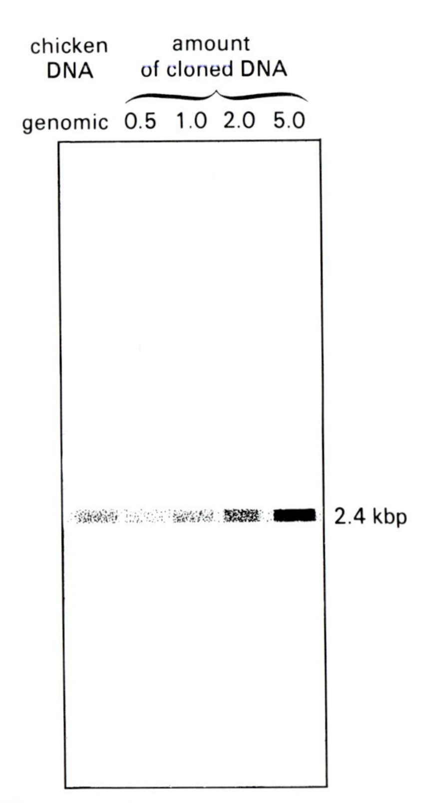

Figure 7.26
Measuring copy number. One lane contains an *Eco*RI digest of 30 μg of chicken DNA. The other lanes contain the cloned 2.35 kbp ovalbumin gene segment in amounts equivalent to 0.5, 1, 2, and 5 copies per haploid genome of the ovalbumin gene in a total of 30 μg of chicken DNA. Electrophoresis of the samples was carried out on an agarose gel. The figure is a schematic drawing of the autoradiogram obtained after annealing a DNA blot with the ^{32}P-labeled 2.35 kbp fragment. The intensity of the band in the chicken DNA corresponds to a haploid copy number of one.

a. *Estimating Copy Number Using DNA Blots*

The degree of darkening on an autoradiogram affords an estimate of the copy number. More precise values are obtained if known amounts of unlabeled cloned DNA are included in separate tracks on the same electrophoresis gel as the genomic DNA sample (Figure 7.26). The extent of darkening of the autoradiogram is a linear function of the amount of DNA on the gel so long as the probe is present in excess. Comparison between the degree of darkening of the genomic DNA and that of the standards gives the copy number.

b. *Estimating Copy Number from the Kinetics of DNA Annealing*

More precise information about copy number can be obtained by kinetic analyses. Kinetic methods antedate those described in Section 7.5a by more than a decade and provided the first indications that many repeated sequences are present in eukaryotic genomes. In order to understand how kinetic analysis is used, a few equations must be derived.

The rate of annealing of complementary single strands in solution depends on DNA concentration and follows second order kinetics. If C_o and C are, respectively, the total concentrations of denatured DNA at zero time and at time t after the start of annealing, then

$$dC/dt = -kC^2 \tag{1}$$

and integration and substitution give

$$C/C_o = \frac{1}{kC_ot + 1}. \tag{2}$$

When 50 percent of the denatured DNA has reannealed,

$$C/C_o = 0.5 \qquad (3)$$

and

$$C_ot_{0.5} = {}^1/k. \qquad (4)$$

C is expressed as moles of mononucleotide equivalents per liter and t as seconds. The product $\mathbf{C_ot}$ is usually used as a word that is spelled and pronounced "Cot."

The second order annealing of several pure and homogeneous DNAs is illustrated in Figure 7.27. As in many annealing experiments, the DNA was sheared into pieces about 400 bp in length before denaturation and annealing. Note that the Cot values on the abscissa are plotted on a log scale to facilitate data comparison. The fit of the curves to a simple second order reaction is clear from the fact that essentially 90 percent of the reaction occurs over two orders of magnitude of Cot value.

The $Cot_{0.5}$ values in Figure 7.27 vary with the total size of the genome because C is expressed as the total concentration of DNA, although the larger the total genome, the lower the actual concentration of any particular annealing fragment. In other words, in the larger genome, each annealing fragment is a smaller percentage of the total and therefore anneals at a higher Cot. Cot curves thus provide a way to estimate genome size so long as a genome of known size is available as a standard and the curves are a reasonable fit to second order kinetics. If N_a and N_b are the lengths of two genomes a and b, then

$$\frac{[Cot_{0.5}]_a}{[Cot_{0.5}]_b} = \frac{N_a}{N_b} \qquad (5)$$

The straightforward second order Cot curves for SV40, T4, and *E. coli* DNAs indicate that each genomic segment is represented only once in the total. However, similar analysis with eukaryotic genomic DNA yields more complicated Cot curves, as illustrated by the data for *Drosophila* DNA (Figure 7.28). Annealing takes place over a wide range of Cot values because many DNA segments are repeated, some many times, others less frequently, and others only a very few times. Those segments that occur

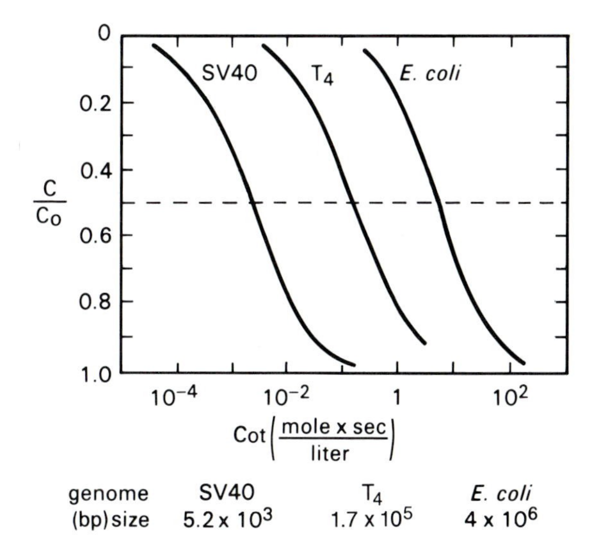

Figure 7.27
Schematic plot showing the reannealing of denatured SV40, T4 phage, and *E. coli* DNAs. See the text (Section 7.5b) for a description of the symbols. Adapted from R. J. Britten and D. E. Kohne, *Science* 161 (1968) p. 529.

only once in the genome are present in lowest concentration and anneal at the highest Cot values. The most highly repeated sequences, being present at the highest concentration, anneal most rapidly and have the lowest Cot values. Assuming ideal second order kinetics for all single copy DNA segments, a least squares analysis of such composite Cot curves allows resolution of a separate curve for single copy sequences and an estimation of the corresponding $Cot_{0.5}$.

The $Cot_{0.5}$ for single copy *Drosophila* sequences is 28.6 mole·sec/liter under the conditions used in Figure 7.28. This value provides a standard against which to compare the $Cot_{0.5}$ value for a cloned *Drosophila* fragment and thereby a means to calculate its copy number:

$$\frac{28.6}{Cot_{0.5}\ \text{clone}} = \text{copy number} \qquad (6)$$

The second curve in Figure 7.28 shows the kinetics of annealing of a cloned fragment in the presence of total genomic *Drosophila* DNA. The cloned fragment is distinctively labeled with a radioisotope so that its annealing is separately measurable. Note that the concentration of the fragment is kept low so that none of it will anneal to itself. The effective concentration of the cloned sequence is provided by the corresponding sequences in the genomic DNA. For this reason, the C_0t on the abscissa used to obtain Cot for the fragment is calculated from the concentration of genomic *Drosophila* DNA, but the C and C_0 on the ordinate refer to the fraction of the cloned DNA that annealed. This is why the relevant C_0t values (abscissa) are indeed those for genomic DNA. The $Cot_{0.5}$ for the cloned probe (Figure 7.28) is 1.06, and the copy number is $28.6/1.06 = 27$.

As a practical matter, obtaining copy numbers of the order of one to ten by this method is difficult because very long times or very high initial DNA concentrations are required to reach the high Cot values. The problem is exacerbated when the genome being studied is very large because the $Cot_{0.5}$ for single copy segments increases. The $Cot_{0.5}$ for single copy sequences in the *Drosophila* genome of 1.7×10^8 bp is 28.6, but the corresponding $Cot_{0.5}$ in mammalian genomes (approximately 3×10^9 bp) is at least tenfold higher.

Not surprisingly, the analysis of annealing kinetics is somewhat more complicated than outlined here. Annealing rates depend not only on DNA concentration but also on the temperature (which must be below the melting temperature of the duplex [Section 1.1f]), the salt concentration, the solution viscosity, and the annealing fragments' size. All these parameters must be carefully held constant or controlled.

Several options are available for measuring the percentage of DNA that becomes double stranded. Advantage can be taken of the fact that duplex DNA but not single strand DNA binds to hydroxyapatite under certain conditions. In another procedure, the percentage of DNA that becomes resistant to digestion by single-strand-specific nuclease (S1) is measured at each time point. The measurement of reassociation kinetics by these two methods yields different values with complex eukaryotic genomes (Figure 7.29). The S1 nuclease method determines only the amount of duplex DNA formed in the reassociation reaction; the hydroxyapatite procedure measures the duplex DNA as well as any unpaired single strand tails

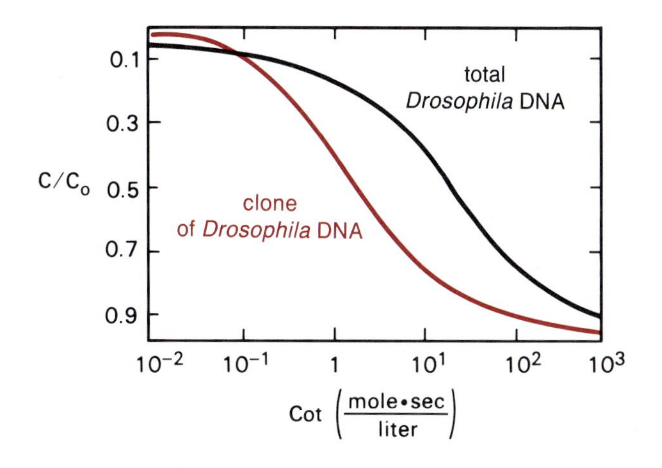

Figure 7.28
The reannealing of *D. melanogaster* total genomic DNA and of a cloned *Drosophila* DNA fragment in the presence of excess *D. melanogaster* DNA. The cloned fragment was present at 4.5×10^{-9} M. See the text (Section 7.5b) for a description of the experiment. Adapted from J. Lis, L. Prestidge, and D. S. Hogness, *Cell* 14 (1978), p. 901.

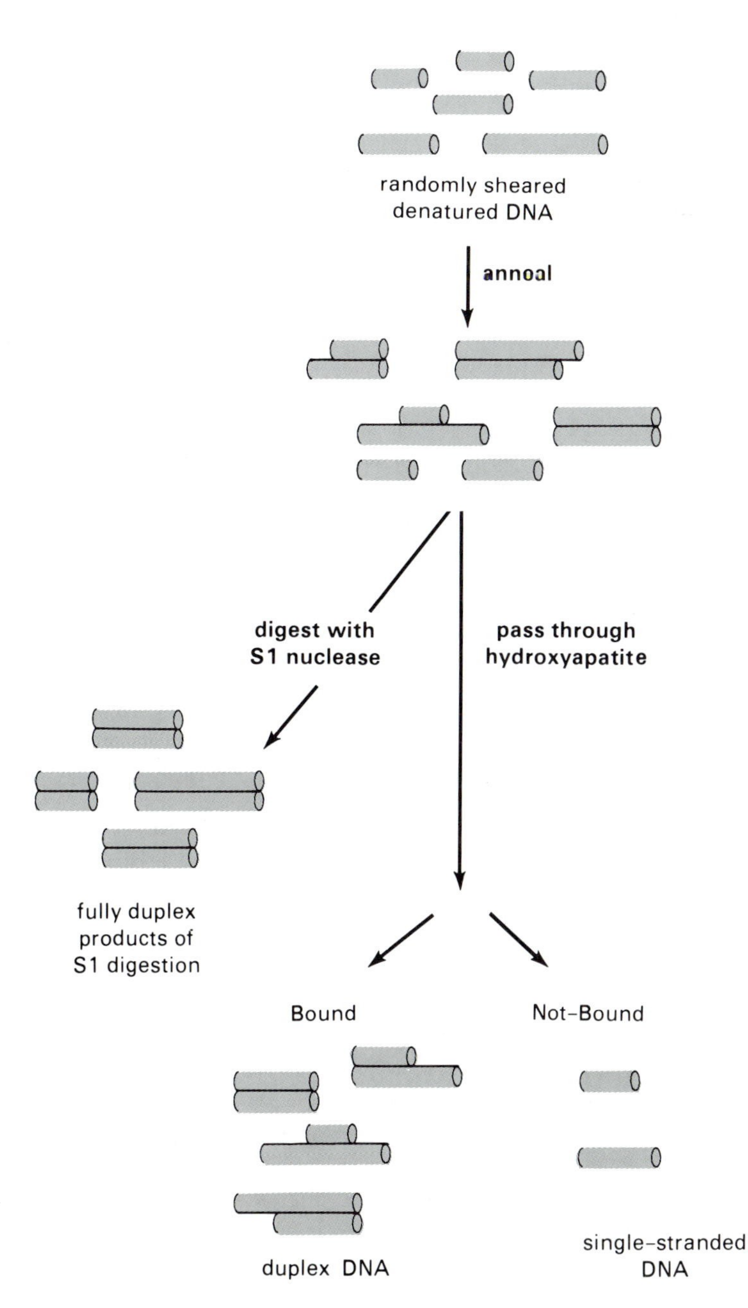

Figure 7.29
The kinetics of reannealing of denatured DNA can be measured by the S1-nuclease technique or by binding duplex strands to hydroxyapatite. At various times, samples are removed, and the portion of the total DNA that became double stranded is determined. The diagram compares the two methods as applied to a sample obtained at one time after the start of reannealing.

linked to the base paired segments. Duplexes with single strand tails form partly because the chains are broken at random into convenient sizes prior to reannealing and partly because of the presence of repeated sequences. S1 nuclease digests all the strands that remain entirely single stranded as well as any single strand tails on otherwise duplex strands. Hydroxyapatite binds all duplex strands, including those with single strand tails, thus giving a larger estimate of the percentage that has annealed than does the S1 procedure. Consequently, the hydroxyapatite assay suggests that the DNA reassociates to a greater extent and at an apparently greater rate than does the S1 nuclease measurement. This difference is minimized when relatively short DNA segments (e.g., 400 bp) are used.

c. Estimating Copy Number by Annealing to Saturation

Another way to obtain copy number is to determine the total amount of a cloned DNA segment that will anneal to a known amount of genomic DNA. Note that in this procedure the cloned segment is present in excess; for the kinetic analysis described in Section 7.5b, the genomic DNA is in excess and drives the annealing. Saturation hybridization is especially useful when the segment in question has a very low copy number, and kinetic analysis is difficult.

In practice, increasing amounts of radioactive cloned segment are incubated with a fixed amount of denatured and fragmented genomic DNA, and the amount of the cloned segment that anneals is measured (Figure 7.30). The reactions are carried out for a long enough time to ensure completion, and the labeled fragment is preferably a single strand to avoid self-annealing at the relatively high concentrations required to obtain saturation. When a plateau is reached, the entire complementary sequence in the genome is saturated with the probe, and the plateau value represents the total amount of homologous sequence in the genome. The copy number can be calculated from the specific radioactivity of the labeled fragment. To improve accuracy, the system is calibrated by separate control reactions in which known amounts of unlabeled cloned fragment are added to genomic DNA samples. In the experiment shown in Figure 7.30, the amount of radioactive fragment that annealed at saturation is equivalent to two copies of the gene per haploid genome. A simple way to carry out the experiment is to fix the denatured genomic DNA samples as dots on nitrocellulose. In a further refinement, the saturating concentration of radioactive probe is determined first and then incubated with a nitrocellulose strip containing several dots, each with a different amount of genomic DNA.

7.6 Altering Cloned Segments: Constructing Mutants

a. General Considerations

Classical genetic analysis depends on the chance occurrence of mutations that are observed by heritable phenotypic variation. Molecular genetics has allowed the development of a **reverse genetics**. Thus, characterized

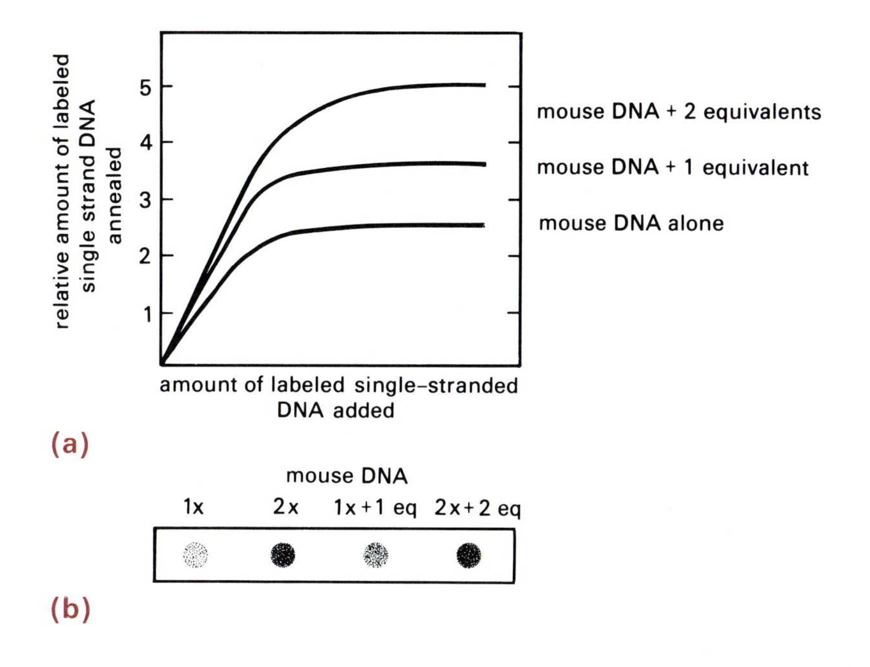

Figure 7.30
Estimating copy number by annealing to saturation. (a) Annealing increasing amounts of a ^{3}H-labeled single strand of a cloned DNA segment to a fixed amount of fragmented, denatured genomic DNA (in solution). In the two upper curves, the equivalent of one or two copies, respectively, of the cloned segment (not radioactive) was added to the reactions to provide standards for comparison. (b) The measurement can also be carried out by fixing dots of the genomic DNA on nitrocellulose (Figure 7.39). After incubating the nitrocellulose with a solution of the ^{32}P-labeled DNA, the matrix is washed and either autoradiographed (as shown here) or cut into individual dots and counted.

cloned genes can be deliberately and precisely mutated to study the effect of specific sequence alterations on phenotype or to locate regulatory elements. For example, if a change is made in the coding region of a gene, a variant polypeptide may be produced, and the effect of the new amino acid sequence on protein conformation or enzyme activity can be studied. Similarly, altering the base pairs in a suspected regulatory region may eliminate or stimulate transcription, thereby confirming the identification.

There are various means to achieve mutagenesis of cloned fragments or small genomes, although only a few are illustrated here. The success of all depends on two considerations. First, the structure of the starting DNA must be well known. A detailed restriction site map is required at a minimum, and knowing the entire sequence of the fragment is preferable. Second, because all the methods yield a mixture of products, the desired construction must be purified by cloning, amplified by replication, and characterized.

Some of the mutagenesis procedures yield randomly placed mutations. Others, however, are directed to particular sites within the DNA; these are referred to as **site-specific** (or **site-directed**) mutations.

b. *Deletion Mutants*

Using restriction endonucleases A neat deletion is possible if the cloned DNA fragment has two neighboring and unique restriction endonuclease sites within the region of interest (Figure 7.31a). After cleavage, flush ends are produced with a single-strand-specific nuclease, and the molecule is religated. This simple approach is, however, only rarely feasible. Alternative procedures permit the formation of deletions around a single unique restriction endonuclease site (Figure 7.31b). After endonuclease cleavage, sequences around the cleavage site are deleted by treatment with either *Bal* 31 or an exonuclease such as *E. coli* exoIII followed by

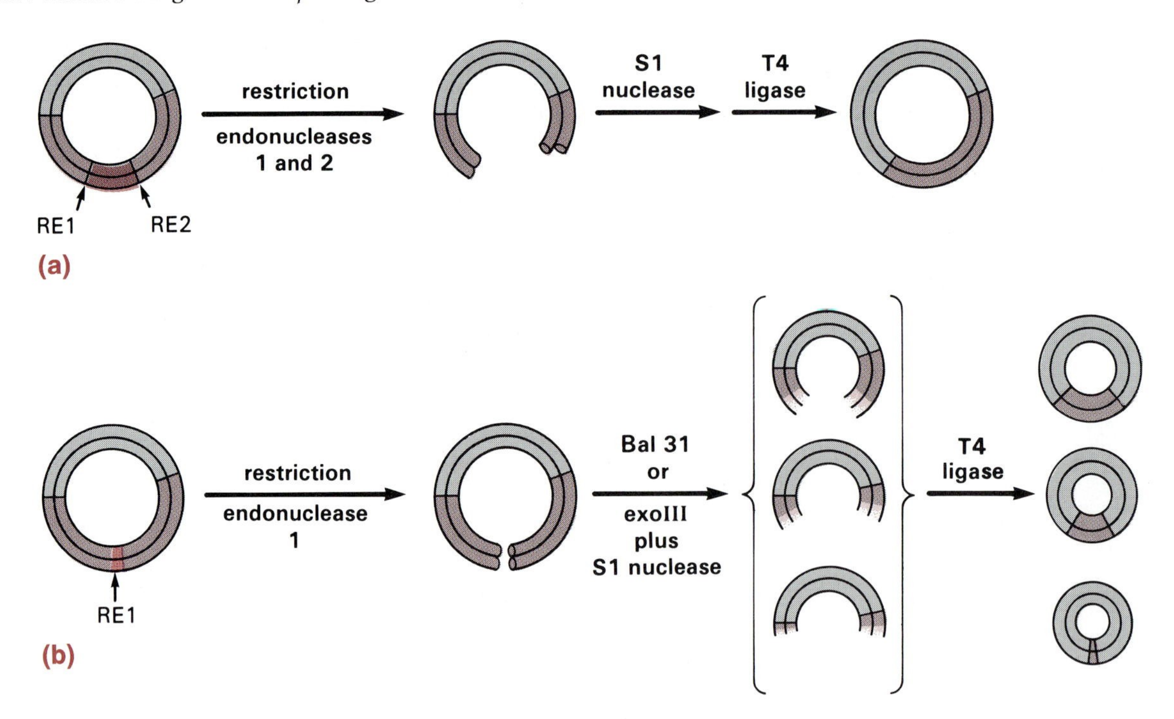

Figure 7.31
Constructing deletion mutants with restriction endonucleases. The grey region is the plasmid vector; the colored is the insert. (a) Using two neighboring and unique restriction endonuclease sites, RE1 and RE2. (b) Using a single unique restriction endonuclease site.

S1 nuclease. Religation yields a family of deletions centered around the original incision, and the different deletion mutants are separated and purified by molecular cloning. In one variation of this method, restriction endonuclease linkers are joined to the ends before recircularization; this has the advantage of providing a useful cleavage site for characterization, sequencing, or subsequent remodeling.

By somewhat more complex manipulations, a series of varying size deletions that go in one direction from a common start site can be constructed (Figure 7.32). The initial recombinant molecule is treated in two ways. From one portion (A), a large segment including the target for deletion is removed by cleavage with two single site restriction endonucleases (RE1 and RE2). A second portion (B) is linearized by treatment with one of the two enzymes (RE1) and then digested with *Bal* 31. Thereafter, cleavage with the second enzyme (RE2) produces a set of fragments of decreasing length (B′, B″, B‴ in Figure 7.32). Ligation of these to A followed by cloning provides a set of deletions all of which start at RE1.

Using nonspecific endonucleases Single cleavages can also be introduced into supercoiled circles by nonspecific endonucleases such as DNase I. A whole set of linear molecules results, each broken at a different site. Once cleavage is achieved, deletions are extended and religated by the methods previously described for restriction endonuclease incision. Several interesting properties of DNase I are involved in the cleavage reaction. In the presence of Mn^{2+}, the enzyme makes a double strand cleavage in duplex DNA, although in Mg^{2+}, it hydrolyzes only one strand at a time, creating single strand nicks. Moreover, the enzyme tends to cleave supercoiled DNA faster than linear duplex DNA, presumably because of perturbations in the B form of DNA in the supercoiled molecules. As a result, after

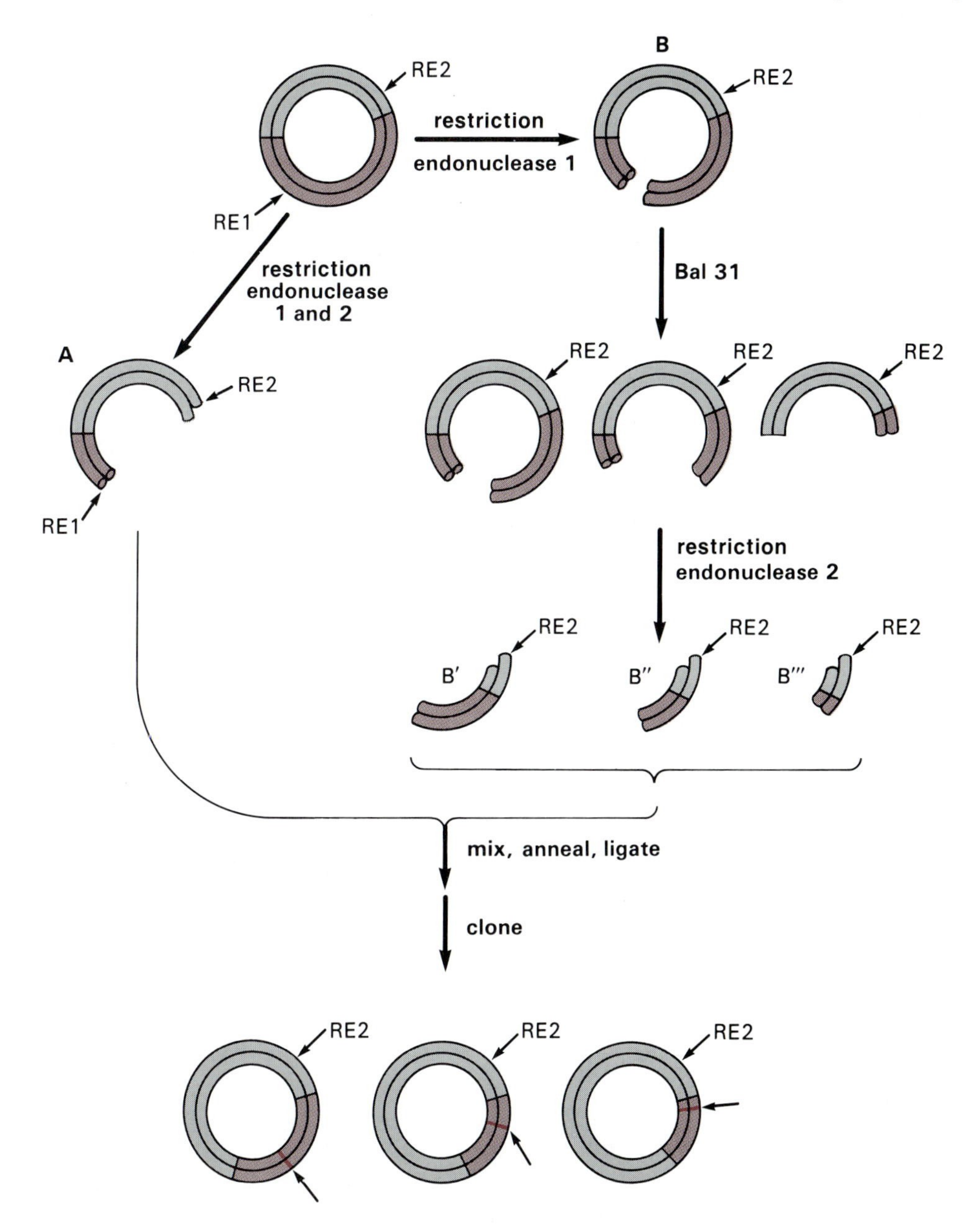

Figure 7.32
Constructing a series of varying size deletions that go in one direction from a common site in the insert (color). The cleavage site for restriction endonuclease 1 (RE1) is not regenerated during the remodeling; its position is shown by the arrows on the products.

brief digestion of supercoiled DNA with DNase I in the presence of Mn^{2+}, the products include a substantial proportion of full length linear duplexes. Treatment with *Bal* 31 or an exonuclease followed by S1 nuclease extends the gap. After religation, the individual products are separable by molecular cloning.

c. *Insertion Mutants*

The ideas behind the construction of insertion mutants are similar to those described for deletion mutants. A single cleavage is made within a cloned DNA segment with either a restriction endonuclease or a nonspecific endonuclease. If necessary, the ends of the now linear molecule are prepared for ligation by filling in or single strand nuclease treatment, and the molecule is then religated in the presence of the segment being inserted. In a simple example, the insertion might be a synthetic fragment containing

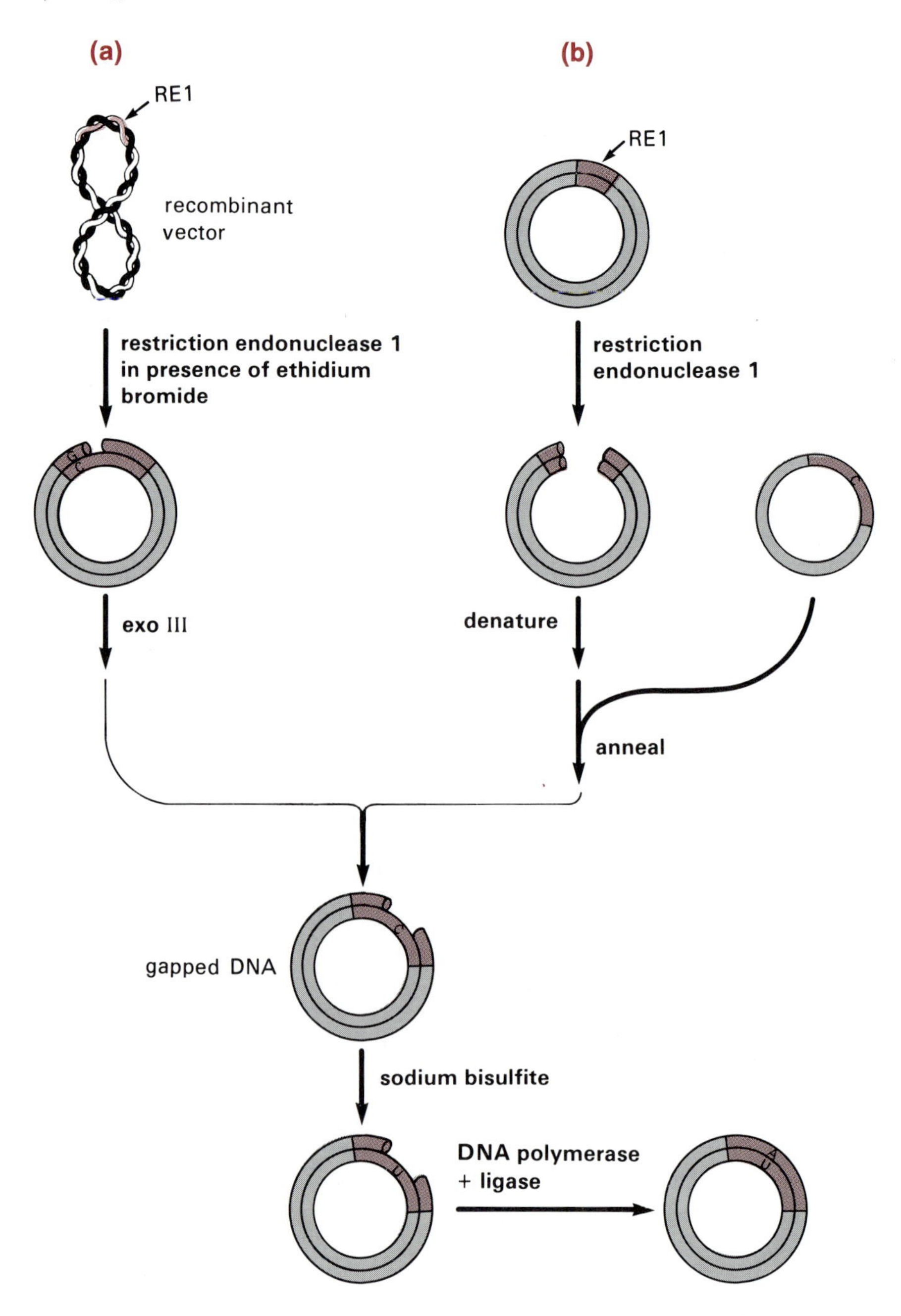

Figure 7.33
Site-specific introduction of a point mutation by the action of sodium bisulfite on a single strand gap in a duplex DNA. The single strand gap in the insert is constructed by (a) restriction endonuclease cleavage in the presence of ethidium bromide followed by exonuclease III digestion or (b) heteroduplex formation between an intact and a deleted insert, each cloned in M13.

multiple restriction endonuclease sites—a polylinker. When the original cleavage is nonspecific (e.g., with DNase I in the presence of Mn^{2+}), the new restriction endonuclease sites in the set of cloned mutants provide a handy way to construct a physical map of the resulting mutations.

d. Point Mutations

Chemical mutagenesis A widely used procedure for generating point mutations in a limited region starts with a duplex circular DNA containing a short single strand region. One way to generate such site-specific gaps is to treat supercoiled DNA with an appropriate restriction endonuclease in the presence of ethidium bromide, which distorts the duplex by intercalation between base pairs (Figure 7.33a). Under these conditions, many restriction endonucleases make only a single strand nick at their recognition sites. Apparently, the intercalation of ethidium into fully duplex DNA (not supercoiled) inhibits cleavage, but in the supercoiled DNA, the single cleavage is allowed. Although not all restriction endonucleases behave in this way, enough do to make this a useful procedure. Once the nick is in place, an exonuclease is used to enlarge it into a small single strand gap in the duplex. Another way to prepare a gapped duplex takes advantage of the properties of the M13 vector system (Figure 7.33b). A single strand M13 recombinant containing the segment of interest is prepared. Another recombinant containing the same segment but with a deletion in the chosen region is prepared in the M13 double strand form, cleaved into a linear molecule, denatured, and annealed with the single strand. The resulting heteroduplex contains a single strand gap.

When DNA containing a single strand gap is treated with sodium bisulfite, the reagent deaminates cytosine residues to form uracil, but only in the single strand region. Deamination is therefore restricted to cytosines within the gap in the duplex. If the gap is relatively small, and the number of deaminated cytosines is limited by a short reaction time and a low concentration of bisulfite, very specific and limited mutations occur. Finally, the gap is filled in with DNA polymerase, and the molecule is religated. In place of the original C·G base pair, the new DNA molecule has a U·A base pair, which is converted to a T·A pair after replication in a cell.

Mutagenic copying Other kinds of localized mutations can be introduced by omitting the deamination but using a mutagenic analogue of a normal deoxyribonucleoside triphosphate during the refilling of a single strand gap. DNA polymerase I utilizes various analogues of the normal deoxyribonucleoside triphosphates as substrates. Some of these are mutagenic. For example, N^6-hydroxydeoxycytidine-5′-triphosphate (HO-dCTP) can be incorporated opposite either an A or a G on the template, depending on whether it is in the imino or amino form, respectively (Figure 7.34). If a gap in duplex DNA is filled in using HO-dCTP instead of dTTP in the reaction, then HO-dC will be inserted opposite A on the template (Figure 7.35). After transfection and intracellular replication, some of the original form will be produced along with some mutated genomes with transitions from T·A to C·G base pairs. Appropriate selection techniques allow cloning of the mutants. Similarly, substitution of dCTP by HO-dCTP yields transitions from C·G to T·A base pairs.

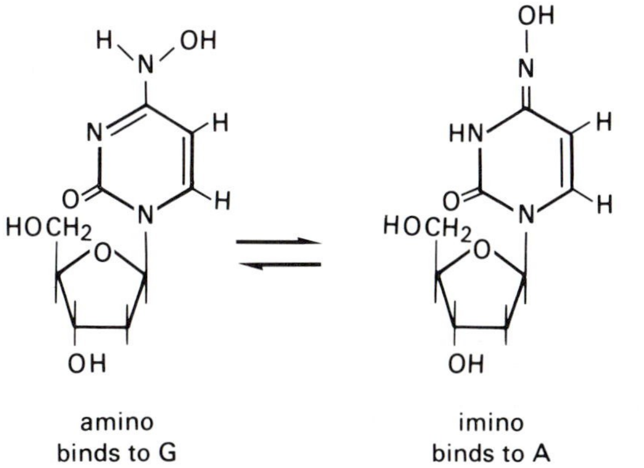

H N OH
N H
O N H
$HOCH_2$ O
OH
amino
binds to G

OH
N
HN H
O N H
$HOCH_2$ O
OH
imino
binds to A

Figure 7.34
Tautomeric forms of *N^6*-hydroxycytidine.

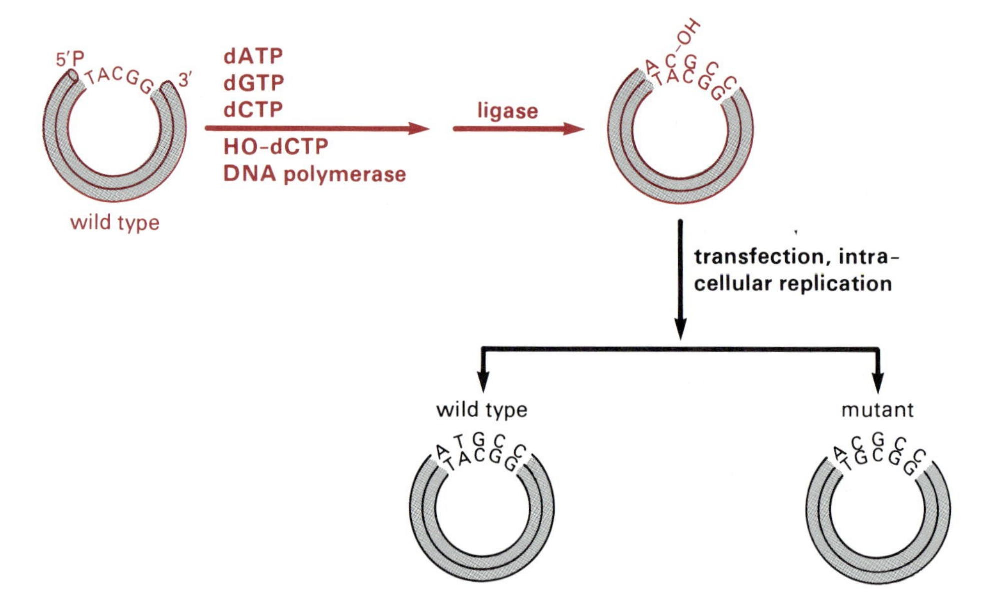

Figure 7.35
Mutagenesis by filling in gaps with DNA polymerase and HO-dCTP. The result of omitting dTTP from the reaction is a transition from a T·A to a C·G base pair.

Site-specific mutagenesis using synthetic oligodeoxynucleotides The fact that oligodeoxynucleotides can be synthesized in good yield makes possible general and precise methods for constructing site-specific point mutations in cloned DNA segments. One such method depends on the formation of a heteroduplex between a single strand synthetic oligodeoxyribonucleotide containing the mutant sequence and the complementary single strand of a recombinant DNA vector containing the wild type segment (Figure 7.36). For example, a gene to be mutated is cloned in M13, and single strand circular recombinant viral DNA is obtained. A synthetic oligodeoxyribonucleotide as short as 8 to 20 nucleotides and containing the mutant sequence is annealed to the circle. This oligodeoxyribonucleotide serves as primer and the remaining single strand as template for *in vitro* DNA synthesis by DNA polymerase I. After the full circle has been copied, the new strand is ligated to form a duplex M13 vector containing a mismatch at the mutated sequence. Note that the progeny of a single infected cell segregates as a mixed population of wild type and mutant recombinant phages that can be resolved in subsequent cloning.

In a second approach, an oligodeoxyribonucleotide in a cloned wild type DNA insert is replaced by a synthetic duplex oligodeoxyribonucleotide containing the mutant sequence—cassette mutagenesis (Figure 7.37). Here, the cloned insert is present in a duplex vector such as pBR322. In the simplest case, unique restriction endonuclease sites that occur in the insert but not in the vector are used to excise the region to be mutated; when such sites are unavailable, additional manipulations are necessary. The replacement oligonucleotide is prepared by annealing two separate synthetic and complementary single strands, each of which contains the appropriate base change. In addition, the synthetic chains are constructed so that the duplex they form will have appropriate cohesive ends. Unlike the heteroduplex method, no wild type genomes are produced upon transfection and cloning of the altered recombinant. However, if a mixture of synthetic oligodeoxyribonucleotides that have alternative bases at the mutated position is used, then the mixture of mutants segregates after transfection and can be separated by cloning. This is a useful

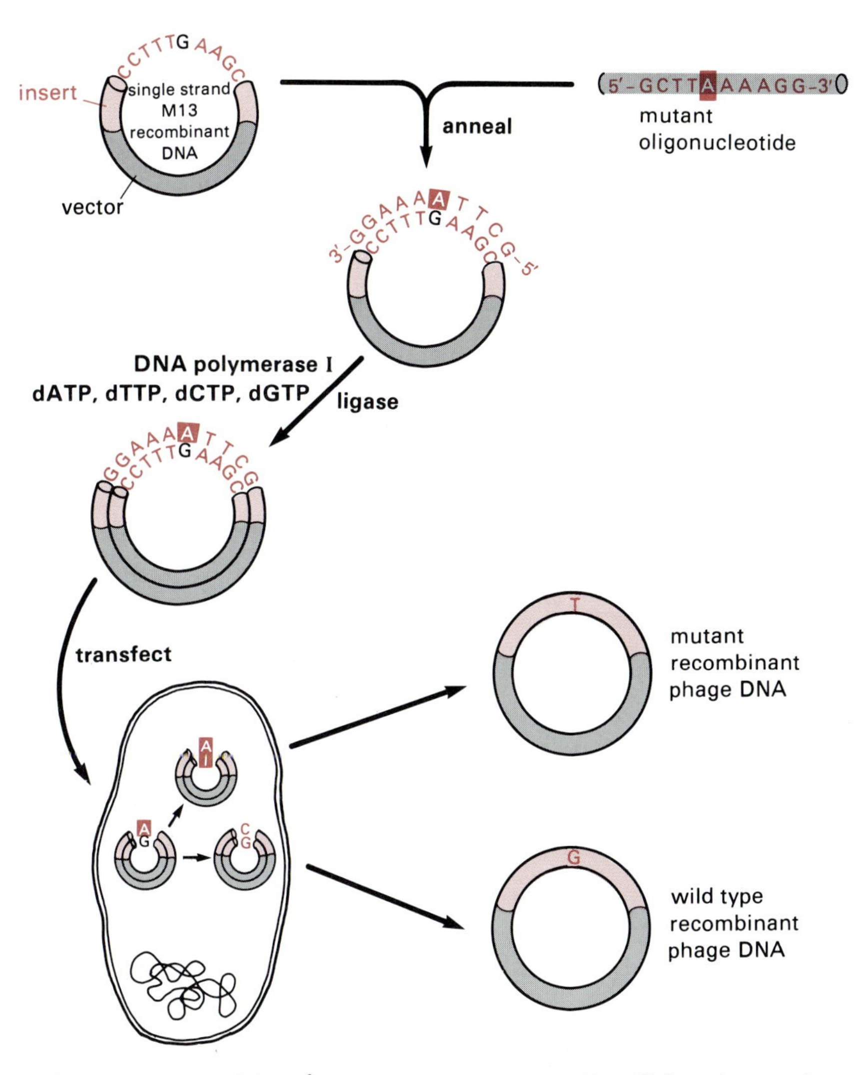

Figure 7.36
Site-specific mutagenesis with synthetic oligonucleotides: the heteroduplex procedure. A mutated synthetic oligonucleotide serves as primer for the synthesis of a mutant insert in a recombinant M13 vector.

way to prepare a variety of mutations at one time. Recall that the synthesis of such mixed oligodeoxyribonucleotides is straightforward; a mixture of mononucleotides rather than a single one is introduced at the desired step in the chemical synthesis.

7.7 Analyzing the Function of Cloned DNA Segments

A frequent reason for cloning genomic DNA segments or cDNAs is to elucidate the function of their intracellular counterparts. Structural analysis of cloned sequences can suggest, by the presence of open reading frames, that they encode genes. The presence of specific promoter sequences or other regulatory elements can often be recognized. But to confirm the deductions from structural analysis, direct functional studies must be performed. Among the questions of interest about a cloned sequence are the following: Is the sequence transcribed in one or more cell types? Are the transcripts mRNAs? Is transcription subject to modulation by changes in the cellular environment? Does the cloned segment include promoters, terminators, or other regulatory signals? If so, how do they work? What is the relation between the structure of the cloned segment and that of

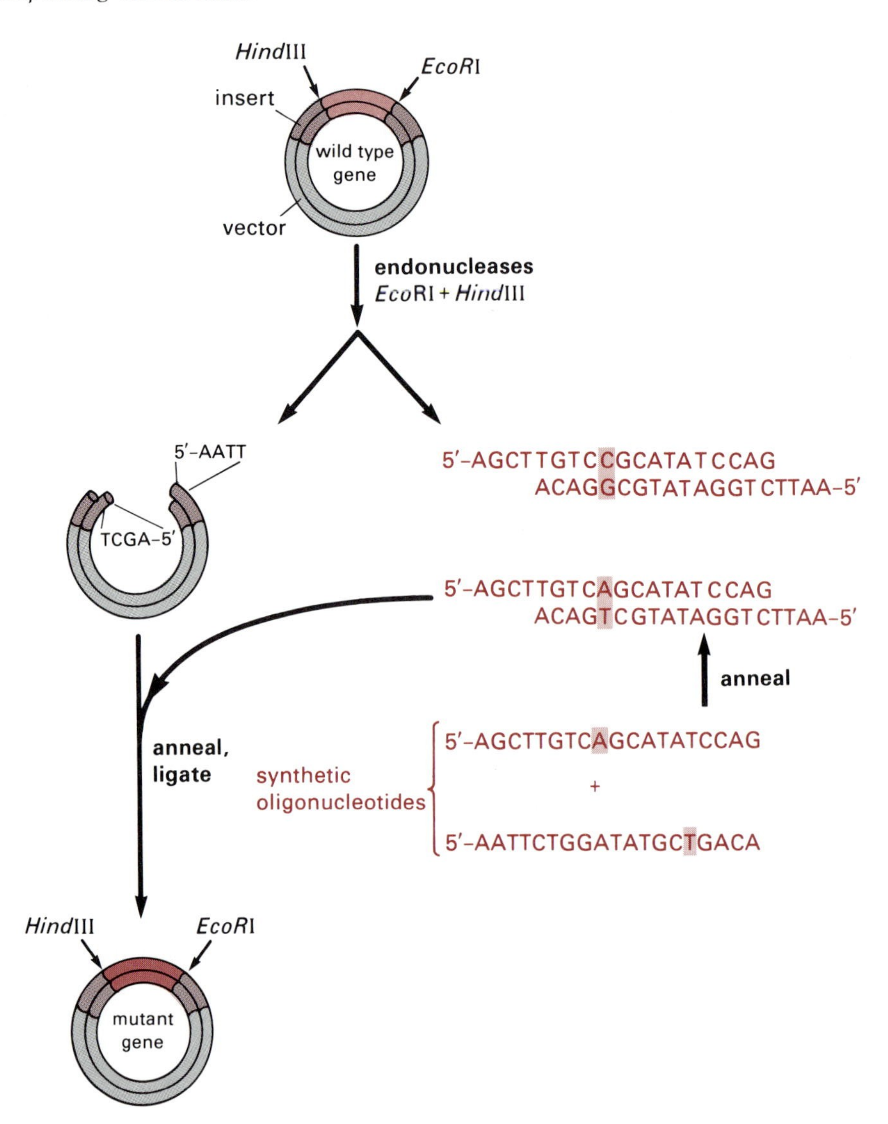

Figure 7.37
Cassette mutagenesis. Two unique restriction endonuclease sites that bracket the region to be mutated are used to excise that region from a cloned recombinant vector. A synthetic duplex oligonucleotide, which contains an altered base pair and cohesive ends that match those on the linear vector, is then inserted by ligation. The two strands of the synthetic oligonucleotide are synthesized separately and annealed together.

the intracellular transcripts? Can the transcripts be translated into polypeptide? A variety of experimental methods is used to study these questions, depending on the particular system being analyzed. Many of them will be described in subsequent chapters. However, certain techniques, fundamental to most of the methods and widely used, are described in the following sections.

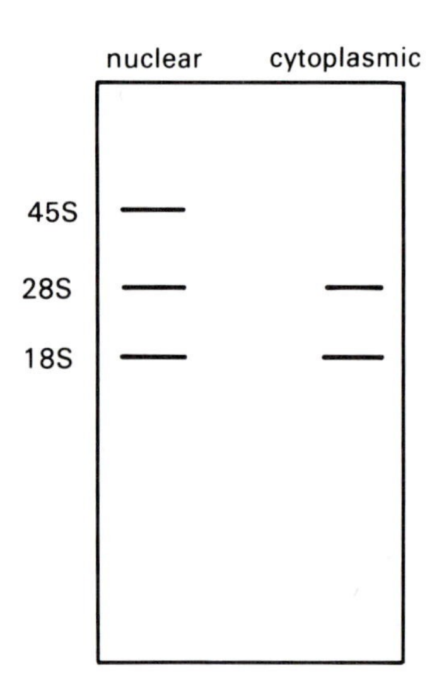

Figure 7.38
Schematic diagram of an RNA blot showing two lanes of electrophoresis on a gel. One lane contained nuclear and the other cytoplasmic RNA. After electrophoresis, the gel was blotted and annealed with ^{32}P-labeled ribosomal RNA and placed in contact with X-ray film. The three bands on the autoradiogram of nuclear RNA correspond to 18S rRNA, 28S rRNA, and the 45S precursor to both (Chapter 8). The 45S RNA is missing in cytoplasmic RNA.

a. Characterizing Intracellular Transcripts Corresponding to Cloned DNA Segments

Among the most important questions about any cloned genomic segment are those related to its transcription *in vivo*. Also, knowing the structural relation between a cloned cDNA and related intracellular transcripts is critical to understanding the significance of the cloned sequence. Three methods form the basis for all such analyses: RNA blotting, analysis with single strand-specific nucleases, and copying of the RNA isolated from cells with reverse transcriptase. All involve prior isolation and purification of RNA from whole cells or specific intracellular compartments such as nuclei or cytoplasm. All depend critically on RNA's ability to form duplex hybrids with complementary cloned DNA.

RNA blotting RNA blotting (Northern blotting) is analogous to DNA blotting (Section 6.1b). Briefly, isolated RNA is separated according to size by electrophoresis through agarose gels (Figure 7.38). Generally, the electrophoresis is performed under conditions that denature the RNA so that the effects of RNA secondary structure on the electrophoretic mobility of the RNAs can be minimized. Alkaline conditions are unsuitable for this purpose because the intranucleotide phosphodiester bonds in RNA are alkali labile; therefore, agents such as glyoxal, formaldehyde, or urea are used. The RNA is then transferred by blotting to an immobilizing matrix without disturbing the RNA distribution along the gel. A labeled DNA (e.g., the clone being characterized) is then used as a probe to find the position on the blot of RNA molecules corresponding to the probe. The blot is incubated with the DNA under conditions that foster annealing. After washing away the excess DNA, the position of the probe is detected by exposing an X-ray film to the blot. Dark bands indicate the position of the homologous RNA on the original electrophoretic gel. In this way, transcripts that derive from a cloned DNA segment are detected. The size of the transcripts can also be estimated if a set of RNA or single strand DNA molecules of known size is separated in a parallel lane on the gel slab. Mobility is approximately inversely proportional to the size of the RNA molecules. In addition, RNA blotting permits estimates of the amount of RNA synthesized in the cells from which it was derived. The method is similar to that used to measure genomic copy number on DNA blots (Figure 7.26). The intensity of the band on the X-ray film is proportional to the amount of homologous RNA present. Similarly, the procedures described in Figure 7.30 and Section 7.5c can also be used to quantitate the amount of a specific RNA in a mixture (Figure 7.39). As in the analogous methods for DNA, it is preferable that the labeled probe be a single strand to avoid having it reanneal with its complementary DNA strand rather than the RNA.

With these simple techniques, a substantial amount can be learned about the functional properties of a cloned DNA segment. Aside from knowing whether or not it is transcribed, variations in the amount of transcript as a function of cell type or cellular environment can be estimated. By analyzing RNA from purified cellular compartments, the location of a transcript in the nucleus, the cytoplasm, or polysomes can be ascertained. Often it is important to know if the homologous RNA is polyadenylated because polyadenylation is a hallmark of most eukaryotic mRNAs. Fractionation of RNA into polyadenylated (polyA$^+$) and nonpolyadenylated (polyA$^-$) populations is readily accomplished. Polyadenylated RNA anneals, under appropriate conditions, to polydT or polyrU. The polymers are themselves usually fixed to inert solid supports, simplifying recovery of the polyA$^-$ RNA that does not bind. Thereafter, the polyA$^+$ fraction can be eluted by using denaturing conditions (Figure 7.40). The RNA in each fraction is then analyzed by electrophoresis and blotting, and the cloned DNA's ability to anneal to each is tested. If the RNA detected by the cloned DNA is cytoplasmic and polyadenylated, it is likely to be an mRNA. If, in addition, it is bound to polysomes, the identification becomes more certain.

The size of an mRNA that anneals to a cloned DNA segment is another revealing fact. Sometimes the RNA is longer than the cloned segment, which shows that only a portion of the gene can be present in the

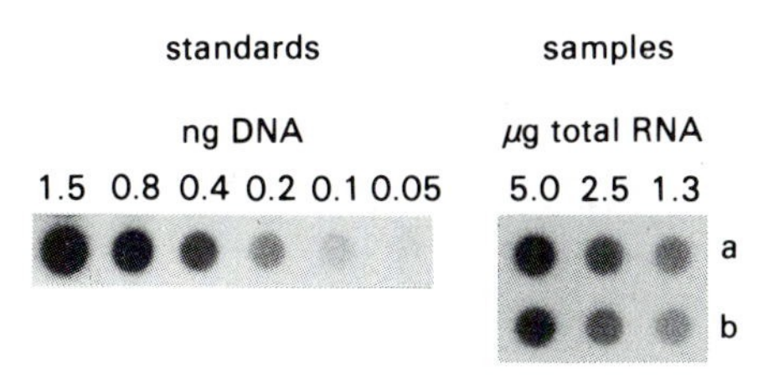

Figure 7.39
Estimating the amount of a specific RNA in a mixture of RNAs by "dot blotting." Three quantities (5.0, 2.5 and 1.3 μg) of two RNA samples (*a* and *b*) were fixed to nitrocellulose and then annealed with an excess of a ^{32}P-labeled single strand DNA probe (cloned in an M13 vector). The amount of the probe that anneals to each dot was detected by autoradiography with X-ray film. The quantity of RNA in the sample that is complementary to the probe is estimated by comparing the intensity of the dots with that obtained with known amounts of the cloned DNA segment. The standards were prepared from duplex, denatured DNA. Courtesy of J. Skowronski.

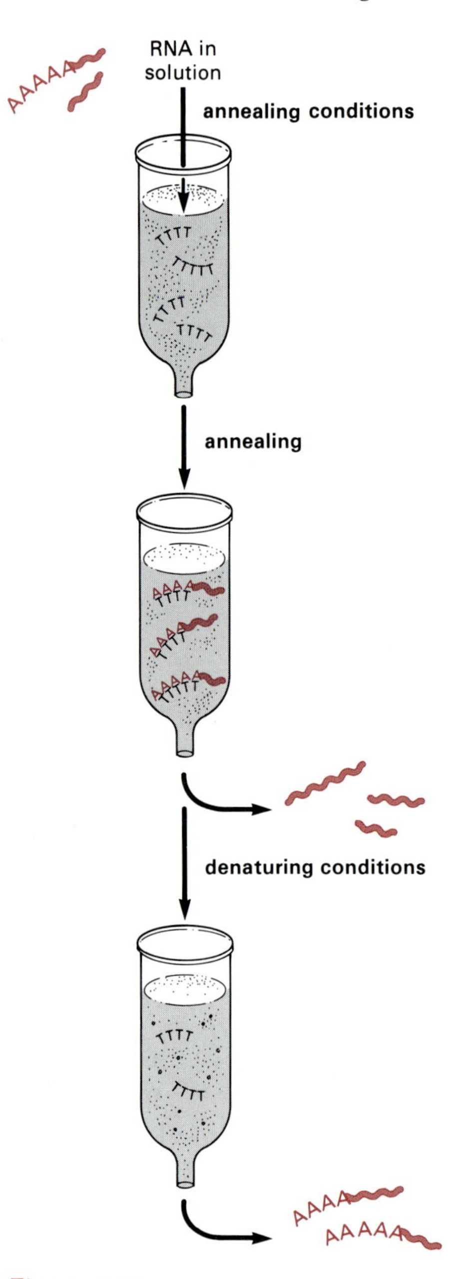

Figure 7.40
Fractionating polyA$^+$ and polyA$^-$ RNA. A sample of RNA is passed through a column containing oligodT bound to cellulose under conditions that favor annealing of the oligoT with the polyA tails found on RNA. The RNA with no polyA tails passes right through. Then the column is eluted under denaturing conditions, and the polyA$^+$ RNA is collected in the eluate. Similar fractionations are obtained with polyU bound to a solid matrix.

clone. In other instances, the RNA is shorter than the cloned segment, indicating that the cloned sequence must contain sequences other than those in the transcript. They may, for example, be genomic sequences that flank the transcribed region or noncoding sequences that interrupt the coding region and were spliced out during mRNA maturation (Section 7.1d and Chapter 8). Such relationships between a cloned DNA segment and the homologous cellular RNA can be more precisely defined by using single strand nucleases or by copying with reverse transcription.

Analyzing the relation between a cloned DNA segment and an intracellular RNA using DNA-RNA hybrids and single strand-specific nucleases This type of analysis has three steps (Figure 7.41). First, the cloned DNA fragment is labeled with a radioactive isotope (e.g., by nick translation, as described in Figure 4.18) and denatured. Alternatively, and preferably, a single strand of DNA can be used. Then the DNA is annealed to isolated cellular RNA in solution under conditions that favor formation of RNA-DNA hybrids. Any DNA sequences present in the cloned fragment that are not represented in the RNA remain single stranded. They are removed by digesting the hybrids with a nuclease specific for single strand DNA. Finally, the sample is denatured and analyzed by gel electrophoresis. The size of the resulting radioactive DNA is determined by exposing the gel to an X-ray film and measuring the position of the dark band relative to the position of DNA molecules of known size.

There are many versions of this basic experiment, and each reveals different features of the RNA structure. They depend on what type of nuclease is used, whether the RNA is total RNA or is, for example, a purified cytoplasmic polyA$^+$ mRNA. One variation is illustrated in Figure 7.41, where a single contiguous region on the DNA is complementary to the RNA, and the only unmatched regions are at the ends of the DNA. However, the method can detect any region of discontinuity between the RNA and DNA, including intervening sequences such as those illustrated in Figure 7.3. Other variations are described in the following chapters. Many of these experiments use the single strand-specific endonuclease called S1 (Section 4.1b); therefore, the procedure is often called **S1 mapping**.

One of the frequent uses of S1 mapping experiments is to align the beginning of an RNA molecule with a specific position on a cloned DNA segment. In many cases, this site corresponds to the site at which transcription was initiated. The precision of the experiment is greatly enhanced by using a small DNA probe so that the length of the protected DNA segment can be measured to a single nucleotide. For this reason, carefully chosen regions of a long DNA fragment are usually subcloned and used as probes. Analysis of the size of the protected DNA fragment is determined on the same kind of electrophoresis gel used to determine primary nucleotide sequence.

Primer extension: analyzing the relation between intracellular RNA and a cloned DNA fragment with reverse transcriptase This analysis requires a relatively short DNA probe that pairs with a region within the RNA and, preferably, not much more than about 100 bases from the suspected 5′ end of the RNA (Figure 7.42). The DNA probe anneals to the RNA and then serves

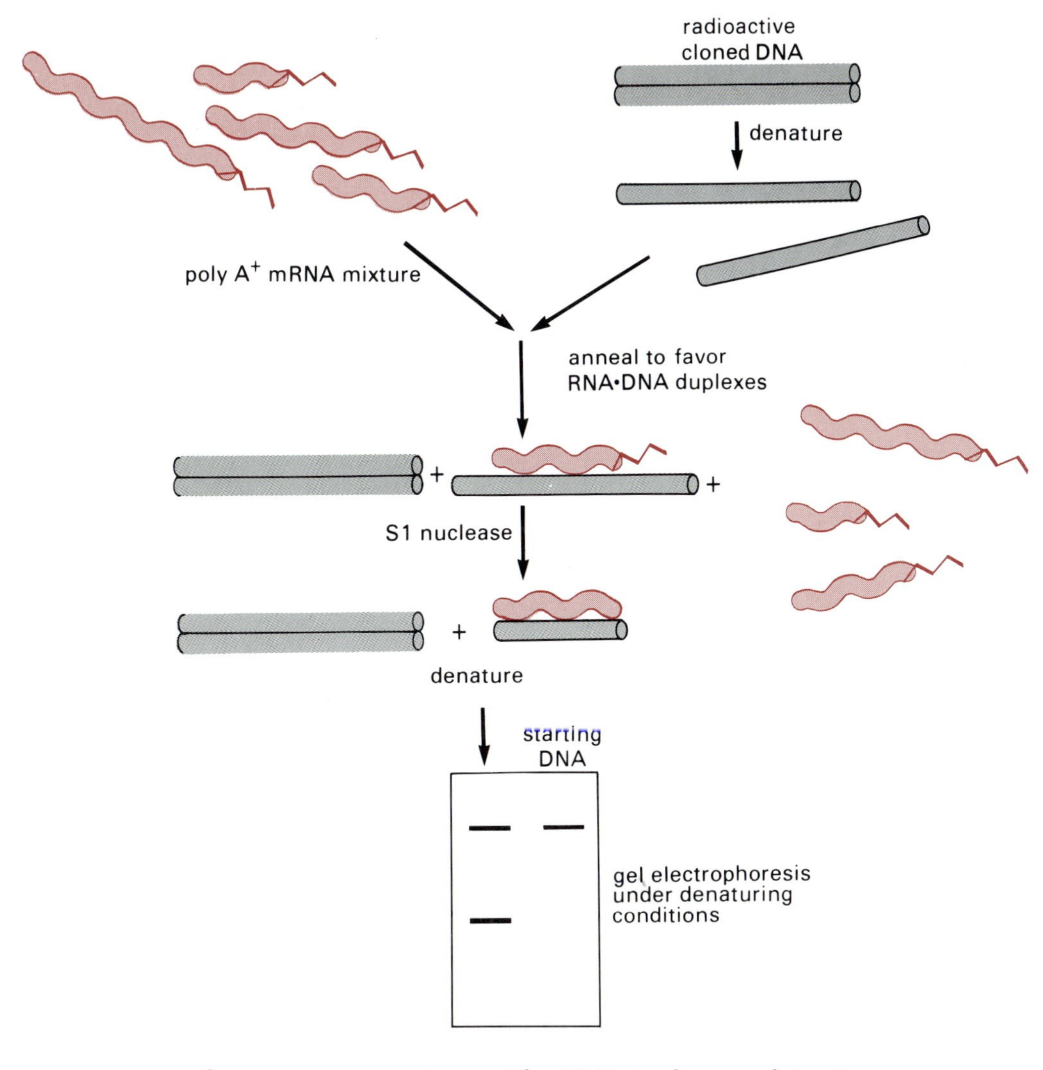

Figure 7.41
Determining the structural relation between a cloned DNA segment and homologous cellular RNA using single strand–specific nuclease S1. In this example, a mixture of cytoplasmic polyA$^+$ RNAs is annealed with a denatured sample of radioactive cloned DNA under conditions that favor formation of RNA · DNA duplexes. Treatment with the S1 nuclease removes those portions of the DNA that do not form a heteroduplex with homologous RNA present in the mixture. The sample is then electrophoresed through a gel under denaturing conditions, and the gel is placed in contact with an X-ray film. Two dark bands (DNA) appear. The larger is the size of the initial DNA strand. The smaller is the size of the RNA that annealed. If marker DNAs of known sizes are electrophoresed in a parallel lane on the gel, the size of the RNA can be calculated.

as a primer for reverse transcription. The RNA is the template. Because the reverse transcriptase stops copying when it comes to the 5′ end of the RNA, the size of the product indicates where the RNA starts. The primer is usually tagged with a radioactive isotope so that the product's position on a gel is readily determined. As long as the priming DNA segment is specific for the RNA being analyzed, the RNA preparation used can be a mixture.

b. Testing Cloned DNAs for Function

Cloned DNA segments can often be transcribed and translated by eukaryotic systems either *in vitro,* using cell extracts or purified enzymes, or *in vivo* after introducing the cloned segments into cells.

In vitro *systems* A variety of eukaryotic cells is routinely used to prepare cell-free extracts containing RNA polymerases I, II, or III. Appropriate regulatory sequences must be present on the DNA segment to promote transcription, and a variety of accessory proteins (or factors) that facilitate transcription is also usually required. Fractionation of such extracts yields purified or partly purified polymerases and factors with which detailed transcriptional mechanisms are studied (Chapter 8). Other types of cell

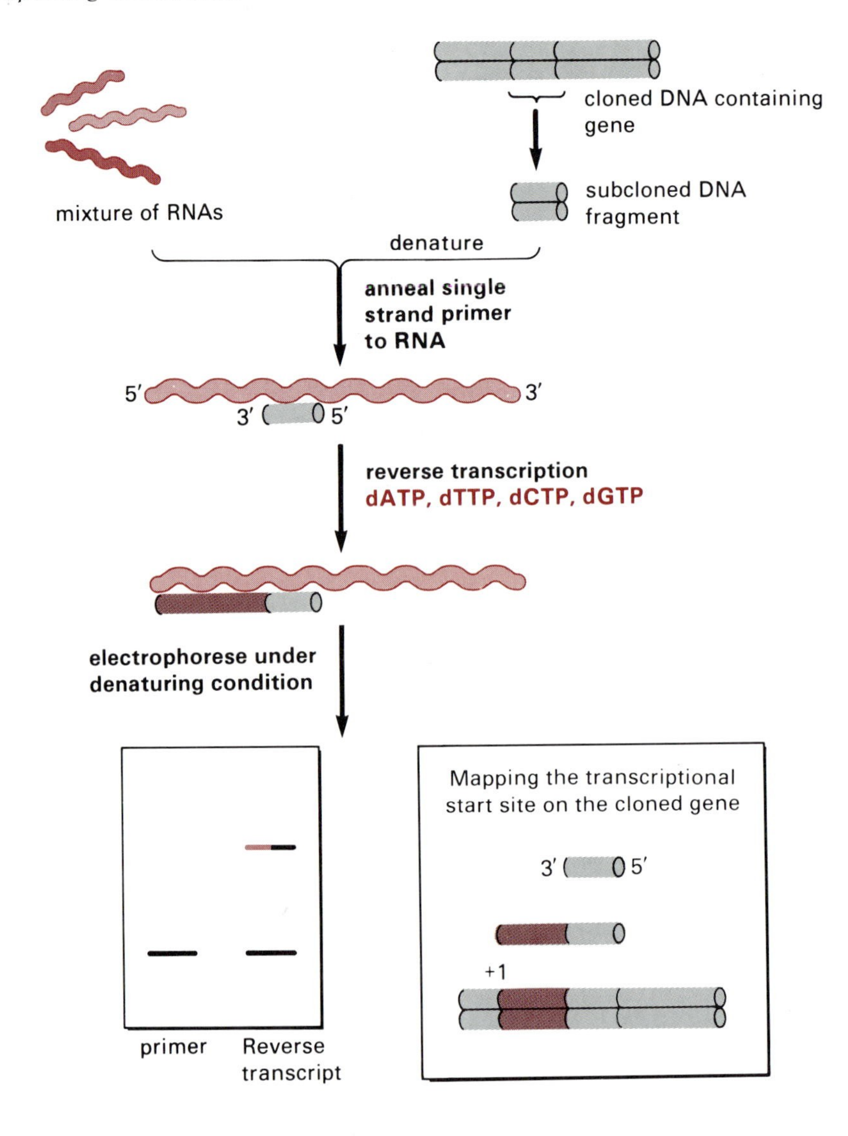

Figure 7.42
Identifying the beginning (5′ end) of an RNA molecule by primer extension. A relatively short synthetic or subcloned DNA fragment homologous to a region near the 5′ end of the RNA of interest is annealed to RNA. Reverse transcriptase is then used to copy the RNA using the DNA fragment as a primer. Copying of the RNA template stops when its 5′ end is reached. Gel electrophoresis under denaturing conditions permits estimation of the size of the extended primer DNA and calculation of the starting point of the RNA. Generally, the priming DNA fragment is radiolabeled to permit detection of the product.

extracts can translate mRNA into corresponding polypeptides, as described in Chapter 6.

In vivo *systems* Whole cells are particularly useful vessels for studying the function of a cloned DNA segment because both transcription and translation can be monitored. The oocytes of frogs (*Xenopus*) are among the best studied biological test tubes. The oocytes are large cells (about 1 μliter in volume) and can be repeatedly collected from single frogs. DNA samples are injected directly into the nucleus (called a germinal vesicle), as little as 5 ngrams generally sufficing. Oocyte histones package the DNA into a chromatin structure, the DNA is transcribed and translated, and the products of gene expression—mRNA and polypeptide—can be detected and analyzed within a few hours. The amount of RNA and polypeptide synthesized by a single oocyte is often sufficient for analysis. In many respects, expression of injected cloned genes follows a normal pathway within the oocytes. The initial nuclear transcripts of both RNA polymerase II and III genes are properly processed, and mature mRNA and tRNA are, for example, passed into the cytoplasm.

Recombinant DNA molecules can also be introduced into eukaryotic cells grown in tissue culture. The molecules may be injected directly into single cells, but a simpler process is to introduce cloned molecules into a large population of cells simultaneously. This is accomplished by the same transfection techniques described in Chapter 5 in connection with cloning in eukaryotic cells. At least some of the molecules introduced in this way arrive in the nucleus.

7.8 Synthesizing Polypeptides Encoded by Cloned Eukaryotic DNA Segments

Examples of the translation, both *in vivo* and *in vitro,* of polypeptides encoded by the inserts in recombinant vectors have been mentioned in a variety of contexts in previous sections. Phenotypic screening and immunological screening techniques both depend on gene expression in a foreign host cell (Chapters 5 and 6). Selection of desired clones by methods such as HART and hybrid selection depend on *in vitro* translation (Section 6.4). Functional analysis of a cloned segment after it is introduced back into the cell of origin depends on both transcription and translation (Section 7.7b). Moreover, the aim of many cloning experiments is to produce a polypeptide. Sometimes the purpose is to produce physiologically significant levels for a phenotypic analysis. Sometimes it is directed toward preparing antibodies. In other cases, the goal is to produce large quantities of a protein. This section describes several host-vector systems that have been designed specifically for protein synthesis.

The production of proteins that are important for research, for clinical medicine, and for manufacturing was one of the earliest goals of recombinant DNA research. Today, several of the commercially available enzymes used to construct recombinant DNA molecules are themselves synthesized from genes cloned into *E. coli* plasmid vectors. These preparations, which are of high purity and reasonable price, include DNA polymerase I, *E. coli* DNA ligase, and reverse transcriptase. Other proteins and mutant forms thereof prepared after site-specific mutagenesis of the corresponding cloned genes are being used to analyze the relation between protein structure and function. These include alcohol dehydrogenase, β-lactamase, cytochrome C, and tyrosyl-tRNA synthetase. Clinically important proteins that were otherwise difficult to obtain have been synthesized for therapeutic use; these include the interferons, human insulin, and human growth hormone. And proteins encoded by animal (including human) viruses such as hepatitis B virus and protozoa such as *Plasmodium falciparum,* the human malaria parasite, are being synthesized for use as antigens in vaccines.

a. *The Choice of an Expression System*

Choosing an appropriate host-vector system is just as important when the aim is to express an already cloned gene as in the initial cloning. *E.*

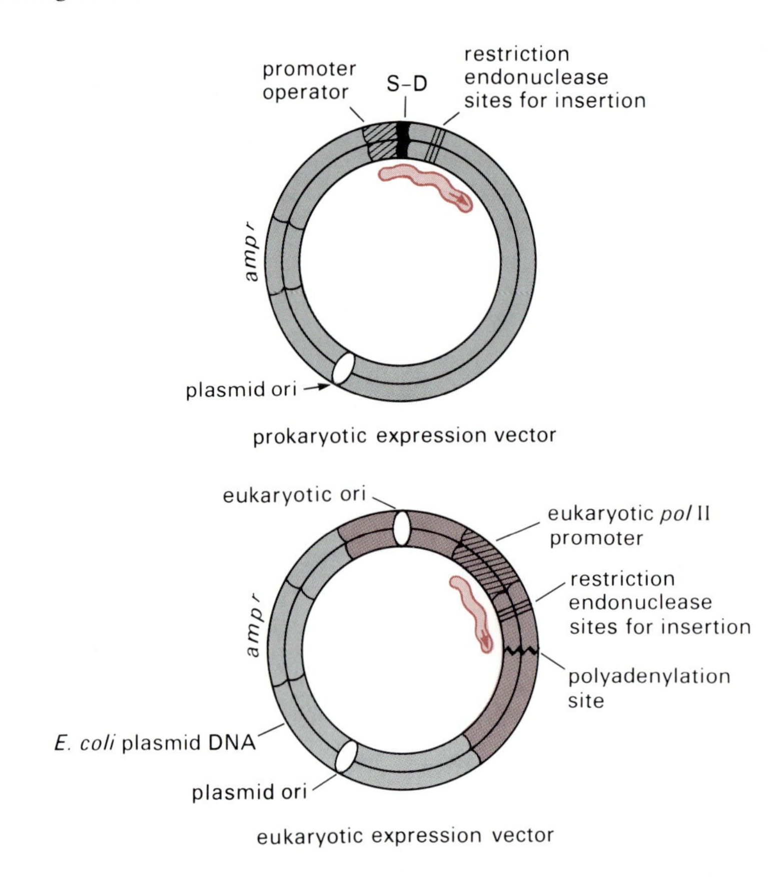

Figure 7.43
General features of expression vectors. Even when the vector is designed for expression in eukaryotic cells, it is usually a shuttle vector and contains sequences required for replication in *E. coli* so that the recombinant molecule can be prepared in quantity. The initiation point and direction of transcription from the promoter driving expression of an insert are indicated inside the circles. S-D is the Shine-Delgarno sequence.

coli is often the host cell of choice for protein production because of the ease with which it can be manipulated and grown on a large scale. Once the existence of introns that interrupt the coding regions in many eukaryotic genes was recognized, cloned cDNAs rather than genes were adopted for use in **expression vectors**. Other problems still preclude synthesis of active forms of certain eukaryotic proteins in *E. coli.* The primary translation products of some eukaryotic genes must undergo specific posttranslational modifications to yield functional molecules. Common modifications include proteolysis, phosphorylation, acetylation, and glycosylation. The enzymes involved in catalyzing these reactions are usually lacking in *E. coli.* Partly to overcome this problem, efficient eukaryotic expression systems have been developed. It is not essential to use cDNAs as coding sequences in animal host-vector systems because the cells are able to remove the intervening sequences.

Vectors designed for the efficient transcription and translation of a cloned segment have elements designed to maximize vector replication as well as gene expression (Figure 7.43). Replication is not itself required for gene expression, but by assuring the formation of a high concentration of templates for transcription, it boosts the level of mRNA and polypeptide produced. Thus, expression vectors generally contain an appropriate origin of replication and encode whatever other functions are required for abundant intracellular replication that are not supplied by the host cells. Transcription depends on the presence in the vector of a promoter that is matched to the RNA polymerase present in the cells: an *E. coli* promoter for expression in the bacteria, an RNA polymerase II promoter for expression in eukaryotic cells. The promoter must be positioned so that a cloned gene can be inserted in the proper orientation and position downstream of the promoter. In eukaryotic expression vectors, it is also necessary to provide a site for the cleavage and polyadenylation of the transcript that is the precursor to functional mRNA (Chapter 8). Correct translation requires the presence of an ATG codon at the start of the coding region and one of the three stop codons at the end. The latter is generally present in the cloned segment, but the start codon is often missing from cDNA inserts and must then be supplied by the vector. For translation in *E. coli,* a ribosome binding site, a Shine-Dalgarno (S-D) sequence, must be properly positioned about ten nucleotides 5′ to the initiator codon (Section 3.6a).

Additional elements are frequently incorporated into expression vectors. These are usually designed to increase the yield of the polypeptide or to simplify the purification of the protein from other cellular proteins. Chief among the factors that tend to decrease yields is the instability of many proteins in a foreign cellular environment.

The expression of each gene (or cDNA) usually presents unique experimental problems. Thus, many expression vectors, while sharing common features, have specific modifications. The systems described in the following sections illustrate some of the different and ingenious expression vectors that have been designed for specific purposes.

b. Expression Vectors Used in E. coli

Vectors designed for efficient protein synthesis in *E. coli* are generally derived from pBR322 or another high copy number plasmid in order to maximize the number of templates for transcription.

Promoters Most *E. coli* expression vectors use one of the following three strong promoters, all of which are described in Section 3.11: the *lac* promoter with its accompanying operator, the *trp* promoter and operator, and the P_L promoter of λ bacteriophage. Generally, a mutant form of the *lac* promoter region called UV5 (Figure 3.7) is used because it permits transcription at a greater rate than the wild type promoter and is, in addition, active even in the absence of the positive effector, CAP-cAMP (Section 3.11c). Another useful promoter, *tac,* is a synthetic construction; its -35 region has the sequence of the *trp* promoter, and its Pribnow box (-10 region) has the sequence of the *lac* UV5 promoter. Thus, the *tac* promoter is identical to the optimal *E. coli* promoter that was deduced from comparing the sequences of many natural promoters (Figure 3.6). It is, as expected, highly efficient.

In addition to their efficiency, these four promoter regions permit temporal control of the initiation of transcription and translation because they are associated with regulatory elements. This is important because the accumulation of large amounts of a foreign protein can debilitate *E. coli* and inhibit growth, thereby limiting the protein's final yield. However, if cells can be grown to a high population density and then induced to begin optimal expression of the cloned gene, yields can often be maximized. Moreover, if the protein being synthesized is unstable in *E. coli,* molecules synthesized early during the growth cycle are likely to be

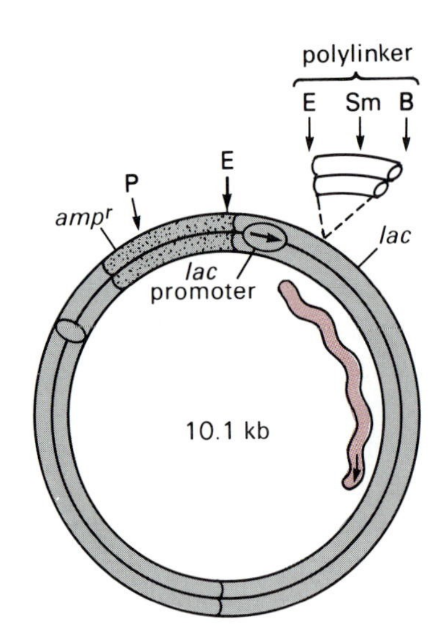

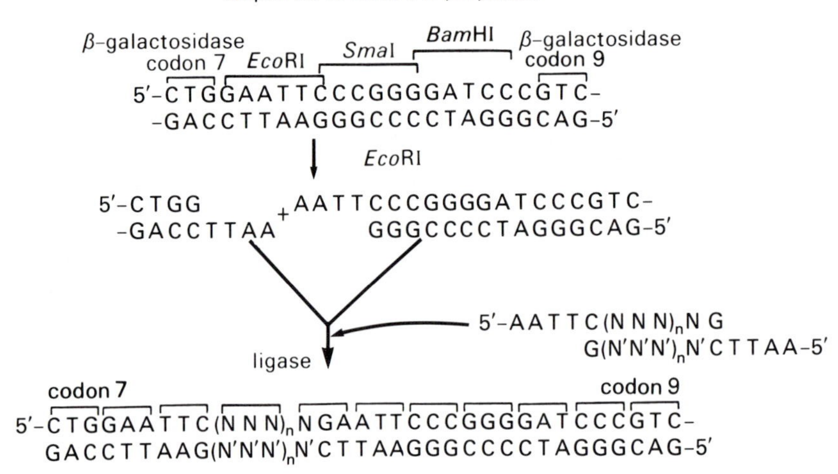

Figure 7.44
An *E. coli* expression vector that utilizes the *lac*UV5 promoter and S-D ribosome binding site. Inserts can be ligated into any of the restriction endonuclease sites in the polylinker that is placed downstream of the seventh codon in the β-galactosidase gene. Inserts with the correct reading frame link the two portions of the β-galactosidase gene into one long contiguous reading frame. The product is a fusion protein. In the example, the insert was added at the *Eco*RI site, N and N′ are paired bases.

degraded by the time the culture reaches a high population density. Again, postponing expression until the culture is well grown is advantageous. When vectors that depend on these promoters for expression of a cloned gene are used, the amount of the foreign protein commonly represents between 1 and 10 percent of the total protein in a cell extract.

An expression vector with the lac *promoter* Plasmid vectors that utilize the *lac*UV5 promoter frequently also contain the S-D sequence of the *lac* operon as well as the coding sequences for β-galactosidase (Figure 7.44). Because the expression from the *lac* operator-promoter segment is subject to the normal regulatory control of the *lac* operon, transcription can be turned on by the addition of the inducer, isopropyl-β-D-thiogalactoside (IPTG). In the example shown, a polylinker segment was inserted in place of the eighth codon of the β-galactosidase gene, and the coding frame was thereby disrupted. *E. coli* cells carrying the vector itself do not synthesize active β-galactosidase (assuming that the cellular genome is *lacZ*$^-$). The two portions of the reading frame are realigned if a fragment with a precise open reading frame is inserted at one of the restriction endonuclease sites in the polylinker. Translation of the mRNA can then begin at the start of the β-galactosidase coding region, continue through the insert and the remaining β-galactosidase coding region, stopping at the normal stop codon at the end of the latter gene. Thus, cells containing a plasmid with insert produce active β-galactosidase because the first eight amino acids are not required for activity, and the foreign amino acids do not interfere with the enzyme's activity. The hybrid polypeptide is known as a **fusion protein** and can be purified by assaying for an increase in the specific activity of the β-galactosidase. Sometimes it is possible to cleave the fusion protein and remove the long carboxy portion of the β-galactosidase. However, fusion proteins are themselves useful because they elicit antibodies to the foreign protein (as well as to β-galactosidase).

An expression vector with the trp *promoter* In the vector shown in Figure 7.45, the *trp* promoter, operator, S-D sequence, and attenuator, as well as about 300 codons of the *trp*E gene, precede a polylinker. Transcription can be controlled by starving *E. coli* cells containing the plasmid for tryptophan after they have grown to a high cell density and then adding 3-β-indolylacrylic acid to the medium. The indolylacrylic acid competes with any remaining tryptophan for binding to the *trp* repressor, but it forms an inactive complex, thereby causing derepression of the promoter. As with the *lac* expression vector, a foreign gene placed in frame into one of the restriction endonuclease sites in the polylinker yields a fusion protein as a product. Translation stops at a fortuitous in-frame stop codon in the vector, just downstream from the end of the insert.

In the example in Figure 7.45, the insert is the reverse transcriptase coding sequence from a retrovirus. An enzymatically active fusion protein is synthesized in *E. coli* cells after transfection and derepression. To increase the stability of the reverse transcriptase, most of the excess *trp*E sequences were deleted by a series of manipulations and reclonings. The few *trp*E amino acids left at the amino terminus of the resulting polypeptide do not interfere significantly with reverse transcriptase activity.

An expression vector that utilizes the P_L promoter of λ bacteriophage The vector illustrated in Figure 7.46 provides for the synthesis of eukaryotic proteins that are not fused to extraneous polypeptide. Transcription from this promoter is repressed by the λ repressor, which is supplied in *trans* when λ lysogens are used as host cells. By using cells lysogenized by phage encoding a temperature-sensitive repressor (cI^{ts}), gene expression can be turned on after the cells are grown at 30° by raising the temperature to 42°. Besides the promoter, the vector contains a segment of DNA derived from the S-D sequence of the λ *c*II gene. Between the promoter and the S-D segment is a group of phage DNA regulatory elements that function in *cis* to relieve transcriptional polarity, the decrease in extent of transcription that sometimes occurs as the distance from the promoter increases. These antipolarity sequences function when the product of the λ *N* gene is supplied in *trans*, as it is in the lysogenized host cells routinely used with this type of vector.

In this vector, the segment that contains the S-D ribosome binding site also contains the ATG codon of the *c*II gene of λ phage. Moreover, the ATG overlaps a *Bam*HI site derived from pBR322 (the starting molecule

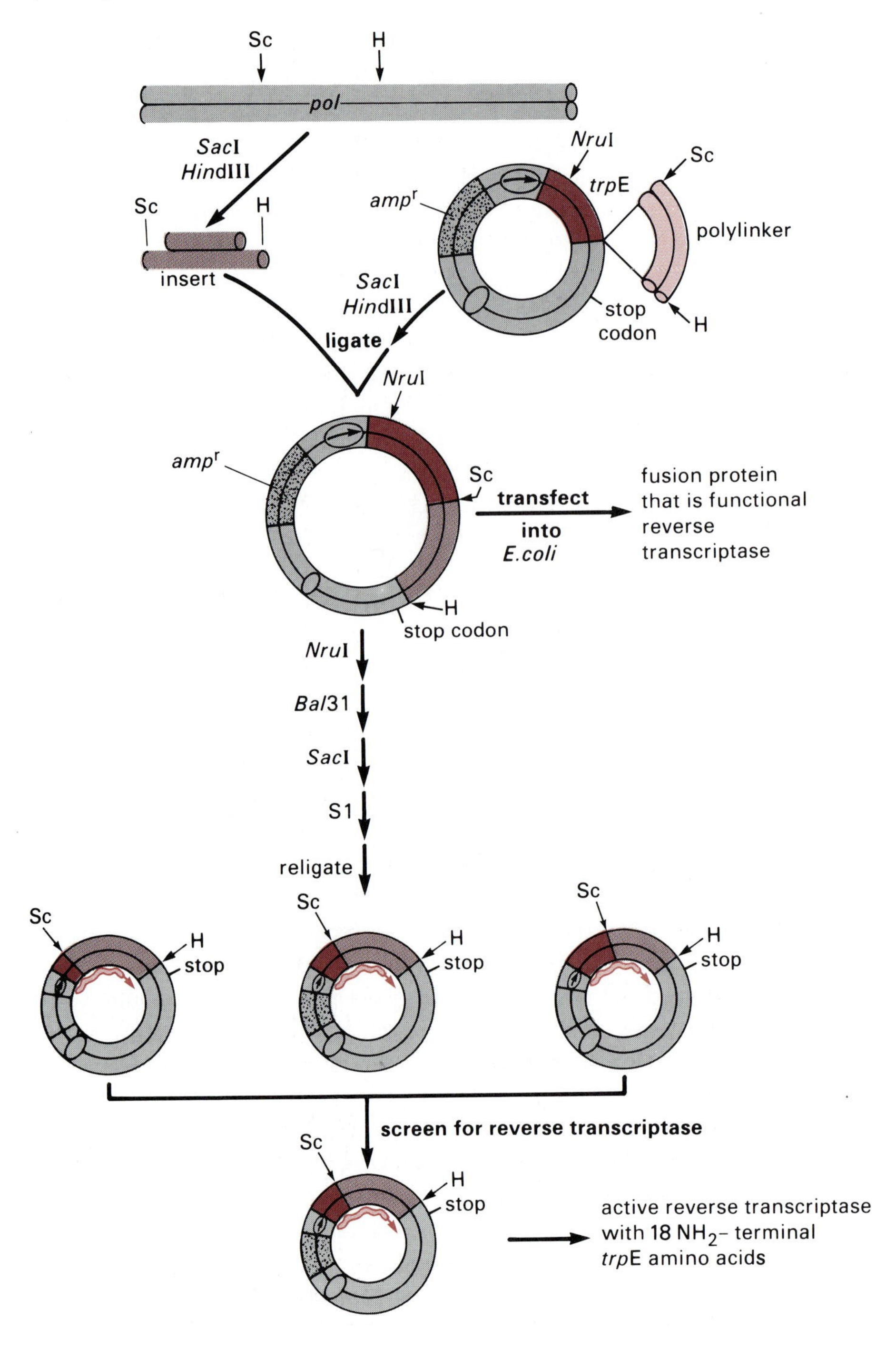

Figure 7.45
An *E. coli* expression vector that utilizes the *trp* promoter and regulatory signals. In this example, the reverse transcriptase (*pol*) coding region of a retrovirus (Section 5.7d) is cloned into a polylinker at the end of the *trp*E coding region. Although the resulting fusion protein was a functional reverse transcriptase, a more stable enzyme was obtained by deleting most of the *trp*E coding sequences by the manipulations outlined at the bottom of the figure. After N. Tanese, M. Roth, and S. P. Goff, *Proc. Natl. Acad. Sci. U.S.A.* 82 (1985), p. 4944.

for vector construction). Eukaryotic coding regions are inserted into the *Bam*HI site in a variety of ways, depending on the particular coding sequence. The critical consideration is that the coding sequence must be inserted so that it is in frame with the ATG. Thus, the vector supplies all the regulatory regions required for transcription and translation except for a stop codon, which is usually found within cloned eukaryotic cDNAs.

c. Expression Vectors Used in Yeast

The most efficient yeast expression vectors are constructed from the yeast plasmid, the 2 μm circle (Section 5.6b). This assures the presence of a high

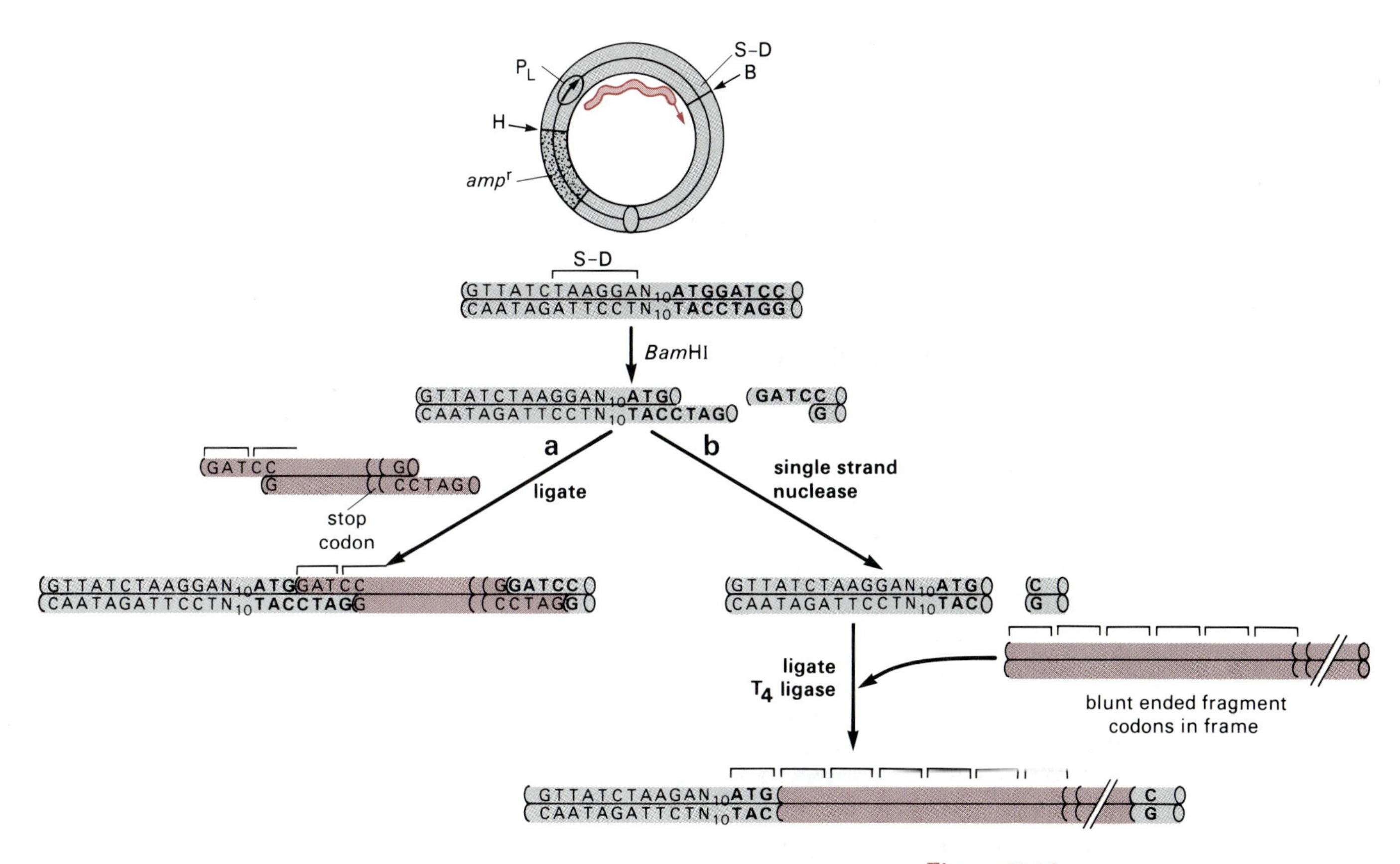

Figure 7.46
An *E. coli* expression vector that utilizes the P_L promoter of λ bacteriophage. The regulatory region of this vector was constructed by ligating various elements from the λ genome to a portion of pBR322. There is a *Bam*HI site into which eukaryotic coding sequences can be inserted just downstream of the ATG supplied by the λ phage sequences. A variety of tricks has been devised that allow the joining of the coding sequence to the vector so that the new codons are in frame with the ATG; two of these are illustrated in (a) and (b). F. M. Rosenberg, Y-S. Ho, and A. Shatzman, *Methods in Enzymology* 101 (1983), p. 123.

copy number of the recombinant within the yeast cells so long as the *REP*3 segment is present in *cis* and functional *REP*1 and *REP*2 genes are present either in *cis* or in *trans* (e.g., on another plasmid). Several different regulatable yeast promoters have been utilized; in the example in Figure 7.47, it is the promoter of the *CYC*1 gene that encodes iso-1-cytochrome C. This promoter and its associated regulatory signals encompass several hundred base pairs, as is typical of the complex eukaryotic regions that regulate transcription by RNA polymerase II (Chapter 8). DNA segments to be translated are inserted, in frame, in the single *Bam*H1 site. This places them directly adjacent to the ATG initiation codon of the *CYC*1 gene. No polyadenylation site is supplied by the vector, but cDNAs usually carry such a site downstream of the translational stop codon.

At least some of the modifications that accompany or follow translation in more complex eukaryotes also occur in yeast. Thus, mRNAs that are translated by ribosomes associated with the endoplasmic reticulum and whose products are normally secreted are also secreted when expressed off a yeast expression vector in yeast cells (Section 3.10c).

d. Expression Vectors for Use in Animal Cells

Two types of vectors, one based on the simian virus 40 genome and one on the bovine papillomavirus (BPV) genome, have been developed for expression of cloned genes in animal cells. (Section 5.7 describes the basic properties of these two viral vector systems.) The papillomavirus vectors are particularly useful for large scale protein synthesis. SV40 vectors are used in many experimental protocols.

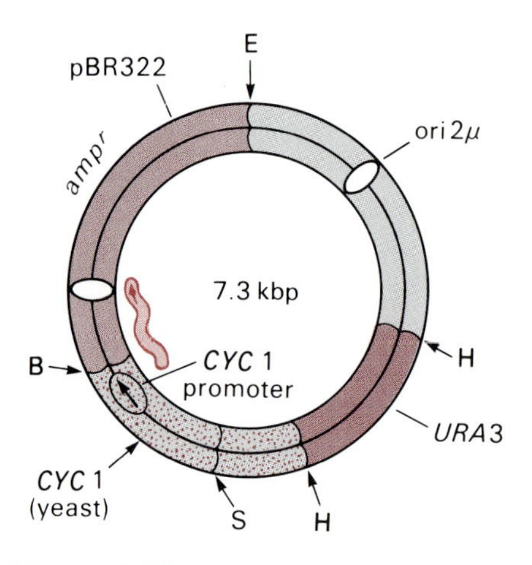

Figure 7.47
A yeast expression vector. This construction is related to the yeast plasmid vector shown in Figure 5.25. It contains, in addition to pBR322 segments, sequences from the 2 μm plasmid that permit replication to high copy number in yeast, a selectable marker (the yeast *URA3* gene), the yeast *CYC1* gene promoter and regulatory signals, and a convenient *Bam*H1 cloning site within the second codon of the *CYC1* coding sequence.

Typical constructions contain either the entire BPV genome or the portion required for stable transformation of hosts such as mouse cells in culture (Figure 7.48). Such vectors also contain a eukaryotic gene (or a portion of a gene) that supplies a promoter, a polyadenylation signal, and other regulatory regions. A convenient restriction endonuclease site is situated between the position at which transcription initiates, downstream of the promoter, and the translational start codon. When a coding sequence is inserted at this site, it will be included in an mRNA that extends from the transcriptional start site to either its own polyadenylation signal or to that supplied in the vector. The coding sequence must supply its own start and stop codons. One of the advantages of BPV expression vectors is that cells transformed by the recombinants can be maintained for months in cell culture. The cells divide continually, providing a constant source of the desired protein. In one example, a previously cloned coding region for the surface antigen of hepatitis B virus was inserted into a BPV vector. The transformed cells secreted as much as 10 mg of antigen per liter of culture per 24 hours.

7.9 Enzymatic Amplification of DNA and RNA Segments

When molecular cloning was first introduced, alternate ways to amplify specific DNA or RNA segments buried within large populations of polynucleotide chains seemed unlikely. Yet such an alternative is now routinely used. The method is called the **polymerase chain reaction** or **PCR**. It provides microgram quantities of DNA copies of either DNA or RNA segments present in amounts as small as a single molecule in a nucleic acid preparation.

PCR is carried out entirely *in vitro* using DNA polymerase and oligonucleotide primers complementary to the two 3′ borders of the duplex segment to be amplified. Thus, PCR development depended on the prior innovations that permit rapid and inexpensive chemical synthesis of oligonucleotides (Section 6.1c). To carry out the procedure, the sequence of the region to be amplified needs to be known so that appropriate oligonucleotide primers can be prepared. Nevertheless, imaginative application of the PCR method has made it an important addition to the field of molecular genetics. Indeed, if Part II of this book had not been virtually

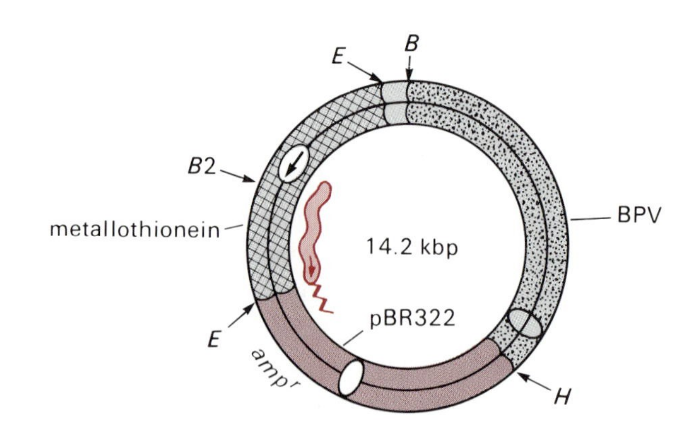

Figure 7.48
A BPV expression vector. The vector contains pBR322 sequences for shuttling through *E. coli*, the transforming region of the BPV genome, and the mouse gene encoding the protein metallothionein. A coding sequence, including a start and stop codon, is inserted into the *Bgl*II restriction endonuclease site between the transcriptional initiation site and the first codon of the metallothionein gene. Mouse cells transformed by the recombinant vector produce an mRNA (inner arrow) that includes the insert and translate the mRNA to produce the corresponding polypeptide.

complete before PCR was fully developed, the method would have been presented as an integral part of the material in Chapters 5, 6, and 7.

The polymerase chain reaction To amplify a DNA segment, two oligonucleotide primers that are each complementary to one of the two 3′ borders of the duplex segment to be amplified are synthesized. The objective of PCR is to copy the sequence of each strand between the regions at which the oligonucleotide primers anneal. Thus, after annealing the primers to a denatured DNA containing the segment to be amplified, the primers are extended using DNA polymerase and the four deoxynucleoside triphosphates (Figure 7.49). Each primer is extended toward the other. The resulting duplex DNAs are then denatured and annealed again with the primers, and the DNA polymerase reaction is repeated. This cycle of steps (denaturation, annealing, and synthesis) may be repeated as many as 60 times. At each cycle, the amount of duplex DNA segment doubles because both new and old DNA molecules anneal to the primers and are copied. In principle, and virtually in practice, 2^n copies of the duplex segment bordered by the primers are produced, where n is the number of cycles.

Using a heat stable DNA polymerase isolated from the thermophilic bacteria *Thermus aquaticus* allows multiple cycles to be carried out after a single addition of the enzyme. The DNA, an excess of primer molecules, the deoxynucleoside triphosphates, and the polymerase are mixed together at the start. Cycle 1 is initiated by heating to a temperature that assures DNA denaturation, followed by cooling to a temperature appropriate for primer annealing. Thereafter, the temperature is adjusted so that DNA synthesis can occur. The second and subsequent cycles are initiated by again heating to the denaturation temperature. Thus, cycling can be automated by using a computer-controlled variable temperature heating block. The whole process takes only a few hours.

Besides permitting automation, use of the *T. aquaticus* DNA polymerase has another advantage. This enzyme is most active between 70° and 75° centigrade. At this temperature, base pairing between the oligonucleotide primers (about 20 residues long) and the DNA is more specific than at 37°, the optimal temperature for *E. coli* DNA polymerase. Consequently, primers are less likely to anneal to imperfectly matched DNA segments, thereby minimizing amplification of unwanted DNA segments, especially when an entire genome's worth of DNA is present. Correct annealing of primers is also fostered by stringent annealing conditions, which are obtained by selecting appropriate conditions of temperature and ionic strength for primer length and base composition. Specificity can be stringent enough to permit simultaneous amplification of two different genomic segments in the same DNA sample, in the presence of two pairs of primers.

An RNA segment is amplified into a duplex DNA segment by essentially the same procedure except that, prior to cycle 1, a cDNA copy of the RNA must be made. If the RNA is an mRNA, then a short polydT can serve as one of the primers for the PCR as well as for the reverse transcriptase step.

Variations on the polymerase chain reaction As described thus far, PCR requires prior knowledge of the sequences at both borders of the segment

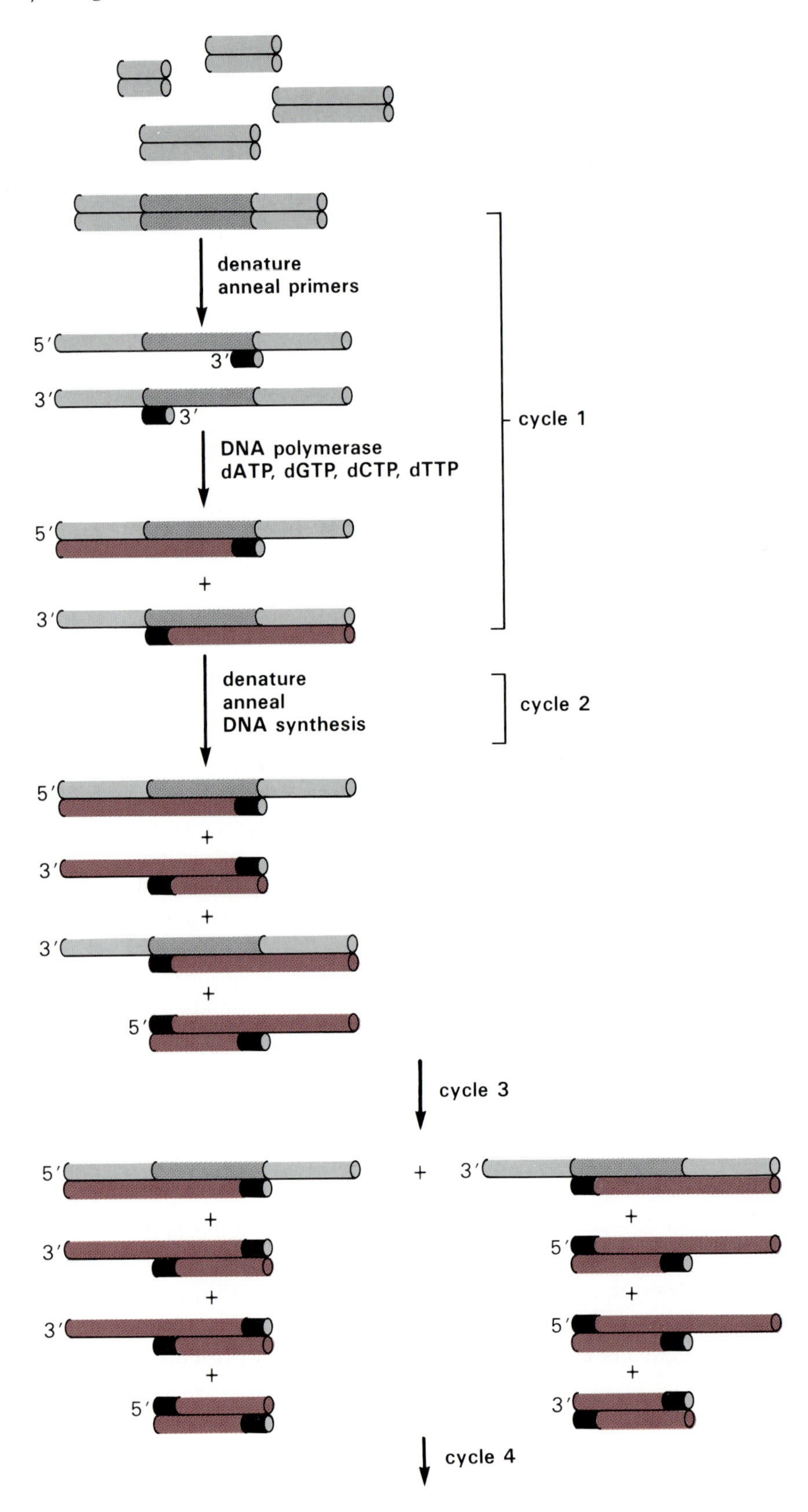

Figure 7.49
The polymerase chain reaction (PCR) amplifies a selected segment of DNA within a large mixture of DNA chains. Two oligonucleotide primers, one to anneal to each strand, define the limits of the amplified segment. The PCR is initiated by denaturing the DNA; then the primers are annealed, allowing DNA polymerase to catalyze chain extension in the presence of the four nucleoside triphosphates (cycle 1). The second and following cycles are initiated by increasing the temperature to denature the duplexes. At each cycle, the amount of the chosen DNA segment doubles. After a few cycles, duplexes representing the selected segment (the short DNAs) predominate. The reaction works efficiently if the amplified segment is less than about 2 kbp. (This figure is continued on the facing page.)

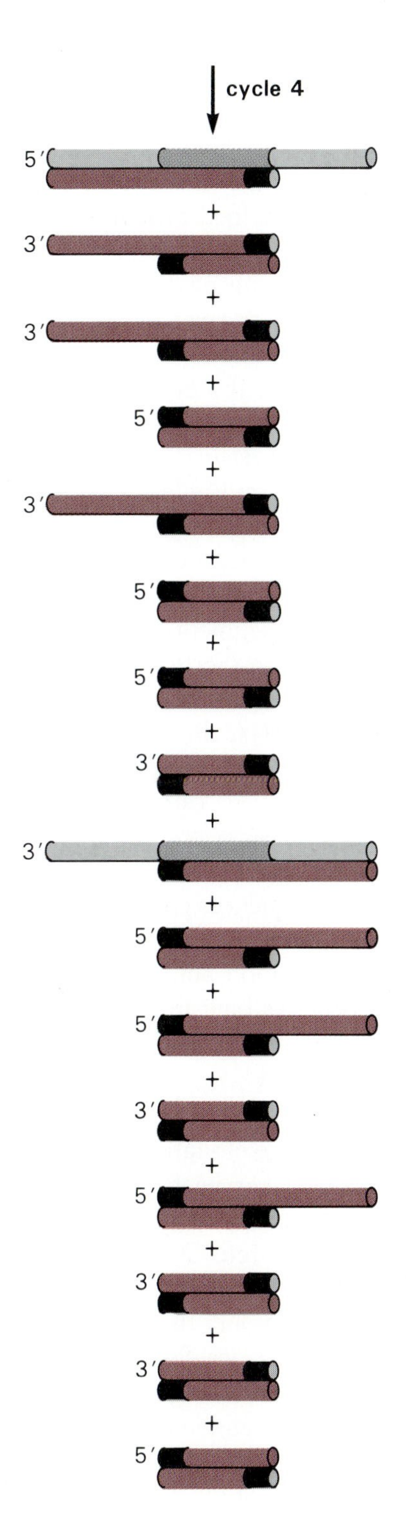

Figure 7.49, continued

to be amplified. This suggests that the method is applicable only to previously cloned and sequenced DNA segments. However, relatively simple modifications expand the opportunities PCR provides. In one such variation, a gene can be isolated if the amino acid sequence of only a short region of the corresponding purified protein is known. For example, by synthesizing 20 base pair long primers predicted by the sequences at the two ends of a 20 amino acid long peptide segment, a 60 bp long genomic fragment can be amplified. Because of the degeneracy of the genetic code, mixtures of primers with alternative bases at required positions are used (Section 6.4b). After PCR, the 60 bp amplified segment is purified by gel electrophoresis or by cloning and used as a probe against a genomic or cDNA library to isolate the desired coding sequence.

Another example of PCR versatility leads to amplification of a complete cDNA copy of an mRNA starting only with knowledge of the sequence of a single small region, from either a peptide or an mRNA. Total mRNA is copied into a single cDNA strand using a short polydT primer. Then a tail (for example, a polydG segment) is added onto the cDNA at the 3′ end with terminal transferase (Section 4.8). The second DNA strand is synthesized using a short polydC primer. Up to this point, there is no specificity for a particular mRNA. In the next step, an oligonucleotide primer of the same sequence as a short region near the 3′ end of the desired mRNA is used. PCR can be carried out using this primer and the polydC primer; specificity is sufficient to give a nearly pure product representing the whole mRNA sequence.

In yet another variation, sequences corresponding to a restriction endonuclease recognition site are included at the 5′ ends of the primers. They remain as single strand tails upon annealing to the initial template, but the amplified fragment acquires built-in terminal restriction endonuclease sites. These can be useful if subsequent cloning or incorporation into an expression vector is anticipated.

Applications of PCR The full exploitation of the PCR method still needs to be realized. But already it has been applied to obtain novel insights into the structure of RNAs, genes, and genomes and into genomic events. Only a few examples can be cited. One example of great significance is the determination of the structure of mutations. Once a gene has been cloned, a pair of primers will amplify the corresponding genomic segment from various individuals in a species. Sequence analysis reveals the nature of a mutation or polymorphism in the gene, be it a single base pair change or a deletion, insertion, or rearrangement. The ease with which this can be done provides diagnostic tools for genetic diseases and data applicable to population genetics. Classes of mutations can also be tracked by using primers that anneal specifically to particular mutant forms of a gene.

Infectious agents such as viruses and bacteria can be detected in the presence of an enormous excess of host cell DNA by using probes complementary to the agents' genomes. Diagnosis of infection by the AIDS virus (human immunodeficiency virus type 1 or HIV-1) is already a reality. In yet another application, the structure of the products of recombination reactions is being analyzed.

One of the most interesting and potentially most powerful applications of PCR is amplification of DNA segments within single cells. This has been demonstrated with both diploid cells and human sperm. The use of sperm has a special significance for genetic analysis in species that cannot be bred experimentally, including humans. With this technique, meiotic recombination rates can be measured directly in large populations of sperm using pairs of primers specific for polymorphic or mutant forms of linked genes.

References for Part II

PERSPECTIVE

The following classical articles and books describe the experiments that provided the theoretical and technical basis for the recombinant DNA method. The listing is chronological.

F. D'Herelle. 1926. The Bacteriophage and Its Behavior. Williams and Wilkins, Baltimore.

O. T. Avery, C. M. Macleod, and M. McCarty. 1944. Studies on the Chemical Nature of the Substance Inducing Transformation of Pneumococcal Types. I. Induction of Transformation by a Desoxyribonucleic Acid Fraction Isolated from *Pneumococcus* Type III. *J. Exp. Med.* 79 137–158.

M. Delbruck. 1946. Experiments with Bacterial Viruses (Bacteriophages). *Harvey Lectures* 41 161–187.

J. Lederberg and E. L. Tatum. 1946. Gene Recombination in *E. coli. Nature* 158 558. (This paper describes conjugation.)

N. Zinder and J. Lederberg. 1952. Genetic Exchange in *Salmonella. J. Bacteriology* 64 679–699. (This paper describes transduction.)

L. L. Cavalli-Sforza, J. Lederberg, and E. M. Lederberg. 1953. An Infective Factor Controlling Sex Compatibility in *Bacterium coli. J. Gen. Microbiol.* 8 89–103.

A. Lwoff. 1954–1955. Control and Interrelations of Metabolic and Viral Diseases of Bacteria. In Harvey Lectures, series L, pp. 92–111. Academic Press, New York.

F. Jacob, P. Schaeffer, and E. L. Wollman. 1960. Episomic Elements in Bacteria. *Symp. Soc. Gen. Microbiol.* 10 67–91.

A. D. Kaiser and D. S. Hogness. 1960. The Transformation of *Escherichia coli* with Deoxyribonucleic Acid Isolated from Bacteriphage λ-dg. *J. Mol. Biol.* 2 392–415.

W. Arber and D. Dussoix. 1962. Host specificity of DNA Produced by *Escherichia coli.* I. Host Controlled Modification of Bacteriophage λ. *J. Mol. Biol.* 5 18–36 and 37–49.

T. Watanabe. 1963. Infective Heredity of Multiple Drug Resistance in Bacteria. *Bacteriological Reviews* 27 87–115.

W. Arber. 1965. Host Controlled Modification of Bacteriophage. *Annu. Rev. Microbiol.* 19 365–378.

T. J. Kelly, Jr., and H. O. Smith. 1970. A Restriction Enzyme from *Hemophilus influenzae.* II. Base Sequence of the Recognition Site. *J. Mol. Biol.* 51 393–409.

H. O. Smith and K. W. Wilcox. 1970. A Restriction Enzyme from *Hemophilus influenzae.* I. Purification and Chemical Properties. *J. Mol. Biol.* 51 371–391.

K. J. Danna and D. Nathans. 1971. Sequence Specific Cleavage of Simian Virus 40 DNA by Restriction Endonucleases of *Hemophilus influenzae. Proc. Natl. Acad. Sci.* U.S.A. 68 2913–2917.

J. Hedgpeth, H. M. Goodman, and H. W. Boyer. 1972. DNA Nucleotide Sequence Restricted by the RI Endonuclease. *Proc. Natl. Acad. Sci.* U.S.A. 69 3448–3452.

D. A. Jackson, R. H. Symons, and P. Berg. 1972. Biochemical Method for Inserting New Genetic Information into DNA of Simian Virus 40: Circular SV40 DNA Molecules Containing λ Phage Genes and the Galactose Operon of *E. coli. Proc. Natl. Acad. Sci.* U.S.A. 69 2904–2909.

J. E. Mertz and R. W. Davis. 1972. Cleavage of DNA by RI Restriction Endonuclease Generates Cohesive Ends. *Proc. Natl. Acad. Sci.* U.S.A. 69 3370–3374.

H. Temin and D. Baltimore. 1972. RNA Directed DNA Synthesis and RNA Tumor Viruses. *Advances in Virus Research* 17 129–186.

S. N. Cohen, A. C. Y. Chang, H. W. Boyer, and R. B. Hellings. 1973. Construction of Biologically Functional Bacterial Plasmids *in vitro. Proc. Natl. Acad. Sci.* U.S.A. 70 3240–3244.

P. E. Lobban and A. D. Kaiser. 1973. Enzymatic End-to-End Joining of DNA Molecules. *J. Mol. Biol.* 78 453–471.

J. F. Morrow, S. N. Cohen, A. C. Y. Chang, H. W. Boyer, H. M. Goodman, and R. B. Helling. 1974. Replication and Transcription of Eukaryotic DNA in *Escherichia coli. Proc. Natl. Acad. Sci.* U.S.A. 71 1743–1747.

The historical development of the recombinant DNA techniques is described in the following works.

J. Cairns, G. S. Stent, and J. D. Watson (eds.). 1966. Phage and the Origins of Molecular Biology. Cold Spring Harbor Laboratory, Cold Spring Harbor, New York.

S. N. Cohen. 1975. The Manipulation of Genes. *Sci. American* 233 25–33.

M. F. Singer. 1979. Introduction and Historical Background. In J. K. Setlow and A. Hollaender (eds.), Genetic Engineering, vol. 1, pp. 1–13. Plenum, New York.

General Reading for Chapters in Part II

Several monographs and laboratory manuals provide basic and practical details regarding the recombinant DNA method and its tools.

P. D. Boyer (ed.). 1981, 1982. The Enzymes, 3rd ed., vols. 14 and 15. Academic Press, New York.

T. Maniatis, E. F. Fritsch, and J. Sambrook. 1982. Molecular Cloning, a Laboratory Manual. Cold Spring Harbor Laboratory, Cold Spring Harbor, New York.

S. M. Linn and R. J. Roberts (eds.). 1983. Nucleases. Cold Spring Harbor Laboratory, Cold Spring Harbor, New York.

J. D. Watson, J. Tooze, and D. T. Kurtz. 1983. Recombinant DNA, a Short Course. Scientific American Books, New York.

D. M. Glover (ed.). 1985, 1987. DNA Cloning, vols. 1, 2, and 3. IRL Press, Oxford, England, and Washington, D.C.

R. W. Old and S. B. Primrose. 1985. Principles of Gene Manipulation, 3rd ed. Blackwell Scientific Publications, Oxford, England.

P. H. Pouwels, B. E. Enger-Valk, and W. J. Brammer. 1985. Cloning Vectors. Elsevier, Amsterdam.

F. M. Ausubel, R. Brent, R. E. Kingston, D. D. Moore, J. A. Smith, J. G. Seidman, and K. Struhl (eds.). 1987. Current Protocols in Molecular Biology. Greene Publishing Associates and Wiley, New York.

S. L. Berger and A. R. Kimmel (eds.). 1987. Guide to Molecular Cloning Techniques. Academic Press, San Diego.

References for Indicated Chapter Sections.

4.1

R. J. Crouch and M. L. Dirksen. 1983. Ribonuclease H. In S. M. Linn and R. J. Roberts (eds.), Nucleases, pp. 211–241. Cold Spring Harbor Laboratory, Cold Spring Harbor, New York.

4.2

D. Nathans. 1979. Restriction Endonucleases, Simian Virus 40 and the New Genetics. *Science* 206 903–910.

S. E. Halford. 1983. How does *Eco*RI Cleave Its Recognition Site on DNA? *Trends Biochem. Sci.* 8 455–460.

C. Kessler and H. J. Holtke. 1986. Specificity of Restriction Endonucleases and Methylases—A Review. *Gene* 47 1–52.

J. A. McClarin, C. A. Frederick, B. C. Wang, P. Greene, H. W. Byer, J. Grable, and J. M. Rosenberg. 1986. Structure of the DNA-*Eco*RI Endonuclease Recognition Complex at 3 Angstrom Resolution. *Science* 234 1526–1541. (A minireview by T. J. Richmond that describes this work appears in *Nature* 326 18–19 [1987].)

M. Nelson and M. McClelland. 1987. The Effect of Site-Specific Methylation on Restriction-Modification Enzymes. *Nucleic Acids Res.* 15 (supplement) r219–r230.

B.-Q. Qiang and I. Schildkraut. 1987. *Not*I and *Sfi*I: Restriction Endonucleases with Octanucleotide Recognition Sequences. *Methods in Enzymology* 155 260–301.

R. J. Roberts. 1987. Restriction Enzymes and Their Isoschizomers. *Nucleic Acids Res.* 15 (supplement) r189–r217. (This is a comprehensive list of restriction endonucleases that is updated annually.)

J. M. Rosenberg, J. A. McClarin, C. A. Frederick, B.-C. Wang, J. Grable, H. W. Boyer, and P. Greene. 1987. Structure and Recognition Mechanism of *Eco*RI Endonuclease. *Trends Biochem. Sci.* 12 395–398.

4.3

G. Chaconas and J. H. van de Sande. 1980. 5'-^{32}P Labeling of RNA and DNA Restriction Fragments. *Methods in Enzymology* 65 75–85.

4.5

I. R. Lehman. 1974. DNA Ligase: Structure, Mechanism and Function. *Science* 186 790–797.

4.6

P. W. J. Rigby, M. Dieckmann, C. Rhodes, and P. Berg. 1977. Labeling DNA to High Specific Activity *in vitro* by Nick Translation *in vitro* with DNA Polymerase I. *J. Mol. Biol.* 113 237–251.

4.8

R. Roychoudhury and R. Wu. 1980. Terminal Transferase-Catalyzed Addition of Nucleotides to the 3' Termini of DNA. *Methods in Enzymology* 65 43–64.

5.1

M. Mandel and A. Higa. 1970. Calcium-Dependent Bacteriophage DNA Infection. *J. Mol. Biol.* 53 159–162.

B. J. Bachmann. 1983. Linkage Map of *Escherichia coli* K12, Edition 7. *Microbiol. Rev.* 47 180–230.

R. Wu and L. Grossman (eds.). 1987. *Methods in Enzymology* 153. (The entire volume is relevant to the material in Chapter 5.)

5.2

S. Falkow. 1975. Infectious Multiple Drug Resistance. Pion, London. (This is a book about the discovery and biology of plasmids.)

F. Bolivar and K. Backman. 1979. Plasmids of *E. coli* as Cloning Vectors. *Methods in Enzymology* 68 245–267.

R. P. Novick. 1980. Plasmids. *Sci. American* 243 102–127.

5.3

J. Messing and J. Vieira. 1982. A New Pair of M13 Vectors for Selecting Either DNA Strand of Double-Digest Restriction Fragments. *Gene* 19 269–276.

A. Campbell. 1983. Bacteriophage λ. In J. A. Shapiro (ed.), Mobile Genetic Elements, pp. 65–103. Academic Press, New York.

M. Feiss. 1986. Terminase and the Recognition, Cutting, and Packaging of λ Chromosomes. *Trends in Genetics* 2 100–104.

5.4

J. Collins and B. Hohn. 1978. Cosmids: A Type of Plasmid Gene-Cloning Vector That Is Packageable *in vitro* in Bacteriophage λ Heads. *Proc. Natl. Acad. Sci.* U.S.A. 75 4242–4246.

J. Collins. 1979. *E. coli* Plasmids Packageable *in vitro* in λ Bacteriophage Particles. *Methods in Enzymology* 68 309–326.

B. Hohn. 1979. *In vitro* packaging of λ and Cosmid DNA. *Methods in Enzymology* 68 299–309.

S. Brenner, G. Cesareni, and J. Karn. 1982. Phasmids: Hybrids Between Co1E1 Plasmids and *E. coli* Bacteriophage λ. *Gene* 17 27–44.

5.5

P. S. Lovett and K. M. Keggins. 1979. *Bacillus subtilis* as a Host for Molecular Cloning. *Methods in Enzymology* 68 342–357.

D. A. Hopwood, M. J. Bibb, K. F. Chater, and T. Kieser. 1987. Plasmid and Phage Vectors for Gene Cloning and Analysis in *Streptomyces*. *Methods in Enzymology* 153 116–166.

5.6

C. Ilgen, P. J. Farabaugh, A. Hinnen, J. M. Walsh, and G.-R. Fink. 1979. Transformation of Yeast. In J. K. Setlow and A. Hollaender (eds.), Genetic Engineering, vol. 1, pp. 117–132. Plenum, New York.

D. T. Stinchcomb, K. Struhl, and R. W. Davis. 1979. Isolation and Characterization of a Yeast Chromosomal Replicator. *Nature* 282 39–43.

L. Clarke and J. Carbon. 1980. Isolation of a Yeast Centromere and Construction of Functional Small Circular Chromosomes. *Nature* 287 504–509.

J. R. Broach. 1982. The Yeast Plasmid 2 μ Circle. *Cell* 28 203–204.

R. J. Rothstein. 1983. One-Step Gene Disruption in Yeast. *Methods in Enzymology* 101 202–211.

K. Struhl. 1983. The New Yeast Genetics. *Nature* 305 391.

R. K. Mortimer and D. Schild. 1985. Genetic Map of *Saccharomyces cerevisiae*, Edition 9. *Microbiol. Rev.* 49 181–213.

J. Campbell. 1988. Eukaryotic DNA Replication: Yeast Bares Its ARSs. *Trends Biochem. Sci.* 13 212–217.

5.7

S. P. Goff and P. Berg. 1976. Construction of Hybrid Viruses Containing SV40 and λ Phage DNA Segments and Their Propagation in Cultured Monkey Cells. *Cell* 9 695–705.

C. Shih, B.-Z. Shilo, M. P. Goldfarb, A. Dannenberg, and R. A. Weinberg. 1979. Passage of Phenotypes of Chemically Transformed Cells via Transfection of DNA and Chromatin. *Proc. Natl. Acad. Sci.* U.S.A. 76 5714–5718.

D. H. Hamer. 1980. DNA Cloning in Mammalian Cells with SV40 Vectors. In J. K. Setlow and A. Hollaender (eds.), Genetic Engineering, vol. 2, pp. 83–101. Plenum, New York.

R. C. Mulligan and P. Berg. 1980. Simian Virus 40 Mediated Expression of a Bacterial Gene in Mammalian Cells. *Science* 209 1422–1427.

A. Pellicer, D. Robins, B. Wold, R. Sweet, J. Jackson, I. Lowy, J. M. Roberts, G. K. Sim, S. Silverstein, and R. Axel. 1980. Altering Genotype and Phenotype by DNA-Mediated Gene Transfer. *Science* 209 1414–1422.

R. C. Mulligan and P. Berg. 1981. Selection for Animal Cells That Express the *E. coli* Gene Coding for Xanthine-Guanine-Phosphoribosyl Transferase. *Proc. Natl. Acad. Sci.* U.S.A. 78 2072–2076.

B. H. Howard. 1983. Vectors for Introducing Genes into Cells of Higher Eukaryotes. *Trends Biochem. Sci.* 8 209–212.

C. L. Cepko, B. E. Roberts, and R. C. Mulligan. 1984. Construction and Applications of a Highly Transmissible Murine Retrovirus Shuttle Vector. *Cell* 37 1053–1062.

J. Yates, N. Warren, D. Reisman, and B. Sugden. 1984. A *cis*-Acting Element from the Epstein-Barr Viral Genome That Permits Stable Replication of Recombinant Plasmids in Latently Infected Cells. *Proc. Natl. Acad. Sci.* U.S.A. 81 3806–3810.

H. M. Temin. 1986. Retrovirus Vectors for Gene Transfer: Efficient Integration into and Expression of Exogenous DNA in Vertebrate Ćell Genomes. In R. Kucherlapati (ed.), Gene Transfer, pp. 149–158. Plenum, New York.

D. DiMaio. 1987. Papillomavirus Clonirg Vectors. In N. P. Salzman and P. M. Howley (eds.), The Papovaviridae, vol 2, The Papillomaviruses, pp. 293–320. Plenum, New York.

C. Cepko. 1988. Retrovirus Vectors and Their Application in Neurobiology. *Neuron.* 1 345–353.

5.8

M. W. Bevan and M.-D. Chilton. 1982. T-DNA of the *Agrobacterium* TI and RI Plasmids. *Annu. Rev. Genet.* 16 357–384.

M.-D. Chilton. 1983. A Vector for Introducing New Genes into Plants. *Sci. American* 249 50–60.

P. Zambryski, H. M. Goodman, M. Van Montagu, and J. Schell. 1983. *Agrobacterium* Tumor Induction. In J. A. Shapiro (ed.), Mobile Genetic Elements, pp. 505–535. Academic Press, New York.

R. B. Horsch, J. E. Fry, N. L. Hoffmann, D. Eichholtz, S. G. Rogers, and R. T. Fraley. 1985. A Simple and General Method for Transferring Genes into Plants. *Science* 227 1229–1231.

R. Deblaere, A. Reynaerts, H. Hofte, J.-P. Hernalsteens, J. Leemans, and M. Van Montagu. 1987. Vectors for

Cloning in Plant Cells. *Methods in Enzymology* 153 277–292.

M. Fromm, J. Callis, L. P. Taylor, and V. Walbott. 1987. Electroporation of DNA and RNA into Plant Protoplasts. *Methods in Enzymology* 153 351–366.

A. Weissbach and H. Weissbach (eds.). 1988. Methods for Plant Molecular Biology. Academic Press, Orlando.

P. Zambryski, J. Tempe, and J. Schell. 1989. Transfer and Function of T-DNA Genes from *Agrobacterium* Ti and Ri Plasmids in Plants. *Cell* 56 193–201.

6.1

E. M. Southern. 1974. Detection of Specific Sequences Among DNA Fragments Separated by Gel Electrophoresis. *J. Mol. Biol.* 98 503–517.

H. Okayama and P. Berg. 1982. High Efficiency Cloning of Full Length cDNA. *Mol. Cell. Biol.* 2 161–170.

V. Pirotta. 1984. Chromosome Microdissection and Microcloning. *Trends Biochem. Sci.* 9 220–221.

M. H. Caruthers. 1985. Gene Synthesis Machines: DNA Chemistry and Its Uses. *Science* 230 281–285.

R. Wu and L. Grossman (eds.). 1987. Chemical Synthesis and Analysis of Oligodeoxynucleotides. *Methods in Enzymology* 154 221–326.

6.4

M. Grunstein and D. S. Hogness. 1975. Colony Hybridization: A Method for the Isolation of Cloned DNAs That Contain a Specific Gene. *Proc. Natl. Acad. Sci.* U.S.A. 72 3961–3965.

R. A. Kramer, J. R. Cameron, and R. W. Davis. 1976. Isolation of Bacteriophage λ Containing Yeast Ribosomal RNA Genes: Screening by *in situ* RNA Hybridization to Plaques. *Cell* 8 227–232.

W. D. Benton and R. W. Davis. 1977. Screening λgt Recombinant Clones by Hydridization to Single Plaques *in situ. Science* 196 180–182.

B. M. Paterson, B. E. Roberts, and E. L. Kuff. 1977. Structural Gene Identification and Mapping by DNA-mRNA Hybrid-Arrested Cell-Free Translation. *Proc. Natl. Acad. Sci.* U.S.A. 74 4370–4374.

M. Grunstein and J. Wallis. 1979. Colony Hybridization. *Methods in Enzymology* 68 379–389.

R. Wu (ed.). 1979. Screening and Selection of Cloned cDNA. *Methods in Enzymology* 68 389–454.

R. A. Young and R. W. Davis. 1983. Efficient Isolation of Genes Using Antibody Probes. *Proc. Natl. Acad. Sci.* U.S.A. 80 1194–1198.

6.5

T. Maniatis, R. C. Hardison, E. Lacy, J. Lauer, C. O'Connell, D. Quon, G. K. Sim, and A. Efstratiadis. 1978. The Isolation of Structural Genes from Libraries of Eucaryotic DNA. *Cell* 15 687–701.

A. Efstratiadis and L. Villa-Komaroff. 1979. Cloning of Double-Stranded cDNA. In J. K. Setlow and A. Hollaender (eds.), Genetic Engineering, vol. 1, pp. 15–36. Plenum, New York.

T. D. Sargent and I. B. Dawid. 1983. Differential Gene Expression in the Gastrula of *Xenopus laevis. Science* 222 135–139.

G. F. Carle and M. V. Olson. 1984. Separation of Chromosomal DNA Molecules from Yeast by Orthogonal-Field-Alternation Gel Electrophoresis. *Nucleic Acids Res.* 12 5647–5664.

D. C. Schwartz and C. R. Cantor. 1984. Separation of Yeast Chromosome-Sized DNAs by Pulsed Field Gradient Gel Electrophoresis. *Cell* 37 67–75.

G. Chu, D. Vollrath, and R. W. Davis. 1986. Separation of Large DNA Molecules by Contour-Clamped Homogeneous Electric Fields. *Science* 234 1582–1585.

R. Wu and L. Grossman (eds.). 1987. cDNA Cloning. *Methods in Enzymology 154 3–83.*

6.6

E. M. Tobin and J. Silverthorne. 1985. Light Regulation of Gene Expression in Higher Plants. *Annu. Rev. Plant Physiol.* 36 569–593.

R. J. Konopka. 1987. Genetics of Biological Rhythms in *Drosophila. Annu. Rev. Genet.* 21 227–236.

M. Heinsenberg. 1988. Designing Heat-Shock Clocks. *Nature* 333 19–20.

7.1

M. Thomas, R. L. White, and R. W. Davis. 1976. Hybridization of RNA to Double-Stranded DNA: Formation of R Loops. *Proc. Natl. Acad. Sci.* U.S.A. 73 2294–2298.

R. L. White and D. S. Hogness. 1977. R Loop Mapping of the 18S and 28S Sequences in the Long and Short Repeating Units of *Drosophila melanogaster* rDNA. *Cell* 10 177–192.

L. Grossman and K. Moldave (eds.). 1980. Determination of Fragment Ordering. *Methods in Enzymology* 65 429–494.

R. C. Parker. 1980. Determination of DNA Fragment Sizes. *Methods in Enzymology* 65 415–426.

M. M. Gottesman (ed.). 1987. *Methods in Enzymology* 151. (The volume contains many articles that pertain to the material in Chapter 7.)

R. Wu (ed.). 1987. *Methods in Enzymology* 155. (The volume contains many articles that pertain to the material in Chapter 7.)

7.2

A. M. Maxam and W. Gilbert. 1977. A New Method for Sequencing DNA. *Proc. Natl. Acad. Sci.* U.S.A. 74 560–564.

F. Sanger, S. Nicklen, and A.R. Coulsen. 1977. DNA Sequencing with Chain-Terminating Inhibitors. *Proc. Natl. Acad. Sci.* U.S.A. 74 5463–5467.

F. Sanger. 1980. Determination of Nucleotide Sequences in DNA. Nobel lecture, December 8. *Bioscience Reports* 1 3–18.

G. M. Church and W. Gilbert. 1984. Genomic Sequencing. *Proc. Natl. Acad. Sci.* U.S.A. 81 1991–1995.

K. B. Mullis and F. A. Faloona. 1987. Specific Synthesis of DNA *in vitro* via a Polymerase-Catalyzed Chain Reaction. *Methods in Enzymology* 155 335–350.

L. M. Smith, R. J. Kaiser, J. Z. Sanders, and L. E. Hood. 1987. The Synthesis and Use of Fluorescent Oligonucleotides in DNA Sequence Analysis. *Methods in Enzymology* 155 260–301.

D. R. Engelke, P. A. Hoener, and F. S. Collins. 1988. Direct Sequencing of Enzymatically Amplified Human Genomic DNA. *Proc. Natl. Acad. Sci.* U.S.A. 85 544–548.

R. J. Saiki, D. H. Gelfand, S. Stoffel, S. J. Scharff, R. Higuchi, G. T. Horn, K. B. Mullis, and H. A. Erlich. 1988. Primer-Directed Enzymatic Amplification of DNA with a Thermostable DNA Polymerase. *Science* 239 487–491.

7.3

T. R. Gingeras and R. J. Roberts. 1980. Steps Toward Computer Analysis of Nucleotide Sequences. *Science* 209 1322–1328.

C. L. Queen and L. J. Korn. 1980. Computer Analysis of Nucleic Acids and Proteins. *Methods in Enzymology* 65 595–609.

D. Söll and R. J. Roberts (eds.). 1984. The Applications of Computers to Research. Nucleic Acids, parts I and II. IRL Press, Oxford, England.

M. J. Bishop and C. J. Rawlings. 1987. Nucleic Acid and Protein Sequence Analysis: A Practical Approach. IRL Press, Oxford, England.

R. F. Doolittle. 1987. Of URFS and ORFS: A Primer on How to Analyze Derived Amino Acid Sequences. University Science Books, Mill Valley, California.

G. von Heijne. 1987. Sequence Analysis in Molecular Biology: Treasure Trove or Trivial Pursuit? Academic Press, New York.

C. DiLisi. 1988. Computers in Molecular Biology: Current Applications and Emerging Trends. *Science* 240 47–52.

Nucleic Acids Research 16 (5A) 1988, whole issue.

7.4

M. E. Harper and G. F. Saunders. 1981. Localization of Single Copy DNA Sequences on G-Banded Human Chromosomes by *in situ* Hybridization. *Chromosoma* 83 431–439.

F. H. Ruddle. 1982. A New Era in Mammalian Gene Mapping: Somatic Cell Genetics and Recombinant DNA Methodologies. *Nature* 294 115–119.

P. D'Eustachio and F. H. Ruddle. 1983. Somatic Cell Genetics and Gene Families. *Science* 220 919–928.

F. M. Rosenburg, Y.-S. Ho, and A. Shatzman. 1983. *Methods in Enzymology* 101 123.

7.5

J. G. Wetmur and N. Davidson. 1968. Kinetics of Renaturation of DNA. *J. Mol. Biol.* 31 349–370.

S. S. Longacre and B. Mach. 1979. A Precise Quantitation of Gene Number by Saturation Hybridization Using Cloned DNA. *Nucleic Acids Res.* 6 1241–1258.

7.6

E. Melgar and D. A. Goldthwait. 1968. Deoxyribonucleic Acid Nucleases II. The Effects of Metals on the Mechanism of Action of Deoxyribonuclease I. *J. Biol. Chem.* 243 4409–4416.

T. E. Shenk, J. Carbon, and P. Berg. 1976. Construction and Analysis of Viable Deletion Mutants of Simian Virus 40. *J. Virology* 18 664–671.

C. A. Hutchison III, S. Phillips, M. H. Edgell, S. Gillam, P. Jahnke, and M. Smith. 1978. Mutagenesis at a Specific Position in a DNA Sequence. *J. Biol. Chem.* 253 6551–6560.

R. M. Myers, L. S. Lerman, and T. Maniatis. 1985. A General Method for Saturation Mutagenesis of Cloned DNA Fragments. *Science* 229 242–247.

D. Shortle and D. Botstein. 1985. Strategies and Application of *in vitro* Mutagenesis. *Science* 229 1193–1201.

M. Smith. 1985. *In vitro* Mutagenesis. *Annu. Rev. Genet.* 19 423–462.

R. Wu and L. Grossman (eds.). 1987. Site-Specific Mutagenesis and Protein Engineering. *Methods in Enzymology* 154 329–533.

7.7

A. J. Berk and P. A. Sharp. 1978. Structure of the Adenovirus 2 Early mRNAs. *Cell* 14 695–711.

D. R. Parks, V. M. Bryan, V. T. Oi, and L. A. Herzenberg. 1979. Antigen-Specific Identification and Cloning of Hybridomas with a Fluorescence-Activated Cell Sorter. *Proc. Natl. Acad. Sci.* U.S.A. 76 1962–1966.

J. B. Gurdon and M. P. Wickens. 1983. The Use of *Xenopus* Oocytes for the Expression of Cloned Genes. *Methods in Enzymology* 101 370–386.

P. Kavathas, V. P. Sukhatme, L. A. Herzenberg, and J. R. Parnes. 1984. Isolation of the Gene Encoding the Human T-Lymphocyte Differentiation Antigen Leu-2(T8) by Gene Transfer and cDNA Subtraction. *Proc. Natl. Acad. Sci.* U.S.A. 81 7688–7692.

7.8

W. Gilbert and L. Villa-Komaroff. 1980. Useful Proteins from Recombinant Bacteria. *Sci. American* 243 74–95.

R. Wu and L. Grossman (eds.). 1987. Expression Vectors. *Methods in Enzymology* 153 385–563.

7.9

H. A. Erlich, D. H. Gelfand, and R. K. Saiki. 1988. Specific DNA Amplification. *Nature* 331 461–462.

M. A. Frohman, M. K. Dush, and G. R. Martin. 1988. Rapid Production of Full-Length cDNAs from Rare Transcripts: Amplification Using a Single Gene-Specific Oligonucleotide Primer. *Proc. Natl. Acad. Sci.* U.S.A. 85 8998–9002.

H. Li, U. B. Gyllensten, X. Cui, R. K. Saiki, H. A. Erlich, and N. Arnheim. 1988. Amplification and Analysis of DNA Sequences in Single Human Sperm and Diploid Cells. *Nature* 335 414–417.

N. Gautam, M. Baetscher, R. Aebersold, and M. I. Simon. 1989. A G Protein Gamma Subunit Shares Homology with ras Proteins. *Science* 244 971–974.

E. Y. Loh, J. F. Elliott, S. Cwirla, L. L. Lanier, and M. M. Davis. 1989. Polymerase Chain Reaction with a Single-Sided Specificity: Analysis of T Cell Receptor δ Chain. *Science* 243 217–220.

The Molecular Anatomy, Expression, and Regulation of Eukaryotic Genes

PART III

Molecular cloning has made essentially every segment of any genome accessible, allowing us to overcome the impasse posed by the complexity of eukaryotic genomes and the difficulty in their genetic analysis. Now, specific, pure DNA segments from any genome can be obtained in sufficient quantities to allow complete chemical description; the use of restriction endonucleases and nucleotide sequence analysis makes such descriptions almost routine. Techniques for deducing the arrangement of overlapping cloned DNA segments have already defined the nucleotide sequences of regions containing more than 150 kbp in mammalian genomes. Although the size and complexity of eukaryotic genomes remain formidable (Table III.1), there is no longer a substantive barrier to knowing the molecular details of their organization and structure. The remaining obstacles are the realities of time and resources. It took about five person-years to determine the sequence of the 50 kilobase pairs in the λ phage genome. Using already available improvements in nucleotide sequencing methods, substantial progress has been made in determining the 4 million base pairs of the *E. coli* chromosome. Moreover, plans are under way to analyze a small human chromosome that is only about ten times larger.

The nucleotide sequence of a gene and its flanking segments alone do not tell us how the gene works or how its expression is regulated during development and differentiation or in response to en-

Table III.1 The Size of Some Eukaryotic Genomes

Species Name		Approximate Size of Haploid Genome (bp)	Haploid Chromosome Number
Common	Proper		
Yeast	*Saccharomyces cerevisiae*	1.35×10^7	16
Slime mold	*Dictyostelium discoides*	7×10^7	7
Trypanosome	*Trypanosoma brucei*	8×10^7	Unknown
Nematode	*Caenorhabditis elegans*	8×10^7	11/12
Silk moth	*Bombyx mori*	5×10^8	28
Vinegar fly	*Drosophila melanogaster*	1.65×10^8	4
Sea urchin	*Strongylocentrotus purpuratus*	8×10^8	21
Toad	*Xenopus laevis*	3×10^9	18
Salamander	*Necturus maculosis*	5×10^{10}	19
Chicken	*Gallus domesticus*	1.2×10^9	39
Mouse	*Mus musculus*	3×10^9	20
Cow	*Bovis domesticus*	3.1×10^9	60
Human	*Homo sapiens*	2.9×10^9	23
Corn	*Zea mays*	5×10^9	10
Onion	*Allium cepa*	1.5×10^{10}	8
Mouse ear cress	*Arabidopsis thaliana*	7×10^7	5

vironmental changes. Neither does the sequence reveal how gene expression is coordinated to ensure the intricate physiological balance characteristic of healthy cells and organisms. Nor can we deduce how the production of a faulty gene product or an altered rate of gene expression will disrupt normal function or produce disease. To comprehend the physiological significance of the molecular details requires biological analysis. Here, too, recombinant DNA techniques provide a powerful experimental approach. When cloned genes are introduced into cells by transfection or injection, they are often expressed and even respond to specific regulatory influences. This provides a way to analyze the relative activities of mutant and normal genes and regulatory signals; both naturally occurring mutant alleles and experimentally constructed mutants can be assessed. Thus, the really special consequence of the recombinant DNA development is that studies of the interdependence of structure and function can now be undertaken. This analytic approach, **reverse genetics,** replaces many of the methods of classical genetics and expands the potential of genetic research in extraordinary ways.

The application of classical genetic and cytogenetic approaches to such organisms as humans was necessarily limited, but now such genomes can be studied at the same fundamental level as had been possible only with bacteria. A few micrograms of DNA isolated from lymphocytes and converted into a library can be used to determine the relation between genotype and phenotype for a large group of functions. Moreover, the universality of recombinant DNA techniques means that the genomes of virtually all obtainable contemporary organisms (and even ancient or extinct organisms whose remains can yield DNA) are accessible, and their similarities and differences can be compared. Important clues about the history of genome evolution can be deduced from such comparisons, adding a new dimension to the classical tools of the evolutionary biologist.

The grand and sweeping statements just made are not wishful thinking. Admittedly, the promise of recombinant DNA techniques has been only partially realized, and many, perhaps most, features of eukaryotic genetic systems remain to be discovered. But even in the short span of about 15 years since the methods were developed, several truly unique aspects of eukaryotic genomes have been discovered.

Our earliest notions about the molecular structure of genes and their genomic organization were predicated on studies with prokaryotes. The major difference between the two classes of organisms appeared to be the sequestering of eukaryotic chromosomes in a nucleus. Some biologists were so bold as to predict that what was true for *E. coli* genes would be equally true of elephants and humans! Now we know that prokaryotes differ from eukaryotes in profound ways. Detailed analysis of many genes and genomes has revealed major and remarkable differences between the two genetic systems. Thus, although eukaryotic genes share certain structural features with their prokaryotic counterparts, the former possess a complexity that drastically alters the rules governing their expression and regulation. Furthermore, both the diversity of DNA segments and their intricate arrangement in eukaryotic genomes have amazed even those who were certain that eukaryotic genomes were more complex than those of prokaryotes. And rearrangement of DNA, once thought to occur only very rarely, is now known to be an inherent, even essential, process. In a few instances, programmed rearrangements are critical features of developmental regulation, and random rearrangements play an important role in evolutionary change. The implications of these complex rearrangements for gene function and for maintaining genome integrity are staggering, but still largely enigmatic.

Overall, the regulation of eukaryotic gene expression is complex and far from being understood. When the whole story is known, it is sure to include mechanisms that have not yet been imagined. Even from the present perspective, it is clear that evolution has yielded a wide variety of mechanisms to control genomic activity.

The Structure and Expression of Eukaryotic Genes

Long before eukaryotic gene structure could be studied directly, it was known that the genetic systems of prokaryotes and eukaryotes

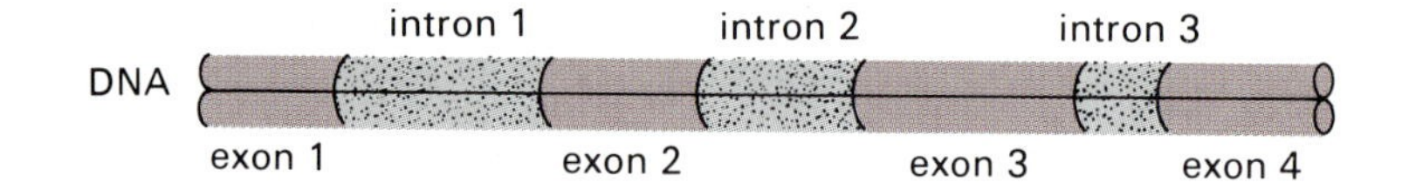

Figure III.1
The organization of exons (coding) and introns (noncoding) in a eukaryotic gene.

share certain fundamental properties. In both, the genetic material is DNA, DNA replication is semiconservative, genetic information flows from DNA to RNA to protein, and the genetic code is the same. It seemed reasonable to expect that eukaryotic organisms, especially multicellular ones, would have more genes than prokaryotes and that they would possess additional control mechanisms to regulate their developmental processes and to integrate their many interrelated functions. But it was widely assumed that the basic structure of the genes and the organization of genetic information in the two types of genomes would be similar. Yet both of these assumptions were shattered almost as soon as the first eukaryotic genes and genomes were analyzed.

Interrupted Genes

By the late 1970s, it was apparent that protein coding sequences in a mammalian gene are not necessarily contained in a single contiguous stretch of DNA, as they are in virtually all prokaryotic genes. Instead, the coding regions are discontinuous, being interrupted by stretches of noncoding DNA; such noncoding DNA segments are called **intervening sequences** or **introns**, and the coding segments of genes are referred to as **exons** (Figure III.1). Subsequently, it became evident that introns are present in most mammalian genes and are also abundant in the genomes of other eukaryotes and certain animal viruses.

The first inklings of the existence of introns emerged from studies of animal viruses. These found that the nucleotide sequences of various messenger RNAs were not colinear with the corresponding nucleotide sequences in the virus DNA from which they were transcribed; indeed, the messenger RNAs seemed to be composed of nucleotide sequences present in discontinuous stretches in the viral DNA. Also, the sequences in β-globin messenger RNA were found to anneal to a collection of genomic restriction endonuclease fragments that together were considerably larger than the size of the messenger RNA. These findings suggested that the polynucleotide segments constituting the messenger RNA were scattered over several segments of the corresponding genome.

In order for the coding sequences of such split genes to produce a functional segment of genetic information, the intervening sequences must be removed, and the coding sequences joined. Intervening sequences can occur between individual codons or within a single codon, and intron removal must be precise so that neither the sequence of codons nor the reading frame is destroyed. It is now clear that the primary (or initial) RNA transcripts of split genes contain the intron sequences and that the introns are removed from the

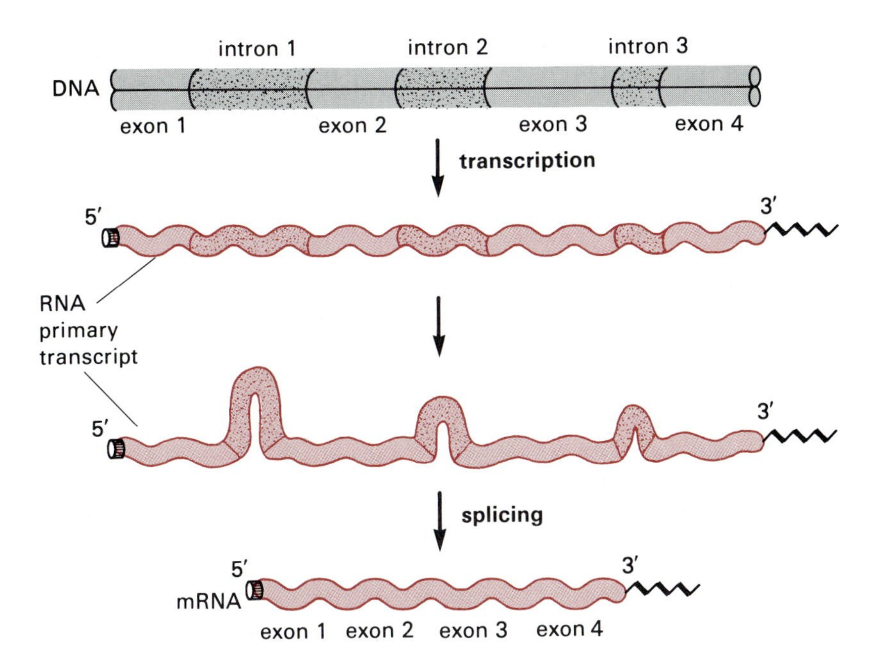

Figure III.2
Splicing a primary transcript removes intron sequences and joins the exons to yield mature messenger RNA.

RNA by a process called **splicing** (Figure III.2). In this way, the interrupted version of the gene is preserved in the DNA for repeated expression and faithful replication. Thus, unlike prokaryotic messenger RNAs, the mature eukaryotic messenger RNAs produced from interrupted genes differ from the nucleotide sequences of the corresponding genes by the absence of the intron segments.

Most eukaryotic genes that code for polypeptides have at least one intervening sequence, and some have very many more (Table III.2). However, intervening sequences are not obligatory, and some polypeptide genes have none at all. Genes that encode functional RNA molecules like transfer RNA or ribosomal RNA may also have intervening sequences, but interruptions in these coding sequences are less common. Generally, the amount of DNA in the introns far exceeds the amount within the exons. The existence of intervening sequences and the enormous variation in their number and size explains several oddities about eukaryotic nuclear RNA, namely, its considerable length and heterogeneity. The **heterogeneous nuclear RNA**, or **hnRNA** for short, is a mixture of transcripts of many nuclear genes. Some of them are primary transcripts and are as long as the genes from which they were copied, but others are partly processed and lack varying numbers of introns. It is a wonder that functional cytoplasmic messenger RNAs are produced from such a complex mixture of RNA precursors.

Regulation of Gene Expression in Eukaryotes

The analysis of eukaryotic gene structure and function has revealed that prokaryotic and eukaryotic genes differ strikingly in the complexity of the short DNA sequences (**motifs**) that regulate transcription (Figure III.3). Three different kinds of regulatory motifs have been identified in prokaryotic genes. One kind includes the sequences that

Table III.2 Intervening Sequences in a Few Representative Genes

		Exons	*Introns*	
Gene Product	Organism	Total bp	Number	Total bp
Adenosine deaminase	Human	1500	11	30,000
Apolipoprotein B	Human	14,000	28	29,000
β-globin	Mouse	432	2	762
Cytochrome b	Yeast (mitochondria)	2200	6	5100
Dihydrofolate reductase	Mouse	568	5	31,500
Erythropoietin	Human	582	4	1562
Factor VIII	Human	9000	25	177,000
Fibroin (silk)	Silk worm	18,000	1	970
Hypoxanthine phosphoribosyl transferase	Mouse	1307	8	32,000
α-interferon	Human	600	0	0
Low density lipoprotein receptor	Human	5100	17	40,000
Phaseolin	French bean	1263	5	515
Thyroglobulin	Human	8500	>40	100,000
$tRNA^{Tyr}$	Yeast	76	1	14
Uricase subunit	Soybean	300	7	4500
Vitellogenin	Toad	6300	33	20,000
Zein	Maize	700	0	0

determine where transcription begins; in *E. coli*, the promoters are defined by two DNA regions that occur about 10 and 35 base pairs before the site of transcriptional initiation. A second kind of element marks the end of a gene or group of genes and triggers termination of transcription. And, finally, there are DNA sequences adjacent to, overlapping, or just beyond the promoter that are recognized by specific effectors such as repressors, activators, and antiterminators to modulate transcription. All these specific regulatory sequences depend upon interactions with proteins to influence the expression of neighboring coding sequences.

The DNA sequences that regulate eukaryotic gene expression occur in a variety of locations and at various distances and directions relative to the transcriptional start and stop sites. Moreover, three different RNA polymerases, I, II and III, transcribe three distinct classes of genes, and each class is associated with different and specific transcriptional control and terminator signals. Molecular genetic and biochemical studies revealed the themes governing transcriptional regulation and provided a basis for understanding differential gene expression. Thus, regulatory sequences consist of complex arrays of relatively short DNA sequence motifs. Each motif is a binding site for a specific protein, a **transcription factor**. For example, genes that are uniquely transcribed in lymphoid cells contain an array of motifs that are recognized by cognate transcription factors restricted to lymphoid cells. Similarly, genes expressed only at certain times or

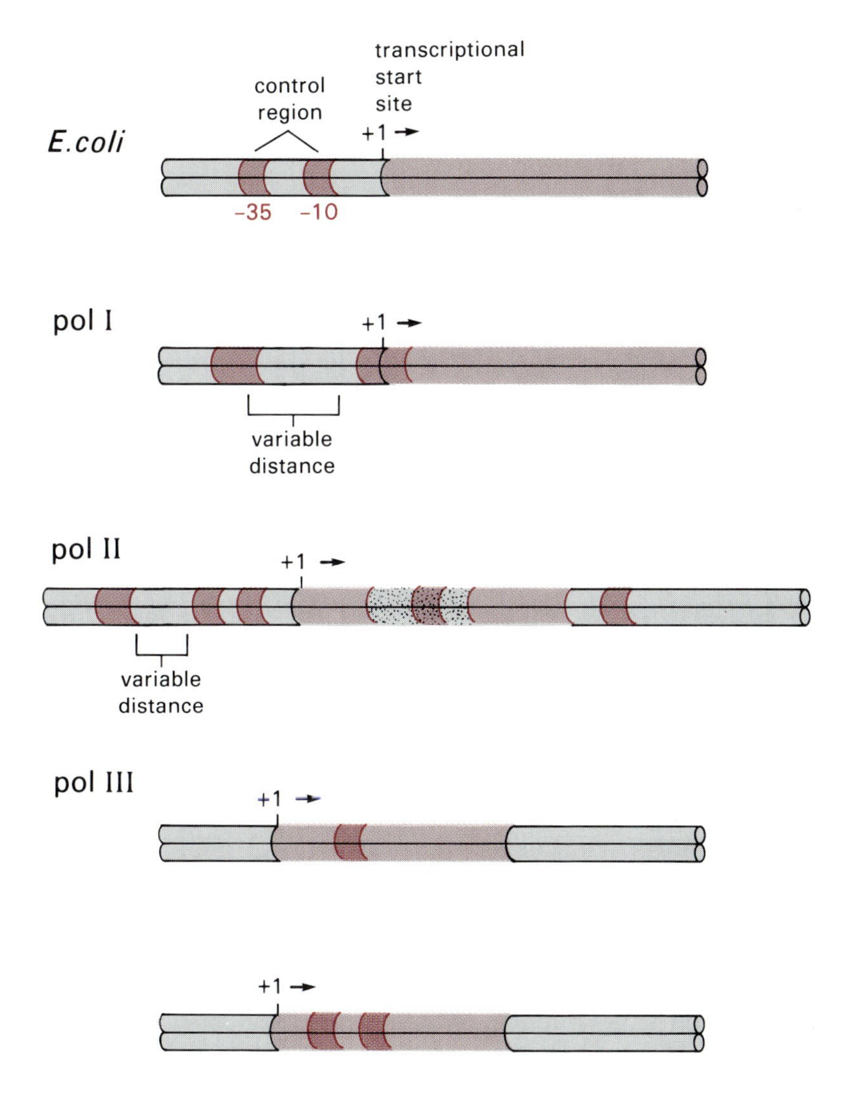

Figure III.3
Schematic drawing showing transcriptional control regions governing the initiation of transcription by *E. coli* RNA polymerase (Chapter 3) and by the three eukaryotic RNA polymerases. RNA polymerase I (pol I) transcribes ribosomal RNA genes. It depends on nucleotide sequences that both surround the transcriptional start site (nucleotide + 1) and occur several hundred to several thousand base pairs in front of the gene. RNA polymerase II (pol II) transcribes genes encoding polypeptides and some RNAs. It depends on complex and variable regulatory regions, including sequences that occur within 100 nucleotides of the transcription start site and others that are as far as 300 to 20,000 bp upstream of the transcription start site; in some instances, important regulatory sequences also occur within or following the gene (note that the exons and an intron are displayed as in Figure III.1). Essential regulatory regions for the RNA genes (transfer RNAs, 5S rRNAs, and other small RNAs) transcribed by RNA polymerase III (pol III) occur within the coding regions, a short distance downstream from the transcription initiation site, as well as upstream of the start site.

only under certain environmental conditions contain regulatory motifs that interact with cognate proteins that are present or active only at those times or under those conditions. Consequently, transcriptional regulation of each gene reflects the particular assortment and arrangement of sequence motifs, the availability of the cognate transcription factors, and the way in which the protein factors influence transcription initiation. Binding of multiple transcription factors in the regulatory region facilitates either the assembly of the relevant RNA polymerase into a transcription complex, the activation of such a complex, or both.

An extraordinary feature of these mechanisms is their universality. Remarkably, corresponding transcription factors from such diverse sources as yeast, *Drosophila*, and mammals are interchangeable. Thus, a yeast or *Drosophila* transcription factor may interact with mammalian sequence motifs and with the other proteins in a mammalian transcription complex as well as does the corresponding mammalian factor. Often, the reverse is also true. This implies a notable conservation of structure in evolution.

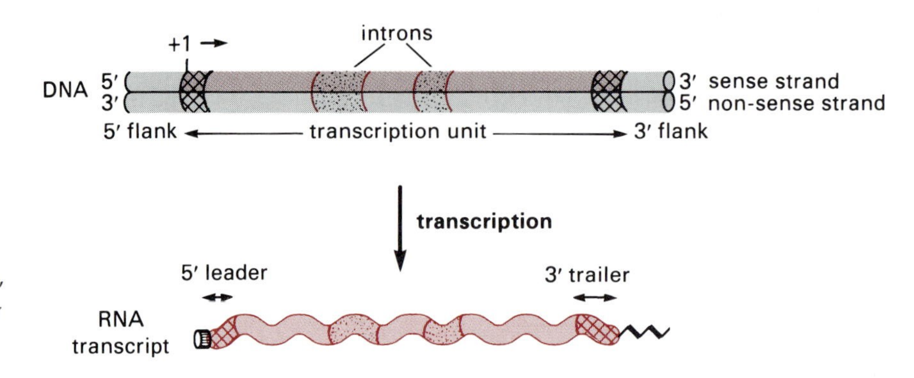

Figure III.4
The DNA segments that comprise a gene include (1) the transcription unit, which contains the 5′ leader and 3′ trailer segments, and the introns and (2) the control regions, which may occur outside of or within the transcription unit. As a general convention, genes are represented diagrammatically from left to right in the direction of transcription. The DNA strand with the same sequence as the messenger RNA is called the sense strand, and its 5′ end is shown at the left and its 3′ end at the right. For convenience, this is often the only strand depicted. The DNA strand that is complementary to the messenger RNA is called the nonsense or template strand. The terms **5′ flank** and **3′ flank** refer to the nucleotide sequences that, respectively, precede and follow transcribed regions.

The mechanism of transcription termination differs with each of the three RNA polymerases; each enzyme relies on sequence motifs at or near the end of the transcription unit. Here, too, the motifs interact with cognate proteins, **termination factors**. Regulation of termination is a consequence of the arrangement of the DNA sequence motifs and the formation of multiprotein assemblies that facilitate termination and modification of the end of the RNA.

The expression of some genes is affected by **epigenetic** changes, which alter a gene's ability to be expressed without altering the basic nucleotide sequence. One form of epigenetic change that has been identified is associated with the level of methylation of the cytosines in 5′-C-G- sequences that flank the 5′ region of a gene. Increased methylation in this region correlates with reduced or no expression of the neighboring gene, and reduced methylation correlates with high levels of expression. Structural changes in chromatin itself, although as yet poorly characterized, can also influence the expression of nearby genes, and high levels of expression may occur only in "unfolded" regions of chromatin.

The unexpected features of eukaryotic genes have stimulated discussion about how a gene, a single unit of hereditary information, should be defined. Several different possible definitions are plausible, but no single one is entirely satisfactory or appropriate for every gene. For the purposes of this book, we have adopted a molecular definition (Figure III.4). A eukaryotic gene is a combination of DNA segments that together constitute an expressible unit, expression leading to the formation of one or more specific functional gene products that may be either RNA molecules or polypeptides. The segments of a gene include (1) the transcribed region (the **transcription unit**), which encompasses the coding sequences, intervening sequences, any 5′ leader and 3′ trailer sequences that surround the ends of the coding sequences, and any regulatory segments included in the transcription unit, and (2) the regulatory sequences that flank the transcription unit and are required for specific expression.

In prokaryotes, transcriptional on-off switches appear to be the primary motif for controlling gene expression, although other mechanisms, such as attenuation, termination, antitermination, translational control, and regulation of the turnover of messenger RNAs and proteins, come into play in special cases. Similarly, transcriptional on-off switches are critical elements in regulating gene expression in eukaryotes. Splicing, particularly the pattern of splicing, influences the kinds of messenger RNAs that are produced and, thereby, the kinds of proteins that are made. There is also suggestive evidence for regulation by transcription attenuation or termination, and there are indications that translation and RNA turnover are also used to regulate protein production. In addition, the unique features of eukaryotic cells and the structure of their genes and genomes confer special opportunities for controlling the flow of genetic information. For example, a complete gene transcript is functionless unless its introns are properly spliced. Similarly, transcripts destined to become messenger RNAs must be modified at their 5′ and 3′ ends; moreover, messenger RNAs must cross the nuclear membrane into the cytoplasm before they can be translated. Each of these events provides a juncture at which regulation may occur.

The Structure of Eukaryotic Genomes

Just as the structure of eukaryotic genes is dramatically different from that of prokaryotic genes, so too is the organization of genetic information distinctive. In bacterial genomes, genes are closely spaced along the DNA, and in some instances, their sequences may overlap. Genes encoding enzymes whose functions are part of the same metabolic pathway, or whose activities are otherwise related, are frequently linked as a single transcription unit. Such an arrangement makes for an efficient use of DNA sequence. But economy seems to have been less important during the evolution of eukaryotes. Quite apart from the large amount of DNA consumed by introns, eukaryotic genes are separated by long stretches of noncoding DNA sequences. As already mentioned, multiple genes in a single transcription unit are rare.

Fundamental to any consideration of eukaryotic genome organization is the fact that the DNA is divided among multiple chromosomes. Cellular chromosomes probably always contain linear duplex DNA, although loops of duplex DNA that create circular domains occur along the otherwise linear backbone. Each interphase chromosome contains a single double helix of DNA, as do each of the two sister chromatids of a metaphase chromosome. The genomes of many eukaryotic viruses, as well as the chromosomes of chloroplasts and many mitochondria, are circular DNAs.

Chromosome Maps

There are no obvious principles governing the distribution of genes among chromosomes. This was already suspected from the extensive mapping of *D. melanogaster* genes by genetic and cytogenetic techniques. These maps, however, accounted for only a small amount of the total DNA, and genetic maps of mammalian genomes were even more primitive. In 1973, only about 60 human genes had been located

Table III.3 The Location of Some Genes on Human Chromosomes

Genes	Chromosome
α-globin cluster	16
β-globin cluster	11
Immunoglobulin	
κ (light chain)	2
λ (light chain)	22
Heavy chain	14
Pseudogenes	9, 32, 15, 18
Viral oncogene homologues	
c-*sis*	22
c-*mos*	8
c-*myb*	6
c-Ha-*ras*-1	11
c-*fes*	15
S-adenosylhomocysteine hydrolase	20
Adenosine deaminase	20
Interferons	
α-cluster	9
β-cluster	9
γ	12
Growth hormone gene cluster	17
Prolactin	6
Major histocompatibility complex (I, II, III)	6
Sarcomeric myosin heavy chains	17
Thymidine kinase	17
Galactokinase	17
Insulin	11
Parathyroid hormone	11
Nerve growth factor, β-subunit	1

on autosomal chromosomes, and by 1981, the number was 400. Now, because of the availability of cloned DNA segments, the number has increased to more than 1000 (Table III.3). However, even the greatly expanded maps fail to imply any systematic placement of genes. Related genes or genes that are present in more than one copy may be clustered together, or they may be dispersed on different chromosomes. For example, genes encoding human cardiac and skeletal actins are on chromosomes 15 and 1, respectively. Also, although the genes encoding the five different human β-globin type polypeptides are clustered in tandem on chromosome 11, those specifying the α-globin type polypeptides are clustered on chromosome 16, even though all the genes are related, and both types of polypeptide are required to form the globin protein of hemoglobin (Figure III.5).

Cloned genes are often obtained with extensive stretches of their flanking DNA intact. An overlapping segment can be cloned from a genomic library if a single-copy DNA segment from the flanking DNA

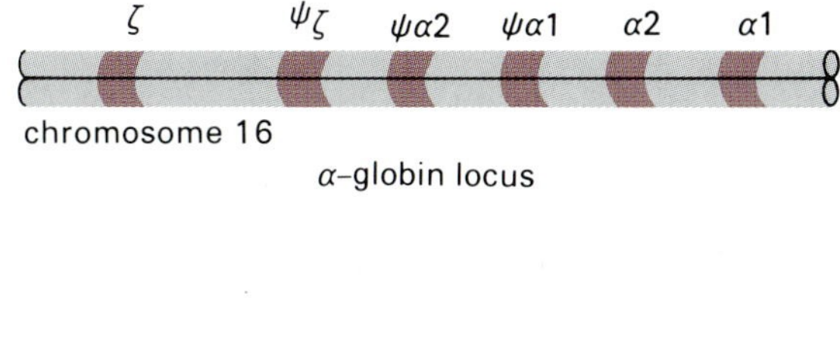

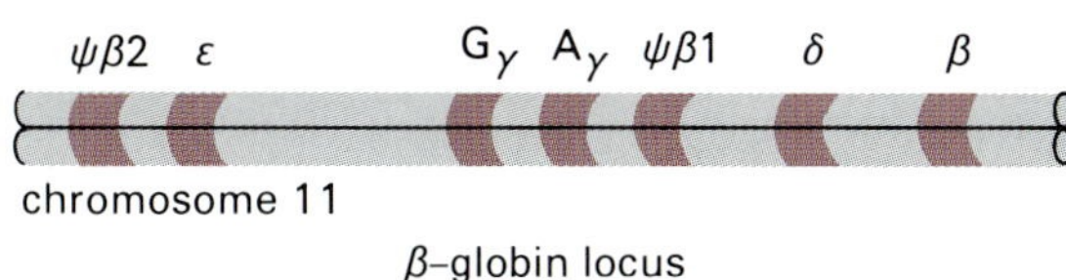

Figure III.5
Maps of the human α- and β-globin gene clusters. Pseudogenes (ψ) occur in both clusters.

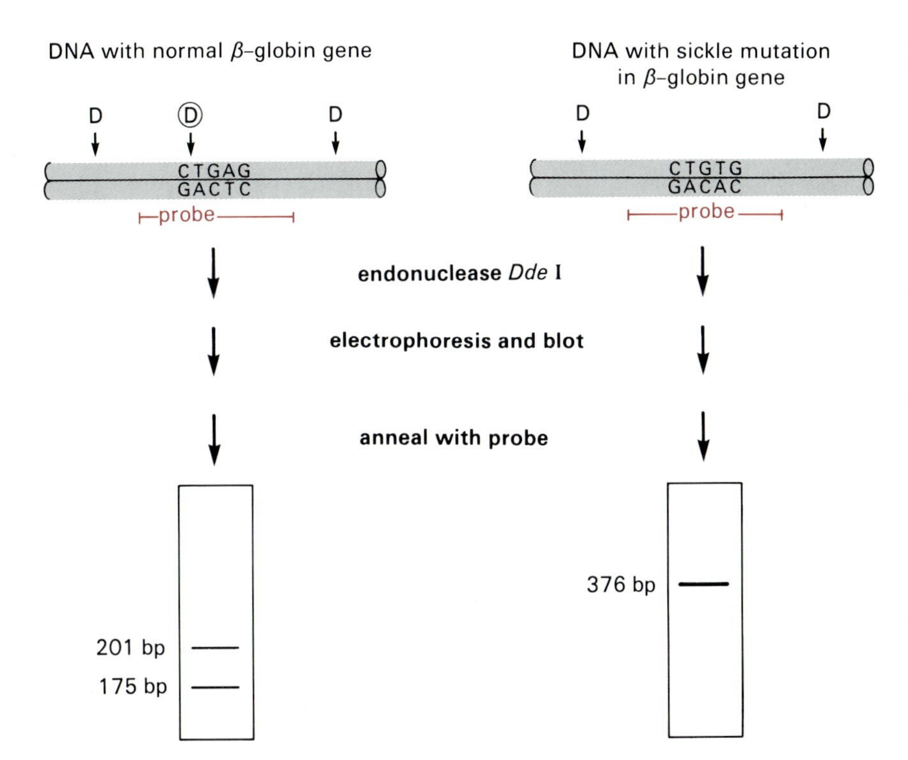

Figure III.6
Restriction endonuclease digests of human DNA detect the mutation in the β-globin gene that is associated with sickle cell disease. The diagram shows normal and mutant DNA. The same base pair change that causes the mutant phenotype eliminates a recognition site for the restriction endonuclease *DdeI*. After digestion with endonuclease *DdeI*, gel electrophoresis, and DNA blotting using a radiolabeled probe that covers this region of the β-globin gene, normal DNA shows two radioactive bands, and the mutant DNA shows one.

is subcloned and used as a probe. And by repeating this procedure, we can obtain extensive chromosomal regions and detailed maps. Both structural and functional information are embodied in these maps. For example, the 65 kbp containing the five human β-globin type genes has been cloned in overlapping segments, and the entire nucleotide sequence is known, thus defining the locations of the coding, regulatory, and intervening sequences.

Maps of the restriction endonuclease sites within a genomic region are extraordinarily useful for detecting variations in the sequence resulting from altered alleles. For example, using a cloned segment from within the human β-globin gene cluster as a hybridization probe, the structure of the β-globin region in DNA from many individuals can be examined by restriction endonuclease digestion and DNA blotting (Figure III.6). A change in the size of a genomic fragment that anneals with the probe indicates an alteration in the sequence either at a particular restriction site or because of a deletion or insertion between two restriction sites; the absence of any fragment that anneals with the probe signifies a deletion of the entire region. Such alterations or **restriction fragment length polymorphisms**, even if they lack phenotypic consequences, provide genetic markers that are as useful as classical markers for genetic analyses on large human families and populations. Indeed, this approach has been used to determine patterns of inheritance for hemoglobinopathies and other genetic diseases. Moreover, a series of polymorphic restriction endonuclease fragments from different positions on a given chromosome can be used as markers for standard genetic linkage analysis, thereby making it possible to construct genetic maps even in the absence of phenotypic markers. This type of analysis has been used to order several genes on the short arm of human chromosome 11: centromere–catalase–parathyroid hormone–ϵ-globin–β-globin–insulin–pepsinogen.

One of the surprising findings emerging from these analyses is the very high frequency of nucleotide sequence differences between individuals. By definition, classical genetic analysis of mutations utilized genomic changes with evident phenotypic consequences. Assuming that the genomes of normal or "wild type" individuals had very few differences seemed reasonable. This simplistic picture is frequently reinforced by emphasizing the existence of dominant and recessive alleles, thereby implying that there may be only two or very few alleles of a particular gene. Yet, restriction fragment length polymorphisms in both coding and noncoding regions are quite frequent, indicating considerable diversity among the genomes of individuals of a species, without any identified phenotypic consequence. There may be subtle effects stemming from some of these polymorphisms, but many will probably turn out to be inconsequential.

The Size of Eukaryotic Genomes

The molecular analysis of eukaryotic genomes is also explaining the apparent paradox that eukaryotic genomes seemed to contain much more DNA than necessary. More than 20 years ago it was argued that, if eukaryotes contain as many as 100,000 genes, and the average size of a gene coding sequence is 1200 bp, then the genome need contain only on the order of 12×10^7 bp. Although some fungal genomes are smaller than this, most mammalian and plant genomes contain 20 to 50 times that amount of DNA (Table III.1). Now, the "excess" DNA can be accounted for in part by the DNA contained in intervening sequences (Table III.2) and by the extensive stretches of DNA between genes.

The fact that some genes appear many times in a eukaryotic genome also contributes to the large size. The existence of multiple copies of a particular gene is not an exclusive property of eukaryotes; *E. coli*, for example, has about seven genes for ribosomal RNA, probably to meet the need for the rapid synthesis of large amounts of ribosomes. Eukaryotes also have multiple gene copies to solve such supply problems; for example, there are several hundred ribosomal RNA genes in mammals. A surprising amount of "excess" DNA is also contributed by nonfunctional copies of genes. These extra copies are called **pseudogenes**, and they are usually unable to express useful information because of deletions or other alterations in their DNA sequence. The number of pseudogenes for a particular gene varies a great deal and is different from one organism to another. Thus, in eukaryotic genomes, a gene may be represented only once, or it may be part of a family of repeated sequences, the members of which may be clustered or dispersed. Members of a family may include several functional genes that are closely related but expressed at different times in development or in different tissues or cells, or they may include pseudogenes, or both (Figure III.5).

Another contribution to the excessive size of eukaryotic genomes is the presence of very large families of repeated DNA sequences whose functions are unknown. Such families often constitute 10 percent and in some instances nearly 50 percent of a genome. More than 15 years ago, analyses of the kinetics of reassociation of denatured eukaryotic DNA showed that large amounts of the DNA reannealed much more rapidly than expected for unique DNA sequences. This high rate of reannealing indicated that such genomes must contain segments that are repeated hundreds of thousands to millions of times. Molecular cloning and sequencing have now confirmed the existence of such highly repeated DNA sequences, which, like repeated genes, are arranged in two different ways. In one arrangement, they occur as tandem arrays, and in the other, they are dispersed among many unlinked genomic loci.

The Relation Between Particular DNA Sequences and Chromosome Morphology

Molecular analysis has revealed that certain kinds of DNA sequences are associated with several distinctive morphological features of chromosomes. The chromosomal regions referred to as nucleolar organizers contain the genes for ribosomal RNA. Telomeric and centromeric regions of most eukaryotic chromosomes contain tandemly repeated nucleotide sequences. Although these repeated nucleotide sequences frequently account for 10 percent or more of the total genomic DNA, their function is largely unknown. Such sequences may not be essential for centromeric function because they are absent from yeast (*Saccharomyces cerevisiae*) chromosomes, where the centromeres function normally in both mitosis and meiosis.

The Rearrangement of DNA Sequences

The arrangement of genes and most other DNA sequences in chromosomes is virtually stable, as is implicit in the establishment of chromosome maps. Genetic information can be reshuffled between homologous chromosomes by reciprocal crossing-over during both meiosis and mitosis. However, although these events bring together new combinations of alleles, they do not cause a reordering of the DNA sequences. Indeed, many gene linkages are much older than individual species, and genes that are syntenic (i.e., on the same chromosome) in one mammal are often syntenic even among mammals in distantly related orders. However, superimposed on the basically stable organization of genomes is a capacity for rearranging genetic information. Rearrangement can occur at random, often producing deleterious mutations. But some rearrangements are normal processes that alter gene expression in an orderly and programmed manner.

Chromosomal aberrations such as breakage, translocation of portions of one chromosome to a nonhomologous chromosome, or formation of dicentric and acentric chromosomes have been known for a long time. At the molecular level, all these morphological alterations are manifested in changes of DNA structure. All organisms encode in their genetic programs the enzymatic machinery to create rearrangements (Section 2.4). These processes include (1) nonreciprocal and nonallelic homologous crossing-over and gene conversion; (2) transposition of DNA segments from one genomic locus to another; (3) formation of additional copies of DNA sequences either in tandem or dispersed to new genomic positions, a process called **amplification**; (4) deletion of DNA sequences; and (5) nonhomologous recombination. Recognition of the fluidity of genomic organization is one of the major achievements of molecular genetics and is especially remarkable because rearrangements of DNA sequences are very infrequent events. This achievement is, to a large extent, a consequence of combining the precision of molecular genetics with the highly selective tools of classical genetic analysis.

Random Genomic Rearrangements

Unprogrammed rearrangements are usually stochastic with respect to time and/or to the nucleotide sequences that are altered. Such rearrangements often produce damaging mutations. However, they may also provide, and may have provided in the past, opportunities for novel responses to environmental conditions and thereby facilitated evolutionary change. Of course, for random rearrangements to have evolutionary significance, they must occur in germ line cells or their precursors; similar events occurring in an occasional somatic cell may be inconsequential to an individual organism. If the somatic cell is a stem cell, however, random rearrangements may prevent the formation of functional, differentiated progeny, or the alteration may be oncogenic and lead to tumor formation.

Among the known types of random genomic rearrangements, one in particular, the transposition of DNA segments from one place to another in a genome, has captured the imagination of both scientists and interested amateurs. "Jumping genes" have had the limelight, even in the popular press. The existence of movable DNA elements was first deduced in the mid-1940s from studies of the genetics and cytogenetics of maize. Twenty years later, bacterial geneticists obtained experimental results that led to the same conclusion; certain DNA segments move about. Ultimately, it was established that many different types of movable elements occur in eukaryotic as well as prokaryotic genomes. Molecular cloning has made it possible to determine the structures of many of these movable elements and to begin to elucidate the actual mechanisms by which they move to new genomic positions. Movable elements are genetic troublemakers and cause mutations of various kinds in both prokaryotes and eukaryotes. When they enter the coding sequence of a gene, the gene's function is lost. When the transposed element lands in the neighborhood of a gene, the expression of that gene can be influenced in complex ways. In fact, many of the mutations that proved so useful early in this century for studying the genetics of *D. melanogaster* are now known to have been caused by insertions of movable elements into genes.

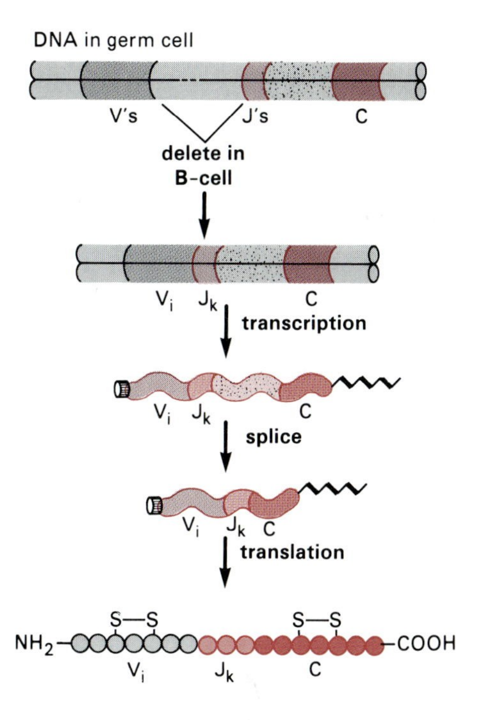

Figure III.7
Genes that encode antibodies are assembled in B lymphocytes by joining DNA segments that are far apart in the DNA of germ line and other types of somatic cells. The example shows, schematically, assembly of a gene for an immunoglobulin light chain from three DNA segments (V, J, and C) that together encode the polypeptide. One each of the multiple Vs and Js in the genome are joined to a single C. Immunoglobulin genes differ, depending on which V and which J is used. In this example, V_i and J_k were joined to C. The diversity of antibody proteins is specified in part by the amino acids encoded by the different Vs and Js (variable region).

Programmed Rearrangements

It is now clear that the expression of genes can be profoundly modified by changing their genomic environments. A gene may be silent with one set of neighboring sequences but expressed if the flanking DNA is changed. Such rearrangements are utilized in a variety of organisms to regulate gene expression and thereby establish specific functions in differentiated cells. Thus, there are genetic programs that cause particular DNA segments to be rearranged in a timely or a cell-specific manner.

Long before it was possible to analyze genomes and genes at the molecular level, geneticists suspected that specific rearrangements might explain certain puzzling biological phenomena. One of these was the alternation between two distinct types of flagellae in individual *Salmonella typhimurium* bacteria. Another, with even more profound implications, was the extraordinary diversity of possible antibody responses in mammals. Molecular analysis has now made it clear that both these phenomena, and others as well, depend on programmed rearrangements of DNA sequences (the immunoglobulin example is illustrated schematically in Figure III.7). In both of these examples, genome reorganization is used to protect the organism from danger. By changing flagellae, the *Salmonella* escapes the immunological offensive of a host organism. By creating new antibodies, an organism defends itself against foreign agents.

Evolutionary Implications

The evolutionary histories of organisms are written in their genomes. Mutations and rearrangements provide genetic variation upon which the evolutionary process depends. In the past, evolutionary relations could be studied only by reference to the consequences of genomic information—the phenotypic properties of organisms. Thus, biological history was initially documented by studying the anatomy of fossils and the comparative anatomy, physiology, and embryology of con-

primate									NH_2-	gly	asp	val	glu	lys	gly	lys	lys	ile	phe	ile	met	lys
wheat	NH_2-	ala	ser	phe	ser	glu	ala	pro	pro	gly	asn	pro	asp	ala	gly	ala	lys	ile	phe	lys	thr	lys
yeast				NH_2-	thr	glu	phe	lys	ala	gly	ser	ala	lys	lys	gly	ala	thr	leu	phe	lys	thr	arg

primate	cys	ser	gln	cys	his	thr	val	glu	lys	gly	gly	lys	his	lys	thr	gly	pro	asn	leu	his	gly	leu	phe
wheat	cys	ala	gln	cys	his	thr	val	asp	ala	gly	ala	gly	his	lys	gln	gly	pro	asn	leu	his	gly	leu	phe
yeast	cys	leu	gln	cys	his	thr	val	glu	lys	gly	gly	pro	his	lys	val	gly	pro	asn	leu	his	gly	ile	phe

primate	gly	arg	lys	thr	gly	gln	ala	pro	gly	tyr	ser	tyr	thr	ala	ala	asn	lys	asn	lys	gly	ile	ile	trp
wheat	gly	arg	gln	ser	gly	thr	thr	ala	gly	tyr	ser	tyr	ser	ala	ala	asn	lys	asn	lys	ala	val	glu	trp
yeast	gly	arg	his	ser	gly	gln	ala	glu	gly	tyr	ser	tyr	thr	asp	ala	asn	ile	lys	lys	asn	val	leu	trp

primate	gly	glu	asp	thr	leu	met	glu	tyr	leu	glu	asn	pro	lys	lys	tyr	ile	pro	gly	thr	lys	met	ile
wheat	glu	glu	asn	thr	leu	tyr	asp	tyr	leu	leu	asn	pro	lys	lys	tyr	ile	pro	gly	thr	lys	met	val
yeast	asp	glu	asn	asn	met	ser	glu	tyr	leu	thr	asn	pro	lys	lys	tyr	ile	pro	gly	thr	lys	met	ala

primate	phe	val	gly	ile	lys	lys	lys	glu	glu	arg	ala	asp	leu	ile	ala	tyr	leu	lys	lys	ala	thr	asn	glu-COOH
wheat	phe	pro	gly	leu	lys	lys	pro	gln	asp	arg	ala	asp	leu	ile	ala	tyr	leu	lys	lys	ala	thr	ser	ser-COOH
yeast	phe	gly	gly	leu	lys	lys	glu	lys	asp	arg	asn	asp	leu	ile	thr	tyr	leu	lys	lys	ala	cys	glu-COOH	

Figure III.8
The structure of cytochrome C is similar in yeast, wheat, and primates. The sequences are read left to right on each line. Amino acids that are identical in all three species are colored. The likelihood that such identity would occur by chance is extremely low.

temporary species. In the middle of this century, comparative protein chemistry was introduced as another phenotypic measure of evolutionary relationships. The discovery that the structure of certain proteins is highly conserved over a broad spectrum of organisms brought many phyla into a common evolutionary history in a way that the fossil record could not achieve. Cytochrome C, for example, is ubiquitous in the mitochondria of plants and animals. It is critical to respiration and must have been among the first essential protein molecules. In the course of evolution, cytochrome C has changed very little (Figure III.8); only 35 out of the 104 amino acids are different in cytochrome C of yeast and humans (discounting the extra amino acids at the amino terminus of the yeast protein).

Cytochrome C and other highly conserved proteins are not useful in analyzing the evolutionary relationships between very closely related species because they may not vary at all among kindred organisms; for example, cytochrome C is identical in humans and chimpanzees. But other proteins evolve at faster rates. Carbonic anhydrase is an example of such a protein; the differences in its structure among primates are sufficient to allow construction of phylogenetic trees. This method is based on the observation that the number of amino acid differences is roughly proportional to the time elapsed since two organisms branched from a common ancestor, as determined from fossil data.

Finally, however, all methods of evolutionary analysis that deal with phenotype are indirect; it is genes and genomes that must change when new phenotypes occur in evolution. Although the study of ancient forms will continue to rely on the morphology of fossils, the comparison of DNA structure is rapidly supplanting other methods for deducing relationships between contemporary species. Thus, the recombinant DNA technique has already had a major impact on evolutionary biology, and its influence in this area is likely to be enormous in the future.

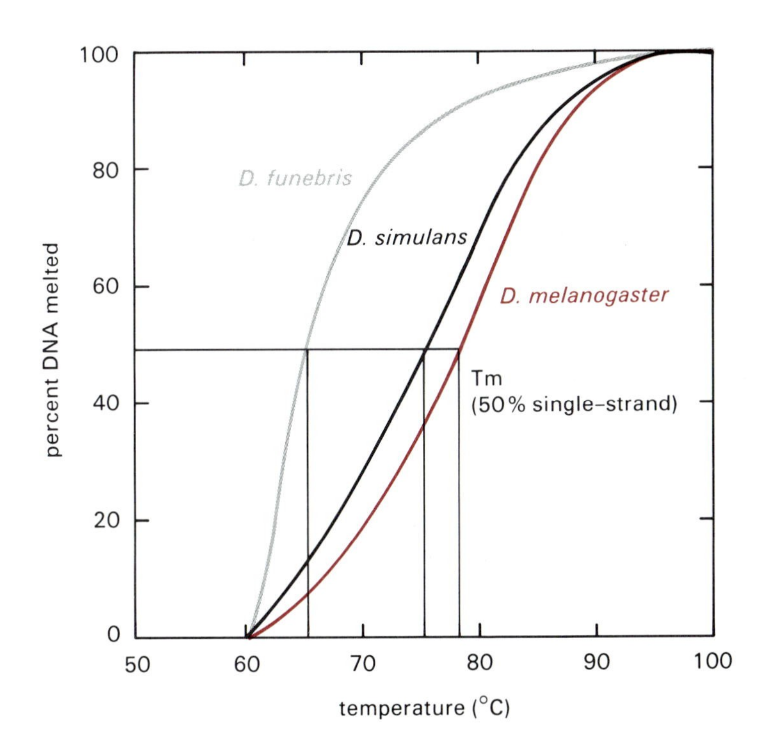

Figure III.9
Melting curves of duplex DNA in which both strands are *Drosophila melanogaster* DNA or in which one strand is *D. melanogaster* and the other is from a closely related species in the same genus. A decrease of 1 degree in melting temperature (Tm) of the heteroduplex compared to that of the homoduplex indicates a mismatch between about 1 percent of the base pairs. Thus, *D. melanogaster* is much more closely related to *D. simulans* than to *D. funebris*. After C. O. Laird and B. J. McCarthy, *Genetics* 60 (1968), p. 303.

Comparative Molecular Genetics

Even before DNA segments could be cloned and sequenced, the importance of DNA sequence comparisons to evolutionary studies was apparent, and methods were developed for assessing the degree to which particular genomes are related. These techniques involve comparing the stability of duplex DNA formed between single strands of two different species (heteroduplex) with the stability of a duplex in which both strands are from a single species (homoduplex). Figure III.9 shows the kind of results obtained with several species of *Drosophila*. Useful as they are, studies of this kind give average values for the divergence between two DNAs. They cannot approach the rich detail of even comparative anatomy and protein chemistry. But comparisons based on direct analysis of the nucleotide sequences in the basic evolving component, the DNA, provide another dimension of detail and, in addition, illuminate previously unapproachable aspects of evolution. Moreover, it is much easier technically to clone and sequence a group of homologous genes than to isolate the polypeptide products and determine their amino acid sequences.

With respect to coding regions, DNA sequences are more informative than polypeptide sequences because the degeneracy of the genetic code allows gene structure to be altered without any change in the corresponding amino acid sequence. And even knowing that an amino acid has been changed does not necessarily reveal how many base pair changes occurred; it allows only an estimate of the minimum number of base pair changes required to substitute one amino acid for another. For example, in replacing an alanine with a valine at position 11 in cytochrome C (Figure III.8), a change of one (e.g., GCU to GUU) or two (e.g., GCU to GUA) base pairs might occur.

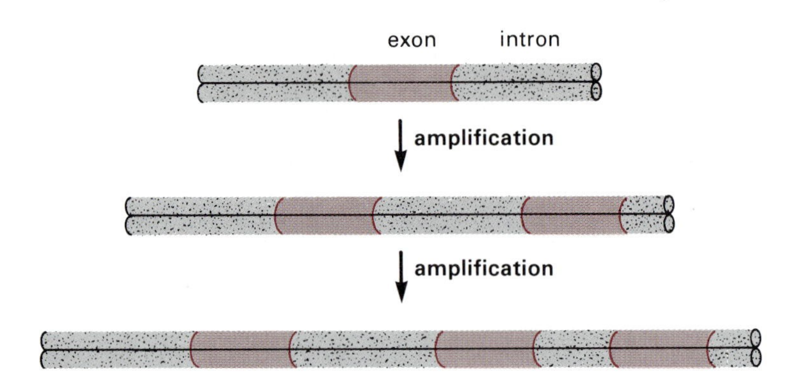

Figure III.10
Some genes evolved by the tandem amplification of a single exon. Examples of such genes include mammalian collagen and α-fetoprotein (Section 9.4a).

A unique advantage of DNA sequences is that noncoding regions, including regulatory sequences, can be compared. This is likely to be of great significance and pertinent to the long evident fact that protein evolution and morphological evolution proceed at very different rates in different groups of organisms. Important phenotypic changes may depend not only on protein structure but on, for example, the relative abundance of individual proteins or the timing of gene expression. DNA sequence comparison can, by analysis of regulatory elements or genomic neighborhoods, elucidate the role of gene regulation in the evolution of species. These changes may, in some cases, involve single base pair changes, but they are also likely to involve genomic rearrangements and the transposition of movable elements.

The sequences of ribosomal RNA, and in particular the 16S to 18S RNA of the small ribosomal subunits, are especially useful for determining evolutionary relationships. Because ribosomes are abundant in all species, the functional RNAs can be isolated and sequenced directly after copying into DNA with reverse transcriptase. Although the function of small subunit RNA is the same in all living things, regions of the molecule are conserved to varying extents. The more highly conserved regions can be used to compare distantly related organisms, and the more rapidly evolving segments provide useful data for closely related species. Thus, all species can be compared through a single type of molecule with a highly conserved function. Representative species of virtually all taxonomic classes have been analyzed. Consequently, detailed phylogenetic relationships have been inferred, providing new insights into evolutionary processes.

The Evolutionary History of Genes

Thus far, we have discussed evolution as it relates to the history of different species. But another kind of biological history, the origin of genes themselves, is also written in DNA structure. Two or more related genes (or DNA sequences) within a single species often derive from a single ancestral sequence. If the ancestral DNA segment was amplified (as a consequence of random events), one copy may still have been sufficient to supply the original function. The additional copies would then be free from the constraints that limit the accumulation of mutations in a functional gene. Mutations in an extra copy could lead to a new functional capacity that, in turn, may be preserved

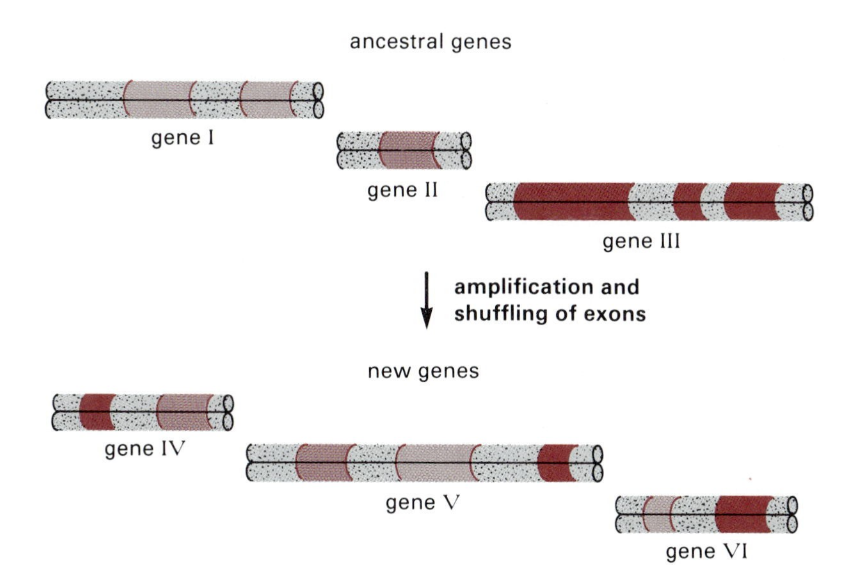

Figure III.11
Some genes evolved by amplifying and shuffling exons. The gene encoding growth hormone and the cell membrane receptor protein for circulating low density lipoprotein appear to have evolved in this way (Sections 9.1c and 9.4a, respectively).

by selection. Alternatively, the additional copies may acquire new or altered regulatory signals and be expressed at different developmental times or in particular differentiated tissues. Still another possibility is that the extra copies could acquire crippling mutations and remain as pseudogenes. The globin gene family includes examples of all three possibilities (Figure III.5).

Some genes seem to have been constructed by joining copies of coding sequences (i.e., of exons). In some cases, this involved the tandem amplification of a single exon, and the repetitious structure of the gene is reflected in the multiple repeats of related peptide segments in the encoded protein (Figure III.10). In other cases, genes appear to be mosaics constructed by shuffling copies of individual exons recruited from different genes (Figure III.11). Thus, exons with the same ancestor occur in several, apparently unrelated genes. Exons that are shared by several proteins are likely to encode polypeptide domains that endow the disparate proteins with related structural or functional properties, such as particular type of secondary structure or the capacity to bind a particular ligand (e.g., ATP).

An interesting hypothesis concerning the assembly of genes by exon repetition and exon shuffling proposes that recombinations within introns served to link exons, thereby leaving the coding segments intact. This view implies that introns are ancient structures and were present in the earliest genes and cells, a view that predicts that introns were lost during the evolution of prokaryotes. A contrasting hypothesis holds that introns represent insertions into preexisting coding regions by a mechanism more prevalent in eukaryotes than in prokaryotes. The idea that introns are ancient avoids several problems inherent in the insertion model, including the mutational consequences of widespread, random insertions of noncoding segments and the need to evolve, simultaneously, a splicing mechanism. In contrast, if introns were present in the earliest genomes, then splicing too might be a very ancient process. The discovery that some introns catalyze their own removal by splicing, without the aid of proteins

(Section 8.5d), has enhanced the plausibility of the very ancient existence of at least some introns. However, the movement of introns from one DNA segment to another is well documented in several instances. The known examples include introns in yeast mitochondrial genes and slime mold ribosomal RNA genes, and all involve a special class of introns (group I). Insertion appears to occur by site-specific gene conversion. Thus, introns may have arisen in several different ways.

The comparison of protein structures provided enough data to permit hypotheses about the role of amplification in genome evolution. But with the availability of cloned genes, these hypotheses have been greatly expanded and strengthened. Most important, molecular analysis reveals that the processes of amplification, divergence, and exon repetition and shuffling were not rare or occasional contributors to genome evolution. The history of substantial portions of eukaryotic DNA, coding and noncoding sequences alike, reflects such processes. The variety of rearrangements seen in contemporary genomes reflects ancient processes that were, and presumably still are, major contributors to evolutionary change.

The Origin of Genetic Systems

What does the proposed great age of introns imply about the nature of the earliest genetic systems? In particular, how does it bear on the question: were the first informational molecules DNA or RNA? Several lines of evidence are consistent with the idea that RNA came first and provided the basis for the earliest coding systems. For example, ribosomal RNA, transfer RNA, and messenger RNA are central elements in the translational apparatus of all organisms and in the operation of the genetic code. This suggests that these molecules predated the evolutionary divergence of prokaryotes and eukaryotes and must have been present in the earliest genetic systems. Moreover, short RNA molecules can be synthesized on an RNA template by purely chemical reactions, in the absence of any proteins. It seems possible then that RNA molecules might have been self-replicating and were transcribed and translated by primitive mechanisms. RNA molecules can also act as enzymes to modify RNAs. The RNA moiety of *E. coli* RNase P catalyzes the site-specific cleavage of transfer RNA precursors (Section 3.3). And, as already mentioned, introns in precursors to ribosomal RNA in certain protozoa and fungi can be spliced out without the intervention of proteins. In contrast, DNA is not known to carry out any of these reactions.

The ancestor of the complex contemporary genetic systems may then have utilized informational RNA molecules (Figure III.12). In these RNAs, short, fortuitously formed coding regions (primitive exons) would have alternated with noncoding regions (ancestral introns). In time, short polypeptides that facilitated RNA processing and replication would have evolved. One of these might have had the properties of a primitive reverse transcriptase and catalyzed the formation of DNA molecules that later became the fundamental depository of cellular genetic information. Such DNA molecules would have retained the exon-intron pattern of the RNA. The genetic system containing the

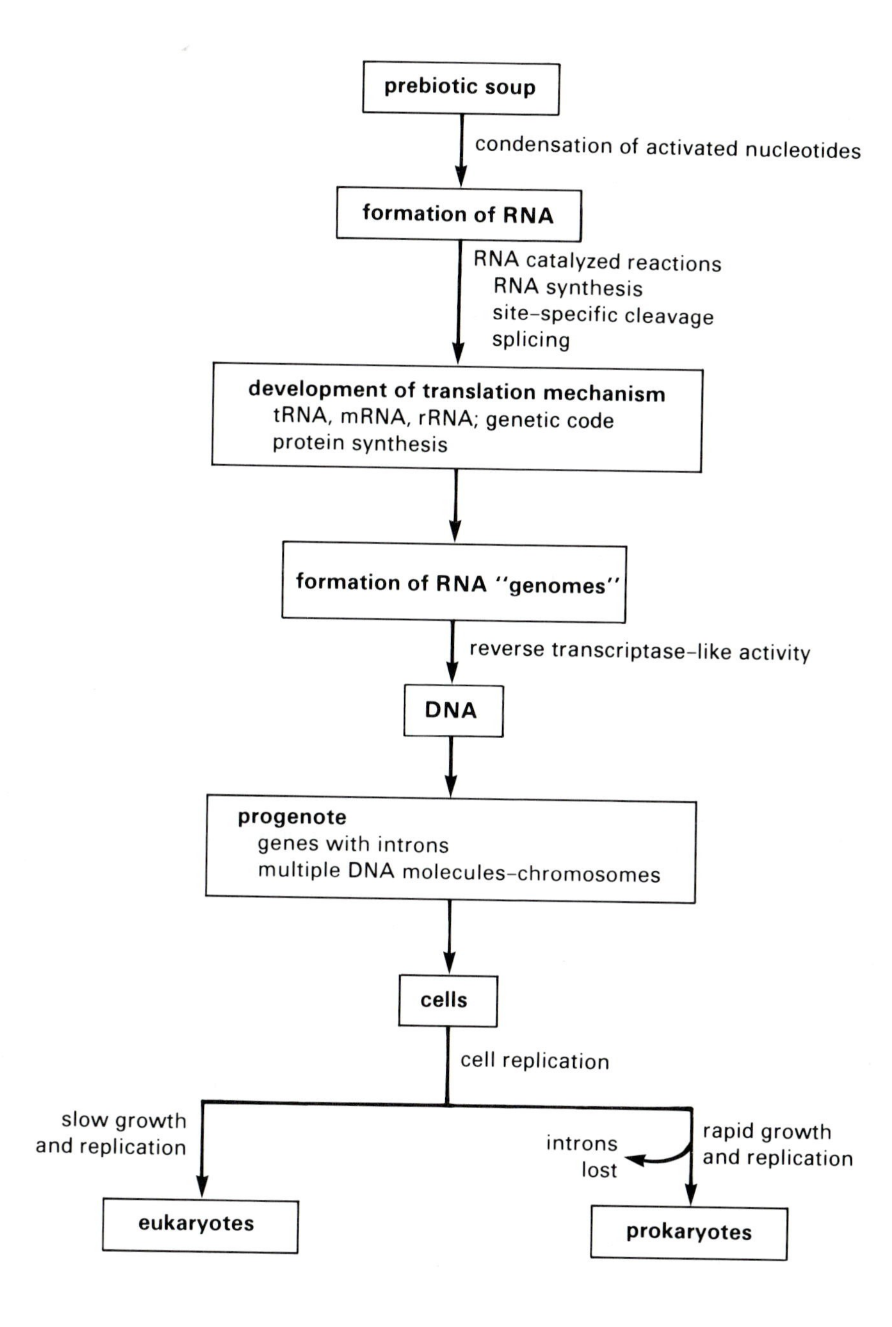

Figure III.12
A speculative scheme for the early molecular and cellular events in the evolution of genetic systems. After J. E. Darnell and W. F. Doolittle, *PNAS* 83 (1986), p. 1271.

DNA, a **progenote**, would become the ancestor of subsequent cellular organisms. Divergent evolution could have given rise to eukaryotic and prokaryotic lines in which retention or loss of introns, respectively, was favored. According to this model, it would not be too surprising to find an occasional intron remaining in a prokaryotic genome. And, indeed, introns occur in polypeptide genes in the *E. coli* bacteriophage T_4 genome as well as in transfer RNA genes in an archaebacterium.

One of the most provocative aspects of the model outlined in Figure III.12 is its implication of independent origins, from ancestral cells, of prokaryotes and eukaryotes. Previously, most speculations regarding early evolution assumed that eukaryotes evolved from prokaryotes. The assumption is based on the idea that the relatively simple contemporary prokaryotic cells are likely to resemble the earliest cells.

Larger and more complex eukaryotic cells are seen to be the product of evolution from the simpler, prokaryotic types. The discovery of introns, splicing, cellular reverse transcriptases, and the catalytic properties of RNA have forced a reevaluation of such concepts. Thus, our ability to analyze the molecular aspects of biological information systems is shaping new approaches to major questions in evolutionary biology.

Conclusion

The recombinant DNA technique and the related technologies of nucleotide sequencing and restriction endonuclease analysis have affected every facet of biological research. Many long-standing puzzles are now resolved, such as the excessive amounts of DNA in eukaryotic organisms and the origin of heterogeneous nuclear RNA. These questions have inevitably been replaced with new ones. However, we can face today's questions optimistically because biological systems no longer seem hopelessly complex.

The three chapters in Part III—Chapters 8, 9, and 10—summarize our present understanding of the structure and expression of eukaryotic genes, the organization of genetic elements, and the kind of rearrangements that occur in eukaryotic genomes, respectively. They take a general rather than a comprehensive approach and strive to show how the new methodologies of molecular genetics provide us with a fresh way of thinking about complex organisms.

CHAPTER 8

The Structure and Regulated Expression of Eukaryotic Genes

The successes of molecular genetics during the 1950s and 1960s engendered a widespread confidence that prokaryotic and eukaryotic genes would be substantially the same in their basic design and mode of expression. After all, it was already evident that both types of organisms shared the same genetic code and used RNA polymerases for transcription and mRNAs, ribosomes, and tRNAs to translate their genetic messages. But even the scanty information about eukaryotic gene regulation that was available by 1970 indicated that these organisms used more complex and possibly wholly new mechanisms for gene expression and regulation. For example, there was clear evidence that eukaryotic mRNAs differ from those in prokaryotes by having unusual modifications of their 5′ termini—so-called "cap" structures instead of the triphosphate group associated with the first nucleotide at the ends of prokaryotic mRNA chains. Moreover, eukaryotic mRNAs have long polyadenylate extensions at their 3′ ends. Aside from the structural distinctions, there were indications that the mechanisms of RNA biogenesis in the two types of organisms differ in fundamental ways. For example, eukaryotic cells use three different kinds of RNA polymerase to synthesize the assorted species of RNA, but prokaryotes make do with one. Moreover, only about 5 percent of the RNA synthesized in the nucleus exits to the cytoplasm. Especially puzzling was the existence of very long nuclear RNA molecules containing the same nucleotide sequences and 5′ and 3′ modifications that occur in smaller cytoplasmic mRNAs.

The speculations generated by these unanticipated findings were difficult to assess in the absence of reliable information about gene structure and methods for studying gene expression. But the advent of molecular cloning techniques and the ensuing structural and functional studies of many eukaryotic genes revealed unanticipated and novel features of the structure of eukaryotic genes and of the complexities of their regulation. This chapter deals with these discoveries and their implications for understanding the complex properties and behavior of eukaryotic organisms.

Molecular cloning and DNA sequencing provide the basic information about a gene's design. But another approach—**reverse genetics**—has been key to identifying the nucleotide sequences relevant to a gene's proper expression and regulation. Reverse genetics utilizes cloned genes as substrates for the *in vitro* construction of variant genes whose functions can be evaluated both *in vitro* and *in vivo* (Section 7.6).

The most useful modifications in a gene's structure include changes in single or groups of nucleotides, deletions or insertions of a few nucleotides or extensive DNA segments, and rearrangements within the gene. Later in this chapter, we discuss how each of these types of modifications has been used to identify the regulatory sequences that ensure a gene's proper expression and account for its characteristic temporal and tissue-specific regulation. In addition, assays of modified, novel genes, created by fusing parts of different genes, improve greatly our ability to identify those sequences that are essential for correct expression. For example, fusions of an SV40 promoter (and constructed variants thereof) to the coding sequences for readily measurable bacterial or eukaryotic cellular proteins provide convenient assays for determining which promoter sequences ensure the proper initiation, efficiency, and regulation of SV40 gene transcription. Analogous chimeric genes containing, for example, the insulin or elastase gene promoters fused to the SV40 T antigen coding region help identify the sequence motifs that restrict insulin or elastase gene expression exclusively to pancreatic β-islet or acinar cells, respectively.

Applying reverse genetics requires an assay, preferably several, to measure the phenotype of the altered gene. Appropriate cell-free systems that can measure the transcriptional activity of normal and modified genes, or the processing and translation of the RNA, offer the greatest opportunity for analyzing the function and consequences of a particular alteration. Transfections into cultured cells of normal and modified genes carried on suitable vectors help identify gene sequences that are important for expression and regulation in their own or related host cells *in vivo*: for example, immunoglobulin genes introduced into lymphoid cells and globin genes into erythroid cell precursors. The introduction of wild type and modified genes into somatic or germ cells or very early embryos creates **transgenic** plants and animals that contain the gene in all their somatic and germ cells. Transgenic organisms are particularly informative for identifying those structural elements that contribute to the tissue-specific regulation of a gene's expression during development.

The eukaryotic genes chosen for illustration in this chapter originate in a wide variety of organisms. Genes from yeast, generally *Saccharomyces cerevisiae*, are important for several reasons. They have certain features characteristic of bacterial, plant, invertebrate, and vertebrate genes. Moreover, the DNA binding proteins that mediate many aspects of yeast transcriptional control are interchangeable with, or work in conjunction with, corresponding signals and proteins from other organisms, including mammals. Also, the extensive knowledge of yeast genetics and the ability to replace normal genes with modified cloned genes (Section 5.6c) enhance the potential of the reverse genetics approach. Mammalian virus genes also figure prominently in our presentation because they frequently reflect structural and functional characteristics of their hosts' genes. Indeed, many novel features of eukaryotic gene structure and regulation were first discovered during studies of viral genomes. Selected genes from plants, invertebrates (e.g., sea urchins and *Drosophila*), and vertebrates (e.g., amphibia, birds, and mammals, including primates) are described because they help explain the complex developmental programs in multicellular organisms.

8.1 Comparative Structural Features of Prokaryotic and Eukaryotic Genes

a. Prokaryotic Genes

The 5′ ends of prokaryotic genes that encode protein or stable RNA products have a distinctive organization of regulatory elements. These sequences include the promoter sequences to which RNA polymerase binds and sequences that influence the rate of transcription initiation by their ability to bind specific repressor or activator proteins (Figure 8.1 and Section 3.2c). Because only a single RNA polymerase, acting in conjunction with sigma (σ) factors, transcribes all of a prokaryotic cell's genes, the sequence organization at different promoters is similar (Figure 3.8). Gen-

(a) a single protein coding gene

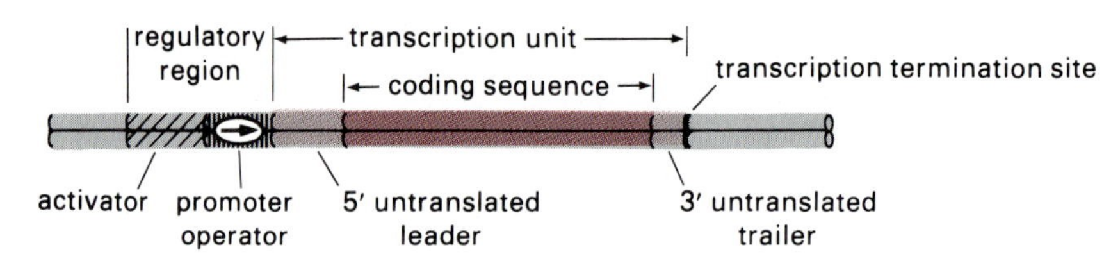

(b) protein coding genes organized into an operon

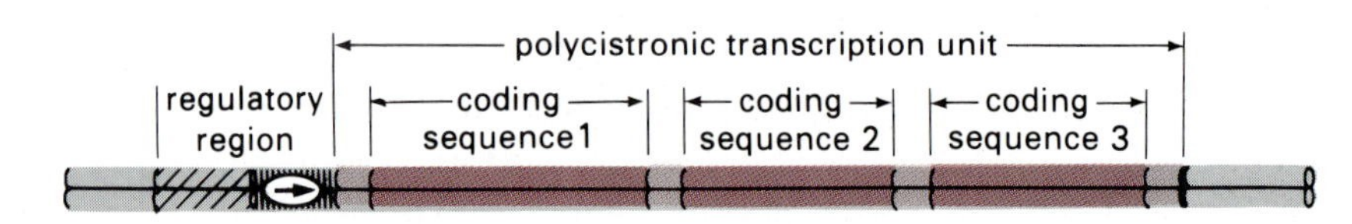

(c) transcription units encoding rRNAs and tRNAs

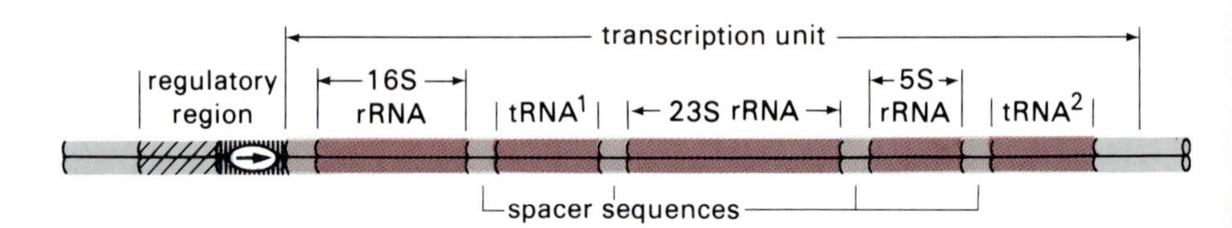

Figure 8.1
Structural features of prokaryotic genes. DNA segments that are transcribed into RNA (mRNAs, rRNAs, or tRNAs) are called transcription units. Protein coding sequences are generally flanked by 5′ untranslated leader and 3′ trailer segments. RNA coding regions have interspersed spacer sequences that are eliminated during the processing of rRNAs and tRNAs.

erally, the sequences comprising the entire regulatory region lie within 50 to 75 base pairs upstream of where transcription begins. Only rarely do positive and negative regulatory sequences occur farther upstream or within or beyond the 3′ end of the transcribed segment.

The nucleotide sequences that cause termination and release of the transcript mark the 3′ end of the gene. The stretch of DNA between the start and finish of transcription defines the **transcription unit**. Most transcripts that encode polypeptides have a variable number of nucleotides that precede (**5′ leader**) and follow (**3′ trailer**) the protein coding sequence. In some instances, the 5′ leader contains sequences that permit additional control of transcription and translation. (Sections 3.11d and 3.11f discuss attenuation and translational regulation, respectively.) Almost invariably, the nucleotide sequences of protein coding genes are colinear with the sequences in their respective mRNAs, and those of RNA genes are colinear with the rRNAs and tRNAs they encode; a thymidylate synthetase gene of T4 phage and a $tRNA^{Ser}$ gene in an *Acanthamoeba* species are known exceptions in that they each contain an intron.

Monocistronic transcription units contain only a single coding sequence for a protein or stable RNA (Figure 8.1a). Most prokaryotic transcription units, however, are **polycistronic** and contain coding sequences for more than one type of protein or RNA (Figure 8.1b and c). Generally, all the coding segments in a polycistronic transcription unit rely on the same 5′ and 3′ regulatory sequences, permitting the multiple gene products to be expressed and regulated coordinately. Transcripts that encode one or more polypeptide sequences are translated directly without any requirement for modification or processing of the primary transcript to smaller RNAs. By contrast, single transcripts that encode multiple RNAs, such as the three species of rRNA or multiple species of tRNA, are cleaved in specific ways posttranscriptionally to yield the mature stable RNA products (Section 3.3 and Figure 8.7).

b. Eukaryotic Genes

Transcription signals The structure and organization of eukaryotic transcription units are considerably more complex than those of prokaryotes. Part of this complexity stems from the use of three discrete transcription systems (exclusive of the ones used to transcribe mitochondrial and chloroplast genes [Sections 9.7a and 9.7c]). Each transcription system is considered in detail in subsequent sections of this chapter, but it is useful to compare them briefly at this point.

Class I genes, which encode the 5.8S, 18S, and 28S rRNAs, are transcribed by RNA polymerase I. All mRNAs and a variety of small nuclear RNAs (snRNAs) are derived from transcription of class II genes by RNA polymerase II. The tRNAs, 5S rRNA, and certain small cytoplasmic RNAs (scRNAs) are transcribed by RNA polymerase III from class III genes. Operationally, transcription by the three eukaryotic RNA polymerases can be distinguished by their relative sensitivities to α-amanitin, a poisonous bicylic octapeptide derived from the *Amanita* mushroom. RNA polymerase II is inactivated by a very low concentration (<0.1 μg/ml) of α-amanitin; a higher level (20 μg/ml) is needed to block RNA polymerase III transcription. RNA polymerase I remains active with even 200 μg/ml.

As anticipated, the three different RNA polymerases require different regulatory sequences in order to initiate transcription, and the typical locations of the regulatory sequences relative to the transcription start sites are distinctive for each enzyme. Also, each type of RNA polymerase requires different accessory proteins (**transcription factors**) that bind to these sequences. Transcription by RNA polymerases I and II depends in part on nucleotide sequences located within about 100 bp surrounding the 5′ end of their respective transcription units. However, these polymerases almost invariably also require transcription factors that bind at specific sequences located nearby or frequently several kilobase pairs distant from the transcription start sites (Sections 8.2c and 8.3d). In class II genes, transcription factors may bind to sites within the transcription unit or even beyond its 3′ end (Section 8.3d). Critical sequences required for transcription of most class III genes are located within the coding region, the sequences 5′ to the transcribed segment generally being dispensable or only marginally involved. In a few cases, however, upstream sequences are also required, and in others, the upstream sequences are all that is required for regulating the gene's transcription (Section 8.4b).

The sequences that specify the 3′ termini of the respective functional RNA products are also distinctive for each of the RNA polymerase systems. Endonucleolytic cleavages of long transcripts made by RNA polymerase I rather than transcription termination account for the 3′ termini of the rRNAs. RNA polymerase II generally does not terminate transcription at unique locations. Instead, the 3′ ends of almost all mature mRNAs

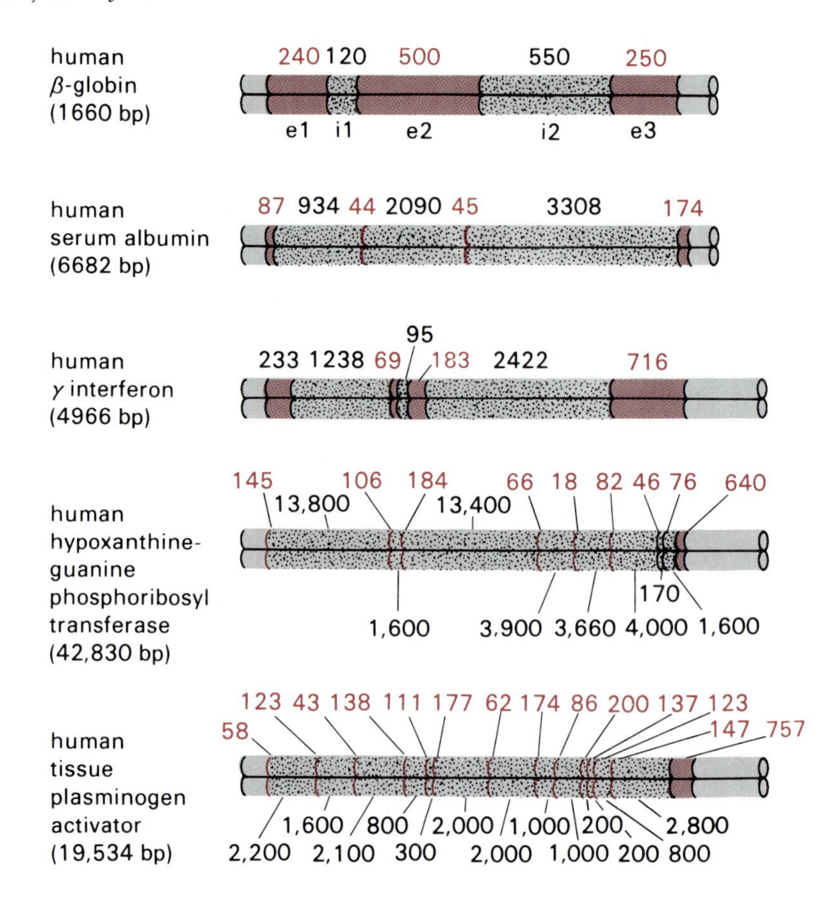

Figure 8.2
Comparative sizes and numbers of exons and introns in various human genes. Introns appear stippled; exons are in color. In the last two examples, most of the exons are shown as colored lines because they are so small relative to the size of the genes. Lengths of DNA are shown in base pairs.

arise by specific cleavage at characteristic sequences and subsequent polyadenylation of the 3′ end. The 3′ ends of RNAs transcribed by RNA polymerase III correspond mainly to the sites at which transcription terminates, although posttranscriptional processing and modification of the 3′ ends also occur.

Mosaic transcription units By contrast with the almost invariable colinearity that exists between prokaryotic genes and the respective RNAs they encode, many eukaryotic genes have mosaic structures. In this context, mosaicism refers to the interspersion of coding (**exon**) and noncoding (**intervening** or **intron**) sequences within the transcription unit (Figure 8.2). Introns occur most frequently in genes that encode proteins and tRNAs and more rarely in rRNAs. With the exception of some genes encoding the five histones, the α and β interferons, and several mammalian virus proteins, all known vertebrate protein coding genes contain introns. By contrast, introns are rare in the protein coding genes of *S. cerevisiae*, although they occur more frequently in another yeast, *Schizosaccharomyces pombe*. Most *Drosophila* genes lack introns, but several of the clusters of genes that regulate the development of the organism's body structures are punctuated by many introns. Introns also occur in many plant genes—maize alcohol dehydrogenase and soybean leghemoglobin being examples.

Introns vary in size, number, and location from one gene to another. Nevertheless, the same genes in different species often have the same number of introns at analogous positions, although the length of the introns and their nucleotide sequences may differ markedly. Generally, the number of introns per gene increases proportionally to the length of the protein coding sequence, and exon sizes tend to be 300 bp, on average. Overall, the total length of intron sequences exceeds the total length in exons, frequently by two to ten times, but occasionally by much more than ten times (Table III.2). Introns are not located at random within transcription units. In tRNA genes, introns occur adjacent to the anticodon loops, and in protein coding genes, the introns are often located between segments that encode discrete structural or functional domains in the fully folded protein (Sections 9.1c and 9.3b).

Monocistronic transcripts As mentioned earlier, most prokaryotic genes are organized into operons, and their expression is mediated by polycistronic mRNAs (Section 3.11b). But protein coding genes in eukaryotic cells yield mRNAs with only a single coding sequence. Even when a gene yields more than one type of mRNA, as by alternate patterns of intron removal (Section 8.5), each mature mRNA has only one translatable coding sequence. Although the 5S rRNA genes are arranged in tandem arrays, each gene is transcribed from its own promoter to yield RNAs with only a single 5S rRNA sequence per chain. By contrast, RNA polymerase I transcripts of the 18S, 5.8S, and 28S rRNA gene clusters yield long transcripts containing one each of the rRNA sequences and spacer sequences, and posttranscriptional cleavages generate the smaller mature rRNAs.

Definition of a eukaryotic gene Prokaryotic and eukaryotic genes share the same logic in their basic design, but the differences in molecular detail are substantial. A revised definition of a eukaryotic gene may help summarize the essence of these differences. Admittedly, no single definition of a eukaryotic gene could be satisfactory to everyone or to every example. The one we have adopted emerges from the molecular structure of genes having a wide range of functions in diverse eukaryotic organisms. The definition takes account of the differing locations and kinds of DNA sequence elements that influence gene expression. It also recognizes, implicitly, that phenotypically observable mutations arise from changes in regulatory signals as well as coding sequences.

We define a gene as a combination of DNA segments that together comprise an expressible unit, a unit that results in the formation of a specific functional gene product that may be either an RNA molecule or a polypeptide. The DNA segments that define the gene include the following:

1. The transcription unit refers to the contiguous stretch of DNA that encodes the sequence in the primary transcript; this includes (a) the coding sequence of either the mature RNA or protein product, (b) the introns, and (c) the 5′ leader and 3′ trailer sequences that appear in mature mRNAs as well as the spacer sequences that are removed during the processing of primary transcripts of RNA coding genes.
2. The minimal sequences needed to initiate correct transcription (the **promoter**) and to create the proper 3′ terminus of the mature RNA.

3. The sequence elements that regulate the rate of transcription initiation; this includes sequences responsible for the inducibility and repression of transcription and the cell, tissue, and temporal specificity of transcription. These regions are so varied in their structure, position, and function as to defy a simple inclusive name. Among them are **enhancers** and **silencers**, sequences that influence transcription initiation from a distance, irrespective of their orientation relative to the transcription start site (Sections 8.2c and 8.3d).

Neither the DNA sequences that influence a gene's configuration within chromatin nor those that regulate its topology are included in our definition of the gene. DNA sequences that encode proteins or RNAs that modulate a particular transcription unit's expression are also excluded from our definition.

The essential features of a prototypical eukaryotic protein coding gene are depicted in Figure 8.3. In this representation, the beginning of the gene is shown at the left and the end of the gene at the right, consistent with the widespread convention for diagramming transcription from left to right. This means that the DNA strand with the same sequence as the RNA transcript, the sense strand, is 5′-to-3′ from left to right and is generally shown as the upper strand. The template for transcription, the nonsense strand, is 3′-to-5′ from left to right (lower strand). For convenience, the sense strand is often the only one shown. The position of the first nucleotide in the transcript is designated +1, and those downstream of +1 (i.e., within the transcription unit) are given positive numbers (e.g., +16). Nucleotides that are upstream of +1 (nontranscribed sequences) are assigned negative numbers (e.g., −25).

A convenient way to examine the relationship between the novel structural features of eukaryotic genes and their expression and regulation is to consider representative genes in the context of the RNA polymerase that initiates their expression. Thus, we discuss, in turn, the structural features of class I (Section 8.2), class II (Section 8.3), and class III (Section 8.4) genes; the characteristics of their respective RNA polymerases; and the distinctive mechanisms of their transcriptional regulation. The processing of primary RNA transcripts, particularly splicing, to produce mature mRNAs, rRNAs, and tRNAs follows (Section 8.5). Section 8.6 examines some of the structural motifs that enable protein transcription factors to recognize and interact with specific DNA sequences and with each other. The chapter concludes with an examination of more global mechanisms of regulation, such as those that influence the rate of mRNA turnover, and the ways

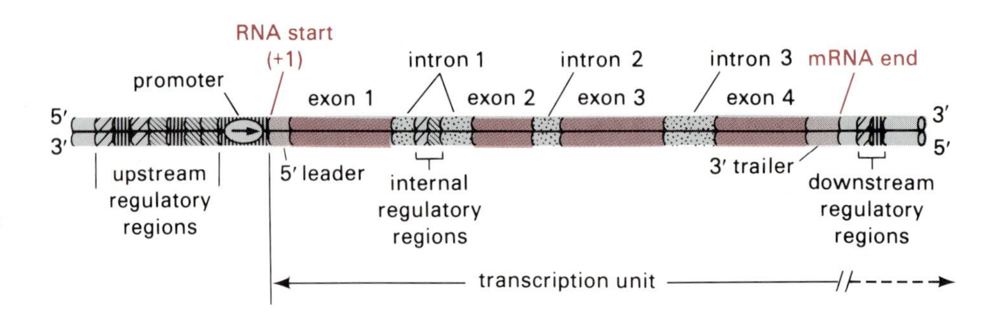

Figure 8.3
Structural features of a prototypic eukaryotic protein coding gene. The gene's transcription unit is defined by the transcription start site (base pair +1) and the region within which transcription terminates. The latter occurs beyond the end of the mature mRNA sequence, but only rarely is it a specific site. Sequences required for transcription initiation generally lie upstream within 100 bp of +1 but often extend for hundreds or even thousands of base pairs. Besides a promoter and upstream regulatory sequences, many genes possess regulatory sequences within introns or beyond the 3′ end of the mature mRNA. The regulatory sequences are shown interspersed with different symbols and different shades of color to indicate possible arrangements of distinctive types of regulatory sequences.

in which large blocks of genes are regulated by modification of the DNA or of the chromatin in which the genes are packaged (Section 8.7).

8.2 Structure and Expression of Class I Genes

The transcription of class I genes by RNA polymerase I accounts for almost half the transcriptional activity of most eukaryotic cells. The sole product of this transcriptional process is a precursor of ribosomal RNA (**pre-rRNA**) that is processed by sequential cleavages to the mature 5.8S, 18S, and 28S rRNA species. (The fourth rRNA species, 5S rRNA, is encoded by a class III gene [Section 8.4].) The number of genes encoding rRNA ranges between a hundred and several thousand, depending upon the eukaryotic species (Section 9.2a–c). They are located in one or a few specific chromosomes at the morphologically distinctive regions called **nucleolar organizers** (Figure I.8). During interphase, these regions are incorporated into **nucleoli**, structures in which rRNAs are actively transcribed, pre-rRNAs are processed, and ribosomes are assembled.

The rRNA genes of virtually all eukaryotic organisms are arranged in long "head-to-tail" repeats (Section 9.2a). Electron micrographs of a segment of nucleolar DNA show several of the rRNA gene clusters being transcribed (Figure 8.4). Note that the "arrowheads," formed by the increasing lengths of the transcripts along the rDNA region, point in the same direction. This indicates that the transcription units in this repeated array are oriented in the same 5′-to-3′ direction. The picture also suggests that there are transcribed and nontranscribed regions regularly spaced along the DNA, but this is more apparent than real.

a. *Transcription Units*

In all species, each rRNA transcription unit encodes 18S, 5.8S, and 28S rRNAs, arranged in that order starting from the 5′ end (Figure 8.5). The three rRNA coding sequences are both flanked and separated by transcribed spacer segments. These are called, respectively, the **external transcribed spacer** (5′ and 3′ **ETSs**) and **internal transcribed spacers** (5′ and

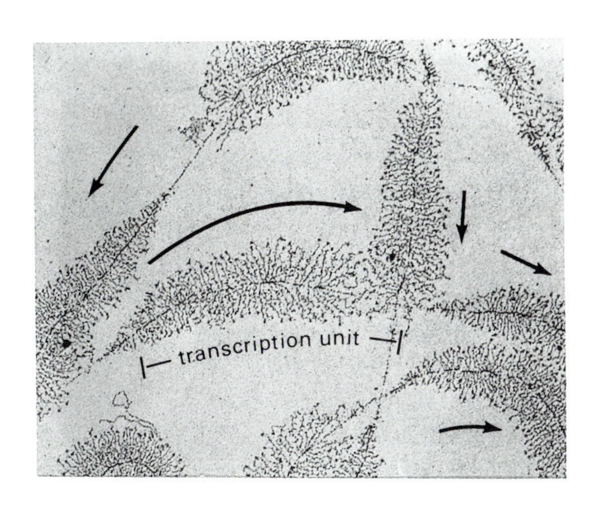

Figure 8.4
Electron micrograph showing transcription of *Xenopus* oocyte rDNA. The long continuous strands are the rDNA, and the featherlike structures are the nascent pre-rRNAs. The arrows indicate the direction of transcription. The shortest transcripts occur at the 5′ end of the transcription units and the longest toward the 3′ end of the transcription units. Note the absence of transcripts between the region of clustered nascent pre-rRNAs. These were originally thought to be nontranscribed spacers, but now it is known that these regions are transcribed (see text) and that the transcripts are too short to be visualized. Photograph courtesy of O. L. Miller, Jr. From O. L. Miller, Jr., and B. R. Beatty, *Science* 164 (1969), p. 955.

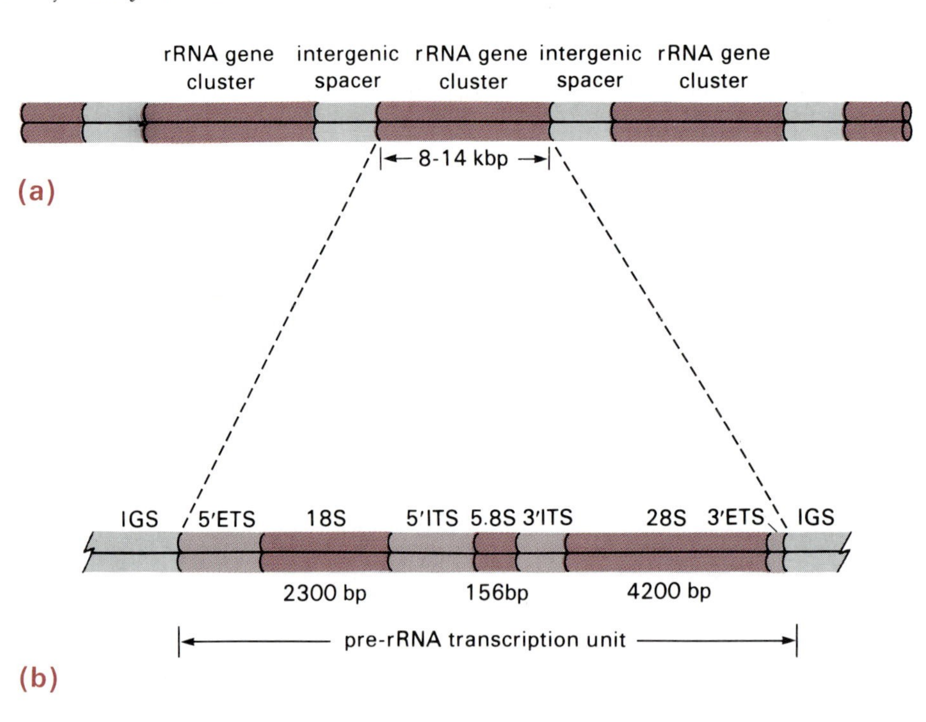

Figure 8.5
Arrangement of eukaryotic rRNA gene clusters. (a) Each rRNA gene cluster about 8 to 14 kbp in length, depending on the species, is separated from another by an intergenic spacer (IGS), which is highly variable in length but relatively constant within a species. (b) Each rRNA gene cluster constitutes a single transcription unit consisting of the 5′ external transcribed spacer (5′ ETS), 18S rRNA coding sequence, 5′ internal transcribed spacer (5′ ITS), 5.8S rRNA coding sequence, 3′ internal transcribed spacer (3′ ITS), 28S rRNA coding sequence, and 3′ ETS. The same intergenic control sequence flanks each rRNA gene cluster. The pre-rRNA transcription unit extends from the IGS's 5′ ETS junction to variable sites in the IGS beyond the end of the 28S rRNA coding sequence.

3′ **ITSs**). The region between tandemly repeated transcription units ranges between a few kbp to nearly 30 kbp in length and was originally referred to as the **nontranscribed spacer** (**NTS**). Because it is now evident that the so-called nontranscribed spacers are transcribed and that transcriptions in this region are a relevant if not an essential feature of the regulation of rDNA expression (Section 8.2c), it is more appropriate to use the term **intergenic spacer** (**IGS**) for this region.

The size and nucleotide sequences of the rRNA coding regions vary only slightly among a wide variety of eukaryotic species; indeed, the extent of nucleotide sequence homology between comparable rRNAs is a useful measure of their evolutionary relationships. The length of the ITSs, however, varies widely from species to species; therefore, the size of the transcription units differs accordingly (Figure 8.6). Thus, the rDNA transcription unit in human DNA is about 14 kbp long, close to the longest known, and the size of the corresponding transcription unit in *S. cerevisiae* is one of the smallest, between 6 and 7 kbp. Nearly half the 28S rRNA coding regions of *Drosophila* rDNA have variable sized insertions; these are not introns, and, indeed, rDNA transcription units with such insertions are not transcribed. The 28S rRNA coding sequences in *Tetrahymena* and several other species are interrupted by 400–1000 bp introns that are spliced out of the pre-rRNA transcripts before they are processed further to the three mature rRNAs (Section 8.5b).

During transcription, the nascent pre-rRNA continuously associates with the corresponding ribosomal proteins and becomes methylated at specific base and ribose residues (Figure 8.7). After completion of the transcript, nucleolar endonucleases cleave the pre-rRNA in the nucleoprotein complex, first at the 5′ end of the 5.8S rRNA sequence and subsequently at the 5′ end of the 18S rRNA and the 3′ end of the 28S rRNA, respectively. Subsequent endonucleolytic cleavages produce the three mature sized rRNAs. The RNA segments corresponding to the transcribed spacers

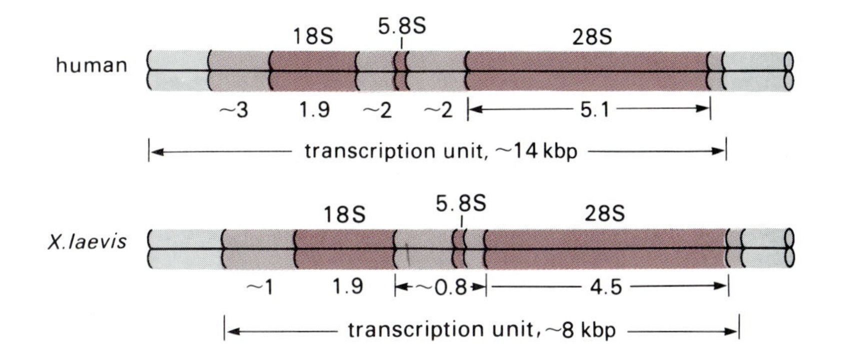

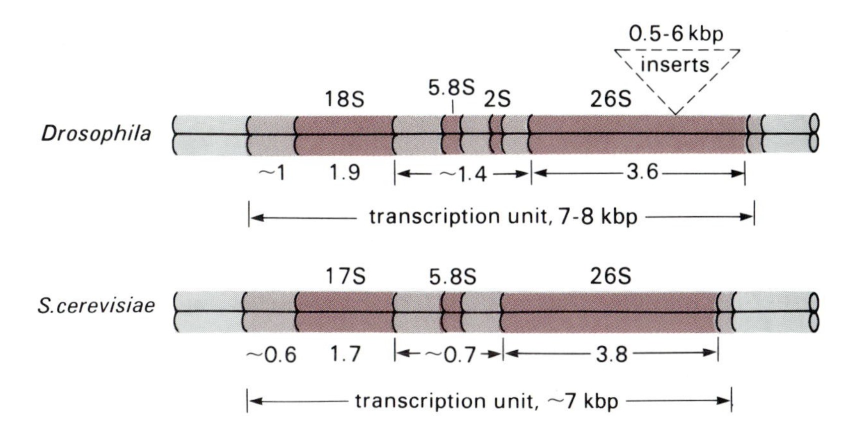

Figure 8.6
Arrangements and sizes of segments in various rRNA gene clusters. The numbers shown below each segment are their approximate lengths in kilobase pairs. Some *Drosophila* 26S rRNA coding sequences contain an insert of 0.5–6 kbp that does not appear in mature 26S rRNA.

are rapidly destroyed and do not accumulate. The cleavages appear to be extremely specific because there is no evidence that exonucleolytic trimming is required to produce the 5′ and 3′ ends of the mature rRNAs. The small nucleolar ribonucleoprotein containing U3 RNA (U3 snRNP) is thought to participate in one of the early steps in processing the pre-rRNA. The 5′ end of 5.8S rRNA can base pair to internal sequences in the 28S rRNA, suggesting that the cleavages between the 5.8S and 28S rRNA segments probably occur in an already base paired structure.

b. Transcription Machinery

RNA polymerases have been purified from a wide variety of cells by standard protein fractionation methods. In contrast to RNA polymerases II and III, which are localized in the nucleoplasm, RNA polymerase I is associated with and isolated from nucleoli. Only RNA polymerase I catalyzes the transcription of the rRNA gene cluster. The specificity of RNA polymerase I for the rRNA genes is inferred from the insensitivity of the enzyme and rRNA synthesis to the inhibitor α-amanitin (Section 8.1b).

The molecular weights of RNA polymerase I, irrespective of its source, are in the range 500 to 600 kDaltons. Moreover, all the enzymes contain 9 to 14 different polypeptide subunits, depending on their source and the methods and assays used to purify them. Generally, the enzymes contain two large subunits, each greater than 100 kDaltons. The large subunits of RNA polymerase I share amino acid sequence similarities with the corresponding two large subunits of RNA polymerases II and III and cross react with certain antibodies to those polymerases. Surprisingly, the

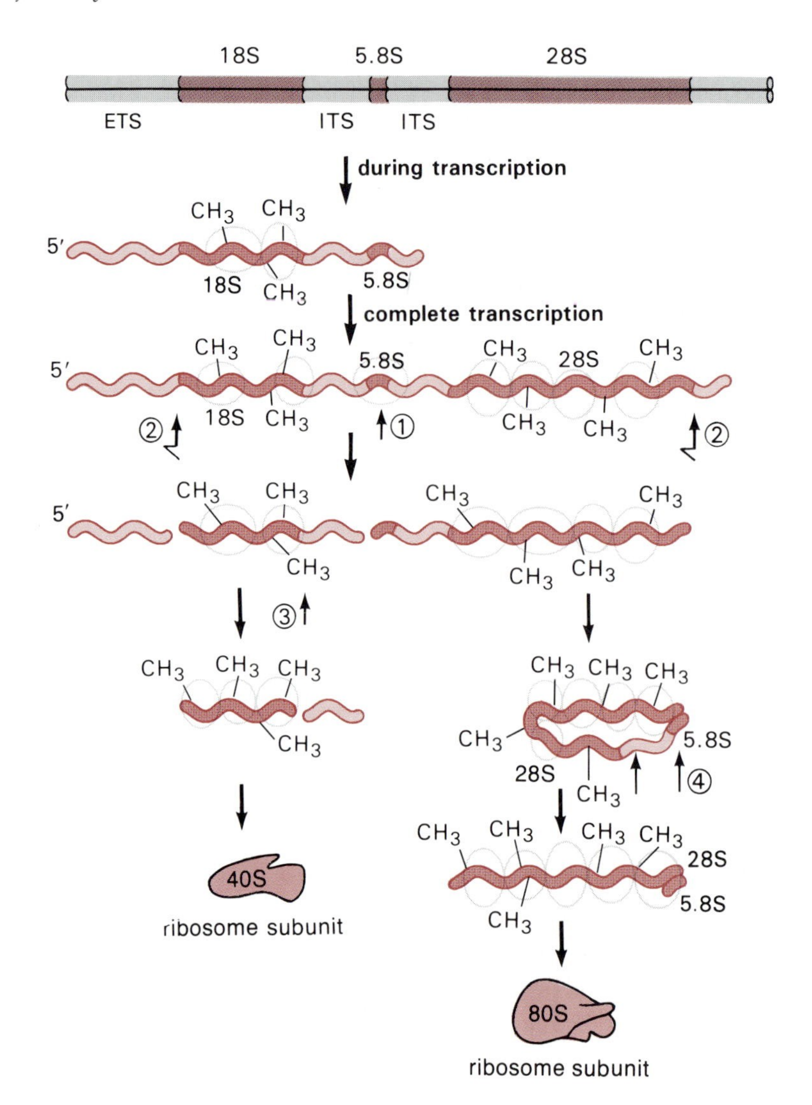

Figure 8.7
Schematic representation of pre-rRNA processing into the mature rRNA substituents of ribosomes. Methylation and association with ribosomal proteins occur on nascent pre-rRNA and continue throughout transcription. The initial cleavage occurs at the 5′ end of the 5.8S rRNA and subsequently at the junctions of the 18S and 28S rRNA sequences with the spacers. Cleavages are shown as arrows pointing upward and are numbered to indicate the order in which they are believed to occur. Binding of additional ribosomal proteins characteristic of each subunit completes ribosome assembly.

RNA polymerase I large subunits also share sequence homology with the two large subunits of *E. coli* RNA polymerase. The other subunits range in size from 15 to 50 kDaltons. Three of the smaller polypeptides are common to all three RNA polymerases, two others are shared by RNA polymerase I and III, and the rest are unique to each enzyme. No functions have yet been assigned to any of the subunits. The isolated RNA polymerase I probably provides the minimal structure necessary for transcribing the rRNA gene cluster because specific accessory proteins are needed for accurate and efficient transcription initiation and termination of an rRNA gene cluster (Section 8.2c). Whether these additional proteins function independently or in transient association with the polymerase remains to be determined. There are clear indications that RNA polymerase I exists in active and inactive forms—the difference between them being either a reversible covalent modification of one of the known subunits or the transient addition and removal of a subunit needed for effective rDNA transcription (recall the role of σ in *E. coli* RNA polymerase; Section 3.2b). Cells not transcribing the rRNA gene cluster have only the inactive form of RNA polymerase I, but after reinitiation of rRNA synthesis, the active form reappears bound to DNA. In the protozoan *Acanthamoeba castellanii*, for example, active enzyme is recovered from the growing trophozoites but not from nongrowing cysts.

c. rDNA Transcription Control Regions

Several different experimental approaches have contributed to our present views about the mechanism and regulation of rDNA transcription. First, DNA sequence comparisons between cloned rRNA genes from the same genome and from genomes of related and unrelated species reveal characteristic nucleotide arrangements that recur at particular locations relative to the transcription unit. Second, measurements of the transcriptional activity of wild type and structurally modified cloned rRNA genes using template-dependent cell extracts, or partially purified RNA polymerase I preparations and purified transcription factors, identify sequences that influence transcription under *in vitro* conditions. Third, the conclusions obtained from *in vitro* studies are reinforced and extended by assaying transcription after transfection or microinjection of variously altered rDNAs into suitable cells.

A striking feature of the transcription signals in rRNA genes is that they are species-specific. Thus, human cell extracts are capable of transcribing cloned human rRNA genes, but do not use mouse, rat, yeast, *Drosophila*, or *Xenopus* rDNA templates. Similarly, mouse cell extracts fail to transcribe rRNA genes from human. But, curiously, *Xenopus* rDNA is transcribed by the mouse system and vice versa. The rRNA genes of even two closely related species of *Drosophila* (*D. melanogaster* and *D. virilis*) can be distinguished by the specificity of their transcription machinery for their own rDNAs. Associated with this species specificity is a lack of apparent DNA sequence conservation around the beginning of the transcription units in human, mouse, *Xenopus*, *Drosophila*, and yeast rRNA genes. This is distinctly different from the situation with class II or class III genes, where the genes within a class, even from widely divergent species, have highly conserved or analogous sequence elements in their transcription control regions. Moreover, the RNA polymerase II and III systems from one species can transcribe the corresponding genes from an unrelated species (Sections 8.3 and 8.4).

The design of rDNA control regions from different species have certain features in common, but there are also striking differences in the nature and arrangement of their regulatory elements. Given the present state of our knowledge, these differences may be more apparent than real. In spite of significant differences in the transcription signals in the DNA, the structural features and activities of the complexes assembled to initiate, elongate, and terminate pre-rRNA transcription seem similar.

Mammalian Two regions of the nearly 30 kbp IGS have been implicated in regulating mammalian rDNA transcription (Figure 8.8). One segment, between base pairs -45 and $+20$ in human rDNA and between base pairs -39 and $+9$ in mouse rDNA, spans the site of transcription initiation ($+1$). Deletion of all or parts of these segments, replacement of short successive sequence intervals by nearly equal length unrelated oligonucleotides **(linker insertions)**, or base pair changes at selected positions abolish or drastically reduce rDNA transcription. Because of their location

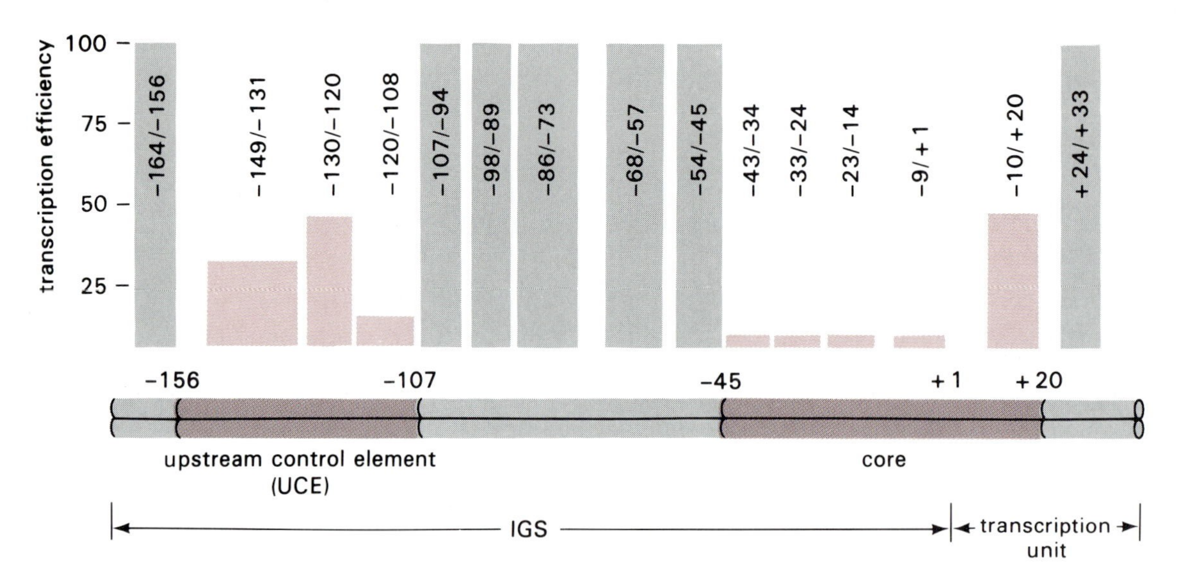

Figure 8.8
The effects of linker insertions mutagenesis at the 5′ end of human rDNA on transcription *in vitro*. rDNAs with substitutions in place of the indicated base pairs were assayed for their ability to be transcribed in human cell extracts. The histogram shows how the substitutions within the indicated intervals spanning base pairs −164 to +33 affect the transcription efficiency relative to wild type rDNA. Adapted from M. M. Haltiner et al., *Mol. Cell. Biol.* 6 (1986), p. 227.

and intrinsic activity for promoting accurate, albeit low levels of transcription (<5 percent), the essential sequence in this region has been termed the **core promoter**.

The second region, which includes sequences between about base pairs −156 and −107, significantly increases transcription from the core promoter both *in vitro* and *in vivo*. This is apparent in experiments involving transfection and injections of modified rDNAs into suitable cells. Deletions or linker insertions that disrupt the sequence in this region reduce rDNA transcription up to 100-fold. Using a more exacting assay than the ones that provided the results shown in Figure 8.8, the essential sequence extends to base pair −186. Although the nucleotide sequence between this **upstream control element** and the core promoter appears not to be critical, changing the distance between them or their orientation relative to one another reduces transcriptional activity. This suggests that the two regulatory sequences operate in concert to effect efficient transcription by RNA polymerase I. The upstream control element probably contains subelements whose sequences serve different functions.

What do these two regulatory regions do and how do they do it? The discovery of stable rDNA transcription complexes provided a clue. If two rDNAs that yield two distinguishable rRNA products are added to an appropriate *in vitro* transcription system, both rRNAs are produced (Figure 8.9). However, if the addition of one rDNA precedes the other by even a few minutes, only the first added rDNA is transcribed, irrespective of which rDNA template is added first. No such preference exists if the DNA added first lacks the rDNA transcription control region or if deletions or single base pair changes have inactivated the upstream control element. The inference is that rDNA reacts with one or more factors in the extract, making them unavailable for reaction with another rDNA even after repeated cycles of transcription. This is confirmed by fractionation of the cell extracts into several protein factors that by themselves or even together do not promote pre-rRNA synthesis. Instead, these factors permit the active form of RNA polymerase I to initiate rDNA transcription.

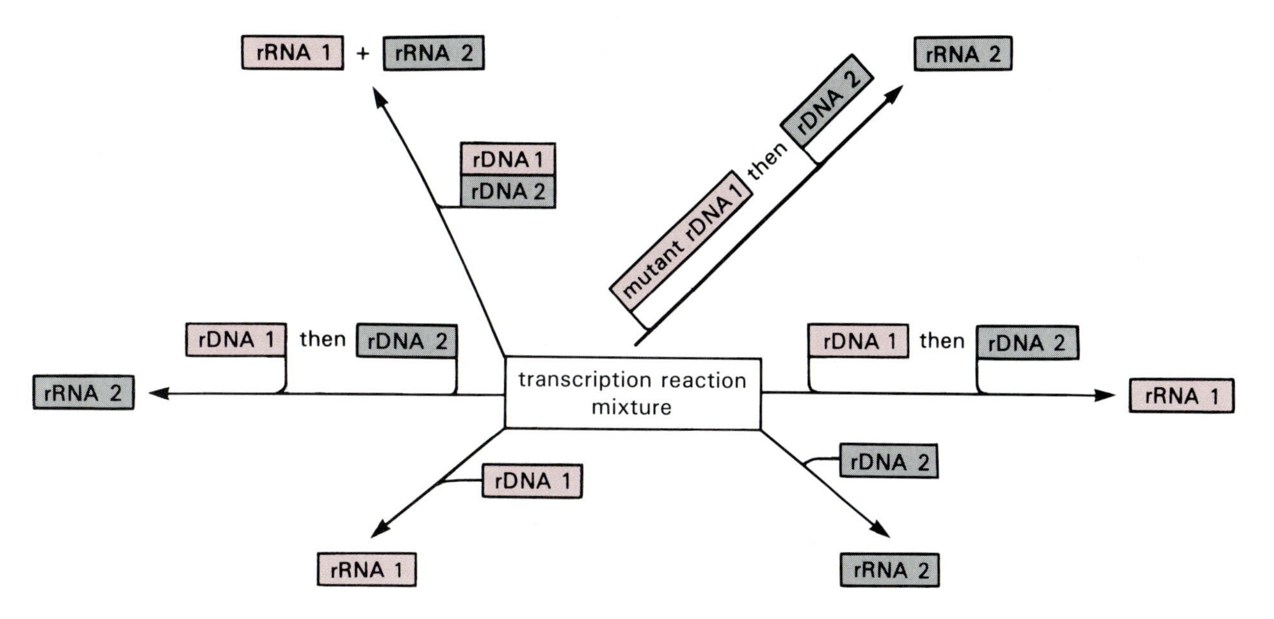

Figure 8.9
Transcription of rDNA involves formation of stable complexes. rDNA 1 and rDNA 2 are equally efficient but distinguishable templates for rRNA synthesis *in vitro*. When both rDNAs are added together, the two rRNAs are made in about equal quantities. However, if one rDNA is added before the other, the first added rDNA is preferentially transcribed. If the first rDNA has a mutation in its promoter region, it fails to prevent transcription of the second rDNA.

The detection and isolation of transcription factors from both human and mouse cells have relied on two kinds of assays: (1) the ability of protein fractions to restore transcription activity to extracts depleted of factors and (2) the ability of such factors to leave a **footprint** on the DNA sequence to which they bind. Footprinting is a general procedure that can detect specific protein binding to DNA by comparing the fragmentation patterns of bound versus free DNA after digestion with DNase I (Figure 8.10). If the bound protein protects against DNase I digestion, the position of the protein on the DNA can be deduced by the loss of characteristic fragments from the array.

A generally consistent view of rDNA transcription initiation is emerging from both *in vitro* and *in vivo* studies with human and rodent transcription systems. Each system relies on the recognition of DNA sequence elements adjacent to the core promoter by specific transcription factors. Human cell extracts are the source of two proteins, UBF-1 and SL-1, which together with RNA polymerase I catalyze transcription from the human rRNA gene promoter. Both proteins have been obtained in nearly pure form, and the mechanism of their action has been examined. UBF-1 by itself binds to and protects the 3′ region of the upstream control element, the sequence between base pairs −75 and −115 (Figures 8.10 and 8.11). The bound UBF-1 also makes contacts with a sequence in the core promoter, the region around base pair −21. SL-1 by itself does not bind to the rRNA gene promoter region. However, in the presence of both SL-1 and UBF-1, the region protected against DNase I digestion extends to base pair −165. SL-1 also improves the binding of UBF-1 to the core promoter element, extending the binding in this region close to base pair +1. Thus, SL-1 and UBF-1 acting together form protein DNA contacts that span the core and upstream control elements. It is not known if SL-1 makes direct contact with the DNA in the presence of UBF-1 or if it alters the recognition properties of UBF-1 to include the additional region. Mutational alterations in the upstream control element prevent the binding by UBF-1 alone, even in the presence of SL-1. Mutations in

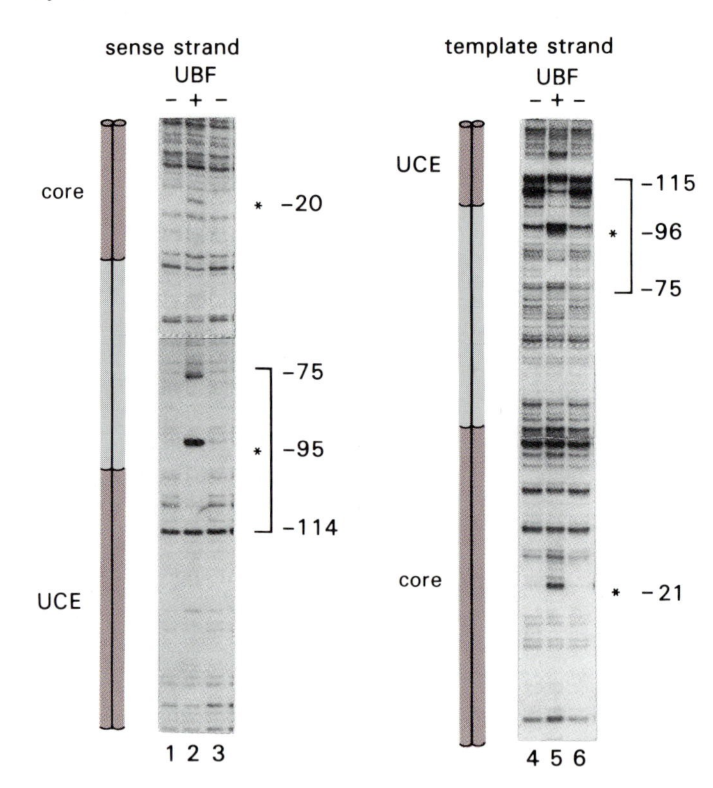

Figure 8.10
DNase I footprint of mammalian rDNA by UBF-1 transcription factor. A 297 bp segment of the human rDNA promoter (between base pairs −199 and +78, labeled with a 5′ ^{32}P at base pair −199) and a 525 base pair segment (from base pair −500 to +24, 5′ labeled at base pair +24) were used to detect the binding of UBF-1 to the sense and template strands, respectively. Lanes 1, 3, 4, and 6 represent the digests produced in the absence of UBF-1. Lanes 2 and 5 are from digests made in the presence of saturating amounts of purified UBF-1. Protected regions (brackets) and enhanced cleavage sites (asterisks) are shown. From S. P. Bell et al., *Science* 241 (1988), p. 1192.

the core promoter element do not block binding of UBF-1 to the upstream element, but they do impair UBF-1's and SL-1's ability to augment transcription by RNA polymerase I.

Present data suggest that protein-protein interactions between UBF-1 and SL-1 result in the formation of a complex that binds to the two rDNA promoter elements, thereby facilitating the binding of RNA polymerase I (Figure 8.12). A partially purified transcription factor for the mouse rRNA gene promoter (variously named TIF-1B or factor D) binds to the analogous core promoter region at around base pairs −40 to −15. The relationship of this factor to UBF-1 is not clear, nor is it known how the binding of this factor facilitates transcription by RNA polymerase I.

Additional sequences farther upstream of base pair −165 also have a novel and important function in facilitating transcription initiation. Be-

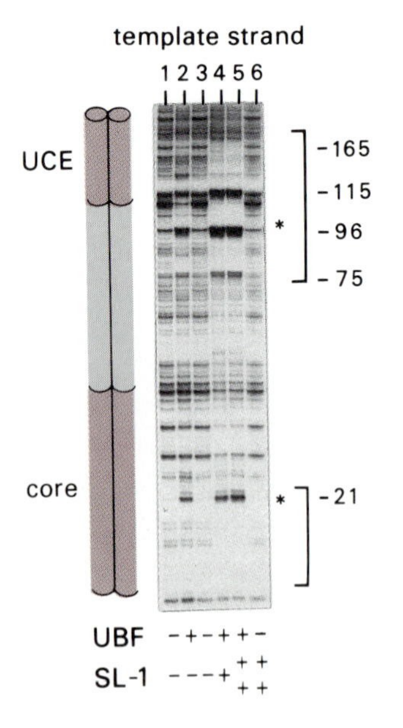

Figure 8.11
Effect of SL-1 on binding of UBF-1 to rDNA promoter region. The same 525 bp end labeled fragment described in Figure 8.10 was digested with DNase I in the presence of either UBF-1 or SL-1 alone or with mixtures of the two proteins. Pluses and minuses indicate addition and omission of the indicated factors, respectively; two pluses represent twice the level of factor. The relative position of the promoter elements are to the left of the digest patterns. Protected regions (brackets) and enhanced cleavage sites (asterisks) are shown. Note that in the presence of both proteins, the footprints are more distinct, and the protected region extends well into the UCE.

cause that role is intimately involved with the termination of rDNA transcription, its discussion is deferred until Section 8.2d.

We noted earlier that the rDNA and the transcription machinery must be from the same or closely related species for transcription to occur. Thus, extracts from rodent cells do not transcribe primate rDNA and vice versa. It is possible to understand the basis of this species specificity by constructing hybrid transcription control regions that have the core promoter from mouse rDNA fused to the human upstream control element or the human core promoter joined to the mouse upstream control region and testing these DNAs in primate extracts (Figure 8.13). Because the mouse promoter does not bind the primate UBF-1/SL-1 combination, its failure to be transcribed is readily explained. The hybrid promoter made from the mouse upstream regulatory element and the human core promoter is inactive for the same reason. But the promoter composed of the human upstream element and the mouse core element is not transcribed by primate RNA polymerase I even though the footprint patterns show that the human UBF-1/SL-1 combination binds to the upstream control region. Therefore, although the binding of UBF-1 and SL-1 to the upstream control segment is necessary to promote transcription in the primate transcription system, the bound factors cannot stimulate transcription if that region is associated with the unrelated core promoter. SL-1 may have to associate with its own species-specific core promoter sequence in order to facilitate RNA polymerase I binding and/or transcription initiation. The species specificity of rDNA transcription appears to reside in the dual requirement for binding the transcription factors to a matched set of an upstream segment and the core promoter.

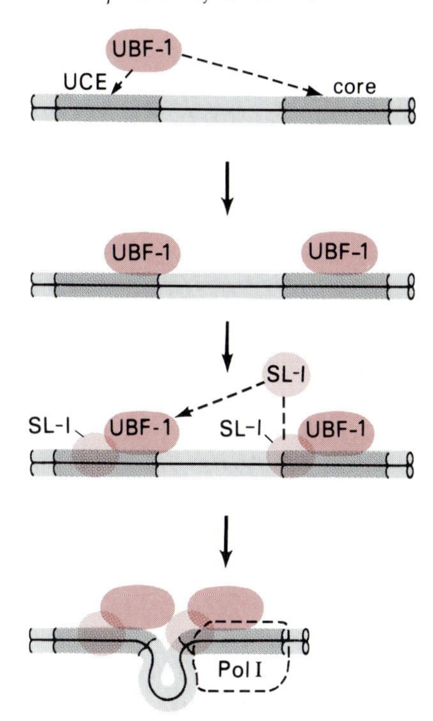

Figure 8.12
Speculative interactions of UBF-1 and SL-1 proteins with rDNA promoter elements in the activation of transcription by RNA polymerase I. UBF-1 and SL-1 bind to both the core and UCE and through protein interactions are presumed to contact each other and either activate or facilitate the binding of RNA polymerase I.

Drosophila Mention has already been made of the fact that rRNA genes from *D. melanogaster* and *D. virilis* are not interchangeable as templates for each other's transcription machinery, either *in vitro* or *in vivo*. Moreover, there is little or no sequence homology between the sequences immediately 5′ to their sites of transcription initiation (i.e., in the IGS). Nevertheless, the two species have similar sequence arrangements in this region. Both contain between five and twelve repeats of about 240 bp segments in the region immediately upstream of base pair +1. Within a spe-

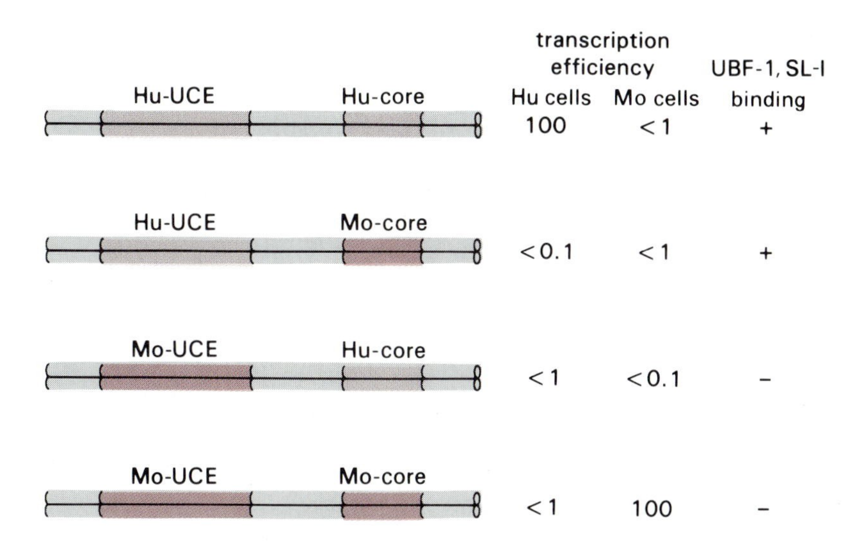

Figure 8.13
The upstream control element and core promoter must be from the same species for transcription but not for binding of UBF-1 and SL-1. The transcription efficiency of the human (Hu) and mouse (Mo) rDNA promoter and of each hybrid promoter was obtained by transfecting the plasmid containing the promoter into cultured primate or mouse cells and measuring the production of the respective rRNAs. Transcription efficiency is expressed as a percentage of the value obtained with the normal arrangements. DNA binding was assessed by footprint analysis, as indicated in Figure 8.10.

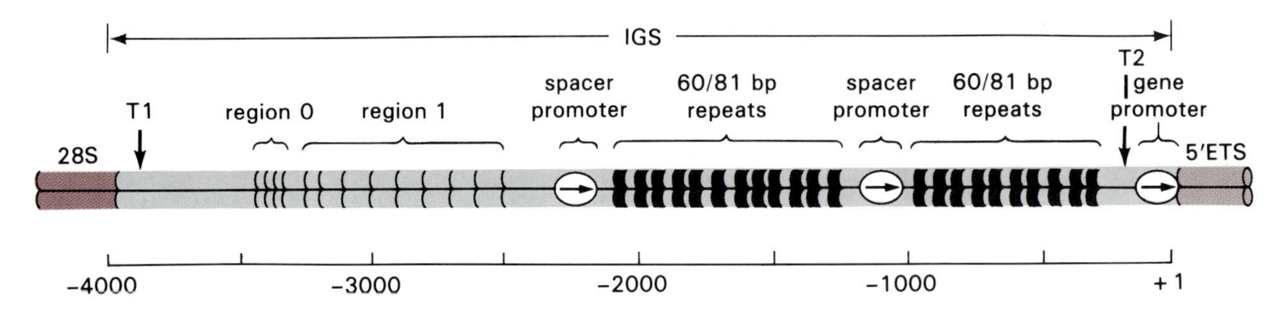

Figure 8.14
A typical intergenic spacer of *X. laevis* rDNA. The region of DNA labeled IGS represents the intergenic spacer. It spans the sequence between the 3′ end of the 28S rRNA coding sequence at the left and the site at which pre-rRNA transcription begins at the start of the 5′ ETS. The ellipses with arrows represent promoter sequences; the small filled-in regions indicate the homologous 60/81 bp repeats. The symbols T1 and T2 represent weak and strong transcription terminators, respectively. Regions 0 and 1 are distinctive sequence arrangements of unknown significance. Adapted from P. Leibhart and R. H. Reeder, *Cell* 45 (1986), p. 431.

cies, the repeats have a nearly identical sequence, but the sequence differs from species to species.

Examinations of the nuclear and cytoplasmic transcripts of *Drosophila* rRNA genes in both species have led to the following conclusions. Analysis of genes with deletions in their control regions, such as the ones described for the mammalian rDNA promoter, has located a core promoter that spans the region between base pairs −43 and +20. Each of the 240 bp repeat elements in the IGS can also promote a low level of transcription, beginning a few base pairs beyond its 3′ end and terminating about 175 bp downstream within the adjacent 240 bp repeat. However, more than 95 percent of the transcription initiations occur just beyond the most downstream repeat (at base pair +1) and continue through the entire transcription unit. Because the short transcripts are rare and absent from the cytoplasmic RNA, their function is obscure. But the organization and maintenance of the repeats in the two previously mentioned *Drosophila* species, as well as their presence in *Xenopus* rRNA genes, suggest that the clustering of promoters and terminators in the IGS upstream of the "true" promoter has a role in regulating the efficiency of rDNA transcription.

Xenopus A typical arrangement of the novel repeats within an IGS that separates rDNA transcription units of *Xenopus laevis* is summarized in Figure 8.14. The core promoter, located between base pairs −147 and +4, is essential for proper transcription initiation of the rDNA. Depending upon which of the many intergenic spacer regions is examined, the 150 bp core promoter sequence is duplicated two to six times within the spacer. Separating these duplications are regions of intermingled and related 60 and 81 bp repeated elements. Interestingly, each of the 60 and 81 bp repeated segments shares a sequence of about 40 bp with the core promoter and its copies in the IGS. Two additional types of repetitive elements, different from the 60 and 81 bp sequences, occur within the farthest upstream portion of the intergenic spacer region (region 0 and region 1). Most of the variation in the spacing between the rDNA transcription units can be ascribed to variable numbers and arrangements of these different repeated segments within individual spacer sequences (Section 9.2a).

Functional roles for each type of repeated sequence element are still only speculative. However, studies of rDNA transcription *in vitro* with RNA polymerase I and its relevant factors, as well as *in vivo* experiments in which suitable cloned rDNA templates are injected into *Xenopus*

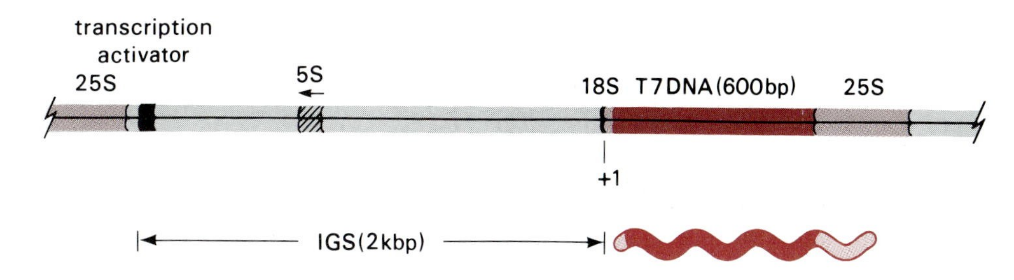

Figure 8.15
A recombinant psuedo-rRNA gene for identifying transcription signals in yeast rRNA genes. The recombinant pseudo-rRNA gene contains 2 kbp of IGS DNA, 130 bp of the 5′ coding sequence of 18S rRNA, 600 bp of T7 DNA, a 600 bp segment containing 500 bp of 25S rRNA coding sequence, and 100 bp of 3′ flanking DNA. Although not relevant to this experiment, the position of the 5S rDNA in the IGS is also shown. The hybrid gene with, or lacking, the 190 bp putative transcription activator at the 5′ most end of the intergenic spacer region was introduced into the yeast chromosome or via a replicating plasmid. Transcription from the hybrid gene was monitored by following the production of a T7-like rRNA.

oocytes, have provided several clues. (1) The 150 bp element that straddles the transcription start site most probably determines the initiation of rDNA transcription because such a repeated element is present in this region of the rDNA in all species. (2) The upstream 150 bp type repeats promote transcription initiation within the spacer, but these transcripts do not continue into the transcription unit. (3) The 40 bp sequences that are common to the 150 bp core promoter and that occur in the 60/81 repeat sequences probably do not function as promoters. They appear instead to enhance the efficiency of transcription initiation. This enhancement is independent of the orientation and distance of the sequence relative to the core promoter. (See Section 8.3d for a discussion of enhancers.) (4) The 60 and 81 bp repeats provide a multiplicity of enhancer sequences and, by their number and spacing, modulate the efficiency of transcription initiation by the core promoter. The function of region 0 and region 1 is unclear. Thus, *Xenopus* rDNA transcription appears to be regulated by clusters of promoter-enhancer elements within the intergenic spacer region.

Yeast Yeast rRNA genes are also organized into tandem transcription units encoding the 18S, 5.8S, and 25S sequences separated by intergenic spacer regions. The yeast clusters are unique, however, in having a 5S rRNA gene, a class III gene, associated with each rDNA repeat (Section 9.2b); but the 5S rRNA gene is transcribed from the opposite DNA strand and by RNA polymerase III (Section 8.4) rather than by RNA polymerase I.

The sequences modulating the expression of the yeast rDNA transcription unit have been delineated using both *in vitro* transcription systems and by transforming yeast cells with plasmids containing a pseudo-rRNA gene. One such pseudo-rRNA transcription unit was constructed by replacing most of an rDNA transcription unit with a segment of bacteriophage T7 DNA. The construct placed the T7 DNA 130 bp downstream of the pre-rRNA transcriptional start site and 500 bp upstream of the end of the 25S rRNA coding sequence (Figure 8.15). This pseudo-rRNA gene was incorporated into an *ARS* and centromere containing plasmid that was maintained as an autonomously replicating plasmid after transformation into yeast cells (Section 5.6b). Transcripts from this chimeric transcription unit could be distinguished from those emanating from the endogenous yeast rRNA genes because of the T7 DNA sequences.

Three discrete regions of the intergenic spacer that affect the rate and accuracy of transcription initiation were identified using pseudo-rRNA gene derivatives having deletions and rearrangements within the IGS. First, the sequence between base pairs −149 and +15 is sufficient to support low level transcription of the pseudo-rRNA. The sequence between base pairs −149 and −133 is especially important for efficient transcription initiation. Second, the presence of the region from base pairs −210 to −2230 restricts initiation to the correct site while increasing the efficiency about twofold. Third, adding the 190 bp region spanning base pairs −2230 to −2420 at the 5′ end of the IGS (Figure 8.15) stimulates transcription at the correct site by about tenfold. However, as with the 60/81 sequences in the *Xenopus* IGS, this element also functions in either orientation—that is, its stimulatory effect is produced irrespective of its orientation relative to the start site of transcription. This suggests that the 190 bp segment also acts as an enhancer.

Because no transcripts corresponding to the intergenic spacer are discernable *in vivo*, the 190 bp upstream element at the most 5′ end of the IGS does not seem to promote transcription initiation. But *in vitro* transcription studies with yeast extracts show clearly that new RNA chains are initiated within this 190 bp segment. The essential sequence for promoting RNA synthesis in this region is contained within a 22 bp segment (base pairs −32 to −11) upstream of the ectopic transcription start site. Thus, in yeast, transcription of the rRNA gene cluster may initiate about 2 kbp upstream of the mature 5′ end of the pre-rRNA. But the transcripts initiated at the 5′ end of IGS are probably terminated near the beginning of the pre-rRNA sequence, and reinitiation of transcription very likely creates the pre-rRNA's correct 5′ terminus. Recall that transcripts are also initiated and terminated within the intergenic spacer regions of *Drosophila* and *Xenopus* rRNA genes. How terminations in this region influence transcription initiation is considered in the next section.

d. rDNA Transcription Termination

Because pre-rRNA transcripts end at or near the 3′ terminus of the 28S rRNA sequence, it was believed initially that the sequence spanning this terminus represented an RNA polymerase I transcription terminator. But subsequent experiments involving *in vitro* transcription of specially constructed rRNA genes show that transcription continues well beyond the 3′ end of the mature 28S rRNA. *Xenopus* transcription continues through the entire intergenic spacer region, terminating at a site about 200 bp upstream of the core promoter for the next 35S pre-rRNA; endonucleolytic cleavage at a site just beyond the end of the 28S rRNA coding region accounts for its mature 3′ end. In yeast, transcripts initiated at the core promoter terminate in the intergenic spacer, beyond the 5S rRNA gene and about 300 bp 5′ to the start of the next pre-rRNA sequence.

In mouse rDNA, transcription termination occurs between 550 and 600 bp beyond the 3′ end of the 28S rRNA sequence. The signal for terminating RNA polymerase I transcription in this system resides in a highly conserved 18 bp sequence on the sense strand (5′-AGGTCGACCAG$^{\mathrm{TA}}_{\mathrm{AT}}$NTCCG-3′, termed the Sal I box) generally preceded by one or more T-rich pyrimidine clusters. Eight such repeats occur between 500 and 1200 bp from the 3′ end of the 28S rDNA (Figure

```
CTCTGCGGGC TTTCCCGTCG CACGCCCGCT CGCTCGCACG CGACCGTGTC GCCGCCCGGG CGTCACGGGG GCGGTCGCCT CGGCCCCCGC GCGGTTGCCC
GAACGACCGT GTGGTGGTTG GGGGGGGGAT CGTCTTCTCC TCCGTCTCCC GAGGACGGTT CGTTTCTCTT TCCCCTTCCG TCGCTCTCCT TGGGTGTGGG
AGCCTCGTGC CGTCGCGACC GCGGCCTGCC GTCGCCTGCC GCCGCAGCCC CTTGCCCTCC GGCCTTGGCC AAGCCGGAGG GCGGAGGAGG GGGATCGGCG
GCGGCGGCGA CCGCGGCGCG GTGACGCACG GTGGGATCCC CATCCTCGGC GCGTCCGTCG GGGACGGCCG GTTGGAGGGG CGGGAGGGGT TTTTCCCGTG
AACGCCGCGT TCGGCGCCAG GCCTCTGGCG GCCGGGGGGG CGCTCTCTCC GCCCGAGCAT CCCCACTCCC GCCCCTCCTC TTCGCGCGCC GCGGCGGCGA
CGTGCGTACG AGGGGAGGAT GTCGCGGTGT GGAGGCGGAG AGGGTCCGGC GCGGCGCCTC TTCCATTTTT TCCCCCCCAA CTTCGGAGGT CGACCAGTAC
                                                                                                          1
TCCGGGCGAC ACTTTGTTTT TTTTTTTTCC CCCGATGCTG GAGGTCGACC AGATGTCCGA AAGTGTCCCC CCCCCCCCCC CCCCCCGGCG CGGAGCGGCG
                                                    2
GGGCCACTCT GGACTCTTTT TTTTTTTTTT TTTTTTTTTT TTAAATTCCT GGAACCTTTA GGTCGACCAG TTGTCCGTCT TTTACTCCTT CATATAGGTC
                                                                           3
GACCAGTACT CCGGGTGGTA CTTTGTCTTT TTCTGAAAAT CCCAGAGGTC GACCAGATAT CCGAAAGTCC TCTCTTTCCC TTTACTCTTC CCCACAGCGA
    4                                                     5
TTCTCTTTTT TTTTTTTTTT TTTGGTGTGC CTCTTTTTGA CTTATATACA TGTAAATAGT GTGTACGTTT ATATACTTAT AGGAGGAGGT CGACCAGTAC
                                                                                                          6
TCCGGGCGAC ACTTTGTTTT TTTTTTTTTT TCCACCGATG ATGGAGGTCG ACCAGATGTC CGAAAGTGTC CCGTCCCCCC CCTCCCCCCC CCGCGACGCG
                                                     7
GCGGGCTCAC TCTGGACTCT TTTTTTTTTT TTTTTTTTTT TTTAAATTTC TGGAACCTTA AGGTCGAC
                                                                      8
```

8.16), but generally, termination occurs within runs of T preceding the first Sal I box. The 18 bp conserved sequence can, by itself, cause transcription termination, but more efficient, accurate termination occurs when there are T-rich stretches both before and after. Small deletions, insertions, and base pair changes within the 18 bp segment or inversion of the segment abolish its terminator activity. A somewhat similar arrangement of conserved repeats (5′-GGGTTGACCA-3′) probably provides the transcription termination signal for the human pre-rRNA. Conserved repeated sequences (5′-GACTTGC-3′ preceded by runs of T) occur in the intergenic spacer regions of *Xenopus* species, termination occurring most frequently at the repeat just before the promoter at the 3′ end of the spacer region.

Indirect evidence initially suggested that termination at these sites depends on sequence-specific binding proteins. These indications were confirmed by the demonstration of one or more protein factors in nuclear extracts that bind to the 18 bp termination sequence. In this instance, binding of the protein ligand to the termination sequence was monitored by the protection of a 5′ end labeled fragment containing that signal against digestion with exonuclease III (Figure 8.17). The principle of this assay is that digestion of the fragment by the enzyme proceeds until it encounters the bound protein. Recall that exonuclease III digests double strand DNA fragments from their 3′ ends (Table 4.1). As a consequence, 5′ end labeled strands will be produced, and their length is determined by the location at which exonucleolytic digestion is blocked by the bound protein. Because each strand in the fragment can be labeled independently at its 5′ end, the two boundaries of the blocked sequence can be ascertained. When such experiments were carried out with segments containing two of the putative termination sequences (the Sal I boxes and their surrounding sequences that occur between base pairs 583 and 665 downstream of the 28S rDNA 3′ end), the exonuclease III fragment pattern was altered by

Figure 8.16
Sequence of the mouse rDNA termination region. The sequence extends over 1168 nucleotides from the first position downstream of the 3′ end of mature 28S rRNA. The 3′ transcribed spacer (565 bp) is delineated by the large grey box. The 18 bp sequence motifs (encompassing *Sal* I sites), which are repeated in the intergenic spacer immediately downstream of the termination site, are shown in color and numbered. Pyrimidine-rich tracts (at least ten nucleotides long with a pyrimidine content of 80 percent) are underlined by dashes. Figure taken from I. Grummt et al., *Cell* 43 (1985), p. 801.

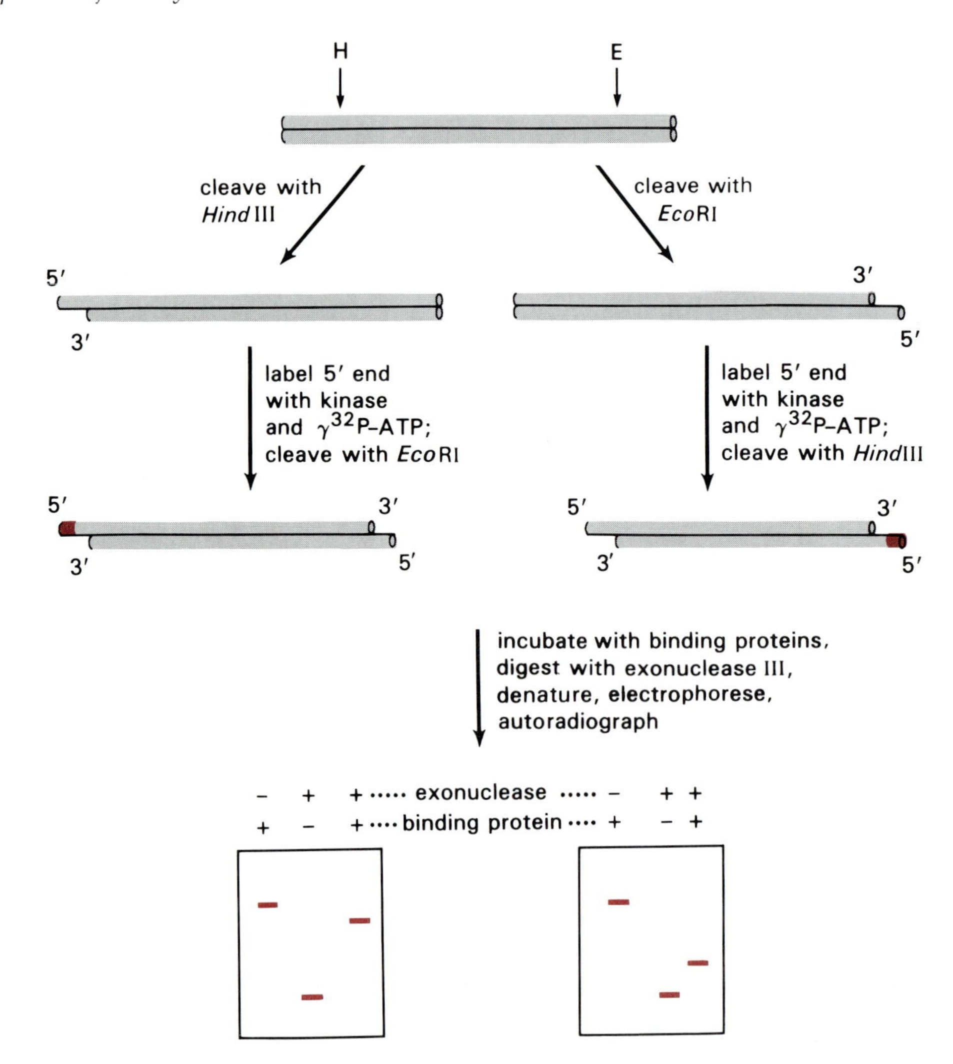

Figure 8.17
Detecting protein binding to DNA by protection of the DNA against degradation by exonuclease III. A segment of DNA of defined length is labeled with ^{32}P at either 5′ end, usually created at a known restriction site. Following incubation of such labeled substrates with cell or nuclear extracts or purified DNA binding proteins, the putative complexes are digested with exonuclease III under conditions in which the complexes are stable. After denaturation, gel electrophoresis, and radioautography, the position of the bands signifies the extent to which each strand was degraded and therefore marks the outer boundaries of the protein binding site. In the example shown, the protein binds closer to the *Eco*RI site than to the *Hind*III site.

prior incubation with a nuclear extract (Figure 8.18). The altered digestion pattern indicates that a protein (or proteins) is bound to a Sal I box sequence. Alterations within a conserved Sal I box sequence or changes in the distance between the two most conserved stretches prevent the binding and impair termination. Moreover, a synthetic oligonucleotide corresponding to the Sal I box consensus sequence promotes transcription termination *in vitro* when introduced into an RNA polymerase I transcription unit. The protein that binds to the Sal I box has been obtained in pure form, although the mechanism of its action is still unknown.

e. Coupling Transcription Termination and Initiation

A novel feature of terminator sequences is that they frequently occur just upstream of the promoters for the next rRNA gene (Figure 8.19). Such terminator sequences have been found at about base pair −170 relative to the start of mouse rDNA transcription, about base pair −185 in the human rDNA upstream control region, and about base pair −200 upstream of the *Xenopus* rDNA transcription unit. Surprisingly, removing or altering these terminator sequences markedly reduces transcription from the +1 site in the next core promoter. Initially, it was surmised that a

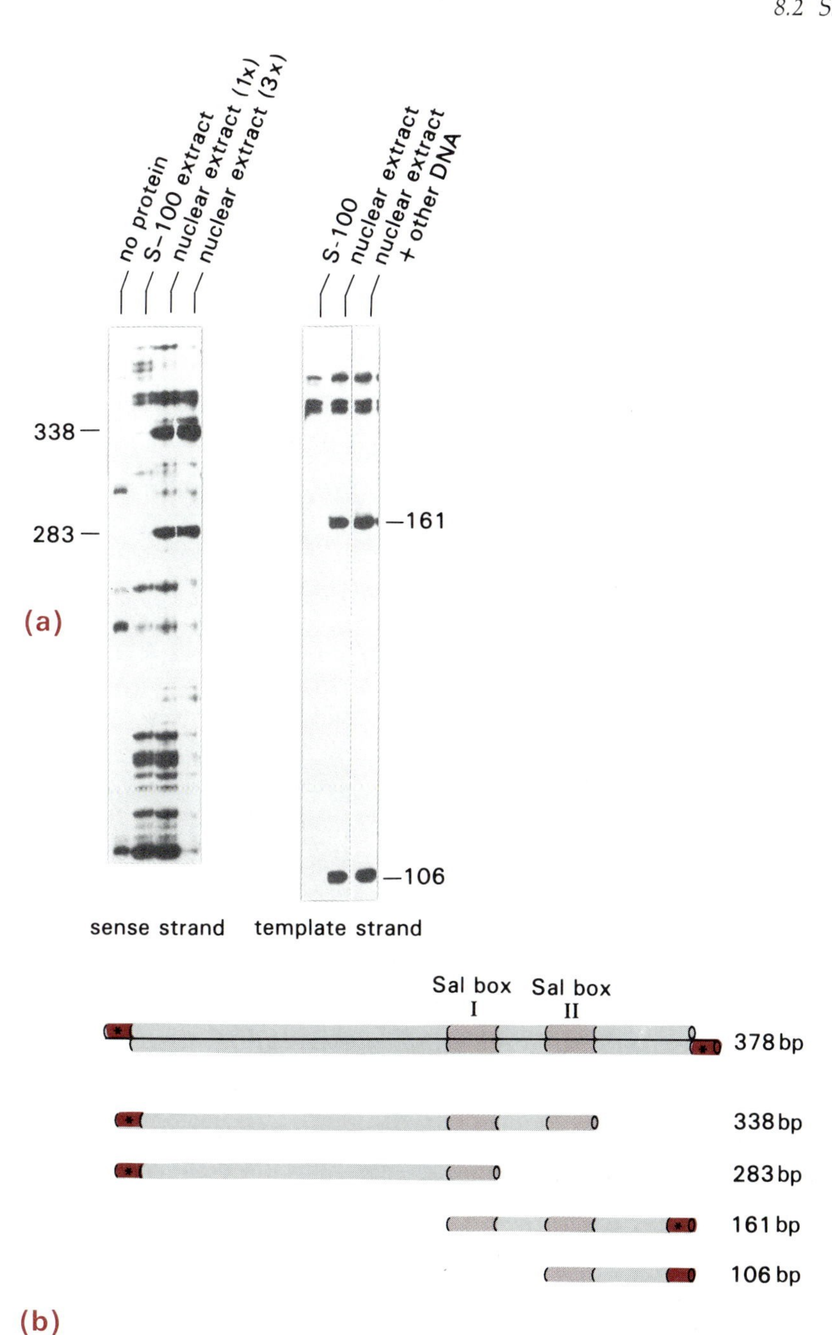

Figure 8.18

Binding of a nuclear termination factor to Sal I boxes. (a) A 378 bp DNA fragment encompassing two Sal I box sequence motifs beyond the end of the mouse 28S rRNA coding sequence was labeled at the 5′ end of either the sense strand or the template strand. The labeled fragment was digested with an excess of exonuclease III in the presence or absence of cell fractions: an S-100 extract (which lacks termination factor activity) or a nuclear extract (which promotes transcription termination at or near the Sal I box sequences). After digestion, the remaining DNA was denatured, electrophoresed, and radioautographed. (b) The relevant features of the labeled (asterisk) starting fragment and the expected exonuclease III digest products assuming binding of a factor at the Sal I box are shown. The lengths of labeled fragments are diagnostic of the locations at which exonuclease III digestion is blocked by the bound factor. The data are taken from I. Grummt et al., *Cell* 45 (1986), p. 837. The picture of the electrophoretic pattern was supplied by I. Grummt.

termination factor might interact with an initiation factor or RNA polymerase I or that activation results from the favorable localization of the enzyme at closely juxtaposed sites near the rDNA promoter. But another explanation is that termination prevents a transcribing polymerase from disrupting transcription complexes that are assembled in the vicinity of the next promoter.

There are numerous puzzles about rRNA gene transcription, particularly as to how the expression of rRNAs, 5S rRNA, and ribosomal protein genes are coordinated. Especially intriguing are the mechanistic and regulatory implications, if any, of the various promoter, enhancer, and terminator sequences in what was initially supposed to be a silent, intergenic spacer region. These findings also raise the question of whether

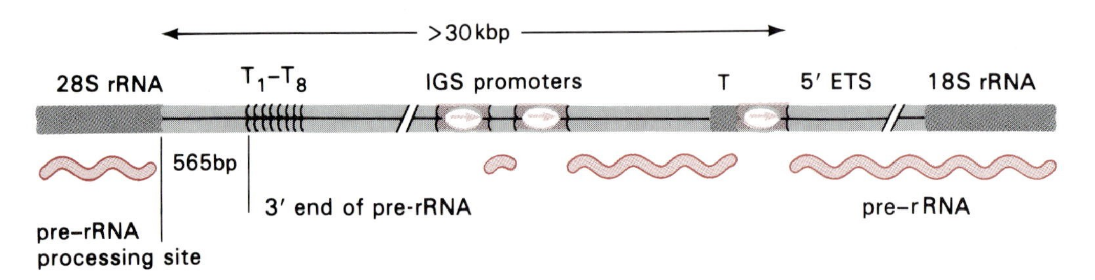

Figure 8.19
The intergenic spacer (IGS) promoters and terminators (T) of the mouse rRNA gene cluster. Transcription from the left through the 28S rRNA sequence terminates within the cluster of terminator sequences (T1–T8), but this RNA is processed to create the mature 3′ end of the 28S rRNA. Transcription initiates at IGS promoters but terminates at T, just upstream of the promoter for the rightward transcription unit. Drawings adapted from I. Grummt et al., *Cell* 47 (1986), p. 901.

coordinated termination and initiation of transcription is a feature in the expression of class II and III genes. From the biochemical viewpoint, it will be important to isolate and characterize the RNA polymerase I related sequence-specific transcription and termination factors so that the molecular basis of their actions and interactions can be elucidated.

8.3 Structure and Expression of Class II Genes

a. *General Considerations*

Features of class II RNAs Class II genes encode all cytoplasmic mRNAs and all but one of the U RNAs (U6) in snRNPs. All transcripts of class II genes have characteristic modifications (caps) at their 5′ ends: 7-methyl guanosine in mRNAs and 2,2,7-trimethyl guanosine in U RNAs. In both instances, the methylated guanosine in the cap is linked to the first nucleotide of the transcript by a triphosphate bridge between their respective 5′ positions (Figure 3.53). Almost all mRNAs have characteristic polyadenylate (poly A) tails beginning varying distances beyond the end of their coding sequences; the poly A stretch is not encoded in the genes from which the mRNAs are transcribed but is added in a separate posttranscriptional reaction. In yeast and other lower eukaryotes, the length of the poly A tails is in the range 75 to 125 residues. The nuclear mRNAs of most vertebrates have poly A tails between 200 and 300 nucleotides long, but, interestingly, the poly A extensions of cytoplasmic mRNA species are smaller and characteristic of the particular mRNA. By contrast, many histone mRNAs and the U RNAs lack poly A tails. We discuss in Section 8.3c how the 5′ cap structures, 3′ ends, and poly A tails are created on each of these RNAs. Some mRNAs contain N^6-methylated adenines, and all U RNAs are distinctive in having a large number of variously modified uracil residues. These modifications are probably made cotranscriptionally or posttranscriptionally, but their physiological function is unknown.

Except in rare instances, mRNAs and U RNAs are derived from primary transcripts whose length and sequence are colinear with the DNA from which they are transcribed. Aside from capping and cleavage at the 3′ end preliminary to polyadenylation, the principal processing step needed for the formation of functional mRNAs is splicing. Although the absence of introns from pre-U RNAs obviates the need for splicing, the primary transcripts require endonucleolytic and exonucleolytic cleavages to create the mature 3′ ends.

Levels of class II gene regulation An organism's or a differentiated cell's phenotype is in large measure determined by the quantity and variety of proteins it contains. These generally reflect the abundances and kinds of mRNAs being synthesized. Thus, the principal mechanism for creating different phenotypes is differential expression of class II genes. This fundamental relationship between gene expression and phenotype was established by comparative measurements of proteins and mRNAs at different stages of organismal development in different differentiated tissues of the same organism and even between identical cells in different physiological states.

In principle and in actuality, the regulation of gene expression may occur at any stage in mRNA biogenesis or during translation of mRNA into protein. A large body of evidence indicates that the predominant determinant of mRNA levels is transcription initiation. This process is regulated mainly by interactions between specific sequence elements that occur at, but often variable distances from, the site of initiation (the ***cis* elements**) and a collection of specific proteins, many but not all of which are sequence-specific DNA binding proteins (***trans* acting** transcription factors). Thus, both the temporal and tissue specificity of transcriptional regulation are governed by the availability and abundance of the different *trans* acting factors. The rate and pattern of pre-mRNA splicing is the next most significant contributor to regulating the level of an mRNA (Section 8.5). Section 8.7 illustrates how differential breakdown of mRNA is also an important regulatory influence in the expression of at least some genes. Transport of mRNAs out of the nucleus to the cytoplasmic sites of translation may also control the expression of protein coding genes, but there is a paucity of data supporting that possibility. In addition, translation of mRNA is regulatable, and several well-documented examples are known (Section 8.7). Selective protein degradation contributes to the regulation of proper protein levels, but the extent to which this occurs is unclear. Where protein turnover influences the steady state levels of critical components of the transcription machinery, for example, *trans* acting transcription factors, it could have a profound effect in regulating mRNA biogenesis.

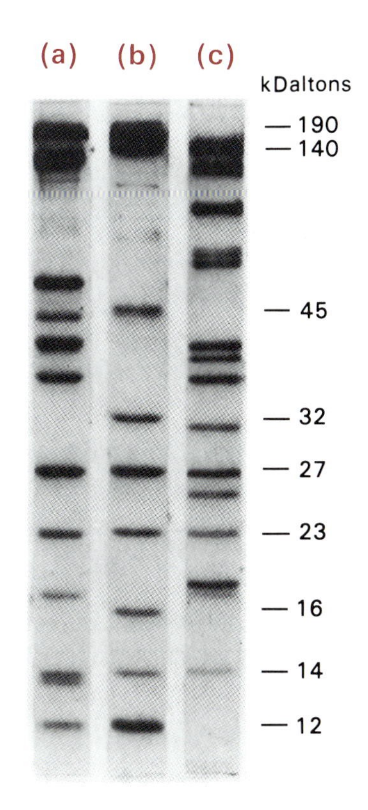

Figure 8.20
Separated protein substituents of mammalian nuclear RNA polymerases. Purified preparations of mammalian RNA polymerase I (a), RNA polymerase II (b), and RNA polymerase III (c) were electrophoresed under denaturing conditions in polyacrylamide gels. The bands represent individual polypeptides that comprise the respective RNA polymerases. Photograph supplied by R. Young.

b. *Transcription Machinery*

RNA polymerase II Class II genes are transcribed in the nucleus by RNA polymerase II. The enzyme has been isolated from a wide variety of organisms, including yeast, *Physarum*, *Aspergillis*, *Acanthamoeba*, higher plants, insects, *Xenopus*, and mammals. The subunit makeup of the enzymes in different organisms is very similar, often being indistinguishable by physical criteria (Figure 8.20). Nine or ten subunits seem to be the consensus. All the enzymes have two large subunits. One, about 220 kDalton in size, is the largest of all the polypeptides in the three kinds of RNA polymerases. The other is about 140 kDalton. RNA polymerase II shares with the other two RNA polymerases three smaller subunits; the remaining four or five subunits are unique to RNA polymerase II. A close structural and chemical similarity between the two largest subunits of all three eukaryotic RNA polymerases was initially suspected from their immunological cross reactivities and common peptides; it is now established by comparisons of the coding sequences in their cloned genes. Moreover, the two largest subunits in each of the three eukaryotic RNA polymerases have amino acid sequences that are strikingly similar to the two large subunits of *E. coli* RNA polymerase.

RNA polymerase II is the most sensitive of the three eukaryotic RNA polymerases to α-amanitin. The toxin permits initiation of RNA chains but blocks their elongation. The α-amanitin binding site appears to be associated with the largest subunit, and α-amanitin resistant mutants of *Drosophila* and *S. cerevisiae* map to the gene encoding the 220 kDalton subunit. The largest RNA polymerase II subunit is also the target for physiological phosphorylation-dephosphorylation reactions, there being multiple sites at which this occurs. The highly phosphorylated form, found in actively transcribing RNA polymerase II *in vivo*, is considerably more active than the unphosphorylated enzyme *in vitro* when tested with the adenovirus major late region promoter.

The genes encoding the largest RNA polymerase II subunit of the yeast, *Drosophila*, and mammalian enzymes encode carboxy terminal polypeptide domains containing multiple tandem repeats of a heptapeptide, Tyr-Ser-Pro-Thr-Ser-Pro-Ser. The number of repeats varies from about 26 in yeast to 52 in mice. This repeated sequence is unique and essential for RNA polymerase activity. Alteration of this sequence in the yeast gene shows that subunits with less than 12 or 13 such repeats are nonfunctional *in vivo*. The abundance of serines and threonines in this region may provide sites for the previously mentioned phosphorylation, but the precise role for the repeated heptamer sequence is unknown.

Transcription factors RNA polymerase II, like the other RNA polymerases (Sections 8.2 and 8.4), requires an array of additional proteins to form a functional transcription complex. Many are DNA binding proteins that recognize one or more of the sequence elements that together constitute the gene's promoter. Some transcription factors appear to function through protein-protein interactions with other factors and thereby modify their binding specificity and affinity. By their interactions with the different DNA sequence motifs and with each other, the various transcription factors form complex protein assemblies that regulate the ability of RNA polymerase II to initiate transcription. Most of the complexes act positively to increase transcription initiation, but some are known to behave negatively and thereby act as repressors. There are even factors that stimulate transcription of one gene while inhibiting that of another. Many of the factors are tissue- or cell-specific or act only at certain intervals in development; these permit genes to be differentially transcribed in different tissues or at different times. In a sense, RNA polymerase II provides the transcribing machine; and the interactions of the factors with the regulatory signals in the DNA, and with each other, determine where, when, and how fast the machine operates.

An ever increasing number of protein factors has been implicated as *trans* acting factors in RNA polymerase II transcription systems, although not all have been thoroughly characterized. The involvement of any one transcription factor is a function of the types and arrangement of the DNA sequence motifs, the *cis* elements, that constitute a gene's regulatory region. The complexity is such that some transcription factors bind to more than one kind of DNA sequence motif, and some DNA sequences are binding sites for more than one kind of factor. Moreover, tandemly arranged or overlapping sequence motifs often create additional binding sites not present in either motif alone or favor the binding of one transcription factor over another. The existing complexity of these interactions, as well as the substantial number of unresolved details, makes a comprehensive description of this important area unattainable. Moreover, different and often unrelated nomenclatures have been used to designate transcription factors. Generally, these are based on their tissue or cell type origin, the principal sequence characteristic of their binding sites, or simply an idiosyncratic choice by the investigator. Therefore, we have elected to concentrate on a description of the characteristic *cis* acting elements in a few model regulatory regions (Sections 8.3d–h) and to mention those factors whose interaction with these regions is best understood. Later, in Section 8.6, we discuss in more detail the distinctive structural properties of selected polypeptide transcription factors that account for their ability to interact with DNA and to activate transcription.

c. *Maturation of mRNAs and U RNAs*

Capping With only one known exception (picornavirus mRNAs), all eukaryotic cellular and viral mRNAs possess caps at their 5′ termini. The cap structures in both nuclear and cytoplasmic RNA polymerase II transcripts are associated with specific cap binding proteins, but the role of the caps or the proteins bound to them in mRNA metabolism is unclear. It is well established, however, that the cap structure interacts specifically with translation initiation factors in the assembly of the translation complex.

Capping of pre-mRNA occurs at the 5′ end of RNA polymerase II transcripts almost immediately after they are initiated; even the shortest RNA chains analyzed are capped. One of the substrates for the capping enzyme (guanylyl transferase) is GTP; the other is the diphosphate end created by removing the terminal phosphate from the triphosphate of the first nucleotide in the nascent RNA (Figure 8.21). In the capping reaction, the terminal phosphate of the nascent pre-mRNA's triphosphate group is replaced by the guanylyl group of GTP, with the accompanying loss of GTP's two terminal phosphates. No reaction occurs if the pre-mRNA has only a 5′ monophosphate end. Thus, caps mark the first nucleotide at which transcription starts. The added guanine residue is methylated at position 7 of the purine ring by guanosine-7-methyl transferase; the 2′ hydroxyl groups of the first and second nucleotides in the nascent RNA are methylated subsequently by 2′-0-methyl transferase. 7-methyl GTP itself is not a substrate for capping. Presumably, U RNAs are capped in an analogous way; but in this instance, subsequent methylation yields 2,2,7-trimethyl guanosine caps as well as 2′-0-methyl substituents on the transcript's first few nucleotides. Because pre-mRNA transcripts made in cell extracts or with purified RNA polymerase II are capped correctly, the responsible enzymatic activities are probably associated with the polymerase itself. The RNA polymerase II mediated transcription initiation and capping reactions are probably obligatorily coupled because the nuclear

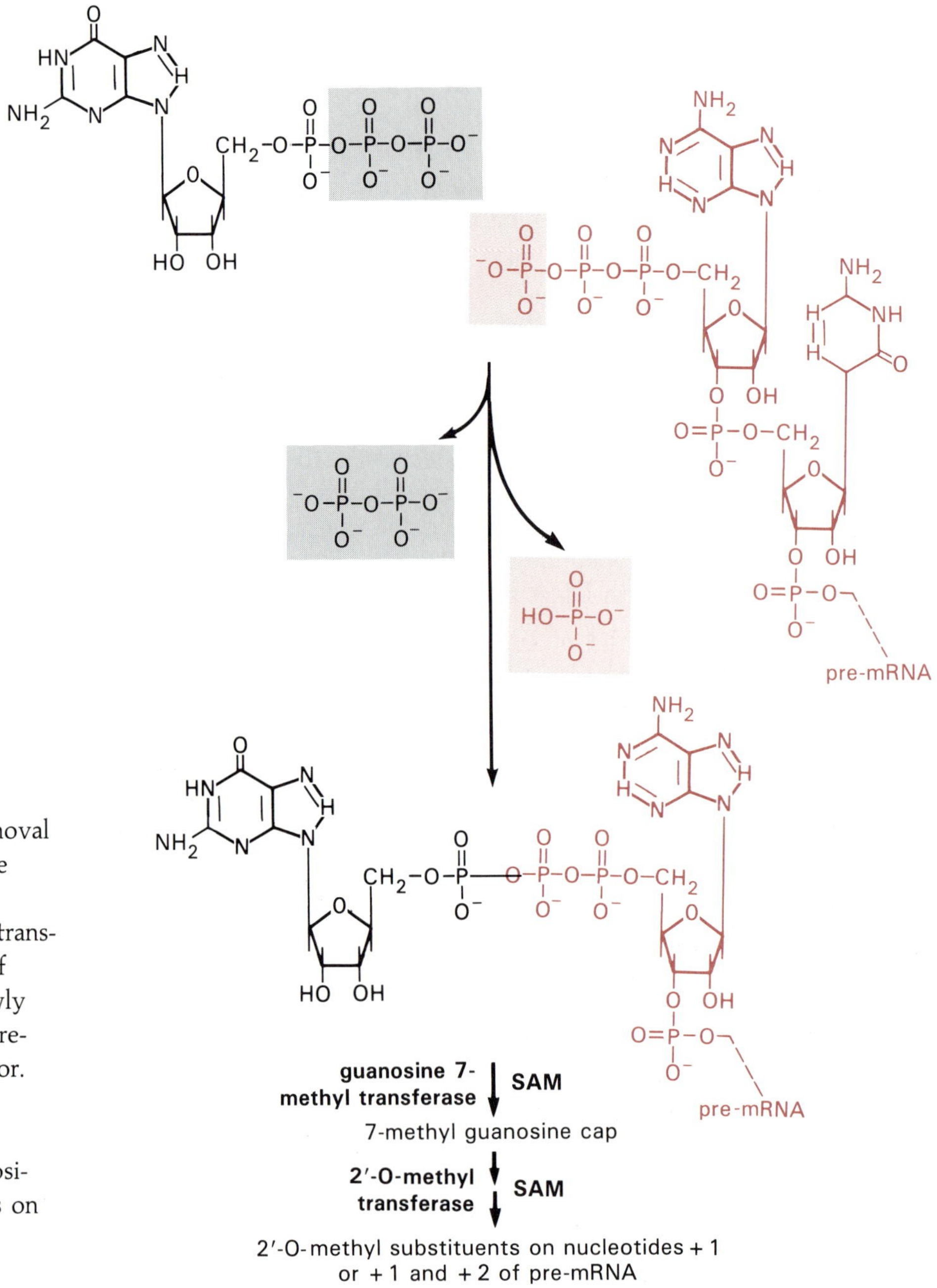

guanosine 7-methyl transferase ↓ SAM

7-methyl guanosine cap

2′-O-methyl transferase ↓ SAM

2′-O-methyl substituents on nucleotides + 1 or + 1 and + 2 of pre-mRNA

Figure 8.21
Creation of 5′ cap structures on eukaryotic mRNAs. Following removal of the terminal phosphate from the 5′ triphosphate end of the nascent pre-mRNA, the enzyme guanylyl transferase links the guanylyl moiety of guanosine triphosphate to the newly created diphosphoryl end of the pre-mRNA to form the 5′ cap precursor. Specific methyl transferases utilize S-adenosyl methionine (SAM) to methylate the guanosine cap at position 7 and the 2′ hydroxyl groups on the first and second riboses in the pre-mRNA chain.

transcripts made by RNA polymerase III, which have triphosphate ends (e.g., U6 RNA, 5S rRNA, and pre-tRNAs), are not capped. One wonders, therefore, what determines the different kinds of caps for pre-mRNAs and pre-U RNAs.

The role of caps during transcription and in the maturation of primary transcripts into mRNA is unclear. For example, the evidence regarding cap involvement in splicing is conflicting. The likelihood that caps have a role in forming the nuclear ribonucleoprotein complexes (nRNPs) in which pre-mRNAs are sequestered during mRNA transport out of the nucleus or in the assembly of cytoplasmic ribonucleoprotein particles (cRNPs) remains unexplored. There are indications that caps and their associated cap binding proteins protect mRNAs against degradation from their 5′ ends. Better known, though, is the clear requirement of caps for the attachment of ribosomes to mRNA prior to translation initiation. One of the initiation factors, eIF-4E, functions as a specific cap recognition factor during the association of a 40S ribosomal subunit with an mRNA 5′ end (Section 3.8b). Removal of the cap structure from an otherwise functional mRNA prevents its translation. The only known exception to the requirement for a 5′ cap in mRNA translation is the picornavirus (e.g., polio virus) genomic RNA, which functions as the mRNA for all viral encoded proteins. Moreover, polio virus infected cells turn off host protein synthesis because a virus encoded protein inactivates the cellular cap binding protein.

Transcription termination and polyadenylation Transcription of many eukaryotic protein genes continues well past the site at which the RNA is cleaved to produce the 3′ end of the mature mRNA. Termination occurs at multiple sites extending over hundreds or even thousands of DNA base pairs. The region in which transcription of the vertebrate β-globin gene ends, for example, is more than 1 kbp beyond the polyadenylation site. Similarly, transcription of the first of the two tandem mammalian α-globin genes stops within the nearly 2 kbp spacer region that separates the two genes. An even more striking example is provided by the transcription of the adenovirus late region. In this instance, transcription continues past five different polyadenylation sites before it terminates at the end of the viral DNA.

By contrast, some eukaryotic protein genes appear to have authentic transcription termination signals. In the human gastrin gene, for example, a termination signal is located 192 bp downstream of the site at which the gastrin mRNA is polyadenylated. This sequence, which terminates transcription both *in vivo* and *in vitro*, is a T-rich stretch, 5′-$T_9A_2T_5AT_4AT_4AT_5$-3′. Transcription of a DNA having this sequence anywhere along its length terminates just 5′ to the signal. A similar sequence with a few mismatches is found both downstream and upstream of several other vertebrate genes. This putative termination signal, and certain T-rich sequences in yeast genes, 5′-CAATCTTG-3′ and 5′-T_5ATA-3′, are reminiscent of ρ-dependent termination signals in *E. coli* (Section 3.8d).

Irrespective of whether there are specific termination signals, transcription most frequently passes through polyadenylation sites. Consequently, the 3′ polyadenylated ends must be created by endonucleolytic scission followed by polymerization of adenylate residues onto the 3′ hydroxyl group created at the cleavage site. Two relatively closely spaced sequences are needed to specify the cleavage-polyadenylation site. One is the virtually invariant AATAAA sequence located 10 to 30 bp upstream of a CA dinucleotide, within or near the site at which cleavage and polyadenylation frequently occur. Changing the T in the AATAAA sequence to G markedly reduces the efficiency of polyadenylation, but the few ends that are formed are polyadenylated. Mutation of the last A to a G in the AATAAA sequence at the end of the first of the two tandem human α-globin genes impairs the production of normal α-globin mRNA. These and other experiments indicate that the requirement for the AATAAA sequence is exacting. However, because AATAAA sequences also exist in the coding regions of many genes, sites at which

transcript	
human α-globin	TCTTTGAATAAAGTCTGAGTGGGCGGCAGCCTGTGTGTGCCTGGGTTCTCTCTGTCCCGGAATGTGCCAACAATGGA
human β-globin	CTGCCTAATAAAAAACATTTATTTTCATTGCAATGATGTATTTAAATTATTTCTGAATATTTTACTAAAAAGGGAAT
human HLA	AGACAAAATAAATGGAACACATGAGAACCTTCCAGAGTCCATGTGTTTCTTGTGCTGATTTGTTGCAGGGG
human α-IFN	TTTTTCATTAAATTTTTACTATACAAAATTTCTTGTGTTTGTTTATTTTTTAAGATTAAATGCCAAGCCTGACTGTA
human γ-IFN	TATATGAATAAAGTGTAAGTTCACAACTACTTATGCTGTGTTGGACTTTTTCTAAGTGAGACCTGGAGTGAAAGAAC
human growth hormone	TGTCCTAATAAAATTAAGTTGCATCATTTTGTCTGACTAGGTGTCCTCTATAATATTATGGGGTGGAGGGGGGTGGT
human preproinsulin	AGATGGAATAAAGCCCTTGAACCAGCCCTGCTGTGCCGTCTGTGTGTCTTGGGGGCCCTGGGCCAAGCCCCACTTCC
mouse metallothionein	GTTAATAATAAAAGCCTGTTTGAGTCTAACTCTGGTTTCTTGGTGTGGTTTGGCAATAAGAAACTGGGGTGACTTGA
SV40 early	TTCACAAATAAAGCATTTTTTTCACTGCATTCTAGTTGTGGTTTGTCCAAACTCATCAATGTATCTTATCATGTCTG
SV40 late	AGCTGCAATAAACAAGTTAACAACAACAATTGCATTCATTTTATGTTTCAGGTTCAGGGGGAGGTGTGGGAGGTTTT
Ad12 Ela	ATGTATAATAAAACTGGTTTCGGTTGAAGTGTCTTGTTAATGTTTGTTTGGGCGTGGTTAAACAGGGATATAAAGCT
Ad2 L2	AATCAAAATAAAAAGTCTGGAGTCTCACGCTCGCTTGGTCCTGTAACTATTTTGTAGAATGGAAGACATCAACTTTG
HSV-1 TK	GACGGCAATAAAAAGACAGAATAAAACGCACGGGTGTTGGGTCGTTTGTTCATAAACGCGGGGTTCGGTCCCAGGGC
ASV	CGATACAATAAACGCCATTTGACCATTCACCACATTGGTGTGTGCACCTGGGTTCATGGCTGGACCGTCGATTCCCT
A-MuLV	GTATCCAATAAACCCTCTTGCAGTTGCATCCGACTTGTGGTCTCGCTGTTCCTTGGGAGGGTCTCCTCTGAGTGCGT

Figure 8.22
Nucleotide sequences at mRNA 3′ termini. The sequences shown are those of the sense strand encoding the 3′ ends of selected mammalian and eukaryotic viral mRNAs. The 5′ most and 3′ most underlined sequences indicate the invariant AATAAA and the GT-rich sequences, respectively. The arrows indicate the principal site at which polyadenylation occurs. The abbreviations correspond to HLA, one of the genes encoding a major histocompatibility antigen; α-IFN and γ-IFN, α- and γ-interferons, respectively; Ad12 E1A and Ad2 L2, the transforming protein E1A of adenovirus 12 and a structural protein of adenovirus 2, respectively; HSV-1 TK, the thymidine kinase gene of *Herpes simplex* virus type 1; ASV, the proviral sequence encoding the 3′ end of the avian sarcoma virus RNA; and A-MuLV, the proviral sequence encoding the Abelson murine leukemia virus. Compilation taken from I. McLauchlan et al., *Nucleic Acids Res.* 13 (1985), p. 1347.

polyadenylation does not occur, a second sequence must be needed to trigger the reactions leading to cleavage and polyadenylation. The location of the second signal was studied by examining the effect on polyadenylation of deletions in the vicinity of the polyadenylation site. The experiments showed that the precise sequence and position of the essential signal varies with the gene. The most consistent signal(s) in many vertebrate protein genes is a GT- or T-rich segment, 5′-YGTGTGYY (Y is either pyrimidine), within about 25 bp downstream, often followed at variable distances by short runs of T (Figure 8.22).

Even authentic polyadenylation sites fail to function in some instances. In the processing of adenovirus 2 late mRNAs, any one of five potential polyadenylation sites, but only one, is used in the processing of each primary transcript (Figure 8.23). The relative abundance of each of the five classes of adenovirus 2 late mRNAs reflects in part the frequency with which the primary transcript is cleaved and polyadenylated at its characteristic terminus. In this instance, the relative abundances of the different mRNAs produced from the single transcript change during the course of the infection, indicating that the choice of the polyadenylation site is regulatable. A second example occurs in the switch from the production of membrane bound immunoglobulin chains to their corresponding secreted forms (Figure 10.72). The two forms differ in a portion of their immunoglobulin heavy chains. In these instances, polyadenylation at the first pair of signals in the pre-mRNA, the one needed to make the mRNA encoding the secreted immunoglobulin chain, is suppressed. Instead, the mRNA is processed at the next polyadenylation signal, yielding a pre-mRNA that can be spliced to produce the mRNA encoding the membrane bound immunoglobulin chain. Thus, the switch from the formation of membrane bound to secreted immunoglobulin chains occurs by regulating the use of a polyadenylation site. The mechanism of differential regulation of the polyadenylation reaction in these two examples, as well as in numerous others, is unknown.

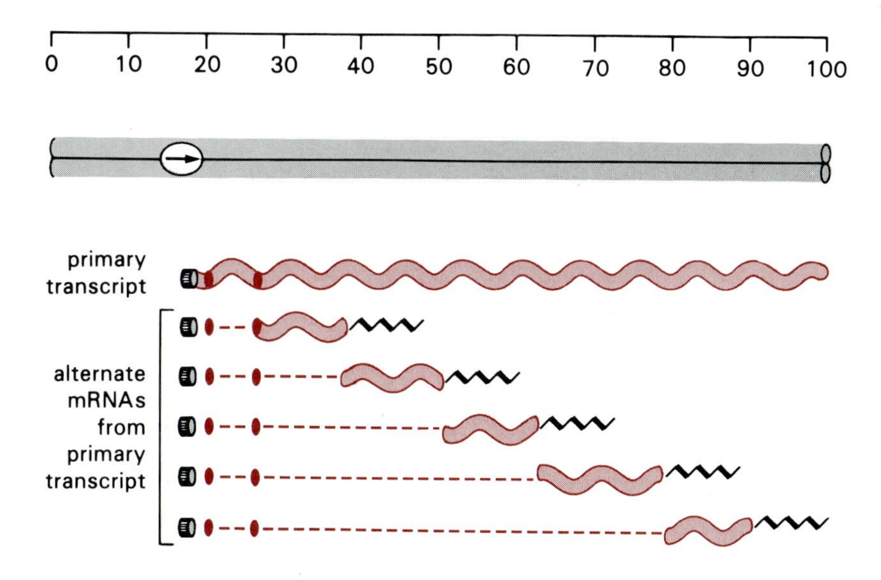

Figure 8.23
Alternately polyadenylated adenovirus mRNAs. The DNA structure represents the adenovirus 2 genome; the numbers above indicate the percentage of the DNA's total length. The late region's primary transcript begins at the major late promoter (about 19 percent), is capped, and continues to the end of the DNA. The pre-mRNAs are cleaved at any one of the indicated locations, and poly A tails (ʌʌʌ) are added. Splicing of each of the polyadenylated forms removes the regions shown by dashes.

The AATAAA sequence has also been implicated in transcription termination. Specifically, the AATAAA sequence at the 3′ end of the mouse major β-globin coding sequence is required for transcription termination that occurs about 1.5 kbp beyond. Furthermore, the same deletions and base changes that block proper polyadenylation also impair the efficiency of the downstream termination. This suggests that recognition of the polyadenylation site must occur before transcription can terminate at the appropriate downstream sequence. One way to account for the coupling of polyadenylation and transcription termination is to suppose that the transcribing RNA polymerase II complex acquires the ability to terminate only after transcribing the polyadenylation signal. This could be accomplished if the transcription complex includes a factor that suppresses inadvertent terminations as it traverses the template but loses that antitermination property as it passes the polyadenylation signal (Figure 8.24). According to this view, if the RNA polymerase II transcription complex lacks this factor, it "falls off" the template at T-rich sequences that it encounters subsequently. This view stems from studies with purified RNA polymerase II *in vitro*, which show that transcription of a variety of DNA templates frequently terminates at T-rich stretches within genes. Alternatively, cleavage of the transcript at the polyadenylation signal could allow an RNA exonuclease or ρ-like protein to "catch" the polymerase and induce termination at downstream T-rich sequences (see Figure 3.12).

Cleavage and polyadenylation occur in various kinds of cell extracts, provided the RNA substrates carry the appropriate signals. Studies of such *in vitro* systems clarified several aspects of the cleavage and polyadenylation reactions. Thus, cleavage and polyadenylation can occur independently of one another. In the presence of ATP analogues (e.g., cordycepin, which is 3′-deoxyadenosine triphosphate), cleavage occurs at the correct site, but polyadenylation does not ensue. Polyadenylation of RNA can occur at 3′ hydroxyl ends that are preceded by an AAUAAA sequence. Fractionation of Hela cell extracts capable of cleaving and polyadenylating RNA substrates yields several protein fractions that together catalyze the cleavage and polyadenylation reactions. One component is a poly A polymerase, a second contains one or more small nuclear ribonucleopro-

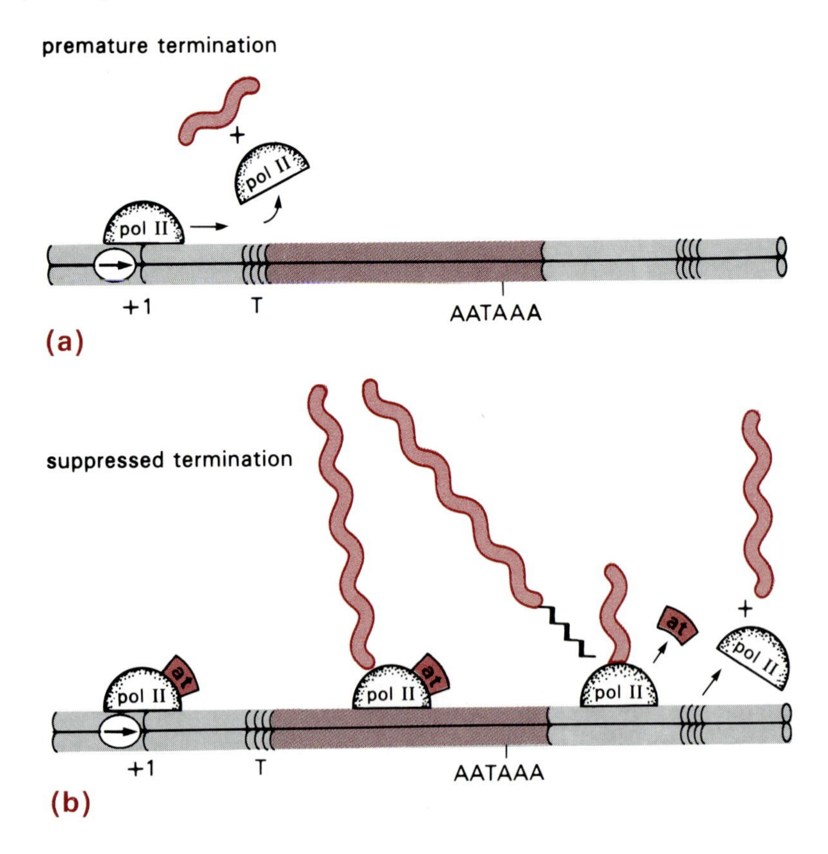

Figure 8.24
A model for coupling polyadenylation and transcription termination. (a) Transcription initiation by RNA polymerase II (pol II) is presumed to terminate prematurely at T-rich sequences designated T located randomly throughout the transcription unit. (b) Interaction of RNA polymerase with an antitermination factor (at) is presumed to suppress terminations at T-rich sequences, allowing the enzyme to transcribe the entire gene sequence. The model further supposes that the at factor dissociates from the polymerase at the polyadenylation signal and that transcription continues until the enzyme terminates transcription at nonspecific T-rich terminator sequences. After J. Logan et al., *Proc. Natl. Acad. Sci.* U.S.A. 84 (1987), p. 8306.

tein particles (snRNPs), and a third is a protein (64 kDaltons) that binds to the region containing the AAUAAA sequence. All three components are needed for the coupled cleavage and polyadenylation, although the poly A polymerase alone can polyadenylate RNA chains that have been cleaved correctly. Although the mechanism is still unclear, the rare U11 snRNP, in conjunction with the 64 kDalton protein and possibly other snRNPs, may form a complex that cleaves the RNA at or near the AAUAAA sequence and permits the poly A polymerase to polyadenylate the 3′ end. The implication of snRNPs is based on finding fragments containing AAUAAA in immunoprecipitates made with antibodies to snRNPs and the inhibition of polyadenylation *in vitro* by antibodies to snRNP. The role of the snRNPs in polyadenylation is consistent with the involvement of other snRNPs in RNA processing: U7 snRNP in creating the 3′ ends of histone mRNAs (discussed later in this section), U3 snRNP in the production of the correct 3′ terminus of 28S rRNA (Section 8.2d), and U1, U2, U5, and U4/U6 snRNPs in pre-mRNA splicing (Section 8.5d).

Because yeast genes and possibly genes in some other small eukaryotes lack the canonical AATAAA polyadenylation signal, the creation of their polyadenylated mRNAs may be coupled to termination. Perhaps sequences like 5′-T_5ATAT, which promote transcription termination, help create the 3′ end for polyadenylation. Alternatively, 3′ end formation may result from endonucleolytic cleavage of extended transcripts followed by polyadenylation with poly A polymerase. Whether the cleavage in yeast involves snRNPs remains to be established.

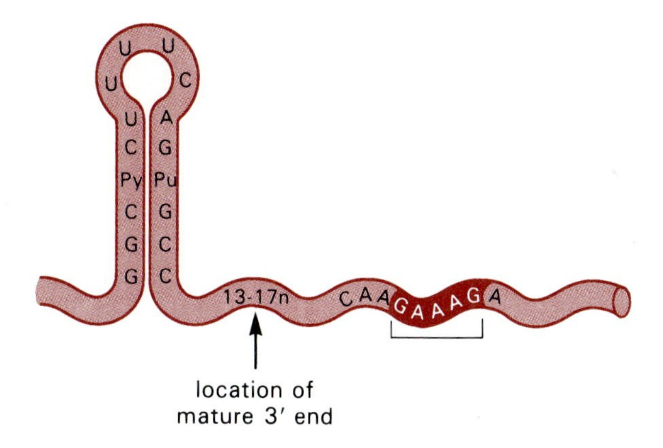

Figure 8.25
Structural and sequence features of the 3′ end of histone mRNAs. The 3′ ends of histone pre-mRNAs contain a characteristic spacing of a sequence capable of forming a 6 base paired stem and a 4 nucleotide loop, separated by about 15 nucleotides from a highly characteristic purine-rich sequence. The mature 3′ end of histone mRNAs is formed by endonuclease cleavage between the stem-loop structure and the purine-rich sequence.

Maturation of histone mRNAs lacking poly A tails Many histone mRNAs are not polyadenylated, although mRNAs encoding some yeast and *Tetrahymena* histones and those mammalian histones whose expression is independent of cell cycle control have poly A tails. Although the histone mRNAs in amphibian oocytes have short poly A tails, these are replaced by nonpolyadenylated histone mRNAs during the organism's maturation.

All nonpolyadenylated histone mRNAs, irrespective of the organism, have a characteristic sequence arrangement spanning the mature 3′ end. This arrangement begins near the end of the coding region with a terminal palindrome that can form a 6 bp double strand stem with a loop containing 4 nucleotides (Figure 8.25). The stem-loop is followed, 13 to 17 nucleotides farther along, by a purine-rich consensus sequence. In sea urchin histone mRNAs, the downstream sequence is absolutely conserved (5′-CAAGAAAGA-3′), but this sequence is less highly conserved in vertebrate histone mRNAs (the consensus sequence is 5′-PuAAAGAGCUG-3′). The nucleotide that becomes the 3′ end of the mature histone mRNA occurs between the stem-loop structure and the purine-rich stretch. Measurements of correct 3′ end formation in a variety of mutant RNAs indicate that changes that prevent formation of the duplex in the stem-loop structure block normal processing, whereas base pair changes that do not disrupt the RNA's ability to form the hairpin do not affect 3′ end formation. Similarly, mutational changes in the purine-rich sequence or even small alterations in its distance from the palindrome drastically reduce 3′ processing. Thus, two closely spaced regions, one preceding the cleavage site and the other downstream of it, are needed to create the proper 3′ terminus. Unexpectedly, histone mRNAs with capped 5′ ends are processed more readily at their 3′ ends than are uncapped mRNAs.

Studies of the 3′ endonucleolytic cleavage of sea urchin and mammalian histone mRNAs *in vivo* and *in vitro* reveal that proper 3′ end formation requires a relatively rare snRNP containing U7 RNA. Alterations of the U7 snRNP's RNA sequence have identified the nucleotides that are essential for promoting the cleavage at the 3′ end of the histone mRNA; these are located close to the U7 RNA's 5′ end. Another factor besides the U7 snRNP, as yet uncharacterized, is also required for the processing reaction. This factor, which is insensitive to nuclease digestion (and thus less likely to be an snRNP), binds to the stem-loop region and promotes interaction between the U7 snRNP and the conserved purine-rich sequence downstream of the mature 3′ end. Comparison of the histone pre-mRNA and

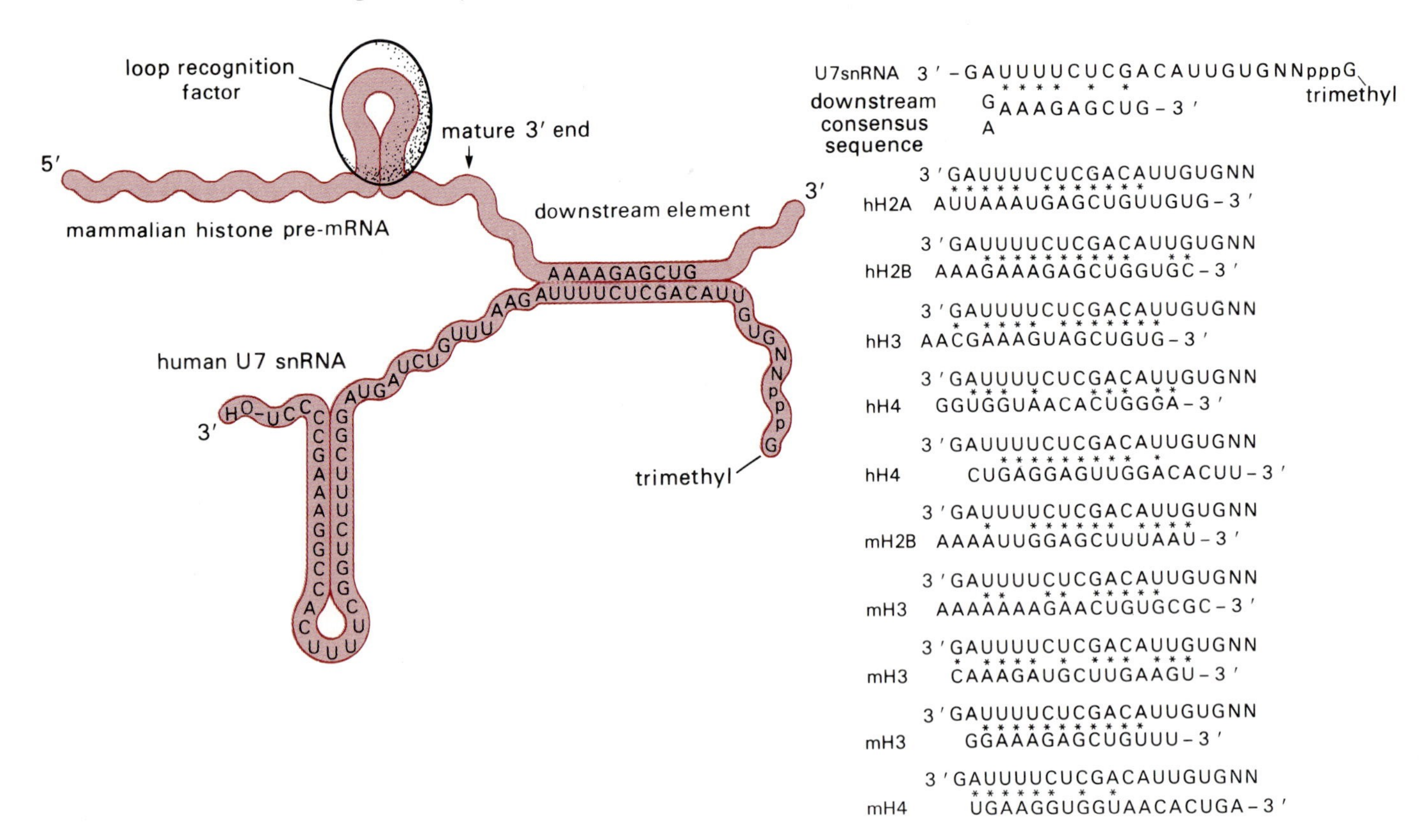

Figure 8.26
Structural features of histone mRNA 3′ ends and their base pairing with U7 snRNA. Extensive base pairing is possible between a discrete nucleotide sequence near the 5′ end of U7 snRNA and the purine-rich sequence downstream of the stem-loop structure. Creation of the mature histone mRNA 3′ terminus occurs by an endonucleolytic cleavage at the indicated position. The pairings between various human (h) and mouse (m) histone mRNA 3′ ends and the U7 snRNA 5′ end are shown at the right of the drawing. The consensus pairings are given at the top. Modified from K. L. Mowry and J. A Steitz, *Trends Biochem. Sci.* 13 (1988), p. 447.

U7 RNA sequences suggests that the purine-rich stretch forms base pairs with the 10-nucleotide sequence (pyrimidine rich) that lies between 10 and 20 nucleotides from the U7 RNA's 5′ end (Figure 8.26). How the cleavage that creates the 3′ ends occurs and is regulated are not known.

Creation of 3′ termini of U RNAs All the U RNAs transcribed by RNA polymerase II lack poly A tails. Their 3′ hydroxyl ends are generated in at least two steps. In the first, either transcription termination or cleavage of a long transcript produces a precursor with a small 3′ extension; the mature 3′ end is then produced by shortening, probably by exonuclease action. An analysis of the efficiency of correct 3′ end formation following various structural modifications in U1 and U2 snRNA genes identifies an essential short conserved sequence on the sense strand, GTTTN$_{0-3}$AAAPuNNAGA (Pu is any purine, and N is any base). This sequence is located 9 to 19 nucleotides beyond the mature ends of U1, U2, and U3 snRNAs from different species. An inverted repeat sequence that can form a stem-loop structure in the pre-snRNA seems not to be essential. Whether the required sequence in the snRNA precursor promotes transcription termination or endonucleolytic cleavage is not known.

A curious feature of U RNA biogenesis is that although the U RNA coding sequence and trimethyl guanosine cap are dispensable for 3′ end formation, proper 3′ ends are formed only if RNA transcription is initiated by a U RNA promoter. If the U RNA promoter is replaced by an mRNA promoter, proper 3′ end formation on the U RNA is impaired. Moreover, transcripts originating at snRNA gene promoters are not polyadenylated even if the templates contain an authentic polyadenylation signal. Con-

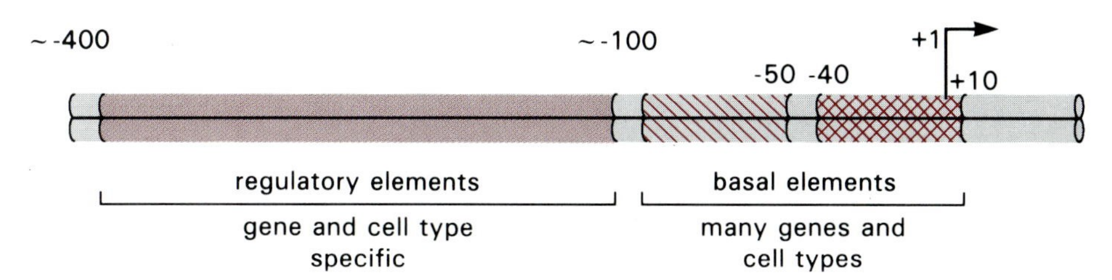

Figure 8.27
Generic sequence elements characteristic of class II gene promoters. Two, or a few, discrete sequence stretches that lie about 100 bp upstream and occasionally overlapping the transcription start site (+1) comprise the basal promoter. Additional regulatory sequence motifs, which generally confer cell- and stage-specific control, occur in the region between base pairs −400 and −100.

versely, transcripts made from mRNA promoters are polyadenylated at polyadenylation signals even if the U RNA processing sequence intervenes. This suggests that specific processing enzymes or transcription termination factors may become associated with transcription complexes at the time they initiate transcription at mRNA or U RNA gene promoters. Similarly, different capping methylases may be included in the transcription complexes for mRNA and U RNA genes.

d. Regulated Expression of Viral Genes

Many of the methods and insights that form the basis for our understanding of the regulation of cellular gene expression were spawned by the analysis of viral transcriptional regulation. As far as we know now, cells and DNA viruses regulate their mRNA levels predominately by varying the rate and location of transcription initiation by RNA polymerase II. Moreover, transcriptional regulation of viral genes, like cellular genes, relies on specific interactions between a complex array of *cis* acting sequence motifs and a wide assortment of *trans* acting transcription factors. Some viruses also encode novel transcription factors that coordinate the expression of groups of viral genes or influence the transcription of selected cellular genes. Indeed, the interplay of viral and cellular *trans* acting factors can affect the transcription of cellular and viral genes differently and even produce positive or negative effects on different genes

The regulation of mammalian virus gene expression recapitulates the principal themes and mechanisms used for the expression of cellular genes. As in most eukaryotic class II genes, viral promoters consist of proximal elements, generally spanning the sequence immediately surrounding the transcriptional start site and extending to about base pair −100, and "distal" regulatory elements, most often between base pairs −100 and −300, that permit cell- and stage-specific gene expression (Figure 8.27). Regulatory sequences that occur within or beyond the transcription unit have not yet been implicated in viral promoters.

Here we describe in detail two viral transcription systems that have been and continue to be especially informative: (1) the SV40 early region and (2) the herpes simplex thymidine kinase (HSTK) gene.

SV40 early gene transcription The general features of SV40's infectious cycle in various hosts are summarized in Section 5.7b. The location of the sequences encoding the mRNAs and proteins as well as the regions governing transcription and replication are shown on the physical map of SV40's genome (Figure 5.34). The early region spans about half the genome's 5243 bp circular length. A single pre-mRNA is transcribed from

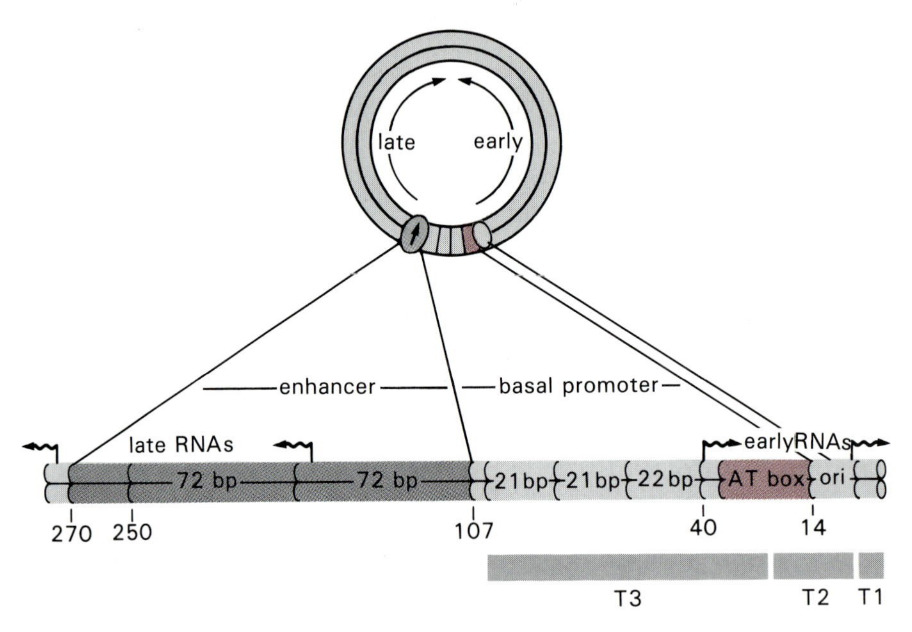

Figure 8.28
Location and organization of the SV40 early and late region promoters. The SV40 DNA genome indicating the position of the regulatory sequences governing replication and early and late region transcription is shown at the top. The locations of the transcriptional start sites for early and late mRNAs are shown as arrows pointing in opposite directions. The origin of DNA replication (ori), the two constituent elements of the basal promoter (the AT box and 21/22 bp repeats), and the tandemly repeated 72 bp segments constituting the enhancer are shown in the expanded version below. The blocks labeled T_1, T_2, and T_3 represent the locations at which T antigen binds, the binding affinities being $T_1 > T_2 \gg T_3$.

this region soon after infection. Differential splicing of this transcript produces two nearly equal length mRNAs that encode two related but distinctive proteins: large T and small T antigens. Large T antigen (Tag) is essential for viral DNA replication and for regulating early and late region transcription, but the function of small T (tag) in the virus life cycle remains problematic. The concomitant increase in the amount of Tag and of progeny viral DNA molecules activates late region transcription. Differential splicing of this region's primary transcript generates two classes of late mRNA that encode the three structural proteins (VP1, VP2, VP3) constituting the virion's capsid.

Thus, the viral genome contains two oppositely oriented and temporally regulated transcription units, each occupying about half the DNA sequence. In the present discussion, we focus on the approximately 300 bp region that separates the two transcription start sites. The *cis* acting signals that govern early and late region transcription as well as replication are interspersed within this segment. They were identified by the study of a wide variety of modifications (e.g., deletions, insertions, base pair substitutions, and rearrangements) within the 300 bp region coupled with both *in vitro* and *in vivo* assays of both transcription and replication. These studies of the interactions between transcription, replication factors, and specific DNA sequences revealed some of the complexities and subtleties that are relevant to SV40's multiplication, as well as some fundamental properties of eukaryotic transcription regulation.

The organization of SV40's *cis* acting signals for regulating replication and transcription are diagrammed in Figure 8.28 and in more detail in Figure 8.29. Transcription initiation of the oppositely oriented early and late regions occurs at the positions indicated by the arrows. Replication, which we consider first, begins within a 27 bp palindromic sequence designated ori. Deletions, insertions, or base pair changes within the palindrome block the initiation of replication. The integrity of the adjacent stretch of 17 A·T base pairs is also essential for efficient replication to

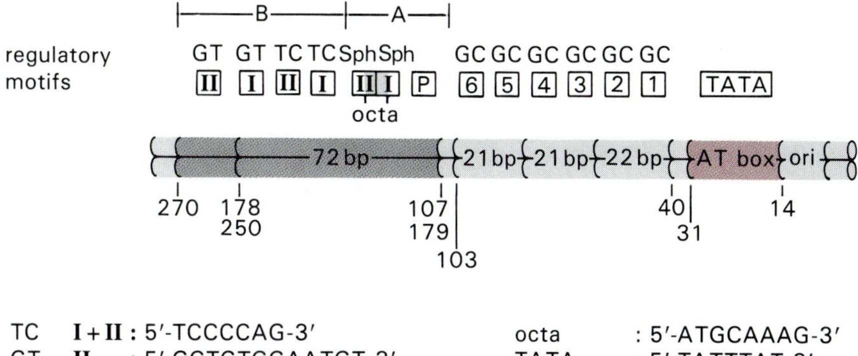

TC	I + II : 5′-TCCCCAG-3′		octa	: 5′-ATGCAAAG-3′
GT	II : 5′-GCTGTGGAATGT-3′		TATA	: 5′-TATTTAT-3′
GT	I : 5′-GGTGTGGAAAGT-3′		GC	: 5′-CCGCCC-3′
Sph	II : 5′-AAGTATGCA-3′		P	: 5′-TTAGTCA-3′
Sph	I : 5′-AAGCATGCA-3′			

Figure 8.29
Sequence motifs constituting the SV40 early region basal promoter and enhancer elements. Sequence motifs, their designations, and their locations within the basal promoter and enhancer elements are indicated above and below the DNA. The numbers below the DNA represent the distance in base pairs upstream of the transcription start site, which is located within ori. The double numbering beneath the 72 bp segment is intended to indicate the tandemly repeated arrangement. The stretches marked A and B represent enhancer domains, which appear to function cooperatively.

begin in ori. Although not essential, the 21/22 bp repeats to the late region side of the A·T sequence augment ori function. We shall see later in this section that the GC-rich 21/22 bp repeats and the adjacent A·T sequence are important elements of the early region transcriptional signals. Their role in promoting replication may stem from their function in transcription.

For replication to begin, the SV40 encoded protein, Tag, must bind at ori. Investigation of this binding reaction indicated that Tag also binds at sequences adjacent to ori. These were first identified as sequences protected against DNase I digestion by binding of Tag. At the initially low levels of Tag that exist early after infection, binding occurs predominately at the sequence encoding the 5′ end of the early pre-mRNA (site T1). As the Tag level rises, binding occurs within ori (site T2). Mutations in site site T1 that impair binding of Tag have little effect on replication. By contrast, altering the palindromic sequence in a variety of ways prevents Tag binding at ori and blocks replication. (Mutations in either site T1 or site T2 also impair Tag's repressor activity on early region transcription.)

Once bound at site T2, Tag functions as an ATP dependent helicase that, in conjunction with a cellular DNA primase and DNA polymerase (and perhaps other proteins as well), promotes the initiation of new DNA chains. The AT- and GC-rich motifs may influence replication because of their proximity per se or because increased transcription initiation in this region facilitates the start of replication.

Each of the Tag binding sites contains at least two copies of a pentanucleotide consensus sequence 5′-GAGGC-3′. The Gs in this sequence are most likely involved in the interaction because they are protected against methylation by dimethylsulfate when Tag is bound. A mutational analysis of the critical nucleotides in site T1 indicates that the Tag binding signal consists of 17 bp containing two pentanucleotide sequences separated by no more or less than 6 to 7 A·T base pairs. This arrangement is thought to permit a Tag dimer to interact with the two appropriately oriented pentanucleotide motifs.

Now we return to a consideration of the different sequence elements that influence early region transcription initiation (Figures 8.28 and 8.29).

One of the elements comprising the basal promoter is associated with the same stretch of 17 A·T base pairs needed for ori function. Mutational alterations or even deletions of the entire region affect the rate of early region transcription only slightly but do impair the accuracy of initiation, resulting in the formation of mRNAs with different capped 5′ ends. Thus, this TATA-like sequence plays a role in directing initiation to a limited number of sites, but it does not have a significant influence on the transcription rate. This contrasts with the situation in many eukaryotic genes, where a somewhat shorter TATA consensus (TATAAA) sequence at about position base pairs −25 to −35 is required for transcription initiation. At the other extreme, there are some eukaryotic promoters that lack an identifiable TATA motif (e.g., the SV40 late region promoter).

SV40's early region promoter also includes the three direct repeats of the 21/22 bp GC-rich sequence that augments ori function. They are located between base pairs −45 to −110 relative to the early RNA start sites (Figures 8.28 and 8.29). Each of the 21/22 bp repeats contains two copies of the six base pair sequence $\begin{smallmatrix}5'\text{-GGGCGG-}3' \\ 3'\text{-CCCGCC-}5'\end{smallmatrix}$. Deletion of this entire 65 bp stretch of SV40 DNA prevents transcription both *in vivo* and *in vitro* even if the surrounding sequence is conserved. However, even one copy of the 21/22 base pair repeat element can support some early region transcription, and two copies are sufficient for maximum transcription rates. Mutational alterations of each base in a 21 base pair repeat segment show clearly that the hexamer sequence is the functionally important motif. Whether the nearly sixfold redundancy of the GC element in SV40's control region is physiologically relevant is unknown.

The SV40 type GC-rich repeat occurs in many other viral and cellular regulatory regions, including HIV-1, herpes simplex thymidine kinase, dihydrofolate reductase, metallothionein-I_A, hydroxymethyl glutarylCoA reductase, and hypoxanthine-guanine phosphoribosyl transferase. This sequence motif functions as a binding site for a specific transcription factor, Spl, which occurs at high levels in many cell types and is described later in this section.

Although necessary, the 21/22 bp elements are by themselves insufficient to promote transcription. Sequences contained within and adjacent to two copies of a 72 bp sequence located between base pairs −110 and −300 are required in addition (Figure 8.29). Some strains of SV40 have only one copy of the 72 bp repeat sequence and its flanking sequence (as shown in Figure 8.29), and that is sufficient for normal transcription activity. Mutational dissection of the 72 bp segment and its upstream flanking sequence demonstrates that this approximately 90 bp region contains several different kinds of subelements (Figure 8.29). Most of the subelement motifs are only weakly stimulatory by themselves, but collectively, in different combinations, they promote high levels of transcription.

Although the 21/22 bp GC-rich elements and the approximately 90 bp upstream segment are both required for efficient early region transcription, their orientation relative to one another and to the transcriptional start sites is not critical. Thus, the entire 21/22 bp repeat segment can be inverted without impairing transcription efficiency. Similarly, the 72 bp upstream element enhances correct transcription initiation in either possible orientation. However, unlike the 21/22 bp repeat element, which functions only in its customary position, the 72 bp segment's location is not critical.

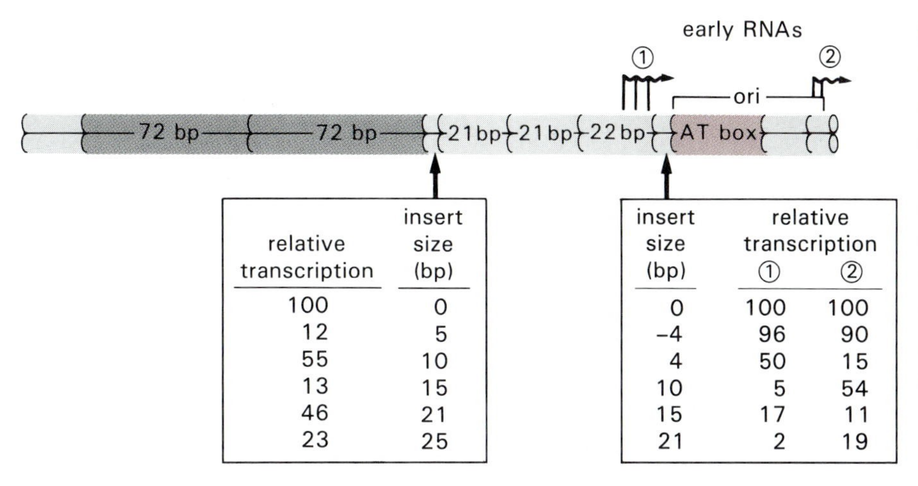

relative transcription	insert size (bp)
100	0
12	5
55	10
13	15
46	21
23	25

insert size (bp)	relative transcription ①	relative transcription ②
0	100	100
−4	96	90
4	50	15
10	5	54
15	17	11
21	2	19

Figure 8.30
Optimal spacing between regulatory elements of SV40 early region promoter. Inserts of different length DNA segments were introduced at either of the positions indicated by arrows, and transcription beginning at positions 1 and 2 was monitored. The relative activities for each modification are expressed as a percentage of the unmodified promoter's activity.

Normal levels of early region transcription occur from the proper start sites even if the 72 bp segment is moved to the early region's intron or several hundred to thousands of base pairs 5′ or 3′ of the transcription unit, and in either orientation. Importantly, however, the 72 bp element alone does not promote transcription, nor is transcription enhanced when it and the 21/22 bp element are on different DNA molecules.

Although the 72 bp and 21/22 bp elements function efficiently even if they are separated by hundreds or thousands of base pairs, there is an optimal arrangement of the two over short distances. Thus, if short DNA segments corresponding in lengths to multiples of a full turn of the DNA helix (10 bp equals a full turn) are inserted between the 72 bp and 21/22 bp repeat segments, the transcriptional activity is only marginally affected. In contrast, if the inserted length corresponds to odd multiples of a half turn of the helix (i.e., 5 or 15 bp), the transcription activity is impaired (Figure 8.30). Similarly, the effect of insertions between the 21/22 bp repeats and the 17 bp A·T segment depends on their length; in this case, the effects are complex and depend on which of the different transcriptional start sites is measured. These results indicate that efficient transcription depends on stereospecific alignments between the 21/22 bp repeat sequence, elements in the 72 bp segment, and other sequences close to the transcription start site. The effective alignments are related to the binding of specific proteins to the various sequence motifs; interactions between the proteins are important for promoting transcription initiation and depend on their proper mutual spatial orientation.

Enhancers The 72 bp SV40 DNA segment also enhances transcription of other genes. For example, a plasmid borne mammalian β-globin gene, which normally is not transcribed after transfection into cultured fibroblasts, is activated if the 72 bp segment is linked to it at varying distances in either orientation. Because of its ability to enhance transcription from a variety of regulatory regions, irrespective of the distance or orientation relative to the start sites, this element is referred to as a transcriptional **enhancer**. Found initially in the SV40 genome, enhancers with very much the same properties but different sequences are now known to be essential features of the transcriptional regulatory signals of many class II genes in all eukaryotic organisms so far examined (Sections 8.3e–h). Enhancer elements also occur in the regulatory regions of the rRNA genes transcribed by RNA polymerase I (Section 8.2c). The existence of enhancers, often kilobase distances 5′ or 3′ to the start of transcription, complicates the identification of a gene's promoter elements.

Commonly, enhancer elements of quite different class II genes are interchangeable. For example, the enhancer element in the murine leukemia virus's regulatory region and the SV40 enhancer segment can be substituted for one another. Furthermore, the SV40 enhancer acts in a wide variety of differentiated mammalian tissues and also stimulates transcription in amphibia, plant cells, and even the yeast *S. pombe*. The enhancer often confers the temporal and cell specificity to a gene's transcriptional activation. For example, as we shall see in the discussion of transcriptional regulation of cellular genes, enhancers associated with some genes are active only at certain developmental stages (Sections 8.3e and 8.3g). Also, the enhancer needed for polyoma virus transcription fails to function in undifferentiated embryonic stem cells but is activated following differentiation. And enhancers in some genes act only in the presence of certain inducers (Section 8.3f).

At a superficial level, enhancers appear to act by increasing the frequency of initiation by RNA polymerase II. This was found by measuring the incorporation of labeled precursors into a gene-specific RNA in isolated nuclei. This method, referred to as **run-on transcription**, is based on the fact that only already transcribing polymerase molecules continue to make RNA in isolated nuclei. In this assay, a transduced β-globin gene produces more β-globin–specific RNA if it is linked to an SV40 enhancer than if the gene lacks the enhancer.

The enhancer is the most complex sequence element governing SV40 transcription (Figure 8.29). Several different kinds of motifs within the 72 bp repeat and 25 bp immediately upstream have been implicated as part of the enhancer. By itself, each sequence motif has very low enhancer activity. The precise identification of the enhancer's motifs is complicated because their activity can be assayed only by the contribution they make to transcription in a particular host and in a particular assay. Some motifs contribute equally well to the enhancer's activity, irrespective of the host cell or transcription system, but others function only in some cell types. Furthermore, mutations in any one motif generally lower the enhancer activity only fivefold to tenfold, compared to the several hundred–fold reduction in transcription observed upon removal of the entire enhancer. This suggests that there is considerable redundancy and only partial dependence on the signals that create a functional enhancer. The extent and type of redundancy may, however, have authentic physiological significance that is not detected by the assays used.

Three of the motifs, designated P, Sph, and octamer (octa), constitute one domain (A), and the paired GT and TC motifs form a second domain (B). The two domains can be inverted relative to one another or separated from each other by up to about 100 bp without appreciably diminishing the enhancer activity. However, deletions in the Sph or octamer motifs of the A domain or in the GT and TC motifs of the B domain impair the enhancer's activity to varying extents, depending on the cells in which transcription is measured. The P motif is not required for enhancer function, but its presence makes transcription responsive to stimulation by phorbol esters, compounds that activate a cellular protein kinase.

The Sph motif consists of a nearly perfect tandem repeat of the 9 bp sequence 5′-AAG$^{T}_{C}$ATGCA-3′; an *Sph* I restriction site in one of the 9 bp repeats accounts for its name. The octamer motif (5′-ATGCAAAG-3′) is formed by the juxtaposition of the two 9 bp Sph repeats. The compound Sph-octamer elements permit the enhancer to function in a wide variety of cells, including lymphocytes. Changes in the sequences of the two Sph motifs that are not also included in the octamer sequence impair enhancer activity in fibroblast or epithelial cells without affecting the activity in lymphocytes. Conversely, mutational alterations affecting the octamer motif eliminate enhancer activity in lymphoid cells but do not affect the enhancer's ability to activate transcription in other cells. Thus, these motifs regulate transcription in both tissue-specific and tissue-nonspecific ways.

Two adjacent repeats of 5′-TCCCCAG-3′, the TC repeats, and a repeated GT-rich sequence, 5′-G$^{C}_{G}$TGTGGA$^{A}_{T}$TGT-3′, occur further upstream in the enhancer. GT-I and GT-II contribute to the enhancer activity in the same cells in which the Sph motifs function. Mutations in either of the GT motifs or in TC-II, as well as insertions between them, reduce enhancer activity, implying that the two motifs function cooperatively. Overlapping sequences in the GT and TC sequence motifs (and minor variations thereof) are sometimes referred to as the core enhancer (5′-GTGG$^{AAA}_{TTT}$G-3′) because it occurs in many viral and cellular enhancers. However, the core motif alone fails to function as an enhancer.

A curious feature of certain of the enhancer's motifs is that although no single one or even pair of them has enhancer activity, multimers are active. For example, tetramers of the paired GT motifs are nearly as effective as the complete enhancer. Multimerization of the octamer motif produces a structure that can substitute for the immunoglobulin enhancer in lymphoid cells but is inactive in fibroblasts. Similarly, multimers of the core sequence can function as an enhancer.

Each of the indicated SV40 enhancer motifs, as well as other *cis* acting regulatory sequences, is a binding site for a transcriptional factor or factors. Collectively, these DNA bound proteins and additional proteins that are acquired through protein-protein interactions are presumed to form a complex that activates RNA polymerase II for transcription initiation. The composition and structure of such transcription complexes are probably not absolutely fixed because, as mentioned previously, combinations of motifs that are inactive by themselves are active after multimerization. Furthermore, mutations that impair the binding of a transcription factor to any one motif diminish but rarely eliminate enhancer activity completely. Clearly, before we understand how the transcription complex is assembled and functions in transcription initiation, we need to identify and characterize the various *trans* acting transcription factors and define their interactions with the various sequence motifs and with each other.

Polypeptide transcription factors A large number of transcription factors presumed to be involved in SV40 early region transcription has been identified and isolated from cell extracts that can transcribe this promoter. Their occurrence, purification, and partial characterization have been monitored in a variety of ways. Some bind specifically to a particular sequence motif, as evidenced both by their ability to protect specific bases in the motif against cleavage by DNase I (footprinting, Section 8.2c) and by decreased electrophoretic mobility of DNAs containing the motif upon binding the protein (**band shifting**). The restoration of transcriptional activity in factor depleted extracts and the failure to do so when their binding sites are destroyed provides evidence that binding of the factors activates transcription. In some instances, the genes encoding a transcription factor have been cloned and sequenced. Their expression in *E. coli* and the ability of these products to activate transcription *in vitro* have provided a way to identify the important regions of the proteins.

Besides the factors needed for enhancer function, which differ from one gene to another, the transcription of most cellular and viral class II genes requires a set of proteins needed to orient RNA polymerase II to the correct initiation sites. Whether these proteins are among the subunits usually found in association with RNA polymerase II or accessory transcription factors that associate and dissociate at the time of initiation is unresolved. The relative efficiency of forming the initiation complex and its stability depend on the presence or absence of a TATA-like sequence at about base pair −25 and on the sequence surrounding the TATA element from base pair −40 to +10. The protein designated TFIID is a ubiquitous mammalian "TATA sequence binding factor." It binds to the AT-rich region in SV40's promoter, as evidenced by its ability to protect the AT block against DNase I cleavages (Figure 8.31). A second factor, TFIIA, binds to this initial complex and stabilizes the association of TFIID

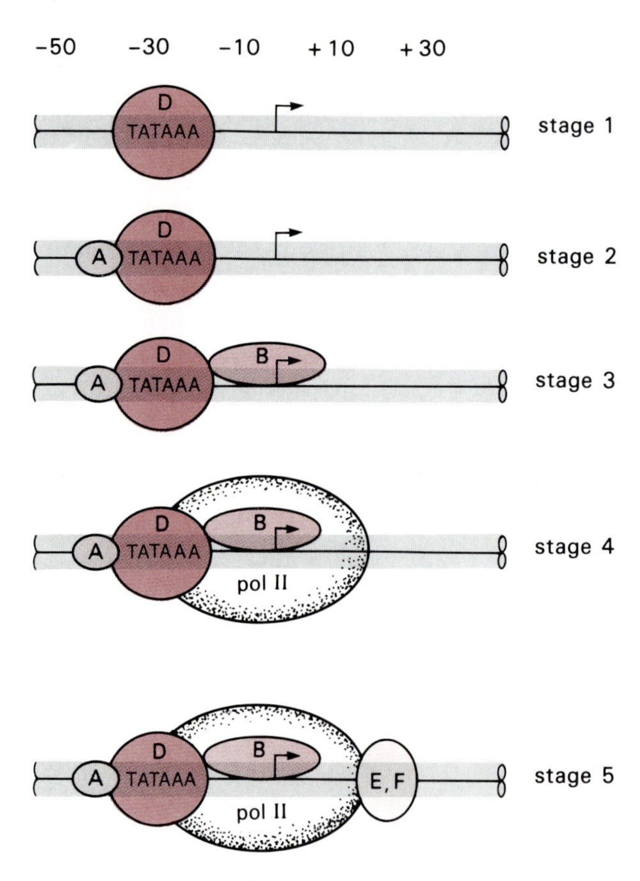

Figure 8.31
Assembly of a transcription complex at the site of transcription initiation. The transcription complex is assembled from the individual transcription factors (TFIIA, B, D, E, F) and RNA polymerase II. TFIID is known to bind to the TATA sequence first, and bindings of TFIIA and TFIIB follow. The subsequent order is speculative. The transcription assembly spans the transcription initiation site and the stretch indicated by the numbers shown above. Adapted from S. Buratowski et al., *Cell* 56 (1989), p. 549.

with the TATA sequence. Formation of the DNA-TFIID-TFIIA complex leads successively to the binding of TFIIB and RNA polymerase II. The preinitiation complex is completed by the binding of two additional factors, TFIIE and TFIIF. In the presence of the four ribonucleoside triphosphates, the complex initiates and elongates the transcript. TFIIA, TFIIB, TFIIE, TFIIF, or RNA polymerase II do not bind to this region of the promoter independent of TFIID; nevertheless, these proteins probably make contact with neighboring DNA sequences. Thus, there is a progressive broadening of the DNase I footprints to include sequences from base pairs -45 to $+30$ as the preinitiation complex is assembled and transcription begins.

Yeast factors analogous to mammalian TFIID and TFIIA have been identified. A TATA binding factor resembling the yeast and mammalian counterparts has also been isolated from *Drosophila*. Remarkably, the yeast TFIID and TFIIA proteins can replace the corresponding mammalian proteins in the assembly of a transcription complex that promotes initiation at the correct start sites. This interchangeability of transcription factors from widely divergent organisms, with little change in the system's efficiency or specificity, is a recurring feature of the eukaryotic transcription machinery. Evidently, the structure and function of proteins involved in multicomponent complexes have been conserved in evolution.

Besides the proteins comprising the preinitiation complex just described, maximum rates of transcription require binding of the transcription factor Sp1 at the 21/22 bp GC-rich motifs as well as a battery of additional transcription factors (Figure 8.32). The latter vary, depending on the particular cell type and conditions in which the SV40 early promoter is expressed.

The transcription factor Sp1 binds as a monomer in the region spanning the three 21/22 bp GC-rich repeats. This binding occurs in the absence of other proteins or nucleoside triphosphates and requires only a single hexamer repeat. When Sp1 is bound to the SV40 early promoter region, the hexamer sequences 5′-GGGCGG-3′ present in each repeat are protected against cleavage by DNase I, and the G residues fail to be methylated by dimethylsulfate (Figure 8.33). The contacts between Sp1 and the GC-rich repeat are restricted to the DNA's major groove and to the Gs on the template strand only. The binding affinity of Sp1 for each of the hexamer repeats varies, possibly because of variations in their spacing and sequence (Figure 8.33). Elimination of this region or mutational changes

Figure 8.32
Transcription factors and their cognate binding sites in the SV40 early region promoter. The locations of the various regulatory sequence motifs are shown in various shades of color and by brackets spanning the designated motif. The proteins that bind to the designated sequences are shown in matching colors below the DNA. The transcription complex consisting of the indicated factors spans the AT stretch, ori, and the start of transcription. T antigen is shown below its two strongest binding sites.

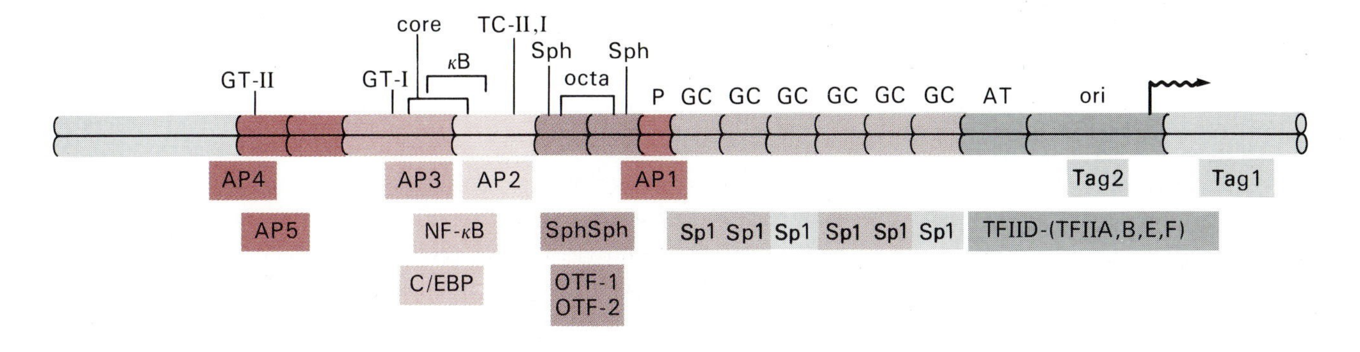

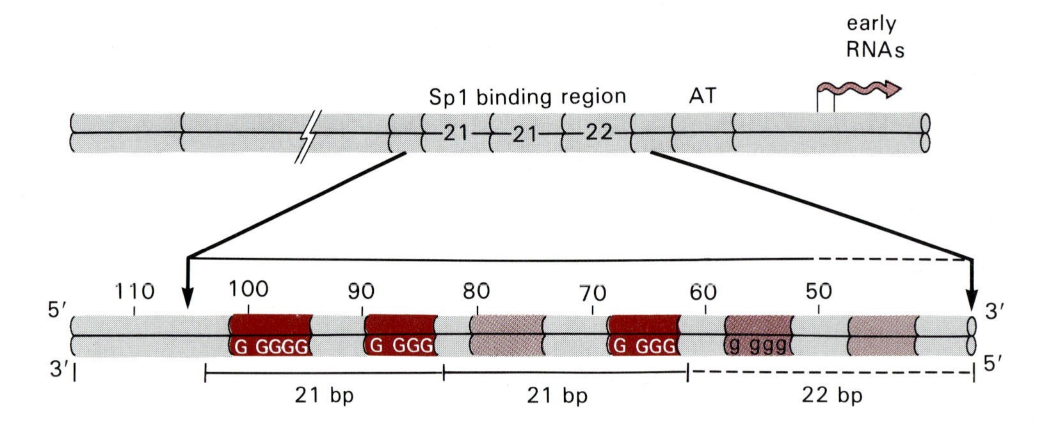

Figure 8.33
Molecular contacts between Sp1 and the GC-rich clusters in the 21/22 base pair repeats. Segments of DNA containing the 21/22 bp repeats were incubated with purified Sp1 protein, and the DNA was reacted with dimethylsulfate. The G residues in the three most highly colored GC-rich clusters were completely protected against methylation. The G residues (g) contained in the next most highly colored segment were partially protected, and those in the most lightly colored segment were not protected. All the protected G residues occur on the same DNA strand.

of specific hexamer sequences alter both the binding and the efficiency of transcription in parallel ways. Thus, binding of Sp1 to this sequence motif is essential for transcription.

Judging from the frequency with which Sp1 binding sites occur in mammalian genes, it is not surprising that Sp1 is a relatively abundant transcription factor. For example, there are between 5000 and 10,000 Sp1 molecules per cell in Hela cells, a commonly used cell line derived from a human cervical tumor. The high Sp1 concentration simplified the task of obtaining the protein in nearly homogeneous form and of cloning its cDNA. The size inferred from the sequence of the cloned cDNA is 80 kDaltons, but the isolated form of Sp1 has an apparent molecular weight of about 90 to 100 kDaltons. The higher molecular weight of cellular Sp1 is probably due to its glycosylation *in vivo*, a modification that is unusual for a nuclear protein. Mutational alterations of different parts of the Sp1 coding sequence identify one of its domains as being required for DNA binding and others as essential for transcriptional activation (Figure 8.34). The DNA binding domain is near the carboxy terminus and has three motifs. Each motif contains characteristically arranged cysteine and histidine residues that can chelate to Zn^{2+}, the so-called **zinc fingers** (Section 8.6a). Alterations within a zinc finger or in the number and spacing of the fingers reduce binding and transcriptional activation. Besides the DNA binding domains, the two glutamine-rich stretches in the amino terminal half of the polypeptide and short sequences that flank the zinc fingers are required for optimal transcriptional activation (Figure 8.34). No role for the serine- and threonine-rich regions has yet been found.

The motif located at the 3′ ends of both 72 bp repeats (designated P in Figure 8.29) serves as a binding site for a protein called AP1 (Figure 8.32). However, because mutations in the P motif that inhibit the binding of AP1 do not affect the enhancer's activity in several *in vivo* transcription assays, the significance of AP1 binding to this sequence is unclear. Their interaction may be important for SV40 early region transcription under only certain circumstances, for example, when other enhancer binding factors are absent or after certain extracellular stimuli.

AP1 is one of a family of closely related proteins of about 40 kDaltons that bind specifically to the consensus sequence 5′-TGANTCA-3′ (where N is any base). In mammals, the AP1 type proteins are encoded by multi-

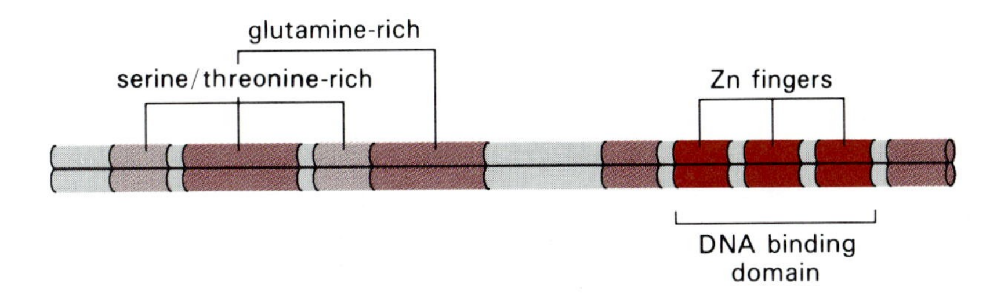

Figure 8.34
The Sp1 coding sequence predicts characteristic protein domains. The DNA segment shown encodes the Sp1 transcription factor polypeptide. The DNA binding domain is confined to the three zinc finger motifs near the carboxy terminus of the protein. Two glutamine-rich domains near the amino terminus and the two short regions flanking the zinc fingers are important for the factor's transcriptional activation. The serine/threonine-rich domains are not essential for DNA binding or activation but may modulate either or both by providing sites for protein modification.

ple genes, several of which have been cloned. The deduced amino acid sequences of several of the AP1 proteins show that their DNA binding domains have been evolutionarily conserved, although other parts of their sequences are dissimilar. Two recent discoveries have focused additional interest on the physiological role of AP1. An oncogene found in a transforming chicken retrovirus, v-*jun*, encodes an AP1-like protein. The cellular counterpart of v-*jun*, c-*jun*, turns out to be one of the AP1 gene family. Each of the AP1-like proteins may have a distinctive regulatory role, possibily dependent upon the context in which its cognate binding motif appears in a gene's regulatory region. Yeast contains a gene, *GCN4*, whose protein product is homologous to the AP1 family in mammals. Just as the yeast TATA binding protein can replace its mammalian analogue, TFIID, GCN4 can activate mammalian genes that utilize AP1, and mammalian AP1 protein can replace GCN4 in the transcription of yeast genes that require that factor. Besides GCN4, yeast also contains additional genes encoding a family of AP1-like proteins. Their characteristics and transcriptional activator properties are still being investigated.

By themselves, AP1 proteins bind only poorly to their recognition motif. For strong binding to occur, AP1 needs to be associated with another protein, called Fos. In Section 8.6b, we discuss the novel structural features of Fos and AP1 and the way they interact with each other to influence transcriptional activity.

The repeated Sph sequence motifs (Figure 8.29) responsible for enhancer function in nonlymphoid cells (e.g., Hela cells) contribute to the enhancer's activity by serving as binding sites for an as yet uncharacterized protein. Mutations in the Sph sequence motif lead to parallel reductions in enhancer activity and protein binding activity. Interestingly, the Sph motif binding protein is absent from lymphoid cells where these motifs do not contribute to enhancer function. Curiously, the same partially purified protein also binds upstream to the GTII motif, a sequence unrelated to the Sph motif. How or if this dual binding specificity influences enhancer activity is unclear. This finding reveals a curious but recurring feature of some transcription factors: an ability to bind to two different and unrelated sequence motifs.

As pointed out earlier, the overlap of the Sph sequence motifs generates an octamer sequence 5′-ATGCAAAG-3′, one whose integrity is essential for enhancer function in lymphoid tissues. The octamer motif, or relatively minor variations of it, occurs frequently in the transcriptional regulatory regions of a wide variety of genes in many different organisms. Its function stems from its ability to bind members of two classes of proteins, OTF-1 and OTF-2. These DNA binding proteins recognize the consensus sequence 5′-ATGCAAAT-3′ or closely related sequences, generally in the context of a variety of other *cis* acting sequences. OTF-1 (about 90 kDaltons and known by a variety of names in the literature) is expressed ubiquitously and has been implicated in transcriptional activation of a variety of cellular and viral promoters. OTF-2 (about 60 kDaltons), however, is present only in cells of lymphoid origin, where it interacts with the octamer motifs present in the promoter-enhancer regions of immunoglobulin and other lymphoid-specific genes to promote their transcription (Section 8.3e). Because SV40 enhancer function is mediated by the Sph binding protein in nonlymphoid cells, OTF-1 and its binding sequence are dispensable. But in lymphocytes, which lack the Sph binding factor, transcription from the SV40 early promoter relies on the interaction of OTF-2 with the octamer sequence motif.

Even though the sequences of the TC-I and TC-II motifs are identical, only the TC-II serves as a binding site for the protein AP2. Mutational changes in the TC-II motif block AP2 binding and lower the enhancer activity. The adjacent TC-II and GT-I motifs together form the core motif that serves as the binding site for the transcription factor AP3 and, in liver, of C/EBP (Figure 8.32). Here, too, alterations in the core sequence motif block AP3 binding and impair enhancer function. It is not clear if enhancer function requires binding of both AP2 and AP3 or whether each alone can activate, depending on cell type or physiological status. The partially purified AP2 and AP3 transcription factors show the anticipated sequence-specific binding properties and can enhance transcription from promoters containing the cognate sequence motifs.

A family of proteins, collectively termed GT binding proteins, has been identified in a variety of different cell extracts by their ability to bind to the GT-II motif and to adjoining sequences (5′-CTGTGGAATGT-3′). The factors are distinguished from each other by their cell specificity, their precise binding sites as defined by DNase I footprinting, methylation protection, and the effect of sequence changes on binding. One of the implicated factors, AP5, is present in Hela cells but not in lymphocytes. This could explain why an intact GT-II motif is essential for enhancer function in Hela cells but is dispensable in lymphocytes. A second factor, AP4, binds to a sequence (5′-CAGCTGTGG-3′) that overlaps the GT-II motif. Its role in enhancer activity is unclear. Although the GT-I and GT-II motifs share 10 of their 12 base pairs, they do not bind the same proteins. AP3, a Hela cell nuclear factor, binds to GT-I but not to GT-II. Mutations in GT-I that block binding of AP3 result in reduced enhancer activity.

The nature of the supramolecular complex created by the binding of the assorted binding factors to their juxtaposed sequence motifs and its role in promoting SV40 early transcription are presently unknown. However, this system has already contributed significant insights into the logic of transcriptional regulation and provides a model for learning the detailed mechanisms of transactivation and differential gene expression.

T antigen: a multipurpose transcriptional regulator Following translation in the cytoplasm from one of the two early mRNAs, Tag is transported to the nucleus, where it promotes initiation of viral DNA replication and represses and activates early and late region transcription, respectively. Tag accomplishes these essential regulatory roles by binding to specific DNA sequences and by interacting with other proteins.

Tag binds to SV40 DNA at the three sites designated T1, T2, and T3 in Figure 8.28. Site T1, which has the highest affinity for Tag, occurs in the DNA segment that specifies the 5′ untranslated sequence of the early transcripts. T2, which binds Tag with a lower affinity, spans the 27 bp palindrome at ori. The 22 bp GC-rich repeats immediately adjacent to the AT block constitute T3, the lowest affinity binding site for Tag. The middle and carboxy terminal domains of Tag appear to be the principal determinants for binding the protein to these sites because the amino terminus is dispensable for this function. However, protein phosphorylations and oligomerizations of the Tag polypeptide influence the binding reactions in as yet unexplained ways.

Both *in vivo* and *in vitro* studies indicate that the binding of Tag to sites T1 and T2 blocks transcription initiation from the most downstream start sites of the early region. Most likely, Tag binding to these sites prevents the assembly of the preinitiation complex between base pairs −45 to +30 (Figure 8.31). Tag may also influence expression from the downstream sites by binding to AP1, thereby blocking the transcription factor's interaction with its cognate enhancer sequence. The interaction of Tag with AP1 could also influence transcription of cellular genes requiring AP1.

Transcription of the early mRNAs is initiated from sites within the ori palindrome in the absence of Tag (the immediate early mRNAs). The delayed early RNAs initiate farther upstream, at or near the junction of the AT element with the 22 bp GC-rich repeat. Curiously, when Tag is bound to sites T1 and T2, transcription of the immediate early mRNAs is repressed, but transcription of the delayed early mRNAs is activated. Transcription from these more upstream sites depends on the integrity of the Sp1 binding sites but, surprisingly, is little affected if the enhancer sequences are deleted. Moreover, mutations in Tag or in ori that prevent replication but do not affect transcription of the immediate early mRNA prevent the shift in transcription to the more distal initiation sites. Although the physiological consequences of the shift in transcriptional start sites are not understood, the less efficient translation of Tag from the longer mRNA may be relevant.

What is accomplished by this complex transcriptional strategy? Very early after infection, Tag is produced at a high rate, and its accumulation in the cell fosters DNA replication and thus an increase in the number of templates for transcription. The rate of Tag accumulation is modulated by the binding of Tag to sites T1 and T2 and possibly by the interaction between Tag and AP1. With the onset of DNA replication, the overall rate of early region transcription is reduced, and there are 5 to 10 times more transcripts emanating from upstream compared to downstream start sites. But, because translation of the longer mRNAs is inefficient, the accumulation of Tag is reduced even more. Concomitant with the down regulation of early region transcription, late region transcription increases so that nearly 20 times more late than early region transcripts are produced; these mRNAs are needed for the synthesis of virion capsid proteins. Thus, the complex and interlaced activities of the viral and cellular transcription factors ensure an orderly and efficient progression of the infectious process.

Transcription of the late region occurs in the direction opposite to that

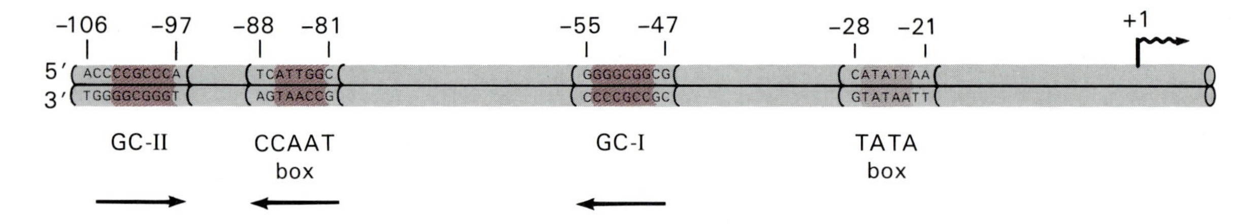

Figure 8.35
Sequence motifs that are essential for transcriptional activity of the Herpes simplex I thymidine kinase promoter. Besides the TATA sequence about 25 bp upstream of the transcription start site (+1), a CCAAT box on the template strand flanked by GC-rich sequences in opposite orientation to each other are required for the HSV-TK promoter function.

of early transcription. Initiations occur at various sites, predominantly within the GC-rich repeats and enhancer. The relatively undefined late promoter lacks a TATA-like sequence but includes the Sp1 binding sites and as yet unspecified regions in the enhancer. The mechanism of Tag's activation of late region transcription is unclear. One view is that activation involves protein-protein interactions rather than DNA binding. Perhaps Tag's association with AP1 or some as yet unidentified transcription factor is responsible for activating late region transcription.

Herpes simplex virus thymidine kinase gene promoter As noted earlier, the SV40 early region is transcribed immediately after infection and without the need for any virus coded proteins. By contrast, transcription of the immediate early protein (IEP) genes of herpes simplex virus (HSV) requires a viral coded transcription factor that is included in the infecting virion. The IEPs are regulatory proteins that activate the transcription of a set of delayed early genes related to DNA replication. Only after the onset of replication can the IEPs activate transcription of the late genes, many of which encode the structural proteins of the virion.

Optimal transcription of the HSV thymidine kinase gene (TK), one of the delayed early genes, requires one of the five IEPs (IEP3) that encodes a large phosphorylated protein (about 175 kDaltons). This was established by showing that a transfected TK gene is expressed at 10 to 100 times higher levels in cells expressing a transfected IEP3 gene than in cells lacking IEP3. The HSV TK gene promoter therefore provides another attractive model for analyzing the interplay of cellular and viral factors in regulating viral gene transcription.

The various sequence motifs that define the HSV TK gene's promoter were identified by introducing deletions, substitutions, and base changes systematically throughout the first hundred base pairs immediately upstream of the transcription start site. The transcriptional competence of the altered promoters was assayed after their injection into *Xenopus* oocytes (Section 7.7b), by transfection of cultured mammalian cells, or by transcription measurements in extracts from uninfected cells. Four essential regions were identified between base pair −105 and the cap site (Figure 8.35). One region, located nearest the transcription start site, is the ubiquitous TATA-like sequence; the other three are more distal, extending to about base pair −105. The distal signals consist of three sequence motifs, two of which are GC-rich hexanucleotides of the type identified in the SV40 early region promoter's 21/22 bp repeats. These flank a sequence that contains a 5′-CCAAT-3′ motif.

Altering the TATA homology disrupts the accuracy and efficiency of transcription initiation while modifications in the three distal motifs cause

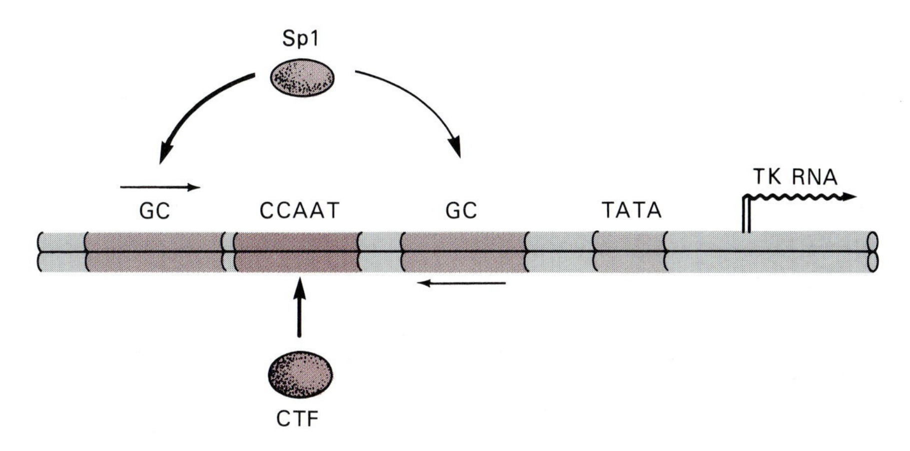

Figure 8.36
Elements of the Herpes simplex I virus thymidine kinase promoter. The HSV-TK promoter comprises four sequence motifs: the TATA homology, two copies of the GC-hexanucleotide oriented in opposite directions (the oppositely pointed arrows), and the CCAAT motif. The stronger binding of Sp1 to the more distal GC motif is indicated by the intensity of the arrows pointing to the two sites. CTF represents a CCAAT transcription factor, and the arrow points to its strong binding site.

a more severe impairment of transcription. Notably, sequence changes in the more upstream GC motif, GC-II, depress transcription to a greater extent than equivalent mutations in GC-I; furthermore, deleting GC-I is less deleterious to transcriptional activity than removing GC-II. Replacing GC-I with the segment containing the GC-II and CCAAT sequences (approximately base pair −75 to −105) in inverted orientation yields an even more active promoter. A likely reason for the difference in activity of the two GC motifs is the greater proximity of the CCAAT sequence to GC-II, although local sequence differences surrounding the GC motifs could also contribute to their different effectiveness.

Fractionation of the proteins necessary for TK mRNA synthesis *in vitro* reveals a requirement for at least two cellular transcription factors (Figure 8.36). One of these is Sp1, which binds to both GC-I and GC-II. The other, CTF, binds to the CCAAT motif as judged by DNAse I footprinting analyses. As expected, mutations in the GC or CCAAT sequence motifs block both the binding of the respective factors and transcription to comparable extents. Initiation of TK gene transcription almost certainly also depends on the proteins that assemble around the TATA sequence between base pair −40 and +35. Thus, as in the case of the SV40 early region promoter, multiple transcription factors are needed to activate the signal contributed by the TATA region. Whether and how Sp1 and CTF interact with each other and influence the formation or activity of the preinitiation complex is not known.

The experiments described thus far analyze the TK promoter in the absence of IEP3. Are there additional sequence motifs involved in mediating transcriptional activation by this protein? The answer appears to be no because transfection experiments with the same mutants used to define the promoter elements show that only the region between base pair −105 and the cap site is needed for transcriptional activation by IEP3 *in vivo*. IEP3 does not bind to the DNA in this region; therefore, it probably activates transcription through protein-protein interactions. The nature of these interactions and how they influence transcription initiation are unknown. IEP3 may stabilize already formed complexes or contribute to the transcription complex's activation because it is stimulatory rather than obligatory.

Many eukaryotic cell and virus promoters contain a CCAAT motif on either of the two DNA strands at about 80 to 120 bp upstream of their transcription start sites. In all instances, the integrity of the sequence and the closely surrounding nucleotides are essential for promoter function. There is now convincing evidence that the CCAAT sequence is part of a motif that activates transcription by serving as a binding site for a transcription factor. But there are multiple—at least three and possibly more—different CCAAT binding proteins. On the basis of DNase I footprints, methylation protection patterns, gel retardation, and binding competition studies, several CCAAT binding proteins have been identified in a variety of mammalian cells. Thus, the CTF that binds to the CCAAT sequence in the TK promoter is a member of the CTF/NF-1 family of CCAAT binding proteins, which also activates the α-globin and murine sarcoma virus promoters. A different protein, present in mammalian liver, binds to the SV40 enhancer core and CCAAT motifs. Hela cells also contain two different CCAAT binding proteins (CP-1 and CP-2) whose promoter binding properties clearly distinguish them from each other and from CTF which is present in the same cells. It would be surprising if there are not still other forms of CCAAT binding proteins because there are growing indications that the CCAAT motif is an important element in a variety of regulatory contexts, including promoters that are developmentally regulated.

The genetic origin of the various CCAAT binding proteins is unclear. Some may be products of separate genes, but three forms of CTF in Hela cells are encoded by alternatively spliced mRNAs originating from the same gene (Section 8.5f). Differential splicing reactions that produce alternate forms of a transcription factor may yield a set of proteins that binds to a common sequence motif but functions in different regulatory contexts. Yeast also possesses CCAAT motifs and related binding proteins. One protein, a dimer of two different subunits (HAP2 and HAP3), activates transcription of the gene encoding iso-2-cytochrome c. The HAP2/HAP3 complex makes the same DNA contacts as mammalian CP-1 protein when it binds to the CCAAT sequence in the yeast cytochrome c promoter. Moreover, the binding of both is affected to the same extent by mutations in the CCAAT element. Astonishingly, components of CP-1 and HAP2/HAP3 are functionally interchangeable. Thus, complexes formed between polypeptide constituents of CP-1 and either HAP2 or HAP3 are fully functional for binding to CCAAT and activating transcription. This finding illustrates further the evolutionary conservation of proteins involved in transcriptional regulation.

e. Tissue- and Stage-Specific Gene Regulation

Multicellular plants and animals are constructed of a wide variety of cell types, each one having a characteristic morphology and specialized function. These specialized features arise at various stages during embryonic development. They are maintained either autonomously or in response to specific cell-cell contacts or extracellular stimuli. Whether autonomous or induced, the diversity of cellular phenotypic properties reflects distinctive patterns of gene expression and the consequent distinctive accumulation and distribution of gene products. One of the central issues in developmental biology is to understand how differential gene expression is achieved in cells that contain a virtually identical genome. We know that transcriptional regulation accounts in part for the establishment and maintenance of the differentiated state. Thus, within cell types, sets of genes become transcriptionally active or inactive in an exquisitely coordinated and specific way throughout life, beginning in the developing embryo (Section 8.3g). Genes that are actively transcribed in one tissue (e.g., in brain or lymphocytes) are not transcribed in others. Moreover, transcription of different sets of genes may be activated in different cell types or tissues in response to the same stimulus. And the same signal may activate transcription of some genes while inhibiting others. In this section, we consider several examples where differential transcriptional regulation is the means for achieving tissue- and stage-specific gene expression.

Immunoglobulin genes: cis *elements* Immunoglobulin (Ig) and T cell receptor (TCR) genes are novel in that their *cis* acting regulatory elements are widely separated in the germ line genome and are brought together by recombination at specific stages during lymphoid cell development (Section 10.6c). Such rearrangements bring transcription signals located at the 5′ end of each Ig gene into relatively close proximity to enhancer segments located within introns for both Ig heavy (IgH) and Ig light (IgL) chain genes (Figures 10.67 and 10.68). Similar rearrangements during T cell ontogeny juxtapose comparable promoter and enhancer sequences that regulate transcription of the genes encoding the TCR's two polypeptide chains (Figure 10.74). Thus, the developmental and cell specificity of Ig and TCR gene expression is regulated by two mechanisms. One operates by determining when and in which cells the rearrangements occur, this step being an essential prerequisite for transcriptional activation. The second mechanism depends on interactions between the rearranged sequence elements and transcription factors whose availability and activity are differentially regulated during B and T cell maturation. Here we examine the latter mode of regulation for the IgH and IgL(κ) genes in B cells. We consider the nature and organization of the sequence motifs governing Ig gene transcription, the transcription factors that bind to these motifs, and those molecular features of the factors that account for their ability to activate transcription. Although omitted from our discussion, transcriptional regulation of the TCR's α and β chain genes in T cells involves comparable sequence motifs and transcription factors.

Several distinctive sequence motifs define an IgH gene's basal promoter (Figure 8.37): a TATA-like sequence situated within 20 to 30 bp of the transcription start site, a conserved octamer (octa) sequence (5′-ATGCAAAT-3′) located about 30 to 60 bp farther upstream, and a consensus heptamer (hepta) sequence (5′-CTCATGA-3′) about 25 bp upstream of the octamer. The most important of the basal promoter's motifs for transcription and cell specificity is the octamer sequence. Its removal destroys the promoter's transcriptional activity. Moreover, inserting the octamer segment into a promoter not normally transcribed in lymphocytes permits transcription in B cells. Deleting the heptamer reduces the IgH gene's basal expression about fivefold, but its contribution to B cell specificity is not known. An IgL(κ) gene's basal promoter re-

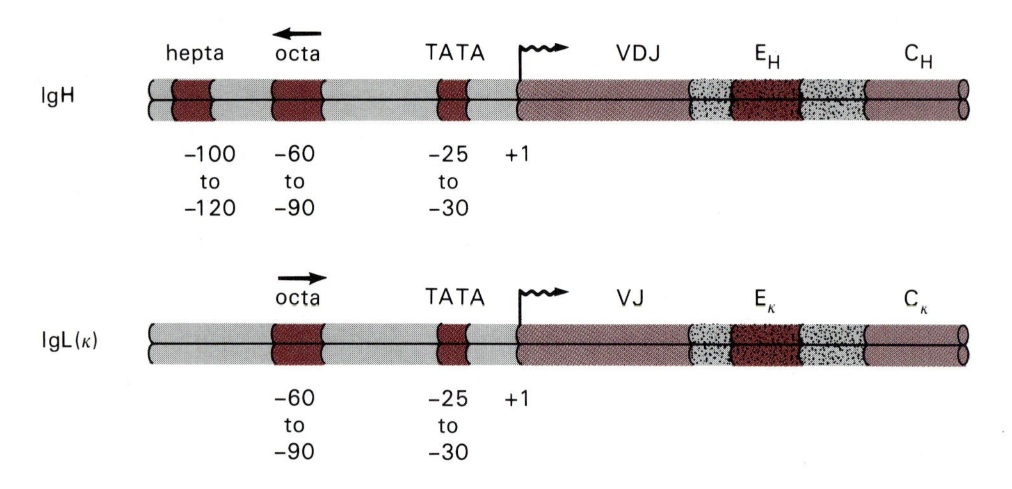

Figure 8.37
Sequence motifs for transcriptional regulation of the immunoglobulin heavy and light (κ) chain genes. The IgH promoter consists of the TATA, octamer (octa), and heptamer (hepta) motifs upstream of the transcription start site and the enhancer element (E_H) in the intron between the variable (VDJ) and constant (C_H) regions. The IgL(κ) gene promoter contains a TATA and an oppositely oriented octamer sequence upstream of the transcription start site and its enhancer E_κ in the intron between the variable (VJ) and constant (C_κ) regions.

sembles the IgH promoter in the arrangement of the TATA-like and octamer motifs, but the octamer sequence in the IgL(κ) promoter has the opposite orientation. All IgL(κ) gene promoters also contain a conserved pentadecanucleotide at about base pair −100. The TATA-like and octamer sequences are essential for IgL(κ) gene transcription, but the role of the pentadecanucleotide is unclear because its removal does not impair transcription in a variety of assays.

Optimum transcription of the IgH and IgL(κ) genes also relies on the integrity of their respective enhancers. The IgH enhancer is located wholly within the intron that precedes the 5′ most exon encoding the IgH constant region (Figures 10.67 and 10.68). Two enhancers have been found in the IgL(κ) gene, one within the intron upstream of the constant region and another 9 kbp beyond the 3′ end of the transcription unit. These enhancers share features characteristic of this class of regulatory element. Each is functional in either orientation and at varying distances and locations relative to the transcriptional start site, and each can activate transcription of heterologous promoters. An important feature of the IgH and IgL enhancers is their preferential activity in B lymphocytes. The IgH enhancer appears to be active at all stages of B cell development, indicating that its action is cell-specific but not stage-specific. The IgL enhancers may be cell-specific as well as stage-specific because induction following the gene rearrangement seems to be needed for initiating IgL gene transcription.

The initial indications that the enhancers contain discrete sequence motifs that serve as protein binding sites were based on the identification of sequences that are protected *in vivo* against methylation by dimethylsulfate. The premise of this approach is that if proteins are bound to a DNA sequence *in vivo,* that sequence will be protected against chemical modifications. Because the protection occurred preferentially on expressed genes in B cells, it was inferred that the bound proteins were cell-specific transcription factors. The four protected sequences in the IgH enhancer and three protected sequences in the IgL(κ) enhancer are referred to in Figure 8.38 as E_H and E_κ motifs, respectively; each is a variant of the consensus sequence 5′-CAGGTGGC-3′. The IgH enhancer also contains three repeats of the core sequence (C_H 1 − 3) previously described in the SV40 early region enhancer (Figure 8.29) and the same octamer sequence

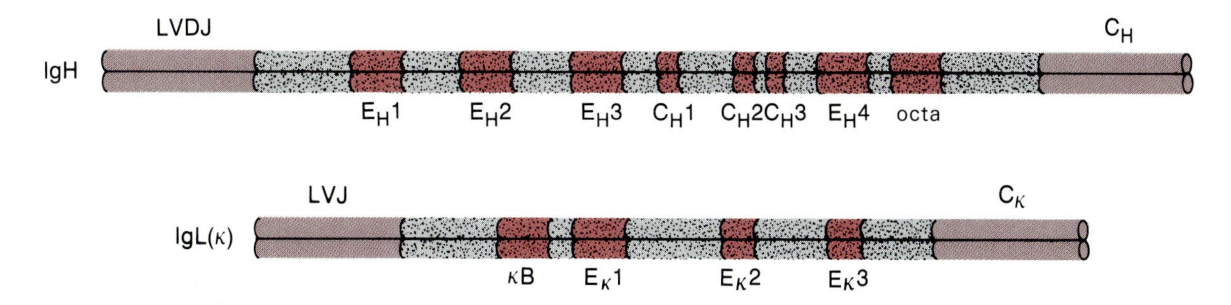

Figure 8.38
The discrete segments that constitute the immunoglobulin heavy and light chain gene enhancers. The introns of the immunoglobulin heavy and light (κ) chain genes are shown in stipple. The various subelements of the heavy chain gene enhancer are designated E_H 1–4 and C_H 1–3, the latter indicating core motifs found in many viral enhancers (Section 8.3d). The subelements of the light chain (κ) enhancer are designated κB and E_κ 1–3.

found in the basal promoter. There are also sequences in the enhancer region that inhibit transcription in nonlymphoid cells because some deletions in the enhancer region increase transcription in fibroblasts while leaving the activity relatively unchanged in lymphoid cells. Although **silencer** regions probably contribute to differential gene expression, their characterization is presently fragmentary and inconsistent.

Assays of the activity of a variety of mutational alterations in the IgH enhancer indicate that the octamer motif is the single most important sequence for B cell-specific enhancement. By this approach, individual E_H or C_H motifs do not appear to contribute significantly to enhancer function. However, this conclusion is probably an oversimplification because alterations in both the octamer and certain E_H motifs cause more drastic impairments in enhancer function than do alterations in either alone. Moreover, multimers of only the E_H or C_H motifs confer preferential enhancer activity in B lymphocytes. At this stage in the analysis of the IgH enhancer motifs, we need to recognize that transcription is often measured in transient transfection assays, conditions that may not be relevant for normal conditions under which these elements function. Therefore, assigning specialized functions to specific regions may be premature.

The IgL(κ) enhancer contains two kinds of sequence motifs. One, the E_κ motifs (Figure 8.38), was identified by methylation protection *in vivo*. The other, dubbed the B motif (5′-GGAAAGTCCCC-3′), was identified as a binding site for a B lymphocyte nuclear protein (NF-κB). Enhancer activity appears to depend most on the B motif, although the $E_\kappa2$ sequence also contributes. An important feature of IgL(κ) gene expression is its inducibility by bacterial lipopolysaccharide, cycloheximide, and phorbol esters, a property that is lost by mutations in the B motif. Thus, the B sequence is critical for both enhancer function and inducibility of IgL(κ) gene transcription. Interestingly, the B motif also occurs in promoters and enhancers that function in nonlymphoid tissues. And, in some instances, these promoters are inducible by the same compounds that activate IgL(κ) gene expression in lymphocytes. These observations can be rationalized by the properties of NF-κB.

The proximal promoters and enhancers of Ig genes each contribute to B lymphoid-specific transcription. The octamer motifs in both the IgH genes' basal promoter and enhancer are the most critical sequence determinants of that cell specificity. Cells lacking a B cell-specific octamer binding protein are unable to use either the IgH promoter or its enhancer. Similarly, the B cell specificity of the IgL(κ) enhancer depends on the basal promoter sequence, and its stage-specific activation late in B cell development depends on the interaction of a κB motif with the protein, NF-κB.

Immunoglobulin genes: trans *acting factors* We have already discussed the protein factors needed to assemble the transcription machinery in the vicinity of the TATA sequence and transcriptional start site. In addition, the IgH and IgL(κ) basal promoters require activation by a lymphocyte-specific protein—an octamer binding protein. Recall that there are two types of octamer binding proteins: one type occurs in most mammalian cells (OTF-1), and the other is found only in lymphocytes (OTF-2) (Section 8.3d). The latter is essential for activating lymphoid-specific transcription from both the IgH and IgL(κ) basal promoters. Both types of proteins bind approximately equally well to the octamer sequence and protect the same residues against DNase I cleavages and methylation by dimethylsulfate. Because lymphoid cells contain both OTF-1 and OTF-2 transcription factors, it is not surprising that they can express a wide variety of genes containing the octamer motif in their promoters. It is paradoxical, however, that the IgH and IgL(κ) promoters fail to function in cells containing OTF-1 but not OTF-2. Perhaps the answer to this puzzle is that OTF-2 binding to the Ig octamer motifs is by itself sufficient to activate transcription initiation, whereas binding by OTF-1 alone is insufficient. Activation by OTF-1 may require interaction with one or more additional factors bound to neighboring motifs. For example, the SV40 early promoter and others contain several transcription factor binding site motifs adjacent to the octamer motif. It is also conceivable that the spacing between the TATA sequence and the octamer sequence in the Ig basal promoters differs from that occurring in the ubiquitously expressed promoters.

cDNAs encoding both the OTF-1 and OTF-2 proteins have been cloned. Using the cDNAs as probes, the ubiquitous expression of OTF-1 and the lymphoid-specific expression of OTF-2 at all stages of lymphocyte development have been confirmed by RNA blotting analysis of cellular mRNAs. If fibroblasts are transformed by plasmids that can express OTF-2, the cells can also express a cointroduced IgH gene. If the OTF-2 plasmid is omitted, the IgH gene is not expressed. This indicates that the presence of OTF-2 is the limiting factor in expressing the IgH gene in nonlymphoid cells. Besides binding to the octamer sequence, OTF-2 also binds to the heptamer sequence in the IgH basal promoter. This added interaction may account for the preferential expression of the IgH promoter compared to IgL expression during early stages of B cell development.

Examination of the cloned cDNA sequences encoding the OTF-1 and OTF-2 proteins shows that they share nearly identical amino acid sequences in their DNA binding domains but diverge elsewhere. The diverged domains may reflect a need for these proteins to interact with different gene-specific or cell-specific transcription factors. Furthermore, the influence of sequences flanking the octamer on the binding affinity of OTF-1 suggests that OTF-1 possesses considerable versatility in its ability to influence transcription of many genes. Further details of the molecular nature of OTF-1 and OTF-2 and their relation to the proteins' DNA binding and transcriptional activation properties are considered in Section 8.6.

Each of the E motifs in the IgH enhancer interacts with distinct nuclear proteins. In each case, binding is prevented by sequence alterations in the motifs. However, the contributions of these interactions to the cell-specific transcriptional activation are unclear because the proteins occur in many different tissues. On balance, the available data suggest that the E_H motifs do not directly influence cell specificity but do influence the level of enhancer function. The E_H motif binding proteins also bind to other promoters, suggesting that they contribute quantitatively rather than qualitatively to Ig gene transcription. Consequently, the principal known determinant of the IgH enhancer's lymphoid specificity is the octamer motif alone.

Transcriptional activation of IgL(κ) genes is mediated by proteins that bind to the E and B motifs in the enhancer. These proteins, NF-κE2, NF-κE3, and NF-κB, were detected in nuclear extracts by the band shift assay (i.e., their ability to retard the electrophoretic mobility of radiolabeled short DNA fragments containing the particular sequence motif, Section 8.3d). The proteins binding to the $E_\kappa 2$ and $E_\kappa 3$ motifs may be the same ones that bind to and influence the IgH enhancer. Because the enhancer's κB site is quantitatively the most significant determinant for IgL(κ) transcription and responsible for the gene's inducibility and tissue- and stage-specific expression, our discussion will focus on the properties of NF-κB and its interactions with the κB site.

Initially, protein complexes involving the κB motif were detected only in nuclear extracts of mature B lymphocytes. Moreover, mutations of the κB site that eliminate IgL(κ) expression and inducibility block the binding of NF-κB. Thus, it seemed most likely that NF-κB is a conventional tissue-specific transcription factor. But this conclusion came into question when NF-κB was detected in extracts of early B lymphocytes, which do not express their IgL(κ) genes. Now we know that NF-κB exists in an inactive complex in the cytosols of a wide variety of cells. The complex is disrupted, and NF-κB appears in the nucleus when such cells are exposed to the same compounds that induce IgL(κ) expression in B cells (e.g., lipopolysaccharide, cycloheximide, and phorbol esters). NF-κB appears to be associated with a labile inhibitor protein (I-κB) in the cytosol. Indeed, fully active NF-κB can be inactivated by mixing with I-κB. Presently, NF-κB activation by the inducers *in vivo* is not fully understood. Quite possibly, cycloheximide prevents synthesis of the labile inhibitor, and lipopolysaccharide and phorbol esters activate enzymes that induce modifications in I-κB, causing the complex to dissociate and NF-κB to enter the nucleus. But the natural mechanism that activates NF-κB and initiates IgL(κ) gene expression in mature B cells is not known.

NF-κB has been purified to homogeneity from bovine and human lymphocytes and probably exists as a heterodimer containing 50 and 65 kDalton polypeptides. The binding activity resides in the 50 kDalton subunit while the 65 kDalton subunit is needed for the inactivation by I-κB. Irrespective of the cell type in which it occurs, NF-κB binds identically to the consensus sequence 5′-GGGR$^{C}_{T}$TYY$^{C}_{T}$C-3′ (where R and Y are purine or pyrimidine, respectively). NF-κB has the unusual property that nucleoside triphosphates stimulate its binding to the κB sequence. A detailed understanding of the NF-κB structure, function, and expression awaits the isolation of its gene.

Many mammalian cell and virus promoters contain the κB sequence motif or minor variations of it. Consequently, the functional status of NF-κB and its inducibility are important considerations for their expression. For example, activation of NF-κB accompanies the interaction of T helper cells with antigen and triggers the production of the growth factor IL2 as well as of the IL2 receptor, both of which are needed for T cell proliferation during normal immune responses. The enhanced transcription of the IL2 and IL2 receptor genes is mediated by NF-κB binding to the κB sequence motifs in their respective promoters. Transcription of the human immunodeficiency virus-type I (HIV-1) genes depends on the binding of NF-κB to a κB sequence in the viral promoter. Viral expression and multiplication, particularly the stability of the latent state in which HIV-1 gene expression is silent, is probably influenced by the intracellular state of NF-κB and thus by antigen stimulation of such latently infected cells. NF-κB activation is not intrinsically lymphoid-specific. It can, in fact, be induced in many different kinds of cells by different stimuli. Thus, the induction of β-interferon production following virus infection of fibroblasts involves activating and binding NF-κB and the resulting stimulation of transcription.

The reversible activation-inactivation of NF-κB allows for rapid regulation of cellular responses to a variety of signals. The few known circumstances under which NF-κB is activated suggest that the full implications of NF-κB as a switch for regulating the expression of many different genes are yet to come.

Liver-specific genes Liver cells make many enzymes and secrete proteins that are not made in other cells. This tissue specificity is regulated mainly at the level of transcription initiation. Here we focus on how this is achieved for two genes: one encoding albumin (*ALB*) and the second, α-fetoprotein (*AFP*). These genes form a small evolutionarily related multigene family. The genes diverged 300 to 500 million years ago; yet both have remained tightly linked, with the same arrangement in both the human and mouse genomes (Section 9.4a). During embryonic development, both genes are expressed in the visceral endoderm of the yolk sac, the fetal liver, and the fetal gastrointestinal tract. Throughout fetal life, the *ALB* and *AFP* genes are expressed about equally in liver and gut; but shortly after birth, *AFP* is shut off in both tissues, and the level of AFP mRNA drops by 10,000-fold within three to four weeks. By contrast, *ALB* gene expression and mRNA levels remain high in the liver throughout adult life and decline only in the gut. The questions posed here are how is tissue-specific transcription of these two genes achieved, and what accounts for their striking differential transcription after birth?

The mouse *ALB* and *AFP* genes are arranged in tandem, the intergenic distance being about 15 kbp (Figure 8.39). Each gene is about 20 to 22 kbp in length and consists principally of introns, each having 15 relatively short exons. Transient expression assays in cultured cells using wild type and selectively modified versions of the *ALB* and *AFP* regulatory regions fused to suitable reporter genes (e.g., chloramphenicol acetyl transferase [CAT] and growth hormone) located the *cis* acting sequences responsible for liver-specific transcription and for the disparate developmental regulation of the two genes.

Tissue-specific expression of the *ALB* and *AFP* genes requires only sequences within 150 bp upstream of their transcription start sites. But tran-

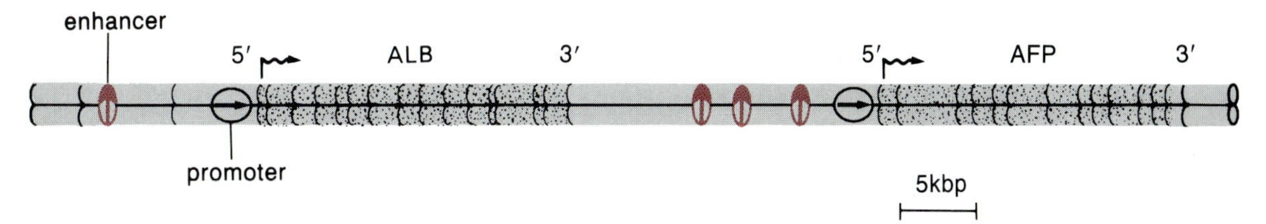

Figure 8.39
Genomic organization and structure of the mammalian albumin (ALB) and alpha-fetoprotein (AFP) genes. The exons, being small relative to the size of the genes, are shown as lines within the stippled introns. The promoter and enhancer elements are indicated, and the transcription start sites are indicated by wavy arrows at the first exons.

scription of the *AFP* gene is increased 20- to 50-fold by sequences located farther upstream. This distal region contains three distinct enhancer elements 6.5, 5, and 2.5 kbp distant from the gene's 5′ end (Figure 8.39). Experiments with transgenic mice containing the variously modified forms of the *AFP* promoter confirm the necessity of the 5′ proximal and distal sequence elements for ensuring the proper timing and levels of tissue-specific transcription. Comparable studies using transgenic mice containing various lengths of DNA surrounding *ALB* show that, besides the 5′ proximal sequence, normal levels of transcription require an element located 9 to 12 kbp upstream as well as the three *AFP* enhancers downstream of the *ALB* gene.

A high resolution mutational dissection of the region immediately 5′ of the rat *ALB* gene and measurements of the effects of these alterations on transcriptional efficiency identify six conserved sequence elements that define the basal promoter and explain its tissue specificity (Figure 8.40). Two sequence elements constitute the basal promoter: the TATA sequence, which defines the transcription start site but does not influence the rate of transcription, and the CCAAT motif, whose alteration markedly impairs transcription in all cell types tested. Liver specificity depends on the four additional sequences: a sequence element (PE) between the TATA and CCAAT motifs and three distal elements (DEI, DEII, and DEIII). The most drastic impairments of the *ALB* promoter's liver specificity occur upon modification of the PE. The next most important motif for liver-specific transcription is DEI. Alterations in either DEII or DEIII produce only moderate reductions in liver-specific transcription, but the removal of both lowers the promoter's activity 25-fold.

Figure 8.40
Comparison of the regulatory sequence motifs responsible for the liver specificity of mammalian albumin gene promoters. The sequence alignments are made with nucleotides representing the transcription start site (+1) as the reference. Dashes indicate gaps introduced to line up the extended regions of identity. See text for relevance of the sequence motifs shown in various color shades.

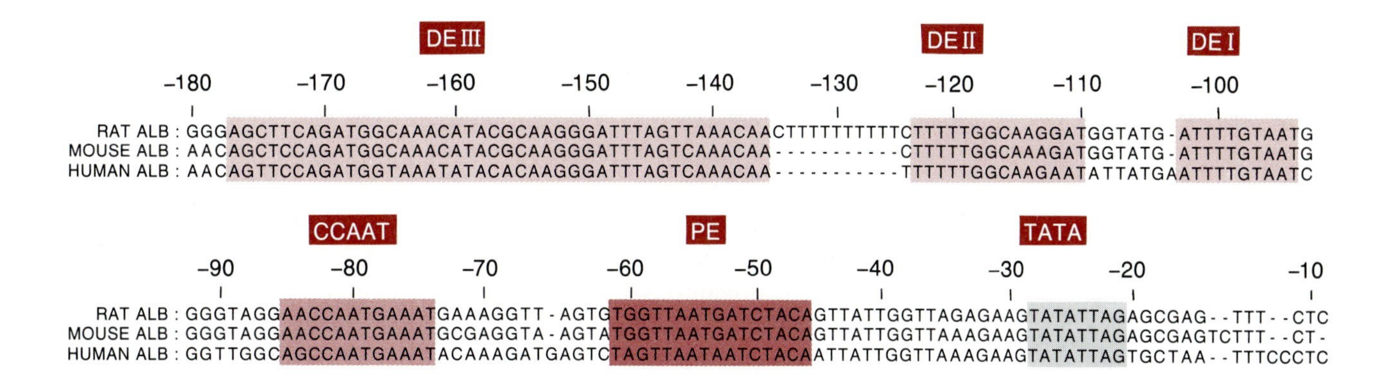

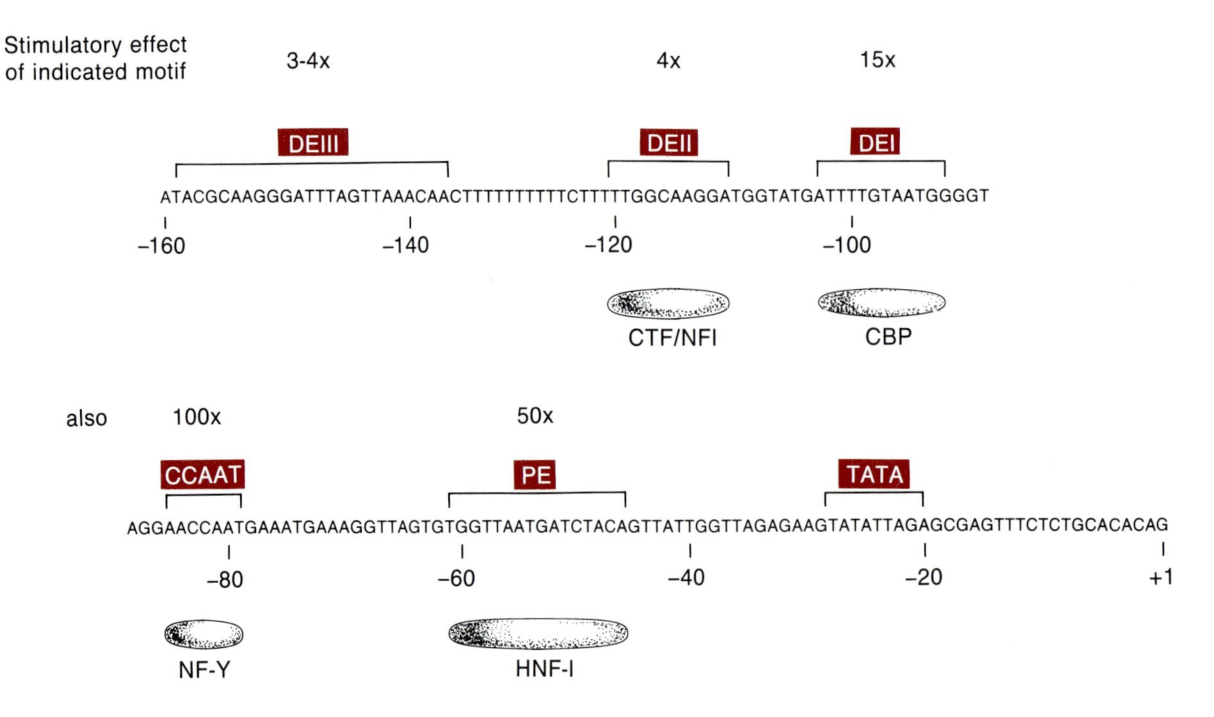

Figure 8.41
Contributions to liver-specific transcription of mammalian albumin gene promoters by various sequence motifs and their cognate transcription factor binding proteins. Each of the sequence motifs and the cognate binding proteins for the most important motifs are indicated. The contribution of each motif to the tissue-specific expression indicated above the motif is derived from the effect of mutational disruptions of that sequence.

Each of the *cis* acting sequence elements contributes a binding site for a specific transcription factor (Figure 8.41). A sequence within the mouse PE element, 5′-GTTAATGATCTAC-3′, is closely related to a sequence motif common to many genes expressed only in liver (Table 8.1). This sequence provides a binding site for a protein HNF-1, which itself is restricted to hepatocytes. Purified HNF-1 (88 kDaltons) binds to and footprints the PE region of the *ALB* promoter as well as the homologous sequence in the *AFP* gene. Moreover, a wide variety of other liver-specific genes and the hepatitis virus promoter, which contain minor variations of the PE sequence motif, also bind HNF-1. Thus, the PE's contribution to the *ALB* gene's tissue-specific expression can be accounted for by the exclusive presence of its ligand, HNF-1. The other regulatory elements in the *ALB* promoter, the CCAAT and TATA motifs, each promotes transcription by binding proteins that recognize those sequences (Section 8.3d); however, neither the TATA nor the CCAAT binding proteins are liver-specific. The proteins that bind to DEI and DEII, C/EBP and NF-1, respectively, have both been implicated in transcriptional activation of a variety of cellular and viral promoters; they too are not restricted to hepatocytes (Section 8.3d).

How can ubiquitous DNA binding proteins contribute to liver-specific expression of the *ALB* promoter? One explanation of the paradox is that the context in which the ubiquitous factors bind is decisive; in this instance, the ubiquitous factors bind in the vicinity of the hepatocyte-specific protein HNF-1. Another possible answer is that the DE binding proteins are hepatocyte-specific members of the C/EBP and NF-1 transcription factor families. In this connection, recall the existence of both ubiquitous and tissue restricted forms of the OTFs.

Thus, *ALB* gene expression is governed by a relatively compact sequence within 150 bp immediately adjacent to the transcription start site. This region contains a multiplicity of specific protein binding sites, some

Table 8.1 Distribution of HNF-1 Binding Sites in Liver-Specific Genes

Promoter	Species	Sequence	Base Pair
Fibrinogen-α chain	rat	G G T G A T G A T T A A C	−47
Fibrinogen-β chain	rat	G T C A A A T A T T A A C	−84
Albumin	rat	G T T A A T G A T C T A C	−53
Albumin	human	G T T A A T A A T C T A C	−51
α-Fetoprotein	rat	G T T A C T A G T T A A C	−49
α-Fetoprotein	rat	G T T A A T T A T T G G C	−115
α-Fetoprotein	human	G T T A C T A G T T A A C	−47
α-Fetoprotein	human	G T T A A T T A T T G G C	−118
Transthyretin	rat	G T T A C T T A T T C T C	−116
Transthyretin	human	G T T A C T T A T T C T C	−116
α_1-Antitrypsin	mouse	G T T A A T — A T T C A T	−63
α_1-Antitrypsin	human	G T T A A T — A T T C A C	−63
Hepatitis virus (pre S1)	human	G T T A A T C A T T A C T	−75
Consensus sequence		G T T A A T N A T T A A C (→ over G T T A A T; ← over A T T A A C)	

Note: The 13 bp sequence shared by the indicated promoters beginning at the indicated base pair upstream of the transcription start site is shown. The palindromic sequence is highly conserved except for the base pair at the center of the dyad symmetry.

SOURCE: Data taken from G. Courtois et al., *Proc. Nat. Acad. Sci.* 85 (1988), p. 7937.

of which contribute to the promoter's restricted tissue expression and others of which regulate the level of expression. Precisely how this aggregate of bound proteins influences the rate of transcription is unknown. Nor is it clear how the enhancers located about 10 kbp upstream and more than 30 kbp downstream of the transcription start site activate the basal promoter.

Similar experimental approaches identified a single sequence element responsible for liver-specific transcription in the *AFP* promoter. That sequence is within an HNF-1 binding site between base pairs −61 and −48. High levels of expression in transfected hepatocytes and in transgenic mice also require three separate enhancer elements located at considerable distance from the basal promoter. No one of the enhancer elements is dispensable, so all contribute to the overall activity. The enhancer elements range in size between 200 and 300 bp, and each is replete with unique and shared sequence motifs. Some of these motifs occur in or resemble those found in other viral and cellular enhancers. Although the identification of the relevant enhancer motifs is still incomplete, there clearly are redundant elements as well as interspersed sequences that permit positive and negative transcription regulation.

What causes the shutoff of *AFP* after an animal's birth? This question was examined in transgenic mice carrying various structurally altered *ALB* and *AFP* genes. For example, transgenic mice carrying an artificial gene containing the three *AFP* enhancers joined to the *ALB* gene and its basal promoter did not shut down *ALB* expression after birth or later in life. Transgenes in which the *ALB* enhancer was fused upstream of an *AFP* gene containing only 1 kbp of its own 5′ upstream sequence were appropriately expressed during fetal development and repressed quickly after birth. This result indicates that the information for stage-specific regulation of *AFP* expression resides in the region 5′ proximal to or in the basal promoter or within the transcription unit itself.

Tissue-specific transcription factor functions in concert with nonspecific factors For previously discussed genes, interactions between *cis* acting sequence motifs and *trans* acting nuclear transcription proteins are the most relevant parameters governing their tissue- and stage-specific expression. The same principle underlies the transcriptional patterns of many and possibly all genes. One might have supposed that one or a few unique transcription factors interacting with a comparable number of specific sequence motifs might be operative in each specific tissue and stage and account for each specialized transcription pattern. Instead, the elements responsible for the particular transcriptional patterns of eukaryotic genes consist of a relatively limited number of sequence motifs and a correspondingly diverse collection of ubiquitous transcription proteins that occur in many different tissues and stages. The sequence motifs are dispersed among the 100 to 400 bp immediately 5′ to the transcription start site or in distal enhancer elements, but usually in both. Generally, one or possibly two transcription factors present in an exclusively tissue- or stage-specific fashion are critical. But it is the combinatorial interactions between such unique and the other ubiquitous proteins and their specific binding sites that give rise to the differential expression of genes. The IgL(κ) genes′ octamer and κB sequence motifs are each represented in many promoters; yet the unique presence of OTF-2 and the inducible activation of NF-κB in lymphocytes accounts for their tissue- and stage-specific expression. The *ALB* gene promoter relies on interactions between several ubiquitous binding sequences and equally widespread transcription factors, but the binding of HNF-1, a hepatocyte-specific transcription factor, confers the liver specificity. Besides the examples detailed here, the expression of many genes likewise depends on the binding of a transcription protein whose presence or activity is specific for a certain cell type, a stage of development, or a time in the cell cycle.

f. Inducible and Repressible Transcription

The transcriptional patterns of selected genes are often altered by external signals that cells receive. Thus, changes in nutritional supplies (e.g., carbon, nitrogen, and energy sources), interactions with various hormones, virus infections, or exposure to metals, light, high temperature, and radiation can either induce or repress the transcription of specific genes or gene sets. In many cases, regulation of transcription up or down is mediated by alterations in the availability of essential transcription factors or by changes in the factor's ability to bind to specific regulatory sequence motifs. Several examples, discussed in this section, illustrate this mode of transcriptional regulation.

Regulation of yeast GAL *genes* The pathway and genes involved in the utilization of galactose by *S. cerevisiae* are summarized in Figure 8.42. The *GAL2* gene specifies a permease required for the uptake of extracellu-

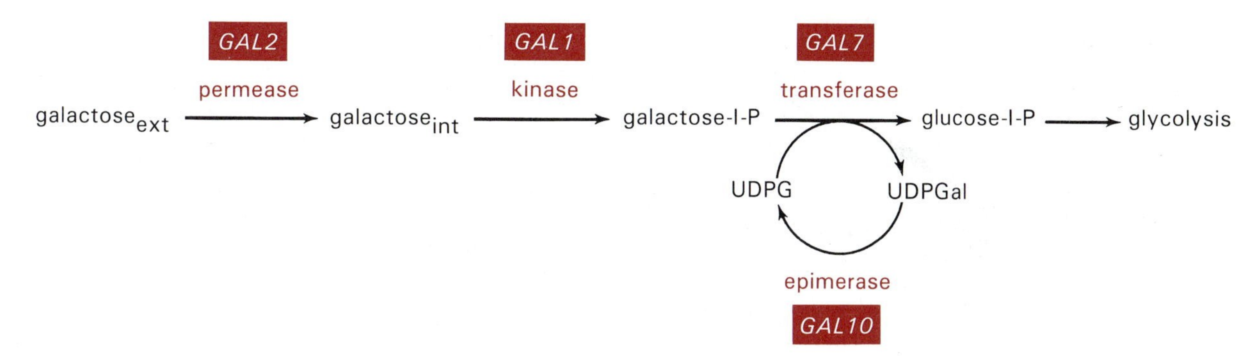

lar galactose. *GAL1* encodes a galactokinase that converts galactose to galactose-1-phosphate. *GAL7* codes for a galactose-1-phosphate uridyl transferase that forms uridine diphosphogalactose (UDPGal) from galactose-1-phosphate and uridine diphosphoglucose (UDPG). The epimerase that regenerates UDPG from UDPGal is encoded by *GAL10*. *GAL1*, *GAL7*, and *GAL10* are clustered on chromosome II, and *GAL2* occurs on chromosome XII (Figure 8.43). The four genes, which are separately transcribed from individual promoters, are unexpressed in cells grown without galactose and are coordinately induced about 1000-fold after growth in galactose.

The induction and repression of the GAL enzymes are regulated at the level of transcription initiation by an unknown inducer derived from galactose and proteins encoded in two unlinked genes, *GAL4* and *GAL80* (Figure 8.43). Mutations in *GAL4* prevent induction of *GAL1*, *GAL2*, *GAL7*, and *GAL10* transcription in the presence of galactose, indicating that *GAL4* encodes a positive regulator of these genes. Many mutations in *GAL80* make the transcription of *GAL1*, *GAL2*, *GAL7*, and *GAL10* constitutive (i.e., independent of inducer), but some render *GAL* gene transcription uninducible. These mutant phenotypes are explainable if the inducer overcomes the inhibition by the *GAL80* product of the GAL4 mediated transcriptional activation. By this view, transcriptional activa-

Figure 8.42
Metabolic pathway for utilizing galactose in *S. cerevisiae*. Four genes (named in the colored boxes) control the ability of *S. cerevisiae* to utilize galactose supplied in the culture medium. The trivial names of the enzymes they encode are shown above or below the gene designation. UDPG and UDPGal correspond to uridine diphosphoglucose and uridine diphosphogalactose, respectively.

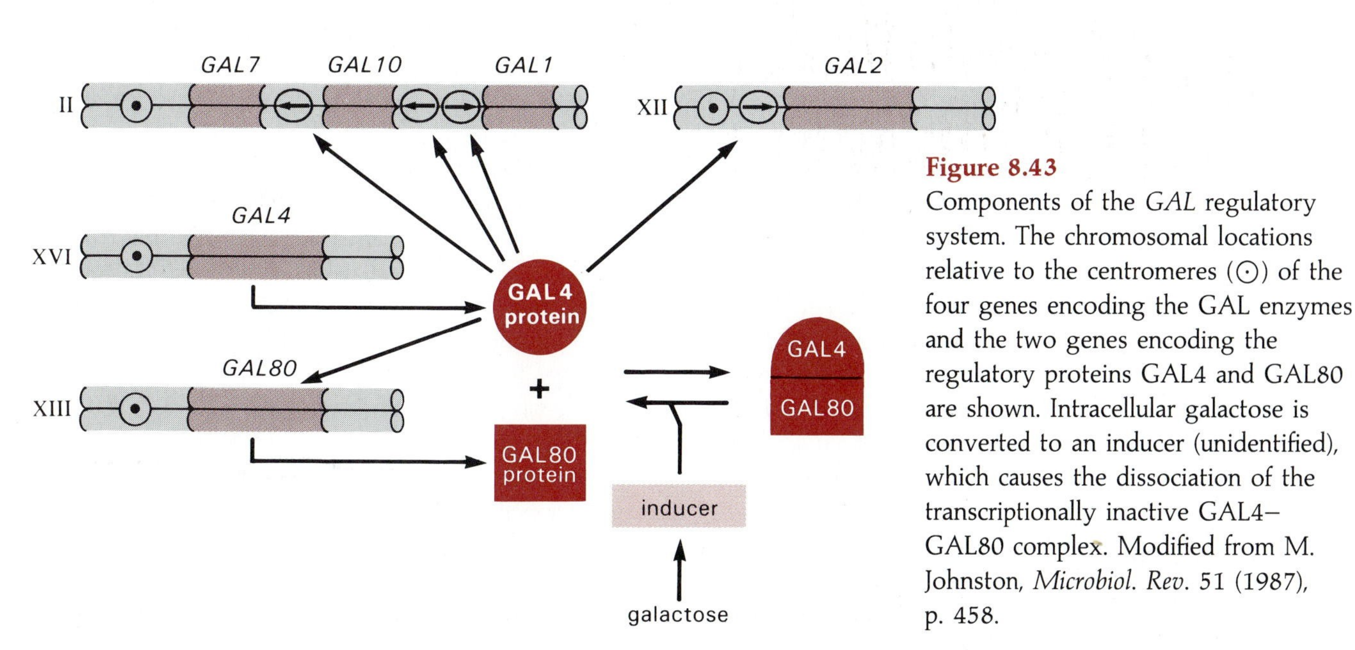

Figure 8.43
Components of the *GAL* regulatory system. The chromosomal locations relative to the centromeres (⊙) of the four genes encoding the GAL enzymes and the two genes encoding the regulatory proteins GAL4 and GAL80 are shown. Intracellular galactose is converted to an inducer (unidentified), which causes the dissociation of the transcriptionally inactive GAL4–GAL80 complex. Modified from M. Johnston, *Microbiol. Rev.* 51 (1987), p. 458.

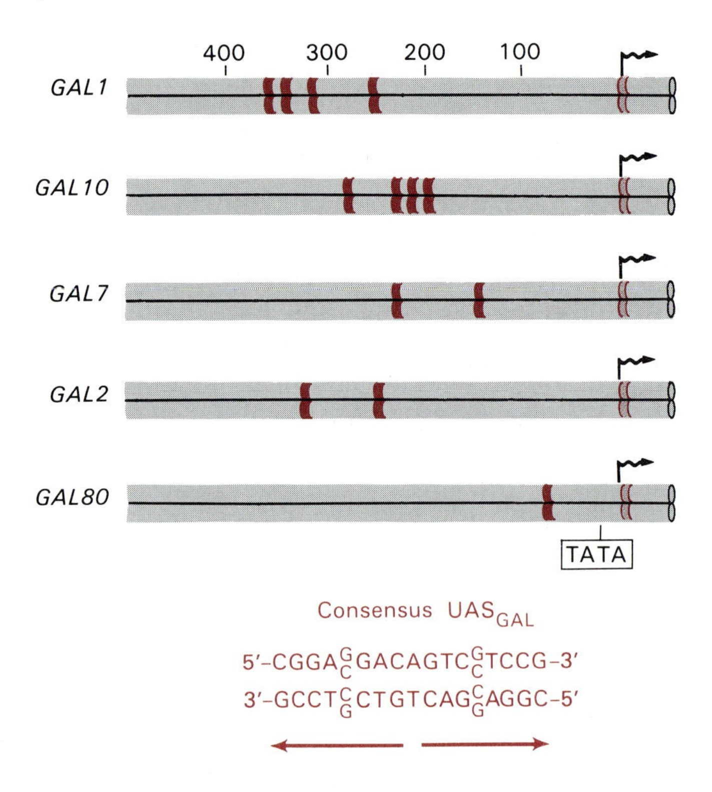

Figure 8.44
Location of the upstream activating sequences (UAS) in the *GAL* gene promoters. The common TATA motif, the transcriptional start sites (wavy arrows), and the UAS motifs are shown. The consensus for the 13 UAS$_{GAL}$ sequences is indicated at the bottom. The divergent arrows indicate the palindromic nature of the sequence. Modified from M. Johnston, *Microbiol. Rev.* 51 (1987), p. 458.

tion of each of the *GAL* genes by the GAL4 protein is blocked by its association with the GAL80 protein. The inducer either dissociates the inactive complex or prevents it from forming. The key to unraveling this regulatory system lies in understanding the GAL4 protein and its interactions with the *GAL* promoters, the transcriptional machinery, and the GAL80 protein.

The GAL4 dependent activation of *GAL* gene transcription requires specific sequences in the *GAL*1, *GAL*2, *GAL*7, *GAL*10, and *GAL*80 promoters. These *cis* regulatory elements are called **upstream activating sequences (UAS)**, a generic term given to such elements in yeast genes. They resemble enhancers in that their function is independent of their orientation and relatively independent of distance. The UAS$_{GAL}$ was first identified on a DNA segment between the *GAL*1 and *GAL*10 genes by its ability to confer *GAL*4 dependent transcriptional activation on the *GAL*1 promoter. A deletion analysis of this region identified multiple copies of a 17 bp palindromic sequence as the essential motif (Figure 8.44). Between one and four of these sequence elements occurs within 200 to 400 bp upstream of the *GAL*1, *GAL*2, *GAL*7, *GAL*10, and *GAL*80 genes, although two are sufficient to provide maximal activation. Any one of the UAS$_{GAL}$ sequences or a synthetic oligonucleotide containing the consensus sequence confers high level GAL4 dependent expression when placed upstream of a variety of yeast basal promoters.

*GAL*4 encodes a 100 kDalton protein that is present at very low levels in both induced and uninduced cells. Protein expressed from the cloned gene in *E. coli* or yeast binds specifically to the UAS$_{GAL}$ motif and protects its 17 bp palindromic sequence against both digestion by DNase I and methylation by dimethylsulfate. Because of the dyad symmetry of the UAS$_{GAL}$ sequence, the GAL4 protein probably binds as a dimer. Examina-

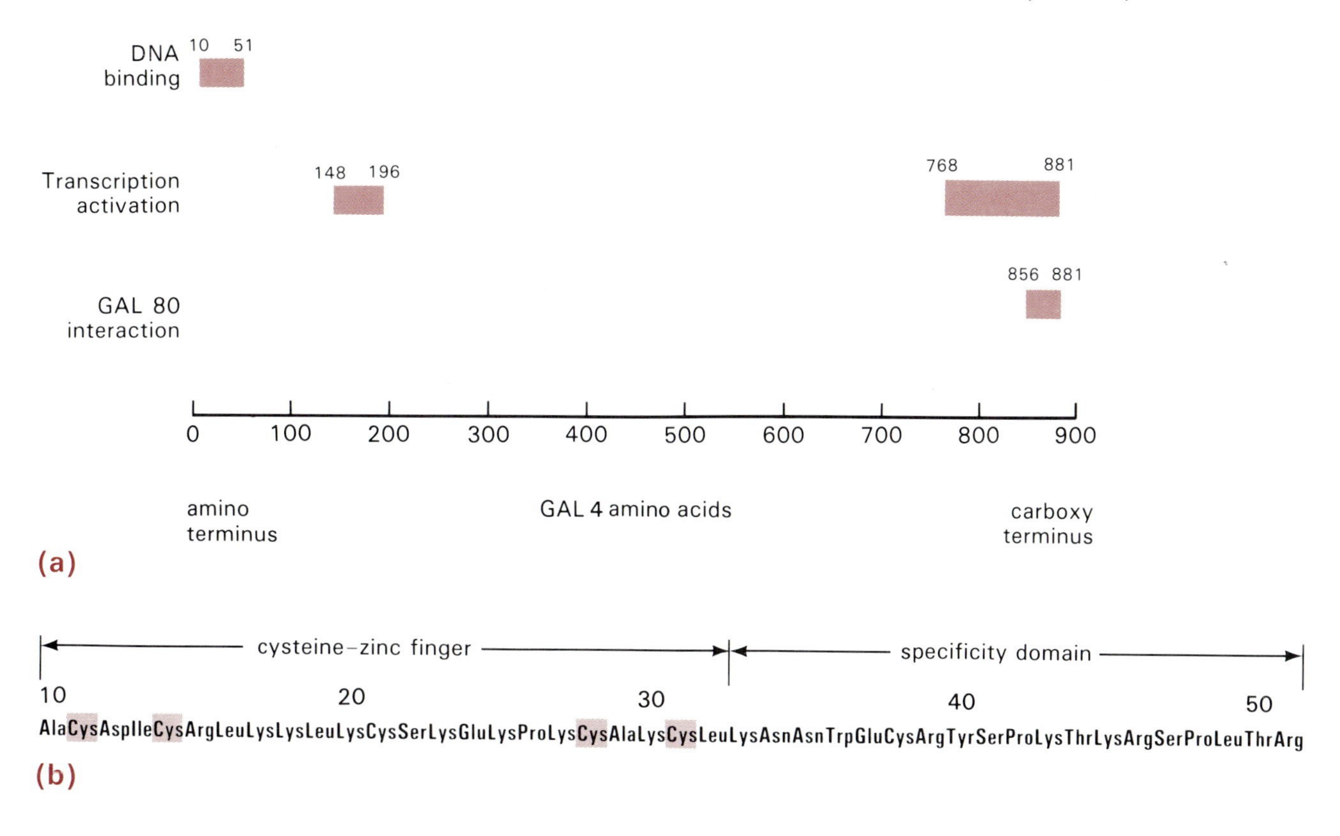

Figure 8.45
Functional domains in GAL4 protein. (a) The regions of GAL4 protein responsible for DNA binding, transcription activation, and interaction with GAL80 are shown along the 881 amino acid long polypeptide chain. The numbers surrounding each region correspond to the amino acid position beginning at the amino terminus. (b) The amino acid sequence of the DNA-binding region reveals a single cysteine-type zinc finger (Section 8.6a) and an adjacent domain responsible for specific binding to UAS_{GAL}. Adapted from M. Johnston, *Microbiol. Rev.* 51 (1987), p. 458.

tion of the GAL4 protein's deduced amino acid sequence and analysis of mutationally altered versions of the protein have identified the protein domains responsible for binding to UAS_{GAL}, transcriptional activation, and interaction with GAL80 protein (Figure 8.45).

The amino acids spanning residues 10 to 51 from the amino terminus are sufficient for specific recognition and binding to the UAS_{GAL} sequence motif; however, this segment by itself is incapable of transcriptional activation. The residues between 10 and 32 contain a spacing of cysteine and other residues characteristic of zinc finger motifs in other DNA binding proteins (Figure 8.45b) (Section 8.6a). In this instance, the zinc finger motif may contribute to both the stability of DNA binding and to its specificity because the ability to recognize UAS_{GAL} requires the adjacent 18 amino acids. Transcriptional activation relies on two regions, one adjacent to the DNA binding domain and one within 120 amino acids of the carboxyl terminus (Figure 8.45a). Although the latter is essential for transcriptional activation, its activity cannot be ascribed to any specific sequence. Instead, the activation function is a consequence of the high density of acidic amino acids in this region, a property characteristic of other transcriptional activator proteins (e.g., the λ phage repressor and the yeast transcription activator, GCN4 [Section 8.6b]). Indeed, variants of such transcription factors with increased numbers of acidic amino acids in these regions are more effective activators. Furthermore, proteins containing the GAL4 transcriptional activator region fused to unrelated DNA binding domains activate transcription from promoters having the sequence motifs corresponding to the DNA binding domain.

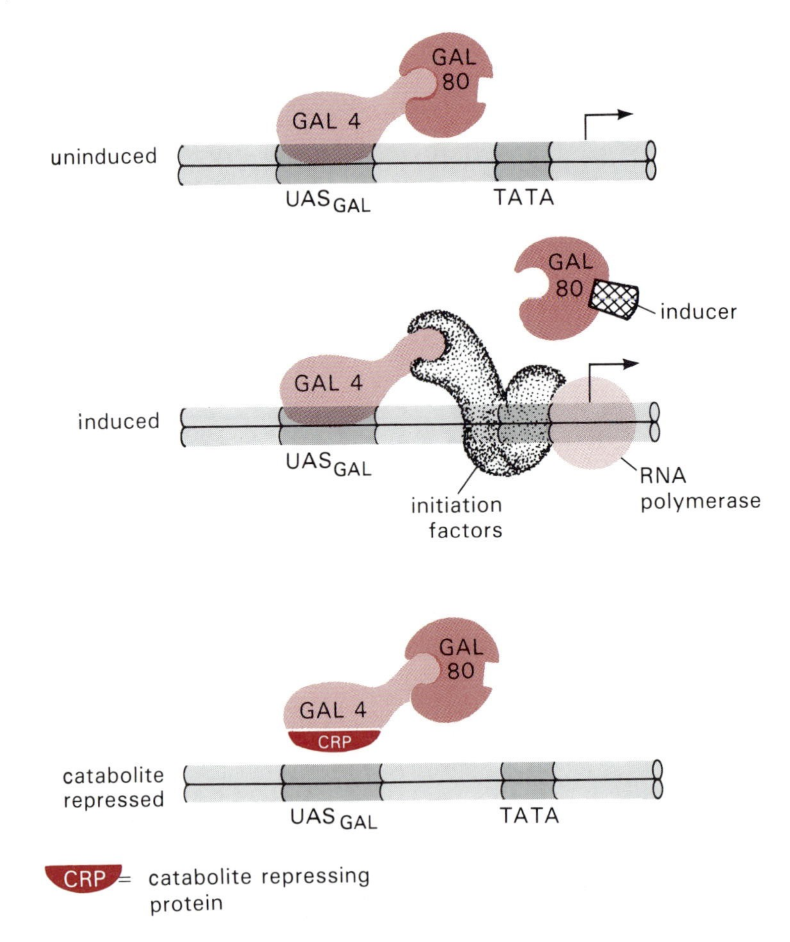

Figure 8.46
Speculative model for GAL4–GAL80 regulation of *GAL* gene transcription. In the uninduced state, GAL4's ability to promote the assembly (or activity) of the transcription complex at TATA is blocked by its association with GAL80. When inducer is present, the GAL80 protein dissociates from its complex with GAL4, allowing GAL4 to interact with (or activate) the transcription complex. In the presence of glucose, a catabolite repressing protein (CRP) binds to GAL4, preventing it from binding to UAS_{GAL}. Modified from M. Johnston, *Microbiol. Rev.* 51 (1987), p. 458.

The mechanism of transcriptional activation by bound GAL4 is not known. The localization of the GAL4 protein's acidic domains near the TATA sequence may influence the assembly of an active initiation complex at the transcription start site. This could be through interactions with the TATA binding protein itself or with other proteins in the complex. Support for this idea comes from the fact that GAL4 alters the DNase I footprint created at the transcription start site by the assembly of the yeast transcription complex (Figure 8.46). If UAS_{GAL} is inserted into the regulatory region of a mammalian promoter (e.g., the mouse mammary tumor virus promoter), transcription from that promoter is activated by the presence of the GAL4 protein *in vivo*. Similarly, placing the UAS_{GAL} within 50 bp of the adenovirus E4 promoter's TATA sequence makes that promoter responsive to GAL4 activation *in vitro*. In both cases, the transcriptional activation by GAL4 is synergistic with the activation produced by the binding of the mammalian transcription factors to their cognate sites in these promoters. Altered forms of the GAL4 protein, which can bind to UAS_{GAL} but are defective in transcriptional activation, fail to stimulate the heterologous promoters. These findings indicate that the GAL4 protein can interact with mammalian transcription proteins to enhance transcription, again emphasizing the universality of the activation mechanisms.

The regulatory role of GAL80 protein is expressed through its interaction with GAL4 protein. Such interactions have been proven by coprecipitation of GAL4 and GAL80 proteins with antibodies to GAL4. Furthermore, although GAL80 by itself does not bind to the UAS_{GAL} sequence, it does bind to the GAL4-UAS_{GAL} complex and alters its electrophoretic mobility. Interaction with the GAL80 protein requires only the 25 amino acids at the GAL4 protein's carboxyl terminus. Indeed, expression of a peptide containing only that sequence causes yeast cells to express the *GAL* genes constitutively, ostensibly because that peptide ties up GAL80 protein and releases transcriptionally active GAL4. By contrast, if GAL4 protein lacking the carboxyl terminal sequence is overproduced, the expression of GAL proteins still requires the inducer; in this instance, the overproduced protein is unable to sequester the GAL80 associated with intact GAL4 protein.

GAL80 protein probably blocks the GAL4 mediated transcriptional activation by preventing the required interaction between GAL4 and the transcription complex and not by altering the binding of GAL4 to UAS_{GAL}. The fact that GAL4 binds to UAS_{GAL} in the absence of inducer explains why the *in vivo* footprint of UAS_{GAL} is the same in the presence or absence of inducer. An as yet unidentified inducer formed from galactose is presumed to bind to GAL80 protein, causing it to dissociate from GAL4. Disruption of the GAL4-GAL80 complex uncovers the GAL4 activation domain, thereby activating transcription. This mechanism of induction contrasts with the induction mechanism used to regulate the *E. coli lac* or *gal* operons, where binding of the inducer prevents the repressor from binding to its DNA operator (Sections 3.11c and d).

The expression of the *GAL* genes is also regulated by the presence of other metabolizable sugars. Thus, *S. cerevisiae* grown on glucose, even in the presence of galactose, fails to produce the GAL proteins. Moreover, the addition of glucose to cells growing on galactose causes a rapid repression of *GAL* gene expression. The nature of the glucose derived repressor of the *GAL* genes is not known; however, several sites of action have been identified. One involves *cis* elements between UAS_{GAL} and TATA, but how this region exerts its inhibitory activity is not clear. Because the DNase I footprint of UAS_{GAL} is lost when cells are grown in glucose, it seems possible that **catabolite repression** is mediated by a protein (CRP) which prevents GAL4 binding to UAS_{GAL}.

Steroid responsive transcription Many aspects of vertebrate development and physiology are regulated by a variety of steroid and similarly acting hormones. The adrenal glucocorticoid and mineralocorticoid steroids influence glycogen and mineral metabolism, responses to stress, and growth and differentiation of many cells and tissues. Control of sexual differentiation in embryos and development of secondary sexual characteristics and of reproductive behavior are mediated by the sex steroids: estrogen, progesterone, and testosterone. Normal bone development and maintenance require the steroidlike vitamin D. And retinoic acid, whose structure slightly resembles those of steroids, functions as a key morphogen and as a regulator of differentiation during embryogenesis. The thyroid hormone, triiodothyronine, although completely unrelated to steroids in structure, initiates its regulatory function in energy metabolism by binding to a receptor that resembles the ones employed for steroids.

The physiological effects of each of these hormones follows from their entry into cells and association with intracellular receptors. The unbound

Figure 8.47
Nucleotide sequences at several hormone responsive elements (HREs). The numbering flanking each sequence refers to the positions relative to the transcription start sites. The letter N refers to any one of the four nucleotides, the arrow over each consensus sequence illustrates its dyad symmetry, and the dot indicates the position of the dyad's axis. The glucocorticoid responsive elements (GRE) are mouse mammary tumor virus (MMTV), human growth hormone (hGH), human metallothionein (hMTIIA), tyrosine oxidase (TO), and tyrosine aminotransferase (TAT). The estrogen responsive elements (ERE) are *Xenopus* vitellogenin (X. vit.) and chicken ovalbumin (ovalbumin). The thyroid hormone responsive element (TRE) is rat growth hormone (rGH).

GRE

MMTV	− 134	TGGTTTGGTATCAAATGTTCTGATCTG	− 108
hGH	+ 87	CCTTTGGGCACAATGTGTCCTGAGGGG	+ 113
hMTIIA	− 268	GCACCCGGTACACTGTGTCCTCCCGCT	− 242
TO	− 439	CTCATATGCACAGCGAGTTCTAGTGAG	− 413
TAT	−2420	TACGCAGGACTTGTTTGTTCTAGTCTT	−2446

consensus GGTACANNNTGTTCT

ERE

X. vit	− 338	AAAGTCAGGTCACAGTGACCTGATCA	− 315
ovalbumin	− 173	TTATTCAGGTAACAATGTGTTTTCTG	− 199

consensus GGTCANNNTG(A/T)CC

TRE

rGH	− 185	GTAAGATCAGGGACGTGACCGCAGG	− 161

consensus GATCANNNNNNTGACC

receptors are localized to different cell compartments: The glucocorticoid receptor is mainly cytoplasmic and moves to the nucleus following ligand binding. The estrogen and progesterone receptors are predominantly nuclear. The thyroid hormone receptor is confined exclusively to the nucleus. However, all receptors undergo structural alterations on associating with their cognate ligands, a change that promotes the binding of the receptor-ligand complex to specific DNA sequence motifs called **hormone response elements (HREs)**. HREs have the properties of enhancers in that they can influence the transcriptional activity of promoters with which they are associated in a position and orientation independent manner. HREs therefore constitute inducible enhancers because they regulate transcription only when they are bound to a functional receptor-hormone complex.

What are the structural features of the receptors and of their modified liganded forms, and how do receptors recognize and bind to their respective HREs? Another central issue is how the association of a receptor-ligand complex with its HRE influences transcription. Our present knowledge concerning these questions, though incomplete, relies on several experimental approaches, all of which stem from the availability of cloned and sequenced cDNAs encoding receptors. Comparisons of the amino acid sequences deduced from the cDNAs identify similar and distinctive regions that point to functionally relevant parts of the receptors' structure. However, the most important and productive approach to correlating receptor structure and function has been the analysis of the effects of mutational alterations in the receptor cDNAs on their receptor activity. Similar strategies applied to hormone responsive promoters identified and eventually characterized the HREs. The construction of chimeric receptor genes encoding hybrid receptors provides additional and reassuring confirmation for assigning receptor functions to specific protein regions and even to specific amino acids.

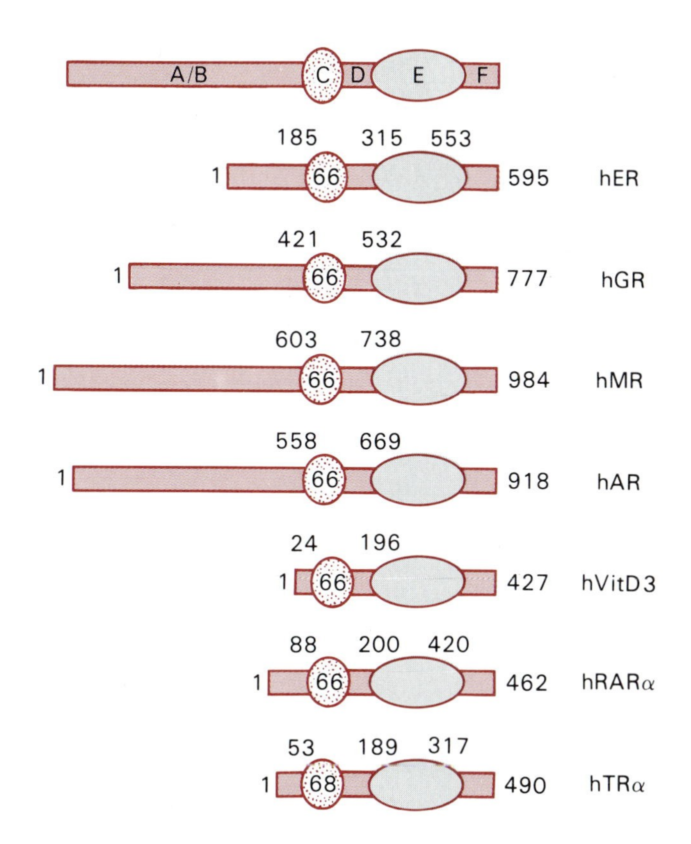

Figure 8.48
Functional domains in steroid and thyroid hormone receptors. The diagram at the top portrays the protein domains of a generic receptor. The individual receptors are hER, human estrogen receptor; hGR, human glucocorticoid receptor; hMR, human mineralocorticoid receptor; hAR, human androgen receptor; hVitD3, human vitamin D3 receptor; hRARα, human α retinoic acid receptor; and hTRα, human α thyroid receptor. The various domains are named A/B, C, D, E, and F, beginning at the protein's amino terminus. The C domain of each receptor contains 66 or 68 amino acids, but the size of the other domains is variable and can be deduced from the amino acid numbering.

Both sequence comparisons and mutational analyses of hormone responsive promoters identify short palindromic sequences located varying distances upstream of the transcription start site that are required for their hormonal regulation. Genes whose transcription is influenced by the same hormone have similar or identical HREs. Receptor binding to the genetically defined HRE sequences has been confirmed by DNase I footprinting with purified glucocorticoid, estrogen, progesterone, and thyroid hormone receptors. Although they differ in their location in the promoter, all HREs share common sequence characteristics: they range in size between 12 and 16 bp with a variable number of nonspecific base pairs flanked by five specific palindromic base pairs (Figure 8.47). Indeed, because of similarities in the HRE sequences, it is possible to change an estrogen responsive element into a glucocorticoid responsive element by changing only two base pairs in the palindrome. The dyad symmetry of HREs suggests that they are bound by receptor dimers.

Examination of the amino acid sequences deduced from cDNA clones encoding the estrogen and glucocorticoid receptors identifies a highly conserved short stretch of amino acids near the middle of each receptor's sequence (region C in Figure 8.48). Screening genomic DNA digests with oligonucleotide probes designed from this conserved sequence indicates that this region is characteristic of a large gene family, ostensibly encoding other receptors. Not surprisingly, the use of the same probes allowed the isolation of additional receptor cDNA clones, suggesting that this domain is important for receptor function. The proteins encoded by some of these cDNA clones are only tentatively classed as receptors because their ligands or promoter HREs are unknown. Several steroid receptor

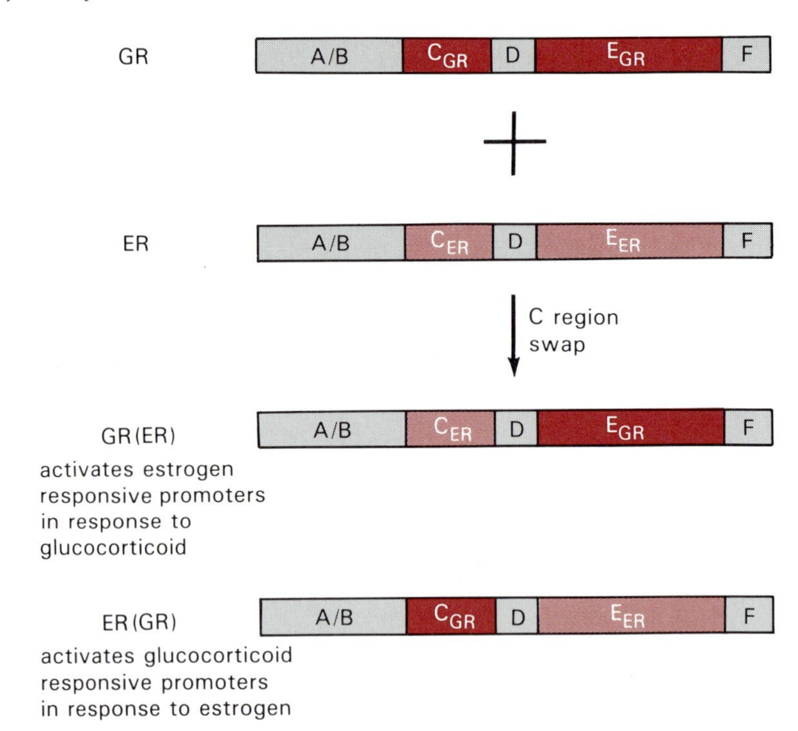

Figure 8.49
Switching DNA binding domains between steroid receptors. The segments encoding the C domains of the glucocorticoid receptor (GR) and the estrogen receptor (ER) were switched in their respective cDNAs so that two chimeric receptors could be produced: GR(ER), a glucocortoid receptor with an estrogen receptor DNA binding domain, and ER(GR), an estrogen receptor with a glucocorticoid receptor DNA binding domain. Transfections with each of these cDNAs into appropriate cells led to the indicated result.

genes have been cloned, partially sequenced, and mapped to their respective chromosomes (Figure 8.48).

Several regions with different degrees of shared homology were uncovered by comparing the amino acid sequences in different receptors. These are designated A/B, C, D, E, and F in Figure 8.48. Region A/B is the least homologous among different receptors, and even comparing the same receptor from different species. By contrast, sequences in region C are highly conserved in all receptors, and further analysis identifies it as the DNA binding site, the one function common to all receptors. Regions D and F differ in both length and sequence in different receptors and even comparing the same receptor in unrelated species. Deletion and mutational analyses of these regions indicate that they are unessential for receptor function. Even though the E regions differ among different receptors, their sequences in the human and chicken estrogen receptors are strikingly similar. This suggests that region E provides the ligand binding domains. The surmises regarding regions C and E, although initially based only on the sequence comparisons, proved to be correct upon more detailed analyses.

The 66 to 68 bp long stretch of amino acids conserved in all receptors (region C) is responsible for binding to the cognate HREs. Mutations affecting this region permit hormone binding but eliminate binding to the HREs. Moreover, truncated receptors lacking regions A/B, D, E, and F still bind to the cognate HRE, although transcriptional regulation is eliminated. The most direct demonstration of the autonomous action of region C in DNA binding derives from experiments in which region C from the glucocorticoid receptor replaces the corresponding region in another receptor (e.g., the estrogen receptor). The chimeric receptor activates glucocorticoid responsive genes in the presence of estrogens (Figure 8.49). Interestingly, intact receptors do not bind to their HREs in the absence of

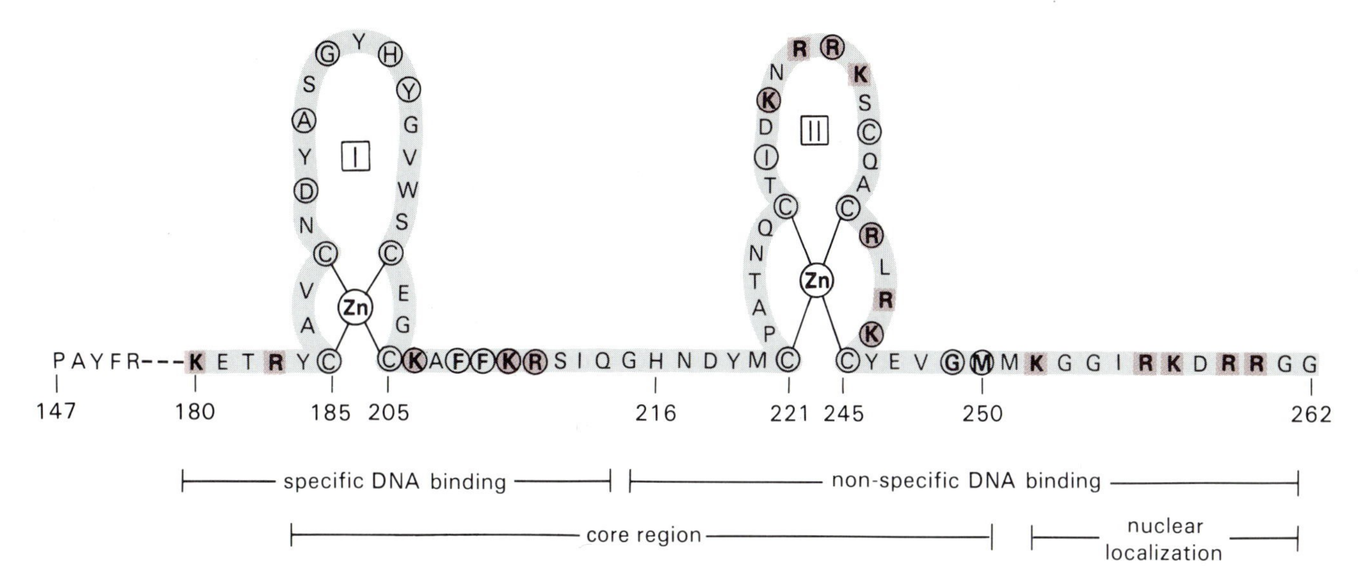

Figure 8.50
Zinc fingers in the DNA binding domain of the human estrogen receptor. The amino acid sequence of the receptor's DNA binding domain (core region) and its surrounding sequence are given, using the single letter designations for the amino acids (Figure 1.24). The basic residues are shown in color, and the most highly conserved amino acids among all receptors are circled. The region between amino acid residues 252 and 262 is presumed to be involved in localizing the receptor to the nucleus.

their ligand but can if the ligand binding domain is removed. This suggests that the receptor's DNA binding domain is unavailable for interaction with its HRE if the ligand binding domain is present and unoccupied; somehow, binding of ligand or removal of the ligand binding domain permits the DNA binding region to interact with its HRE.

Examinations of many receptor C regions reveal an arrangement of cysteine, hydrophobic, and basic amino acids consistent with two zinc fingers (Figure 8.50). Two zinc ions are associated with this region in a purified glucocorticoid receptor. Each of the four cysteine residues implicated in the zinc finger structure is essential for DNA binding. This suggests that, in this instance, each finger's zinc ion is coordinated to four cysteines instead of to cysteines and histidines, as occurs in several other transcription factors (Section 8.6a). Region C autonomy is further supported by the fact that its coding sequence is separated from those for regions A/B and D by introns; interestingly, the two zinc fingers are encoded by separate exons.

Both putative zinc finger motifs are necessary for high affinity binding to HRE. However, if each zinc finger is swapped between different receptors, the chimeric receptors have the target gene specificity characteristic of the receptor's zinc finger I. Mutational alterations of the sequence preceding or in zinc finger I do not alter the receptor's DNA binding specificity. But several sequence changes immediately following the last cysteine of the estrogen receptor's zinc finger I impair binding to the estrogen responsive element and permit the receptor to bind to the glucocorticoid responsive element. Thus, the discrimination by the estrogen receptor between the estrogen and glucocorticoid responsive elements, whose palindromic sequences differ by only two base pairs, is strongly influenced by three amino acids of the estrogen receptor's zinc finger I. It seems, therefore, that specific interactions between the receptor and its HRE depend on the arrangement of hydrophobic residues in zinc finger I. Contacts between the DNA backbone and the zinc finger II domain and basic amino acids also contribute to the interaction's stability.

The palindromic nature of HREs suggests that receptors bind as dimers. Receptor dimerization involves two of the receptor's domains. The stronger interactions that appear after ligand binding involve sequences in region E. A set of weaker interactions relies on sequences in the second zinc finger. Ostensibly, the receptor binds more strongly and specifically as a dimer than as a monomer. The mechanism of hormone induced dimerization is unknown, but it may be part of the same process that allows the receptor to bind to its HRE once it has bound hormone. Both HRE binding and dimerization may be the consequence of a conformational change in the receptor induced by hormone binding. There are also indications that in some instances hormone binding may cause dissociation of an inhibitor that prevents spontaneous dimerization of the receptor.

Hormone binding is essential for both specific and efficient binding of receptors to their respective HREs and for transcriptional regulation of hormone responsive genes. A receptor's ability to bind hormone requires the integrity of its E region because deletions, insertions, or base changes in E eliminate binding. Interestingly, the receptor's ability to bind its ligand is unimpaired if regions A/B, C, or D are lacking, presumably because region E can bind ligand independently of the rest of the protein. This autonomously functioning characteristic of region E is further illustrated by swapping E regions between different receptors, as was done with the DNA binding domains (Figure 8.49). Thus, if the estrogen receptor's region E is replaced by the glucocorticoid receptor's region E, the chimeric receptor activates transcription from estrogen responsive genes in response to glucocorticoids but not to estrogens.

Although binding of intact receptors to their respective HREs does not occur in the absence of bound hormone, receptors that have region E deleted bind readily to their HREs in the absence of hormone. This indicates that the structural change accompanying hormone binding either unmasks or activates the receptor's DNA binding domain, possibly by promoting the required receptor dimerization. The structural change induced in the glucocorticoid family of receptors by hormone binding also triggers the ligand-receptor complex's transfer from the cytoplasm to the nucleus. The ligand binding domain also contributes to the receptors' ability to activate transcription after binding to their respective HREs. Thus, in yeast expressing a fusion protein containing the glucocorticoid receptor's hormone binding domain and the yeast GAL4 protein's UAS_{GAL} binding domain, *GAL* gene expression is inducible by the addition of glucocorticoids. Because binding this truncated region of GAL4 protein to the UAS_{GAL} does not activate transcription, this finding means that region E bound to hormone can activate transcription in yeast.

Antihormones, compounds that antagonize the physiologic response to hormones, may act by binding to the receptor without triggering the structural changes needed for dimerization, for binding to the HRE, or for influencing transcription from the responsive genes.

The end result of binding the activated receptor-hormone complexes to their respective HREs is transcriptional regulation, which, depending on the promoter, may be the activation or inhibition of transcription initiation. One approach to learning how the bound receptor produces these positive and negative transcriptional activities is to identify the receptor regions responsible for their transcriptional effects. Here, too, mutational alterations in cDNAs encoding steroid, thyroid hormone, vitamin D3, and retinoid receptors have been used to explore transcriptional regulation by the cognate HREs. Binding of the receptor-hormone complex is prerequisite for transcriptional regulation, so the issue becomes what regions of the receptor structure mediate this regulatory effect. Taken together, the results show that multiple receptor regions are involved, the most important being regions A/B and E. However, the relative importance of the two regions varies for different receptors and among different promoters containing the same HRE. For example, if the estrogen receptor's E region is removed, the receptor can be bound, but it fails to activate transcription normally from several estrogen responsive promoters. By contrast, although deletions within several receptors' E regions prevent hormone, but not DNA binding, the effect on transcriptional activation is variable, depending on the promoter and, in some cases, on the cells in which the receptor is tested. Mutations in region A/B reduce transcriptional activation by the bound receptor, but the magnitude of the effect varies among receptors and target genes. Thus, an estrogen receptor lacking the entire region A/B still binds to estrogen responsive elements of two test promoters, but transcriptional activation is nearly normal with one and is lost with the other.

A dramatic demonstration of the influence of regions A/B and E on transcriptional activation stems from the use of chimeric receptors containing only the DNA binding domains of the GAL4 regulatory protein fused to either the A/B or E domain of several steroid receptors. When chimeric genes encoding fusion proteins containing either the A/B-*GAL*4 or E-*GAL*4 regions are transfected into mammalian cells together with reporter (β-galactosidase) genes whose promoters contain a UAS_{GAL} near the TATA sequence, transcription of the β-galactosidase gene is activated. Mutations in either region A/B or E that impair the receptor's transcriptional activity also prevent the activations by the chimeric proteins. It seems, therefore, that if a steroid receptor's region A/B or E is bound in the vicinity of a mammalian promoter, it can activate transcription.

The steroid receptor's transcriptional activation domains can also transactivate yeast promoters in yeast. If the *GAL*1 promoter, including the UAS_{GAL}, is fused to the β-galactosidase coding sequence, β-galactosidase expression is induced by galactose addition (Figure 8.51a). Replacement of the UAS_{GAL} by an estrogen responsive element permits a test of whether the estrogen receptor can activate the yeast transcription machinery and whether the activation is estrogen dependent. In fact, transcription of β-galactosidase from that construct does not occur even in the presence of estrogen because the estrogen receptor is lacking (Figure 8.51b). But when a functional estrogen receptor is expressed from its cDNA in cells containing the estrogen responsive element between the *GAL*1 promoter and the β-galactosidase gene, transcription depends completely on the addition of estrogen and does not respond to galactose (Figure 8.51c). Thus, the estrogen receptor clearly can interact with and influence the yeast transcription apparatus.

A model embodying the various structural features of the hormone receptors and how they might be involved in transcriptional activation is summarized in Figure 8.52. Two forms of the free receptor are indicated. In one (I), domain E is free and uncomplexed. The second, form II,

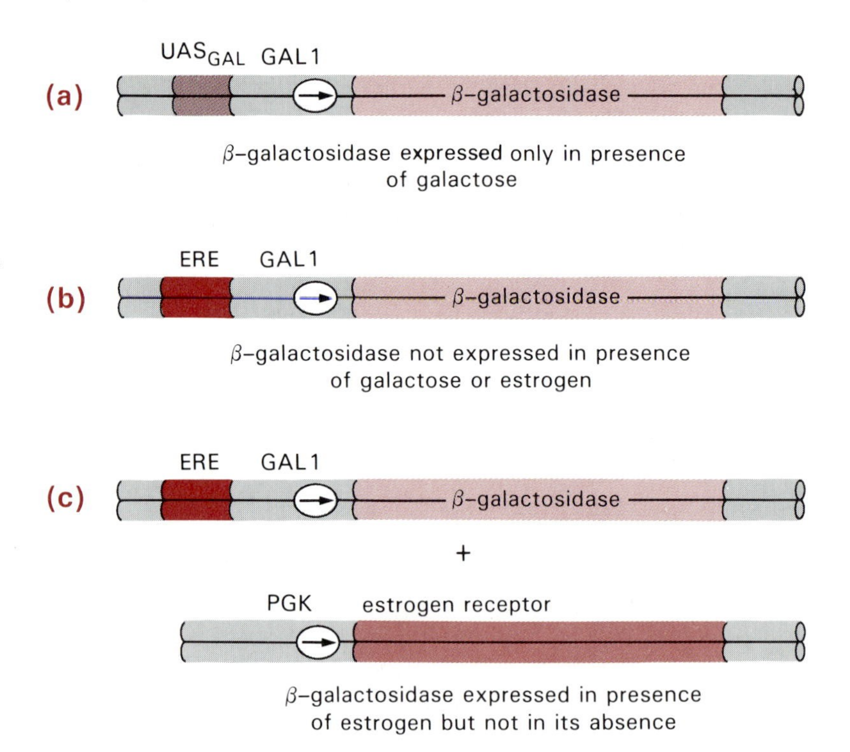

Figure 8.51
The human receptor can activate transcription in *S. cerevisiae.* (a) A β-galactosidase gene transcribed from a *GAL1* promoter. Transfection of this gene into yeast leads to galactose inducible expression of β-galactosidase. (b) The *GAL1* promoter's UAS$_{GAL}$ replaced by an estrogen responsive element (ERE). Transfection of this gene into yeast does not promote β-galactosidase expression when either estrogen or galactose is added. (c) Cotransfections of the gene shown in (b) with a gene encoding the human estrogen receptor transcribed from the yeast glycerophosphate kinase (GPK) promoter leads to estrogen dependent expression of β-galactosidase.

represents a putative complex involving the receptor and an inhibitor. Both forms I and II are presumed to be unable to bind to hormone responsive elements. According to the model, binding of hormone to form I alters the conformation of domain E and allows the receptor to dimerize (form III) and bind cooperatively to the HRE (form IV). When hormone is bound to form II, the inhibitory protein dissociates from the receptor, permitting dimerization (form III) and binding to the HRE. The model suggests that dimerization results from interactions between the liganded E domains and the two DNA binding C domains.

In considering how the bound receptor influences transcription at hormone responsive promoters, it should be emphasized that the HREs are not the exclusive regulatory motifs in these promoters. On the contrary, most of these promoters contain many different kinds of regulatory motifs that bind different kinds of transcription factors. Indeed, if a glucocorticoid or estrogen responsive element is introduced into a promoter ordinarily not activated by the hormone, the promoters become responsive to the steroid if the corresponding receptor is present. Although generally synergistic, the magnitude of the transcriptional response is influenced by the kinds of neighboring sequence motifs and their location relative to the hormone binding sequence.

Although we have so far emphasized that the binding of receptors to their cognate HREs activates the promoter's transcription, in some instances, the binding causes transcriptional repression. For example, the prolactin gene promoter contains numerous binding sites for the receptor-hormone complex, but binding inhibits rather than activates transcription. The sequence of the negative glucocorticoid responsive element differs from elements that activate transcription. Thus, the glucocorticoid receptor's DNA binding domain can bind to both negative and positive gluco-

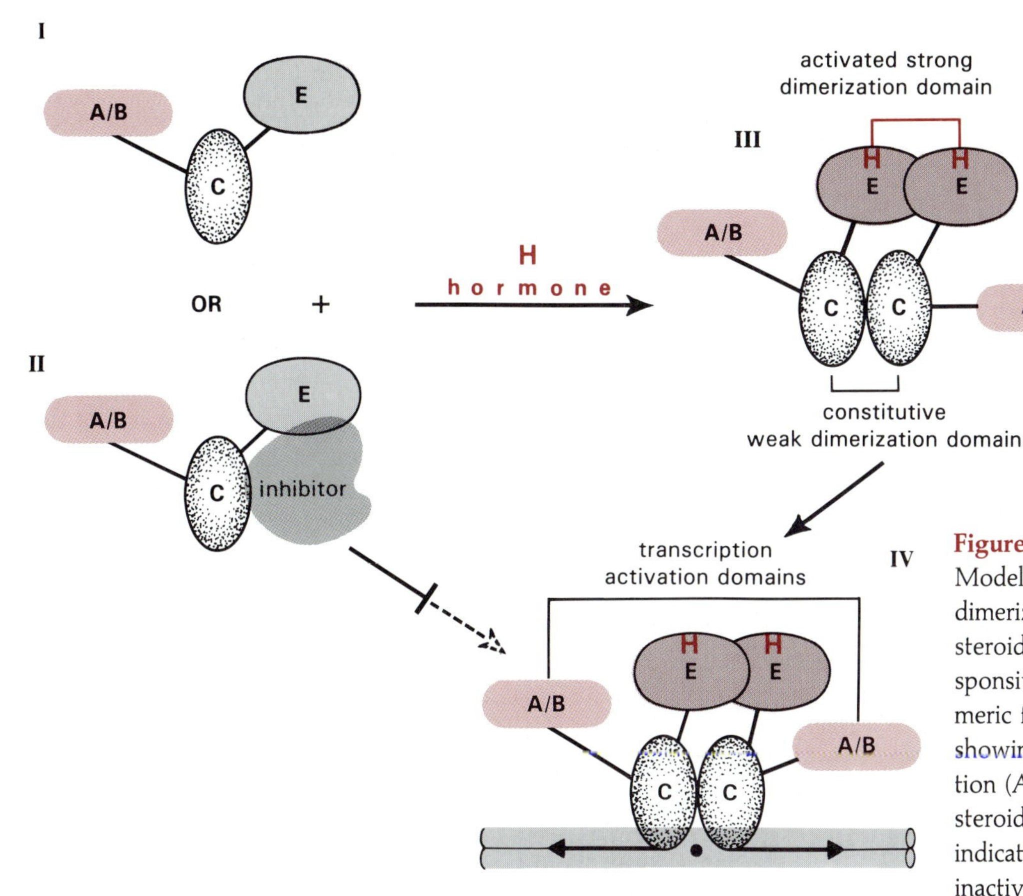

Figure 8.52
Model for hormone induced dimerization and binding of steroid receptors to steroid responsive elements. The monomeric form of a steroid receptor showing the transcription activation (A/B), DNA binding (C), and steroid binding (E) domains is indicated as form I. A putative inactive cytoplasmic complex, form II, is presumed to be formed by association of the monomeric steroid receptor with a protein inhibitor; the nature of the association is unknown. Binding of the hormone (H) to the E domain induces receptor dimerization via strong interactions between the hormone binding domains and weaker constitutive interactions between the DNA binding domains. Alternatively, hormone binding to the inactive form II leads to dissociation of the inhibitor, followed by dimerization. Form III can enter the nucleus and bind to the cognate hormone responsive element. The diverging arrows in the DNA indicate the dyad symmetry of the binding site's sequence. Adapted from S. Green and P. Chambon, *Trends Genet.* 4 (1988), p. 309.

corticoid responsive elements. The analysis to date suggests that binding of the receptor-hormone complex to a negative element interferes with the binding of other, positively acting transcription factors.

There is a remarkable similarity in the regulation of the yeast *GAL* genes and the mammalian steroid responsive genes. The *GAL* genes are not transcribed unless the GAL4 protein is bound to the UAS_{GAL} motifs. Similarly, transcription of hormone inducible genes requires that the receptor be bound to its cognate HRE. In each case, binding of the transcriptional activator to its DNA site is triggered by an inducer, the appropriate hormone in the case of the receptor, and a compound derived from galactose where *GAL4* is involved. Furthermore, activator binding to the promoter is blocked in the absence of the inducer by the association of GAL80 with GAL4 in the case of yeast, and probably by the receptor's domain E in the other. An important distinction between the two inducible systems is that the negative and positive regulatory components of the *GAL* system are separate polypeptides (GAL4 and GAL80 proteins), and the corresponding components regulating transcription of hormone responsive genes are contained in the same protein, the receptor. Although we do not know how the transcriptional machinery is activated in the two systems, it is striking that the activating domains of the two regulatory regions contain clusters of acidic amino acids of relatively undefined composition and sequence. Equally astonishing is the interchangeability of the activation domains between yeast and mammals.

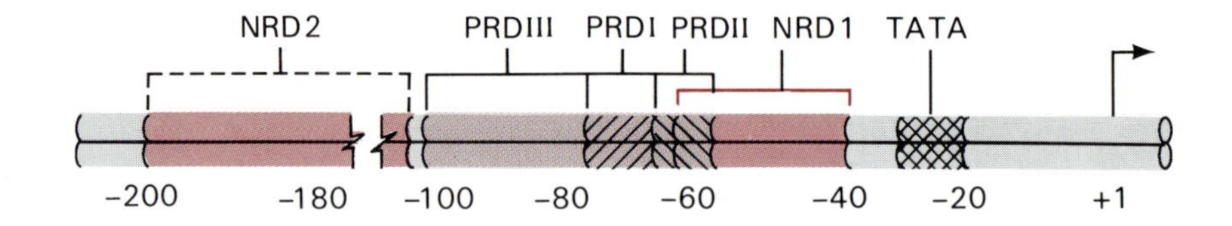

Figure 8.53
Organization of positive and negative regulatory elements in the IFN-β promoter. The locations of the two negative regulatory domains, NRD1 and NRD2, as well as three positive regulatory domains, PRDI, PRDII, and PRDIII, are shown relative to the TATA sequence and the transcription start site (+1).

β-interferon gene regulation Interferons (IFNs) are secreted polypeptides that are able to protect cells of the same species from infection by viruses. Different kinds of IFNs occur in vertebrate leukocytes (α), fibroblasts (β), and cells of the immune system (γ). Although not detectable in normal fibroblasts, large amounts of IFN-β mRNA are induced within four to eight hours after virus infection or, in some cells, after exposure to double strand RNA; such double strand RNAs are presumed to arise in the course of virus multiplication. The IFN-β mRNA accumulation is due in large part to increased transcription induced by the infection.

The regulatory sequences governing IFN-β transcription span the region between base pairs −204 and +1 (Figure 8.53). The ubiquitous TATA element occupies its usual location around base pair −25, and the regions governing IFN-β inducibility (IRE) reside between base pairs −200 and −37. The IRE consists of five discernable subelements, two of which act negatively (NRD1 and NRD2) and three of which mediate the induced transcriptional activation (PRD I, II, and III). Note that NRD1 overlaps PRD II, and NRD2 either abuts or overlaps PRD III. The NRD1 and NRD2 motifs were identified because mutational alterations of their sequences reduce or eliminate the need for induction (i.e., IFN-β transcription becomes constitutive). By contrast, changes in either of the three PRD sequences impair the viral induced transcriptional response. Thus, the regulation of IFN-β expression involves both negative and positive mechanisms.

Methylation protection experiments carried out *in vivo* indicate that both NRD regions are protected in uninfected cells but become reactive to methylation when the cells are infected or induced. However, the three PRD regions are readily methylated in the uninfected cells and become fully protected after infection. These findings indicate that these regions are probably bound by nuclear factors in a reciprocal fashion; those involved in repressing transcription are occupied in the uninduced state, and the mediators of the induced response are bound after induction (Figure 8.54).

Nuclear extracts from infected or induced cells contain distinct proteins that bind to each of the PRD elements. These proteins are most likely present in an inactive form in uninduced extracts because induction of IFN-β transcription does not require new protein synthesis. PRD II contains a sequence closely resembling the κB site in the IgL(κ) enhancer and, as expected, binds NF-κB. Mutations that prevent NF-κB binding to PRD II also impair the promoter's inducible response. Recall that NF-κB is a ubiquitous protein existing mainly in the cytoplasm as an inactive complex but is activatable by a variety of treatments, including virus infection. Thus, activation of NF-κB by virus infection is probably an important contributor to IFN-β induction. Neither the inducible factors that

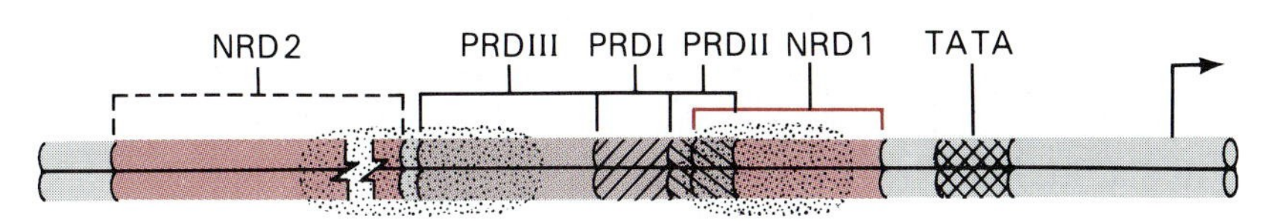

induced

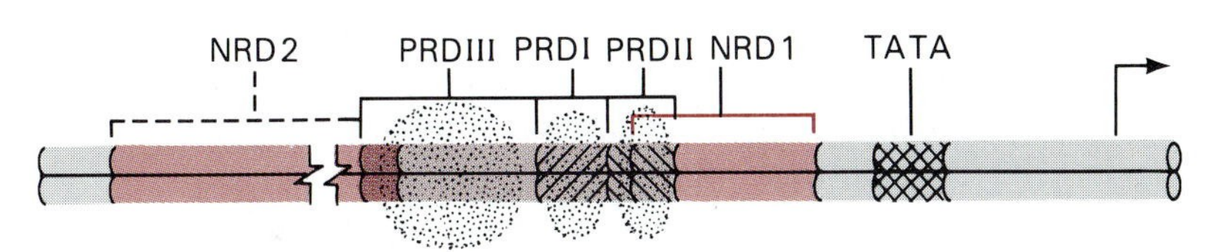

Figure 8.54
Binding sites for regulatory proteins in the IFN-β gene promoter. In the uninduced state, proteins are presumed to bind to sites in NRD1 and NRD2 and overlap sites in PRDII and PRDIII, respectively. After induction, the NRD binding proteins no longer bind, and proteins occupy the three PRD sites. The latter state activates transcription.

bind to PRD III and PRD I nor the proteins that bind to the NRD elements have been identified. A plausible but still unproven model for IFN-β regulation is that the bound NRD factors restrict the binding of the PRD II and III proteins to their cognate sequences, thereby keeping the promoter off (Figure 8.54). Induction may inactivate the NRD factors and cause them to dissociate from their binding sites, or the NRD proteins could be displaced by the binding of activated PRD proteins.

Transcriptional regulation by heat, light, and metals Heat shock and similar environmental stresses induce a typical physiological reaction called the **heat-shock** response. Every eukaryotic and prokaryotic organism can mount a heat-shock response. In most instances, normal protein synthesis is shut down, and a specific set of proteins—heat-shock proteins (HSPs)—is induced. The heat-shock response may be essential for survival because mutations in yeast that prevent it are lethal. The mechanism for inducing heat-shock genes is remarkably conserved among *Drosophila, Xenopus,* yeast, and mammals. For example, the *Drosophila HSP70* gene (which encodes a 70 kDalton heat-shock protein) introduced into *Xenopus,* mouse, and monkey cells undergoes heat-induced transcriptional activation as efficiently as the endogenous heat-shock genes.

The appearance of HSPs depends on transcriptional activation of the heat-shock genes. This induction is immediate and increases the amount of HSP mRNA from less than a molecule per cell to thousands per cell within an hour of heat shock. Some heat-shock genes are also developmentally regulated and do not require heat shock for their transcriptional activation. Cloned and sequenced heat-shock genes from many species have helped define the key features of their transcriptional response to heat shock. Generally, the promoter sequences responsible for transcriptional regulation in response to heat shock consist of several repeats of a relatively short sequence motif (heat-shock elements—HSEs) scattered within 200 bp upstream of TATA. The consensus sequence obtained from the regulatory regions of many heat-shock promoters of diverse species is 5′-CNGAANTTCNG-3′, where the Ns indicate any one of the four bases. Such HSE sequences when placed upstream of TATA sequences of heterologous promoters (e.g., the herpes TK or β-globin promoters) make their transcription heat inducible.

As with *cis* acting motifs in other promoters, HSEs are binding sites for transcription regulatory proteins (HSTFs). HSTFs from various species are similar in their physical and functional characteristics, and, indeed, the yeast and *Drosophila* HSTFs are identical in size and DNA binding properties. Many experiments indicate that HSTFs are tightly bound to the HSEs during, but not before, heat shock. Thus, the HSTFs are present in an inactive state in unshocked cells, and heat shock activates their binding to the HSEs of all heat inducible promoters. The gene encoding the yeast HSTF has been cloned, thereby paving the way for an analysis of its binding to HSEs and the mechanism of transcriptional activation of various heat-shock promoters.

Higher plants have multiple photoreceptors that mediate responses to light. Light of different wavelengths serves as a signal for regulating complex developmental and regulatory processes. The idea that one of the photoreceptors' functions is to activate or repress gene expression is an old one. Recent investigations using molecular genetic approaches to dissect the regulatory elements of plant gene promoters provide experimental support for transcriptional activation. The light induced activation of a nuclear gene that encodes the small subunit of ribulose 1,5-bisphosphate carboxylase (*rbc*S), has been analyzed in this way (Figure 6.36). One kbp of the gene's 5′ upstream sequence fused to the chloramphenicol transacetylase gene was used to assay the promoter's activity. This chimeric gene, and others with varying size deletions and rearrangements, were introduced into tobacco plants using Ti plasmid mediated transformation (Section 5.8c). Promoter function was assessed in light and dark kept tobacco calli. Low levels of transcription from an *rbc*S promoter is mediated by a sequence located within 90 bp upstream of the TATA sequence. However, maximum levels of light activated transcription require multiple upstream enhancer sequences between base pairs −975 and −90. Shortening that sequence leads to progressively lower levels of light induction. A similar arrangement of a TATA proximal light response element and an upstream enhancer occur in the promoter of a gene that encodes a light activated, light harvesting, chlorophyll associated protein. In each case, the specific sequence motifs responsible for the light responsiveness have been delineated, but the transcription factors with which they interact have not been identified. Moreover, the mechanism by which the photoreceptors transduce the light signal to the transcription regulatory machinery is unknown.

Just as organisms can respond to heat, light, and a variety of physical stimuli (e.g., radiation) by altering the transcriptional patterns of sets of genes, so too are there homeostatic adjustments in response to metals. Some of the metals are toxic (e.g., cadmium and mercury); others are required nutrients but toxic in high concentrations (e.g., copper and zinc). A principal contributor to the response is transcriptional activation of the metallothionein genes. These genes encode small cysteine-rich polypeptides that bind both essential and toxic heavy metals.

Mammalian species contain several distinctive but structurally related metallothionein genes, all of which are tandemly linked on the same chromosome. Each contains the same arrangement of exons and introns. Metal-

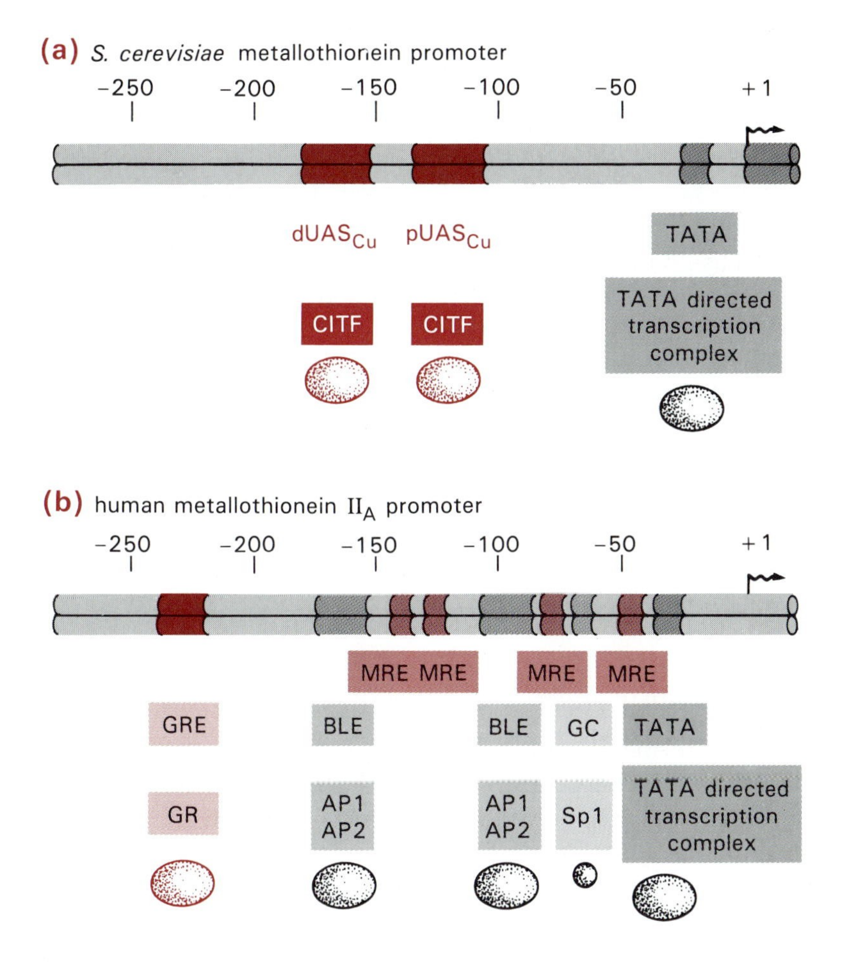

Figure 8.55
The organization of the mammalian and yeast metallothionein gene promoter regulatory motifs. (a) *S. cerevisiae* metallothionein gene promoter: Besides the TATA motif, proximal and distal sequence motifs (UAS_{Cu} and $dUAS_{Cu}$, respectively) are essential for copper inducible transcription. The TATA sequence directs the assembly of the transcription complex, and the two UAS_{Cu} motifs bind a copper inducible transcription factor (CITF). (b) Human metallothionein II_A gene promoter: Basal activity of this promoter (i.e., in the absence of metals) relies on the TATA motif and its assembled transcription complex (an Sp1 protein bound to a GC-rich hexamer, and the AP1 and AP2 transcription factors bound to the two basal level enhancers [BLEs]). Metal inducible transcription requires a cluster of metal responsive elements (MREs), which bind a MRE binding protein (MRE-BP) for their activity. Glucocorticoid responsivenes relies on the binding of the glucocortocoid receptor (GR) to a glucocorticoid responsive element (GRE).

lothionein gene transcription is transiently induced by the same heavy metals that bind to the protein, thereby providing a mechanism for cells to protect themselves against metal toxicity. The inducible response to metals occurs in all cells, and each of the several different metallothionein genes that can be present in a single organism is activated. The mammalian genes encoding metallothioneins are also induced by stress and a variety of poisons, the induction being mediated by glucocorticoid release.

S. cerevisiae contains only a single metallothionein gene. It is not essential for the yeast's survival, but it does protect the organism against copper toxicity. Dissection of the promoter identified two copper responsive elements located between base pairs −200 and −100 and the usual TATA sequence at about base pair −30 relative to the transcription start site (Figure 8.55a). The two upstream copper activating sequences (UAS_{Cu}), which are imperfect repeats of a 32 to 34 bp sequence, are binding sites for a copper inducible transcription factor. This protein (CITF), whose gene (*ACE-1*) has been cloned, sequenced, and expressed, binds to UAS_{Cu} only when copper is bound to the protein. The copper binding and DNA binding domains of CITF are localized to the amino terminus of the protein. A likely explanation for the copper dependent binding to UAS_{Cu} is that the proper binding domain is induced only when a cluster of copper ions is chelated to the cysteine residues located in that region of the protein. Interestingly, the carboxy terminal half of the CITF is highly acidic, a feature characteristic of other transcriptional activating factors (Section 8.6b).

The regulatory elements controlling transcription of the human metallothionein II promoter have been identified by mutational analysis and transcription assays of the modified promoters *in vivo* (Figure 8.55b). Several regions distributed within 175 bp upstream of the transcription start site are necessary for basal levels of transcription (i.e., in the absence of inducers); these include TATA and GC motifs and two basal level enhancers (BLEs). Besides the TATA factor and the accessory transcription factors that bind over the TATA motif and adjacent sequences, basal transcription requires Sp1 binding to the GC motif and AP1 and AP2 binding to the BLEs. Metal induction requires, in addition, the metal responsive elements (MREs) and a protein whose binding to the MRE is metal dependent. Thus, the amount of metal coordinated MRE binding factor, which mirrors the level of free metal, regulates the promoter's transcriptional activity. The transcriptional response to glucocorticoids, which is independent of the presence of metal inducers, is a consequence of the binding of the liganded glucocorticoid receptor to the promoter's glucocorticoid response element.

g. Transcriptional Regulation of Morphogenesis

The principal theme emerging from our discussion so far is that transcriptional regulation depends on the combinatorial association of a variable number of distinctive transcription proteins with an array of cognate sequence motifs in promoters and enhancers. Ostensibly, the efficiency of transcription initiation is regulated by interactions between the DNA bound proteins (and possibly by proteins associated with them) and one or more parts of the basic transcription machinery, although the molecular details of the regulation are unknown. According to this model, the extent of a gene's expression will depend on the availability, concentration, and activity of one or more transcription factors involved in its transcription. This mechanistic logic probably accounts in large part for the wide variety of differentiated cell and tissue phenotypes and their ability to respond to environmental changes. This same logic also appears to be used to direct embryonic development along an orderly path in space and time. Thus, the generation of complex spatial and temporal patterns of expression during embryogenesis involves the coordinated interplay of both ubiquitously expressed and spatially restricted specific transcription factors.

An extended discussion of the involvement of transcriptional regulators as the determinants of embryonic development is beyond the scope of this section. That will be dealt with in Volume II. For now, it is sufficient to point out how transcription factors can create a hierarchical cascade of gene expression that can have both temporal and spatial dimensions, depending on when and where the initiator of the cascade is generated.

The most highly developed system for analyzing the molecular events governing the early stages of embryo development is *Drosophila*. Such insects are metameric (i.e., composed of serially repeating body segments, each of which differentiates into particular structures and patterns according to their position in the animal). The generation and subsequent diversification of such segments during embryogenesis depend first on

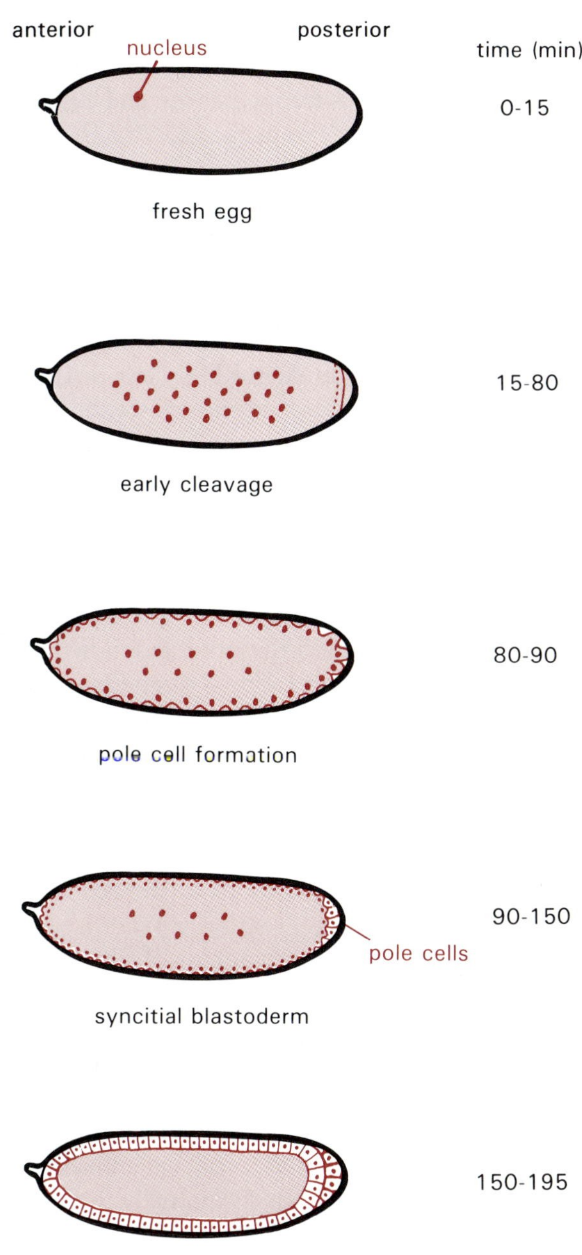

Figure 8.56
Morphological changes during the development of the *Drosophila* egg to the cellular blastoderm. The freshly laid egg contains a single nucleus, which divides rapidly during the first hour after hatching to produce a large number of evenly distributed nuclei (syncitial nuclei). At about 90 minutes, many of the nuclei migrate to the periphery of the egg, and pole cell differentiation begins at the posterior end of the egg. Nuclear migration to the periphery continues for about another hour, at which time cellularization ensues to produce the cellular blastoderm.

subdividing the embryo into repeating units and then specifying their differentiation pathways. Embryogenesis begins with rapid nuclear divisions, followed by migration of the nuclei to the egg's perimeter and then the formation of a membrane around each nucleus. The resulting structure is the cellular blastoderm (Figure 8.56). However, the framework for the subsequent elaboration of the body plan and the fate of each cell are already established even before cellularization of the blastoderm has occurred. This contrasts with mammalian embryogenesis, in which the body architecture and differentiated structures are not yet established in the multicellular blastocyst.

The principal determinants for establishing the *Drosophila* embryo's anterior-posterior pattern are the products of the *bicoid* (*bcd*) and *nanos*

(*nos*) genes. Transcription of the *bcd* gene occurs in special ovarian cells that cluster around one end of the developing oocyte. The bcd mRNA passes into the oocyte, where it is trapped at the end destined to become the anterior pole of the embryo. Products made by the *nos* gene group are localized at the posterior pole of the egg. Consequently, by the time the egg is laid, it is both morphologically and molecularly polarized (Figure 8.57). As translation of the bcd mRNA proceeds, the protein diffuses away from the anterior pole and becomes gradually distributed over about half the egg's length. Concomitantly, proteins directed by nos mRNA move anteriorly. Thus, even before cellularization has occurred, two different products directed by maternal mRNAs are distributed inhomogeneously along the embryo's future anterior-posterior axis.

One consequence of the overlapping gradients of the bcd and nos products is the regional activation of a group of genes referred to as the **gap genes**, which are responsible for generating the embryo's initial seven segments. The three principal members of the gap gene group, *hunchback* (*hb*), *kruppel* (*Kr*), and *knirps* (*kni*), are also expressed in a region-specific fashion. Expression of *hb* occurs in only two regions: one is localized to the posterior quarter of the egg, and the other extends from the anterior pole to about the middle of the egg. The *Kr* gene is initially expressed in a single broad region in the middle of the embryo, and the kni product appears both anteriorly and posteriorly of the Kr band. Studies of mutants that are unable to make the bcd or nos products indicate that the expression of these three gap genes is regulated by the products of the *bcd* and *nos* genes and that the geographic localization of the gap gene products is determined by the relative concentrations of the bcd and nos products along the anterior-posterior axis.

The bcd protein acts as a positive regulator of *hb* expression, causing hb to accumulate at the anterior end of the embryo. The *nos* gene product appears to inhibit the function of the hb protein. Moreover, the *bcd* and *nos* gene products block *Kr* gene expression in the anterior and posterior regions of the embryo, respectively, so that Kr mRNA accumulates in the central region (Figure 8.57). With time, the boundaries of the gap gene transcripts narrow because their products influence each other's transcriptions: thus, the hb and Kr products mutually repress one another's transcription, and the kni product represses *Kr* expression. Therefore, the establishment of stable regions of gap gene expression occurs in two stages. At first, gap gene expression is differentially regulated by the distribution of maternal determinants. Later it is regulated by mutual repression.

The spatial localization of the gap gene products results in position-specific regulation of the next set of genes in the hierarchy, the **pair rule** class. Their transient expression is localized to seven or eight stripes. These periodic patterns of expression lead to localized expression of the **segment polarity** genes, several of which (*fushi terazu* [*ftz*], *even skipped* [*eve*], and *engrailed* [*eng*]) are involved in further subdivision of the segments into compartments (Figure 8.57). Each gene class is clearly influenced by the action of earlier acting genes and by some members of the same class. Thus, the stages during which *Drosophila*'s body plan is established during embryogenesis are created by a regulatory hierarchy in which the products of maternal genes influence the expression of the

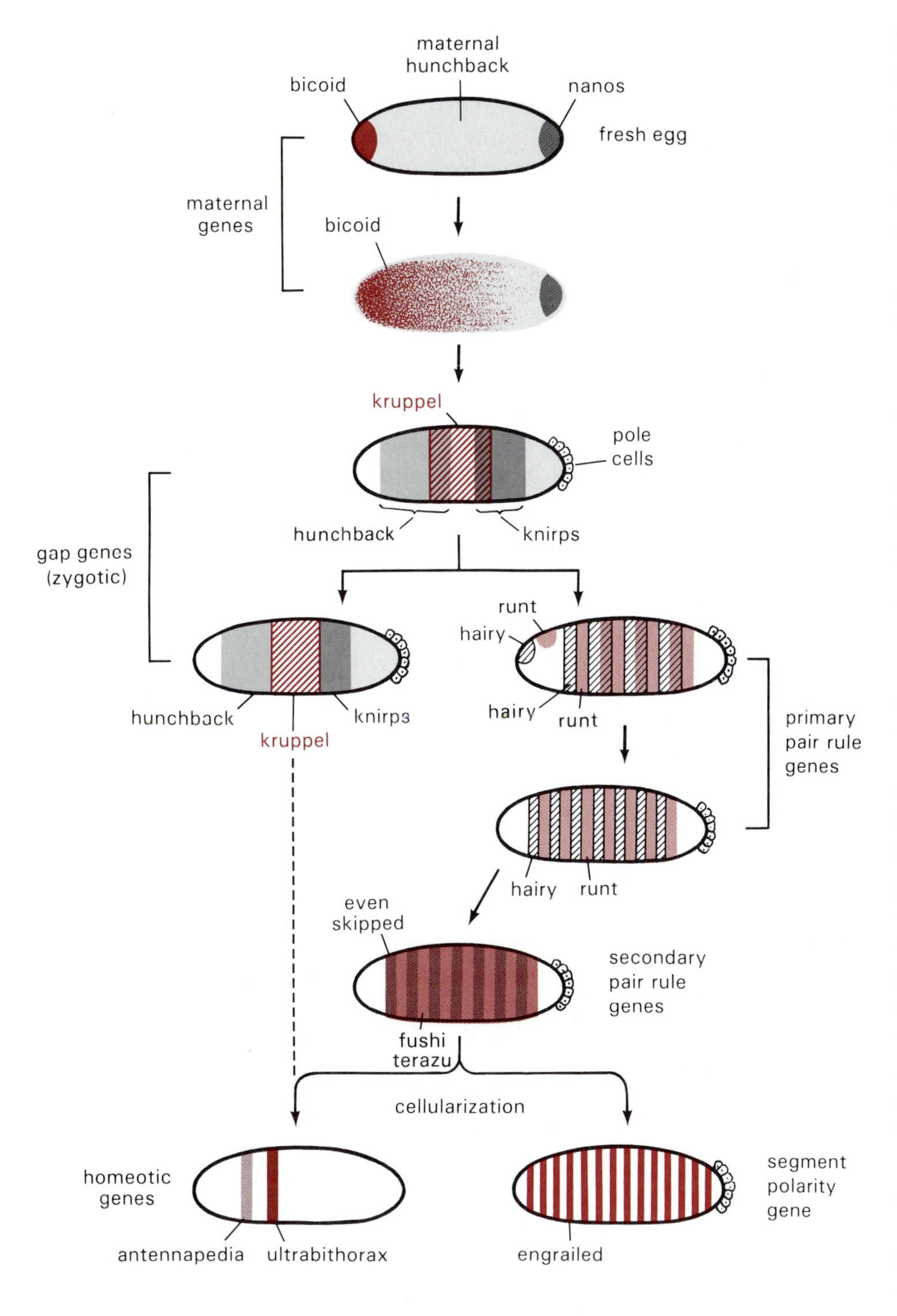

Figure 8.57
Patterns of gene expression during *Drosophila* early development. The diagram summarizes the temporal and spatial expression of some of the genes involved in establishing the anterior-posterior organization of the *Drosophila* embryo in the three hours following fertilization. Cellularization of the embryo occurs as *antennapedia* and *ultrabithorax* gene products are expressed. Expression of the gap genes *hunchback, kruppel,* and *knirps* is influenced by maternal gene products in the egg (e.g., *bicoid* and *nanos*). The pattern of gap gene expression reflects the decreasing concentration gradient of bicoid mRNA and protein from the anterior to posterior regions of the egg and the influence of the gap gene products on each other's expression. By the time defined patterns of hunchback, kruppel, and knirps products are established, several of the pair rule genes *hairy* and *runt* are expressed, initially in poorly defined stripes, but subsequently in a well-ordered alternating array of seven stripes each. Slightly later, additional pair rule genes, *evenskipped* and *fushi terazu,* appear in seven alternating stripes each. This collection of early gene products induces the appearance of a characteristic pattern of *engrailed* gene expression and the appearance of *antennapedia* and *ultrabithorax* gene products, predominantly in two bands in the anterior half of the egg.

first expressed zygotic genes, whose products influence each other's expression, as well as sets of genes expressed subsequently in an interdependent fashion. A consequence of this network of interactions is that by the time cellularization of the embryo is complete, with all the cells appearing morphologically similar, the fate of cells in different parts of the embryo is set because they express different combinations of the pair-rule and segment-polarity genes. The various combinations of these gene products in each segment dictate the expression of the **homeotic** genes, *ultrabithorax* (*ubx*) and *antennapedia* (*antp*), which regulate later stages in the fly's development and tissue differentiation. Homeotic genes are those in which mutations transform one part of the adult body into another quite different structure. For example, the homeotic mutation *Antenna pedia* transforms the antenna on the fly's head into a normal leg structure. Thus, one of the products of the *Antp* gene probably acts in the determination of the developmental fate of certain cells.

How is this hierarchical pattern of spatially precise gene expression created and maintained, especially at a stage when the embryo consists only of free nuclei? Although most of the details of how this happens still need to be filled in, transcriptional regulation clearly accounts for part of the differential gene expression during the early stages of embryonic development. Cloning and sequencing of the genes that are transcribed at various levels in the hierarchy indicate that some are sequence-specific DNA binding proteins that either activate or repress the transcription of genes within their class or genes below them in the hierarchy.

As mentioned earlier, the zygotically induced *hb* gene transcripts are concentrated in the most anterior portion of the egg, the region with the highest levels of the bcd protein. Because most *bcd* mutants fail to express the *hb* gene, it was surmised that the bcd product is needed to activate *hb* transcription. The availability of the cloned *bcd* gene and its protein product, as well as the cloned *hb* gene and its transcriptional regulatory region, made it possible to test that hypothesis. The bcd protein binds with varying affinity to multiple regions of the *hb* promoter near the *hb* transcription start site (Figure 8.58). The consensus sequence for three strong binding sites, which are located 100 bp apart between base pairs −293 and −55, conforms to 5′-TCTAATCC-3′. The involvement of the bcd protein in regulating *hb* transcription is supported by *in vivo* experiments. Thus, the injection of a chimeric gene containing the *hb* promoter fused to the coding sequence for chloramphenicol transacetylase into early *Drosophila* embryos leads to the gene's expression in wild type embryos but not in *bcd* mutant females. Successive deletions of the *hb* promoter sequences identify the stretch between base pairs −300 and +1 as being sufficient for *bcd* dependent and region-specific expression. This region contains the three closely spaced copies of the strong *bcd* binding sites. Two additional but weaker bcd binding sites occur farther upstream in the region about base pair −2300. These sites promote the formation of a larger hb mRNA.

When chimeric genes containing the strong bcd binding sequences fused to the coding sequence for β-galactosidase are introduced into embryos by P element transformation (Section 10.2b), β-galactosidase is expressed in the anterior portion extending to about the middle of the embryo. When the weak bcd binding sequences are used to activate the marker gene's promoter, expression is limited to the 25 percent most anterior region of the embryo. This indicates that the *hb* promoter can sense the bcd gradient in two ways to yield two kinds of mRNA.

The bcd protein's DNA binding domain is encoded in a stretch of about 60 amino acids referred to as the **homeo domain** because it was first recognized in *Drosophila* homeotic genes. This DNA binding domain (Section 8.6a) occurs in proteins encoded by a large variety of regulatory genes in invertebrate as well as vertebrate genomes. They almost invari-

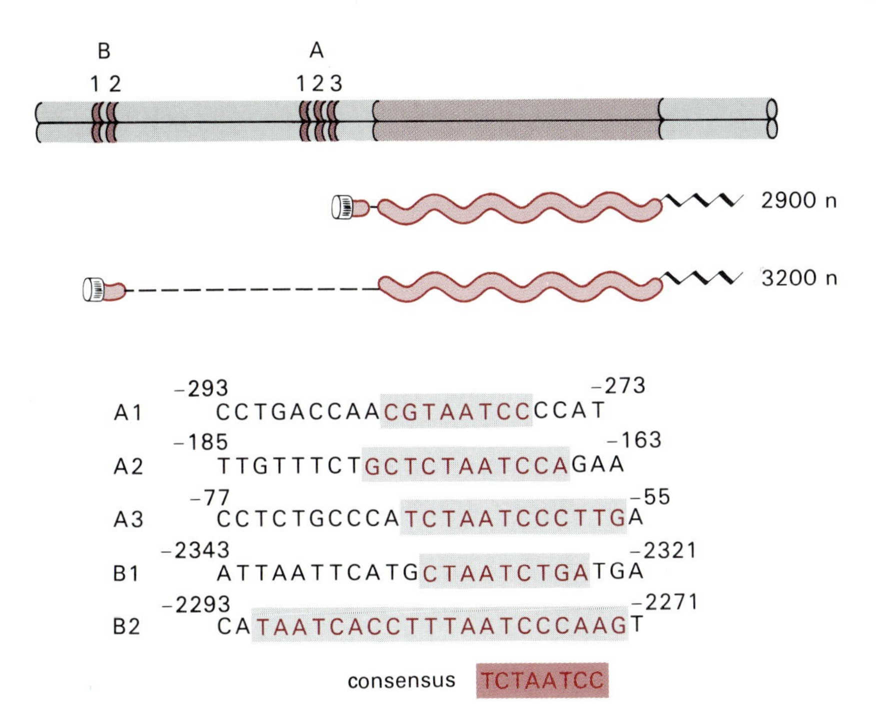

Figure 8.58
Regulatory elements in the *Drosophila hunchback (hb)* gene promoter. The first 300 bp upstream of the *hb* transcription unit contain three closely spaced strong bcd protein binding sites (A1, A2, and A3); the sequences responsible for binding are shown in color. These are essential for transcription of the 2900 nucleotide long mRNA encoding the hb protein. Two weak bcd binding sites occur about 2300 nucleotides upstream of the transcription start site, and their sequences (B1 and B2) are shown highlighted in color. B1 and B2 promote the formation of the 3200 nucleotide long mRNA (after splicing).

ably specify transcriptional regulators. Many of the *Drosophila* genes mentioned so far in connection with the regulatory hierarchy encode homeo domains. Some also contain the zinc finger motif and therefore may be able to bind to different kinds of DNA sequences.

h. *Transcriptional Regulation of U RNA Genes*

The uracil-rich small nuclear RNAs (U snRNAs) of all species are encoded in a family of highly expressed multicopy genes (Section 9.2e). At least 200,000 transcripts of each major type of U RNA (U1, U2, U4, U5, and U6) are present in the vertebrate nucleus. These RNAs are components of snRNPs that participate in the processing of pre-mRNAs (Section 8.5d). Other U RNAs, U3 and U7, are part of less abundant snRNPs that may function in the processing of pre-rRNA and 3′ ends of histone mRNAs, respectively (Section 8.3c). RNA polymerase II has been implicated in the transcription of U1, U2, U4, and U5 RNA genes because of the transcriptional sensitivity to α-amanitin (Section 8.1b). Furthermore, U RNA transcripts initially have the typical 7-methyl guanosine cap, which is then converted to the 2,2,7-trimethyl guanosine cap characteristic of U RNAs. By contrast, the U6 RNA gene is transcribed by RNA polymerase III, and the mature RNA lacks a cap structure. Nevertheless, certain features of the U6 RNA gene's promoter are shared with typical class II gene promoters (Section 8.4b). For example, the region upstream of the U6 RNA start site contains GC boxes, an octamer, and TATA sequence, all regulatory motifs characteristic of RNA polymerase II promoters. The ambivalence of the transcriptional signals is further illustrated by the fact that the *Xenopus tropicalis* U6 promoter is transcribed in *Xenopus* oocytes by either RNA polymerase II or III. And with only a few alterations in the 50 bp upstream of the U2 RNA start site, the promoter originally

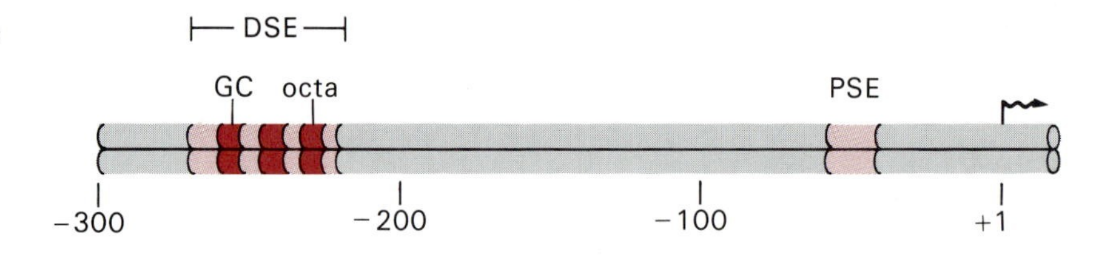

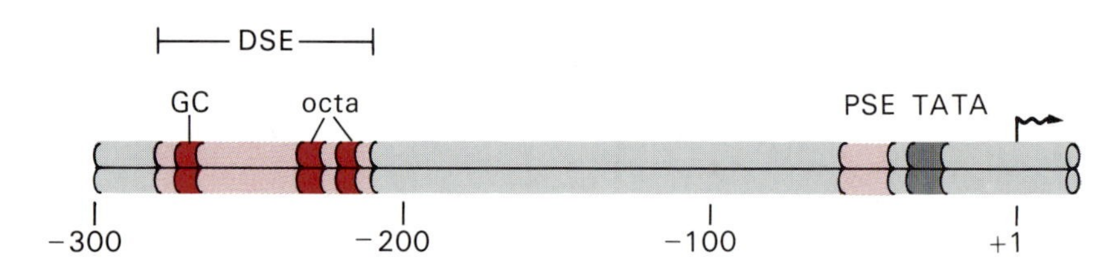

Figure 8.59
Transcription regulating elements of U RNA promoters. (a) RNA polymerase II is responsible for transcription from the U1, U2, U4, and U5 RNA promoters. These promoters lack a TATA sequence but require a proximal sequence element (PSE) and a distal sequence element (DSE) consisting of the GC-rich Sp1 binding site and the octamer motif (octa) characteristic of immunoglobulin gene promoters. (b) Transcription from the U6 RNA promoter is mediated by RNA polymerase III. This promoter contains a characteristic TATA motif and a required proximal sequence element (PSE), as well as a distal sequence element (DSE) containing clustered GC and octamer motifs.

specific for RNA polymerase II is changed to one specific for RNA polymerase III. Clearly, the distinction between class II and III promoters is becoming blurred.

The regulatory elements of the U RNA promoters normally transcribed by RNA polymerase II are all organized similarly (Figure 8.59a). Each contains at least two elements that contribute to promoter activity. One, the proximal sequence element (PSE), constitutes the basal promoter and is located between base pairs -70 and -40. The other, a distal regulatory element (DSE), is contained in the region between base pairs -300 and -200. Neither region alone is able to promote transcription of U RNAs. The PSE is essential for transcription and for specifying the RNA start site. The DSE functions in either orientation and in a relatively distance independent fashion. It therefore has the properties of a conventional enhancer.

Although the PSE element identifies the transcriptional start site for U RNA synthesis, it lacks a recognizable TATA motif. Furthermore, with the exception of the U6 gene promoter (Figure 8.59b), all the PSE sequences are similar and unique to the major U RNA gene promoters. So far, however, the sequence motif and the binding proteins relevant to the PSE's function in transcription are unknown. By contrast, the sequence motifs and binding proteins responsible for the DSE's enhancer function are characterized. Two types of sequences have been found: the ubiquitous GC box (5′-GGGCGG-3′) and the octamer motif (5′-ATGCAAAT-3′). Mutations in either sequence inactivate the enhancer, so the two must be functionally interdependent. The GC box binds Sp1, and the octamer sequence binds the ubiquitous octamer binding protein OTF-1. DNase I footprints and mutational studies with the U RNA promoter's GC box and octamer motif identify the same critical residues that have been implicated in mRNA gene promoters. Some DSEs also bind AP1 and AP2.

The regulatory sequences required for transcription of the U6 RNA gene are entirely upstream of the coding sequence, even though the gene is transcribed by RNA polymerase III (Section 8.4b). Equally puzzling is the fact that the *cis* sequences required for the RNA polymerase III transcription machinery are the same as those used by the transcription factors that function with RNA polymerase II (Figure 8.59b). Thus, the U6 RNA promoter resembles the U1, U2, U4, and U5 RNA promoters in that it contains a PSE close to the transcription start site and a DSE element, comprised of a GC-hexanucleotide and two octamer sequences between base pairs −300 and −200. However, the U6 RNA promoter also contains a TATA sequence near the transcription start site, and the PSE differs from the one in the U1, U2, U4, and U5 RNA promoters. It is not clear how RNA polymerase III utilizes these *cis* regulatory sequences to promote transcription of the U6 RNA gene.

A curious feature of U RNA promoters is that when they are fused to an mRNA gene's coding region, the transcripts are not polyadenylated at their 3′ ends (Section 8.3c). Similarly, the 3′ ends of U RNAs transcribed from mRNA gene promoters are not processed correctly. Thus, there appears to be a need for coordination between the mechanism of transcription initiation and the events that mature the 3′ ends of mRNAs and U RNAs. Another paradox is that the DSEs from U RNA promoters do not enhance transcription from either the TK gene's basal promoter or the β-globin gene promoter, although enhancers from heterologous mRNA promoters do both efficiently. This suggests that the transcription factors bound to the DSE are unable to interact with or activate TATA directed complexes. Perhaps the transcription complex assembled at the PSE motif is different from and not interchangeable with the one at the TATA sequence. This distinct type of complex at the start of transcription may influence the nature of the events at the 3′ end of the transcription unit.

8.4 Structure and Expression of Class III Genes

a. General Features

RNAs encoded by class III genes The RNAs transcribed from class III genes do not encode proteins. Instead, their products are small RNAs that are involved in protein synthesis (tRNAs and 5S rRNA), in intracellular protein transport (7SL RNA), and in posttranscriptional processing (U6 RNA). Class III genes also encode several other RNAs whose physiological roles are unknown (7SK RNA and Alu RNAs). Adenovirus genomes encode two class III genes, VA1 and VA2, whose RNA products (157 and 140 nucleotides long, respectively) divert an infected cell's translational machinery for efficient synthesis of viral structural proteins. Two small RNAs, EBERI and EBERII (166 and 172 nucleotides in length, respectively) are also expressed from class III genes encoded in the Epstein-Barr Herpes virus genome.

Class III genes fall into two groups. In one, which encodes the 5S rRNA, the 7SL, 7SK, U6, and the VA RNAs, the primary transcript has the same length as the mature functional RNA; therefore, processing of the transcript is not required. The 5′ triphosphate end group that results from transcription initiation is the 5′ end, and the last residue to be transcribed is the 3′ end of the mature RNA. The second group of class III genes encodes tRNAs whose mature forms are produced by removing nucleotides from both the 5′ and 3′ ends of the primary transcripts by specific endonucleases. The 3′ terminal –CCA sequence, which is required for the function of all tRNAs, is not encoded in the genes and is added by a special enzyme. Because some tRNA genes contain an intron, splicing is required to produce their mature tRNA products (Section 8.5).

Class III genes' transcription signals The signals governing transcription initiation and termination of the various kinds of class III genes are highly conserved among eukaryotic organisms as diverse as yeast, *Drosophila, Xenopus,* and mammals. Consequently, the tRNA, 5S rRNA, and other small cellular and viral RNA genes from various sources are correctly transcribed by heterologous transcription systems.

A distinctive feature of the sequences that control transcription initiation of many class III genes is their location within the coding region. These **internal control regions** (ICRs) vary somewhat from one class III gene to another, but overall, they have a striking resemblance to one another. With some of these class III genes, additional regulatory sequences are located within 30 bp upstream of the transcription start site. Other class III genes (e.g., the mammalian U6 RNA and the 7SK RNA) lack an internal control region and rely instead on regulatory sequences located wholly 5′ to the transcription start site. The nucleotide sequences specifying transcription termination at the 3′ end of class III genes are all similar, minor variations causing different efficiencies of termination.

RNA polymerase III transcription machinery RNA polymerase III, which is localized in the nucleoplasm of eukaryotic cells, has been purified extensively from yeast, insects, amphibians, plants, and several mammalian sources; near homogeneous forms of the enzyme are available from yeast, *Xenopus,* and mammals. Each of the enzymes is large, about 650 kDaltons, and contains 10 to 15 subunits, depending on the enzyme's source and the isolation procedure. Two of the subunits have molecular weights in the range of 140 to 170 kDaltons, and their amino acid sequences resemble those of the corresponding large subunits of RNA polymerases I and II as well as those of the β and β' subunits of *E. coli* RNA polymerase. The remaining subunits appear to be present in equimolar quantities and range in molecular weights from 10 to 90 kDaltons. The approximately 90 kDalton subunit is likely to be unique to RNA polymerase III, but three of the smaller subunits are closely related to comparable size subunits in purified RNA polymerase I and II as judged by amino acid sequence similarities and immunological cross reactions.

Purified RNA polymerase III preparations require several additional proteins for accurate and efficient transcription initiation on class III genes. The particular assortment of required transcription factors varies with different class III genes, specifically with the location and kind of sequence elements that govern the gene's transcriptional activity. In most cases, the required transcription factors are unique to class III gene expression. But some class III genes that depend on regulatory elements in the 5′ upstream sequence (e.g., tRNAs and the small nuclear and cytoplasmic RNAs) require additional transcription factors that may be related to pro-

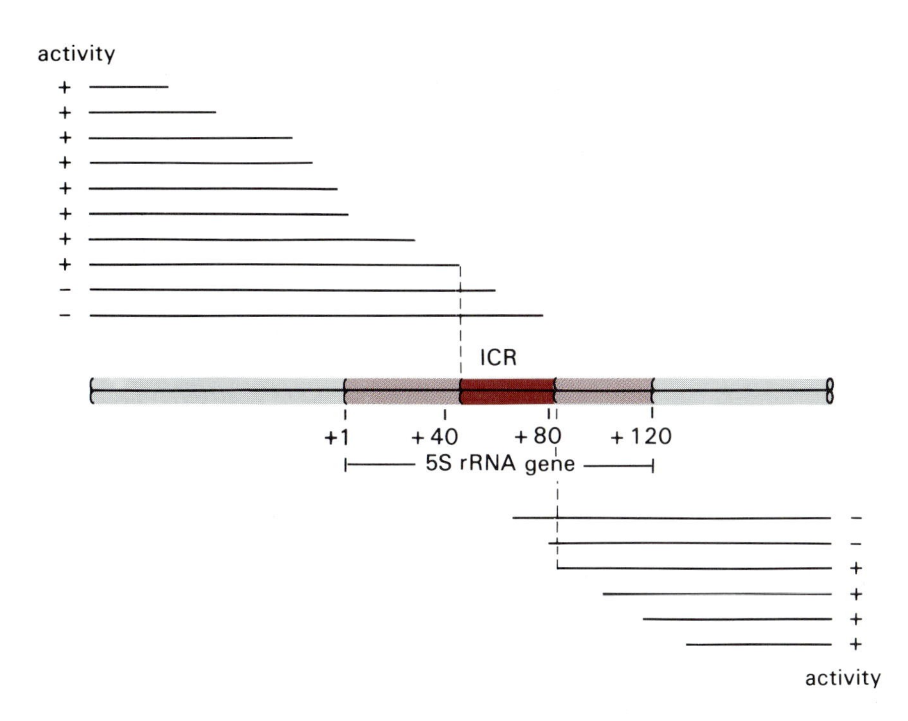

Figure 8.60
The effect of deletions around and within a 5S rRNA gene on transcription initiation. The positions marked base pairs +1 and +120 are the normal initiation and termination sites, respectively. Deletions, which are shown as lines starting from either the 5′ or 3′ side of the cloned gene, were prepared by *in vitro* mutagenesis and recloning. Transcriptional activity of each construct was determined in *Xenopus* oocyte nuclei or cell-free extracts.

teins previously thought to be specific to RNA polymerase II transcription systems (Section 8.3h). Transcription termination of 5S rRNA genes appears to be mediated by RNA polymerase III alone, although associated transcription factors are probably needed for other class III genes.

b. *Sequences Regulating Transcription Initiation*

Xenopus *5S rRNA genes' internal control regions* Cloned *Xenopus* DNA containing a 5S rRNA coding region and its 5′ and 3′ flanking regions is accurately transcribed after injection into *Xenopus* oocyte nuclei or incubation with unfractionated oocyte nuclear extracts. The product is predominantly the 120 nucleotide long 5S rRNA. This key finding opened the way for a reverse genetic analysis of the structural requirements relevant for accurate and efficient transcription.

When 5S rRNA genes with varying size deletions at the 5′ end were tested for transcriptional activity, it became evident that the sequence flanking the 5′ end of the transcription unit and more than the first third of the coding region are dispensable for proper transcription initiation (Figure 8.60). Transcription of such truncated genes yields RNA products whose lengths are close to 120 nucleotides; because of the deletions, transcription actually begins in the plasmid sequence that is fused to the 5′ end of the residual coding sequence. Similarly, accurate and essentially normal levels of transcription initiation occur when the 3′ flanking sequence and the 3′ terminal third of the coding sequence are removed from the gene. However, when the deletions extend more than about 50 bp into the 5′ end or about 30 bp into the 3′ end of the coding region, transcription is abolished. This type of experiment, which provided the first demonstration of an intragenic transcription signal, located the boundaries of the *Xenopus* 5S rRNA gene's internal control region (ICR) be-

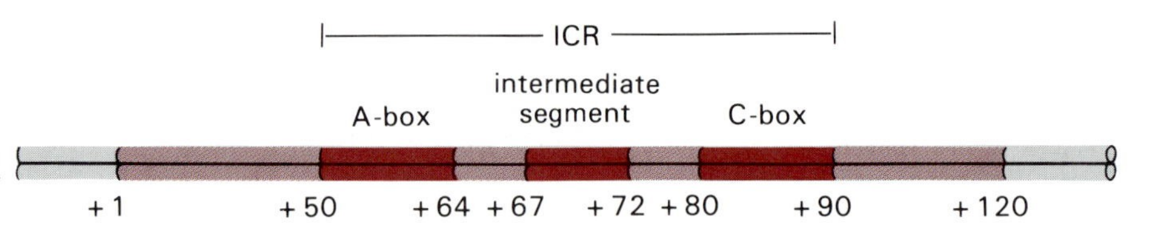

Figure 8.61
Fine structure of the *X. laevis* 5S rRNA gene's internal control region (ICR). The internal control region is composed of an A-box, an intermediate segment, and a C-box. Their positions in the coding sequence are indicated by the most intense color, and the 5S rRNA coding sequence is shown in lighter color.

tween about base pairs +50 and +90. Indeed, a cloned 5S rDNA segment spanning base pairs +41 and +87 is sufficient for transcription by RNA polymerase III *in vitro*, initiation occurring about 50 bp upstream of the junction between the vector and the cloned segment.

A more refined definition of this gene's ICR was made by analyzing the effects of small internal deletions, insertions in place of deletions, or specific base pair changes on transcription *in vitro* and *in vivo*. These studies revealed that the ICR consists of three functionally distinct sequence domains (Figure 8.61). One spans base pair +50 to +64 and resembles in both sequence and function the so-called A-box, now known to be characteristic of the ICRs in all class III genes. A second domain encompasses base pair +80 to +90 and is unique to the ICRs of 5S rRNA genes (the C-box). The segment between base pairs +67 and +72 also contributes to the ICR function. Although sequence changes in this region cause only small effects on transcription, variations in the spacing between the A-box and C-box segments have more drastic consequences. Thus, sequences within the *Xenopus* 5S rRNA genes' coding region are necessary and sufficient for regulating transcription initiation. How these signals fulfill that function is considered in Section 8.4c.

Ubiquity of class III genes' internal control regions Mutational studies of the type mentioned previously have identified ICRs in the 5S rRNA and tRNA genes of all eukaryotes that have been examined. For example, linker insertions in the *Drosophila* 5S rRNA gene between base pairs +46 and +65 and between base pairs +76 and +98 prevent its transcription; by contrast, insertions between base pairs +3 and +20 reduce transcription efficiency only slightly. ICRs with nearly the same boundaries have been identified in the 5S rRNA genes of *Bombyx*, *Neurospora*, and *Saccharomyces* species.

Similar sorts of analyses with modified tRNA genes, using *in vitro* transcription systems or the *Xenopus* oocyte assay, have established that their transcription also requires an ICR. The nucleotide sequences at the 5′ and 3′ ends of the coding sequence are not essential for transcription initiation. Deletions and linker insertions have identified two essential regions whose positions vary only slightly from one tRNA gene to another, regardless of the species (Figure 8.62). One, the A-box, is located about base pairs +10 to +20, and the other, the B-box, lies between base pairs +50 and +65. The nucleotide sequences of the A- and B-boxes are not invariant, their consensus sequences being 5′-TGGCNNAGTGG-3′ and 5′-GGTCGANNC-3′ for A- and B-boxes, respectively. The nucleotide sequence between the A- and B-boxes is not critical, but the distance between them is important. If the distance between the A- and B-boxes is shortened to less than 25 base pairs or increased to more than about

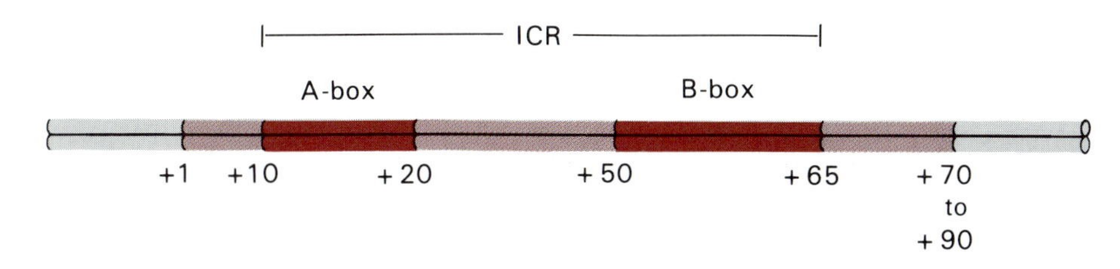

Figure 8.62
Location of the internal control region (ICR) in a tRNA gene. The positions of characteristic A-box and B-box motifs are shown here in a $tRNA^{Met}$ gene. The colors used to designate the two boxes and the tRNA coding sequence are those used in Figure 8.61. The precise positions of the A- and B-boxes differ slightly in different tRNAs. The transcription start site is designated +1. A similar bipartite arrangement of A- and B-boxes occurs in the VA RNAs, EBER RNAs, and 7SL RNA. The consensus sequence for A-boxes is 5′-TGGCNNAGTGG-3′ and for B-boxes, 5′-GGTTCGANNC-3′, where N represents any nucleotide.

60 base pairs, transcriptional activity is reduced. Transcription is prevented if the distance exceeds 100 base pairs or if the A- and B-boxes are fused. The same kind of bipartite, A- and B-box sequence arrangement occurs in the coding sequences of 7SL RNA genes and the four small RNA genes encoded in the adenovirus and Epstein-Barr virus genomes.

Besides their essential role in transcription, the sequences in the A- and B-boxes contribute significantly to tRNA secondary structure; the A-box forms part of the D-stem and loop, and the B-box contains the TΨUG stem and loop structure (Figure 8.63). The dual requirement of these sequences for transcription of tRNA genes and for proper folding into tRNAs that are functional in protein synthesis has interesting ramifications for the evolution of these genes. Indeed, if a C·G base pair at position 56 in the yeast $tRNA^{Tyr}$ gene is changed to a G·C base pair, a change that replaces a C with a G in the TΨUG loop, the gene fails to be transcribed.

Interchangeability of A-box regions of different internal control regions Do the similar elements of the ICRs imply similar functions in different class III genes? The answer is yes with respect to the A-box sequences in the ICRs of 5S rRNA and tRNA genes. Thus, recombinants constructed by replacing the first 20 base pairs of the 5S rRNA gene's ICR with the A-box sequence from a tRNA gene are efficiently transcribed (Figure 8.64). The reciprocal replacement of a tRNA A-box with the first 20 base pairs of the 5S rRNA gene's ICR is also active. These findings suggest that although the ICRs in different class III genes are distinctive, they are most likely variations of a fundamental regulatory motif.

Upstream sequences' effect on transcription regulation of some class III genes Following the discovery that the ICR was sufficient for promoting transcription initiation of *Xenopus* oocyte 5S rRNA genes, it was presumed that all class III genes rely on this strategy. That proved to be a premature conclusion. Other 5S rRNA genes (e.g., from *Xenopus* [somatic], *Bombyx*,

Figure 8.63
The internal control region (ICR) in $tRNA^{Leu}$ of *Xenopus laevis*. The 100 bp $tRNA^{Leu}$ gene sequence is shown. The sites for transcription initiation and termination are indicated above the DNA. The 83 bp mature tRNA sequence and the various domains typical of tRNAs are shown below the DNA (see also Figure 3.29). The locations of the A- and B-boxes are indicated by color. After G. Galli et al., *Nature* 294 (1981), p. 626.

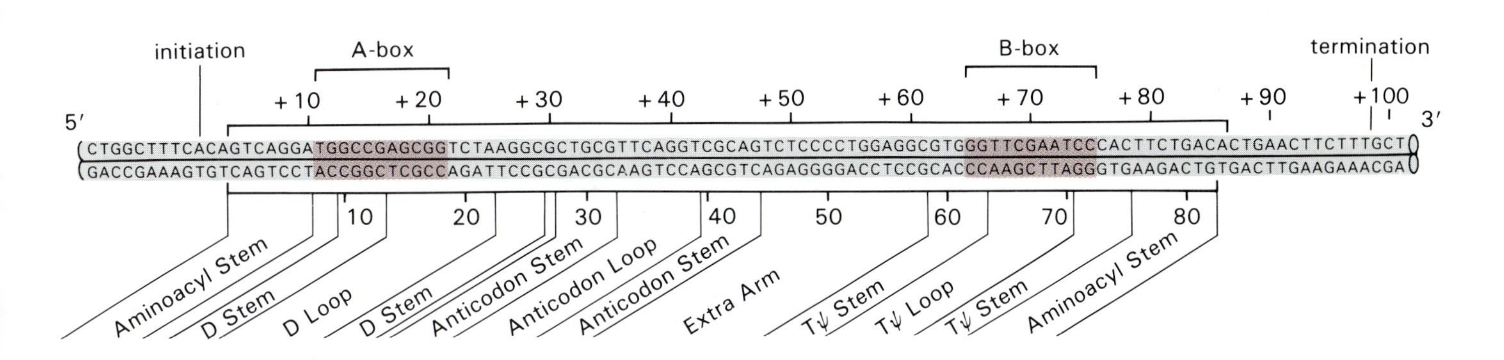

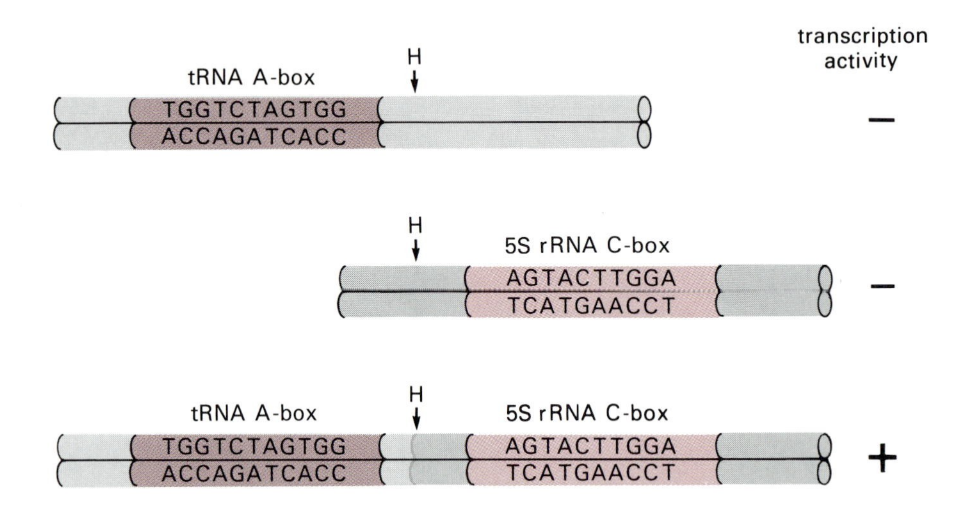

Figure 8.64
Exchangeability of 5S rRNA and tRNA gene transcriptional control sequences. Recombinant DNAs were prepared in plasmid vectors and tested for transcription after injection into *X. laevis* oocyte nuclei. The $tRNA^{Pro}$ gene segment was cloned from the nematode *C. elegans*. The 5S rRNA gene segment was from *X. borealis*. The vector sequences and the *Hind*III sequence at which the two regulatory segments were joined are indicated in two shades of grey and by an H, respectively. With the active hybrid control region, transcription initiation occurs about 50 bp upstream of the start of the $tRNA^{Pro}$ A-box. Because transcription of the tRNA gene ordinarily begins about 10 bp upstream of the A-box, the 5S rRNA gene's C-box must also contribute to the specification of the initiation site.

Drosophila, Neurospora, Saccharomyces, and mammalian species) require sequences located 5′ to the transcription initiation site. Thus, deleting or replacing the sequences between base pair −34 and −11 in a *Bombyx mori* 5S rRNA gene and between base pair −39 and −26 in a *Drosophila* 5S rRNA gene results in loss of their ability to be transcribed. With the exception of *Xenopus*, most tRNA genes, from whatever species examined, also require 5′ upstream sequences for efficient and accurate transcription initiation, and these sequences appear to be characteristic of the species rather than of the gene. Thus, many of the *Bombyx* tRNA genes share homology in the 5′ upstream region with each other and with 5S rRNA genes. Other class III genes that rely on both 5′ upstream sequences and an ICR motif for accurate and efficient transcription are the viral VA and EBER RNAs and the 7SL RNA gene.

Transcription control of some class III genes by upstream sequences alone Perhaps the most surprising turnabout from the initial conclusion that ICRs are the sole determinants for regulating RNA polymerase III transcription is the finding that transcription of 7SK, U6 RNA, and several small RNA genes (EBER) in the Epstein-Barr virus genome utilize only sequences upstream of their coding regions for transcriptional regulation. The coding region of the U6 RNA gene contains an A-box characteristic of tRNA genes and the 5′ region of the 5S rRNA gene's ICR, but mutations within that segment or its complete removal do not affect transcription promoted by the 5′ upstream sequence. The 7SK RNA gene lacks the A-box analogue entirely. Not surprisingly, therefore, a fusion of a 243 bp segment of the 7SK RNA gene's upstream sequence (base pair −245 to −3) to a completely unrelated sequence is transcribed by RNA polymerase III *in vitro* and *in vivo* as efficiently as the complete 7SK RNA gene; analogous experiments with the upstream sequences taken from U6 RNA genes produce the same result. By contrast, optimal transcription of the EBER RNAs requires both the A- and B-boxes of the ICR and several 5′ sequence motifs characteristic of RNA polymerase II promoter-enhancers.

Transcriptional activity with the 7SK RNA upstream element ceases when sequences between base pair −37 and −3 are removed (Figure

```
CTGCAGTATTTAGCATGCCCCACCCATCTGCAAGGCATTCTGGATAGTGTCAAAACAGCCGGAAATCAA
|
-243
(100%)

GTCCGTTTATCTCAAACTTTAGCATTTTGGGAATAAATGATATTTGCTATGCTGGTTAAATTAGATTTT
                    |
                  -154
                  (63%)

AGTTAAATTTCCTGCTGAAGCTCTAGTACGATAAGCAACTTGACCTAAGTGTAAAGTTGAGACTTCCTT
                                             |             |       |
                                            -59           -45     -37
                                          (25.3%)        (24%)   (21%)

CAGGTTTATATAGCTTGTGCGCCGCTTGGGTACCTCG
       |             |      |    | |
      -26           -15     -8  -3 +1
     (<1%)         (<1%)  (<1%) (<1%)
```

Figure 8.65
Sequence of the 5′ flanking region of the 7SK RNA gene from base pair −243 to the transcription start site. Deletions in the gene were made by *Bal*31 endonuclease digestion and extend to the indicated position relative to the transcription start site (the single boxed base marked +1). The negative number denoting each mutant indicates the position at which the deletion ends; thus, mutant −243 contains 243 bp 5′ to the start site, and mutant −154 contains only 154 bp 5′ to the start site. The values shown below the deletion end points are the transcription efficiencies relative to the longest 5′ end. After S. Murphy et al., *Cell* 51 (1987), p. 81.

8.65). Also, at least 79 bp of the U6 RNA gene's upstream region are needed to promote transcription of the adjacent sequence *in vitro*. In both instances, sequences farther upstream, between base pair −243 and −37 in the 7SK RNA gene's 5′ flanking sequence and between base pair −315 and −79 in the U6 RNA's 5′ flanking sequence, are needed for wild type levels of transcription *in vivo* (Section 8.3g). Interestingly, the 5′ flanking sequences of the U6 and 7SK RNA genes are interchangeable functionally. However, the 5′ flanking region of the 7SL RNA gene, which uses both internal and 5′ flanking sequences, does not replace the corresponding region of the 7SK gene even though both are transcribed by RNA polymerase III.

Tantalizing features of the U6 and 7SK RNA genes' upstream regions are sequences characteristic of many class II genes (Figure 8.65). Thus, both genes have TATA sequences at base pair −29 to −24, and these are essential for efficient transcription initiation. Moreover, the U6 and 7SK RNA promoters share similar sequence motifs between base pair −40 and −50 with those of the other U RNA promoters. Additionally, within several hundred base pairs upstream of the U6 and 7SK RNA promoters are motifs (e.g., the GC [5′-GGGCGG-3′] and octamer [5′-ATGCAAAT-3′] sequences) that are characteristic of class II gene promoters and enhancers (Section 8.3h). The physiological role of the proximal and distal promoter elements or the necessity for both is not known. Nor is the physiological rationale for the striking similarity between the 7SK and U6 RNA gene upstream promoter sequences understood. Also puzzling is the considerable homology between the regulatory sequence upstream of the U6 RNA gene—a class III gene—and the regulatory sequences of the genes encoding U1 through U5 RNAs—all class II genes (Section 8.3h). The existence of transcription factors capable of interacting with regulatory sequences common to both class II and class III genes may permit coregulation of diverse sets of genes.

Another hint suggesting a physiological connection between RNA polymerases II and III transcription signals is the presence of an RNA polymerase III promoter in the DNA encoding the untranslated leader sequence of a class II gene (the c-*myc* gene). Furthermore, both the RNA polymerase II and III promoter signals cause transcription to begin at the same site. RNA polymerase III transcription is blocked, however, at the end of the first exon while RNA polymerase II transcribes through to the end. The c-*myc* gene's first exon contains an A-box but no B-box, and there are 5′ upstream sequences reminiscent of those needed for transcription of the *Bombyx mori* 5S rRNA and tRNA genes. Whether the cell cycle control of c-*myc* transcription is mediated by both RNA polymerase II and RNA polymerase III transcription signals remains to be established, as does the role of premature termination by the latter.

c. *Transcription Factor Modulation of Transcription Initiation by RNA Polymerase III*

Additional proteins are needed for RNA polymerase III to initiate transcription, and three have been identified so far. Transcription of all class III genes that depend on an ICR requires factors TFIIIB and TFIIIC (now known to be a mixture of TFIIIC-1 and TFIIIC-2). TFIIIA is uniquely required for transcription of all 5S rRNA genes. These transcription factors occur in all eukaryotes, none of them showing a strong species specificity for the source of the template. Thus, human 5S rRNA genes are transcribed as efficiently and accurately as the *Xenopus* genes after injection into *Xenopus* oocyte nuclei; similarly, human cell extracts containing RNA polymerase III and factors transcribe the two 5S rDNAs equally well. The factors, rather than RNA polymerase III itself, interact first with the control regions and form stable preinitiation complexes. Such complexes have been demonstrated by template competition experiments (described in Section 8.2c), by footprinting, nitrocellulose filter binding, and other procedures.

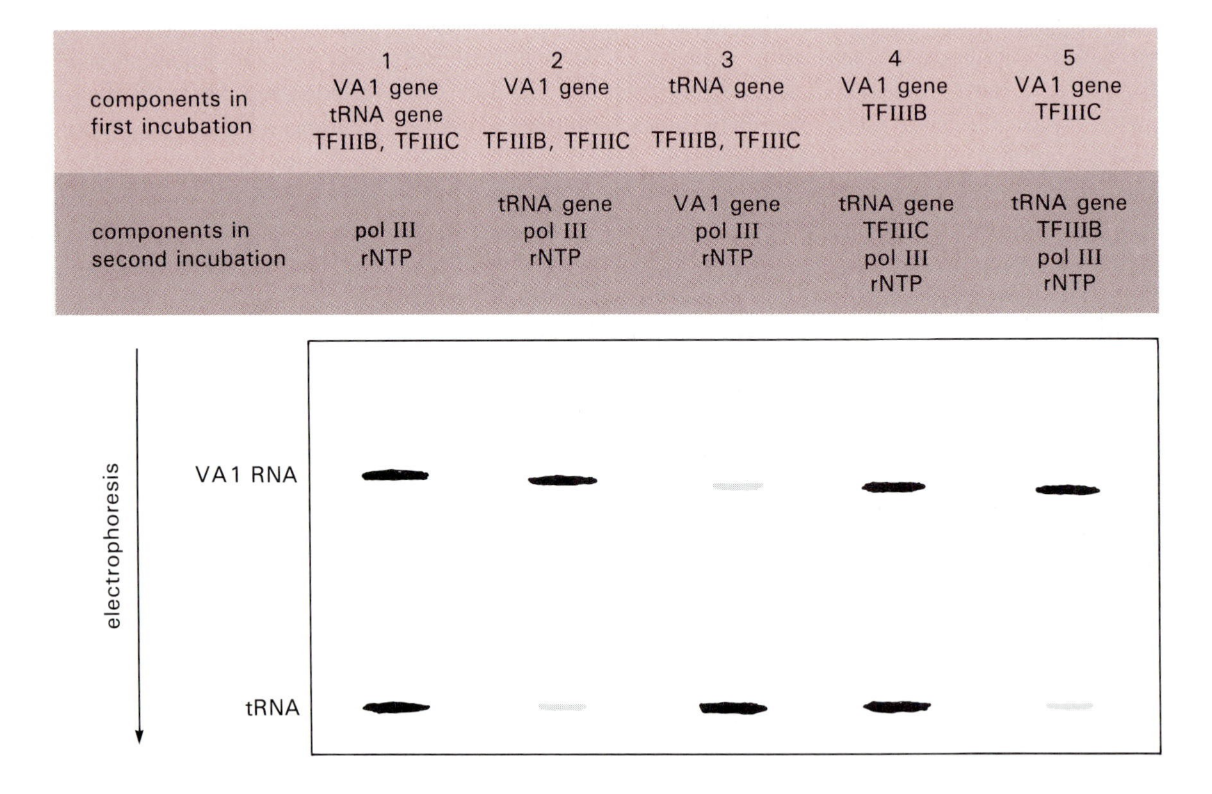

Figure 8.66
Formation of stable complexes between transcription factors and class III gene control regions. Preincubation of a tRNA or VA RNA gene with TFIIIB and TFIIIC leads to their transcription after the addition of ribonucleoside triphosphates (rNTP) and RNA polymerase III (pol III). However, preincubation of either the VA1 RNA gene (lane 2) or tRNA gene alone (lane 3) with TFIIIB and TFIIIC, followed by the addition of the other gene along with the polymerase and substrates, results in transcription of only the preincubated RNA gene. Lanes 4 and 5 show that TFIIIC can form a preinitiation complex with the VA1 RNA gene (or the tRNA gene) in the absence of TFIIIB. After A. B. Lasser et al., *Science* 222 (1983), p. 740.

TFIIIB and TFIIIC interactions with A- and B-boxes Template competition experiments have helped reveal some features of complex formation between TFIIIB and TFIIIC and the A- and B-boxes in class III genes. When a VA1 RNA and tRNA gene are incubated together in a transcription system *in vitro*, both RNA products are produced (Figure 8.66). However, if prior to adding RNA polymerase III, one of the DNAs is incubated with TFIIIB and TFIIIC, then that template is transcribed preferentially when the second gene is added later (compare lanes 1–3 in Figure 8.66). Such preferential transcription indicates that the factors form a tight complex with the first gene and are not available for transcription of the gene added later. Note also that it makes a difference as to which factor is preincubated with the template first (compare lanes 4 and 5 in Figure 8.66). This indicates that TFIIIC alone can form a complex with the VA DNA. Moreover, at least one but possibly both factors seem to remain bound to the template even after multiple rounds of transcription.

Because TFIIIC binding blocks cleavage at a restriction site in the VA DNA's B-box, its binding site is probably at or near that site. Footprinting studies with the separated TFIIIC-1 and TFIIIC-2 indicate that TFIIIC-1 alone fails to bind strongly to the DNA. However, TFIIIC-2 does. TFIIIC-2 protects the 3′ portion of the control region, including the B-box, and the binding of TFIIIC-1 extends the footprint in the 5′ direction over the A-box to the transcription start site. Both TFIIIB and TFIIIC are needed to form a tight complex with the B- and A-boxes of tRNA genes. Thus, the two TFIIICs interact first with the B-box to form a complex whose stability and extent of binding depend on the particular B-box sequence, and binding of TFIIIB stabilizes the preinitiation complex. RNA polymerase III is presumed to interact with this complex and initiates transcription at the appropriate position. RNA polymerase III may bind to the complex exclusively by protein-protein interactions, by specific contacts with parts of the ICR, or possibly with transcription factors bound to regulatory motifs upstream of the transcription start sites.

One inference drawn from *in vitro* experiments is supported by the finding that a tRNA gene contained in a chromatin structure can be transcribed properly by RNA polymerase III alone; the addition of TFIIIB and TFIIIC is not required. In the same reaction, a VA RNA gene, present as free DNA, is not transcribed unless the two required transcription factors are added. In chromatin, therefore, the TFIIIB and TFIIIC proteins are most likely assembled on the ICRs in a stable complex, and the factors are dissociated when the histones are removed from the DNA.

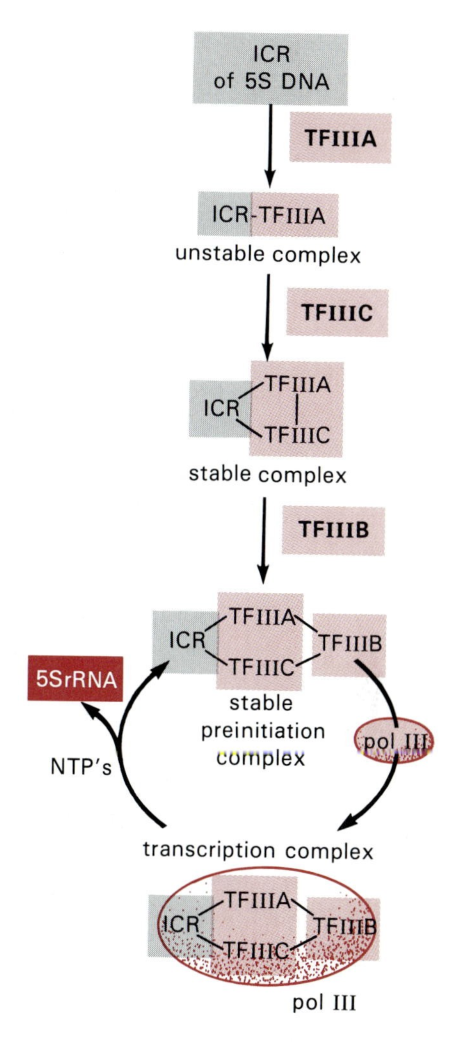

Figure 8.67
The ordered binding of transcription factors to the 5S rRNA gene's internal control region (ICR). Although shown as separate steps, the binding of TFIIIB and RNA polymerase III (pol III) may occur together. The diagram indicates that the preinitiation complex is conserved during each transcription cycle.

Preinitiation complexes containing 5S rRNA genes and transcription factors The participation of transcription factors in promoting transcription initiation of 5S rRNA genes was examined using the methods mentioned previously. The results of such studies indicate specific requirements and ordered binding (Figure 8.67). TFIIIA alone forms a weak complex with the ICR. Although TFIIIC does not bind to the ICR by itself, it binds to the TFIIIA-5S rDNA complex and stabilizes it significantly. Whether one or both of the TFIIICs is required for this stabilization is unclear. The principal determinants of the TFIIIA-5S rDNA interaction are in the C-box between base pairs +80 and +90, although some contacts are also made with nucleotides between base pairs +50 and +64 (i.e., the A-box). Stabili-

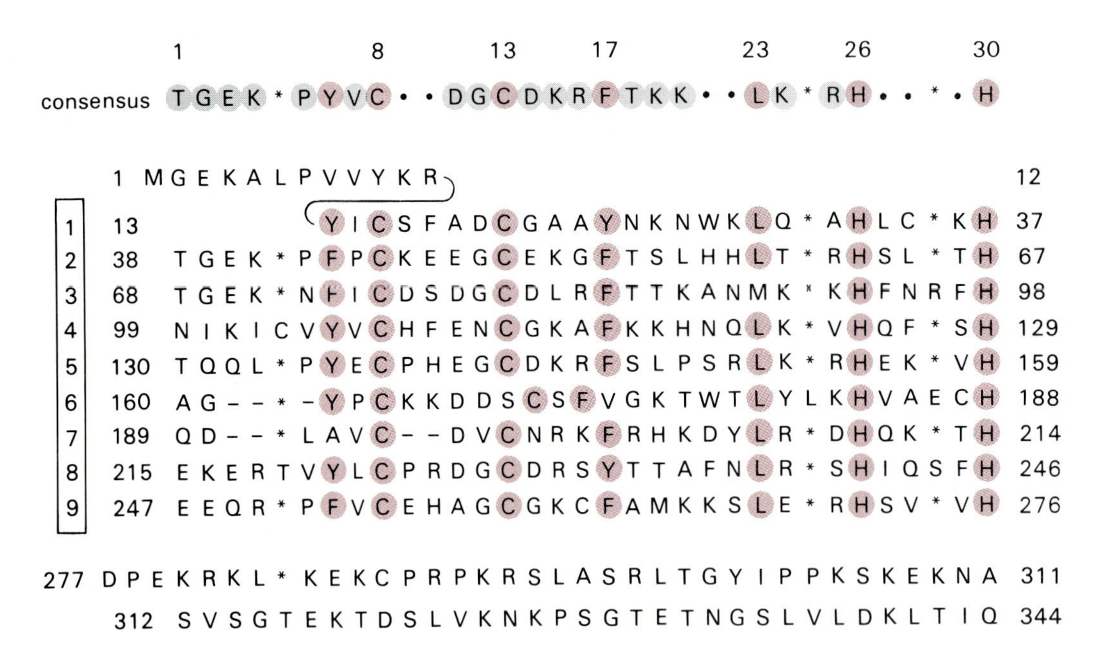

Figure 8.68
Amino acid sequence of TFIIIA from *X. laevis*. The sequence, which is aligned to emphasize the repeated units, uses the one-letter code for amino acids (see Figure 1.24). The molecule contains an amino terminal region (residues 1–12), the 9 repeat units (residues 13–276), a lysine-rich region (residues 277–309) near the carboxyl end, and a short tail region. The repeat units are numbered 1–9 on the left side of the diagram. The encircled consensus sequence shown at the top presents the characteristic features of a typical repeat unit, numbered as for a length of 30 residues. The end point of each unit has been chosen arbitrarily after H30. The most highly conserved residues are in color. C8, C13, H26, and H30 are believed to bind to a Zn^{2+} ion. Y6, F7, and L23 are hydrophobic residues that may form an inner core for the proposed multiple-domain structure. Asterisks (*) mark positions where an insertion sometimes occurs in the normal pattern, and a dash (—) in the main body of the repeats indicates an alignment gap. Dots (· · ·) mark variable positions in the sequence. Taken from J. Miller et al., *EMBO J.* 4 (1985), p. 1609.

zation of the TFIIIA-5S rDNA complex by TFIIIC probably involves interactions between the two transcription factors and contacts between TFIIIC and the A-box. TFIIIB does not bind to 5S rDNA alone, nor does it alter the footprint produced by the binding of TFIIIA and TFIIIC. However, TFIIIB is required for transcription by RNA polymerase III. It probably acts by creating a multiprotein complex with which RNA polymerase III associates to begin transcription. Because TFIIIA is the only transcription factor capable by itself of recognizing and binding to the 5S rDNA's ICR, considerable effort has been made to learn the mechanism and specificity of its interactions.

Molecular structure of TFIIIA *Xenopus* TFIIIA is a zinc metalloprotein of 40 kDaltons whose 344 amino acid sequence is known from the nucleotide sequence of its cDNA. The cloned *Xenopus* TFIIIA gene is 11 kbp in length and contains eight introns. Nearly two-thirds of the amino acids at the amino terminus (residues 13–276) form nine copies of an imperfectly repeated sequence of about 30 amino acids; each copy contains two invariant pairs of cysteine and histidine residues (Figure 8.68). Many of the TFIIIA gene's eight introns separate the coding regions of the 30 amino acid repeated segments. The entire carboxy terminal domain, which lacks the repeat motif, is encoded in a single exon.

Atomic absorption spectroscopy shows that TFIIIA contains nine (±one) zinc atoms associated with each molecule of purified protein. Extended X-ray absorption fine structure spectroscopy confirms that each zinc atom is chelated through the pairs of cysteine and histidine residues within each 30 amino acid repeat (Figure 8.69). The currently favored structure for TFIIIA is one in which the protein has nine repeating zinc finger domains, each of which contains a zinc atom tetrahedrally coordinated to the invariant pairs of cysteines and histidines (Figure 8.69). The sequence between the paired cysteines and histidines contains an invariant phenylalanine (or tyrosine) and a leucine (or methionine) and a dispropor-

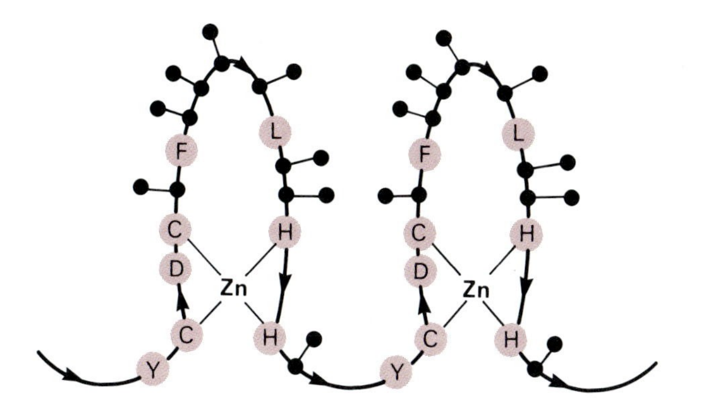

Figure 8.69
Zinc fingers: a folding scheme for the linear arrangement of repeated domains in TFIIIA. Each domain is centered on a tetrahedral arrangement of zinc ion ligands. Only the significant residues of a typical repeat are indicated. Colored circles mark the conserved amino acids, which include the C(cys) and H(his) residues to which the zinc ions are liganded, the negatively charged D(asp)11, and the three hydrophobic groups that may form a structural core. Small black circles mark side chains that may interact with the DNA. In this model, the zinc ion draws the ends of each repeating unit together, allowing the 14 to 25 central residues to form a potential DNA-binding loop or "finger." Adapted from J. Miller et al., *EMBO J.* 4 (1985), p. 1609.

tionate number of basic amino acids (Figures 8.68 and 8.69). A tyrosine or phenylalanine residue invariably precedes each finger, and an acidic residue is generally between the two cysteines. The basic and other residues in the fingers constitute the DNA binding sequences in TFIIIA.

TFIIIA probably binds to the approximately 50 bp constituting the 5S rRNA gene's ICR with about 5 bp interacting with each of the nine zinc fingers. Studies that measure the accessibility of the DNA in the protein-DNA complex to dimethylsulfate and micrococcal nuclease suggest that the contacts made between TFIIIA and the DNA are spaced about 5 bp apart or half a helical turn. However, except for a possible structural periodicity of guanine residues in the ICR, there is no apparent sequence repeat that would provide similar binding sites for each of the nine fingers. The currently favored model for binding proposes that TFIIIA interacts with guanine clusters on one face of the double helix (Figure 8.70). Such extended contacts probably explain the apparent stability of the TFIIIA-TFIIIC-ICR complex and the persistence of the entire complex through multiple rounds of transcription. The complex could remain stably bound as RNA polymerase III traverses the DNA by disengaging one or a few fingers at a time. Several RNA polymerase II transcription factors also rely on the zinc finger motif for binding to their cognate regulatory sequences (Sections 8.3e and f and 8.6a).

Limited proteolysis experiments and an analysis of the activity of variously truncated forms of TFIIIA protein reveal that the region responsible for activating transcription of 5S rRNA genes is localized to about 20 amino acids near the carboxyl end of the protein; this region has no role in the binding of TFIIIA to the ICR. The remaining 75 percent of the protein, which contains the nine zinc fingers and the amino terminus, is

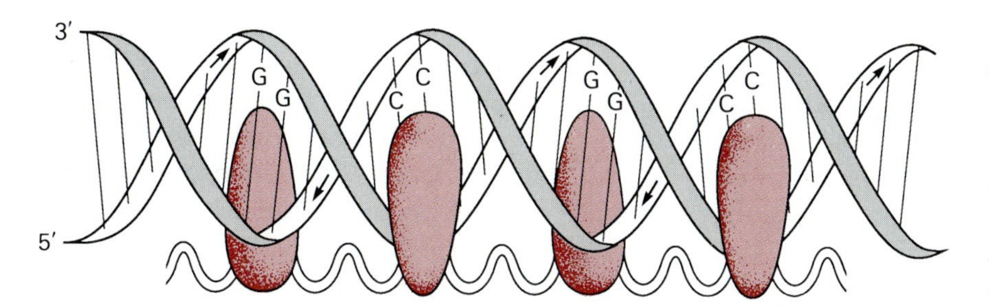

Figure 8.70
Schematic model for the interaction of the TFIIIA zinc fingers with the double helix. In the model, the protein lies on one face of the DNA helix, successive fingers pointing into the major groove alternately in "front" and "behind." Every other finger makes structurally equivalent contacts located 10 bp apart, and the N7 atoms of guanine residues spaced 5 nucleotides apart are approached from different directions. In this idealized drawing, the axis of the DNA helix is shown to be straight and the fingers are regularly arranged. Adapted from L. Fairall et al., *J. Mol. Biol.* 192 (1986), p. 577.

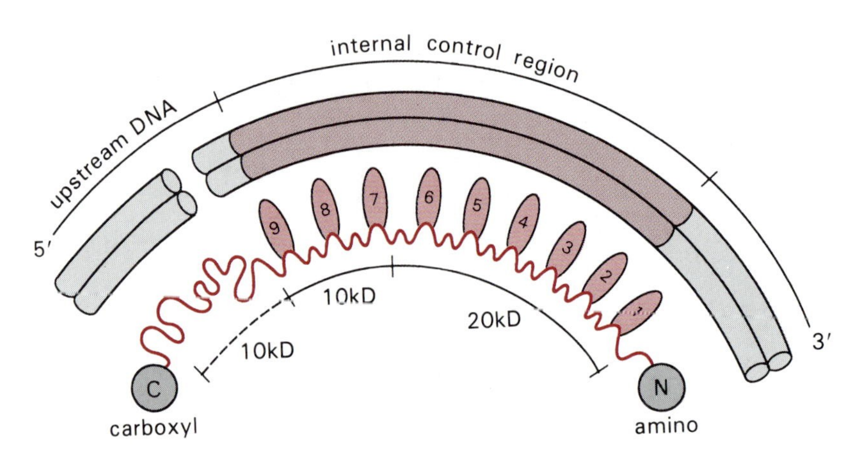

Figure 8.71
Structural features of TFIIIA and a model for its interactions with the 5S rRNA gene's internal control region (ICR). The ICR, which is contained between base pairs +50 and +90, is drawn with the 5′ end of the sense strand at the left. The protein sequence runs from right to left; the evidence for this orientation is given in the text. The amino terminus is followed by nine repeat units (residues 1–276) in contact with the control region. These units together account for 30 kDaltons of the protein. The numbered repeats represent the zinc fingers linked by flexible joints. Residues 277–344 include a lysine-rich region, followed by sequences at the carboxyl end. Together these parts form a separate domain of 10 kDaltons. The relationship between the grooves of the DNA and the positions of the protein units is as suggested in Figure 8.70. Adapted from J. Miller et al., *EMBO J.* 4 (1985), p. 1609.

responsible for the strong interactions with the ICR; this region is unable to activate transcription. A model summarizing these findings is shown in Figure 8.71.

Each of the zinc fingers contributes to the overall binding efficiency, although the largest contribution to the binding energy stems from the interaction of the two amino terminal fingers with the C-box. The remaining fingers make their contacts with the A-box and the quantitatively less significant region between the two boxes (Figure 8.61). Precisely which amino acids are responsible for recognizing the various sequence elements of the ICR is not known.

The region near the carboxy terminus may activate transcription by influencing RNA polymerase III directly or by interacting with TFIIIB and/or TFIIIC. Although it is known that TFIIIB and TFIIIC stabilize the binding of TFIIIA to the ICR, it is not clear whether these factors also have a part in promoting transcription initiation by the polymerase. The fact that TFIIIB and TFIIIC are sufficient to promote transcription of all other class III genes suggests that they participate in transcription initiation with RNA polymerase III in a direct way.

Is the complex formed by interaction of the three factors and the ICR stable to the passage of RNA polymerase III as it transcribes past the control region? This question was examined with a 5S rRNA gene fused to a bacterial promoter so that the 5S rDNA sequence could be transcribed *in vitro* with a purified preparation of a bacterial RNA polymerase. It was found that under conditions where RNA polymerase III can accurately transcribe the 5S rRNA gene, the bacterial RNA polymerase transcribes the 5S rDNA as efficiently in the presence or absence of the three transcription factors. Following transcription by the bacterial RNA polymerase, 5S rRNA could still be synthesized after addition of only RNA polymerase III. This result indicates that the preinitiation complex containing the 5S rDNA, TFIIIA, TFIIIB, and TFIIIC is not disrupted during transcription by bacterial RNA polymerase. It is important that the conditions used for the transcription by bacterial RNA polymerase were not appropriate for the reassembly of the preinitiation complex if it had dissociated. If the complex had dissociated, the inability to re-form prior to or after the addition of RNA polymerase III would have prevented 5S rRNA synthesis. Although it remains unproven, it is inferred that the complex of 5S rDNA and the transcription factors is also stable to repeated rounds of transcription by RNA polymerase III. Given the extended contacts between the nine repeating zinc finger domains and the 50 bp of the ICR (Figure 8.71), some DNA-protein contacts may be maintained while others are disrupted during transcription.

A different outcome results when the DNA in the complex is replicated. This was shown by introducing an SV40 origin of DNA replication into a plasmid containing a 5S rRNA gene. After assembling the same kind of preinitiation complex, the ingredients needed for replicating the plasmid *in vitro* were added. Following replication, the DNA was tested for its ability to support 5S rRNA synthesis when RNA polymerase III was added. In this case, 5S rRNA was made only if the three transcription factors were added and preinitiation complexes could be re-formed. Furthermore, digestion of the replicated DNA with DNase I showed the same fragment pattern as uncomplexed DNA, indicating that TFIIIA and TFIIIC were not bound. Apparently, the complex of 5S rDNA with the three transcription factors is disrupted during replication but not during transcription of the DNA. The instability during DNA replication may result from the need to disrupt large regions of the duplex to accommodate the replication machinery at forks.

d. Transcription Termination by RNA Polymerase III

Although initiation of transcription by RNA polymerase III requires complex regulatory sequences and several proteins besides the polymerase itself, transcription termination requires only the enzyme and a cluster of deoxyadenylate residues on the template strand. Termination occurs accurately *in vitro* with highly purified RNA polymerase III and a duplex DNA template. Accurate and efficient termination occurs even on class III genes that lack an ICR, thus eliminating the possibility that association of the template with TFIIIA, TFIIIB, or TFIIIC plays a part in the termination process. Specifically designed cloned DNA transcription templates have been used to define the termination signal. These experiments show that termination depends on the number of deoxyadenylate residues clustered on the template strand and on the sequence surrounding the deoxyadenylate cluster (Table 8.2). For example, four deoxyadenylate residues surrounded by deoxythymidylate residues in the template strand do not promote efficient termination, but four deoxyadenylate residues surrounded on one side by CG and on the other by a GC dinucleotide (5′-GCAAAAGC-3′), as occurs in a *Xenopus borealis* somatic 5S rRNA gene, is an efficient termination signal. Mutations that alter the number of adenylate residues in the cluster reduce termination efficiency. RNA synthesis stops with a uridylate residue that is complementary to one of the deoxyadenylate residues in the run. Neither distant sequences nor RNA secondary structure seems to influence termination. If the 3′ end of a 5S rRNA gene is deleted from a cloned gene, transcription continues until the RNA polymerase III meets an appropriate fortuitous signal in the flanking vector sequence. And if runs of adenylate residues are introduced at an inappropriate site in the template strand of a class III gene, termination often occurs at that altered site.

Table 8.2 Sequences at Which RNA Polymerase III Terminates Transcription

Efficient Termination Sites	
Xenopus borealis (somatic)	5′-TAGGCTTTTGCACT-3′
Xenopus borealis (oocyte-type 1)	-TAGGCTTTTAGACT
Xenopus borealis (oocyte-type 3)	-TAGTCTTTTTCCAG
Adenovirus VA1	-GCTCCTTTTGGCTT
Bombyx mori tRNA$^{\text{A1a}}$	-ACGATTTTGTTAT
Weak Termination Sites	
Xenopus laevis (oocyte-type 1)	-TAGGCTTTTCAAAG
Very Weak Termination Sites	
Xenopus laevis (oocyte-type 2)	-TAGGCTTTCCAAAG
Xenopus laevis (oocyte pseudogene)	-AGAAGTTTTCAAAG

Note: RNA polymerase III termination signals occur at the 3′ end of 5S rRNA coding sequences and beyond the 3′ end of tRNA coding sequences. Termination occurs within the T cluster (underlined) but generally not at a unique T. The relatively inefficient termination sequences contain short runs of A residues following or preceding the T clusters. The 5S rRNA transcript requires no processing at the 3′ end, but nucleotides must be removed from the 3′ end of the tRNA gene transcripts before the 3′ CCA end can be added to produce a mature tRNA.

SOURCE: From D. F. Bogenhagen and D. D. Brown, *Cell* 24 (1981), p. 261.

e. Control of 5S rRNA Gene Expression During Development

The two kinds of Xenopus *5S rRNA genes* Dramatic changes in the expression of 5S rRNA genes occur during development and differentiation of *Xenopus* species (Table 8.3). During the early phases of oogenesis, oocytes grow rapidly and synthesize a large amount of 5S rRNA to meet the large requirement for ribosomes during early embryogenesis. By contrast, the mature egg does not synthesize RNA of any kind. Synthesis of all RNAs, including 5S rRNA, resumes at or just before formation of the blastula (4000 cells) and continues thereafter into adult life.

Two distinct sets of 5S rRNA genes are expressed during the several stages of development. One is expressed almost exclusively during the oocyte stage (the **oocyte 5S rRNA** genes); the second is expressed in both oocytes and in somatic cells of the developing embryo and adult tissues (the **somatic 5S rRNA** genes). Thus, although there are about equal numbers of the oocyte and somatic 5S rRNA transcripts in the blastula, more than 98 percent of the 5S rRNA is of the somatic type by the tadpole stage and beyond. This differential expression is especially dramatic because there are about 20,000 oocyte type 5S rRNA genes and only about 400 of the somatic type in the *Xenopus* genome (Section 9.2b). What accounts for the preferential expression of the oocyte 5S rRNA genes in oocytes and their repression in somatic cells? How do the somatic 5S rRNA genes take over and become the nearly exclusive template for 5S rRNA synthesis in somatic cells? We know now that this differential gene function is mediated by transcriptional regulation. Furthermore, this

Table 8.3 Changing Cellular Proportions of Oocyte and Somatic 5S rRNA and Levels of TFIIIA During *Xenopus* Development

Stage of Development	Predominant Type of 5S rRNA Being Made	Molecules of TFIIIA per 5S rRNA Gene
Immature oocyte	Oocyte	5×10^7
↓ growth		
Mature oocyte	Oocyte	5×10^6
↓ meiosis I		
Unfertilized egg	Neither RNA made	4×10^5
↓ meiosis II, fertilization, cell division		
Blastula embryo (4000 cells)	Equal somatic and oocyte	10
↓ cell division, differentiation		
Gastrula embryo	Somatic	2
↓ cell division, differentiation		
Swimming tadpole	Somatic	0.4

SOURCE: Data from W. M. Wormington et al., *Cold Spring Harbor Symposium* 47 (1982), p. 879.

regulation seems to depend on several parameters. Some of those implicated are the changing levels of transcription factors during development, notably TFIIIA but possibly others, the differential stability of the oocyte and somatic genes' complexes with factors, the state of the chromatin structure in which the two types of 5S rDNA are packaged, and the time at which each type of 5S rDNA is replicated during S phase.

Differential gene expression in vitro In distinction to the disparity in their numbers, the oocyte and somatic 5S rRNA genes differ only slightly in structure, primarily in three to six base pairs (depending on the species) in the sequence near the 5′ end of the ICR (Figure 8.72). Transcription of cloned somatic 5S rRNA genes is about five times more efficient than cloned oocyte genes when the two are used as templates with either oocyte or somatic cell extracts or with purified RNA polymerase III and TFIIIA, TFIIIB, and TFIIIC. This slight difference in transcription probably reflects the somewhat greater binding affinity of TFIIIA for the somatic genes' ICR. But because there are 50 times more oocyte than somatic 5S rRNA genes in cells, it seems unlikely that the marginally greater binding affinity of TFIIIA for somatic genes' ICRs can explain the differential expression observed *in vivo*.

The discovery that the pattern of differential gene expression typical of development can be reproduced using purified chromatin as a template for RNA polymerase III has made it possible to investigate the switch in expression *in vitro*. Unfertilized eggs or cells in the early stages of blastula

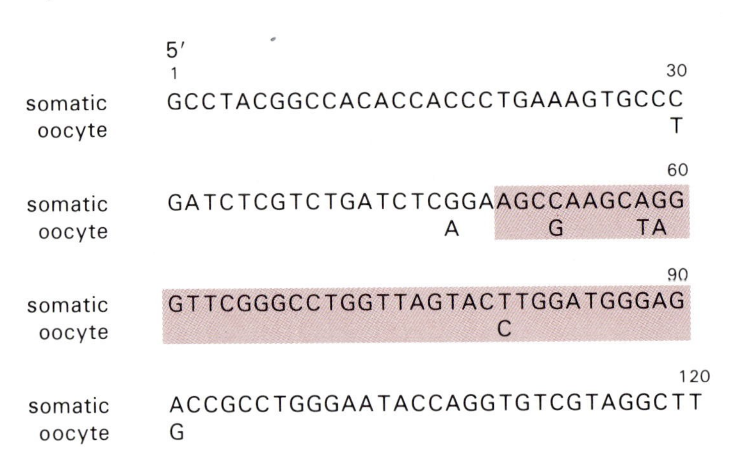

Figure 8.72
Nucleotide sequence differences between *X. laevis* somatic and oocyte 5S rRNA genes. Only the sequence of the first 120 nucleotides of the sense strands and the differences between the two types of genes are shown. The boxed segment includes the internal control regions. From P. J. Ford and R. D. Brown, *Cell* 8 (1976), p. 485.

development do not make any RNA, and chromatin isolated from such cells fails to support 5S rRNA synthesis when incubated with RNA polymerase III. By contrast, transcription of chromatin purified from oocytes yields mainly oocyte 5S rRNA, but chromatin from somatic cells yields principally somatic 5S rRNA under the same conditions. Only RNA polymerase III needs to be added to synthesize oocyte or somatic 5S rRNA with oocyte or somatic cell chromatin as the template, respectively. Thus, the genes are probably associated with the three transcription factors in the form of a preinitiation complex in their respective chromatins. But even if the three transcription factors are added along with the polymerase, the oocyte genes in the somatic cell chromatin are not transcribed. However, the oocyte 5S rRNA genes in somatic cell chromatin can be transcribed if the chromatin is first extracted with 0.6 *M* NaCl and then the three transcription factors are added along with the polymerase. Transcription of the somatic 5S rRNA genes is unaffected by the salt wash and still requires only RNA polymerase III addition. One explanation of this result is that the oocyte genes in the repressed chromatin lack the needed transcription factors and are inaccessible to added factors or that a potentially active complex exists in the repressed chromatin but is inaccessible to the polymerase. Disrupting the chromatin by salt extraction might open up the chromatin or remove unidentified proteins, making the oocyte genes available for association with the added factors and RNA polymerase III with restoration of transcription activity.

Although the nucleosome histones H2A, H2B, H3, and H4 are not removed by the salt extraction, histone H1 is efficiently removed from chromatin by such a treatment. One possibility is that histone H1 maintains the oocyte type 5S rRNA genes in a repressed condition in somatic cell chromatin and that removing histone H1 makes the oocyte type 5S rRNA genes accessible for transcription. The fact that adding purified histone H1 to the histone H1 depleted somatic cell chromatin blocks transcription of the oocyte type 5S rRNA genes supports this deduction. This suggests that the differential expression of 5S rRNA genes in somatic cells depends on histone H1 mediated condensation of the chromatin containing the oocyte 5S rRNA genes.

A clue to how histone H1 could restrict the binding of the transcription factors to the ICR emerges from an analysis of the competition between the binding of nucleosomes (histone octamers) and TFIIIA. Footprinting

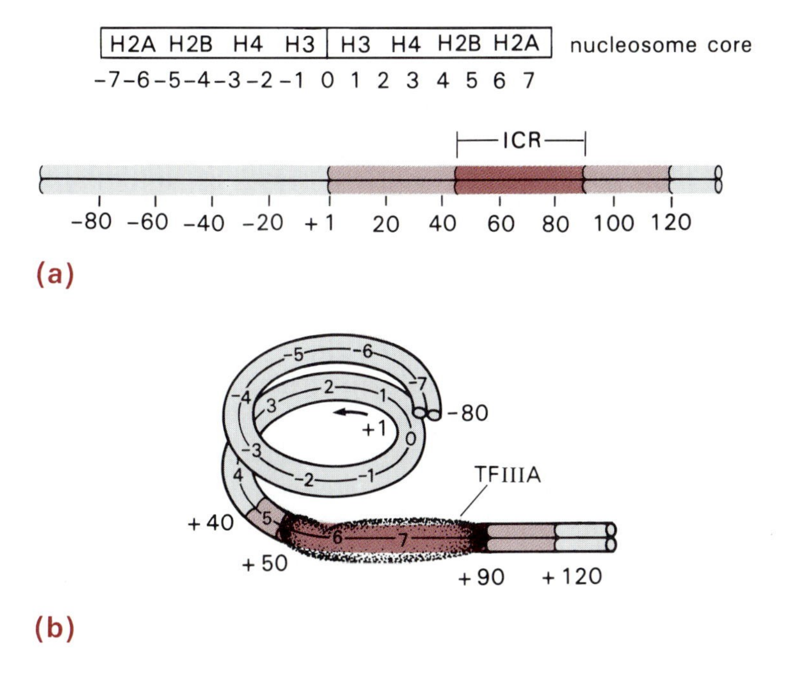

Figure 8.73
TFIIIA competes with nucleosomes for binding to the 5S rDNA ICR. (a) The *X. laevis* 5S rRNA gene (colored) and its flanking sequences indicating the position of the ICR. The position at which the histone octamer binds to the gene and its flanking sequence are indicated by the box enclosing the designations for the individual histones. (b) The diagram shows how the DNA in (a) would be wound around the nucleosome core (not shown) and the relative position of the ICR. Because histone H1 binds where the DNA enters and leaves the nucleosome, its binding would be expected to block binding of TFIIIA to the ICR. Adapted from D. Rhodes, *EMBO J.* 4 (1985), p. 3473.

experiments show that the nucleosome core binds to the 5′ end of the 5S rRNA gene, extending about two-thirds of the way into the ICR (Figure 8.73). But TFIIIA can still bind to the exposed portion of the ICR and eventually displace the nucleosome from contact with the 5′ end of the control region. This is presumably sufficient for TFIIIB, TFIIIC, and RNA polymerase III to bind to the gene and transcription to begin at the proper site. Because histone H1 binds to nucleosomes where the DNA enters and leaves the octamer (Figure 1.17), its presence in the chromatin probably prevents the initial binding of TFIIIA and thereby blocks access of TFIIIC, TFIIIB, and RNA polymerase III to the template.

Switching 5S rRNA gene expression during Xenopus *development* If histone H1 binding to the oocyte genes accounts for the differential expression in differentiated somatic cells, we need to ask when and how during development the switch occurs from predominantly oocyte to nearly exclusively somatic gene expression. A large amount of TFIIIA is synthesized concomitant with the activation of oocyte 5S rRNA genes in maturing oocytes (Table 8.3). This is not surprising because TFIIIA is required for transcription. In addition to binding to 5S rRNA genes, TFIIIA also forms a 7S ribonucleoprotein particle with the newly synthesized 5S rRNA. The particle appears to be a storage depot for the 5S rRNA prior to its incorporation into ribosomes later in oogenesis. By forming such particles, the 5S rRNA competes with its own genes for TFIIIA, thereby establishing a feedback regulation for the RNA's production.

After the early burst of TFIIIA synthesis, its synthesis slows down and continues to decrease after fertilization and the beginning of embryogenesis. The gradual repression of oocyte 5S rRNA gene expression after the mid-blastula stage correlates with a further decrease in the intracellular concentration of TFIIIA (Table 8.3). Thus, immature oocytes have a large excess of TFIIIA over the total number of 5S rRNA genes and produce a large quantity of oocyte type 5S rRNA; somatic cells have only about one molecule of TFIIIA for every 2–3 5S rRNA genes. The competition for TFIIIA by oocyte and somatic type 5S rRNA genes is strongly influenced by a greater instability of oocyte 5S rRNA gene transcription complexes in fertilized eggs, a phenomenon also observed in extracts of early blastulas. Consequently, the transcription complexes involving oocyte 5S rRNA genes break down, but complexes involving somatic 5S rRNA genes are preferentially retained. As the embryo develops, lowered synthesis of TFIIIA and its binding to 5S rRNA limit its availability for complex formation. The preferential lability of oocyte 5S rRNA gene transcription complexes, together with a preferential binding of TFIIIA and TFIIIC to somatic 5S rRNA genes, causes a net transfer of transcription factors from oocyte to somatic 5S rRNA genes. As the transcription factors are removed from the oocyte 5S rRNA genes, that region may be "shut down" by a histone H1 mediated condensation of the chromatin. Similarly, the persistent association of the somatic 5S rRNA gene with the three transcription factors may prevent such a condensation in somatic cells.

But how is this preferential state of somatic 5S rRNA gene activation maintained during the DNA replication that occurs in successive cell divisions? DNA replication disrupts the transcription complexes, so they must be re-formed before transcription can begin again. Interestingly, in *Xenopus* somatic cells, the somatic 5S rRNA genes replicate during the first quarter of the S phase, and the oocyte genes replicate late in the S phase. Such a staging of chromosome replication could allow the somatic 5S rRNA genes to re-form functional transcription complexes without competition from the oocyte genes.

Our understanding of the developmental regulation of 5S rRNA genes in *Xenopus* is presently incomplete. However, several mechanisms appear to be relevant: (1) hyperproduction of TFIIIA during oogenesis, which permits the expression of both types of 5S rRNA genes in proportion to their abundance; (2) cessation of TFIIIA formation during egg maturation and subsequent stages of embryonic development and only a low level maintenance in somatic cells; (3) the influence during development of TFIIIA level on the interaction of TFIIIA with 5S rRNA; (4) preferential breakdown of oocyte 5S rRNA gene transcription complexes and, because of the competitive advantage in forming transcription complexes, the preferential accumulation of somatic 5S rRNA gene transcription complexes; (5) preferential condensation of the chromatin containing "empty" oocyte 5S rRNA genes by histone H1; and (6) earlier replication of somatic 5S rRNA genes during S phase, which permits preferential re-formation of active somatic 5S rRNA gene transcription complexes. Additional controls that could be influential are (1) the differential regulation of TFIIIA gene expression during early stages of oogenesis compared to later stages of embryogenesis and in somatic cells, (2) the availability of functional TFIIIC and TFIIIB, and (3) the influence of transcription factors that operate at regulatory sequences upstream of the genes.

8.5 Dealing with Introns

The coding sequences of eukaryotic genes (**exons**) are frequently interrupted by noncoding stretches of DNA (**introns**). This astonishing revelation raised several intriguing fundamental questions. (1) Where, when, and how did introns appear in genes, and what role did they serve in biological evolution? (2) How are the primary transcripts of such mosaic genes processed into mature RNAs that lack introns? (3) What implications does this requirement and the mechanism of splicing have for the regulation of gene expression? (4) What function, if any, do introns serve in the myriad transactions of current day genomes? For example, are introns vestigial remains of the evolutionary history of genes, or do they have an essential role in the life of contemporary organisms? Some of the issues raised in the first question are discussed in Perspective III and Chapter 9. Our principal aim here is to consider what we know about the structure of introns, the mechanisms by which their sequences are removed during the biogenesis of mature functional RNAs, and the relevance of splicing in eukaryotic gene expression.

a. *The Frequency of Introns*

The frequency of split genes varies greatly among eukaryotic species. They are most prevalent in plants and animals and the viruses that infect them. Introns are rarer in invertebrate genes (e.g., *Drosophila*, nematodes, and sea urchins). Nevertheless, several genes that guide morphogenesis in *Drosophila* (e.g., the *ultrabithorax* and *antennapedia* clusters) have complex arrangements of introns and exons. Most nuclear genes in *S. cerevisiae* lack introns (actin genes and some tRNA genes are exceptions), but split genes are abundant in the distantly related *S. pombe*. Although introns are rare in *S. cerevisiae* nuclear genes, they are common in the organism's mitochondrial genes. At the time of their discovery, split genes were thought to be characteristic of eukaryotic genes only. But the discovery of an intron in a prokaryotic gene-thymidylate synthetase of T4 bacteriophage proved this presumption wrong. Since then, introns have also been discovered in a $tRNA^{Ser}$ gene and the large rRNA gene of *Archaebacteria*. It would be surprising if wider ranging searches among prokaryotes did not turn up more genes with introns.

Introns occur in nuclear genes encoding mRNAs, rRNAs (so far only the large rRNA genes of lower eukaryotes), and a subset of tRNA species. Split genes are also frequently found in mitochondrial and chloroplast genes that specify messenger, ribosomal, and transfer RNAs used in these organelles. However, in mammalian genes, where introns are almost always present, there are exceptions. Most genes encoding the five histones lack introns, as do the two families of genes encoding the α and β interferons; by contrast, the single γ interferon gene has multiple introns (Figure 8.2). There are no reported instances of introns in 5.8S and 5S rRNA genes or in any of the U RNA, 7SL RNA, and 7SK RNA genes.

b. *Different Kinds of Introns*

All introns are transcribed as part of precursor RNAs and subsequently removed by a cleavage-ligation process called **splicing**. Structural features of introns and the underlying splicing mechanisms form the basis of a classification for different kinds of introns (Table 8.4).

Table 8.4 Consensus Sequences at Different Kinds of Splice Sites

Intron Type	5′ Splice Junction	Near 3′ Splice Junction	3′ Splice Junction
Nuclear pre-mRNA (general)	CRG↓$\underline{\mathrm{GU}}{}^{\mathrm{A}}_{\mathrm{G}}$AGU	A	$\mathrm{Y_n}\underline{\mathrm{AG}}$↓N
Nuclear pre-mRNA (yeast)	↓$\underline{\mathrm{GU}}$AUGU	$\underline{\mathrm{UACUAAC}}$	$\mathrm{Y_n}$AG↓N
tRNA	N↓N		N↓N
Group I (nuclear rRNA, mitochondrial mRNA and rRNA)	$\underline{\mathrm{U}}$↓		$\underline{\mathrm{G}}$↓
Group II (mitochondrial mRNA)	↓$\underline{\mathrm{GUGCG}}$		$\mathrm{Y_n}\underline{\mathrm{AU}}$↓
Chloroplast mRNA *(Euglena)*	↓$\underline{\mathrm{GUG^{C}_{U}G}}$		$\mathrm{Y_n}\underline{\mathrm{A}}{}^{\mathrm{U}}_{\mathrm{C}}$↓

Note: The bases shown are as they occur in RNA. Y represents a pyrimidine, and the subscript n indicates multiple pyrimidines. R indicates a purine. The underlined letters indicate the invariant bases, and the arrows represent the splice sites.
SOURCE: After T. Cech, *Cell* 44 (1986), p. 207.

Introns in nuclear mRNA genes The introns in nuclear protein coding genes, the first kind to be discovered, vary in size from 100 bp to well over 10 kbp. Introns from corresponding genes in vertebrate species can be as dissimilar from each other in length and sequence as two introns from unrelated genes. Exon lengths seem to cluster around 52, 140, 223, and 299 bp, with the latter being the most prevalent. Nevertheless, exons of as few as 15 to 30 bp or as many as several hundred or a thousand base pairs are known, but these are rarer. The most distinctive common features associated with introns are the sequences at their 5′ (upstream or donor) and 3′ (downstream or acceptor) borders (i.e., the exon-intron junctions or **splice sites**).

The nucleotide sequences at each exon-intron junction are remarkably conserved in virtually all nuclear mRNA genes of almost all species that have been studied (Table 8.4). The exon sequence most frequently flanking the 5′ splice site is CRG (where R is a purine); that bordering 3′ splice sites is most often a single G. Nevertheless, there are wide variations in the sequences bordering introns, and mutations at these locations never abolish splicing, although they may affect the rate. The most invariant structural features lie within the intron. The first two nucleotides at the 5′ end of the intron in RNA are virtually always GU (two exceptions having GC); the next four are not invariant, although the sequence ${}^{\mathrm{A}}_{\mathrm{G}}$AGU appears to be the consensus. Alterations of either the G or U residue at the junction generally abolish splicing; changes at the adjacent positions affect splicing variably. The six nucleotides at the 5′ end of the intron are sufficient for 5′ splice site function. Even cryptic splice junctions (Section 8.5f), those used only rarely or when the predominant splice sites are abolished or impaired, have the GU sequence at the 5′ end of the sequence that is spliced out. The 3′ terminus of the intron invariably ends with AG, most often preceded in mammalian introns by a pyrimidine-rich stretch ($\mathrm{Y_n}$NYAG). Here, too, mutations that replace the invariant A or G with another base block splicing at that junction.

An A residue near the 3′ end of the intron is an essential participant in the splicing of nuclear pre-mRNAs (Section 8.5d). In mammalian introns, this A residue does not occur at an invariant position or sequence because any one of several As within 18 to 37 nucleotides upstream of the 3′ splice site may be used. However, mutational alterations in the sequence neighboring the A residue cause a substantial decrease in splicing efficiency *in vitro*; thus, although the essential A does not occur within an invariant context, the surrounding sequence influences its utilization in splicing. By contrast, yeast nuclear pre-RNAs rely on a discrete heptanucleotide sequence, 5′-UACUAAC-3′, located between 6 and 59 nucleotides upstream of the 3′ splice site, to provide the reactive A residue.

The occurrence of consensus splice site sequences does not always predict that an intron can be spliced. Sometimes either or both of these splice site sequences occur within exons and introns at positions where splicing normally does not take place. However, such cryptic splice sites do function under certain circumstances (e.g., when the authentic sites are altered or missing).

Occasionally, introns have more than one 5′ or 3′ splice junction, thereby making it possible to have alternate splicing patterns. For example, there are two introns within the SV40 early region, each of which has the same 3′ splice site but different 5′ junctions. The outcome of such alternative splices are two different mRNAs from the same primary transcript. Under certain circumstances, the choice of alternative splice sites is regulated temporally and in a tissue-specific manner (Section 8.5f). Sometimes existing splice sites are not used. For example, retroviral genomes are derived from the transcripts of the proviral DNA without splicing, but production of the mRNAs that encode certain viral proteins requires splicing. In this instance, the same transcripts may be spliced or remain unspliced.

As previously mentioned, introns may harbor other genetic elements, such as enhancers, other genes, possibly replication and chromosomal packaging signals, or sequences needed for packaging pre-mRNAs into ribonucleoprotein particles.

Introns in tRNA genes The introns in tRNA genes vary in size from 14 to nearly 60 nucleotides, but they invariably occur at the same location: one nucleotide to the 3′ side of the anticodon (Figure 8.74). Generally, if a tRNA gene has an intron, all the other genes encoding the same tRNA in the species have the same intron. However, the sequences within the introns and at their junctions differ markedly among genes that encode different tRNA species. Given the importance and location of the intragenic control region for RNA polymerase III mediated transcription of tRNA genes (Section 8.4b), a relevant issue is whether the introns in some tRNA genes influence their transcription. However, removal of a yeast $tRNA^{Tyr}$ suppressor gene's intron by site-directed mutagenesis does not impair its ability to be expressed after introduction into cells. Nevertheless, although the tRNA sequence is transcribed and trimmed normally, the U residue in the anticodon fails to be modified to the customary Ψ. Whether this finding presages a role for introns in posttranscriptional modifications of tRNA or whether this effect is unique to $tRNA^{Tyr}$ is not clear.

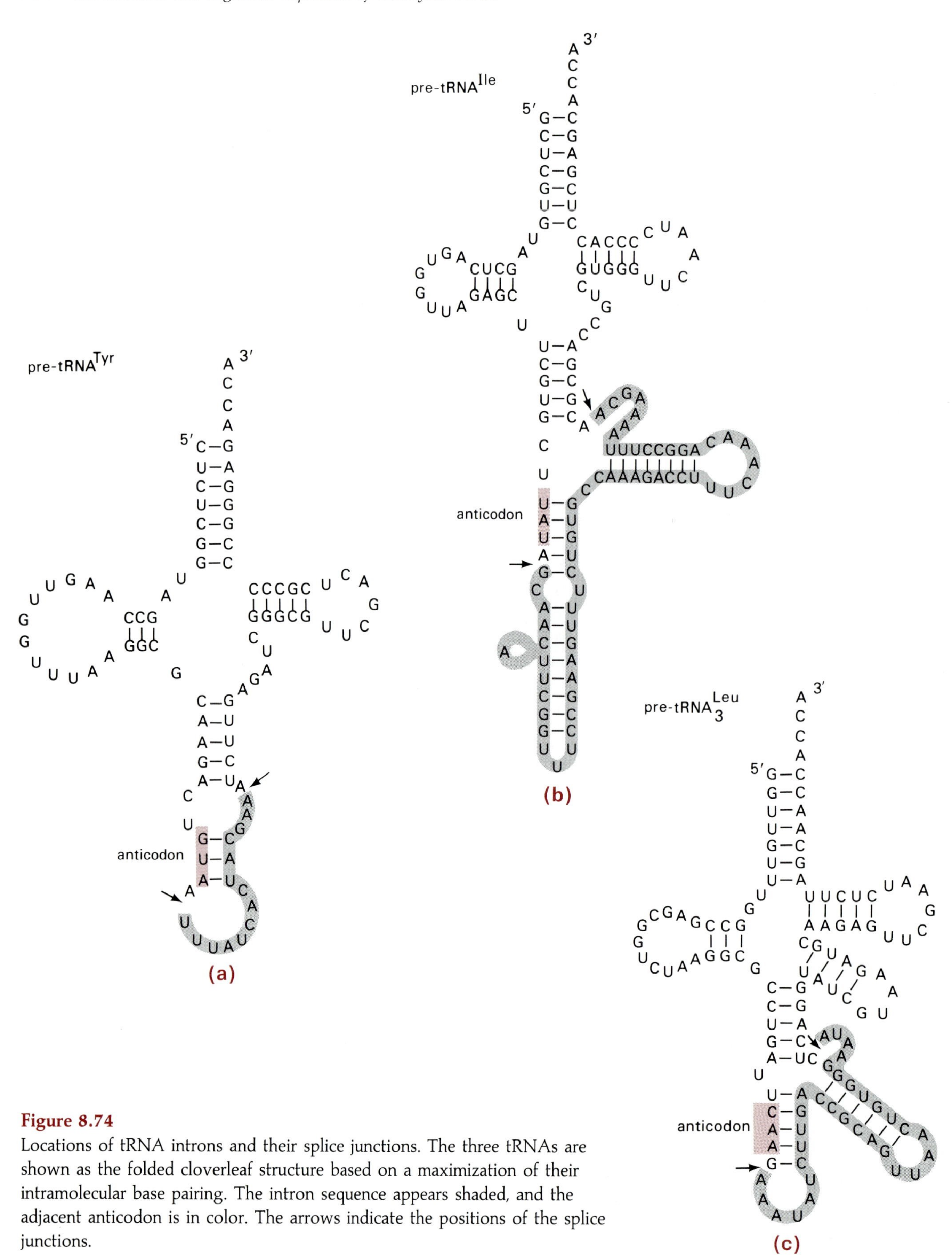

Figure 8.74
Locations of tRNA introns and their splice junctions. The three tRNAs are shown as the folded cloverleaf structure based on a maximization of their intramolecular base pairing. The intron sequence appears shaded, and the adjacent anticodon is in color. The arrows indicate the positions of the splice junctions.

Special types of introns: group I The nuclear rRNA genes of certain lower eukaryotes (e.g., *Tetrahymena thermophila* and *Physarum polycephalum*) have a special kind of intron and a unique splicing mechanism. This kind of intron, called group I, also occurs in chloroplast, yeast, and fungal mitochondrial rRNA genes; in certain yeast and fungal mitochondrial mRNA genes; and in several chloroplast tRNA genes in higher plants. The paradoxical intron in the T4 phage thymidylate synthetase gene is also a group I intron. All in all, many genes containing group I introns have been identified, but none has been found in vertebrate genes.

Although group I introns vary in size—the *Tetrahymena* pre-rRNA intron is about 400 residues long, and some of the mitochondrial pre-mRNA introns are more than a thousand nucleotides long—they share a number of characteristics: (1) Splicing is self-catalyzed and can occur *in vitro* in the absence of any proteins. (2) The essential information for splicing is encoded within the intron in multiple, relatively short internal sequences that permit folding into a characteristic three-dimensional structure. (3) Splicing is initiated by free guanosine or any of its 5′-phosphorylated derivatives. (4) Spliced and ligated rRNA and a linear RNA somewhat smaller than the intron are the products.

The critical structural features common to all class I introns reside in four separate nucleotide sequences, each about ten nucleotides long (Table 8.5). The relative locations of these sequence elements in the *Tetrahymena* rRNA intron and in the yeast mitochondrial cytochrome b intron and the way they are paired are shown in Figure 8.75. The presumption is that the intramolecular base pairing organizes the intron into a configuration favorable for the self-catalyzed splicing events (Section 8.5c). Both chemical modifications and site directed substitutions within these conserved pairing sequences prevent or impair the intron's splicing potential. Comparable sequences with the same intramolecular base pairing capabilities exist in the group I intron of the yeast mitochondrial cytochrome oxidase gene. Here, too, mutational alterations of nucleotides involved in creating the folded secondary structure prevent splicing and expression of the oxidase enzyme. Further comments on the structural features of group I introns are reserved for the discussion of the splicing mechanism (Section 8.5c).

Table 8.5 Sequence Elements Conserved in Many Group I Introns

Element	Synonym	Consensus Sequence	Complementary to
9R′	E	Not conserved	Box9R
A	P	AUGCUGGAAA	B
B	Q	AAUCAGCAGG	A
Box9L	R	UCAGAGACUACA	Box2
Box9R	E′	Not conserved	9R′
Box2	S	AAGAUAUAGUCC	Box9L

Note: The portion of each consensus sequence that is proposed to be involved in base pairing is underlined.

SOURCE: After T. Cech, *Int. Rev. of Cytology* 93 (1985), p. 3.

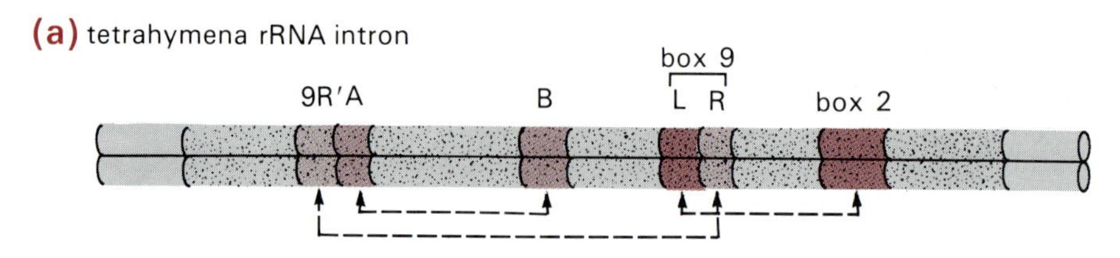

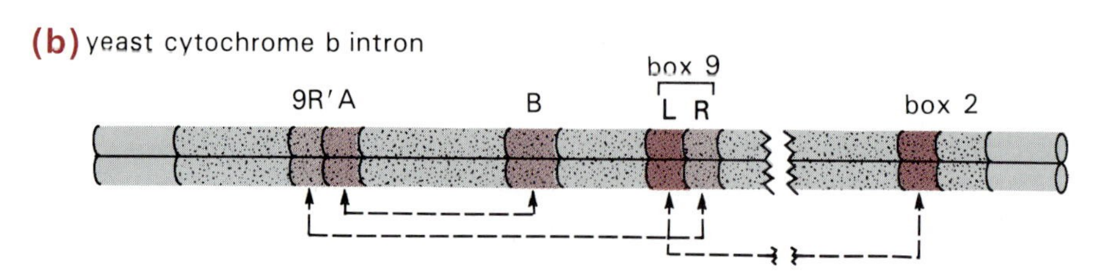

Figure 8.75
Organization of conserved sequences in mitochondrial and nuclear class I introns. Introns are shown stippled and exons without stipple. The similarly colored regions within each intron are complementary and are presumed to base pair with each other as indicated by the dashed double-headed arrows as in Figure 8.77. (a) The interactions are based on direct biochemical analysis of the excised RNA and on computer predictions. (b) These interactions are based on phylogenetic sequence comparisons and genetic data. Adapted from T. Cech, *Intl. Rev. Cytol.* 93 (1985), p. 3.

Special types of introns: group II Group II introns are less widespread than those of group I. They have been identified in two yeast mitochondrial genes—those encoding a cytochrome oxidase subunit and cytochrome b; interestingly, these genes also contain group I introns. Group II introns lack the group I consensus sequences, but they too have a distinctive secondary structure that relies on intramolecular base pairing. The group II introns also undergo self-splicing reactions *in vitro*, but in this instance, a residue within the intron, rather than added guanosine, initiates the reaction. The third key difference between group II and group I introns is in the structure of the excised intron. Rather than the circular intermediates and linear products formed during the splicing of group I introns, spliced group II introns occur as **lariats**, structures in which the 5′ phosphoryl end of the intron RNA is linked through a phosphodiester bond to the 2′ hydroxyl group of an internal nucleotide (Figure 8.76).

c. *Autocatalytically Spliced Introns*

Self-splicing seemed astonishing when it was first discovered, and the studies of its mechanism have proved to be no less remarkable. Moreover, the mechanism of self-splicing provides a basis for understanding how and why accessory proteins and a special class of ribonucleoproteins (**snRNPs**) are needed for splicing nuclear pre-mRNAs. Self-splicing occurs via an ordered sequence of self-promoted phosphodiester exchanges (**transesterifications**); in a sense, the RNA functions as an enzyme, a **ribozyme**, by providing the intramolecular environment and reactive groups to promote a series of ordered, specific transesterifications.

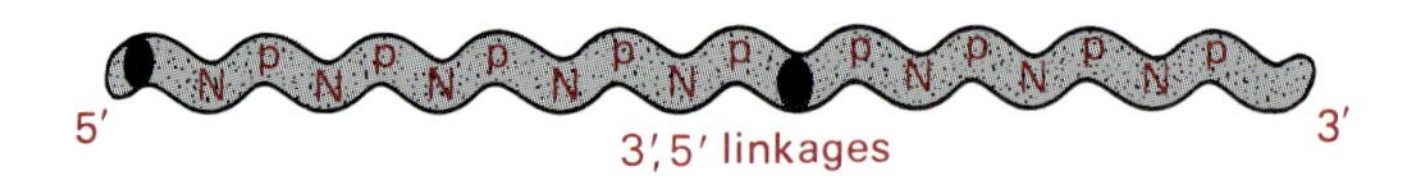

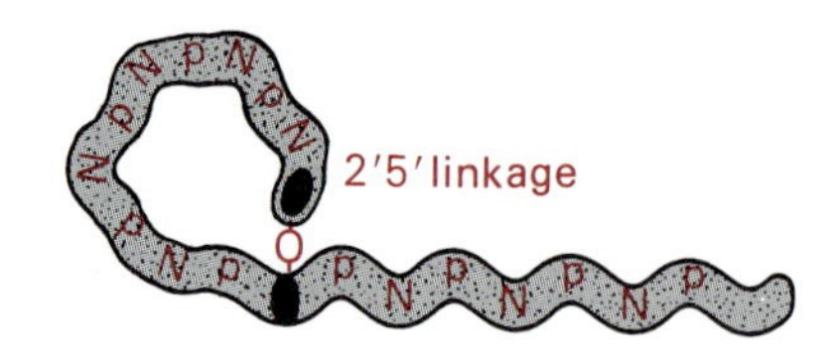

Figure 8.76
Formation of 2′, 5′ phosphodiester linkages connecting distant nucleotides in the chain. Connection between a 5′ phosphate and a 2′ hydroxyl of an internal nucleotide forms a 2′, 5′ linkage and a lariat structure.

Self-splicing group I introns Splicing of the *Tetrahymena* pre-rRNA intron, a prototypic group I intron, proceeds by a series of concerted transesterification reactions during which phosphate esters are exchanged without intermediary hydrolysis. Except for the initiation step, promoted by free guanosine or its nucleotide, all reactive groups involved in the transesterification reactions are contained within the intron sequence. Furthermore, the specificity of the bond exchanges is a consequence of a three-dimensional intron organization that depends on intramolecular base pairing between the distant but conserved sequence blocks in the intron (Figures 8.75 and 8.77).

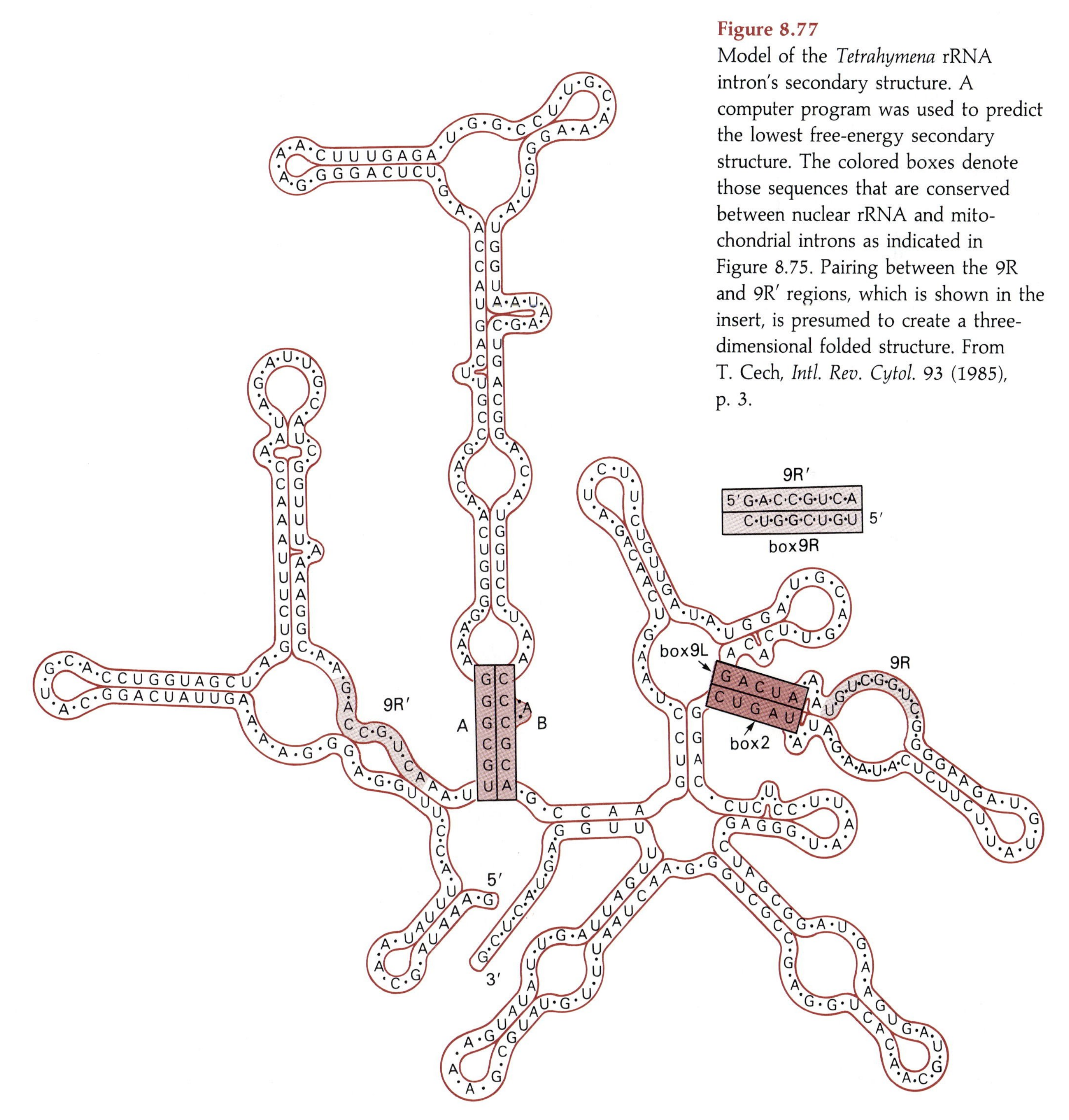

Figure 8.77
Model of the *Tetrahymena* rRNA intron's secondary structure. A computer program was used to predict the lowest free-energy secondary structure. The colored boxes denote those sequences that are conserved between nuclear rRNA and mitochondrial introns as indicated in Figure 8.75. Pairing between the 9R and 9R′ regions, which is shown in the insert, is presumed to create a three-dimensional folded structure. From T. Cech, *Intl. Rev. Cytol.* 93 (1985), p. 3.

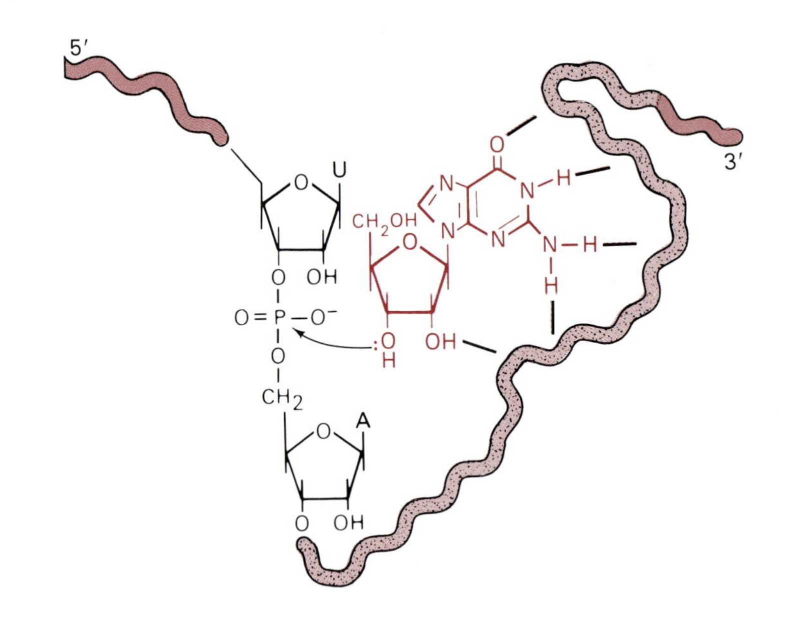

Figure 8.78
The guanosine binding site in the *Tetrahymena* rRNA intron. The intron, which is shown stippled, includes the adenylate residue at its 5′ junction, and the exon (unstippled) includes the uridylate at its 3′ junction. The proposed mode of binding guanosine via hydrogen bonds could involve different bases, sugars, or phosphates that are brought into close proximity by the RNA secondary and tertiary structure. The arrow indicates the nucleophilic attack by the 3′ hydroxyl of guanosine's ribose at the phosphorus atom at the 5′ splice site. This transesterification is presumed to be the first step of RNA self-splicing. Adapted from B. L. Bass and T. Cech, *Nature* 308 (1984), p. 820.

The first step in the splicing cascade is the binding of guanosine to an intron sequence (Figure 8.78). The unshared pair of electrons in the 3′ hydroxyl group of the bound guanosine can make a nucleophilic attack on the phosphate group at the 5′ exon-intron junction (-UpA-), causing cleavage at that site (Figure 8.79). The 3′ hydroxyl group created at the cleavage site (the end of the 5′ exon) reacts with the phosphate at the 3′ exon-intron junction, causing the two exons to be ligated together with release of the 413 nucleotide long linear intron. The linear intron segment then undergoes two successive intramolecular transesterifications and subsequent hydrolysis reactions, resulting in the removal of first 15 and then 4 nucleotides to yield the final processed intron. Note that there is essentially no net energy change accompanying the overall reaction; each phosphodiester bond disruption is compensated by the concomitant formation of a different phosphodiester bond.

Guanosine or one of its 5′ phosphorylated derivatives is specific for initiating the splicing reaction. Modifications of either the 2′ or 3′ hydroxyl groups or the guanosine's base pairing potential to cytosine residues impairs its initiating capacity. Ostensibly, base pairing of the guanosine with a complementary residue in the intron is important for properly positioning the guanosine 3′ hydroxyl group for its reaction with the phosphate at the 5′ splice junction.

The two exon-intron junctions are probably brought into proximity to enable the two exons to be ligated following the initial cleavage by guanosine. This is facilitated by folding the intron sequence in such a way as to juxtapose the two junctions, presumably in the vicinity of the guanosine binding site. One possibility is to have a sequence in the intron that can base pair with the exon sequences at both splice junctions (Figure 8.80). Such an "internal guide sequence" could enable the 3′ hydroxyl group created by the guanosine mediated cleavage of the 5′ exon-intron junction to react with the phosphate at the 3′ splice site to form the spliced product. The G residue that terminates the intron and the adjacent

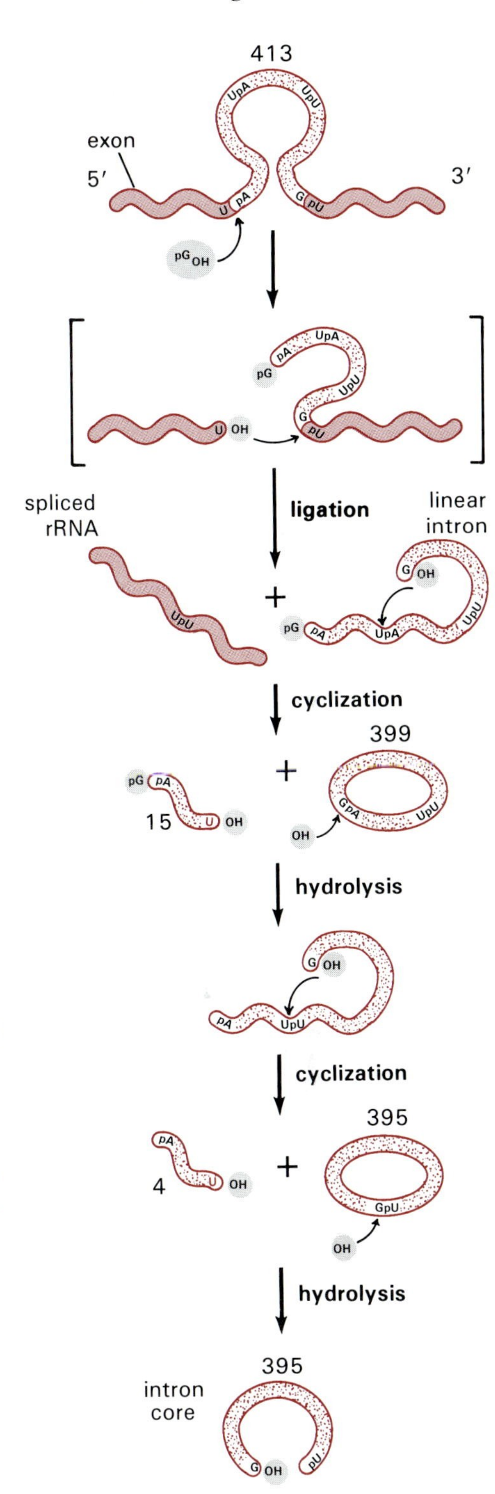

Figure 8.79
Reactions in the self-splicing of a group I intron. The example shown is for the rRNA intron of *Tetrahymena thermophila*. Stippled segments are intron, and unstippled segments are exon. Important reactants are highlighted in deep color. All the indicated structures, except for the bracketed initial cleavage product, have been isolated. Where indicated, the numbers refer to the lengths of the RNA segments.

intronic nucleotides specify the 3′ splice site. Conformational changes in the released linear intron are presumed to facilitate the concurrent circularizations and release of the short oligonucleotides that follow (Figure 8.81).

The mechanism of self-splicing elucidated with the *Tetrahymena* rRNA gene's intron probably serves for the other group I introns as well. All share the same four conserved sequences called A, B, 9L, and 2 and the two nonconserved but complementary sequences called 9R and 9R′ (Table 8.5 and Figure 8.75). The six sequence elements always occur in the same order in the intron, 5′-9R′-A-B-9L-9R-2-3′. Moreover, most group I introns also have a sequence that can act as an internal guide. Mutations that prevent splicing are known to occur in a variety of yeast and fungal mitochondrial mRNA and rRNA genes. Such mutations generally disrupt the potential for base pairing. Thus, for example, the inactivation of splicing due to alterations in the sequence of either 9R or 9R′ is reversed by compensating changes in the other. The catalytic or enzymelike properties and behavior of the self-splicing introns are considered further in Section 8.5d.

The *in vivo* splicing of some group I introns in yeast mitochondrial pre-mRNA genes depends on the same conserved sequences needed for *Tetrahymena* pre-rRNA splicing, and the products of splicing are also analogous. Yet these pre-mRNAs do not undergo self-splicing *in vitro*. Splicing of these introns depends on transacting proteins encoded within the same or related genes, generally by open reading frames that span

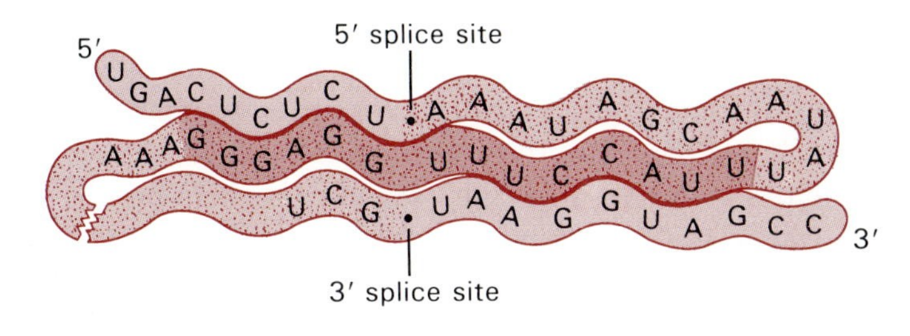

Figure 8.80
An internal guide sequence may bridge the two exon junctions. Intron sequences are shown stippled and exons unstippled. The intron's internal guide sequence is in deep color and shown base paired to exon sequences at the 5′ and 3′ splice sites.

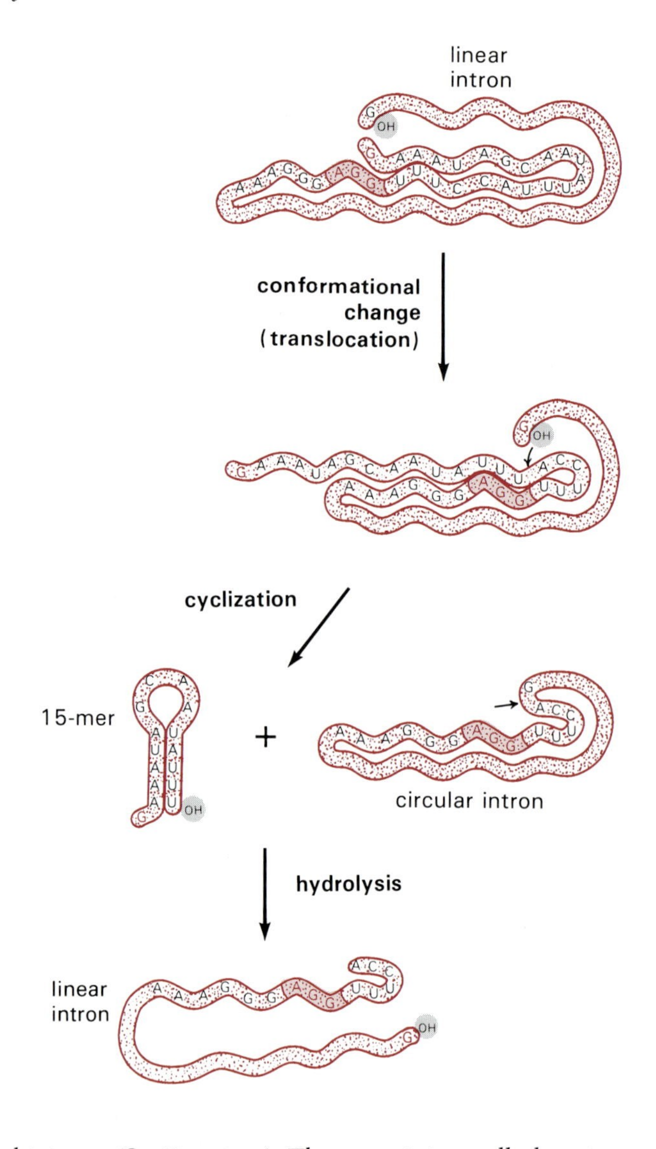

Figure 8.81
Alternative base pairing within the excised intron promote further reactions, leading to shortening and cyclization. Once released, the linear intron may undergo conformational alterations due to changes in the intramolecular base pairing pattern. Thus, pairing of the GGG (colored segment) with a run of 3 Us may allow the 3′ hydroxyl group of the terminal G to cleave the AU linkage.

exons and introns (Section 9.7a). These proteins, called **maturases**, persist only fleetingly and apparently help fold the pre-mRNA's intron into the proper conformation for RNA directed splicing to proceed. This explains why nonsense or frameshift mutations in the intronic sequence coding for the maturase prevent splicing of that intron as well as splicing of the group I introns in other mitochondrial genes. This model also rationalizes why suppressor mutations or compensating frameshift mutations that restore production of the maturase also restore proper splicing. Thus, correct folding of some group I introns may occur spontaneously and be stable *in vitro*, but others may require proteins to direct and/or stabilize the autocatalytically active structure. Whether protein-protein interactions play a role in the folding is not known.

Another sort of self-splicing Group II introns in yeast mitochondrial pre-mRNAs also have conserved sequences and probably defined secondary structures, but these are distinct from those of group I. An intron of this type also undergoes self-splicing, although less is known about the secondary and tertiary structural features of this substrate and consequently about the molecular details of the splicing. Two features of the

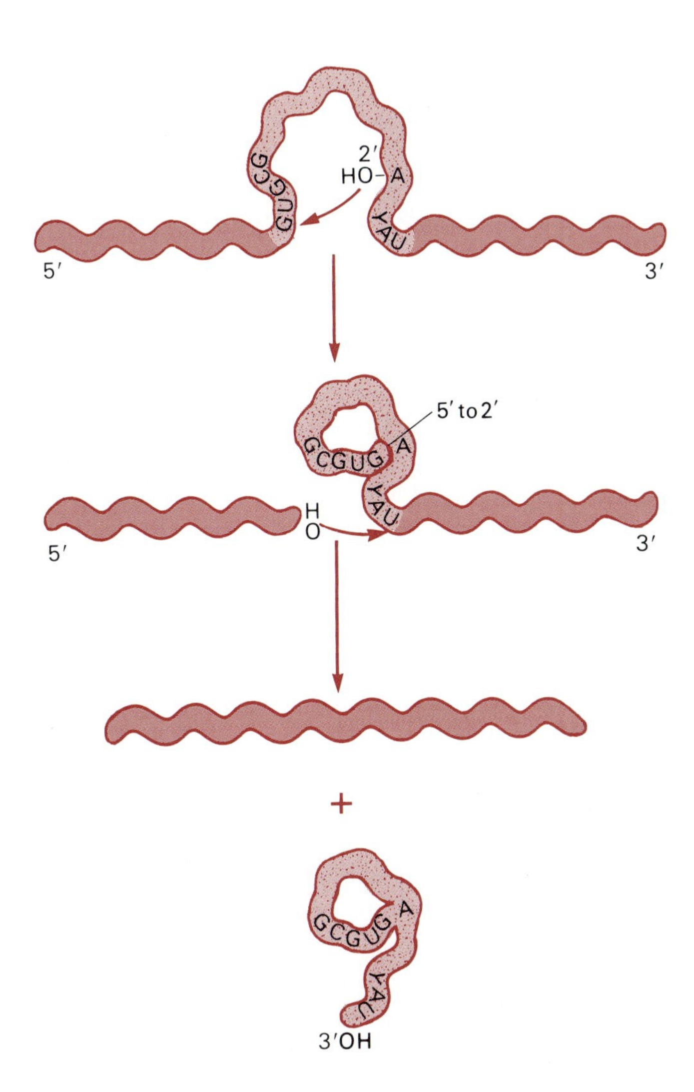

Figure 8.82
Splicing group II introns. A 2′ hydroxyl group on an adenylate residue in the intron acts as the nucleophile to cause cleavage at the 5′ splice site. Joining of the 5′ exon's 3′ hydroxyl group to the 3′ exon's phosphate completes the splicing, thereby releasing the linear lariat RNA. The numbers identify the phosphate residues at the intron junctions and their fates in the spliced product and the intron lariat.

mechanism are clear: (1) Unlike the self-splicing reaction of group I introns, splicing of group II introns proceeds without the need for a nucleoside initiator. (2) The product of the splicing reaction is a lariat structure (Figure 8.76).

How does a group II intron splice itself out, and what accounts for the novel RNA lariat product? As with group I introns, the process is presumed to occur via two transesterification steps, one involving cleavage of the 5′ splice site and the second cleavage of the 3′ splice site and ligation of the two exons. The fundamental difference, however, is the nature of the attacking nucleophile: guanosine with group I introns and the 2′ hydroxyl group of a nucleotide somewhere in the intron with the group II introns (Figure 8.82). The lariat results from the creation of the novel 2′, 5′ phosphodiester linkage in the midst of the RNA sequence.

The three-dimensional structure of the intron, whether acquired spontaneously or facilitated by bound proteins, must bring the lariat forming 2′ hydroxyl group into close proximity to the 5′ splice site and activate the transesterification. The choice of which internucleotide bonds are cleaved probably involves the group II–specific sequence, GUGCG, at the 5′ end of the intron and the YAU sequence at the intron's 3′ end (Table 8.4). As shown in Section 8.5d, this mechanism of intron splicing closely resembles that employed for splicing the introns in nuclear pre-mRNAs. The need for *trans* acting ribonucleoprotein particles to mediate the splicing of the latter type of introns may reflect the need for more sophisticated and specific ways to fold the many different kinds of pre-mRNAs and to ensure splicing between the proper exon-intron junctions.

d. Splicing Nuclear Pre-mRNA Introns

Much of what has been learned about the mechanism of splicing introns from nuclear pre-mRNAs was facilitated by the availability of cloned split genes. Particularly useful were natural or deliberately altered forms of these genes. Specially constructed substrates with convenient fusions of exons and introns also helped identify the structural requirements for faithful splicing. But the most informative approach proved to be biochemical, the use of cell extracts competent to carry out splicing, and, subsequently, the purification of the components of the splicing machinery. Especially important were capped RNA substrates, prepared by transcribing specially designed DNA templates that were inserted downstream of the phage SP6 promoter with the specific SP6 RNA polymerase (Figure 6.23).

General features Splicing of nuclear pre-mRNAs occurs in the nucleus, perhaps concomitantly with the transcription of some genes and only after completion of the transcript with other genes. There are indications that capping of the transcript's 5′ end is involved, but in what way is not clear; polyadenylation of the RNA is clearly not required. A principal problem, both conceptually and practically, is how pre-mRNAs with multiple introns are spliced accurately so that neighboring exons are spliced to each other. Because it is known that the 5′ splice site of one intron can be spliced to the 3′ splice site of another, it is important to understand how this is prevented in the normal course of splicing introns and how it is promoted to permit alternative modes of splicing (Section 8.5f).

Pathway of splicing reactions We focus first on the cleavages and ligations that constitute the splicing reactions. The initial step in splicing is the assembly of a splicing complex. The earliest products detected during splicing *in vitro* result from a cleavage precisely at the 5′ splice site; one product contains the 5′ exon, and the other contains the intron and the 3′ exon (Figure 8.83). Because no RNA species other than the completely spliced product is altered at the 3′ splice site, cleavage at the 5′ splice site must precede cleavage at the 3′ splice site. As the reaction proceeds, two products accumulate: the correctly spliced, ligated exons and the released complete intron. Both the initial cleavage product and the excised intron contain lariat structures of the type described for the self-splicing of group II mitochondrial introns (Figure 8.83). The lariat structure was initially inferred from the excised intron's anomolous electrophoretic mobility, its inability to be reverse transcribed beyond a point near the 3′ end of the intron,. and its resistance to complete digestion by several ribonucleases. Subsequently, the structure was confirmed by the isolation of the branched nucleotides. The branch point always contains the 5′ phosphate of the guanosine at the intron junction joined in diester linkage

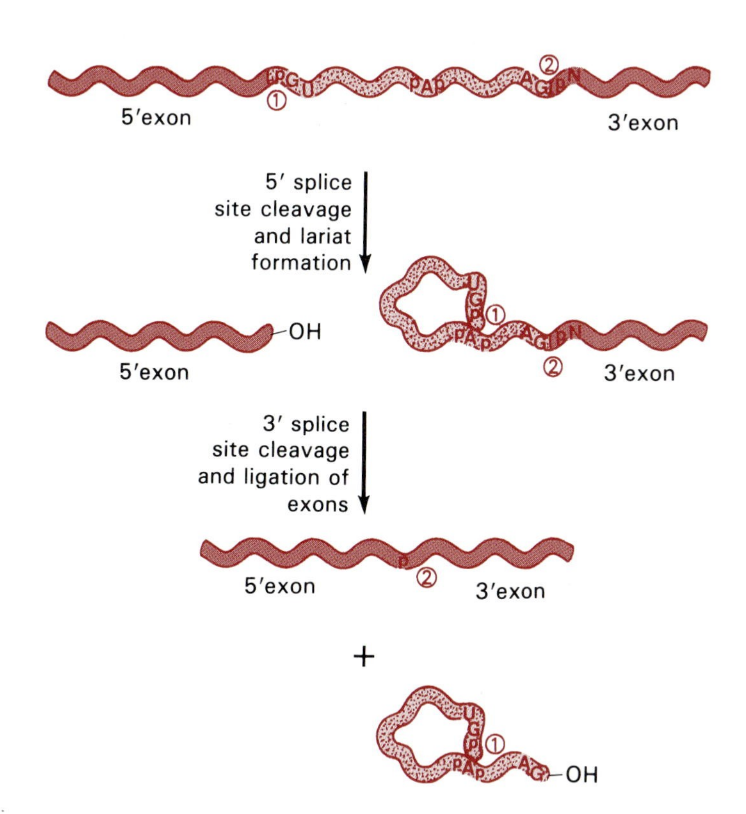

Figure 8.83
Splicing pathway for nuclear pre-mRNA introns. Introns are shown stippled, and the flanking exons are unstippled. The overall splicing reaction involves cleavage of the 5′ splice site and concomitant formation of the 5′, 2′ phosphodiester linkage. Cleavage of the 3′ splice site and ligation of the exon ends complete the process. The nucleotide sequences at the relevant sites are indicated.

to the 2′ hydroxyl group of the adenosine residue at the branch point irrespective of the intron spliced. The puzzling existence of nuclear RNAs with occasional branches emanating from the 2′ hydroxyl groups of adenylate residues was at last explained.

Pre-mRNA splicing in spliceosomes In an overall sense, splicing of nuclear pre-mRNA and group II introns have similar outcomes: removal of the intron as a lariat and ligation of the two exon junctions. They differ, however, in that the splicing of nuclear pre-mRNAs requires a multitude of nuclear factors—proteins and ribonucleoprotein complexes (**snRNPs**). The large multisubunit complex that catalyzes splicing has been termed a **spliceosome**. Spliceosomes, which assemble on the intron in an apparently ordered sequence prior to the initial cleavage at the 5′ splice site, contain the intron bound with at least five distinct snRNPs and several additional proteins not normally associated with the snRNPs. The formation of spliceosomes relies on base pairing between RNAs, associations of proteins with RNA, and associations of proteins with proteins.

snRNPs The identification and characterization of snRNP particles preceded recognition of their involvement in nuclear splicing. More recently, sophisticated fractionation procedures that depend on the availability of antibodies directed against the unique trimethyl guanosine cap structure of snRNAs have improved the purification of snRNPs. Individual snRNPs have also been obtained using antibodies present in sera from patients with various autoimmune disorders. These sera specifically precipitate one or a few different snRNPs. The snRNPs in mammalian spliceosomes contain U1, U2, U4, U5, and U6 RNAs (Table 8.6). These RNAs are called U RNAs because of their unusually high content of uracil and its modified

Table 8.6 Size of U RNAs

U RNA	Size in Nucleotides	
	Yeast	Mammalian
U1	568	164
U2	1175	187
U3	—	217
U4	160	145
U5	214	116
U6	112	106
U7	—	65
U11	—	131

Note: Dashes indicate where data are not available.

forms. U1, U2, and U5 RNAs each occur in a distinctive snRNP; U4 and U6 RNAs, which are extensively base paired to each other, constitute the RNA part of another snRNP. snRNPs containing U3 RNA participate in the processing of pre-rRNA in the nucleolus (Section 8.2d), the snRNP containing U7 RNA is involved in creating the 3′ ends of histone mRNAs (Section 8.3c), and U11 RNA has been implicated in the polyadenylation of mRNAs (Section 8.3c). There are additional snRNPs in mammalian nuclei, but their molecular makeup and functions are unknown.

The nucleotide sequences of corresponding U RNAs from all vertebrates so far examined are about 95 percent identical. This is even true for the U1 RNA from humans and *Drosophila*. snRNPs containing analogous small U-rich RNAs also occur in cockroach, sea urchin, *Tetrahymena, Dictyostelium*, yeast, and dinoflagellates. Even among such widely divergent species, the U RNAs share regions of homology with each other, although substantial parts of their sequences are unique to each species. The extraordinary conservation of these molecules and their participation in specific steps in RNA processing indicate that this is a fundamental reaction, probably originating early in evolution.

The snRNPs involved in spliceosomes (U1, U2, U5, and U4/U6) all contain a common set of seven proteins (Table 8.7). U1, U2, and U5 snRNPs also contain several proteins distinctive of their type. The autoimmune sera commonly used to isolate and classify snRNPs are often specific for one or a few proteins in the particle. U3 snRNP, which functions in rRNA processing, contains at least six proteins not present in snRNPs involved in splicing. The proteins in the lower abundance U snRNPs, U7 and U11, are not as well known.

snRNP interaction with intron sequences The first hint about how snRNPs are involved in splicing came from the recognition that the nucleotide sequence at the 5′ end of U1 RNA is complementary to the consensus sequence at all intron's 5′ splice sites (Figure 8.84). Furthermore, U1 snRNP binds specifically to the 5′ splice junction *in vitro*, where it protects a sequence of about 17 nucleotides at the intron's 5′ splice site from RNase digestion; the RNA is neither cut nor modified by the binding. The U2 snRNP binds to and protects a 40 nucleotide region that contains the A

Table 8.7 Proteins in Mammalian snRNPs

Proteins	snRNPs Containing Indicated U RNA: U1	U2	U3	U5	U4/U6	Molecular Weight (kDaltons)
B, B′, D, D′, E, F, G	+	+	−	+	+	28, 29, 16, 15.5, 12, 11, 9
70k, A, C	+					70, 34, 22
A′, B″		+				33, 28
25K, 1BP				+		25, 100 or 70
(34K, 74K, 59K			+			34, 74, 59
30K, 13K, 125K)			+			30, 13, 125

Note: Because of their low abundance, U7 to U11 snRNP protein constituents are not identified. Proteins are currently named by letters or by their apparent molecular weights. The molecular weights of the proteins common to U1, U2, U5, U4/U6 are taken from the amino acid composition deduced from their cloned cDNAs.

nucleotide involved in forming the branched lariat structure. The molecular basis for the binding of the U2 snRNP to this region of mammalian introns is not entirely clear because the sequence is not highly conserved. However, in yeast, where the sequence containing the lariat's A residue is stringently conserved, the requirement for complementarity between the 5′ end of U2 RNA and the intron sequence is clearly established by genetic experiments. A stretch of about 15 nucleotides at the 3′ end of the intron is bound by U5 snRNP. However, physiologically, this site may be bound and activated by a complex of U5 snRNP and the U4/U6 snRNP. The binding of both U2 and U5 snRNPs to their respective target sequences requires additional proteins. One protein (AF) binds to the intron's 3′ junction (the AG- and pyrimidine-rich tract) and facilitates the binding of the U2 snRNP nearby. Binding the complex of U4/U6 and U5 snRNPs also requires a protein that stabilizes the association of the complex with the 3′ intron junction. Most likely, protein-protein and protein-RNA interactions play a part in the formation of these complexes.

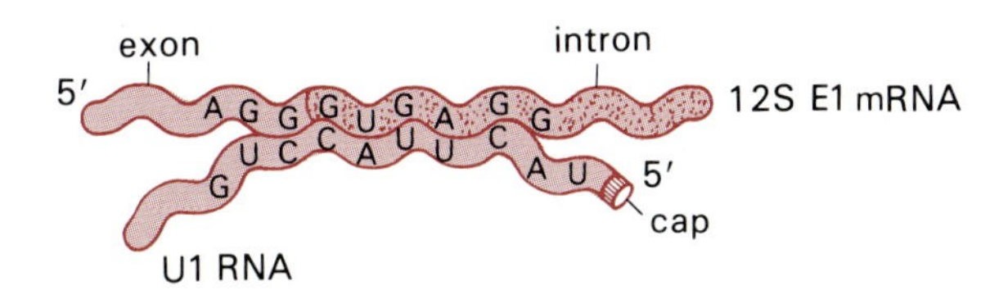

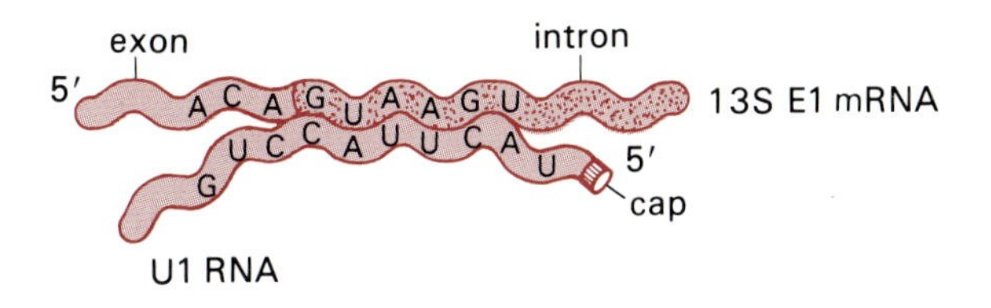

Figure 8.84
Association of U1 RNA with exon-intron junctions in mRNA. Intron (stippled)-exon (unstippled) junctions in two adenovirus E1 mRNAs can base pair with the 5′ end of U1 RNA.

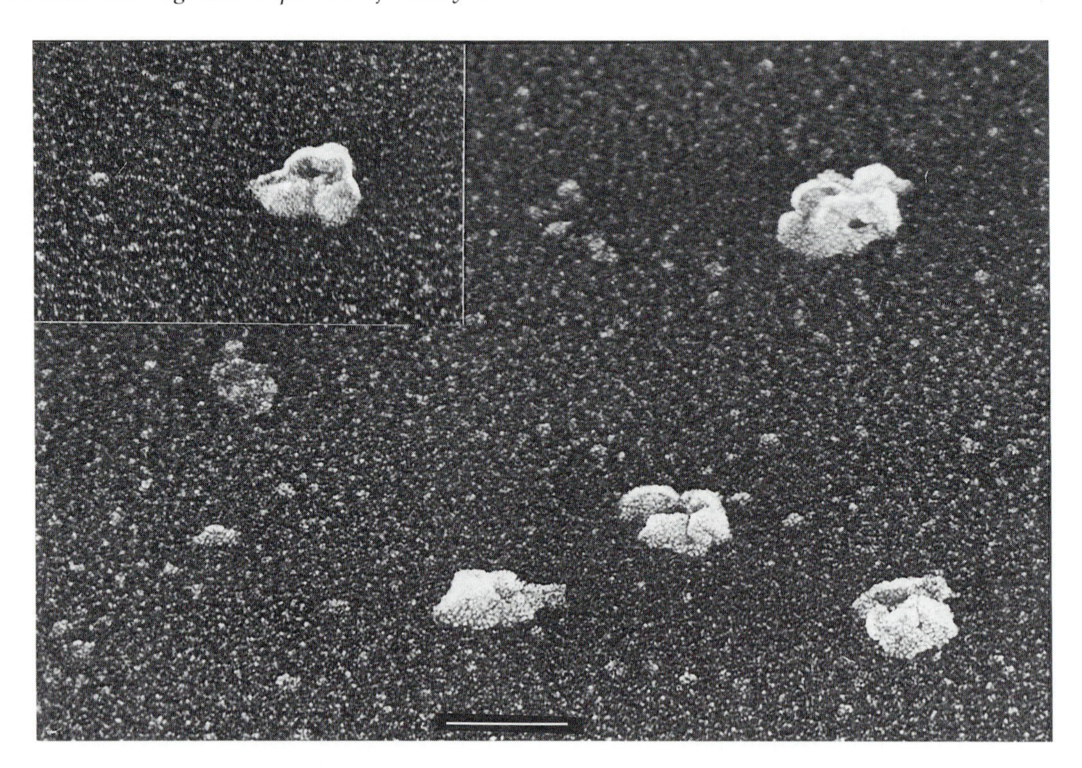

Figure 8.85
Electron micrograph of spliceosomes. The picture shown is Figure 10 in R. Reed et al., *Cell* 53 (1988), p. 949. It was kindly provided by J. Griffith.

The actual involvement of U1 and U2 snRNPs in the splicing reaction is inferred from the fact that immunoprecipitation of U1 and U2 snRNPs by their respective antibodies from extracts that catalyze pre-mRNA splicing inhibits splicing. Moreover, appropriately fractionated splicing extracts are dependent on the addition of U1 and U2 snRNPs. Similarly, exogenous snRNPs must be added to the splicing reaction when the extract's own snRNPs are destroyed by targeted endonuclease (RNaseH) cleavage using short oligodeoxynucleotides complementary to each of the U RNAs.

Spliceosome assembly Splicing occurs in spliceosomes; therefore, their structure, assembly, and mechanism of action need to be considered. Spliceosomes isolated from a reaction in which the 3′ cleavage step is inactivated appear to be ellipsoid-shaped particles whose dimensions are about 25 by 50 μm (Figure 8.85). Within the particles, the intron is associated with one each of the U1, U2, U4/U6, and U5 snRNPs. Functional spliceosomes also contain other proteins, particularly those that are involved in binding the snRNPs to their target sites or are normally bound to nuclear pre-mRNA.

Only minimal amounts of exon sequence seem to be needed for snRNP binding to an intron. In our present picture of spliceosome assembly, U1 snRNP binds first to the 5′ splice junction irrespective of the existence of functional U2 snRNP and U5 snRNP binding sites (Figure 8.86); but binding alone is not sufficient to promote the subsequent cleavage at the 5′ splice site. Association of U2, U5, and the U4/U6 snRNP complexes occurs subsequently and requires ATP. U2 snRNP may rely on base pairing for binding to its site in the intron; however, incorporation of the

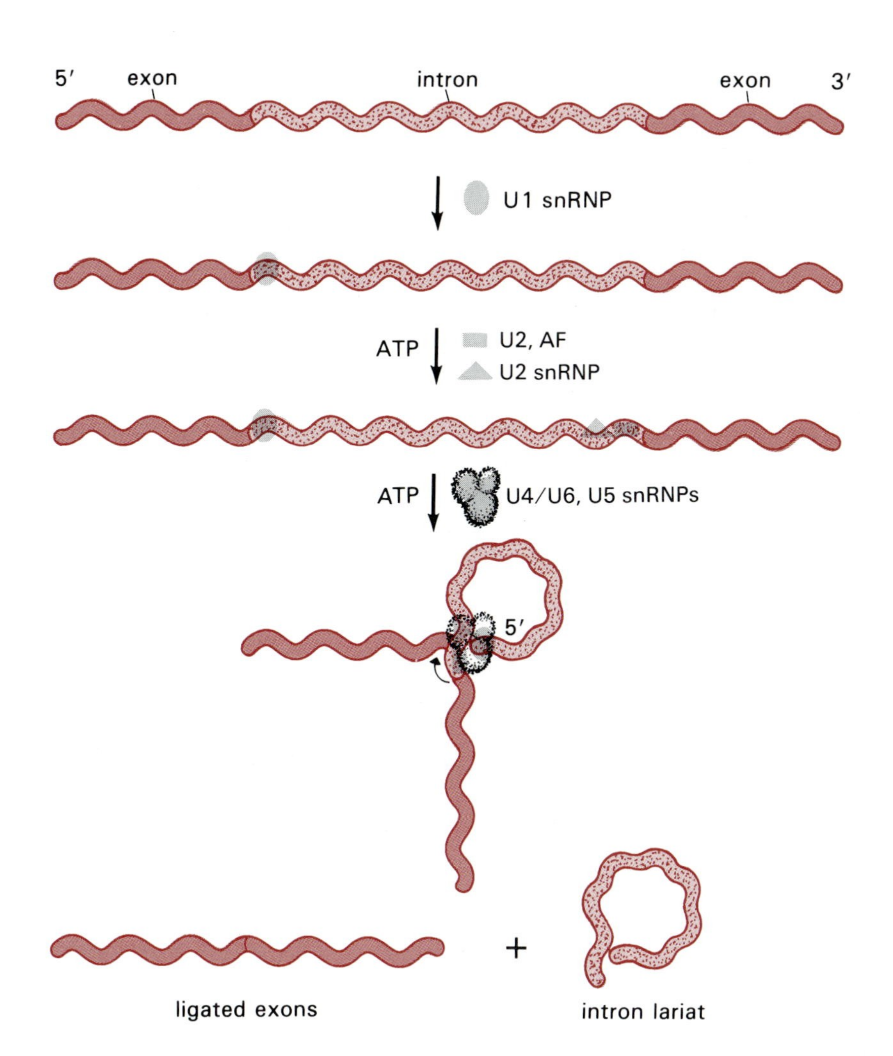

Figure 8.86
Ordered assembly of spliceosomes during pre-mRNA splicing. The fate of the spliceosome following ligation of the two exons and release of the intron lariat is not shown.

complex containing the U4/U6 and U5 snRNPs into the spliceosome depends on interactions between snRNPs. A similar scheme accounts for spliceosome assembly during splicing of yeast pre-mRNA. The fate of the spliceosome following exon ligation is uncertain, but individual or subassemblies of snRNPs are probably recycled to other introns as the intron RNA is degraded. These reaction paths represent a possible, but not obligatory, order of spliceosome assembly. To establish the pathway, an analysis of the precursor-product relationships is needed. Also, the model lacks an explanation for how the 3′ splice site is cleaved, how ligation of the two exons occurs, and how the joinings are restricted to the proper two exons.

Self-splicing and spliceosome mediated splicing The end result of splicing group II and pre-mRNA introns is the same: the intron is excised as a lariat structure, and the two flanking exons are joined. Moreover, the chemistry of the two processes is similar. In both, a 2′ hydroxyl group within the intron serves as the nucleophile to promote cleavage at the 5′ splice site, and the 3′ hydroxyl group of the upstream exon is the nucleophile that cleaves the 3′ splice site by forming the exon-exon bond. The key difference is that at least some group II introns are self-splicing *in vitro*, but nuclear pre-mRNA splicing requires an elaborate machinery of multiple snRNPs and accessory proteins. Yet some group II introns do require maturases for their splicing *in vivo,* although it seems likely that both the nucleotides and catalytic centers involved in the autocatalytic and maturase aided splicing are the same. Accepting that autocatalytic splicing requires a highly specific folded RNA structure, maturases may be required for folding or stabilizing a structure that can undergo the correct, sequential transesterifications. Thus, the only way maturase aided and autocatalytic splicing reactions may differ is in how the intron achieves the catalytically active conformation.

The same considerations may hold for spliceosome mediated splicing. Perhaps snRNPs and the associated factors create a "scaffold" upon which the intron can be properly folded so that the two intron junctions and the branch point residue are juxtaposed and activated. In that sense, the spliceosome's involvement in nuclear pre-mRNA splicing is analogous to that of the maturases for splicing some group II introns. One might wonder why such a complex array of snRNPs is needed to fold nuclear pre-mRNA introns when a single maturase protein suffices for those group II introns that required the protein. The answer may lie in the complexity of the splicing challenge. Mitochondrial group II introns are similar to one another, each having several highly conserved sequence patches. Perhaps only one or two proteins are needed to fold and maintain the intron's active conformation. But the extraordinary diversity of intron sizes and sequences and the large number of different introns in nuclear pre-mRNAs poses additional problems. No single protein is likely to be able to fold all introns into a self-reactive form. However, highly conserved snRNPs that can base pair with the only two or three conserved sequences in all introns and associate with one another by interactions via snRNPs may be able to fold any intron into the conformation that favors the two RNA catalyzed transesterifications.

Trans *splicing* Our consideration of splicing thus far has focused on intramolecular or *cis* reactions, but one may ask whether intermolecular or ***trans* splicing** also occurs. More specifically, can two exons located on separate RNA molecules be ligated together with the concomitant elimination of their flanking introns? Intermolecular splicing has in fact been demonstrated *in vitro* with specially designed RNA substrates. More significantly, *trans* splicing has been established as an essential step in the cellular production of all *Trypanosome* mRNAs and of three of the four actin gene mRNAs in *C. elegans*. In each case, the splice sites are of the canonical type, GU at the 5′ junction and AG at the 3′ junction. Moreover, the product in each case is a branched structure.

Two experimental designs have been used to demonstrate *trans* splicing *in vitro*. In the first, the 5′ exon and flanking intron segment are on one RNA, and the 3′ exon and associated intron are on a second RNA (Figure 8.87a). Complementary base pairing between sequences in the two intron segments holds the two RNAs together. Proper joining of the exons and elimination of the intron segments occurs in cell extracts about 15 percent as efficiently as with molecules whose introns are in the same RNA molecule. Thus, neither the existence of secondary structure nor a covalent interruption in the intron prevents splicing. The second experiment also relies on complementary sequences in two introns to join the two splice-

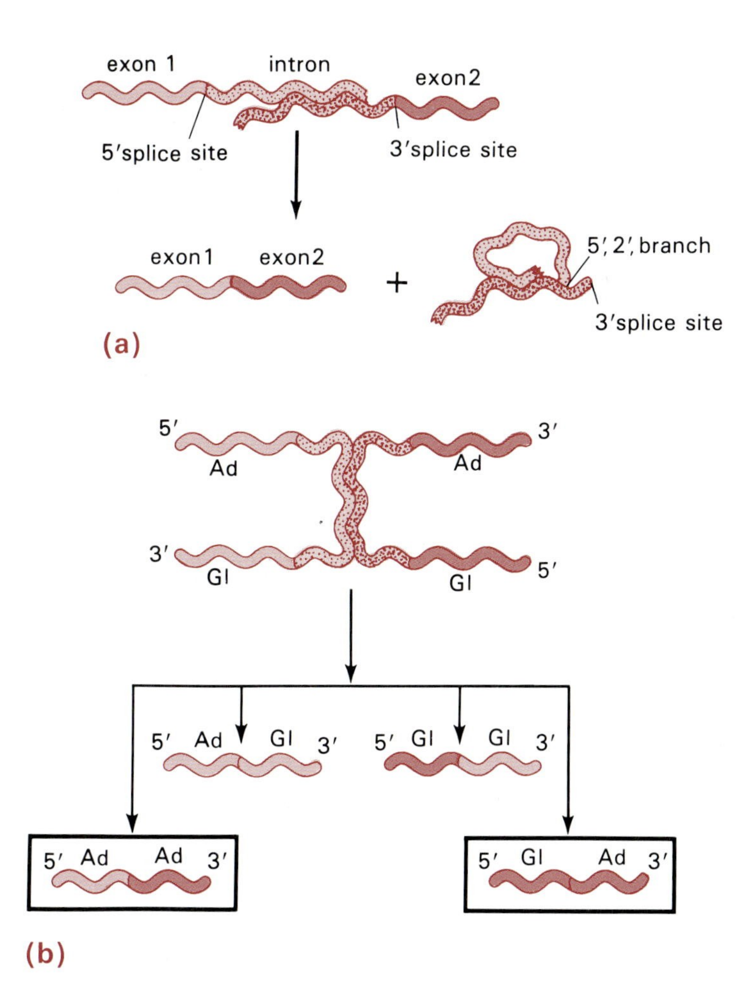

Figure 8.87
Trans splicing of mRNAs *in vitro*. (a) Splicing between exons located on separate RNAs was tested in an extract that catalyzes splicing of pre-mRNAs. One RNA contains an exon and an intron joined by a 5′ splice site. The other RNA consists of an exon and an intron joined by a 3′ splice site. Complementary sequences in the two introns hold the two RNAs together. The excised introns have a modified lariatlike structure. (b) Two separate mRNAs whose introns are complementary over much of their lengths were used as substrates for *in vitro* splicing. Ad refers to 5′ and 3′ exon sequences from adenovirus 2 mRNAs, and Gl refers to 5′ and 3′ exon sequences from globin mRNA. Introns are shown stippled. Four possible spliced products, two formed by *trans* splicing and two by *cis* splicing, are shown below. The boxes indicate the predominant spliced products.

able RNAs (Figure 8.87b). In this design, four spliced products are possible: two formed by *trans* splicing and two by *cis* splicing. The outcome of *in vitro* splicing with this substrate is to form the *trans* spliced 5′ adeno-adeno 3′ and the *cis* spliced 5′ globin-adeno 3′ products. The *cis* spliced 5′ adeno-globin 3′ and *trans* spliced 5′ globin-globin 3′ products are also produced, but in reduced quantities, possibly because of the proximity of the base paired stem to the intron's branch point.

The cellular formation of functional actin mRNA and several other mRNAs in *C. elegans* occurs by *trans* splicing. In the case of actin, the cap and first 22 nucleotides at the 5′ termini of the muscle-specific actin mRNAs are derived from the 5′ terminus of a 100 nucleotide long RNA transcribed from the 5S rDNA repeat located on a different chromosome. However, the best studied example of cellular *trans* splicing is the formation of *Trypanosome* mRNAs (Section 10.6d). All *Trypanosome* mRNAs contain an identical 35 nucleotide 5′ untranslated sequence preceding an uninterrupted coding sequence. These 5′ **miniexons** are derived from the 5′ terminal stretch of a 137 nucleotide RNA that is transcribed from hundreds of tandem copies of the corresponding 137 bp DNA segment in the genome (Figure 8.88). *Trans* splicing joins the 35 nucleotide 5′ miniexon from one RNA to the protein coding exon on another RNA. First a branched RNA molecule is formed by joining the two introns. In the next step, the 3′ splice site is cleaved, and the two exons are joined. The branched segment containing "intron" sequences from the two separate

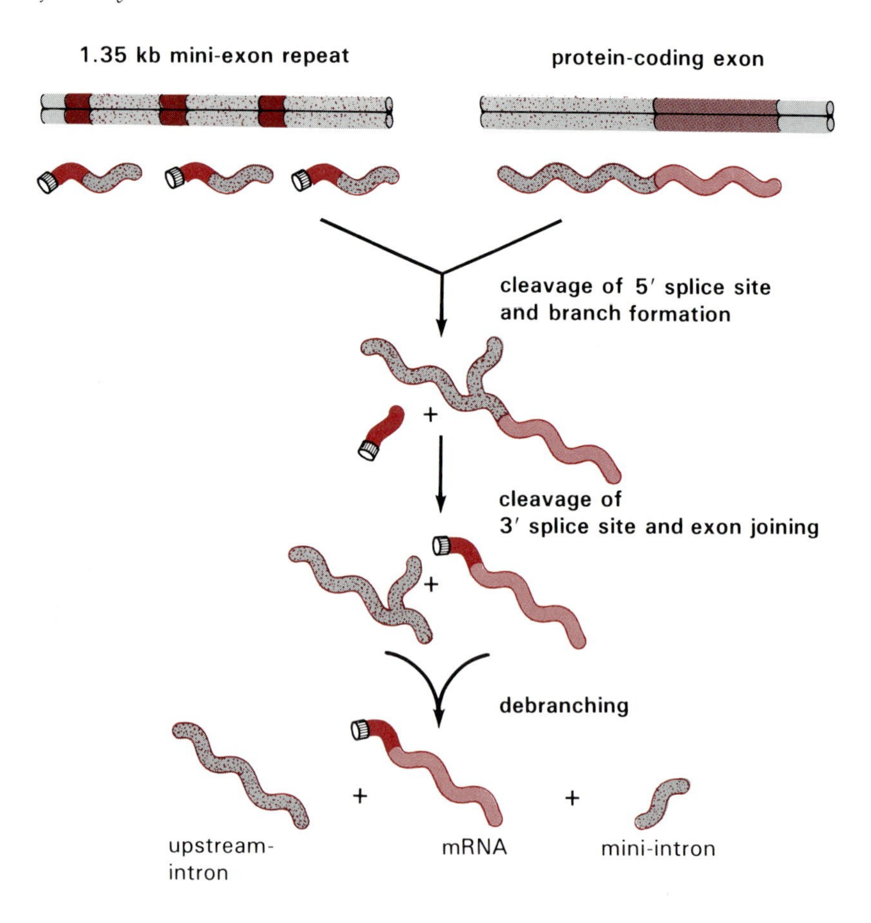

Figure 8.88
Trans splicing of *Trypanosome* mRNAs. The 5′ mini-exon of each *Trypanosome* mRNA derives from short RNAs transcribed from tandemly repeated transcription units clustered at a specific chromosomal location. These 137 nucleotide transcripts consist of a 35 nucleotide 5′ mini-exon and the remainder as intron (stippled). Protein coding exons transcribed from individual genes undergo *trans* splicing as indicated.

RNAs is then cleaved by a debranching enzyme, resulting in the separation of the introns that were associated with the two RNAs. Particularly noteworthy are the facts that (1) the sequence immediately downstream of the *Trypanosome* miniexon conforms to a consensus 5′ splice site sequence (GUAUGA), (2) the intron junction associated with the coding exon is similarly analogous to sequences at 3′ splice sites ($[C/U]_n$NNAG), and (3) an intron nucleotide provides the branch point, suggesting that the *trans* and *cis* splicing mechanisms are analogous. The mystery is how the splicing complex is established from two separate RNAs.

e. *Splicing tRNAs*

The way in which tRNA introns are spliced is best understood in yeast, with some information having been derived from studies with other lower eukaryotes and plants. All the enzymes are known and highly purified, and all the intermediates are characterized, even if their existence and rationale is puzzling (Figure 8.89a). A specific tRNA endonuclease cleaves the pre-tRNA at the 5′ intron junction, creating a 2′, 3′ cyclic phosphodiester at the end of the 5′ half of the tRNA. The same enzyme cleaves the other intron junction, producing a 5′ hydroxyl group at the end of the tRNA's 3′ half. These cleavages are reminiscent of cleavages made by some RNAses. The cyclic phosphodiester at the end of the 5′ exon is then cleaved to a 2′ phosphomonoester by a cyclic phosphodiesterase. After phosphorylation of its 5′ hydroxyl end, the 3′ exon is adenylylated by a

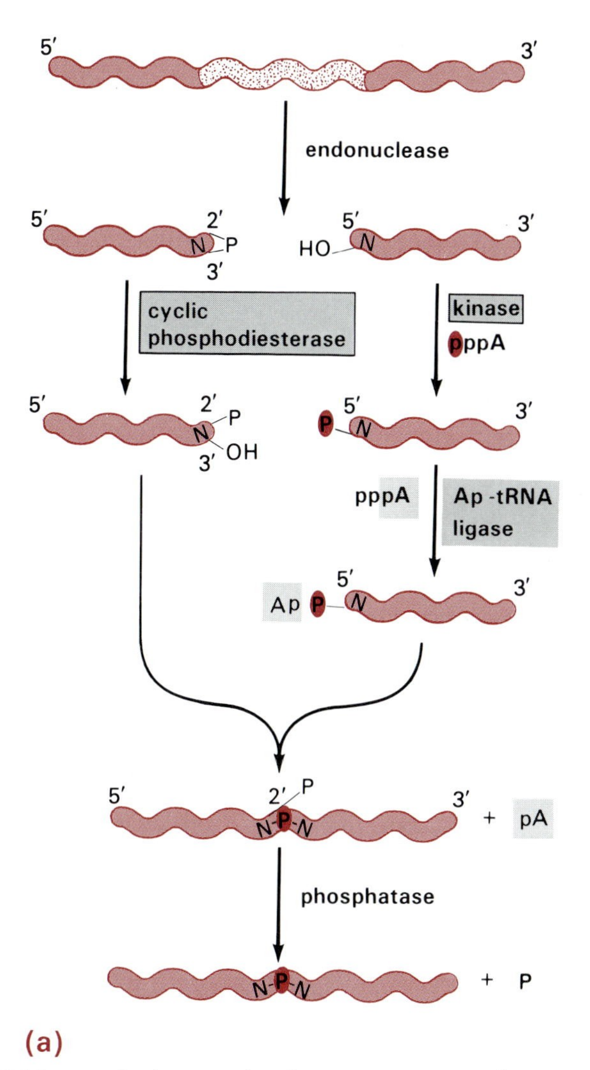

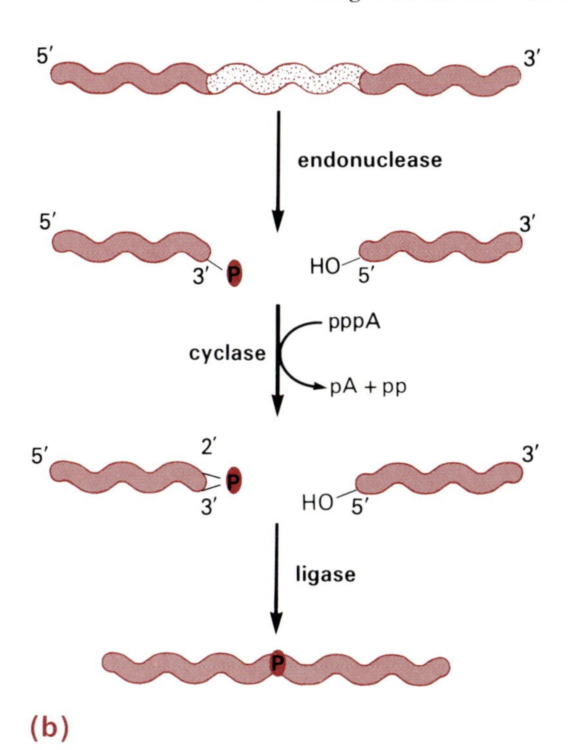

Figure 8.89
Pathway for splicing tRNAs. (a) The splicing reactions in yeast involve three proteins: an endonuclease, which cleaves the pre-tRNA at both splice junctions to produce the indicated ends; a multifunctional protein, which catalyzes all the reactions up to the phosphatase; and a 2′ phosphatase. In this case, the phosphate residue at the junction of the two exons is derived from the terminal phosphate of ATP. (b) In vertebrates, discrete enzymes catalyze the three reactions shown. Each of the enzymes is specific for tRNA splicing. Note that the phosphate at the junction of the two exons is derived from the phosphate at the exon-intron junction.

tRNA-specific ligase; this latter reaction is the same as that catalyzed by DNA ligases (Section 2.1e). Ligation of the two half-molecules through their respective activated ends creates an unusual 2′ phosphate, 3′, 5′ phosphodiester linkage. Removal of the 2′ phosphate by a phosphatase yields the mature spliced tRNA. Note that the phosphate in the newly created diester linkage is derived from ATP, a feature that distinguishes the yeast tRNA splicing reaction from the analogous process in vertebrates (Figure 8.89b). In that series of reactions, the initial endonuclease cleavage generates 5′ hydroxyl and 3′ phosphomonoester ends, and the latter end is converted to a 2′, 3′ cyclic phosphodiester by an ATP dependent cyclase. Then a tRNA-specific ligase joins the two halves without the necessity for further activation of the ends.

Fractionation of the yeast splicing system has yielded highly purified preparations of two enzymes. One catalyzes the endonucleolytic cleavage at the intron junctions, and the other provides the cyclic phosphodiesterase, kinase, adenylylation, and ligation activities, all embodied in a single polypeptide. Both enzymes are highly specific for these reactions in tRNA splicing; however, they act indiscriminately on any tRNA introns and join any two tRNA segments. Indeed, some interesting hybrid tRNAs have been constructed from half-molecules derived from different tRNAs.

f. Alternative Splicing: Multiple Proteins from One Gene

Most pre-mRNAs are spliced by a pathway that excises each intron at its respective 5′ and 3′ splice sites. Consequently, each one of the transcript's exons is conserved in its original order as a continuous sequence in a single mature mRNA (**constitutive splicing**). However, some pre-mRNAs are spliced in more than one way, thereby yielding a family of structurally related mRNAs, each with a subset of exons and each encoding one member of a family of protein **isoforms**. This form of RNA processing is called **alternative splicing**. An increasing number of genes in organisms ranging from *Drosophila* to humans and in the viruses they harbor, are known to use alternative splicing in the maturation of their pre-mRNAs. These genes encode a wide variety of proteins, including some involved in the cytoskeleton, muscle contraction, membrane receptors, peptide hormones, intermediary metabolism, and DNA transposition.

Alternative splicing provides the means to diversify the output from a single gene without altering its genomic organization. It represents an efficient way to generate a variety of mRNAs encoding structurally related proteins in which the shared amino acid sequences are specified by the common exons and the individually distinctive sequences derive from the alternatively spliced exons. In some instances, alternatively spliced mRNAs are produced concurrently, and the several protein isoforms may perform the same or different functions. The four myelin basic protein isoforms are all components of central nervous system myelin sheaths; the two isoforms of immunoglobulins are membrane bound or secreted (Section 10.6c). Some gene transcripts are spliced differently in specific tissues; for example, the single calcitonin gene expresses calcitonin in the thyroid and its isoform—calcitonin gene related protein—in the brain, each from its own distinctively spliced mRNA. And in some gene systems (e.g., the mammalian troponin T gene and the *Drosophila ultrabithorax* complex), the relative abundances and types of alternatively spliced mRNA are developmentally regulated. Alternative splicing has an important role in sex determination in *Drosophila* embryogenesis. Thus, the different activities of regulatory genes in males and females are due largely to sex-specific differences in RNA splicing that lead to the production of functionally different transcripts in the two sexes. Moreover, the activity of individual genes in this regulatory hierarchy are not only themselves controlled at the level of splicing, but they specify the splicing pattern of the transcripts that function subsequently in the hierarchy.

Alternative splicing patterns Three general classes of alternative splicing have been differentiated (Figure 8.90). The type I pattern results from the use of different promoters to generate related pre-mRNAs with distinctive 5′ proximal regions of varying lengths. The mouse α-amylase (Section 9.3b) and vertebrate myosin light chain (MLC) genes typify this class. The MLC gene is transcribed from two promoters located about 10 kbp apart (Figure 8.91). The longer pre-mRNA contains all the exons, but the shorter one lacks the first exon. Alternative splicing generates two nearly equal sized mRNAs, each containing the five 3′ proximal constitutive exons but differing in their 5′ exons. During splicing of the longer pre-mRNA, exon 1 is joined to exon 4, ignoring the 3′ splice

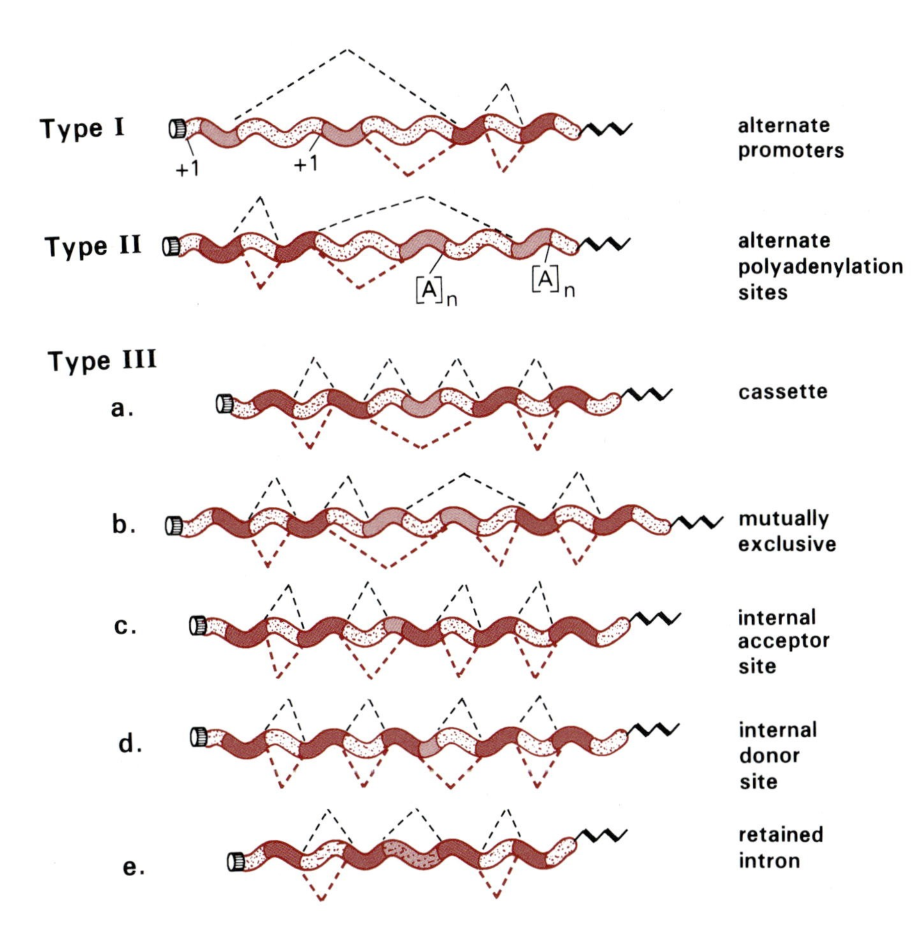

Figure 8.90
Alternate patterns of splicing pre-mRNAs. The dashed lines connecting exons above and below the RNAs indicate various patterns of splicing, depending upon whether the pre-mRNAs have alternate 5′ ends, alternate 3′ ends, or identical 5′ and 3′ ends.

sites of exons 2 and 3. By contrast, exons 2, 3, and 5 are joined during splicing of the shorter pre-mRNA, but the exon 3 to exon 4 splice fails to occur. Sequence analysis has shown that the 3′ splice site flanking exon 2 is defective, accounting for the failure to join exons 1 and 2. But the failure to splice exons 1 and 3 and exons 3 and 4 must be attributed to causes discussed later in this section.

The second type of alternative splicing involves pre-mRNAs that have different and variable length 3′ proximal sequences, usually because the transcript is polyadenylated at different locations (Figure 8.90). The five

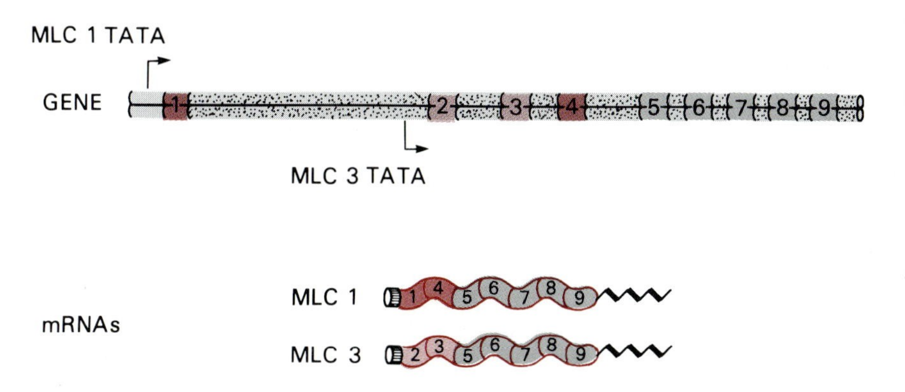

Figure 8.91
Alternately spliced mRNAs resulting from alternative transcription start sites. The myosin light chain gene (MLC 1) with its nine exons is shown above; introns are stippled. The two arrows represent alternate transcription initiation sites. The two kinds of primary transcripts are spliced to include common and distinctive exons.

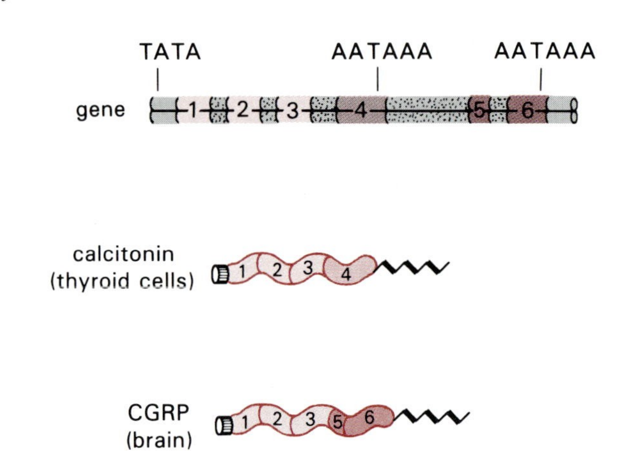

Figure 8.92
Alternately spliced mRNAs from pre-mRNAs with alternative polyadenylated 3′ ends. The calcitonin gene's exons (numbered) and introns (stippled), its promoter's TATA, and its polyadenylation signals (AATAAA) are shown. Two different spliced mRNAs, one encoding calcitonin and the other encoding a calcitonin related protein (CGRP), are formed from transcripts that are polyadenylated at the first or second polyadenylation signal, respectively.

different groups of distinctive adenovirus 2 late mRNAs arise by alternative splicing made possible by the use of five different polyadenylation sites. The same explanation accounts for the origin of two species of immunoglobulin heavy chain mRNAs (Section 10.6c). The single mammalian calcitonin gene generates two mRNAs (Figure 8.92). One is derived from the shorter pre-mRNA, and the other encodes the calcitonin gene related peptide. All four exons in the shorter calcitonin pre-mRNA are retained. The longer pre-mRNA has two additional exons, 5 and 6; these are retained in the calcitonin gene related peptide mRNA, but exon 4 is lost. The 3′ splice site in intron 4 is the preferred partner for the 5′ splice site of the third intron. In this instance, the tissue-specific use of different polyadenylation signals determines which splices are made.

In the third and most diverse and intriguing type of alternative splicing, identical pre-mRNAs give rise to different functional mRNAs (Figure 8.90). All the potential exons are present in the pre-mRNA, and splicing selections are made between the existing introns. The α and β splicing patterns in Figure 8.90 are found in many genes, the fast skeletal muscle troponin T gene being a good example (Figure 8.93). The α and β groups of troponin T mRNAs contain either exon 16 or 17, respectively. This results from mutually exclusive splicing of exons 15, 16, and 18 or 15, 17, and 18. The choice of which exon is used is developmentally regulated. Thus, exon 16 is specific to adult muscle troponin T mRNAs, and exon 17 is used in both adult and embryonic mRNAs. Each of the 32 possible combinatorial alternative splices involving exons 4 to 8 are represented in the α and β groups; the pattern ranges from complete exclusion to retention of all five exons in the mature mRNAs.

Splicing involving 5′ to 3′ splice sites that are internal to exons or introns also occurs. Thus, although an authentic 5′ splice site usually splices to the normal 3′ splice site at the end of the same intron, it may use an alternate 3′ splice site within the intron, resulting in elongation of the next exon (Figure 8.90c). Alternatively, use of an alternate 3′ splice site in the next exon leads to truncation of that exon. Similar outcomes ensue if alternate 5′ splice sites within introns or exons are used (Figure 8.90d). The existence and use of such "cryptic" splice junctions necessitates that the alternate splices generate consistent translational reading frames. This latter mode of alternative splicing is used by SV40 and polyoma viruses to express multiple mRNAs from a single early region transcript (Figure

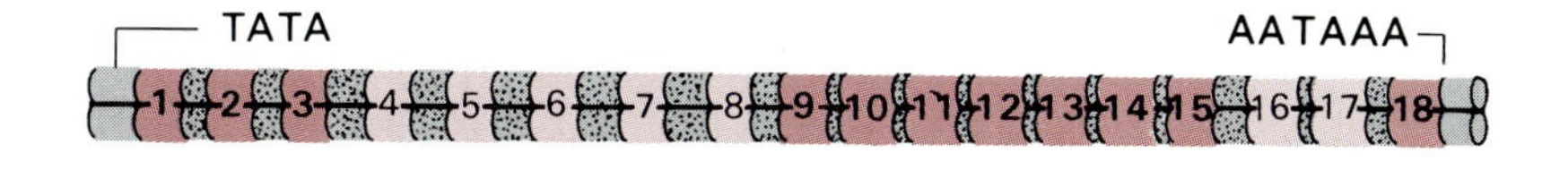

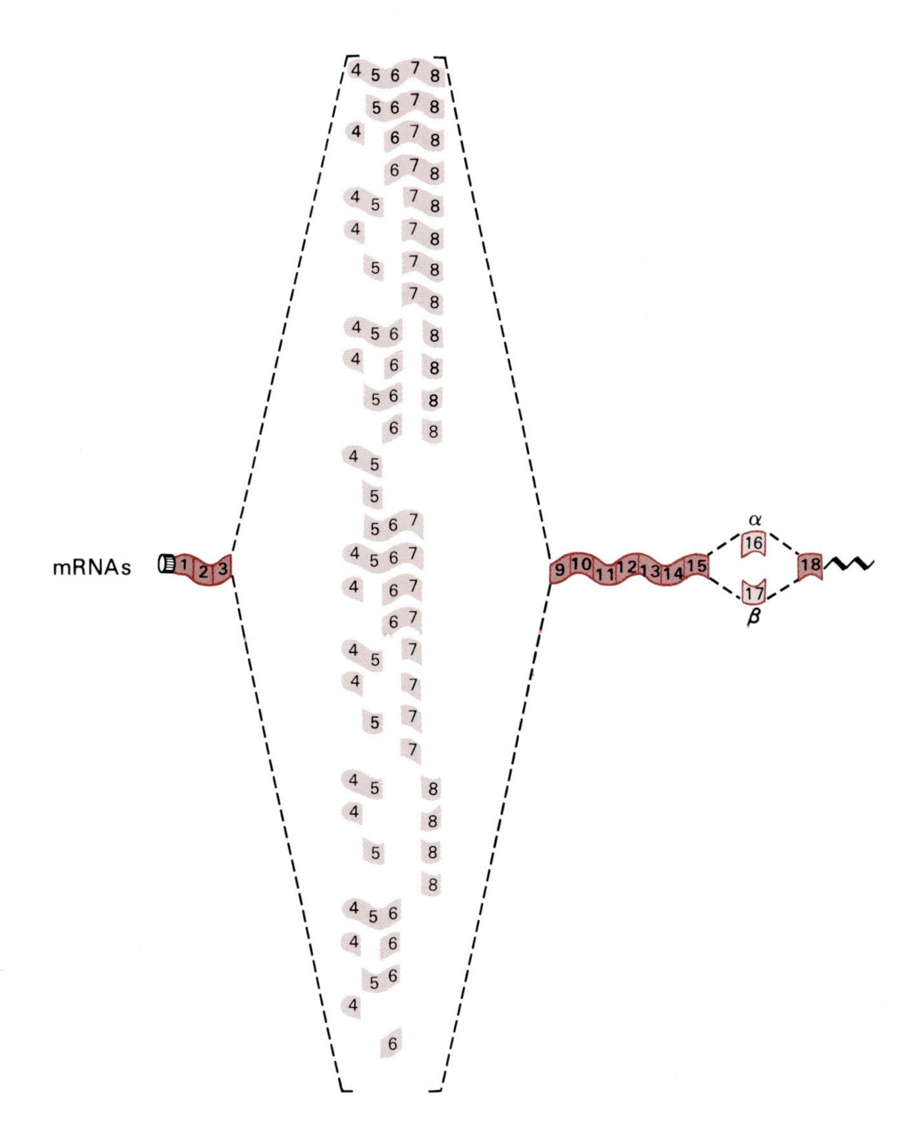

Figure 8.93
Alternately spliced troponin T mRNAs from pre-mRNAs with identical 5′ and 3′ ends. The fast skeletal troponin T gene's exons (numbered segments) and introns (stippled segments), as well as the TATA and AATAAA transcription initiation and polyadenylation signals, respectively, are shown above. There are two groups of spliced mRNAs, resulting from alternative inclusions of exons 16 (α) or 17 (β). Each of these has 32 different mRNAs, depending on which of exons 4–8 are retained in the splicing.

8.94). This type of splicing muddles the distinction between the coding properties of exons and introns.

A particularly interesting example of tissue-specific alternative splicing involves the gene encoding the *Drosophila* P element transposase (Section 10.2b). There the failure to splice an intron containing a translational termination codon prevents expression of the transposase in somatic tissue. Expression of the transposase in germ line tissue depends on the ability of that tissue to splice out the intron, thereby extending the reading frame to an exon required for transposase activity.

An extreme example of differential splicing occurs in retroviral gene expression. The integrated provirus is transcribed into a complete copy of viral genomic RNA, which is ultimately packaged into virions; no splicing occurs in the RNA destined for virions (Section 5.7d). The same full length transcript is also an mRNA that encodes both the core and

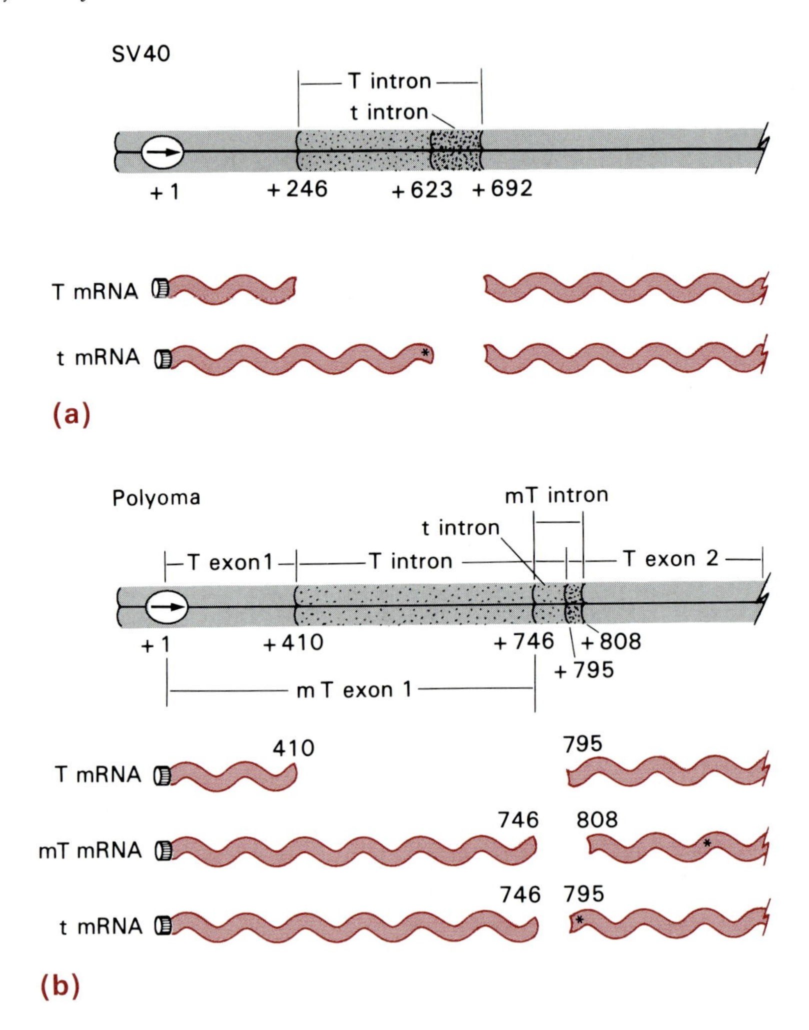

Figure 8.94
Alternative splicing generates multiple mRNAs from the SV40 and polyoma virus early regions. (a) The 5′ proximal portion of SV40's early region indicates the locations of large T and small t introns. Shown below are the two alternately spliced mRNAs encoding large T and small t proteins. The asterisk indicates the position of a termination codon that is absent from the large T mRNA but is present in the small t mRNA and ends the small t coding sequence. (b) There are three overlapping introns in the polyoma early region. Alternate splicing removes one of the three, resulting in the formation of three different early mRNAs whose retained exons are indicated in the mRNAs.

reverse transcriptase proteins. However, the same transcript is, in addition, a pre-mRNA that is spliced to yield the envelope mRNA. And in the cases of the human T cell leukemia and human immunodeficiency retroviruses, multiple modes of splicing the primary transcript yield a host of other mRNAs, each encoding a specialized protein needed for viral multiplication and pathogenesis (Figure IV.6).

Splice site selection The principal conundrum of both constitutive and alternative pre-mRNA splicing is how the correct pairs of exon junctions are chosen for joining. In considering this problem, two broad categories of possible explanations exist. One invokes structural elements intrinsic to the pre-mRNA itself or to the intermediates formed along the path toward a completely spliced mRNA: these are referred to as *cis* factors. The second involves *trans* acting factors, proteins, or other RNAs that influence the choice of introns to be spliced. Almost surely there is a hierarchy of splice site efficiencies, and that hierarchy may reflect the relative competitive advantage of the most favored 5′ and 3′ splice sites during spliceosome formation. In the case of the splicing of SV40's two T antigen mRNAs, the two alternative splicing reactions are determined by the choice of branch points in the intron. There are several persuasive examples in which mutations at one splice junction allow a normally unused site to act in its place. And there is considerable evidence to suggest that *trans* factors also influence the outcome of splicing reactions. This is especially true where presumably identical pre-mRNAs are spliced differentially, depending on physiological conditions or on the particular tissue. Special snRNPs or maturaselike proteins could be expressed and function in a tissue-specific or developmentally programmed manner, thereby influencing the splicing pattern. The next few years should clarify some of these issues.

8.6 Novel Structural Motifs in Transcription Factors

Transcription factors must be able to perform at least two essential functions: binding to specific DNA sequences and influencing transcription initiation. Studies with a variety of DNA binding proteins establish that specific DNA binding is the result of cooperative and additive interactions between a DNA sequence and characteristic structural motifs in the protein. One such motif contributes an array of amino acids that make contacts with sequence independent structural features of the DNA duplex (e.g., the sugar-phosphate backbone of the B helix). Specificity is generally achieved by an associated structural motif that makes Van der Waals and hydrogen bond contacts between selected amino acid side chains and the edges of base pairs in the DNA binding site's major groove. The two kinds of motifs and the amino acids responsible for specific strong binding are known in several instances: in the binding of the lambda (λ) phage repressor and Cro proteins, and in the *E. coli trp* repressor and CAP proteins. It already seems unlikely that the structural parameters for permissible interactions define an "amino acid code" in the way that complementarity limits the nature of purine-pyrimidine pairing in nucleic acids. In fact, studies on the protein-DNA interactions mentioned previously suggest that a particular amino acid can contact more than one kind of base, and a given base is able to interact with more than one kind of amino acid side chain.

Although obligatory, the binding of transcription factors to DNA is insufficient for transcriptional regulation. The bound protein must also possess a transcription regulatory domain. The binding and regulatory domains may be part of the same or different proteins; in the latter case, the two proteins interact in order to create a functional, promoter-specific, transcription regulatory factor.

Studies with eukaryotic transcription factors are in their infancy. Nevertheless, several structural domains that create the architectural framework for specific amino acid–base pair interactions to occur have already been identified. In each protein, the structural domains responsible for specific DNA binding and for transcriptional regulation are distinctive and separate. These two are not interdependent, because the DNA binding and regulatory domains from different proteins can be swapped to create transcription factors with properties characteristic of each domain. Moreover, provided that the cognate DNA binding sequence is placed near a TATA sequence, such chimeric proteins can bind and regulate transcription in a variety of eukaryotes (e.g., yeast, *Drosophila*, frogs, and mammals).

a. DNA Binding Domains

Helix-turn-helix The best characterized structural motif for DNA binding is the **helix-turn-helix**, first identified in the λ Cro and repressor proteins. This type of motif allows the yeast *MAT* α2 protein, which represses transcription of genes involved in determining mating types (Section 10.6b) to bind to specific sequences in a variety of promoters.

The distinctive feature of a helix-turn-helix structure is its geometry rather than its amino acid sequence: two adjacent α helices separated by a turn of several amino acids contribute the most important contacts with the DNA binding site. X-ray crystallographic analysis of several proteins and their protein-DNA complexes suggests a model for the proteins' DNA binding specificity. The structural data indicate that both Cro and repressor form dimers, the dimerization resulting from interactions between the antiparallel β3 sheets in Cro and between the α5 helices of the repressor (Figure 8.95). The DNA sequences to which Cro and repressor bind have dyad symmetry, permitting the helix-turn-helix motifs of the two monomers to bind side by side and to each half of the dyad, thereby stabilizing the interactions with the DNA (Figure 8.96). Contacts between amino acids in one of the monomer's helices, the "recognition helix" (the α3 helix in the Cro and repressor proteins), and a stretch of six bases in the binding site's major groove are the most important determinants of the binding specificity. A second helix along the backbone (the α2 helix in each protein) makes contacts with the sugar-phosphate backbone and possibly bases adjoining the recognition sequence, thereby stabilizing the protein DNA interaction. This view of the binding is supported by experiments that alter the binding specificity by swapping α3 helices between proteins with different DNA binding specificities. The assignment of recognition and stabilization functions to particular amino acids in the two helices relies on mutational alterations of key residues in each domain.

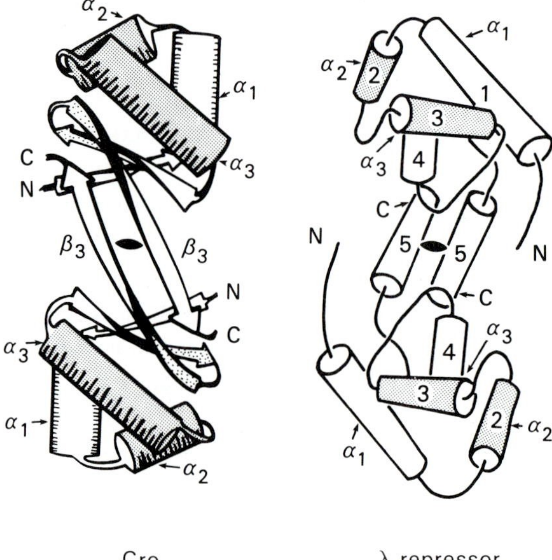

Figure 8.95
Schematic drawing of dimers of uncomplexed Cro protein and λ repressor amino terminal domains. The views shown are down the dimers' respective twofold axes of symmetry. Cro consists of three α helices, shown as barrels, and a three-stranded antiparallel β sheet, illustrated by arrows. The λ repressor contains five α helices with no β sheet. The amino terminal domains make contact via the fifth α helix in each subunit. Binding to the DNA involves each repressor subunit's α2 and α3 helices, which are shown stippled. Adapted from Y. Takeda et al., *Science* 221 (1983), p. 1020.

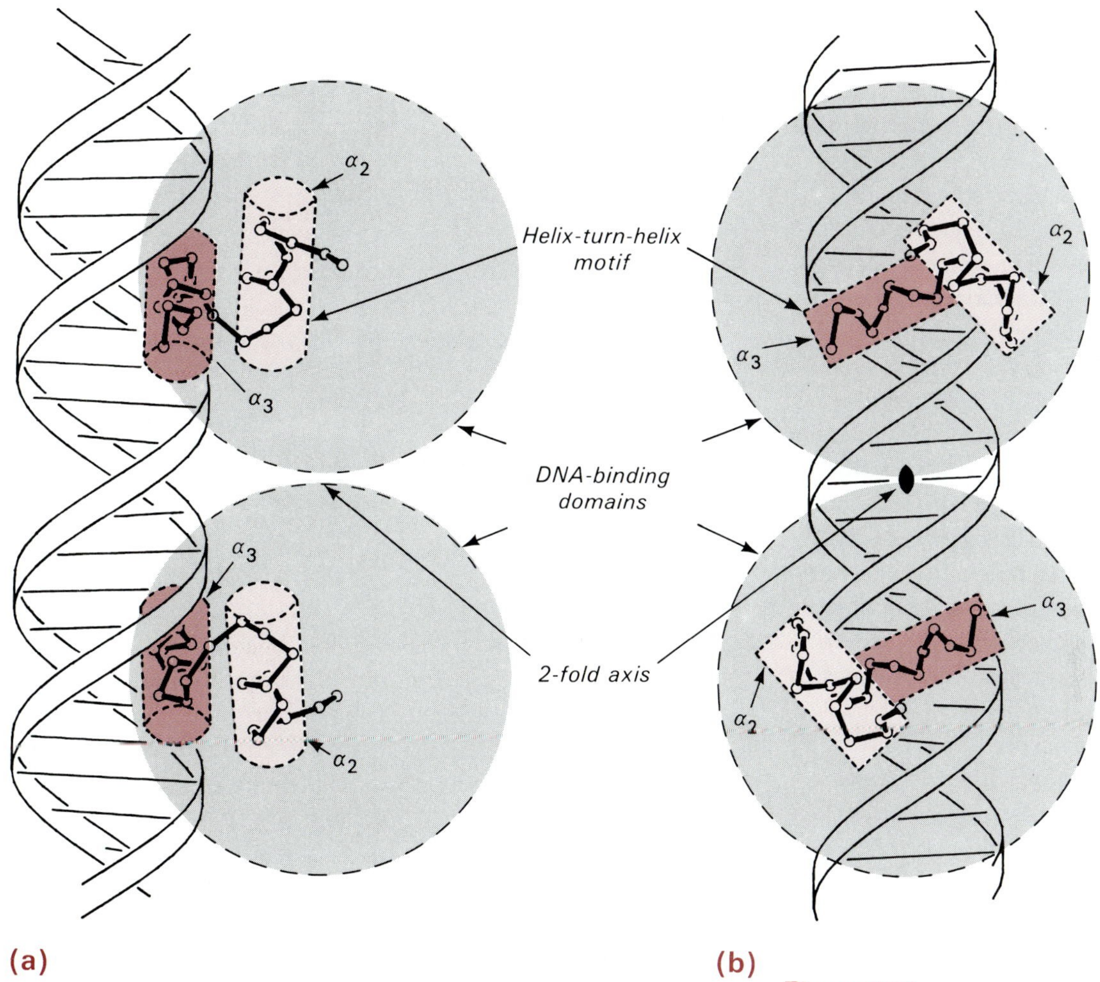

Figure 8.96
Mode of DNA binding by proteins with the helix-turn-helix structural motif. The protein's overall DNA binding domain is represented by the large grey circular field. The α3 recognition helix and the α2 stabilization helix are shown in deeper and lighter colors, respectively. At the left (a) is a side view, and at the right (b) is the face-on projection. The twofold axis of symmetry of the protein dimer is indicated for each view. Adapted from D. H. Ohlendorf, W. F. Anderson, and B. W. Mathews, *J. Mol. Evol.* 19 (1983), p. 109.

Homeodomains An increasing number of known genes concerned with regulating a variety of developmental processes in nearly every eukaryotic species contains a characteristic 180 bp sequence within their protein coding region. This sequence is called the **homeobox** because it was initially found in *Drosophila* homeotic genes. Expression of these genes, which specify the segmental identities that give rise to the head, thoracic, and abdominal regions, follows the expression of the maternal, zygote, and segmentation genes but precedes the differentiation of individual cell types within each segment (Section 8.3g). Since its discovery, the homeobox has been found in many other *Drosophila* genes, specifically those that commit cells to particular pathways of early embryonic development. Virtually every one of the *Drosophila* homeobox genes is expressed in a characteristic subset of embryonic cells, and every cell type contains a unique combination of homeobox gene products. Particular homeobox genes appear to encode transcription regulatory factors that influence the expression of other homeobox genes as well as of other nonregulatory structural proteins. Since their discovery in *Drosophila*, homeoboxes have also been implicated in genes involved in developmental or related regulation in worms, frogs, mice, and humans. These genes may be viewed as lineage determining functions. Thus, the early embryo is initially divided on the basis of homeobox gene expression into anterior and posterior areas of different developmental potential prior

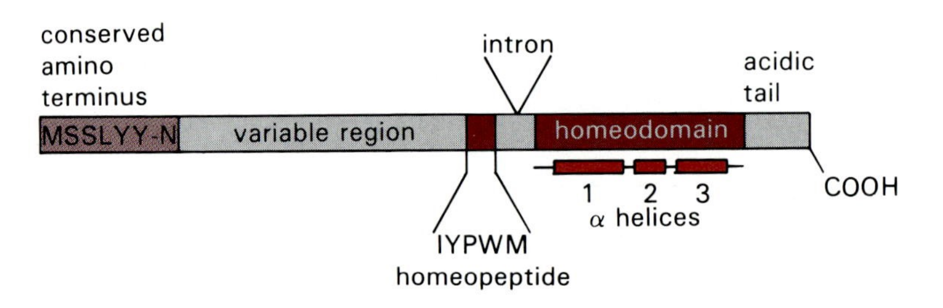

1 ... 20
Arg Lys Arg Gly Arg Gln Thr Tyr Thr Arg Tyr Gln Thr Leu Glu Leu Glu Lys Glu Phe

21 ... 40
His Phe Asn Arg Tyr Leu Thr Arg Arg Arg Arg Ile Glu Ile Ala His Ala Leu Cys Leu

41 ... 60
Thr Glu Arg Gln Ile Lys Ile Trp Phe Gln Asn Arg Arg Met Lys Trp Lys Lys Glu Asn

Figure 8.97
Amino acid sequence organization of proteins with homeodomains. The top of the figure summarizes characteristic features of homeodomain proteins. Many have a characteristic amino terminal amino acid sequence containing methionine (M), serine (S), leucine (L), tyrosine (Y), a variable amino acid (-), and asparagine (N). This is followed by a region of variable length and sequence and a short invariant sequence of five amino acids: isoleucine (I), tyrosine (Y), proline (P), tryptophan (W), and methionine (M). The strongly basic homeodomain consists of a 60 amino acid stretch near the highly acidic (rich in glutamic acids) carboxy terminus. The homeodomain sequence can fold into three α helical regions to form a helix-turn-helix DNA binding motif. The sequence below corresponds to the homeodomain of the protein encoded by the *Drosophila antennapedia* gene. The basic residues are highlighted in color. Information for the figure was adapted from C. V. E. Wright et al., *Trends Biochem. Sci.* 14 (1989), p. 52, and W. J. Gehring, *Sci. American* 253 (October 1985), p. 153.

to the formation of individual organs. Wherever it is found, the 180 bp homeobox specifies a 60 amino acid **homeodomain**, now known to be the DNA binding motif for this class of proteins. Aside from their almost invariant location in the proteins in which they occur, perhaps the most significant feature of homeodomains is their potential to form a helix-turn-helix motif. Indeed, X-ray crystallography has established the existence of a helix-turn-helix structure in a homeodomain peptide.

The analysis of many homeoprotein sequences suggests that there are similarities and some differences between the vertebrate and *Drosophila* homeoproteins. One striking similarity is the high content of basic amino acids (about 30 percent of the residues are lysine or arginine) that is relevant to their DNA binding function. Among the vertebrate homeoproteins, several other features stand out (Figure 8.97). Most have a conserved amino terminal sequence consisting of Met, Ser, Ser, Leu, Tyr, Tyr, and Asn, followed by a variable sequence region, which, nevertheless, has weak or short patches of homology among different homeoprotein classes. A conserved pentapeptide of the composition Ile, Tyr, Pro, Trp, Met occurs just in front of the homeodomain, which is followed by a sequence rich in acidic amino acids at the carboxyl end.

Several lines of evidence indicate that the homeodomain is important for the DNA binding activity of this class of proteins. Thus, point mutations and frameshift mutations in the homeodomain completely impair the protein's DNA binding activity, whereas mutations elsewhere in the protein have less striking effects. Swapping homeodomains among various homeoproteins results in binding characteristics consistent with the source of the homeodomain. The DNA sequences to which homeoproteins bind generally do not have dyad symmetry; consequently, only monomers are presumed to bind. An important feature of some homeoproteins is their ability to bind different but related DNA sequences, although with different binding affinities. Thus, differential binding specificity may enable a single homeoprotein to regulate transcription of more than one kind of gene.

Details about how homeodomain proteins bind to DNA specifically

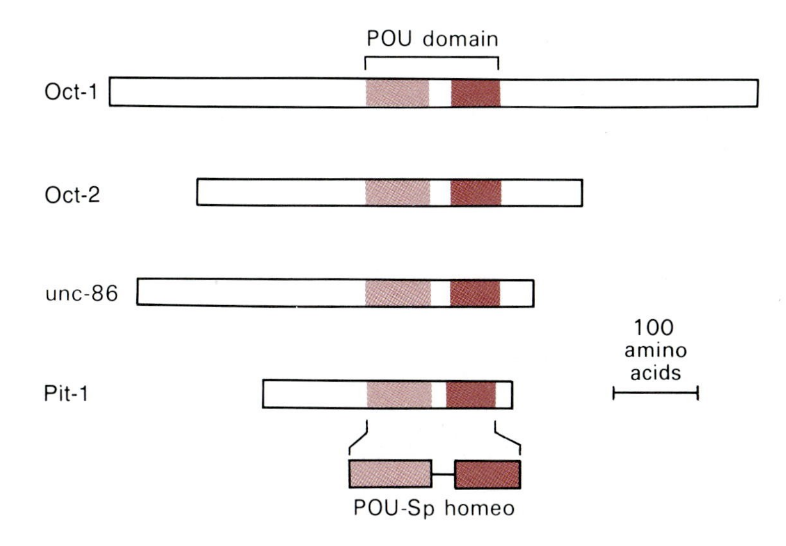

Figure 8.98
Location and organization of the POU domains of several transcription factors. The colored regions identify the location and organization of the POU-specific (POU-SP) and homeodomains (homeo) of transcription factors: the octamer binding proteins Oct-1 (OTF-1) and Oct-2 (OTF-2), the transcription factor encoded by the *C. elegans* gene *Unc-86*, and a pituitary-specific transcription factor, Pit-1. The POU domain is about the same size in each protein, although its position relative to the ends differs among this group of proteins.

are lacking. The most widely held view is that the amino acid arrangement in and around the helix-turn-helix motif is crucial for the protein's DNA binding specificity. Other conserved protein features may influence DNA binding, but these contributions are likely to be minor. Conceivably, the DNA binding and/or transcription regulation properties of these proteins are modulated by small molecules. More likely, these proteins possess regions that permit them to interact with other proteins so that a complex of proteins may be the functional entity. Once bound to DNA, the homeoproteins influence transcription from their target promoters, but the mechanism is unknown. The acidic nature of their carboxy termini and the clustered prolines and glutamines in their amino-terminal variable regions may influence that critical function (Section 8.6b).

POU domains Several transcription factors, whose functions involve tissue-specific and developmental regulation, have a complex regulatory region of about 160 amino acids, which includes the characteristic homeodomain. This region is highly conserved in a transcription factor (Pit-1) needed for transcriptional activation of the rat prolactin and growth hormone genes, in both the ubiquitous and lymphoid-specific octamer binding proteins OTF-1 and OTF-2, and in a *C. elegans* transcriptional regulator of neuronal development (Unc-86). The acronym **POU** refers to the proteins in which it was first detected.

The location and size of the two conserved motifs that together constitute the POU domain in the four proteins are shown in Figure 8.98, and the extent of the sequence identity in these domains is summarized in Figure 8.99. The sequences of the three homeodomains are closely related, and as is the case for other homeoproteins, the sequence allows for a helix-turn-helix structure. The highly conserved carboxy terminal third of the homeodomain, referred to as the WFC region because of the presence of tryptophan (W), phenylalanine (F), and cysteine (C), probably conforms to the recognition helix of that motif. This is inferred from the fact that alterations in the WFC region impair the binding of the two OTF proteins to the octamer sequence. Whether these proteins bind to their cognate DNA sequences via the helix-turn-helix structure of the

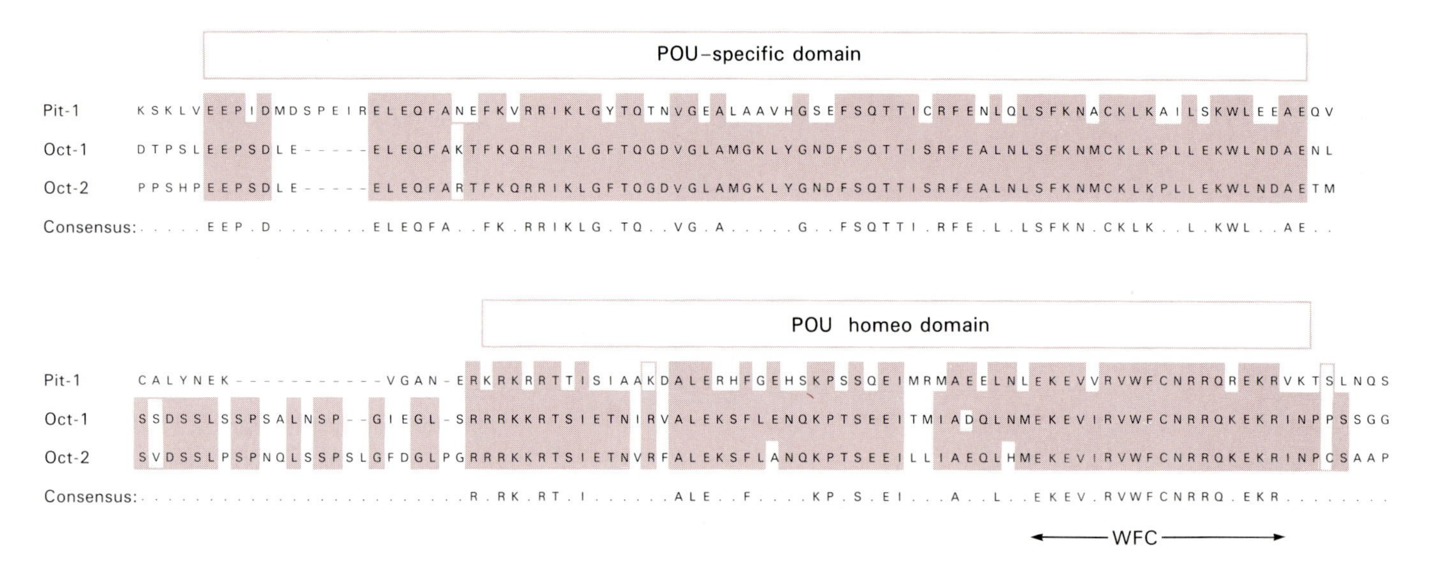

Figure 8.99

Amino acid sequence similarities in the POU domains of several transcription factors. The amino acid sequence of the POU domains in the Pit-1, Oct-1 (OTF-1), and Oct-2 (OTF-2) transcription factors are shown in the single letter designations (see Figure 1.24). The sequences have been aligned to maximize the number of identical positions. The color screen indicates where the same amino acid occurs in at least two of the three proteins. The consensus sequence indicates where all three proteins have the identical amino acid. Note that the consensus does not include instances where the dissimilarities involve amino acids of like charge or structure. The label WFC at the carboxy terminus of the POU homeodomain specifies a highly conserved sequence containing the amino acids, tryptophan (W), phenylalanine (F), and cysteine (C). As discussed in the text, this region is essential for the activity of the transcription factors containing the POU domains. Adapted from R. G. Clerc et al., *Gene and Devel.* 2 (1988), p. 1513.

homeodomain and what role the POU-specific domain has in that binding and/or transcription regulation are not known. Conceivably, the POU-specific region provides for interactions of these proteins with others to modify or refine their regulatory actions.

Zinc fingers An examination of the amino acid sequence in TFIIIA, one of the factors required for 5S rRNA gene transcription, turned up the zinc finger motif in its DNA binding domain (Section 8.4c). Since then, a large number of proteins with structurally related sequence arrangements capable of chelating zinc ion has been discovered. The yeast GAL4 protein (Section 8.3f), the mammalian Sp1 factor (Section 8.3d), and the steroid receptors (Section 8.3f) all contain varying numbers of zinc fingers that are involved in DNA binding. The *Drosophila* gap gene *hunchback* encodes a protein that contains multiple zinc finger motifs (Section 8.3g). A mammalian sterol inducible protein, which regulates both a group of genes involved in sterol biosynthesis as well as the gene encoding the low density lipoprotein receptor, contains six zinc fingers for binding to the sterol responsive elements in these genes' promoters. The correlation between transcriptional regulation and the occurrence of the finger structure is becoming so high that finding zinc finger structures encoded in unknown gene sequences is strong reason to consider a transcription regulatory role for the gene product.

At least three kinds of zinc fingers have been identified. They differ in the number and location of the zinc coordinated cysteine and histidine residues. The basic structural feature of one zinc finger type (the TFIIIA type) consists of a loop formed by the coordination of a zinc atom with pairs of cysteines and histidines spaced 12 amino acids apart (Figure 8.69). Another type has variable numbers of cysteines for coordination to the zinc atom. Some, like the GAL4 zinc finger, contain six cysteines, only four of which are coordinated to zinc; the steroid receptor's DNA binding domains have four cysteines in their zinc fingers. A third motif utilizes three cysteines and one histidine to form the finger. But it is possible that any combination of cysteines and histidines is capable of forming a finger by coordination to zinc. Other metals may also associate with these residues to form the finger motif, and in cases where there are more than four potential coordination sites, the metal may coordinate with alternative pairs of sites.

The existence of the zinc finger structure has been established by a variety of physical means: circular dichroism and nuclear magnetic resonance spectroscopy and, most definitively, a three-dimensional structure analysis of a peptide containing a single zinc finger. The zinc atom is buried in the interior of the molecule surrounded by α and other types of helical stretches connected by β strands (Figure 8.100).

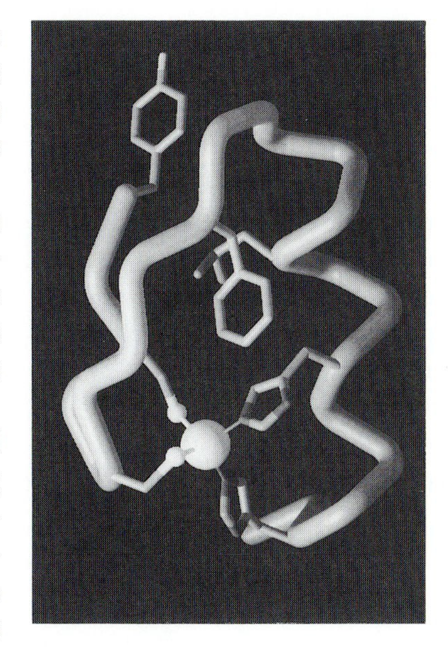

Figure 8.100
Computer graphic representation of a single zinc finger structure deduced from nuclear magnetic resonance analysis. The Cα backbone chain is represented as a tube. The zinc atom (shown as a sphere) is coordinated by two cysteine and two histidine ligands. Invariant phenylalanine and leucine residues adjacent to and within the finger are also shown. Adapted from M. Lee et al., *Science* 245 (1989), p. 635. M. E. Pique and P. E. Wright supplied the photographic image.

The ability of TFIIIA to bind to the 5S rRNA gene's internal control region correlates with the integrity of the zinc fingers, and the binding of Sp1 to its cognate DNA sequence requires zinc in the protein. Furthermore, mutations around the GAL4 or glucocorticoid receptor finger regions inactivate their binding to the respective target sequences. Two striking facts that support the DNA binding activity of zinc finger domains are (1) that truncated segments of either TFIIIA or Sp1 proteins, containing only the finger regions, still bind specifically to DNA and (2) that swapping the estrogen receptor's finger regions for the corresponding fingers from the glucocorticoid receptor creates a hybrid receptor that binds to the glucocorticoid responsive element and not to the estrogen receptor's binding site.

The precise way in which the zinc fingers bind to DNA is not entirely clear. It is known that binding occurs in the DNA's major groove, but the residues that contribute to binding specificity and stability are not entirely clear. In some instances, residues in the finger's helix contribute to the binding specificity. But in the estrogen receptor, one of the two fingers is the principal determinant of the receptor's DNA binding specificity; moreover, several residues that contribute to that finger's binding specificity occur at its base (Section 8.3f). Similarly, the single zinc finger in the GAL4 protein can be replaced by a finger from an unrelated gene, but the modified protein still binds to UAS_{GAL}. Thus, the zinc fingers create a loop structure that permits local residues to interact specifically with their DNA binding sites. The fingers may also contribute nonspecific interactions that increase the stability of the protein-DNA complex; this is reminiscent of the role of the second helix in the helix-turn-helix motif.

Leucine zippers Several transcription factors, notably the C/EBP, the AP1-Jun family of mammalian DNA binding transcriptional regulators, and the GCN4 related factors in yeast, share an unusual bipartite structural motif called the **leucine zipper** (Figure 8.101). Several mammalian transcription factors, notably those encoded by the c-*myc* and c-*fos* genes, also possess the leucine zipper motif even though they do not by themselves bind DNA.

A characteristic feature of the leucine zipper is the periodic repetition of leucines in a relatively long α helical segment (Figure 8.102). When the amino acid sequence of the α helix is displayed as a helical wheel, which portrays the amino acid side chains viewed down the helical axis, the leucines appear on the same side of the helix every other turn. Further-

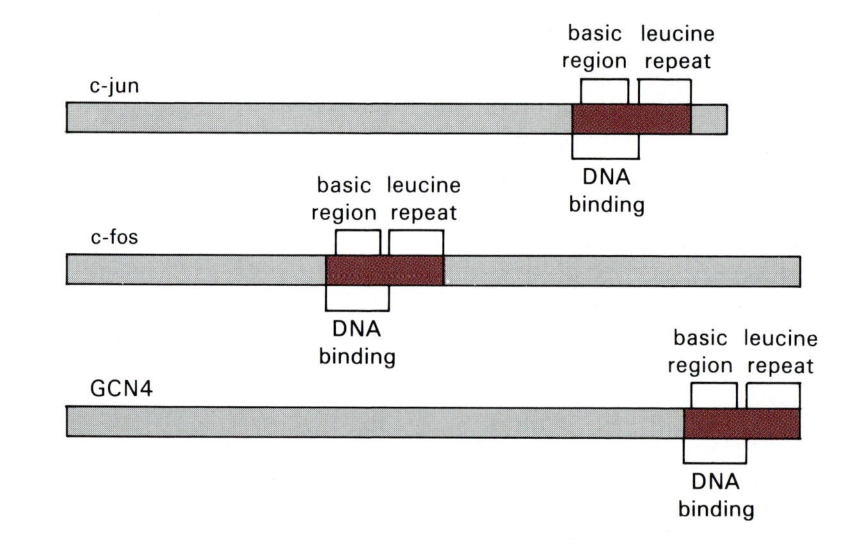

Figure 8.101
Location and organization of the leucine zipper domains in mammalian and yeast transcription factors. Leucine zipper domains consist of the characteristic helical region containing leucines every seventh residue (see Figure 8.103) and a strongly basic region by which the proteins bind to their recognition sequence in DNA. The proteins are shown with the amino termini at the left and the carboxy termini at the right.

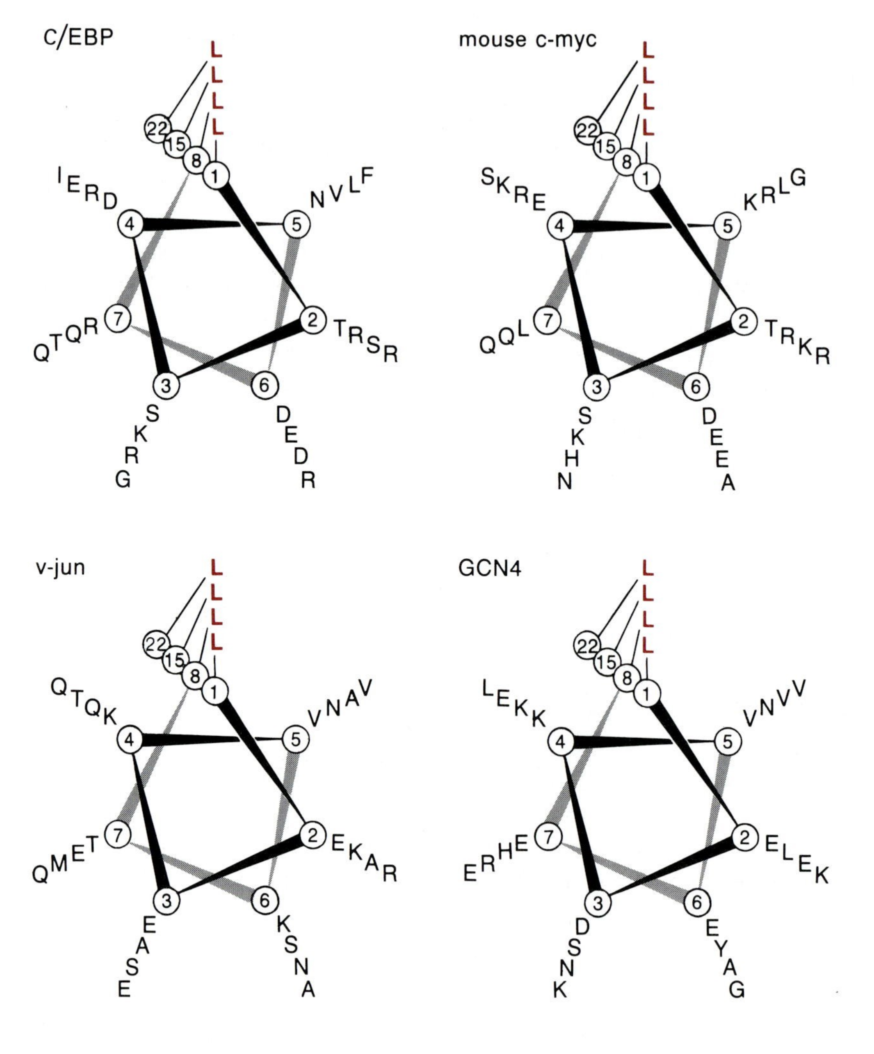

Figure 8.102
Representations of α helical regions in leucine zipper motifs as helical wheels. The helical wheel representation shows how the leucine residues (L) are spaced seven residues apart in the leucine zippers of the enhancer binding protein (C/EBP), the murine myc protein (c-myc), the protein encoded by the viral oncogene v-jun, and the yeast GCN4 protein. This representation also shows the other amino acids in the single letter designations (Figure 1.24) constituting the α helical regions of the leucine zipper. The amino acids appear in the order they would appear as the helix descends into the paper.

```
                      basic region                        leucine repeat
c-jun   ESQERIKAERKRMRNRIAASKCRKRKLERIARLEEKVKTLKAQNSELASTANMLREQVAQL
c-fos   SPEEEEKRRIRRERNKHAAAKCRNRRRELTDTLQAETDQLEDEKSALQTEIANLLKEKEKL
GCN4    VPESSDPAALKRARNTEAARRSRARKLQRMKQLEDKVEELLSKNYHLENEVARLKKLVGER
```

Figure 8.103
Amino acid sequences of several leucine zipper basic regions and leucine repeats. DNA binding by proteins with the leucine zipper structural motif occurs via the basic region adjacent to the leucine repeats. The basic amino acids within this sequence are highlighted in color.

more, there is an unusually large number of negatively charged residues on the opposite side of the helix. Their potential for forming ion pairs, as well as the novel stacking of the hydrophobic residues, probably contribute to the stability of these long helices. An important feature of the leucine zipper motif is the highly basic amino acid sequence adjacent to the helically stacked leucines (Figures 8.101 and 8.103).

Proteins with a leucine zipper motif are able to form dimers, either homodimers or heterodimers. This occurs by hydrophobic interactions between residues arrayed on the surface of parallel coiled coils (Figure 8.104). Interactions between the cognate DNA sequences and amino acids in and around the proteins' basic domains probably account for the specificity and relative affinities of DNA binding.

Proteins possessing a leucine zipper motif invariably bind DNA as dimers. Indeed, dimer formation is a prerequisite to efficient binding. If dimer formation is impaired by mutations that alter the sequence or the secondary structure of the leucine zipper, DNA binding is prevented even though the basic region is left intact. Alterations in the basic region also prevent DNA binding, but without affecting the protein's ability to dimerize. Thus, the binding of these proteins to their specific sites on DNA involves an interesting interplay of two structural features: the leucine zipper provides a dimerization motif, and the basic region contributes the DNA binding motif. However, the basic region's function depends on dimerization even though dimerization is independent of the basic region.

The transcription factors GCN4 and C/EBP bind to DNA only as homodimers; by contrast, homodimers of the AP1-Jun family of transcription factors bind only poorly. However, the AP1-Jun factors form heterodimers with members of the fos protein class, resulting in complexes that bind DNA considerably more efficiently. Even though the fos proteins

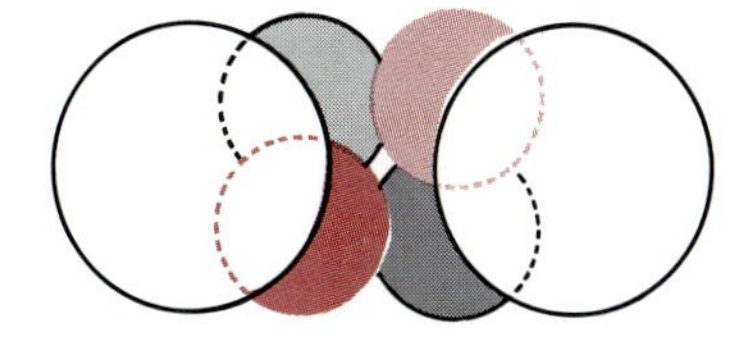

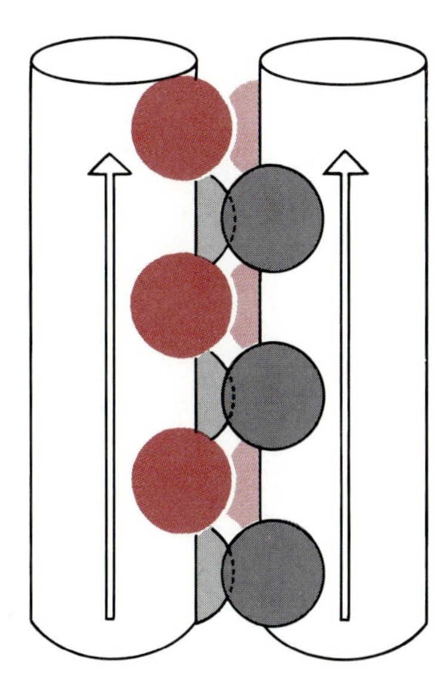

Figure 8.104
Diagrammatic representation of the coiled-coil structure for the leucine zipper. A schematic representation of the leucine zipper's coiled-coil structure is shown in top and side views. In the model the parallel helices are represented by white cylinders with arrows showing their orientation. Leucines in the heptad repeat are represented by dark spheres, and the repeated hydrophobic residues are represented by lightly shaded spheres. Adapted from Figure 4 in T. G. Oas et al., *Biochemistry* 29 (1990), p. 2891. P. S. Kim kindly provided the figure before publication.

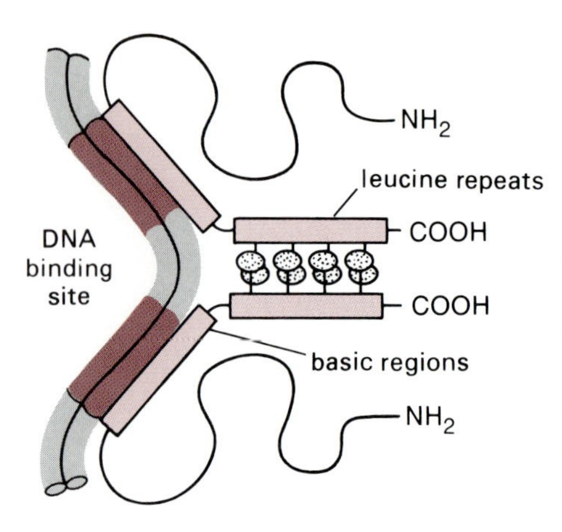

Figure 8.105
Dimers of leucine zipper transcription factors bind to their cognate DNA sequence via the basic regions. The model presumes that dimers are held together by interactions between each monomer's leucine repeats and associated hydrophobic residues. Contacts are made between residues in the protein's symmetrically arranged basic regions and the DNA's symmetrical dyad binding site. Adapted from W. H. Landschulz et al., *Science* 243 (1989), p. 1689.

have the characteristic leucine zipper and basic domains, they neither form homodimers nor bind DNA. The inability of fos to bind DNA may be due to its inability to form homodimers, which makes its ability to form DNA binding heterodimers with other factors especially intriguing.

Dimerization dependent DNA binding probably stems from the close apposition of the two basic regions that results from pairing of the two proteins through their leucine zippers (Figure 8.105). This suggests that leucine zipper proteins bind to DNA cooperatively, particularly to sites that have dyad symmetry. In that case, each basic region can contribute to the binding by interactions with the half-site. Heterodimeric factors might also be able to bind cooperatively to such palindromic sites, but it is conceivable that one of the subunits modulates the specificity or stability of the interactions. Heterodimers also possess the potential for dual or multiple binding specificities. This permits them to bind differentially at closely related sequences or to regulate promoters having distinctive arrangements of several signals. Given the very clear indications that transcriptional regulation relies on combinatorial interactions between proteins and their cognate DNAs, the existence of heteromeric factors creates another dimension for fine-tuning transcriptional control.

b. Transcription Regulatory Domains

Binding to a specific DNA sequence is only one of the two essential functions of transcription regulatory proteins. The bound proteins must also interact with the transcription complex to activate or repress initiation. We know now that each of these functions—binding and regulation—can be assigned to discrete protein domains. Irrespective of the transcription factor, the binding domain is relatively small yet highly specific; moreover, the binding domain functions autonomously (i.e., it recognizes and binds its cognate sequence even in the absence of the rest of the protein [Sections 8.3 and 8.6a]). By contrast, the regions responsible for transcription regulation are less discrete and relatively nonspecific. They can regulate transcription from a wide variety of promoters, provided that the regulatory domains are bound in the vicinity of the TATA sequence. Thus, hybrid proteins containing one transcription factor's regulatory region fused to another factor's DNA binding region will regulate transcription from promoters that contain the latter's DNA binding sequence. This occurs even if the regulatory region and the transcription machinery are from different species. In this section, we examine the characteristics of various transcription regulatory domains and possible mechanisms by which they act.

Transcriptional activation at a distance via DNA bending Throughout the descriptions of the different eukaryotic promoters (Sections 8.2–8.4), we stressed the fact that transcription is most often regulated by *cis* acting sequences located at considerable distances from the transcription initiation sites. Because we also suppose that transcription is regulated by proteins bound at these distant sites, we need to consider how these proteins' regulatory regions influence the transcription complex's activity. Much of the experimental evidence points to protein-protein interactions between the transcription factors' regulatory regions and components of

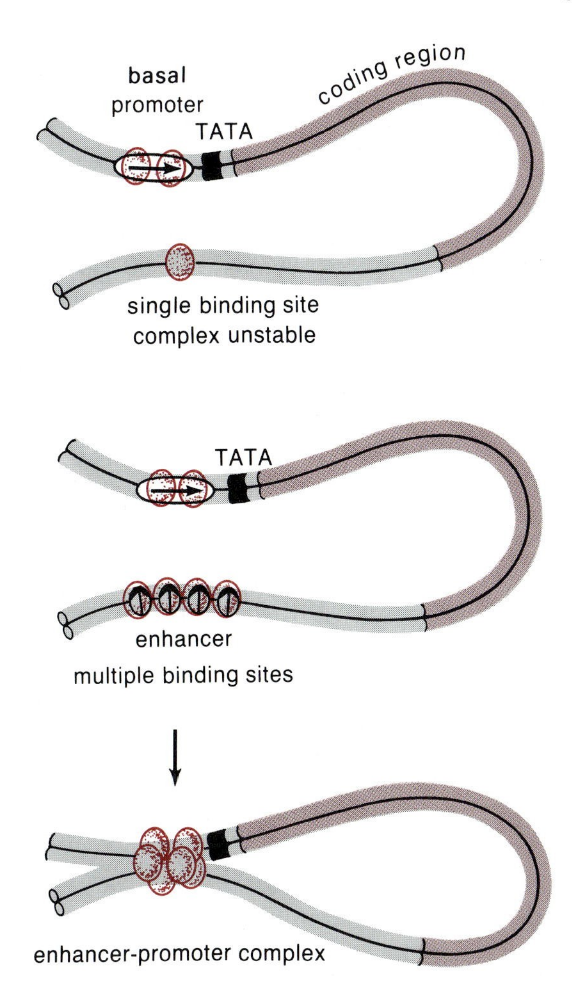

Figure 8.106
Multiple proteins bound at distal enhancers influence the assembly of transcription complexes. Multiple binding sites for the same or different proteins promote or stabilize the assembly of the transcription machinery in the region of the TATA sequence. DNA bending permits proteins bound at a distance, here shown downstream of the transcription unit, to interact with proteins bound upstream of the transcription start site.

the transcription complex (Figure 8.106). The necessary contacts between distantly bound proteins are made possible by looping out the DNA between the two binding sites. In complex promoters, those that can accommodate the binding of many different proteins, alternative, cooperative, or even negative interactions between the bound proteins permit differential regulation of the promoter's activity.

Prokaryotic transcription systems provided the first experimental support for the DNA looping model. In those instances, the activators generally bind close to the RNA polymerase binding site, and amino acids on one face of the regulatory protein's activation domain interact with the bound polymerase. Mutational alterations of these amino acids generally prevent transcriptional activation without affecting DNA binding. Also, there are clear indications that DNA looping permits such protein-protein interactions when the binding sites are farther apart. Not surprisingly, DNA flexibility and topological constraints strongly influence the efficiency of the protein interactions over short separations but appear to be less important over kilobase distances.

Acidic activation domains We have already emphasized that specific DNA binding can be achieved using several different structural motifs (Section 8.6a). In each instance, the amino acid sequences responsible for the specific contacts between these motifs and their cognate base pairs are relatively invariant, and even minor substitutions affect the efficiency of binding. By contrast, although the capacity for transcriptional activation can also be attributed to certain regions of each protein, the boundaries and sequences of most activation domains are considerably less well defined. A mutational analysis of the GAL4 protein localizes its activation domains to two regions: residues 148–238 near the amino terminus and amino acids 768–881 at the carboxy terminus (Figure 8.45). The most characteristic feature of these regions is the relative abundance of negatively charged amino acids: a net charge of −7 and −9 at the amino and carboxy terminal regions, respectively. Either of these regions can activate transcription when it is fused to the *GAL4* DNA binding domain; moreover, either of these regions activates transcription if it is attached to any binding sequence that brings the GAL4 protein region near a promoter's TATA sequence.

The yeast transcriptional activator GCN4 positively regulates the expression of genes involved in amino acid biosynthesis and is closely related to and interchangeable with the mammalian AP1-Jun family of transcription factors (Section 8.7d). GCN4 has a single activation region consisting of a stretch of 88 amino acids with a net charge of −16 near the center of the 281 amino acid sequence. The glucocorticoid receptors also have acidic activating domains. In the case of the rat glucocorticoid receptor, there is a single cluster of negatively charged amino acids near the amino terminus. The human glucocorticoid receptor has an additional acidic domain in the carboxy terminal region. An acidic activation domain also exists in the *bcd* encoded transcription regulator (Section 8.3g), thus extending the generality of this feature to *Drosophila*. Transcriptional activators that do not bind to DNA (e.g., VP16, the virion associated transcription factor of *Herpes simplex* that triggers expression of viral encoded genes after infection [Section 8.3d], and E1a, the essential transcriptional activator in adenovirus infections) have acidic activation domains. Although these proteins alone do not activate transcription, fusion proteins containing their acidic domains joined to a DNA binding domain activate the promoters to which they bind.

The acidic character of transcription activation domains came to light in the observation that activation is enhanced proportionately by increasing the net negative charge in one such domain. Indeed, even segments of *E. coli* proteins containing a high proportion of acidic amino acids can activate *GAL* gene transcription when fused to the GAL4 DNA binding domain. For example, GAL80 protein can be changed from an inhibitor to an activator of GAL4-dependent transcription if it is joined to an *E. coli* acidic segment. In this instance, the *E. coli* segment provides the activation function while the GAL4 activation domain remains blocked by its association with GAL80.

Although transcriptional activation does not depend on a defined sequence of acidic amino acids, there appears to be a preference for sequences that can form amphipathic α helices. Such helices have negatively charged residues along one surface and hydrophobic amino acids along the other, such as exists in leucine zippers (Section 8.6a). Thus, a synthetic peptide containing 15 amino acids designed so that it can assume an amphipathic helical structure activates transcription when joined to an appropriate DNA binding domain; the same 15 amino acids assembled in a scrambled order do not.

The critical issue of how the acidic activation domains influence transcription remains unanswered. Because the activating domains already mentioned function in a wide variety of heterologous systems (e.g., in *Drosophila*, plant, yeast, and mammalian cells), it is likely that they interact with a common component of the transcription machinery. Thus, activation could result from interactions of the acidic domains with the basic histones in nucleosomes so as to influence their behavior in the promoter region. Alternatively, the activation domains might contact a basic region in RNA polymerase II or in one of the proteins in the transcription complex and thereby facilitate the assembly or activity of the complex. The DNase I footprint formed by binding the transcription complex in the TATA region is altered when mammalian transcription factors are bound nearby. The synthetic amphipathic helical segment joined to the GAL4 DNA binding domain produces a similar alteration in that footprint if the UAS_{GAL} sequence is positioned near TATA.

Irrespective of which part of the transcription machinery is affected by the localized binding of proteins with acidic domains, that interaction appears to rely on electrostatic effects rather than on stereo-specific or complementary contacts. Because transcription factors from yeast or mammals function equally well with the other's transcription machinery, this relaxed sequence requirement appears to have been conserved over long evolutionary history. Inasmuch as the magnitude of transcriptional activation is directly proportional to the net negative charge, changes in that quantity could provide a way to modulate the regulatory response. Thus, the net negative charge in the activation domain could be altered by amidation of glutamic or aspartic acid residues.

Other activation domains Some transcription factors possess functional activation domains that lack an obvious clustering of acidic amino acids. Sp1, for example, contains activation regions rich in glutamine, and in serine and threonine. How these regions activate transcription is unknown. But serine hydroxyl groups could serve as sites for phosphorylation and dephosphorylation, providing for flexible levels of acidic residues. Also, reversible deamidations and amidations of glutamines would increase or decrease the negative charge in these regions and thereby modulate their transcriptional activity. The recently discovered glycosylation of Sp1 and other transcription factors might also influence their transcriptional activity by adding regions whose electrostatic charge can be altered in a regulated way.

8.7 Global Influences on Gene Expression

So far, our consideration of gene expression has focused on the transcriptional regulation of single genes and the processing of their RNA transcripts into mature RNA species. The rate, time, and location of these processes are governed by proteins that recognize relatively short DNA sequence motifs and/or the surfaces of other proteins. These interactions influence transcription initiation, termination, or processing. The availability or functional state of these proteins is often the principal determinant of a gene's expression. But there is also compelling evidence that reversible modifications in the state of the DNA can influence gene expression, particularly transcriptional competence. For example, alterations in the extent of DNA's superhelicity, that is, its topological status and the consequent changes in conformation, e.g., from B to Z or other forms of DNA (Section 1.1c) can influence transcriptional activity strikingly. The state of the chromatin in which the gene is embedded, specifically, the higher order packing of nucleosomes (Section 1.1g), adds another level of complexity affecting gene expression. Generally, transcriptional activation correlates with transitions from highly condensed chromatin structures to more unfolded, "open" forms. Transcriptional activity is also influenced by the extent of cytosine methylation in the dinucleotide base pairs $\substack{5'\text{-CG-}3' \\ 3'\text{-GC-}5'}$. Transcriptional repression is often associated with hypermethylation in and around genes, and activation with hypomethylation.

In this section, we examine some of these modifications and their consequences for gene expression. Because these regulatory mechanisms are not gene-specific but affect the expression of many genes, we refer to them as being global. This section concludes with two illustrations of how nontranscriptional mechanisms contribute to the regulation of mRNA levels. One operates posttranscriptionally by selective destruction of mRNAs. The other involves a novel mode of translational regulation of a yeast transcription factor.

a. DNA Packing

The basic unit of chromatin organization in eukaryotic cells is the nucleosome, a repeating subunit in which 145 bp of DNA are wrapped into two left-handed superhelical turns around an octamer of histone proteins containing two molecules each of H2A, H2B, H3, and H4 (Section 1.1g). Nucleosomes are organized into 10 nm diameter filaments by interactions with histone H1, which is bound to DNA entering and leaving the nucleosome (Figure 1.17). A third level of organization is achieved by winding the 10 nm filament into a helix with six nucleosomes per turn: the 30 nm diameter solenoid (Figure 1.18). Still more complex chromatin structures are formed by condensing the 30 nm filament, but details of these structures are unknown.

There is strong electron microscopic evidence that eukaryotic chromatin is organized into discrete regions, loops, or domains of about 30–300 kbp, each anchored to a protein-rich matrix (Figure 8.107). Each loop appears to have a single replication origin and to behave as a unit of replication. The loops are independent units of supercoiling, their topological structure being independent of the state of the surrounding loops. This is probably achieved by having the ends of the loops anchored to the matrix. Although each loop contains multiple transcription units, the activity of the entire region may be coordinated, being either transcriptionally repressed or potentially active.

Repressed chromatin Many genes in multicellular eukaryotic organisms are expressed only at certain times in development and/or in a tissue-specific manner. This means that cells must have an efficient means for repressing transcription of all genes other than those that are appropriate

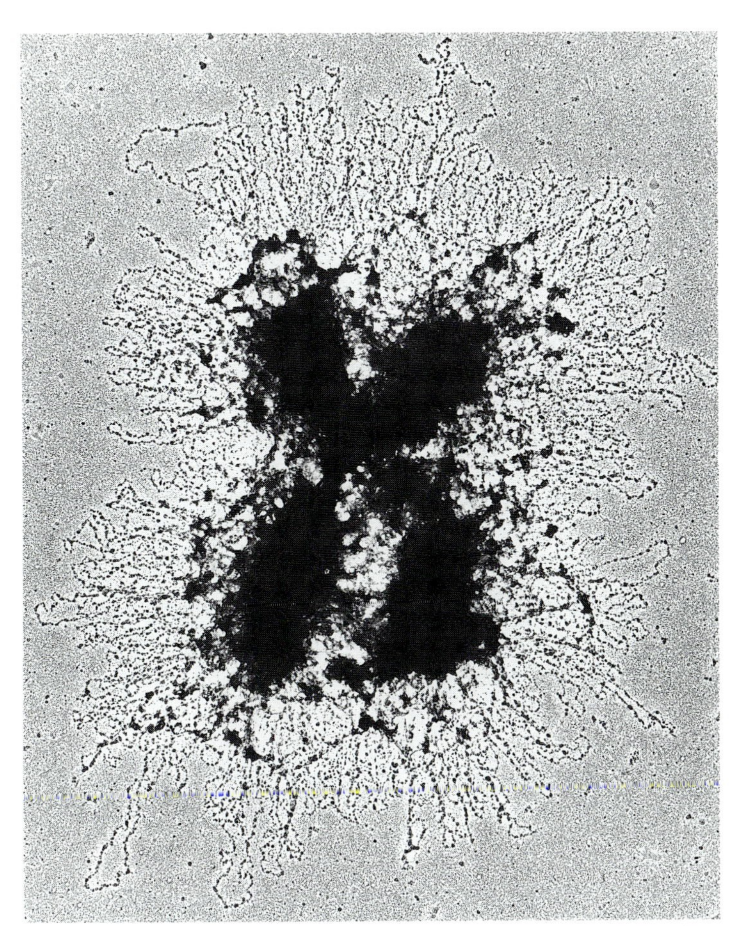

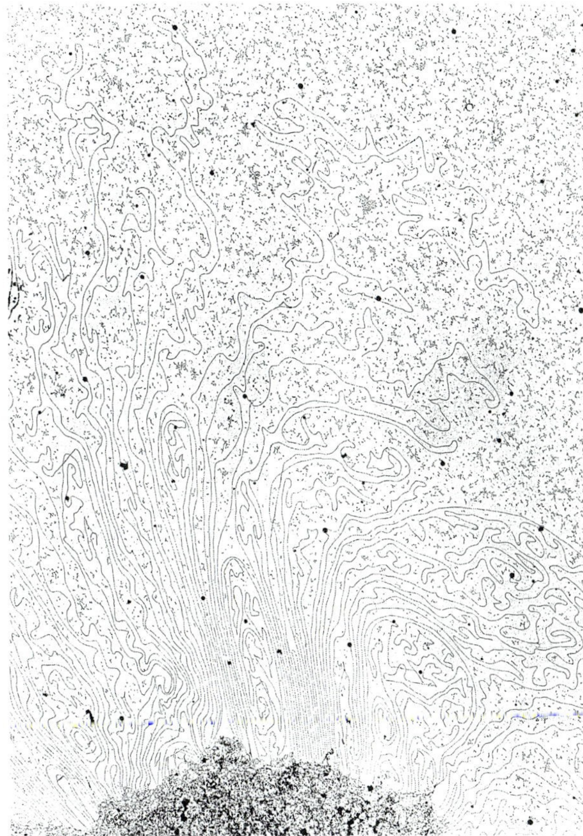

Figure 8.107
Eukaryotic chromatin contains variable length loops, the ends of which are anchored to a protein matrix. The micrograph at the left shows a metaphase chromosome from which chromatin fibers have been extruded. The looplike organization of the chromatin fibers is evident, although precise points of the loop attachments are not. Photograph appeared in W. C. Earnshaw and U. K. Laemmli, *J. Cell Biol.* 96 (1983), p. 84. The electron micrograph at the right was obtained from chromosomes treated with dextran sulfate and heparin, purified and spread with cytochrome. Note that both ends of a DNA loop appear to emanate from adjacent points in the protein matrix at the bottom of the photograph. Photograph appeared in J. R. Paulson and U. K. Laemmli, *Cell* 12 (1977), p. 817. Both photographs were supplied by U. K. Laemmli.

to the particular cell type. Repressing large blocks of genes by limiting the availability of required transcription factors seems an unlikely mechanism for this kind of regulation. Support for this view comes from experiments in which genes not ordinarily expressed by a particular cell type can be expressed when introduced by transfection. For example, transfected globin genes are expressed about 1000 times more efficiently in fibroblast cells than are the endogenous chromosomal globin genes; moreover, this difference persists through successive cell divisions even though the transfected genes are integrated into chromosomal sites. Indeed, it is common, though not universal, for the transcriptional activity of a transfected gene to be orders of magnitude higher than that of the corresponding endogeneous gene. A widely held explanation is that such repression results from the sequestration of the silent genes into higher order chromatin structures, making them unavailable for interaction with the transcription machinery.

In the absence of detailed information about such higher order chromatin structures, it is impossible to describe the transition that occurs in going from repressed to expressed chromatin. One clue comes from correlations between the time at which a particular gene is replicated and its transcriptional competence. The ubiquitously expressed housekeeping genes (genes that are expressed in all tissues) such as dihydrofolate reductase, cytoplasmic actin, and glucose 6-phosphate dehydrogenase replicate during the first half of the S-phase; unexpressed genes tend to replicate late. Moreover, genes tend to replicate early in tissues in which they are expressed and late in tissues in which they are silent. Similarly, in mammals, the inactive X-chromosome replicates later than the active X-chromosome. These correlations suggest that replication early in S-phase may create states that are more accessible to the transcription machinery or more efficient at binding essential transcription factors.

Nuclease sensitivity Judging from their increased sensitivity to digestion with DNase I or micrococcal nuclease, transcriptionally active chromatin regions are less tightly packed than chromatin containing transcriptionally inactive genes. This general nuclease sensitivity often extends for several thousand base pairs flanking a transcription unit. The structural basis for this differential nuclease sensitivity is not settled, but biochemical studies suggest that nucleosomes along active genes are low in histone H1, enriched for acetylated and ubiquinated histones, and complexed with several kinds of highly acidic proteins (HMG proteins). These modifications may alter the histone association with DNA and may foster the irregular spacing of nucleosomes, rendering the DNA more sensitive to nuclease digestion. Thus, the normal periodicity of nuclease cleavage sites of the heat-shock protein genes in chromatin is altered after heat induction. The same kind of altered nuclease cleavage patterns occurs in the oviduct's but not the liver's ovalbumin gene and in the lymphocyte's IgL(κ) gene after its rearrangement to the transcriptionally active form.

Nuclease hypersensitivity Superimposed on the regional nuclease sensitivity, actively transcribed genes have hypersensitive sites within and surrounding their transcription units. Such DNase I hypersensitive cleavage sites do not occur in the corresponding naked DNA. Generally, hypersensitive sites are located immediately upstream of a gene's transcription start site, but in several instances, they occur within or at the 3′ end of the transcription unit; recall that these sites can contain enhancer and transcription termination sequences. Furthermore, certain genes have DNase I hypersensitive sites at considerable distances upstream of the cap site (e.g., several kilobase pairs upstream of the avian ovalbumin gene cluster and more than 20 kbp upstream of the human β-globin gene). In most instances, the occurrence of DNase I hypersensitive sites is restricted to the tissue or cell that is either expressing or programmed to express the particular gene. The hypersensitive site at the 5′ end of the preproinsulin gene, for example, is found in pancreatic β-islet cells but not in liver. Similarly, hypersensitive sites are found associated with rearranged immunoglobulin genes in B cells but not in nonhematopoietic tissues.

Nucleosome-free regions Nuclease hypersensitive sites probably arise where there are discontinuities in the arrangement of nucleosomes. This was first noted in the correlation between the location of the hypersensitive sites and a nucleosome-free zone in the SV40 genome. Electron micrographs of isolated SV40 minichromosomes reveal a nucleosome-free gap in the DNA (Figure 8.108). The gap spans the region between the early gene's transcription start site and the most upstream end of the enhancer. When this 400 bp segment is translocated to other regions of

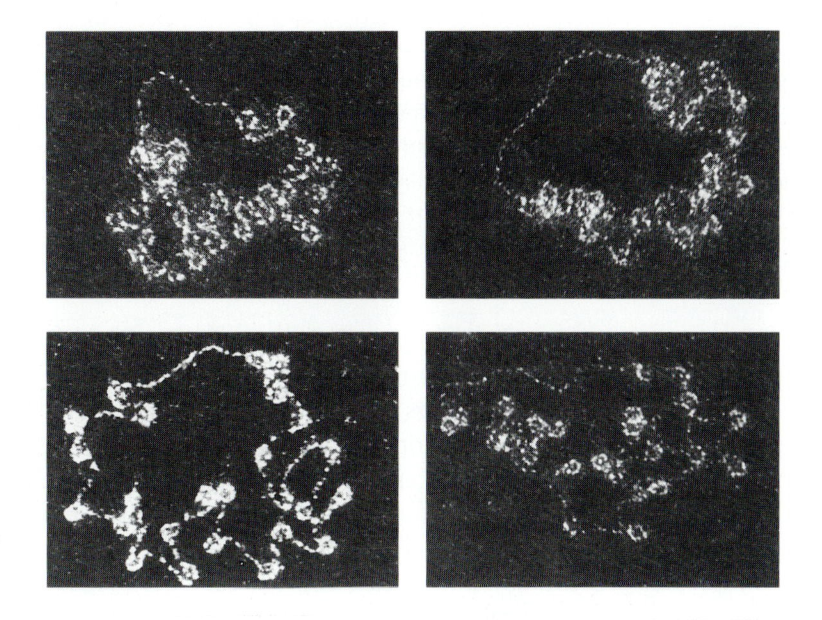

Figure 8.108
SV40 minichromosomes with a stretch of DNA devoid of nucleosomes. SV40 minichromosomes were isolated from infected cell nuclei, stained with aqueous uranylacetate, and visualized by dark field electron microscopy. Nucleosomes are clearly absent from a specific region of the DNA. Photographs from Figure 2 in S. Saragosti, G. Moyne, and M. Yaniv, *Cell* 20 (1980), p. 65. They were supplied by M. Yaniv.

the SV40 genome or into unrelated plasmid DNAs, nucleosome-free gaps and DNase I hypersensitive sites are created at these locations. This indicates that the two properties are most likely a consequence of the nucleotide sequence in this stretch of DNA. However, the nuclease sensitivity of the gapped region is probably modulated by nonhistone proteins bound in this region, possibly Z-DNA binding proteins (Section 8.7b) and transcription factors bound to the enhancer (Section 8.3d).

Similar nuclease hypersensitive sites and nucleosome gaps occur in the enhancer regions of the IgH and IgL(κ) genes (Section 8.3e) and at the UAS sequences associated with the 5′ end of the *GAL* and metallothionein genes (Section 8.3f). Nuclease hypersensitive cleavage sites appear in the glucocorticoid responsive element (Section 8.3f) within minutes after the addition of glucocorticoid to cells and disappear soon after removal of the hormone. DNase I hypersensitive sites have also been linked to transcriptional activation of rRNA genes, for such sites occur both upstream of the transcription initiation sites and around the intergenic spacer regions of rRNA genes. Indeed, nuclease hypersensitivity has become a surrogate indicator for transcriptional regulatory regions.

Because the location of transcriptional regulatory sequences corresponds to both the nuclease hypersensitive and nucleosome-free sites, there is reason to believe that all are related. The binding of transcription factors to their cognate binding sequences possibly displaces nucleosomes or alters their packing, the result being visible gaps in the chromatin array or even more subtle perturbations that permit nuclease cleavages. If, however, densely packed nucleosomes are stabilized by, for example, histone H1 or by folding into higher structures, access of essential transcription proteins to their cognate sequence signals may be restricted, thereby preventing transcription initiation.

Much remains to be discovered about the structural differences between expressed and repressed chromatin and the conditions governing the transition from one to the other. At present, we know only that chromatin regions may remain in the condensed unexpressed state through many rounds of DNA replication and become disassembled and expressible after appropriate exogeneous or endogeneous stimuli. Whether the activation of specific chromatin regions involves the intervention of positive-acting proteins or the removal of repressing proteins or both remains to be learned.

b. DNA Topology and Conformation

Superhelicity in transcriptional regulation There is considerable evidence that prokaryotic gene expression is modulated by the superhelical state of the DNA. Mutations or inhibitors of topoisomerases, which influence DNA's superhelicity, activate transcription of some genes while inhibiting others. Because the binding of proteins to negatively supercoiled DNA is aided by a favorable free energy change, it is surmised that changes in DNA superhelicity influence the ability of RNA polymerase or activator and repressor proteins to bind to their cognate regulatory sequences. Alterations in the superhelical state or the extent of supercoiling could also affect DNA bending and thereby the ability of distantly bound transcription proteins to interact with each other.

In Section 8.7a, we noted that eukaryotic chromatin is organized into a series of variable length loops, the ends of which are anchored to a protein matrix (Figure 8.107). Potentially, this organization allows for the superhelical density of one loop to be independent of the torsional state of the surrounding loops. Interestingly, DNA topoisomerase II, an enzyme responsible for relaxing supercoiled DNA (Section 2.1f), is one of the major proteins in the chromosomal scaffold to which the loops are anchored. Moreover, the DNase I hypersensitive sites associated with transcriptional control regions are frequently sensitive to single strand-specific nucleases, suggesting that these regions are associated with negatively supercoiled domains. In addition, topoisomerase I appears to be bound at the nuclease hypersensitive sites. Both positive and negative effects on transcription by compounds that intercalate into DNA or inhibit topoisomerases, both of which alter the DNA's superhelicity, have also fueled the notion that DNA topology is an important parameter governing transcriptional regulation. However, the lack of a transcriptional phenotype of yeast mutants deficient in DNA topoisomerase I or II suggests that if a eukaryotic DNA topoisomerase is required for transcription, either of the two probably suffices.

Most of the evidence implicating topoisomerases and their effects on DNA topology in transcriptional regulation in eukaryotes is indirect and inconclusive. Indeed, there is no direct evidence either for or against the view that changes in superhelical density within chromosomal loops influence transcriptional activity of the entire region or of genes located within the region.

B- and Z-DNA conformers and transcription DNA can assume both right-handed (B) and left-handed (Z) conformations (Section 1.1c and Figure 1.8). The relative proportions of the two forms at equilibrium are influenced by the DNA sequence, superhelicity, and a variety of environmental conditions such as ionic strength, temperature, and metals. Under appropriate circumstances, right-handed and left-handed regions may be interspersed within a given stretch of DNA. In these instances, sequences at or near the junctions behave anomolously as substrates for various enzymes.

Although the formation of Z-DNA was established *in vitro* by a variety of physical methods, its existence *in vivo* was, until recently, problematic. Now its presence in both prokaryotic and eukaryotic cells has been verified. As anticipated, the level of Z-DNA is strongly influenced *in vivo* by the extent of the DNA's superhelicity. Its formation is also favored in stretches of nucleotide sequences containing alternating purines and pyrimidines. For example, stretches of the middle repetitive sequences, $d(CA)_n \cdot d(GT)_n$ are prone to adopt the Z conformation. In addition, methylation of C residues in stretches of alternating Cs and Gs (i.e., 5′-CGCGCG-3′) promote Z-DNA conformations. And there are indications that sequences able to adopt the Z conformation correspond to the nuclease hypersensitive sites in the transcription regulatory regions of some genes.

Proteins that bind specifically to Z-DNA have been identified and purified. For example, several Z-DNA binding proteins are associated with the enhancer of the SV40 early region promoter. Nevertheless, aside from the existence of Z-DNA forming segments in or near transcription regulatory regions and of proteins that recognize Z-DNA conformations, the involvement of Z-DNA in transcriptional regulation is at present speculative. In one instance, a Z-DNA binding protein has been implicated in DNA recombination, but here too the precise involvement of the DNA conformation or the binding protein in the recombinational process is unknown.

c. DNA Methylation

There is considerable evidence that vertebrate gene expression is correlated with the methylation state of specific cytosine residues within and around a gene's regulatory regions. With some notable exceptions, the presence of methyl groups (**hypermethylation**) is associated with diminished gene expression and often with total inactivity. Conversely, gene activity is generally associated with the absence of or reduced methylation (**hypomethylation**) at the same sites.

One of the intriguing features of DNA methylation is its stability through many cell generations. Thus, genes that are expressed in a tissue-specific manner remain hypermethylated throughout embryonic life in all the somatic tissues save the one in which the gene is expressed. Conversely, regions around "housekeeping" genes, those that are expressed throughout development in all tissues, remain hypomethylated. Moreover, transfected unmethylated genes are generally expressed transiently, even after integration into chromosomal DNA and subsequent cell division. Conversely, transfected methylated DNAs fail to be expressed transiently or after integration. Significantly, integrated unmethylated DNA remains unmethylated and is expressed throughout many cell generations even though the corresponding cellular gene remains methylated and unexpressed.

Another indication of the relation between gene expression and methylation stems from studies of murine retrovirus. The virus fails to propagate or express its genes after infection of embryonic mouse cells.

However, both propagation and expression occur if the infection occurs after the embryonic cells are induced to differentiate. One correlative distinction between the two cellular states is that the integrated proviral DNA becomes methylated in the undifferentiated cells and remains so, whereas it fails to be methylated when introduced following differentiation. The inverse correlation between gene activity and methylation and the maintenance of these parameters in a tissue-specific and temporal fashion have stimulated considerable research into this mechanism of gene regulation.

Methylation at special cytosine residues Almost all methyl groups added to eukaryotic DNA occur at the 5 position of cytosine residues (5-meC) in the dinucleotide 5′-CG-3′. (In higher plants, the 5′-most cytosine of the trinucleotide 5′-CNG-3′, where N is any nucleotide, is occasionally methylated.) Recall that in vertebrates, the frequency of the CG dinucleotide is about one-fifth of that predicted from the DNA's base composition (Section 1.1). Thus, based on the content of G + C in mammalian DNAs (40 percent) and assuming an unbiased arrangement of nearest base neighbors, the frequency of the CG dinucleotide should be about 0.04, the same as that found for the GC dinucleotide. However, the actual frequency of CG is 0.008.

The underrepresentation of CG dinucleotides and their frequent methylation suggest that the two characteristics are related. This correlation probably stems from the consequences of spontaneous deamination of cytosines in DNA. Deamination of C residues is readily repaired by a mechanism that recognizes uracil in DNA (Section 2.3a). But deamination of 5-meC results in the formation of a T residue, and T residues are not repaired. Consequently, unessential CG dinucleotides that become methylated can be lost. Therefore, the retention of sequences containing methylated CGs in regulatory regions probably reflects their special function in gene expression.

Between 50 and 90 percent of the C residues in CG dinucleotides are methylated in vertebrates. The actual extent, however, varies from tissue to tissue, at various times in development, and from region to region in the DNA. Direct analysis of the amount of 5-meC relative to the total amount of C residues measures the overall extent of methylation. The more meaningful information concerns the methylation patterns of specific DNA regions. This can be assessed with pairs of restriction endonucleases that differ in their ability to cleave DNA depending on whether or not their recognition sites contain a C or 5-meC (Section 4.2e). For example, *Hpa*II and *Msp*I endonucleases both cleave the sequence 5′-CCGG-3′, but only *Msp*I cleaves at 5′-CmeCGG-3′. In contrast, *Hpa*II cuts DNA at 5′-meCCGG-3′, but *Msp*I does not. A comparison of the *Hpa*II and *Msp*I cleavage patterns at various regions of DNA identifies methylated sites. 5-meC containing CG dinucleotides that are not part of a restriction site or that cannot be distinguished by available isoschizomeric restriction endonuclease pairs can be identified by sequencing genomic DNA directly using chemical procedures that do not cleave at 5-meC residues. In this type of analysis, fragments produced by cleavages at Cs are missing from the sequencing ladders (Section 7.2c).

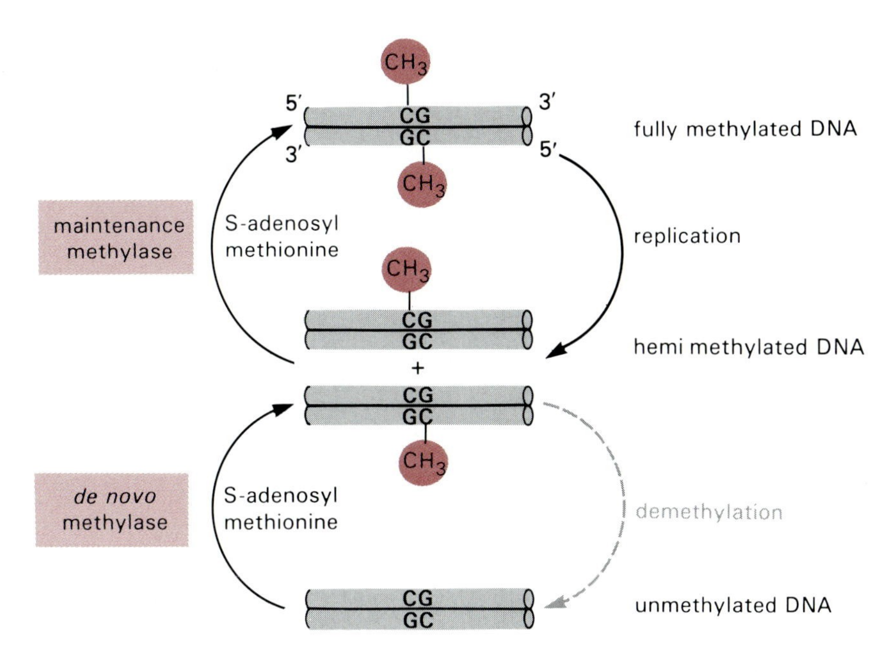

Figure 8.109
Maintenance and *de novo* methylation of DNA. Replication of fully methylated DNA results in formation of hemimethylated DNA, the newly synthesized strands being unmethylated. Maintenance methylases rapidly methylate the unmethylated C residues opposite the methylated Cs to regenerate fully methylated DNA. The broken arrow labeled demethylation connotes the existence of a pathway for complete loss of methyl groups, although the mechanism of that loss is unknown. *De novo* methylases are presumed to exist to account for the ability of cells to methylate unmethylated DNA. All DNA methylases are known to use S-adenosyl methionine as the methyl donor.

Mechanisms of de novo *and maintenance methylation and loss of meCs* The variable methylation patterns among different tissues and during differentiation raise the question of how the patterns are established, maintained, and altered within a cell lineage. Acquiring and maintaining a methylation pattern present one aspect of the problem; loss of meC residues presents another. *De novo* methylation involves methylation at both opposing CG dinucleotides; maintenance methylation requires methylation of one of the two CG dinucleotides in hemimethylated sequences (Figure 8.109). *De novo* methylation establishes new patterns, and maintenance methylation modifies newly synthesized DNA strands to assure the inheritance of the parental methylation pattern in the descendants.

The mechanism for specific or global loss of meCs is unknown, although various models have been considered. One proposes direct replacement of methylated with unmethylated C residues. Supporting this idea is the observation that *in vitro* differentiation of certain leukemia cells induced by chemical agents is accompanied by a massive demethylation of DNA. During this process, radiolabeled deoxycytidine, but not deoxyadenosine, is incorporated into unreplicated cellular DNA (Figure 8.110). An alternative explanation is that the association of tissue-specific or stage-specific proteins with a sequence containing a methylation site blocks the action of the maintenance methylase. The site will become completely unmethylated following the next round of replication. A third model posits that methylation occurs at selective CG sites in highly condensed chromatin, thereby stabilizing the transcriptionally inactive form of the DNA, and that loss of meC occurs following the unfolding to the active state.

Only a beginning has been made in characterizing the enzymes responsible for DNA methylation. Several mammalian DNA methyl transferases have been purified, and the gene encoding one has been cloned and sequenced; their molecular weights range between 135 and 175 kDaltons. S-adenosyl methionine serves as the methyl donor. The isolated

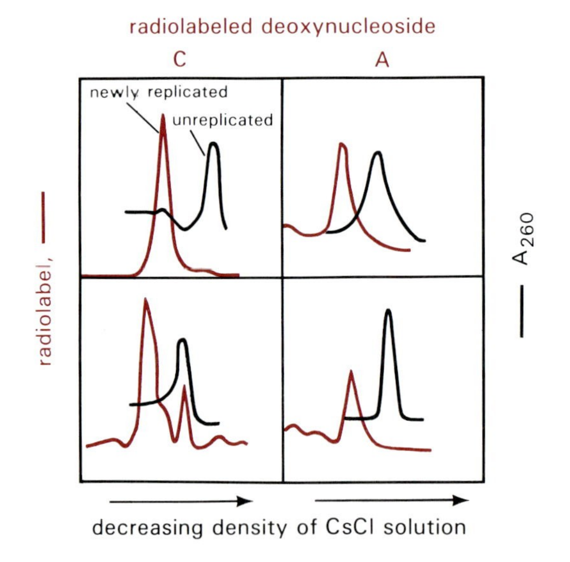

Figure 8.110
Extensive demethylation of DNA can occur without new DNA synthesis. Mouse erythroleukemia cells synchronized in the G1 phase of the cell cycle were induced to differentiate by the addition of hexamethylenebisacetamide. After 16 hours, 5-bromodeoxyuridine and 5-fluorodeoxyuridine were added to the medium. The 5-bromodeoxyuridine is phosphorylated to the corresponding deoxynucleoside triphosphate and incorporated into newly synthesized DNA. Consequently, unreplicated DNA has a light density (neither strand has the heavy bromodeoxy nucleoside), but newly synthesized DNA will be heavier (one strand has the heavy nucleoside). One hour later, radiolabeled deoxycytidine (C, left side of figure) or deoxyadenosine (A, right side of figure) was added to separate portions of cells. Radiolabeled deoxycytidine or deoxyadenosine is incorporated into the newly replicated DNA in both the induced (the upper two curves) and the uninduced cells (the lower two panels). By contrast, in the induced cells, which are known to be depleted of 5-meC, only radiolabeled C is incorporated into the unreplicated DNA. Thus, C can replace 5-meC in DNA by a mechanism that does not involve extensive DNA synthesis. Adapted from A. Razin et al., *Proc. Nat. Acad. Sci.* U.S.A. 83 (1986), p. 2827.

enzymes methylate hemimethylated DNA at a considerably greater rate than unmethylated DNA, suggesting that they are maintenance methylases. Use of part of the mouse β globin gene promoter as a substrate revealed the specificity of a purified mammalian DNA methyl transferase to be quite complex. The promoter region contains ten CG dinucleotides in a 368 bp stretch. Of the ten, only five or six, which lie in a cluster of 29 bp about 100 bp upstream of the transcription start site, are methylated, and even these are methylated to different extents and at different rates. Apparently, the sequences surrounding the CGs influence the methylation pattern. *In vivo*, the state of the chromatin may do so as well. Other factors, particularly those that regulate methylation and demethylation, remain to be discovered.

Methylation and gene expression Numerous studies with viral as well as nonviral eukaryotic genomes support the notion that hypermethylation correlates with inhibited or inactivated promoters, but actively transcribed promoters are frequently hypomethylated or unmethylated. Studies with adenovirus 12 transformed cells illustrate this correlation. Adenovirus 12 DNA is unmethylated. After infecting hamster cells, the viral DNA is integrated into the cellular chromosomes. Certain of the viral genes are expressed; others are not. Generally, but not invariably, the unexpressed genes are methylated, and the expressed genes remain unmethylated. A detailed examination of the adenovirus E2A region reveals that in the transformed cell line HE1, the E2A gene is transcribed and translated, and none of the 14 5′-CCGG-3′ sequences in the E2A region is methylated (Table 8.8). However, in transformed cell lines HE2 and HE3, the E2A gene is silent, and all 14 of these CG dinucleotides are methylated. Using the genomic sequencing technique referred to earlier, one can focus on all the E2A promoter's CG dinucleotides, irrespective of whether they are included in *Hpa*II sites. In cell line HE1, where the E2A promoter is active, the 13 CG dinucleotides between bp +24 and −160 are unmethylated. In cell lines HE2 and HE3, the same promoter is inactive, and these same CG sequences are methylated on both strands. Thus, the

Table 8.8 Adenovirus E2A Gene Function and Methylation in Transformed Cells

Cell Line	Presence of Intact E2A Promoter	Expression of E2A Gene[1]	Methylation of 14 5′-CCGG-3′ sites[2]	Methylation of All 13 CGs Between bp +24 and −160[3]
HE1	+	+	−	−
HE2	+	−	+	+
HE3	+	−	+	+

[1] E2A mRNA and protein formation.
[2] 5-meC is detected by differential cleavages with *Hpa*II and *Msp*I endonucleases.
[3] 5-meC in all CG dinucleotides is detected by DNA sequencing.

inverse correlation is perfect over a region of about 180 bp of the E2A promoter.

Methylation of the E2A gene *in vitro* with the *Hpa*II methylase almost invariably inactivates its transcriptional activity upon transient transfection or stable transformation. In contrast, the unmethylated promoter is functional in both assays (Figure 8.111). Furthermore, once integrated, the DNA retains the same methylated pattern as the transfecting DNA. Similarly, methylation with different CG-specific methylases *in vitro* causes the inactivation of the adenovirus E1A, α globin, thymidine kinase, adenine phosphoribosyl transferase, and IgL(κ) gene promoters. In general, the sites of inactivating methylations are localized to promoter regions, but differential methylation of distantly located regulatory sequences also occurs. The activity of hemimethylated E2A promoter DNA, made by annealing unmethylated single strands with oligonucleotides containing methylated 5′-CCGG-3′ sequences, is markedly reduced but not completely abolished.

The correlation between methylation and repression is often more complex than implied by the illustrations discussed previously. For example, the pattern of methylation in the approximately 30 kbp of human DNA

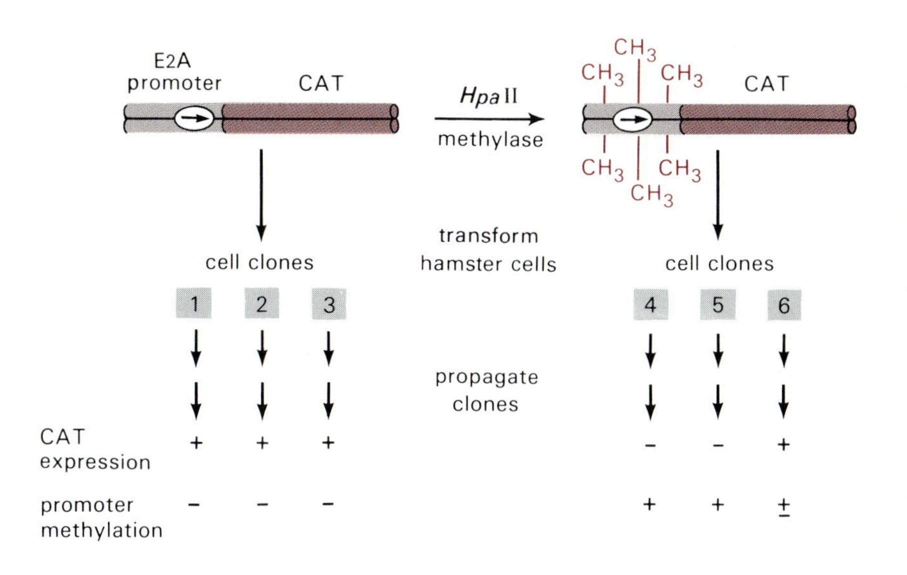

Figure 8.111
Promoters that are methylated *in vitro* remain methylated and inactive *in vivo*. A construct containing the adenovirus E2A promoter fused to the chloramphenicol transacetylase (CAT) coding sequence was methylated *in vitro* with purified *Hpa*II methylase. After transfection of hamster cells with the unmethylated and methylated DNAs, CAT expression and the methylation state of the promotor were examined in independent clones. The results of that analysis are shown as + or − to indicate either the presence or absence of methyl groups or the presence or absence of CAT expression, respectively.

that includes the human γ-, δ-, and β-globin genes (Figure III.5) has been examined in cells that express one or another of these genes. This region contains 17 recognition sites for restriction endonucleases whose activity depends on the methylation state of a CG dinucleotide: *Hpa*II, *Hha*I, *Xho*I, and *Sal*I (Sections 4.2c and e). Genomic DNA digests from the various cell types were analyzed by DNA blotting with appropriate probes cloned from the region. The data indicate that the γ-globin gene region is largely unmethylated in a leukemic cell line that expressed the γ-globin genes and hypomethylated in erythroid cells from midterm fetal liver, the major site of γ-globin gene expression in human fetuses. The sites neighboring the other globin genes are more highly methylated. In contrast, the whole globin region is hypermethylated in adult liver, lymphocyte, sperm, and fetal brain cell DNAs. More complex details emerge when individual cleavage sites are distinguished. Thus, some sites remain unmethylated even in adult liver, but other sites appear to be methylated in at least some fetal liver erythoid cells. In placenta, which does not express the globin genes, the entire region is hypomethylated. This analysis suggests that hypomethylation can be associated with gene expression and may even be required, but it cannot by itself be determining.

The correlation between inhibition of transcription and methylation of the corresponding regulatory regions holds for a variety of genes, although not all that have been tested. Inhibition might result from interference with the binding of transcription factors by 5-meC residues. For example, the methylation of a CG dinucleotide in the promoter inhibits the binding of an adenovirus major late transcription factor and *in vitro* transcription. Methylation of another CG dinucleotide a few bases downstream has no effect. The binding of appropriate transcription factors to the adenovirus E2 and mammalian tyrosine aminotransferase promoters is also blocked by methylation of CG sequences in their transcription factor binding sites. The cAMP-responsive element (CRE), present in many inducible and tissue-specific genes, contains a CG dinucleotide (5′-TGACGTCA-3′). Methylation of promoters that depend on one or paired CREs for their transcriptional activity clearly blocks the binding of the CRE transcription factor and prevents transcription of a linked gene both *in vitro* and *in vivo*. Taken together, these observations indicate that methylation of CG dinucleotides within transcription factor binding sequences can block binding of the factor and inhibit the promoter's function.

However, methylation of the CG dinucleotide in the Sp1 binding site, 5′-GGGCGG-3′ (Section 8.3d), appears to be an exception. Neither Sp1 binding nor Sp1 dependent transcription is inhibited by methylation either *in vitro* or *in vivo*. Sp1 binding sites may be exceptional because of the way the protein and DNA interact. Perhaps the binding and activation characteristics of Sp1 have been selected to be insensitive to methylation at its cognate sequence motif because of its widespread role in the transcription of many essential and ubiquitously expressed genes.

Additional evidence supporting a role for methylation in regulating specific gene expression comes from studies with 5-azacytidine (5-azaC). This cytosine analogue, which contains a nitrogen atom instead of carbon at position 5 in the pyrimidine ring, can be incorporated into DNA during replication but cannot be methylated. Exposure of cells to 5-azaC is followed by extensive loss of 5-meC residues. The mechanism by which this occurs is unclear, but some evidence suggests that 5-azaC residues in DNA inhibit DNA methyl transferase. The extensive loss of meC is often transient, but in some cell types, certain genes become stably hypomethylated.

The most remarkable property of 5-azaC is its ability to change the phenotype of certain cells, including altering the state of differentiation and inducing specific gene expression. Where the repression of endogenous or transfected genes has been attributed to their hypermethylation, treatment of the cells with 5-azaC can cause either a transient or stable derepression. Those genes that become stably hypomethylated are most likely to be transcriptionally activated. However, some genes that are demethylated by the 5-azaC treatment are not activated or become activated only after additional treatments. Thus, although 5-azaC causes a transient genomewide loss of meCs, it is not clear that this alteration is directly responsible for the activation process. Loss of meC residues could occur after events that directly activate expression of certain genes.

Confounding the generalization that methylated CG dinucleotides have a critical part in regulating transcription is the lack of meCs in other eukaryotic genomes. Thus, meC is absent from yeast and other fungal genomes and from invertebrates such as *Drosophila* and the nematode *C. elegans*. Yet these organisms maintain their genomes in complex chromosomes and express their genes differentially according to developmental programs and environmental stimuli. The need for and function of DNA methylation in vertebrates therefore remains an enigma.

Unmethylated CG clusters Although most CG dinucleotides in vertebrate DNA are highly methylated, clusters of stably unmethylated CGs exist throughout the genome. Thus, about 1 percent of vertebrate DNA appears as small fragments (average length 120 bp) after digestion with *Hpa*II endonuclease, the enzyme that cleaves unmethylated 5′-CCGG-3′ sequences. These fragments are referred to as *Hpa*II tiny fragments (HTF). When blots of genomic DNA digests are probed with cloned HTFs, most are localized in "islands" of unmethylated CGs, usually 1–2 kbp long. Altogether there appear to be about 30,000 such islands, spaced on average about 100 kbp apart. Such clusters of unmethylated CGs are referred to as **CG** or **HTF islands**. The shared features of CG islands are (1) a higher average G + C content than bulk DNA, (2) a CG dinucleotide frequency that is predicted by the islands' G + C content, and (3) the lack of methyl groups on most of the CG dinucleotides.

About 70 percent of CG islands are associated with housekeeping genes; others occur with tissue-specific genes (Table 8.9). Examples of genes not associated with CG islands are listed in Table 8.10. CG islands generally occur within regions where transcription begins (Figure 8.112). Thus, most, if not all, housekeeping genes transcribed by RNA polymerase II have CG islands at their 5′ ends. The lengths of the CG islands are about the same irrespective of the lengths of the gene (Figure 8.113). Island sequences can extend into the transcribed region, often into the first exon and intron.

At present, neither the origins of CG islands nor their function are known, but interesting speculations on both abound. For example, en-

Table 8.9 Genes Associated with CG Islands

Genes	Species
Genes for metabolic enzymes	
Hypoxanthine-guanosine phosphoribosyl transferase	Mouse, Human
Adenine phosphoribosyl transferase	Mouse, Hamster
Dihydrofolate reductase	Mouse, Human
Glucose-6-phosphate dehydrogenase	Human
Hydroxy methyl glutaryl CoA reductase	Human
Glyceraldehyde-3-phosphate dehydrogenase	Chicken
Triose phosphate isomerase	Human
Adenosine deaminase	Human
Phosphoglycerate kinase	Human
Cellular inducible genes	
Metallothionien I	Mouse
Superoxide dismutase	Human
Heat-shock	Human
Cellular versions of viral oncogenes	
c-*int*1	Human
c-*myc*	Human, Mouse
c-*src*	Chicken
c-*fos*	Mouse
c-*Ha-ras*	Human
Cellular structural protein genes	
Ribosomal protein L-30	Mouse
Ribosomal protein L-32	Mouse
β-Actin	Rat
β-Tubulin	Rat, Human
Class I histocompatibility	Man, Mouse
Genes for tissue-specific proteins	
α2(I) collagen	Chicken
Retinol binding protein	Human
α-Globin	Human, Goat
Thy-1	Human, Mouse
Class II histocompatability	Human

Note: This partial list includes extensively sequenced genes that show HTF-like regions. The hypoxanthine-guanosine phosphoribosyl transferase and glucose-6-phosphate dehydrogenase are X-chromosome linked genes whose islands are methylated on the inactive X-chromosome. The 5′ end of the glucose-6-phosphate dehydrogenase gene has not been identified, but there are HTF-like sequences near the 3′ end.
SOURCE: Adapted from A. Bird, *Nature* 321 (1986), p. 209.

richment of CG dinucleotides at the start of many genes may stem from their protection against methylation by bound transcription factors. In comparison, CG dinucleotides in other genomic regions are unprotected and can be lost by methylation and deamination to TGs. This view accounts for the clustering of unmethylated CGs at the 5′ ends of genes and their predicted abundance based on G + C content. In this model,

Table 8.10 Genes Showing No Evidence of CG Islands

Genes	Species
β-Globin family	Human, Mouse, Rabbit, Goat
Placental lactogen	Human
Nerve growth factor	Human
α-Amylase I and II	Mouse
Growth hormone	Mouse
Insulin	Mouse
α-Interferon	Mouse
Fibrinogen	Rat
δ-Crystallin	Mouse
Myoglobin	Seal

Note: This partial list includes extensively sequenced genes that are deficient in CGs throughout the gene and its 5′ and 3′ flanks.

SOURCE: A. Bird, *Nature* 321 (1986), p. 209.

the islands need not have a special function: their methylation is prevented only because they bind one or several proteins. In an alternative speculation, the densely clustered unmethylated CGs derive from a special property related to chromatin structure, permitting CG-rich regions to be constitutively available for interaction with elements of the transcription machinery. This would mean that methylation excludes protein factors from an inactive island and that the factors exclude methylation from active islands. Both speculations presume that the CG islands are essential for the expression of their linked genes. Rigorous tests of these predictions have not been made, but some relevant facts are known. When genes with CG islands are transfected into cells, the CGs in the islands remain unmethylated, but CG dinucleotides in the flanking regions of the gene become methylated. Moreover, if the CG dinucleotides in the islands are methylated *in vitro* and then transfected into cells, the gene's expression is impaired, and the islands remain methylated thereafter.

As mentioned earlier, CG islands are found associated with some genes that are expressed in a tissue-specific fashion (Table 8.9). Moreover, these DNA regions are unmethylated in cells that express the gene as well as

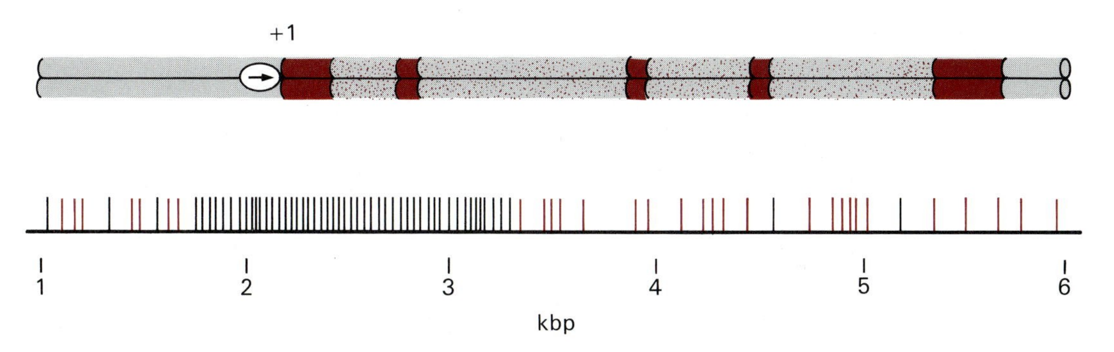

Figure 8.112
Distribution of methylated and unmethylated CpGs at a model vertebrate gene. A typical vertebrate gene containing exons (in color) and introns (stippled) and a promoter upstream of the transcription unit are shown at the top. The vertical lines below the gene indicate the positions of CpGs in and surrounding the gene. Methylated CpGs are shown in color and unmethylated CpGs in black. The region of closely spaced unmethylated CpGs clustered over the promoter and 5′ end of the gene constitutes a CpG island. Although all CpGs can be located by genomic sequencing, only a fraction of them can be detected with restriction endonucleases because not all sequences containing CpGs can be cleaved by the available enzymes. Adapted from A. P. Bird, *Trends Genet.* 3 (1987), p. 342.

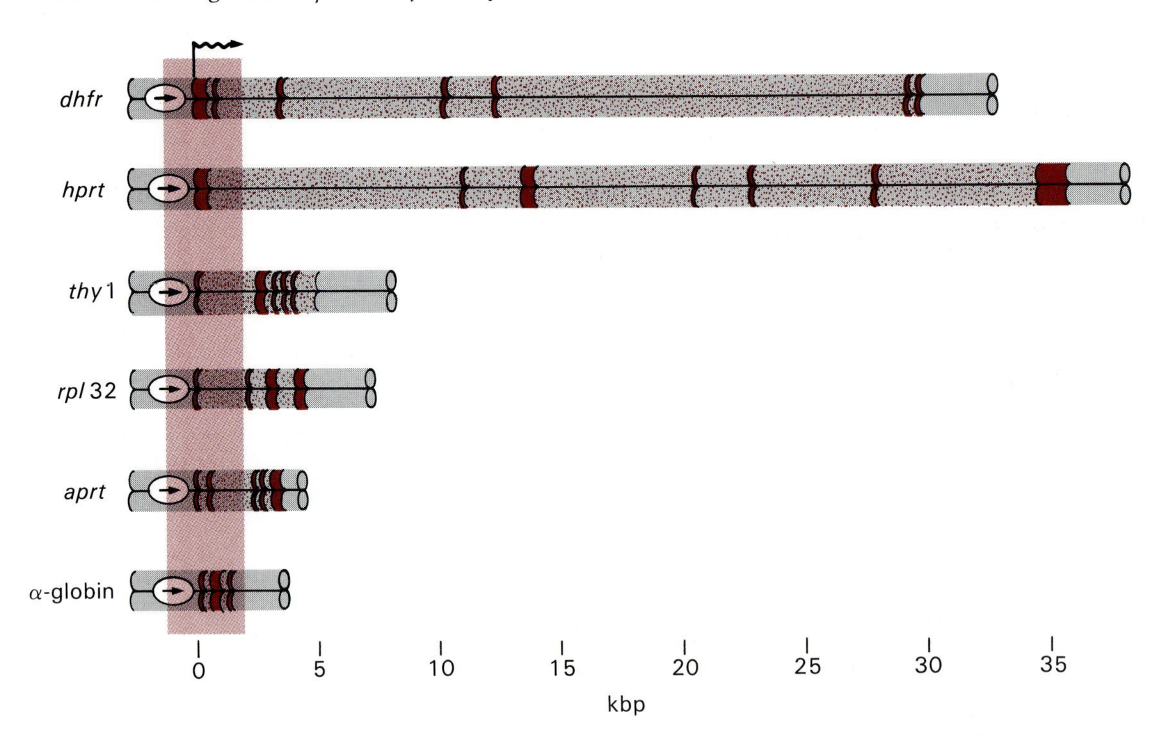

Figure 8.113
Similar location and size of CG islands at six mammalian genes. The promoters, transcription start sites (wavy arrows), exons (colored solid regions), and introns (colored stippled regions) of six genes are illustrated. The colored bar indicates the region within which CG dinucleotides are generally unmethylated; the so-called CG islands. Methylated CG dinucleotides generally occur outside this region. The genes are *dhfr*, mouse dihydrofolate reductase; *hprt*, mouse hypoxanthine phosphoribosyl transferase; *thy-1*, mouse thy-1 surface glycoprotein; *rpl*, mouse ribosomal protein L-32; *aprt*, mouse adenine phosphoribosyl transferase; and α-globin, human α-globin. Data taken from A. Bird, *Trends Genet.* 3 (1987), p. 342.

in those where the gene is silent. This strengthens the view that the occurrence of unmethylated CG clusters in a gene's promoter region is by itself insufficient to account for preferential transcription in appropriate cells. Tissue-specific positive factors or ubiquitous inhibitory factors that bind to 5-meC are needed to account for the differential tissue expression.

Methylation and developmental processes Two very striking correlations link methylation with critical mammalian developmental processes: **X-chromosome inactivation** in females and differential **imprinting** of autosomal chromosomes by male and female parents.

In placental mammals, inactivation of one of the two X-chromosomes in female cells assures equivalent contributions of X-chromosome information in XX females and XY males. The inactivation occurs in three steps during early embryogenesis (Figure 8.114). Initially, one of the two active X-chromosomes is selected for inactivation. Then the inactivation spreads progressively throughout the chromosome. Finally, the modified state is stabilized so that the selected X-chromosome remains inactive during subsequent cell generations. X-chromosome inactivation does not occur in all embryonic cells at the same time. At about the fourth day following implantation, the paternally derived X-chromosome is preferentially inactivated in cells that contribute to the extraembryonic tissues. After the sixth day of gestation, either of the two X-chromosomes, chosen at random, is inactivated in each cell destined for the fetus. Thus, the female fetus develops as a mosaic in which cells express either the paternal or maternal X-chromosome. Aneuploid individuals having more than two X-chromosomes inactivate all but one.

The inactive X appears cytologically in interphase nuclei as a highly condensed, heterochromatic structure (**Barr body**). In contrast to the active homologue, the inactive X-chromosome replicates late in S-phase. Virtu-

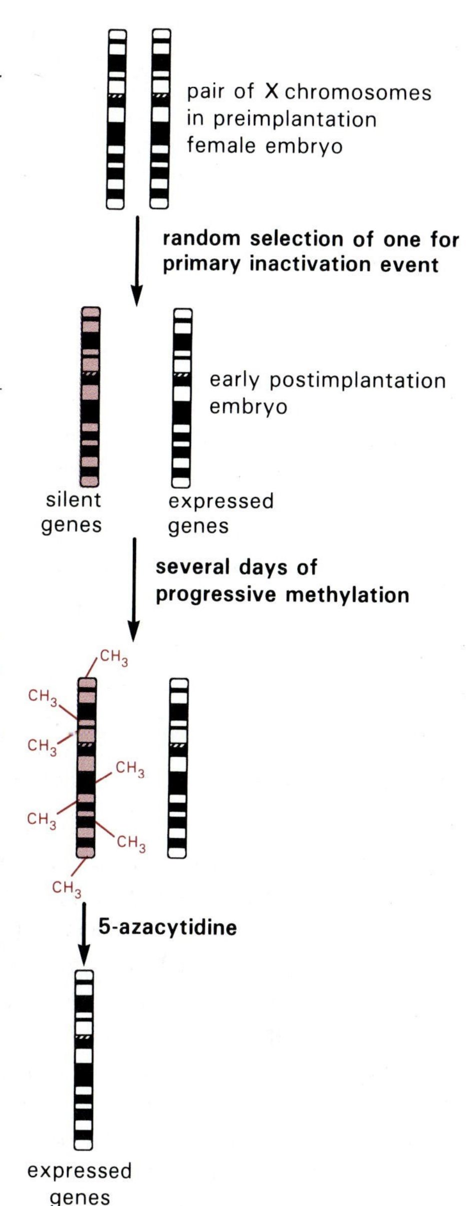

Figure 8.114
Mode of inactivation of one of two mammalian X-chromosomes. Both X-chromosomes in preimplantation embryo tissues are active and unmethylated. Gene expression in one of the two X-chromosomes is silenced in the post-implantation embryonic tissues while the other remains transcriptionally active. Inactivation appears to precede extensive methylation, but methylation follows soon after. Treatment of cells containing an active and inactive X-chromosome with azacytidine leads to extensive loss of 5-meC residues from the inactive X-chromosome and activation of that chromosome's gene expression.

ally all the genes encoded on the inactive X-chromosome fail to be expressed. This inactivity is remarkably stable, and only rare reactivations of discrete regions have been noted. In normal development, reactivation occurs only in germ cells just prior to meiosis. However, activation of X-chromosome genes can be induced by treating cells with 5-azaC (Figure 8.114). This finding raises the possibility that X inactivation is a consequence of DNA methylation. Several lines of indirect evidence support this notion, but convincing proof that methylation is the cause rather than an effect of inactivation is lacking.

One line of evidence stems from the fact that only DNA from mouse cells whose *hprt* gene is on an active X-chromosome (aX) can transform *hprt*$^-$ cells to an *hprt*$^+$ phenotype by DNA transfer. However, after treatment with 5-azaC, DNA from the formerly inactive X-chromosome (iX) is active in transforming *hprt*$^-$ cells to *hprt*$^+$. Additional support for the methylation hypothesis derives from an examination of the methylation sites in the *hprt* gene from aX- and iX-chromosomes. Overall, the methylation pattern of CG containing sequences in the aX-chromosome's *hprt* gene is similar in all tissues and characteristic of other housekeeping genes in the aX-chromosome and in autosomes. Thus, the sites in a region surrounding the *hprt* gene's first exon are completely unmethylated in the aX-chromosome. In contrast, these sites are nearly completely methylated in the iX-chromosome. Treatment of cells with 5-azaC alters the methylation pattern of the iX-chromosome to resemble that of the aX-chromosome. Nevertheless, when both the kinetics of X inactivation and the alteration of the methylation pattern of the *hprt* gene are examined during various stages of postimplantation development, the two processes are not coordinate. Methylation of the specific sites in the *hprt* gene does not occur in most of the embryo's cells until several days after X-chromosome inactivation. Consequently, methylation may not have a causal role in either selecting the chromosome for inactivation or in propagating the inactivation throughout the chromosome at these early stages (Figure 8.114). Instead, methylation may serve as a secondary mechanism for maintaining the inactive state.

Genomic imprinting refers to the differential modification of the maternal and paternal genetic contributions to the zygote leading to differential expression of parental alleles during development and in the adult. In the

Table 8.11 Dependence of a Transgene's Activity and Methylation Pattern on Parental Source of Transgene

Parent Contributing Transgene	Number of Mice with Indicated Properties: Gene Expressed	Gene Not Expressed	Methylation: Hypo	Hyper
Mother	0	42	0	59
Father	20	0	50	0

Note: A description of the experiment appears in the text. The numbers of animals used to test expression were different from the numbers of animals used to test for DNA methylation levels.
SOURCE: Adapted from J. L. Swain et al., *Cell* 50 (1987), p. 719.

mouse, the distinctive modifications of maternal and paternal genomes, and presumably specific genes, are essential for normal development of the embryo and fetus.

The nature of the imprinting modifications and the mechanisms by which they arise, are maintained, and disappear are not known. However, there is a difference in the 5-meC content in sperm and egg DNAs. This suggests that sex-specific differences in the extent of DNA methylation imposed at the time of gametogenesis may contribute to the imprinting process. Consistent with this notion are experiments indicating that transgenes are differentially expressed and methylated depending on the sex of the parental donor. Thus, a transgene under the transcriptional control of a viral promoter is expressed if it is acquired from the father, and then only in heart tissue; when the same gene is inherited from the mother, it remains silent in all tissues (Table 8.11). Measurements of the transgene's relative susceptibility to cleavage at its 5′-CCGG-3′ sites with the *Hpa*II and *Msp*1 endonucleases indicate a strong correlation between the transgene's expression and its degree of methylation. When the transgene is inherited from the father, it is hypomethylated and expressed; when the transgene derives from the mother, it is hypermethylated and silent.

Imprinting is established prior to or during gametogenesis and persists stably through many generations of somatic cell division. It is erased in the germ line, only to be differentially reestablished in the sperm and the egg (Figure 8.115). The egg genome is decidedly hypomethylated, and the sperm genome is relatively hypermethylated. Both sets of chromosomes become partly demethylated during the embryonic preimplantation state. Thereafter, throughout the development of the fetus, the parental chromosomes are progressively remethylated, albeit differentially. Nothing is known about the mechanisms, consequences, or regulation of the methylation-demethylation modifications in the two sexes. Here, too, methylation may be secondary to some prior imprinting step and serve only to stabilize the imprinted signal.

d. *Regulated Utilization of mRNA*

Relatively little is known about the mechanisms regulating mRNA stability and translatability, although each of these two parameters influences gene

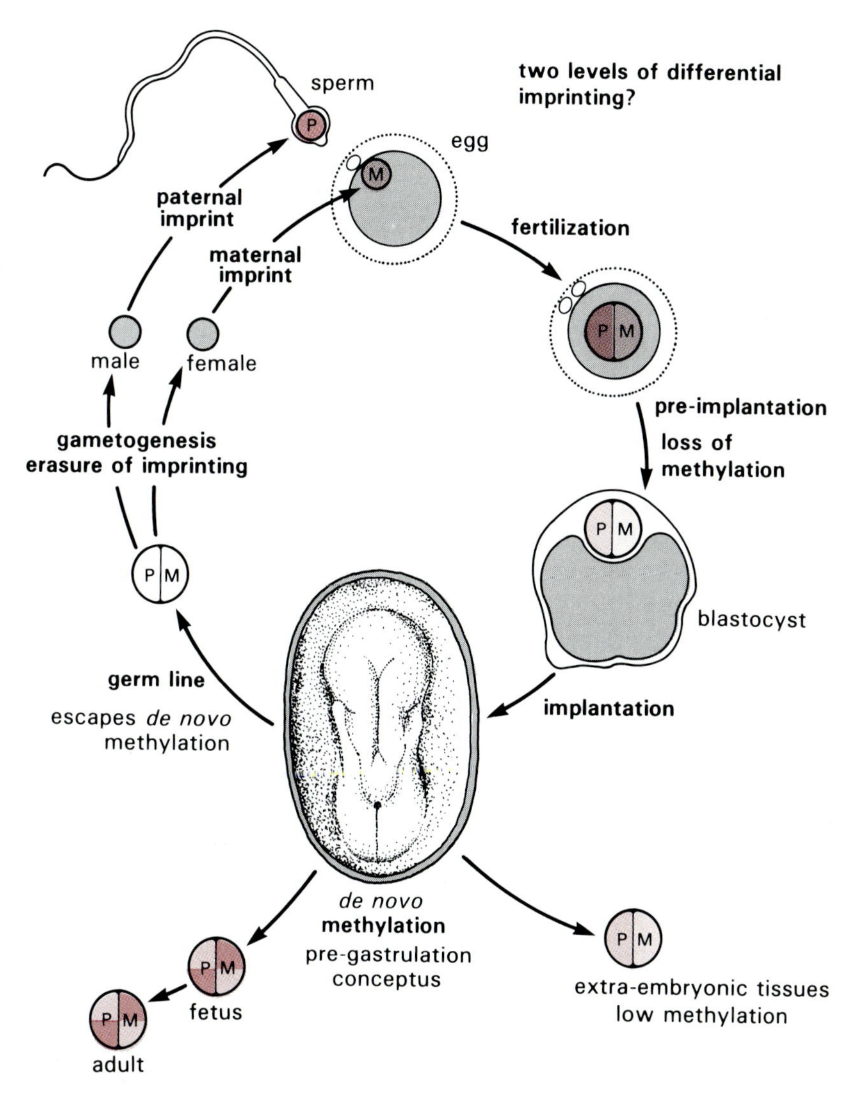

Figure 8.115
Model indicating differential methylation of gametes and somatic cells during mammalian development. The extent of DNA methylation at various stages of development is shown in various shades of color: the deeper the color, the more extensive the methylation. P and M refer to paternal and maternal origins. The uniform coloring of the P or M regions is not intended to indicate that each is uniformly hypermethylated or hypomethylated. The DNA in cells destined for the germ line is largely unmethylated, but during gametogenesis, distinctive and different methylation patterns are established for the sperm and egg nuclei. The fertilized ovum therefore contains two types of DNA: one having a methylation pattern characteristic of the sperm and the other resembling that of the egg. Methyl groups are lost from both the paternal and maternal DNA contributions during the preimplantation and implantation stages. Further development of the conceptus to the fetal and adult stages is accompanied by the establishment of distinctive methylation patterns for the paternally and maternally inherited chromosomes. Adapted from M. Monk, *Genes and Devel.* 2 (1988), p. 921.

expression in significant ways. Several well-studied examples illustrate the importance of these processes.

Influence on stability of sequences at mRNA ends Special sequences in the untranslated regions of eukaryotic mRNAs are important elements in determining mRNA stability. The poly A sequence, in association with the poly A binding protein, contributes to mRNA stability by protecting the 3′ ends against exonucleolytic attack. Thus, mutations in the polyadenylation signal that prevent polyadenylation (Section 8.3c) result in highly unstable transcripts. Also, removal of the poly A tail from mature mRNAs causes their rapid destruction after injection into *Xenopus* oocytes or mammalian cells; restoration of the poly A ends restores the characteristic mRNA stability. There are some indications that progressive shortening of the poly A tails accompanies successive rounds of mRNA translation and that at some stage, the mRNA's 3′ end becomes susceptible to exonuclease degradation. Thus, the translatability of an mRNA may influence its stability. The stability of histone mRNAs, most of which lack

GM-CSF	UAAUAUUUAUAUAUUUAUAUUUUUAAAAUAUUUAUUUAUUUAUUUAUUUAA
IFNα	UAUUUAUUUAUUUAA
IFNβ	UUUUGAAAUUUUUAUUAAAUUAUGAGUUAUUUUUAUUUAUUUAAAUUUUAUUUUGGAAAA
IFNγ	UAUUUAUUAAUAUUUAACAUUAUUUAUAU
IL1	UUAUUUUUUAAUUAUUAUUUAUAUAUGUAUUUAUAAAUAUAUUUAAGAUAAUUAUAAUAU
IL2	UAUUUAUUUAAAUAUUUAAAUUUUAUAUUUAU
fos	GUUUUUAAUUUAUUUAUUAAGAUGGAUUCUCAGAUAUUUAUAUUUUUAUUUUAUUUUUUU
myc	UAAUUUUUUUUAUUUAAGUACAUUUUGCUUUUUAAAGUUGAUUUUUUUCUAUUGUUUUUA

Figure 8.116
Destabilizing AU-rich sequences occur frequently in the 3′ untranslated region of several human lymphokine and nuclear transcription factor mRNAs. In some of the mRNAs, the AUUUA sequences (shown underlined) are overlapping (shown overlined). The human mRNAs shown encode GM-CSF, granulocyte macrophage colony stimulating factor; IFNα, IFNβ, and IFNγ correspond to α, β, and γ interferons, respectively; IL1 and IL2 designate interleukin 1 and 2, respectively; fos and myc represent human c-fos and c-myc. Data taken from G. Shaw and R. Kamen, *Cell* 46 (1986), p. 659.

poly A modifications, is attributed to the resistance to exonuclease digestion afforded by stem-loop structures at their 3′ ends (Section 8.3c).

Recently, a specific sequence promoting destruction has been identified in the 3′ untranslated region of some eukaryotic mRNAs. This sequence, 5′-AUUUA-3′, appears multiple times, occasionally in tandem overlaps, upstream of the polyadenylation signal of several short-lived mRNAs (Figure 8.116). These include a variety of mRNAs encoding cellular proteins that regulate cellular growth and differentiation, as well as mRNAs transcribed from the normal versions of several cellular and viral oncogenes. One feature these mRNAs share is their relative metabolic instability, a property that accounts for the relatively short-lived physiological action of their encoded proteins.

Deletion of the AU-rich region from the normal cellular c-*fos* gene increases the steady state level of its mRNA and converts it to an oncogene. Conversely, the viral oncogene v-*fos* can be made nononcogenic by exchanging its 3′ end for the corresponding region of the nononcogenic c-*fos* gene. Presumably, the ability to produce tumors results from increased levels of the longer-lived mRNA's translation product. Further support for the AU-rich region's destabilizing activity comes from experiments in which ordinarily stable mRNAs, such as β-globin mRNA, become unstable after one or several repeats of the AU-rich sequences are introduced into their 3′ ends (Figure 8.117).

How the AU-rich sequences induce mRNA instability is not known. Possibly, a protein binds to the sequence, making it susceptible to endonuclease action and subsequent exonuclease digestion. There are indications, however, that the AU-rich sequences enhance mRNA degradation by increasing the rate of poly A tail shortening.

There is only fragmentary evidence concerning the contribution of 5′ untranslated sequences to mRNA stability. The steady state level of histone mRNAs and the synthesis of most histone proteins are closely correlated with DNA synthesis during the S-phase of the cell cycle. At the end of S-phase, histone mRNA levels decrease. This selective destabilization following cessation of DNA synthesis is posttranscriptional and for histone H3 is mediated by a sequence in the 5′ untranslated region of the mRNA. The mechanism of the differential destabilization of histone mRNAs is unknown but may depend on a protein made only after DNA synthesis ends.

Iron as a regulator of transferrin receptor mRNA stability A striking example of regulated mRNA degradation is the way iron governs the level of

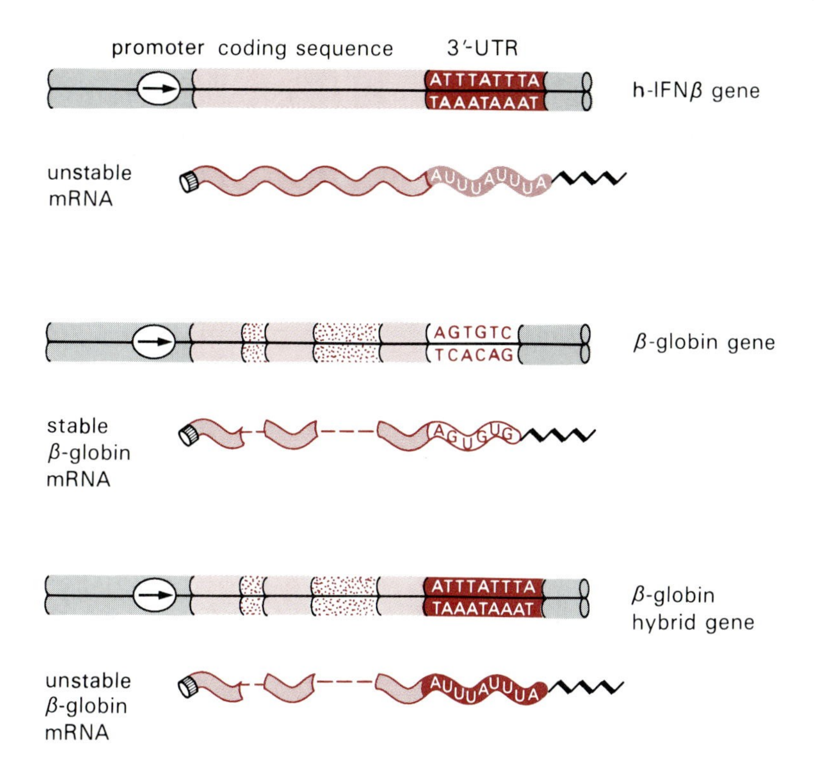

Figure 8.117
AUUUA sequences in an mRNA's 3′ untranslated region targets it for rapid degradation. mRNAs that contain one or more AUUUA sequences in their 3′ untranslated region are destroyed rapidly. This was demonstrated by an experiment in which the AUUUA sequence from the highly unstable human interferonβ (h-IFN-β) mRNA was introduced into the 3′ end of the ordinarily stable β-globin mRNA. As a consequence, the hybrid β-globin mRNA is as unstable as the IFN-β mRNA. The moderately colored solid regions indicate exon sequences, and the stippled segments are introns. The origins of the gene segments and RNA destabilization sequence are shown by their respective colors. Dashed lines in the mRNAs indicate the position of the spliced out introns.

transferrin receptor (TfR) mRNA. In proliferating mammalian cells, iron availability modulates the synthesis of at least two proteins critical to iron metabolism. The TfR is the principal means of iron uptake. Its synthesis is decreased when iron is abundant and increased when iron is scarce (Table 8.12). Ferritin, which serves as an intracellular storage protein for iron, is made at a higher rate when iron is available than when it is in short supply. Although the formation of both proteins is regulated by iron, the mechanisms are different. Iron controls TfR synthesis by altering the degradation rate of TfR mRNA. By contrast, ferritin mRNA levels remain unchanged in response to iron; instead, iron regulates the rate at which ferritin mRNA is translated. Nevertheless, although iron regulation in the two instances is achieved by two different processes, they share common regulatory elements. Here we discuss the iron mediated regulation of TfR mRNA levels and, later in this section, how iron regulates the translation of ferritin mRNA.

After a cloned TfR cDNA linked to an SV40 promoter is transfected into mammalian cells, the level of TfR mRNA is regulated by iron, even though transcription is under the control of the SV40 promoter, which

Table 8.12 Regulation of Transferrin Receptor and Ferritin by Iron

	Transferrin Receptor		Ferritin	
Conditions	Protein	mRNA	Protein	mRNA
Limiting iron	↑	↑	↓	No change
Sufficient iron	↓	↓	↑	No change

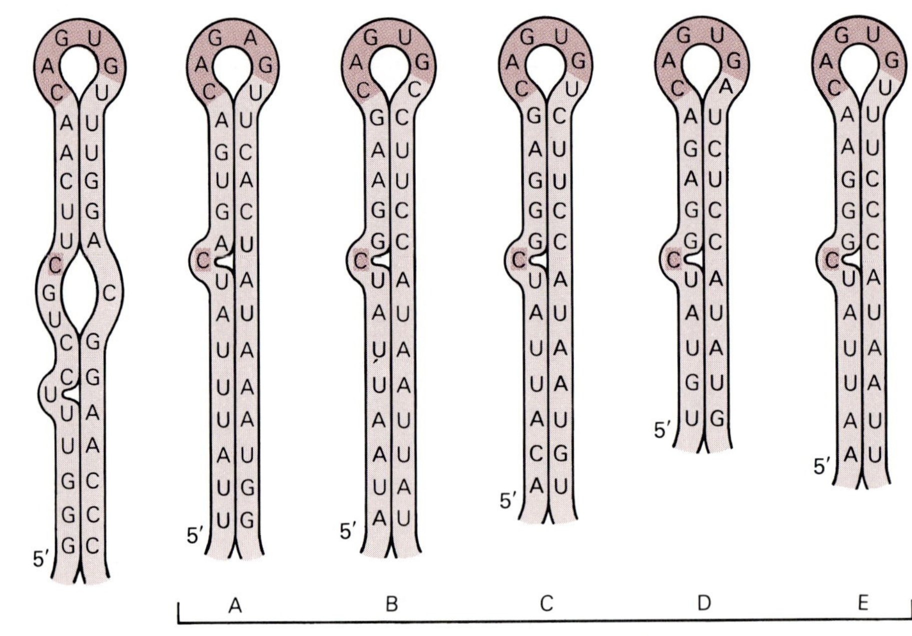

Figure 8.118
Stem-loop structures in the ferritin H chain and transferrin receptor mRNAs. The single stem-loop at the left occurs in the 5′ untranslated region (5′ UTR) of the human ferritin H chain mRNA, and the five stem-loops (A–E) at the right occur arranged in tandem in the 3′ untranslated region (3′ UTR) of the human transferrin receptor (TfR) mRNA. The residues essential for the regulatory activity—five in the loops and a single unpaired cytosine residue in the stems—are shown in deep color.

does not respond to iron. When the cDNA's 3′ untranslated region is deleted, the mRNA level no longer responds to iron. Moreover, if the TfR's 3′ untranslated region is introduced into the 3′ untranslated regions of several different cDNAs, their mRNA levels become regulatable by iron. Thus, the TfR mRNA's 3′ untranslated region is both necessary and sufficient for its regulation by iron.

The TfR mRNA's iron responsive element (IRE) spans about a third (680 nucleotides) of the 3′ untranslated region. Its sequence has the potential to form five similar stem-loop structures (Figure 8.118). These stem-loop structures have in common (1) a loop of $CAG^{A}/_{U}GN$, (2) an upper stem of five base pairs, (3) an unpaired C separated by five base pairs from the 5′ end of the loop, and (4) a lower stem of variable length. This characteristic structural domain occurs in the 3′ untranslated region of TfR mRNA from all species that contain the gene. Neither any of the stem-loop structures individually nor the first two or last three together confers iron regulation on the TfR mRNA levels. This suggests that co-operative interactions among several of the stem-loop structures are needed for the iron mediated regulation.

Cytosolic extracts of cells deprived of iron contain a protein that binds specifically to the IRE (IRE-BP). When iron is abundant, the amount of functional IRE-BP is decreased. One plausible model for IRE regulation of TfR mRNA degradation in response to iron is that the binding of IRE-BP to the clustered stem-loop structures, either alone or in association with other proteins, prevents the action of an endonuclease in or near the IRE (Figure 8.119). The initial endonuclease cleavage is presumed to make the rest of the mRNA susceptible to rapid exonuclease degradation. This scheme allows for continuing synthesis of TfR when cells need to sequester limited amounts of iron from the medium.

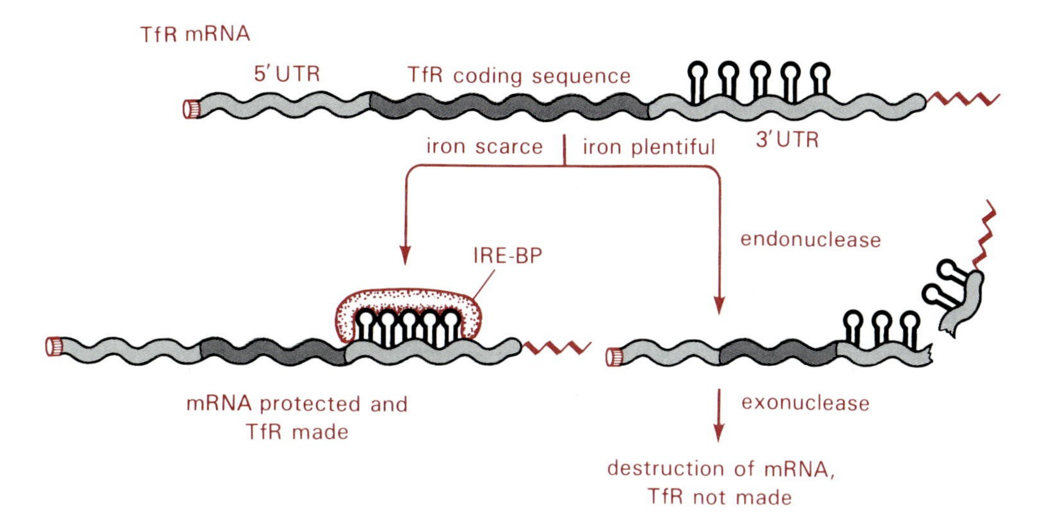

Figure 8.119
Iron availability regulates transferrin receptor (TfR) formation by modulating the stability of the TfR mRNA. The metabolic stability of the TfR mRNA is determined by the interaction or lack of interaction of a specific protein, the iron responsive element binding protein (IRE-BP), with the five stem-loop structures (IREs) in the 3′ untranslated region (3′ UTR). The availability or activity of the IRE-BP is determined by the level of ferric iron. An abundance of iron leads to inactivation of the IRE-BP's ability to bind to the IREs, thereby allowing the 3′ UTR to be cleaved by an RNA endonuclease. A scarcity of iron, however, permits IRE-BP to bind to the IREs and to protect the 3′ UTR from endonucleolytic cleavage.

Studies with the purified IRE-BP indicate that its ability to bind to the TfR mRNA requires a free sulfhydryl group. This suggests that high levels of iron favor a disulfide or inactive form of IRE-BP and that low amounts of iron promote the reduction of the protein's disulfides to free sulfhydryl groups and an active state (Figure 8.120). Thus, the rate of TfR mRNA degradation would reflect the amount of IRE-BP bound to the IRE, which in turn is determined by the iron regulated state of oxidation of the IRE-BP cysteine residues. In effect, the intracellular levels of iron operate a "sulfhydryl switch," which determines whether or not the IRE-BP binds and protects TfR mRNA from degradation.

Iron as a regulator of ferritin mRNA translation Ferritin synthesis in mammalian cells is regulated posttranscriptionally by iron. Transcription of the ferritin gene is independent of iron concentration, as is its mRNA stability, but translation of the mRNA is subject to iron regulation (Table 8.12). Deletion analyses identify a 43 bp sequence within the 5′ untranslated leader region of ferritin mRNA that is responsible for iron regulation of ferritin synthesis. Computer modeling of this region's secondary structure predicts the existence of a single stem-loop that closely resembles the consensus stem-loop in the TfR mRNA's IRE (Figure 8.118). Indeed, genes whose expression is not normally regulated by iron acquire iron responsive expression when a synthetic oligonucleotide capable of assuming a similar stem-loop structure, or any one of the stem-loops in the TfR's IRE, is introduced into their 5′ leader regions. Each of these stem-loops is as effective as the ferritin gene's 5′ stem-loop in conferring translational regulation by iron. Here too, then, the stem-loop structural motif is necessary and sufficient for iron regulated translation.

The same protein that binds to the TfR mRNA's IRE also binds to the stem-loop structure at the ferritin mRNA's 5′ end, even though there is only a single stem-loop. A reasonable, but not yet proven, way for the bound IRE-BP to regulate translation is by blocking ribosome access to the start of the ferritin coding sequence. Thus, ferritin mRNA levels are unaffected by iron, but synthesis of ferritin is repressed. In this way, the cell turns off ferritin synthesis when iron is limiting while the stability of the ferritin mRNA allows a rapid response to an increase in iron.

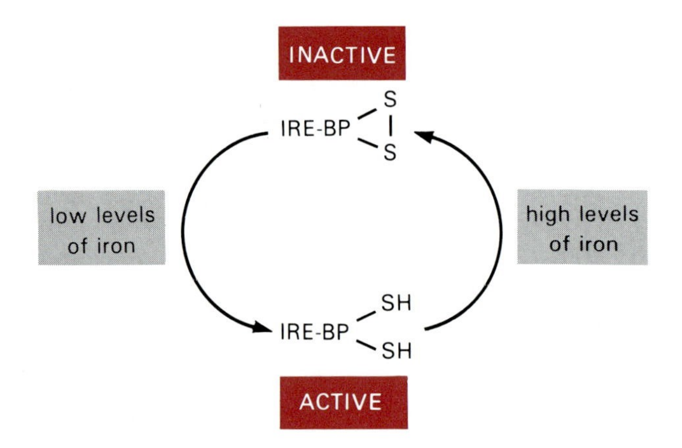

Figure 8.120
The IRE-BP's regulatory activity is determined by the oxidation state of the protein's sulfhydryl groups. Free sulfhydryl groups (SH) in the IRE-BP, the state favored by a scarcity of iron, are required for binding of the IRE-BP to the IREs. With an abundance of iron, the sulfhydryl groups exist in the oxidized form (S-S). Consequently, the IRE-BP is unable to bind to the IREs, and the TfR mRNA is degraded.

Translational regulation of mRNA stability The expression of α- and β-tubulins, the two principal constituents of microtubules, is regulated in complex eukaryotes by two mechanisms. One involves transcriptional regulation of a multigene family that encodes the two subunits (Section 9.3b). The amount of tubulin is also regulated in most animal cells by tubulin heterodimers ($\alpha\beta$), which modulate the stability of tubulin mRNAs.

The signal for the tubulin induced instability of tubulin mRNAs resides in the first four codons of the coding sequence. Changes throughout the mRNA sequence permit normal mRNA turnover, whereas alterations of the nucleotide sequence that change the amino terminal sequence Met Arg Glu Ile (MREI) stabilize the mRNA. Changes in these four codons that do not alter the amino acid sequence have no effect on tubulin mRNA stability. Also, the MREI coding sequence must be at the amino terminus because mRNAs encoding this peptide sequence internally are not destabilized. Moreover, the MREI codons must be followed by a length of RNA sufficient to encode a polypeptide that is long enough for the MREI portion to emerge from the ribosome.

One model to account for this behavior asserts that as the level of tubulin heterodimers rises, they or some associated factor interact with the nascent amino terminal tubulin tetrapeptide just after it emerges from the ribosome (Figure 8.121). This binding could cause translation to stall, allowing an endonuclease to cleave the mRNA and initiate its destruction. Much remains to be defined about the nature of the protein-tetrapeptide interaction and the entities involved in the associated destruction of the mRNA. Nevertheless, this feedback mechanism permits a protein to regulate the lifetime of its own mRNA.

Translational control of a global transcription factor Many genes encoding enzymes in multiple amino acid biosynthetic pathways are under a common control in *S. cerevisiae*. Thus, starvation of yeast cells for any one of ten amino acids leads to elevated transcription of as many as 30 genes involved in amino acid biosynthesis. Cloning and sequencing of many genes that respond to amino acid deprivation indicate that general amino acid control depends on a short nucleotide sequence. Often, the sequence is repeated several times in either orientation in the promoter regions of coregulated genes. The placement and function of these motifs resemble

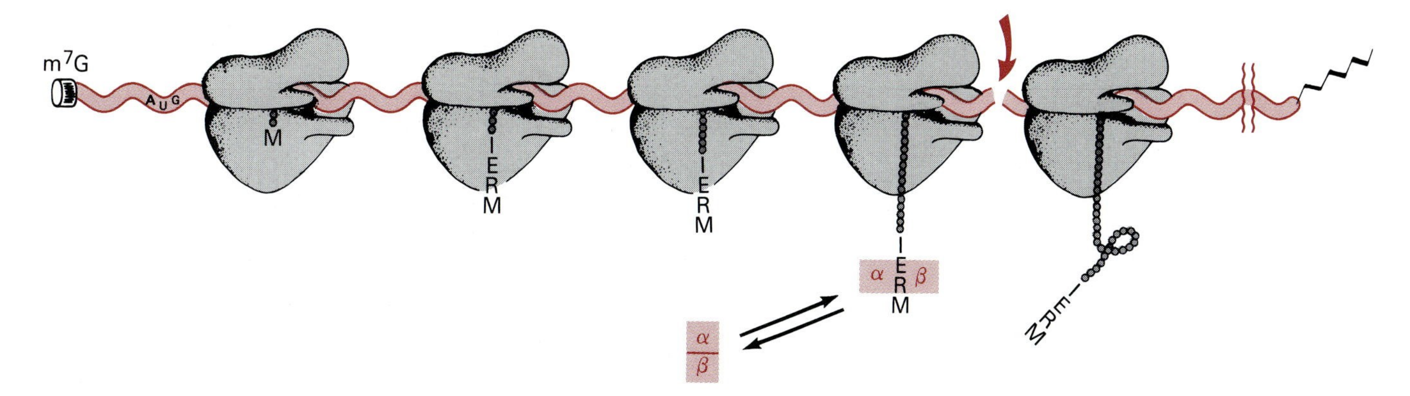

Figure 8.121
Model for autoregulated instability of β-tubulin mRNA. The diagram shows successive stages in the translation of β-tubulin mRNA (from the 5′ cap site at the left toward the poly A tail at the 3′ end). The growing polypeptide chain is shown as a string of small spheres. As the amino terminus of the nascent polypeptide emerges from the ribosome, free α and β tubulin subunits associate with the terminal MREI (methionine, arginine, glutamine, isoleucine) sequence. This activates an endonuclease that cleaves the mRNA nearby, prevents completion of newly initiated protein chains, and triggers degradation of the remainder of the mRNA. The source of the endonuclease, free or ribosome-bound, is not known. Adapted from T. J. Yen et al., *Nature* 334 (1988), p. 500.

the UAS_{GAL} elements mentioned in Section 8.3f. The consensus for the general control sequence motifs is 5′-RRTGACTCATTT-3′, where R represents either of the two purines. The invariant segment, TGACTC, is a binding site for the positive transcription factor GCN4. Recall that GCN4 is highly homologous to the mammalian transcription factor AP1 and its oncogenic counterpart, Jun, and that both of these bind the same hexamer motif (Sections 8.3d and 8.6a). The general control response hinges on the availability of GCN4, its binding to the TGACTC motifs, and activation of the transcriptional machinery. We have already discussed the latter two properties in Sections 8.3 and 8.6. Here we focus on the regulation of GCN4 formation.

The principal determinant of GCN4 expression is a novel mode of translational control. The key to understanding the regulation is in the sequence arrangement in the GCN4 mRNA. The GCN4 coding sequence begins about 600 nucleotides downstream of the mRNA's 5′ end. This unusually long 5′ leader differs from that in most yeast mRNAs by containing four short open reading frames (ORFs), each complete with a start and stop codon (Figure 8.122). Such upstream AUG codons generally inhibit translation initiation at more distal start codons (Section 3.8). The novel feature of GCN4 expression is that the inhibitory effect can be modulated in response to amino acid starvation.

Mutational alterations in this region reveal that the amino acids encoded by the four upstream ORFs are not essential for GCN4 function *in vivo*, nor are they needed for translation of the GCN4 coding sequence *in vitro* or *in vivo*. However, removal of the four upstream AUGs or modifications that convert them to other codons prevent the normal increase in GCN4 formation in response to amino acid deprivations. Interestingly, the four ORFs are not equivalent because their consecutive removal beginning with ORF4 results in a stepwise loss in the ability to repress GCN4 translation when the supply of amino acids is adequate.

The amino acid mediated control of GCN4 translation depends on complex interactions between the products of at least three *gcn* genes and ten or more *gcd* genes. The GCN1, GCN2, and GCN3 proteins are needed to promote GCN4 synthesis during amino acid deprivation. The collection of GCD proteins acts to block GCN4 formation when the supply of amino acids is adequate. The various outcomes of mutations in the *gcd* and *gcn* genes indicate that the GCD proteins enhance the negative effect of the leader AUGs on translation of GCN4 and that the GCN proteins function

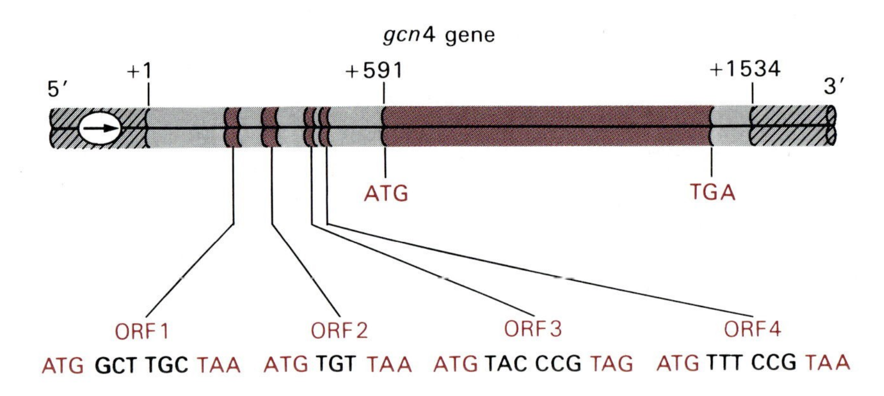

Figure 8.122
Organization of open reading frames (ORFs) in the *S. cerevisiae gcn4* gene. The *gcn4* gene transcription unit begins and ends at residues +1 and +1534, respectively. GCN4 is encoded by the sequence beginning at the ATG (bp +591) and ending at the TGA. The location of the four ORFs in the 5′ untranslated region of the GCN4 mRNA are shown in the same color as the GCN4 coding sequence, and their sequences are indicated below.

indirectly as positive regulators of GCN4 formation by overcoming the GCD protein's negative effects at these AUGs (Figure 8.123).

The precise details of how the GCD and GCN proteins combine to monitor the amino acid supply and convert that signal into either inhibitory or stimulatory responses by the translation machinery are not known. The most favored explanation is that these regulatory actions affect the 40S ribosomal subunit's ability to scan past the upstream start codons. Eukaryotic ribosomes, in contrast to prokaryotic ribosomes, normally initiate translation at the first AUG downstream of the 5′ cap (Section 3.8). When they encounter a termination codon, translation is terminated, and reinitiation at the next AUG codon occurs inefficiently. Multiple and successive start and stop codons, as occur in the GCN4 mRNA's leader sequence, serve as a strong barrier to the translation of the GCN4 coding sequence. The four successive initiation-termination reactions within a relatively short distance are an efficient way to repress GCN4 translation. Perhaps the GCD proteins act on some component of the translation machinery to reduce the likelihood of reinitiation even further. The GCN proteins may antagonize the action of the GCD proteins and create alternate modifications that enhance the ribosome's ability to reinitiate translation following termination. How the GCN proteins sense the status of the amino acid supply is not known.

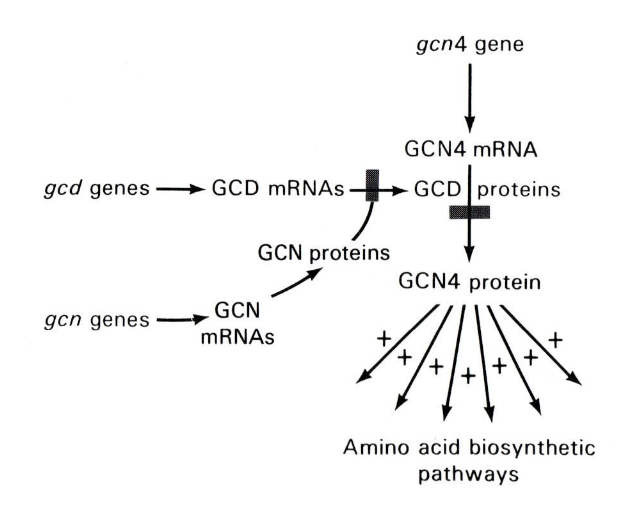

Figure 8.123
GCN4 formation is regulated by *gcd* and *gcn* gene products. Passage of the ribosomes or translation per se of the upstream open reading frames in GCN4 mRNA is impaired by GCD proteins. This block is alleviated by the GCN1, GCN2, and GCN3 proteins, which inhibit translation of the GCD mRNAs.

CHAPTER 9

The Molecular Anatomy of Eukaryotic Genomes

9.1 Architectural Elements

Two provocative questions motivate the study of the molecular architecture of genomes: To what extent does the arrangement of genetic information influence its expression and thus the major biological phenomena of development, differentiation, and disease? What do the arrangements indicate about how genomes and organisms evolve?

We are concerned with three levels of organization. One involves individual genes and their associated regulatory elements, as described in Chapter 8. At this level, substantial information and some general principles have been established, the fruits of merging the study of structure and function. A second level, the main subject of this chapter, is the number and location of genes and other DNA segments relative to one another. A large amount of detail is available, but comprehensive rules about the organization of genomes have not yet emerged. Curiously, however, long stretches of GC- or AT-rich sequences (**isochores**) occur in warm-blooded vertebrates and are associated with R and G chromosome bands, respectively. In only a few known cases does the placement of genes on individual chromosomes have a recognizable pattern. For example, the polycistronic operons common in prokaryotes are not typical of eukaryotes. The available data suggest a high degree of organizational freedom, although the current picture could eventually prove misleading. The third organizational level refers to the way in which chromatin structure influences the underlying linear arrangement of genetic information and gene expression. Although gene expression is clearly related to chromatin structure, we know very little about this interdependence (Section 8.7a).

In this chapter, after reviewing some definitions and evolutionary considerations, we describe the arrangements on chromosomes of various types of genes and other DNA segments. The discussion is simultaneously complex and oversimplified. Each DNA segment, whether gene or not, has special and interesting properties. In addition, the genome of each species is unique, although closely related species share many common properties. We point out recurring organizational patterns when they are discernable and give specific examples to illustrate the extraordinarily diverse arrangements that exist. But the abundant diversity apparent in this chapter is probably infinitesimal compared to what occurs in nature because only a very few DNA segments and species have been studied.

a. Several Classifications of DNA Segments

Genes As previously explained, we use a molecular definition for a gene. A eukaryotic gene is a combination of DNA segments that together constitute an expressible unit. Expression leads to the formation of one or more specific functional gene products that may be either RNA molecules or polypeptides. Each gene includes one or more DNA segments that regulate the transcription of the gene and thus its expression. Coding regions are DNA segments that encode a polypeptide or a functional RNA, or portions thereof. Those DNA segments whose sequences are not reflected in a gene product are called noncoding regions. Some noncoding regions, such as the regulatory signals that flank coding regions and the intervening sequences that interrupt coding regions, are parts of genes. Other noncoding regions, such as segments concerned with replication and segments of unknown significance, are found between genes and in special locations.

Pseudogenes **Pseudogenes** are genomic segments that are structurally similar to specific functional genes, but are neither allelic nor able to yield functional gene products. Some pseudogenes contain debilitating mutations in coding or regulatory regions or both. Others have premature stop codons or are truncated and contain only part of the coding or regulatory regions. Thus, pseudogenes may be silent, or they may be transcribed and even translated into aberrant polypeptides. They are often found closely linked to the corresponding functional gene and may be flanked by sequences homologous to those that flank the corresponding functional gene. Such pseudogenes usually contain introns and appear to be the result of tandem reiterations of DNA segments.

Processed pseudogenes constitute a subclass of pseudogenes that is distinctive in both position and structure. They are not generally linked to the corresponding functional gene but instead are dispersed to distant genomic positions, including different chromosomes. Their structures typically appear more like DNA copies of mRNA than like genes. For example, processed genes usually lack intervening sequences found in the functional gene, and the 5′ and 3′ flanking sequences around a processed gene are usually different from those around the functional gene. Typically, there is a stretch of A residues at the 3′ end of processed genes. Processed pseudogenes appear to be generated by reverse transcription and transposition (Section 10.4).

Special sequence arrangements In earlier chapters, we introduced terms, as needed, to describe special kinds of nucleotide sequence arrangements. Those terms are reviewed here because we use them frequently in the following discussions of the anatomy of eukaryotic genomes (Figure 9.1).

Repeated sequences that are oriented in the same direction along a DNA molecule are called direct repeats and are also referred to as "head-to-tail." Such direct repeats can occur in uninterrupted arrays called tandem repeats, or they may be separated by unrelated sequences. The length of the repeated unit can be anywhere from two to thousands of base pairs.

Sequences that are repeated in opposite directions are called inverted repeats (or dyad symmetries or head-to-head or tail-to-tail). With inverted repeats, a given sequence (5′ to 3′) and its Watson-Crick complementary sequence (3′ to 5′) each occur on both DNA strands. A single DNA (or RNA) strand containing such an inverted repeat can form an intramolecular double helical segment if there are at least three or four residues separating the repeats. If the distance between the sequence and its inversion is rather long, the entire length is called a **stem-and-loop** because the segment between the inverted repeats can loop out as a single strand when the intrachain duplex (the stem) forms. If the sequence and the inverted sequence are separated by very few nucleotides, the double helical single strand is called a **hairpin**.

Sometimes a sequence and its inversion occur within the same group of base pairs (e.g., the recognition sites of the restriction endonucleases shown in Figure 4.7); the sequence is then called a palindrome. Palin-

Figure 9.1
Nucleotide sequence arrangements. The symbol N represents any nucleotide. Note that at least three or four nucleotide residues are required to form the loop in a hairpin.

direct repeats

tandem direct repeats

5′-AAGAGAAGAGAAGAGAAGAG-3′
3′-TTCTCTTCTCTTCTCTTCTC-5′

inverted repeats

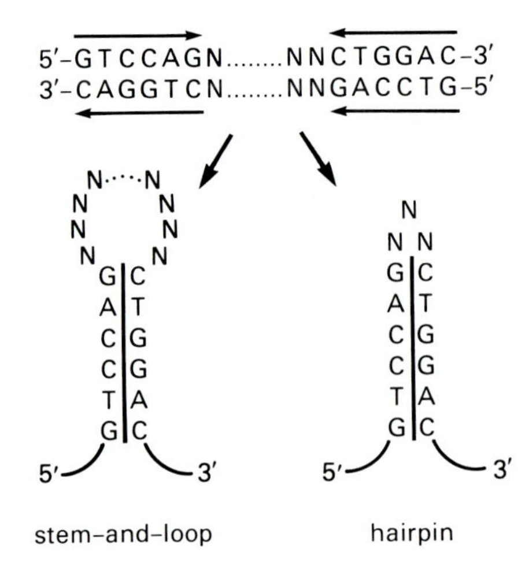

palindrome

5′-GAATTC-3′
3′-CTTAAG-5′

true palindrome (symmetry)

5′-GTCAATGAAGTAACTG-3′
3′-CAGTTACTTCATTGAC-5′

dromes are distinguished from true palindromes or symmetrical sequences; the latter are DNA segments in which the sequence of bases on a single strand is the same regardless of whether it is read in the 5′-to-3′ or 3′-to-5′ direction. In a double helical molecule, a true palindrome on one strand is always mirrored by a complementary true palindrome on the other.

b. The Repetition of DNA Sequences

A very large proportion of most eukaryotic genomes is composed of DNA segments that are repeated either precisely or in variant form more than once. Copy numbers vary from two to millions. A gene and its cognate pseudogenes, for example, constitute one kind of repeated sequence **family**. In this instance, the family contains both coding and noncoding members. Other families, like the repeated histone genes, are primarily composed of coding members; still others, like **satellite** DNAs, contain exclusively noncoding units. Sequence repetition is not unique to eukaryotes. Prokaryotes such as *E. coli* and *S. typhimurium* contain short (20 to 40 bp) palindromic sequences that are repeated 500 or more times in intergenic regions. They make up as much as 0.5 percent of the genomes. Although the full significance of the repeats is not known, they may be sites for DNA gyrase action because gyrase binds to them with high affinity. In addition, prokaryotic DNA may contain repeated units several kilobase pairs in length. Some of these are genes for rRNA; others are transposable elements (Section 10.2a). These repeats, as well as fortuitous and quite short sequence duplications, foster the spontaneous formation of deletions and tandem repetitions by unequal homologous crossing-over. Even considering the repetitious elements in prokaryotes, however, the situation in many eukaryotes is quite distinctive. Fifty percent or more of these genomes are frequently composed of reiterations, and there may in fact be very few truly unique DNA segments. It is important to distinguish between a unique DNA segment (or sequence) and a unique gene. A specific functional gene that is present only once per haploid genome and is thus unique may belong to a family of repeated sequences that includes related genes as well as pseudogenes.

Some repeated sequence families take the form of uninterrupted tandem arrays of the repeating units (Figure 9.1). In others, the family members are clustered but have the directly or inversely repeated units separated by other kinds of sequence. The members of still other families are dispersed to distant places in the genome, including different chromosomes; in some cases, tandem arrays are themselves dispersed.

Variability of repeated sequence family members The nucleotide sequences of members of a particular repeat family may be identical or may vary in one or more base pairs. In the latter case, the family is said to be **polymorphic**. In some families, the divergence is considerable. The terms **subfamilies** and **superfamilies** are used to describe sets of family members with greater and lesser similarity, respectively. The extent of divergence between different family members is estimated with increasing accuracy by four methods: (1) comparing restriction endonuclease maps, (2) measuring their ability to anneal together, (3) measuring the thermal stability of hybrid duplexes formed between family members, and (4) comparing actual nucleotide sequences. Stringent annealing conditions (usually relatively high temperature, about 65°C, and relatively low salt, about 0.45 M NaCl) generally allow duplex formation only between DNA segments whose nucleotide sequences are complementary in at least 85 percent of the nucleotide residues. Lowering the stringency by decreasing the temperature or increasing the salt concentration permits annealing of more distantly related segments, but the lower limit of detectability is about 70 percent complementarity. The stability of the heteroduplexes formed is estimated by measuring their conversion to single strands as a function of temperature (Section 1.1f). As a general rule of thumb, the T_m is lowered 1°C for every 1 percent of mismatched base pairs.

Duplex DNA segments are said to be homologous if their sequences are identical. However, the term **homologous** is often also used to describe similar but not identical DNA segments. Frequently, a qualitative modifier is applied to indicate the extent of identity (e.g., highly homologous meaning highly similar). When the actual sequences are known, quantitative statements can be made. Thus, two DNA segments are described as being 70 percent homologous (or 70 percent similar) if 70 percent of their base pairs (or nucleotide residues) are identical and colinear. This usage derives from the concept of homologous chromosomes, but it is different from that typical of the literature on biological evolution. There, homologous generally means that morphological structures or DNA sequences share a common ancestor, and the idea of percent homology is thus meaningless. Although DNA sequences that are 70 or even 30 percent similar are likely to be derived from a common ancestral DNA sequence, they may not be so related. Thus, the common use of percent homology in the molecular genetics literature and in this book is somewhat ambiguous. The term percent similarity, which we also use, avoids the ambiguity but is only rarely found in the literature.

The size of repeated sequence families The number of related units in different eukaryotic sequence families varies from two to millions, and the **copy number** is frequently different in the genomes of different species. Although the actual copy number is the most informative description, many families can now only be described as highly, moderately (or middle), or infrequently repeated. Historically, these classifications were associated with ranges of Cot values (Section 7.5b). For example, mammalian highly repetitive DNA (10^5 copies or more) was defined by Cot values less than 0.1, middle repetitive (between 10^2 and 10^5 copies) by Cot values between 0.1 and 100, and infrequently repetitive (fewer than 100 copies) by Cot values greater than 100.

Consensus sequences When a repeated sequence family has many members, each of which differs from the others at only a few positions in the nucleotide sequence, it is often convenient to define an average, or **consensus**, sequence for all members of the family. A consensus sequence represents the most frequent base at each position in the population of family members and is not necessarily identical to the sequence of any single member. There are two ways to obtain a consensus sequence. Sometimes it is possible to isolate a large population of the repeating units free from unrelated sequences and subject the mixture to sequence analysis. Even if each molecule in the population differs from the others at a few positions, a unique nucleotide sequence is often obtained because the sequencing methods do not detect the small percentage of divergent nucleotides at any single position (Section 7.2). In the second approach, the sequences of a number of individually cloned family members are compared, and the most frequent base at each residue is calculated.

It is helpful to remember that the definition of a family of repeated sequences is pragmatic. In the absence of actual sequence data, a family is generally taken to be all sequences that anneal to one another, or to a single cloned family member, under stringent conditions. Generally, this means that sequences that diverge 15 to 20 percent from a consensus sequence are barely detectable, and other, more divergent sequences are not counted as family members. Also, the identification of repeated sequence families by reannealing systematically omits families whose members are too short to reanneal. For example, the 5′-ACT(A or T)$_n$TA family members that precede tRNA genes (Section 9.2d) are not detected. Often, nucleotide sequence data show that two DNA sequences that do not anneal with one another are sufficiently similar (e.g., 30 to 50 percent) to be grouped within a family (or superfamily).

c. Evolutionary Implications of Repetition

The role of DNA sequence amplification in genome evolution Related sequences within a genome could, in principle, have arisen either independently or by copying a single, original DNA segment—a "founder" sequence. The likelihood that two sequences originated independently is more remote the greater their similarity and the length of the homologous region. There is no doubt that the multiplication of ancestral founder sequences accounts for the establishment of multisequence families and, thus, for extensive portions of contemporary genomes. Increases in the number of copies of a DNA segment, either in evolutionary or experimental time, is called **amplification**. A variety of mechanisms may account for the amplification of particular DNA sequences either in clustered arrays or dispersed to new genomic loci (Chapter 10). If a single founder sequence is physiologically sufficient, then no particular advantage follows from accumulating additional copies. All but one copy are free to collect mutations, including base pair changes, deletions, and insertions. An altered copy may then become nonfunctional, it may come to encode a distinctive new function, or it may respond to different regulatory signals. If altered sequences do prove useful or adaptive, the new variants may themselves be conserved by selective pressure. Otherwise they are classified as pseudogenes. In this way, the amplification of DNA sequences provides raw material for evolution.

It is instructive to compare the structures of particular genes and multigene families among different species as well as within individual species. The interspecies differences in sequences, copy number, intervening sequences, and organization are important additions to the more classical investigations of evolutionary relationships and mechanisms. To illustrate these concepts, we consider the vertebrate multigene family that includes genes for growth hormone and for prolactin. This superfamily provides a good example of how gene structure informs evolutionary considerations both within and among species.

The growth hormone gene superfamily in vertebrates Although growth hormone and prolactin (which has different functions in mammals than in other vertebrates) differ in about 65 percent of their amino acid residues, they are about the same size (190–199 amino acids, depending on the species) and have common amino acids in certain analogous positions. The genes encoding the two hormones are each interrupted by four introns at essentially the same sites, but the size of the introns is very different in the two hormone genes, as are the sequences in the 5′ and 3′ flanking regions. Each gene is expressed in a different type of pituitary cell and is regulated by distinctive regulatory signals. Most vertebrates have only one prolactin gene and one growth hormone gene per haploid genome. Placental mammals have at least one additional family member, a gene for the hormone placental lactogen, and probably several other, more distantly related family members as well. Growth hormone and placental lactogen genes have identically positioned introns and are about 92 percent similar in coding sequences (85 percent in amino acid sequences). They are also highly homologous in their introns and for at least 500 bp in their 5′ flanking regions; the 3′ flanking regions vary more.

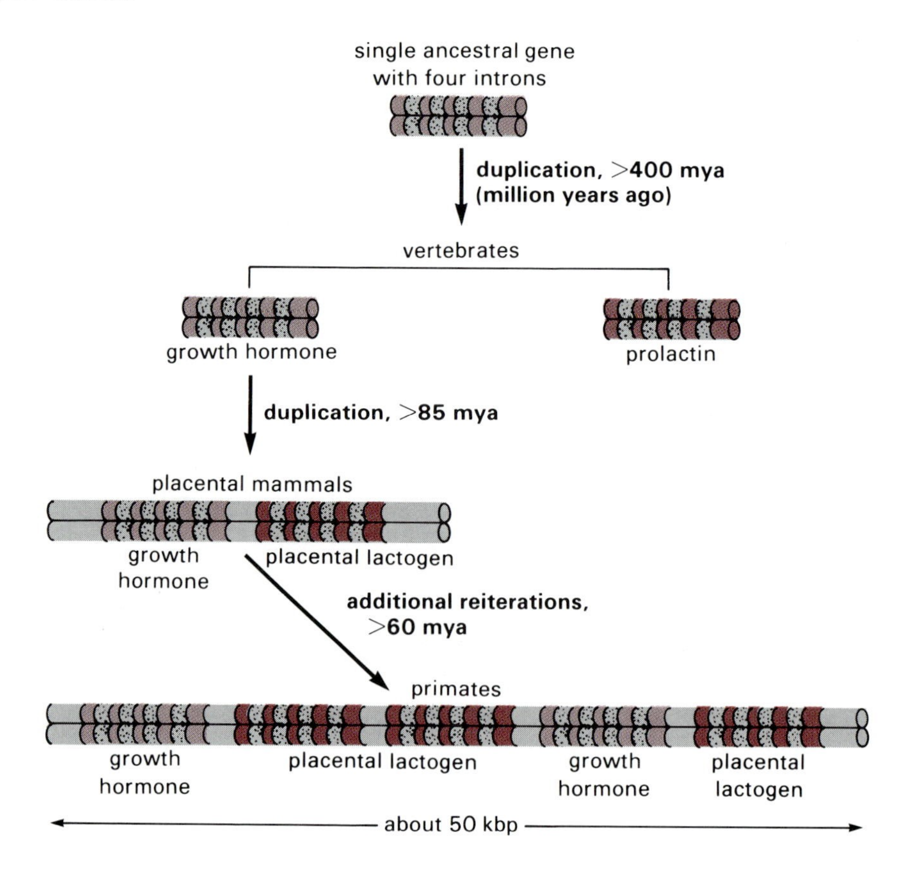

Figure 9.2
Schematic diagram showing the evolutionary history of the human growth hormone multigene family (not to scale). Variations in intron sizes are not shown. After G. S. Barsch, P. H. Seeburg, and R. E. Gelinas, *Nucleic Acids Res.* 11 (1983), p. 3939.

The similarities are interesting because growth hormone and placental lactogen are produced in different tissues (anterior pituitary and placenta, respectively), and expression of the genes is subject to different controls. Small proximal sequence variations or more distant DNA segments may account for the differential regulation.

Because all vertebrates have the two distantly related genes for prolactin and growth hormone, it is likely that a single ancestral gene was duplicated before amphibians and fish diverged about 400 million years ago (Figure 9.2). Thereafter, the two genes evolved independently in each species. Genes related to one another in this way are called **paralogous.** There is no way to know if the original duplication produced linked genes or dispersed copies. In contemporary primates, the genes are dispersed. In humans, for example, the prolactin gene is on chromosome 6, and the growth hormone genes are on chromosome 17. Paralogous genes are distinguished from **orthologous** genes, which are structurally and functionally homologous genes in different species. Thus, for example, the growth hormone genes in all vertebrates are orthologous; chicken, human, bovine, and rat growth hormone genes are 75 percent homologous in their coding regions.

All placental mammals have a placental lactogen gene, but other vertebrates do not, so a duplication of the growth hormone gene is likely to have occurred more than 85 to 100 million years ago, when many mammalian lines were established. These paralogous genes have remained together and maintained linkage. Morever, primates have multiple linked

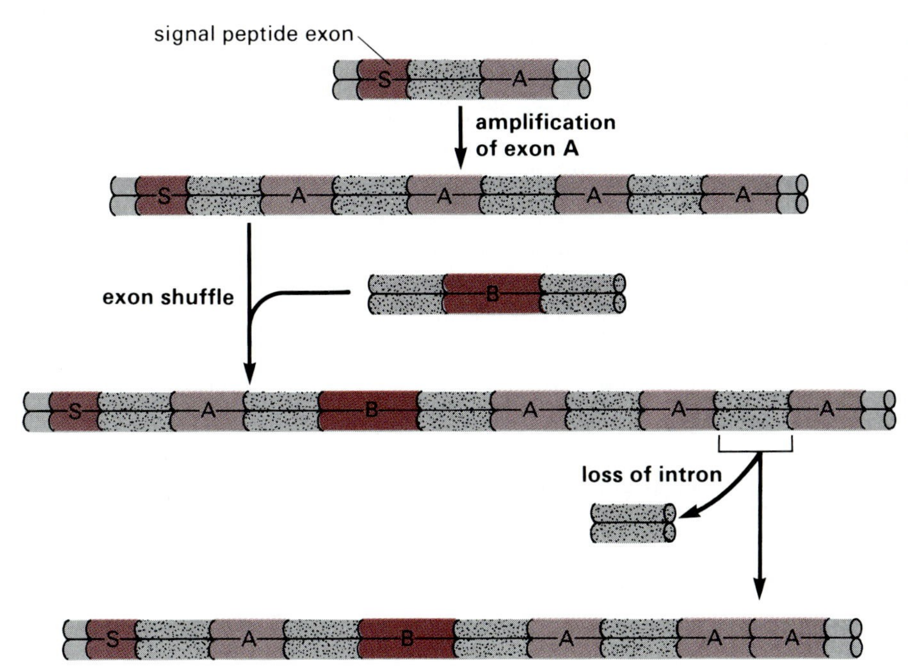

Figure 9.3
A schematic model that can explain the evolution of the growth hormone (and related) genes. The steps involve (a) tandem repetition of an ancestral exon, (b) insertion of an exon (exon shuffling), and (c) loss of an intron. After A. Barta et al., *Proc. Natl. Acad. Sci. USA* 78 (1981), p. 4367.

genes for both growth hormone and placental lactogen: in humans, for example, two growth hormone genes and three placental lactogen genes occur on chromosome 17. This indicates that additional amplifications took place before the primate line developed, some 60 million years ago. The alternative to these evolutionary models is to assume independent origins for very similar genes, a process that is highly unlikely, given the marked similarities within the superfamily.

The role of exon repetition and exon shuffling in growth hormone gene evolution The structure of the growth hormone gene (and of other genes in the superfamily) indicates that exon amplification contributed to its present structure. Cloning and sequencing of the rat growth hormone gene demonstrated the presence of five exons (Figure 9.3). The first exon and part of the second encode a polypeptide signal sequence (Section 3.10b). The second, fourth, and fifth exons are similar to one another, suggesting that they arose by a fourfold tandem amplification of an ancestral exon (exon 5 contains two copies of the homologous region). Exon 3 is not similar to the others and could be the result of exon shuffling—the insertion of an exon copied from another gene. Indeed, although the peptide encoded by exon 3 has antiinsulin activity, it does not promote growth; growth hormone itself shows both activities.

Molecular clocks Detailed comparisons of gene structures in the growth hormone superfamily illustrate additional aspects of evolution at the molecular level. The fundamental hypothesis regarding molecular evolution states that mutations accumulate at a more or less constant rate over time. The percentage of sequence similarity between contemporary related coding regions reflects (1) the time since the genes diverged from a common ancestral gene, (2) the rate at which mutations occur, and (3) the rate at which mutations become fixed in a population. These ideas should be applicable whether the homologous genes are orthologous or paralogous. Data for many protein sequences are consistent with this general hypothesis. However, the evolutionary "clock" runs at different rates for different proteins, presumably mainly because of differences in the rate of fixation. Thus, the **unit evolutionary period (UEP)**, the time in millions of years during which two coding sequences diverge 1 percent, is 15 for cytochrome C and 400 for histone H4, a moderately and a very highly conserved protein, respectively. UEP is calculated by comparing the percentage of divergence between orthologous coding sequences in two species with the time since the two species diverged according to the fossil record.

The UEP for both growth hormone and prolactin genes in humans and rats is about 4.5. The clock hypothesis can be tested by using this UEP value and comparing the sequences for human growth hormone and human placental lactogen genes. The divergence is about 10 percent, which gives a value of about 45 million years (10×4.5) for the amplification leading to the placental lactogen gene. This is a rather surprising result because the paleontological data sets 85–100 million years ago as the time when the mammalian line diverged extensively. One explanation for the discrepancy is that the same duplication occurred independently in individual mammalian species, a not impossible but unlikely set of events. Another is that the UEP for the placental lactogen gene is different from that of the growth hormone gene in humans. Still another explanation is that the simple clock model is insufficient, and more has happened during evolution than the accumulation and fixation of single base pair mutations; some additional mechanisms may limit the independent evolution of the two genes. Two additional facts are important to these considerations. First, the human growth hormone genes are more similar to the human placental lactogen genes than they are to the orthologous growth hormone genes in rats and cows. Second, the introns within the human growth hormone and placental lactogen genes are almost as similar as the coding sequences; normally, analogous introns in paralogous genes diverge extensively in length and sequence. Evidently, the evolution of the two paralogous genes is not independent, a phenomenon termed **concerted evolution** or **homogenization**. It has been noted in several other multigene families (e.g., globin genes, Section 9.3b) and in dispersed repeated sequences (Section 9.5e).

What processes can account for the homogenization of gene structure in a pair of related, but distinctive genes? Several alternatives exist. One is unequal crossing-over, which accounts for the homogenization of rRNA genes in at least some genomes (Section 9.2a). Another possibility involves recent deletions of genes, followed by new amplifications. A third is gene conversion (Section 2.4). At present, it is usually impossible to choose among these alternatives for particular homogenized families. Nevertheless, several different mechanisms clearly contribute to molecular evolution. While the evolutionary clock no doubt keeps on ticking, it can be influenced by recombinational events as well as by mutation.

The evolution of introns Are introns as ancient as the primordial coding regions, or were they inserted later during evolution? In the growth hormone family and many other multigene families, the position and number of introns is the same in both orthologous and paralogous genes, although intron lengths and sequences are not conserved. This situation is consistent with the idea that the interruptions are as old as the common ancestral genes but are relatively free from selective pressure. The structures of mammalian insulin genes, for example, support this view. Vertebrates generally have a single (haploid) insulin gene with two introns. But some rodents have a second functional gene with only one intron. It is easier to construct models for the derivation of the occasional second gene by duplication and the loss of an intron than to start by assuming that the ancestral gene for all vertebrates had only one intron.

A similar argument can be made in the globin gene family, which is described later in this chapter. Plant genes for leghemoglobin have three introns, two of which correspond in position to the two introns in vertebrate hemoglobin genes. The positions of introns in the globin family also illustrate another general finding: exons frequently encode distinctive structural and functional protein domains. These observations all point to the presence of introns in the earliest genes. The self-splicing reactions described in Section 8.5d may reflect the mechanisms by which primitive genes were assembled into functional units. An early evolutionary origin of introns does not, however, preclude the possibility that some introns were later inserted into preexisting coding regions. It has, for example, been suggested that some introns in the gene family encoding serine proteases (e.g., thrombin, trypsin, chymotrypsin) represent insertions. Recent experiments provide models for intron insertion. Thus, group I and II introns (Sections 8.5 and 9.7a) insert into target sites by gene conversion or by reversal of self-splicing reactions. The group I gene conversions depend on proteins encoded by the intron.

9.2 Genes Encoding RNA

An extensive inventory of RNA molecules is used in translating mRNAs and in processing RNA precursors. Many of these RNA molecules are required in relatively large amounts. Unlike protein coding genes, where each gene transcription yields multiple gene products through repeated translation of the mRNA, each RNA gene transcription yields only one copy of the final RNA product. The demand for many RNA molecules is often met by having a large number of redundant genes.

a. Genes for 18S, 5.8S, and 28S rRNA

A great deal was learned about the genes for rRNA (rDNA) before recombinant DNA techniques were developed because the rRNA obtained from purified ribosomes is pure enough to be used as a hybridization probe without cloning. Moreover, the average G + C content of rDNA is higher than that of many entire genomes, and the many copies of rDNA that are typical of most eukaryotic genomes are repeated in long tandem arrays. This means that high molecular weight DNA molecules containing multiple rRNA genes have a relatively high buoyant density compared to the rest of genomic DNA and can thus be isolated as a

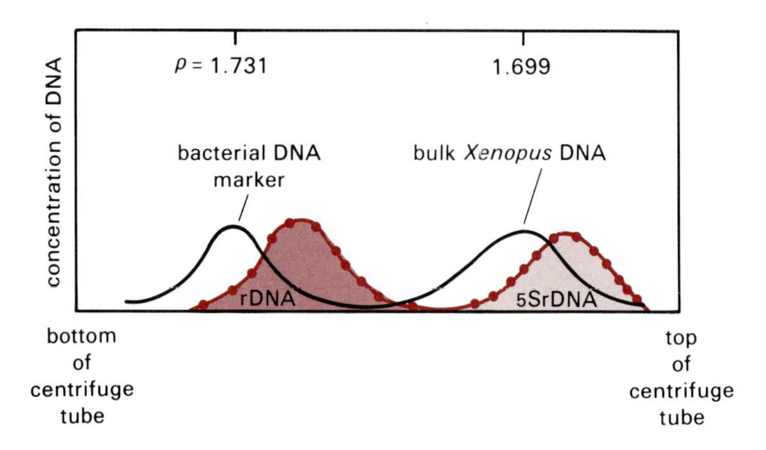

Figure 9.4
Isopycnic centrifugation of total *Xenopus* genomic DNA. DNA is centrifuged at high speed in a solution containing a density gradient formed from a concentrated solution of a salt such as CsCl. After centrifugation to equilibrium, DNA molecules form a band at the position in the salt gradient that corresponds to their own buoyant density (ρ). Because the density of duplex DNA increases in direct proportion to the percentage of G · C base pairs, long arrays of sequence with a G · C content that differs significantly from that of the bulk of the genome appear as separate or "satellite" bands. After D. D. Brown and C. S. Weber, *J. Mol. Biol.* 34 (1968), p. 681.

distinct fraction upon centrifugation of total genomic DNA to equilibrium in a density gradient (**isopycnic centrifugation**) (Figure 9.4). Through a variety of techniques, including *in situ* hybridization with radioactive rRNA probes, rDNA was found to occur at the discrete chromosomal areas called nucleolar organizer (NO) regions, the regions at which nucleoli form (Perspective I). The transcription of rDNA and the processing of rRNA precursors occur within these organelles (Section 8.2).

In contrast to the hundreds of copies of rRNA genes found in most eukaryotes, genes for ribosomal proteins have low copy numbers. Genomic clones and cDNAs encoding ribosomal proteins have been isolated from several organisms, including yeast, *Drosophila*, *Xenopus*, mouse, and humans. Functional genes for particular ribosomal proteins occur between one and ten times per genome, and processed pseudogenes are found.

The linkage of genes for three rRNAs in a single transcription unit Typically, coding regions for eukaryotic 18S, 5.8S, and 28S rRNAs are grouped in that order in a single RNA polymerase I transcription unit (Section 8.2). The long transcript is the precursor from which all three RNAs are produced by a series of processing steps. These include endonucleolytic cleavages and methylation both on bases and on 2′-hydroxyl groups. The entire transcribed unit includes two noncoding spacer DNA segments (called **internal transcribed spacers, ITS**) that separate the three coding regions from one another and noncoding regions that precede the first (18S rRNA) gene and follow the last (28S rRNA) (called **external transcribed spacers, ETS**) gene. Because the size and sequence of rRNAs vary only slightly among eukaryotes, the differences in transcription unit size from one species to another (Table 9.1) reflect different sizes of the spacer regions.

DNA segments on both sides of the transcriptional start site at the 5′ end of the ETS are essential for transcription by RNA polymerase I. These sequences are markedly different from one species to another, in contrast to the high degree of similarity among the coding regions. Their differences are probably related to the species specificity of RNA polymerase I transcription, which appears to require interactions between these sequences and species-specific accessory transcription factors for the enzyme (Section 8.2c).

Table 9.1 Location, Copy Number, and Size of rRNA Genes

Species	Located on Chromosome	Approx. Copy Number (per haploid genome)	Approx. Size of Transcription Unit (kb)
C. elegans	1	55	6
D. discoides			42
S. cerevisiae	12	100	7
T. pyriformis		1	
D. melanogaster	X, Y	200	8
X. laevis	(1 location)	600	8
G. domesticus		200	10.5
*M. musculus**	15, 18, 19 12, 16, 18	100	13
H. sapiens	13, 14, 15, 21, 22	150–200	14
Z. mays	6	3000–9000	

* In two different inbred strains.

Tandem repetition of rRNA genes The copy number of rDNA is known in many organisms from saturation hybridization experiments using pure rRNA as a probe (Table 9.1). By and large multiple copies of the rDNA transcription unit are clustered in long, direct tandem arrays and clusters often occur on several different chromosomes. The rDNA transcription units within the clusters are separated by **intergenic spacer** segments called **IGS** (Figure 9.5). IGS lengths vary from about 2 kbp in yeast to 30 kbp in mammals and the overall length of the repeated unit, including the transcription unit plus IGS, varies accordingly. The sequences within the IGS are also markedly different from one species to another (Section 8.2).

Unequal crossing-over among tandem repeats of rDNA The copy number of the repeated rDNA units can vary among individuals in a given species. In the human genome, where the rDNA units are found at five chromosomal loci each locus contains a different number of rDNA units, and the number on each chromosome varies from one individual to another. In some organisms, the rDNA copy number even varies markedly among somatic cells during the lifetime of a single individual. Thus, the expansion and contraction of copy number is a common property of rDNA, as it seems to be of other long tandem arrays (Section 9.4c). Several mechanisms might account for such dynamic genomic properties. One of these, unequal homologous recombination between nonallelic tandem repetitions (Figures 2.53 and 2.54), can explain the changing rDNA copy number as well as another remarkable feature of rDNA. The nucleotide

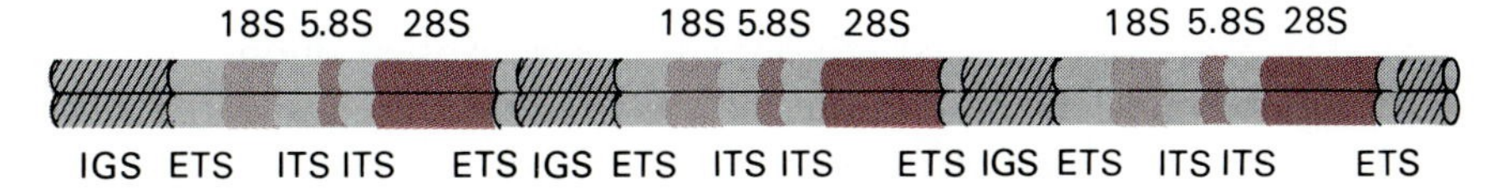

Figure 9.5
The organization of rDNAs. rDNA transcription units are typically clustered in long tandem arrays. The diagram shows the coding regions for 18S, 5.8S, and 28S rRNA as well as the three types of spacers: external transcribed spacer (ETS), internal transcribed spacer (ITS), and intergenic spacer (IGS).

Figure 9.6 (on opposite page)
Demonstrating unequal crossing-over between rDNA units in yeast. (a) A plasmid containing yeast rDNA and a yeast *LEU2* gene transfects *leu2*$^-$ yeast cells. Transformants that are *LEU*$^+$ are selected. Analysis of restriction endonuclease digests of the chromosomal DNA by DNA blotting and annealing shows that the entire recombinant plasmid recombined into a genomic rDNA repeat unit. (b) Haploid yeast cells transformed as described in (a) were mated with a wild type strain containing the *LEU2* gene. The two strains are distinguished by the presence of different restriction endonuclease sites in their respective rDNA repeats (and by the color of their chromosomes in the diagram). The bars indicate rDNA repeats. The diploids were induced to undergo meiosis and sporulation, and the individual haploid cells in the tetrads (derived from single diploid cells) were analyzed for *LEU* phenotype, the presence of *LEU2* sequence (by DNA blotting), and the parental rDNA type (A or wild type) by restriction endonuclease analysis of the rDNA repeats. About 10 percent of the tetrads contained a cell that was *leu*$^-$. DNA from the *leu*$^-$ cells did not hybridize with the *LEU2* sequence probe. DNA from one of the *LEU*$^+$ cells in such tetrads had twice as much *LEU2* DNA as the initial haploid cells. Thus, crossing-over was frequent. Both the *leu*$^-$ cells and those with the double dose of *LEU2* sequences had A type rDNA, indicating that recombination was between sister chromatids. Tetrad spores that contain both A type and wild type rDNA are rare. Thus, crossing-over is much more frequent between sister chromatids than between homologous chromosome pairs. After T. Petes, *Cell* 19 (1980), p. 765.

sequences of all human rDNA units, for example, are surprisingly similar, even though the tandem clusters reside on different chromosomes. If the rDNA units on each chromosome were isolated from one another, they should collect different random mutations. In time, they should diverge from one another, especially in relatively unimportant regions of the IGS. The uniformity of the human rDNA units therefore speaks for some interaction between them. Unequal homologous recombination between rDNA units on different chromosomes as well as between non-allelic units on sister chromatids would put all the copies into a common network and thereby account for the similarity of all the rDNA sequences. At the same time, it would explain the ongoing change in copy number at individual chromosomal clusters.

Crossing-over within rDNA units is much more easily studied in yeast than in mammals. When a recombinant DNA plasmid containing marker yeast sequences is transfected into yeast, some of the recipient cells incorporate the marker gene into chromosomal DNA (Section 5.6c). The transformation occurs as a result of recombination between the sequences of the marker gene and homologous genomic segments. A plasmid containing both yeast rDNA and a *LEU2* gene recombines preferentially with genomic rDNA sequences, presumably because of their abundance (Figure 9.6). The transformants, which are selectable because the *LEU2* gene is expressed, contain a tandem array of rDNA units interrupted at one position by the *LEU2* gene (and other vector sequences). When the transformants are mated with wild type yeast cells, they yield diploid cells that can be induced to undergo meiosis and then sporulation. The four possible products of meiosis are represented in the spores as four separate haploid cells called **tetrads**; these cells can be separately dissected, grown, and analyzed. The results show that unequal crossing-over occurs between DNA strands of sister chromatids at some point during meiosis. The diagram in Figure 9.6 emphasizes that the crossing-over leads to varying numbers of repeat units within an rDNA cluster.

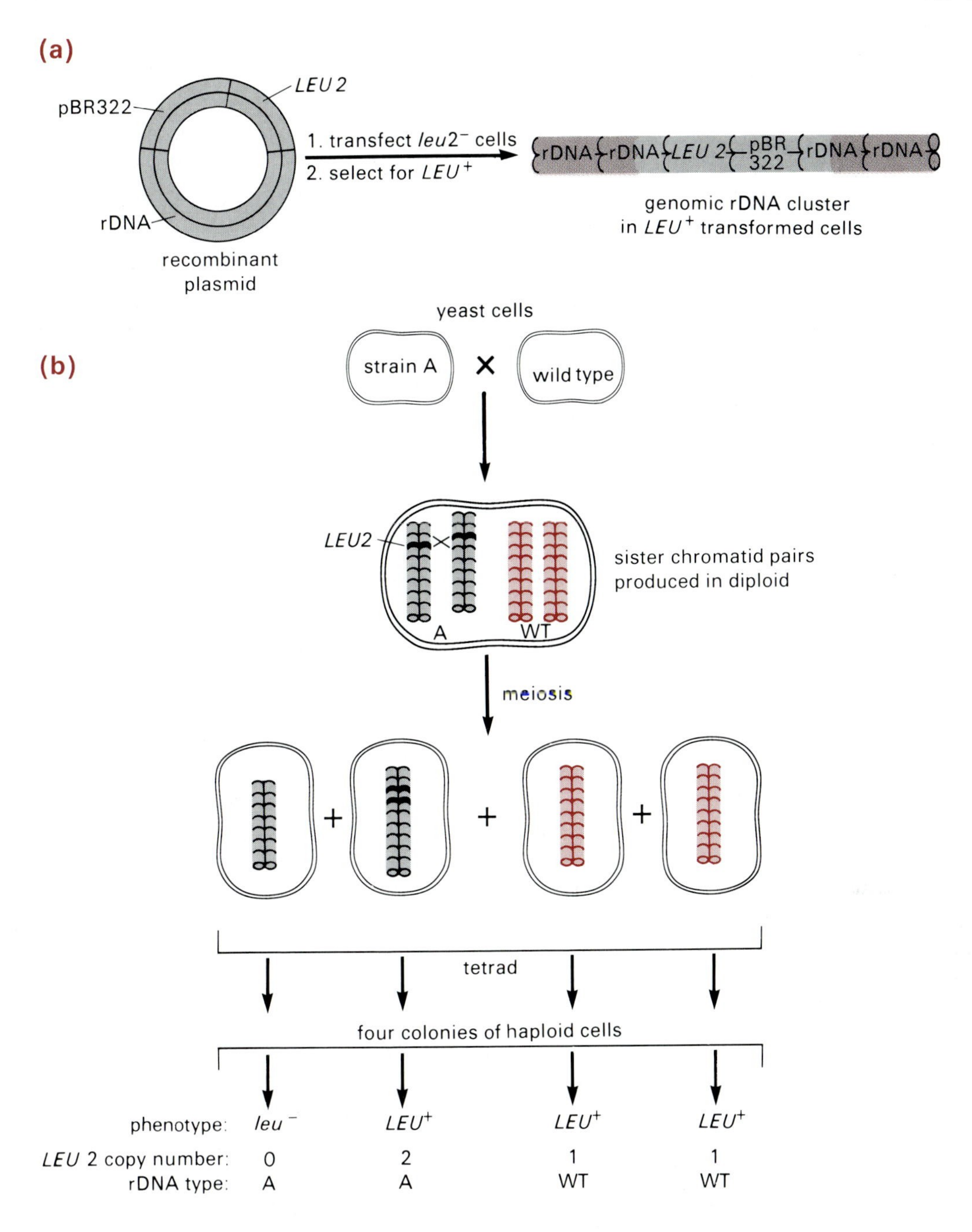

Divergence of IGS spacer region sequences The structures of the IGSs are complex and differ both in sequence and in organization from one species to another (Section 8.2c). This is in marked contrast to the highly conserved rDNA coding regions. The one feature IGSs often have in common is clusters of direct tandem repeats, though the repeated sequences are different in different organisms. Except in mammals, the repeated units are frequently similar to the sequences around the transcription start sites, and thus to RNA polymerase I promoter signals. Some of these repeated sequences serve as minor transcriptional start sites and thereby account for the presence of a low level of RNA homologous to the IGS in some species. Some contain enhancers of rDNA expression. The ubiquitous presence of tandem repeats within IGSs suggests that their presence,

Table 9.2 Localization, Copy Number, and Repeat Unit Size of 5S rRNA Genes

Species	Located on Chromosome	Approx. Copy Number (per haploid genome)	Approx. Size of Repeat Unit (including IGS) (bp)
S. cerevisiae	12	150	
D. melanogaster	2R, band 56EF	160	375 ±5
X. laevis	Many or all, telomeric	400 somatic	850–900
		20,000 oocyte	600–1000
X. borealis		700 somatic	850–900
		9000 oocyte	1000–2000
H. sapiens	1	2000	

though not their specific sequence, is related to some advantageous effect on rDNA transcription.

Unequal crossing-over between tandem repeat units within IGSs leads to different numbers of repeat units, just as it leads to different numbers of complete rDNA units. Thus, the IGS evolves both in concert with and independent of the long rDNA unit within which it resides. Moreover, as a result of the crossing-over, the overall length of rDNA repeat units varies among closely related species and even within a single species.

b. Genes for 5S rRNA

Genes for 5S rRNA are usually not linked to the genes for 18S, 5.8S, and 28S rRNA; the only known exceptions are in some fungi and protozoa. In accordance with the fact that 5S rRNA genes are transcribed by RNA polymerase III, portions of the internal coding sequences function as transcriptional regulatory elements (Section 8.4b). Genes for the 120 nucleotide long 5S rRNA are not interrupted by introns.

Tandem repetition of 5S rRNA genes As with the clusters of genes for the other rRNAs, the 5S rRNA genes of many eukaryotes are repeated many times in long tandem arrays (Table 9.2). Isopycnic centrifugation of bulk *Xenopus* DNA, for example, yields 5S rDNA as a separate band even though it represents less than 1 percent of the genome. Electron micrographs of partially denatured tandem arrays of 5S rDNA revealed a structure of A·T-rich, easily melted regions, alternating with more G·C-rich segments. The 120 base pair long coding region is the G·C-rich portion and represents less than 20 percent of the total repeat length. The remainder is an A·T-rich, nontranscribed spacer. The presence of variant 5S rRNA genes was deduced from the heterogeneity of purified 5S rDNA upon density gradient centrifugation and from its multicomponent melting curve. The copy number of 5S rDNA in various genomes has been determined by saturation hybridization using purified 5S rRNA.

In *D. melanogaster* the approximately 160 5S rRNA genes are repeated in tandem on one region of chromosome 2. Each repeat unit contains the

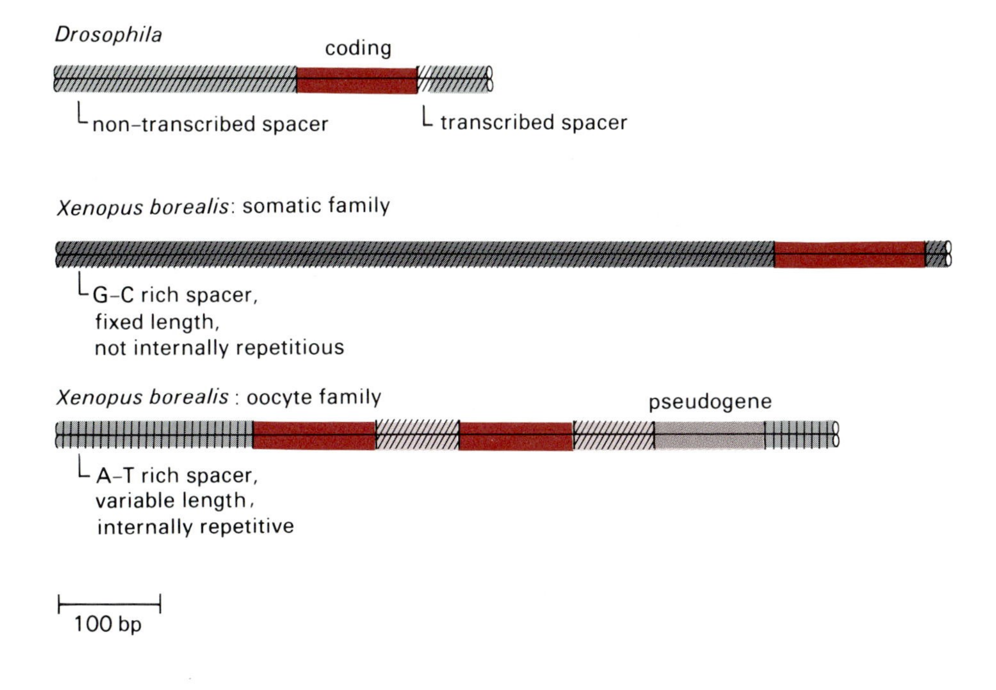

Figure 9.7
Structure of some 5S rDNA repeat units. Within the units, spacer segments and one or more genes or pseudogenes may alternate. In the *X. borealis* oocyte 5S rDNA, the repeated units contain several but varying numbers of genes and pseudogenes separated from one another by about 80 base pairs of spacer. In turn, these units are separated by long, variable length spacers that are AT-rich and internally repetitious. This spacer contains multiple tandem repeats of a 21 bp unit (5′-CGTCGCGTCG-TTTTTGTCGCG-3′) and changes length by alterations in the copy number of this repeated segment; cleavage of the 5S rDNA with the restriction endonuclease *HhaI* (recognition site GCGC) facilitates analysis of the structure because it yields the 21 bp repeat units as a separate fragment. Few transcripts of the pseudogene are found in the cell; it differs in 15 positions from the sequence of the functional gene.

120 bp long coding region, about 15 bp of transcribed spacer at the 3′ end of the transcription unit, and about 250 bp of nontranscribed spacer (Figure 9.7). All 160 repeat units appear to be in a single cluster, uninterrupted by other sequences. Over 20 tandem repeats have been cloned together in a single segment more than 8 kbp in length. The A·T-rich region in the spacer has internal sequence repeats and is variable in length (± about 5 bp).

The organization of Xenopus *5S rRNA genes* Although *D. melanogaster* and other organisms have relatively simple arrays of 5S rRNA genes, different *Xenopus* species (and other amphibia and fish) each have two families of 5S rRNA genes (Section 8.4e, Table 9.2, and Figure 9.7). One family of 5S rRNA genes includes thousands of gene copies and is expressed only during oogenesis. The other family contains many fewer copies and is expressed both in oocytes and in somatic cells at most stages of development. Each family has a distinctive coding sequence, but within each family, the coding sequences are homogeneous. Both families are arranged as tandem repeats. The repeating unit of the somatic 5S rRNA genes is simple. A long spacer region alternates with the short transcription unit. But the oocyte 5S repeat units are complicated and have different forms in different, even related, species. Within a species, there is some variation in the length and composition of the unit.

c. The Linkage of all Four Yeast rRNA Genes

The rRNA genes of some fungi and protozoa, including the yeast *S. cerevisiae*, are distinctive in that all the genes, including the 5S rRNA gene, are linked in a single repeating unit (Figure 9.8). However, transcription of the 5S rRNA gene is from the DNA strand opposite to that used for the long precursor to 18S, 5.8S, and 25S rRNA. The two distinctive tran-

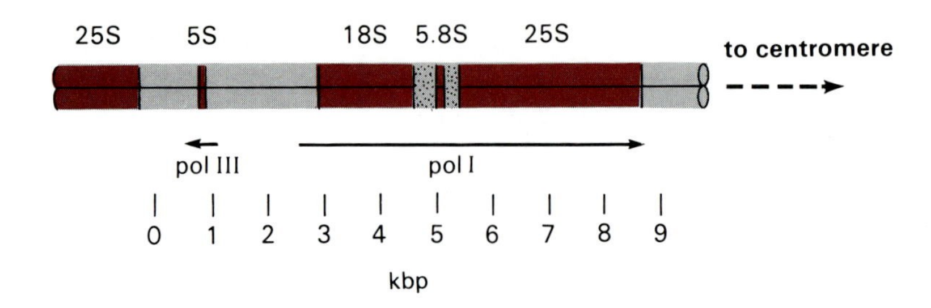

Figure 9.8
The yeast rDNA repeat unit. The arrows show the direction of transcription.

scription units use different RNA polymerases: RNA polymerase I for the long polycistronic transcript and RNA polymerase III for 5S rRNA. In another yeast, *Schizosaccharomyces pombe*, and in the fungus *Neurospora crassa*, the 5S rRNA genes are neither part of the rDNA repeat nor organized in tandem arrays; they are dispersed in the genome. Genetic experiments mapped most if not all the *S. cerevisiae* rDNA units to a single tandem cluster on chromosome 12. The map of the region is as follows: -*URA*4-*rRNA*-*PEP*3-*GAL*2-*ASP*5-centromere. (The orientation of the rDNA units relative to the centromere is shown in Figure 9.8)

d. Genes Encoding tRNAs

The structure of tRNA genes Genes for tRNA, like those for 5S rRNA, are transcribed by RNA polymerase III, and promoter sequences are similarly located internally, within the coding sequence (Section 8.4b). There is no fixed rule about intervening sequences; some tRNA genes have them, others do not. Within a given species, such as yeast, *Drosophila*, or *Xenopus*, both interrupted and noninterrupted genes occur. As a rule, introns in tRNA genes are short, 45 base pairs or less, and they occur just to the 3′ side of the nucleotides encoding the anticodon region.

The repetition and varied arrangements of tRNA genes Like the genes for rRNA, those for each tRNA are usually present in multiple copies in eukaryotic genomes (Table 9.3). But in contrast to the orderly arrangement of rDNA, tRNA gene organization seems haphazard. Different tRNA genes may or may not be in close proximity to one another.

Table 9.3 Copy Number and Distribution of tRNA Genes

Species	Clustered (c) or Dispersed (d)	Approx. Total Copy Number (per haploid genome)
S. cerevisiae	d	360
Tetrahymena pyroformis		800
D. melanogaster	c, d	800 (~12 of each tRNA)*
X. laevis	c, d	7000 (~200 of each tRNA)
Rattus norvegicus	c, d	6500 (~100 of each tRNA)*
H. sapiens	c, d	1300 (10–20 of each tRNA)

* Some of these genes are known to be pseudogenes.

Multiple genes for a single tRNA may or may not be clustered. For example, annealing of tRNA probes to *Drosophila* chromosomes *in situ* revealed about 50 different loci at which many genes are clustered. The individual loci contain genes for a mixture of different tRNAs (Figure 9.9). Although genes for particular tRNAs may reside at more than one cluster site, some identical tRNA genes occur in tandem repeats. In other words, all possible arrangements are found.

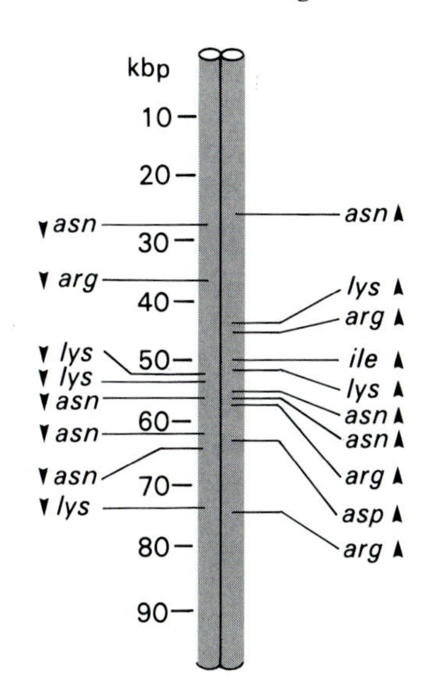

Figure 9.9
The gross anatomy of a cluster of tRNA genes at chromosome region 42A of *Drosophila*. The arrow heads show the direction of transcription. After P. H. Yen and N. Davidson, *Cell* 22 (1980), p. 137.

Specific repeated DNA sequences preceding tRNA genes Functional tRNA genes are found in the midst of diverse flanking sequences. Nevertheless, several specific DNA sequences often occur close to tRNA genes. The arrangement is interesting, although no functional significance is associated with the recurring units as yet. The sequence 5′-ACT(A or T)$_n$TA (where n is from 0 to 5) often occurs within the 100 bp preceding tRNA genes in *Drosophila*, yeast, and other organisms. A 340 bp long sequence called sigma occurs about 30 times in the yeast genome, each time within 18 bp of the start of a tRNA gene. Both possible orientations of sigma relative to the direction of transcription of the tRNA gene are found. However, most yeast tRNA genes are not associated with a sigma segment, and alleles of a tRNA^{Tyr} gene both with and without a neighboring sigma show no detectable phenotypic difference. Sigma itself is a copy of the long terminal repeat (LTR) found at the two ends of a yeast retrotransposon called Ty3 (Section 10.3). Such "solo" LTRs are common remnants of retrotransposons. The localization of sigma in front of tRNA genes suggests a positional specificity for Ty3 insertion.

e. Genes for Small Nuclear and Cytoplasmic RNAs

All eukaryotic cells contain a variety of small, stable RNA molecules, many of which reside in nucleoprotein particles in either the nucleus or the cytoplasm. Several of the small nuclear RNAs (snRNAs) are known to perform critical functions in the processing of primary transcripts to mature mRNAs and rRNA (Section 8.5). The function of small cytoplasmic RNAs (scRNAs) is generally unknown, expect for the 7SL RNA of the signal recognition particle (Section 3.10b).

Multiple genes and pseudogenes for snRNAs Metazoan genomes contain multiple DNA segments that anneal with U1, U2, U3, U4, U5, U6, and U7 snRNAs or to cDNAs made from these molecules (Table 9.4). Within each genome, some of the homologous regions are dispersed, but others are clustered, frequently in tandem arrays. As with the 5S rRNA genes, the coding region can be buried in a long DNA segment that constitutes the repeat unit. As a general but not necessarily fixed rule, functional genes tend to be clustered, and pseudogenes are dispersed. The functional genes have coding sequences that are identical with the sequence of the respective snRNAs. Functional genes for a particular snRNA usually have very similar sequences flanking the coding region; these contain transcriptional regulatory elements (Secton 8.3). Pseudogenes diverge from the sequence of the snRNA by extensive deletions, insertions, or single base pair changes and have highly divergent flanking sequences. The actual organization and number of both genes and pseudogenes vary from one

Table 9.4 Genes for Some snRNAs in Metazoans

			Coding Genes		
Name	Known in	RNA Length in Bases	Approx. Copies of Functional Gene	Dispersed (d) or Clustered (c)	Polymerase
U1	All metazoans	164	30 (human)	c	II
			5 (chicken)	c	
			10 (mouse)	c (head to head)	
			500 (*Xenopus*)	c (tandem) d	
			1–2 (*Phaseolus vulgaris*)		
			3–4 (*Drosophila*)	d	
U2	All metazoans	188–189	20–40 (human)	c (tandem)	II
			10 (mouse)		
			500 (*Xenopus*)	c (tandem)	
			(sea urchin)	c (tandem)	
			5 (*Drosophila*)	c, d	
U3	All metazoans (nucleoli)	216	6 (mouse)	c, d	II
U4	All metazoans	142–146	4 (*Drosophila*)	d	II
			2 (chicken)	c	
U5	All metazoans	116–118	*Xenopus*	tandem	II
U6	All metazoans	107–108	200 (mouse)	d	III
			3 (*Drosophila*)	c	III
U7	Sea urchins	58	5	c	II
$4.5S_I$	Rodents	98–99	10^4 (rat)	d	III

Note: There are at least two classes of U1 *Xenopus* genes: class I, about 500 in number, are tandemly repeated and expressed in early embryos; class II, a few in number, are dispersed and expressed in oocytes. In yeast, sequences homologous to U4 and U6 are in a single gene.

snRNA to another and from species to species. Genes for the less abundant snRNAs such as U11 have not yet been studied. Notably, yeast has only one copy of each of its snRNA genes and no pseudogenes. Yeast and metazoan snRNAs have comparable roles in the processing of primary transcripts to form functional RNAs (Section 8.5).

The complexity of the sequence families in many species is illustrated by the U1 family in human DNA. The copy number of U1 snRNA genes in the human genome is about 30, and these are clustered in a tandem array on the short arm of chromosome 1. Each functional gene in the array is embedded in a repeat unit that is at least 44 kbp long. Pseudogenes occur ten times more frequently and in various genomic loci.

scRNA genes The 7SL RNA is a critical constituent of the signal recognition particle that functions in the translocation of newly synthesized secretory and membrane polypeptides across the lipid bilayer of the endoplasmic reticulum. The nucleotide sequences of rodent and primate 7SL RNAs are virtually identical, and that of *Drosophila* 7SL RNA is 64 percent homologous to the mammalian sequences. In human DNA, there are three genes for 7SL RNA and several hundred pseudogenes (Table 9.5). But there are, in addition, almost 10^6 scattered repeats of sequences that are

Table 9.5 Genes for Some scRNAs

RNA	Known in	Approximate RNA Length (in bases)	Haploid Copy Number of Functional Genes	RNA Polymerase
7SL	Vertebrates, invertebrates	300	3–4 (human)	III
			2 (*Drosophila*)	
		254	1 (*S. pombe*)	
7SK	Vertebrates	330	≤10 (human)	III
4.5S	Rodent	90–94	850 (mouse)	III
			690 (rat)	

homologous to about half of the 7SL RNA sequence. The half that is so abundantly repeated is a combination of the 5′ end and the 3′ end with the center portion deleted and is called an **Alu** sequence (Section 9.5c). A similar situation exists in rodents.

The functions of the other scRNAs are not known, but most of them are associated in mammals with large sequence families containing both genes and pseudogenes.

9.3 Genes Encoding Polypeptides

a. *Some General Considerations*

Genes for polypeptides typically have coding regions (exons) interrupted by one or more introns. However, intervening sequences are not obligatory, and some polypeptide genes (e.g., for some histones and most interferons) have none. The number of intervening sequences varies widely from one gene to another. Moreover, the amount of genomic DNA taken up by intervening sequences frequently exceeds that in the coding sequences by a large amount (Table III.2). Overall, there may be as much as tenfold more DNA in introns than in coding sequences in some organisms. Generally, analogous exons belonging to paralogous genes within a particular genome or to orthologous genes in the genomes of different species are more highly conserved in nucleotide sequence than are the introns. The position of the introns tends to be conserved, but the length and sequence tend to vary.

All known polypeptide genes are transcribed by RNA polymerase II and are therefore frequently associated with similar promoter elements and polyadenylation signals. Many polypeptide genes are also associated with more specific regulatory sequences that modulate gene expression in response to developmental, hormonal, or environmental conditions (Section 8.3).

In contrast to genes encoding RNA, genes encoding specific polypeptides are generally represented only once in a genome. However, even when there is only a single, functional gene for a particular polypeptide, there are frequently several or many segments within the genome that are homologous to the gene sequence. Thus, single copy genes are often members of a family of closely related sequences (e.g., the growth hormone gene family). The family members may include several genes that

Table 9.6 Some Families of Related Genes and Proteins

Closely Homologous in Structure and Function
β-globins
α-globins
Actins
Tubulins
Zeins
Growth hormone/placental lactogen
Steroid receptors
Opsins
Related, But Divergent in Structure and Function
α-globins/β-globins
Prolactin/growth hormone
Serum albumin/α-fetoprotein
α-interferons/β-interferons
RNA polymerases
Distant Cousins
Haptoglobulin/serum proteases (chymotrypsin)
Antithrombin III/ovalbumin/angiotensinogen
Ceruloplasmin/factor V/factor VIII

encode slightly variant proteins (such as **isozymes**). Alternatively, family members may be associated with different regulatory signals to permit expression in different tissues or at different times in development (like the growth hormone and placental lactogen genes). (Recall that the definition of a gene used in this book includes certain control sequences that are not found in the transcription unit.) Also, some family members may be pseudogenes. As a general convention, two nonallelic genes are considered identical only if they encode virtually identical polypeptides and are subject to the same regulation and modulation. But whether they are identical or not, closely related genes are considered members of a multigene family (Table 9.6).

Single copy genes may also belong to a large family of distantly related sequences. Genes in such a superfamily encode quite different polypeptides, although the similarity of gene structure is marked (e.g., the previously discussed prolactin and growth hormone genes). The classification of individual genes into superfamilies is not always clear-cut. Family members often show a gradient of nucleotide sequence similarity from more than 95 percent to 50 percent or less. Only DNA segments with matches of better than about 80 percent anneal with one another under stringent conditions, so distantly related family members may be identifiable only by comparison of primary nucleotide sequence. Furthermore, the similarity between two proteins is not a simple function of the percentage of similarity between the corresponding genes. Given the degeneracy of the genetic code, extensive changes in the third positions of codons may result in only minimal changes in amino acid sequence, but the same number of changes in the first or second positions of codons yields radical changes in polypeptide structure.

Table 9.7 Some Polypeptide Multigene Families

Gene Family	Organism	Approximate Family Size (number of genes)	Linked (l) or Dispersed (d)	Characteristics
Acid phosphatase	Yeast	≥4	l, d	Individual genes are regulated in diverse ways (e.g., one is repressed by Pi, one is induced by Pi, another is constitutive)
α-amylase	Mouse	≥3	l	Multiple genes are expressed in pancreas, one in liver and salivary gland. Different mouse strains have variable numbers of pancreatic α-amylase genes
	Rat	≥9	?	4 encode pancreatic amylase
	Barley	≥7	?	At least 3 functional genes
Collagens (interstitial)	Human	4	d	Interstitial collagens, types I (a heterotrimer), II, and III (homotrimers), are encoded by 4 genes dispersed to chromosomes 2, 7, 12, and 17
Myosin heavy chain	Rat	≥10		At least 1 gene for each type of muscle; at least 1 nonmuscle gene
Ovalbumin	Chicken	3	l	All 3 regulated by estrogen; the function of 2 is unknown
Serum albumin	Mouse	2	l	Serum albumin and α-fetoprotein
Vitellogenin	Toad	4	?	Pairs of genes diverge 20% from one another; within pairs, the genes are 5% divergent; all 4 are expressed
Ribulose-1,5-bisphosphate carboxylase, small subunit	Tomato	5	l, d	Expressed in different tissues at different times in development, dependent on light
Zein	Maize	100	d, l	Encode a family of kernel storage proteins

Note: Pi stands for inorganic phosphate. In this and the following tables, linked (l) means that two or more genes are clustered in one locus, and dispersed (d) means that family members are in different loci. In some families, there are both linked and dispersed copies.

Although many genes belong to multigene or multisequence families and superfamilies, some are probably truly unique sequences in certain genomes. Among these are the human thyroglobulin gene and the yeast actin gene.

b. Examples of Multigene Families

Each multigene family encoding polypeptides is distinctive, and characteristics often vary from one species to another (Table 9.7). Several ex-

Table 9.8 Actin Gene Families

Organism	Number of Proteins Known	Gene Number (including pseudogenes)	Linked (l) or Dispersed (d)	Actin Type: Muscle (m) or Cytoplasmic (c)
S. cerevisiae	1	1		c
Dictyostelium	>2	17	l, d	c
C. elegans		4	l, d	c
Oxytricha	>2	3		c
Soybean		≥2		c
Drosophila	3	6	d	c
S. purpuratus		8	l, d	m, c
Chicken	6	8–10	d	m, c
Rat	6	≥8	d	m, c
Human	6	20–30	d	m, c

Note: The number of functional genes for different types of actins can be estimated with probes from the 3′ noncoding regions (see text). Linkage is ascertained by the presence of more than one gene in a cloned genomic segment. The various *Dictyostelium* genes differ up to 10 percent in sequence, and at least eight are expressed; some of the genes may encode identical proteins. The four actin genes in the nematode (*C. elegans*) are functional; three function in muscular contraction. In *Drosophila*, all six genes are expressed; one protein, although similar to mammalian cytoplasmic actin, is specific to muscle tissue. In the sea urchin (*S. purpuratus*), at least six are different functional genes. Annealing of appropriate probes to DNA from human-mouse somatic cell hybrids indicates that there are single functional genes for β-actin, γ-actin, and cardiac and skeletal α-actins on human chromosomes 7, 17, 15, and 1, respectively.

amples selected to illustrate the different ways that multigene families provide for a diversity of physiological capabilities and the fine-tuning of cellular processes are described here.

Actin genes: a multigene family conserved in evolution Actin is involved in various kinds of cellular motion including intracellular motility and muscle contraction. Accordingly, eukaryotes contain several different kinds of actin and corresponding distinct genes. The several genes are clustered, dispersed, or both in different genomes (Table 9.8). Yeast is the only organism known to have but a single actin gene.

Cytoplasmic actins, those that are involved in cellular motility, are similar in all eukaryotes; all the actins of invertebrates and plants are in this class. In *Drosophila* and presumably other invertebrates, certain cytoplasmic actins function in muscle contraction, although in vertebrates, a distinctive group, the α-actins play this role. There are six known actin proteins in birds and mammals. Two α-actins are found in skeletal and cardiac muscle, respectively, and are associated with muscle contraction. Two actins, an α and a γ, occur in smooth muscle. The cytoplasms of virtually all mammalian and avian cell types contain cytoplasmic β- and γ-actin.

Although the several kinds of actins are structurally distinct, they have very similar amino acid sequences and differ mainly near their NH_2 termini. For example, human β- and γ-actins differ in four amino acid residues. The different actin genes comprise multigene families in most organisms,

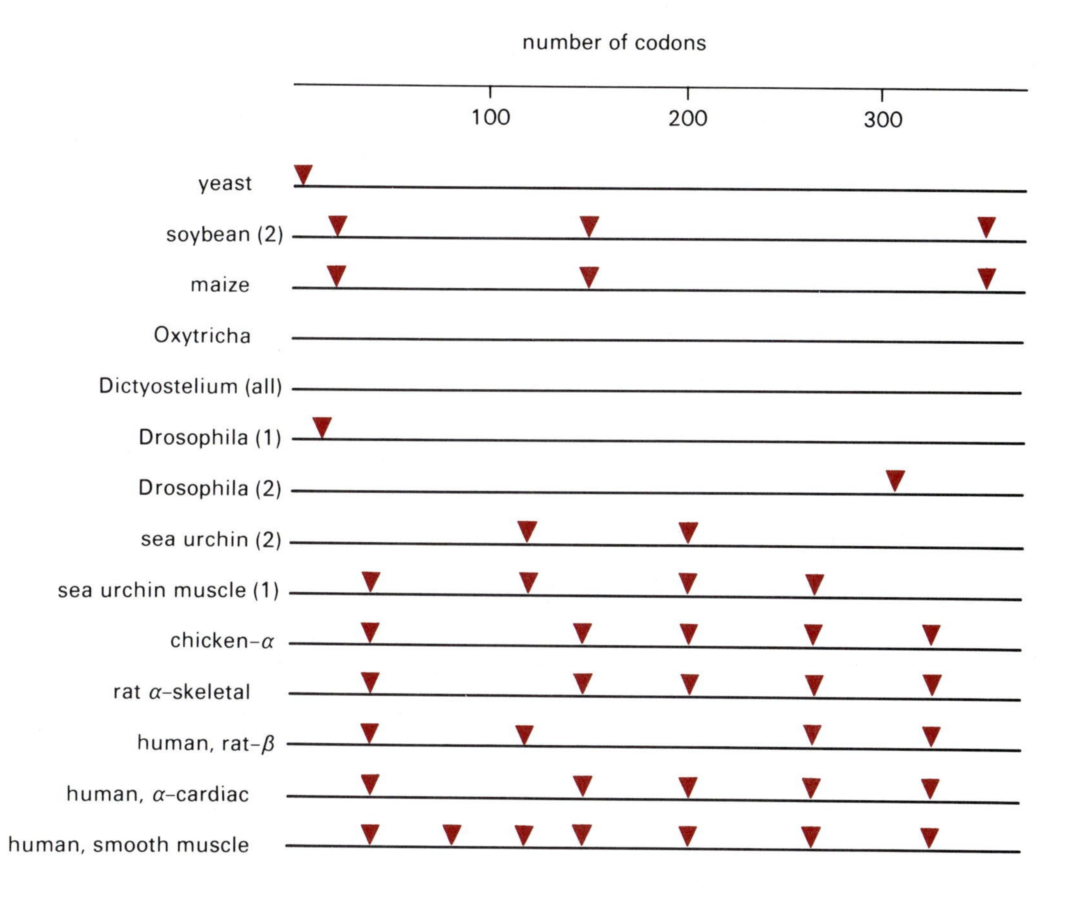

Figure 9.10
The number and location of introns in some actin gene coding regions. The scale is the number of codons. The numbers in parentheses show the number of different genes known to have the indicated arrangement. Several actin genes have an additional intron in the untranslated 5′ leader region (not shown).

and their coding regions are homologous from one organism to another. Skeletal muscle α-actin, for example, has exactly the same amino acid sequence in chickens as in mammals, and actin mRNA isolated from vegetative *Dictyostelium* cells works well as a probe to select actin gene clones from such diverse DNAs as yeast, maize, soybean, and mammals.

The 5′ and 3′ flanking regions of the various actin genes diverge more than do the coding regions. However, equivalent actin genes such as the mammalian β-actins tend to have similar flanking regions in different species, and these are markedly different from the regions flanking the α-actin genes in the same species. Frequently, the flanking regions of the several types of actin genes are so distinctive that they can be used as hybridization probes to identify a specific type of gene and its transcripts. For example, cloned probes representing each of the six different functional sea urchin actin gene 3′ trailer sequences can be used to measure the differential transcription of the corresponding genes during development and differentiation. Thus, although all the actin coding regions and actins are very closely related, each type of actin appears to be associated with characteristic noncoding sequences, sequences that are likely to have regulatory significance. This situation illustrates the twofold demands on the evolutionary process. One is concerned with the functional properties of the gene product and the second with the regulation of gene expression in differentiated cells.

Another unusual aspect of the actin multigene family is the variation in the number and location of introns (Figure 9.10). This is evident comparing both paralogous genes such as the β-actin and skeletal α-actin gene of the rat or orthologous genes like the plant and invertebrate actin genes. The situation is very different from that of most orthologous genes and multigene families, in which the several genes have identically placed introns. The evolution of actin genes seems unlikely to have included multiple insertions of new introns. Each such insertion would have to provide for appropriate splice junction sequences and preserve the reading frame on either side. An alternative interpretation is that the most ancient actin genes had introns at all the positions found in contemporary genes and that different ones were lost during the evolution of different phyla. According to the latter model, *Dictyostelium* and *Oxytricha* were very efficient at purging introns because none is left in the coding regions.

Individual actin genes tend to be dispersed among chromosomes. In the mouse, for example, the genes for skeletal and cardiac muscle α-actins are on chromosomes 3 and 17, respectively, while the gene for cytoplasmic β-actin is on chromosome 5.

Most organisms studied seem to have more actin genes than known actin proteins. Although unrecognized forms of actin may be encoded by some of these "extra" sequences, many of them are probably pseudogenes, including processed pseudogenes. In humans, at least 2 of the 20 actinlike sequences are processed β-actin pseudogenes; they lack introns and contain many different mutations within the coding region. All six genes in *Drosophila* and six of the sea urchin genes are functional and are expressed according to a specific developmental program. In *Drosophila,* two genes for cytoplasmic actins are expressed at all developmental stages, two genes predominantly in larval, pupal, and adult intersegmental muscles, and two genes predominantly in adult thoracic and leg muscles. Presumably, differential gene expression is governed by specific regulatory sequences that are associated with the flanking regions of the several genes.

Tubulin genes: a multigene family that includes genes for the two different subunits of a heterodimeric protein Microtubules are associated with a variety of processes fundamental to all eukaryotic cells. These include meiosis, mitosis, cell motility, and secretion. Not surprisingly, then, the structure of tubulin, the protein from which microtubules are constructed, is conserved among all eukaryotic organisms, as are the DNA sequences encoding tubulin. cDNA clones prepared from chicken tubulin mRNA anneal to tubulin genes from such distantly related organisms as yeast and mammals.

Tubulin itself is a heterodimer composed of two polypeptides, α and β. The α- and β-subunits contain 450 or 451 and 445 amino acids, respectively, and are about 40 percent similar. Genes for both subunits belong to the same multigene superfamily, although they are different enough not to anneal.

Typically, families of genes encode multiple **isotypes** of both the α- and β-subunits in each eukaryotic species (Table 9.9). There are usually only minor structural differences (less than 10 percent) in the amino acid sequences of the various α- or β-subunits encoded by these genes. Much of the divergence occurs near the carboxyl terminus of the polypeptides. The variations in tubulin structure appear to relate to the construction of somewhat different kinds of microtubules that are associated with dif-

Table 9.9 Tubulin Gene Families

Organism	Number of known Functional Genes	Approximate Number of Pseudogenes for α or β	Linked (l) or Dispersed (d)
S. cerevisiae	2α, 1β	0	d
Chlamydomonas	2α, 2β	0	d
Trypanosoma	About 15α, 15β	0	l (tandem)
Drosophila	4α, 4β	0	d
Sea urchin	2α, 3β	10	l, d
Chicken	2α, 7β	<5	d
Mammals	About 5α, 5β	15	d
Arabidopsis thaliana	1α (possibly 4)		

ferent cellular processes, different cell types, and different developmental stages. Thus, various α- or β-isotypes may be specialized for specific microtubule functions, although specialization may not be absolute, and isotypes may be fungible. Additional alterations in tubulin structure are made by posttranslational processes, which also contribute to the functional specialization of microtubules. The concept of functionally specialized isotypes is supported by a high degree of conservation of specific isotypes in vertebrates. For example, the carboxyl terminal sequences of the predominant neural β-tubulin are virtually identical in several different vertebrates.

The 5′ and 3′ flanking regions of members of the α- and β-tubulin multigene families diverge much more than do the coding sequences. This probably relates to independent regulation of expression. Thus, one of the mouse α-tubulin genes is expressed only in testis. In *Drosophila,* the concentrations of mRNAs transcribed from the various tubulin genes vary from tissue to tissue and in different developmental stages. One of the α-tubulin genes is expressed specifically in adult male flies, one in ovaries, and two are constitutively expressed. In yeast, the two α-tubulin genes appear to encode functionally equivalent polypeptides, although one gene is more efficiently expressed than the other. The single β-tubulin gene of yeast is essential for growth. A yeast mutant characterized as being blocked in the cell division cycle (a *cdc* mutation) was subsequently identified as a mutation in the β-tubulin gene.

In most species, the tubulin genes are dispersed in the genome. For example, although all the *Drosophila* tubulin genes reside on chromosome 3, they are not close to one another. Some protozoa provide striking exceptions to the general distribution pattern of tubulin genes. In *Tetrahymena brucei,* the approximately ten gene copies are clustered in a single tandem array of alternating α- and β-tubulin genes; in *Leishmania enrietti,* the same number of copies is clustered in separate α- and β-tubulin gene arrays.

Some sequences that are homologous to tubulin genes are pseudogenes, and many of these are of the processed type. A rat α-tubulin pro-

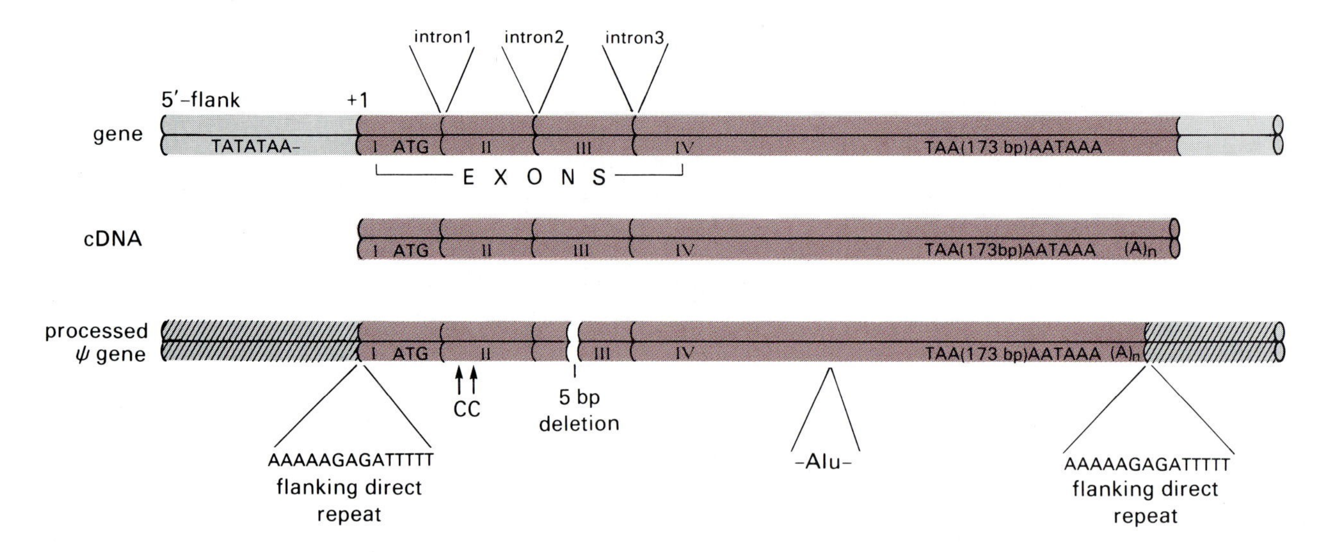

Figure 9.11
The structural relationships between a rat α tubulin gene, its messenger RNA (cDNA), and a processed pseudogene. The pseudogene has acquired two additional C residues, a 5 bp deletion, and an insert of an Alu sequence (see Section 9.5c). The drawing is not to scale. After I. Lemishka and P. A. Sharp, *Nature* 300 (1982), p. 330.

cessed pseudogene illustrates the typical features mentioned in Section 9.1a (Figure 9.11). The sequence of the functional gene, including portions of the 5′ and 3′ flanking regions, are known. There are about 120 bp between the TATA signal and the ATG initiation codon, and the gene has three introns (the first comes immediately after the ATG), as is typical of vertebrate tubulin genes. The tubulin mRNA sequence was determined from a cloned cDNA; it starts about 20 bp downstream from the TATA signal and continues for 100 bp to the ATG. Thereafter, the sequence is identical to that of the exons with the intron sequences spliced out. The cDNA sequence ends with a series of A residues that starts 16 bp after the polyadenylation signal. In contrast to the gene, the processed gene looks more like the cDNA. On the 5′ side of the ATG, it is identical to both the gene and the cDNA for 99 bp (i.e., throughout the 5′ leader). Further in the 5′ direction, the sequences are not homologous to the corresponding region of the gene itself. At the 3′ end, the sequence of the processed gene is homologous to the cDNA through the series of A residues. Thereafter, the sequence is completely different from that in the 3′ flanking region of the gene itself. Moreover, the introns are all missing. The processed gene would be a perfect copy of the cDNA except that there are several changes in the coding region that preclude making a good α-tubulin. Frameshifts are caused by two C residues inserted in exon 2, a 5 bp deletion in exon 3, and the insertion of an unrelated segment of 110 bp (marked Alu on Figure 9.11) into exon 4 (Section 9.5c discusses Alu). The α-tubulin processed gene is flanked on either side by direct repeats of a short DNA sequence, another common feature of processed pseudogenes. Such surrounding direct repeats are typical of DNA segments that have been inserted by transposition into new genomic loci. They represent reiterations of the sequence that preexisted at the insertion site and are called target site duplications (Section 10.1b).

The globin gene superfamily The vertebrate hemoglobins are heterotetramers containing two each of an α- and a β-globin polypeptide. Multiple genes and pseudogenes for the α- and β-chains occur in all vertebrates, and these are part of a multigene superfamily that includes the coding

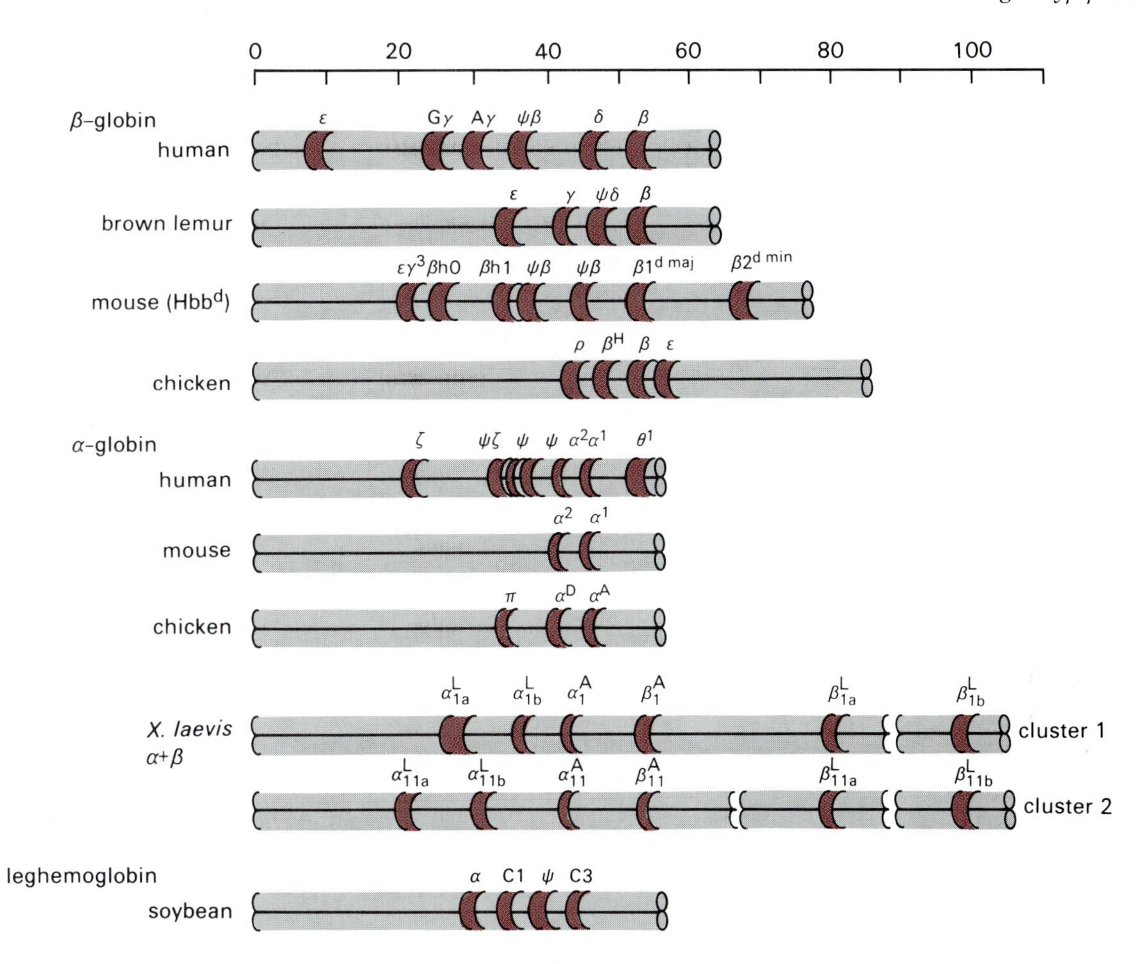

Figure 9.12
The organization of β- and α-globin-like genes in several vertebrates and of leghemoglobin genes in soybeans. The genes are shown without introns. The scale at the top indicates kbp. Direction of transcription is from left to right in all cases. The function, if any, of human α-$\theta 1$ gene is unknown.

sequences for invertebrate globins, vertebrate myoglobin, and plant leghemoglobin (Figure 9.12). All these proteins are functionally similar in that they contain heme and associate reversibly with oxygen.

Globin mRNAs and genes were among the very first targets of recombinant DNA research. More than 90 percent of the soluble protein in red blood cells is hemoglobin, and the bulk of the mRNA in reticulocytes and nucleated red cells is globin mRNA. No other cell types transcribe the globin genes to a significant extent. The availability of relatively pure globin mRNA, the large number of known human globin mutations, and extensive knowledge about the globin proteins together stimulated the study of the molecular genetics of these genes. This work provided some of the earliest evidence for the presence of introns in cellular genes as well as for the nature of RNA polymerase II promoters. And because so much primary sequence data are available for various human alleles and also for globin genes in other species, many ideas concerning evolution at the level of DNA sequence originated in studies of the globins.

Typically, mammals have multiple α- and β-globin genes and pseudogenes. In humans, the β-globin genes are clustered within a 65 kbp stretch of DNA on chromosome 11 and the α-globin genes are arranged in a 25 kbp cluster on chromosome 16 (Figure 9.12). The β-globin cluster contains five genes (ε, ${}^{G}\gamma$, ${}^{A}\gamma$, δ, and β) and the α-cluster three (ζ, α_1, and α_2). Both clusters also include several pseudogenes.

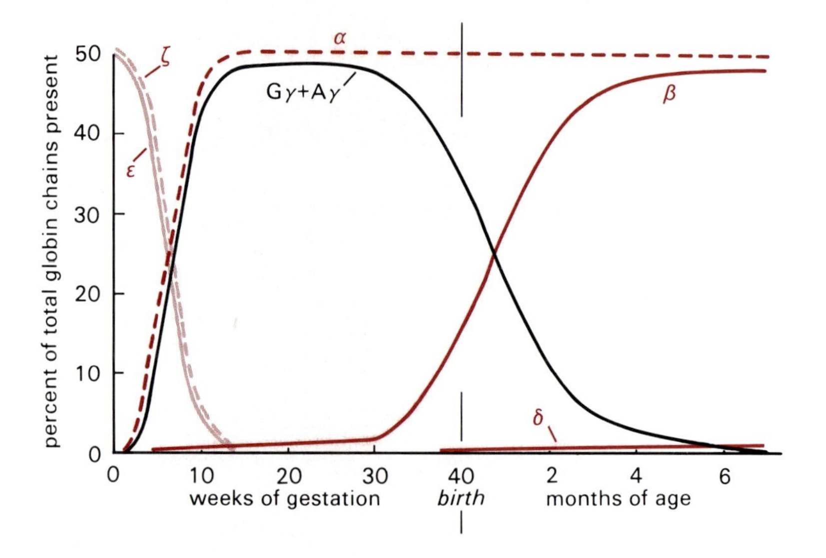

Figure 9.13
Time course of expression of members of the globin gene family in human development.

In humans and other mammals, the different globin genes are expressed at different times during development (Figure 9.13). ζ and ε are expressed in very early embryos: the embryonic human hemoglobin is ζ_2/ε_2. By ten weeks of gestation, the α- and γ-genes are activated to produce an α_2/γ_2 globin that circulates during much of fetal life. Soon after birth, the β-like γ-chain is replaced by the β-globin polypeptide to form the α_2/β_2 globin that is the predominant hemoglobin protein during the remainder of the organism's lifetime. The δ-globin gene is also expressed in adult organisms, but only produces a very small portion of the total β-globin-like chains. Two identical polypeptides (α_1 and α_2) are encoded by the two α-globin genes. Although α_1 and α_2 are expressed at the same time, α_2 is more efficiently transcribed and contributes the bulk of the α-polypeptide. Not in all, but in many species, including humans, the order of genes in the α- and β-clusters is the same as the order in which they are expressed during development (going in the direction of transcription, left to right on the diagram in Figure 9.12).

Because of the extensive data on genes and proteins in the globin family, the evolutionary history of human globins can be traced in some detail. A single "globin" gene is assumed to have existed in a remote animal ancestor of contemporary species. An amplification of the gene is postulated more than 400 million years ago because, although the cyclostome fishes have a single globin chain, the jawed fishes, the gnathostomes, encode both α- and β-globin chains. In amphibia (e.g., *Xenopus laevis*), the α- and β-globin genes are closely linked in tandem, as they presumably are in the fishes (Figure 9.12). Note that there are two different clusters of linked globin genes in *Xenopus*; this apparently reflects the facts that the genomes are actually tetraploid and that the two sets of chromosomes evolved separately. In birds and mammals, the α- and β-globin genes are not linked. It is not known whether their separation was by division of a cluster, such as is found in *Xenopus*, or by independent amplification of individual genes after transposition. The avian and mammalian hemoglobin genes are assumed to have evolved separately starting about 250 to 300 million years ago because the fossil record places the divergence of the organisms at this time.

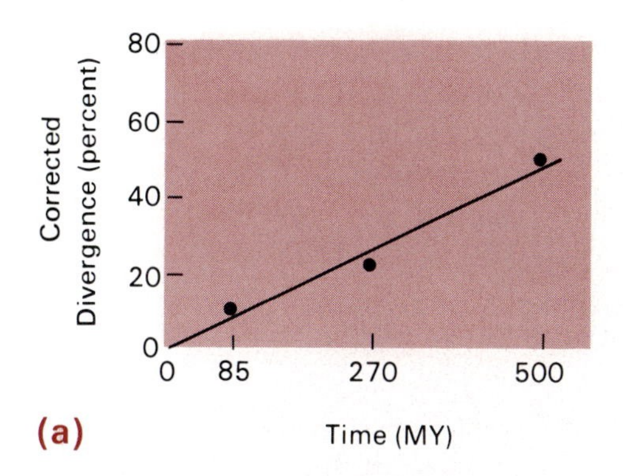

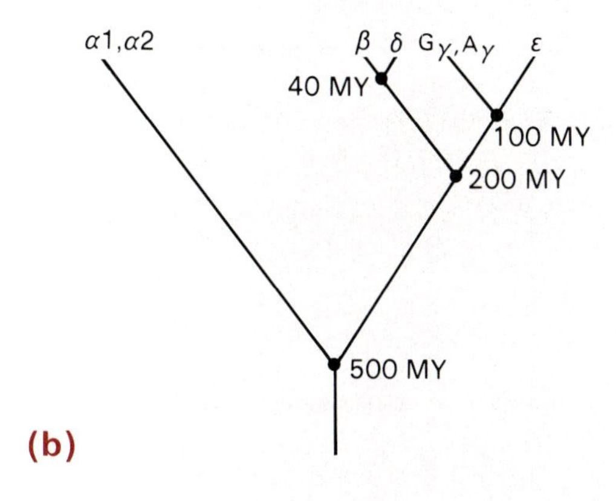

Figure 9.14
The evolution of the globin gene superfamily. (a) Calibration of the evolutionary clock. The nucleotide divergence at sites where a change in base pair leads to a codon for a new amino acid (replacement sites) was calculated for pairs of genes. The average percentage of divergence (ordinate) is plotted against the divergence time as deduced from the fossil record (abscissa). The comparisons used were as follows: (1) 500 million years, α- and β-globin genes in individual species; (2) 270 million years, α- and β-globin genes of chickens with α- and β-globin genes of mammals, respectively; (3) 85 million years, α- and β-globin genes of different mammals, respectively. The slope gives a unit evolutionary period (UEP) of about 10. (b) An evolutionary tree for the human β-globin-like genes. The percentage of divergence at replacement sites of the indicated pairs of globin genes was multiplied by the UEP (10) to give divergence times, and the tree was constructed from these. The branch points represent the time at which pairs of genes began to diverge and is equivalent to either the time of duplication or the time since the last homogenization of the pair. Courtesy of A. Efstratiadis, as published by A. Efstratiadis et al., *Cell* 21 (1980), p. 653. Copyright Cell Press.

The first tandem duplication of the β-gene is likely to have occurred as much as 200 million years ago, well before the evolution of modern mammals began. The regulatory components of two β-genes then diverged so that one was expressed early and the other late in development. Subsequent tandem duplications varied from one mammalian species to another, leading to the present complex arrangements of genes and pseudogenes. In the line leading to the primates, each of the two β-globin genes duplicated again (although not at the same time). This series of events established the cluster of four β-globin-like genes that is the basic arrangement in all primates; in humans, ε and γ are expressed early in development, and δ and β are expressed late. $^{G}\gamma$ and $^{A}\gamma$ derive from a duplication of the γ-genes that is typical of Old World monkeys and may have occurred some 20 to 40 million years ago. Similar types of events can be traced for the α-globin cluster.

The entire nucleotide sequence of the human β-globin gene cluster and large parts of the β-clusters in other species are known. These data permit a comparison of the phylogeny based on the number of genes and the fossil record with that based on the molecular clock hypothesis (Figure 9.14). The basic agreement is reasonable, but the detailed comparison is informative. For example, the time estimated for the divergence of the β- and δ-gene is about 40 million years, which fits well with the time estimated for establishment of a line leading to the New World and Old World monkeys—35 to 40 million years ago. In contrast, comparison of the $^{A}\gamma$ and $^{G}\gamma$ genes indicates that they are identical over the first two exons and diverge 1.2 percent in the third exon. Yet this duplication is typical of all Old World monkeys, and so the sequences should have had at least 6 to 10 million years to diverge at a UEP of 10 (Figure 9.14). The comparison suggests that the first two exons of $^{A}\gamma$ and $^{G}\gamma$ were corrected to uniformity quite recently. Their very recent divergence time estimated from the nucleotide sequence data defines the time of homogenization, not the time of duplication.

The structural similarity of myoglobin and hemoglobin has been known for a long time (Figure 9.15). Determination of the amino acid sequence of leghemoglobin revealed about 15 percent similarity to the vertebrate globins, suggesting that all three proteins might belong to the same family. The related structures and functions stimulated speculation about a common ancestry for the genes encoding these proteins. Before these various genes had been cloned and sequenced, convergent evolution seemed a likely alternative possibility, but the sequence data strongly confirm the idea of common ancestry. Not only do the similarities among the genes include the encoded polypeptides and the size of the coding regions but, most notably, the placement of the introns.

All the vertebrate globin and myoglobin genes have two introns at highly conserved positions (Figure 9.16). The first is at around codon 30

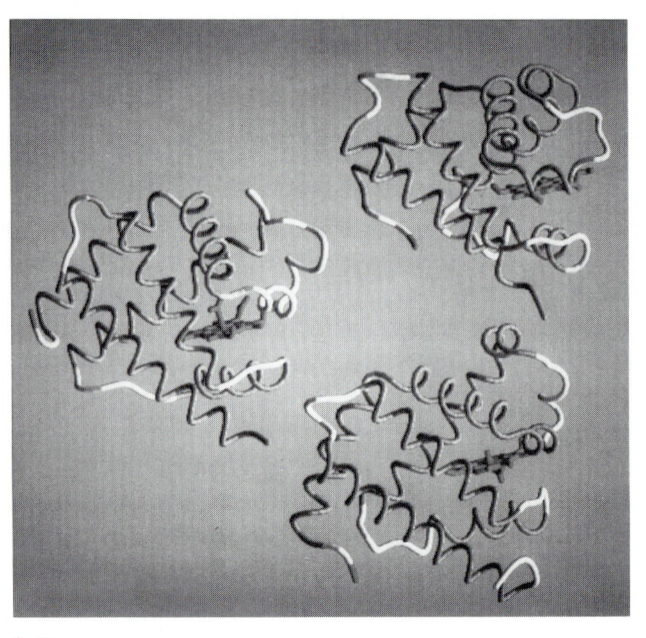

(a)

```
myoglobin         V — L S E G E W Q L V L H V W A K V E A D V A G H G Q D I L I R L F K S H P E T L E K F D R F K H L K T E A
α-globin          V — L S P A D K T N V K A A W G K V G A H A G E Y G A E A L E R M F L S F P T T K T Y F P H F — D L S H — —
β-globin          V H L T P E E K S A V T A L W G K V — — N V D E V G G E A L G R L L V V Y P W T Q R F F E S F G D L S T P D
leghemoglobin-a   V A F T E K Q D A L V S S S F E A F K A N I P Q Y S V V F Y T S I L E K A P A A K D L F S F L A N G V D P —

myoglobin         E M K A S E D L K K H G V T V L — T A L G A I L K K K G H — H E A E L K P L A Q S H A T K H K I P I K Y L E
α-globin          — — — G S A Q V K G H G K K V A — D A L T N A V A H V D D — M P N A L S A L S D L H A H K L R V D P V N F K
β-globin          A V M G N P K V K A H G K K V L — G A F S D G L A H L D N — L K G T F A T L S E L H C D K L H V D P E N F R
leghemoglobin-a   — — — T N P K L T G H A E K L F A L V R D S A G Q L K A S G T V V A D A A L G S V H A Q K A V T D P Q — F V

myoglobin         F I S E A I I H V L H S R H P G N F G A D A Q G A M N K A L E L F R K D I A A K Y K E L G Y Q G-153
α-globin          L L S H C L L V T L A A H L P A E F T P A V H A S L D K F L A S V S T V L T S K Y R-141
β-globin          L L G N V L V C V L A H H F G K E F T P P V Q A A Y Q K V V A G V A N A L A H K Y H-146
leghemoglobin-a   V V K E A L L K T I K A A V G D K W S D E L S R A W E V A Y D E L A A A I K — K A-144
```

(b)

Figure 9.15
Globin structures. (a) Computer-generated three-dimensional structures of leghemoglobin (left), β subunit of hemoglobin (upper right), and myoglobin (lower right). Courtesy of Richard Feldman. The heme moieties are seen almost perpendicular to the page. (b) The amino acid sequences of myoglobin (sperm whale), α- and β-globin (human), and leghemoglobin a (soybean).The single-letter amino acid code is used (Figure 1.24), and a dash (–) means that no amino acid occurs. Amino acids that are identical in all four polypeptides are boxed. Dotted boxes indicate nonidentical but similar amino acids (conservative changes) such as serine/threonine or valine/leucine/isoleucine. The amino acids that are identical in α- and β-globin are shaded. The total number of amino acids in each sequence is given at the end.

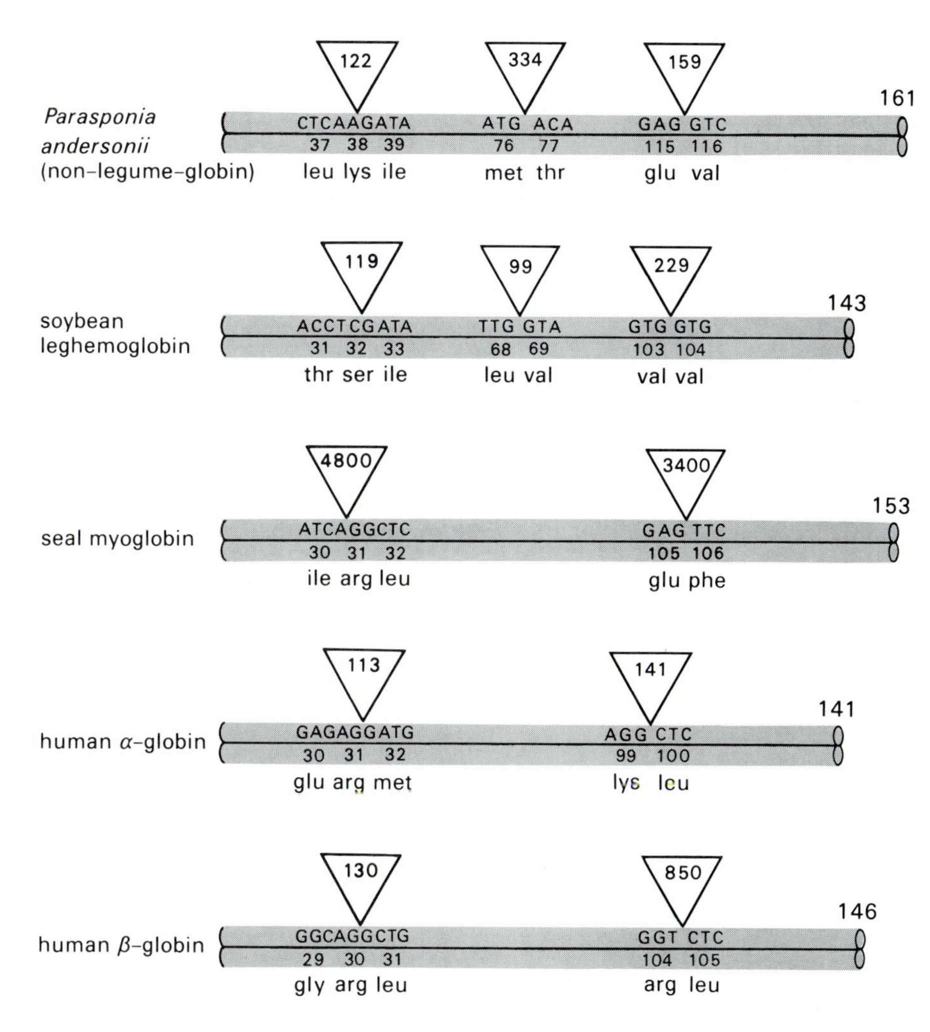

Figure 9.16
The placement of introns in genes of the globin superfamily. The number of amino acids (codons) in each protein is shown at the right. The ATG initiation codon is not counted because the methionine is cleaved off the mature polypeptide. Introns are triangles with the size in bp inside. The codon numbers and corresponding amino acids surrounding the introns are shown (refer to Figure 9.15b). Note that the first intron always interrupts a codon, and the other introns are between codons.

from the NH_2 terminus and always interrupts a codon. The second intron is between two codons at about codon 100. Two of the three introns in the plant leghemoglobins are at virtually the same positions as those of the vertebrate globin genes. A third is at about the middle of the coding region.

Analysis of intron distribution in the globin genes strongly supported the idea that individual exons encode distinctive structural and functional polypeptide domains. The three-dimensional structures of the vertebrate globin subunits show several compact regions that are relatively distant from one another. Two of these compact regions were recognized as corresponding to exons 1 and 3 (residues 1–30 and 105–146 in the β-globin chain in Figure 9.15). But another pair of distant but compact regions (residues 40–60 and 70–100) did not correspond to two exons. If the relation between exons and structural polypeptide domains was real, then, it was argued, an intron ought to occur between these regions. Soon after this hypothesis was made, the structure of a soybean leghemoglobin gene was determined, and the extra intron was demonstrated at the predicted position. The common ancestor of all contemporary globin family genes most likely contained at least three introns. The domain-exon relation can also be perceived in functional aspects of the globins. For example, the region between about amino acids 65 and 96 binds heme and also contains residues involved in forming an $\alpha\beta$ globin dimer. Similarly, residues required for the formation of the $\alpha_2\beta_2$ tetramer are clustered in exon 3.

Discerning if the ancestral gene existed prior to the formation of separate plant and animal lineages is difficult. Only a limited number of dicotyledonous plants, those that fix N_2 in symbiotic association with microorganisms like *Rhizobium*, have leghemoglobin genes; such plants include legumes and nonlegumes. These species may have acquired globin genes by horizontal transfer from another genome (e.g., viral infection) rather than by a vertical evolutionary line. In plants, as in animals, an ancestral gene multiplied to form clustered multigene families that include functional genes and pseudogenes (Figure 9.12). In animals, a duplication that evolved into the paralogous genes for myoglobin and hemoglobin is estimated to have occurred more than 500 million years ago because the cyclostome fish (e.g., lamprey) have both genes.

The α-amylase family: two genes from one coding sequence Different strains of mice contain varying numbers of functional genes for pancreatic α-amylase. In some mouse strains, two essentially identical, nonallelic genes encode mouse pancreatic amylase, and a third gene is expressed in salivary glands and liver (Table 9.7). The pancreatic and salivary amylase proteins differ in 12 percent of their amino acids, and all three genes have similar 5′ but not 3′ flanking regions. The introns diverge markedly in sequence and length. All three genes are located on mouse chromosome 3, and chromosome "walking" experiments showed that the salivary amylase and one of the pancreatic amylase genes are only 22 kbp apart. Mouse DNA contains additional segments that are homologous to α-amylase coding regions, and at least one of these is a pseudogene that is linked to the cluster on chromosome 3.

The unique mouse α-amylase coding region expressed in salivary glands and liver is preceded by two different transcriptional control regions that are spaced almost 3 kbp apart (Figure 9.17). There is a relatively low and equal level of expression from the control unit closest to the coding sequence in both liver and parotid gland. However, in the adult parotid gland, the upstream unit is activated and produces about 30 times as many transcripts as does the downstream one. This accounts for the high level of α-amylase that is typical of parotid glands. Each control region is associated with its own TATA signal. The two primary transcripts produced have a different 5′ exon, each of which is later spliced to the same exon, the one that contains the AUG initiation codon. Thus, the 5′ leaders of the two resulting mRNAs differ. The exon that includes the transcription initiation signal for low level expression is part of the large intron that is spliced out from the parotid-specific mRNA.

Rat α-amylase genes are dissimilar. Distinct genes encoding different polypeptides are expressed in parotid glands, pancreas, and liver, and at least four different genes may encode pancreatic amylase. Pseudogenes or processed genes also occur. Like the mouse *Amy-1* gene, a rat gene expressed in liver and salivary glands has two independent promoters.

Interferon genes Cells from many vertebrates secrete polypeptides known as interferons (IFN) in response to a variety of external stimuli (Table 9.10). Viral infections and double strand RNAs, for example, elicit a set

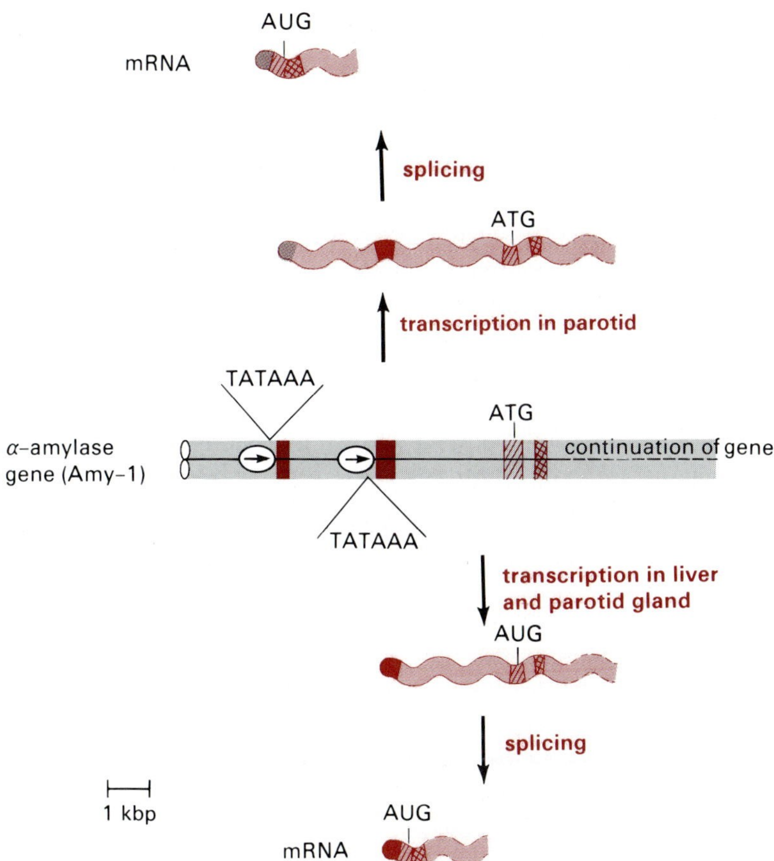

Figure 9.17
A mouse α-amylase coding sequence (*Amy*-1) has two different promoters. One yields low levels of transcripts in liver and parotid glands. The other specifically promotes high levels of mRNA in parotid glands. After U. Schibler et al., *Cell* 33 (1983), p. 501.

of IFN-αs in leukocytes and a related set of IFN-βs in fibroblasts. Within each set, the various proteins have very similar structures. Quite a different protein, the single IFN-γ (or immune interferon), is produced by lymphocytes upon stimulation of DNA replication by mitogens. All the interferons elicit in turn a range of cellular responses, including the inhibition of viral infection, the modulation of immune responses, and antitumor activity.

Table 9.10 Human Interferon Genes

	IFN-α	IFN-β	IFN-γ
Name	Leukocyte	Fibroblast	Immune
Number of nonallelic genes	≥15	≥2	1
Pseudo-(or processed) gene	>3	?	?
Chromosome location	9p	9p; 2; others	12 (q24.1)
Introns	None	None	3
Homology with IFN-α genes (%)	85–95	50	Little
Homology with IFN-α polypeptides (%)	80	30	<10
Size of protein (# amino acids)	155–166	166	146

Note: The protein sizes are those of the IFNs themselves. The corresponding genes also encode 20–23 amino acid long signal peptides that are removed during translation.

The potential importance of the interferons as therapeutic agents for viral diseases and cancer has provided a strong impetus for studying them. Initially, progress was limited by the very small amounts of the proteins produced by stimulated cells. Recombinant DNA techniques circumvented these problems because the polypeptides can be produced in good yield in *E. coli* using cloned cDNAs. However, the cloning itself was not simple because the mRNAs, like the polypeptides, are not abundant. The mRNA for IFN-γ, for example, is an estimated 0.04 percent of the polyadenylated RNA of stimulated lymphocytes. Special approaches are required to clone cDNAs for such rare messengers. In this case, crude RNA was first fractionated by size and the fractions tested for their ability to be translated into IFN-γ after injection into *Xenopus laevis* oocytes. The active fraction was then converted to double strand cDNA for cloning in a plasmid vector. The resulting colonies were subjected to differential screening (Section 6.5b). IFN-γ coding sequences were found in colonies that annealed with cDNA probes from stimulated lymphocytes but not from normal lymphocytes. The cDNA clones were then used as probes to isolate the *IFN-*γ gene. This type of approach led to the isolation of full length cDNA clones for the various interferons and, by inference, to the amino acid sequence of the polypeptides. Moreover, it also gave a detailed picture of the organization of an interesting multigene family.

Among the cloned human DNA segments containing *IFN-*α genes, some contain more than one gene, and others represent overlapping genomic regions. This information, plus the identification of a single chromosomal locus for *IFN-*α by *in situ* hybridization, showed that the genes encoding the various *IFN-*αs are closely clustered. Genetic analysis indicates that *IFN-*β genes are near neighbors of the *IFN-*αs on human chromosome 9. The single *IFN-*γ gene is on another chromosome. Neighboring genes have similar structures. For example, the *IFN-*αs and *IFN-*β on chromosome 9 are unusual in having no introns and are similar both in coding and 5′ flanking sequence. In contrast, the *IFN-*γ gene on chromosome 12, which has very little similarity to the *IFN-*αs or *IFN-*β at either the gene or polypeptide level, has three introns. The interferon gene families in other mammals have similar general properties.

c. *Histone Genes: Conservation of Coding Sequences and Divergence of Organization*

General properties of histone genes The primary structures of the histones are highly conserved among widely different eukaryotes. This is not surprising, given their fundamental and common role in chromatin structure (Section 1.1g). Still, they do vary. The sequences of H1 histones tend to diverge the most and those of H3 and H4 the least. Within a single species, slightly different histone molecules may be synthesized at different stages of development or of the cell cycle or in different tissues. For example, expression of most histone genes correlates with the S-phase of the cell cycle and thus with DNA replication. Expression of others, called replacement type genes, occurs at a low level throughout the cell cycle. In contrast to the conservation of the polypeptide sequences, the numbers and arrangements of histone genes differ remarkably among species (Figure

species	arrangement of histone genes	approximate cluster length kbp	approximate copy number
yeast *S. cerevisiae*	H2A · H2B H3 · H4	6 and 13	2 2
Drosophila	H3 · H4 · H2A · H2B · H1	5	100
sea urchin *S. purpuratus*	H1 · H4 · H2B · H3 · H2A scattered, some in pairs	6–7	500 (early) 10 (late)
newt *N. viridescens*	H4 · H2A · H2B · H3 · H1	9	700
Xenopus	H3 · H4 · H2A · H2B · H1B · H3 · H4 H3 · H4 · H2A · H2B	16 6	25
chicken	H3 · H2A · H4 · H1 · H2A · H2B H2A · H4 · H2A · H2B H3 · H4 H5 (expressed in erythrocytes in place of H1)	14 >10 >10	10 1
human	H3 · H4 · H4 · H3 · H2A · H2B H4 · H3 · H1 · H2B · H2A	20 15	10–20 5
rice	H2A · H2B · H4	6	unknown

Figure 9.18
Typical examples of linked groups of histone genes in various species. Not all groups within a species are necessarily the same. In humans, and perhaps in some other species, some copies are pseudogenes. Where known, the direction of transcription is indicated by an arrow above the gene. After C. C. Hentschel and M. L. Birnstiel, *Cell* 25 (1981), p. 301.

9.18). This emphasizes, as does the organization of tRNA genes, that conservation of coding sequences need not be associated with conservation of copy number or genomic organization.

Genes for the different histones tend to be linked. But unlike the rRNA genes, three of which occur in a single transcription unit, each histone gene is transcribed as a separate unit, often from opposite DNA strands. Also in contrast to rRNA genes, the order of the linked histone genes is not necessarily the same from one species to another or even within a single species. Moreover, although linked groups of histone genes are repeated in tandem arrays in some organisms, they are dispersed in birds and mammals. Most replication type histone genes lack introns, and their mRNAs lack 3′ polyA tails.

Two experimental approaches allowed certain basic observations to be made about histone gene organization before recombinant DNA techniques were available. First, the separation of histone genes as satellite bands after density gradient centrifugation of total DNA established their high copy number and tandem organization in sea urchins. Second, histone mRNA was partly purified from an enriched source.

Cloning histone genes In the first ten hours or so after fertilization, sea urchin eggs undergo 2×10^9 to 2×10^{10} cell divisions with a concomitant rapid DNA replication and histone synthesis. During this period, about 30 percent of the proteins synthesized are nuclear proteins, and the histone mRNAs are abundant. Because of their characteristic size, they are enriched in the fraction of mRNA that sediments around 9S. In addition, electrophoresis on polyacrylamide gels separates the mRNAs for the indi-

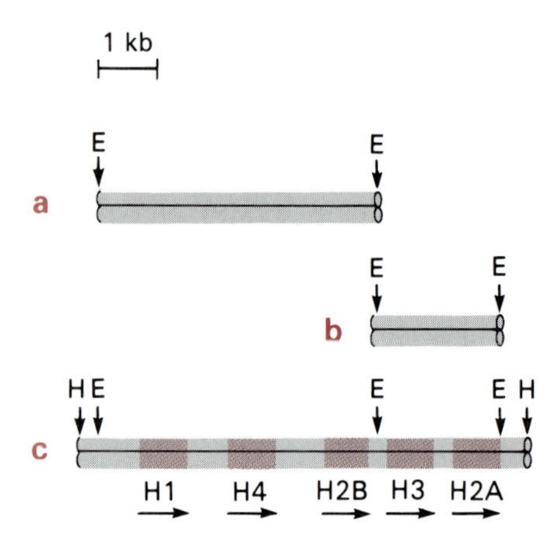

Figure 9.19
A linked group of five sea urchin early histone genes (*S. purpuratus*). Two cloned fragments (a and b) that anneal to total histone mRNA but not to each other were isolated from an *Eco*RI genomic library. Both a and b anneal to a single 7 kbp fragment generated by cleavage of genomic DNA with endonuclease *Hind*III (c). After double digestion of genomic DNA with endonuclease *Eco*RI plus *Hind*III, no 7 kbp fragment annealed with either a or b. Instead, two fragments the size of a and b, respectively, annealed with a mixture of a and b. Thus, the two fragments are linked within a 7 kbp segment of genomic DNA.

vidual histones, which can then be identified by translation *in vitro* into the corresponding polypeptides. The sea urchin mRNAs can also be used as probes for cloning histone genes in plants, fungi, and animals.

Histone genes in sea urchins A special set of genes is responsible for the rapid synthesis of histones in very early sea urchin embryos. After blastulation, expression of a different set of histone genes is stimulated and is responsible for the bulk of histone synthesis throughout the remainder of development and adulthood. The two sets are called the early and late sea urchin histone genes. Their organization varies from one species of sea urchin to another, and that found in *Strongylocentrotus purpuratus* is used here as an example.

Genomic segments containing the early histone genes were cloned from a library of *Eco*RI endonuclease fragments using total histone mRNA as a probe (Figure 9.19). Two cloned fragments that did not cross-hybridize to each other were obtained. Genomic mapping indicated, however, that the two are linked within 7 kbp of one another on the genome. The presence and order of all five histone genes in the two clones was established by annealing their DNA with the electrophoretically separated mRNAs and analyzing the heteroduplex products by electron microscopy. The individual mRNAs were also annealed to the separated strands of the cloned segments; all the RNAs hybridized to only one of the two strands, showing that each of the five genes was transcribed in the same direction. Later, nucleotide sequence analysis confirmed all these conclusions and established the precise sizes of the coding regions and the spacers between them. The several hundred copies of the linked group of early histone genes occur in long tandem arrays. Similar arrangements are found in several different genera of sea urchin.

The late histone genes are repeated only about ten times, in marked contrast to the 500-fold repetition of the early genes. Besides the decreased copy number, the "late" genes differ from the "early" ones in several ways. For one thing, they are neither linked in groups of five nor arranged in tandem repeats. Some pairs of genes are neighbors (e.g., H3 and H4), residing within 1 kbp of each other, but others are "loners," at least 6 kbp away from any other histone gene. Neither the order of the genes nor their transcriptional direction has any relation to the arrangement of the early genes. Moreover, the organization of the "late" genes is variable from one genus of sea urchins to another.

The cloned genes for cognate "early" and "late" histones do not cross-hybridize when stringent annealing conditions are used, even though their gene products are very similar (or even identical) in structure. The differences in the coding sequences reflect the use of alternate codons for the same amino acid. The sequences flanking cognate "early" and "late" genes are even more divergent, except for the maintenance of those oligonucleotide sequences essential for transcription by RNA polymerase II.

Histone genes in other species With *Drosophila, in situ* hybridization of cloned genes to polytene chromosomes located all the histone genes at one region of chromosome 2. The five different genes are linked within a unit that is repeated about 100 times; the gene order and transcriptional directions are completely different from those of the sea urchin genes (Figure 9.18). The switch from high to low levels of histone synthesis during *Drosophila* embryogenesis is apparently not associated with the expression of alternate genes, as it is in sea urchins, but rather involves the modulation of expression from the same gene set.

In yeast (*S. cerevisiae*) DNA, there are two copies of a segment containing one gene for H2A and one for H2B and two copies of a segment containing one gene for H3 and one for H4 (Figure 9.18). Homologous genes within the pairs of clusters may encode proteins with slightly different (i.e., the H2As) or identical (i.e., the H3s) amino acid sequences. It is not known whether the variations have any functional significance. No H1 histone (or gene) has been identified in yeast.

Vertebrates, like invertebrates, have a diversity of histone gene arrangements (Figure 9.18). In *Xenopus*, for example, several linked groups of the five different histone genes occur, but they have different gene orders. At least two of the groups are distinguished by variant H1 genes. Each of the two H1 genes hybridizes to a distinctive oocyte mRNA, and the two messengers yield different H1 proteins upon translation, *in vitro*. The *Xenopus* genome also has several tandem arrays of groups of histone genes, but unlike the case of the "early" histone genes of sea urchins, neighboring groups within a tandem array show restriction endonuclease site polymorphism.

The location of the approximately 700 histone genes in the newt *Notophthalmus viridescens* emphasizes the irrelevance of position to at least some aspects of gene expression. DNA segments containing genes for the five histones are 9 kbp long, and each segment is embedded within a long array of tandem repeats of an unrelated sequence on a chromosome arm. Typically, there are 50 to 100 kbp of the unrelated tandem repeats between each group of histone genes. The same tandemly repeated sequence appears also in centromeres (Section 9.4c).

Warm-blooded animals have relatively few histone genes. They are grouped neither uniformly together nor in tandem arrays, but neighboring genes are frequently found.

9.4 Tandem Repetition of DNA Sequences: A Common Characteristic of Eukaryotic Genomes

One of the most striking lessons learned from studying genome organization is how frequently related DNA sequences are arranged in tandem. Sometimes entire linked groups of identical genes are repeated in tandem, as with rRNA and histone genes. In other instances, related but distinctive genes appear in tandem, as in the globin, ovalbumin, and human growth hormone multigene families discussed earlier. Tandem repetition is not limited to complete genes, however; it is evident even within coding regions and in the noncoding areas of DNA.

a. *Genes Constructed of Tandem Repeats of DNA Segments*

The internal structures of some coding regions look as though they were constructed from tandem, though divergent, repeats. Thus, amplification mechanisms appear to have contributed, during evolution, to the construction of genes. Four examples indicate the widespread occurrence of internally repetitive coding sequences.

Serum albumin and α-fetoprotein genes The predominant protein in mammalian fetal serum is α-fetoprotein. After birth, serum albumin takes over (Section 8.3e). Both proteins are synthesized in the liver; each is transcribed from a single gene. In mice, the two genes are linked in tandem, 15 kbp apart, on chromosome 5, where the albumin gene precedes the α-fetoprotein gene, in the direction of transcription. Although the genes are so divergent that they do not cross-hybridize, their structures are similar enough to indicate their origin as tandem reiterations that have diverged in time. For example, the positions of the 14 intervening sequences and lengths of the 15 exons are closely similar in both genes (the structure of the α-fetoprotein gene is shown in Figure 9.20a). In humans, two similar genes and a third member of the family, the gene for the vitamin D–binding plasma protein, group-specific component (Gc), are linked on the long arm of chromosome 4.

The two proteins are immunologically cross-reactive and match in about 35 percent of their approximately 580 amino acids. Each of them comprises three similarly folded domains. Figure 9.20b shows this structure and the disulfide bonds that hold it together. The striking thing about this structure is that the three domains themselves are very similar to one another both in the placement of the cysteine residues and in the numbers of amino acids between neighboring cysteines. Each protein molecule appears to be constructed from three repetitions of related polypeptides. Analysis of the mouse α-fetoprotein gene confirms this relation. In Figure 9.20a, the DNA segments corresponding to the three protein domains are aligned. The number and position of the intervening sequences within each domain are nearly identical. Moreover, the coding sequences in domain I are 39 and 35 percent homologous to those in domains II and III, respectively. Thus, the gene represents three tandem copies of a DNA segment. The three repeats have diverged over time, but their common ancestry is clear in the contemporary structure. The albumin genes of *X. laevis* have a similarly repetitive internal structure, suggesting a common precursor to the amphibian and mammalian genes. However, there is no α-fetoprotein in *Xenopus*. Instead, independent reiterations gave rise to additional albumin genes.

The gene encoding the receptor for low density lipoprotein The principle form of circulating cholesterol in vertebrates is a low density lipoprotein, LDL. LDL itself is composed of phospholipid membrane, a core of protein called apoprotein B, and an interior of cholesterol esters. LDL enters cells by interaction with a specific receptor that occurs on the plasma membrane of many cell types, followed by receptor mediated endocytosis. LDL and LDL receptors dissociate in the endosomes. The receptor is recycled to the plasma membrane, and the LDL is directed to lysosomes, where the apoprotein B is degraded. Freed cholesterol is used by the cells for new membrane synthesis. In addition, the intracellular cholesterol regulates the rate of cholesterol biosynthesis by suppressing transcription of the gene encoding 3-hydroxy-3-methylglutaryl coenzyme A reductase, the first enzyme unique to the synthetic pathway. Thus, the LDL receptor serves

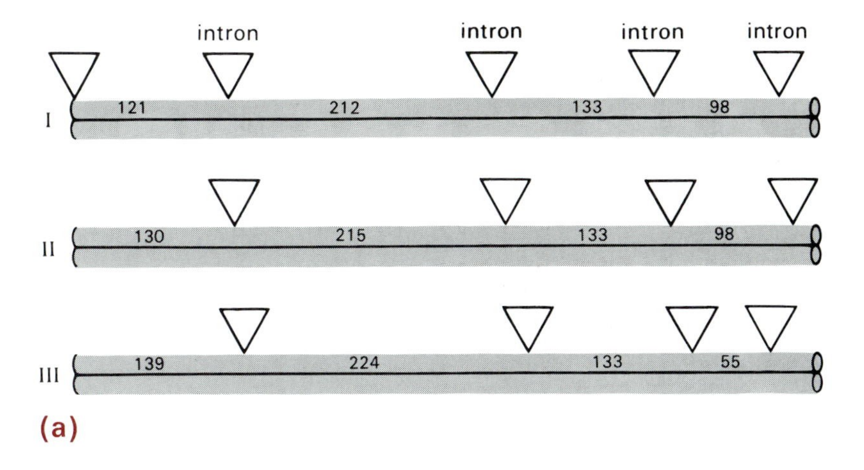

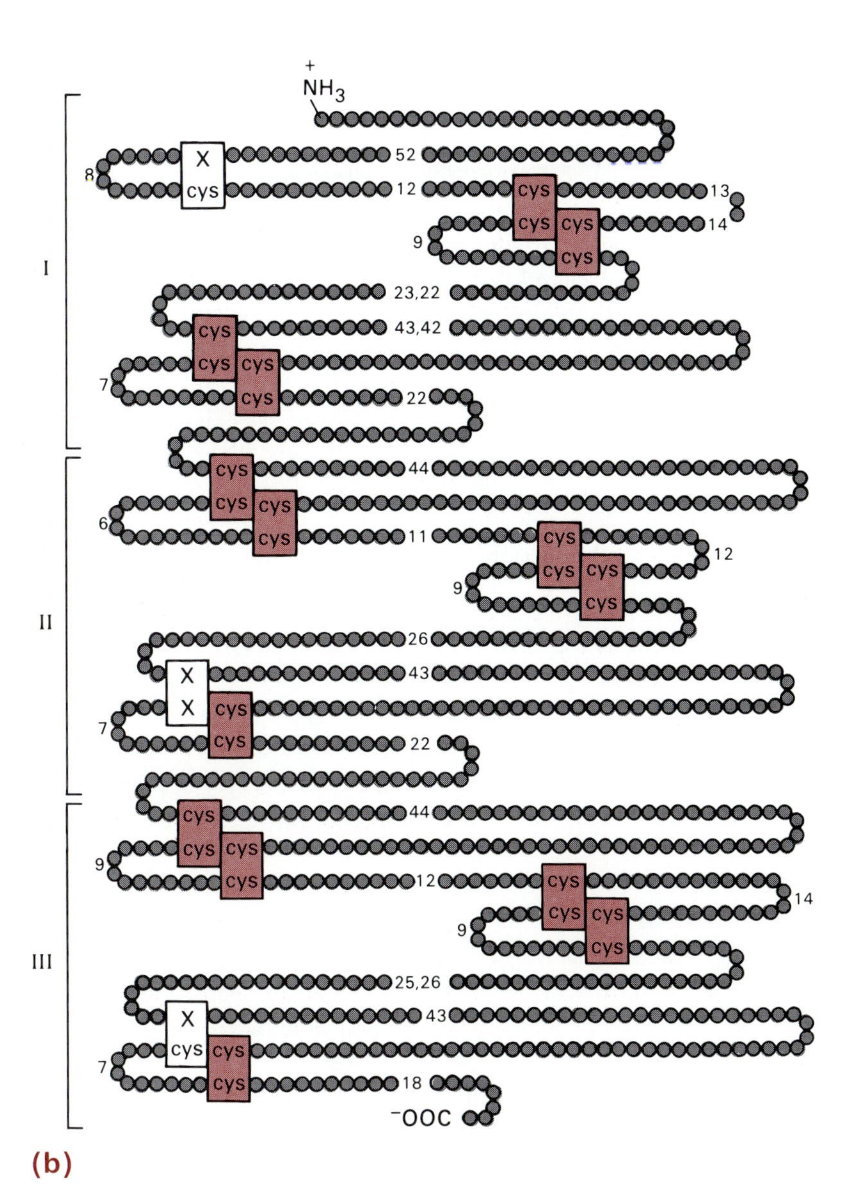

Figure 9.20

(a) The positions of intervening sequences in the mouse α-fetoprotein gene are shown by inverted triangles. Numbers indicate exon size (bp). The entire gene is displayed on three lines: I, II, and III (see text). A fourteenth intron is in the 5′ leader region (not shown). (b) Diagram showing the overall similarity of mouse α-fetoprotein and human serum albumin. The three similarly folded domains in each protein are labeled I, II, and III, corresponding to the three segments of the gene shown in (a). The Xs mark positions where a cysteine occurs in serum albumin but not in α-fetoprotein. The numbers show how many amino acids occur between cysteine residues. Where two numbers are given, the two proteins have different numbers of amino acids between cysteines. Note that the small circles do not show the number of amino acids, but are a schematic representation of the polypeptide. After M. B. Gorin et al., *J. Biol. Chem.* 256 (1981), p. 1954.

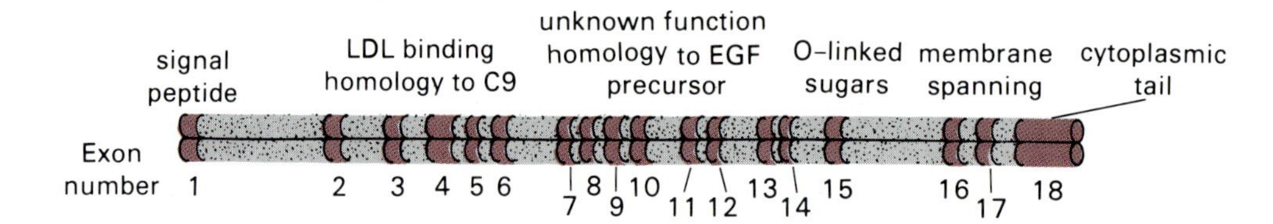

Figure 9.21
A schematic diagram of the human LDL receptor gene showing the 18 exons and the functional domains that they encode. C9 is the protein complement factor C9, and EGF is epidermal growth factor. After M. Brown and J. Goldstein, *Science* 232 (1986), p. 34.

two functions: it is essential for regulating the rate of cholesterol formation and for the clearance of cholesterol from plasma. Defects in LDL receptor function cause the disease familial hypercholesterolemia (Section 9.5d).

The LDL receptor itself is a complex protein with 839 amino acids organized into several domains (Figure 9.21): a domain responsible for LDL binding, which extends outside the cell; another external domain of unknown function; a small, highly glycosylated domain that is located just outside the plasma membrane; a small hydrophobic domain that spans the plasma membrane; and a domain that extends into the cytoplasm and is essential for proper endocytosis.

The single human LDL receptor gene has 18 exons, which correlate with the several protein domains (Figure 9.21). Exon 1 encodes a signal sequence that is removed in the lumen of the endoplasmic reticulum. Exons 2 through 6 encode the LDL binding domain and include seven tandem repeats of a sequence that encodes a 40 amino acid long cysteine-rich peptide. The repeats are divergent, but clearly have the same ancestor. Moreover, the seven related polypeptide sequences are also homologous to a 40 amino acid long unit that is repeated in one of the proteins of the serum complement complex—complement factor C9. Thus, both exon shuffling and exon repetition were involved in the formation of the LDL binding domain of the receptor. These two processes also were important in the formation of the coding sequences (exons 7 through 14) for the domain of unknown function in the middle of the protein. Exons 7, 8, and 14 each encode 40 residue long polypeptide segments that are similar to sequences repeated in another protein, epidermal growth factor.

Drosophila *"glue" protein gene* During the larval stage, *Drosophila* salivary glands excrete a group of proteins that glue the pupae to external surfaces. The mRNA for one of these proteins, called sgs 4, was cloned as a cDNA and used as a probe to isolate the *Sgs*4 gene from a *Drosophila* library. Near the 5′ end of the *Sgs*4 coding region, there are 19 direct tandem repeats of a sequence 21 base pairs long that encodes 7 amino acids (Figure 9.22). Altogether, the tandem array accounts for almost half the length of the coding sequence. The repeats are not perfect, and thus the repeating heptapeptide unit also diverges. In several strains of *Drosophila*, the lengths of the *Sgs*4 gene differ by multiples of the 21 base pair repeat. One strain has 19 repeats; others have 22 and 27. Such changes can be readily explained by unequal homologous crossing-over.

The collagen gene Collagen is the major extracellular stuctural protein in vertebrates. Although various different collagens exist, they are all characterized by a triple helix formed of three polypeptide chains that

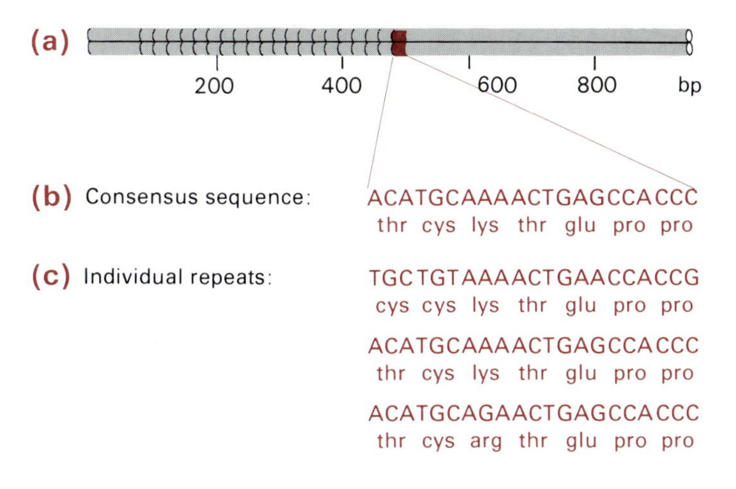

Figure 9.22
(a) Structure of the *Sgs4* gene in a *Drosophila* strain with 19 tandem repeats of the 21 bp long unit. The consensus sequence of the repeat unit (b) as well as several individual repeats (c) are shown. After M. A. T. Muskavetch and D. S. Hogness, *Cell* 29 (1982), p. 1041.

are not necessarily all identical (Figure 1.30). The collagen polypeptides are unique in having a glycine every third amino acid (gly-X-Y); they are also rich in proline and lysine residues, many of which are hydroxylated (in a posttranslational modification).

Genes for the α2(I) type of collagen polypeptide are similar in chicken, mouse, and human (Figure 9.23a). Electron microscopic views of collagen mRNA annealed to cloned regions of a 38 kbp long chicken gene revealed a remarkable organization. The gene comprises at least 52 exons. The intervening sequences vary in length from about 80 bp to more than 2 kbp. However, all the exon lengths are multiples of 9 base pairs, and most of them are 54 or 108 bp long (Figure 9.23b). The structure can be seen as built from tandem amplifications of a 9 bp unit, thereby explaining the gly-X-Y repeating tripeptide of the polypeptide chain.

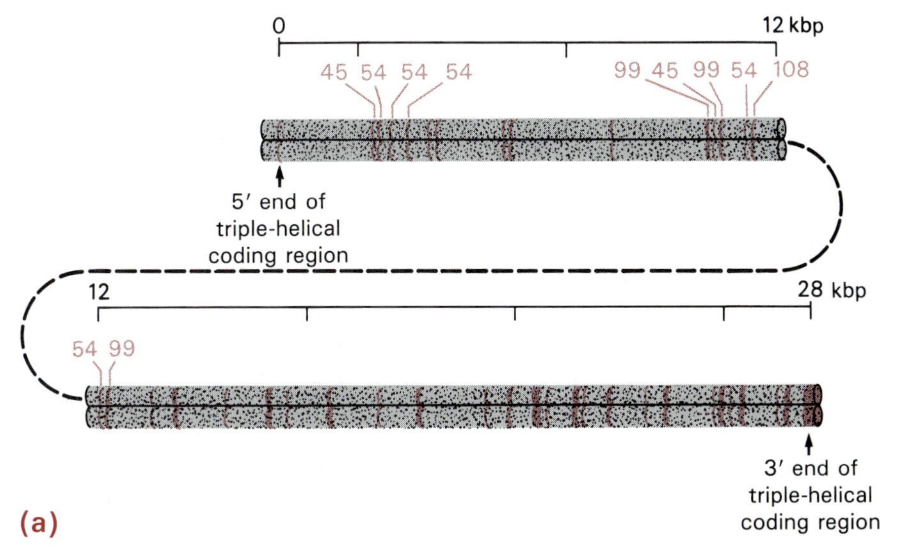

a typical exon	GGT	GAA	CCA	GGC	CCT	GCT	GGT	GCC	AAG	GGA	GAA	AGT	GGT	AAC	AAG	GGT	GAG	CCT
	gly	glu	pro	gly	pro	ala	gly	ala	lys	gly	glu	ser	gly	asn	lys	gly	glu	pro

(b)

Figure 9.23
(a) Structure of the coding region for the triple-helix portion of a chicken α2(I) collagen gene. The region is shown in two segments. Exons are dark bars, and introns are stippled. The size of some exons (in base pairs) is shown. After J. Wozney et al., *Nature* 294 (1981), p. 129. (b) The structure of one exon and the encoded polypeptide.

b. Tandem Repetitions Outside of Coding Regions

Tandem repetitions turn up often between genes and in introns. The length of the repeated units varies from two nucleotides up. A few examples are listed in Table 9.11. Although such repeats are ubiquitous and represent many different sequences, such as $(CCG)_n$, $(AT)_n$, and $(TAACCC)_n$, we cannot at present ascribe any function to most of them. Among the simpler repeated units, one has attracted considerable attention.

DNA probes composed of alternating CAs (5′-CACA . . .) (or GTs) anneal with many restriction endonuclease bands in the DNA of all eukaryotes. The sequence is thus repeated many times in most genomes. The ability of such arrays to form Z-DNA may contribute to their functional role.

Associations with polymorphisms As a general matter, tandem repetitions, wherever they appear, seem to be associated with intraspecies polymorphisms. At present, unequal crossing-over between related nonallelic repeat units or slippage during replication (Section 10.1c) seem likely explanations for these alterations. The spacer regions between oocyte 5S rRNA genes in *X. laevis* and between tandem rRNA transcription units usually contain multiple internal tandem repeats (Section 9.2). These are associated with size polymorphism of the spacer regions. The size variation is accounted for by the gain or loss of one or more complete repeat units in the same way that the copy number of the tandem repetition in the *Drosophila Sgs4* gene changes from one strain to another.

Expression of immunoglobulin genes requires substantial rearrangements of DNA (Section 10.6c). The recombinations that form active heavy chain genes take place within special noncoding sequences called "switch" regions that contain multiple tandem repeats of the sequence 5′-

Table 9.11 Some Tandem Repetitions That Have Been Observed Interspersed in a Variety of Eukaryotic Genomes

Repeated Unit	Known in	Examples of Locations
5′-CAAAGTTTGAGTTTT	*Xenopus laevis*	Spacer in oocyte 5S RNA genes
5′-ACAGGGGTGTGGGG	Human	Upstream of insulin gene
5′-CA	Many eukaryotes	Many genomic positions and yeast telomeres
5′-GA	Many eukaryotes	Spacer between H2A and H1 histone genes in sea urchin
5′-TCTCC	Chicken	Intron of X gene in ovalbumin family
5′-GGAAG	Chicken	2.5 kbp 5′ of conalbumin gene
5′-GGAAA	Pheasant	2.5 kbp 5′ of conalbumin gene
5′-AT	Many eukaryotes	Many locations

(a) (AGGGCTGGAGGAGGGCTGGAGGAGGGCTGGAGG)$_{18}$

(b) (AGAGGTGGGCAGGTGG)$_{29}$

(c) (GGGAGG$\overset{\mathrm{C}}{\underset{\mathrm{T}}{\text{or}}}$GGGCAGGAGG)$_{14}$

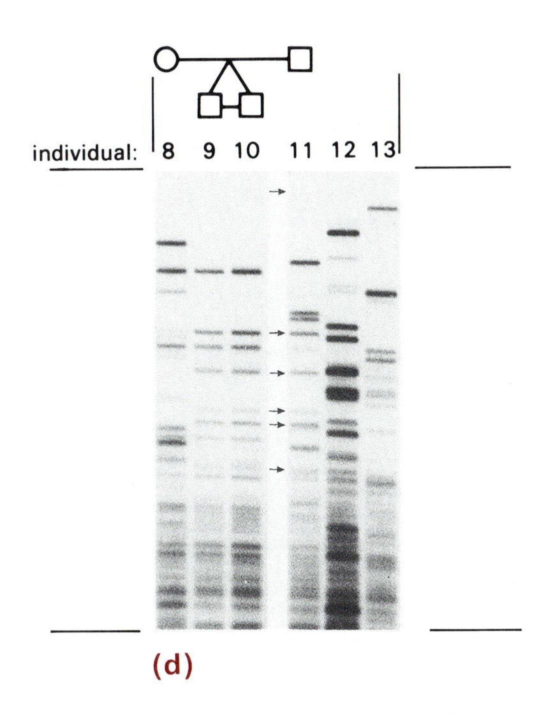

Figure 9.24
Probes of cloned minisatellite sequences detect polymorphic patterns of restriction endonuclease fragments in human DNA. The sequences of three related human minisatellites (a, b, c) are shown at the top. (d) A DNA blot obtained by annealing a cloned minisatellite b probe to *Hinf*I digested DNA. Four DNAs are from a single family: mother, father, and identical twin offspring. A circle indicates a female, a square, a male. The last two lanes are DNA from two unrelated males. The paternal bands in the twins are marked with arrows. After A. Jeffreys et al., *Nature* 316 (1985), p. 76. Photo courtesy of A. Jeffreys.

(GAGCT)$_n$GGGGT-3′, where n equals from one to five. This genomic region too is polymorphic.

About 500 bp upstream from the 5′ end of the human insulin gene, there is a tandem array of variants of the sequence 5′-ACAGGGGTGTGGGG-3′. The segment does not occur elsewhere in the genome. This tandem array is the site of extensive polymorphism; many different alleles have been noted, and the majority of individuals are heterozygous. The polymorphism was observed initially as variant lengths of restriction endonuclease fragments that hybridized with a cloned probe from this region. It is caused by different numbers of repeat units, from 26 to 200 or so, as well as by the occasional alterations in the repeat sequence itself. The function of the sequence is unknown; it is probably unrelated to its position near the insulin gene because not a single copy of the sequence is found in the analogous neighborhood near a rat insulin gene.

Minisatellites The tendency of tandem repeats to be polymorphic is dramatically illustrated by so-called **minisatellite** sequences. Several versions of such relatively simple tandem repeats are dispersed in mammalian genomes (Figure 9.24). When cloned probes containing a minisatellite segment are annealed with DNA blots containing restriction endonuclease digests of DNA, multiple bands hybridize. The pattern of bands varies from one individual of a species to another, but it is the same when DNA from several tissues of a single individual is examined. The bands are inherited in Mendelian fashion, and it is possible to identify those bands inherited from each parent. Because the DNA isolated from a few drops of blood is sufficient for the DNA blot procedure, the minisatellites can be used in forensic medicine for positive identification of individual humans.

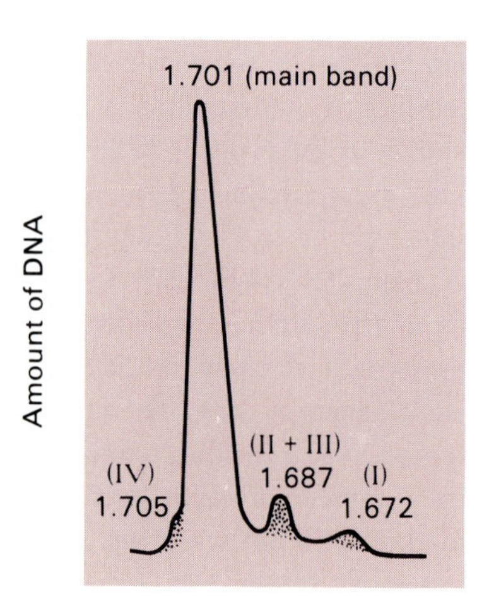

Figure 9.25
Resolution of centromeric satellite bands upon centrifugation of total *Drosophila* DNA to equilibrium in a CsCl density gradient (see Figure 9.4). The numbers give the densities (ρ). Repeated centrifugation is required to resolve satellite IV and satellites II and III (see Table 9.13). After S. A. Endow, M. L. Polan, and J. G. Gall, *J. Mol. Biol.* 96 (1975), p. 665.

c. *Tandem Repetitions at Centromeres and Telomeres*

The greatest concentrations of tandem repetitions typically occur at the centromeres and telomeres of animal and plant chromosomes. Hundreds or even thousands of copies of a repeating unit may be found within one chromosomal region. They often constitute a substantial portion of the total genomic DNA (Table 9.12). As a result, they are generally among the sequences that reanneal most rapidly (i.e., at very low Cot values) when total genomic DNA is denatured and allowed to reassociate. Like other long tandem arrays of repeated sequences, these elements frequently form separate "satellite bands" when total DNA is centrifuged to equilibrium on density gradients (Figure 9.25). All such satellite bands, including those comprising rDNA or histone gene repeats, have been called **satellite DNA** at one time or another. The term is now generally used more restrictively to refer to long arrays of tandem repeats of centromeric and telomeric origin, regardless of whether they do or do not form bands on density gradients.

Many organisms have multiple distinctive satellites. Sometimes these appear as individual bands upon density gradient centrifugation, as do the *Drosophila* satellites shown in Figure 9.25. In other organisms, some or all of the centromeric and telomeric tandem repeats fail to show up as density satellites. Such **cryptic satellites** are found in the main density band of DNA or are not abundant enough to be detected as a band.

The various satellites are distinguished by the length and the nucleotide sequence of the repeat unit. The length of satellite repeat units varies

Table 9.12 Percentage of Genomes That Are Centromeric Satellite DNA

Species	Percent (approx.)
Gecarcinus lateralis (land crab)	21
Drosophila melanogaster	16
Mus musculus	8
Rattus norvegicus	1–3
Bovis domesticus	>23
Cercopithecus aethiops (African green monkey)	13–20
Homo sapiens	Probably about 5

Note: Precise species names are used because satellite content and sequence varies from species to species, even within a genus.

from two base pairs, as in a crab satellite, whose structure is simply alternating deoxyadenylate and thymidylate residues, $d(AT)_n \cdot d(TA)_n$, to several thousand base pairs.

General properties of centromeric and telomeric satellites RNA that is homologous to satellites is only very rarely detectable. Apparently, this DNA is not transcribed to any significant extent and is most likely noncoding. Centromeres and telomeres are generally heterochromatic, that is, they stain very densely with the dye Giemsa 11 (see the "C"-bands in Figure 7.23), in contrast to the less densely staining, euchromatic regions of chromosomes. Satellite DNAs are thus associated with heterochromatin, and the two terms are frequently considered synonymous, although the precise relation between tandem repeats and the staining habits is not understood. Other chromosomal regions that contain long tandem arrays are also heterochromatic, including the nucleolar organizer region, which contains the rDNA repeats. Similarly, other genetically silent chromosome regions, such as the inactivated X-chromosome, are also heterochromatic. Centromeric satellites are among the last DNA segments to be replicated in S-phase and are often underreplicated in polytene chromosomes compared to other DNA sequences.

Analysis of satellite DNA with restriction endonucleases The very complex nature of satellite DNAs, both those separated by isopycnic centrifugation and those that are cryptic, was elucidated by restriction endonuclease digestion coupled with DNA blotting and annealing. Cloning and sequence analysis further clarified these unusual genomic segments.

If a site for a particular restriction endonuclease occurs in each repeat of a repetitious tandem array, then the array is digested to unit size fragments by that enzyme (Figure 9.26a). Very frequently, the number of unit length fragments in total genomic DNA is sufficient to be seen as a distinct band on electrophoresis gels after staining with ethidium. After elution of the band from the gel, the DNA can be used directly for primary sequence analysis. The sequence obtained is a consensus sequence and not necessarily the sequence of any particular repeat unit because the repeat units in centromeric satellite arrays are rarely identical. Rather, they represent a family whose members differ at one or more residues. Some of the sequence divergence represents random changes in base pairs in an otherwise relatively homogeneous array. In other instances, the same base pair change(s) may recur within many or all of the repeat units in a portion of the array. When part of the array differs consistently from the overall satellite sequence, it is said to be a subarray or satellite domain. Primary sequence analysis of cloned satellite arrays, domains, and monomeric repeat units has confirmed that a great deal of variation occurs within what once appeared to be very homogeneous DNA sequences.

Note that if sequence divergence occurs within a restriction endonuclease site in the repeated units, then digestion with the enzyme produces, in addition to the monomeric repeat unit, DNA fragments 2, 3, 4, and so forth times the size of the monomer. A ladder of segments is observed on an electrophoretic gel and is a telltale sign of the presence of a satellite (Figure 9.26b).

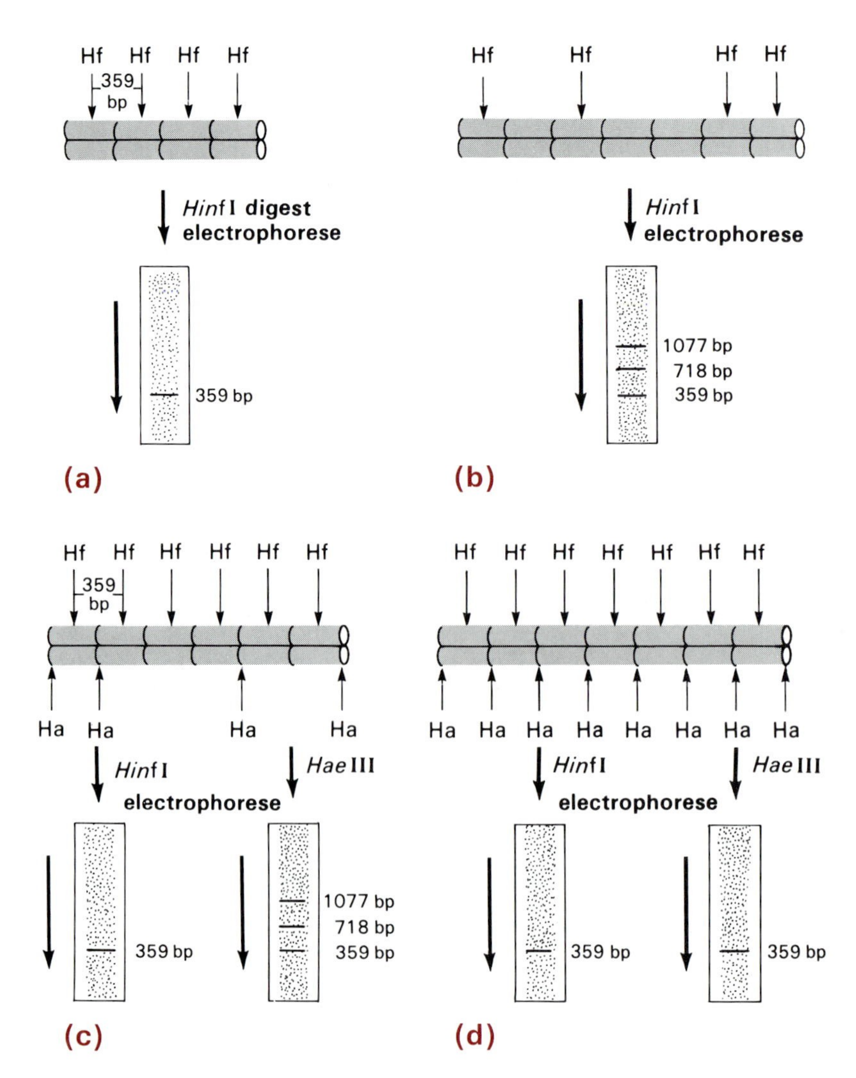

Figure 9.26
Digestion of *Drosophila* centromeric satellite III by restriction endonucleases. Different domains (a, b, c, and d) yield distinctive fragment patterns upon electrophoresis of restriction endonuclease digests. (a) The basic repeat unit is 359 bp long and contains one endonuclease *Hinf*I site. Digestion with *Hinf*I converts most of the satellite to a set of 359 bp long fragments that are abundant enough to be seen as a band against the smear of other genomic fragments after gel electrophoresis and staining with ethidium bromide. (b) In some domains of satellite III, the *Hinf*I site is missing from some repeat units. Digestion with *Hinf*I yields a ladder of DNA fragments that are 359, 2 × 359, 3 × 359 (and so forth) base pairs long. (c) Endonuclease *Hae*III sites occur frequently in the repeat units within some domains of satellite III. Digestion of satellite III with endonuclease *Hae*III yields a ladder of fragments that are multiples of 359 bp in length. Fragments up to 15 × 359 bp occur (not shown). (d) A satellite III domain that has both *Hinf*I and *Hae*III sites in each repeat unit.

Sometimes a satellite repeat unit contains sites for several restriction endonucleases. Digestion with any of these enzymes will produce the same unit size fragments from the satellite DNA. However, alternate enzymes may instead yield different size fragments from distinctive satellites in the same DNA sample. Thus, analysis of a genome with a battery of enzymes affords an initial survey of its multiple satellite components. More detailed and quantitative characterizations are possible when satellite segments, cloned or even uncloned, are used as probes on DNA blots. Total genomic DNA or bands isolated after density gradient centrifugation can be used. Annealing of the same probes to chromosomes *in situ* can locate the satellite to centromeres or telomeres. More often than not such analyses reveal the presence of multiple centromeric satellites in a particular species. Moreover, the different satellites in a species may all have distinctive repeat lengths and nucleotide sequences.

Species-specific satellite DNAs Each species has a distinctive set of centromeric satellites, even when compared with closely related species in the same genus. This is a remarkable fact in view of the high degree of sequence conservation and synteny elsewhere in genomes and raises difficult problems for any speculations on the function of satellites. We de-

Table 9.13 The Centromeric Satellites of *D. melanogaster* and *D. virilis*

		D. melanogaster
Satellite	% of Genome	Repeat Unit
I	2	5′-[AATAT]AATATAATAT etc.
	1	5′-[AATATAT]AATATATAAT etc.
II	3	5′-[AATAACATAG]AATAAC etc.
III	5	254 and 359 bp repeat units
IV	4	5′-[AAGAG]AAGAGAA etc.
	0.5	5′-[AAGAGAG]AAGAGAG etc.

		D. virilis
Satellite	% of Genome	Sequence
I	25	5′-[ACAAACT]ACAAACTAC etc.
II	8	5′-[ATAAACT]ATAAACTATAA etc.
III	8	5′-[ACAAATT]ACAAATTAC etc.
Ic	0.1	5′-[AATATAG]AATATAGAA etc.

scribe the satellites of several organisms to illustrate the complexity that has been observed.

Centromeric satellites in D. melanogaster There are four different satellites in *D. melanogaster* DNA (Figure 9.25 and Table 9.13). Together they constitute about 16 percent of the genome. Satellite I contains two simple and closely related domains, one with a pentameric and one a heptameric repeat unit. About 80 percent of satellite II is composed of tandem repeats of a decamer. Satellite IV, like I, includes both a pentameric and a heptameric domain, and these have an obvious relation to I. Satellite III has a unique structure. Its repeat length is 359 base pairs, and most repeat units have a recognition site for endonuclease *Hinf*I (Figure 9.26a). The approximately 16,000 repeats of the satellite III monomer are polymorphic. Many of them differ from one another and from the consensus sequence in a few base pairs. This is apparent when satellite III or long cloned segments of it are digested with restriction endonuclease *Hae*III. Separation of the digestion products by gel electrophoresis yields a "ladder" of fragments (Figure 9.26c). This means that many of the units in the tandem array have a *Hae*III recognition site, but others do not. Cloning of long tandem arrays of satellite III has demonstrated the occurrence of distinctive domains within what appeared to be a single homogeneous satellite. Some domains contain both *Hae*III and *Hinf*I sites in each repeating unit (Figure 9.26d). Others contain a *Hinf*I site in each unit but no *Hae*III sites at all (Figure 9.26a). Sequence analysis of cloned repeat units shows that other differences also occur in family members. For example, one domain of satellite III is made up of a tandem array of repeat units containing only 254 of the usual 359 base pairs. Thus, although some *Drosophila* centromeric satellites are quite simple sequence arrays, satellite III is quite complex.

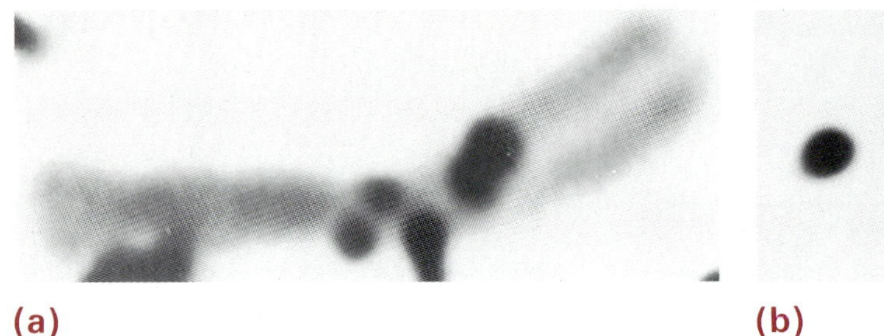

(a)

(b)

Figure 9.27
The annealing of a centromeric satellite I (ρ = 1.672) probe to *Drosophila* mitotic chromosome 3 (a) and Y (b), *in situ.* (c) A schematic diagram showing the distribution of satellites I and IV (ρ = 1.705) on chromosome 3. The relative positions of the two satellites near the junction between centromeric heterochromatin and euchromatin on the 3R (right) arm were determined by electron microscopy of the annealed preparations. After D. M. Steffenson, R. Appels, and W. J. Peacock, *Chromosoma* (Berl.) 82 (1979), p. 525.

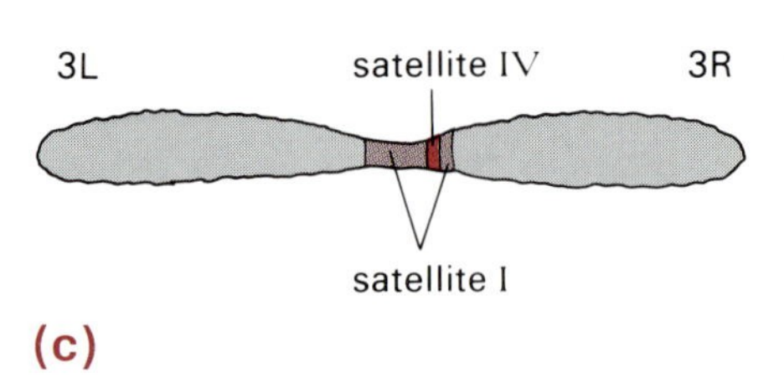

(c)

When *Drosophila* polytene chromosomes are annealed *in situ* with radioactive probes of satellites I and IV, hybridization occurs over all the centromeres (Figure 9.27). However, the amount of hybridization varies markedly from one chromosome to another, indicating that the number of repeat units varies greatly. Also, each satellite hybridizes to characteristic regions within each centromere, indicating a specific location. Satellite III hybridizes only to the X- and Y-chromosomes. Thus, individual centromeric satellites have specific loci. Moreover, the *Drosophila* centromeric satellites are species specific. Compare the satellites of *Drosophila virilis* with those of *D. melanogaster* (Table 9.13). Although the two flies are members of the same genus, their centromeric satellites are remarkably different. Only Ic of *D. virilis* is similar to any of the *D. melanogaster* repeats. Note, too, that three of the *D. virilis* satellites are similar enough to one another to preclude distinguishing their chromosomal locations by *in situ* hybridization. If they had not been separable on the basis of density, they might have been termed domains of a single satellite.

Bovine and mouse satellites There are eight different bovine centromeric satellites (Figure 9.28). Together they comprise over 23 percent of the genome, and, in kinetic analysis, they are partitioned between highly repeated and moderately repeated sequences (see Section 9.1b for definitions). The bovine repeat units are a complex set of interrelated sequences. Primary sequence analysis revealed that at least five of the eight satellites share one or more common subsequences, although they are sufficiently divergent that neither cross-annealing experiments nor restriction endonuclease analysis hinted at the relationship. The repeated units in most of the bovine satellites are more than 1 kbp long. Within these long units, shorter periodic repetitions also occur.

These complex structures are not unique to cows. Satellites in other species, although of characteristic, species-specific sequence, also display both short range and long range periodicities. The major centromeric satellite of *Mus musculus*, for example, has an overall repeat unit of 234 bp. *Sau*96I, *Eco*RII, and *Ava*II endonuclease sites appear once in each typical unit. Sequence analysis shows the 234 bp repeat to include four (imperfect)

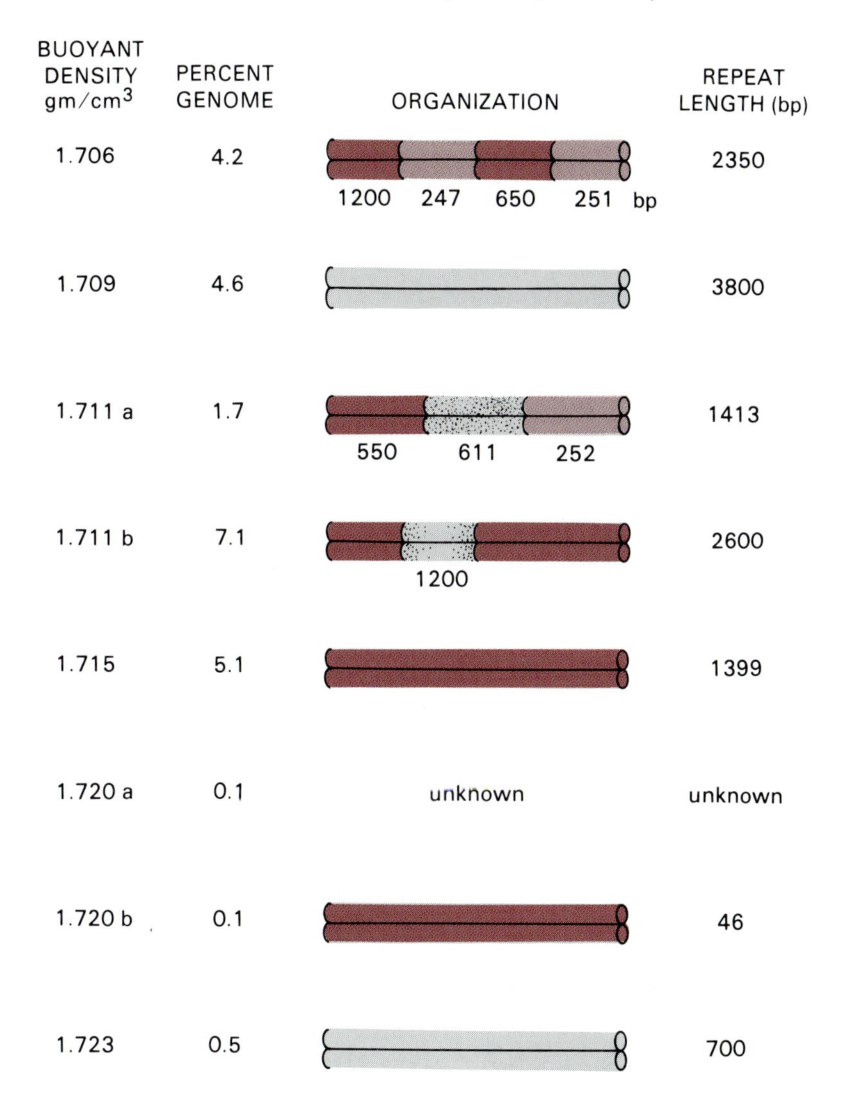

Figure 9.28
The bovine satellites were isolated by density gradient centrifugation and are named according to their densities. Homologous portions of the different satellites are indicated by similar coloring. The 1.711b is similar to 1.715 except that about 60 percent of the 1399 bp long repeat units are interrupted by an extra 1200 base pairs, which shares some sequence homology with a region of 1.711a. The sequences of the 1.723 and 1.709 satellites are unrelated to the others. The related sequences of 1.715 and 1.711b are themselves constructed from a series of internal tandem repeats, each 31 bp long. These satellites then have a long-range periodicity superimposed on a short-range periodicity. Similarly, the related segments in the long repeat units of 1.706 and 1.711a contain multiple internal tandem repeats of variants of a common 23 bp long sequence, and the 46 bp repeat unit of 1.720b is a dimer of the same 23 bp long segment.

internal tandem repeats 58, 60, 58, and 58 base pairs in length. Each of these can be further divided into two related but variant segments 28 and 30 bp long, respectively. Both of these are also related, and each could have evolved from a common nonanucleotide.

The α-satellites of primates Primates contain satellites with both long and short repeat units. At least one class of centromeric satellite in every primate that has been investigated includes a species-specific variant of an approximately 170 base pair long repeat unit called alpha (α) (Figure 9.29). Although the α-satellites of different species do not always cross-hybridize, their close relation is apparent from their primary nucleotide (consensus) sequences. For example, the approximately 340 bp repeat units of baboon and some human α-satellites are dimers of the 172 base pair long African green monkey repeat unit. The two monomers that make up the human dimer differ about 30 percent from one another and from the monkey sequence. The human and baboon sequences are also different from one another. A recognition site for endonuclease *Bam*H1 appears once every 343 base pairs in the baboon α-satellite; an *Eco*RI endonuclease recognition site occurs once in every 340 base pairs in one domain of

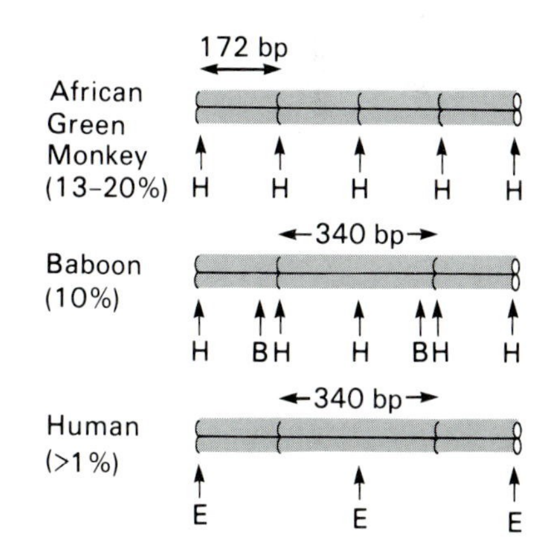

Figure 9.29
Primate α-satellites and their genomic abundance. The monkey and baboon structures are typical of the various primate species-specific α-satellites. The human structure shown represents one domain of the total human α-satellite. Although α-satellite sequences may represent 5 percent or more of human DNA, this domain is less than 1 percent.

human α-satellite. Within a species, individual repeat units are divergent. In monkey α-satellite, the sequences of cloned 172 bp long units diverge an average of 3 percent from the consensus sequence. The variations are of two types. First, there are randomly distributed changes in base pairs that could have arisen from single base pair mutations. Second, there are changes that are common to many repeat units. These probably arose by single base pair mutation followed by amplification of the divergent repeat unit. For example, many monomer units have a uniquely positioned *Eco*RI site and others a specific *Hae*III endonuclease site, neither of which appears in the consensus sequence. Those units that have either the *Eco*RI or the *Hae*III sites tend to occur in distinct tandem arrays that define satellite domains.

Some human α-satellite domains have repeat lengths that are much longer than dimers of the monomeric 170 bp. Domains that are pentameric or dodecameric have been identified. Often, the divergence from the consensus sequence of the monomers in these lengthy repeats is as much as 15 percent. Cloned probes representing such domains then fail to anneal to other domains under stringent conditions. Using such specific probes in conjunction with somatic cell hybrid lines or *in situ* hybridization, domains are localized to the centromeres of specific chromosomes. For example, a domain with a dodecameric repeat has a repeat length of 12 × 171 or about 2.1 kbp (Figure 9.30). Single *Bam*H1 and *Pst*I sites occur every 2.1 kbp. Annealing experiments indicate that there are about 5000 repeats of the dodecamer, and they occur on the X-chromosome. Several different domains may occur on one chromosome, but each chromosome is associated with specific long range domains of α-satellite. Such chromosomal specificity occurs in other primates and may be a general feature of satellite DNAs.

d. *Speculation on the Function of Tandem Repeats*

As yet, no function has been associated with many of the tandem repeats so prevalent in eukaryotic genomes. There are exceptions, of course, such as the tandem repeats of some genes, the tandem repeats within genes, and special sequences like the switch regions associated with immunoglobulin heavy chain genes. But the puzzle of the many scattered tandem repeats and the huge arrays at centromeres and telomeres remains. Perhaps this only reflects our ignorance, and, in time, their function will be clarified. But it is also possible that some DNA sequences have no particular function at all. Tandem repeats may be generated and eliminated by ongoing processes such as DNA replication and recombination. If their coming and

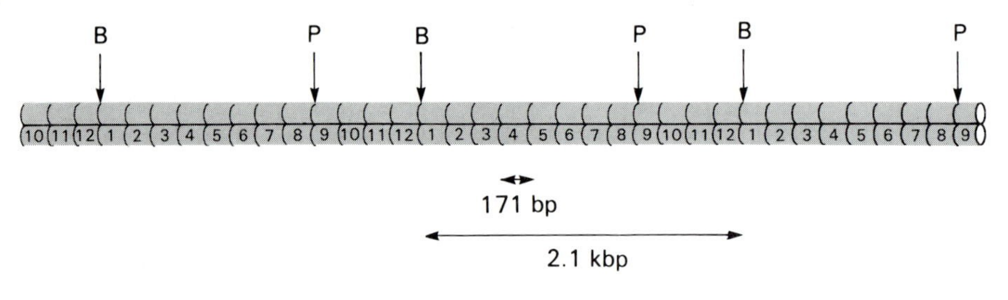

Figure 9.30
A domain of α-satellite that is located on the human X-chromosome. Each 2.1 kbp repeat unit contains 12 copies of the α-satellite monomer (170 bp). The 12 repeats diverge 10 to 20 percent from one another, but within each higher order repeat, the corresponding monomers (i.e., 1, 2, 3, etc.) are well conserved. Thus, the *Bam*H1 and *Pst*I sites occur in the same monomer unit within each 12-mer. After J. S. Waye and H. F. Willard, *Nucleic Acids Res.* 13 (1985), p. 2731.

going does not disturb the function of important coding or regulatory segments, they will accumulate and disappear unaffected by the pressures of natural selection. The polymorphism of some tandemly repeated elements fits this speculation. Nevertheless, the suggestion is unsettling, given the notion that much of genome evolution is driven by natural selection. Genomes also appear to evolve by neutral changes in DNA structure, mutations that do not affect the phenotype either positively or negatively. However, the concept of **neutral molecular evolution** is generally applied to single base pair changes or short deletions and reiterations. It could be applicable to short dispersed tandem repeats of no obvious function. It is intuitively more difficult to apply it to centromeric and telomeric satellites that often constitute a very high percentage of total genomic DNA.

The ubiquity and enormous bulk of centromeric and telomeric satellites, as well as their interesting positions on chromosomes, seem to suggest association with important function(s). Cytogenetic analysis of heterochromatin, which is often taken to be synonymous with satellite DNA, has been used for many years to attempt to elucidate the function of these chromosomal regions. A very large number of possible cellular functions has been considered. Much speculation has centered on mitotic or germ line processes, including chromosome rearrangement, pairing, segregation, and recombination. Most hypotheses assume that the sequence of the satellite DNA, like the sequence of the rest of the genome, is important to its function. However, in spite of a great deal of experimental work, none of the hypotheses has been proven, and many are eliminated by the experimental data. No phenotypic effects, and thus no function, can now be ascribed to satellites. Indeed, a strong case can be made for a lack of essential function and for the absence of important informational content in the DNA sequence. For example, functional *S. cerevisiae* centromeres have no centromeric satellite (Section 9.6a). *Drosophila* mutants that lack most, or all, of their centromeric heterochromatin function well. The freedom of satellite DNAs to adopt quite different primary sequences even in closely related species suggests that there is little reason to preserve any particular sequence. Moreover, recent data indicate that the amount and sequence organization of a centromeric satellite can vary extensively even among the individuals of a single species. This is true of α-satellite sequences in humans. Also, individual variation has been observed in a minor centromeric satellite of African green monkeys. The data suggest that neither the amount nor the precise sequence of the tandem repeats matters for centromeric function.

The only really compelling evidence in favor of some role for centromeric satellite DNA is its nearly ubiquitous occurrence. Even the fission yeast, *Schizosaccharomyces pombe,* has long tandem repeats at its centromeres, in contrast to the budding yeast, *Saccharomyces cerevisiae.* As indicated in Section 9.4e, the same processes that can lead to changes in and expansion of satellites can also lead to their elimination, but do not. Perhaps functional centromeres benefit in some way from the presence of tandem repeats themselves, regardless of their actual sequence. There is one observation on maize chromosomes that is interesting and perhaps relevant in this regard. Knobs of heterochromatin are visible at about 23 noncentromeric locations and are associated with a remarkable property.

During meiotic divisions, the knobs can form extra functional centromeres. Like maize centromeric heterochromatin, knob heterochromatin is replicated late in S-phase, and the knobs contain long tandem arrays of a DNA sequence that is distinct from the sequence of centromeric satellites. Thus, the tandem arrays themselves, not their specific sequence, appear to be associated with centromere function.

e. *Mechanisms for the Formation and Evolution of Tandem Repeats*

General recombination between nonallelic, homologous regions on DNA can yield reiterations of DNA sequence (Section 2.4). Repeated cycles of such unequal crossing-over lead to both the expansion and contraction of long tandem arrays. The initial multiplications can involve rather short homologous regions (Figure 9.31), such as the nonanucleotide that occurs with a short range periodicity in the mouse satellite. Although other mechanisms, such as gene conversion or slippage during DNA replication, may also be involved, unequal homologous recombination can account for most of the known plasticity of tandem repeats.

The evolution of satellite DNAs Most satellites are restricted to one species or a group of closely related species, which suggests that they may be of relatively recent origin or change more in evolution than many other genomic regions. The changes leading to species-specific satellites are likely to involve alternating cycles of amplification (and deletion) by unequal crossing-over and sequence change (including single base pair changes, deletions, and insertions), as outlined in Figure 9.32. This model explains interspecific variation, the presence of distinctive intraspecies domains, and contemporary polymorphism within a species. Observed variations will depend on the frequency of recombinational events compared to the rate at which sequence changes occur and are fixed by natural selection or by **genetic drift** (i.e., chance fixation). The model also accounts for the construction of long range repeat units from short oligonucleotide stretches. The model does not, however, suggest how or why particular sequences happen to be amplified into satellites. One widely discussed hypothesis holds that, within related species, a common special library of sequences is available for amplification to satellite DNA. However, neither the nature of the proposed library nor the kinds of sequences deposited in it have been described. In short, the question of whether diverse satellite sequences have any common property has no more definitive answer than the question of satellite function. Many different genomic segments, including unique genes, can undergo amplification in the genome and even be stably maintained in amplified arrays under special selective conditions (Section 10.8). The amplifications are not usually stable, however, under normal conditions. Satellite sequences are also not stable in the sense that they change rapidly. Nevertheless, the location, pattern, and abundance of the tandem repeats are retained. Is there something special about centromeres and telomeres that allows or even fosters the presence of tandem repeats, regardless of the particular primary nucleotide sequence?

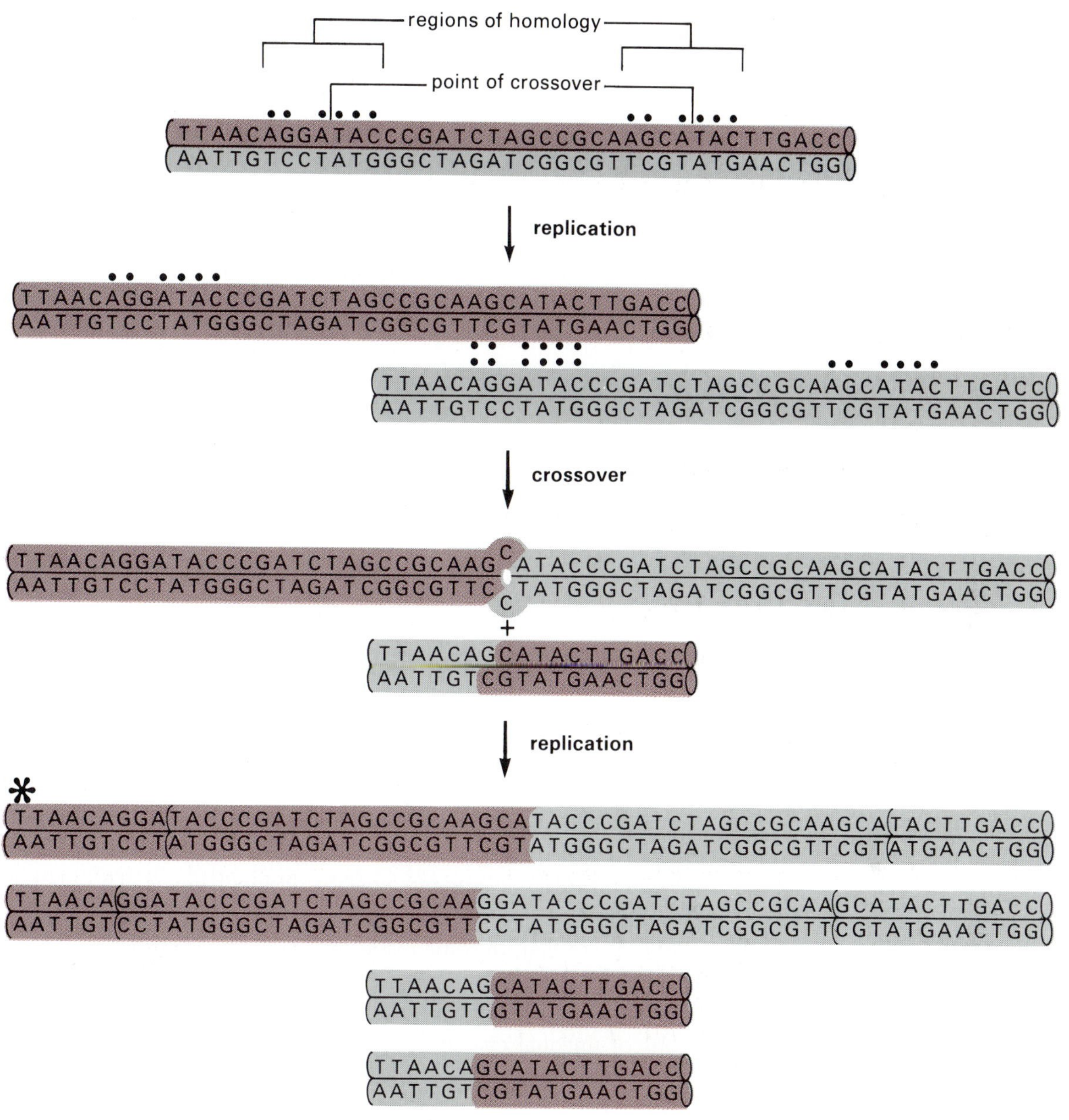

Figure 9.31
The expansion and contraction of tandem arrays by unequal crossing-over. Note that recombination may be between segments of a single DNA molecule, between sister chromatids, or between homologous chromosomes. The example is given for sister chromatids formed by the first replication. The DNA single strands that derive from the upper strand of the starting molecule are colored, and those that derive from the lower strand are grey. The two molecules produced by the crossover are shown with short heteroduplex regions, in which one strand derives from one of the parental molecules and the other strand from the other. Mismatched base pairs, such as that shown in one of the recombinant molecules here, are possible in these heteroduplex regions. Despite the complexity of the two molecules produced by crossover, each of the four molecules resulting from subsequent replication has the pattern expected if it were produced by a breakage and reunion at single points of crossover in the recombining molecules. The point of crossover for the molecule marked with an asterisk is indicated in the starting sequence. The other three molecules would be produced if the crossover point were shifted to other sites within the region of homology. The two longer final molecules carry tandem duplications of that part of the starting sequence lying between their respective points of crossover; the duplicated sequences are indicated. The two shorter final molecules have deletions. Taken from G. P. Smith, *Science* 191 (1976), p. 528.

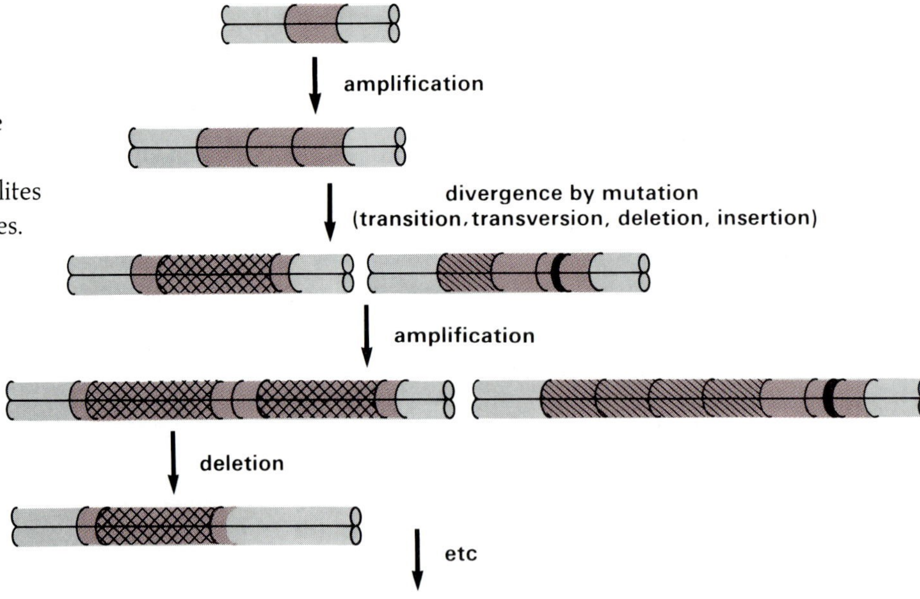

Figure 9.32
Schematic diagram suggesting how satellite DNAs might change over rather short evolutionary times. The scheme can be applied to the generation of both species-specific satellites and satellite domains within a species. After E. M. Southern, *Nature* 227 (1970), p. 794.

The evolution of tandem repetitions of coding sequences Earlier in this chapter, we described several examples of tandem arrays of genes. In some of these, like rDNA, the various coding segments are identical. In others, like the globins and the human growth hormone/placental lactogen cluster, the several coding segments specify different genes. The overall process of amplification, followed by mutation of one copy, followed by selection is by now familiar. In some instances, the detailed structure of the repeated array gives specific hints as to how the reiterations may have occurred at the molecular level. The resulting models often invoke amplification by means of unequal crossing-over between members of repeated sequence families that are scattered in the genome (Figure 9.33).

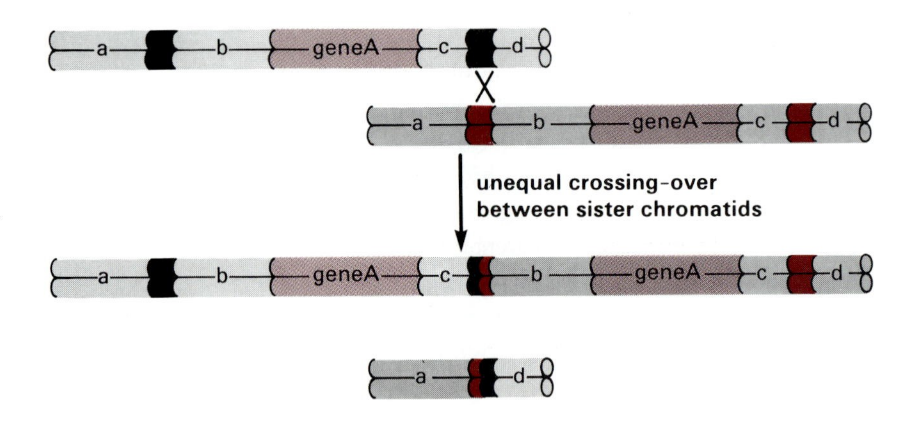

Figure 9.33
A model for the evolution of tandemly repeated copies of genes. Homologous but unequal crossing-over occurs between two copies of a dispersed, repeated sequence (shown in black on the sister chromatid at the top).

9.5 Repeated Sequences Dispersed in Genomes

Members of multigene families often recur in unlinked genomic locations, either as functional genes or as processed pseudogenes. With few exceptions, however, the copy number of these sequences is rather low, smaller than 50. The 1000 pseudogenes for U1 RNA in human DNA is the largest number yet recorded. However, many eukaryotic DNAs also contain other, very large families of repeated sequences whose members are scattered throughout genomes in as many as 10^6 copies. Although some members of some families may be functional genes, the role of the bulk of these sequences remains an enigma.

A notable feature of these abundant sequence families is their diversity from one group of organisms to another. The very large interspersed sequence families of rodents, for example, are different from those of primates, although related families with characteristic sequence variations do occur in different species. In addition, the relative abundance of related families can vary more than tenfold from one species to another. These complex relationships are characteristic of plant, many invertebrate, and vertebrate genomes. Moreover, the plasticity of interspersed repeats is not limited to sequence and copy number. Many interspersed repeats are movable genetic elements. Their locations are not necessarily the same, even in closely related species. Also, otherwise identical alleles in a particular species can differ by the presence or absence of a repeat unit. The mechanisms by which interspersed repeats are amplified and transposed into new genomic locations are discussed in Chapter 10. Here, we concentrate on their overall organizational patterns, bearing in mind that the very nature of these sequences defies generalization.

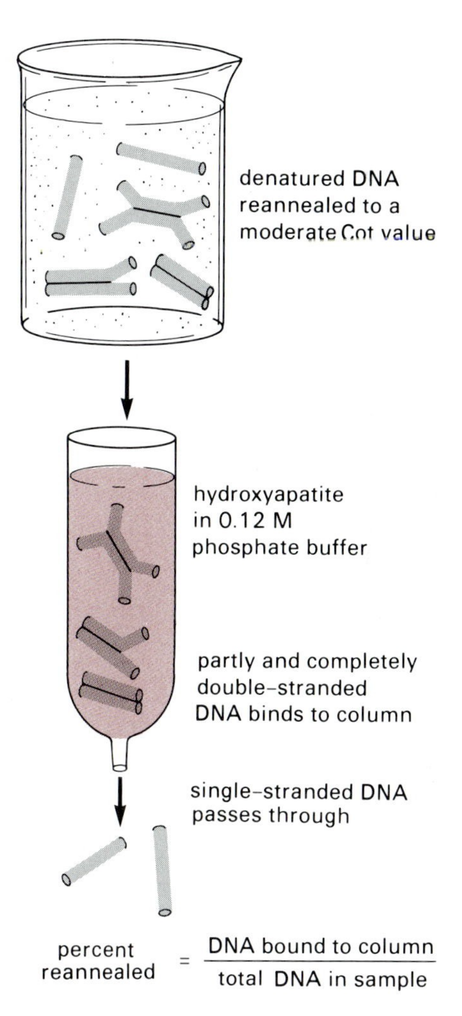

Figure 9.34
Measuring the extent of reannealing of denatured DNA with hydroxyapatite. Partly and completely double-stranded DNA chains bind to hydroxyapatite in 0.12M phosphate buffer. Single-stranded chains pass through the column.

a. *Patterns of Interspersion*

The presence of dispersed repeated sequence families was first indicated by analysis of reannealing kinetics. A procedure that came to be called **interspersion analysis** demonstrated that highly repeated sequences frequently interrupt other DNA sequences.

Interspersion analysis In genomic Cot curves, repeated DNA sequences reanneal at low to moderate Cot values and single copy sequences at high Cot values (Chapter 7.5b). When hydroxyapatite is used to measure the percentage of DNA that reanneals at a particular Cot value, both partly and completely double strand molecules are counted as double stranded (Figure 9.34). As a result, the percentage of a genome that reanneals at low and moderate Cot values often appears to increase with the length of the DNA fragments. The apparent increase in percentage reannealed occurs because repeated sequences that anneal at relatively low Cot values are interspersed in the genome among sequences of low copy number (Figure 9.35). In interspersion analysis, the percentage of DNA that registers as double stranded on hydroxyapatite is measured as a function of the length of the DNA fragments. To obtain DNA samples of different lengths, total genomic DNA is fragmented by random shearing and then fractionated into different size classes by gradient centrifugation. The

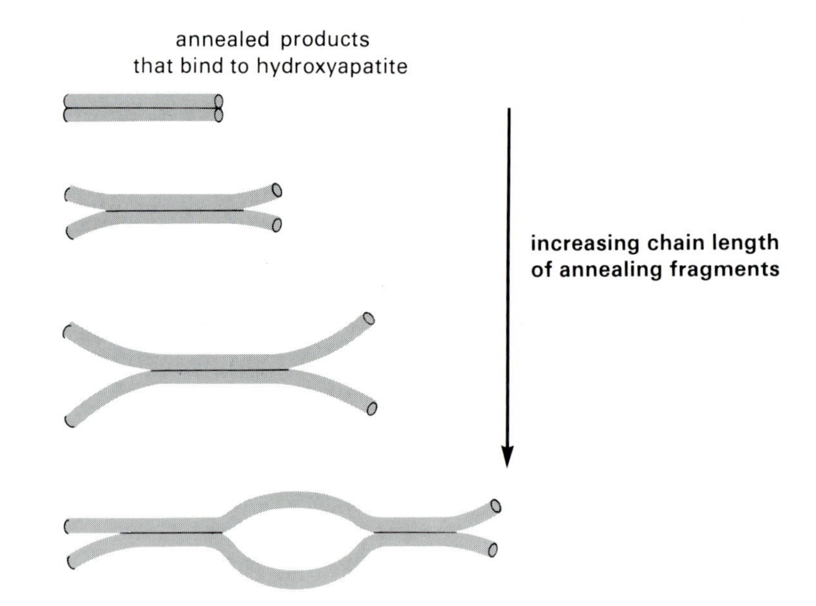

Figure 9.35
At a given moderate cot value, the percentage of DNA bound to hydroxyapatite increases with chain length. The duplex regions will be members of interspersed, repeated sequence families.

denatured DNA samples are then annealed at a fixed Cot value that is low enough to ensure that single copy sequences themselves do not reanneal.

Different patterns of interspersion Some typical interspersion curves are shown in Figure 9.36. In each case, the percentage of DNA that binds to hydroxyapatite increases with chain length, but the shapes of the curves differ dramatically. Fully 60 percent of human DNA binds to hydroxyapatite when the fragments reach about 2.2 kbp in length. Because only about 20 percent of human DNA is moderately or highly repetitive (the intercept on the ordinate in Figure 9.36), low copy number sequences must frequently flank repeated segments. Further, the break in the slope at about 2.2 kbp suggests that, on the average, 2.2 kbp of relatively rare sequences separate the repeated sequences in 60 percent of the genome. In contrast to these results, the curve obtained with *Drosophila* DNA rises continuously. The average distance between repeated sequences in *Drosophila* is at least 12 kbp, the longest fragments used in the experiment. Electron micrographs of the reannealed molecules demonstrate that most of the repeated sequences in human DNA are about 300 bp in length, and those in *Drosophila* are between 5 and 6 kbp long.

The human and *Drosophila* genomes define two different interspersion patterns (Figure 9.37). In the human or short range pattern, repeated seg-

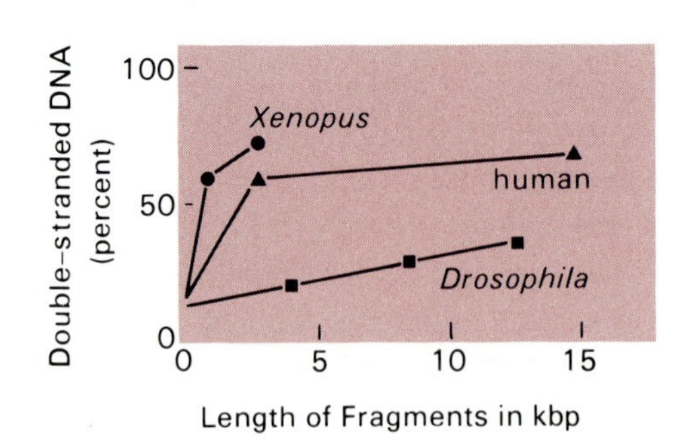

Figure 9.36
Interspersion curves for *Drosophila*, *Xenopus*, and human DNA.

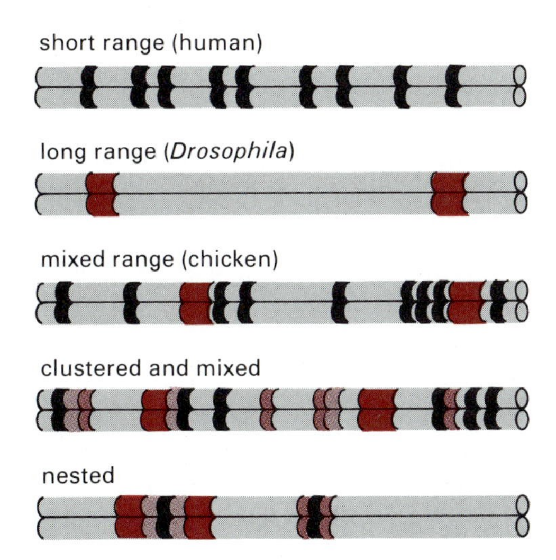

Figure 9.37
Schematic diagram illustrating different types of interspersion of repeated sequences. Different colors represent members of different families of repeats.

ments a few hundred base pairs in length are separated by up to a few thousand base pairs of low copy sequence. In the *Drosophila* or long range pattern, repeated segments several kilobase pairs in length are separated by tens of kilobase pairs of low copy sequence. In addition to *Drosophila*, a few other insects and fungi show long range interspersion, but most organisms, including peas, corn, *Xenopus* (Figure 9.36), and primates, show short range interspersion. Some (e.g., chicken) show interspersion patterns between the two extremes.

Molecular cloning has revealed that many organisms have both short and long range interspersion patterns superimposed on one another. Kinetic analysis reveals only the preponderant type. Also, different repeated elements can be clustered together or even nested within one another (Figure 9.37). Members of interspersed repeat families are found between genes, in introns, within satellite DNA, and even within coding and regulatory regions, where they cause mutations.

Fold-back DNA The percentage of DNA that reanneals increases with chain length, even at very low DNA concentrations (i.e., at Cot values around 10^{-6}). Such reannealing must be intramolecular, indicating that repeated sequences sometimes occur in opposite orientations along DNA chains (Figure 9.38). The fraction of DNA that reanneals under these conditions is termed **snap-back** or **fold-back** DNA, and the characteristic stem-loop structures are visible in the electron microscope. Molecular cloning demonstrated clearly that members of interspersed repeated families occur in both orientations, even within rather short genomic segments.

b. *Interspersed Repeats in Invertebrates*

Different invertebrates illustrate the variety of pattern and abundance typical of interspersed repeated sequences. A comparison of the *Drosophila* and sea urchin genomes emphasizes the observed diversity.

Drosophila Although interspersed repeats constitute about 9 percent of the *Drosophila* genome, no very highly repeated sequences have been

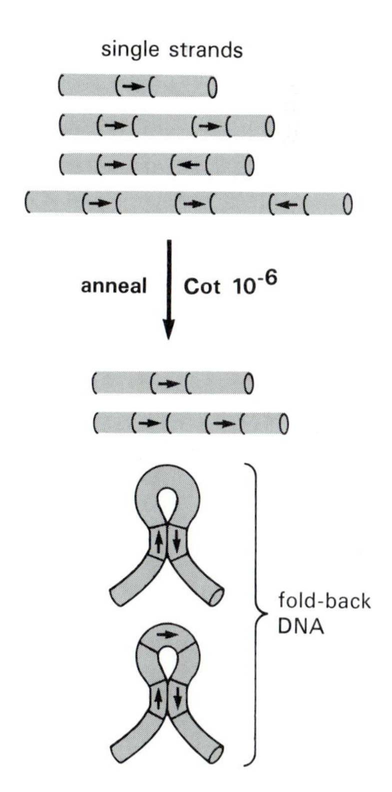

Figure 9.38
The formation of "fold-back" DNA.

described. Instead, there are from 50 to 100 families, each with only 10 to a few hundred members. Many of these are movable elements and are several kilobase pairs long (Sections 10.2 and 10.3). Some regions of the *Drosophila* genome contain clusters of members of several different repeated sequence families. As many as 17 different families can contribute multiple copies to a single cluster as long as 15 kbp. The arrangement of the repeated units within a cluster appears to be scrambled. Clustered scrambled repeats are also found in the chicken genome and some portions of sea urchin DNA.

Sea urchins Sea urchins have almost 1000 distinct families of interspersed repeated sequences, most of which are about 300 bp in length. The size of the families ranges from fewer than 100 to 10^4 members, and there is a total of at least 5×10^5 such segments in the genome. In some families, the different members are very homogeneous in sequence; in others, the family members diverge extensively. Any particular family may occur in several different sea urchin species, but its relative abundance can vary as much as 20-fold even in closely related species. When species in different orders are compared, the relative copy number of different families changes even more dramatically, although the total percentage of the genome occupied by such sequences remains remarkably constant. Furthermore, although a family may be identified in several species by cross-hybridization, it often has a species-specific consensus sequence. For example, particular sequence alterations, revealed by distinctive restriction endonuclease sites, may characterize all family members in one species but not in a related species.

Many of the repeat sequences in sea urchins are transcribed as integral parts of unrelated long transcription units, as when the family member is within an intron. Such transcripts are abundant in nuclei and include single copy sequences as well as the repeated elements. Thus, the interspersion pattern of the genome is preserved in nuclear RNA. Both strands of the repeated elements are represented in the transcripts, although usually not on the same RNA molecule.

The function, if any, of the repeated DNA sequences and their RNA counterparts is unknown. Except in sea urchin oocytes, the repeated sequence transcripts are largely confined to the nucleus and are not present in cytoplasmic polyadenylated mRNA. Even in the oocyte cytoplasmic RNA, there is no proof that the repetitive sequences are associated with functional mRNA. The sequences of those transcripts that have been studied are not likely to encode proteins because they are punctuated by frequent termination codons in all three reading frames.

c. *The Very Large Families of Interspersed Repeats in Mammalian Genomes*

Classical interspersion analysis indicated that human and other mammalian genomes have short range interspersion patterns. The molecular details provided by restriction endonuclease analysis and by cloning and sequencing confirm this conclusion. Although different families and subfamilies of interspersed repeats occur, a very small number of families has

rat ID
(10^5 copies per genome)

```
GGGGCTGGGG ATTTAGCTCA G-TGGTAGAG CGCTTGCCTA GGAAGCGCAA GGCCCTGGGT TCGGTCCCCA GCTCCGAAAA AAAAAAAAAA
      GGGG GCGTAGCTCA GATGGTAGAG CGCTCGCTTA GCATGCGAGA GGTACCGGGA TCGATACCCG GCGCCTCCA
```

ala-tRNA

mouse type II
(8×10^4 copies per genome)

```
GGGGCTGGAG AGATGGCTCA GTGGTTAAGA GCACCTGACT GCTCTTCCGA AGGT------ ---CCTGAGT TCAATTCCCA GCAACCACAT GGTGGCTCAC
      GACG AGGTGGCCGA GTGGTTAAG- GCGATGGACT GCTAATCCAT TGTGCTCTGC ACGCGTGGGT TCGAATCCCA TCCTCGTCG
```

ser-tRNA

Figure 9.39
Sequence similarity between members of SINE families and tRNAs. From G. R. Daniels and P. L. Deininger, *Nature* 317 (1985), p. 819.

very high copy numbers and dominates the genomes. A particular dominant family may account for more than 5 percent of the total DNA, and together the several families may add up to as much as 20 percent. Some of these families have members that are between 100 and 500 bp long and are called SINES, for *s*hort *in*terspersed repeats. Others, which have members that may be 6 kbp or more in length, are called LINES, for *l*ong *in*terspersed repeats. Organisms other than mammals, including slime molds, locusts, echinoderms, fishes, and amphibians, are now known to contain interspersed repeats that are like the mammalian SINES. Similarly, sequences like mammalian LINES occur in plants and invertebrates (Section 10.3f).

SINES Short, highly repeated, and interspersed mammalian sequences were first classified together as SINES simply on the basis of length and abundance. Recently, a more significant common property was established. SINES appear to be processed pseudogenes derived from class III genes that encode small cytoplasmic RNAs, including tRNAs (Figure 9.39) and 7SL RNA (Figure 9.40). Thus, the sequences of SINES are homologous to the sequences of such RNAs (and their genes). Often, SINES are surrounded by direct repeats that are target site duplications (Section 10.1b) and have a 3′ terminal A-rich stretch.

Each species (or order) has its own typical set of SINE families. A few examples will suffice to illustrate the diversity and complexity of these

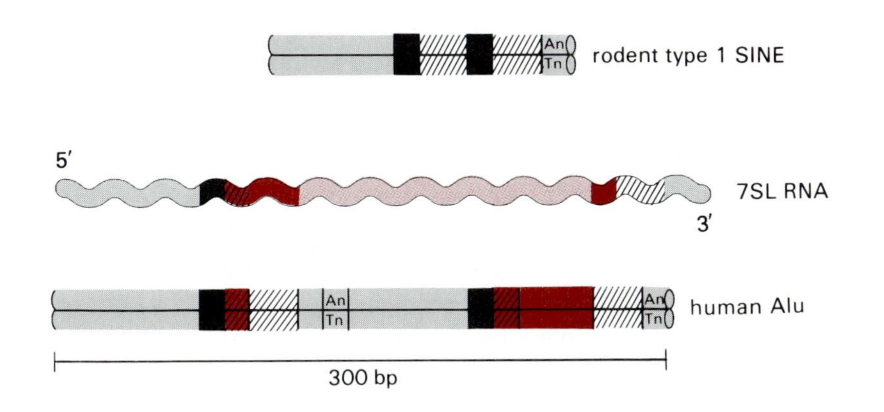

Figure 9.40
The relations between 7SL RNA and the consensus sequences of rodent type I SINEs and primate Alus are indicated by similar markings. A_n indicates the A-rich sequence at the 3′ end of one strand of each monomer (and T_n the complement). The human 7SL RNA genes have the same structure as the RNA. The central portion of the 7SL RNA sequence is not related to Alu or type I SINES but is independently repeated in the human genome. From E. Ullu and C. Tschudi, *Nature* 312 (1984), p. 171.

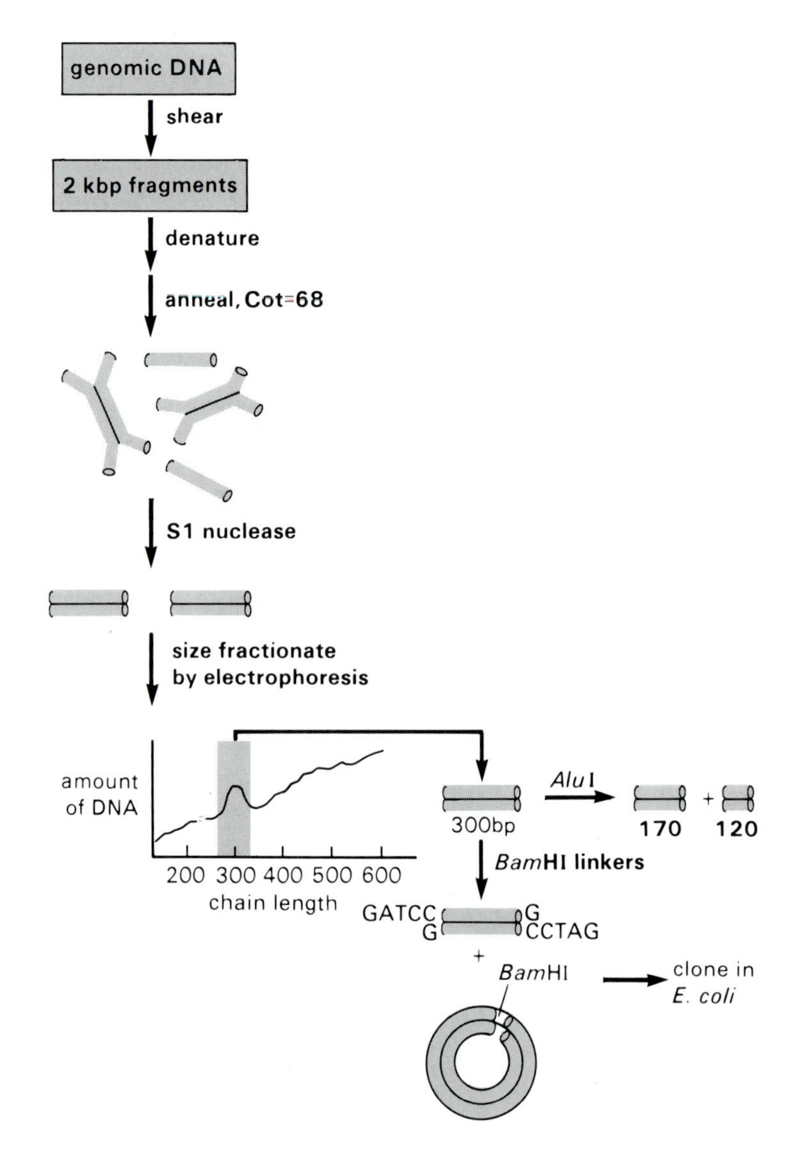

Figure 9.41
The isolation of human Alu sequences. Total human DNA is sheared to 2 kbp long fragments, denatured, and reannealed to a cot value of 68, at which low copy number sequences remain single stranded, and moderately and highly repetitive sequences form duplexes. Treatment with a single strand-specific nuclease eliminates all but the duplex segments. When these are fractionated by size, a substantial proportion of the DNA is about 300 bp in length, and the majority of the 300 bp fraction is cleaved by the restriction endonuclease *Alu*I into fragments of 170 and 120 bp. These experiments show the frequent repetition of a sequence about 300 bp long that typically includes a single *Alu*I site. Representative Alu family members were cloned from the 300 bp fraction using *Bam*H1 linkers and pBR322 as vector. After P. L. Deininger et al., *J. Mol. Biol.* 151 (1981), p. 17.

sequences. One of the best studied SINE families is the Alu family of Old World primates, named for an *Alu*I restriction endonuclease site typical of the sequence. Alu units are found flanking genes, in introns, within satellite DNA, and in clusters with other interspersed repeated sequences. Within any particular species, the Alu sequences diverge from one another up to about 15 percent. When Old World primates are compared with one another, the sequences of Alu family members vary to about the same extent as within a single species.

Typical primate Alu sequences are dimeric, composed of two head-to-tail repeats of a sequence about 130 bp long (Figure 9.40). The two monomeric repeats are not identical, and the most striking difference is an extra 31 bp long segment within the second repeat. Each monomer ends (at the 3′ terminus in the top strand shown in Figure 9.40) in a stretch of A residues of variable length. Alu monomers contain sequences that are homologous to the 5′ and 3′ portions of 7SL RNA; the central 155 bp of the 7SL RNA is deleted in Alu sequences.

Alu family sequences were first isolated and cloned (at random) from human DNA (Figure 9.41). A consensus sequence was derived using sev-

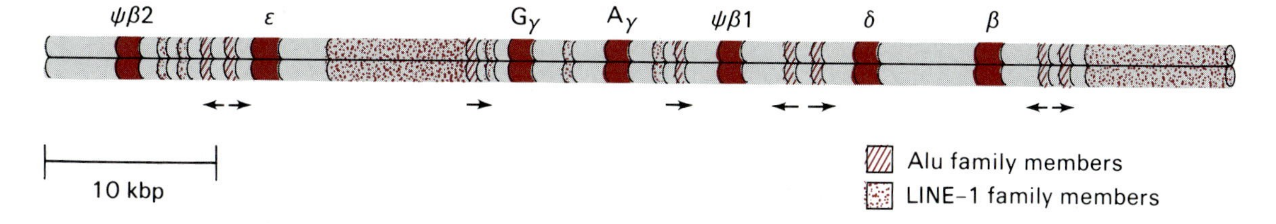

Figure 9.42
Interspersed repeated sequences pepper the human β-globin region.

eral of these. Cloned Alu family members were then used as probes to measure copy number by analysis of reannealing kinetics. The results indicate the presence of as many as 9×10^5 copies per haploid genome (or about 9 percent of human DNA). On the average, there is one Alu sequence every 5 kbp in the genomes of humans and other Old World primates. More than 90 percent of the phage in typical human genomic libraries anneals with an Alu probe, and most cloned segments of 15–20 kbp in length contain more than one family member. Neighboring Alu sequences can occur in either direction, and some are therefore included in the fold-back fraction of genomic DNA. For example, of the eight Alu sequences within the 65 kbp stretch of DNA that contains the human β-globin multigene family, five go in one direction and three in the other (Figure 9.42).

Considering that there are only three functional 7SL RNA genes in human DNA, the ubiquity of the Alu sequences is remarkable. Moreover, the generation of all these primate Alus must have taken place after the major radiation of mammalian orders in evolution. This is because the rodent SINE family that is homologous to 7SL RNA is markedly different from the primate Alu families (Figure 9.40). Thus, the approximately 10^5 copies of the rodent type I family are about 130 bp long, essentially equivalent to one of the two head-to-tail monomers typical of primate Alus. Other rodent SINEs, such as the Type II and ID families, are homologous to various tRNA genes (Figure 9.39).

The homology between SINE family members and their parental class III genes includes the internal promoter elements, the A and B boxes, and SINE units contained in cloned DNA segments can often be transcribed *in vitro* by RNA polymerase III (Figure 9.43). Transcription begins precisely at the 5′ end of the unit and terminates at the first stretch of four or more A residues on the template strand. However, in spite of the fact that many SINE units can be efficiently transcribed *in vitro*, most of them

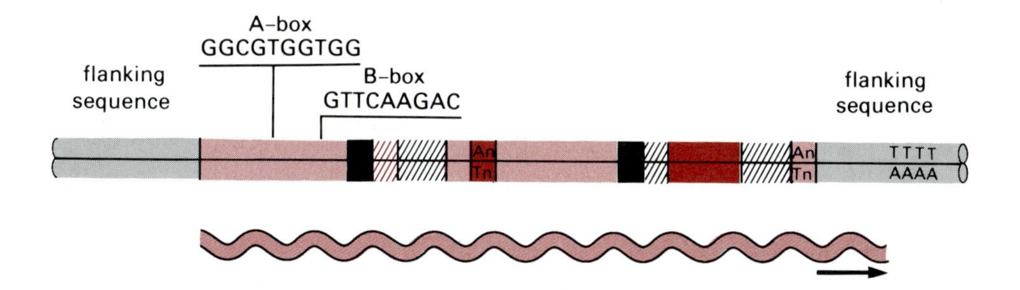

Figure 9.43
In vitro transcription of a SINE unit by Pol III. A typical cloned primate Alu segment is used as an example. The bipartite transcriptional control regions (the A- and B-box) are shown (Section 8.3), as is the initiation site corresponding to the beginning of the repeat unit.

are not active templates for RNA polymerase III *in vivo*. Nevertheless, the most interesting current models that are invoked to explain the amplification of SINES and their insertion into DNA depend on the reverse transcription of RNA polymerase III transcripts (Section 10.4).

SINE sequences are also transcribed by RNA polymerase II. This occurs when a SINE is in an intron, 5′ leader or 3′ trailer of an unrelated, class II transcription unit. For this reason, HnRNA contains abundant transcripts of SINE sequences interspersed among other sequences. Upon processing of a primary transcript into mRNA, the SINE is generally eliminated, and cytoplasmic RNA contains about ten times fewer SINE sequences than does nuclear RNA.

LINES In addition to multiple SINE families, mammalian genomes all have one abundant family of LINES, the LINE-1 family, whose members can be as long as 6 to 7 kbp (Figure 9.44). Additional LINE families occur in some of these genomes, but they are much less frequent. The LINE-1 families in different mammals are all related but divergent; the more closely related the species, the more similar they are. Many LINE-1 family members are much less than 6 or 7 kbp in length. Such truncated segments usually lack sequences from the left or 5′ end (as drawn in Figure 9.44). Sometimes they are missing internal sequences (i.e., they carry deletions compared to long family members). Moreover, not all family members are colinear. In some, portions are inverted or rearranged relative to one another. There are many more truncated than full length family members in both mouse and primate genomes. As a result, the genomic copy number of subregions from within the long LINE-1 sequences varies and increases from the far left to the far right end. In humans, for example, there are about 3.5×10^3 copies of the far left end sequences and almost 10^5 copies of the far right end sequences.

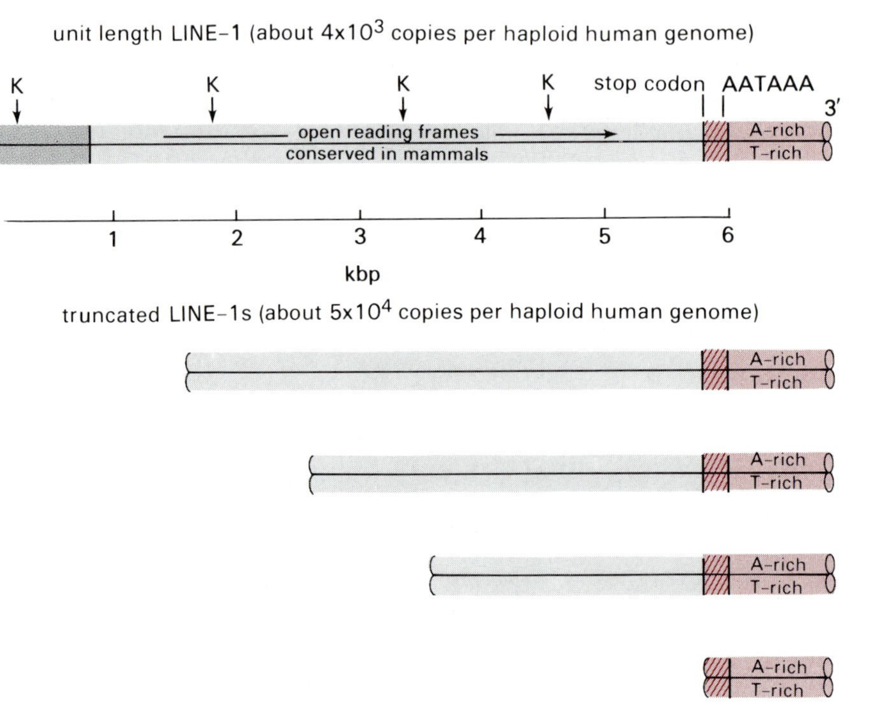

Figure 9.44
The full length human LINE-1 family member at the top is similar to those found in all mammals. A few of the characteristic (primate) restriction endonuclease sites are indicated. For example, the 1.5 and 1.2 kbp long *KpnI* segments within the primate LINE-1 are abundant enough to be visualized by ethidium bromide staining after gel electrophoresis of genomic digests. The light grey region (about 5 kbp) is similar in all mammals and contains extensive regions of open reading frame (5′ to 3′, left to right on the upper strand). The segments at the 5′ and 3′ ends are markedly different in length and sequence from one mammalian order to another. They do not contain open reading frames. As shown below, most family members are truncated at the 5′ end.

LINE-1 sequences are found flanking genes, in introns, and inserted into satellite DNA repeats. Analysis of LINE-1 family members in cloned genomic segments, as well as *in situ* annealing of LINE-1 sequence probes to intact chromosomes, show that they are dispersed to many chromosomal sites. Whether the dispersal is random or not is unknown. At some sites, there may be clusters of LINE-1s. For example, there is a surprising number of family members in the 65 kbp stretch of human DNA that contains the β-globin genes (Figure 9.42).

Like SINES, LINE-1 units are often surrounded by short direct repeats that appear to be target site duplications (Section 10.1b), and they have a 3′ terminal A-rich stretch. However, unlike SINES, LINE-1 units do not appear to be pseudogene versions of any common gene. Rather, some of the longest LINE-1 units may be movable elements that are amplified and transposed into new genomic sites by a mechanism involving transcription into an RNA intermediate and reverse transcription into DNA (Section 10.3f). The reverse transcriptase may be encoded within an approximately 4 kbp portion of the LINE-1 unit, a portion that is relatively well conserved among mammalian species and contains a long open reading frame on one strand (Figure 9.44). Truncated and rearranged LINE-1 units could arise from incomplete or aberrant transcripts or reverse transcripts, or from recombinational events accompanying insertion.

LINE-1 family sequences are transcribed when RNA polymerase II reads through a gene that has a family member in its noncoding region. As a result, RNAs with homology to both strands of LINE-1 sequences are relatively abundant in HnRNA. Discrete transcripts have not been detected in polyadenylated cytoplasmic RNA except in a human teratocarcinoma cell line and in mouse lymphocytes, where full length transcripts of the strand with the open reading frame occur. No proteins encoded by these RNAs have been identified.

d. The Function of Interspersed Repeats

The existence of hundreds of families of repeated sequences scattered in tens to hundreds of thousands of places in eukaryotic genomes is puzzling. SINES and LINES, for example, make up at least 10 percent of mammalian genomes. Aside from the possibility that a few members of a few families may contain functional genes, what roles do the bulk of the interspersed repeats play? There are two interesting hypotheses about their function. The first suggests that repeated DNA sequences, their transcripts, or both are involved in regulating gene expression. The second proposes that most such sequences have no function at all.

Repeated sequence families as regulators of gene expression Recall that, unlike prokaryotes, where related genes for a particular metabolic pathway are often linked and regulated within a single operon (Section 3.11), related eukaryotic genes are usually widely dispersed. For example, the α- and β-globin genes must be expressed coordinately, although they are on different chromosomes. In other instances, whole sets of dispersed genes must be expressed at the same time during development or in a particular tissue. Such coordinate regulation could be achieved if all members of the gene set were associated with a common regulatory element. A single effector molecule interacting with the dispersed identical regulatory elements could then turn the whole set on or off. Control of gene expression might operate at the DNA level by regulating transcription or at the RNA level. For example, HnRNA appears to maintain the interspersed repeat pattern of genomic DNA, so coordinate regulation could operate by controlling the maturation of primary transcripts into mRNA. Alternatively, copies of the repeat sequence families might be maintained in mature mRNA and regulate gene expression at the level of translation.

What evidence is available to support this general hypothesis or any of its alternative forms? In fact, several families of repeated sequences are known to be involved in regulating specific groups of genes. Typically, these families have very short members, less than 10 bp, and are therefore not included in the mass of interspersed repeat families identified by reannealing kinetics or readily detected by cloning. For example, certain common sequences are part of the regulatory regions associated with genes that are transcribed by one or the other of the distinctive RNA polymerases; they are, in a formal sense, both repetitive and dispersed.

When yeast cells are starved for amino acids, they respond by expressing those genes required for amino acid biosynthesis. In the 5′ flanking region of at least some of these genes is the sequence 5′-TGACTC-3′. The sequence occurs once or more at irregular intervals between 33 to 330 bp before the start site for transcription amid otherwise nonhomologous sequences. A single copy of 5′-TGACTC-3′ is necessary and sufficient for a gene to be expressed in response to starvation, as long as the remainder of the transcriptional control elements is also intact. Stimulation of transcription is mediated by the DNA binding protein GCN4 that specifically recognizes the hexanucleotide (Section 8.3f). Another short repeated sequence is always found in the 5′ flanking region of *Drosophila* genes that are activated in response to elevated temperature. The (consensus) segment 5′-CTNGAANNTTCNAG-3′ appears within 150 bp of the start of transcription of these heat-shock genes, and it is required for the response to heat, both in *Drosophila* and in mammalian cells in culture. There is also a set of sequences that are associated with introns and are required for efficient and correct splicing. They include the dinucleotides 5′-GT-3′ and 5′-AG-3′, which occur as the first two and last two residues in eukaryotic introns (Section 8.5c). In yeast, the sequence 5′-TACTAAC-3′ appears in all known introns, between 20 and 55 residues upstream of the AG at the 3′ (acceptor) splice site (Section 8.5c). For example, in the single intron in the yeast actin gene (Figure 9.10), it precedes the 3′ splice site by 47 nucleotides. When the sequence is deleted from genes cloned in a yeast vector, the vector no longer yields mRNA in yeast because the transcripts are not spliced.

What about the very abundant families whose members are considerably longer than 10 bp, including SINES and LINES? Is there any evidence that they have regulatory functions of any kind? In sea urchins and in *Dictyostelium*, for example, different repeated sequence families occur in the different sets of transcripts that appear in embryos or in adult tissues. Also, the families of repeated sequences represented in nuclear transcripts differ markedly from one type of differentiated cell to another. Could these RNAs contribute to the regulation of sets of functional genes whose expression is similarly timed? In an analogous way, certain SINE sequences occur in transcripts isolated from specific mammalian cell types. A mouse type II SINE family sequence, B2, is common within mRNAs that are present at elevated concentrations in oncogenic cells and in mouse embryos at particular stages of development. It has also been identified in the 3′ trailer of mRNAs for several genes in the major histocompatibility complex of mice. Such findings may suggest a specific role for repeated elements, but additional evidence is needed before the repeats can be assigned a causal role in the regulation of gene expression.

Repeated sequence families as nonfunctional entities The notion that some genomic segments might be without genetic function was introduced in Section 9.4d. Pseudogenes and processed genes seem likely to represent such sequences. To the extent that many interspersed families include a large number of pseudo- and processed genes, they may turn out to be largely functionless. Such DNA has been called **selfish** because its abundance may simply reflect its own ability to multiply itself and disperse. Possible mechanisms for amplification and dispersion are described in Chapter 10. For now, it is pertinent to recognize that such events are not themselves innocent; interspersion can lead to insertional mutations in important regulatory or coding sequences. Moreover, multiple interspersed repeats can foster deletions of functional segments by homologous recombination. Thus, unlimited selfishness could be devastating to functional genomes.

Alu sequences foster deletions The potential of interspersed repeats to cause mutational deletions is illustrated by the human LDL receptor gene (Section 9.4a). Homozygotes for mutant LDL receptor genes accumulate very high levels of LDL cholesterol in plasma, resulting in premature atherosclerosis and heart attacks even in childhood. Estimates indicate that there is one mutant allele in every 500 individuals. Heterozygotes suffer from a milder form of **familial hypercholesterolemia**. Cloning and structural comparison of genes from several individuals with familial hypercholesterolemia shows that some defective LDL receptors are encoded by genes that have suffered extensive deletions. There are many Alu sequences within the 45 kbp of DNA that contains the LDL receptor gene. Several of these are in the 3′ half of the gene, including 3 Alus in the 3′ untranslated portion of the last exon. Two characterized and independent deletion mutants were apparently caused by homologous recombination between such Alu units (Figure 9.45). As a consequence, the exons that encode the membrane spanning region and cytoplasmic tail of the receptor are deleted. Mutant proteins are produced and secreted from the cell, rather than taking up the normal position in the plasma membrane.

e. *The Remarkable Uniformity of Dispersed Repeated Sequences*

How can the intraspecies homogeneity of members of an interspersed repeated sequence family be explained? Tandem repeats, such as the arrays of rDNA units on different human chromosomes, are also similar in sequence. Reciprocal homologous crossing-over between nonallelic family members is one way to achieve such homogenizations and is

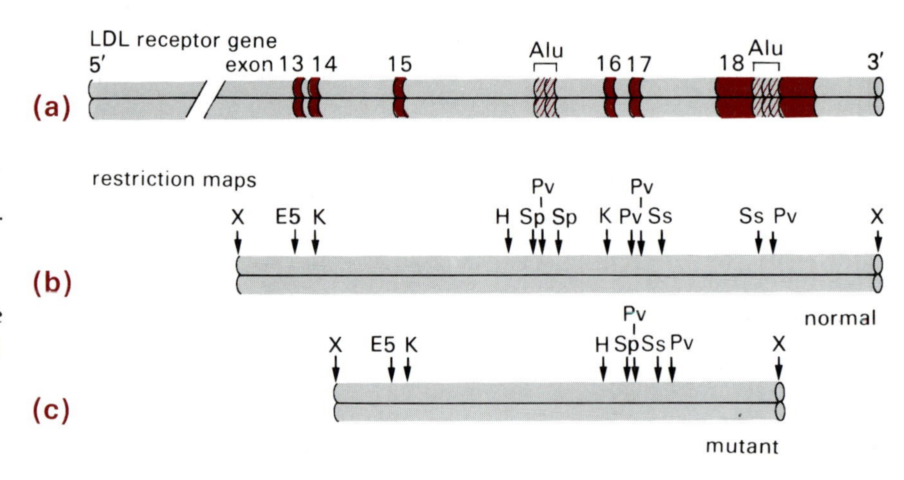

Figure 9.45
A deletion mutant in the LDL-receptor gene that is generated by recombination between Alu sequences. (a) The 3′ portion of the normal gene with exons numbered as in Figure 9.21. The Alu sequences in one intron and in the 3′ terminal exon (exon 18) are shown. (b) The restriction endonuclease map of the same portion of the normal gene. (c) The restriction endonuclease map of a cloned mutant gene. The restriction endonuclease sites that are missing in the mutant allow an estimate of the end points of the deletion and place them in the Alu sequences. This was confirmed by comparing the nucleotide sequences of cloned normal and mutant genes. After M. A. Lehrman et al., *Science* 227 (1985), p. 140.

known to occur between rDNA units in yeast (Section 9.2a). Analogous reactions can also explain the intraspecies homogeneity of other tandem repeats, such as centromeric satellite DNAs and spacer sequences. However, the homogenization mechanisms typical of tandem arrays cannot be applied to interspersed repeats without severely jeopardizing the integrity of unique sequences that surround the repeats.

Possible mechanisms for homogenization of dispersed repeats It makes sense to assume that members of homologous SINE or LINE families in different organisms are all derived from a common ancestral sequence. In each evolutionary line, distinctive versions of these sequences arose and became very abundant. For example, the primate Alu sequences are all similar. Mice do not have such dimeric Alu sequences. Instead, they have many copies of a monomeric relative of Alu, the type I SINE. Yet both of these families are likely to have originated in the rearrangement, amplification, and dispersal of sequences derived from ancestral and conserved 7SL RNA genes.

It is difficult to imagine that intraspecies homogenization of these families is the result of natural selection acting on each family member. If multiple copies are functional, then their redundancy would tend to minimize the importance of any single mutant copy. Indeed, most interspersed repeats diverge from the species-specific consensus sequence at between 5 and 15 percent of the nucleotide residues. However, if most of the copies are nonfunctional, then homogenization to a species-specific version is even less likely to be influenced by selective pressure. It is therefore most reasonable to consider that each intraspecies copy evolves in some sort of communication with the others, a communication that tends to make them all similar.

Consider one SINE family as an example for speculation. At some point prior to the evolution of vertebrates, amplification and dispersal of a rearranged 7SL RNA gene occurred. Thus, *Xenopus*, but not *Drosophila*, has such a SINE family, although both have 7SL RNA genes. Within each distinctive vertebrate species, an abundant family was maintained, but the ancestral family members were replaced by a single species-specific version serving as a new **founder** sequence. All or most of the ancestral copies were lost from the new species, and the family was

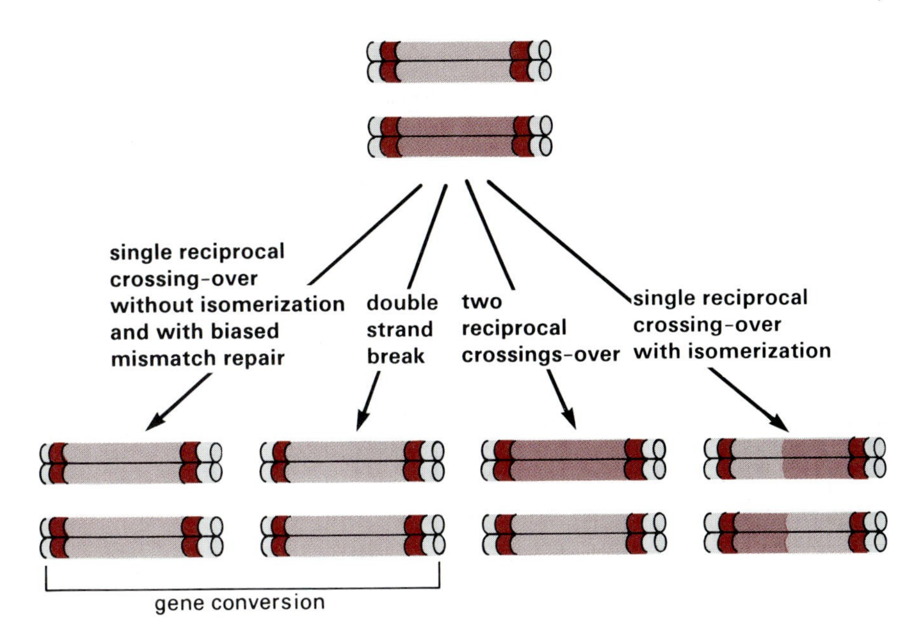

Figure 9.46
Recombination between nonallelic repeated units (shown by two shades of color) yields different products, depending on the mechanism. All the mechanisms depicted involve homologous recombination within the repeat units themselves.

regenerated *in toto* from the founder sequence. If this happened all at once, the event would be cataclysmic and destroy the genome. But it could proceed by the gradual replacement of existing SINES by the founder sequence over relatively long evolutionary times. In the 70 million years since primates and rodents diverged, assuming selective neutrality, a replacement rate of about 0.01 per year could homogenize a family as big as the primate Alu.

One possible replacement mechanism involves recombinational events between nonallelic members of a repeated sequence family such that only regions within the repeated elements, and not the flanking sequences, crossover. This avoids causing a major rearrangement. Thus, a single crossing-over leaves flanking sequences intact and gives nonreciprocal results if the cross-strand intermediate is resolved by breakage and reunion without isomerization and if repair of the heteroduplex region is biased toward one of the interacting copies (Figures 2.61 and 9.46). Alternatively, gene conversion according to the double strand break model described in Figures 2.60 and 9.46 gives the same result. By these mechanisms, part or all of the two homologous sequences can become identical (i.e., be homogenized). If there exists some bias such that one family member is favored over others in gene conversions, then that member will establish itself as a founder. In time, individual gene conversions will tend to homogenize the whole family. The correcting copy need not be chromosomal DNA; it could be a cDNA or even an RNA. This type of model suggests one possible source of directional bias. Thus, the founder element(s) would be those loci that are most abundantly transcribed in germ line cells (or their early embryonic progenitors). Note that neither two reciprocal crossovers within the repeat nor a single crossover resolved after isomerization results in conversion (Figure 9.46).

Gene conversion in yeast In yeast, where gene conversion can be studied genetically, a directional bias exists. Some loci are more efficient tem-

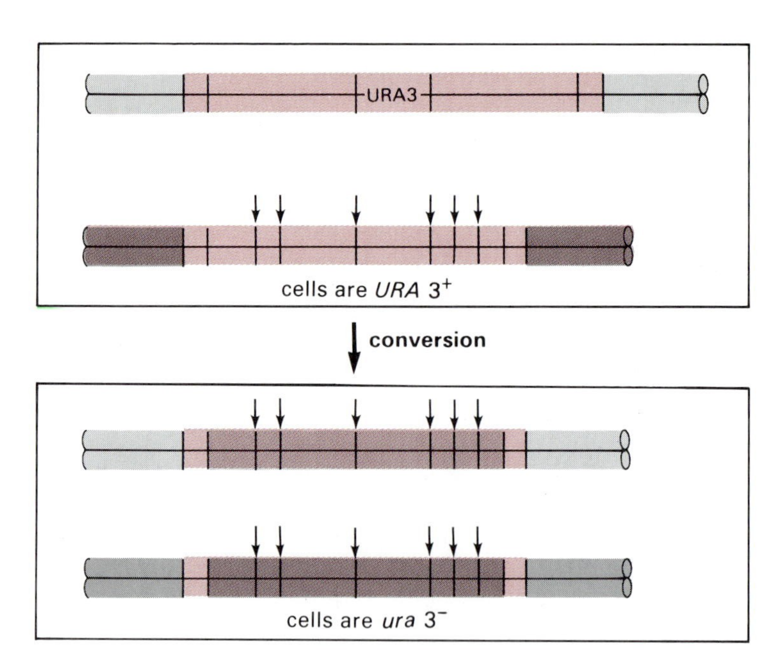

Figure 9.47
A yeast Ty element undergoes gene conversion. Ty sequences (colored) are repeated, interspersed, and transposable in yeast (Section 10.3b). The vertical bars on one of the Tys represent restriction endonuclease sites, and these vary some from one Ty element to another. A strain was constructed with the following properties. The normal *URA3* gene was deleted. Another *URA3* gene, embedded in a Ty element, was introduced, making the cells *URA3*$^+$. Several *ura3*$^-$ derivatives were isolated from the *URA3*$^+$ strain. The *URA3* sequence was missing, but the Ty element remained in place, as shown by its unaltered flanking sequence. The loss of the *URA3* DNA was not the only change in the Ty. The pattern of restriction endonuclease cleavage sites was altered and was similar to the patterns of Ty elements elsewhere in the genome.

plates than others. The reason for the bias is unknown, but the rate of conversion in yeast is high enough so that even a small advantage allows one locus to homogenize a multisequence family. Direct evidence for the conversion of one member of a repeat sequence family to a sequence like that of another member comes from the study of the yeast interspersed repeats called Ty. (Ty is a transposable element described in Section 10.3b.) In yeast, exchange or conservation of flanking sequences is readily scored genetically by reference to mutations in flanking marker genes. In the cases analyzed, loss of a particular Ty was detected by virtue of the simultaneous loss of an essential gene, *URA3*, that had been inserted within it (Figure 9.47); the strain used had no other functional *URA3* gene. Loss of *URA3* was accompanied by conversion of its surrounding Ty sequence to a different Ty sequence as indicated by characteristic restriction endonuclease sites. The new form of Ty was derived from another genomic copy. Because this type of conversion involves the loss of extensive extra sequences (*URA3*), it is more simply explained by the double strand break model for gene conversion than by the single crossing-over and repair model. It is also unlikely that two reciprocal crossings-over occurred within the Ty because the frequency of the replacement is ten times higher than that of even single reciprocal recombinations at a particular locus.

Besides replacement by gene conversion, deletion of progenitor type SINES and insertion of new copies is also likely to contribute to homogenization. Otherwise we might expect that the copy number of a dispersed repeated family would be relatively constant from one species to another, and this is not so. There are, for example, about 9×10^5 and 3×10^5 Alu sequences in humans and chimpanzees, respectively. Definitive evidence for insertion (and/or deletion) comes from the identification of alleles of single genes that differ by the presence of a SINE in a flanking region or intron. Possible mechanisms for insertion or deletion are described in Section 10.4.

9.6 Sequences at Centromeres and Telomeres

Centromeres and telomeres are the most sharply defined morphological features of chromosomes. It was suspected for many years that their gross anatomy and specialized functions would be associated with special DNA sequences. However, for a long time, the only relations established at the molecular level were the presence of satellite DNA at centromeres and telomeres. This information revealed nothing about function. Because molecular cloning makes it possible to study the relationship between structure and function, speculation is now being replaced by definitive information.

a. *Sequences at Centromeres*

Gross anatomy of centromeres In mammals, centromeres have a complex trilaminar disklike structure called a **kinetochore**. There is one kinetochore disk on each side of the chromosome. During mitosis, the microtubules of the spindle fibers appear to attach directly to the dense outer layer of the kinetochore, where loops of chromatin fibers can be seen (Figure 9.48). All mammalian kinetochores must have some structural similarities because they all form complexes with specific antibodies present in the sera of humans suffering from systemic scleroderma, a rare autoimmune disease. Only the kinetochores react with the antibody in mitotic cells, but complexes are also formed at well-defined regions of interphase chromosomes. The latter presumably identify the otherwise obscure centromeres in interphase cells. The scleroderma antibody does not interact with microtubules or with other proteins that are associated with microtubules.

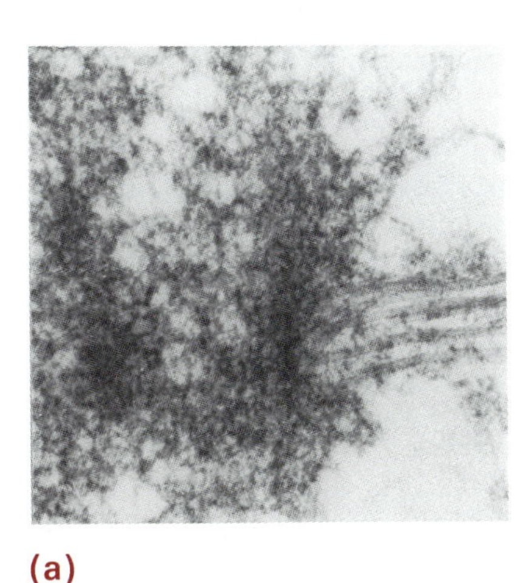

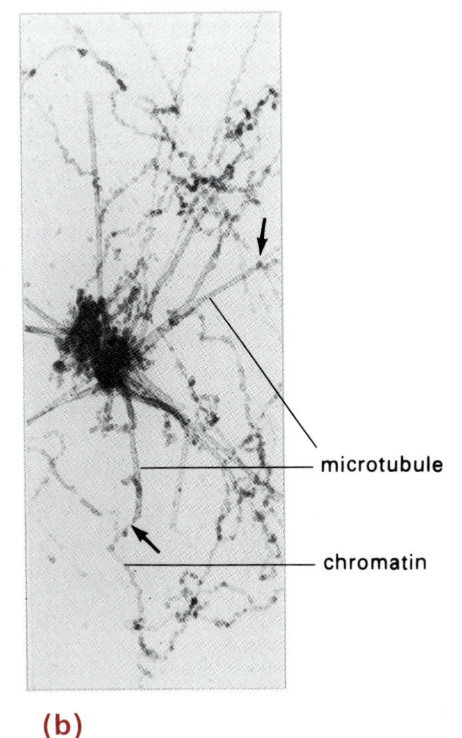

Figure 9.48
Attachment of microtubules to the centromeric region of mitotic chromosomes. (a) Microtubules are directly attached to chromatin fibers in the centromeric (kinetochore) region of Chinese hamster cells (magnification: 40,000X). From H. Ris and P. L. Witt, *Chromosoma* 82 (1981), p. 153. (b) In the yeast, *S. cerivisiae*, a single microtubule joins the chromosomes to the spindle fiber. There is no visible kinetochorelike structure at the attachment site (arrows) of the 20 nm diameter microtubule to the 20 nm chromatin fiber (magnification: 48,000X). From J. B. Peterson and H. Ris, *J. Cell Science* 22 (1976), p. 219.

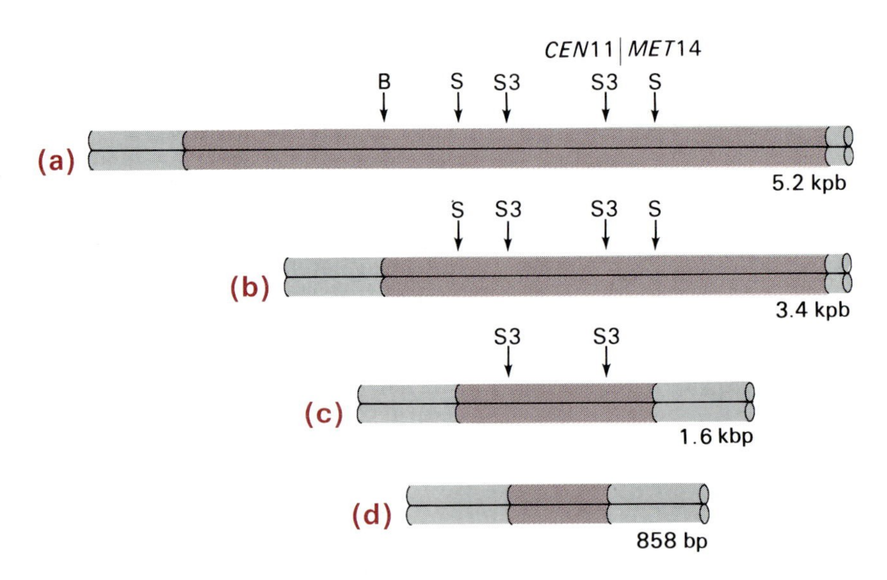

Figure 9.49
Subcloning the *CEN*11 region from a cloned yeast DNA fragment. A yeast synthetic plasmid vector that contains the yeast *TRP*1 gene and an ARS sequence was used to prepare a genomic library. After transfecting *met*14$^-$ and *trp*1$^-$ yeast cells with the library, viable colonies were selected in growth medium devoid of methionine and tryptophan. Cells in one such colony contained a plasmid that had the *MET*14 gene within a 5.2 kbp yeast DNA fragment (a) that also contained a functional centromere. Successive subcloning in the same type of vector (b, c, and d) yielded an 858 bp piece of DNA that still had centromere function but not the *MET*14 gene. Vector sequences are shown in grey.

Yeast (*S. cerevisiae*) chromosomes are small, and only a single microtubule joins each mitotic centromere (in contrast to the large bundles of microtubules seen in vertebrates). Uniquely among known eukaryotes, centromeric satellite DNA has not been detected in *S. cerevisiae*. However, small DNA segments (*CEN*) from yeast centromeres do supply both mitotic and meiotic centromeric function in recombinant vectors (Section 5.6b).

Cloning yeast CEN *regions* Detailed genetic maps of *S. cerevisiae* chromosomes place certain genes close to centromeres. For example, the gene *MET*14, which is required for methionine biosynthesis, is closely linked to the centromere of chromosome 11 (*CEN*11). This linkage provided a way to clone the centromeric sequence itself. A library of yeast genomic sequences was prepared in a yeast synthetic plasmid vector, and a colony containing a plasmid that carries the *MET*14 gene was isolated (Figure 9.49). Genetic analysis suggested that the plasmid also contains a functional centromere because it is stable in yeast cells, its copy number is restricted to about one per cell, and it segregates like a proper "minichromosome" during mitosis and meiosis. Subcloning permitted the isolation of a fragment that maintains centromeric function and is only 858 bp long. Centromeric segments from other yeast chromosomes have also been cloned, and the nucleotide sequences of several are compared in Figure 9.50. The minimum size for full centromere function is estimated to be about 150 base pairs.

	element I		element II		element III
CEN3	ATAAGTCACATGAT	----	83 bp 95% A+T	----	GTATTTGATTTCCGAAAGT•TAAAA
CEN4	AAAGGTCACATGCT	----	79 bp 94% A+T	----	GTTTATGATTACCGAAACA•TAAAA
CEN6	TTTCATCACGTGCT	----	84 bp 95% A+T	----	GTTTTTGTTTTCCGAAGATGTAAAA
CEN11	ATAAGTCACATGAT	----	84 bp 93% A+T	----	GTTCATGATTTCCGAACGTATAAAA
CEN14	GTTAGTCACGTGCA	----	84 bp 94% A+T	----	GTATTTGTCTTCCGAAAAG•TAAAA

Figure 9.50
The conserved nucleotide sequences in several yeast *CEN* regions. Highly conserved bases are boxed. A dot is a missing base. Note that region III has an internal palindrome.

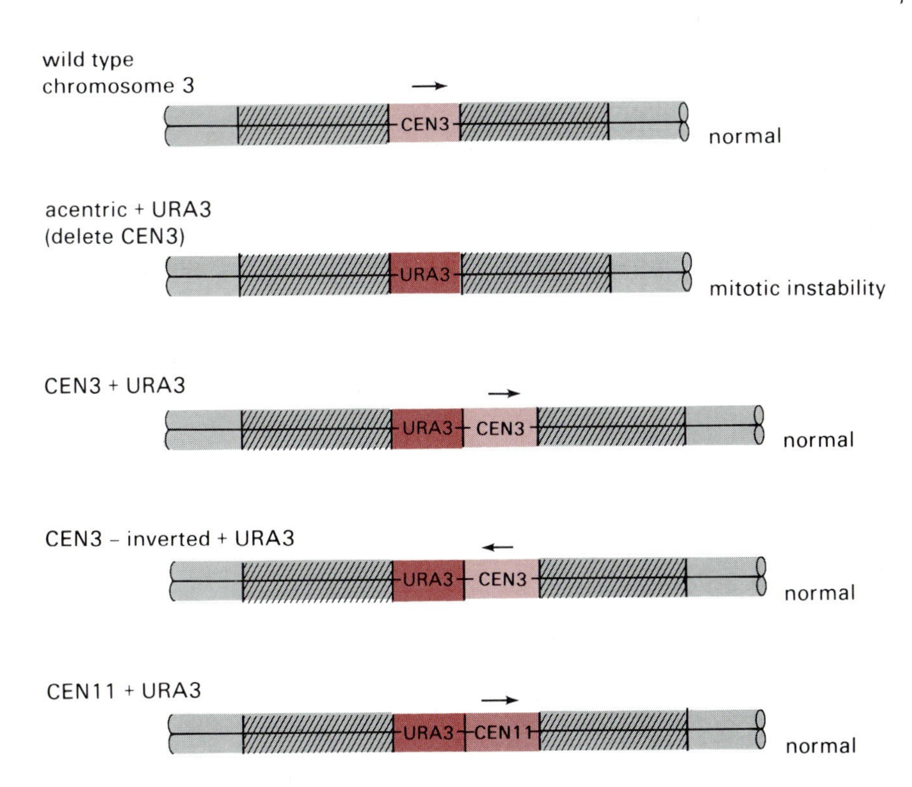

Figure 9.51
Effects of altered *CEN* sequences on centromere function. Wild type *CEN*3 is shown at the top. Site-specific transformation (Figure 5.29) of one of the two chromosome 3s in diploid cells was used to construct the altered chromosomes depicted below. The *URA*3 gene provided a selectable marker for isolation of transformed yeast colonies. The hatched regions represent *CEN*3 flanking sequences that were used in the transforming vectors to provide for homologous recombination. The phenotype associated with each altered chromosome 3 is shown on the right. L. Clark and J. Carbon, *Nature* 305 (1983), p. 23.

Sequence at yeast centromeres The common and distinctive structural features of several *CEN* regions are summarized in Figure 9.50. The flanking sequences at the centromeres are all different. Although the *CEN* regions do not anneal with one another under stringent conditions, they do share two short homologous regions (marked elements I and III in Figure 9.50). I and III are, in each instance, separated by a region (II) that is rich in A·T base pairs. Deletion of the segment from I through III eliminates the mitotic and meiotic centromere function of the plasmids. More specific mutations and deletions within these elements indicate that mitotic centromeric function depends strongly on the integrity of region III and less so on that of I and II, whereas meiotic *CEN* function depends upon the integrity of all three elements. In other experiments, a 627 bp segment that contains *CEN*3 was deleted from the chromosome itself (Figure 9.51). This acentric chromosome is unstable in yeast. However, when the 627 bp segment is inverted within the chromosome, it functions normally. Perhaps most surprisingly, the *CEN*3 region of chromosome 3 can be completely replaced by the cloned *CEN*11 with no untoward effects. This experiment shows that functional centromeres need not be chromosome specific. Moreover, it very clearly indicates that the pairing of homologous chromosomes in meiosis does not depend on homologous centromeres but is more likely to involve similarities between chromosome arms.

The segments of yeast chromatin that include regions I through III of *CEN* are more resistant than the surrounding sequences to the action of DNase I and micrococcal nuclease. The protection against DNase digestion is maintained even when *CEN* is in minichromosome vectors. Thus, these segments must have an unusual chromatin structure. Nuclease resis-

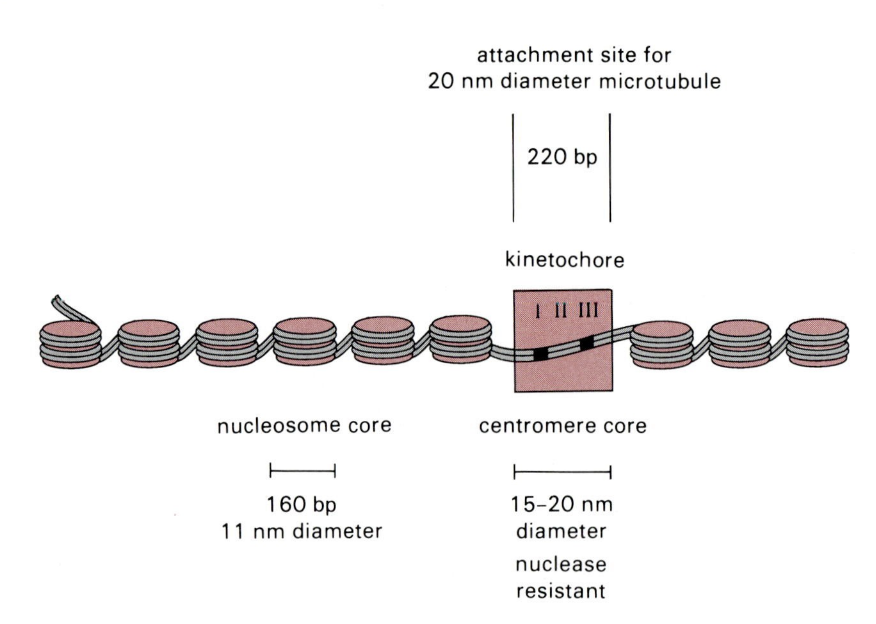

Figure 9.52
Speculative representation of a yeast centromere, based on data for *CEN*11. The functional centromeric sequences containing regions I, II, and III (220 bp) are surrounded by chromatin containing typical nucleosomes. The protection of the 220 bp from digestion by DNases is caused by interaction with centromere-specific DNA binding proteins (box). A microtubule attaches to the core region, which may be the functional kinetochore. After J. S. Bloom and J. Carbon, *Cell* 29 (1982), p. 305.

tance could be caused by special, centromere-specific DNA binding proteins, by the attachment of a microtubule, or by both. Recently, such binding proteins have been detected in yeast cell extracts, using exonuclease III protection and DNA mobility shift assays. Figure 9.52 shows a speculative model for the structure of a yeast centromere. In the model, the *CEN* region itself, along with specific DNA binding proteins, is the kinetochore to which a microtubule attaches in preparation for segregation. Although this gives a possible explanation for one function of the *CEN* sequences, it does not address the question of how *CEN* segments control the copy number of minichromosomes and chromosomes. One possibility is that replication of the *CEN* segment is not completed until mitotic anaphase or the second anaphase in meiosis. This implies that sequences on either or both sides of *CEN* might block the passage of DNA replication forks until a specific signal allowed completion of replication in anaphase. The number of chromosomes would thus be limited to one per daughter cell.

b. Sequences at Telomeres

Telomeres, the ends of eukaryotic chromosomes, are also the ends of linear duplex DNA. Two important molecular puzzles are associated with these structures. The first concerns replication; how are the 5′ termini of chromosomal duplex DNA completed if DNA polymerases do not initiate new chains (Section 2.1h)? Do eukaryotic chromosomes solve this problem with strategies similar to those adopted by the linear duplex adenovirus genome, or do they use alternative mechanisms? The second puzzle concerns the stability of DNA termini in eukaryotic cells. Generally, if breaks in DNA duplexes are not rapidly repaired by ligation, they invite degradation by exonucleases or recombination. Yet the ends of chromosomes are stable, and individual chromosomes are not normally joined to one another end to end. Both puzzles suggest that telomeres have special molecular properties, and the application of molecular cloning techniques has demonstrated this to be true. Certain peculiar properties of ciliated protozoa, in particular *Tetrahymena*, provided the opening experimental wedge.

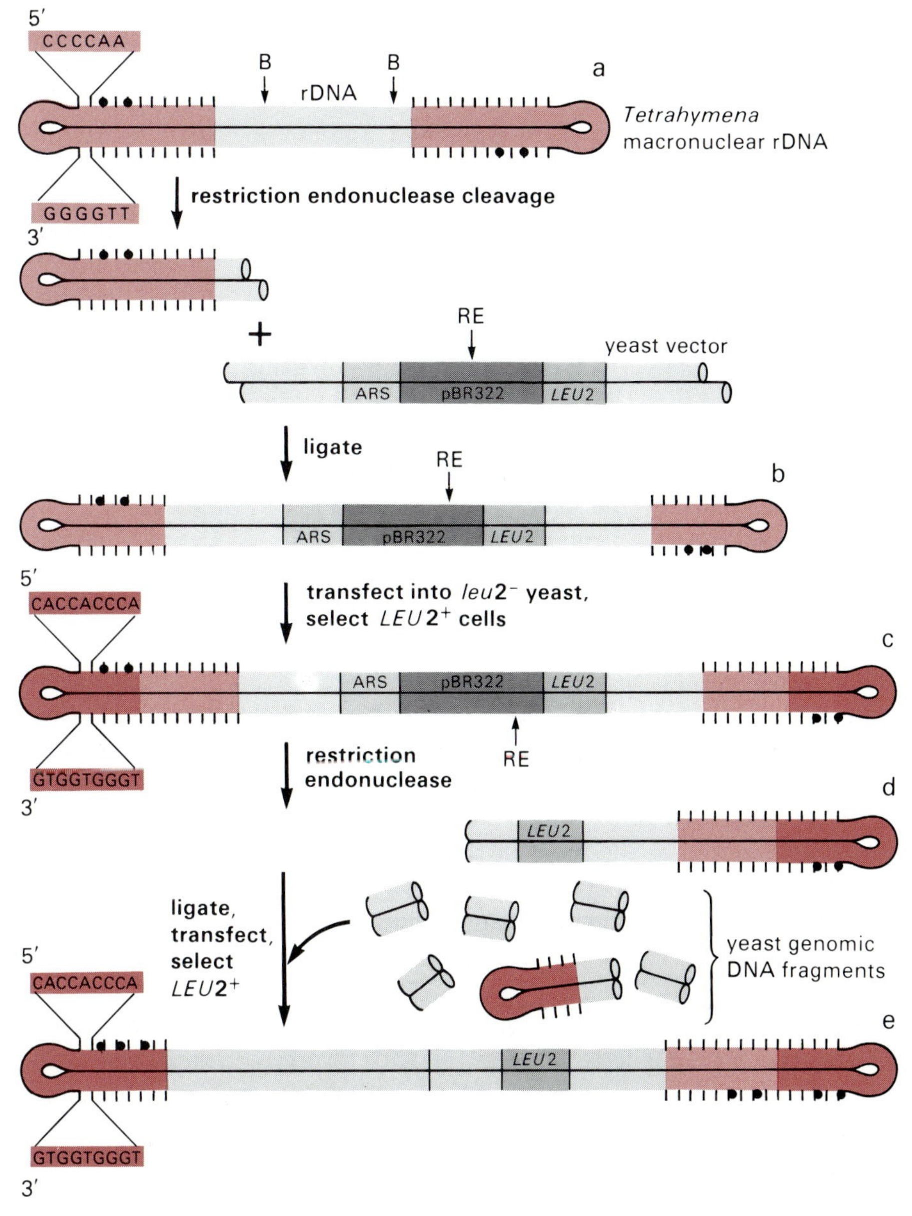

Figure 9.53
The ends of *Tetrahymena* macronuclear DNA fragments (a) act like telomeres in yeast. Dots indicate nicks in the DNA. Restriction endonuclease fragments from the ends of the rDNA fragments were ligated to a linearized yeast–*E. coli* shuttle vector. After transfection, the linear recombinant molecules (b) replicate in yeast. The linear duplex isolated after replication in yeast (c) was then used to generate a vector to clone yeast telomeres, assuming that its ends were *Tetrahymena* telomeres. After cleavage with a restriction endonuclease (RE), the fragments were ligated to restriction endonuclease fragments representing the entire yeast genome. Stable linear duplexes that replicate in yeast were isolated after transfection and selection for *LEU2*$^{+}$ cells (e). The duplexes contained two yeast telomeres, one of which was added onto the *Tetrahymena* telomere present on the vector fragment (d). Further analysis showed that after replication of (b) in yeast, the molecule had already acquired typical yeast telomeres (c), with repetitive terminal sequences, C_nA, where n equals 1, 2, or 3. In the diagram, yeast telomeres are colored more intensely than *Tetrahymena* telomeres. Note that (e) can replicate without the ARS sequence present in (b), probably because (e) acquired an ARS sequence within the joined yeast telomere segment.

Cloning telomeric sequences Some protozoa sequester their germ line genomes in a **micronucleus** and form a "somatic" **macronucleus** from which genes are expressed. Macronuclear DNA contains only some of the sequences present in the micronucleus, is fragmented, and contains many copies of short linear DNA molecules, each of which contains only a limited number of genes. In *Tetrahymena,* those short macronuclear linear duplexes that contain the rRNA genes are abundant enough to be isolated and characterized without cloning. They have, at their ends, between 20 and 70 tandem repeats of the hexanucleotide sequence 5′-CCCCAA-3′ on one strand and the complement 5′-TTGGGG-3′ on the other (Figure 9.53a). Nicks occur at specific positions within some of the C-rich tandem repeats, and there is a hairpin loop at the terminus. Thus, the "ends" of the rDNA are unlike normal termini of DNA strands. These structures are functional telomeres and behave as such when they are introduced into yeast on linear vectors (Figure 9.53b). Thus, linear DNA fragments containing a yeast ARS replication sequence and the termini from the

Tetrahymena rDNA molecules replicate stably in yeast. Normally, in the absence of such special ends, linear fragments are very unstable.

The replicated linear recombinant molecules (Figure 9.53c) isolated from yeast were used as vectors for the cloning of yeast telomeres themselves. One of the unusual ends was first removed by cleavage of the duplex with an appropriate restriction endonuclease (Figure 9.53d). A total digest of yeast DNA was ligated onto the cut end. After transfection, cells containing molecules that replicate in yeast as stable linear DNAs were selected (Figure 9.53d). The functional yeast telomeres were purified by subcloning and used to estimate the copy number of the sequences in the yeast genome. The results indicated the presence of 30–40 copies per haploid, just the number expected if homologous segments occur on each end of the 16 yeast chromosomes.

The structure of the yeast telomere is remarkably like that of *Tetrahymena*. A sequence of the general structure 5′-C_nA-3′ (n equals 1, 2, or 3) occurs in tandem on one strand and contains single strand breaks at specific locations. Additional experiments showed that many simple eukaryotes have similar kinds of tandem repeats at telomeres (e.g., 5′-CCCTAA-3′ in trypanosomes). The very short units rich in Cs and As are generally at the very ends of chromosomes. Other tandem arrays with longer repeat units may occur very close to the termini in several species (e.g., *S. cerevisiae*). They may be involved in the various telomeric interactions identified by cytogenetic analysis, including association between telomeres and between telomeres and the nuclear envelope. Sequence-specific binding proteins that bind to the yeast or the ciliate telomeric repeats have been found in the respective organisms. These proteins may contribute to the stability of chromosome ends by protecting the DNA from degradation and ligation.

Elongation of telomeres during replication When linear DNAs with *Tetrahymena* telomeres replicate in yeast, the ends of newly synthesized molecules are altered (compare b and c in Figure 9.53). They are elongated about 200 bp by the addition of the tandem 5′-C_nA-3′ repeats typical of yeast telomeres but not those of *Tetrahymena*. Moreover, the single strand nicks are positioned like those of yeast, not *Tetrahymena* telomeres. Similar observations are made when telomeres from several other species are tested in heterologous host cells. Such elongation of chromosome ends appears to be a normal part of replication. In trypanosomes, for example, cloned telomeric probes were annealed to restriction endonuclease fragments from DNA isolated at various stages during synchronized multiplication starting with a single cell. In the course of a few hundred generations, the fragments that annealed with the probe grew longer at the rate of 7 to 10 bp per generation (and subsequently lost most of the added base pairs). Similar observations have been made with *Tetrahymena*, and, in other organisms, restriction endonuclease fragments that contain telomere sequences are heterogeneous in length.

The clues offered by the unusual properties of telomeres in yeast and ciliated protozoa resolved into a coherent, if incomplete, picture with the discovery of the enzyme, **telomerase**. Telomerase is a terminal deoxynucleotidyl transferase. The *Tetrahymena* enzyme adds 5′-TTGGGG-3′ repeats, one nucleotide at a time, onto the 3′ ends of specific oligonu-

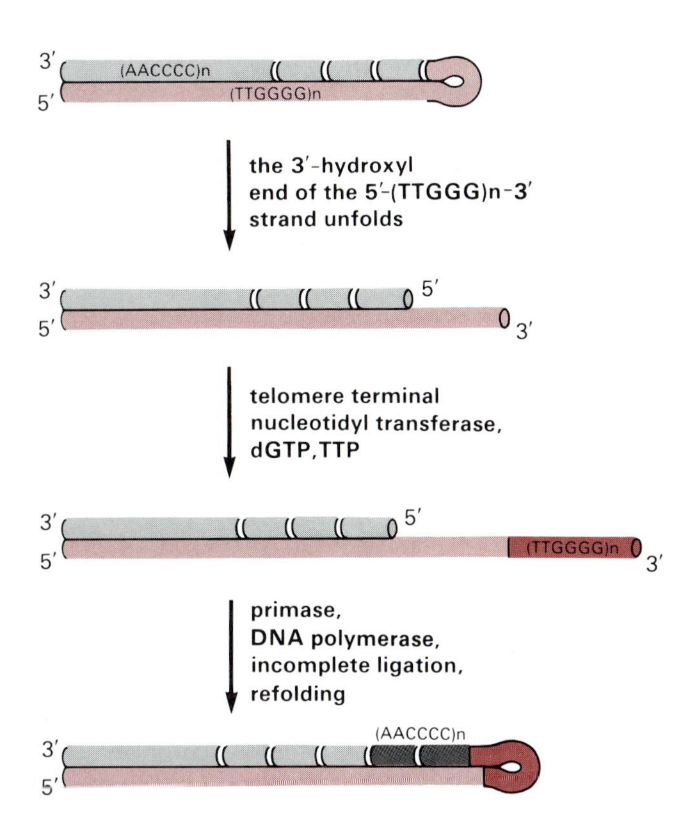

Figure 9.54
A model for the synthesis of *Tetrahymena* telomeres. The folded back 5′-$(TTGGGG)_n$-3′ strand and the nicks in the 5′-$(CCCCAA)_n$-3′ strand are shown schematically in the top diagram. The 3′ end of the 5′-$(TTGGGG)_n$-3′ strand becomes accessible for elongation by the telomere terminal nucleotidyl transferase (telomerase), which adds, one nucleotide at a time, multiple 5′-TTGGGG-3′ units. Primase and DNA polymerase can then copy the 5′-$(TTGGGG)_n$-3′ strand, forming new 5′-$(CCCCAA)_n$-3′ units. Incomplete ligation would leave nicks on the C-rich strand. The folded back loop at the 3′ end of the 5′-$(TTGGGG)_n$-3′ strand re-forms, stabilized by G · G interactions. After C. W. Greider and E. H. Blackburn, *Cell* 43 (1985), p. 405.

cleotide primers: $(TTGGGG)_n$ (*Tetrahymena*-like) and $(TGTGTGGG)_n$ (yeast-like). Thus, telomerase can build the telomeres; the parental DNA duplex does not provide a template (Figure 9.54). Telomerase itself is a large ribonucleoprotein complex; the RNA as well as the proteins are essential for enzymatic activity. Indeed, the telomerase RNA contains the sequence 5′-CAACCCCAA-3′, which appears to provide the template for TTGGGG synthesis. These recent results do not clarify the mechanism by which the C-rich telomeric strand is synthesized. It is possible that the G-rich repeats serve as template and primer. The hairpin loops at the termini involve the G-rich strand and are stabilized by unusual interactions between pairs of G residues.

These mechanisms, elucidated with single cell organisms, are likely to be common in most eukaryotes. Cloning and sequencing of telomeric sequences from the plant *Arabidopsis* and from humans show that they are composed of oligonucleotide repeats very similar in sequence to those of the ciliates and yeast. Telomerase has been demonstrated in human cells.

Formation of multiple telomeres in the macronuclei of protozoa Reconsider now the *Tetrahymena* rDNA molecules. There are only two genomic rDNA genes in the diploid *Tetrahymena* micronuclear DNA. Amplification in the macronucleus gives hundreds of linear molecules, each of which includes the rDNA and the telomeric structure. Yet the rDNA is not telomeric in the micronucleus. Furthermore, most of the 2 to 100 kbp linear duplex DNA molecules typical of the *Tetrahymena* macronucleus contain the same type of terminal sequences. These molecules include many genes besides rDNA and are produced as part of a general program of DNA fragmentation (and elimination), leading to the development of

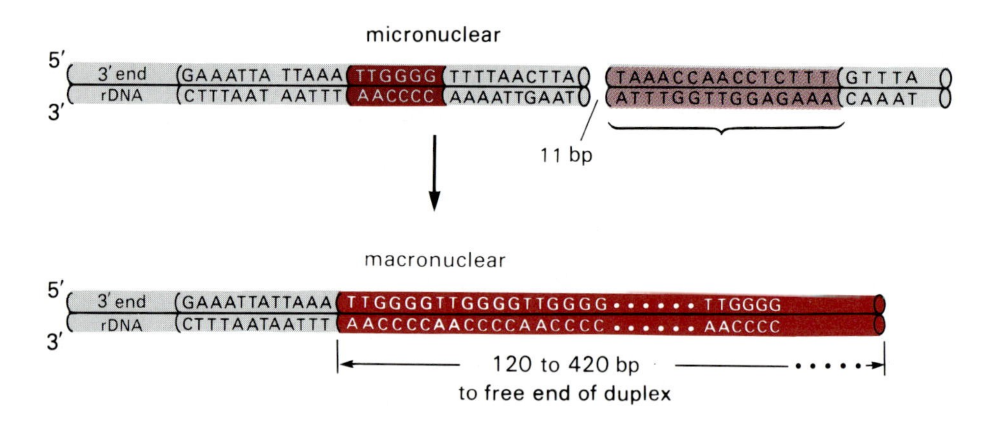

Figure 9.55
Sequences at the 3′ end of rDNA in the micro- and macronuclear DNA of *Tetrahymena*. The 15 bp sequence at which cleavage of micronuclear DNA occurs is bracketed. A single copy of 5′-CCCCAA occurs in the micronuclear locus, compared to the many tandem copies in the macronuclear rDNA. After B. O. King and M-C. Yao, *Cell* 31 (1982), p. 177, and M-C. Yao, K. Zheng, and C-H. Yao, *Cell* 48 (1987), p. 779.

the somatic macronucleus. A similar situation holds in a different ciliated protozoan, *Oxytricha*, except that the terminal repeated sequence is 5′-CCCCAAAA-3′, as are the repeats at the telomeres of *Oxytricha*'s micronuclear chromosomes. For example, the macronuclear actin gene of *Oxytricha* is contained in 1.6 kbp fragments flanked by four copies of 5′-CCCCAAAA-3′ (or 5′-TTTTGGGG-3′). How are such terminal segments formed? In *Tetrahymena*, a conserved 15 bp sequence occurs at the sites where micronuclear DNA is cleaved to form macronuclear DNA fragments (Figure 9.55). The sequence is likely to provide a recognition site for a specific endonuclease. After (or concomitant with) cleavage, some of the surrounding DNA is eliminated, and the 5′-GGGGTT-3′ ends are added, presumably by a terminal nucleotidyl transferase such as that described in the previous paragraph. Note that the micronuclear progenitors of macronuclear fragments in both *Tetrahymena* and *Oxytricha* need not have any flanking 5′-CCCC$(A)_n$-3′ repeats. Altogether, the formation of macronuclear DNA is associated with the excision of segments from micronuclear DNA, loss of some noncoding flanking DNA, addition of the oligonucleotide repeat, formation of a hairpin, and construction of functional telomeres. Hairpin formation may contribute to the stabilization of the DNA fragments by inhibiting degradation and inappropriate ligation to other duplexes.

c. Yeast Artificial Chromosomes

All the segments required to construct a functional yeast chromosome are now available. Cloned fragments containing an ARS sequence for replication, a *CEN* sequence for centromere function, and telomeric segments can be ligated together *in vitro* in a suitable arrangement. Other DNA segments can be included in the linear duplex to supply one or more selectable marker genes such as *LEU2* or *HIS3* (Figure 9.56). When the total DNA in the linear molecule is less than about 20 kbp, the synthetic chromosomes are unstable in mitotic division, and the centromere does not function to control copy number. Thus, many molecules accumulate in each cell. However, stability, control of copy number, and accuracy of segregation all improve if the molecules are longer than 50 kbp. Those that are 150 kbp behave more or less properly in both mitosis and meiosis. Evidently, a minimum length is critical to chromosomal move-

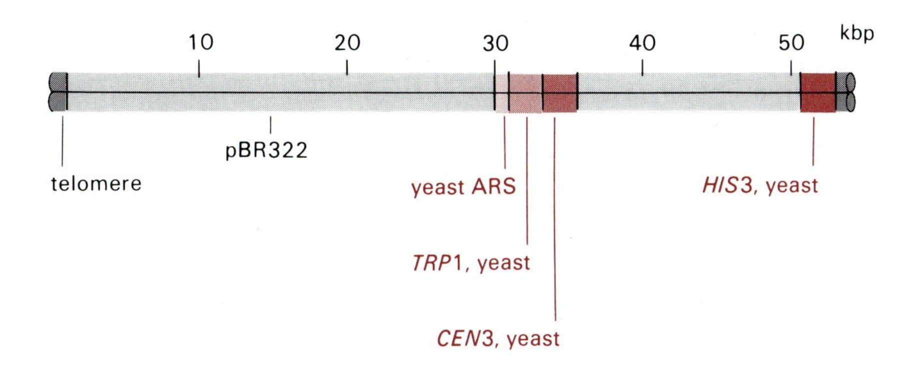

Figure 9.56
A functional synthetic yeast chromosome. After A. W. Murray and J. W. Szostak, *Nature* 305 (1983), p. 189.

ment during cell division. Such yeast artificial chromosomes (**YACs**) are used as vectors in the cloning of very long DNA segments for mapping complex eukaryotic genomes.

9.7 Genomes of Eukaryotic Organelles: The DNA of Mitochondria and Chloroplasts

The packaging of essential genetic information within extrachromosomal genomes is a distinctive feature of eukaryotic cells. There are two such types of cytoplasmic DNA molecules; one in the mitochondria of all eukaryotes and the other in the chloroplasts of green plants and algae. As might be expected for cytoplasmic elements, both are inherited maternally and in non-Mendelian fashion.

Most of the proteins that make up the functional and structural components of mitochondria and chloroplasts are encoded in chromosomal DNA, synthesized on cytoplasmic ribosomes, and transported to the organelle (Section 3.10c). But a few are encoded in the extranuclear DNA and are synthesized on special organellar ribosomes. Thus, the construction of the organelles is a joint effort between two genomes and two translational apparatuses. The RNA components of the organellar ribosomes are themselves encoded within the mitochondrial and chloroplast genomes, as are the tRNAs used in translation.

The interplay between chromosomal and both mitochondrial and chloroplast gene products is apparent genetically. For example, mitochondrial mutations in yeast and maize can be suppressed by secondary chromosomal mutations. Certain types of male sterility (production of infertile pollen) that are widespread in plants are associated with mitochondrial DNA mutations and governed by non-Mendelian, maternal inheritance. Mutations in chromosomal genes restore fertility to these plants, and, indeed, the various types of cytoplasmic male sterility are distinguishable because each is restored by a different chromosomal gene.

Table 9.14 lists the genomic provenance of some mitochondrial constituents. Even individual enzymes are composed of polypeptides contributed by the two separate genomes. In addition, the expression of some mitochondrial genes depends on functions provided by the products of nuclear genes (e.g., the processing of yeast mitochondrial DNA transcripts

Table 9.14 The Genomic Provenance of Some Mitochondrial Constituents

Chromosomal (nuclear)	Mitochondrial
Enzymes + factors for replication	
RNA polymerase + factors for transcription	
RNA splicing enzymes (yeast)	Maturases for splicing (yeast)
Ribosomal proteins	1 ribosomal protein (yeast)
Aminoacyl tRNA synthetases	tRNA
Factors for translation	rRNA
Subunits of cytochrome oxidase	Subunits of cytochrome oxidase
Subunits of ATPase	Subunits of ATPase
Subunits of NADH-dehydrogenase	Subunits of NADH-dehydrogenase

Note: The distribution varies some from one organism to another. The respiratory chain NADH dehydrogenase is also known as complex I or the NADH-ubiquinone oxidoreductase.

to functional mRNAs). Although Table 9.14 summarizes the general situation, different organisms have somewhat different arrangements. A gene that is mitochondrial in one organism can be genomic in another, and sometimes both genomes contain a gene for a given polypeptide. A very similar situation occurs with chloroplasts.

a. *Mitochondrial Genomes*

Many mitochondrial genomes are closed, circular, duplex, superhelical DNA molecules. Most of them encode a very similar set of functions. But beyond this, it is difficult to generalize about mitochondrial DNAs. Even those of closely related species may have very different restriction endonuclease maps and only partly similar sequences. The order of homologous genes around the DNA circle also varies. In plants, individual organisms sometimes harbor mitochondrial genomes of different sizes, and both linear and circular molecules occur. Other organisms, including some fungi and protozoa, have linear mitochondrial genomes.

Mitochondrial genomes vary enormously in size. In animals, they are relatively small, usually less than 20 kbp. But yeast mitochondrial DNA is about 80 kbp long, and that of plants ranges from several hundred to several thousand kilobase pairs. An interesting common feature of all mitochondrial DNAs is the absence of detectable methylation. In the very large mitochondrial genomes, a great proportion of the extra DNA is taken by up noncoding sequences of unknown function. In contrast, the mammalian versions are very efficiently organized, with little or no DNA between genes.

The DNA in mitochondria evolves much more rapidly than does nuclear DNA. Mutations collect about ten times as fast. As a result, there is extensive intraspecies polymorphism, and measurable changes are detectable even within a few generations. Because of the rapidity of change, restriction endonuclease site polymorphism in mitochondria is a conve-

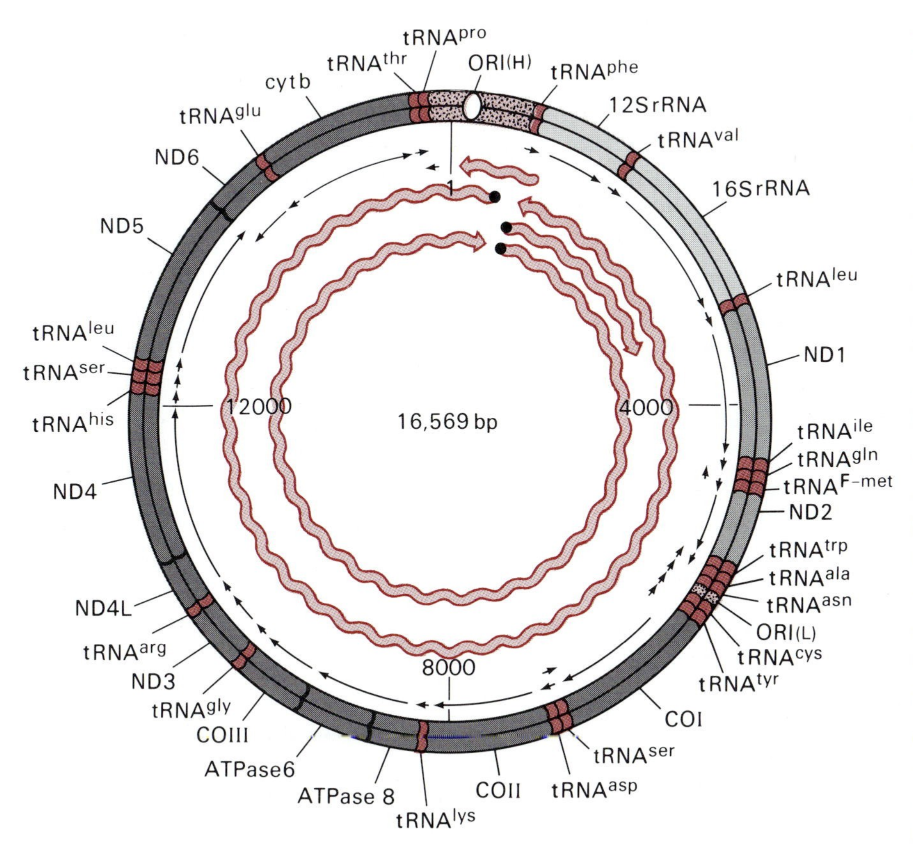

Figure 9.57
Human mitochondrial DNA. Protein-coding genes are darker grey; they encode the three large subunits of cytochrome C oxidase (*CO*-I, *CO*-II, *CO*-III), cytochrome b (*cyt*b), subunits 6 and 8 of the ATPase complex, and subunits of the NADH dehydrogenase complex (*ND*). Genes encoding the various rRNAs are light grey; tRNA genes are brightly colored. Each gene has been given a systematic name, but for simplicity, they are indicated here and in Figure 9.58 by abbreviations for the gene products. The dots show the approximate location of transcriptional start sites on the two strands. Note that two RNAs are transcribed from the H strand template (clockwise). The short one is the source of the bulk of the rRNA; the longer one gives rise to the mRNAs and most of the tRNAs. Only a single, full length L strand transcript is made. A 7S RNA is transcribed in the counter-clockwise direction very close to the initiation site. This RNA may play a role in transcript regulation or maturation. Stippled regions are origins of DNA replication. Small arrows show the 5′ to 3′ direction of the sense strand for the various genes.

nient parameter for studying intraspecies population genetics in plants and animals, including humans.

Mammalian mitochondrial DNA The entire nucleotide sequence of several mammalian mitochondrial DNAs is known (Figure 9.57). All these genomes are very similarly organized, as is *Drosophila* and *Xenopus* mitochondrial DNA. Nevertheless, the similarity of the nucleotide sequences ranges from only 20 percent to 90 percent at different loci, and large swings in extent of homology occur even among single coding regions.

The nucleotide defined as residue 1 is in the region of the origin of mitochondrial DNA replication, a segment that may also be important to the regulation of mitochondrial gene expression (Figure 9.57). Proceeding clockwise, the first gene encountered in human mitochondrial DNA is for a $tRNA^{Phe}$, followed closely by the gene for 12S rRNA, the RNA of the small ribosomal subunit of mitochondrial ribosomes, and so forth. Between most of the RNA or protein coding regions, there are one or more tRNA genes. The spaces between genes are generally less than 25 base pairs long, with most of them fewer than 3. In some cases (e.g., between $tRNA^{Ile}$ and $tRNA^{Gln}$ and between ATPase subunits 6 and 8), the end of one gene overlaps the start of the next.

The transcription of mammalian mitochondrial DNA Unlike most eukaryotic genes, each of which constitutes an independent transcription unit, both strands of the entire mitochondrial genome are transcribed into single

RNA chains starting from sites near the origin of H strand DNA replication, ORI-H (H for heavy because it bands to a higher density in isopycnic centrifugation in CsCl than does the other, light, or L strand). In addition, shorter transcripts of the H strand that stop at the 3′ end of the 16S rRNA gene are made. About ten times more of the short H strand transcripts are made than the long H strand transcripts. The short H strand transcripts can be processed to yield 12S rRNA, 16S rRNA, $tRNA^{Phe}$ and $tRNA^{Val}$. Their abundance provides for a high rate of accumulation of the rRNAs. The longer H strand transcript and the L strand transcript are processed to give mRNAs and the other tRNAs. Both strands give rise to functional RNAs, although one, the L strand, is the predominant template strand and hybridizes to most of the mature RNAs (Figure 9.57).

The two mitochondrial DNA strands are transcribed by a special mitochondrial RNA polymerase (encoded in nuclear DNA). Neither the mechanism of transcription initiation nor how the large primary transcripts are converted into discrete RNAs is understood in detail. The promoter elements are contained in segments that are 40 bp or less in length and bracket the transcriptional start sites. These differ from one mammalian species to another; accordingly, the mitochondrial RNA polymerases are species specific. Current models ascribe a unique role in the maturation of the long RNAs to the tRNA coding regions that flank almost all genes. Briefly, the tRNA regions of the long transcripts might assume folded, intramolecular, hydrogen bonded structures. Specific ribonucleases, perhaps similar to *E. coli* RNase P, would recognize these folded regions and cleave on either side, thereby generating both the tRNAs and rRNAs or mRNAs. An RNase P-like activity has indeed been purified from mammalian mitochondria. The processed transcripts then undergo typical maturation reactions; the mRNAs are polyadenylated, and the tRNAs acquire the customary 3′ termini, 5′ CCA. However, the 5′ termini of the mRNAs are unusual for eukaryotes in that they are not capped. No splicing is necessary because none of the genes contains an intervening sequence.

The L strand transcripts are also important for the initiation of DNA synthesis at ORI-H. Some of the transcripts appear to be cleaved within a region of about 200 nucleotides from the site of transcription initiation. These short RNAs then serve as primers for DNA replication.

The unusual mammalian mitochondrial translational apparatus The structure of mitochondrial mRNAs raises several problems regarding translation. Not only are typical cap structures missing, but there are no significant lengths of 5′ leader sequences upstream from the initiation codons. Thus, mitochondrial ribosomes must recognize and bind to the mRNAs in an unusual fashion. Another oddity of mitochondrial translation is the use of unusual codons. In addition to AUG, three other triplets, AUA, AUU, and AUC, can act as initiation codons (Table 9.15) and signify methionine. The universal stop codons UAA and UAG are used in mammalian mitochondria, but the other normal stop codon, UGA, encodes tryptophan. The universal arginine codons AGA and AGG are used as stop codons rather than coding for amino acids.

Some mitochondrial genes have no stop codon at all at the end of the coding sequence. One example is the gene for ATPase subunit 6. In these

Table 9.15 Human Mitochondrial Codons and Anticodons

Codon		*Anti*	*Codon*		*Anti*	*Codon*		*Anti*	*Codon*		*Anti*
UUU	phe	AAG	UCU	ser	AGU	UAU	tyr	AUG	UGU	cys	ACG
UUC			UCC			UAC			UGC		
UUA	leu	AAU	UCA			UAA	stop		UGA	trp	ACU
UUG			UCG			UAG			UGG		
CUU	leu	GAU	CCU	pro	GGU	CAU	his	GUG	CGU	arg	GCU
CUC			CCC			CAC			CGC		
CUA			CCA			CAA	gln	GUU	CGA		
CUG			CCG			CAG			CGG		
AUU	ile	UAG				AAU	asn	UUG	AGU	ser	UCG
AUC			ACU	thr	UGU	AAC			AGC		
AUU	met	UAC	ACC								
AUC			ACA			AAA	lys	UUU	AGA	stop	
AUA			ACG			AAG			AGG		
AUG											
GUU	val	CAU	GCU	ala	CGU	GAU	asp	CUG	GGU	gly	CCU
GUC			GCC			GAC			GGC		
GUA			GCA			GAA	glu	CUU	GGA		
GUG			GCG			GAG			GGG		

Note: The codons are shown 5′-to-3′ and the anticodons 3′-to-5′. In some mammals, AUA, AUU, and AUC can act as initiator codons, being recognized by N-formyl-methionine-tRNA anticodon: 3′-UAC-5′.

cases, processing of the primary transcript leaves either a terminal U or UA. The subsequent polyadenylation of the 3′ terminus of the messenger creates the UAA stop codon.

Correlated with the unusual genetic code is an unconventional family of tRNAs. Each of the 22 human mitochondrial tRNAs is indicated on the map (Figure 9.57) and also shown on Table 9.15, where it is represented by its anticodon. Unlike the universal code, which utilizes separate tRNAs for different redundant codons, each mitochondrial tRNA is used for a set of two or four redundant codons. Each tRNA that interacts with a two-member set of codons utilizes G·U base pairings in the codon-anticodon interaction for one member of the set. Each tRNA that interacts with a four-member set of codons has a U at the 5′ terminus of the anticodon. Either the U can base pair with any of the four bases, or only two base pairs operate in the codon-anticodon interaction.

Yeast mitochondrial DNA Although it contains just about the same number of genes as does human mitochondrial DNA, the yeast genome is almost five times larger—about 80 kbp (Figure 9.58). Most of the excess is consumed by large AT-rich segments of unknown function and by intervening sequences that interrupt coding regions, a feature that is absent from mammalian mitochondrial genomes. To a large extent, the extreme

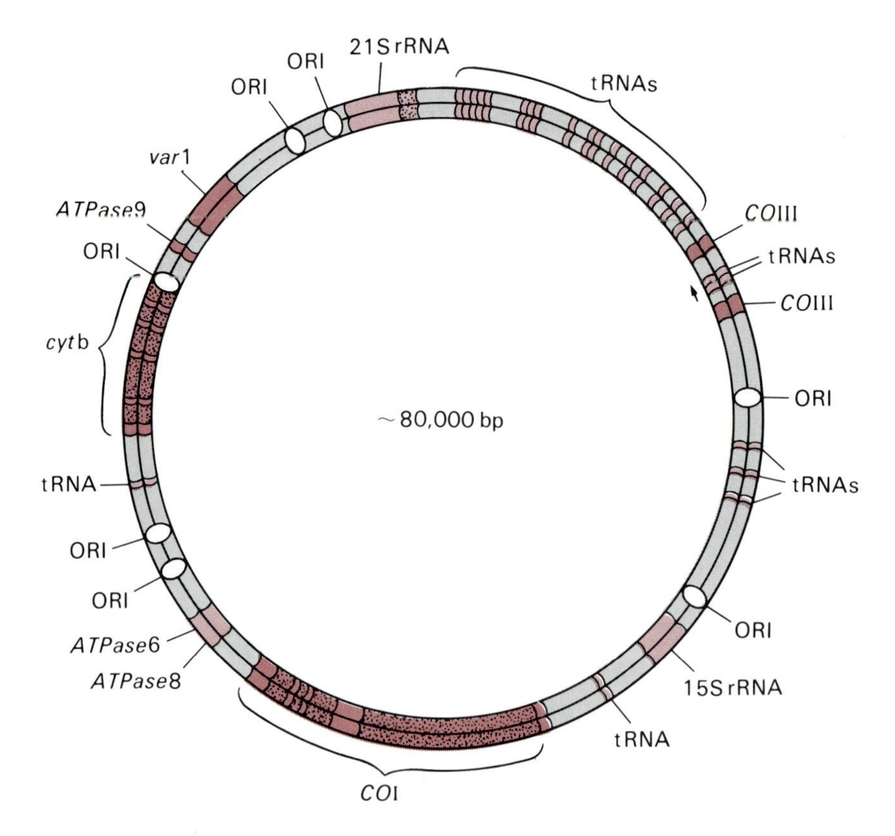

Figure 9.58
Some features of yeast (*S. cerevisiae*) mitochondrial DNA. Genes encoding proteins are darkly colored. Genes encoding RNAs are lightly colored. Stipples are intervening sequences (several of which contain open reading frames, only some of which are shown). Grey shows regions that are uncharacterized; some of these include origins of replication. Transcription is mainly clockwise; one known exception is the tRNA at about 2 o'clock (see arrow).

variability of mitochondrial DNA structure from one strain of yeast to another is accounted for by changes in the noncoding regions. However, remarkable changes also occur within coding regions without severely damaging the function of the corresponding gene products.

Transcription and translation of yeast mitochondrial DNA Whatever the constraints on the size of mammalian mitochondrial genomes (and those of flat worms, sea urchins, insects, *Chlamydomonas*, and so forth), they are absent in yeast. The freedom to be larger in yeast is associated with fewer peculiarities in structure and function. The tRNA genes do not serve as punctuation marks that separate other genes. Transcription starts from at least 19 independent promoters, not from a single one; it is catalyzed by a specific mitochondrial RNA polymerase that is encoded in chromosomal DNA. A short sequence that serves as a promoter (consensus 5′-A/TTATAAGTA-3′) and occupies positions from −9 to +1 relative to the initiation of transcription at +1 is found at the 5′ end of each transcription unit. The long transcripts are processed by enonuclease action into individual, functional RNAs. Although the yeast mitochondrial genetic code also departs from the universal rules, it does so less than its mammalian counterparts. Similarly, rRNAs are a more customary size, 3200 and 1660 nucleotides in length, compared to 1559 and 954 for the human ones. Although the coding sequences are clearly homologous to those in the human mitochondrial genome, the mRNAs have 5′ terminal leaders and 3′ terminal trailers, like most mRNAs. Compare, for example, the human and yeast cytochrome b genes (Figure 9.59).

Nevertheless, the yeast mitochondrial genome demonstrates some idiosyncrasies of its own. It is the only genome known in which the rRNA

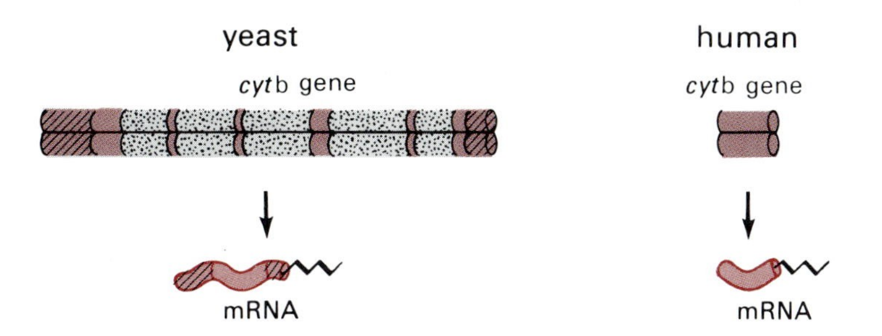

Figure 9.59
The cytochrome b genes (*cytb*) and messenger RNAs of yeast and human mitochondria. Exons are colored; yeast introns are stippled and grey. After P. Borst and L. A. Grivell, *Nature* 290 (1981), p. 443.

genes are separated by unrelated genes. Also, it encodes one ribosomal protein at the *var*-1 locus, although the others are encoded by chromosomal DNA, as are all the ribosomal proteins of animal mitochondria. And it provided the first examples of the presence of coding regions within intervening sequences.

The genes inside yeast mitochondrial DNA introns Introns occur in the 21S rRNA, cytochrome b (*cyt* b), and cytochrome oxidase subunit I (*CO*-I) genes in *S. cerevisiae* mitochondrial DNA. Some of these introns are optional and are not present in all yeast strains. Others appear to be essential because mutants deficient in respiration map both to mitochondrial exons and introns. Genetic analysis also suggested that intervening sequences in, for example, *cyt* b are required for cytochrome b synthesis. Some of the intron mutations are, in addition, **pleiotropic**; not only is the synthesis of cytochrome b inhibited, but the expression of *CO*-I is altered. Such a property suggests that the intron might encode a *trans* acting protein, a concept that was confirmed by the discovery of open reading frames in many yeast mitochondrial introns. Other intron mutations interfere with the formation of the specialized secondary structures that are required for splicing the introns (Section 8.5d).

Cloning and sequencing demonstrated that, within the *cyt* b gene, two of the intervening sequences, I2 and I4, contain long open reading frames capable of encoding proteins (Figure 9.60). Maturation of the primary transcript begins with removing I1 and joining exons 1 and 2 (splicing of such group II introns is described in Section 8.5c). This RNA serves as the mRNA for synthesis of the I2 maturase, a protein involved in subsequent RNA processing. The coding sequences for I2 maturase start in the first exon of the cytochrome b gene and continue into the open coding region of I2. In the next step, the I2 maturase assists in the removal of I2, thereby destroying its own mRNA, a unique regulatory mechanism. Genetic and molecular studies also indicate that the long open reading frame in I4 encodes another maturase that participates in the splicing out of I4. And the I4 maturase is responsible for the pleiotropic affects of the previously mentioned mutations. Mutations in *cyt* b I4 block the splicing of an intron in transcripts of the *CO*-I gene. Similarly, the first intron of the *CO*-I gene itself encodes a maturase.

The 21S rDNA intron is optional. This 1.1 kbp DNA segment is not required for splicing, either of itself or of other mitochondrial introns. The intron has nevertheless been studied extensively because of its unusual properties. When yeast strains carrying the 21S rDNA intron (ω^+) are crossed with strains that lack the intron (ω^-), almost all the progeny mitochondria, including those derived from the ω^- parent, are ω^+. This results from the insertion of a copy of the 1.1 kbp intron into the ω^-

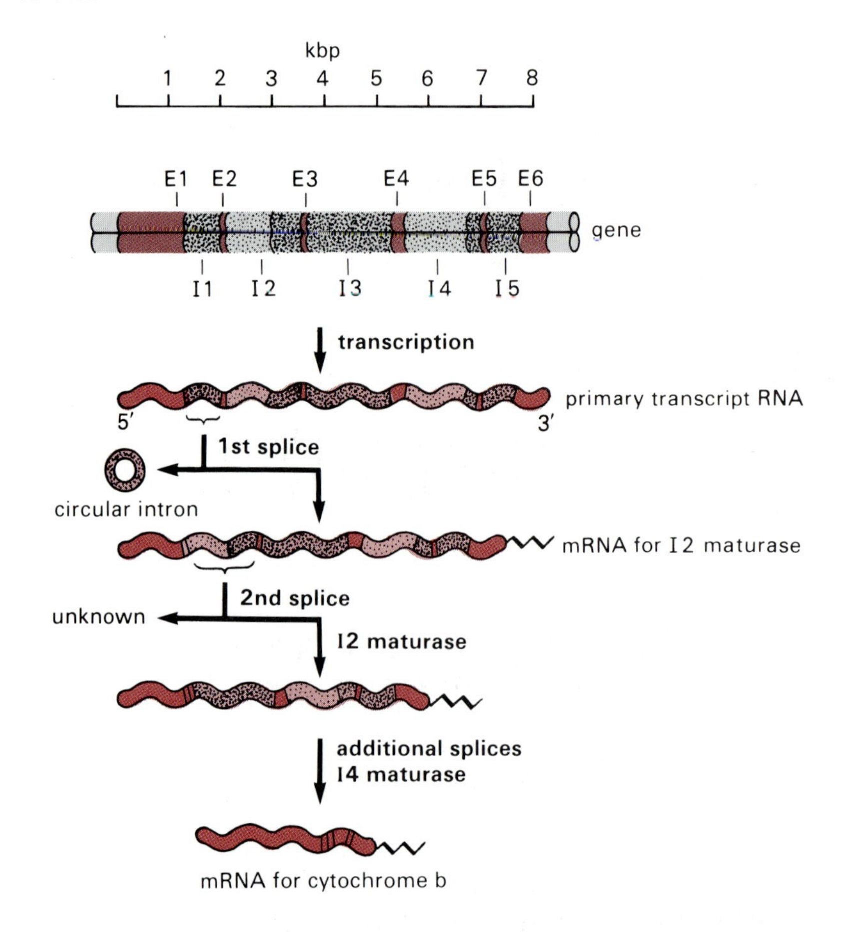

Figure 9.60
The yeast cytochrome b gene. Exons are colored; introns are stippled and grey. Lightly stippled intron regions include coding regions for the I2 and I4 maturases. The details of the final splicing reactions are unknown.

21S rDNA. Insertion is always at a precise position in the 21S rDNA: 5′-GATAACAG-3′. Mutations within this target sequence (ω^n) abolish insertion. Mutations within the 1.1 kbp intron also abolish insertion if they disrupt the 235 codon long open reading frame contained therein. The intron encodes a site-specific endonuclease that is required for insertion into an ω^- 21S rDNA. Thus, the intron is a type of transposable element (Chapter 10). Note that this system provides a model for explaining how introns might arise through insertion into a preexisting gene.

The replication of mitochondrial DNA Cultured mammalian cells contain thousands of mitochondrial DNA molecules. The actual numbers vary from one organism to another, and there does not seem to be any precise number, even in a particular cell type. Although mitochondrial DNA replication is not coupled to the cell cycle, it usually produces enough new molecules to supply each daughter cell fully.

Replication of mouse mitochondrial DNA is best understood (Figure 2.5). Recall that each strand of the duplex has its own replication origin, Ori-H and Ori-L. In mouse cells, but not those of all animals, most mitochondrial genomes include a short segment of replicated H strand (550–650 bp long) and consequently a displacement loop (**D loop**) in the Ori-H region. The molecules thus appear to be poised to start replication in earnest. Another unusual feature of the mature genomes is the scattered presence of ribonucleotides in place of deoxyribonucleotides. Presumably, these are the remains of RNA primers of DNA replication and the insertion of ribonucleotides during chain extension. However, why mitochondrial DNA synthesis should be so sloppy in this regard is unknown. In contrast to mammals, yeast mitochondrial DNA has seven or more scattered origins of replication (Figure 9.58). Each origin region contains three short GC-rich segments separated by two long regions that are almost exclusively AT.

With one exception, there is no information on the enzymes or protein factors involved in replication. It is known that, of the three typical eukaryotic DNA polymerases, the γ-polymerase participates in mitochondrial DNA synthesis.

b. The Unusual Mitochondrial DNA of Trypanosomes

Cells of the order of protozoans called kinetoplastidae, which includes the trypanosomes, have as a distinguishing characteristic a peculiar disklike structure within their single mitochondrion. The disks are called **kinetoplasts**, and they contain thousands of circular DNA molecules. In *Trypanosoma brucei*, a typical example, 45 of the circles are about 21 kbp in circumference (**maxicircles**); the remaining 5500 are about 1 kbp long (**minicircles**). The circular DNA molecules form an interlocking network (Figure 9.61). Closed circular DNA molecules linked in this way are called **catenated**. No other example of massive catenated networks is known. Isolated kinetoplast DNA networks are about 10 μm in diameter, but within cells, they are condensed into disks tenfold smaller. Invariably, the disk is located at the base of the trypanosome's flagellum.

Except for a conserved segment a few hundred base pairs long, the minicircles within a single network often vary in sequence, and even the conserved segment is not necessarily the same from one species to the next. All known minicircles do, however, have within the conserved region the segment 5′-GGGGTTGGTGTAA-3′, and this appears to be an origin of replication. Moreover, in the vicinity of the conserved region, minicircles contain a region of bent helix (Section 1.1e). The function of the minicircles is unknown, and it is not clear whether they encode any genes. One small 240 nucleotide long transcript that includes part of the conserved region has been identified, but it, like the entire minicircle, contains only relatively short open reading frames of unknown significance. The maxicircles represent functioning mitochondrial genomes similar to those already described. Unlike minicircles, all the maxicircles in a network are very similar, if not identical in sequence. Like the yeast mitochondrial genome, they are very rich in A·T base pairs (80 percent).

The replication of the complex kinetoplast DNA network occurs once in each cell generation. It starts with unlinking the catenates to yield

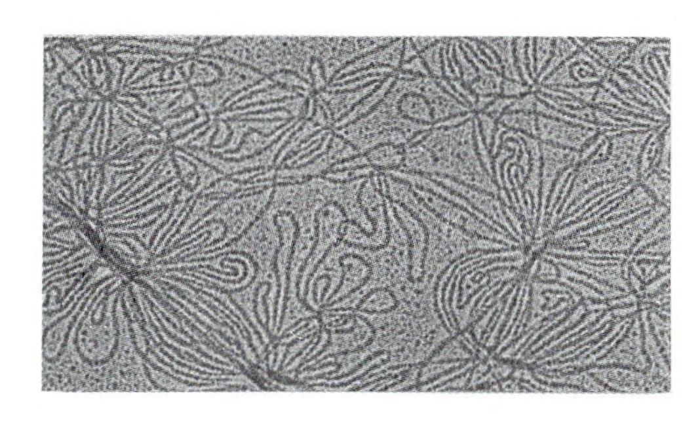

Figure 9.61
Trypanosome (*Crithidia fasciculata*) kinetoplasts: an electron micrograph showing part of a catenated network of mini- and maxicircular DNA. Each loop is a minicircle of 2.5 kbp. Courtesy of P. T. Englund. See P. T. Englund, *Ann. Rev. Biochem.* 51 (1982), p. 701.

individual minicircles, presumably through the action of a type II topoisomerase. The minicircles are replicated to form free duplex circles with single strand gaps, and these are then reattached to the matrix. Thereafter, the gaps are repaired to form closed duplexes. Finally, the matrix divides in two, and the two networks segregate into the daughter cells. In contrast, maxicircles remain attached to the network, where they appear to replicate by a rolling circle mechanism. The new linear molecules cleaved from the rolling circle then attach to the network in a circular form.

RNA editing The kinetoplastidae have revealed a surprising departure from the virtually universal pathway of genetic informational flow. Some maxicircle coding sequences differ substantially from the sequences found in the corresponding functional mRNAs. For example, the sense strand of the *T. brucei CO*-II gene includes the segment

5′-AAAGTAGAGAACCTGGTAGGTGTAATGAAATAA-3′

but the sequence of the *CO*-II mRNA (deduced from the cDNA sequence) is

5′AAAGUAGA**UUGU**A**U**ACCUGGUAGGUGUAAUGAAAUAA-3′

(In each case, the illustrated portion of the reading frame starts with the 5′ AAA codon for the amino acid lysine.) The U residues shown in boldface were inserted into the primary transcript. The insertions (+4) shift the reading frame, thereby allowing complete translation of the RNA into the *CO*-II polypeptide. This process, **RNA editing**, indicates that DNA is not always the sole source of coding information. Indeed, proper expression of five different kinetoplastidae mitochondrial genes, in three species, is now known to involve RNA editing. In addition to shifting reading frames, some of the U insertions form essential start (AUG) or stop (UAA) codons. In one case, the *CO*-III gene of *T. brucei*, more than 50 percent of the residues in the mRNA are posttranscriptionally inserted U residues. In other kinetoplastidae, mitochondrial DNA coding sequences homologous to the *T. brucei CO*-III cDNA, but not the U-deficient *T. brucei CO*-III gene, are found.

Besides insertions, RNA editing can involve deletions of U residues encoded by the gene. In some instances, functional mRNA appears to be produced in several steps, including inserting multiple Us at given sites followed by trimming out of all except those Us required for the final functional mRNA. Many aspects of these insertion reactions are not understood, including the source of the information specifying the sequence of the proper, final mRNA or how a primary transcript is recognized to be in need of editing. A particular gene may require RNA editing in one species but be intact in another. It is also likely that RNA editing can be regulated. Thus, with *T. brucei*, some RNAs (e.g., *CO*-II) are edited in the parasitic form that inhabits insect hosts, but not in the form typically found in mammalian blood. This seems likely to be related to the fact that in the insect, the protozoa depend on oxidative phosphorylation to produce ATP; but in the mammalian bloodstream, ATP is produced by glycolysis.

How general is RNA editing? Is it confined to kinetoplastidae and to mitochondria? Two recently observed examples suggest that editing may occur in many types of organisms and on nuclear transcripts. Thus, the mammalian apolipoprotein-B gene produces two mRNAs; in one, a genomically encoded C is converted to a U, with the concomitant formation of a stop codon. Also, two mRNAs derived from a single gene of simian Paramyxovirus SV4 differ by the insertion in one of them of two (uncoded) G residues. Apparently, RNA editing is not confined to the insertion and deletion of U residues. Whether the mechanisms used by mammalian systems are related to those that operate in kinetoplastidae remains to be discovered.

c. The DNA of Chloroplasts

The genomes of chloroplasts are similar in some respects to those of mitochondria. They are closed circular duplex DNAs that encode some, but not the majority, of the genes used in chloroplast structure and function. Chloroplast DNAs run to enormous sizes, generally between 120 and 180 kbp. Each chloroplast itself may have tens of copies of the genome. The complete DNA sequences of chloroplast DNAs from a liverwort (a bryophyte), tobacco (a dicot), and rice (a monocot) are known.

Structure of chloroplast DNA Restriction endonuclease mapping indicates that, unlike mitochondrial DNA, the structure of chloroplast DNA is quite stable within a species. The spinach chloroplast genome (Figure 9.62) is typical of that of many angiosperms—both monocots and dicots. Four rRNA genes (16S, 23S, 5S, and 4.5S RNAs) occur within an approximately 25 kbp segment, and this segment is repeated, in inverted orientation, after a separation of about 15 kbp. Other genes have been located, including that for the large subunit of ribulose 1,5-bisphosphate carboxylase, *rbc*L (the gene for the small subunit is chromosomal), that for a 32 kDalton constituent of the thylakoid membrane (photogene 32, *psb*A, expressed only in the light), and several encoding subunits of the chloroplast RNA polymerase (*rpo*). Altogether, chloroplast DNA encodes about 100 polypeptides and 35 RNAs (rRNA and tRNA). The gene products of many of the open reading frames have yet to be identified. The only known exceptions to the spinachlike organization in angiosperms are in a few legumes (like peas), which have lost one copy of the inverted repeat. Indeed, the inverted repeat feature is found even in the green alga *Chlamydomonas reinhardii*, although not in *Euglena*. The latter is distinctive in having three tandem repeats of a segment (5.6 kbp) that includes the rRNA genes. In contrast to mitochondrial genomes, chloroplast DNA appears to use the universal genetic code. There are many tRNA genes, and most of the universal codons appear in those polypeptide genes that have been analyzed.

No clear picture of chloroplast DNA replication or of the locations of origins of replication is yet available.

Structure of chloroplast genes The structures of chloroplast proteins and their genes are highly conserved over a wide range of species. The rate of evolution of coding sequences appears to be much slower than that of nuclear genes.

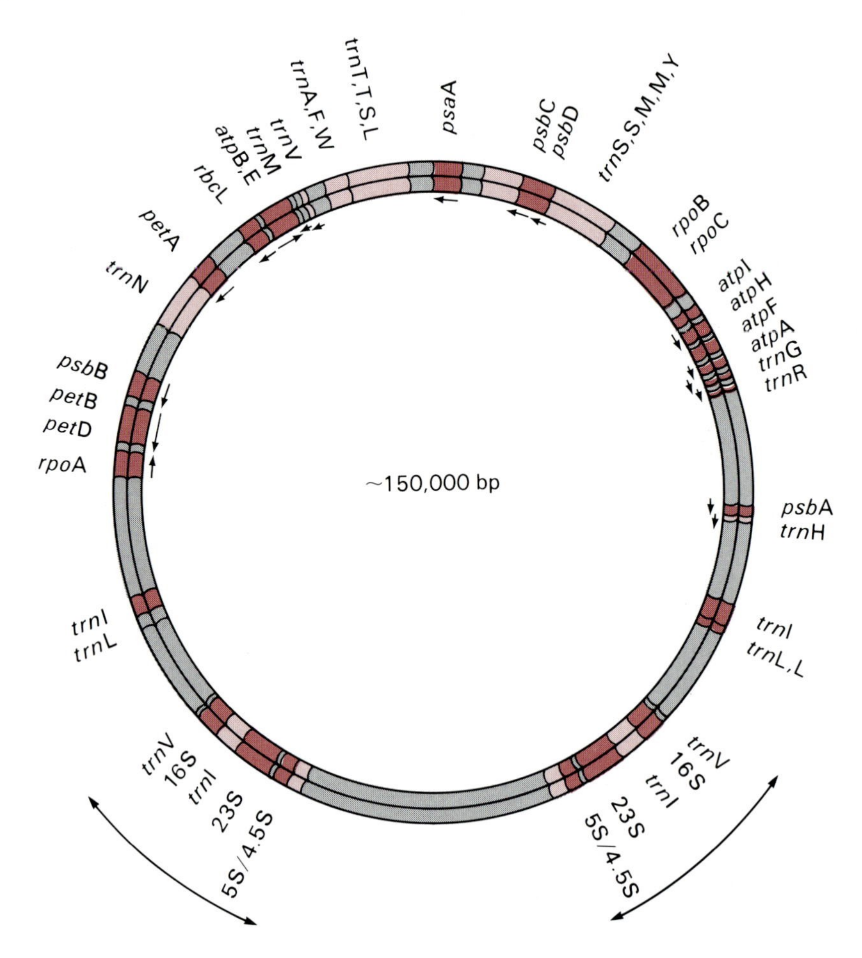

Figure 9.62
Schematic diagram of spinach chloroplast DNA showing the location of some genes. The double arrows on the outside show the inverted repeats that contain genes for ribosomal RNAs. Small arrows show transcription direction. The gene designations are *trn*, tRNAs (identified by single letter abbreviation for the amino acid, as is customary in the field); *psa*, components of photosystem I; *psb*, components of photosystem II; *pet*, polypeptides of the cytochrome complex; *rbc*L, large subunit of ribulose 1, 5-bisphosphate carboxylase/oxygenase; *atp*, subunits of the H^+-ATPase complex; *rpo*, subunits of RNA polymerase. Genes not shown include those for ribosomal proteins. Courtesy of W. Gruissem. After W. Gruissem and G. Zurawski, *EMBO J.* 4 (1985), p. 1637, and G. Sijben-Müller et al., *Nucleic Acids Res.* 14 (1986), p. 1029.

Introns occur in the tRNA, rRNA, and polypeptide genes of various chloroplasts, but their distribution is complex. For example, although introns occur in the chloroplast tRNA genes of many organisms, they are lacking in the tRNA genes of *Euglena*. Introns are typical in *Euglena* polypeptide genes, although many of the homologous mRNAs in cyanobacteria and higher plant chloroplasts may not require splicing. Even among higher plant species, individual chloroplast polypeptide genes may or may not contain introns. *Chlamydomonas reinhardii*, unlike all the other organisms analyzed, has an interrupted 23S rRNA gene. Overall, the data do not, at least as yet, lend themselves to coherent interpretation regarding the significance of the presence or absence of introns.

Chloroplast introns appear similar to either the group I or group II introns described in Section 8.5c. Although not yet rigorously proven, splicing probably involves related mechanisms. Most chloroplast introns do not have open reading frames of significant length. However, a 509 codon long open reading frame occurs in an intron of a tobacco chloroplast $tRNA^{Lys}$ gene and a 163 codon long one in the intron in the *C. reinhardii* 23S rRNA gene. The predicted proteins have regions that are homologous to portions of maturases encoded in yeast mitochondrial introns. Thus, it may be that in chloroplasts, as in mitochondria, functional coding regions can be included in introns.

Several mRNAs in both algae and plants depend for their formation on *trans* splicing (Section 8.5d). For example, the 5′ exon of the gene encoding the chloroplast ribosomal protein s1z is transcribed from one DNA strand, and the other two exons are transcribed from a region 30 kbp away on the other strand. *Trans* splicing joins the transcripts into a functional mRNA.

d. The Origin of Organelle DNA

A widely accepted hypothesis states that mitochondria and chloroplasts evolved from prokaryotic endosymbionts that lived within the cytoplasm of the ancestors of eukaryotes. Functions supplied by the prokaryotes, such as aerobic metabolism and photosynthesis, became essential to the hosts. The genes for these functions and related structures have been maintained in contemporary organisms while genes encoding other functions and structures of the endosymbiont prokaryotes were lost during evolution. Various aspects of chloroplast gene expression support this hypothesis. Thus, promoters for chloroplast gene expression resemble the typical -35 and -10 sequences of prokaryotic promoters (Figure 9.63). Moreover, chloroplast ribosomes are quite similar to those of prokaryotes. The rRNAs are similar in size and sequence, and several chloroplast ribosomal proteins are immunologically related to specific *E. coli* ribosomal proteins. Translation on chloroplast ribosomes is inhibited by chloramphenicol, which also inhibits prokaryotic translation. The case is not as strong when gene expression in mitochondria is examined for similarities to prokaryotes. The peculiar translational apparatus of mitochondria, particularly that of animals, bears only token resemblance to that of known contemporary prokaryotes. Nevertheless, careful analysis of plant mitochondrial rRNA suggests an ancestry in common with the α type of purple bacteria, a group that includes known plant symbionts like *Agrobacteria* and *Rhizobacteria*. It may be that the ribosomes and genetic code of mitochondria evolved from a now extinct (or unknown) prokaryote.

The genes required for the construction and function of mitochondria and chloroplasts are divided between nuclear and organellar DNA. Moreover, in different organisms, the same gene may be either organellar or nuclear. For example, the gene encoding subunit 9 of the ATPase complex is mitochondrial in *S. cerevisiae* but is nuclear in other fungi and humans. In its simplest form, the endosymbiont hypothesis predicts that all the genes for mitochondrial and chloroplast structure and function should reside on the organelle's genome, the descendant of the original prokaryotic DNA. Why then are many of these genes found on nuclear chromosomes in contemporary organisms? One explanation with experimental foundation is that there was in the past, and continues to be, considerable traffic of DNA sequences from organelle to nucleus and also between the organellar genomes themselves (Figure 9.64).

For example, the critical enzyme in photosynthesis, ribulose 1,5-bisphosphate carboxylase/oxygenase, is a heteromultimer composed of eight identical large (L) and eight identical small (S) subunits. In photosynthetic eukaryotes, L is encoded by *rbc*L in the chloroplast genome (Figure 9.62) and S, like most chloroplast polypeptides, in the nucleus (Figure 6.36). However, in the protist, *Cyanophora paradoxa*, both L and S are encoded in the extranuclear genome that is part of the chloroplast-

Figure 9.63
Promoters for transcription of spinach chloroplast genes compared to *E. coli* promoters. From W. Gruissem and G. Zurawski, *EMBO J.* 4 (1985), p. 1637.

	−35 sequence	−10 sequence
E. coli consensus sequence	TCTTGACAT	TATAAT
spinach chloroplast gene $tRNA_2^{met}$	TATTGCTTA	TATAAT
*rbc*L	GGTTGCGCC	TATACA
*psb*A	GGTTGACAC	TATACT

like structure (cyanelle) typical of cyanobacteria. This location is consistent with an endosymbiont origin for eukaryotic chloroplasts. The suggested scenario is that, in a common ancestor to contemporary photosynthetic eukaryotes, a DNA segment including the S gene moved from an endosymbiont or its derivative into nuclear DNA.

Additional evidence for intergenomic exchange is the finding that the nuclear genome of spinach contains sequences (presumably noncoding) that anneal with cloned probes representing almost the entire chloroplast DNA, including sequences from *rbc*L. DNA sequences homologous to *rbc*L and other parts of chloroplast DNA also occur in the mitochondria of some plants. Some of these sequences appear to result from relatively recent transfers because the mitochondrial DNA is more similar to the chloroplast DNA of the same species than to that of other species. Thus, chloroplast DNA segments can move to the nucleus and to the mitochondria. There are also multiple examples of the transfer of mitochondrial DNA sequences. Annealing is detected between mitochondrial gene segments and nuclear DNA in such diverse organisms as yeasts, locusts, sea urchins, and rats. These examples illustrate that DNA rearrangements are not limited to the nuclear genome or barred by organelle or nuclear membranes. However, the mechanisms by which organellar DNA moves about and integrates into new genomes are in general unknown. The mechanisms for rearrangement of nuclear DNA that are described in the next chapter may provide useful models.

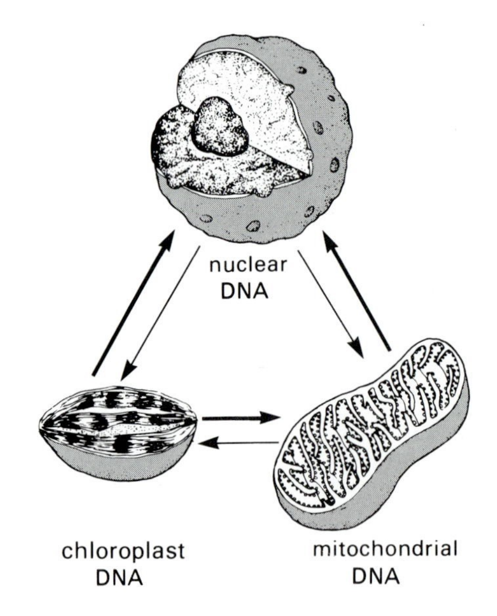

Figure 9.64
DNA segments have moved from organelle to nuclear DNA. The heavy arrows indicate documented movements. The transfers shown with thin arrows may also have occurred, but are not proven. After R. Lewin, *Science* 224 (1984), p. 970.

CHAPTER 10

Genomic Rearrangements

The description in Chapter 9 of the organization of contemporary genomes illustrates the profound evolutionary significance of the rearrangement and duplication of DNA sequences. Moreover, several different reorganizing strategies can be inferred from comparative genomic anatomy. One is the movement of DNA segments from their original locations to different sites. Another is the duplication of DNA segments at a particular locus giving rise to tandem reiterations. In still other instances, DNA segments are duplicated and moved to distant genomic locations by both DNA mediated processes and reactions involving RNA intermediates and reverse transcription. Finally, additional copies of essential segments can mutate and be adapted for new or modified functions. Together these strategies lead to an increase in genome size and phenotypic capacity and, in combination with natural selection, to flexibility and evolution.

In this chapter, we summarize the experimental systems, both genetic and biochemical, that illuminate the ways in which genomic rearrangements can occur. These observations confirm the strategies inferred from the structure of present-day genomes. To the extent that they are understood, the mechanisms described here provide credible models for the events that must have occurred during evolution. Among these mechanisms, one in particular—reverse transcription—demands special emphasis because it appears to be universally important in eukaryotic genomes. The incorporation into genomes of DNA copies of RNA molecules accounts for a significant share of eukaryotic genomic fluidity. Recent evidence suggests that reverse transcription also occurs in prokaryotes; the full significance of this is not yet known.

All the reactions described here involve an alteration in the position of a DNA segment relative to neighboring sequences. These events are ubiquitous and diverse, but most of them occur relatively rarely. Many other genomic rearrangements arise from homologous recombination between allelic sequences and do not alter the neighborhood of a DNA segment. Overall, genome organization is quite static. It could not be otherwise if the viability of individuals and species is to be maintained.

Many genomic rearrangements are **unprogrammed**. Such processes are not associated with specific effects on gene expression and involve some element of randomness. There may be a random component in the frequency of the events, in the particular DNA segments that participate in the reactions, or in both. One example of such partly random events is the transposition of DNA sequences from one genomic locus to another. Another is the duplication and further amplification of DNA segments. However, similar transpositions and amplifications are also associated with decidedly nonrandom or **programmed** changes. Such programmed events play a critical role in the control of expression of some genes during differentiation and development in certain cell types.

10.1 General Features of Unprogrammed Transpositions

Some DNA segments are inherently mobile. They can move into new genomic locations where they may or may not significantly alter neighboring gene expression. For example, insertion of a mobile segment into a new genomic site is mutagenic if a coding or critical regulatory segment is disrupted. In the simplest case, expression of the gene is eliminated. But movable elements frequently include regulatory regions themselves. In such cases, expression of genes in the neighborhood of the insertion (or **target**) site may be influenced in complex ways, including the imposition of new regulatory modes.

Several different types of mobile DNA elements are distinguishable. It would be desirable to group the elements according to related transposition mechanisms. However, the transposition mechanism is still conjectural for many eukaryotic and prokaryotic elements. Therefore, the classifications used here are primarily based on common structural features. Although it is reasonable to presume that common morphological details relate to common transposition mechanisms, such a conclusion is not certain.

a. *Different Types of Mobile Elements*

Transposable elements The first mobile DNA segments to be studied at the molecular level were the **transposable elements** found in prokaryotes and in their phages and plasmids. Although several different types are now known, they all share two properties: first, the DNA within the element encodes a gene or genes that are required for transposition, and, second, specific DNA sequences are repeated in inverted orientation at the two ends of the element and are required for transposition. The prokaryotic transposable elements themselves encode no functions essential for the organisms that harbor them, although they often contain special genes, sucn as those encoding antibiotic resistance. The movement of these elements often has profound mutagenic effects.

Many eukaryotic mobile elements are like their prokaryotic kin in that their transposition requires genes encoded within the element as well as specific terminal DNA sequences. If their transposition does not appear to involve reverse transcription, these elements are classified here as transposable elements. The **controlling elements** of maize, which were the first movable units discovered, fall into this class, as do several elements in *Drosophila*. Similar elements are known to occur in other invertebrates and in some flowering plants, where they are responsible for the dramatic variegation of petal color.

Retrotransposons Eukaryotic movable elements whose transposition depends on transcription and reverse transcription are termed **retrotransposons**. They contain a central DNA segment that encodes, among other proteins, a reverse transcriptase. In some retrotransposons, called here class I, the central segment is surrounded by long direct repetitions of sequence at both ends of the movable unit—**long terminal repeats** or **LTRs**. Class I retrotransposons also have short inverted repeats at either end. They resemble retroviral proviruses in structure, in transcriptional properties, and in transposition mechanism, hence the name. The distinctive difference between class I retrotransposons and retroviruses is the absence of any viable extracellular form of the former. Families of such retrotransposons are known in various invertebrates, including yeast and *Drosophila*, in plants, and in several mammals.

Class II retrotransposons have no terminal repeats, and one end often has a stretch rich in A·T base pairs. Thus, they are not like retroviral proviruses. Although less well understood than class I, the class II elements also occur in a wide variety of eukaryotes. Notably, neither class of retrotransposons has a known prokaryotic counterpart.

Retrogenes Diverse DNA segments that lack the specific structural and coding properties of transposable elements or retrotransposons also move about in genomes. Unlike the other classes of movable elements, they are very heterogeneous in size and structure. They have no terminal repeats, and an A·T-rich stretch frequently occurs at one end. This class of elements, called **retrogenes**, includes processed pseudogenes and SINES and is found in different eukaryotes but is most abundant in mammals. Transposition of retrogenes is likely to proceed by the intermediate formation of RNA followed by reverse transcription. However, the process appears to be passive in that the elements do not encode functions required for transposition (e.g., reverse transcriptase).

b. The Formation of Target Site Duplications

Regardless of type, insertion of a mobile element into a new genomic locus is generally accompanied by duplication of a short DNA sequence at the target site. These duplications flank the two ends of the insertion (Figure 10.1). The virtually universal formation of **target site duplications** upon insertion suggests that, regardless of other mechanistic differences, most insertions involve staggered single strand nicks in potential target sites, as shown in the model in Figure 10.1. Conventionally, a target site in a genome is called "filled" if it contains a mobile element and "empty" if it does not.

The size of the duplicated DNA sequence varies from one mobile element to another. Transposable elements and class I retrotransposons are generally associated with specific size target site duplications. For example, insertions of the *E. coli* transposon Tn3 into a new location always cause 5 bp target size duplications. A maize element called *Ac* causes 8 bp duplications and the *Drosophila* retrotransposon called gypsy, 4 bp. Class II retrotransposons and retrogenes, however, cause variable size target site duplications. The target site duplications caused by insertion of an F element of *Drosophila* (a class II retrotransposon), for example, vary from 8 to 13 bp. Fixed or variable size target site duplications are likely to reflect the specificity of the enzyme responsible for making the staggered nicks in the target site. The fixed characteristic sizes seen with transposable elements and class I retrotransposons are probably the result of nicking by specific endonucleases encoded within each element.

c. Mobile Elements as Dispersed Repeated Sequences

Each type of mobile element generally occurs many times in the genome. Thus, the elements constitute families of dispersed repeated sequences. As such, they foster further genomic rearrangements because they provide sites for nonallelic, homologous recombinations. These yield dele-

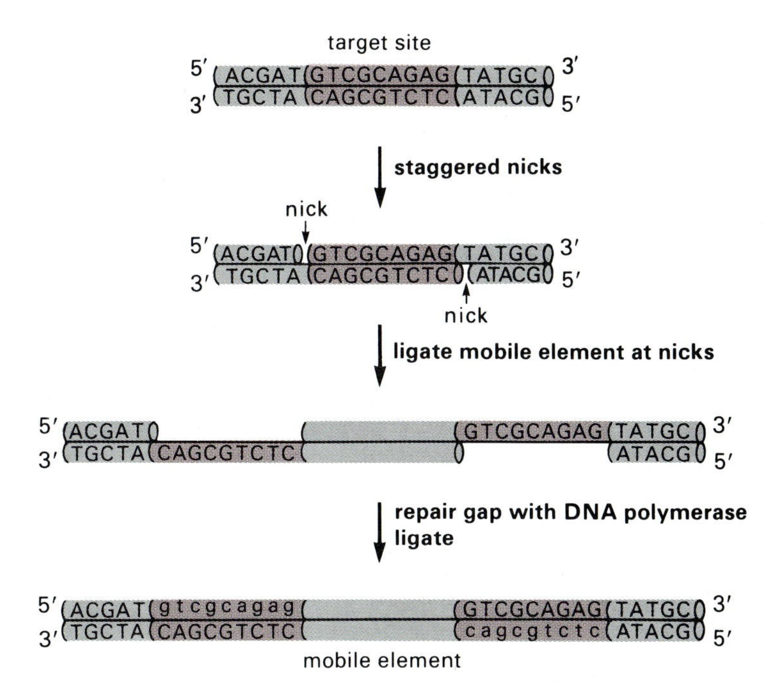

Figure 10.1
A model for the duplication of target site sequences upon insertion of a mobile element. In the first step, both DNA strands are cleaved by endonuclease. The two cleavages are several base pairs apart (nine in this model), yielding staggered nicks. In this example, the nicks generate overhangs on each strand in the 5′ direction. The 3′ ends of the mobile element are ligated to the 5′ termini at the nicks. Repair of the resulting gaps yields two copies of the target site sequence, one at either end of the insertion. The distance between the staggered nicks determines the length of the duplication. A similar series of reactions can be written if the initial nicks generate overhangs in the 3′ direction and the 5′ ends of the mobile element are ligated to the 3′ ends of the nicked target.

tions, duplications, inversions, and translocations. The ability of mobile elements to foster such rearrangements has been extensively documented with the class I retrotransposon Ty of yeast (Section 10.3). When such rearrangements occur in single somatic cells in multicellular organisms, they have only a small probability of influencing the function of a complex tissue. But in germ line cells, important mutations and chromosomal aberrations can result. Such rearrangements, if fixed in populations by natural selection or genetic drift, can also contribute significantly to evolutionary change.

Homologous recombinations between nonallelic dispersed repeats of a particular family are frequently invoked to explain the repetition of DNA segments (Section 2.4a). Also, some of the differences between homologous loci in different mammals (e.g., those encoding β-globin-like genes) are explicable in terms of deletions between direct repeats. Unequal homologous crossing-over between dispersed repeats on sister chromatids or on homologous chromosomes, intrastrand recombination, and "slipped mispairing" (Figure 10.2) are all possible mechanisms leading to new genomic arrangements. In *Drosophila,* combined genetic and molecular analysis confirms that frequent unequal crossing-over between certain pairs of distinct alleles results from the presence, in each member of the pair, of the same movable element at a different position. The anticipated reciprocal products of recombination, corresponding deletions and duplications, occur. Gross chromosomal alterations of the kind well known from cytogenetic analysis of mammalian cells can also be generated by reciprocal homologous recombination between dispersed repeated sequences on different chromosomes.

Given the high frequency of interspersed repeated sequences in most eukaryotic genomes, it is almost surprising that chromosome structure is as stable as it is. Perhaps recombination between these segments is actively suppressed by some unknown mechanism. In yeast, where such recombinations are readily measured by genetic techniques, translocations

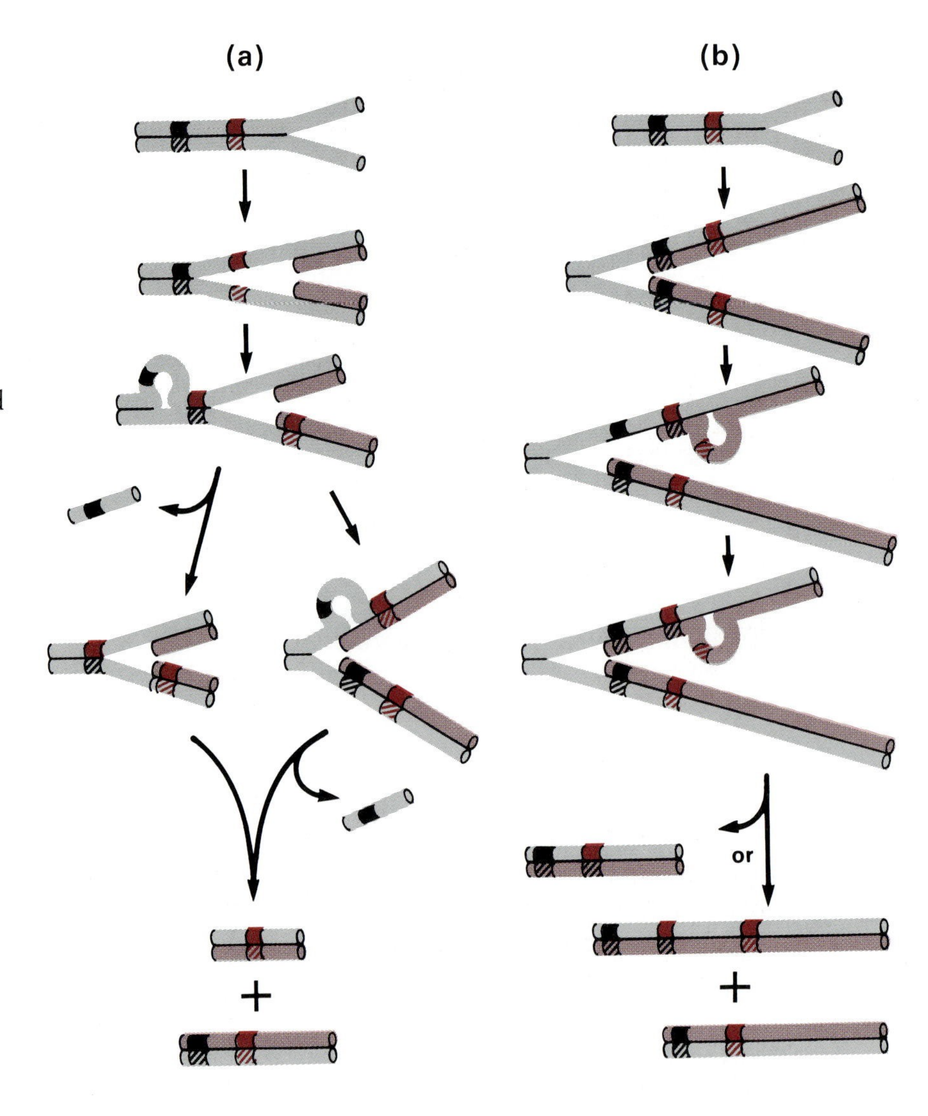

Figure 10.2
Deletions and amplifications by "slipped mispairing" can occur if two direct repeats (boxes) on one DNA duplex are assumed to mispair out of register close to a replication fork. (a) Slippage of a template strand and excision of the loop or replication across the resulting loop, followed by repair, leads to deletion. (b) Slippage of a strand being synthesized followed by repair can yield amplification.

and other rearrangements are surprisingly rare. For example, in cells containing two copies of the *HIS3* gene on nonhomologous chromosomes, reciprocal crossovers accounted for only about 10 percent of the recombinational events scored for the gene; the others are gene conversions.

10.2 Transposable Elements

a. Prokaryotic Transposable Elements

Transposable elements were described briefly in Chapter 5, when plasmid vectors were discussed. The R-plasmid first described in Figure 5.1 is shown again in Figure 10.3, but the several kinds of transposable elements are indicated in more detail. One type, designated IS for **insertion sequence**, occurs either independently (e.g., IS2) or at the ends of longer units called **composite transposons** (e.g., two IS10s surround Tn10). Other **transposons** (e.g., Tn3) are not associated with IS elements. All three classes (ISs, composite transposons, and transposons) have inverted repeated DNA sequences at their two ends. These repeats, as well as one or more genes encoded within the element, are required for transposition.

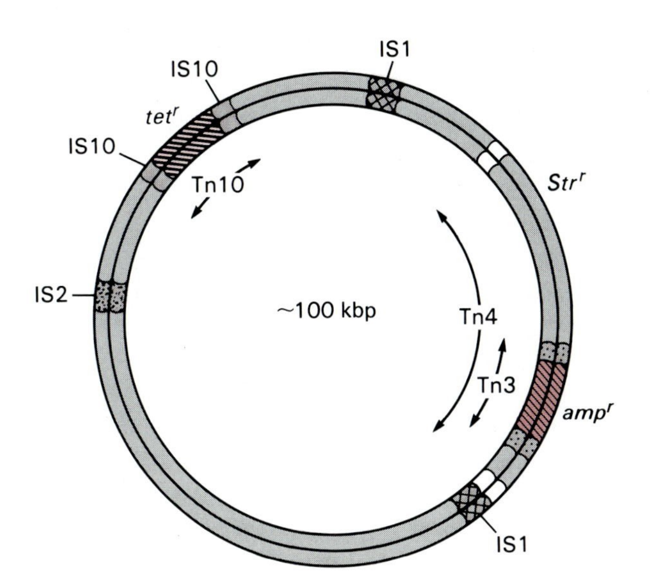

Figure 10.3
The genome of an R-plasmid showing insertion sequences (IS) and transposons (Tn). Antibiotic resistance genes are defined in Table 5.3.

Moreover, insertion of any of these transposable elements is generally associated with target site duplications of a length specific to the particular element.

Insertion sequences (IS) IS elements, such as IS1, IS2, and IS10, have defined lengths and unique nucleotide sequences (Table 10.1). Their movement into new genomic locations often causes mutations by interrupting essential regulatory or coding regions. The frequency of transposition varies, depending on the element, and is about 10^{-5} to 10^{-7} per generation. When an IS is transposed to a new position, the original IS remains in place; thus, insertion involves the precise synthesis of a second copy and depends on the host replication functions. Insertion does not depend on the major recombinational apparatus of *E. coli*, and there need not be any homology between the IS and its genomic target site.

The length and sequence of the various *E. coli* ISs are different from one another (Table 10.1). However, the two ends of each IS are always inverted repetitions (sometimes imperfect) of each other (Figure 10.4), and the repetitions are required for transposition to occur. The nucleotide

Table 10.1 Typical IS Elements of *E. coli*

IS	Total Length (bp)	Length of Terminal Inverted Repeats (bp)	Length of Target Site Duplication (bp)	Typical Occurrence
IS1	768	23	9	4–19 copies on *E. coli* chromosome 2 copies on plasmid R6 >40 copies on *Shigella dysenteriae*
IS2	1327	41	5	0–12 copies on *E. coli* chromosome 1 copy on R6 1 copy on F-plasmid
IS4	1426	18	11	1–2 copies on *E. coli* chromosome
IS10	1329	23	9	On R-plasmids

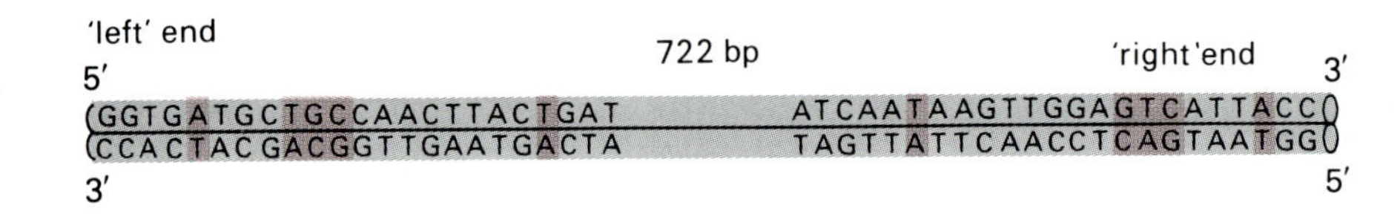

Figure 10.4
The inverted terminal repeats of IS1. Mismatches are boxed.

sequences of the inverted repeats are different for each IS, and they vary from 10 to 40 bp in length. Another characteristic property of each IS is the size of the target site duplication it causes upon insertion (Table 10.1). Most of the DNA in each IS is occupied by one or more structural genes and accompanying regulatory signals. The gene products are **transposases** that are required for transposition. In contrast to transposons, ISs do not usually encode any genes other than those required for transposition.

The presence of an IS element is often associated with a variety of mutational events besides the insertional inactivation of coding or regulatory sequences at the target site. For example, an IS can increase the rate of formation of deletions and inversions in neighboring genomic regions by 100- to 1000-fold. Furthermore, promoter elements within the IS itself can modulate the expression of nearby genes.

Composite transposons ISs, which are about 1 kbp long, frequently occur as components of larger, more complex composite transposons (Figures 10.3 and 10.5 and Table 10.2). These elements are up to thousands of base pairs long, and their central regions include a variety of genes, such as those encoding antibiotic resistance. The central region is flanked on both sides by complete IS units of the same type (i.e., either IS1s or IS10s, etc.), which are termed IS-L and IS-R (for left and right, respectively). In different composite transposons, IS-L and IS-R are found in either the same or the inverted direction, relative to one another. Because the ISs themselves contain inverted terminal repeats, the transposons also have such repeats, regardless of whether the flanking ISs are direct or inverted repetitions.

All the information required for movement of a composite transposon is supplied by the IS portion of its structure and is the same as that utilized by the independent IS itself: inverted terminal repeats and one or more structural genes for transposase. The requirement for inverted terminal repeats means that the IS-L and IS-R must be identical at the outside borders. However, one copy of the structural genes required for transposition is sufficient, so IS-L and IS-R need not be identical in sequence throughout; one of them can be altered by mutations that cause an attenuation or loss of function. In the case of Tn10 (Figure 10.5), for example, either IS10-L or IS10-R alone provides transposase, but IS10-R is much more efficient. Also, IS10-R can act as an independent IS element,

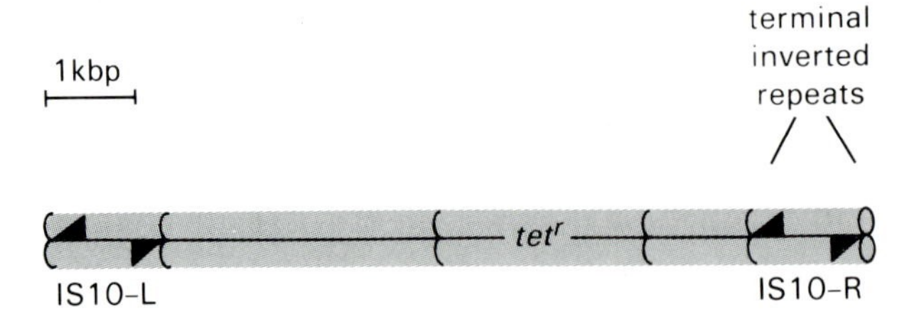

Figure 10.5
The composite transposon Tn10 has two copies of IS10 flanking a central DNA segment. The central region encodes tetracycline resistance.

Table 10.2 Typical *E. coli* Transposons

Transposon	Size (bp)	IS Termini	IS Orientation	Terminal Inverted Repeat (bp)	Target Site Duplication (bp)	Genes Carried
Tn3	4957	None	—	38	5	*amp*^{R}
Tn9	2638	IS1	Direct	23	9	*cam*^{R}
Tn10	9300	IS10	Inverted	1329	9	*tet*^{R}

whereas IS10-L cannot. Presumably, the few sequence differences between these almost identical IS10s are responsible for the differential activity of their respective gene products. In some composite transposons, one of the IS elements is completely nonfunctional (except for the terminal inverted repeats).

Noncomposite transposons Some transposons are not associated with IS sequences at all (Table 10.2). Rather, the functions required for transposition are encoded within the body of the element. Tn3 is a good example (Figure 10.6); its structure includes inverted 38 bp long terminal repeats, IR-L and IR-R, that are required for transposition. In between the repeats are sequences encoding three genes—two for transposition proteins (*tnp*A, *tnp*R) and one for β-lactamase (*amp*). In addition, there is a 170 bp long noncoding region that contains the promoters for *tnp*A and *tnp*R, as well as a special region called *res*, which is required for transposition. One important difference between an IS and a noncomposite transposon is the presence of genes like *amp*, which are unrelated to transposition.

Target sites Thus far, the description of prokaryotic transposable elements has concentrated on the elements themselves. What of the DNA segments into which they move—the target sites? Some transposable elements are quite fastidious and insert most frequently within short genomic regions that are homologous to the termini of the element itself. Others are more promiscuous and have less or even no obvious preference for particular target sites, although they may show a tendency to move into regions rich in A·T base pairs. Several of the elements demonstrate a specificity for a particular palindromic target site (e.g., the consensus sequence is 5′-CTAG for IS5 and 5′-GCTNAGC for Tn10). Because each transposable element makes a characteristic size duplication, it is likely that the spacing of the target site nicks is determined, at least partly, by functions encoded within the elements.

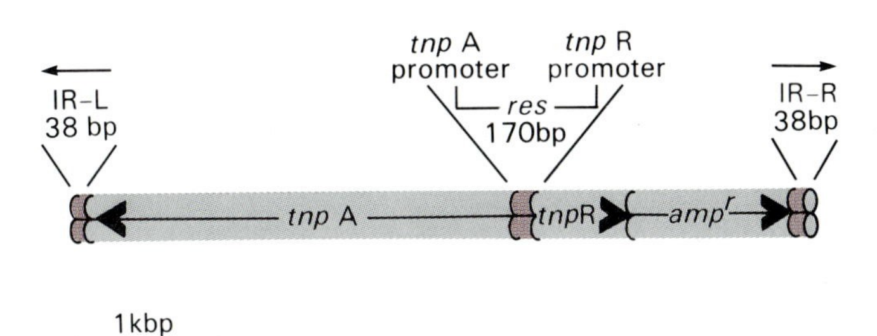

Figure 10.6
The noncomposite transposon Tn3. The arrow heads show the direction of transcription of the three genes.

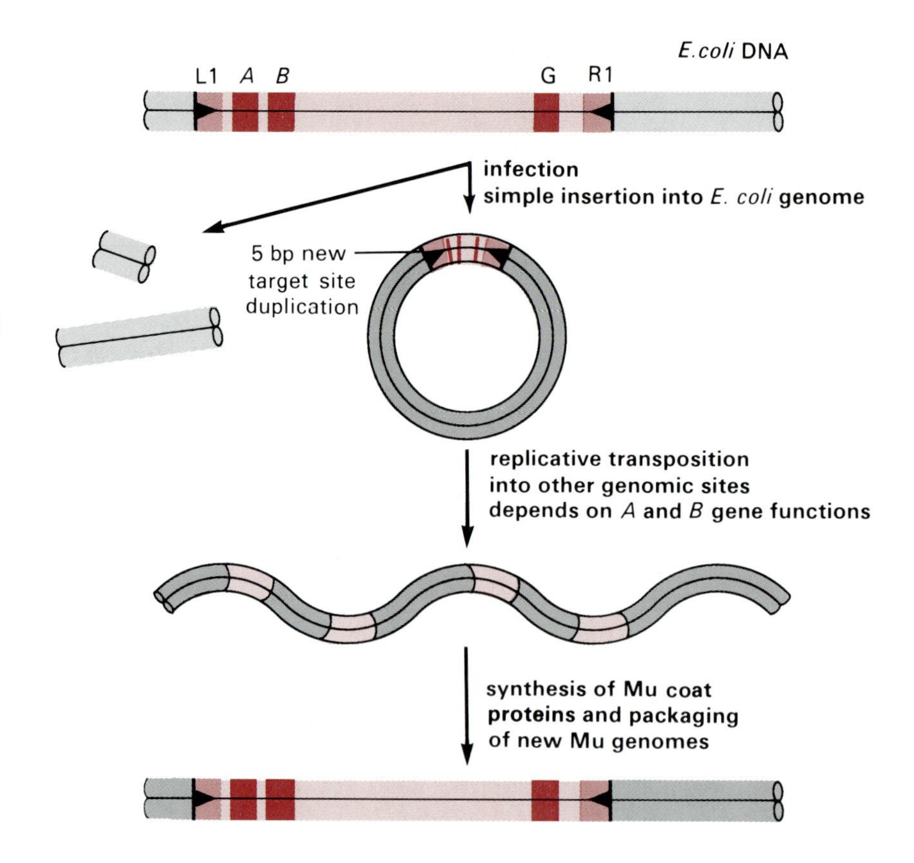

Figure 10.7
A schematic diagram showing the structure of the genome of Mu bacteripohage and its life cycle. The drawing is not to scale. The left and right terminal repeats (L1 and R1) are indicated, as are genes encoding the transposase (*A*) and functions required for replication (*B*). The segment marked G is described in Section 10.6a.

Bacteriophage Mu An *E. coli* temperate bacteriophage called Mu behaves like an IS or transposon. The Mu genome is a linear duplex DNA of about 39 kbp with approximately 20 bp long terminal, inverted (imperfect) repeats that are flanked by *E. coli* DNA segments (Figure 10.7). Upon infection, Mu DNA integrates into the *E. coli* chromosome, forming 5 bp target site duplications and creating a stable lysogen. Such Mu lysogens can be induced to replicate Mu DNA and undergo a lytic cycle. However, unlike the events that occur upon induction of λ phage lysogens, the replicating Mu DNA remains integrated. New phage genomes are generated through *in situ* replication of phage DNA. The newly synthesized Mu DNA, which is not joined to the integrated copy, is then inserted into new chromosomal positions. Ultimately, the host genome is destroyed, and Mu DNA, joined at both ends to fragments of *E. coli* DNA, is packaged into phage particles.

The presence of an integrated Mu DNA prophage in *E. coli* DNA promotes a variety of rearrangements that are related to its transposition. These rearrangements are similar to those fostered by ISs and transposons. Because Mu DNA transposes at a relatively high rate and Mu transposition can be studied in cell-free systems, Mu has served as a model for the elucidation of bacterial transposition mechanisms at the molecular level.

Transposition mechanisms Many model reaction mechanisms have been proposed to explain transposition, and different elements probably transpose in somewhat different fashions. However, the models generally fall into one of two classes because two different kinds of transposition

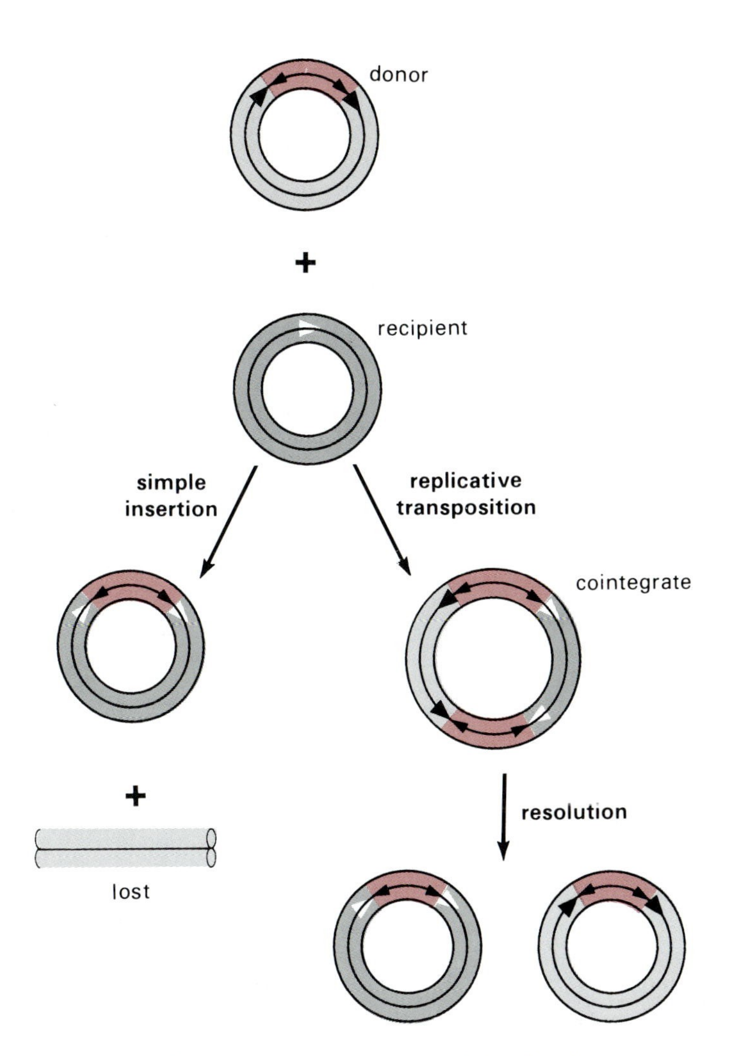

Figure 10.8
Two possible modes for inserting a transposable element into a new location. The transposable element donor and the recipient DNA are shown as two separate circles, although both donor and recipient sites could reside on the same DNA duplex. The element (colored) is depicted with its terminal inverted repeats and flanking target site duplications (thick arrow heads). One mode involves the intermediate formation of a cointegrate with duplication of the element followed by resolution of the cointegrate into two products, each of which contains a copy of the transposable element. In the second mode, simple insertion, the transposable element is moved from the donor to the recipient with no element replication. The donor circle loses the element, and its subsequent fate is unknown. Regardless of mode, the target site on the recipient is duplicated.

have been observed (Figure 10.8). To simplify the discussion of these models, we consider the case of transposition from one genome to another (e.g., from a plasmid to a bacterial genome or the reverse). (Note that an integrated Mu DNA can also transpose to a plasmid DNA.) The situation when transposition is intramolecular is more complex and is described briefly at the end of this section.

In one class, termed **cointegration**, the donor genome that carries the transposable element fuses with the recipient DNA molecule. The **cointegrate** product contains all the DNA of the donor and all the DNA of the recipient, with a copy of the transposable element at each junction between donor and recipient DNA. The process involves breaking preexisting phosphodiester bonds and forming new ones, duplicating the entire element, and forming a target site duplication. The cointegrate structure can then be **resolved** into the two separate original DNAs, each of which now carries a copy of the element. In addition to functions encoded on the element, transposition by cointegration utilizes the cell's replicative functions. Because a complete duplication of the element occurs, this type of mechanism is sometimes referred to as **replicative transposition**.

The second class of transposition mechanisms is called **simple insertion**. The transposable element is moved into a new genomic position with no other rearrangement of the target DNA save the formation of

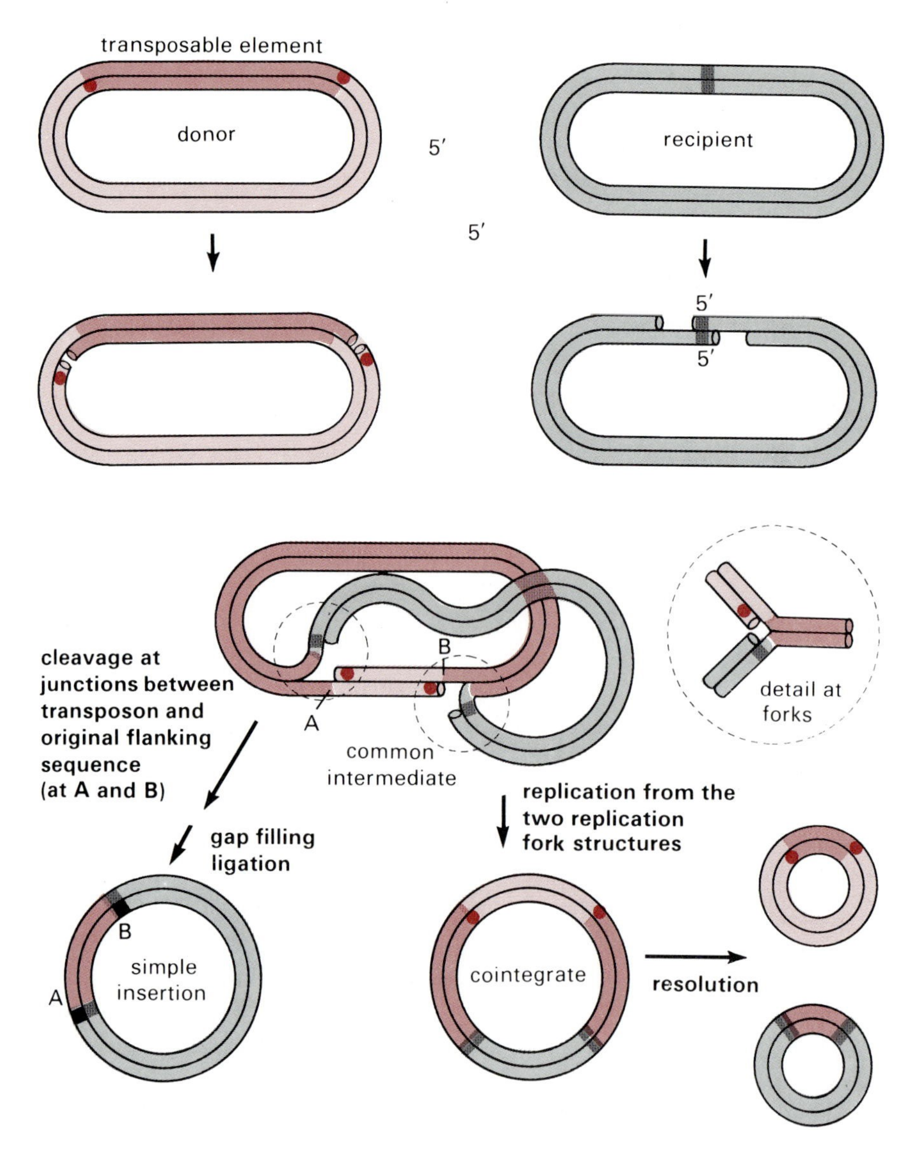

Figure 10.9
A model showing replicative and nonreplicative transposition as alternative outcomes from a common intermediate. Both donor and recipient DNAs are shown as circular. Dots represent 5′ termini on the donor element. Squares represent the target site on the recipient (target site duplications surrounding the donor transposon are omitted for convenience). The first step is the formation of staggered single strand nicks in the donor and recipient. In the donor, they occur at either end of the element such that the element is at the 3′ terminus of the nicked strand. In the recipient, the nicks are spaced by a distance characteristic of the element and yield single strand overhangs of a 5′ terminus, as shown in Figure 10.1. In the next step, the free 5′ ends in the recipient are ligated to the free 3′ ends of the donor to yield the common intermediate. After K. Mizuuchi, *Cell* 39 (1984), p. 395.

a target site duplication. This type of transposition is sometimes called **conservative** (or **nonreplicative**) because there is no net duplication of the transposable element. Some transposable elements take part in both cointegrate and simple insertion events; Mu DNA is one example.

In one very attractive scheme, the two types of transposition are alternative outcomes following the initial formation of a common intermediate structure (Figure 10.9). Mu transposition follows this model, and other transposable elements may as well. The reaction scheme is depicted with circular donor and recipient DNAs because the *in vitro* experiments in support of the model used circular plasmid DNA substrates; a modified Mu was the transposable element on the donor. *In vivo*, donors and recipients such as plasmids or the *E. coli* chromosome are also circular.

According to this scheme, transposition starts with the formation of single strand nicks in the donor and recipient circles. In the donor, the nicks occur at both 3′ ends of the element. In the recipient, the nicks are staggered at the target site and generate overhangs in the 5′ direction. The nicked ends of the transposable element are then joined to the termini in the recipient to form the common intermediate. A simple insertion

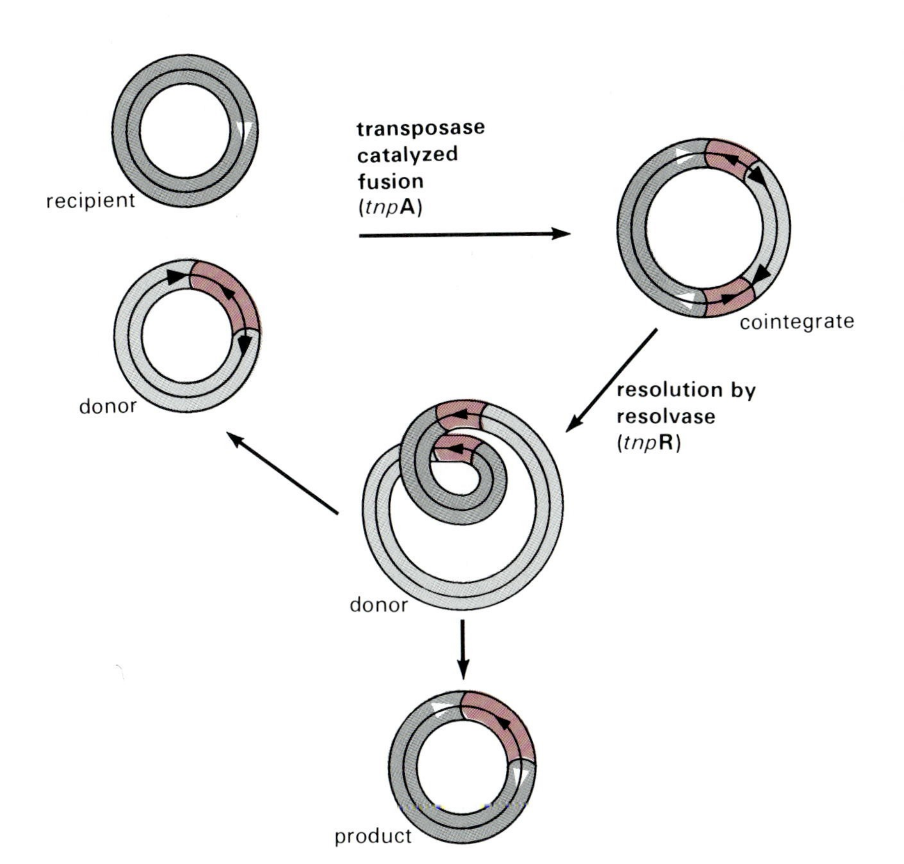

Figure 10.10
Replicative transposition of Tn3 by the intermediate formation and resolution of a cointegrate. Only the donor target site duplications are shown.

occurs if there are additional cleavages on the opposite strands of the transposable element from those initially nicked. Gap repair and ligation generate the target site duplications. The nonreplicative transposition of Tn10 may proceed in this way. The reaction is facilitated by the Tn10 transposase, which promotes strand breaks at Tn10 termini and also seems to foster the joining of broken termini. For replicative transposition, the common intermediate undergoes a completely different set of reactions. Notice that the intermediate structure is analogous to a circular DNA with two replication forks. If replication is initiated at each fork, the final product is a circle that contains the complete donor and recipient DNAs as well as two semiconservatively replicated copies of the transposable element, a cointegrate. Replication of the transposable element is an inherent part of cointegrate formation. If resolution of the cointegrate occurs, the net reaction is the transposition of a new copy of the element into a new target site. Resolution of cointegrates can be through homologous recombination between the two copies of the transposable element mediated by the recombination functions of *E. coli* (Figure 10.10). However, more efficient resolution can occur through processes encoded by transposable elements themselves.

The transposition of Tn3 and related noncomposite transposons is replicative and involves cointegrate formation (Figure 10.10). The cointegrate contains direct repeats of Tn3 at the two junctions between the donor and recipient DNAs. Fusion requires transposase, the gene product of *tnpA*, as well as IR-L and IR-R; it also requires host replicative functions. Resolution involves intramolecular site-specific recombination between the two *res* sites in the two Tn3 elements (Figure 10.6) and is

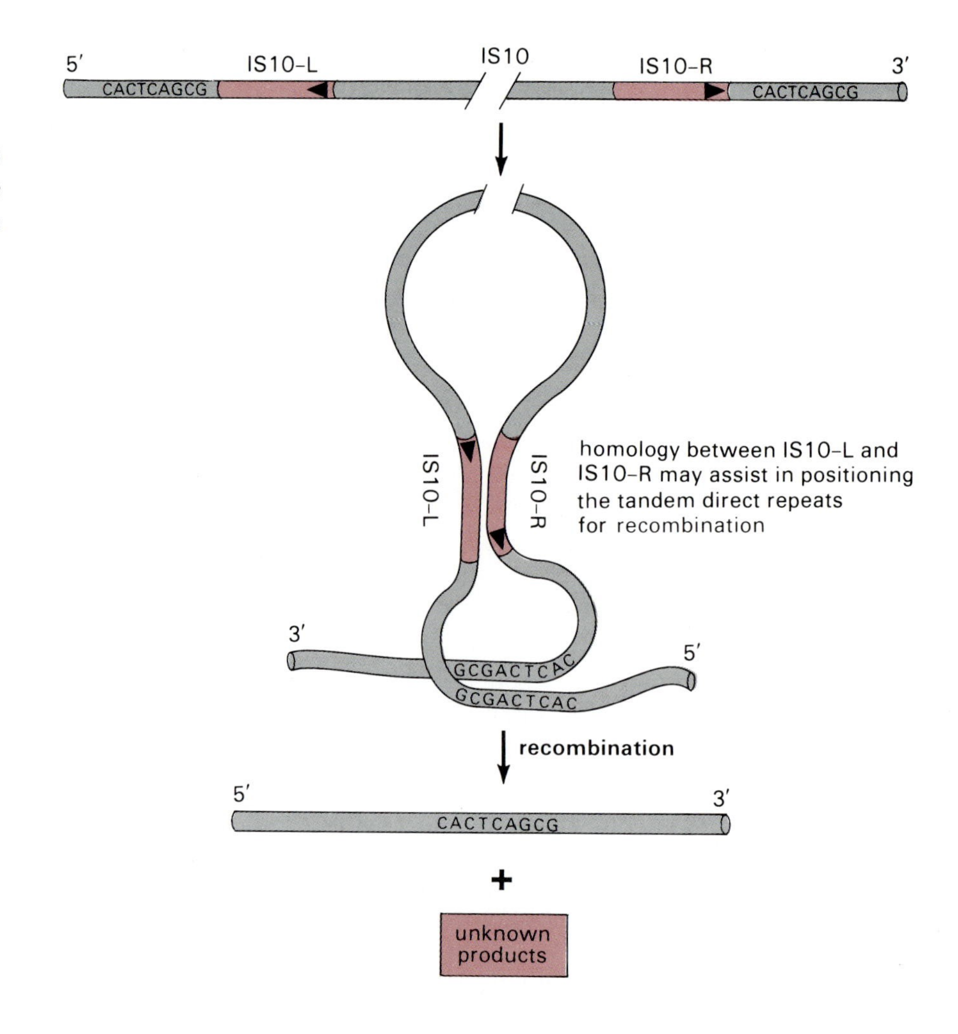

Figure 10.11
Homologous recombination between target site duplications excises the transposable element and regenerates the original target site. Only a single strand is shown, and Tn10 is used as an example.

mediated by the product of the *tnp*R gene, the **resolvase** protein. If a Tn3 is *tnp*R$^-$, then the general recombination functions of *E. coli* allow resolution of the cointegrates, but at considerably lower efficiency. The mechanism of Tn3 transposase action is unknown. However, the reactions catalyzed by purified resolvase have been studied *in vitro*. Its duplex DNA substrate must be circular and negatively supercoiled and must contain two *res* sites, both oriented in the same direction along the cointegrate. The site-specific resolvase binds to the two *res* sites and catalyzes a topoisomeraselike exchange of strands that leads to crossing-over and recombination. The reaction involves four DNA strands, four phosphodiester bond breaks, and four new joins. No DNA replication occurs during resolution. Note that the promoters for both *tnp*A and *tnp*R are within *res* (Figure 10.6); when resolvase binds to *res*, it represses expression of both genes, thereby limiting additional transpositions.

Excision Insertion and excision of transposable elements are very different processes, and precise excision does not depend on functions encoded in the elements. Many excisions are in fact unrelated to insertion elsewhere; the excised element is lost, and its fate is unknown. These events are likely to be the result of homologous recombination between the target site duplications (Figure 10.11). By this means, the entire element and one of the flanking repeats may be eliminated, and the original empty target site may be regenerated. Thus, excision can correct a mutation caused by an insertion. The frequency of excision is generally several orders of magnitude lower than that of insertion. For some elements, these reactions depend on the host *rec*A functions. For others, *rec*A is not required, but the frequency of excision is affected by other host recombination and repair functions.

Associations between transposable elements and a variety of mutations Although insertion of a transposable element into a critical regulatory region or a coding sequence often eliminates expression of the gene, more complex outcomes are frequent (Figure 10.12). Sometimes an appropriately placed promoter within the element mediates the expression of a neighboring gene that would otherwise be silent. For example, the IS10-R of Tn10 has a strong promoter about 100 bp from one end; transcription is initiated in the direction away from the body of the element and into the adjoining DNA (Figure 10.12a). Other kinds of effects are observed when an element inserts into one gene of a polycistronic operon. For example, there may be a polar effect on the expression of later genes in the string either by stopping transcription at a termination site within the element (Figure 10.12b) or by inhibiting translation of the distal coding sequences on the mRNA (Figure 10.12c).

A different kind of mutagenesis results from the propensity of transposable elements to induce rearrangements in neighboring DNA when a single genome serves as both donor and recipient for cointegrate formation (intramolecular transposition). All these events, which include deletions and inversions, are associated with the formation of a junction between one or more ends of the element and a new DNA sequence (Figure 10.13); the detailed mechanisms are not well understood. Finally, homologous recombination between two identical elements of the same orientation on one DNA duplex will result in the deletion of the sequences in between.

b. The P Elements of *Drosophila*

Hybrid dysgenesis is a genetic peculiarity associated with some strains of *D. melanogaster*. As we shall see, the properties of dysgenic systems are consistent with the activity of transposable elements. Several such systems are known in this species, and one, the P system, is particularly well characterized both at the genetic and the molecular levels.

Transposition of P elements About 40 members of the transposable P element family are found dispersed throughout the (haploid) genome of some strains of *D. melanogaster*. Other strains and some other species of *Drosophila* have no P elements whatsoever. The longest P elements are 2.9 kbp and highly conserved in sequence; others range in size from 0.5 to 2.9 kbp, having suffered internal deletions relative to the 2.9 kbp full length structure. All P elements, regardless of size, have 31 bp long inverted terminal repeats.

The ability of P elements to transpose is observed in several ways. First, the number of copies of P varies from one strain of flies to another. Second, the genomic location of P elements is different in diverse strains.

Figure 10.12
Some mutagenic effects of transposable elements.

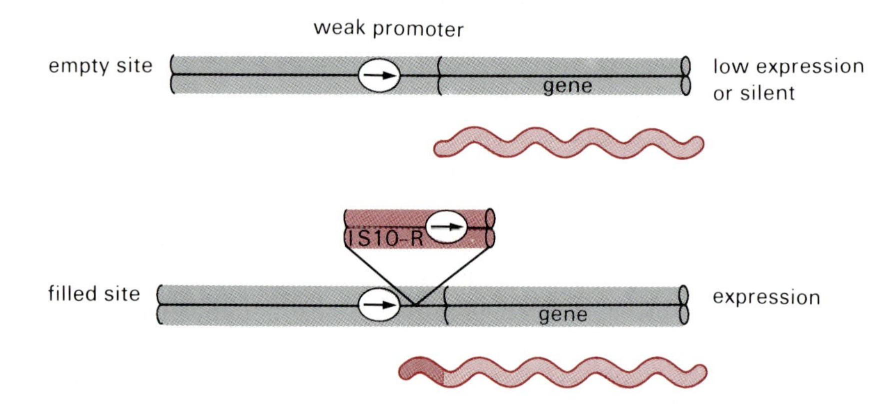

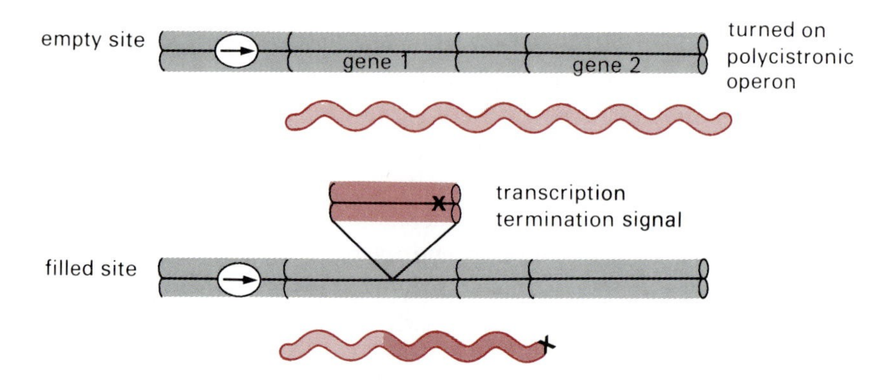

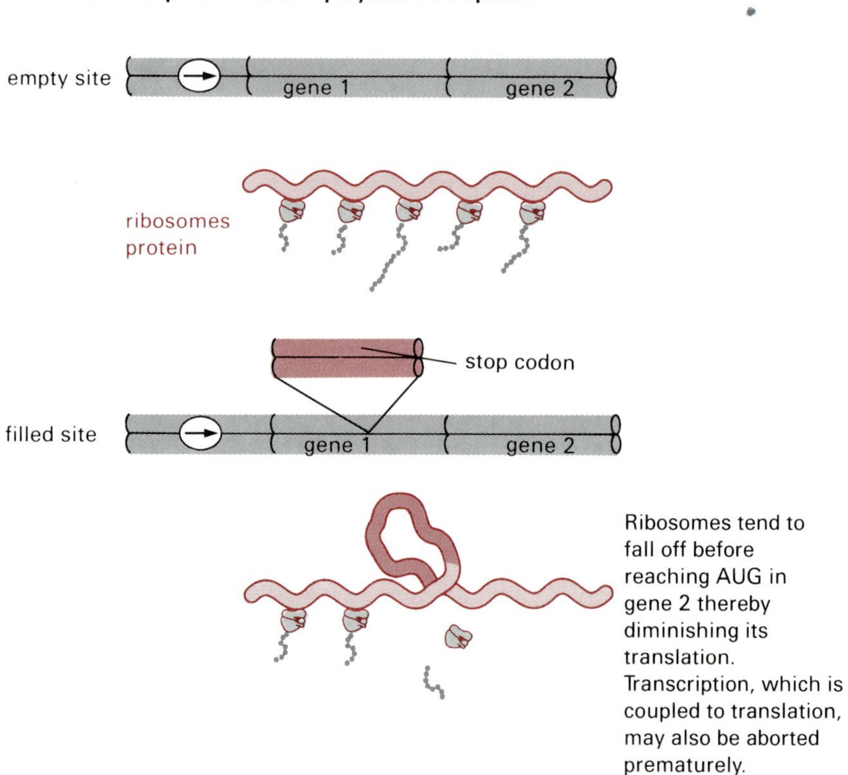

For example, the DNAs from various strains produce distinctive patterns of bands when digested with a specific restriction endonuclease and analyzed by DNA blotting using a cloned, P sequence probe. Third, and most definitively, insertion and excision of P elements at specific loci correlate with mutations and reversions, respectively, in specific genes.

The *white* locus of *D. melanogaster* is associated with the deposition of pigment in the eye and has been studied extensively for many years. Typically, mutants in this locus have white rather than the normal red eyes. DNA-blotting experiments using cloned DNA probes from the region around wild type and mutant *white* loci, demonstrated that one group of *white* mutants is caused by the insertion of a P element into the locus (Figure 10.14). Proof that the P element actually caused the mutation came from analogous studies on DNA from a red-eyed revertant fly. There are no detectable differences between the mutant and revertant DNA other than the P element is gone and only a single copy of the target site that was duplicated at the ends of P in the mutant remains in the revertant.

P elements display a marked preference for certain target sites. The consensus structure of the favored site, 5′-GGCCAGAC, was deduced by comparing the target site duplications around a number of P elements. Insertion is accompanied by the formation of an 8 bp long target site duplication.

Other types of *D. melanogaster* movable elements also vary in copy number and location and cause insertional mutations (Section 10.3). How-

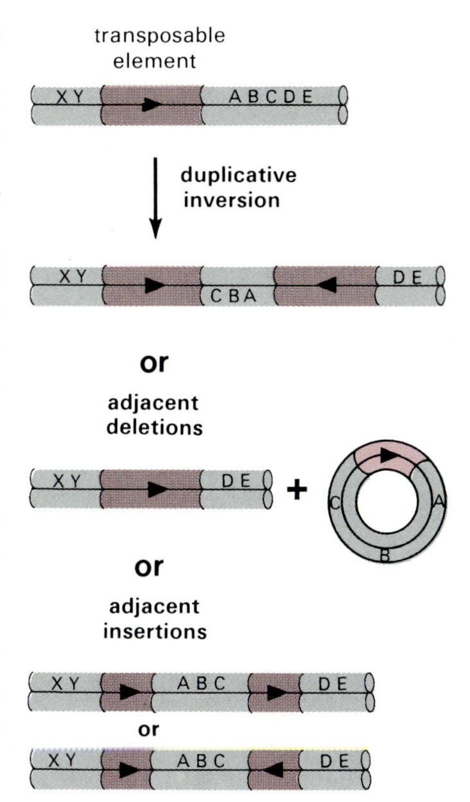

Figure 10.13
Transposable elements are associated with rearrangements of DNA sequences. After N. Kleckner, *Ann. Rev. Gen.* 15 (1981), p. 341.

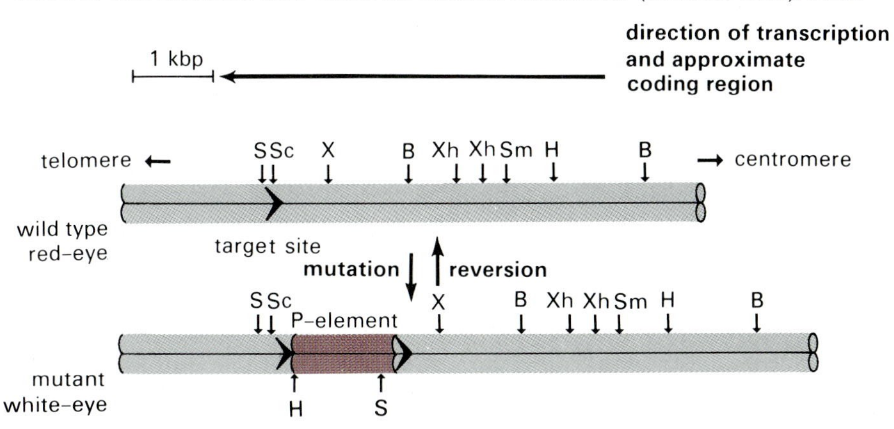

Figure 10.14
A mutation to white eye is caused by inserting a P element into the *Drosophila white* locus on the X-chromosome. Overlapping cloned segments representing more than 25 kbp around the wild type *white* locus were used to construct the regional genomic map. Using these segments as probes, fragments covering the homologous region were cloned from a recombinant library constructed with DNA from a white mutant. A map of the mutant locus was constructed. The mutant locus contains 1.4 kbp of extra DNA compared to the wild type, and the extra DNA anneals to cloned P element DNA. Similarly, a map of the locus was constructed for a red-eyed revertant of the white-eyed mutant fly. In the revertant, the P element sequence is lost, and only a single copy remains of the 8 bp target site that was duplicated at the ends of the P element. Thus, the revertant and wild type loci have the same structure.

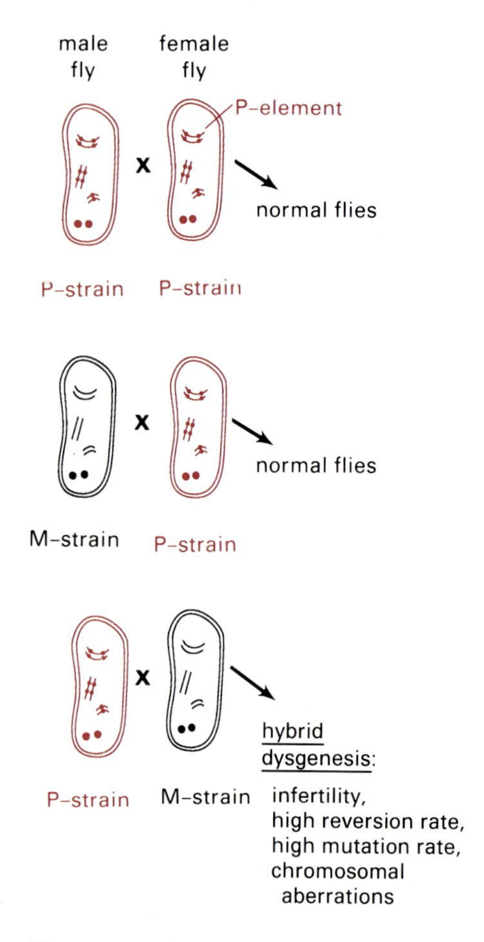

Figure 10.15
Hybrid dysgenesis. *Drosophila* chromosome pairs are shown schematically; the small dots on the chromosomes represent P elements.

ever, their structures and apparent transposition mechanisms are quite different from that of P. The properties of P, including the inverted terminal repeats and the inclusion of genes encoding one or more functions required for transposition, suggest a similarity to bacterial transposable elements. Moreover, like Tn3, P directs the repression of P-encoded functions, including transposition, although this is a complex phenomenon that depends on other genetic components.

Hybrid dysgenesis in strains harboring P elements The expression of genes encoded in P elements exerts profound effects in the germ cells of flies that are progeny of a cross between males of a strain containing P (P strain) and females of a strain lacking P (M strain). This complex set of effects is what is called **hybrid dysgenesis**. Dysgenic flies are often sterile. When they are fertile, their offspring show high levels of mutations and chromosomal rearrangements, traits that are passed to successive generations. Furthermore, some of the mutations are quite unstable in dysgenic flies and revert to wild type or altered alleles at a high rate. Another characteristic of the syndrome is that recombination occurs in dysgenic males, unlike normal *Drosophila* males.

Hybrid dysgenesis is caused by the introduction of the P elements carried on the paternal chromosomes into the environment of an M strain egg. There is no hybrid dysgenesis when P males cross with P females or when M males cross with P females (Figure 10.15). Thus, both P elements and a P-free maternal environment are required to yield hybrid dysgenesis. A useful explanatory model states that the presence of P in maternal chromosomes leads to repression of P transposition, thereby preventing hybrid dysgenesis.

The high rates of mutation and reversion in dysgenic flies are largely the result of the insertion and excision of P elements. Insertion appears to occur only in germ line cells and at a rate as high as about one insertion per chromosome per generation of flies (estimated by *in situ* hybridization to polytene chromosomes). In revertants, excision is often precise, as illustrated by the reversion of the white locus mutation described previously (Figure 10.14). The frequency of excision can be estimated genetically from reversion rates. It occurs in as many as four out of a thousand flies, a rate much higher than that found with bacterial transposons. In contrast to these rates, both insertional mutations and excisional revertants are much less frequent in nondysgenic flies (e.g., offspring of a P male and a P female).

Hybrid dysgenesis requires the presence of a full length P element. In the shorter P elements, sequences required to promote transposition and excision are deleted. However, full length elements can supply P functions in *trans* to short elements in the same cell. For example, the sn^w mutation of the *singed* locus was caused by insertion of a defective short P element. Mutant flies have unusual bristles. It is possible by genetic manipulation to obtain flies carrying the sn^w mutation, but no other P element. Male flies of this type do not produce hybrid dysgenesis upon mating with M strain females, and the sn^w mutation is stable. However, it becomes unstable when a full length P element is introduced into the genome; thus, a complete P element provides the excision function in *trans* for a defective element (Figure 10.16).

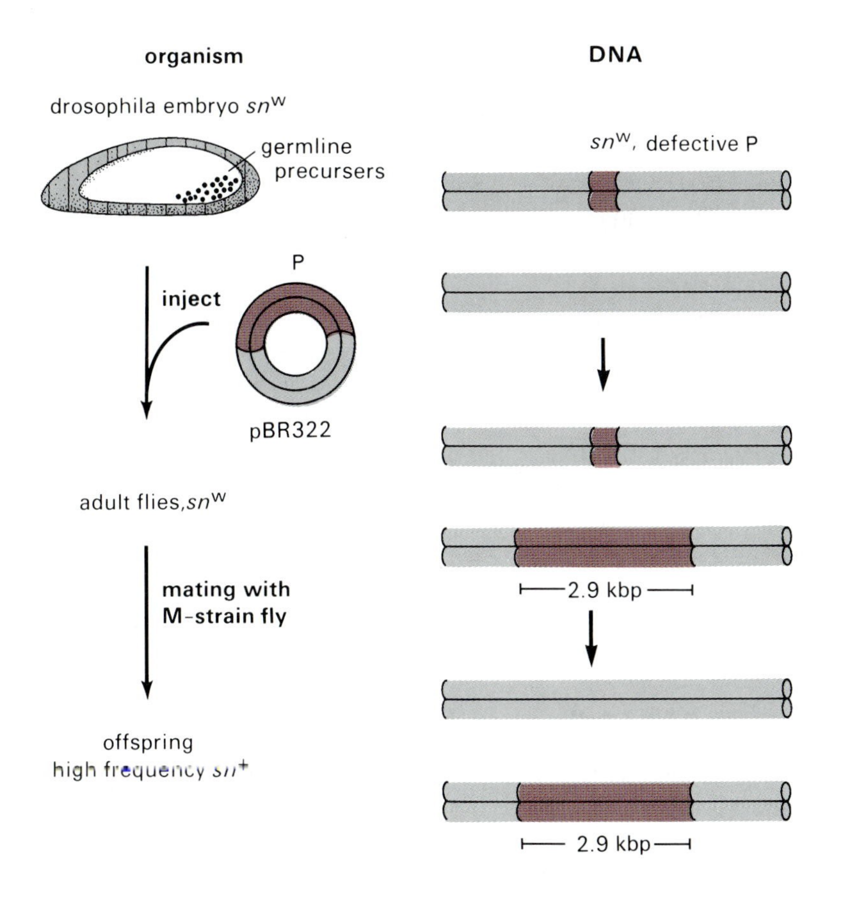

Figure 10.16
A single 2.9 kbp long P element encodes functions required for transposition. The experiment and the phenotype of the flies are shown on the left, and two regions of the genomic DNA are on the right. A complete 2.9 kbp P element was cloned in pBR322 and injected into very early sn^{w} embryos that had no other chromosomal P elements except for the short defective P associated with the sn^{w} mutation. The adult flies that grew from the embryos had a stable sn^{w} phenotype. However, the sn^{w} mutation was unstable in the offspring of some of these adult flies. Furthermore, the DNA of those adults and their offspring contained the complete new P element in their genomic DNA. Thus, some of the injected P elements were inserted into the DNA of embryonic germ line cells, were replicated faithfully during differentiation and development, and were expressed (in germ cells) to provide typical P functions. The pBR322 sequences that were covalently joined to P in the injected DNA were absent in the DNA of the progeny flies. Thus, P did not insert through an unspecific recombinational event but was most probably transposed intact from the vector into the *Drosophila* genome by the action of its own specific transposition system. This indicates that the P-encoded transposition system must have been expressed upon the initial injection. After A. C. Spradling and G. M. Rubin, *Science* 218 (1982), p. 341.

Functions encoded in P elements The hybrid dysgenesis phenotype implies that P elements encode a number of functions. One or more of these promote (1) transposition (e.g., a transposase) and thus a high rate of mutational insertion, (2) excision, and (3) sterility, presumably the result of the high mutation rate. Transposase activity appears to be limited to germ line cells. In addition, the suppression of transposition, and consequently the other phenotypic properties of P females, suggests that P elements are likely to encode a repressor of P-functions. The absence of repressor in the cytoplasm of M strain females would then allow for expression of P element genes and hybrid dysgenesis.

The nucleotide sequence of a full length P element contains four open reading frames (exons) on one strand (Figure 10.17). Two major polyadenylated P element transcripts with the sequence of that strand occur in somatic cells of dysgenic and P strain embryos. Both RNAs start close to the 5′ end of the P element sense strand. The shorter RNA, 2.5 kb, ends with a poly A tail close to residue 2700; and the longer 3 kb RNA terminates near the end of P, perhaps in flanking genomic sequences. Both are spliced in the same way. The first two introns are removed to form a single long open reading frame (ORF); the fourth ORF remains separated and unlikely to be translated. This observation was puzzling because analysis of mutations in full length P elements showed that all four ORFs are required for transposase. One possible explanation for the apparent discrepancy was that the splicing of the last intron to yield an mRNA for transposase is confined to germ line cells. Such a regulatory model could explain why P element transposition occurs only in such cells.

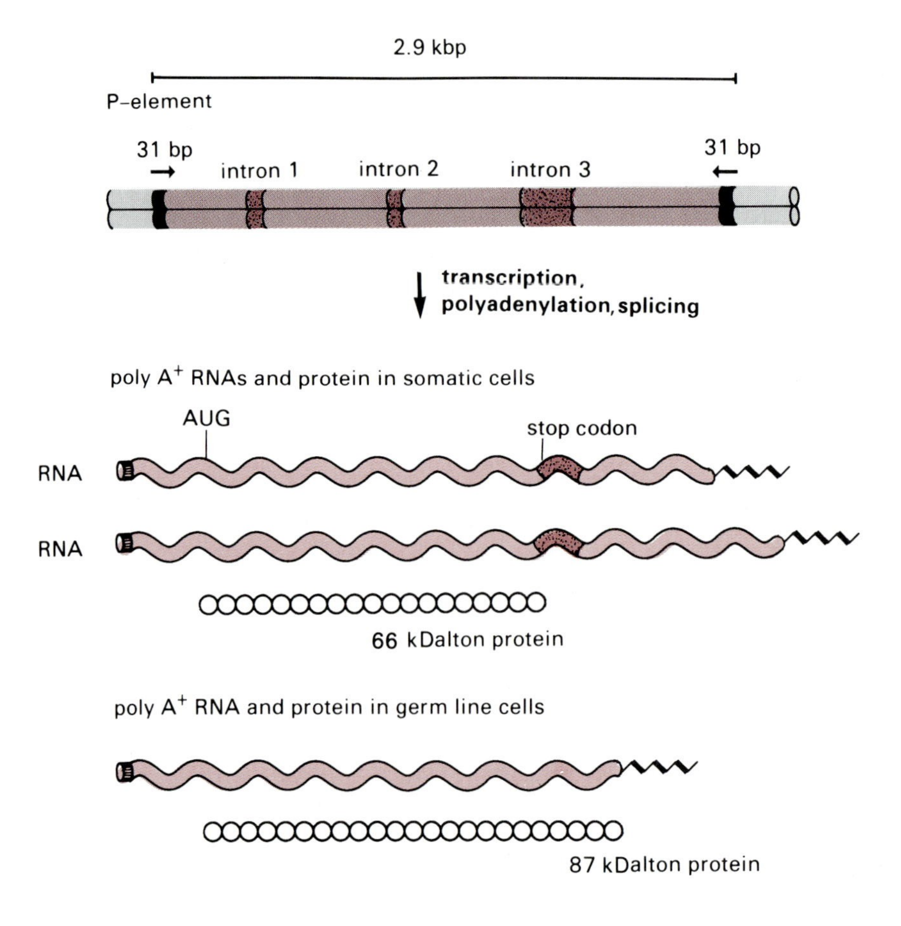

Figure 10.17
Structure of a full length, active P element is at the top. The 31 bp inverted terminal repeats, the four exons, and the three introns are shown. In the middle are the structures of two polyadenylated P RNAs found in somatic cells and the 66 kDalton protein they encode. Transcription initiates at base pair 87. At the bottom is the presumed structure of the polyadenylated RNA that occurs only in germ line cells and encodes an 87 kDalton transposase; in this RNA, all three introns have been removed by splicing. After K. O'Hare and G. M. Rubin, *Cell* 34 (1983), p. 25, and R. E. Karess and G. M. Rubin, *Cell* 38 (1984), p. 135.

It is not possible to get enough RNA from the *D. melanogaster* germ line cells to test this hypothesis directly. Instead, reverse genetic experiments were carried out using a protocol like that shown in Figure 10.16. P elements were mutated by site directed mutagenesis (Section 7.6) at the 5′ or 3′ splice junctions between the third and fourth exon. In another construction, the intron between the two exons was precisely removed. The mutated elements were injected into flies carrying the sn^w mutation but no other P element. Offspring were scored according to whether sn^w was stable or unstable, instability indicating an active P element. P elements with single base pair mutations at either splice junction failed to destabilize sn^w, consistent with the model. However, injection of the P element lacking the intron did destabilize sn^w in the same way as a normal P element does. Moreover, the intronless P even promoted transposition and excision in somatic cells, confirming the idea that germ line cells are uniquely able to carry out the specific splicing reaction. Consistent with this, P RNA containing intron 3 is found in all somatic cells and throughout development. Thus, transcriptional regulation is not involved in restricting transposase activity to germ line cells. Rather, regulation of gene expression occurs by control of primary transcript processing.

Somatic *D. melanogaster* cells containing the modified P element that lacks the third intron, but not those containing normal P elements, synthesize an 87 kDalton polypeptide. This is the size encoded by the fully spliced P RNA and is likely to be the transposase. Another, shorter polypeptide is translated from P RNAs that retain intron 3, such as the somatic cell 2.5 and 3 kb RNAs. This 66 kDalton P encoded polypeptide appears to supply the repressor of transposition that operates in somatic cells and P strain eggs.

The mechanisms by which P elements are inserted or excised are not known. Besides transposase, an approximately 65 kDalton *Drosophila* polypeptide which binds to the 31 bp terminal repeats may be involved. The nucleotide sequence of target sites left empty by P element excision are complex. Some show a clean excision of P and complete loss of one copy of the target site duplication (Figure 10.14). In others, the entire second copy of the target site sequence or remnants of it, and even remnants of the P element, remain behind.

Implications The experiments described in Figure 10.16 indicated that P elements could be used as vectors to introduce essentially any gene into *D. melanogaster* (Perspective IV). Moreover, the P element system is also interesting from an evolutionary standpoint. The emergence of a P element can lead to the reproductive isolation of fly populations that inhabit the same environmental niche. In this way, P elements might be a factor in speciation, even though they probably confer no inherent selective advantage because M strain flies are perfectly viable in their absence. Because the P family is only one of several families of movable elements that cause hybrid dysgenesis in *Drosophila*, the evolutionary potential of transposable elements in this genus is extensive. It is important, in this connection, to note that these other elements need not have the properties of transposable elements. At least one, the I element, is a class II retrotransposon (Section 10.3f). Also, there are transposable elements in *Drosophila* that are similar to P but are not associated with hybrid dysgenesis (e.g., an element called hobo).

c. The Controlling Elements of Maize

Transposable genetic elements were first recognized in studies of maize genetics, where they were termed **controlling elements**. At least three different families of controlling elements were identified; current evidence suggests that there are more than a dozen. Similar elements occur in other plant species. Each family in maize has two genetically defined membership categories: **autonomous** members are inherently unstable and capable of independent transposition and excision; **nonautonomous** members are themselves stable and are capable of transposition only when an autonomous family member coexists in the same genome. Autonomous members allow transposition only of nonautonomous members of the same family; indeed, this is the way families are defined. Because autonomous members appear to encode functions required for transposition and because of their structural features, maize controlling elements are grouped here with the prokaryotic transposable elements and the P element of *Drosophila*. The functional distinction between autonomous and nonautonomous controlling elements is, in fact, analogous to the functional differences between full length and shortened P elements. Nevertheless, there are important differences between the two systems. For example, the P elements are active only in germ line cells as displayed by hybrid dysgenesis, but the controlling elements cause somatic as well as germ line mutations. In maize, transposition occurs at reproducible frequencies at particular times during development; the effects on gene expression thus have the aspect of a regular, timed pattern. Although

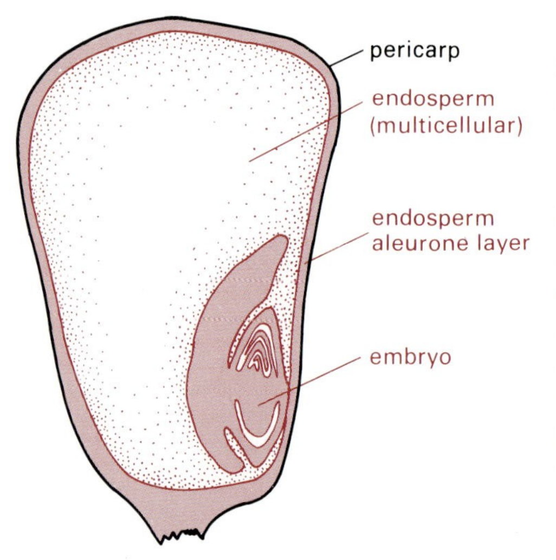

Figure 10.18
A maize kernel. The pericarp (black) is made up of maternal cells. The rest of the kernel has the genetic constitution of the embryo.

these effects appear to be superimposed on the inherent developmental pattern specified by the fixed maize genome, it is not hard to imagine how they might evolve to play an essential role in development itself.

The genetic system Without some appreciation of the special characteristics of maize development, it is very difficult to understand how classical genetic and cytogenetic analysis led to the discovery of controlling elements. Each kernel contains an embryo and a bulky multicellular endosperm that feeds the growing plant during germination (Figure 10.18). Embryo and endosperm arise from separate fusions of haploid nuclei, but they usually have identical genetic information. The pairs of male and female nuclei that form them are sisters, produced by mitotic division of haploid precursors. Therefore, germ line mutations that give rise to observable phenotypic changes in the endosperm are conveniently preserved in the companion embryo for further study. Each of the several hundred kernels on an ear is the result of a separate mating, so rare germ line mutations are observable.

Somatic mutations, too, are readily studied. During endosperm differentiation, single precursor cells yield localized clones of cells. A mutation occurring in a single cell at a specific time during development is later observed as a patch of cells with the mutant phenotype in a specific area of the endosperm. Many variant phenotypes with little or no effect on viability are visually evident in endosperm and are convenient experimental tools (Table 10.3). Consider, for example, the *C* locus. Plants homozygous for a recessive mutation in *C* are defective in the synthesis of anthocyanin pigment and have almost colorless kernels (Figure 10.19a).

Properties of the controlling elements established by genetic analysis Homozygous plants with an unstable recessive allele in the *C* locus give pale kernels with darkly pigmented patches (Figure 10.19b–e). Each such patch is a clone of cells derived from one that reverted to the dominant, wild

Table 10.3 Some Experimentally Useful Mutations in Maize

Genetic Locus	Mutant Phenotype	Wild Type Phenotype	Chromosome	Function Encoded
C	Colorless aleurone	Pigmented aleurone	9	Regulation of anthocyanin production
Bz (*bronze*)	Pale aleurone	Pigmented aleurone	9	Glycosylation of anthocyanin by UDP-glucose:flavonol 3-0-glycosyl transferase
Wx (*waxy*)	Lack of amylose in endosperm; stains pale brown with I_2-KI solution	Contains amylose; stains blue-black with I_2-KI	9	Glucosyltransferase for synthesis of amylose
Sh (*shrunken*)	Insufficient starch yields shrunken kernel	Plump kernel	9	Sucrose synthetase for starch biosynthesis
*Adh*1	Lacks alcohol dehydrogenase		1	1 subunit of alcohol dehydrogenase

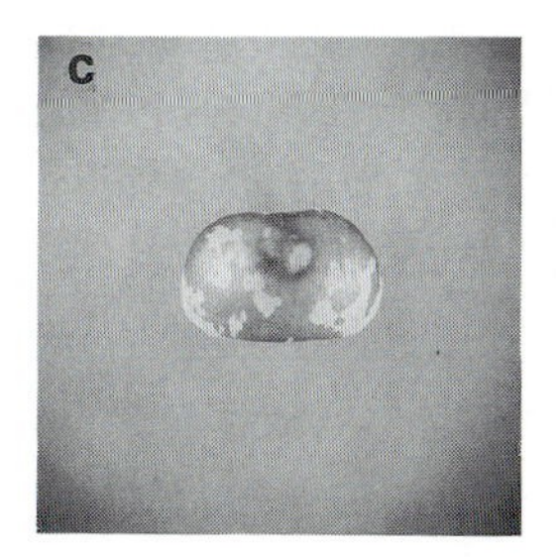

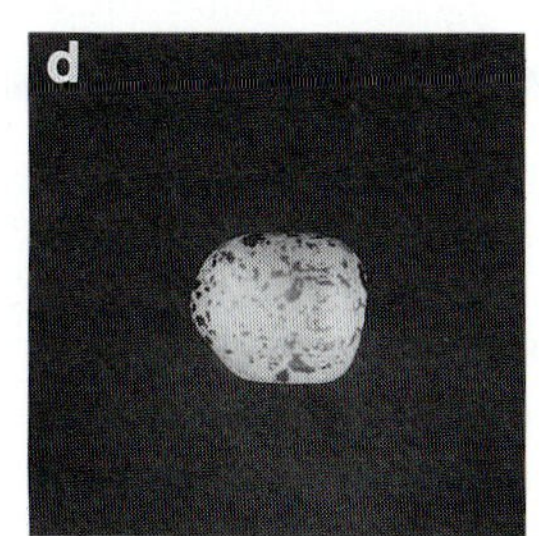

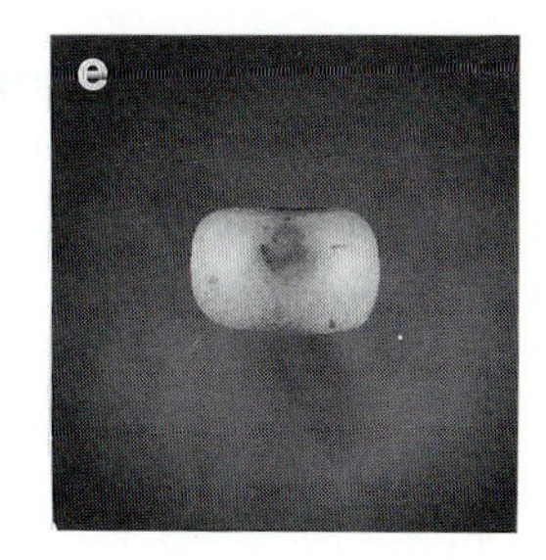

Figure 10.19
Phenotypic properties associated with the *C* locus. (a) A corn cob with some wild type (purple) and some mutant (colorless) kernels. The colorless kernels have stable mutation in a gene (*C*) required for anthocyanin synthesis. (b) Some of the cob's kernels are colorless, as in (a); others are variegated because the mutation, which is caused by insertion of a transposable element, is unstable. Clones of cells in which the element has transposed away have a functional gene and synthesize anthocyanin. Also shown are details of kernels in which the transposable element was excised from the *C* locus early (c), at an intermediate time (d), and late and infrequently (e) during kernel development. Courtesy of Nina Fedoroff.

type allele and can synthesize anthocyanin. Unstable, recessive mutations are caused by insertion of a controlling element. Reversion is associated with secondary rearrangements, such as the excision of the element and a return to the dominant phenotype. In addition to these properties, the presence of the controlling element fosters deletions, duplications, inversions, translocations, chromosome breaks, fusions, and rearrangements in neighboring chromosomal loci. This is reminiscent of the rearrangements in neighboring sequences that are associated with intramolecular cointegrate formation by bacterial transposons. However, it is not known whether the mechanism in plants is related to that in bacteria.

Autonomous controlling elements cause inherently unstable insertional mutations (Figure 10.20). Nonautonomous elements are usually associated with mutations that are inherently stable; however, they become unstable when an autonomous element of the same family is present elsewhere in the genome (Figure 10.20). It is helpful to keep the properties of the *D. melanogaster* P element in mind; using the language of maize genetics, a full length P element is autonomous and fosters its own transposition and excision, but a short defective P element is nonautonomous and is mobile only when a full length P provides the necessary functions.

In one family of controlling elements called *Ac-Ds, Ac* (for *activator*) is the autonomous element that provides for transposition and excision of

Figure 10.20
Families of maize controlling elements contain both autonomous and non-autonomous elements. (a) Insertion of nonautonomous elements into genomes free of autonomous elements yields stable mutations. Unstable mutations are caused by inserting autonomous elements. (b) Insertional mutations caused by nonautonomous elements are unstable if an autonomous member of the same family exists in the genome.

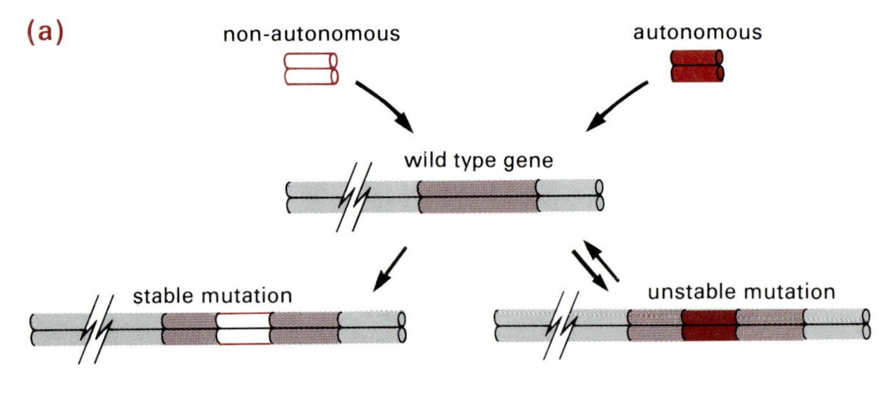

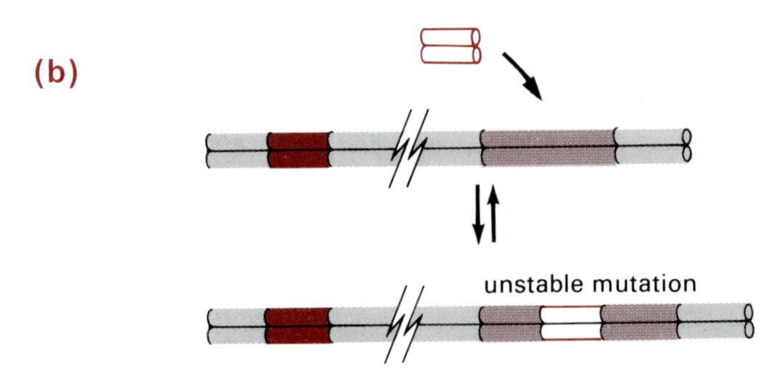

itself and of related nonautonomous elements termed *Ds* (for *dissociation*). *Ds* elements are unstable only when the genome contains one or more *Ac* elements. Moreover, *Ds* elements are produced by alterations in an *Ac*, as might be expected if the analogy with P is pertinent. For example, independent mutant strains with *Ds* at the *Bz* locus arise from strains with an *Ac* at the same position (Figure 10.21). They retain the mutant *Bz*

Figure 10.21
Ds elements are produced by alterations in an *Ac*. *Ds* is activated if the genome acquires an autonomous *Ac*.

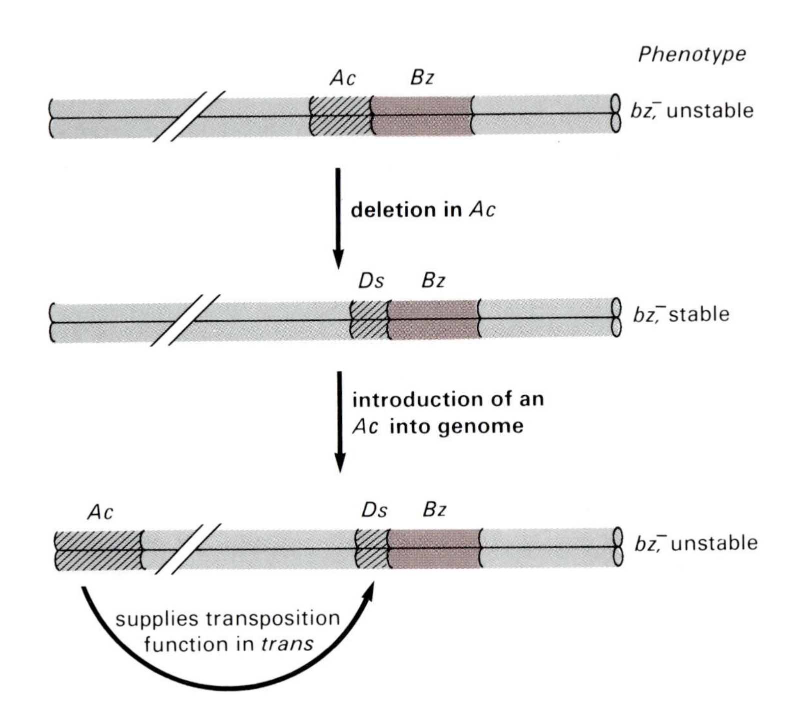

phenotype but are stable unless an additional *Ac* is introduced into the genome in an appropriate cross (just as the sn^w mutation in *D. melanogaster* is stable unless a full length P element is present elsewhere in the genome).

A large variety of phenotypes results from insertion of *Ac* or *Ds* elements near a particular gene. The level of transcription may be altered, the normal control of gene expression at specific developmental times or in specific tissues may be disrupted, or a mutant protein may be produced. Still other phenotypic properties are related to the instability of regions around an *Ac* or a *Ds* (in the presence of an *Ac*). Notable among them are (1) chromosome breaks that result in the splitting off and loss of acentric chromosome fragments and the disappearance of all functions encoded on those fragments (Figure 10.22) and (2) the appearance of additional mutations in the vicinity of the inserted *Ds*. Curiously, these events are much more frequent around *Dss* than around *Acs*. Not all *Ds* elements display the full range of effects even in the presence of an *Ac*. Some produce frequent chromosome breaks; others do so only rarely. Some have lost the ability to transpose but still revert and elicit neighboring mutations. The various properties of *Ds* elements are associated with different alterations in *Ac*.

Analysis of Ac-Ds mutants at the molecular level Cloned DNA segments containing either *Ac* or *Ds* have been isolated. A set derived from the *waxy* locus (Table 10.3) illustrates the general approach. Wild type *waxy* encodes a 65 kDalton protein, a 58 kDalton derivative of which is tightly bound to starch granules. Poly(A)$^+$ mRNA was isolated from endosperm and used to prepare plasmid clones containing the corresponding cDNAs. The cDNA clones in the library were screened by hybrid selection (Figure 6.26) for those that encoded mRNA for the 65 kDalton protein. The positive cDNA clones were used in turn to select homologous genomic clones from maize DNA libraries. Five such cloned genomic segments are illustrated in Figure 10.23. One is derived from a wild type *waxy* locus, one from an *Ac* mutant, one (*Ds*-1) from a *Ds* mutant that arose from the *Ac* mutant itself, and one from a phenotypically wild type revertant of the *Ds*-1 mutant. The fifth segment is from a separately derived *Ds* mutant (*Ds*-2). The *Ac* induced mutation is associated with a 4.5 kbp DNA insertion near the 3′ end of the *waxy* transcription unit; this 4.5 kbp segment is the *Ac* element itself. The *Ds*-1 mutant derived from the *Ac* mutant has an insertion in exactly the same position, but it is missing several hundred bp from within the element. The wild type revertant DNA lacks the insertion and is like the original wild type DNA. As already seen in the case of bacterial transposable elements and the P element of *Drosophila*, this set of relations is a hallmark of a transposable unit. The independent *Ds*-2 mutant also has an insertion (2.0 kbp) in the *waxy* transcription unit, but, as might be anticipated of an independent event, it is in a different position from the others. Analysis in the electron microscope of heteroduplex DNA formed between *Ds*-2 and *Ac* revealed that *Ds*-2 contains about 1 kbp of DNA from each end of *Ac* but is missing about 2.3 kbp from the interior. Some other *Ds* units are even shorter. For example, similar experiments with a mutant *Adh*1 locus revealed a 405 bp *Ds* insertion in the 5′ leader region (Figure 10.24). In

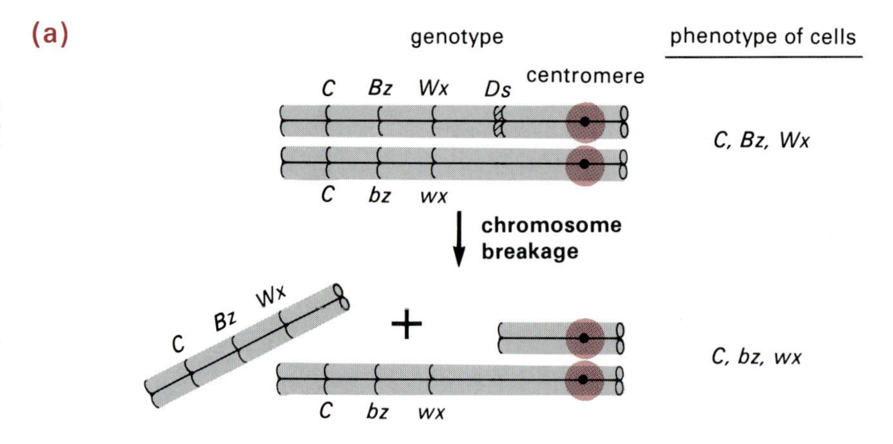

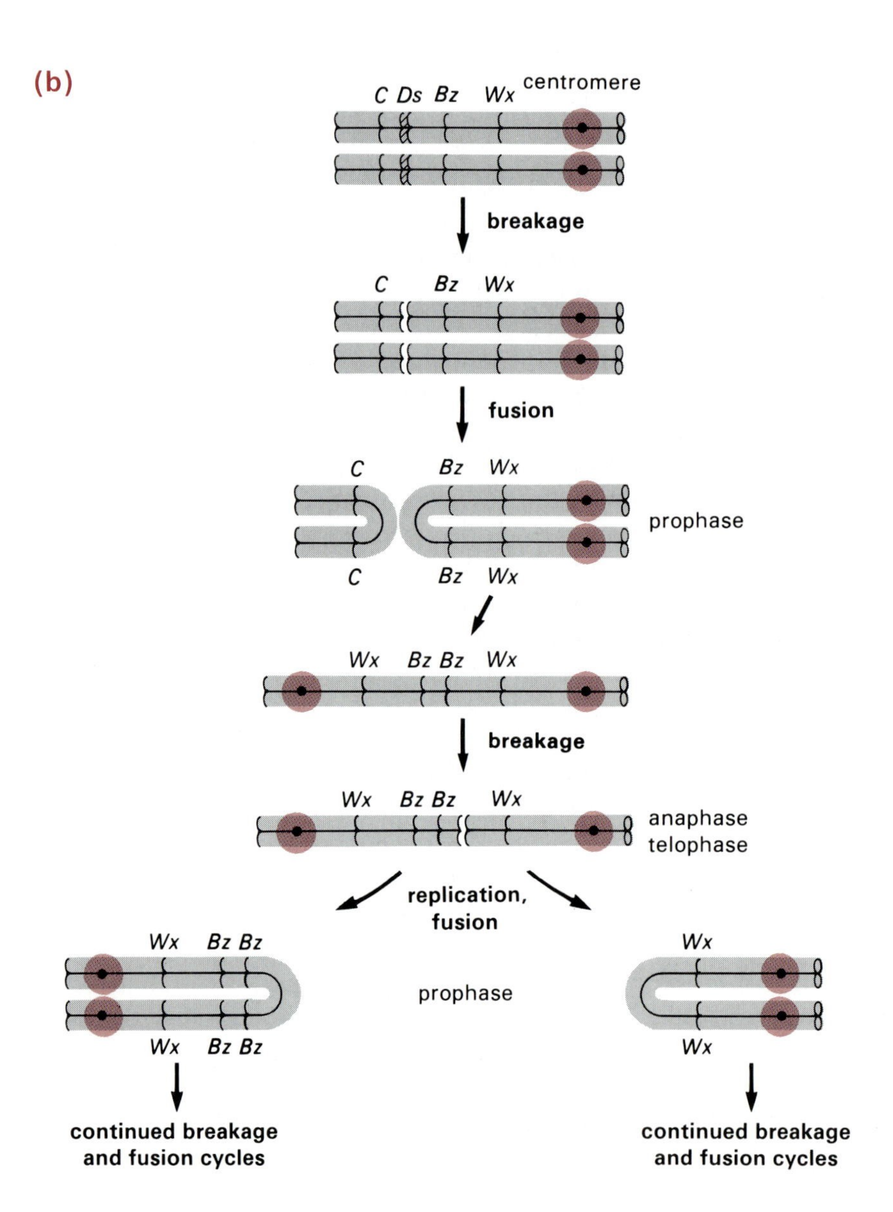

Figure 10.22
Chromosome breakage and rearrangement in the vicinity of members of the *Ac-Ds* family. In (a), one homologue of chromosome 9 contains a *Ds* insertion proximal (relative to centromere) to the *waxy* locus. Chromosome breakage at the *Ds* leads to loss in daughter cells of an acentric fragment carrying the dominant wild type *C*, *Bz*, and *Wx* loci. In (b), after breakage at *Ds*, sister chromatids fuse to yield a U-shaped acentric fragment and a U-shaped dicentric fragment that may be pulled apart and broken at subsequent mitoses. Again, fusion of the newly formed ends can occur, and a continuing cycle of chromosome instability ensues. In contrast to the broken ends, chromosome fusions do not occur at telomeres (Section 9.6b).

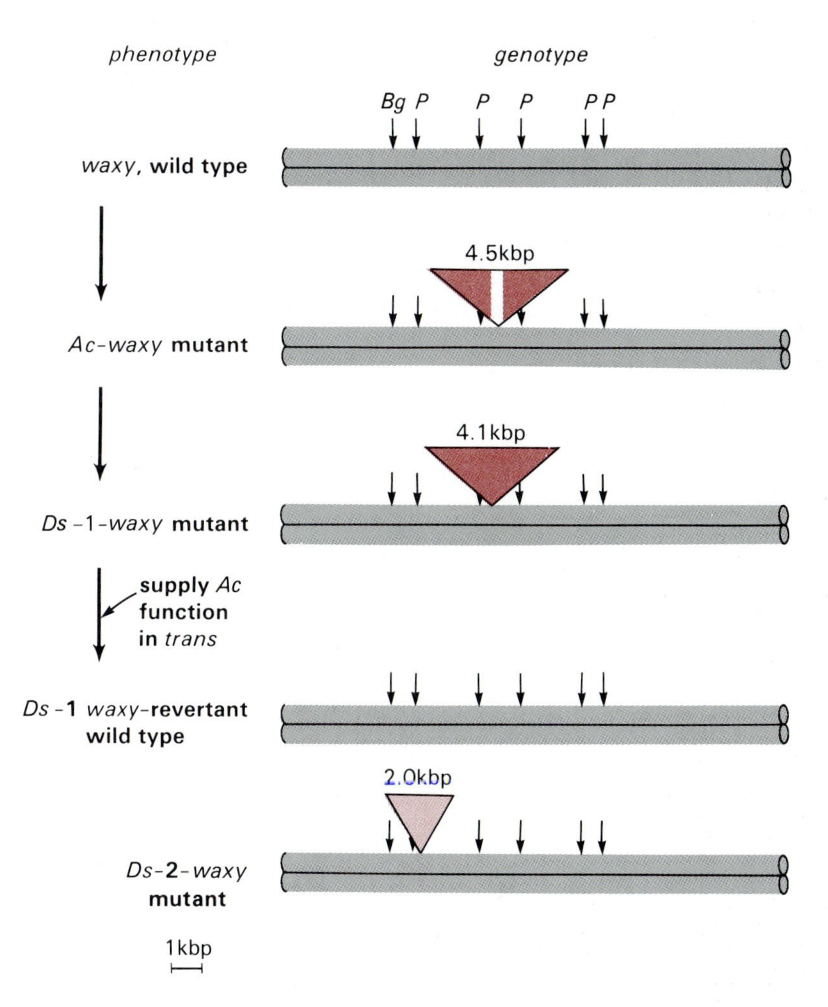

Figure 10.23
Genomic clones containing *Ac* and *Ds* mutations at the *waxy* locus of maize.

this case, sequences unrelated to *Ac* are found between termini typical of *Ac*. Thus, *Ds* elements are likely to be quite heterogeneous internally.

The nucleotide sequence of *Ac* is known (Figure 10.25). At the ends of the 4.6 kbp element, there are 11 bp terminal inverted repeats. There are also several sets of separate inverted repeats close to both ends. The terminal repeats are conserved in *Ds* elements. Five open reading frames that can encode polypeptide occur in *Ac*. Deletions in several different exons render the elements incapable of transposition. A discrete 3.5 kb *Ac* transcript that lacks the four intron sequences is found in plant tissue known to encode active *Ac* elements. It is likely that this is an mRNA for an *Ac* transposase and possibly other *Ac* specified functions.

Insertion of *Ds* or *Ac* is associated with a target site duplication of a set size, 8 bp. However, the target site duplication is not usually lost upon loss of the element and reversion. For example, in a revertant of the *Adh*1 mutant described previously, the extra copy of the duplicated 8 bp remains as a telltale ghost of the prior insertion (Figure 10.24). Frequently, one or both of the remaining copies of the duplicated sequence is slightly altered by, for example, a point mutation or deletion of a base pair. Thus, excision is imprecise and may or may not lead to phenotypic reversion. Similar observations have been made with other plant transposable elements.

The mechanisms by which *Ac* and *Ds* elements are inserted or excised are unknown. Genetic and molecular evidence suggests a nonreplicative

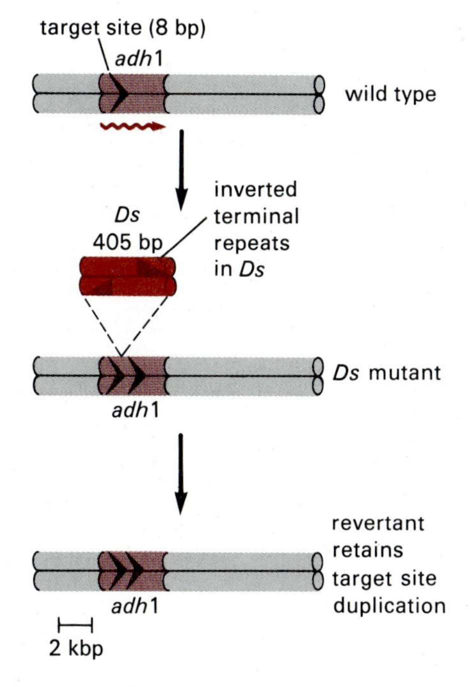

Figure 10.24
Genomic clones containing the *Adh*1 gene of maize. The wavy arrow shows the direction of transcription.

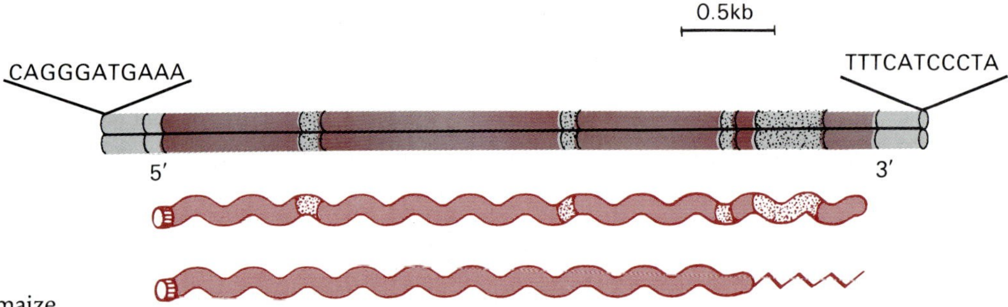

Figure 10.25
The structure of the 4.6 kbp maize transposable element, *Ac*. The imperfect 11 bp terminal repeats are shown, as are the extent and direction of the single identified transcription unit. Five exons together encode a long open reading frame in the 3.5 kb mRNA shown below.

insertion, involving excision of an *Ac* (or *Ds*) from one genomic site and insertion of the same DNA molecule in a new position. The unusual remains of the target site duplications that are left behind in previously filled sites suggest that *Ac* (or *Ds*) excision may be similar to P elements excision.

Repetition of Ac-Ds sequences Subcloned segments from within *Ac* and *Ds* anneal to multiple bands in restriction endonuclease digests of maize DNA. Altogether, about 35 copies are estimated to occur in the genomes of several strains. However, the patterns of annealing are complex, indicating that many different kinds of *Ds* elements occur and perhaps several different kinds of *Ac* as well. This variability is already apparent from the *Ds* elements described previously; they are, respectively, 4.1, 2.0, and 0.405 kbp long. Moreover, a *Ds* insertion mutation in an intron of the sucrose synthase gene (the *shrunken* locus, *Sh*, Table 10.3) is associated with a complex element that is about 30 kbp long and contains several repeats of *Ds* sequences as well as unrelated sequences (Figure 10.26). The ends of the 30 kbp element are imperfect inverted repeats of *Ds*, each of which is more than 3 kbp long. This is not the only peculiarity in the vicinity of the mutant shrunken locus on chromosome 9; on the 5′ side of the gene, there is a duplication of a portion of the mutant locus. The duplication includes parts of the gene and of the *Ds* sequences. The structure reinforces deductions made from genetic analysis, namely, that the presence of an *Ac-Ds* element encourages secondary rearrangements in neighboring sequences. Similarly, molecular analysis of a revertant of the same mutant confirms the conclusions of genetic analyses made 30 years ago; revertants to Sh^+ must retain a *Ds* element because the locus still displays frequent chromosome breakage. Thus, although the cloned revertant Sh^+ gene has lost the entire 30 kbp insert and is very much like the wild type gene in structure, it retains the neighboring partial duplication of gene and *Ds* element sequences. The early genetic analysis of maize has proved to be a remarkably accurate predictor of the presence of movable elements and of the complex molecular structures associated with *Ac-Ds* controlling elements.

10.3 Retrotransposons

a. *General Characteristics of Class I Retrotransposons*

A typical class I retrotransposon has a central DNA segment of several kilobase pairs flanked by LTRs that are several hundred to a thousand base pairs in length (Figure 10.27). A few base pairs at the beginning and end of each LTR are inverted repeats of each other. Insertion of a retrotransposon into a new location on DNA is accompanied by formation of a target site duplication, and the size but not the sequence of the duplication is always the same for a given element. Like all the movable elements, these retrotransposons constitute multisequence families in their respective genomes.

Typically, class I retrotransposons are actively transcribed and translated, although transcription is often modulated during development and differentiation. Some transcripts are a good deal smaller than the full length of the elements themselves. However, some transcripts are initiated within one LTR and end within the other, thus representing all the DNA sequences within the element. These long transcripts have direct repeats at their ends (i.e., they are terminally redundant). Such transcripts are intermediates in transposition. They appear to be reverse transcribed much as retrovirus RNA genomes are reverse transcribed to yield DNA proviruses (Section 2.2a), although the details of retrotransposon reverse transcription have not yet been studied. Retrotransposons encode several proteins. Of these, one is a reverse transcriptase, and others form an intracellular ribonucleoprotein particle with the specific transcripts.

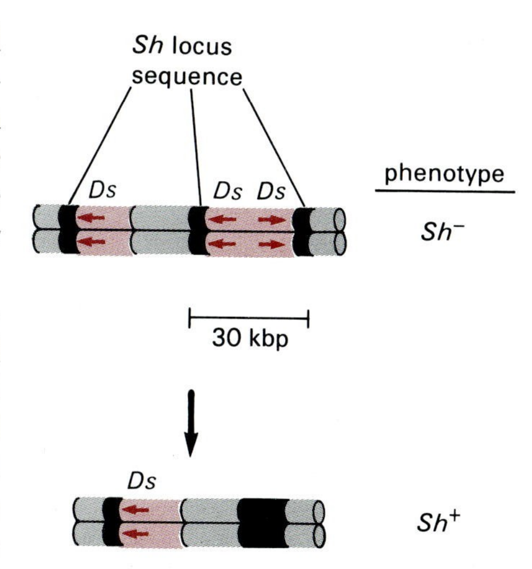

Figure 10.26
Complex *Ds* insertions in the maize sucrose synthase (*Sh*) gene. The structures of the mutant (Sh^-) and revertant (Sh^+) loci were determined from cloned genomic DNA segments. In *Sh*, the gene is interrupted by inverted duplications of *Ds* sequence that surround uncharacterized, non-*Ds* DNA. On the 5′ side of the gene (to the left), more than 20 kbp of sequence is duplicated, including part of the gene and part of the complex insertion. In the Sh^+ revertant, the insertion in the gene is lost, but the upstream duplication remains. The drawing is not to scale. After U. Courage-Tebbe et al., *Cell* 34 (1983), p. 383.

b. The Ty Elements of Yeast

The structure of Ty elements Ty elements have a typical class I retrotransposon structure (Figure 10.28). They are about 5.9 kbp long, and the central region, called ε, is flanked by 330 bp LTRs that are called δ. The inverted repeats at either end of the δs are only 2 bp long. Other notable structural features are the 5 bp long direct repeats within ε at its junctions with the two δs. Ty insertions are always accompanied by 5 bp long target site duplications.

The 35 or so Ty elements in the haploid genome of laboratory strains of *S. cerevisiae* account for about 1 percent of the genome. Other yeast strains contain fewer copies of Ty. Besides complete Ty elements, *S. cerevisiae* contains about 100 solo δ segments that are not attached to complete Tys. Annealing of cloned ε probes to restriction endonuclease digests of genomic DNA from different yeast strains yields varying numbers and sizes of homologous segments, as expected if the location of the elements is not fixed.

The multiple copies of Ty in a single genome are not identical in sequence, though they are similar enough to anneal to one another. Polymorphism is apparent, for example, in the different restriction endonuclease maps determined for individual elements. However, the two δs associated with a single Ty are usually identical in sequence, in contrast to the divergence observed between the IS-L and IS-R of some bacterial composite transposons.

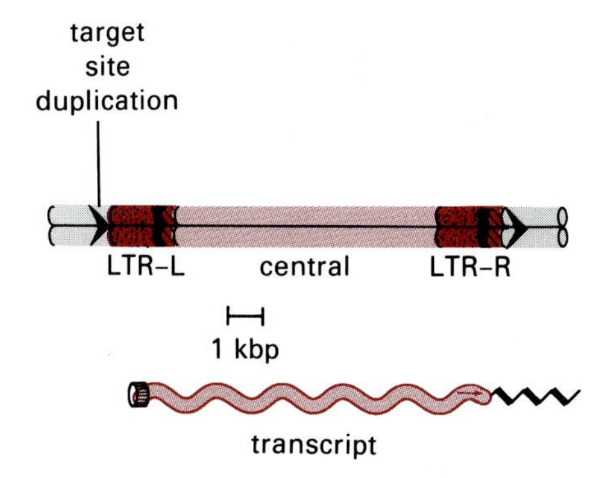

Figure 10.27
A typical class I retrotransposon and its transcript.

Ty moves The ability of Ty to transpose from one genomic locus to another was demonstrated in several ways. First, the variability from strain to strain in the sizes and numbers of restriction endonuclease fragments that anneal to Ty sequences is very suggestive. Second, molecular cloning shows that allelic genomic regions from different yeast strains differ by

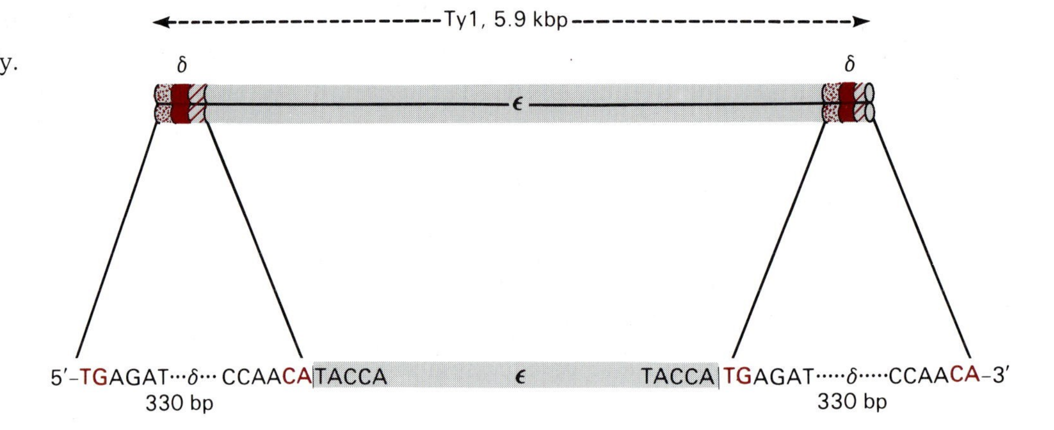

Figure 10.28
The yeast retrotransposon Ty.

the presence or absence of Ty. Ty was in fact discovered because a yeast suppressor $tRNA^{Tyr}$ annealed to different size *Eco*RI fragments in two different strains. When the two different fragments were cloned, they proved to be identical in the region immediately surrounding the tRNA gene, but to diverge dramatically on one side of the gene (Figure 10.29). The divergence is caused by the presence, in strain 2, of a Ty. The primary nucleotide sequence of this region shows a clean insertion of Ty. The only other alteration is the target site duplication: a pentanucleotide (5′-GAAAC) on either side of the Ty. Note that there are several solo δs at identical sites in this region in both strains.

Third, the mobility of Ty was demonstrated directly by dividing a single yeast colony into five cultures and allowing each to grow for a month, transferring a small inoculum to a fresh culture every day. At the end of the month, DNA from each culture was analyzed by the DNA blot procedure using a cloned Ty (ε) probe. Two of the five samples had a new hybridizing band that was not seen with DNA from the initial culture. The new bands were different in the two samples. Thus, rearrangement of Ty occurred.

Finally, the isolation of mutations caused by insertion of a Ty element conclusively demonstrates that transposition is an ongoing process. Numerous such events have been observed, although Ty transposes into any

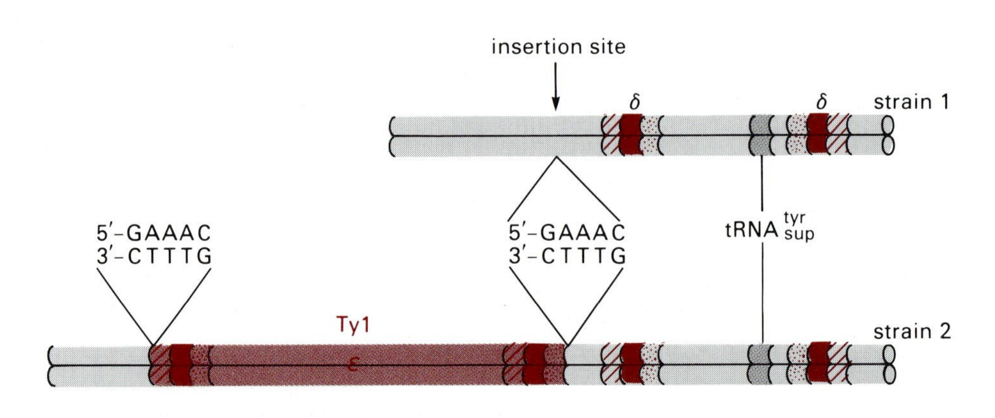

Figure 10.29
Allelic regions of two different yeast strains differ by the absence (strain 1) and presence (strain 2) of a Ty element. After J. R. Cameron, E. Y. Loh, and R. W. Davis, *Cell* 16 (1979), p. 739, and J. Gafner and P. Philippsen, *Nature* 286 (1980), p. 414.

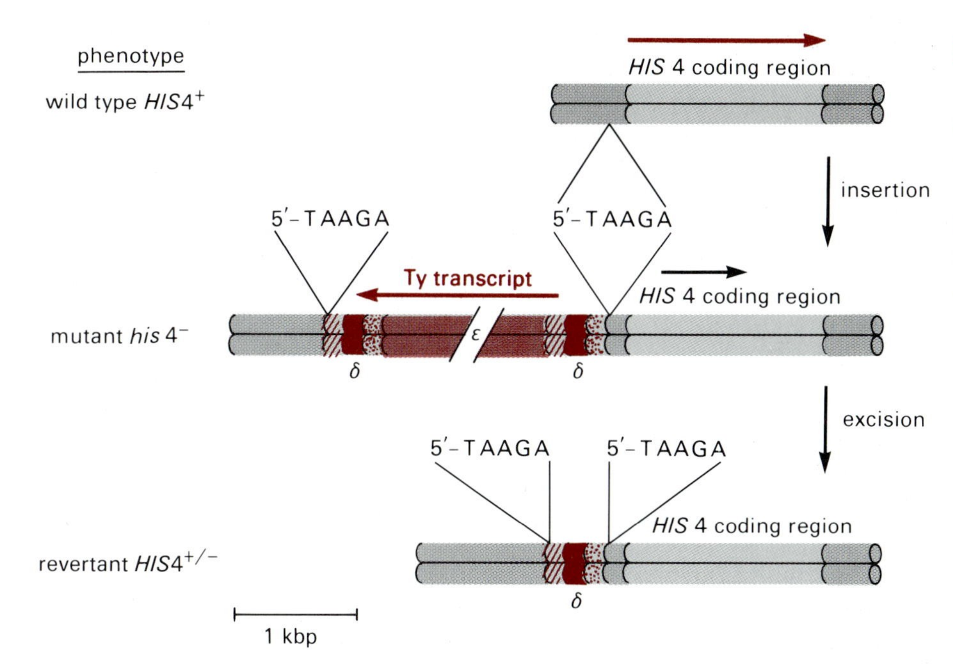

Figure 10.30
Insertion of a Ty element in the promoter region of *HIS4* creates a *his4*$^-$ mutant. Partial reversion to *HIS4*$^+$ involves excising the Ty, leaving a solo δ remnant. After P. J. Farabough and G. R. Fink, *Nature* 286 (1980), p. 352.

particular locus in only 1 out of 10^8 cells. Some Ty insertional mutations interrupt coding regions, but many of them are caused by entry of the element into the 5′ regulatory region of the affected transcription unit. Apparently, such sites are especially favorable Ty recipients, for reasons that are not yet understood. These mutations have several different phenotypes. In some, transcription of the neighboring gene is prevented; here the Ty may be in either orientation with reference to its own transcription (Figure 10.30). Others cause as much as a 50-fold overproduction of the mRNA and gene product of the neighboring gene. In overproducers, Ty is always inserted so that its own transcription is in the direction away from the gene (the direction shown in Figure 10.30). Overproduction depends on transcription of the Ty itself and thus on a haploid cell because Ty is poorly transcribed in diploid yeast (Section 10.5b). Moreover, the effect of Ty insertion on neighboring genes differs from one Ty to another, and some Ty elements have no obvious effect on nearby genes.

These differences depend, at least in part, on the sequence of a short DNA segment about 600 bp from the 5′ end of Ty. A change in one base pair from C · G to T · A in this segment markedly decreases neighboring gene activation. Activation appears to be mediated by a host-encoded *trans* acting protein factor that binds to the regulatory segment. In addition, various effects of Ty on the expression of a neighboring gene are suppressed by at least three other unlinked yeast mutations called *spt* for **suppressor of Ty**. *SPT* genes' products are required for abundant Ty transcription.

Having established that Ty moves, the next logical question is how it moves. However, in order to describe the mechanism, we must first discuss the transcription and translation of Ty.

The transcription of Ty Haploid yeast cells typically contain a very large number of transcripts of Ty: 5 to 10 percent of the polyadenylated RNA

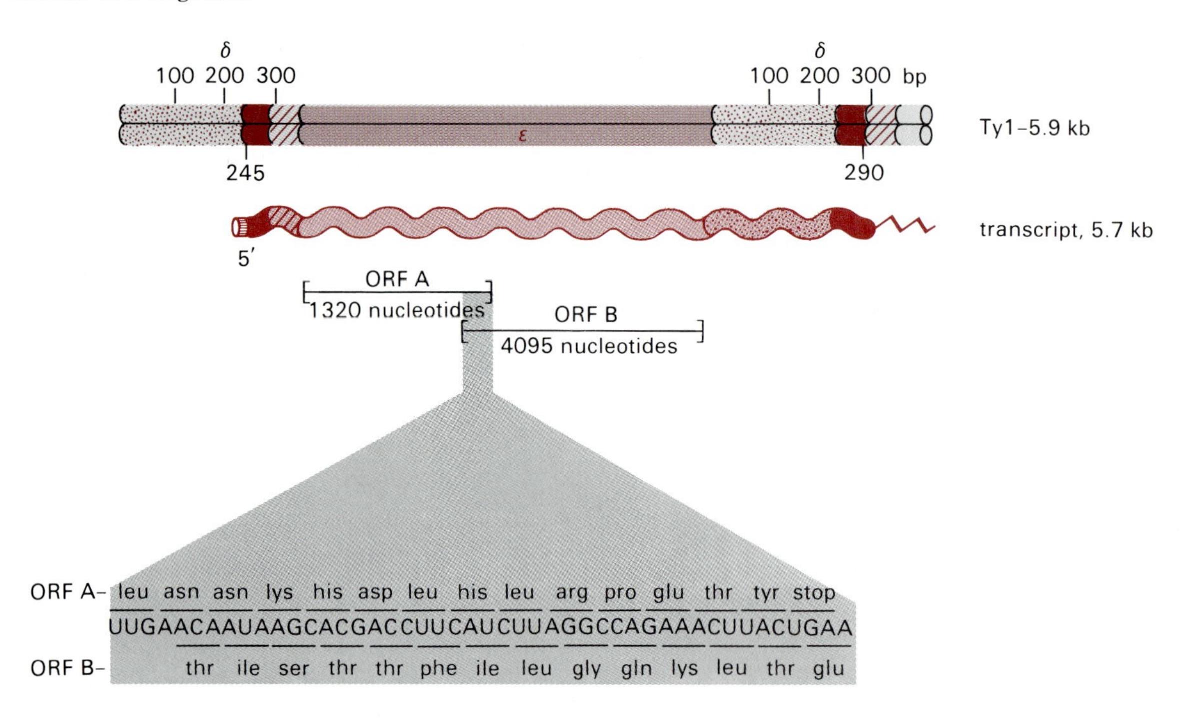

Figure 10.31
Transcription and translation of Ty. The 5.7 kb RNA is the most abundant Ty transcript found in yeast cells. The transcript has two overlapping open reading frames on one strand, ORF A and ORF B. The region of overlap between ORF A and ORF B is shown with the two reading frames. Upstream and downstream of this overlap ORF A and ORF B, respectively, are uninterrupted. The diagram is not to scale. After R. T. Elder et al., *Proc. Natl. Acad. Sci. USA* 80 (1983), p. 2432, and J. Mellor et al., *Nature* 313 (1985), p. 243.

is Ty RNA. The bulk of this is in a 5.7 kb long molecule that covers almost the entire length of the element. Individually cloned Ty cDNAs have slightly different sequences, so it is likely that many of the individual Ty elements are transcribed.

The 5′ end of the 5.7 kb long transcript is 245 bp from the beginning of the left-hand δ, and the transcript continues through the remaining 95 nucleotides of the δ, through the entire ε, and into the second δ, stopping about 40 bp from the end of the right-hand δ (Figure 10.31). A poly A tail that is not encoded in the element completes the 3′ end of the transcript. Several important points emerge from this structure, and essentially similar ideas are applicable to other class I retrotransposons and to retroviruses.

First, although the Ty transcripts omit the beginning and end of the Ty DNA, all of the nucleotide sequence of the element is represented in the transcript. This is because the omitted segments in the left-hand δ are transcribed from the right-hand δ, and, similarly, those omitted from the far end of the right-hand δ are transcribed from the left-hand δ. Note too that about 50 nucleotides are duplicated in the transcript. The first 50 nucleotides at the 5′ end of the RNA are the same as the last 50 at the right end, just before the polyadenylate segment. The transcript then has 50 bp long direct repeats at its two ends. Assuming that the promoter sequences for Ty transcription lie within the element itself, which is consistent with the fact that all Ty elements are similarly transcribed, regardless of the surrounding genomic DNA, then both δ sequences might be expected to initiate transcripts because they are essentially identical. However, no transcripts that initiate in the distal δ have been found. Similarly, if sequences within the distal δ dictate either termination of transcription or polyadenylation, or both, then why do the transcripts not stop in the homologous sequences just 50 bp downstream from the initiation site? These unexplained puzzles remind us that transcription is subject to many levels of control; some of them are only dimly perceived, and others are

Table 10.4 Examples of Similar Regions in Retroviral and Retrotransposon Polypeptides

Retroviruses			
Rous sarcoma virus	V L **P Q G** M	· 24 ·	C M L H **Y** M **D D** L L **L** A A T
Moloney murine leukemia virus	R L **P Q G** F	· 25 ·	I L L Q **Y V D D** L L **L** A A T
Retrotransposons—class I			
Ty (ORF B)	R P **P** P H L	· 52 ·	T I C L **F V D D** M V **L** F S K
Copia	R L **P Q G** I	· 58 ·	Y V L L **Y V D D** V V **I** A T G
IAP	V L **P Q G** M	· 25 ·	I L L L **Y** M **D D** I L **L** C H K
Retrotransposons—class II			
F	G I **P Q G** S	· 20 ·	T V S T **F A D D** T A **I** L S R
I	G V **P Q G** S	· 24 ·	K F N A **Y A D D** F F **L** I I N
LINE-1 (human)	G T R **Q G** C	· 34 ·	K L S L **F A D D** M I **V** Y L E

Note: Only two of several such similar regions in each element are illustrated. In retroviruses, these segments are known to be in the part of the *pol* genes that encode reverse transcriptase; the same is likely to be true for the retrotransposons. Boldface amino acids are either identical or very similar in most examples. Substitutions by chemically similar amino acid residues (e.g., tyrosine for phenylalanine or isoleucine for leucine) are termed **conservative replacements**. The numbers indicate the number of amino acids separating the two polypeptide segments.

surely not yet recognized at all. One control on Ty transcription relates to the mating type locus (Section 10.6b); diploids that are heterozygous for a and α contain only low levels of Ty transcripts compared to haploids. In addition, Ty transcription requires the *SPT* genes' products.

The translation of Ty The complete nucleotide sequence of several Ty elements is known. Two long open reading frames occur on the strand that has the sequence of the 5.7 kb transcript. The 3′ end of the first frame (ORF A) overlaps the 5′ end of the second (ORF B) by about 40 nucleotides (the number varies in different Tys), and the triplets in ORF A and ORF B are read in different reading frames (Figure 10.31). Trans lation of these frames is at a very low level in yeast cells, in spite of the abundance of Ty transcripts. However, when an extra Ty is introduced into the cells on a plasmid vector that attains a high copy number, Ty polypeptides are abundantly produced. The 1.3 kb long ORF A is translated in a straightforward manner to yield a polypeptide of about 55 kDaltons. Remarkably, an additional polypeptide corresponding to the translation of both open reading frames, about 190 kDaltons, is also synthesized at about 5 percent of the level of the ORF A product. To accomplish synthesis of this long polypeptide, there is an unusual shift in the coding frame during translation because, as already mentioned, the two frames are not in phase. This frameshift occurs in the region of overlap between the frames. The long polypeptide is processed by proteolysis to form several different proteins. One of these appears to correspond to some or all of the product of ORF A. A second protein, representing a translation product of ORF B, includes the reverse transcriptase. Note that an alternative mechanism for producing an mRNA that could be translated into the ORF B polypeptide might have involved splicing out the first frame. Surprisingly, this very generally utilized process is not used by Ty. Instead, the unusual frameshift occurs.

The amino acid sequence of Ty's ORF B contains regions of homology to the polypeptides encoded by retroviral *pol* genes (Table 10.4). In retroviruses, these segments yield several enzymatic functions, including protease, endonuclease (also called integrase), and reverse transcriptase (RT). When a Ty element is transcribed and translated efficiently in yeast cells, active reverse transcriptase is found in cell extracts. Partly purified fractions from such extracts synthesize Ty DNA in the presence of added deoxynucleoside triphosphates and Mg^{2+}. Introduction of a mutation in ORF B eliminates the enzymatic activity, thus proving that the reverse transcriptase is encoded by Ty. Cells with detectable reverse transcriptase also contain a 90 kDalton protein that binds to antibodies raised to a chemically synthesized nonapeptide whose structure was deduced from ORF B. The 90 kDalton protein copurifies with reverse transcriptase activity. Note that, besides the 90 kDalton molecule, the 190 kDalton fusion protein produced by the frameshift yields several additional ORF B polypeptides by posttranslational proteolysis. One of these is the protease responsible for posttranslational processing, and another is likely to be an integrase required for transposition.

The appearance of Ty reverse transcriptase activity in yeast cells is accompanied by the occurrence of cytoplasmic particles. These viruslike particles are about 60 nm in diameter and contain the enzyme, Ty RNA, and, in at least some particles, full length, linear duplex Ty DNA. A 48 kDalton polypeptide derived from proteolysis of the ORF A product is the main structural protein of the particles. Thus, ORF A is analogous to the *gag* gene of retroviruses. However, unlike retroviruses, Ty elements do not produce infectious particles.

Neither reverse transcriptase nor Ty particles are detectable in normal yeast cells, in spite of the presence of about 35 chromosomal Ty elements and a large amount of polyadenylated Ty RNA. The experiments described in the previous paragraphs depended on very high levels of Ty transcription and translation from a Ty element on a specially constructed plasmid (an example of such a plasmid is described in the following section). Most of the chromosomal Tys seem to be either defective or repressed. They do not yield detectable reverse transcriptase or particles, and, in contrast to the Ty on the plasmid, they foster only a very low level of Ty transposition. However, transposition of otherwise non-self-transposable chromomal Ty elements is elevated when a functional Ty is expressed to high levels, for example, with Tys on a multicopy plasmid or with Tys transcribed from an inducible heterologous promoter (see following discussion).

RNA as an intermediate in Ty transposition Ty transposes by reverse transcription of an RNA copy of the element in much the same way as retrovirus DNA, the provirus, is synthesized from retroviral genomic RNA. This mechanism was demonstrated by the following direct analysis (Figure 10.32).

Two plasmid vectors were introduced together into *his3*$^-$ yeast cells. One contained the Ty and was the retrotransposon donor. The second plasmid was the target for transposition and contained a *his3* gene from which the promoter was deleted. Many of the *HIS*$^+$ revertants isolated from such a yeast population contain the Ty transposed into the 5′ flanking region of the mutant *his3* gene on the target plasmid.

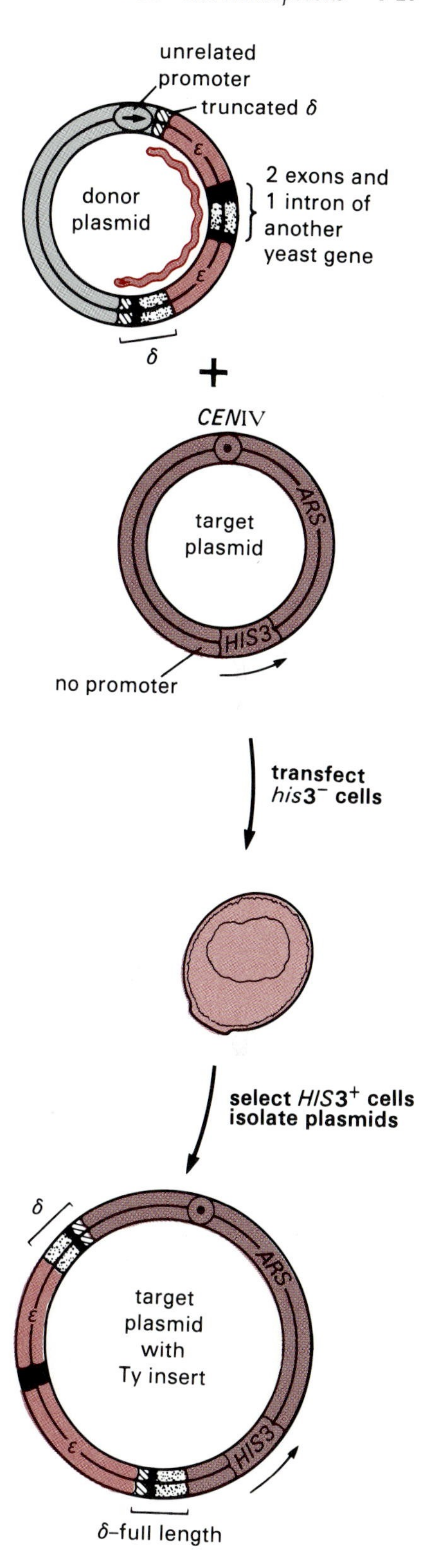

Figure 10.32
Experiments demonstrating that RNA is an intermediate in Ty transposition. Two plasmids were introduced into *his3*$^-$ yeast cells. The donor plasmid contained a Ty element that lacks the first 243 bp of the 5′ δ; an unrelated promoter replaces the Ty promoter, and a normal Ty transcript is produced (see Figure 10.31). In addition, a yeast intron (stippled) along with its neighboring exon sequences (black) is inserted into the central ε part of the Ty. The target plasmid contains a *HIS3* gene that lacks its promoter, the centromere region of chromosome 4 (to assure stability), and an *ARS* sequence (for replication). *HIS3*$^+$ colonies were isolated from the cell population, and the target plasmid was isolated and analyzed. Ty had transposed from the donor into the 5′ flanking region of the *his3* gene, providing the missing promoter function (note that the orientation of the inserted Ty is not known). The transposed Ty has two complete δs, and the intron was removed by precise splicing. After J. D. Boeke et al., *Cell* 40 (1985), p. 491.

For these experiments, the Ty donor was modified in several ways. The first 243 bp of the 5′ δ were deleted, and the missing Ty promoter was replaced by a promoter that is inducible by the addition of galactose (the yeast *GAL1* promoter, Section 8.3f). Ty transcription can initiate at the correct position (base pair number 245, Figure 10.31) and yield a correct Ty RNA, though under control of the inducible promoter. Transposition occurred normally, albeit at an unusually high rate. The Tys transposed into the target plasmids contained two complete flanking δs although the donor Ty lacked the 5′ portion of the 5′ δ. Two conclusions stem from these observations. First, a full 5′ δ is not required for transposition, which is very different from the situation with transposable elements. Second, the truncated Ty regenerates a full Ty element. Thus, the 3′ δ must provide a template for reconstruction of the missing portions of the 5′ δ. This can be accomplished using the RNA as intermediate because all the necessary sequences are available at least once in the donor. The proposed mechanism is virtually the same as that established for reverse transcription of retroviral RNA (Figure 2.41).

The participation of an RNA intermediate is even more persuasively demonstrated by a second modification of the donor Ty. An intron from an unrelated yeast gene along with short regions of its flanking exons was inserted into the central ε region. Normally, Tys do not contain introns. The transposed Tys in the target plasmids did not contain the intron. It was precisely spliced out, leaving behind its original flanking exons, which are foreign to Ty. Because such splicing reactions occur only on RNA, the role of RNA and reverse transcription in Ty transposition is clearly implicated. Note also that the continuing presence of the foreign exons proves that it was the Ty in the donor plasmid that was transposed rather than some other Ty in genomic DNA. Additional support for the role of Ty transcripts in transposition stems from the fact that *spt3* mutants fail to transpose chromosomal Ty elements but do transpose the plasmid borne Ty elements that are under transcriptional control of the non-Ty, inducible promoter. Recently, transposition was demonstrated in a cell-free system composed of purified Ty particles and an appropriate target DNA. It is likely that the linear duplex Ty DNA produced by reverse transcription and found within particles is the immediate precursor of the transposed element. Evidence suggesting that other class I retrotransposons transpose in a similar way is described in the next section.

Excision and other rearrangements associated with Ty In analogy with bacterial transposable elements, Ty is occasionally excised; the frequency is about 1 in 10^5 to 10^6 cells per generation. In the case of the *his*4 mutants, excision is accompanied by at least a partial reversion to the HIS4$^+$ phenotype (Figure 10.30). Excision generally leaves a trace of the Ty insertion in the form of a solo δ flanked by the same 5 bp repeats seen in the mutant. Homologous reciprocal recombination between the directly repeated δs flanking the Ty is the most likely mechanistic explanation (Figure 10.33a). It is also reasonable to assume that many of the solo δs originated from similar excision of complete Tys. The second product of such excisions, a circular Ty with one δ, is most likely lost. Although similar reciprocal recombinations between the 5 bp target site duplications are possible in principle, such events, which would regenerate an unaltered target site, are not observed.

Other genomic rearrangements arise from recombinations between dispersed Tys or solo δs. Some of these are known only from their genetic consequences, but others are described in molecular terms. Such events between any two particular Tys occur in as many as 1 percent of all cells undergoing meiosis; in dividing haploid cells, recombination between the same two Tys occurs at a frequency of only 10^{-7}. Deletions arise from reciprocal recombinations between Tys or solo δs that are in the same orientation on the same chromosome (Figure 10.33b); this occurs at a frequency of about 10^{-7} at the wild type locus shown in Figure 10.29. Inversions arise from reciprocal recombinations between Tys that are in opposite orientations on one chromosome (Figure 10.33c). When the interacting Tys are on different chromosomes, reciprocal homologous recombination results in chromosome translocations (Figure 10.33d). When unequal crossing-over occurs between Ty elements at different loci on homologous chromosomes in diploids, one chromosome has a deletion and the other a duplication (Figure 10.33e). Note that, in the reactions shown in Figure 10.33d and e, dicentric and acentric chromosomes will

---

Figure 10.33 (PAGE 749)
Recombinations between Tys and δs yield different rearrangements. A, B, C, D, E, and F are genetic markers in the yeast genomic DNA. (a) Deletion of Ty with retention of one δ due to homologous reciprocal recombination between flanking δs. (b) Deletion due to homologous reciprocal recombination between two Tys arranged in the same direction on one chromosome. (c) Inversion due to homologous reciprocal recombination between two Tys arranged in opposite directions on one chromosome. (d) Reciprocal recombination between two Tys on nonhomologous chromosomes. (e) Unequal crossing-over due to recombination between two nonallelic Tys on homologous chromosomes. (f) Homogenization of Tys on nonhomologous chromosomes by gene conversion. See also Figure 9.47.

(a) Homologous reciprocal recombination between flanking δ's

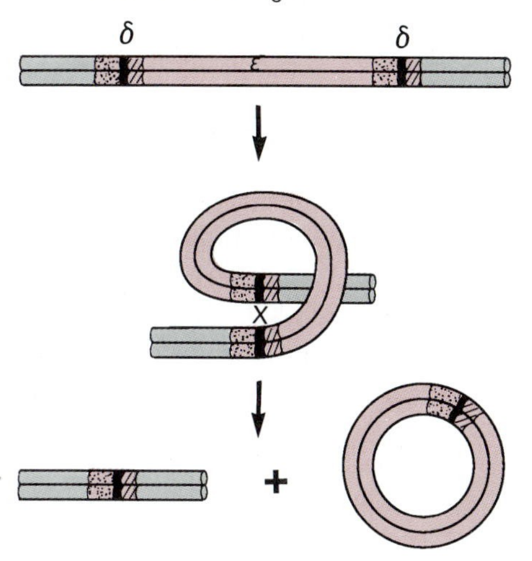

(b) Reciprocal recombination between two Tys in the same orientation on one chromosome

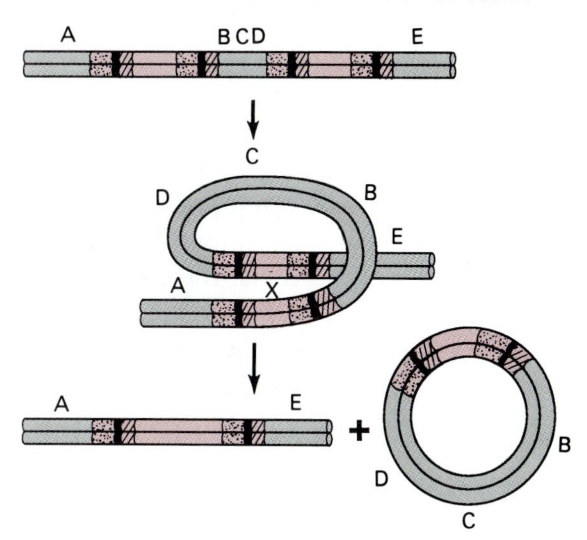

(c) Reciprocal recombination between two Tys in opposite orientations on one chromosome

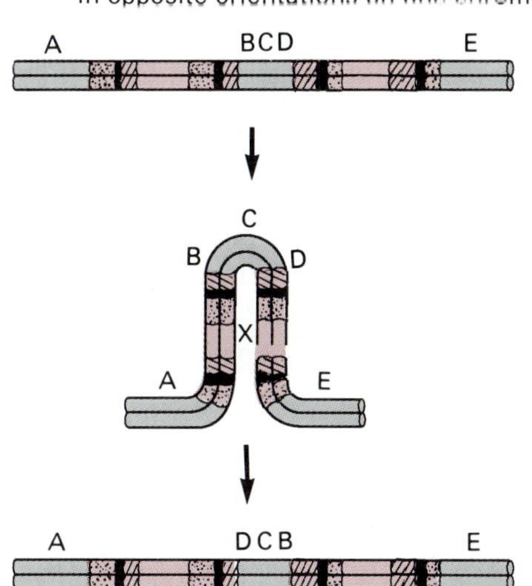

(d) Reciprocal recombination between two Tys on non homologous chromosomes

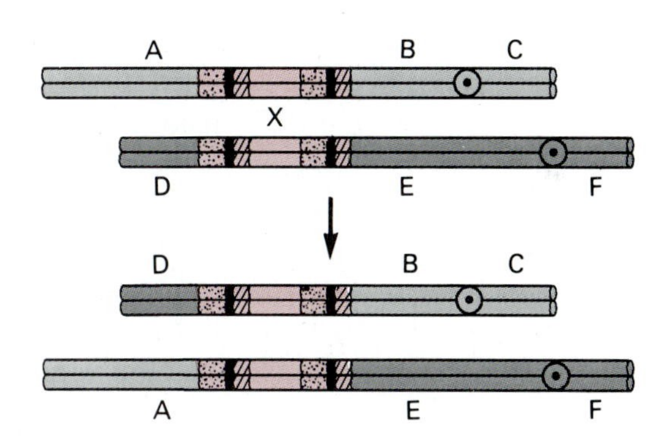

(e) Unequal crossing over between two non-allelic Tys on homologous chromosomes

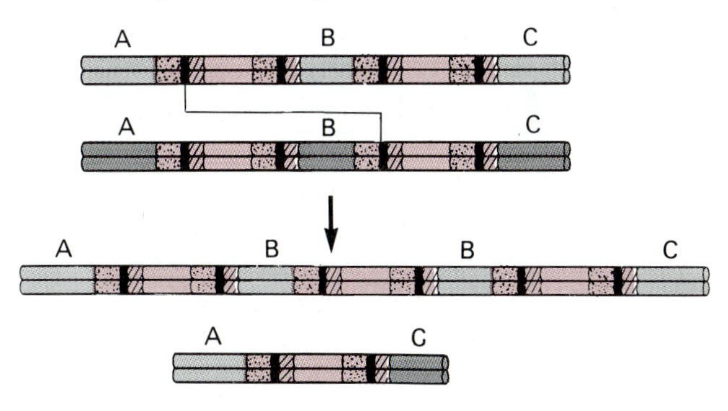

(f) Gene conversion

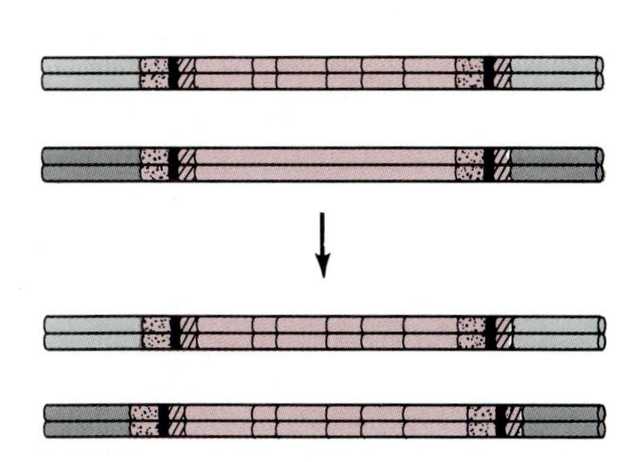

result if the two centromeres are on opposite sides of the crossover point. Finally, gene conversion between two Ty elements can alter the sequence of a Ty element without causing a change in flanking DNA sequences (Figure 10.33f).

c. *The Copialike Elements of Drosophila*

The panoply of *D. melanogaster* class I retrotransposon families are called, as a group, **copialike** elements after one of the first discovered families, copia. The many copialike elements, as well as the class II retrotransposons and transposable elements like P, together contribute to the characteristic long range interspersion pattern of the *D. melanogaster* genome (Section 9.5a).

The copia group Copious amounts of RNA that hybridize to a family of interspersed repeated DNA sequences gave the name to the prototypical element of this type. As many as 20 distinctive copialike families may occur in *D. melanogaster*, accounting for as much as 10 percent of the genome (5 to 100 copies per family). Some of the better characterized ones are listed in Figure 10.34. Like Ty, family members are several kilobase pairs in length and have direct terminal repeats several hundred base pairs long (LTRs) surrounding a central region. Also, like the δs flanking Tys, the LTRs bounding each individual family member are identical to one another but diverge somewhat from those flanking other family members. And again like Ty, the two ends of each LTR are short inverted repeats of one another. Unlike the situation with Ty, isolated copies of the LTR analogous to solo δs are rarely observed. Insertion of each element into a new genomic site is accompanied by a target site duplication of characteristic size; the actual sequence of the duplicated segment varies. The individual copialike families do not cross-hybridize and do not share marked sequence similarity. Within each family, members do cross-hybridize but are often polymorphic as a result of sequence divergence and deletions. Other members of the *Drosophila* genus (e.g., *D. simulans*) have distinctive sets of copialike families. Some share limited homology with the *D. melanogaster* elements; others appear to be species specific.

family	target site duplication	approx. copy no.	kbp		
copia	5 bp	30	276	5.1	276
412	4 bp	30	500	7.0	500
B104	5 bp	100	429	8.5	429
mdg1	4 bp	25	440	7.2	440
297	4 bp	30	412	7.0	412
gypsy	4 bp	10	500	7.3	500

Figure 10.34
Some of the families in the copialike group of retrotransposons in *Drosophila*.

Transposition The position of family members within genomes varies from one strain or stock of *D. melanogaster* to another, as is dramatically demonstrated by *in situ* hybridization. Figure 10.35 shows the annealing of a cloned member of the 412 family to paired homologous chromosomes of a fly that was the offspring of a mating between two strains of *D. melanogaster*. In one region, the polytene chromosomes are separated, and the different locations of 412 are easily seen by the different positions of the silver grains. Restriction endonuclease analysis like that described for Ty also shows variation in the position of family members from one strain to another. Changes in both position and total number of elements occurs in *Drosophila* cells grown in culture; the number of elements can increase as much as fivefold over that in fly DNA. The rate at which individual copialike elements rearrange could be as high as 10^{-3} to 10^{-4} per generation in cell culture, although it is much lower in flies themselves. In fact, the constancy in the overall number of elements per genome in the organisms is striking and suggests that some active process may limit the spread of each family.

Copialike elements, like other movable elements, are mutagenic when they insert into genes. Many examples have been characterized at the molecular level, including insertions of copia at the *white* locus that cause either apricot (w^a) or white eyes. As already described for Ty insertion, the mutant phenotypes associated with insertion may have several different explanations. Moreover, the effect of the insertions is sometimes modulated by unlinked *Drosophila* genes, perhaps reflecting the control by those genes of functions encoded within the element itself. As with Ty, mutations caused by copialike elements can revert when the element is excised by homologous recombination between the two LTRs; a single LTR and the 5 bp target site duplication remain. However, in contrast to Ty, such events appear to be rare because there are very few solo LTRs in the genome.

Transcription of the copia group Transcripts homologous to some copia group members amount to between about 0.5 and 3 percent of the total RNA of *Drosophila* cells. The RNA is found in both nucleus and cytoplasm. Some of it is polyadenylated, and some is not. The abundance of the transcripts varies, depending on the developmental stage and the particular element involved. Transcripts of B104, for example, are not detected in cells in culture but are abundant during embryogenesis. Copia transcripts are present in cultured cells and throughout development but are more

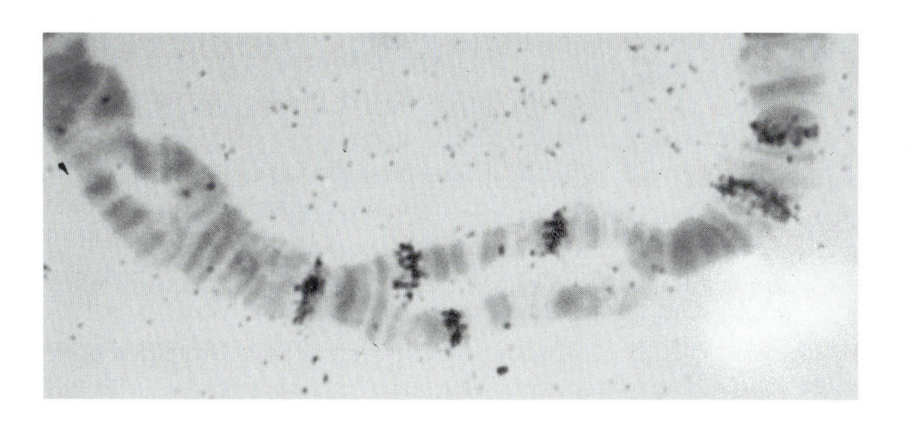

Figure 10.35
In situ annealing of a retrotransposon family 412 probe to a *Drosophila* polytene chromosome. The individual was a cross between parents of two different *Drosophila* strains. In this asynaptic region of chromosome 2, the varying location of 412 sequences in the two chromatid types is apparent in the silver grain distribution. Courtesy of G. Rubin, *Cold Spring Harbor Symp. Quant. Biol.* 45 (1980), p. 620.

abundant in larvae and adults than in embryos. The 412 transcripts appear in embryos and decrease in adults.

The most abundant transcripts of the copia elements are similar to the RNA produced from Ty in yeast. Typically, they cover almost the entire length of the element (about 5 kb), starting and ending within the terminal direct repeats and thus terminally redundant, and are complementary to only one DNA strand. Transcription appears to be unidirectional. Smaller size transcripts are also detectable in tissue culture cells (e.g., a 2 kb RNA from copia), but their significance is problematic because they are very minor RNAs in flies themselves.

Translation of the copia group Much of the copia RNA isolated from tissue culture cells is capped and is translated *in vitro* into several polypeptides between 18 and 51 kDaltons in length. These must be encoded by mRNAs that include the left half of copia because subcloned segments from the left half arrest translation (Section 6.4b). Much of this mRNA activity may reside in the 2 kb RNA.

A long open reading frame covers almost the entire region between the LTRs in copia and in the 5 kb copia RNA. In other copialike elements, the analogous region has several long open frames, similar to the situation with Ty. The predicted gene products contain regions of homology to Ty and to retroviral gene products, including regions associated with protease, integrase, and reverse transcriptase (Table 10.4). Thus, the copialike elements probably yield one or more polyproteins that can be cleaved by proteolysis to form the typical proteins and enzymes.

Ribonucleoprotein particles containing long transcripts of copialike elements occur in *Drosophila* cells. Particles containing 5 kb copia transcripts include about six polypeptides and four additional RNAs that are less than 300 nucleotides long, as well as reverse transcriptase activity. Both the polypeptides synthesized *in vitro* from copia RNA and those within the particles are precipitated by antibodies made against the particles. Thus, the copia RNA encodes the proteins with which it is associated. Moreover, although not proven conclusively, it is likely that the copia RNA in the particles is an intermediate in replication and transposition of copia by reverse transcription, as is Ty RNA. DNA·RNA hybrids of copialike retrotransposons have been detected and could be intermediates in this process.

d. The IAP Sequences of Mice

Elements that strongly resemble class I retrotransposons have been observed in various vertebrates. However, only one family, the murine IAP elements, is well characterized.

Intracisternal A-particles The cytoplasmic ribonucleoprotein complexes of mouse cells called **intracisternal A-particles (IAPs)** have an unusual natural history. Particles appear very early in mouse embryo cells and in many different kinds of mouse tumor cells, but except in the thymus (and even there in only certain strains), they are seen only rarely in normal tissues. Both the molecular components of the particles and their visual appearance are like those of some retroviruses, although they are not infectious. The donut shaped particles form by budding off from the endoplasmic reticulum membrane and then remain in the cisternae; they are about 80 nm in diameter.

Purified IAP particles contain a major structural protein of 73 kDalton, reverse transcriptase, and several specific RNA molecules, most of which are homologous to one another and are about 7.2 kb in size (35S). In some mouse myeloma cells, nearly 8 percent of the total cytoplasmic polyadenylated RNA is IAP RNA. This RNA is translated, in a cell-free system, into the 73 kDalton major structural protein.

IAP RNA is not homologous to that of most murine retroviruses; one exception is a partial similarity to the genome of a retrovirus endogenous to two Asian mouse species.

Genes encoding intracisternal A-particles Sequences that hybridize with IAP RNA occur in about 1000 dispersed positions on all the chromosomes of *Mus musculus;* other mouse species have only 20 to 25 copies, but Syrian hamsters have about 900 copies. The repeated units are as long as 7.1 kbp, although shorter family members exist. Like all interspersed repeated elements, individual family members diverge from one another. The elements contain a central region flanked by 338 bp long LTRs, each of which has short terminal inverted repeats (4 bp). Thus, IAP DNA is typical of class I retrotransposons (Figure 10.27). The target site duplications associated with insertion are 6 bp. The 5′ ends of the long RNAs found in IAP particles correspond to sequences in the LTR-L; thus, transcription is like that of Ty and copia.

Transposition and mutagenesis by IAP DNA IAP elements move to previously empty genomic positions and cause mutations in both germ line and somatic cells. For example, mutant cell lines that do not produce immunoglobulin light chains have been obtained from cells that are abundant producers. In some of the defective cell lines, IAP elements have been found inserted in the introns of the light chain gene; the IAP element may occur in either direction relative to the direction of transcription of the immunoglobulin gene. In another example, insertion of an IAP element increases expression of a neighboring gene, as is typical of many transposable elements. Thus, an IAP element inserted into the 5′ portion of the coding region of the murine oncogene c-*mos* is associated with an increased level of transcription of the c-*mos* sequences downstream of the insertion; as in similar examples with Ty, the IAP is inserted so that its normal direction of transcription is away from the c-*mos* sequences. S1 nuclease analysis of the aberrant c-*mos* transcripts suggested that LTR-L contains an additional control element that fosters transcription away from the body of the IAP element. The presence of bidirectional promoters and an enhancer in LTR-L were directly demonstrated using specially constructed expression vectors.

e. Comparison with Retroviruses

Structural similarities The structural similarities between class I retrotransposons and the integrated double strand DNA form of retrovirus genomes (provirus) are by now familiar (Figure 10.36). All of these DNA units have the same overall form: long terminal direct repeats (LTRs) surrounding

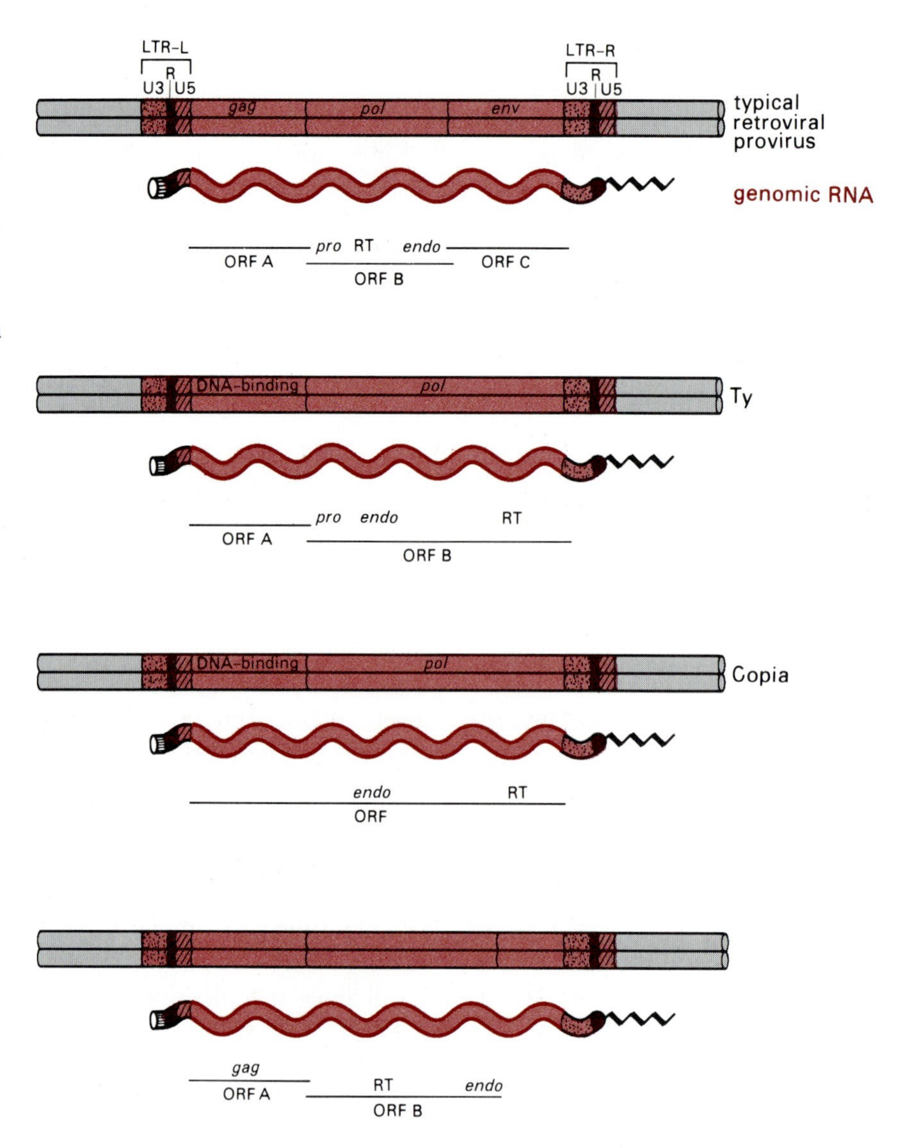

Figure 10.36
Structure of the DNA form of a retrovirus genome inserted into host cell DNA—a provirus, compared with the structures of Ty, copia, and IAP. The provirus *pol* region encodes a protease (*pro*) reverse transcriptase (*RT*), and endonuclease (*endo*). The retrotransposon ORFs encode polypeptides with regions of homology to the retroviral proteins. Overlapping ORFs are indicated.

a longer central region. Although the sequences of the LTRs differ from one family to another, they all begin and end in short inverted repeats; consequently, the complete units also begin and end with short inverted repeats. Moreover, the first two nucleotides of the inverted repeats are the same, 5′-TG, in many of the different units (Figure 10.37). In some cases, the similarity between the inverted repeats extends for several additional base pairs. There are also similarities between the U3, R, and U5 regions of provirus LTRs and sequences in the LTRs of retrotransposons. Signals associated with RNA polymerase II transcription, including the promoter sequence TATAAA and the polyadenylation signal AATAAA, occur within the LTRs in similar alignments. Moreover, many retrotransposons have a sequence similar to the retroviral site (the P sequence, Figure 2.39) that anneals with the tRNA primer for the initiation of reverse transcription (Figure 10.37). As in the retroviral genomes, the P sequence is at the border between LTR-L and the central DNA segment. Similarly,

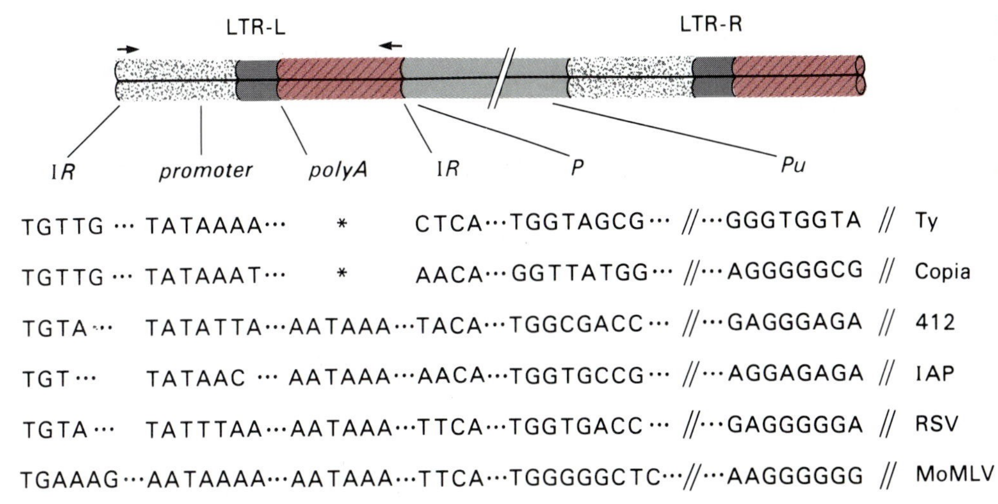

Figure 10.37
Common structural features in Ty, copia, 412, IAP DNA, and integrated forms of retroviruses (proviruses). An asterisk indicates that no equivalent sequences have been identified.

the region in the central segment of retrotransposons that is immediately adjacent to LTR-R is often purine-rich, as is the analogous region in retrovirus sequences (Figure 10.37).

Transcriptional and coding similarities The major transcripts of Ty, copia, and IAP DNA are also very much like transcripts of the proviruses. The RNAs that have been characterized begin within LTR-L and end within LTR-R in such a way that the transcript (1) contains at least one copy of all the sequences in the element and (2) has direct terminal repeats (Figure 10.31). As in retroviruses, the central regions of the class I retrotransposons are almost completely occupied by open reading frames. The coding region closest to LTR-L encodes protein that interacts with nucleic acid (analogous to *gag* in the retroviruses). The next coding region (*pol* for polymerase) specifies reverse transcriptase and also probably proteins homologous to the protease (*pro*), integrase (*int*), and RNase H of retroviruses. Retroviruses contain a third coding region (*env*) toward the 3′ end of their genomes; this encodes an envelope protein that encases the extracellular virion. Most retrotransposons appear to lack a functional *env* region, which may explain why they do not produce infectious particles. IAP DNA, unlike that of other retrotransposons, has a region long enough to be an *env* gene. However, the region is not a functional coding sequence because it contains multiple stop codons.

In retroviruses, as well as those class I retrotransposons that have been studied, the *gag* or *gag*-like reading frame (ORF A) is translated normally, though it is often then subject to proteolytic processing to yield multiple products. Copia has only a single reading frame, and all the proteins (e.g., *gag*-like, reverse transcriptase) appear to be the products of posttranslational processing. In other retrotransposons, and in many retroviruses, the *gag*-like and *pol* reading frames overlap (Figure 10.36) but utilize alternate sets of triplets, as in Ty (Figure 10.31). Some percentage of the translations starting at the beginning of the *gag*-like region continue through the discontinuity, shifting reading frames, to produce a fusion protein containing both *gag*-like and *pol* sequences. The fusion protein is then processed to form the individual proteins.

Localization of transcripts in ribonucleoprotein particles The analogy between retroviruses and retrotransposons extends to the packaging of retrotransposon transcripts into ribonucleoprotein particles. Besides the IAPs in mice, copialike element RNAs and Ty RNAs are found in particles in *Drosophila* and yeast, respectively. The particles contain protein components that are encoded by the *gag*-like and *pol* regions of the associated RNA.

Transposition of retrotransposons Given all the other similarities, it is not surprising that class I retrotransposons move by a mechanism similar to that used by retroviral genomes when they integrate into cellular DNA. All the details are not firmly established for each retrotransposon family, but the following general model is pertinent. A transcript of the retrotransposon is converted by the reverse transcriptase encoded in the element into a duplex DNA copy, which is then inserted into a new genomic locus. The actual details of the complex set of reactions involved in reverse transcription of retroviral RNA are described in Sections 2.2a and 5.7d.

One consequence of the reverse transcriptase model is that both LTRs of a new element arise from the same sequence because the RNA itself contains only one complete set of the LTR sequences. A common source for the two LTRs explains why those surrounding a single retrotransposon are virtually identical while the LTRs associated with other family members diverge. It is important to recognize that many, if not most, of the members of a retrotransposon family are probably incapable of independent transposition. Base pair changes and other sequence modifications in both regulatory and coding regions are common among family members in a genome and can limit transcription and translation into functional gene products. Some incompetent elements may be transposable if reverse transcriptase and other functions are supplied in *trans.*

Three duplex DNA structures are produced by reverse transcription of retroviral RNA, one linear and two circular forms that differ by having one or two copies of the LTR sequence (Figure 2.40). Rare circular DNA forms of Ty, IAP, and copialike elements have been detected and amplified by cloning. For example, two different size circles of copia are seen, 5.0 and 4.7 kbp (Figure 10.38); they differ in having either one or two copies of the LTRs, just as do the circular retrovirus DNAs. The Ty circles, however, contain only a single LTR. These circles could be formed by reverse transcription of the RNA in a manner similar to the formation of retroviral circles. However, they can also be generated by familiar genomic recombinational processes involving only DNA (Figure 10.39). In any case, the unintegrated circular retrotransposon DNAs may be dead ends. Recent *in vitro* experiments indicate that for both Ty and retrovirus, linear duplex DNAs are the transposition intermediates.

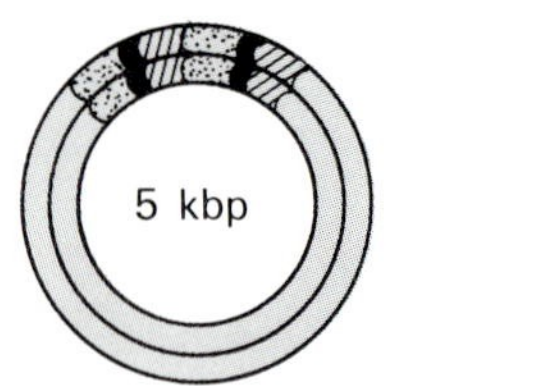

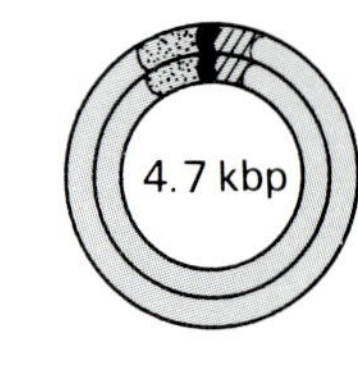

Figure 10.38
Two circular forms of copia differ in having one or two copies of the LTRs.

The final step in transposition is the insertion of the duplex DNA into the recipient genome. The reaction has been shown by *in vitro* experiments to be similar for retrotransposons and retroviruses holds. Of the

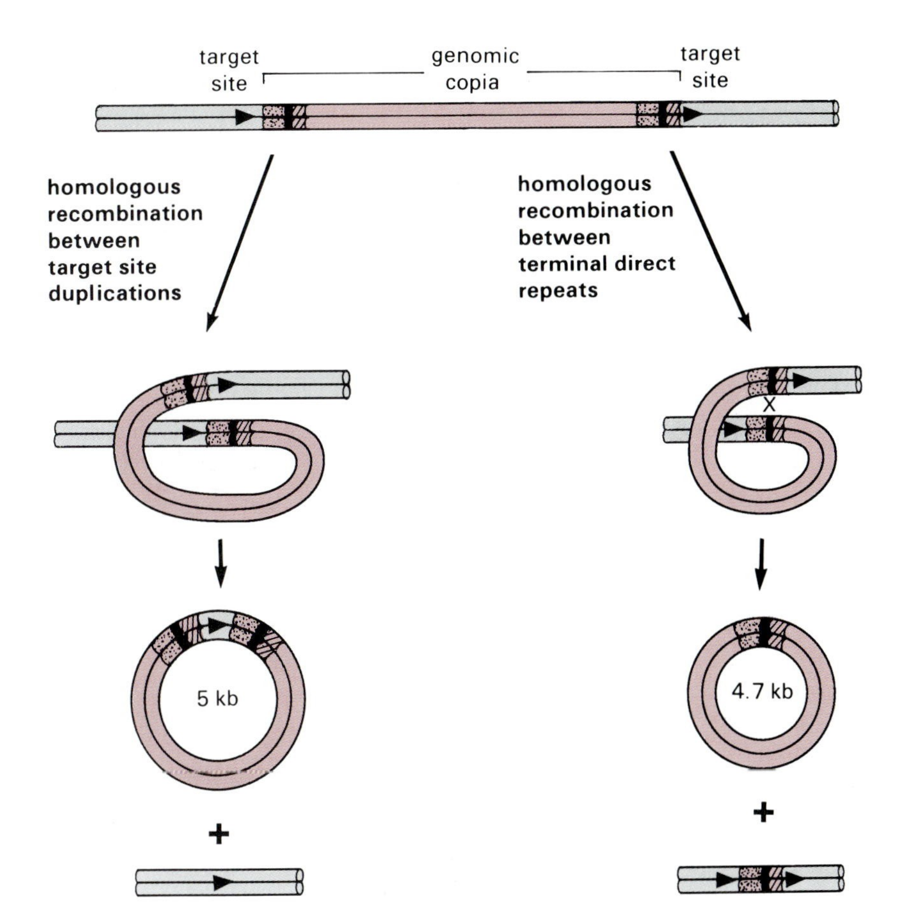

Figure 10.39

Model for origin of copia DNA circles through excision.

two models in Figure 10.40, that utilizing linear duplex DNA actually occurs. Note that these reactions are similar to those described for a simple insertion of a prokaryotic transposable element on Figure 10.9. To initiate insertion, four phosphodiester bonds are broken, two in the donor and two (staggered) in the recipient, and two new phosphodiester bonds are formed between 3′ termini on the donor and 5′ termini on the recipient.

Because so many features of class I retrotransposons and their transcripts are similar to those of retroviruses, it seems likely that the two had common ancient ancestors. Apparently, evolution yielded retroviruses as viable extracellular forms and confined retrotransposons to an exclusively intracellular habitat. Both types of elements are important because they provide for the flow of information from RNA into the genome, as may the class II retrotransposons (Section 10.3f). In this way, genomes can expand, and sequences can be rearranged. A continuing source of fresh material is assured for the evolutionary process.

f. Class II Retrotransposons

A typical class II retrotransposon has no long terminal repeats and is several kbp long (Figure 10.41). Target site duplications occur, but their

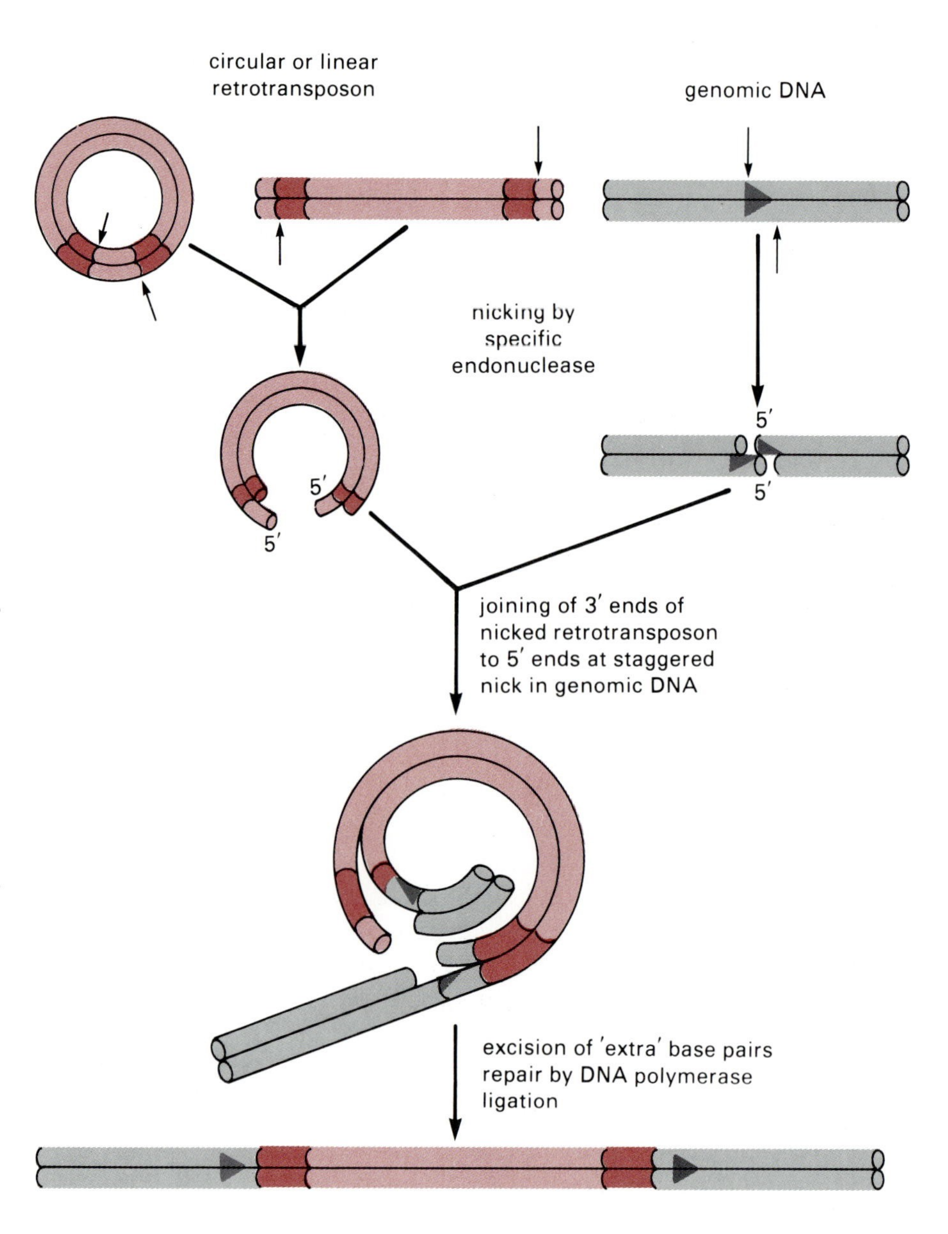

Figure 10.40
Model depicting the insertion of circular or linear duplex retroviral or retrotransposon DNA into a recipient genome. Staggered cuts are made in the recipient genome (arrows), presumably by the retroviral or retrotransposon encoded endonuclease. The number of base pairs between the nicks is likely to be specified by the endonuclease because that number sets the characteristic size of the target site duplication for each element. Similarly, the donor DNA is cleaved (arrows). Retroviral duplex circles typically contain 4 bp of extra DNA (hatched) between the U5 and U3 regions of the two joined LTRs (colored); presumably, the linear DNA has two of these bases at each end. Cuts are proposed to occur on either side of these base pairs in either the linear or circular donor. Retrotransposon duplexes may have similar extra base pairs. After joining the 3′ ends of the donor to the 5′ ends of the target DNA, the extra base pairs are removed, the gaps are filled by DNA polymerase, and ligation occurs. After P. O. Brown et al., *Cell* 49 (1987), p. 347.

lengths vary from one element to another in a genome. On one strand, the 3′ end contains a stretch rich in A residues. The central region of the strand with the A-rich terminus contains open reading frames, portions of which predict polypeptide segments similar to conserved regions of the *pol* and *gag* segments of retroviruses and class I retrotransposons (Table 10.4). Thus, although no direct demonstration of enzymatic activity or transposition mechanism has yet been made, class II retrotransposons are likely to encode their own reverse transcriptase and transpose by reverse transcription of an RNA intermediate. Compared to the wealth of information about retroviruses and class I retrotransposons, a great deal remains to be learned about the class II elements.

The F and G elements of Drosophila There are about 50 members of the F family scattered in the euchromatin and chromocenter of the *D. melanogaster* genome. The longest and most common F elements are 4.7 kbp, but others are as short as 3.4 kbp. At the 3′ end of all F elements is a

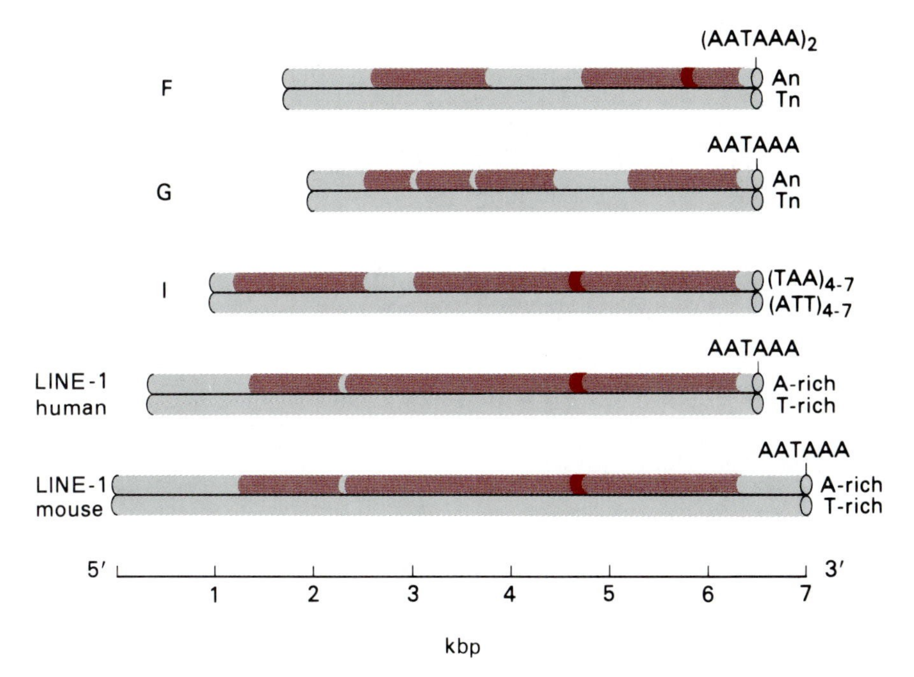

Figure 10.41
Structure of class II retrotransposons. Long open reading frames are colored, and the regions around which there is homology to known reverse transcriptases (Table 10.4) are marked.

stretch of 12 to 30 adenine residues (Figure 10.42). Just before the A-rich stretch, there are two overlapping polyadenylation signals, AATAAA. The short F elements are truncated at the 5′ end and are thus similar to the mammalian LINE-1 sequences.

At least one long open reading frame occurs in F elements that have been sequenced (Figure 10.41). Because it is not known whether these are active or disabled F elements, it may be that the open reading frame(s) in functional Fs will be somewhat differently organized. The common open frame predicts a protein with regional similarity to portions of reverse transcriptase (Table 10.4). Cellular transcripts of F are very rare, and their significance is unknown. No cellular polypeptides encoded by F have been detected.

The mobility of F was first suggested by its variable location in individual flies and by the existence of target site duplications. *In situ* hybridization of cloned F probes to polytene chromosomes of different *D. melanogaster* strains yields silver grains at dispersed positions. Even within a strain, the location of F varies among individual flies, giving different patterns of hybridization upon DNA blot analysis of restriction endonuclease digests of genomic DNA. The interstrain differences were exploited to clone and compare corresponding empty and filled sites as described in Section 10.2b for P elements. These experiments defined the

Figure 10.42
Three different F elements are homologous but have different 5′ ends. They are flanked by target site duplications of 8–13 bp, the sequences of which are different around each copy of F. The polyadenylation sites are boxed: AATAAA.

5′-CAGTTGCCGACCAATGAAGCATTTCGATCGCC----4.7 kb--------CAATAAATAAAAGTAAAGT$(A)_{18}$GCAGTTGCCGACCA-

5′-ATGTTTGAGATCGCC----4.7 kb--------CAATAAATAAAAGCAAAGC$(A)_{26}$ATGTTTGA-

5′-AATAATGCGGTCTGC -4.2 kb- CGATAAATAAAAGCAAAGTAAAAT$(A)_{22}$AATAATGCGG

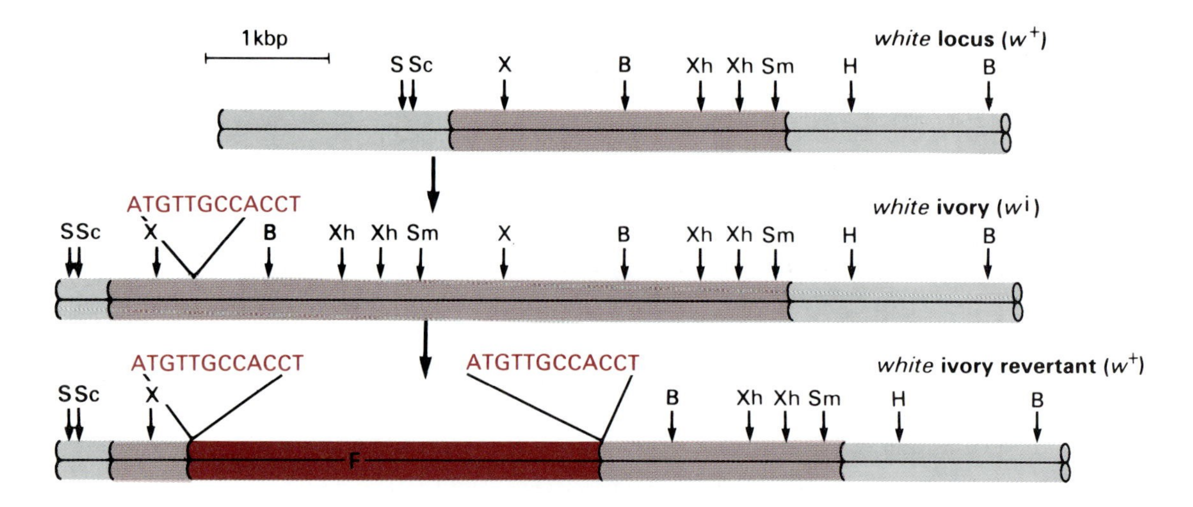

Figure 10.43
Evidence demonstrating the transposition of an F element. *White* ivory (ω^i) is a *Drosophila* mutation associated with a 3 kbp duplication of a portion of the wild type *white* (eye color) locus. In a wild type revertant of ω^i (ω^+), the duplication was lost, and an F element, surrounded by target site duplications, was inserted. After R. E. Karess and G. M. Rubin, *Cell* 30 (1982), p. 63.

borders of the full length F element itself and showed that the target site duplication length is different around various copies of F, ranging from 8 to 13 bp. The white locus of *Drosophila* provided the opportunity to demonstrate directly that F moves. White-ivory is a mutation caused by the duplication of 3 kbp of DNA in the white locus (Figure 10.43). In one revertant of white-ivory, the duplication is deleted, and an F element is inserted.

G elements are very similar to F elements in structure (Figure 10.41), but their organization in the *D. melanogaster* genome is quite different. They occur in tandem arrays that are inserted into the intergenic spacers (IGS) of ribosomal RNA gene clusters. These rDNA units appear to be localized in the chromocenters of polytene chromosomes because cloned G element sequences anneal exclusively to that chromosomal region.

There are several open reading frames in sequenced G elements, one of which predicts a polypeptide with similarities to reverse transcriptase. As with F, no transcripts or protein gene products of G have been identified.

The I element of Drosophila Distinct from the hybrid dysgenesis caused by P elements, a second hybrid dysgenesis system called **I-R** occurs in certain *Drosophila* strains. Progeny of crosses between so-called **inducer** males (I) and **reactive** (R) females are dysgenic. As with P, other types of crosses (e.g., R males with I females) yield normal progeny. (Figure 10.15 illustrates the corresponding crosses in the P-M system.) The phenotypic consequences of I-R dysgenesis are similar to those of P-M.

The working model of the I-R system is analogous to that for P-M. A transposable I element is activated when it enters the cytoplasm of an R strain egg upon fertilization. If it enters an I strain egg, activation is repressed in some (unknown) way. Consistent with this, several independent mutations that occur in the *white* locus of I-R dysgenic flies are known to be caused by insertions of 5.4 kbp DNA segments, I elements. Target site duplications of variable length (10 to 14 bp) flank individual elements.

The structure of an I element is similar to that of F and G (Figure 10.41). There are no terminal repeats, either direct or inverted. The A-rich end is a tandem array of between four and seven repeats of the sequence TAA, but there is no polyadenylation signal. Two long open reading frames (ORFs), separated by 471 bp, occur on the strand with the A-rich 3′ end. No potential splice sites are apparent. The ORF closest to the 3′ terminus of I can encode a polypeptide with similarities to reverse transcriptase (Table 10.4). Genetic data indicate that I elements encode a function required for transposition as well as a function that regulates transposition. The former could well be the reverse transcriptase. I element transcripts have been identified; they are confined exclusively to *Drosophila* ovaries.

The I element depicted in Figure 10.41 is derived from an element that is known, from genetic experiments, to be transposable. Moreover, its 5′ end is the same as that of the several other I elements that have been sequenced. These considerations suggest that the sequenced I element is a self-contained unit including everything necessary for transposition. Assuming that the transposition mechanism involves an RNA intermediate representing the entire unit, then transcriptional regulatory elements are probably present within the element itself. The same may be true of other class II retrotransposons. Indeed, recent work with another *Drosophila* class II retrotransposon called Jockey has demonstrated promoter activity within the element in the region preceding the first (5′) ORF. Transcription does not occur if the 5′terminal 13 bp are deleted. Thus, these elements appear to utilize unusual, internal RNA polymerase II regulatory sequences.

The LINE-1 family of mammals The most abundant mammalian long interspersed repeated sequence, LINE-1 (Section 9.5c), is a class II retrotransposon family similar to the F, G, Jockey and I elements of *Drosophila* (Figure 10.41). Their mobility is known from newly generated mutations associated with LINE-1 insertions. For example, insertion of a LINE-1 sequence into the X chromosome gene for factor VIII (a protein in the blood clotting system) was observed in hemophiliac boys whose mothers had wild type alleles (i.e., no insertion).

Although many of the 10^4 to 10^5 copies of LINE-1 in mammalian genomes are truncated at the 5′ end, several thousand are full length, between 6 and 7 kbp, depending on the species. Variable length target site duplications (9 to 19 bp) may or may not be present. The full length elements have no terminal repeats, either direct or inverted. Their 3′ A-rich ends vary from almost uninterrupted A stretches to tandem repeats of TAAA or similar units, reminiscent of the A-rich ends of I elements. Comparison of LINE-1 elements from several mammalian orders indicates a conserved 5 kbp central region. Long open reading frames (ORFs) covering the 5 kbp conserved region occur on one strand. In most cloned and sequenced LINE-1s, the ORFs are broken by stop codons and frameshifts caused by short deletions and insertions. However, the structures of consensus sequences and a few individual elements suggest that all LINE-1s have diverged from a basic element with two ORFs. Thus, although most LINE-1s do not include functional genes, one or more of the full length elements in each species may encode one or more polypeptides. The shorter, 5′ ORF (ORF I) is less well conserved than the longer, 3′ ORF (ORF II), which encodes a polypeptide with similarities to reverse transcriptase (Table 10.4 and Figure 10.41). Flanking the two ORFs at the 5′ and 3′ sides are segments that vary in length and sequence from one mammalian species to another.

Nuclear RNA in many mammalian cell types contains large amounts of heterogeneously sized RNA that anneals with LINE-1 probes. Both strands of the element are represented. Many of these RNAs probably represent the primary transcripts of unrelated genes that contain LINE-1 insertions (e.g., in introns). In contrast, discrete polyadenylated RNA containing the entire LINE-1 sequence has been detected in one human cell type, a human teratocarcinoma cell line, and in mouse lymphocytes, and these RNAs represent the strand containing the ORFs. (Similar Jockey RNAs are found in most *Drosophila* cells.). Thus, these RNAs could, in principle, be mRNA or intermediates in transposition. They are not homogeneous. cDNAs synthesized and cloned from the 6.5 kb human RNA differ from one another in about 2 percent of the residues. Moreover, many of them contain base pair changes that interrupt the two reading frames, so that only a few are likely to be functional mRNAs. However, the break between the two ORFs shown in Figure 10.41 is conserved in the human cDNAs (and thus the RNAs) and in rodent and human genomic LINE-1s.

General characteristics of class II retrotransposons Putting aside the absence of long terminal repeats, class II retrotransposons have several striking similarities to retroviruses and class I retrotransposons. One is the potential to encode reverse transcriptase. A second is the frequent presence of two ORFs on one strand. This organization is similar to the relation between *gag* (and *gag*-like) and *pol* coding regions in retroviruses and class I retrotransposons. However, although the ORF IIs of all the class II elements may encode reverse transcriptases, their ORF Is are not similar (except for closely related species). And as with the elements that contain LTRs, the break between the two ORFs differs in various class II families (Figure 10.41). In F, G, and I elements, several hundred bases separate the two frames. In the human LINE-1, they are separated by two stop codons that are in the same frame as the two ORFs. In the murine LINE-1, the two ORFs overlap, as they do in some retroviruses and in Ty. If ORF II of the class II retrotransposons is to be translated, then either the RNA must be appropriately spliced, or the break between frames must somehow be suppressed or overcome during translation.

Because of the lack of LTRs, class II elements must be reverse transcribed and transposed by mechanisms that differ from those used by retroviruses and class I retrotransposons. The processes have not yet been studied. However, circular F and LINE-1 DNAs have been detected and could be intermediates in transposition.

10.4 Retrogenes

A diverse group of elements here called **retrogenes** (the term **retroposons** has also been used) also appear to transpose through an RNA

Table 10.5 Estimates of the Copy Number of Some Processed Polypeptide Pseudogenes

		Estimated Copy Number	
Gene	Organism	Total Genes	Functional Genes
Actin	Human	20–30	6
α-tubulin	Rat	4	1
β-tubulin	Human	15–20	3
IFN-α	Human	15	10
U1-RNA	Human	1100	100
Glyceraldehyde 3-phosphate dehydrogenase	Chicken	1	1
	Human	10–30	1
	Mouse	>200	1

intermediate, although the mechanism of retrogene transposition is unproven. Retrogenes include a very diverse group of sequences that, notably, do not themselves encode reverse transcriptase. The common structural features of retrogenes are (1) the absence of terminally redundant sequences, either direct or inverted, (2) the presence, at the 3′ end of one strand, of a segment that is rich in A residues, and (3) variable size target site duplications. By convention, the sequences of these elements are displayed as the strand containing the 3′ A-rich end. The evidence that retrogenes are mobile is largely indirect. It includes the dispersal of family members to different genomic loci, the presence of target site duplications, and the existence of alleles that differ by the presence or absence of a retrogene.

Transposition of retrogenes appears to be passive; RNA may be reverse transcribed by enzymes supplied by retrotransposons or retroviruses. Thus, processed pseudogenes are retrogenes that may be copies of mRNAs (processed polypeptide pseudogenes) or of small RNAs like tRNA or 7SL RNA. Indeed, most retrogenes appear to be processed pseudogenes (i.e., noncoding copies of functional genes). Altogether, retrogenes can account for about 10 percent of the total size of some genomes, thus contributing significantly to the reverse flow of information from RNA back into DNA. The copy numbers of particular processed polypeptide pseudogenes tend to be relatively low, and the number of processed forms for any specific gene may vary markedly from one species to another (Table 10.5). Notably, however, the processed pseudogenes of the 7SL RNA (e.g., the Alu SINE family in primates and the type I SINE family in mice) are extraordinarily frequent, more than 5×10^5 copies in human DNA.

a. *Processed Polypeptide Pseudogenes*

Structure Processed polypeptide pseudogenes constitute a distinctive class of pseudogenes (Section 9.1a). Although they have variable forms, they are generally more like copies of mRNAs than like the corresponding

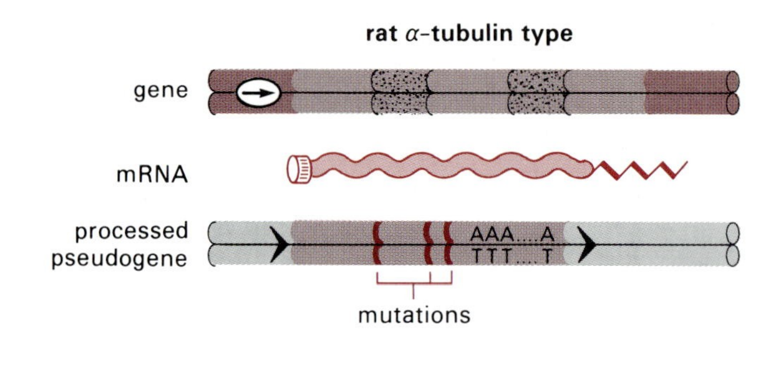

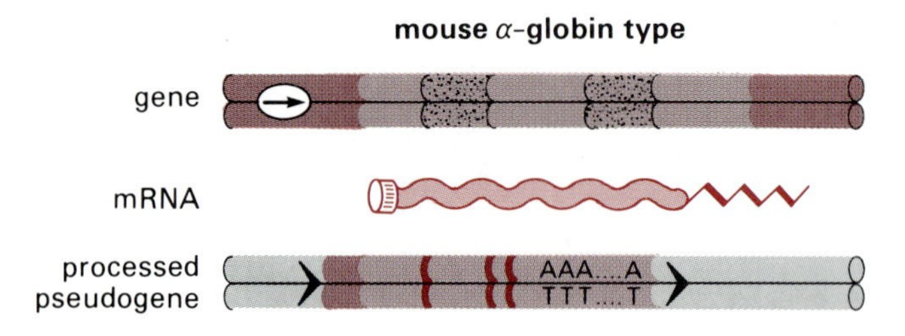

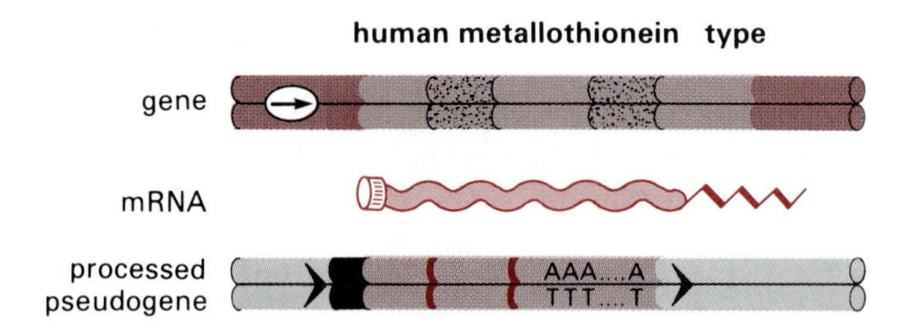

Figure 10.44
Comparison of the structures of polypeptide genes and their processed pseudogenes.

genes themselves (Figure 10.44). Moreover, they are located at genomic sites that are remote from the locus of the corresponding functional gene, often on a different chromosome. Curiously, processed polypeptide pseudogenes are most abundant in mammals, rare in other vertebrates and probably plants, and not yet detected in invertebrates.

A rat α-tubulin processed gene is prototypical; it begins where the mRNA begins, lacks all intervening sequences, is flanked by DNA sequences very different from those surrounding the gene itself and thus lacks transcriptional control signals, carries several debilitating mutations, and has a 3′ terminal poly A stretch (Figures 9.11 and 10.44). Other processed polypeptide genes have only some of the prototypical features. For example, a mouse α-globin processed gene on chromosome 15 lacks the two introns of the α-globin gene but includes sequences that precede the transcriptional start site of the functional gene on chromosome 11 and does not have an A-rich 3′ end. Still another type of processed gene has, upstream of the sequence corresponding to the 5′ end of the mRNA, a sequence that is not homologous to the region preceding the actual gene but appears to be part of the transposed region because it is within the segment flanked by the presumed target site duplications (the metallothionein processed gene in Figure 10.44).

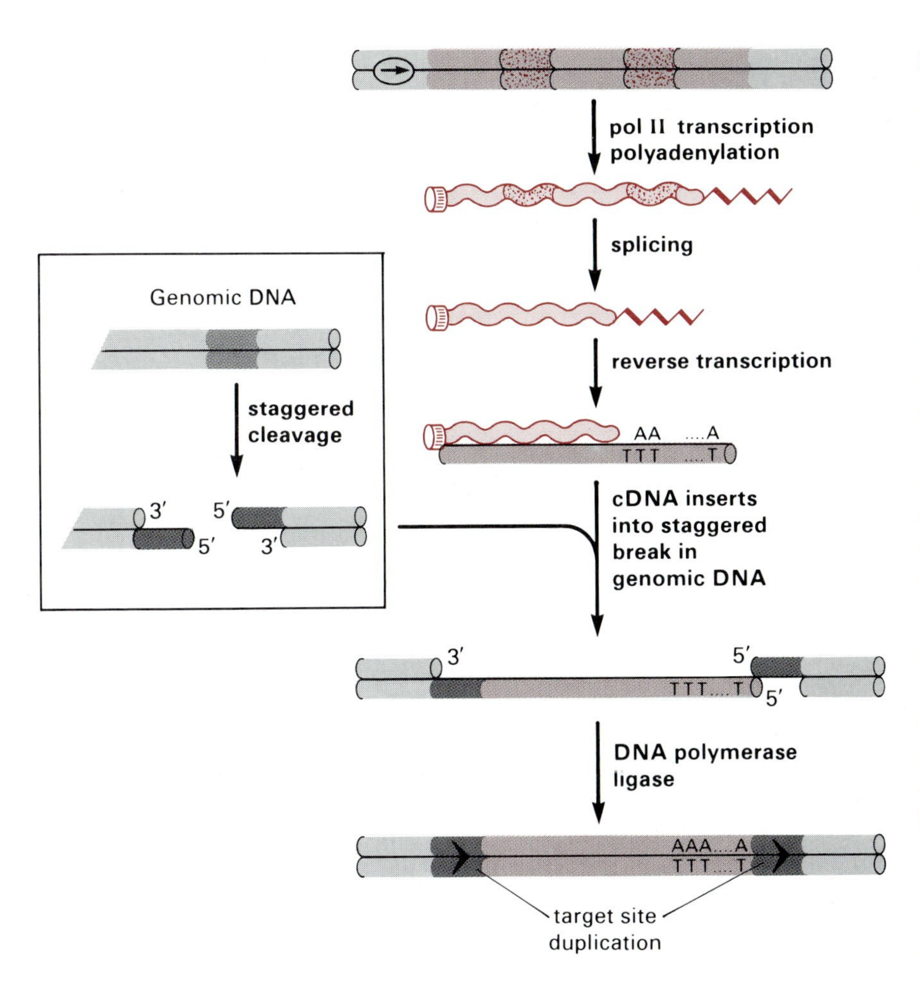

Figure 10.45
A general model for transposing retrogenes by means of an RNA intermediate. The formation of a processed polypeptide pseudogene is used as an example. A functional messenger RNA is produced by transcription and processing, including splicing out introns. Reverse transcription yields one cDNA strand, and the RNA template is digested away by, for example, RNase H. As depicted here, the single DNA strand is ligated through its 3′ hydroxyl end to the 5′ end of a staggered break in genomic DNA, and the second DNA strand is synthesized by DNA polymerase, which also fills in the remaining gaps. Alternatively, a double strand cDNA may be synthesized prior to insertion, or an RNA strand may be joined to the break with both cDNA strands being synthesized in place. Ligation finally seals the retrogene into place. The target site duplication forms as a consequence of these reactions. Note that the transposition is a conservative process because the original gene is still in place.

Models for the generation of processed polypeptide pseudogenes The structure of processed polypeptide genes suggests that they are DNA copies of the corresponding mRNA. A model reaction for their formation and transposition involves the reverse transcription of the RNA and the subsequent insertion of the cDNA into a staggered break in the genome (Figure 10.45). The target site duplication formed as a result of the initial staggered break in the empty site is variable in size. This presumably reflects an imprecise nicking, in contrast to the fixed size of target site duplications that occur with transposable elements and class I retrotransposons. There are two serious and unresolved problems with this model. The first is that the availability of reverse transcriptase in normal early embryo or germ line cells remains to be demonstrated. The reaction must proceed in such cells if a heritable insertion is to be made. The second is that an appropriate primer must be found. It is not immediately obvious what serves as a primer for reverse transcription of RNA polymerase II transcripts unless oligo-dTs are available in the cell; then the reaction would be similar to those carried out with reverse transcriptase *in vitro* (Figure 4.20). The primer problem aside, the model is attractive for processed genes of the rat α-tubulin type that are essentially perfect copies of mRNAs. Modifications are necessary in order to adapt the model to the formation of processed genes of the mouse α-globin or human metallothionein types.

The model in Figure 10.45 omits important details, and different mechanisms can be written for several of the steps. For example, synthesis of the second cDNA strand could occur prior to insertion into a new genomic locus. Alternatively, the RNA itself might be joined to a DNA break followed by synthesis of the first cDNA strand *in situ*. This whole class of models is attractive, and supporting evidence continues to accumulate, although other possibilities are consistent with much of the data. Thus, a variety of mechanisms may disperse sequences to distant genomic loci. One interesting idea postulates that retroviruses serve as natural recombinant vectors. Retrovirus genomes do acquire cellular genes and, in the course of one or more life cycles, can deliver them back into new genomic locations. During their life cycle, retroviral genomes alternate between DNA and RNA forms. In the RNA form, splicing could remove introns from the cellular segments, thereby generating a processed gene structure in the next DNA stage. This model can explain the presence within processed genes of unexpected sequences on the 5′ side of mRNA initiation sites and the absence of A-rich 3′ termini. One prediction of this model is that some processed genes should be neighbors of genomic retroviral sequences. Another model proposes the packaging of cellular mRNAs in retroviral (or retrotransposon) particles, followed by reverse transcription and insertion. Such reactions have been demonstrated, although at extremely low frequencies, in model systems. A quite different possible route for dispersal of genomic sequences is one that does not involve RNA at all but rather the direct transposition of DNA sequences. For example, a segment including a gene might be duplicated in tandem, followed by direct transposition of one copy to a distant location in the genome. Transposition might be mediated by neighboring transposable elements or inverted repeats. This kind of mechanism can explain the origin of dispersed pseudogenes, which, unlike the examples mentioned so far, retain introns; a distinctive mouse α-globin pseudogene on chromosome 17 has such a structure.

b. Processed RNA Pseudogenes

Structure The snRNA (small nuclear RNA) gene families (Section 9.2e) include different kinds of pseudogenes that diverge from the sequence of the functional RNA and its genes in one or a few scattered base pairs. Some of these appear to be processed genes because they are dispersed in the genome and because their flanking sequences are not homologous to the sequences neighboring true genes. Like processed polypeptide genes, processed snRNA genes differ from the structure of corresponding true genes in diverse ways (Figure 10.46). They are often truncated and lack sequences from the 3′ end of the RNA. All the snRNA processed genes have 5′ ends that coincide precisely with the start of the RNA. Sometimes, but not always, the processed genes have 3′ terminal A-rich stretches and the short direct repeats that indicate a target site duplication.

There are interesting differences between processed pseudogenes of snRNA genes and those of scRNA (small cytoplasmic RNA) genes. First,

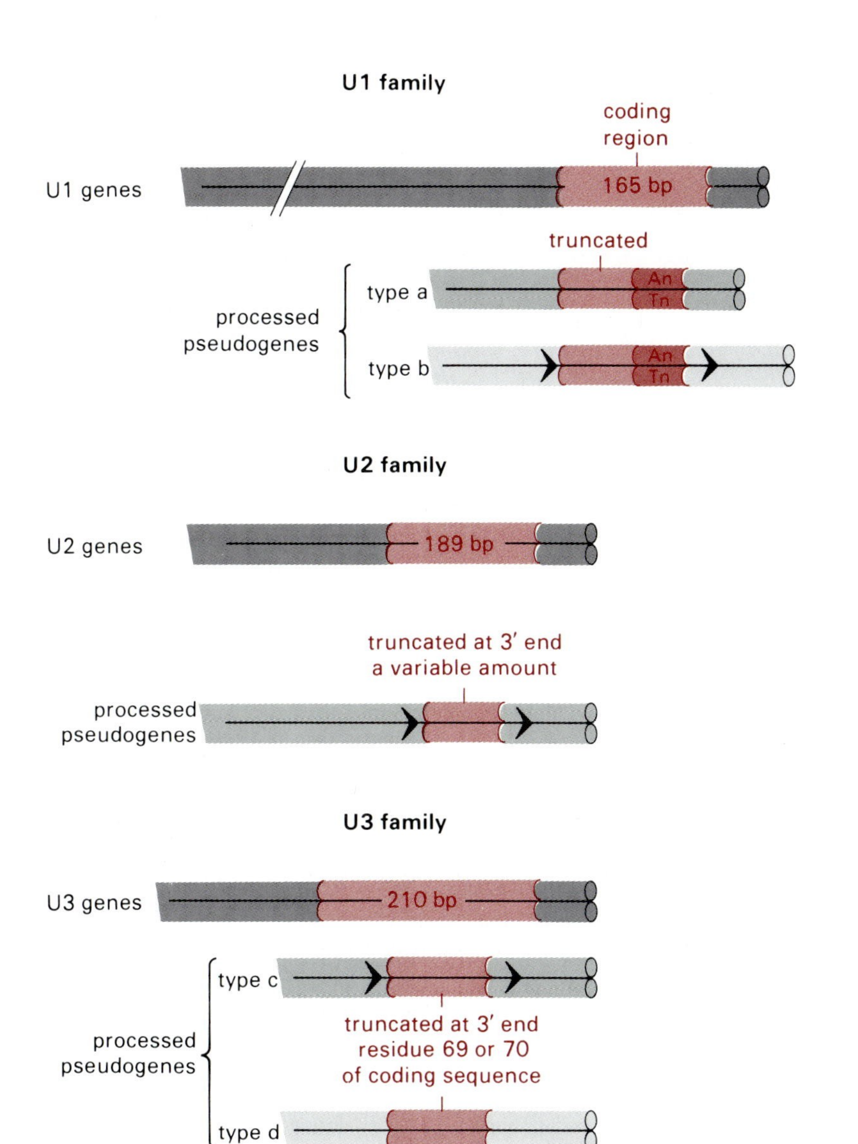

Figure 10.46

Processed pseudogenes of snRNAs have a variety of structures. All the examples are taken from mammalian genomes. Type a U1 processed genes lack sequences found at the 3′ end of functional U1 genes and U1 RNA itself. Bordering the truncated 3′ end of type a sequences is a short A-rich stretch, although U1 RNA itself does not end in a series of A residues. Type b U1 processed genes are not truncated, terminate in an A-rich stretch, and have flanking direct repeats of 20 bp or less. U2 processed genes are truncated to varying extents at their 3′ end and are flanked by about 20 bp long direct repeats. U3 processed genes are truncated at the 3′ end, particularly at position 69 or 70 of the 210 nucleotide long RNA. Some have flanking direct repeats 16–20 bp in length (type c); others do not (type d).

the latter are among the most abundant dispersed repeats in mammalian genomes. They include processed versions of tRNA genes and the Alu-SINES of primates and the type I SINES of mice, both of which are processed versions of the 7SL RNA genes. Second, the scRNA processed pseudogenes preserve the typical polymerase III internal promoter elements of the functional genes and are themselves transcribed, at least *in vitro.* (Most of the snRNA genes are class II genes with external promoters.) Finally, the structure of scRNA processed pseudogenes is markedly different from that of the functional genes and RNAs and varies from one order of mammals to another (Section 9.5c). For example, in Alu-SINES, the central portion of the 7SL RNA sequence is deleted, the two outer segments are joined, and the resulting segment (approximately 130 bp) is repeated, imperfectly, in tandem. Rodent type I SINES also lack the central portion of the 7SL RNA, but here, only a small part of the remainder is duplicated in tandem. In spite of these differences, the dispersed location and almost universal association of these segments with target site duplications clearly indicates that they are processed pseudogenes.

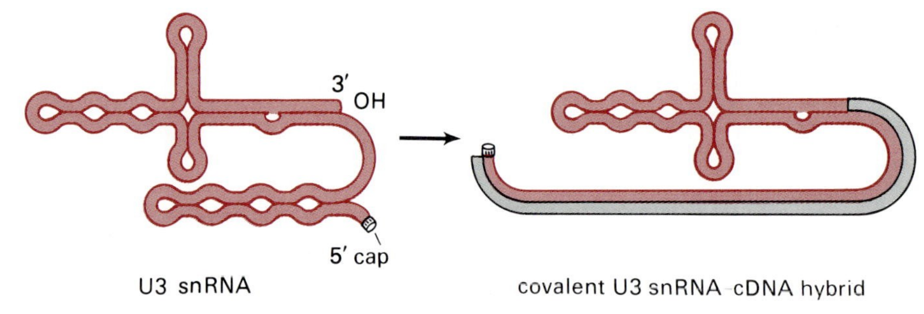

Figure 10.47
U3 RNA provides its own primer for reverse transcriptase. The 3′ end of the RNA anneals to a complementary internal sequence. cDNA synthesis involves the addition of dNTP residues to the 3′ terminal hydroxyl (residue 213). The cDNA product is complementary to residues 1 to 74, consistent with the structure of type c and type d U3 processed genes (Figure 10.46). After L. B. Bernstein, S. M. Mount, and A. M. Weiner, *Cell* 32 (1983), p. 461.

Models for the formation of processed RNA pseudogenes The reactions in Figure 10.45 also serve as the basis of models for production and transposition of processed RNA pseudogenes. Modifications of the model are required to account for specific properties of snRNA pseudogenes, such as the appearance of poly A tails (they are not present on the RNAs themselves), the frequent absence of sequences from the 3′ end of the RNA, and the occasional absence of flanking direct repeats (Figure 10.46). Moreover, as with the processed polypeptide genes, a primer must be found for first strand cDNA synthesis. In some cases, the primer could be provided by the RNA itself through duplex formation between sequences at the 3′ terminus and complementary internal sequences. Such a reaction has been demonstrated *in vitro* using U3 RNA and reverse transcriptase from a retrovirus (Figure 10.47). The secondary structure of a thermodynamically favored form of U3 RNA involves the hydrogen bonding of the 3′ end to the region around residue 70, thereby providing the appropriate intramolecular template-primer. Note that this model explains the 3′ end truncation of the U3 pseudogene.

Because scRNAs are transcribed by RNA polymerase III, the model actually works quite well for sequences such as Alu and type I SINES. Transcripts of these SINE sequences start at the beginning of the element and terminate at the first T-rich segment encountered by the polymerase. Normally, this would be some distance beyond the end of the SINE; thus, the transcripts would contain some unrelated sequences between the A-rich stretch at the end of the element and the U-rich stretch at the end of the transcript. The U-rich stretch at the 3′ end might bend around and pair with the A-rich region to provide a primer for reverse transcription. Degradation of the RNA template and insertion of the cDNA into a staggered break in the genome would complete transposition.

The new element contains the entire sequence of the RNA, including some but not necessarily all of the original A-rich stretch, but it contains no extraneous sequences. Because the newly transposed retrogene also contains the internal promoter elements of the parent gene, it can itself be the source of new family members. This "autocatalytic" effect might account for the high abundance of Alu elements. Note that these reactions account for the amplification and transposition of SINE sequences, but they do not explain the origin of the SINEs themselves. One or more of the many genomic SINEs must have arisen directly from the 7SL RNA or tRNA genes or from 7SL RNA or tRNA themselves. The process may have depended on reverse transcription, as shown in Figure 10.45. But if

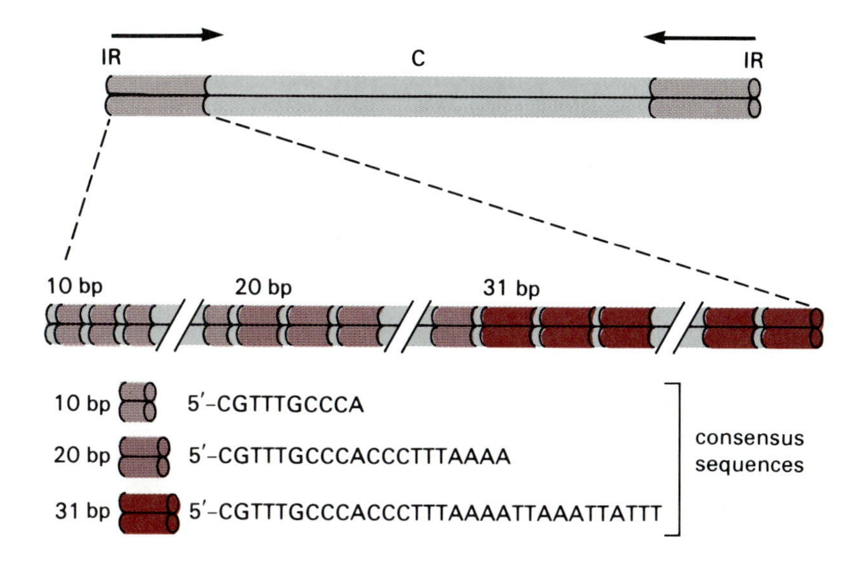

Figure 10.48
Schematic diagram of a movable fold-back (FB) element of *Drosophila*. Both the inverted terminal repeats (IR) and the central region (C) are variable in length. IR is between 400 and 3400 bp; C may be from less than 100 to 4000 bp. Different FB elements have different sequences in C. The diagram at the bottom shows the complex arrangement of tandem repeats typical of the IR region.

so, some complex alterations in the normal structures were made, including the addition of the A-rich stretch at the 3′ end (which does not appear in the RNA) and, in the case of 7SL RNA, deletion of the central portion and tandem duplications (Figure 9.40).

10.5 Other Unusual Movable Elements

Many eukaryotic movable elements can be classified, at least tentatively, as transposable elements, retrotransposons, or retrogenes. But others have unique structures and unusual properties. At the present time, most of the unusual elements that have been described are from invertebrates, but it seems likely that a great variety of movable elements will eventually be found in all eukaryotes. A detailed description of each of these distinctive elements is beyond the scope of this book. Instead, we briefly mention a few here to illustrate their diverse forms.

a. The Fold-Back Elements of Drosophila

The **FB** (for fold-back) elements of *Drosophila* (and similar elements in *Xenopus*) have long terminal repetitions in inverted configurations relative to one another (Figure 10.48). The inverted terminal repeats surround a central region. Two other distinguishing features occur. First, the sequence in the central region is not fixed; different FB elements have central regions of differing size and sequence. Second, the lengths of the inverted terminal repeats vary and are complex arrays of direct tandem repeats. Deletions (or additions) of the tandem repeats form through unequal crossing-over, as is typical of such arrays. The two terminal repeats within a single family member are more similar in length and sequence than repeats found in different family members, but they are not identical. However, the outer edges of the inverted repeats are highly conserved, even from one FB element to another. Upon insertion, the *Drosophila* FB elements cause mutations, and excision of the inserted FB can be accompanied by reversion to wild type. Besides moving about as an independent unit, FB is responsible for the transposition of long segments of the *Drosophila* genome. Members of one class of large transposable elements called TE are composed of two FB elements surrounding a long stretch of unrelated sequence. In one example, a total of several hundred kilobase pairs of DNA from the X chromosome moved to chromosome 2 and thence to multiple distant sites; FB elements were found at either end of the transposed unit. Transposition is accompanied by a variety of mutations and rearrangements both within the TE and in adjacent genes. These TE properties are reminiscent of the features typical of complex transposons in prokaryotes.

b. Insertions in Drosophila rDNA

A surprisingly high percentage of the regions encoding 28S rRNA are interrupted in some *Drosophila* species, but not by introns. The interrupted rDNA units appear to be nonfunctional and are transcribed at very low levels; the interruptions themselves serve no known function. One kind of insertion sequence, called type I, occurs in tandem arrays at the chromocenter and at a few scattered euchromatin regions, as well as within the 28S rRNA coding region. Of the several hundred type I family members in *Drosophila* DNA, about half are inserted in a specific site in the 28S rDNA units on the X chromosome. The type I insertions in rDNA are mainly 5 kbp long, but partial units 0.55 and 1 kbp also exist, and those at the chromocenter are even more heterogeneous. The evidence that type I units are mobile is circumstantial. None of the usual structural features of well-characterized movable elements is present, nor has transposition been seen experimentally. The nucleotide sequences immediately surrounding the type I elements are often similar, even though the elements are in very different genomic loci. Thus, insertion may favor particular target sequences and could involve site-specific recombination. This would also help explain why type I insertions are always found at one specific position in rDNA.

10.6 Programmed Rearrangements and the Modulation of Gene Expression

Programmed reorganizations with specific effects on gene expression have been recognized in prokaryotes, yeast, tetrahymena, trypanosomes, and mammals. Because the distribution of these systems is so diverse, many organisms probably have genes that are modulated by analogous rearrangements. We describe here several prokaryotic site-specific rearrangements that regulate gene expression by inverting DNA segments in order to introduce some general principles before discussing eukaryotic systems.

a. Prokaryotic Models—Translocation Through Flip-Flop Inversions

Alternating expression of two genes encoding Salmonella flagellin Flagellin, the major protein constituent of *Salmonella* flagellae, has two antigenically

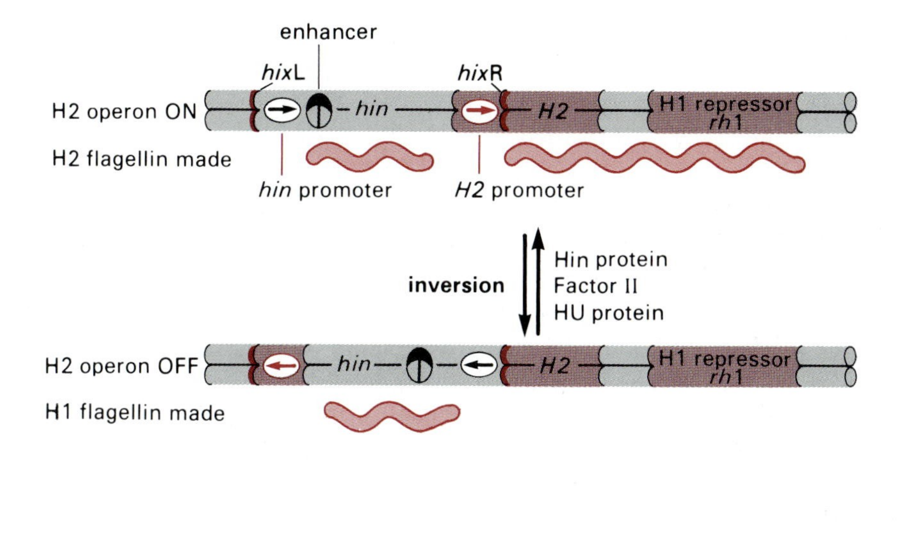

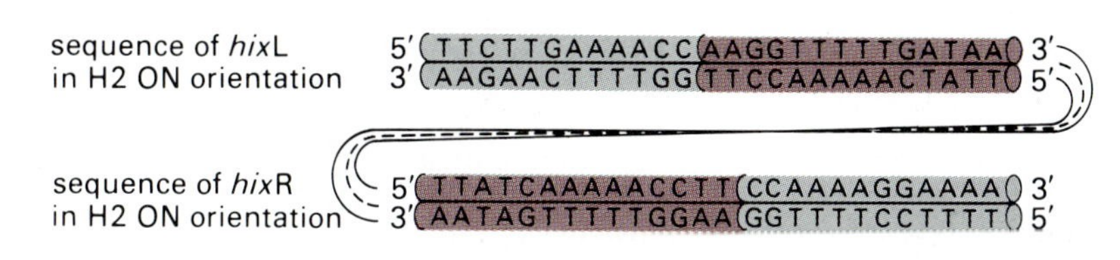

Figure 10.49
The invertible region that controls phase variation in the expression of *Salmonella* flagellin genes. At the top, the invertible segment is in the direction of transcription of the *H2* and *rh1* genes. Below, it is turned around, and *H2* and *rh1* are silent. The *hin* gene is within the invertible segment; its product is required in *trans* for inversion. *Hix*L and *hix*R (sequences shown at bottom) are required in *cis*, and the enhancer, when present in *cis*, stimulates the rate of inversion. The *hix* sequences are shown as they occur in the top (H2 ON) orientation of the segment. The 14 bp sequences that are repeated in *hix*L and *hix*R are highlighted.

distinguishable forms, H1 and H2. They are encoded by two nonallelic genes that are related but unlinked on the *Salmonella* chromosome. Only one of the two is expressed at any one time, and different *Salmonella* strains switch between H1 and H2 at frequencies between 10^{-5} and 10^{-3} per bacterium in a single generation. Presumably, this flexibility permits the *Salmonella* some protection against the immune response of an infected host.

An explanatory model for the switching between H1 and H2, or **phase variation**, as it is called, was derived from extensive genetic experiments. The *H1* gene is a passive element. The controlling factors are the *H2* gene and closely linked genetic determinants. When *H2* is expressed, a neighboring gene *rh*1 that encodes a repressor of *H1* is also expressed; no *H1* protein is made, and the flagellin is the *H2* type. The switch mechanism turns off expression of both *H2* and *rh*1, and then *H1* flagellin is produced. As described in the following paragraphs, more recent molecular analysis fully confirms this model.

The *H2* and *rh*1 genes are part of an operon that is controlled by a single promoter about 100 bp upstream of the *H2* gene (Figure 10.49). This promoter is within a 1 kbp long segment that can be inverted. In one position, the promoter is functional, and *H2* and *rh*1 are expressed. Upon inversion, the promoter is far removed from the start of the *H2* gene and is oriented in the wrong direction; expression of *H2* and *rh*1 is switched off, and *H1* expression is derepressed. When the 1 kbp invertible segment changes back to the original direction, *H2* and *rh*1 are again expressed, and *H1* shuts off.

Genetic analysis suggested that the invertible region contains, in addition to the promoter, three elements that are required for inversion; two *cis* acting segments at either end (*hix*L and *hix*R) and a gene called *hin* (for H-inversion), whose product acts in *trans*. Again, the genetic interpretation is now confirmed at the molecular level. The 1 kbp invertible segment contains sequences that correspond to the genetically defined functions. First, *hix*L and *hix*R are 26 bp long sequences at the two ends of the 1 kbp long unit, each of which includes a copy of a 14 bp segment (in inverted orientations). The 26 bp units themselves are imperfect palindromes. Second, there is a sequence within the 1 kbp unit that is the *hin* gene and encodes a 190 amino acid long polypeptide. *In vitro* experiments with purified Hin protein and recombinant plasmids containing 2 *hix* sites (but not the rest of the 1 kbp segment) revealed that Hin protein binds specifically to the *hix* sites and that a third, 60 bp, sequence within the *hin* coding region is also required in *cis* for efficient inversion. This element remains active even if it is placed outside of the inverting DNA segment or if it is present in an orientation opposite to its normal one. Thus, it shares properties with transcriptional enhancers. *In vitro* experiments show that two additional proteins are required for inversion. One, a small basic protein called factor II (or Fis), binds to the enhancer. The other is the prokaryotic histonelike protein, HU. Together, factor II, HU, and Hin protein assemble the DNA in a configuration that fosters the rearrangement.

The nucleotide sequence of the entire 1 kbp invertible element and its flanking sequences is known, and the sequences at the two ends were determined independently for both orientations of the invertible piece. Although the promoter for *hin* has not been located precisely, it is likely to start within the terminal repeat; the ATG start codon for *hin* is within 100 bp from the start of the repeat. Note that inversion does not destroy the *hin* promoter because of the similarity of the two ends. Thus, *hin* can be expressed in either orientation, as expected from the bidirectional invertibility of the segment. Sequence analysis also identified the promoter for the *H2* operon within and close to one end of the 1 kbp segment. The H2 coding sequences themselves begin 16 bp outside of the invertible segment.

Control of host specificity by inversion in the phage Mu genome The phage Mu also modulates gene expression by inversion and by a mechanism similar to that of the flagellin system of *Salmonella*. Within the 37 kbp Mu genome, a 3 kbp region called G is invertible (Figures 10.7 and 10.50). The borders of G are 34 bp inverted repeats, and 26 out of the 34 bp are, like the repeats flanking the *Salmonella* segment, required for the flip-flop. However, a functional gene *gin* that is required for inversion lies outside the invertible segment, not inside, as does *hin* in the *Salmonella* segment. A recombinational enhancer analogous to the one in the *Salmonella* system occurs within the *gin* coding sequences, and additional cellular proteins stimulate inversion *in vitro* in the Mu system. Inversion of the G segment occurs in about 1 out of 10^6 phage in a lytic cycle and alters the host specificity of phage Mu. In one orientation, called G(+), the phage infects *E. coli* K12. When G is in the opposite G(−) orientation, Mu infects *E. coli* C and certain other bacteria as well. G itself encodes proteins for phage tail fibers, and different proteins are expressed, depending on the orientation of G. Thus, the change in host range is likely to reflect altered absorption to recipient cells.

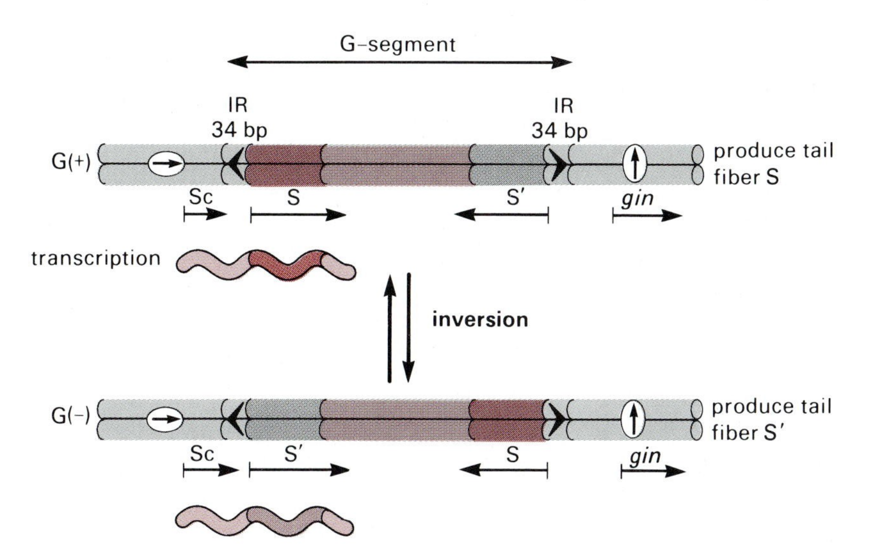

Figure 10.50
Schematic map of the phage Mu G segment and flanking regions in the G(+) and G(−) configurations.

G contains coding sequences for two alternate sets of tail fiber proteins, which occur in opposite orientations within G (S and S′ in Figure 10.50). In the G(+) position, the S operon is expressed, and in G(−), the S′ operon. The promoter for the two operons S and S′ is actually external to G (in a region termed Sc), as are the sequences encoding the 5′ ends of the mRNAs and the NH_2 terminal ends of the first protein in each operon. Regardless of whether S or S′ proteins are synthesized, this NH_2 terminus is constant and encoded by the Sc region; only the carboxy portion of the protein varies. Moreover, because the promoter for S is external to G, the orientation of G has no effect on the regulation of S expression, but only on which kind of tail fiber protein is synthesized.

Relation of Hin and Gin The strikingly similar *Salmonella* and Mu inversion systems very likely represent two adaptations of a common ancestral genetic program because the Hin and Gin proteins are structurally related and can substitute for one another. *Salmonella hin*$^-$ cells that are lysogenic for phage Mu undergo phase variation. Conversely, a functional *hin* gene permits inversion of the Mu G segment in a *gin*$^-$ phage. Because both Hin and Gin act in conjunction with the inverted repeats that flank the flip-flop segment, it is not surprising that the inverted repeats bordering the Mu G region are partly homologous to the *Salmonella* inverted repeats. There are additional members in this family of proteins. For example, an invertible segment in the genome of certain *E. coli* strains is associated with homologous flanking repeated sequences and a protein called Pin that complements *gin*$^-$ Mu phage. The amino acid sequences of the Hin, Gin, and Pin proteins are identical in more than 60 percent of the positions. They are also structurally related to the Tn3 resolvase.

Mechanism of site-specific flip-flop inversions A detailed analysis of the mechanism of prokaryotic flip-flop inversions is possible because (1) the recombinations can be carried out *in vitro* with purified components and (2) recombinant DNA techniques provide precisely designed substrates, including ones with site-specific mutations. DNA segments bounded by two inverted repeats (*hix* elements or the homologous inverted repeats

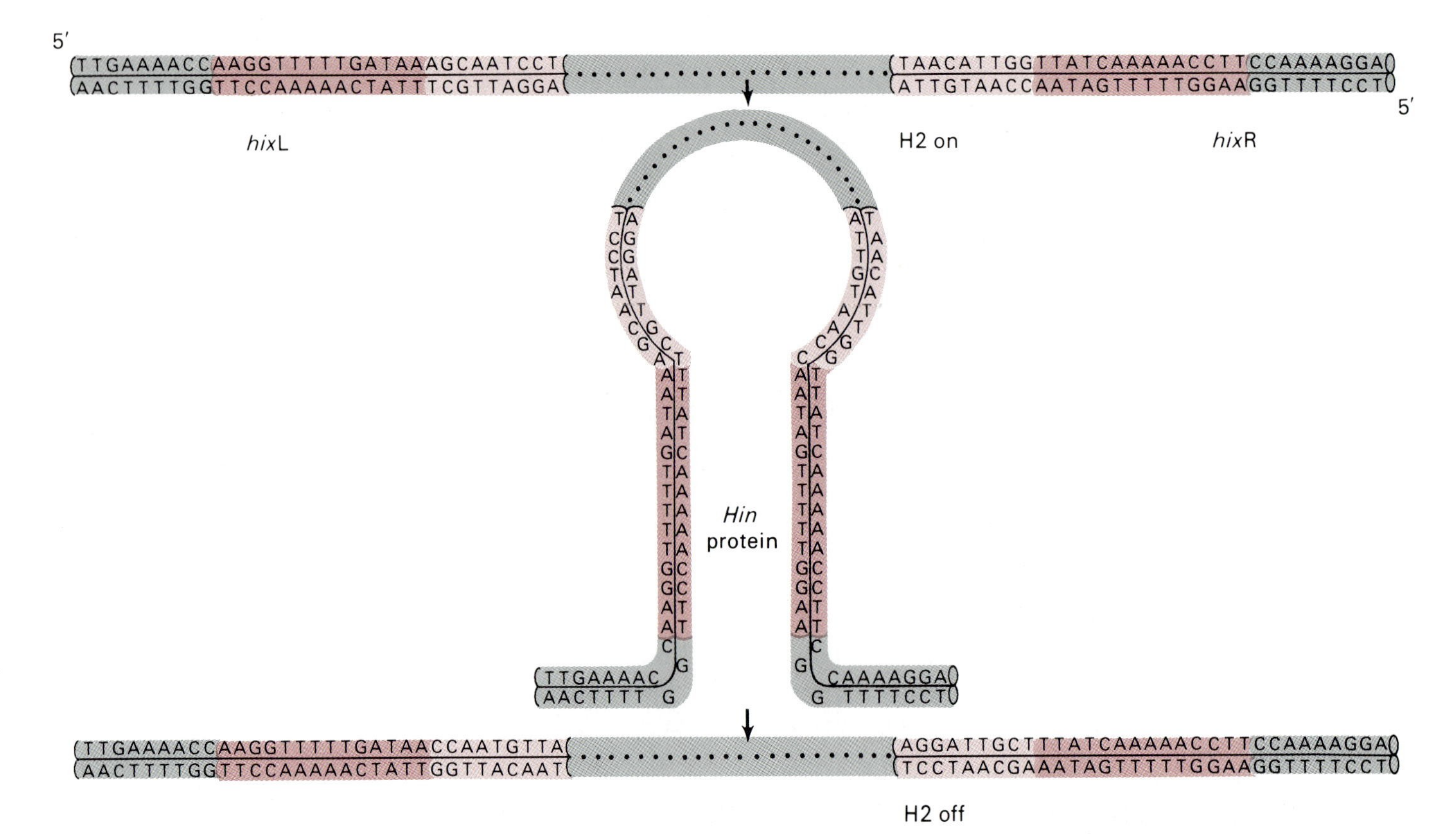

Figure 10.51
Site-specific recombination between *hix*L and *hix*R inverts the segment between them. After J. R. Zieg and M. Simon, *Proc. Natl. Acad. Sci. USA* 77 (1980), p. 4196.

of phage Mu G segment) and present in a supercoiled plasmid are inverted in the presence of purified Hin or Gin. The distance between the repeats can vary from hundreds to thousands of base pairs. Hin and Gin are recombinases, and they also are DNA binding proteins that bind to the inverted repeats at the boundary of the invertible segment. They are required in excess. The repeats are the actual site of strand exchange leading to inversion. Inversion is greatly stimulated by the addition of factor II and HU protein and by the presence of an enhancer element, in *cis*. Factor II binds specifically to the enhancer.

Inversion appears to involve formation of a complex containing appropriately aligned DNA segments and the recombinase (Hin or Gin). Factor II bound to the enhancer may facilitate complex formation. HU may further facilitate complex formation; there are hints that it might promote the bending of the DNA that is required to bring the two recombination sites into alignment. The inversion systems require that the homologous recombination sites be present on the same DNA molecule and in inverted orientation. This contrasts with the Tn3 resolvase, which prefers that the two sites be in the same orientation. However, like the Tn3 resolvase, the recombination leading to inversion involves four DNA strands, four phosphodiester bond cleavages, and four new joins (Figure 10.51).

b. Yeast Mating Types—A Cassette Mechanism

Mating between haploid cells of different mating types Budding *S. cerevisiae* cells may be either haploid or diploid. Diploids are heterozygous at a

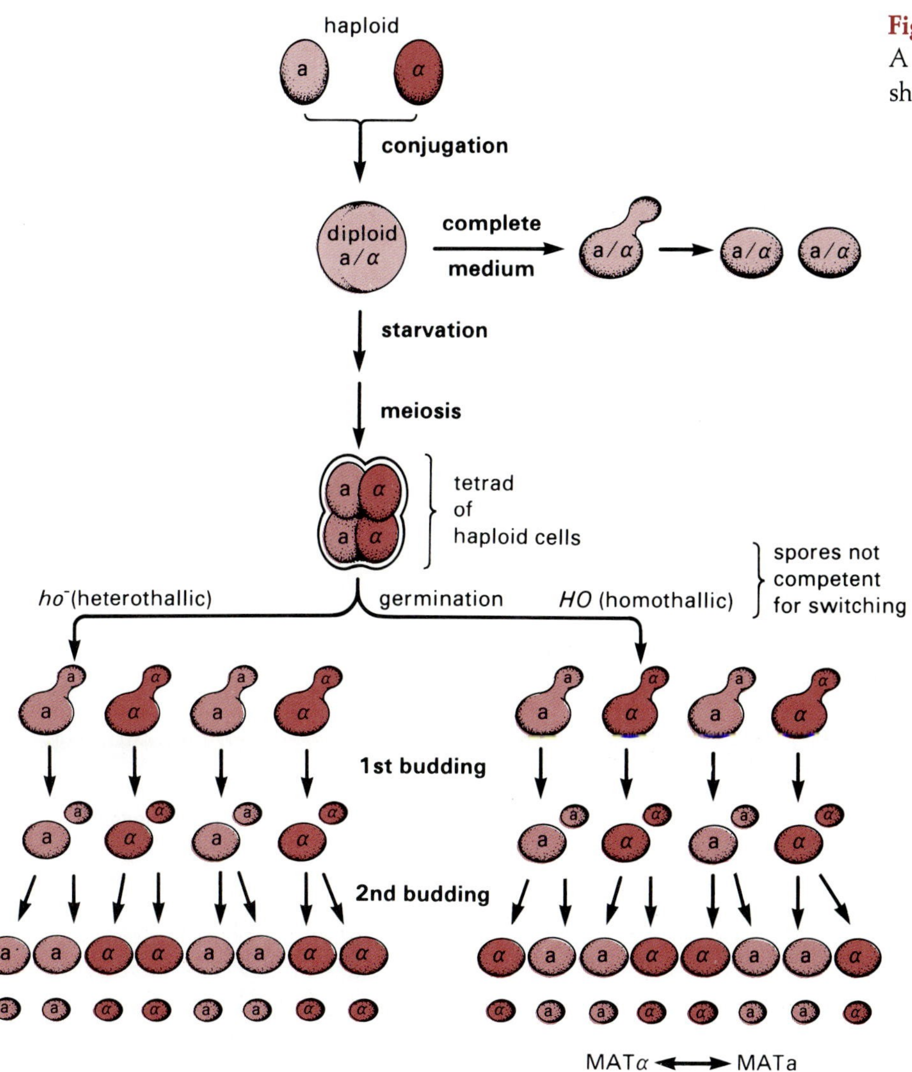

Figure 10.52
A summary of yeast cell division showing mating types.

locus called the **mating type locus** (*MAT*); the two wild type and codominant alleles are called *MAT*a and *MAT*α. Thus, haploid cells are either *MAT*a or *MAT*α. Successful conjugation of two haploid cells to form a diploid requires that they be of opposite mating types, one *MAT*a and one *MAT*α. A *MAT*a/*MAT*α diploid cell is unable to mate by conjugation, but it can be induced by starvation to undergo meiosis and sporulate to form four haploid daughter cells constituting a tetrad (Figure 10.52 and Section 9.2a).

Some *S. cerevisiae* strains, called **heterothallic**, produce haploids with a stable mating type. Others fall in the **homothallic** group and are characterized by instability of the *MAT* locus in haploids. In homothallic haploids, the allele at *MAT* readily interconverts between *MAT*a and *MAT*α. *MAT*a cells give rise to *MAT*α progeny with over 80 percent efficiency at every cell division (budding) after the first; only cells that have budded at least once interconvert *MAT*a and *MAT*α. Within a few cell divisions, a single homothallic cell yields a population capable of conjugation and diploid formation (a self-mating thallophyte, thus homothallic). But a clone of heterothallic yeast must encounter a population of the opposite mating type before the cells can conjugate.

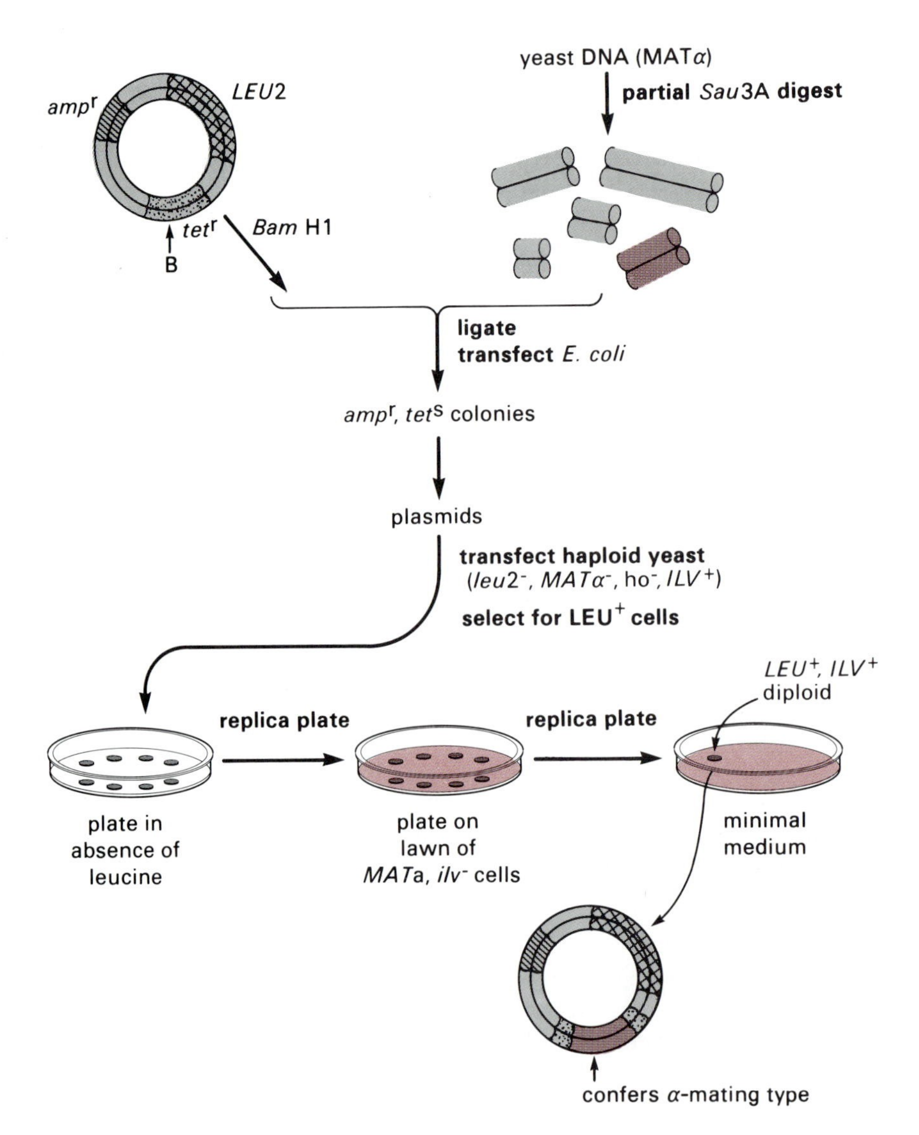

Figure 10.53
Cloning of a yeast segment that confers α-mating phenotype to *MAT*α⁻, *ho*⁻ cells (*MAT*α⁻ implies *HML*α⁻ as well). A synthetic yeast-*E. coli* shuttle vector (Section 5.8c) containing the yeast *LEU2* gene was cleaved at the endonuclease *Bam*H1 site. Yeast DNA fragments generated by endonuclease *Sau*3A digestion of yeast DNA from *MAT*α cells were ligated to the vector. After transfection into *E. coli* K12, all the amp^r, tet^s clones were pooled and grown. Plasmids isolated from this population were transfected into haploid yeast (*leu2*⁻, *MAT*α⁻, *ho*⁻, ILV⁺), and all the *LEU2*⁺ colonies were screened by replica plating for their ability to mate with a *MAT*a strain. The *MAT*a strain was also *ilv*⁻, so only diploids containing a functional *ILV* gene from the other parent grow on minimal medium. A single positive clone was detected, and the corresponding plasmid was isolated. After K. Nasmyth and K. Tatchell, *Cell* 19 (1980), p. 753.

Whether a yeast is heterothallic or homothallic is determined by a single gene, *HO*, whose product is required for the homothallic phenotype. Haploid cells that contain a mutant allele of *HO* (*ho*⁻) are heterothallic and unable to interconvert between *MAT*a and *MAT*α. Besides *MAT* and *HO*, which are unlinked and reside on chromosomes 3 and 4, respectively, two other genetic loci, *HML*α and *HMR*a, are required for interconversion. Genetic experiments showed that (1) *HML*α is required for the conversion of *MAT*a to *MAT*α, and *HMR*a is needed for the reverse conversion, and (2) *HML*α and *HMR*a are located on the left and right arms of chromosome 3, with *MAT* somewhere in between on the right arm.

The cassette model On the basis of these experiments, a model was proposed that describes a molecular mechanism for the switch from *MAT*a to *MAT*α and back. *HML*α and *HMR*a contain silent copies of genes associated with the α and a mating types, respectively. Replicas of either the α or a genes are translocated from *HML*α or *HMR*a to *MAT*, where they are then expressed. At the same time, the previous occupant of the *MAT* locus is replaced. According to this model, switching from one mating type to another is like changing a cassette on a tape player (*MAT*), except that the storage copies at *HML*α and *HMR*a remain in place while replicas enter *MAT*. Molecular cloning and characterization of *HML*α, *HMR*a, and *MAT* fully support the cassette model and provide rich details about this example of cell differentiation.

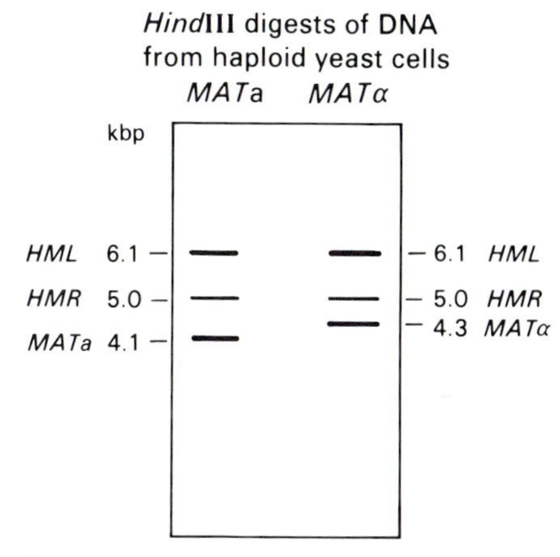

Figure 10.54
A cloned yeast segment that imparts α-mating type phenotype anneals to four different segments of yeast DNA that correspond, respectively, to *MAT*α, *MAT*a, *HML*α, and *HMR*a. The cloned segment described in Figure 10.53 is the probe. Haploid *MAT*a or haploid *MAT*α (heterothallic) cells each yield two common bands, 6.1 and 5.0 kbp, and one unique band, 4.3 kbp for *MAT*α and 4.1 kbp for *MAT*a. When two such strains are mated and tetrad spores isolated and grown, the 4.3 kbp *Hind*III fragment segregates with the *MAT*α phenotype and the 4.1 with *MAT*a. Thus, these two fragments include the *MAT*α and *MAT*a loci, respectively. A second inference is that the other two bands at 6.1 and 5.0 kbp represent the silent copies at *HML*α and *HMR*a and furthermore that the two loci share extensive homology. The identity of the four restriction fragments with *MAT*a, *MAT*α, *HML*α, and *HMR*a was confirmed by cloning all three fragments from the genomic DNA of both *MAT*a and *MAT*α yeast strains, using the original plasmid as a probe. From K. Nasmyth and K. Tatchell, *Cell* 19 (1980), p. 753.

*The structure of HML*α*, HMR*a*, and MAT* Characterization of these loci began by cloning a yeast fragment that confers the α-mating phenotype to *ho*$^-$ haploid yeast cells with a defective *MAT*α locus (Figure 10.53). The cloned fragment was used as a probe to screen restriction endonuclease digests of various genomic yeast DNAs by DNA blotting (Figure 10.54). This analysis showed four different hybridizing fragments produced by endonuclease *Hind*III that correspond, respectively, to segments containing *MAT*α, *MAT*a, *HML*α, and *HMR*a. Each of these was cloned. All possible pairs of the cloned segments were then denatured and reannealed together, and the resulting heteroduplexes analyzed in the electron microscope (Figure 10.55). As expected from the cassette hypothesis, annealing of *MAT*α to *HML*α or *MAT*a to *HMR*a gives an uninterrupted homologous region; the *MAT*α/*HML*α duplex region is about 800 bp longer than the 1.6 kbp long *MAT*a/*HML*a duplex. A hybrid is formed between HMLα and *HMR*a but is interrupted by an approximately 800 bp long nonhomologous stretch. The heteroduplexes also show that the sequences surrounding *HML*, *HMR*, and *MAT* are not homologous, although the flanking regions around *MAT* are the same, regardless of whether they surround *MAT*a or *MAT*α. Finally, the heteroduplex structures show that the extra 800 bp associated with *MAT*α and *HML*α occur also in *MAT*a, but not in *HMR*a. All these data are summarized in the maps of the three loci shown in Figure 10.56. The maps show that the *MAT* locus contains either α-type sequences such as those at *HML*α or a-type sequences such as those at *HMR*a.

Mating type switch The switch in mating type is effected by the transposition of a copy of *HML*α or *HMR*a into *MAT*. Genetic and structural studies have established certain properties of the switch: (1) The product of the *HO* locus is required. (2) Copies of sequences reserved at *HML*α or *HMR*a occur at *MAT* in *MAT*α or *MAT*a cells, respectively. (3) The sequences at *HML*α and *HMR*a are unchanged upon their appearance at *MAT*. (4) When switching occurs, the a or α cassette that was at *MAT* is lost. The reactions must be asymmetric because a or α sequences at *MAT* do not normally move into *HML* or *HMR*. How does this occur?

The *HO* locus encodes an endonuclease that recognizes and makes a staggered double strand cleavage at a specific site in *MAT* (arrows in Figure 10.56 and Figure 10.57). The *HO* endonuclease is not present in heterothallic yeast (*ho*$^-$) or diploid (a/α) homothallic yeast, and sequences at *MAT* do not switch in such cells. A significant (and detectable) proportion of the genomes in a switching population do in fact have a break at the *HO* endonuclease cleavage site. However, the identical segments at *HML*α and *HMR*a remain intact; somehow these loci are protected from cleavage. Assuming that the cleavage is an early critical step in switching,

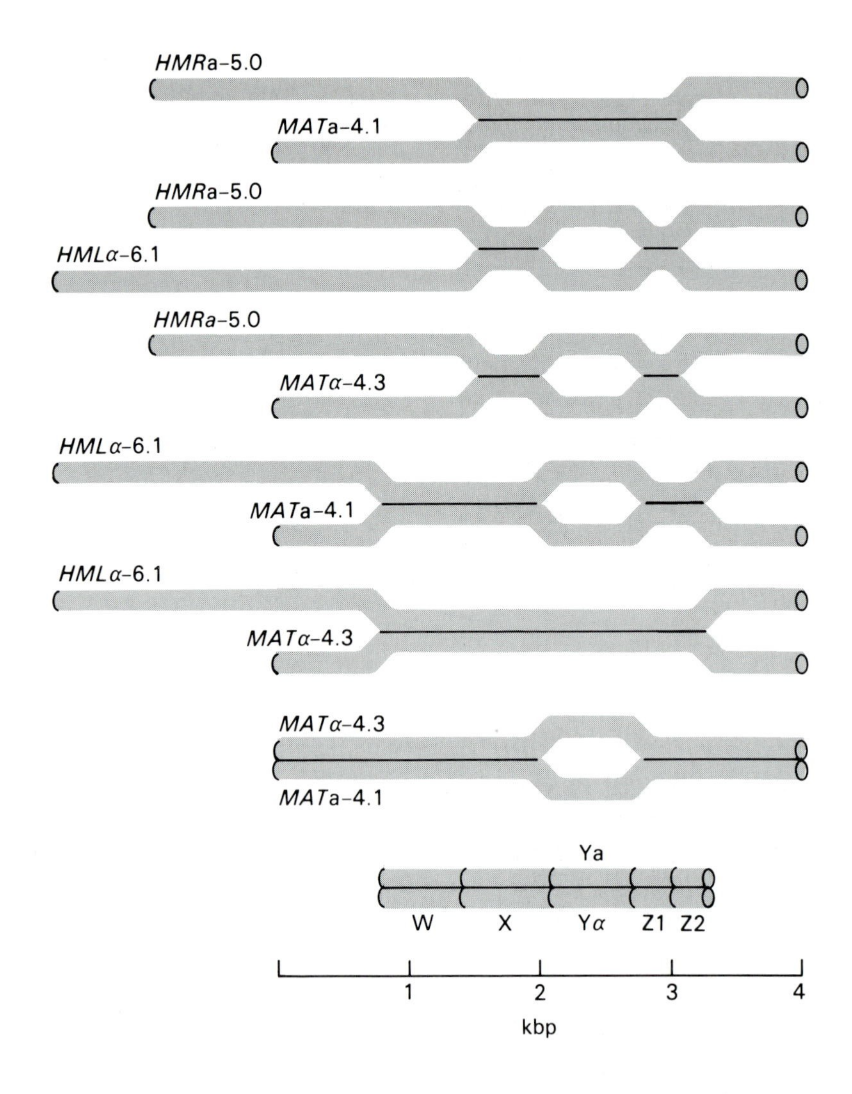

Figure 10.55
Schematic diagram of heteroduplex structures between cloned *HMLα*, *HMR*a, *MATα*, and *MAT*a loci. At the left, each strand in the heteroduplexes is identified, and its length in kb is given. The single strand loops (region Y) are not equal in length, although they are drawn that way for convenience. After K. Nasmyth and K. Tatchell, *Cell* 19 (1980), p. 753.

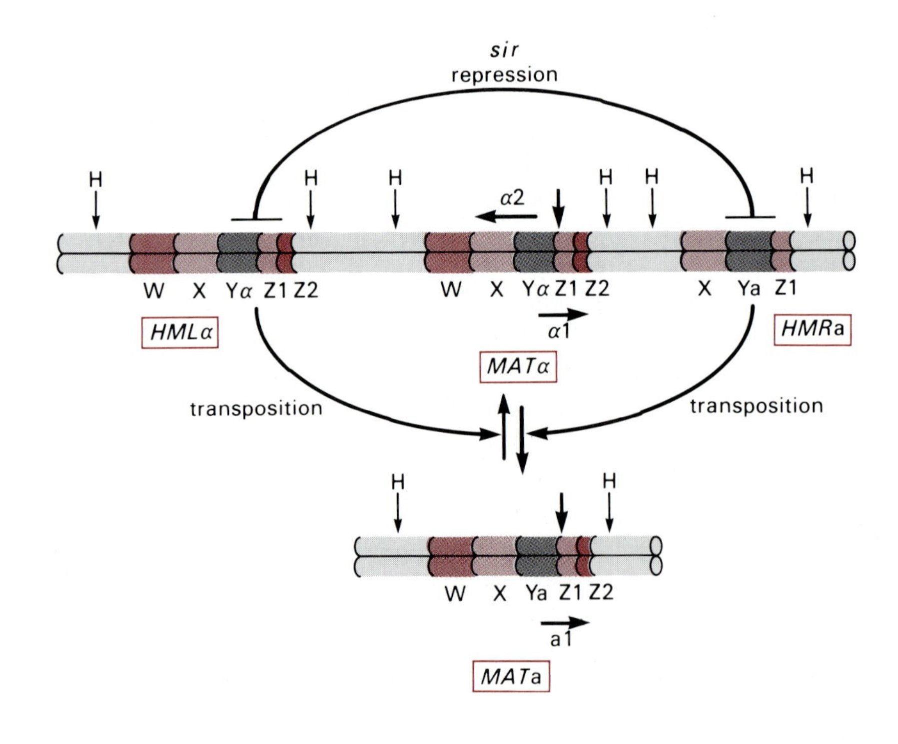

Figure 10.56
A map showing the three mating type loci on yeast chromosome III. In the drawing on the top line (not to scale), the active *MATα* configuration is shown. The *MAT*a configuration is shown below. The *MAT* locus is divided into several segments called W (723 bp), X (704 bp), Yα (747 bp) or Ya (642 bp), Z1 (239 bp), and Z2 (89 bp). *HMLα* has the same segments, but *HMR*a lacks W and Z2. With *MATα*, the two genes α1 and α2 are transcribed in opposite directions from a central promoter in Yα. With *MAT*a, gene al is expressed from a promoter in Ya. The bold vertical arrows show the cleavage sites of the HO endonuclease.

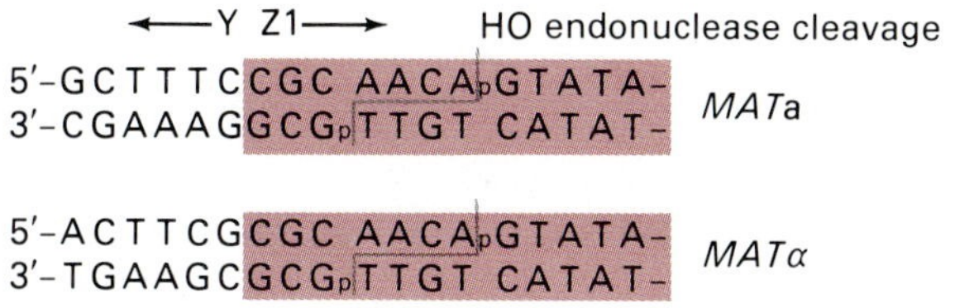

```
←Y Z1→        HO endonuclease cleavage
5'-GCTTTCCGC AACApGTATA-
3'-CGAAAGGCGpTTGT CATAT-   MATa

5'-ACTTCGCGC AACApGTATA-
3'-TGAAGCGCGpTTGT CATAT-   MATα
```

Figure 10.57
The HO endonuclease makes a staggered double strand cleavage at a specific sequence close to the Y-Z1 junction at *MAT*a and *MAT*α.

the protection of *HML*α and *HMR*a accounts at least in part for the asymmetry of transposition.

The *HO* gene itself has been cloned; it was selected from a yeast library in a plasmid vector by virtue of its ability to convert heterothallic cells into homothallic ones. A debilitating mutation appears to account for the absence of *HO* endonuclease in heterothallic cells. Using the cloned gene as a probe, levels of mRNA for *HO* gene product were measured under various conditions and in cells mutant in specific cell cycle functions. Diploid (a/α) cells make no *HO* mRNA, thereby explaining the fact that no switching occurs. In homothallic haploid cells, both *HO* mRNA and *HO* endonuclease are produced late in G1, but by the time of mitosis, the mRNA and the unstable enzyme activity are undetectable. Also, neither *HO* mRNA nor *HO* endonuclease appears in G1 of new daughter cells; this is consistent with the curious fact that only cells that have budded at least once are competent to switch (Figure 10.52). Competence to produce *HO* mRNA is acquired in the course of the cell cycle. Sequences dispersed over at least 1400 bp in the 5′ flanking region of the *HO* gene as well as several proteins are required for the complex regulation of transcription.

A good deal more information will be required before we know the details of the transposition mechanism. The most interesting models envisage a nonreciprocal homologous crossing-over (gene conversion) somewhat similar to that described in Figure 2.60. The reaction takes place between two homologous sequences on chromosome 3; either *HML*α and *MAT*a or *HMR*a and *MAT*α (Figure 10.58). Extra steps are required to account for the loss of the preexisting cassette in the *MAT* locus. Note that the double strand cleavage made by *HO* endonuclease at *MAT* fits well with this model. Moreover, mating type switching also requires the products of other genes that are involved in DNA repair and recombination.

Expression of mating type genes Mating type genes are expressed in *MAT* but not in *HML*α or *HMR*a. So far we have described a single phenotypic difference between a (*MAT*a) and α (*MAT*α) yeast cells, namely, the ability of a cells to mate with α cells but not a cells with a cells or α cells with α cells. What accounts for this, and what other phenotypic properties are associated with *MAT*a and *MAT*α? The a segment encodes one gene, *a*1, and the α segment two, α1 and α2. The products of α1, α2, and *a*1 are regulatory proteins that alter the expression of other genes. The effects of these regulatory proteins as deduced from genetic experiments are summarized in Figure 10.59. In *MAT*α cells, the α1 gene product acts as a positive regulator, turning on transcription of a set of genes that is otherwise silent and is required for the α cell phenotype. These

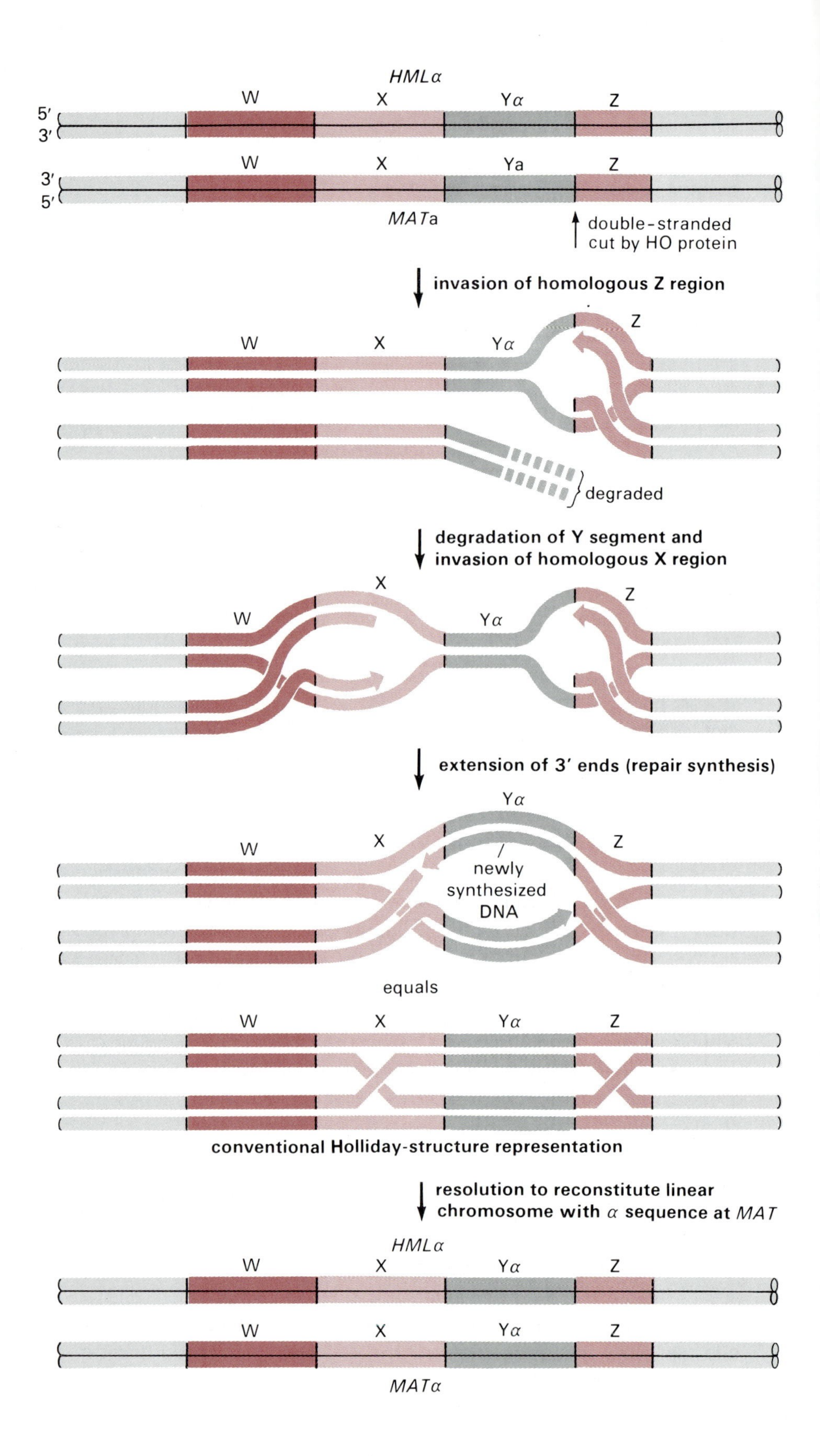

Figure 10.58
A gene conversion model for the mechanism of the mating type switch. Adapted from J. D. Watson et al., *Molecular Biology of the Gene* (Menlo Park, CA: Benjamin/Cummings, 1987), p. 580.

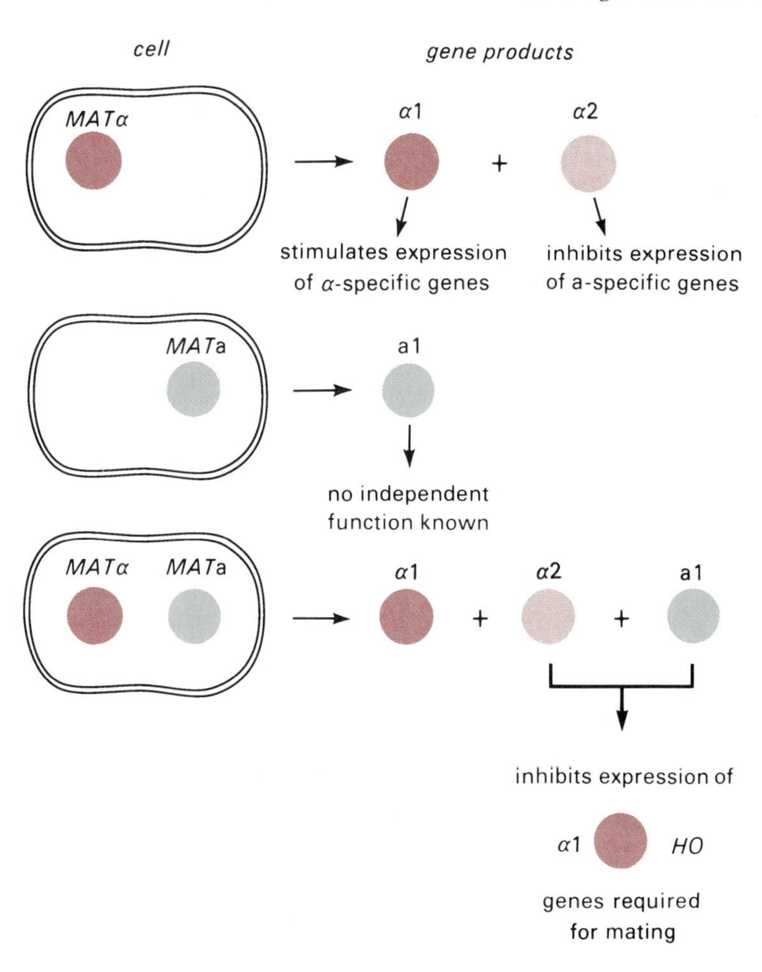

Figure 10.59
The effects of *MAT*α and *MAT*a gene products on gene expression.

include the gene *MF*α1 that encodes α-factor, a secreted oligopeptide **pheromone** recognized by a cells. However, α2 is a negative regulator, turning off the expression of genes that are expressed in *MAT*a cells. Both α1 and α2 affect their respective target genes' transcription through interaction with specific DNA sequences in the 5′ upstream regions. The activity of α2 depends also on its cooperative interaction with another protein that is ubiquitous in yeast cells. In addition, in *MAT*a/*MAT*α diploids, where both *MAT* loci are expressed, α2 acts in conjunction with *a*1 to (1) inhibit the expression of α1 and thus all the genes that are positively regulated by α1, (2) inhibit the expression of a variety of other genes that are required for mating, as well as *HO* and the Ty genes (Section 10.3b), and (3) make the cells competent for meiosis and sporulation if other requirements (i.e., starvation) are met. The coordinate repression oof the several genes by the combination of the α2 and *a*1 gene proteins depends on a common specific 20 bp long sequence that occurs in the 5′ flanking region of all the affected genes, including α1. This *cis*-acting sequence is different from the sequence recognized by α2 itself. The interaction with *a*1 protein alters the binding specificity of α2 protein. This complex scheme explains the inability of *HO* diploid cells to switch and to mate. *MAT* is the master control locus; it controls switching through regulating *HO*, and it controls mating by regulating the expression of sets of mating type specific genes. Note, too, that *MAT*a provides no genes necessary for the a mating type; an a mating type results simply from the absence of α gene expression. The whole system operates to assure an alternation between sexual and asexual reproduction.

The transcription units for the three genes are shown in Figure 10.56. The assignments were made by *in vitro* mutagenesis of cloned *MAT* DNA and reinsertion by transformation into the yeast genome; the mutation site was then correlated with a mutant gene product. Transcription of all three genes is outward from promoter regions in the center of the region labeled Y.

The silent a and α genes at HML and HMR Although the *a* and α genes at *HML* and *HMR* contain complete 5′ and 3′ flanking sequences, including promoters, they are not transcribed until transposed into MAT. What keeps them quiet? The answer is, at least in part, a set of four gene products, SIR 1, 2, 3, and 4 (for **silent information regulator**). The existence of the SIR proteins was initially inferred from four yeast mutations, each of which results in expression of α or *a* genes at *HML* or *HMR*. *SIR* genes are unlinked to *HML*, *HMR*, or *MAT*. Thus, *HML*α and *HMR*a contain all the elements required for the expression of α and *a* genes, but the SIR proteins repress that expression. Repression is mediated through interaction of the SIR proteins with regions that flank *HML*α and *HMR*a (Figure 10.60). These regions are outside the transposable units and more than 1000 bp from the sequences that control transcription of α1, α2, and *a*1.

The targets of SIR protein repression at *HML*α were defined by construction of a series of deletion mutants in cloned *HML* segments; each mutant clone was used to transform yeast cells bearing various SIR mutations (Figure 10.60). All deletions that traversed a 130 bp segment (E) to the left of *HML* lost sensitivity to SIR repression. Similar results were obtained for *HMR*. The E segments may be regarded as silencer elements (Section 8.3e); silencers have properties characteristic of transcriptional enhancers, except that they serve to repress rather than stimulate transcription. When the E segment is inverted, or moved to an alternate site still within 2.5 kbp of the mating type genes, repression by *SIR* gene products still occurs. Thus, the E segments, like transcriptional enhancers, function bidirectionally and at a distance. Similar deletion studies identified other regions (I) to the right of *HML* and *HMR*, which contribute to SIR-mediated repression.

The net effect of transposition from *HML* or *HMR* to *MAT* is to remove coding sequences from an environment that represses gene ex-

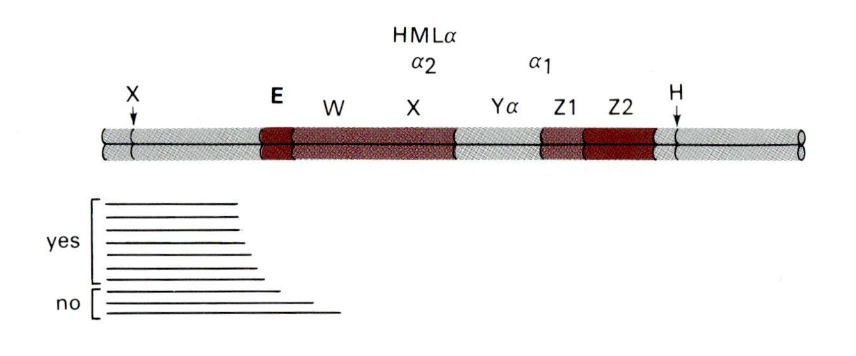

Figure 10.60
In vitro mutagenesis defined the site of action of SIR proteins at *HML*. The horizontal bars indicate the length of individual deletions constructed by cleavage of a cloned *HML* segment at the *Xho* I site followed by digestion with BAL31 (see Section 7.6b). Each mutant was used to transform yeast cells bearing various SIR mutations. The ability of the various deletions to be repressed by SIR proteins is summarized in the column at the far left (yes or no). The region (E) that is sensitive to SIR action as shown by these and other experiments is indicated.

pression. Just how the interaction of SIR proteins with sequences so far removed from the promoters inhibits transcription remains to be determined. It may be related to the presence of *ARS* sequences (Section 5.6b) at the SIR interaction sites because derepressed cells must go through an S phase before repression can be renewed. An altered chromatin structure may also be involved. If the proposed alteration in chromatin structure were to account for the resistance of *HML*α or *HMR*a to cleavage by *HO*, it would also explain why *MAT* is normally the recipient rather than the donor in switching. Consistent with this hypothesis, both *HML* and *HMR* become recipients of transposition in sir^- strains.

c. Genes Encoding Vertebrate Immune Proteins

The vertebrate immune response generates a nearly limitless variety of specific proteins—secreted **antibodies** and **receptors** in the plasma membranes of lymphocytes. Their role is to defend the organism against infections by viruses, bacteria, parasites, and probably the proliferation of tumor cells. Unlike genes for other cellular proteins, the genes encoding the antibody and receptor proteins are not present in the zygote. Instead, precursors of the antibody and receptor genes exist as separate and discrete DNA segments in germ line DNA, and functional genes are assembled by several specific rearrangements that occur only during the development of **B** and **T** lymphocytes (also called B and T cells).

There seems to be no limit to the types of chemical structures that can elicit an immune response. Virtually all naturally occurring molecules (e.g., proteins, carbohydrates, lipids, and nucleic acids) are **antigenic**. Even chemicals that do not exist in nature can provoke an immune reaction. Each distinctive structural feature of an **antigen**—an **epitope**—elicits different antibodies during an immune response. Even a single epitope frequently generates distinctive but related antibodies.

Two remarkable properties influence the immune system's versatility and specificity. First, there is the immune system's ability to distinguish between **self** (the chemical structures indigenous to an organism) and **nonself** (those structures that are foreign). Rejection of transplanted cells and tissues is a consequence of this discriminatory ability. **Autoimmunity** (the breaking of tolerance to self antigens) occurs rarely, usually in connection with certain pathologic states. Second, the immune response has a memory—that is, the organism mounts a more effective and rapid antibody response when it is confronted with the antigen a second time, even many years later.

Cellular and humoral immunity There are two distinctive but related immune systems, each associated with a different class of lymphocyte (Figure 10.61): **cellular immunity**, provided by T cells, and **humoral immunity**, dependent upon both B and T lymphocytes. Cellular immunity, which is mediated by a special class of T cell, the **cytotoxic** or "killer" **T lymphocytes** (**CTL**), is directed against cells carrying foreign antigens on their surface (e.g., tumor cells expressing novel proteins on their surface or virus particles on the surface of virus infected cells). Killing the target cell requires cell-to-cell interaction with the CTL, ostensibly to allow one or more specialized proteins produced by the CTL to act

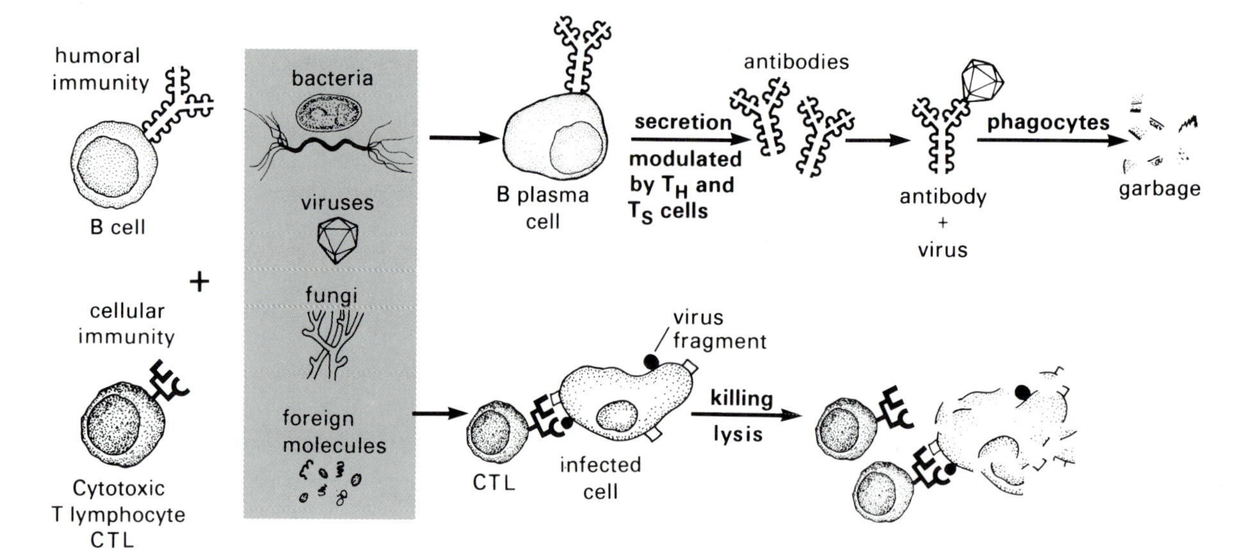

Figure 10.61
The two immune systems. In humoral immunity, circulating antibodies secreted by B cells interact with foreign antigens, including bacteria, viruses, fungi, and proteins. Cellular immunity is mediated by T cells, which interact directly with cells carrying both foreign and specific, normal surface proteins.

at the points of contact. Humoral immunity is mediated by circulating antibodies produced by B cells under the regulatory control of two kinds of T cells—**T helper (T_H)** and **T suppressor** (T_S) cells. Antibodies circulate in the serum and lymph, but certain types are localized to specific duct and intestinal secretions. Wherever they occur, antibodies bind offending antigens that are free in solution or on the surface of an infecting cell. The antibody-antigen complexes are rapidly removed and destroyed by **phagocytic** cells such as **macrophages**. Combination of some antibodies with cellular antigens (e.g., bacteria) triggers a proteolytic system, **complement**, that causes lysis of the foreign cells.

Antibodies Antibodies are extraordinarily specific. They can distinguish between closely related epitopes, for example, between two stereo (*d*- and *l*-enantiomers) or geometric (*cis*- and *trans*-) isomers. Differences in only a few amino acids between two proteins (e.g., between bovine and human insulin) can be detected by antibodies raised against either antigen. The remarkable specificity of each antibody, as well as their functional similarities, can best be understood by examining their structures (Figures 1.37, 10.62, and 10.63). Each immunoglobulin molecule is a distinctive heterotetramer composed of two identical pairs of polypeptide chains.

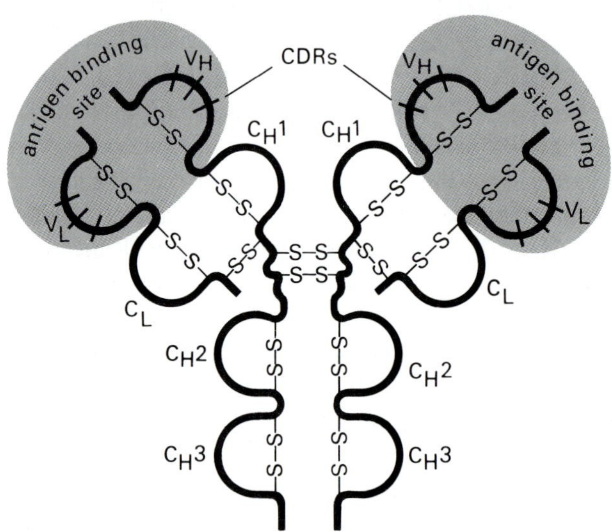

Figure 10.62
Schematic diagram of an antibody. The variable and constant regions of the light (V_L, C_L) and heavy (V_H, C_H) chains are marked. Several domains in the heavy chain constant regions (C_H1, C_H2, and C_H3) are indicated, as are the antigen binding sites in the variable regions and the complementarity determining regions (CDRs).

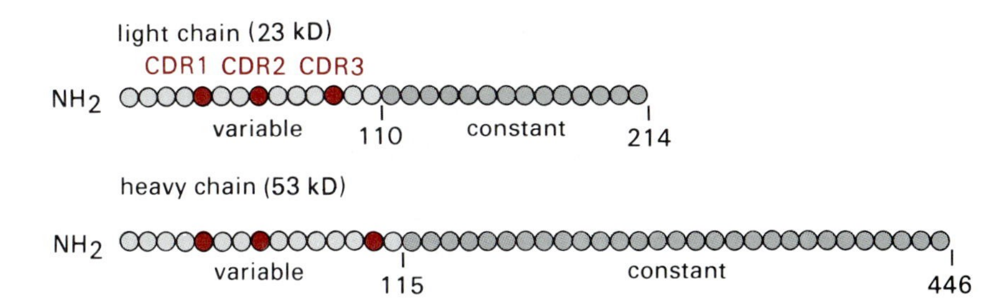

Figure 10.63
Schematic diagram of light and heavy chains of an immunoglobulin.

Each pair consists of a large or **heavy chain** (**H** chain) and a small or **light chain** (**L** chain). The H and L chains of each dimer are held together by disulfide bonds, as are the two H chains of each tetramer. The two H chains in an antibody molecule have identical amino acid sequences, but each antibody's *H chains* are distinctive in the approximately 115 residues at the amino terminal end; this region is referred to as the H chain **variable region**, $\mathbf{V_H}$. The corresponding approximately 110 amino terminal amino acids in the two identical L chains, $\mathbf{V_L}$, are also distinctive for each antibody. The amino acid sequence of the carboxy terminal half of L chains (about 110 amino acids) is referred to as the **constant region**, $\mathbf{C_L}$, because it is either one of two sequences, the $\boldsymbol{\kappa}$ or $\boldsymbol{\lambda}$ type, in all L chains, irrespective of the V_L sequence. H chains also have a constant region sequence, $\mathbf{C_H}$, that includes the approximately 330 amino acids in the carboxy terminal portion of the chain. There are several characteristic types of H chain constant regions, and these define the class and potential effector functions of their antibodies.

Only two types of light chains are known, kappa, κ and lambda, λ. Each type is defined by the amino acid sequence of its C_L segment and includes L chains with many different V_L sequences. Heavy chains are more diverse. Five major types of C_H regions are known: μ, δ, γ, ε, and α. These define the five classes of immunoglobulins, the so-called **isotypes**, IgM, IgD, IgG, IgE, and IgA, respectively. Antibodies of the five classes play specific roles in humoral immunity. For example, IgGs are the common circulating antibodies, IgEs are involved in allergic responses, and IgAs, which are concentrated in fluids surrounding epithelial cells, appear to protect against microbial infections. Each of the five heavy chain classes may be associated with either κ or λ L chains.

The structure formed by the interaction of each V_H and associated V_L region creates a unique combining site for interaction with specific antigens and thereby defines the antibody's specificity; thus, each antibody is bivalent because it contains two identical combining sites for antigen. Specificity is determined by the fit, or **complementarity**, of antigen and antibody (Figure 10.64). Three loops of amino acids in V_L and V_H are primarily responsible for the specific binding; these are referred to as the **hypervariable** or **complementarity-determining regions**, CDR1, CDR2, and CDR3. It is in the amino acid sequences of the CDRs that V_L and V_H segments of different antibodies differ most from one another.

Long before eukaryotic genes could be studied directly, the amino acid sequences of a variety of immunoglobulins were known. These studies demonstrated the existence of variable and constant sequences in the polypeptides. Many different hypotheses were advanced to try to explain the enormous diversity of V regions and the seemingly unlimited numbers of different antigenic specificities that can be produced. At a fundamental

Figure 10.64
(a) Schematic diagram showing binding of antigen to a specific immunoglobulin binding site. The dark spots represent the CDRs. (b) The computer generated structures at the bottom are the heavy and light chain variable regions of a lysozyme antibody (left) and lysozyme (right). The structure at the top shows how the two molecules "fit" together. Binding is stabilized by Van der Waals and electrostatic interactions and by hydrogen bonding. Courtesy of S. Sheriff. See S. Sheriff et al., *Proc. Natl. Acad. Sci. USA* 84 (1987), p. 8075.

(a)

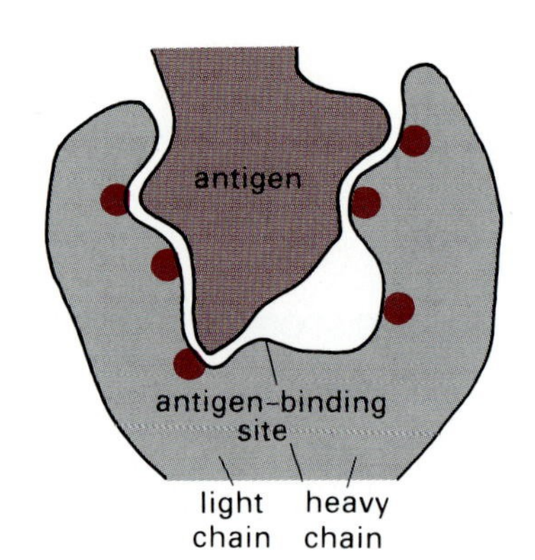

(b)

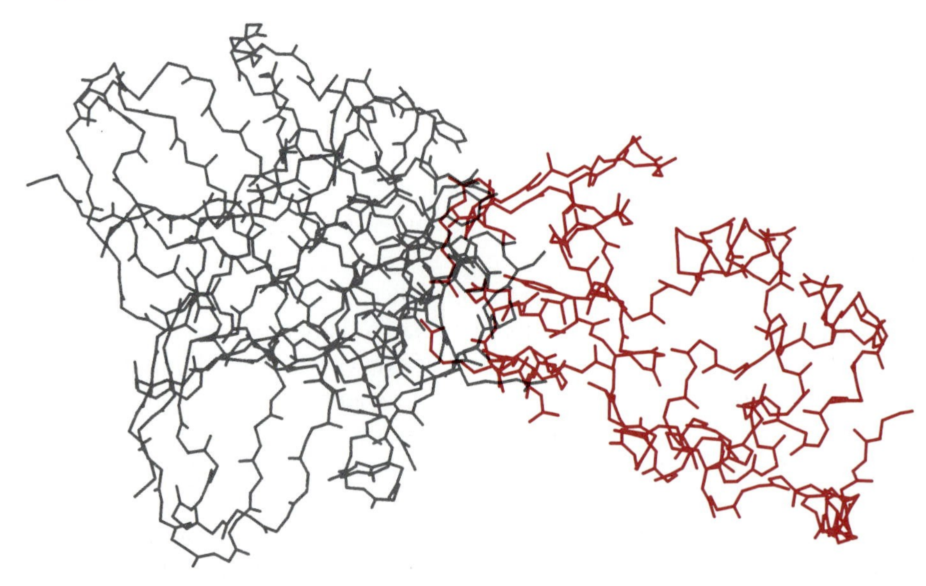

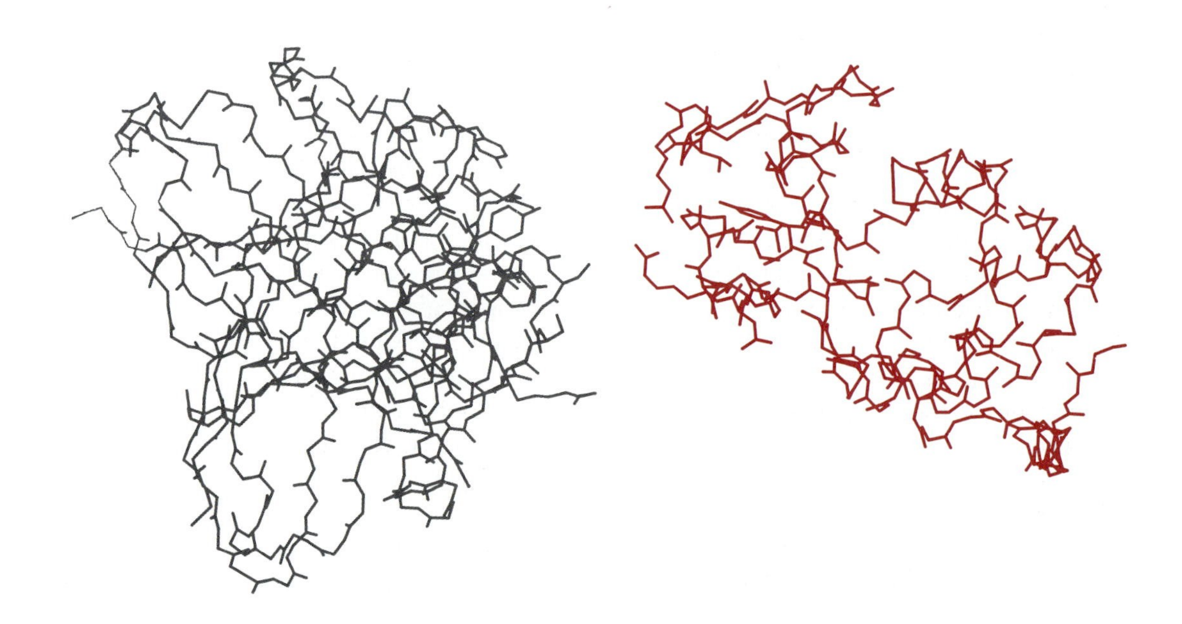

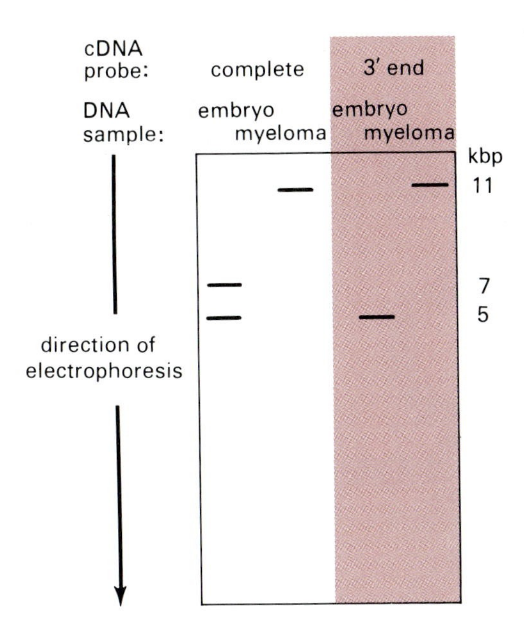

Figure 10.65
κ light chain coding regions rearrange during B cell maturation. Genomic DNA was isolated from a mouse embryo and from a mouse myeloma. Myelomas are B cell tumors, all of whose cells represent a clone and secrete a single homogeneous antibody. The DNAs were digested with restriction endonuclease *Bam*H1, electrophoresed, blotted, and annealed with cDNA probes representing the complete myeloma immunoglobulin mRNA or the 3′ end (C_κ) portion of the mRNA. Both probes anneal to a single restriction endonuclease fragment in myeloma DNA, but they anneal to different fragments in embryo DNA (or DNA from adult somatic cells other than B cells). After N. Hozumi and S. Tonegawa, *Proc. Natl. Acad. Sci. USA* 73 (1976), p. 3628.

level, the problem is how to generate a virtually infinite number of genes that encode different V regions but the same C regions. As it turned out, no single hypothesis was correct. Several different mechanisms are utilized to generate V region diversity.

Light chain genes The first demonstration that complete L chain genes are formed by DNA rearrangements during B cell differentiation came from comparisons between the restriction patterns of L chain coding sequences in cells that make antibodies (e.g., myeloma cells) and those that do not (e.g., liver or sperm) (Figure 10.65). This analysis showed that the V and C regions of an expressed L chain gene are joined on a single restriction endonuclease fragment in myeloma cell DNA, but occur on separate fragments in comparable digests of germ line or nonlymphoid cell DNAs. Since the original experiments, both germ line and B cell coding sequences for many different κ and λ light chain genes have been cloned and sequenced. Besides confirming that rearrangements occur, the sequence data provided insight into the mechanisms involved in the rearrangement.

Functional λ or κ L chain genes are formed in much the same way, and the processes are similar in most vertebrates. Figure 10.66 illustrates how a functional human κ L chain gene is constructed. Three DNA segments are used: V_κ, J_κ, and C_κ. All three are found on human chromosome 2, where there are about 300 tandemly arranged V_κ segments, 5 tandem J_κ segments, and 1 C_κ segment. In human germ line and most somatic cells, the V_κ array is 23 kbp from the J_κ segments. But during the differentiation of each B cell, one V_κ region is joined by recombination to any one of the J_κ segments. During the joining reaction, the DNA between the V_κ and J_κ regions in the germ line is either lost or conserved, depending on the relative orientations of the V_κ and J_κ segments in the germ line.

Each V_κ segment has three functional parts, a promoter followed by two exons, L_κ and V_κ (Figure 10.67). L_κ encodes a signal sequence that directs

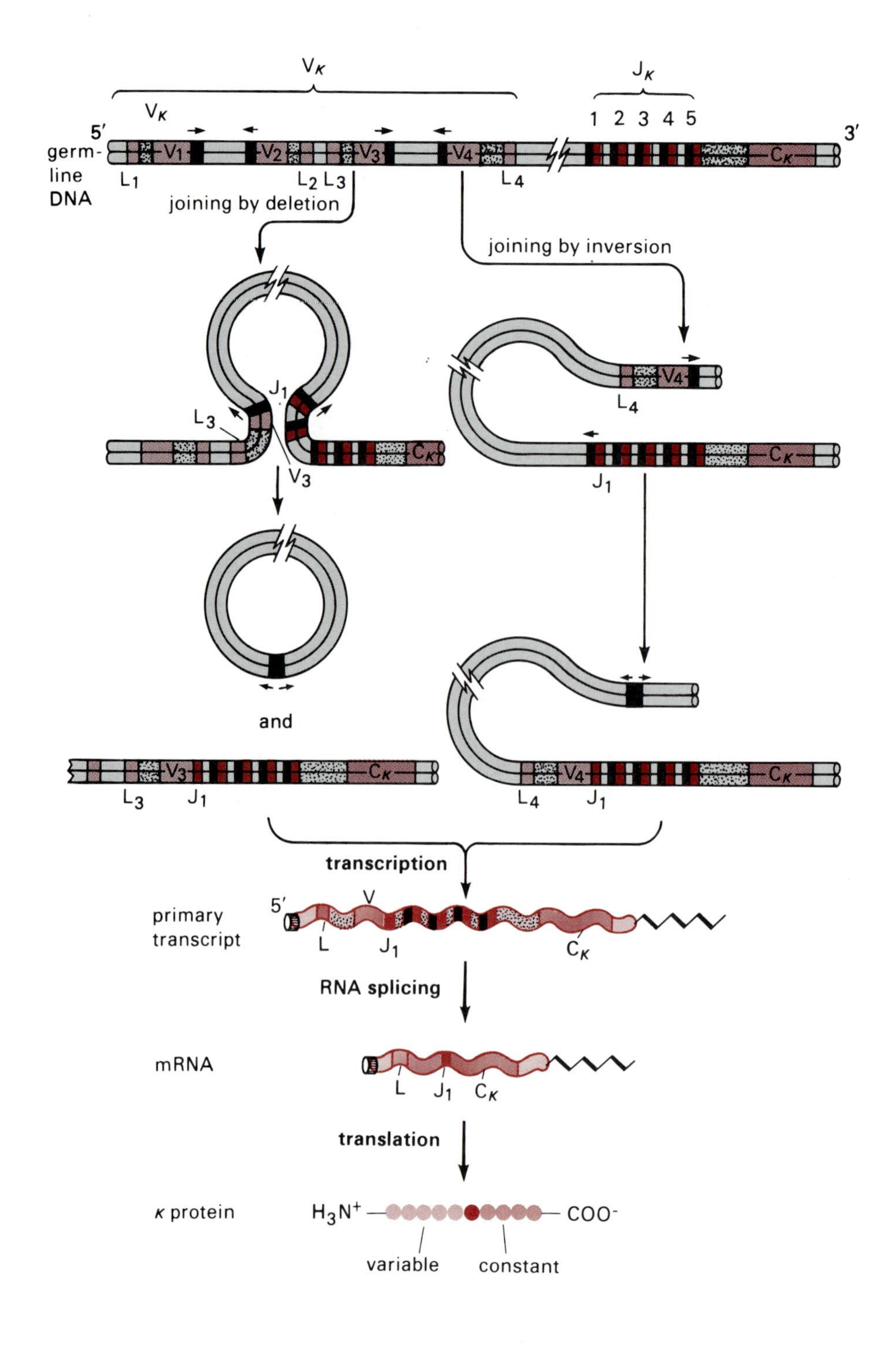

Figure 10.66
The rearrangements of germ line DNA regions required to form a functional κ light chain gene. The multiple V_κ regions occur in both orientations relative to the J_κ and C_κ segments. During B cell differentiation, a single V_κ region is joined to a single J_κ region. After J. Darnell, H. Lodish, and D. Baltimore, *Molecular Cell Biology* (New York: Scientific American Books, 1986), p. 1104.

the nascent polypeptide to the endoplasmic reticulum, and V_κ encodes about 80 amino acids of the κ chain's V region. The remainder of the L chain's V region, about 30 amino acids, is encoded by the joined J_κ segment. After the joining, the rearranged and functional κ L chain gene contains a promoter, three exons, and two introns. One intron separates the L_κ and V_κ-J_κ exons, and the other occurs between the V_κ-J_κ and C_κ exons. One of the L chains's CDR regions, CDR3, is encoded by the sequence created by the joining of V_κ to J_κ.

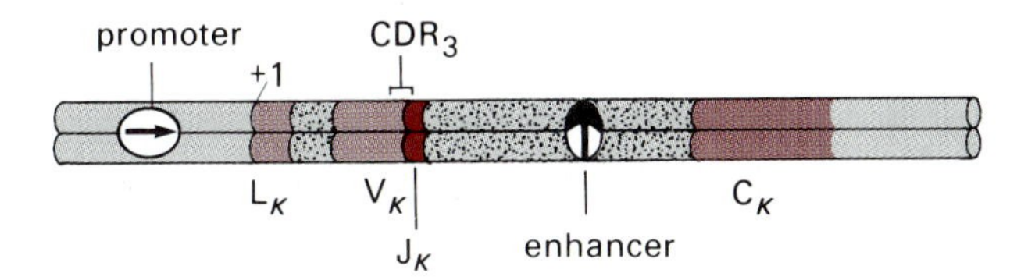

Figure 10.67
The functional elements of a rearranged L_κ gene.

Besides the alternative use of λ or κ chains, three mechanisms account for the diversity of L chains in antibody molecules. First, any one of the hundreds of V_κ segments can be rearranged and joined to any one of the distant J_κ segments, which are associated with C_κ. Thus, the potential specificity of the L chain is determined in part by which V_κ and J_κ segments are joined. Analogous sets of V_λ, J_λ, and C_λ coding regions combine to form λ light chains with different specificities. Second, the reaction joining the V_L and J_L segments together is imprecise, giving rise to several alternative coding sequences. Third, following formation of a functional L chain gene, the V_L region undergoes unusually frequent mutational alterations during the subsequent proliferation of the B cells. Such mutations further enlarge the diversity of the L chain's antigen binding region.

The site-specific recombinations that join a V_L region to a J_L region depend on a specific **recombinase** system. Recombination depends on bipartite oligonucleotide **recognition signals** that recur at the 3′ end of V_L and the 5′ end of J_L (Figure 10.68). These consist of 7 and 9 bp long sequences separated by 11 bp (about one helical turn) at V_L and 23 bp (about two helical turns) at J_L. Recombination depends on both the integrity of the heptamer and nonamer sequences and on the spacing, but not the sequence in the spacer. Although the specificity for these recognition signals is high, the actual joining of the V_L and J_L segments is rather sloppy; a variable small number of base pairs is either lost or gained at the junction between V_L and J_L. The consequence of this imprecision is that some joinings produce in-phase, translatable junctions between the V_L and J_L reading frames, but many do not. Moreover, the in-phase junctions are different from one joining reaction to another. This introduces variations in the coding sequence of the V_L region's CDR3 domain and contributes to the variability and diversity of V_L region sequences.

The mechanism of L chain gene assembly evolved to maximize the diversity of antibody chains, even at the expense of producing a large

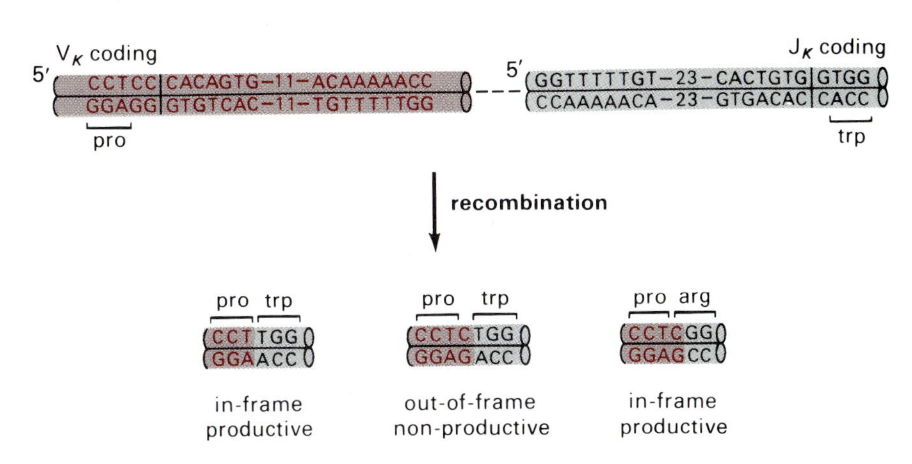

Figure 10.68
The joining of V_κ to J_κ depends on recombination between specific homologous recognition signals. Each signal has a 7 and a 9 bp long segment. These are separated by 11 and 23 bp in V_κ and J_κ, respectively. The differential spacing between heptamer and nonamer is required for recombination. The joining reaction is imprecise and results in the loss or gain of a variable number of base pairs from the V_κ and J_κ coding regions. Some joints put the two regions out of frame so that no light chain can be translated. Different in-frame joints occur and contribute to the variability of the CDR3 region of the light chain.

number of defective L chain genes. The joining of any one of about 300 V_κ regions to any one of the five J_κ segments can yield about 1500 different L_κ chain genes. Imprecision in the joining reaction may yield as many as 10 different in-phase junctions and thus expands the total to about 1.5×10^4 different L_κ chains. A similar abundance of possible λ light chains makes the total number of possible light chains very high. In humans, which have several different λ constant regions, the number of different kinds of L chains is considerably larger.

Besides the generation of diversity, the construction of a light chain gene also activates the gene's expression (Section 8.3e). The rearrangement brings the promoter preceding each V_κ segment into the vicinity of an enhancer within the intron separating J_κ from C_κ (Figure 10.67).

Heavy chain genes The same general recombinational mechanisms that generate L chain diversity are also used to create an enormous variety of H chain genes encoding distinctive V_H amino acid sequences. But the V regions of H chains are even more diverse than those of L chains because the V_H coding region is constructed by joining three rather than two segments: V_H, D_H, and J_H (Figure 10.69). These segments all occur

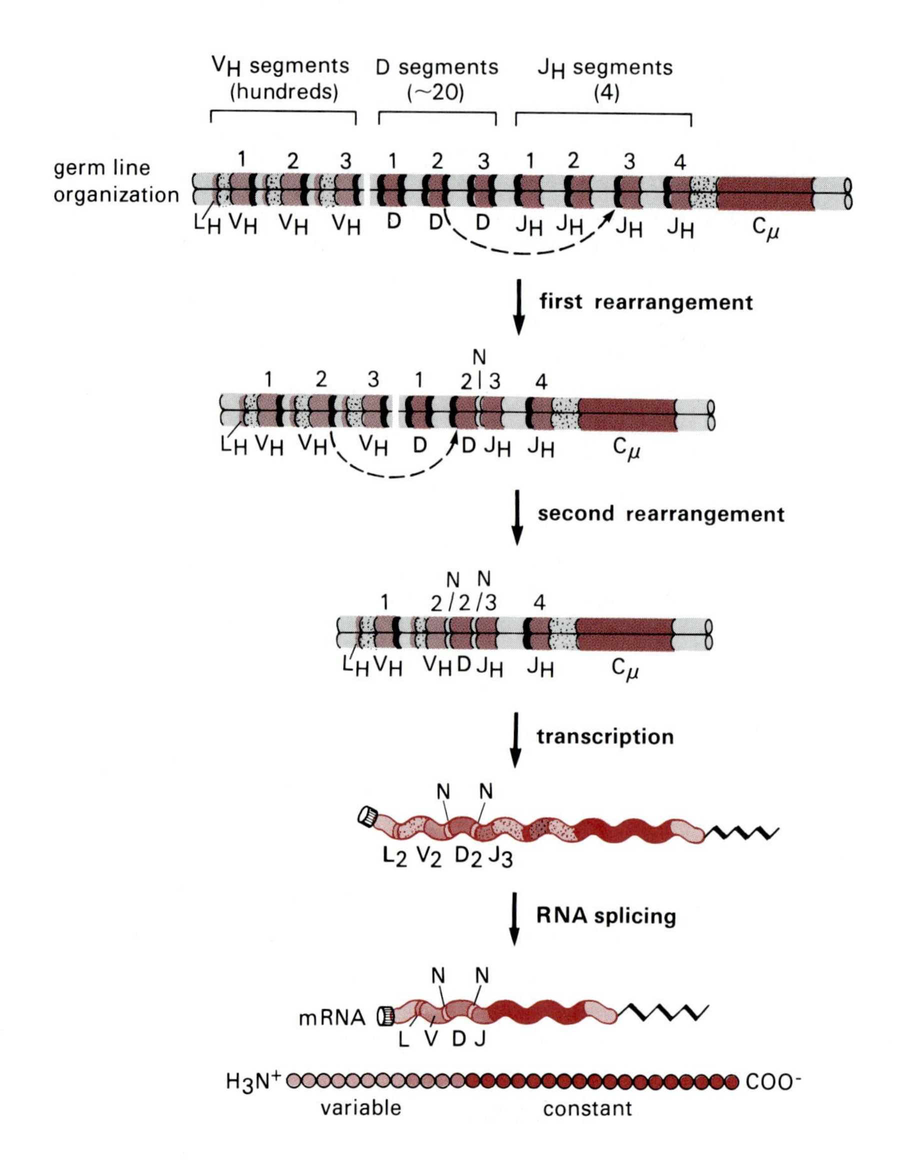

Figure 10.69
The rearrangements of germ line DNA that yield functional heavy chain genes in B cells.

on human chromosome 14, but the clusters of V_H's, D_H's, and J_H's are separated by long distances. Random combinations of one of several hundreds of V_H's, with one of 20 D_H's, and one of 5 J_H's yield between 10^4 and 10^5 different chains. The same 7 and 9 bp long sequences that promote joining of V_L and J_L regions also serve as the recombination signals for H chain rearrangement: these occur at the 3′ end of V_H, at the 5′ and 3′ ends of D_H, and at the 5′ end of J_H. Here, too, the sequence independent spacings of 11 and 23 bp between the heptamer-nonamer pairs of each recombining segment are critical. The two joining reactions occur in sequence. The first joins one of the D_H elements to one of the J_H's, and the second links a V_H segment to the already recombined D_HJ_H segment. Both the V_HD_H and D_HJ_H rearrangements are imprecise and are frequently accompanied by deletions and insertions of new nucleotides at either or both of the joining boundaries. The inserted nucleotides, termed **N regions**, probably result from the action of terminal deoxynucleotidyl transferase (Section 4.8), an enzyme that was known for more than 25 years before a physiological role could be identified. These insertions may encode as many as ten new amino acids in an H chain. The combined effect of deletions and insertions on either side of D_H is to increase the variability of V_H chains several hundredfold. The selection of different V_H, D_H, and J_H segments provides for about 10^5 combinations. N region variability expands the number of different H chains to about 10^7 to 10^8. Different combinations between H and L chains can readily account for 10^{11} to 10^{12} possible immunoglobulin combining sites.

The constant regions of H chains, C_H, also vary, but these differences do not influence the nature of the variable antigen binding sites (CDRs). Thus, a particular rearranged V_H segment can be joined to a μ, δ, γ, ε, or α C_H segment; which C_H region is used defines the antibody isotype: IgM, IgD, IgG, IgE, or IgA, respectively. The nature of the H chain determines, in part, the particular physiological effect produced by the antigen-antibody interaction. Before explaining how H chains with different C_H regions are created, we need to consider how B cells differentiate. This description also explains how the immune system produces large amounts of specific antibodies in response to an antigen.

B cell differentiation The **stem** cell progenitors of B cells, and indeed of all hematopoietic cells, originate first in fetal liver and spleen; but after birth, they arise in the bone marrow. The initial steps leading to the formation of a circulating but nonproliferative **virgin B cell** occur in the absence of antigen (Figure 10.70). Nevertheless, virgin B cells already have rearranged functional H and L chain genes and express antibodies of the membrane bound IgM and IgD classes. Some virgin B cells produce only IgM; others produce both IgM and IgD. The approximately 10^4 surface IgM molecules associated with each virgin B cell are all identical. But each cell's IgMs differ from those on other cells in structure and antigenic specificity because of the distinctive V_H and V_L regions of their respective H and L chains. Thus, to a large extent, antibody diversity and specificity are determined before antigens are even encountered. In virgin B cells producing both IgM and IgD, both immunoglobulins have the same V_H and V_L regions and differ only in the C_H region, $C\mu$ or $C\delta$.

Four different L chains could conceivably result from independent rearrangements of the κ and λ L chain alleles on each of the pairs of chromo-

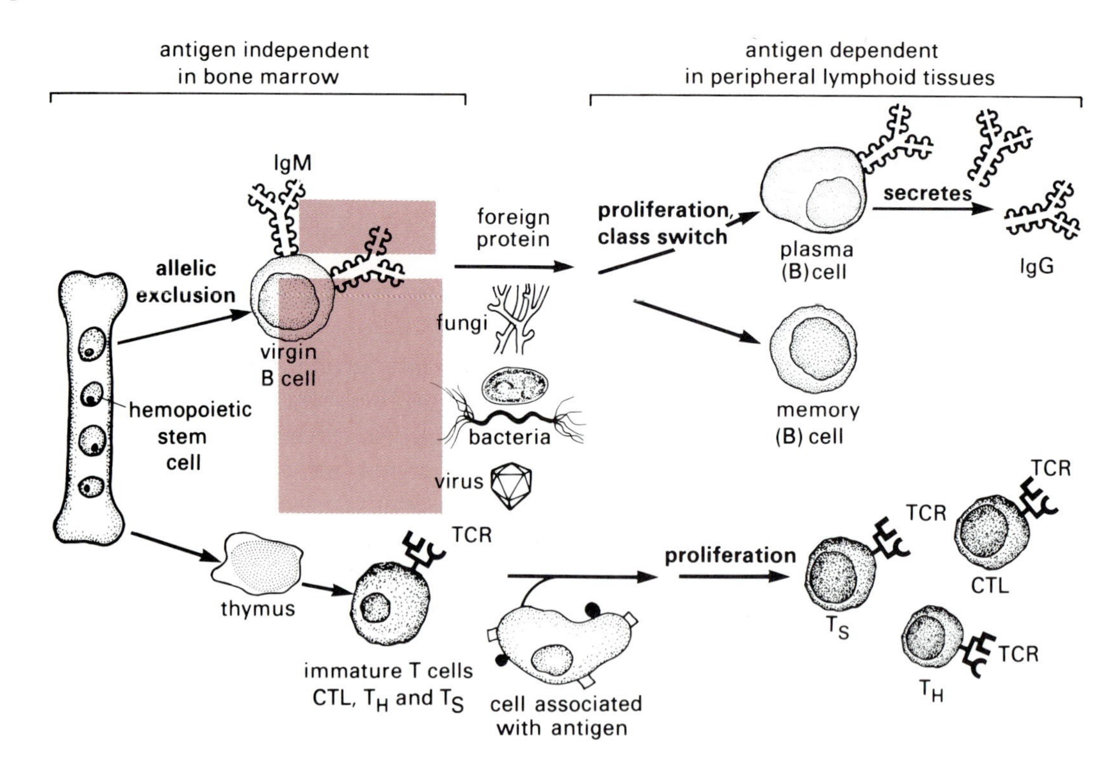

Figure 10.70
T and B lymphocytes differentiate from stem cells that originate in bone marrow. T lymphocyte precursors must spend some time in the thymus to become immunologically active, thus, T cells. B cells derive their name from their original identification in the avian organ, the bursa of Fabricius.

some 2 and 22, respectively. Similarly, two different H chains could arise from two different rearrangements in the two chromosomes 14. How is it, then, that a virgin B cell makes only one kind of H and one kind of L chain? Rearrangements taking place during development of the virgin B cells occur in a nonrandom order. H chain gene rearrangement occurs first on one of the two chromosomes. If the recombinational joining is grossly aberrant or if the initial D_HJ_H or subsequent $V_HD_HJ_H$ joinings create an unexpressible gene, rearrangement reactions will be initiated on the second chromosome. Successful completion of an H chain rearrangement triggers successive attempts to create a functional κ L chain gene on either chromosome 2. Failure of these recombinations initiates rearrangements at the λ L chain loci. Cells that fail to make both functional H and functional L chain genes do not survive. Although the mechanism for achieving this ordered and exclusionary process is not understood, it seems that once a functional H chain or L chain can be made by a proper genome rearrangement, further rearrangements of the alternative alleles stop—a phenomenon called **allelic exclusion**.

The further differentiation of virgin B cells to mature, short-lived, antibody secreting **plasma cells** or, alternatively, to long-lived, nonproducing **memory cells** occurs outside the bone marrow. Both transformations depend on the interaction of the B cell's membrane bound IgM with a cognate antigen (Figure 10.70). Failure to bind antigen dooms the virgin B cell to death within a few days. Interaction of the surface IgM with a well-fitting antigen triggers a panoply of physiological responses. These include active growth and proliferation, a change from the production of membrane bound antibody to secretable antibody, and a switch in the class of H chain being expressed. Most such induced B cells have limited life spans, but some differentiate into memory cells that, for many years,

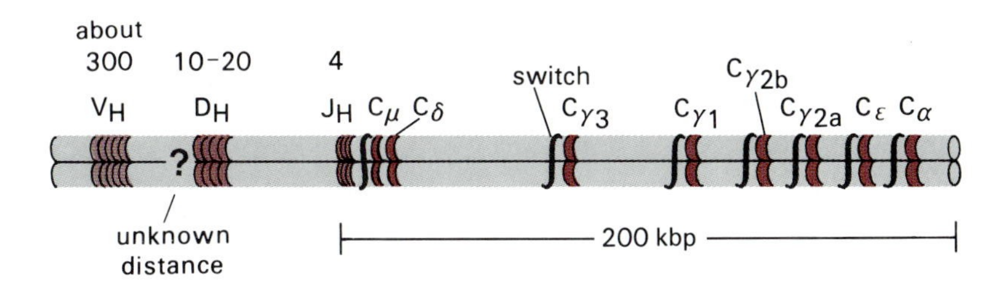

Figure 10.71
Schematic diagram of the cluster of C_H coding regions in the mouse genome.

retain the ability to respond to the original antigen by a renewed production of the antibody.

The mechanisms of the growth and proliferative responses to antigen binding require complex interactions with antigen processing cells (macrophages), helper T cells, and growth and differentiation factors called **lymphokines**; a discussion of these inputs is beyond the scope of this chapter. The mechanism of the changeover to the production of secretable antibody and the switch in expression from μ (or δ) chains to one of the other isotypes, γ or ε or α, involve differential mRNA processing and additional genomic rearrangements, respectively. The following discussion illustrates the general principles.

The structure and expression of C_H coding regions The coding sequences for the H chain constant regions, μ, δ, γ, ε, and α, occur in that order in a tandem array covering about 200 kbp in both mouse and human genomic DNA (Figure 10.71). The array, which consists of one each of C_μ, C_δ, C_ε, and C_α and multiple C_γ's, begins downstream of the cluster of J_H segments. Each C has multiple exons and introns.

In virgin B cells, transcription of the expressed membrane bound μ chain starts upstream of the leader segment, L_H, and continues through the first intron, the rearranged V_H region, the second intron, and the C_μ introns and exons; polyadenylation occurs within the last C_μ exon (Figure 10.72). The sequences encoding the typical H chain are in the first four of the six C_μ exons. The last two exons encode a polypeptide segment that anchors the μ chains of the antibody in the membrane. The membrane bound form of the μ chain is called μ_m, and the corresponding immunoglobulin is IgM_m. The antigen binding site of the membrane bound IgM is located on the outside of the B cell, capable of binding its cognate antigen. Binding of properly presented antigen to the surface IgM is the initiating signal for several events. An early one is a change from the production of IgM_m to IgM_s, a secreted variant of IgM that has the same antigen binding specificity, but whose μ chain carboxy terminal amino acid sequence is altered. Concomitantly, proliferation of the B cells produces a clone of plasma cells that secrete the same IgM_s. The changeover from μ_m to μ_s production is accomplished by use of an alternate polyadenylation signal for μ chain mRNA (Figure 10.72). This signal occurs in the intron between exons 4 and 5 of C_μ. Its use eliminates from the primary transcript the coding sequences for the μ_m membrane anchor and leads to the retention in the μ_s-mRNA of a coding segment at the end of exon 4 that encodes the carboxy terminus of μ_s. This extra sequence is removed by splicing from μ_m-mRNA.

Synthesis of IgD also depends on the use of alternate polyadenylation

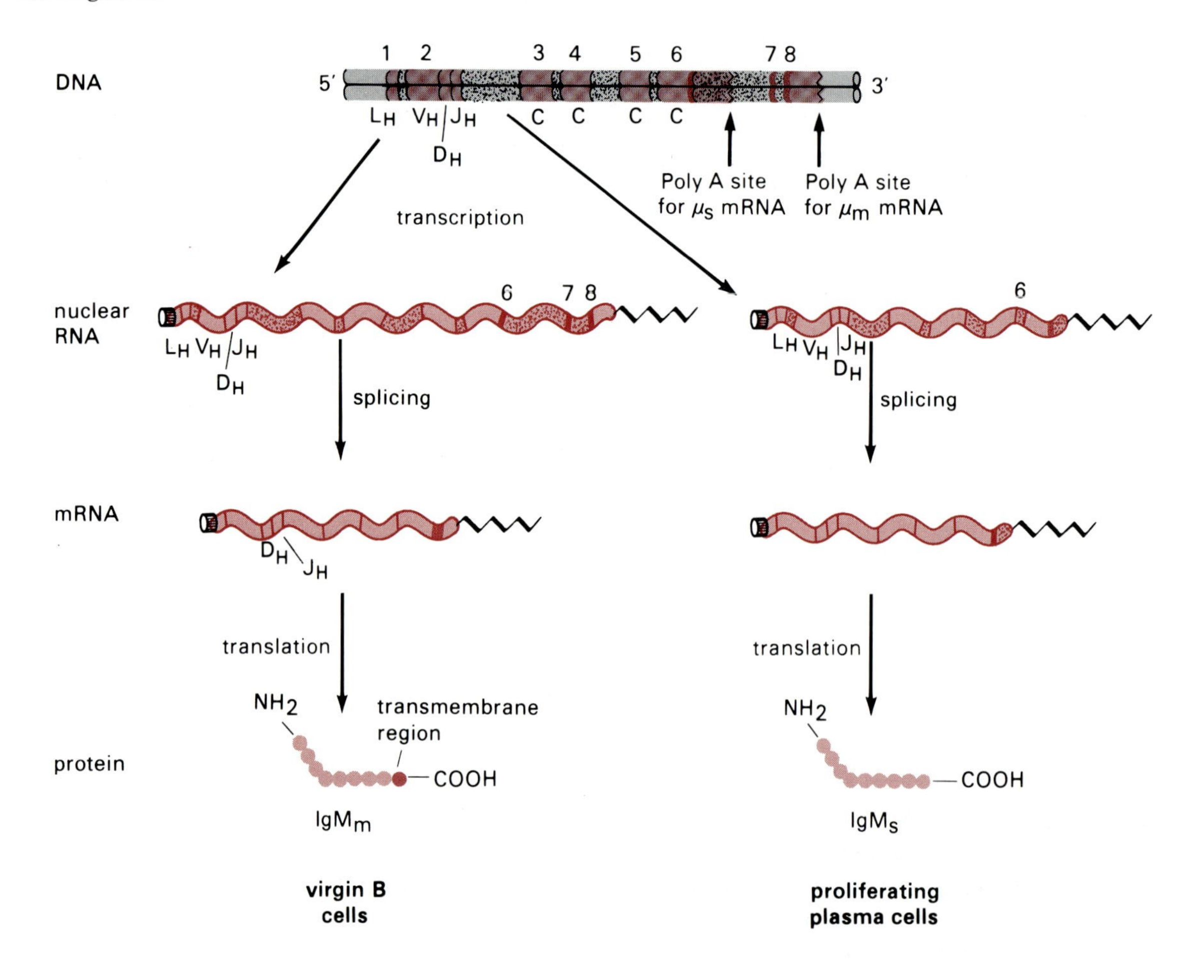

Figure 10.72
Alternative polyadenylation signals in the C_μ coding region yield heavy chains that are either membrane bound ($C\mu_m$) or secreted ($C\mu_s$). The six C_μ exons are shown here as exons 3 to 8. After J. Darnell, D. Baltimore, and H. Lodish, *Molecular Cell Biology* (New York: Scientific American Books, 1986), p. 1113.

signals and splicing. If the heavy chain primary transcript continues through C_δ, the second C_H region in the tandem array, and is polyadenylated beyond the C_δ coding region, then differential splicing from the rearranged VDJ of some transcripts can eliminate C_μ and lead to IgD_m synthesis in some virgin B cells (Figure 10.71). Differential use of polyadenylation sites analogous to those used in the IgM case explains the changeover from IgD_m to IgD_s.

Antigen induced differentiation of virgin B cells into plasma cells or memory B cells is also accompanied by **isotype switching** (e.g., from expression of IgM to IgG). Initially, this appears to result from alternate splicing of extremely long pre-mRNAs that extend from the normal transcriptional start site 5′ of the V region leader sequence to the polyadenylation signals at the end of the C_γ, C_ε, or C_α coding regions (Figure 10.71). Splicing from the 3′ end of the $V_HD_HJ_H$ segment to the 3′ intron junction upstream of the alternate C regions yields mRNAs that encode the same V_H sequence but different C_H sequences. However, this mechanism of isotype switching is supplanted by one that involves recombinational joining of the rearranged $V_HD_HJ_H$ segment to one of the C_H regions downstream of C_μ and C_δ, with concomitant loss of the DNA sequence between (Figure 10.73). For example, deletion of C_μ, C_δ, $C_{\gamma 3}$, and $C_{\gamma 1}$ results in the expression of a heavy chain with a $C_{\gamma 2b}$ constant region and thus of IgG_{2b}. Deletion of C_μ, C_δ, $C_{\gamma 3}$, $C_{\gamma 1}$, $C_{\gamma 2b}$, $C_{\gamma 2a}$, and C_ε

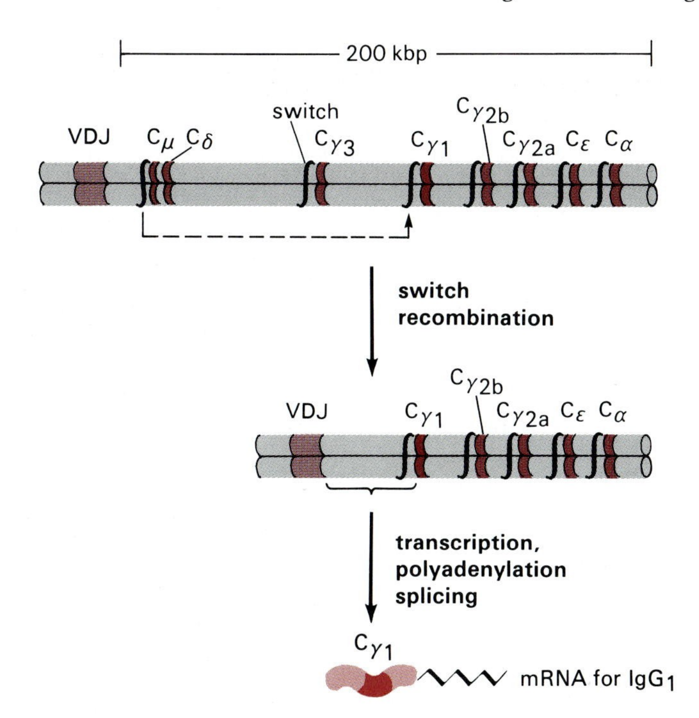

Figure 10.73
The rearrangements that occur during switch recombination. In this example, the immunoglobulin switches from the IgM class to IgG_1. The details of the C region exons and introns are omitted.

leads to IgA synthesis. The light chains and the V_H regions, as well as the antigen specificity, remain the same.

Heavy chain switching occurs by homologous recombination between DNA sequences within the introns, about 2 to 3 kbp in front of each of the C_H regions except C_δ. These sequences are called **S**, for switch. The switch sequences in front of the mouse C_H regions are composed of multiple copies of tandemly repeated sequences, generally about 40 to 80 nucleotides long but as short as 4 to 6 nucleotides in the case of the S_μ region. The nucleotide sequences GAGCTG and TGGG are common to all the S regions in mice.

Clonal selection Many proposals were made to account for the origin of the nearly limitless variety of antibodies and their extraordinary combining specificities. One of these, **clonal selection**, in conjunction with the explanation of diversity offered by DNA rearrangement, proved to be the most important. The idea of clonal selection can be summarized as follows. The early, antigen independent, steps in B cell maturation yield a large variety of virgin B cells (Figure 10.70). Each of these carries a specific antibody in the form of a surface IgM. The specificity lies in the uniqueness of the V_L and V_H amino acid sequences and the contributions they make to the CDRs. When a virgin B cell interacts by chance with a closely fitting antigen, it is stimulated to proliferate and secrete immunoglobulin. Thus, after antigen binding, the virgin B cell gives rise to a circulating clone of plasma cells that secrete immunoglobulins with the same antigen binding specificity as the initially stimulated B cell.

This description of clonal selection omits many interesting details, particularly the critical role of accessory cells and growth factors in facilitating and modulating the transformations of virgin B cells. One additional important feature is that not all the progeny of an activated virgin B cell become short-lived, antibody secreting plasma cells. Some become quiescent and immortal memory B cells. Memory B cells retain a history of the animal's previous encounters with antigens. The record is stored as circulating B cells with membrane bound antibodies (IgM) whose V_L and V_H regions were selected for their close fit to a foreign antigen. On encountering the same antigen later in life, the organism's memory B cell is stimulated to proliferate and secrete the same antibody as its progenitor virgin B cell.

The clonal selection model, by itself, does not account fully for the range of antibody diversity, nor for the finding that the quality of fit between antibody and antigen changes during subsequent encounters with antigen. These characteristics of the immune system can be explained by another generator of V_H and V_L diversity. During their quiescent phase, memory B cells accumulate mutations in and closely surrounding their rearranged V_H regions at nearly a millionfold greater rate than occurs spontaneously in other cellular genes. The amino acid changes caused by these somatic mutations accumulate in the CDR regions of the antibody and alter the combining specificities of already rearranged V regions. Some of the changes may so drastically alter the specificity of the antibody's combining site as to create a new or inactive antibody. But in others, the mutational alterations improve the fit between an antibody and its antigen. In that case, a second exposure to antigen preferentially activates those B cells that are capable of producing the tightest binding antibodies. Because this system creates cells that have "learned" to recognize an offending organism more efficiently on reinfection, the organism's defenses are substantially improved.

The role of T cells As mentioned earlier, both cellular and humoral immunity rely on T lymphocytes. These cells arise in the bone marrow but differentiate to their mature forms in the thymus (Figure 10.70). Cytotoxic T cells (CTLs) are the main actors in cellular immunity, and T helper (T_H) and T suppressor (T_S) cells modulate the humoral immune response. All the physiological functions attributed to T cells stem from the interaction of their **T cell surface receptor (TCR)** with cell associated antigens. In the case of CTLs, this interaction kills the cells bearing the antigen. T_H cells act as helpers by facilitating the proliferation of virgin B cells when they encounter their cognate antigen. The interaction between the T_H cell's TCR and the B cell's bound antigen triggers the T cell to release specific B cell growth and differentiation proteins (lymphokines) that stimulate B cell proliferation and secretion of antibody. The precise role of T_S cells is unknown, but they may down-regulate the B cell response and participate in the maintenance of immunological tolerance to self-antigens.

TCRs resemble membrane bound antibodies in their structure and antigen recognition function (Figure 10.74). However, there are a number of notable differences. TCRs remain fixed in the cell's plasma membrane and are never secreted. Moreover, whereas the surface bound antibodies on B cells bind both soluble and cell associated antigens, TCRs bind only cell associated antigens, Also, TCRs generally recognize foreign antigens only when they are associated with specific proteins located on the antigen presenting cell's surface—the products of the so-called **major histocompatibility complex (MHC)** genes. How the TCR is able to recognize

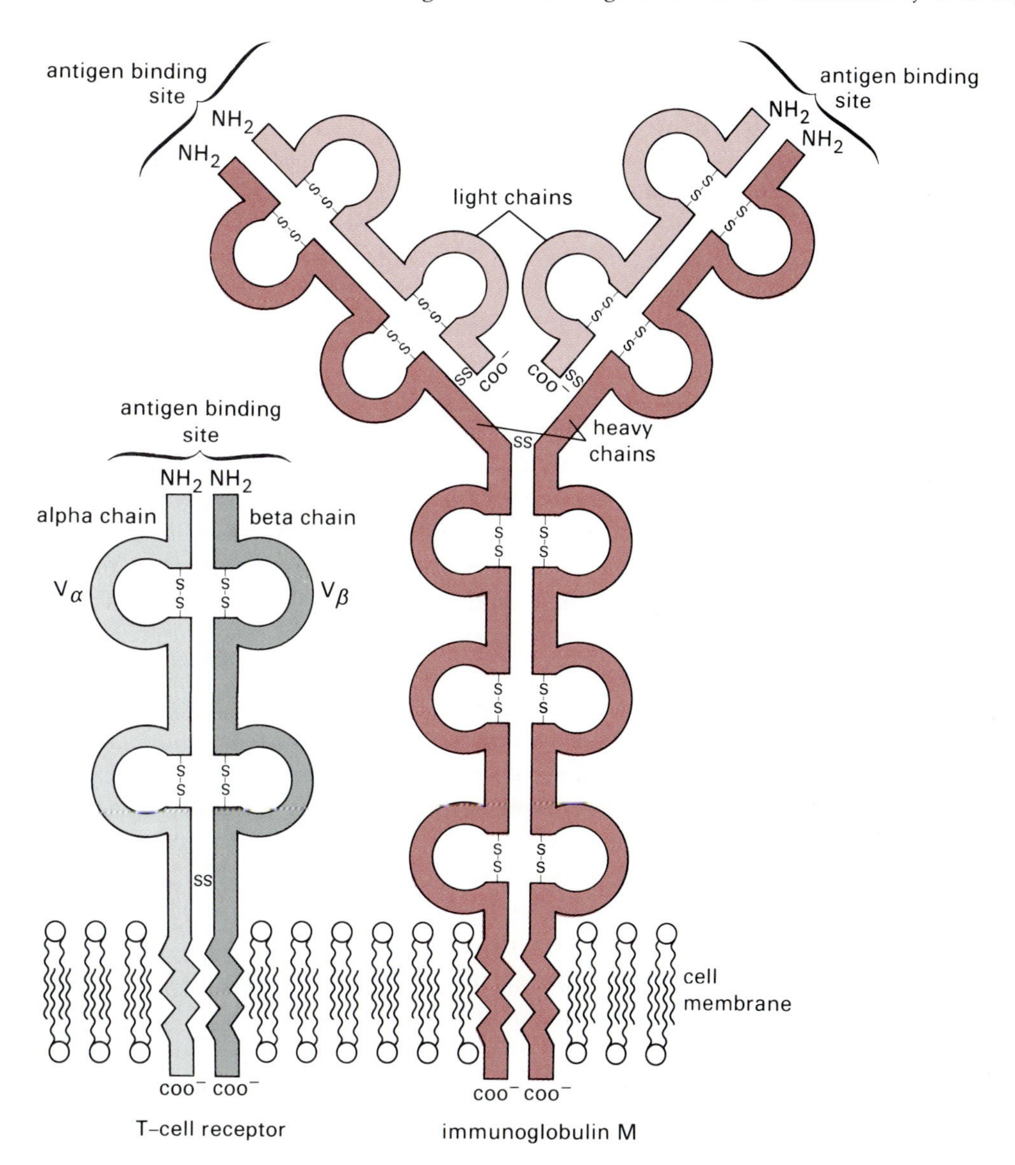

Figure 10.74
A schematic diagram comparing the structures of a T cell receptor and a membrane bound immunoglobulin. The T cell receptor has two chains, α and β, each of which is about 40 kDalton and has a membrane binding, carboxy terminal portion. The variable regions and antigen binding site are in the amino half of the chains. After T. Honjo and S. Habo, *Ann. Rev. Biochem.* 54 (1985), p. 820.

and bind the foreign antigen and MHC proteins, and why this dual recognition and binding function is necessary, are questions that have made the TCR an intriguing target for study. But in the context of this chapter, we shall focus on the structure of the TCR and the genomic rearrangements that lead to the expression of the genes encoding its constituent polypeptides.

All TCRs are plasma membrane bound heterodimers (Figure 10.74). Most consist of two chains, called α and β, but a small subset of TCRs contains two other chains, γ and δ. The α, β, and δ chains are about 45 kDaltons in size; γ is about 35 kDaltons. Different types of T cells carry either $\alpha\beta$ or $\gamma\delta$ TCRs. The two chains of a pair are held together by disulfide linkages. As with the H and L chains of antibodies, each of the TCR chains contains an amino terminal variable region that is largely extracellular and a carboxy terminal constant region, part of which anchors the chain in the membrane. A TCR's antigen binding specificity is attributed to the three-dimensional structure created by the juxtaposition of the V_α and V_β regions or the V_γ and V_δ regions.

The similarity between TCRs and antibodies is continued in the way the diversity of, for example, α and β chain V regions is acquired at the genetic level. Information for the V region of the β chain is split into 3 separate DNA regions on chromosome 6 in mice and chromosome 7 in humans: approximately 20 V_β, 2 D_β, and about 6 J_β segments are associated with each of 2 C_β segments. The α chain gene segments, about 50 V_α, 50 J_α, and 1 C_α, are localized to chromosome 14 in mice and humans. The coding segments for γ chains are also split, but these are arranged in four clusters, each having several V, J, and C segments that can be joined to produce a functional γ chain. Although not studied as thoroughly, the rearrangement mechanisms resemble those of antibody genes. The same motif of heptamers and nonamers separated by nonconserved spacers of 11–12 and 22–23 base pairs provides the recognition signals. The same recombinase probably catalyzes the rearrangements in B and T cells. The recombinational event in T cells is also imprecise, thereby creating considerable diversity by the variability in the joints. Rearrangement of the β chain locus precedes that of the α chain, and successful rearrangement of one of the β or α chain alleles prevents rearrangement at the other allele (allelic exclusion). So far as is known, somatic mutation of TCR V regions does not occur; therefore, the diversity of TCR specificities derives from the multiplicity of the various gene segments and the sloppiness of the joining reaction. A significant corollary of the presence of both α chain genes and heavy chain (immunoglobulin) genes on human chromosome 14 is the potential for aberrant joining of V_H regions with J_α regions mediated by the common recognition signals and recombinase. Such reactions are responsible for several chromosome 14 rearrangements associated with human T cell tumors.

As with B cells, the antigenic specificity of individual T cells is established during the preliminary stage of T cell maturation. This takes place in the thymus and in the absence of antigen (Figure 10.70). Later, interaction of a T cell with a cell carrying a foreign surface antigen that can bind to the TCR fosters proliferation of a T cell clone with that specificity. A clone of cytotoxic T cells is formed, for example, in response to interaction with an infected cell carrying viral antigens. Similarly, it is likely that a clone of helper T cells forms after an immature T cell interacts with a macrophage that has ingested the protein antigen and displays the proteolytic digestion products on its surface.

For a cell carrying surface antigens to interact with the TCR on a CTL or T_H requires that it also display an MHC protein on its surface (Figure 10.75). These proteins are encoded by multigene families called H_2 (in mice) and HLA (in humans). The human HLA gene complex, for example, spans at least 3×10^6 bp on chromosome 6. A ternary complex of MHC, antigen, and T cell receptor appears to be required for a T cell response, but the molecular details of these interactions are only beginning to be clarified. One group of MHC proteins, called class I MHC, occurs on the surface of most somatic cells. The recognition and binding by a CTL of a foreign antigen on a target cell requires that the antigen be associated with a class I MHC protein. For a T_H cell to bind a foreign antigen, it must occur associated with a class II MHC protein, the kind known to occur on both macrophages and B cells. MHC genes are highly polymorphic among individuals of a species; therefore, the class I and II proteins are highly variable among individuals. Differences in the class I

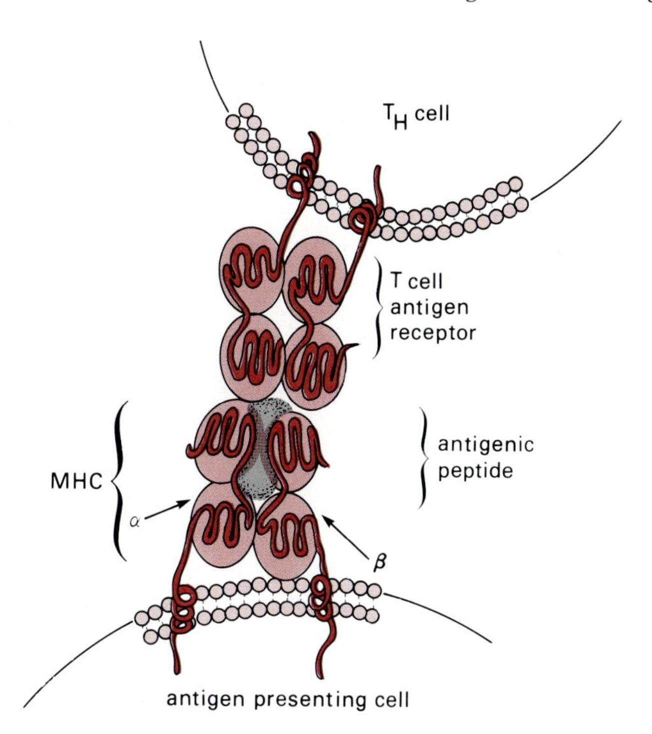

Figure 10.75
Diagram showing T cell receptor binding to both antigen and MHC surface protein. Courtesy of H. McDevitt.

proteins between individuals is what accounts for the failure of one person's cells and tissues to grow in most other members of the species, a phenomenon referred to as **graft rejection**. These differences also influence the probability of autoimmune reactions against an individual's own antigens: a breakdown of tolerance. Differences in class II proteins may account for some of the variation in an individual's immune responsiveness to certain antigens.

d. The Variable Surface Antigens of Trypanosomes

Trypanosomiasis In parts of Africa, the protozoan *Trypanosoma brucei* shuttles between the tsetse fly (*Glossina*) and mammals, causing African sleeping sickness in humans and related diseases in important domestic animals. The parasite divides by fission and reaches high population densities in connective tissues and extracellular fluids, whereupon more than 99.9 percent of the parasites are destroyed by the host's immune response. Seven to 10 days later, the few organisms that were resistant to the prevailing antibodies yield another population explosion. The process repeats itself with serologically distinct serial episodes occurring at seven- to ten-day intervals for months (Figure 10.76). With each episode, another **variant antigen type** (or **VAT**) of the trypanosome multiplies. Infection may finally be overcome in wild animals, but in humans and domestic animals, untreated trypanosomiasis is usually fatal. Death cannot be attributed to any single symptom, but is associated with complex immunological aberrations and infection of the central nervous system.

The essential feature of the trypanosome's success is its uncanny ability to escape immunological surveillance. Within each massive population, a few organisms (at a rate of about 10^{-6} per cell division) switch to a new variant antigen, and these survive to multiply anew.

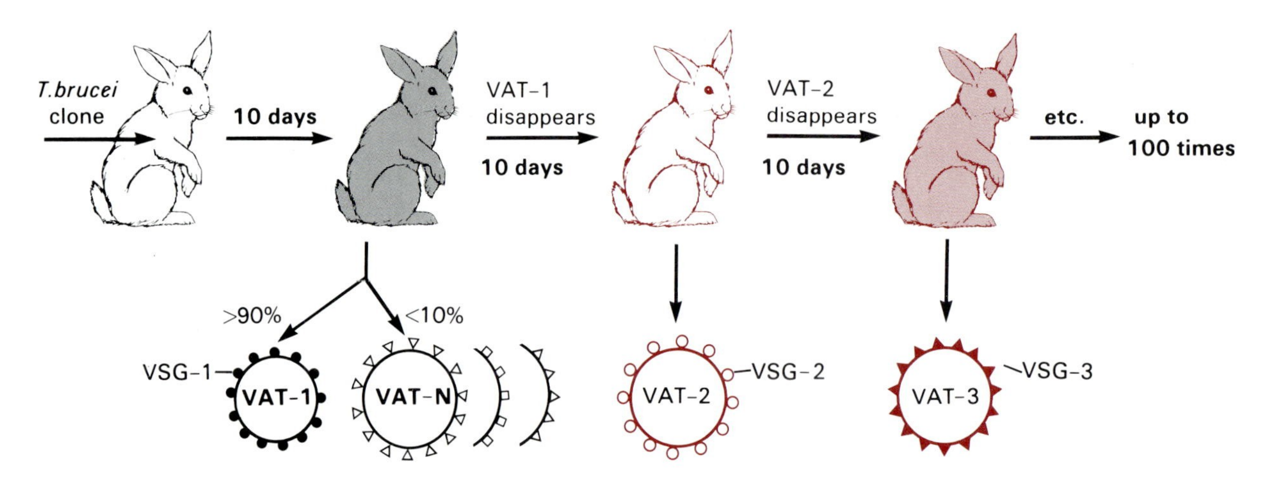

Figure 10.76
The cyclical course of infection of rabbits by *T. brucei*. VAT is the variable antigen type of the protozoan and VSG, the corresponding variable surface glycoprotein. In the first cycle, a significant proportion of the organisms carry VSGs other than the predominant one (VAT-N). Each VAT can be stably maintained by transfer from the rabbit to irradiated mice or rats in which the cyclical infection is severely retarded. Trypanosome counts are as high as 10^9 per ml of rat blood.

Variant antigen types (VATs) The single antigenic determinant that is exposed to the host is the trypanosome's surface antigen, a glycoprotein. About 10^7 molecules of the antigen are densely packed over the plasma membrane. Each VAT has a new and different surface antigen (**variable surface glycoprotein** or **VSG**), and 100 or more different VSGs can be produced by the serial offspring of a single cloned trypanosome (Figure 10.76). When individual host organisms are infected with the same trypanosome strain, the same set of VATS occurs. Also, protozoa in the salivary glands of tsetse flies infected with that trypanosome strain produce a subset of the same antigens. Unlike the mammalian hosts, which produce predominantly one VAT at a time, the trypanosomes in the fly are an antigenically heterogeneous population. Thus, there is a repertoire of genes encoding the different surface antigens in the trypanosome genome. At each switch to a new VAT, the expression of one gene is turned off, and a new one is turned on; in the fly, different cells express different surface antigen genes. Different *T. brucei* strains have distinctive repertoires of surface antigens. The repertoire produced by a single trypanosome clone is called a **serodeme**.

Not only does each *T. brucei* strain express a particular serodeme, but the repertoire of surface antigens usually appears in a particular if imprecise order. Thus, certain VATs tend to appear early in the infectious cycle, others later. Moreover, when the parasites are passed to previously uninfected organisms, the typical sequence is reinitiated. Usually, the very first mass population in a newly infected host contains multiple VATs, and the following cycles tend to be more homogeneous (Figure 10.76).

Variable surface glycoproteins (VSGs) Although each is distinctive, all VSGs have a similar basic structure; they are homodimers of glycosylated polypeptides of about 500 amino acids (Figure 10.77). Certain positions are glycosylated in all VSGs; other positions may or may not be modified by a carbohydrate moiety. The polypeptides themselves can be divided into a variable and a relatively constant region. The variable regions encompass about 400 amino acids, starting at the amino terminus. The constant region includes the 100 residues preceding the carboxy terminus and is similar, but not identical, from one VSG to another. Two subclasses

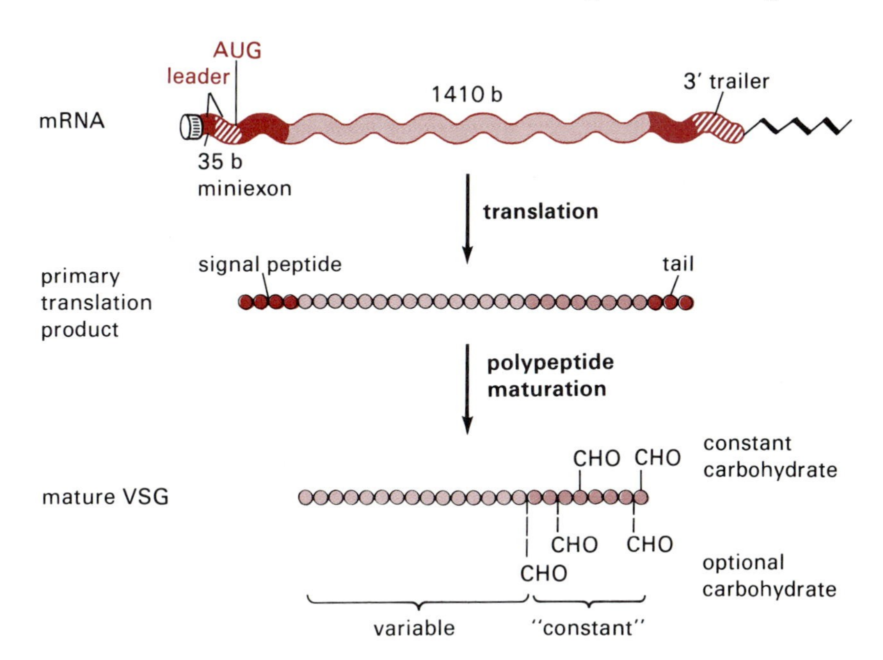

Figure 10.77
A typical *T. brucei* variable surface antigen (VSG), its corresponding mRNA (as determined from the structure of a cloned cDNA), and the primary translation product.

of *T. brucei* VSGs are recognized, based on the extent of homology at the carboxy terminus. The carboxy terminal homology is apparent even in VSGs from different serodemes. For this reason, and in particular because of the constant glycosylation sites associated with the carboxy end of VSGs, antibodies made to different *purified* VSGs, but not to the trypanosomes, cross-react. The primary antigenic determinant on the trypanosome itself is the NH_2 terminus of the VSG. This is because the carboxy terminal portions of the VSGs face inward toward the trypanosome interior while the variable NH_2 portions face outward.

Mature VSGs isolated from trypanosomes are derived from precursor polypeptides that are extended at both the NH_2 and COOH terminal ends. This is illustrated by comparing the structure of a VSG polypeptide with its mRNA and primary translation product (Figure 10.77). The structures of VSG mRNAs are known from cDNA clones; the VSGs are about 10 percent of total trypanosome protein, and a corresponding mRNA abundance facilitates cDNA cloning. Hybrid arrest translation (Figure 6.25) in conjunction with the specific antibodies proved the identity of the cDNA clones. A typical cDNA includes about 100 nucleotides of a 5′ nontranslated leader followed by 99 nucleotides encoding a typical signal peptide; the 33 corresponding amino acids do not appear in the VSG. The signal peptide coding region is followed by the sequences encoding the 50 kDalton mature VSG and thereafter by another 60–70 nucleotides that encode a C terminal region that, like the signal peptide, is missing from the mature VSG. The C terminal extension is very hydrophobic; it may facilitate the anchoring of the VSG on the trypanosome surface and be concomitantly removed.

The VSG multigene family The distinctive serodemes associated with different trypanosome clones are each encoded by a set of genes. Regardless of which VSG is being expressed, genes for all the VSGs in the serodeme are present in each organism. There is evidence that at least some VSG genes are clustered in trypanosome genomes. Some of the approximately 10^3 VSG genes in the genome may be quite similar or even identical, but others diverge considerably. The extent of divergence generally decreases toward the 3′ end of the genes, as expected from the conservation of the VSG carboxy terminus. This gradient of divergence is apparent when different regions within a cDNA clone are subcloned and used as probes against genomic digests in DNA blotting experiments. Subclones from the 5′ end of the cDNA may anneal to one or a few bands, probes from the middle of the same coding sequence often yield a greater number of bands, and probes from the 3′ end hybridize to as many as 50 bands. Few of these are perfect repetitions because increasingly stringent annealing conditions eliminate many of them. Notably, probes that represent the 3′ untranslated region of the mRNA anneal to the greatest number of genomic fragments.

So far we have been discussing multigene families in individual *T. brucei* clones. What is the relation between the gene families encoding different serodemes? Some VSG genes and proteins recur in many but not all serodemes. Others may be unique to a particular serodeme, indicating that, unlike most genes and gene families, at least some VSG genes evolve quite rapidly.

Switching expression from one VSG gene to another Whenever a trypanosome clone switches from one VAT to another, it turns off one VSG gene and turns on another. Each VAT makes only one kind of VSG mRNA. Earlier in this chapter, we described three different switch mechanisms for rapid on-off regulation of transcription. In the case of *Salmonella* flagellin, recurrent inversions of the promoter turn the *H2* operon on and off. When the G segment of Mu is inverted, the promoter remains in place, but alternate genes move into a position downstream from the promoter. In the yeast mating type switch, segments containing genes and promoters move into an active genomic region. Trypanosomes have adopted the latter mechanism for switching VSG genes, but they have added some tricks of their own as well. The situation is complex, and much about the mechanisms remains to be clarified. But one thing is plain; originality and versatility are hallmarks of trypanosomes and other protozoa as well.

Association of expressed VSG genes with telomeric regions of trypanosome chromosomes Expressed VSG genes are identified by annealing cDNA probes to genomic DNA blots under stringent conditions and also by their hypersensitivity, within chromatin, to cleavage to DNase I (Section 8.7a). The maps of these segments, and especially the sequences flanking the expressed VSG gene on both sides, are peculiar (Figure 10.78).

Bordering the expressed genes at the 5′ side, there is often a region barren of restriction endonuclease sites (Figure 10.78b). Farther away from the genes, the maps take on more normal aspects. The distance between the VSG coding region and the 5′ distal endonuclease sites can vary, even if DNAs from two different trypanosome clones expressing the same VSG are compared (compare Figure 10.78b and c). This is so, regardless of the fact that the arrangement of endonuclease sites in the 5′ distal cluster may be identical, indicating that, in both clones, the VSG gene is in the same genomic locus. The variable length of the barren region is related

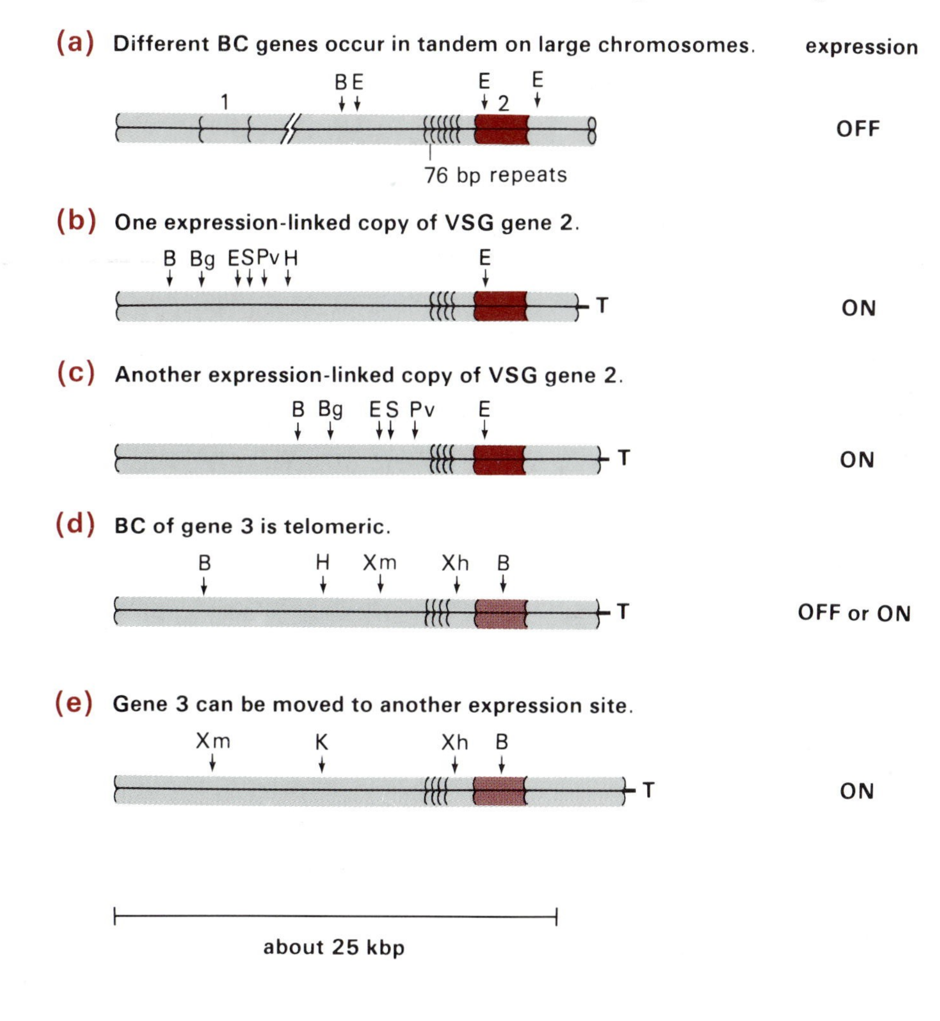

Figure 10.78
Schematic drawings showing the relation between the basic copy (BC) and expressed copies (ELCs) of VSG genes. The scale is not precise. Coding regions are in color. Transposed units include sequences on the 5′ side of the coding region. The 76 bp long repeats at the 5′ side of the coding region are indicated by vertical lines, but the number of repeats is only diagramatic. There are usually more repeats associated with ELCs than BCs. The breakpoint of the transposed unit can be, but is not necessarily, within the repeats. (a) A tandem array of BCs (1 and 2) in the internal region of a large chromosome. (b) Structure of one ELC of gene 2 after transposition. (c) Structure of another ELC of gene 2. The distance between the gene and the cluster of restriction sites on the far side of the telomere (T) is different in (b) and (c). It is not clear whether these two ELCs are on different telomeres or on the same telomere with variant flanking regions. (d) Structure of a telomeric BC for gene 3. At this site, the gene can be silent or expressed. (e) Another ELC of gene 3 after movement to a different telomere.

to the presence of a variable number of tandem copies of a 76 bp repeated sequence.

The genomic region 3′ to the expressed genes also has remarkably few restriction endonuclease sites (Figure 10.78b, c, and e). Moreover, the maps cannot be extended more than 5 to 10 kbp on the 3′ side of the VSG gene; the actual distance varies from one expressed gene to another. Regardless of what restriction endonuclease is used, it produces a fragment with one terminus at this "end" position. Also, those fragments are shortened if the DNA preparation is treated with *Bal* 31 (Section 4.1c) prior to restriction endonuclease cleavage. The explanation for these observations is that "end" is the natural, preexisting end of a DNA duplex. Expressed VSG genes are always quite close to a telomere. Invariably, the 3′ end of the sense strand is proximal to the telomere.

T. brucei has about 100 chromosomes. Many of them are quite small, less than 10^6 bp. Expressed VSG genes have been identified in at least three different telomeric regions, but the actual number of potentially active telomeric loci is unknown. However, it is clear that although a telomeric location is required for expression, it is not sufficient. VSG genes other than the one being expressed at a particular time may also be in telomeric locations (Figure 10.78d). Such silent telomeric VSG genes can be activated without changing their location. Still other silent VSG genes are not telomeric, but are in internal chromosomal loci (Figure 10.78a).

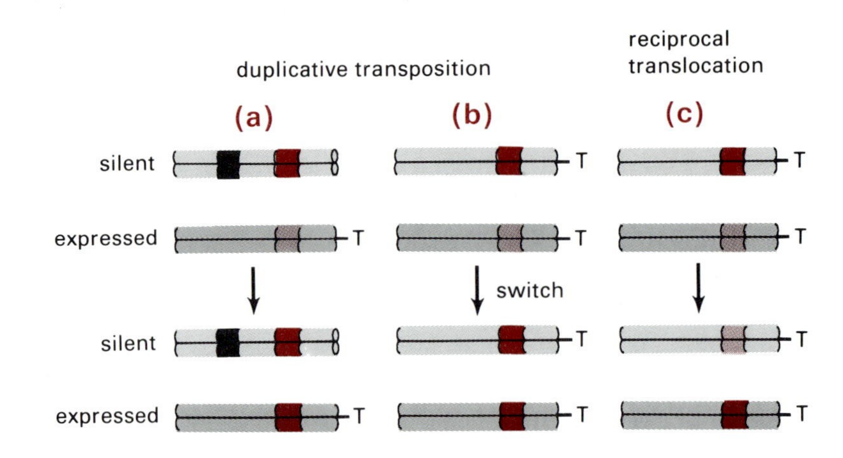

Figure 10.79
Three mechanisms for inserting VSG genes into potentially expressible telomeric regions. Duplicative transpositions from an internal chromosomal site (a) or from another telomere (b) are likely to involve gene conversion. At each new insertion, the previously active gene is either lost (duplicative transposition) or moved to a new site (reciprocal translocation [c]).

Prior to being expressed, copies of the internal genes are transposed to telomeric sites. Similarly, some telomerically located but silent VSG genes undergo duplicative transposition to a new telomere prior to expression (Figure 10.78d and e). Thus, expression of a single VSG gene requires (1) that it be in an appropriate telomeric region, (2) that the telomeric region be activated for gene expression, and (3) that VSG genes in other potentially active telomeric regions be silenced. Although several routes for insertion of VSG genes into potentially expressible telomeric regions are known, the way in which gene expression in those regions is activated or deactivated is still mysterious. Some recent experiments suggest that regional (uncharacterized) modifications of C residues may be associated with lack of gene expression. Also, *Pvu*II and *Pst*I restriction endonuclease sites in silent telomeric regions, but not in transcribed ones, are resistant to cleavage.

VSG gene insertion into potentially active telomeric sites Several ways by which a new VSG gene can be placed in a potentially active telomeric site are known. A copy of a silent, **basic copy** (**BC**) VSG gene can be inserted, by a duplicative transposition, into a potentially active telomeric site. The basic copy VSG may be from either an internal chromosomal locus or from a silent telomere (Figure 10.79a or b). Duplicative transposition, most likely by a gene conversion type event such as occurs in yeast mating type switches, is the most frequent mechanism for VSG gene switching. Consequently, a strain synthesizing a particular VSG often has one more copy of that gene than one that is not. When the next switch is made, the extra copy is frequently but not always lost, and another VSG gene can take its place. An alternative to duplicative transposition, reciprocal recombination, can occur when a VSG gene moves from a silent telomere to a potentially active one (Figure 10.79c). In this instance, the previously active gene is not lost and can be reactivated by a subsequent exchange. All these types of switches can be followed by pulsed-field gel electrophoresis and DNA blotting (Figure 10.80).

Figure 10.78 (a, b, and c) compares the structures of the basic copy of a VSG gene in an internal chromosomal location and two transposed expressed copies (**ELCs**, for **expression-linked copies**) of the gene from two different trypanosome clones. The structure of the BC region was obtained from phage and cosmid clones selected from genomic libraries by

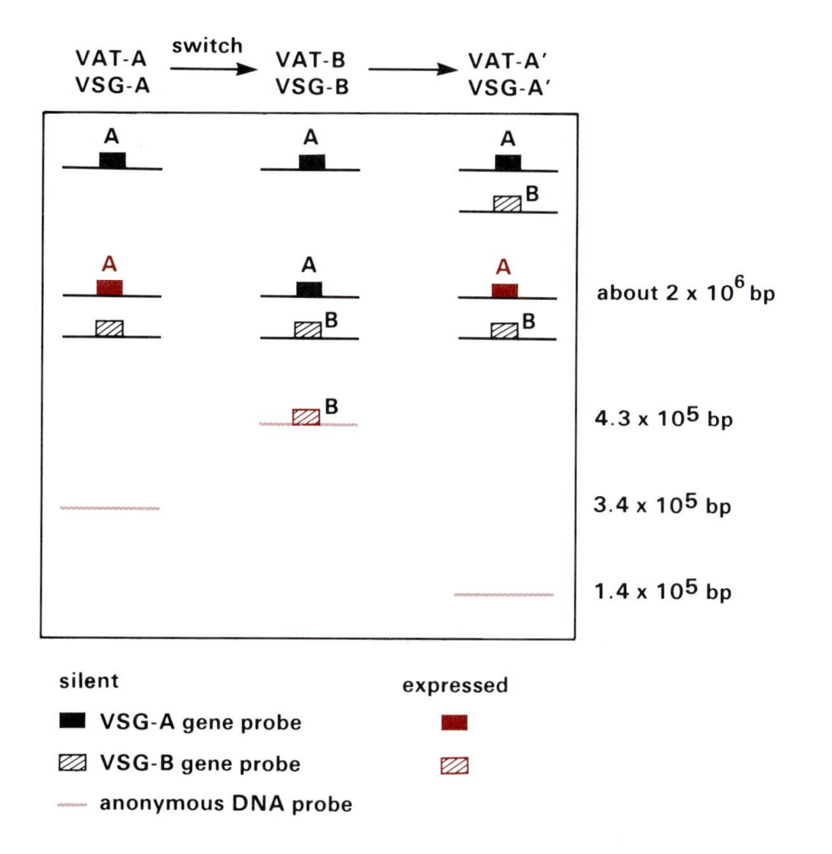

Figure 10.80
Genomic changes accompanying the switch from expression of VSG-A to VSG-B and back to VSG-A (A′) in one trypanosome clone. Each column summarizes the results of DNA blotting experiments on undigested trypanosome DNA separated into its constituent chromosomal DNAs by pulsed-field gel electrophoresis (Section 6.5). Each bar represents a DNA segment that annealed with a specific probe (see key). VSG genes are represented by boxes and are colored if being expressed in the VAT being analyzed. VAT-A expresses a VSG-A gene present in the telomere of an approximately 2000 kbp chromosome. A second, silent, basic copy of the VSG-A gene is present in a large chromosome that remains at the top of the electrophoresis gel. VAT-A also has a silent basic copy of a VSG-B gene in a telomeric location. Upon switching to VAT-B, the telomeric VSG-A gene is silenced, but it remains in the once active telomere. A copy of the telomeric VSG-B gene is inserted into an active telomere on a 340 kbp chromosome; the duplicative transposition appears to involve about 90 kbp of DNA because the recipient chromosome increases in size to 430 kbp (the recipient is identified by annealing to an anonymous DNA probe). Subsequently, a relapse to VSG-A expression occurs (VAT-A′). The previously active VSG-B gene is removed from the expressible site by recombination into a large chromosome (top of the gel) and silenced; the residue of the previously active chromosome is 140 kbp. The VSG-A gene is reactivated without changing its location. Adapted from C. Shea et al., *J. Biol. Chem. 261* (1986), p. 6056.

annealing with cDNA probes; it is the same regardless of whether the DNA is cloned from a VAT that is expressing the gene or from one that is not. The ELCs of transposable VSG genes are difficult to clone in *E. coli*, presumably because the sequences are unstable. Therefore, the structural features of ELCs were initially deduced partly from DNA blotting analysis of genomic DNA using subcloned regions of BC as probes and partly from sequence analysis of short subcloned regions of the ELC. The sizes of the transposable BC units are variable (1 to 40 kbp), but they are frequently about 3 kbp. They include sequences that appear in mature mRNA as well as variable lengths of 5′ flanking sequences that do not. The 3′ end of the transposable unit is close to the 3′ end of the gene. Outside of the transposable unit itself, the ELC and BC regions are not homologous. Note that the region around the 5′ end of the BC transposable unit includes a few tandem repeats of the same 76 bp unit that occurs in larger numbers in tandem at the 5′ side of the telomeric ELC. Thus, transposition into the expression site could involve homologous pairing between the 76 bp repeats. In such recombinations, different copies of the repeat units can participate, at random, and variable positions within the repeats can be involved. As a result, even transpositions into identical expression sites can result in the variable distances between the VSG gene and the upstream endonuclease sites that are found.

The situation at the 3′ end of the transposed unit is at least as complex and interesting as that at the 5′ end. Recall the unusual conservation of 3′ terminal coding sequences among otherwise quite divergent VSG genes. Even more remarkably, the conservation extends into the 3′ untranslated region of various BCs, ELCs, and cDNAs. In many other multigene families, homologous coding regions are associated with quite divergent 3′

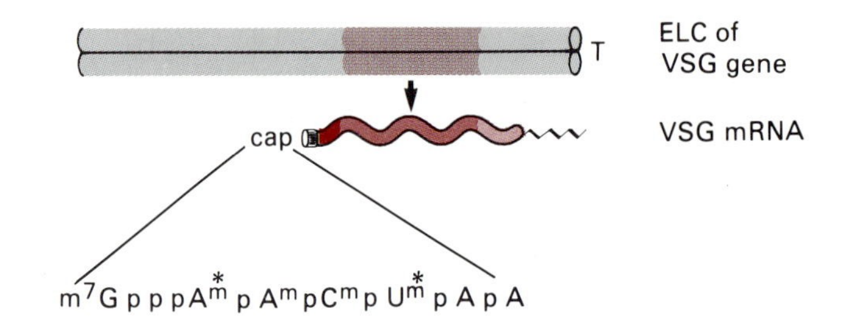

Figure 10.81
The 5′ end of a VSG mRNA (and other trypanosome mRNAs) is a capped, 35 nucleotide miniexon that is not present on the ELC gene. (T is telomere.) The cap structure itself is unusual; besides the terminal, added 7-methylguanosine cap, the first four residues of the 35 nucleotide segment are modified. Two, an A and a C, are 2′-0-methylated. The detailed structures of the other two, an A and a U (asterisks), are unknown. After K. L. Perry, K. P. Watkins, and N. Agabian, *Proc. Natl. Acad. Sci. USA* 84 (1987), p. 8190.

nontranslated sequences. What is the significance of the unusual conservation of the 3′ untranslated regions? Some clues come from a detailed comparison of the 3′ nontranslated region of mRNAs (and the ELCs) with the corresponding BCs. Basically, the cDNAs and the BCs are similar. However, they are usually not precisely identical; single base pair changes and small deletions occur, although rare cDNAs are actually identical in sequence to their BC. These observations could be explained by assuming that the duplicative transposition of a copy of a BC involves, in addition to the pairing between 76 bp tandem repeats at the 5′ side, homologous recombination at the 3′ end of the transposing unit and the 3′ end of the VSG it is replacing. Because of branch migration, the crossover point in the recombination could, in principle, occur anywhere in the homologous region (Section 2.4b), which extends from within the 3′ coding region through the end of the untranslated region. Thus, any particular ELC could be composed of a patchwork of sequences, some derived from the old gene and some from the newly arrived one. Notice, too, that if this were the case, there would be two elements contributing to the selective pressure that conserves the 3′ end: (1) maintenance of the structure of the carboxy terminus of VSGs and (2) maintenance of a sequence that promotes homologous recombination.

VSG gene transcription VSG genes need to be in particular telomeric regions to be expressed. But, as already pointed out, an appropriate position is only one requirement because some VSG genes are silent even when they are in potentially expressible telomeric sites (Figure 10.80). What kind of switch turns expression on and off? The answer is uncertain, but several relevant facts are known. For example, the length of DNA between the VSG coding region and the telomere is generally longer for an expressing gene than for a silent one. More interesting, however, are the unusual properties of the 5′ ends of the mRNAs and the length of the transcription units.

The 5′ flanking sequence of a VSG mRNA is quite different from the sequence in the corresponding region of the ELC from which it is transcribed (Figure 10.81). First, it starts with 35 untranslated bp unlike anything close to the ELC, and, second, some of the sequences in the ELC are missing. The same 35 bp are found at the start of all VSG mRNAs as well as at the start of most, if not all, trypanosome mRNAs. These 35 bp are found in a 1.4 kbp unit that is repeated about 200 times in the trypanosome genome, unlinked to VSG genes. Most of the units are in direct tandem arrays, but a few are scattered. The repeats are transcribed *in vivo* to yield capped RNAs that are 140 nucleotides long (mini exon derived or medRNA) and have the 35 bp segment at the 5′ end. Tran-

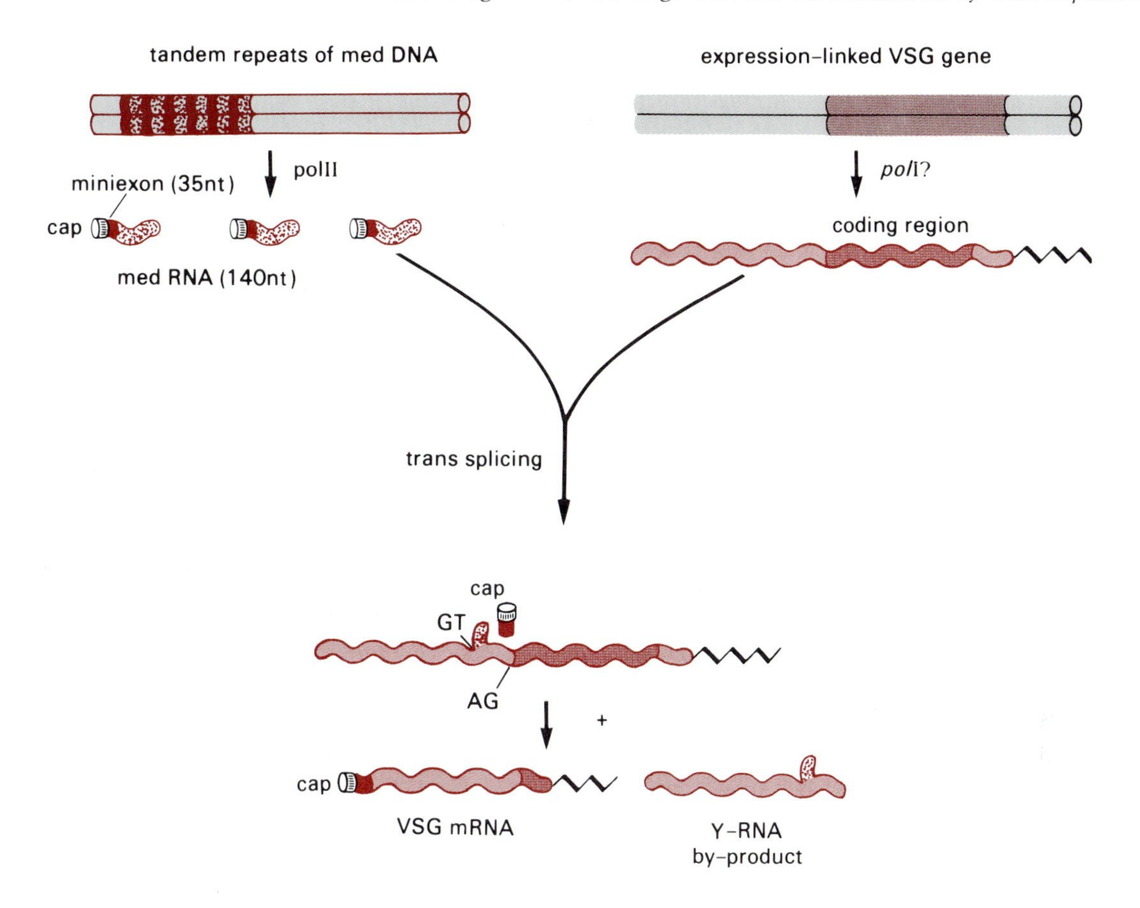

Figure 10.82
Transcription and processing of VSG mRNA. medRNA is transcribed from the tandem genomic repeats, and the VSG gene is transcribed as part of a long primary transcript. The medRNA and long VSG transcript are directly joined by *trans* splicing with the intermediary formation of a Y-RNA. At the junction of the miniexon and the rest of the medRNA, the sequence contains a typical donor splice junction, 5′-GT-. Similarly, the sequence preceding the 5′ end of the VSG gene coding region is a typical acceptor splice junction, 5′-AG-. After L. H. T. Van der Ploeg, *Cell* 47 (1986), p. 479.

scription appears to be by RNA polymerase II. The formation of mRNA in trypanosomes, whether of VSG or other protein genes, involves transfer of the 7-methylguanosine capped 35 bp 5′ segment from medRNA to the transcript of the coding region. Because trypanosomes generally have a high concentration of medRNA, it is unlikely to be important in the transcriptional regulation of VSG gene expression sites. Therefore, the regulation is likely to be associated with transcription at the expression site itself.

Expressible telomeric VSG gene loci can be either transcriptionally active or inactive, and only very rarely (during the switching process itself) is more than one such locus simultaneously active in a single cell. Transcription can initiate as much as 60 kbp upstream of the VSG gene, although the distance varies. The long primary transcript contains multiple, unrelated coding regions, in addition to the VSG region near its 3′ end. Formation of functional VSG mRNA requires the joining of the medRNA with the region of the primary transcript containing the VSG coding sequence (Figure 10.82). The two RNAs are joined by a *trans* splicing (Section 8.5d) that directly joins the 35 bp exon to the VSG coding regions (Figure 10.82). The 100 nucleotide 3′ end of medRNA is released from the polyadenylated fraction of trypanosome RNA by a debranching enzyme from human cells. This is the result predicted from the Y structured by-product that forms during *trans* splicing (Section 8.5d). Functional mRNAs for the unrelated coding sequences that are upstream of the VSG gene in the primary transcript could be formed in the same way.

In spite of the detailed description of VSG mRNA formation, the mechanism whereby a particular expression site is turned on or off is still not understood. Transcription of VSGs is notably resistant to even very high α-amanitin concentrations, and sequences similar to trypanosome rDNA promoters are found in the promoter regions. Thus, RNA polymerase I may be involved. One interesting speculation is that active transcription takes place in the nucleolus, known to be the location of most, if not all, RNA polymerase I. If so, then entry of the expression site into the nucleolus may regulate transcription.

10.7 Programmed Amplification and the Modulation of Gene Expression

The regulation of gene expression during development sometimes depends on a programmed amplification of specific DNA segments. Well-documented examples occur in protozoa, invertebrates, and vertebrates. In each case, amplification yields a large number of gene copies at a specific time and in a specific cell type, thereby assuring an abundant supply of a critical gene product. However, very different molecular strategies are used to effect amplification in the several known examples.

a. Disproportionate Replication: The Drosophila Chorion Genes

Meeting the high demand for eggshell proteins For a brief period during insect oogenesis, the ovarian follicle cells that surround a maturing oocyte secrete large amounts of the various proteins that form the chorion (eggshell). In some insects, this short-term supply problem is solved by timely and abundant transcription from highly repetitive families of chorion genes that are part of the germ line and somatic genome. *Drosophila*, however, maintains only single copies of the various chorion protein genes in its normal haploid genome. About 18 hours before chorion protein synthesis starts, the chorion genes in the follicle cells begin to be amplified. Then, in a 5 hour long critical period, the amplified genes are prolifically expressed.

Of the 20 to 50 different *Drosophila* chorion genes, a few are found within two clusters: one tandem cluster of 4 genes on chromosome 3 and one of 6 genes on the X-chromosome (Figure 10.83). Amplification of these genomic regions in follicle cells was demonstrated directly by measuring the change in genomic copy number of sets of restriction endonuclease fragments derived from the cloned genes. Such experiments show that, on the X-chromosome, for example, the DNA is amplified about 15-fold in the region containing the genes themselves. The flanking sequences are amplified to a lesser extent, with the copy number falling off gradually on either side of the genes. Sequences about 40 kbp away from the gene cluster are amplified only about 2-fold. Another characteristic of the amplification is that it does not generate any new restriction fragments that anneal to probes from the amplified region. Thus, no new combinations of DNA are formed. Together, the gradient of amplification

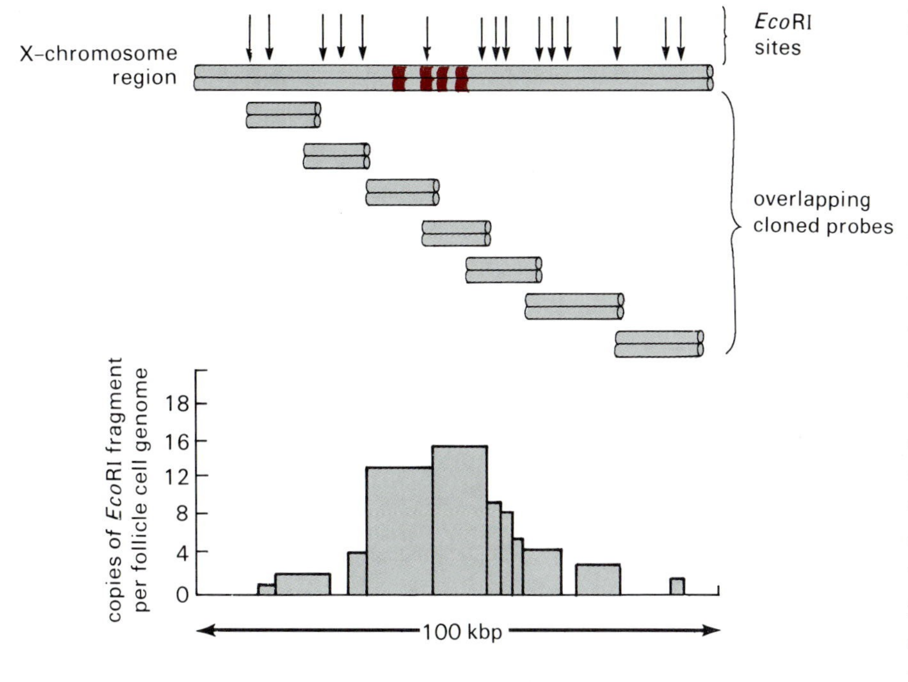

Figure 10.83
DNA amplification in follicle cells in the region of the chorion gene cluster on the *Drosophila* X-chromosome. At the top is a map of a 100 kbp region that contains four chorion protein genes (in color). The arrows indicate endonuclease *Eco*RI sites. A set of overlapping subclones of the region was constructed and is shown below the map. The subclones were each used as probes against separate genomic DNA blots of both embryo and follicle cell DNA digested with endonuclease *Eco*RI. The amount of each radioactive probe that annealed to each genomic *Eco*RI fragment was measured. The relative copy number of each *Eco*RI fragment in follicle cells was deduced from the ratio of the amount of radioactivity annealed to follicle cell DNA to that annealed to embryonic DNA (bottom). The copy number of each fragment in embryonic cells is assumed to be 1. After A. C. Spradling, *Cell* 27 (1981), p. 193.

and the lack of recombination suggest a mechanism by which the expansion of chorion gene copy number occurs.

Amplification by disproportionate replication Imagine that an origin of replication exists near the cluster of genes and that DNA synthesis begins bidirectionally from this origin. After a while, another round of replication begins at the same origin. Multiple successive initiations of bidirectional replication will generate multiple copies of the region (Figure 10.84). Moreover, because each successive replication fork must stop short of the preceding one, a decreasing gradient of copy numbers is produced on both sides of the origin. The newly generated copies are linked covalently neither to one another nor to the original duplex, so no new combinations of DNA sequences are formed. This model assumes that the normal controls prohibiting multiple initiations of replication at a single origin are relaxed. This seems to be so because chorion gene amplification continues even after normal DNA synthesis ceases in follicle cells. This hypothesis, sometimes called the **onion-skin** model after the similarity of the amplified region to the layers of an onion, is supported by striking electron micrographs of amplifying follicle cell DNA (Figure 10.85). The picture looks exactly like one-half of the model shown schematically in Figure 10.84; the origin of replication is at the top, and multiple replication forks occur at varying distances from the origin.

Competence for amplification is conferred by short (less than 5 kbp) DNA segments within the chorion gene regions on chromosomes 3 and X. When these short segments are inserted in other genomic sites, the new sites can also be amplified in a timely way in follicle cells. Insertion of the extra copies of the **amplification control elements** is accomplished by first inserting them into P elements (by *in vitro* recombination and cloning) and then injecting the P element containing vectors into early *Drosophila* embryos (Section 10.2b).

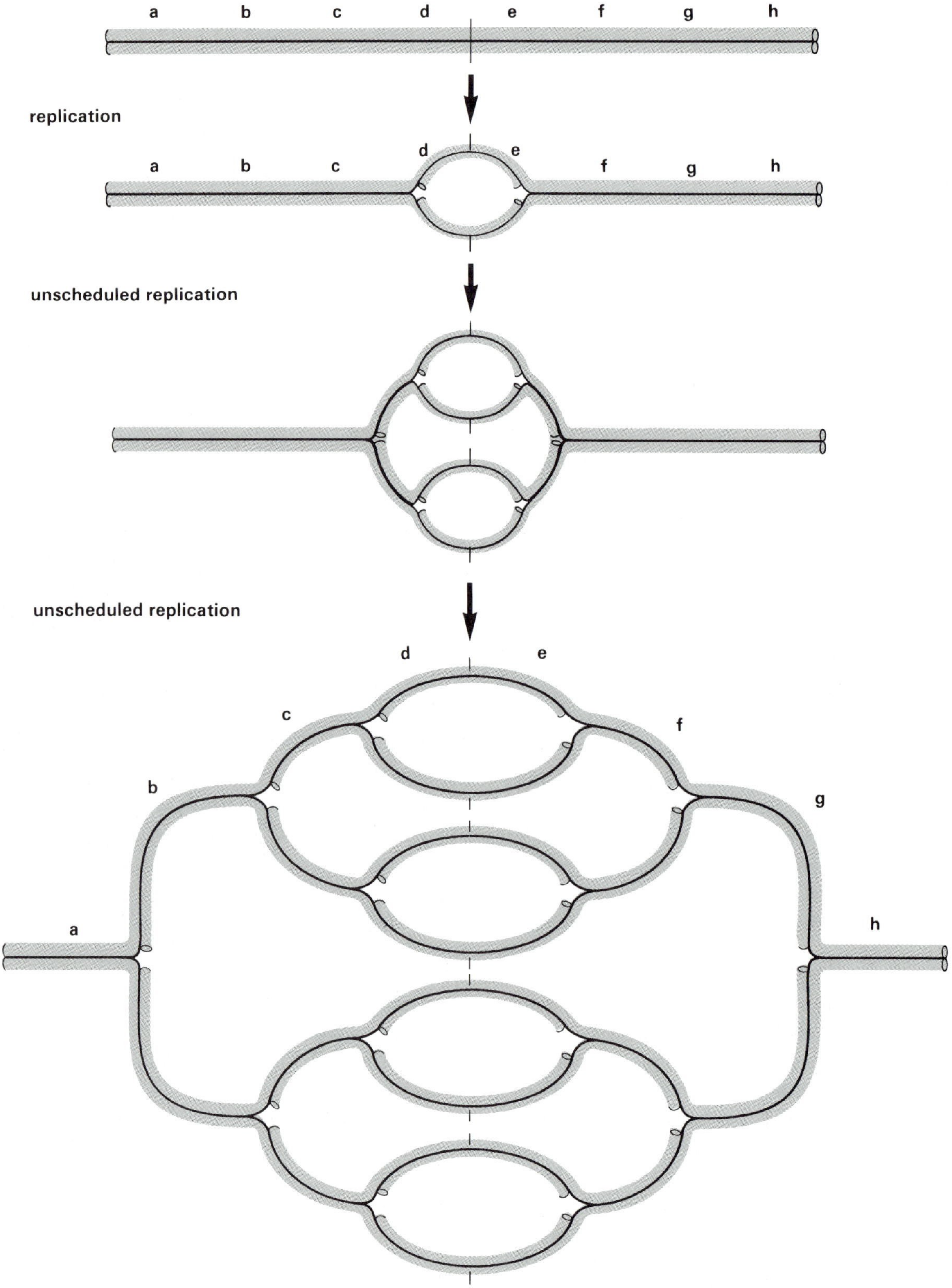

a b c d e f g h

replication

a b c d e f g h

unscheduled replication

unscheduled replication

d e

c f

b g

a h

Figure 10.84

Model showing amplification by multiple reinitiations of bidirectional DNA synthesis at a single origin: the "onion-skin" model for disproportionate replication. Note that only the two original DNA strands are ligated together. After G. Stark and G. M. Wahl, *Ann. Rev. Biochem.* 53 (1984), p. 478.

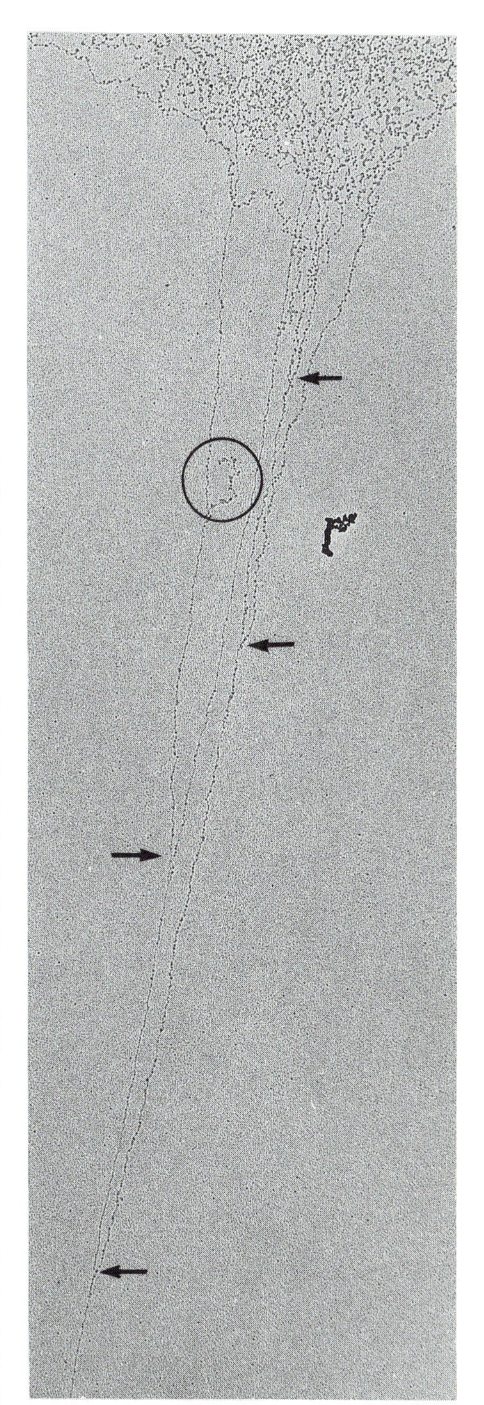

Figure 10.85
Electron micrograph of *Drosophila* follicle cell DNA undergoing amplification in a region including chorion genes. The arrows point to some of the replication forks. Multiple overlapping replication forks are apparent. Nucleosomes are visible as small dots along the DNA. The ring encircles a region that is being transcribed (an RNA strand dotted with ribonucleoprotein particles). Courtesy of Y. N. Osheim and O. L. Miller, Jr., as published in *Cell* 33 (1983), p. 543. Copyright Cell Press.

b. *Xenopus* rDNA: Rolling Circle Amplification

Although most organisms contain multiple copies of rDNA in their genomes (Section 9.2a). these are sometimes amplified still further within oocytes to assure an adequate supply of ribosomes for embryogenesis. *Xenopus* is the best studied example. The 600 (haploid) genomic rDNA copies are amplified over a thousandfold, and these are contained within about a thousand extrachromosomal nucleoli.

Amplification occurs in two stages; a small number of extrachromosomal rDNA copies are generated in the primordial germ cell, and then massive amplification takes place during meiosis by a rolling circle mechanism of DNA replication (Section 2.1g). The final products are large circular DNA molecules, each of which contains as many as 100 rDNA repeat units. Unlike the regular genomic rDNA repeats, in which the length of the intergenic spacer (IGS) varies, most of the IGSs in the extrachromosomal rDNA circles in one oocyte are the same length. It seems that only one or a few chromosomal genes give rise to the abundant circles in each oocyte. However, different oocytes, even in a single organism, amplify different genomic rDNA repeats and have different length IGSs. The total number of copies of rDNA seems to be controlled, and no further amplification occurs after fertilization. Just how a single rDNA repeat unit is incorporated into a rolling circle remains a puzzle.

c. Amplification of *Tetrahymena* rDNA in Macronuclei

Germ line DNA of ciliated protozoa is split into hundreds of fragments during the formation of the transcriptionally active macronucleus (Section 9.6b). Concomitantly, some of the micronuclear DNA is eliminated, telomeric sequences are added to the macronuclear segments, and most of the segments are amplified about 45-fold. However, the fragments containing rDNA are amplified preferentially. They attain copy numbers as high as 10^4 and are packaged within multiple nucleoli. The macronuclear DNA structures and copy numbers are maintained during the amitotic vegetative replication of the macronucleus. In contrast, micronuclear DNA contains only a single (haploid) copy of ribosomal RNA genes. Little is understood about the mechanism underlying preferential replication of the rDNA. Each macronuclear rDNA fragment contains two entire rDNA transcription units as well as about 4 kbp of nontranscribed DNA (Figure 10.86). The two halves of the molecules are inverted repeats of one another; thus, the whole fragment is palindromic. The fragments replicate from origins of replication in the central IGS regions. DNA

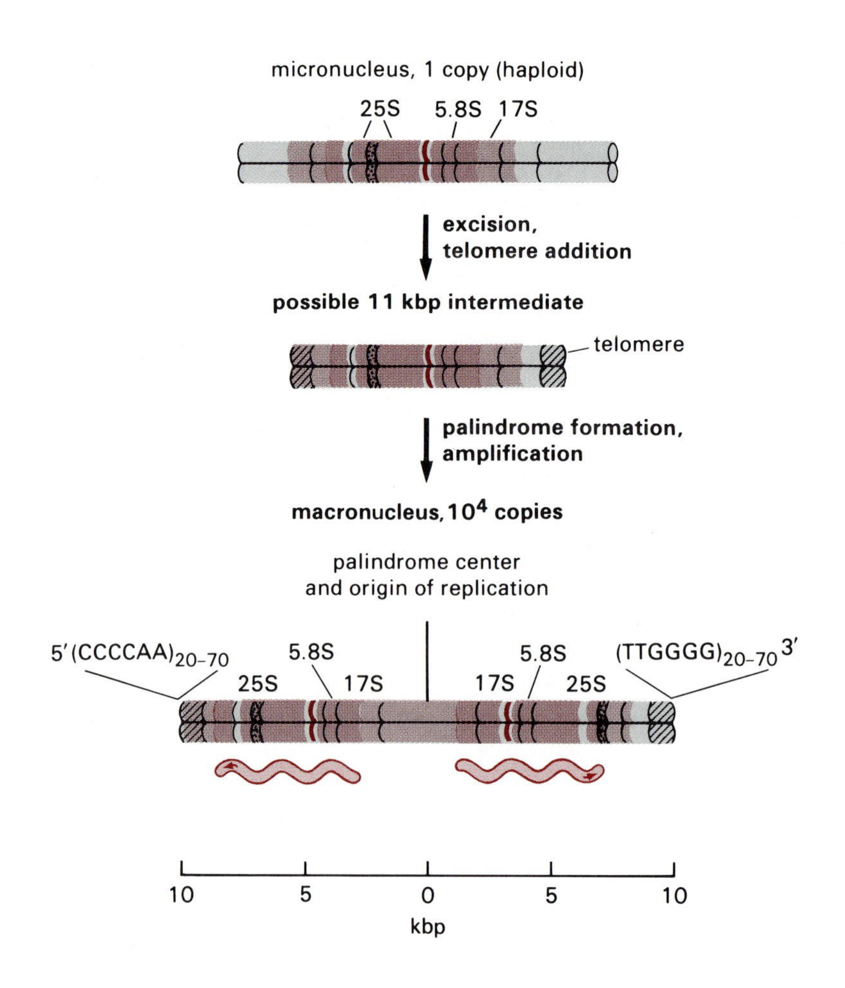

Figure 10.86
The single micronuclear rRNA gene locus of *Tetrahymena* and a possible pathway for its excision and amplification into the macronuclear rDNA fragments. After D. D. Larson et al., *Cell* 47 (1986), p. 229.

synthesis proceeds bidirectionally to the fragment ends. Sequences within the IGS regions that are required for efficient rDNA replication overlap those that function as transcriptional enhancers (Section 8.2), suggesting that DNA replication is linked in some way to transcription regulation.

10.8 Unprogrammed Tandem Amplifications

When the activity of an essential gene product becomes insufficient, either by mutation or because of the presence of a specific inhibitor, mutant cells that overcome the deficiency can often be found. Among such mutant cells, a substantial proportion are viable because they overproduce the gene product. Many have acquired multiple tandem repeats of the deficient gene, and the multiple gene copies produce a superabundance of mRNA and of the corresponding polypeptide. Both bacterial cells and a variety of eukaryotic cells in culture undergo such gene amplification. Amplifications also occur in tumor cells. By and large, these events are rare, and it remains to be seen how frequently amplification occurs in normal somatic cells *in vivo*.

a. Amplifications that Correct Enzyme Deficiencies

Drug resistance and gene amplification Treatment of cells with inhibitors (drugs) whose action is specifically directed toward a single essential enzyme kills almost all the cells but allows the growth of resistant mutants. Two such systems in which the inhibitor is tightly bound to the enzyme are shown in Figure 10.87. Methotrexate (amethopterin), an analogue of folic acid, inhibits dihydrofolate reductase (DHFR), which catalyzes the synthesis of tetrahydrofolate, a metabolite essential to the formation of, for example, dTMP. N-(phosphonacetyl)-L-aspartate (PALA) inhibits aspartate transcarbamoylase, an enzyme that participates in pyrimidine nucleotide biosynthesis. When cells in culture are exposed to low concentrations of these inhibitors (e.g., 10^{-8} M methotrexate), a few resistant cells (one in 10^6 or 10^7) can be selected and grown. If these are then grown at an even higher concentration of inhibitor, again, only a few colonies form. Proceeding in this stepwise manner, cells that tolerate very high drug concentrations (e.g., 10^{-3} M methotrexate) are obtained. Many of the drug resistant cells overcome the inhibition by producing high levels of both the inhibited enzyme and the corresponding mRNA. The underlying cause of the increase in mRNA and enzyme is the amplification of the single gene for dihydrofolate reductase or aspartate transcarbamoylase.

Cells that are highly resistant to the drugs are a rich source from which mRNA can be purified and cloned as cDNA. In turn, the cloned cDNA provides a tool for precise measurement of mRNA concentration and gene copy number. An example of the stepwise selection of cells that are resistant to high levels of PALA is shown in Table 10.6. After each successive selection step, the cells are resistant to a higher concentration of PALA. This resistance is accompanied by increasing amounts of aspartate transcarbamoylase. Kinetic analysis using a cloned cDNA probe and either RNA or DNA isolated from the resistant cells (to drive the annealing of the probe) shows increasing numbers of copies of both the mRNA and the corresponding gene.

Is this phenomenon specific to these two genes, or is amplification a more general phenomenon? Table 10.7 lists several other systems in which

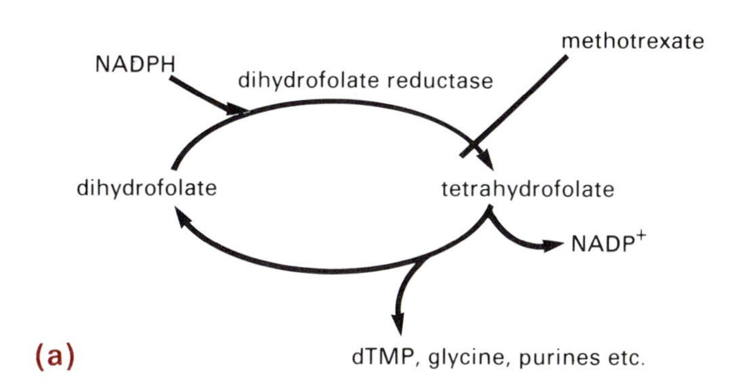

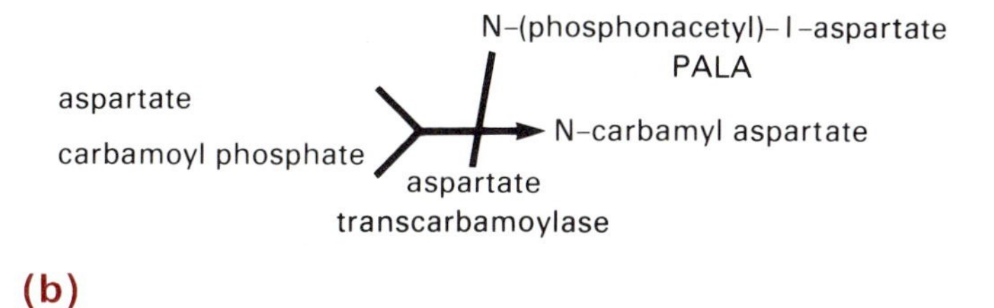

Figure 10.87
Inhibition of (a) dihydrofolate reductase by methotrexate and (b) aspartate transcarbamoylase by N-(phosphonacetyl)-L-aspartate (PALA).

Table 10.6 Development of Resistance to PALA in Syrian Hamster Cells

	PALA Conc.		*Relative Level of Aspartate Transcarbamoylase*			
Cell Line	Cells Selected in	For 50% Inhibition of Growth	mRNA	Enzyme	$Cot_{0.5}$*	Relative Gene Number
Wild type	0 mM	6 μM	1	1	2980	1
A	0.1 mM	1.6 mM	6.6	6.4	225	13
B	10 mM	17 mM	42	25	54	55
C	25 mM	34 mM	68	73	25	190

* $Cot_{0.5}$ for Syrian hamster DNA using a cloned cDNA of aspartate transcarbamoylase mRNA as probe (Section 7.5b). After G. M. Wahl, R. A. Padgett and G. R. Stark, *J. Biol. Chem.* 254 (1978), p. 8679.

cells with amplified genes have been selected. Although the number of examples is small, the result has been the same in every system tested. Note that, in one of the examples, selection of cells with amplified genes does not depend on the use of a drug. Altogether, the experiments suggest that the amplification of many different genomic regions occurs spontaneously, albeit at low frequencies (10^{-4} to 10^{-7}, depending on the system), in many cultured cells. Spontaneous amplification implies that selective pressure is simply a tool that allows these rare events to be observed.

Reversion of a mutant phenotype by gene amplification Analysis of secondary mutations that relieve a mutant phenotype provides additional evidence for the conclusion that random DNA amplifications can occur in eukaryotic cells. Hypoxanthine-guanine phosphoribosyl transferase (HPRTase) allows the salvage and reuse of purines in animal cells (Figure 5.32). Cells with a mutant form of the *hprt* gene that encodes a debilitated HPRTase do not grow in medium that contains hypoxanthine, aminopterin, and thymidine (HAT medium) because the aminopterin (a close relative of methotrexate) blocks the *de novo* synthesis of purine nucleotides

Table 10.7 Gene Amplifications that Correct Enzyme Deficiencies (selected examples)

Genes Encoding	Cells	Selective Tool
Dihydrofolate reductase	Mouse, Chinese hamster	Methotrexate (MTX) resistance
Aspartate transcarbamoylase	Syrian hamster	N-(phosphonacetyl)-L-aspartate (PALA) resistance
Metallothionein	Mouse, monkey	Cd^{2+} resistance
Asparagine synthetase	Chinese hamster	β-aspartylhydroxamate resistance
Hypoxanthine-guanine phosphoribosyl transferase	Mouse, Chinese hamster	Reversion of $hprt^-$ cells to $hprt^+$
Adenosine deaminase	Rat	Deoxycoformycin resistance
*ADH4**	Yeast	Antimycin A resistance

* The *ADH4* gene of yeast encodes either an isozyme of alcohol dehydrogenase or a protein regulating alcohol dehydrogenase gene expression.

(Figure 10.87), and the hypoxanthine cannot be efficiently utilized. Revertant cells that grow well in HAT medium frequently have amplified the mutant *hprt* gene, in one case, about 50-fold. They synthesize an increased level of mutant HPRTase mRNA and up to 50-fold more HPRTase protein than do wild type cells (Table 10.7). That the excess protein is the mutant form is indicated by its unusual kinetic properties and relative instability. To summarize, cloned cells with a single mutated *hprt* gene that encodes a variant and unstable enzyme yield revertants that overcome the enzyme deficiency by amplifying the mutant gene, thereby permitting synthesis of a large amount of the defective enzyme. What the enzyme lacks in quality is made up for by quantity.

b. The Structure of Amplified Genes

Changes in DNA sequence The amplified unit includes the entire gene as well as other DNA sequences. Amplified segments typically contain the regulatory sequences that flank the transcription unit on both the 5′ and 3′ borders, the coding region, and the introns, as well as extensive additional sequences. The repeated units are arranged in tandem, and some repeat units contain inverted duplications of the gene. Consider first aspartate transcarbamoylase. In animals, this enzymatic activity is part of a single polypeptide that has two additional enzymatic activities. The other two are carbamoylphosphate synthetase and dihydroorotase, and their actions precede and follow, respectively, aspartate transcarbamoylase activity in the pathway leading to pyrimidine nucleotide biosynthesis. The triple-headed enzyme, called CAD, is encoded by a single gene. In Syrian hamsters, the gene is 25 kbp long, and most of it is taken up by at least 37 intervening sequences; the mRNA is only 7.9 kb long and contains about 6300 coding bases. However, the amplified unit may be more than 500 kbp long. Similarly, the mouse dihydrofolate reductase gene is 31 kbp in length, contains at least five introns, and encodes a protein 186 amino acids long in an mRNA that is approximately 1.6 kb long. The length of the reiterated unit varies from one resistant cell line to another but can be more than 100 kbp.

The simplest explanation for the fact that the repeated units are so much longer than the genes themselves is that long genomic segments including the genes are amplified. However, the structure of the repeated units suggests that the process is much more complex. In some cases, the amplified units include sequences from unlinked genomic loci that do not normally neighbor the gene in question. In addition, individual repeat units are not always identical. Repeat units of different length and sequence content occur even in cloned cell lines. Moreover, rearrangements within the repeated units can occur continually during extended growth of the cells under conditions that maintain the selective pressure favoring amplification.

Changes in chromosome structure Often, but not always, gene amplification is accompanied by marked morphological changes in chromosomes. These are of several types, and which type is observed depends on what kind of cell and which gene is being studied. Sometimes, very

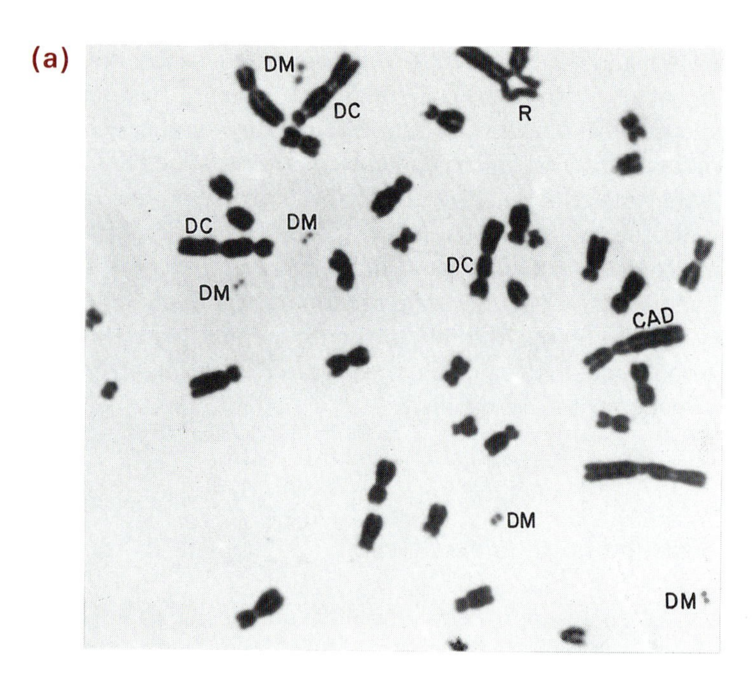

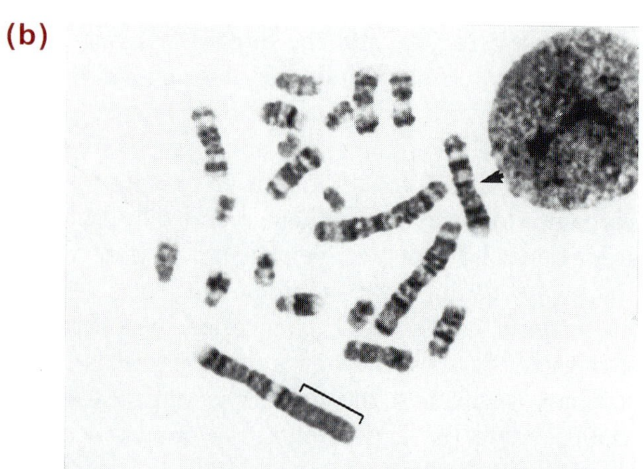

Figure 10.88
Chromosomal aberrations that are evident on metaphase spreads accompany gene amplification. (a) Double minute (DM) supernumerary chromosomes, dicentric chromosomes (DC), and amplified chromosomal regions (CAD) are found in PALA resistant Syrian hamster cells. From G. M. Wahl et al., *Mol. Cell. Biol.* 2 (1982), p. 308. (b) A long, homogeneously staining region (bracket) is found on one chromosome 2 in methotrexate-resistant Chinese hamster fibroblasts. The arrow points to the normal chromosome 2. Courtesy of J. L. Biedler. PAGE 817: (c) The top shows *in situ* hybridization of a CAD gene probe to a metaphase spread from PALA resistant Syrian hamster cells. The bottom shows a karyotype of the cells. The elongated chromosome B9 contains the site of CAD gene amplification. Some of the other chromosome pairs are also aberrant. From G. M. Wahl et al., *Mol. Cell. Biol.* 2 (1982), p. 308.

small supernumerary chromosomes called **double minutes** collect (Figure 10.88a), and these probably contain the amplified sequences. The double minutes are acentric and segregate unequally at cell division. Unequal segregation, together with the fact that cells containing substantial numbers of double minutes grow less well than normal cells, means that drug resistant cells tend to be lost in a nonselective (drug-free) medium.

A second kind of morphological change is the appearance on otherwise normal chromosomes of regions that fail to give typical banding patterns upon staining (Figure 7.23); they are referred to as **homogeneously staining regions**, or **HSRs**. Frequently, HSRs are long, and the chromosome carrying an HSR is thus longer than its homologue (Figure 10.88b). Clusters of silver grains appear over the HSRs of methotrexate resistant cells upon annealing *in situ* with radioactive probes containing the cloned DHFR sequences (and exposure of photographic emulsions). Thus, the amplified sequences are within the HSRs. In Syrian hamster cells containing amplified aspartate transcarbamoylase genes, extra long chromosome regions also occur. Because they show banding patterns, they are

(c)

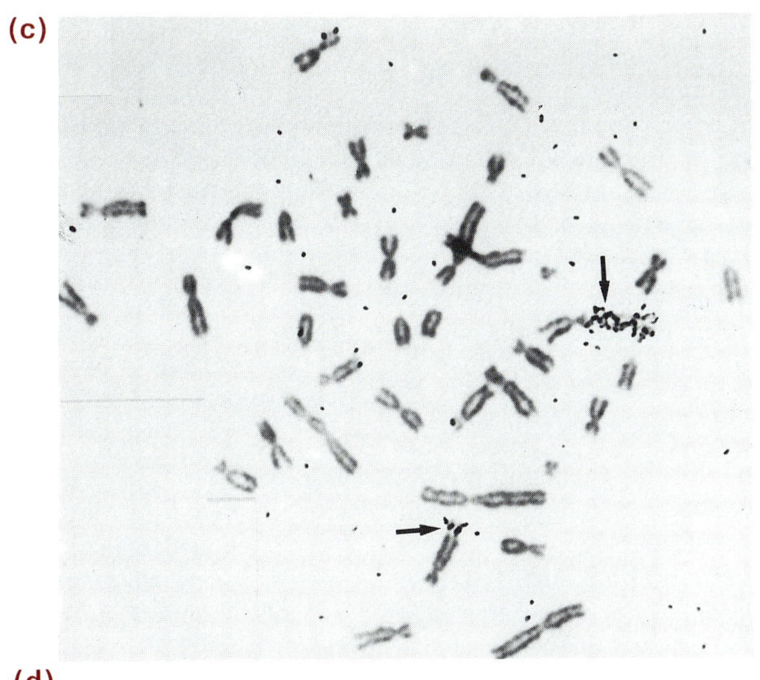

(d)

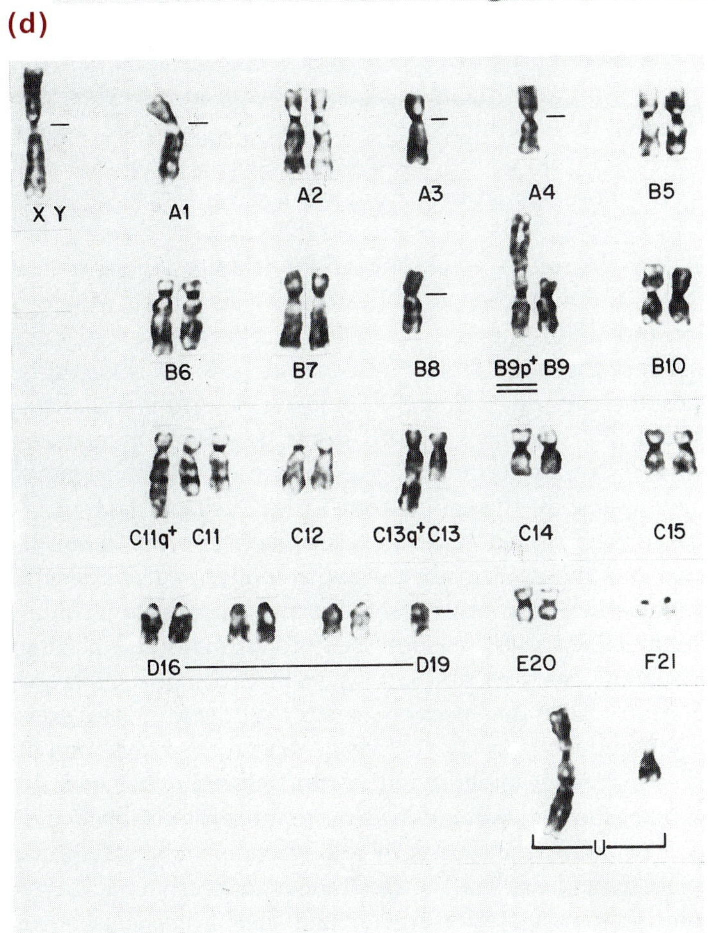
X Y
A1
A2
A3
A4
B5
B6
B7
B8
B9p+ B9
B10
C11q+ C11
C12
C13q+ C13
C14
C15
D16
D19
E20
F21
U

not HSRs, but they do nevertheless contain the amplified DNA (Figure 10.88c). In some instances, expansion of the *CAD* gene is known to occur at the normal Syrian hamster *CAD* locus (chromosome B9p), although other chromosomes may also have multiple gene copies as well as a variety of chromosomal abnormalities (Figure 10.88d). Double minutes can also occur. In each case examined, amplification is associated with only one of a pair of homologous chromosomes. In contrast with the instability of amplified genes on double minutes, amplified genes within otherwise normal chromosomes tend to be stable even after extended growth in nonselective medium.

c. *Amplification Mechanism*

Models for the amplification process must account for the following facts: (1) Expansion occurs at the normal locus of the gene, at distant loci, or in disconnected double minutes. (2) Genomic segments much longer than the gene itself are amplified. (3) Amplified units include rearrangements of the initial genomic sequences. (4) The amplified units are not all identical in structure and are changeable. The models should also explain the instability of some amplified segments upon removal of selective pressure, although the mechanism for deletion may not be a simple reversal of amplification. Moreover, the models need to consider whether the amplification process itself is influenced by the imposed selective pressure or whether the only role of the selective pressure is to allow maintenance and thus detection of ongoing rare genomic events.

Implications of the cytology During continued growth of methotrexate resistant cells in the presence of the drug, cells with a stable resistant phenotype and HSRs emerge from cell populations that start out with an unstable phenotype and double minutes. These observations suggest that, in some instances, double minute chromosomes form first and that extrachromosomal DNA is then inserted from the double minutes back into chromosomal DNA by recombination. Thereafter, additional amplification may occur. Support for this model comes from the fact that transformation of cells by foreign cloned genes (e.g., mouse cells with Chinese hamster *DHFR* gene or Chinese hamster cells with mouse *DHFR* gene), followed by the stepwise selection procedure, results in amplification of the foreign gene. The amplified unit includes the foreign gene and accompanying vector sequences, such as pBR322, as well as the host DNA segments that happen to flank the inserted foreign segments. Such experiments indicate that amplification of a native gene can involve a complex series of reactions including amplification *in situ* at the original gene locus followed by excision and formation of an extrachromosomal intermediate like a double minute, reinsertion into the genome, and additional amplification. This would explain the fact that amplification can occur at loci other than the normal site for a gene, and such a mechanism would also provide opportunities for linkage by recombination with otherwise distant genomic sequences. One or another of these processes may be more frequent with particular cell types or particular genes than others, thus accounting for the diverse manifestations of amplification in different systems. The complexities apparent from the cytological observations also imply diversity and complexity at the molecular level.

Molecular mechanisms Programmed amplifications utilize a diversity of molecular modes, and any or all of them may be involved in the random amplification of genes. Moreover, different mechanisms may operate at the various stages of stepwise amplifications. Both disproportionate replication (Section 10.7a) and rolling circle amplification (Section 10.7b) can yield long tandem arrays. The onion-skin model for disproportionate replication described in Figure 10.84 does not, by itself, yield linked tandem copies. It is possible, however, to imagine that opposite ends of the several copies of a DNA segment in the onion skin might be brought in close proximity and joined by recombination to form tandem arrays, including those with inverted duplications of the gene within the repeat units (Figure 10.89). The presence of members of an interspersed repeat family at opposite ends of the amplified unit could, for example, facilitate linkage. Alternatively, opposite ends of one new copy might join up to form an extrachromosomal circle that could then be amplified as a rolling circle. Note that these models predict that the amplified unit must contain an origin of replication. Support for this class of models comes from the identification of approximately 250 kbp extrachromosomal closed circular duplex DNA in one clone of PALA resistant hamster cells. These molecules, too small to be scored as double minutes, contain the *CAD* gene and replicate autonomously. They could represent intermediates between the original chromosomal genes and double minutes.

Successive unequal crossings-over between repeated sequences on sister chromatids or homologous chromosomes can also yield long tandem arrays (Sections 2.4a and 9.4e). This type of mechanism requires multiple rounds of cell division to generate arrays with many tandem copies and is thus consistent with the process of stepwise amplification. However, this model does not explain amplification at chromosomal loci that are distant from the position of the original gene.

One consequence of all these models is that a new DNA segment, one that is not present in the unamplified state, will be formed at the junction between tandem repeats. Contrast this with the *Drosophila* chorion gene amplification described in Section 10.7a in which the extra DNA copies are not covalently linked to the rest of the genome, and no new segment is formed. The formation of new junction sequences has been demonstrated in the *CAD* gene system (Figure 10.90). Note that the junction sequence itself is amplified together with the whole unit. Cell lines that are independently isolated after *CAD* amplification have different junction fragments, suggesting that there is no fixed position at which the repeated unit begins. Moreover, at least some amplified segments contain inverted repeats of the amplified gene. In addition, as noted previously, the structures of amplified units differ even within a cloned cell line. Thus, several different junction segments may be observed; the diagram in Figure 10.90 is therefore simplified and schematic.

Instability of amplified genes The loss of amplified genes upon removal of selective pressure suggests that the tandem arrays are dynamic and subject to continual deletion and reexpansion. Such fluidity was demonstrated directly using specially constructed transformed cells. Mouse cells carrying mutations in both thymidine kinase (tk^-) and adenine phosphoribosyl transferase ($aprt^-$) genes were transfected with a recombinant

Figure 10.89
Adapting the onion-skin model to produce long tandem arrays or a rolling circle. After G. Stark and G. M. Wahl, *Ann. Rev. Biochem.* 53 (1984), p. 478.

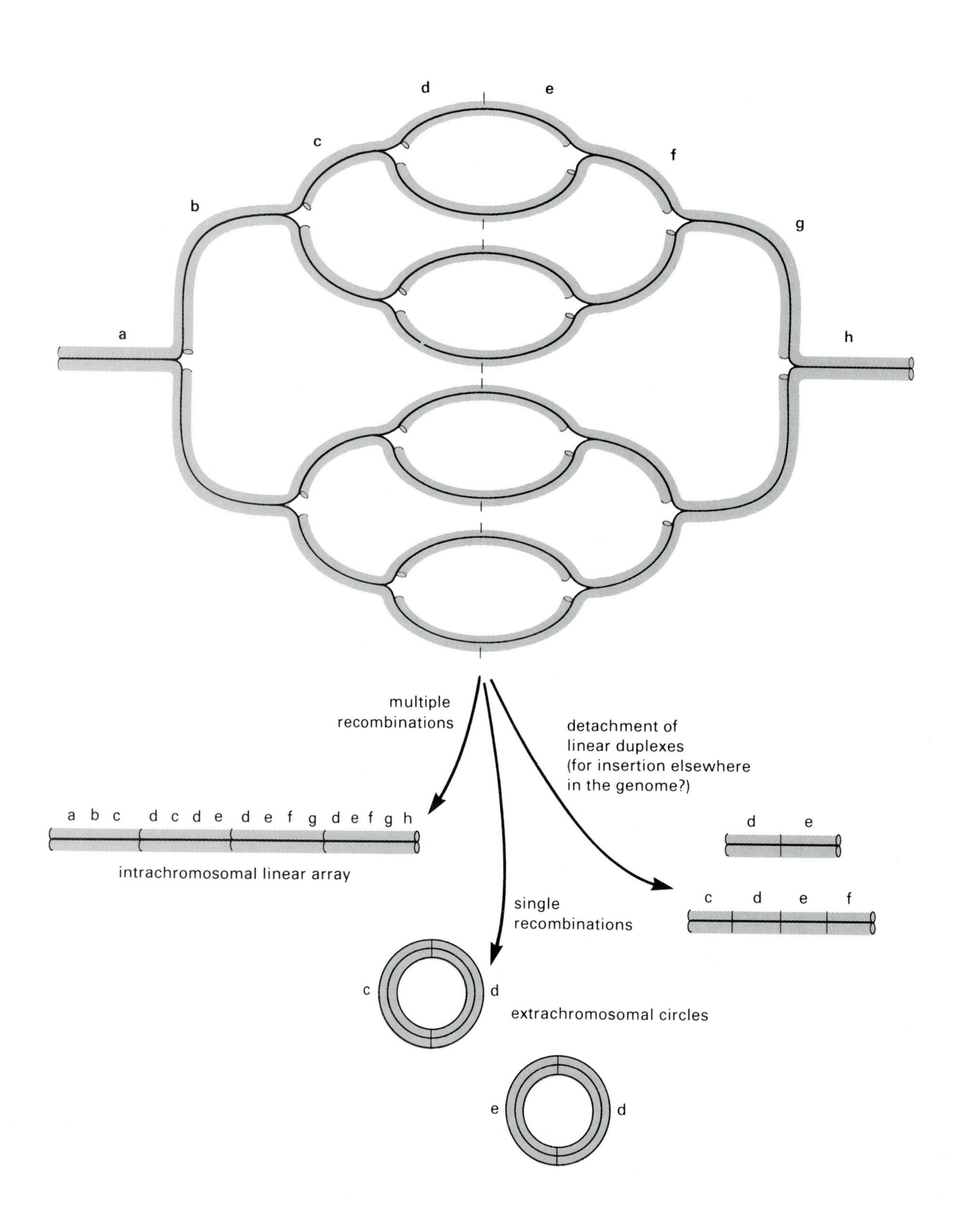

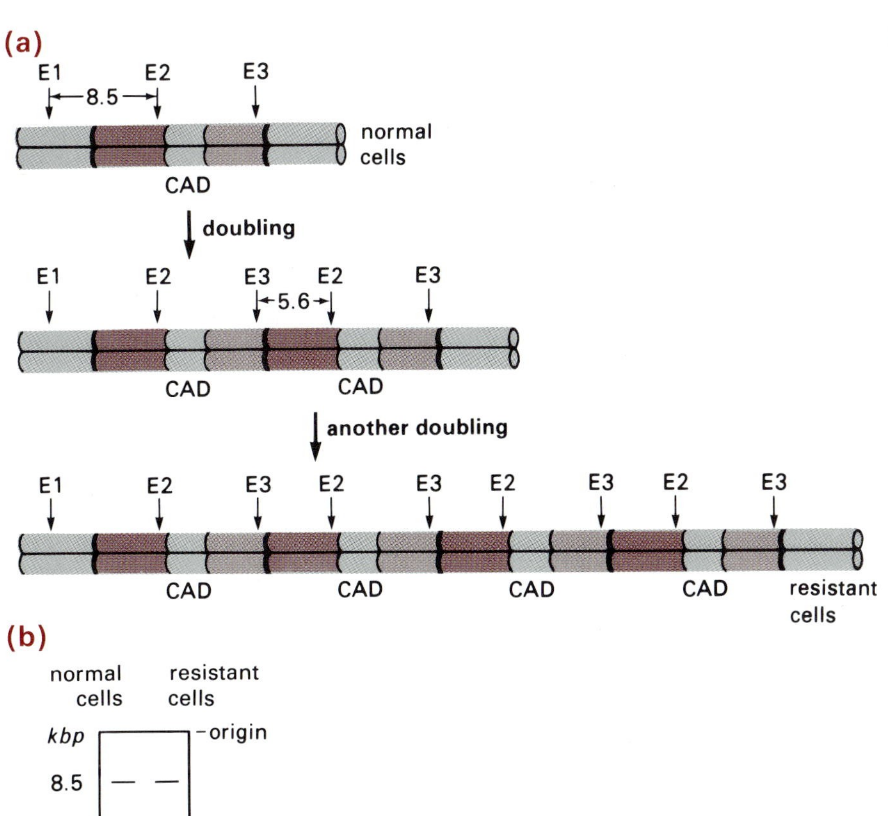

Figure 10.90
The generation of tandem repeats of the CAD gene results in the formation and amplification of a new junction segment. (a) Schematic diagram of amplification. A segment from the stippled region was subcloned to use as a probe. (b) The annealing of the probe to DNA blots containing endonuclease *Eco*RI digests of DNA from normal cells and from cell line C (Table 10.6). The normal DNA has one fragment (8.5 kbp) that anneals to the probe and is present at the level of one copy per haploid genome. The 8.5 kbp band is present at the same level in DNA from cell line C. However, cell line C also produces a band at the position of the junction fragment (5.6 kbp); and its high abundance, as revealed by the intensity of the autoradiographic response, is consistent with the 190-fold amplification of the CAD gene.

vector containing both a *tk* gene from Herpes simplex virus and an *aprt* gene from hamsters. Cell lines that were both tk^+ and $aprt^+$ and contained tandem amplifications of a DNA segment that includes both genes were obtained by growth in a suitable selection medium (Figure 10.91). When selective pressure for amplified *tk* is maintained but the cells are placed in growth medium that selects for an $aprt^-$ phenotype, $aprt^-$ mutants appear at about the same frequency as they do in normal cell populations that have a single wild type *aprt* gene. This is quite surprising because it suggests that all the amplified copies of $aprt^+$, 20 in number in these cells, are mutated simultaneously. Indeed, that is exactly what happens. In some cloned $aprt^-$, tk^+ cells, all the *aprt* genes are deleted, thereby accounting for the $aprt^-$ phenotype. In others, all 20 copies of the $aprt^-$ gene have the same mutation. The particular mutation varies from one cell line to another, but the overall picture is always the same; all 20 *aprt* genes are changed in the same way. Similarly, in those instances where the $aprt^-$ phenotype is due to deletion, all the deletions are in exactly the same position in every copy in the tandem array. Because identical random mutations in all copies are highly unlikely, an alteration must have occurred in one copy, and then all the other copies must have been replaced by that one. Some mechanism operates to maintain identity

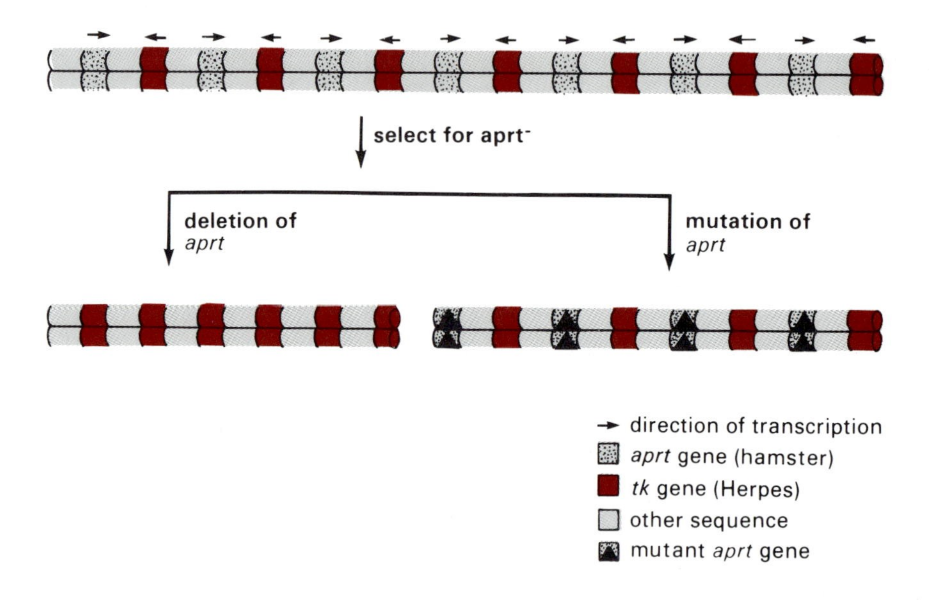

Figure 10.91
Changes in an amplified tandem array spread throughout the array.

among the multiple copies in a tandem array. Just as it is presently impossible to choose among the various possible mechanisms for spontaneous amplification itself, no specific mechanism, can be associated with the tendency to homogenize the members of a tandem array. In addition to unequal crossing-over, which can contract and reexpand tandem arrays, cycles of deletion and reamplification through, for example, disproportionate replication and gene conversion are possible mechanistic explanations.

Stimulation of amplification by chemical agents Current information is consistent with infrequent, spontaneous, and random amplification of many genomic segments. Estimates of spontaneous frequency in cultured cells (in the absence of selective pressure) range, in the best cases, from 10^{-3} to 10^{-4} times per cell per generation for each amplifiable unit. Note that, in a genome of about 10^9 bp in which any segment can amplify, and assuming an average amplified unit of 10^5 bp (and thus 10^4 such units in the genome), most newly divided cells would contain one new amplification. Moreover, particular conditions designed to select for amplification, like the stepwise increase in methotrexate concentration that yields amplified dihydrofolate reductase genes, may themselves influence the frequency of amplification. Some compounds, including inhibitors of DNA replication and many carcinogens, are known to increase the amplification frequency of random genomic segments. Homogeneous staining regions and double minute chromosomes are frequently observed in tumor cells that have been either induced by carcinogens or have occurred spontaneously. These and other chromosomal aberrations in tumors are associated with the amplification of DNA sequences, including those encoding known oncogenes.

It may be that a variety of agents that interfere with orderly DNA replication foster extra, unprogrammed replication in random genomic regions. Under appropriate selective conditions, cells with extra copies of the critical gene will be at an advantage and will outgrow normal sister cells. After several rounds of cell division, especially if increasingly stringent selective pressure is applied, massive gene amplification can result.

References for Part III

PERSPECTIVE

The following articles provide introductions to or overviews of the concepts emphasized in this Perspective essay. The references are grouped according to the subheadings in the essay. Within each section, references are given chronologically.

THE STRUCTURE AND EXPRESSION OF EUKARYOTIC GENES

P. Chambon. 1981. Split Genes. *Sci. American* 244(5) 60–71.

W. Gilbert. 1981. DNA Sequencing and Gene Structure. Nobel Lecture, December 8, 1980. *Science* 214 1305–1312.

A. B. Lassar, P. L. Martin, and R. G. Roeder. 1983. Transcription of Class III Genes: Formation of Preinitiation Complexes. *Science* 222 740–748.

E. J. Milner-White. 1984. Isozymes, Isoproteins and Introns. *Trends Biochem. Sci.* 9 517–519.

E. Serfling, M. Jasin, and W. Schaffner. 1985. Enhancers and Eukaryotic Gene Transcription. *Trends Genet.* 1 224–230.

W. Gilbert, M. Marchionni, and G. McKnight. 1986. On the Antiquity of Introns. *Cell* 46 151–154.

G. Dreyfuss, M. S. Swanson, and S. Pinol-Roma. 1988. Heterogeneous Nuclear Ribonucleoprotein Particles and the Pathway of mRNA Formation. *Trends Biochem. Sci.* 13 86–89.

E. P. Geiduschek and G. P. Tocchini-Valentini. 1988. Transcription by RNA Polymerase III. *Annu. Rev. Biochem.* 57 873–914.

T. W. Traut. 1988. Do Exons Code for Structural or Functional Units in Proteins? *Proc. Natl. Acad. Sci.* U.S.A. 85 2944–2948.

B. Wasylyk. 1988. Transcription Elements and Factors of RNA Polymerase B Promoters of Higher Eukaryotes. *Crit. Rev. Biochem.* 23 77–120.

J. A. Witkowski. 1988. The Discovery of Split Genes. *Trends Biochem. Sci.* 13 110–113.

M. Belfort. 1989. Bacteriophage Introns: Parasites Within Parasites. *Trends Genet.* 5 209–213.

P. N. Benfey and N. Chua. 1989. Regulated Genes in Transgenic Plants. *Science* 244 174–181.

R. G. Brennan and B. W. Matthews. 1989. Structural Basis of DNA-Protein Recognition. *Trends Biochem. Sci.* 14 286–290.

W. S. Dynan. 1989. Modularity in Promoters and Enhancers. *Cell* 58 1–4.

R. B. Goldberg, S. J. Barker, and L. Perez-Grau. 1989. Regulation of Gene Expression During Plant Embryogenesis. *Cell* 56 149–160.

N. D. Hastie and R. C. Allshire. 1989. Human Telomeres: Fusion and Interstitial Sites. *Trends Genet.* 5 326–331.

S. Murphy, B. Moorefield, and T. Pieler. 1989. Common Mechanisms of Promoter Recognition by RNA Polymerases II and III. *Trends Genet.* 5 123–126.

M. Ptashne. 1989. How Gene Activators Work. *Sci. American* 260(1) 40–47.

J. H. Rogers. 1989. How Were Introns Inserted into Nuclear *Genes? Trends Genet.* 5 213–216.

THE STRUCTURE OF EUKARYOTIC GENOMES

E. O. Long and I. B. Dawid. 1980. Repeated Genes in Eukaryotes. *Annu. Rev. Biochem.* 49 727–764.

J. G. Gall. 1981. Chromosome Structure and the C-Value Paradox. *J. Cell Biol.* 91 3s–14s.

A. C. Spradling and G. M. Rubin. 1981. *Drosophila* Genome Organization: Conserved and Dynamic Aspects. *Annu. Rev. Genet.* 15 219–264.

D. Baltimore. 1985. Retroviruses and Retrotransposons: The Role of Reverse Transcriptase in Shaping the Eukaryotic Genome. *Cell* 40 481–482.

E. F. Vanin. 1985. Processed Pseudogenes: Characteristics and Evolution. *Annu. Rev. Genet.* 19 253–272.

J. H. Nadeau. 1989. Maps of Linkage and Synteny Homologies Between Mouse and Man. *Trends Genet.* 5 82–86.

THE REARRANGEMENT OF DNA SEQUENCES

B. McClintock. 1951. Chromosome Organization and Gene Expression. *Cold Spring Harbor Symp. Quant. Biol.* 16 13–57.

J. Lederberg and T. Iino. 1956. Phase Variation in Salmonella. *Genetics* 41 743–757.

M. P. Calos and J. H. Miller. 1980. Transposable Elements. *Cell* 20 579–595.

R. T. Schimke. 1980. Gene Amplification and Drug Resistance. *Sci. American* 243(5) 60–69.

P. Leder. 1982. The Genetics of Antibody Diversity. *Sci. American* 246(5) 102–115.

N. V. Federoff. 1984. Transposable Genetic Elements in Maize. *Sci. American* 250(6) 84–98.

D. Baltimore. 1985. Retroviruses and Retrotransposons: The Role of Reverse Transcriptase in Shaping the Eukeryotic Genome. *Cell* 40 481–482.

S. Tonegawa. 1985. The Molecules of the Immune System. *Sci. American* 253(4) 122–131.

P. Borst and D. R. Greaves. 1987. Programmed Gene Rearrangements Altering Gene Expression. *Science* 235 658–667.

H. Varmus. 1987. Reverse Transcription. *Sci. American* 257(3) 56–64.

D. J. Finnegan. 1989. Eukaryotic Transposable Elements and Genome Evolution. *Trends Genet.* 5 103–107.

A. M. Lambowitz. 1989. Infectious Introns. *Cell* 56 323–326.

H. E. Varmus. 1989. Reverse Transcription in Bacteria. *Cell* 56 721–724.

EVOLUTIONARY IMPLICATIONS

F. Ayala. 1978. The Mechanisms of Evolution. *Sci. American* 239(3) 56–69.

J. E. Darnell, Jr. 1978. Implications of RNA-RNA Splicing in Evolution of Eukaryotic Cells. *Science* 202 1257–1260.

W. Gilbert. 1978. Why Genes in Pieces? *Nature* 271 501.

M. Kimura. 1979. The Neutral Theory of Molecular Evolution. *Sci. American* 241(5) 98–126.

D. Reanney. 1984. Genetic Noise in Evolution? *Nature* 307 318–319.

J. E. Darnell, Jr. 1985. RNA. *Sci. American* 253(4) 68–78.

R. F. Doolittle. 1985. The Geneology of Some Recently Evolved Vertebrate Proteins. *Trends Biochem. Sci.* 10 233–237.

R. F. Doolittle. 1985. Proteins. *Sci. American* 253(4) 88–99.

W. Gilbert. 1985. Genes in Pieces Revisited. *Science* 228 823–824.

A. C. Wilson. 1985. The Molecular Basis of Evolution. *Sci. American* 253(4) 164–173.

T. R. Cech. 1986. Biological Catalysis by RNA. *Annu. Rev. Biochem.* 55 599–629.

J. E. Darnell and W. F. Doolittle. 1986. Speculations on the Early Course of Evolution. *Proc. Natl. Acad. Sci.* U.S.A. 83 1271–1275.

G. A. Dover. 1986. Molecular Drive in Multigene Families: How Biological Novelties Arise, Spread, and Are Assimilated. *Trends Genet.* 2 159–165.

N. R. Pace, G. J. Olsen, and C. R. Woese. 1986. Ribosomal RNA Phylogeny and the Primary Lines of Evolutionary Descent. *Cell* 45 325–326.

W. Gilbert. 1987. The RNA World. *Cold Spring Harbor Symp. Quant. Biol.* 52 901–905.

A. M. Weiner and N. Maizels. 1987. tRNA-like Structures Tag the Ends of Genomic RNA Molecules for Replication: Implications for the Origin of Protein Synthesis. *Proc. Natl. Acad. Sci.* U.S.A. 84 7383–7387.

U. Wintersberger and E. Wintersberger. 1987. RNA Makes DNA: A Speculative View of the Evolution of DNA Replication Mechanisms. *Trends Genet.* 3 198–202

L. L. Cavalli-Sforza, A. Piazza, P. Menozzi, and J. Mountain. 1988. Reconstruction of Human Evolution: Bringing Together Genetic, Archeological, and Linguistic Data. *Proc. Natl. Acad. Sci.* U.S.A. 85 6002–6006.

C. R. Woese. 1987. Bacterial Evolution. *Microbiol. Rev.* 51 221–271.

K. G. Field, G. J. Olsen, D. J. Lane, S. J. Giovannoni, M. T. Ghiselin, E. C. Raff, N. R. Pace, and R. A. Raff. 1988. Molecular Phylogeny of the Animal Kingdom. *Science* 239 748–753.

R. F. Doolittle. 1989. Similar Amino Acids Revisited. *Trends Biochem. Sci.* 14 244–245.

G. F. Joyce. 1989. RNA Evolution and the Origins of Life. *Nature* 338 217–224.

S. Pääbo, R. G. Higuchi, and A. C. Wilson. 1989. Ancient DNA and the Polymerase Chain Reaction. The Emerging Field of Molecular Archaeology. *J. Biol. Chem.* 264 9709–9712.

General Reading for Chapters in Part III

J. Darnell, H. Lodish, and D. Baltimore. 1986. Molecular Cell Biology. Scientific American Books, New York.

The Molecular Biology of *Homo sapiens*. 1986. *Cold Spring Harbor Symp. Quant. Biol.* 51 whole issue.

D. Friefelder. 1987. Molecular Biology, 2nd ed. Jones and Bartlett, Boston.

J. D. Watson, N. H. Hopkins, J. W. Roberts, J. A. Steitz, and A. M. Weiner. 1987. Molecular Biology of the Gene, 4th ed. Benjamin/Cummings, Menlo Park, Ca.

B. John and G. Miklos. 1988. The Eukaryote Genome in Development and Evolution. Allen and Unwin, London.

L. Stryer. 1988. Biochemistry, 3rd ed. W. H. Freeman, San Francisco.

Evolution of Catalysis. 1988. *Cold Spring Harbor Symp. Quant. Biol.* 52 whole issue.

B. Alberts, D. Bray, J. Lewis, M. Raff, K. Roberts, and J. D. Watson. 1989. Molecular Biology of the Cell, 2nd ed. Garland Publishing, New York and London.

D. E. Berg and M. M. Howe (eds). 1989. Mobile DNA. American Society for Microbiology, Washington, D.C.

B. Lewin. 1990. Genes, 4th ed. Wiley, New York.

J. D. Rawn. 1988. Biochemistry. Carolina Biological Supply Co., Burlington, North Carolina.

The following books and articles cover subjects in indicated chapter sections.

8.2a

R. H. Reeder. 1985. Mechanisms of Nucleolar Dominance in Animals and Plants. *J. Cell Biol.* 101 2013–2016.

B. Sollner-Webb and J. Tower. 1986. Transcription of Cloned Eukaryotic Ribosomal RNA Genes. *Annu. Rev. Biochem.* 55 801–830.

R. G. Worton, J. Sutherland, J. E. Sylvester, H. F. Willard, S. Bodrug, I. Dube, C. Duff, V. Kean, P. N. Ray, and R. D. Schmickel. 1988. Human Ribosomal RNA Genes: Orientation of the Tandem Array and Conservation of the 5′ End. *Science* 239 64–67.

8.2b

P. Chambon. 1974. Eucaryotic RNA Polymerases. In P. D. Boyer (ed.), The Enzymes, pp. 261–331. Academic Press, New York.

L. A. Allison, M. Moyle, M. Shales, and C. J. Ingles. 1985. Extensive Homology Among the Largest Subunits of Eukaryotic and Prokaryotic RNA Polymerases. *Cell* 42 599–610.

J. Biggs, L. L. Searles, and A. L. Greenleaf. 1985. Structure of the Eukaryotic Transcription Apparatus: Features of the Gene for the Largest Subunit of *Drosophila* RNA Polymerase II. *Cell* 42 611–621.

A. Sentenac. 1985. Eukaryotic RNA Polymerases. *Crit. Rev. Biochem.* 18 31–90.

E. Bateman and M. R. Paule. 1986. Regulation of Eukaryotic Ribosomal RNA Transcription by RNA Polymerase Modification. *Cell* 47 445–450.

P. Bucher and E. N. Trifonov. 1986. Compilation and Analysis of Eukaryotic POL II Promoter Sequences. *Nucleic Acids Res.* 14 10009–10026.

J. L. Buhler, M. Riva, C. Mann, P. Thuriaux, S. Menet, J. Y. Micovin, I. Threich, S. Mariotte, and A. Sentenac. 1987. Eukaryotic RNA Polymerases, Subunits and Genes. In W. S. Reznikoff et al. (eds.), RNA Polymerase and the Regulation of Transcription, pp. 25–36. Elsevier, Amsterdam.

D. L. Cadena and M. E. Dahmus. 1987. Messenger RNA Synthesis in Mammalian Cells Is Catalyzed by the Phosphorylated Form of RNA Polymerase II. *J. Biol. Chem.* 262 12468–12474.

M. Nonet, D. Sweetser, and R. A. Young. 1987. Functional Redundancy and Structural Polymorphism in the Large Subunit of RNA Polymerase II. *Cell* 50 909–915.

J. Tower and B. Sollner-Webb. 1987. Transcription of Mouse rDNA Is Regulated by an Activated Subform of RNA Polymerase I. *Cell* 50 873–883.

8.2c

E. A. Elion and J. R. Warner. 1984. The Major Promoter Element of rRNA Transcription in Yeast Lies 2 kb Upstream. *Cell* 39 663–673.

R. Miesfeld and N. Arnheim. 1984. Species-Specific rDNA Transcription Is Due to Promoter-Specific Binding Factors. *Mol. Cell. Biol.* 4 221–227.

R. Miesfeld, B. Sollner-Webb, C. Croce, and N. Arnheim. 1984. The Absence of a Human-Specific Ribosomal DNA Transcription Factor Leads to Nucleolar Dominance in Mouse > Human Hybrid Cells. *Mol. Cell. Biol.* 4 1306–1312.

R. H. Reeder. 1984. Enhancers and Ribosomal Gene Spacers. *Cell* 38 349–351.

I. Grummt, H. Sorbaz, A. Hofmann, and E. Roth. 1985. Spacer Sequences Downstream of the 28S RNA Coding Region Are Part of the Mouse rDNA Transcription Unit. *Nucleic Acids Res.* 13 2293–2304.

R. M. Learned, S. Cordes, and R. Tjian. 1985. Purification and Characterization of a Transcription Factor That Confers Promoter Specificity to Human RNA Polymerase I. *Mol. Cell. Biol.* 5 1358–1369.

K. G. Miller, J. Tower, and B. Sollner-Webb. 1985. A Complex Control Region of the Mouse rRNA Gene Directs Accurate Initiation by RNA Polymerase I. *Mol. Cell. Biol.* 5 554–562.

V. L. Murtif and P. M. M. Rae. 1985. *In Vivo* Transcription of rDNA Spacers in *Drosophila*. *Nucleic Acids Res.* 13 3221–3224.

M. E. Swanson, M. Yip, and M. J. Holland. 1985. Characterization of an RNA Polymerase I-Dependent Promoter Within the Spacer Region of Yeast Ribosomal Cistrons. *J. Biol. Chem.* 260 9905–9915.

R. F. J. DeWinter and T. Moss. 1986. Spacer Promoters Are Essential for Efficient Enhancement of *X. laevis* Ribosomal Transcription. *Cell* 44 313–318.

P. Labhart and R. H. Reeder. 1986. Characterization of Three Sites of RNA 3′ End Formation in the *Xenopus* Ribosomal Gene Spacer. *Cell* 45 431–443.

R. M. Learned, T. K. Learned, M. M. Haltiner, and R. T. Tjian. 1986. Human RDNA Transcription Is Modulated by the Coordinate Binding of Two Factors to an Upstream Control Element. *Cell* 45 847–857.

J. Windle and B. Sollner-Webb. 1986. Upstream Domains of the *Xenopus laevis* rDNA Promoter Are Revealed in Microinjected Oocytes. *Mol. Cell. Biol.* 6 1228–1234.

P. Kownin, E. Bateman, and M. R. Paule. 1987. Eukaryotic RNA Polymerase I Promoter Binding Is Directed by Protein Contacts with Transcription Initiation Factor and Is DNA Sequence-Independent. *Cell* 50 693–699.

E. Bateman and M. R. Paule. 1988. Promoter Occlusion During Ribosomal RNA Transcription. *Cell* 54 985–992.

S. P. Bell, R. M. Learned, H. M. Jantzen, and R. Tjian. 1988. Functional Cooperativity Between Transcription Factors UBF1 and SL1 Mediates Human Ribosomal RNA Synthesis. *Science* 241 1192–1197.

M. H. Jones, R. M. Learned, and R. Tjian. 1988. Analysis of Clustered Point Mutations in the Human Ribosomal RNA Gene Promoter by Transient Expression *in Vivo*. *Proc. Natl. Acad. Sci.* U.S.A. 85 669–673.

8.2d

I. Grummt, U. Maier, A. Ohrlein, N. Hassouna, and J. P. Bachellerie. 1985. Transcription of Mouse rDNA Terminates Downstream of the 3′ End of 28s RNA and Involves Interaction of Factors with Repeated Sequences in the 3′ Spacer. *Cell* 43 801–810.

S. M. Baker and T. Platt. 1986. Pol I Transcription: Which Comes First, the End or the Beginning? *Cell* 47 839–840.

I. Grummt, H. Rosenbauer, I. Niedermeyer, U. Maier, and A. Ohrlein. 1986. A Repeated 18 bp Sequence Motif in the Mouse rDNA Spacer Mediates Binding of a Nuclear Factor and Transcription Termination. *Cell* 45 837–846.

P. Labhart and R. H. Reeder. 1986. Characterization of Three Sites of RNA 3′ End Formation in the *Xenopus* Ribosomal Gene Spacer. *Cell* 45 431–443.

I. Bartsch, C. Schoneberg, and I. Grummt. 1987. Evolutionary Changes of Sequences and Factors That Direct Transcription Termination of Human and Mouse Ribosomal Genes. *Mol. Cell. Biol.* 7 2521–2529.

S. L. Henderson, K. Ryan, and B. Sollner-Webb. 1989. The Promoter-Proximal rDNA Terminator Augments Initiation by Preventing Disruption of the Stable Transcription Complex Caused by Polymerase Read-In. *Genes and Devel.* 3 212–223.

A. Kuhn and I. Grummt. 1989. 3′-End Formation of Mouse Pre-rRNA Involves Both Transcription Termination and a Specific Processing Reaction. *Genes and Devel.* 3 224–231.

8.2e

I. Grummt, A. Kuhn, I. Bartsch, and H. Rosenbauer. 1986. A Transcription Terminator Located Upstream of the Mouse rDNA Initiation Site Affects rRNA Synthesis. *Cell* 47 901–911.

S. Henderson and B. Sollner-Webb. 1986. A Transcriptional Terminator Is a Novel Element of the Promoter of the Mouse Ribosomal RNA Gene. *Cell* 47 891–900.

B. McStay and R. H. Reeder. 1986. A Termination Site for *Xenopus* RNA Polymerase I Also Acts as an Element of an Adjacent Promoter. *Cell* 47 913–920.

A. Kuhn, A. Normann, I. Bartsch, and I. Grummt. 1988. The Mouse Ribosomal Gene Terminator Consists of Three Functionally Separable Sequence Elements. *EMBO J.* 7 1497–1502.

8.3c

K. S. Zaret and F. Sherman. 1982. DNA Sequence Required for Efficient Transcription Termination in Yeast. *Cell* 28 563–573.

C. Montell, E. F. Fisher, M. H. Caruthers, and A. J. Berk. 1983. Inhibition of RNA Cleavage but Not Polyadenylation by a Point Mutation in mRNA 3′ Consensus Sequence AAUAAA. *Nature* 305 600–605.

A. Gil and N. J. Proudfoot. 1984. A Sequence Downstream of AAUAAA Is Required for Rabbit β-Globin mRNA 3′-End Formation. *Nature* 312 473–474.

S. Henikoff and E. H. Cohen. 1984. Sequences Responsible for Transcription Termination on a Gene Segment in *Saccharomyces cerevisiea*. *Mol. Cell. Biol.* 4 1515–1520.

I. Pettersson, M. Hinterberger, T. Mimori, E. Gottlieb, and J. A. Steitz. 1984. The Structure of Mammalian Small Nuclear Ribonucleoproteins. *J. Biol. Chem.* 259 5907–5914.

B. M. Bhat and W. S. M. Wold. 1985. AATAAA as Well as Downstream Sequences Are Required for RNA 3′-End Formation in the E3 Complex Transcription Unit of Adenovirus. *Mol. Cell. Biol.* 5 3183–3193.

M. L. Birnstiel, M. Busslinger, and K. Strub. 1985. Transcription Termination and 3′ Processing: The End Is in Site! *Cell* 41 349–359.

L. Conway and M. Wickens. 1985. A Sequence Downstream of A-A-U-A-A-A Is Required for Formation of Simian Virus 40 Late mRNA 3′ Termini in Frog Oocytes. *Proc. Natl. Acad. Sci.* U.S.A. 82 3949–3953.

R. P. Hart, M. A. McDevitt, H. Ali, and J. R. Nevins. 1985. Definition of Essential Sequences and Functional Equivalence of Elements Downstream of the Adenovirus E2A and the Early Simian Virus 40 Polyadenylation Sites. *Mol. Cell. Biol.* 5 2975–2983.

J. L. Manley, H. Yu, and L. Ryner. 1985. RNA Sequence Containing Hexanucleotide AAUAAA Directs Efficient mRNA Polyadenylation *in Vitro*. *Mol. Cell. Biol.* 5 373–379.

J. McLauchlan, D. Gaffney, J. L. Whitton, and J. B. Clements. 1985. The Consensus Sequence YGTGTTYY Located Downstream from the AATAAA Signal Is Required for Efficient Formation of mRNA 3′ Termini. *Nucleic Acids Res.* 13 1347–1368.

C. L. Moore and P. A. Sharp. 1985. Accurate Cleavage and Polyadenylation of Exogenous RNA Substrate. *Cell* 41 845–855.

K. H. Baek, K. Sato, R. Ito, and K. Agarwal. 1986. RNA Polymerase II Transcription Terminates at a Specific DNA Sequence in a HeLa Cell-Free Reaction. *Proc. Natl. Acad. Sci.* U.S.A. 83 7623–7627.

G. Ciliberto, N. Dathan, R. Frank, L. Philipson, and I. W. Mattaj. 1986. Formation of the 3′ End on U snRNAs Requires at Least Three Sequence Elements. *EMBO J.* 5 2931–2937.

C. Hashimoto and J. A. Steitz. 1986. A Small Nuclear Ribonucleoprotein Associates with the AAUAAA Polyadenylation Signal *in Vitro*. *Cell* 45 581–591.

N. Hernandez and A. M. Weiner. 1986. Formation of the 3′ End of U1 snRNA Requires Compatible snRNA Promoter Elements. *Cell* 47 249–258.

M. A. McDevitt, R. P. Hart, W. W. Wong, and J. R. Nevins. 1986. Sequences Capable of Restoring Poly(A) Site Function Define Two Distinct Downstream Elements. *EMBO J.* 5 2907–2913.

T. Platt. 1986. Transcription Termination and the Regulation of Gene Expression. *Annu. Rev. Biochem.* 55 339–372.

K. Sato, R. Ito, K. H. Baek, and K. Agarwal. 1986. A Specific DNA Sequence Controls Termination of Transcription in the Gastrin Gene. *Mol. Cell. Biol.* 6 1032–1043.

H. E. N. de Vegvar, E. Lund, and J. E. Dahlberg. 1986. 3′ End Formation of U1 snRNA Precursors Is Coupled to Transcription from snRNA Promoters. *Cell* 47 259–266.

E. Whitelaw and N. Proudfoot. 1986. α-Thalassaemia Caused by a Poly(A) Site Mutation Reveals That Transcriptional Termination Is Linked to 3′ End Processing in the Human α2 Globin Gene. *EMBO J.* 5 2915–2922.

R. A. Ach and A. M. Weiner. 1987. The Highly Conserved U Small Nuclear RNA 3′-End Formation Signal Is Quite Tolerant to Mutation. *Mol. Cell. Biol.* 7 2070–2079.

R. L. Dedrick, C. M. Kane, and M. J. Chamberlin. 1987. Purified RNA Polymerase II Recognizes Specific Termination Sites During Transcription *in Vitro*. *J. Biol. Chem.* 262 9098–9108.

A. Gil and N. J. Proudfoot. 1987. Position-Dependent Sequence Elements Downstream of AAUAAA Are Required for Efficient Rabbit β-Globin mRNA 3′ End Formation. *Cell* 49 399–406.

J. Logan, E. Falck-Pedersen, J. E. Darnell, Jr., and T. Shenk. 1987. A Poly(A) Addition Site and a Downstream Termination Region Are Required for Efficient Cessation of Transcription by RNA Polymerase II in the Mouse β-Globin Gene. *Proc. Natl. Acad. Sci. U.S.A.* 84 8306–8310.

K. L. Mowry and J. A. Steitz. 1987. Both Conserved Signals on Mammalian Histone Pre-mRNAs Associate with Small Nuclear Ribonucleoproteins During 3′ End Formation *in Vitro*. *Mol. Cell. Biol.* 7 1663–1672.

K. L. Mowry and J. A. Steitz. 1987. Identification of the Human U7 snRNP as One of Several Factors Involved in the 3′ End Maturation of Histone Premessenger RNA's. *Science* 38 1682–1687.

D. Reines, D. Wells, M. J. Chamberlin, and C. M. Kane. 1987. Identification of Intrinsic Termination Sites *in Vitro* for RNA Polymerase II Within Eukaryotic Gene Sequences. *J. Mol. Biol.* 196 299–312.

J. S. Butler and T. Platt. 1988. RNA Processing Generates the Mature 3′ End of Yeast CYC1 Messenger RNA *in Vitro*. *Science* 242 1270–1274.

G. Christofori and W. Keller. 1988. 3′ Cleavage and Polyadenylation of mRNA Precursors *in Vitro* Requires a Poly(A) Polymerase, a Cleavage Factor, and a snRNP. *Cell* 54 875–889.

G. M. Gilmartin, F. Schaufele, G. Schaffner, and M. L. Birnstiel. 1988. Functional Analysis of the Sea Urchin U7 Small Nuclear RNA. *Mol. Cell. Biol.* 8 1076–1084.

K. L. Mowry and J. A. Steitz. 1988. snRNP Mediators of 3′ End Processing: Functional Fossils? *Trends Biochem. Sci.* 13 447–451.

Y. Takagaki, L. C. Ryner, and J. L. Manley. 1988. Separation and Characterization of a Poly(A) Polymerase and a Cleavage/Specificity Factor Required for Pre-mRNA Polyadenylation. *Cell* 52 731–742.

J. Wilusz, D. I. Feig, and T. Shenk. 1988. The C Proteins of Heterogeneous Ribonucleoprotein Complexes Interact with RNA Sequences Downstream of Polyadenylation Cleavage Sites. *Mol. Cell. Biol.* 8 4477–4483.

J. Wilusz and T. Shenk. 1988. A 64 kd Nuclear Protein Binds to RNA Segments That Include the AAUAAA Polyadenylation Motif. *Cell* 52 221–228.

N. Levitt, D. Briggs, A. Gil, and N. J. Proudfoot. 1989. Definition of an Efficient Synthetic Poly(A) Site. *Genes and Devel.* 3 1019–1025.

N. J. Proudfoot. 1989. How RNA Polymerase II Terminates Transcription in Higher Eukaryotes. *Trends Biochem. Sci.* 14 105–110.

M. D. Sheets and M. Wickens. 1989. Two Phases in the Addition of a Poly(A) Tail. *Genes and Devel.* 3 1401–1412.

8.3d

W. S. Dynan and R. Tjian. 1983. The Promoter-Specific Transcription Factor Sp 1 Binds to Upstream Sequences in the SV40 Early Promoter. *Cell* 35 79–87.

M. Fromm and P. Berg. 1983. Simian Virus 40 Early- and Late-Region Promoter Functions Are Enhanced by the 72-Base-Pair Repeat Inserted at Distant Locations and Inverted Orientations. *Mol. Cell. Biol.* 3 991–999.

M. Fromm and P. Berg. 1983. Transcription *in Vivo* from SV40 Early Promoter Deletion Mutants Without Repression by Large T Antigen. *J. Mol. Appl. Genet.* 2 127–135.

J. Brady, M. Radonovich, M. Thoren, G. Das, and N. P. Salzman. 1984. Simian Virus 40 Major Late Promoter: An Upstream DNA Sequence Required for Efficient *in Vitro* Transcription. *Mol. Cell. Biol.* 4 133–141.

A. R. Buchman and P. Berg. 1984. Unusual Regulation of Simian Virus 40 Early-Region Transcription in Genomes Containing Two Origins of DNA Replication. *Mol. Cell. Biol.* 4 1915–1928.

A. R. Buchman, M. Fromm, and P. Berg. 1984. Complex Regulation of Simian Virus 40 Early-Region Transcription from Different Overlapping Promoters. *Mol. Cell. Biol.* 4 1900–1914.

D. Gidoni, W. S. Dynan, and R. Tjian. 1984. Multiple Specific Contacts Between a Mammalian Transcription Factor and Its Cognate Promoters. *Nature* 312 409–413.

S. L. McKnight, R. C. Kingsbury, A. Spence, and M. Smith. 1984. The Distal Transcription Signals of the Herpesvirus *tk* Gene Share a Common Hexanucleotide Control Sequence. *Cell* 37 253–262.

F. Weber, J. de Villiers, and W. Schaffner. 1984. An SV40 "Enhancer Trap" Incorporates Exogenous Enhancers or Generates Enhancers from Its Own Sequences. *Cell* 36 983–992.

W. S. Dynan and R. Tjian. 1985. Control of Eukaryotic Messenger RNA Synthesis by Sequence-Specific DNA-Binding Proteins. *Nature* 316 774–778.

D. Gidoni, J. T. Kadonaga, H. Barrera-Saldana, K. Takahashi, P. Chambon, and R. Tjian. 1985. Bidirectional SV40 Transcription Mediated by Tandem SP1 Binding Interactions. *Science* 230 511–517.

K. A. Jones, K. R. Yamamoto, and R. Tjian. 1985. Two Distinct Transcription Factors Bind to the HSV Thymidine Kinase Promoter *in Vitro*. *Cell* 42 559–572.

J. M. Keller and J. C. Alwine. 1985. Analysis of an Activatable Promoter: Sequences in the Simian Virus 40 Late Promoter Required for T-Antigen-Mediated *trans* Activation. *Mol. Cell. Biol.* 5 1859–1869.

K. Ryder, E. Vakalopoulou, R. Mertz, I. Mastrangelo, P. Hough, P. Tegtmeyer, and E. Fanning. 1985. Seventeen Base Pairs of Region I Encode a Novel Tripartite Binding Signal for SV40 T Antigen. *Cell* 42 539–548.

E. Serfling, M. Jasin, and W. Schaffner. 1985. Enhancers and Eukaryotic Gene Transcription. *Trends Genet.* 1 224–230.

R. Treisman and T. Maniatis. 1985. Simian Virus 40 Enhancer Increases Number of RNA Polymerase II Molecules on Linked DNA. *Nature* 315 72–77.

M. R. Briggs, J. T. Kadonaga, S. P. Bell, and R. Tjian. 1986. Purification and Biochemical Characterization of the Promoter-Specific Transcription Factor, Sp1. *Science* 234 47–52.

W. Herr and J. Clarke. 1986. The SV40 Enhancer Is Composed of Multiple Functional Elements That Can Compensate for One Another. *Cell* 45 461–470.

J. T. Kadonaga, K. A. Jones, and R. Tjian. 1986. Promoter-Specific Activation of RNA Polymerase II Transcription by Sp1. *Trends Biochem. Sci.* 11 20–23.

J. T. Kadonaga and R. Tjian. 1986. Affinity Purification of Sequence-Specific DNA Binding Proteins. *Proc. Natl. Acad. Sci.* U.S.A. 83 5889–5893.

S. McKnight and R. Tjian. 1986. Transcriptional Selectivity of Viral Genes in Mammaliam Cells. *Cell* 46 795–805.

S. E. Plon and J. C. Wang. 1986. Transcription of the Human β-Globin Gene Is Stimulated by an SV40 Enhancer to Which It Is Physically Linked but Topologically Uncoupled. *Cell* 45 575–580.

P. Sassone-Corsi and E. Borrelli. 1986. Transcriptional Regulation by *trans*-Acting Factors. *Trends Genet.* 2 215–219.

K. Takahashi, M. Vigneron, H. Matthes, A. Wildeman, M. Zenke, and P. Chambon. 1986. Requirement of Stereospecific Alignments for Initiation from the Simian Virus 40 Early Promoter. *Nature 319* 121–126.

M. Zenke, T. Grundstrom, H. Matthes, M. Wintzerith, C. Schatz, A. Wildeman, and P. Chambon. 1986. Multiple Sequence Motifs Are Involved in SV40 Enhancer Function. *EMBO J.* 5 387–397.

A. Dorn. J. Bollekens, A. Staub, C. Benoist, and D. Mathis. 1987. A Multiciplicity of CCAAT Box-Binding Proteins. *Cell* 50 863–872.

R. D. Everett. 1987. The Regulation of Transcription of Viral and Cellular Genes by Herpesvirus Immediate-Early Gene Products. *Anticancer Res.* 7 589–604.

L. Guarente. 1987. Regulatory Proteins in Yeast. *Annu. Rev. Genet.* 21 425–452.

W. Lee, P. Mitchell, and R. Tjian. 1987. Purified Transcription Factor AP-1 Interacts with TPA-Inducible Enhancer Elements. *Cell* 49 741–752.

E. May, F. Omilli, M. Ernoult-Lange, M. Zenke, and P. Chambon. 1987. The Sequence Motifs That Are Involved in SV40 Enhancer Function Also Control SV40 Late Promoter Activity. *Nucleic Acids Res.* 15 2445–2461.

P. J. Mitchell, C. Wang, and R. Tjian. 1987. Positive and Negative Regulation of Transcription *in Vitro*: Enhancer-Binding Protein AP-2 Is Inhibited by SV40 T Antigen. *Cell* 50 847–861.

R. Rosales, M. Vigneron, M. Macchi, I. Davidson, J. H. Xiao, and P. Chambon. 1987. *In Vitro* Binding of Cell-Specific and Ubiquitous Nuclear Proteins to the Octamer Motif of the SV40 Enhancer and Related Motifs Present in Other Promoters and Enhancers. *EMBO J.* 6 3015–3025.

S. Schirm, J. Jiricny, and W. Schaffner. 1987. The SV40 Enhancer Can Be Dissected into Multiple Segments, Each with a Different Cell Type Specificity. *Genes and Devel.* 1 65–74.

J. H. Xiao, I. Davidson, D. Ferrandon, R. Rosales, M. Vigneron, M. Macchi, F. Ruffenach, and P. Chambon. 1987. One Cell-Specific and Three Ubiquitous Nuclear Proteins Bind *in Vitro* to Overlapping Motifs in the Domain B1 of the SV40 Enhancer. *EMBO J.* 6 3005–3013.

S. M. Abmayr, J. L. Workman, and R. G. Roeder. 1988. The Pseudorabies Immediate Early Protein Stimulates *in Vitro* Transcription by Facilitating TFIID: Promoter Interactions. *Genes and Devel.* 2 542–553.

T. J. Bos, D. Bohmann, H. Tsuchie, R. Tjian, and P. K. Vogt. 1988. v-*jun* Encodes a Nuclear Protein with Enhancer Binding Properties of AP-1. *Cell* 52 705–712.

S. Buratowski, S. Hahn, P. A. Sharp, and L. Guarente. 1988. Function of a Yeast TATA Element-Binding Protein in a Mammalian Transcription System. *Nature* 334 37–42.

B. Cavallini, J. Huet, J-L. Plassat, A. Sentenac, J-M. Egly, and P. Chambon. 1988. A Yeast Activity Can Substitute for the HeLa Cell TATA Box Factor. *Nature* 334 77–80.

L. A. Chodosh, A. S. Baldwin, R. W. Carthew, and P. A. Sharp. 1988. Human CCAAT-Binding Proteins Have Heterologous Subunits. *Cell* 53 11–24.

L. A. Chodosh, J. Olesen, S. Hahn, A. S. Baldwin, L. Guarente, and P. A. Sharp. 1988. A Yeast and a Human CCAAT-Binding Protein Have Heterologous Subunits That Are Functionally Interchangeable. *Cell* 53 25–35.

L. Clark, R. M. Pollock, and R. T. Hay. 1988. Identification and Purification of EBP1: A HeLa Cell Protein That Binds to a Region Overlapping the "Core" of the SV40 Enhancer. *Genes and Devel.* 2 991–1002.

A. J. Courey and R. Tjian. 1988. Analysis of Sp-1 *in Vivo* Reveals Multiple Transcriptional Domains, Including a Novel Glutamine-Rich Activation Motif. *Cell* 55 887–898.

C. Fromental, M. Kanno, H. Nomiyama, and P. Chambon. 1988. Cooperativity and Hierarchical Levels of Functional Organization in the SV40 Enhancer. *Cell* 54 943–953.

S. Hahn and L. Guarente. 1988. Yeast HAP2 and HAP3: Transcriptional Activators in a Heteromeric Complex. *Science* 240 317–321.

K. D. Harshman, W. S. Moye-Rowley, and C. S. Parker. 1988. Transcriptional Activation by the SV40 AP-1 Recognition Element in Yeast Is Mediated by a Factor Similar to AP-1 That Is Distinct from GCN4. *Cell* 53 321–330.

N. C. Jones, P. W. J. Rigby, and E. B. Ziff. 1988. *Trans*-Acting Protein Factors and the Regulation of Eukaryotic Transcription: Lessons from Studies on DNA Tumor Viruses. *Genes and Devel.* 2 267–281.

J. T. Kadonaga, A. J. Courey, J. Ladika, and R. Tjian. 1988. Distinct Regions of SP1 Modulate DNA Binding and Transcriptional Activation. *Science* 242 1566–1569.

S. N. Maity, P. T. Golumbek, G. Karsenty, and B. de Crombrugghe. 1988. Selective Activation of Transcription by a Novel CCAAT Binding Factor. *Science* 241 582–584.

M. Raymondjean, S. Cereghini, and M. Yaniv. 1988. Several Distinct "CCAAT" Box Binding Proteins Coexist in Eukaryotic Cells. *Proc. Natl. Acad. Sci.* U.S.A. 85 757–761.

B. Wasylyk. 1988. Enhancers and Transcription Factors in the Control of Gene Expression. *Biochimica et Biophysica Acta* 951 17–35.

S. Buratowski, S. Hahn, L. Guarente, and P. A. Sharp. 1989. Five Intermediate Complexes in Transciption Initiation by RNA Polymerase II. *Cell* 56 549–561.

P. J. Mitchell and R. Tjian. 1989. Transcriptional Regulation in Mammalian Cells by Sequence-Specific DNA Binding Proteins. *Science* 245 371–378.

W. S. Moye-Rowley, K. D. Harshman, and C. S. Parker. 1989. Yeast *YAP1* Encodes a Novel Form of the Jun Family of Transcriptional Activator Proteins. *Genes and Devel.* 3 283–292.

A. G. Saltzman and R. Weinmann. 1989. Promoter Specificity and Modulation of RNA Polymerase II Transcription. *FASEB J.* 3 1723–1733.

8.3e

M. Mercola, X. Wang, J. Olsen, and K. Calame. 1983. Transcriptional Enhancer Activity in the Immunoglobulin Heavy Chain Locus. *Science* 221 663–665.

F. G. Falkner and H. G. Zachau. 1984. Correct Transcription of an Immunoglobulin κ Gene Requires an Upstream Fragment Containing Conserved Sequence Elements. *Nature* 310 71–74.

D. Picard and W. Schaffner. 1984. A Lymphocyte-Specific Enhancer in the Mouse Immunoglobulin κ Gene. *Nature* 307 80–82.

F. W. Alt, T. K. Blackwell, and G. D. Yancopoulos. 1985. Immunoglobulin Genes in Transgenic Mice. *Trends Genet.* 1 231–236.

K. Calame. 1985. Mechanisms Which Regulate Immunoglobulin Gene Expression. *Annu. Rev. Immunol.* 3 159–195.

R. Grosschedl and D. Baltimore. 1985. Cell-Type Specificity of Immunoglobulin Gene Expression Is Regulated by at Least Three DNA Sequence Elements. *Cell* 41 885–897.

R. Krumlauf, R. E. Hammer, S. M. Tilghman, and R. L. Brinster. 1985. Developmental Regulation of α-Fetoprotein Genes in Transgenic Mice. *Mol. Cell. Biol.* 5 1639–1648.

M. Mercola, J. Goverman, C. Mirell, and K. Calame. 1985. Immunoglobulin Heavy-Chain Enhancer Requires One or More Tissue-Specific Factors. *Science* 227 266–270.

D. Picard and W. Schaffner. 1985. Cell-Type Preference of Immunoglobulin and Gene Promoters. *EMBO J.* 4 2831–2838.

R. Godbout, R. Ingram, and S. M. Tilghman. 1986. Multiple Regulatory Elements in the Intergenic Region Between the α-Fetoprotein and Albumin Genes. *Mol. Cell. Biol.* 6 477–487.

C. L. Peterson, K. Orth, and K. L. Calame. 1986. Binding *in Vitro* of Multiple Cellular Proteins to Immunoglobulin Heavy-Chain Enhancer DNA. *Mol. Cell. Biol.* 6 4168–4178.

R. Sen and D. Baltimore. 1986. Inducibility of κ Immunoglobulin Enhancer-Binding Protein NF-κB by a Posttranslational Mechanism. *Cell* 47 921–928.

R. Sen and D. Baltimore. 1986. Multiple Nuclear Factors Interact with the Immunoglobulin Enhancer Sequences. *Cell* 46 705–716.

H. Singh, R. Sen, D. Baltimore, and P. A. Sharp. 1986. A Nuclear Factor That Binds to a Conserved Sequence Motif in Transcriptional Control Elements of Immunoglobulin Genes. *Nature* 319 154–158.

L. M. Staudt, H. Singh, R. Sen, T. Wirth, P. A. Sharp, and D. Baltimore. 1986. A Lymphoid-Specific Protein Binding to the Octamer Motif of Immunoglobulin Genes. *Nature* 323 640–643.

M. L. Atchison and R. P. Perry. 1987. The Role of the κ Enhancer and Its Binding Factor NF-κB in the Development Regulation of κ Gene Transcription. *Cell* 48 121–128.

L. E. Babiss, R. S. Herbst, A. L. Bennett, and J. E. Darnell, Jr. 1987. Factors That Interact with the Rat Albumin Promoter Are Present Both in Hepatocytes and Other Cell Types. *Genes and Devel.* 1 256–267.

S. Cereghini, M. Raymondjean, A. G. Carranca, P. Herbomel, and M. Yaniv. 1987. Factors Involved in Control of Tissue-Specific Expression of Albumin Gene. *Cell* 50 627–638.

S. Eaton and K. Calame. 1987. Multiple Elements Are Necessary for the Function of an Immunoglobulin Heavy Chain Promoter. *Proc. Natl. Acad. Sci.* U.S.A. 84 7634–7638.

J. M. Heard, P. Herbomel, M. O. Ott, A. Mottura-Rollier, M. Weiss, and M. Yaniv. 1987. Determinants of Rat Albumin Promoter Tissue Specificity Analyzed by an Improved Transient Expression System. *Mol. Cell. Biol.* 7 2425–2434.

J. L. Imler, C. Lemaire, C. Wasylyk, and B. Wasylyk. 1987. Negative Regulation Contributes to Tissue Specificity of the Immunoglobulin Heavy-Chain Enhancer. *Mol. Cell. Biol.* 7 2558–2567.

M. Lenardo, J. W. Pierce, and D. Baltimore. 1987. Protein-Binding Sites in Ig Gene Enhancers Determine Transcriptional Activity and Inducibility. *Science* 236 1573–1577.

S. Lichsteiner, J. Wuarin, and U. Schibler. 1987. The Interplay of DNA-Binding Proteins on the Promoter of the Mouse Albumin Gene. *Cell* 51 963–973.

C. Peterson and K. Calame. 1987. Complex Protein-DNA Interactions on the Mouse Immunoglobulin Heavy Chain Enhancer. *Mol. Cell. Biol.* 7 4194–4203.

C. Scheidereit, A. Heguy, and R. G. Roeder. 1987. Identification and Purification of a Human Lymphoid-Specific Octamer-Binding Protein (OTF-2) That Activates Transcription of an Immunoglobulin Promoter *in Vitro*. *Cell* 51 783–793.

K. Araki, H. Maeda, J. Wang, D. Kitamura, and T. Watanabe. 1988. Purification of a Nuclear *Trans*-Acting Factor Involved in the Regulated Transcription of a Human Immunoglobulin Heavy Chain Gene. *Cell* 53 723–730.

A. S. Baldwin, Jr., and P. A. Sharp. 1988. Two Transcription Factors, NF-κB and H2TF1, Interact with a Single Regulatory Sequence in the Class I Major Histocompatibility Complex Promoter. *Proc. Natl. Acad. Sci.* U.S.A. 85 723–727.

K. Calame and S. Eaton. 1988. Transcriptional Controlling Elements in the Immunoglobulin and T Cell Receptor Loci. *Adv. Immunol.* 43 235–275.

S. Cereghini, M. Blumenfeld, and M. Yaniv. 1988. A Liver-Specific Factor Essential for Albumin Transcription Differs Between Differentiated and Dedifferentiated Rat Hepatoma Cells. *Genes and Devel.* 2 957–974.

R. H. Costa, E. Lai, D. R. Grayson, and J. E. Darnell, Jr. 1988. The Cell-Specific Enhancer of the Mouse Transthyretin (Prealbumin) Gene Binds a Common Factor at One Site and a Liver-Specific Factor(s) at Two Other Sites. *Mol. Cell. Biol.* 8 81–90.

G. Courtois, S. Baumhueter, and G. R. Crabtree. 1988. Purified Hepatocyte Nuclear Factor 1 Interacts with a Family of Hepatocyte-Specific Promoters. *Proc. Natl. Acad. Sci.* U.S.A. 85 7937–7941.

R. Godbout, R. S. Ingram, and S. M. Tilghman. 1988. Fine-Structure Mapping of the Three Mouse α-Fetoprotein Gene Enhancers. *Mol. Cell. Biol.* 8 1169–1178.

R. Godbout and S. M. Tilghman. 1988. Configuration of the α-Fetoprotein Regulatory Domain During Development. *Genes and Devel.* 2 949–956.

D. R. Grayson, R. H. Costa, K. G. Xanthopoulos, and J. E. Darnell, Jr. 1988. A Cell-Specific Enhancer of the Mouse α1-Antitrypsin Gene Has Multiple Functional Regions and Corresponding Protein-Binding Sites. *Mol. Cell. Biol.* 8 1055–1066.

D. R. Grayson, R. H. Costa, K. G. Xanthopoulos, and J. E. Darnell. 1988. One Factor Recognizes the Liver-Specific Enhancers in α1-Antitrypsin and Transthyretin Genes. *Science* 239 786–788.

R. Grosschedl and M. Marx. 1988. Stable Propagation of the Active Transcriptional State of an Immunoglobulin Gene Requires Continuous Enhancer Function. *Cell* 55 645–654.

K. Kawakami, C. Scheidereit, and R. G. Roeder. 1988. Identification and Purification of a Human Immunoglobulin-Enhancer-Binding Protein (NF-κB) That Activates Transcription from a Human Immunodeficiency Virus Type 1 Promoter *in Vitro*. *Proc. Natl. Acad. Sci.* U.S.A. 85 4700–4704.

H. Ko, P. Fast, W. McBride, and L. M. Staudt. 1988. A Human Protein Specific for the Immunoglobulin Octamer DNA Motif Contains a Functional Homeobox Domain. *Cell* 55 135–144.

J. H. LeBowitz, T. Kobayashi, L. Staudt, D. Baltimore, and P. A. Sharp. 1988. Octamer-Binding Proteins from B or HeLa Cells Stimulate Transcription of the Immunoglobulin Heavy-Chain Promoter *in Vitro*. *Genes and Devel.* 2 1227–1237.

M. M. Muller, S. Ruppert, W. Schaffner, and P. Matthias. 1988. A Cloned Octamer Transcription Factor Stimulates Transcription from Lymphoid-Specific Promoters in Non-B Cells. *Nature* 336 544–551.

J. Perez-Mutul, M. Macchi, and B. Wasylyk. 1988. Mutational Analysis of the Contribution of Sequence Motifs Within the IgH Enhancer to Tissue Specific Transcriptional Activation. *Nucleic Acids Res.* 16 6085–6096.

C. Scheidereit, J. A. Cromlish, T. Gerster, K. Kawakami, C. Balmaceda, R. A. Currie, and R. G. Roeder. 1988. A Human Lymphoid-Specific Transcription Factor That Activates Immunoglobulin Genes Is a Homeobox Protein. *Nature* 336 551–557.

H. Singh, J. H. LeBowitz, A. S. Baldwin, Jr., and P. A. Sharp. 1988. Molecular Cloning of an Enhancer Binding Protein: Isolation by Screening of an Expression Library with a Recognition Site DNA. *Cell* 52 415–423.

L. M. Staudt, R. G. Clerc, H. Singh, J. H. LeBowitz, P. A.

Sharp, and D. Baltimore. 1988. Cloning of a Lymphoid-Specific cDNA Encoding a Protein Binding the Regulatory Octamer DNA Motif. *Science* 241 577–579.

B. Wasylyk. 1988. Transcription Elements and Factors of RNA Polymerase B Promoters of Higher Eukaryotes. *Crit. Rev. Biochem.* 23 77–120.

K. L. Calame. 1989. Immunoglobulin Gene Transcription: Molecular Mechanisms. *Trends Genet.* 5 395–399.

S. A. Camper and S. M. Tilghman. 1989. Postnatal Repression of the α-Fetoprotein Gene Is Enhancer Independent. *Genes and Devel.* 3 537–546.

A. D. Friedman, W. H. Landschulz, and S. L. McKnight. 1989. CCAAT/Enhancer Binding Protein Activates the Promoter of the Serum Albumin Gene in Cultured Hepatoma Cells. *Genes and Devel.* 3 1314–1322.

M. A. Garcia-Blanco, R. G. Clerc, and P. A. Sharp. 1989. The DNA-Binding Homeo Domain of the Oct-2 Protein. *Genes and Devel.* 3 739–745.

P. Herbomel, A. Mottura, F. Tronche, M. O. Ott, M. Yaniv, and M. Weiss. 1989. The Rat Albumin Promoter Is Composed of Six Distinct Positive Elements Within 130 Nucleotides. *Mol. Cell. Biol.* 9 4750–4758.

I. Kemler, E. Schreiber, M. M. Muller, P. Matthias, and W. Schaffner. 1989. Octamer Transcription Factors Bind to Two Different Sequence Motifs of the Immunoglobulin Heavy Chain Promoter. *EMBO J.* 8 2001–2008.

M. J. Lenardo and D. Baltimore. 1989. NF-κB: A Pleiotropic Mediator of Inducible and Tissue Specific Gene Control. *Cell* 58 227–229.

K. B. Meyer and M. S. Neuberger. 1989. The Immunoglobulin κ Locus Contains a Second, Stronger B-Cell-Specific Enhancer Which Is Located Downstream of the Constant Region. *EMBO J.* 8 1959–1964.

C. Murre, P. S. McCaw, and D. Baltimore. 1989. A New DNA Binding and Dimerization Motif in Immunoglobulin Enhancer Binding, Daughterless, MyoD, and myc Proteins. *Cell* 56 777–783.

R. H. Scheuermann and U. Chen. 1989. A Developmental-Specific Factor Binds to Suppressor Sites Flanking the Immunoglobulin Heavy-Chain Enhancer. *Genes and Devel.* 3 1255–1266.

8.3f

R. I. Richards, A. Heguy, and M. Karin. 1984. Structural and Functional Analysis of the Human Metallothionein-1_A Gene: Differential Induction by Metal Ions and Glucocorticoids. *Cell* 37 263–272.

C. Wu. 1984. Activating Protein Factor Binds *in Vitro* to Upstream Control Sequences in Heat Shock Gene Chromatin. *Nature* 311 81–84.

M. Bienz. 1985. Transient and Developmental Activation of Heat-Shock Genes. *Trends Biochem. Sci.* 10 157–161.

J. Simpson, M. P. Timko, A. R. Cashmore, J. Schell, M. VanMontagu, and L. Herrera-Estrella. 1985. Light-Inducible and Tissue-Specific Expression of a Chimaeric Gene Under Control of the 5′-Flanking Sequence of a Pea Chlorophyll a/b-Binding Protein Gene. *EMBO J.* 4 2723–2729.

M. P. Timko, A. P. Kausch, C. Castresana, J. Fassler, L. Herrera-Estrella, G. Van den Broeck, M. Van Montagu, J. Schell, and A. R. Cashmore. 1985. Light Regulation of Plant Gene Expression by an Upstream Enhancer-Like Element. *Nature* 318 579–582.

M. Bienz and H. R. B. Pelham. 1986. Heat Shock Regulatory Elements Function as an Inducible Enhancer in the *Xenopus* hsp70 Gene and When Linked to a Heterologous Promoter. *Cell* 45 753–760.

S. Green, P. Walter, V. Kumar, A. Krust, J. Bornert, P. Argos, and P. Chambon. 1986. Human Oestrogen Receptor cDNA: Sequence, Expression and Homology to v-erb-A. *Nature* 320 134–139.

D. H. Hamer. 1986. Metallothionein. *Annu. Rev. Biochem.* 55 913–951.

R. Klemenz and W. J. Gehring. 1986. Sequence Requirement for Expression of the *Drosophila melanogaster* Heat Shock Protein hsp22 Gene During Heat Shock and Normal Development. *Mol. Cell. Biol.* 6 2011–2019.

R. D. Andersen, S. J. Taplitz, S. Wong, G. Bristol, B. Larkin, and H. R. Herschman. 1987. Metal-Dependent Binding of a Factor *in Vivo* to the Metal-Responsive Elements of the Metallothionein-1 Gene Promoter. *Mol. Cell. Biol.* 7 3574–3581.

V. Giguere, E. S. Ong, P. Segui, and R. M. Evans. 1987. Identification of a Receptor for the Morphogen Retinoic Acid. *Nature* 330 624–629.

G. Gill and M. Ptashne. 1987. Mutants of GAL4 Protein Altered in an Activation Function. *Cell* 51 121–126.

S. Green and P. Chambon. 1987. Oestradiol Induction of a Glucocorticoid-Responsive Gene by a Chimaeric Receptor. *Nature* 325 75–78.

M. Johnston. 1987. A Model Fungal Gene Regulatory Mechanism: The GAL Genes of *Saccharomyces cerevisiae*. *Microbiol. Rev.* 51 458–476.

M. Karin, A. Haslinger, A. Heguy, T. Dietlin, and T. Cooke. 1987. Metal-Responsive Elements Act as Positive Modulators of Human Metallothionein-II_A Enhancer Activity. *Mol. Cell. Biol.* 7 606–613.

R. E. Kingston, T. J. Schuetz, and Z. Lavin. 1987. Heat-Inducible Human Factor That Binds to a Human *hsp* 70 Promoter. *Mol. Cell. Biol.* 7 1530–1534.

D. Kuhl, J. de la Fuente, M. Chaturvedi, S. Parimoo, J. Ryals, F. Meyer, and C. Weissmann. 1987. Reversible Silencing of Enhancers by Sequences Derived from the Human IFN-α Promoter. *Cell* 50 1057–1069.

C. Kuhlemeier, P. J. Green, and N. Chua. 1987. Regulation of Gene Expression in Higher Plants. *Annu. Rev. Plant Physiol.* 38 221–257.

V. Kumar, S. Green, G. Stack, M. Berry, J. Jin, and P. Chambon. 1987. Functional Domains of the Human Estrogen Receptor. *Cell* 51 941–951.

W. Lee, A. Haslinger, M. Karin, and R. Tjian. 1987. Activation of Transcription by Two Factors That Bind Promoter and Enhancer Sequences of the Human Metallothionein Gene and SV40. *Nature* 325 368–372.

N. F. Lue, D. I. Chasman, A. R. Buchman, and R. D. Kornberg. 1987. Interaction of GAL4 and GAL80 Gene Regulatory Proteins *in Vitro*. *Mol. Cell. Biol.* 7 3446–3451.

J. Ma and M. Ptashne. 1987. Deletion Analysis of GAL4 Defines Two Transcriptional Activating Segments. *Cell* 48 847–853.

M. Petkovich, N. J. Brand, A. Krust, and P. Chambon. 1987. A Human Retinoic Acid Receptor Which Belongs to the Family of Nuclear Receptors. *Nature* 330 444–450.

G. Wiederrecht, D. J. Shuey, W. A. Kibbe, and C. S. Parker. 1987. The *Saccharomyces* and *Drosophila* Heat Shock Transcription Factors Are Identical in Size and DNA Binding Properties. *Cell* 48 507–515.

V. Zimarino and C. Wu. 1987. Induction of Sequence-Specific Binding of *Drosophila* Heat Shock Activator Protein Without Protein Synthesis. *Nature* 327 727–730.

I. E. Akerblom, E. P. Slater, M. Geato, J. D. Baxter, and P. L. Mellon. 1988. Negative Regulation by Glucocorticoids Through Interference with a cAMP Responsive Enhancer. *Science* 241 350–353.

N. Brand, M. Petkovich, A. Krust, P. Chambon, H. de The, A. Marchio, P. Tiollais, and A. Dejean. 1988. Identification of a Second Human Retinoic Acid Receptor. *Nature* 332 850–853.

R. M. Evans. 1988. The Steroid and Thyroid Hormone Receptor Superfamily. *Science* 240 889–895.

P. Furst, S. Hu, R. Hackett, and D. Hamer. 1988. Copper Activates Metallothionein Gene Transcription by Altering the Conformation of a Specific DNA Binding Protein. *Cell* 55 705–717.

V. Giguere, N. Yang, P. Segui, and R. M. Evans. 1988. Identification of a New Class of Steroid Hormone Receptors. *Nature* 331 91–94.

S. Green and P. Chambon. 1988. Nuclear Receptors Enhance Our Understanding of Transcription Regulation. *Trends Genet.* 4 309–314.

S. Green, V. Kumar, I. Theulaz, W. Wahli, and P. Chambon. 1988. The N-Terminal DNA-Binding "Zinc Finger" of the Oestrogen and Glucocorticoid Receptors Determines Target Gene Specificity. *EMBO J.* 7 3037–3044.

S. M. Hollenberg and R. M. Evans. 1988. Multiple and Cooperative Trans-Activation Domains of the Human Glucocorticoid Receptor. *Cell* 55 899–906.

M. Horikoshi, M. F. Carey, H. Kakidani, and R. G. Roeder. 1988. Mechanism of Action of a Yeast Activator: Direct Effect of GAL4 Derivatives on Mammalian TFIID-Promoter Interactions. *Cell* 54 665–669.

H. Kakidani and M. Ptashne. 1988. GAL4 Activates Gene Expression in Mammalian Cells. *Cell* 52 161–167.

V. Kumar and P. Chambon. 1988. The Estrogen Receptor Binds Tightly to Its Responsive Element as a Ligand-Induced Homodimer. *Cell* 55 145–156.

Y. Lin, M. F. Carey, M. Ptashne, and M. R. Green. 1988. GAL4 Derivatives Function Alone and Synergistically with Mammalian Activators *in Vitro*. *Cell* 54 659–664.

T. Maniatis. 1988. Mechanisms of Human β-Interferon Gene Regulation. Harvey Lectures, Series 82 71–104.

D. Metzger, J. H. White, and P. Chambon. 1988. The Human Oestrogen Receptor Functions in Yeast. *Nature* 334 31–36.

A. E. Oro, S. M. Hollenberg, and R. M. Evans. 1988. Transcriptional Inhibition by a Glucocorticoid Receptor-β-Galactosidase Fusion Protein. *Cell* 55 1109–1114.

A. E. Oro, E. S. Ong, J. S. Margolis, J. W. Posakony, M. McKeown, and R. M. Evans. 1988. The *Drosophila* Gene *Knirps*-Related Is a Member of the Steroid-Receptor Gene Superfamily. *Nature* 336 493–496.

M. Ponglikitmongkol, S. Green, and P. Chambon. 1988. Genomic Organization of the Human Oestrogen Receptor Gene. *EMBO J.* 7 3385–3388.

D. D. Sakai, S. Helms, J. Carlstedt-Duke, J. Gustafsson, F. M. Rottman, and K. R. Yamamoto. 1988. Hormone-Mediated Repression: A Negative Glucocorticoid Response Element from the Bovine Prolactin Gene. *Genes and Devel.* 2 1144–1154.

M. Schena and K. R. Yamamoto. 1988. Mammalian Glucocorticoid Receptor Derivatives Enhance Transcription in Yeast. *Science* 241 965–967.

R. Schule, M. Muller, C. Kaltschmidt, and R. Renkawitz. 1988. Many Transcription Factors Interact Synergistically with Steroid Receptors. *Science* 242 1418–1420.

L. Tora, H. Gronemeyer, B. Turcotte, M. Gaub, and P. Chambon. 1988. The N-Terminal Region of the Chicken Progesterone Receptor Specifies Target Gene Activation. *Nature* 333 185–188.

S. Y. Tsai, J. Carlstedt-Duke, N. L. Weigel, K. Dahlman, J. Gustafsson, M. Tsai, and B. W. O'Malley. 1988. Molecular Interactions of Steroid Hormone Receptor with its Enhancer Element: Evidence for Receptor Dimer Formation. *Cell* 55 361–369.

K. Umesono, V. Giguere, C. K. Glass, M. G. Rosenfeld, and R. M. Evans. 1988. Retinoic Acid and Thyroid Hormone Induce Gene Expression Through a Common Responsive Element. *Nature* 336 262–265.

M. L. Waterman, S. Adler, C. Nelson, G. L. Greene, R. M. Evans, and M. G. Rosenfeld. 1988. A Single Domain of the Estrogen Receptor Confers Deoxyribonucleic Acid Binding and Transcriptional Activation of the Rat Prolactin Gene. *Mol. Endocrin.* 2 14–21.

N. Webster, J. R. Jin, S. Green, M. Hollis, and P. Chambon. 1988. The Yeast UAS_G Is a Transcriptional Enhancer in Human HeLa Cells in the Presence of the GAL4 Trans-Activator. *Cell* 52 169–178.

G. Wiederrecht, D. Seto, and C. S. Parker. 1988. Isolation of the Gene Encoding the *S. cerevisiae* Heat Shock Transcription Factor. *Cell* 54 841–853.

P. N. Benfey and N. Chua. 1989. Regulated Genes in Transgenic Plants. *Science* 244 174–181.

E. A. Craig. 1989. Essential Roles of 70kDa Heat Inducible Proteins. *BioEssays* 11 48–52.

K. Damm, C. C. Thompson, and R. M. Evans. 1989. Protein Encoded by v-erbA Functions as a Thyroid-Hormone Receptor Antagonist. *Nature* 339 593–596.

R. B. Goldberg, S. J. Barker, and L. Perez-Grau. 1989. Regulation of Gene Expression During Plant Embryogenesis. *Cell* 56 149–160.

M. J. Lenardo, C. Fan, T. Maniatis, and D. Baltimore. 1989. The Involvement of NF-κB in β-Interferon Gene Regulation Reveals Its Role as a Widely Inducible Second Messenger. *Cell* 57 287–294.

S. Mader, V. Kumar, H. deVerneuil, and P. Chambon. 1989. Three Amino Acids of the Oestrogen Receptor Are Essential to Its Ability to Distinguish an Oestrogen from a Glucocorticoid-Responsive Element. *Nature* 338 271–274.

M. Meyer, H. Gronemeyer, B. Turcotte, M. Bocquel, D. Tasset, and P. Chambon. 1989. Steroid Hormone Receptors Compete for Factors That Mediate Their Enhancer Function. *Cell* 57 433–442.

S. Y. Tsai, M. Tsai, and B. W. O'Malley. 1989. Cooperative Binding of Steroid Hormone Receptors Contributes to Transcriptional Synergism at Target Enhancer Elements. *Cell* 57 443–448.

8.3g

W. J. Gehring. 1985. The Molecular Basis of Development. *Sci. American* 253 153–162.

W. J. Gehring. 1987. Homeo Boxes in the Study of Development. *Science* 236 1245–1252.

M. P. Scott and S. B. Carroll. 1987. The Segmentation and Homeotic Gene Network in Early *Drosophila* Development. *Cell* 51 689–698.

W. Driever and C. Nusslein-Volhard. 1988. The *Bicoid* Protein Determines Position in the *Drosophila* Embryo in a Concentration-Dependent Manner. *Cell* 54 95–104.

W. Driever and C. Nusslein-Volhard. 1988. A Gradient of *Bicoid* Protein in *Drosophila* Embryos. *Cell* 54 83–93.

P. W. Ingham. 1988. The Molecular Genetics of Embryonic Pattern Formation in *Drosophila*. *Nature* 335 25–34.

P. A. Lawrence. 1988. Background to *Bicoid*. *Cell* 54 1–2.

M. D. Biggin and R. Tjian. 1989. Transcription Factors and the Control of *Drosophila* Development. *Trends Genet.* 5 377–383.

W. Driever and C. Nusslein-Volhard. 1989. The *Bicoid* Protein Is a Positive Regulator of *Hunchback* Transcription in the Early *Drosophila* Embryo. *Nature* 337 138–143.

W. Driever, G. Thoma, and C. Nusslein-Volhard. 1989. Determination of Spatial Domains of Zygotic Gene Expression in the *Drosophila* Embryo by the Affinity of Binding Sites for the *Bicoid* Morphogen. *Nature* 340 363–367.

M. J. Pankratz, M. Hoch, E. Seifert, and H. Jackie. 1989. *Kruppel* Requirement for *Knirps* Enhancement Reflects Overlapping Gap Gene Activities in the *Drosophila* Embryo. *Nature* 341 337–340.

D. Stanojevic, T. Hoey, and M. Levine. 1989. Sequence-Specific DNA-Binding Activities of the Gap Proteins Encoded by *Hunchback* and *Kruppel* in *Drosophila*. *Nature* 341 331–335.

J. Treisman and C. Desplan. 1989. The Products of the *Drosophila* Gap Genes *Hunchback* and *Kruppel* Bind to the *Hunchback* Promoters. *Nature* 341 335–337.

A. S. Wilkins. 1989. Organizing the *Drosophila* Posterior Pattern: Why Has the Fruit Fly Made Life So Complicated for Itself? *BioEssays* 11 67.

C. V. E. Wright. 1989. Vertebrate Homeodomain Proteins: Families of Region-Specific Transcription Factors. *Trends Biochem. Sci.* 14 52–56.

8.3h

M. Ares, Jr., J. Chung, L. Giglio, and A. M. Weiner. 1987. Distinct Factors with Sp1 and NF-A Specificities Bind to Adjacent Functional Elements of the Human U2 snRNA Gene Enhancer. *Genes and Devel.* 1 808–817.

P. Carbon, S. Murgo, J. Ebel, A. Krol, G. Tebb, and I. W. Mattaj. 1987. A Common Octamer Motif Binding Protein Is Involved in the Transcription of U6 snRNA by RNA Polymerase III and U2 snRNA by RNA Polymerase II. *Cell* 51 71–79.

J. T. Murphy, J. T. Skuzeski, E. Lund, T. H. Steinberg, R. R. Burgess, and J. E. Dahlberg. 1987. Functional Elements of the Human U1 RNA Promoter. *J. Biol. Chem.* 262 1795–1803.

G. Das, D. Henning, D. Wright, and R. Reddy. 1988. Upstream Regulatory Elements Are Necessary and Sufficient for Transcription of a U6 RNA Gene by RNA Polymerase III. *EMBO J.* 7 503–512.

G. R. Kunkel and T. Pederson. 1988. Upstream Elements Required for Efficient Transcription of a Human U6 RNA Gene Resemble Those of U1 and U2 Genes Even Though a Different Polymerase Is Used. *Genes and Devel.* 2 196–204.

I. W. Mattaj, N. A. Dathan, H. D. Parry, P. Carbon, and A. Krol. 1988. Changing the RNA Polymerase Specificity of U snRNA Gene Promoters. *Cell* 55 435–442.

M. Tanaka, U. Grossniklaus, W. Herr, and N. Hernandez. 1988. Activation of the U2 snRNA Promoter by the Octamer Motif Defines a New Class of RNA Polymerase II Enhancer Elements. *Genes and Devel.* 1 1764–1778.

P. Weller, C. Bark, L. Janson, and U. Pettersson. 1988. Transcription Analysis of a Human U4C Gene: Involvement of Transcription Factors Novel to snRNA Gene Expression. *Genes and Devel*. 2 1389–1399.

S. Murphy, B. Moorefield, and T. Pieler. 1989. Common Mechanisms of Promoter Recognition by RNA Polymerases II and III. *Trends Genet*. 5 123–126.

H. D. Parry, D. Scherly, and I. W. Mattaj. 1989. "Snurpogenesis": The Transcription and Assembly of U snRNP Components. *Trends Biochem. Sci*. 14 15–19.

8.4a

G. Ciliberto, L. Castagnoli, and R. Cortese. 1983. Transcription by RNA Polymerase III. Genome Function, Cell Interactions, and Differentiation. *Current Topics in Dev. Biol*. 18 59–88.

A. L. Bak and A. L. Jorgensen. 1984. RNA Polymerase III Control Regions in Retrovirus LTR, Alu-Type Repetitive DNA, and Papovavirus. *J. Theor. Biol*. 108 339–348.

E. P. Geiduschek and G. P. Tocchini-Valentini. 1988. Transcription by RNA Polymerase III. *Annu. Rev. Biochem*. 57 873–914.

8.4b

D. F. Bogenhagen, S. Sakonju, and D. D. Brown. 1980. A Control Region in the Center of the 5S RNA Gene Directs Specific Initiation of Transcription: II. The 3′ Border of the Region. *Cell* 19 27–35.

S. Sakonju, D. F. Bogenhagen, and D. D. Brown. 1980. A Control Region in the Center of the 5S RNA Gene Directs Specific Initiation of Transcription: I. The 5′ Border of the Region. *Cell* 19 13–25.

G. Galli, H. Hofstetter, and M. L. Birnstiel. 1981. Two Conserved Sequence Blocks Within Eukaryotic tRNA Genes Are Major Promoter Elements. *Nature* 294 626–631.

D. Larson, J. Bradford-Wilcox, L. S. Young, and K. U. Sprague. 1983. A Short 5′ Flanking Region Containing Conserved Sequences Is Required for Silkworm Alanine tRNA Gene Activity. *Proc. Natl. Acad. Sci.* U.S.A. 80 3416–3420.

D. F. Bogenhagen. 1985. The Intragenic Control Region of the *Xenopus* 5 S RNA Gene Contains Two Factor A Binding Domains That Must Be Aligned Properly for Efficient Transcription Initiation. *J. Biol. Chem*. 260 6466–6471.

K. C. Raymond, G. J. Raymond, and J. D. Johnson. 1985. *In Vivo* Modulation of Yeast tRNA Gene Expression by 5′-Flanking Sequences. *EMBO J*. 4 2649–2656.

S. J. Sharp, J. Schaack, I. Cooley, D. J. Burke, and D. Soll. 1985. Structure and Transcription of Eukaryotic tRNA Genes. *Crit. Rev. Biochem*. 19 107–144.

E. Ullu and A. M. Weiner. 1985. Upstream Sequences Modulate the Internal Promoter of the Human 7SL RNA Gene. *Nature* 318 371–375.

G. R. Kunkel, R. L. Maser, J. P. Calvet, and T. Pederson. 1986. U6 Small Nuclear RNA Is Transcribed by RNA Polymerase III. *Proc. Natl. Acad. Sci*. U.S.A. 83 8575–8579.

M. J. Morry and J. D. Harding. 1986. Modulation of Transcriptional Activity and Stable Complex Formation by 5′-Flanking Regions of Mouse $tRNA^{His}$ Genes. *Mol. Cell. Biol*. 6 105–115.

J. Chung, D. J. Sussman, R. Zeller, and P. Leder. 1987. The c-*myc* Gene Encodes Superimposed RNA Polymerase II and III Promoters. *Cell* 51 1001–1008.

A. D. Garcia, A. M. O'Connell, and S. J. Sharp. 1987. Formation of an Active Transcription Complex in the *Drosophila melanogaster* 5S RNA Gene Is Dependent on an Upstream Region. *Mol. Cell. Biol*. 7 2046–2051.

S. Murphy, C. diLiegro, and M. Melli. 1987. The *in Vitro* Transcription of the 7SK RNA Gene by RNA Polymerase III Is Dependent Only on the Presence of an Upstream Promoter. *Cell* 51 81–87.

T. Pieler, J. Hamm, and R. G. Roeder. 1987. The 5S Gene Internal Control Region Is Composed of Three Distinct Sequence Elements, Organized as Two Functional Domains with Variable Spacing. *Cell* 48 91–100.

S. J. Sharp and A. D. Garcia. 1988. Transcription of the *Drosophila melanogaster* 5S RNA Gene Requires an Upstream Promoter and Four Intragenic Sequence Elements. *Mol. Cell. Biol*. 8 1266–1274.

E. T. Wilson, D. P. Condliffe, and K. U. Sprague. 1988. Transcriptional Properties of BmX, a Moderately Repetitive Silkworm Gene That Is an RNA Polymerase III Template. *Mol. Cell. Biol*. 8 624–631.

8.4c

A. B. Lassar, P. L. Martin, and R. G. Roeder. 1983. Transcription of Class III Genes: Formation of Preinitiation Complexes. *Science* 222 740–748.

R. E. Baker and B. D. Hall. 1984. Structural Features of Yeast tRNA Genes Which Affect Transcription Factor Binding. *EMBO J*. 3 2793–2800.

D. D. Brown. 1984. The Role of Stable Complexes That Repress and Activate Eucaryotic Genes. *Cell* 37 359–365.

D. R. Smith, I. J. Jackson, and D. D. Brown. 1984. Domains of the Positive Transcription Factor Specific for the *Xenopus* 5S RNA Gene. *Cell* 37 645–652.

T. Enver. 1985. A Pulling Out of Fingers. *Nature* 317 385–386.

W. K. Hoeffler and R. G. Roeder. 1985. Enhancement of RNA Polymerase III Transcription by the E1A Gene Product of Adenovirus. *Cell* 41 955–963.

J. Miller, A. D. McLachlan, and A. Klug. 1985. Repetitive Zinc-Binding Domains in the Protein Transcription Factor IIIA from *Xenopus* Oocytes. *EMBO J*. 4 1609–1614.

D. R. Setzer and D. D. Brown. 1985. Formation and Stability of the 5S RNA Transcription Complex. *J. Biol. Chem*. 260 2483–2492.

M. J. Taylor and J. Segall. 1985. Characterization of

Factors and DNA Sequences Required for Accurate Transcription of the *Saccharomyces cerevisiae* 5S RNA Gene. *J. Biol. Chem.* 260 4531–4540.

L. Fairall, D. Rhodes, and A. Klug. 1986. Mapping of the Sites of Protection on a 5S RNA Gene by the *Xenopus* Transcription Factor IIIA. *J. Mol. Biol.* 192 577–591.

S. C. Harrison. 1986. Fingers and DNA Half-Turns. *Nature* 322 597–598.

D. Rhodes and A. Klug. 1986. An Underlying Repeat in Some Transcriptional Control Sequences Corresponding to Half a Double Helical Turn of DNA. *Cell* 46 123–132.

A. P. Wolffe and D. D. Brown. 1986. DNA Replication *in Vitro* Erases a *Xenopus* 5S RNA Gene Transcription Complex. *Cell* 47 217–227.

A. Klug and D. Rhodes. 1987. "Zinc Fingers": A Novel Protein Motif for Nucleic Acid Recognition. *Trends Biochem. Sci.* 12 464–469.

G. A. McConkey and D. F. Bogenhagen. 1987. Transition Mutations Within the *Xenopus borealis* Somatic 5S RNA Gene Can Have Independent Effects on Transcription and TFIIIA Binding. *Mol. Cell. Biol.* 7 486–494.

W. R. Folk. 1988. Changing Directions in Pol III Transcription. *Genes and Devel.* 2 373–375.

W. K. Hoeffler, R. Kovelman, and R. G. Roeder. 1988. Activation of Transcription Factor IIIC by the Adenovirus E1A Protein. *Cell* 53 907–920.

B. Sollner-Webb. 1988. Surprises in Polymerase III Transcription. *Cell* 52 153–154.

K. E. Vrana, M. E. A. Churchill, T. D. Tullius, and D. D. Brown. 1988. Mapping Functional Regions of Transcription Factor TFIIIA. *Mol. Cell. Biol.* 8 1684–1696.

8.4d

N. R. Cozzarelli, S. P. Gerrard, M. Schlissel, D. D. Brown, and D. F. Bogenhagen. 1983. Purified RNA Polymerase III Accurately and Efficiently Terminates Transcription of 5S RNA Genes. *Cell* 34 829–835.

J. Hess, C. Perez-Stable, G. J. Wu, B. Weir, I. Tinoco, Jr., and C. K. J. Shen. 1985. End-to-End Transcription of an Alu Family Repeat. *J. Mol. Biol.* 184 7–21.

8.4e

W. M. Wormington, M. Schlissel, and D. D. Brown. 1982. Developmental Regulation of *Xenopus* 5S RNA Genes. *Cold Spring Harbor Symp. Quant. Biol.* 47 879–884.

W. M. Wormington and D. D. Brown. 1983. Onset of 5S RNA Gene Regulation During *Xenopus* Embryogenesis. *Devel. Biol.* 99 248–257.

M. S. Schlissel and D. D. Brown. 1984. The Transcriptional Regulation of *Xenopus* 5S RNA Genes in Chromatin: The Roles of Active Stable Transcription Complexes and Histone H1. *Cell* 37 903–913.

D. D. Brown and M. S. Schlissel. 1985. A Positive Transcription Factor Controls the Differential Expression of Two 5S RNA Genes. *Cell* 42 759–767.

M. T. Andrews and D. D. Brown. 1987. Transient Activation of Oocyte 5S RNA Genes in *Xenopus* Embryos by Raising the Level of the Trans-Acting Factor TFIIIA. *Cell* 51 445–453.

L. J. Korn, D. R. Guinta, and J. Y. Tso. 1987. TFIIIA Mediated Control of 5S RNA Gene Expression in *Xenopus*. In R. A. Firtel and E. H. Davidson (eds.), Molecular Approaches to Developmental Biology. Proceedings of the Dupont-Genetech UCLA Symposium, Keystone, Colorado, March 30–April 6, 1986 (UCLA Symposia on Molecular and Cellular Biology, Vol. 51), pp. 25–37. Alan R. Liss, New York.

L. J. Peck, L. Millstein, P. Eversole-Cire, J. M. Gottesfeld, and A. Varshavsky. 1987. Transcriptionally Inactive Oocyte-Type 5S RNA Genes of *Xenopus laevis* Are Complexed with TFIIIA *in Vitro*. *Mol. Cell. Biol.* 7 3503–3510.

A. P. Wolffe and D. D. Brown. 1987. Differential 5S RNA Gene Expression *in Vitro*. *Cell* 51 733–740.

G. A. McConkey and D. F. Bogenhagen. 1988. TFIIIA Binds with Equal Affinity to Somatic and Major Oocyte 5S RNA Genes. *Genes and Devel.* 2 205–214.

A. P. Wolffe and D. D. Brown. 1988. Developmental Regulation of Two 5S Ribosomal RNA Genes. *Science* 241 1626–1632.

8.5b and c

T. R. Cech. 1983. RNA Splicing: Three Themes with Variations. *Cell* 34 713–716.

F. K. Chu, G. F. Maley, F. Maley, and M. Belfort. 1984. Intervening Sequence in the Thymidylate Synthase Gene of Bacteriophage T4. *Proc. Natl. Acad. Sci. U.S.A.* 81 3049–3053.

G. Garriga and A. M. Lambowitz. 1984. RNA Splicing in Neurospora Mitochondria: Self-Splicing of a Mitochondrial Intron *in Vitro*. *Cell* 38 631–641.

H. F. Tabak, G. Van der Horst, K. A. Osinga, and A. C. Arnberg. 1984. Splicing of Large Ribosomal Precursor RNA and Processing of Intron RNA in Yeast Mitochondria. *Cell* 39 623–629.

M. Belfort, J. Pedersen-Lane, D. West, K. Ehrenman, G. Maley, F. Chu, and F. Maley. 1985. Processing of the Intron-Containing Thymidylate Synthase (td) Gene of Phage T4 Is at the RNA Level. *Cell* 41 375–382.

T. R. Cech. 1985. Self-Splicing RNA: Implications for Evolution. *Intl. Rev. Cytol.* 93 3–22.

J. V. Price and T. R. Cech. 1985. Coupling of *Tetrahymena* Ribosomal RNA Splicing to β-Galactosidase Expression in *Escherichia coli*. *Science* 228 719–722.

F. J. Schmidt. 1985. RNA Splicing in Prokaryotes: Bacteriophage T4 Leads the Way. *Cell* 41 339–340.

G. van der Horst and H. F. Tabak. 1985. Self-Splicing of Yeast Mitochondrial Ribosomal and Messenger RNA Precursors. *Cell* 40 759–766.

A. C. Arnberg, G. Van der Horst, and H. F. Tabak. 1986. Formation of Lariats and Cicles in Self-Splicing of the Precursor to the Large Ribosomal RNA of Yeast Mitochondria. *Cell* 44 235–242.

T. R. Cech. 1986. The Generality of Self-Splicing RNA: Relationship to Nuclear mRNA Splicing. *Cell* 44 207–210.

T. R. Cech. 1986. RNA as an Enzyme. *Sci. American* 255 64–75.

F. K. Chu, G. F. Maley, D. K. West, M. Belfort, and F. Maley. 1986. Characterization of the Intron in the Phage T4 Thymidylate Synthase Gene and Evidence for Its Self-Excision from the Primary Transcript. *Cell* 45 157–166.

G. Garriga and A. M. Lambowitz. 1986. Protein-Dependent Splicing of a Group I Intron in Ribonucleoprotein Particles and Soluble Fractions. *Cell* 46 669–680.

J. M. Gott, D. A. Shub, and M. Belfort. 1986. Multiple Self-Splicing Introns in Bacteriophage T4: Evidence from Autocatalytic GTP Labeling of RNA *in Vitro*. *Cell* 47 81–87.

C. L. Peebles, P. S. Perlman, K. L. Mecklenburg, M. L. Petrillo, J. H. Tabor, K. A. Jarrell, and H.-L. Cheng. 1986. A Self-Splicing RNA Excises an Intron Lariat. *Cell* 44 213–223.

C. Schmelzer and R. J. Schweyen. 1986. Self-Splicing of Group II Introns *in Vitro*: Mapping of the Branch Point and Mutational Inhibition of Lariat Formation. *Cell* 46 557–565.

H. F. Tabak and L. A. Grivell. 1986. RNA Catalysis in the Excision of Yeast Mitochondrial Introns. *Trends Genet.* 2 51–55.

R. van der Veen, A. C. Arnberg, G. van der Horst, L. Bonen, H. F. Tabak, and L. A. Grivell. 1986. Excised Group II Introns in Yeast Mitochondria Are Lariats and Can Be Formed by Self-Splicing *in Vitro*. *Cell* 44 225–234.

R. B. Waring, P. Towner, S. J. Minter, and R. W. Davies. 1986. Splice-Site Selection by a Self-Splicing RNA of *Tetrahymena*. *Nature* 321 133–139.

R. A. Akins and A. M. Lambowitz. 1987. A Protein Required for Splicing Group I Introns in *Neurospora* Mitochondria Is Mitochondrial Tyrosyl-tRNA Synthetase or a Derivative Thereof. *Cell* 50 331–345.

T. R. Cech. 1987. The Chemistry of Self-Splicing RNA and RNA Enzymes. *Science* 236 1532–1539.

D. H. Hall, C. M. Povinelli, K. Ehrenman, J. Pedersen-Lane, F. Chu, and M. Belfort. 1987. Two Domains for Splicing in the Intron of the Phage T4 Thymidylate Synthase (td) Gene Established by Nondirected Mutagenesis. *Cell* 48 63–71.

J. Kjems and R. A. Garrett. 1988. Novel Splicing Mechanism for the Ribosomal RNA Intron in the Archaebacterium *Desulfurococcus mobilis*. *Cell* 54 693–703.

J. V. Price and T. R. Cech. 1988. Determinants of the 3′ Splice Site for Self-Splicing of the *Tetrahymena* Pre-rRNA. *Genes and Devel.* 2 1439–1447.

S. A. Woodson and T. R. Cech. 1989. Reverse Self-Splicing of the *Tetrahymena* Group I Intron: Implication for the Directionality of Splicing and for Intron Transposition. *Cell* 57 335–345.

S. Augustin, M. W. Müller, and R. J. Schweyen. 1990. Reverse Self-Splicing of Group II Intron RNAs *in Vitro*. *Nature* 343 383–386.

M. Mörl and C. Schmelzer. 1990. Integration of Group II Intron bI1 into a Foreign RNA by Reversal of the Self-Splicing Reaction *in Vitro*. *Cell* 60 629–636.

8.5d

H. Busch, R. Reddy, L. Rothblum, and Y. C. Choi. 1982. SnRNAs, SnRNPs, and RNA Processing. *Annu. Rev.* Biochem. 51 617–654.

H. Domdey, B. Apostol, R. J. Lin, A. Newman, E. Brody, and J. Abelson. 1984. Lariat Structures Are *in Vivo* Intermediates in Yeast Pre-mRNA Splicing. *Cell* 39 611–621.

C. Hashimoto and J. A. Steitz. 1984. U4 and U6 RNAs Coexist in a Single Small Nuclear Ribonucleoprotein Particle. *Nucleic Acids Res.* 12 3283–3293.

W. Keller. 1984. The RNA Lariat: A New Ring to the Splicing of mRNA Precursors. *Cell* 39 423–425.

J. R. Rodriguez, C. W. Pikielny, and M. Rosbash. 1984. *In Vivo* Characterization of Yeast mRNA Processing Intermediates. *Cell* 39 603–610.

D. L. Black, B. Chabot, and J. A. Steitz. 1985. U2 as Well as U1 Small Nuclear Ribonucleoproteins Are Involved in Premessenger RNA Splicing. *Cell* 42 737–750.

M. M. Konarska, R. A. Padgett, and P. A. Sharp. 1985. Trans Splicing of mRNA Precursors *in Vitro*. *Cell* 42 165–171.

A. J. Newman, R. J. Lin, S. C. Cheng, and J. Abelson. 1985. Molecular Consequences of Specific Intron Mutations on Yeast mRNA Splicing *in Vivo* and *in Vitro*. *Cell* 42 335–344.

P. A. Sharp. 1985. On the Origin of RNA Splicing and Introns. *Cell* 42 397–400.

D. Solnick. 1985. Trans Splicing of mRNA Precursors. *Cell* 42 157–164.

M. Ares, Jr. 1986. U2 RNA from Yeast Is Unexpectedly Large and Contains Homology to Vertebrate U4, U5 and U6 Small Nuclear RNAs. *Cell* 47 49–59.

T. R. Cech. 1986. The Generality of Self-Splicing RNA: Relationship to Nuclear mRNA Splicing. *Cell* 44 207–210.

V. Gerke and J. A. Steitz. 1986. A Protein Associated with Small Nuclear Ribonucleoprotein Particles Recognizes the 3′ Splice Site of Premessenger RNA. *Cell* 47 973–984.

M. R. Green. 1986. Pre-mRNA Splicing. *Annu. Rev. Genet.* 20 671–708.

C. Guthrie. 1986. Finding Functions for Small Nuclear RNAs in Yeast. *Trends Biochem. Sci.* 11 430–434.

W. J. Murphy, K. P. Watkins, and N. Agabian. 1986. Identification of a Novel Y Branch Structure as an Intermediate in Trypanosome mRNA Processing: Evidence for *Trans* Splicing. *Cell* 47 517–525.

R. A. Padgett, P. J. Grabowski, M. M. Konarska, S. Seiler, and P. A. Sharp. 1986. Splicing of Messenger RNA Precursors. *Annu. Rev. Biochem.* 55 1119–1150.

C. W. Pikielny and M. Robash. 1986. Specific Small Nuclear RNAs Are Associated with Yeast Spliceosomes. *Cell* 45 869–877.

R. E. Sutton and J. C. Boothroyd. 1986. Evidence for *Trans* Splicing in Trypanosomes. *Cell* 47 527–535.

J. Tazi, C. Alibert, J. Temsamani, I. Reveillaud, G. Cathala, C. Brunel, and P. Jeanteur. 1986. A Protein That Specifically Recognizes the 3′ Splice Site of Mammalian Pre-mRNA Introns Is Associated with a Small Nuclear Ribonucleoprotein. *Cell* 47 755–766.

L. H. T. Van der Ploeg. 1986. Discontinuous Transcription and Splicing in Trypanosomes. *Cell* 47 479–480.

Y. Zhuang and A. M. Weiner. 1986. A Compensatory Base Change in U1 snRNA Suppresses a 5′ Splice Site Mutation. *Cell* 46 827–835.

A. Bindereif and M. R. Green. 1987. An Ordered Pathway of snRNP Binding During Mammalian Pre-mRNA Splicing Complex Assembly. *EMBO J.* 6 2415–2424.

S. C. Cheng and J. Abelson. 1987. Spliceosome Assembly in Yeast. *Genes and Devel.* 1 1014–1027.

J. R. Couto, J. Tamm, R. Parker, and C. Guthrie. 1987. A *Trans*-Acting Suppressor Restores Splicing of a Yeast Intron with a Branch Point Mutation. *Genes and Devel.* 1 445–455.

M. M. Konarska and P. A. Sharp. 1987. Interactions Between Small Nuclear Ribonucleoprotein Particles in Formation of Spliceosomes. *Cell* 49 763–774.

M. Krause and D. Hirsh. 1987. A *Trans*-Spliced Leader Sequence on Actin mRNA in *C. elegans*. *Cell* 49 753–761.

M. Lossky, G. J. Anderson, S. P. Jackson, and J. Beggs. 1987. Identification of a Yeast snRNP Protein and Detection of snRNP-snRNP Interactions. *Cell* 51 1019–1026.

R. Luhrmann. 1987. snRNP Proteins. In M. L. Birnstiel (ed.), Structure and Function of Major and Minor Small Nuclear Ribonucleoprotein Particles, pp. 71–99. Springer Verlag, Heidelberg.

T. Maniatis and R. Reed. 1987. The Role of Small Nuclear Ribonucleoprotein Particles in Pre-mRNA Splicing. *Nature* 325 673–678.

R. Parker, P. G. Siliciano, and C. Guthrie. 1987. Recognition of the TACTAAC Box During mRNA Splicing in Yeast Involves Base Pairing to the U2-like snRNA. *Cell* 49 229–239.

P. A. Sharp. 1987. *Trans* Splicing: Variation on a Familiar Theme? *Cell* 50 147–148.

S. Bektesh, K. Van Doren, and D. Hirsh. 1988. Presence of the *Caenorhabditis elegans* Spliced Leader on Different mRNAs and in Different Genera of Nematodes. *Genes and Devel.* 2 1277–1283.

C. Guthrie and B. Patterson. 1988. Spliceosomal snRNAs. *Annu. Rev. Genet.* 22 387–419.

D. H. Kedes and J. A. Steitz. 1988. Correct *in Vivo* Splicing of the Mouse Immunoglobulin k Light-Chain Pre-mRNA Is Dependent on 5′ Splice-Site Position Even in the Absence of Transcription. *Genes and Devel.* 2 1448–1459.

A. Kramer. 1988. Presplicing Complex Formation Requires Two Proteins and U2 snRNP. *Genes and Devel.* 2 1155–1167.

R. Reed, J. Griffith, and T. Maniatis. 1988. Purification and Visualization of Native Spliceosomes. *Cell* 53 949–961.

R. Reed and T. Maniatis. 1988. The Role of the Mammalian Branchpoint Sequence in Pre-mRNA Splicing. *Genes and Devel.* 2 1268–1276.

B. Ruskin, P. D. Zamore, and M. R. Green. 1988. A Factor, U2AF, Is Required for U2 snRNP Binding and Splicing Complex Assembly. *Cell* 52 207–219.

P. G. Siliciano and C. Guthrie. 1988. 5′ Splice Site Selection in Yeast: Genetic Alterations in Base-Pairing with U1 Reveal Additional Requirements. *Genes and Devel.* 2 1258–1267.

J. A. Steitz. 1988. "Snurps." *Sci. American* 258(6) 56–63.

8.5e

C. L. Peebles, R. C. Ogden, G. Knapp, and J. Abelson. 1979. Splicing of Yeast tRNA Precursors: A Two-Stage Reaction. *Cell* 18 27–35.

E. M. De Robertis, P. Black, and K. Nishikura. 1981. Intranuclear Location of the tRNA Splicing Enzymes. *Cell* 23 89–93.

C. L. Greer, C. L. Peebles, P. Gegenheimer, and J. Abelson. 1983. Mechanism of Action of a Yeast RNA Ligase in tRNA Splicing. *Cell* 32 537–546.

C. L. Peebles, D. Gegenheimer, and J. Abelson. 1983. Precise Excision of Intervening Sequences from Precursor tRNAs by a Membrane-Associated Yeast Endonuclease. *Cell* 32 525–536.

M. I. Baldi, E. Mattoccia, S. Ciafre, D. G. Attardi, and G. P. Tocchini-Valentini. 1986. Binding and Cleavage of pre-tRNA by the *Xenopus* Splicing Endonuclease: Two Separable Steps of the Intron Excision Reaction. *Cell* 47 965–971.

C. L. Greer. 1986. Assembly of a tRNA Splicing Complex: Evidence for Concerted Excision and Joining Steps in Splicing *in Vitro*. *Mol. Cell. Biol.* 6 635–644.

E. M. Phizicky, R. C. Schwartz, and J. Abelson. 1986. *S. cerevisiae* tRNA Ligase: Purification of the Protein and Isolation of the Structural Gene. *J. Biol. Chem.* 261 2978–2986.

C. L. Greer, D. Soll, and I. Willis. 1987. Substrate Recognition and Identification of Splice Sites by the tRNA-Splicing Endonuclease and Ligase from *Saccharomyces cerevisiae*. *Mol. Cell. Biol.* 7 76–84.

V. M. Reyes and J. Abelson. 1988. Substrate Recognition and Splice Site Determination in Yeast tRNA Splicing. *Cell* 55 719–730.

N. Stange, H. J. Gross, and H. Beier. 1988. Wheat Germ Splicing Endonuclease Is Highly Specific for Plant Pre-tRNAs. *EMBO J.* 7 3823–3828.

M. Winey, I. Edelman, and M. R. Culbertson. 1989. A Synthetic Intron in a Naturally Intronless Yeast Pre-tRNA Is Spliced Efficiently *in Vivo*. *Mol. Cell. Biol.* 9 329–331.

8.5f

M. G. Rosenfeld, C. R. Lin, S. G. Amara, L. Stolarsky, B. A. Roos, E. S. Ong, and R. M. Evans. 1982. Calcitonin mRNA Polymorphism: Peptide Switching Associated with Alternative RNA Splicing Events. *Proc. Natl. Acad. Sci.* U.S.A. 79 1717–1721.

M. Kress, D. Glaros, G. Khoury, and G. Jay. 1983. Alternative RNA Splicing in Expression of the H-2K Gene. *Nature* 306 602–604.

D. W. Chung and E. W. Davie. 1984. γ and γ' Chains of Human Fibrinogen Are Produced by Alternative mRNA Processing. *Biochemistry* 23 4232–4236.

F. de Ferra, H. Engh, L. Hudson, J. Kamholz, C. Puckett, S. Molineaux, and R. A. Lazzarini. 1985. Alternative Splicing Accounts for the Four Forms of Myelin Basic Protein. *Cell* 43 721–727.

A. R. Kornblith, K. Umezawa, K. Vibe-Pedersen, and F. E. Baralle. 1985. Primary Structure of Human Fibronectin: Differential Splicing May Generate at Least 10 Polypeptides from a Single Gene. *EMBO J.* 4 1755–1759.

D. Solnick. 1985. Alternative Splicing Caused by RNA Secondary Structure. *Cell* 43 667–676.

R. Zamoyska, A. C. Vollmer, K. C. Sizer, C. W. Liaw, and J. R. Parnes. 1985. Two Lyt-2 Polypeptides Arise from a Single Gene by Alternative Splicing Patterns of mRNA. *Cell* 43 153–163.

Y. Ben-Neriah, A. Bernards, M. Paskind, G. Q. Daley, and D. Baltimore. 1986. Alternative 5′ Exons in *c-abl* mRNA. *Cell* 44 577–586.

S. E. Leff and M. G. Rosenfeld. 1986. Complex Transcriptional Units: Diversity in Gene Expression by Alternative RNA Processing. *Annu. Rev. Biochem.* 55 1091–1117.

J. I. Paul, J. E. Schwarzbauer, J. W. Tamkun, and R. O. Hynes. 1986. Cell-Type-Specific Fibronectin Subunits Generated by Alternative Splicing. *J. Biol. Chem.* 261 12258–12265.

R. Reed and T. Maniatis. 1986. A Role for Exon Sequences and Splice-Site Proximity in Splice-Site Selection. *Cell* 46 681–690.

A. Andreadis, M. E. Gallego, and B. Nadal-Ginard. 1987. Generation of Protein Isoform Diversity by Alternative Splicing: Mechanistic and Biological Implications. *Annu. Rev. Cell Biol.* 3 207–242.

R. E. Breitbart, A. Andreadis, and B. Nadal-Ginard. 1987. Alternative Splicing: A Ubiquitous Mechanism for the Generation of Multiple Protein Isoforms from Single Genes. *Annu. Rev. Biochem.* 56 467–495.

R. E. Breitbart and B. Nadal-Ginard. 1987. Developmentally Induced, Muscle-Specific *Trans* Factors Control the Differential Splicing of Alternative and Constitutive Troponin T Exons. *Cell* 49 793–803.

J. C. S. Noble, C. Prives, and J. L. Manley. 1988. Alternative Splicing of SV40 Early Pre-mRNA Is Determined by Branch Site Selection. *Genes and Devel.* 2 1460–1475.

J. C. S. Noble Z. Q. Pan, C. Prives, and J. L. Manley. 1987. Splicing of SV40 Early Pre-mRNA to Large T and Small t mRNAs Utilizes Different Patterns of Lariat Branch Sites. *Cell* 50 227–236.

B. S. Baker. 1989. Sex in Flies: The Splice of Life. *Nature* 340 521–524.

8.6a

C. O. Pabo and R. T. Sauer. 1984. Protein-DNA Recognition. *Annu. Rev. Biochem.* 53 293–321.

J. E. Anderson, M. Ptashne, and S. C. Harrison. 1985. A Phage Repressor-Operator Complex at 7 Å Resolution. *Nature* 316 596–605.

W. J. Gehring. 1985. The Homeo Box: A Key to the Understanding of Development? *Cell* 40 3–5.

J. M. Berg. 1986. More Metal-Binding Fingers. *Nature* 319 264–265.

J. M. Berg. 1986. Potential Metal-Binding Domains in Nucleic Acid Binding Proteins. *Science* 232 485–487.

U. B. Rosenberg, C. Schroder, A. Preiss, A. Kienlin, S. Cote, I. Riede, and H. Jackle. 1986. Structural Homology of the Product of the *Drosophila kruppel* Gene with *Xenopus* Transcription Factor IIIA. *Nature* 319 336–339.

A. Klug and D. Rhodes. 1987. "Zinc Fingers": A Novel Protein Motif for Nucleic Acid Recognition. *Trends Biochem. Sci.* 12 464–469.

K. Struhl. 1987. The DNA-Binding Domains of the Jun Oncoprotein and the Yeast CN4 Transcriptional Activator Protein Are Functionally Homologous. *Cell* 50 841–846.

J. M. Berg. 1988. Proposed Structure for the Zinc-Binding Domains from Transcription Factor IIIA and Related Proteins. *Proc. Natl. Acad. Sci.* U.S.A. 85 99–102.

M. Bodner, J. Castrillo, L. E. Theill, T. Deerinck, M. Ellisman, and M. Karin. 1988. The Pituitary-Specific Transcription Factor GHF-1 Is a Homeobox-Containing Protein. *Cell* 55 505–518.

R. Chiu, W. J. Boyle, J. Meek, T. Smeal, T. Hunter, and M. Karin. 1988. The c-Fos Protein Interacts with c-Jun/AP-1 to Stimulate Transcription of AP-1 Responsive Genes. *Cell* 54 541–552.

R. G. Clerc, L. M. Corcoran, H. H. LeBowitz, and D. Baltimore. 1988. The B Cell Specific Oct-2 Protein Contains POU-Box and Homeo Box Type Domains. *Genes and Devel.* 2 1570–1581.

T. Curran and B. R. Franza, Jr. 1988. Fos and Jun: The AP-1 Connection. *Cell* 55 395–397.

C. Desplan, J. Theis, and P. H. O'Farrell. 1988. The Sequence Specificity of Homeodomain-DNA Interaction. *Cell* 54 1081–1090.

R. M. Evans and S. M. Hollenberg. 1988. Zinc Fingers: Gilt by Association. *Cell* 52 1–3.

T. D. Halazonetis, K. Georgopoulos, M. E. Greenberg, and P. Leder. 1988. c-Jun Dimerizes with Itself and with c-Fos, Forming Complexes of Different DNA Binding Affinities. *Cell* 55 917–924.

W. Herr, R. A. Sturm, R. G. Clerc, L. M. Corcoran, D. Baltimore, P. A. Sharp, H. A. Ingraham, M. G. Rosenfeld, M. Finney, G. Ruvkun, and H. R. Horvitz. 1988. The POU Domain: A Large Conserved Region in the Mammalian Pit-1, Oct-1, Oct-2, and *Caenorhabditis elegans Unc*-86 Gene Products. *Genes and Devel.* 2 1513–1516.

H. A. Ingraham, R. Chen, H. J. Mangalam, H. P. Elsholtz, S. E. Flynn, C. R. Lin, D. M. Simmons, L. Swanson, and M. G. Rosenfeld. 1988. A Tissue-Specific Transcription Factor Containing a Homeodomain Specifies a Pituitary Phenotype. *Cell* 55 519–529.

T. Kouzarides and E. Ziff. 1988. The Role of the Leucine Zipper in the Fos-Jun Interaction. *Nature* 336 646–651.

W. H. Landschulz, P. F. Johnson, and S. L. McKnight. 1988. The Leucine Zipper: A Hypothetical Structure Common to a New Class of DNA Binding Proteins. *Science* 240 1759–1764.

M. Levine and T. Hoey. 1988. Homeobox Proteins as Sequence-Specific Transcription Factors. *Cell* 55 537–540.

Y. Nakabeppu, K. Ryder, and D. Nathans. 1988. DNA Binding Activities of Three Murine Jun Proteins: Stimulation by Fos. *Cell* 55 907–915.

F. J. Rauscher III, P. J. Voulalas, B. R. Franza, Jr., and T. Curran. 1988. Fos and Jun Bind Cooperatively to the AP-1 Site: Reconstitution *in Vitro*. *Genes and Devel.* 2 1687–1699.

M. Robertson. 1988. Homeo Boxes, POU Proteins and the Limits to Promiscuity. *Nature* 336 522–524.

R. T. Sauer, D. L. Smith, and A. D. Johnson. 1988. Flexibility of the Yeast a2 Repressor Enables It to Occupy the Ends of Its Operator, Leaving the Center Free. *Genes and Devel.* 2 807–816.

R. A. Sturm, G. Das, and W. Herr. 1988. The Ubiquitous Octamer-Binding Protein Oct-1 Contains a POU Domain with a Homeo Box Subdomain. *Genes and Devel.* 2 1582–1599.

R. G. Brennan and B. W. Matthews. 1989. Structural Basis of DNA-Protein Recognition. *Trends Biochem. Sci.* 14 286–290.

W. H. Landschulz, P. F. Johnson, and S. L. McKnight. 1989. The DNA Binding Domain of the Rat Liver Nuclear Protein C/EBP is Bipartite. *Science* 243 1681–1687.

M. S. Lee, G. P. Gippert, K. V. Soman, D. A. Case, and P. E. Wright. 1989. Three-Dimensional Solution Structure of a Single Zinc Finger DNA-Binding Domain. *Science* 245 635–637.

A. R. Oliphant, C. J. Branch, and K. Struhl. 1989. Defining the Sequence Specificity of DNA-Binding Proteins by Selecting Binding Sites from Random-Sequence Oligonucleotides: Analysis of Yeast GCN4 Protein. *Mol. Cell. Biol.* 9 2944–2949.

E. K. O'Shea, R. Rutkowski, W. F. Stafford III, and P. S. Kim. 1989. Preferential Heterodimer Formation by Isolated Leucine Zippers from Fos and Jun. *Science* 245 646–648.

T. B. Rajavashisth, A. K. Taylor, A. Andalibi, K. L. Svenson, and A. J. Lusis. 1989. Identification of a Zinc Finger Protein That Binds to the Sterol Regulatory Element. *Science* 245 640–643.

M. P. Scott, J. W. Tamkun, and G. W. Hartzell III. 1989. The Structure and Function of the Homeodomain. *Biochim. Biophys. Acta. Ser. Rev. Cancer* 989(1) 25–48.

W. Shaffner. 1989. How Do Different Transcription Factors Binding the Same DNA Sequence Sort Out Their Jobs? *Trends Genet.* 5 37–38.

K. Struhl. 1989. Helix-Turn-Helix, Zinc-Finger, and Leucine-Zipper Motifs for Eukaryotic Transcriptional Regulatory Proteins. *Trends Biochem. Sci.* 14 127–140.

C. V. E. Wright, K. W. Y. Cho, G. Oliver, and E. M. DeRobertis. 1989. Vertebrate Homeodomain Proteins: Families of Region-Specific Transcription Factors. *Trends Biochem. Sci.* 14 52–56.

8.6b

I. A. Hope and K. Struhl. 1986. Functional Dissection of a Eukaryotic Transcriptional Activator Protein, GCN4 of Yeast. *Cell* 46 885–894.

L. Keegan, G. Gill, and M. Ptashne. 1986. Separation of DNA Binding from the Transcription-Activating Function of Eukaryotic Regulatory Protein. *Science* 231 699–704.

J. Ma and M. Ptashne. 1987. A New Class of Yeast Transcriptional Activators. *Cell* 51 113–119.

K. Struhl. 1987. Promoters, Activator Proteins, and the Mechanism of Transcriptional Initiation in Yeast. *Cell* 49 295–297.

L. Guarente. 1988. UASs and Enhancers: Common Mechanism of Transcriptional Activation in Yeast and Mammals. *Cell* 52 303–305.

I. A. Hope, S. Mahadevan, and K. Struhl. 1988. Structural and Functional Characterization of the Short Acidic Transcriptional Activation Region of Yeast GCN4 Protein. *Nature* 333 635–640.

S. P. Jackson and R. Tjian. 1988. O-Glycosylation of Eukaryotic Transcription Factors: Implications for Mechanisms of Transcriptional Regulation. *Cell* 55 125–133.

M. Ptashne. 1988. How Eukaryotic Transcriptional Activators Work. *Nature* 335 683–689.

K. Struhl. 1988. The JUN Oncoprotein, a Vertebrate Transcription Factor, Activates Transcription in Yeast. *Nature* 332 649–650.

J. C. Wang and G. N. Giaever. 1988. Action at a Distance Along a DNA. *Science* 240 300–304.

J. W. Lillie and M. R. Green. 1989. Transcription Activation by the Adenovirus Ela Protein. *Nature* 338 39–44.

C. V. E. Wright, K. W. Y. Cho, G. Oliver, and E. M. DeRobertis. 1989. Vertebrate Homeo Domain Proteins: Families of Region-Specific Transcription Factors. *Trends Biochem. Sci.* 14 52–56.

8.7a

J. R. Paulson and U. K. Laemmli. 1977. The Structure of Histone-Depleted Metaphase Chromosomes. *Cell* 12 817–826.

M. P. F. Marsden and U. K. Laemmli. 1979. Metaphase Chromosome Structure: Evidence for a Radial Loop Model. *Cell* 17 849–858.

C. Wu, P. M. Bingham, K. J. Livak, R. Holmgren, and S. C. R. Elgin. 1979. The Chromatin Structure of Specific Genes: I. Evidence for Higher Order Domains of Defined DNA Sequence. *Cell* 16 797–806.

C. Wu, Y.-C. Wong, and S. C. R. Elgin. 1979. The Chromatin Structure of Specific Genes: II. Disruption of Chromatin Structure During Gene Activity. *Cell* 16 807–814.

S. Saragosti, G. Moyne, and M. Yaniv. 1980. Absence of Nucleosomes in a Fraction of SV40 Chromatin Between the Origin of Replication and the Region Coding for the Late Leader RNA. *Cell* 20 65–73.

H. Weintraub. 1985. Assembly and Propagation of Repressed and Derepressed Chromosomal States. *Cell* 42 705–711.

M. Yaniv and S. Cereghini. 1986. Structure of Transcriptionally Active Chromatin. *Crit. Rev. Biochem.* 31 1–26.

Y. Lorch, J. W. LaPointe, and R. D. Kornberg. 1987. Nucleosomes Inhibit the Initiation of Transcription but Allow Chain Elongation with the Displacement of Histones. *Cell* 49 203–210.

M. A. Goldman. 1988. The Chromatin Domain as a Unit of Gene Regulation. *BioEssays* 9 51–55.

Y. Lorch, J. W. LaPointe, and R. D. Kornberg. 1988. On the Displacement of Histones from DNA by Transciption. *Cell* 55 743–744.

8.7b

M. Gellert. 1981. DNA Topoisomerases. *Annu. Rev. Biochem.* 50 879–910.

A. Rich, A. Mondheim, and A. H.-J. Wang. 1984. The Chemistry and Biology of Left-Handed Z-DNA. *Annu. Rev. Biochem.* 53 791–846.

J. C. Wang. 1985. DNA Toposiomerases. *Annu. Rev. Biochem.* 54 665–697.

A. Jaworski, W.-T. Hsieh, J. A. Blaho, J. E. Larson, and R. D. Wells. 1987. Left-Handed DNA *in Vivo*. *Science* 238 773–777.

F. Lancillotti, M. C. Lopez, P. Arias, and C. Alonso. 1987. Z-DNA in Transcriptionally Active Chromosomes. *Proc. Natl. Acad. Sci.* U.S.A. 84 1560–1564.

R. D. Wells. 1988. Unusual DNA Structures. *J. Biol. Chem.* 263 1095–1098.

W. Zacharias, A. Jaworski, J. E. Larson, and R. D. Wells. 1988. The B- to Z-DNA Equilibrium *in Vivo* Is Perturbed by Biological Processes. *Proc. Natl. Acad. Sci.* U.S.A. 85 7069–7073.

H. Zhang, J. C. Wang, and L. F. Liu. 1988. Involvement of DNA Topoisomerase I in Transcription of Human Ribosomal RNA Genes. *Proc. Natl. Acad. Sci.* U.S.A. 85 1060–1064.

A. R. Rahmouni and R. D. Wells. 1989. Localized Supercoiling Stabilized Short (CG) Sequences in the Z-Structure *in Vivo*. *Science* 246 358–363.

B. Wittig, T. Dorbic, and A. Rich. 1989. The Level of Z-DNA in Metabolically Active, Permeabilized Mammalian Cell Nuclei Is Regulated by Torsional Strain. *J. Cell Biol.* 108 755–764.

8.7c

L. H. T. van der Ploeg and R. A. Flavell. 1980. DNA Methylation in the Human $\gamma\delta\beta$-Globin Locus in Erythroid and Nonerythroid Tissues. *Cell* 19 947–958.

M. Busslinger, J. Hurst, and R. A. Flavell. 1983. DNA Methylation and the Regulation of Globin Gene Expression. *Cell* 34 197–206.

W. Doerfler. 1983. DNA Methylation and Gene Activity. *Annu. Rev. Biochem.* 52 93–124.

A. Bolden, C. Ward, J. A. Siedlecki, and A. Weissbach. 1984. DNA Methylation. Inhibition of *de Novo* and Maintenance Methylation *in Vitro* by RNA and Synthetic Polynucleotides. *J. Biol.* Chem. 259 12437–12443.

A. Bird, M. Taggart, M. Frommer, O. J. Miller, and D. Macleod. 1985. A Fraction of the Mouse Genome That Is Derived from Islands of Nonmethylated, CpG-Rich DNA. *Cell* 40 91–99.

G. L. Cantoni and A. Razin (eds.). 1985. Biochemistry and Biology of DNA Methylation, vol. 198, Progress in Clinical and Biological Research. Alan R. Liss, New York.

P. A. Jones. 1985. Altering Gene Expression with 5-Azacytidine. *Cell* 40 485–486.

A. P. Bird. 1986. CpG-Rich Islands and the Function of DNA Methylation. *Nature* 321 209–213.

M. Monk. 1986. Methylation and the X Chromosome. *BioEssays* 4 204–208.

A. Razin, M. Szyf, T. Kafri, M. Roll, H. Giloh, S. Scarpa, D. Carotti, and G. L. Cantoni. 1986. Replacement of 3-Methylcytosine by Cytosine: A Possible Mechanism for Transient DNA Demethylation During Differentiation. *Proc. Natl. Acad. Sci.* U.S.A. 83 2827–2831.

A. P. Bird. 1987. CpG Islands as Gene Markers in the Vertebrate Nucleus. *Trends Genet.* 3 342–347.

R. Holliday. 1987. The Inheritance of Epigenetic Defects. *Science* 238 163–170.

L. F. Lock, N. Takagi, and G. R. Martin. 1987. Methylation of the *Hprt* Gene on the Inactive X Occurs After Chromosome Inactivation. *Cell* 48 39–46.

M. Monk. 1987. Genomic Imprinting: Memories of Mother and Father. *Nature* 328 203–204.

J. P. Sanford, H. J. Clark, V. M. Chapmen, and J. Rossant. 1987. Differences in DNA Methylation During Oogenesis and Spermatogenesis and Their Persistence During Early Embryogenesis in the Mouse. *Genes and Devel.* 1 1039–1046.

J. L. Swain, T. A. Stewart, and P. Leder. 1987. Parental Legacy Determines Methylation and Expression of an Autosomal Transgene: A Molecular Mechanism for Parental Imprinting. *Cell* 50 719–727.

G. Theiss, R. Schleicher, G. Schimpff-Weiland, and H. Follman. 1987. DNA Methylation in Wheat. Purification and Properties of DNA Methyltransferase. *J. Eur. Biochem.* 167 89–96.

H. Cedar. 1988. DNA Methylation and Gene Activity. *Cell* 53 3–4.

M. Holler, G. Westin, J. Jiricny, and W. Schaffner. 1988. Sp1 Transcription Factor Binds DNA and Activates Transcription Even When the Binding Site Is CpG Methylated. *Genes and Devel.* 2 1127–1135.

R. Khan, X.-Y. Zhang, P. C. Supakar, K. C. Ehrlich, and M. Ehrlich. 1988. Human Methylated DNA-Binding Protein. *J. Biol. Chem.* 263 14374–14383.

M. Monk. 1988. Genomic Imprinting. *Genes and Devel.* 2 921–925.

W. Doerfler. 1989. Complexities in Gene Regulation by Promoter Methylation. *Nucleic Acids and Mol. Biol.* 3 92–119.

R. Holliday. 1989. A Different Kind of Inheritance. *Sci. American* 260(6) 60–70.

S. M. M. Iguchi-Ariga and W. Schaffner. 1989. CpG Methylation of the cAMP-Response Enhancer/Promoter Sequence TGACGTCA Abolishes Specific Factor Binding as Well as Transcriptional Activation. *Genes and Devel.* 3 612–619.

8.7d

G. R. Fink. 1986. Translational Control of Transcription in Eukaryotes. *Cell* 45 155–156.

T. Morris, F. Marashi, L. Weber, E. Hickey, D. Greenspan, J. Bonner, J. Stein, and G. Stein. 1986. Involvement of the 5′-Leader Sequence in Coupling the Stability of a Human H3 Histone mRNA with DNA Replication. *Proc. Natl. Acad. Sci.* U.S.A. 83 981–985.

P. P. Mueller and A. G. Hinnebusch. 1986. Multiple Upstream AUG Codons Mediate Translational Control of *GCN4*. *Cell* 45 201–207.

G. Shaw and R. Kamen. 1986. A Conserved AU Sequence from the 3′ Untranslated Region of GM-CSF mRNA Mediates Selective mRNA Degradation. *Cell* 46 659–667.

G. Brawerman. 1987. Determinants of Messenger RNA Stability. *Cell* 48 5–6.

M. W. Hentze, S. W. Caughman, T. A. Roualt, J. G. Barriocanal, A. Dancis, J. B. Harford, and R. D. Klausner. 1987. Identification of the Iron-Responsive Element for the Translational Regulation of Human Ferritin mRNA. *Science* 238 1570–1573.

M. W. Hentze, T. A. Rouault, S. W. Caughman, A. Dancis, J. B. Harford, and R. D. Klausner. 1987. A *cis*-Acting Element Is Necessary and Sufficient for Translational Regulation of Human Ferritin Expression in Response to Iron. *Proc. Natl. Acad. Sci.* U.S.A. 84 6730–6734.

J. S. Pachter, T. J. Yen, and D. W. Cleveland. 1987. Autoregulation of Tubulin Expression Is Achieved Through Specific Degradation of Polysomal Tubulin mRNAs. *Cell* 51 283–292.

T. A. Rouault, M. W. Hentze, A. Dancis, W. Caughman, J. B. Harford, and R. D. Klausner. 1987. Influence of Altered Transcription on the Translational Control of Human Ferritin Expression. *Proc. Natl. Acad. Sci.* U.S.A. 84 6335–6339.

J. L. Casey, B. Di Jeso, K. Rao, R. D. Klausner, and J. B. Harford. 1988. Two Genetic Loci Participate in the Regulation by Iron of the Gene for the Human Transferrin Receptor. *Proc. Natl. Acad. Sci.* U.S.A. 85 1787–1791.

J. L. Casey, M. W. Hentze, D. M. Koeller, S. W. Caughman, T. A. Rouault, R. D. Klausner, and J. B. Harford. 1988. Iron-Responsive Elements: Regulatory RNA Sequences That Control mRNA Levels and Translation. *Science* 240 924–927.

S. W. Caughman, M. W. Hentze, T. A. Rouault, J. B. Harford, and R. D. Klausner. 1988. The Iron-Responsive Element Is the Single Element Responsible for Iron-Dependent Translational Regulation of Ferritin Biosynthesis. *J. Biol. Chem.* 263 19048–19052.

M. W. Hentze, S. W. Caughman, J. L. Casey, D. M. Koeller, T. A. Rouault, J. B. Harford, and R. D. Klausner. 1988. A Model for the Structure and Functions of Iron-Responsive Elements. *Gene* 72 201–208.

A. G. Hinnebusch. 1988. Mechanisms of Gene Regulation in the General Control of Amino Acid Biosynthesis in *Saccharomyces cerevisiae*. *Microbiol. Rev.* 52 248–273.

R. D. Klausner. 1988. From Receptors to Genes—Insights from Molecular Iron Metabolism. *Clin. Res.* 36 494–500.

T. A. Rouault, M. W. Hentze, S. W. Caughman, J. B. Harford, and R. D. Klausner. 1988. Binding of a Cytosolic Protein to the Iron-Responsive Element of Human Ferritin Messenger RNA. *Science* 241 1207–1210.

T. Wilson and R. Treisman. 1988. Removal of Poly(A) and Consequent Degradation of c-*fos* mRNA Facilitated by 3′ AU-Rich Sequences. *Nature* 336 396–399.

T. J. Yen, D. A. Gay, J. S. Pachter, and D. W. Cleveland. 1988. Autoregulated Changes in Stability of Polyribosome-Bound β-Tubulin mRNAs Are Specified by the First 13 Translated Nucleotides. *Mol. Cell. Biol.* 8 1224–1235.

T. J. Yen, P. S. Machlin, and D. W. Cleveland. 1988. Autoregulated Instability of β-Tubulin mRNAs by Recognition of the Nascent Amino Terminus of β-Tubulin. *Nature* 334 580–585.

M. W. Hentze, T. A. Rouault, J. B. Harford, and R. D. Klausner. 1989. Oxidation-Reduction and the Molecular Mechanism of a Regulatory RNA-Protein Interaction. *Science* 244 357–359.

C. J. Paddon, E. M. Hannig, and A. G. Hinnebusch. 1989. Amino Acid Sequence Similarity Between GCN3 and GCD2, Positive and Negative Translational Regulators of *GCN4*: Evidence for Antagonism by Competition. *Genetics* 122 551–559.

D. Tzamarias, I. Roussou, and G. Thireos. 1989. Coupling of GCN4 mRNA Translational Activation with Decreased Rates of Polypeptide Chain Initiation. *Cell* 57 947–954.

R. C. Wek, B. M. Jackson, and A. G. Hinnebusch. 1989. Juxtaposition of Domains Homologous to Protein Kinases and Histidyl-tRNA Synthetases in GCN2 Protein Suggests a Mechanism for Coupling *GCN4* Expression to Amino Acid Availability. *Proc. Natl. Acad. Sci.* U.S.A. 86 4579–4583.

9.1a

J. Battey, E. E. Max, W. O. McBride, D. Swan, and P. Leder. 1982. A Processed Human Immunoglobulin ε Gene Has Moved to Chromosome 9. *Proc. Natl. Acad. Sci.* U.S.A. 79 5956–5959.

E. F. Vanin. 1985. Processed Pseudogenes: Characteristics and Evolution. *Annu. Rev. Genet.* 19 253–272.

G. Bernardi. 1989. The Isochore Organization of the Human Genome. *Annu. Rev. Genet.* 23 637–641.

9.1b

E. O. Long and I. B. Dawid. 1980. Repeated Genes in Eukaryotes. *Annu. Rev. Biochem.* 49 727–764.

E. Gilson, J. M. Clement, D. Perrin, and M. Hofnung. 1987. Palindromic Units: A Case of Highly Repetitive DNA Sequences in Bacteria. *Trends Genet.* 3 226–230.

Y. Yang and G. F-L. Ames. 1988. DNA Gyrase Binds to the Family of Prokaryotic Repetitive Extragenic Palindromic Sequences. *Proc. Natl. Acad. Sci.* U.S.A. 85 8850–8854.

9.1c

A. C. Wilson, S. S. Carlson, and T. J. White. 1977. Biochemical Evolution. *Annu. Rev. Biochem.* 46 573–639.

D. Baltimore. 1981. Gene Conversion: Some Implications for Immunoglobulin Genes. *Cell* 24 592–594.

F. A. Perler, A. Efstratiadis, P. Lomedico, W. Gilbert, R. Kolodner, and J. Dodgson. 1981. The Evolution of Genes: The Chicken Preproinsulin Gene. *Cell* 20 555–566.

V. J. Kidd and G. F. Saunders. 1982. Linkage Arrangement of Human Placental Lactogen and Growth Hormone Genes. *J. Biol. Chem.* 157 10673–10680.

G. S. Barsh, P. H. Seeburg, and R. E. Gelinas. 1983. The Human Growth Hormone Gene Family: Structure and Evolution of the Chromosomal Locus. *Nucleic Acids Res.* 11 3939–3958.

G. A. Dover and R. B. Flavell. 1984. Molecular Coevolution: DNA Divergence and the Maintenance of Function. *Cell* 38 622–623.

G. R. Fink and T. D. Petes. 1984. Gene Conversion in the Absence of Reciprocal Recombination. *Nature* 310 728–729.

C. C. F. Blake. 1985. Exons and the Evolution of Proteins. *Intl. Rev. Cytol.* 93 149–185.

D. I. H. Linzer and D. Nathans. 1985. A New Member of the Prolactin-Growth Hormone Gene Family Expressed in Mouse Placenta. *EMBO J.* 4 1419–1423.

D. Strauss and W. Gilbert. 1985. Genetic Engineering in the Precambrian. Structure of the Chicken Triosephosphate Isomerase Gene. *Mol. Cell. Biol.* 5 3497–3506.

G. A. Dover. 1986. Molecular Drive in Multigene Families: How Biological Novelties Arise, Spread and Are Assimilated. *Trends Genet.* 2 159–165.

N. Maeda and O. Smithies. 1986. The Evolution of Multigene Families: Human Haptoglobulin Genes. *Annu. Rev. Genet.* 20 81–108.

M. Marchionni and W. Gilbert. 1986. The Triosephosphate Isomerase Gene from Maize: Introns Antedate the Plant-Animal Divergence. *Cell* 46 133–141.

G. L. McKnight, P. J. O'Hara, and M. L. Parker. 1986. Nucleotide Sequence of the Triosephosphate Isomerase Gene from *Aspergillus nidulans*: Implications for a Differential Loss of Introns. *Cell* 46 143–147.

Molecular Evolutionary Clock. 1987. *J. Mol. Evol.* 26 (1 and 2) special issue.

S. J. O'Brien, H. N. Seuanez, and J. E. Womack. 1988. Mammalian Genome Organization: An Evolutionary View. *Annu. Rev. Genet.* 22 323–351.

A. C. Wilson, H. Ochman, and E. M. Prager. 1987. Molecular Time Scale for Evolution. *Trends Genet.* 3 241–247.

J. Felsenstein. 1988. Phylogenies from Molecular Sequences: Inference and Reliability. *Annu. Rev. Genet.* 22 521–565.

M. R. Green. 1988. Mobile RNA Catalysts. *Nature* 336 716–718.

E. Y. Chen, Y-C. Liao, D. H. Smith, H. A. Barrera-Saldana, R. E. Gelinas, and P. H. Seeburg. 1989. The Human Growth Hormone Locus: Nucleotide Sequence, Biology, and Evolution. *Genomics* 4 479–497.

N. J. Dibb and A. J. Newman. 1989. Evidence That Introns Arose at Proto-Splice Sites. *EMBO J.* 8 2015–2021.

P. S. Perlman and R. S. Butow. 1989. Mobile Introns and Intron-Encoded Proteins. *Science* 246 1106–1109.

J. Rogers. 1989. How Were Introns Inserted into Nuclear Genes? *Trends Genet.* 5 213–216.

9.2a

I. B. Dawid, D. D. Brown, and R. H. Reeder. 1970. Composition and Structure of Chromosomal and Amplified Ribosomal DNAs of *Xenopus laevis*. *J. Mol. Biol.* 51 341–360.

N. V. Federoff. 1979. On Spacers. *Cell* 16 697–710.

J. R. Bedbrook and W. L. Gerlach. 1980. Cloning of Repeated Sequence DNA from Cereal Plants. In J. K. Setlow and A. Hollaender (eds.), Genetic Engineering, vol. 2, pp. 1–19. Plenum Press, New York.

E. O. Long and I. B. Dawid. 1980. Repeated Genes in Eukaryotes. *Annu. Rev. Biochem.* 49 727–764.

P. D'Eustachio, O. Meyuhas, F. Ruddle, and R. Perry. 1981. Chromosomal Distribution of Ribosomal Protein Genes in the Mouse. *Cell* 24 307–312.

D. M. Glover. 1981. The rDNA of *Drosophila melanogaster*. *Cell* 26 297–298.

M. Krystal, P. D'Eustachio, F. H. Ruddle, and N. Arnheim. 1981. Human Nucleolus Organizers on Non-Homologous Chromosomes Can Share the Same Ribosomal RNA Gene Variants. *Proc. Natl. Acad. Sci.* U.S.A. 78 5744–5748.

O. L. Miller, Jr. 1981. The Nucleolus, Chromosomes, and Visualization of Genetic Activity. *J. Cell Biol.* 91 15s–27s.

J. Cortadas and M. C. Pavon. 1982. The Organization of Ribosomal Genes in Vertebrates. *EMBO J.* 1 1075–1080.

S. A. Gerbi, C. Jeppesen, B. Stebbins-Boaz, and M. Ares, Jr. 1987. Evolution of Eukaryotic rRNA: Constraints Imposed by RNA Interactions. *Cold Spring Harbor Symp. Quant. Biol.* 52 709–719.

R. S. Hawley and C. H. Marcus. 1989. Recombinational Controls of rDNA Redundancy in *Drosophila*. *Annu. Rev. Genet.* 23 87–120.

9.2b

D. D. Brown and K. Sugimoto. 1973. 5S DNAs of *Xenopus laevis* and *Xenopus mulleri*: Evolution of a Gene Family. *J. Mol. Biol.* 78 397–415.

R. C. Peterson, J. L. Doering, and D. D. Brown. 1980. Characterization of Two *Xenopus* Somatic 5S DNAs and One Minor Oocyte-Specific 5S DNA. *Cell* 20 131–141.

J. Mao, B. Appel, J. Schaak, S. Sharp, H. Yamada, and D. Soll. 1982. The 5S RNA Genes of *Schizosaccharomyces pombe*. *Nucleic Acids Res.* 10 487–500.

9.2c

G. I. Bell, L. J. DeGennaro, D. H. Gelfand, R. J. Bishop, P. Valenzuela, and W. J. Rutter. 1977. Ribosomal RNA Genes of *Saccharomyces cerevisiae*. I. Physical Map of the Repeating Unit and Location of the Regions Coding for 5S, 5.8S, 18S, and 25S Ribosomal RNA. *J. Biol. Chem.* 252 8118–8125.

T. D. Petes. 1979. Yeast Ribosomal Genes Are Located on Chromosome XII. *Proc. Natl. Acad. Sci.* U.S.A. 76 410–414.

T. D. Petes. 1980. Unequal Meiotic Recombination Within Tandem Arrays of Yeast Ribosomal DNA Genes. *Cell* 19 765–774.

J. W. Szostak and R. Wu. 1980. Unequal Crossing Over in the Ribosomal DNA of *Saccharomyces cerevisiae*. *Nature* 284 426–430.

T. J. Zamb and T. D. Petes. 1982. Analysis of the Junction Between Ribosomal RNA Genes and Single-Copy Chromosomal Sequences in the Yeast *Saccharomyces cerevisiae*. *Cell* 28 355–364.

9.2d

P. H. Yen and N. Davidson. 1980. The Gross Anatomy of a tRNA Gene Cluster at Region 42A of the *D. melanogaster* Chromosome. *Cell* 22 137–148.

E. Kubli. 1982. The Genetics of Transfer RNA in *Drosophila*. *Adv. Genet.* 21 123–172.

S. G. Clarkson. 1983. Transfer RNA Genes. In N. MacLean, S. P. Gregory, and R. A. Flavell (eds.), Eukaryotic Genes: Their Structure, Activity and Regulation, pp. 239–261. Butterworth, London.

9.2e

H. Busch, R. Reddy, L. Rothblum, and C. Y. Choi. 1982. SnRNAs, SnRNPs, and RNA Processing. *Annu. Rev. Biochem.* 51 617–654.

H. P. Saluz, T. Schmidt, R. Dudler, M. Altwegg, E. Stumm-Zollinger, E. Kubli, and P. S. Chen. 1983. The Genes Coding for 4 snRNAs of *Drosophila melanogaster*: Localization and Determination of Gene Numbers. *Nucleic Acids Res.* 11 77–90.

J. A. Wise, D. Tollervey, D. Maloney, H. Swerdlow, E. J. Dunn, and C. Guthrie. 1983. Yeast Contains Small Nuclear RNAs Encoded by Single Copy Genes. *Cell* 35 743–751.

E. Lund and J. E. Dahlberg. 1984. True Genes for Human U1 Small Nuclear RNA: Copy Number, Polymorphism, and Methylation. *J. Biol. Chem.* 259 2013–2021.

S. W. Van Arsdell and A. M. Weiner. 1984. Human Genes

for U2 Small Nuclear RNA Are Tandemly Repeated. *Mol. Cell. Biol.* 4 492–499.

R. Zeller, M-T. Carri, I. W. Mattaj, and E. M. DeRobertis. 1984. *Xenopus laevis* U1 snRNA Genes: Characterization of Transcriptionally Active Genes Reveals Major and Minor Repeated Families. *EMBO J.* 3 1075–1081.

L. B. Bernstein, T. Manser, and A. M. Weiner. 1985. Human U1 Small Nuclear RNA Genes: Extensive Conservation of Flanking Sequences Suggests Cycles of Gene Amplification and Transposition. *Mol. Cell. Biol.* 5 2159–2171.

C. Guthrie. 1986. Finding Functions for Small Nuclear RNAs in Yeast. *Trends Biochem. Sci.* 11 430–434.

K. A. Montzka and J. A. Steitz. 1988. Additional Low-Abundance Human Small Nuclear Ribonucleoproteins: U11, U12, etc. *Proc. Natl. Acad. Sci.* U.S.A. 85 8885–8889.

9.3b

A. Efstratiadis, J. W. Posakony, T. Maniatis, R. M. Lawn, C. O'Connell, R. A. Spritz, J. K. DeRiel, B. G. Forget, S. M. Weissman J. L. Slightom, A. E. Blechl, F. E. Baralle, C. C. Shoulders, and N. J. Proudfoot. 1980. The Structure and Evolution of the Human β-Globin Gene Family. *Cell* 21 653–668.

E. F. Fritsch, C. K. J. Shen, R. M. Lawn, and T. Maniatis. 1980. Molecular Cloning and Characterization of the Human β-like Globin Gene Cluster. *Cell* 19 959–972.

I. Lemischka and P. A. Sharp. 1982. The Sequences of an Expressed Rat α-Tubulin Gene and a Pseudogene with an Inserted Repetitive Element. *Nature* 300 330–335.

C. D. Wilde, L. T. Chow, F. C. Wefald, and N. J. Cowan. 1982. Structure of Two Human α-Tubulin Genes. *Proc. Natl. Acad. Sci.* U.S.A. 79 96–100.

M. B. Buckingham and A. J. Minty. 1983. Contractile Protein Genes. In N. MacLean, S. P. Gregory, and R. A. Flavell (eds.), Eukaryotic Genes: Their Structure, Activity, and Regulation, vol. 2, pp. 365–395. Butterworth, London.

H. Czosnek, U. Nudel, Y. Mayer, P. E. Barker, D. D. Pravtcheva, F. H. Ruddle, and D. Yaffe. 1983. The Genes Coding for the Cardiac Muscle Actin, and Skeletal Muscle Actin, and the Cytoplasmic β-Actin Are Located on Three Different Mouse Chromosomes. *EMBO J.* 2 1977–1979.

S. Pestka. 1983. Interferon Genes. *Arch. Biochem. Biophys.* 221 1–37.

F. S. Collins and S. M. Weissman. 1984. The Molecular Genetics of Human Hemoglobin. *Prog. Nucleic Acid Res. and Mol. Biol.* 31 317–462.

R. Garcia, B. Paz-Aliaga, S. G. Ernst, and W. R. Crain, Jr. 1984. Sea Urchin Actin Genes Show Different Patterns of Expression: Muscle Specific, Embryo Specific, and Constitutive. *Mol. Cell. Biol.* 4 840–845.

A. D. Sagar, P. B. Sehgal, L. T. May, M. Inouye, D. L. Slate, L. Shulman, and F. H. Ruddle. 1984. Interferon-β-Related DNA Is Dispersed in the Human Genome. *Science* 223 1312–1315.

R. J. Shott, J. J. Lee, R. J. Britten, and E. H. Davidson. 1984. Differential Expression of the Actin Gene Family of *Strongylocentrotus purpuratus*. *Devel. Biol.* 101 295–306.

D. W. Cleveland and K. F. Sullivan. 1985. Molecular Biology and Genetics of Tubulin. *Annu. Rev. Biochem.* 54 331–365.

S. Karlsson and A. W. Nienhuis. 1985. Developmental Regulation of Human Globin Genes. *Annu. Rev. Biochem.* 54 1071–1108.

S-Y. Ng, P. Gunning, R. Eddy, P. Ponte, J. Leavitt, T. Shows, and L. Kedes. 1985. Evolution of the Functional Human β-Actin Gene and Its Multi-Pseudogene Family: Conservation of Noncoding Regions and Chromosomal Dispersion of Pseudogenes. *Mol. Cell. Biol.* 5 2720–2732.

M. Ohlsson, J. Feder, L. L. Cavalli-Sforza, and A. von Gabain. 1985. Close Linkage of α and β Interferons and Infrequent Duplication of β Interferon in Humans. *Proc. Natl. Acad. Sci.* U.S.A 82 4473–4476.

A-C. Pittet and U. Schibler. 1985. Mouse α-Amylase Locus: Amy-1a and Amy-2a Are Closely Linked. *J. Mol. Biol.* 182 359–365.

P. Shaw, B. Sordat, and U. Schibler. 1985. The Two Promoters of the Mouse α-Amylase Gene Amy-1a Are Differentially Activated During Parotid Gland Differentiation. *Cell* 40 907–912.

K. Wiebauer, D. L. Gumucio, J. M. Jones, R. M. Caldwell, H. T. Hartle, and M. H. Meisler. 1985. A 78-Kilobase Region of Mouse Chromosome 3 Contains Salivary and Pancreatic Amylase Genes and a Pseudogene. *Proc. Natl. Acad. Sci.* U.S.A. 82 5446–5449.

S. Pestka, J. A. Langer, K. C. Zoon, and C. E. Samuels. 1987. Interferons and Their Actions. *Annu. Rev. Biochem.* 56 727–777.

H. P. Erba, R. Eddy, T. Shows, L. Kedes, and P. Gunning. 1988. Structure, Chromosomal Location, and Expression of the Human γ-Actin Gene: Differential Evolution, Location, and Expression of the Cytoskeletal β- and γ-Actin Genes. *Mol. Cell. Biol.* 8 1775–1789.

S. Esteal. 1988. Rate Constancy of Globin Gene Evolution in Placental Mammals. *Proc. Natl. Acad. Sci.* U.S.A. 85 622–7626.

G. Romeo, G. Fiorucci, and G. B. Rossi. 1989. Interferons in Cell Growth and Development. *Trends Genet.* 5 19–24.

9.3c

R. H. Cohn, J. C. Lowry, and L. H. Kedes. 1976. Histone Genes of the Sea Urchin (*S. purpuratus*) Cloned in

E. coli: Order, Polarity, and Strandedness of the Five Histone-Coding and Spacer Regions. *Cell* 9 147–161.

C. C. Hentschel and M. L. Birnstiel. 1981. The Organization and Expression of Histone Gene Families. *Cell* 25 301–313.

A. Ruiz-Carrillo, M. Affolter, and J. Renaud. 1983. Genomic Organization of the Genes Coding for the Six Main Histones of the Chicken: Complete Sequence of the H5 Gene. *J. Mol. Biol.* 170 843–859.

M. M. Smith and K. Murray. 1983. Yeast H3 and H4 Histone Messenger RNAs Are Transcribed from Two Non-Allelic Gene Sets. *J. Mol. Biol.* 169 641–661.

P. C. Turner and H. R. Woodland. 1983. Histone Gene Number and Organization in *Xenopus*: *Xenopus borealis* Has a Homogeneous Major Cluster. *Nucleic Acids Res.* 11 971–986.

L. P. Woudt, A. Pastink, A. E. Kempers-Veenstra, A. E. M. Jansen, W. H. Mager, and R. J. Planta. 1983. The Genes Coding for Histone H3 and H4 in *Neurospora crassa* Are Unique and Contain Intervening Sequences. *Nucleic Acids. Res.* 11 5347–5360.

R. W. Old and H. R. Woodland. 1984. Histone Genes: Not So Simple After All. *Cell* 38 624–626.

P. Tripputi, B. S. Emanuel, C. M. Croce, L. G. Green, G. S. Stern, and J. L. Stern. 1986. Human Histone Genes Map to Multiple Chromosomes. *Proc. Natl. Acad. Sci.* U.S.A. 83 3185–3188.

D. Schumperli. 1988. Multilevel Regulation of Replication-Dependent Histone Genes. *Trends Genet.* 4 187–191.

9.4a

F. A. Eiferman, P. R. Young, R. W. Scott, and S. M. Tilghman. 1981. Intragenic Amplification and Divergence in the Mouse α-Fetoprotein Gene. *Nature* 294 713–718.

S. Ohno. 1981. Original Domain for the Serum Albumin Family Arose from Repeated Sequences. *Proc. Natl. Acad. Sci.* U.S.A. 78 7657–7661.

A. Dugaiczk, S. W. Law, and O. E. Dennison. 1982. Nucleotide Sequence and the Encoded Amino Acids of Human Serum Albumin mRNA. *Proc. Natl. Acad. Sci.* U.S.A. 79 71–75.

M. A. T. Muskavitch and D. S. Hogness. 1982. An Expandable Gene that Encodes a *Drosophila* Glue Protein Is Not Expressed in Variants Lacking Remote Upstrem Sequences. *Cell* 29 1041–1051.

F. Alexander, P. R. Young, and S. M. Tilghman. 1984. Evolution of the Albumin: α-Fetoprotein Ancestral Gene from the Amplification of a 27 Nucleotide Sequence. *J. Mol. Biol.* 173 159–176.

M. S. Brown and J. L. Goldstein. 1986. A Receptor-Mediated Pathway for Cholesterol Homeostasis. *Science* 232 34–47.

9.4b

P. Gill, A. J. Jeffreys, and D. J. Werrett. 1985. Forensic Application of DNA Fingerprints. *Nature* 318 577–579.

A. J. Jeffreys, V. Wilson, and S. L. Thein. 1985. Individual-Specific "Fingerprints" of Human DNA. *Nature* 316 76–79.

A. P. Jarman and R. A. Wells. 1989. Hypervariable Minisatellites: Recombinators or Innocent Bystanders? *Trends. Genet.* 5 367–372.

9.4c

W. J. Peacock, D. Brutlag, E. Goldring, R. Appels, C. W. Hinton, and D. L. Lindsey. 1973. The Organization of Highly Repeated DNA Sequences in *Drosophila melanogaster* Chromosomes. *Cold Spring Harbor Symp. Quant. Biol.* 38 405–416.

W. J. Peacock, A. R. Lohe, W. L. Gerlach, P. Dunsmuir, E. S. Dennis, and R. Appels. 1977. Fine Structure and Evolution of DNA in Heterochromatin. *Cold Spring Harbor Symp. Quant. Biol.* 42 1121–1135.

D. Brutlag. 1980 Molecular Arrangement and Evolution of Heterochromatic DNA. *Annu. Rev. Genet.* 14 121–144.

A. J. Hilliker, R. Appels, and A. Schalet. 1980. The Genetic Analysis of *D. melanogaster* Heterochromatin. *Cell* 21 607–619.

M. F. Singer. 1982. Highly Repeated Sequences in Mammalian Genomes. *Intl. Rev. Cytol.* 76 67–112.

M. J. M. Pages and G. P. Roizes. 1984. Nature and Organization of the Sequence Variations in the Long-Range Periodicity Calf Satellite DNA I. *J. Mol. Biol.* 173 143–157.

H. F. Willard and J. S. Waye. 1987. Hierarchical Order in Chromosome-Specific Human Alpha Satellite DNA. *Trends Genet.* 3 192–198.

R. S. Verma (ed.). 1988. Heterochromatin. Cambridge University Press, Cambridge.

S. J. Durfy and H. F. Willard. 1989. Patterns of Intra- and Interarray Sequence Variation in Alpha Satellite from the Human X Chromosome: Evidence for Short-Range Homogenization of Tandemly Repeated DNA Sequences. *Genomics* 5 810–821.

R. Wevrick and H. F. Willard. 1989. Long Range Organization of Tandem Arrays of α-Satellite DNA at the Centromeres of Human Chromosomes: High Frequency Array-Length Polymorphism and Meiotic Stability. *Proc. Natl. Acad. Sci.* U.S.A. 86 9394–9398.

9.4d

A. Pryor, K. Faulkner, M. M. Rhoades, and W. J. Peacock. 1980. Asynchronous Replication of Heterochromatin in Maize. *Proc. Natl. Acad. Sci.* U.S.A. 77 6705–6709.

9.4e

E. M. Southern. 1970. Base Sequence and Evolution of Guinea Pig α-Satellite DNA. *Nature* 227 794–798.

G. P. Smith. 1976. Evolution of Repeated DNA Sequences by Unequal Crossover. *Science* 191 528–535.

9.5a

R. J. Britten and D. E. Kohne. 1968. Repeated Sequences in DNA. *Science* 161 529–540.

J. E. Manning, C. W. Schmid, and N. Davidson. 1975. Interspersion of Repetitive and Nonrepetitive DNA Sequences in the *Drosophila melanogaster* Genome. *Cell* 4 141–155.

C. W. Schmid and P. L. Deininger. 1975. Sequence Organization of the Human Genome. *Cell* 6 345–358.

A. C. Spradling and G. M. Rubin. 1981. *Drosophila* Genome Organization: Conserved and Dynamic Aspects. *Annu. Rev. Genet.* 15 219–264.

9.5b

D. E. Graham, B. R. Neufeld, E. H. Davidson, and R. J. Britten. 1974. Interspersion of Repetitive and Nonrepetitive DNA Sequences in the Sea Urchin Genome. *Cell* 1 127–137.

W. H. Klein, T. L. Thomas, C. Lai, R. H. Scheller, R. J. Britten, and E. H. Davidson. 1978. Characteristics of Individual Repetitive Sequence Families in the Sea Urchin Genome Studied with Cloned Repeats. *Cell* 14 889–900.

J. W. Posakony, R. H. Scheller, D. M. Anderson, R. J. Britten, and E. H. Davidson. 1981. Repetitive Sequences of the Sea Urchin Genome. Nucleotide Sequences of Cloned Repeat Elements. *J. Mol. Biol.* 149 41–67.

J. W. Posakony, C. N. Flytzanis, R. J. Britten, and E. H. Davidson. 1983. Interspersed Sequence Organization and Developmental Representation of Cloned Poly(A) RNAs from Sea Urchin Eggs. *J. Mol. Biol.* 167 361–389.

9.5c

P. L. Deininger, D. J. Jolly, C. M. Rubin, T. Friedmann, and C. W. Schmid. 1981. Base Sequence Studies of 300 Nucleotide Renatured Repeated Human DNA Clones. *J. Mol. Biol.* 151 17–33.

P. Jagadeeswaran, B. Forget, and S. M. Weissman. 1981. Short Interspersed Repetitive DNA Elements in Eucaryotes: Transposable DNA Elements Generated by Reverse Transcription of RNA Pol III Transcripts? *Cell* 26 141–142.

W. R. Jelinek and C. W. Schmid. 1982. Repetitive Sequences in Eukaryotic DNA and Their Expression. *Annu. Rev. Biochem.* 51 813–844.

E. Ullu and C. Tschudi. 1984. Alu Sequences Are Processed 7SL RNA Genes. *Nature* 312 171–172.

G. R. Daniels and P. L. Deininger. 1985. Several Major Mammalian SINE Families Are Derived from tRNA Genes. *Nature* 317 819–822.

H. R. Hwu, J. W. Roberts, E. H. Davidson, and R. J. Britten. 1986. Insertion and/or Deletion of Many Repeated DNA Sequences in Human and Higher Ape Evolution. *Proc. Natl. Acad. Sci.* U.S.A. 83 3875–3879.

K. Matsumoto, K. Murakami, and N. Okada. 1986. Gene for Lysyl tRNA1 May Be a Progenitor of the Highly Repetitive and Transcribable Sequences Present in the Salmon Genome. *Proc. Natl. Acad. Sci.* U.S.A. 83 3156–3160.

R. J. Britten, W. F. Baron, D. B. Stout, and E. H. Davidson. 1988. Sources and Evolution of Human Alu Repeated Sequences. *Proc. Natl. Acad. Sci.* U.S.A. 85 4770–4774.

R. Dornburg and H. M. Temin. 1988. Retroviral Vector System for the Study of cDNA Gene Formation. *Mol. Cell. Biol.* 8 2328–2334.

T. G. Fanning and M. F. Singer. 1988. LINE-1: A Mammalian Transposable Element. *Biochem. Biophys. Acta* 910 203–212.

J. R. Korenberg and M. C. Rykowski. 1988. Human Genome Organization: Alu, LINES, and the Molecular Structure of Metaphase Chromosome Bands. *Cell* 53 391–400.

T. L. Chen and L. Manuelidis. 1989. SINEs and LINEs Cluster in Distinct DNA Fragments of Giemsa Band Size. *Chromosoma* (Berl) 98 309–316.

P. L. Deininger. 1989. SINEs: Short Interspersed Repeated DNA Elements in Higher Eucaryotes. In D. E. Berg and M. M. Howe (eds.), Mobile DNA, pp. 619–636. American Society of Microbiology, Washington, D.C.

C. A. Hutchison III, S. C. Hardies, D. D. Loeb, W. R. Shehee, and M. H. Edgell. 1989. LINES and Related Retroposons: Long Interspersed Repeated Sequences in the Eucaryotic Genome. In D. E. Berg and M. M. Howe (eds.), Mobile DNA, pp. 593–618. American Society of Microbiology, Washington, D.C.

R. K. Moyzis, D. C. Torney, J. Meyne, J. M. Buckingham, J. R. Wu, C. Burks, K. M. Sirotkin, and W. B. Goad. 1989. The Distribution of Interspersed Repetitive DNA Sequences in the Human Genome. *Genomics* 4 273–289.

9.5d

R. J. Britten and E. H. Davidson. 1969. Gene Regulation in Higher Cells: A Theory. *Science* 165 349–358.

F. del Rey, T. F. Donahue, and G. R. Fink. 1983. The Histidine tRNA Genes of Yeast. *J. Biol. Chem.* 258 8175–8182.

G. S. Roeder. 1983. Unequal Crossing-Over Between Yeast Transposable Elements. *Mol. Gen. Genet.* 190 117–121.

M. Kress, Y. Barra, J. G. Seidman, G. Khoury, and G. Jay. 1984. Functional Insertion of an Alu Type 2 (B2 Sine) Repetitive Sequence in Murine Class I Genes. *Science* 226 974–977.

A. Hinnebusch, G. Lucchini, and G. R. Fink. 1985. A Synthetic HIS4 Regulatory Element Confers General Amino Acid Control on the Cytochrome c Gene (*CYC*1) of Yeast. *Proc. Natl. Acad. Sci.* U.S.A. 82 498–502.

M. A. Lehrman, W. J. Schneider, T. C. Sudhof, M. S. Brown, J. L. Goldstein, and D. W. Russell. 1985. Mutation in LDL Receptor: Alu-Alu Recombination Deletes Exons Encoding Transmembrane and Cytoplasmic Domains. *Science* 227 140–146.

J. H. Rogers. 1985. The Origin and Evolution of Retroposons. *Intl. Rev. Cytol.* 93 187–279.

D. Tautz, M. Trick, and G. A. Dover. 1986. Cryptic Simplicity in DNA Is a Major Source of Genetic Variation. *Nature* 322 652–656.

9.5e

T. Nagylaki and T. D. Petes. 1982. Intrachromosomal Gene Conversion and the Maintenance of Sequence Homogeneity Among Repeated Genes. *Genetics* 100 315–337.

A. M. Weiner, P. L. Deininger, and A. Efstratiadis. 1986. Nonviral Retroposons: Genes, Pseudogenes and Transposable Elements Generated by the Reverse Flow of Genetic Information. *Annu. Rev. Biochem.* 55 631–661.

9.6a

H. Ris and P. L. Witt. 1981. Structure of the Mammalian Kinetochore. *Chromosoma* (Berl.) 82 153–170.

M-C. Ya and C-H. Yao. 1981. Repeated Hexanucleotide C-C-C-C-A-A Is Present Near Free Ends of Macronuclear DNA of *Tetrahymena. Proc. Natl. Acad. Sci.* U.S.A. 78 7436–7439.

D. T. Stinchcomb, C. Mann, and R. W. Davis. 1982. Centromeric DNA from *Saccharomyces cerevisiae. J. Mol. Biol.* 158 157–179.

L. Clarke and J. Carbon. 1983. Genomic Substitutions of Centromeres in *Saccharomyces cerevisiae. Nature* 305 23–28.

J. Carbon. 1984. Yeast Centromeres: Structure and Function. *Cell* 37 351–353.

L. Clarke and J. Carbon. 1985. The Structure and Function of Yeast Centromeres. *Annu. Rev. Genet.* 19 29–56.

P. Hieter, C. Mann, M. Snyder, and R. W. Davis. 1985. Mitotic Stability of Yeast Chromosomes: A Colony Color Assay that Measures Nondisjunction and Chromosome Loss. *Cell* 40 393–403.

P. Hieter, D. Pridmore, J. H. Hegemann, M. Thomas, R. W. Davis, and P. Philippsen. 1985. Functional Selection and Analysis of Yeast Centromeric DNA. *Cell* 42 913–921.

D. Koshland, J. C. Kent, and L. H. Hartwell. 1985. Genetic Analysis of the Mitotic Transmission of Minichromosomes. *Cell* 40 381–392.

J. H. Hegemann, J. H. Shero, G. Cottarei, P. Philippsen, and P. Hieter. 1988. Mutational Analysis of Centromere DNA from Chromosome VI of *Saccharomyces cerevisiae. Mol. Cell. Biol.* 8 2523–2535.

R. Ng and J. Carbon. 1988. Mutational and *in Vitro* Protein Binding Studies on Centromere DNA from *Saccharomyces cerevisiae. Mol. Cell. Biol.* 7 4522–4534.

M. Saunders, M. Fitzgerald-Hayes, and K. Bloom. 1988. Chromatin Structure of Altered Yeast Centromeres. *Proc. Natl. Acad. Sci.* U.S.A. 85 175–179.

R. E. Baker, M. Fitzgerald-Hayes, and T. C. O'Brien. 1989. Purification of the Yeast Centromere Binding Protein CP1 and a Mutational Analysis of Its Binding Site. *J. Biol. Chem.* 264 10843–10850.

M. Cai and R. W. Davis. 1989. Purification of a Yeast Centromere-Binding Protein That Is Able to Distinguish Single Base-Pair Mutations in its Recognition Site. *Mol. Cell. Biol.* 9 2544–2550.

G. Cottarel, J. H. Shero, P. Hieter, and J. H. Hegemann. 1989. A 125-Base-Pair *CEN6* DNA Fragment Is Sufficient for Complete Meiotic and Mitotic Centromere Functions in *Saccharomyces cerevisiae. Mol. Cell. Biol.* 9 3342–3349.

9.6b

J. G. Gall, K. Karrier, and M-C. Yao. 1977. The Ribosomal DNA of *Tetrahymena*. In P. Ts'o (ed.), The Molecular Biology of the Mammalian Genetic Apparatus, pp. 80–85. Elsevier-North Holland Biomedical Press, Amsterdam.

E. H. Blackburn. 1984. Telomeres: Do the Ends Justify the Means? *Cell* 37 7–8.

E. H. Blackburn and J. W. Szostak. 1984. The Molecular Structure of Centromeres and Telomeres. *Annu. Rev.* Biochem. 53 163–194.

C. W. Greider and E. H. Blackburn. 1987. The Telomere Transferase of *Tetrahymena* Is a Ribonucleoprotein Enzyme with Two Kinds of Primer Specificity. *Cell* 51 887–898.

E. H. Henderson, C. C. Hardin, S. K. Walk, I. Tinoco, Jr., and E. H. Blackburn. 1987. Telomeric DNA Oligonucleotides Form Novel Intramolecular Structures Containing Guanine·Guanine Base Pairs. *Cell* 51 899–908.

M-C. Yao, K. Zheng, and C-H. Yao. 1987. A Conserved Nucleotide Sequence at the Sites of Developmentally Regulated Chromosomal Breakage in *Tetrahymena*. Cell 48 779–788.

R. C. Allshire, J. R. Gosden, S. H. Cross, G. Cranston, D. Rout, N. Sugawara, J. W. Szostak, P. A. Fantes, and N. D. Hastie. 1988. Telomeric Repeat from *Tetrahymena* Cross-Hybridizes with Human Telomeres. *Nature* 332 656–659.

T. R. Cech. 1988. G-Strings at Chromosome Ends. *Nature* 332 777–778.

R. M. McCarroll and W. L. Fangman. 1988. Time of Replication of Yeast Centromeres and Telomeres. *Cell* 54 505–513.

R. K. Moyzis, J. M. Buckingham, L. S. Cram, M. Dani, L. L. Deaven, M. D. Jones, J. Meyne, R. L. Ratliff, and J-R. Wu. 1988. A Highly Conserved Repetitive DNA Sequence, (TTAGGG)n, Present at the Telomeres of Human Chromosomes. *Proc. Natl. Acad. Sci.* U.S.A. 85 6662–6626.

E. J. Richards and F. Ausubel. 1988. Isolation of a Higher Eukaryotic Telomere from *Arabidopsis thaliani. Cell* 53 127–136.

J. Shampay and E. H. Blackburn. 1988. Generation of Telomere-Length Heterogeneity in *Saccharomyces cerevisiae. Proc. Natl. Acad. Sci.* U.S.A. 85 534–538.

A. M. Weiner. 1988. Eukaryotic Nuclear Telomeres: Molecular Fossils of the RNP World? *Cell* 52 155–157.

J. Meyne, R. L. Ratliff, and R. K. Moyzis. 1989. Conservation of the Human Telomere Sequence (TTAGGG)n Among Vertebrates. *Proc. Natl. Acad. Sci.* U.S.A. 86 7049–7053.

M. K. Raghuraman and T. R. Cech. 1989. Assembly and Self-Association of *Oxytricha* Telomeric Nucleoprotein Complexes. *Cell* 59 719–728.

J. W. Szostak. 1989. The Beginning of the Ends. *Nature* 337 303–304.

V. A. Zakian. 1989. Structure and Function of Telomeres. *Annu. Rev. Genet.* 23 579–604.

V. Lundblad and E. H. Blackburn. 1990. RNA-Dependent Polymerase Motifs in EST1: Tentative Identification of a Protein Component of an Essential Yeast Telomerase. *Cell* 60 529–530.

9.6c

A. W. Murray. 1985. Chromosome Structure and Behavior. *Trends Biochem. Sci.* 10 112–115.

D. T. Burke, G. F. Carle, and M. V. Olson. 1987. Cloning of Large Segments of Exogenous DNA into Yeast by Means of Artificial Chromosome Vectors. *Science* 236 806–812.

A. W. Murray and J. W. Szostak. 1987. Artificial Chromosomes. *Sci. American* 257(5) 62–68.

9.7a

M. L. Claisse, P. P. Slonimski, J. Johnson, and H. R. Mahler. 1980. Mutations Within an Intron and Its Flanking Sites: Patterns of Novel Polypeptides Generated by Mutants in One Segment of the *cob-box* Region of Yeast Mitochondrial DNA. *Mol. Gen. Genet.* 177 375–387.

H. Bechmann, A. Haid, R. J. Schweyen, S. Mathews, and F. Kaudewitz. 1981. Expression of the Split Gene *COB* in Yeast mtDNA: Translation of Intervening Sequences in Mutant Strains. *J. Biol. Chem.* 256 3525–3531.

B. Weiss-Brummer, G. Rodel, R. J. Schweyen, and F. Kaudewitz. 1982. Expression of the Split Gene *cob* in Yeast: Evidence for a Precursor of a "Maturase" Protein Translated from Intron 4 and Preceding Exons. *Cell* 29 527–536.

G. Attardi. 1985. Animal Mitochondrial DNA: An Extreme Example of Genetic Economy. *Intl. Rev. Cytol.* 93 93–145.

T. K. Biswas, J. C. Edwards, M. Rabinowitz, and G. S. Getz. 1985. Characterization of a Yeast Mitochondrial Promoter by Deletion Mutagenesis. *Proc. Natl. Acad. Sci.* U.S.A. 82 1954–1958.

M. Douglas and M. Takeda. 1985. Nuclear Genes Encoding Mitochondrial Proteins in Yeast. *Trends Biochem. Sci.* 10 192–194.

J. E. Hixson and D. A. Clayton. 1985. Initiation of Transcription from Each of the Two Human Mitochondrial Promoters Requires Unique Nucleotides at the Transcriptional Start Sites. *Proc. Natl. Acad. Sci.* U.S.A. 82 2660–2664.

B. Jacquier and B. Dujon. 1985. An Intron-Encoded Protein Is Active in a Gene Conversion Process that Spreads an Intron into a Mitochondrial gene. *Cell* 41 383–394.

I. G. Macreadie, R. M. Scott, A. R. Zinn, and R. A. Butow. 1985. Transposition of an Intron in Yeast Mitochondria Requires a Protein Encoded by that Intron. *Cell* 41 395–402.

T. W. Wong and D. A. Clayton. 1985. *In Vitro* Replication of Human Mitochondrial DNA: Accurate Initiation at the Origin of Light-Strand Synthesis. *Cell* 42 951–958.

M. de Zamaroczy and G. Bernardi. 1985. Sequence Organization of the Mitochondrial Genome of Yeast—A Review. *Gene* 37 1–17.

G. Attardi, 1986. The Elucidation of the Human Mitochondrial Genome: A Historical Perspective. *BioEssays* 5 34–39.

R. L. Cann, M. Stoneking, and A. C. Wilson. 1986. Mitochondrial DNA and Human Evolution. *Nature* 325 31–36.

T. D. Fox. 1986. Nuclear Gene Products Required for Translation of Specific Mitochondrially Coded mRNAs in Yeast. *Trends Genet.* 2 97–100.

A. L. Greenleaf, J. L. Kelly, and I. R. Lehman. 1986. Yeast *RPO41* Gene Product Is Required for Transcription and Maintenance of the Mitochondrial Genome. *Proc. Natl. Acad. Sci.* U.S.A. 83 3391–3394.

R. M. Mulligan and V. Walbot. 1986. Gene Expression and Recombination in Plant Mitochondrial Genomes. *Trends Genet.* 2 263–266.

K. O'Hare. 1986. Genes Within Genes. *Trends Genet.* 2 33.

A. Tzagaloff and A. M. Myers. 1986. Genetics of Mitochondrial Biogenesis. *Annu. Rev. Biochem.* 55 249–285.

R. Bordonne, G. Dirheimer, and R. P. Martin. 1987. Transcription Initiation and RNA Processing of a Yeast Mitochondrial tRNA Gene Cluster. *Nucleic Acids Res.* 15 7381–7394.

K. J. Netwon. 1988. Plant Mitochondrial Genomes: Organization, Expression and Variation. *Annu. Rev. Plant Physiol. and Mol. Biol.* 39 503–532.

L. A. Grivell and R. J. Schweyen. 1989. RNA Splicing in Yeast Mitochondria: Taking Out the Twists. *Trends Genet.* 5 39–41.

A. Delahodde, V. Goguel, A. M. Becam, F. Creusot, J. Perea, J. Banroques, and C. Jacq. 1989. Site-Specific DNA Endonuclease and RNA Maturase Activities of Two Homologous Intron-Encoded Proteins from Yeast Mitochondria. *Cell* 56 431–441.

C. S. Levings III and G. G. Brown. 1989. Molecular Biology of Plant Mitochondria. *Cell* 56 171–179.

J. M. Wenzlau, R. J. Saidanha, R. A. Butow, and P. S. Perlman. 1989. A Latent Intron-Encoded Maturase Is also an Endonuclease Needed for Intron Mobility. *Cell* 56 421–430.

9.7b

P. Englund et al. 1982. Kinetoplasts. *Annu. Rev. Biochem.* 51 695–726.

S. L. Hajduk, V. A. Klein, and P. T. Englund. 1984. Replication of Kinetoplast DNA Maxicircles. *Cell* 36 483–492.

J. Griffith, M. Bleyman, C. A. Rauch, P. A. Kitchin, and P. T. Englund. 1986. Visualization of the Bent Helix in Kinetoplast DNA by Electron Microscopy. *Cell* 46 717–724.

H. Eisen. 1988. RNA Editing: Who's on First? *Cell* 53 331–332.

A. I. Lamond. 1988. RNA Editing and the Mysterious Undercover Genes of Trypanosomatid Mitochondria. *Trends Biochem. Sci.* 13 283–284.

N. Maizels and A. Weiner. 1988. In Search of a Template. *Nature* 334 469–470.

K. A. Ryan, T. A. Shapiro, C. A. Rauch, J. D. Griffith, and P. T. Englund. 1988. A Knotted Free Minicircle in Kinetoplast DNA. *Proc. Natl. Acad. Sci.* U.S.A. 85 5844–5848.

P. S. Covello and M. W. Gray. 1989. RNA Editing in Plant Mitochondria. *Nature* 341 662–666.

J. M. Gualberto, L. Lamattina, G. Bonnard, J-H. Weil, and J-M. Grienenberger. 1989. RNA Editing in Wheat Mitochondria Results in the Conservation of Protein Sequences. *Nature* 341 660–662.

K. A. Ryan and P. T. Englund. 1989. Replication of Kinetoplast DNA in *Trypanosoma equiperdum*. Minicircle H Strand Fragments Which Map at Specific Locations. *J. Biol. Chem.* 264 823–830.

L. Simpson and J. Shaw. 1989. RNA Editing and the Mitochondrial Cryptogenes of Kinetoplastid Protozoa. *Cell* 57 355–366.

9.7c

L. Bogorad. 1981. Chloroplasts. *J. Cell. Biol.* 91 256s–270s.

J. C. Gingrich and R. B. Hallick. 1985. The *Euglena gracilis* Chloroplast Ribulose-1,5-Bisphosphate Carboxylase Gene. *J. Biol. Chem.* 260 16162–16168.

S. D. Kung and C. M. Lin. 1985. Chloroplast Promoters from Higher Plants. *Nucleic Acids Res.* 13 7543–7549.

J. D. Palmer. 1985. Comparative Organization of Chloroplast Genomes. *Annu. Rev. Genet.* 19 325–354.

G. Van den Broeck, M. P. Timko, A. P. Kausch, A. R. Cashmore, M. Van Montagu, and L. Herrera-Estrella. 1985. Targeting of a Foreign Protein to Chloroplasts by Fusion to the Transit Peptide from the Small Subunit of Ribulose 1,5-Bisphosphate Carboxylase. *Nature* 313 358–363.

J. Gray. 1986. Wonders of Chloroplast DNA. *Nature* 322 501–502.

G. W. Schmidt and M. L. Mishkind. 1986. The Transport of Proteins into Chloroplasts. *Annu. Rev. Biochem.* 55 879–912.

K. Shinozaki, M. Ohme, M. Tanaka, T. Wakasugi, N. Hayashida, T. Matsubayashi, N. Zaita, J. Chunwongse, J. Obokata, K. Yamaguchi-Shinozaki, C. Ohto, K. Torazawa, B. Y. Meng, M. Sugita, H. Deno, T. Kamogashira, K. Yamada, J. Kusuda, F. Takaiwa, A. Kato, N. Tohdoh, H. Shimada, and M. Sugiura. 1986. The Complete Nucleotide Sequence of the Tobacco Chloroplast Genome: Its Gene Organization and Expression. *EMBO J.* 5 2043–2049.

M. Wu, J. K. Lou, D. Y. Chang, C. H. Chang, and Z. Q. Nie. 1986. Structure and Function of a Chloroplast DNA Replication Origin of *Chlamydomonas reinhardtii*. *Proc. Natl. Acad. Sci.* U.S.A. 83 6761–6765.

K. Umesono and H. Ozeki. 1987 Chloroplast Gene Organization in Plants. *Trends Genet.* 3 281–287.

J. E. Mullet. 1988. Chloroplast Development and Gene Expression. *Annu. Rev. Plant Physiol. and Plant Mol. Biol.* 39 475–502.

K. Ohyama, T. Kohchi, T. Sano, and Y. Yamada. 1988. Newly Identified Groups of Genes in Chloroplasts. *Trends Biochem. Sci.* 13 19–22.

W. Gruissem. 1989. Chloroplast Gene Expression: How Plants Turn Their Plastids On. *Cell* 56 161–170.

9.7d

L. Margulis. 1970. Origin of Eukaryotic Cells. Yale University Press, New Haven, Conn.

R. Lewin. 1983. Promiscuous DNA Leaps All Barriers. *Science* 219 478–479.

D. Yang, Y. Oyaizu, G. J. Olsen, and C. R. Woese. 1985. Mitochondrial Origins. *Proc. Natl. Acad. Sci.* U.S.A. 82 4443–4447.

M-C. Shih, G. Lazar, and H. M. Goodman. 1986. Evidence in Favor of the Symbiotic Origin of Chloroplasts: Primary Structure and Evolution of Tobacco Glyceraldehyde-3-Phosphate Dehydrogenase. *Cell* 47 73–80.

M-C. Shih, P. Heinrich, and H. M. Goodman. 1988. Intron Existence Predated the Divergence of Eukaryotes and Prokaryotes. *Science* 242 1164–1166.

M. W. Gray. 1989. The Evolutionary Origins of Organelles. *Trends Genet.* 5 294–299.

10.1a

D. J. Finnegan and D. H. Fawcett. 1986. Transposable Elements in *Drosophila melanogaster*. *Oxford Surv. Eukar. Genes* 3 1–62.

A. J. Kingsman, K. F. Chater, and S. M. Kingsman (eds.). 1988. Transposition. Society for General Microbiology Symposium 43, Cambridge University Press, Cambridge.

D. E. Berg and M. M. Howe. 1989. Mobile DNA. *Amer. Soc. Microbiol.*, Washington, D.C.

10.1c

M. L. Goldberg, J-Y. Sheen, W. J. Gehring, and M. M. Green. 1983. Unequal Crossing-Over Associated with Asymmetrical Synapsis Between Nomadic Elements in the *Drosophila menalogaster* Genome. *Proc. Natl. Sci.* U.S.A. 80 5017–5021.

G. Levinson and G. A. Gutman. 1987. Slipped-Strand Mispairing: A Major Mechanism for DNA Sequence Evolution. A Review. *Mol. Biol. Evol.* 4 203–221.

10.2a

N. Kleckner. 1981. Transposable Elements in Prokaryotes. *Annu. Rev. Genet.* 15 341–404.

N. Grindley. 1983. Transposition of Tn3 and Related Transposons. *Cell* 32 3–5.

S. Iida, J. Meyer, and W. Arber. 1983. Prokaryotic IS Elements. In J. A. Shapiro (ed.), Mobile Genetic Elements, pp. 159–221. Academic Press, New York.

A. Toussaint and A. Resibois. 1983. Phage Mu: Transposition as a Life Style. In J. A. Shapiro (ed.), Mobile Genetic Elements, pp. 105–158. Academic Press, New York.

R. Craigie and K. Mizuuchi. 1985. Mechanism of Transposition of Bacteriophage Mu: Structure of a Transposition Intermediate. *Cell* 41 867–876.

K. M. Derbyshire and N. D. F. Grindley. 1986. Replicative and Conservative Transposition in Bacteria. *Cell* 47 325–327.

K. Mizuuchi and R. Craigie. 1986. Mechanism of Bacteriophage Mu Transposition. *Annu. Rev.* Genet. 20 385–429.

D. Brunier, B. Michel, and S. D. Ehrlich. 1988. Copy Choice Illegitimate DNA Recombination. *Cell* 52 883–892.

D. J. Galas and M. Chandler. 1989. Bacterial Insertion Sequences. In D. E. Berg and M. M. Howe (eds.), Mobile DNA, pp. 109–162. American Society for Microbiology, Washington, D.C.

N. Kleckner. 1989. Transposon Tn10. In D. E. Berg and M. M. Howe (eds.), Mobile DNA, pp. 163–184. American Society for Microbiology, Washington, D.C.

M. L. Pato. 1989. Bacteriophage Mu. In D. E. Berg and M. M. Howe (eds.), Mobile DNA, pp. 23–52. American Society for Microbiology, Washington, D.C.

D. Sheratt. 1989. Tn3 and Related Transposable Elements: Site Specific Recombination and Transposition. In D. E. Berg and M. M. Howe (eds.), Mobile DNA, pp. 163–184. American Society for Microbiology, Washington, D.C.

W. M. Stark, M. R. Boocock, and D. J. Sherratt. 1989. Site-Specific Recombination by Tn3 Resolvase. *Trends Genet.* 5 304–309.

10.2b

P. M. Bingham, M. G. Kidwell, and G. M. Rubin. 1982. The Molecular Basis of P-M Hybrid Dysgenesis: The Role of the P Element, a P-Strain-Specific Transposon Family. *Cell* 29 995–1004.

G. M. Rubin and A. C. Spradling. 1982. Genetic Transformation of *Drosophila* with Transposable Element Vectors. *Science* 218 348–353.

A. C. Spradling and G. M. Rubin. 1982. Transposition of Cloned P Elements into *Drosophila* Germ Line Chromosomes. *Science* 218 341–347.

R. E. Karess and G. M. Rubin. 1984. Analysis of P Transposable Element Functions in *Drosophila*. *Cell* 28 135–146.

K. O'Hare. 1985. The Mechanism and Control of P Element Transposition in *Drosophila melanogaster*. *Trends Genet.* 1 250–254.

D. C. Rio and G. M. Rubin. 1988. Identification and Purification of a *Drosophila* Protein that Binds to the Terminal 31-Base-Pair Inverted Repeats of the P Transposable Element. *Proc. Natl. Acad. Sci.* U.S.A. 85 82–833.

M. Snyder and W. F. Doolittle. 1988. P Elements in *Drosophila*: Selection at Many Levels. *Trends Genet.* 4 147–149.

W. R. Engels. 1989. P Elements in *Drosophila melanogaster*. In D. E. Berg and M. M. Howe (eds.), Mobile DNA, pp. 437–484. American Society for Microbiology, Washington, D.C.

10.2c

B. McClintock. 1951. Chromosome Organization and Gene Expression. *Cold Spring Harbor Symp. Quant. Biol.* 16 13–47.

B. Burr and F. A. Burr. 1982. Ds Controlling Elements of Maize at the Shrunken Locus Are Large and Dissimilar Insertions. *Cell* 29 977–986.

H-P. Doring and P. Starlinger. 1984. Barbara McClintock's Controlling Elements: Now at the DNA Level. *Cell* 39 253–259.

M. Muller-Neumann, J. I. Yoder, and P. Starlinger. 1984. The DNA Sequence of the Transposable Element Ac of *Zea mays* L. *Mol. Gen. Genet.* 198 19–24.

W. D. Sutton, W. L. Gerlach, D. Schwartz, and W. J. Peacock. 1984. Molecular Analysis of Ds Controlling Element Mutations at the Adh Locus of Maize. *Science* 223 1265–1268.

J. W. Schiefelbein, V. Rabox, N. V. Fedoroff, and O. E. Nelson, Jr. 1985. Deletions Within a Defective Suppressor-Mutator Element in Maize Affect the Frequency and Developmental Timing of Its Excision from the Bronze Locus. *Proc. Natl. Acad. Sci.* U.S.A. 82 4783–4787.

B. Baker, J. Schell, H. Lorz, and N. Fedoroff. 1986. Transposition of the Maize Controlling Element "Activator" in Tobacco. *Proc. Natl. Acad. Sci.* U.S.A. 83 4844–4848.

H. Dooner, J. English, E. Ralston, and E. Weck. 1986. A Single Genetic Unit Specifies Two Transposition Functions in the Maize Element Activator. *Science* 234 210–211.

H-P. Doring and P. Starlinger. 1986. Molecular Genetics of Transposable Elements in Plants. *Annu. Rev. Genet.* 20 175–200.

S. R. Wessler. 1988. Phenotypic Diversity Mediated by the Maize Transposable Elements Ac and Spm. *Science* 242 399–405.

G. Coupland, C. Plum, S. Chatterjee, A. Post, and P. Starlinger. 1989. Sequences Near the Termini Are Required for Transposition of the Maize Transposon *Ac* in Transgenic Tobacco Plants. *Proc. Natl. Acad. Sci.* U.S.A. 86 9385–9388.

N. V. Fedoroff. 1989. Maize Transposable Elements. In D. E. Berg and M. M. Howe (eds.), Mobile DNA, pp. 375–412. American Society for Microbiology, Washington, D.C.

10.3b

R. T. Elder, E. Y. Loh, and R. W. Davis. 1983. RNA from the Yeast Transposable Element Tyl Has Both Ends in the Direct Repeats, a Structure Similar to Retrovirus RNA. *Proc. Natl. Acad. Sci.* U.S.A. 80 2432–2436.

C. E. Paquin and V. M. Williamson. 1984. Temperature Effects on the Rate of Ty Transposition. *Science* 226 53–55.

G. S. Roeder, M. Smith, and E. J. Lambie. 1984. Intrachromosomal Movement of Genetically Marked *S. cerevisiae* Transposons by Gene Conversion. *Mol. Cell. Biol.* 4 703–711.

S. J. Silverman and G. R. Fink. 1984. Effects of Ty Insertions on *HIS*4 Transcription in *Saccharomyces cerevisiae*. *Mol. Cell. Biol.* 4 1246–1251.

J. D. Boeke, D. J. Garfinkel, C. A. Styles, and G. R. Fink. 1985. Ty Elements Transpose Through an RNA Intermediate. *Cell* 40 491–500.

J. Clare and P. Farabaugh. 1985. Nucleotide Sequence of a Yeast Ty Element. Evidence for an Unusual Mechanism of Gene Expression. *Proc. Natl. Acad. Sci.* U.S.A. 82 2829–2833.

D. J. Garfinkel, J. D. Boeke, and G. R. Fink. 1985. Ty-Element Transposition: Reverse Transcriptase and Virus-Like Particles. *Cell* 42 507–517.

J. D. Boeke, C. A. Styles, and G. R. Fink. 1986. *Saccharomyces cerevisiae* SPT3 Gene Is Required for Transposition and Transpositional Recombination of Chromosomal Ty Elements. *Mol. Cell. Biol.* 6 3575–3581.

G. R. Fink, J. D. Boeke, and D. J. Garfinkel. 1986. The Mechanism and Consequences of Retrotransposition. *Trends Genet.* 2 118–123.

S. E. Adams, J. Mellor, K. Gull, S. B. Sim, M. F. Tuite, S. M. Kingsman, and A. J. Kingsman. 1987. The Functions and Relationships of Ty-VLP Proteins in Yeast Reflect Those of Mammalian Retroviral Proteins. *Cell* 49 111–119.

L. R. Coney and G. S. Roeder. 1988. Control of Yeast Gene Expression by Transposable Elements: Maximum Expression Requires a Functional Ty Activator Sequence and a Defective Ty Promoter. *Mol. Cell. Biol.* 8 4009–4017.

D. J. Eichinger and J. D. Boeke. 1988. The DNA Intermediate in Yeast Tyl Element Transposition Copurifies with Virus-Like Particles: Cell-Free Tyl Transposition. *Cell* 54 955–966.

A. Goel and R. E. Pearlman. 1988. Transposable Element-Mediated Enhancement of Gene Expression in *Saccharomyces cerevisiae* Involves Sequence-Specific Binding of a Trans-Acting Factor. *Mol. Cell. Biol.* 8 2572–2580.

A. J. Kingsman and S. M. Kingsman. 1988. Ty: A Retroelement Moving Forward. *Cell* 53 333–335.

J. D. Boeke. 1989. Transposable Elements in *Saccharomyces cerevisiae*. In D. E. Berg and M. M. Howe (eds.), Mobile DNA, pp. 335–374. American Society for Microbiology, Washington, D.C.

J. D. Boeke and V. G. Corces. 1989. Transcription and Reverse Transcription of Retrotransposons. *Annu. Rev. Microbiol.* 43 403–434.

K. Yu and R. T. Elder. 1989. A Region Internal to the Coding Sequences Is Essential for Transcription of the Yeast Ty-D15 Element. *Mol. Cell. Biol.* 9 3667–3678.

10.3c

A. J. Flavell, S. W. Ruby, J. J. O'Toole, B. E. Roberts, and G. M. Rubin. 1980. Translation and Developmental Regulation of RNA Encoded by the Eukaryotic Transposable Element Copia. *Proc. Natl. Acad. Sci.* U.S.A. 77 7107–7111.

N. A. Tchurikv, Y. V. Ilyin, K. G. Skryabin, E. V. Ananiev, A. A. Bayev, Jr., A. S. Krayev, E. S. Zelentsova, V. V.

Kulguskin, N. V. Lyubomirskaya, and G. P. Georgiev. 1981. General Properties of Mobile Dispersed Genetic Elements in *Drosophila melangaster*. *Cold Spring Harbor Symp. Quant. Biol.* 45 655–665.

A. J. Flavell and D. Ish-Horowicz. 1983. The Origin of Extrachromosomal Circular Copia Elements. *Cell* 34 415–419.

T. Shiba and K. Saigo. 1983. Retrovirus-Like Particles Containing RNA Homologous to the Transposable Element Copia in *Drosophila melanogaster*. *Nature* 302 119–124.

S. M. Mount and G. M. Rubin. 1985. Complete Nucleotide Sequence of the *Drosophila* Transposable Element Copia: Homology Between Copia and Retroviral Proteins. *Mol. Cell. Biol.* 5 1630–1638.

A. Flavell. 1986. Transposon Tricks Revealed. *Nature* 320 397.

T. Hazelrigg. 1987. The *Drosophila* White Gene: A Molecular Update. *Trends Genet.* 3 43–47.

P. M. Bingham and Z. Zachar. 1989. Retrotransposons and the FB Transposon from *Drosophila melanogaster*. In D. E. Berg and M. M. Howe (eds.), Mobile DNA, pp. 485–502. American Society for Microbiology, Washington, D.C.

10.3d

S. H. Wilson and E. L. Kuff. 1972. A Novel DNA Polymerase Activity Found in Association with Intracisternal A-Type Particles. *Proc. Natl. Acad. Sci*, U.S.A. 69 1531–1536.

E. L. Kuff, L. A. Smith, and K. K. Lueders. 1981. Intracisternal A Particle Genes in *Mus musculus*: A Conserved Family of Retrovirus-Like Elements. *Mol. Cell. Biol.* 1 216–277.

M. Horowitz, S. Luria, G. Rechavi, and D. Givol. 1984. Mechanism of Activation of the Mouse c-*mos* Oncogene by the LTR of an Intracisternal A-Particle Gene. *EMBO J.* 3 2937–2941.

L. Piko, M. D. Hammons, and K. D. Taylor. 1984. Amounts, Synthesis, and Some Properties of Intracisternal A Particle Related RNA in Early Mouse Embryos. *Proc. Natl. Acad. Sci.* U.S.A. 81 488–492.

E. L. Kuff and K. K. Lueders. 1988. The Intracisternal A-Particle Gene Family: Structure and Functional Aspects. *Adv. Cancer Res.* 51 183–276.

10.3e

T. Fujiwara and K. Mizuuchi. 1988. Retroviral DNA Integration: Structure of an Integration Intermediate. *Cell* 54 497–504.

R. F. Doolittle, D-F. Feng, M. S. Johnson, and M. A. McClure. 1989. Origins and Evolutionary Relationships of Retroviruses. *Quart. Rev. Biol.* 64 1–30.

D. P. Grandgenett and S. R. Mumm. 1990. Unraveling Retrovirus Integration. *Cell* 60 3–4.

10.3f

P. P. DiNocera, M. E. Digan, and I. B. Dawid. 1983. A Family of Oligo-Adenylate-Terminated Transposable Sequences in *Drosophila melanogaster*. *J. Mol. Biol.* 168 715–727.

A. Bucheton, R. Paro, H. M. Sang, A. Pelisson, and D. J. Finnegan. 1984. The Molecular Basis of I-R Hybrid Dysgenesis in *Drosophila melanogaster*: Identification, Cloning, and Properties of the I Factor. *Cell* 38 153–163.

H. M. Sang, A. Pelisson, A. Bucheton, and D. J. Finnegan. 1984. Molecular Lesions Associated with White Gene Mutations Induced by I-R Hybrid Dysgenesis in *Drosophila melanogaster*. *EMBO J.* 3 3079–3085.

D. H. Fawcett, C. K. Lister, E. Kellett, and D. J. Finnegan. 1986. Transposable Elements Controlling I-R Hybrid Dysgenesis in *D. melanogaster* Are Similar to Mammalian Lines. *Cell* 47 1007–1015.

S. Antonarakis and H. H. Kazazian, Jr. 1988. The Molecular Basis of Hemophilia A in Man. *Trends Genet.* 4 233–237.

P. P. DiNocera. 1988. Close Relationship Between Non-Viral Retroposons in *Drosophila melanogaster*. *Nucleic Acids Res.* 16 4041–4052.

L. J. Mizrokhi, S. G. Georgieva, and Y. V. Ilyin. 1988. Jockey, a Mobile *Drosophila* Element Similar to Mammalian LINEs, Is Transcribed from the Internal Promoter by RNA Polymerase II. *Cell* 54 685–691.

J. Skowronski, T. G. Fanning, and M. F. Singer. 1988. Unit-Length Line-1 Transcripts in Human Teratocarcinoma Cells. *Mol. Cell. Biol.* 8 1385–1397.

D. J. Finnegan. 1989. F and Related Elements in *Drosophila melanogaster*. In D. E. Berg and M. M. Howe (eds.), Mobile DNA, pp. 519–522. American Society for Microbiology, Washington, D.C.

D. J. Finnegan. 1989. The I Factor and I-R Hybrid Dysgenesis in *Drosophila melanogaster*. In D. E. Berg and M. M. Howe (eds.), Mobile DNA, pp. 503–518. American Society for Microbiology, Washington, D.C.

C. A. Hutchison III, S. C. Hardies, D. D. Loeb, W. R. Shehee, and M. H. Edgell. 1989. LINES and Related Retroposons: Long Interspersed Repeated Sequences in the Eucaryotic Genome. In D. E. Berg and M. M. Howe (eds.), Mobile DNA, pp. 593–618. American Society for Microbiology, Washington, D.C.

A. Bucheton. 1990. I Transposable Elements and I-R Hybrid Dysgenesis in *Drosophila*. *Trends Genet.* 6 16–21.

10.4a

J. Battey, E. E. Max, W. O. McBride, D. Swan, and P. Leder. 1982. A Processed Human Immunoglobulin ε Gene Has Moved to Chromosome 9. *Proc. Natl. Acad. Sci.* U.S.A. 79 5956–5960.

S. Lewis and M. Gellert. 1989. The Mechanism of Antigen Receptor Gene Assembly. *Cell* 59 585–588.

S. G. Lutzker and F. W. Alt. 1989. Immunoglobulin Heavy-Chain Class Switching. In D. E. Berg and M. M. Howe (eds.), Mobile DNA, pp. 693–714. American Society for Microbiology, Washington, D.C.

H. G. Zachau. 1989. Immunoglobulin Light-Chain Genes of the κ Type in Man and Mouse. In T. Honjo, F. W. Alt, and T. Rabbitts (eds.), The Immunoglobulin Genes, pp. 91–109. Academic Press, London.

10.6d

P. Borst and G. A. M. Cross. 1982. Molecular Basis for Trypanosome Antigenic Variation. *Cell* 29 291–303.

P. T. Englund, S. L. Hajduk, and J. C. Marini. 1982. The Molecular Biology of Trypanosomes. *Annu. Rev. Biochem.* 51 695–726.

R. G. Nelson, M. Parsons, P. J. Barr, K. Stuart, M. Selkirk, and N. Agabian. 1983. Sequences Homologous to the Variant Antigen mRNA Spliced Leader Are Located in Tandem Repeats and Variable Orphons in *Trypanosoma brucei*. *Cell* 34 901–909.

E. Pays, S. Van Assel, M. Laurent, M. Darville, T. Vervoort, N. Van Meirvenne, and M. Steinert. 1983. Gene Conversion as a Mechanism for Antigenic Variation in Trypanosomes. *Cell* 34 371–381.

J. R. Young, J. S. Shah, G. Matthyssens, and R. O. Williams. 1983. Relationship Between Multiple Copies of a *T. brucei* Variable Surface Glycoprotein Gene Whose Expression Is Not Controlled by Duplication. *Cell* 32 1149–1159.

D. A. Campbell, M. P. van Bree, and J. C. Boothroyd. 1984. The 5′-Limit of Transposition and Upstream Barren Region of a Trypanosome VSG Gene: Tandem 76 Base-Pair Repeats Flanking $(TAA)_{90}$. *Nucleic Acids Res.* 12 2759–2773.

D. M. Dorfman and J. E. Donelson. 1984. Characterization of the 1.35 Kilobase DNA Repeat Unit Containing the Conserved 35 Nucleotides at the 5′-Termini of Variable Surface Glycoprotein mRNAs in *Trypanosoma brucei*. *Nucleic Acids Res.* 12 4907–4920.

P. A. M. Michels, L. H. T. Van der Ploeg, A. Y. C. Lium, and P. Borst. 1984. The Inactivation and Reactivation of an Expression-Linked Gene Copy for a Variant Surface Glycoprotein in *Trypanosoma brucei*. *EMBO J.* 3 1345–1351.

M. Milhausen, R. G. Nelson, S. Sather, M. Selkirk, and N. Agabian. 1984. Identification of a Small RNA Containing the Trypanosome Spliced Leader: A Donor of Shared 5′ Sequences of Trypanosomatid mRNAs? *Cell* 38 721–729.

L. H. T. Van der Ploeg, D. C. Schwartz, C. Cantor, and and P. Borst. 1984. Antigenic Variation in *Trypanosoma brucei* Analyzed by Electrophoretic Separation of Chromosome-Sized DNA Molecules. *Cell* 37 77–84.

F. E. G. Cox. 1985. Chromosomes of Malaria Parasites and Trypanosomes. *Nature* 315 280–281.

J. E. Donelson and A. C. Rice-Ficht. 1985. Molecular Biology of Trypanosome Antigenic Variation. *Microbiol. Rev.* 49 107–125.

R. M. Klug. 1985. The Role of RNA Priming in Viral and Trypanosomal mRNA Synthesis. *Cell* 41 651–652.

P. Borst. 1986. Discontinuous Transcription and Antigenic Variation in Trypanosomes. *Annu. Rev. Biochem.* 55 701–732.

E. Pays. 1986. Variability of Antigen Genes in African Trypanosomes. *Trends Genet.* 2 21–26.

L. Van der Ploeg. 1986. Discontinuous Transcription and Splicing in Trypanosomes. *Cell* 47 479–480.

L. H. T. Van der Ploeg. 1987. Control of Variant Surface Antigen Switching in Trypanosomes. *Cell* 51 159–161.

E. Pays and M. Steinert. 1988. Control of Antigen Gene Expression in African Trypanosomes. *Annu. Rev. Genet.* 22 107–126.

R. E. Sutton and J. C. Boothroyd. 1988. The Cap of Both Mini-Exon-Derived RNA and mRNA of Trypanosomes Is 7-Methylguanosine. *Mol. Cell. Biol.* 8 494–496.

J. E. Donelson. 1989. DNA Rearrangements and Antigenic Variation in African Trypanosomes. In D. E. Berg and M. M. Howe (eds.), Mobile DNA, pp. 763–782. American Society for Microbiology, Washington, D.C.

E. Pays, P. Tebabi, A. Pays, H. Coquelet, P. Revelard, D. Salmon, and M. Steinert. 1989. The Genes and Transcripts of an Antigen Gene Expression Site from *T. brucei*. *Cell* 57 835–845.

C. Roth, F. Bringaud, R. E. Layden, T. Baltz and H. Eisen. 1989. Active Late-Appearing Variable Surface Antigen Genes in *Trypanosoma equiperdum* Are Constructed Entirely from Pseudogenes. *Proc. Natl. Acad. Sci. U.S.A.* 86 9375–9379.

10.7a

A. C. Spradling and A. P. Mahowald. 1981. A Chromosome Inversion Alters the Pattern of Specific DNA Replication in *Drosophila* Follicle Cells. *Cell* 27 203–209.

T. H. Eickbush and F. C. Kafatos. 1982. A Walk in the Chorion Locus of *Bombyx mori*. *Cell* 29 633–643.

D. V. de Cicco and A. C. Spradling. 1984. Localization of a Cis-Acting Element Responsible for the Developmentally Regulated Amplification of *Drosophila* Chorion Gene. *Cell* 38 45–54.

F. C. Kafatos, W. Orr, and C. Delidakis. 1985. Developmentally Regulated Gene Amplification. *Trends Genet.* 1 301–306.

B. Wakimoto, L. J. Kalfayan, and A. C. Spradling. 1986. Developmentally Regulated Expression of *Drosophila* Chorion Genes Introduced at Diverse Chromosomal Positions. *J. Mol. Biol.* 187 33–45.

A. Spradling and T. Orr-Weaver. 1987. Regulation of DNA Replication During *Drosophila* Development. *Annu. Rev. Genet.* 21 373–403.

10.7b

I. B. Dawid, D. D. Brown, and R. H. Reeder. 1970. Composition and Structure of Chromosomal and Amplified Ribosomal DNAs of *Xenopus laevis*. *J. Mol. Biol.* 51 341–360.

S. A. Endow and K. C. Atwood. 1988. Magnification: Gene Amplification by an Inducible System of Sister Chromatid Exchange. *Trends Genet.* 4 348–351.

10.7c

M. A. Truett and J. G. Gall. 1977. The Replication of Ribosomal DNA in the Macronucleus of *Tetrahymena*. *Chromosoma* (Berl.) 64 295–303.

M-C. Yao and J. G. Gall. 1977. A Single Integrated Gene for Ribosomal RNA in a Eucaryote, *Tetrahymena pyriformis*. *Cell* 12 121–132.

E. H. Blackburn and J. G. Gall. 1978. A Tandemly Repeated Sequence at the Termini of the Extrachromosmal Ribosomal RNA Genes in *Tetrahymena*. *J. Mol. Biol.* 120 33–53.

E. H. Blackburn and K. M. Karrer. 1986. Genomic Reorganization in Ciliated Protozoans. *Annu. Rev. Genet.* 20 501–521.

D. D. Larson, E. H. Blackburn, P. C. Yaeger, and E. Orias. 1986. Control of rDNA Replication in *Tetrahymena* Involves a *Cis*-Acting Upstream Repeat of a Promoter Element. *Cell* 47 229–240.

M-C. Yao, K. Zheng, and C-H. Yao. 1987. A Conserved Nucleotide Sequence at the Sites of Developmentally Regulated Chromosomal Breakage in *Tetrahymena*. *Cell* 48 779–788.

E. K. Robinson, P. D. Cohen, and E. H. Blackburn. 1989. A Novel DNA Deletion-Ligation Reaction Catalyzed *in Vitro* by a Developmentally Controlled Activity from *Tetrahymena* Cells. *Cell* 58 887–900.

M-C. Yao. 1989. Site-Specific Chromosome Breakage and DNA Deletion in Ciliates. In D. E. Berg and M. M. Howe (eds.), Mobile DNA, pp. 715–734. American Society for Microbiology, Washington, D. C.

10.8a

T. D. Kempe, E. A. Swyryd, M. Brulet, and G. R. Stark. 1976. Stable Mutants of Mammalian Cells That Overproduce the First Three Enzymes of Pyrimidine Nucleotide Biosynthesis. *Cell* 9 541–550.

G. M. Wahl, R. A. Padgett, and G. R. Stark. 1979. Gene Amplification Causes Overproduction of First Three Enzymes of UMP Synthesis in N-(Phosphonacetyl)-1-Aspartate-Resistant Hamster Cells. *J. Biol. Chem.* 254 8679–8689.

J. L. Hamlin, J. D. Milbrandt, N. H. Heintz, and J. C. Azizkhan. 1984. DNA Sequence Amplification in Mammalian Cells. *Intl. Rev. Cytol.* 90 31–82.

R. T. Schimke. 1984. Gene Amplification in Cultured Animal Cells. *Cell* 37 705–713.

G. Stark and G. M. Wahl. 1984. Gene Amplification. *Annu. Rev. Biochem.* 53 447–491.

K. Alitalo. 1985. Amplification of Cellular Oncogenes in Cancer Cells. *Trends Biochem. Sci.* 10 194–197.

R. T. Schimke. 1988. Gene Amplification in Cultured Cells. *J. Biol. Chem.* 263 5989–5992.

10.8b

R. J. Kaufman, P. C. Brown, and R. T. Schimke. 1979. Amplified Dihydrofolate Reductase Genes in Unstably Methotrexate-Resistant Cells Are Associated with Double Minute Chromosomes. *Proc. Natl. Acad. Sci.* U.S.A. 76 5669–5673.

J. K. Cowell. 1982. Double Minutes and Homogeneously Staining Regions: Gene Amplification in Mammalian Cells. *Annu. Rev. Genet.* 16 21–59.

M. Montoya-Zavala and J. L. Hamlin. 1985. Similar 150 Kilobase DNA Sequences Are Amplified in Independently Derived Methotrexate Resistant Chinese Hamster Cells. *Mol. Cell. Biol.* 5 619–627.

M. Ford and M. Fried. 1986. Large Inverted Duplications Are Associated with Gene Amplification. *Cell* 45 425–430.

E. Giulotto, I. Saito, and G. R. Stark. 1986. Structure of DNA Formed in the First Step of CAD Gene Amplification. *EMBO J.* 5 2115–2121.

I. Saito, R. Groves, E. Giulotto, M. Rolfe, and G. R. Stark. 1989. Evolution and Stability of Chromosomal DNA Coamplified with the CAD Gene. *Mol. Cell. Biol.* 9 2445–2452.

10.8c

M. Botchan, W. Topp, and J. Sambrook. 1978. Studies on Simian Virus 40 Excision from Cellular Chromosomes. *Cold Spring Harbor Symp. Quant. Biol.* 43 709–719.

M. Wigler, M. Perucho, D. Kurtz, S. Dana, A. Pellicer, R. Axel, and S. Silverstein. 1980. Transformation of Mammalian Cells with an Amplifiable Dominant-Acting Gene. *Proc. Natl. Acad. Sci.* U.S.A. 77 3567–3570.

J. M. Roberts and R. Axel. 1982. Gene Amplification and Gene Correction in Somatic Cells. *Cell* 29 109–119.

J. M. Roberts, L. B. Buck, and R. Axel. 1983. A Structure for Amplified DNA. *Cell* 33 53–63.

K. Alitalo. 1985. Amplification of Cellular Oncogenes in Cancer Cells. *Trends Biochem. Sci.* 10 194–197.

S. M. Carroll, M. L. DeRose, P. Gaudray, C. M. Moore, D. R. Needham-Vandevanter, D. D. Von Hoff, and G. M. Wahl. 1988. Double Minute Chromosomes Can

Be Produced from Precursors Derived from a Chromosomal Deletion. *Mol. Cell. Biol.* 8 1525–1533.

D. D. Von Hoff, D. R. Needham-VanDevanter, J. Yucel, B. E. Windle, and G. M. Wahl. 1988. Amplified Human *MYC* Oncogenes Localized to Replicating Submicroscopic Circular DNA Molecules. *Proc. Natl. Acad. Sci. U.S.A.* 85 4804–4808.

J. C. Ruiz, K. Choi, D. D. Von Hoff, I. B. Roninson, and G. M. Wahl. 1989. Autonomously Replicating Episomes Contain *mdrl* Genes in a Multidrug-Resistant Human Cell Line. *Mol. Cell. Biol.* 9 109–115.

G. R. Stark, M. Debatisse, E. Giulotto, and G. M. Wahl. 1989. Recent Progress in Understanding Mechanisms of Mammalian DNA Amplification. *Cell* 57 901–908.

Understanding and Manipulating Biological Systems

PART IV

Preceding chapters tried to show, through well-studied examples, how modern experimental techniques provide rich molecular detail about gene structure and function. Through its broad overview of the structure, organization, and expression of eukaryotic genes and genomes, Part III stressed the unity of genetic processes among different organisms. However, that overview tends to mask the enormous variety of biological mechanisms used to convert genotypes into successful organisms. In fact, the number of genetic strategies living organisms employ seems to be limitless. The principal generalization is that anything that can work is likely to occur somewhere in nature. Molecular analysis has only begun to reveal the varied genetic tactics that account for the diversity of organismal form, habitat, behavior, and function. Here we introduce the ways in which such complexity is being analyzed. The presentation is both a summary of what has already been discussed and a preface to Volume II of this book.

Volume II examines in detail the impact of advances in eukaryotic molecular genetics on our understanding of several complex, interactive genetic systems. Viral, cellular, and organismal models are included. Volume II also describes how the new concepts and techniques are used to alter the phenotypes of individual gene products, of cells, and of whole organisms. The potential of these novel experimental opportunities extends beyond increased understanding by encompassing the genetic modification of plants and animals. Such modification is not an entirely new enterprise. Humans have long been successful at selecting genetically modified organisms of other species. All our agriculturally significant plants and animals are the result of thousands of years of manipulative breeding. Some of these are virtually man-made species. Others are hybrids of originally independent species. Debilitated viruses, bred and selected for their altered phenotypes, are the basis of the successful vaccines for polio and other viral diseases. Naturally arising mutants of various microorganisms, selected because they are efficient producers of antibiotics, are grown in large quantities. Modern techniques now allow us to design and select for modified organismal phenotypes with more precision and efficiency. Moreover, the new methods reduce the chance that undesirable and undetected modifications will accompany the selected phenotype, a serious drawback of the nonspecific breeding techniques.

Understanding Complex Biological Systems

Biochemistry and genetics epitomize what is called the **reductionist approach** to biology; single biological elements (for example, a protein, an enzyme, or a gene) are examined in great detail through a series of narrowly defined questions and experimental protocols. This approach can illuminate many phenotypic characteristics of complex systems, including whole organisms, if those phenotypes are the direct consequence of a single gene's function. An example is the common red-green color blindness phenotype of human males.

The Multigene Family for Visual Pigments

The cone cells of the retina are the color-sensitive components of vertebrate eyes. Other, more abundant, rod cells are sensitive to dim light and give monochromatic perception. Rod cells' ability to absorb light resides in the membrane-bound visual pigment rhodopsin, a protein containing a bound chromophore, 11-*cis*-retinal, a vitamin A derivative. When light strikes the retina, the absorbed energy is transformed into an ion flow (electrical signal). The large amounts of pure rhodopsin that can be isolated from bovine eyes permitted its extensive biochemical characterization. After more than 40 years of work, we now have a detailed picture of how rod cells absorb light and, through a cascade of enzymatic reactions, amplify and transmit the excitation to neurons. The reactions involve signal transduction proteins and GTP. The bovine gene for opsin, the protein portion of rhodopsin, was cloned, characterized, and used as a probe to isolate the human opsin gene (Figure 6.38).

Less was known about the color-sensitive components of cone cells. Some basic ideas about the role of pigments in color perception had been deduced as early as the late eighteenth century by John Dalton, who was himself color-blind. And the association of genes determining red-green color blindness with the X-chromosome was established as early as 1911 because of the prevalence of red-green color blindness in males. But the paucity of cone cells made isolating their visual pigments and understanding the molecular basis of color vision difficult. Recently, by exploiting a partial sequence homology

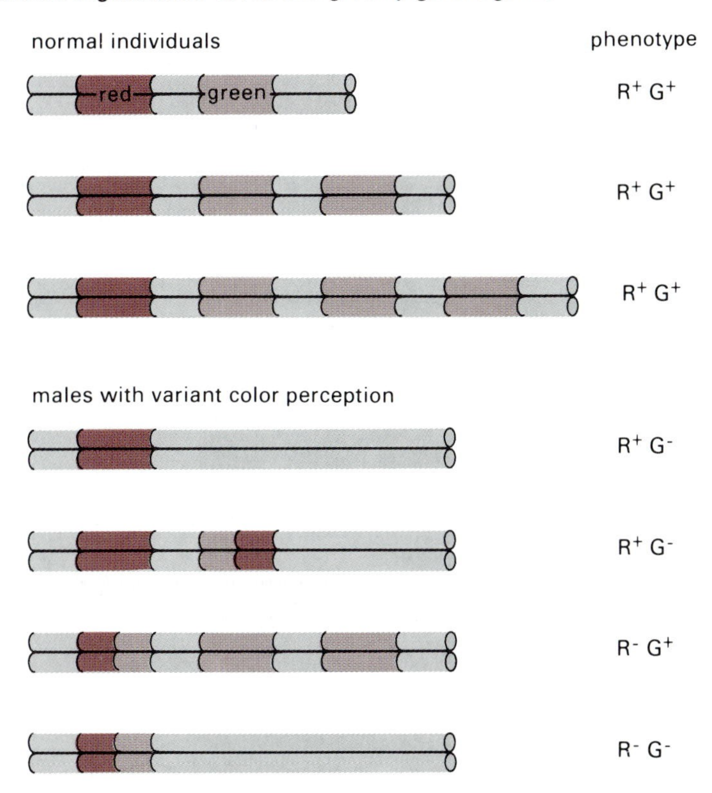

Figure IV.1
The organization of the human multigene family that encodes visual pigments. All the genes are related to one another. Genes for the red and green pigments occur in tandem on the X-chromosome. The close homology between the two genes fosters recombination. The predicted amino acid sequences of the red and green pigments are identical at 96 percent of the residues; all other pairwise comparisons (e.g., green versus blue) show 40 to 45 percent identity. Adapted from J. Nathans, *Annu. Rev. Neurosci.* 10 (1987), p. 163.

between opsin genes and human genes encoding the protein moieties of the pigments for red, green, and blue perception, the human color pigment genes were cloned (Figure 6.38).

The genes for all four visual pigment proteins make up a multigene family, apparently derived from a single, ancestral gene by amplification and mutation of both coding and regulatory sequences (Figure IV.1). With the genes in hand, we can now study the mechanism whereby individual neighboring cells synthesize only one of the pigments: rhodopsin in rod cells and red-, green-, or blue-sensitive pigment in different cones. Gene cloning has also provided an explanation for the high frequency of human color blindness. In normal males, a single gene for the red-sensitive pigment and varying numbers (one to three) of genes for the green-sensitive pigment are arranged in tandem on the long arm of the X-chromosome. DNA blotting experiments using genomic DNA from males with different red-green color-blind phenotypes indicate that aberrant color perception is often associated with mutations caused by recombinations such as unequal crossing-over or gene conversion. The high level of homology between the color vision genes and their proximity to one another fosters multiple recombinational events leading to a high frequency of mutations.

Coordinating Interactions Between Many Gene Products

Unlike color blindness, the vast majority of organisms' phenotypic characteristics arise from complex interactions between the products of many genes. To understand these phenotypes, we must study the properties of physiological systems (e.g., the nervous system), of organelles, of whole tissues, of whole organisms, of populations of organisms, and of the interaction of organisms with their environments. The alternative to reductionist biology, the **holistic** approach, has long maintained that to understand organismal behavior and function fully, we must study phenotypes in the context of the whole organism. Now, however, initial efforts suggest that the expanding resources of molecular genetics are applicable to the analysis of such complex multigenic systems. Thus, like single gene phenotypes, complex phenotypes can be described in molecular terms. Examples described earlier in this book include the regulation of serum cholesterol levels by low density lipoprotein (LDL) and the LDL receptor, the vertebrate immune systems, and the control of mating type in yeast. Another example, whose molecular features are becoming discernable, is the formation and assembly of ribosomes, organelles that are essential to all cellular functions.

Ribosome production involves many different genes encoding a variety of RNAs and proteins. Moreover, the expression of these many genes is regulated by independent as well as coordinated interactions. Thus, shortly after the completion of mitosis, the initiation of transcription of ribosomal RNA genes triggers the formation of a nucleolus and with it the synthesis and localization of about 80 ribosomal proteins, several enzymes, several small nuclear RNAs required for ribosomal RNA maturation, and other proteins that contribute to the complex nucleolar structure. Although the transcription of ribosomal protein genes is carried out by RNA polymerase II, the formation of ribosomal RNAs requires RNA polymerases I and III. In contrast to ribosomal RNA genes, ribosomal protein genes are not highly repeated. The output of these different transcription systems must be coordinated with one another and with the translation of the ribosomal and other proteins. In yeast, for example, there is evidence that ribosomal RNA maturation may be linked to levels of ribosomal proteins. Finally, the components of the ribosome must be assembled in an orderly manner to form the particle.

Mastering the intricacies of isolated genetic elements is only a first step in solving the higher order complexities of living organisms. As the molecular designs of complex systems become comprehensible, the already arid reductionist-holist debate will become irrelevant. We can anticipate a continuum of biological understanding starting with the fundamental properties of genes and genomes and extending to the complex, hierarchical interactions fundamental to living things. The tasks will be challenging because many different organisms and diverse kinds of systems need to be considered (for example, the way viruses act out their specialized life cycles, the coordinated regulation of genes during development and differentiation, and the mechanisms by which organisms achieve homeostasis and stability).

The Life Cycles of Viruses

To date, the goal of understanding the relation between a complete genome and the consequent phenotype has been most closely approached with certain viruses. Virion particles generally consist of a genome enclosed in a protein coat (or capsid). In some more complex virions, a lipid bilayer impregnated with proteins (an envelope) surrounds the capsid (Figure IV.2). When a virion infects a new cell, the coat is discarded, and the genome begins to direct the synthesis of new virion particles. Viral genomes are relatively small compared to those of free living eukaryotes, and they encode only a limited number of proteins, generally including those that will form the capsid and one or more proteins required for replication and expression of the viral genome. Host cells provide energy and metabolites. To varying extents, viruses also co-opt host cell mechanisms for transcription, translation, and replication. The detailed molecular descriptions of several viruses and their life cycles now available are the consequence of two technical developments: recombinant DNA techniques and sophisticated cell culture methods that permit the study of viruses *in vitro*, thereby avoiding the complications that accompany the infection of whole animals.

Viral genomes may be either DNA or RNA and either linear or circular molecules (Figure IV.3). Some viruses kill their host cells in the course of replicating themselves. Others insert their genomes into the host cell DNA, thereby transforming the cells and becoming permanent residents of the infected cell and all its progeny; in many cases, the integrated viral genomes retain the ability to produce new viral particles. This diversity of form and function illustrates the general notions, expressed before, that evolution tries out many different genetic mechanisms and that the only requirement for adoption, or evolutionary fixation, is that the strategy work. In this, viruses are no different from more complex organisms; they are only smaller and less independent.

Some viral DNAs, including those of papovaviruses, adenoviruses, and herpesviruses, are replicated by DNA polymerase. Others, such as those of hepadnaviruses (e.g., hepatitis B virus) and cauliflower mosaic virus, replicate through the intermediary formation of an RNA by the cell's RNA polymerase II followed by reverse transcription to yield a new DNA genome; the reverse transcriptase is encoded in the viral genome. Retroviruses also encode reverse transcriptase, which copies the single strand RNA genome into duplex DNA prior to its insertion into the cellular genome as a provirus; new viral genomes are transcribed from the provirus by the cellular RNA polymerase (Sections 2.2a and 5.7d). Other RNA viruses, such as polio and foot-and-mouth, replicate by copying RNA directly into RNA through the mediation of viral-encoded RNA polymerases (Section 2.5).

Recombinant DNA techniques have been especially important in the study of hepatitis B virus, a serious cause of disease worldwide. In this case, alternative approaches were extremely limited because until very recently no *in vitro* cell culture system was known to sup-

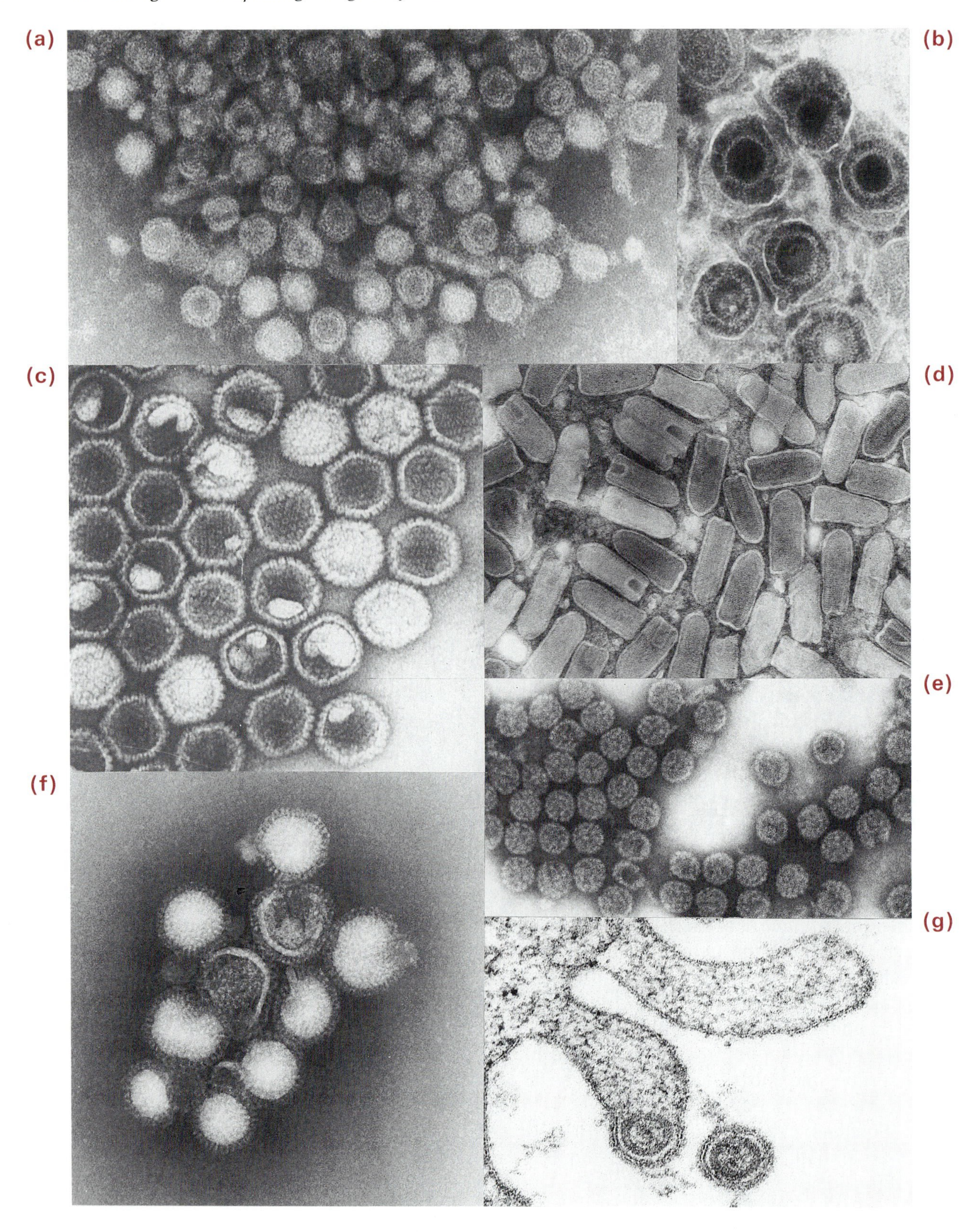
(a)
(b)
(c)
(d)
(e)
(f)
(g)

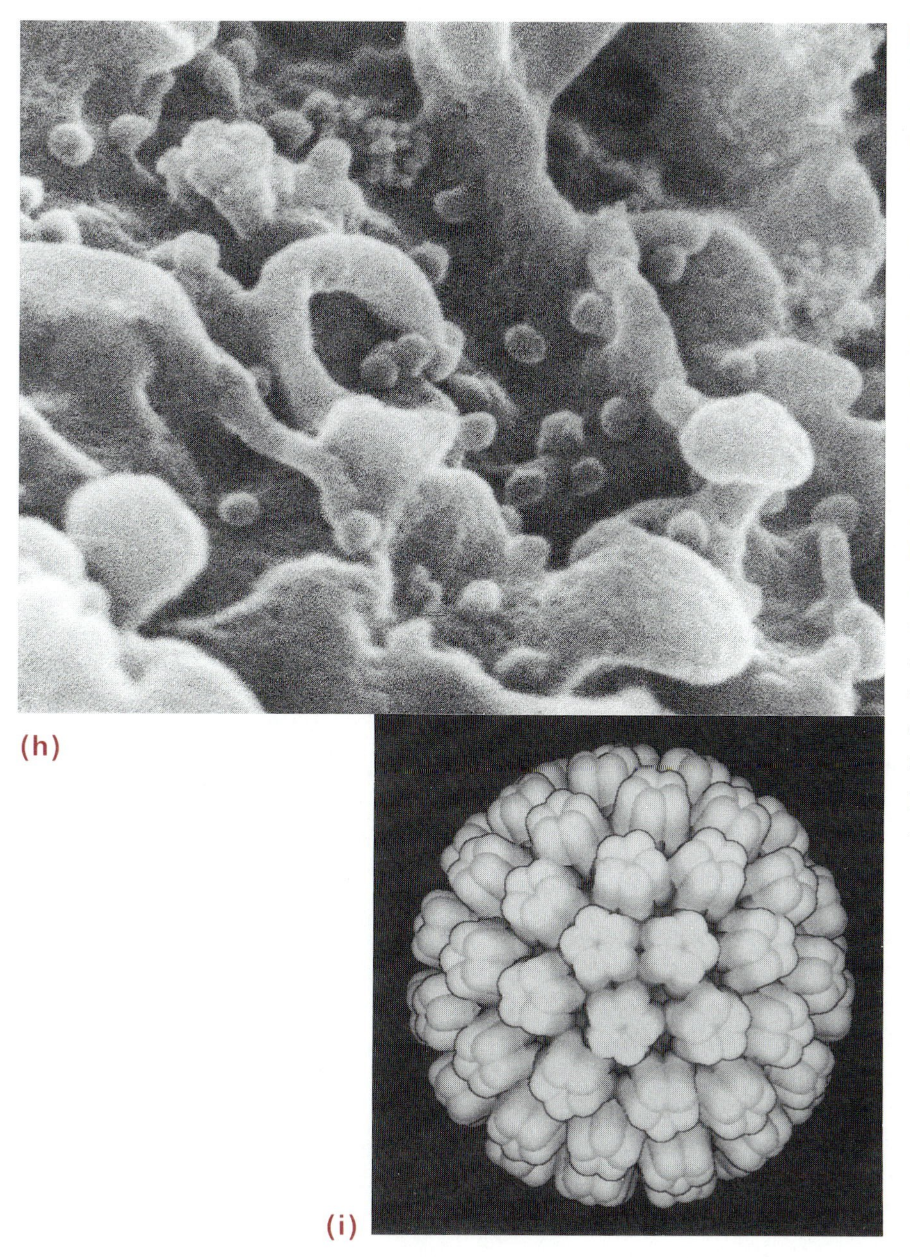

Figure IV.2
Electron micrographs of several types of virus. The numbers in parentheses in the caption give the magnification. See also Figure I.14. (a) Hepatitis B virus (170,000X); (b) Herpes Simplex 1 (114,000X); (c) Herpes Simplex 1, showing purified capsids without envelopes (127,000X); (d) Vesicular Stomatitis virus (107,000X); (e) Human papillomavirus (94,000X); (f) Influenza virus (128,000X); (g) Budding murine type C retrovirus (100,000X); (h) HIV-1 (AIDS virus) shown in a scanning electronmicrograph of the surface of human T4 lymphocytes (74,000X); (i) Computer graphics representation of the polyomavirus capsid, diameter about 500Å. Photos (a)–(e) and (h) are courtesy of F. A. Murphy. Photo (g) is courtesy of J. E. Dahlberg. Photo (i) is courtesy of D. M. Salunke and is as published by D. M. Salunke et al., *Cell* 46 (1986), p. 895. Copyright Cell Press.

port the replication of the virus. Moreover, only small quantities of its circular DNA genome can be isolated from the blood of infected humans. The amount is, however, sufficient to permit cloning the DNA in *E. coli* systems, thereby providing ample viral DNA for chemical characterization and for the production of immunogenic proteins for vaccine development. Hepatitis B surface protein synthesized in yeast cells was the first vaccine produced by recombinant DNA techniques to be licensed for human use in the United States. Similar approaches have also yielded clones encoding a surface protein from the infectious agent responsible for hepatitis C, a particularly serious human pathogen. This advance may lead to an effective vaccine for this type of infection.

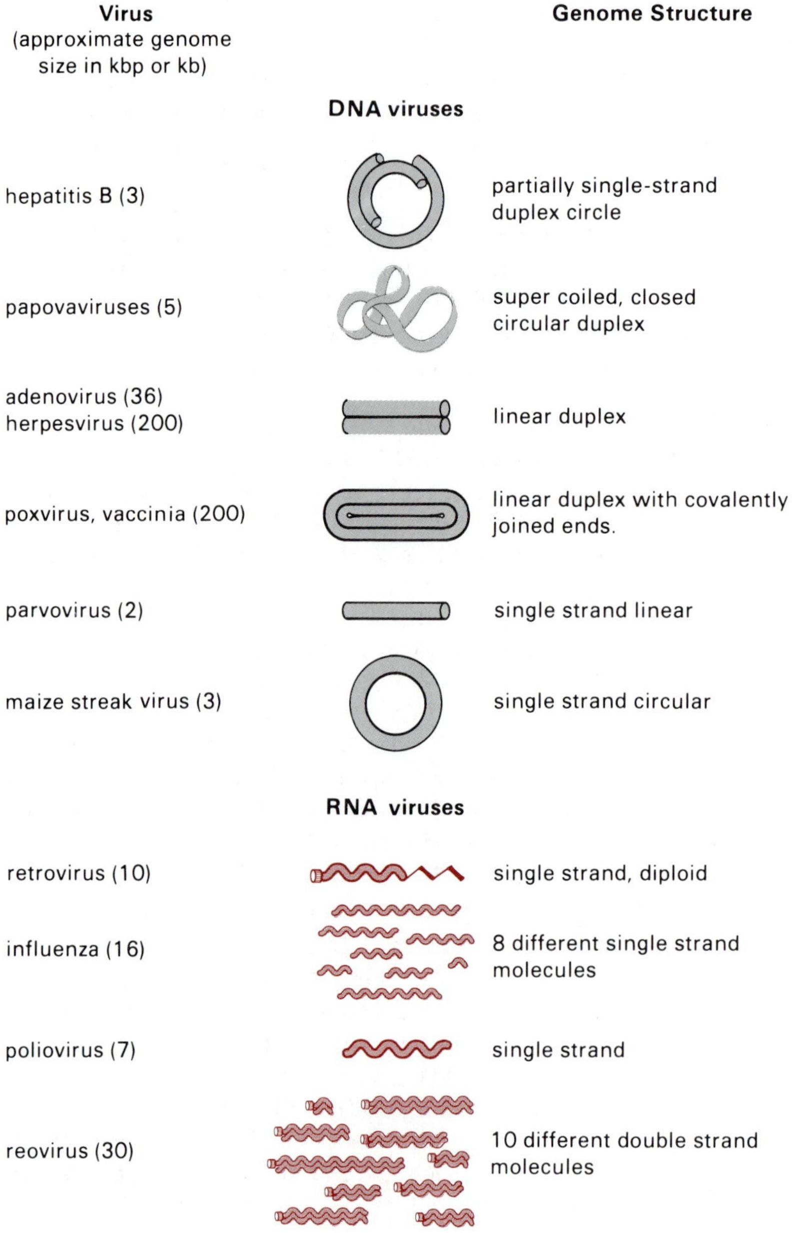

Figure IV.3
Viral genomes may be DNA or RNA, single or double strand, circular or linear.

Virus (approximate genome size in kbp or kb)	Genome Structure
DNA viruses	
hepatitis B (3)	partially single-strand duplex circle
papovaviruses (5)	super coiled, closed circular duplex
adenovirus (36) herpesvirus (200)	linear duplex
poxvirus, vaccinia (200)	linear duplex with covalently joined ends.
parvovirus (2)	single strand linear
maize streak virus (3)	single strand circular
RNA viruses	
retrovirus (10)	single strand, diploid
influenza (16)	8 different single strand molecules
poliovirus (7)	single strand
reovirus (30)	10 different double strand molecules

The papovavirus simian virus 40 (SV40) and adenovirus have served as prototypical eukaryotic genomes in studies on gene structure and on the nature of transcriptional regulatory signals. Many of the fundamental regulatory signals governing transcription by RNA polymerase II were first elucidated using these genomes, including the roles of TATA units, of GC-rich units, of enhancers, and of the polyadenylation signal AATAAA (Section 8.3). Studies with these viruses also led to the discovery of introns and splicing (Section 8.5) and of proteins such as Sp 1 that regulate transcription (Section 8.3d). Today these viruses remain attractive model systems for analyzing additional questions about eukaryotic molecular genetics and the source of critical elements for constructing eukaryotic vectors. For example,

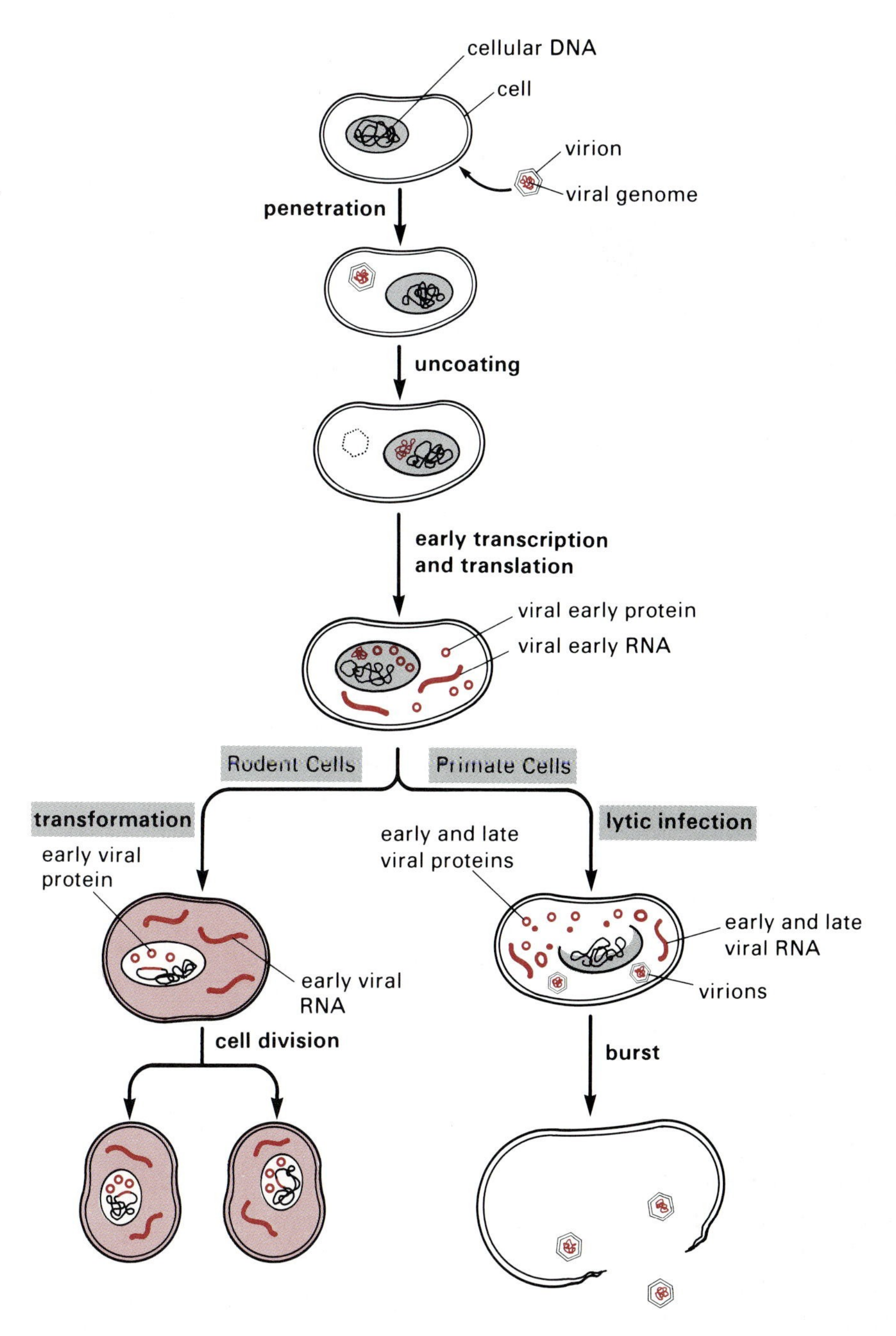

Figure IV.4
Schematic diagram of the SV40 life cycle showing uncoating and expression of early genes followed by the alternate pathways of lytic infection (replication, expression of late functions, and assembly of new virion particles) or transformation (insertion of viral DNA into host genome) in primate or rodent cells, respectively.

the mechanisms by which these viral genomes replicate are being studied, and *in vitro* systems that permit replication have been developed. These and other DNA viruses are also models for studying differentiation and development in complex organisms because their life cycles involve an orderly temporal series of genetic events, and they also display alternative life-styles. SV40, for example, multiplies in certain primate cells but is unable to replicate in rodent cells, where its genome is inserted into cellular DNA, causing an oncogenic transformation (Figure IV.4).

In contrast to eukaryotic cellular genomes, there is very little or no space between genes on the SV40 and adenovirus genomes (Figure IV.5). Coding and regulatory sequences can overlap on the DNA,

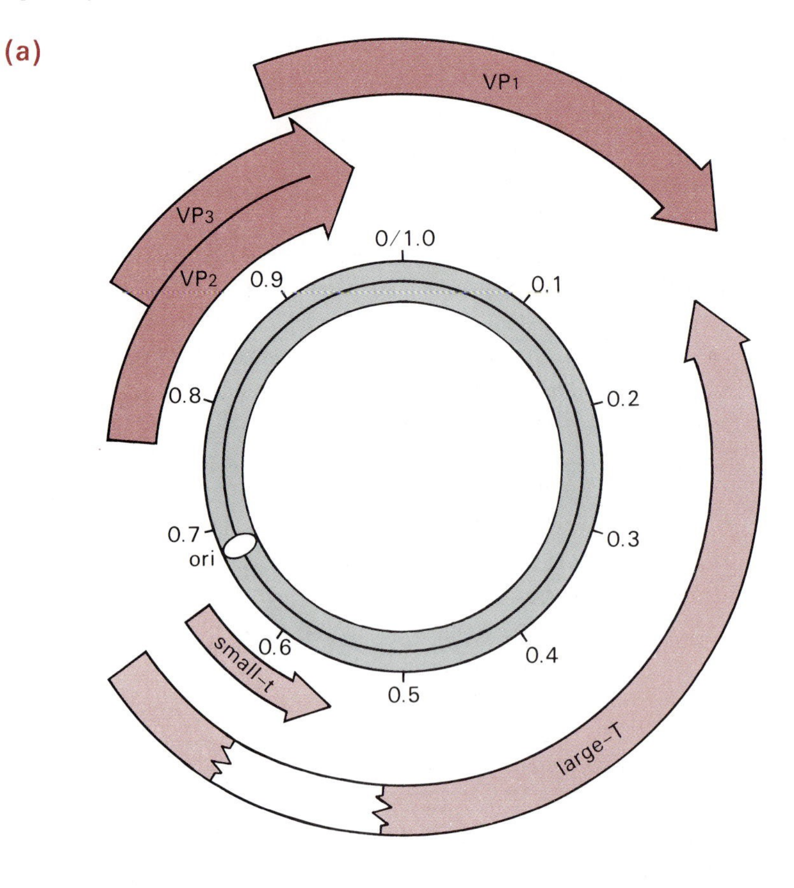

Figure IV.5
Schematic diagrams of (a) the simian virus 40 genome and (b) the adenovirus genome. Early (E) and late (L) mRNAs are shown as bars whose arrow heads indicate the direction of transcription. The lengths are not precise, and many details, including introns, are omitted.

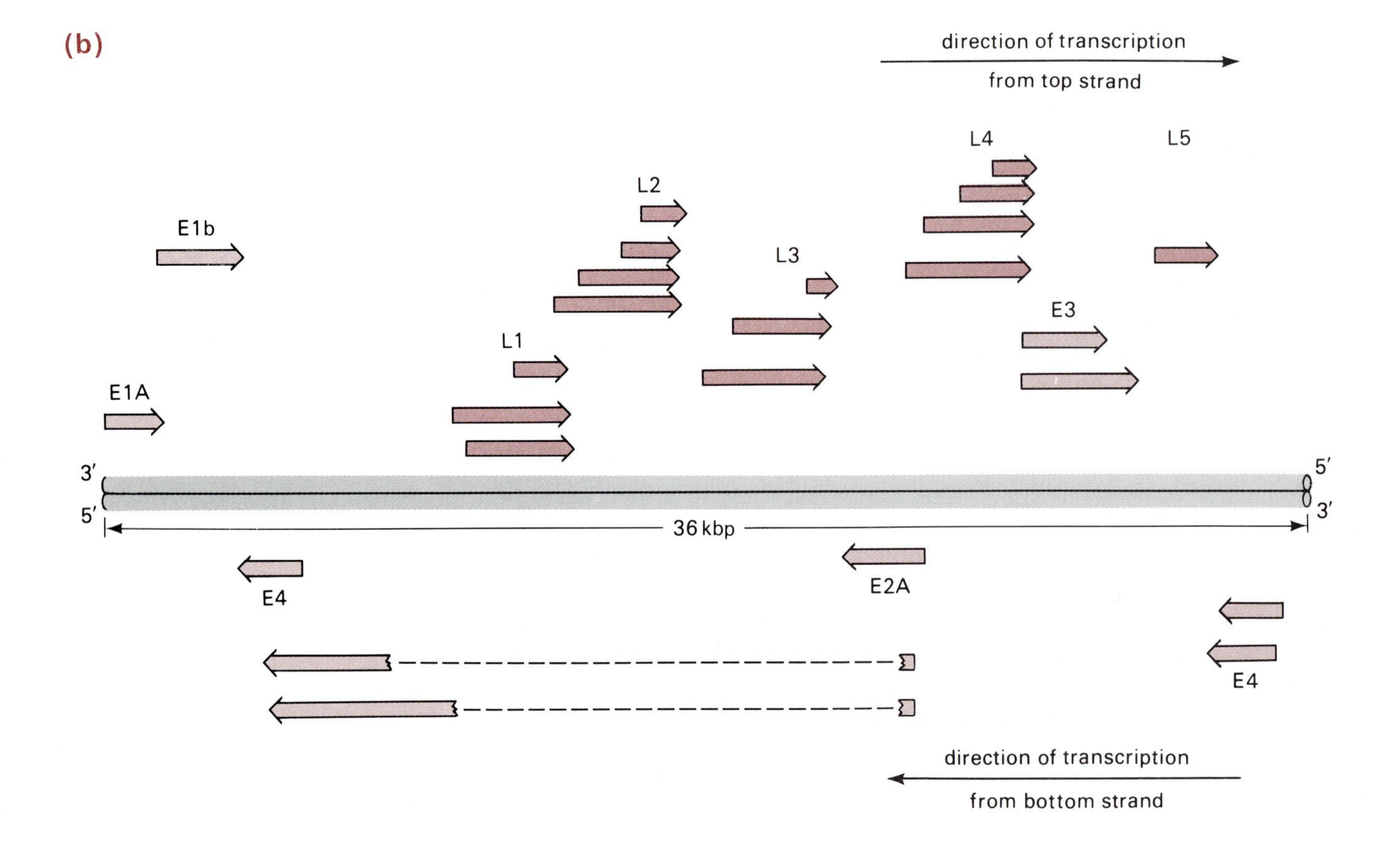

and, in the adenoviruses, important genetic information even occurs on both DNA strands in certain regions. Adenoviral DNA is seven times the size of papovavirus DNA. The enlarged coding capacity permits a more complex life cycle and virion structure, although the adenovirions are still rather simple: the genome surrounded by a capsid composed of proteins. In contrast, in the large family of herpesviruses, the capsids are surrounded by a lipid bilayer that is studded with viral-encoded glycoproteins. The complex life cycles of herpesviruses are encoded in linear duplex DNA genomes as large as 250 kbp, and the virions contain about 30 different viral-encoded proteins, compared to 3 in SV40 and about 10 in adenoviruses. The herpesvirus family includes several viruses that are associated with serious medical and veterinary problems. For example, herpes simplex viruses 1 and 2 cause recurring facial and genital lesions, cytomegalovirus causes often fatal disease in developing fetuses, Epstein-Barr virus is associated with Burkitt's lymphoma and infectious mononucleosis, and Marek's disease virus causes tumors in birds.

Molecular cloning has also been of special significance to the study of RNA viruses because the nucleotide sequences of large RNA molecules are virtually impossible to determine directly. Moreover, the DNA form of cloned viral genes, when introduced by appropriate recombinant vectors into bacterial and animal cells, can be used to produce viral proteins for both biochemical studies and the preparation of antibodies. For example, the foot-and-mouth disease virus, a member of the **picornavirus** family ("pico" for small and "rna" for an RNA genome) is a formidable experimental challenge because its genome is an 8 kb long single strand of RNA and because it is highly contagious among young, cloven-hoofed animals. Its importation into the United States is strictly prohibited. However, cDNAs encoding antigenic viral coat proteins have been synthesized and cloned, and can be safely manipulated in the laboratory. Expression of the cloned foot-and-mouth virus sequence in *E. coli* provides polypeptides that may be effective vaccines. Also, because the cloning provides the amino acid sequence of the coat protein, synthetic polypeptides can be investigated for their ability to provoke immunity.

The class of viruses now called retroviruses was first discovered because some retroviruses are tumorigenic. In 1911, Peyton Rous demonstrated that a virus, Rous Sarcoma virus (RSV), elicits the formation of sarcomas in chickens. Subsequently, enormous effort was devoted to understanding vertebrate RNA tumor viruses on the assumption that the studies would lead to an understanding of human cancer even though no human RNA tumor virus had been identified when the work began. Now we know that many of the oncogenic retroviral genomes carry coding segments other than the usual viral genes. These segments are modified cellular genes (Figure IV.6) that account for the oncogenicity of many retroviruses; they are called **viral oncogenes**, or v-*onc*. Each of the several classes of v-*onc*s encodes proteins that, when produced in the cell, impart a tumor cell phenotype. The normal cellular genes from which the v-*onc*s are derived are called **protooncogenes**. Such protooncogenes do not pro-

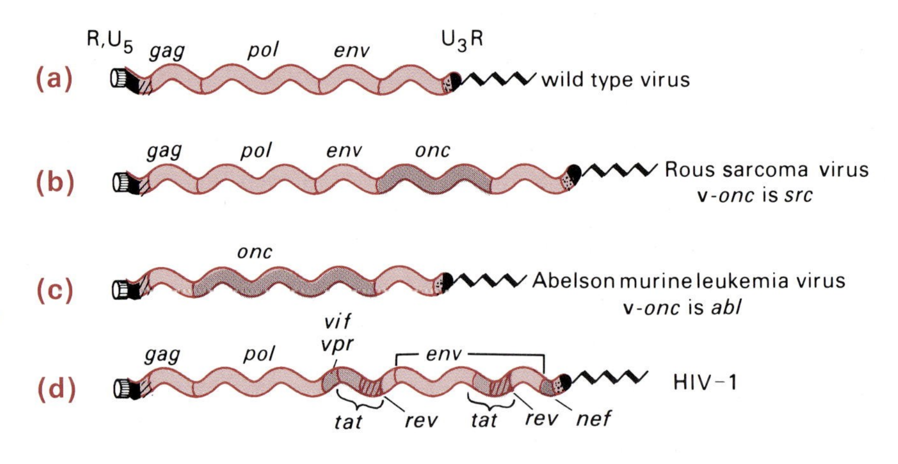

Figure IV.6
Schematic diagrams showing (a) a typical "wild type" retroviral genome with genes for nucleocapsid (*gag*), reverse transcriptase (*pol*), and envelope protein (*env*); (b) a nondefective retroviral genome carrying the v-*onc* gene, *src*; (c) a defective retroviral genome that carries the v-*onc* gene *abl* but cannot replicate independently because it lacks *pol* and *env*; and (d) the human immunodeficiency retrovirus, HIV-1, which causes AIDS, showing schematically the several genes that regulate viral gene expression (note that some of these have interrupted or overlapping reading frames). (d) is after R. Gallo et al., *Nature* 333 (1988), p. 504.

duce cancer cells under normal circumstances but can do so if they are modified or changed by their incorporation into viral genomes.

Thus, the tumorigenic retroviruses are transducing viruses analogous to a λ bacteriophage. They must have recombined, in the past, with a cellular sequence that can now be introduced into the genome of newly infected cells as part of a provirus (Figure IV.7a). Usually, the structure of the v-*onc* gene is altered compared to that of the normal cellular counterpart. Also, transcription of a proviral v-*onc* is controlled by the proviral regulatory signals, not by the signals

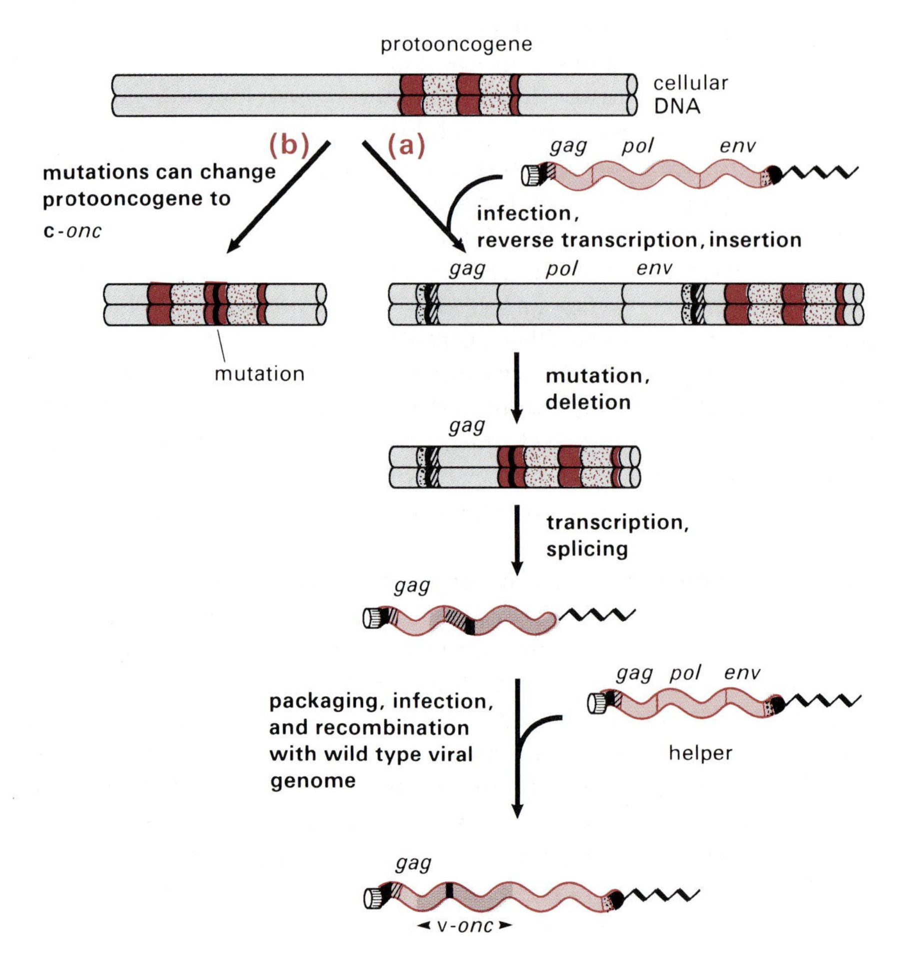

Figure IV.7
Hypothetical schemes for the formation of a v-*onc* (a) and a c-*onc* (b) from a cellular protooncogene.

Table IV.1 The Properties of Some Oncogenes

Gene	Occurrence of v-*onc*	Occurrence of Protooncogene	Characteristics of Gene Product
src	Rous sarcoma virus (avian)	vertebrates *Drosophila*	A plasma membrane protein kinase that phosphorylates tyrosine residues
abl	Abelson murine leukemia virus	vertebrates *Drosophila*	Same as *src*
*erb*B	Avian erythroblastosis	vertebrates *Drosophila*	Protein kinase that phosphorylates tyrosine residues; normal gene encodes plasma membrane receptor for epidermal growth factor
H-*ras*	Harvey murine sarcoma virus	vertebrates *Drosophila*, yeast	A plasma membrane GDP/GTP binding protein; normal protein is a GTPase
myc	Avian MC29 myelocytomatosis virus	vertebrates	Located on nuclear matrix

normally associated with the protooncogene in the cell's genome. In complex and not yet well understood ways, the abnormal regulation and (or) the abnormal product of v-*onc* expression yields the tumor phenotype. And just as cancer itself is not a single disease but a disparate set of disorders, so too different oncogenes encode different kinds of functions, and their tumorigenicity follows different mechanisms (Table IV.1).

Identification of the proteins encoded by some of the 50 or so known protooncogenes indicates that several of them may be involved in the control of normal cell growth. For example, some encode growth factors and others the plasma membrane receptors for growth factors. Genes and proteins homologous to some oncogenes are even known to occur in yeast and *D. melanogaster*, where they are associated with important, conserved functions. For example, two genes related to the H-*ras* protooncogene are found in yeast. Yeast cells cannot grow if the two H-*ras*-like genes are eliminated, but a human H-*ras* gene inserted into yeast DNA corrects the defect. The proteins encoded by viral and cellular *ras* genes and the *ras*-like yeast genes bind GTP and GDP specifically and tightly, and are localized in the plasma membrane. In yeast, the *ras*-like proteins interact with the system that regulates the levels of cyclic-AMP, the **second messenger** that mediates responses of all eukaryotic cells to altered extracellular environments (i.e., hormones).

Protooncogenes can become oncogenic even without an association with a retrovirus; such abnormal cellular genes are called c-*onc*s (Figure IV.7b). Again, mutations, rearrangements, and amplifications can alter the regulation or gene product of a protooncogene to yield a c-*onc* and a tumor phenotype. For example, a mutation that alters a single amino acid in the normal human H-*ras* gene product is sufficient to change the protooncogene into a c-*onc*. Such altered H-*ras* genes are found in several human tumors (e.g., bladder carcinomas). Thus, a progression of experiments starting with vertebrate retroviruses has led to the discovery of normal genes, protooncogenes, whose c-*onc* counterparts are instrumental in inducing human tumors.

Yet another extension of this work is the recognition that anti-oncogenes also exist. Proteins encoded by anti-oncogenes can prevent tumorigenic transformation of cells. Examples include the gene associated with Wilms' tumor of the kidney and the retinoblastoma gene. Thus, humans who inherit a single mutated retinoblastoma gene (hemizygotes) produce the gene product from the remaining, functional allele and are normal. However, if the functional allele mutates in a somatic cell, the cell can initiate a tumor. Insight into this phenomenon came from study of the heritable childhood tumor of the eye, retinoblastoma. The retinoblastoma gene has now been cloned and characterized. It encodes a nuclear, DNA- binding protein. Transfection of the gene into tumor cells in culture reverses the tumor phenotype. Thus, the anti-oncogene product appears to function in some (unknown) manner to suppress growth, in distinction to many oncogene products, which stimulate cell proliferation.

There has been a direct, if unpredictable, track from fundamental research on the RNA viruses that cause tumors in birds and nonhuman mammals to urgent clinical problems such as human cancer and the acquired immune deficiency syndrome, AIDS. AIDS is caused by a retrovirus called HIV (previously HTLV-III) that infects macrophages and a class of T-lymphocytes, causing a devastating destruction of the immune response. Other human lymphotropic viruses, HTLV-I and HTLV-II, also infect T-cells but are tumorigenic and cause leukemias. HTLV-I was in fact the first human retrovirus to be identified. Because of the prior intensive study of vertebrate retroviruses and the availability of recombinant DNA techniques, the structures of the HTLVs and of HIV (Figure IV.6) and their biological properties were rapidly elucidated. This permitted the cloning of viral genes, their expression in *E. coli*, and diagnostic tests for the viruses and for antibodies they induce in infected individuals. In the case of HIV, these tools allow the screening of blood and the exclusion of contaminated blood from use in transfusions, an important element in controlling the disease's spread. The detailed knowledge of HIV is also suggesting rational experimental approaches to designing therapeutic protocols for AIDS.

The study of retroviruses has had important ramifications for understanding the oncogenicity of some DNA viruses. For example, SV40 is oncogenic in newborn rodents, and some papillomaviruses, adenoviruses, and herpesviruses may be oncogenic in humans. The oncogenicity of these viruses is also associated with viral genes, although as far as is known, the v-*onc*s of DNA viruses have no normal cellular homologues. Recent evidence suggests that at least some DNA virus v-*oncs* may influence cell growth indirectly by interaction with the products of anti-oncogenes like the retinoblastoma gene. Presumably, the interaction inhibits the normal growth-inhibiting function of the anti-oncogene product, thereby fostering cell proliferation and tumorigenesis.

Another important consequence of the study of retroviruses is their adaptation as recombinant vectors (Section 5.7d). These vectors have proven to be particularly useful for introducing new genes into early mammalian embryos and thus into essentially all the cells of complete experimental animals, including germ line cells; the resulting animals are called **transgenic**. The importance of transgenic animals to the study of tissue-specific expression of genes and of the activity of *onc* genes is described in Chapter 8 later in this essay.

Mapping Genomes

Underlying the fruitful analysis of viral genetic systems are correlations between physical and genetic maps, detailed information about genome structure, and an interplay between structural, genetic, and biochemical data. Similar information is critical for the eventual understanding of more complex genetic systems from bacteria to humans.

Physical and Genetic Maps

The first step in establishing a complete physical map is to describe a genome (or its individual chromosomes) as an ordered set of restriction endonuclease fragments. Ultimately, the nucleotide sequence of the genome provides the complete structure. A restriction fragment map of the approximately 4.5×10^3 kbp *E. coli* chromosome has already been constructed (Figure IV.8). It consists of 21 DNA fragments produced by digestion with the enzyme *Not*I, which cleaves at an 8 bp recognition site, and thus relatively rarely. The fragments' positions on the circular genome were determined by a combination of methods: DNA blotting using previously cloned and characterized *E. coli* genes as probes; analysis of partial *Not*I digests; comparison of fragments obtained from wild type DNA with those from *E. coli* strains carrying λ prophage insertions at different, known loci or known large deletions or inversions. Because over 1000 genes have been located on the *E. coli* chromosome, a detailed correlation of genetic and physical maps is in hand.

Similar approaches can be used to establish physical maps for larger, more complex, eukaryotic genomes. In such instances, libraries of cloned fragments making up the entire genome will be required because even enzymes like *Not*I produce too many fragments to be totally resolvable by current techniques. The cloned segments need to be large to facilitate their ordering; thus, cosmids, or even artificial yeast chromosomes, YACs (Section 9.6c), are likely to be the vectors of choice. Restriction endonuclease site maps (i.e., physical maps) of extensive regions of two small eukaryotic genomes, *Caenorhabditis elegans* and *S. cerevisiae*, are already available in computerized data banks. To construct such maps, large numbers of long cloned fragments were randomly selected from genomic libraries in λ phage or cosmid vectors. The sizes of the fragments generated from each cloned segment by cleavage with a small set of restriction endonucleases were catalogued. Specially designed computer programs compared the individual fragment sets and thereby indicated which clones are likely to contain overlapping genomic regions. About 95 percent of the nematode genome (haploid size 8×10^7 bp) is covered by overlapping cosmid clones at the time we write. More

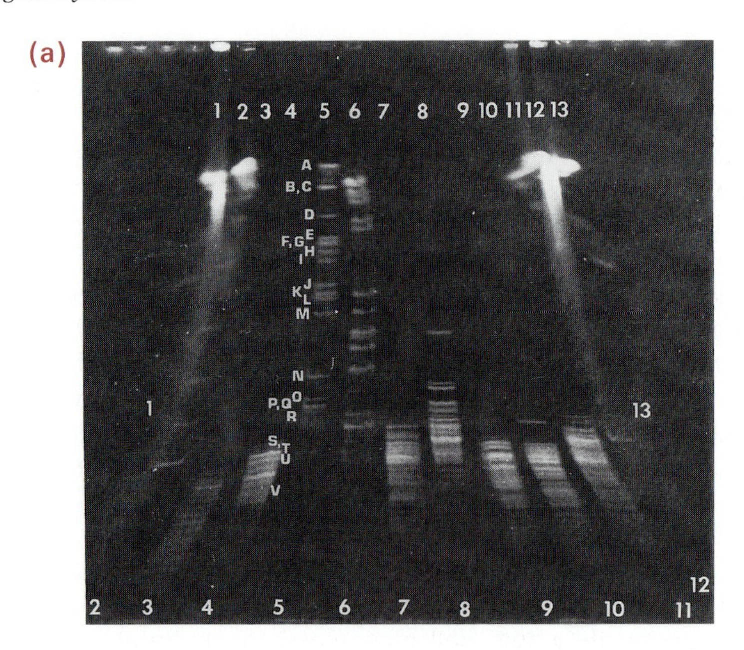

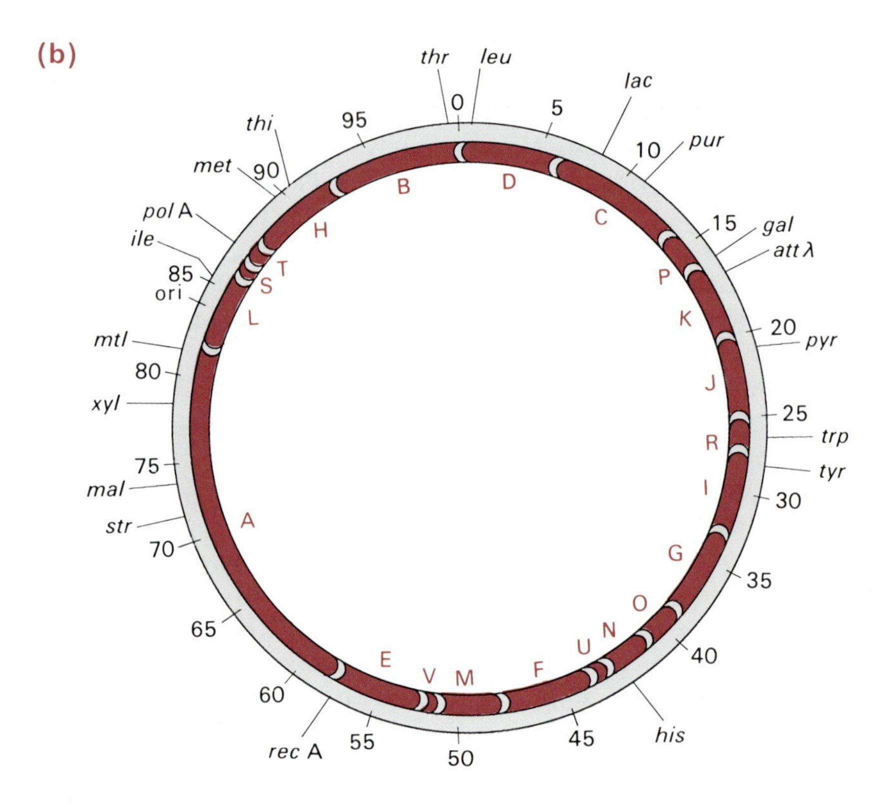

Figure IV.8
The correlated physical and genetic map of *E. coli.* (a) Photograph showing the separation of the 21 DNA fragments (20 to 1000 kbp in length) produced by *Not*I digestion of *E. coli* (strain EMG2 DNA) after separation by pulsed-field gel electrophoresis (Figure 6.31). Fragment Q is F plasmid DNA that was present in the *E. coli* strain used and linearized by cleavage at its single *Not*I site. Courtesy of C. L. Smith. (b) A correlation of the genetic map of *E. coli* (100 units) with the map of *Not*I fragments. The *Not*I fragments are lettered as in (a). Some of the genes shown on the map are also seen in Figure II.3. Cloned segments of these genes were used as probes to determine which of the *Not*I fragments carried which gene. For example, the *gal* gene is on *Not*I fragment P. After C. L. Smith et al., *Science* 236 (1987), p. 1448.

than 95 percent of the yeast genome (haploid size 1.4×10^7 bp) is accounted for in 5000 15 kbp segments that have been melded in clusters up to 100 kbp. Experience being acquired with these small genomes will be applicable to mapping large plant and animal genomes. One problematic aspect has emerged already. Even carefully designed genomic libraries are not random assortments of all possible genomic segments. Some segments are more and some less abundant in the library than expected from their genomic representation. Some are very rare or absent. The larger the genome being analyzed, the more such breaches will complicate map construction. Moreover, the highly repeated sequence families in many eukaryotic genomes will add to the difficulties.

The overlapping clones obtained in mapping projects will eventually be used to determine the entire nucleotide sequence of the genomes. Meanwhile, the sequences of genomic segments that have been cloned for specific purposes are accumulating. Altogether, more than 2×10^7 bp of sequence are stored. As of summer 1990, about 7×10^6 bp of sequence data from scattered regions of the human genome are recorded in central computerized data banks, and extensive data are also available for *S. cerevisiae*, *C. elegans*, *D. melanogaster*, and mice. Plans are currently being made to accelerate data collection using improved DNA sequencing methodologies, with the eventual aim of obtaining complete sequences for certain of these species' genomes.

The physical and genetic maps can be merged, as with *E. coli*, using cloned genes that have been genetically mapped by recombinational analysis. For organisms like yeast, the nematode, and *D. melanogaster*, for which relatively rich genetic maps are available, informative correlations can be made. With *D. melanogaster*, the correlations are facilitated by extensive cytogenetic data. The combined molecular-genetic maps then make possible the cloning of genes that are defined by phenotype and locus but whose gene product is unknown (for example, the cloning of the *D. melanogaster per* gene, Figure 6.39).

Mapping Large Genomes

Extension of mapping techniques to the very large, complex genomes of plants and vertebrates, including humans, is a great challenge. Problems arise from the combination of large genomic size with a paucity of genetic markers, as well as the difficulty of carrying out experimental crosses. With humans, such crosses are of course impossible. In the past, human geneticists depended on rare, large families carrying specific mutant phenotypes to perform linkage analysis and establish genetic maps. Known loci were sparsely spread on the human genome. Besides hampering fundamental analysis of human genetics, the inadequate human genetic maps thwarted early diagnosis of genetic diseases. Until recently, reliable fetal diagnostic procedures were available for only a limited number of diseases—those associated with either grossly altered chromosomes (e.g., Down's syndrome, which is associated with trisomy of chromosome

21) or with phenotypic consequences that can be detected early in fetal development (e.g., Tay-Sachs disease, which is caused by a lack of hexosaminidase A, a lysosomal hydrolase). The situation began to change as larger numbers of cloned human DNA segments became available and as larger numbers of human genes were mapped to specific chromosomal locations. Section 7.4b describes the two methods that, in conjunction with cloned DNA probes, make the mapping of single-copy DNA sequences on human chromosomes an almost routine procedure: *in situ* hybridization and analysis by DNA blotting of human/rodent somatic cell hybrids. Neither of these methods depends on mutations, linkage analysis, or family trees.

In spite of the large increase in the number of identifiable loci, the human genetic map remained sparsely populated until very recently. Consider, for example, that 1000 genes with an average size of 10 kbp (exons plus introns) account for only 10^4 out of the 3×10^6 kbp in the haploid human genome; each of the thousand could be separated by millions of base pairs, well beyond the range amenable to chromosomal walking or even adequate recombinational analysis, in view of the small number of genetic crosses, represented by family pedigrees, available for analysis. Moreover, as far as diagnosis is concerned, the utility of these procedures is limited by a lack of information about the mutant genes and defective gene products responsible for many genetic diseases. Fortunately, these problems are now being overcome by a completely new approach that is applicable to generational analysis of genes in a family, to fetal diagnosis, to analysis of gene distribution in populations, to recombinational linkage analysis, and to mapping. The new method is also applicable to other organisms. Maize chromosomes, for example, are being mapped in this way, an endeavor of importance to both biology and agriculture.

Polymorphic Restriction Endonuclease Fragments

The new approach depends on the fact that allelic genomic regions frequently differ by single base pair changes, deletions, insertions, or rearrangements that alter a restriction endonuclease recognition site or the spacing between two sites for a particular enzyme. As a consequence, the size of the DNA fragment(s) that anneals with a regional probe may be different on homologous chromosomes, and the hybridization pattern can change from one individual in a species to another (for example, the alleles of the ovalbumin gene described in Section 7.4a and the polymorphic minisatellite regions described in Section 9.4b). Thus, such **restriction fragment length polymorphisms** (or RFLPs, pronounced riflip) constitute markers for alleles at a particular locus. The analyses can be carried out on the small amounts of DNA available from adult lymphocytes or from cultured fetal cells derived from extraembryonic fluid or the placental chorionic villi.

In the simplest case, a RFLP results from the same sequence change that causes the mutant phenotype; this is the situation in the mutant β-globin gene that causes sickle cell disease (Figure III.6). Although RFLPs of this type are rare, analysis of restriction endonuclease digests of human (and other mammalian) DNA with a variety of randomly selected cloned probes has revealed that allelic positions on homologous chromosomes frequently give rise to polymorphic fragments even among normal individuals. The number of variable allelic positions is much larger than was anticipated from genetic and gross structural data. It has been estimated, for example, that one out of every 100 base pairs in the 65 kbp that contain the human β-globin gene family may be polymorphic.

RFLPs are usually inherited in Mendelian fashion; thus, each one constitutes a genetic marker essentially analogous to a classical phenotypic marker. Even random single-copy DNA segments of unknown significance, so-called anonymous DNA probes, are as useful markers for mapping, as are segments of known function; they need not be known genes or genes at all. Thus, RFLPs provide an enormous number of human genetic markers. They may be used alone or in conjunction with phenotypic markers or chromosomal abberrations for analyzing recombination rates and determining linkage. They have already been used to help locate the genes and to provide diagnostic tools for diseases whose molecular defects were not even identified, including Duchenne's muscular dystrophy, Huntington's disease, and cystic fibrosis. In the case of Duchenne's muscular dystrophy, identification of the molecular defect followed. Siting of the gene on the short arm of the X-chromosome led to cloning of the gene and corresponding cDNAs. The gene product, synthesized in *E. coli*, yielded antibodies that identified a skeletal muscle protein present in normal humans but not in Duchenne's patients.

A further refinement of this general mapping technique allows the detection of previously known sequence polymorphisms that are not associated with RFLPs. Oligonucleotides as short as 16 bases anneal specifically to complementary genomic segments. Because duplex formation between the oligomer and DNA depends on relatively few base pairs, even a single base pair mismatch significantly decreases the melting temperature of the duplex. Thus, annealing of an oligomer probe to a genomic DNA blot under carefully defined conditions can distinguish one allele from another.

Differential Gene Expression in Specific Cells and Tissues

A fertilized egg yields, through successive cell divisions, thousands of kinds of cells and many highly differentiated tissues. The manifold shapes and functions of differentiated cells are phenotypes caused by the selective expression of specific genes in particular cell types or tissues or at precise and limited times during development. Other genes appear to be expressed in most if not all cell types. These so-called "housekeeping" genes encode such ubiquitous products as ribosomes, transfer RNAs, actin, and the enzymes required for nucleotide biosynthesis and energy production.

The positional and temporal regulation of cell- and tissue-specific gene expression depends primarily on controlling the initiation of transcription. This is achieved by regulating the assembly of transcription complexes near the site at which transcription is initiated. Interactions between specific DNA binding proteins (transcription factors) and their cognate short DNA sequence motifs are presumed to influence either that assembly process or the transcriptional activity of the assembled complex itself. Some transcription factors act through protein-protein interactions rather than by binding to DNA. Various combinations of a relatively limited number of such DNA sequence motifs, each binding a unique constellation of proteins, can influence the transcription machinery in myriad ways. The DNA binding or transcriptional activating properties of some transcription factors are often influenced by their association with small molecules (e.g. steroids or metals). Covalent modifications made at specific amino acids in a transcription factor's sequence may even serve to fine tune its ability to influence transcription. This can provide for an enormous variety of cell-, tissue-, and stage-specific regulatory responses. Other specialized mechanisms, for example, rearrangements of genomic DNA, serve to both form and activate the expression of immunoglobulin and T cell receptor genes in mammals. Although the transcriptional on/off switches are probably not absolute, experimental evidence suggests that rates of expression may be regulated over many orders of magnitude. Besides the regulation of transcription initiation, polypeptide levels are modulated by controls on messenger RNA processing, stability, and translation, on posttranslational modification, transport, and destruction of polypeptides.

Differential gene expression in specific cells and tissues involves not just single genes, but sets of genes whose expression is turned on and off in a coordinated manner (Table IV.2). Also, the several members of some multigene families encode related or identical proteins that are expressed in different tissues or at different times during development. In the murine α-fetoprotein/serum albumin multigene family, for example, the two genes are arranged in tandem on chromosome 5, and both are expressed in embryos. Shortly after birth, expression of the α-fetoprotein gene decreases at least 10^4-fold, a change that depends on the *trans* active product of another mouse gene (Sections 8.3e and 9.4a).

One approach to understanding differential gene expression during development is to analyze developing organisms, including early embryos. Later in this essay, we outline some recent work of this kind. Other aspects of the problem are illuminated by studying already differentiated systems. For example, expression of different α- and β-globin genes in vertebrate red blood cell precursors at different developmental stages was demonstrated by such an approach (Section 9.3b). The ontogeny of lymphocytes is investigated by studying the immunoglobulin and T cell receptor gene rearrangements at different stages of B and T cell development, respectively. Tumor cells that represent particular stages in B cell differentiation are especially useful.

Once the genes that are specifically expressed in a particular cell type are identified, the DNA motifs and *trans* acting elements that govern transcription can be analyzed. Similarly, other regulatory

Table IV.2 Examples of Sets of Genes That Are Coordinately Regulated in Specific Cell Types or at Specific Times During Development

Genes	Cells or Location
Chymotrypsin, trypsin, pancreatic amylase: simultaneous expression.	Exocrine cells of pancreas
α-fetoprotein and serum albumin expressed in fetal yolk sac and liver. After birth, serum albumin expressed in liver.	Fetal yolk sac and liver (mouse)
α-skeletal actin, α-cardiac actin, α-vascular smooth muscle actin, α-enteric smooth muscle actin	Muscle: skeletal, cardiac, vascular smooth, enteric smooth
Collagen: type I, type II	Extracellular matrix: tendon, bone, skin, smooth muscle, cartilage
α- and β-type globins are simultaneously expressed. Different α- and β-type globins are expressed in embryonic, fetal, and adult cells.	Red blood cells
Enzymes for biosynthesis of various amino acids expressed upon starvation.	Yeast (*S. cerevisiae*)
Heat-shock proteins synthesized at elevated temperatures.	Many organisms
Multiple different myosin gene family members are expressed in body wall, muscle, or pharynx muscle and in fixed ratios.	*C. elegans*

targets such as splicing or posttranslational modifications can be identified and studied. Fortuitously identified mutations have also contributed to an understanding of development. For example, the human genetic diseases called thalassemias are caused by a failure to synthesize adequate amounts of globin polypeptides. In several thalassemias, functional globin messenger RNA cannot be produced because of mutations in splice junction positions. Other thalassemias are caused by the deletion of one or more globin genes. In some cases, where the adult β-globin gene is deleted, the expression of the fetal genes continues into adulthood, suggesting the existence of interconnected regulatory systems.

Currently, the application of reverse genetics is obviating the dependency on chance events and expanding the range of questions that can be studied. Specific types of mutations in selected genes can be designed and introduced into cells or into the early embryos of certain experimental organisms. Here, we describe a few examples from the broad range of current studies on control of gene expression in differentiated cells. Work on embryos themselves and on differentiated tissues in adults derived from such embryos is being carried out, as described later.

Differential Gene Expression in the Pancreas

The pancreas synthesizes and excretes a group of highly specialized proteins. Insulin synthesis is detectable only in β-type endocrine cells of the pancreatic islets; pancreatic exocrine cells synthesize chymo-

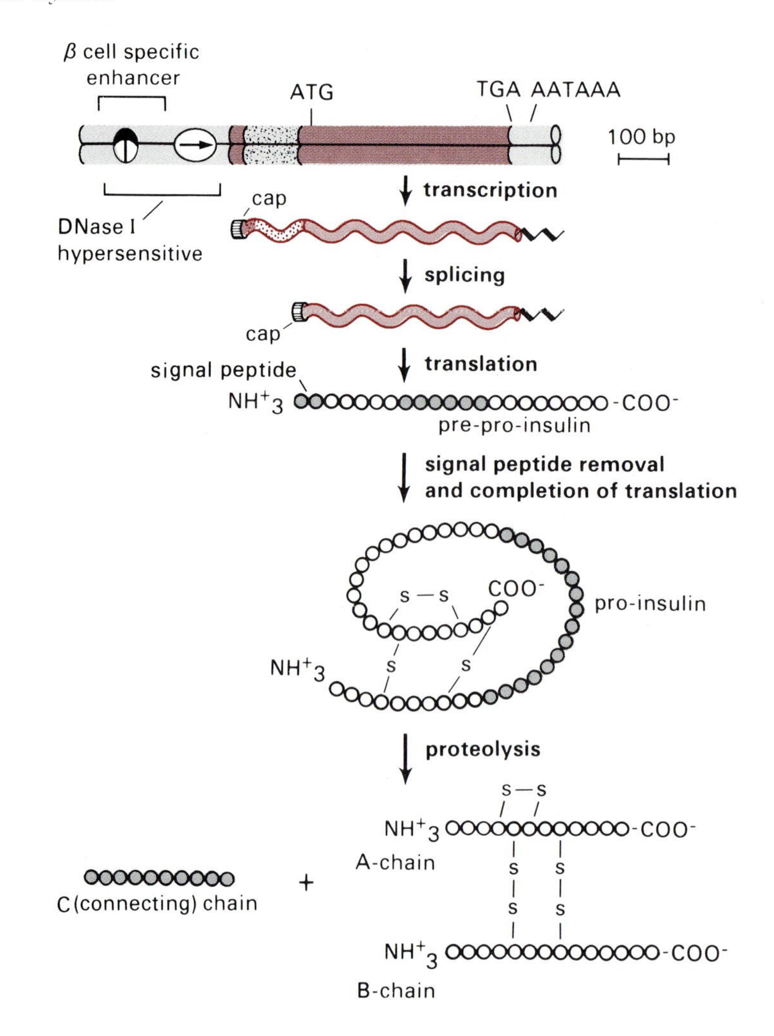

Figure IV.9
A typical mammalian insulin gene showing its intron, coding sequences, pancreatic *β* cell-specific enhancer region and the TATA box (promoter). Also shown are the gene transcripts and the primary translation product and its relation to the active, heterodimeric insulin molecule (see also Figure 1.27).

trypsin, trypsin, and pancreatic amylase. Nevertheless, all the genes are present in all cells. What processes and elements are responsible for this exquisitely selective gene expression? And how, during development, are these precise and limited synthetic capacities acquired? The 5′ flanking region of the insulin gene transcription unit contains a complex enhancer region that is active only in insulin-producing cells and not, for example, in fibroblasts (Figure IV.9). Two major determinants of enhancer activity occur about 100 bp apart and share a common 8 bp core sequence. Each of these core elements binds to a *β*-cell-specific *trans* activating protein. Moreover, a negative *trans* acting factor appears to repress the enhancer in cells that do not synthesize insulin. Similarly, *cis* acting elements and corresponding *trans* activating proteins direct the synthesis of the exocrine cell-specific enzymes. Thus, during differentiation, the characteristic synthetic capacities of exocrine and endocrine cells must be established by the switching on (and off) of sets of regulatory elements. Much more work will be needed to understand how the switches operate during development.

Still other controls on insulin synthesis, for example, can occur during the posttranslational processing that is required to produce active hormone from the longer polypeptide precursor encoded by insulin genes. Indeed, such proteolytic processing of polypeptide precursors plays an important role in modulating the concentrations of many hormones and other intercellular polypeptide messengers.

Posttranslational Modifications as Regulators of the Flow of Gene Products

Opportunities for regulating the flow of gene products by posttranslational events occur if the nascent polypeptides require modification before they are active. Such modifications are especially common in the myriad polypeptides that serve as intercellular messengers in multicellular organisms. Polypeptide endocrine hormones like insulin circulate in the blood and coordinate the activities of distantly positioned cells. Other peptides act over shorter distances to influence the activity of cells in the vicinity of the secreting cell. For example, peptide neurotransmitters carry messages between synapsing nerve cells. Basically, all these peptide messengers work in similar fashions, regardless of whether they are wide ranging like insulin or confined to small spaces like the neurotransmitter, enkephalin. In the initial step, they bind to highly specific receptors on the surfaces of particular target cells. This interaction then triggers a precise response, the nature of which depends on the properties of the receptor cell: growth, secretion of another polypeptide, expression of a particular gene, firing of a neuron, initiation of specific fixed behavioral patterns, and so forth.

Although all cells encounter many different peptide messengers, each cell type can respond only to those messengers for which it has specific receptors. Any single peptide messenger can, however, elicit different types of responses, if receptors for that messenger occur on different cell types. The peptide messenger cholesystokinin is a good example. Cholesystokinin is secreted by certain intestinal cells and travels through the bloodstream. It bypasses many cells completely, but in the gall bladder and in the brain, cholesystokinin meets cells with matching receptors. Gall bladder cells respond by increasing the bile flow into the intestines. In the brain, cholesystokinin acts as a neurotransmitter. Many different intercellular peptide messengers are known, and new ones are regularly discovered. These, together with other intercellular messengers that are not peptides (e.g., acetylcholine, epinephrine, norepinephrine, and the lipid-soluble prostaglandins) coordinate an extraordinary number of subtle physiological events.

The genes that encode the various polypeptide intercellular messengers generally encode proteins that are several hundred amino acids long. The peptide messenger itself, the product of posttranslational processing, may be as small as a pentapeptide, as are, for example, the two enkephalins that act as natural opiates in the brain and regulate peristaltic contractions in the gut (Table IV.3). Typical of all proteins destined to be secreted (Figure IV.10), primary translation products of these genes contain an amino-terminal signal peptide that is cleaved off as the nascent polypeptide enters the endoplasmic reti-

Table IV.3 Enkephalins: Peptide Messengers

tyr-gly-gly-phe-met	met-enkephalin
tyr-gly-gly-phe-leu	leu-enkephalin

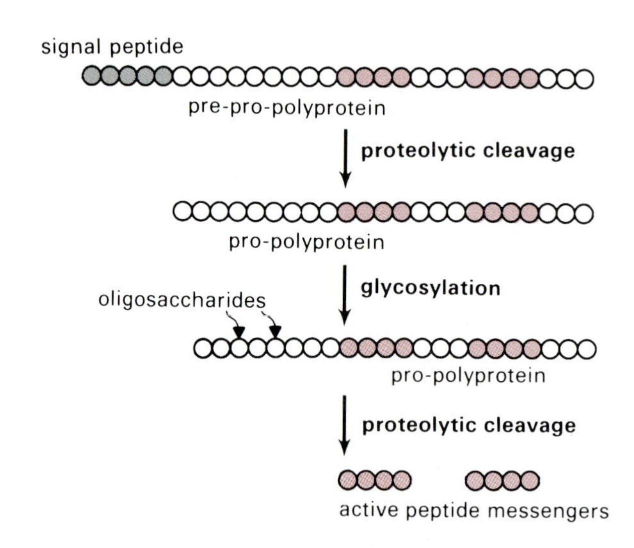

Figure IV.10
Genes encoding peptide messengers encode a long pre-pro-polyprotein. The signal peptide is cleaved off in the endoplasmic reticulum, where the propolyprotein is glycosylated. Proteolytic processing of the polyprotein can produce one or more functional peptide messengers.

culum. Frequently, the remaining portion of the polypeptide is glycosylated. Finally, proteolysis at specific sites trims the long precursor down to yield one or more active peptide messengers. For example, the 263 amino acid long protein called pre-pro-enkephalin A is encoded by a single gene in mammals and produces six molecules of met-enkephalin and one of leu-enkephalin.

The characteristic properties of mammalian genes encoding intercellular messengers originated early in evolution. The same general strategies are used to encode and synthesize the small peptides called pheromones that permit individual yeast cells to communicate their mating types to one another. And small peptides that elicit and coordinate fixed behavioral patterns in invertebrates such as the marine slug *Aplysia* are produced in a similar way. In this system, as in mammals, a cascade of peptide messengers brings many different cell types into communication with one another, providing the manifold elements that contribute to coordinated activities.

Specific proteases catalyze the cleavage of the long precursors to yield active peptide messengers. Their concentration in the cellular compartments (e.g., the Golgi apparatus and secretory granules) that contain the precursors could therefore control the rate of peptide messenger formation. Unfortunately, definitive identification and characterization of such proteases have proven difficult. Among the problems is that of distinguishing specific proteases from nonspecific proteolytic enzymes such as those present in lysosomes. More definitive experiments have been done in yeast, where a mutation has identified a gene encoding a protease required for specific cleavages of the precursor to the pheromone, α-factor. The same protease is responsible for the processing of other preproteins to their mature forms and appears to have a mammalian counterpart.

Control of Gene Expression by DNA Sequence Rearrangement

Differential gene expression that depends on genomic rearrangement is widespread in nature, although it appears to operate with a very limited number of genes in any particular organism. Several examples of such programmed reorganizations were described in Chapter 10. One is the construction of the genes for immune proteins (Section 10.6c). It depends on *cis* recombinational signals and proteins that are encoded in the genome and occur specifically in T and B cells. The rearrangements have three functional consequences. They lead to formation of complete genes by joining DNA segments that are not contiguous in the germ line genome (e.g., V, J, and C regions to form an immunoglobulin light chain gene). They allow efficient transcription initiation by bringing promoter and enhancer into proximity. They generate diversity in the segments encoding immune protein variable regions and thereby provide for an enormous repertoire of antigenic responses.

Other scheduled reorganizations include mechanisms by which prokaryotes respond to new environments, yeast cells switch their mating types, and trypanosomes evade the immune response of host organisms. In other systems, such as the *Xenopus* ribosomal RNA genes and the *D. melanogaster* genes encoding the chorion proteins, specific genes are extensively amplified in order to meet a large demand for their gene products. In still other cases, massive losses of DNA occur. Certain protozoa, such as *Tetrahymena*, sequester their germ line genomes in a micronucleus and form a "somatic" macronucleus from which genes are expressed. As much as 90 percent of the genome may be lost because virtually all repetitive DNA sequences are eliminated during development of the macronucleus from the germinal micronucleus. A large number of multicellular invertebrates, including certain nematodes, insects, and crustaceans, discard large proportions of their highly repeated DNA sequences from somatic stem cells but reserve the sequences in germ line cells. The phenomenon was first observed microscopically in 1887 as the diminution of chromosome size in developing nematodes. Thus, the statement that every cell in a complex organism has the same DNA as the fertilized egg from which it arose is not precisely true. Nevertheless, the extent to which specific DNA rearrangements contribute to the differentiation of somatic cells appears to be small; the overwhelming number of already cloned genes have the same structure in germ line and somatic cells.

Modifying Biological Systems

The potential of molecular genetics extends beyond the increased ability to understand biological systems. It includes a novel competence to alter the genotype and phenotype of individual gene products, of cells, and of whole organisms. The influence over nature that comes from an ability to alter the genotype of whole organisms is qualitatively different from what was previously possible and inevitably involves political, social, and cultural issues in addition to scientific questions. Human beings have of course manipulated the genotypes of plants and animals by selective breeding since initiating organized agriculture millennia ago. And it is not only humans that have power over other species. Organisms of all types influence the genotypes of other species, including humans: subtly, by applying selective pressure for certain traits (e.g., ability to use alternate food supplies), and boldly, by killing off large portions of populations that lack the genetic makeup to protect themselves. But there is no question that the new biology gives human beings more precise and more specific tools for manipulating both themselves and other species. For this reason, biologists' euphoria is tempered by awe and an acceptance that society at large will participate in decisions concerning the genetic modification of living organisms.

The concepts and techniques of molecular genetics will be used to modify biological molecules and systems in many, presently unknown ways. Currently, the approaches range from synthesizing altered proteins to introducing new genes into plants and animals. The remainder of this essay gives an overview of the current experimental systems; more detailed descriptions of some will be found in Volume II of this book.

Synthesizing Normal and Modified Eukaryotic Proteins

The earliest practical goal of recombinant DNA research was to produce medically and economically important proteins such as vaccines and intercellular peptide messengers (e.g., insulin, growth hormone, and oxytocin), which are vital tools of clinical medicine. The idea was to clone the gene encoding the polypeptide, insert it into a plasmid that replicates in *E. coli*, arrange the insertion so that an *E. coli* promoter would regulate transcription, and then allow the *E. coli* ribosomes to synthesize large quantities of the protein. What appeared to be a relatively straightforward scheme turned out to be more complicated than expected (Section 7.8). First, most eukaryotic genes have introns, but *E. coli* genes do not; the bacteria lack splicing mechanisms and thus are unable to produce the correct messenger RNA from a eukaryotic gene. Second, the primary translation products of many eukaryotic genes, and in particular the polypeptide hormone precursors, require specific posttranslational processing to form an active gene product, and these processes are not carried out by *E. coli*. Finally, success in obtaining good yields of many eukaryotic proteins in *E. coli* is thwarted by the toxicity of the polypeptide to the bacterial cells, by the degradation of the polypeptide by bacterial proteases, or by the insolubility of the product in bacterial cells.

The solution to the first problem is to use a cloned cDNA rather than a cloned gene because the introns are absent in a cDNA clone. The second problem can sometimes be solved if, rather than the naturally occurring cDNA, a cDNA encoding the final processed product is used. Insulin, for example, is a heterodimeric protein, but its two polypeptide chains are contained within a single long precursor that is cleaved to yield the two polypeptides (Figure IV.10). *E. coli* can be adapted to synthesize human insulin by using two different recombinant plasmids, each containing a cDNA corresponding to one of the two chains of the active insulin molecule. Each of the polypeptides is synthesized in a different population of *E. coli*, and then the two purified polypeptides are mixed to form insulin (Figure IV.11).

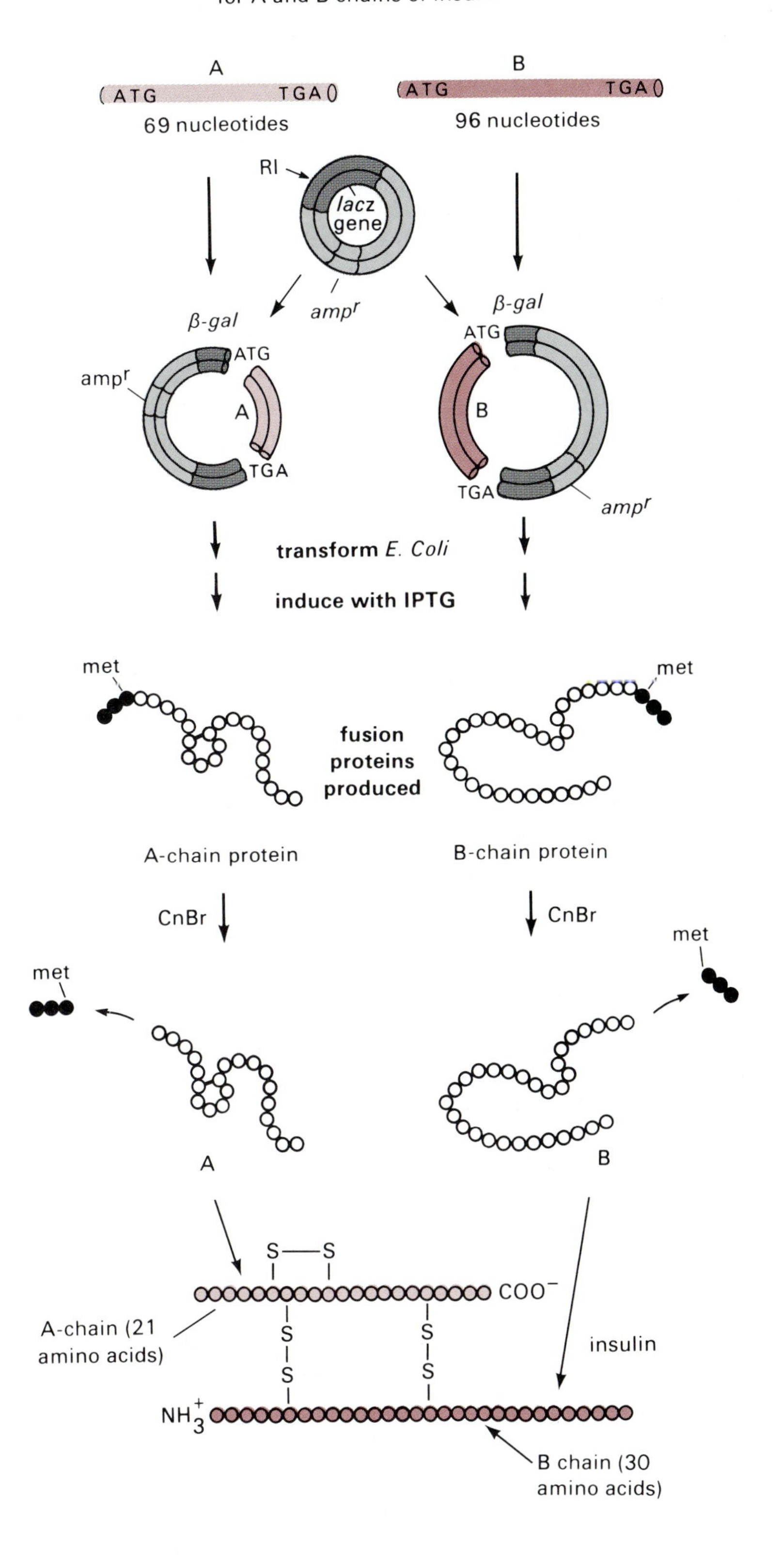

Figure IV.11

Schematic diagram showing the production of insulin in *E. coli.* DNA segments encoding the two insulin chains, A and B, each with an ATG start and TGA stop codon, were synthesized (only a single strand is shown) and inserted into *E. coli* plasmid vectors downstream of the *β*-galactosidase (*lacZ*) promoter. *E. coli* cells were transfected with one or the other plasmid and cloned. Induction of cell populations with isopropyl-thiogalactoside (IPTG) results in synthesis of the two polypeptides. A short region of fused *β*-galactosidase polypeptide including the single methionine (from the ATG) is removed from the two purified polypeptides by cyanogen bromide treatment. The complete A and B chains are then mixed under conditions that permit proper disulfide bond formation to yield active insulin. Note that the success of the strategy depends on the absence of methionine residues in insulin. After J. D. Watson, J. Tooze, and D. T. Kurtz, *Recombinant DNA: A Short Course* (San Francisco: W. H. Freeman, 1983), Figure 18.2.

Because such cDNAs could not be made by copying natural messenger RNA, they were synthesized chemically, using the known amino acid sequence of insulin to determine the nucleotide sequence. The third problem has been addressed by using inducible *E. coli* promoters to control the synthesis of the foreign protein; the cells are grown to a high population density in the absence of inducer and then induced to initiate transcription and translation of the desired gene. Yields can be enhanced by using strains of *E. coli* that carry mutations in protease genes. In some cases, expression of the recombinant gene in a yeast system or in animal cells rather than in *E. coli* proves more useful. In a eukaryotic environment, appropriate posttranslational modifications (e.g., glycosylation) of the nascent polypeptide can occur, and expression vectors can even be designed to permit the secretion of the polypeptide through the normal secretory pathways. Although it has taken longer than the optimistic predictions made in the mid 1970s, important polypeptides produced by recombinant DNA methods are now in use. The hepatitis B virus vaccine was already mentioned. Clinical trials are being carried out with a vaccine against malaria, a disease that affects more than 100 million people each year. Pigs are being protected from disease by a vaccine against the causative bacterial toxin. Human insulin and growth hormone are available. Lymphokines are being tested as therapeutic agents against cancer, and tissue plasminogen activator is a powerful agent in combating heart disease.

Quite apart from its application to the synthesis of therapeutic polypeptides, the ability to synthesize substantial quantities of specific polypeptides has impressive implications for the study of protein structure and function. The three-dimensional structure and consequently the biological activity of each protein depends largely on its unique amino acid sequence. Chemical studies have shown that modification of individual amino acid side chains or groups of side chains strikingly alters a protein's ability to renature and form a fully active secondary, tertiary, or quaternary structure. Analysis of proteins produced by genes containing mutations in coding regions tells the same story. The genetic and phenotypic defects in sickle cell anemia described in Section 1.3 were among the first and remain one of the most dramatic illustrations of this principle.

Rather than depending on random mutations, the amino acid sequence of a protein can now be systematically changed by site-specific mutagenesis of its cloned gene or cDNA. The physical, chemical, and functional properties of the altered protein can now be studied after synthesizing and purifying them from *E. coli* extracts. We can delineate regions of the protein that are important for forming the proper secondary, tertiary, and quaternary structure and define those amino acids that contribute to the active site of enzymes and those that form the binding site for substrates. Enzymes and proteins can be redesigned to change, for example, the pH optima of enzymatic reactions, an enzyme's substrate specificity or affinity, and its temperature dependence. Thus, the ramifications of the recombinant DNA technique have been extended to chemistry and foster an understanding of proteins at a previously impossible level of sophistication. A few examples illustrate the potential of these investigations.

Changing the histidine to aspartate at position 10 of the insulin B chain yields a hormone that binds the plasma membrane insulin receptor almost five times more efficiently than normal insulin. This change mimics a mutation found in patients with the disease familial hyperproinsulinemia. Cytochrome *c* contains an evolutionarily conserved phenylalanine: residue 87 in yeast cytochrome *c* (Figure III.8). When the phenylalanine codon (TTT) is mutated to a glycine (GGT), the resulting cytochrome *c* is still active; but when reduced, it has a markedly diminished ability to transfer electrons to cytochrome *c* peroxidase. The glycine at residue 226 of trypsin is part of the substrate binding pocket of this proteolytic enzyme. Structural considerations suggested that replacing it with a larger amino acid, alanine, might favor lysine over arginine peptide substrates because of competition for space within the pocket; the longer arginine side chain extends deeper into the pocket and fills the normal space more than does lysine. The experimental data confirm the prediction (Figure IV.12). Though the mutant enzyme is less efficient overall than the wild type, the relative effect on both the enzyme-substrate affinity (K_m) and the catalytic rate (k_{cat}) is greater for an arginine peptide substrate than for a lysine peptide substrate.

Genetic Engineering

Protein chemists utilize *in vitro* mutagenesis to produce altered pure proteins in bulk. Biologists are interested in studying the effect of altered protein structure, regulatory regions, or genomic architecture on the phenotypes of whole cells and organisms in order to understand normal and abnormal states. To achieve this aim, altered genes are introduced into cells and whole organisms. Being able to accomplish such alterations accounts for the most profound implications of the recombinant DNA techniques. It is here that biology changes from its traditional pursuit of understanding how living things are constructed and function to a manipulative science that can make permanent, heritable changes in organisms. The term **genetic engineering**, which we have thus far eschewed, is apt. Moreover, it is here that the reductionism inherent in molecular genetics yields to the study of whole cells and organisms and the pleiotropic effects of single genes on physiological, morphological, and developmental systems.

The transformation of eukaryotic cells in culture by recombinant vectors containing genes, cDNAs, and specifically mutagenized versions of natural DNA segments is one of the basic techniques in molecular genetics. It is an essential part of many of the experiments described in this book. When the cells in question are bacteria or yeast, then of course whole organisms, albeit single-celled organisms, are being manipulated. When an experiment involves correcting a yeast or bacterial mutation, then it can be described as gene therapy. The extension of such reverse genetics and gene therapy to diploid organisms that reproduce sexually involves major conceptual and experimental issues.

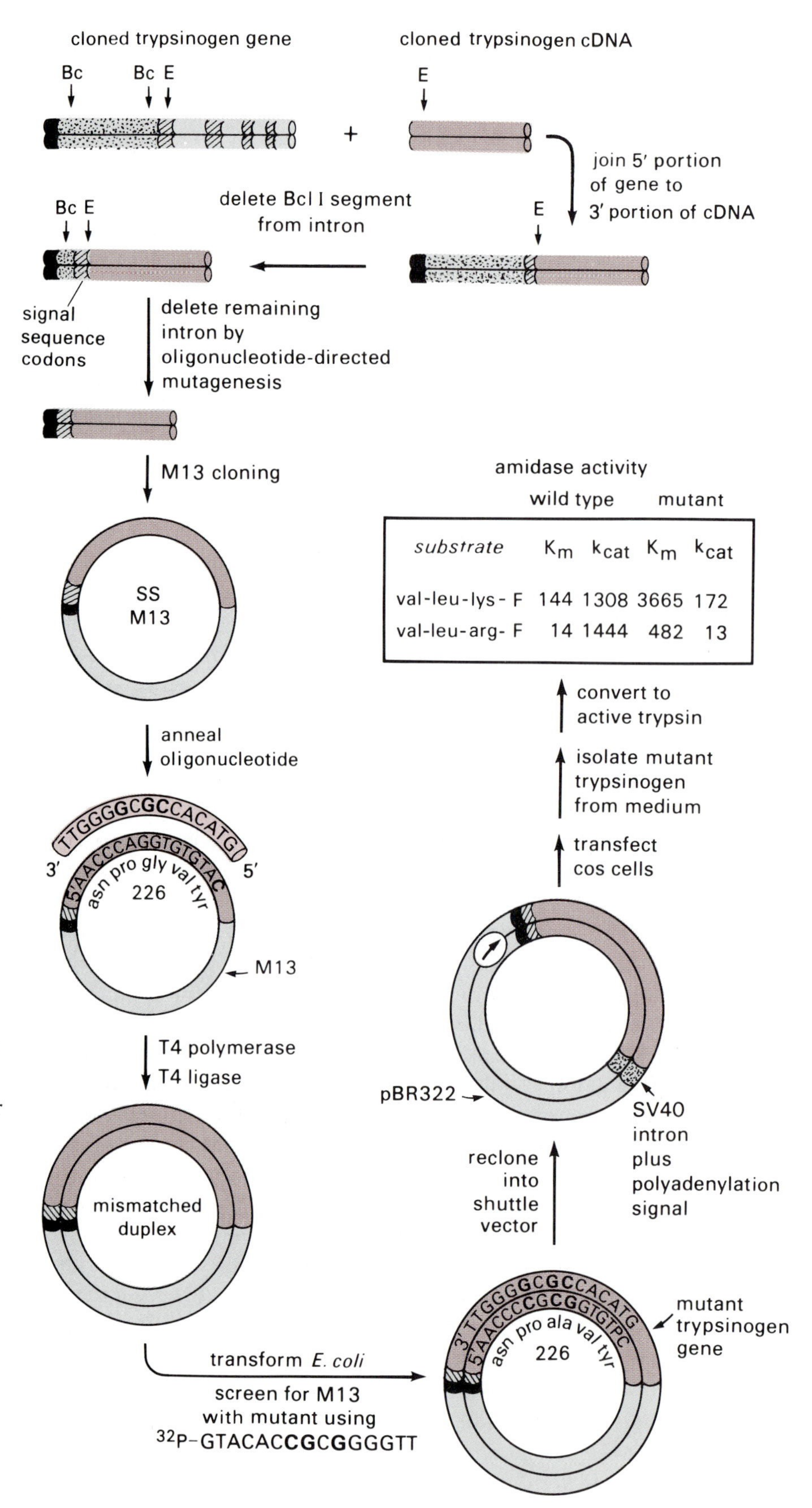

substrate	wild type K_m	wild type k_{cat}	mutant K_m	mutant k_{cat}
val-leu-lys-F	144	1308	3665	172
val-leu-arg-F	14	1444	482	13

Figure IV.12

Altering the substrate binding pocket of trypsin. Glycine 226 is believed to permit entry of large amino acid side chains into the trypsin binding pocket. To alter glycine 226, a full trypsinogen gene lacking introns was constructed by combining the 5′ part of a cloned gene with the 3′ part of a cloned cDNA. This provided the coding region for the NH_2 terminal signal sequence that was missing in the cDNA. After inserting the reconstructed gene into an M13 vector, oligonucleotide-directed mutagenesis was carried out (Figure 7.36), changing the codon for residue 226 from GGT (glycine) to GCG (alanine). The mutated gene was placed in an SV40-pBR322 shuttle vector so that the SV40 early promoter would transcribe the trypsinogen gene in cos cells (Section 5.7b). Trypsinogen was purified from the medium into which it was secreted and converted by proteolysis to active trypsin. The relative abilities of the mutant enzyme and normal trypsin to hydrolyze peptide arginine- and lysine-amides was measured using a fluorescent amide group, F. The relative specificity of the mutant is shifted in favor of the lysine substrate, as indicated by the changes in both K_m (units, μM) and k_{cat} (units, minutes^{-1}). The K_m change may reflect a crowding of the arginine side chain in the specificity pocket when the larger alanine replaces glycine. After C. S. Craik et al., *Science* 228 (1985), p. 291.

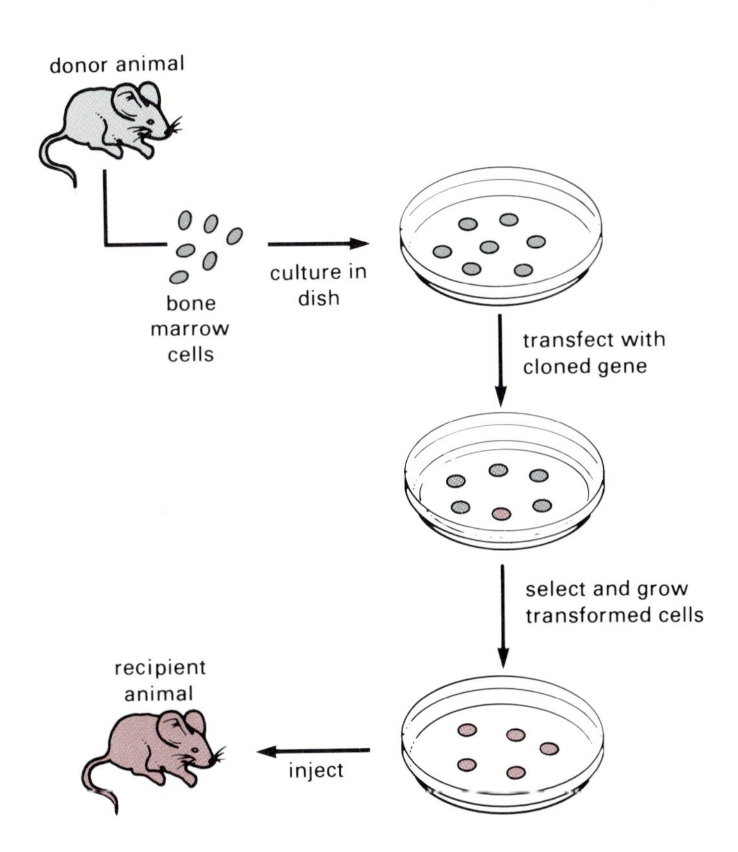

Figure IV.13
The steps involved in altering the genome of bone marrow cells and reintroducing the cells into a whole animal. The recipient animal is X irradiated to improve the likelihood that the injected cells will be established in its bone marrow.

Modifying Somatic Cells

One approach to the genetic modifications of multicellular organisms is to change the genotype of only one type of differentiated cell: somatic cell modification. The current experimental approach requires several steps: removing cells from the organism, placing them in a culture medium in which they survive and divide, transforming them with a recombinant vector containing the gene of interest (or the corresponding cDNA), and reintroducing the cells into the individual from which they were obtained. This is the basic protocol presently being developed for therapeutic intervention in some human genetic diseases. The experiments utilize mammalian bone marrow cells, which include the stem cells, the progenitors of circulating blood cells (Figure IV.13). Transformation is being attempted in a variety of ways, mainly using retroviral vectors carrying functional mammalian genes of special interest. At this writing, the steps in the overall scheme are being tested in experimental mammals, and it remains to be seen if the results are sufficiently promising to warrant trials designed to treat debilitating human genetic diseases. Several diseases would be ameliorated by the presence of a functional gene in blood cells. For example, humans who are homozygous for nonfunctional mutations in the adenosine deaminase gene suffer from severe immune system deficiencies.

Modifying Germ Cells

Somatic cell alteration does not introduce heritable changes into multicellular organisms. To do so requires modifying germ line cells,

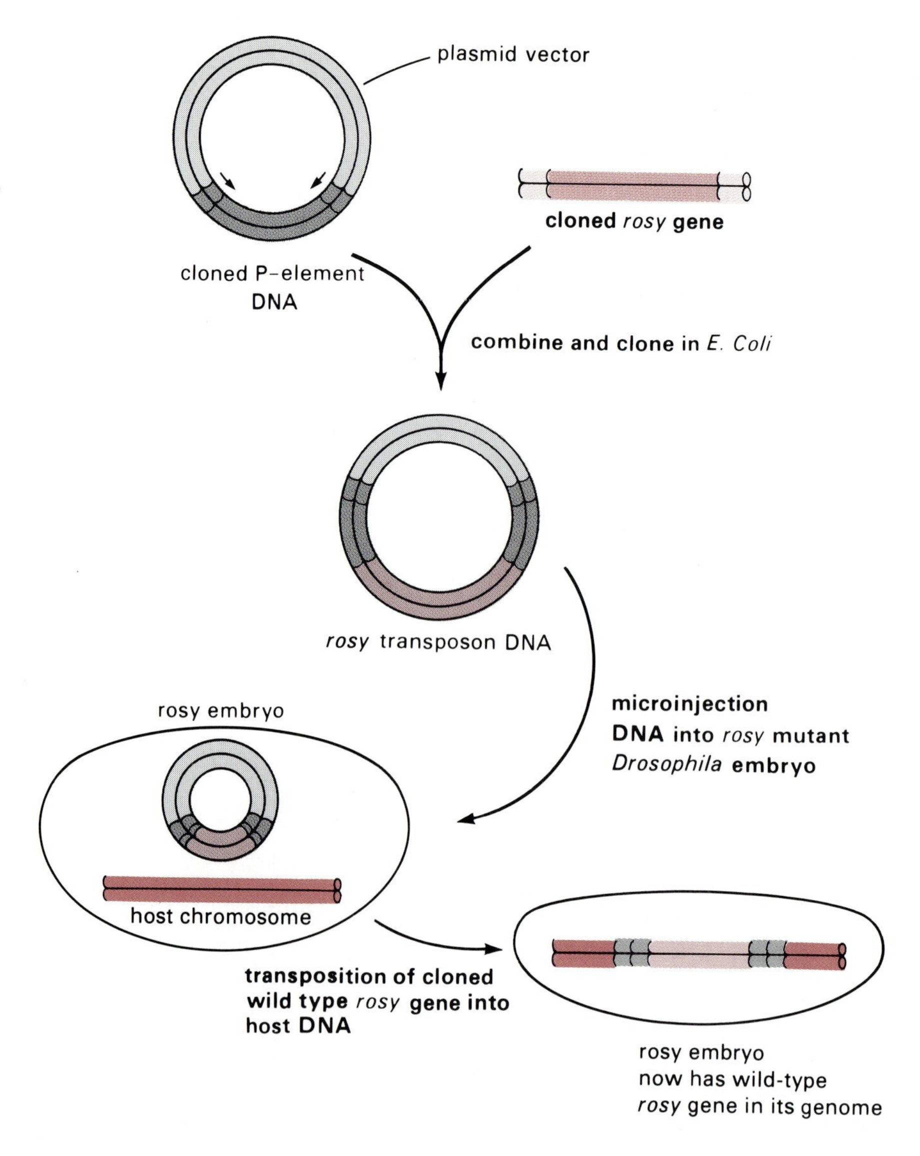

Figure IV.14
Schematic outline showing the modification of *Drosophila* by P element "transduction." The *rosy* locus encodes xanthine dehydrogenase, and the wild-type *rosy* gene gives a wild-type eye color. After G. M. Rubin and A. Spradling, *Science* 218 (1982), p. 348.

an approach currently being investigated in three types of organisms: *D. melanogaster*, experimental mammals, and plants. In *D. melanogaster*, germ line cells can be modified by using the transposable P element that was described in Section 10.2b. A recombinant vector carrying a gene of interest inserted into a functional P element is introduced into very early *D. melanogaster* embryos by microinjection. The P element, along with the extra gene, is transposed from the vector into genomic DNA (Figure IV.14). Adult flies that develop from such embryos frequently carry the P element in their germ line DNA and express the new gene. This technique provides extraordinary opportunities for studying the regulation of gene expression during development and differentiation. Among the questions that can be answered are the following. Is the gene expressed in all cells or only in the cell in which the normal gene would be expressed? What regulatory DNA segments must accompany the newly introduced gene if expression is to mimic that of the normal gene both temporally and as to cell type? Is the chromosomal position of the newly introduced gene important for its appropriate regulation during development and differentiation?

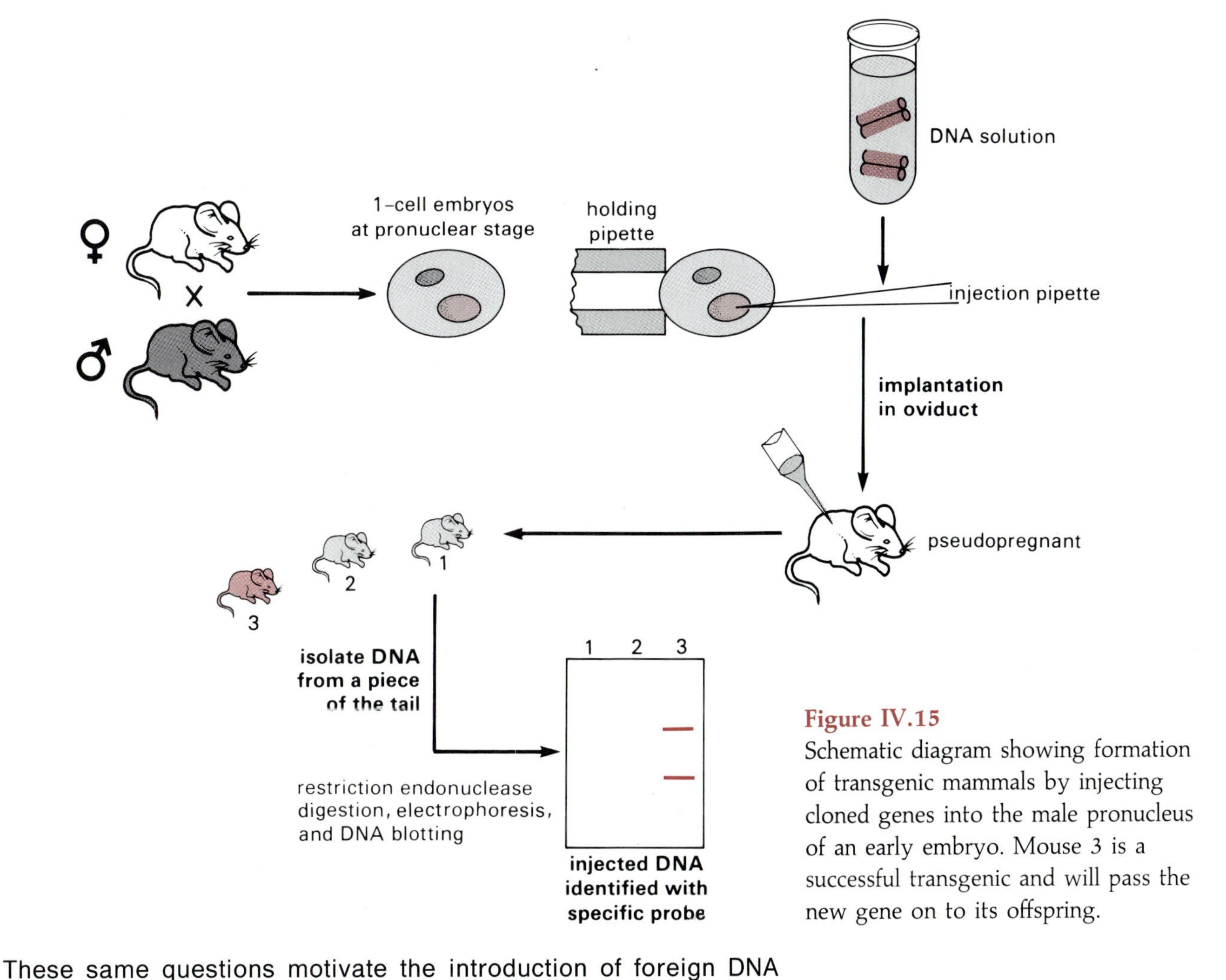

Figure IV.15
Schematic diagram showing formation of transgenic mammals by injecting cloned genes into the male pronucleus of an early embryo. Mouse 3 is a successful transgenic and will pass the new gene on to its offspring.

These same questions motivate the introduction of foreign DNA sequences into the germ line cells of experimental mammals. Up until now, mice have been the preferred experimental tools for producing such **transgenic** animals. Sheep, rabbits, and pigs have also been studied, with the eventual aim of improving the production of these important sources of food. Transgenic animals can be constructed in several ways, but one method has been most successful. Cloned genes are injected into the pronucleus of a fertilized egg cell, which is implanted into a foster mother (Figure IV.15). The foreign DNA integrates into the recipient's genome early enough in development to transform both germ line and somatic cell lineages and is inherited in Mendelian fashion. Transgenic offspring and their progeny can be inspected for abnormalities and analyzed for the time and place of gene expression. Generally, the cloned segment (the **transgene**) contains various promoter elements and sequences important for tissue-specific gene expression. In such experiments, it is often convenient to attach the relevant regulatory elements to a sequence encoding an easily detectable product. In other instances, the effects of a specific gene product in a transgenic animal may be of more interest than the role of regulatory signals. Here, the proper coding sequence can be joined to a heterologous promoter to assure gene expression.

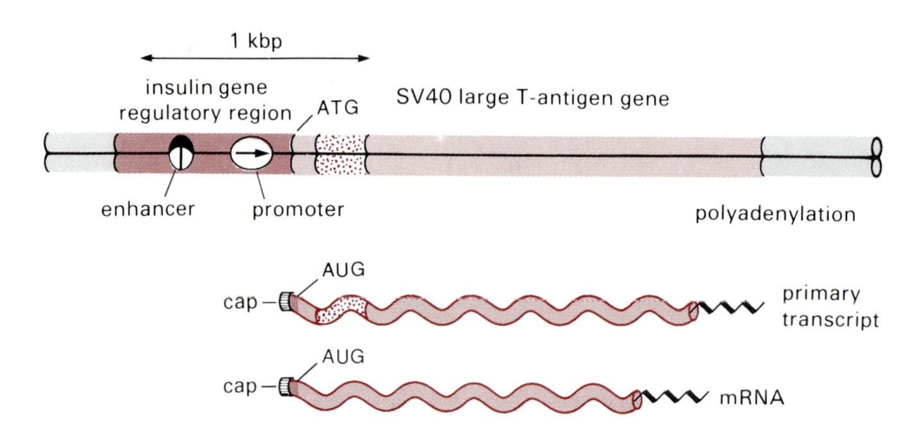

Figure IV.16
A recombinant hybrid gene containing the SV40 oncogene (large T-antigen) under the control of transcriptional regulatory sequences derived from the rat insulin gene (see Figure IV.9) was used to form transgenic mice. All such mice and their offspring develop insulinomas (tumors of the pancreatic β-cells). After D. Hanahan, *Nature* 315 (1985), p. 115.

The applicability of transgenic mice to a variety of biological questions can be illustrated by experiments with insulin genes. As mentioned earlier, the restriction of insulin gene expression to the β cells of the pancreas is governed by *cis* acting transcriptional regulatory segments in the 5′ flanking region of the insulin gene (Figure IV.9). When the 5′ flanking region of the rat insulin gene is joined to the coding region of the SV40 oncogene (large-T antigen) and the recombinant molecules are introduced into early mouse embryos, the progeny genomes carry the integrated recombinant DNA (Figure IV.16). Large-T antigen is synthesized exclusively in the mice's pancreatic β cells. Its oncogenic potential is manifest in the formation of β cell tumors, insulinomas. This experiment confirms the interpretation of *in vitro* experiments by demonstrating that the insulin gene regulatory signals dictate tissue-specific gene expression. A similar experiment in which this oncogene is linked to the transcriptional control region of the elastase gene, normally expressed in pancreatic exocrine acinar cells, provokes pancreatic adenomas rather than insulinomas. Such experiments provide a remarkable and widely applicable model system for the study of tumorigenesis and the development of precise therapeutic tools. For example, introducing a different oncogene, *myc*, under the control of a mouse mammary tumor virus promoter, yields mammary tumors.

The experimental modification of plant genotypes shares with the animal experiments the goal of understanding fundamental biological processes. With plants, the process of tissue differentiation and the mechanism by which light induces specific gene expression are important current questions. In addition, such research is motivated by a desire to improve the quality of agriculturally important crops. Thus, genes that increase crop resistance to herbicides and insect pests are a major research focus (Figure IV.17). Many of these experiments depend on recombinant vectors derived from the Ti plasmid of *Agrobacteria* and were, until recently, restricted to dicotyledonous plants (Section 5.8). Newer techniques involving manipulation of the Ti system, electroporation, intranuclear microinjection, or the transformation of plant cell protoplasts have been successfully used to transform cells from monocotyledonous species, an important advance because the major cereal crops are monocots. However, the generation of

Figure IV.17
The effect of a *Bacillus thuringiensis* toxin gene on caterpillars' appetites for tomato plant leaves. The plant on the right contains the transgene. Courtesy of H. Schneiderman and the Monsanto Co.

whole fertile plants from transformed monocot protoplasts is still problematic. In contrast, whole dicotyledonous plants and fertile seeds carrying expressible recombinant genes have been produced from totipotent cells transformed by Ti vectors.

New Directions

Neither this essay nor the more detailed material in Volume II comprehensively illustrate the many ways whereby molecular genetics is being applied to the analysis of complex systems. Regrettably, important topics have been omitted. Nevertheless, it is apparent from even the few experiments summarized here that many of the curiosities about biological phenomena are giving way to experimental analysis. One such phenomenon is the development of a complex multicellular organism from a zygote, a process that involves differentiating primordial cells to yield the diverse cell types and tissues found in adult organisms. Molecular genetics has opened new windows on this marvelous process. Of the many interesting experiments being performed on plants, invertebrates, and vertebrates, including mammals, we can mention only a few.

Embryological Development and Differentiation of Complex Organisms

During the cell divisions that follow fertilization, individual cells develop different sizes, shapes, functions and potentials, and specific spatial relationships to one another. Moreover, all these events occur within defined temporal patterns. In some species, the mouse, for example, the cells formed in the first few cell divisions all have the same potential for further development into any part of the adult. As cell division proceeds, different cells become destined to give rise to

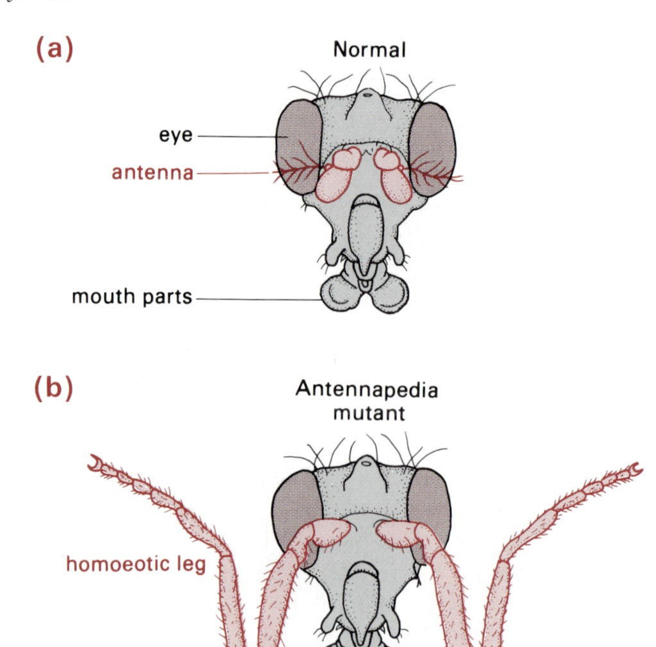

Figure IV.18
The homeotic mutation called *Antennapedia* results in the growth of legs instead of antennae on the head segment of *Drosophila*. After B. Alberts et al., *Mol. Biol. Cell* (1983), p. 840. Garland Publishing Co., New York.

only a limited number of adult cell types, and they are then said to be determined. Determined cells are committed to differentiate in a specific manner, and their daughter cells inherit their restricted program. Some cells, for example, are determined for extraembryonic tissue, such as the membranes that surround the fetus; others will contribute to the formation of the gut lining (endoderm), the nervous system and epidermis (ectoderm), or muscles, blood vessels, and other organs (mesoderm). More restrictive specializations to the individual cell types of various organ systems are programmed at subsequent cell divisions. Determination is followed by differentiation, the acquisition of distinctive phenotypes, be it shape or biochemical properties. Eventually, more than 200 cell types are distinguishable in vertebrates. A large body of evidence indicates that differential gene expression, in both time and place, is the underlying determinant of development and differentiation. Carefully controlled regulatory networks function in very early embryos and are sensitive to the position of the cell, the nature of the neighboring cells, and the time. One aim of a molecular approach to understanding determination and differentiation is to correlate spatial and temporal changes of early embryonic cells with differential gene expression. This approach has been particularly successful with *D. melanogaster*, largely because mutants with defects in early morphological development are available.

The segments seen in both developing and adult flies are normally associated with specific morphological features such as eyes, antennae, wings, or legs. Classical genetics identified a group of *D. melanogaster* mutants (e.g., *Antennapedia*, *bithorax*, etc.) in which particular morphological features develop in abnormal positions. In *Antennapedia* mutants, for example, legs appear where antennae

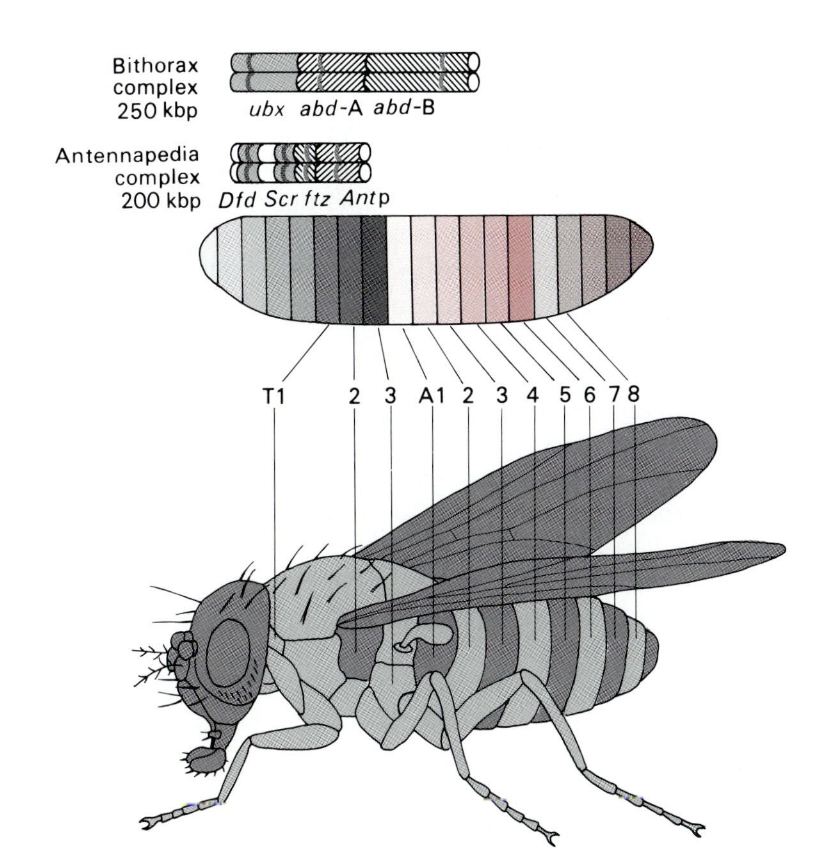

Figure IV.19
Schematic representation of the relation between homeotic genes and *Drosophila* segments. *Drosophila* larvae are segmented, and each segment gives rise to a distinct region of the adult fly. Large clusters of homeotic genes (e.g., *ubx, abd, Antp*) on chromosome 3, the *Antennapedia* and *bithorax* complexes, influence the morphological development of anterior (head and thorax) and posterior (thorax and abdomen) segments, respectively. The thoracic (T) and abdominal (A) segments of the larva are matched to corresponding regions of the adult fly. The gene complexes are shown (not to scale) with the homeo boxes indicated in full color. After K. Harding et al., *Science* 229 (1985), p. 1236, and W. J. Gehring, *Science* 236 (1987), p. 1245.

normally grow (Figure IV.18). The genes involved in this type of morphological transformation are called, as a group, **homeotic** genes, and the consequences of specific homeotic mutations are generally localized to one or more of the segments (Figure IV.19). Homeotic genes control the development of the central nervous system components and muscles within the segments as well as the external structures. Homeotic genes do not determine the number of segments but only the morphological development; mutations in other *D. melanogaster* genes affect the number of segments.

Two enormous regions of *D. melanogaster* chromosome 3, each covering 250 kbp or more, contain clusters of either *Antennapedia* or *bithorax* genes. Multiple coding regions that are associated with the proper development of specific segments occur within each cluster. Some regions seem to control the function of all the genes in a cluster, for example, the *ubx* region of the *bithorax* cluster. The functioning of particular homeotic genes in specific segments depends at least in part on their cell-specific transcription. Thus, transcripts of a homeotic gene are most abundant within cells of the segmental locus that is modified by mutations in that same gene. Homeotic gene products are DNA-binding proteins specific for regulatory motifs in other genes. Altogether, these experiments demonstrate that, like biochemical specificity, morphological development is controlled by differential gene expression.

Genes that are expressed at specific times in development are also being identified in organisms as diverse as corn, nematode worms, sea urchins, frogs, and mice. cDNA libraries prepared from cells at particular developmental stages are used to obtain clones representing messenger RNA expressed at that stage but not others. An example of the isolation of cDNAs that represent genes transcribed in *Xenopus* gastrulae but not in eggs was described in Section 6.5b. However, the potential for rapid progress with many of these organisms is not high, primarily because only rudimentary genetic information is available. The nematode *Caenorhabditis elegans* is an exception; not only has extensive genetic analysis been carried out, but the line of descent from the zygote is known for every one of the 2000 cells in the adult worms. It is possible to identify which *C. elegans* genes are expressed and when as particular cell lineages branch off in the course of determination and differentiation. Successful techniques for introducing cloned genes (and reverse mutations) into *C. elegans* germ line cells are permitting even more precise experiments. Among mammals, mice are the most promising animals for experimental work. Extensive genetic information exists, and many mouse mutations are known. Moreover, cells from early mouse developmental stages are accessible. With mice and to some extent with other mammals, there are also available useful cultured cell lines that originate from early embryos or from teratocarcinomas, tumors representing abnormally proliferating embryonic cells. Most promising, however, are two quite new avenues for studying mouse development. The first of these originates in the study of *D. melanogaster* homeotic genes. The second depends on transgenic mice.

All the coding regions of homeotic genes contain homologous DNA sequences about 180 bp long, known as **homeo boxes** (Figure IV.19). These segments give rise to related polypeptide domains in the various homeotic gene products, and these domains impart the DNA binding property to the proteins. Even more significantly, the characteristic homeo box domain is highly conserved in evolution. Sequences homologous to *D. melanogaster* homeo boxes occur in yeast, invertebrates, and vertebrates, including mammals. In each species studied, homeo box proteins are DNA binding proteins that interact with a variety of sequence motifs and with other proteins to contribute to complex regulatory responses. Moreover, the homeo box sequences are transcribed in embryonic mouse cells starting at about ten days of gestation. This circumstantial evidence has excited interest in the possibility that distantly related organisms such as *D. melanogaster* and mammals utilize similar developmental strategies. Perhaps the identification of the gene products associated with homeo boxes will lead to substantial progress in understanding mammalian morphogenetic development.

Insertional Mutagenesis

The second new experimental approach to understanding gene function in mammalian determination and differentiation derives from the use of transgenic mice. This technique was first developed to study the timing and location of gene expression and to experiment with genetic therapy (Figure IV.15). The methods are such that the newly introduced genes enter the host genome at nonspecific and unpredictable sites. This lack of specificity can frustrate gene expression studies and gene therapy. However, it has an advantage for the study of development. When newly introduced DNA is inserted by chance within coding regions or important regulatory regions, it is mutagenic; expression of the target gene may be modulated or eliminated. An altered gene product may be produced. This is equivalent to the technique called **insertional mutagenesis**, a standard tool of prokaryote, yeast, *D. melanogaster*, and maize genetics. With the latter organisms, transposable elements are used as insertional mutagenic agents. In mice, both recombinant retroviral vectors and DNA segments play the role of transposable elements when injected into zygotes. Quite fortuitously then, striking heritable modifications in the phenotype of early embryos and adults are sometimes observed in transgenic mice and their offspring. First generation transgenic mice are heterozygous for the insertional mutation. Selective inbreeding supplies both heterozygous and homozygous offspring and thus the opportunity to observe the phenotypic consequences of recessive mutations. Because the inserted segment generally includes a sequence that is foreign to the mouse genome (i.e., the vector), it can readily be identified in DNA blots or in genomic libraries and can be cloned along with the surrounding sequences of the target gene. In this way, a lethal insertional mutation that arrests mouse development after about 11 days of gestation was associated with a collagen gene. Another insertional mutation resulted in a severe defect in mouse limb bone development. The target gene proved to be an allele of a mutant gene first detected by chance more than 20 years ago (Figure IV.20). Cloning of the gene and identification of the gene product should provide clues into mammalian morphogenesis.

Targeted and Untargeted Modification of Mammalian Genomes

Virtually all the approaches described above rely on the ability of eukaryotic cells to integrate DNA at random sites in their genomes. But for many purposes it is most useful and, in some cases, imperative to introduce new genetic information at specific chromosome sites or to modify existing genes in directed ways. Up until recently, this was possible only with yeast (Section 5.6c). Now, considerable progress along these lines has been made with mammalian cells by designing special vectors which recombine with their homologous sequences in chromosomes. Using these methods, and special techniques for preferentially recovering *the desired cells and introducing them into early mouse embryos*, it is possible to produce animals with specific mutations in selected genes or to eliminate mutations from defective genes.

Future Research

Molecular genetics has made a beginning. A wealth of detail about many biological systems is already available, but the successes do not amount to a complete or even a very profound understanding.

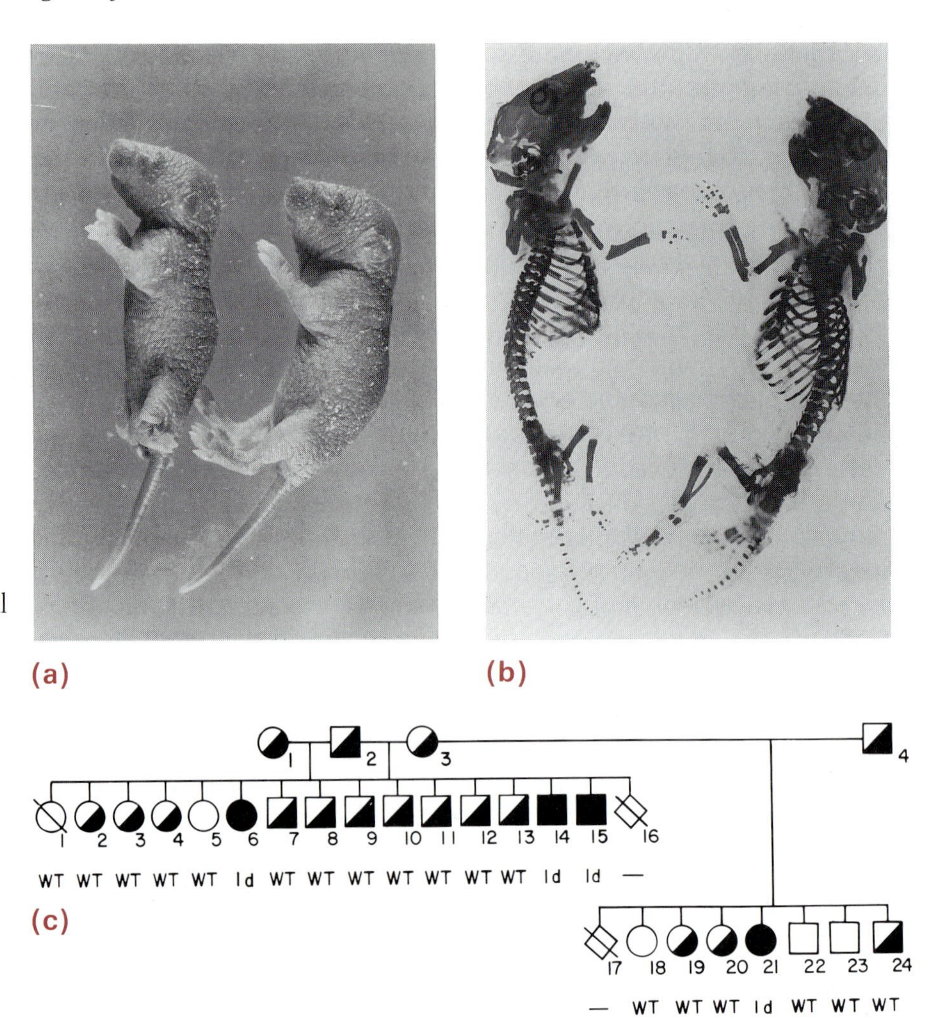

Figure IV.20
Insertional mutagenesis in a transgenic mouse creates an inherited limb deformity. Transgenic mice (Figure IV.15) were obtained after injecting an embryo with a DNA segment carrying a retroviral LTR as promoter and the mouse protooncogene *myc* on a pBR322 vector and breeding and inbreeding the progeny. The resulting mouse strain carries a recessive autosomal mutation. The phenotype of homozygous mice is a limb deformity as shown on the left in the photographs of four-day-old homozygous mutant mice (a) and skeletons (b). On the right in the photographs are normal four-day-old mice. (c) The limb abnormality (Id) segregated during breeding with DNA sequences that anneal with the LTR-*myc* sequences as shown in the pedigree. Circles are females; squares are males. Black symbols designate animals that are homozygous for the LTR-*myc* insertion, as determined by DNA blotting. Half-black symbols designate individuals heterozygous for the insertion. Animals designated by open symbols contained no LTR-*myc* insertion in their DNA. The letters below the symbols show the phenotype, wild type or Id. A slash (\) through the symbol indicates that the animal died before data could be collected. Using mouse/hamster somatic cell hybrids, the LTR-*myc* sequences were localized on mouse chromosome 2. Similar spontaneous (or radiation-induced) mouse mutations yielding limb deformities were previously mapped to chromosome 2. In one of these mutants, *ld*, the phenotype is virtually the same as that observed in the transgenic mice, and the affected genes are probably alleles. After R. P. Woychik et cl., *Nature* 318 (1985), p. 36.

On the contrary, current ignorance is vaster than current knowledge. Nothing in the man-made world rivals the complexity and diversity of living things. No man-made information system approaches, in content, the amount of data encoded in genomes or, in complexity, the intricate regulatory networks that control gene expression. There remain to be discovered mechanisms and concepts that no one has yet even imagined. In some instances, we have learned enough at least to identify important areas of ignorance. Certain or these concern long-standing questions, like development and differentiation, or the molecular basis of mind. Others are new questions, raised by the very achievements of molecular genetics itself. And, of course, we should be wary; some things that we now think we know may become less clear or even prove wrong in the years to come.

Another area that is attracting a large measure of attention is oncogenes: *v*-oncs, *c*-oncs, and protooncogenes. Considerable progress has already been made in identifying the normal biochemical functions of protooncogenes, although many still remain mysterious. We are only beginning to scratch the surface of the important question of the critical differences accounting for the normal function of a protooncogene and the oncogenicity of its *c*-onc counterpart. At this time, there is no complete description of how a *c*-onc or a *v*-onc turns a normal cell into a tumorigenic one. Activation of protein phosphorylation and the permanent turning on of transcriptional regulatory switches are likely mechanisms in some cases. How many cellular aberrations other than *v*-onc or *c*-onc expression are required to yield a tumorigenic cell? And, most important, can the process of tumorigenesis be interrupted by precise molecular tools?

Genetic recombination was first described early in the twentieth century. At that time, recombination meant homologous, meiotic crossing-over. Now we know that there are many different kinds of recombination, and they occur in both meiotic and mitotic cells: homologous crossing-over, gene conversion, transposition of movable elements, genome reorganization, and so forth. What are the molecular mechanisms? Which type of recombination predominates in different organisms, and when (in development) and where (in what cell types)? Are there mechanisms to modulate recombination? These questions now engage many active investigators.

Future Concerns

The fast pace of modern biology is driven by technology, by major challenges in health, agriculture, and industry, and, ultimately, by curiosity. Because we and our offspring are the products of functional genetic systems, our intellectual curiosity is buttressed by personal involvement. This same combination of intellectual curiosity and personal relevance is responsible for the intense public interest in the recombinant DNA revolution and its product, genetic engineering. And it is why these achievements will identify twentieth-century biology as a great historical landmark. Thus far, the revolution has been a positive endeavor. Our understanding of ourselves and other organisms has deepened. Important products (such as hormones and vaccines and enzymes for research and commerce) are being produced and developed. Harmless microorganisms are being manipulated to make them beneficial environmental agents. Still, there are disquieting aspects to the revolution. Care is required to assure that no unexpected detrimental properties accompany the beneficial characteristics introduced into altered microorganisms. Deep thought and cautious analysis are required if we are to avoid unwise use of the sophisticated diagnostic techniques afforded by the new methods or of somatic cell gene therapy. The challenges are great, particularly because the course of future research is inherently unpredictable in outcome.

It is sobering to us that almost nothing recorded in this book was known when we completed our formal educations over 30 years ago. Our chief regret is that we are not likely to see the new perspective 30 years hence. We can be certain of only one thing: from future research will emerge major new concepts, concepts that will be as unexpected in their time as were introns and movable elements to the present generation of biologists. A changing perspective is the history of science, and molecular genetics will not be immune to that imperative.

References for Part IV

PERSPECTIVE

The following articles and books provide an introduction to or an overview of biological systems that are being illuminated by molecular genetics and are discussed in this Perspective essay. The listing is divided into sections that correlate with the sections in the essay. Within each section, references are given chronologically.

UNDERSTANDING COMPLEX BIOLOGICAL SYSTEMS

M. S. Brown and J. L. Goldstein. 1986. A Receptor-Mediated Pathway for Cholesterol Homeostasis. *Science* 232 34–47.

A. W. Murray and M. W. Kirschner. 1989. Dominoes and Clocks: The Union of Two Views of the Cell Cycle. *Science* 246 614–621.

The Multigene Family for Visual Pigments

J. Nathans. 1987. Molecular Biology of Visual Pigments. *Annu. Rev. Neurosci.* 10 163–194.

T. Piantanida. 1988. The Molecular Genetics of Color Vision and Color Blindness. *Trends Genet.* 4 319–323.

D. Vollrath, J. Nathans, and R. W. Davis. 1988. Tandem Array of Human Visual Pigment Genes at Xq28. *Science* 240 1669–1672.

J. Nathans. 1989. The Genes for Color Vision. *Sci.* American 260(2) 42–49.

Coordinating Interactions Between Many Gene Products

J. Sommerville. 1985. Organizing the Nucleolus. *Nature* 318 410–411.

G. Jordan. 1987. At the Heart of the Nucleolus. *Nature* 329 489–490.

R. J. Planta and H. A. Raue. 1988. Control of Ribosome Biogenesis in Yeast. *Trends Genet.* 4 64–68.

S. Rose. 1988. Reflections on Reductionism. *Trends Biochem. Sci.* 13 160–162.

F. Amaldi, I. Bozzoni, E. Beccari, and P. Pierandrei-Amaldi. 1989. Expression of Ribosomal Protein Genes and Regulation of Ribosome Biosynthesis in *Xenopus* Development. *Trends Biochem. Sci.* 14 175–178.

The Life Cycles of Viruses

J. M. Bishop. 1982. Oncogenes. *Sci. American* 246(3) 80–92.

J. M. Bishop. 1983. Cellular Oncogenes and Retroviruses. *Annu. Rev. Biochem.* 52 301–354.

R. A. Weinberg. 1983. A Molecular Basis of Cancer. *Sci. American* 249(5) 126–143.

H. E. Varmus. 1984. Molecular Genetics of Cellular Oncogenes. *Annu. Rev. Genet.* 18 553–612.

R. DiMarchi, G. Brooke, C. Gale, V. Cracknell, T. Doel, and N. Mowat. 1986. Protection of Cattle Against Foot-and-Mouth Disease by a Synthetic Peptide. *Science* 232 639–641.

J. M. Bishop. 1987. The Molecular Genetics of Cancer. *Science* 235 305–311.

B-Z. Shilo. 1987. Proto-Oncogenes in *Drosophila melanogaster*. *Trends Genet.* 3 69–72.

A. S. Fauci. 1988. The Human Immunodeficiency Virus: Infectivity and Mechanisms of Pathogenesis. *Science* 239 617–622.

D. I. H. Linzer. 1988. The Marriage of Oncogenes and Anti-Oncogenes. *Trends Genet.* 4 245–247.

P. C. Nowell and C. M. Croce. 1988. Chromosomal Approaches to Oncogenes and Oncogenesis. FASEB J. 2 3054–3060.

R. A. Weinberg. 1988. Finding the Anti-Oncogene. *Sci. American* 259(3) 4–51.

MAPPING GENOMES

Physical and Genetic Maps

A. Coulson, J. Sulston, S. Brenner, and J. Karn. 1986. Toward a Physical Map of the Genome of the Nematode, *Caenorhabditis elegans*. *Proc. Natl. Acad. Sci.* U.S.A. 83 7821–7825.

M. V. Olson, J. E. Dutchik, M. Y. Graham, G. M. Brodeur, C. Helms, M. Frank, M. MacCollin, R. Scheinman, and T. Frank. 1986. Random-Clone Strategy for Genomic Restriction Mapping in Yeast. *Proc. Natl. Acad. Sci.* U.S.A. 83 7826–7830.

H. Cooke. 1987. Cloning in Yeast: An Appropriate Scale for Mammalian Genomes. *Trends Genet.* 3 173–175.

Y. Kohara, K. Akiyama, and K. Isono. 1987. The Physical Map of the Whole *E. coli* Chromosome: Application of a New Strategy for Rapid Analysis and Sorting of a Large Genomic Library. *Cell* 50 495–508.

C. L. Smith and C. R. Cantor. 1987. Preparation and Manipulation of Large DNA Molecules. *Trends Biochem. Sci.* 12 284–287.

C. L. Smith, J. G. Econome, A. Schutt, S. Klco, and C. R. Cantor. 1987. A Physical Map of the *Escherichia coli* Genome. *Science* 236 1448–1453.

D. Vollrath, R. W. Davis, C. Connelly, and P. Hieter. 1988. Physical Mapping of Large DNA by Chromosome Fragmentation. *Proc. Natl. Acad. Sci.* U.S.A. 85 6027–6031.

V. Knott, D. J. Blake, and G. G. Brownlee. 1989. Completion of the Detailed Restriction Map of the *E. coli* Genome by the Isolation of Overlapping Cosmid Clones. Nucleic Acids Res. 17 5901–5912.

Mapping Large Genomes

V. A. McKusick. 1986. The Gene Map of *Homo sapiens*: Status and Prospects. *Cold Spring Harbor Symp. Quant. Biol.* 51 15–27.

D. T. Burke, G. F. Carle, and M. V. Olson. 1987. Cloning of Large Segments of DNA into Yeast by Means of Artificial Chromosome Vectors. *Science* 236 806–812.

H. Donis-Keller, P. Green, C. Helms, S. Cartinhour, B. Weiffenbach, K. Stephens, T. P. Keith, D. W. Bowden, D. R. Smith, E. S. Lander, D. Botstein, G. Akots, K. S. Rediker, T. Gravius, V. A. Brown, M. B. Rising, C. Parker, J. A. Powers, D. E. Watt, E. R. Kauffman, A. Bricker, P. Phipps, H. Muller-Kahle, T. R. Fulton, S. Ng, J. W. Schumm, J. C. Braman, R. Knowlton, D. F. Barker, S. M. Crooks, S. E. Lincoln, M. J. Daly, and J. Abrahamson. 1987. A Genetic Linkage Map of the Human Genome. *Cell* 51 319–327.

T. Helentjaris. 1987. A Genetic Linkage Map for Maize Based on RFLPs. *Trends Genet.* 3 217–221.

Human Gene Mapping 9. 1987. *Cytogenet. Cell Genet.* 46, Nos. 1–4.

V. A. McKusick and F. H. Ruddle. 1987. Editorial: Toward a Complete Map of the Human Genome. *Genomics* 1 103–106.

Y. Nakamura, M. Lathrop, P. O'Connell, M. Leppert, D. Barker, E. Wright, M. Skolnick, S. Kondoleon, M. Litt, J-M. Lalouel, and R. White. 1988. A Mapped Set of DNA Markers for Human Chromosome 17. *Genomics* 2 302–309.

C. M. Rick and J. I. Yoder. 1988. Classical and Molecular Genetics of Tomato: Highlights and Perspectives. *Annu. Rev. Genet.* 22 281–300.

J. Schmidtke and D. N. Copper. 1989. A Comprehensive List of Cloned Human DNA Sequences. *Nucleic Acids Res.* 17 (suppl.) r173–r281.

R. White and J-M. Lalouel. 1988. Linked Sets of Genetic Markers for Human Chromosomes. *Annu. Rev. Genet.* 22 259–279.

Polymorphic Restriction Endonuclease Fragments

A. J. Jeffreys. 1979. DNA Sequence Variants in the $^{G}\gamma$-, $^{A}\gamma$-, δ- and β-Globin Genes of Man. *Cell* 18 1–10.

S. H. Orkin and H. H. Kazazian, Jr. 1984. The Mutation and Polymorphism of the Human Beta-Globin Gene and Its Surrounding DNA. *Annu. Rev. Genet.* 18 131–171.

J. F. Gusella. 1986. DNA Polymorphism and Human Disease. *Annu. Rev. Biochem.* 55 831–854.

C. T. Caskey. 1987. Disease Diagnosis by Recombinant DNA Methods. *Science* 236 1223–1228.

A. P. Monaco and L. M. Kunkel. 1987. A Giant Locus for the Duchenne and Becker Muscular Dystrophy Gene. *Trends Genet.* 3 33–37.

M. Namanùra, M. Leppert, P. O'Connell, R. Wolff, T. Holm, M. Culver, C. Martin, E. Fujimoto, M. Hoff, E. Kumlin, and R. White. 1987. Variable Number of Tandem Repeat (VNTR) Markers for Human Gene Mapping. *Science* 235 1616–1622.

M. Dean. 1988. Molecular and Genetic Analysis of Cystic Fibrosis. *Genomics* 3 93–99.

J. A. Witkowski. 1988. The Molecular Genetics of Duchenne Muscular Dystrophy: The Beginning of the End? *Trends Genet.* 4 27–30.

R. G. Woron and M. W. Thompson. 1988. Genetics of Duchenne Muscular Dystrophy. *Annu. Rev. Genet.* 22 601–629.

B-S. Karem, J. M. Rommens, J. A. Buchanan, D. Markiewicz, T. K. Cox, A. Chakravarti, M. Buchwald, and L-C. Tsui. 1989. Identification of the Cystic Fibrosis Gene: Genetic Analysis. *Science* 245 1073–1080.

J. L. Mandel. 1989. Dystrophin: The Gene and Its Product. *Nature* 339 584–586.

B. Martin, J. Nienhuis, G. King, and A. Schaefer. 1989. Restriction Fragment Length Polymorphisms Associated with Water Use in Tomato. *Science* 243 1725–1728.

J. R. Riordan, J. M. Rommens, B-S. Karem, N. Alon, R. Rozmahel, Z. Grzelzak, J. Zielenski, S. Lok, N. Plavsic, J-L. Chou, M. L. Drumm, M. C. Iannuzzi, F. S. Collins, and L-C. Tsui. 1989. Identification of the Cystic Fibrosis Gene: Cloning and Characterization of Complementary DNA. *Science* 245 1066–1073.

J. M. Rommens, M. C. Iannuzzi, B-S. Karem, M. L. Drumm, G. Kelmer, M. Dean, R. Rozmahel, J. L. Cole, D. Kennedy, N. Hidaka, M. Zsiga, M. Buchwald, J. R. Riordan, L-C. Tsui, and F. S. Collins. 1989. Identification of the Cystic Fibrosis Gene: Chromosome Walking and Jumping. *Science* 245 1059–1065.

DIFFERENTIAL GENE EXPRESSION IN SPECIFIC CELLS AND TISSUES

Differential Gene Expression in the Pancreas

T. Edlund, M. D. Walker, P. J. Barr, and W. J. Rutter. 1985. Cell-Specific Expression of the Rat Insulin Gene: Evidence for Role of Two Distinct 5′ Flanking Elements. *Science* 230 912–916.

O. Karlsson, T. Edlund, J. B. Moss, W. J. Rutter, and M. D. Walker. 1987. A Mutational Analysis of the Insulin Gene Transcription Control Region: Expression in Beta Cells Is Dependent on Two Related

Sequences Within the Enhancer. *Proc. Natl. Acad. Sci.* U.S.A. 84 8819–8823.

L. G. Moss, J. B. Moss, and W. J. Rutter. 1988. Systematic Binding Analysis of the Insulin Gene Transcription Control Region: Insulin and Immunoglobulin Enhancers Utilize Similar Transactivators. *Mol. Cell. Biol.* 8 2620–2627.

H. Ohlsson, O. Karlsson, and T. Edlund. 1988. A Beta-Cell-Specific Protein Binds to the Two Major Regulatory Sequences of the Insulin Gene Enhancer. *Proc. Natl. Acad. Sci.* U.S.A. 85 4228–4231.

Posttranslational Modifications as Regulators of the Flow of Gene Products

J. Douglas, O. Civelli, and E. Herbert. 1984. Polyprotein Gene Expression: Generation of Diversity of Neuroendocrine Peptides. *Annu. Rev. Biochem.* 53 665–715.

S. H. Snyder. 1985. The Molecular Basis of Communication Between Cells. *Sci. American* 253(4) 132–141.

T. T. Puck and F-T. Kao. 1982. Somatic Cell Genetics and Its Application to Medicine. *Annu. Rev. Genet.* 16 225–271.

J. Stanbury, J. B. Wyngaarden, D. S. Fredrickson, J. L. Goldstein, and M. S. Brown. 1983. The Metabolic Basis of Inherited Disease, 5th ed. McGraw-Hill, New York.

K. E. Davies (ed.). 1986. Human Genetic Diseases: A Practical Approach. IRL Press, Oxford and Washington, D.C.

MODIFYING BIOLOGICAL SYSTEMS

Synthesizing Normal and Modified Eukaryotic Proteins

J. F. Young, W. T. Hockmeyer, M. Gross, W. R. Ballou, R. A. Wirtz, J. H. Trosper, R. L. Beaudoin, M. R. Hollingdale, L. H. Miller, C. L. Diggs, and M. Rosenberg. 1985. Expression of *Plasmodium falciparum* Circumsporozoite Proteins in *E. coli* for Potential Use in a Human Malaria Vaccine. *Science* 228 958–962.

A. J. Clark, P. Simons, I. Wilmut, and R. Lathe. 1987. Pharmaceuticals from Transgenic Livestock. *Trends Biotech.* 5 20–24.

C. S. Craik, S. Roczniak, S. Sprang, R. Fletterick, and W. Rutter. 1987. Redesigning Trypsin Via Genetic Engineering. *J. Cell. Biochem.* 33 199–211.

J. R. Knowles. 1987. Tinkering with Enzymes: What Are We Learning? *Science* 236 1252–1258.

N. Liang, G. J. Pielak, A. G. Mauk, M. Smith, and B. M. Hoffman. 1987. Yeast Cytochrome *c* with Phenylalanine or Tyrosine at Position 87 Transfers Electrons to (Zinc Cytochrome *c* Peroxidase)$^{+}$ at a Rate Ten Thousand Times That of the Serine-87 or Glycine-87 Variants. *Proc. Natl. Acad. Sci.* U.S.A. 84 1249–1252.

D. Shortle. 1989. Probing the Determinants of Protein Folding and Stability with Amino Acid Substitutions. *J. Biol. Chem.* 264 5315–5318.

Modifying Somatic Cells

T. Friedmann. 1983. Gene Therapy: Fact and Fiction. Cold Spring Harbor Laboratory, Cold Spring Harbor, New York.

W. F. Anderson. 1984. Prospects for Human Gene Therapy. *Science* 226 401–409.

J. Ellis and A. Bernstein. 1988. Gene Targeting with Retroviral Vectors. Cold Spring Harbor Laboratory, Cold Spring Harbor, New York.

Y. Gluzman and S. H. Hughes. 1988. Viral Vectors. Cold Spring Harbor Laboratory, Cold Spring Harbor, New York.

T. Friedmann. 1989. Progress Toward Human Gene Therapy. *Science* 244 1275–1281.

Modifying Germ Cells

G. M. Rubin and A. C. Spradling. 1982. Genetic Transformation of *Drosophila* with Transposable Element Vectors. *Science* 218 348–353.

A. C. Spradling and G. M. Rubin. 1982. Transposition of Cloned P Elements into *Drosophila* Germ Line Chromosomes. *Science* 218 341–347.

R. L. Chisholm. 1983. Gene Therapy in *Drosophila*. *Trends Biochem. Sci.* 8 191–193.

R. Jaenisch, K. Harbers, A. Schnieke, J. Lohler, I. Chumakov, D. Jahner, D. Grotkopp, and E. Hoffmann. 1983. Germline Integration of Moloney Murine Leukemia Virus at the *Mov*13 Locus Leads to Recessive Lethal Mutation and Early Embryonic Death. *Cell* 32 209–216.

T. A. Stewart, P. K. Pattengale, and P. Leder. 1984. Spontaneous Mammary Adeno-Carcinomas in Transgenic Mice that Carry and Express MTV/*myc* Fusion Genes. *Cell.* 38 627–637.

J. W. Gordon and F. H. Ruddle. 1985. DNA-Mediated Genetic Transformation of Mouse Embryos and Bone Marrow—A Review. *Gene* 33 121–136.

D. Hanahan. 1985. Heritable Formation of Pancreatic Beta-Cell Tumours in Transgenic Mice Expressing Recombinant Insulin/Simian Virus 40 Oncogenes. *Nature* 315 115–123.

R. L. Brinster and R. D. Palmiter. 1986. Introduction of Genes into the Germ Line of Animals. In The Harvey Lectures, pp. 1–38, series 80. Alan R. Liss, New York.

D. Hanahan. 1986. Oncogenesis in Transgenic Mice. In T. Graf and P. Kahn (eds.), Oncogenes and Growth Control, pp. 349–363. Spring-Verlag, Heidelberg.

R. D. Palmiter and R. L. Brinster. 1986. Germ-Line Transformation of Mice. *Annu. Rev. Genet.* 20 465–499.

R. B. Church. 1987. Embryo Manipulation and Gene Transfer in Domestic Animals. *Trends Biotech.* 5 13–19.

E. C. Cocking and M. R. Davey. 1987. Gene Transfer in Cereals. *Science* 236 1259–1262.

R. M. Goodman, H. Hauptili, A. Crossway, and V. C. Knauf. 1987. Gene Transfer in Crop Improvement. *Science* 236 48–54.

D. Hanahan. 1988. Dissecting Multistep Tumorigenesis in Transgenic Mice. *Annu. Rev. Genet.* 22 479–519.

C. A. Rhodes, D. A. Pierce, I. J. Mettler, D. Mascarenhas, and J. J. Detmer. 1988. Genetically Transformed Maize Plants from Protoplasts. *Science* 240 204–207.

L. Willmitzer. 1988. The Use of Transgenic Plants to Study Plant Gene Expression. *Trends Genet.* 4 13–18.

M. R. Capecchi. 1989. Altering the Genome by Homologous Recombination. *Science* 224 1288–1292.

Cell 56(2). 1989. Entire issue is devoted to reviews of plant systems.

C. S. Gasser and R. T. Fraley. 1989. Genetically Engineering Plants for Crop Improvement. *Science* 244 1293–1299.

D. Hanahan. 1989. Transgenic Mice as Probes into Complex Systems. *Science* 246 1265–1275.

S. Thompson, A. R. Clarke, A. M. Pow, M. L. Hooper, and D. W. Melton. 1989. Germ Line Transmission and Expression of a Corrected HPRT Gene Produced by Gene Targeting in Embryonic Stem Cells. *Cell* 56 313–321.

NEW DIRECTIONS

Embryological Development and Differentiation of Complex Organisms

P. W. Sternberg and H. R. Horvitz. 1984. The Genetic Control of Cell Lineage During Nematode Development. *Annu. Rev. Genet.* 18 489–524.

W. Bender. 1985. Homeotic Gene Products as Growth Factors. *Cell* 43 559–560.

W. J. Gehring. 1985. The Molecular Basis of Development. *Sci. American* 253(4) 152B–162.

E. H. Davidson. 1986. Gene Activity in Early Development, 3rd ed. Academic Press, Orlando.

W. J. Gehring and Y. Hiromi. 1986. Homeotic Genes and the Homeobox. *Genetics* 20 147–173.

I. Duncan. 1987. The Bithorax Complex. *Annu. Rev. Genet.* 21 285–319.

W. J. Gehring. 1987. Homeo Boxes in the Study of Development. *Science* 236 1245–1252.

M. Robertson. 1987. Towards a Biochemistry of Morphogenesis. *Nature* 330 420–421.

P. W. Ingham. 1988. The Molecular Genetics of Embryonic Pattern Formation in *Drosophila. Nature* 335 25–34.

M. Akam. 1989. Hox and Hom: Homologous Gene Clusters in Insects and Vertebrates. *Cell* 57 347–349.

G. R. Dressler. 1989. An Update on the Vertebrate Homeobox. *Trends Genet.* 5 129–131.

Insertional Mutagenesis

R. P. Woychik, T. A. Stewart, L. G. Davis, P. D'Eustachio, and P. Leder. 1985. An Inherited Limb Deformity Created by Insertional Mutagenesis in a Transgenic Mouse. *Nature* 318 36–40.

T. Gridley, P. Soriano, and R. Jaenisch. 1987. Insertional Mutagenesis in Mice. *Trends Genet.* 3 162–166.

L. Cooley, R. Kelley, and A. Spradling. 1988. Insertional Mutagenesis of the *Drosophila* Genome with Single P Elements. *Science* 239 1121–1128.

FUTURE CONCERNS

Office of Technology Assessment, U.S. Congress. 1981. Impacts of Applied Genetics: Microorganisms, Plants, and Animals. U.S. Government Printing Office, Washington, D.C.

President's Commission for the Study of Ethical Problems in Medicine and Biomedical and Behavioral Research. 1982. Splicing Life: The Social and Ethical Issues of Genetic Engineering with Human Beings. U.S. Government Printing Office, Washington, D.C.

D. J. Weatherall. The New Genetics and Clinical Practice. 1982. Nuffield Provincial Hospitals Trust, London.

J. G. Perpich (ed). 1986. Biotechnology in Society: Private Initiatives and Public Oversight. Pergamon Press, New York.

Introduction of Recombinant DNA-Engineered Organisms into the Environment: Key issues. 1987. National Academy Press, Washington, D.C.

Commission on Life Sciences, National Research Council. 1988. Mapping and Sequencing the Human Genome. National Academy Press, Washington, D.C.

E. K. Nichols. 1988. Human Gene Therapy. Institute of Medicine, National Academy of Sciences. Harvard University Press, Cambridge, Mass.

Office of Technology Assessment, U.S. Congress. 1988. Field-Testing Engineered Organisms: Genetic and Ecological Issues. U.S. Government Printing Office, Washington, D.C.

Office of Technology Assessment, U.S. Congress. 1988. Mapping Our Genes. Genome Projects: How Big? How Fast? U.S. Government Printing Office, Washington, D.C.

L. Walters. 1989. Report to the Biomedical Ethics Advisory Committee of the Biomedical Ethics Board. Washington, D.C. Available through National Reference Center for Bioethics Literature, Kennedy Institute of Ethics, Georgetown University, Washington, D.C.

Index

Note: Page numbers followed by *t* and *f* indicate tables and figures, respectively.

A

B

C

D

E

F

G

H

I

J

K

L

M

N

O

P

R

T

U

V

W

X

Y

Z